TODAY'S TECHNICIAN ™

Shop Manual for
Automotive
Electricity & Electronics

Fourth Edition

TODAY'S TECHNICIAN ™

Shop Manual for
Automotive
Electricity & Electronics

Fourth Edition

Barry Hollembeak

THOMSON

DELMAR LEARNING

Australia   Canada   Mexico   Singapore   Spain   United Kingdom   United States

THOMSON
DELMAR LEARNING

Today's Technician/Classroom Manual for Automotive Electricity & Electronics, 4th Edition

Barry Hollembeak

Vice President, Technology and Trades ABU:
David Garza

Director of Learning Solutions:
Sandy Clark

Senior Acquisitions Editor:
David Boelio

Product Manager:
Matthew Thouin

Marketing Director:
Deborah Yarnell

Channel Manager:
William Lawrensen

Marketing Coordinator:
Mark Pierro

Project Editor:
Toni Hansen

Art/Design Specialist:
Cheri Plasse

Technology Project Manager:
Kevin Smith

Technology Project Specialist:
Linda Verde

Editorial Assistant:
Andrea Domkowski

Library of Congress Cataloging-in-Publication Data
Hollembeak, Barry.
    Shop manual for automotive electricity and electronics / Barry Hollembeak. — 4th ed.
    p. cm. — (Today's technician)
Includes index.
ISBN 1-4180-1267-X
    1. Automobiles—Electric equipment—Maintenance and repair. 2. Automobiles—Electronic equipment—Maintenance and repair. I. Title. II. Title: Automotive electricity and electronics. III. Series.

TL272.H63 2006
629.25'40288—dc22

2006044533

NOTICE TO THE READER

CONTENTS

Preface   xiii

CHAPTER 1   **Safety**   1

Introduction 1 ● Personal Safety 1 ● Tool and Equipment Safety 7 ● Lifting the Vehicle 8 ● Running the Vehicle While in the Shop 13 ● Fire Hazards and Prevention 14 ● Hazardous Materials 18 ● Electrical System Safety 21 ● Summary 26 ● Terms to Know 26 ● ASE-Style Review Questions 27 ● Job Sheets 29

CHAPTER 2   **Special Tools and Procedures**   33

Introduction 33 ● Basic Electrical Troubleshooting 34 ● Test Equipment 35 ● Multimeters 38 ● Lab Scopes and Oscilloscopes 49 ● Scan Tools 56 ● Static Strap 59 ● Memory Keepers 60 ● Service Information 61 ● Working as an Electrical Systems Technician 66 ● Diagnostics 69 ● Summary 70 ● Terms to Know 71 ● ASE-Style Review Questions 71 ● ASE Challenge Questions 72 ● Job Sheets 73

CHAPTER 3   **Basic Electrical Troubleshooting and Service**   87

Introduction 87 ● Testing Circuit Protection Devices 88 ● Testing and Replacing Electrical Components 92 ● Testing Diodes 99 ● Testing Transistors 101 ● Testing for Circuit Defects 103 ● Digital Storage Oscilloscope (DSO) 111 ● Terms to Know 116 ● ASE-Style Review Questions 117 ● ASE Challenge Questions 118 ● Job Sheets 119

CHAPTER 4

Wiring Repair and Reading Circuit Diagrams   **133**

Introduction 133 ● Wire Repair 133 ● Replacing Fusible Links 141 ● Repairing Connector Terminals 142 ● Reading Wiring Diagrams 147 ● Terms to Know 150 ● ASE-Style Review Questions 150 ● ASE Challenge Questions 151 ● Job Sheets 153

CHAPTER 5

Battery Diagnosis and Service   **171**

Introduction 171 ● General Precautions 172 ● Battery Inspection 173 ● Charging the Battery 175 ● Battery Test Series 178 ● Battery Drain Test 188 ● Battery Removal and Cleaning 190 ● Maintaining or Restoring Memory Functions 194 ● Jumping the Battery 194 ● Terms to Know 196 ● ASE-Style Review Questions 196 ● ASE Challenge Questions 197 ● Job Sheets 199

CHAPTER 6

Starting System Diagnosis and Service   **209**

Introduction 209 ● Starting System Service Cautions 210 ● Starting System Troubleshooting 210 ● Testing the Starting System 214 ● Starter Motor Removal 225 ● Starter Motor Disassembly 227 ● Starter Motor Component Tests 232 ● Starter Reassembly 235 ● Terms to Know 238 ● ASE-Style Review Questions 238 ● ASE Challenge Questions 240 ● Job Sheets 241

CHAPTER 7

Charging System Testing and Service   **249**

Introduction 249 ● Charging System Service Cautions 251 ● AC Generator Noises 251 ● Charging System Troubleshooting 252 ● Voltage Output Testing 252 ● Voltage Drop Testing 253 ● Field Current Draw Test 254 ● Current Output Testing 256 ● Full-Field Test 256 ● Regulator Test 262 ● Diode/Stator Test 263 ● Diode Pattern Testing 263 ● Charging System Requirement Test 266 ● AC Generator Removal and Replacement 266 ● AC Generator Disassembly 268 ● AC Generator Component Testing 271 ● AC Generator Reassembly 274 ● Terms to Know 276 ● ASE-Style Review Questions 276 ● ASE Challenge Questions 277 ● Job Sheets 279

| CHAPTER 8 | **Lighting Circuits Repair and Diagnosis** | **287** |

Introduction 287 ● Headlights 288 ● Concealed Headlights 300 ● Headlight Switch Testing and Replacement 301 ● Dimmer Switch Testing and Replacement 304 ● Taillight Assemblies 308 ● Interior Lights 314 ● Terms to Know 318 ● ASE-Style Review Questions 318 ● ASE Challenge Questions 319 ● Job Sheets 321

| CHAPTER 9 | **Electrical Accessories Diagnosis and Repair** | **335** |

Introduction 335 ● Horn Diagnosis 335 ● Wiper System Service 339 ● Windshield Washer System Service 348 ● Blower Motor Service 349 ● Electric Defogger Diagnosis and Service 352 ● Power Window Diagnosis 355 ● Power Seat Diagnosis 357 ● Power Door Lock Diagnosis 358 ● Terms to Know 359 ● ASE-Style Review Questions 359 ● ASE Challenge Questions 360 ● Job Sheets 361

| CHAPTER 10 | **Body Computer System Diagnosis** | **375** |

Introduction 375 ● Electronic Service Precautions 377 ● Diagnosis of Computer Voltage Supply and Ground Circuits 378 ● Trouble Codes 383 ● Visual Inspection 383 ● Entering Diagnostics 384 ● Testing Actuators 387 ● Testing Sensors 389 ● PROM Replacement 394 ● Flashing the BCM 396 ● Terms to Know 398 ● ASE-Style Review Questions 398 ● ASE Challenge Questions 399 ● Job Sheets 401

| CHAPTER 11 | **Vehicle Multiplexing Diagnostics** | **409** |

Introduction 409 ● Communication Fault Codes 410 ● ISO 9141-2 Bus System Diagnostics 410 ● ISO-K Bus System Diagnostics 412 ● Class A Bus System Diagnostics 413 ● J1850 Bus System Diagnostics 415 ● Controller Area Network Bus Diagnostics 418 ● Local Interconnect Network Bus Diagnostics 422 ● Enhanced Scan Tool Diagnostics 423 ● Terms to Know 425 ● ASE-Style Review Questions 425 ● ASE Challenge Questions 426 ● Job Sheets 427

CHAPTER 12

Advanced Lighting Systems Diagnosis and Service   **451**

Introduction 451 • Computer-Controlled Concealed Headlight Diagnosis 451 • Testing Computer-Controlled Headlight Systems 456 • Automatic Headlight System Diagnosis 458 • Illuminated Entry System Diagnosis 472 • Fiber-Optics Diagnosis 477 • Terms to Know 477 • ASE-Style Review Questions 477 • ASE Challenge Questions 479 • Job Sheets 481

CHAPTER 13

Instrumentation and Warning Lamp System Diagnosis and Repair   **491**

Introduction 491 • Instrument Panel and Printed Circuit Removal 492 • BCM Diagnostics 493 • Self-Diagnostics 493 • Speedometer Diagnosis and Repair 499 • Tachometer 504 • Conventional Instrument Cluster Gauge Diagnosis 506 • Electronic Gauge Diagnosis 509 • Gauge Sending Units 510 • Warning Lamps 512 • Trip Computers 514 • Terms to Know 515 • ASE-Style Review Questions 515 • ASE Challenge Questions 516 • Job Sheets 517

CHAPTER 14

Diagnosis of Electronic Chassis and Accessory Systems   **527**

Introduction 527 • Automatic Temperature Control System Diagnosis 528 • Diagnosis and Service of Electronic Cruise Control Systems 544 • Memory Seat Diagnosis 556 • Climate-Controlled Seat Diagnosis and Service 560 • Electronic Sunroof Diagnosis 561 • Antitheft System Troubleshooting 562 • Immobilizer System Service 565 • Automatic Door Lock System Troubleshooting 567 • Remote Keyless Entry Diagnosis 569 • Electronically Heated Windshield Service 570 • Speed-Sensitive Steering Diagnostic 574 • Antilock Brake Service 575 • Audio and Video Entertainment 583 • Terms to Know 586 • ASE-Style Review Questions 586 • ASE Challenge Questions 587 • Job Sheets 589

CHAPTER 15 **Servicing Passive Restraint Systems** **611**

Introduction 611 ● Automatic Seat Belt Service 611 ● Air Bag Safety and Service Warnings 616 ● Diagnostic System Check 617 ● Retrieving Fault Codes 618 ● Air Bag System Testing 621 ● Inspection After an Accident 623 ● Cleanup Procedure After Deployment 623 ● Component Replacement 623 ● Occupant Classification System Service 626 ● Terms to Know 631 ● ASE-Style Review Questions 631 ● ASE Challenge Questions 632 ● Job Sheets 633

CHAPTER 16 **Hybrid and High-Voltage System Service** **645**

Introduction 645 ● Safety Precautions 645 ● High-Voltage Service Plug 647 ● Self-Diagnosis Capabilities 648 ● High-Voltage Battery Service 650 ● Inverter/Converter Assembly 652 ● System Main Relay 654 ● Terms to Know 655 ● ASE-Style Review Questions 655 ● ASE Challenge Questions 656 ● Job Sheets 657

Appendix A **ASE Practice Examination** **663**

Appendix B **Metric Conversions** **681**

Appendix C **Electrical and Electronic Special Tool Suppliers** **682**

Appendix D **J1930 Terminology List** **683**

Glossary **689**

Index **702**

Photo Sequences

1. Using a Dry Chemical Fire Extinguisher **18**

2. Performing a Voltage Drop Test **41**

3. Identifying Bipolar Transistors **102**

4. Voltage Drop Test to Locate High Circuit Resistance **110**

5. Soldering Copper Wire **137**

6. Performing a Battery Capacity Test **186**

7. Removing and Cleaning the Battery **191**

8. Typical Procedure for Delco Remy Starter Disassembly **228**

9. Typical Procedure for Disassembly of Gear Reduction Starter **230**

10. Performing the Current Output Test **257**

11. Typical Procedure for IAR Generator Disassembly **269**

12. Replacing a Sealed-Beam Headlight **289**

13. Removal and Testing of the Multifunction Switch **305**

14. Wiper Motor Removal **347**

15. Typical Procedure for Grid Wire Repair **354**

16. Typical Procedure for Scan Tester Diagnosis **385**

17. Typical Procedure for Replacing the PROM Chip **395**

18. Flashing a BCM **397**

19. Advanced Scan Tool Functions **424**

20. Typical Procedure for Replacing a Photocell Assembly **461**

21. Automatic High-Beam Camera Calibration **467**

22. Bench Testing the Fuel Level Sender Unit **511**

23. Typical Procedure for Replacing the Cruise Control Servo Assembly **537**

24. Typical Procedure for Testing the Antitheft Relay **566**

25. Typical Procedure for Pump and Motor Assembly Replacement **581**

26. Typical Procedure for Removing the Air Bag Module **624**

27. Occupant Classification Validation Use '05 Durango, 300 Liberty, or Chrysler Minivan **629**

Job Sheets

1. Shop Safety Survey **29**

2. Fire Hazard Inspection **31**

3. Using Ohm's Law to Calculate Electrical Properties **73**

4. Meter Symbol Interpretation **79**

5. Using a Test Light **81**

6. Use of a Voltmeter **83**

7. Use of a Ohmmeter **85**

8. Using a DSO on Sensors and Switches **119**

9. Testing Circuit Protection Devices **121**

10. Testing Switches **123**

11. Testing for an Open Circuit **125**

12. Testing for a Short to Ground **127**

13. Testing a Diode **129**

14. Part Identification on a Wiring Diagram **153**

15. Using a Component Locator **155**

16. Using a Wiring Diagram **157**

17. Soldering Copper Wires **167**

18. Inspecting and Cleaning a Battery **199**

19. Testing the Battery's Capacity **201**

20. Three-Minute Charge Test **203**

21. Removing, Cleaning, and Installing a Battery **205**

22. Current Draw Test **241**

23. Testing the Starting System Circuit **243**

24. Starter Free-Speed Testing **245**

25. Inspecting Drive Belts **279**

26. Testing Charging System Output **281**

27. Testing the Charging System Circuit **283**

28. Diagnosing Lighting Systems **321**

29. Checking a Headlight Switch **323**

30. Headlight Aiming **325**

31. Diagnosing a Windshield Wiper Circuit **361**

32. Testing the Windshield Washer Circuit **363**

33. Testing a Rear-Window Defogger **365**

34. Using a Scan Tool **401**

35. Using a Lab Scope to Tool Sensors and Switches **403**

36. Testing an Ambient Temperature Sensor **405**

37. Flashing a BCM **407**

38. ISO 9141-2 Bus System Diagnosis **427**

39. ISO-K Bus System Diagnosis **431**

40. J1850 Bus System Diagnosis **433**

41. CAN Bus System Diagnosis **439**

42. LIN Bus System Diagnosis **445**

43. Testing an Automatic Headlight System **481**

44. Testing the BCM-Controlled Headlight System **483**

45. Diagnosing Automatic High-Beam Systems **487**

46. Checking a Fuel Gauge **517**

47. Testing an Oil Pressure Warning Light Circuit **519**

48. Testing the Electronic Instrument Cluster **521**

49. ATC Diagnostics **589**

50. Testing the Electronic Cruise Control System **591**

51. Testing the Antitheft System Operation **593**

52. Testing an ABS Wheel Speed Sensor **595**

53. Programming an Immobilizer Key **597**

54. Working Safely Around Air Bags **633**

55. Diagnosing "Driver Squib Circuit Open" Fault **635**

56. Clockspring Replacement and Centering **639**

57. Occupant Classification System Validation Procedure **641**

58. Hybrid Safety **657**

59. Hybrid System DTCs **659**

PREFACE

Thanks to the support the Today's Technician™ series has received from those who teach automotive technology, Thomson Delmar Learning, the leader in automotive-related textbooks, is able to live up to its promise to provide new editions of the series regularly. We have listened and responded to our critics and our fans and present this new updated and revised fourth edition. By revising our series regularly, we can and will respond to changes in the industry, changes in technology, changes in the certification process, and to the ever-changing needs of those who teach automotive technology.

The Today's Technician™ series, by Thomson Delmar Learning, features textbooks that cover all mechanical and electrical systems of automobiles and light trucks (whereas the Heavy-duty Trucks portion of the series does the same for heavy-duty vehicles). Principally, the individual titles correspond to the main areas of ASE (National Institute for Automotive Service Excellence) certification. Additional titles include remedial skills and theories common to all of the certification areas and advanced or specific subject areas that reflect the latest technological trends. Each text is divided into two volumes: a Classroom Manual and a Shop Manual.

Unlike yesterday's mechanic, the technician of today and for the future must know the underlying theory of all automotive systems and be able to service and maintain those systems. Dividing the material into two volumes provides the reader with the information needed to begin a successful career as an automotive technician without interrupting the learning process by mixing cognitive and performance learning objectives into one volume.

The design of Thomson Delmar Learning's Today's Technician™ series was based on features that are known to promote improved student learning. The design was further enhanced by a careful study of survey results, in which the respondents were asked to value particular features. Some of these features can be found in other textbooks, whereas others are unique to this series.

Each Classroom Manual contains the principles of operation for each system and subsystem. The Classroom Manual also contains discussions on design variations of key components used by the different vehicle manufacturers. This volume is organized to build on basic facts and theories. The primary objective of this volume is to allow the reader to gain an understanding of how each system and subsystem operates. This understanding is necessary to diagnose the complex automobiles of today and tomorrow. Although the basics contained in the Classroom Manual provide the knowledge needed for diagnostics, diagnostic procedures appear only in the Shop Manual. An understanding of the basics is also a requirement for competence in the skill areas covered in the Shop Manual.

A spiral-bound Shop Manual covers the "how-to's." This volume includes step-by-step instructions for diagnostic and repair procedures. Photo Sequences are used to illustrate some of the common service procedures. Other common procedures are listed and are accompanied with line drawings and photos that allow the reader to visualize and conceptualize the finest details of the procedure. This volume also contains the reasons for performing the procedures as well as when that particular service is appropriate.

The two volumes are designed to be used together and are arranged in corresponding chapters. Not only are the chapters in the volumes linked together, but also the contents of the chapters are linked. This linking of content is evidenced by marginal callouts that refer the reader to the chapter and page in which that the same topic is addressed in the other volume. This feature is valuable to instructors. Users of other two-volume textbooks without this feature must search the index or table of contents to locate supporting information in the other volume. This is not only cumbersome but also creates additional work for an instructor when planning the presentation of material and when making reading assignments. It is also valuable to the students; with page references, they also know exactly where to look for supportive information.

Both volumes contain clear and thoughtfully selected illustrations, many of which are original drawings or photos specially prepared for inclusion in this series. This means that the art is a vital part of each textbook and not merely inserted to increase the numbers of illustrations.

The page layout used in the series is designed to include information that would otherwise break up the flow of information presented to the reader. The main body of the text includes all of the "need-to-know" information and illustrations. In the wide side margins of each page are many of the special features of the series. Items that are truly "nice-to-know" information include simple examples of concepts just introduced in the text, explanations or definitions of terms that will not be defined in the glossary, examples of common trade jargon used to describe a part or operation, and exceptions to the norm explained in the text. Many textbooks attempt to include this type of information and insert it in the main body of text; this tends to interrupt the thought process and cannot be pedagogically justified. By placing this information off to the side of the main text, the reader can select when to refer to it.

Classroom Manual

Features of this manual include:

Cognitive Objectives
These objectives define the contents of the chapter and define what the student should have learned upon completion of the chapter.
Each topic is divided into small units to promote easier understanding and learning.

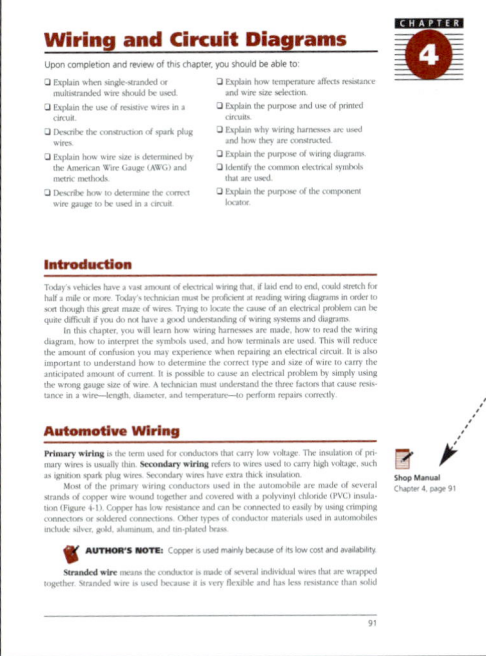

Cross-References to the Shop Manual
Reference to the appropriate page in the Shop Manual is given whenever necessary. Although the chapters of the two manuals are synchronized, material covered in other chapters of the Shop Manual may be fundamental to the topic discussed in the Classroom Manual.

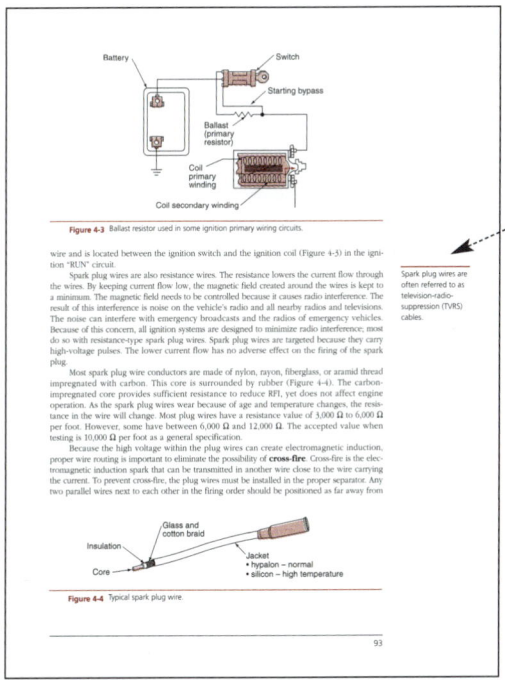

Marginal Notes
These notes add "nice-to-know" information to the discussion. They may include examples or exceptions, or may give the common trade jargon for a component.

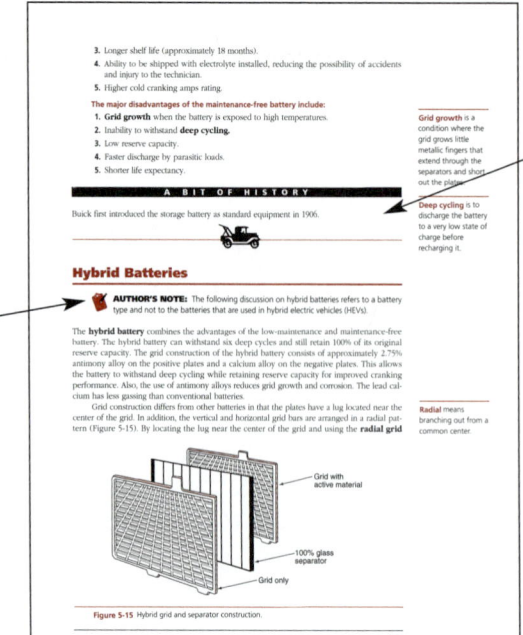

A Bit of History

This feature gives the student a sense of the evolution of the automobile. This feature not only contains nice-to-know information, but also should spark some interest in the subject matter.

Author's Notes

This feature includes simple explanations, stories, or examples of complex topics. These are included to help students understand difficult concepts.

Summaries

Each chapter concludes with a summary of key points from the chapter. These are designed to help the reader review the chapter contents.

Terms to Know List

A list of new terms appears next to the Summary.

Review Questions

Short-answer essay, fill-in-the-blank, and multiple-choice questions are found at the end of each chapter. These questions are designed to accurately assess the student's competence in the stated objectives at the beginning of the chapter.

Shop Manual

To stress the importance of safe work habits, the Shop Manual dedicates one full chapter to safety. Other important features of this manual include:

Performance-Based Objectives

These objectives define the contents of the chapter and define what the student should have learned upon completion of the chapter. These objectives also correspond with the list of required tasks for ASE certification. *Each ASE task is addressed.*

Although this textbook is not designed to simply prepare someone for the certification exams, it is organized around the ASE task list. These tasks are defined generically when the procedure is commonly followed and specifically when the procedure is unique for specific vehicle models. Imported- and domestic-model automobiles and light trucks are included in the procedures.

Marginal Notes

These notes add "nice-to-know" information to the discussion. They may include examples or exceptions, or may give the common trade jargon for a component.

Special Tools List

Whenever a special tool is required to complete a task, it is listed in the margin next to the procedure.

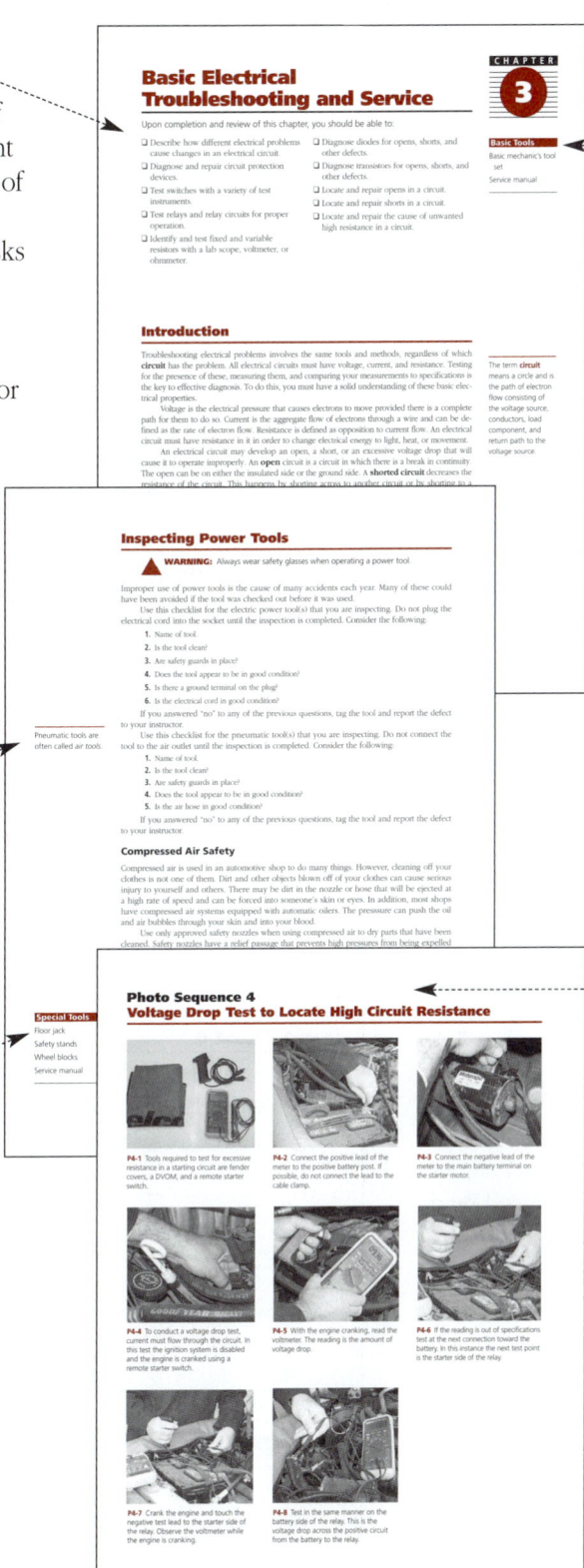

Basic Tools List

Each chapter begins with a list of the basic tools needed to perform the tasks included in the chapter.

Photo Sequences

Many procedures are illustrated in detailed Photo Sequences. These detailed photographs show the students what to expect when they perform particular procedures. They also can provide the student a familiarity with a system or type of equipment, which the school may not have.

Cautions and Warnings

Throughout the text, warnings are given to alert the reader to potentially hazardous materials or unsafe conditions. Cautions are given to advise the student of things that can go wrong if instructions are not followed or if a nonacceptable part or tool is used.

Service Tips

Whenever a short-cut or special procedure is appropriate, it is described in the text. These tips are generally those things commonly done by experienced technicians.

Cross-References to the Classroom Manual

Reference to the appropriate page in the Classroom Manual is given whenever necessary. Although the chapters of the two manuals are synchronized, material covered in other chapters of the Classroom Manual may be fundamental to the topic discussed in the Shop Manual.

Customer Care

This feature highlights those little things a technician can do or say to enhance customer relations.

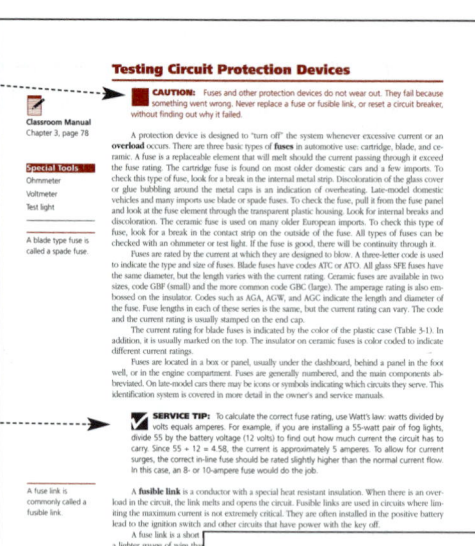

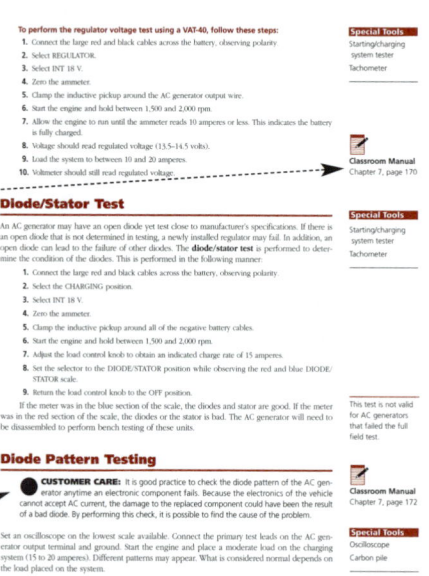

Job Sheets

Located at the end of each chapter, the Job Sheets provide a format for students to perform procedures covered in the chapter. A reference to the ASE task addressed by the procedure is included on the Job Sheet.

Case Studies

Case Studies concentrate on the ability to properly diagnose the systems. Beginning with Chapter 3, each chapter ends with a case study in which a vehicle has a problem, and the logic used by a technician to solve the problem is explained.

Terms to Know List

Terms in this list can be found in the Glossary at the end of the manual.

ASE-Style Review Questions

Each chapter contains ASE-style review questions that reflect the performance-based objectives listed at the beginning of the chapter. These questions can be used to review the chapter as well as to prepare for the ASE certification exam.

ASE Challenge Questions

Each technical chapter ends with five ASE challenge questions. These are not more review questions; rather, they test the students' ability to apply general knowledge to the contents of the chapter.

ASE Practice Examination

A 50-question ASE practice exam, located in Appendix A, is included to test students on the contents of the Shop Manual.

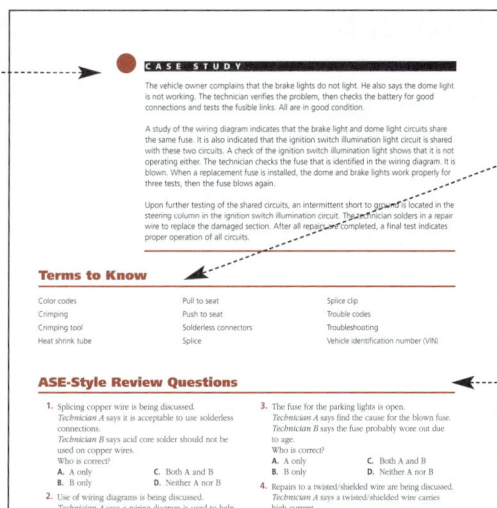

Reviewers

The author and publisher would like to extend a special thanks to the following instructors for their contributions to this text:

Steve Bertram
Palomar College, San Marcos, CA

Terry Enyart
University of Northwestern Ohio, Lima, OH

C. Neel Flannagan
Aiken Technical College, Graniteville, SC

Robert Gibbens
North Central Kansas Technical College, Beloit, KS

Chris Hadfield
College of Lake County, Grayslake, IL

Don Lumsdon
Ivy Tech State College, Terre Haute, IN

Kent McCleary
University of Northwestern Ohio, Lima, OH

Dick Rogers
Lincoln Land Community College, Springfield, IL

Shane Sampson
Western Iowa Tech Community College, Sioux City, IA

Dan Wilson
Westwood College, Denver, CO

John Wood
Ranken Technical College, St. Louis, MO

Safety

Upon completion and review of this chapter, you should be able to:

❑ Explain how safety is a part of professionalism.

❑ List and describe personal safety responsibilities.

❑ List the different types of eye protection devices and explain the proper application of each.

❑ Lift heavy objects properly.

❑ Inspect power tools before use.

❑ Raise a vehicle using a floor jack and safety stands.

❑ Raise a vehicle using a hoist.

❑ Demonstrate the ability to properly run the engine in the shop.

❑ Classify fires and fire extinguishers.

❑ Locate, identify, and inspect fire extinguishers in the shop.

❑ Explain the proper use of the fire extinguisher.

❑ Define hazardous materials.

❑ Explain the right-to-know law or workplace hazardous materials information systems (WHMIS).

❑ Describe the responsibilities of the employer and the employee concerning hazardous materials.

❑ Determine what constitutes hazardous waste and how to properly dispose of it.

❑ Describe the basic safety rules of servicing electrical systems.

❑ Work around batteries safely.

❑ Explain the safety precautions associated with charging and starting systems.

❑ List the safety precautions associated with servicing the air bag system.

❑ Explain the safety precautions that are necessary when servicing the antilock brake system.

Introduction

Being a professional technician is more than being knowledgeable about automotive systems, it is also an attitude. Being a professional technician includes having an understanding of all the hazards that may exist in the workplace. One of the most obvious traits of a professional is the ability to work productively and safely. This is where knowledge becomes very important. You need it to be productive and you need it to ensure your own safety and the safety of others. This chapter discusses the safety concerns associated with working in an automotive repair shop and the safety concerns associated with the vehicle's fuel and emission systems. In addition to basic shop safety, working on the vehicle's fuel and emission systems presents many special concerns.

Safety is everyone's business. However, never assume the person working next to you is as conscientious as you are. You must be aware of what is going on around you at all times. As a professional technician, you must perform your work in a manner that protects not only you but others in the workplace as well.

Personal Safety

Personal safety encompasses all aspects of preventing injury, including awareness, attitudes, and dress. All three of these are manifested through neat work habits. Cleaning up spills, keeping tools clean, and organizing the tools and materials in the shop all help to prevent accidents. Rushing to complete a job may result in a lack of consideration for personal safety and may ultimately cause an accident. Taking time to be neat and safe is rewarded by fewer accidents, higher customer satisfaction, and better pay.

Dress and Appearance

Nothing displays professional pride and a positive attitude more than the way you dress (Figure 1-1). Customers demand a professional atmosphere in the service shop. Your appearance instills customer confidence, as well as expresses your attitude toward safety. Wearing proper and neat clothing can prevent injuries.

Loose-fitting clothing, or clothing that hangs out freely, can cause serious injury. Long-sleeve shirts should have their cuffs buttoned or rolled up tightly. Shirttails should be tucked in at all times. Some job positions within an automotive repair facility may require the employee to wear a necktie. If a necktie is worn in the shop area, it should be tucked inside your shirt. Clip-on ties are recommended if you must wear a tie.

Long hair is a serious safety concern. Very serious injury can result if hair becomes caught in rotating machinery, fan belts, or fans. If your hair is long enough to touch the bottom of your shirt collar, it should be tied back and tucked under a hat.

Jewelry has no place in the automotive shop. Rings, watches, bracelets, necklaces, earrings, and so forth can cause serious inj. . The gold, silver, and other metals used in jewelry are excellent **conductors** of electricity. Your body is also a good conductor. When electrical current flows through a conductor, it generates heat. The heat can be great enough to cause severe burns. Jewelry can also become caught in moving parts, causing serious cuts. Necklaces can cause serious injury or even death if they become caught in moving equipment.

You should wear shoes or b. that will protect your feet in the event that something falls or you stumble into something. It is a good idea to wear safety shoes or boots with steel toes and shanks. Most safety shoes also have slip-resistant soles. Tennis and jogging shoes provide little protection and are not satisfactory footwear in the automotive shop.

Figure 1-1 Wearing clean and properly fitting clothes is an indication of how serious you are about safety.

Smoking, Alcohol, and Drugs in the Shop

Due to the potential hazards, never smoke when working in the shop. A spark from a cigarette or lighter may ignite flammable materials in the workplace. If the shop has designated smoking areas, smoke only in those areas. As a courtesy to your customers, do not smoke in their vehicles. Nonsmokers may not appreciate cigarette odor in their vehicles.

The use of drugs or alcohol must be avoided while working in the shop. Even a small amount of drugs or alcohol affects reaction time. In an emergency situation, slow reaction time may cause personal injury. If a heavy object falls of the workbench and your reaction time is slowed by drugs or alcohol, you may not get out of the way quickly enough to prevent injury. Also you are a hazard to your coworkers if you are not performing at your best.

Eye Protection

The importance of wearing proper eye protection cannot be overemphasized. Every working day there are more than 1,000 eye injuries. Many of these injuries result in blindness. Almost all of these are preventable. The safest and surest way to protect your eyes is to wear the proper eye protection any time you enter the shop. At a minimum, wear eye protection when grinding, using power tools, hammering, cutting, chiseling, or performing service under the car. In addition, wear eye protection when doing any work that can cause sparks, dirt, or rust to enter your eyes, and when you are working around chemicals. Remember, just because you are not doing the work yourself does not mean you cannot suffer an eye injury. Many eye injuries are caused by a coworker. Wear eye protection any time you are near an eye hazard.

The key to protecting your eyes is the use of *proper* eye protection. Prescription glasses do not provide adequate protection. Regular glasses are designed to impact standards that are far below those that are required in the workplace. The lens may stop a flying object, but the frame many allow the lens to pop out and hit your face, causing injury. In addition, regular glasses do not provide side protection.

There are many types of eye protection (Figure 1-2). One of the best ways to protect your eyes is to wear **occupational safety glasses.** These glasses are light and comfortable. They are constructed of tempered glass or safety plastic lens and have frames that prevent the lens from being pushed out upon impact. They have side shields to prevent the entry of objects from the side. Occupational safety glasses are available in prescription lens so they can be worn instead of regular corrective lens glasses.

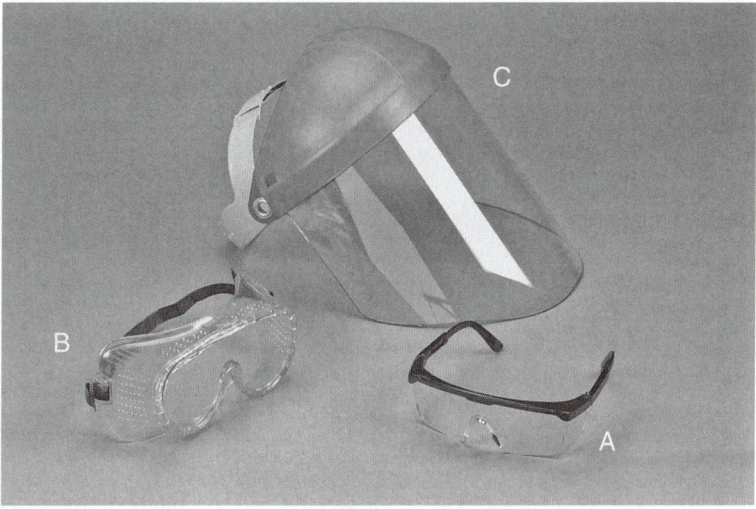

Figure 1-2 Different types of eye protection: (A) safety glasses with side shields; (B) safety goggles that may be worn over prescription glasses; and (C) a face shield that is worn over safety glasses or goggles and protects the face.

Safety goggles fit snugly around the area of your eyes to prevent the entry of objects and to provide protection from liquid splashes. The force of impact on the lens is distributed throughout the entire area where the safety goggles are in contact with your face and forehead. Safety goggles are designed to fit over regular glasses.

Face shields are clear plastic shields that protect the entire face. They are used when there is potential for sparks, flying objects, or splashed liquids, which can cause neck, facial, and eye injuries. The plastic is not as strong or impact resistant as occupational safety glasses or safety goggles. If there is a danger of high-impact objects hitting the face shield, it is a good practice to wear safety glasses under the face shield.

Safety glasses provide little or no protection against chemicals. When working with chemicals, such as battery acid, refrigerants, cleaning solutions, and so forth, safety goggles should be worn. Full-face shields are not intended to provided primary protection for your eyes. They are designed to provide primary protection for your face and neck and should be worn in addition to eye protection.

Before removing your eye protection, close your eyes. Pieces of metal, dirt, or other foreign material may have accumulated on the outside. These could fall into your eyes when you remove your glasses or shield.

Eyewash Fountains

Eye injuries may occur in various ways in an automotive shop. The following are some common types of eye injuries:

1. Thermal burns from excessive heat.
2. Irradiation burns from excessive light, such as from an arc welder.
3. Chemical burns from strong liquids such as gasoline or battery electrolyte.
4. Foreign material in the eye.
5. Penetration of the eye by a sharp object.
6. A blow from a blunt object.

Wearing safety glasses and observing shop safety rules will prevent most eye accidents. If a chemical gets in your eyes, it must be washed out immediately to prevent a chemical burn. An eyewash fountain is the most effective way to wash the eyes, and every shop should be equipped with some eyewash facility (Figure 1-3). Be sure you know the location of the eyewash fountain in the shop.

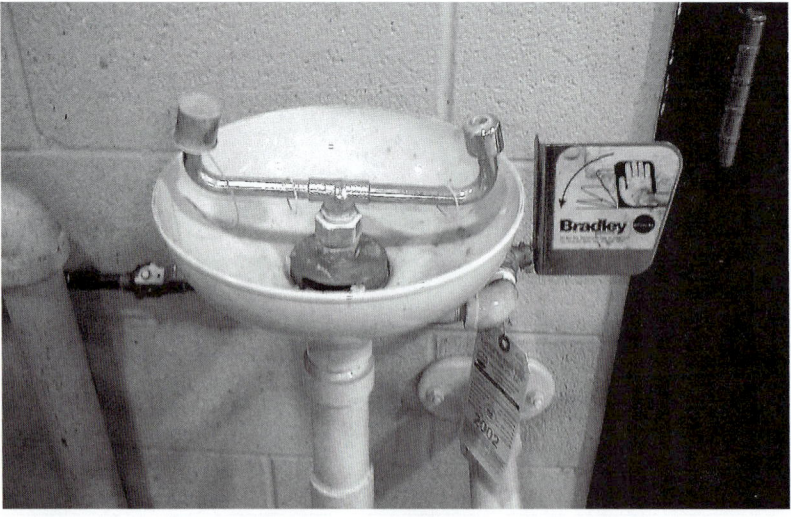

Figure 1-3 An eyewash fountain is used to remove chemicals and dirt from the eyes.

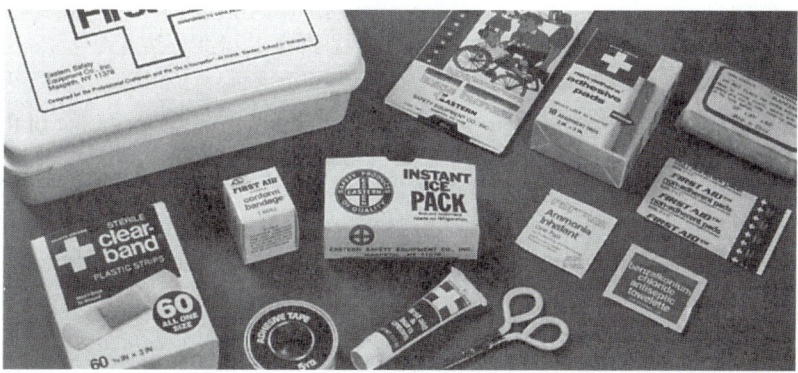

Figure 1-4 First-aid kit.

First-aid Kits

First-aid kits should be clearly identified and conveniently located (Figure 1-4). These kits contain such items as bandages and ointment required for minor cuts. All shop personnel must be familiar with the location of first-aid kits. At least one of the shop personnel should have basic first-aid training, and this person should be in charge of administering first aid and keeping first-aid kits filled.

Hand Protection

Good hand protection is often overlooked. A scrape, cut, or burn can seriously impair your ability to work for many days. A well-fitting pair of heavy work gloves should be worn while grinding, welding, or when handling chemicals or high-temperature components. Special rubber gloves are recommended for handling **caustic** chemicals. Caustic chemicals have the ability to destroy or eat through something and are considered extremely corrosive.

Many technicians wear latex, vinyl, or nitrile gloves to help protect their hands and to keep them clean. These are similar to the type of gloves worn by doctors and dentists during examinations. Latex gloves are inexpensive and provide good hand protection, however some people are allergic to latex. If you wear latex gloves and develop a rash or redness on your hands, discontinue use. Vinyl gloves are also available and provide good resistance to tears and many nonaggressive liquids. Also, vinyl gloves are latex-free, so those who are allergic to latex can wear them. At a higher cost, nitrile gloves are latex-free synthetic rubber gloves that are superior to latex or vinyl in puncture resistance. In addition, nitrile gloves resist a wide range of chemicals that are harmful to either latex or vinyl.

Latex, vinyl, or nitrile gloves should be worn if you have an open cut or other injury on your hand, to prevent infection and the spread of diseases. In addition, these gloves should be worn if you are required to render first aid or medical assistance to someone who is injured. Because of the serious nature of blood-borne pathogens (disease- and infection-causing microorganisms carried by blood and other potentially infectious materials), it is important that you take every precaution to protect yourself regardless of the perceived status of the individual you are assisting. In other words, whether or not you think the blood/body fluid is infected with blood-borne pathogens, you treat it as if it is.

Rotating Belts and Pulleys

Many times the technician must work around rotating parts such as generators, power steering pumps, air pumps, water pumps, and air conditioner (A/C) compressors. Other rotating equipment or components of concern include tire changers, spin balancers, drills, bench grinders, and drive shafts. Always think before acting. Be aware of where you are placing your hands and fingers at all times. Do not place rags, tools, or test equipment near moving parts. In addition, make sure you are not wearing any loose clothing or jewelry that can become caught.

Electric Cooling Fans

Be very cautious around electric cooling fans. Some of these fans will operate even if the ignition switch is turned off. They are controlled by a temperature-sensing unit in the engine block or radiator and may turn on any time the coolant temperature reaches a certain temperature. Before working on or around an electric cooling fan, you should become familiar with its operation, and, if necessary, you should disconnect the electrical connector to the fan motor or the negative battery cable.

Lifting

Back injuries are one of the most crippling injuries in the industry, yet most of them are preventable. Most occupational back injuries are caused by improper lifting practices. These injuries can be avoided by following a few simple lifting guidelines:

1. Do not lift a heavy object by yourself. Seek help from someone else.

2. Do not lift more than you can handle. If the object is too heavy, use the proper equipment to lift it.

3. Do not attempt to lift an object if there is not a good way to hold onto it. Study the object to determine the best balance and grip points.

4. Do not lift with your back. Your legs have some of the strongest muscles in your body. Use them.

5. Place your body close to the object. Keep your back and elbows straight (Figure 1-5).

6. Make sure you have a good grip on the object. Do not attempt to readjust the load once you have lifted it. If you are not comfortable with your balance and grip, lower the object and reposition yourself.

7. When lifting, keep the object as close to your body as possible. Keep your back straight and lift with your legs.

8. While carrying the object, do not twist your body to change directions. Use your feet to turn your whole body in the new direction.

9. To set the load down, keep the object close to your body. Bend at the knees and keep your back straight. Do not bend forward or twist.

10. If you need to place the object onto a shelf or benchtop, place an edge of the object on the surface and slide it into place. Do not lean forward.

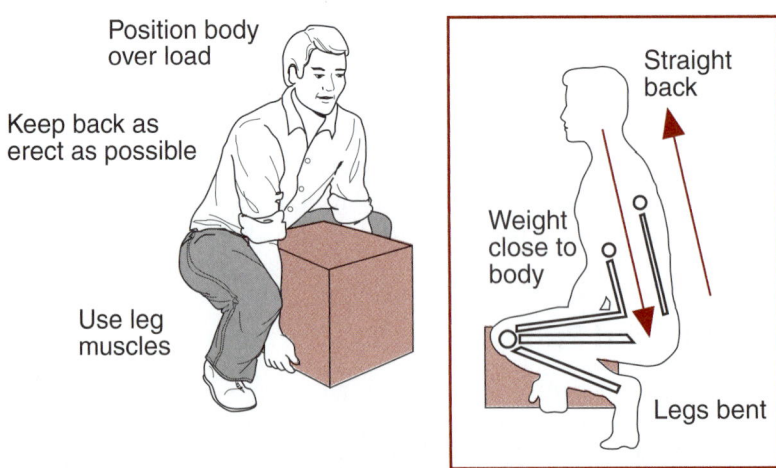

Figure 1-5 When lifting a heavy object, keep your back straight and lift with your legs.

Tool and Equipment Safety

Technicians would not be able to do their job without tools and equipment. Most injuries caused by tools and equipment are the result of improper use, improper maintenance, and/or carelessness.

Hand Tools

Hand tools use only the force generated from the body to operate. They multiply the force received through leverage to accomplish the work. Here are some very simple steps that you can take to help assure safe hand tool use:

1. Do not use tools that are worn out or broken.
2. Do not use a tool to do something that it was not designed to do. Use the proper tool for the job.
3. Keep your tools clean and in good condition.
4. Point sharp edges of tools away from yourself.
5. Do not hold small components in your hands while using tools such as screwdrivers. The tool may slip and cause injury to your hand.
6. Examine your work area for things that can cause injury if a tool slips or a fastener breaks loose quickly. Readjust yourself or the tool to avoid them.
7. Do not put sharp tools in your pockets.

Power Tool Safety

 WARNING: Always wear safety glasses when operating a power tool.

Power tools use forces other than those generated by the body. They can use compressed air, electricity, or hydraulic pressure to generate and multiply force. Many times a technician will be required to use power tools when performing electrical service. Drills and hole saws will be used to install new accessories onto the vehicle or to drill holes for wiring to pass through. Grinders, drill presses, and hydraulic presses may be used to help fabricate or modify components. **Pneumatic tools** are powered by compressed air and are used to remove or fasten components. All of these tools can cause injury if not used properly. Use the following guidelines when working with power tools:

Pneumatic tools are often called *air tools*.

1. Ask your instructor if you are not sure of the correct operation of a tool.
2. Always wear proper eye protection when using power tools.
3. Check that all safety guards and safety equipment are installed on the tool.
4. Before using an electrical tool, check the condition of the plug and cord. The plug should be a three-prong plug. Never cut off the grounding prong. Do not use the tool if the wires are frayed or broken. Plug the tool only into a grounded receptacle.
5. Before using an air tool, check the condition of the air hose. Do not use the tool if the hose shows signs of weakness such as bulges or fraying. Also, the tool should be properly oiled.
6. Before using a hydraulic tool, check the condition of all hoses and gauges. Do not use the tool if any of these are defective.
7. Make sure other people are not in the area when you turn the tool on.
8. Do not leave the area with the tool still running. Stay with the tool until it stops. Then disconnect it.
9. Make all adjustments to the tool before turning it on.
10. If the tool is defective or does not pass your safety inspection, put a sign on it and report it to your supervisor.

Inspecting Power Tools

Improper use of power tools is the cause of many accidents each year. Many of these could have been avoided if the tool was checked out before it was used.

Use this checklist for the electric power tool(s) that you are inspecting. Do not plug the electrical cord into the socket until the inspection is completed. Consider the following:

1. Name of tool.
2. Is the tool clean?
3. Are safety guards in place?
4. Does the tool appear to be in good condition?
5. Is there a ground terminal on the plug?
6. Is the electrical cord in good condition?

If you answered "no" to any of the previous questions, tag the tool and report the defect to your instructor.

Use this checklist for the pneumatic tool(s) that you are inspecting. Do not connect the tool to the air outlet until the inspection is completed. Consider the following:

1. Name of tool.
2. Is the tool clean?
3. Are safety guards in place?
4. Does the tool appear to be in good condition?
5. Is the air hose in good condition?

If you answered "no" to any of the previous questions, tag the tool and report the defect to your instructor.

Compressed Air Safety

Compressed air is used in an automotive shop to do many things. However, cleaning off your clothes is not one of them. Dirt and other objects blown off of your clothes can cause serious injury to yourself and others. There may be dirt in the nozzle or hose that will be ejected at a high rate of speed and can be forced into someone's skin or eyes. In addition, most shops have compressed air systems equipped with automatic oilers. The presssure can push the oil and air bubbles through your skin and into your blood.

Use only approved safety nozzles when using compressed air to dry parts that have been cleaned. Safety nozzles have a relief passage that prevents high pressures from being expelled directly out the front. It is best not to use compressed air to dry parts; however, there are instances where air must be used to dry small passages. In these cases, only use air pressure that has been regulated to about 25 psi.

Check the air hoses for signs of wear. Do not use them if they are bulging, frayed, or if the couplers are damaged.

Lifting the Vehicle

Many service procedures require lifting the vehicle. There are two basic methods of lifting the vehicle from the floor: floor jack and safety stands, and hoists. Each requires the technician to follow certain safety rules to prevent injury and vehicle damage.

Floor Jack and Safety Stand Use

Floor jacks are used to lift the vehicle a short distance off the floor or when only a portion of the vehicle needs to be raised (Figure 1-6). Before using a floor jack, check it for signs of hydraulic fluid leaks and for damage that would compromise its safe use. Before lifting the vehicle, place wheel blocks in front of and behind one of the tires that will remain on the ground.

Many jack manufacturers and service manuals provide illustrations for the proper lift points on a vehicle (Figure 1-7). If this information is not available, always place the floor jack on major strength parts. These areas include the frame, cross member, or differential. If you are in doubt about the proper lift point, ask your instructor. Never lift on sheet metal or plastic parts.

The floor jack is to be used only to lift the vehicle off the floor. It is not intended to support the vehicle while someone is under it. Use **jack stands** to support the vehicle

Jack stands are also referred to as floor stands.

Figure 1-6 Floor jacks are used to raise the vehicle a short distance off the floor.

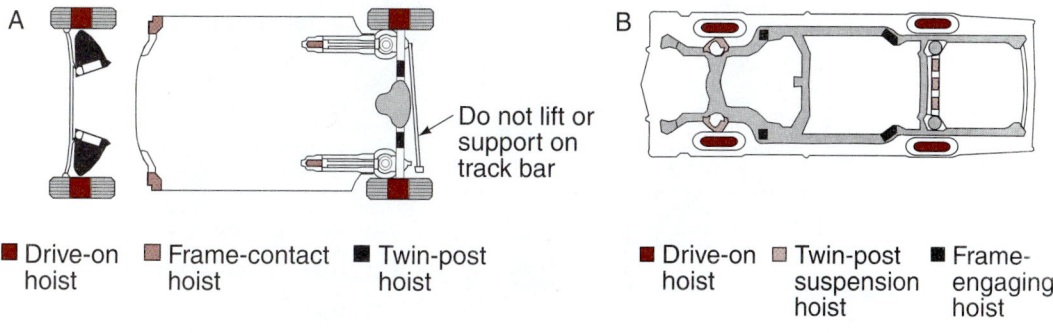

A

Do not lift or support on track bar

■ Drive-on hoist ■ Frame-contact hoist ■ Twin-post hoist

B

■ Drive-on hoist □ Twin-post suspension hoist ■ Frame-engaging hoist

Figure 1-7 Examples of proper lift points for (A) unibody and (B) frame/body vehicles.

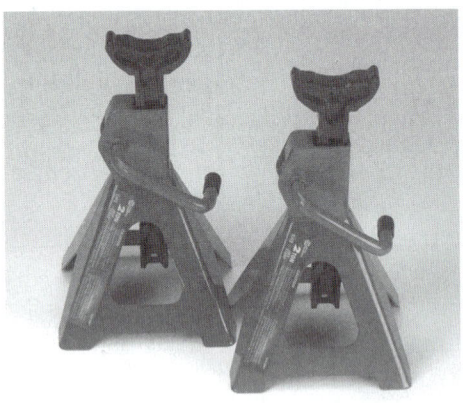

Figure 1-8 Jack stands should be used to support the vehicle after it has been lifted by a floor jack.

Figure 1-9 Support the vehicle by jack stands located at each corner that is lifted. Do not rely on the floor jack to support the vehicle.

(Figure 1-8). Use one jack stand for each quarter of the vehicle that is lifted (Figure 1-9). Place the jack stand under the frame or a major support component of the vehicle. When the vehicle is lowered onto the stands, make sure that they do not tilt.

Before using the floor jack, make sure it has a sufficient rating to lift and sustain the weight of the vehicle. Next, inspect it for proper lubrication and hydraulic fluid level. Check the operation of the jack while looking for signs of hydraulic fluid leaks. If the jack does not pass any one of these inspections, tag it and notify your instructor immediately.

To lift the entire vehicle, begin by placing the vehicle's transmission into PARK. Place it in first gear if the vehicle has a manual transmission. Set the parking brake and place wheel blocks around the rear wheels (Figure 1-10). Position the floor jack under the front of the vehicle at a location strong enough to support the weight. The jack should be centered between the front tires and positioned so that the lift will be straight up and down (Figure 1-11).

Figure 1-10 Before lifting the vehicle with floor jacks, block the wheels to prevent them from rolling.

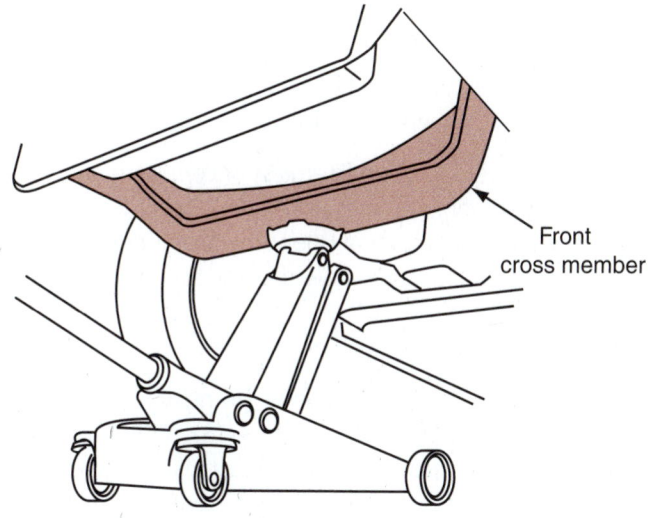

Front
cross member

Figure 1-11 Correct lifting location for raising the front of the vehicle.

Many jack manufacturers and service manuals provide illustrations for the **vehicle lift points** for the type of vehicle being lifted. Vehicle lift points are the areas the manufacturer recommends for safe vehicle lifting. These areas are structurally strong enough to sustain the stress of lifting. If this information is not available, always place the floor jack on major strength parts. These areas include the frame, cross member, or differential. Never lift on sheet metal or plastic parts.

WARNING: If you are lifting only one wheel of the vehicle, be careful not to lift it so high that it can slip off of the jack saddle.

Operate the jack until the jack saddle contacts the vehicle lift point. Check for good contact. If things look good, lift the front of the vehicle a couple of inches off of the floor. Recheck the position of the jack. Continue to check the jack position throughout the lifting procedure. If the vehicle or jack begins to lean, lower the jack and reset it. Lift the vehicle to the required height. Do not lift higher than is necessary.

CAUTION: Never place blocks of wood, and so on, between the vehicle frame and floor jack to obtain additional lift. If additional lift is required, the blocks should be placed under the floor jack.

Do not get under a vehicle that is supported only by a floor jack. Place jack stands under the vehicle in locations that will support the weight. Use two safety stands to support the front of the vehicle. Make sure the safety stands are located where they will not lean or slip. Slowly lower the vehicle onto the stands.

Place the floor jack under the rear of the vehicle (Figure 1-12). Follow the same procedure to raise the rear of the vehicle. Use two safety stands to support the rear of the vehicle.

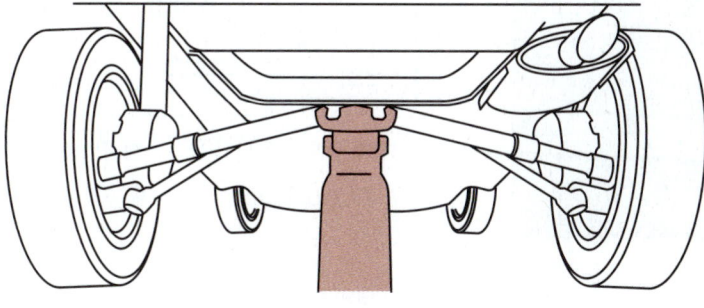

Figure 1-12 Proper lift procedure for raising the rear of the vehicle.

When the vehicle is properly lifted and supported by safety stands, it is safe to work under the vehicle.

Use the same lift points to lower the vehicle. Once one end of the vehicle is on the floor, place wheel blocks around those wheels. Then lower the other end.

CAUTION: Some vehicles with fiberglass or composite body structures may require special lifting considerations. Some of these may include opening the doors, deck lid, and hood before lifting. Others may not allow for opening any doors after the vehicle is lifted. Refer to the service manual and follow the procedures listed.

Hoist Safety

Special Tools

Frame contact hoist

Service manual

Hoists are used when the entire vehicle needs to be raised, usually high enough for the technician to stand underneath the vehicle (Figure 1-13). When a vehicle is placed on the hoist, it must be centered. The balance of the vehicle must be taken into consideration, as well as the effects on balance if a heavy component is removed from the vehicle.

Place the hoist pads under proper lift points of the vehicle. It may be necessary to adjust the pad height in order to lift the vehicle level. Stop raising the vehicle when it is a few inches off the floor to confirm good pad contact. Once the vehicle is raised to the desired height, use the hoist's locking mechanism to prevent accidental lowering of the vehicle.

If the vehicle is not level on the pads, or the pads are not in the proper position, lower the vehicle and readjust as needed. Never get under a vehicle that is not sitting properly on the hoist.

Locate the correct lift points in the service manual for the vehicle you are working on (Figure 1-14). Pay close attention to any warnings or special considerations listed there.

Center the vehicle over the hoist; keep in mind the vehicle's center of gravity and balance point. Locate the hoist pads under the lift points (Figure 1-15). Adjust the pads so that the vehicle will be lifted level.

Lift the vehicle a few inches off the floor. Shake the vehicle while observing for signs of any movement. If the vehicle is not secure on the hoist, or unusual noises are heard while lifting, lower it to the floor and reset the pads.

Once the vehicle is at the desired height, lock the hoist. Do not get under the vehicle until the hoist locks have been set.

Figure 1-13 One type of hoist that may be used in the automotive shop. The hoist is used to raise the entire vehicle.

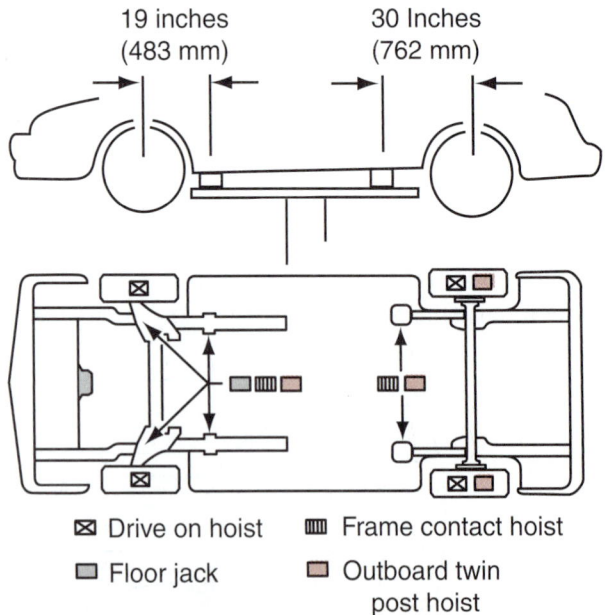

Figure 1-14 Lift point illustrations are usually provided in the service manual.

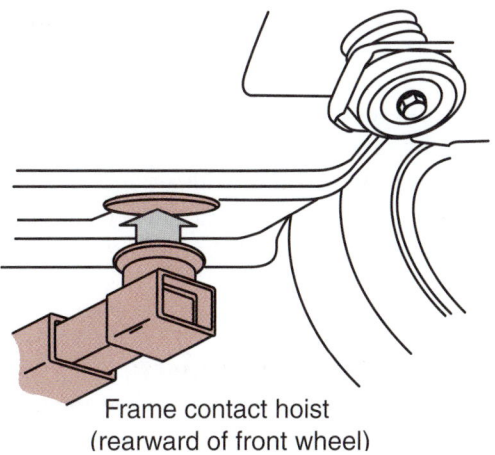

Frame contact hoist
(rearward of front wheel)

Figure 1-15 Locate the hoist pad at the correct lift point for the vehicle you are lifting.

To lower the vehicle, release the locks and put the control valve into the "lower" position. Once it is returned to the floor, push the contact pads out of the path of the tires.

Running the Vehicle While in the Shop

Many times it will be necessary to run the engine while the vehicle is in the shop. This presents the possibility of carbon monoxide poisoning if it is not done safely. **Carbon monoxide** is an odorless, colorless, and toxic gas produced as a result of the combustion process. Most shops are equipped with a ventilation system that will remove the vehicle exhaust. If the shop is not equipped with a ventilation system, a hose must be routed from the vehicle's exhaust to outdoors.

▲ **WARNING:** Some of the early warning signs of carbon monoxide poisoning are headaches, dizziness, and blurred vision. If you experience any of these while in the shop, notify your instructor immediately and get some fresh air. If symptoms persist, seek medical attention.

Check the operation of the shop's ventilation system before connecting to the vehicle. Turn on the ventilation motor and place your hand over the hose (Figure 1-16). You should feel a strong, consistent vacuum. A weak vacuum indicates a restriction or a leak in the ventilation system. If you encounter a problem, notify your instructor.

Connect the ventilation hose to the vehicle's exhaust (Figure 1-17). Place wheel blocks around a tire and place the transmission in PARK or NEUTRAL. Set the parking brake. The engine is now ready to be started.

Figure 1-16 Testing the ventilation system before using it.

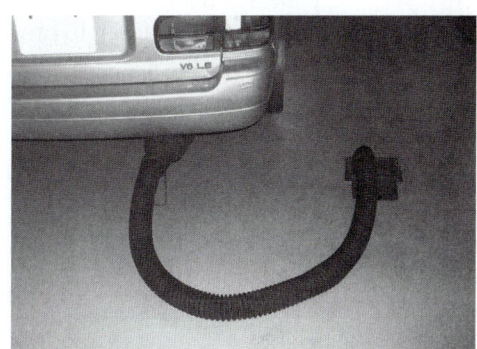

Figure 1-17 Connect the ventilation system hose to the vehicle's exhaust before starting the engine. Make sure the hose is secure and there are no kinks in the hose.

Be aware of the warning signs of carbon monoxide poisoning. Report any symptoms to your instructor immediately.

WARNING: Be careful when working around electric engine cooling fans. These fans are controlled by a thermostat and can come on without warning, even when the engine is not running. Whenever you must work around these fans, disconnect the electrical connector to the fan motor before reaching into the area around the fan. Make sure you re-connect the connector before you return the car to the customer.

Fire Hazards and Prevention

Fires are classified by the types of materials that are involved (Table 1-1). Technicians should be able to locate the correct fire extinguisher to control all the types of fire they are likely to experience (Figure 1-18). Technicians must also be able to fight a fire in an emergency.

Labels on the fire extinguisher will indicate the types of fires that it will put out. Become familiar with the use of a fire extinguisher.

	Class of fire	Typical fuel involved	Type of extinguisher
Class **A** Fires (green)	For ordinary combustibles. Put out a Class A fire by lowering its temperature or by coating the burning combustibles.	Wood Paper Cloth Rubber Plastics Rubbish Upholstery	Multipurpose dry chemical
Class **B** Fires (red)	For flammable liquids. Put out a Class B fire by smothering it. Use an extinguisher that gives a blanketing, flame-interrupting effect; cover whole flaming liquid surface.	Gasoline Oil Grease Paint Lighter fluid	Carbon dioxide Halogenated agent Standard dry chemical Purple K dry chemical Multipurpose dry chemical
Class **C** Fires (blue)	For electrical equipment. Put out a Class C fire by shutting off power as quickly as possible and by always using a nonconducting extinguisher agent to prevent electric shock.	Motors Appliances Wiring Fuse boxes Switchboards	Carbon dioxide Halogenated agent Standard dry chemical Purple K dry chemical Multipurpose dry chemical
Class **D** Fires (yellow)	For combustible metals. Put out a Class D fire of metal chips, turnings, or shavings by smothering or coating with a specially designed extinguisher agent.	Aluminum Magnesium Potassium Sodium Titanium Zirconium	Dry power extinguisher and agents only

Table 1-1 A guide to fire classification and fire extinguisher types.

Figure 1-18 Know the location and types of fire extinguishers that are available in the shop.

Gasoline

Gasoline is so commonly found in automotive repair shops that its dangers are often forgotten. A slight spark or an increase in heat can cause a fire or explosion. Gasoline is a very explosive liquid and is very powerful. One exploding gallon of gasoline has a force equal to fourteen sticks of dynamite. The expanding vapors from gasoline are extremely dangerous, and these vapors are present even in cold temperatures. Gasoline vapors are heavier than air; therefore, when an open container of gasoline is sitting about, the vapors spill out over the sides of the container onto the floor. These fumes are more **flammable** (support combustion) than liquid gasoline and can easily explode.

Never smoke around gasoline since even the droppings of hot ashes can ignite the gasoline. If an engine has a gasoline leak or you have caused a leak by disconnecting a fuel line, stop the leak and clean up the spilled gasoline immediately. While stopping the leak, be extra careful not to cause sparks. If any rags are used in the cleanup of the gasoline, they must be placed in an approved container (Figure 1-19). Due to extreme fire hazards, it is important to immediately clean up any gasoline

Figure 1-19 Dirty rags and towels must be stored in approved containers.

Figure 1-20 Store any type of combustible materials in an approved safety cabinet.

spilled on the floor. Also, many of the compounds in petroleum are toxic, especially if they are in high concentrations.

The chemicals in petroleum that do not evaporate quickly are biodegradable. Optimum degradation occurs if the gasoline is diluted and there is enough air, water, and nutrients for the microbes to "eat up" the chemicals. These properties of gasoline are an advantage in the cleanup and disposal of small spills. Spreading absorbent material such as kitty litter, sand, ground corncobs, straw, sawdust, woodchips, peat, synthetic absorbent pads, or even dirt can stop the flow and soak up the gasoline. Keep in mind that the absorbent does not make gasoline nonflammable.

Brooms can be used to sweep up the absorbent material and put it into buckets, garbage cans, or barrels. Remember to control ignition sources. Be aware of local laws concerning gasoline spills. Some states or municipalities require notification of any gasoline spill larger than 5 gallons.

Gasoline should always be stored in approved containers (Figure 1-20) and never in glass containers. If the glass container is knocked over or dropped, a terrible explosion can occur. Approved gasoline storage cans have a flash-arresting screen at the outlet. These screens prevent external ignition sources from igniting the gasoline within the can while the gasoline is being poured.

Follow these safety precautions regarding gasoline containers:

1. Always use approved gasoline containers that are painted red for proper identification.

2. Do not fill gasoline containers completely. Always leave the level of gasoline at least 1 inch (25 mm) from the top of the container. This action allows expansion of the gasoline at higher temperatures. If gasoline containers are completely full, the gasoline will expand when the temperature increases. This expansion forces gasoline from the can and creates a dangerous spill.

3. If gasoline containers must be stored, place them in a well-ventilated area such as a storage shed. Do not store gasoline containers in your home or in the trunk of a vehicle.

4. When a gasoline container must be transported, be sure it is secured against upsets. Do not transport or fill gasoline containers on plastic truck bed liners. Static electricity can be generated and ignite the vapors.

5. Do not store a partially filled gasoline container for long periods of time because it may give off vapors and produce a potential danger.

6. Never leave gasoline containers open except while filling or pouring gasoline from the container.

7. Do not prime an engine with gasoline while cranking the engine.

8. Never use gasoline as a cleaning agent.

Diesel Fuel

A **volatile** substance easily vaporizes or explodes. Although diesel fuel is not as volatile as gasoline, it should still be stored and handled in the same way as gasoline. It is also not as refined as gasoline and tends to be a dirty fuel. It normally contains impurities, including active microscopic organisms that can be highly infectious. If diesel fuel happens to enter an open cut or sore, thoroughly wash it immediately. If it gets into your eyes, flush them immediately and seek medical help.

Solvents

Cleaning solvents are not as volatile as gasoline, but they are still flammable. They should be treated and stored in the same way as gasoline. Whenever using solvents, wear eye protection.

Rags

Oily and greasy rags can also cause fires. Used rags should be stored in an approved container and never thrown out with normal trash. Like gasoline, oil is a hydrocarbon and can ignite with or without a spark or flame.

Fire Extinguisher Use

Fire extinguishers are portable apparatuses containing chemicals, water, foam, or special gas that can be discharged to extinguish a small fire. Tour the shop area and become familiar with the location of the fire extinguishers. Use a report sheet to record the locations. Also indicate the type of each extinguisher and what kinds of fires it will extinguish. Check the gauge and record the state of charge for each extinguisher.

WARNING: The following is not intended to be used as a lab exercise unless expressly directed by your instructor. Use the photo sequence as a guide to become familiar with fire extinguisher use. You must be willing to fight a fire in the shop, if the occasion arises.

WARNING: Do not risk your life fighting a fire. If it is evident that the fire is out of control, get out. Always be aware of where you are and the location of the nearest exit. Do not open garage doors in the event of a fire because the extra oxygen will intensify the flames.

The proper use of a fire extinguisher is very important. It is possible to deplete an extinguisher and still not put out even the smallest of fires if the extinguisher is used improperly. Procedures vary depending on the agents used. Technicians must become familiar with all the extinguishers equipped in the shop. Photo Sequence 1 illustrates the proper use of a multipurpose dry chemical extinguisher. This type of extinguisher is the most widely used in the automotive shop.

Photo Sequence 1
Using A Dry Chemical Fire Extinguisher

P1-1 Multipurpose dry chemical fire extinguisher.

P1-2 Hold the fire extinguisher in an upright position.

P1-3 Pull the safety pin from the handle.

P1-4 Free the hose from its retainer and aim it at the base of the fire.

P1-5 Stand 8 feet from the fire; do not go any closer to the fire than this distance.

P1-6 Squeeze the lever while sweeping the hose from side to side. Keep the hose aimed at the *base* of the fire.

Hazardous Materials

Hazardous materials are materials that can cause illness, injury, or death or pollute water, air, or land. Many solvents and other chemicals used in an automotive shop have warning and caution labels that should be read and understood by everyone who uses them. Many service procedures generate what are known as hazardous wastes. Examples of hazardous waste are used or dirty cleaning solvents and other liquid cleaners.

Right-to-Know Laws

In the United States, **right-to-know laws** concerning hazardous materials and wastes protect every employee in a workplace. The general intent of these laws is for employers to provide a safe working place as it relates to hazardous materials. The right-to-know laws state that employees have a right to know when the materials they use at work are hazardous. The right-to-know laws started with the **Hazard Communication Standard** published by the Occupational Safety and Health Administration (OSHA) in 1983. This document was originally intended for chemical companies and manufacturers that required employees to handle hazardous materials in their work situation. At the present time, most states have established their own right-to-know laws. Meanwhile, the federal courts have decided to apply these laws to all companies, including automotive service shops.

Under the right-to-know laws, the employer has three responsibilities regarding its employees' handling of hazardous materials. The first responsibility concerns employee training and providing information. All employees must be trained about the types of hazardous materials they will encounter in the workplace. All employees must be informed about their rights under legislation regarding the handling of hazardous materials. In addition, information about each hazardous material must be posted on **material safety data sheets (MSDS)** available from the manufacturer (Figure 1-21). These sheets contain extensive information and facts about hazardous materials. In Canada, MSDS sheets are called **workplace hazardous materials information systems (WHMIS).**

The employer has a responsibility to place MSDS where they are easily accessible by all employees. The MSDS provide extensive information about the hazardous material such as:

1. Chemical name.

2. Physical characteristics.

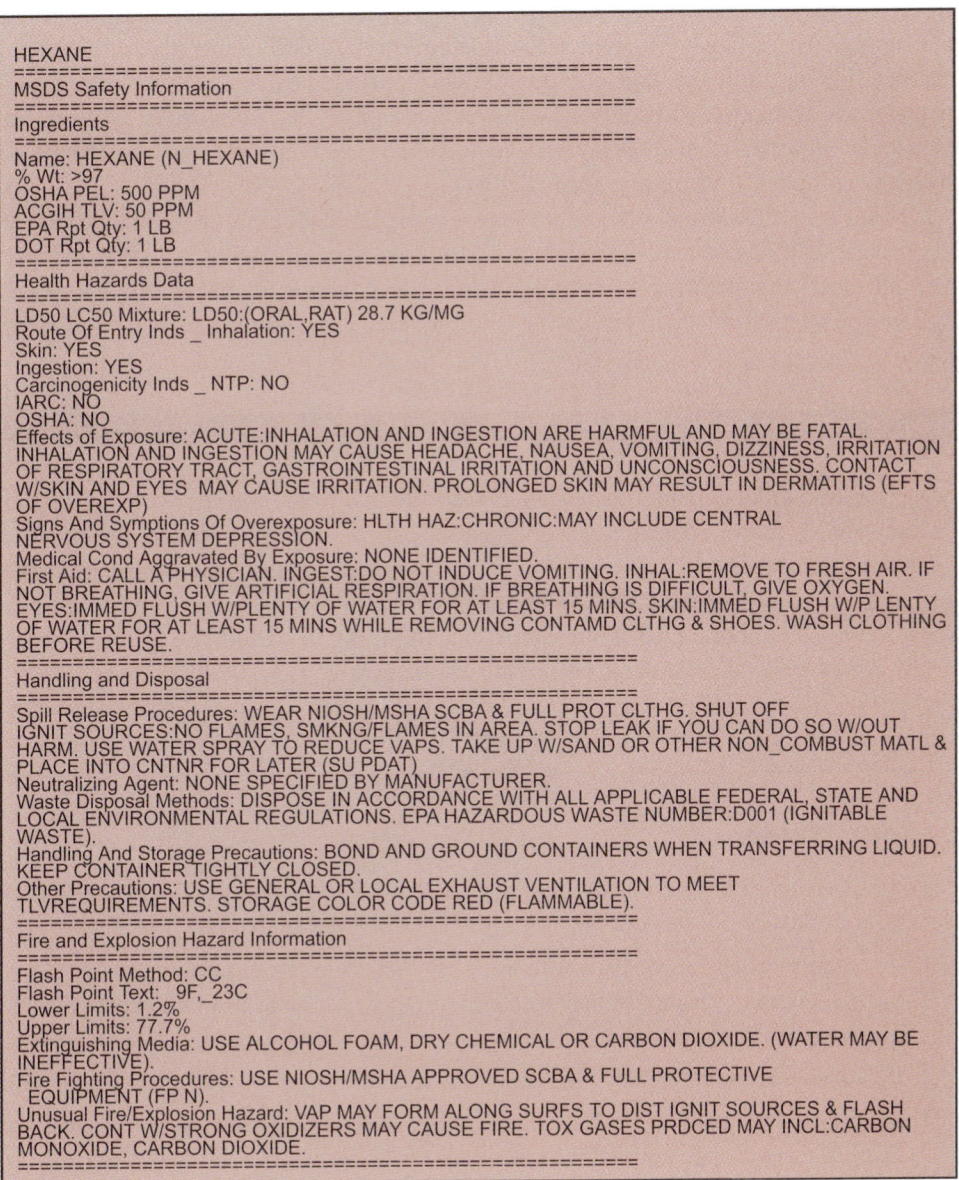

```
HEXANE
===================================================================
MSDS Safety Information
===================================================================
Ingredients
===================================================================
Name: HEXANE (N_HEXANE)
% Wt: >97
OSHA PEL: 500 PPM
ACGIH TLV: 50 PPM
EPA Rpt Qty: 1 LB
DOT Rpt Qty: 1 LB
===================================================================
Health Hazards Data
===================================================================
LD50 LC50 Mixture: LD50:(ORAL,RAT) 28.7 KG/MG
Route Of Entry Inds _ Inhalation: YES
Skin: YES
Ingestion: YES
Carcinogenicity Inds _ NTP: NO
IARC: NO
OSHA: NO
Effects of Exposure: ACUTE:INHALATION AND INGESTION ARE HARMFUL AND MAY BE FATAL.
INHALATION AND INGESTION MAY CAUSE HEADACHE, NAUSEA, VOMITING, DIZZINESS, IRRITATION
OF RESPIRATORY TRACT, GASTROINTESTINAL IRRITATION AND UNCONSCIOUSNESS. CONTACT
W/SKIN AND EYES  MAY CAUSE IRRITATION. PROLONGED SKIN MAY RESULT IN DERMATITIS (EFTS
OF OVEREXP)
Signs And Symptons Of Overexposure: HLTH HAZ:CHRONIC:MAY INCLUDE CENTRAL
NERVOUS SYSTEM DEPRESSION.
Medical Cond Aggravated By Exposure: NONE IDENTIFIED.
First Aid: CALL A PHYSICIAN. INGEST:DO NOT INDUCE VOMITING. INHAL:REMOVE TO FRESH AIR. IF
NOT BREATHING, GIVE ARTIFICIAL RESPIRATION. IF BREATHING IS DIFFICULT, GIVE OXYGEN.
EYES:IMMED FLUSH W/PLENTY OF WATER FOR AT LEAST 15 MINS. SKIN:IMMED FLUSH W/P LENTY
OF WATER FOR AT LEAST 15 MINS WHILE REMOVING CONTAMD CLTHG & SHOES. WASH CLOTHING
BEFORE REUSE.
===================================================================
Handling and Disposal
===================================================================
Spill Release Procedures: WEAR NIOSH/MSHA SCBA & FULL PROT CLTHG. SHUT OFF
IGNIT SOURCES:NO FLAMES, SMKNG/FLAMES IN AREA. STOP LEAK IF YOU CAN DO SO W/OUT
HARM. USE WATER SPRAY TO REDUCE VAPS. TAKE UP W/SAND OR OTHER NON_COMBUST MATL &
PLACE INTO CNTNR FOR LATER (SU PDAT)
Neutralizing Agent: NONE SPECIFIED BY MANUFACTURER.
Waste Disposal Methods: DISPOSE IN ACCORDANCE WITH ALL APPLICABLE FEDERAL, STATE AND
LOCAL ENVIRONMENTAL REGULATIONS. EPA HAZARDOUS WASTE NUMBER:D001 (IGNITABLE
WASTE).
Handling And Storage Precautions: BOND AND GROUND CONTAINERS WHEN TRANSFERRING LIQUID.
KEEP CONTAINER TIGHTLY CLOSED.
Other Precautions: USE GENERAL OR LOCAL EXHAUST VENTILATION TO MEET
TLVREQUIREMENTS. STORAGE COLOR CODE RED (FLAMMABLE).
===================================================================
Fire and Explosion Hazard Information
===================================================================
Flash Point Method: CC
Flash Point Text:  9F,_23C
Lower Limits: 1.2%
Upper Limits: 77.7%
Extinguishing Media: USE ALCOHOL FOAM, DRY CHEMICAL OR CARBON DIOXIDE. (WATER MAY BE
INEFFECTIVE).
Fire Fighting Procedures: USE NIOSH/MSHA APPROVED SCBA & FULL PROTECTIVE
   EQUIPMENT (FP N).
Unusual Fire/Explosion Hazard: VAP MAY FORM ALONG SURFS TO DIST IGNIT SOURCES & FLASH
BACK. CONT W/STRONG OXIDIZERS MAY CAUSE FIRE. TOX GASES PRDCED MAY INCL:CARBON
MONOXIDE, CARBON DIOXIDE.
===================================================================
```

Figure 1-21 An example of a Material Safety Data Sheet (MSDS).

3. Protective equipment required for handling.

4. Explosion and fire hazards.

5. Other incompatible materials.

6. Health hazards such as signs and symptoms of exposure, medical conditions aggravated by exposure, and emergency and first-aid procedures.

7. Safe handling precautions.

8. Spill and leak procedures.

The second responsibility of the employer is to make sure that all hazardous materials are properly labeled. The label information must include health, fire, and reactivity hazards posed by the material, as well as the protective equipment necessary to handle the material. The manufacturer must supply all warning and precautionary information about hazardous materials, and this information must be read and understood by the employee before handling the material. Pay great attention to the information on the label. By doing so, you will use the substance in the proper and safe way, thereby preventing hazardous conditions.

The third responsibility of the employer is for maintaining permanent files regarding hazardous materials. These files must include information on hazardous materials in the shop, proof of employee training programs, and information about accidents such as spills or leaks of hazardous materials. The employer's files must also include proof that employees' requests for hazardous material information, such as MSD, have been met. The employer must maintain a general right-to-know compliance procedure manual.

There are responsibilities for the employees as well. Employees must be familiar with the intended purpose of the substance, the recommended protective equipment, accident and spill procedures, and any other information regarding the safe handling of hazardous materials. This training must be given annually to employees and provided to new employees as part of their job orientation.

A BIT OF HISTORY

During the 1960s, disabling injuries increased 20% and 14,000 workers were dying on the job each year. In pressing for prompt passage of workplace safety and health legislation, Senator Harrison A. Williams Jr. called attention to the need to protect workers against such hazards as noise, cotton dust, and asbestos. Representative William A. Steiger also worked for passage of a bill to protect workers. On December 29, 1970, President Richard M. Nixon signed The Occupational Safety and Health Act of 1970, also known as the Williams-Steiger Act.

Hazardous Waste

Waste is considered hazardous if it is on the Environmental Protection Agency (EPA) list of known and harmful materials or if it has one or more of the following characteristics:

1. Any material that reacts violently with water or other chemicals is considered hazardous. If a material releases cyanide gas, hydrogen sulfide gas, or similar gases when exposed to low-pH acid solutions, it is hazardous.

2. If a material burns the skin or dissolves metals and other materials, it is considered hazardous.

3. Materials are hazardous if they leach one or more of eight heavy metals in concentrations greater than 100 times the primary drinking water standard. These materials are considered toxic.

4. A liquid is hazardous if the temperature at which the vapors on the surface of the fuel will ignite when exposed to an open flame is below 140°F (60°C), and a solid is hazardous if it ignites spontaneously.

A complete list of EPA hazardous wastes can be found in the Code of Federal Regulations. It should be noted that no material is considered hazardous waste until the shop is finished using it and is ready to dispose of it. New oil is not a hazardous waste; however, used oil is. Once you drain oil from an engine, you have generated hazardous waste and now become responsible for its proper disposal. There are many other wastes that need to be handled properly after you have removed them, such as batteries, brake fluid, transmission fluid, and engine coolant.

No fluids drained from a vehicle should be allowed to enter sewage drains. Some fluids, such as coolant, can be captured and recycled in the shop with special equipment. Filters for fluids (transmission, fuel, and oil filters) also need to be handled in designated ways. Used filters need to be drained and then crushed or disposed of in a special shipping barrel. Most regulations demand that oil filters be drained for at least 24 hours before they are disposed of or crushed.

Federal and state laws control the disposal of hazardous waste materials. It is the responsibility of the employer and the employee to ensure that everyone in the shop is familiar with these laws. Hazardous waste disposal laws include the **Resource Conservation and Recovery Act (RCRA).** This law basically states that hazardous waste generators are responsible for the waste from the time it becomes a waste material until the proper waste disposal is completed. Therefore, the user must store hazardous waste material properly and safely and be responsible for the transportation of this material until it arrives at an approved hazardous waste disposal site where it is processed according to the law. A licensed waste management firm normally does the disposal. The hazardous waste coordinator for the shop should have a written contract with the hazardous waste hauler.

The RCRA controls these types of automotive waste:

1. Paint and body repair products waste.

2. Solvents for parts and equipment cleaning.

3. Batteries and battery acid.

4. Mild acids used for metal cleaning and preparation.

5. Waste oil, engine coolants, or antifreeze.

6. Air-conditioning refrigerants.

7. Engine oil filters.

NEVER, under any circumstances, use these methods to dispose of hazardous waste material:

1. Pour hazardous wastes on weeds to kill them.

2. Pour hazardous wastes on gravel streets to prevent dust.

3. Throw hazardous wastes in a dumpster.

4. Dispose of hazardous wastes anywhere but an approved disposal site.

5. Pour hazardous wastes down sewers, toilets, sinks, or floor drains.

6. Bury hazardous wastes in the ground.

Electrical System Safety

There are many safety requirements that must be followed when working on the vehicle's electrical system. In addition to personal safety, there is the concern of damaging the electrical system with improper service techniques. The following are a few of the safety rules.

Battery Safety

Before attempting to do any type of work on or around the battery, you must be aware of certain precautions. To avoid personal injury or property damage, follow these precautions:

1. Battery acid is very corrosive. Do not allow it to come into contact with your skin, eyes, or clothing. If battery acid should get into your eyes, rinse them thoroughly with clean water and seek immediate medical attention. If battery acid comes into contact with your skin, wash with clean

water. Baking soda added to the water will neutralize the acid. If the acid is swallowed, drink large quantities of water or milk followed by milk of magnesia and a beaten egg or vegetable oil.

2. When making connections to a battery, be careful to observe polarity, positive to positive and negative to negative.

3. When disconnecting battery cables, always disconnect the negative (ground) cable first.

4. When connecting battery cables, always connect the negative cable last.

5. Avoid any arcing or open flames near a battery. The vapors produced by the cycling of a battery are very explosive. Do not smoke around a battery.

6. Follow the manufacturer's instructions when charging a battery. Charge the battery in a well-ventilated area. Do not connect or disconnect the charger leads while the charger is turned on.

7. Do not add electrolyte to the battery if it is low. Add only distilled water.

8. Do not wear any jewelry while servicing the battery. These items are excellent conductors of electricity. Severe burns may result if current flows through them by accidental contact with the battery positive terminal and a ground.

9. Never lay tools across the battery. They may come into contact with both terminals, shorting out the battery and causing it to explode. This will damage both the tools and the battery.

10. Wear safety glasses and/or a face shield when servicing the battery.

Starting System Service Safety

Before testing or servicing the starter system, become familiar with these precautions that should be observed:

1. Refer to the recommendations given in the service manual for correct procedures for disconnecting a battery. Some vehicles with on-board computers must be supplied with an auxiliary power source to maintain computer memories.

2. Disconnect the battery ground cable before disconnecting any of the starter circuit's wires or removing the starter motor.

3. Be sure the vehicle is properly positioned on the hoist or on safety jack stands.

4. Before performing any cranking test, be sure the vehicle transmission is in PARK or NEUTRAL and the parking brakes are applied. Put wheel blocks in front of and behind one tire.

5. Follow the service manual procedures for disabling the ignition system.

6. Be sure all test leads are clear of any moving engine components.

7. Never clean any electrical components in solvent or gasoline. Clean with denatured alcohol, or wipe with clean rags only.

Charging System Service Safety

The following are some general rules for servicing the generator and the charging system:

1. Do not run the vehicle with the battery disconnected. The battery acts as a buffer and stabilizes any voltage spikes that may cause damage to the vehicle's electronics.

2. When performing charging system tests, do not allow output voltage to increase over 16 volts.

3. If the battery needs to be recharged, disconnect the battery cables while charging.

4. Do not attempt to remove electrical components from the vehicle with the battery connected.

5. Before connecting or disconnecting any electrical connections, the ignition switch must be in the OFF position, unless directed otherwise in the service manual.

6. Avoid contact with the BAT terminal of the generator while the battery is connected. **BAT** is the terminal identifier for the conductor from the generator to the battery positive terminal. Battery voltage is always present at this terminal.

Batteries are very dangerous components of the vehicle. It is important that you be able to demonstrate the ability to work around the battery in a safe manner. Throughout this manual there will be many instances where you will be required to perform a task involving the battery. Chapter 5 covers the subject of removing and testing the battery. Do not perform any tests or disconnect the battery until you have completed that chapter. The purpose of this section is to assist you in becoming more familiar with the battery to allow you to work safely around it.

Point out the following components of the battery to your instructor:

1. Negative terminal.
2. Positive terminal.
3. Vents.

Answer the following questions concerning battery safety (answer written or orally per your instructor):

1. Why be concerned about battery acid?
2. What should be done if battery acid is splashed into your eyes?
3. What should you do if battery acid gets onto your skin?
4. What is meant by polarity? Why is it a concern when connecting a battery?
5. Which terminal must be disconnected first?
6. When connecting battery cables, which cable is to be connected last?
7. Why is wearing jewelry discouraged around the battery?
8. Why is smoking not allowed around the battery?
9. Why are tools not to be laid across the top of the battery?
10. What safety protection should be worn while servicing or working around the battery?
11. What other safety precautions must be observed?

If you do not understand any of the safety precautions associated with working on or around the battery, ask your instructor.

> **CAUTION:** Always double-check the polarity of the battery charger's connections and leads before turning the charger on. Incorrect polarity can damage the battery or cause it to explode.

Air Bag Safety

The **air bag system** is designed as a supplemental restraint that, in the case of an accident, will deploy a bag out of the steering wheel or passenger-side dash panel to provide additional protection against head and face injuries. An air bag system demands that the technician pay close attention to safety warnings and precautions when working on or *around* it. Most air bags are deployed by an explosive charge. Accidental deployment of the air bag can result in serious injury. When working on or *around* the steering wheel or air bag module, be aware of your hands and arms. Do not place your arm over the module. If the air bag deploys, injury can result.

Air bag systems contain a means of deploying the bag even if the battery is disconnected. This system is needed in the event the battery is damaged or disconnected during an accident. The reserve energy can be stored for over 30 minutes after the battery is disconnected. Follow the service manual procedures for disabling the system.

The **air bag module** is the air bag and inflator assembly together in a single package. When carrying the air bag module, carry it so that the bag and trim are facing up and away from your body. In the event of accidental deployment, the charge will be away from you. Do not face the module toward any other people.

When you place the module on the bench, face the bag and trim up. This provides a free space for the bag to expand if it deploys. If the module will be stored for any period of time, it must be stored in a cool dry place. Store the module with the trim up and do not place anything on top of the module.

While troubleshooting the air bag system, do not use electrical testers such as battery-powered or AC-powered voltmeters, ohmmeters, and so on, or any other equipment except those specified in the service manual. Do not use a test light to troubleshoot the system.

When it is necessary to make a repair or replace a component in the air bag system, always disconnect the negative battery cable before making the repair. It is a good practice to insulate the terminal with tape or a rubber hose to prevent it from coming into contact with the battery post. Some manufacturers recommend that the air bag inflator module(s) be disconnected in addition to the negative battery cable.

 AUTHOR'S NOTE: Be sure to follow all manufacturer's warnings, cautions, and special service notes when working on or around the air bag system.

Although it is unlikely that an air bag module (Figure 1-22) will inflate on its own, it is possible to ignite it while performing service. To prevent injury, be aware of where the module(s) is located in the vehicle. The most common location is in the center of the steering wheel. However, many vehicles also have a passenger-side air bag (Figure 1-23). Some manufacturers are installing air bag systems for the back-seat passengers as well.

WARNING: Obey all of the warnings in the service manual when working on or around the air bag system. Failure to follow these warnings may result in air bag deployment and injury.

To prevent accidental deployment of the air bag system, disconnect the negative battery cable before disconnecting or connecting any electrical connectors in the system. It is important to be able to recognize the components of the air bag system. Most manufacturers place the wiring of the air bag system into a bright yellow harness tube or use bright yellow insulation

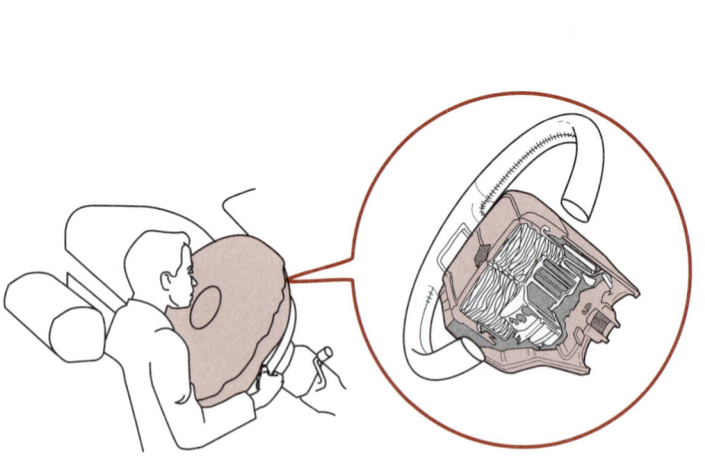

Figure 1-22 Typical air bag module.

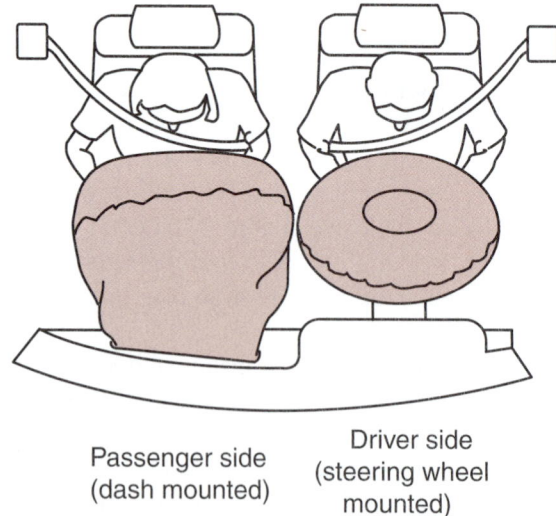

Passenger side
(dash mounted)

Driver side
(steering wheel mounted)

Figure 1-23 An air bag system with both a driver and passenger-side air bag module.

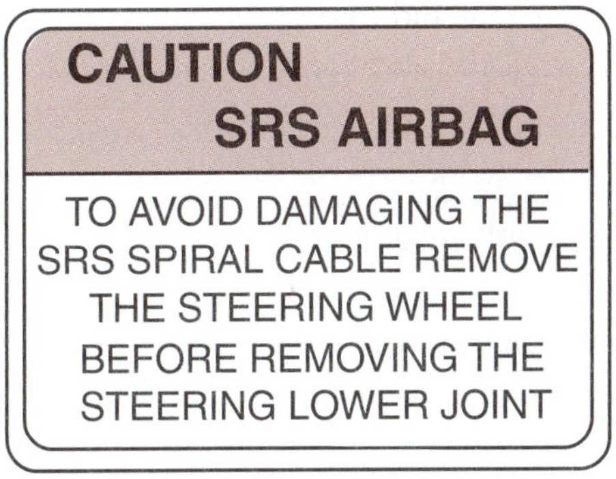

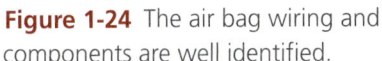

CAUTION
SRS AIRBAG

TO AVOID DAMAGING THE
SRS SPIRAL CABLE REMOVE
THE STEERING WHEEL
BEFORE REMOVING THE
STEERING LOWER JOINT

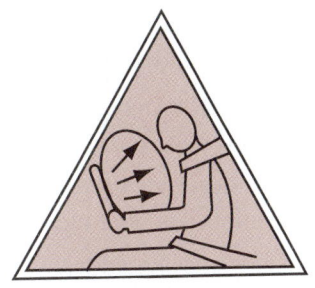

Figure 1-24 The air bag wiring and components are well identified.

Figure 1-25 Some manufacturers identify air bag circuits in their service manual with a warning symbol.

(Figure 1-24). The wires are usually tagged to alert the technician. Walk around an air bag–equipped vehicle with your instructor. Your instructor will point out the components of the system and review the necessary safety precautions.

Some manufacturers denote air bag related–circuitry in their service manuals with a warning symbol (Figure 1-25).

Antilock Brake System Safety

Antilock brake systems (ABS) automatically pulsate the brakes to prevent wheel lockup under panic stop and poor traction conditon. ABS is available on most of today's vehicles. There are many different systems used, and each has its own safety requirements regarding servicing the system. Become familiar with the warnings and cautions associated with the system you are working on by studying the service manual before performing any service.

Certain components of the ABS are not intended to be serviced individually. Do not attempt to remove or disconnect these components. Only those components with approved removal and installation procedures in the service manual should be serviced.

Some operations require that the tubes, hoses, and fittings be disconnected. Many earlier antilock brake systems used high hydraulic pressures and an accumulator to store this pressurized fluid. Before disconnecting any lines or fittings, the accumulator must be fully depressurized. Follow the service manual procedures for depressurizing the system.

Many late-model ABS systems do not use an accumulator; therefore these systems do not require depressurizing. However, always refer to the correct service manual before servicing a brake system.

Some service operations to the antilock brake system require that the tubes, hoses, and fittings be disconnected. The ABS can use hydraulic pressures as high as 2,800 psi and an accumulator to store this pressurized fluid. Before disconnecting any lines or fittings, the accumulator must be fully depressurized. The following is a common method of depressurizing the ABS. However, follow the service manual procedures for the vehicle you are working on.

1. Place the ignition switch in the OFF position.
2. Pump the brake pedal a minimum of 20 times.
3. There should be noticeable change in pedal feel when the accumulator is discharged.

Summary

❏ Being a professional technician means more than having knowledge of vehicle systems. It also requires an understanding of all the hazards in the work area.

❏ As a professional technician, you should work responsibly to protect yourself and the people around you.

❏ Technicians must be aware that it is their responsibility to prevent injuries in the shop, and that their actions and attitudes reflect how seriously they accept that responsibility.

❏ Long-sleeve shirts should have their cuffs buttoned or rolled up tightly, and shirttails should be tucked in at all times. Neckties should be tucked inside your shirt; ideally, only clip-on ties should be worn.

❏ Long hair should be tied back and tucked under a hat.

❏ Jewelry has no place in the automotive shop.

❏ The safest and surest method of protecting your eyes is to wear proper eye protection anytime you enter the shop.

❏ When working around rotating pulleys and belts, be aware of where you are placing your hands and fingers at all times.

❏ Most occupational back injuries are caused by improper lifting practices.

❏ Most injuries caused by tools and equipment are the result of improper use, improper maintenance, and carelessness.

❏ Never use compressed air for cleaning off your clothes.

❏ The floor jack is to be used only to lift the vehicle off the floor. Use jack stands to support the vehicle after it is lifted.

❏ Fires are classified by the types of materials involved. Fire extinguishers are classified by the type of fire they will extinguish.

❏ Batteries can cause serious injury if all safety rules are not followed when working on or around them.

❏ The air bag system demands that the technician pay close attention to safety warnings and precautions when working on or around them.

❏ Air bag systems contain a means of deploying the bag even if the battery is disconnected. The reserve energy can be stored for over 30 minutes after the battery is disconnected.

Terms to Know

Air bag module	Floor jacks	Resource Conservation and Recovery Act (RCRA)
Air bag system	Hand tools	
Antilock brake systems (ABS)	Hazard Communication Standard	Right-to-know laws
BAT	Hazardous materials	Safety goggles
Carbon monoxide (CO)	Hoists	Vehicle lift point
Caustic	Jack stands	Volatile
Conductors	Material safety data sheets (MSDS)	Workplace hazardous materials information systems (WHMIS)
Face shields	Occupational safety glasses	
Fire extinguishers	Pneumatic tools	
Flammable	Power tools	

ASE-Style Review Questions

1. Which of the following is **not** included in proper lifting practices?
   - **A.** Do not lift with your back.
   - **B.** Keep your back and elbows straight.
   - **C.** Lift the object by yourself so others are out of the way.
   - **D.** While carrying the object, do not twist your body to change directions.

2. *Technician A* says the right-to-know laws require employers to train employees regarding hazardous waste materials.
   Technician B says the right-to-know laws have no provisions requiring proper labeling of hazardous materials.
   Who is correct?
   - **A.** A only
   - **B.** B only
   - **C.** Both A and B
   - **D.** Neither A nor B

3. *Technician A* says it is safe to work around the electric cooling fan when the engine is off and the engine temperatures are low.
   Technician B says the electric cooling fans can only come on if the engine is running.
   Who is correct?
   - **A.** A only
   - **B.** B only
   - **C.** Both A and B
   - **D.** Neither A nor B

4. *Technician A* says electrical fires are extinguished with Class A fire extinguishers.
   Technician B says gasoline is extinguished with Class B fire extinguishers.
   Who is correct?
   - **A.** A only
   - **B.** B only
   - **C.** Both A and B
   - **D.** Neither A nor B

5. *Technician A* says material safety data sheets (MSDS) explain employers' and employees' responsibilities regarding handling and disposal of hazardous materials.
   Technician B says material safety data sheets (MSDS) contain specific information about hazardous materials.
   Who is correct?
   - **A.** A only
   - **B.** B only
   - **C.** Both A and B
   - **D.** Neither A nor B

6. *Technician A* says a solid that ignites spontaneously is considered a hazardous material.
   Technician B says a liquid is considered a hazardous material if the vapors on the surface will ignite when exposed to an open flame whose temperature is below 140°F (60°C).
   Who is correct?
   - **A.** A only
   - **B.** B only
   - **C.** Both A and B
   - **D.** Neither A nor B

7. *Technician A* says safety glasses should be worn when working with battery acid and refrigerants.
   Technician B says full-face shields are designed to provide protection for the face, neck, and eyes of the technician.
   Who is correct?
   - **A.** A only
   - **B.** B only
   - **C.** Both A and B
   - **D.** Neither A nor B

8. *Technician A* says the way people dress reflects very little about their attitude toward safety.
   Technician B says jewelry can be a personal safety hazard and should be removed while working in an automotive repair shop.
   Who is correct?
   - **A.** A only
   - **B.** B only
   - **C.** Both A and B
   - **D.** Neither A nor B

9. Which of the following statements concerning gasoline storage is true?
   - **A.** Always use approved gasoline containers that are painted blue for proper identification.
   - **B.** Prevent air from entering the container by filling gasoline containers completely full.
   - **C.** Always transport gasoline containers on plastic bed liners.
   - **D.** Do not store a partially filled gasoline container for long periods of time.

10. *Technician A* says hazardous waste materials may be hauled to an approved hazardous waste disposal site or recycled in the shop.
    Technician B says the disposal of all hazardous waste materials must be done in accordance with federal, state, and local laws.
    Who is correct?
    - **A.** A only
    - **B.** B only
    - **C.** Both A and B
    - **D.** Neither A nor B

Job Sheet 1

Name _____   Date _____

Shop Safety Survey

As a professional technician, safety should be one of your first concerns. This job sheet will increase your awareness of shop safety items. As you survey your shop and answer the following questions, you will learn how to evaluate the safety of any workplace.

Procedure

Your instructor will review your work at each Instructor Response point.

Task Completed

1. Before you begin to evaluate your workplace, evaluate yourself. Are you dressed for work? ☐ Yes ☐ No

   If yes, why? _____

   _____

   If no, what must you correct to be properly dressed? _____

   _____

2. Are your safety glasses OSHA approved? ☐ Yes ☐ No

   Do they have side shields? _____

3. Carefully inspect your shop, noting any potential hazards. _____

   _____

   NOTE: A hazard is not necessarily a safety violation but is an area of which you must be aware.

4. Are there safety areas marked around grinders and other machinery? ☐ Yes ☐ No

5. What is the air pressure in your shop? _____

6. Where are the tools stored in your shop? _____

   _____

   Are they clean and neatly stored? ☐ Yes ☐ No

7. If you could, how would you improve the tool storage? _____

   _____

   _____

8. What kind of hoists are used in your shop? _____

   _____

9. Ask your instructor to demonstrate hoist usage.   ☐

10. Where is the first-aid kit in your shop? _____

    _____

11. Where is the main power shutoff located? _____

    _____

12. List the location of the exits. _____

    _____

13. Describe the emergency evacuation procedures. _____

_____

_____

_____

_____

14. Where are the hazardous materials stored? _____

_____

_____

_____

_____

15. What is the procedure for handling hazardous waste? _____

_____

_____

_____

_____

16. What is the procedure to be followed in your shop in case of an accident? _____

_____

_____

_____

_____

17. Have your instructor supply you with a vehicle make, model, and year. Using the appropriate shop manual, draw an illustration showing the lifting points on the given vehicle.

☐

Instructor's Response _____

_____

_____

_____

Job Sheet 2

Name _____ Date _____

Fire Hazard Inspection

Fire is always a danger in any automotive shop. The very nature of automotive work involves the use of many highly flammable chemicals. Because of this, a technician must be very careful. Watch for and immediately correct all fire hazards.

Procedure

Task Completed

1. Are there any flammable liquids stored in your shop? _____

    Are they stored properly? ☐ Yes ☐ No

    Why or why not? _____

    _____

2. Where are the fire extinguishers located in your shop? _____

    _____

    Against what class fires can each extinguisher be used? _____

    _____

3. Explain to your instructor how to use each fire extinguisher in your shop. ☐

4. Does your shop have a fire blanket? ☐ Yes ☐ No

    If so, where is it kept? _____

    _____

5. Where are the fire alarms located? _____

    _____

6. Where are the fire exits located? _____

    _____

7. Where are the fire escape routes posted? _____

    _____

8. Where are the fireproof cabinets located? _____

    _____

9. Where are dirty shop towels to be disposed? _____

    _____

Instructor's Response _____

_____

_____

Special Tools and Procedures

Upon completion and review of this chapter, you should be able to:

❑ Explain the proper use of jumper wires.

❑ Explain the proper use of a test light.

❑ Explain the proper use of a logic probe.

❑ Explain the proper use of analog volt/amp/ohmmeters.

❑ Explain the proper use of digital volt/amp/ohmmeters.

❑ Describe when to use the different types of multimeters.

❑ Explain the proper use of a digital storage oscilloscope.

❑ Use Ohm's law to determine electrical values in different types of circuits.

❑ Locate service information.

❑ Explain the concepts of working as an electrical systems technician.

Basic Tools

Basic mechanic's tool set

Classroom Manual
Chapter 2, page 26

Introduction

This chapter covers some of the typical shop procedures that the electrical systems technician may encounter. This includes proper troubleshooting procedures, the use of special test equipment, the use of service information, and workplace practices.

To be able to properly diagnose electrical components and circuits, you must be able to use many different types of electrical test equipment. In this chapter, you will learn when and how to use the most common types of test equipment. You will also learn which test instrument is best to use to identify the cause of the various types of electrical problems.

Electrical current is a term used to describe the movement or flow of electricity. The greater number of electrons flowing past a given point in a given amount of time, the more current the circuit has. This current, like the flow of water or any other substance, can be measured. The unit for measuring electrical current is the ampere. The instrument used to measure electrical current flow in a circuit is called an **ammeter.**

When any substance flows, it meets resistance. The resistance to electrical flow can be measured. The resistance to current flow produces heat. This heat can be measured to determine the amount of resistance. A unit of measured resistance is called an ohm. Resistance can be measured by an instrument called an **ohmmeter.**

Voltage is electrical pressure. Voltage is the force developed by the attraction of electrons to protons. The more positive one side of the circuit is, the more voltage is present in the circuit. Voltage does not flow; rather it is the pressure that causes current flow.

To have electricity, some force is needed to move the electrons between atoms. This force is the pressure that exists between a positive and negative point within an electrical circuit. This force, also called electromotive force (EMF), is measured in units called volts. One volt is the amount of pressure (force) required to move 1 ampere of current through a resistance of 1 ohm. Voltage is measured by an instrument called a **voltmeter.**

The amount of current that flows in a circuit is determined by the resistance in that circuit. As resistance goes up, the current goes down. The energy used by a load is measured in volts. Amperage stays constant in a circuit, but the voltage is dropped as it powers a load. Measuring voltage drop determines the amount of electrical energy changed to another form of energy by the load.

Also, this chapter covers what it means to work as an electrical systems technician. This includes compensation programs, the importance of workplace and customer relations, communication, and certification.

Basic Electrical Troubleshooting

Troubleshooting electrical problems involves using meters, test lights, and jumper wires to determine if any part of the circuit is open or shorted, or if there is unwanted resistance.

To troubleshoot a problem, always begin by verifying the customer's complaint. Then operate the system and others, to get a complete understanding of the problem. Often there are other problems, which are not as evident or bothersome to the customer, that will provide helpful information for diagnostics. Obtain the correct wiring diagram for the car and study the circuit that is affected. From the diagram, you should be able to identify testing points and probable problem areas. Then test and use logic to identify the cause of the problem.

An ammeter and a voltmeter connected to a circuit at the different locations shown in Figure 2-1 should give readings as indicated when there are no problems in the circuit. An open exists whenever there is not a complete path for current flow. If there is an open anywhere in the circuit, the ammeter will read zero current. If the open is in the 1-ohm resistor, a voltmeter connected from C to ground will read zero. However, if the resistor is open and the voltmeter is connected to points B and C, the reading will be 12 volts. The reason is that the battery, ammeter, voltmeter, 2-ohm resistor, and 3-ohm resistor are all connected together to form a series circuit. Because of the open in the circuit, there is only current flow in the circuit through the meter, not the rest of the circuit. This current flow is very low because the meter has such high resistance. Therefore, the voltmeter will show a reading of 12 volts, indicating no voltage drop across the resistors.

To help you understand this concept, look at what happens if the 2-ohm resistor is open instead of the 1-ohm resistor. A voltmeter connected from point C to ground would indicate 12 volts. The 1-ohm resistor in series to the high resistance of the voltmeter would have little effect on the circuit. If an open should occur between point E and ground, a voltmeter connected from points B, C, D, or E to ground would read 12 volts. A voltmeter connected across any one of the resistors, from B to C, C to D, or D to E, would read zero volts, because there will be no voltage drops if there is no current flow.

A short would be indicated by excessive current and/or abnormal voltage drops. These examples illustrate how a voltmeter and ammeter can be used to check for problems in a circuit. An ohmmeter also can be used to measure the values of each component and compare

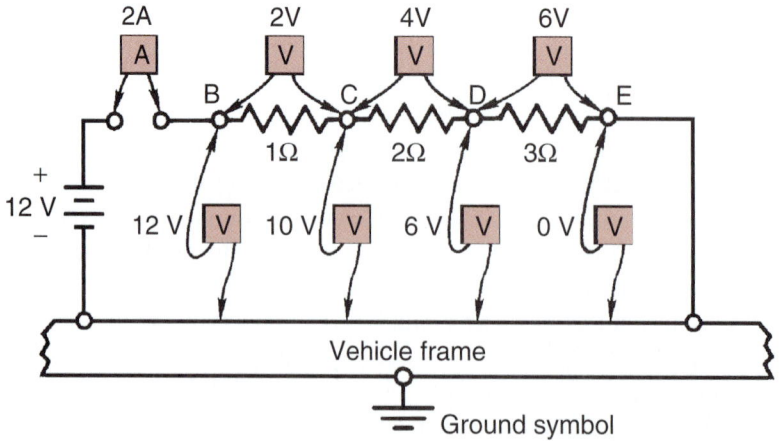

Figure 2-1 A basic circuit being tested with an ammeter and a voltmeter.

these measurements to specifications. If there is no continuity across a part, it is open. If there is more resistance than called for, there is high internal resistance. If there is less resistance than specified, the part is shorted.

 CAUTION: Any broken, frayed, or damaged insulation material requires replacement or repair to the wire. Exposed conductor material from damaged insulation can result in a safety hazard and damage to circuit components.

WARNING: Because the human body is a conductor of electricity, observe all safety rules when working with electricity.

Test Equipment

Since electricity is an invisible force, the proper use of test tools will permit the technician to "see" the flow of electrons. Knowing what is being looked at and being able to interpret various meter types will assist in electrical system diagnosis. To diagnosis and repair electrical circuits correctly, a number of common tools and instruments are used. The most common tools are jumper wires, test lights, voltmeters, ammeters, and ohmmeters.

Jumper Wires

One of the simplest types of test equipment is the **jumper wire.** Connecting one end of the jumper wire to battery positive will provide an excellent 12-volt power supply for testing a component. If the component does not operate when it is in its own circuit, but does operate when battery voltage is jumped to it, the component is good and the fault is within the circuit (Figure 2-2).

Jumper wires can be used to bypass individual wires, connectors, components, or switches (Figure 2-3). Bypassing a component or wire helps to determine if that part is faulty. If the problem is no longer evident after the jumper wire is installed, the part being bypassed is usually faulty.

 CAUTION: Jumper wires can never be used to jump across the load of the circuit.

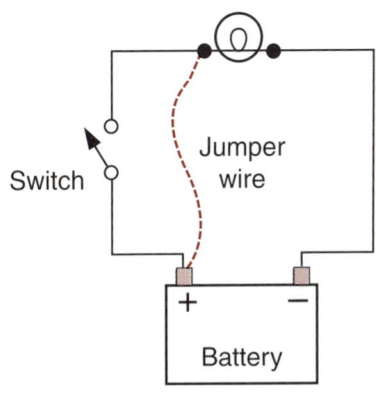

Figure 2-2 Using a jumper wire to bypass the switch.

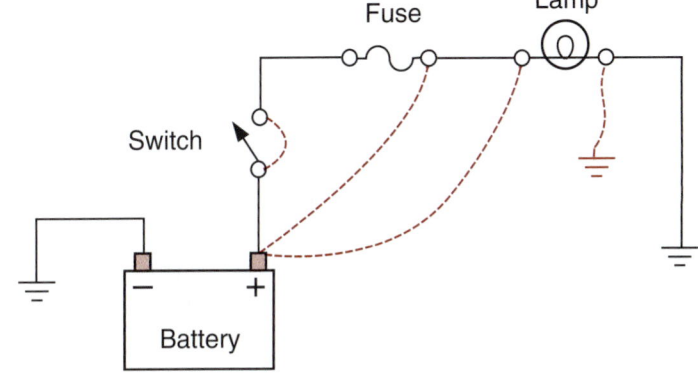

Figure 2-3 Examples of locations where a jumper wire can be used to bypass a portion of the circuit and to test the ground circuit. Remember, if you bypass the circuit fuse, the jumper wire should be fitted with a fuse. Never bypass the load component.

To protect the circuit being tested, it is recommended the jumper wire be fitted with an in-line fuse holder. This will allow the quick changing of fuses to correctly protect the circuit. Using a fused jumper wire will help prevent damage to the circuit if the jumper wire is connected improperly.

 WARNING: Never connect a jumper wire across the terminals of the battery. The battery could explode, causing serious injury.

Test Lights

There are two types of test lights commonly used in diagnosing electrical problems: non-powered and self-powered. A **test light** is used when the technician needs to "look" for electrical power in the circuit. A typical non-powered test light has a transparent handle that contains a light bulb. A sharp probe extends from one end of the handle while a ground wire with a clamp extends from the other end (Figure 2-4). If the circuit is operating properly, clamping the lead of the test light to ground and probing the circuit at a point of voltage should light the lamp (Figure 2-5).

A test light is limited because it does not display how much voltage is at the point of the circuit being tested. However, by understanding the effects of voltage drop, the technician will be able to interpret the brightness of the test light and relate the results to the expectations of a good circuit. If the lamp is connected after a voltage drop, the lamp will light dimly. Connecting the test lamp before the voltage drop should light the lamp brightly. The light should not illuminate at all if it is probing for voltage after the last resistance.

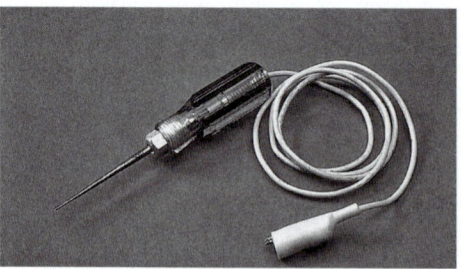

Figure 2-4 A typical test light used to probe for voltage in a circuit.

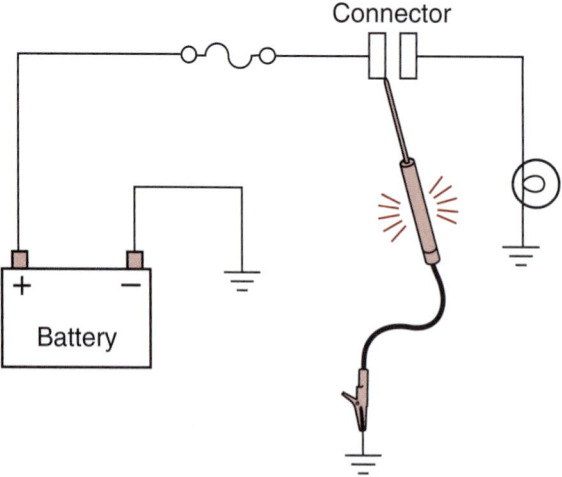

Figure 2-5 If voltage is present, the test light will illuminate.

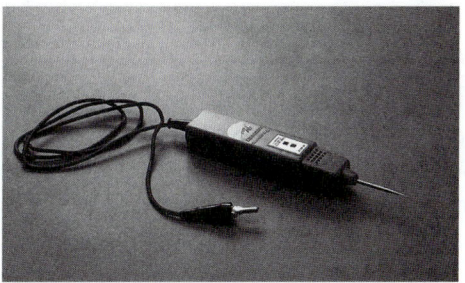

Figure 2-6 Typical self-powered continuity tester.

CAUTION: Do not use a test light to probe for power in an electronic circuit. The increased current draw of the test light may damage the system components.

Another type of circuit tester is the self-powered continuity tester (Figure 2-6). The continuity tester has an internal battery that powers a light bulb. With the power in the circuit turned off or disconnected, the ground clip is connected to the ground terminal of the load component. By probing the feed wire, the light will illuminate if the circuit is complete (has continuity). If there is an open in the circuit, the lamp will not illuminate.

WARNING: Never use a self-powered test light to test the air bag system. The battery in the tester can cause the air bag to deploy.

CAUTION: Do not connect a self-powered test light to a circuit that is powered. Doing so will damage the test light.

Logic Probes

Many computer-controlled systems use a pulsed voltage to transmit messages or to operate a component. A standard or self-powered test light should not be used to test these circuits since they may damage the computer. However, a **logic probe** (Figure 2-7) can be used. A logic probe looks something like a test light except it contains three different-colored LEDs. The red LED will light if there is high voltage at the point in the circuit being tested. The green LED will light to indicate the presence of low voltage. The yellow LED lights to indicate the presence of voltage pulses. If the voltage is a pulsed voltage from a high level to a low level, the yellow LED will be on and the red and green LEDs will cycle, indicating the change in voltage.

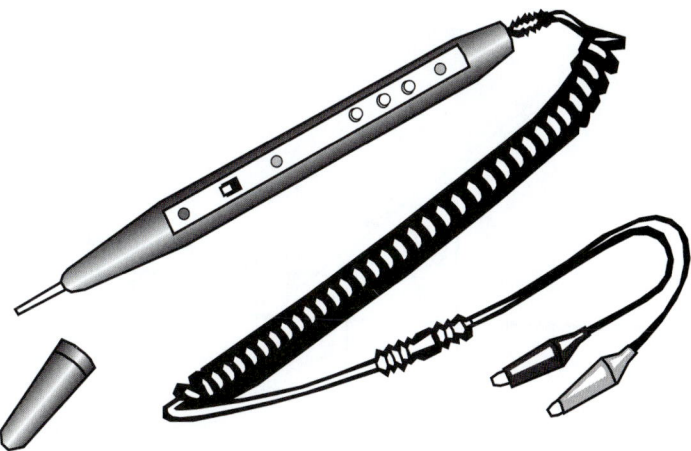

Figure 2-7 A typical logic probe.

Multimeters

The **multimeter** is one of the most versatile tools used to diagnose electrical systems. It can be used to measure voltage, resistance, and amperage. In addition, some types of multimeters are designed to test diodes and measure frequency, duty cycle, temperature, and rotation speed. Multimeters are available in analog (swing needle) and digital display.

Analog Meters

Analog meters use a sweeping needle and a scale to display test values (Figure 2-8). All analog meters use a **D'Arsonval movement** (Figure 2-9). A D'Arsonval movement is a small coil of wire mounted in the center of a permanent horseshoe-type magnet. A pointer or needle is mounted to the coil. When taking a measurement, current flows through the coil and creates a magnetic field around the coil. The coil rotates within the permanent magnet as its magnetic field interacts with the magnetic field of the permanent magnet. The amount of rotation is determined by the strength of the magnetic field around the coil. The direction of the rotation is determined by the direction of the current flow through the coil. Since the needle moves with the coil, it reflects the amount of coil movement and its direction. To give accurate readings, analog meters must be calibrated. This is done through a zero adjustment knob or adjusting screw in the meter. These adjustments move the coil to its base point.

Special Tools

Analog volt or ohm meter

Figure 2-8 An analog meter.

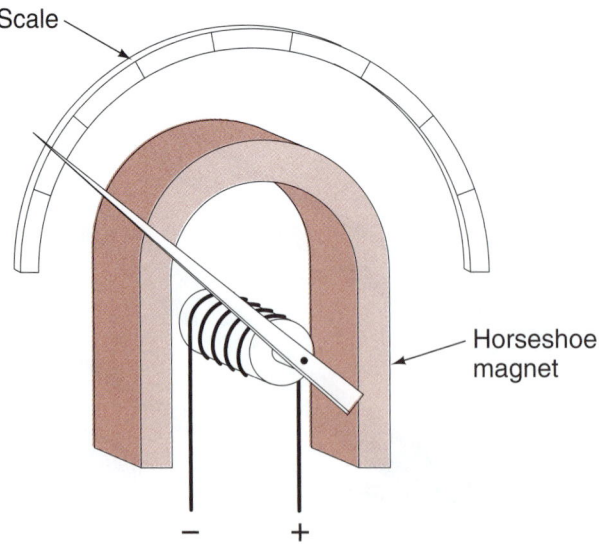

Figure 2-9 D'Arsonval movement is the basis for the movement of an analog meter

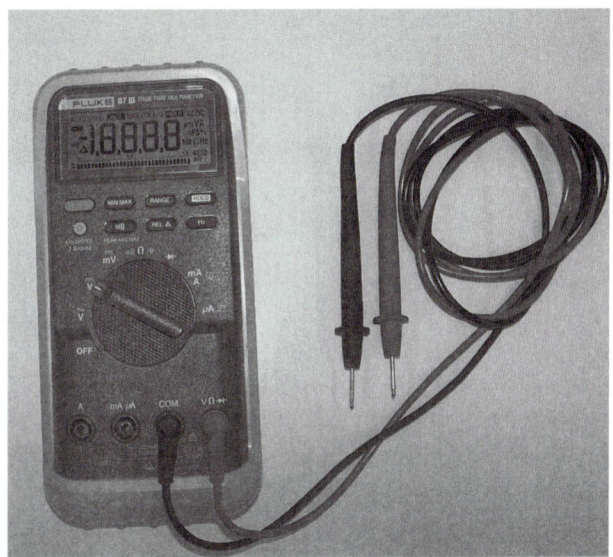

Figure 2-10 Digital multimeter.

There are some service procedures that specify that an analog meter be used instead of a digital meter. Always follow the manufacturer's recommendations while performing diagnostic and/or service routines.

Digital Meters

With modern vehicles incorporating computer-controlled systems, the need for **digital multi-meters (DMM)** is required (Figure 2-10). Digital multimeters (also called DVOMs) display values using liquid crystal displays instead of a swinging needle. They are basically a computer that determines the measured value and displays it for the technician. Computer systems have integrated circuits that operate on very low amounts of current. Analog meters can overload computer circuits and burn out the IC chips since they allow a large amount of current to flow through the circuit. On the other hand, most digital multimeters have very high input resistance (impedance) which prevents the meter from drawing current when connected to a circuit. Most DMMs have at least 10 megohms (10 million ohms) impedance. This reduces the risk of damaging computer circuits and components.

Digital multimeters (DMM) are also referred to as DVOMs (digital volt/ohmmeter).

> **CAUTION:** Not all DMMs are rated at 10 megohms of impedance. Be sure of the meter you are using to prevent electronic component damage.

Digital meters rely on electronic circuitry to measure electrical values. The measurements are displayed with LEDs or on a liquid crystal display (LCD). Digital meters tend to give more accurate readings and are certainly much easier to read. Rather than reading a scale at the point where the needle lines up, digital meters simply display the measurement in a numerical value. This also eliminates the almost certain error caused by viewing an analog meter at an angle.

Voltmeters

A voltmeter can be used to measure the voltage available at the battery. It can also be used to test the voltage available at the terminals of any component or connector. In addition, a voltmeter tests voltage drop across an electrical circuit, component, switch, or connector.

A voltmeter has two leads: a red positive lead and a black negative lead. The red lead should be connected to the positive side of the circuit or component. The black lead should be

Classroom Manual
Chapter 2, page 27

Special Tools
DMM

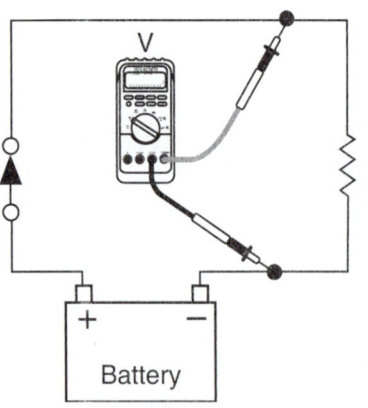

Figure 2-11 Connecting a voltmeter in parallel to the circuit.

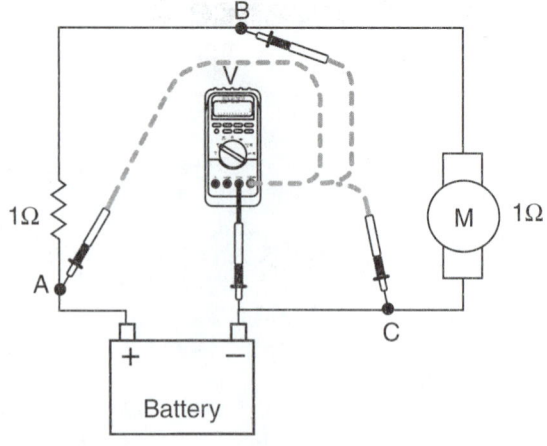

Figure 2-12 Checking voltage in a closed circuit.

connected to ground or to the negative side of the component. A voltmeter is connected in parallel with a circuit (Figure 2-11).

Figure 2-12 shows how to check for voltage in a closed circuit. The voltage at point A is 12 volts positive. There is a drop of 6 volts over the 1-ohm resistor and the reading is 6 volts positive at point B. The remaining voltage drops in the motor load and the voltmeter reads 0 at point C. indicating normal motor circuit operation.

When reading voltage in the same circuit that has an open (Figure 2-13A), 12 volts will be indicated at any point ahead of the open. This is indicated at points A, B, and C, but not through X. Since the circuit is open, and there is no electrical flow, there is no voltage drop across a resistor or load.

SERVICE TIP: Using a voltmeter to measure voltage drop is one of the easiest ways of testing a circuit's ability to carry current under load.

Classroom Manual
Chapter 2, page 30

The loss of voltage due to resistance in wires, connectors, and loads is called **voltage drop.** Voltage drop is the amount of electrical energy converted to another form of energy. For example, to make a lamp light, electrical energy is converted to heat energy. It is the heat that makes the lamp light.

To measure voltage drop across each load, it must be determined what point is the most positive and what point is the most negative in the circuit. A point in the circuit can be either positive or negative, depending on what is being measured. Referring to Figure 2-13B, if voltage

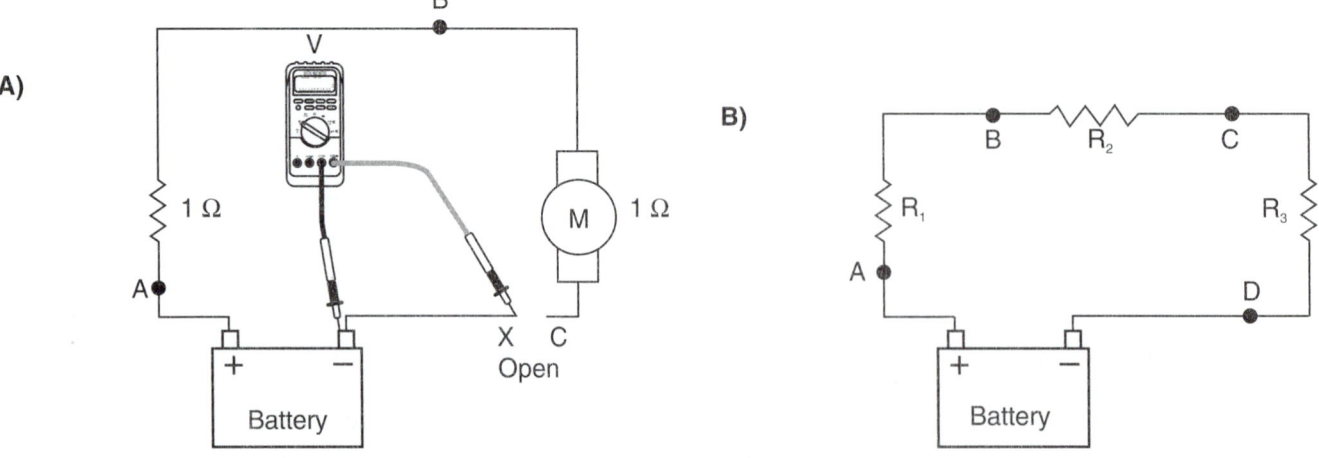

Figure 2-13 Using a voltmeter to (A) find an open circuit and (B) to check voltage drop.

Photo Sequence 2
Performing a Voltage Drop Test

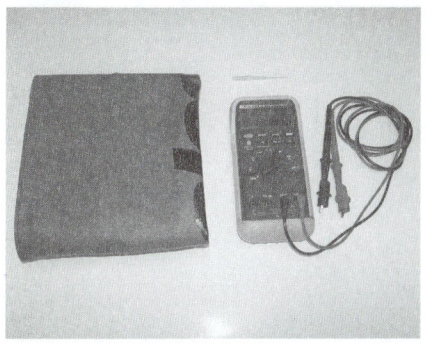

P2-1 The tools required to perform this task: voltmeter, backprobing tools, and fender covers.

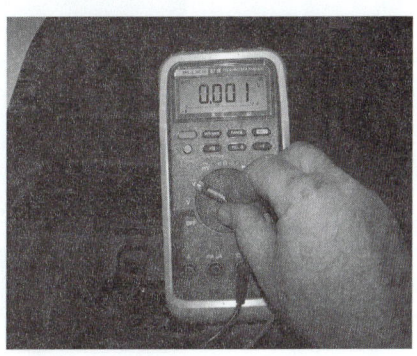

P2-2 Set the voltmeter to its lowest DC voltage scale.

P2-3 To test the voltage drop of the whole insulated side of the circuit, connect the red (positive) voltmeter test lead to the battery positive (+) terminal.

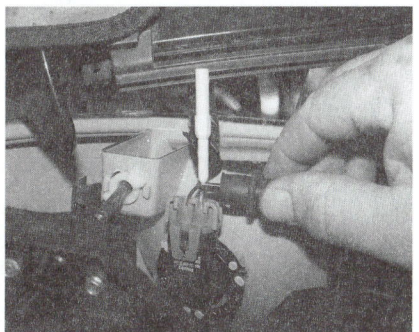

P2-4 Use the backprobing tools to connect the black (negative) voltmeter test lead into the low-beam terminal of the headlight socket. Make sure that you are connected to the input side of the headlight.

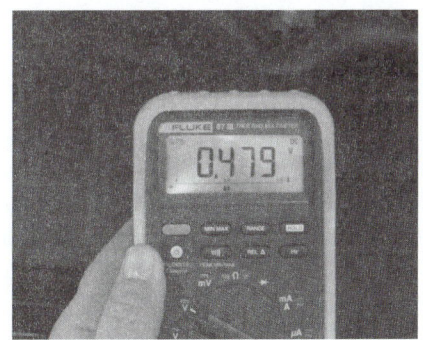

P2-5 Turn on the headlights (low beam) and observe the voltmeter readings. The voltmeter will indicate the amount of voltage that is dropped between the battery and the headlight.

drop across R_1 is being measured, the positive meter lead is placed at point A and the negative lead at point B. The voltmeter will measure the difference in volts between these points. However, to measure the voltage drop across R_2 the polarity of point B is positive and point C is negative. The reason being that point B is the most negative for R_1, yet it is the most positive point for R_2. To measure voltage drop across R_3, point C is positive and point D is negative. The positive lead of the voltmeter should be placed as close as possible to the positive side of the battery in the circuit. The procedure for measuring a voltage drop is shown in Photo Sequence 2.

Often, voltage drop testing requires the technician to **backprobe** a connector to access the terminal. Backprobing requires the use of a backprobing tool (Figure 2-14) that will slide into the back side of a connector. The tool should be inserted between the wire's insulation and the

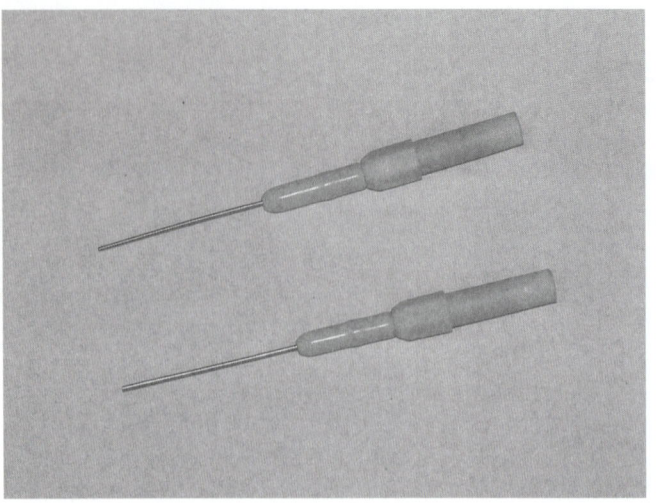

Figure 2-14 Backprobing tools.

Figure 2-15 Backprobing tool properly inserted into the back of a connector.

connector's seal until it contacts the terminal (Figure 2-15). When performed properly the wire, connector, or terminal will not be damaged. It is important to use as small a backprobing tool as possible.

SERVICE TIP: A "T" pin that is used for sewing pant and suit hems can be used as a backprobing tool.

All wiring must have resistance values low enough to allow enough voltage to the load for proper operation. The maximum allowable voltage loss due to voltage drops across wires, connectors, and other conductors in an automotive circuit is 10% of the system voltage. Therefore, in a 12-volt automotive electrical system, this maximum total loss is 1.2 volts.

Figure 2-16 shows two headlights connected to a 12-volt battery by two wires. Each wire has a resistance of only 0.05 ohms. Each headlight has a resistance of 2 ohms. The two headlights are wired in parallel and have a total resistance of 1 ohm. If the wires had no resistance, the total resistance of the circuit would be 1 ohm and the current flowing through the circuit would be 12 amps. However, the wires have resistance and the total resistance of the circuit is 1.1 ohms (1 ohm + 0.05 ohms + 0.05 ohms). Therefore the circuit current is 10.9 amps ($I = E/R$, $I = 12$ V/1.1 ohms). The voltage drop across the bulbs would be 12 volts if there was no resistance in the wires. Now the voltage drop across the bulbs is only 10.9 volts ($E = I \times R$, $E = 10.9$

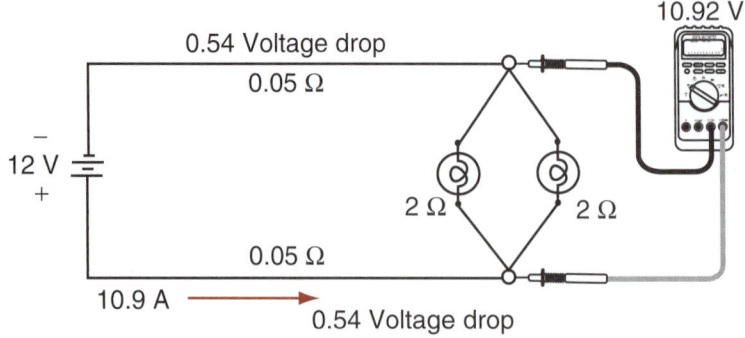

Figure 2-16 Measuring the voltage drop across a lamp. Notice the wires offer some resistance.

amps × 1 ohm). Although it may not be very noticeable, the light bulbs will not be as bright as they should be because of the decreased current and the decreased voltage drop.

A voltmeter can also be used to check for proper circuit grounding. If the voltmeter reading indicates full voltage at the lights, but the bulbs are not illuminated, the bulbs or sockets could be bad or the ground connection is faulty. An easy way to check for a defective bulb is to replace it with a known good bulb.

If the bulbs are good, the problem lies in either the light sockets or ground wires. Connect the voltmeter to the ground wire and a good ground. If the light socket is defective, the voltmeter will read 0 volts. If the socket is not defective, but the ground wire is broken or disconnected, the voltmeter will read very close to battery voltage. In fact, any voltage reading would indicate a bad or poor ground circuit. The higher the voltage, the greater the problem.

Thus far, we have discussed using a voltmeter on a direct current (DC) circuit. The voltmeter can also measure alternating current (AC). There are two methods used to display AC voltage: **root mean square (RMS)** and **average responding.** If the AC voltage signal is a true **sine wave,** both methods would display the same value. However, most automotive sensors do not produce a pure sine wave signal. The technician must know how the different meters will display the AC voltage reading under these circumstances. The type of display the voltmeter uses can be found in the meter's operating manual.

Ohmmeters

In contrast to the voltmeter, which uses the voltage available in the circuit, an ohmmeter is battery powered. The circuit being tested must be open. If the power is on in the circuit, the ohmmeter may be damaged. Most analog ohmmeters use a multiplier to determine higher resistance readings. A multiposition switch is used to select these multipliers or ranges. Normal ranges are labeled R × 1, R × 10, R × 100, and R × 1K. The reading on the ohmmeter must be multiplied by the value indicated by the range. If the reading is 22 ohms and the selected range is R × 1, the resistance value is 22 ohms. However, if the meter reads 22 ohms and the selected range is R × 100, the resistance value is 2,200 (22 × 100) ohms. The ranges are selected according to the component being tested. If the resistance of the component is unknown, start measuring with the lowest range and move up through the ranges until a good reading can be made. This will prevent misinterpreting infinity readings.

The two leads of the ohmmeter are placed across or in parallel with the circuit or component being tested. The red lead is placed on the positive side of the circuit and the black lead is placed on the negative side of the circuit (Figure 2-17). The meter sends current through

Root mean square (RMS) meters convert the AC signal to a comparable DC voltage signal.

Average responding meters display the average voltage peak.

A **sine wave** is a waveform that shows voltage changing polarity from positive to negative.

Classroom Manual Chapter 2, page 29

Special Tools
DMM

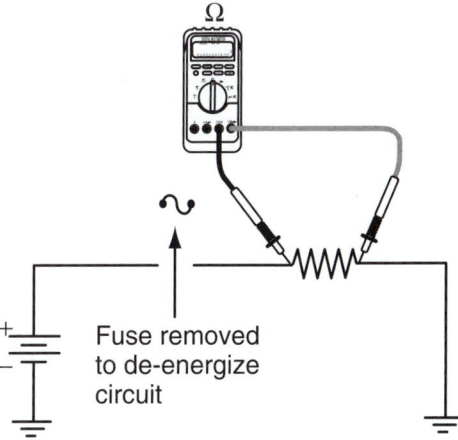

Figure 2-17 Measuring resistance with an ohmmeter. After power is removed to the circuit the meter is connected in parallel with the component being tested.

the component and determines the amount of resistance based on the voltage dropped across the load. The scale of an ohmmeter reads from 0 to infinity (∞). A 0 reading means there is no resistance in the circuit and may indicate a short in a component that should show a specific resistance. For example, a coil winding should have a high resistance value, a 0-ohm reading would indicate the coil windings are being bypassed. An infinity reading indicates a number higher than the meter can measure. This usually is an indication of an open circuit.

The test chart, shown in Figure 2-18, illustrates the readings that may be expected from an ohmmeter or voltmeter under different conditions. It is important to become familiar with these examples in order to analyze circuits.

> **CAUTION:** Since the ohmmeter is self-powered, never use an ohmmeter on a powered circuit.

Ohmmeters are also used to trace and check wires or cables. Assume that one wire of a four-wire cable is to be found. Connect one probe of the ohmmeter to the known wire at one end of the cable and touch the other probe to each wire at the other end of the cable. Any evidence of resistance, such as meter needle deflection, indicates the correct wire. Using this same method, you can check a suspected defective wire. If low resistance is shown on the meter, the wire is sound. If infinite resistance is measured, the wire is defective (open). If the wire is okay, continue checking by connecting the probe to other leads. Any indication of resistance indicates that the wire is shorted to one of the other wires and that the harness is defective.

Type of Defect	Test Unit	Expected Results
Open	Ohmmeter	∞ infinite resistance between conductor ends
	Test light	No light after open
	Voltmeter	∅ volts at end of conductor after the open
Short to Ground	Ohmmeter	∅ resistance to ground
	Test light	Lights if connected across fuse
	Voltmeter	Generally not used to test for ground
Short	Ohmmeter	Lower than specified resistance through load component
		∅ resistance to adjacent conductor
	Test light	Light will illuminate on both conductors
	Voltmeter	A voltage will be read on both conductors
Excessive Resistance	Ohmmeter	Higher than specified resistance through circuit
	Test light	Light illuminates dimly
	Voltmeter	Voltage will be read when connected in parallel over resistance

Figure 2-18 Circuit test chart.

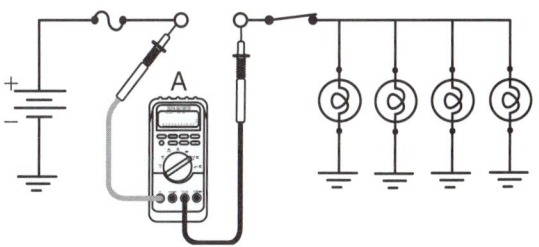

Figure 2-19 Measuring current flow with an ammeter. The meter must be connected in series with the circuit.

Ammeters

An ammeter measures current flow in a circuit. Circuit problems can also be identified by using an ammeter. An ammeter must be placed into series with the circuit being tested (Figure 2-19). Normally, this requires disconnecting a wire or connector from a component and connecting the ammeter between the wire or connector and the component. The red lead of the ammeter should always be connected to the most positive side of the connector and the black lead should be connected to the least positive side.

Classroom Manual
Chapter 2, page 28

Special Tools
DMM

CAUTION: Never place the leads of an ammeter across the battery or a load. This puts the meter in parallel with the circuit and will blow the fuse in the ammeter or possibly destroy the meter.

Most hand-held multimeters have a 10-ampere protection device. This is the highest amount of current flow the meter can read. When using the ammeter, start on a high scale and work down to obtain the most accurate readings.

It is much easier to test current using an ammeter with an inductive pickup (Figure 2-20). The pickup clamps around the wire or cable being tested. The inductive clamp has an arrow on it to indicate proper attachment to the conductor. The arrow indicates current flow direction using the conventional flow theory of positive to negative. These ammeters measure amperage based on the magnetic field created by the current flowing through the wire. This type of pickup eliminates the need to separate the circuit to insert the meter leads.

Figure 2-20 An ammeter with an inductive pickup. The inductive pickup eliminates the need to connect the meter in series with the circuit.

Because ammeters are built with very low internal resistance, connecting them in series does not add any appreciable resistance to the circuit. Therefore, an accurate measurement of the current flow can be taken.

For example, assume that a circuit normally draws 5 amps and is protected by a 6-amp fuse. If the circuit constantly blows the 6-ampere fuse, a short exists somewhere in the circuit. Mathematically, each light should draw 1.25 amperes ($5 \div 4 = 1.25$). To find the short, disconnect all lights by removing them from their sockets. Then, close the switch and read the ammeter. With the load disconnected, the meter should read 0 amperes. If there is any reading, the wire between the fuse block and the socket is shorted to ground.

If zero amps was measured, reconnect each light in sequence; the reading should increase 1.25 amperes with each bulb. If, when making any connection, the reading is higher than expected, the problem is in that part of the light circuit.

> **CAUTION:** When testing for a short, always use a fused jumper wire. Never bypass the fuse with a wire. The fuse should be rated at no more than 50% higher capacity than specifications. This offers circuit protection and provides enough amperage for testing. After the problem is found and corrected, be sure to install the specified rating of fuse for circuit protection.

Additional DVOM Functions

Some multimeters feature additional functions besides measuring AC or DC voltage, amperage, and ohms. Many multimeters are capable of measuring engine revolutions per minute (rpm), ignition dwell, diode condition, distributor condition, frequency, and temperature.

Some meters have a MIN/MAX function that displays the minimum, maximum, and average values received by the meter during the time the test was being recorded. This feature is valuable when checking sensors, output commands, or circuits for electrical **noise.** Noise is an unwanted voltage signal that rides on a signal. Noise is usually the result of radio frequency interference (RFI) or electromagnetic induction (EMI). The noise causes slight increases and decreases in the voltage signal to, or from, the computer. Another definition of noise is an AC signal riding on a DC voltage. The computer may attempt to react to the small changes in the signal as a result of the noise. This means the computer is responding to the noise and not the voltage signal, resulting in incorrect component operation.

Also, some multimeters may have the capabilities to measure **duty cycle,** pulse width, and frequency. Duty cycle is the measurement of the amount of "on" time as compared to the time of a **cycle** (Figure 2-21). A cycle is one set of changes in a signal that repeats itself several times. The duty cycle is displayed in a percentage. For example, a 60% duty cycle means the device is on 60% of the time and off 40% of the time of one cycle.

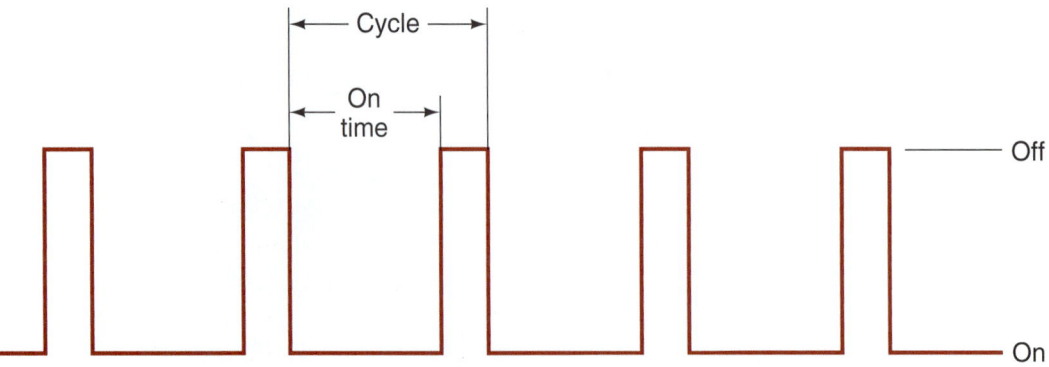

Figure 2-21 Duty cycle.

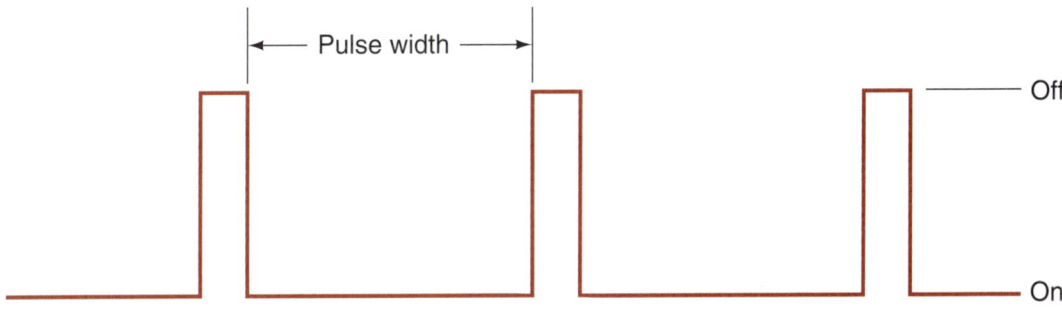

Figure 2-22 Pulse width.

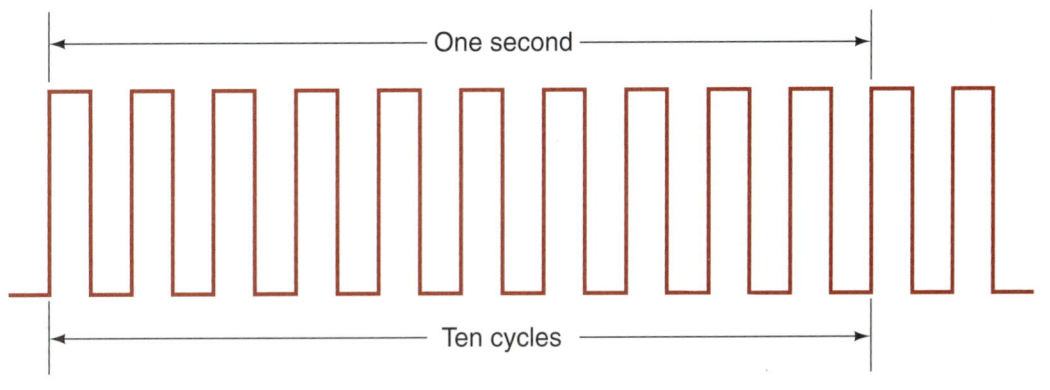

Figure 2-23 Frequency signal.

Pulse width is similar to duty cycle except that it is a measurement of time the device is turned on within a cycle (Figure 2-22). Pulse width is usually measured in milliseconds.

As previously mentioned, some multimeters can measure **frequency** (Figure 2-23). Frequency is a measure of the number of cycles that occur in one second. The higher the frequency, the more cycles that occur in a second. Frequency is measured in **hertz.** If the cycle occurs once per second, the frequency is one hertz. If the cycle is 300 time per second, the frequency is 300 hertz.

To accurately measure duty cycle, pulse width, and frequency, the meter's trigger level must be set. The trigger level tells the meter when to start counting. Trigger levels can be set at certain voltage levels or at a rise or fall in the voltage. Normally, meters have a built-in trigger level that corresponds with the voltage setting. If the voltage does not reach the trigger level, the meter will not begin to recognize a cycle. On some meters, you can select between a rise or fall in voltage to trigger the cycle count. A rise in voltage is a positive increase in voltage. This setting is used to monitor the activity of devices whose power feed is controlled by a computer. A fall in voltage is negative voltage. This setting is used to monitor ground-controlled devices.

Reading the DVOM

With the increased use of the digital volt/ohmmeter (DVOM), it is important for the technician to be able to accurately read the meter. By becoming proficient in the use of the DVOM, technicians will have confidence in their conclusions and recommended repairs. This will eliminate the replacement of parts that were not faulty. There are deviations between the different DVOM manufacturers as to the way the display is presented, but most follow the method described here.

PREFIX	SYMBOL	RELATION TO BASIC UNIT
Mega	M	1,000,000
Kilo	K	1,000
Milli	m	0.001 or $\dfrac{1}{1,000}$
Micro	μ	0.000001 or $\dfrac{1}{1,000,000}$
Nano	n	0.000000001
Pico	p	0.000000000001

Figure 2-24 Common prefixes used on digital multimeters.

Multimeters either have an "auto range" feature, in which the appropriate scale is automatically selected by the meter, or they must be manually set to a particular scale. Either way, to designate particular ranges and readings, meters display a prefix before the reading or range. Meters use the prefix because they cannot display long numbers. Values such as 20,200 Ω cannot be displayed as a whole number. As a result, scales are expressed as a multiple of tens or use the prefix K, M, m, and μ (Figure 2-24). The prefix K stands for kilo and represents 1,000 units. For example, a reading of 10K equals 10,000. An M stands for mega and represents 1,000,000 units. A reading of 10M would represent 10,000,000. An m stands for milli and represents 0.001 of a unit. A reading of 10m would be 0.010. The symbol μ stands for micro and represents 0.000001 of a unit. In this case a reading of 125.0μ would represent 0.000125.

If the display has no prefix before the unit being measured (V, A, Ω), the reading displayed is read directly. For example, if the reading was 1.243 V, the actual voltage value is 1.243. However, if there is a prefix displayed, then the decimal point will need to be floated to determine actual readings. If the prefix is M (mega), then the decimal is floated six places to the right. For example, a reading display of 2.50 MΩ is actually 2,500,000 ohms. A reading display of 0.250 MΩ is actually 250,000 ohms.

A prefix of K (kilo) means the decimal point needs to move three places to the right. For example, a display reading of 56.4 KΩ is actually 56,400 ohms. A reading of 1.264 KΩ is actually 1,264 ohms.

If the prefix is m (milli), the decimal is floated three places to the left. For example, a reading of 25.4 mA is representing 0.0254 amperes. A display of 165.0 mA is actually 0.165 amperes.

Finally, if the prefix is a μ (micro), the decimal is floated six points to the left. A reading displayed as 125.3μ would represent 0.0001253 amperes while a reading of 4.6 mA is actually 0.0000046 amperes.

When using the ohmmeter function of the DVOM, make sure power to the circuit being tested is turned off. Also, be sure to calibrate the meter before taking measurements. This is done by holding the two test leads together. Most DVOMs will self-calibrate while others will need to be adjusted by turning a knob until the meter reads zero. Connect the DVOM in parallel to the portion of the circuit being tested. If continuity is good, the DVOM will read zero or close to zero even on the lowest scale. If the continuity is very poor, the DVOM will display an infinite reading. This reading is usually shown as a "1.000," a "1," or an "OL."

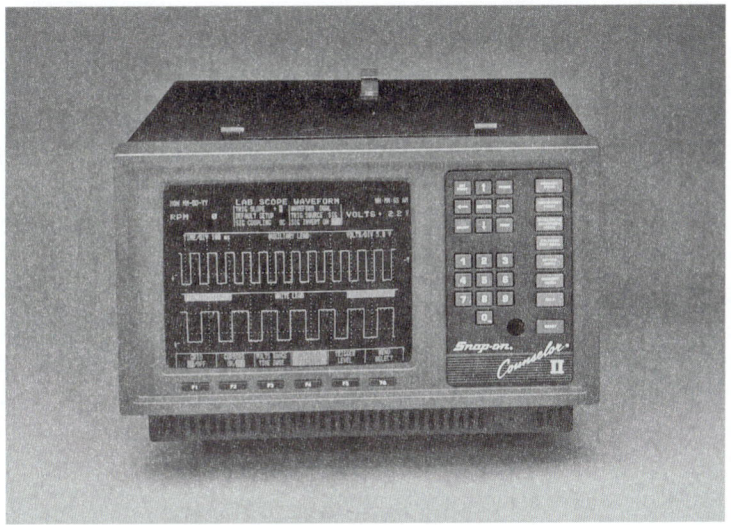

Figure 2-25 Oscilloscope.

Lab Scopes and Oscilloscopes

An oscilloscope is a visual voltmeter (Figure 2-25). For many years, technicians have used scopes to diagnose ignition, fuel injection, and charging systems. These scopes, called "tune-up scopes," were normally part of a large diagnostic machine, although some were stand-alone units. In recent years, an electronic scope, referred to as a lab scope, has become the diagnostic tool of choice for many good technicians.

The oscilloscope is very useful in diagnosing many electrical problems quickly and accurately. Digital and analog voltmeters do not react fast enough to read systems that cycle quickly. The oscilloscope may be considered as a very-fast-reacting voltmeter that reads and displays voltages. The scope allows the technician to view voltage over time (Figure 2-26).

Oscilloscopes are commonly called scopes.

Special Tools

Lab scope

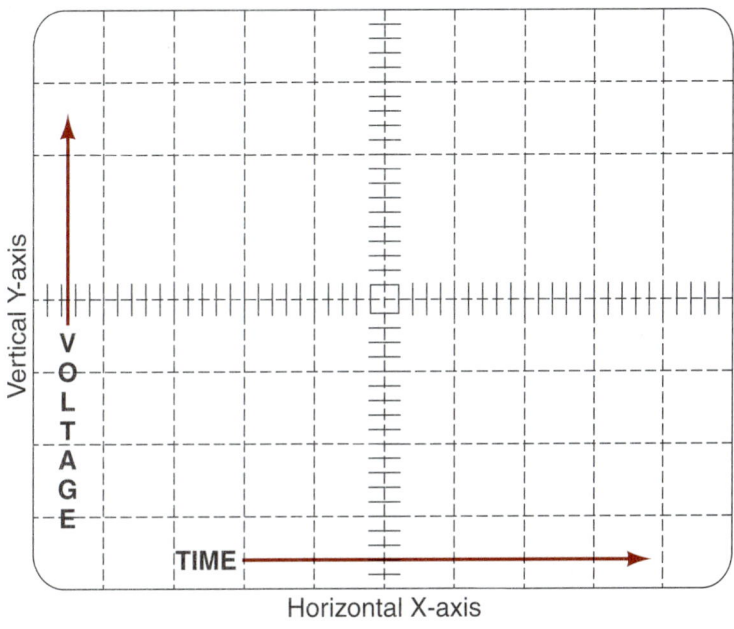

Figure 2-26 Grids on a scope screen serve as a time and voltage reference.

These voltage readings appear as a voltage trace on the oscilloscope screen. Some smaller oscilloscopes use liquid crystal displays (LCD). However, most larger screens are cathode ray tubes (CRT), which are very similar to the picture tube in a television set. High voltage from an internal source is supplied to an electron gun in the back of the CRT when the oscilloscope is turned on. The screen of an analog scope is actually an anode. It has a positive charge and attracts negatively charged particles from the cathode or electron gun. This electron gun emits a continual beam of electrons against the front of the CRT. The external leads on the oscilloscope are connected to deflection plates above, below, and on each side of the electron beam. When a voltage signal is supplied from the external leads to the deflection plates, the electron beam is distorted and strikes the front of the screen in different locations to indicate the voltage signal from the external leads. As the beam of electrons strikes the inside of the screen, it causes the phosphorous coating on the screen to glow.

An upward movement of the voltage trace on an oscilloscope screen indicates an increase in voltage, and a downward movement of this trace represents a decrease in voltage. As the voltage trace moves across an oscilloscope screen, it represents a specific length of time. Most oscilloscopes of this type are referred to as analog scopes or real-time scopes. This means the voltage activity is displayed without any delay.

The squares formed by the grid are called major divisions. These are further defined by smaller divisions called minor divisions.

The waveform on a scope is commonly called a trace.

Today, most technicians use a variation of the oscilloscope called a lab scope. These divisions set up a grid pattern on the screen. Time is represented by the horizontal movement of the waveform. Voltage is measured with the vertical position of the waveform. Since the scope displays voltage over time, the waveform moves from left (the beginning of measured time) to the right (the end of measured time). The value of the divisions can be adjusted to improve the view of the voltage waveform. For example, the vertical scale can be adjusted so that each division represents 0.5 volts, and the horizontal scale can be adjusted so that each division equals 0.005 (5 milliseconds). This allows the technician to view small changes in voltage that occur in a very short period of time. The grid serves as a reference for measurements.

Since a scope displays actual voltage, it will display any electrical noise or disturbances that accompany the voltage signal (Figure 2-27). This noise can cause intermittent problems

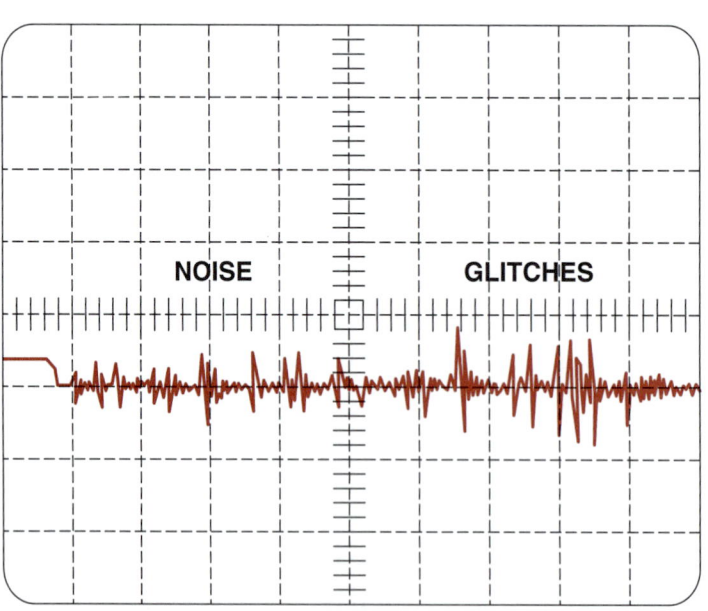

Figure 2-27 RFI noise and glitches showing as voltage signals.

with unpredictable results. When a computer receives a voltage signal with noise, it will try to react to the minute changes. As a result, the computer responds to the noise rather than the voltage signal.

Electrical disturbances or **glitches** are momentary changes in the signal. These can be caused by intermittent shorts to ground, shorts to power, or opens in the circuit. These problems can occur for only a moment or may last for some time. A lab scope is handy for finding these and other causes of intermittent problems. By observing a voltage signal and wiggling or pulling a wiring harness, any looseness can be detected by a change in the voltage signal. This type of testing is commonly referred to as a "wiggle test."

The digital scope, commonly called a digital storage oscilloscope or DSO, converts the voltage signal into digital information and stores it into its memory. Some DSOs send the signal directly to a computer or a printer, or save it to a disk. To help with diagnosis, a technician can "freeze" the **captured signal** for closer analysis. DSOs also have the ability to capture low-frequency signals. Low-frequency signals tend to flicker when displayed on an analog screen. To have a clean waveform on an analog scope, the signal must be repetitive and occurring in real time. The signal on a DSO is not quite real time. Rather, it displays the signal as it occurred a short time before.

This delay is actually very slight. Most DSOs have a sampling rate of one million samples per second. This is quick enough to serve as an excellent diagnostic tool. This fast sampling rate allows slight changes in voltage to be observed. Slight and quick voltage changes cannot be observed on an analog scope.

Both an analog and a digital scope can be dual-trace scopes (Figure 2-28). This means they both have the capability of displaying two traces at one time. By watching two traces simultaneously, you can watch the cause and effect of a sensor, as well as compare a good or normal waveform to the one being displayed.

Waveforms

A waveform represents voltage over time. Any change in the amplitude of the trace indicates a change in the voltage. When the trace is a straight horizontal line, the voltage is constant

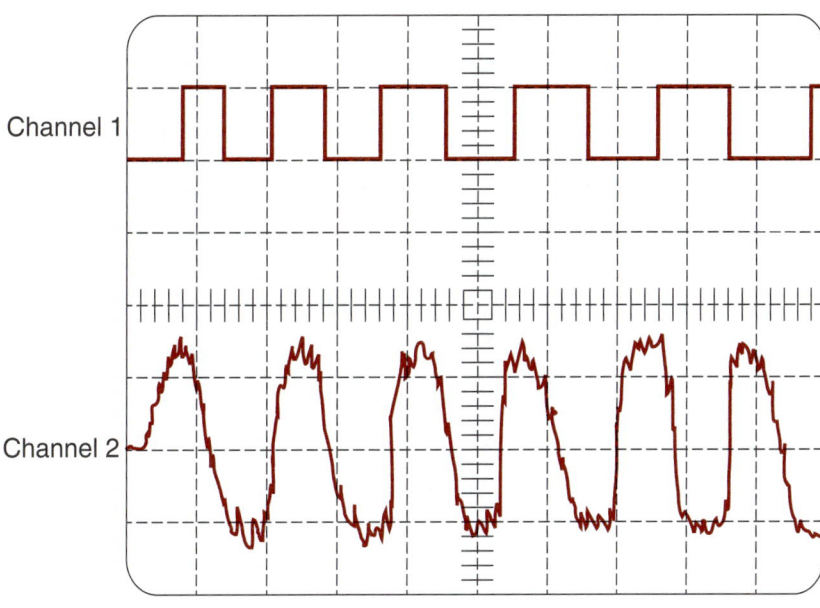

Figure 2-28 A dual-trace scope can show two patterns at the same time.

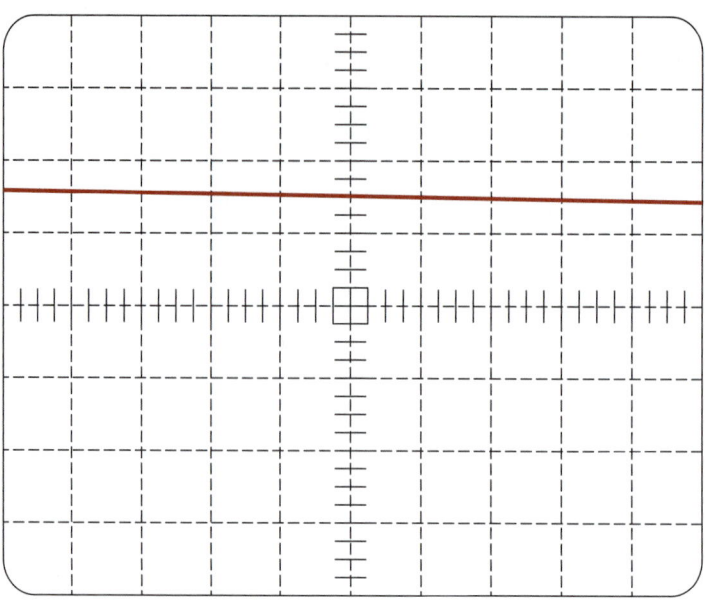

Figure 2-29 A waveform showing a constant voltage.

(Figure 2-29). A diagonal line up or down represents a gradual increase or decrease in voltage. A sudden rise or fall in the trace indicates a sudden change in voltage.

Scopes can display AC and DC, one at a time or both, as in the case of noise caused by RFI. The consistent change of polarity and amplitude of the AC signal causes slight changes in the DC voltage signal. A normal AC signal changes its polarity and amplitude over a period of time. The waveform created by AC voltage is typically a sine wave (Figure 2-30). One complete sine wave shows the voltage moving from zero to its positive peak, moving down through zero to its negative peak, and returning to zero. If the rise and fall from positive and negative is the same, the wave is said to be **sinusoidal.** If the rise and fall are not the same, the wave is nonsinusoidal.

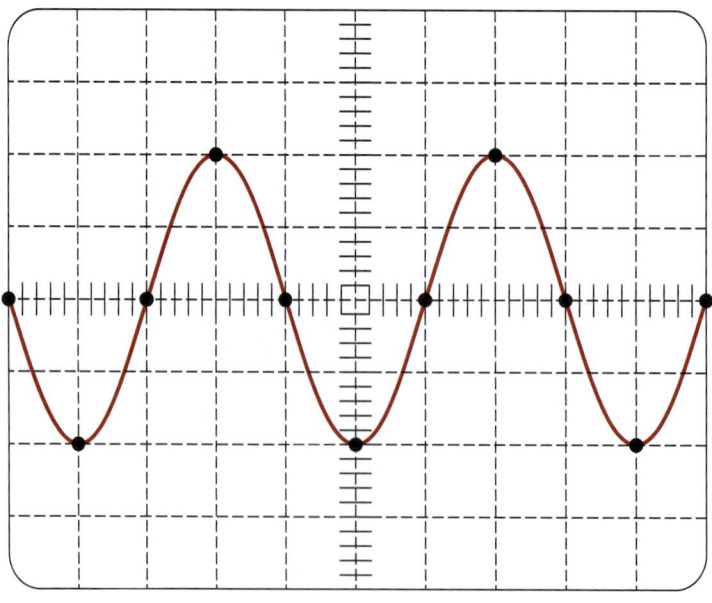

Figure 2-30 An AC voltage sine wave.

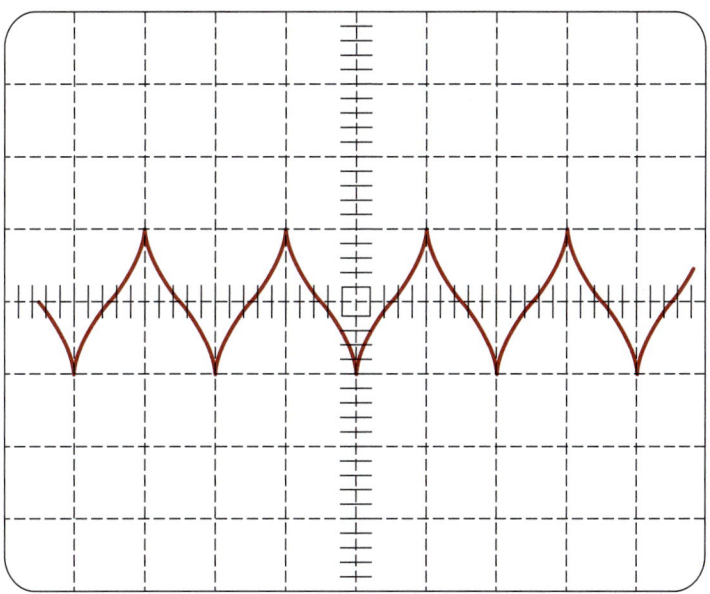

Figure 2-31 An AC voltage trace from a typical permanent magnetic generator-type sensor.

One complete sine wave is a cycle. The number of cycles that occur per second is the frequency of the signal. Checking frequency or cycle time is one way of checking the operation of some electrical components. Input sensors are the most common components that produce AC voltage. Permanent magnet voltage generators produce an AC voltage that can be checked on a scope (Figure 2-31). AC voltage waveforms should also be checked for noise and glitches. These may send false information to the computer.

DC voltage waveforms may appear as a straight line or a line showing a change in voltage. Sometimes a DC voltage waveform will appear as a **square wave,** which shows voltage making an immediate change (Figure 2-32). This type of wave represents voltage being applied (circuit being turned on), voltage being maintained (circuit remaining on), and no voltage applied (circuit is turned off). Of course, a DC voltage waveform may also show gradual voltage changes.

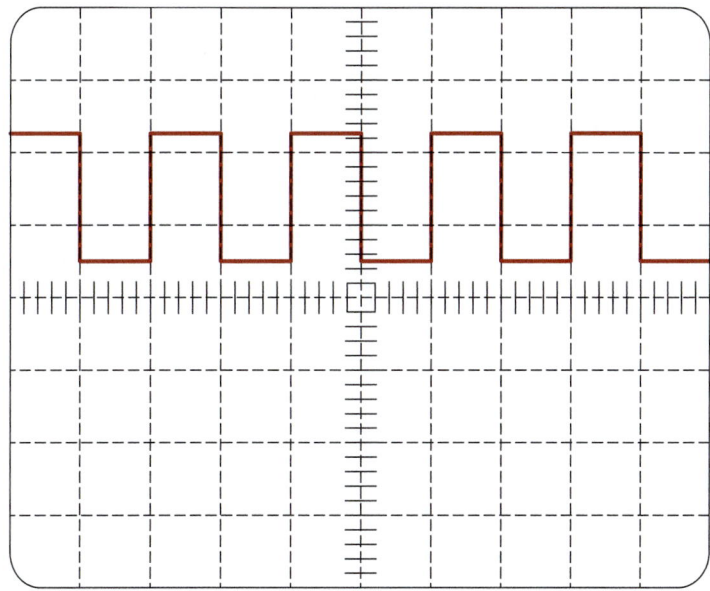

Figure 2-32 Typical square wave.

Scope Controls

Depending on the manufacturer and the model of the scope, the type and number of its controls will vary. However, nearly all scopes have these controls: intensity, vertical (Y-axis) adjustments, horizontal (X-axis) adjustments, and trigger adjustments. The intensity control is used to adjust the brightness of the trace. This allows for clear viewing regardless of the light around the scope screen.

The vertical adjustment controls the voltage displayed. The voltage setting of the scope is the voltage that will be shown per division (Figure 2-33). If the scope is set at 0.5 (500 milli) volts, a 5-volt signal will need 10 divisions. Likewise, if the scope is set to 1 volt, 5 volts will need only 5 divisions. While using a scope, it is important to set the vertical so that voltage can be accurately read. Setting the voltage too low may cause the waveform to move off the screen, while setting it too high may cause the trace to be flat and unreadable. The vertical position control allows the vertical position of the trace to be moved anywhere on the screen.

The horizontal position control allows the horizontal position of the trace to be set on the screen. The horizontal control is actually the time control of the trace (Figure 2-34). Setting the horizontal control is setting the time base of the scope's sweep rate. If the time per division is set too low, the complete trace may not show across the screen. Also, if the time per division is set too high, the trace may be too crowded for detailed observation. The time per division (TIME/DIV) can be set from very short periods of time (millionths of a second) to full seconds.

Trigger controls tell the scope when to begin a trace across the screen. Setting the trigger is important when trying to observe the timing of something. Proper triggering will allow the trace to repeatedly begin and end at the same points on the screen. There are usually numerous trigger controls on a scope. The trigger mode selector has a NORM and AUTO position. In the NORM setting, no trace will appear on the screen until a voltage signal occurs within the set time base. The AUTO setting will display a trace regardless of the time base.

Slope and level controls are used to define the actual trigger voltage. The slope switch determines whether the trace will begin on a rising or falling of the voltage signal (Figure 2-35). The level control determines where the time base will be triggered according to a certain point on the slope.

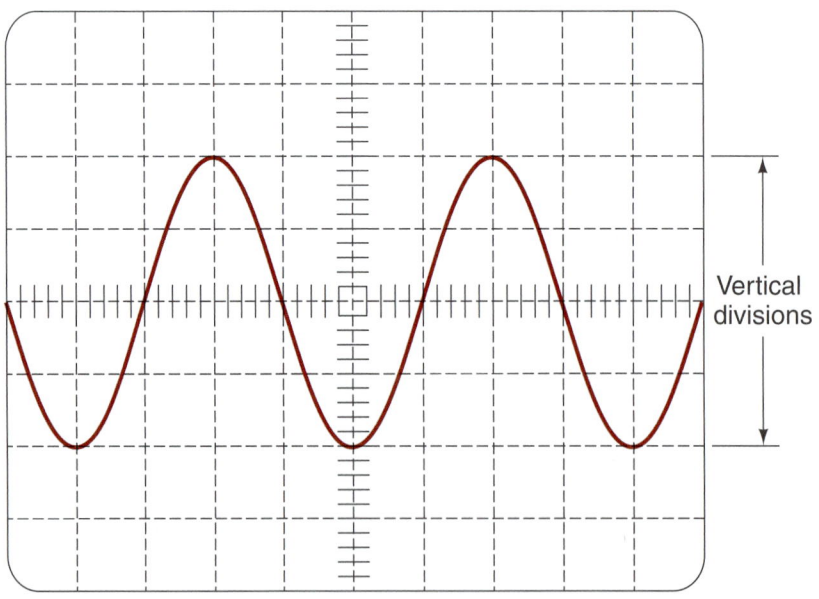

Figure 2-33 Vertical divisions represent voltage.

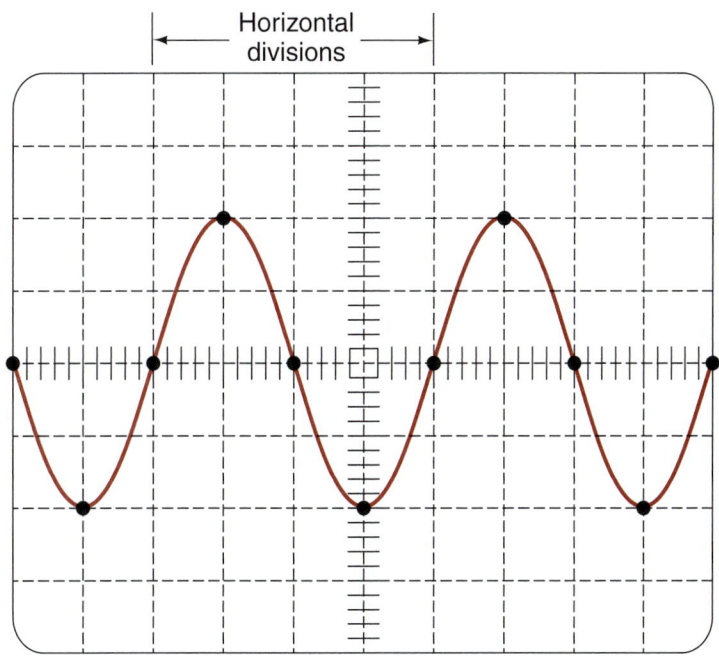

Figure 2-34 Horizontal divisions represent time.

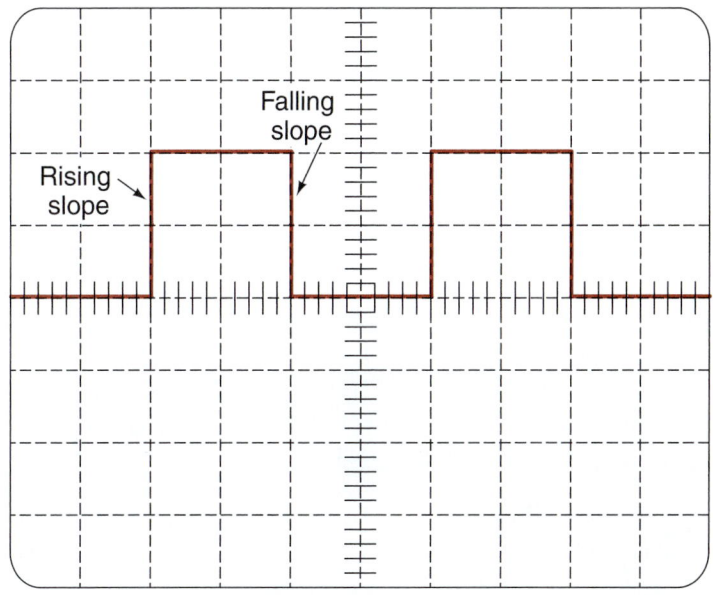

Figure 2-35 A trigger can be set to start the trace with a rise or fall of voltage.

A trigger source switch tells the scope which input signal to trigger on. This can be Channel 1, Channel 2, line voltage, or an external signal. External signal triggering is very useful when observing a trace of a component that may be affected by the operation of another component. An example of this would be observing fuel injector activity when changes in throttle position are made. The external trigger would be voltage change at the throttle position sensor. The displayed trace would be the cycling of a fuel injector. Channel 1 and Channel 2 inputs are determined by the points of the circuit being probed. Some scopes have a switch that allows inputs from both channels to be observed at the same time or alternately.

Scan Tools

The introduction of power train computer controls brought with it the need for tools capable of troubleshooting electronic control systems. There are a variety of computer **scan tools** available today that do just that (Figure 2-36). Connected to the computer through diagnostic connectors, a scan tool can access diagnostic trouble codes (DTCs), run tests to check system operations, and monitor the activity of the system. Trouble codes and test results are displayed on an LED screen or printed out on the scanner printer.

A scan tool receives its testing information from one of several sources. Some scan tools have a programmable read-only memory (PROM) chip that contains all the information needed to diagnose specific model lines. This chip is normally contained in a cartridge, which is plugged into the tool. The type of vehicle being tested determines the appropriate cartridge that should be inserted. These cartridges contain the test information for that particular car. A cartridge is typically needed for each make and model vehicle. As new systems are introduced on new car models, a new cartridge is made available.

Scan testers have the capability to test many onboard computer systems such as engine computers, antilock brake computers, air bag computers, and suspension computers, depending on the year and make of vehicle and the type of scan tester. In many cases, the technician must select the computer system to be tested with the scan tester after the tester is connected to the vehicle.

The scan tester is connected to specific diagnostic connectors on various vehicles. Some manufacturers have one diagnostic connector and they connect the data wire from each onboard computer to a specific terminal in this connector. Other vehicle manufacturers have several different diagnostic connectors on each vehicle and each of these connectors may be connected to one or more onboard computers. A set of connectors is supplied with the scan tester to allow tester connection to various diagnostic connectors on different vehicles.

The scan tester must be programmed for the model year, make of vehicle, and type of engine. With some scan testers, this selection is made by pressing the appropriate buttons on the tester as directed by the digital tester display. On other scan testers, the appropriate memory card must be installed in the tester for the vehicle being tested. Some scan testers have a built-in printer to print test results, while other scan testers may be connected to an external printer.

Figure 2-36 A typical handheld scan tester.

As automotive computer systems become more complex, the diagnostic capabilities of scan testers continue to expand. Many scan testers now have the capability to store, or "freeze," data into the tester during a road test and then play back this data when the vehicle is returned to the shop.

Some scan testers now display diagnostic information based on the fault code in the computer memory. Service bulletins published by the scan tester manufacturer may be indexed by the tester after the vehicle information is entered in the tester. Other scan testers will display sensor specifications for the vehicle being tested.

Scan Tester Features

Scan testers display data and **diagnostic trouble codes (DTCs)** on computer systems and perform many other diagnostic functions. Diagnostic trouble codes (DTCs) are fault codes that represent a circuit failure in a monitored system. On many vehicles, scan testers have the capability to diagnose various computer systems such as engine, transmission, antilock brake system (ABS), suspension, and air bag. Scan testers vary depending on the manufacturer, but many scan testers have the following features:

1. A display window displays data and messages to the technician. Messages are displayed from left to right. Most scan testers display at least four readings at the same time.

2. A memory cartridge that plugs into the scan tester. These memory cartridges are designed for specific vehicles and electronic systems. For example, a different cartridge may be required for the transmission computer and the engine computer. Most scan tester manufacturers supply memory cartridges for domestic and imported vehicles.

3. A power cord that connects from the scan tester to the battery terminals or cigarette lighter socket.

4. An adapter cord that plugs into the scan tester and connects to the data link connector (DLC) on the vehicle (Figure 2-37). A special adapter cord is supplied with the tester for the diagnostic connector on each make of vehicle.

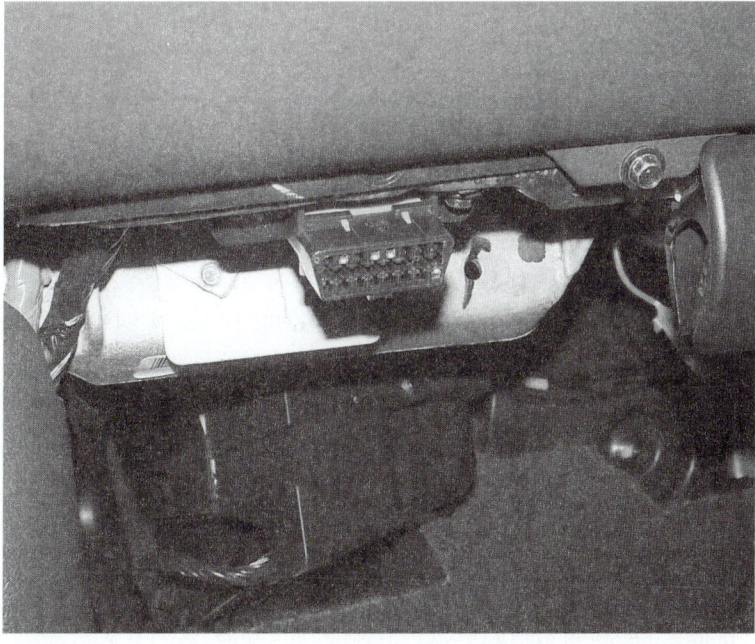

Figure 2-37 The scan tester connects to the diagnostic link connector (DLC).

Figure 2-38 Scan tester button controls.

5. A serial interface for connecting optional devices, such as a printer, terminal, or personal computer.

6. A keypad that allows the technician to enter data and reply to tester messages.

Typical keys that may be on a scan tester include (Figure 2-38):

1. Numbered keys covering digits 0 through 9.

2. Horizontal or vertical arrow keys that allow the technician to move backward and forward through test modes and menus.

3. ENTER keys to enter information into the tester.

4. PAGE BACK key that allows the technician to interrupt the current procedure and go back to the previous modes.

5. "F" keys to allow the technician to perform special functions described in the scan tester manufacturer's manuals.

6. MORE key that allows the technician to obtain additional diagnostic information from the scan tester software.

7. YES and NO keys to allow the technician to select or reject specific procedures.

⚠️ **WARNING:** The tester manufacturer's and vehicle manufacturer's recommended scan tester diagnostic procedures must be followed while diagnosing computer systems. Improper test procedures may result in scan tester damage and computer system damage.

■ **CAUTION:** Always keep scan tester leads away from rotating parts such as belts and fan blades. Personal injury or property damage may result if scan tester leads become tangled in rotating parts.

The advantages offered by the use of a scan tool include:

1. A scan tester provides quick access to data from various on-board computers. Some vehicles have several diagnostic link connectors (DLCs) to which the scan tester must be connected to access data from a specific computer. For example, some vehicles have separate DLCs to access the powertrain control module (PCM) and antilock brake system (ABS) computer data. Vehicles that have on-board diagnostic (OBD II) systems have a central DLC and data links from the various on-board computers to this DLC. Accessing this computer data greatly reduces diagnostic time.

2. A wide variety of modules are available for many scan testers. These modules allow the same scan tester to display data from many vehicles, including imported vehicles. Some scan tester modules access service bulletin information related to engine and transmission problems. This information is available in a book published by the scan tester manufacturer.

3. The vehicle can be driven on a road test with the scan tester connected to the DLC. This allows the technician to observe computer data during various operating conditions when a specific problem may occur. Most scan testers have a snapshot capability which freezes computer data into the scan tester memory for a specific period of time. This data may be played back after the technician returns to the shop.

4. Most scan testers can be connected to a printer and a copy of the scan tester data can be printed. This allows improved communication between the customer, service writer, and the technician.

5. A scan tester can be connected to a personal computer (PC). This connection allows data to be transferred from the scan tester to the PC. This data may be saved and recalled at a future time. With a computer modem, this information may be transferred to an off-site diagnostic center for analysis.

Static Strap

You are probably familiar with static charges in one form or another. The most common experience with static electricity is when you slide your feet across a carpeted floor and then touch something. You might feel and see a slight static discharge. The action of sliding your feet across the carpet placed a slight electrical charge on you. A change in the number of electrons on you puts you at a different charge level than the objects around you. When you touch them, there is a discharge between you and the object. Although this discharge generally does nothing to you other than to perhaps surprise you, it can do potentially great damage to electronic circuitry. Today's technicians must realize that static electricity will have to be discharged safely before they begin working on an electronic component or processor.

To effectively work on these circuits, some precautions are necessary. Generally these can be summarized by the statement that you must be at the same electrical potential as the component you are working on and the vehicle you are working in. Many manufacturers suggest the use of static straps that connect the technician, the component, and the ground system of the vehicle together (Figure 2-39). The theory behind this is to place all things that will touch at the same electrical potential so that a discharge will not take place. Even if you do not have all the special static straps, run jumper ground wires between the components and the vehicle, and ground yourself to the vehicle before you begin working.

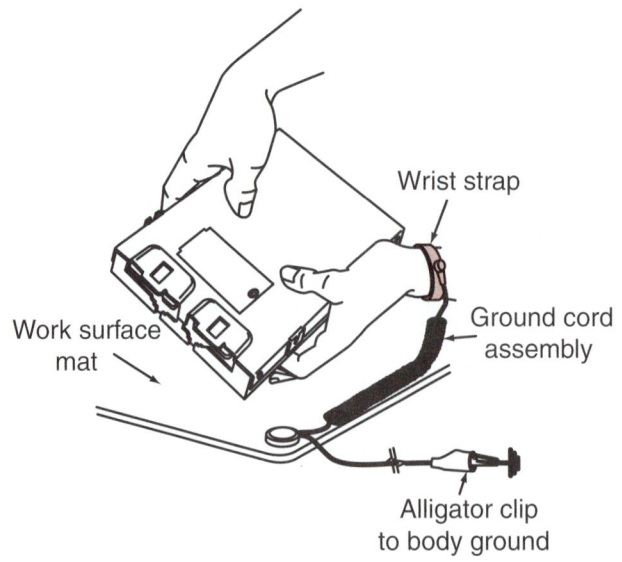

Wrist strap

Ground cord assembly

Work surface mat

Alligator clip to body ground

Figure 2-39 Static straps are needed to protect electronic devices.

Memory Keepers

Memory keepers, or battery backups, maintain the computer's memory when the battery is disconnected for service. Such components as radio presets, clock settings, memory seat positions, and computer functions are lost when the battery is disconnected. A simple memory keeper (Figure 2-40) plugs into the power outlet (cigarette lighter) socket, and a 9-volt battery will keep the settings for up to 4 hours. These devices should not be used if service is being performed to the air bag system.

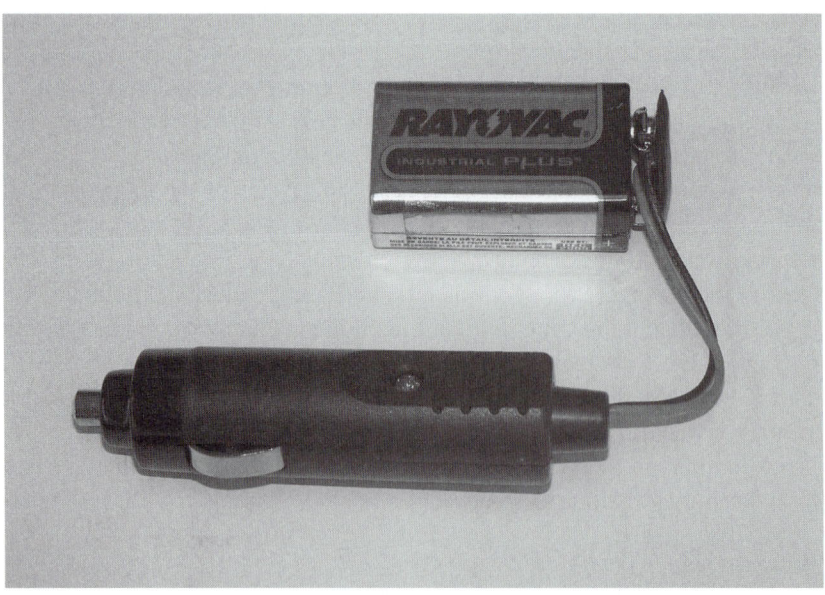

Figure 2-40 A memory keeper will keep the computer memories operational while the battery is disconnected.

Service Information

With today's complex electrical systems, it would be impossible to repair every customer concern that is brought into the service bay without having the proper service information. The service manual (in either paper form or electronic format) is one of the most important and valuable tools for today's technician. It provides information concerning system description, service procedures, specifications, and diagnostic information. In addition, the service manual provides information concerning wiring harness connections and routing, component location, and fluid capacities. Service manuals can be supplied by the vehicle manufacturer or through aftermarket suppliers.

It is also important to stay current with updated service information. This is usually published as Technical Service Bulletins (TSBs). These documents may provide information concerning fixes for a problem, new part numbers to replace a defective unit, corrections to service manual information, and general information of system operation.

To obtain the correct information, you must be able to identify the engine you are working on. This may involve using the vehicle's VIN number. This number has a code for model year and engine. Which numbers are used varies between manufacturers, but the service manual will provide instructions for proper VIN usage.

In the past, each manufacturer and manual publisher used its own method of organizing its manuals. Recent guidelines now require manufacturer service manuals to have a standard organization (Figure 2-41).

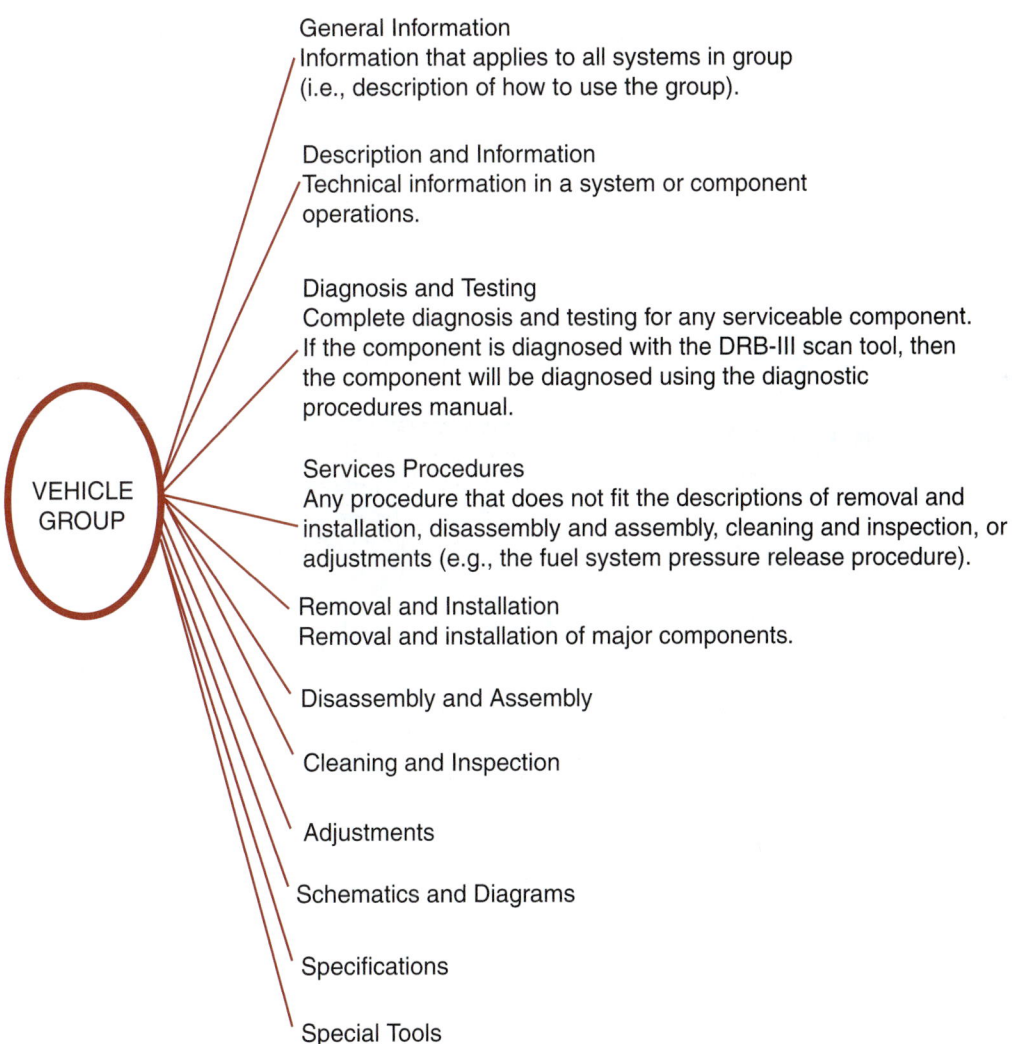

Figure 2-41 Service manuals use a uniform layout.

Procedural information provides the steps necessary to perform the task. Most service manuals provide illustrations to guide the technician through the task (Figure 2-42). To get the most out of the service manual, you must use the correct manual for the vehicle and system being worked on, and follow each step in order. Some technicians lead themselves down the wrong trail by making assumptions and skipping steps.

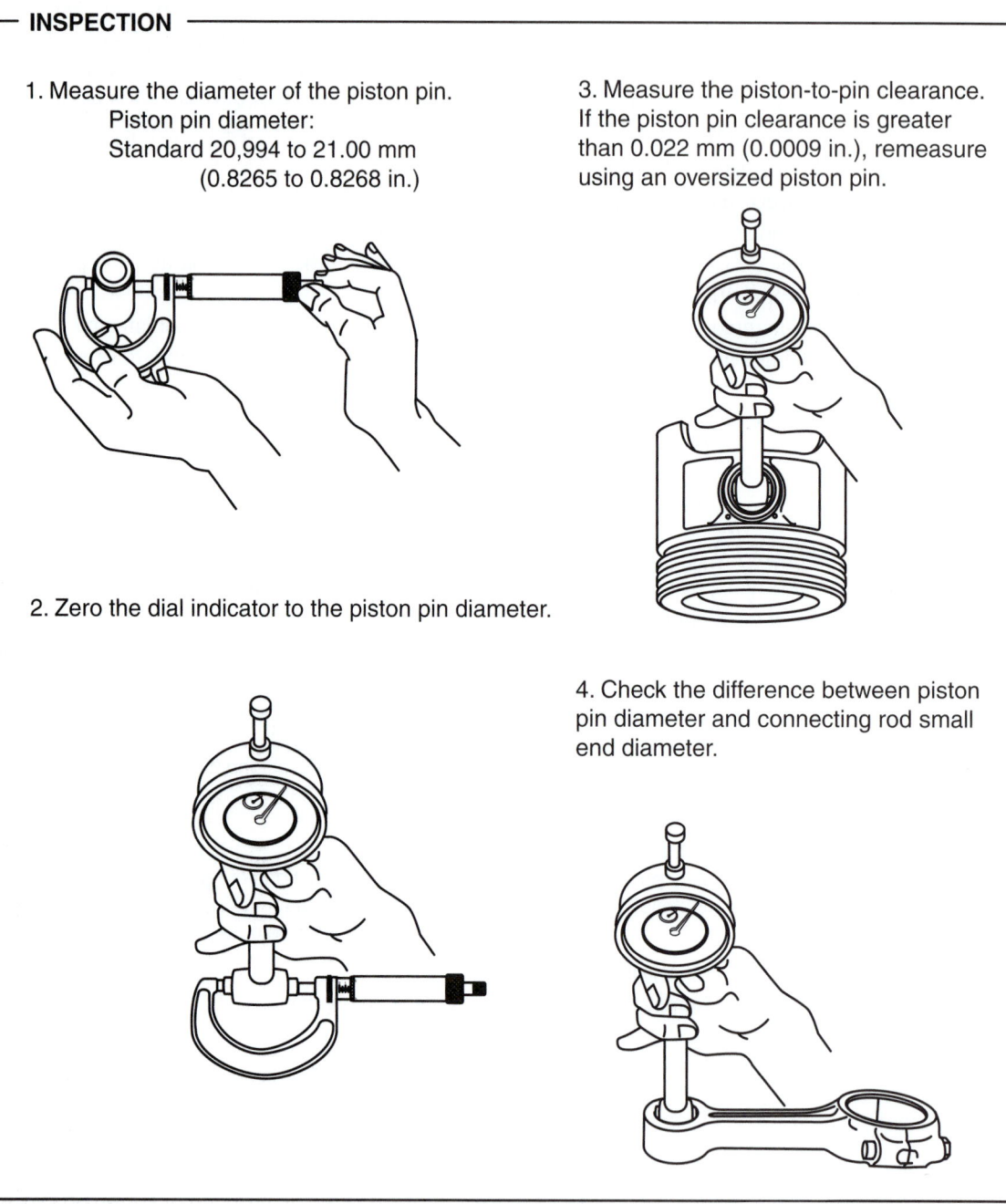

INSPECTION

1. Measure the diameter of the piston pin.
   Piston pin diameter:
   Standard 20,994 to 21.00 mm
   (0.8265 to 0.8268 in.)

2. Zero the dial indicator to the piston pin diameter.

3. Measure the piston-to-pin clearance. If the piston pin clearance is greater than 0.022 mm (0.0009 in.), remeasure using an oversized piston pin.

4. Check the difference between piston pin diameter and connecting rod small end diameter.

Figure 2-42 The service manual may use illustrations to guide the technician through the task.

TORQUE SPECIFICATIONS

Fan Assembly Motor to Fan	3.3 Nm (29 lbs. in.)
Fan to Radiator Support Bolt	9 Nm (80lbs. in.)
Hose Clamps	
Heater Hose	1.7 Nm (15 lbs. in.)
Radiator Hose	3.4 Nm (30 lbs. in.)
Lower Air Deflector to Impact Bar	2 Nm (18 lbs. ft.)
Throttle Body Inlet Pipe Bolt 3.1 L (VIN T)	25 Nm (18 lbs. ft.)
Transmission Oil Cooler Fittings at Radiator	27 Nm. (20 lbs. ft.)
Trans. Oil Cooler Pipe Connections (Alum. Radiator)	20 Nm (15 lbs. ft.)
Radiator Outlet Pipe to Block 2.01 (VIN K)	27 Nm (20 lbs. ft.)
Radiator to Radiator Support Bolts	10 Nm (90 lbs. in.)
Thermostat Housing Bolts	
2.0 L (VIN K) & 3.1 L (VIN T)	27 Nm (20 lbs. ft.)
Coolant Pump to Block Bolts	
2.0 L (VIN K)	25 Nm (19 lbs. ft.)
3.1 L (VIN T)	24 Nm (18 lbs. ft.)
Coolant Pump to Front Cover Bolts	
3.1 L (VIN T)	10 Nm (90 lbs. in.)
Coolant Pump Pulley to Pump Bolts	
2.0 L (VIN K)	24 Nm (17 lbs. ft.)
3.1 L (VIN T)	21 Nm (15 lbs. ft.)
Surge Tank Bolts	4 Nm (15 lbs. ft.)
Surge Tank Pipe to Block Bolt 3.1 L (VIN T)	8 Nm(70 lbs. in.)
Temperature Sending or Gauge Unit	27 Nm (20 lbs. ft.)

Figure 2-43 Specifications can be found in chart form.

Torque, end play, and clearance specifications may be located within the text of the procedural information. In addition, specifications may be provided in a series of tables (Figure 2-43). The heading above the table provides a quick reference to the type of specification information being provided.

Diagnostic procedures are often presented in a chart form or a tree (Figure 2-44). The tree guides you through the process as system tests are performed. The result of a test then directs you to a branch. Keep following the steps until the problem is isolated.

Since the service manual is divided into several major component areas, a table of contents is provided for easy access to the information. Each component area of the vehicle is covered

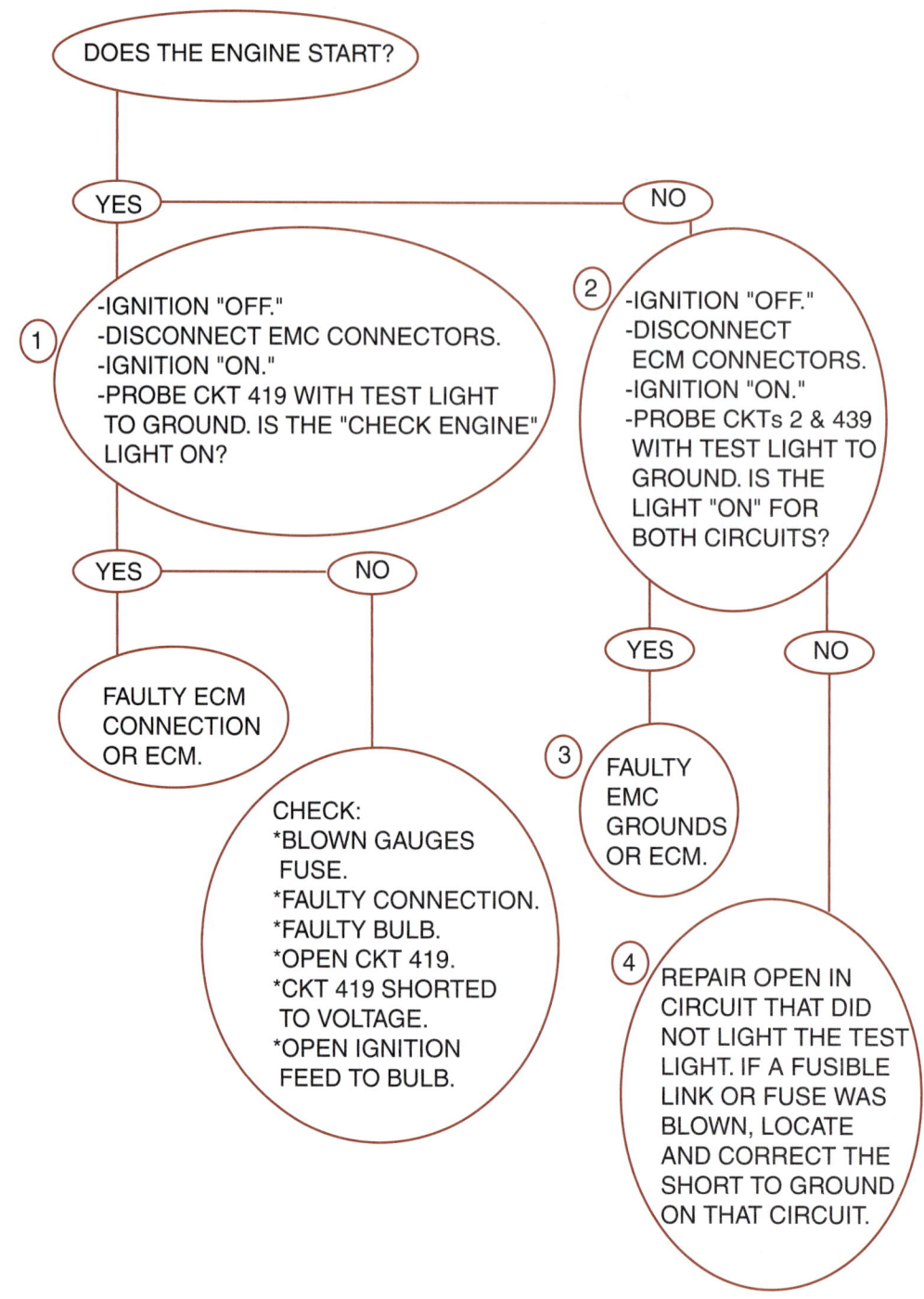

Figure 2-44 A diagnostic tree leads the technician to the cause of the fault.

under a section in the service manual (Figure 2-45). Using the table of contents identifies the section to turn to. Once in the appropriate section, a smaller, more specific table of contents will direct you to the page on which the information is located. Due to the extensive amount of information provided in service manuals and the number of component areas, the manual may be divided into several volumes.

TABLE OF CONTENTS	SECTION NUMBER
GENERAL INFO. AND LUBE	
General Information	0A
Maintenance and Lubrication	0B
HEATING AND AIR COND.	
Heating and Vent.(nonA/C)	1A
Air Conditioning System	1B
V-5 A/C Compressor Overhaul	1D3
BUMPERS AND FRONT BODY PANELS	
Bumpers (See 10-4)	
Fr. End Body Panels (See 10-5)	
STEERING, SUSPENSION, TIRES, AND WHEELS	
Diagnosis	3
Wheel Alignment	3A
Power Steering Gear & Pump	3B1
Front Suspension	3C
Rear Suspension	3D
Tires and Wheels	3E
Steering Col. On-Vehicle Service	3F
Steering Col. - Std. Unit Repair	3F1
Steering Col. - Tilt, Unit Repair	3F2
DRIVE AXLES	
Drive Axles	4D
BRAKES	
General Info. - Diagnosis and On-Car Service	5
Compact Master Cylinder	5A1
Disc Brake Caliper	5B2
Drum Brake - Anchor Plate	5C2
Power Brake Booster Assembly	5D2
ENGINES	
General Information	6
2.0 Liter I-4 Engine	6A1
3.1 Liter V6 Engine	6A3
Cooling System	6B
Fuel System	6C
Engine Electrical - General	6D
Battery	6D1
Cranking System	6D2
Charging System	6D3
Ignition System	6D4
Engine Wiring	6D5
Driveability & Emissions - Gen.	6E
Driveability & Emissions - TBI	6E2
Driveability & Emissions - PFI	6E3
Exhaust System	6F

TABLE OF CONTENTS	SECTION NUMBER
TRANSAXLE	
Auto.Transaxel On-Car Serv.	7A
Auto. Trans. - Hydraulic Diagnosis	3T40-HD
Auto. Trans. - Unit Repair	3T40
Man. Trans. On-Car Service	7B
5-Sp. 5TM40 Man. Trans. Unit Repair	7B1
5-Sp. Isuzu Man. Trans. Unit Repair	7B2
Clutch	7C
CHASSIS ELECTRICAL, INSTRUMENT PANEL & WASHER WIPER	
Electrical Diagnosis	8A
Lighting and Horns	8B
Instrument Panel and Console	8C
Windshield Wiper/Washer	8E5
ACCESSORIES	
Audio System	9A
Cruise Control	9B
Engine Block Heater	9C
BODY SERVICE	
General Body Service	10-1
Stationary Glass	10-2
Underbody	10-3
Bumpers	10-4
Body Front End	10-5
Doors	10-6
Rear Quarters	10-7
Body Rear End	10-8
Roof & Convertible Top	10-9
Seats	10-10
Safety Belts	10-11
Body Wiring	10-12
Unibody Collision Repair	11-1
Welded Panel Replacement	11-2
INDEX	
Alphabetical Index	

Figure 2-45 The table of contents directs the technician to the major components areas of the service manual.

Figure 2-46 Computerized service information and Web-based retrieval systems are replacing paper service manuals.

Service and parts information can also be provided through computer services (Figure 2-46). This is a popular method since libraries of service manuals can take large areas of space. Computerized systems can have the information stored on disks, or the computer can be connected to a central database. The computer system helps the technician find the required information quicker and easier than in a book-type manual. Using the computer keyboard, light pen, touch-sensitive screen, or mouse, the technician makes choices from a series of menus on the monitor. If needed, the information can be printed to paper.

Another tool that can be of assistance to the technician is hotline services. There are many companies and organizations that provide online assistance to technicians. The hotline assistants use database information, along with factory service manuals, TSBs, and other sources to help the technician repair the vehicle.

Also, some companies and organizations provide a service of tracking the service history of vehicles. Manufacturers will generally do this for any of their vehicles that is serviced in one of their dealerships. This information lets the technician know if service has been performed on the vehicle that may be related to the present problem. Also, some Web site organizations will track if the vehicle has ever been "totaled" in an accident, stolen, or caught in a flood.

Working as an Electrical Systems Technician

To be a successful automotive technician, you need to have good training, a desire to succeed, and a commitment to becoming a good technician and a good employee. A good employee works well with others and strives to make the business successful. The required training is not just in the automotive field. Good technicians need to have good reading, writing, and math skills. These skills will allow you to better understand and use the material found in service manuals and textbooks, as well as provide you with the basics for good communications with customers and others.

Compensation

Technicians are typically paid according to their abilities. Most often, new or apprentice technicians are paid by the hour. While being paid they are learning the trade and the business. Time is usually spent working with master technicians or doing low-skill jobs. As apprentices learn more, they can earn more and take on more complex jobs. Once technicians have demonstrated a satisfactory level of skills, they can go on **flat rate.**

Flat rate is a pay system in which technicians are paid for the amount of work they do. Each job has a flat rate time. Pay is based on that time, regardless of how long it took to complete the job. To explain how this system works, let's look at a technician who is paid $15.00 per flat rate hour. If a job has a flat rate time of 3 hours, the technician will be paid $45.00 for the job, regardless of how long it took to complete. Experienced technicians beat the flat rate time nearly all of the time. Their weekly pay is based on the time "turned in," not on the time spent. If the technician turns in 60 hours of work in a 40-hour workweek, he or she actually earned $22.50 each hour worked. However, if he or she turned in only 30 hours in the 40-hour week, the hourly pay is $11.25.

The flat rate system favors good technicians that work in a shop that has a large volume of work. The use of flat rate times allows for more accurate repair estimates to the customers. It also rewards skilled and productive technicians.

Workplace Relationships

When you begin a job, you enter into a business agreement with your employer. When you become an employee, you sell your time, skills, and efforts. In return, your employer pays you for these resources. As part of the employment agreement, your employer also has certain responsibilities:

- Instruction and Supervision—You should be told what is expected of you. A supervisor should observe your work and tell you if it is satisfactory and offer ways to improve your performance.
- A Clean, Safe, Place to Work—An employer should provide a clean and safe work area as well as a place for personal cleanup.
- Wages—You should know how much you are to be paid, what your pay will be based on, and when you will be paid before accepting a job.
- Fringe Benefits—When hired, you should be told what benefits to expect, such as paid vacations and employer contributions to health insurance and retirement plans.
- Opportunity—You should be given a chance to succeed and possibly advance within the company.
- Fair Treatment—All employees should be treated equally, without prejudice or favoritism.

On the other side of this business transaction, employees have responsibilities to their employers. Your obligations as an employee to the employer include the following:

- Regular Attendance—A good employee is reliable.
- Following Directions—As an employee, you are part of a team, and doing things your way may not serve the best interests of the company.
- Responsibility—You must be willing to answer for your behavior and work. You need to also realize that you are legally responsible for the work you do.
- Productivity—Remember, you are paid for your time as well as your skills and effort.
- Loyalty—Loyalty is expected; by being loyal you will act in the best interests of your employer, both on and off the job.

Customer Relations

Another responsibility you have is good customer relations. Learn to listen and communicate clearly. Be polite and organized, particularly when dealing with customers. Always be as honest as you possibly can.

Look like and present yourself as a professional, which is what automotive technicians are. Professionals are proud of what they do and they show it. Always dress and act appropriately and watch your language, even when you think no one is near.

Respect the vehicles you work on. They are important to the lives of your customers. Always return the vehicle to the owner in a clean, undamaged condition. Remember, a vehicle is the second-largest expense a customer has. Treat it that way. It doesn't matter if you like the car. It belongs to the customer; treat it respectfully.

Explain the repair process to the customer in understandable terms. Whenever you are explaining something to a customer, make sure you do this in a simple way without making the customer feel stupid. Always show the customer respect and be courteous. Not only is this the right thing to do but it also leads to loyal customers.

Communicating with the Customer

Depending on the size of the service center that you are working in, you may or may not talk directly with the customer. A service advisor or manager might be in between the consumer and you, the technician. In either case, getting the correct information from the customer cannot be overemphasized. The customer is likely the person who was driving the vehicle when the problem showed up. The conversation that someone has with the customer can be extremely useful and save hours of fruitless work. A repair ticket that states "drivability problem" will require the technician to figure out the driving conditions that were present when the problem occurred. It is possible that the technician will not be able to duplicate the conditions and not observe the problem. The vehicle is returned to the consumer with the note of "no problem found. " Think about how frustrating this could be, especially if the costumer experiences the problem the very next day. The conversation with the customer should have revealed important information for the technician. When did the problem occur? What specifically did the vehicle do or not do? What was the outside temperature? Was the engine warm or cold? What driving conditions produced the problem? Can you duplicate the problem with the technician in the vehicle?

Think about how much more information the technician has in the second example. It is likely that less time will be necessary to fix the vehicle because the technician has a starting point. Once you get out in the field, try to develop your communication skills and especially your ability to listen to the customer. The customer's information, if you have heard it, will save you countless hours of frustration and no doubt result in a better repair. Better repairs bring customers back to the service center the next time they need service.

ASE Certification

An obvious sign of your knowledge and abilities, as well as your dedication to the trade, is ASE certification. The National Institute for Automotive Service Excellence (ASE) has established a voluntary certification program for automotive, heavy-duty truck, auto body repair, and engine machine shop technicians. In addition to these programs, ASE also offers individual testing in some specialty areas. This certification system combines voluntary testing with on-the-job experience to confirm that technicians have the skills needed to work on today's vehicles (Figure 2-47). ASE recognizes two distinct levels of service capability—the automotive technician and the master automotive technician.

Figure 2-47 ASE certification communicates your pride in your profession to the customers.

After passing at least one exam and providing proof of two years of hands-on work experience, you become ASE certified. Retesting is necessary every five years to remain certified. A technician who passes one examination receives an automotive technician shoulder patch. The master automotive technician patch is awarded to technicians who pass all eight of the basic automotive certification exams. You may receive credit for one of the two years by substituting relevant formal training in one, or a combination, of the following:

- High school training. Three years of training may be substituted for one year of experience.
- Post–high school training. Two years of post–high school training in a trade school, technical institute, or community college may be counted as one year of work experience.
- Short courses. For shorter periods of post–high school training you may substitute two months of training for one month of work experience.
- Apprenticeship programs. You may receive full credit for the experience requirement by satisfactorily completing a three- or four-year apprenticeship program.

Each certification test consists of 40 to 80 multiple-choice questions. The questions are written by a panel of technical service experts, including domestic and import vehicle manufacturers, repair and test equipment and parts manufacturers, working automotive technicians, and automotive instructors. All questions are pretested and quality checked on a national sample of technicians before they are included in the actual test. Many test questions force the student to choose between two distinct repair or diagnostic methods. The test questions focus on basic technical knowledge, repair knowledge and skill, and testing and diagnostic knowledge and skill.

Diagnostics

The true measure of a good technician is an ability to find and correct the cause of problems. Service manuals and other information sources will guide you through the diagnosis and repair of problems, but these guidelines will not always lead you to the exact cause of the problem. To do this you must use your knowledge and take a logical approach while troubleshooting. Diagnosis is not guessing, and it's more than following a series of interrelated steps in order to find the solution to a specific problem. Diagnosis is a way of looking at systems that are not functioning the way they should and finding out why. It is knowing how the system should work and deciding if it is working correctly. Through an understanding of the purpose and operation of the system, you can accurately diagnose problems.

Most good technicians use the same basic diagnostic approach. Simply because this is a logical approach, it can quickly lead to the cause of a problem. Logical diagnosis follows these steps:

1. Gather information about the problem.
2. Verify that the problem exists.
3. Thoroughly define what the problem is and when it occurs.
4. Research all available information and knowledge to determine the possible causes of the problem.
5. Isolate the problem by testing.
6. Continue testing to pinpoint the cause of the problem.
7. Locate and repair the problem.
8. Verify the repair.

Summary

❏ A test light is used when the technician needs to "look" for electrical power in the circuit. The test light allows the technician to see if current is at a point in the circuit by lighting the light.

❏ A logic probe provides a means for testing voltages on electronic circuits without damaging the circuit.

❏ Digital multimeters (DMM) display values using liquid crystal displays instead of a swinging needle. They are computers that determine the measured value and display it for the technician.

❏ A voltmeter measures the voltage potential between two points in a circuit.

❏ An ohmmeter is used to measure the resistance of a circuit or part of a circuit.

❏ An ammeter is a special meter used to measure current flow in a circuit.

❏ Electrical noise is an unwanted voltage signal that rides on a signal. Noise is usually the result of radio frequency interference (RFI) or electromagnetic induction (EMI).

❏ Duty cycle is the percentage of on time the circuit component is turned on as compared to the total time of the cycle.

❏ A cycle is one set of changes in a signal that repeats itself several times.

❏ Pulse width is the amount of time, measured in milliseconds, that a component is turned on.

❏ Frequency is a measure of the number of cycles that occur in one second.

❏ Hertz is the measurement of frequency.

❏ A lab scope provides a visual display of electrical waves.

❏ Glitches may be the result of momentary shorts to ground, shorts to power, or opens in the circuit.

❏ A sine wave is a waveform that shows voltage-changing polarity from positive to negative.

❏ Having straight vertical sides and a flat top indicating a fast-acting on-off voltage state identifies a square wave pattern.

❏ Scan tools interface with the vehicle's computer system to allow the technician to "talk" with the computers.

❏ The service manual (in either paper form or electronic format) is one of the most important and valuable tools for today's technician.

❏ The service manual provides information concerning engine identification, service procedures, specifications, and diagnostic information.

❏ Technical Service Bulletins (TSBs) may provide information concerning fixes for a problem, new part numbers to replace a defective unit, corrections to service manual information, and general information of system operation.

❏ Service manual procedural information provides the steps necessary to perform the task.

❏ Since the service manual is divided into several major component areas, a table of contents is provided for easy access to the information.

❏ Service and parts information can also be provided through computer services.

❏ Technicians are typically paid according to their abilities. New or apprentice technicians are paid by the hour. Once technicians have demonstrated a satisfactory level of skills, they can go on flat rate.

❏ When you begin a job, you enter into a business agreement with your employer. When you become an employee, you sell your time, skills, and efforts. In return, your employer pays you for these resources.

- ❏ As part of the employment agreement, your employer also has certain responsibilities: instruction and supervision; a clean, safe place to work; wages; fringe benefits; opportunity; and fair treatment.

- ❏ Your obligations as an employee to the employer include regular attendance, following directions, responsibility, productivity, and loyalty.

- ❏ When communicating with customers, be polite, respectful, organized, and honest.

- ❏ An obvious sign of your knowledge and abilities, as well as your dedication to the trade is ASE certification.

- ❏ The true measure of a good technician is an ability to find and correct the cause of problems.

- ❏ Diagnosis is not guessing, and it is more than following a series of interrelated steps in order to find the solution to a specific problem.

Terms to Know

Ammeter	Flat rate	Pulse width
Average responding	Frequency	Root mean square (RMS)
Backprobe	Glitches	Scan tool
Captured signal	Hertz	Sine wave
Cycle	Jumper wire	Sinusoidal
D'Arsonval movement	Logic probe	Square wave
Diagnostic trouble codes (DTCs)	Multimeter	Test light
Digital multimeter (DMM)	Noise	Voltage drop
Duty cycle	Ohmmeter	Voltmeter

ASE-Style Review Questions

1. *Technician A* says a test light is ideal for checking for voltage on a low-current, low-power circuit. *Technician B* says to use an analog multimeter to test computer-controlled circuits. Who is correct?
   - **A.** A only
   - **B.** B only
   - **C.** Both A and B
   - **D.** Neither A nor B

2. The use of an ammeter is being discussed. *Technician A* says the ammeter is used to measure current flow. *Technician B* says the ammeter must be connected in parallel to the circuit being tested. Who is correct?
   - **A.** A only
   - **B.** B only
   - **C.** Both A and B
   - **D.** Neither A nor B

3. A DVOM is being used to measure current flow. The meter is displaying 85.5 mA. *Technician A* says this represents 0.0855 amperes. *Technician B* says the decimal point needs to be moved six points to the left. Who is correct?
   - **A.** A only
   - **B.** B only
   - **C.** Both A and B
   - **D.** Neither A nor B

4. *Technician A* says a 10% duty cycle indicates that the load device is turned on most of the time. *Technician B* says the pulse width is measured in degrees. Who is correct?
   - **A.** A only
   - **B.** B only
   - **C.** Both A and B
   - **D.** Neither A nor B

5. A vehicle is being tested for a draw against the battery with the ignition switch in the OFF position. The specifications state the draw should be between 10 and 30 milliamps. The DVOM reads 0.251 amps.

Technician A says this draw is within the specification range.

Technician B says the draw is too high.

Who is correct?

A. A only   **C.** Both A and B

B. B only   **D.** Neither A nor B

6. *Technician A* says that after an individual passes a particular ASE-certification exam, he or she is certified in that test area.

Technician B says that all of the questions on an ASE-certification exam are written as *Technician A* and *Technician B* questions.

Who is correct?

A. A only   **C.** Both A and B

B. B only   **D.** Neither A nor B

7. *Technician A* says a voltmeter that is connected in parallel to the load device will indicate the voltage drop across the device.

Technician B says that a ohmmeter reading of 0.00 Ω when connected in parallel to the coil of an A/C compressor indicates the coil is shorted.

Who is correct?

A. A only   **C.** Both A and B

B. B only   **D.** Neither A nor B

8. *Technician A* says the upward voltage traces on an oscilloscope screen indicate a specific length of time. *Technician B* says the cathode ray tube (CRT) in an oscilloscope is like a very-fast-reacting voltmeter.

Who is correct?

A. A only   **C.** Both A and B

B. B only   **D.** Neither A nor B

9. The best way of determining if a switch is faulty is being discussed.

Technician A says to use a jumper wire to bypass the switch and to connect the various circuits controlled by the switch.

Technician B says to use an ammeter to test continuity through the switch.

Who is correct?

A. A only   **C.** Both A and B

B. B only   **D.** Neither A nor B

10. *Technician A* says the service manual (in either paper form or electronic format) is one of the most important and valuable tools for today's technician. *Technician B* says service bulletins (TSBs) provide information concerning fixes for a problem, new part numbers to replace a defective unit, corrections to service manual information, and general information of system operation.

Who is correct?

A. A only   **C.** Both A and B

B. B only   **D.** Neither A nor B

ASE Challenge Questions

1. *Technician A* says a voltmeter measures the electrical potential between two points of the circuit.

Technician B says that an ammeter reading of 0.00 when connected in series to a circuit indicates a short to ground.

Who is correct?

A. A only   **C.** Both A and B

B. B only   **D.** Neither A nor B

2. A scanner allows a technician to:

A. measure circuit resistance.

B. determine load voltage drop.

C. view computer inputs.

D. load test the battery.

3. All of the following are true concerning the use of an ohmmeter EXCEPT:

A. Connect the ohmmeter with the circuit power off.

B. An "OL" indicates that the reading is over limits.

C. 0.00 indicates no resistance.

D. Connect the test leads of the meter in series to the load.

4. The signal-output of a throttle position sensor is being checked (from idle to the wide-open throttle position) with an oscilloscope. The test lead is set to the 10x position. Which of the following represents the correct position of the vertical adjustment selector?

A. 0.1 Volts   **C.** 2 Volts

B. 0.5 Volts   **D.** 5 Volts

5. *Technician A* says that logic probes can be very helpful in testing analog signals.

Technician B says the nonpowered test lights should not be used to test most computer circuits.

Who is correct?

A. A only   **C.** Both A and B

B. B only   **D.** Neither A nor B

Job Sheet 3

Name _____   Date _____

Using Ohm's Law to Calculate Electrical Properties

Using Ohm's law, solve the following problems:

Exercise 1—Series Circuit

Refer to the circuit presented below.
Use Ohm's law to calculate the following values, when R_1 = 2 ohms and R_2 = 4 ohms:

Total circuit resistance = _____ ohms

Circuit current = _____ amps

Current through R_1 = _____ amps

Current through R_2 = _____ amps

Voltage drop across R_1 = _____ volts

Voltage drop across R_2 = _____ volts

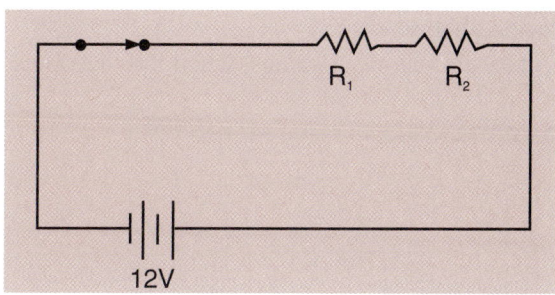

If the resistance of R_1 increases to 8 ohms, what are the new values?

Total circuit resistance = _____ ohms

Circuit current = _____ amps

Current through R_1 = _____ amps

Current through R_2 = _____ amps

Voltage drop across R_1 = _____ volts

Voltage drop across R_2 = _____ volts

Exercise 2—Series Circuit

Refer to the circuit provided below.
Use Ohm's law to calculate the following values, when R_1 = 3 ohms and R_2 = 6 ohms:

Total circuit resistance = _____ ohms

Circuit current = _____ amps

Current through R_1 = _____ amps

Current through R_2 = _____ amps

Voltage drop across R_1 = _____ volts

Voltage drop across R_2 = _____ volts

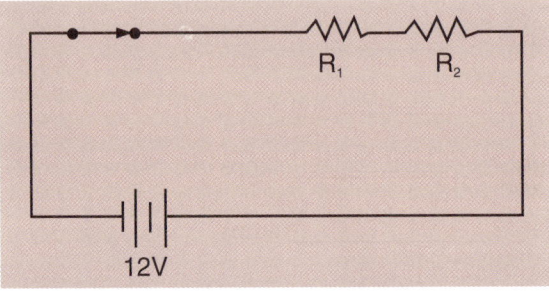

Exercise 3—Parallel Circuit

Refer to the circuit provided below.
Use Ohm's law to calculate the following values, when R_1 = 12 ohms and R_2 = 12 ohms:

Total circuit resistance = _____ ohms

Circuit current = _____ amps

Current through R_1 = _____ amps

Current through R_2 = _____ amps

Voltage drop across R_1 = _____ volts

Voltage drop across R_2 = _____ volts

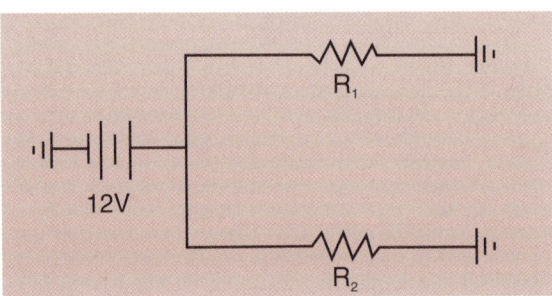

Exercise 4—Parallel Circuit

Refer to the circuit provided below.
Use Ohm's law to calculate the following values, when R_1 = 1 ohm, R_2 = 3 ohms, R_3 = 2 ohms, and R_4 = 2 ohms:

Total circuit resistance = _____ ohms

Circuit current = _____ amps

Current through R_1 = _____ amps

Current through R_2 = _____ amps

Current through R_3 = _____ amps

Current through R_4 = _____ amps

Voltage drop across R_1 = _____ volts

Voltage drop across R_2 = _____ volts

Voltage drop across R_3 = _____ volts

Voltage drop across R_4 = _____ volts

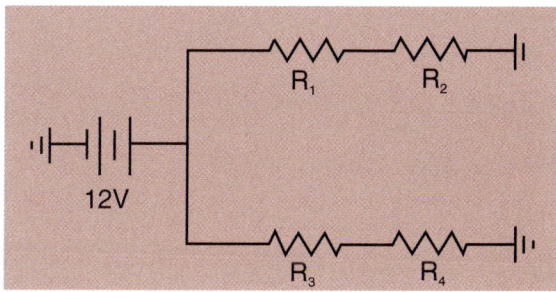

Exercise 5—Parallel Circuit

Refer to the circuit on page 76.
Use Ohm's law to calculate the following values, when R_1 = 1 ohm, R_2 = 3 ohms, R_3 = 2 ohms, and R_4 = 10 ohms:

Total circuit resistance = _____ ohms

Circuit current = _____ amps

Current through R_1 = _____ amps

Current through R_2 = _____ amps

Current through R_3 = _____ amps

Current through R_4 = _____ amps

Voltage drop across R_1 = _____ volts

Voltage drop across R_2 = _____ volts

Voltage drop across R_3 = _____ volts

Voltage drop across R_4 = _____ volts

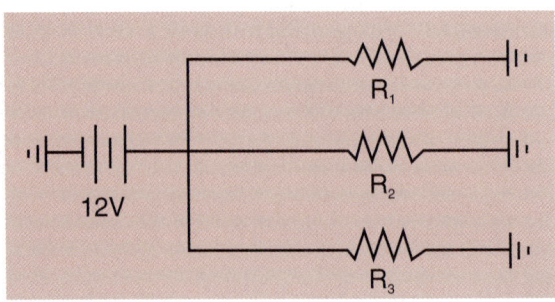

Exercise 6—Parallel Circuit

Refer to the circuit provided below.
Use Ohm's law to calculate the following values, when R_1 = 2 ohms, R_2 = 3 ohms, and R_3 = 6 ohms:

Total circuit resistance = _____ ohms

Circuit current = _____ amps

Current through R_1 = _____ amps

Current through R_2 = _____ amps

Current through R_3 = _____ amps

Voltage drop across R_1 = _____ volts

Voltage drop across R_2 = _____ volts

Voltage drop across R_3 = _____ volts

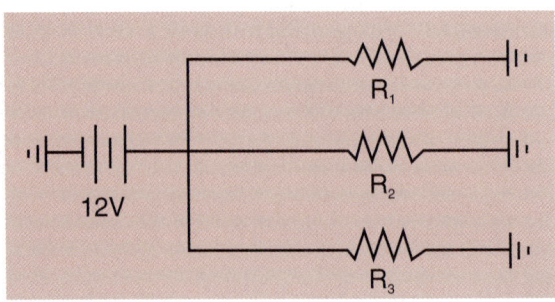

Exercise 7—Series-Parallel Circuit

Refer to the circuit provided below.

Use Ohm's law to calculate the following values, when R_1 = 1 ohm, R_2 = 2 ohms, R_3 = 3 ohms, and R_4 = 6 ohms:

Total circuit resistance = _____ ohms

Circuit current = _____ amps

Current through R_1 = _____ amps

Current through R_2 = _____ amps

Current through R_3 = _____ amps

Current through R_4 = _____ amps

Voltage drop across R_1 = _____ volts

Voltage drop across R_2 = _____ volts

Voltage drop across R_3 = _____ volts

Voltage drop across R_4 = _____ volts

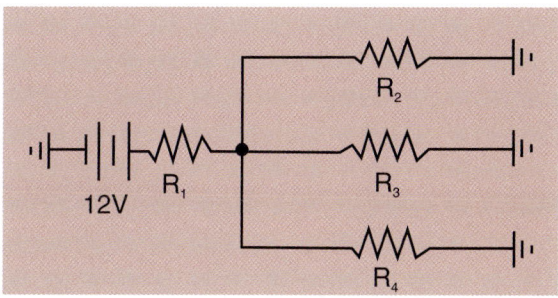

Exercise 8—Series-Parallel Circuit

Refer to the circuit provided on page 78.

Use Ohm's law to calculate the following values, when R_1 = 6 ohms, R_2 = 3 ohms, and R_3 = 2 ohms:

Total circuit resistance = _____ ohms

Circuit current = _____ amps

Current through R_1 = _____ amps

Current through R_2 = _____ amps

Current through R_3 = _____ amps

Voltage drop across R_1 = _____ volts

Voltage drop across R_2 = _____ volts

Voltage drop across R_3 = _____ volts

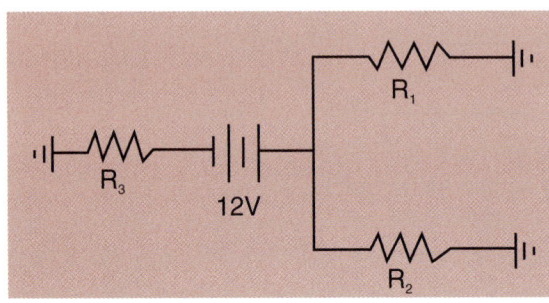

Instructor's Response _____

_____

_____

_____

Job Sheet 4

Name _____ Date _____

Meter Symbol Interpretation

Upon completion of this job sheet, you should be able to convert commonly found symbols into numeric values.

Convert the following values into the electrical units noted:

1. 2.4 K Ω   = _____ Ω

2. 954 mV   = _____ V

3. 5.76 K Ω   = _____ Ω

4. 2 mA   = _____ A

5. 22 K Ω   = _____ Ω

6. 4.5 mA   = _____ A

7. 456 mA   = _____ A

8. 1024 mV   = _____ V

9. 0.786 K Ω   = _____ Ω

10. 32 K Ω + 112 Ω   = _____ Ω

11. 1400 Ω   = _____ K Ω

12. 0.000235 A   = _____ mA

13. 0.987 V   = _____ mV

14. 5 K V   = _____ mV

15. 123955 Ω   = _____ K Ω

16. 144,000 mA   = _____ A

17. 126 mV + 11.874 V   = _____ V

18. 320,000 Ω   = _____ Ω

19. 0.000045 A   = _____ mA

20. 12,600 mV   = _____ V

Instructor's Response _____

_____

_____

_____

Job Sheet 5

Name _____ Date _____

Using a Test Light

Upon completion of this job sheet, you should be able to properly use a test light to check continuity in electrical circuits.

ASE Correlation

This job sheet is related to the ASE Electrical/Electronic System Certification Exam's content area: *General Electric/Electronic System Diagnosis;* task: Check continuity in electrical circuits with a test light and determine needed repairs.

Tools and Materials

A vehicle
A test light
Wiring diagrams for the vehicle

Describe the vehicle being worked on.

Year _____ Make _____ Model _____

VIN _____ Engine type and size _____

Procedure

Task Completed

1. Using the proper wiring diagram and your instructor's assistance, determine which wire is the positive feed and which wire is the ground for one of the low-beam headlights on your assigned vehicle. Identify the color of each wire:

   Positive _____ Negative _____

2. Test proper operation of the test light by connecting the negative lead to the negative post of the battery and touching the probe tip to the positive lead of the battery. If the test light is working properly, the light should illuminate. If the light does not turn on, inform your instructor. Did the light work properly? ☐ Yes ☐ No

3. With the headlight switch in the OFF position, disconnect the headlight connector. ☐

4. Connect the negative lead of the test light to a good ground. Locate the probe of the test light into the connector cavity identified as positive feed. Did the test light come on? ☐ Yes ☐ No

   Explain your results: _____

   _____

   _____

5. Turn the headlight switch into the ON position and retest the circuit again. Did the test light come on? ☐ Yes ☐ No

6. Turn the headlight switch OFF.

7. Move the wire clip of the test light to the positive post of the battery and locate the probe into the cavity identified as the ground. Does the test light come on?
   ☐ Yes   ☐ No

   Explain the reason: _____

   _____

   _____

Instructor's Response _____

_____

_____

_____

Job Sheet 6

Name _____ Date _____

Use of a Voltmeter

Upon completion of this job sheet, you should be able to measure available voltage and voltage drop.

ASE Correlation

This job sheet is related to the ASE Electrical/Electronic System Certification Exam's content area: *General Electrical/Electronic System Diagnosis;* task: Check applied voltages and voltage drops in electrical/electronic circuits and components with a voltmeter; determine needed repairs.

Tools and Materials

A vehicle	Wiring diagram for vehicle
A DMM	Basic hand tools

Describe the vehicle being worked on.

Year _____ Make _____ Model _____

VIN _____ Engine type and size _____

Procedure

Task Completed

1. Set the DMM to the appropriate scale to read 12 volts DC.   ☐

2. Connect the meter across the battery (positive to positive and negative to negative).

   What is your reading on the meter? _____ volts

3. With the meter still connected across the battery, turn on the headlights of the vehicle.

   What is your reading on the meter? _____ volts

4. Keep the headlights on. Connect the positive lead of the meter to the point on the vehicle where the battery's ground cable attaches to the frame. Keep the negative lead where it is.

   What is your reading on the meter? _____ volts

   What is being measured? _____

5. Disconnect the meter from the battery and turn off the headlights.   ☐

6. Refer to the correct wiring diagram and determine what wire at the right headlight delivers current to the lamp when the headlights are on and low beams selected.

   Color of the wire _____

7. From the wiring diagram, identify where the headlight is grounded.

   Place of ground _____

☐   **8.** Connect the negative lead of the meter to the point where the headlight is grounded.

☐   **9.** Connect the positive lead of the meter to the power input of the headlight.

10. Turn on the headlights.

What is your reading on the meter? _____ volts

What is being measured? _____

11. What is the difference between the reading here and the battery's voltage?
_____ volts

12. Explain why there is a difference.

_____

_____

_____

_____

Instructor's Response _____

_____

_____

_____

Job Sheet 7

Name _____ Date _____

Use of an Ohmmeter

Upon completion of this job sheet, you should be able to check continuity of a circuit and measure resistance on a variety of components.

ASE Correlation

This job sheet is related to the ASE Electrical/Electronic System Certificate Exam's content area: *General Electrical/Electronic System Diagnosis;* task: check continuity and resistances in electrical/electronic circuits and components with an ohmmeter; determine needed repairs.

Tools and Materials

A vehicle   An analog ohmmeter
A DMM   Wiring diagram for the vehicle

NOTE: An ohmmeter works by sending a small amount of current through the path to be measured. Because of this, all circuits and components being tested must be disconnected from power. An ohmmeter must never be connected to an energized circuit; doing so may damage the meter. The safest way to measure ohms is to disconnect the negative battery cable before taking resistance readings.

Describe the vehicle being worked on.

Year _____ Make _____ Model _____

VIN _____ Engine type and size _____

Procedure

Task Completed
☐

1. Locate the fuse panel or power distribution box.
2. With no power to the fuses, check the resistance of each fuse. Summarize your findings:

   _____
   _____
   _____

3. Now connect the leads of the analog meter across the negative battery cable.
   Your reading is: _____ ohms
   Now connect the leads of the digital meter across the negative battery cable.
   Your reading is: _____ ohms
4. Disconnect the primary wires leading to the ignition coil.
   Now connect the leads of the analog meter across the terminals of the coil.
   Your reading is: _____ ohms
   Now connect the leads of the digital meter across the terminals of the coil.
   Your reading is: _____ ohms
5. Reconnect the wires to the coil.
   Carefully remove one spark plug wire from the spark plug and ignition coil
   or distributor cap.
   Now connect the leads of the analog meter across the wire.
   You reading is: _____ ohms
   Now connect the leads of the digital meter across the wire.
   Your reading is: _____ ohms

6. Carefully reinstall the spark plug wire.

   Locate the cigar lighter inside the vehicle.

   Now connect the leads of the analog meter from the heating coil to its case.

   Your reading is: _____ ohms

   Now connect the leads of the digital meter from the heating coil to its case.

   Your reading is: _____ ohms

7. Refer to the service manual and find out how to remove the bulb in the dome light. Remove it.

   Now connect the leads of the analog meter across the bulb.

   Your reading is: _____ ohms

   Now connect the leads of the digital meter across the bulb.

   Your reading is: _____ ohms

8. Reinstall the bulb.

   Remove the rear brake light bulb.

   Now connect the leads of the analog meter across the bulb.

   Your reading is: _____ ohms

   Now connect the leads of the digital meter across the bulb.

   Your reading is: _____ ohms

9. Reinstall the bulb.

   Disconnect the wire connector to one of the headlights.

   From the wiring diagram, identify which terminals are for low-beam operation.

   Now connect the leads of the analog meter across the low-beam terminals.

   Your reading is: _____ ohms

   Now connect the leads of the digital meter across the low-beam terminals.

   Your reading is: _____ ohms

10. From the wiring diagram, identify which terminals are for high-beam operation.

    Now connect the leads of the analog meter across the high-beam terminals.

    Your reading is: _____ ohms

    Now connect the leads of the digital meter across the high-beam terminals.

    Your reading is: _____ ohms

11. You should have gotten slightly different readings from the digital meter and the analog meter. Explain why:

    _____

    _____

    _____

    _____

12. You measured the resistance across several different light bulbs. On each you should have read a different amount of resistance. Based on your findings, which light bulb would be the brightest and which would be the dimmest? Explain why.

    _____

    _____

    _____

Instructor's Response _____

_____

_____

_____

Basic Electrical Troubleshooting and Service

Upon completion and review of this chapter, you should be able to:

❏ Describe how different electrical problems cause changes in an electrical circuit.

❏ Diagnose and repair circuit protection devices.

❏ Test switches with a variety of test instruments.

❏ Test relays and relay circuits for proper operation.

❏ Identify and test fixed and variable resistors with a voltmeter, ohmmeter, or lab scope.

❏ Diagnose diodes for opens, shorts, and other defects.

❏ Diagnose transistors for opens, shorts, and other defects.

❏ Locate and repair opens in a circuit.

❏ Locate and repair shorts in a circuit.

❏ Locate and repair the cause of unwanted high resistance in a circuit.

❏ Set up a digital storage oscilloscope (DSO) and identify wave form.

Basic Tools

Basic mechanic's tool set

Service manual

_____

Introduction

Troubleshooting electrical problems involves the same tools and methods, regardless of which **circuit** has the problem. All electrical circuits must have voltage, current, and resistance. Testing for the presence of these, measuring them, and comparing your measurements to specifications is the key to effective diagnosis. To do this, you must have a solid understanding of these basic electrical properties.

Voltage is the electrical pressure that causes electrons to move provided there is a complete path for them to do so. Current is the aggregate flow of electrons through a wire and can be defined as the rate of electron flow. Resistance is defined as opposition to current flow. An electrical circuit must have resistance in it in order to change electrical energy to light, heat, or movement.

An electrical circuit may develop an open, a short, or an excessive voltage drop that will cause it to operate improperly. An **open circuit** is a circuit in which there is a break in continuity. The open can be on either the insulated side or the ground side. A **shorted circuit** decreases the resistance of the circuit. This happens by shorting across to another circuit or by shorting to a ground. When there is a circuit-to-circuit short, one of the circuits is not controlled by its switch. Since the shorted circuit becomes a new parallel leg to the circuit, the entire parallel circuit will turn on and off with the switch controlling the other circuit. With this type of problem, many strange things can happen. When a circuit is shorted to ground, a new parallel leg is present. This new leg has very low resistance and causes the current in the circuit to increase drastically.

High-resistance problems can occur anywhere in the circuit. However, the effect of high resistance is the same regardless of where it is. Additional or unwanted resistance in series with a circuit will always reduce the current in the circuit and will reduce the amount of voltage drop by the component in the circuit.

The term **circuit** means a circle and is the path of electron flow consisting of the voltage source, conductors, load component, and return path to the voltage source.

Testing Circuit Protection Devices

Classroom Manual
Chapter 3, page 78

Special Tools

Ohmmeter

Voltmeter

Test light

A blade type fuse is called a spade fuse.

■ **CAUTION:** Fuses and other protection devices do not wear out. They fail because something went wrong. Never replace a fuse or fusible link, or reset a circuit breaker, without finding out why it failed.

A protection device is designed to "turn off" the system whenever excessive current or an **overload** occurs. There are three basic types of **fuses** in automotive use: cartridge, blade, and ceramic. A fuse is a replaceable element that will melt should the current passing through it exceed the fuse rating. The cartridge fuse is found on most older domestic cars and a few imports. To check this type of fuse, look for a break in the internal metal strip. Discoloration of the glass cover or glue bubbling around the metal caps is an indication of overheating. Late-model domestic vehicles and many imports use blade or spade fuses. To check the fuse, pull it from the fuse panel and look at the fuse element through the transparent plastic housing. Look for internal breaks and discoloration. The ceramic fuse is used on many older European imports. To check this type of fuse, look for a break in the contact strip on the outside of the fuse. All types of fuses can be checked with an ohmmeter or test light. If the fuse is good, there will be continuity through it.

Fuses are rated by the current at which they are designed to blow. A three-letter code is used to indicate the type and size of fuses. Blade fuses have codes ATC or ATO. All glass SFE fuses have the same diameter, but the length varies with the current rating. Ceramic fuses are available in two sizes, code GBF (small) and the more common code GBC (large). The amperage rating is also embossed on the insulator. Codes such as AGA, AGW, and AGC indicate the length and diameter of the fuse. Fuse lengths in each of these series is the same, but the current rating can vary. The code and the current rating are usually stamped on the end cap.

The current rating for blade fuses is indicated by the color of the plastic case (Table 3-1). In addition, it is usually marked on the top. The insulator on ceramic fuses is color coded to indicate different current ratings.

Fuses are located in a box or panel, usually under the dashboard, behind a panel in the foot well, or in the engine compartment. Fuses are generally numbered, and the main components abbreviated. On late-model cars there may be icons or symbols indicating which circuits they serve. This identification system is covered in more detail in the owner's and service manuals.

■ **SERVICE TIP:** To calculate the correct fuse rating, use Watt's law: watts divided by volts equals amperes. For example, if you are installing a 55-watt pair of fog lights, divide 55 by the battery voltage (12 volts) to find out how much current the circuit has to carry. Since 55 ÷ 12 = 4.58, the current is approximately 5 amperes. To allow for current surges, the correct in-line fuse should be rated slightly higher than the normal current flow. In this case, an 8- or 10-ampere fuse would do the job.

A fuse link is commonly called a **fusible link**.

A **fusible link** is a conductor with a special heat-resistant insulation. When there is an overload in the circuit, the link melts and opens the circuit. Fusible links are used in circuits where limiting the maximum current is not extremely critical. They are often installed in the positive battery lead to the ignition switch and other circuits that have power with the key off.

A fusible link is a short length of small-gauge wire installed in a conductor. Since the fusible link is a lighter gauge of wire than the main conductor, it melts and opens the circuit before damage can occur in the rest of the circuit. Fusible link wire is covered with a special insulation that bubbles when it overheats, indicating that the link has melted. If the insulation appears good, pull lightly on the wire. If the link stretches, the wire has melted. Of course, when it is hard to determine if the fusible link is burned out, check for continuity through the link with a test light or ohmmeter.

To replace a fusible link, cut the protected wire where it is connected to the fusible link. Then, tightly crimp or solder a new fusible link of the same rating as the original link. Since the insulation

TABLE 3-1 Typical color coding of protection devices.

Blade Fuse Color Coding	
Ampere Rating	**Housing Color**
4	pink
5	tan
10	red
15	light blue
20	yellow
25	natural
30	light green
Fuse Link Color Coding	
Wire Link Size	**Insulation Color**
20 GA	blue
18 GA	brown or red
16 GA	black or orange
14 GA	green
12 GA	gray
Maxi-fuse Color Coding	
Ampere Rating	**Housing Color**
20	yellow
30	light green
40	amber
50	red
60	blue

on the manufacturer's fusible links is flameproof, never fabricate a fusible link from ordinary wire because the insulation may not be flameproof.

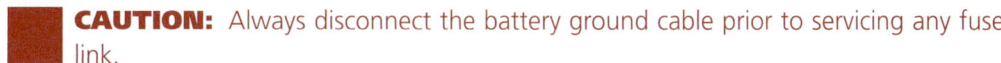

CAUTION: Always disconnect the battery ground cable prior to servicing any fuse link.

Many late-model vehicles use maxi-fuses instead of fusible links. Maxi-fuses look and operate like two-prong, blade or spade fuses, except they are much larger and can handle more current. (Typically, a maxi-fuse is four to five times larger.) Maxi-fuses are located in a fuse box in the engine compartment and/or passenger compartment, under the dash or the rear seat.

Maxi-fuses are easier to inspect and replace than are fuse links. To check a maxi-fuse, look at the fuse element through the transparent plastic housing. If there is a break in the element, the maxi-fuse has blown. To replace it, pull it from its fuse box or panel. Always replace a blown maxi-fuse with a new one having the same ampere rating.

Some circuits are protected by circuit breakers. Like fuses, they are rated in amperes. There are two types of circuit breakers: cycling or those that must be manually reset.

In the cycling type, the bimetal arm will begin to cool once the current to it is stopped. Once it returns to its original shape, the contacts are closed and power is restored. If the current is still too high, the cycle of breaking the circuit will be repeated.

A circuit breaker is typically abbreviated c.b. in a fuse chart of a service manual.

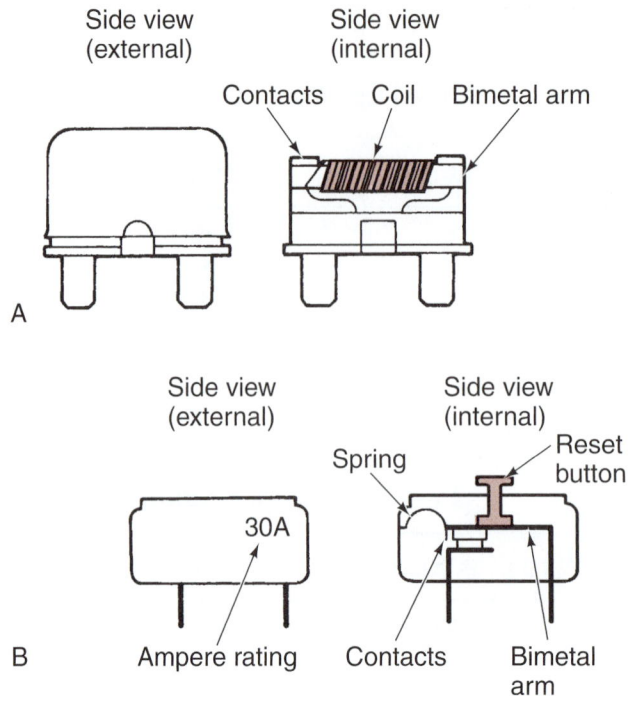

Figure 3-1 The two basic types of circuit breakers.

Two types of noncycling or resettable breakers are used. One is reset by removing the power from the circuit. There is a coil wrapped around a bimetal arm (Figure 3-1A). When there is excessive current and the contacts open, a small current passes through the coil. This current through the coil is not enough to operate a load, but it does heat up both the coil and the bimetal arm. This keeps the arm in the open position until power is removed. The other type is reset by depressing a reset button. A spring pushes the bimetal arm down and holds the contacts together (Figure 3-1B). When an overcurrent condition exists and the bimetal arm heats up, the bimetal arm bends enough to overcome the spring and the contacts snap open. The contacts stay open until the reset button is pushed, which snaps the contacts together again.

A visual inspection of a fuse or fusible link will not always determine if it has an open (Figure 3-2). To accurately test a circuit protection device, use an ohmmeter, voltmeter, or test light.

With the fuse or circuit breaker removed from the vehicle, connect the ohmmeter's test leads across the protection device's terminals (Figure 3-3). On its lowest scale, the ohmmeter should read 0 to 1 ohms. If it reads infinite, the protection device is open. Test a fusible link in the same way (Figure 3-4). Before connecting the ohmmeter across the fusible link, make sure there is no current flow through the circuit. To be safe, disconnect the negative cable of the battery.

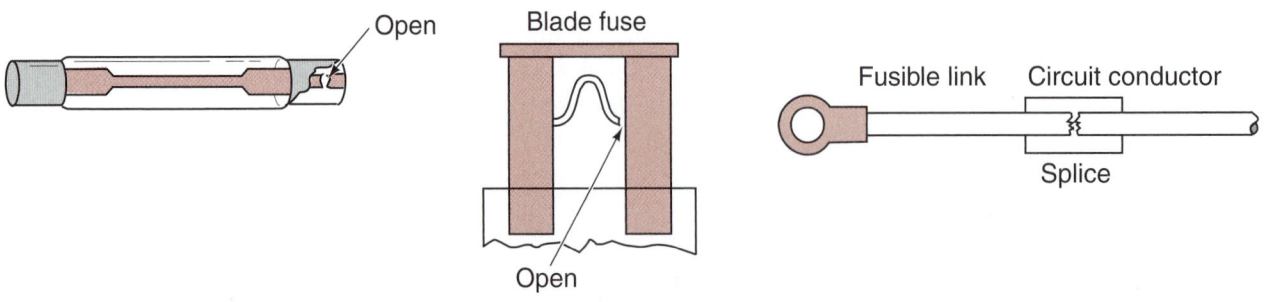

Figure 3-2 A fuse can have a hidden fault that the technician cannot see.

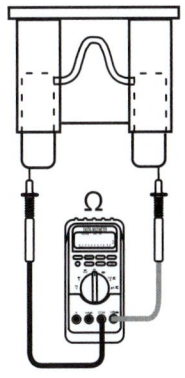
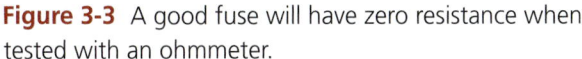
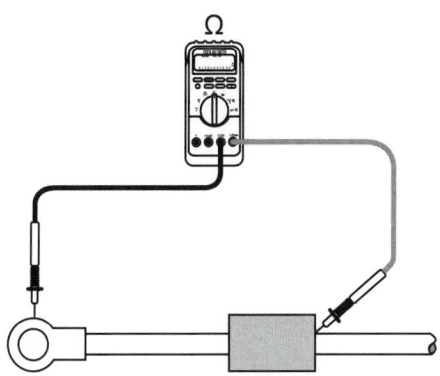

Figure 3-3 A good fuse will have zero resistance when tested with an ohmmeter.

Figure 3-4 A fusible link can be tested with an ohmmeter, once it is disconnected from power.

To test a circuit protection device with a voltmeter, check for available voltage at both terminals of the unit (Figure 3-5). If the device is good, voltage will be present on both sides. A test light can be used in place of a voltmeter. The lamp should illuminate when each test terminal is touched with the lamp's probe.

 SERVICE TIP: Before using a test light, it is good practice to check the tester's lamp. To do this, simply connect the test light across the battery. The light should come on.

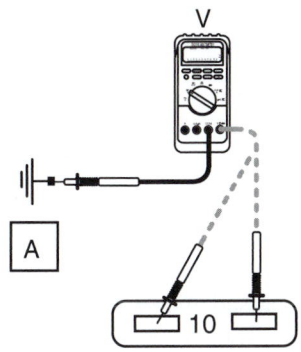

A

Top view of mini fuse

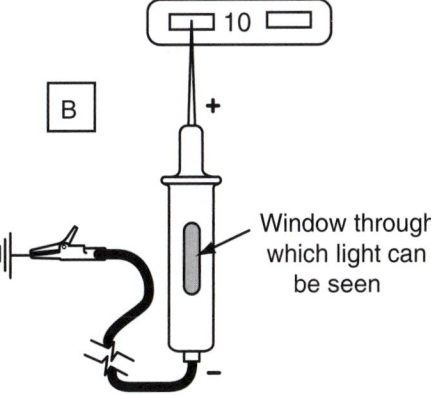

B

Window through which light can be seen

Figure 3-5 (A) Voltmeter test of a circuit protection device. Battery voltage should be present on both sides. (B) The test light will illuminate on both terminals if the fuse is good.

Measuring voltage drop across a fuse or other circuit protection device will tell you more about its condition than whether or not it is open. If a fuse, a fuse link, or circuit breaker is in good condition, a voltage drop of zero will be measured. If 12 volts is read, the fuse is open. Any reading between zero and 12 volts indicates some voltage drop. If there is voltage drop across the fuse, it has resistance and should be replaced. Make sure you check the fuse holder for resistance as well.

CAUTION: Fuses are rated by amperage and voltage. Never install a larger rated fuse into a circuit than the one that was designed by the manufacturer. Doing so may damage or destroy the circuit. Also do not replace a fusible link with a resistor wire or vice versa.

CAUTION: Do not use an unfused jumper wire to bypass the protection device. Circuit damage may result.

CUSTOMER CARE: Any time you install additional electrical accessories for customers, provide them with information concerning the type and size of fuses installed so they can put this information with their owner's manual.

Testing and Replacing Electrical Components

All electrical components can fail. Testing them is the best way of determining if they are good or bad. For the most part, the proper way to check electrical components is determined by what the component is supposed to do. If we think about what something is supposed to do and how it does it, we can figure out how to test it. Often, removing the component and testing it on a bench is the best way to check it.

Switches

The easiest method of testing a **normally open (NO)** switch is to use a fused jumper wire to bypass the switch (Figure 3-6). An NO switch will not allow current flow when it is in its rest position. The contacts are open until they are acted on by an outside force that closes to complete the circuit. If the circuit operates with the switch bypassed, the switch is defective. Voltage drop across switches should also be checked. Ideally, when the switch is closed, there should be no voltage drop. Any voltage drop indicates resistance, and the switch should be replaced.

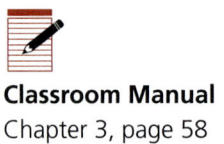

Classroom Manual
Chapter 3, page 58

Special Tools

Jumper wires

Voltmeter

Test light

Ohmmeter

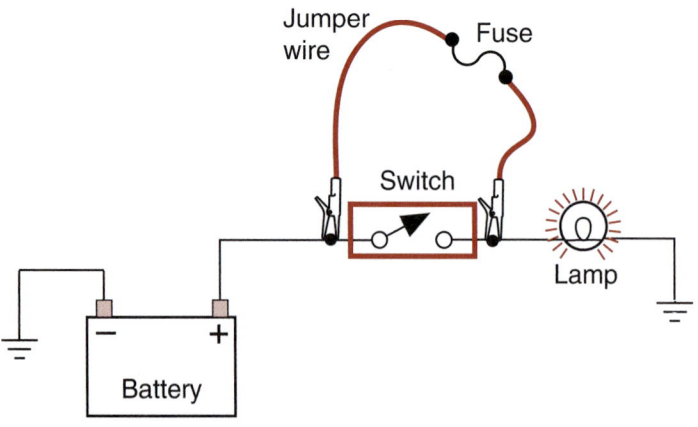

Figure 3-6 Using a fused jumper wire to bypass the switch.

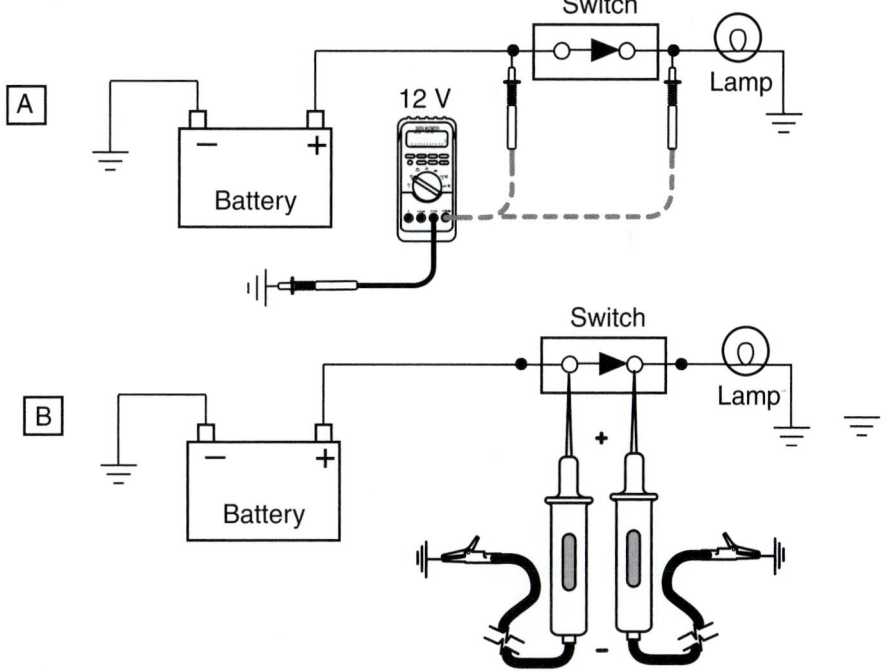

Figure 3-7 (A) Using a voltmeter to test a switch; the same voltage should be on both sides of the switch. (B) Using a test lamp to test a switch; the lamp should illuminate on both sides of the switch.

A voltmeter or test light can be used to check for voltage on both sides of the switch (Figure 3-7). A faulty NO switch would have voltage present at the input side of the switch but not on the output side when in the ON position.

CAUTION: Use a jumper wire to bypass nonresistive portions of the circuit. Do not use the jumper wire to bypass the load component. The high current will damage the circuit. The wire size and fuse of the jumper wire must be appropriate for the current in the circuit.

If the switch is removed, it can be tested with an ohmmeter. With the switch contacts open, there should be no continuity between the terminals (Figure 3-8). When the contacts are closed,

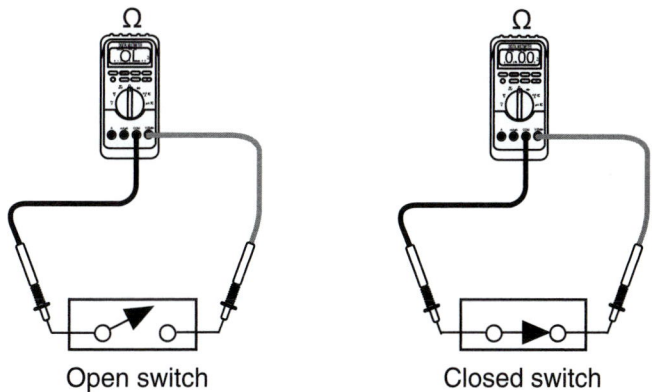

Open switch   Closed switch

Figure 3-8 The continuity through a switch can be checked with an ohmmeter. With the switch closed, there should be zero resistance. With the switch open, there should be infinite resistance.

Headlamp switch connector
(connector end view)

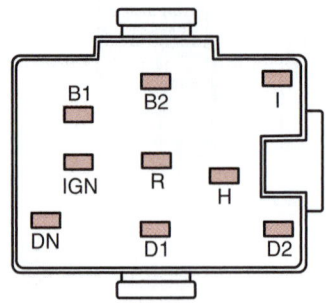

Pin number	Circuit	Circuit function
B1	38 (BK/O)	Power supply to battery
B2	195 (T/W)	Tail lamp switch feed
I	19 (LB/R)	Instrument panel lamp feed
IGN		Not used
R	14 (BR)	Tail lamp and side marker lamps
H	15 (R/Y)	Headlamp dimmer switch feed
DN		Not used
D1	54 (LG/Y)	Interior lamp switch feed
D2	706 (GY)	Battery saver door switch feed

Figure 3-9 Headlight switch continuity checks.

Classroom Manual
Chapter 3, page 61

Special Tools

Jumper wires

Test light

Voltmeter

Ohmmeter

If the circuit being tested is powered through the ignition switch, it must be in the RUN position.

To obtain accurate voltmeter test results, the battery must be fully charged and in good condition.

there should be zero resistance through the switch contacts. On complex ganged-type switches, the technician should consult the service manual for a continuity diagram (Figure 3-9). If there is no continuity chart, use the wiring diagram to make your own chart.

Relays

A **relay** is a device that uses low current to control a high-current circuit. The relay can be either a normally open or normally closed design. The relay can be checked using a jumper wire, voltmeter, ohmmeter, or test light. If the terminals are easily accessible, the jumper wire and test light may be the fastest method.

Check the wiring diagram for the relay being tested to determine if the control is through an insulated or ground switch. Use the illustration (Figure 3-10) as a guide to test a ground-switch-controlled relay. Follow these steps:

1. Use a voltmeter to check for available voltage to the battery side of the relay (terminal A). If voltage is not present at this point, the fault is in the circuit from the battery to the relay. If voltage is present, continue testing.

2. Probe for voltage at control terminal B. If voltage is not present at this terminal, then the fault is in the relay coil. If voltage is present, continue testing.

3. Use a jumper wire to connect terminal B to a good ground. If the horn sounds, the fault is in the control circuit from terminal B to the horn switch ground. If the horn does not sound, continue testing.

4. Connect the jumper wire from the battery positive to terminal C. If the horn did not sound, there is a fault in the circuit from the relay to the horn ground. If the horn sounded, the fault is in the relay.

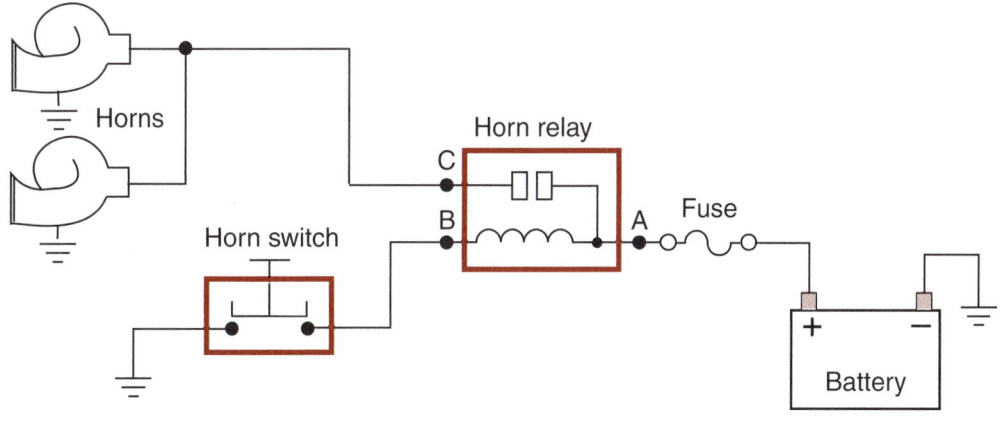

Figure 3-10 A relay circuit with a ground control switch.

If the relay is controlled by the computer, it is not recommended that a test light be used. The test light may draw more current than the circuit is designed to carry and damage the computer. Refer to the illustration (Figure 3-11) for procedures using a voltmeter to test a relay. Use a digital volt-ohm meter (DVOM) set as follows:

1. Connect the negative voltmeter test lead to a good ground.

2. Connect the positive voltmeter test lead to the output wire (terminal B). Turn on the ignition switch. If no voltage is present at this terminal, go to step 3. If the voltmeter reads 10.5 volts or higher, turn off the control circuit. The voltmeter should then read 0 volts. If it does, then the relay is good. If the voltmeter still reads any voltage, the relay is not opening and needs to be replaced.

3. Connect the positive voltmeter test lead to the power input terminal (terminal A). The voltmeter should indicate at least 10.5 volts. If below this value, the circuit from the battery to the relay is faulty. If the voltage value is correct, continue testing.

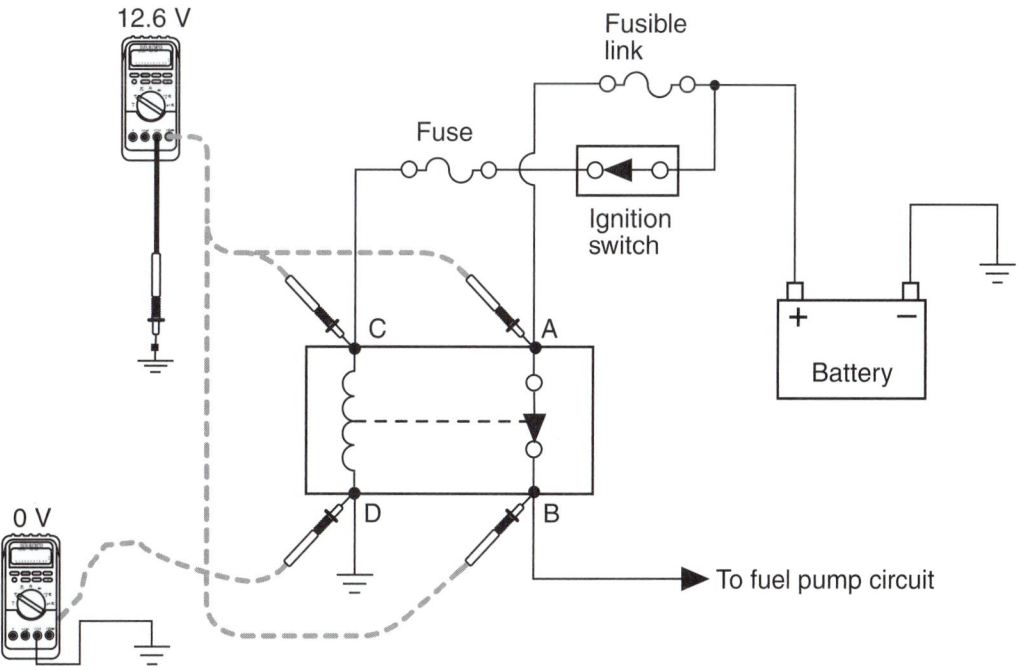

Figure 3-11 Testing relay operation with a voltmeter.

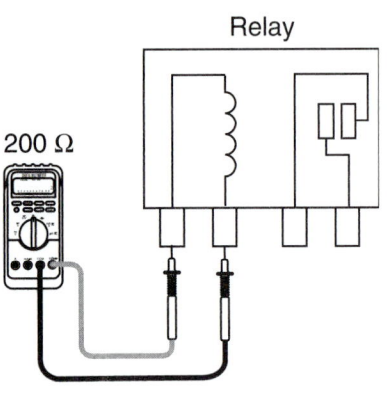

Relay

200 Ω

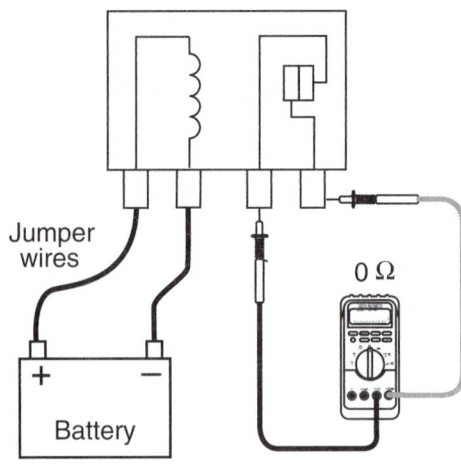

Jumper wires

0 Ω

+ −

Battery

Figure 3-12 Testing the resistance of the relay coil.

Figure 3-13 Bench testing a relay.

4. Connect the positive voltmeter test lead to the control circuit terminal (terminal C). The voltage should read 10.5 volts or higher. If not, check the circuit from the battery to the relay (including the ignition switch). If the voltage is 10.5 volts or higher, continue testing.

5. Connect the positive voltmeter test lead to the relay ground terminal (terminal D). If more than 1 volt is indicated on the meter, there is a poor ground connection. If the reading is less than 1 volt, replace the relay.

> **CAUTION:** It is not recommended that a test light be used to probe for power in a computer-controlled circuit. The increased draw of the test light may damage the system components.

Be careful not to touch the coil terminals with the ohmmeter test leads while the coil is energized.

Special Tools

Ohmmeter

If the relay terminals are not accessible, remove the relay from its holding fixture and bench test it. Use an ohmmeter to test for continuity between the relay coil terminals (Figure 3-12). If the meter indicates an infinite reading, replace the relay. If continuity is indicated, use a pair of jumper wires to energize the coil (Figure 3-13). Check for continuity through the relay contacts. If the meter indicates an infinite reading, the relay is defective. If there is continuity, the relay is good and the circuits will have to be checked.

Be sure to check your service manual for resistance specifications and compare the relay to them. It is easy to check for an open coil. However, a shorted coil will also prevent the relay from working. Low resistance across a coil would indicate that it is shorted. Too low of resistance may also damage the transistors and/or driver circuits because of the excessive current that would result.

> **SERVICE TIP:** The procedures presented above can be used to test the relay to determine the type of fault it has. However, the easiest way to test a relay is to substitute it with a *known good* relay of the same type. If the circuit operates with the substitute relay, the old relay is the faulty component.

Testing Stepped Resistors

A **stepped resistor** has two or more fixed resistor values (Figure 3-14). The best method of testing a stepped resistor is to use an ohmmeter. To obtain accurate test results, it is a good practice to remove the resistor from the circuit. Connect the ohmmeter leads to the two ends of the resistor (Figure 3-15). Compare the results with manufacturer's specifications. Be sure to place the ohmmeter on the correct scale to read the anticipated amount of resistance.

Classroom Manual
Chapter 3, page 65

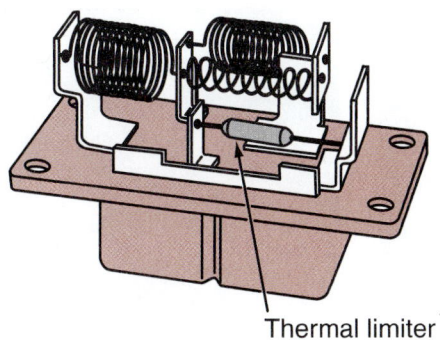

Thermal limiter

Figure 3-14 A stepped resistor used in the heater blower motor circuit.

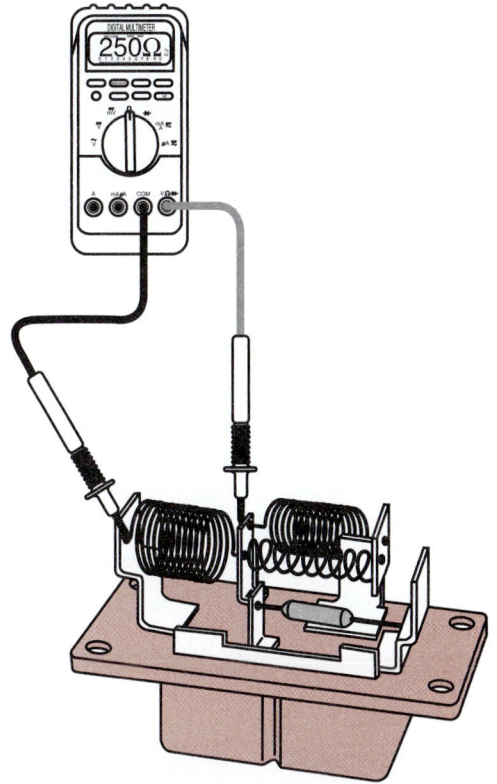

Figure 3-15 Ohmmeter testing of a stepped resistor.

A stepped resistor can also be checked with a voltmeter or DSO. By measuring the voltage after each part of the resistor block and comparing the readings to specifications, you can tell if the resistor is good or not.

Testing Variable Resistors

A **variable resistor** provides for an infinite number of resistance values within a range. As with the stepped resistor, the best method of testing a variable resistor is with an ohmmeter. However, it is possible to use a voltmeter, DSO, or test light.

A **rheostat** is a two-terminal variable resistor used to regulate strength of an electrical circuit. To test a rheostat, locate the input and output terminals and connect the test leads to them. Rotate the resistor knob slowly while observing the ohmmeter. The resistance value should remain

Special Tools

Ohmmeter

Voltmeter

Test light

DSO

Classroom Manual
Chapter 3, page 65

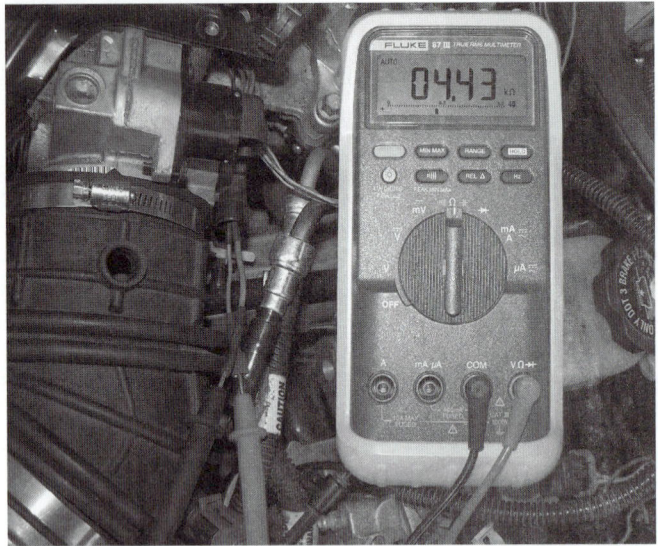

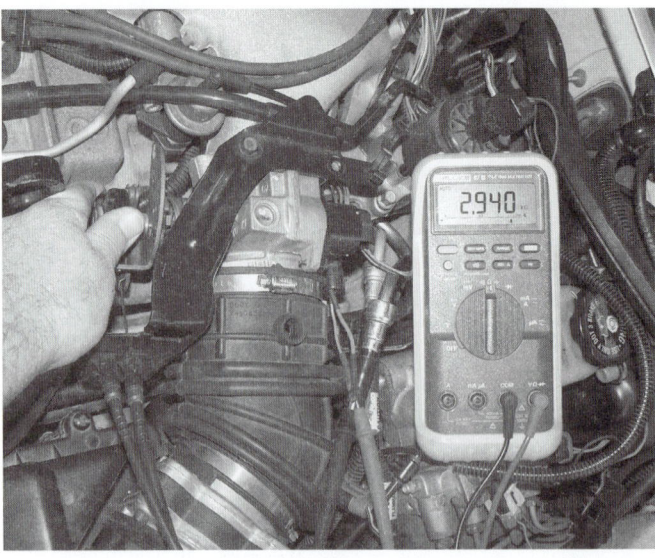

Figure 3-16 Using an ohmmeter to test the continuity between terminals A and C of a potentiometer.

Figure 3-17 Testing continuity between terminals A and B of a potentiometer while the wiper is being moved.

within the specification limits and change in a smooth and constant manner. If the resistance values are out of limits or the resistance value jumps as the knob is turned, replace the rheostat.

If a voltmeter is used, the readings should be smooth and consistent. A test light should change in brightness as the knob is turned; the rheostat is defective if the light blinks at any point.

A **potentiometer** is a three-wire variable resistor that acts as a voltage divider to produce a continuously variable output signal proportional to a mechanical position. To test a potentiometer, connect the ohmmeter test leads to terminals A and C (Figure 3-16). Check the results with specifications. Next connect the ohmmeter test leads to terminals A and B (Figure 3-17). Check the resistance at the stop and observe the ohmmeter as the wiper is moved to the other stop. The resistance values should be within specification and smooth and constant.

A voltmeter can be used in the same manner. However, jumper wires may need to be used to gain access to the test points (Figure 3-18). Because potentiometers are primarily used in computer-controlled circuits, it is not recommended that a test light be used.

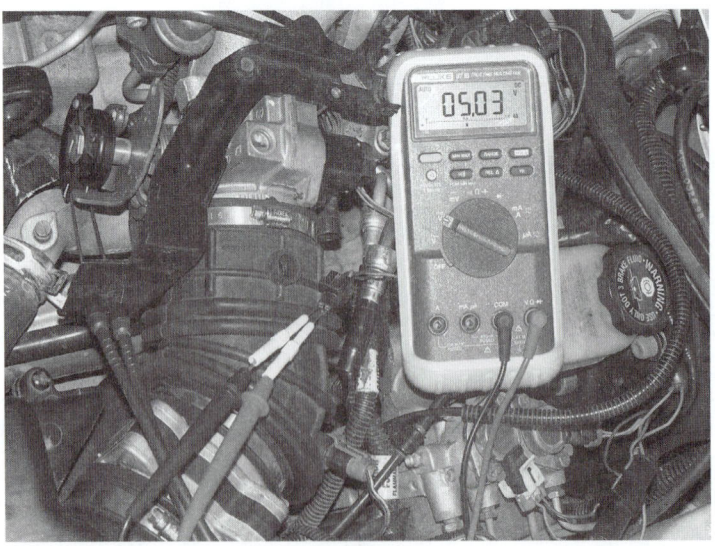

Figure 3-18 It may be necessary to use jumper wires to connect the wire connector to the sensor in order to measure voltage.

CAUTION: Do not pierce the insulation to test the potentiometer. These circuits usually operate on 5 to 9 volts. Piercing the insulation may break some of the wire strands, resulting in a voltage drop that will give errant information to the computer. Even if the conductor is not broken, moisture can enter and cause corrosion.

Testing Diodes

A **diode** is an electrical one-way check valve that will allow current to flow in one direction only. Regardless of the bias of the diode, it should allow current flow in one direction only. To test a diode, use an analog ohmmeter. Connect the meter's leads across the diode (Figure 3-19). Observe the reading on the meter. Then reverse the meter's leads and observe the reading on the meter. The resistance in one direction should be very high or infinite and in the other direction, the resistance should be close to zero. If any other readings are observed, the diode is bad. A diode that has low resistance in both directions is shorted. A diode that has high resistance or an infinite reading in both directions is open.

You may run into problems when checking a diode with a high-impedance digital ohmmeter. Since many diodes won't allow current flow through them unless the voltage is at least 0.6 volts, a digital meter may not be able to forward bias the diode. This will result in readings that indicate the diode is open, when in fact it may not be. Because of this problem, many multimeters are equipped with a diode-testing feature. This feature allows for increased voltage at the test leads. The value displayed is the voltage required to forward bias the diode. A silicon diode should read between 400 and 800 mv in the forward direction and open in reverse. For a germanium diode, it is between 200 and 400 mv in the forward direction. If the diode is open, the meter will display "OL" or another reading to indicate infinity or out-of-range. Some meters during diode check will make a beeping noise when there is continuity.

Diodes may also be tested with a voltmeter. Using the same logic as when testing with an ohmmeter, test the voltage drop across the diode. The meter should read low voltage in one direction and near source voltage in the other direction.

Testing Zener Diodes

If the Zener diode is out of the circuit and you need to diagnose it for an open or short, then test the Zener as described for the standard diode. However, if you desire to measure its Zener voltage

Classroom Manual
Chapter 3, page 68

Special Tools

Analog ohmmeter

An analog ohmmeter should be used to test a **diode**. The test current of a digital ohmmeter can pass both ways through a good diode, giving false indications to the technician.

The voltage displayed on the meter is referred to as turn on voltage or diode drop.

Classroom Manual
Chapter 3, page 69

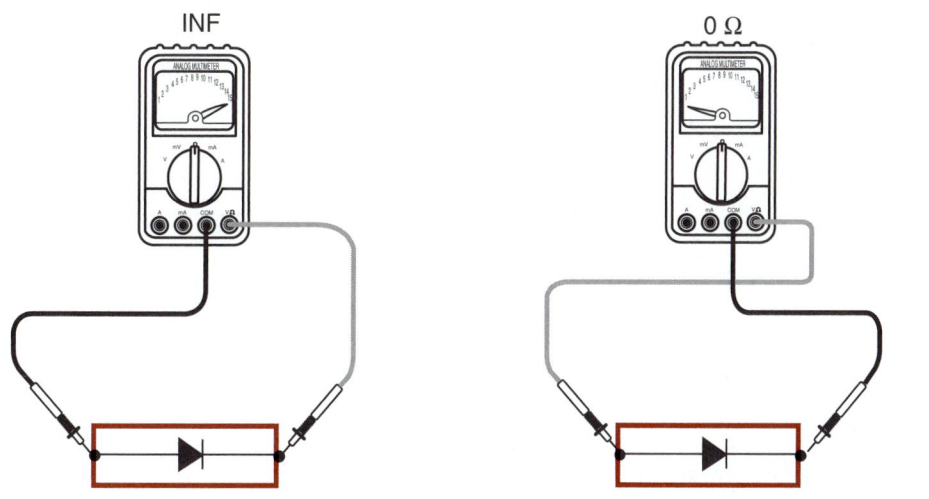

Figure 3-19 Use an analog ohmmeter to test a diode for an open and short.

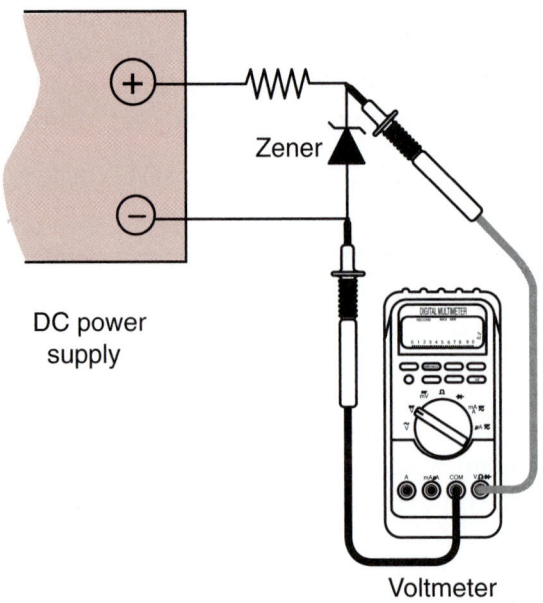

Figure 3-20 Making a test power supply to measure Zener voltage.

level, you will have to build a test circuit (Figure 3-20). The power supply voltage should be set to a value slightly higher than the Zener value. For example, for a 12-volt diode, the supply voltage should be about 15 volts. This can be made using many styles of "project boxes" from most electronic stores. The value of the resistor R should limit the current to about one milliamp. For example, using 15 volts with a 12-volt Zener, use a 3.3 K resistor.

Once the circuit is built, read the Zener voltage using a digital voltmeter. If the voltmeter indicates 600 mv, the diode is reverse biased and will need to be reinstalled into the circuit.

Testing LEDs

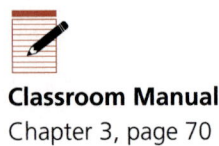

Classroom Manual
Chapter 3, page 70

The turn-on voltage of an LED is usually between 1.5 to 2.5 volts. If your DMM has a diode test function, then the LED can be tested in the same manner as a standard diode. The difference will be that the meter will read 1,600 or 50 when the diode conducts instead of the 600 you read on a standard diode.

It is possible to test an LED without the use of a DMM. To do this, first build the test circuit as shown (Figure 3-21). By plugging the LED into the circuit, it should light. If the LED doesn't light, then reverse the polarity on the diode. If it still doesn't light, then the LED is faulty.

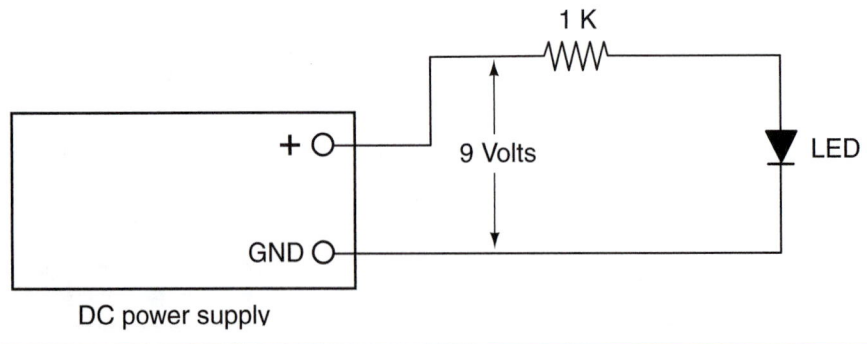

Figure 3-21 Testing an LED with a test circuit.

Testing Zener Diodes and LEDs in a Circuit

It is not necessary to remove a Zener diode or the LED from the circuit to test it. To test a Zener, use a voltmeter and measure the voltage across it. Connect the negative lead to the anode and the positive lead to the cathode. The meter should read the Zener voltage. If you read zero volts, the Zener is shorted. This is true if the power and ground circuits to the Zener are confirmed as being good. If the voltmeter reads a voltage that is higher than the Zener's rated voltage, the diode is open.

For an LED that is supposed to be lit but isn't, use a voltmeter to measure the voltage across it. If you measure more than 3 volts, the LED is open.

Testing Transistors

Although replacing a transistor that is part of an integrated circuit board is not often done in the automotive repair industry, some bipolar transistors can be easily tested and replaced. In order to test a bipolar transistor, it is first necessary to identify its type (NPN or PNP) and lead arrangement. To perform this test, it is first necessary to identify the base leg of the transistor. There are several configurations of the transistor legs (Figure 3-22). This can be done using a multimeter. Since transistors behave as back-to-back diodes, the collector and emitter can be identified based on the fact that the doping for the base-to-emitter junction is always much higher than for the base-to-collector junction. Therefore, the forward voltage drop will be a couple of milli volts higher on the DVOM reading when set on the diode test function. Follow Photo Sequence 3 to identify the type of bipolar transistor.

> **CAUTION:** Do not hold the transistor in your hand while testing it. For every degree the transistor increases in temperature, the base-emitter diode drop decreases by 2 mV. This is a significant amount when determining the base-to-emitter and base-to-collector junctions.

If none of the six possible lead connection combinations indicates a pair of low readings, or if more than one combination results in a pair of low readings, the transistor is probably faulty. Keep in mind that the base-to-collector junction voltage drop is always slightly lower than the emitter-to-base junction drop.

Once the type of transistor is identified, it can be tested using the diode function of the DVOM. Connect the red meter lead to the base of the transistor and the black lead to the emitter. A good NPN transistor will read a voltage of between 450 and 900 mv. A good PNP transistor will read open. With the red lead still on the base, move the black lead to the collector. The reading should be the same as the previous test.

Classroom Manual
Chapter 3, page 71

This discussion is for testing transistors that are out of the circuit only.

Special Tools

DVOM

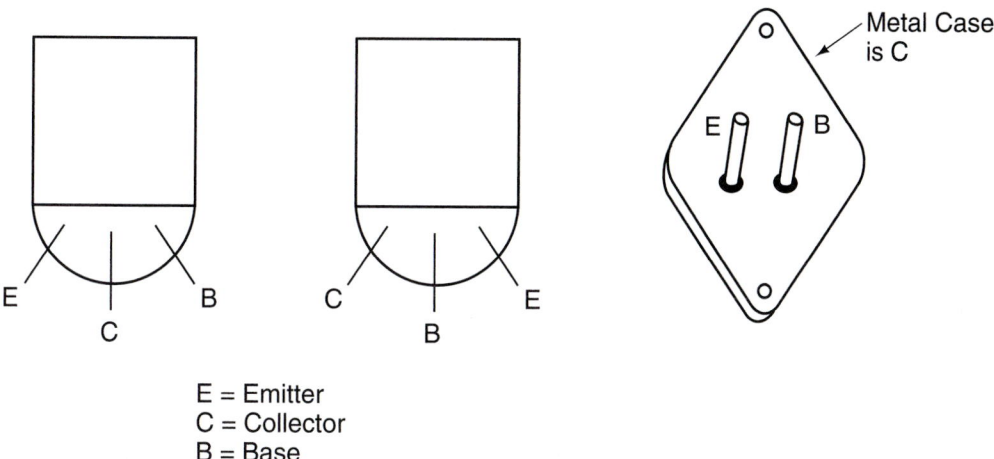

E = Emitter
C = Collector
B = Base

Figure 3-22 There are several configurations of the transistor legs.

Photo Sequence 3
Identifying Bipolar Transistors

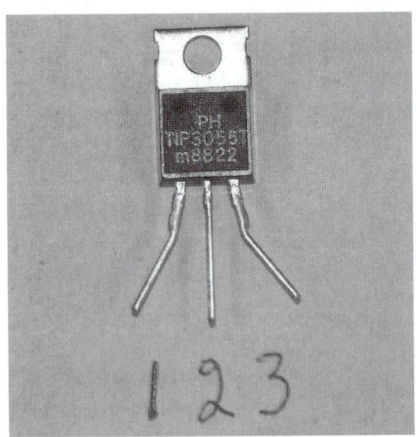

P3-1 Label the pins on the unknown device 1, 2, and 3.

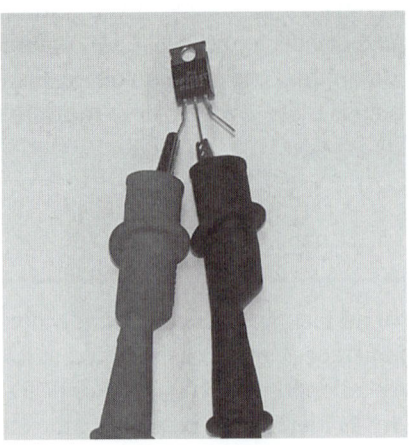

P3-2 Put the positive probe of the DVOM on pin 1 and measure the diode drop to pins 2 and 3.

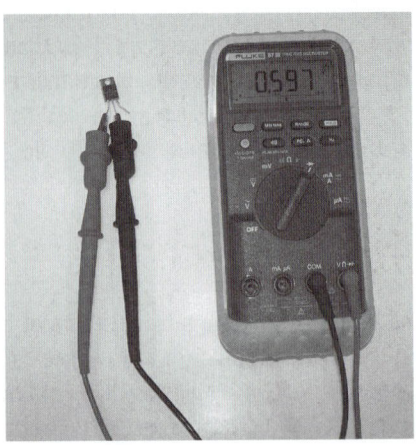

P3-3 If the positive probe is on the base of a good NPN transistor, you should read a low diode drop to pins 2 and 3. The base to collector diode drop will be slightly lower than the base to emitter reading.

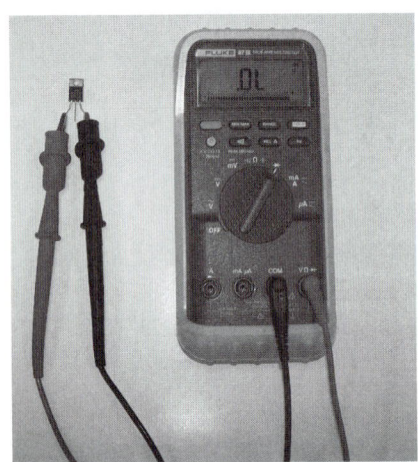

P3-4 If one or both measurements to pins 2 and 3 is high, put the positive probe on pin 2 and retest. If still high, put the positive probe on pin 3 and retest.

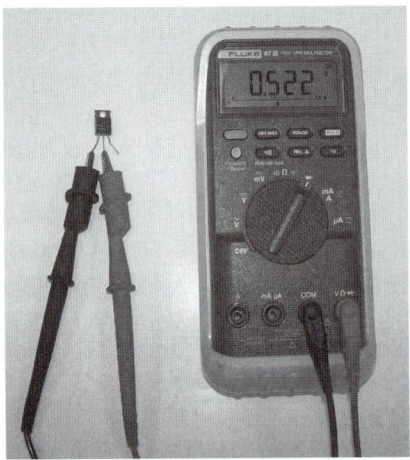

P3-5 If the reading is high when all three pins are tested, repeat the tests with the negative probe as the common pin. A pair of low readings now indicates a PNP transistor.

Next, reverse the meter leads and repeat the test. With the black lead connected to the base of the transistor and the red lead to the emitter, a good PNP transistor will read a voltage of between 450 and 900 mv. A good NPN transistor will read open. Leave the black lead on the base and move the red lead to the collector. The reading should be the same as the previous test.

Finally, place one meter lead on the collector, the other on the emitter. The meter should read open. Reverse your meter leads and the meter should read open. This is the same for both NPN and PNP transistors.

 SERVICE TIP: Some power transistors have built-in diodes that are reverse biased across the collector-to-emitter junction and resistors between the base-to-emitter junction. The resistor is usually 50 ohms. If not aware of this, you can be confused by the reading if testing the transistor as a standard bipolar transistor. You will need to know the specifications of the power transistor in order to properly test it. Power transistors without internal damper diodes test just about like bipolar transistors.

Testing Darlington Transistors

A **Darlington** is a special type of configuration usually consisting of two transistors fabricated on the same chip or mounted in the same package.

Testing is basically similar to that of normal bipolar transistors except that in the forward direction the base-to-emitter reading will be 0.2 to 1.4 volts when reading the DVOM on the diode function. This higher voltage is due to the pair of junctions that are in series.

Testing for Circuit Defects

Electrical circuits may develop an open, a short, a ground, or an excessive voltage drop that will cause the circuit to operate improperly.

Testing for Opens

It is possible to test for opens using a voltmeter, DSO, test light, self-powered test light, ohmmeter, or a jumper wire. The test equipment used will depend on the circuit being tested and the accessibility of the components.

The technician must determine the correct operation of the circuit before attempting to determine what is wrong. The illustration (Figure 3-23) shows the voltmeter readings that should be obtained in a properly operating parallel circuit.

Classroom Manual
Chapter 3, page 75

A reading of 1.2 V may be too high for some DVOMs. This will result in a good **Darlington** testing as open. Confirm that the open circuit reading on your DVOM is higher than 1.4 V.

Classroom Manual
Chapter 3, page 83

Special Tools

Voltmeter

Test light

Self-powered test light

Ohmmeter

Jumper wire

DSO

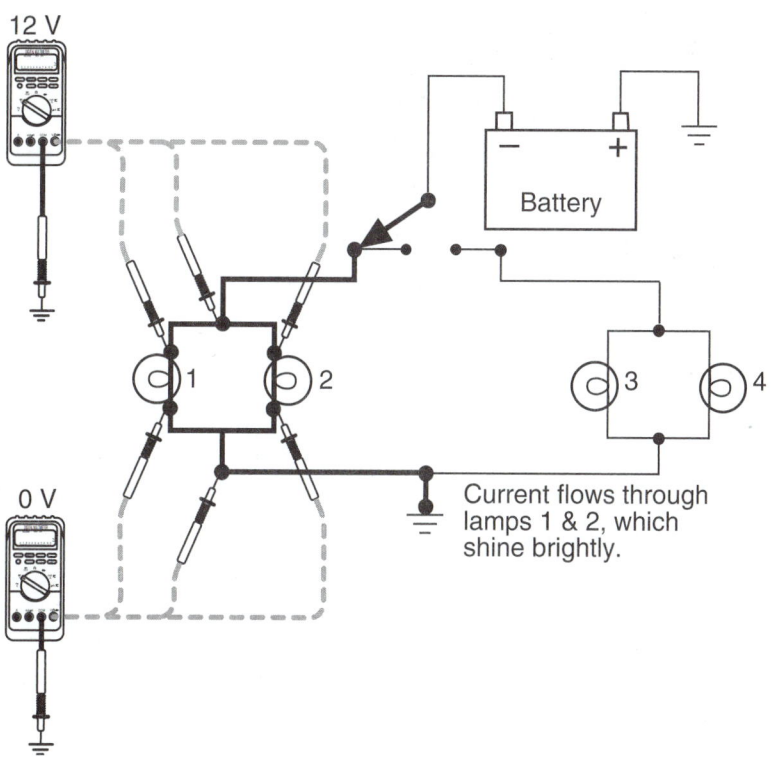

Current flows through lamps 1 & 2, which shine brightly.

Figure 3-23 Voltmeter readings that would be expected in a properly operating parallel circuit.

The easiest method of testing a circuit is to start at the most accessible place and work from there. If the load component is easily accessible, test for voltage at the input to the load (Figure 3-24). Use the following procedure for locating the open:

1. Check for voltage at point A. If voltage is 10.5 volts or higher, check the ground side (point B). If less than 10.5 volt is present, there is excessive resistance or an open in the ground circuit. If the voltage at point A is less than 10.5 volts, continue testing.

2. Work toward the battery. Test all connections for voltage. If voltage is present at a connection, then the open is between that connection and the previously tested location (Figure 3-25). Use a jumper wire to bypass that section to confirm the location of the open.

3. If battery voltage is present at point B, the open is in the ground circuit. Use a jumper wire to connect the ground circuit. Then retest the component.

In more complex circuits, the open may have very different results. In a normally operating circuit, the voltmeter readings would be as indicated in the illustration (Figure 3-26). If an open occurs in the ground side of the circuit, the circuit converts to a series circuit (Figure 3-27).

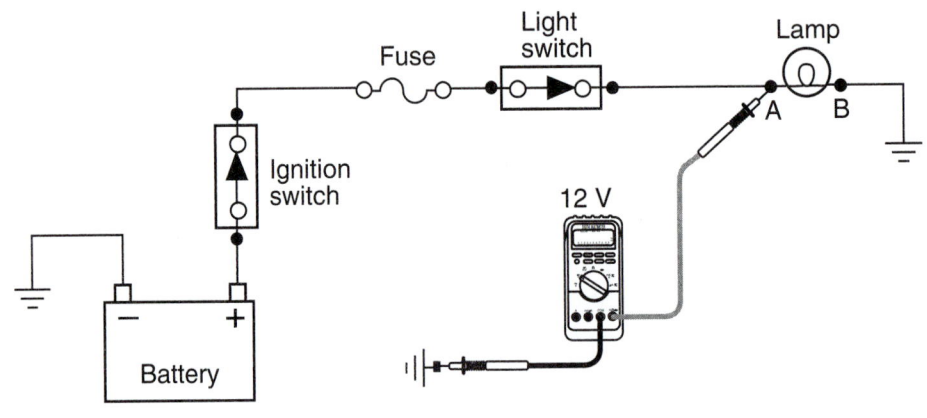

Figure 3-24 Locating an open by testing for voltage.

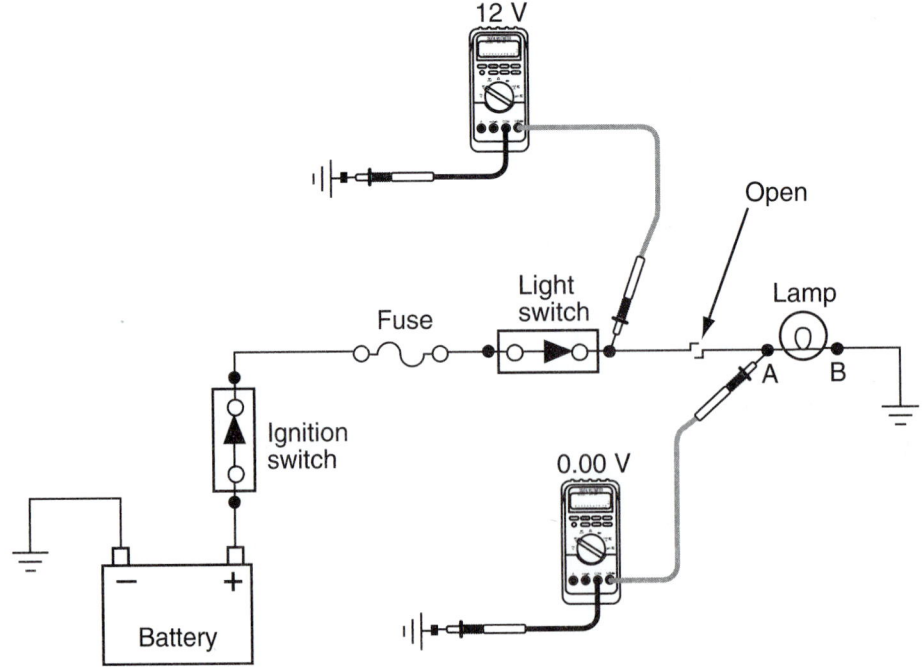

Figure 3-25 An open is present between the point where voltage was measured and where it was not.

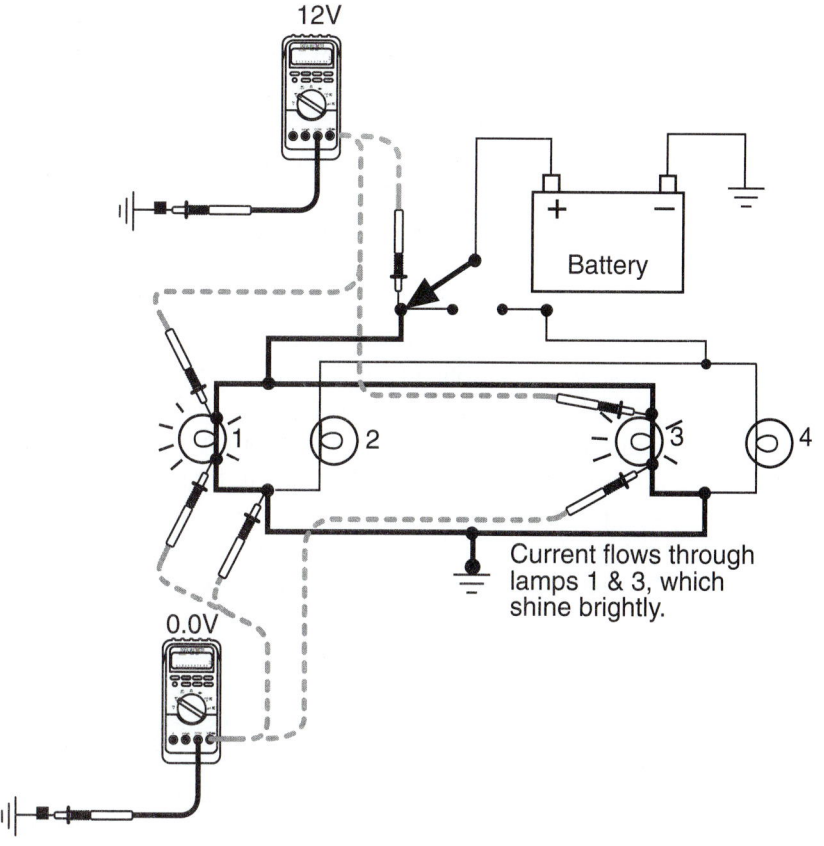

Figure 3-26 Properly operating complex parallel circuit.

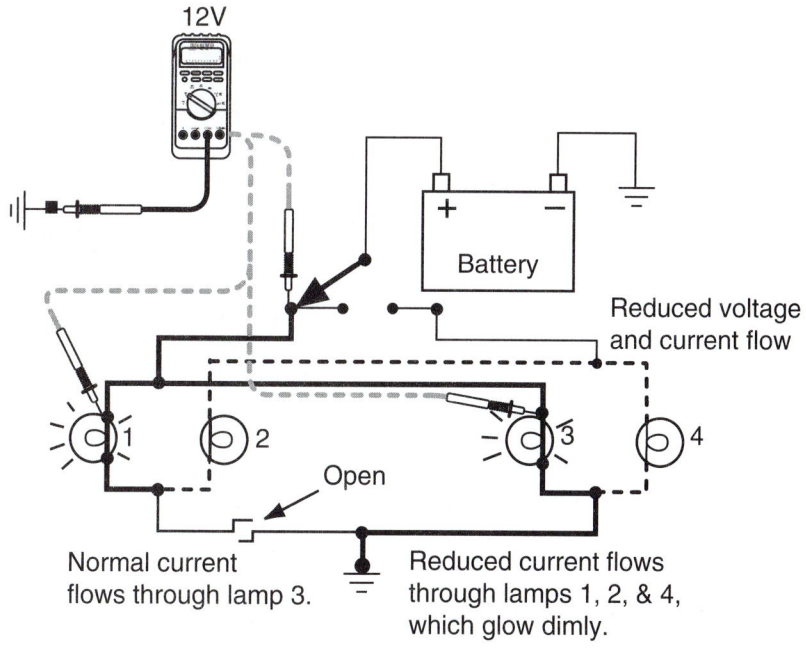

Figure 3-27 An open in the ground circuit can convert the circuit to a series circuit. The dashed line represents the resulting path to ground.

This is a form of **feedback** that results in lamps coming on that are not intended to. If the electrons cannot find a path to ground through the intended circuit, they will attempt to find an alternate path to ground. This may result in turning on components that are in the path. This type of defect is referred to as feedback. Normal voltage is applied to lamp 3, but lamps 1, 2, and 4 are in series and will illuminate dimmer than normal. The voltmeter will read 12 volts at the locations illustrated in Figure 3-27. However, the voltmeter will not indicate 0 volts on the ground side of bulb 1.

Testing for Shorts

Locating a copper-to-copper short can be one of the most difficult tasks for a technician. If the short is within a component, the component will operate at less than optimum or not at all. An ohmmeter can be used to check the resistance of the component. If there is a short, then the amount of resistance will be lower than specified. If specifications for the component are not available, it may be necessary to replace the component with a known good unit. Do this only after it has been determined that the insulated and ground side circuits are in good condition.

If the short is between circuits, the result will be components operating when not intended (Figure 3-28). Visually check the wiring for signs of burned insulation and melted conductors that will indicate a short. Also check common connectors that are shared by the two affected circuits. Corrosion can form between two terminals of the connector and result in the short.

If the visual inspection does not isolate the cause of the copper-to-copper short, remove one of the fuses for the affected circuits. (If the affected circuits share a common fuse, remove it.) Install a buzzer that has been fitted with terminals across the fuse holder terminals (Figure 3-29). Activate the circuit that the buzzer is connected to. In Figure 3-28, if the buzzer is connected to fuse B, then switch 1 would be turned on. Disconnect the loads that are supposed to be activated by this switch (lamp 1). Disconnect the wire connectors in the circuit from the load back to the switch. If the buzzer stops when a connector is disconnected, the short is in that portion of the circuit.

Classroom Manual
Chapter 3, page 84

Special Tools

Ohmmeter

Buzzer fitted with
  terminals and
  in-line fuse

A short may not
blow a fuse
depending on the
amount of current
flowing.

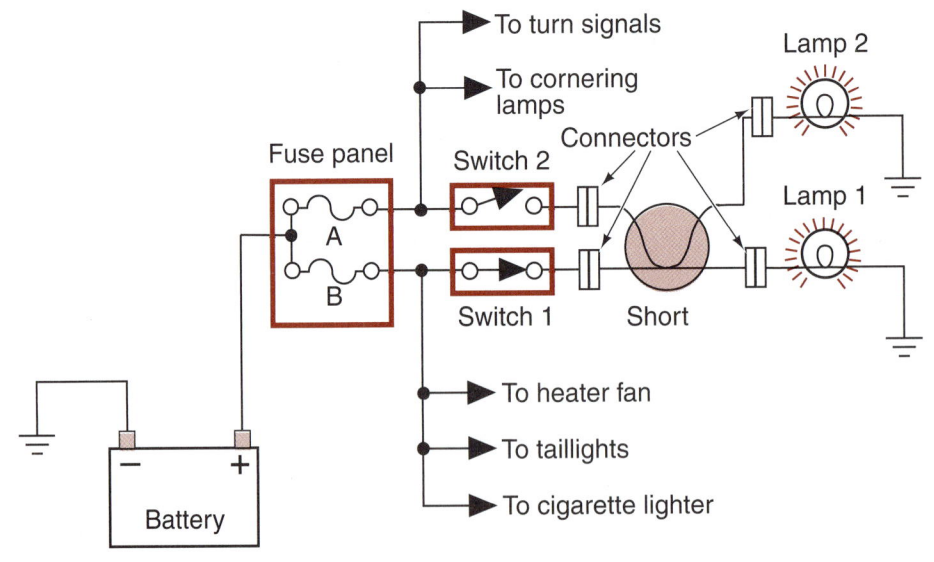

Figure 3-28 A copper-to-copper short between two circuits.

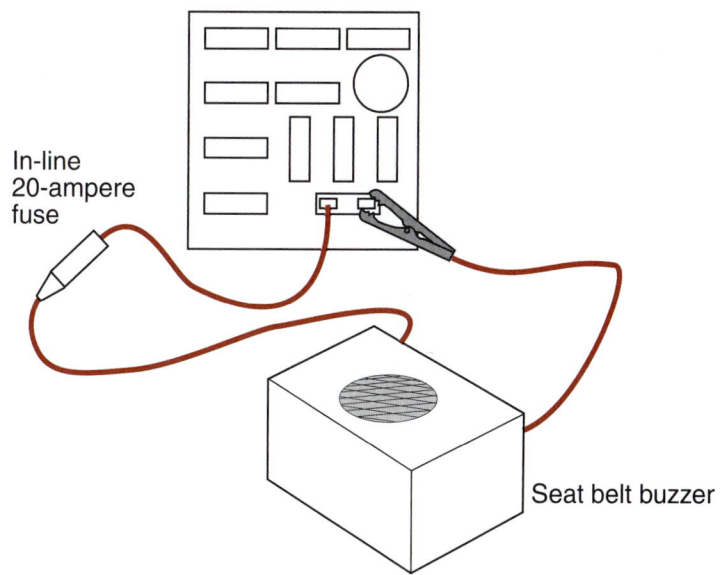

In-line 20-ampere fuse

Seat belt buzzer

Figure 3-29 The buzzer will sound until the cause of the short is found and corrected.

Testing for a Short to Ground

A fuse that blows as soon as it is installed indicates a **short to ground.** This condition allows current to return to ground before it has reached the intended load component. If the circuit is unfused, the insulation and conductor will melt. Not all shorts will blow the fuse, however. If the short to ground is on the ground side of the load component but before a grounding switch, the component will not turn off (Figure 3-30). If the short to ground is after the load and grounding switch (if applicable), circuit operation will not be affected.

To confirm that the circuit has a ground before the load, remove the fuse and connect a test light in series across the fuse connections. If the lamp lights, the circuit has a short to ground.

It is difficult to test for shorts to ground with a test light or voltmeter because the fuse blows before any testing can be conducted. To prevent this, connect a cycling circuit breaker that is fitted

Classroom Manual
Chapter 3, page 84

Special Tools

Test light
Circuit breaker fitted
  with terminals
Gauss gauge
  or compass
Ohmmeter

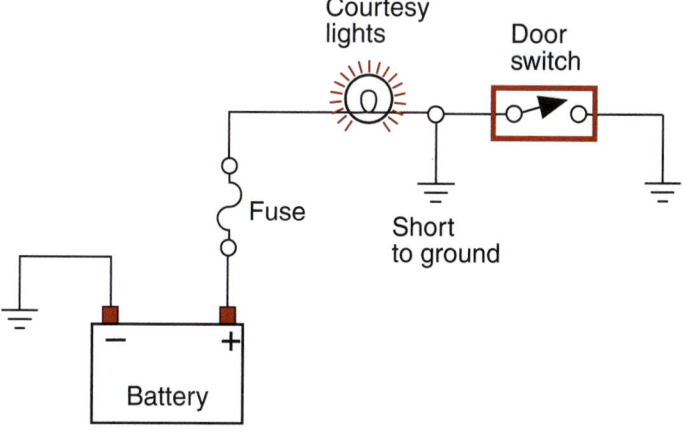

Courtesy lights

Door switch

Fuse

Short to ground

Battery

Figure 3-30 A ground in this location will cause the lamp to remain on.

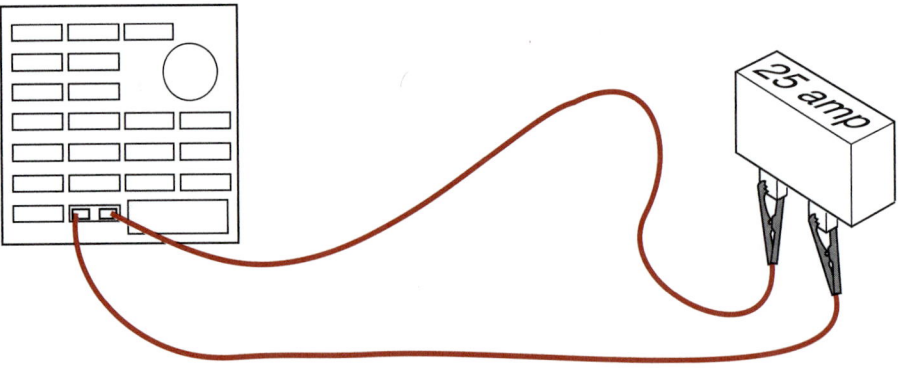

Figure 3-31 Use a circuit breaker to protect the circuit while checking for the short to ground.

A buzzer can be substituted for the test light.

with alligator clips across the fuse holder (Figure 3-31). The circuit breaker will continue to cycle open and closed, allowing the technician to test for voltage.

CAUTION: Use a circuit breaker that is rated between 25 and 30 amperes. The use of a circuit breaker rated too high will damage the circuit.

Testing for shorts may be complicated if there are several circuits protected by a single fuse and if the ground is located in a section of wire that is not accessible. There are a couple of methods that can be used to locate the fault.

One method is to connect a test light, in series with a cycling circuit breaker, across the fuse holder (Figure 3-32). While observing the test light, disconnect individual circuits one at a time until the light goes out. The fault is in the circuit that was disconnected when the light went off.

A second method is to use a **Gauss gauge** or a compass to locate the short to ground. A Gauss gauge is a meter that is sensitive to the magnetic field surrounding a wire conducting current. The gauge or compass works on the principle that a magnetic field is developed around a conductor that is carrying current. With a cycling circuit breaker bypassing the blown fuse, trace the path of the circuit with the gauge or compass. The needle will fluctuate as long as the gauge is over the conductor. The needle will stop fluctuating when the point of the short to ground is passed (Figure 3-33). This method will work even through the vehicle's trim. It will be necessary to follow all of the circuits protected by the fuse. Consult the wiring diagram for this information.

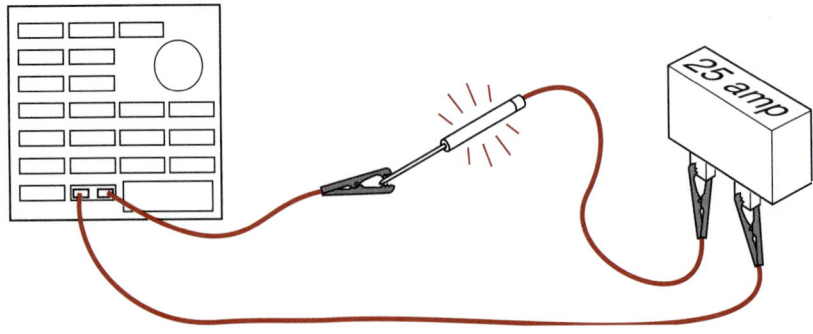

Figure 3-32 The test light will allow the technician to "see" when he has located the faulty circuit.

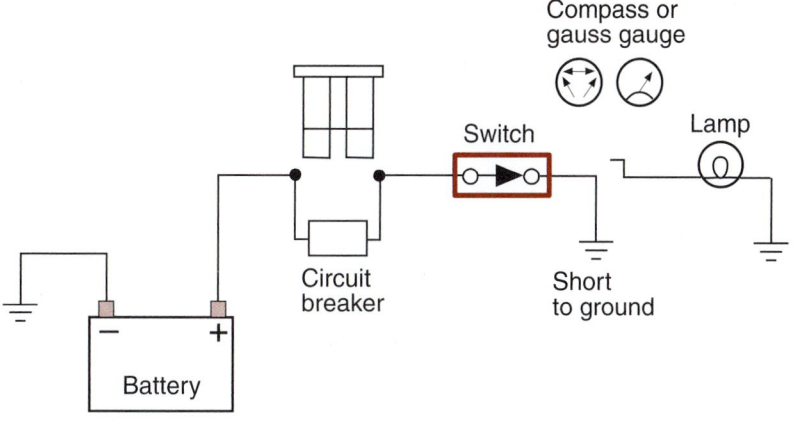

Figure 3-33 The needle of a compass or Gauss gauge will fluctuate over the portion of the circuit that has current flowing through it. Once the ground has been passed, the needle will stop fluctuating.

Testing for Voltage Drop

Voltage drop, when considered as a defect, defines the portion of applied voltage that is used up in other points of the circuit rather than that used by the load component. It is a resistance in the circuit that reduces the amount of electrical pressure available beyond the resistance. Excessive voltage drop may appear on either the insulated or ground return side of a circuit. To test for voltage drop, the circuit must be active (current flowing). The source voltage must be as specified before voltage drop readings can be valid. Whenever voltage drop is suspected, both sides of the circuit must be checked.

Excessive voltage drop caused by high resistance can be identified by dim or flickering lamps, inoperative load components, or slower-than-normal electrical motor speeds. Excessive resistance will not cause the fuse to blow.

To perform a voltage drop test on any circuit, the positive voltmeter lead must be connected to the most positive portion of the circuit. Follow the instructions shown in Photo Sequence 4 to conduct a voltage drop test. Consult the service manual for the maximum amount of voltage drop allowed. When the voltage drop decreases to within specifications, the cause of the excessive resistance has been located. In Photo Sequence 4, a faulty relay was the cause of the excessive resistance.

When testing the ground side of the circuit, the ground connection terminal of the load component is the most positive location and the battery negative post is the most negative (Figure 3-34). Usually more than 0.1 volt indicates excessive resistance in the ground circuit.

Classroom Manual
Chapter 3, page 86

Special Tools

Voltmeter

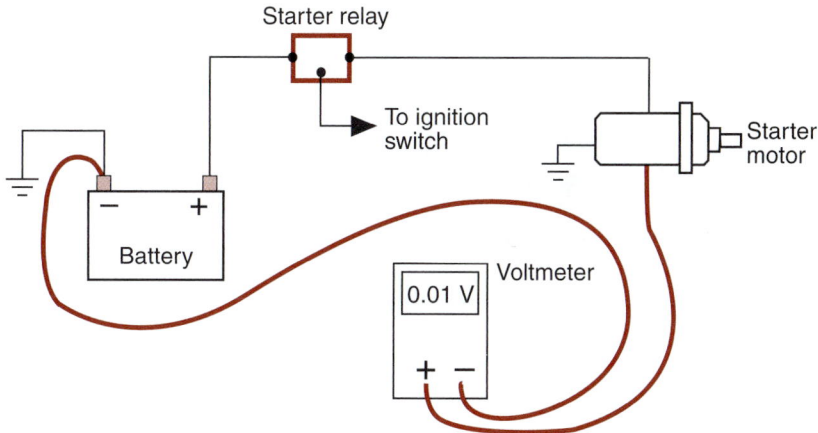

Figure 3-34 Testing the ground side of the starter motor circuit for high resistance by measuring voltage drop. Notice the voltmeter connections.

Photo Sequence 4
Voltage Drop Test to Locate High Circuit Resistance

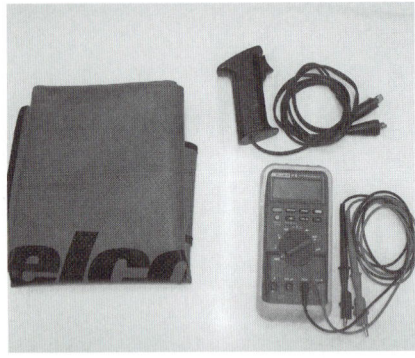

P4-1 Tools required to test for excessive resistance in a starting circuit are fender covers, a DVOM, and a remote starter switch.

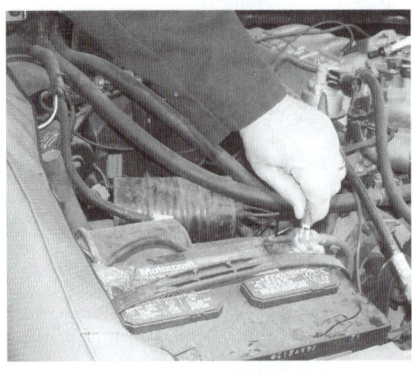

P4-2 Connect the positive lead of the meter to the positive battery post. If possible, do not connect the lead to the cable clamp.

P4-3 Connect the negative lead of the meter to the main battery terminal on the starter motor.

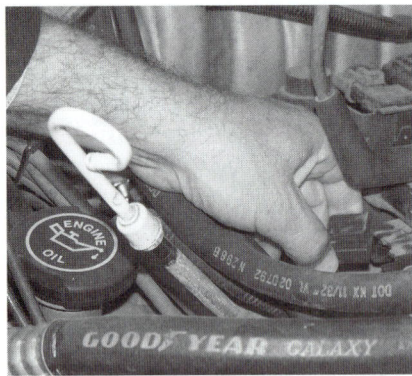

P4-4 To conduct a voltage drop test, current must flow through the circuit. In this test the ignition system is disabled and the engine is cranked using a remote starter switch.

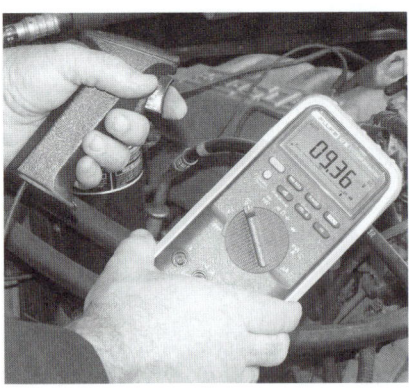

P4-5 With the engine cranking, read the voltmeter. The reading is the amount of voltage drop.

P4-6 If the reading is out of specifications test at the next connection toward the battery. In this instance the next test point is the starter side of the relay.

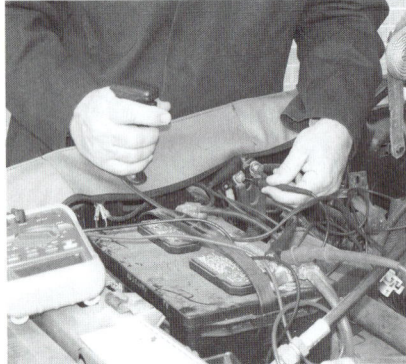

P4-7 Crank the engine and touch the negative test lead to the starter side of the relay. Observe the voltmeter while the engine is cranking.

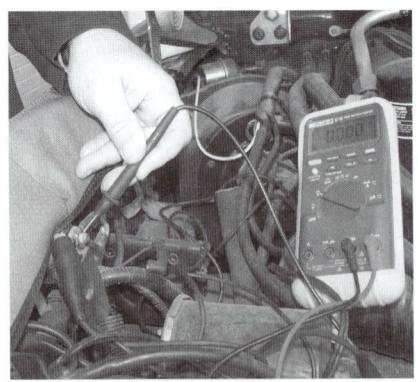

P4-8 Test in the same manner on the battery side of the relay. This is the voltage drop across the positive circuit from the battery to the relay.

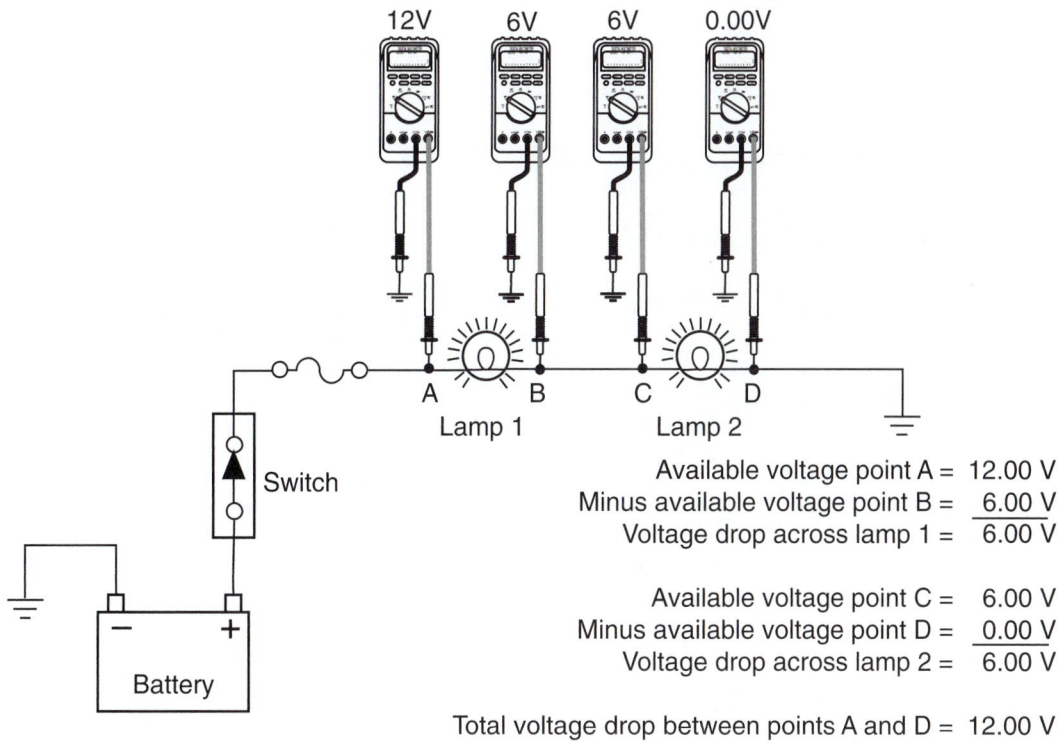

Available voltage point A = 12.00 V
Minus available voltage point B = 6.00 V
Voltage drop across lamp 1 = 6.00 V

Available voltage point C = 6.00 V
Minus available voltage point D = 0.00 V
Voltage drop across lamp 2 = 6.00 V

Total voltage drop between points A and D = 12.00 V

Figure 3-35 Using available voltage to calculate voltage drop over a component. This method is used if the wires of the circuit are too long to test with standard test leads.

According to many manuals, the maximum allowable voltage drop for an entire circuit, except for the drop across the load, is 10% of the source voltage. Although 1.2 volts is the maximum acceptable amount, it is still too much. Many good technicians use 0.5 volts as the maximum allowable drop. However, there should be no more than 0.1 volts dropped across any one wire or connector. This is the most important specification to consider and remember.

It is possible to calculate voltage drop by testing for available voltage. Use Ohm's law to determine the correct amount of voltage drop that should be across a component. Test for available voltage on both sides of the load component (Figure 3-35). Subtract the available voltage readings to obtain the amount of voltage drop across the component.

Digital Storage Oscilloscope (DSO)

The greatest advantage of a DSO is the speed at which it samples electrical signals. Mechanical switching speed of switches and relays is measured in thousands of a second or milliseconds. Electrical/electronic switching speed is measured in millionths of a second or microseconds. Radio frequency interference (RFI) is measured in billionths of a second. The DSO operates at 25 million samples per second. Another method of expressing this operating speed is to say the DSO is capable of sampling a signal in 40 billionths of a second. DSO sampling speed is at least 47,000 times faster than automotive testers such as other engine analyzers.

This sampling speed allows the DSO to provide an extremely accurate, expanded display of input sensors and output actuators compared to multimeters or other analog scopes. Such increased speed allows the DSO to display glitches or momentary defects in input sensors and output actuators. The extremely fast sampling of the DSO allows this scope to display a graph of input sensor and output actuator operation. Some DSOs have the capability to display two voltage

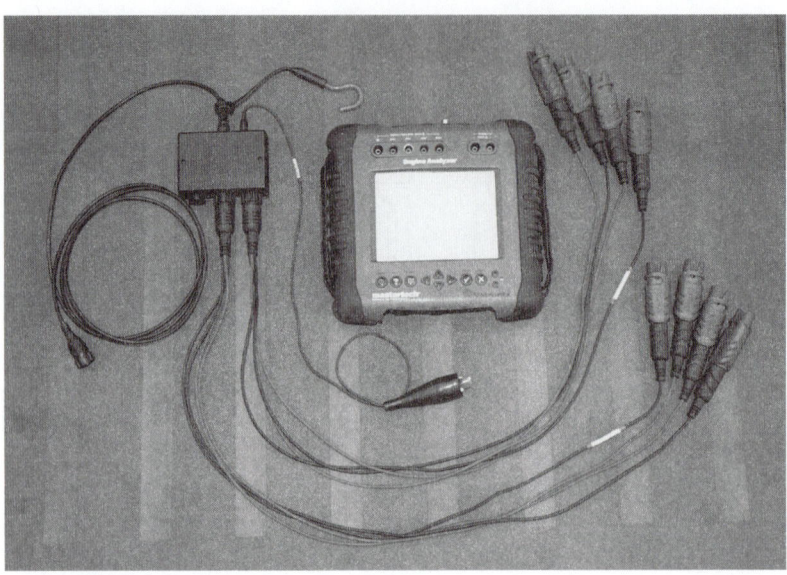

Figure 3-36 Digital storage oscilloscope.

traces across the screen (Figure 3-36). Other DSOs, such as the Simu-Tech, display six voltage traces simultaneously.

DSO Screen

CAUTION: While diagnosing computer systems, always place test equipment, such as DSOs or scan testers, in a secure position where they will not fall on the floor or into rotating components. Severe meter damage may occur if the DSO or scan tester is dropped.

On the DSO screen, voltage is displayed vertically. Voltage change is shown as a vertical movement. Vertical grids on the screen provide a voltage measurement, and the voltage level between the grids is adjustable (Figure 3-37). The technician must know the voltage in the circuit being tested in order to select the voltage per division on the DSO that provides the most detail with the signal remaining on the screen.

Horizontal movement on the screen represents time. The milliseconds per division on the horizontal grid are adjustable with the time button on the DSO. Each time a DSO samples a voltage

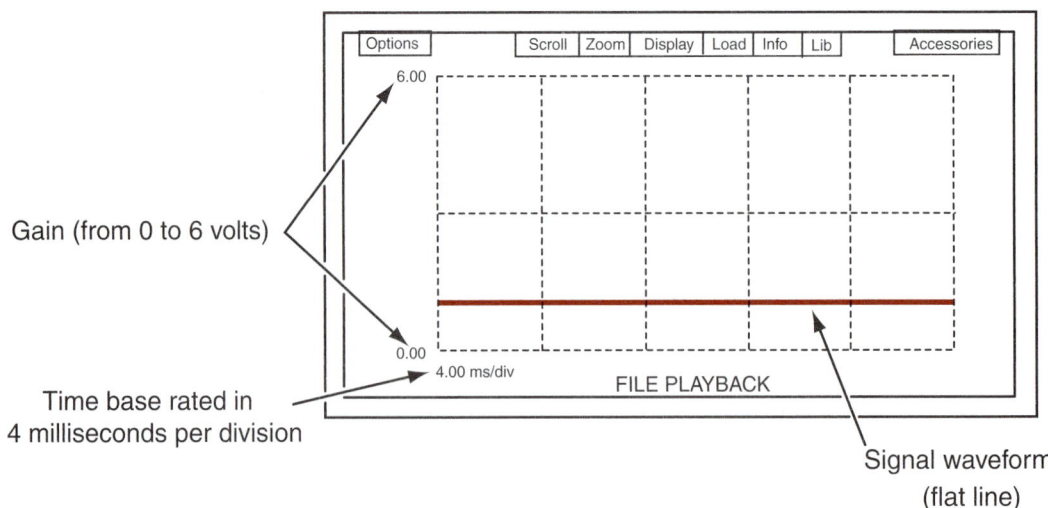

Figure 3-37 DSO vertical and horizontal screen grids.

signal, it displays a dot on the screen. The DSO then connects the dots to provide a waveform. When a faster signal is being read, a shorter time base should be selected on the DSO.

If the time base selected is too long and the voltage too high, the waveform is too small to read. Conversely, if the time base is too short and the voltage scale too low, the waveform is too large for the screen. The technician must select the proper time base for the voltage signal being measured so the waveform is displayed on the screen.

Peak, Average, and Root Mean Square Related to an AC Voltage Waveform

The term **peak** represents the highest point in one cycle of an AC voltage waveform. When both the highest and lowest peaks are considered in an AC voltage waveform, the term **peak-to-peak voltage** is the total voltage measured between these peaks. For example, an AC voltage waveform with a 60-volt peak would have a 120-volt peak-to-peak. The average voltage in an AC voltage waveform is calculated by multiplying 0.637 $\times$ peak voltage. The average voltage on a 60-volt peak would be 0.637 $\times$ 60 = 38.2 volts.

In many cases, root mean square (RMS) is used to describe AC voltage. For example, if one cycle of an AC voltage waveform from a 120-volt household electrical outlet is divided into four parts at 90° intervals, the instantaneous voltage and current are recorded for each degree in a 90° interval and then averaged. The square root of the average may be calculated by multiplying 0.7071 $\times$ the peak voltage. The peak voltage for the average 120-volt household outlet is about 170 volts at 60 hertz (Hz). Therefore, 0.7071 $\times$ 170 = 120.207 RMS.

Selecting DSO Voltage and Time Base

To display a waveform for a 120-volt household electrical outlet, round off the peak voltage of 170 volts to 200 volts. There are eight vertical voltage divisions on the DSO screen with four divisions above and below the centerline. Select 50 volts per division to display the high and low peaks on the waveform. Assuming the 0-volt position in the waveform is positioned in the center of the screen, the 50 volts per division selection provides 200 volts above and below the screen centerline to display the 170-volt peak voltage above and below the centerline. If the volts per division setting is increased, the peaks appear shorter on the screen.

In a 60-Hz AC voltage, one cycle occurs in approximately 18 milliseconds (ms). Displaying one complete AC cycle requires about 20 ms. The average DSO has 10 horizontal divisions. Since 20 ÷ 10 = 2 ms per division, this time base selection displays one AC voltage waveform. If 4 ms per division is selected, 2 AC voltage waveforms are displayed. Increasing the time base displays more AC voltage cycles on the screen. Conversely, decreasing the time base displays fewer AC voltage cycles on the screen, and the waveform appears expanded.

Each time a DSO takes a voltage sample, it displays a dot on the screen and then connects these dots to display a waveform. When the ms time base is too low, the waveform is expanded horizontally and a reduced number of dots are used in the waveform display. This may result in an altered and incomplete waveform display. Ideally, one to three cycles should be displayed on the screen for the best display.

When the DEFAULT button is pressed on some DSOs, a baseline volts per division and ms per division is automatically selected internally. If the volts per division and ms per division selected by the technician are incorrect for the voltage signal being tested, this default mode baseline should provide settings to display a waveform on the screen. Then the volts per division and ms per division may be adjusted to provide the desired display.

Trigger and Trigger Slope

The trigger selection tells the DSO when to begin displaying a waveform. Until the DSO has a trigger level, it doesn't know when to begin the waveform display. When testing an input sensor that operates in a 0-volt to 5-volt range, select a trigger level of one-half this range.

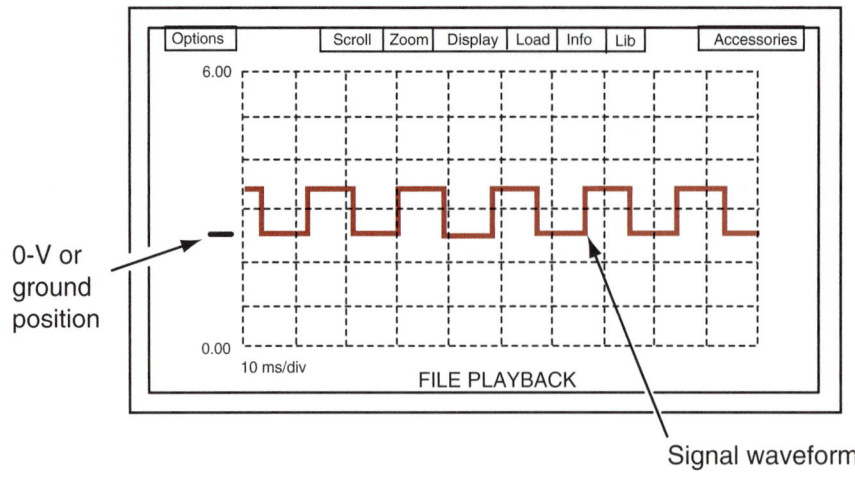

0-V or ground position

Signal waveform

Figure 3-38 Ground level reference marker on the left side of the screen.

Trigger slope informs the DSO whether the voltage signal is moving upward or downward when it crosses the trigger level. When a negative trigger slope is selected, the voltage signal is moving downward as it crosses the trigger level. Selecting a positive trigger level results in an upward voltage signal trace when it crosses the trigger level. A marker on the left side of the screen indicates the 0-volt or ground voltage position. This marker may be moved with the DSO controls (Figure 3-38). A second marker at the top of the screen indicates the trigger location. Since the control buttons on DSOs vary, the technician must spend some time to become familiar with a particular DSO.

Types of Voltage Signals

An **analog signal** is a varying voltage within a specific range over a period of time (Figure 3-39). A throttle position sensor (TPS) produces an analog voltage signal each time the throttle is opened (Figure 3-40).

A **digital signal** is either on or off. It may be described as one that is always high or low. The leading edge of a digital signal represents an increasing voltage, while the trailing edge of the signal represents a decreasing voltage. If the component is turned on by insulated side switching,

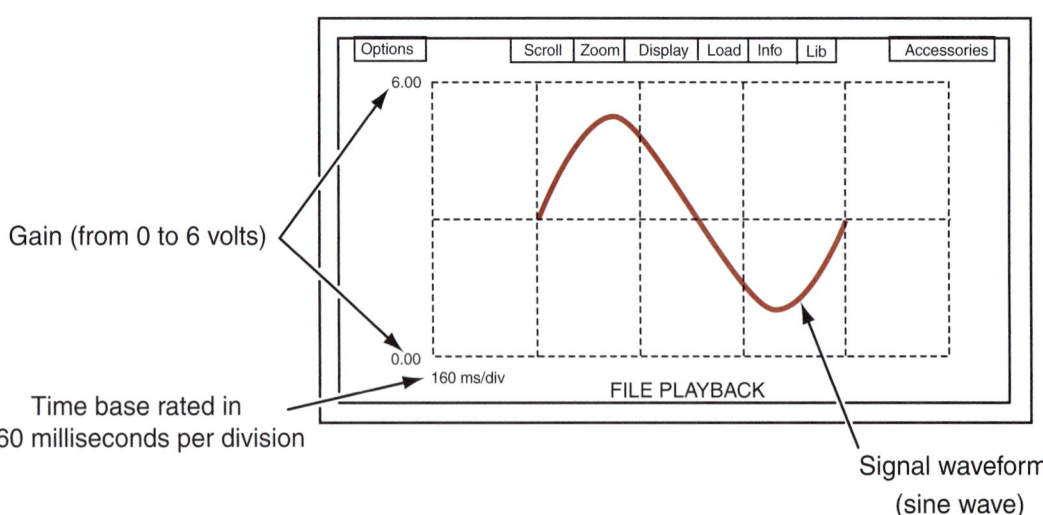

Gain (from 0 to 6 volts)

Time base rated in 60 milliseconds per division

Signal waveform (sine wave)

Figure 3-39 Analog voltage signal.

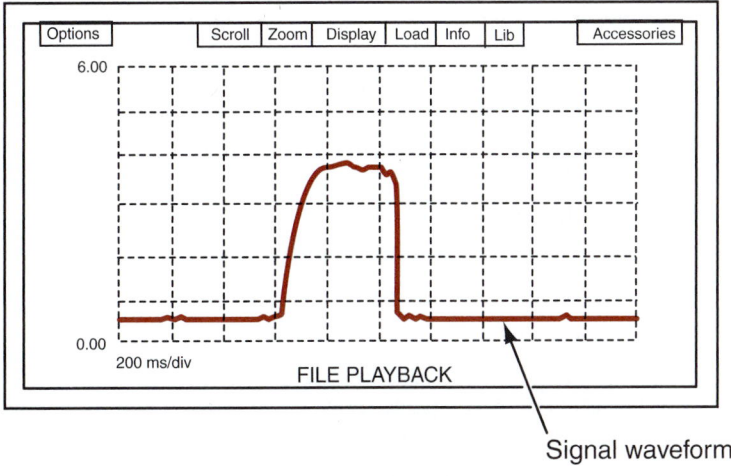

Signal waveform

Figure 3-40 TPS analog voltage signal.

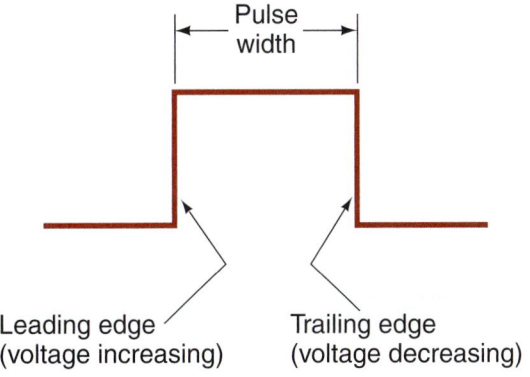

Figure 3-41 The component on time in a cycle is called pulse width.

the line across the top of the leading and trailing edge signals represents the length of component on time, which is called pulse width (Figure 3-41).

The computer measures the distance between the leading edge of a digital signal and leading edge of the next signal to determine the frequency of the waveform. The distance between the leading edge of one digital signal and the leading edge of the next digital signal is referred to as one cycle (Figure 3-42). The computer counts the number of cycles over a period of time

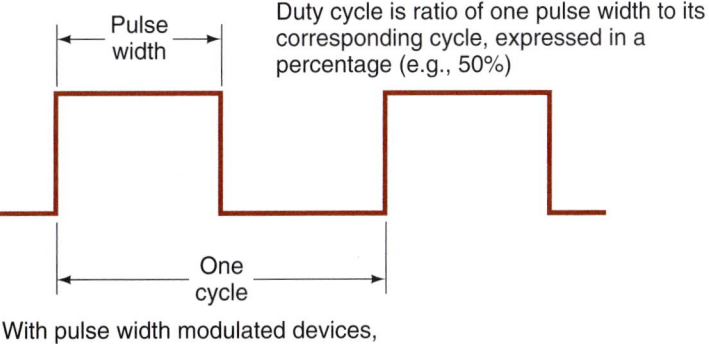

Figure 3-42 The distance between leading edges is one cycle.

to establish the frequency. For example, if 92 cycles are occurring per second, the frequency is 92 hertz (Hz).

The relationship between the on time and off time in a digital signal is called duty cycle. For example, if the component on time and off time are equal, the component has a 50% duty cycle. When the component has a 90% duty cycle, the component is on for 90% of the time in a cycle and off for 10% of the time in a cycle. The computer controls some outputs, such as a carburetor mixture control solenoid, by varying the pulse width while the frequency remains constant. This type of computer control is referred to as **pulse width modulation (PWM).** Other outputs, such as fuel injectors, are controlled by varying the frequency and the pulse width.

User-Friendly DSOs

DSOs with simplified, user-friendly controls have recently been introduced to the automotive service industry. In these DSOs, the auto-range function automatically selects the proper voltage and time base for the signal being received. The technician may use the DSO controls to turn off the auto-range function, and manually select the voltage range and time base.

When the menu key is pressed, various menus are displayed and the vertical arrow keys allow the technician to scroll through the menu to select a specific test. Digital readings are displayed on the screen with most waveforms. For example, minimum, average, and maximum millivolt (mV) readings are provided with an O2 sensor waveform. The DSO automatically adjusts for zirconia or titania O2 sensors. This DSO has multimeter and ignition waveform capabilities.

Some user-friendly DSOs, such as the OTC Vision, have a removable application module to help prevent scope obsolescence. A software program card plugs into the bottom of the DSO. This DSO also sets the voltage range and time base automatically for the signal being tested. Four voltage waveforms or six multimeter functions may be displayed on the DSO screen.

CASE STUDY

A customer brings his vehicle to the shop because the dash lights are not illuminating. The technician checks the fuses and all are good. She then substitutes a bulb, known to be good, in the printed circuit, but it still does not illuminate. Next she uses a voltmeter and checks for applied voltage to the panel light circuit. The test indicates that 12.6 volts are present.

The technician then performs a voltage drop test on the ground side of the circuit and the voltmeter indicates 12.6 volts. She concludes that the printed circuit has an open in the ground side of its circuit. Upon receiving written approval from the customer to perform repairs, she replaces the printed circuit. Verification of the repair confirms her diagnosis.

Terms to Know

Analog signal	Fusible link	Peak-to-peak voltage	Short to ground
Darlington	Gauss gauge	Potentiometer	Shorted circuit
Digital signal	Normally open (NO)	Pulse width modulation (PWM)	Stepped resistor
Diode	Open circuit		Variable resistor
Feedback	Overload	Relay	
Fuse	Peak	Rheostat	

ASE-Style Review Questions

1. Circuit defects are being discussed.
   Technician A says an open can only be on the ground side of the circuit.
   Technician B says an unwanted resistance can result from a corroded connector.
   Who is correct?
   - **A.** A only
   - **B.** B only
   - **C.** Both A and B
   - **D.** Neither A nor B

2. Testing the fuse is being discussed.
   Technician A says sometimes a visual inspection of a fuse or fusible link does not reveal that it is open.
   Technician B says to use a jumper wire to bypass the fuse in order to test the circuit.
   Who is correct?
   - **A.** A only
   - **B.** B only
   - **C.** Both A and B
   - **D.** Neither A nor B

3. Testing of a switch is being discussed.
   Technician A says a switch can be tested with a voltmeter.
   Technician B says to use an ohmmeter to test a switch.
   Who is correct?
   - **A.** A only
   - **B.** B only
   - **C.** Both A and B
   - **D.** Neither A nor B

4. *Technician A* says a relay can only be tested while it is connected to the circuit.
   Technician B says a stepped resistor can be tested while it is disconnected from the circuit.
   Who is correct?
   - **A.** A only
   - **B.** B only
   - **C.** Both A and B
   - **D.** Neither A nor B

5. An LED does not light. A voltmeter indicates 0 volts across the LED. All of the following can be the cause EXCEPT:
   - **A.** Current-limiting resistor is open.
   - **B.** The LED is open.
   - **C.** Short to ground between the current-limiting resistor and the LED.
   - **D.** Open ground circuit.

6. Voltage drop testing is being discussed.
   Technician A says it is possible to calculate voltage drop by testing for available voltage on both sides of the component.
   Technician B says excessive voltage drop can be on either the power side or the ground side of the circuit.
   Who is correct?
   - **A.** A only
   - **B.** B only
   - **C.** Both A and B
   - **D.** Neither A nor B

7. The results of copper-to-copper shorts are being discussed.
   Technician A says if there is a short in an electrical motor then the amount of resistance will be higher than specified.
   Technician B says a short between circuits can result in both circuits operating by closing one switch.
   Who is correct?
   - **A.** A only
   - **B.** B only
   - **C.** Both A and B
   - **D.** Neither A nor B

8. The testing of a shorted circuit is being discussed.
   Technician A says if a fuse blows as soon as it is installed, this indicates a short to ground.
   Technician B says if the short to ground is on the ground side of the load component but before a grounding switch, the component will not turn off.
   Who is correct?
   - **A.** A only
   - **B.** B only
   - **C.** Both A and B
   - **D.** Neither A nor B

9. *Technician A* says test the base-collector junction and the base-emitter junction as if they were standard diodes.
   Technician B says the resistance between the collector and emitter should read open circuit.
   Who is correct?
   - **A.** A only
   - **B.** B only
   - **C.** Both A and B
   - **D.** Neither A nor B

10. Testing of a potentiometer is being discussed.
    Technician A says a voltmeter can be used if jumper wires are connected to the wire connector and sensor to gain access to the test points.
    Technician B says the wires can be pierced to test for voltage.
    Who is correct?
    - **A.** A only
    - **B.** B only
    - **C.** Both A and B
    - **D.** Neither A nor B

ASE Challenge Questions

1. The fuse for an A/C blower motor circuit fails after a short period of time.
   Technician A says that the blower motor may be binding internally.
   Technician B says that the blower motor ground circuit may have excessive resistance.
   Who is correct?
   A. A only
   B. B only
   C. Both A and B
   D. Neither A nor B

2. The brake lights of a vehicle are inoperative. A voltmeter that is connected across the terminals of the brake light switch indicates system voltage (12 volts) regardless of whether the brake pedal is depressed.
   Technician A says that there is no power available at the brake light bulbs.
   Technician B says that the brake light switch may be open.
   Who is correct?
   A. A only
   B. B only
   C. Both A and B
   D. Neither A nor B

3. Relay testing is being discussed.
   Technician A says that it is acceptable during the test sequence to bypass the relay control coil terminals with a fused jumper wire.
   Technician B says that the voltage drop across the relay load contact terminals is checked with the relay control circuit de-energized.
   Who is correct?
   A. A only
   B. B only
   C. Both A and B
   D. Neither A nor B

4. The troubleshooting of a parallel circuit that contains three dimly lit bulbs is being discussed. A voltmeter that is placed across each of the bulbs indicates 7.2 volts.
   Technician A says that the power supply that is common to all three bulbs may be faulty.
   Technician B says that the ground terminal that is common to all three bulbs may have excessive resistance.
   Who is correct?
   A. A only
   B. B only
   C. Both A and B
   D. Neither A nor B

5. An A/C compressor clutch coil spike suppression diode is being tested with an analog ohmmeter. The meter indicates infinite resistance in both directions.
   Technician A says that the diode is electrically open.
   Technician B says that the use of this diode would result in the immediate failure of the circuit fuse.
   Who is correct?
   A. A only
   B. B only
   C. Both A and B
   D. Neither A nor B

Job Sheet 8

Name _____ Date _____

Using a DSO on Sensors and Switches

Upon completion of this job sheet, you should be able to connect a DSO and observe the activity of various sensors and switches.

Tools and Materials

A vehicle with accessible sensors and switches
Service manual for the above vehicle
Component locator manual for the above vehicle
A DSO
A DMM

Describe the vehicle being worked on:

Year _____ Make _____ Model _____

VIN _____ Engine type and size _____

Procedure

Task Completed

1. Connect the DSO across the battery. Make sure the scope is properly set. Observe the trace on the scope. Is there evidence of noise? Explain.

   _____

   _____

   _____

   _____

2. Locate the A/C compressor clutch control wires. Start and run the engine. Connect the DMM to read available voltage. Observe the meter, then turn the compressor on. What happened on the meter?

   _____

   _____

   Now connect the DSO to the same point with the compressor turned off. Observe the waveform, then turn the compressor on. What happened to the trace?

   _____

   _____

3. Turn off the engine but keep the ignition on. Locate the TP sensor and identify the purpose of each wire to it. List each wire and describe the purpose of each.

   _____

   _____

4. Connect the DMM to read reference voltage at the TP sensor. What do you read?

_____

_____

Now move the leads to read the output of the sensor. Starting with the throttle closed, slowly open the throttle until it is wide open. Watch the voltmeter while doing this. Describe your readings.

_____

_____

5. Now connect the DSO to read reference voltage at the TP sensor. What do you see on the trace?

_____

_____

Now move the leads to read the output of the sensor. Starting with the throttle closed, slowly open the throttle until it is wide open. Watch the trace while doing this. Describe your readings.

_____

_____

6. Now run the engine. Locate the oxygen sensor and identify the purpose of each wire to it. Connect the DMM to read voltage generated by the sensor. (To do this, you may use an electrical connector for the O_2 sensor that is positioned away from the hot exhaust manifold.) Watch the meter and describe what happened below.

_____

_____

_____

_____

7. Now connect the DSO to read voltage output from the sensor. Watch the trace and describe what happened below.

_____

_____

_____

_____

8. Explain what you observed as the differences between testing with a DMM and a DSO.

_____

_____

_____

_____

Instructor's Response _____

_____

_____

Job Sheet 9

Name _____ Date _____

Testing Circuit Protection Devices

Upon completion of this job sheet, you should be able to test circuit protection devices for opens.

ASE Correlation

This job sheet is related to the ASE Electrical/Electronic System Certification Exam's content area: *General Electrical System Diagnosis;* task: Inspect, test, and replace fusible links, circuit breakers, and fuses.

Tools and Materials

A vehicle equipped with fusible links
Fender covers
A DVOM
Test light

Describe the vehicle being worked on:

Year _____ Make _____ Model _____

VIN _____ Engine type and size _____

Procedure:

Task Completed

1. Locate the fuse panel or power distribution center.   ☐

2. Check that the test light is working properly by connecting it across the battery.   ☐

3. Connect the negative lead of a test light to a good ground.   ☐

4. Turn the ignition switch to the RUN position.   ☐

5. Touch the probe of the test light onto the metal test tabs on each side of the fuse.   ☐

6. Did the test light illuminate on each side of the fuse?

   Why or why not?

   _____

   _____

   _____

7. Repeat for all fuses in the fuse box. Record your findings.

   _____

   _____

   _____

☐

8. Remove any fuses that failed the test.

9. Visually inspect the fuse. Record your findings.

_____

_____

_____

10. Use an ohmmeter and test the fuses. Record your findings.

_____

_____

_____

☐

11. Locate a fusible link on the vehicle.

12. Use a voltmeter and measure the voltage drop over the link. Record your results.

_____

_____

_____

13. Disconnect power to the fusible link and use an ohmmeter to test the link. Record your results.

_____

_____

_____

Instructor's Response _____

_____

_____

Job Sheet 10

Name _____ Date _____

Testing Switches

Upon completion of this job sheet, you should be able to test a switch and properly determine needed repairs.

ASE Correlation

This job sheet is related to the ASE Electrical/Electronic System Certification Exam's content area: *General Electrical System Diagnosis;* task: Check continuity and resistance in electrical/electronic circuits and components with an ohmmeter and determine needed repairs. Check electrical/ electronic circuits with jumper wires and determine needed repairs.

Tools and Materials

A vehicle
DVOM
Test light
Jumper wires
Wiring diagram for the vehicle

Describe the vehicle being worked on:

Year _____ Make _____ Model _____

VIN _____._____ Engine type and size _____

Procedure

Task Completed

1. Locate the brake light switch (or other switch as directed by your instructor). ☐

2. Disconnect the switch from the wire harness. ☐

3. Use an ohmmeter to measure the resistance of the switch with the brake pedal released. Record your results. _____

    _____

4. With the ohmmeter still connected across the switch terminal, press the brake pedal and record the ohmmeter reading: _____

    _____

5. Based on your results, is the switch operating properly?   ☐ Yes ☐ No

    Why? _____

    _____

6. With the electrical connector to the switch still unplugged, connect a jumper wire across the battery feed and brake light circuits. ☐

7. Do the brake lights come on?   ☐ Yes ☐ No

What is the faulty component if the brake lights did not come on with the switch connected and the brakes depressed, but they do come on when jumped across the terminals? _____

_____

Instructor's Response _____

_____

_____

Job Sheet 11

11

Name _____ Date _____

Testing for an Open Circuit

Upon completion of this job sheet, you should be able to test a circuit and locate the open.

ASE Correlation

This job sheet is related to the ASE Electrical/Electronic System Certification Exam's content area: *General Electrical System Diagnosis;* task: Find shorts, grounds, opens, and high-resistance problems in electrical/electronic circuits and determine needed repairs.

Tools and Materials

A vehicle
DVOM
Test light
Wiring diagram for the vehicle

Describe the vehicle being worked on:

Year _____ Make _____ Model _____

VIN _____ Engine type and size _____

Procedure

Task Completed

1. What is the customer's complaint? _____
_____

2. Can the complaint be verified?   ☐ Yes ☐ No
(If no, consult your instructor.)

3. Are there any other related symptoms?   ☐ Yes ☐ No
If yes, describe the symptom. _____
_____

4. Following the wiring diagram, use a voltmeter or test light to trace the circuit.   ☐

5. Describe the location of the open.

_____
_____
_____

Instructor's Response _____

_____
_____

Job Sheet 12

Name _____ Date _____

Testing for a Short to Ground

Upon completion of this job sheet, you should be able to test a circuit and locate the short to ground.

ASE Correlation

This job sheet is related to the ASE Electrical/Electronic System Certification Exam's content area: *General Electrical System Diagnosis;* task: Find shorts, grounds, opens, and high-resistance problems in electrical/electronic circuits and determine needed repairs.

Tools and Materials

A vehicle
DVOM
Test light
Circuit breaker fitted with alligator clips
Gauss gauge
Wiring diagram for the vehicle

Describe the vehicle being worked on.

Year _____ Make _____ Model _____

VIN _____ Engine type and size _____

Procedure

Task Completed

1. Which circuit is affected by the short to ground? _____
   _____

2. Are there any other related symptoms?   ☐ Yes ☐ No
   If yes, describe the symptom(s). _____
   _____

3. Pull the fuse or circuit protection for the affected circuit from the fuse box.   ☐

4. With ignition switch in the RUN position, connect the test light across the fuse terminals in the fuse box. Does the test light illuminate?   ☐ Yes ☐ No

5. What does this test indicate?

   _____
   _____
   _____

6. With the test light connected across fuse terminals of the fuse box, disconnect components and connectors in the affected circuit that are identified in the wiring diagram.   ☐

7. Did the test light go out when a component was disconnected?   ☐ Yes ☐ No

8. What can be concluded thus far? _____

_____

_____

_____

☐

9. Connect the test circuit breaker across the fuse terminals of the fuse box and use the Gauss gauge to find the location of the short to ground.

10. Describe the location of the short to ground. _____

_____

_____

_____

Instructor's Response _____

_____

_____

_____

Job Sheet 13

Name _____ Date _____

Testing a Diode

Upon completion of this job sheet, you should be able to test a diode for a short or open.

ASE Correlation

This job sheet is related to the ASE Electrical/Electronic System Certification Exam's content area: *General Electrical System Diagnosis;* task: Check continuity and resistance in electrical/electronic circuits and components with an ohmmeter and determine needed repairs.

Tools and Materials

An assortment of diodes
Analog ohmmeter
DVOM

Procedure

Task Completed

1. Which side is the stripe around the diode on?
   ☐ Anode   ☐ Cathode

2. Using the analog ohmmeter, measure the resistance through the diode by connecting the red test lead to the anode and the black lead to the cathode side of the diode.   ☐

3. Record your reading. _____

4. Reverse the test leads so the red test lead is on the cathode and the black lead is on the anode side of the diode.   ☐

5. Record your reading. _____

6. What is your conclusion concerning the condition of this diode?

   _____

   _____

7. Use the ohmmeter function of the DVOM and measure the resistance in both directions through the diode. Record your results.

   Forward biased _____

   Reverse biased _____

8. Are the readings the same as with the analog ohmmeter?   ☐ Yes   ☐ No

9. Explain why a DVOM ohmmeter is not recommended for testing a diode.

   _____

   _____

   _____

10. Use a DVOM to test a second diode using the diode test function. Connect the red test lead to the anode and the black lead to the cathode side of the diode. Record the meter readings. _____

11. What does this reading represent?

_____

_____

12. Reverse the test leads across the diode. What is the reading?

_____

_____

13. What does this reading represent?

_____

_____

14. What is your conclusion concerning this diode?

_____

_____

Instructor's Response _____

_____

_____

DIAGNOSTIC CHART 3-1

PROBLEM AREA:	Electrical opens.
SYMPTOMS:	Electrical component will not operate.
POSSIBLE CAUSES:	**1.** Broken conductor. **2.** Defective switch. **3.** Defective relay. **4.** Blown fuses. **5.** Burned fusible links. **6.** Burned or defective circuit breakers.

DIAGNOSTIC CHART 3-2

PROBLEM AREA:	Excessive resistance resulting in lowered electrical output.
SYMPTOMS:	Electrical components fail to operate or operate at reduced efficiency.
POSSIBLE CAUSES:	**1.** Excessive resistance. **2.** Broken conductor. **3.** Blown circuit protection device. **4.** Open or excessive resistance in the switch. **5.** Open or excessive resistance in the relay. **6.** Improper stepped resistor values. **7.** Defective variable resistor. **8.** Open or shorted diodes.

DIAGNOSTIC CHART 3-3

PROBLEM AREA:	Copper-to-copper short or short to ground.
SYMPTOMS:	No electrical component operation or operation when another control switch is activated.
POSSIBLE CAUSES:	**1.** Broken or burned insulation and/or connectors causing copper-to-copper short. **2.** Broken or burned insulation and/or connectors causing a short to ground.

Wiring Repair and Reading Circuit Diagrams

Upon completion and review of this chapter, you should be able to:

❏ Perform repairs to copper wire using solderless connections.

❏ Solder splices to copper wire.

❏ Repair aluminum wire according to manufacturer's requirements.

❏ Repair twisted/shielded wire.

❏ Replace fusible links.

❏ Repair and/or replace the terminals of a hard-shell connector.

❏ Repair and/or replace the terminals of weather-pack and metri-pack connectors.

❏ Read a wiring diagram to correctly determine the operation of the circuit.

❏ Use the wiring diagram to diagnose possible causes for the system fault.

Basic Tools

Basic mechanic's tool set

Service manual

Introduction

Many electrical repairs will involve the replacement or repairing of a damaged conductor. To locate the problem area, today's technician must be capable of reading and understanding electrical diagrams and schematics. Once it is determined that the battery is operating correctly, the schematic should always be the starting point in **troubleshooting** an electrical system. Troubleshooting is the diagnostic procedure of locating and identifying the cause of the fault. It is a step-by-step process of elimination by use of cause and effect. By using the schematic, the technician is able to understand how the circuit should work. This is essential before attempting to figure out why it does not work.

The component locator will assist the technician in finding the location of the electrical components shown in the schematic. Many times the component locator will also list the locations of connectors, grounds, and splices.

The process of troubleshooting an electrical complaint is as follows:

1. Confirm the complaint. Perform a check of the system to gain an understanding of what is wrong. If the faulty system is monitored by the onboard computer, enter diagnostics to retrieve any **trouble codes.** Trouble codes are the output of the self-diagnostics program in the form of alpha/numeric codes that indicate faulty circuits or components.

2. Study the electrical schematic. This will indicate any shared circuits. Trying to operate the shared circuits will help direct the technician to the problem area. If the shared circuits operate correctly, the problem is isolated to the wiring or components of the problem system. If the shared circuits do not operate, the problem is usually in the power or ground circuit.

3. Locate and repair the fault. By narrowing down the possible causes and taking measurements as required, the fault is located. Before replacing any components, check the ground and power leads. If these are good, then the component is bad.

4. Test the repair. Repeat a check of the system to confirm that it is operating properly.

Wire Repair

Not all electrical repairs involve removing and replacing a faulty component. Many times the cause of the malfunction is a damaged conductor. The technician must make a repair to the circuit

Classroom Manual
Chapter 4, page 91

that will not increase the resistance. It should also be a permanent repair. There are many methods to repair a damaged wire. The type of repair used will depend on factors such as:

1. Type of repair required.
2. Ease of access to the damaged area.
3. Type of conductor.
4. Size of wire.
5. Circuit requirements.
6. Manufacturer's recommendations.

The most common methods of wire repair include wrapping damaged insulation with electrical tape or tubing, crimping the connections with solderless connectors, and soldering splices.

Copper Wire Repairs

Classroom Manual
Chapter 4, page 91

Copper wire is the most commonly used primary wire in the automobile. The insulation may break down, or the wire may break due to stress or excessive motion. The wire may also be damaged due to excessive current flow through the wire. Any of these conditions require that the wire be repaired.

WARNING: Repair of air bag wiring must be done to manufacturer's specifications.

CAUTION: A solid copper wire may be used in low-voltage, low-current circuits where flexibility is not required. Do not use solid wire where high voltage, high current, or flexibility is required, unless solid wire was used by the manufacturer.

In some instances, it may be necessary to bypass a length of wire that is not accessible. In this case, cut the wire before it enters the inaccessible portion and at the other end where it leaves the area. Install a replacement wire and reroute it to the load component (Figure 4-1). Be sure to protect the wire by using straps, hangers, and grommets as needed.

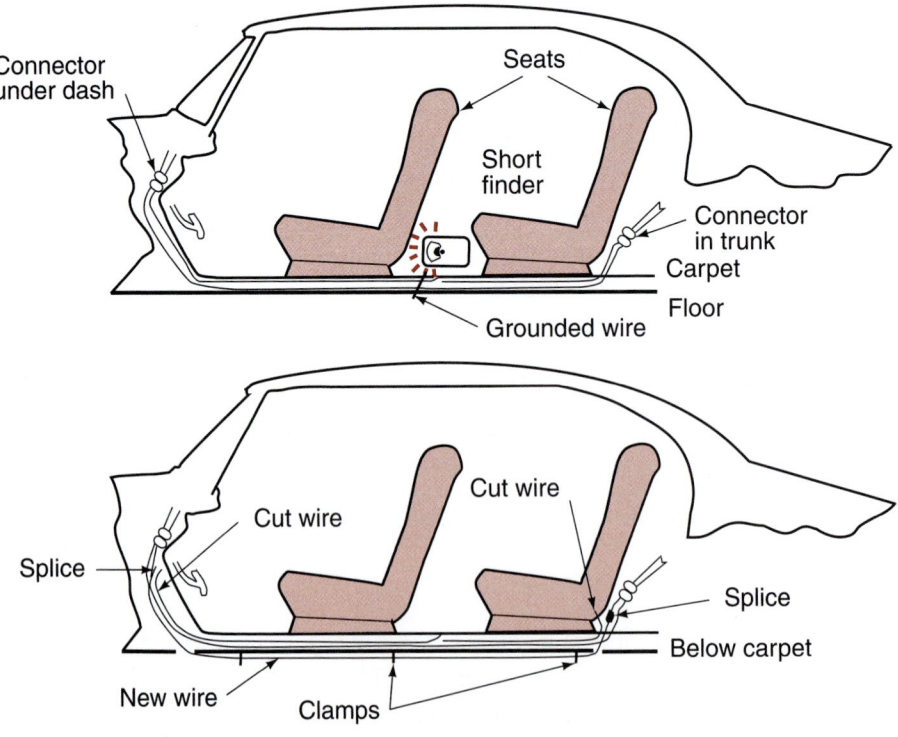

Figure 4-1 Routing a new wire to replace a damaged wire that is in an inaccessible location.

The two most common methods of splicing copper wire are with solderless connectors or by soldering.

Crimping. Crimping of **solderless connectors** is an acceptable method to **splice** wires that are not subjected to weather elements, dirt, corrosion, or excessive movement. Also, do not use crimped connections in electronic circuits. A poor connection, or corrosion over time, can result in improper electronic control operation of the system. Do the following to make a splice using solderless connections:

1. Use the correct size of stripping opening on the **crimping tool** to remove enough insulation to allow the wire to completely penetrate the connector. The crimping tool has different areas for performing several functions (Figure 4-3). This single tool will cut the wire, strip the insulation, and crimp the connector.

2. Place the wire into the connector and crimp the connector (Figure 4-4). To get a proper crimp, place the open area of the connector facing toward the anvil. Be sure the wire is compressed under the crimp.

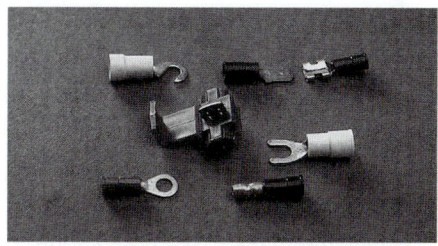

Figure 4-2 Types of solderless connectors and terminals commonly used for wire repair.

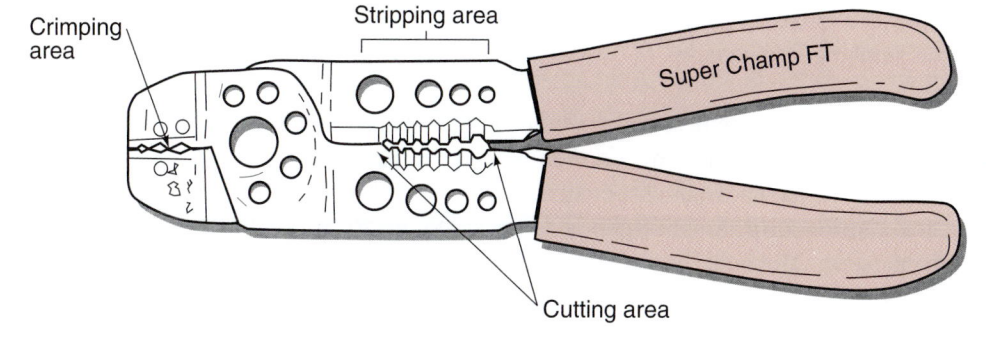

Figure 4-3 A typical crimping tool used for making electrical repairs.

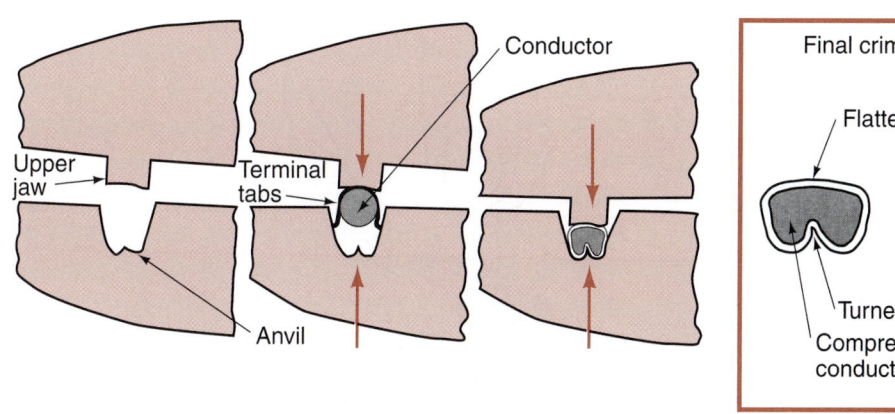

Figure 4-4 Properly crimping a connector.

Special Tools

Crimping tool

Electrical tape or heat shrink tube

Solderless connector

Safety glasses

Fender covers

Crimping means to bend, or deform by pinching, a connector so that the wire connection is securely held in place.

Solderless connectors are hollow metal tubes covered with insulating plastic. They can be butt connectors or terminal ends (Figure 4-2).

Splice is a term used to mean joining of single wire ends or the joining of two or more electrical conductors at a single point.

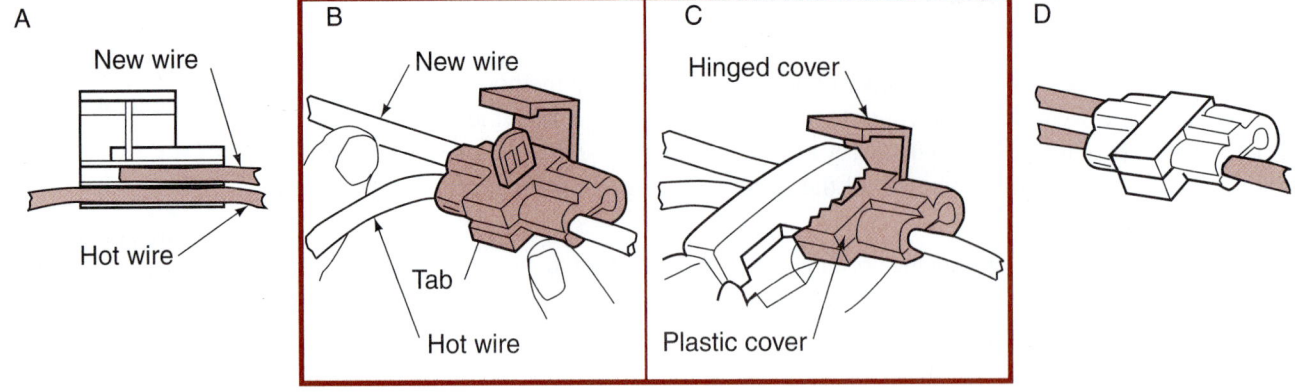

A

New wire

Hot wire

B

New wire

Tab

Hot wire

C

Hinged cover

Plastic cover

D

Figure 4-5 Using the tap connector to splice in another wire. (A) Place wires in position in the connector, (B) close the connector around the wires, (C) use pliers to force the tab into the conductors, (D) close the hinged cover.

3. Insert the stripped end of the other wire into the connector and crimp in the same manner.

4. Use electrical tape or a piece of **heat shrink tube** to provide additional protection. Heat shrink tube is plastic tubing that shrinks in diameter when exposed to heat.

Another type of crimping connector is the tap splice connector. This type of connector allows for adding an additional circuit to an existing feed wire without stripping the wires (Figure 4-5). Although tap connectors make connecting wires easy, these should not be used to provide power to critical components. Tap connectors cannot be used on electronic circuits. They are unreliable for making and maintaining a good connection, and their use is discouraged. Also, make sure the fuse of the circuit being tapped into has a large enough capacity before adding the circuit. Tap connectors add a circuit in parallel with another circuit. This causes circuit resistance to decrease and circuit amperage to increase.

Classroom Manual
Chapter 4, page 96

The process of applying solder to the tip of the iron is called *tinning*.

Soldering. Soldering is the best way to splice copper wires. Solder is an alloy of tin and lead. It is melted over a splice to hold the wire ends together. Soldering may be a splicing procedure, but it is also an art that takes much practice. Photo Sequence 5 illustrates the soldering process when using a **splice clip.** A splice clip is a special connector used along with solder to assure a good connection. The splice clip is different from a solderless connection in that it does not have insulation. Some have a hole provided for applying solder (Figure 4-6).

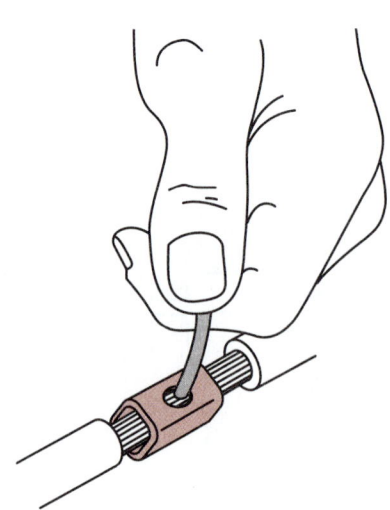

Figure 4-6 Splice clip. Some have a hole for applying solder.

Photo Sequence 5
Soldering Copper Wire

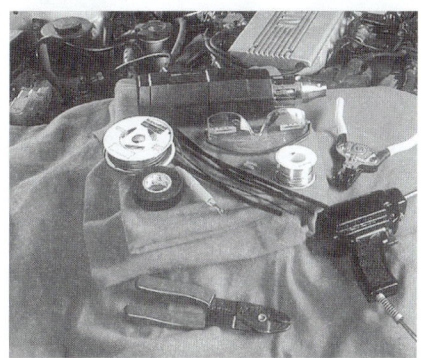

P5-1 Tools required to solder copper wire: 100-watt soldering iron, 60/40 rosin core solder, crimping tool, splice clip, heat shrink tube, heating gun, safety glasses, sewing seam ripper, electrical tape, and fender covers.

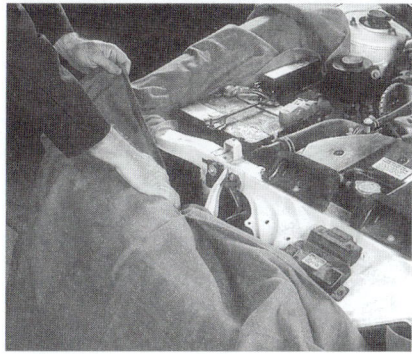

P5-2 Place the fender covers over the vehicle fenders.

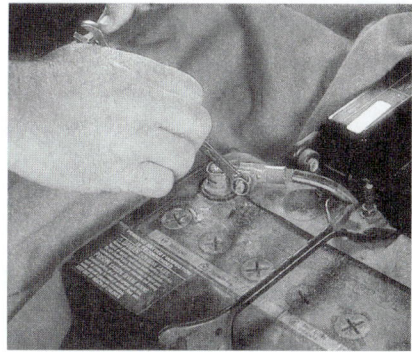

P5-3 Disconnect the fuse that powers the circuit being repaired. Note: If the circuit is not protected by a fuse, then disconnect the battery.

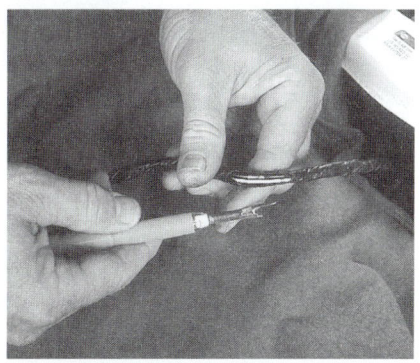

P5-4 If the wiring harness is taped, use a seam ripper to open the wiring harness.

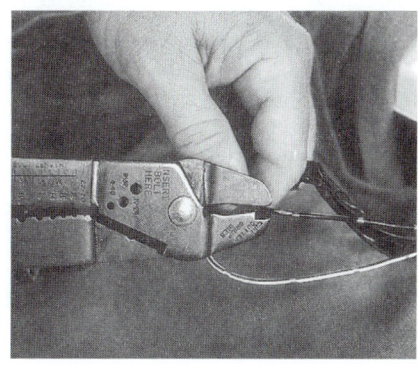

P5-5 Cut out the damaged wire using the wire cutters on the crimping tool.

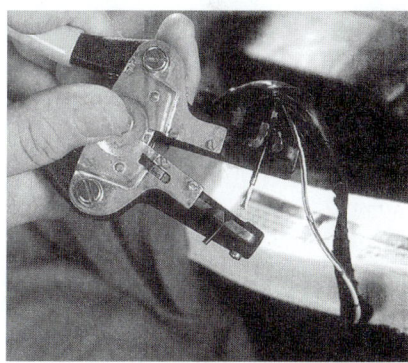

P5-6 Using the correct size stripper, remove about 1/2 inch (12 mm) of the insulation from both wires. Be careful not to nick or cut any of the wires.

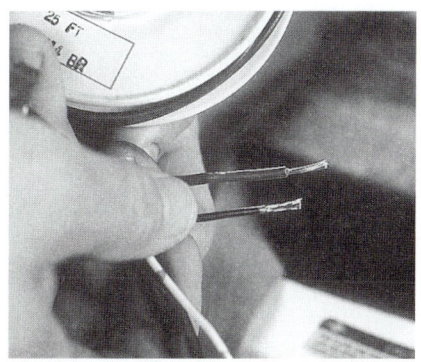

P5-7 Determine the correct gauge and length of replacement wire.

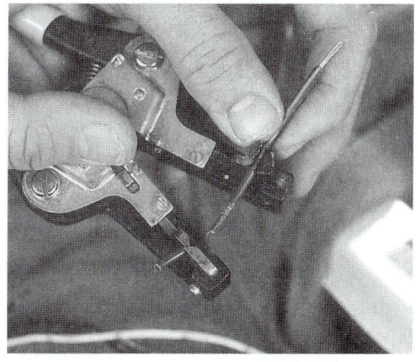

P5-8 Using the correct size stripper, remove 1/2 inch (12 mm) of insulation from each end of the replacement wire.

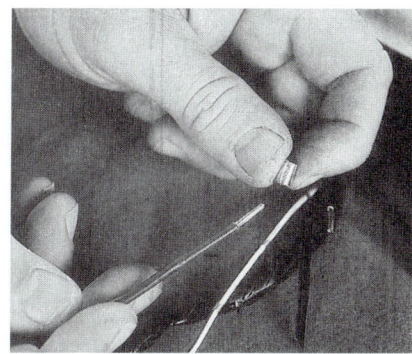

P5-9 Select the proper size splice clip to hold the splice.

Photo Sequence 5
Soldering Copper Wire (continued)

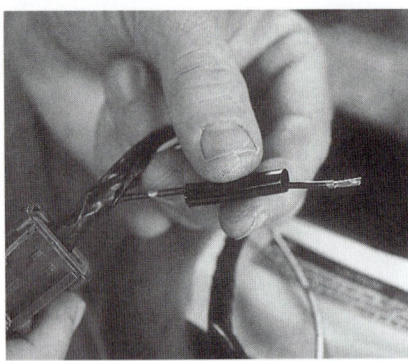

P5-10 Place the correct length and size of heat shrink tube over the two ends of the wire. Slide them far enough away so they are not exposed to the heat of the soldering iron.

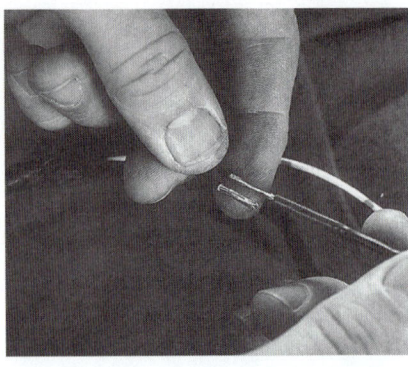

P5-11 Overlap the two splice ends and hold in place with thumb and forefinger.

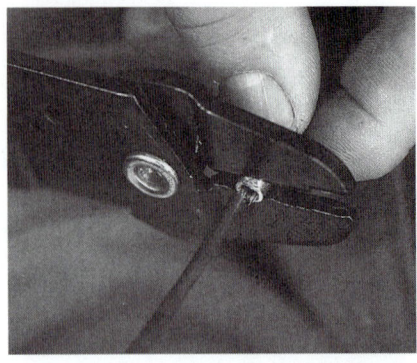

P5-12 Center the splice clip around the wires and crimp in place. Make sure that wires extend beyond the splice clip in both directions. Crimp the clip on both ends.

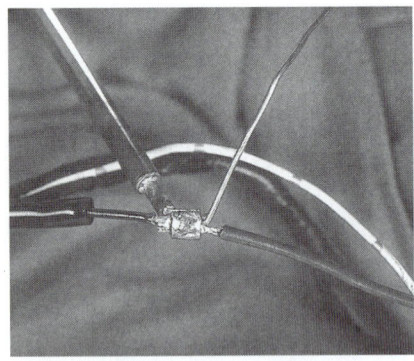

P5-13 Heat the splice clip with the soldering iron while applying solder to the opening in the back of the clip. Do not apply solder to the iron; the iron should be 180 degrees away from the opening of the clip.

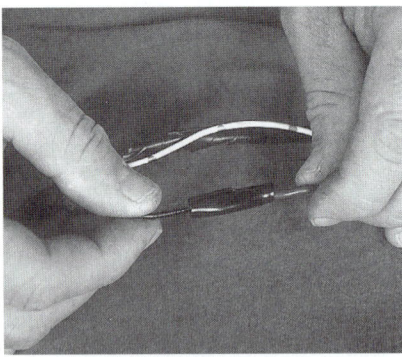

P5-14 After the solder cools, slide the heat shrink tube over the splice.

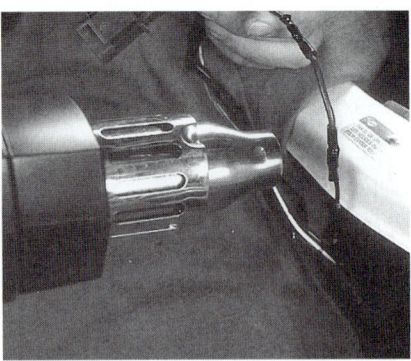

P5-15 Heat the tube with the hot air gun until it shrinks around the splice. Do not overheat the tube.

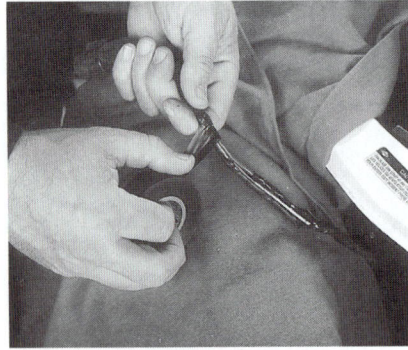

P5-16 Retape the wiring harness.

If a splice clip is not used, the wire ends should be braided together tightly. Then the splice should be heated with the soldering gun. It is important to note that when soldering, the solder should melt by the heat of the wire splice, not the heat of the soldering tool. Always use rosin core solder when making electrical repairs. Acid core solder is used for other purposes than electrical repairs and can cause the wire to corrode, which would lead to high resistance.

WARNING: Before cutting into a wire to make a splice, look for other splices or connections first. Never have two or more splices within 1.5 inches (40 mm) of each other. Also always use wire of the same size or larger than the wire being replaced.

An alternate method of soldering wires together is to use wire joints in place of splice clips. Remove about 1 inch (25 mm) of the insulation from the wires. Join the wires using one of the methods illustrated (Figure 4-7). Heat the twisted connection with the soldering iron. Apply the solder to the strands of wire. Do not apply the solder directly to the soldering iron. The solder should melt and flow evenly among all of the wire strands (Figure 4-8). Insulate the splice with electrical tape or heat shrinking tube.

Special Tools

100-watt soldering iron

60/40 rosin core solder

Crimping tool

Splice clip

Heat shrink tube

Heating gun

Safety glasses

Sewing seam ripper

Electrical tape

Fender covers

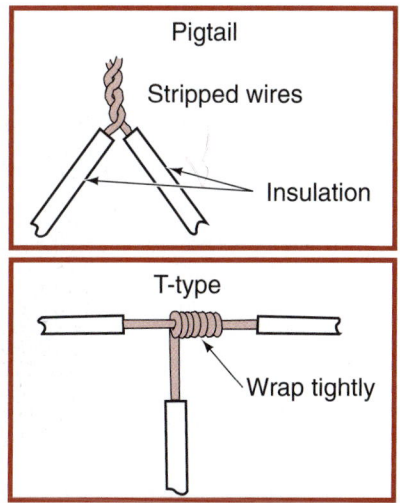
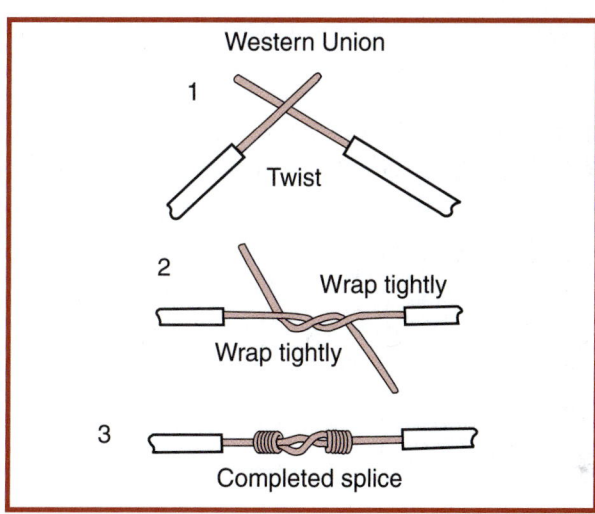

Figure 4-7 Methods for joining wires together.

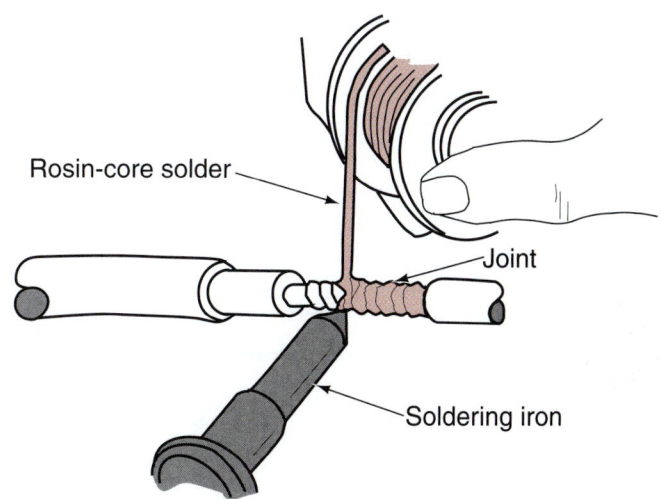

Figure 4-8 When soldering, apply the solder to the joint, not to the tip of the soldering iron.

SERVICE TIP: It is easier to heat the wire with the soldering iron if the solder is first melted onto the tip of the iron. This will transfer the heat from the iron to the wire more quickly.

Repairing Aluminum Wire

⚠️ **WARNING:** Attempting to solder aluminum wire will damage the conductor.

Special Tools

Crimping tool

Petroleum jelly

Safety glasses

Electrical tape or
heat shrink tube

General Motors has used single-stranded aluminum wire in limited applications where no flexing of the wire is expected. This wire usually has a thick plastic insulator and is placed in a brown harness.

After cutting away the damaged wire, strip all wire of the last 1/4 inch (6 mm) of insulation. Be careful not to nick or damage the conductor. Apply a generous coating of petroleum jelly to the wire and connector (Figure 4-9). The petroleum jelly will prevent corrosion from developing in the core.

Crimp the connector in the usual manner. Insulate the splice with heat shrink tube. Do not use electrical tape since it will not stay in place due to the petroleum jelly.

Splicing Twisted/Shielded Wire

Special Tools

Crimping tool

Splice clip

100-watt soldering
gun

Fender covers

Safety glasses

60/40 rosin core
solder

Twisted/shielded wire is used in computer circuits. It protects the circuit from electrical noise that would interfere with the operation of the computer controls (Figure 4-10). These wires may carry as low as 0.1 ampere of current. It is important that the splice made in these wires does not have any resistance. The added resistance may give false signals to the computer or actuator. Do the following to splice this type of wire:

1. Locate and cut out the damaged section of wire.

2. Being careful not to cut into the mylar tape or to cut the drain wire, remove about 1 inch (25 mm) of the outer jacket from the ends of the cable.

3. Unwrap the mylar tape. Do not remove the tape from the cable (Figure 4-11).

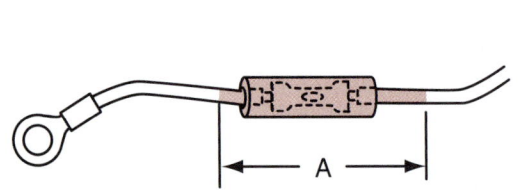

Figure 4-9 Apply petroleum jelly to the areas shown.

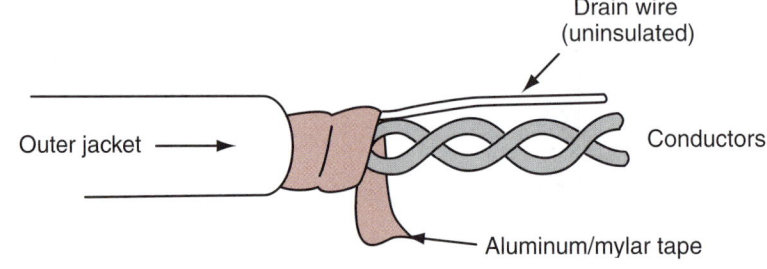

Figure 4-10 Twisted/shielded wire used in computer circuits.

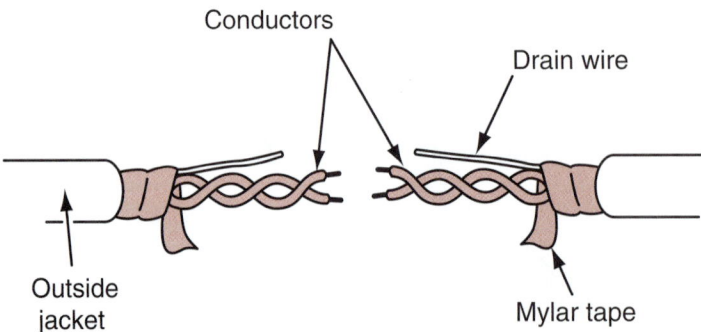

Figure 4-11 Before attempting to repair the wire, unwrap the mylar tape to expose the conductors.

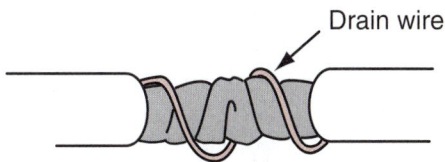

Drain wire

Figure 4-12 The drain wire should be wrapped around the outside of the mylar tape.

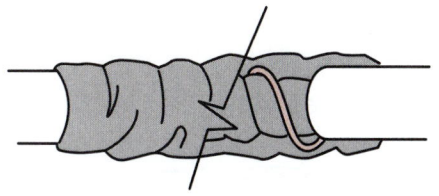

Figure 4-13 The completed repair.

CAUTION: When repairing the wires, stagger the splice connections. This will prevent shorts.

4. Untwist the conductors and remove the insulation from the ends.
5. Use a splice clip to connect the two wires and solder the splice.
6. Wrap the conductors with the mylar tape. Do not wrap the drain wire in the tape.
7. If the drain wire is cut, splice the drain wire and solder the connection.
8. Wrap the drain wire around the conductors and the mylar tape (Figure 4-12).
9. Use electrical tape or heat shrink tube to insulate the cable (Figure 4-13).

Replacing Fusible Links

Special Tools

Crimping tool

100-watt soldering iron

60/40 rosin core solder

Safety glasses

Fender covers

Not all fusible link open circuits are detectable by visual inspection only. Test for battery voltage on both sides of the fusible link to confirm its condition. If the fusible link must be replaced, it is cut out of the circuit and a new fusible link is crimped or soldered into place.

There are two types of insulation used on fusible links: Hypalon and Silicone/GXL. Hypalon can be used to replace either type of link. However, do not use Silicone/GXL to replace Hypalon. To identify the type of insulation, cut the blown link's insulation back. The insulation of the Hypalon link is a solid color all the way through. The insulation of Silicon/GXL will have a white inner core.

CAUTION: Disconnect the battery negative cable before performing any repairs to the fusible link.

When cutting off the damaged fusible link from the feed wire, cut it beyond the splice (Figure 4-14). When making the repair link, do not use a fusible link that is cut longer than 9 inches (228 mm). This length will not provide sufficient overload protection. Splice in the repair link by crimping or soldering.

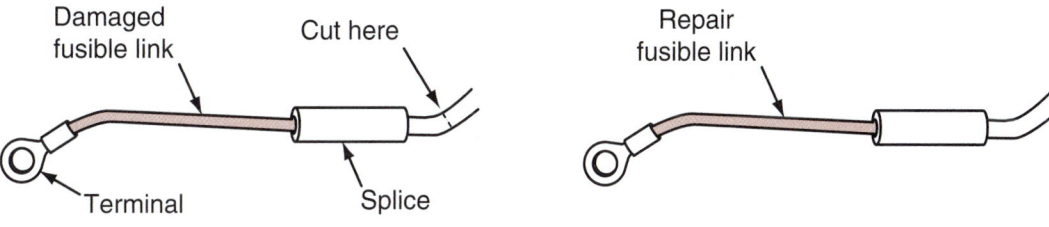

Damaged fusible link · Cut here · Repair fusible link · Terminal · Splice

Figure 4-14 Replacing a fusible link.

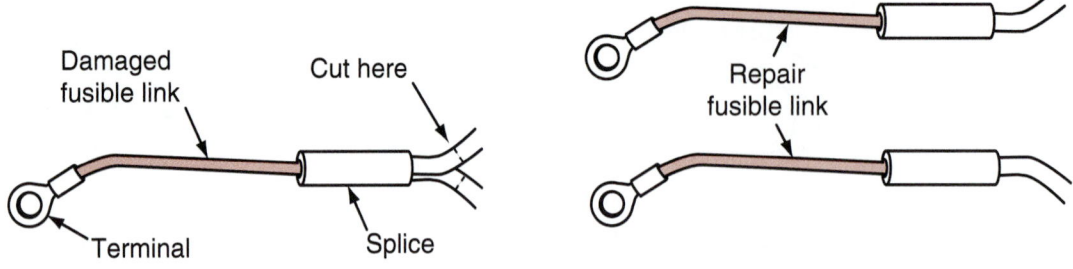

Damaged fusible link

Cut here

Repair fusible link

Terminal

Splice

Figure 4-15 Replacing a fusible link that feeds two circuits.

If the damaged fusible link feeds two harness wires, use two fusible links. Splice one link to each of the harness wires (Figure 4-15).

CAUTION: Use the fusible link gauge required by the manufacturer.

CUSTOMER CARE: If the fusible link or any of the fuses are blown, it is important to locate the cause. Fuses do not wear out. If they blow, it is due to an overload of current in the circuit. By using a cause-and-effect diagnosis approach, the fault can be identified and repaired the first time the vehicle is in the shop. The following illustrates how to use this method:

Effect	*Cause*
Turn signals not operating	Blown fuse
Blown fuse	Short to ground
Short to ground	Broken wire in trunk
Broken wire	Loose jack handle

This method led the technician to fix the fault and find the cause.

Repairing Connector Terminals

The connector terminal is subjected to needed repairs due to abuse, improper disconnecting procedures, and exposure potential to the elements. The method of repair depends on the type of connector.

Classroom Manual
Chapter 4, page 97

Molded Connectors

If the connector is a one-piece, molded-type connector, it cannot be disassembled for repairs (Figure 4-16). Although the connector halves can be separated, the connector itself cannot be disassembled. If the connector is damaged, it must be cut off and a new connector spliced in.

Classroom Manual
Chapter 4, page 97

Hard-Shell Connectors

WARNING: Do not place your fingers or body next to the connector. If excessive force is needed to depress the tang, the pick may be pushed out the back and cause injury.

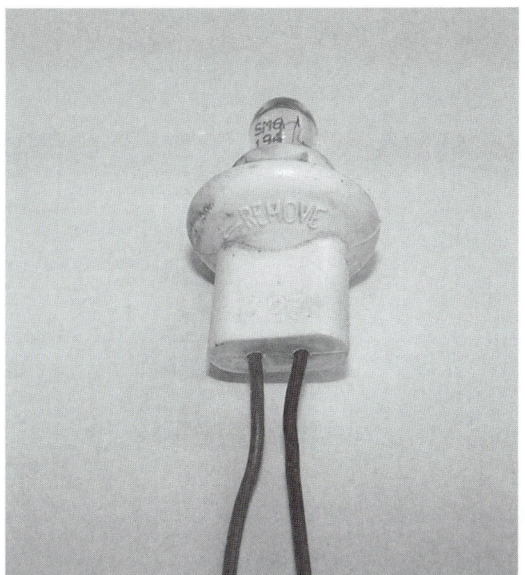

Figure 4-16 Molded connector halves are a one-piece design that cannot be disassembled for repairs.

Hard-shell connectors usually provide a means of removing the terminals for repair. Use a pick, or the special tool, to depress the locking tang of the connector (Figure 4-17). Pull the lead back far enough to release the locking tang from the connector. Remove the pick, then pull the lead completely out of the connector. Make the repair to the terminal using the same procedures for repairing copper or aluminum wire.

Re-form the terminal locking tang to assure a good lock in the connector (Figure 4-18). Use the pick to bend the lock tang back into its original shape. Insert the lead into the back of the connector. A noticeable "catch" should be felt when the lead is halfway through the connector. Gently push back and forth on the lead to confirm that the terminal is locked in place.

Special Tools

Pick

Crimping tool

Safety glasses

Fender covers

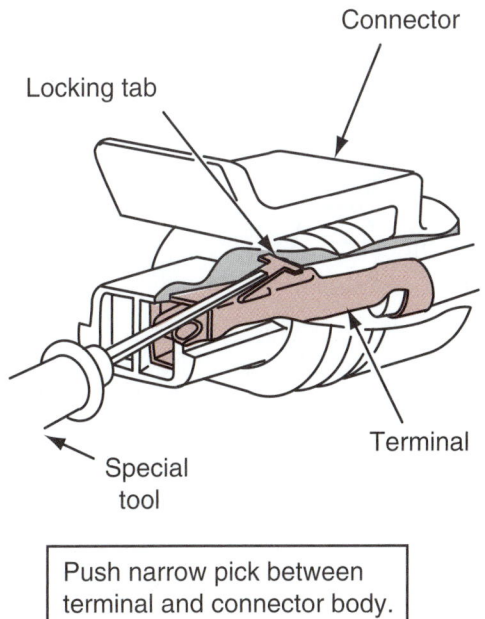

Push narrow pick between terminal and connector body.

Figure 4-17 Depress the locking tang to remove the terminal from the connector.

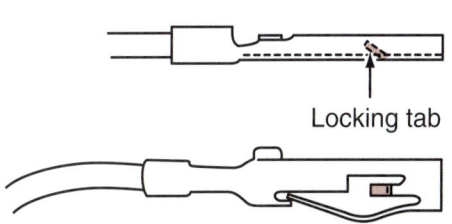

Figure 4-18 Before inserting the terminal back into the connector, re-form the locking tang to its original position.

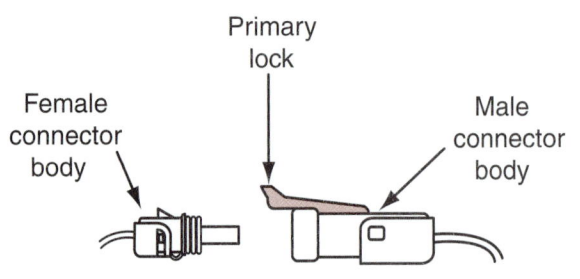

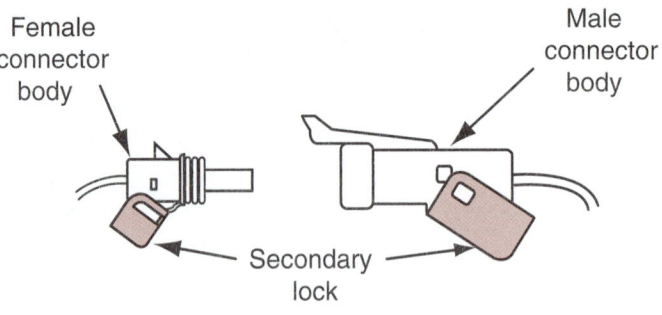

Figure 4-19 The weather-pack connector has two locks. Use the primary lock to separate the halves.

Figure 4-20 Unlock the secondary lock to remove the terminals from the connector.

Classroom Manual
Chapter 4, page 97

Special Tools

Weather-pack tool

Crimping tool

Safety glasses

Fender cover

Weather-Pack Connectors

The terminals of the weather-pack connector are secured by a hinged secondary lock or a plastic terminal retainer. To perform repairs, first disconnect the two halves by pulling up on the primary lock while pulling the two halves apart (Figure 4-19). Unlock the secondary locks and swing them open (Figure 4-20).

Depress the terminal locking tanks using the special weather-pack tool. Push the cylinder of the tool into the terminal cavity from the front until it stops (Figure 4-21). Pull the tool out, then gently pull the lead out of the back of the connector (Figure 4-22).

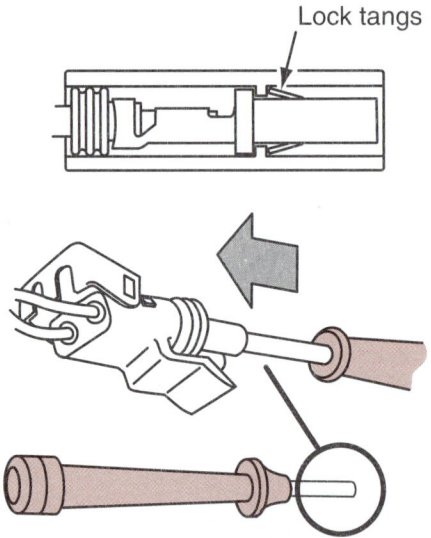

Figure 4-21 Use the recommended special tool to unlock the tang on the terminal.

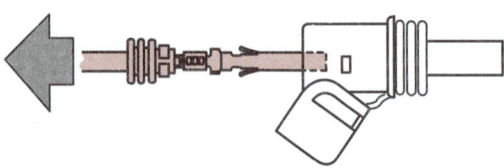

Figure 4-22 After the lock tang has been depressed, remove the lead from the back of the connector.

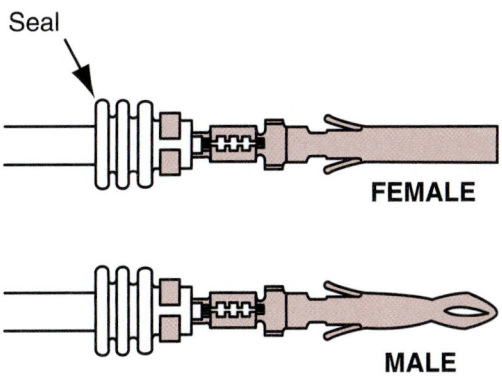

Figure 4-23 Male and female connectors.

The terminal is either a male or female connector (Figure 4-23). Use the correct terminal for the repair. Feed the wire through the seal and connect the repair lead in the normal manner of crimping and soldering (Figure 4-24). Re-form the terminal lock tang by bending it back into its original position (Figure 4-25).

Insert the lead from the back of the connector until a noticeable "catch" is felt. Gently push and pull on the lead to confirm that it is locked to the connector. Close the secondary locks and reconnect the connector halves.

SERVICE TIP: Use a weather-pack repair kit that provides new seals for a complete repair.

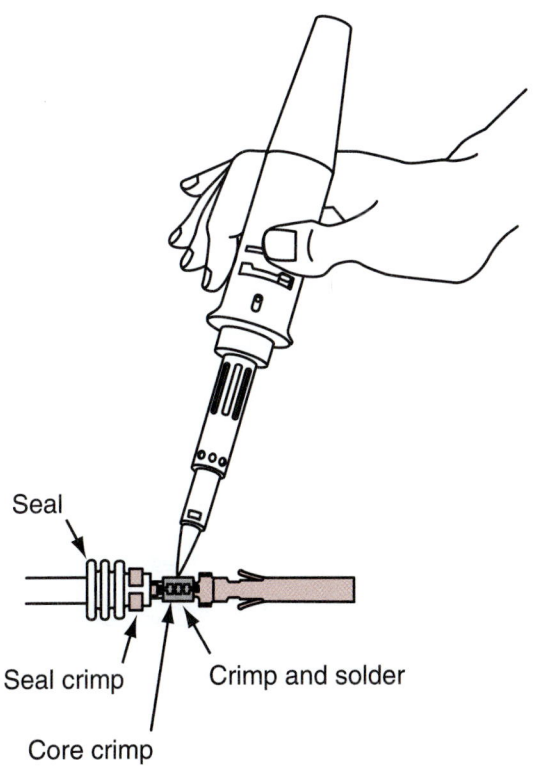

Figure 4-24 Crimp and solder the terminal to the lead.

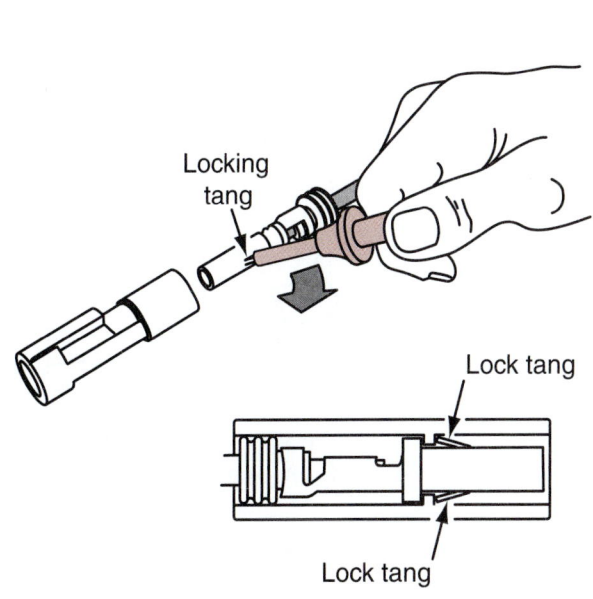

Figure 4-25 Re-forming the locking tangs of the terminal.

Special Tools

Pick

Crimping tool

Safety glasses

Fender covers

The connectors are called **pull to seat** or **push to seat** to depict the method used to install the terminals into the connector.

Metri-Pack Connectors

There are two types of metri-pack connectors: **pull to seat** and **push to seat.** The push-to-seat terminal removal is illustrated (Figures 4-26 and 4-27).

The pull-to-seat terminal is removed by inserting a pick into the connector and under the lock tang (Figure 4-28). Gently pull back on the lead while prying up on the lock tang. When the lock tang is free of the tab in the connector, push the lead through the *front* of the connector.

To make the repairs to the terminal, insert the stripped wire through the seal and the connector body (Figure 4-29). Crimp and solder the terminal to the wire. Pull the wire lead and terminal back into the connector body until the terminal is locked (Figure 4-30).

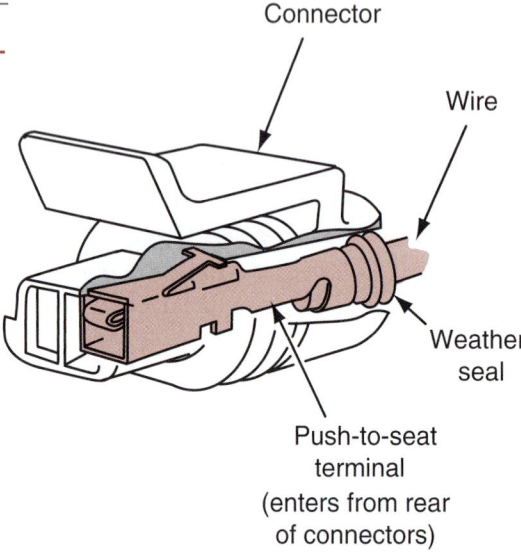

Connector

Wire

Weather seal

Push-to-seat terminal (enters from rear of connectors)

Figure 4-26 Use a wide pick to unlock the male terminal locking tang.

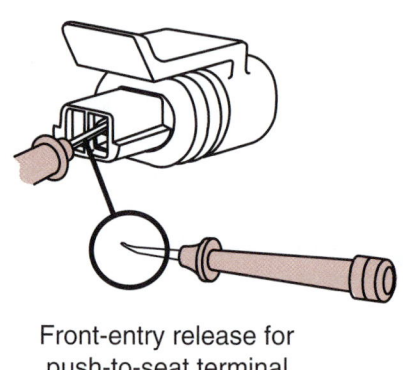

Front-entry release for push-to-seat terminal

Figure 4-27 Use a wide pick to unlock the nib of the terminal retainer for female terminals.

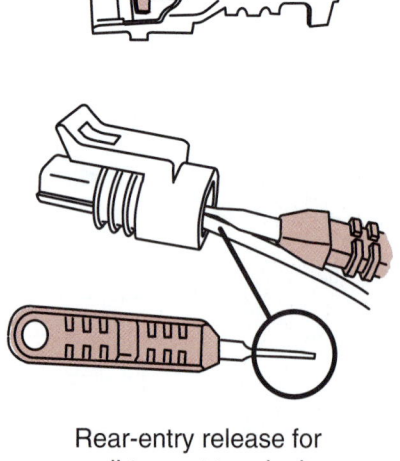

Lock tab

Rear-entry release for pull-to-seat terminal

Figure 4-28 Pull up on the lock tang to release the terminal from the connector.

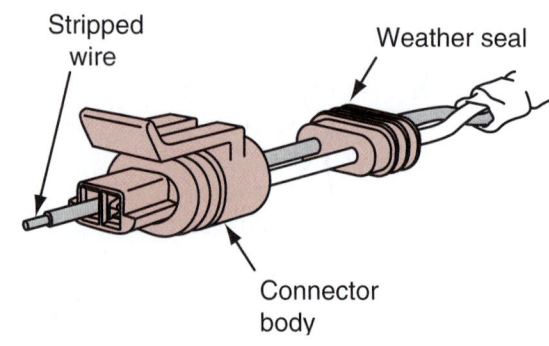

Stripped wire

Weather seal

Connector body

Figure 4-29 The wire lead must be installed into the seal and connector before attaching the terminal.

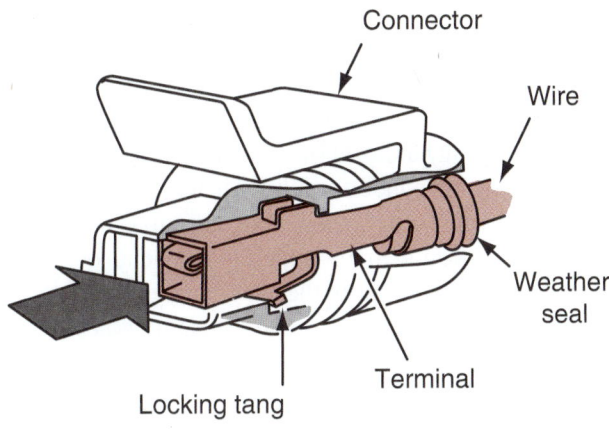

Figure 4-30 Make sure that the terminal locks into the connector body.

Reading Wiring Diagrams

Classroom Manual
Chapter 4, page 103

When attempting to locate a possible cause for system malfunction, it is important to have the correct wiring diagram for the vehicle being worked on. There may be a different diagram for each model and even for the same models equipped with different options. Also, diagrams may differ between two- and four-door models. In some cases, it may be necessary to use the date of manufacture and/or the **vehicle identification number (VIN)** to determine the correct diagram to use. The VIN is assigned to a vehicle for identification purposes. The identification plate is usually located on the cowl, next to the left upper instrument panel. It is visible from the outside of the vehicle. Other locations for the VIN plate are in the door opening, glove box, and engine compartment (Figure 4-31).

Next, study the method used to identify circuits and **color codes.** Usually this information is provided in the service manual. Color codes are used to assist in tracing the wires. In most color codes, the first group of letters designates the base color of the insulation and the second group of letters indicates the color of the tracer. Also become familiar with the electrical symbols used by the manufacturer.

Before you begin to try to use a wiring diagram, you must first have an idea of what you are looking for. It may be a particular component, circuit, or connector. The best way to start the process is by identifying the component or one of the components that doesn't work correctly. Then look in the index for the wiring diagram and find where that component is shown.

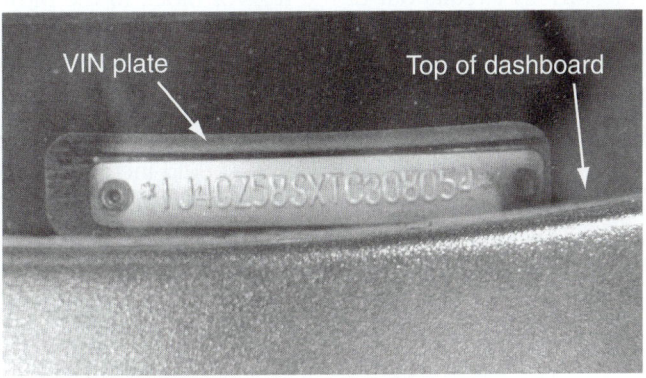

Figure 4-31 VIN plate location.

The electrical section in most service manuals breaks down the electrical system of the automobile into individual circuits. This approach makes it easier to find a particular component. Of course, you still need to use the index to find the page the component and its circuit is on.

If the service manual uses a total vehicle wiring diagram, finding the component may be a little trickier. Wiring diagrams are usually indexed by grids. The diagram is marked into equal sections like a street map. The wiring diagram's index will list a letter and number for each major component and many different connection points. If the wiring diagram is not indexed, you can locate the component by relating its general location in the vehicle to a general location on the wiring diagram. Most system diagrams are drawn so the front of the car is on the left of the diagram.

Once you have found the component or part of the circuit you were looking for, identify all of the components, connectors, and wires that are related to that component. This is done by tracing through the circuit, starting at the component. Tracing does not mean taking a pencil and marking on the wiring diagram. Tracing means taking your pencil and drawing out the circuit on another piece of paper. It doesn't have to be pretty to work; it just needs to be accurate. Tracing may also mean taking your finger and following the wires to where they lead. In order for tracing to have any value, you need to identify the power source for the component and/or for the circuit, all related loads (sometimes this involves tracing the circuit back through other pages of the wiring diagram), and the ground connection for the component and for all of the related loads.

After you have traced the circuit, study it and make sure you know how the circuit is supposed to work. Then describe the problem you are hoping to solve. Ask yourself what could cause this. Limit your answers to those items in your traced wiring diagram. Also, limit your answers to the description of the problem. It is wise to make a list of all probable causes of the problem; then number them according to probability. For example, if no dashlights come on, it is possible that all of the bulbs are burned out. However, it is not as probable as a blown fuse. After you have listed the probable causes in order of probability, look at the wiring diagram to identify how you can quickly test to find out which is the cause. Diagnostics is made easier as your knowledge of electricity grows. It also becomes easier with a good understanding of how the circuit works.

Figure 4-32 is a schematic for a blower control circuit of a heater system. By tracing through the circuit, the technician should be able to determine the correct operation of the system. It cannot be overemphasized that the technician should not attempt to figure out why the circuit is not working until it is understood how it is supposed to work.

In this circuit, battery voltage is applied to the fuse box where the 30-ampere fuse supplies the battery voltage to the motor through circuit number 181. Notice that circuit 181 connects to the blower motor resistor block through connector C606. When it leaves the resistor block (after flowing through the thermal limiter) it goes through connector C001 and attaches to the motor through connector C002. Details of these connectors are shown with the schematic.

The circuit then leaves the motor and enters the resistor block through the same connectors. There is a splice in the resistor block that connects this wire to a series of resistors. Connector C606 directs the various circuits out of the resistor block to the blower switch through connector C613. If the switch is placed in the OFF position, the circuit to ground is opened and the blower motor should not operate. If the switch is placed in the LOW position, the current will flow through all of the resistors and the switch through circuit 260. The switch completes the circuit to ground. As the switch is placed in different speed positions, the amount of resistance and the circuit number change. The motor speed should increase as the amount of resistance decreases.

If the customer complains that the heater motor does not work at all, in any speed position, consider the following possible causes:

1. Open fuse.

2. Open in the lead from the battery to the fuse box.

3. Bad ground connection after the switch.

4. Inoperative blower motor.

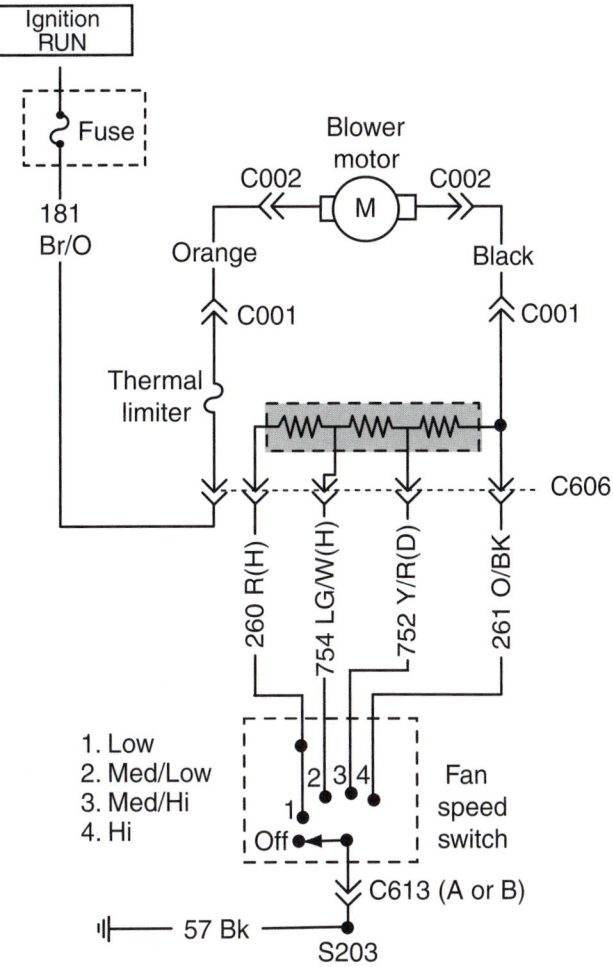

Figure 4-32 Heater system wiring diagram.

5. Open in circuit 181.

6. Open in the orange wire between connectors C001 and C002.

7. Open thermal limiter.

8. Open in the black wire between connectors C002 and C001.

9. Disconnected, damaged, or corroded connector C606 or C613.

10. Faulty switch.

However, if the customer complains that the motor will not operate in the LOW position only, the problem is limited to three possibilities:

1. The third resistor in the series is open.

2. An open in circuit 260 from the resistor block to the blower switch.

3. A faulty switch.

Once the potential problem areas are determined, use the color code to locate the exact wires. Test the leads for expected voltages at different locations in the circuit.

By understanding the way the system is supposed to work, the problem of determining where to look for the problems is simplified. Practice in reading wiring schematics is the only good teacher.

The vehicle owner complains that the brake lights do not light. He also says the dome light is not working. The technician verifies the problem, then checks the battery for good connections and tests the fusible links. All are in good condition.

A study of the wiring diagram indicates that the brake light and dome light circuits share the same fuse. It is also indicated that the ignition switch illumination light circuit is shared with these two circuits. A check of the ignition switch illumination light shows that it is not operating either. The technician checks the fuse that is identified in the wiring diagram. It is blown. When a replacement fuse is installed, the dome and brake lights work properly for three tests, then the fuse blows again.

Upon further testing of the shared circuits, an intermittent short to ground is located in the steering column in the ignition switch illumination circuit. The technician solders in a repair wire to replace the damaged section. After all repairs are completed, a final test indicates proper operation of all circuits.

Terms to Know

Color codes	Pull to seat	Splice clip
Crimping	Push to seat	Trouble codes
Crimping tool	Solderless connectors	Troubleshooting
Heat shrink tube	Splice	Vehicle identification number (VIN)

ASE-Style Review Questions

1. Splicing copper wire is being discussed.
   Technician A says it is acceptable to use solderless connections.
   Technician B says acid core solder should not be used on copper wires.
   Who is correct?
   A. A only
   B. B only
   C. Both A and B
   D. Neither A nor B

2. Use of wiring diagrams is being discussed.
   Technician A says a wiring diagram is used to help find the fault.
   Technician B says the wiring diagram will give the exact location of the components in the car.
   Who is correct?
   A. A only
   B. B only
   C. Both A and B
   D. Neither A nor B

3. The fuse for the parking lights is open.
   Technician A says find the cause for the blown fuse.
   Technician B says the fuse probably wore out due to age.
   Who is correct?
   A. A only
   B. B only
   C. Both A and B
   D. Neither A nor B

4. Repairs to a twisted/shielded wire are being discussed.
   Technician A says a twisted/shielded wire carries high current.
   Technician B says because a twisted/shielded wire carries low current, any repairs to the wire must not increase the resistance of the circuit.
   Who is correct?
   A. A only
   B. B only
   C. Both A and B
   D. Neither A nor B

5. Replacement of fusible links is being discussed.
   Technician A says not all open fusible links are detectable by visual inspection.
   Technician B says to test for battery voltage on both sides of the fusible link to confirm its condition.
   Who is correct?
   A. A only   **C.** Both A and B
   B. B only   **D.** Neither A nor B

6. *Technician A* says troubleshooting is the diagnostic procedure of locating and identifying the cause of the fault.
   Technician B says troubleshooting is a step-by-step process of elimination by use of cause and effect.
   Who is correct?
   A. A only   **C.** Both A and B
   B. B only   **D.** Neither A nor B

7. *Technician A* says a replacement fusible link should be at least 9 inches (25 mm) long.
   Technician B says if the damaged fusible link feeds two harness wires, use one replacement fusible link.
   Who is correct?
   A. A only   **C.** Both A and B
   B. B only   **D.** Neither A nor B

8. Repairing connectors is being discussed.
   Technician A says molded connectors are a one-piece design and cannot be separated for repairs.
   Technician B says to replace the seals when repairing the weather-pack connector.
   Who is correct?
   A. A only   **C.** Both A and B
   B. B only   **D.** Neither A nor B

9. *Technician A* says all metri-pack connectors use male terminals.
   Technician B says the connectors of the metri-pack are called pull to seat or push to seat to depict the method used to install the terminals into the connector.
   Who is correct?
   A. A only   **C.** Both A and B
   B. B only   **D.** Neither A nor B

10. *Technician A* says there may be a different wiring diagram for each model of a vehicle.
    Technician B says it may be necessary to use the VIN number to get the correct wiring diagram for the vehicle.
    Who is correct?
    A. A only   **C.** Both A and B
    B. B only   **D.** Neither A nor B

ASE Challenge Questions

1. *Technician A* says that acid core solder should be used whenever copper wires are to be soldered.
   Technician B says that solderless connectors should not be used if a weather-resistant connection is desired.
   Who is correct?
   A. A only   **C.** Both A and B
   B. B only   **D.** Neither A nor B

2. Which of the following electrical troubleshooting routines is in the correct sequence?
   A. Study the electrical diagram, confirm the complaint, locate and repair the fault, test the repair.
   B. Study the electrical diagram, locate and repair the fault, confirm the complaint, test the repair.
   C. Confirm the complaint, study the electrical diagram, test the repair, locate and repair the fault.
   D. Confirm the complaint, study the electrical diagram, locate and repair the fault, test the repair.

3. Wire repair procedures are being discussed.
   Technician A says the aluminum wire should never be soldered; it should only be crimped whenever repairs are necessary.
   Technician B says that damaged twisted/shielded wires should be connected with splice clips and then the connections should be soldered.
   Who is correct?
   A. A only   **C.** Both A and B
   B. B only   **D.** Neither A nor B

4. *Technician A* says that when replacing a fusible link the gauge size of the replacement fusible link can be decreased but never increased.

Technician B says that a 14-gauge fusible link can be replaced with an equivalent length of 14-gauge stranded wire.

Who is correct?

A. A only

B. B only

C. Both A and B

D. Neither A nor B

5. *Technician A* says that a wire identified as "754 LG/W (H)" on a wiring diagram refers to a wire located in circuit 754 that is light green with white hash marks.

Technician B says that a wire that is labeled 181 BR/O refers to an 18-gauge wire that is brown with an orange stripe.

Who is correct?

A. A only

B. B only

C. Both A and B

D. Neither A nor B

Job Sheet 14

Name _____ Date _____

Part Identification on a Wiring Diagram

Upon completion of this job sheet, you should be able to locate different parts on a wiring diagram.

ASE Correlation

This job sheet is related to the ASE Electrical/Electronic System Certification Exam's content area: *General Electrical System Diagnosis;* task: Read and interpret electrical schematic diagrams and symbols.

Tools and Materials

A wiring diagram in a service manual (assigned by the instructor)

Describe the vehicle being worked on.

Year _____ Make _____ Model _____

VIN _____ Engine type and size _____

Service Manual used _____

Wiring diagram is found on pages _____ to _____ .

Procedure

Task Completed

Study the wiring diagram; then answer the following questions.

1. Are all circuit grounds clearly marked in the wiring diagrams? _____

   _____

2. What is represented by lines that cross each other? _____

   _____

   _____

3. Are most switches shown in their normally open or normally closed position?

   _____

4. Do all wires have a color code listed by them? _____

5. Is the internal circuitry of all components shown in their schematic drawing?

   _____

6. List the page and figure numbers (the location) where the following electrical components are shown in the wiring diagram. Then **draw the schematical symbol** used by this wiring diagram to represent the part.

Component	Location	Drawing
Windshield wiper motor	_____	_____
Dome (courtesy) light	_____	_____
A/C compressor clutch	_____	_____
Turn signal flasher unit	_____	_____
Fuse	_____	_____
Fuel gauge sending unit	_____	_____

Instructor's Response _____

_____

_____

_____

Job Sheet 15

Name _____ Date _____

Using a Component Locator

Upon completion of this job sheet, you should be able to identify components and their locations indicated on the wiring diagram that is used for electrical system diagnosis.

Tools and Materials

Service manual with a component locator section (supplied by instructor)

Describe the vehicle being worked on.

Year _____ Make _____ Model _____

VIN _____ Engine type and size _____

Service Manual used _____

Procedure

Task Completed

Using the component locator for the vehicle that has been assigned, the student will find the locations for the taillight, headlight, and stoplight circuits, connectors, connections, plugs, and switches.

1. Taillight circuit:

    Location of source of power: _____

    Location of switch: _____

    Number of connectors: _____ ; their location: _____

    _____

    _____

    Location of taillight bulb: _____ ; how you get to them: _____

    _____

    _____

2. Headlight circuit:

    Location of source of power: _____

    Location of switch: _____

    Number of connectors: _____ ; their location: _____

    _____

    _____

    Location of taillight bulb: _____ ; how you get to them: _____

    _____

    _____

Task Completed

3. Stoplight circuit:

Location of source of power: _____

Location of switch: _____

Number of connectors: _____ ; their location: _____

_____

_____

Location of taillight bulb: _____ ; how you get to them: _____

_____

_____

Instructor's Response _____

_____

_____

Job Sheet 16

Name _____ Date _____

Using a Wiring Diagram

Upon completion of this job sheet, you should be able to find the power source, ground connection, and controls for electrical circuits using a wiring diagram.

ASE Correlation

This job sheet is related to the ASE Electrical/Electronic System Certification Exam's content area: *General Electrical System Diagnosis;* task: Read and interpret electrical schematic diagrams and symbols.

Exercise 1

Using the diagram in Figure 4-33, answer the following (list wire colors):

    1. How is the windshield wiper motor circuit powered? _____

       _____

       _____

       _____

       _____

    2. How is the circuit grounded?

       _____

       _____

       _____

       _____

    3. How is the circuit controlled?

       a. If it is controlled by a switch, is the switch normally open or closed?
       b. Is the circuit power switched or ground switched?
       c. Is the circuit controlled by a variable resistor?
       d. Is the circuit controlled by a mechanical or vacuum-operated device?

       _____

       _____

       _____

       _____

    4. How is the circuit protected (identify all fuses)?

       _____

       _____

       _____

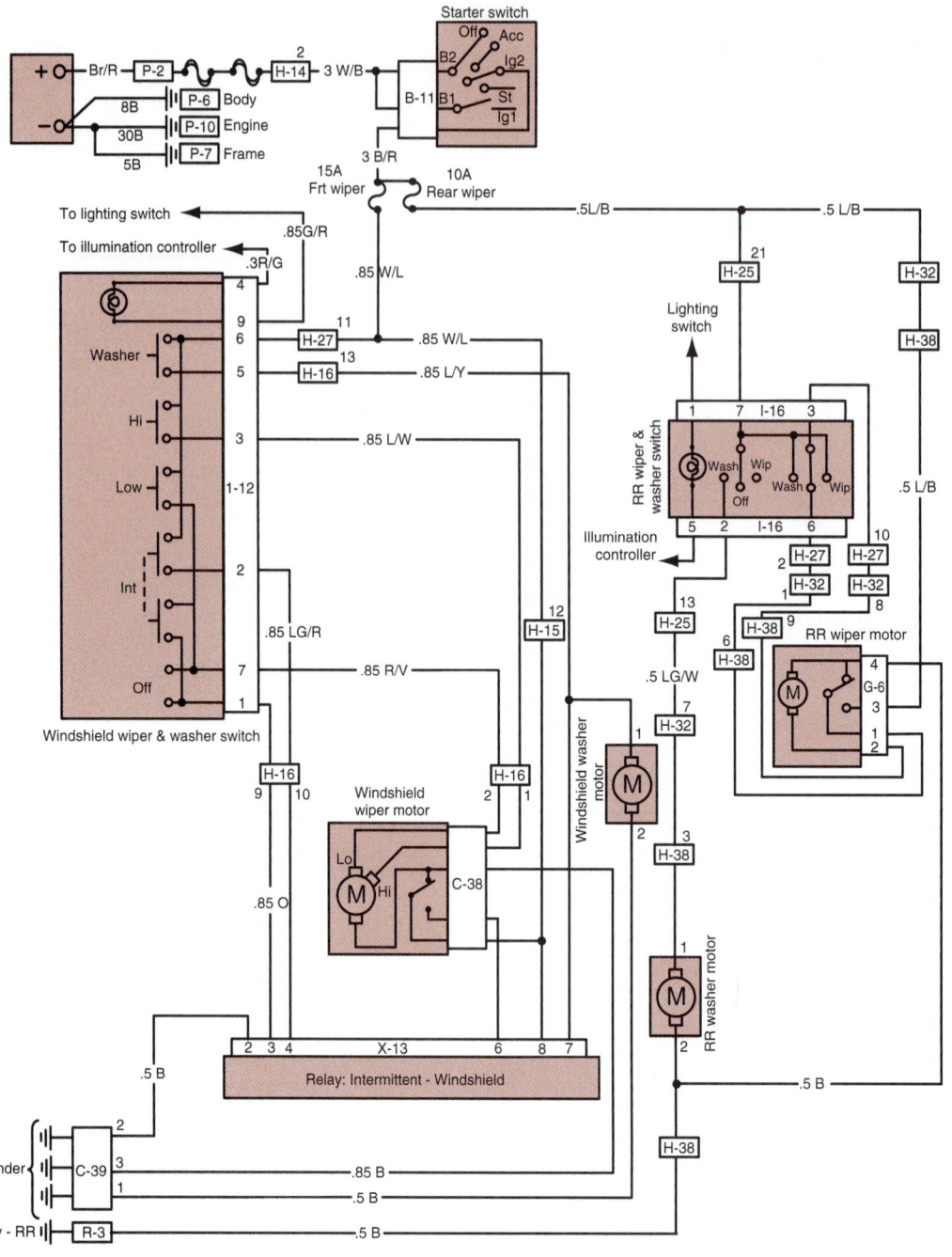

Figure 4-33

Exercise 2

Using the wiring diagram in Figure 4-34, answer the following:

1. How is the circuit powered?_____

_____

_____

_____

_____

2. How is the circuit grounded?

_____

_____

_____

_____

3. How is the circuit controlled?

    a. If it is controlled by a switch, is the switch normally open or closed?
    b. Is the circuit power switched or ground switched?
    c. Is the circuit controlled by a variable resistor?
    d. Is the circuit controlled by a mechanical or vacuum-operated device?

_____

_____

_____

_____

4. How is the circuit protected?

_____

_____

_____

_____

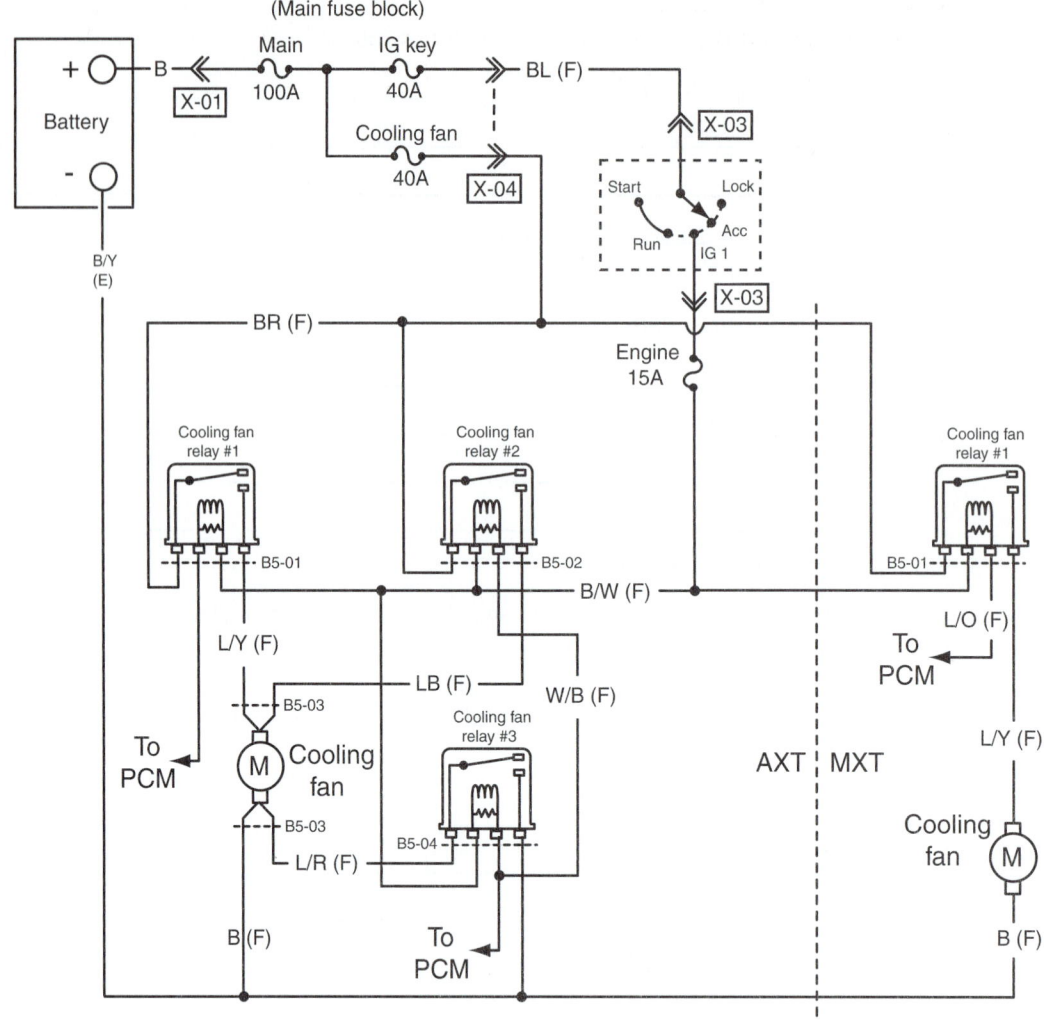

Figure 4-34

Exercise 3

Using the wiring diagram in Figure 4-35, answer the following:

1. How is the circuit powered?

2. How is the circuit grounded?

3. How is the circuit controlled?
   a. If it is controlled by a switch, is the switch normally open or closed?
   b. Is the circuit power switched or ground switched?
   c. Is the circuit controlled by a variable resistor?
   d. Is the circuit controlled by a mechanical or vacuum-operated device?

_____

_____

_____

_____

4. How is the circuit protected?

_____

_____

_____

_____

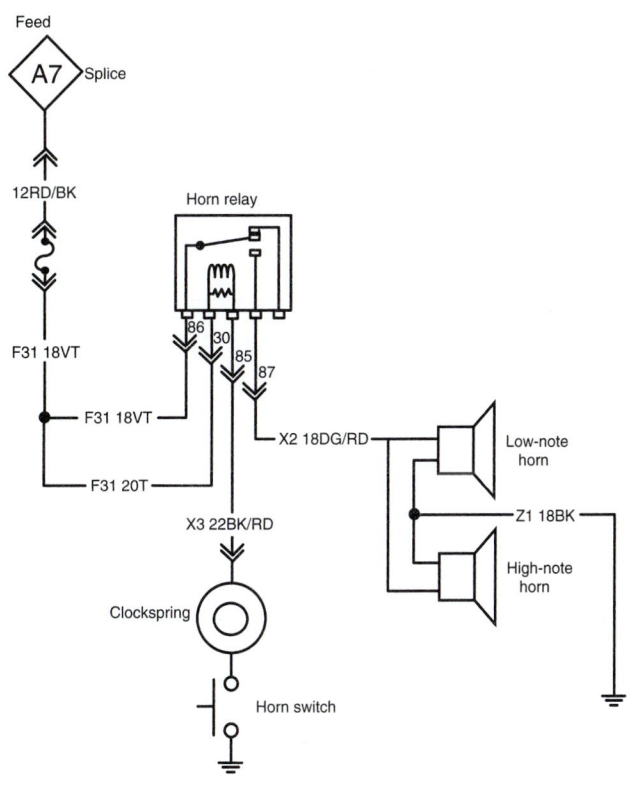

Figure 4-35

Exercise 4

Using the wiring diagram in Figure 4-36, answer the following:

1. How is the circuit powered?_____

_____

_____

_____

_____

2. How is the circuit grounded?

_____

_____

_____

_____

3. How is the circuit controlled?
 a. If it is controlled by a switch, is the switch normally open or closed?
 b. Is the circuit power switched or ground switched?
 c. Is the circuit controlled by a variable resistor?
 d. Is the circuit controlled by a mechanical or vacuum-operated device?

_____

_____

_____

_____

4. How is the circuit protected?

_____

_____

_____

_____

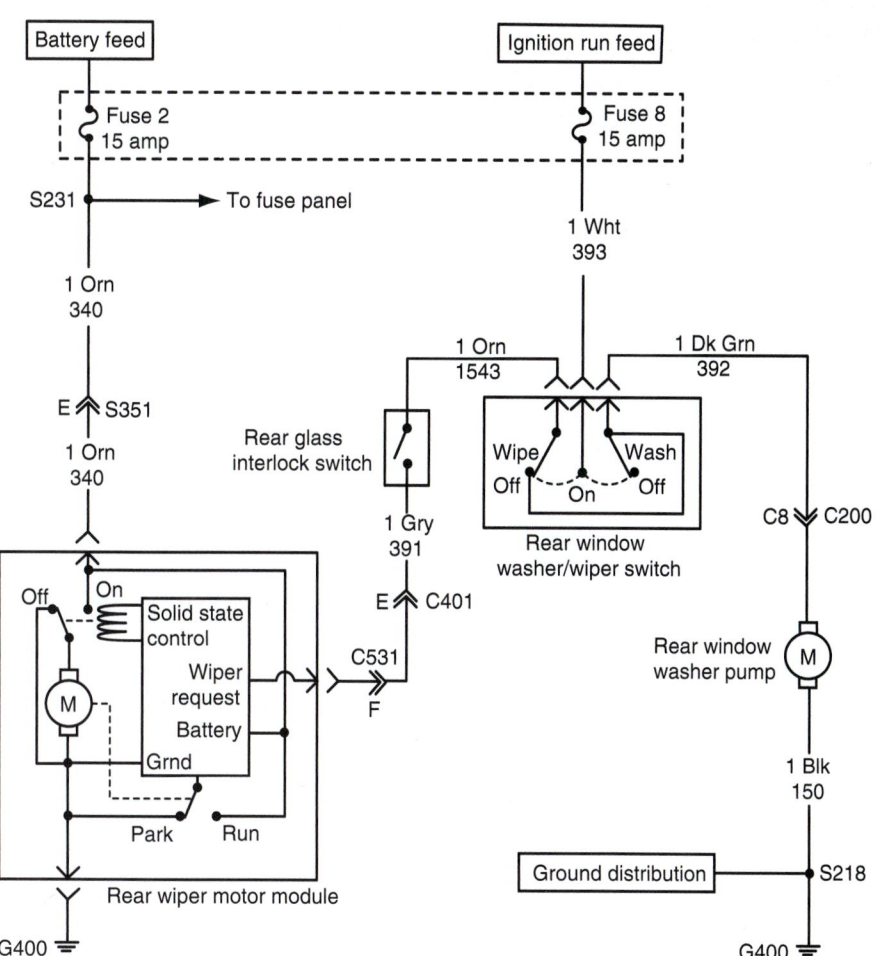

Figure 4-36

Exercise 5

Using the wiring diagram in Figure 4-37, answer the following:

1. How is the circuit powered?_____

_____

_____

_____

_____

2. How is the circuit grounded?

_____

_____

_____

_____

3. How is the circuit controlled?
   a. If it is controlled by a switch, is the switch normally open or closed?
   b. Is the circuit power switched or ground switched?
   c. Is the circuit controlled by a variable resistor?
   d. Is the circuit controlled by a mechanical or vacuum-operated device?

_____

_____

_____

_____

4. How is the circuit protected?

_____

_____

_____

_____

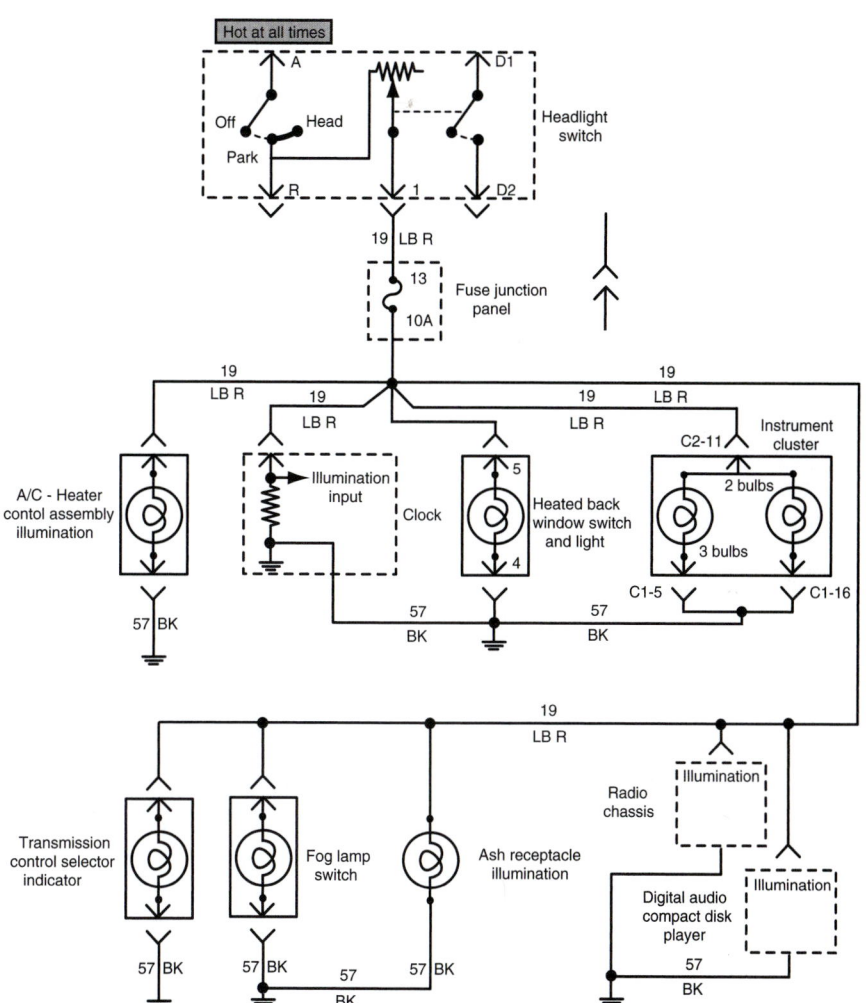

Figure 4-37

Instructor's Response _____

_____

_____

_____

Job Sheet 17

17

Name _____ Date _____

Soldering Copper Wires

Upon completion of this job sheet, you should be able to properly solder two copper wires.

Tools and Materials

Two pieces of copper wire
100-watt soldering iron
60/40 rosin core solder
Crimping tool
Splice clip
Heat shrink tube
Heating gun
Safety glasses

Procedure:

Task Completed

1. Using the correct size stripper, remove about 1/2 inch (12 mm) of the insulation from both wires. Be careful not to nick or cut any of the wires. ☐

2. Select the proper size splice clip to hold the splice. ☐

3. Place the correct length and size of heat shrink tube over the wire. Slide the tube far enough away so the wires are not exposed to the heat of the soldering iron. ☐

4. Overlap the two splice ends and hold in place with thumb and forefinger. While the wire ends are overlapped, center the splice clip around the wires and crimp into place. ☐

5. Heat the splice clip with the soldering iron while applying solder to the opening in the back of the clip. Apply only enough solder to make a good connection. The solder should travel through the wire. ☐

6. After the solder cools, slide the heat shrink tube over the splice and heat the tube with the hot air gun until it shrinks around the splice. Do not overheat the tube. ☐

Instructor's Response _____

_____

_____

_____

DIAGNOSTIC CHART 4-1

PROBLEM AREA:	Burned fusible link.
SYMPTOMS:	Several electrical components fail to operate.
POSSIBLE CAUSES:	**1.** Burned or defective fusible link.

DIAGNOSTIC CHART 4-2

PROBLEM AREA:	Burned, broken, or defective connectors, insulation, or conductors.
SYMPTOMS:	Open or short circuits that prevent component operation.
POSSIBLE CAUSES:	**1.** Defective wiring. **2.** Improper or no wire protectors. **3.** Defective connectors.

Battery Diagnosis and Service

Upon completion and review of this chapter, you should be able to:

❑ Demonstrate all safety precautions and rules associated with servicing the battery.

❑ Perform a visual inspection of the battery, cables, and terminals.

❑ Test a conventional battery's specific gravity.

❑ Perform an open circuit test and accurately interpret the results.

❑ Test the capacity of the battery to deliver both current and voltage and to accurately interpret the results.

❑ Perform a 3-minute charge test to determine if the battery is sulfated.

❑ Correctly slow and fast charge a battery, in or out of the vehicle.

❑ Describe the differences between slow and fast charging and determine when either method should be used.

❑ Jump-start a vehicle by use of a booster battery and jumper cables.

❑ Perform a battery drain test and accurately determine the causes of battery drains.

❑ Perform a battery leakage test and determine the needed corrections.

❑ Do a battery terminal test and accurately interpret the results.

❑ Remove, clean, and reinstall the battery properly.

Basic Tools

Basic mechanic's tool set
Service manual

Introduction

Classroom Manual
Chapter 5, page 114

A discharged or weak battery can affect more than just the starting of the engine. The battery is the heart of the electrical system of the vehicle. It is important that it is not overlooked when servicing most electrical problems. The function of the battery is to:

1. Operate the starting motor, ignition system, electronic fuel injection, and other electrical devices for the engine during cranking and starting.

2. Supply all the electrical power for the vehicle accessories whenever the engine is not running or at low idle.

3. Supplement current for a limited time whenever electrical demands exceed charging system output.

4. Act as a stabilizer of voltage for the entire automotive electrical system.

5. Store energy for extended periods of time.

6. Keep power to computer memory.

Because of its importance, the battery should be checked whenever the vehicle is brought into the shop for service. A battery test series will show the state of charge and output voltage of the battery, which determine if it is good, is in need of recharging, or must be replaced.

There are many different manufacturers of battery test equipment. For years, the most popular tester was the Sun Electrical Corporation's VAT-40 (Figure 5-1). Most modern testers are

Figure 5-1 A Sun VAT-40 battery, starting, and charging system tester.

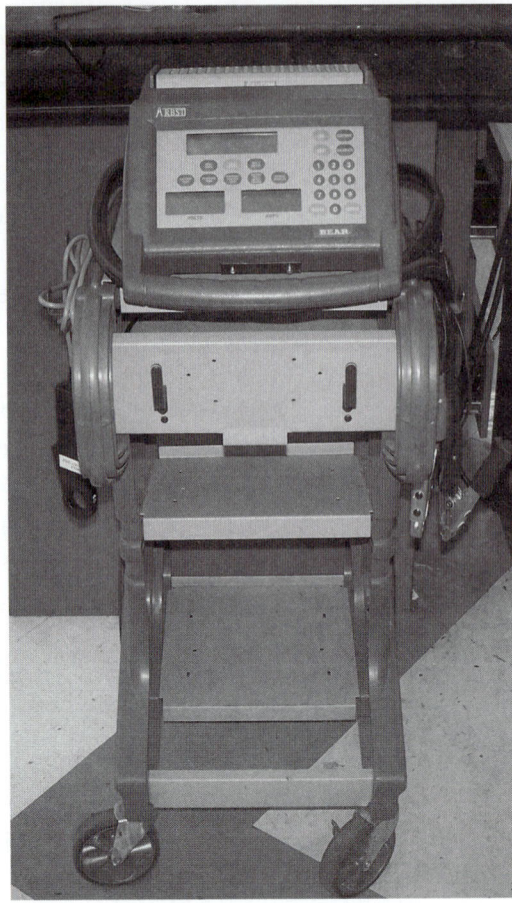

Figure 5-2 A computer-based generator, regulator, battery, and starter tester.

computer based and conduct the tests automatically after a particular test is selected (Figure 5–2). Always follow the procedures given by the tester's manufacturer.

General Precautions

Before attempting to do any type of work on or around the battery, the technician must be aware of certain precautions. To avoid personal injury or property damage, take the following precautions:

1. Battery acid is very corrosive. Do not allow it to come in contact with skin, eyes, or clothing. If battery acid gets into your eyes, rinse them thoroughly with clean water and receive immediate medical attention. If battery acid comes in contactwith skin, wash with clean water. Baking soda added to the water will help to neutralize the acid. If the acid is swallowed, drink large quantities of water or milk followed by milk of magnesia and a beaten egg or vegetable oil.

2. When making connections to a battery, be careful to observe polarity, positive to positive and negative to negative.

3. When disconnecting battery cables, always disconnect the negative (ground) cable first.

4. When connecting battery cables, always connect the negative cable last.

5. Avoid any arcing or open flames near a battery. The vapors produced by the battery cycling are very explosive. Do not smoke around a battery.

6. Follow manufacturer's instructions when charging a battery. Charge the battery in a well-ventilated area. Do not connect or disconnect the charger leads while the charger is turned on.

7. Do not add additional electrolyte to the battery if it is low. Add only distilled water.

8. Do not wear any jewelry or watches while servicing the battery. These items are excellent conductors of electricity. They can cause severe burns if current flows through them by accidental contact with the battery positive terminal and ground.

9. Never lay tools across the battery. They may come into contact with both terminals, shorting out the battery and causing it to explode.

10. Wear safety glasses or face shield when servicing the battery.

11. If the battery's electrolyte is frozen, allow it to defrost before doing any service or testing of the battery. While it is defrosting, look for leaks in the case. Leakage means the battery is cracked and should be replaced.

Battery Inspection

Before performing any electrical tests, the battery should be inspected, along with the cables and terminals. The complete visual inspection of the battery should include the following items:

1. Battery date code: This provides information as to the age of the battery (Figure 5-3).

2. Condition of battery case: Check for dirt, grease, and electrolyte condensation. Any of these contaminants can create an electrical path between the terminals and cause the battery to drain. Also check for damaged or missing vent caps and cracks in the case. A cracked or buckled case could be caused by excessive tightening of the holddown fixture, excessive under-hood temperatures, buckled plates from extended undercharged conditions, freezing, or excessive charge rate.

Classroom Manual
Chapter 5, page 115

Special Tools
Fender covers
Safety glasses
Battery filler bottle

Figure 5-3 The battery sticker will usually have the date the battery was sold, plus additional information.

Figure 5-4 Use a battery filler bottle to top off the cells.

3. Electrolyte level, color, and odor: If necessary, add distilled water to fill to 1/2 inch (12 mm) above the top of the plates. Use a battery filler bottle to fill the battery (Figure 5-4). After adding water, charge the battery before any tests are performed. Discoloration of electrolyte and the presence of a "rotten egg" odor indicate an excessive charge rate, excessive deep cycling, impurities in the electrolyte solution, or an old battery.

4. Condition of battery cables and terminals: Check for corrosion, broken clamps, frayed cables, and loose terminals (Figure 5-5). These conditions result in voltage drop between the battery terminal and the end of the cable.

5. Battery abuse: This includes the use of bungee cords and 2 × 4s for holddown fixtures, too small of a battery rating for the application, and obvious neglect to periodic maintenance. In addition, inspect the terminals for indications that they have been hit by a hammer and for improper cable removal procedures. Finally, check for proper cable length.

6. Battery tray and holddown fixture: Check for proper tightness. Also check for signs of acid corrosion of the tray and holddown unit. Replace as needed.

7. If the battery has a built-in hydrometer, check its color indicator (Figure 5-6).

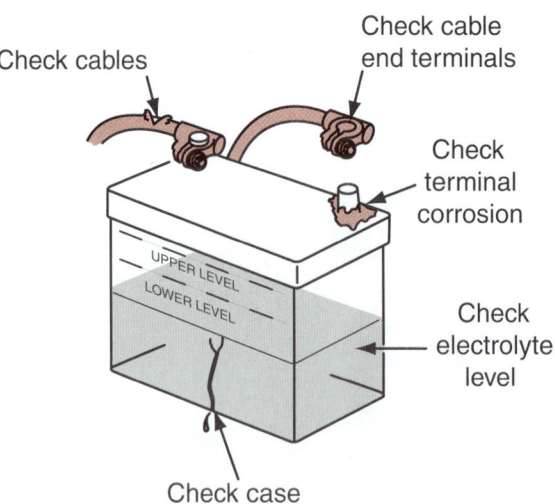

Figure 5-5 Inspect the condition of the battery cables and terminals.

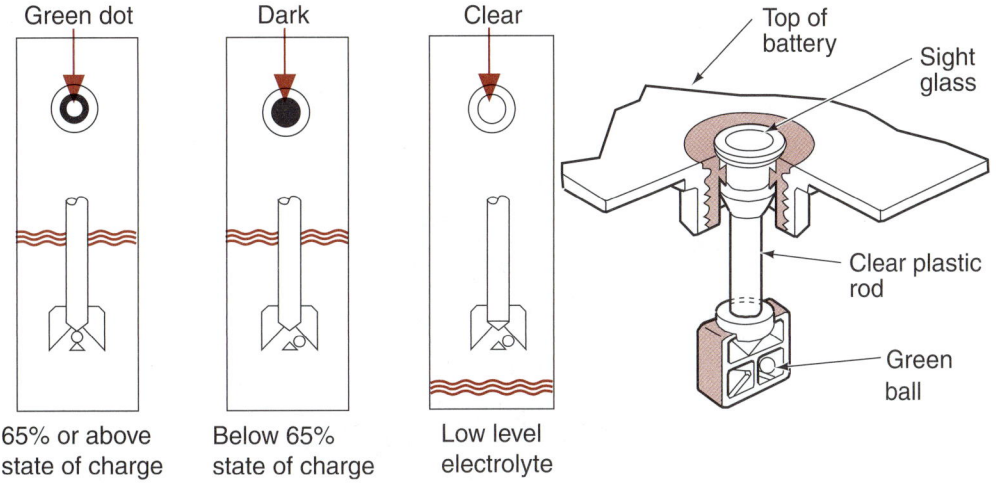

Green dot	Dark	Clear	
65% or above state of charge	Below 65% state of charge	Low level electrolyte	Top of battery / Sight glass / Clear plastic rod / Green ball

Figure 5-6 Built-in hydrometer used to indicate a battery's state of charge.

SERVICE TIP: Grid growth can cause the battery plate to short out the cell. If there is normal electrolyte level in all cells but one, that cell is probably shorted and the electrolyte has been converted to hydrogen gas.

Charging the Battery

WARNING: There are many safety precautions associated with charging the battery. The hydrogen gases produced by a charging battery are very explosive. Exploding batteries are responsible for over 15,000 injuries per year that are severe enough to require hospital treatment. Keep sparks, flames, and lighted cigarettes away from the battery. Also, do not use the battery to lay tools on. They may short across the terminals and result in the battery exploding. Always wear eye protection and proper clothing when working near the battery. Also, most jewelry is an excellent conductor of electricity. Do not wear any jewelry when performing work on or near the battery. Do not remove the vent caps while charging. Do not connect or disconnect the charger leads while the charger is turned on.

CAUTION: Before charging a battery that has been in cold weather, check the electrolyte for ice crystals. **Do not attempt to charge a frozen battery.** Forcing current through a frozen battery may cause it to explode. Allow it to warm at room temperature for a few hours before charging.

To **charge** the battery means to pass an electric current through the battery in an opposite direction than during discharge. If the battery needs to be recharged, the safest method is to remove the battery from the vehicle. The battery can be charged in the vehicle, however. If the battery is to be charged in the vehicle, it is important to protect any vehicle computers by removing the negative battery cable.

CAUTION: If the battery is to be removed from the vehicle, disconnect the negative battery cable first. Lift the battery out with a carrying tool (Figure 5-7).

When connecting the charger to the battery, make sure the charger is turned off. Connect the cable leads to the battery terminals, observing polarity. Attempting to charge the battery while the cables are reversed will result in battery damage. For this reason, many battery chargers have

Classroom Manual
Chapter 5, page 118

Special Tools
Safety glasses
Battery charger
Voltmeter
Fender covers

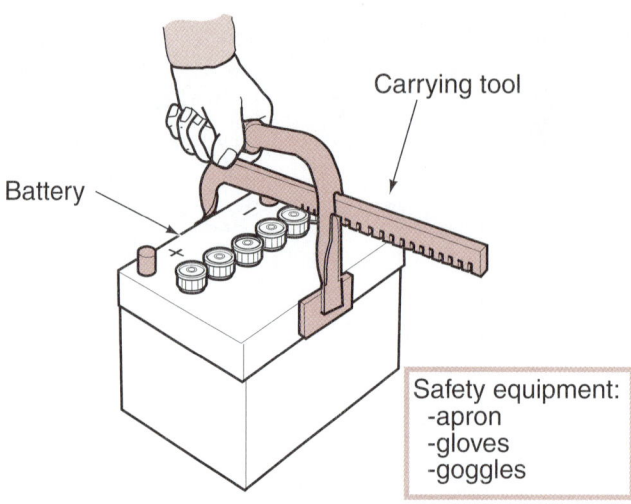

Figure 5-7 Always use a battery carrier to lift the battery.

a warning system to alert the technician that the cables are connected in reverse polarity. Rotate the clamps slightly on the terminals to assure a good connection.

Depending on the requirements and amount of time available, the battery can be either slow or fast charged. Each method of charging has its advantages and disadvantages.

Slow Charging

Slow charging means the charge rate is between 3 and 15 amperes for a long period of time. Slow charging the battery has two advantages: It is the only way to restore the battery to a fully charged state and it minimizes the chances of overcharging the battery. Slow charging the battery causes the lead sulfate on the plates to convert to lead peroxide and sponge lead throughout the thickness of the plate.

Slow charging is often referred to as "trickle charging."

Fast Charging

Fast charging uses a high current for a short period of time to boost the battery. Fast charging the battery will bring the state of charge up high enough to crank the engine. However, fast charging is unable to recharge the battery as effectively as slow charging. Fast charging the battery converts only the lead sulfate on the outside of the plates. The conversion does not go through the plates. After the battery has been fast charged to a point that it will crank the engine, it should then be slow charged to a full state.

CAUTION: Fast charging the battery requires that the battery be monitored at all times and the charging time must be controlled. Do not fast charge a battery for longer than 2 hours. Excessive fast charging can damage the battery. Do not allow the voltage of a 12-volt battery to exceed 15.5 volts. Also, don't allow the temperature to rise above 125°F.

Charge Rate

The **charge rate** is the speed at which the battery can safely be recharged at a set amperage. The charge rate required to recharge a battery depends on several factors:

1. Battery capacity. High-capacity batteries require longer charging time.
2. State of charge.

Open Circuit Voltage	Battery Specific Gravity	State of Charge	Charging Time to Full Charge at 80°F (267°C)					
			at 60 amps	at 50 amps	at 40 amps	at 30 amps	at 20 amps	at 10 amps
12.6	1.265	100%	Full Charge					
12.4	1.225	75%	15 min.	20 min.	27 min.	35 min.	48 min.	90 min.
12.2	1.190	50%	35 min.	45 min.	55 min.	75 min.	95 min.	180 min.
12.0	1.155	25%	50 min.	65 min.	85 min.	115 min.	145 min.	280 min.
11.8	1.120	0%	65 min.	85 min.	110 min.	150 min.	150 min.	370 min.

Figure 5-8 Table showing the rate and time of charging a battery. Electrolyte temperatures should not exceed 125°F (51.7°C) during charging.

3. Battery temperature.

4. Battery condition.

Slow chargering is the easiest on the battery. However, slow charging requires a long period of time. The basic rule of thumb for slow charging the battery is 1 ampere for each positive plate in one cell.

SERVICE TIP: If a battery is severely discharged and will not take a slow charge, connect a good battery in parallel (with jumper cables). Fast charge for 30 minutes, then disconnect the good battery and slow charge the discharged battery.

Slow charging of the battery may not always be practical due to the time involved. In these cases, fast charging is the only alternative. To determine the charging time for a full charge based on charge rate amperes, use the illustration (Figure 5-8).

An alternative method is to connect a voltmeter across the battery terminals while it is charging. If the voltmeter reads fewer than 15 volts, the charging rate is low enough. If the voltmeter reads over 15 volts, reduce the charging rate until voltage reads below 15 volts. Keeping the voltage at 15 volts will ensure the quickest charge and a safe rate for the battery.

CAUTION: To prevent damage to the AC generator and computers, disconnect the negative battery cable before fast charging the battery in the vehicle.

CAUTION: Fast charging at rates over 30 amperes for longer than 2 hours can result in permanent battery damage.

There are three methods of determining if the battery is fully charged:

1. Specific gravity holds at 1.264 or higher after the battery is stabilized.

2. An open circuit voltage test indicates 12.68 or higher after the battery has been stabilized.

3. The ammeter on the battery charger falls to approximately 3 amperes or less and remains at that level for 1 hour.

Recharging Gel Cell Batteries

Most gel cell batteries will accept being recharged very well. This is due, in part, to their low internal resistance. However, overcharging is very harmful to gel cell batteries. Since gel cell batteries use a special sealing design, overcharging will dry out the electrolyte by forcing the oxygen and hydrogen from the battery through the safety valves.

If a battery is continually undercharged, a layer of sulfate will build up on the positive plate. The sulfate then acts to resist the flow of electrons. This may also result in plate shedding, reducing performance and shorting battery life.

When recharging the gel cell battery, it is critical that the charger being used will properly limit the voltage to no more than 14.1 volts and no less than 13.8 volts at 68°F (20°C). This requires special charging equipment designed to recharge gel cell batteries. Older-type battery chargers use higher voltages and charging rates and usually cannot be used to recharge gel cell batteries. A gel cell battery that is charged at too high of a voltage may experience shorter battery life because of damage resulting from the battery temperature increasing to over 100°F (37.8°C).

Most modern battery chargers are compatible with gel cell–type batteries and have a switch to cycle between flooded and gel batteries. Be sure to check the documentation supplied with the charger you are using to recharge a gel cell battery.

Classroom Manual
Chapter 5, page 125

Special Tools

Safety glasses

Voltmeter

Terminal pliers

Terminal puller

Terminal and clamp cleaner

Fender covers

The battery must be fully charged to perform the battery terminal test.

Battery Test Series

When the battery and cables have been completely inspected and any problems have been corrected, the battery is ready to be tested further. For the tests to be accurate, the battery must be fully charged.

Battery Terminal Test

The **battery terminal test** checks for poor electrical connections between the battery cables and terminals. Use a voltmeter to measure voltage drop across the cables and terminals. It is good practice to perform the battery terminal test anytime the battery cable is disconnected and reconnected to the terminals. By performing this test, comebacks, due to loose or faulty connections, can be reduced.

Connect the negative voltmeter test lead to the cable clamp and connect the positive meter lead to the battery terminal (Figure 5-9). Disable the ignition system to prevent the vehicle from starting. This may be done by removing the ignition coil secondary wire from the distributor cap and putting it to ground (Figure 5-10). Many systems require the removal of the fuel pump or electronic fuel injection (EFI) relay in order to prevent the engine from starting.

CAUTION: Always refer to the manufacturer's service manual for the correct procedure for disabling the ignition system.

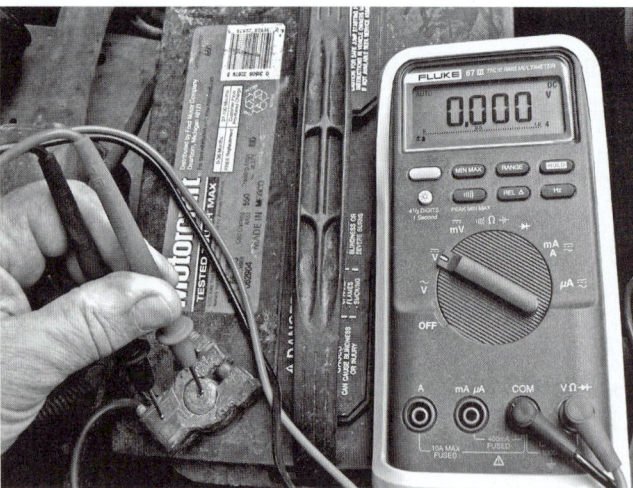

Figure 5-9 Voltmeter connections for the battery terminal test.

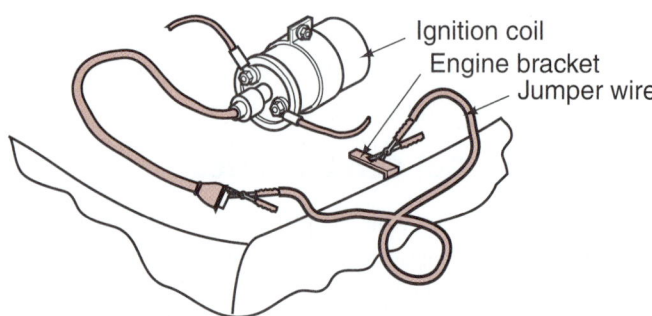

Figure 5-10 Grounding the coil's secondary cable to prevent the engine from starting and to protect the coil.

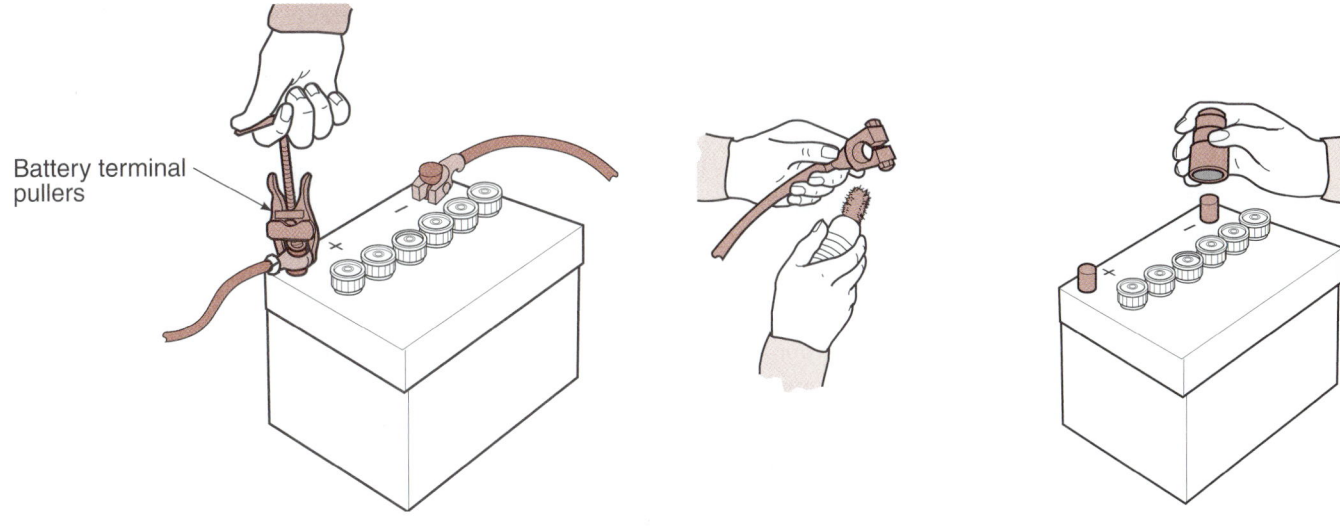

Battery terminal pullers

Figure 5-11 Use battery clamp pullers to remove the cable end from the terminal. Do not pry the clamp off.

Figure 5-12 A terminal cleaning tool is used to clean the battery's terminals and the cables' clamps.

Crank the engine and observe the voltmeter reading. If the voltmeter shows over 0.3 volt, there is a high resistance at the cable connection. Remove the battery cable using the clamp puller (Figure 5-11). Clean the cable ends and battery terminals (Figure 5-12).

Battery Leakage Test

Battery drain can be caused by a dirty battery. The dirt can actually allow current flow over the battery case. This current flow can drain a battery as quickly as leaving a light on. A **battery leakage test** is conducted to see if current is flowing across the battery case. To perform a battery leakage test, set a voltmeter to a low DC volt scale. Connect the negative test lead to the negative terminal of the battery. Move the red test lead across the top and sides of the battery case (Figure 5-13). If the meter reads voltage, a current path from the negative terminal of the battery to its positive terminal is being completed through the dirt. Keep in mind that you should not measure voltage anywhere on the case of the battery. If voltage is present, remove the battery. Then use a baking

Classroom Manual
Chapter 5, page 131

Special Tools

Voltmeter
Fender covers
Safety glasses

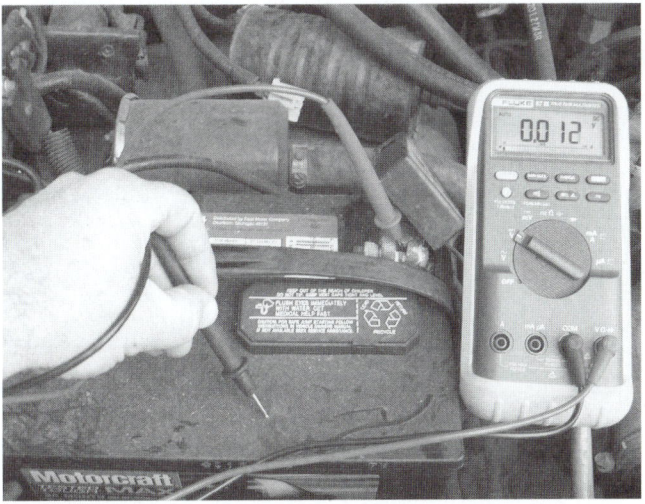

Figure 5-13 Using a voltmeter to perform the battery leakage test.

soda and water mixture to clean the case of the battery. When cleaning the battery, don't allow the baking soda and water solution to enter its cells. After the case is clean, rinse it off with clean water.

Classroom Manual
Chapter 5, page 118

Special Tools

Hydrometer

State of Charge Test

Measuring the **state of charge** is a check of the battery's electrolyte and plates. The state of charge test uses the measurement of the specific gravity to determine the charge of the battery.

To use the hydrometer to test the battery's state of charge:

1. Remove all battery vent caps.

2. Check the electrolyte level. It must be high enough to withdraw the correct amount of solution into the **hydrometer.** A hydrometer measures the specific gravity of a liquid (Figure 5-14).

3. Squeeze the bulb and place the pickup tube into the electrolyte of a cell (Figure 5-15).

4. Slowly release the bulb. Draw in enough solution until the float is freely suspended in the barrel. Hold the hydrometer in a vertical position.

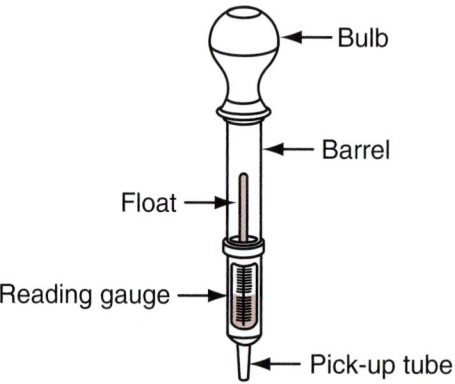

Figure 5-14 A temperature correction hydrometer is used to measure the specific gravity of the electrolyte solution, providing an indication of the battery's state of charge.

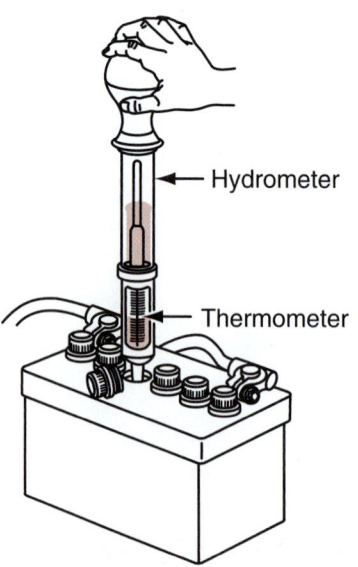

Figure 5-15 Drawing electrolyte out of the battery's cell and into the hydrometer.

The float rises and the specific gravity is read where the float scale intersects the top of the solution (Figure 5-16). The reading must also be corrected by compensating for temperatures (Figure 5-17).

If the electrolyte is too low to perform the test, add distilled water to the cell. Do not take hydrometer readings until the water and electrolyte have been mixed by charging the battery.

Test Results. As a battery becomes discharged, its electrolyte has a larger percentage of water. Thus, a discharged battery's electrolyte will have a lower specific gravity number than that of a fully charged battery.

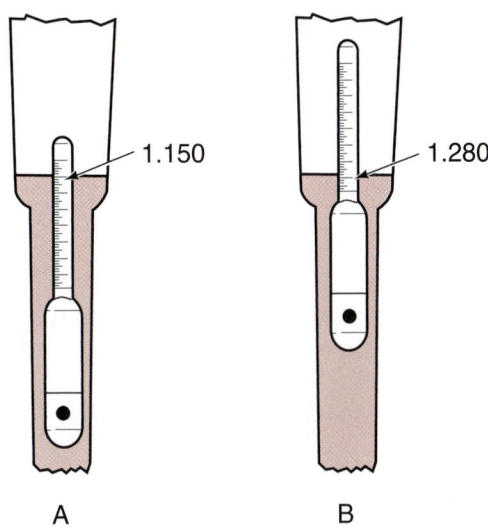

A   B

Figure 5-16 The specific gravity of the electrolyte is read at the point where the electrolyte intersects the float. (A) shows a low reading. (B) shows a high reading.

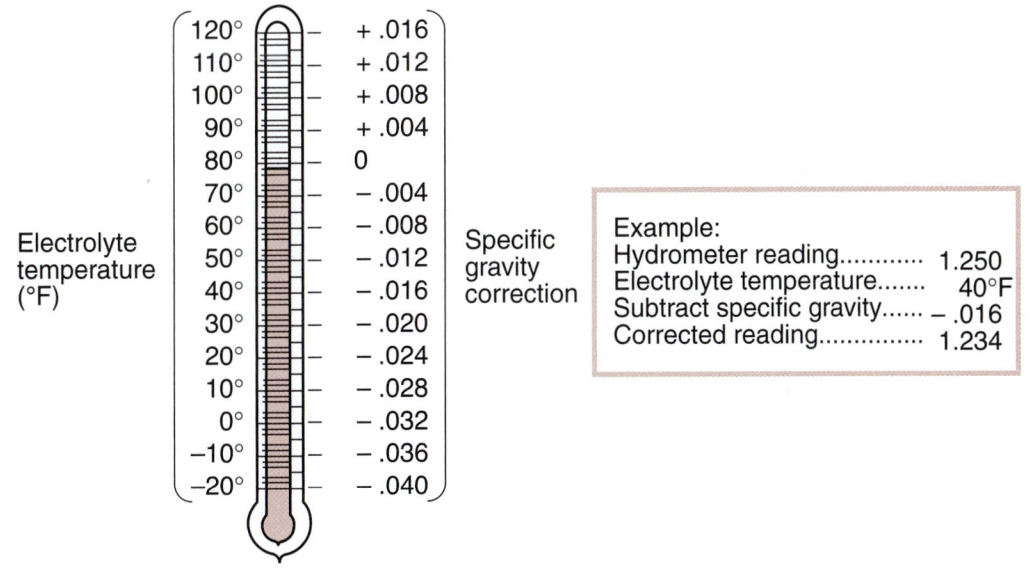

Figure 5-17 Correct the specific gravity reading according to the temperature of the electrolyte.

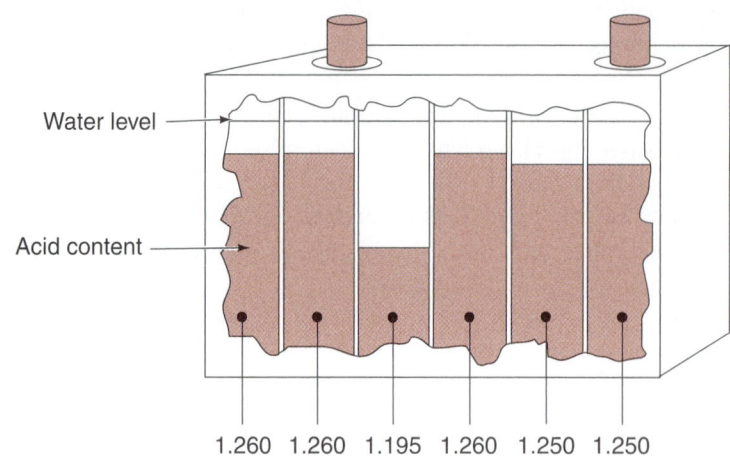

Figure 5-18 A defective cell can be determined by the specific gravity readings.

1.260 1.260 1.195 1.260 1.250 1.250

A fully charged battery will have a hydrometer reading near 1.265. Remember, the specific gravity is also influenced by the temperature of the electrolyte and the readings must be corrected to the temperature. If the corrected hydrometer reading is below 1.265, the battery needs recharging or it may be defective.

A defective battery can be determined with a hydrometer by checking every cell. If the specific gravity has a 0.050-point variation between the highest and lowest cell readings, the battery is defective (Figure 5-18). When all the cells have an equal gravity, even if all are low, the battery can usually be regenerated by recharging.

Specific gravity tests should not be used as the sole determinant of battery condition. If the cells of the battery do not have the same specific gravity, the battery should be replaced. When the specific gravity of all the cells is good or bad, the voltage of the battery must be considered before coming to a conclusion about the battery's condition. A battery with low specific gravity and acceptable voltage is normally only discharged, perhaps due to a charging system problem. However, a battery with good specific gravity readings but low voltage readings is always bad and needs to be replaced.

Optical Refractometer

A **refractometer** uses the refractions of light to determine and display a very accurate specific gravity reading. All that is required is a couple of drops of electrolyte to be placed on the glass slide of the refractometer (Figure 5-19). While you hold the lens of the meter up to your eye, the light will refract through the sample and display the specific gravity in the window (Figure 5-20).

Open Circuit Voltage Test

The **open circuit voltage test** is used to determine the battery's state of charge. It is used when a hydrometer is not available or cannot be used. To obtain accurate open circuit voltage test results, the battery must be stabilized. To **stabilize** the battery means the surface charge is removed by placing a large load on the battery for 15 seconds. If the battery has just been recharged, perform the capacity test, then wait at least 10 minutes to allow battery voltage to stabilize. Connect a voltmeter across the battery terminals, observing polarity (Figure 5-21). Measure the open circuit voltage. Take the reading to the 1/10 volt.

Figure 5-19 Placing a sample of electrolyte onto the refractometer's slide.

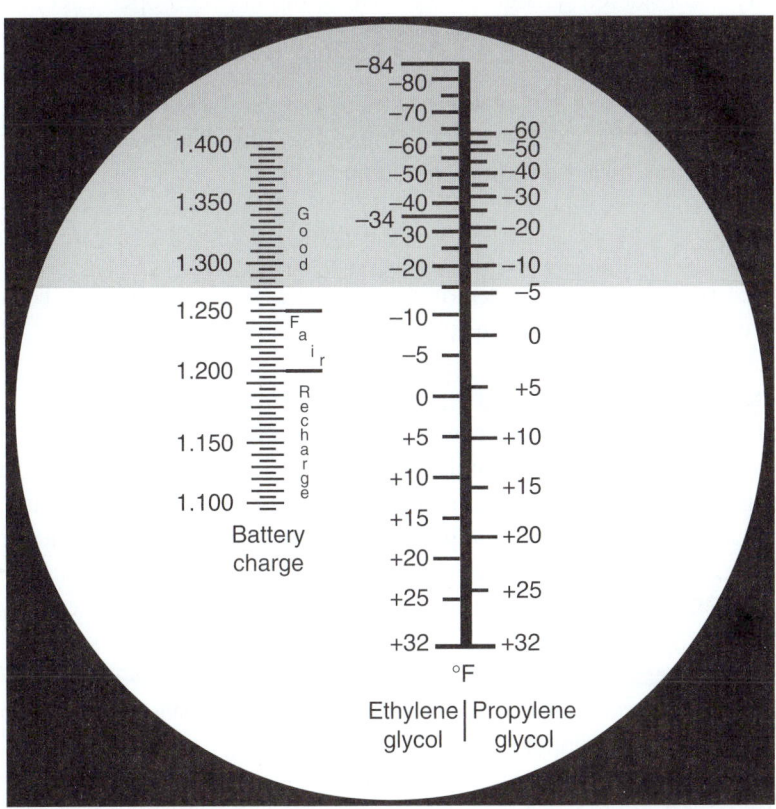

Figure 5-20 The window will indicate the specific gravity.

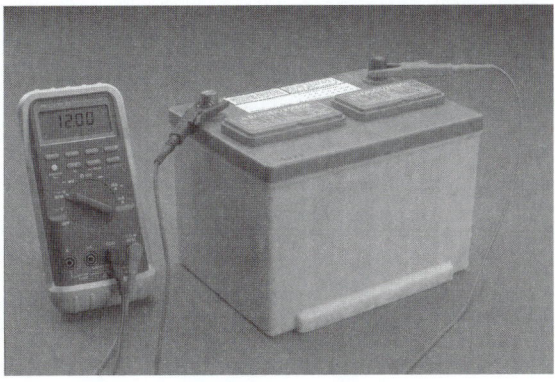

Figure 5-21 Open circuit voltage test using a voltmeter.

Open Circuit Voltage Table	
Open Circuit Voltage	Charge Percentage
11.7 volts or less	0%
12.0 volts	25%
12.2 volts	50%
12.4 volts	75%
12.6 volts or more	100%

Figure 5-22 Open circuit voltage test results relate to the specific gravity of the battery's cells.

To analyze the open circuit voltage test results, consider that a battery at a temperature of 80°F (26.7°C), in good condition, should show about 12.4 volts. If the state of charge is 75% or more, the battery is considered "charged." The relationship between open circuit voltage test results and charge percentage is illustrated (Figure 5-22).

> **SERVICE TIP:** If the vehicle has many circuits that place a constant drain on the battery (computer, clock, memory radios, etc.), the negative battery cable should be disconnected before taking a voltmeter reading.

Capacity Test

Classroom Manual
Chapter 5, page 126

Special Tools

VAT-60

Electrolyte thermometer

Safety glasses

Fender covers

The capacity test is also called the "load test."

The **capacity test** provides a realistic determination of the battery's condition. For this test to be accurate, the battery must pass the state of charge or open circuit voltage test. If it does not, recharge the battery and test it again.

In the capacity test, a specified load is placed on the battery while the terminal voltage is observed. A good battery should produce current equal to 50% of its cold-cranking rating (or three times its ampere-hour rating) for 15 seconds and still provide 9.6 volts to start the engine.

Depending on the equipment used, certain steps need to be followed. If the tester uses a carbon pile to load the battery, follow the general steps outlined in Photo Sequence 6.

> **CAUTION:** While performing the load test, do not load the battery longer than 15 seconds.

Once the load test is completed and you have recorded the voltage value, the readings need to be corrected to electrolyte temperature (Figure 5-23).

If voltage level is below the specifications listed in Figure 5-23, observe the battery voltage for the next 10 minutes. If the voltage raises to 12.45 volts or higher, the battery must be replaced. This means that the battery can hold a charge but has insufficient cold-cranking amperes.

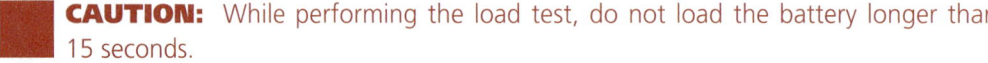

Electrolyte Temperature								
°F	70+	60	50	40	30	20	10	0
°C	21+	16	10	4	−1	−7	−12	−18
Minimum Voltage (12-Volt Battery)	9.6	9.5	9.4	9.3	9.1	8.9	8.7	8.5

Figure 5-23 Correcting the readings of the capacity test to temperature readings.

Photo Sequence 6
Performing a Battery Capacity Test

P6-1 Charge the battery to at least 75% and allow the battery to stabilize.

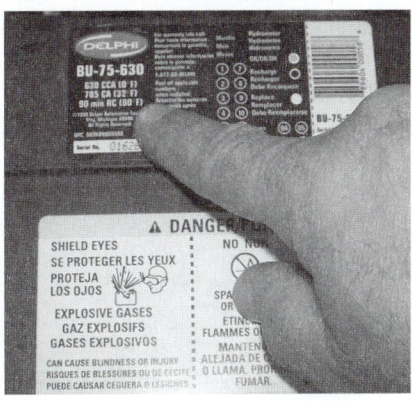

P6-2 Determine the rating on the battery's label and figure the load test specification.

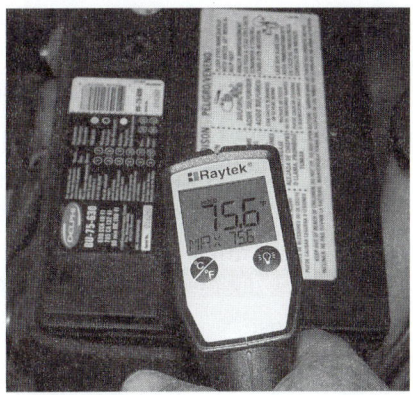

P6-3 Determine the temperature of the electrolyte. If the case is sealed, measure the temperature of the case.

P6-4 Connect the large load leads across the battery terminals, observing polarity.

P6-5 Confirm that the meter reads zero amps. If needed, zero the ammeter.

P6-6 Connect the ammeter's inductive pickup around the negative load tester lead (not the battery's negative cable).

P6-7 Turn the load control knob slowly until the ammeter indicates the amperage amount determined in step 2.

P6-8 Read and record the voltmeter while applying the load for 15 seconds.

If the voltage does not return to 12.4 volts, recharge the battery until the open circuit test indicates a voltage of 12.66 volts. Repeat the capacity test. If the battery fails again, replace the battery.

If the capacity test readings of a clean and fully charged battery are equal to or above specification, the battery is good. If the battery tests are borderline, perform the 3-minute charge test.

Three-minute Charge Test

If a conventional battery fails the load test, it is not always the fault of the battery. It is possible that the battery has not been receiving an adequate charge from the charging system. The **3-minute charge test** determines the battery's ability to accept a charge, and for **sulfation.** Sulfation is a chemical action within the battery that interferes with the ability of the cells to deliver current and accept a charge (Figure 5-24). A battery must have failed the load test to get accurate results from a 3-minute charge test.

Classroom Manual
Chapter 5, page 119

Special Tools

Battery charger
Voltmeter
Safety glasses
Fender covers

To conduct the 3-minute test:

1. Remove the ground cable. The battery must be disconnected from the vehicle's electrical system since the high voltage that is possible during this test can damage the computers.
2. Connect a battery charger to the battery, observing polarity.
3. Connect a voltmeter across the battery terminals, observing polarity.
4. Turn on the battery charger to 40 amperes (20–25 for maintenance-free batteries).
5. Maintain this rate of charge for 3 minutes.
6. Check the voltage reading at 3 minutes. If fewer than 15.5 volts, the battery is not sulfated. If the voltmeter reading is above 15.5 volts, the battery is sulfated or there is a poor internal connection.
7. If the battery passes the 3-minute test, slowly recharge the battery and do the load test again.
8. If the battery passes the load test this time, test the charging system.

⚠️ **WARNING:** Some battery manufacturers, such as Delco, do NOT recommend the 3-minute charge test.

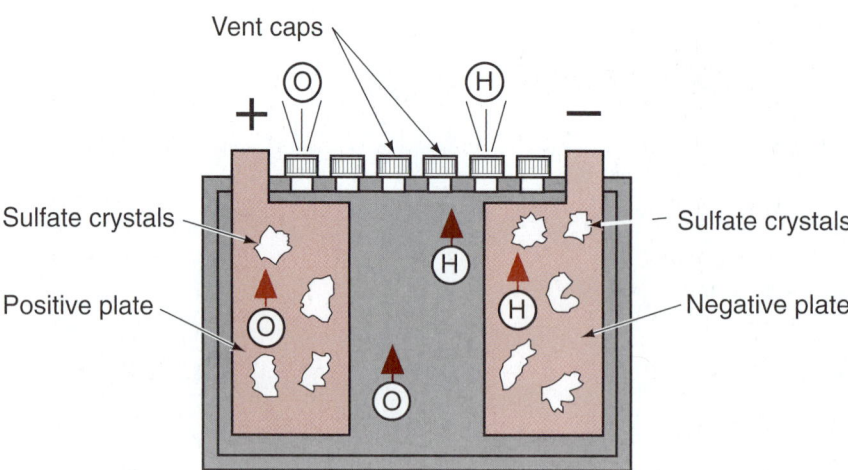

Figure 5-24 A sulfated battery is the result of sulfate crystals that penetrate the plates. The crystals become insoluble and will not allow the battery cell to deliver current nor accept a charge.

CUSTOMER CARE: One of the best things you can do for your customers is to assist them in choosing the correct battery. Battery selection needs to be based on the make of the vehicle, electrical options on the vehicle, driving habits, and climatic conditions. The largest current capacity rating that can be achieved in a given battery group may benefit some customers but may be a waste of money for others.

Computerized Load-Testing Equipment

Recently, automotive equipment manufactures have developed battery load testers that use computer technology. With this type of equipment, the technician types in the CCA rating of the battery and the tester automatically performs the test. The results on the test are printed out or displayed for the technician. Some systems will plot the voltages on a graph as the test is performed. The tester compares the results from the load test with its data for the CCA rating and makes a determination for the proper recommendation to the technician.

Battery Conductance Testing

Conductance is a measurement of the battery's plate surface that is available for chemical reaction. This determines how much power the battery can supply and describes the ability of a battery to conduct current. Conductance testing of the battery has proven to be a reliable test of the battery's capacity. The higher the conductance value (or lower internal resistance), the better the performance potential of the battery.

Conductance testing is performed by sending a low-frequency AC signal through the battery. A portion of the AC current's returned pulse is then captured and a conductance measurement is calculated. Conductance battery testers have the ability to accurately test batteries that are not fully charged. Battery internal damage is also detected without having to charge the battery. An on-screen display directs the technician through the steps required to perform the battery test (Figure 5-25).

Figure 5-25 Conductance tester.

Special Tools

Test iight

VAT-60 or equivalent

Multiplying coil

Terminal pliers

Terminal puller

Cable clamp
 spreader

Safety glasses

Fender covers

The open circuit
voltage reading must
be 11.5 volts or
higher to perform
the battery drain
test.

If the vehicle is
equipped with
computer-controlled
air suspension
systems, it may be
necessary to
disconnect the
module to eliminate
it from the test.

Battery Drain Test

If a customer complains that the battery is dead every time he attempts to start the vehicle after it has not been used for a short while, the problem may be a current drain from one of the electrical systems. The most common cause for this type of drain is a light that is not turning off—such as glove box, trunk, or engine compartment illumination lights.

These **parasitic drains** on the battery can cause various driveability problems. With low-battery voltage several problems can result; for example:

1. The computer may go into backup mode or "limp-in" mode of operation.
2. The computer may set false trouble codes.
3. To compensate for the low-battery voltage, the computer may raise the engine speed.

The procedure for performing the battery drain test may vary according to the manufacturer. However, battery drain can often be observed by connecting an ammeter in series with the negative battery cable or by placing the inductive ammeter pickup lead around the negative cable. If the meter reads 30 milliamps or more, there is excessive drain. Visually check the trunk, glove box, and under-hood lights to see if they are on. If they are, remove the bulb and watch the battery drain. If the drain is now within specifications, find out why the circuit is staying on and repair the problem. If the cause of the drain is not the lights, go to the fuse panel or distribution center and remove one fuse at a time while watching the ammeter. When the drain decreases, the circuit protected by the fuse you removed last is the source of the problem.

The following is a typical procedure for determining and locating parasitic drains against the battery. First, it is necessary to determine if a draw is occurring. This can be done by connecting an inductive ammeter clamp around the negative battery cable. The meter must be capable of reading less than 1 amp accurately. With all accessories turned off, the ammeter reading indicates the current draw against the battery. It is normal for some vehicles to have higher than 30 milliamps draw for up to an hour after the ignition switch is turned off. This allows the computers to perform their "administrative" tasks. Be sure to confirm the normal time-out period with the proper service information. If the ammeter reads higher than allowed current, the cause of the drain must be determined.

If an inductive clamp is not used, then the ammeter needs to be connected in series with the negative battery cable and the battery terminal. In this case it may be advisable to use a test light in series with the negative battery cable and the battery terminal first (Figure 5-26). This is done to prevent the vehicle's computers from powering down and then powering back up again when

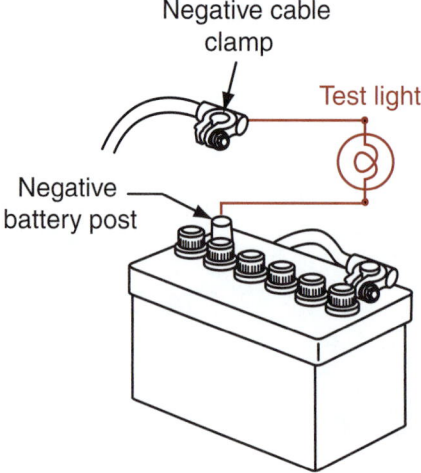

Negative cable
clamp

Test light

Negative
battery post

Figure 5-26 Using a test light to prevent the vehicle's computers from powering down.

the ammeter is connected. The additional amperage of the computers powering back up may blow the fuse in the ammeter. Prior to disconnecting the clamp from the terminal, connect the test lamp. As the clamp is removed, maintain connection of the lamp with the clamp and terminal. If the test light is on, there is a drain against the battery that is sufficient to light the lamp. After the time period has expired, the test light should go out. If not, then connect the ammeter leads in series between the cable and terminal (with the test light still connected). Once the ammeter is connected, remove the test light and read the amount of current draw.

An alternate method of isolating the circuit that has the excessive drain is to measure the voltage drop across the fuse. Since a fuse may protect several circuits, it is possible to mask the problem when a fuse is pulled. For example, if a fuse protects the power supply circuits to three different computers, then pulling that fuse will cause them all to totally turn off. Usually a computer does not turn off completely when it powers down, but the draw is very slight. If one of the three computers was failing to power down, then excessive drain would be against the battery. If the fuse is pulled and all three computers power down, it is possible that the defect computer will reset and when the fuse is plugged in again there no longer will be a parasitic load. Since pulling the fuse caused the load reading on the ammeter to drop, the circuit is identified. However, once the fuse was plugged back in, the load is no longer indicated and isolating the cause is made more difficult. It would be advantageous to identify the circuit first, then unplug the computers one at a time until the draw is identified.

To perform this method of circuit identification, use the test tabs on the top of the fuse (maxifuses will require removing the lens). Connect a voltmeter set on the millivolt scale across the test points and observe the meter's reading (Figure 5-27). A fuse has resistance so that it can get hot and burn through if excessive current passes through it; this resistance will provide a voltage drop. If there is no (or very little) current flowing through the circuit, then the voltmeter will read 0 volts.

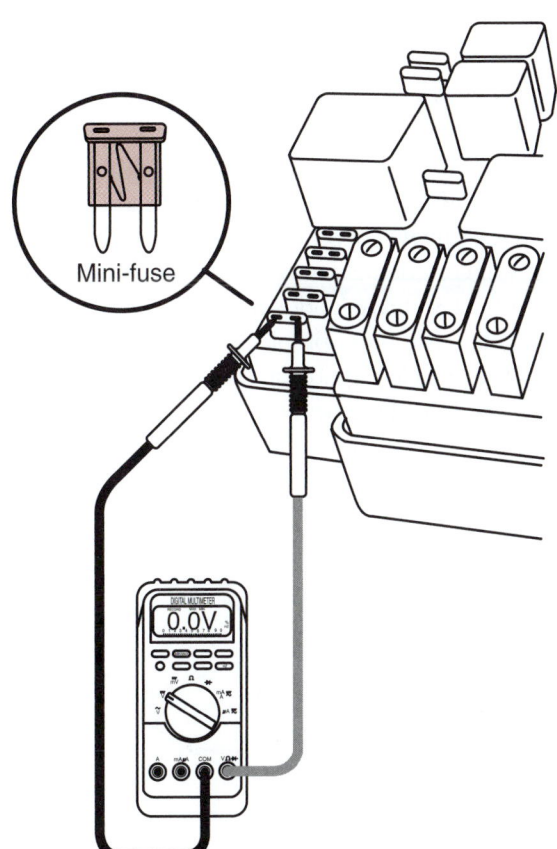

Figure 5-27 Using a voltmeter across a fuse to determine current draw.

Fuse Value	Fuse Type	Divide by
5	Mini	16.5
10	Mini	7.5
15	Mini	4.5
20	Mini	3.5
25	Mini	2.5
30	Mini	2.0
20	Cartridge	1.0
30	Cartridge	1.5
40	Cartridge	1.0
50	Cartridge	0.5

Figure 5-28 Determining how much current is flowing through a fuse.

However, if current is flowing through the circuit, then the voltmeter will provide a reading. The reading will indicate how much current is flowing in the circuit, which is a function of the rating of the fuse (Figure 5-28). For example, if the voltmeter reads 20 millivolts when connected across a 15-amp fuse, then the current draw is $20 \div 4.5 = 4.4$ amps. It is not important to determine the amount of current flow. You are not looking for 1 or 2 milliamps but more than 30 milliamps; therefore, any reading displayed will indicate excessive current flow.

If the ammeter indicates a constant current drain, check all of the under-hood, glove box, trunk, and courtesy lights to confirm they are shutting off properly. If the cause of the drain is not located, remove the fuses one at a time until the amperage drops to the acceptable level. When this happens, the fault lies in that fuse's circuit and it must be traced.

If the faulty circuit is not located by pulling the fuses, disconnect the leads at the starter relay one at a time until the amperage drops. Trace the faulty circuit.

Battery Removal and Cleaning

It is natural for dirt and grease to collect on the top of the battery. If allowed to accumulate, the dirt and grease can form a conductive path between the battery terminals. This may result in a drain on the battery. Also, normal battery gassing will deposit sulfuric acid as the vapors condense. Over a period of time, the sulfuric acid will corrode the battery terminals, cable clamps, and holddown fixtures. As the corrosion builds on the terminals, it adds resistance to the entire electrical system.

Periodic battery cleaning will eliminate these problems. To be able to clean the battery correctly, it is best to remove it from the vehicle. Removing the battery protects the vehicle's finish and other under-hood components. Follow Photo Sequence 7 for the procedure for removing the battery from the vehicle and cleaning it. Consult the manufacturer's service manual for precautions concerning the vehicle's computer controls.

Classroom Manual
Chapter 5, page 130

Special Tools

Baking soda

Cleaning brushes

Terminal pliers
 (Figure 5-29)

Cable clamp
 spreader

Terminal puller

Terminal and clamp
 cleaning tool

Battery-lifting strap

Putty knife

Protective terminal
 coating

Safety glasses

Heavy rubber gloves

Fender covers

Figure 5-29 Battery terminal pliers.

Photo Sequence 7
Removing and Cleaning the Battery

P7-1 Tools needed to remove the battery from the vehicle: rags, baking soda, pan, terminal pliers, cable clamp spreader, terminal puller, assorted wrenches, terminal and clamp cleaning tool, battery-lifting strap, safety glasses, heavy rubber gloves, rubber apron, and fender covers.

P7-2 Place the fender covers on the vehicle to protect the finish.

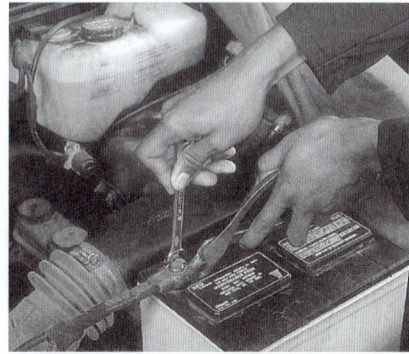

P7-3 Loosen the clamp bolt for the negative cable using terminal pliers and wrench of correct size. Be careful not to put excessive force against the terminal.

P7-4 Use the terminal puller to remove the cable from the terminal. Do not pry the cable off of the terminal.

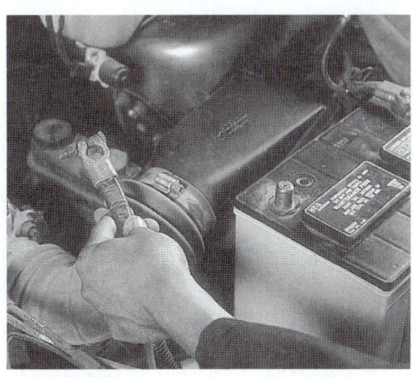

P7-5 Locate the negative cable away from the battery.

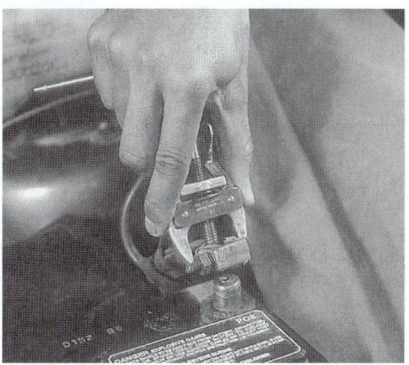

P7-6 Loosen the clamp bolt for the positive cable and use the terminal puller to remove the cable. If the battery has a heat shield, remove it.

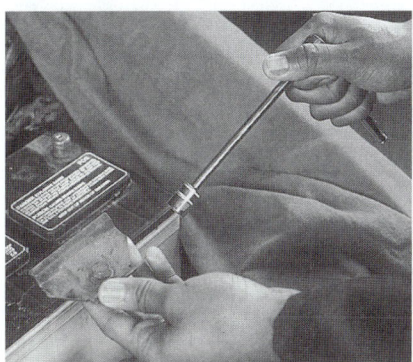

P7-7 Disconnect the holddown fixture.

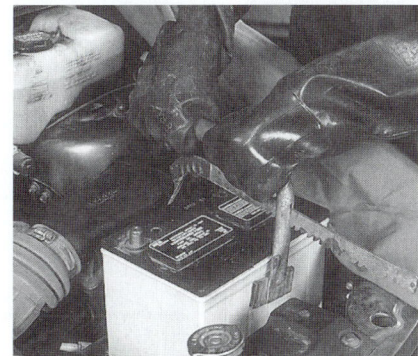

P7-8 Using the battery-lifting clamp, remove the battery out of the tray. Keep the battery away from your body. Wear protective clothing to prevent acid spills on your hands.

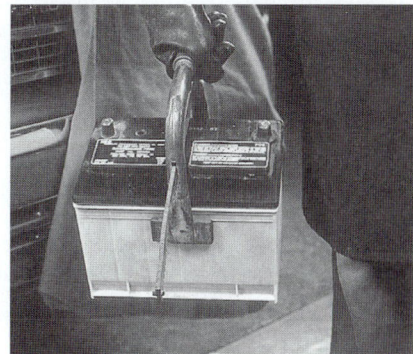

P7-9 Transport the battery to the bench. Keep it away from your clothes.

Removing and Cleaning the Battery (continued)

P7-10 Mix a solution of baking soda and water.

P7-11 Brush the solution over the battery case. Be careful not to allow the solution to enter the cells of the battery.

P7-12 Flush the solution off with water.

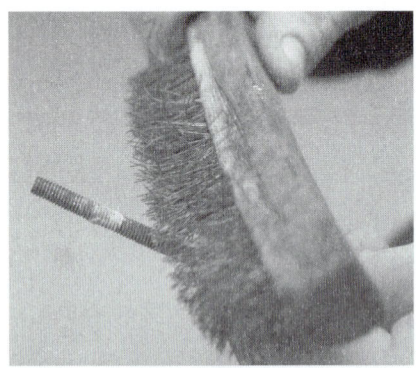

P7-13 Use a wire brush to remove corrosion from the holddown brackets.

P7-14 Use the terminal cleaning tool to clean the cable.

P7-15 Use the cleaning tool to clean the battery posts.

P7-16 Install the battery into the tray and install the holddown hardware. The cables are installed positive first, followed by the negative cable.

P7-17 Coat the battery terminals with corrosion preventive spray. Perform a terminal test to assure tight connections.

CAUTION: Do not allow the baking soda solution to enter the battery cells. This will neutralize the acid in the electrolyte and destroy a battery.

CUSTOMER CARE: Before installing the battery back into the vehicle, it is a good practice to clean the battery tray. First scape away any heavy corrosion with a putty knife. Next, clean the tray with a baking soda and water solution (Figure 5-30). Flush with water and allow to dry. After the tray has dried, paint it with rust-resistant paint. After the paint has completely dried, coat the tray with silicone base spray. These extra steps will protect the tray from corrosion.

When replacing the battery into the vehicle, make sure it is properly seated in the tray. Connect the holddown fixture and secure. Do not overtighten the holddown fixture. Install the positive cable and secure, then install the negative cable. Be sure to observe polarity. Perform a battery terminal test to confirm good connections.

WARNING: Be careful not to touch the positive terminal with the wrench when tightening the negative cable clamp.

Spray the cable clamps with a protective coating to prevent corrosion. A little grease or petroleum jelly will prevent corrosion as well. Also, protective pads are available that go under the clamp and around the terminal (Figure 5-31).

The baking soda and water solution consists of 1 tablespoon per quart of water.

Make sure the cell caps are securely in place before cleaning the battery.

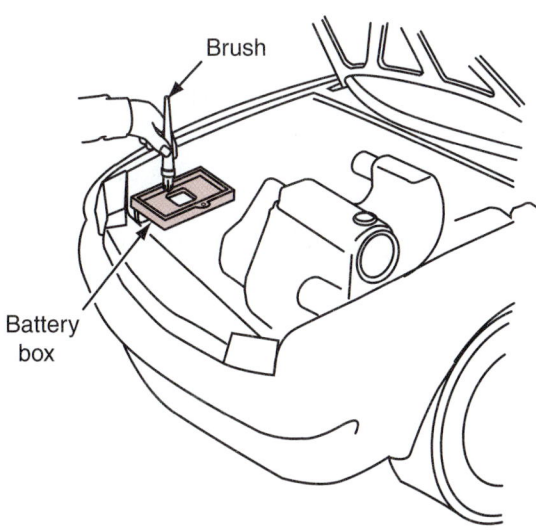

Figure 5-30 After scraping the battery tray clean with a putty knife, use the baking soda mixture to remove all electrolyte residue.

Figure 5-31 A protective pad under the battery clamp prevents corrosion of the clamp and terminal.

Maintaining or Restoring Memory Functions

Today's vehicles are equipped with several computer control modules that use memories to store values and presets. These include radio station selection, memory seat positions, transmission shift schedule learning, adaptive fuel strategies, and so on. These memories require battery voltage to be maintained and will reset if the battery is disconnected. If the battery is disconnected to perform battery tests, battery service, or to be charged, then a memory keeper discussed in Chapter 2 of this text can be used to maintain the memories. However, if the battery is to be disconnected to perform electrical circuit tests or repairs, then the memory keeper should not be used and the memory function should be restored.

Restoring the memory presets for memory seats and radio station recalls, as well as other preset functions, can be done by first using a scanner to determine and record the preset positions. After the battery is connected, reset the presets to the recorded values.

Jumping the Battery

Special Tools

Booster battery

Jumper cables

Fender covers

Safety glasses

WARNING: Before charging a battery that has been in cold weather, check the electrolyte for ice crystals. Do not attempt to charge a frozen battery. Forcing current through a frozen battery may cause it to explode. Allow it to warm at room temperature for a few hours before charging.

There will be times when you will have to use a boost battery and jumper cables to jump start a vehicle (Figure 5-32). It is important that all safety precautions be followed. Jump-starting a dead battery can be dangerous if it is not done correctly. The following steps should be followed to safely jump-start most vehicles:

1. Make sure the two vehicles are not touching each other. The excessive current flow through the vehicles' bodies can damage the small ground straps that attach the engine block to the frame. These small wires are designed to carry only 30 amperes. If the vehicles are touching, as much as 400 amperes may be carried through them.
2. For each vehicle, engage the parking brake and put the transmission in NEUTRAL or PARK.

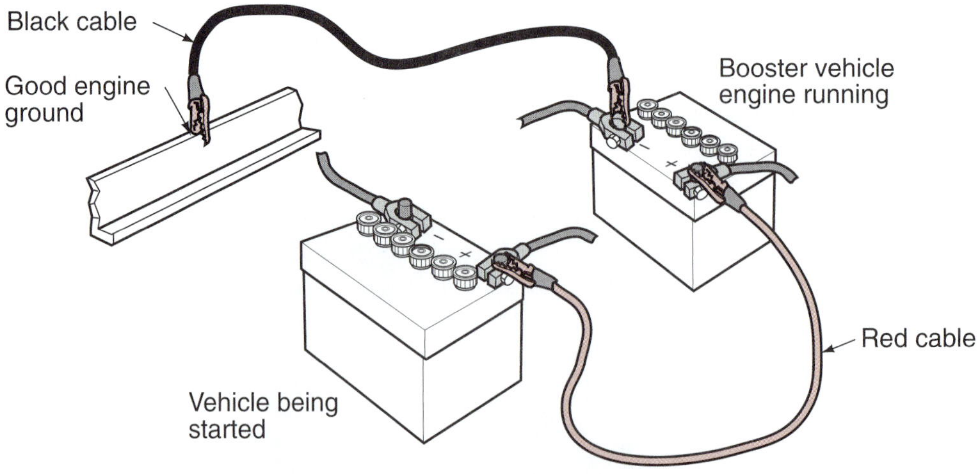

Figure 5-32 Proper jumper cable connections for jump-starting a vehicle.

3. Turn off the ignition switch and all accessories, on both vehicles.

4. Attach one end of the positive jumper cable to the disabled battery's positive terminal.

5. Connect the other end of the positive jumper cable to the booster battery's positive terminal.

6. Attach one end of the negative jumper cable to the booster battery's negative terminal.

7. Attach the other end of the negative jumper cable to an engine ground on the disabled vehicle.

WARNING: Do not connect this cable end to the battery negative terminal. Doing so may create a spark that will cause the battery to explode.

8. Attempt to start the disabled vehicle. If the disabled vehicle does not start readily, start the jumper vehicle and run at fast idle to prevent excessive current draw.

9. Once the disabled vehicle starts, disconnect the ground-connected negative jumper cable from its engine block.

10. Disconnect the negative jumper cable from the booster battery.

11. Disconnect the positive jumper cable from the booster battery, then from the other battery.

CAUTION: Do not use more than 16 volts to jump-start a vehicle that is equipped with an engine control module. The excess voltage may damage the electronic components.

WARNING: A battery that has been rapidly discharged will create hydrogen gas. Do not attach jumper cables to a weak battery if starting the vehicle has been attempted. Wait for at least 10 minutes before connecting the jumper cable and attempting to start the vehicle.

CASE STUDY

A customer complains that the vehicle will not start without having to jump the battery. The technician learns that this happens every time the customer attempts to start the vehicle. The customer also says the voltmeter in the dash has been reading higher than normal.

The technician verifies the complaint. The engine turns over very slowly for a few seconds then does not turn. After jumping the battery to get the vehicle into the shop, the technician makes a visual inspection of the battery and cables. The open circuit voltage test shows a voltage of 12.5 volts across the terminals. When the battery is subjected to the capacity test, the voltage drops to 7.8 volts at 80 degrees. After 10 minutes, the open circuit voltage is back up to 12.5 volts. The technician determines that the battery is sulfated. The battery can't handle the load of cranking the engine, which is why it always needs to be jumped. The higher-than-normal voltmeter readings indicate the charging system is trying to keep the battery charged. The technician calls the customer with a price quote. The customer agrees to have the battery replaced. While replacing the battery, the technician cleans the battery tray, the cable clamps, and sprays the clamps with a corrosion protector.

Terms to Know

Battery leakage test	Fast charging	Slow charging
Battery terminal test	Hydrometer	Stabilize
Capacity test	Open circuit voltage test	State of charge
Charge	Parasitic drains	Sulfation
Charge rate	Refractometer	3-minute charge test

ASE-Style Review Questions

1. Battery terminal connections are being discussed. *Technician A* says when disconnecting battery cables, always disconnect the negative cable first. *Technician B* says when connecting battery cables, always connect the negative cable first. Who is correct?
   - **A.** A only
   - **B.** B only
   - **C.** Both A and B
   - **D.** Neither A nor B

2. A customer's battery is always dead when she attempts to start her car in the morning. After jumping the battery one time in the morning, the car will start throughout the day with no problems. All of the following can be the cause EXCEPT:
   - **A.** The starter motor is drawing too much current.
   - **B.** The glove box light is staying on.
   - **C.** The computer is not powering down.
   - **D.** A relay contact is stuck closed.

3. The specific gravity of a battery has been tested. All cells have a corrected reading of about 1.200. *Technician A* says the battery needs to be recharged before further testing. *Technician B* says the battery is sulfated and needs to be replaced. Who is correct?
   - **A.** A only
   - **B.** B only
   - **C.** Both A and B
   - **D.** Neither A nor B

4. When charging a battery:
   - **A.** Connect a voltmeter across the battery terminals while the battery is charging and keep the charge rate at fewer than 10 volts.
   - **B.** Disconnect the negative battery cable before charging to prevent damage to the alternator and computers.
   - **C.** Charge at an amperage rate of 50% of the CCA rating.
   - **D.** All of the above.

5. Which statement concerning the battery leakage test is correct?
   - **A.** The test is used to determine if the battery can provide current and voltage when loaded.
   - **B.** A voltmeter reading of 0.05 when performing the test is acceptable.
   - **C.** Both A and B.
   - **D.** Neither A nor B.

6. The open circuit test is being discussed. *Technician A* says the battery must be stabilized before the open circuit voltage test is performed. *Technician B* says a test result of 12.4 volts is acceptable. Who is correct?
   - **A.** A only
   - **B.** B only
   - **C.** Both A and B
   - **D.** Neither A nor B

7. A maintenance-free battery has failed the capacity test. *Technician A* says if the voltage recovers to 12.45 volts, the battery is still good. *Technician B* says if the voltage level does not return to 12.4 volts, recharge the battery and repeat the capacity test. Who is correct?
   - **A.** A only
   - **B.** B only
   - **C.** Both A and B
   - **D.** Neither A nor B

8. The results of a 3-minute charge test are being discussed. *Technician A* says if the voltmeter indicates fewer than 15.5 volts, the battery must be replaced. *Technician B* says if the voltmeter reading is above 15.5 volts the battery is good. Who is correct?
   - **A.** A only
   - **B.** B only
   - **C.** Both A and B
   - **D.** Neither A nor B

9. While jump-starting a vehicle, a puff of smoke is observed and the engine ground cable is burned.
Technician A says this happened because the two vehicles were touching.
Technician B says this was caused by connecting the negative jumper cable to the disabled vehicle's engine ground.
Who is correct?
   A. A only
   B. B only
   C. Both A and B
   D. Neither A nor B

10. Ice crystals are found in the electrolyte. This can be caused by:
   A. A discharged battery.
   B. Use of tap water.
   C. Reversed battery connections.
   D. Improper holddowns.

ASE Challenge Questions

1. Charging a battery with a battery charger is being discussed.
Technician A says that battery-charging voltage should never exceed 15 volts.
Technician B says that a battery can be considered fully charged when its open circuit voltage exceeds 12.1 volts after it has been stabilized.
Who is correct?
   A. Technician A
   B. Technician B
   C. Both A and B
   D. Neither A nor B

2. *Technician A* says that a battery terminal test is performed by placing voltmeter leads between the positive battery post and the negative battery terminal.
Technician B says that the total amount of voltage drop between the negative battery post and the negative battery terminal should not exceed 300 mV.
Who is correct?
   A. Technician A
   B. Technician B
   C. Both A and B
   D. Neither A nor B

3. *Technician A* says that an ammeter is used when performing a battery leakage test.
Technician B says that a fully charged battery will have a specific gravity of at least 1.265.
Who is correct?
   A. Technician A
   B. Technician B
   C. Both A and B
   D. Neither A nor B

4. *Technician A* says that the 3-minute charge test is performed after a battery has failed a capacity test.
Technician B says that if battery voltage is below 15.5 volts at the end of the 3-minute charge test the battery is probably sulfated.
Who is correct?
   A. Technician A
   B. Technician B
   C. Both A and B
   D. Neither A nor B

5. The battery current drain test is being discussed.
Technician A says that an ammeter that can read as low as 20 mA should be used.
Technician B says that a current drain caused by an internally shorted battery could not be measured with an ammeter.
Who is correct?
   A. Technician A
   B. Technician B
   C. Both A and B
   D. Neither A nor B

Job Sheet 18

18

Name _____ Date _____

Inspecting and Cleaning a Battery

Upon completion of this job sheet, you should be able to visually inspect a battery.

ASE Correlation

This job sheet is related to the ASE Electrical/Electronic System Certification Exam's content area: *Battery Diagnosis and Service;* task: Inspect, clean, fill, or replace battery.

Tools and Materials

A vehicle with a 12-volt battery	Basic tool kit	Battery cable puller
A DMM	Baking soda	Wire brush
Safety glasses	Terminal cleaning brush	

Describe the vehicle being worked on:

Year _____ Make _____ Model _____

VIN _____ Engine type and size _____

Procedure

Task Completed

1. Describe the general appearance of the battery.

_____

_____

_____

_____

2. Describe the general appearance of the cables and terminals.

_____

_____

_____

_____

3. Check the tightness of the cables at both ends. Describe their condition.

_____

_____

_____

4. Connect the positive lead of the meter (set on DC volts) to the positive terminal of the battery. ☐

5. Put the negative lead on the battery case, and move it all around the top and sides of the case. What readings do you get on the voltmeter?

_____

_____

_____

6. What is indicated by the readings?

_____

_____

_____

7. Measure the voltage of the battery. Your reading was: _____ volts.

8. What do you know about the condition of the battery based on the visual inspection and the tests that you did?

_____

_____

_____

_____

_____

Instructor's Response _____

_____

_____

Job Sheet 19

19

Name _____ Date _____

Testing the Battery's Capacity

Upon completion of this job sheet, you should be able to test a battery's capacity.

ASE Correlation

This job sheet is related to the ASE Electrical/Electronic System Certification Exam's content area: *Battery Diagnosis and Service;* task: Perform battery capacity (load, high-rate discharge) test; determine needed service.

Tools and Materials

A vehicle with a 12-volt battery
Service manual
Starting charging system tester (VAT-60 or similar)
Safety glasses

Describe the vehicle being worked on:

Year _____ Make _____ Model _____

VIN _____ Engine type and size _____

Procedure

Task Completed

1. Perform a battery state of charge test:

   a. Record the specific gravity readings for each cell:

      (1) _____ (2) _____ (3) _____ (4) _____ (5) _____ (6) _____

   b. If the battery is a maintenance-free-type battery, what is the open circuit voltage?

      _____ volts

2. Summarize the battery's state of charge from the above.

   _____

   _____

   _____

   _____

3. Connect the starting charging system tester to the battery.   ☐

4. Locate the rating of the battery. What is the rating? _____

5. Based on the rating, how much load should be put on the battery during the capacity

   test? _____ amps

6. Conduct a battery load test.

   Battery voltage decreased to _____ volts after _____ seconds.

7. What is the specification for the vehicle being serviced? _____

8. Describe the results of the battery load (capacity) test. Include in the results your service recommendations and the reasons for them.

_____

_____

_____

_____

_____

_____

Instructor's Response _____

_____

_____

_____

Job Sheet 20

20

Name _____ Date _____

Three-Minute Charge Test

Upon completion of this job sheet, you should be able to perform the 3-minute charge test.

Tools and Materials

DMM
Battery charger

Describe the vehicle being worked on:

Year _____ Make _____ Model _____

VIN _____ Engine type and size _____

Procedure

Task Completed

1. Describe the purpose of this test.

   _____

   _____

   _____

2. Determine if the battery manufacturer recommends this test. ☐

3. Disconnect the negative battery cable. Why is this step important?

   _____

   _____

4. Connect the battery charger and DMM leads across the battery. At what rate should the battery charger be set?

5. Turn on the battery charger and record the voltmeter reading after 3 minutes. _____

6. Describe your results and provide recommendations.

   _____

   _____

Instructor's Response _____

_____

_____

_____

Job Sheet 21

21

Name _____ Date _____

Removing, Cleaning, and Installing a Battery

Upon completion of this job sheet, you should be able to remove, clean, and reinstall a battery.

ASE Correlation

This job sheet is related to the ASE Electrical/Electronic System Certification Exam's content area: *Battery Diagnosis and Service;* task: Inspect, clean, fill, or replace a battery.

Tools and Materials

A vehicle with a 12-volt battery	Battery cable puller
A wash pan	Battery terminal cleaner and brushes
A box of clean baking soda	Cable end spreader
Water	Safety glasses
Cleaning brush	Rubber gloves
Hand tools	

Describe the vehicle being worked on:

Year _____ Make _____ Model _____

VIN _____ Engine type and size _____

Procedure

Task Completed

NOTE: Before disconnecting the battery, record the stations of all preset buttons on the radio. Also, if the vehicle has memory seats or other like accessories, record those settings as well.

1. Disconnect the negative cable of the battery. Use the cable puller if the cable is difficult to remove. ☐

2. Carefully remove the battery holddown strap. Describe the condition of the holddown.

   _____

   _____

   _____

3. Disconnect the positive cable of the battery. Use the cable puller if the cable is difficult to remove. Move anything that may interfere with the removal of the battery. ☐

4. With a terminal cleaner, clean off the terminals of the battery. ☐

5. With a battery-carrying strap, remove the battery from the vehicle and place it on a workbench or on the floor close to a drain. ☐

6. Inspect the battery tray. Describe its condition and your recommendations.

_____

_____

_____

_____

☐ 7. Make a mixture of baking soda and water. The mixture should be like a paste.

☐ 8. With the brush, scrub the top, sides, and bottom of the battery with the baking soda paste. Be careful not to allow the paste to get inside the battery.

☐ 9. After all of the battery has been scrubbed, wash the paste off with clean water.

☐ 10. Allow the battery to drip dry; then wipe the water off the battery with a clean rag.

☐ 11. Use the leftover paste to clean the battery tray and the holddown assembly.

☐ 12. Rinse these off with water.

☐ 13. Use a clean, dry cloth to wipe the tray dry. Then reinstall the battery.

☐ 14. Clean the battery cable terminals and spread them slightly with the spreader tool.

☐ 15. Install the battery holddown assembly; make sure it is tight and that the battery is positioned properly.

☐ 16. Connect the positive cable to the battery.

☐ 17. Connect the negative cable to the battery.

☐ 18. Reset all preset stations on the radio and other accessories with memory.

Instructor's Response _____

_____

_____

_____

DIAGNOSTIC CHART 5-1

PROBLEM AREA:	Battery
SYMPTOMS:	No-start condition after vehicle is parked overnight.
POSSIBLE CAUSES:	**1.** Glove box, trunk, interior illumination lights not turning off due to defective switches. **2.** Circuit short to ground. **3.** Circuit copper-to-copper shorts. **4.** Contaminated battery case.

DIAGNOSTIC CHART 5-2

PROBLEM AREA:	Battery
SYMPTOMS:	Battery requires jump-starting after engine shutdown.
POSSIBLE CAUSES:	**1.** Low specific gravity. **2.** Defective battery cell(s).

DIAGNOSTIC CHART 5-3

PROBLEM AREA:	Battery
SYMPTOMS:	Inadequate current to start engine under heavy load conditions and battery will not accept a charge.
POSSIBLE CAUSES:	**1.** Contaminated battery case resulting in constant current draw. **2.** Contaminated battery terminals. **3.** Undercharged battery. **4.** Defective battery. **5.** Sulfated battery. **6.** Damaged battery.

DIAGNOSTIC CHART 5-4

PROBLEM AREA:	Battery
SYMPTOMS:	Starter cranks engine slowly or fails to turn engine; low state of charge.
POSSIBLE CAUSES:	**1.** Discharged battery. **2.** Contaminated terminal clamps. **3.** Defective battery cables.

Starting System Diagnosis and Service

Upon completion and review of this chapter, you should be able to:

❏ Perform a systematic diagnosis of the starting system.

❏ Determine what can cause slow-crank and no-crank conditions.

❏ Perform a quick check test series to determine the problem areas in the starting system.

❏ Perform and accurately interpret the results of a current draw test.

❏ Perform and accurately interpret the results of an insulated circuit resistance test and a ground circuit test.

❏ Perform the solenoid test series and accurately diagnose the solenoid.

❏ Perform the no-crank test and recommend needed repairs as indicated.

❏ Diagnose the starter motor condition by use of the free speed test.

❏ Remove and reinstall a starter motor.

❏ Disassemble, clean, inspect, repair, and reassemble a starter motor.

Basic Tools

Basic mechanic's tool set

Service manual

_____

Introduction

Perhaps one of the most aggravating experiences to a car owner is to have an engine that will not start. However, not all starting problems are caused by the starting system. The ignition and fuel systems must also be in proper condition to perform their functions. In addition, the internal condition of the engine must be such that compression, correct valve timing, and free rotation are all obtained.

The starter motor must be capable of rotating the engine fast enough to start and run under its own power. The starting system is a combination of mechanical and electrical parts that work together to start the engine. The basic starting system includes the following components (Figure 6-1):

1. Battery.

2. Cable and wires.

Classroom Manual
Chapter 6, page 136

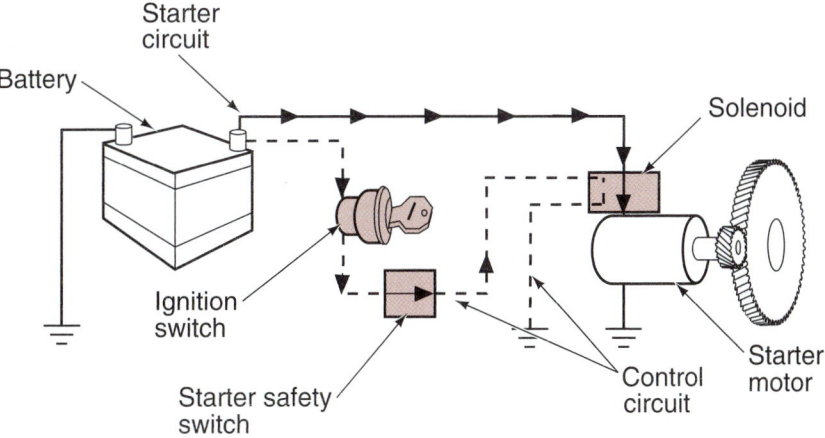

Figure 6-1 Major components of the starting system. The solid line represents the starting circuit, while the dashed line represents the starter control circuit.

3. Ignition switch.

4. Starter solenoid or relay.

5. Starter motor.

6. Starter drive and flywheel ring gear.

7. Starting safety switch.

In this chapter you will perform the required tests to make a decision concerning the condition and operation of these components. You will also remove, disassemble, reassemble, and reinstall the starter motor.

Starting System Service Cautions

Before beginning any service on the starter system, some precautions must be observed. Along with the precautions outlined in Chapter 1, when servicing the battery, several other precautions should be followed:

1. Refer to the manufacturer's manuals for correct procedures for disconnecting a battery. Some vehicles with on-board computers must be supplied with an auxiliary power source.

2. Disconnect the battery ground cable before disconnecting any of the starter circuit's wires or removing the starter motor.

3. Be sure the vehicle is properly positioned on the hoist or on safety jack stands.

4. Before performing any cranking test, be sure the vehicle is in PARK or NEUTRAL and the parking brakes are applied.

5. Follow the manufacturer's directions for disabling the ignition system.

6. Be sure the test leads are clear of any moving engine components.

7. Never clean any electrical components in solvent or gasoline. Clean with compressed air, denatured alcohol, or wipe with clean rags only.

Starting System Troubleshooting

Customer complaints concerning the starting system generally fall into four categories: **no crank, slow cranking,** starter spins but does not turn engine, and excessive noise. As with any electrical system complaint, a systematic approach to diagnosing the starting system will make the task easier. First, the battery must be in good condition and fully charged. Perform a complete battery test series to confirm the battery's condition. Many starting system complaints are actually attributable to battery problems. If the starting system tests are performed with a weak battery, the results can be misleading. The conclusions may be erroneous and costly.

Before performing any tests on the starting system, first begin with a visual inspection of the circuit. Repair or replace any corroded or loose connections, frayed wires, or any other trouble sources. The battery terminals must be clean and the starter motor must be properly grounded.

The diagnostic chart shows a logical sequence to follow whenever a starting system complaint is made (Figure 6-2). What tests are performed is determined by whether or not the starter will crank the engine.

If the customer complains of a no-crank situation, attempt to rotate the engine by the crankshaft pulley bolt. Rotate the crankshaft two full rotations in a clockwise direction, using a large socket wrench. If the engine does not rotate, it may be seized due to its being operated with no oil, broken engine components, or **hydrostatic lock.** Since liquid cannot be compressed, if there is a leak that allows antifreeze from the cooling system to enter the cylinder, the cylinder can fill

Classroom Manual
Chapter 6, page 137

No crank means that when the ignition switch is placed in the START position, the starter does not turn the engine. This may be accompanied by a buzzing noise, which indicates the starter motor drive has engaged the ring gear, but the engine does not rotate. There may also be no clicking sounds from the starter motor or solenoid.

Slow cranking means the starter drive engages the ring gear, but because the engine turns slowly it cannot start. Some manufacturers provide specifications for engine cranking speed.

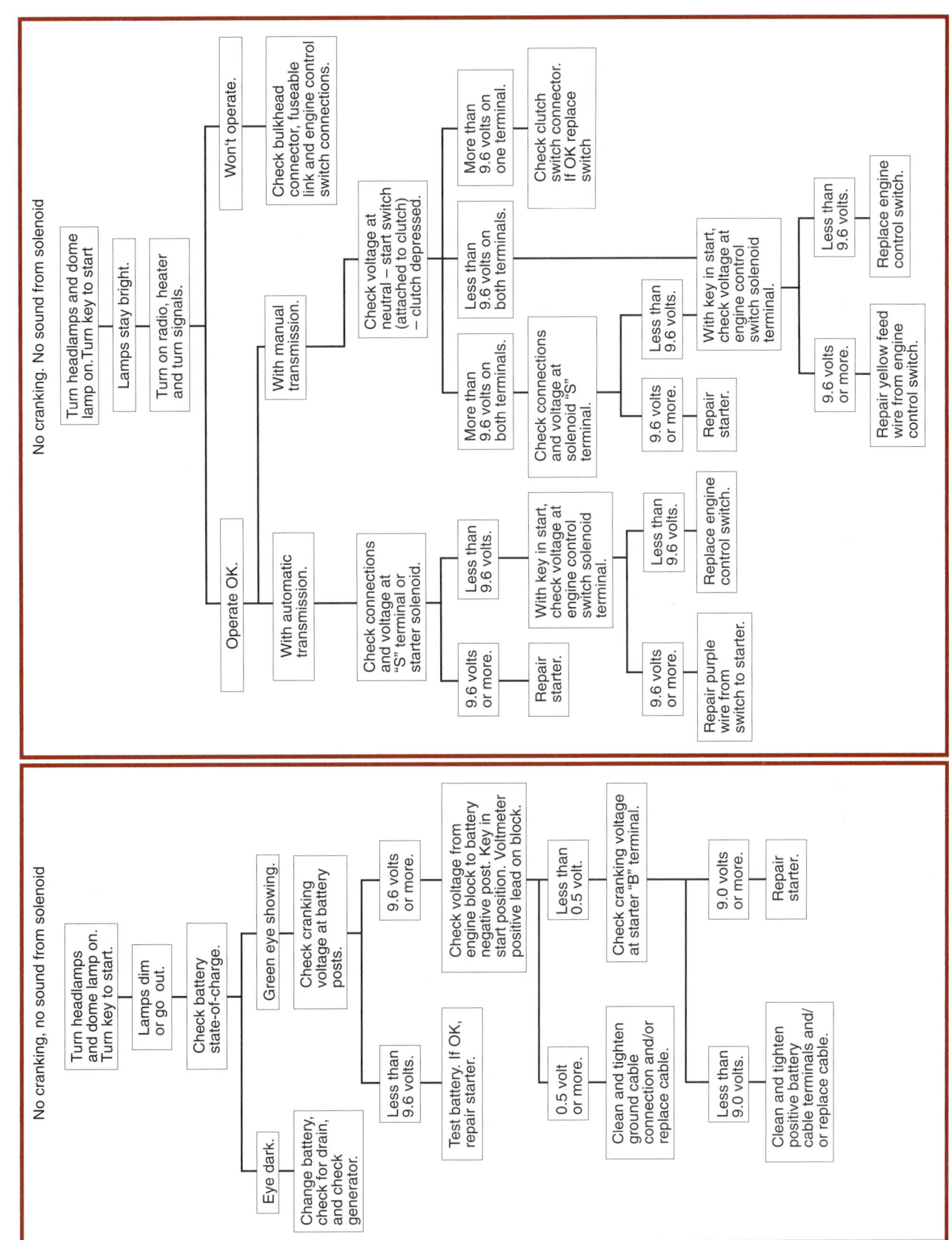

Figure 6-2 Starting system diagnostic chart.

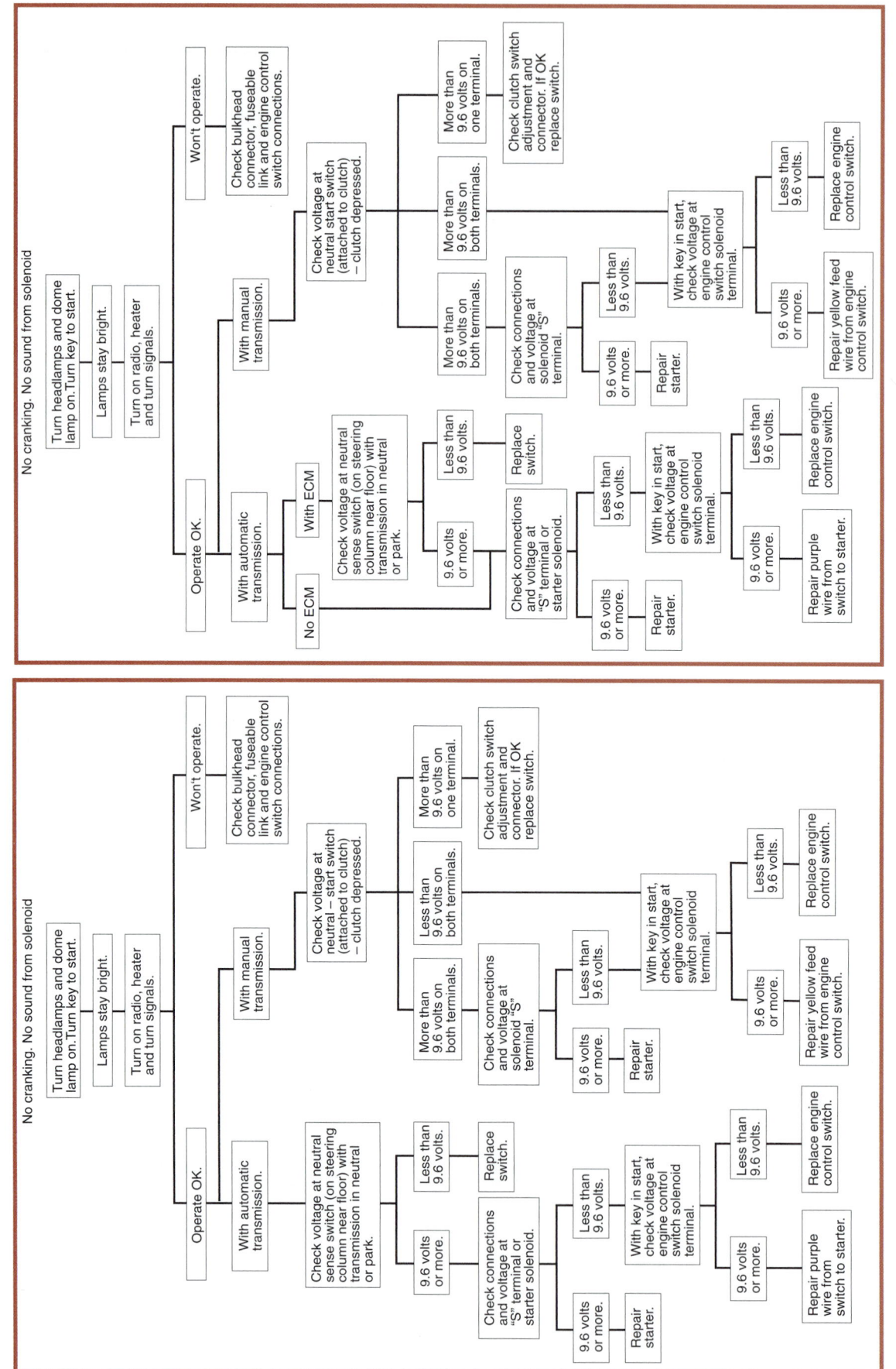

Figure 6-2 (*continued*)

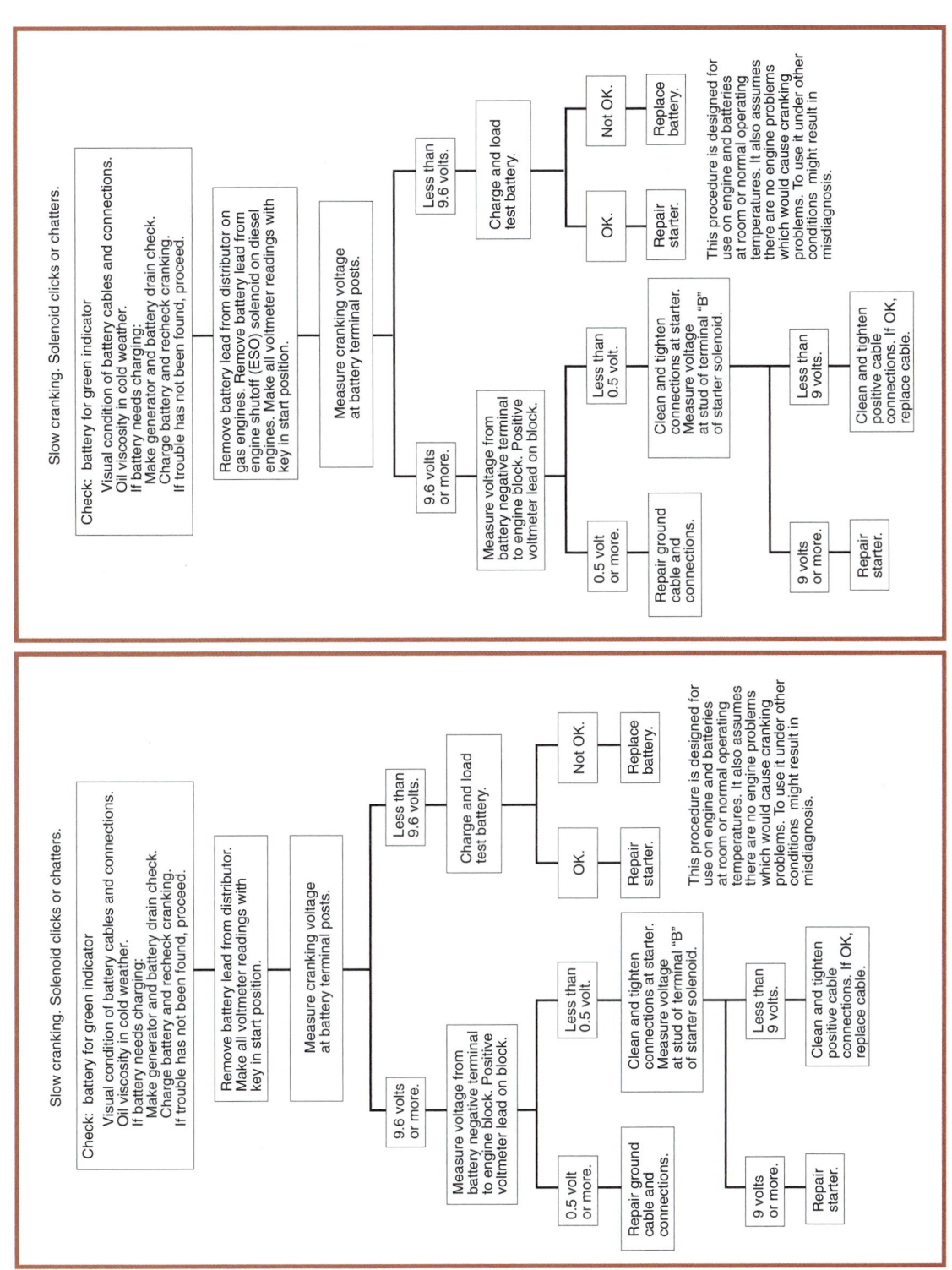

Figure 6-2 (continued)

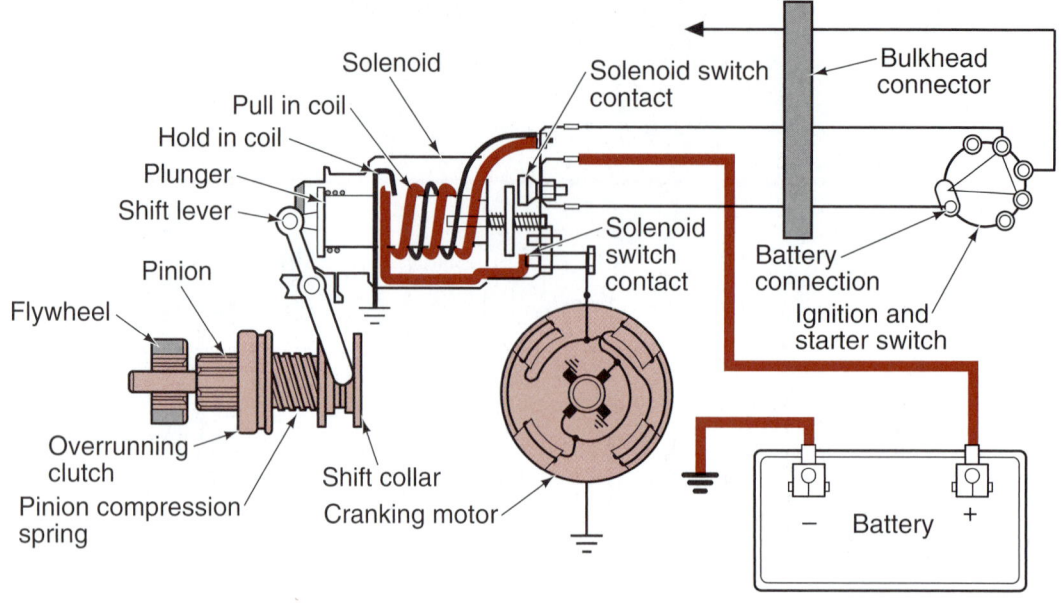

Figure 6-3 Excessive wear, loose electrical connections, or excessive voltage drop in any of these areas can cause a slow-crank or no-crank condition.

to such a level that the piston is unable to move upward. This condition is referred to as hydrostatic lock.

Several potential trouble spots in the circuit can cause a slow-crank or no-crank complaint. (Figure 6-3). Excessive voltage drops in these areas will cause the starter motor to operate slower than required to start the engine. The speed that the starter motor rotates the engine is important to engine starting. If the speed is too slow, compression is lost and the air/fuel mixture draw is impeded. Most manufacturers require a speed of approximately 250 rpm during engine cranking.

If the starter spins but the engine does not rotate, the most likely cause is a faulty starter drive. If the starter drive is at fault, the starter motor will have to be removed to install a new drive mechanism. Before faulting the starter drive, also check the starter ring gear teeth for wear or breakage, and for incorrect gear mesh of the ring gear and starter motor pinion gear.

Most noises can be traced to the starter drive mechanism. The starter drive can be replaced as a separate component of the starter.

CUSTOMER CARE: Always treat the customer's car with respect. Place fender covers over the fenders when performing tasks under the hood. Do not lay tools on the vehicle's finish. Clean your hands before entering the vehicle. Place a seat protector over the seats and paper mats on the floor boards. Give the car back to the customer at least as clean as when you received it.

Testing the Starting System

As with the battery testing series, the tests for the starting system are performed with a starting/charging system tester (Figure 6-4). The starter performance and battery performance are so closely related that it is important for a full battery test series to be done before trying to test the starter system. If the battery fails the load test and is fully charged, it must be replaced before doing any other tests.

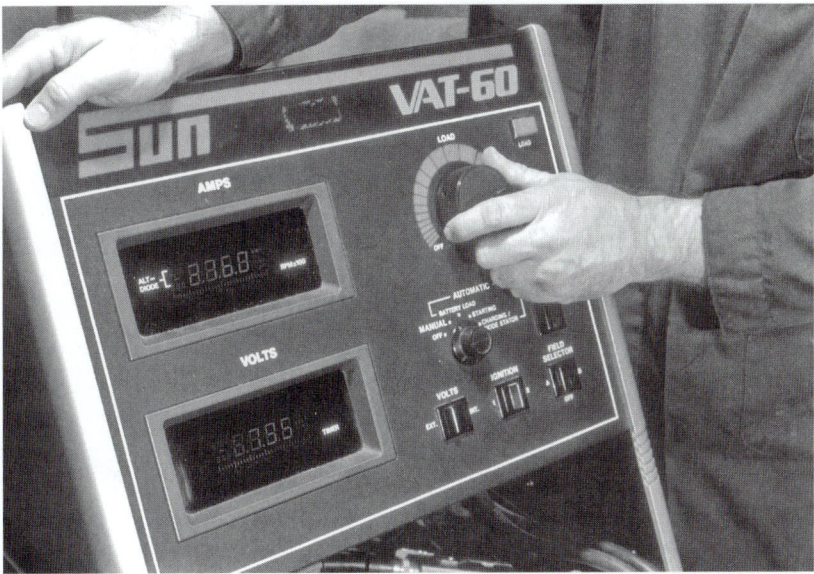

Figure 6-4 Starter tests can be performed with any tester capable of measuring high current, such as the VAT-60.

Quick Testing

The **quick test** will isolate the problem area and determine whether the starter motor, solenoid, or control circuit is at fault. If the starter does not turn the engine at all, and the engine is in good mechanical condition, the quick test can be performed to locate the problem area. To perform this test, make sure the transmission is in neutral and set the parking brake. Turn on the headlights. Next, turn the ignition switch to the START position while observing the headlights.

 SERVICE TIP: Check the fusible link if the engine does not crank and the headlights do not come on.

There are three things that can happen to the headlights during this test:

1. They will go out.
2. They will dim.
3. They will remain at the same brightness level.

If the lights go out completely, the most likely cause is a poor connection at one of the battery terminals. Check the battery cables for tight and clean connections. It will be necessary to remove the cable from the terminal and clean the cable clamp and battery terminals of all corrosion.

If the headlights dim when the ignition switch is turned to the START position, the battery may be discharged. Check the battery condition. If it is good, then there may be a mechanical condition in the engine that is preventing it from rotating. If the engine rotates when turning it with a socket wrench on the pulley nut, the starter motor may have internal damage. A bent starter armature, worn bearings, thrown armature windings, loose pole shoe screws, or any other worn component in the starter motor that will allow the armature to drag can cause a high current demand.

● **CUSTOMER CARE:** If the starter windings are thrown, this indicates several different problems. The most common is that the driver is keeping the ignition switch in the START position too long after the engine has started. Other causes include the driver

Special Tools

Fender covers

Jumper cables

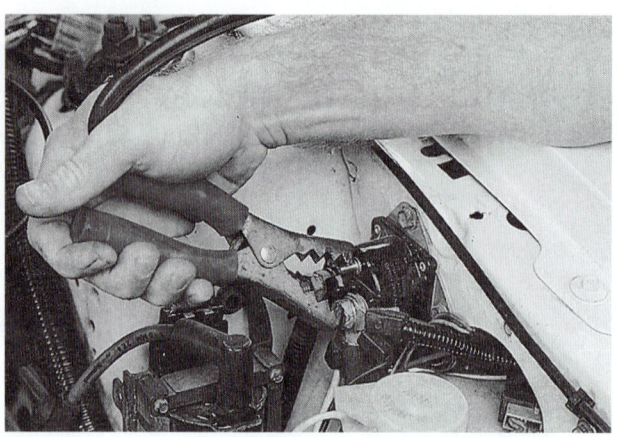

Figure 6-5 Bypassing the solenoid to determine if the solenoid or the control circuit is faulty.

opening the throttle plates too wide while starting the engine, which results in excessive armature speeds when the engine does start. Also, the windings can be thrown because of excessive heat buildup in the motor. The motor is designed to operate for very short periods of time. If it is operated for longer than 15 seconds, heat begins to buildup at a very fast rate. If the engine does not start after a 15-second crank, the starter motor needs to cool for about 2 minutes before the next attempt to start the engine.

If the lights stay brightly lit and the starter makes no sound (listen for a deep clicking noise), there is an open in the circuit. The fault is in either the solenoid or the control circuit. To test the solenoid, bypass the solenoid by bridging the BAT and S terminals (Figure 6-5).

 WARNING: A starter can draw up to 400 amperes. The tool used to jump the terminals must be able to carry this high current and must have an insulated handle.

If the starter rotates with the solenoid bypassed, the control circuit is at fault. If the starter does not rotate and the lights do not dim, the solenoid is at fault. (Also, listen for the starter drive engaging.) If the starter rotates slowly and the headlights dim, there is excessive current draw and the system will have to be tested further.

Lab Scope Testing

A lab scope or graphing multimeter is a good method of testing a starting system. Prior to disabling the ignition system to prevent the engine from starting, use the ammeter probe to view the starter signal. Most current probes convert amps to millivolts to be read on a lab scope. The amp probe is connected over either the positive or negative battery cable. Be sure to orient the arrow on the current probe with the direction of current flow. Set the scope settings of 0.5 volt/division and 100 ms/division to start with and adjust later as needed. Using the meter's record function will make interpretation of the pattern results easier.

To get the engine to rotate over requires high starter current. Once the engine is rotating, lower current is required. A typical pattern will indicate this higher current draw at initial cranking (Figure 6-6). During this time, both the pull-in and hold-in windings of the starter solenoid are energized. Shortly after the high current spike, the pull-in windings are de-energized and current will drop to a lower value. After the engine starts, the current draw drops to zero as the ignition switch is placed in the RUN position. Excessively high or low current indicates a problem with the circuit, such as a bad starter or resistance from corrosion or loose connections. At this time, current draw and circuit resistance testing will be required.

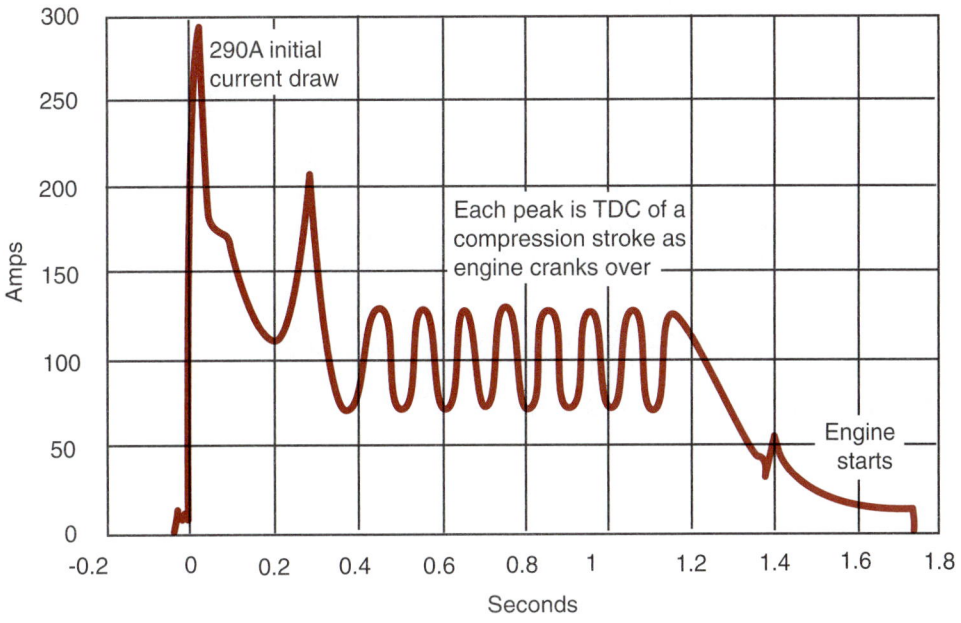

Figure 6-6 When the starter is energized, the current will spike; then it drops as starter speed increases.

Current Draw Test

The **current draw test** measures the amount of current that the starter draws when actuated. It determines the electrical and mechanical condition of the starting system. If the starter motor cranks the engine, the technician should perform the current draw test. The following procedure uses a typical starting/charging system tester and is similar to the procedure for other starting and charging system testers:

1. Connect the large red and black test leads across the battery, observing polarity.
2. Follow the manufacturer's instructions to zero the ammeter.
3. Connect the amps inductive probe around the battery ground cable. If more than one ground cable is used, clamp the probe around all of them (Figure 6-7).

Special Tools

Fender covers
Starting/charging
  system tester
Jumper wires

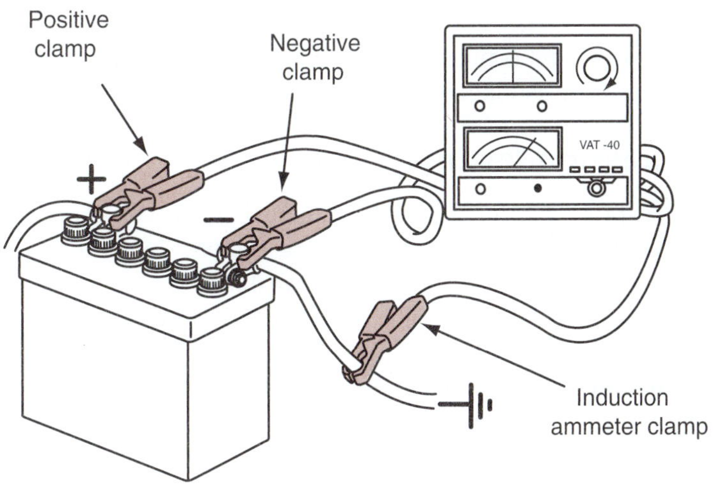

Figure 6-7 Test lead connections to perform the starter current draw test.

4. Make sure all loads are turned off (lights, radio, etc.).

5. Select STARTING TEST.

6. Follow the service manual procedure to disable the ignition system to prevent the vehicle from starting. Some systems may require that the fuel pump or EFI relay be removed or the ignition module be disconnected in order to prevent the engine from starting.

> **CAUTION:** Always refer to the manufacturer's service manual for the correct procedure for disabling the ignition system.

7. Crank the engine for 10–15 seconds and note the voltmeter and ammeter readings.

> **CAUTION:** Do not operate the starter motor for longer than 15 seconds. Allow the motor to cool between cranking attempts.

The specification for current draw is the maximum allowable; the specification for cranking voltage is the minimum allowable.

After recording the readings from the current draw test, compare them with the manufacturer's specifications. If specifications are not available, correctly functioning systems, as a rule, will crank at 9.6 volts or higher. Current draw is dependent on engine size. Most V-8 engines will have a current draw of about 200 amperes, six-cylinder engines about 150 amperes, and four-cylinder engines about 125 amperes.

If the readings obtained from the current draw test are out of specifications, then additional testing will be required to isolate the problem. If the readings were on the borderline of the specifications or there is an intermittent problem, then detailed testing for bad components will pinpoint potential failures.

The following provides an interpretation of the current draw test:

❑ Voltage is 9.6 volts or more and amperage is higher than specified—indicates impedance to rotation of the starter motor. This includes worn bushings, a mechanical blockage, and excessive advanced ignition timing. Shorted starter motor windings can cause high current draw.

❑ Voltage is 9.6 volts or more and amperage is lower than specified—indicates high electrical circuit resistance.

❑ Voltage is less than 9.6 volts and current is higher than specified—indicates the amperage is draining the battery. Perform the battery test series to confirm the battery is good. If the battery passes this test series, then see the first point above.

❑ Voltage is less than 9.6 volts and current is lower than specified—indicates a faulty battery.

Because the voltage reading obtained from the current draw test was taken at the battery, this reading may not be an exact representation of the actual voltage delivered to the starter. Voltage losses due to bad cables, connections, and relays (or solenoids) may diminish the amount of voltage to the starter. These should be tested before removing the starter from the vehicle.

Relative Compression Testing

Although usually a test procedure for determining internal engine condition, compression testing is also used to identify hard starting conditions. Good compression and a properly functioning starting system are critical for fast starts and good driveability. Due to the increased difficulty of accessing spark plugs, normal compression testing by removing the spark plugs and installing a compression gauge is being replaced with a nonintrusive diagnostic routine referred to as **relative compression testing.** Relative compression testing uses current draw during cranking to determine if a cylinder has lower compression relative to the other cylinders. This test also provides a good indication of the starting system.

Every time a piston is forced up during the compression stroke, it takes work. Since current represents work, the lab scope provides a method of observing current on a small timeline scale;

with the scope you can visualize the effect that each cylinder's compression stroke has on starter current while the engine is cranking.

To perform this test, clamp the lab scope's amp probe around the starter cable between the battery and the starter motor. With the ignition system disabled, crank the engine while monitoring the lab scope pattern. As discussed earlier, the initial current should go high and peak, followed by a ripple pattern (Figure 6-8). As the pistons approach top dead center compression stroke, the resistance increases and causes the current trace to also increase. This should result in a trace with even peaks across the screen. If compression of a cylinder is lower than that of other cylinders, then the current peak of that cylinder will also be lower (Figure 6-9).

By using an external trigger placed around a spark plug, it is possible to identify which cylinder has the low compression. Once the trigger is installed onto the spark plug, adjust the scope to trigger off this signal. For example, if the lab scope is set to trigger off of the rpm pickup that was placed on the number one plug wire, then every time the screen is updated, the number one cylinder is the first one on the screen. By following the engine's firing order, each cylinder number can be determined.

Most engine analyzers have a variation of this test that will test the relative current comparison of each cylinder during cranking. These analyzers provide a readout of the results and may suggest the faulty cylinder.

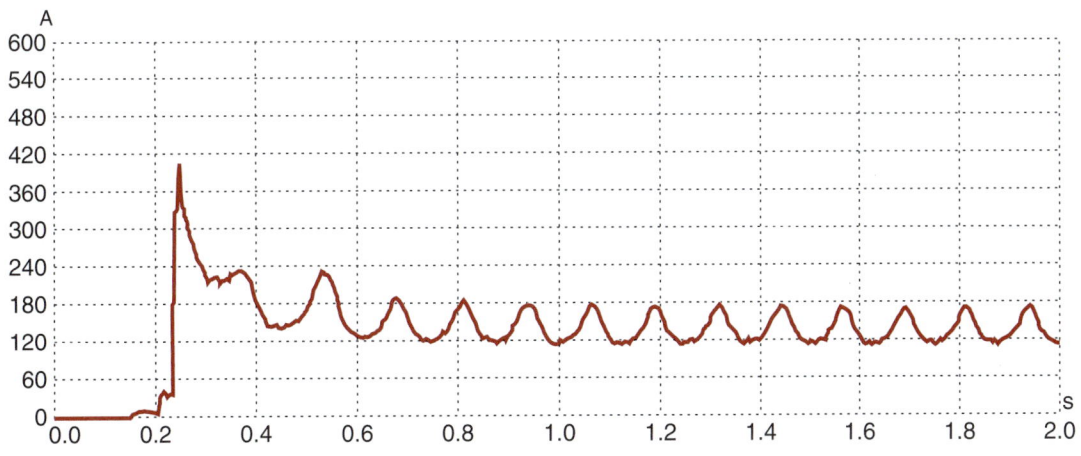

Figure 6-8 Lab scope pattern showing good relative compression between cylinders.

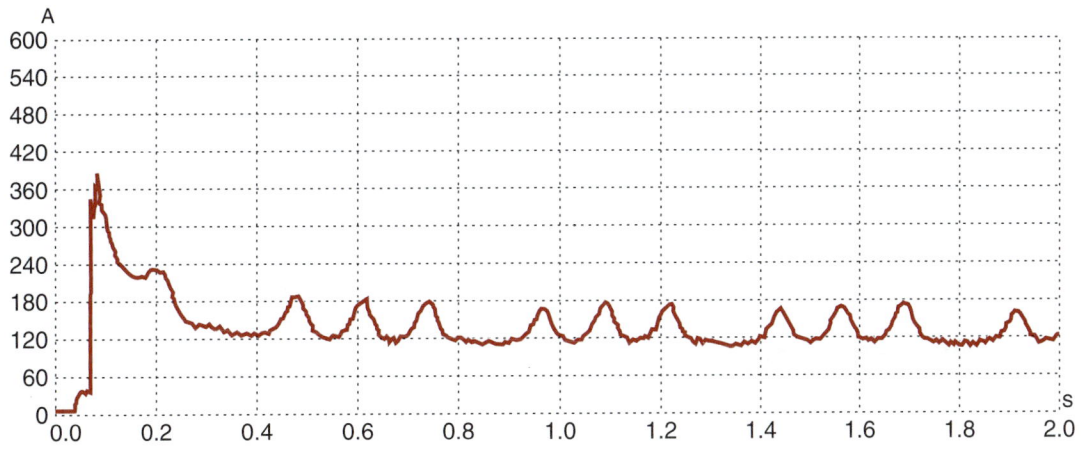

Figure 6-9 Lab scope pattern indicating a cylinder with a lower relative compression.

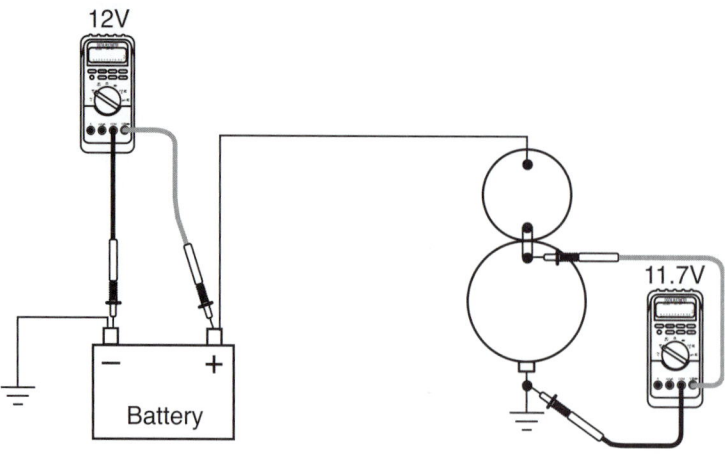

12V

Battery

11.7V

Figure 6-10 Voltage drop testing to identify sources of excessive resistance.

Classroom Manual
Chapter 6, page 147

Special Tools

Voltmeter

Fender covers

Starting/charging
   system tester

Insulated Circuit Resistance Test

An electrical resistance will have a different pressure or voltage on each side of the resistance. Voltage is dropped when current flows through resistance. Most manufacturers design their starting systems to have very little resistance to the flow of current to the starter motor. Most have less than 0.2 volt dropped on each side of the circuit. This means the voltage across the starter input terminal to the starter ground should be within 0.4 volt of battery voltage (Figure 6-10).

Voltage drops are measured by connecting a voltmeter in parallel with the circuit section being tested. In order to obtain a voltage drop reading, a load on the circuit must be applied. The following is a typical test procedure:

1. Connect the test leads as shown (Figure 6-11), depending on the type of system being tested. Usually the positive lead of the voltmeter is connected to the positive battery post and the negative lead is connected to the starter battery (BATT) terminal.

2. Disable the ignition system as discussed in current draw testing.

3. Crank the engine for 15 seconds and observe the voltmeter scale.

This tests for voltage drop in the entire circuit, so if voltage drop is excessive, the cause of the drop must be located. To locate the cause of the excessive voltage drop, move the voltmeter lead on the starter toward the battery. Check each connection while moving toward the battery. With each move of the test lead, crank the engine while observing the voltmeter reading. Continue to test each connection until a noticeable decrease in voltage drop is detected. The cause of the excessive voltage drop will be located between that point and the preceding point.

SERVICE TIP: As a general rule, allow up to 0.2 volt per cable and 0.1 volt per connection to be dropped. Switches can be as high as 0.3 volt. Use the wiring diagram for the vehicle to determine the number of conductors and connections used in the circuit. This will provide a specification for you if no other specifications are available.

Special Tools

Voltmeter

Fender covers

Ground Circuit Test

A ground circuit test is performed to measure the voltage drop in the ground side of the circuit (Figure 6-12). If the starter motor connection to ground is broken or loose, the circuit would be opened. This could cause an intermediate starter system problem or a starter motor that will not crank the engine. To perform the ground circuit test, connect the voltmeter positive lead to

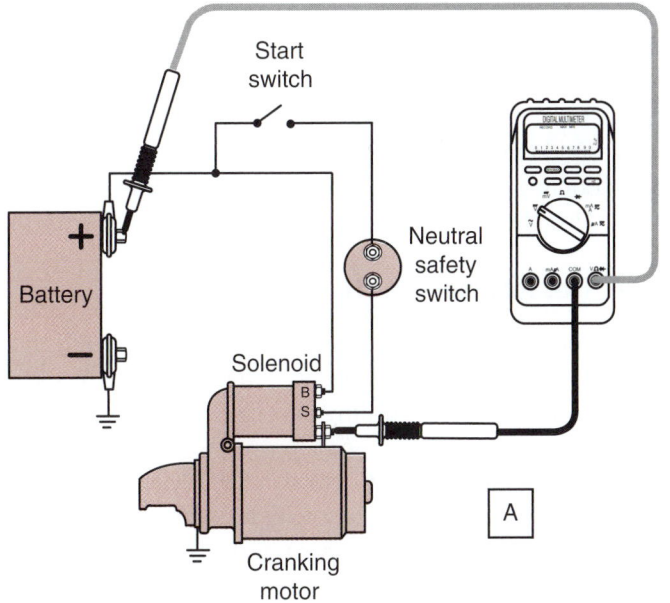

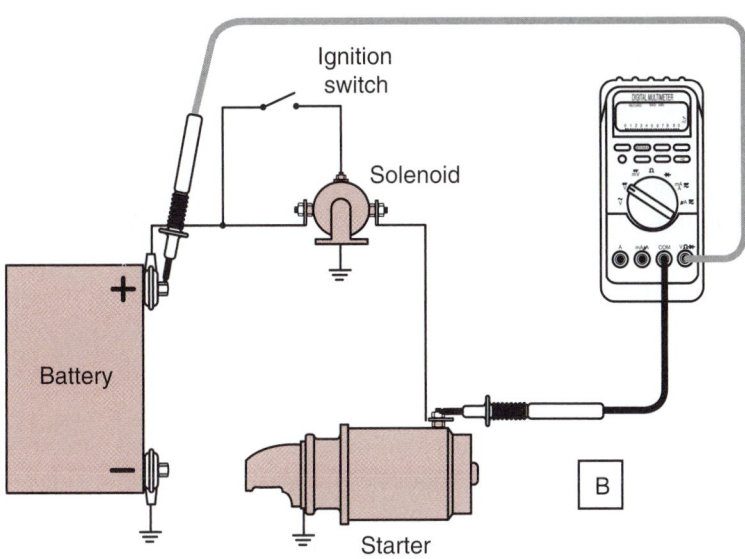

Figure 6-11 (A) Test lead connections for a starter-mounted solenoid, (B) test lead connections for relay-controlled systems.

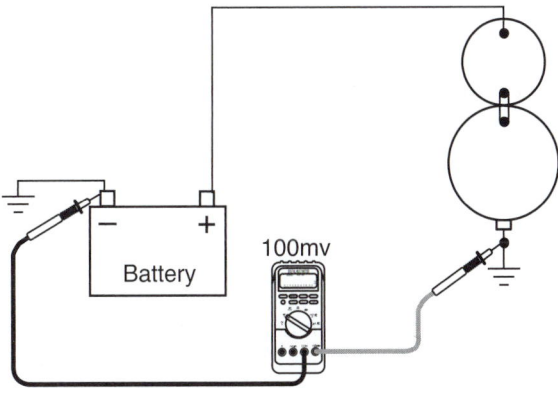

Figure 6-12 Voltage drop testing of the ground-side circuit of the starting system.

the starter motor case and the negative test lead to the ground battery terminal. Make sure any paint is removed from the area where the lead is connected to the case. Crank the engine while observing the voltmeter.

Less than 0.2 volt indicates the ground circuit is good. If more than 0.2 volt is observed, then there is a poor ground circuit connection. A poor ground circuit connection could be the result of loose starter mounting bolts, paint on the starter motor case, or a bad battery ground terminal post connection. Also check the ground cable for high resistance or for being undersized.

Solenoid Circuit Resistance Testing

Classroom Manual
Chapter 6, page 147

Special Tools

Voltmeter

Fender covers

Starting/charging system tester

High resistance in the solenoid will reduce the current flow through the solenoid windings and cause the solenoid to function improperly. If the solenoid has high resistance in the windings, it may result in the contacts burning and causing excessive resistance to the starter motor.

The **solenoid circuit resistance test** determines the electrical condition of the solenoid and the control circuit. To perform the solenoid circuit resistance test, first disable the ignition system. Using a voltmeter or starting/charging system tester, connect the positive voltmeter lead to the BAT terminal of the solenoid. Connect the negative test lead to the field coil terminal (M terminal). Crank the engine while observing the voltmeter reading. If the voltmeter reading indicates a voltage drop of greater than 0.2 volt, then the solenoid is defective. If this test proves the starter solenoid is good, then the solenoid switch circuit should be tested. Follow these steps:

1. Disable the ignition system.
2. Connect the voltmeter leads to both solenoid switch terminals, observing polarity.
3. Crank the engine and observe the voltmeter reading.

The total voltage drop should be less than 0.5 volt. If the indicated voltage drop is in excess of 0.5 volt, move the voltmeter leads up the circuit and test each component. The voltage drop across each component should be less than 0.1 volt.

Continue to move the voltmeter leads to test for voltage drop through the wires, starter relay, neutral safety switch, and ignition switch.

Special Tools

Test light

Jumper wires

Fender covers

Open Circuit Test

In order to perform voltage drop tests, a load must be placed on the circuit. If there is a no-crank complaint, a voltage drop test cannot be performed. Most no-crank problems are the result of opens in the circuit. The easiest way to diagnose this problem is with the use of a test light. On a system that uses a starter motor–mounted solenoid, the M terminal is the end of the circuit (Figure 6-13). Connecting the positive lead of the test light to the M terminal and the negative lead to a good ground, the light should be on when the ignition switch is located in the START position (Figure 6-14). If the light comes on, then the complete insulated circuit (including the ignition switch, wires, neutral safety switch, solenoid, and all connections) is operating properly. The open is in either the starter motor or the ground circuit.

If the test light comes on very dim, then there is very high resistance in the circuit. By working the test light through the circuit and back to the battery, the reason for the high resistance should be found. If the test light did not come on, follow the same procedure of backtracking the circuit toward the battery until the open is found. Also, check for voltage at the B+ terminal (Figure 6-15). Connecting the test light as shown should light the bulb with the key off. The light should stay on when the ignition switch is turned to the START position. If the light goes out in the START position, repair the cable or end connections.

 SERVICE TIP: A voltmeter can be used in place of the test light. At the M terminal there should be more than 9.6 volts present when the ignition switch is in the START position.

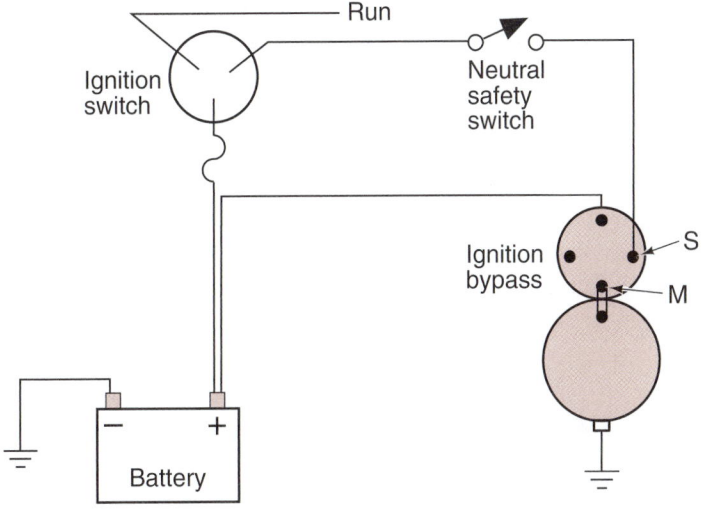

Figure 6-13 Starting system using a solenoid shift.

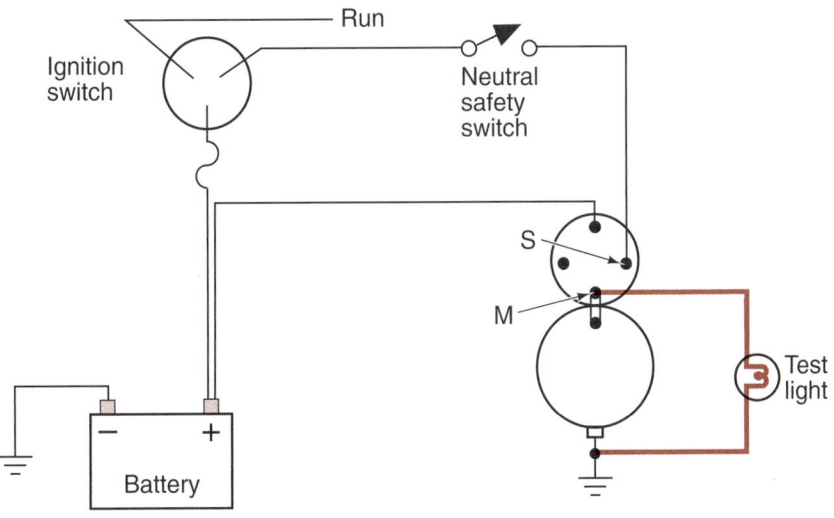

Figure 6-14 Test light connections for testing the solenoid and control circuit.

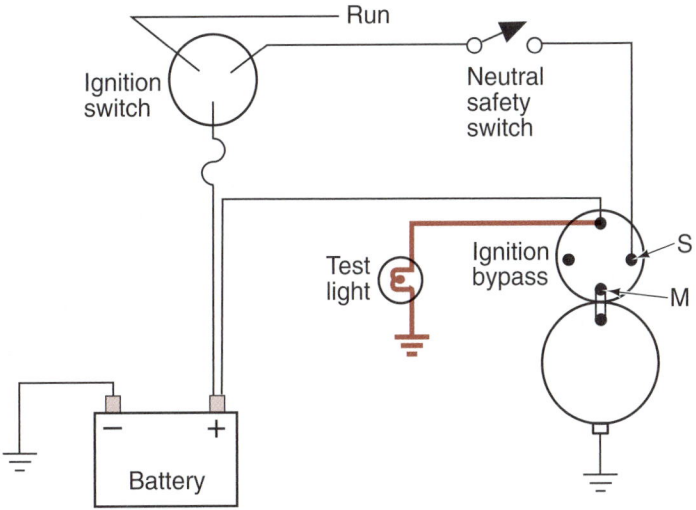

Figure 6-15 Test light connections for checking voltage at the BAT terminal.

Once the open has been found, it can be verified by using jumper wires to jump across the defective component or connection. If jumping across the solenoid, for example, and the starter spins, then there is an open in the solenoid. The same procedure can be used to jump across the ignition switch, neutral safety switch, or open wires.

If the test light did not come on when connected to the M terminal, then make a simple test of the ground circuit. This is done by connecting the ground lead of the test light onto the starter body and the positive lead to the M terminal of the solenoid. The light should come on bright with the ignition switch in the START position. If the ground circuit is good, then the starter is suspect and should be bench tested.

The Ford starting system is tested in the same manner, except there is an additional battery cable to test.

 SERVICE TIP: Most starter safety switches are adjustable. Sometimes a no-start problem can be corrected by checking and adjusting (or replacing) the starter safety switch.

Free Speed Test

In the event that the starter has failed the previous tests, or a new starter is going to be installed, a **free speed test** should be performed. The free speed test determines the free rotational speed of the armature. Some manufacturers recommend this test procedure over the current draw test. The starter must be removed from the vehicle, as described in the next section. With the starter removed from the vehicle, perform the test as follows (Figure 6-16):

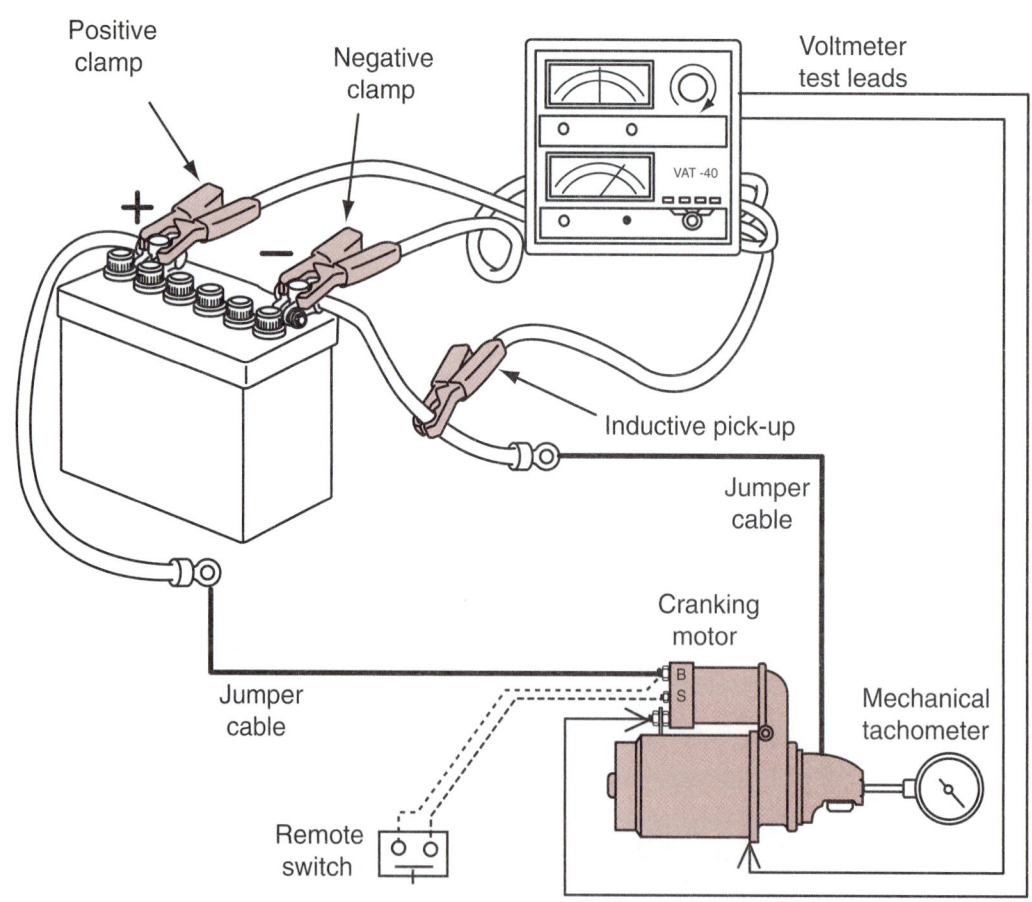

Figure 6-16 Starting/charging system tester connections for a free speed test.

> **CAUTION:** Do not overtighten the vise against the starter frame assembly. It is possible to crack the frame or the pole shoes.

1. Place the starter motor into a secure vise.

2. Attach an rpm indicator to the armature shaft at the drive housing end.

3. Connect a remote starter switch between the BAT and S terminals of the solenoid.

4. Connect the jumper cables, as shown in Figure 6-16.

5. Connect the large red and black test leads of the tester across the battery, observing polarity.

6. Follow the manufacturer's instructions for zeroing the ammeter.

7. Connect the amp inductive probe around the jumper cable from the battery negative terminal to the starter frame.

8. Place the test selector to the STARTING position.

9. Load the battery by rotating the load control knob until a voltage reading of 10 volts is obtained.

> **CAUTION:** Failure to load the battery to this level can result in the armature windings being thrown. Because there is no load on the starter, the rpm's will be excessive if more than 10 volts are used.

10. Close the remote starter switch while reading the ammeter, voltmeter, and tachometer scales.

Compare the test results with the manufacturer's specifications. General specifications will be about 6,000 to 12,000 rpm with a current draw of 60 to 85 amperes. Voltage should remain at 10 volts. If the test results are within specifications, the starter motor is ready to be installed into the vehicle.

If the current draw was excessive and rpm slower than specifications, there is excessive resistance to rotation. This could be caused by:

1. Worn bushings or bearings.

2. Shorted armature.

3. Grounded armature.

4. Shorted field windings.

5. Bent armature.

If there was no current draw, and the starter did not rotate, this could be caused by one of the following:

1. Open field windings.

2. Open armature coils.

3. Broken brush or brush spring.

Low armature speed with low current draw indicates excessive resistance. There may be a poor connection between the commutator and the brushes. Also, any connection in the starter and to the starter may be faulty.

If the armature speed and current draw readings are high, check for a shorted field winding.

Starter Motor Removal

If the tests indicate the starter motor must be removed, the first step is to disconnect the battery from the system.

Special Tools

Fender covers

Battery cable puller

It may be necessary to place the vehicle on a lift to gain access to the starter motor. Before lifting the vehicle, disconnect all wires, fasteners, and so forth, that can be reached from the top of the engine compartment.

Disconnect the wires leading to the solenoid terminals. To prevent confusion, it is a good practice to use a piece of tape to identify the different wires.

On some vehicles, it may be necessary to disconnect the exhaust system to be able to remove the starter motor. Spray the exhaust system fasteners with a penetrating oil to assist in removal. Loosen the starter mounting bolts and remove all but one. Support the starter motor; remove the remaining bolt. Then remove the starter motor.

Reverse the procedure to install the starter motor. Be sure all electrical connections are tight. If you are installing a new or remanufactured starter, remove any paint that may prevent a good ground connection. Be careful not to drop the starter. Make sure it is properly supported.

The shims are used to provide proper pinion-to-ring clearance.

Some General Motors starters use shims between the starter motor and the mounting pad (Figure 6-17). To check this clearance, insert a flat-blade screwdriver into the access slot on the

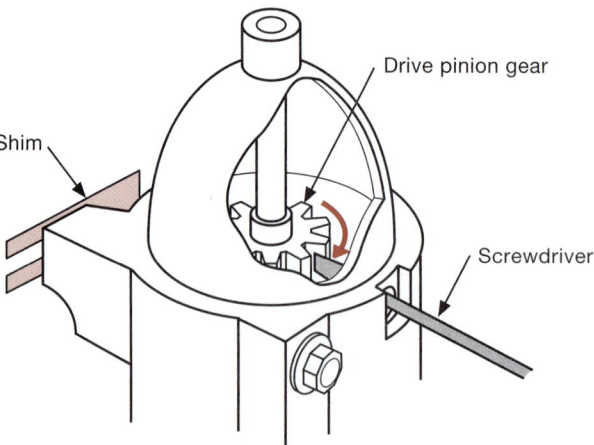

Drive pinion gear

Shim

Screwdriver

One shim will increase clearance by approximately 0.005".
More than one shim may be required.

Figure 6-17 Shimming the starter to obtain proper pinion to ring gear clearance.

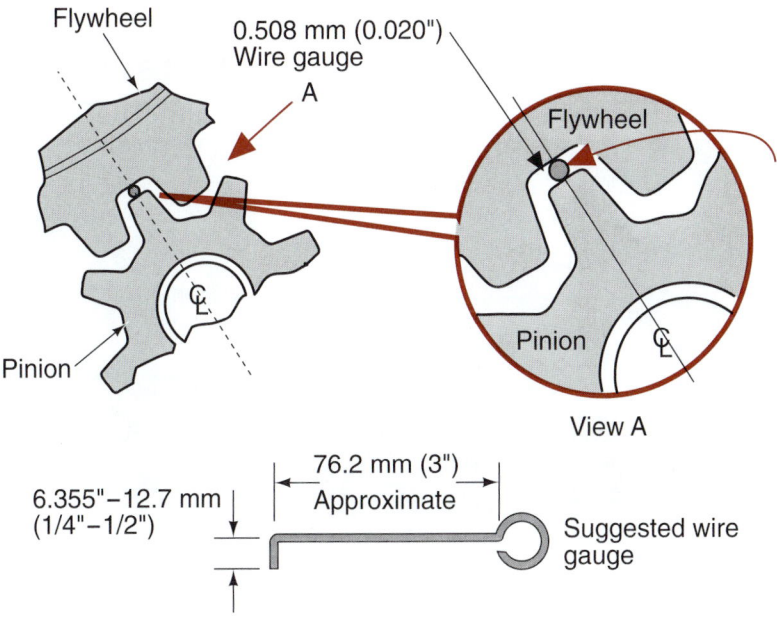

Figure 6-18 Checking the clearance between the pinion gear and ring gear.

side of the drive housing. Pry the drive pinion gear into the engaged position. Use a piece of wire that is 0.020" in diameter to check the clearance between the gears (Figure 6-18).

If the clearance between the two gears is excessive, the starter will produce a high-pitched whine while the engine is being cranked. If the clearance is too small, the starter will make a high-pitched whine after the engine starts and the ignition switch is returned to the RUN position.

> ✓ **SERVICE TIP:** The major cause of drive housing breakage is due to too small of a clearance between the pinion and ring gears. It is always better to have a little more clearance than too small of a clearance.

Starter Motor Disassembly

If it is determined that the starter is the defective part, it can be disassembled, bench tested, and rebuilt. To reduce vehicle downtime to a minimum, many repair facilities do not rebuild starters. They replace them instead. However, many shops will replace the starter drive mechanism, which may require several of the following disassembling steps. The decision to rebuild or replace the starter motor is based on several factors:

1. What is best for the customer.
2. Shop policies.
3. Cost.
4. Time.
5. Type of starter.

If the starter is to be rebuilt, the technician should study the manufacturer's service manual to become familiar with the disassembly procedures for the particular starter. Photo Sequence 8 illustrates a typical procedure for disassembly of a Delco Remy starter. Always refer

Classroom Manual
Chapter 6, page 156

Photo Sequence 8
Typical Procedure for Delco Remy Starter Disassembly

P8-1 Always have a clean and organized work area. Tools required to disassemble the Delco Remy starter: rags, assorted wrenches, snap ring pliers, flat-blade screwdriver, ball-peen hammer, plastic-head hammer, punch, scribe, safety glasses, and arbor press.

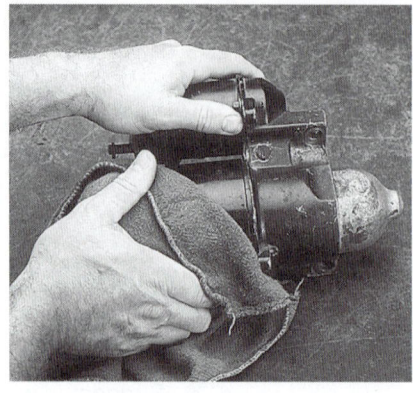

P8-2 Clean the case.

P8-3 Scribe reference marks at each end of the starter end housings and the frame.

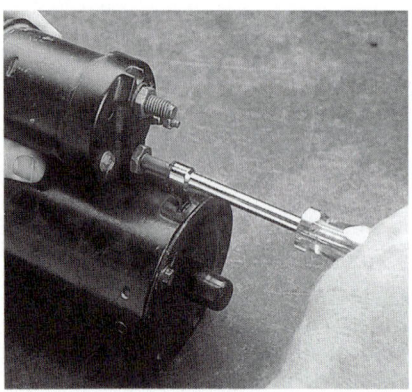

P8-4 Disconnect the field coil connection at the solenoid's M terminal.

P8-5 Remove the two screws that attach the solenoid to the starter drive housing.

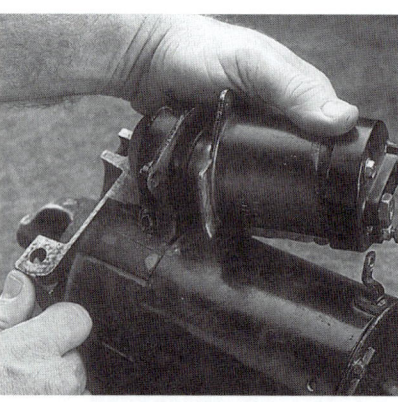

P8-6 Rotate the solenoid until the locking flange of the solenoid is free. Then remove the solenoid.

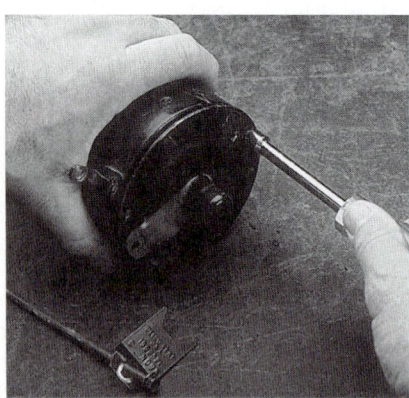

P8-7 Remove the through bolts from the end frame.

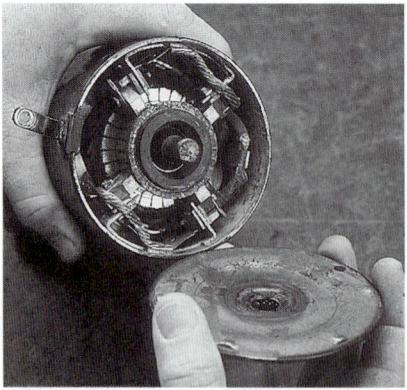

P8-8 Remove the end frame.

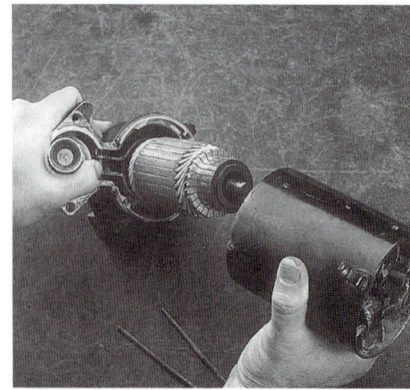

P8-9 Separate the armature from the frame.

Typical Procedure for Delco Remy Starter Disassembly (Continued)

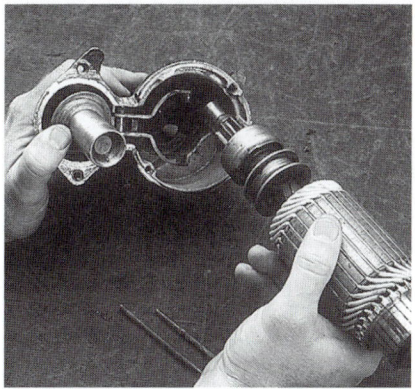

P8-10 Remove the armature from the drive housing. Note: On some units it may be necessary to remove the shift lever from the drive housing before removing the armature.

P8-11 Place a 5/8"-deep socket over the armature shaft until it contacts the retaining ring of the starter drive.

P8-12 Tap end of socket with a plastic hammer to drive the retainer toward the armature. Move it only far enough to access to the snap ring.

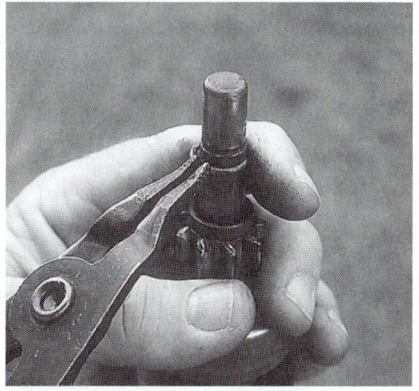

P8-13 Remove the snap ring.

P8-14 Remove the retainer from the shaft and remove the clutch and spring from the shaft. Press out the drive housing bushing and the end frame bushing.

to the specific manufacturer's service manual for the starter motor you are working on. The disassembled view of this starter is shown in Figure 6-19.

Photo Sequence 9 illustrates a typical procedure for disassembling a permanent magnet, gear reduction starter. Again, be sure to refer to the specific manufacturer's service information for the starter motor you are working on.

Classroom Manual
Chapter 6, page 158

CAUTION: Do not clean the starter motor components in solvent or gasoline. The residue left can ignite and destroy the starter. Use compressed air that is regulated to 25 psi, wipe with clean rags, or use denatured alcohol to clean the starter components.

The starter motor can be cleaned and inspected when it is disassembled. Inspect the end frame and drive housing for cracks or broken ends. Check the frame assembly for loose pole shoes and broken or frayed wires. Inspect the drive gear for worn teeth and proper overrunning clutch operation. The commutator should be free of flat spots and should not be excessively burned. Check the brushes for wear. Replace them if worn past manufacturer's specifications.

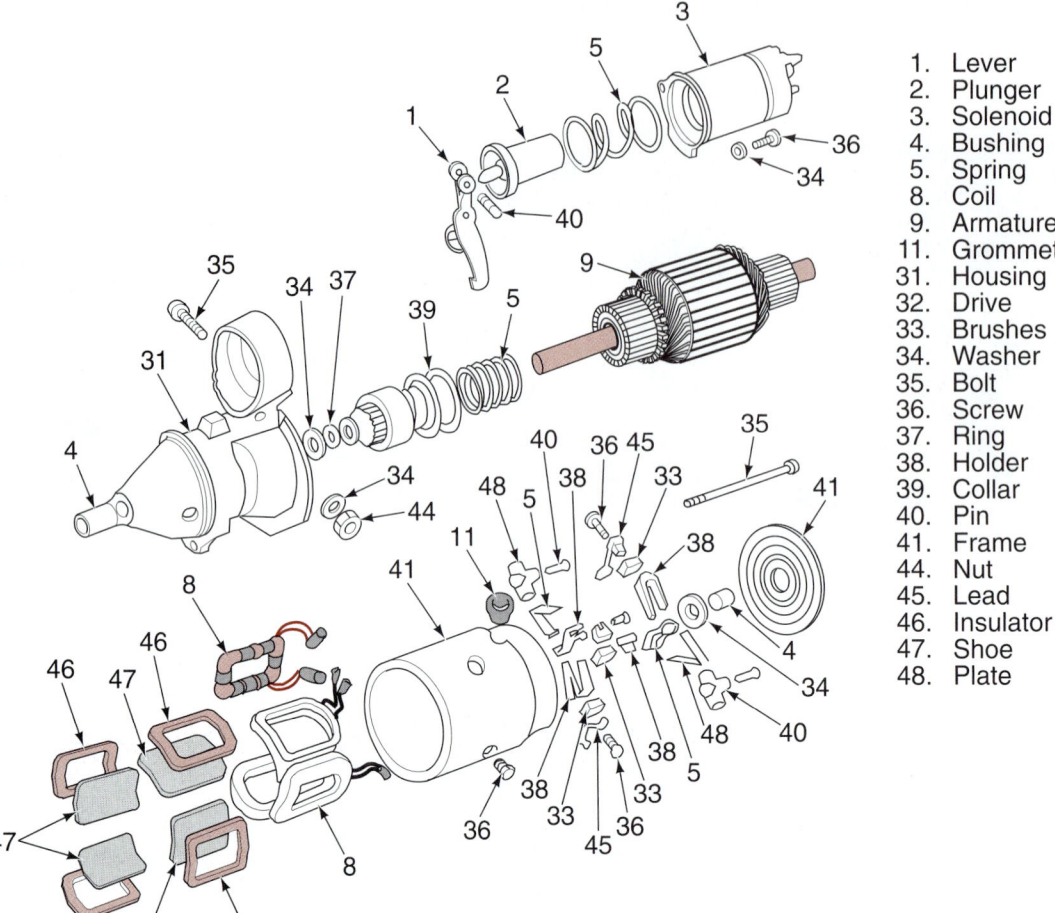

1. Lever
2. Plunger
3. Solenoid
4. Bushing
5. Spring
8. Coil
9. Armature
11. Grommet
31. Housing
32. Drive
33. Brushes
34. Washer
35. Bolt
36. Screw
37. Ring
38. Holder
39. Collar
40. Pin
41. Frame
44. Nut
45. Lead
46. Insulator
47. Shoe
48. Plate

Figure 6-19 Delco Remy 10MT starter.

Photo Sequence 9
Typical Procedure for Disassembly of Gear Reduction Starter

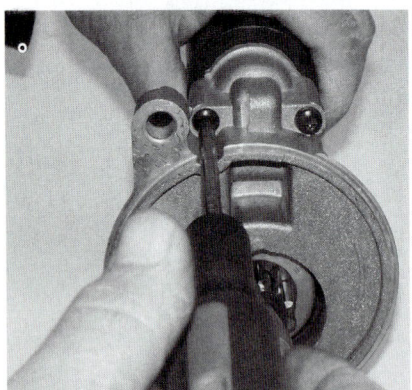

P9-1 Disconnect the lead to the field coil and remove the solenoid retainer screws.

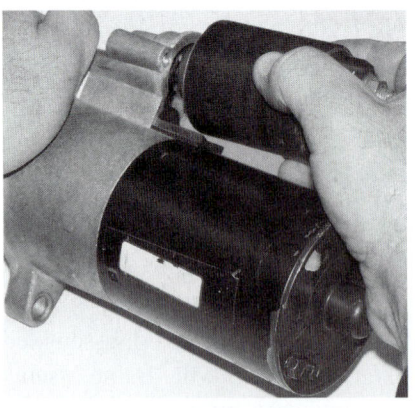

P9-2 Remove the solenoid housing while working the plunger off of the drive lever.

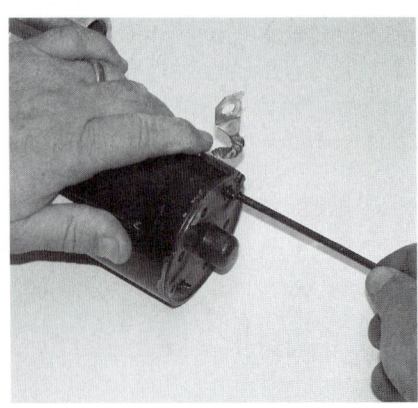

P9-3 Remove the frame through bolts.

Typical Procedure for Disassembly of Gear Reduction Starter (Continued)

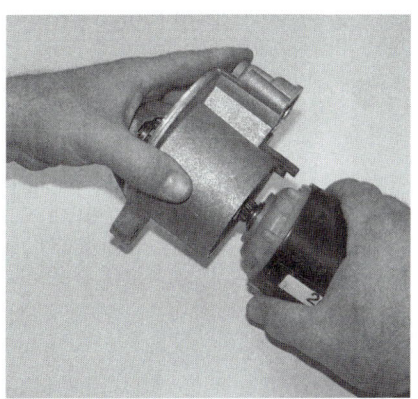

P9-4 Separate the drive end frame from the body. Remove the seal also.

P9-5 Remove the O-ring from the end of the drive gear and then remove the retainer ring and C-clip.

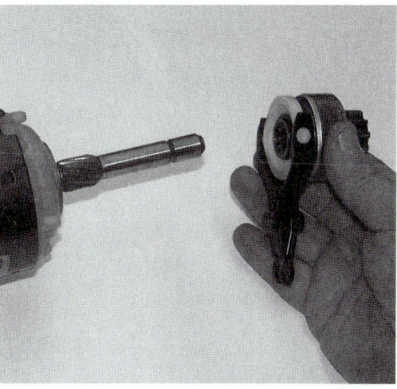

P9-6 Remove the drive from the output shaft.

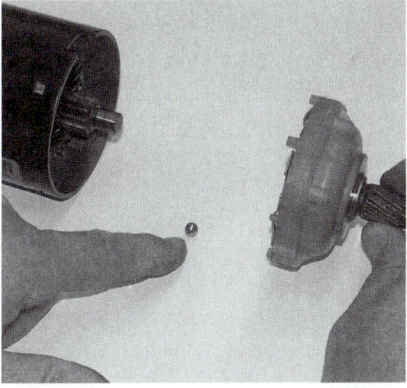

P9-7 Separate the output shaft and stationary gear assembly from the armature. Be sure to locate and retain the thrust ball located in a seat in the output shaft.

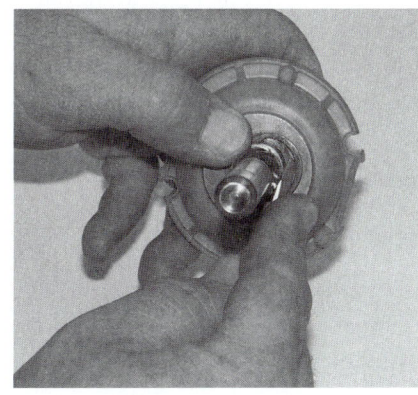

P9-8 Remove the lock ring from the output shaft and remove the stationary gear from the shaft.

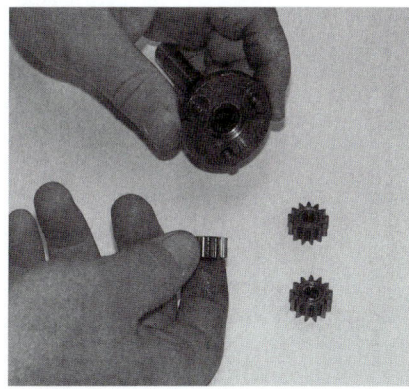

P9-9 Remove the planetary gears from the output shaft.

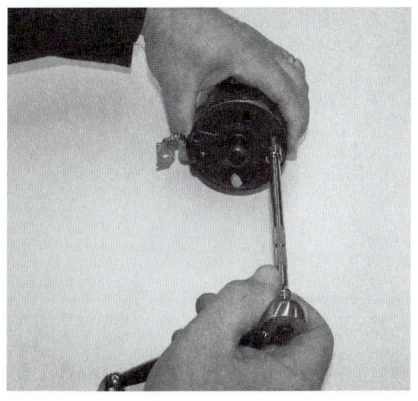

P9-10 Remove the fasteners that attach the end plate to the brush plate.

P9-11 Remove the armature and brush assembly from the body.

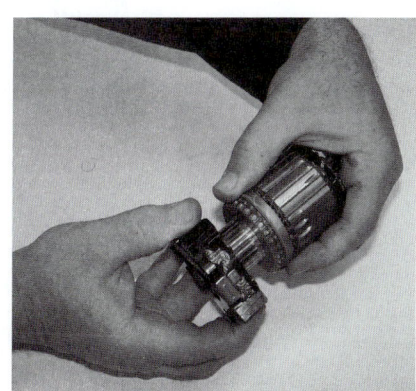

P9-12 Separate the brush assembly from the armature.

Starter Motor Component Tests

With the starter motor disassembled and the components cleaned, you are ready to perform tests that will isolate the reason for the failure. The armature and field coils are checked for shorts and opens. In most cases, the whole starter motor assembly is replaced if the armature or field coils are bad.

Classroom Manual
Chapter 6, page 142

Special Tools

Ohmmeter

Field Coil Testing

The field coil and frame assembly should be tested for opens and shorts to ground. In most cases, if one of these conditions is found, the starter is considered unrebuildable in the field and will need to be replaced with a new unit.

Field coils can be wired in a number of different ways. The most effective testing of the coils for opens and shorts is determined by how the coils are wired. There are two things to do to determine the best way to check the field coils: refer to a service manual for specific instructions and/or refer to the wiring diagram for the starting circuit. By looking at the wiring diagram, you will be able to tell where the coils get their power and where they ground. Knowing these things is critical to testing the coils. The following procedure is valid for many, but not all, vehicles.

Using an ohmmeter, place one lead on the starter motor input terminal. Connect the other lead to the insulated brushes (Figure 6-20). The ohmmeter should indicate zero resistance. If there is resistance in the field coil, replace the coil and/or the frame assembly.

Place one lead of the ohmmeter on the starter motor input terminal and the other lead to the starter frame (Figure 6-21). An infinite reading should be obtained. If the ohmmeter indicates continuity, there is a short to ground in the field coil.

Classroom Manual
Chapter 6, page 140

Special Tools

Growler

Hacksaw blade

Armature Short Test

A growler produces a very strong magnetic field that is capable of inducing a current flow and magnetism in a conductor. It is used to test the armature for shorts and grounds (Figure 6-22).

To test the armature for shorts, place the armature in the growler and hold a thin steel blade parallel to the core (Figure 6-23). Slowly rotate the armature and observe the steel blade. If the blade begins to vibrate or pull toward the core, the armature is shorted and in need of replacement.

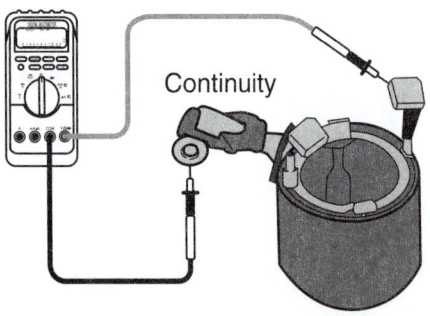

Figure 6-20 Testing the field coils for opens.

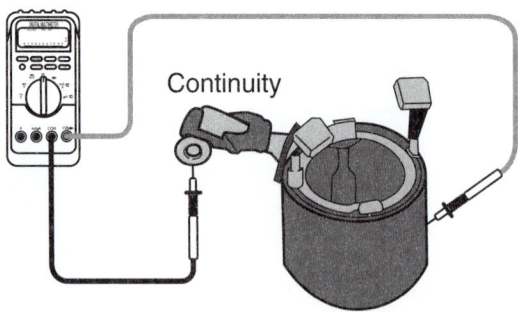

Figure 6-21 Testing the field coils for shorts to ground.

Figure 6-22 A growler is used to test the armature for shorts.

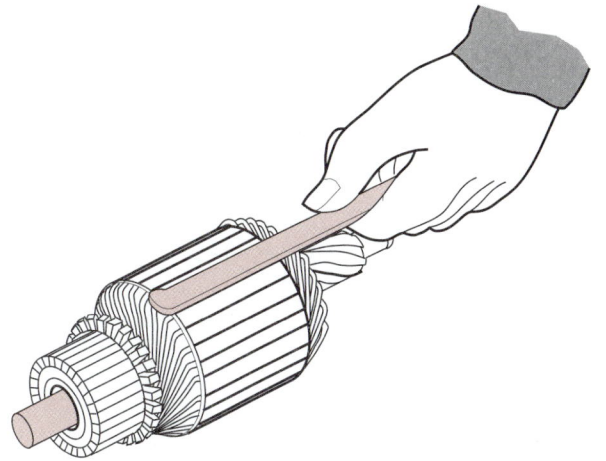

Figure 6-23 The growler generates a magnetic field. If there is a short, the hacksaw blade will vibrate over the area of the short.

Armature Ground Test

With the armature placed in the growler, use a continuity tester or ohmmeter to check for continuity between the armature core and any bar of the commutator (Figure 6-24). If there is continuity, then the armature is grounded and in need of replacement.

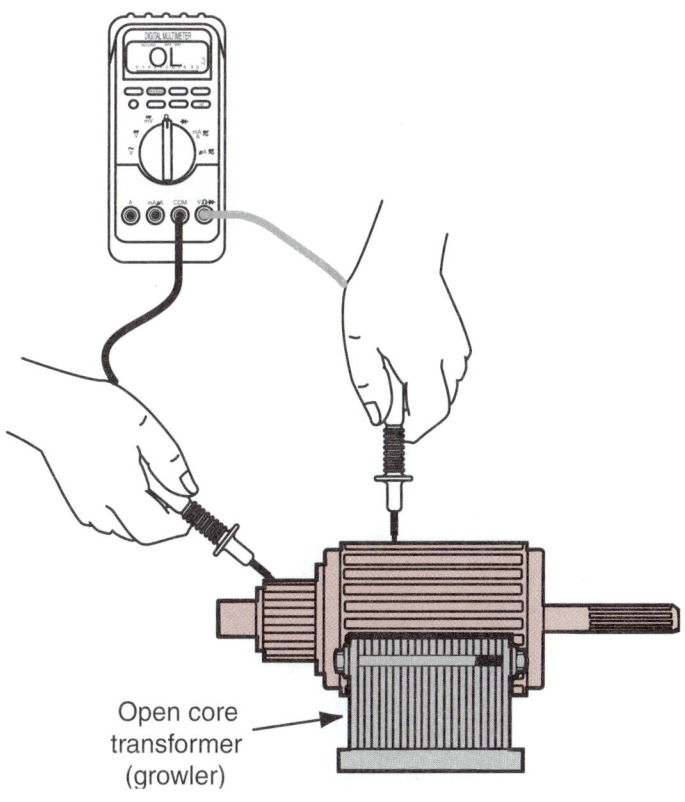

Open core transformer (growler)

Figure 6-24 Checking an armature for a short.

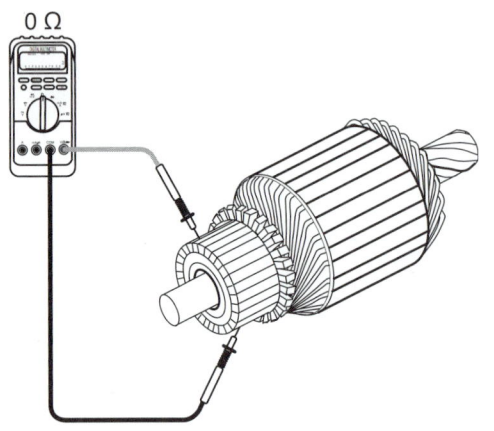

Figure 6-25 Testing the commutator for opens. There should be zero resistance between the segments.

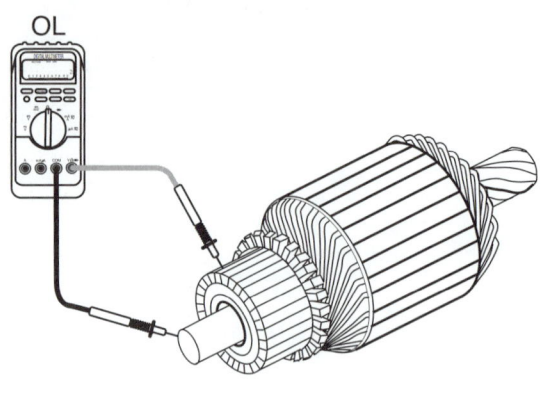

Figure 6-26 Testing the armature for short to ground. The meter should read infinite when placed on the shaft and the different segments of the commutator.

Special Tools

Ohmmeter

Crocus cloth

Classroom Manual
Chapter 6, page 139

Crocus cloth is used to polish metals. While polishing, it removes very little metal.

Commutator Tests

If a growler is not available, the armature commutator can be tested for opens and grounds using an ohmmeter. The commutator should be cleaned with **crocus cloth.** To check for continuity, place the ohmmeter on the lowest scale. Connect the test leads to any two commutator sectors (Figure 6-25). There should be zero ohms of resistance. The armature will have to be replaced if there is resistance.

To test the armature for short to ground, use an ohmmeter and connect one of the test leads to the armature shaft. Connect the other lead to the commutator segments (Figure 6-26). Check each sector. There should be no continuity to ground. The armature will have to be replaced if there is continuity.

Brush Inspection

Use an ohmmeter to test continuity through the brush holder (Figure 6-27). Connect one of the test leads to the positive brush and the other test lead to the negative brush. There

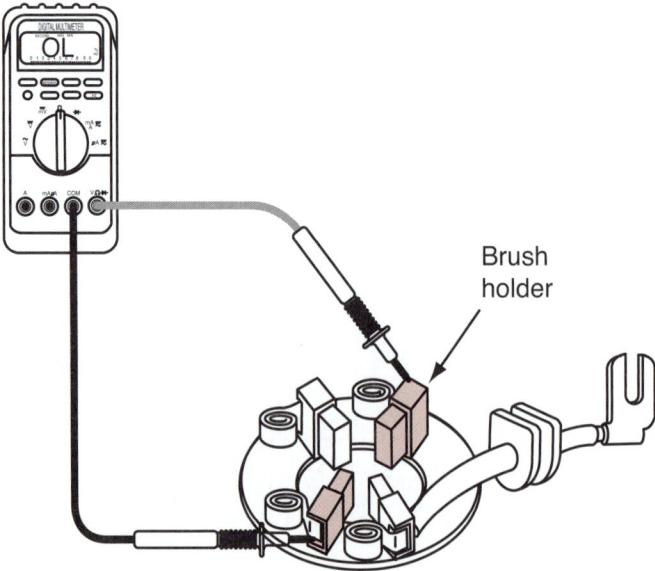

Brush holder

Figure 6-27 Typical brush holder.

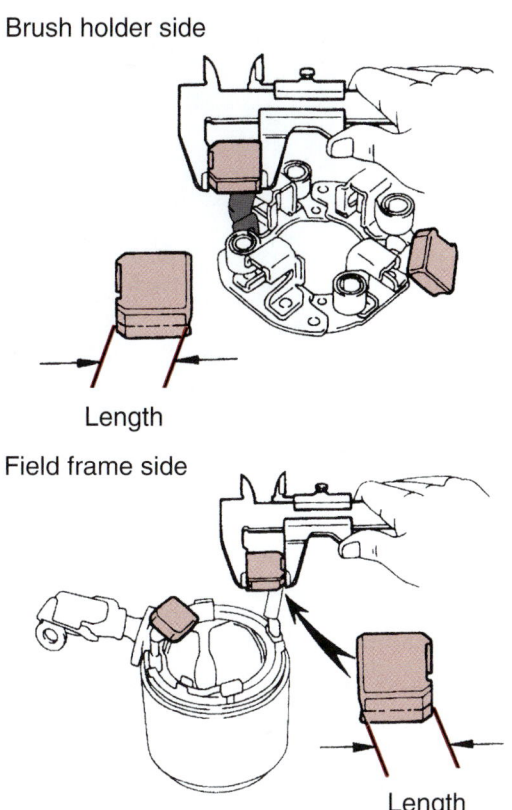

Brush holder side

Length

Field frame side

Length

Figure 6-28 Measure the length of the brushes to determine if they are worn.

should be no continuity between the brushes. If the ohmmeter indicated continuity, replace the brush holder.

Another check of the brush assembly requires checking spring tension. To do this, install the brushes into the brush holder and slide the assembly over the commutator. Use a spring scale to measure the spring tension of the holders at the point where the spring lifts off the brush. If the spring tension is below specifications, replace the springs or the brush holder assembly.

Finally, measure the length of each brush (Figure 6-28). If they are shorter than specifications, replace the brushes. Some starters may not have serviceable brushes, thus the brush holder assembly will need to be replaced.

Overrunning Clutch Inspection

The overrunning clutch may be inspected by sliding it onto the armature shaft. Attempt to rotate the clutch in both directions. If it is working properly, the clutch should rotate smoothly in one direction and lock in the other. If it fails to lock, or locks in both directions, replace the overrunning clutch.

Starter Reassembly

If the brushes are worn beyond specifications, they must be replaced. Manufacturers use two methods of connecting the brushes; they are either soldered to the coil leads or screwed to terminals.

Classroom Manual
Chapter 6, page 145

Figure 6-29 Removing the worn brushes.

Figure 6-30 Soldering a new brush to the field coil lead.

If the brushes are soldered to the coil leads, cut the old leads (Figure 6-29). Place a piece of heat-shrink tube over the brush connector. Crimp the new brush lead connector to the coil leads. Solder the brush connector to the coil lead with rosin core solder (Figure 6-30). Slide the heat-shrink tube over the soldered connection and use a heating gun to shrink the tube.

SERVICE TIP: To seat the new brushes to the commutator, slide the brushes into their holders and then place the assembly onto the commutator. Slide a piece of fine sandpaper between the commutator and the brushes with the grain facing the brushes. Rotate the armature to sand down the face of the brushes so their contour matches the commutator.

To reassemble the starter motor, basically reverse the disassembly procedures. Additional steps are listed here:

1. With a high-temperature grease, lubricate the splines on the armature shaft that the drive gear rides on.
2. To install the snap ring onto the armature shaft, stand the commutator end of the armature on a block of wood. Position the snap ring onto the shaft and hold in place with a block of wood. Hit the block of wood with a hammer to drive the snap ring onto the shaft (Figure 6-31).

Special Tools

Feeler gauge set

100-watt soldering iron

Rosin core solder

Heat-shrink tube

High-temperature grease

Two blocks of wood

Jumper cables

Jumper wire

Starting/charging system tester

Remote starter switch

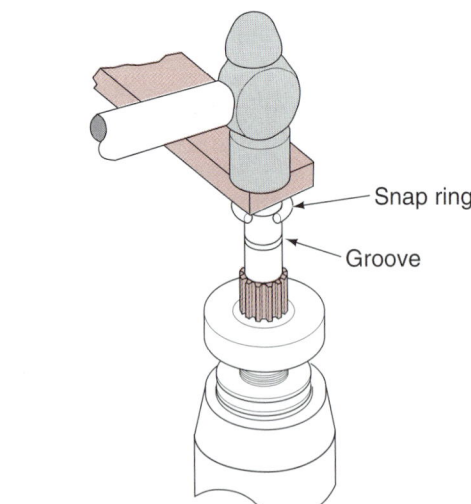

Snap ring

Groove

Figure 6-31 Once the snap ring is centered on the shaft, a hammer and block of wood can be used to install the ring onto the shaft.

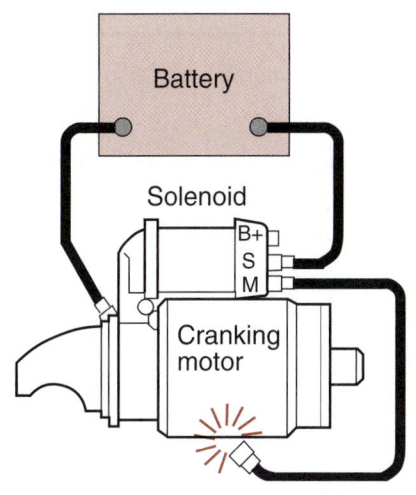
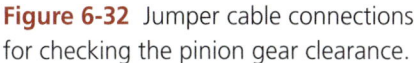

Figure 6-32 Jumper cable connections for checking the pinion gear clearance.

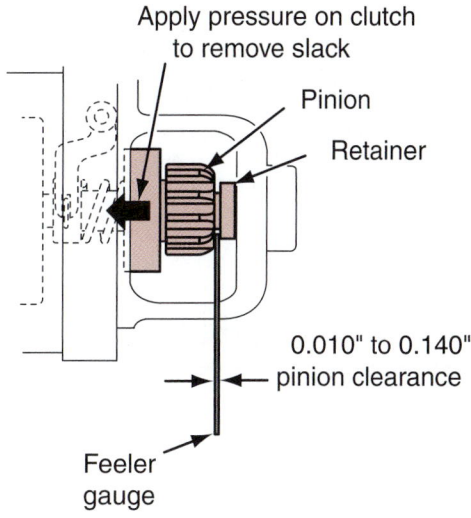

Figure 6-33 Checking the pinion gear to drive housing clearance.

3. Lubricate the bearings with high-temperature grease.

4. Apply sealing compound to the solenoid flange before installing the solenoid to the frame.

5. Use the scribe marks to locate the correct position of the frame-to-frame end and drive housing.

6. Check the pinion gear clearance. Disconnect the M terminal to the starter motor's field coils. Connect a jumper cable from the battery positive terminal to the S terminal of the solenoid. Connect the other jumper cable from the battery negative terminal to the starter frame (Figure 6-32). Connect a jumper wire from the M terminal and momentarily touch the other end of the jumper wire to the starter motor frame. This will shift the pinion gear into the cranking position and hold it there until the battery is disconnected. Once the solenoid is energized, push the pinion back toward the armature; this removes any slack. Check the clearance with a feeler gauge (Figure 6-33). Compare clearance with specifications; normally, specifications call for a clearance of 0.010 to 0.140 inches (0.25 to 0.35 mm).

SERVICE TIP: There is no provision on most starters to adjust the pinion clearance. However, if the clearance is excessive, it may indicate excessive wear of the solenoid linkage or shift lever.

7. Perform the free spin test before installing the starter into the vehicle.

CASE STUDY

A customer has a vehicle towed to the service center because it does not start. The technician verifies the complaint and observes that the starter drive engages the flywheel, but the engine does not rotate. By turning the crankshaft, the technician checks that the engine is able to rotate. The engine turns freely through two complete revolutions.

Next, the technician performs a visual inspection of the starting system. All connections are cleaned and tightened. A complete battery service is performed to confirm good battery condition. The battery passes all tests. The technician performs the solenoid circuit-resistance test and it indicates an excessive voltage drop as a result of burned contacts. The solenoid is replaced and the repair verified before returning the vehicle to the owner. The extra resistance caused by the burned contacts prevented sufficient current flow to the starter motor. This did not allow enough torque to be generated to rotate the engine.

Terms to Know

Crocus cloth

Current draw test

Free speed test

Hydrostatic lock

No crank

Quick test

Relative compression testing

Slow cranking

Solenoid circuit-resistance test

ASE-Style Review Questions

1. The starter circuit shown in Figure 6-34 has a fully charged battery. The starter relay and solenoid do not operate when the ignition switch is placed in the START position.
   Technician A says this could be caused by a grounded circuit at terminal 85 of the starter relay.
   Technician B says this could be caused by an open to terminal 86 of the starter relay.
   Who is correct?
   - **A.** A only
   - **B.** B only
   - **C.** Both A and B
   - **D.** Neither A nor B

2. Pinion gear to ring gear clearance is being discussed.
   Technician A says if the clearance is excessive, the starter will produce a high-pitched whine while the engine is being cranked.
   Technician B says if the clearance is too small, the starter will make a high-pitched whine after the engine starts.
   Who is correct?
   - **A.** A only
   - **B.** B only
   - **C.** Both A and B
   - **D.** Neither A nor B

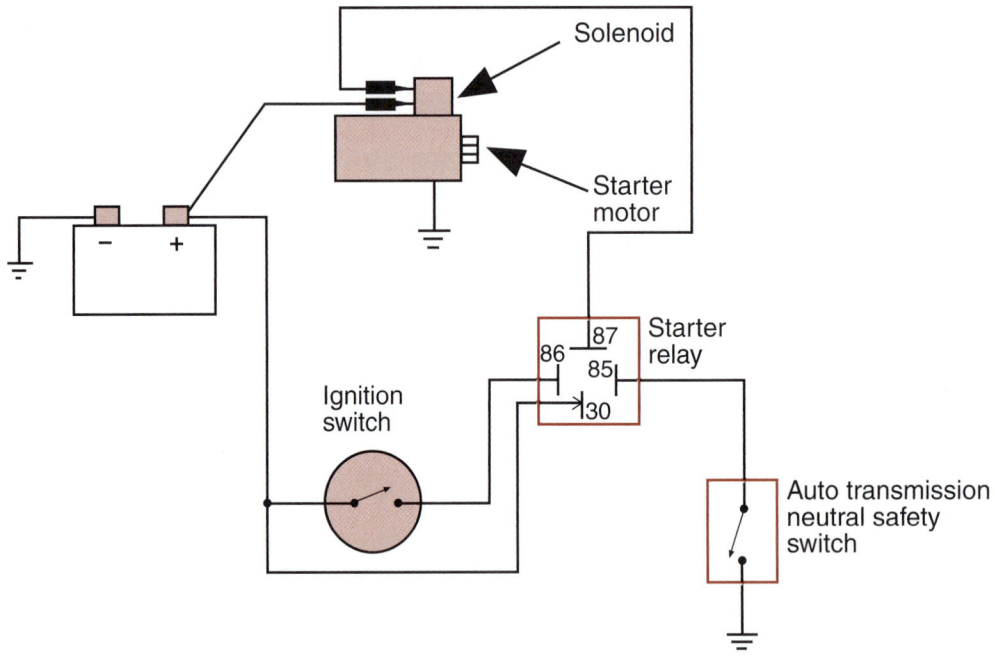

Figure 6-34 Starter circuit.

3. A 600-millivolt drop is measured across the starter motor solenoid while the engine is being cranked. What is the repair?
   A. Replace the battery.
   B. Replace the starter motor.
   C. Confirm cable connections.
   D. All of the above.

4. In the starter system shown in Figure 6-34, a voltmeter indicates battery voltage to the input of the neutral safety switch and battery voltage on the output terminal with the transmission in neutral and the ignition switch in the START position. This indicates:
   A. Normal operation.
   B. The neutral safety switch is working properly.
   C. A faulty starter relay coil winding.
   D. An open neutral safety switch ground.

5. During a starter current draw test, the voltage is 10.2 volts and amperage is above specifications. This could be caused by all of the following EXCEPT:
   A. Excessive circuit resistance.
   B. Shorted starter windings.
   C. Worn starter motor bushings.
   D. Internal engine failure.

6. Armature testing is being discussed.
   Technician A says to test for shorts, place the armature in the growler and hold a thin steel blade parallel to the core and watch for blade vibrations that would indicate a short.
   Technician B says there should be zero resistance between the commutator sectors and the armature shaft.
   Who is correct?
   A. A only
   B. B only
   C. Both A and B
   D. Neither A nor B

7. *Technician A* says it is important that a full battery test series be done before trying to test the starter system.
   Technician B says the internal condition of the engine has little effect on the operation of the starting system.
   Who is correct?
   A. A only
   B. B only
   C. Both A and B
   D. Neither A nor B

8. The starter motor has been rebuilt and is ready to install in the vehicle.
   Technician A says to perform the free spin test before installing the starter into the vehicle.
   Technician B says to remove the M terminal connector before installing the starter motor.
   Who is correct?
   A. A only
   B. B only
   C. Both A and B
   D. Neither A nor B

9. Voltage drop testing of the solenoid control circuit is being discussed.
   Technician A says the maximum amount of voltage drop allowed is 0.9 volt.
   Technician B says the voltage drop across each wire should be less than 0.1 volt.
   Who is correct?
   A. A only
   B. B only
   C. Both A and B
   D. Neither A nor B

10. Technician A says most starter noises come from the armature.
    Technician B says the starter drive cannot be replaced on most starters.
    Who is correct?
    A. A only
    B. B only
    C. Both A and B
    D. Neither A nor B

ASE Challenge Questions

1. A vehicle is towed into the shop due to a no-start (no-crank) condition. The headlights are turned on, and when the ignition key is placed in the START position, the lights remain bright but the starter does not crank the engine.
   Technician A says that the starter may be binding internally.
   Technician B says that the battery may be discharged.
   Who is correct?
   A. A only   **C.** Both A and B
   B. B only   **D.** Neither A nor B

2. A vehicle is being tested for a slow-crank condition. The starter current draw is 525 amps.
   Technician A says that the field windings in the starter may be shorted.
   Technician B says that the starter drive gear may be slipping.
   Who is correct?
   A. A only   **C.** Both A and B
   B. B only   **D.** Neither A nor B

3. The starter draw of a vehicle with a slow-crank condition is 75 amps. The battery has a good state of charge and it passed a capacity test.
   Technician A says that a starting circuit voltage drop test should be performed.
   Technician B says that the neutral safety switch may have excessive resistance.
   Who is correct?
   A. A only   **C.** Both A and B
   B. B only   **D.** Neither A nor B

4. A voltmeter that is connected across a starter solenoid's battery and motor terminals indicates 12 volts when the ignition key is turned to the START position. A distinct click is heard from the solenoid when this occurs but the engine does not crank.
   Technician A says that there is excessive voltage drop in the circuit between the battery and the starter solenoid.
   Technician B says that the starter solenoid probably needs to be replaced.
   Who is correct?
   A. A only   **C.** Both A and B
   B. B only   **D.** Neither A nor B

5. *Technician A* says that incorrect pinion gear to ring gear clearance can result in noisy cranking and high starter draw.
   Technician B says that starting circuit voltage drop should be checked before performing a starter draw test.
   Who is correct?
   A. A only   **C.** Both A and B
   B. B only   **D.** Neither A nor B

Job Sheet 22

22

Name _____ Date _____

Current Draw Test

Upon completion of this job sheet, you should be able to measure the current draw of a starter motor and interpret the results of the test.

ASE Correlation

This job sheet is related to the ASE Electrical/Electronic System Certification Exam's content area: *Starting System Diagnosis and Repair;* task: Perform starter current draw test; determine needed repairs.

Tools and Materials

A vehicle with a 12-volt battery
Starting/charging system tester
Service manual

Describe the vehicle being worked on:

Year _____ Make _____ Model _____

VIN _____ Engine type and size _____

Procedure

Task Completed

1. Perform a battery test series to confirm the battery is good. If necessary, refer to the job sheets in Chapter 5. ☐

2. Disable the ignition or fuel injection systems to prevent the engine from starting. ☐

3. Expected starter current draw is: _____ amps.
   Voltage should not drop below: _____ volts.

4. Connect the starting charging system tester cables to the vehicle. ☐

5. Zero the ammeter on the tester. ☐

6. Be prepared to observe the amperage when the engine begins to crank and while it is cranking. Also, note the voltage when you stop cranking the engine.
   The initial current draw was: _____ amps.
   After _____ seconds, the current draw was _____ amps and the voltage dropped to _____ volts.

7. What is indicated by the test results? Compare your measurements to the specifications.

_____

_____

_____

_____

_____

☐

8. Reconnect the ignition or fuel injection system and start the engine.

Instructor's Response _____

_____

_____

Job Sheet 23

23

Name _____ Date _____

Testing the Starting System Circuit

Upon completion of this job sheet, you should be able to visually inspect the starting circuit and perform voltage drop testing for excessive resistance.

ASE Correlation

This job sheet is related to the ASE Electrical/Electronic System Certification Exam's content area: *Starting System Diagnosis and Repair;* task: Perform starter circuit voltage drop tests; determine needed repairs.

Tools and Materials

A vehicle with a 12-volt battery
Wiring diagram for the vehicle assigned
A DMM

Describe the vehicle being worked on:

Year _____ Make _____ Model _____

VIN_____ Engine type and size _____

Procedure

Task Completed

1. Draw a simple starter circuit for the vehicle. Use the wiring diagram as a guide.

☐

2. Disable the ignition or fuel injection systems to prevent the engine from starting.

☐

3. Connect the voltmeter across the battery's negative cable.
   Crank the engine with the starter and observe the readings on the meter.
   Your reading was _____ volts.

4. What does this indicate?

   _____

   _____

   _____

5. Connect the voltmeter across the battery's positive cable (from battery to starter motor). Crank the engine with the starter and observe the readings on the meter.
   Your reading was _____ volts.

6. This test measured the voltage drop across everything in the positive side of the circuit. What is included in this circuit?

_____

_____

_____

7. What do the test results suggest? What are your recommendations?

_____

_____

_____

☐   8. Reconnect the ignition or fuel injection systems to allow the engine to start.

Instructor's Response _____

_____

_____

Job Sheet 24

24

Name _____ Date _____

Starter Free Speed Testing

Upon completion of this job sheet, you should be able to perform a free speed test and interpret the results.

ASE Correlation

This job sheet is related to the ASE Electrical/Electronic System Certification Exam's content area: *Starting System Diagnosis and Repair;* task: Perform starter free-running (bench) tests and determine needed repairs.

Tools and Materials

Starter motor
Appropriate service manual

Procedure

Task Completed

1. Identify the starter motor assigned to you. _____

   _____

   _____

2. What page in the service manual is this test procedure found on? _____

3. What is the rpm specification? _____

   _____

4. What is the current draw specification? _____

   _____

5. List the steps necessary to perform the free speed test: _____

   _____

   _____

6. List the precautions associated with this test: _____

   _____

   _____

   _____

7. Perform the free speed test and record the results:

   Starter rpm _____ Starter current draw _____

8. What do the test results indicate? What are your recommendations?

   _____

   _____

Instructor's Response _____

_____

_____

DIAGNOSTIC CHART 6-1

PROBLEM AREA:	Starting system operation
SYMPTOMS:	Starter fails to turn engine or operates at reduced efficiency. Excessive current draw.
POSSIBLE CAUSES:	**1.** Shorted armature. **2.** Worn starter bushings. **3.** Bent armature. **4.** Thrown armature windings. **5.** Loose pole shoes. **6.** Grounded armature. **7.** Shorted field windings.

DIAGNOSTIC CHART 6-2

PROBLEM AREA:	Starting system operation
SYMPTOMS:	No or reduced starter operation. Current draw too low.
POSSIBLE CAUSES:	**1.** Worn brushes. **2.** Excessive circuit voltage drop.

DIAGNOSTIC CHART 6-3

PROBLEM AREA:	Starting system operation
SYMPTOMS:	Starter fails to rotate the engine or operates at reduced efficiency. Excessive starter circuit resistance.
POSSIBLE CAUSES:	**1.** Excessive starter circuit voltage drop. **2.** Improper connections. **3.** Corroded ground connections. **4.** High resistance in solenoid or relay.

DIAGNOSTIC CHART 6-4

PROBLEM AREA:	Starting system operation
SYMPTOMS:	Starter fails to rotate engine or rotates too slowly. Failed starter test series.
POSSIBLE CAUSES:	**1.** Defective or worn starter. **2.** Worn brushes. **3.** Shorted field coils. **4.** Open field coils. **5.** Shorted armature. **6.** Open armature.

DIAGNOSTIC CHART 6-5

PROBLEM AREA:	Starter control circuit
SYMPTOMS:	Starter does not operate when ignition switch is located in the START position. No sounds.
POSSIBLE CAUSES:	**1.** Open circuit. **2.** Faulty ignition switch. **3.** Park/neutral switch faulty or misadjusted. **4.** Faulty starter relay/solenoid. **5.** Faulty starter.

DIAGNOSTIC CHART 6-6

PROBLEM AREA:	Starter control circuit
SYMPTOMS:	Starter does not operate when ignition switch is located in the START position. Relay/solenoid clicks.
POSSIBLE CAUSES:	**1.** High resistance in starter control circuit. **2.** High resistance in started circuit. **3.** Faulty starter relay/solenoid. **4.** Faulty starter.

DIAGNOSTIC CHART 6-7

PROBLEM AREA:	Starter drive
SYMPTOMS:	Starter spins but does not rotate the engine.
POSSIBLE CAUSES:	**1.** Defective one-way clutch. **2.** Broken teeth on ring gear. **3.** Faulty starter motor.

DIAGNOSTIC CHART 6-8

PROBLEM AREA:	Starter drive fails to disengage after engine starts.
SYMPTOMS:	Excessive noise after engine starts.
POSSIBLE CAUSES:	**1.** Faulty ignition switch. **2.** Faulty relay/solenoid. **3.** Faulty starter motor. **4.** Improper starter mounting.

Charging System Testing and Service

Upon completion and review of this chapter, you should be able to:

- ❏ Diagnose charging system problems that cause an undercharge or no-charge condition.
- ❏ Diagnose charging system problems that cause an overcharge condition.
- ❏ Inspect, adjust, and replace generator drive belts, pulleys, and fans.
- ❏ Perform charging system output tests and determine needed repairs.
- ❏ Perform charging system circuit voltage drop tests and determine needed repairs.

- ❏ Perform voltage regulator tests and determine needed repairs.
- ❏ Test and replace AC generator diodes and/or rectifier bridge.
- ❏ Remove and replace the AC generator.
- ❏ Disassemble, clean, and inspect AC generator components.
- ❏ Inspect and replace AC generator brushes and brush holders.
- ❏ Test and diagnose the rotor.
- ❏ Test and diagnose the stator.

Basic Tools

Basic mechanic's tool set

Service manual

Introduction

 WARNING: Always wear safety glasses when performing charging system tests.

Whenever there is a charging system problem, make sure the battery is thoroughly checked first. The battery supplies the electrical power for the charging system. If the battery is bad, the charging system cannot be expected to work its best. AC generators are designed to maintain the charge of a battery, not to charge a dead battery.

There are many different types of testers that can be used to test the charging system and AC generators. Some handheld multimeters have the ability to conduct many tests. However, the best testers to use are those designed to test the entire system. These testers are commonly referred to as starting/charging system testers. Often, in this chapter, a reference is made to using a VAT-40. This tester, made by Sun Electric Corporation, is commonly found in service departments. Although a VAT-40 is mentioned in the text, this does not mean that this is the only tester that can be used. Any starting/charging system tester can be used (Figure 7-1). Always follow the operating procedures for the specific tester being used.

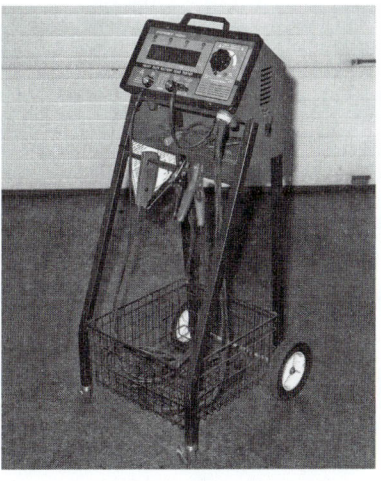

Figure 7-1 A typical battery-starting charging system tester.

When performing the tests, be sure of the connections you are making. Refer to the service manual for identification of the various terminals for the AC generator and regulator. Connecting a test lead to the wrong terminal can result in AC generator damage, as well as damage to other electrical and electronic components.

Before attempting to test the charging system, first check the battery. The state of charge must be considered before faulting any of the charging system components. If the battery passes the state of charge test, a load test should be performed to determine the capacity. If the battery fails this test, use a battery charger to fully recharge the battery and test again. It is important that the battery be in good condition in order to obtain accurate charging system test results. In addition, the battery must be fully charged before proceeding with the diagnosis of the charging system.

It is also important to perform a preliminary inspection. Many problems can be detected during this simple step. Check the following items:

1. Condition of the drive belt (Figure 7-2). If the drive belt is worn or glazed, it will not allow enough rotor rpm's to produce sufficient current.

2. Drive belt tension (Figure 7-3).

3. Electrical connections to the AC generator.

4. Electrical connections to the regulator.

5. Ground connections at the engine and chassis.

6. Battery cables and terminals.

7. Fuses and fusible links.

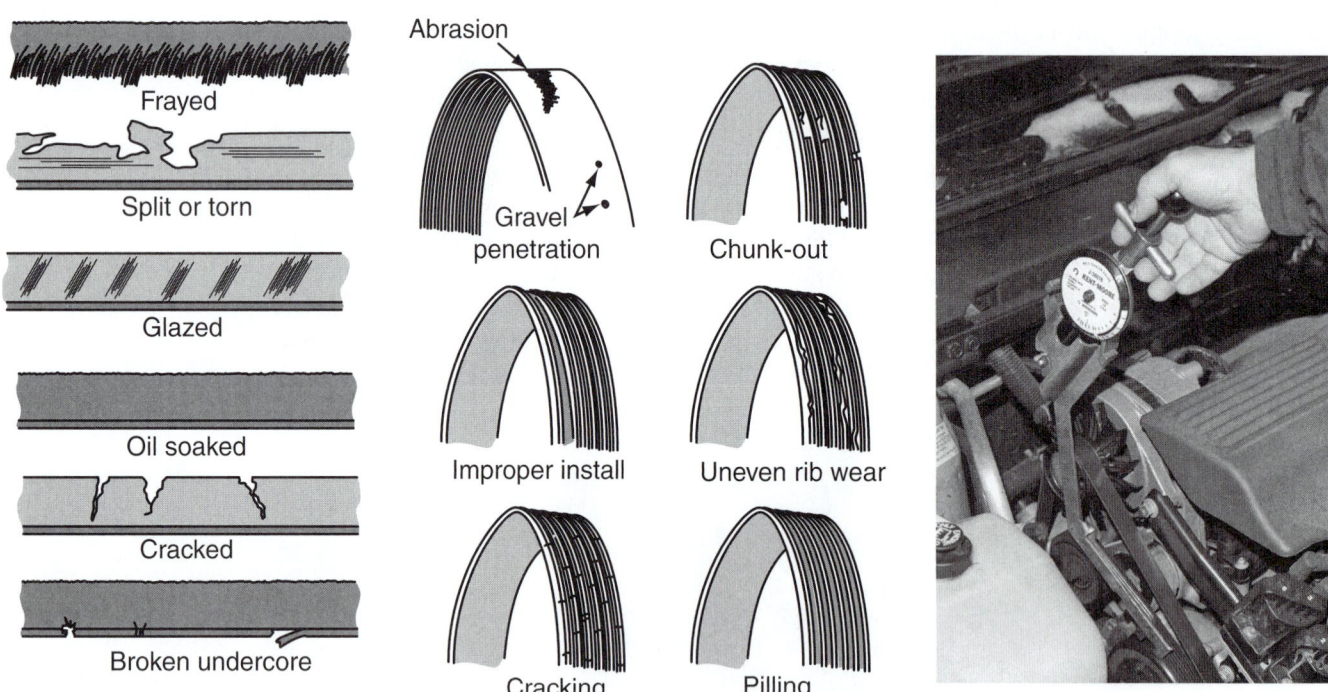

Figure 7-2 The generator drive belt must be replaced if any of these conditions exist.

Figure 7-3 Checking belt tension using a tension gauge.

8. Excessive current drain caused by a light or other electrical component remaining on after the ignition switch is turned off.

9. Symptoms of undercharging. These include slow-cranking, discharged battery, low instrument panel ammeter or voltmeter readings, and charge indicator lamp on.

10. Symptoms of overcharging. These include high ammeter and voltmeter readings, battery boiling, and charge indicator lamp on.

CAUTION: Do not overtighten the drive belt. Early bearing failure can occur if the belt is tightened beyond manufacturer's specifications.

SERVICE TIP: To check the fusible link to the AC generator, use a voltmeter and test for voltage at the BAT terminal. If the battery is good, voltage should be present. If there is no voltage, the fusible link is probably burned out. A better test would be to measure the voltage drop across the link. This will identify any high resistance in the circuit.

The manufacturer of the vehicle you are working on may have several additional tests to perform. It is important to always follow the procedures outlined by the manufacturer for the vehicle being tested.

WARNING: Many charging system tests require that the vehicle be operated in the shop area. Always place wheel blocks against the drive wheels. Be sure there is proper ventilation of the vehicle's exhaust. Also, be aware of the drive belts and cooling fan. Be sure of where your hands and tools are at all times.

Charging System Service Cautions

The following are some of the general rules when servicing the charging system:

1. Do not run the vehicle with the battery disconnected. The battery acts as a buffer and stabilizes any voltage spikes that may cause damage to the vehicle's electronics.

2. Do not allow output voltage to increase over 16 volts when performing charging system tests.

3. If the battery needs to be recharged, disconnect the cables while charging.

4. Do not attempt to remove electrical components from the vehicle with the battery connected.

5. Before connecting or disconnecting any electrical connections, the ignition switch must be in the OFF position.

6. Avoid contact with the BAT terminal of the AC generator while the battery is connected. Battery voltage is always present at this terminal.

AC Generator Noises

Noises that come from the AC generator can be from three sources. The causes of the noises are identifiable by the types of noises they make. A loose belt will make a squealing noise. Check the belt condition and tension. Replace the belt if necessary.

Classroom Manual
Chapter 7, page 167

SERVICE TIP: With the engine off, rub a piece of bar soap on the pulley surface of the drive belts. Do this one belt at a time until the noise stops. This way you will know which belt is the cause of the noise.

A squealing noise can also be caused by faulty bearings. The bearings are used to support the rotor in the housing halves. To test for bearing noises, use a length of hose, a long screwdriver, or a technician's stethoscope. By placing the end of the probe tool close to the bearings and listening on the other end, any bearing noise will be transmitted so you will be able to hear it. Bearing replacement will require disassembly of the AC generator.

 WARNING: This test is performed with the engine running. Use caution around the drive belts, fan, and other moving components.

A whining noise can be caused by shorted diodes or stator, or by a dry rotor bearing. A quick way to test for the cause of a whining sound is to disconnect the wiring to the generator, then start and run the engine. If the noise is not there, the cause of the noise is a magnetic whine due to shorted diodes or stator windings. Use a scope to verify the condition of the diodes and stator. If the noise remains, the cause is mechanical and probably due to worn bearings.

Charging System Troubleshooting

Troubleshooting charts assist the technician in diagnosing the charging system. They give several possible causes for the customer complaint. They also instruct the technician in what tests to perform or what service is required.

Voltage Output Testing

Classroom Manual
Chapter 7, page 178

Special Tools

Voltmeter

Tachometer

Once the visual inspection and preliminary checks are completed, the next step is to perform a **voltage output test.** The voltage output test is used to make a quick determination about whether or not the charging system is working properly. If the charging system is operating correctly, then check for battery drain. The following procedure is for performing the test:

1. Connect the voltmeter across the battery terminals, observing polarity.
2. Connect the tachometer, following the manufacturer's procedure.
3. With the engine off, record the base voltage value across the battery.
4. Start the engine. Because most AC generators do not produce maximum voltage output until 1,500 to 2,000 engine rpm, the engine speed needs to be brought up to this level.
5. Observe the voltmeter reading. It should read between 13.5 and 14.5 volts.

If the charging voltage was too high, there may be a problem in the following areas:

1. Defective voltage regulator.
2. Poor voltage regulator ground connection.
3. Short to ground in the field coil control circuit, causing the AC generator to full field.
4. High resistance in the "sense" circuit between the battery and the PCM or voltage regulator.

If the charging voltage was too low, the fault might be:

1. Loose or glazed drive belt.
2. Defective voltage regulator.
3. Defective AC generator.
4. Discharged battery.
5. Loose or corroded battery cable terminals.

If the voltage reading was correct, perform a load test to check the voltage output under a load condition:

1. With the engine running at idle, turn on the headlights and the heater fan motor to high speed.

2. Increase the engine speed to approximately 2,000 rpm.

3. Check the voltmeter reading. It should increase a minimum of 0.5 volt over the base voltage reading taken previously. Some vehicle manufacturer's specifications require a rise in voltage of 2.5 volts over the base voltage value obtained in step 3 of the test.

4. If the voltage increases, the charging system is operating properly. If the voltage did not increase, perform the following test series to locate the fault.

 SERVICE TIP: If the charging system passes the no-load test but fails the load test, check the condition and tension of the drive belt closely.

Voltage Drop Testing

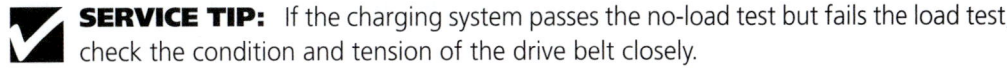

The voltage drop of all wires and connections combined should not exceed 3% of the system voltage. Any particular wire or connection should not exceed 0.2 volt; total system drops should be less than 0.7 volt. The ground side voltage drop should be less than 0.2 volt. The **voltage drop test** determines if the battery, regulator, and AC generator are all operating at the same potential. To perform the voltage drop test using a typical charging system tester, follow these steps:

Special Tools

Voltmeter
Tachometer

1. Connect the large red cable to the battery positive terminal.

2. Connect the large black cable to the battery negative terminal.

3. Select CHARGING.

4. Select EXT 3 V.

5. Zero the ammeter.

6. Clamp the inductive pickup around the AC generator output wire.

7. Using the small red and black test leads, connect at the following locations:

   Insulated circuit: Red lead to AC generator output terminal. Black lead to the battery positive terminal.

   Ground circuit: Red lead to battery negative terminal. Black lead to AC generator housing.

8. Start the engine and hold the engine speed between 1,500 and 2,000 rpm.

9. Using the carbon pile knob, load the system between 9 and 20 amperes.

Some manufacturers recommend measuring voltage drop when the generator is putting out its maximum. Always follow the recommendations of the manufacturer. Remember, if the AC generator is not putting out any current, there will be no voltage drop even if the circuit is very corroded.

 SERVICE TIP: Turning on the headlights may be substituted for the carbon pile.

10. Read voltmeter.

The general specifications for voltage drop testing are:

   Insulated circuit: less than 0.5 volt.
   Ground circuit: less than 0.2 volt.

If a higher voltage drop is observed, work up the circuit to find the fault. Check every wire and connection.

Alternate Procedure

If a starter/charging system tester is not available, the voltage drop test can be performed with just a voltmeter. Simply follow steps 7 through 10 in the previous procedure and turn on the headlights to load the system.

Classroom Manual
Chapter 7, page 182

Special Tools

Starting/charging system tester

Multiplying coil

A **multiplying coil** is made of 10 wraps of wire. This multiplies the ammeter reading so that a starting/charging system tester's scale can be used to read lower current. For example, if the needle is pointing to 25 amperes, when using the multiplying coil, the actual reading is 2.50 amperes.

Field Current Draw Test

Because field current is required to create a magnetic field, it is necessary to determine if current is flowing to the field coil. To perform the **field current draw** test using a typical starting/charging system tester, follow these steps:

1. Connect the large red and black cables across the battery, observing polarity.
2. Select CHARGING.
3. Select INT 18 V.
4. Zero the ammeter.
5. Disconnect the field wire from either the AC generator or the regulator.
6. Connect the **multiplying coil** to the field terminal. Make the connection toward the AC generator.
7. Connect the field lead of the tester to the multiplying coil.
8. Clamp the inductive pickup around the loop of the multiplying coil (Figure 7-4).
9. Move the toggle switch to the proper field-type position (A or B).
10. Read the ammeter while the toggle switch is depressed.
11. Compare results with manufacturer's specifications.

For GM, A circuit systems, steps 1 through 4 are the same. Remove the field plug from the AC generator and connect a Y-type connector between terminals 1 and 2. Connect the multiplying coil to the Y connector. Connect the field lead of the tester to the multiplying coil. The inductive pickup is clamped around the loop of the multiplying coil. Press the toggle switch to B and read the field current draw on the ammeter.

 CAUTION: Be sure to move the toggle switch to B even though an A circuit is being tested.

Figure 7-4 Connecting the multiplying coil and amp pickup to a generator.

If the readings are within the specification limits, then the field circuit is good. If the readings are over specifications, a shorted field circuit or bad regulator may be the problem. If the readings were too low, then there is high electrical resistance that may be caused by worn brushes.

To test Ford's integral alternator/regulator (IAR) system, use a voltmeter as follows:

1. With the ignition switch in the OFF position, connect the negative voltmeter lead to the generator housing.

2. Connect the positive voltmeter lead to the F terminal screw of the regulator (Figure 7-5).

3. Check the voltmeter reading. It should indicate battery voltage. If it reads battery voltage, the field circuit is normal.

4. If the voltmeter reading is less than battery voltage, disconnect the wiring plug from the regulator.

5. Connect the positive voltmeter lead to the I terminal of the plug (Figure 7-6).

Classroom Manual
Chapter 7, page 192

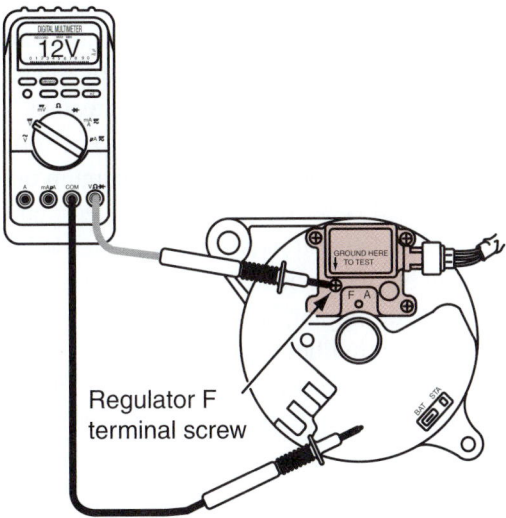

Figure 7-5 Testing the IAR generator field circuit.

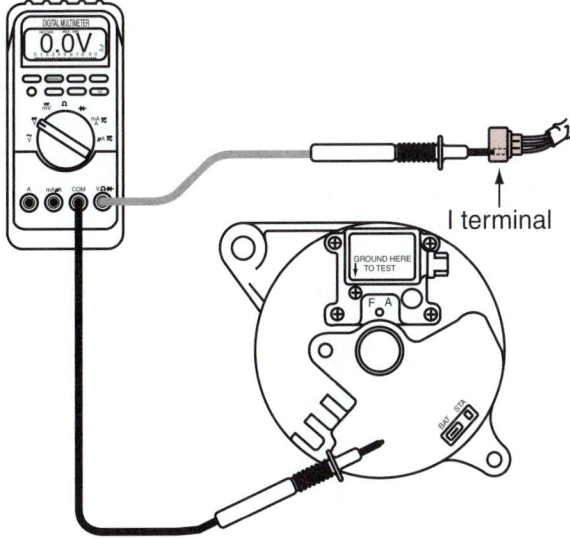

Figure 7-6 Wiring plug (connector) terminal identification.

6. Check the voltmeter reading. It should indicate 0 volts. If there is voltage present, repair the I lead from the ignition switch. The I lead is receiving voltage from another source.

7. If there was no voltage present in step 6, connect the positive voltmeter lead to the S terminal of the regulator wiring plug.

8. Check the voltmeter reading. If there are 0 volts, replace or service the regulator.

9. If voltage is indicated, disconnect the wiring plug from the AC generator.

10. Check for voltage to the regulator wiring plug S terminal.

11. If voltage is still present, repair the S terminal wire lead to the AC generator. The S terminal wire is receiving voltage from another source.

12. Replace the rectifier bridge if no voltage is present.

Current Output Testing

Classroom Manual
Chapter 7, page 165

Special Tools

Starting/charging
system tester
Tachometer

The system must be loaded in order to obtain AC generator current output. By connecting a carbon pile to maintain system voltage at 12 volts, the signal voltage to the regulator will be reduced. When this occurs, the regulator attempts to recharge the battery by full fielding. This will produce the maximum current output to the battery. The **current output test** will determine the maximum output of the alternator. Follow the steps in Photo Sequence 10 to perform the current output test.

Once the maximum current output is known, add the maximum output reading to the reading obtained in step 5. This total should be within 10% of the rated output of the AC generator.

If the ammeter reading indicates that output is 2 to 8 amperes below the specification, then an open diode or slipping belt may be the problem. If the output reading indicates 10 to 15 amps below specifications, a shorted diode or slipping belt may be the problem. If the AC generator output is below specifications, perform the full field test.

When testing General Motors CS-130 and 144 AC generators, first use a voltmeter to test for voltage at the L and I terminals. Battery voltage should be indicated at both terminals with the ignition switch in the RUN position.

Full Field Test

Classroom Manual
Chapter 7, page 181

Special Tools

Various jumper wires

Starting/charging
system tester

Tachometer

The **full field test** will determine if the detected problem lies in the AC generator or the regulator. The full field test needs to be performed only if the charging system failed the output test. This test is performed by manually **full fielding** the AC generator with the regulator bypassed. Full fielding means the field windings are constantly energized with full battery current. Full fielding will produce maximum AC generator output. If this test still produces lower than specified output, the AC generator is the cause of the problem. If the output is within specifications with the regulator bypassed, then the problem is within the regulator.

CAUTION: The full field test is to be performed on AC generator systems only. Do not use this test on vehicles equipped with a DC generator system.

When full fielding the system, the battery should be loaded to protect vehicle electronics and computers. With the voltage regulator bypassed, there is no control of voltage output. The AC generator is capable of producing well over 30 volts. This increased voltage will damage the circuits not designed to handle that high of a voltage.

CAUTION: Not all AC generators can be full fielded. Check the manufacturer's procedures before attempting to full field an AC generator.

Photo Sequence 10
Performing the Current Output Test

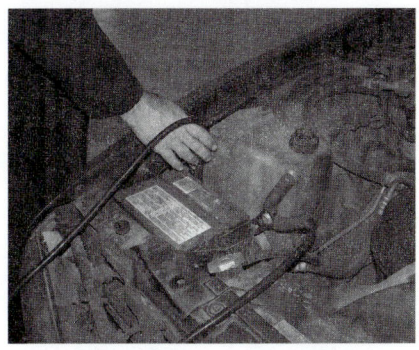

P10-1 Connect the large red and black cables across the battery, observing polarity.

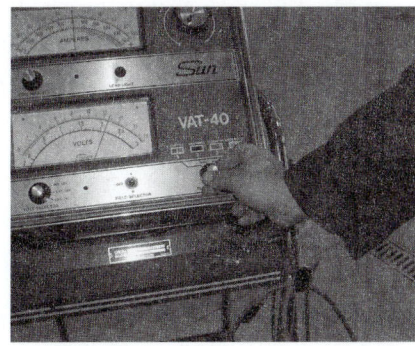

P10-2 Select CHARGING.

P10-3 Zero the ammeter.

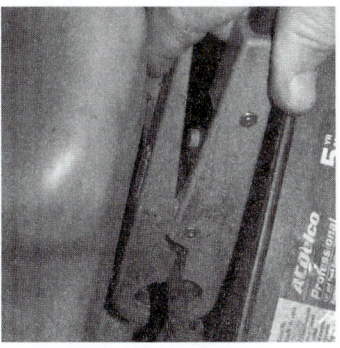

P10-4 Connect the inductive pickup around all battery ground cables.

P10-5 With the ignition switch in the RUN position, engine not running, observe the ammeter reading. This reading indicates how much current is required to operate any full-time accessories.

P10-6 Start the engine and hold between 1,500 and 2,000 rpm.

P10-7 Turn the load knob for the carbon pile slowly, until the highest ammeter reading possible is obtained. Do not reduce battery voltage below 12 volts.

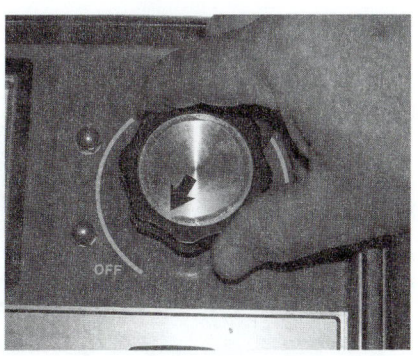

P10-8 Return the load control knob to the OFF position.

P10-9 The highest reading indicates maximum current output.

General Motors Full Field Testing

To perform the full field test on a GM A circuit SI-type AC generator with an internal voltage regulator, insert a screwdriver into the D-shaped test hole (Figure 7-7). This test hole lines up with a small tab that is attached to the negative brush. By inserting a screwdriver into the D hole about 1/2 in. (12.7 mm) and grounding it to the housing, the regulator is bypassed. Perform the output test again with the regulator bypassed. If the output is within specifications, the regulator is at fault.

CAUTION: Do not force the screwdriver into the D hole more than 1 in. (25.4 mm). Damage to several electrical systems can result. Do not full field for longer than 10 seconds.

SERVICE TIP: If the means of loading the battery is not available, do the full field test for a very short time. Do not allow voltage output to increase over 16 volts. Use the vehicle accessories to put a load on the battery.

A variation of the test calls for shorting the negative brush in the D hole while the ignition switch is in the RUN position. If the brushes and rotor are good, then the rear bearing should be magnetized and attract a metal screwdriver (Figure 7-8).

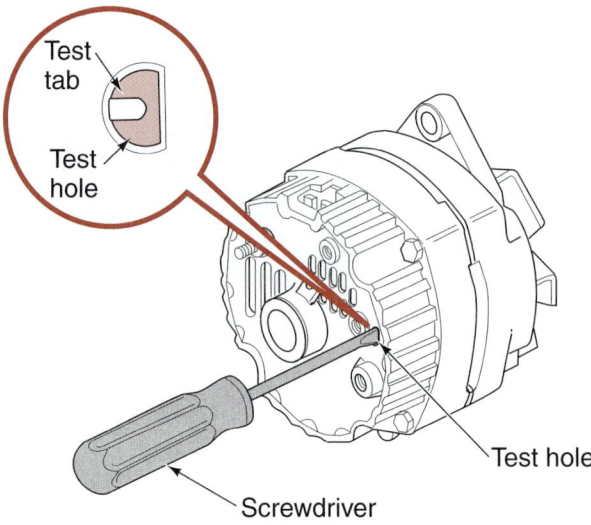

Test tab

Test hole

Test hole

Screwdriver

Figure 7-7 Full fielding the GM 10SI AC generator by grounding the tab in the D test hole.

Figure 7-8 Quick check of the rotor and brushes.

If the vehicle is equipped with a CS or AD series generator, General Motors does not recommend a manual full field test. Instead, use the current output test to confirm proper operation of the generator.

Ford Full Field Testing

Ford Motor Company has utilized different designs of the integral regulator. The early design had one terminal, called the exciter, which was connected to the outside of the regulator (Figure 7-9). The wiring schematic for this type of design is illustrated (Figure 7-10). By removing the protective cover from the field terminal (closest to the rear bearing), the field circuit can be grounded and the regulator bypassed.

Classroom Manual
Chapter 7, page 192

Special Tools
Voltmeter
Ohmmeter
Jumper wires

Ford refers to this test as the under voltage test.

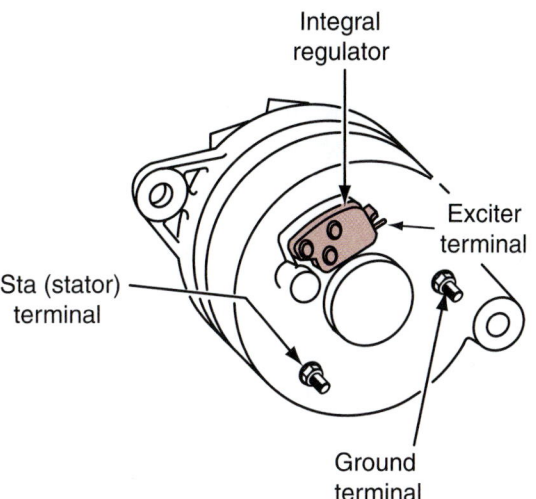

Figure 7-9 Integral regulator with exciter terminal.

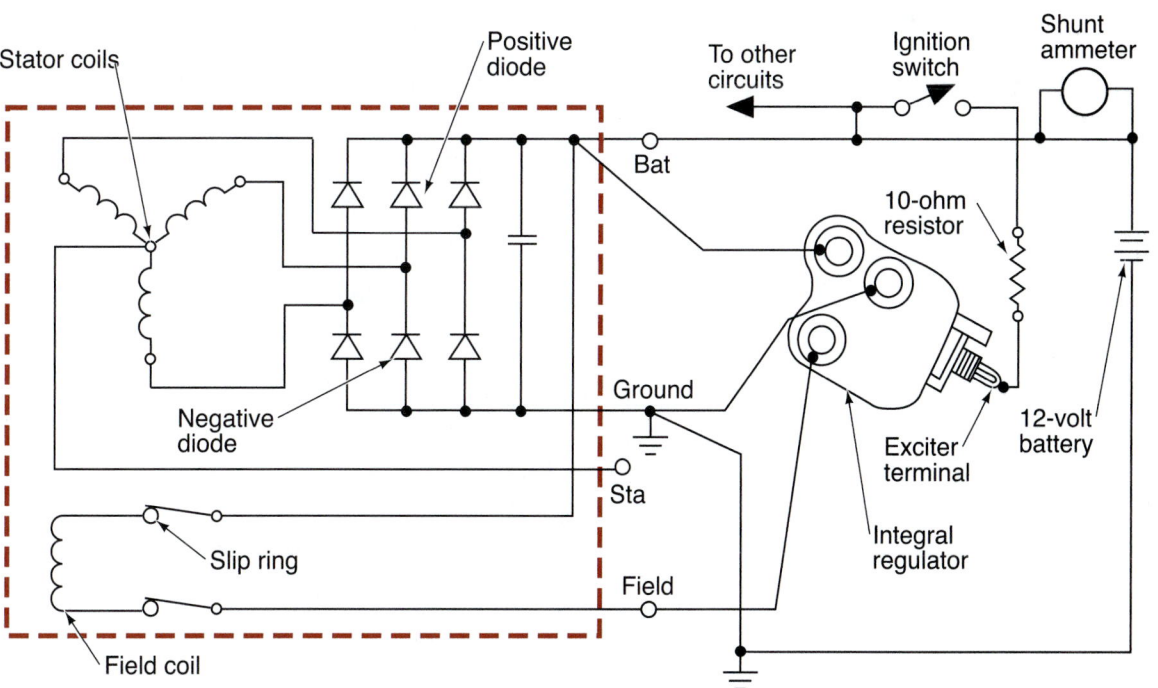

Figure 7-10 Wiring schematic of integral regulator with exciter terminal.

Before full fielding the IAR AC generator, check the rotor and field circuit resistance:

1. Disconnect the wiring plug to the regulator.

2. Connect an ohmmeter between the regulator A and F terminals.

3. Read the ohmmeter. The resistance should not be below 2.4 ohms.

If the resistance is less than 2.4 ohms, there is a short to ground somewhere in the circuit. Check for the following:

1. A failed regulator.

2. A shorted rotor circuit.

3. A shorted field circuit.

CAUTION: Do not replace the regulator without first repairing any shorts in the rotor or field circuits. To do so may damage the new regulator.

The illustration (Figure 7-11) shows the wiring of the IAR system. To full field this system, disconnect the wiring connector to the AC generator and install a twelve-gauge wire jumper

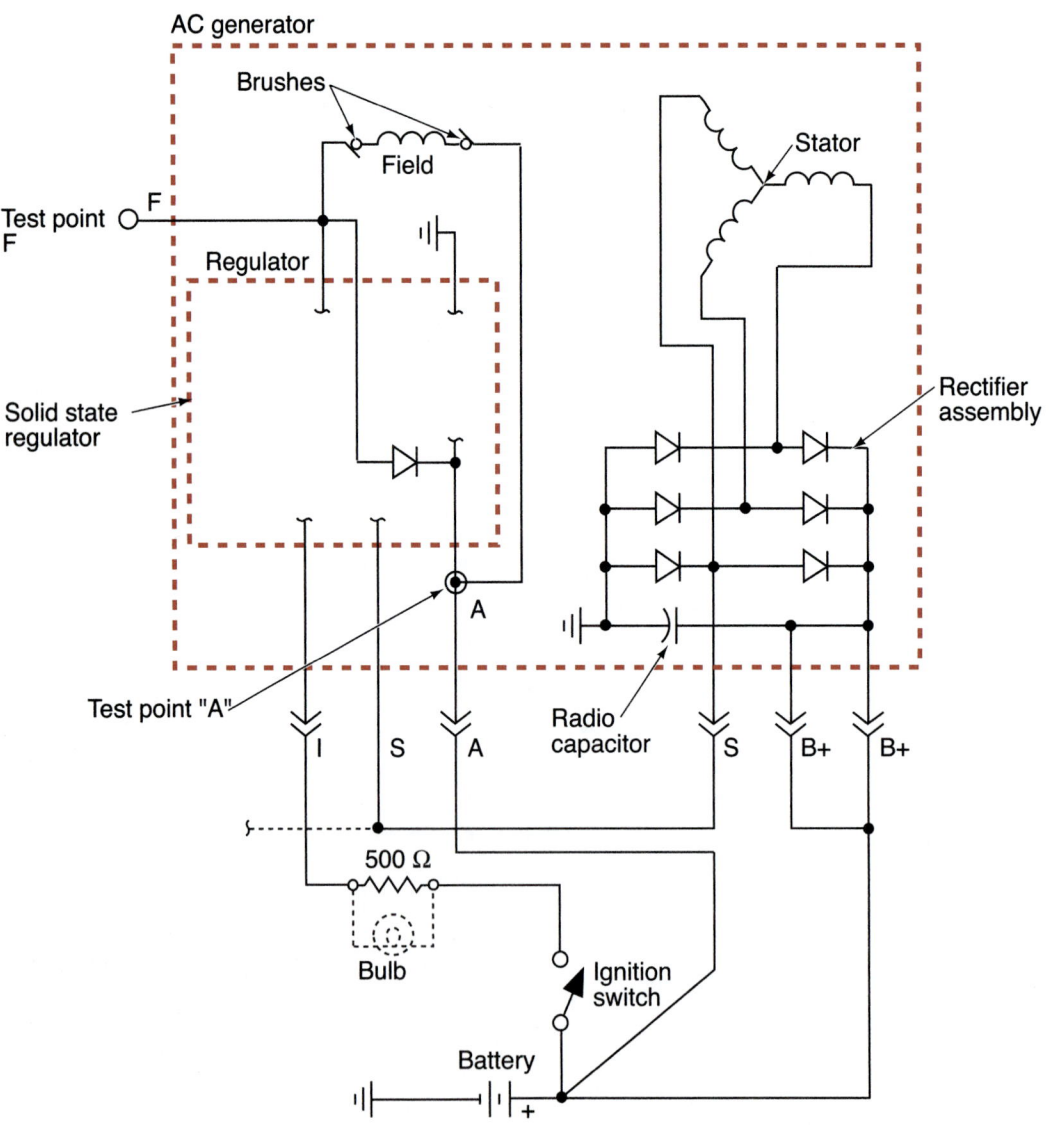

Figure 7-11 Wiring diagram of Ford's IAR charging system.

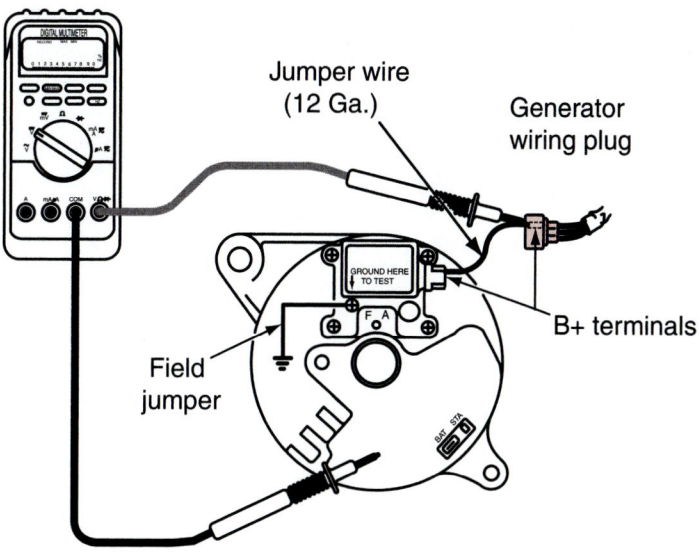

Figure 7-12 Jumper wire connections between B+ terminals.

between the B+ terminal blades (Figure 7-12). Connect another jumper wire from the regulator F terminal screw to ground. Connect a voltmeter with the positive lead connected to one of the B+ jumper wire terminals and the negative test lead to a good ground. Start the engine and perform the load output test. The regulator is faulty if the voltage rises to specifications. If the voltage does not rise to specifications, the AC generator needs to be serviced or replaced.

DaimlerChrysler Full Field Testing

Most late-model DaimlerChrysler vehicles use an A-type field with the voltage regulator inside the powertrain control module. For DaimlerChrysler vehicles with computer-controlled charging systems, refer to the section on "Special Full Field Testing."

Early DaimlerChrysler vehicles used an isolated field with two field leads. To full field these systems, disconnect the green field wire from the AC generator. Then connect a jumper wire from the AC generator field terminal to ground. Start the engine and check the output.

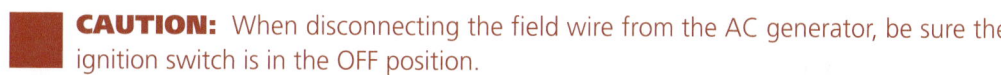 **CAUTION:** When disconnecting the field wire from the AC generator, be sure the ignition switch is in the OFF position.

Special Full Field Testing

Some AC generators that use internal regulators, or computer-controlled regulators, do not provide for a means of full fielding. This can be determined by looking at the wiring diagram for the charging system. In fact, by studying the circuit in the wiring diagram, you should be able to full field any AC generator that can be full fielded. The following procedure uses a starting/charging system tester to full field an AC generator while observing the output of the system. If the AC generator fails this test, it should be repaired or replaced.

CAUTION: Check the manufacturer's service manual to see if there are instructions that prohibit full fielding the AC generator in this manner. If the AC generator cannot be full fielded, it will have to be disassembled and bench tested.

Classroom Manual
Chapter 7, page 194

Special Tools

VAT-40 or similar tester

Some manufacturers with computer-controlled charging systems may provide for on-board diagnostics of the system.

When using the charging system tester, the full field test is performed as follows:

1. Connect the large red and black leads across the battery, observing polarity.

2. Disconnect the field wire from the AC generator.

3. Connect the blue field test lead from the charging system tester to the field terminal. Make the connection toward the AC generator.

 CAUTION: To prevent AC generator damage, do not use the blue test lead while the regulator is still connected to the system.

4. Select CHARGING.

5. Select INT 18 V.

6. Zero the ammeter.

7. Clamp the inductive pickup around the AC generator output wire.

8. Turn the ignition switch to the RUN position. Do not start the engine. Record the ammeter reading.

9. Start the engine and hold the speed between 1,500 and 2,000 rpm.

10. Turn the load control knob until the voltmeter reads the voltage specified for maximum output.

11. Read the current output on the 100-amp scale and compare to specifications. If the reading is within specifications, the regulator and AC generator are fine. If the reading is below specifications, continue testing.

12. Release the load control knob and allow the engine to run at 1,500 to 2,000 rpm.

13. Turn the load control knob until the voltmeter indicates 2 volts less than system voltage.

14. Full field the AC generator using the field switch. Use the load control knob to prevent system voltage from exceeding 14 volts.

15. Adjust the load control knob to the voltage reading that is required in the manufacturer's specifications.

16. Read the output on the 100-ampere scale while depressing the toggle switch.

17. Release the field selector switch and return the load knob to the OFF position.

18. Add the reading obtained in step 8 to the reading in step 16. Compare to manufacturer's specifications.

 CAUTION: Always refer to the manufacturer's service manual for the correct location of the AC generator terminals. Incorrect test connections can damage the AC generator and the vehicle's electrical system components.

Some manufacturers may provide a method of full fielding the generator using a scan tool. The scan tool will direct the PCM to full field the generator and display the sensed battery voltage. While full fielding, the voltage should increase. However, if the voltage does not increase, you still must determine if the fault is in the generator, the insulated side of the field circuit, the field circuit between the generator and the PCM, or the PCM itself.

Classroom Manual
Chapter 7, page 181

If the circuit type is unknown, test in both the A and B positions.

Regulator Test

The regulator test is used to determine if the regulator is maintaining the correct voltage output under different load demands.

To perform the regulator voltage test using a VAT-40, follow these steps:

1. Connect the large red and black cables across the battery, observing polarity.
2. Select REGULATOR.
3. Select INT 18 V.
4. Zero the ammeter.
5. Clamp the inductive pickup around the AC generator output wire.
6. Start the engine and hold between 1,500 and 2,000 rpm.
7. Allow the engine to run until the ammeter reads 10 amperes or less. This indicates the battery is fully charged.
8. Voltage should read regulated voltage (13.5–14.5 volts).
9. Load the system to between 10 and 20 amperes.
10. Voltmeter should still read regulated voltage.

Diode/Stator Test

An AC generator may have an open diode yet test close to manufacturer's specifications. If there is an open diode that is not determined in testing, a newly installed regulator may fail. In addition, an open diode can lead to the failure of other diodes. The **diode/stator test** is performed to determine the condition of the diodes. This is performed in the following manner:

1. Connect the large red and black cables across the battery, observing polarity.
2. Select the CHARGING position.
3. Select INT 18 V.
4. Zero the ammeter.
5. Clamp the inductive pickup around all of the negative battery cables.
6. Start the engine and hold between 1,500 and 2,000 rpm.
7. Adjust the load control knob to obtain an indicated charge rate of 15 amperes.
8. Set the selector to the DIODE/STATOR position while observing the red and blue DIODE/STATOR scale.
9. Return the load control knob to the OFF position.

If the meter was in the blue section of the scale, the diodes and stator are good. If the meter was in the red section of the scale, the diodes or the stator is bad. The AC generator will need to be disassembled to perform bench testing of these units.

Diode Pattern Testing

● **CUSTOMER CARE:** It is good practice to check the diode pattern of the AC generator anytime an electronic component fails. Because the electronics of the vehicle cannot accept AC current, the damage to the replaced component could have been the result of a bad diode. By performing this check, it is possible to find the cause of the problem.

Set an oscilloscope on the lowest scale available. Connect the primary test leads on the AC generator output terminal and ground. Start the engine and place a moderate load on the charging system (15 to 20 amperes). Different patterns may appear. What is considered normal depends on the load placed on the system.

Special Tools

Starting/charging
  system tester
Tachometer

_____

Classroom Manual
Chapter 7, page 170

Special Tools

Starting/charging
  system tester
Tachometer

_____

This test is not valid for AC generators that failed the full field test.

Classroom Manual
Chapter 7, page 172

Special Tools

Oscilloscope
Carbon pile

_____

The diode pattern (Figure 7-13) illustrates a good pattern. However, the second pattern shown (Figure 7-14) is also a good pattern if the AC generator is under a full load. The third pattern shown (Figure 7-15) is a good pattern for some AC generators.

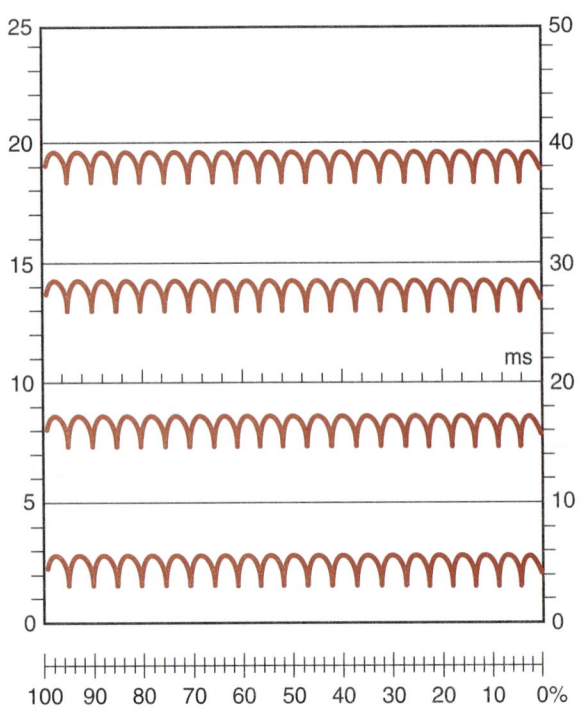

Figure 7-13 Good diode waveform.

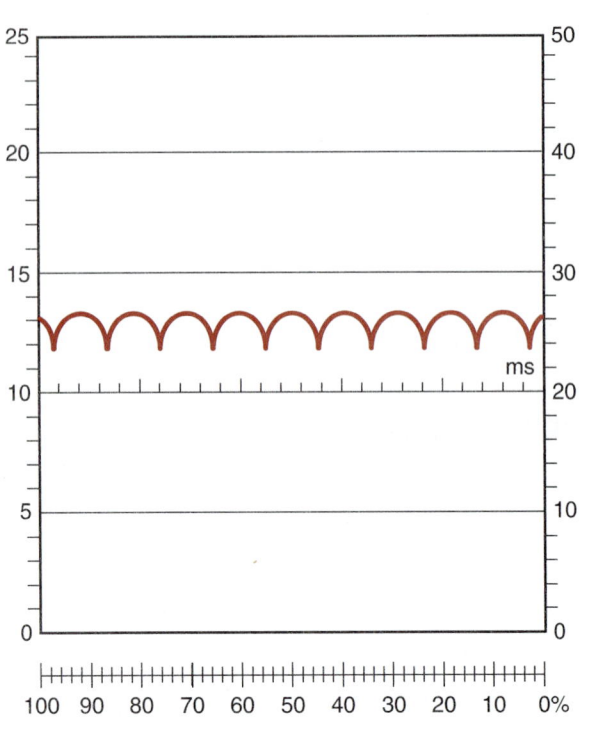

Figure 7-14 Good diode waveform when the AC generator is operating under full load.

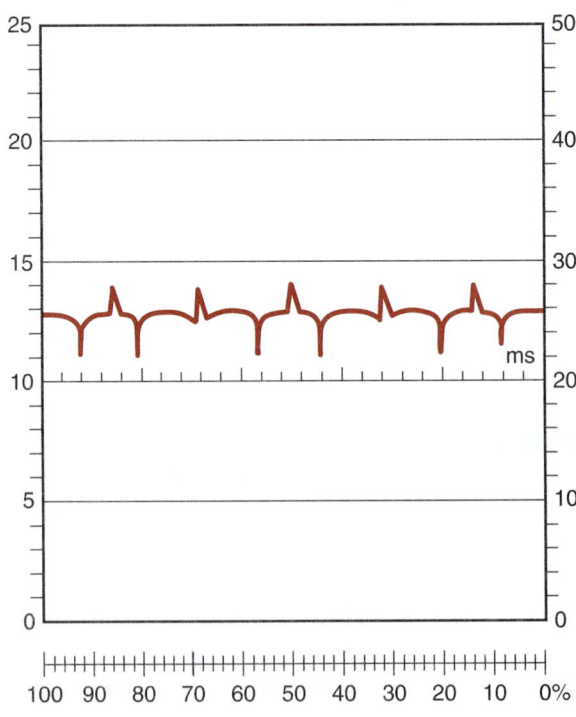

Figure 7-15 Good diode waveform when AC generator is operating under no load demands.

Patterns that have high-resistance, open, and shorted diodes are illustrated (Figures 7-16 through 7-19). Remember to check the waveforms for noise. If the diodes don't rectify all of the AC voltage, some will ride on the DC output.

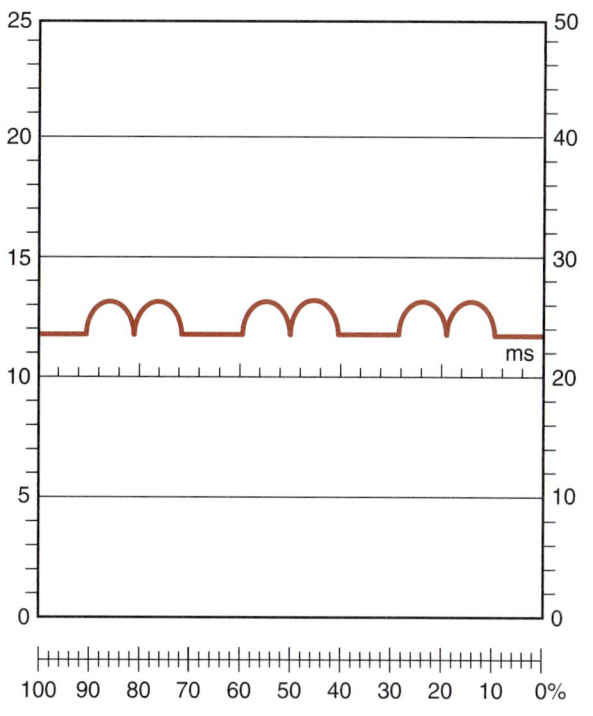

Figure 7-16 Shorted diodes or shorted stator winding when the AC generator is placed under full load.

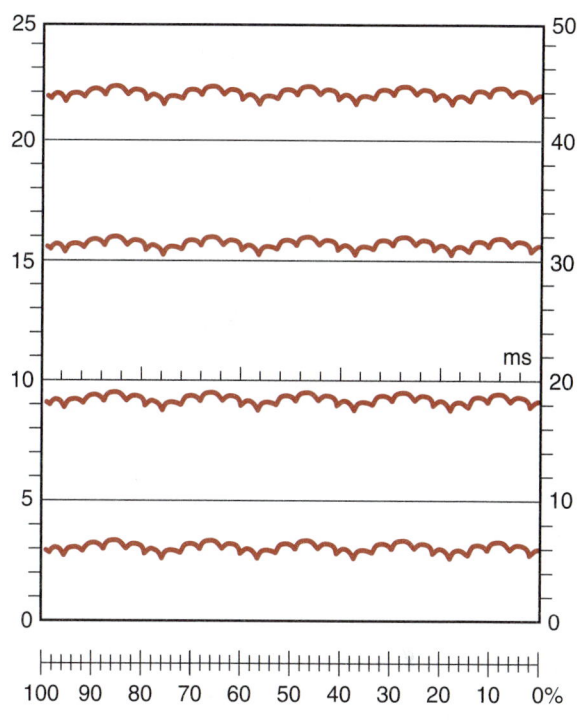

Figure 7-17 A waveform indicating high resistance.

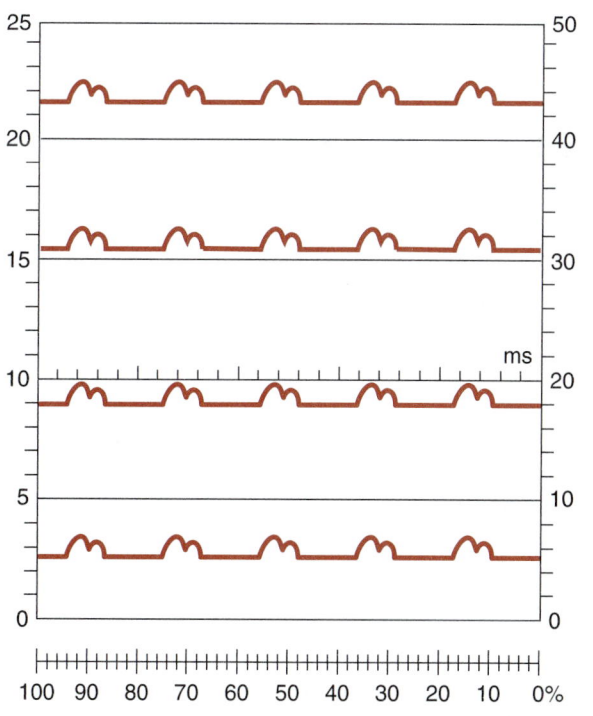

Figure 7-18 Waveform pattern indicating one open and one shorted diode.

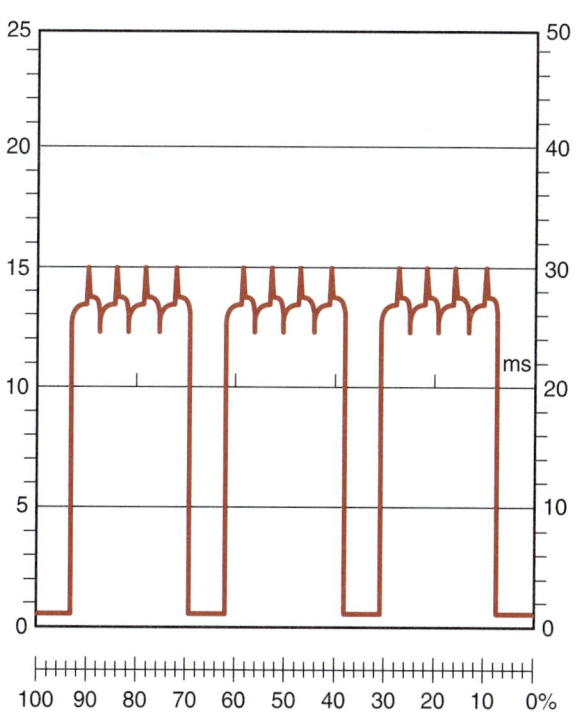

Figure 7-19 Waveform indicating an open diode in the diode trio.

SERVICE TIP: Instead of using a carbon pile, it is possible to place a moderate load on the charging system by turning on the headlights and a few other electrical accessories.

An alternate way to check the action of the diodes is to check for AC voltage at the battery. Do this with a DMM set to a low AC voltage scale. Connect the meter across the battery; load the charging system by turning on the headlights. Ideally there should be zero volts AC at the battery. A voltage reading of more than 0.5 VAC is excessive and indicates the diodes are not rectifying the AC output of the AC generator.

Special Tools

Starting/charging system tester

Charging System Requirement Test

It is possible to have a charging system that is working properly, yet not meet the requirements of the vehicle's electrical system. If an AC generator is installed on the vehicle without sufficient output to meet the demands of the vehicle, the customer may have complaints that are identical to those of a charging system that is not functioning at all. The actual AC generator output should be at least 10% to 20% greater than the load demand. The charging system requirement test is used to determine the total electrical demand of the vehicle's electrical system.

To determine the vehicle's electrical requirement:

1. Connect the large red and black leads across the battery, observing polarity.
2. Select the CHARGING position.
3. Select INT 18 V.
4. Zero the ammeter.
5. Clamp the inductive pickup around all of the negative battery cables.
6. Turn the ignition switch to the RUN position. Do not start the engine.
7. Turn on all accessories to their highest positions.
8. Read the ammeter. The indicated amperage is the total load demand of the vehicle.

AC Generator Removal and Replacement

Special Tools

Fender covers

Pry bar

Belt tension gauge

Battery terminal puller

AC generator removal varies according to the engine size, engine placement, and vehicle accessories (such as power steering and air conditioning). The following is the typical procedure for removal and replacement of the AC generator:

1. Place fender covers over the fenders.
2. Disconnect the battery ground cable.

WARNING: Never attempt to remove the AC generator or disconnect any wires to the generator without first disconnecting the battery negative cable. Always wear safety glasses when working around the battery.

3. Disconnect the wiring harness connections to the AC generator.
4. Loosen the drive belt tensioner (Figure 7-20).
5. Remove the drive belt.
6. Remove the upper bolt that attaches the AC generator to the mounting bracket.
7. Remove the generator lower bolt while supporting the generator.
8. Remove the generator from the vehicle.

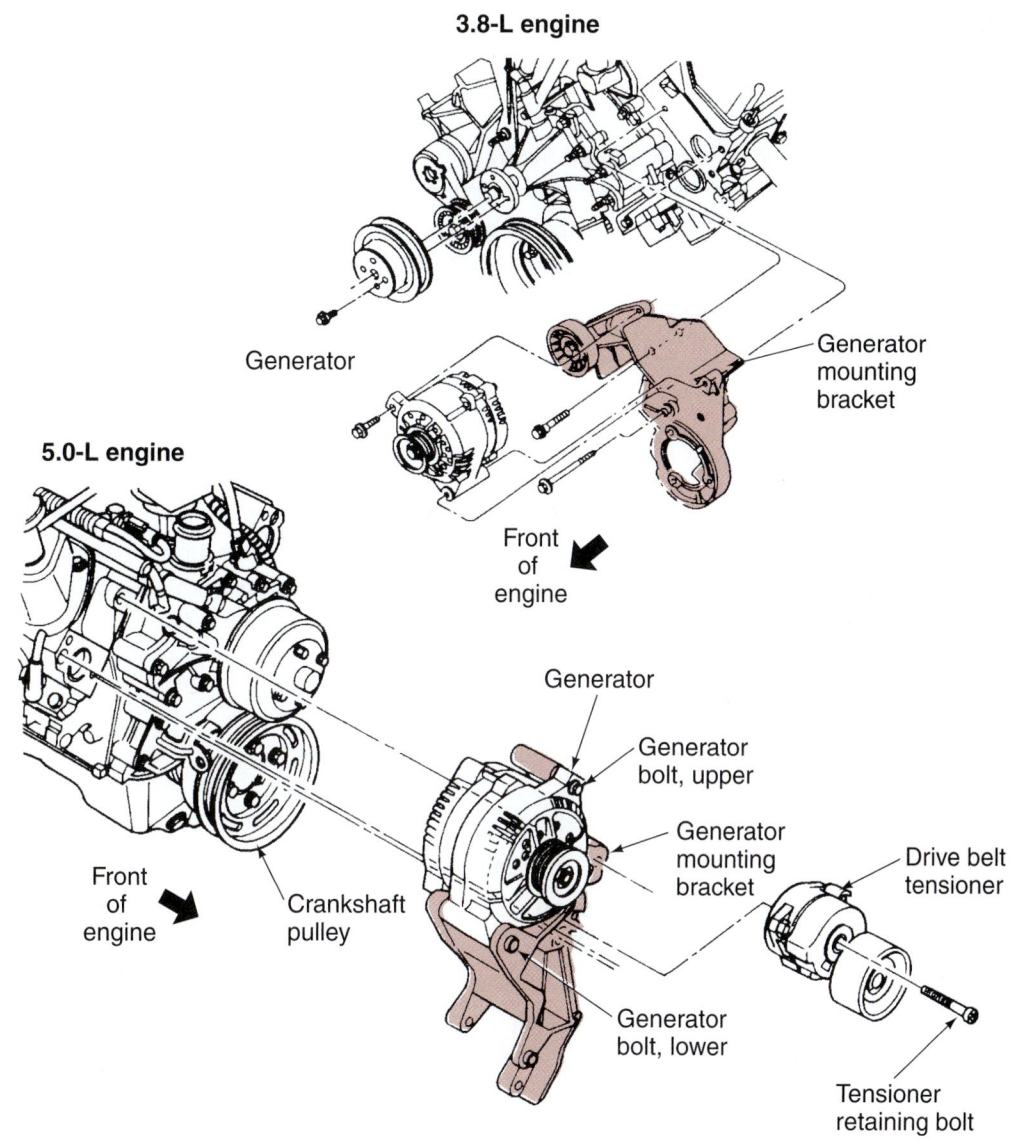

3.8-L engine

Generator

Generator mounting bracket

5.0-L engine

Generator

Front of engine

Generator bolt, upper

Generator mounting bracket

Generator bolt, lower

Drive belt tensioner

Tensioner retaining bolt

Front of engine

Crankshaft pulley

Figure 7-20 Examples of AC generator mounting brackets and fasteners.

Reverse the removal procedure to install the AC generator. To adjust the belt tension, leave the pivot and adjusting arm bolts loose. Look up the correct belt tension specification for the vehicle you are working on. Install a belt tension gauge on the belt and apply pressure to the AC generator front housing only when adjusting the belt tension. Once the correct tension reading is obtained, tighten the bolts to specified torque value. Manufacturers may provide different specifications for new and used belts. A belt is considered used if it has more than 15 minutes of run time.

CAUTION: Do not pry on the rear housing of the generator. It may crack the housing.

Many manufacturers now give belt tension specifications in frequency. A frequency-measuring tool equipped with a special microphone probe is used to measure the frequency of the belt to determine proper tension. The end of the microphone probe is placed about 1 in. (25 mm) from the belt in the center of the span between pulleys. Using your finger, pluck the belt a minimum of three times. The tool will display the frequency in hertz (Hz). Adjust the belt until proper frequency is obtained.

Also, many manufacturers use a self-adjusting mechanism to maintain proper belt tension. Follow the service procedures for the vehicle being worked on to properly set initial tension.

AC Generator Disassembly

Special Tools

Ohmmeter

100-watt soldering iron

Resin core solder

Arbor press

Soft jaw vise

Scribe

High-temperature bearing grease

400-grain emery cloth

Pulley puller

Clean rags

Heat sink grease

If the AC generator fails the previous tests, the technician must decide whether to rebuild or replace the AC generator. This decision is based on several factors:

1. What is best for the customer.
2. Shop policies.
3. Cost.
4. Time.
5. Type of AC generator.

Once the decision is made to disassemble the AC generator, the technician should study the manufacturer's service manual and become familiar with the procedure for the particular AC generator being rebuilt. Photo Sequence 11 shows the procedure for disassembling the Ford IAR AC generator. A disassembled view of this AC generator is shown (Figure 7-21).

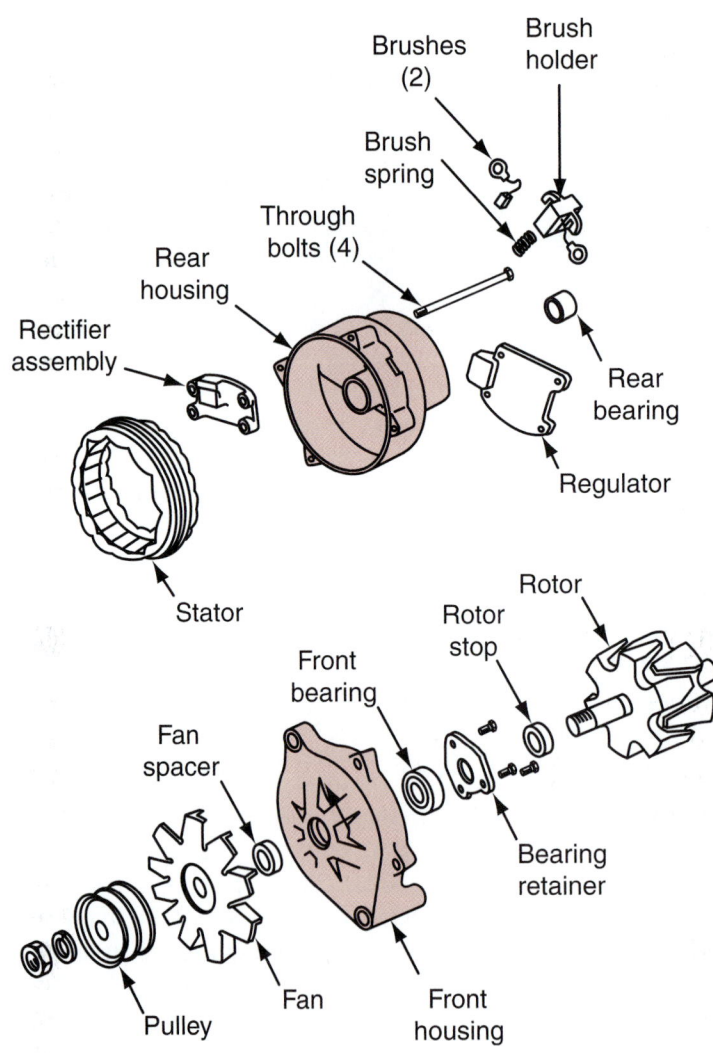

Figure 7-21 Disassembled view of Ford's IAR generator.

Photo Sequence 11
Typical Procedure for IAR Generator Disassembly

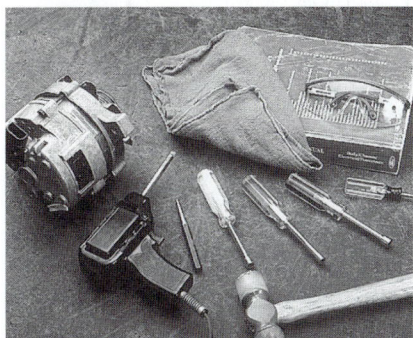

P11-1 Always have a clean and organized work area. Tools required to disassemble the Ford IAR AC generator: rags, T20 TORX wrench, plastic hammer, arbor press, 100-watt soldering iron, soft jaw vise, safety glasses, and assorted nut drivers.

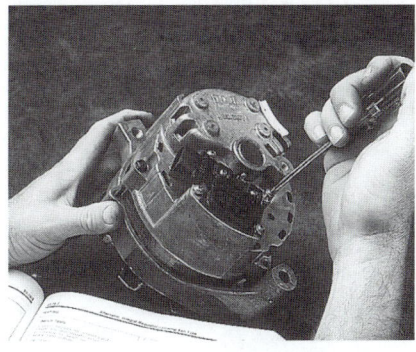

P11-2 Using a T20 TORX, remove the four attaching screws that hold the regulator to the AC generator rear housing.

P11-3 Remove the regulator and brush assembly as one unit.

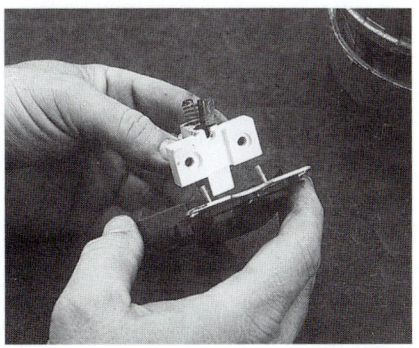

P11-4 Using a T20 TORX, remove the two screws that attach the regulator to the brush holder. Separate the regulator from the brush holder. Remove the A terminal insulator from the regulator.

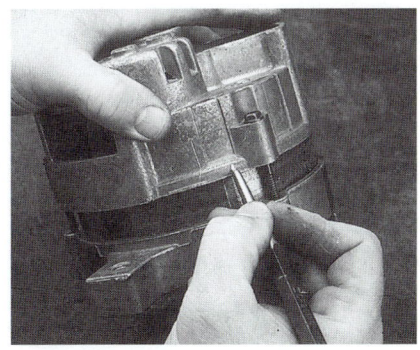

P11-5 Scribe or mark the two housing ends and the stator core for reference during assembly.

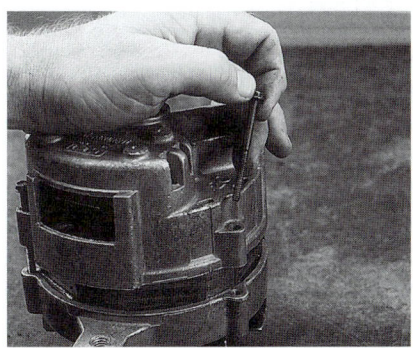

P11-6 Remove the three through bolts that attach the two housings.

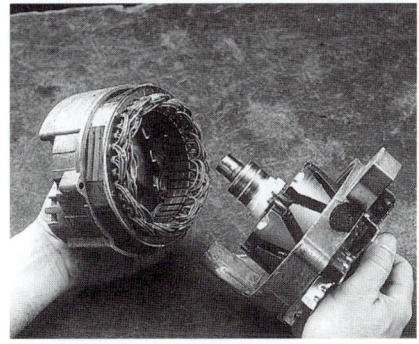

P11-7 Separate the front housing from the rear housing. The rotor will come out with the front housing, while the stator will stay with the rear housing. NOTE: It may be necessary to tap the front housing with a plastic or dead weight hammer to get the halves to separate.

P11-8 Separate the three stator lead terminals from the rectifier bridge.

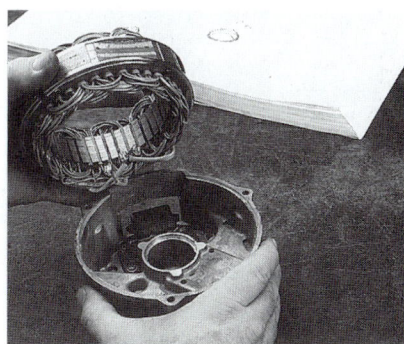

P11-9 Remove the stator coil from the housing.

Typical Procedure for IAR Generator Disassembly (Continued)

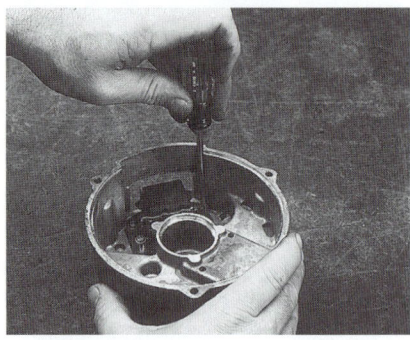

P11-10 Using a T20 TORX, remove the four attaching bolts that hold the rectifier bridge.

P11-11 Remove the rectifier bridge from the housing.

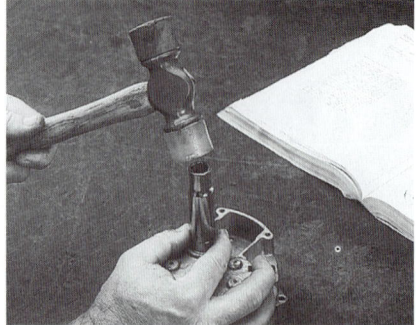

P11-12 Use a socket to tap out the bearing from the housing.

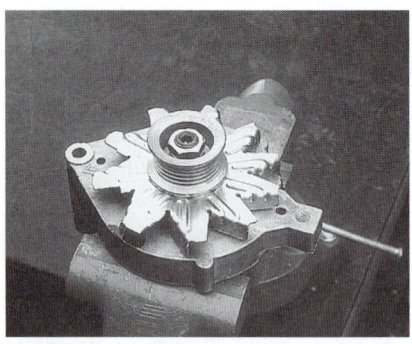

P11-13 Clamp the rotor in a soft jaw vise.

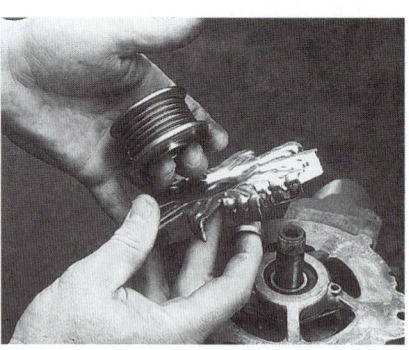

P11-14 Remove the pulley attaching nut, flatwasher, drive pulley, fan, and fan spacer from the rotor shaft.

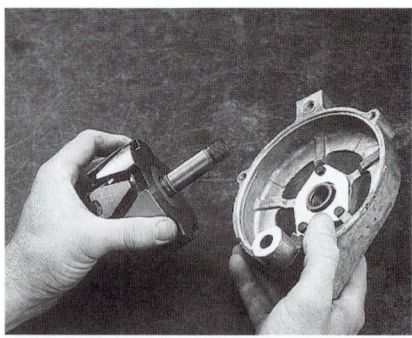

P11-15 Separate the front housing from the rotor. If the stop ring is damaged, remove it from the rotor; if not, leave it on the rotor shaft.

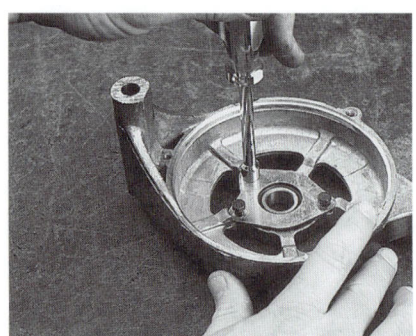

P11-16 Remove the three screws that hold the bearing retainer to the front housing.

P11-17 Remove the bearing retainer.

P11-18 Remove the front bearing from the housing. NOTE: It may be necessary to use an arbor press to remove the bearing if it does not slide out.

Once the AC generator is disassembled, the components must be cleaned and inspected. Using a clean cloth, wipe the stator, rotor, and front bearing. Inspect the front and rear bearings by rotating them on the rotor shaft. Check for noises, looseness, or roughness. Replace the defective bearing if any of these conditions are present.

CAUTION: Do not use solvent cleaners on these components.

Check the rotor shaft rear bearing surface. If the surface is not smooth, the rotor will have to be replaced. Visual inspection of the rotor includes checking the slip rings for smoothness and roundness. If the rings are discolored, dirty, scratched, nicked, or have burrs, they may be cleaned with fine-grit emery cloth. Caution must be observed to prevent creating flat spots while polishing the slip rings. The minimum diameter of the slip rings is 1.22 in. (31 mm).

Inspect the terminals and wire leads of the rotor and stator. Also, check both the rotor and stator for signs of burnt insulation of the windings. If there is damage to the insulation, replace the component.

Inspect the housing halves for cracks. Also check the fan and pulley for looseness on the rotor shaft and for cracks. Replace any part that does not pass inspection. Remove the heat transfer grease that is in the rectifier mounting area with a clean cloth.

The AC generator's brushes should be inspected and tested anytime the unit is disassembled. Brushes should be replaced whenever they are worn shorter than 1/4 in. (6.35 mm) in length. Also, the brush springs must be checked for sufficient strength to keep constant contact of brushes with slip rings. Brush continuity may be checked using an ohmmeter. There should be zero resistance through the brush path. Replace the brushes if there is any resistance indicated.

Classroom Manual
Chapter 7, page 169

AC Generator Component Testing

Once the AC generator is disassembled and cleaned, the individual components can be tested. The chart (Figure 7-22) illustrates the test connections and results for the major components.

COMPONENT	TEST CONNECTION	NORMAL READING	IF READING WAS:	TROUBLE IS:
Rotor	Ohmmeter from slip ring to rotor shaft	Infinite resistance	Very low	Grounded
	Test lamp from slip ring to shaft	No light	Lamp lights	Grounded
	Test lamp across slip rings	Lamp lights	No light	Open
Stator	Ohmmeter from any stator lead to frame	Infinite resistance	Very low	Grounded
	Test lamp from lead to frame	No light	Lamp lights	Grounded
	Ohmmeter across any pair of leads	Less than $^1/_2\,\Omega$	Any very high reading	Open
Diodes	Ohmmeter across diode, then reverse leads	Low reading one way; high reading other way	Both readings low Both readings high	Shorted Open
	12-V test lamp across diode, then reverse leads	Lamp lights one way, but not other way	No light either way Lamp lights both ways	Open Shorted

Figure 7-22 Guidelines for bench testing a generator.

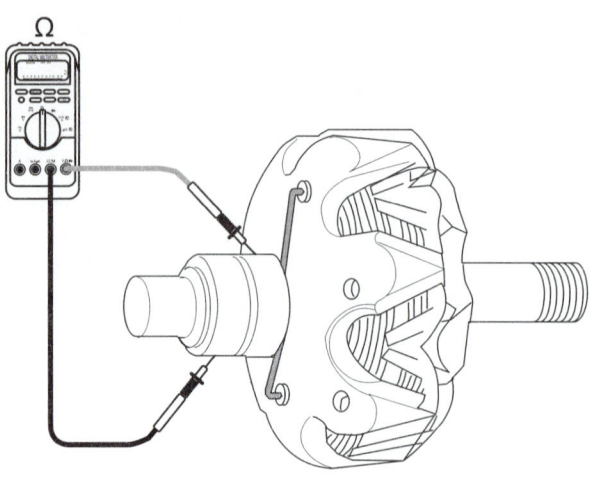

Figure 7-23 Test connections for checking for opens in the rotor windings.

Figure 7-24 Testing the rotor for shorts to ground.

Classroom Manual
Chapter 7, page 167

Special Tools

Ohmmeter

110-volt test light

Classroom Manual
Chapter 7, page 170

Special Tool

Ohmmeter

Remove the stator leads from the diodes before testing the stator windings.

Testing the Rotor

The most important test of a rotor is a complete visual inspection. Carefully check the rotor windings for signs of discoloration or overheating. If these signs are present, the rotor is no good. Also carefully inspect the slip rings; they should be flat, smooth, and free of damage. If the rotor passes the visual inspection, proceed to test it with an ohmmeter.

An ohmmeter can be used to measure the resistance between the slip rings (Figure 7-23). Always check the manufacturer's service manual for the correct specification for the unit you are working on. If specifications are not available, the following are some typical values:

GM	2.4 to 3.5 Ω
Ford	3.0 to 5.5 Ω
DaimlerChrysler	3.0 to 6.0 Ω

If the resistance reading is below specifications, then the rotor is shorted. If the resistance is high, the rotor connections are badly corroded or open. The rotor must be replaced if any of these conditions are found.

Connecting the ohmmeter from each of the slip rings to the rotor shaft should show infinite resistance (Figure 7-24). If the reading was very low, the field coil is grounded and the rotor will have to be replaced.

SERVICE TIP: In many instances, it is less expensive to replace the AC generator than to replace the rotor only.

Testing the Stator

When testing a stator, a visual inspection is the most productive test. Look for discoloration or other damage to the windings. Often the assembly will look fine but will actually be damaged due to excessive heat. One quick way of checking for this is to take the blade of a knife and scrape the windings. If the coating or varnish flakes off, the windings overheated and the varnish is baked. Pay special attention to the connectors. Any signs of damage or breakage indicate the stator should not be reused. If the stator passes the visual inspection, it should be checked for opens and shorts to ground with an ohmmeter. To test for an open, connect the ohmmeter test leads to any pair of stator leads (Figure 7-25). Continue to test the stator until all combinations of pair connections are completed. On all of these connections, the resistance should be less than 0.5 volt. If the ohmmeter reads infinity between any two leads, the stator has an open and it must be replaced.

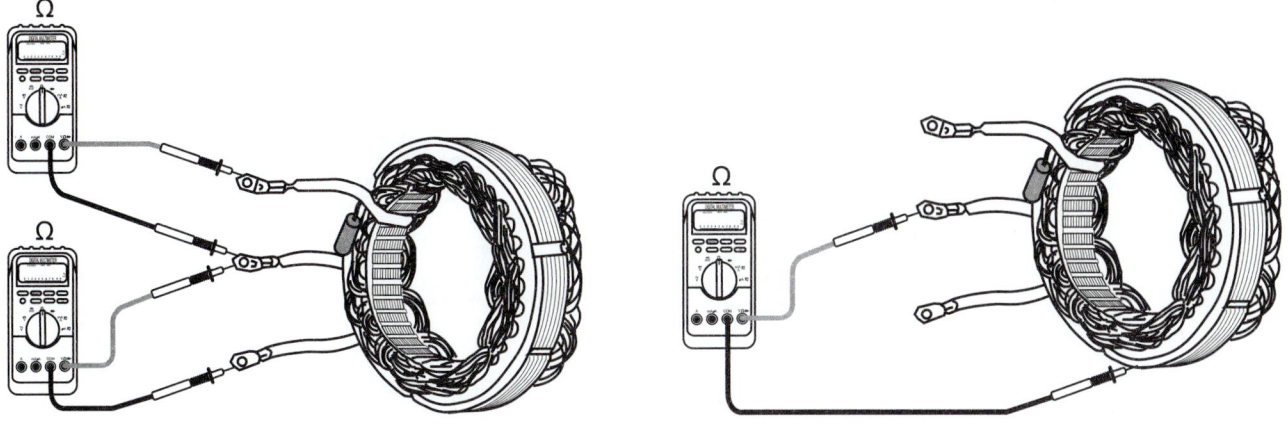

Figure 7-25 Testing the stator for opens.

Figure 7-26 Testing the stator for shorts to ground.

To test the stator for a short to ground, connect the ohmmeter to the stator leads and stator frame (Figure 7-26). The ohmmeter should read infinity on all three stator leads. If the reading is less than infinity, the stator is shorted to ground and it must be replaced.

 SERVICE TIP: A shorted stator is difficult to test for because the resistance is very low for a normal stator. If all other components test okay, but output was low, a shorted stator is the probable cause.

Testing the Diodes (Rectifier Bridge)

Because a diode should allow current to flow in only one direction, it must be tested for continuity in both directions. Using the analog ohmmeter or the diode test function of the DMM, connect the test leads to the diode lead and case (Figure 7-27). Read the ohmmeter scale. If the diode is good, it will show high resistance in one direction and low resistance in the opposite direction. If both

Classroom Manual
Chapter 7, page 172

The diodes must be separated from all other electrical components before testing.

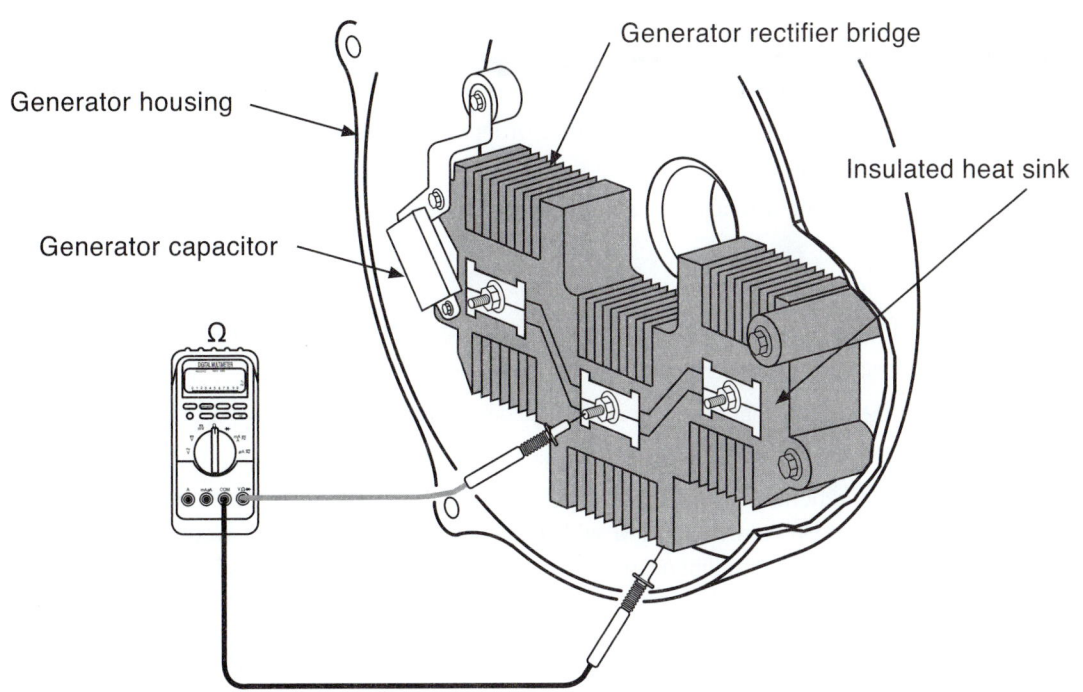

Generator rectifier bridge

Generator housing

Insulated heat sink

Generator capacitor

Figure 7-27 Testing the rectifier bridge.

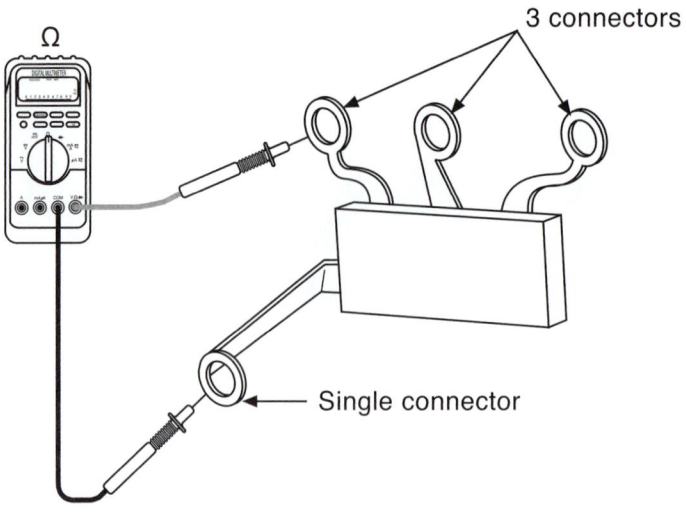

Figure 7-28 Diode trio test connections.

Classroom Manual
Chapter 7, page 177

Special Tool

DMM ohmmeter

readings are low, the diode is shorted. If there is high resistance in both directions, the diode is open. Test all six diodes and replace any that are defective.

Testing the Diode Trio

If the AC generator is equipped with a diode trio, it must be tested for opens and shorts. The procedure is much the same as with the rectifier bridge test. Connect one of the DMM test leads to the signal connector. Connect the other test lead to one of the three connectors (Figure 7-28). Test each of the three connectors in diode test mode. Record your readings.

Reverse the ohmmeter leads and record the readings obtained on each of the three connectors. The ohmmeter should read "OL" in one direction and above 600 mV in the other direction. These results should be obtained on all three connections. If not, replace the diode trio.

AC Generator Reassembly

The reassembly procedure is basically the reverse order of the disassembly. The following are suggestions to assist in the assembly process:

1. Always check the manufacturer's specifications for the proper torque values of the attaching screws.
2. Use high-temperature grease on the bearings.
3. Check for free rotor rotation after installing and torquing the pulley to the rotor shaft.
4. Apply heat sink grease across the rectifier bridge base plate.
5. Protect the diodes from excess heat while soldering the connections. Use a pair of needle nose pliers as a heat sink (Figure 7-29).

When assembling a generator, always follow the recommendations of the manufacturer. It is critical that all screws and bolts be installed with the insulating washers that were present before disassembly. These insulators maintain proper circuit polarity. If a washer is left out, a short circuit will exist.

When installing the rotor into the brushes, most generators are equipped with a hole that allows a pin or paper clip to be inserted into the brush holder to keep the brushes back and

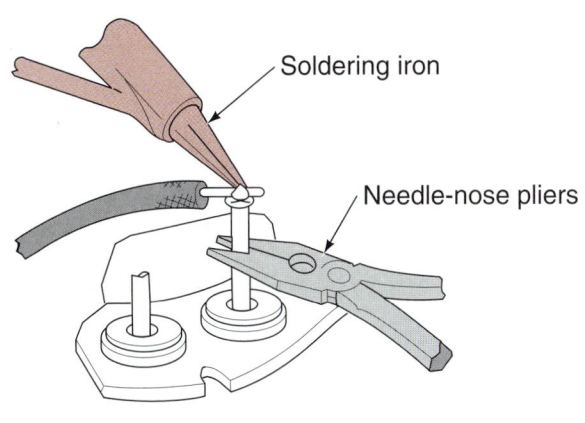

Figure 7-29 When soldering diodes, protect them from excessive heat by providing a heat sink.

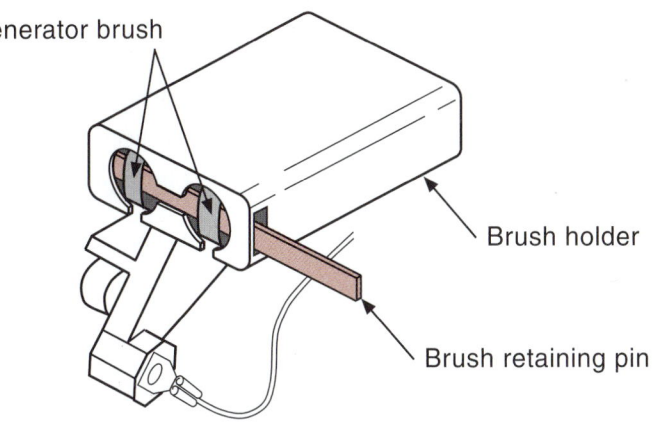

Figure 7-30 When assembling a generator, use a pin to hold the brushes back so the rotor can be inserted into the brush holder.

allow the rotor to fit into them (Figure 7-30). After the rotor is in place, the pin can be removed and the brushes will snap into place on the slip rings. Before installing the pin to hold the brushes, make sure the brush springs are properly positioned behind the brushes.

CASE STUDY

A customer complains that his vechicle's engine "dies" and requires a jump-start. The engine will run for a few minutes, then die again. If the headlights are turned on while the engine is running, the engine dies immediately.

The technician boost starts the engine to confirm the customer's complaint. The engine dies just as reported. The technician slow charges the battery to full capacity. Next, a full battery test series is performed. The battery passes all tests.

The technician then performs a visual inspection of the charging system. During this check, a worn and glazed AC generator drive belt is discovered. The technician does not know for certain yet if the belt is the only problem. However, the charging system tests will not be accurate if the belt is worn. The customer is informed of what the technician has found thus far, and that other tests will still need to be performed to confirm any other problems. The technician gives the customer an estimate for the belt and receives permission to replace it.

After the new belt is installed, the technician performs the voltage output test and the current output test. The requirement test is also performed to confirm that the correct size AC generator is installed. The technician also checks the diode pattern. The charging system passes all tests.

The customer is notified that the car is ready to be picked up. When the customer arrives, the technician tactfully reminds the customer that all belts should be checked every 6 months and replaced per the manufacturer's maintenance schedule.

Terms to Know

Current output test

Diode/stator test

Field current draw test

Full field test

Full fielding

Multiplying coil

Voltage drop test

Voltage output test

ASE-Style Review Questions

1. Charging system testing is being discussed.
   Technician A says before attempting to test the charging system, the battery must be checked.
   Technician B says the state of charge of the battery is not a concern to charging system testing.
   Who is correct?
   - **A.** A only
   - **B.** B only
   - **C.** Both A and B
   - **D.** Neither A nor B

2. AC generator noise complaints are being discussed.
   Technician A says a loose belt will make a grumbling noise.
   Technician B says a whining noise can be caused by a shorted diode.
   Who is correct?
   - **A.** A only
   - **B.** B only
   - **C.** Both A and B
   - **D.** Neither A nor B

3. *Technician A* says the no-load/load voltage output test is used to make a quick determination concerning whether or not the charging system is working properly.
   Technician B says when testing a charging system, the first step is to perform a visual inspection and preliminary checks of the charging system.
   Who is correct?
   - **A.** A only
   - **B.** B only
   - **C.** Both A and B
   - **D.** Neither A nor B

4. Test results of the voltage output test are being discussed.
   Technician A says if the charging voltage is too high, there may be a loose or glazed drive belt.
   Technician B says if the charging voltage is too low, the fault might be a grounded field wire from the regulator (full fielding).
   Who is correct?
   - **A.** A only
   - **B.** B only
   - **C.** Both A and B
   - **D.** Neither A nor B

5. Voltage drop testing is being discussed.
   Technician A says the total system drops should be less than 0.7 volt.
   Technician B says the ground side voltage drop should be less than 0.2 volt.
   Who is correct?
   - **A.** A only
   - **B.** B only
   - **C.** Both A and B
   - **D.** Neither A nor B

6. *Technician A* says the field current draw test determines if there is current available to the field windings.
   Technician B says a slipping belt can cause a low reading when performing the field current draw test.
   Who is correct?
   - **A.** A only
   - **B.** B only
   - **C.** Both A and B
   - **D.** Neither A nor B

7. Full field test procedures are being discussed.
   Technician A says to full field a GM A circuit AC generator, insert a screwdriver into the D-shaped test hole and ground the tab.
   Technician B says check the rotor and field circuit resistance before full fielding the IAR AC generator.
   Who is correct?
   - **A.** A only
   - **B.** B only
   - **C.** Both A and B
   - **D.** Neither A nor B

8. *Technician A* says full fielding means the field windings are constantly energized with full battery voltage.
   Technician B says the full fielding test should be performed only if the charging system passes the output test.
   Who is correct?
   - **A.** A only
   - **B.** B only
   - **C.** Both A and B
   - **D.** Neither A nor B

9. *Technician A* says if full fielding with the regulator bypassed produces lower-than-specified output, the regulator is the cause of the problem.
Technician B says the full field test will isolate whether the detected problem lies in the AC generator or the regulator.
Who is correct?
   A. A only
   B. B only
   C. Both A and B
   D. Neither A nor B

10. *Technician A* says all AC generators have a means of full fielding and bypassing the regulator.
Technician B says many import and domestic AC generators must be disassembled to determine the cause of the charging system failure.
Who is correct?
   A. A only
   B. B only
   C. Both A and B
   D. Neither A nor B

ASE Challenge Questions

1. A vehicle's battery discharges in a very short period of time due to a shorted cell. Before replacing the battery, a charging system test is performed. The engine is started and all accessories are turned off.
Technician A says that the charging system's amperage output will be lower than normal.
Technician B says that the alternator's field current will be high when the engine is running.
Who is correct?
   A. A only
   B. B only
   C. Both A and B
   D. Neither A nor B

2. The charging system voltage of a vehicle equipped with an external voltage regulator is 15.5 volts at 1,500 rpm.
Technician A says that the voltage regulator sensing circuit may have excessive resistance.
Technician B says that the alternator's field circuit may have excessive resistance.
Who is correct?
   A. A only
   B. B only
   C. Both A and B
   D. Neither A nor B

3. Charging system voltage is being measured at two places at the same time.
Connecting the voltmeter's leads across the battery terminals results in a 12.8-volt reading, while connecting the voltmeter to the alternator output terminal and the alternator case results in a 14.2-volt reading.
Technician A says that there may be excessive resistance on the ground side of the charging system.
Technician B says that there may be excessive resistance on the positive side of the charging system.
Who is correct?
   A. A only
   B. B only
   C. Both A and B
   D. Neither A nor B

4. A customer says that his battery completely discharges every few days even though he drives his car daily 50 miles. A test of his charging system is performed and the following measurements are recorded with all accessories turned "on."
At 1,300 rpm the charging system voltage and amperage is 12.4 volts and 60 amps. A replacement alternator produces the same measurements.
Technician A says that there may be excessive electrical demand on the alternator.
Technician B says that the charging system may have excessive voltage drop.
Who is correct?
   A. A only
   B. B only
   C. Both A and B
   D. Neither A nor B

5. Which of the following could result in battery overheating and eventual premature failure?
   A. Open field circuit.
   B. Shorted stator windings.
   C. Open diode.
   D. High-resistance voltage regulator sensing circuit.

Job Sheet 25

25

Name _____ Date _____

Inspecting Drive Belts

Upon completion of this job sheet, you should be able to visually inspect drive belts and check their tightness.

ASE Correlation

This job sheet is related to the ASE Electrical/Electronic System Certification Exam's content area: *Charging System Diagnosis and Repair;* task: Inspect, adjust, and replace alternator drive belts, pulleys, and fans.

Tools and Materials

Two vehicles, one with a serpentine belt and the other with V-belts
Service manuals for the vehicles assigned
Belt tension gauge

Describe the vehicle being worked on:

Year _____ Make _____ Model _____
VIN _____

Year _____ Make _____ Model _____
VIN _____

Procedure

Task Completed

1. On the vehicle with a serpentine belt, carefully inspect the belt and describe the general condition of the belt.

   _____
   _____
   _____

2. With the proper belt tension gauge, check the tension of the belt. Belt tension should be _____. You found _____.

3. Based on the above, what are your recommendations?

   _____
   _____
   _____

4. Describe the procedure for adjusting the tension of the belt.

   _____
   _____
   _____

5. On the vehicle with V-belts, you will find more than one drive belt. List the different belts by their purpose.

_____

_____

_____

6. Carefully inspect the belts and describe the general condition of each one.

_____

_____

_____

7. Check the tension of the AC generator drive belt. Belt tensions should be

_____. You found _____.

8. Based on the above, what are your recommendations?

_____

_____

9. Describe the procedure for adjusting the tension of the belt.

_____

_____

_____

Instructor's Response _____

_____

_____

_____

Job Sheet 26

26

Name _____ Date _____

Testing Charging System Output

Upon completion of this job sheet, you should be able to measure the output of the charging system.

ASE Correlation

This job sheet is related to the ASE Electrical/Electronic System Certification Exam's content area: *Charging System Diagnosis and Repair;* task: Perform charging system output test; determine needed repairs.

Tools and Materials

A vehicle
Service manual for the vehicle assigned
Starting/charging system tester (VAT-60 or similar)

Describe the vehicle being worked on:

Year _____ Make _____ Model _____

VIN _____ Engine type and size _____

Procedure

Task Completed

1. Identify the type and model of AC generator. What type is it?

   _____

   What are the output specifications for this AC generator?

   _____ amps and _____ volts at _____ rpm

2. Connect the starting/charging system tester to the vehicle. ☐

3. Start the engine and run it at the specified engine speed. ☐

4. Observe the output to the battery. The meter readings are:

   _____ amps and _____ volts.

5. Compare readings to specifications and give recommendations.

   _____

   _____

   _____

   _____

6. If readings are outside of the specifications, refer to the service manual to find the proper way to full field the AC generator. Describe the method.

_____

_____

_____

_____

7. Full field the generator and observe the output to the battery. The meter readings are: _____ amps and _____ volts.

8. Compare readings to specifications and give recommendations.

_____

_____

_____

_____

Instructor's Response _____

_____

_____

_____

Job Sheet 27

27

Name _____ Date _____

Testing the Charging System Circuit

Upon completion of this job sheet, you should be able to visually inspect and test the insulated and ground side of the charging system circuit.

ASE Correlation

This job sheet is related to the ASE Electrical/Electronic System Certification Exam's content area: *Charging System Diagnosis and Repair;* task: Diagnose charging system problems that can cause undercharge, a no-charge, or an overcharge condition.

Tools and Materials

A vehicle
Wiring diagram for the vehicle assigned
A DMM

Describe the vehicle being worked on:

Year _____ Make _____ Model _____

VIN _____ Engine type and size _____

Procedure

Task Completed

1. Describe the general appearance of the AC generator and the wires that are attached to it.

   _____

   _____

   _____

2. Measure the open circuit voltage of the battery. Your measurement was _____ volts.

3. From the wiring diagram, identify the output, input, and ground wires for the AC generator. Describe these wires, by color and location, below.

   _____

   _____

   _____

   _____

4. Start the engine and allow it to run. Then turn on the headlights in the high-beam mode. ☐

5. Connect the DMM across the charging system's output wire. Measure the voltage drop and record your readings.

_____

_____

6. Connect the DMM across the charging system's input wire. Measure the voltage drop and record your readings.

_____

7. Connect the DMM across the charging system's ground wire. Measure the voltage drop and record your readings.

_____

8. Based on the test, what is indicated by the results?

_____

_____

_____

Instructor's Response _____

_____

_____

DIAGNOSTIC CHART 7-1

PROBLEM AREA:	Charging system operation
SYMPTOMS:	**1.** Battery is too low to start engine. **2.** Headlight illumination dim. **3.** Electrical accessories do not operate properly. **4.** Charging indicator warning light illuminated.
POSSIBLE CAUSES:	**1.** Slipping or worn drive belt. **2.** Poor battery cable connections. **3.** Faulty voltage regulator. **4.** Shorted stator. **5.** Open stator. **6.** Open diode. **7.** Shorted diode. **8.** Slipping or worn drive belt. **9.** Worn brushes. **10.** Open field coil circuit. **11.** Shorted field coil. **12.** Worn or slipping drive belt. **13.** Worn brushes. **14.** Faulty rectifier bridge. **15.** Faulty diode trio. **16.** Excessive resistance in system circuit. **17.** Open in the circuit. **18.** Improper ground connection. **19.** Loose or corroded connection.

DIAGNOSTIC CHART 7-2

PROBLEM AREA:	Charging system operation (excessive charging)
SYMPTOMS:	Battery electrolyte level constantly low. Bulbs burn out. Brighter than normal bulb illumination. Electrical system component failures.
POSSIBLE CAUSES:	**1.** Faulty voltage regulator. **2.** Grounded field coil circuit. **3.** Excessive resistance in sensing circuit. **4.** Faulty generator.

DIAGNOSTIC CHART 7-3

PROBLEM AREA:	Voltage regulation
SYMPTOMS:	**1.** Charging system overcharging the battery. **2.** Charging system output below specifications.
POSSIBLE CAUSES:	**1.** Defective voltage regulator. **2.** Open or short in sense circuit.

DIAGNOSTIC CHART 7-4

PROBLEM AREA:	Diodes
SYMPTOMS:	**1.** Excessive noises. **2.** Low or no charging system output. **3.** AC generator fails output test.
POSSIBLE CAUSES:	**1.** Open diodes. **2.** Shorted diodes.

DIAGNOSTIC CHART 7-5

PROBLEM AREA:	AC generator pulleys
SYMPTOMS:	**1.** Belts wear prematurely. **2.** Noises.
POSSIBLE CAUSES:	**1.** Pulley bent.

DIAGNOSTIC CHART 7-6

PROBLEM AREA:	AC generator
SYMPTOMS:	Noises.
POSSIBLE CAUSES:	**1.** Bent fan blades. **2.** Worn or dry bearings.

Lighting Circuits Repair and Diagnosis

Upon completion and review of this chapter, you should be able to:

❑ Correctly replace sealed-beam and composite headlights.

❑ Correctly service the high-intensity discharge (HID) lamp and ballast.

❑ Correctly aim sealed-beam and composite headlights.

❑ Diagnose the cause of brighter-than-normal lights.

❑ Diagnose the cause of dimmer-than-normal lights.

❑ Diagnose lighting systems that do not operate.

❑ Determine causes for incorrect concealed headlight operation.

❑ Remove and replace dash-mounted and steering column–mounted headlight switches.

❑ Replace multifunction switches.

❑ Test and determine needed repairs of the dimmer switch and related circuits.

❑ Replace the dimmer switch.

❑ Diagnose incorrect taillight assembly operation.

❑ Diagnose the turn signal system for improper operation.

❑ Replace the turn signal switch.

❑ Diagnose the interior lights, including courtesy, instrument, and panel lights.

Basic Tools

Basic mechanic's tool set

Service manual

Introduction

The lighting system of the vehicle is becoming very complex. There may be over 50 light bulbs and hundreds of feet of wiring in the lighting circuits. The circuits include circuit protectors, switches, lamps, and connectors. Any failure requires a systematic approach to diagnose, locate, and correct the fault in the minimum amount of time.

The importance of a proper operating lighting system cannot be overemphasized. The lighting system should be checked whenever the vehicle is brought into the shop for repairs. Often a customer may not be aware of a light failure. If a lighting circuit is not operating properly, there is a potential danger to the driver and other people. When today's technician performs repairs on the lighting systems, the repairs must assure vehicle safety and meet all applicable laws. Be sure to use the correct lamp type and size for the application (Figure 8-1).

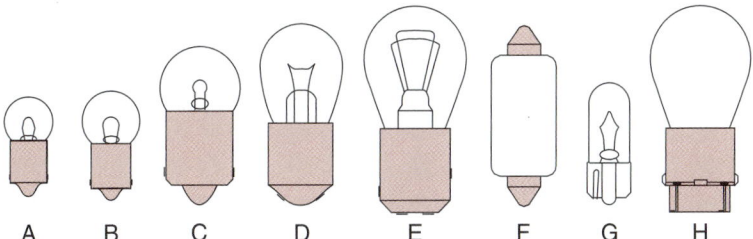

Common automotive bulbs:

A,B   Miniature bayonet for indicator and instrument lights
C –   Single contact bayonet for license and courtesy lights
D –   Double contact bayonet for trunk and underhood lights
E –   Double contact bayonet with staggered indexing lugs for stop, turn, and brake lights
F –   Cartridge type for dome lights
G –   Wedge base for instrument lights
H –   Blade double contact for stop, turn, and brake lights

Figure 8-1 Correct selection of the lamp is important for proper operation of the system.

Before performing any lighting system tests, check the battery for state of charge. Also, be sure all cable connections are clean and tight. Visually check the wires for damaged insulation, loose connections, and improper routing. A troubleshooting chart for the lighting system is located at the end of this chapter.

When troubleshooting the lighting system, if only one bulb is not operating it is usually faster to replace it with a known good unit first. Check the connector for signs of corrosion. When testing the circuit with a voltmeter, ohmmeter, or test light, check the most easily reached components first.

Headlights

There are four basic types of headlights used on automobiles today: standard sealed beam, halogen sealed beam, composite, and HID. The most common service performed on the headlights is lamp replacement and aiming. These procedures vary depending on the type of headlights used.

Because the lighting circuits are largely regulated by federal laws, the systems are similar between the various manufacturers. However, there are variations that exist in these circuits. Consult the service manual for the vehicle you are working on.

Headlight Replacement

One of the most common lighting system repairs is replacing the headlight. After a period of time, the filament may burn through or the lens may be broken. Before the headlight is replaced, however, a voltmeter or test light should be used to confirm that voltage is present. Next, check the ground for proper connections. If these test good, the headlight is probably faulty and needs to be replaced. If there is no voltage present at the connector, work back toward the switch and battery until the fault is located.

SERVICE TIP: If it is necessary to replace the headlight lamp often, check the charging system. A too-high voltage output will cause the filament to burn hotter than it is designed to burn and will shorten the life of the lamp.

The procedure for replacing the headlight differs depending on the type of bulb used. Most conventional sealed-beam and halogen sealed-beam headlights are replaced in the same manner. Composites require different procedures. Always refer to the service manual for the vehicle you are working on.

CAUTION: Because of the construction and placement of the prisms in the lens, it is important that the headlight is installed in its proper position. The lens is usually marked "TOP" to indicate the proper installed position.

Sealed-Beam. Photo Sequence 12 illustrates a common procedure for replacing a sealed-beam headlight.

After confirming proper headlight operation, check headlight aiming as described in the next section.

CAUTION: It is not recommended that halogen sealed-beam and standard sealed-beam headlights be mixed on the vehicle. Also, if the vehicle was equipped with halogen headlights as original equipment, do not replace the headlights with standard sealed beams. Doing so may result in poor light quality.

Classroom Manual
Chapter 8, page 206

Classroom Manual
Chapter 8, page 206

Photo Sequence 12
Replacing a Sealed-Beam Headlight

P12-1 Place fender covers around the work area.

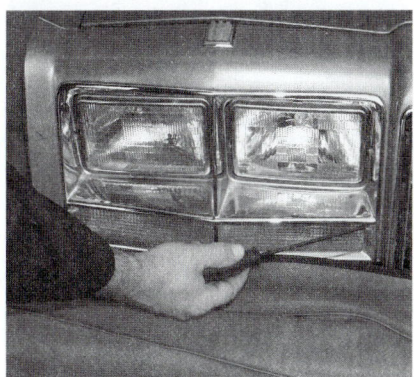

P12-2 Sealed-beam headlight replacement usually requires the removal of the bezel. Some vehicles may require that the turn signal light assembly be removed before the headlight is accessible.

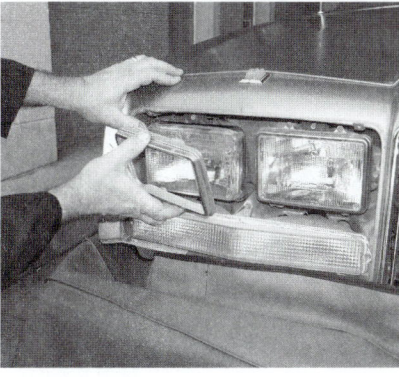

P12-3 Remove the retaining ring screws and the retaining trim. Do not turn the two headlight aiming adjustment screws.

P12-4 Remove the headlight from the shell assembly.

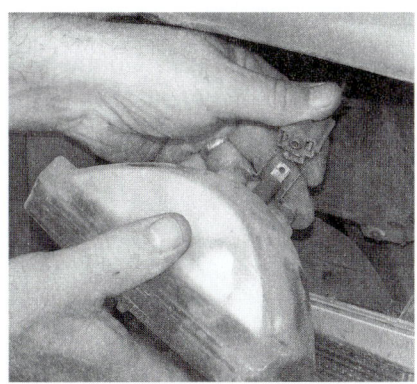

P12-5 Disconnect the wire connector from the back of the lamp.

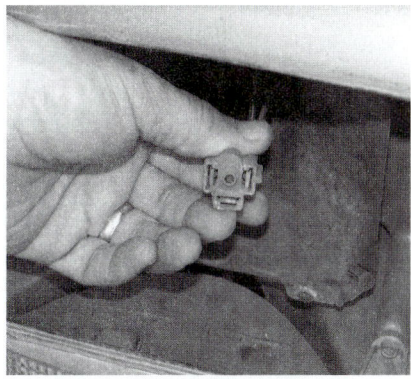

P12-6 Check the wire connector for corrosion or other foreign material. Clean as necessary.

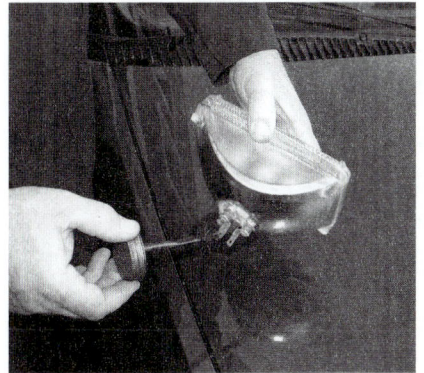

P12-7 Coat the connector terminals and the prongs of the new headlight with dielectric grease to prevent corrosion.

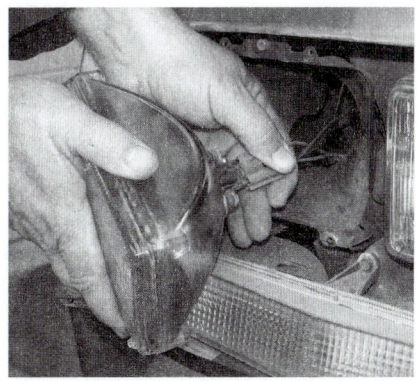

P12-8 Install the wire connector onto the new headlight's connector prongs.

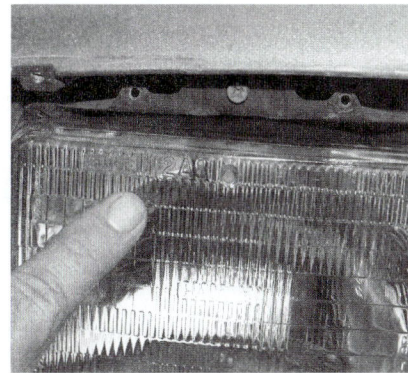

P12-9 Place the headlight into the shell assembly. When positioning the headlight, be sure that the embossed number is at the top. Many headlights are marked "TOP."

Photo Sequence 12
Replacing a Sealed-Beam Headlight (continued)

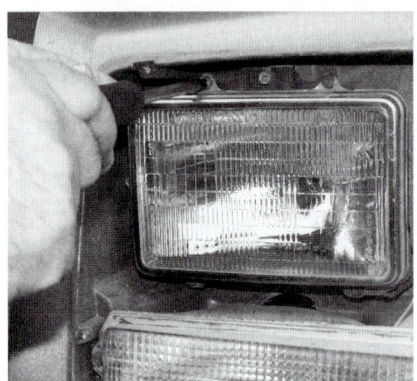

P12-10 Install the retainer trim and fasteners.

P12-11 Install the headlight bezel.

P12-12 Check operation of the headlight.

CUSTOMER CARE: Because the filament of the halogen lamp is contained in its own bulb, cracking or breaking of the lens does not prevent headlight operation. The filament will continue to operate as long as the filament envelope has not been broken. However, a broken lens will result in poor light quality and should be replaced for the safety of the customer.

Classroom Manual
Chapter 8, page 208

Composite. To replace a composite headlight:

1. Place fender covers around the work area.
2. Remove the wire connector from the bulb.
3. Unlock the bulb retaining ring by rotating it 1/8 of a turn (Figure 8-2).
4. Slide off the retaining ring from the base.

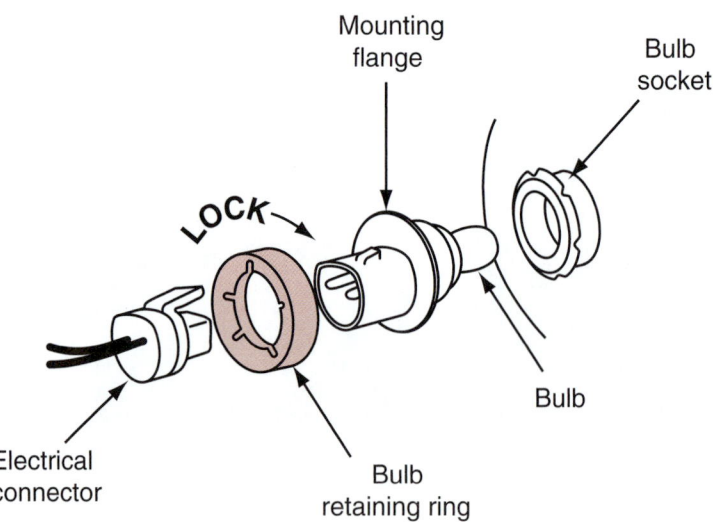

Figure 8-2 Composite headlight bulb replacement.

5. Gently pull the bulb straight back out of the socket. To prevent breaking the bulb and locating tabs, do not rotate it while pulling.

6. Check the wire connector for corrosion or other foreign material. Clean as needed.

7. Coat the connector terminals and the prongs of the new headlight with dielectric grease to prevent corrosion.

CAUTION: Do not get any of the dielectric grease onto the bulb. The bulb's life will be shortened.

8. Place the new bulb in the socket (Figure 8-3). The flat part of the base should face up. It may be necessary to turn the bulb slightly to align the locating tabs.

CAUTION: Whenever technicians replace a composite lamp, care must be taken not to touch the envelope with the fingers. Staining the bulb with normal skin oil can substantially shorten the life of the bulb. Handle the lamp only by its base. Also dispose of the old lamp properly.

9. The mounting flange on the bulb base should contact the rear of the socket.

10. Install the retaining ring over the base and lock the ring into the socket.

11. Reconnect the wire connector to the bulb. It will snap and lock into position when properly installed.

12. Check headlight operation.

CAUTION: To prevent early failure, do not energize the bulb unless it is installed into the socket.

13. Check and adjust headlight aiming as needed.

Because the housings of a composite headlight system are vented, condensation may develop inside the lens assembly. This condensation is not harmful to the bulb and does not affect headlight operation.

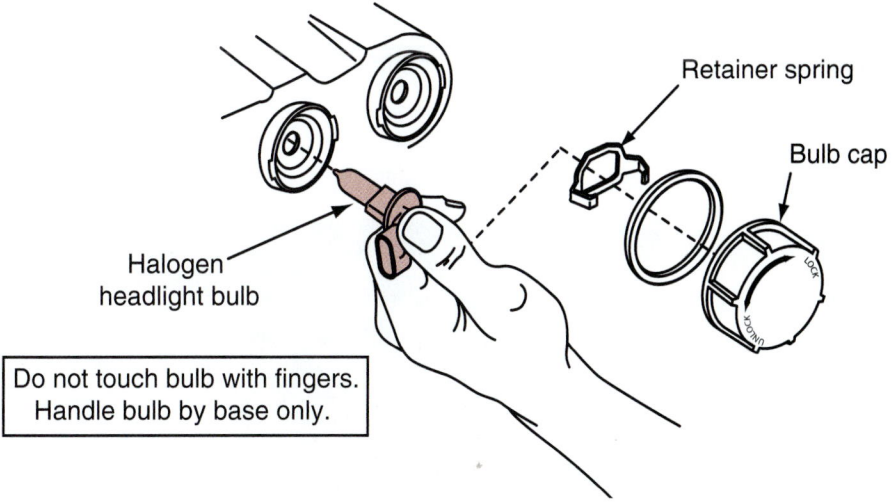

Retainer spring

Bulb cap

Halogen headlight bulb

Do not touch bulb with fingers.
Handle bulb by base only.

Figure 8-3 The correct method of handling the composite bulb during replacement.

⚠️ **WARNING:** Before testing the HID system, remove the headlight assembly from the vehicle. The HID system operates on high voltage and current. If the system is accidentally shorted, personal injury or death may occur.

HID Lamp and Ballast Service. The high-intensity discharge (HID) lamps require battery voltage and ground to operate as any other lamp. The ballast has internal circuit protection to prevent damage resulting from an open or shorted circuit. In addition, the circuit protection will come into effect if an overcharge or undercharge condition exists. If any of these faults are detected, the headlights will not turn on for that key cycle, or until the fault is corrected. If the customer states that both headlights do not operate, recycle the ignition or the headlight switch to see if the condition reoccurs. If the lights are still not coming on, then check battery condition and the charging system. It is unlikely that both lamps failed at the same time, so check the switch input, HID relay, power feed, and ground.

A burned-out lamp will appear black or smoky. If a visual inspection does not indicate a faulty lamp, then check for battery voltage to the ballast. If battery voltage is above specifications, check the ground circuit for the ballast. If neither of these circuits has an open or short, then substitute a known good ballast. If the headlight on the opposite side of the vehicle is operating properly, substitute its ballast with the side that does not operate. If the headlight illuminates, replace the ballast. If the light still does not come on, replace the bulb/ignitor element.

Depending on design, it may be necessary to remove the headlight assembly prior to replacing the bulb and ignitor. Some vehicles provide access that allows the bulb and ignitor to be serviced without removing the headlight assembly. Regardless of how the bulb element is accessed, replacement procedures are very similar.

Prior to servicing the headlight element, disconnect and isolate the battery negative cable or remove the HID relay. This will prevent accidental operation of the headlights.

⚠️ **WARNING:** The HID system produces a high voltage and current. Do not attempt to operate the headlights without all connections properly made.

⚠️ **WARNING:** Never attempt to open the ballast or ignitor assembly. The ignitor is a component of the bulb and cannot be separated.

To gain access to the bulb assembly, remove the rubber boot that protects the element (Figure 8-4). At the ignitor, disconnect the cable coming from the ballast (Figure 8-5). As seen

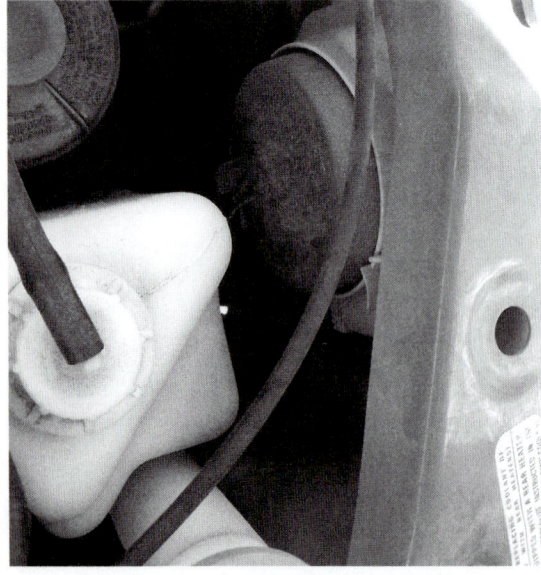

Figure 8-4 The HID element is protected by a rubber boot on the back of the headlight assembly.

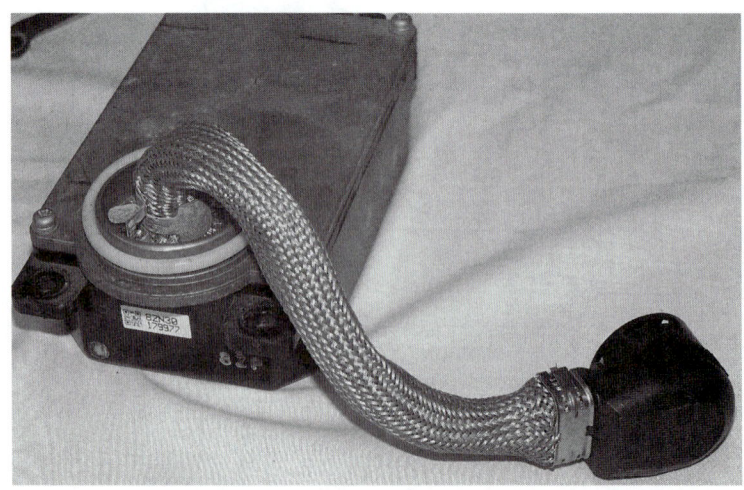

Figure 8-5 The ignitor is connected to the ballast by a cable.

Figure 8-6 HID ballast connectors.

in Figure 8-6, a wire clip will retain the element to the headlight assembly. Release the clip and remove the element by grabbing the ignitor and pulling the unit rearward. It may be necessary to rock the ignitor slightly as it is being pulled.

To install the new element, push the bulb element assembly into the headlight unit. Once properly seated, secure with the retainer clip. Connect the cable from the ballast to the ignitor's connector. Install the rubber protective cover. Make sure the cover is properly seated. If the cover is not seated, water may enter the element and damage the unit. If necessary, install the headlight assembly. Reconnect the battery or connect the relay and test for proper operation.

The ballast can be attached to the bottom of the headlight assembly, to the vehicle's frame, or to the inner fender well. A cable connects the ballast to the ignitor. In addition, another connector attaches the ballast to the vehicle's wiring system (Figure 8-6). It may be necessary to remove the headlight assembly to gain access to the ballast. Prior to removing the ballast, disconnect and isolate the battery negative cable or remove the HID relay.

Remove the screws that retain the ballast to the headlight assembly or frame. Next, disconnect the main electrical connector and the cable to the ignitor from the ballast. The cable may be secured by a wire lever that must be rotated down. Installation of the new ballast is reverse order of the removal procedure. Be sure to properly secure the wire retainer around the cable connector.

Headlight Aiming

The headlights should be checked for proper aiming whenever the lamps are replaced. Proper aiming is important for good light projection onto the road and to prevent discomfort and dangerous conditions for oncoming drivers.

Correct headlight beam position is so critical that government regulations control limits for headlight aiming. For example, a headlight that is misaimed by one degree downward will reduce the vision distance by 156 feet. The following are maximum allowable limits that have been established by all states:

1. Low beam: In the horizontal plane, the left edge of the headlight high-intensity area should be within 4 inches (102 mm) to the right or left of the vertical centerline of the

Classroom Manual
Chapter 8, page 206

Special Tools

Headlight aiming
  unit

Torx driver

Fender covers

Safety glasses

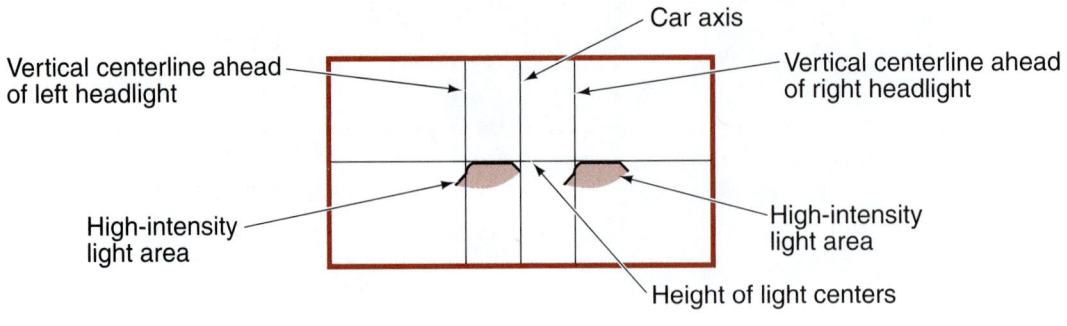

Figure 8-7 Low-beam headlight aiming adjustment pattern.

lamp. In the vertical plane, the top edges of the headlight high-intensity area should be within 4 inches (102 mm) above or below the horizontal centerline of the lamp (Figure 8-7).

2. High beam: In the horizontal plane, the center of the headlight high-intensity area should be within 4 inches (102 mm) to the right or left of the vertical centerline of the lamp. In the vertical plane, the center of the headlight high-intensity area should be within 4 inches (102 mm) above or below the horizontal centerline of the lamp (Figure 8-8).

Before the headlights are aimed, the vehicle must be checked for proper **curb height.** Curb height is the height of the vehicle when it has no passengers or loads and has normal fluid levels and tire pressure. This includes checking the springs; tire inflation; removing any additional load in the vehicle; a half-filled fuel tank; and removing dirt, ice, snow, and so on, from the vehicle.

Sealed-Beam. Most shops use portable mechanical aiming units (Figure 8-9). These are secured to the headlight lens by suction cups (Figure 8-10). The aiming unit should have a variety of adapters to attach to the various styles of headlights. Before using the aiming equipment, be sure to follow the manufacturer's procedure for calibration. Park the vehicle on a level floor area and place fender covers around the work area. It may be necessary to remove the trim and bezel from around the headlight. Using the correct adapter, connect the calibrated aimer units to the headlights. Be sure the adapters fit the headlight aiming pads on the lens (Figure 8-11). Zero the horizontal adjustment dial. Confirm that the split-image target lines are

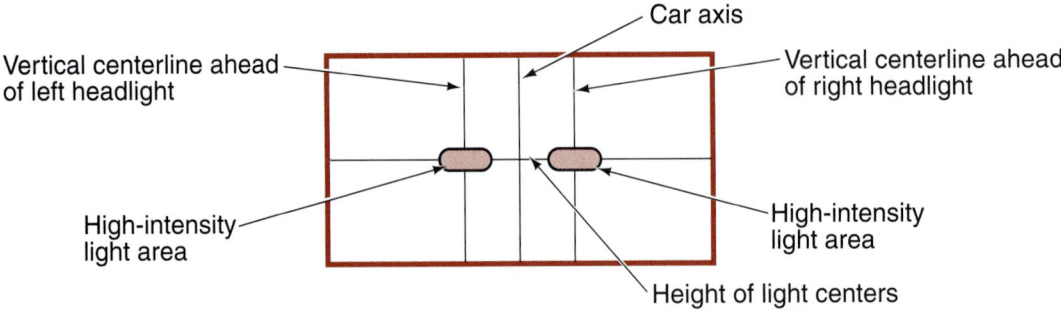

Figure 8-8 High-beam headlight aiming adjustment pattern.

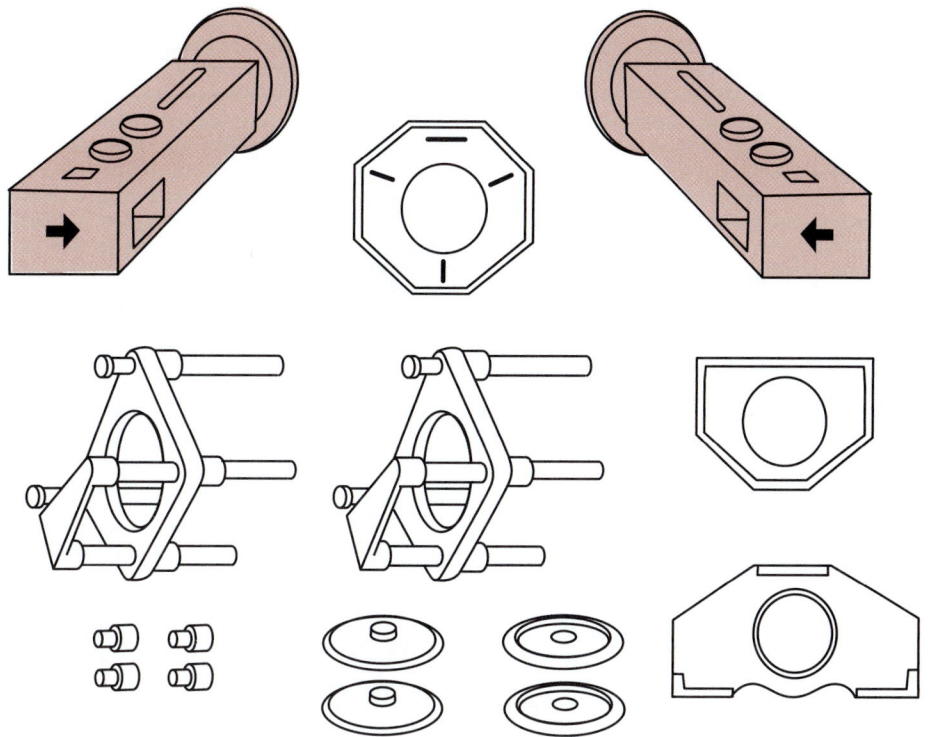

Figure 8-9 Typical portable mechanical headlight aiming equipment and adapters.

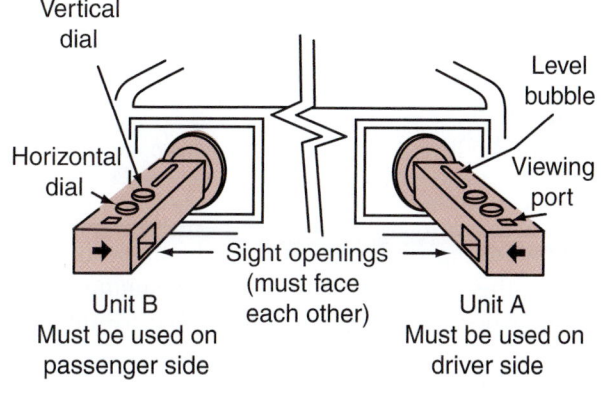

Vertical dial

Horizontal dial

Level bubble

Viewing port

Sight openings (must face each other)

Unit B
Must be used on passenger side

Unit A
Must be used on driver side

Figure 8-10 The aiming units attach to the headlight lens with suction cups.

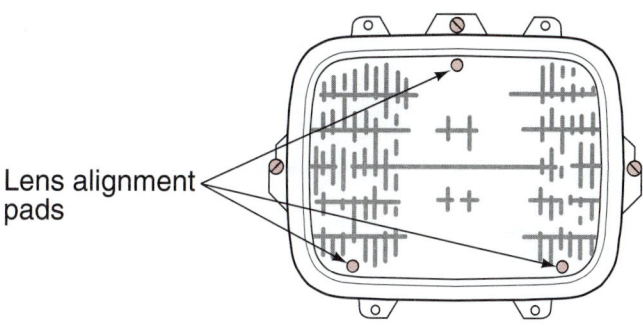

Lens alignment pads

Figure 8-11 Headlight aiming pads.

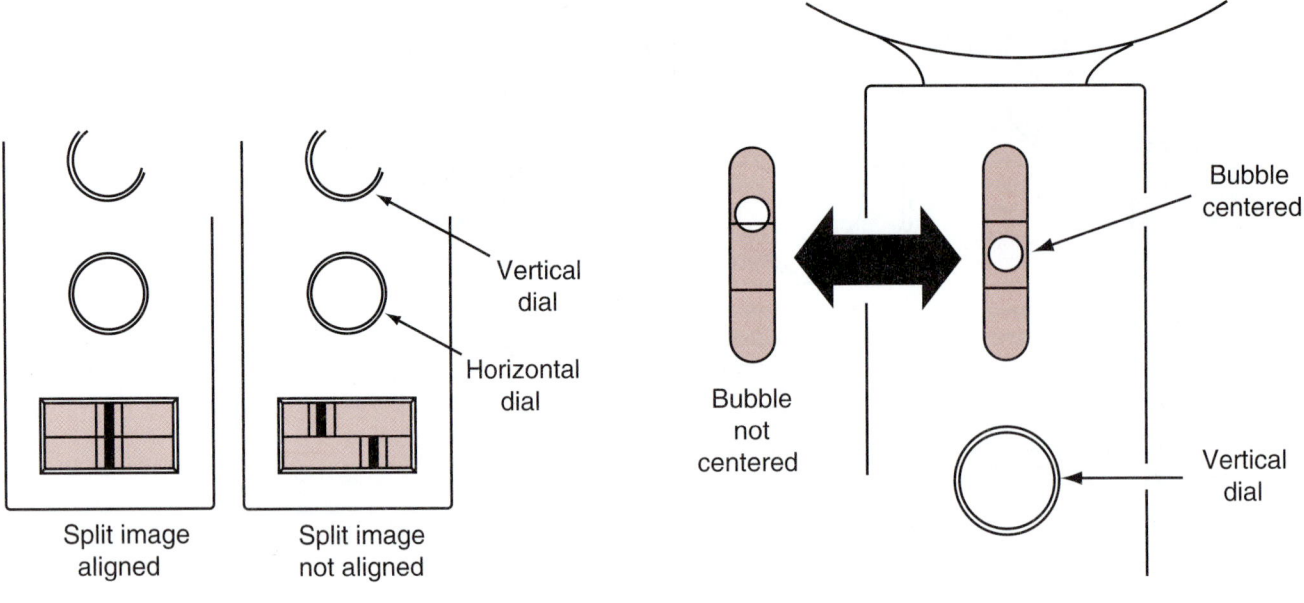

Figure 8-12 Split-image target.

Figure 8-13 Center the spirit level by turning the vertical aiming screw.

visible in the view port (Figure 8-12). If the target lines are not seen, rotate the aimer unit. Turn the headlight horizontal adjusting screw until the split-image target lines are aligned. Repeat for the headlight on the other side.

To set the vertical aim of the headlight, turn the vertical adjustment dial on the aiming unit to zero. Turn the vertical adjustment screw until the spirit level bubble is centered (Figure 8-13). Recheck the horizontal aiming on each headlight. The vertical adjustment may have altered the original adjustments.

If the vehicle is equipped with a four-headlight system, repeat the procedures for the other pair of lamps.

Composite. To adjust some composite and HID headlight designs, special adapters are required (Figure 8-14). Also, the lens must have headlight aiming pads to be able to use a mechanical aiming unit. The headlight assembly will have a number molded on it. The

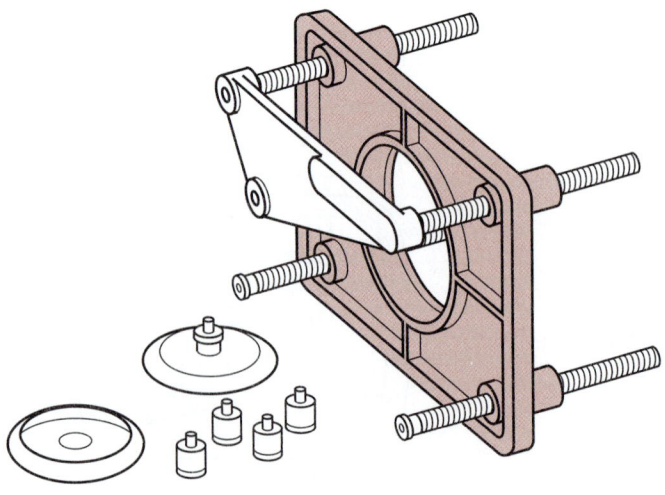

Figure 8-14 Special adapter for aiming composite headlights.

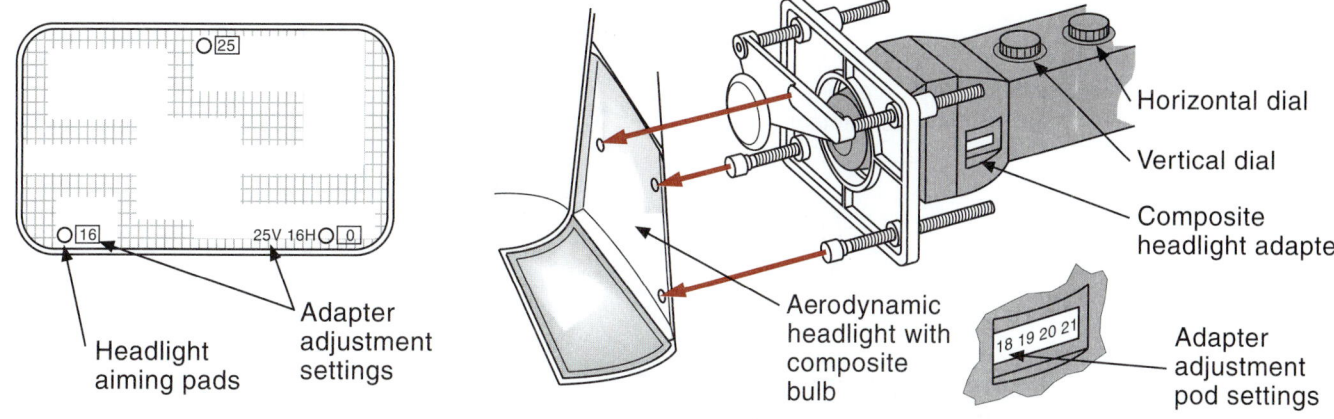

Figure 8-15 Connect the aiming equipment to the headlight lens. The lens must have aiming pads.

adjustment rod setting must be set to that number and locked in place. The aiming unit is attached to the headlight lens in the same manner as with sealed-beam headlights (Figure 8-15).

The adjustment procedure is identical to that of the sealed-beam headlights. The illustration (Figure 8-16) shows the typical location of the headlight adjusting screws.

Many composite and HID headlight designs do not have alignment pads on the lens. These systems usually adjust the beam location by moving the reflector position. Since the lens does not move, conventional headlight aimers are not used. For the most part, these systems are aimed by locating the vehicle 25 feet away from a blank wall with the vehicle on a level surface. The wall is marked, based on the centerline of the vehicle; then the location of the beam on the wall is adjusted to meet manufacturer's specifications. Some manufacturers may require that the headlights be adjusted with the high beams.

Many late-model vehicles have spirit levels built into the headlamp assembly (Figure 8-17). These are not always to be used for initial headlamp adjustment. They are supplied for the driver to be able to adjust his headlights based on the load in the vehicle. For example, if the trunk is loaded, the front of the vehicle is lifted and the light beam is too high. By turning the adjuster wheels until the bubble is in the middle of the level, the beam is returned to its original position. After the load is removed from the trunk, the headlights are adjusted until the bubble is returned to the middle. Whenever the headlights are adjusted, the technician should adjust the spirit level also.

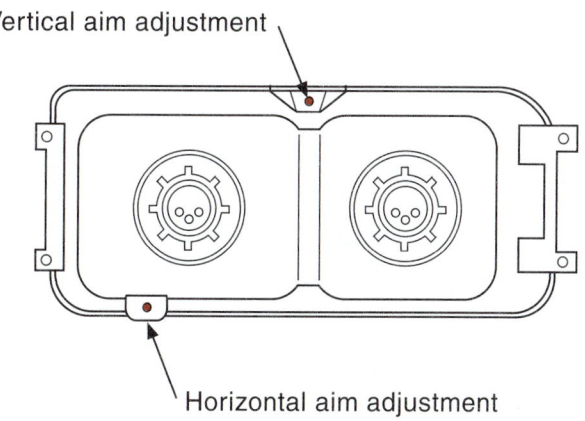

Figure 8-16 Composite headlight aiming screw locations.

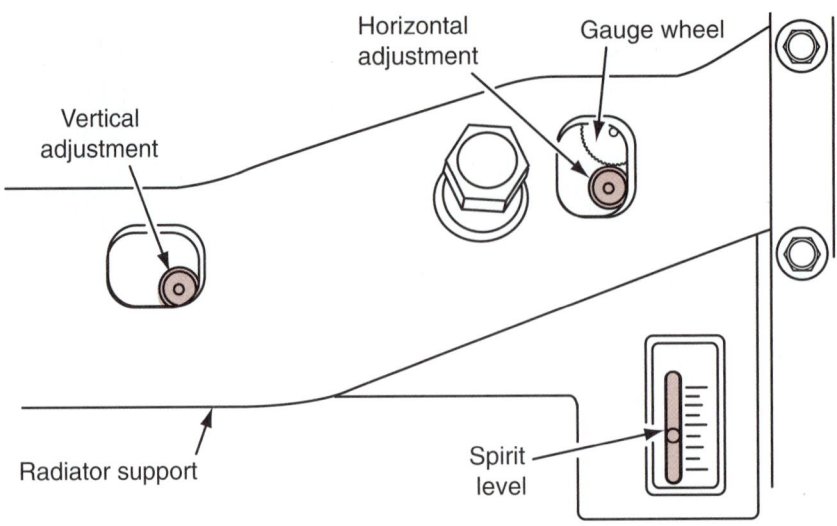

Figure 8-17 Some headlight assemblies are equipped with spirit levels to provide easy adjustments to offset vehicle loads.

Classroom Manual
Chapter 8, page 213

Dimmer- or Brighter-than-Normal Lights

The complete headlight circuit consists of the headlight switch, dimmer switch, high-beam indicator, and the headlights (Figure 8-18). Excessive resistance in these units, or at their connections, can result in lower illumination levels of the headlights.

Special Tools

Voltmeter

Safety glasses

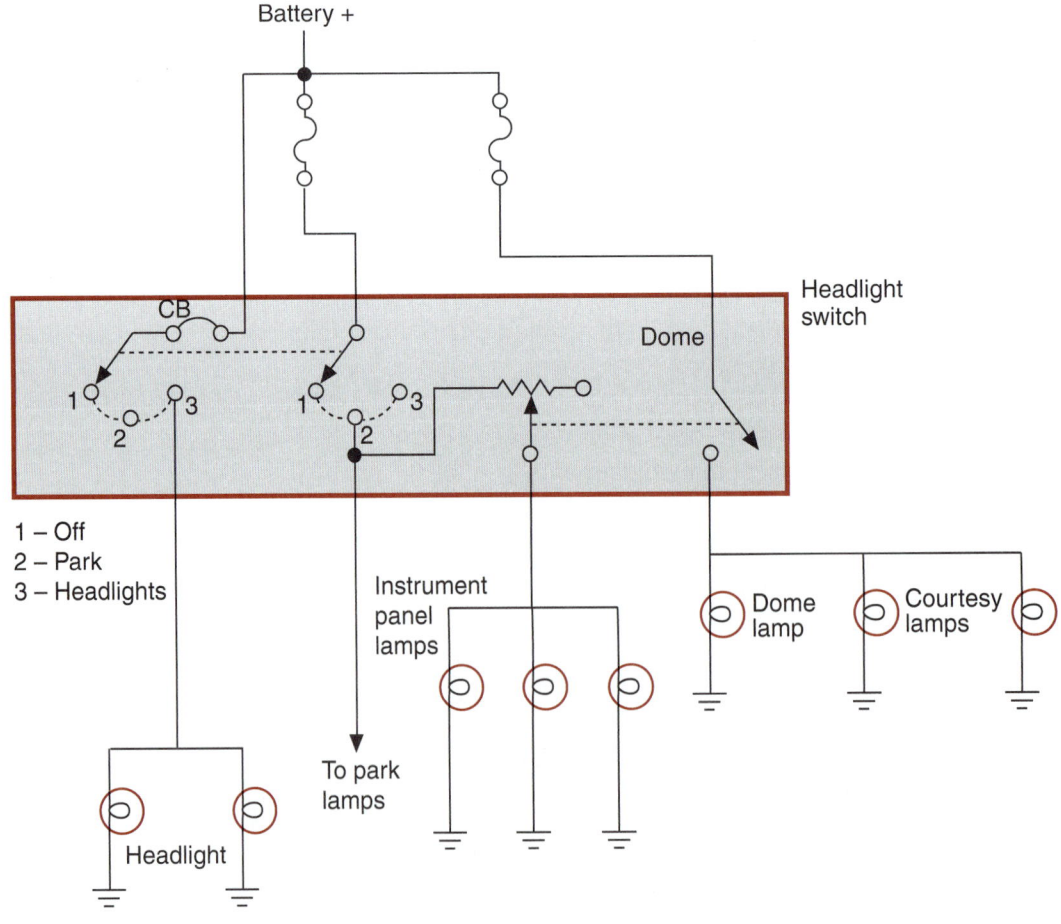

Figure 8-18 Headlight circuit components.

The extra resistance can be on the insulated side or the ground side of the circuit. To locate the excessive resistance, perform a voltage drop test (Figure 8-19). Consult the wiring diagram to determine the number of connectors and switches. This will provide you with the specification for maximum voltage drop. Start at the light and work toward the battery.

All headlight systems are wired in parallel. If both headlights are dim, then the excessive resistance is in the common portions of the circuit. Dim headlights can also be the result of low generator output.

Other causes of dim lights can be the use of the wrong lamps, improper circuit routing, the addition of extra electrical loads to the circuit, and the wrong size conductors.

✓ SERVICE TIP: Headlights do not wear out and get dimmer with age. If one of the headlights is dimmer than the other, there is excessive voltage drop in that circuit. If a new headlight is installed, the breaking and making of the socket connection may clean the terminals enough to make a good contact. Once the new headlight is installed, it may operate properly. Do not be fooled. It was not the headlight that was at fault. It was the connection.

Brighter-than-normal lights can be the result of higher-than-specified generator output or improper lamp application. It is also possible that the dimmer switch contacts are "welded" into the high-beam position.

Although this procedure is being shown for the headlights, it is identical for all lighting systems. The only difference in the test results will be if the circuit uses insulated bulbs or grounded bulbs. If testing the turn signal circuit, bypass the flasher with a jumper wire.

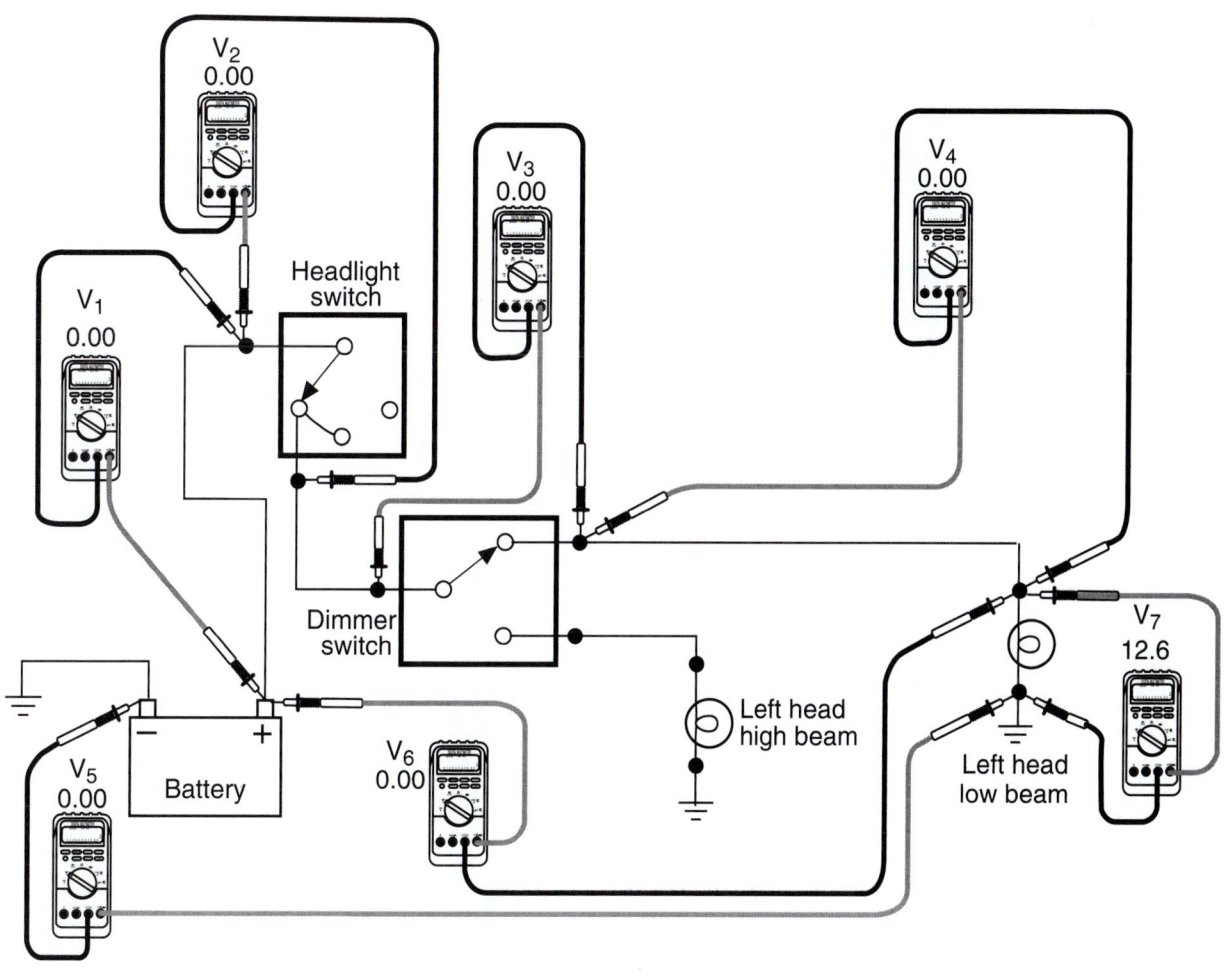

Figure 8-19 Voltage drop testing the headlight system.

Classroom Manual
Chapter 8, page 215

Special Tools

Voltmeter

Vacuum gauge

Vacuum pump

Fender covers

Safety glasses

In recent years, some manufacturers started using the body computer to control the operation of the concealed headlight system. On these vehicles, it may be necessary to enter body computer diagnostics to troubleshoot the system.

Concealed Headlights

A vehicle equipped with a concealed headlight system hides the lamps behind doors when the headlights are turned off. When the headlight switch is turned to the HEADLIGHT position, the headlight doors open. The headlight doors can be controlled either by electric motors or by vacuum. In vacuum-controlled systems, a vacuum distribution valve controls the direction of vacuum to various vacuum motors or to vent. Electrically controlled systems use either a torsion bar to open both headlight doors from a single motor or a separate motor for each headlight door.

If the electrically operated doors fail to operate, check the fuse first. If the fuse is good, check for voltage at the connection to the motor. If voltage is not present, then trace the circuit back to the switch and battery.

If there was voltage at the connector, check the ground circuit. Also, check the operation of the limit switches before condemning the motor.

Vacuum-controlled doors are tested for the presence of vacuum through the distribution valve to the vacuum motors. Use a vacuum gauge to check the amount of vacuum being applied to the motor. If vacuum is present to the motor, use a vacuum pump to check the operation of the motor.

 WARNING: Be careful not to get hands and fingers caught in the door if it should suddenly open.

All concealed headlight systems have a means of manually opening the doors. Check the service manual for the correct procedure for opening the doors. Most electrical doors have a knob that is rotated to open the doors (Figure 8-20).

 WARNING: Be sure the headlight switch is in the OFF position before manually opening the doors. The doors may snap open, catching fingers between the door and the vehicle body.

SERVICE TIP: Most vacuum-controlled doors use the vacuum to close the door. If the headlight doors open while the vehicle is sitting overnight, check for a leak and test the one-way check valve.

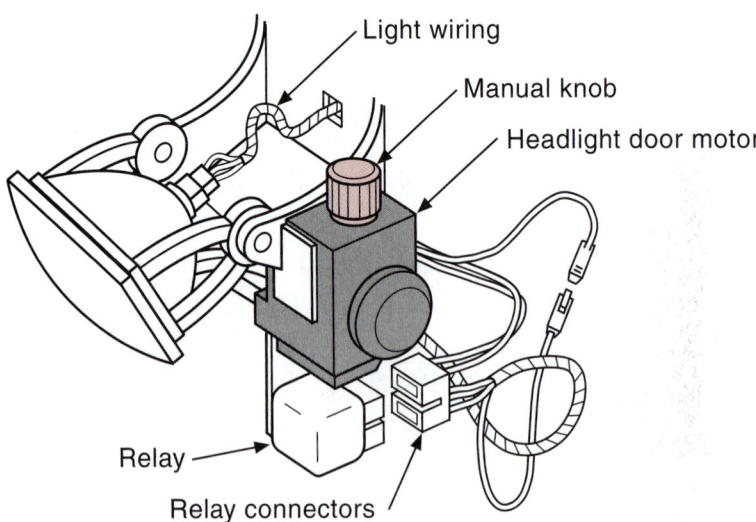

Figure 8-20 Manual control knob is used to open the doors in the event of electrical problems.

Headlight Switch Testing and Replacement

The headlight switch controls most of the vehicle's lighting systems. The headlight switch will generally receive direct battery voltage to two of its terminals. Disconnect the battery before removing the headlight switch.

In the headlight switch circuit, a rheostat is used to control the instrument cluster illumination lamp brightness. Most dash-mounted headlight switches incorporate the rheostat into the switch assembly. Steering column–mounted switches may have the rheostat located on the dash.

Many customer complaints concerning the lighting systems can be the result of a faulty headlight switch. For example, dim or no instrument panel lights, dim or no headlights, dim or no parking lights, and improperly operating dome lights can all be caused by the headlight switch.

☑ **SERVICE TIP:** Headlights that flash on and off as the vehicle goes over road irregularities indicate a loose connection. Headlights that flash on and off at a constant rate indicate that the circuit breaker is being tripped. There is an overload in the circuit that must be traced and repaired.

Dash-Mounted Switches

There are many methods used to retain the headlight switch to the dash. Consult the service manual of the vehicle you are working on. The following is a common method of removing the headlight switch:

⚠ **WARNING:** If the vehicle is equipped with airbags, disable the supplemental restraint system before attempting any component diagnosis or service. Failure to take the proper precautions could result in accidental airbag deployment.

1. Place fender covers on the fenders.
2. Install a memory keeper and disconnect and isolate the battery negative cable.
3. Remove the lower cluster bezel from the instrument panel.
4. Disconnect the wire harness connector from the back of the headlamp switch.
5. Remove the fasteners that secure the headlamp switch to the back of the cluster bezel.
6. Remove the headlamp switch from the cluster bezel.

With the switch removed, it can be tested for continuity and the connector plug will serve as a test point for the lighting circuits. First, test at the connector.

The following is a typical procedure for testing the headlight switch connector on the harness side. A test light and jumper wires are used to test the circuits. This procedure would be very similar for any non-computer-controlled headlight system—just use the service manual to determine the function of each terminal. In this procedure the terminals are identified as follows (Figure 8-21):

- ❏ Terminal B — battery.
- ❏ Terminal A — fuse.
- ❏ Terminal H — headlights.
- ❏ Terminal R — rear park and side marker lights.
- ❏ Terminal I — instrument panel lights.
- ❏ Terminal D1 — dome light feed.
- ❏ Terminal D2 — dome light.

Consult the service manual for terminal identification for the vehicle you are working on. If this is not listed in a separate chart, you should be able to identify the circuits from the wiring diagrams.

Classroom Manual
Chapter 8, page 210

Special Tools

12-volt test light
Ohmmeter or self-powered test light
Jumper wire
Safety glasses
Battery terminal pliers
Terminal pullers

Some headlight switches are retained by spring clips. Remove them by compressing the springs.

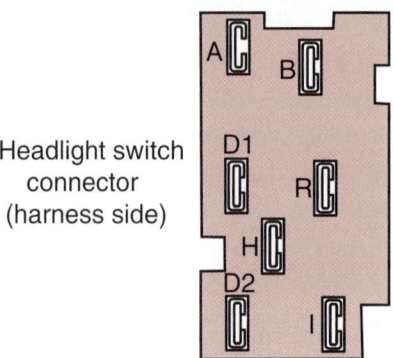

Headlight switch connector (harness side)

Figure 8-21 The harness side connector to the headlight switch provides a good test location.

1. Connect the 12-volt test light across terminal B and ground. The test light should light. If not, there is an open in the circuit back to the battery.

2. Connect the test light across terminal A and ground. The test light should come on. If not, repair the circuit back to the fuse panel.

3. Connect a fused jumper wire between terminals B and H. The headlights should come on. If the headlights fail to turn on, trace the H circuit to the headlights. Also, check the ground circuit side from the headlights.

4. Connect a fused jumper wire between terminals A and R. The rear lamps should illuminate. If not, trace the circuit to the rear lights. Also, check the ground return path.

5. Connect a fused jumper wire between terminals A and I. The instrument panel lights should come on. If not, trace the circuit to the panel lights.

If all the tests performed at the switch connector pass, then the problem is in the headlight switch. The chart (Figure 8-22) indicates the test results that should be obtained when testing the headlight switch for continuity for the system just discussed. Use an ohmmeter or a self-powered test light to test the switch. Most service manuals will provide a table similar to Figure 8-22. If a chart is not available, use the wiring diagram to determine which terminals should or should not have continuity in the different switch positions.

Switch terminals	Switch positions		
	Off	Park	Headlamp
B to H	No continuity	No continuity	Continuity
B to R	No continuity	No continuity	No continuity
B to A	No continuity	No continuity	No continuity
R to H	No continuity	No continuity	No continuity
R to A	No continuity	Continuity	Continuity
H to A	No continuity	No continuity	No continuity
D1 to D2	Continuity only when rheostat fully counterclockwise		
I to R	Continuity only when rheostat fully counterclockwise Then slowly rotate clockwise and test lamp should dim		

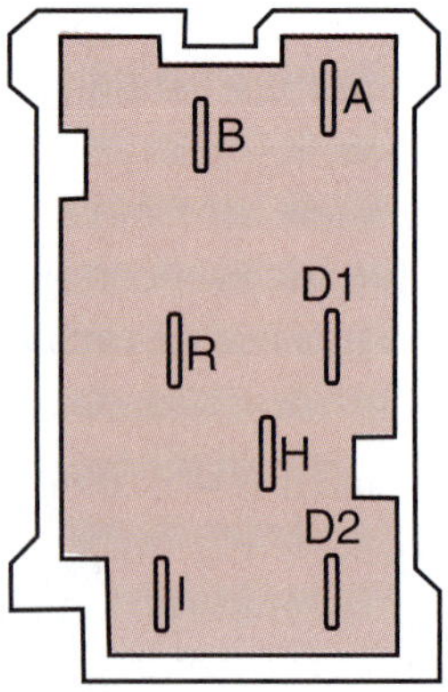

Figure 8-22 Continuity test chart for the headlight switch.

Steering Column–Mounted Switches

On some vehicles, it is possible to test the steering column–mounted switch without removing it. The test is conducted at the connector at the base of the column (Figure 8-23). However, on some models, it is necessary to remove the column cover and/or the steering wheel to gain access.

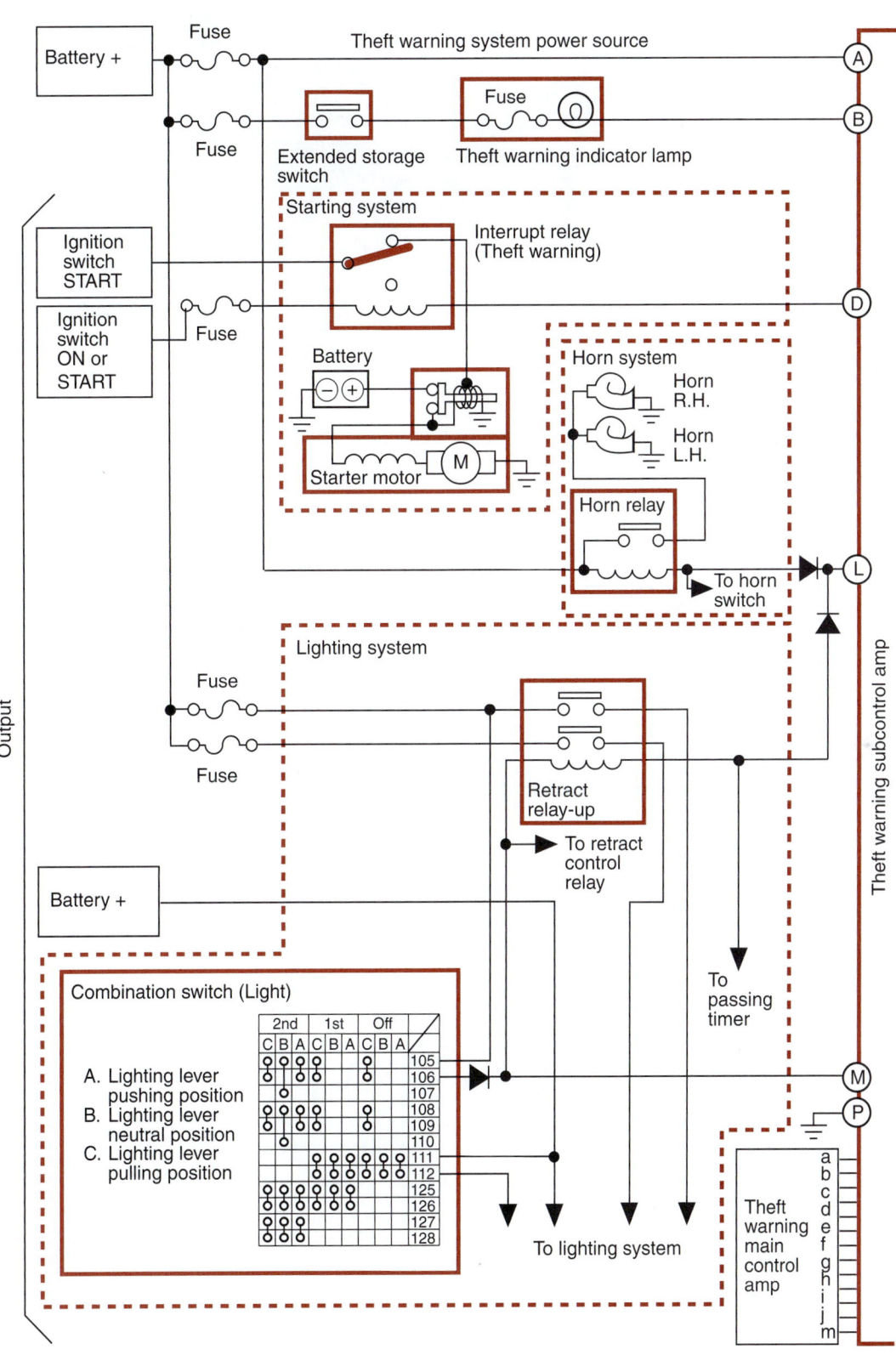

Figure 8-23 Using the connector to test the multifunction switch.

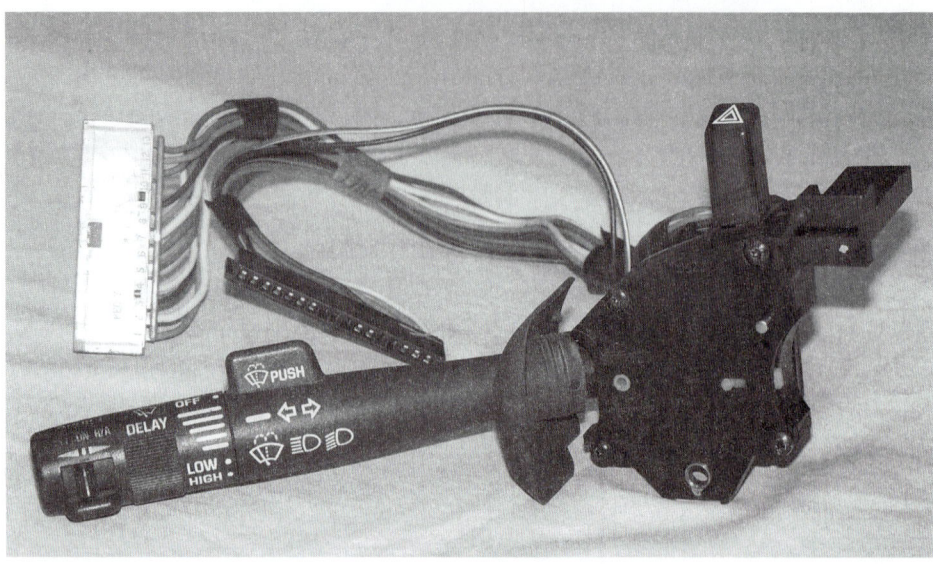

Figure 8-24 Multifunction switch.

A common procedure for removing and testing the **multifunction switch** is shown in Photo Sequence 13. The switch is called a multifunction switch because it can have a combination of any of the following switches in a single unit: headlights, turn signal, hazard, dimmer switch, horn, and flash to pass (Figure 8-24).

Dimmer Switch Testing and Replacement

Classroom Manual
Chapter 8, page 212

Special Tools

Jumper wires

Safety glasses

Battery terminal pliers

Terminal pullers

The **dimmer switch** is connected in series within the headlight circuit and controls the current path for high and low beams. The dimmer switch is located either on the floor board next to the left kick panel or on the steering column. Testing of the switch is done by using a set of jumper wires to bypass the switch (Figure 8-25). If the headlights operate with the switch bypassed, the switch is faulty. This is a common problem with older vehicles that have the dimmer switch on the floor board because the switch is subjected to damage due to rust, dirt, and so on.

CAUTION: Remove the battery negative cable before replacing the dimmer switch.

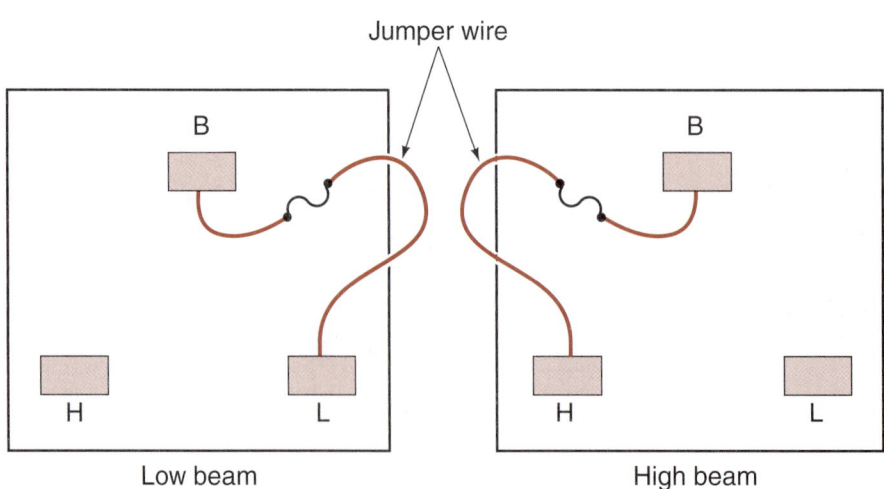

Figure 8-25 Fused jumper wire connections to bypass the dimmer switch.

Photo Sequence 13
Removal and Testing of the Multifunction Switch

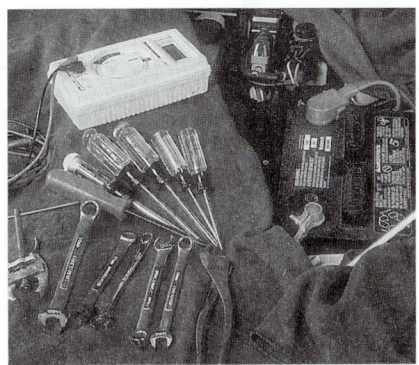

P13-1 Tools required to remove and test the multifunction switch: fender covers, battery terminal pliers, terminal pullers, assorted combination wrenches, torx drivers, and ohmmeter.

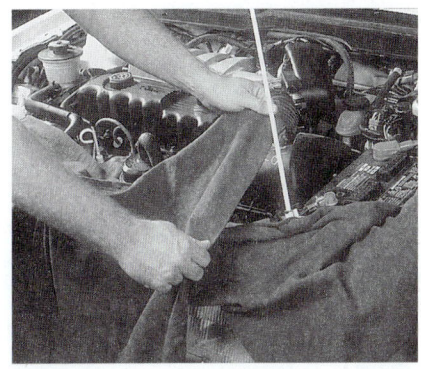

P13-2 Place the fender covers around the battery work area.

P13-3 Loosen the negative battery clamp bolt and remove the clamp using terminal pullers. Place the battery cable where it cannot contact the battery.

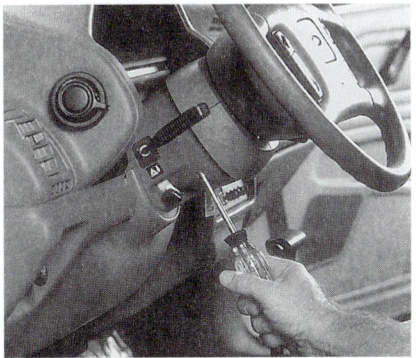

P13-4 Remove the shroud retaining screws and remove the lower shroud from the column.

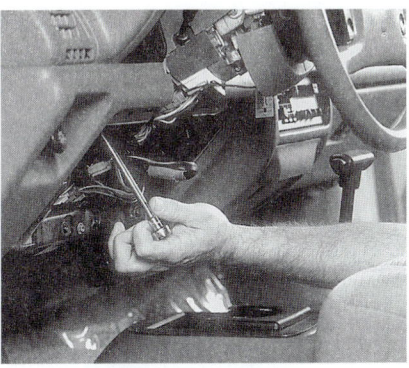

P13-5 If needed to gain access to the upper shroud, loosen the steering column attaching nuts. Do not remove the nuts.

P13-6 Lower the steering column enough to remove the upper shroud.

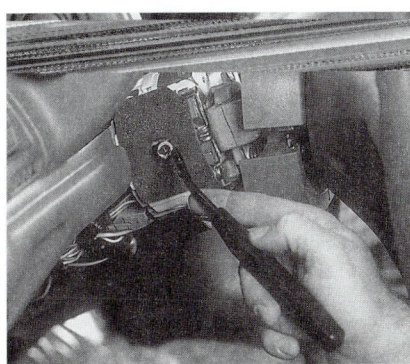

P13-7 Remove the turn signal lever by slightly rotating the outer end of the lever then pulling straight out on the lever. Some levers may be attached with fasteners.

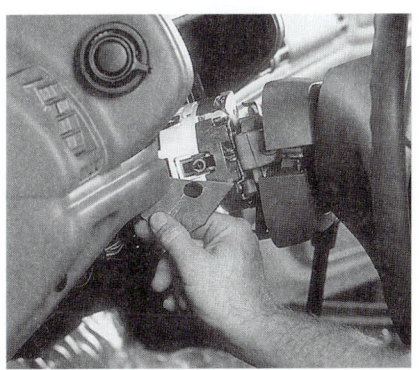

P13-8 Peel back the foam shield from the turn signal switch.

P13-9 Disconnect the turn signal switch electrical connectors.

Removal and Testing of the Multifunction Switch (continued)

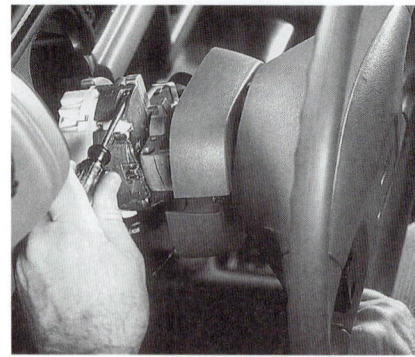

P13-10 Remove the screws attaching the turn signal switch to the lock cylinder assembly.

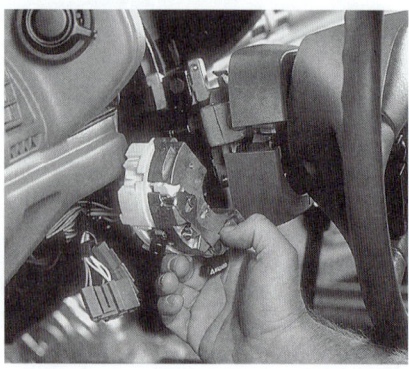

P13-11 Disengage the switch from the lock assembly.

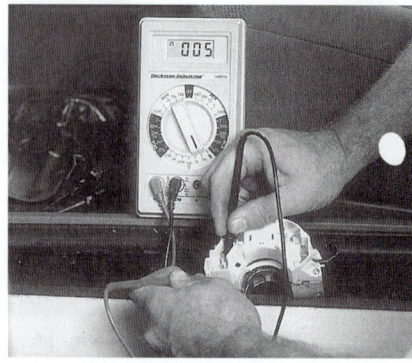

P13-12 Use the ohmmeter to test the switch. Check for continuity from terminal 15 to 13 when the dimmer switch is in the low-beam position.

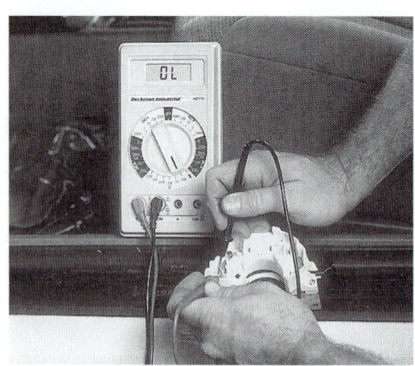

P13-13 With the switch in the low-beam position, the circuit should be open between terminals 15 and 12.

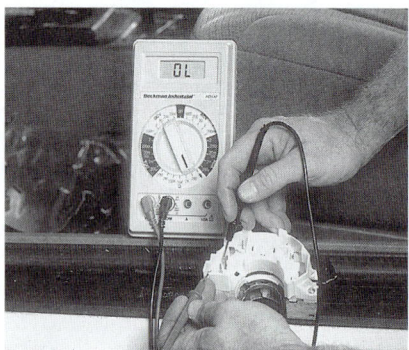

P13-14 The circuits between terminals 196 and 13, and 196 and 12, should be open in the low-beam position.

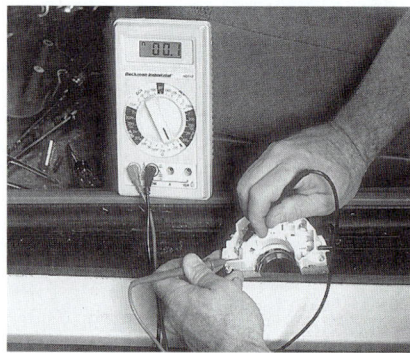

P13-15 With the switch in the high-beam position, continuity should be between terminals 15 and 12. Circuits 15 to 13, 196 to 13, and 196 to 12 should be open.

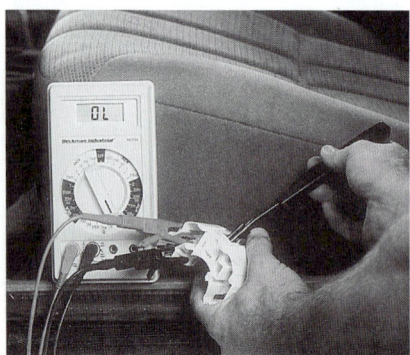

P13-16 When the dimmer switch is placed in the flash-to-pass position, there should be a closed circuit between terminals 196 and 12, and an open circuit between 196 and 13.

Floor-Mounted Switches

Removal and replacement of the floor-mounted dimmer switch is done by first pulling back on the floor mat to expose the switch. Next, disconnect the wire plug and remove the holddown fasteners. Install the new dimmer switch and relocate the mat so it does not interfere with switch operation.

Steering Column–Mounted Switches

The steering column–mounted dimmer switch can be operated by an actuator control rod from the lever to a remotely mounted switch (Figure 8-26). Another style incorporates the dimmer switch into the multifunction switch (Figure 8-27).

To remove the remote switch, first place fender covers on the fenders of the vehicle and disconnect the battery negative cable. Disconnect the wire connector at the switch. Remove the two switch mounting screws and disengage the switch from the actuator rod.

When installing the switch, make sure the actuator rod is firmly seated into the switch. During the installation, adjust the position of the switch so that all actuator rod slack is taken

Use a memory keeper before disconnecting the battery.

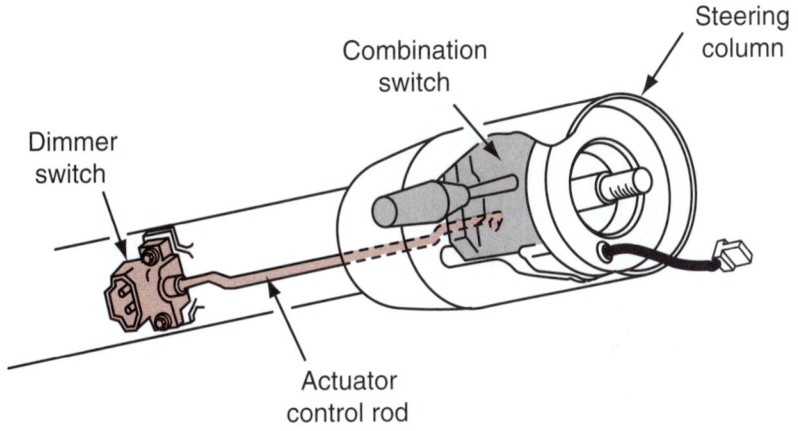

Figure 8-26 Steering column mounted–dimmer switch.

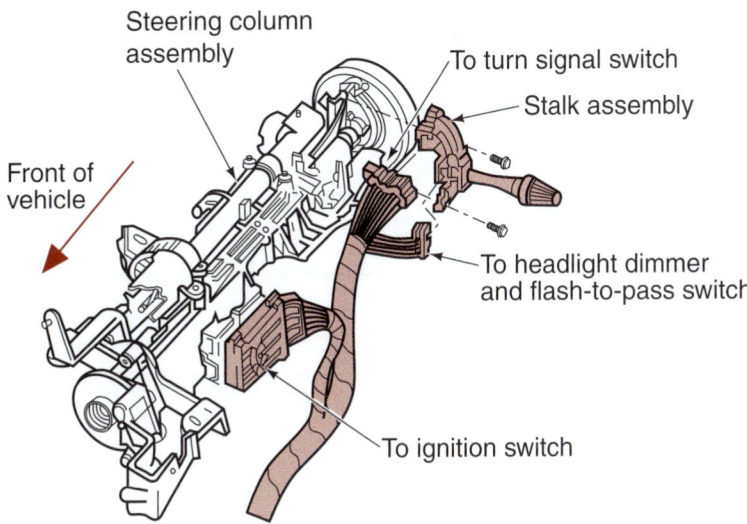

Figure 8-27 Dimmer switch incorporated into the multifunction switch.

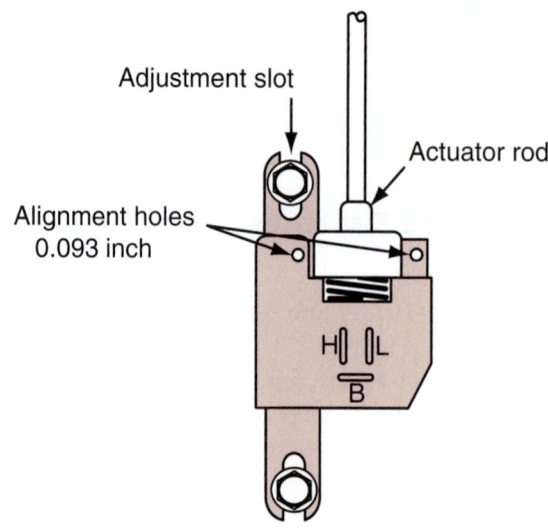

Figure 8-28 Insert a dowel into the alignment holes to adjust the dimmer switch.

up. If the switch has alignment holes, compress the switch until two appropriately sized dowels can be inserted into the alignment holes (Figure 8-28). While applying a slight rearward pressure, install and tighten the mounting bolts.

When the switch is adjusted properly, it will click when the lever is lifted and again when it is returned to its downward position. The second click should occur just before the stop.

If the dimmer switch is a part of the multifunction switch, follow the general procedure shown in Photo Sequence 13.

To reinstall the multiswitch, reverse the procedure. Torque the steering column attaching nuts to the amount specified in the service manual.

Taillight Assemblies

Classroom Manual
Chapter 8, page 219

In a three-bulb taillight system, the brake lights are controlled directly by the brake light switch (Figure 8-29). The brake lights on both sides of the vehicle are wired in parallel. Most brake light systems use dual-filament bulbs that perform multifunctions. In this type of circuit, the brake lights are wired through the turn signal and hazard switches (Figure 8-30).

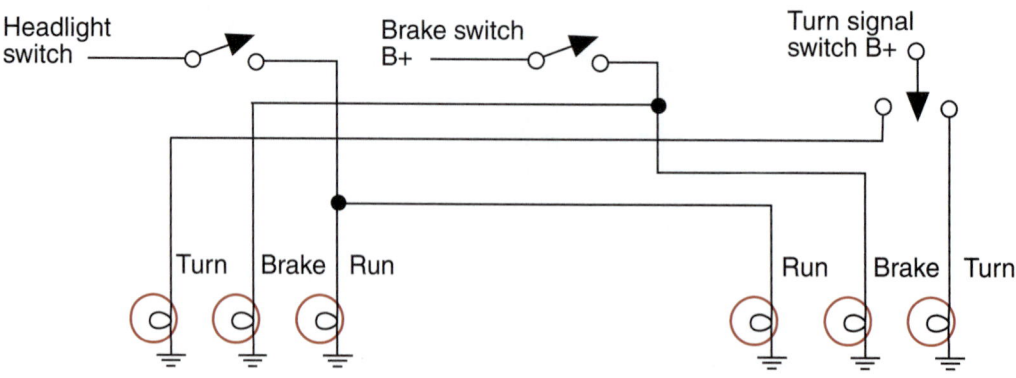

Figure 8-29 Three-bulb taillight circuit has individual control for each bulb.

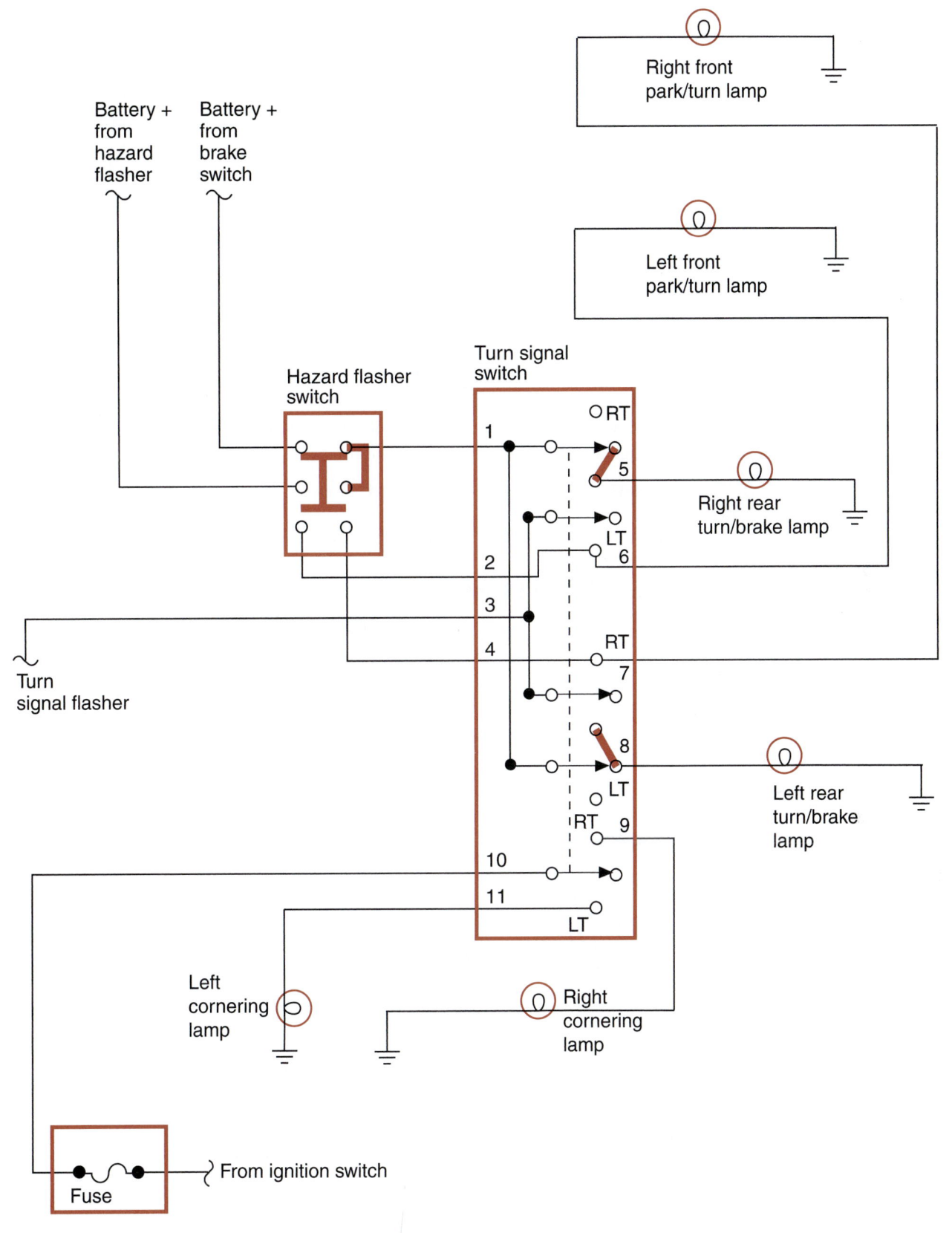

Figure 8-30 Turn signal switch used in a two-bulb taillight circuit.

If all of the taillights do not operate, check the condition of the fuse. If it is good, use a voltmeter to test the circuit. With the headlight switch in the PARK position (first detent), check for voltage at the last common connection between the switch and the lamps. If battery voltage is present, then the problem is in the individual circuits from that connector to the lamps. If no battery voltage is present, test for voltage from the switch terminal. If no voltage is present at this terminal yet the headlights operate when in the ON position, replace the switch. If battery voltage is present, the problem is between the switch and the last common connection. If there was no voltage present at the switch terminal, check for battery voltage into the switch.

Most taillight bulbs can be replaced without removal of the lens assembly. The bulb and socket are removed by twisting the socket slightly and pulling it out of the lens assembly (Figure 8-31). To remove the bulb from the socket, push in on the bulb slightly while turning it. When the lugs align with the channels of the socket, pull the bulb from the socket (Figure 8-32).

If the bulb is a blade base–type bulb, pull it straight out of the socket without twisting. On all bulb types it is a good practice to use an oil-free rag or wear nylon gloves to grasp the bulb. This will keep oil off of the bulb but also will protect you if the bulb should break.

The illustration (Figure 8-33) shows how the lens assembly is fastened to the vehicle body. Remove the attachment nuts from the back of the assembly to remove it.

CAUTION: Using the wrong type of lamp for the socket and application can result in "crazy lights." This is a result of feedback caused by the incorrect bulb (Figure 8-34).

Turn Signals and Brake Lights

Classroom Manual
Chapter 8, page
220, 225

To test the turn signal switches, use a 12-volt test light to probe for voltage into and out of the switch. The ignition switch must be in the RUN position for the circuit to operate. If voltage is present on the input side of the switch but not on the output side, the switch is faulty.

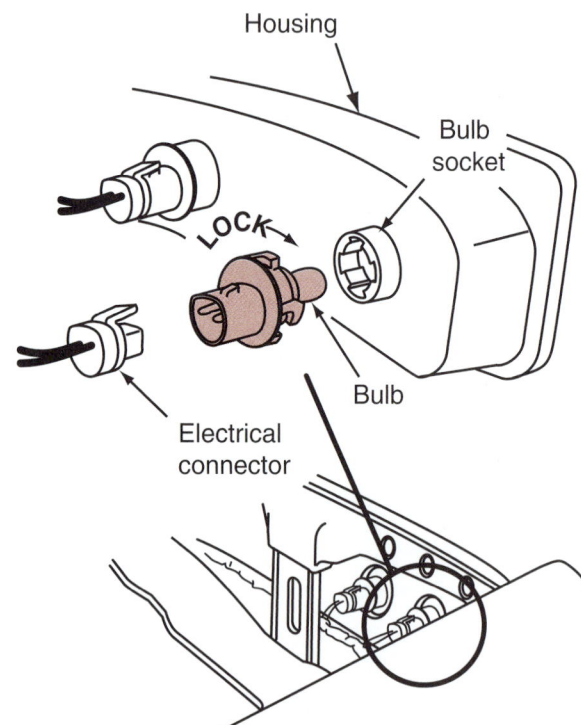

Figure 8-31 Bulb and socket removal from the taillight lens assembly.

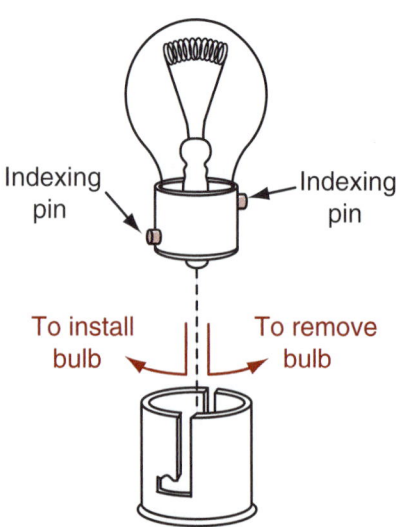

Figure 8-32 Removing the bulb from the socket.

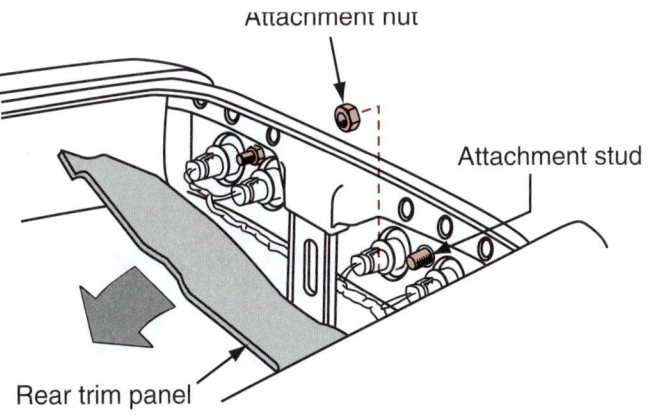

Attachment nut

Attachment stud

Rear trim panel

Figure 8-33 Taillight lens assembly.

Special Tools

12-volt test light

Steering wheel puller

Lock ring compressor

Safety glasses

Most turn signal switches receive their voltage from the ignition switch when it is in the RUN position only. This prevents the turn signals from operating while the ignition switch is in the OFF position.

Check brake light operation through the turn signal switch in the same manner. Also, check the brake light switch for proper adjustment and operations (Figure 8-35).

SERVICE TIP: The turn signal switches used in a two-bulb system also control a portion of the operation of the brake lights; thus they have a complex system of contact points. The technician must remember that many brake light problems are caused by worn contact points in the turn signal switch.

Many early vehicles use a turn signal switch that is separate from the multifunction switch. The steering wheel will have to be removed to gain access to the turn signal switch on these vehicles. The following procedure is a common method of turn signal switch replacement:

1. Place the fender covers on the fenders.

2. Install a memory keeper and disconnect the battery negative cable.

WARNING: If the vehicle is equipped with an air bag system, wait the recommended amount of time before removing any other components. Fifteen minutes or more may be required to discharge the capacitors that are used to fire the air bags.

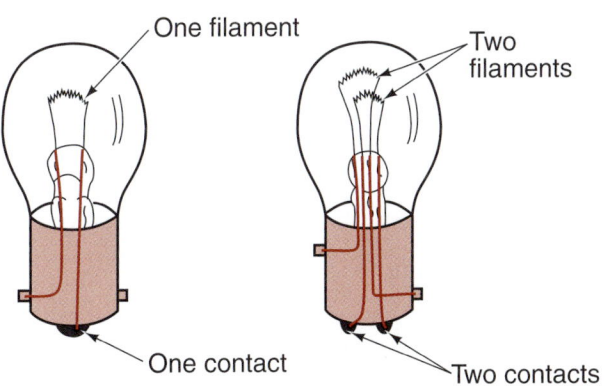

One filament

Two filaments

One contact

Two contacts

Single-filament bulb   Double-filament bulb

Figure 8-34 If the single-filament bulb is mistakenly installed into a socket designed for dual filament, the single contact of the bulb will short across the two contacts of the socket. This will result in lighting circuits operating when they are not supposed to.

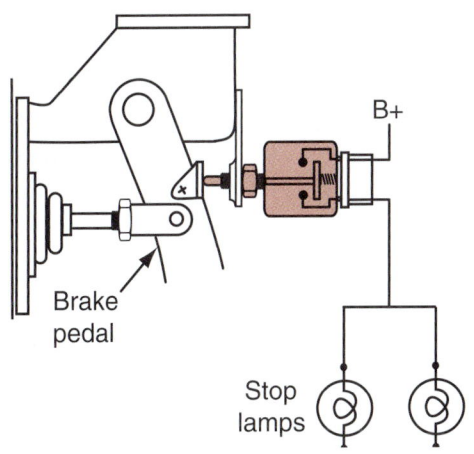

B+

Brake pedal

Stop lamps

Figure 8-35 Typical brake light switch operation. Check for continuity in both positions. Should be open when at rest and closed when the pedal is depressed.

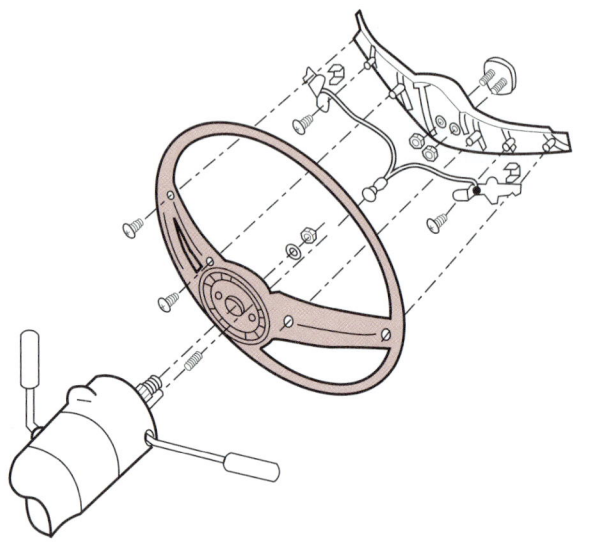

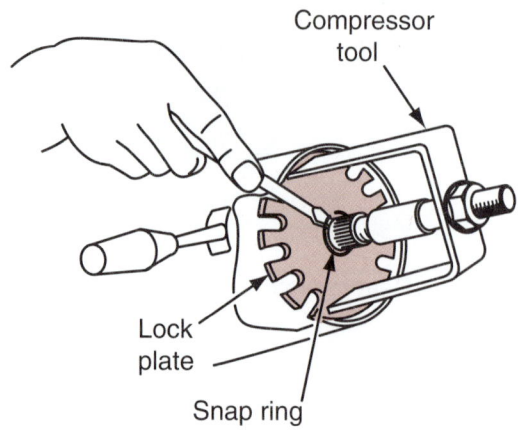

Figure 8-36 Steering wheel attachment.

Figure 8-37 To remove the snap ring, use the compressing tool to relieve the pressure against the snap ring.

3. Remove the steering column trim.

4. Remove the horn pad from the steering wheel (Figure 8-36).

5. Remove the steering shaft nut and horn collar (if equipped).

6. Use a suitable puller to remove the steering wheel.

7. If needed, use a suitable compressor to compress the preload spring to the lock plate (Figure 8-37). Compress the spring only enough to remove the snap ring.

8. Use a pick and a small, flat-blade screwdriver to remove the snap ring.

9. Remove the lock plate, horn contact carrier, and spring.

10. Remove the bolts at the upper steering column support and the upper mounting bracket from the column.

11. Disconnect the turn signal wiring connector.

12. Wrap tape around the wire and connector (Figure 8-38).

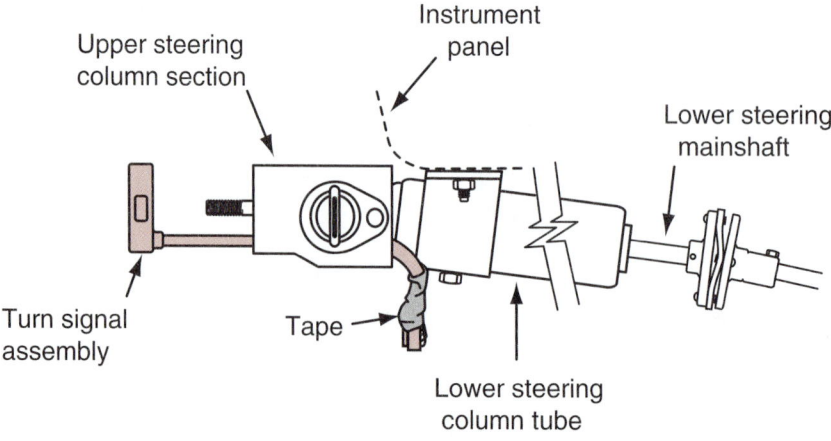

Figure 8-38 Tape the turn signal connector to make removal easier.

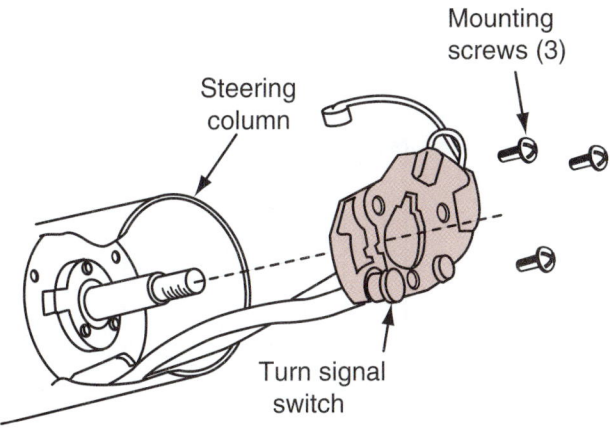

Figure 8-39 Remove the turn signal switch from the column.

13. Remove the hazard warning knob from the column.

14. Remove the switch retaining screws and remove the switch (Figure 8-39).

On vehicles equipped with cornering lights, the turn signal switch has an additional set of contacts that operate the cornering light circuit only. The contacts can receive voltage from either the ignition switch or the headlight switch. If the ignition switch provides the power, the cornering lights will be activated any time the turn signals are used. If the contacts receive the voltage from the headlight switch, the cornering lights do not operate unless the headlight switch is in the PARK or HEADLIGHT position.

Flashers. The flasher uses a bimetallic strip and a heating coil to flash the turn signals (Figure 8-40). The most common complaints that are attributable to the flasher is the speed of the

Classroom Manual
Chapter 8, page 225

Special Tools

Jumper wires
Safety glasses

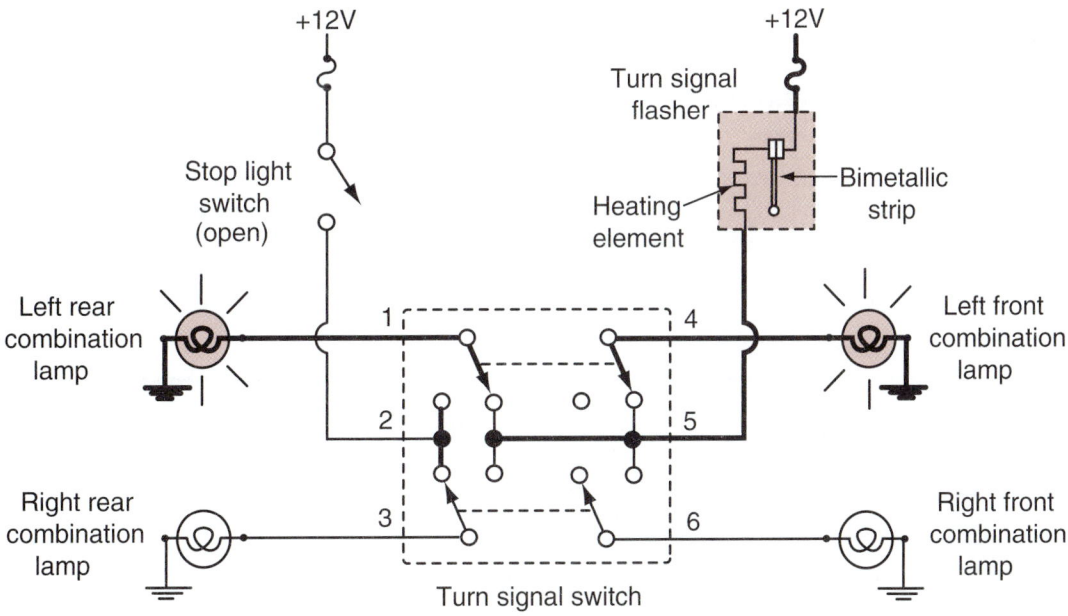

Figure 8-40 As current flows through the heating element, the bimetallic strip will bend and break the circuit. This happens several times a minute to provide for turn signal flashing.

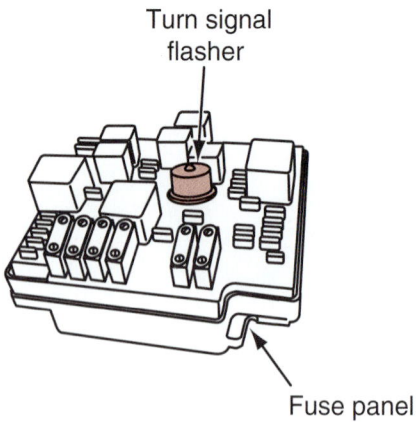

Turn signal flasher

Fuse panel

Figure 8-41 Many manufacturers locate the flasher unit in the fuse panel.

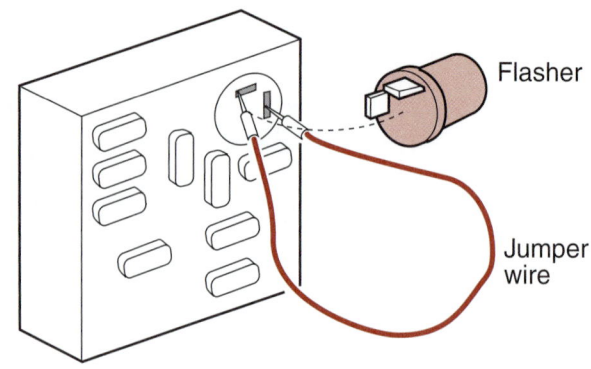

Flasher

Jumper wire

Figure 8-42 Connecting jumper wires across the terminals to bypass the flasher unit.

The ignition switch must be in the RUN position for the turn signal circuit to be powered.

Not all flashers are located in the fuse box. Use the component locator to find where the flasher is installed.

Classroom Manual
Chapter 8, page 236

Special Tools

Voltmeter

Safety glasses

Any time a blown fuse is found, the cause of the circuit overload must be traced.

flashing rate. If the flasher is of the wrong type and rating, the amount of time required to heat the coil will differ from what the manufacturer designed into the circuit. Also, newer flashers that use electronic circuits will flash at an increased speed if one of the turn signal bulbs is burned out or the circuit is defective. If the flasher is rated higher than required, the flashing rate is reduced because it takes longer for the current to heat the coil.

Check the size and type of light bulbs in the circuit. Use only the lamp size recommended by the manufacturer. If these checks do not correct the problem, test the generator output. Voltage output that is higher or lower than specified may cause the flasher rate to be incorrect. If the charging system output is within specifications, check for excessive resistance in the turn signal circuit. Check both sides of the circuit.

If none of the turn signals operate, check the fuse. Next, check the flasher. Remove the flasher from the fuse box (Figure 8-41). Connect a jumper wire across the fuse box terminals (Figure 8-42). If the turn signal lamps come on with the lever in either indicator position, the flasher is faulty. If the lights still do not illuminate, test the turn signal switch.

SERVICE TIP: If the turn signals operate properly in one direction but do not flash in the other, the problem is not in the flasher unit. A burned-out lamp filament will not cause enough current to flow to heat the bimetallic strip sufficiently to cause it to open. Thus the lights do not flash. Locate the faulty bulb and replace it.

Interior Lights

Interior lighting includes courtesy lights, map lights, and instrument panel lights.

Courtesy Lights

Courtesy lights operate from the headlight and door switches. They receive their power source directly from a fused battery connection. The switches can control either the ground circuit (Figure 8-43) or the insulated circuit (Figure 8-44). The courtesy lights may also be activated from the headlight switch by turning the switch knob to the extreme counterclockwise position. The contacts in the switch close and complete the circuit.

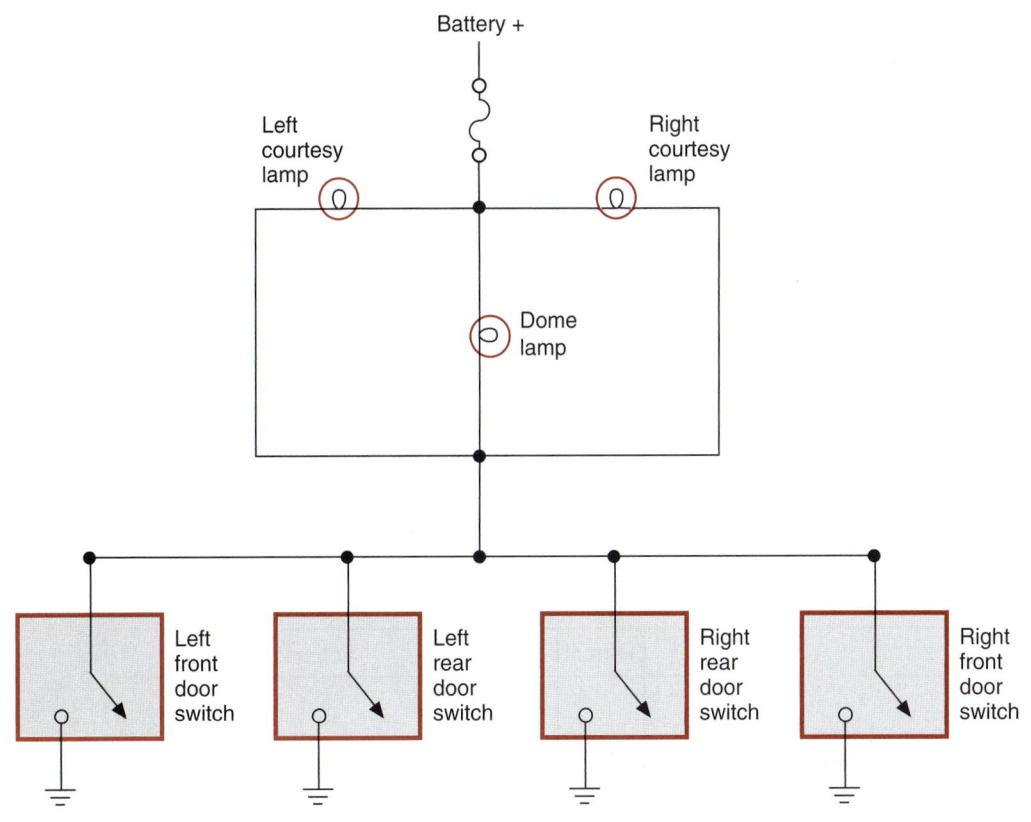

Figure 8-43 Courtesy lights using ground side switches.

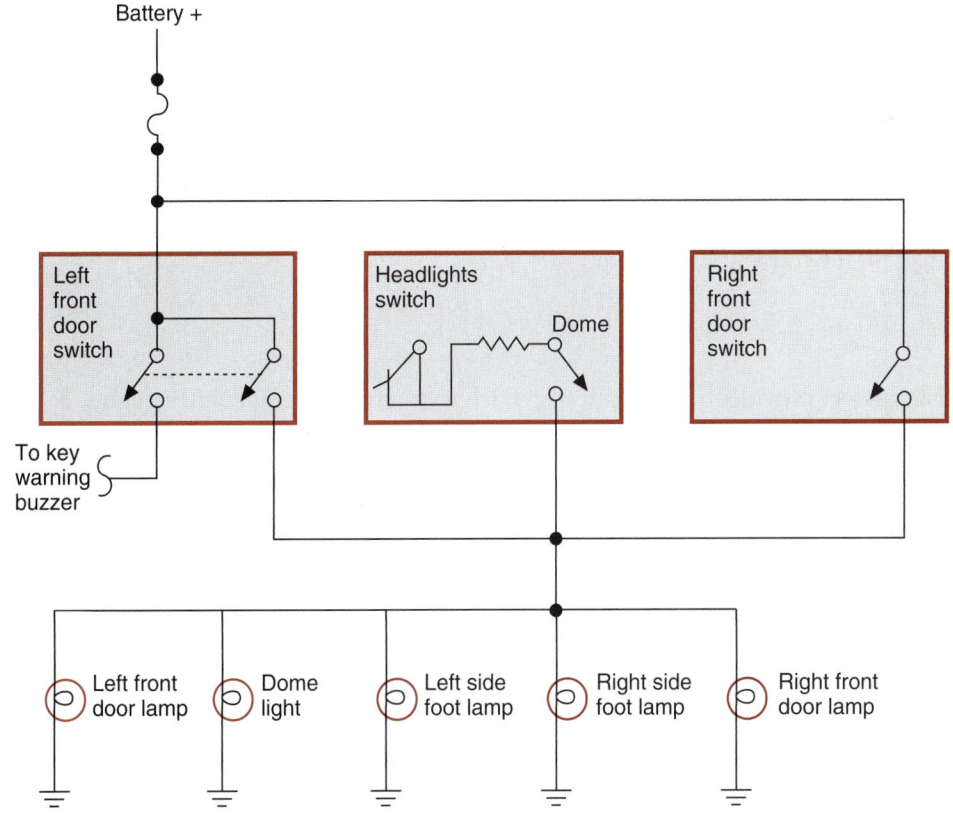

Figure 8-44 Courtesy lights using insulated side switches.

COURTESY LAMPS DO NOT TURN ON WHEN ONE DOOR IS OPENED
OK WHEN OTHER DOORS ARE OPENED

Test step	Result	Action to take
A0 **Verify condition**		Go to A1
A1 **Check power** Check for power at door switch	~~OK~~	Check the power circuit back to fuse
	OK	Go to A2
A2 **Check the door switch** Check the door switch for proper operation	~~OK~~	Replace the switch
	OK	Check circuit from switch to lamp

COURTESY LAMPS DO NOT COME ON WHEN HEADLAMP SWITCH
IS TURNED COUNTERCLOCKWISE TO STOP

Test step	Result	Action to take
B0 **Verify condition**		Go to B1
B1 **Check operation of door switches** Check to see if courtesy lamps operate from door switches	~~OK~~	Go to chart C
	OK	Go to B2
B2 **Check for power** Check for power at headlamp switch	~~OK~~	Check circuit back to fuse
	OK	Go to B3
B3 **Check for continuity** Check continuity of headlamp switch	~~OK~~	Replace headlamp switch
	OK	Check circuit from switch to lamp

Figure 8-45 A diagnostic chart for the courtesy light system.

Courtesy lights and illuminated entry lights usually use the same bulbs. Courtesy lights are lights that come on when the door is open or the switch is turned on. Illuminated entry is a term used for turning on lamps prior to entering the vehicle, such as when the handle is lifted or remote keyless entry is used.

Figure 8-45 provides a systematic approach to troubleshooting courtesy lights. Follow the steps in proper order to locate the fault.

If all the lights of the circuit do not light, begin by checking the fuses. If the fuse is good, then use a voltmeter to check for battery voltage at the last common connection. If voltage is present at the fuse box but not at the common connection, the problem is between these two points. If battery voltage is present at the common connection, trace the individual circuits until the cause(s) for the open is located.

If the courtesy lights do not come on when only one of the doors is opened but do come on when any of the other doors are opened, the problem is in the affected door's switch. In order to check the switch and circuit, bypass the switch with a jumper wire. If the lights come on with the switch bypassed, it is a faulty switch. If the lights do not come on with the switch bypassed, there is a problem in the circuit.

COURTESY LAMP DOES NOT COME ON WHEN ALL DOORS ARE OPENED

Test step		Result	Action to take
C0 **Verify condition**	Vehicle with only one courtesy lamp		Go to C1
	Vehicle with more than one courtesy lamp		Go to C2
C1 **Check operation of fuse circuit** Check operation of other circuits that share the same fuse		⊘OK	Go to C4
		OK	Go to C2
C2 **Check for power** Check for power at the bulb		⊘OK	Replace bulb
		OK	Go to C3
C3 **Check for continuity** Check continuity of bulb		⊘OK	Replace bulb
		OK	Check bulb ground
C4 **Check fuse** Check continuity of fuse		⊘OK	Replace fuse
		OK	Go to C5
C2 **Check for power** Check for power through the fuse		⊘OK	Check power feed circuit
		OK	Check for open circuit between fuse and common point in lamp circuit

Figure 8-45 *(continued)*

Instrument Cluster and Panel Lights

The power source for the instrument panel lights is provided through the headlight switch. The contacts are closed when the headlight switch is located in the PARK or HEADLIGHT position. The current must flow through a variable resistor (rheostat) that is either a part of the headlight switch or a separate dial on the dash. The resistance of the rheostat is varied by turning the knob. By varying the resistance, changes in the current flow to the lamps control the brightness of the lights.

Test for voltage output from the headlight switch to determine if the switch is operating properly. Vary the amount of resistance in the rheostat while observing the voltmeter. The voltage reading from the rheostat should vary as the knob is turned. If voltage is present to the printed circuit, check the ground.

If all connections are good, remove the dash and test the printed circuit board. Use an ohmmeter to check for opens and shorts in the printed circuit board from the connector plug to the lamp sockets. If the printed circuit is bad, it must be replaced. There are no repairs to the board.

Classroom Manual
Chapter 8, page 238

Special Tools

Voltmeter
Ohmmeter
Safety glasses

CAUTION: Be careful when testing the printed circuit. Do not touch the circuit paths with your fingers. Do not scratch the lamination with the test probes. Doing so may destroy a good circuit board.

A customer says the cornering lights on the vehicle are working only part of the time. The technician turns on the turn signals and the cornering lights illuminate. She then shakes the vehicle and notices that the cornering lights stop working. The lights come on when the vehicle is shaken again.

The technician checks all of the connections at the battery and at the light sockets. All are good. The technician then shakes the vehicle until the cornering lights go out. Next, she uses a test light to check for voltage at the light sockets. The test light does not come on. Tracing the circuit backward toward the turn signal switch, the technician discovers a voltage into the switch, but not out of it. Once the turn signal switch is removed from the steering column, the cause of the problem is easily found. The spring that maintains pressure on the contacts of the cornering light switch has been dislodged. The contacts intermittently have continuity until the vehicle goes over a bump in the road.

Terms to Know

Curb height

Dimmer switch

Multifunction switch

ASE-Style Review Questions

1. When the headlights are switched from low beam to high beam all headlights go out. All of the following can cause this EXCEPT:
   A. Faulty dimmer switch.
   B. Defective high-beam relay.
   C. Defective ignition switch.
   D. Open in high-beam circuit.

2. None of the turn signals operate. The LEAST likely cause of this is:
   A. Burned-out bulbs.
   B. Faulty turn signal switch.
   C. Faulty turn signal flasher.
   D. Open circuit from ignition feed to turn signal.

3. A customer states that the headlights are brighter than normal and that she has to replace the lamps regularly.
   Technician A says this can be caused by too-high generator output.
   Technician B says this can be caused by excessive voltage drop in the circuit.
   Who is correct?
   A. A only
   B. B only
   C. Both A and B
   D. Neither A nor B

4. A customer states that none of the external parking and headlights turn on.
   Technician A says the circuit from the battery to the switch may be faulty.
   Technician B says the headlight switch can be at fault.
   Who is correct?
   A. A only
   B. B only
   C. Both A and B
   D. Neither A nor B

5. The taillight assembly is being discussed.
   Technician A says most taillight assemblies require the removal of the lens to replace the bulbs.
   Technician B says if all of the taillights do not operate, probe for voltage at a common connection.
   Who is correct?
   A. A only
   B. B only
   C. Both A and B
   D. Neither A nor B

6. The turn signals of a vehicle operate in the left direction only.
   Technician A says the flasher is bad.
   Technician B says the fuse is blown.
   Who is correct?
   A. A only
   B. B only
   C. Both A and B
   D. Neither A nor B

7. Diagnosis of the instrument panel lights is being discussed.

Technician A says the power source for the instrument panel lights is provided through the headlight switch and rheostat.

Technician B says the printed circuit board must be replaced.

Who is correct?

A. A only   **C.** Both A and B

B. B only   **D.** Neither A nor B

8. Composite headlight bulb replacement is being discussed.

Technician A says care must be taken not to touch the envelope with your fingers.

Technician B says not to energize the bulb unless it is installed into the socket.

Who is correct?

A. A only   **C.** Both A and B

B. B only   **D.** Neither A nor B

9. *Technician A* says the dimmer switch is connected in parallel to the headlight circuit.

Technician B says concealed headlight doors cannot be manually opened.

Who is correct?

A. A only   **C.** Both A and B

B. B only   **D.** Neither A nor B

10. *Technician A* says if only one lamp in the circuit is not operating, the fastest check is to replace the bulb with a known good one.

Technician B says to start the diagnostic tests at the easiest location to access.

Who is correct?

A. A only   **C.** Both A and B

B. B only   **D.** Neither A nor B

ASE Challenge Questions

1. The brake lights of a vehicle equipped with dual-function turn signal/brake lamps are inoperative. The turn signals are functioning properly. However, there is no power at the brake light terminals when the brake pedal is depressed.

Technician A says that the brake light switch may be faulty.

Technician B says that the turn signal switch may be faulty.

Who is correct?

A. A only   **C.** Both A and B

B. B only   **D.** Neither A nor B

2. The left turn signals of a vehicle are flashing very slowly; the right turn signals are operating correctly.

Technician A says that this may be caused by excessive circuit resistance on the left turn signal circuit.

Technician B says that this may be occurring because someone may have installed bulbs with higher-than-normal wattage ratings on the left side of the vehicle.

Who is correct?

A. A only   **C.** Both A and B

B. B only   **D.** Neither A nor B

3. The right back-up lamp on a vehicle is dim; the left back-up lamp is working correctly.

Technician A says that the back-up lamp switch may have excessive resistance.

Technician B says that the back-up lamp fuse may have corroded terminals.

Who is correct?

A. A only   **C.** Both A and B

B. B only   **D.** Neither A nor B

4. All of the following could cause premature failure of a composite bulb except:

A. High-charging system voltage.

B. Excessive bulb circuit ground resistance.

C. Improper bulb handling.

D. Cracked lamp housing.

5. The courtesy lamps of a vehicle equipped with insulated side door plunger switches are remaining on after all of the doors are closed.

Technician A says that one of the door plunger switches may be shorted to ground.

Technician B says that one of the door plunger switches may be remaining electrically "open."

Who is correct?

A. A only   **C.** Both A and B

B. B only   **D.** Neither A nor B

Job Sheet 28

28

Name _____ Date _____

Diagnosing Lighting Systems

Upon completion of this job sheet, you should be able to diagnosis the cause of a no-light operation fault and determine needed repairs.

ASE Correlation

This job sheet is related to the ASE Electrical/Electronic System Certification Exam's content area: *Headlights, Parking Lights, Taillights, Dash Lights, and Courtesy Lights;* task: Diagnose no-light condition and determine necessary action.

Tools and Materials

Vehicle
Wiring diagram for the vehicle
DMM
Test light

Describe the vehicle being worked on:

Year _____ Make _____ Model _____

VIN _____ Engine type and size _____

Procedure

Task Completed

1. Record the concern the customer would have with the lighting system.

   _____
   _____
   _____

2. Confirm the customer's concern by performing a check of the system and record your results.

   _____
   _____

3. Determine if there are any other symptoms that may be related to the customer's concern by testing the operation of other systems. Record your results.

   _____
   _____
   _____
   _____

Task Completed

4. Analyze the symptoms and reference the wiring diagram to determine the most likely location of the fault. Record the component(s) and its (their) location(s).

_____

_____

_____

☐

5. Locate a component from step 4 and return to the vehicle to test it. Describe the test procedure used and record the results.

6. Based on your diagnostic checks, was the cause of the problem located?

☐ Yes   ☐ No

If yes, record your findings: _____

If no, continue to test other components identified in step 4.

7. If the fault is determined, make the repair and verify proper operation.
Was the repair successful?   ☐ Yes   ☐ No

If no, continue testing the system.

Instructor's Response _____

_____

_____

_____

Job Sheet 29

29

Name _____ Date _____

Checking a Headlight Switch

Upon completion of this job sheet, you should be able to check the operation of a headlight switch with an ohmmeter.

ASE Correlation

This job sheet is related to the ASE Electrical/Electronic System Certification Exam's content area: *Headlights, Parking Lights, Taillights, Dash Lights, and Courtesy Lights;* task: Inspect, test, and repair or replace headlight dimmer switches, relays, control units, sensors, sockets, connectors, and wires of headlight circuits.

Tools and Materials

A vehicle
A wiring diagram for the vehicle
A DMM

Describe the vehicle being worked on:

Year _____ Make _____ Model _____

VIN _____ Engine type and size _____

Procedure

Task Completed

1. Put the headlight switch in all possible positions and observe which lights are controlled by each position. List each position and the controlled lights below.

   _____

   _____

   _____

   _____

   _____

   _____

2. Locate the headlight switch in the wiring diagram and draw the switch with each possible connection and possible position. Label the lights controlled by each position of the switch.

3. Remove the fuse to the headlights or disconnect the battery's negative cable. Remove the headlight switch according to the procedures outlined in the service manual. Describe the procedure to your instructor before removing it.
Instructor's OK to move to the next step. _____.

4. Identify the various terminals of the switch and list the different terminals that should have continuity in the various switch positions.

_____

_____

_____

5. Connect the ohmmeter across these terminals, one switch position at a time, and record your readings below:

_____

_____

_____

_____

6. Based on the test, what are your conclusions about the switch?

_____

_____

_____

_____

7. Reinstall the switch and connect the negative battery cable or reinstall the fuse. Then check the operation of the headlights.

Instructor's Response _____

_____

_____

_____

Job Sheet 30

30

Name _____ Date _____

Headlight Aiming

Upon completion of this job sheet, you should be able to adjust the aim of headlights using portable headlight aiming equipment.

ASE Correlation

This job sheet is related to the ASE Electrical/Electronic System Certification Exam's content area: *Headlights, Parking Lights, Taillights, Dash Lights, and Courtesy Lights;* task: Inspect, replace, and aim headlights/bulbs.

Tools and Materials

A vehicle with adjustable headlights
Portable headlight aiming kit
Hand tools

Describe the vehicle being worked on:

Year _____ Make _____ Model _____

VIN _____ Engine type and size _____

Procedure

Task Completed

1. Describe the type of headlights used on the vehicle.

   _____

   _____

2. Park the vehicle on a level floor.

   Install the calibrated aiming units to the headlights. (Make sure the adapters fit the headlight aiming pads on the lens.) ☐

3. Zero the horizontal adjustment dial. Are the split-image target lines visible in the view port? _____ If the lines cannot be seen, what should you do?

   _____

4. Turn the headlight horizontal adjusting screw until the split-image target lines are aligned. Then repeat this for the other headlight. List any problems you may have had doing this.

   _____

   _____

   _____

5. Turn the vertical adjustment dial on the aiming unit to zero.
Turn the vertical adjustment screw until the spirit level bubble is centered.
Recheck your horizontal setting after adjusting the vertical.

6. List any problems you had making the vertical adjustment.

_____

_____

_____

7. If the headlight assembly has four lamp assemblies, repeat steps 4 and 5 on the other two lamps. List below any problems you may have had doing this.

_____

_____

_____

Instructor's Response _____

_____

_____

DIAGNOSTIC CHART 8-1

PROBLEM AREA:	Headlights
SYMPTOMS:	Bright illumination, early bulb failure.
POSSIBLE CAUSES:	**1.** Alternator output too high. **2.** Defective dimmer switch.

DIAGNOSTIC CHART 8-2

PROBLEM AREA:	Headlights
SYMPTOMS:	Intermittent headlight operation, headlights flicker.
POSSIBLE CAUSES:	**1.** Defective circuit breaker. **2.** Overload in circuit. **3.** Improper connection. **4.** Defective switch. **5.** Poor ground. **6.** Excessive resistance.

DIAGNOSTIC CHART 8-3

PROBLEM AREA:	Headlights
SYMPTOMS:	Dim headlight illumination
POSSIBLE CAUSES:	**1.** Poor ground connection. **2.** Corroded headlight socket. **3.** Poor battery cable connections. **4.** Low generator output. **5.** Loose or broken generator drive belt.

DIAGNOSTIC CHART 8-4

PROBLEM AREA:	Headlights
SYMPTOMS:	No or improper headlight operation. No road illumination in front of the vehicle.
POSSIBLE CAUSES:	**1.** Burned-out headlights. **2.** Defective headlight switch. **3.** Open circuit. **4.** Defective circuit breaker. **5.** Overload in circuit. **6.** Improper or poor connection. **7.** Poor ground. **8.** Excessive resistance. **9.** Defective relay. **10.** Blown fuse. **11.** Faulty dimmer switch. **12.** Short in insulated circuit. **13.** Improper bulb application. **14.** Improper headlight aiming.

DIAGNOSTIC CHART 8-5

PROBLEM AREA:	Concealed headlight system
SYMPTOMS:	Headlight doors fail to operate properly.
POSSIBLE CAUSES:	**1.** Loose or broken vacuum connections. **2.** Defective vacuum motors. **3.** Defective switch. **4.** Defective electrical door motors. **5.** Defective limit switch(es). **6.** Open circuit. **7.** Poor ground connection.

DIAGNOSTIC CHART 8-6

PROBLEM AREA:	Park/taillights
SYMPTOMS:	Brighter-than-normal parking/taillights. Early bulb failure.
POSSIBLE CAUSES:	**1.** Too high of alternator output. **2.** Improper bulb application.

DIAGNOSTIC CHART 8-7

PROBLEM AREA:	Park/taillights
SYMPTOMS:	Intermittent park/taillight operation. Parking and/or taillights flicker.
POSSIBLE CAUSES:	**1.** Improper connection. **2.** Defective switch. **3.** Poor ground. **4.** Excessive resistance.

DIAGNOSTIC CHART 8-8

PROBLEM AREA:	Park/taillights
SYMPTOMS:	No parking and/or taillight operation.
POSSIBLE CAUSES:	**1.** Burned-out bulbs. **2.** Defective headlight switch. **3.** Open circuit. **4.** Improper connection. **5.** Poor ground. **6.** Excessive resistance. **7.** Faulty relay. **8.** Improper bulb application.

DIAGNOSTIC CHART 8-9

PROBLEM AREA:	Instrument cluster lighting
SYMPTOMS:	Intermittent brightness control of instrument cluster light circuits. Dash lights flicker.
POSSIBLE CAUSES:	**1.** Improper connection. **2.** Defective headlight switch rheostat. **3.** Poor ground. **4.** Excessive resistance. **5.** Faulty printed circuit.

DIAGNOSTIC CHART 8-10

PROBLEM AREA:	Instrument cluster lighting
SYMPTOMS:	Low-level light intensity from panel illumination lights.
POSSIBLE CAUSES:	**1.** Burned-out bulbs. **2.** Defective headlight switch rheostat. **3.** Improper bulb application. **4.** Improper connection. **5.** Poor ground. **6.** Excessive resistance. **7.** Defective or faulty printed circuit.

DIAGNOSTIC CHART 8-11

PROBLEM AREA:	Instrument cluster lighting
SYMPTOMS:	No bulb illumination.
POSSIBLE CAUSES:	**1.** Blown circuit protection device. **2.** Burned-out bulbs. **3.** Defective headlight switch rheostat. **4.** Open circuit. **5.** Improper connection. **6.** Poor ground. **7.** Excessive resistance. **8.** Improper bulb application. **9.** Defective printed circuit.

DIAGNOSTIC CHART 8-12

PROBLEM AREA:	Instrument cluster lighting
SYMPTOMS:	No dash light brightness control.
POSSIBLE CAUSES:	**1.** Defective headlight switch rheostat.

DIAGNOSTIC CHART 8-13

PROBLEM AREA:	Courtesy lights
SYMPTOMS:	Intermittent courtesy light operation.
POSSIBLE CAUSES:	**1.** Improper connection. **2.** Defective headlight switch. **3.** Defective door jam switch. **4.** Defective or sticking door switch. **5.** Poor ground. **6.** Excessive resistance.

DIAGNOSTIC CHART 8-14

PROBLEM AREA:	Courtesy lights
SYMPTOMS:	Dimmer-than-normal courtesy lights. Battery condition good.
POSSIBLE CAUSES:	**1.** Improper bulb application. **2.** Improper connection. **3.** Poor ground. **4.** Excessive resistance.

DIAGNOSTIC CHART 8-15

PROBLEM AREA:	Courtesy light operation
SYMPTOMS:	No courtesy light illumination.
POSSIBLE CAUSES:	**1.** Blown circuit protection device. **2.** Burned-out bulbs. **3.** Defective headlight switch. **4.** Defective door switches. **5.** Open circuit. **6.** Improper connection. **7.** Poor ground. **8.** Excessive resistance. **9.** Improper bulb application.

DIAGNOSTIC CHART 8-16

PROBLEM AREA:	Courtesy lights.
SYMPTOMS:	Courtesy lights stay on all of the time.
POSSIBLE CAUSES:	**1.** Defective door jamb switch. **2.** Defective headlight switch. **3.** Shorted circuit.

DIAGNOSTIC CHART 8-17

PROBLEM AREA:	Stop (brake) lamp operation
SYMPTOMS:	Intermittent stop lamp operation.
POSSIBLE CAUSES:	**1.** Misadjusted brake light switch. **2.** Poor ground connection. **3.** Excessive resistance. **4.** Faulty sockets. **5.** Poor connections. **6.** Faulty turn signal switch contacts. **7.** Defective brake light switch.

DIAGNOSTIC CHART 8-18

PROBLEM AREA:	Stop (brake) lamp operation
SYMPTOMS:	Dimmer-than-normal stop lights.
POSSIBLE CAUSES:	**1.** Excessive circuit resistance. **2.** Poor ground connection. **3.** Improper bulb application. **4.** Improper connections. **5.** Faulty turn signal switch contacts. **6.** Improper bulb application.

DIAGNOSTIC CHART 8-19

PROBLEM AREA:	Stop (brake) lamp operation
SYMPTOMS:	No stop lamps illuminate. Stop lights fail to illuminate when the brakes are applied.
POSSIBLE CAUSES:	**1.** Faulty brake light switch. **2.** Open in the circuit. **3.** Improper bulb application. **4.** Faulty turn signal switch. **5.** Improper common ground connection. **6.** Burned-out light bulbs.

DIAGNOSTIC CHART 8-20

PROBLEM AREA:	Turn signal operation.
SYMPTOMS:	Turn signals do not operate in either direction.
POSSIBLE CAUSES:	**1.** Blown fuse. **2.** Defective or worn flasher unit. **3.** Defective or faulty turn signal switch. **4.** Open circuit.

DIAGNOSTIC CHART 8-21

PROBLEM AREA:	Turn signals
SYMPTOMS:	Turn signal lamp does not illuminate. Turn signal indicator illuminates but does not flash.
POSSIBLE CAUSES:	**1.** Improper bulb. **2.** Burned-out bulb. **3.** Open circuit. **4.** Failed flasher unit.

DIAGNOSTIC CHART 8-22

PROBLEM AREA:	Hazard light operation
SYMPTOMS:	Hazard lights fail to operate when activated.
POSSIBLE CAUSES:	**1.** Blown fuse. **2.** Defective or worn flasher unit. **3.** Defective or faulty hazard light switch. **4.** Open circuit. **5.** Defective turn signal switch.

DIAGNOSTIC CHART 8-23

PROBLEM AREA:	Back-up light operation.
SYMPTOMS:	Back-up lights fail to operate some of the time.
POSSIBLE CAUSES:	**1.** Misadjusted back-up light switch. **2.** Poor ground connection. **3.** Excessive resistance. **4.** Faulty sockets. **5.** Poor connections.

DIAGNOSTIC CHART 8-24

PROBLEM AREA:	Back-up light operation
SYMPTOMS:	Dimmer-than-normal back-up lights
POSSIBLE CAUSES:	**1.** Excessive circuit resistance. **2.** Poor ground connection. **3.** Improper bulb application. **4.** Improper connections. **5.** Faulty back-up switch contacts.

DIAGNOSTIC CHART 8-25

PROBLEM AREA:	Back-up light operation
SYMPTOMS:	Back-up lights fail to illuminate when the transmission is in reverse.
POSSIBLE CAUSES:	**1.** Faulty back-up light switch. **2.** Misadjusted back-up light switch. **3.** Blown fuse. **4.** Open in the circuit. **5.** Improper bulb application. **6.** Improper common ground connection. **7.** Burned-out light bulbs.

DIAGNOSTIC CHART 8-26

PROBLEM AREA:	Back-up light does not operate
SYMPTOMS:	One back-up light fails to illuminate when the transmission is in reverse.
POSSIBLE CAUSES:	**1.** Burned-out lamp. **2.** Loose connection. **3.** Open circuit to lamp.

Electrical Accessories Diagnosis and Repair

Upon completion and review of this chapter, you should be able to:

- ❏ Identify the causes of no operation, intermittent operation, or constant horn operation.
- ❏ Diagnose the cause of poor sound quality from the horn system.
- ❏ Perform diagnosis and repair of no windshield wiper operation in one speed only or in all speeds.
- ❏ Identify causes for slower-than-normal wiper operation.
- ❏ Determine the cause for improper park operation.
- ❏ Identify causes for continuous wiper operation.
- ❏ Diagnose faulty intermittent wiper system operation.

- ❏ Remove and install wiper motors and wiper switches.
- ❏ Determine the causes for improper operation of the windshield washer system and be able to replace the pump if required.
- ❏ Perform diagnosis of problems associated with blower motor circuits.
- ❏ Diagnose and repair electric rear window defoggers.
- ❏ Diagnose the power window system.
- ❏ Diagnose common problems associated with the power seat system.
- ❏ Perform diagnosis of the power door lock system.

Basic Tools

Basic mechanic's tool set

Service manual

Introduction

The electrical accessories included in this chapter represent the most often performed electrical repairs. Most of the systems discussed do not provide for rebuilding of components. The technician must be capable of diagnosing the fault and then replacing the defective part. As with any electrical system, always use a systematic diagnostic approach to finding the cause. Refer to the service manual to obtain information concerning correct system operation. The fault is easier to locate once you understand how the system is supposed to operate.

Included in this chapter are diagnostic and repair procedures for horn systems, windshield wipers and washer systems, blower motors, electric defogger systems, power seat and window systems, and power door locks. Although the procedures here are typical, always refer to the service manual for the vehicle you are working on.

Horn Diagnosis

Customer complaints associated with the horn system can include no operation, intermittent operation, continuous operation, or poor sound quality. Testing of the horn system varies between systems that do and do not use a relay.

No Horn Operation

Systems with Relay. When a customer complains of no horn operation, first confirm the complaint by depressing the horn button. If it is mounted in the steering wheel, rotate the steering wheel from stop to stop while depressing the horn button. If the horn sounds intermittently while the steering wheel is turned, the problem is probably in the sliding contact

Classroom Manual
Chapter 9, page 246

Special Tools

Jumper wires

Voltmeter

Test light

Ohmmeter

Fender covers

ring in the steering column, or the tension spring is worn or broken. If the horn does not sound during this test, continue to check the system as follows:

1. Check the fuse or fusible link. If defective, replace as needed. After replacement of the fuse, operate all other circuits it protects. It is possible that another circuit is faulty but the customer has not noticed.

2. Connect a fused jumper wire from the battery positive terminal to the horn terminal. If the horn sounds, continue testing; if the horn does not sound, check the ground connection. Replace the horn if the ground is good.

3. Remove the relay from its connector and check for voltage at terminals 30 and 86 (Figure 9-1). If there is no voltage at these points, trace the wiring from the relay to the battery to locate the problem. Continue testing if voltage is present at the power feed terminals.

4. With the relay removed, connect a test light across terminals 85 and 86 of the relay connector (Figure 9-2). Press the horn switch and observe the test light. The test light should illuminate. If it does not, the horn switch or its circuit is faulty—go to step 7. If the light illuminates, continue testing.

5. Use a fused jumper wire and jump terminals 30 and 87 (Figure 9-3). If the horns sound, replace the relay. If the horns do not sound, continue to test.

6. In a multiple-horn system, test for voltage at the last common connection between the horn relay and the horns. On a single-horn system, test for voltage at the horn terminal. Voltage should be present at this connection only when the horn button is depressed. If there is no voltage at this connection, repair the open between the relay and the common connection. If voltage is present, check the individual circuits from the connection to the horns; repair as needed.

7. Check for voltage on the battery side of the horn switch. If there is no voltage at this location, the fault is between the relay and the switch. Continue testing if voltage is present.

8. Check for continuity through the switch. If good, check the ground connection for the switch. Then recheck operation. Replace the horn switch if there is no continuity when the button is depressed.

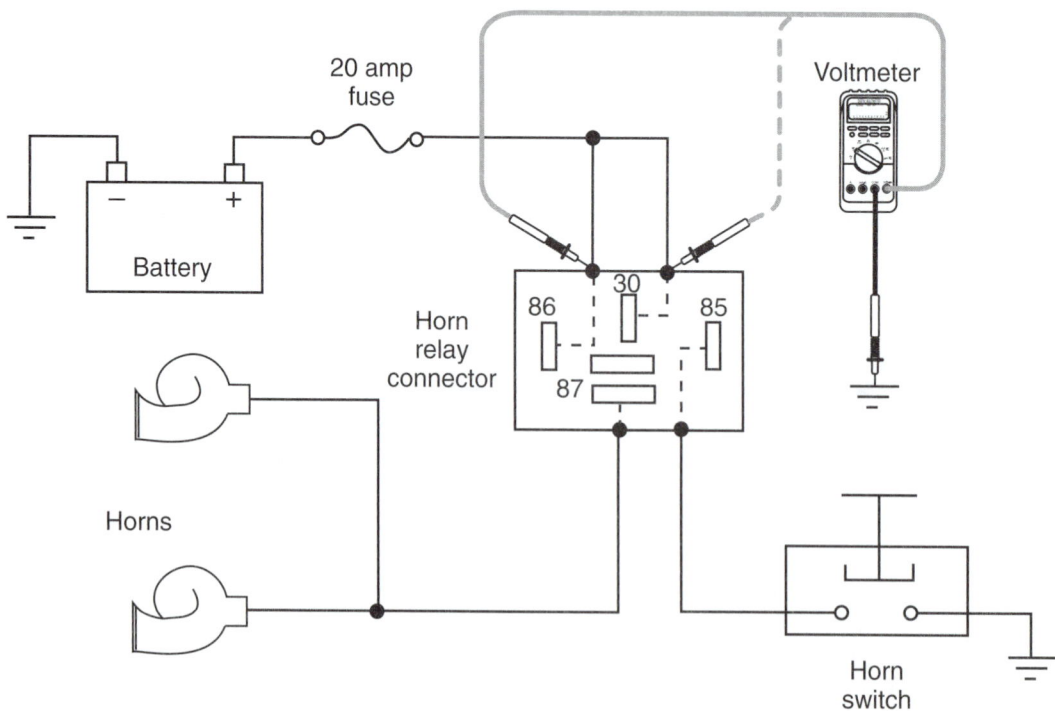

Figure 9-1 Testing for voltage to the relay connector.

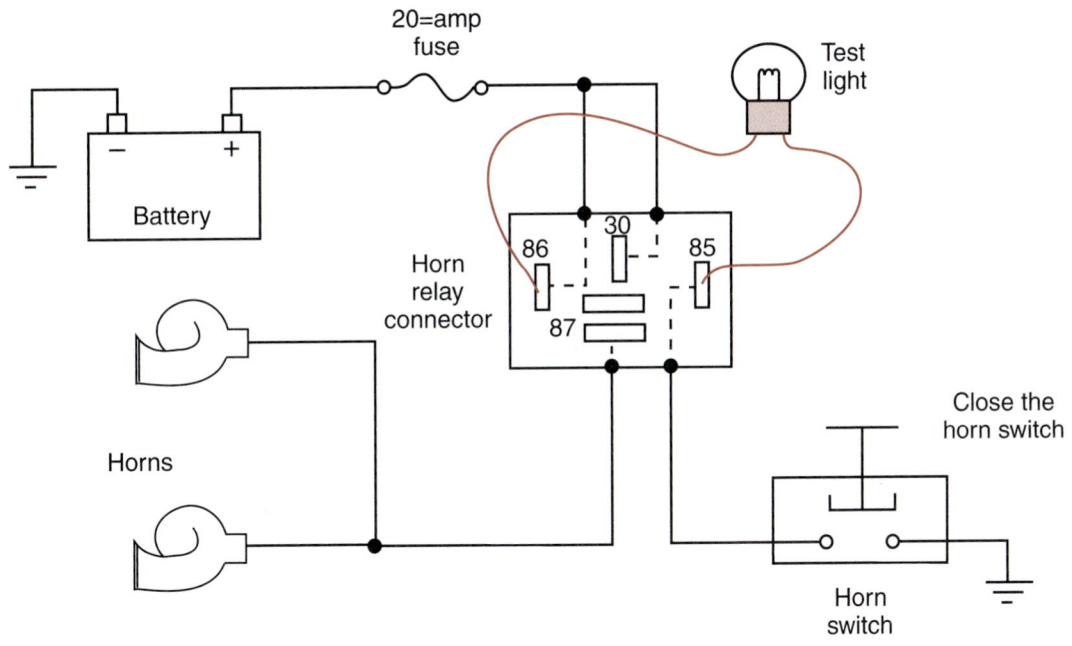

Figure 9-2 Using a test light to test the horn switch circuit.

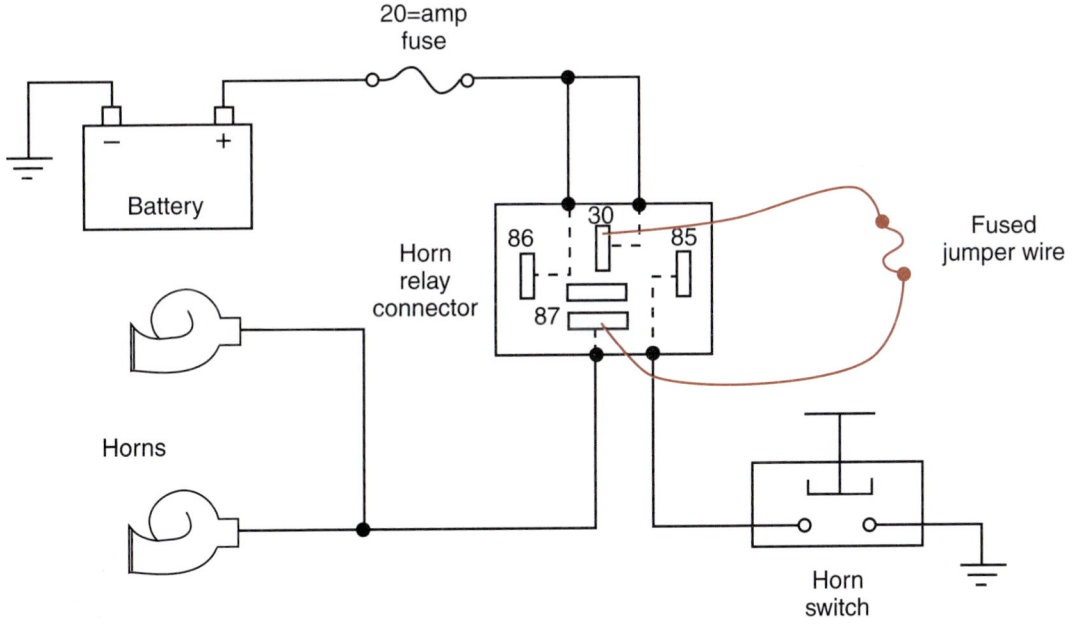

Figure 9-3 Using a fused jumper wire to test the horn circuit. If the horn sounds, the relay is faulty.

Systems Without Relay. Follow the steps described under "Systems with Relay" to confirm the complaint. If this step confirms that the horn is not operational, perform the following tests to locate the fault:

1. Check the fuse or fusible link. If defective, replace as needed. After replacing the fuse, operate all other circuits it protects. It is possible that another circuit is faulty but the customer has not noticed.

2. Connect a jumper wire from the battery positive terminal to the horn terminal. If the horn sounds, continue testing; if the horn does not sound, check the ground connection. Replace the horn if the ground is good.

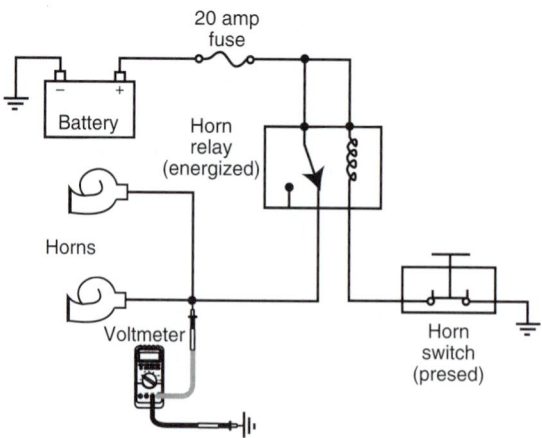

Figure 9-4 Simplified horn circuit without relay.

3. In a multiple-horn system, test for voltage at the last common connection between the horn switch and the horns (Figure 9-4). On a single-horn system, test for voltage at the horn terminal. Voltage should be present at this connection only when the horn button is depressed. Continue testing if there is no voltage at this connection. If voltage is present, check the individual circuits from the connection to the horns; repair as needed.

4. Check for voltage at the horn side of the switch when the button is depressed. If voltage is present, the problem is in the circuit from the switch to the horn(s). Continue testing if there is no voltage at this connection.

5. Check for voltage at the battery side of the switch. If voltage is present, the switch is faulty and must be replaced. If there is no voltage at this terminal, the problem is in the circuit from the battery to the switch.

Poor Sound Quality

Classroom Manual
Chapter 9, page 249

Special Tools

Voltmeter

Jumper wires

Fender covers

Poor sound quality can be the result of several factors. In a multiple-horn system, if one of the horns is not operating, the horn sound quality may suffer. Other reasons include a damaged diaphragm, excessive circuit resistance, poor ground connections, or improperly adjusted horns.

If one horn of a multiple-horn system is not operating, use a jumper wire from the battery positive terminal to the horn terminal to determine whether the fault is in the horn or in the circuit. If the horn sounds, the problem is in the circuit between the last common connection and the affected horn.

If one or all horns are producing poor quality sound, use a voltmeter to measure the voltage at the horn terminal when the horn switch is closed. The voltage should be within 0.5 volt of battery voltage. If the voltage measured is less than battery voltage, there is excessive voltage drop in the circuit. Work back through the circuit measuring voltage drop across connectors, relays, and switches to find the source of the high resistance.

If the applied voltage to the horn terminal is good, connect a jumper wire between the battery positive terminal and the horn terminal to activate the horn. With the horn sounding, turn the adjusting screw counterclockwise one quarter to three eighths of a turn. Replace the horn if the sound quality cannot be improved.

Horn Sounds Continuously

Special Tools

Ohmmeter

Fender cover

A horn that sounds continuously is usually caused by a sticking horn switch or sticking contact points in the relay. To find the fault, disconnect the horn relay from the circuit. Use an ohmmeter to check for continuity from the battery feed terminal of the relay to the horn circuit terminal (Figure 9-5). If there is continuity, the relay is defective. If there is no continuity through the relay, test the switch.

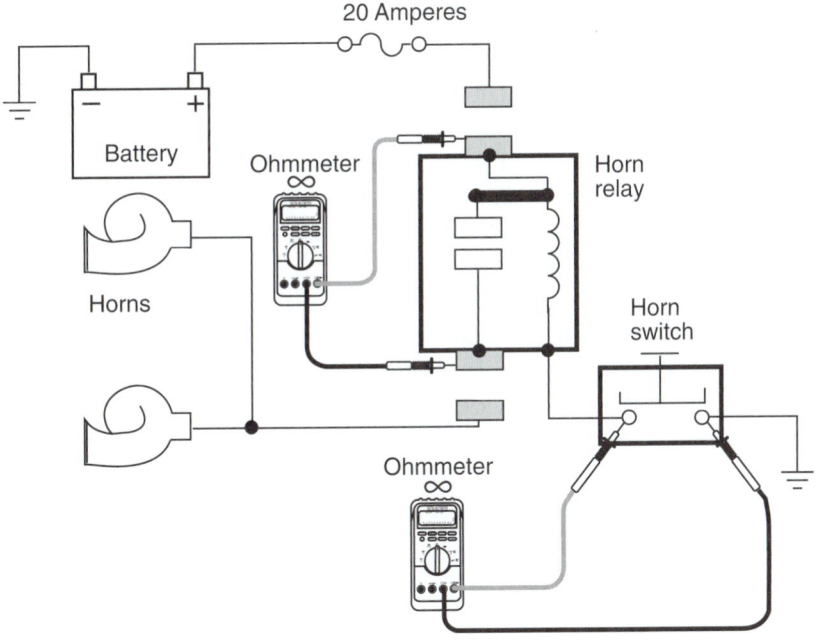

Figure 9-5 Ohmmeter tests to find cause of continuous horn operation.

The easiest way to test the switch circuit is to measure for continuity between the relay connector terminal 85 and ground. With the horn switch released, there should not be continuity to ground. If there is continuity, then the circuit to the horn switch is shorted to ground or the switch contacts are stuck. Use an ohmmeter to test the switch in the normal manner.

Wiper System Service

Customer complaints concerning windshield wiper operation can include no operation, intermittent operation, continuous operation, and wipers will not park. Other complaints have to do with blade adjustment (such as blades slapping the molding or one blade parks lower than the other).

When a customer brings the vehicle into the shop because of faulty windshield wiper operation, the technician needs to determine if it is an electrical or mechanical problem. To do this, simply disconnect the arms to the wiper blades from the motor (Figure 9-6). Turn on the wiper system. Observe operation of the motor. The problem is mechanical if the motor is operating properly.

Classroom Manual
Chapter 9, page 250

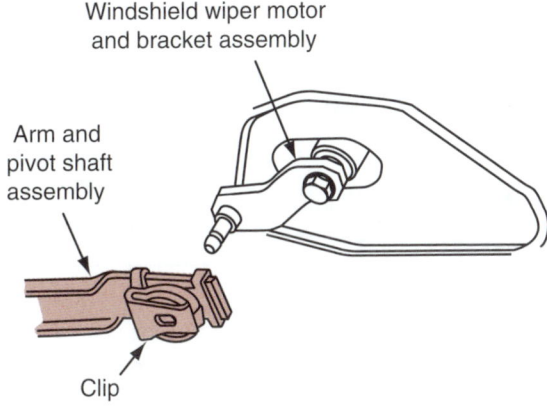

Note: Hand press to install

Figure 9-6 Disconnecting the mechanical arms from the motor.

Special Tools

Voltmeter

Fender covers

This instance is used as an example of troubleshooting the wiper system. Be sure to use the correct schematic for the vehicle you are working on.

No Operation in One Speed Only

Problems that cause the system to not operate in only one switch position are usually electrical. Use the service manual's wiring schematic to determine proper operation. For example, use the three-speed wiper system schematic illustrated (Figure 9-7) to determine the cause of a motor that does not operate in the MEDIUM speed position only. The problem is that the 7-ohm resistor is open. The problem could not be the shunt field in the motor because LOW and HIGH speeds operate; nor could it be in the wiring to the motor because this is shared by all speeds.

An opened resistor can be verified by using a voltmeter to measure voltage at the terminal leading to the shunt field. If it drops to 0 volts in the MEDIUM position, the switch must be replaced. By proper use of the wiring schematic and by understanding the correct operation of the system, you are able to diagnose this problem without having to use any test equipment. The voltmeter is used only to verify your conclusions.

In two-speed systems, the motor operating in only one speed position can be caused by several different faults. It will require the use of wiring schematics and test equipment to locate. Use (Figure 9-8) to step through a common test sequence to locate the reason why the motor does not operate in the HIGH position.

CAUTION: Do not leave the ignition switch in the RUN position without the engine running for extended periods of time. This may result in damage to the ignition system components.

Turn the ignition switch to the ACC position if the wipers will operate in this position. If not, place the ignition switch in the RUN position. Place the wiper switch in the HIGH position. Use a voltmeter to test for voltage at the high-speed connector of the motor. If voltage is present at this point, the high-speed brush is worn or the wire from the terminal to the brush is open. Most shops do not rebuild the wiper motor; replacement is usually the preferred service. If there is no voltage

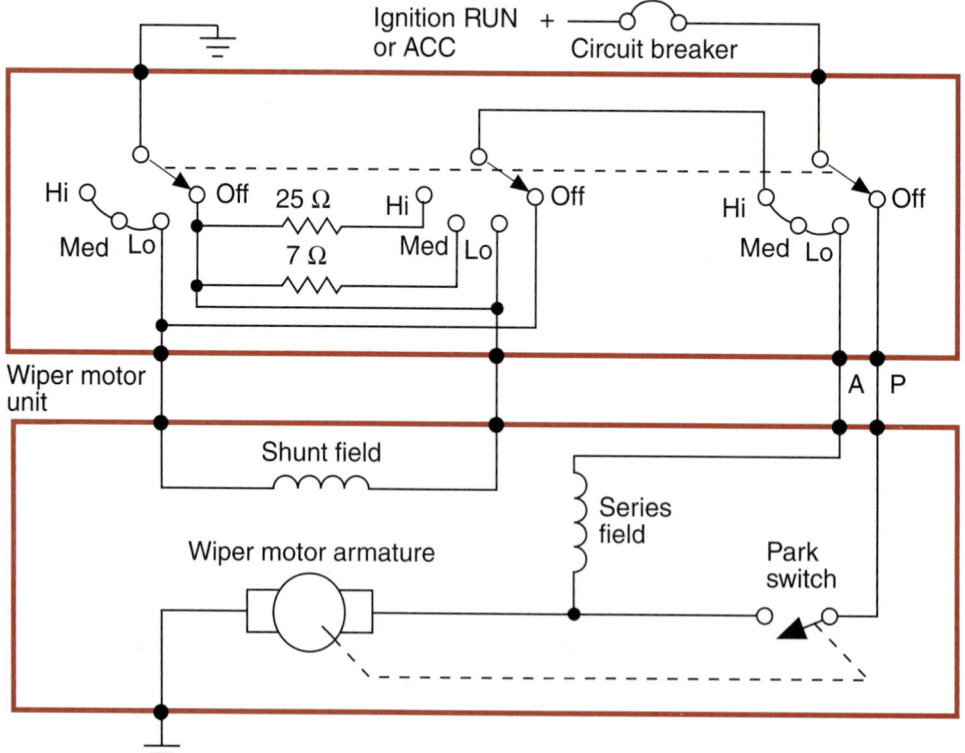

Figure 9-7 Three-speed wiper system schematic.

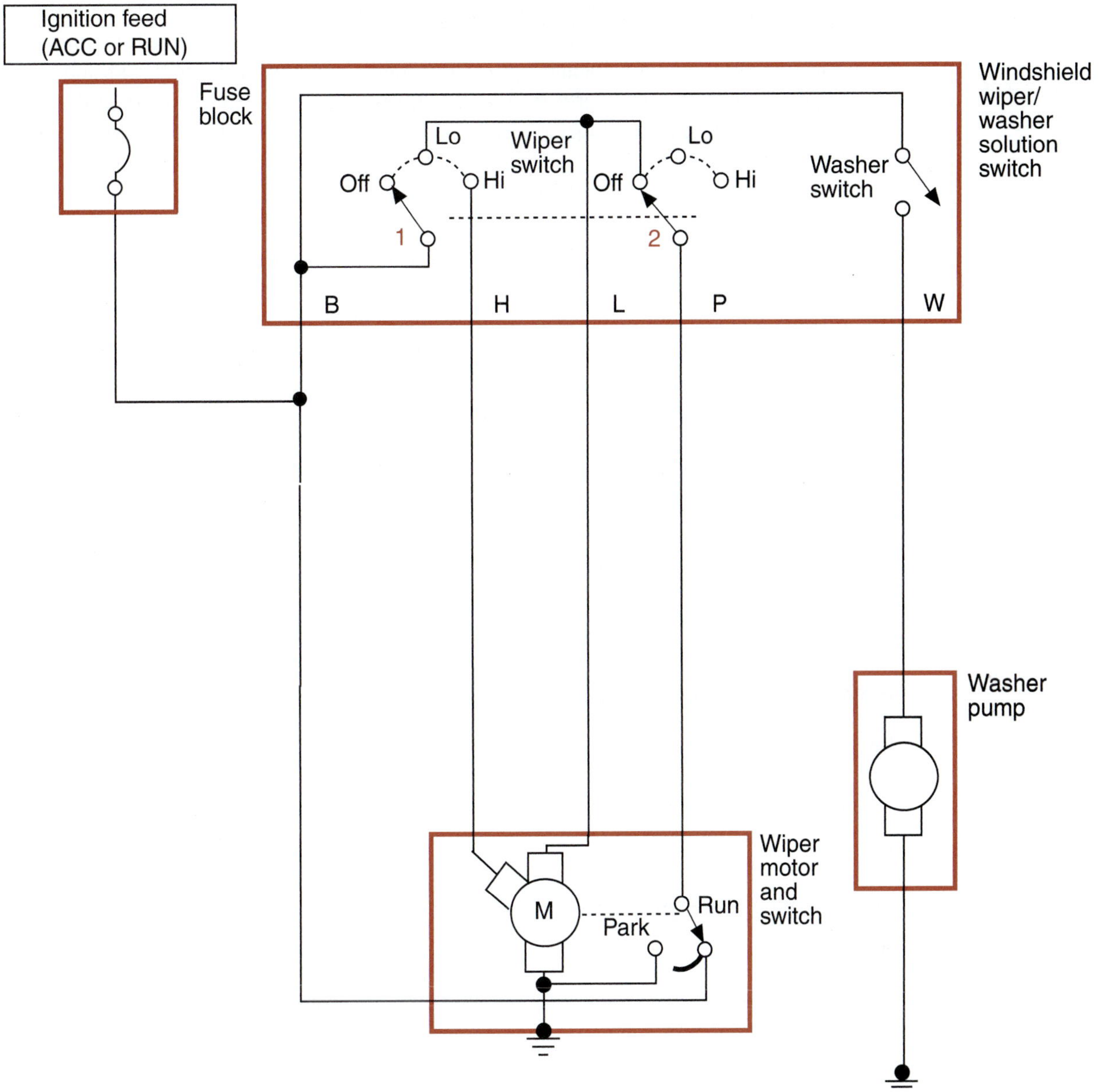

Ignition feed (ACC or RUN)

Fuse block

Wiper switch

Off — Lo — Hi

1

Off — Lo — Hi

2

B   H   L   P

Washer switch

Windshield wiper/ washer solution switch

W

Washer pump

Wiper motor and switch

M   Park — Run

Figure 9-8 Two-speed wiper circuit.

present at the high-speed connector, check for voltage at connector terminal H for the switch. If voltage is present at this point, the fault is in the circuit from the switch to the motor. If there is no voltage at this point, replace the switch.

To test for no LOW speed operation only, use the same procedure to test the low-speed circuit.

No Wiper Operation

If the wiper motor does not operate in any speed position, check the fuse. If the fuse is blown, replace it and test the operation of the motor. Also check for binding in the mechanical portion of the system. This can cause an overload and blow the fuse.

If the fuse is good, check the motor ground by using a jumper wire from the motor body to a good chassis ground. If the motor operates when the ignition switch is in the RUN position and the wiper switch is placed in all speed positions, repair the ground connection. Continue testing if the motor does not operate.

Classroom Manual
Chapter 9, page 250

Special Tools

Digital voltmeter

Fender covers

Use a voltmeter to check for voltage at the low-speed terminal of the motor with the ignition switch in the RUN or ACC position and the wiper switch in LOW position. If there is no voltage at this point, test for voltage on the low-speed terminal of the wiper switch. If the voltmeter indicates battery voltage at this terminal, the fault is in the circuit between the switch and motor. Look for indications of burned insulation or other damage that would affect both the high- and low-speed circuit.

No voltage at the low-speed terminal of the wiper switch indicates the fault may be in the switch or the power feed circuit. Test for battery voltage at the battery supply terminal of the switch. If there is voltage at this point, the switch is faulty and needs to be replaced. If no voltage is at the supply terminal, trace the circuit back to the battery to locate the fault.

If battery voltage was present at the low-speed terminal of the motor, check for voltage at the high-speed terminal. Voltage at both of these terminals indicates the motor is faulty and needs to be replaced. If there is no voltage at the high-speed terminal, use the procedure just described to trace the high-speed circuit.

 SERVICE TIP: No voltage at either terminal means the problem is probably at a shared location. In most systems this would be the power supply portion of the wiper switch and the switch itself. If there is no voltage at either terminal, go directly to the power supply terminal of the switch. If the switch is good and power is through the switch, the problem is in the wiring loom. Check all connectors. If good, remove the harness protector and inspect the wires for burned insulation or breaks.

Slower-than-Normal Wiper Speeds

Classroom Manual
Chapter 9, page 250

Special Tools

Ammeter

Voltmeter

Fender covers

Slower-than-normal wiper speeds can be caused by electrical or mechanical faults. An ammeter can be used to determine the current draw of the motor with and without the mechanical portion connected. If the draw changes substantially when the mechanical portion is removed, the fault is in the arms and/or wiper blades.

Most electrical circuit faults that result in slow wiper operation are caused by excessive resistance. If the complaint is that all speeds are slow, use the voltage drop test procedure to check for resistance in the power feed supply circuit to the wiper switch. If the power supply circuit is good, then check the switch for excessive resistance.

If the insulated side voltage tests fail to locate the problem, check the voltage drop on the ground side of the wiper motor. Connect the voltmeter positive lead to the ground terminal of the motor (or motor body) and the negative lead to the vehicle chassis. The voltage drop should be no more than 0.1 volt. If excessive, repair the ground circuit connections. If voltage drop on both the insulated and ground sides of the motor are within specifications, the fault is in the motor.

Wipers Will Not Park

Classroom Manual
Chapter 9, page 253

Special Tools

Test light

Voltmeter

Fender covers

The most common complaint associated with a faulty park switch is that the wipers stop in the position they are in when the switch is turned off. This may not be the direct fault of the **park switch,** however. The park switch is located inside the motor assembly. It supplies current to the motor after the wiper control switch has been turned to the PARK position. This allows the motor to continue operating until the wipers have reached their PARK position.

The operation of the park switch can usually be observed by removing the motor cover (Figure 9-9). Operate the wipers through three or four cycles while observing the latch arm. When the wiper switch is placed in the OFF position, the park switch latch must be in position to catch the drive pawl. Check to make sure the drive pawl is not bent. If good, replace the park switch.

A faulty wiper switch can also cause the park feature to not operate. Using the illustration (Figure 9-10), if wiper 2 is bent or broken so it does not make an electrical connection with the contacts, the wipers will not park even with the park switch in the PARK position, as shown. To test the switch, check for voltage at the low-speed circuit when the switch is moved from the LOW

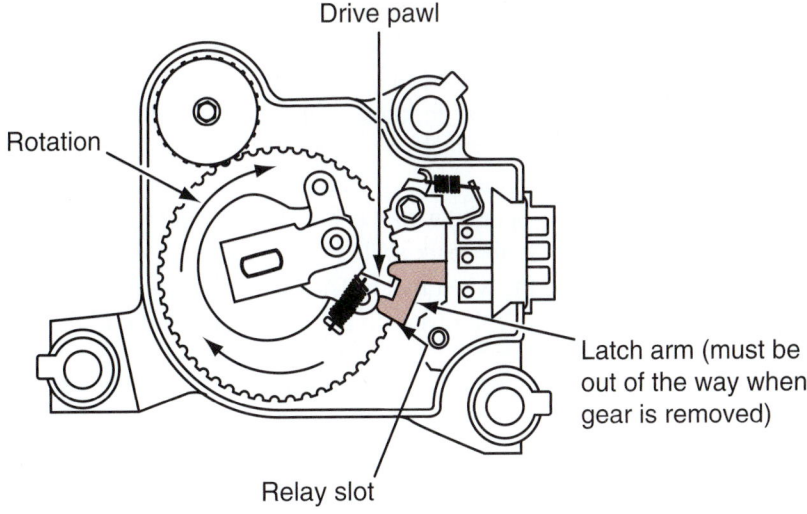

Figure 9-9 Checking the operation of the park switch while the motor is operating.

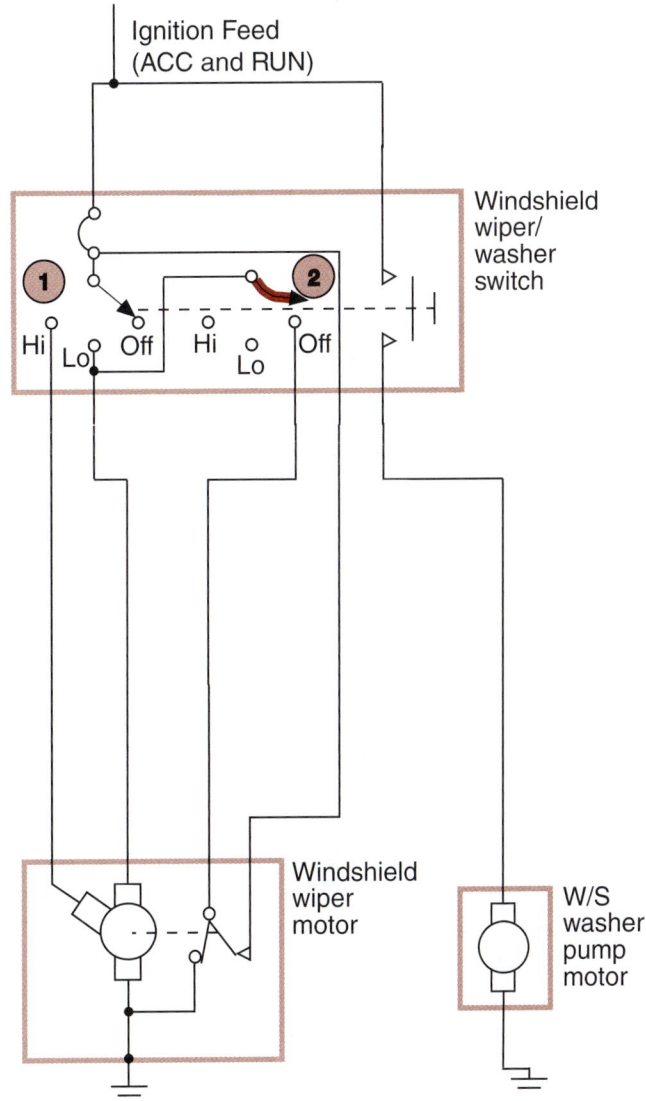

Figure 9-10 A faulty wiper in the switch can cause the park feature to not operate.

to the OFF position. If the switch is operating properly, there should be voltage present for a few seconds after the switch is in the OFF position. No voltage at this circuit when the wiper switch is turned off indicates that the problem may be in the switch.

The park switch operation can also be checked by using a test light to probe for voltage on the park switch circuit when the wiper switch is turned off. Probing for voltage at this circuit should produce a pulsating light when the motor is running.

If the wiper blades continue to operate with the wiper switch in the OFF position, the most probable cause is "welded" contacts in the park switch (Figure 9-11). If the park switch does not open, current will continue to flow to the wiper motor. The only way to turn off the wipers is to turn off the ignition switch, physically remove the wires to the motor, or pull the fuse.

Some motors provide for replacement of the park switch. However, most shops replace the motor.

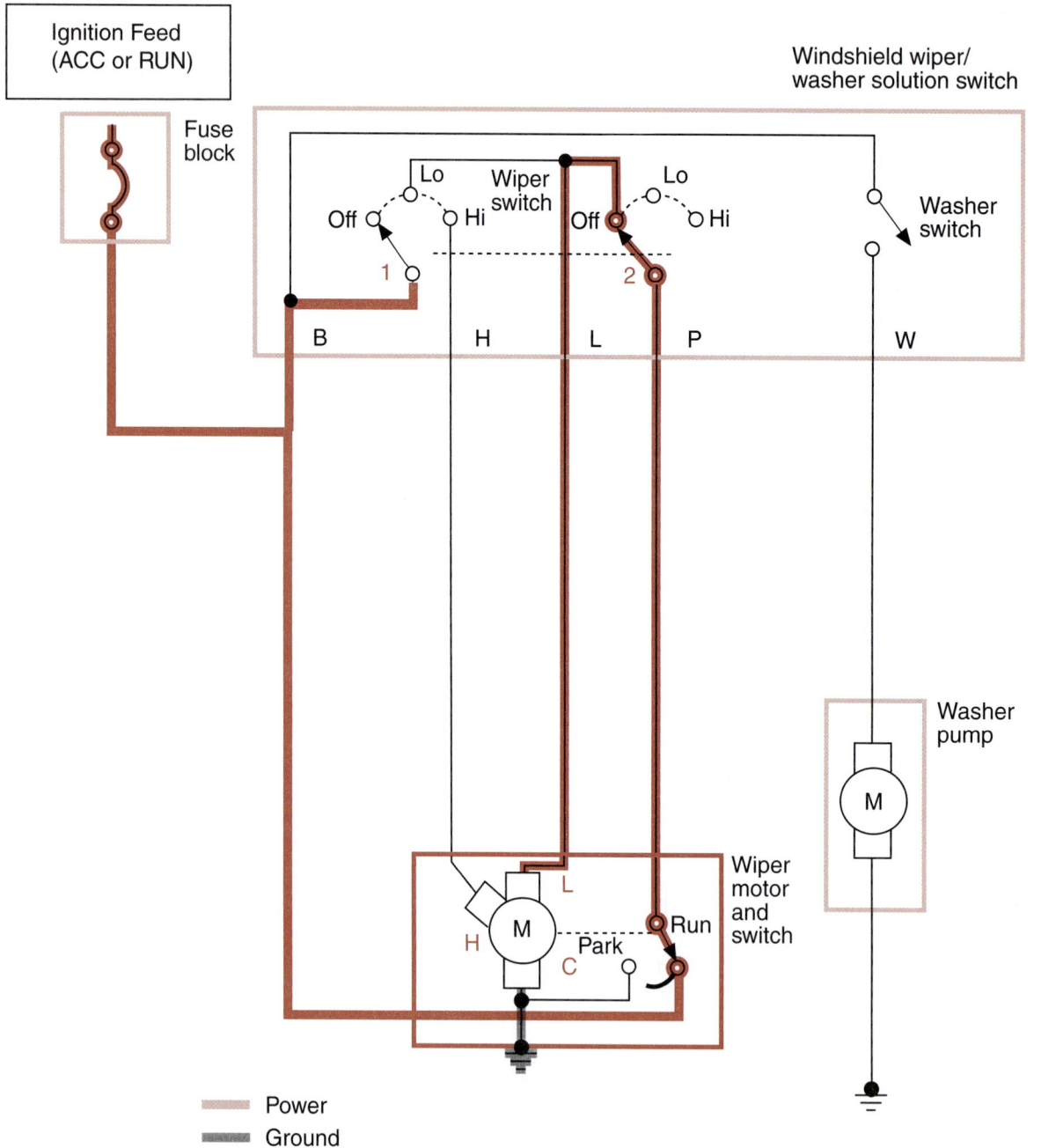

Figure 9-11 Sticking contacts in the park switch can cause the wipers to operate even after the switch is turned off.

Intermittent Wiper System Diagnosis

The illustration shown (Figure 9-12) is a schematic of the intermittent wiper system that Ford uses. If the intermittent function is the only portion of the system that fails to operate properly, begin by checking the ground connection for the timer module. If the ground is good, perform a continuity test of the switch using an ohmmeter (Figure 9-13). If the switch is good and all wires and connections are good, then replace the module.

Classroom Manual
Chapter 9, page 263

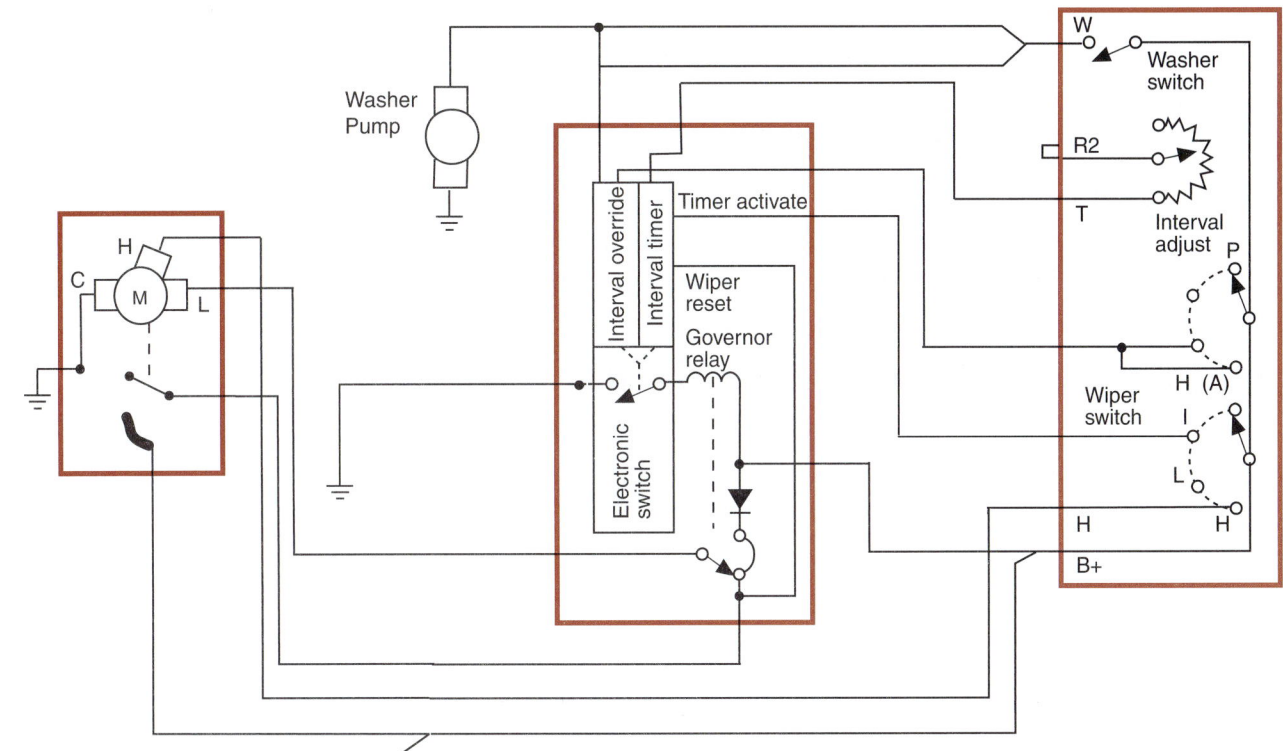

Figure 9-12 Interval wiper system schematic.

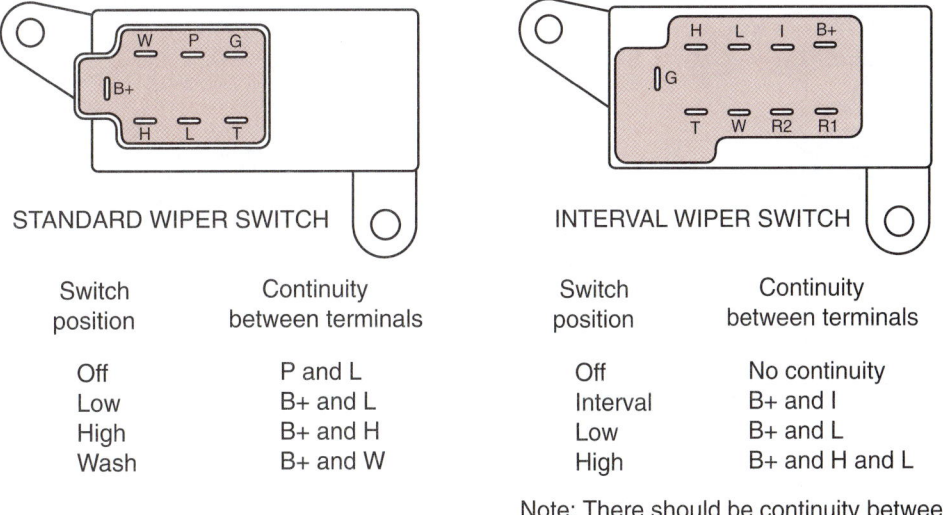

STANDARD WIPER SWITCH

Switch position	Continuity between terminals
Off	P and L
Low	B+ and L
High	B+ and H
Wash	B+ and W

INTERVAL WIPER SWITCH

Switch position	Continuity between terminals
Off	No continuity
Interval	B+ and I
Low	B+ and L
High	B+ and H and L

Note: There should be continuity between terminals R1 and R2 throughout variable resistance range

Figure 9-13 Wiper switch continuity chart.

Special Tools

Fender covers

Battery terminal
  puller

Battery terminal
  pliers

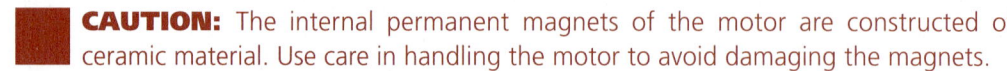

CAUTION: Some manufacturers have incorporated the intermittent wiper system into the body computer. These systems may incorporate self-diagnosis through the body-control module (BCM). Do not measure resistance through the module; damage to the circuits may result.

Wiper Motor Removal and Installation

Removal procedures differ among manufacturers. Some motors are situated in areas that may require the removal of several engine compartment components. Always refer to the correct service manual for the recommended procedure. Photo Sequence 14 provides a common method of removing a wiper motor.

CAUTION: The internal permanent magnets of the motor are constructed of ceramic material. Use care in handling the motor to avoid damaging the magnets.

On some vehicles, the linkage is removed by lifting the locking tab and pulling the clip away from the pin (Figure 9-14). Installation is basically the reverse of the removal procedure. But be sure to attach the ground wire to one of the mounting bolts during installation.

Special Tools

Fender covers

Battery terminal
  puller

Battery terminal
  pliers

Wiper Switch Removal and Installation

The wiper switch removal procedure differs among manufacturers and depends on switch location. The procedures presented here are common. However, always refer to the manufacturer's service manual for correct procedures. Always protect the customer's investment by using fender covers while disconnecting the battery negative cable.

Dash-Mounted Switches. Depending on the location of the switch control, it may be necessary to remove the finish panel (Figure 9-15). Usually the finish panel is held in place by a combination of fasteners and clips.

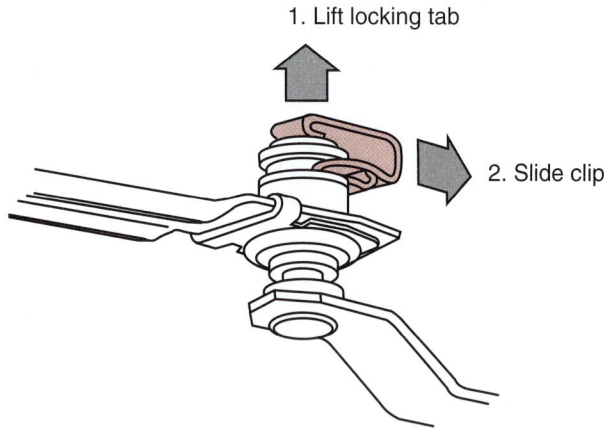

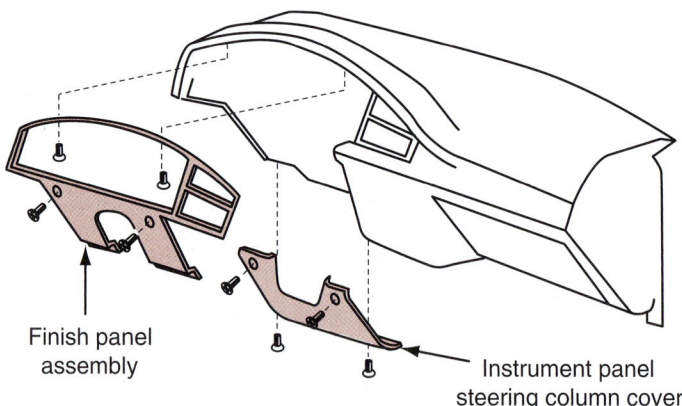

Figure 9-14 To remove the clip, lift up the locking tab and pull the clip.

Figure 9-15 It may be necessary to remove the instrument panel finish trim to gain access to the switch.

Photo Sequence 14
Wiper Motor Removal

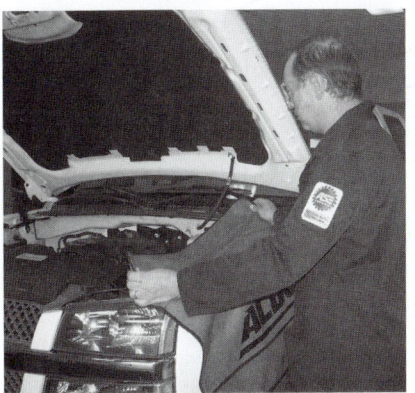

P14-1 Place fender covers over the vehicle's fenders.

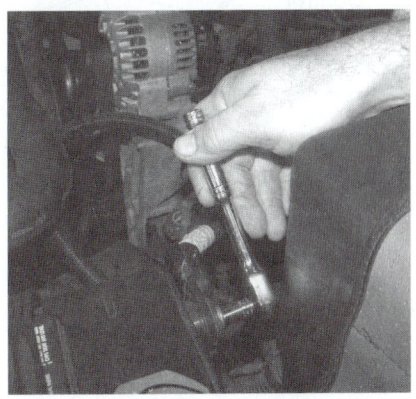

P14-2 Disconnect the battery negative cable.

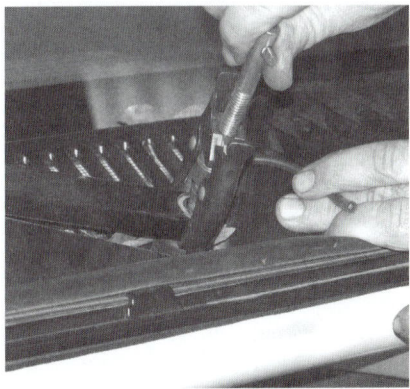

P14-3 Disconnect the wiper arms from the linkage.

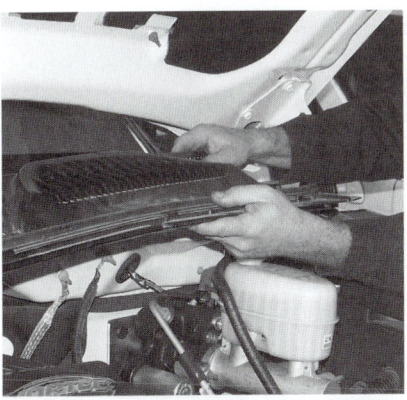

P14-4 To gain access to the motor, remove the shield cover from the crowl.

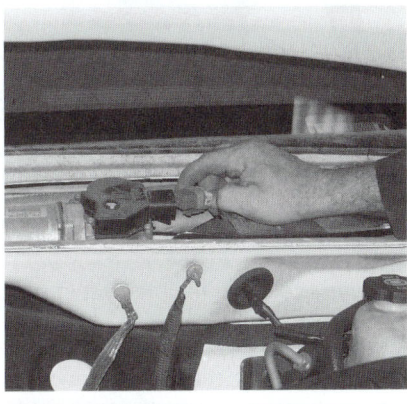

P14-5 Disconnect the wire connector at the motor.

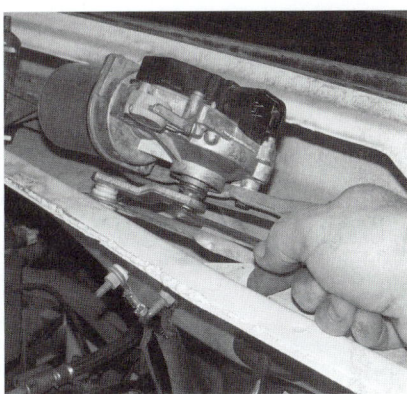

P14-6 Remove the linkage from the motor.

P14-7 Remove the attaching bolts from the motor assembly.

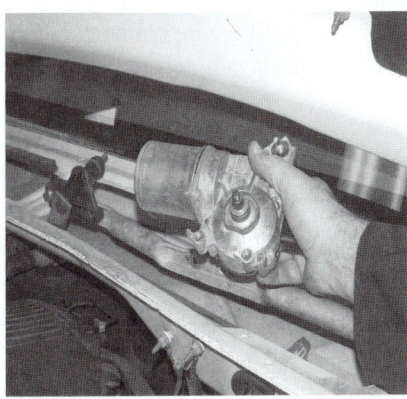

P14-8 Remove the motor.

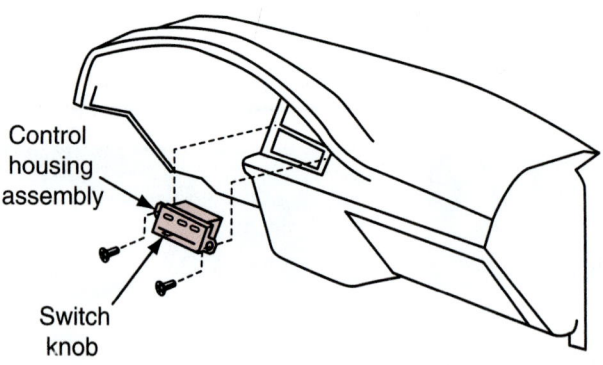

Figure 9-16 Dash-mounted wiper switch removal.

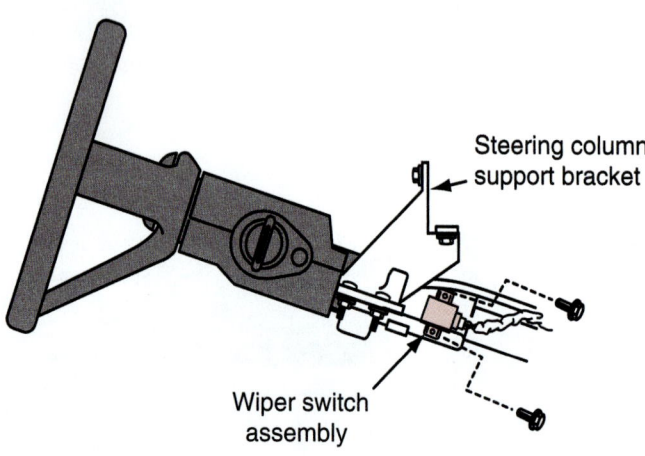

Figure 9-17 Steering column–mounted wiper switch removal.

Remove the switch housing retaining screws. Then remove the housing (Figure 9-16). Pull off the wiper switch knob. Disconnect the wire connectors from the switch. Remove the switch from the dash.

Reverse the procedure to install the switch.

Steering Column–Mounted Switches. Remove the upper and lower steering column shrouds to expose the switch. Disconnect the wire connectors to the switch. It may be necessary to peel back the foam to gain access to the retaining screws. Remove the screws and the switch (Figure 9-17).

Windshield Washer System Service

Classroom Manual
Chapter 9, page 267

Special Tools

Fender covers

Voltmeter

Test light

Ohmmeter

Safety glasses

Many windshield washer problems are due to restrictions in the delivery system. To check for restrictions, remove the hose from the pump and operate the system. If the pump ejects a stream of fluid, then the fault is in the delivery system. If the pump does not deliver a spray of fluid, continue testing using the following procedure:

1. Make checks of obvious conditions such as low fluid level, blown fuses, or disconnected wires.
2. Activate the washer switch while observing the motor. If the motor operates but does not squirt fluid, check for blockage at the pump. Remove any foreign material. If there is no blockage, then replace the motor.

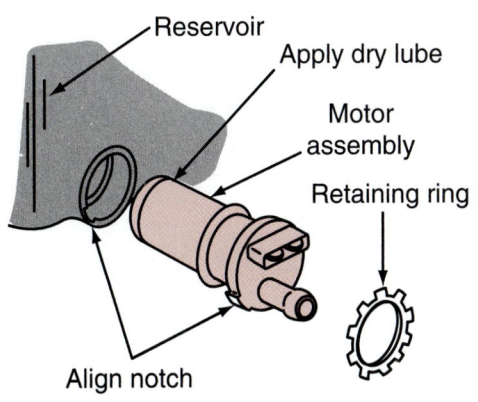

Figure 9-18 Reservoir-mounted washer pump and motor.

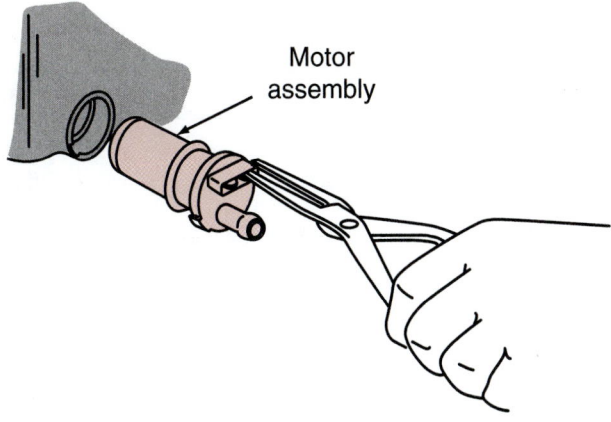

Figure 9-19 Use a pair of pliers to pull the motor out of the reservoir.

3. If the motor does not operate, use a voltmeter or test light to check for voltage at the washer pump motor with the switch closed. If there is voltage, then check the ground circuit with an ohmmeter. If the ground connection is good, then replace the pump motor.

4. If there is no voltage to the pump motor in step 3, trace the circuit back to the switch. Test the switch for proper operation. If there is power into the switch but not out of it to the motor, replace the switch.

If the motor is in need of replacement, follow this procedure for pumps installed in the reservoir (Figure 9-18). Disconnect the wire connector and hoses from the pump. Remove the reservoir assembly from the vehicle. Use a small blade screwdriver to pry out the retainer ring.

WARNING: Wear safety glasses to prevent the ring from striking your eyes. Also be careful to position the palm of your hand so that if the screwdriver slips it will not puncture your skin.

Use a pair of pliers to grip one of the walls that surrounds the terminals (Figure 9-19). Pull out the motor, seal, and impeller.

Before installing the pump assembly, lubricate the seal with a dry lubricant. The lubricant is used to prevent the seal from sticking to the wall of the reservoir. Align the small projection on the motor with the slot in the reservoir and assemble. Make sure the seal seats against the bottom of the motor cavity. Use a 12-point, 1-in. socket to hand press the retaining ring into place.

Replace the reservoir assembly in the vehicle. Reconnect the hose and wires. When refilling the reservoir, do so slowly to prevent air from being trapped in the reservoir. Check system operation while checking for leaks.

CAUTION: Do not operate the washer pump without fluid. Doing so may damage the new pump motor.

Blower Motor Service

Conventional blower motor speed is controlled by sending current through a **resistor block.** The resistor block is a series of resistors with different values. There is usually one less resistor than there are fan speed positions because the high-speed circuit bypasses the resistors. The higher the resistance value, the slower the fan speed. The position of the switch determines which resistor will be added to the circuit. Circuits can use either ground side switches or insulated side switches.

Classroom Manual
Chapter 9, page 269

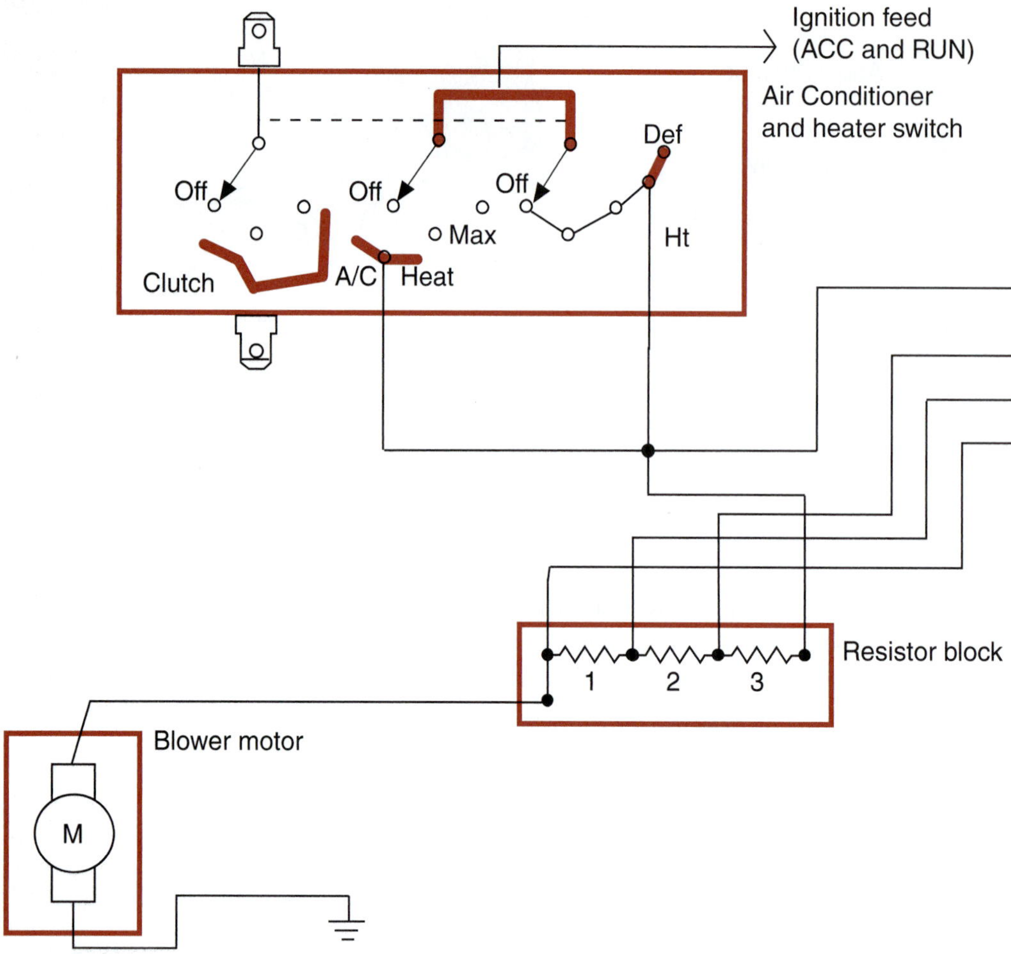

Figure 9-20 Typical wiring for a four-speed fan motor circuit.

If the customer complaint is that the fan operates in only a couple of speed positions, the most likely cause is an open resistor in the resistor block. Using the illustration (Figure 9-20), if resistor 1 is open, the motor will not operate in any position except high speed. If resistor 2 was open, the motor would operate in high and M2 speeds only. If resistor 3 was open, the motor would operate in all speeds except low.

CAUTION: The resistor block is mounted in the heater/air conditioning housing where it is cooled by air flow from the fan. Do not run the fan motor with the resistor block removed from the air flow because it may overheat and burn the coils.

If the motor operates in any one of the speed select positions, the fault is not in the motor. If the motor fails to operate at all, begin by inspecting the fuse. If the fuse is good, use the correct wiring schematic to determine whether ground or insulated side switches are used. The diagnostic procedure used depends on the circuit design.

Inoperable Motor With Insulated Switches

Use a jumper wire to bypass the switch and resistor block to check motor operation. Connect the jumper wire from a battery positive supply to the motor terminal. If the motor does not operate, connect a second jumper wire from the motor body to a good ground. Replace the motor if it still does not operate.

Classroom Manual
Chapter 9, page 269

If the motor operated when the switch and resistor block were bypassed, trace up the circuit toward the switch. Use a voltmeter or test light to check for voltage in and out of the blower speed control switch. The switch is faulty if voltage is at the input terminal but not at any of the output terminals. No voltage at the input terminal indicates an open in the circuit between the battery and the switch.

> ✔ **SERVICE TIP:** Because the high-speed circuit bypasses the resistor block, it is doubtful that no motor operation would be the fault of open resistors. However, an open in the wire from the block to the motor can cause the problem. Most likely the switch is bad and in need of replacement. Always confirm your diagnosis by doing a continuity test on the switch.

Inoperable Motor With Ground Side Switches

Using the illustration (Figure 9-21) as an example of negative side switch blower motor circuit to test the motor, connect the jumper wire from the motor negative terminal to a good ground. This bypasses the switch and resistor block. If the motor does not operate, use a voltmeter or test light to check for voltage at the battery terminal of the motor. If voltage is present, then the motor is defective.

If there is no voltage at the battery terminal of the motor, the problem is in the circuit from the battery to the motor. Be sure to check the circuit breaker.

Operation of the motor when the jumper wire is connected to the ground terminal indicates the problem is in the switch side of the circuit. Use an ohmmeter to check the ground connection for the switch. A jumper wire or test light can also be used to test this connection.

If the ground connection is good, use a voltmeter or test light to probe for voltage at any of the circuits from the resistor block to the switch. Replace the switch if there is power to this point. Replace the resistor block if there is no power at these points.

Special Tools

Jumper wires

Test light

Voltmeter

Fender covers

Classroom Manual
Chapter 9, page 269

Special Tools

Jumper wires

Voltmeter

Ohmmeter

Test light

Fender covers

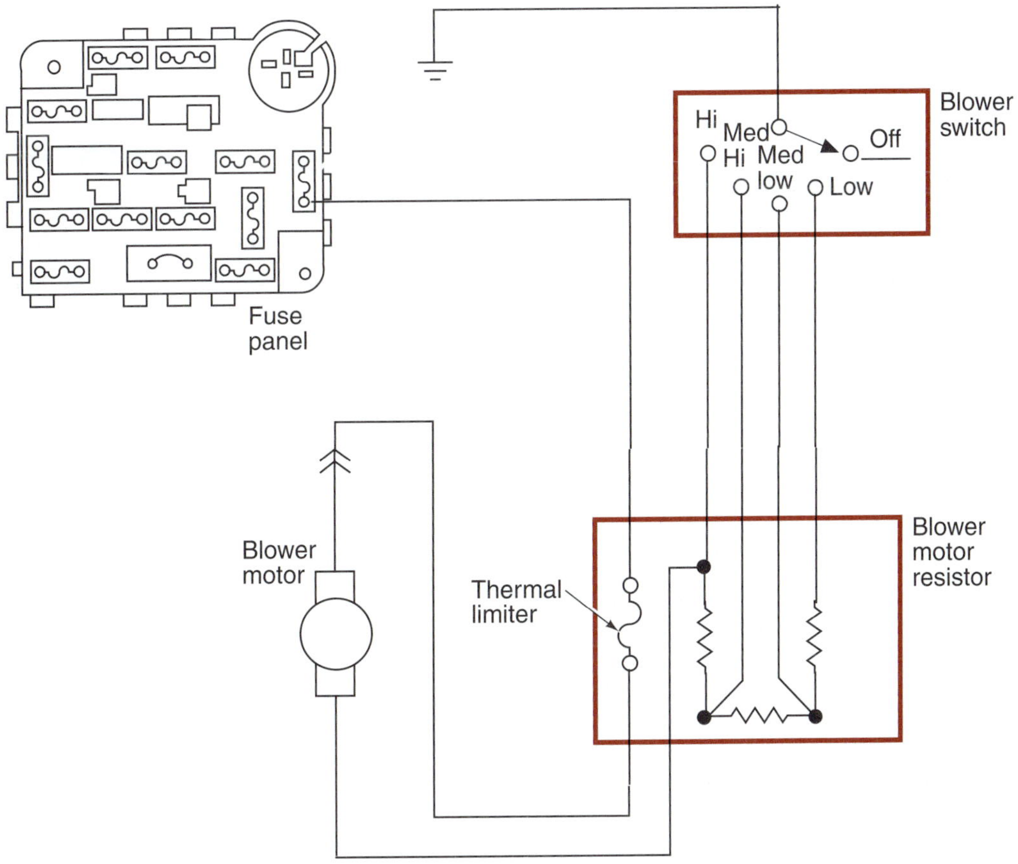

Figure 9-21 Blower motor circuit using negative side switch.

Classroom Manual

Chapter 9, page 269

General Motors designed many of its blower fans to constantly operate. Do not confuse this normal operation with a circuit defect.

Classroom Manual

Chapter 9, page 273

Special Tools

Test light

Grease pencil

Special Tools

Heat gun or blow-dryer

Defogger repair kit

Test light

Masking tape

Clean cloth

Wooden stick

Alcohol

Steel wool

Constantly Operating Blower Fans

Constantly running blower motors are more common in ground side switch systems. A short to ground at any point on the ground side of the circuit will cause the motor to run. Other areas to check include the switch and the circuit between the switch and the resistor block.

In insulated side switch circuits, check for copper-to-copper shorts in the power side of the system. If the motor is receiving power from another circuit, due to a copper-to-copper short, the motor will continue to run whenever current is flowing through that circuit. Some systems may incorporate a relay, and if the contact points fuse together, the motor will continue to operate.

Electric Defogger Diagnosis and Service

If the rear window defogger fails to operate when the switch is activated, use a test light to test the **grids.** The rear window defogger grids are a series of horizontal, ceramic, silver-compounded lines that are baked into the surface of the window. Under normal conditions, the test light should be bright on one side of the grid and off on the other side. If the test light has full brilliance on both sides of the grid, then the ground connection for the grid is broken.

If the test light does not illuminate at any position on the grid, use normal test procedures to check the switch and relay circuits. There may be several fuses involved in the system. Use the correct wiring diagram to determine the fuse identification.

Most rear window defogger complaints will be associated with broken grids. These will generally be complaints that only a portion of the window is cleared while the rest remains foggy. Some grid wire breaks are easily detected by visual inspection. However, many are too small to see. To test the grid lines, start the engine and activate the system. (Remember, the system is controlled by a timer.) Use a test light to check each grid wire to locate the breaks. Test each grid in at least two places—one on each side of the center line. The test results that should be obtained on each grid are illustrated (Figure 9-22).

CAUTION: Be careful not to tear the grid with the test light. Only a light touch on the grid should be required.

If the test light does not indicate normal operation on a specific grid line, place the test light probe on the grid at the left bus bar and work toward the right until the light goes out. The point where the light goes out is the location of a break (Figure 9-23). Mark the location of the break with a grease pencil on the *outside* of the glass.

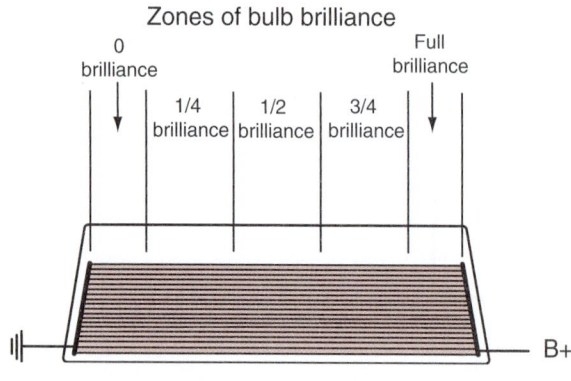

Figure 9-22 Zones of test light brilliance while probing a rear window defogger grid.

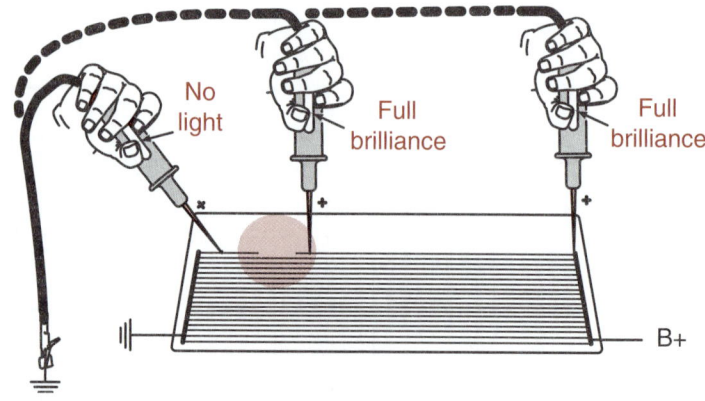

Figure 9-23 Test light brilliance when passed over a break.

The rear window defogger should turn off about 10 minutes after activation. If the circuit fails to turn off, check the ground for the control module (Figure 9-24). If the ground is good, replace the module.

Grid Wire Repair

If the grid wire is broken, it is possible to repair the grid with a special repair kit. Follow the procedures in Photo Sequence 15 to repair the grid.

Special Tools

Soldering iron
Rosin flux paste
3% silver solder
Steel wool

Some manufacturers refer to the rear window defogger system as electric backlight (EBL).

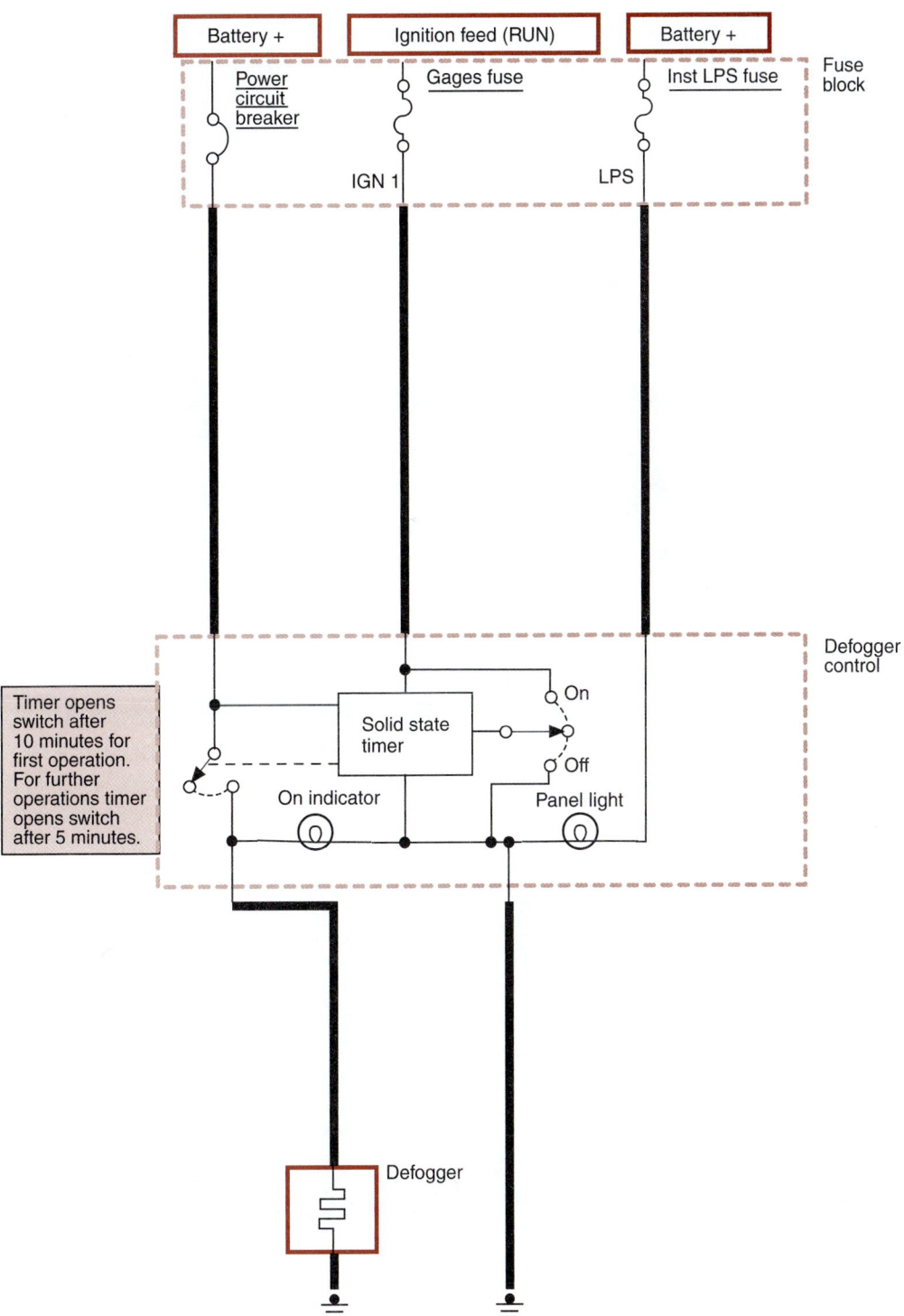

Figure 9-24 The defogger control incorporates a solid-state timer circuit.

Photo Sequence 15
Typical Procedure for Grid Wire Repair

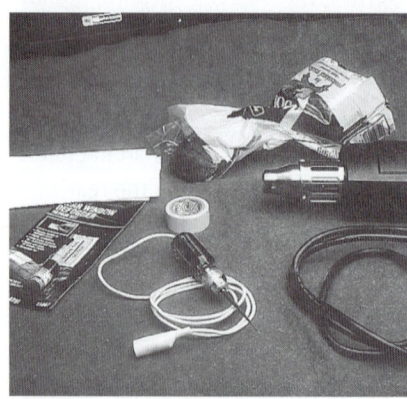

P15-1 Tools required to perform this task include masking tape, repair kit, 500°F heat gun, test light, steel wool, alcohol, and a clean cloth.

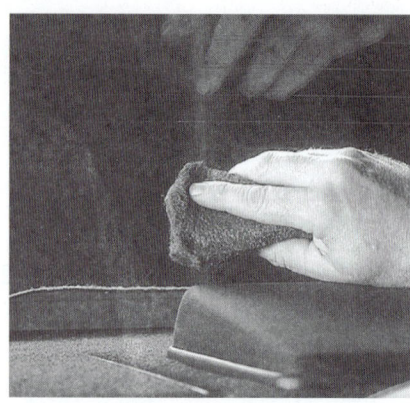

P15-2 Clean the grid line area to be repaired by buffing with steel wool and wiping clean with a cloth dampened with alcohol. Clean an area about 1/4 inch (6 mm) on each side of the break.

P15-3 Position a piece of tape above and below the grid. The tape is used to control the width of the repair.

P15-4 Mix the hardener and silver plastic thoroughly. If the hardener has crystallized, immerse the packet in hot water.

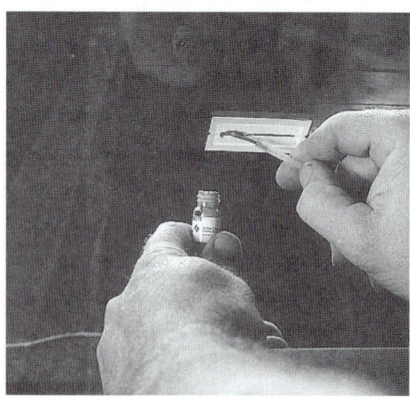

P15-5 Apply the grid repair material to the repair area using a small stick.

P15-6 Remove the tape.

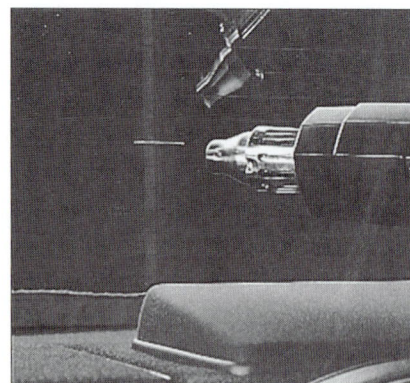

P15-7 Apply heat to the repair area for 2 minutes. Hold heat gun 1 inch (25 mm) from the repair.

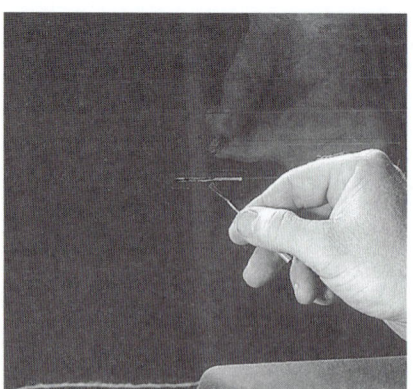

P15-8 Inspect the repair. If it is discolored, apply a coat of tincture of iodine to the repair. Allow to dry for 30 seconds, then wipe off the excess with a cloth.

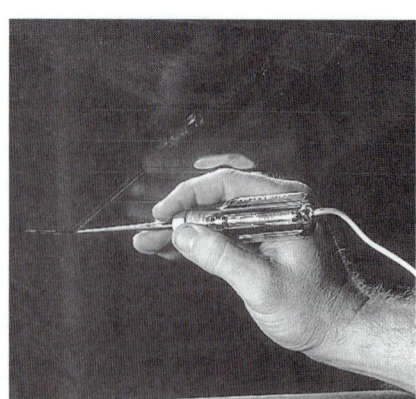

P15-9 Test the repair with a test light. Note: It takes 24 hours for the repair to fully cure.

Bus Bar Lead Repair

The bus bar lead wire can be resoldered using a solder containing 3% silver and a rosin flux paste. Clean the repair area using a steel wool pad. Apply the rosin flux paste in small quantities to the wire lead and bus bar. **Tin** the solder iron tip with the solder. Finish the repair by soldering the wires to the bus bar. Be careful not to overheat the wire.

Power Window Diagnosis

Use the illustration (Figure 9-25) as a guide to diagnosing the power window circuit. If the window does not operate, begin by testing the circuit breaker. Use a test light or voltmeter to test for voltage at both sides of the circuit breaker. If voltage is present at both sides, then the circuit breaker is good. If there is voltage into the circuit breaker but not out of it, the circuit breaker is bad. If there is no voltage into the circuit breaker, then there is an open in the feed from the battery.

SERVICE TIP: It does not matter if the motor or the switches are tested next. Test the unit that is easiest to get to first.

If the circuit breaker is good, use jumper wires to test the motor. The motor is a reversible motor, so connections to the motor terminals are not polarity sensitive. Disconnect the wire connectors to the motor. Connect battery positive to one of the terminals and ground the other. If the motor does not operate, reverse the jumper wire connections. The motor should reverse directions when the polarity is reversed. If the motor does not operate in one or both directions, it is defective and needs to be replaced.

WARNING: Do not place your hands into the window's operating area. Make the final test connections outside of the door where there is no danger of getting caught in the window track.

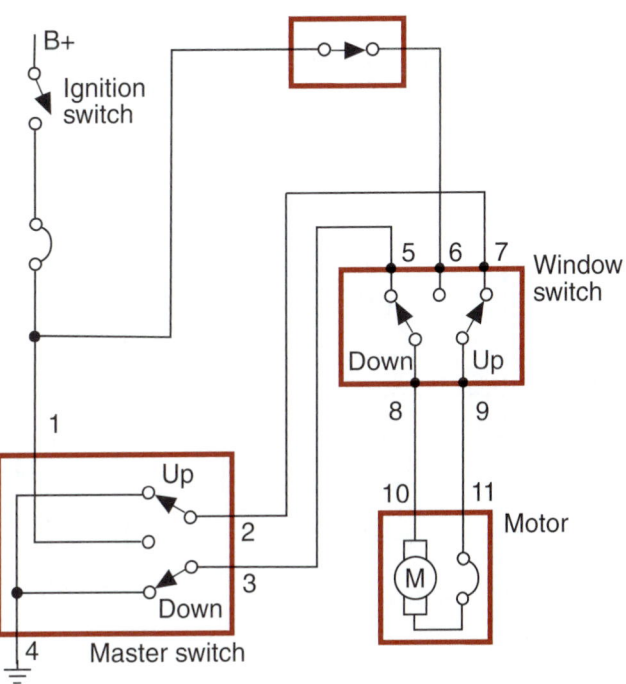

Figure 9-25 Simplified power window circuit.

To **tin** the tip is a process of applying solder to the tip of the soldering iron to provide better heating control.

Classroom Manual
Chapter 9, page 279

Special Tools

Test light
Voltmeter
Jumper wires

The circuit used in this example is typical. However, use the service manual for the vehicle you are working on to get the correct wiring schematic.

If the motor operates when the switches are bypassed, the problem is in the control circuit. To test the master switch, connect the test light between terminals 1 and 2 (Figure 9-26). When the master switch is in the rest (OFF) position, the test light should illuminate. If the light does not glow, there is an open in the circuit to the window switch or from the window switch to ground at terminal 4. Check the ground at terminal 4 for good connections. If good, continue testing.

If the test light illuminates when connected across terminal 1 and 2, place the switch in the UP position. The test light will go out if the switch is good. Repeat the test between terminals 1 and 3. Place the switch in the DOWN position.

If the master switch is good, test the window switch. Battery voltage should be present at terminal 6. If not, check to see if the lockout switch is closed. Check the circuit from terminal 6 to the circuit breaker. Connect the test light across terminals 5 and 6 (Figure 9-27). The light should come on. Move your test light to connect between terminals 6 and 7. Again, the test light should come on. Next, connect the test light between terminals 8 and 9 of the window switch. With the switch in the "at rest" position, the test light should be off. Placing the window switch in either the UP or DOWN positions should illuminate the test light. If the light does not come on, you will need to isolate the problem. It may be the switch, the circuits between the switch and the motor, or the motor itself. Use common test methods to determine the fault.

Slower-than-normal operating speeds are an indication of excessive resistance or of binding in the mechanical linkage. Use the voltage drop test method to locate the cause of excessive resistance. Excessive resistance can be in the switch circuits, the ground circuit, or in the motor. If the problem is mechanical, lubricate the track and check for binding or bent linkage.

> ⚠️ **WARNING:** Follow the manufacturer's recommended procedure when removing the power window motor. The springs used in window regulators can cause serious injury if removed improperly.

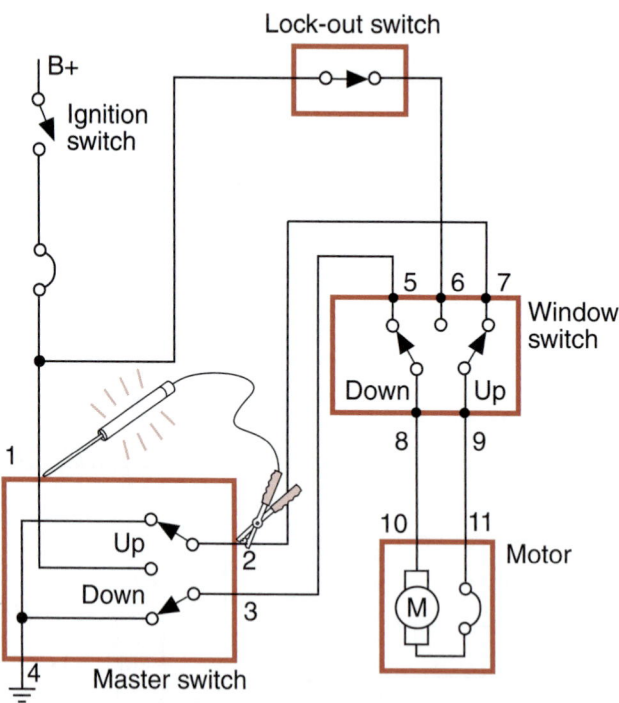

Figure 9-26 Using a test light to check the operation of a power window master switch.

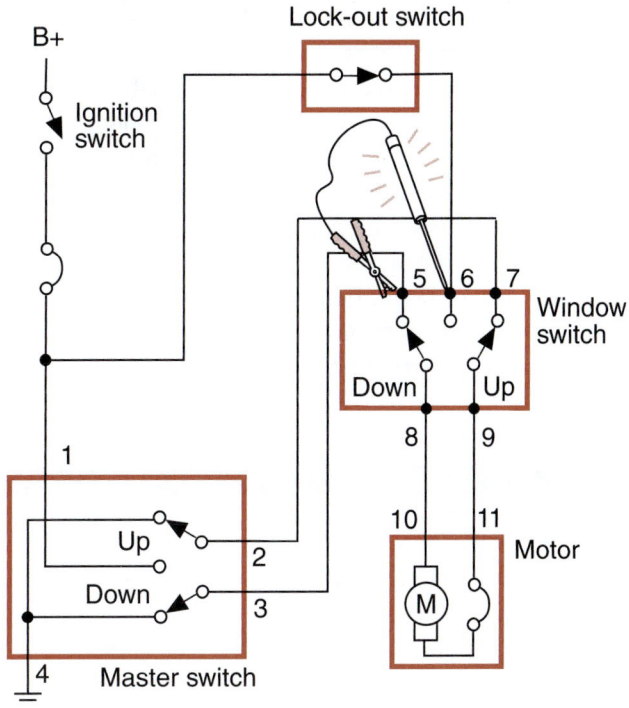

Figure 9-27 Test light connections for testing the window switch.

⬤ **CUSTOMER CARE:** Testing and repair of the power window system will usually require the door panels to be removed. There are several methods used by manufacturers to secure the door panel. Always refer to the proper service information to determine the correct methods of removal and installation of the panel to prevent damage. Also, use new clips (if applicable) to assure a tight connection and eliminate noise from the panel. Most doors will have a sound dampening material behind the panel. Therefore, you must remember to reinstall this material.

Power Seat Diagnosis

The power seat system is usually very simple to troubleshoot. Test for voltage to the input of the switch control. If voltage is available to the switch, remove it from the seat or arm rest. Using a continuity chart from the service manual, test the switch for proper operation. If the switch is operating properly, it may be necessary to remove the seat to test the motors and circuits to the motors.

The power seat motors are tested in the same manner as the power window motor. Be sure to test each armature of the trimotor. If any of the armatures fail to operate, the trimotor must be replaced as a unit.

▲ **WARNING:** Be careful when making the jumper wire connection to test the motor. Do not place your hands in locations where they can become pinched or trapped when the seat moves.

▲ **WARNING:** If the trimotor needs to be replaced, follow the manufacturer's service procedures closely. Improper removal of the springs may result in personal injury.

Noisy operation of the seat can generate from the motor, transmission, or cable. If the motor or transmission is the cause of the noise, it must be replaced. A noisy cable can usually be cured with a dry lubricant, provided the cable is not damaged.

Classroom Manual
Chapter 9, page 284

Special Tools

Jumper wires
Ohmmeter
Test light
Voltmeter

Special Tools

Jumper wires

Ohmmeter

Test light

Power Door Lock Diagnosis

To test the door lock motor, apply 12 volts directly to the motor terminals. The actuator rod should complete its travel in less than 1 second. Reverse polarity to test operation in both directions.

The switch is checked for continuity using an ohmmeter. There should be no continuity between any terminals when the switch is in its neutral position. Use the circuit schematic to determine when there should be continuity between terminals.

If the system uses a relay, use the schematic to determine relay circuit operation. In this example, battery voltage should be present at terminal 4 of the connector. Using an ohmmeter, check the ground connects of terminals 1 and 5 of the connector. To test the relay, connect a test light across terminal 3 and ground. Ground terminal 1 and apply power to terminals 2 and 4. The test light will light if the relay is good.

CASE STUDY

The customer complains that although the fan blower motor operates well in low- and medium-speed positions, it runs slow in the high-speed position. A look at the wiring schematic (Figure 9-28) for the circuit indicates that the resistor block does not control high-speed operation. Because the motor operates properly in low and medium speeds, the fault is not in the motor or the power supply feed. By examining the schematic, the only place that there could be resistance in the high-speed circuit is at the switch wiper or in the circuit between the switch and the resistor block. A voltage drop test confirms the location of the resistance is the connector to the switch.

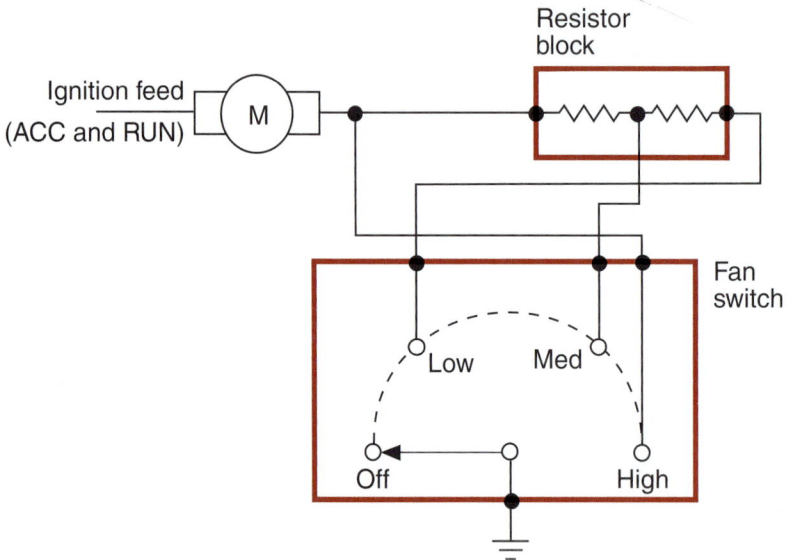

Figure 9-28 Three-speed blower motor schematic.

Terms to Know

Grids

Resistor block

Park switch

Tin

ASE-Style Review Questions

1. All of the following can cause the horn to sound continuously EXCEPT:
   A. Circuit between relay and horn shorted to power.
   B. Short to ground between relay and horn.
   C. Horn switch contacts stuck closed.
   D. Short to ground between the relay coil and the horn switch.

2. The two-speed windshield wiper operates in HIGH position only.
   Technician A says the low-speed brush may be worn.
   Technician B says the motor has a faulty ground connection.
   Who is correct?
   A. A only
   B. B only
   C. Both A and B
   D. Neither A nor B

3. What is the LEAST likely cause of slow windshield wiper operation?
   A. Binding mechanical linkage.
   B. Excessive resistance in the motor's ground circuit.
   C. Short to power at the low-speed brush.
   D. Worn common brush.

4. The wiper motor does not park when the wiper switch is placed in the OFF position.
   Technician A says the park switch is faulty.
   Technician B says the wiper switch is faulty.
   Who is correct?
   A. A only
   B. B only
   C. Both A and B
   D. Neither A nor B

5. The intermittent wipers do not operate. The wipers work fine in low and high speed. What can be the cause?
   A. Open circuit between the windshield wiper switch and the intermittent wiper module.
   B. Excessive resistance in the ground connection for the intermittent wiper module.
   C. Faulty wiper switch.
   D. All of the above.

6. The heater fan motor does not operate in high-speed position.
   Technician A says the cause is a faulty resistor block.
   Technician B says the motor is defective.
   Who is correct?
   A. A only
   B. B only
   C. Both A and B
   D. Neither A nor B

7. The grid of a rear window defogger is only removing some areas of fog from the window.
   Technician A says the timer circuit is faulty.
   Technician B says the grid is damaged.
   Who is correct?
   A. A only
   B. B only
   C. Both A and B
   D. Neither A nor B

8. The passenger-side power window does not operate in either direction, despite which switch is used (master or window).
   Technician A says the problem is a faulty master switch.
   Technician B says the problem is a worn motor.
   Who is correct?
   A. A only
   B. B only
   C. Both A and B
   D. Neither A nor B

9. The six-way power seat is not operating in any switch position.
   Technician A says to check the circuit breaker.
   Technician B says to use a continuity chart to test the switch.
   Who is correct?
   A. A only
   B. B only
   C. Both A and B
   D. Neither A nor B

10. The power door locks will lock the door, but they do not unlock it.
    Technician A says the motor is faulty.
    Technician B says the unlock relay is faulty.
    Who is correct?
    A. A only
    B. B only
    C. Both A and B
    D. Neither A nor B

ASE Challenge Questions

1. The horn of a vehicle equipped with a horn relay is sounding continuously.
   Which of the following could be the cause of this problem?
   A. Shorted horn relay coil.
   B. Open horn switch.
   C. Grounded wire on the switched side of the horn relay.
   D. Open horn relay coil.

2. The high-speed position of a two-speed windshield wiper system is inoperative; the low-speed position is fine. Which of the following could cause this problem?
   A. High resistance on the ground side of the motor.
   B. Open motor resistor.
   C. Faulty park switch.
   D. Worn motor brush.

3. An A/C blower motor is turning slowly. A voltmeter that is connected across the power and ground terminals of the motor is indicating 13.6 volts when the blower speed switch is in the HIGH position and the engine is running; an ammeter indicates that the motor is drawing about 3 amps.
   Technician A says that the blower motor ground circuit may have excessive resistance.
   Technician B says that the blower motor armature may be binding.
   Who is correct?
   A. A only
   B. B only
   C. Both A and B
   D. Neither A nor B

4. The blower motor of a vehicle equipped with a ground-side controlled motor circuit is running at high speed whenever the engine is running; the blower switch has no control of the motor speed.
   Technician A says that ground side of the motor may be shorted to ground.
   Technician B says that the blower switch contacts may have excessive resistance.
   Who is correct?
   A. A only
   B. B only
   C. Both A and B
   D. Neither A nor B

5. A power window motor is inoperative. A voltmeter that is connected across the power and ground terminals of the motor indicates 0 volts when the window switch is moved to the DOWN position; when the ground lead of the voltmeter is connected to a chassis ground (and the switch is in the DOWN position) the voltmeter then indicates 12.6 volts.
   Technician A says that the window motor has an open internal circuit.
   Technician B says that the window switch may have an open contact.
   Who is correct?
   A. A only
   B. B only
   C. Both A and B
   D. Neither A nor B

Job Sheet 31

31

Name _____ Date _____

Diagnosing a Windshield Wiper Circuit

Upon completion of this job sheet, you should be able to diagnose an inoperative or poorly working windshield wiper system.

ASE Correlation

This job sheet is related to the ASE Electrical/Electronic System Certification Exam's content area: *Horn and Wiper/Washer Diagnosis and Repair;* task: Diagnose the cause of wiper problems, including constant, intermittent, poor speed control, parking problems, and no operation of wiper.

Tools and Materials

A vehicle
Wiring diagram for the chosen vehicle
Service manual for the chosen vehicle
A DMM

Describe the vehicle being worked on:

Year _____ Make _____ Model _____

VIN _____ Engine type and size _____

Procedure

Task Completed

1. Describe the general operation of the windshield wipers. Check the operation in all speeds and modes.

   _____

   _____

   _____

   _____

2. Check the mechanical linkages for evidence of binding or breakage. Record your findings.

   _____

   _____

   _____

   _____

3. Draw the windshield wiper circuit below. Include its power source, controls, and ground.

4. Describe how the motor is controlled to operate at different speeds.

_____

_____

_____

_____

5. Connect the voltmeter across the ground circuit; energize the motor. What was your reading on the meter? What does this indicate?

_____

_____

_____

_____

6. Probe the power feed to the motor in the various switch positions. Observe your voltmeter readings. What were they? What is indicated by these readings?

_____

_____

_____

_____

7. Describe the general operation of the windshield wiper motor.

_____

_____

_____

_____

Instructor's Response _____

_____

_____

Job Sheet 32

32

Name _____ Date _____

Testing the Windshield Washer Circuit

Upon completion of this job sheet, you should be able to diagnose problems in the windshield washer circuit.

ASE Correlation

This job sheet is related to the ASE Electrical/Electronic System Certification Exam's content area: *Horn and Wiper/Washer Diagnosis and Repair;* task: Diagnose the cause of constant, intermittent, or no operation of windshield washer.

Tools and Materials

A vehicle
Wiring diagram for the chosen vehicle
A DMM
Hand tools

Describe the vehicle being worked on:

Year _____ Make _____ Model _____

VIN _____ Engine type and size _____

Procedure

Task Completed

1. Describe the general operation of the windshield wipers.

   _____

   _____

   _____

   _____

   ☐

2. Check the fluid level in the washer fluid reservoir.

   Replenish the level, if necessary.

3. Remove the fluid output line from the washer pump.

   Activate the washer pump. Does fluid come out of the pump? _____ What are your conclusions from this?

   _____

   _____

   _____

   _____

4. If fluid comes out of the pump but none sprays on the windshield, disconnect the fluid lines from the washer fluid nozzles. Activate the pump. Does fluid come out of the pump? _____ What are your conclusions from this?

_____

_____

_____

_____

5. If the pump does not operate when switched on, run a jumper wire from the battery to the pump. What happened? What are your conclusions from this test?

_____

_____

_____

_____

6. Connect the voltmeter across the ground of the pump. Activate the pump. Describe what happened. What are your conclusions from this?

_____

_____

_____

_____

7. What are your recommendations about the windshield washer system?

_____

_____

_____

_____

Instructor's Response _____

_____

_____

_____

Job Sheet 33

33

Name _____ Date _____

Testing a Rear Window Defogger

Upon completion of this job sheet, you should be able to diagnose inoperative and poorly working rear window defogger units.

ASE Correlation

This job sheet is related to the ASE Electrical/Electronic System Certification Exam's content area: *Accessories Diagnosis and Repair;* task: Diagnose the cause of poor, intermittent, or no operation of rear window defogger.

Tools and Materials

A vehicle with a rear window defogger
Wiring diagram for the chosen vehicle
A DMM

Describe the vehicle being worked on:

Year _____ Make _____ Model _____

VIN _____ Engine type and size _____

Procedure

Task Completed

1. From the wiring diagram, identify the power feed to the rear window defogger. With the circuit activated, check for voltage at the power terminal to the defogger unit. Was there battery voltage present? _____
   If no voltage was present, continue your testing at the control switch.
   If much less than battery voltage is present at the terminal, check for excessive resistance in the power circuit.

   ☐
   ☐

2. Identify the ground circuit for the defogger; measure the voltage drop across this part of the circuit. Your reading is _____. If more than 0.1 volt, check the wire, connectors, and ground connection for the cause of higher-than-normal resistance.

3. If all of the grids in the defogger circuit are good, applied voltage should be dropped across each grid. To test the condition of each grid, measure the voltage drop across each grid, starting from the top grid and moving down. List your findings below.

   _____

   _____

   _____

   _____

   _____

   _____

4. If one or more of the grids do not drop applied voltage, continue your diagnosis by moving the negative meter lead toward the positive side of the grid and watch the meter. When battery voltage is measured, this indicates there is an open after that point. The grid should be repaired.

5. On a working grid, connect the positive lead of the meter to the power feed terminal and move the negative to the following positions; record your measured voltage.

3/4 of the way across the grid _____

1/2 of the way across the grid _____

1/4 of the way across the grid _____

Describe what is happening.

_____

_____

_____

_____

6. Turn off the circuit, measure the resistance across the following points of the grid, and record your resistance readings.

3/4 of the way across the grid _____

1/2 of the way across the grid _____

1/4 of the way across the grid _____

Describe why you had these readings.

_____

_____

_____

_____

Instructor's Response _____

_____

_____

_____

DIAGNOSTIC CHART 9-1

PROBLEM AREA:	Horn system
SYMPTOMS:	Horn sounds even when switch is not activated
POSSIBLE CAUSES:	**1.** Faulty horn switch **2.** Horn control circuit shorted to ground **3.** Faulty horn relay

DIAGNOSTIC CHART 9-2

PROBLEM AREA:	Horn system
SYMPTOMS:	Intermittent horn operation
POSSIBLE CAUSES:	**1.** Faulty horn switch **2.** Poor clock spring contacts **3.** Faulty horn relay **4.** Poor ground connection at horn **5.** Poor ground connection at switch

DIAGNOSTIC CHART 9-3

PROBLEM AREA:	Horn system
SYMPTOMS:	Horn fails to sound when the horn switch is activated
POSSIBLE CAUSES:	**1.** Blown fuse **2.** Faulty horn switch **3.** Poor clock spring contacts **4.** Faulty horn relay **5.** Poor ground connection at horn **6.** Poor ground connection at switch

DIAGNOSTIC CHART 9-4

PROBLEM AREA:	Wiper system operation
SYMPTOMS:	Wipers operate any time the ignition switch is in the RUN position.
POSSIBLE CAUSES:	**1.** Faulty wiper switch **2.** Defective park switch or activation arm **3.** Shorted control circuit

DIAGNOSTIC CHART 9-5

PROBLEM AREA:	Wiper system operation
SYMPTOMS:	Wipers operate some of the time
POSSIBLE CAUSES:	1. Poor ground connection 2. Poor control circuit connection 3. Faulty switch 4. Worn motor contacts or brushes

DIAGNOSTIC CHART 9-6

PROBLEM AREA:	Wiper system operation
SYMPTOMS:	Wipers fail to function when switch is activated
POSSIBLE CAUSES:	1. Blown fuse 2. Open in the control circuit 3. Short in the control circuit 4. Faulty wiper motor 5. Poor ground connection 6. Mechanical linkage binding 7. Faulty wiper switch

DIAGNOSTIC CHART 9-7

PROBLEM AREA:	Wiper system operation
SYMPTOMS:	1. Wipers only operate on low speed 2. Wipers only operate on high speed
POSSIBLE CAUSES:	1. Faulty wiper switch 2. Worn brushes 3. Poor control circuit connections 4. Open in control circuit

DIAGNOSTIC CHART 9-8

PROBLEM AREA:	Wiper system operation
SYMPTOMS:	1. Wipers stop at on windshield when switch is turned off. 2. Wipers remain on when wiper switch is turned off
POSSIBLE CAUSES:	1. Faulty park switch 2. Activation arm broken or out of adjustment

DIAGNOSTIC CHART 9-9

PROBLEM AREA:	Wiper system operation
SYMPTOMS:	Wipers will not shut off
POSSIBLE CAUSES:	**1.** Faulty switch **2.** Faulty park switch or circuit

DIAGNOSTIC CHART 9-10

PROBLEM AREA:	Intermittent wiper system operation.
SYMPTOMS:	No intermittent wiper operation, low and high speed operate normally
POSSIBLE CAUSES:	Faulty switch

DIAGNOSTIC CHART 9-11

PROBLEM AREA:	Wiper motor operation
SYMPTOMS:	Slow or no wiper operation
POSSIBLE CAUSES:	Faulty wiper motor Binding wiper linkage

DIAGNOSTIC CHART 9-12

PROBLEM AREA:	Washer system operation
SYMPTOMS:	Washer operates without switch activation
POSSIBLE CAUSES:	**1.** Faulty switch **2.** Short in control circuit

DIAGNOSTIC CHART 9-13

PROBLEM AREA:	Washer system operation
SYMPTOMS:	Intermittent washers operation
POSSIBLE CAUSES:	**1.** Faulty switch **2.** Faulty pump **3.** Poor control circuit connections **4.** Poor ground connection

DIAGNOSTIC CHART 9-14

PROBLEM AREA:	Washer system operation
SYMPTOMS:	Windshield washer fails to operate
POSSIBLE CAUSES:	1. Blown fuse 2. Faulty switch 3. Faulty motor 4. Poor ground connection 5. Open or short in control circuit 6. Restriction in hoses

DIAGNOSTIC CHART 9-15

PROBLEM AREA:	Power side window operation
SYMPTOMS:	Power windows operate slower than normal
POSSIBLE CAUSES:	1. Control circuit resistance 2. Faulty motor 3. Poor ground connection 4. Improperly adjusted regulators 5. Binding linkages

DIAGNOSTIC CHART 9-16

PROBLEM AREA:	Power side window operation
SYMPTOMS:	Intermittent power window operation
POSSIBLE CAUSES:	1. Faulty switch 2. Faulty motor 3. Poor ground connections 4. Poor control circuit connections 5. Binding linkage 6. Faulty circuit breaker

DIAGNOSTIC CHART 9-17

PROBLEM AREA:	Power side window operation
SYMPTOMS:	Power windows fail to operate when switch is activated
POSSIBLE CAUSES:	1. Faulty switch 2. Faulty motor 3. Poor ground connections 4. Poor control circuit connections 5. Binding linkage 6. Faulty circuit breaker

DIAGNOSTIC CHART 9-18

PROBLEM AREA:	Power seat operation
SYMPTOMS:	Power seats operate slower than normal
POSSIBLE CAUSES:	**1.** Control circuit resistance **2.** Faulty motor **3.** Poor ground connection **4.** Binding linkages **5.** Faulty motor transmission

DIAGNOSTIC CHART 9-19

PROBLEM AREA:	Power seat operation
SYMPTOMS:	Intermittent power seat operation
POSSIBLE CAUSES:	**1.** Faulty switch **2.** Faulty motor **3.** Poor ground connections **4.** Poor control circuit connections **5.** Binding linkage **6.** Faulty circuit breaker **7.** Faulty transmission

DIAGNOSTIC CHART 9-20

PROBLEM AREA:	Power seat operation
SYMPTOMS:	Power windows fail to operate when switch is activated
POSSIBLE CAUSES:	**1.** Faulty switch **2.** Faulty motor **3.** Poor ground connections **4.** Poor control circuit connections **5.** Binding linkage **6.** Faulty circuit breaker **7.** Faulty transmission

DIAGNOSTIC CHART 9-21

PROBLEM AREA:	Rear-window defogger operation
SYMPTOMS:	Rear-window defogger fails to clear the window
POSSIBLE CAUSES:	**1.** Open in the grid **2.** Excessive circuit resistance

DIAGNOSTIC CHART 9-22

PROBLEM AREA:	Rear window defogger operation
SYMPTOMS:	Intermittent rear window defogger operation
POSSIBLE CAUSES:	**1.** Faulty switch **2.** Poor circuit control connection **3.** Poor ground connection **4.** Loose connection to grid

DIAGNOSTIC CHART 9-23

PROBLEM AREA:	Rear window defogger operation
SYMPTOMS:	Defogger fails to clear window
POSSIBLE CAUSES:	**1.** Faulty switch **2.** Blown fuse **3.** Open in control circuit **4.** Grid connection loose **5.** Broken grid bus **6.** Poor ground connection

DIAGNOSTIC CHART 9-24

PROBLEM AREA:	Power door lock system operation
SYMPTOMS:	Power door locks fail to operate some of the time
POSSIBLE CAUSES:	**1.** Faulty circuit breaker **2.** Faulty switch **3.** Poor control circuit connections **4.** Poor ground connections **5.** Faulty motor or solenoid **6.** Faulty relay

DIAGNOSTIC CHART 9-25

PROBLEM AREA:	Power door lock system operation
SYMPTOMS:	All door locks do not operate in either direction
POSSIBLE CAUSES:	**1.** Faulty circuit breaker **2.** Faulty master switch **3.** Open in control circuit **4.** Short in control circuit **5.** Poor ground connection **6.** Faulty relay **7.** Poor relay control circuit connections

DIAGNOSTIC CHART 9-26

PROBLEM AREA:	Power door lock system operation
SYMPTOMS:	One door lock does not operate in either direction
POSSIBLE CAUSES:	**1.** Obstruction or binding of linkage **2.** Open in control circuit **3.** Short in control circuit **4.** Poor ground connection **5.** Faulty motor or solenoid

DIAGNOSTIC CHART 9-27

PROBLEM AREA:	Power door lock system operation
SYMPTOMS:	All locks work from one switch only
POSSIBLE CAUSES:	**1.** Faulty switch **2.** Open in control circuit **3.** Short in control circuit

Body Computer System Diagnosis

Upon completion and review of this chapter, you should be able to:

❏ Describe the service precautions associated with servicing the BCM.

❏ Diagnose computer voltage supply and ground circuits.

❏ Distinguish between hard and intermittent codes.

❏ Perform flash code retrieval on various vehicles.

❏ Erase fault codes.

❏ Perform a complete visual inspection of the problem system.

❏ Enter BCM diagnostics by use of a scan tool.

❏ Enter BCM diagnostics through the ECC panel.

❏ Perform basic actuator tests.

❏ Perform basic sensor tests.

❏ Replace computer PROM chips properly.

❏ Properly flash the BCM

Basic Tools

Basic mechanic's tool kit

Service manual

Introduction

Because the body control module (BCM) controls many of the functions of the vehicle's electrical systems, it is important for today's technician to be able to properly diagnose problems with this system. The use of body computers has expanded to include the functions of climate control, lighting circuits, cruise control, antilock braking, electronic suspension systems, electronic shift transmissions, and alternator regulation. In some systems, the direction light, the rear window defogger, the illuminated entry, and the intermittent wiper systems are included in the body controller function (Figure 10-1).

As discussed in the Classroom Manual, a computer processes the physical conditions that represent information (data). The operation of the computer is divided into four basic functions:

1. *Input:* A voltage signal that is sent from an input device. This device can be a sensor or a button activated by the driver or technician.

2. *Processing:* The computer uses the input information and compares it to programmed instructions. The logic circuits process the input signals into output demands.

3. *Storage:* The program instructions are stored into an electronic memory. Some of the input signals are also stored for later processing.

4. *Output:* After the computer has processed the sensor input and checked its programmed instructions, it will put out control commands to various output devices. These output devices may be the instrument panel display or a system actuator. The output of one computer can also be used as an input to another computer.

Understanding these four computer functions will help you organize the troubleshooting process. When a system is tested, you are attempting to isolate a problem with one of these functions.

In the process of controlling the various electrical systems, the BCM continuously monitors operating conditions for possible system malfunctions. The computer compares system conditions against programmed parameters. If the conditions fall outside of these limits, the computer detects a malfunction. A **diagnostic and trouble code (DTC)** is set to indicate the portion of the system at fault. The technician can access this code for aid in troubleshooting.

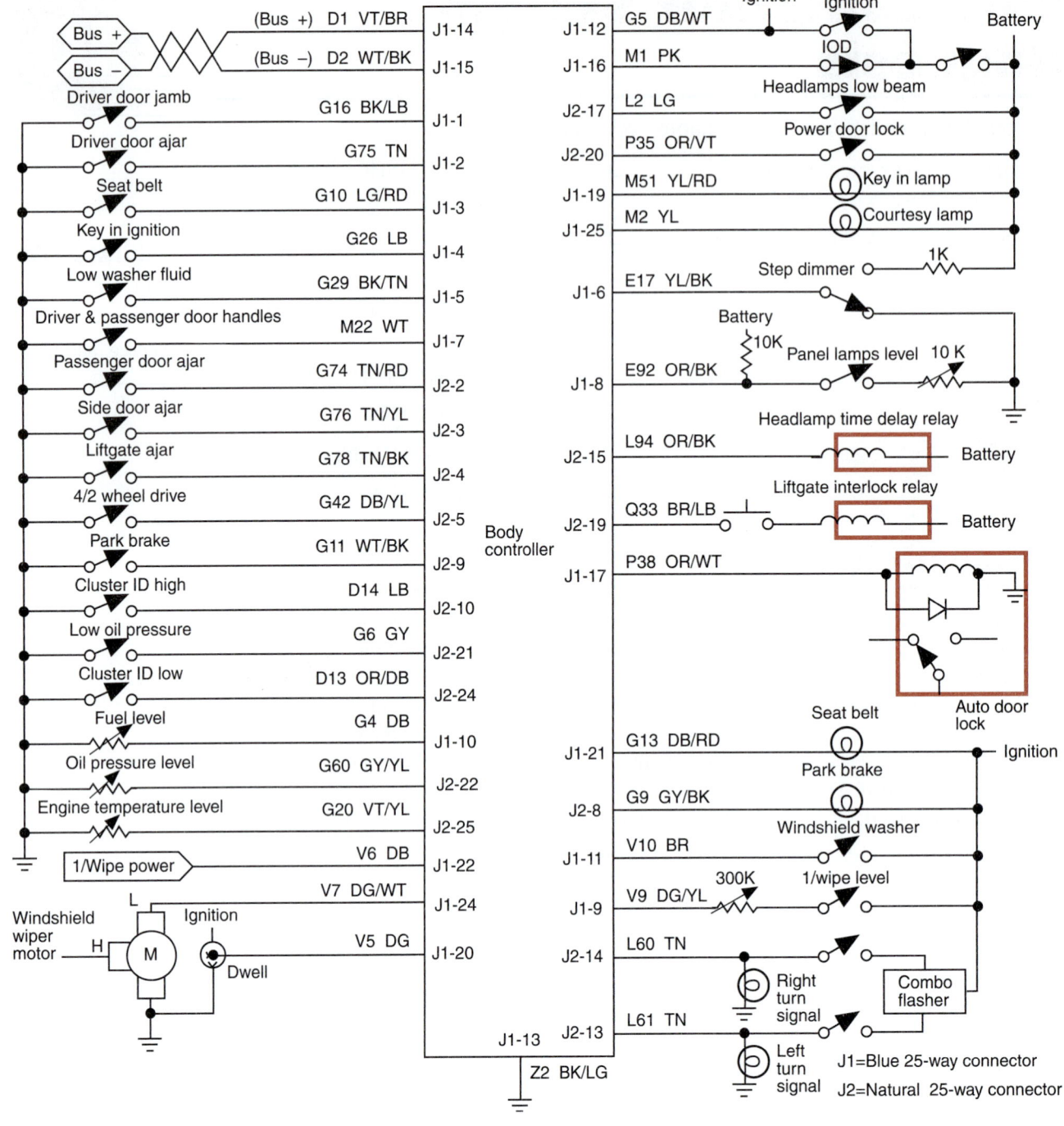

Figure 10-1 The body controller controls many of the vehicle's electrical systems.

If a malfunction results in improper system operation, the computer may minimize the effects by using **fail soft** action. This provides limited system operation by substituting a fixed input value if a sensor circuit should fail. For example, if the automatic temperature control system has a malfunction from the ambient temperature sensor, instead of shutting down the whole system, the computer will provide a fixed value as its own input. This fixed value can be programmed into the computer's memory, or it can be the last received signal from the sensor prior to failure. This allows the system to operate on a limited basis instead of shutting down completely. Some other faults may result in the automatic temperature control system switching to high fan speed, full heat, or defrost mode.

There are several things you need to know before you learn how to access the computer's memory to gain information concerning system operation. You need to become familiar with what you're looking at and you must follow proper precautions when servicing these systems.

Electronic Service Precautions

The technician must take some precautions before servicing the body computer or any of its controlled systems. The BCM is designed to withstand normal current draws associated with normal operation. However, overloading any of the system circuits will result in damage to the BCM. Follow these service precautions to prevent BCM and circuit damage:

1. Do not ground or apply voltage to any controlled circuits unless the service manual instructs you to do so.

2. Use only a high **impedance** multimeter (10 megohm or greater) to test the circuits. Impedance is the combined opposition to current created by the resistance, capacitance, and inductance of the meter. Never use a test light unless specifically instructed to do so in the service manual.

3. Make sure the ignition switch is turned off before making or breaking electrical connections to the BCM.

4. Unless instructed otherwise in the service manual, turn off the ignition switch before making or breaking any electrical connections to sensors or actuators.

5. Turn the ignition switch off whenever disconnecting or connecting the battery terminals. Also turn it off when pulling and replacing the fuse.

6. Do not connect any other electrical accessories to the insulated or ground circuits of the computer-controlled systems.

7. Use only manufacturer's specific test and replacement procedures for the year and model of vehicle being serviced.

8. Wear an **electrostatic discharge (ESD) strap** to ground your body to prevent static discharges that may damage electronic components.

Static electricity can be 25,000 volts or higher.

By following these precautions, plus those listed in the service manual, you can avoid having to replace expensive components.

Electrostatic Discharge

Some manufacturers mark certain components and circuits with a code or symbol to warn technicians that the units are sensitive to electrostatic discharge (Figure 10-2). Static electricity can destroy or render a component useless.

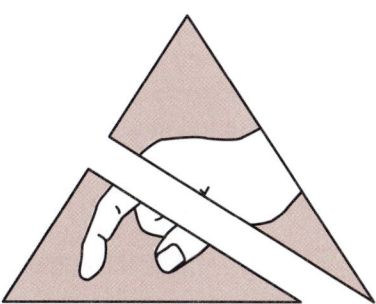

Figure 10-2 One type of electrostatic discharge (ESD) symbol used to warn the technician that the component or circuit is sensitive to static.

When handling any electronic part, especially those that are static sensitive, follow the guidelines below to reduce the possibility of electrostatic buildup on your body and the inadvertent discharge to the electronic part. If you are not sure if a part is sensitive to static, treat it as if it is.

1. Always touch a known good ground before handling the part. This should be repeated while handling the part and more frequently after sliding across a seat, sitting down from a standing position, or walking a distance.

2. Avoid touching the electrical terminals of the part unless you are instructed to do so in the written service procedures. It is good practice to keep your fingers off all electrical terminals as the oil from your skin can cause corrosion.

3. When you are using a voltmeter, always connect the negative meter lead first.

4. Do not remove a part from its protective package until it is time to install the part.

5. Before removing the part from its package, ground yourself and the package to a known good ground on the vehicle.

6. When replacing a PROM, ground your body by putting on an ESD strap and connect the other end to a good ground.

Electromagnetic Interference

Electromagnetic interference (EMI) or radio frequency interference (RFI) can cause problems with the vehicle's on-board computers. Unfortunately, an automobile's spark plug wires, ignition coil, and generator all possess the ability to generate these radio waves. Under the right conditions, RFI can trigger sensors or actuators. The result may be an intermittent drivability problem with system operation.

To minimize the effects of RFI, make sure your visual inspection is thorough. Also check to make sure that sensor wires running to the computer are routed away from potential RFI sources. Rerouting a wire by no more than an inch or two (25 to 50 mm) may keep RFI from falsely triggering or interfering with computer operation. RFI can be present on a voltage signal or on a ground.

Most manufacturers shield their BCM from EMI and RFI. However, this shielding will only work if the BCM is properly grounded. Always confirm a good ground before condemning the BCM. Some BCMs may have up to five grounds.

Classroom Manual
Chapter 10, page 298

Special Tools

DVOM

Lab scope

Diagnosis of Computer Voltage Supply and Ground Circuits

✔ **SERVICE TIP:** Never replace a computer unless the ground wires and voltage supply wires are proven to be in satisfactory condition.

Like any other electrical or electronic component, a computer cannot operate properly unless it has satisfactory voltage supply at the required terminals and proper ground connections. A computer wiring diagram for the vehicle being tested must be available for these tests. First, measure the voltage across the battery terminals as a reference. Now, backprobe the battery feed terminal to the computer and connect the DVOM leads to this terminal and to ground. The voltage at the battery feed terminal should be 0.5 volt of battery voltage with the ignition switch off. If the proper voltage is not present at this terminal, check the computer fuse and related circuit. Turn the ignition switch to the RUN position and check applied voltage to *all* battery and ignition feed terminals of the PCM. The voltage measured at these terminals should be within 0.5 volt of battery voltage. If the specified voltage is not available, test the voltage supply wires to these terminals. These terminals may be connected through fuses, fuse links, or relays. Always refer to the vehicle manufacturer's wiring diagram for the vehicle being tested.

SERVICE TIP: When diagnosing computer problems, it is usually helpful to ask the customer about service work that has been performed lately on the vehicle. If service work has been performed, it is possible that a computer ground wire may be loose or disconnected.

Computer ground wires usually extend from the computer to a ground connection on the engine or battery. Often there is more than one ground wire. In addition, the fasteners used to attach the computer to the vehicle chassis may be used as a ground. With the ignition switch in the RUN position, perform a voltage drop test across the ground wires. Compare your results with specifications. A good circuit should not drop more than 0.2 volt on the ground circuit.

Sensor Return Circuit

Many manufacturers use a single sensor return circuit for all of the BCM sensors. This makes a good ground critical since a sensor return circuit that has high resistance will cause *all* of the sensors to indicate an erroneous voltage reading to the PCM. Consider a temperature-sensing circuit as an example (Figure 10-3). (A) indicates the normal voltage sensed by the PCM with the thermistor at 10K ohms. (B) indicates the voltage sensed by the PCM with the thermistor still at 10K ohms but with an additional 200 ohms of resistance in the sensor's return circuit. The voltage is

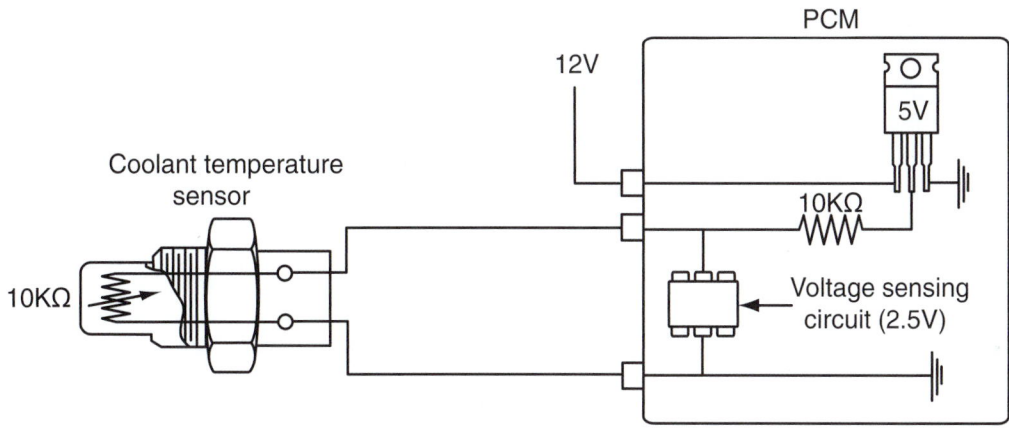

(A) NORMAL SENSOR RETURN CIRCUIT

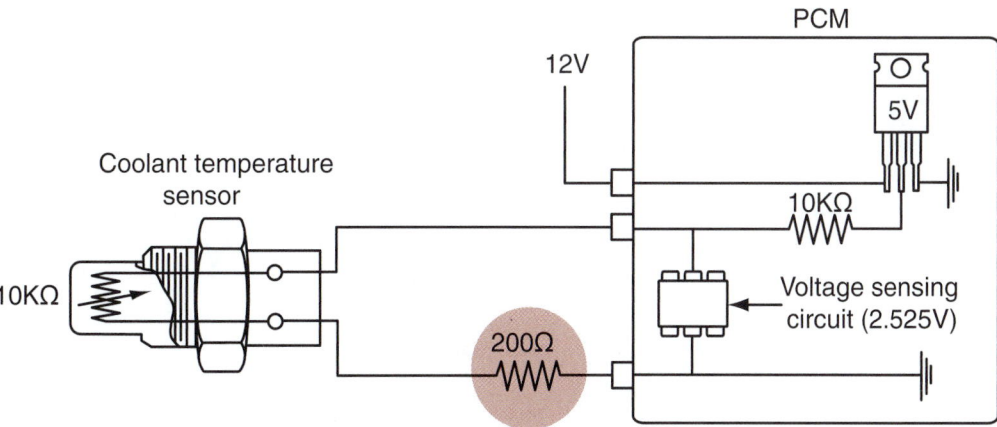

(B) RESISTANCE IN SENSOR RETURN CIRCUIT

Figure 10-3 The effect of resistance on the sensor return circuit. (A) Normal sensor voltage. (B) Voltage reading with ETC at same resistance, but sensor return circuit has additional resistance.

now off by 25 millivolts. Since the 200 ohms of resistance is also affecting all of the other sensors that share this return circuit, they are all off by 25 millivolts also. This may not seem like enough voltage to cause any problems, but remember that most systems are very sensitive to small voltage changes.

In addition to causing some system operation problems, this type of fault may not set a fault code. The voltage values are within the normal operating parameters of the sensor, so continuity faults will not set. Also, since all sensors are off by the same amount, rationality faults are not detected. System operation problems may occur as a result of the computer making bad decisions based on erroneous information.

Poor grounds can also allow electromagnetic interference (EMI) or noise to be present on the reference voltage signal. This noise causes minute changes in the voltage going to the sensor; therefore, the output signal from the sensor will also have these voltage changes. The computer will try to respond to these changes, which can cause a system operation problem. The best way to check for noise is to use a lab scope.

Connect the lab scope between the 5-volt reference signal to the sensor and ground. The trace on the scope should be flat (Figure 10-4). If noise is present, move the scope's negative probe to a known good ground. If the noise disappears, the sensor's ground circuit is faulty or has resistance. If the noise is still present, the voltage feed circuit is faulty or there is EMI in the circuit from another source, such as the AC generator. Find and repair the cause of the noise.

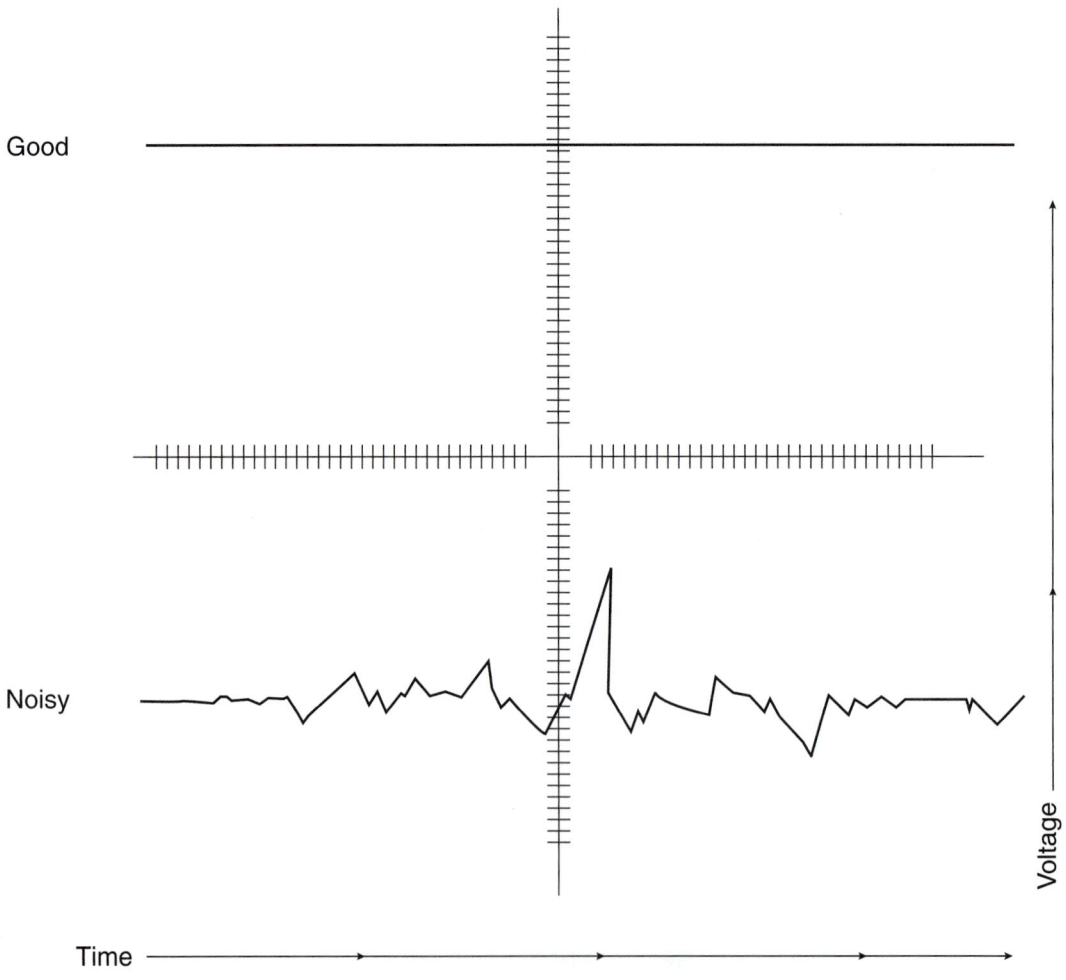

Figure 10-4 (Top) Voltage signal trace of a good circuit. (Bottom) Trace of a circuit with excess noise.

Circuit noise may be present in either the positive side or the negative side of a circuit. It may also be evident by a popping noise from the radio. However, noise can cause a variety of problems in any electrical circuit. The most common sources of noise are electric motors, relays, solenoids, AC generators, ignition systems, switches, and A/C compressor clutches. Typically, noise is the result of an electrical device being turned on and off. Sometimes the source of the noise is a defective suppression device. Some of the commonly used noise suppression devices are resistor-type secondary cables and spark plugs, shielded cables, capacitors, diodes, and resistors. If the source of the noise is not a poor ground or a defective component, check the suppression devices.

Resistors that are used for noise suppression do not eliminate the spikes, but they do limit their intensity. When testing a circuit that uses a resistor to limit noise, if the lab scope trace indicates large voltage spikes, the resistor may be bad. Clamping diodes are used to suppress noise and induced voltages in circuits such as A/C compressor clutches and relays. If the diode is bad, a negative spike will result (Figure 10-5). Capacitors or chokes are used to control noise from motors and generators (Figure 10-6). To avoid much frustration during diagnosis of a computer system (especially the inputs), check the integrity of the ground as one of the first steps.

SERVICE TIP: In light of the total number of vehicles being produced that use computer control systems, very few customer complaints are actually the fault of the computer. The computer should be replaced only if *all* other possible causes have been checked and confirmed as operating properly. Always check the condition of the battery feed and ground circuits of the computer before condemning the computer.

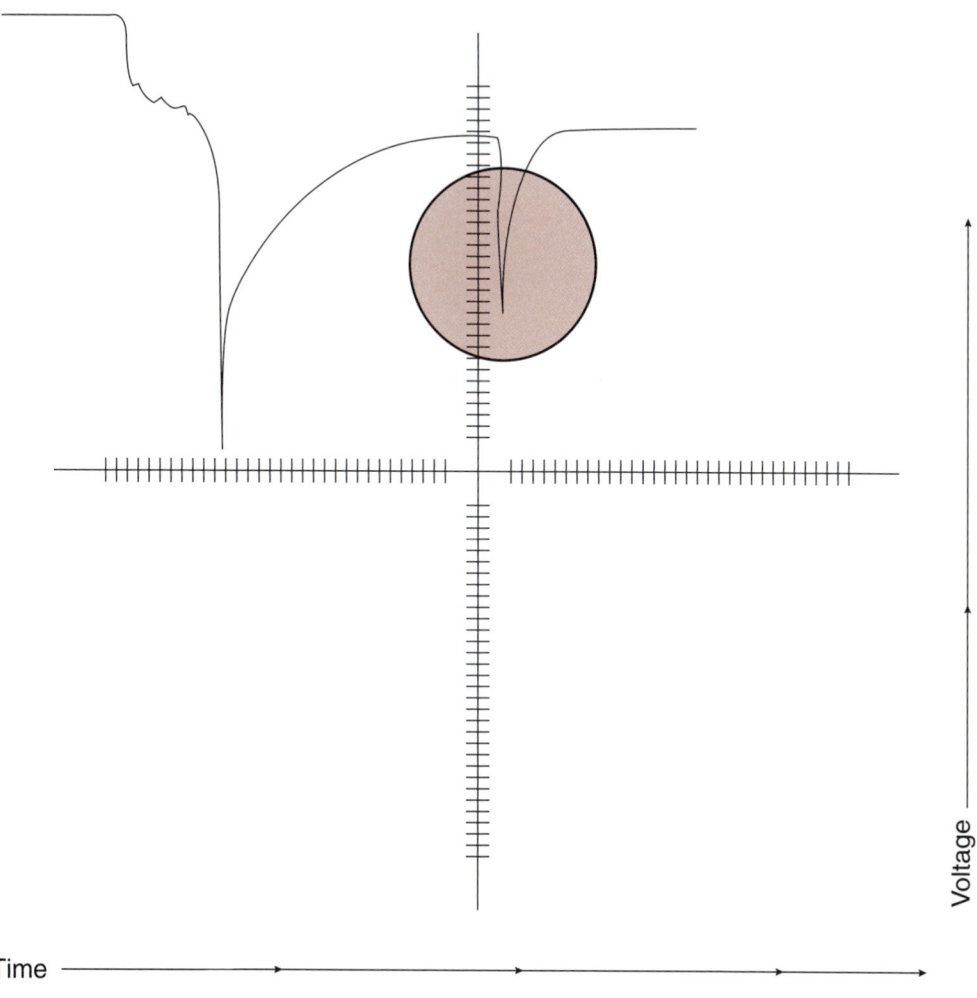

Figure 10-5 A voltage trace of an A/C circuit with a bad diode.

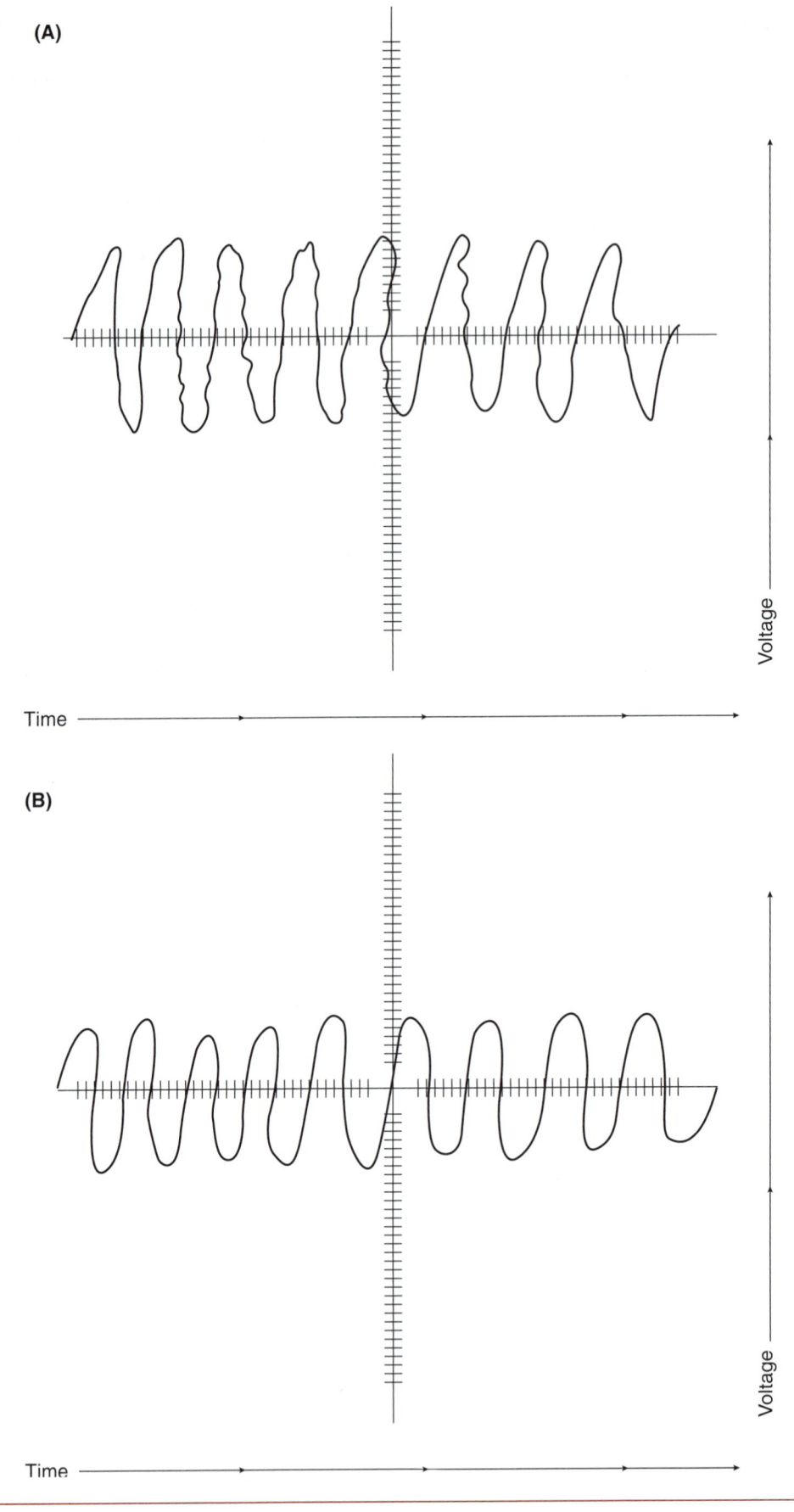

Figure 10-6 (A) Voltage trace of motor without a chock. (B) Voltage trace of a motor with a chock.

Trouble Codes

Most BCMs are capable of displaying the stored faults in memory. Fault codes can be displayed by the scan tool and, in early systems, by a method referred to as **flash codes.** Flash codes are DTCs displayed by flashing a lamp or LED. The method used to retrieve the codes varies greatly, and the technician must refer to the correct service manual for the procedure. Depending on system design, the computer may store codes for long periods of time; some lose the code when the ignition switch is turned off. Systems that do not retain the code when the ignition is turned off require that the technician test drive the vehicle and attempt to duplicate the fault. Once the fault is detected by the computer, the code must be retrieved before the ignition switch is turned off again.

The trouble code does not necessarily indicate the faulty component; it only indicates that circuit of the system that is not operating properly. For example, the code displayed may be F11, indicating an A/C high side temperature sensor problem. This does not mean the sensor is bad; the fault is in that circuit, which includes the wiring, connections, sensor, and BCM. To locate the problem, follow the diagnostic procedure in the service manual for the code received.

Differences Between Hard and Intermittent Codes

Some BCMs will store trouble codes in their memory until they are erased by the technician or until a set amount of engine starts have passed. Some computers will display two sets of fault codes. Usually, the first set of codes displayed represents all codes stored in memory, including both hard and intermittent codes. The second set of codes displayed are only **hard codes.** Hard codes are failures that were detected the last time the BCM tested the circuit. The codes displayed in the first set but not in the second set are **intermittent codes.** Intermittent codes are those that have occurred in the past but were not present during the last BCM test of the circuit.

Most diagnostic charts cannot be used to locate causes of intermittent codes. This is because the testing at various points of the chart depends on the fault being present to locate the problem. If the fault is not present, the technician may be erroneously instructed to replace the BCM module, even though it is not defective.

Many intermittent problems are the result of poor electrical connections. Diagnosis should start with a good visual inspection of the connectors involved with the code. Even on hard codes, visually inspect the circuit before conducting any other tests.

Visual Inspection

Perhaps the most important check to be made before diagnosing a BCM-controlled system is a complete visual inspection. The visual inspection can identify faults that could cause the technician to spend wasted time in diagnostics. In addition, the problem can be pinpointed without any further steps.

Inspect the following:

1. All sensors and actuators for physical damage.
2. Electrical connections into sensors, actuators, and control modules.
3. All ground connections.
4. Wiring for signs of burned or chafed spots, pinched wires, or contact with sharp edges or hot exhaust manifolds.
5. All vacuum hoses for pinches, cuts, or disconnects.

The time spent performing a visual inspection is worthwhile. Put forth the effort to check wires and hoses that are hidden under other components.

Entering Diagnostics

There are as many methods of entering BCM diagnostics as there are vehicle manufacturers. Most require a scan tool. A scan tool is a microprocessor designed to communicate with the BCM. It will access trouble codes and run system operation, actuator, and sensor tests (Figure 10-7). The scan tool is plugged into the diagnostic link connector (DLC) for the system being tested. Some manufacturers provide a single DLC and the technician chooses the system to be tested through the scan tool. Always refer to the correct service manual for the vehicle being serviced. Use only the methods identified in the service manual for retrieving trouble codes. Once the trouble codes are retrieved, consult the appropriate diagnostic chart for instructions on isolating the fault. It is also important to check the codes in the order the manufacturer requires.

Using a Scanner

Special Tools

Scan tool

Service manual

Do not touch the pad of the scanner during the power-up sequence.

CAUTION: The following is given as a guide and is intended to complement the service manual. Improper methods of code retrieval may result in damage to the computer.

Connecting the scan tool into the DLC will access the technician to information concerning the operation of most vehicle systems. Some scanners require the use of adapters and cartridges. Follow Photo Sequence 16 as a typical method used to enter body controller diagnostics.

When the technician has programmed the scan tester by performing the initial entries, some entry options appear on the screen. These entry options vary depending on the scan tester and the vehicle being tested. The technician presses the number beside the desired selection to proceed with the test procedure. In the first four selections, the tester is asking the technician to select the computer system to be tested. If "data line" is selected, the scan tester provides a voltage reading from each input sensor in the system.

When the technician makes a selection from the initial test selection menu, the scan tester moves on to the actual test selections. These selections vary depending on the scan tester and the vehicle being tested.

Since the scan tool provides the voltage value that the BCM is sensing from an input circuit, diagnosis of a circuit fault is made simpler. For example, if you are monitoring the input voltage of a two-wire input sensor such as the ambient temperature sensor and the scan tool indicates 5.0 volts on the circuit, this indicates the circuit has an open. The open could be anywhere in the circuit between the BCM and sensor ground (including the sensor). If the monitored voltage shows 0 volts,

Figure 10-7 A scan tester is used to enter the computer's diagnostic capabilities.

Photo Sequence 16
Typical Procedure for Scan Tester Diagnosis

P16-1 Turn the ignition switch to the OFF position.

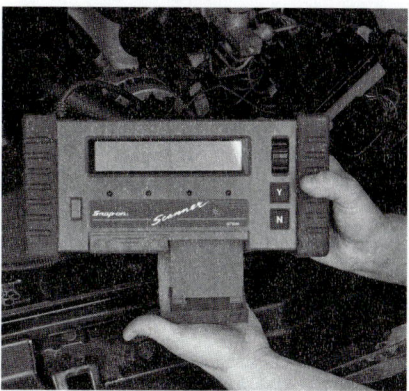

P16-2 Install the proper module into the scan tool.

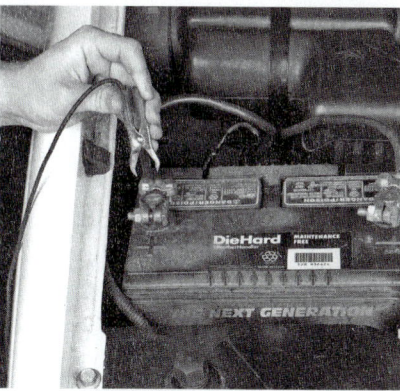

P16-3 Connect the scan tester power cables to the vehicle's battery terminals, observing polarity.

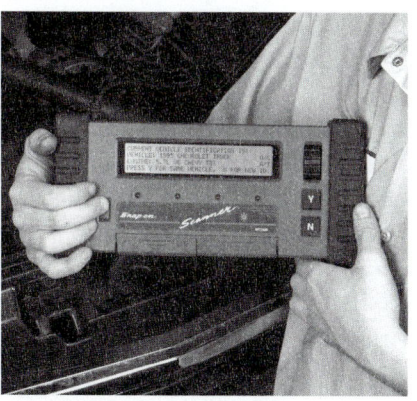

P16-4 Enter the model year and the VIN code in the scan tester.

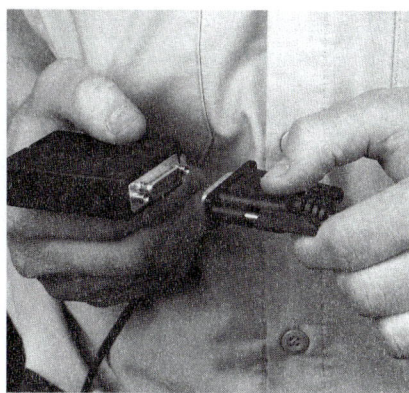

P16-5 Select the proper scan tool data cable for the vehicle being tested.

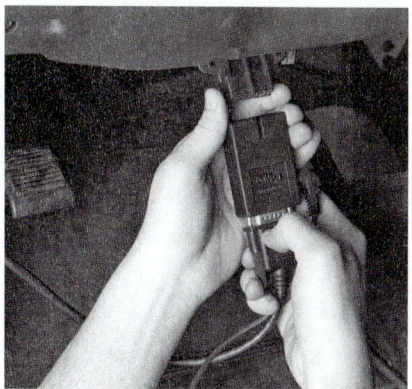

P16-6 Connect the scan tester cable to the DLC and turn the ignition switch to RUN.

P16-7 Select the BODY function and retrieve the DTCs with the scan tester.

P16-8 Operate the system being diagnosed and obtain the input sensor and output actuator data with the scan tool.

P16-9 Compare the input sensor and output actuator data to specifications in the appropriate service manual. Determine data that is not written specifications.

this usually indicates a short to ground. The short could be the sensor or in the circuit between the BCM and the sensor. A reading of 0 volts may also be caused by a faulty BCM not sending the reference voltage to the circuit.

Cadillac BCM Trouble Code Retrieval

CAUTION: The procedures may change between models, years, and the type of instrument cluster installed. Refer to the correct service manual for the vehicle being diagnosed.

Many manufacturers will provide a method of retrieving diagnostic fault codes without using a scanner. The methods used to display DTCs varies greatly between manufacturers and often between different lines of the same manufacturer. Always refer to the proper service information to determine the correct procedure to retrieve DTCs. Also, beginning in 1996, many manufactures have discontinued the practice of displaying DTCs without a scan tool. Make sure that the vehicle you are working on supports this option. The following is an example of retrieving fault codes without a scan tool.

Many Cadillac vehicles allow access to trouble codes, and other system operation information, through the electronic climate control panel (ECC). The BCM and ECM share information with each other. Thus both ECM and BCM codes are retrieved through the ECC. To enter diagnostics, begin by placing the ignition switch in the RUN position. Depress the OFF and WARMER buttons on the ECC panel simultaneously (Figure 10-8). Hold the buttons until all display segments are illuminated.

Cadillac uses the on-board ECC panel to display trouble codes, whereas other GM vehicles use a Tech I or II scan tool. Starting in 1996, the Tech II scan tool is used to retrieve codes on certain models. In 1996, Cadillac switched to the use of the Tech I scan tool to retrieve class 2 data. All General Motors vehicles produced in 1997 or later use the Tech II scanner. When diagnosing GM systems, make sure to follow the procedures designated by GM.

CAUTION: Diagnosis should not be attempted if all segments do not illuminate. It is possible to misdiagnose a problem as a result of not receiving the correct code. For example, if two segments of the display fail to illuminate, a code 24 could look like code 21.

When the segment check is completed, the computer will display any trouble codes in its memory. An "8.8.8" will be displayed for about 1 second. Then an "..E" will appear. This signals the beginning of engine controller trouble codes. The display will show all engine controller trouble codes beginning with the lowest number and progressing through the higher numbers. All codes associated with the engine controller will be prefixed with an "E." If there are no codes, the "..E" will not be displayed. The "E" codes will be displayed twice. The first set of codes are all in memory, both hard and intermittent. The second set will be only hard codes. An ".E.E" is displayed to separate the two sets of codes.

In some GM systems, the body codes are prefixed with a "B."

If there were no trouble codes in memory, the ".7.0" will be displayed after the segment illumination test.

Once all "E" codes are displayed, the computer will display BCM codes. The BCM codes are usually prefixed by an "F." An "..F" will precede the first set of codes displayed. The first set will be all codes stored in memory since the last 100 engine starts. An ".F.F" will appear to signal the separation of the first pass and the second. The second set of codes will be all hard codes.

When all codes are displayed, ".7.0" will be displayed. This indicates the system is ready for the next diagnostic feature to be selected. To erase the BCM trouble codes, press the OFF and LOW buttons together until "F.0.0" appears. Release the buttons and ".7.0" will reappear. Turn off the ignition switch and wait at least 10 seconds before reentering the diagnostic mode.

When in the diagnostic mode, it is possible to exit the system without erasing the trouble codes. Press AUTO on the ECC panel and the temperature will reappear in the display panel. This exits the diagnostic mode.

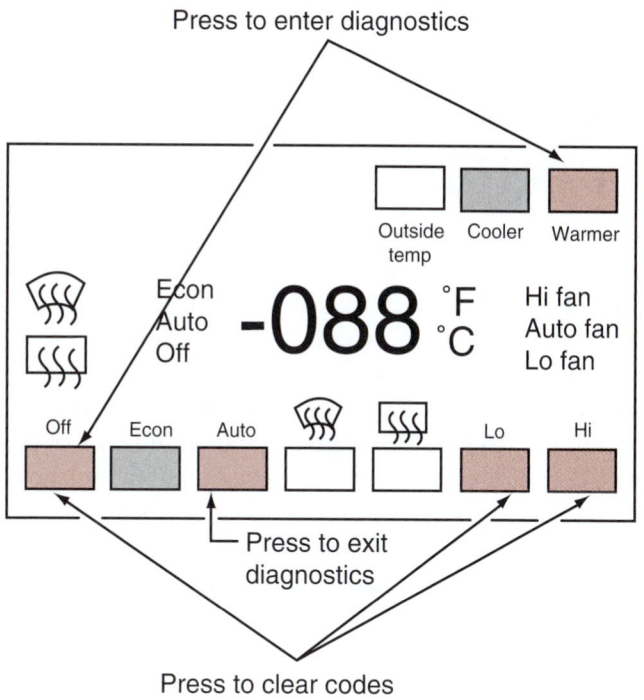

Press to enter diagnostics

Outside temp · Cooler · Warmer

Econ
Auto
Off

-088 °F °C

Hi fan
Auto fan
Lo fan

Off · Econ · Auto · Lo · Hi

Press to exit diagnostics

Press to clear codes

Snapshot: Econ and Cooler

Snapshot review: Econ and Warmer

Increment: Hi

Decrement: Lo

Figure 10-8 The buttons on the ECC panel allow the technician to access information from the computer when in the diagnostic mode.

Testing Actuators

Testing of actuators is included here to orient you to the basic procedures. Specific procedures will be presented throughout this manual for individual systems and types of actuators.

Most computer-controlled actuators are electromechanical devices that convert the output commands from the computer into mechanical action. These actuators are used to open and close switches, control vacuum flow to other components, and operate doors or valves, depending on the requirements of the system.

Most systems allow for testing of the actuator through the scan tool or ECC panel while in the correct mode. Actuators that are duty cycled by the computer are more accurately diagnosed through this method. This will allow the technician to activate selected actuators to test their operation.

 **WARNING:** When performing actuation tests on systems such as power windows and power seats, be sure that you not are in a position that will allow you to be pinched or caught by moving components.

In the General Motors system, the technician can access selected actuator operation. At the .7.0 display (before erasing codes), press the OUTSIDE TEMP button. The display should switch to F.8.0. Use the HIGH and LOW buttons to scroll through the parameters. Use the parameter chart from the service manual. For example, if parameter P.2.3 is selected, the actual air blend door position is displayed in percentage.

If the actuator must be tested by other means than the scanner, follow the manufacturer's procedures very carefully. Because some of the actuators used by the BCM operate with 5 to 7 volts, do not connect a jumper from a 12-volt source unless the service manual directs this. Some

Classroom Manual
Chapter 10, page 315

Special Tools

Scan tool

DMM

Service manual

actuators are easily tested with a voltmeter by testing for input voltage to the actuator. If there is input voltage of the correct level, check for a good ground connection. If both of these are good, then the actuator is faulty. If an ohmmeter needs to be used to measure the resistance of an actuator, disconnect the actuator from the circuit first.

Special Tools

Lab scope
Service manual

Testing Actuators with a Lab Scope

Since most actuators are electromechanical devices, when they fail it is because they are electrically faulty or mechanically faulty. By observing the action of an actuator on a lab scope, you will be able to watch an actuator's electrical activity. Normally, if there is a mechanical fault, it will affect the activator's electrical activity as well. Therefore, you get a good sense of the actuator's condition by watching it on a lab scope.

To test an actuator, you need to know what type it is. Most actuators are solenoids. The computer controls the action of the solenoid by controlling the pulse width of the control signal. By watching the control signal, you can see the turning on and off of the solenoid (Figure 10-9). The voltage spikes are caused by the discharge of the coil in the solenoid.

Some actuators are controlled pulse-width modulated signals (Figure 10-10). These signals show a changing pulse width. These devices are controlled by varying the pulse width, signal frequency, and voltage levels.

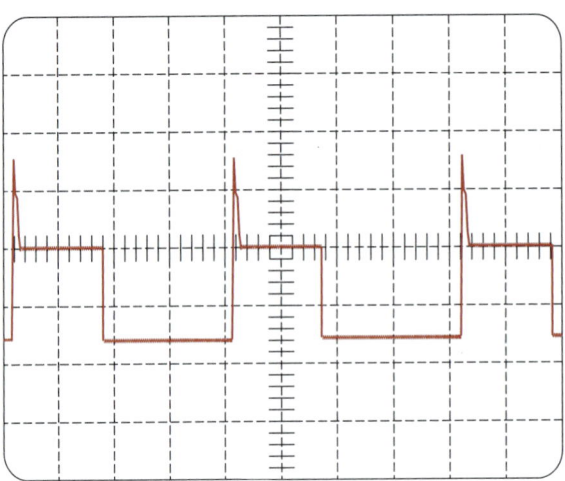

Figure 10-9 A typical solenoid control signal.

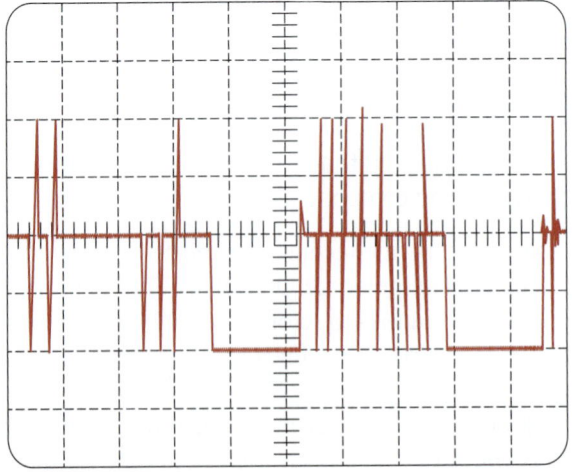

Figure 10-10 A typical pulse-width modulated solenoid control signal.

Both waveforms should be checked for amplitude, time, and shape. You should also observe changes to the pulse width as operating conditions change. A bad waveform will have noise, glitches, or rounded corners. You should be able to see evidence that the actuator immediately turns on and off according to the commands of the computer.

Testing Sensors

Testing of sensors is included here to orient you to the basic procedures. Specific procedures will be presented throughout this manual for individual systems and types of sensors.

There are many different designs of sensors, depending on the operation of the system they monitor. Some sensors are nothing more than a switch that completes the circuit. Others are complex chemical reaction devices that generate their own voltage under different conditions.

Thermistor and Potentiometer Testing

Thermistors are used to sense engine coolant or ambient temperatures. Potentiometers are used to measure linear or rotary movement. These units are tested by measuring the input voltage to the sensor and the feedback voltage to the computer. The feedback voltage to the computer should change smoothly as the resistance value of the sensor changes. To test these voltage signals, a series of jumper wires may be required (Figure 10-11). The jumper wires provide a method of gaining access to the terminals of weather-pack connectors without breaking the wire insulation.

 CAUTION: Do not use a test light to test for voltage. This may damage the system. Also, do not probe for voltage by sticking the wire insulation.

Classroom Manual
Chapter 10, page 307

Special Tools
DMM
Lab scope
Jumper wires
Service manual

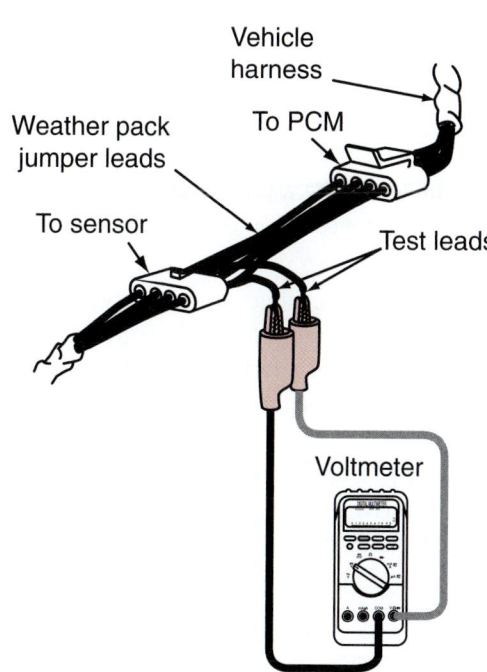

Figure 10-11 Jumper wires connected between the sensor and the harness allow the technician to probe for voltage or test resistance without damaging the wiring.

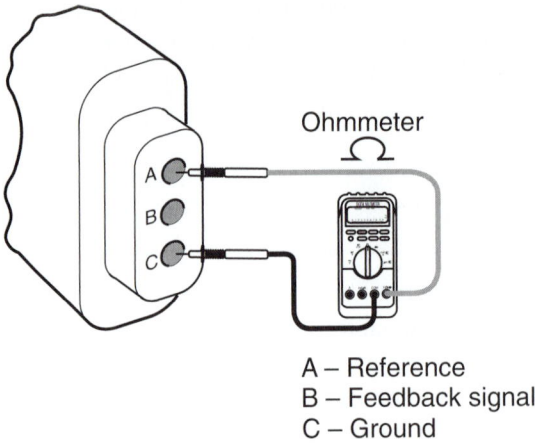

Figure 10-12 Connecting an ohmmeter to test the potentiometer. This will give the fixed resistance value.

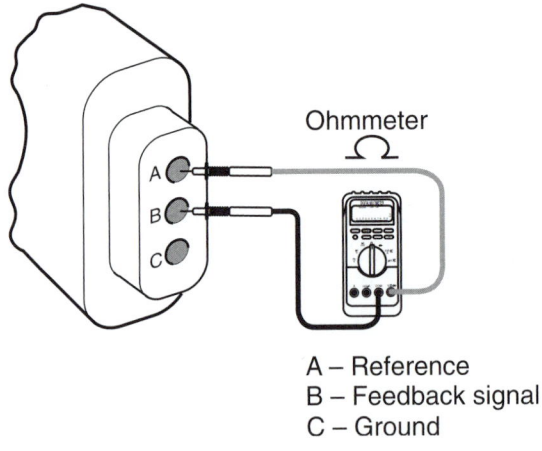

Figure 10-13 Ohmmeter connection to test the wiper movement. As the wiper is moved from one end of its travel to the other, the resistance reading should change smoothly.

Classroom Manual
Chapter 10, page 310

Special Tools

DMM

Lab scope

Service manual

Many vehicle speed sensors (VSSs), wheel-speed sensors, and ignition pickup units are magnetic pulse generators.

An ohmmeter can be used to measure the changes in resistor values. Disconnect the sensor from the system. Connect the ohmmeter leads to the reference and ground terminals (Figure 10-12). Check the results against specifications. If good, connect the leads between the reference terminal and the feedback terminal (Figure 10-13). The resistance should change smoothly and consistently as the wiper position is changed.

Potentiometer and thermistor testing with a lab scope is a good way to watch the sweep of the resistor. The waveform (Figure 10-14) is a DC signal that moves up as the voltage increases. Most potentiometers in computer systems are fed a reference voltage of 5 volts. Therefore, the voltage output of these sensors will range from zero to 5 volts. The change in voltage should be smooth. Look for glitches in the signal. These can be caused by changes in resistance or an open.

Testing Magnetic Pulse Generators

Magnetic pulse generators are commonly used to send data to the computer concerning the speed of the monitored component. This data provides information concerning vehicle speed and individual wheel speed. The signal from the speed sensors is used for computer-driven instrumentation, cruise control, antilock braking, speed sensitive steering, and automatic ride control systems.

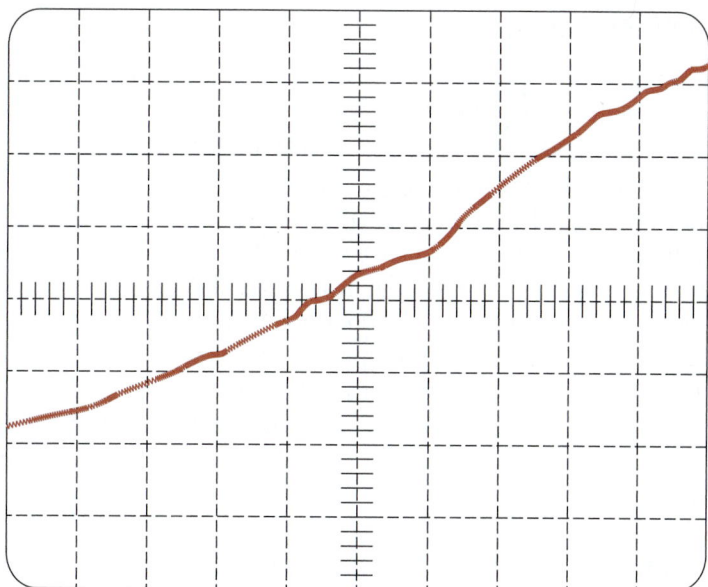

Figure 10-14 A DC signal for a good throttle position sensor, measured from closed throttle to wide-open throttle.

To test the magnetic pulse generator, disconnect it from the system. Use an ohmmeter to test the resistance value of the coil and compare the results with specifications. The voltage generation of the sensor can be tested by connecting a voltmeter across the sensor terminals. The voltmeter must be in the AC position and on the lowest scale. Rotate the shaft while observing the voltage signal. It should increase and decrease with changes in shaft speed.

Magnetic pulse generators can also be tested with a lab scope. Instead of connecting a voltmeter across the sensor's terminals, connect the lab scope leads. The expected pattern is an AC signal that should be a perfect sine wave when the speed is constant. When the speed is changing, the AC signal should change in amplitude and frequency as shown in Figure 10-15.

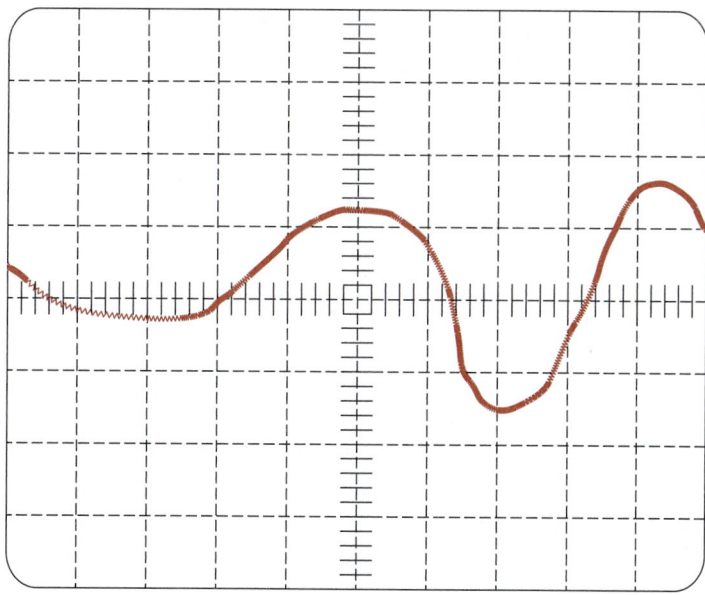

Figure 10-15 An AC voltage signal from a magnetic pulse generator used as a wheel speed sensor. The signal is changing over time because the speed of the wheel is accelerating quickly.

Special Tools

DMM

Lab scope

Feeler gauge set

Jumper wires

Service manual

Testing Hall-Effect Sensors

The best way to test Hall-effect inputs is to use a scope. A DVOM can be used to confirm that the proper voltages and grounds are supplied to the sensor, but it will not be able to indicate the quality of the signal. With a lab scope, the unit can be checked while the monitored component is operating. Connect the positive lead of the scope to the signal wire by back probing the connector. The connector must be plugged into the Hall-effect switch. With the component operating, or being rotated, the trace should show a clean 5-volt or 12-volt square wave pattern (based on design) that increases in frequency as shaft rpm increases and decreases (Figure 10-16). The voltage value should go to a full 5 or 12 volts. If it does not, then there is resistance on the signal circuit between the Hall-effect switch and the computer. The voltage should also return to 0 volts. If it does not, then there is resistance on the ground circuit. In addition, a sloping rising and falling line may indicate that the tone wheel is too far away from the magnet of the Hall-effect switch or that the transistor is faulty (Figure 10-17). Check the trace for glitches and noise that may be the result of RFI or EMI (Figure 10-18).

SERVICE TIP: In many systems that use a Hall-effect switch, it is possible to test the switch by use of a scratch test. For example, if the fuel/ignition system uses a Hall-effect crankshaft sensor, disconnect the sensor connector and use a terminal probe and a jumper wire to momentarily ground the signal wire back to the PCM. If the circuit is good, the PCM should respond by activating the fuel pump and/or the shutdown relays. If the PCM responds now, the problem is the Hall-effect switch.

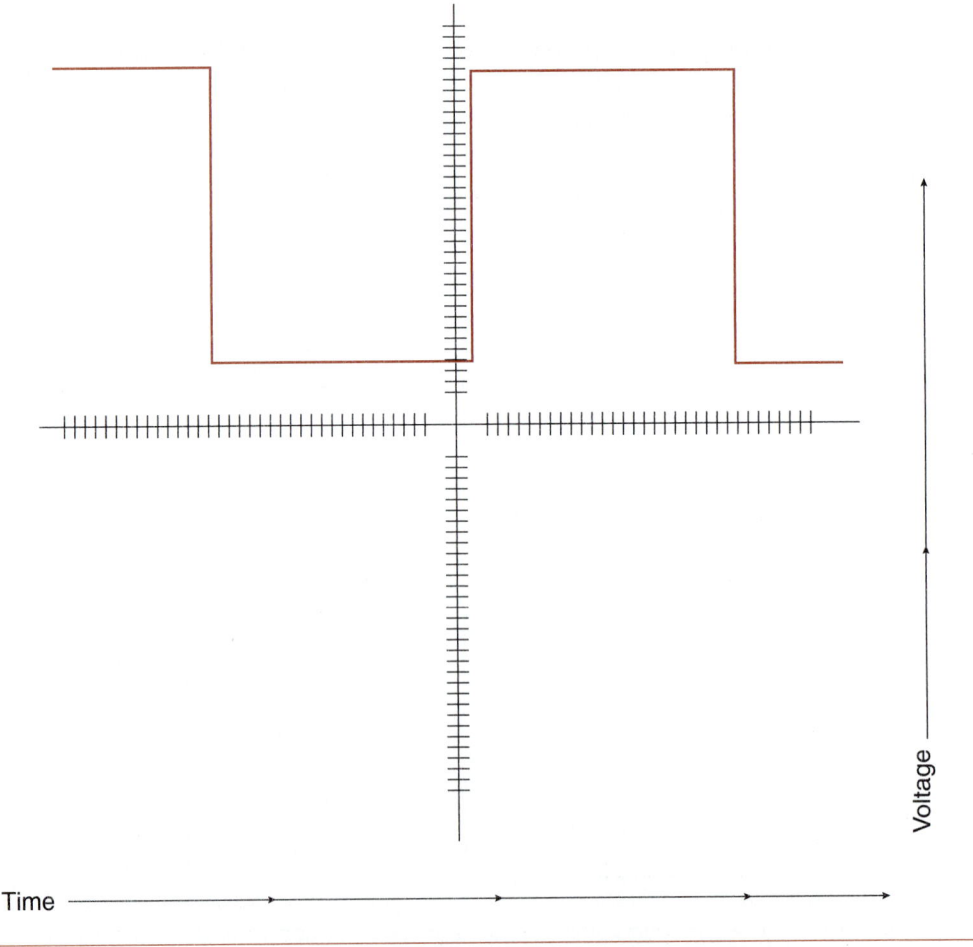

Figure 10-16 A good Hall-effect switch signal should be a square wave pattern.

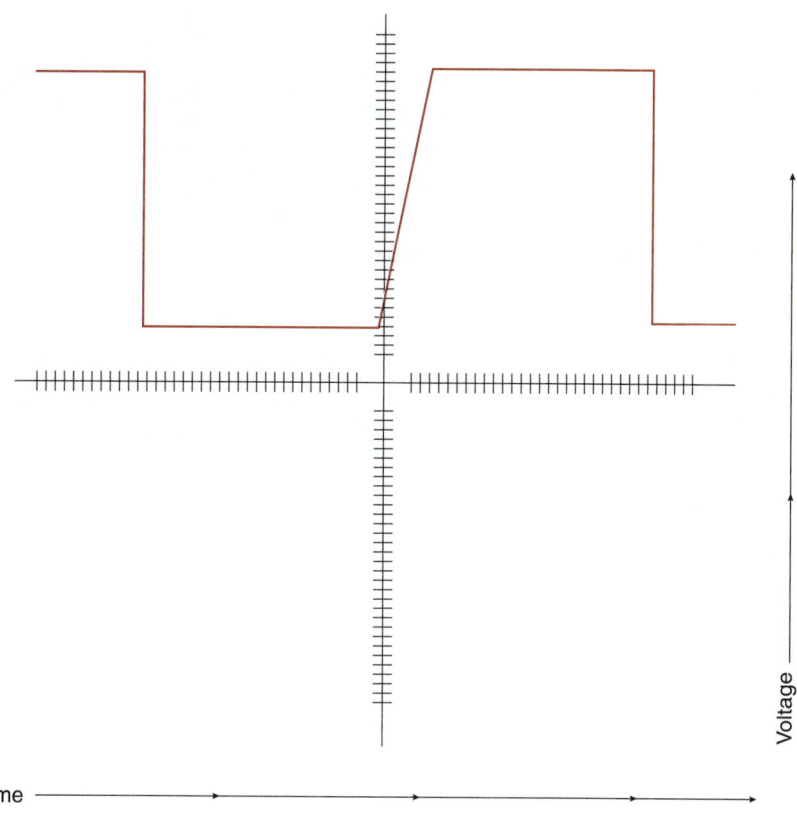

Figure 10-17 A sloping rising or falling line may indicate the air gap is out of adjustment or the transistor in the Hall-effect switch is faulty.

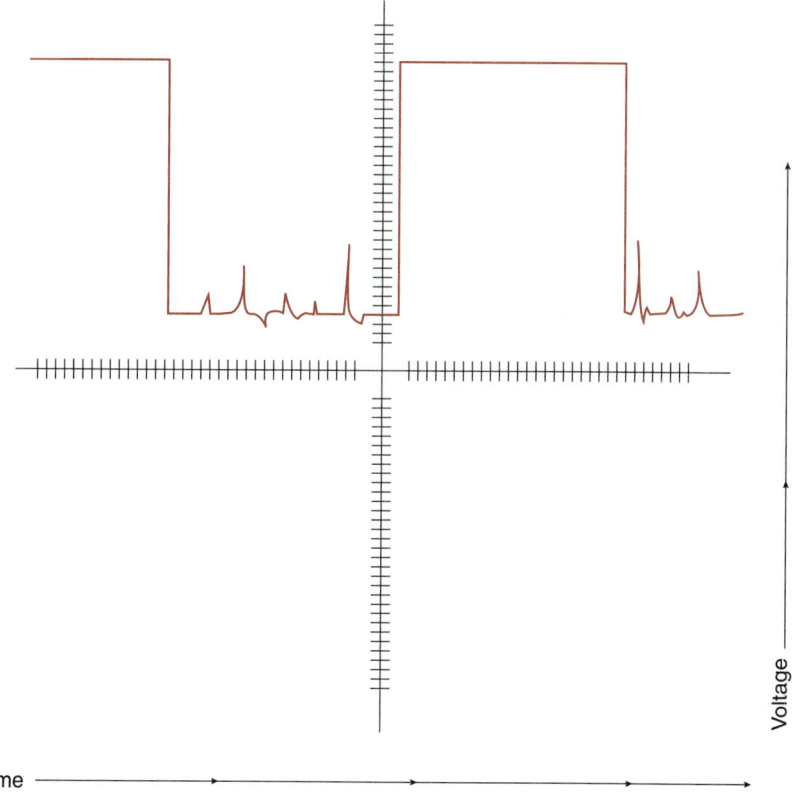

Figure 10-18 The Hall-effect trace pattern should be free of glitches and noise.

Figure 10-19 The breakout box provides test points for voltmeter and ohmmeter connections.

Figure 10-20 Using the breakout box to test the circuit.

⚠ **WARNING:** Since the magnetic pulse generator and Hall-effect switches require movement to generate their respective signal, do not place your hands and fingers around moving components. Also, be sure none of your clothing is hanging loose.

Special Tools

Scanner

Service manual

Sensor Testing with a Scan Tool

Most scan tools will display the voltage values or switch position of many sensors. Access to this information differs depending on the scan tool used. Selecting the SENSOR DISPLAY function will show the values of selected sensors. If the technician requires switch position, select INPUTS/OUTPUTS and the display will indicate the positions of various switches used as inputs to the computer.

Special Tools

Breakout box

DMMl

Breakout boxes are also known as pin testers because they can check the signals in a given circuit at the module pin terminals used for that particular circuit.

Breakout Boxes

The vehicle's wiring harness creates special problems when it comes to diagnosing circuits. A **breakout box** connected between the module and the wiring harness will allow the technician to "see" the exact information the computer is receiving and sending. Breakout boxes (Figure 10-19) allow the technician to test circuits, sensors, and actuators by providing test points.

Once the breakout box is connected into the system, a DMM can be used to measure the voltage signals and resistance values of the circuit (Figure 10-20). The diagnostic manual provided with the breakout box will direct the technician through a series of test procedures. Comparing test results with specifications will lead the technician to the problem area.

The breakout box has the advantage in that it taps directly into the sensor or actuator circuit. This provides the technician with the exact voltage signal being sent or received. The scan tool provides the technician with an interpretation of these values only. In addition, hard-to-get-to components can be tested without having to disconnect them from the circuit.

Classroom Manual
Chapter 10, page 300

Special Tools

PROM removal tool

PROM Replacement

Some manufacturers provide for PROM replacement in their body computers. To replace the PROM (if the diagnostic chart leads you to this step), first remove the BCM from the vehicle. Follow all service precautions while servicing the BCM.

Follow Photo Sequence 17 for the correct method of replacing the PROM.

Photo Sequence 17
Typical Procedure for Replacing the PROM Chip

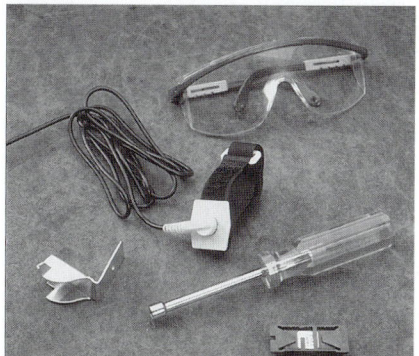

P17-1 Tools required to remove and replace the PROM chip: rocker-type PROM removal tool, ESD strap, safety glasses, and replacement PROM.

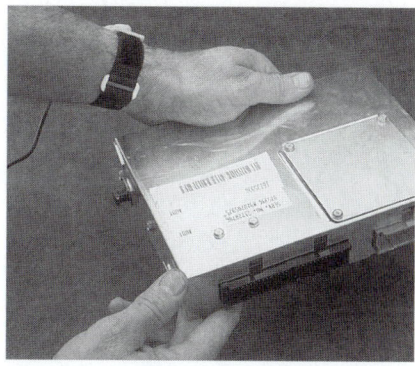

P17-2 Place the BCM onto the work bench with the prom access cover facing up. Be careful not to touch the electrical connectors with your fingers.

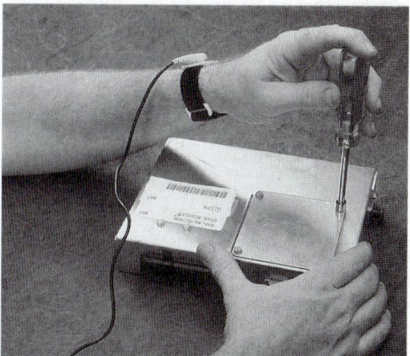

P17-3 Remove the PROM chip access cover.

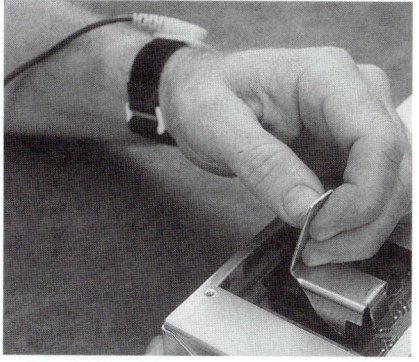

P17-4 Using the rocker-type PROM removal tool, engage one end of the PROM carrier with hook end of tool. Grasp the PROM carrier with the tool only at the narrow ends of the carrier.

P17-5 Press on the vertical bar end of the tool and rock the end of the PROM carrier up as far as possible.

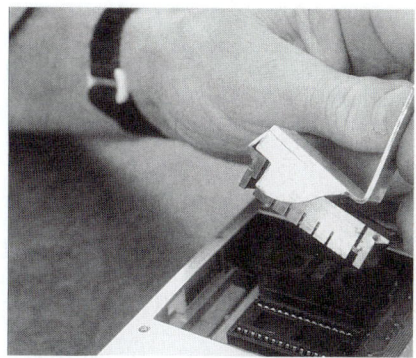

P17-6 Repeat the process on the other end of the carrier until the PROM carrier is removed from the socket.

P17-7 Inspect the replacement PROM part number for proper calibration.

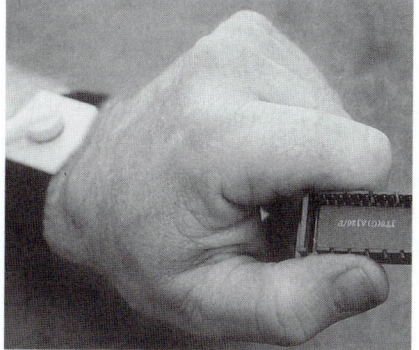

P17-8 Check for proper orientation of the PROM in the carrier. The notch in the PROM should be referenced to the smaller notch in the carrier. If the replacement PROM does not come in its own carrier, it will be necessary to remove the old PROM and install the replacement PROM into the carrier. Be careful not to bend the pins.

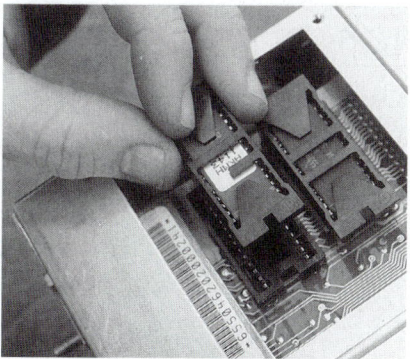

P17-9 Align the PROM carrier with the socket. The small notch of the carrier must be aligned with the small notch in the socket.

Typical Procedure for Replacing the PROM Chip (continued)

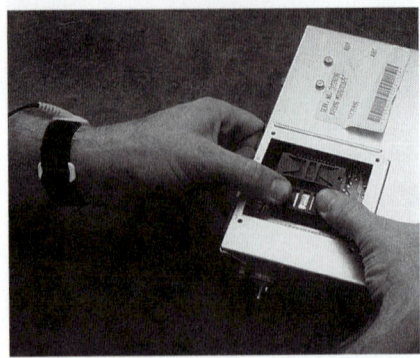

P17-10 Press the PROM carrier until it is firmly seated into the socket. Do not press on the PROM.

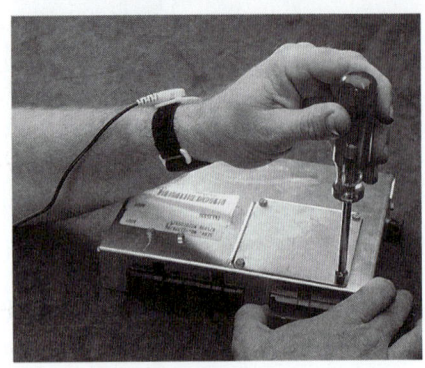

P17-11 Replace the PROM cover. Reinstall the BCM into the vehicle.

CAUTION: Installing a PROM chip in backward will immediately destroy the chip. In addition, electrostatic discharge (ESD) will destroy the chip. There are static straps available that will prevent ESD while you are working on the unit.

CUSTOMER CARE: Usually, removal and replacement of the body computer involves removal of covers and/or trim pieces. Be careful not to damage these components and to properly secure them upon installation. Keep the interior of the vehicle as clean as possible by removing any indications that you were ever in the vehicle before returning it to the customer.

Flashing the BCM

Classroom Manual
Chapter 10, page 299

Special Tools

Scan tool

Many modern BCMs are produced with EEPROM chips that allow the basic programming of the computer to be altered or rewritten. This enables manufacturers to reduce costs by having the technician **flash** the computer instead of replacing it. To flash a computer means to remove the existing programming and overwrite it with new software. New program instructions are downloaded into the scan tool and then downloaded into the BCM over a dedicated circuit.

Each scan tool is a little different in the procedure used to obtain a download and to flash the BCM. Some scan tools are connected to a PC and the software is transferred from a CD-ROM disc or from an Internet site. Some scan tools are connected directly to a LAN and obtain the flash from the Internet site. Usually the automotive shop has to pay a subscription fee to obtain these downloads. Photo Sequence 18 shows a typical procedure for flashing a BCM. It is important to follow the service manual procedures for the vehicle you are working on. Also, connecting a battery charger to keep the battery at 13.5 volts will prevent problems during the download resulting from low battery voltage. Be sure to check and clear any DTCs that may have been set during the flash procedure.

Photo Sequence 18
Flashing a BCM

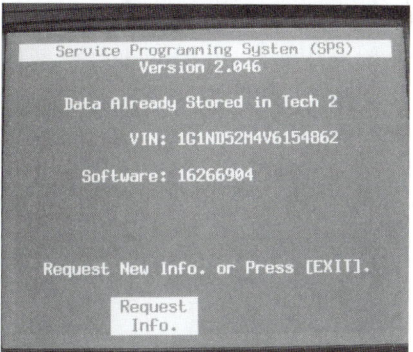

P18-1 Use the scan tool to obtain the BCM's part number or the VIN as displayed and record.

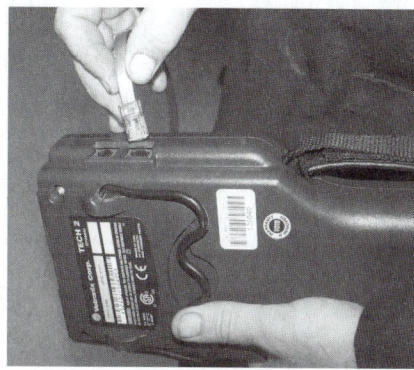

P18-2 Connect the scan tool to a PC that links the scanner to the flash software. Some scan tools will connect directly to an Internet site through a LAN.

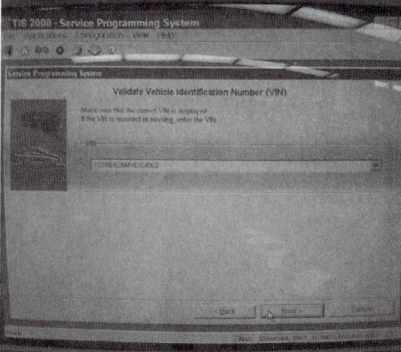

P18-3 Enter the BCM part number in the field and select the "Show Updates" button.

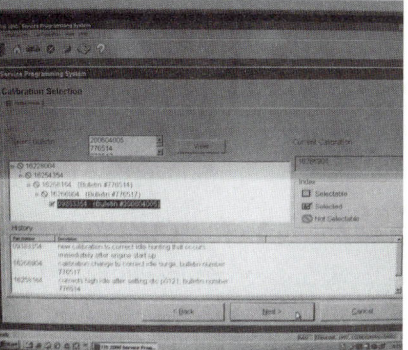

P18-4 Select the desired flash file.

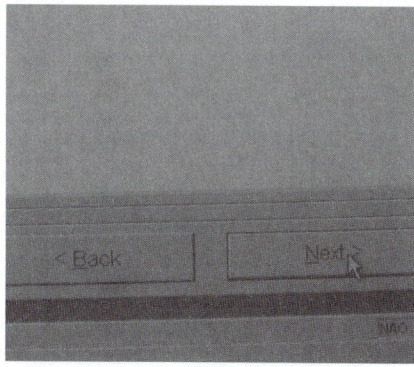

P18-5 Select the "Next" button to begin the download to the scan tool.

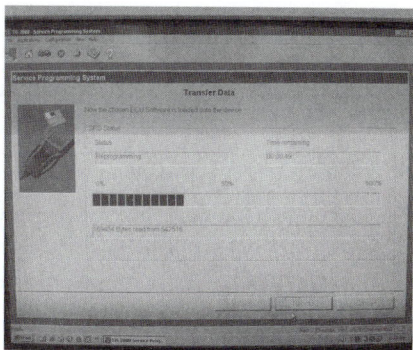

P18-6 Monitor the progress of the download to the scan tool.

P18-7 After successful download, return to the vehicle and connect a battery charger to the vehicle's battery. Maintain about 13.5 volts on the battery.

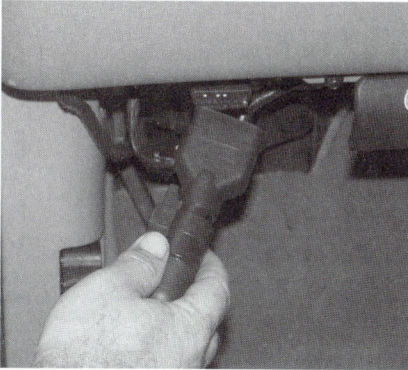

P18-8 Connect the scan tool to the DLC and power the tool on.

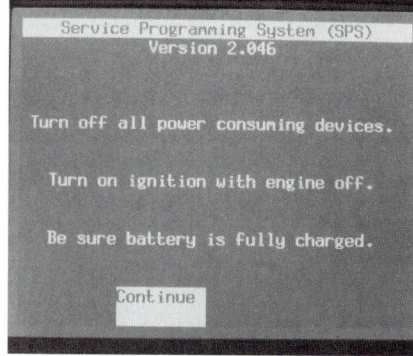

P18-9 Navigate the scan tool to perform the desired flash. Follow any instructions that are displayed on the screen.

A customer brings in a Cadillac Fleetwood because the instantaneous fuel economy constantly reads zero mpg. The technician verifies the complaint during the road test. When diagnostics are entered, a code F32 is displayed as a hard code. A check of the diagnostic chart indicates this problem is associated with a data problem between the ECM and BCM. A visual inspection does not reveal any obvious problems. The diagnostic chart refers the technician to check for a short or open to ground in the data link from the ECM to the BCM; this is located and repaired. The technician confirms the repair by test driving the vehicle.

Terms to Know

Breakout box

Diagnostic and trouble code (DTC)

Electrostatic discharge (ESD) strap

Fail soft

Flash

Flash codes

Hard codes

Impedance

Intermittent codes

ASE-Style Review Questions

1. All of the following statements about servicing the BCM are true EXCEPT:
   A. Always check voltage supply circuits before replacing the BCM.
   B. Analog voltmeters must be used to test circuit voltage.
   C. Always check grounds before replacing the BCM.
   D. Turn the ignitions switch to the OFF position before disconnecting or connecting components.

2. Diagnostic trouble codes are being discussed.
   Technician A says hard code failures are those that have occurred in the past but were not present during the last BCM test of the circuit.
   Technician B says intermittent codes are those that were detected the last time the BCM tested the circuit.
   Who is correct?
   A. A only
   B. B only
   C. Both A and B
   D. Neither A nor B

3. *Technician A* says DTCs will indicate the exact failure in the circuit.
   Technician B says DTCs will direct the technician to the circuit with a fault in it.
   Who is correct?
   A. A only
   B. B only
   C. Both A and B
   D. Neither A nor B

4. *Technician A* says an NTC-type ECT sensor voltage signal will drop in voltage as the monitored temperature increases.
   Technician B says that an NTC-type ECT sensor will increase resistance as monitored temperature increase. who is correct?
   A. A only
   B. B only
   C. Both A and B
   D. Neither A nor B

5. Accessing trouble codes through the ECC on General Motors vehicles is being discussed.
   Technician A says depress the OFF and WARMER buttons simultaneously.
   Technician B says the body codes are prefixed with an "E."
   Who is correct?
   A. A only
   B. B only
   C. Both A and B
   D. Neither A nor B

6. When testing input sensors, which statement is most correct?
   A. Sensors can only be tested with a scan tool.
   B. A breakout box can be used to test sensor signals.
   C. Sensor circuits can be tested with a test light.
   D. All of the above.

7. When flashing a computer, all of the following are true EXCEPT:
   A. A battery charger should be connected to the vehicle's battery.
   B. Special instructions may be displayed on the scan tool screen during the flash.
   C. The flash process loads new software programming into the computer.
   D. The BCM must be removed from the vehicle before it is flashed.

8. Testing of the magnetic pulse generator is being discussed.
   Technician A says use an ohmmeter to test the resistance value of the coil.
   Technician B says the voltage generation of the sensor can be tested by connecting a voltmeter across the sensor terminals.
   Who is correct?
   A. A only
   B. B only
   C. Both A and B
   D. Neither A nor B

9. Replacement of the PROM is being discussed.
   Technician A says the notch in the PROM should be referenced to the smaller notch in the carrier.
   Technician B says the small notch of the carrier must be aligned with the small notch in the socket.
   Who is correct?
   A. A only
   B. B only
   C. Both A and B
   D. Neither A nor B

10. *Technician A* says installing a PROM chip in backward will immediately destroy the chip.
    Technician B says electrostatic discharge will destroy the chip.
    Who is correct?
    A. A only
    B. B only
    C. Both A and B
    D. Neither A nor B

ASE Challenge Questions

1. The radiator/condenser fan of a BCM-equipped vehicle is operating at full speed whenever the engine is running.
   Technician A says that the coolant temperature sensor may be open.
   Technician B says that for some reason the BCM may have initiated a fail soft action.
   Who is correct?
   A. A only
   B. B only
   C. Both A and B
   D. Neither A nor B

2. BCM system troubleshooting is being discussed.
   Technician A says that if the BCM "sees" a problem with any of its input sensors or output devices it will always command the "service engine" lamp to illuminate.
   Technician B says that most diagnostic charts cannot be used to diagnose intermittent problems.
   Who is correct?
   A. A only
   B. B only
   C. Both A and B
   D. Neither A nor B

3. The in-car temperature sensor of a vehicle equipped with a BCM and automatic temperature control (ATC) occasionally shorts internally, resulting in erratic operation of the ATC system. However, when the vehicle is brought into the shop for testing, the ATC system performs correctly.
   Technician A says that an intermittent trouble code will be stored by the BCM.
   Technician B says that an intermittent and a hard trouble code will be stored by the BCM.
   Who is correct?
   A. A only
   B. B only
   C. Both A and B
   D. Neither A nor B

4. The diagnosis of BCM-controlled output actuators is being discussed.

Technician A says that the speed of the radiator fan motor can be controlled and monitored by pressing certain buttons on the ECC display.

Technician B says that some scan tools can control the operation of output actuators.

Who is correct?

A. A only

B. B only

C. Both A and B

D. Neither A nor B

5. Sensor testing is being discussed.

Technician A says that sensor information displayed by a scan tool is "live" data, meaning that the scan tool displays the data the exact instant that it receives it.

Technician B says that it is preferable to use a DMM or lab scope when "sweeping" potentiometer-type sensors.

Who is correct?

A. A only

B. B only

C. Both A and B

D. Neither A nor B

Job Sheet 34

34

Name _____ Date _____

Using a Scan Tool

Upon completion of this job sheet, you should be able to hook up a scan tool and retrieve the codes from a computer.

ASE Correlation

This job sheet is related to the ASE Electrical/Electronic System Certification Exam's content area: *General Electrical System Diagnosis;* task: Use scan tool data to diagnose electronic systems; interpret readings and determine needed repairs.

Tools and Materials

Vehicle equipped with a BCM
Scan tool

Describe the vehicle being worked on:

Year _____ Make _____ Model _____

VIN _____ Engine type and size _____

Procedure

Task Completed

1. Locate the data link connector for the BCM. Where is the connector located?

   _____

   _____

2. Connect the scan tool to the DLC and turn the ignition switch to the RUN position. What version is the scan tool software? _____

3. Select the body computer function. What information is displayed about the body computer?

   _____

   _____

   _____

   _____

4. How is this information useful?

   _____

   _____

   _____

   _____

5. Retrieve any fault codes that may be in the BCM's memory. Record the codes below.

_____

_____

_____

_____

6. Are the fault codes' hard codes intermittent? _____

7. For the circuit with the fault code associated with it, look at all sensors and input/output values. Are any of the values out of specifications? _____

   If so, which sensors and/or actuators:

   _____

   _____

   _____

   _____

8. Follow the diagnostic chart for the fault code. What are your conclusions?

   _____

   _____

   _____

   _____

Instructor's Response _____

_____

_____

Job Sheet 35

35

Name _____ Date _____

Using a Lab Scope to Test Sensors and Switches

Upon completion of this job sheet, you should be able to connect a lab scope and observe the activity of various sensors and switches.

ASE Correlation

This job sheet is related to the ASE Electrical/Electronic System Certification Exam's content area: *General Electrical/Electronic System Diagnosis;* task: Check electronic circuit waveforms using an oscilloscope and determine needed repairs. Also, find shorts, grounds, opens, and high-resistance problems in electrical/electronic circuits; determine needed repairs.

Tools and Materials

A vehicle with accessible sensors and switches   A lab scope
Service manual for the chosen vehicle   A DMM
Component locator manual for the chosen vehicle

Describe the vehicle being worked on:

Year _____ Make _____ Model _____

VIN _____ Engine type and size _____

Procedure

Task Completed

1. Connect the lab scope tools across the battery. Make sure the scope is properly set. Observe the trace on the scope. Is there evidence of noise? Explain.

   _____

   _____

   _____

2. Locate the A/C compressor clutch control wires. Start and run the engine. Connect the DMM to read available voltage. Observe the meter; then turn the compressor on. What happened on the meter?

   _____

   _____

   _____

   Now connect the lab scope to the same point with the compressor turned off. Observe the waveform, then turn the compressor on. What happened to the trace?

   _____

   _____

   _____

3. Turn off the engine but keep the ignition on. Locate the TP sensor and identify the purpose of each wire to it. List each wire and describe the purpose of each.

_____

_____

_____

4. Connect the DMM to read reference voltage at the TP sensor. What do you read?

_____

Now move the leads to read the output of the sensor. Starting with the throttle closed, slowly open the throttle until it is wide open. Watch the voltmeter while doing this. Describe your readings.

_____

_____

5. Now connect the lab scope to read reference voltage at the TP sensor. What do you see on the trace?

_____

Now move the leads to read the output of the sensor. Starting with the throttle closed, slowly open the throttle until it is wide open. Watch the trace while doing this. Describe your readings.

_____

_____

6. Explain what you observed as the differences between testing with a DMM and a lab scope.

_____

_____

_____

_____

Instructor's Response _____

_____

_____

Job Sheet 36

36

Name _____ Date _____

Testing an Ambient Temperature Sensor

Upon completion of this job sheet, you should be able to check the operation of an ambient temperature sensor (ATS).

ASE Correlation

This job sheet is related to the ASE Electrical/Electronic System Certification Exam's content area: *General Electrical System Diagnosis;* task: Check continuity and resistances in electrical/electronic circuits and components with an ohmmeter; interpret readings and determine needed repairs.

Tools and Materials

DMM

Describe the vehicle being worked on:

Year _____ Make _____ Model _____

VIN _____ Engine type and size _____

Procedure

Task Completed

1. Describe the location of the ATS sensor.

_____

_____

_____

2. What color of wires are connected to the sensor?

_____

_____

3. Record the resistance specifications for a normal ATS sensor for this vehicle.

_____

_____

_____

_____

4. Disconnect the electrical connector to the sensor.   ☐

5. Measure the resistance of the sensor: _____ ohms at approximately _____°F.

6. Conclusions:

_____

_____

_____

_____

_____

Instructor's Response _____

_____

_____

Job Sheet 37

37

Name _____ Date _____

Flashing a BCM

Upon completion of this job sheet, you should be able to flash a BCM.

Tools and Materials

Scan tool
Access to flash software
Battery charger

Describe the vehicle being worked on:

Year _____ Make _____ Model _____

VIN _____ Engine type and size _____

Procedure

Task Completed

1. Is there a TSB that addresses flashing the BCM for your assigned vehicle? ☐ Yes ☐ No

   If yes, what is the document number? _____

   If no, go to step 3.

2. According to the TSB, what is the purpose of the flash?

   _____

   _____

   _____

3. Connect the scan tool to the DLC and access the BCM. Obtain the BCM part number and record. _____

4. Describe the procedure used to obtain the flash download into your scan tool.

   _____

   _____

5. Follow the procedure outlined in step 4 and download the flash file to your scan tool. Describe the navigation required to download the file to the scan tool.

   _____

   _____

6. How do you know the download was successful?

   _____

   _____

7. Return to the vehicle and connect a battery charger to the battery. Why is this step important?

_____

_____

8. Perform the flash of the BCM. Record any steps the scan tool directs you to perform during the flash.

_____

_____

_____

9. How do you know if the flash update was successful?

_____

10. After the flash is completed, check for DTCs. Were any DTCs set?   ☐ Yes   ☐ No

   If yes, erase DTCs.

Instructor's Response _____

_____

_____

Vehicle Multiplexing Diagnostics

Upon completion and review of this chapter, you should be able to:

❑ Describe the purpose of U- and B-codes.

❑ Properly diagnose an ISO 9141-2 bus system and determine needed repairs.

❑ Properly diagnose an ISO-K bus system and determine needed repairs.

❑ Properly diagnose a class A bus system and determine needed repairs.

❑ Properly diagnose a J1850 bus system and determine needed repairs.

❑ Properly diagnose a Controller Area Network (CAN) bus system and determine needed repairs.

❑ Properly diagnose a local interconnect network (LIN) bus system and determine needed repairs.

Basic Tools

Hand tools
Fender covers

Introduction

If the vehicle's multiplexing system should fail, the symptoms can range from a single function (such as instrument gauges) not operating to multiple function failures, including engine no-start. Diagnosing the bus system is not much different than diagnosing any other electrical system. Begin by verifying the customer's complaint; determine if there are any related symptoms, and then analyze the symptoms to develop a logical troubleshooting plan. It is important to understand how the bus system you are diagnosing should operate and what are the normal voltages on the system. Bus system failures include circuit opens, shorts, high resistance, and component failures. In addition, do not be quick to condemn the bus system if a module is not communicating. It is possible the module is not powering up due to loss of battery voltage feed or loss of ground. This chapter discusses those items that the technician must be aware of while diagnosing different bus networks. The most common bus systems are discussed here.

Since most bus systems communicate with the scan tool, they will have a point of connection at the data link connector (DLC). To assist in testing of the data bus, a **J1962 breakout box (BOB)** is available (Figure 11-1). Since J1962 is the mandated DLC configuration for OBD II, this tool will work on any OBD II–compliant vehicle. The J1962 BOB provides a pass through test point that connects in series between the DLC and the scan tool. This provides easy testing of voltages and resistance of any of the DLC circuits without risk of damage to the DLC.

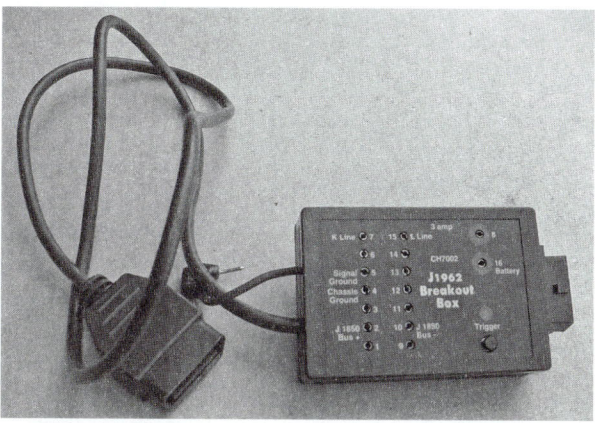

Figure 11-1 The J1962 BOB makes pin-out testing of the DLC easier.

Communication Fault Codes

Diagnostic trouble codes assigned to the vehicle communication network are called **U-codes.** These codes follow the same SAE guideline as the "P-codes" used for powertrain faults. The prefix "U" indicates the fault is associated with network communications.

In addition, DTCs that are assigned to the vehicle's body systems and control modules are called **B-codes.** Typically, these codes refer to a failure of the system the module operates (such as a sensor failure). If a module relies on a bus message from another module but does not receive it, the first module may set both a U-code and a B-code. The U-code would be due to the loss of communication with the second module, and the B-code due to the system not able to perform a function. B-codes also follow the same SAE guideline as the P-code.

Most modules on the bus network are capable of setting U-codes if they detect abnormal conditions. Most bus modules can detect loss of communication conditions with one or more modules or a bus failure. Some modules may also be able to monitor the actual voltage on the bus circuits and set additional trouble codes for conditions such as voltage high, low, shorted, or open bus circuits.

In addition, most modules will report the status of the DTCs. If the conditions currently exist, then the DTC is reported as being active. If the condition no longer exists, then the DTC is reported as being stored.

Classroom Manual
Chapter 11, page 325

Special Tools

J1962 BOB

Scan tool

DMM

Lab scope

ISO 9141-2 Bus System Diagnostics

The ISO 9141-2 standard bus system provides for communication links between the scan tool and the module. Many OBD II vehicles will use this protocol for communication to the powertrain control module (PCM). In addition, some manufacturers will use the system for communication between the scan tool and other modules on the vehicle (Figure 11-2).

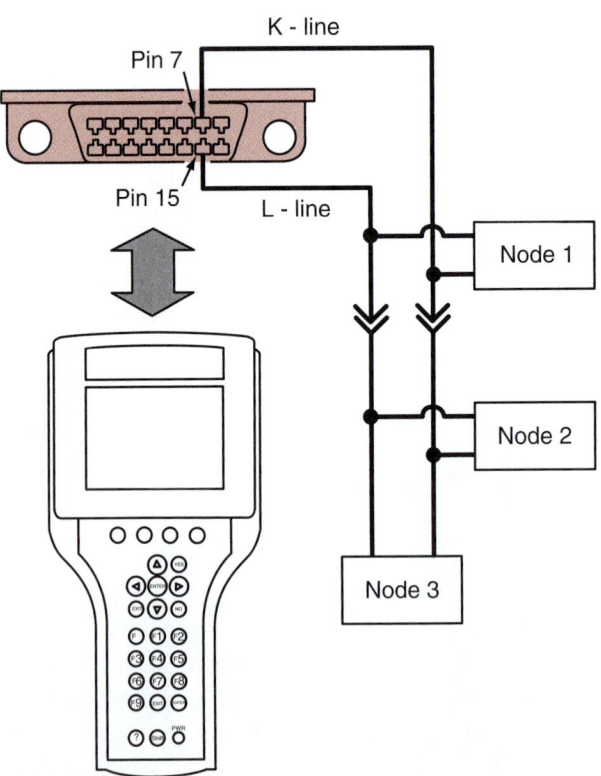

Figure 11-2 ISO 9142-2 bus system used to communicate between the scan tool and modules on the vehicle.

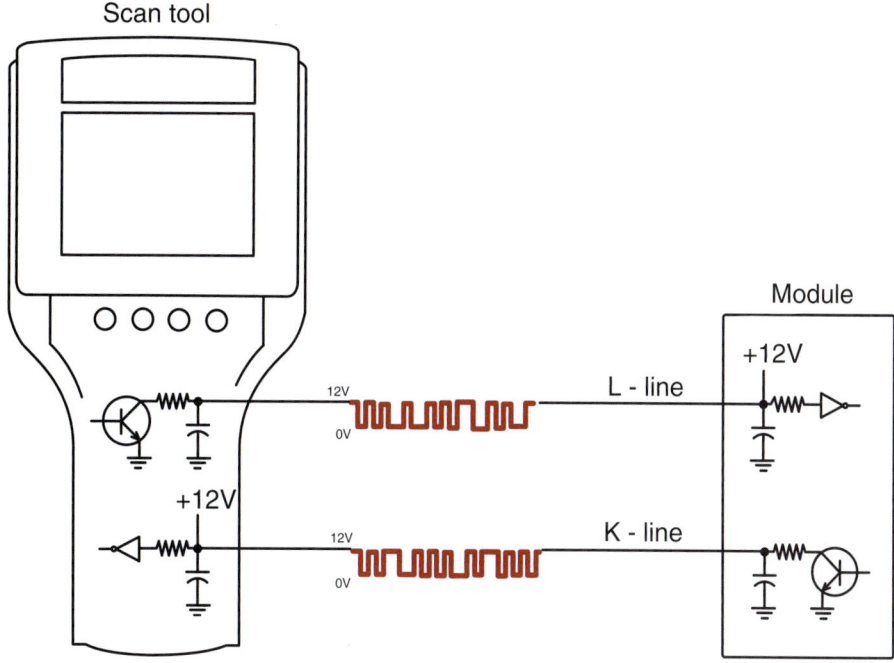

Figure 11-3 The K-line transmits data from the module to the scan tool, and an L-line receives data from the scan tool.

The ISO 9141-2 bus uses a K-line to transmit data from the module to the scan tool and an L-line for the module to receive data from the scan tool (Figure 11-3). The scan tool supplies bias to the module on the K-line, while the module supplies bias to the scan tool on the L-line. Communication occurs when the transmitting node pulls the voltage low. If a failure occurs in this bus system, then communications between the scan tool and the module will not be possible. Since this bus system is not used for communications between modules on the vehicle, the customer may not have any noticeable problems with vehicle or accessory operation.

When diagnosing a failure due to the scan tool not being able to communicate with the PCM, it is important to analyze the symptoms. Review the wiring diagram and bus system information in the proper service manual to determine if other modules on the vehicle are diagnosed using the ISO 9141-2 bus system. If other modules are on the bus, then use the scan tool to attempt to connect with each of these modules. If the scan tool connects to any module using the ISO 9141-2 bus, then it is not a total bus failure and the technician will need to diagnose for a partial bus failure. If no modules respond, then the technician will need to diagnose for a total bus failure.

SERVICE TIP: Some scan tools provide a **vehicle module scan** function that will query all of the modules on the bus to respond and then list those that did reply. This makes it simple to see if any other modules are responding and which ones are not.

Referring to Figure 11-2, if node 1 responded, but nodes 2 and 3 did not, then the first location to check would be the in-line connector. If nodes 1 and 3 only responded, then the problem is in the bus circuit to node 2. Since the circuit is wired in parallel, the problem cannot be a short to ground or voltage. Either the bus has an open between the splice and the node or the node is not powering up due to faulty battery feed or ground circuits.

If no module responds, then check voltages on the K-line and the L-line. Connect the J1962 BOB to the DLC and test for voltages at the proper terminals. Without the scan tool connected, there should be voltage only on the L-line. Voltage on the L-line is supplied by the module. Zero volts here can indicate an open circuit, or short to ground. Disconnect the battery and use an ohmmeter to determine the type of fault. If an open is indicated, the most likely location is between the DLC and the splice to the first node. A short to ground can be anywhere in the circuit. To locate the short, refer to the service information and determine if there is an in-line connector in the circuit. If the in-line connector is accessible, disconnect it and see if the resistance changes. If it does, then the short is downstream of the in-line connector. If the ohmmeter reading does not change, then the fault is between the DLC and the in-line connector. Next disconnect the modules on the side of the in-line connector that the fault was isolated to. After each module is disconnected, check the resistance reading. If disconnecting a module changes the reading, then that module has an internal fault and needs to be replaced. If the ohmmeter still indicates a short to ground after all of the modules are disconnected, then the fault is in the harness.

With the scan tool connected, voltage should be present on the K-line. Since the voltage on the K-line is supplied from the scan tool, zero volts on this line means that either the circuit is shorted to ground or the scan tool is faulty. Use an ohmmeter to determine if the K-line is shorted to ground. If so, follow the same procedure just described for the L-line to locate the short.

Classroom Manual
Chapter 11, page 325

Special Tools
J1962 BOB
Scan tool
DMM
Lab scope

ISO-K Bus System Diagnostics

Vehicles that use ISO-K for the communication connection between the module and the scan tool may have their own dedicated lines from individual terminals in the DLC to the module (Figure 11-4). The K-line from the DLC to the PCM will be from terminal 7 of the DLC. The ISO-K bus is used only for communications between the scan tool and the module on a signal wire; it is not used for intermodule communications. The scan tool supplies the voltage onto the K-line.

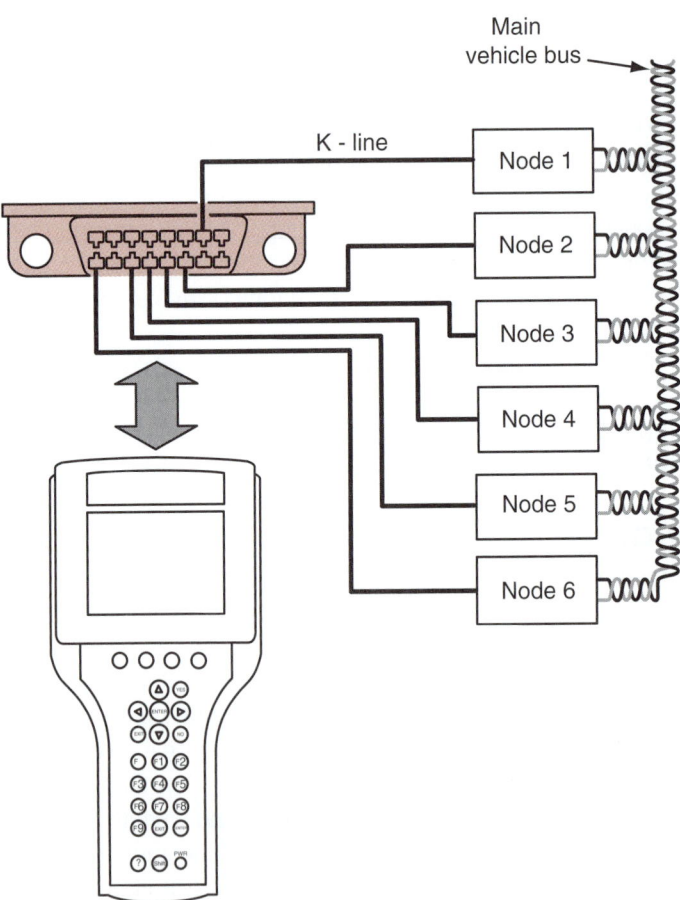

Figure 11-4 An ISO-K bus circuit used to connect several modules to the scan tool. Each module uses its own dedicated circuit from the DLC.

Channel 1

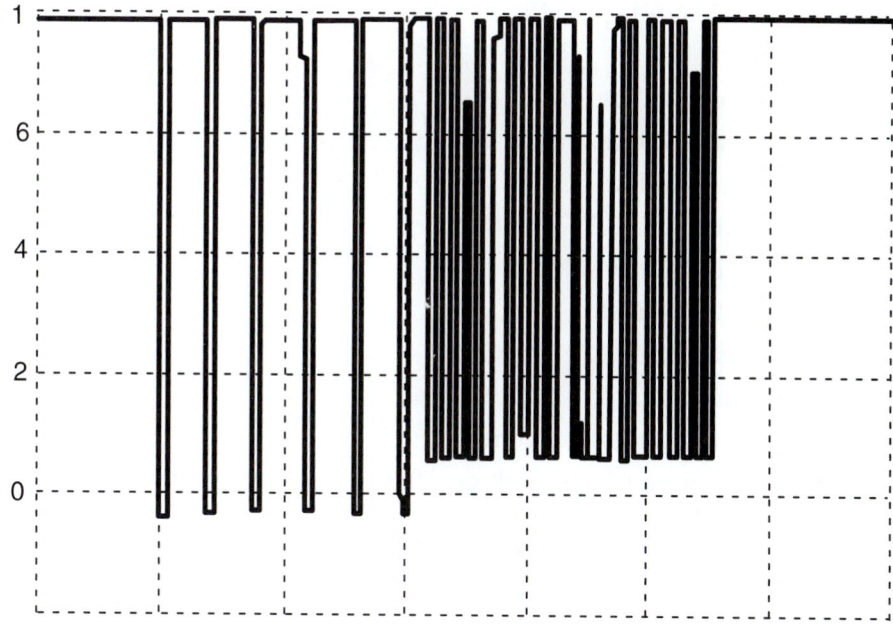

Figure 11-5 ISO-K transmission trace. The negative voltage is from the scan tool attempting to request data from the module; when the module is transmitting data to the scan tool, the voltage is pulled to +0.5 volts.

If communications between the scan tool and module are not possible, begin diagnosis by connecting the J1962 BOB to the DLC. With the ignition key in the OFF position and the scan tool disconnected, there should be zero volts on the K-line. If voltage is present on this circuit at this time, there is a short to voltage between the DLC and the module, or internal of the module.

If zero volts are indicated, test for short to ground with an ohmmeter. If a short to ground is indicated, it may be in the wiring between the DLC and the module, or internal of the module.

Next, connect the scan tool with the ignition switch in the OFF position. It would be best to use a lab scope to observe the voltages. Set the scope to read 10 ms per division on the time frame. At this time, up to 12 volts should be seen on the scope trace. While observing the scope trace, turn the ignition key to the RUN position and attempt to establish communications between the scan tool and the module. The scope trace should indicate a digital signal as communication is established and as data is transmitted (Figure 11-5). Notice in the scope trace that the biased voltage from the scan tool is about 10.5 volts. When the scan tool is attempting to request data from the module, the voltage is pulled low, to −0.5 volts. Finally, when the module is transmitting data to the scan tool, the voltage is pulled to +0.5 volts.

Class A Bus System Diagnostics

As discussed in the Classroom Manual, the class A bus system is a slow-speed bus. This system will have a master module and several slave modules. The master module will supply the bias voltage onto the bus system. If this module should fail, then no bus communications will be possible. In the Classroom Manual, the Chrysler Collision Detection (CCD) multiplexing system was presented as an example of how a class A bus operates. In this chapter, the CCD bus system will be used to provide an example of failure modes and how to diagnose a typical class A bus system.

Depending on scan tool type, the scan tool may attempt to diagnose the bus system if communication fails to occur. If this happens, the scan tool will display the cause of the problem on the screen. The messages displayed can be as follows:

Classroom Manual
Chapter 11, page 327

Special Tools
J1962 BOB
Scan tool
DMM
Lab scope

- ❏ Bus (−) open.
- ❏ Bus (+) and bus (−) open.
- ❏ Bus (+) open.
- ❏ Bus (+) shorted to bus (−).
- ❏ Bus bias level too high.
- ❏ Bus bias level too low.
- ❏ Bus shorted to 5 volts.
- ❏ Bus shorted to battery voltage.
- ❏ Bus shorted to ground.
- ❏ No bus bias.
- ❏ No bus termination.
- ❏ Not receiving bus messages correctly.

The CCD bus network uses the voltage difference between the two wires to transmit data (Figure 11-6). The bus bias on the two wires is about 2.50 volts. The bus can operate with a voltage between 1.5 and 3.5 volts. Voltages that are outside of this range will not transmit data.

Since the CCD bus is a vehicle-wide bus system, failure of this bus will result in electrical systems not operating properly. First, determine if it is a total bus failure or a partial bus failure by attempting to communicate with all of the modules on the vehicle. If it is possible to communicate with a module on the bus system, then approach the diagnostics as a partial bus failure. If no modules can communicate with the scan tool, then a total bus failure is indicated.

SERVICE TIP: The PCM does not communicate with the scan tool on the CCD bus. It uses the K-line for this purpose. Do not be fooled into thinking that the problem is a partial CCD bus failure based on the response of the PCM.

If a total bus failure is indicated, then possible causes include:
- ❏ A faulty master module.
- ❏ Faulty power or ground circuits to the master module.
- ❏ An open in one of the bus circuits from the master module.

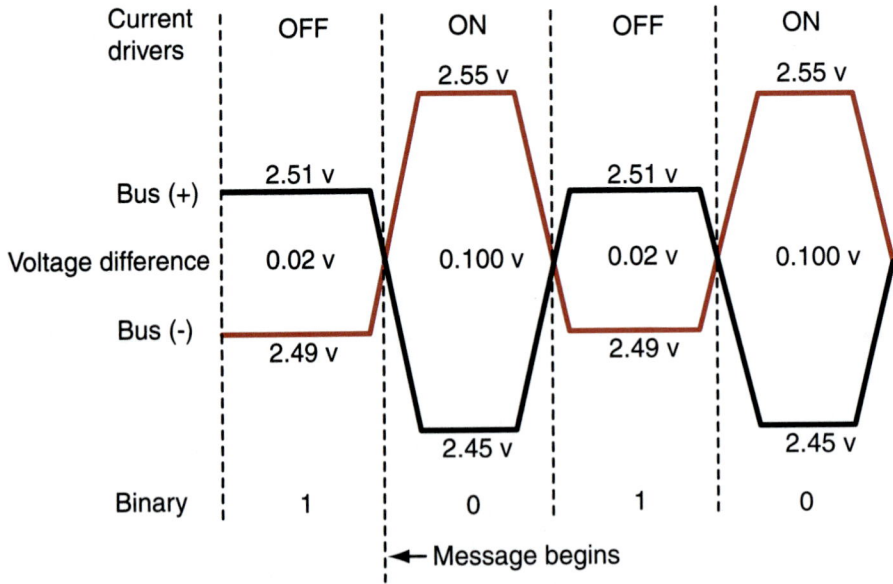

Figure 11-6 Normal CCD bus voltages during message transmission. Voltage difference is used to transmit the bit 1 or bit 0.

❏ A short to ground in one of the bus circuits.

❏ A short to voltage in one of the bus circuits.

❏ The two bus wires are shorted together.

To determine the type of fault that is causing total bus failure, use the J1962 BOB and connect it to the DLC. Measure the voltage on each bus circuit. The CCD system uses pin 3 of the DLC for CCD (+) and pin 11 for CCD (−). CCD (+) should have 2.49 volts with the ignition key in the RUN position. CCD (−) should have 2.51 volts.

If the voltmeter reads 12 volts on either circuit, that circuit is shorted to battery voltage. To isolate the location of the short, turn the ignition switch to the OFF position and watch the voltmeter. If the voltage drops, then the short is to ignition switched voltage. If the voltage is still reading battery voltage, the circuit is shorted to a direct battery feed. Remove fuses and relays one at a time while watching the voltmeter. If the voltage drops after a fuse or relay is pulled, note the identification of the fuse or relay and use the service manual to determine what circuits and components are protected by that fuse or controlled by that relay. Once this is known, return to the vehicle, reinstall the fuse, and begin to unplug each component one at a time while observing the voltmeter. Once a component is unplugged and the voltage drops to normal, that component is the fault. If all components are unplugged and the voltage is still high, the fault is in the wiring of the circuit for those components.

If the voltmeter indicated zero volts on either or both of the bus circuits, then the cause can be an open or a short to ground. To determine the type of fault, use an ohmmeter and measure the termination resistance of the bus system by disconnecting the battery and placing the test leads into pins 3 and 11. Most CCD bus systems will have 60 ohms of termination resistance. However, some vehicles will have 120 ohms of termination resistance, so it is important to check the service information.

If the system should have 60 ohms and the ohmmeter reads 120 ohms, then one of the termination resistors is open or there is an open in the bus circuit to a module that has one of the termination resistors. Usually the termination resistors will reside in the PCM and the BCM. Each of these modules will have a 120-ohm termination resistor (Figure 11-7). To locate the open

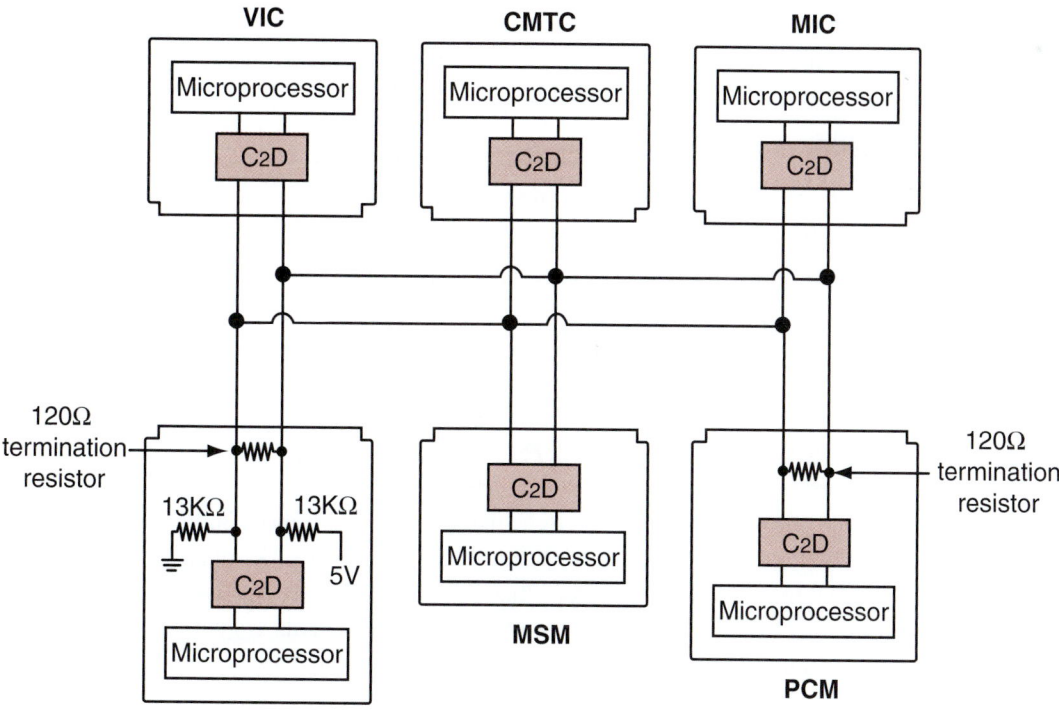

Figure 11-7 Usually BCM two modules will have termination resistors. Total circuit resistance will be 60 ohms.

resistor, disconnect either the PCM or BCM while observing the ohmmeter. If the resistance does not change, then the termination resistor that is open is in the module that was just unplugged. If the resistance changes to infinite, the remaining termination resistor in the circuit was just removed, so the fault is within the termination of the other module. Before replacing the module, check for an open in the bus circuits to that module. If the bus circuits are good, and module power and ground circuits are confirmed, replace the module.

If termination resistance is normal and there is no indication of a short to ground on any of the bus circuits, the fault is probably the master module not powering up and thus not able to supply the bias for the bus. Usually the master module is the BCM in the CCD bus system. Confirm the power and ground circuits are good before replacing the BCM.

If ohmmeter testing indicates a short to ground in one of the bus circuits, isolation of the fault must still be performed. The fault may be a faulty module that has an internal short, or it could be the bus wire is shorted to ground within the wiring harness. The system circuit needs to be broken down into smaller sections to isolate where the fault is. Examine the wiring diagram for the vehicle and determine if there is a connector that would separate the bus network. Identify any connectors that can be used and their location. With the ohmmeter measuring resistance between the affected circuit and chassis ground, unplug the connector. If the ohmmeter still reads low resistance, the fault is between the DLC and the connector. If the ohmmeter reads infinite, the fault is downstream of the connector. Now that this is known, move the ohmmeter test leads (if necessary) to observe resistance readings, as modules on the appropriate side of the connector are unplugged. If the readings change after a module is unplugged, then that module is the fault. If the reading is still low after all modules are unplugged, the fault is in the wiring harness.

If the initial tests indicate a partial bus failure, try to determine which modules are not communicating. If there is more than one, use the wiring diagrams for the vehicle to determine if there is anything common between the affected modules. For example, look for a common connector, common splice, common voltage supply to the modules, or common ground for the modules. Locate this common component on the vehicle and confirm proper function and repair as needed.

> **SERVICE TIP:** If the scan tool cannot communicate with the PCM, and other modules on the CCD bus system have logged a loss of communication fault code against the PCM, check the common voltage supply circuits to the sensors. If these circuits short to ground, the PCM will turn off. This will also result in a no-start, and the PCM may not store any DTCs.

When diagnosing the CCD bus system, remember the following:

1. Normal bus voltages with the ignition switch in the RUN position are approximately 2.5 volts.
2. Normal bus voltages with the ignition switch in the OFF position are zero volts on both circuits.
3. Normal termination resistance is typically 60 ohms.
4. The wires must be twisted at the rate of one twist every 1¾ inches (44 mm).

Classroom Manual
Chapter 11, pages 326, 331

Classroom Manual Chapter 11, pages 326, 331

Special Tools

J1962 BOB
Scan tool
DMM
Lab scope

J1850 Bus System Diagnostics

The J1850 bus system can be either a 10.4 Kb/s variable pulse width modulated (VPWM) system or a 41.6 Kb/s pulse width modulated (PWM) system. In the Classroom Manual, the Programmable Communication Interface (PCI) bus system that Chrysler uses was discussed. This system is similar to the class 2 bus system that General Motors uses as well. This system will be discussed here as an example of failure modes and how to diagnose a typical J1850 bus system.

Since the PCI bus is a vehicle bus system, a failure in it will cause electric systems to not operate properly. Like the other bus system diagnostics just discussed, begin by determining if the problem is a total bus failure or a partial bus failure.

If a total bus failure is indicated, then the system needs to be tested for proper voltage and termination resistance. Use the J1962 BOB to provide a diagnostic test point to measure these values. Total bus failure can only occur if there is a short to voltage or a short to ground in the circuit. Termination resistance will be different on every vehicle since each module provides its own termination. Since vehicles will have different options on them, the resistance will be different between vehicles. However, by measuring the resistance between pin 2 of the DLC (at the J1962 BOB) and chassis ground, a low reading will confirm a short to ground in the system. The procedure to locate this type of fault is presented later.

The normal voltage on the PCI bus is zero volts when it is at rest, and the voltage pulls up to near 7.5 volts when the bus is active (Figure 11-8). The best method of determining proper voltage on this system is with a lab scope. A DMM set to the voltmeter will not be able to give actual voltage values since the activity is too fast, so the meter will display the average. The MIN/MAX feature of the voltmeter will freeze the readings, but they still will not be actual voltages. However, the voltmeter will indicate if the bus is shorted to battery voltage (12 volts) or is a short to ground (0 volts). A reading of zero volts may also indicate an open, but this type of fault will not result in total bus failure.

If a short to battery voltage is indicated, then the location of the fault needs to be isolated. In order to do this, the system needs to be broken down into manageable pieces or smaller circuits. With the voltmeter still measuring PCI voltage, turn the ignition key to the OFF position. Wait for 20 seconds and read the voltmeter again. If the voltmeter now reads zero volts, then the bus network is shorted to ignition switched voltage. If the voltmeter still reads battery voltage, then the bus is shorted to direct battery voltage. In either case, the circuit that is shorted to the bus network can be further isolated by pulling fuses and relays one at a time while watching the voltmeter. If the voltage drops after a fuse or relay is pulled, note the identification of the fuse or relay and use the service manual to determine what circuits and components are protected by that fuse or controlled by that relay. Once this is known, return to the vehicle, reinstall the fuse, and begin to unplug each component one at a time while observing the voltmeter. Once a component is unplugged and the voltage drops to normal, that component is the fault. If all components are unplugged and the voltage is still high, the fault is in the wiring of the circuit for those components.

If a short to ground was indicated while testing the termination resistance as discussed earlier, isolation of the fault must still be performed. The fault may be a faulty module that has an internal short, or it could be the PCI bus wire is shorted to ground within the wiring harness. Again the trick is to break the system down into smaller circuits. Depending on how the vehicle is wired, some may use a common hub for all of the bus circuits. If this is the case, this location provides

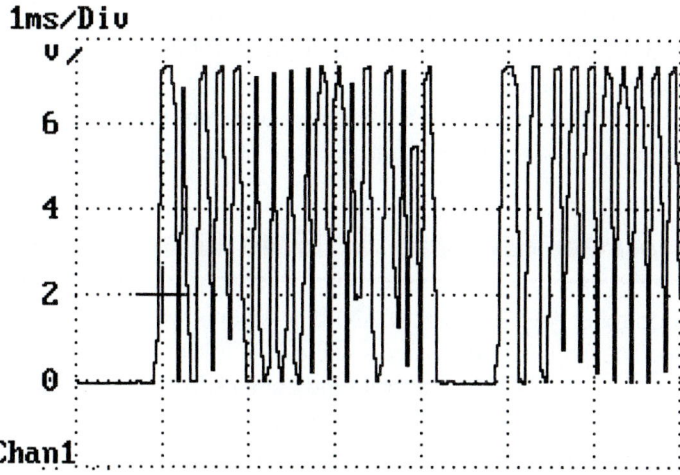

Figure 11-8 The normal voltage on the PCI bus is zero volts when it is at rest. To transmit data, the voltage is pulled up to near 7.5 volts.

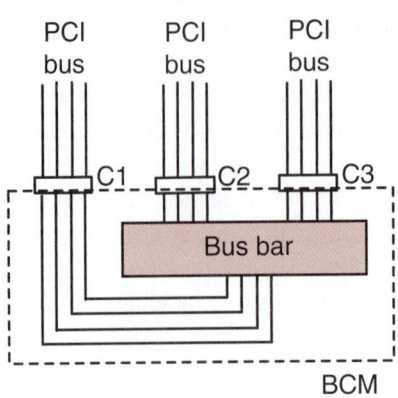

Figure 11-9 Common hub or bus bar within a controller helps make isolating the cause of bus failures easier for the technician.

an excellent place to begin diagnostics. For example, Figure 11-9 shows a PCI bus system that uses the BCM for a common hub for the bus. The BCM has three connectors attached to it that contain the different bus circuits from all of the modules. By unplugging these connectors (one at a time) and observing system operation, the system can be broken down into a few circuits. Once a connector is unplugged and some of the bus systems become active, the fault is within the bus circuits of that connector. It is possible that the connector has several bus wires in it. Use the ohmmeter to determine which bus wire is shorted to ground. Use the service information to determine what module is on that circuit and unplug the module. Again, measure resistance between the BCM connector and chassis ground. If the reading is infinite, the module is the fault. If the reading is still low, the wire is shorted to ground.

If the bus system does not have a common hub, examine the wiring diagram for the vehicle and determine if there is a connector that would separate the bus network. Identify any connectors that can be used and their location. With the ohmmeter measuring resistance between pin 2 of the DLC (using the J1962 BOB) and chassis ground, unplug the connector. If the ohmmeter still reads low resistance, the fault is between the DLC and the connector. If the ohmmeter reads any other value, the fault is downstream of the connector. Now that this is known, move the ohmmeter test leads (if necessary) to observe resistance readings, as modules on the appropriate side of the connector are unplugged. As before, if the readings change after a module is unplugged, then that module is the fault. If the reading is still low after all modules are unplugged, the fault is in the wiring harness.

If the initial tests indicate a partial bus failure, try to determine which modules are not communicating. If there is more than one, use the wiring diagrams for the vehicle to determine if there is anything common between the affected modules. For example, look for a common connector, common splice, common voltage supply to the modules, or common ground for the modules. Locate this common component on the vehicle and confirm proper function and repair as needed.

Controller Area Network Bus Diagnostics

Vehicles that utilize Controller Area Network (CAN) bus systems will generally have two or three different CAN systems. The vehicle body systems will communicate over the medium speed CAN B network. Those systems such as engine controls and antilock brakes that require data at a faster rate will communicate on the CAN C bus. Beginning in the 2005 model year, manufacturers have started migrating to the new requirement for using a CAN bus network for diagnostics (often referred to as diagnostic CAN C).

The first step after verifying the customer's complaint is to determine which bus network is at fault. Usually this can be easily determined by a scan tool. Use the scan tool to access the central gateway module (CGW) for the bus networks. Since this is the common component for all

Classroom Manual
Chapter 11, pages
326, 332

Special Tools

J1962 BOB

Scan tool

DMM

Lab scope

three CAN bus networks, it will usually provide a DTC that will indicate which bus network has the fault. If the scan tool cannot communicate to the central gateway module, then the diagnostic CAN C circuit is faulty.

Once the faulty network is identified, diagnostics of either CAN network can be performed using a DMM or a lab scope. However, methods are different based on the network and the type of fault. The following diagnostic procedure should be followed to isolate the cause:

1. If the fault code indicates an open circuit fault with either the CAN B (+) or the CAN B (−) circuit (Figure 11-10), then determine and locate an easily accessed module on the CAN B bus system. At the connector for this component, create a short to ground on the opposite Can B circuit. For example, if the fault code indicates that the CAN B (+) is open, then short the CAN B (−) circuit

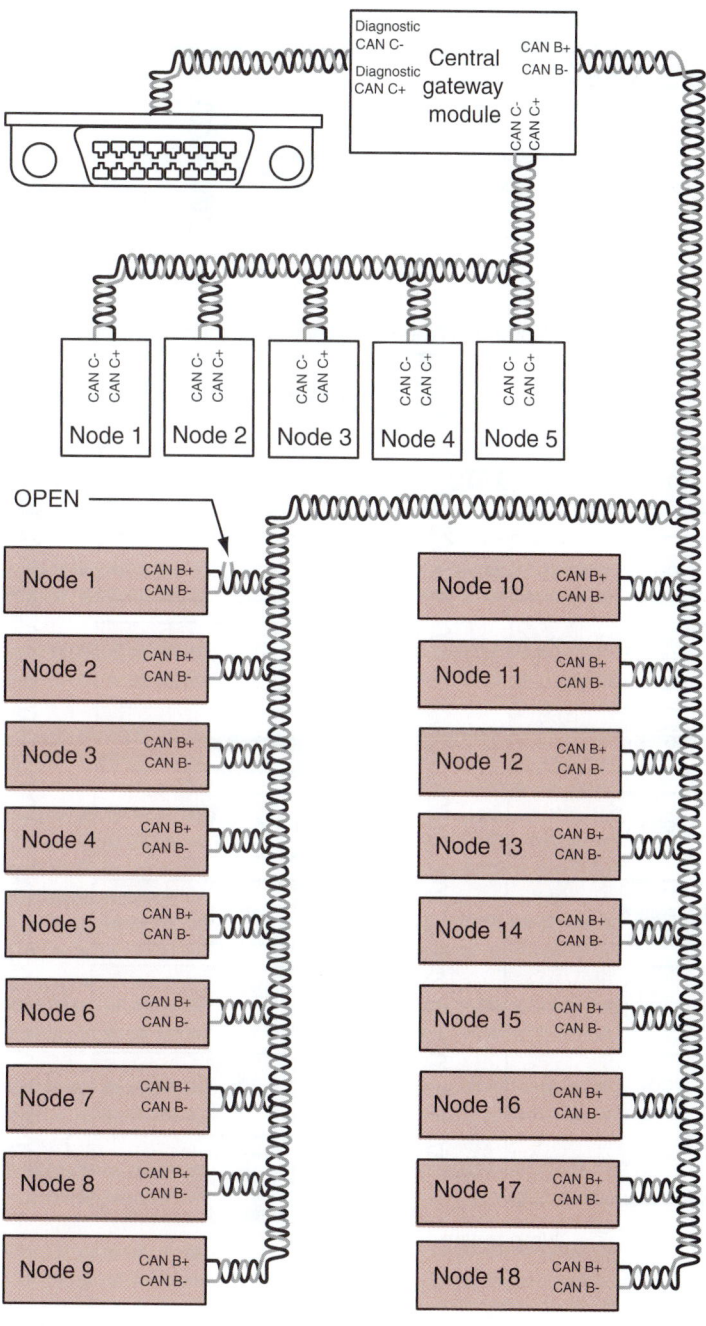

Figure 11-10 An open bus (+) wire at CAN B bus node 2 does not prevent the nodes from being active on the bus.

to ground. Since CAN B buses that operate at speeds up to 125 Kb/s are fault tolerant and will operate in single-wire mode as long as an electrical potential exists between one of the circuits and ground, the open may not cause any noticeable symptoms. However, when the other circuit is shorted to ground, that module can no longer communicate. All other modules will communicate using the single-wire mode (Figure 11-11). With the wire shorted to ground, use the scan tool to determine which module is not communicating on the bus. Locate this module on the vehicle. The fault will be between the splice from the affected bus circuit to the module.

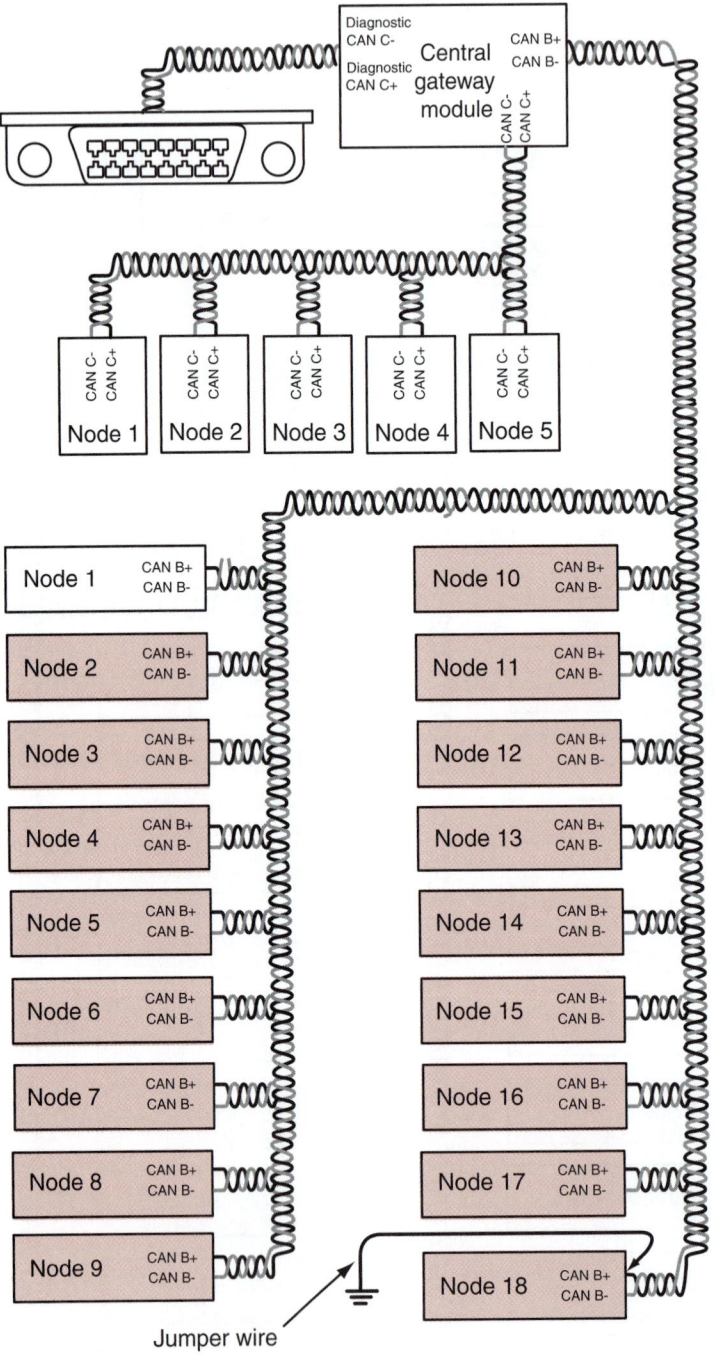

Figure 11-11 When the bus (–) wire at CAN B bus node 18 is shorted to ground, all modules continue to communicate except node 1. Node 1 cannot communicate since it no longer has a circuit.

2. If the fault code indicates one of the CAN B circuits is high or low, then use a DMM to determine if the fault is a short to voltage (12 volts) or shorted to ground (0 volts). If the voltage on both circuits is between 500 mv and 1.0 volts, the two circuits are shorted together. Locating a shorted circuit will be similar to the procedure described earlier for the J1850 VPWM bus. Use the wiring diagram for the vehicle and determine if connectors are used that can break the circuits down into smaller sections. Return to the vehicle and, with a voltmeter connected to the shorted circuit, observe the readings while opening the connectors. Once it is determined which side of the connector has the short, modules or other connectors on that branch of the circuit can then be disconnected to locate the root cause of the failure.

3. If the fault code indicates both CAN B (+) and CAN B (−) circuits are low or high, this indicates that a total bus failure has occurred. Symptoms may include headlamps on, warning indicator lamps on in the instrument panel, instrument panel backlighting at full intensity, gauges inoperative, and possible no-start. To locate the cause of this failure, use the diagnostic procedure for locating a short circuit, as previously described in 2.

4. Some modules do not set specific codes that isolate the type of fault. They will set a generic "CAN B BUS" fault. First determine if the single-wire failure is an open or short by measuring the voltages at an easily accessed module connector. If this fault is set due to an open, short to ground each CAN B circuit one at a time while monitoring the scan tool and observing symptoms. If shorting one of the circuits to ground does not affect operation or set additional DTCs, remove the short and repeat the procedure on the other CAN B circuit to determine where the open exists. The open will be on the CAN B circuit that was *not* shorted to ground. If the tests indicate that the circuit(s) is shorted to ground, shorted to voltage, or shorted to each other, then follow the diagnostic procedure outlined previously in 2.

5. If the fault code indicated that a CAN C circuit is shorted, this would cause a total failure of the CAN C bus. The short can be to voltage, ground, or the two circuits together. Symptoms may include illuminated warning indicator lamps in the instrument panel, inoperative gauges, and possible no-start. To locate the short circuit, determine if it is caused by a defective module by disconnecting the modules one at a time and observing symptoms and ohmmeter readings on the CAN circuit. If disconnecting all of the modules fails to indicate a faulty module, the problem is in the wiring harness.

6. If the fault code indicates a loss of communication with a single module (on either the CAN B or CAN C network), check for proper power and ground circuit at the affected module. If these circuits are confirmed good, then test the CAN circuits for opens between the splice to the network and the module. If the circuits do not have an open, replace the module.

CUSTOMER CARE: Modules can be very expensive and should be replaced only after all power feeds and ground circuits are tested and confirmed good.

When diagnosing the CAN bus networks, remember these facts:

1. A normal voltmeter reading on the CAN B (+) circuit with the ignition key in the RUN position is between 280 and 920 mv.

2. A normal voltmeter reading on the CAN B (−) circuit with the ignition key in the RUN position is between 4.08 and 4.72 volts.

3. A normal voltmeter reading on the CAN B (+) circuit with the ignition key in the OFF position and the bus asleep is zero volts.

4. A normal voltmeter reading on the CAN B (−) circuit with the ignition key in the OFF position and the bus asleep is battery voltage.

5. CAN B termination resistance cannot be measured. However, it should be infinite with the battery disconnected. A low reading indicates a short.

6. The CAN B bus can become active from inputs or from the ignition switch.

7. A normal CAN C bus termination is 60 ohms.

8. A normal diagnostic CAN C termination is 60 ohms.

9. A normal voltmeter reading on the CAN C (+) circuit with the ignition key in the RUN position is about 2.60 volts.

10. A normal voltmeter reading on the CAN C (−) circuit with the ignition key in the RUN position is about 2.4 volts.

11. The CAN C bus is only active with the ignition in the RUN position.

12. The wires must be twisted at a rate of one twist every inch (25.4 mm).

Classroom Manual
Chapter 11, page 337

Special Tools
Scan tool
DMM
Lab scope

Local Interconnect Network Bus Diagnostics

The local interconnect network (LIN) bus is a supplemental bus network that is used along with the vehicle's main multiplexing system. The master module communicates data from the slave modules onto the main bus network. The scan tool will not have access to the slave module. All diagnostics with the scan tool are performed through the master module.

Figure 11-12 is an illustration of a LIN bus system used for steering wheel–mounted radio controls. The steering column module (SCM) is the master module and is also connected to the CAN B bus network. The right steering wheel switch assembly is actually the slave module and is capable of sending data to the SCM and of receiving data from the SCM. The left steering wheel switch is a multiplex switch system that uses different resistance values for each switch position. The different resistances result in a unique voltage drop that is sensed by the slave module. The request action of the driver that is indicated by pressing a switch is received by the slave module and sent to the master module. The master module then puts the data onto the CAN B bus network, and the message is sent to the radio, which performs the requested action.

In this example, the scan tool will communicate with the SCM. By observing the data display information on the scan tool, each switch position input should be indicated. In addition, the master module (the SCM in this case) will store a DTC if a system indicates a fault. For example,

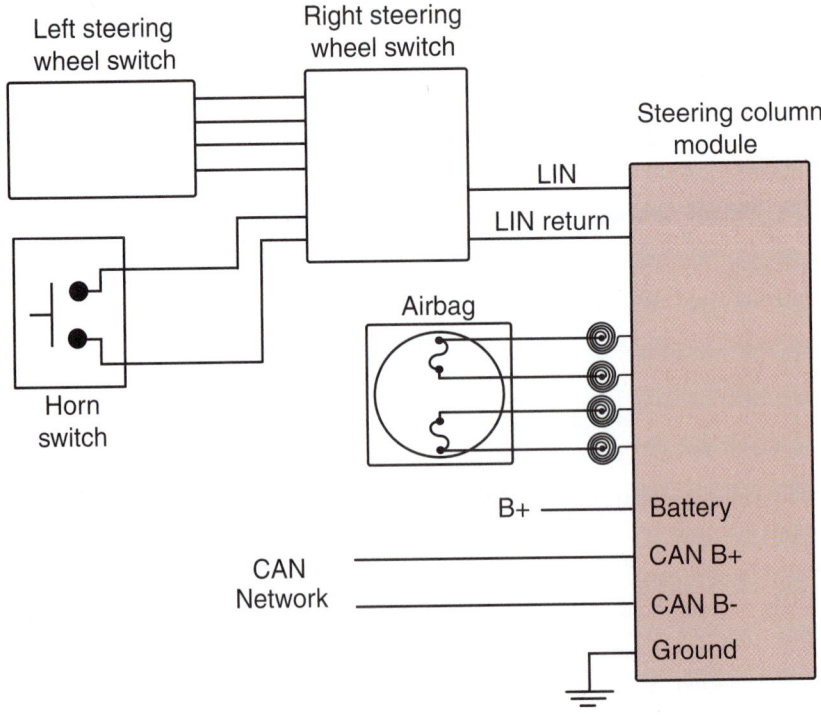

Figure 11-12 LIN bus system used for remote radio controls on the steering wheel.

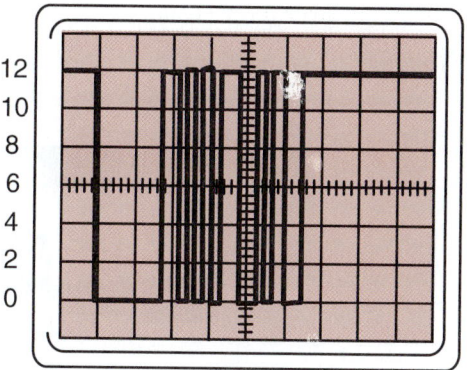

Figure 11-13 Normal LIN bus trace.

if a switch in the left steering wheel switch assembly indicates that it is stuck in the pressed position for an excessive amount of time, the SCM will set a DTC for this condition.

If a customer states that the steering wheel radio controls fail to operate, use the scan tool to see if any of the button pushes are indicated by the master module. If any are indicated, then the LIN bus circuit is working. If none of the switch inputs are indicated on the scan tool, then the LIN bus circuit, the slave module, or the master module has failed.

Normal voltages on the LIN bus is up to 12 volts, which is pulled low during communications (Figure 11-13). However, if the slave module is disconnected from the master module and the LIN bus circuit is tested with a voltmeter with the ignition switch in the RUN position, the reading will be about 9.5 volts. This is due to the master module pulse width modulating the signal in an attempt to communicate with the slave module. The voltmeter is indicating an average reading. If the slave module is plugged in and the voltage is tested, the reading will now be about 8.3 volts since both modules are pulling the circuit low in an attempt to communicate. The most accurate way of determining LIN bus voltages is with a lab scope.

If zero volts are indicated by the voltmeter or lab scope when the slave module is disconnected, the cause could be a faulty master module, an open circuit between the master and slave module connector, or a short to ground in the LIN circuit. Use the voltmeter and ohmmeter to diagnose and locate the fault.

If normal voltage is indicated with the slave module disconnected, but then reads zero volts when it is reconnected, the fault is the slave module. In this case the slave module has an internal short to ground.

> **CUSTOMER CARE:** Many people do not realize the subjectivity of the communication networks to RFI and EMI. These signals may interfere with module communications. If the customer says he is experiencing intermittent problems with the vehicle, it is possible the problem is due to an add-on feature such as a radio, cell phone, entertainment system, and so on. If this is found to be the problem, inform the customer that these systems may require special shielding and location alterations in order to prevent future instances.

Enhanced Scan Tool Diagnostics

Some scan tools provide enhanced diagnostics of the CAN bus networks. In addition to performing the module scan, these units may provide advanced loss of communication screens, network topology screens, and so on. All of these additional features are designed to assist the technician in diagnosing the CAN bus network. Photo Sequence 19 provides an example of these features and how they are accessed. Since scan tool manufacturers differ in their approach to diagnostics and display of information, this Photo Sequence is provided as a guide only. Always refer to the instructions for the scan tool you are using.

Photo Sequence 19
Advanced Scan Tool Functions

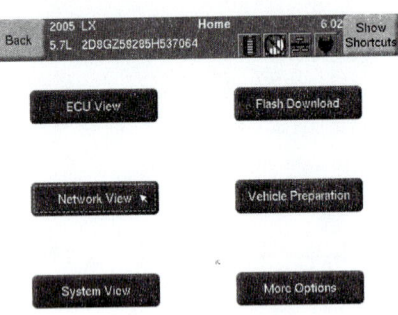

P19-1 The "Home" screen of the scan tool. Select "Network View."

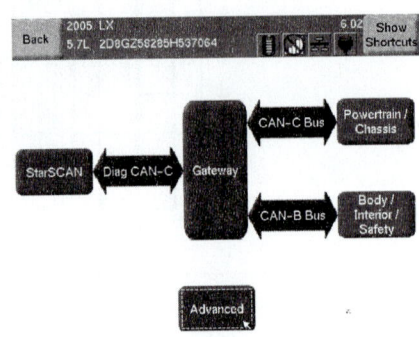

P19-2 The network view provides an icon look of the CAN bus network from the scan tool to the different bus systems. Red-colored icons indicate a fault has been recorder in that system. From this screen select the "Advanced" button.

P19-3 This is the Loss of Communications Summary screen. A green check mark indicates that the module is communicating with the scan tool. The red Xs indicates that those modules are not communicating on the network bus. Under the "Active" column, the 6 indicates that six other modules have logged loss of communication fault codes against the module and the faults are currently present. If any logged against faults were stored, they would be under the "Stored" column. Press the "Network Topology" button.

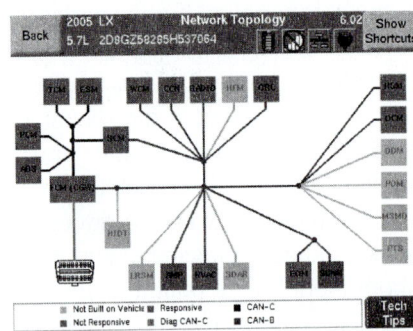

P19-4 This is a "live" layout of the vehicle's communication system. The color codes indicate if the module is active on the bus, if the vehicle is not built with that module, and the different bus networks. This is an actual representation of the vehicle's communication wiring. The red icons indicate modules that are not communicating. In this case, a common connector or splice is not the cause of the problem. The next logical diagnostic step would be to see if the two affected modules share a common power circuit or ground.

P19-5 Another option from the "Loss of Communication" screen is to press the "Advanced LoC" button. This takes the technician to another loss of communications screen.

P19-6 This advanced loss of communications screen indicates the type of fault code as "E" for electrical, "C" for communications, "S" for implausible signal, and "L.A." for logged against. Also columns indicate the number of active or stored codes and which modules have logged the fault.

A customer has had his vehicle towed to the repair facility. Other shops have attempted to repair the problem of an intermittent no-start condition, but the problem persists. The technician attempts to duplicate the problem, but the engine will now start and run fine. The technician connects the scan tool and accesses the PCM to read fault codes. There are no active or stored DTCs. Realizing that the vehicle he is working on is equipped with the CAN bus network, he accesses the central gateway module to see if it has any DTCs stored in it. Here he finds a stored "Loss of communication with PCM" fault. There are no active faults. Realizing that this is an intermittent problem, he checks the connectors at the PCM. Here he finds a loose terminal and makes the necessary repairs. The intermittent no-start was due to the vehicle's security system not being able to communicate with the PCM; thus the PCM would not allow the engine to start. No fault codes were set in the PCM since it was doing what it was supposed to. Other modules did log the loss of communication fault against the PCM, since they could not communicate with the PCM when the terminal connection came loose.

Terms to Know

B-codes

U-codes

J1962 breakout box (BOB)

Vehicle module scan

ASE-Style Review Questions

1. In a J1850 VPWM bus system, how can a short to ground be isolated?
   A. Use a jumper wire to supply battery voltage to pin 2 of the DLC while observing the DMM.
   B. Disconnect in-line connectors one at a time while observing the DMM.
   C. Use a jumper wire to short pin 15 of the DLC to ground while observing the DMM.
   D. All of the above.

2. If one of the CAN C bus circuits were shorted to ground, all of the following could occur EXCEPT:
   A. No-start.
   B. No communication between scan tool and any CAN C modules.
   C. CGW module logs a loss of communication fault code against CAN C modules.
   D. The bus will operate in single-wire mode.

3. All of the following can cause total bus failure of the J1850 VPWM bus system, EXCEPT:
   A. Open bus wire to the BCM.
   B. Bus wire shorted to ground.
   C. Bus wire shorted to battery voltage.
   D. Internal short in a module.

4. The lab scope trace of an ISO-K bus system shows the voltage going to −0.5 volts. This would indicate:
   A. The scan tool bias voltage.
   B. The module attempting to transmit data to the scan tool.
   C. The scan tool attempting to transmit data to the module.
   D. Excessive electrical noise on the circuit.

5. A class A bus that has zero volts on both bus circuits may indicate:
   A. Both circuits are shorted to ground.
   B. The master module's power feed circuit is open.
   C. An open bus connector to the master module.
   D. All of the above.

6. To isolate a bus circuit that is shorted to battery voltage, all of the following methods can be used EXCEPT:

- **A.** Use a jumper wire and connect the faulty circuit to ground.
- **B.** Turn the ignition switch off to see if it is switched voltage.
- **C.** Pull fuses one at a time to determine the shorted circuit.
- **D.** Pull relays one at a time to determine shorted circuits or components.

7. On a LIN bus, zero volts are indicated on the bus circuit with the slave module disconnected. What is the LEAST likely cause?

- **A.** An open circuit between the master module and the slave module connector.
- **B.** A shorted circuit between the master module and the slave module connector.
- **C.** Faulty master module.
- **D.** Faulty slave module.

8. *Technician A* says on the LIN bus system the scan tool accesses the slave modules to display their data. *Technician B* says the LIN slave module can only send data to the master module. Who is correct?

- **A.** A only
- **B.** B only
- **C.** Both A and B
- **D.** Neither A nor B

9. When the scan tool is connected, voltage on the K-line reads zero. *Technician A* says this means the circuit may be shorted to ground. *Technician B* says the scan tool may be faulty. Who is correct?

- **A.** A only
- **B.** B only
- **C.** Both A and B
- **D.** Neither A nor B

10. *Technician A* says normal voltages on the LIN bus is up to 12 volts. *Technician B* says 12 volts on the LIN bus indicates data transmission. Who is correct?

- **A.** A only
- **B.** B only
- **C.** Both A and B
- **D.** Neither A nor B

ASE Challenge Questions

1. *Technician A* says on a CAN B system that if one of the bus circuits is shorted to ground, it may be possible to locate the fault by jumping battery voltage to the faulty circuit. *Technician B* says to locate the short to ground, use a jumper wire and ground the opposite circuit. Who is correct?

- **A.** A only
- **B.** B only
- **C.** Both A and B
- **D.** Neither A nor B

2. *Technician A* says on an ISO 9141-2 bus system, without the scan tool connected, there should be voltage only on the L-line. *Technician B* the module supplies says the voltage on the L-line. Who is correct?

- **A.** A only
- **B.** B only
- **C.** Both A and B
- **D.** Neither A nor B

3. *Technician A* says on CAN B bus systems it may be possible for the bus to operate if there is an electrical potential between one of the circuits and ground. *Technician B* says if the two CAN B circuits are shorted together, total bus failure will result. Who is correct?

- **A.** A only
- **B.** B only
- **C.** Both A and B
- **D.** Neither A nor B

4. One module fails to communicate on the bus. *Technician A* says this can be due to an internal short of one of the bus circuits. *Technician B* says this could be due to the module ground circuit being faulty. Who is correct?

- **A.** A only
- **B.** B only
- **C.** Both A and B
- **D.** Neither A nor B

5. *Technician A* says U-codes identify a failure with a body system function. *Technician B* says B-codes are set to indicate bus network communication failures. Who is correct?

- **A.** A only
- **B.** B only
- **C.** Both A and B
- **D.** Neither A nor B

Job Sheet 38

Name _____ Date _____

ISO 9141-2 Bus System Diagnosis

Upon completion of this job sheet, you will be able to determine normal ISO 9141-2 bus operation. You will also observe faulty bus system operation.

ASE Correlation

This job sheet is related to the ASE Electrical/Electronic System Certification Exam's content area: *Accessories Diagnosis and Repair;* task: Check for module communication errors using a scan tool.

Tools and Materials

J1962 BOB
Scan tool
DMM
Lab scope
Jumper wires
Service manual
Vehicle with an ISO 9141-2 bus system

Describe the vehicle being worked on:

Year _____ Make _____ Model _____

VIN _____ Engine type and size _____

Procedure

Task 1

Task Completed

1. Use the service manual to identify DLC cavities assigned to the ISO 9141-2 circuits. These circuits are usually referred to as the K-line and the L-line.

   a. K-line

   Cavity number _____

   Circuit identification _____

   Color code _____

   b. L-line

   Cavity number _____

   Circuit identification _____

   Color code _____

2. Connect the J1962 BOB to the DLC. Use the voltmeter function of the DMM to measure the voltage on each bus circuit without a scan tool connected to the DLC.

   K-line _____

   L-line _____

3. Connect the scan tool to the DLC and observe and record the voltages for each bus circuit.

K-line _____

L-line _____

4. Establish communications between the scan tool and the PCM and record the voltages on each bus circuit.

K-line _____

L-line _____

5. Remove the scan tool from the DLC and use a lab scope to trace the voltage on each bus circuit with the ignition switch in the OFF position. If available, use the print function to print the trace for each circuit. If not available, describe what was observed.

K-line _____

L-line _____

6. Connect the scan tool, and with the ignition switch in the OFF position, use a lab scope to trace the voltage on each bus circuit. If available, use the print function to print the trace for each circuit. If not available, describe what was observed.

K-line _____

L-line _____

7. Turn the ignition switch to the RUN position, and use a lab scope to trace the voltage on each bus circuit. If available, use the print function to print the trace for each circuit. If not available, describe what was observed.

K-line _____

L-line _____

8. Establish communications between the scan tool and the PCM while observing the lab scope trace. If available, use the print function to print the trace for each circuit. If not available, describe what was observed.

K-line _____

L-line _____

9. Based on your observations of normal bus operation, what can you conclude concerning normal voltage?

_____

_____

_____

Task 2

1. Use a jumper wire and short the K-line from the J1962 BOB to chassis ground.

2. Connect the scan tool, and turn the ignition switch to the RUN position; attempt to establish communications with the PCM. Record the results.

_____

_____

3. Use a lab scope to trace the voltages on both circuits. If available, use the print function to print the trace for each circuit. If not available, describe what was observed.

K-line _____

L-line _____

4. Based on your observations, what is the effect of the short to ground?

_____

_____

Instructor's Response _____

_____

_____

_____

Job Sheet 39

39

Name _____ Date _____

ISO-K Bus System Diagnosis

Upon completion of this job sheet, you will identify normal operating characteristics of the ISO-K bus using a lab scope and a scan tool.

ASE Correlation

This job sheet is related to the ASE Electrical/Electronic System Certification Exam's content area: *Accessories Diagnosis and Repair;* task: Check for module communication errors using a scan tool.

Tools and Materials

J1962 BOB
Scan tool
DMM
Lab scope
Jumper wires
Service manual
Vehicle with an ISO-K bus system

Describe the vehicle being worked on:

Year _____ Make _____ Model _____

VIN _____ Engine type and size _____

Procedure

Task Completed

1. Use the service manual to identify DLC cavity assigned to the ISO-K circuit to the PCM.

   a. K-line

   Cavity number   _____

   Circuit identification _____

   Color code   _____

2. Connect the J1962 BOB to the DLC. Use a lab scope to trace the voltage on the K-line with the ignition switch in the OFF position. If available, use the print function to print the trace for the K-line circuit. If not available, describe what was observed.

   _____

3. Connect the scan tool, and with the ignition switch in the OFF position, use a lab scope to trace the voltage on the K-line. If available, use the print function to print the trace for the K-line circuit. If not available, describe what was observed.

   _____

4. With the ignition switch still in the OFF position, attempt to communicate with the PCM. Use a lab scope to trace the voltage on the K-line. If available, use the print function to print the trace for the K-line circuit. If not available, describe what was observed.

_____

5. Turn the ignition switch to the RUN position. Establish communications between the scan tool and the PCM while observing the lab scope trace. If available, use the print function to print the trace for the K-line circuit. If not available, describe what was observed.

_____

6. What is indicated by each of the voltage values seen on the trace?

_____

_____

_____

Job Sheet 40

40

Name _____ Date _____

J1850 Bus System Diagnosis

Upon completion of this job sheet, you will be able to use the scan tool and a DMM to monitor normal J1850 bus activity. You will also use the lab scope to observe normal bus operation. Finally, you will use the DMM to diagnose full and partial failures of the J1850 bus.

ASE Correlation

This job sheet is related to the ASE Electrical/Electronic System Certification Exam's content area: *Accessories Diagnosis and Repair;* task: Check for module communication errors using a scan tool.

Tools and Materials

J1962 BOB
Scan tool
DMM
Lab scope
Jumper wires
Service manual
Vehicle with J18050 bus
Instructor-installed bugs

Describe the vehicle being worked on:

Year _____ Make _____ Model _____

VIN _____ Engine type and size _____

Procedure

Task Completed

Task 1

1. Use the service manual to identify the DLC cavity number(s), circuit identification, and color code(s). Record your findings.

   a. Cavity number(s) _____

   b. Circuit identification _____

   c. Color code(s) _____

2. Connect the J 1962 BOB to the DLC, and connect the scan tool to the J1962 BOB. Turn the ignition switch to the RUN position. ☐

3. If available, use the scan tool to perform an 1850 module scan. List all active modules on the J1850 bus.

   _____

   _____

   _____

   _____

4. Does the J1850 bus appear to be operating normally?   ☐ Yes   ☐ No

Explain your answer:

_____

_____

5. Connect the DMM positive lead to the J1850 bus circuit of the J 1962 BOB (a jumper wire may be required). Connect the negative lead to a good chassis ground. With the DMM set to read DC volts, observe and record the voltage range:

MIN _____   MAX _____

6. Remove the DMM, and use a lab scope to observe the J1850 bus activity. If available, use the print function to print off the trace. If not available, describe the lab scope display:

_____

_____

_____

☐

7. Turn the ignition switch to the OFF position, and disconnect the vehicle battery.

8. Use the ohmmeter function of the DMM to measure the resistance between the J1962 BOB cavity for the J1850 bus and chassis ground. Observe and record the J1850 bus termination resistance. _____

9. Disconnect a regular node from the J1850 bus and record the bus termination resistance. _____

10. Disconnect the scan tool from the J1850 bus and record the bus termination resistance. _____

11. Disconnect the PCM from the J1850 bus and record the bus termination resistance. _____

12. Based on your observations, are the nodes wired in series or parallel? _____

13. Based on your observations, which module had the greatest impact on bus terminal resistance and why?

_____

_____

☐

14. Reconnect all modules and then reconnect the vehicle battery.

Task 2

1. Start the engine, and use a jumper wire to short the J1850 bus circuit of the J1962 BOB to chassis ground. What are the observable conditions?

_____

_____

_____

2. Using the scan tool, perform a J1850 module scan and record the results.

_____

3. Use the voltmeter function of the DMM to observe and record the J1850 bus voltage.
_____

4. Does the measured voltage indicate normal operation?   ☐ Yes   ☐ No

5. Turn the ignition switch to the OFF position, and disconnect the vehicle battery.   ☐

6. Use the ohmmeter function of the DMM; observe and record the J1850 bus termination resistance. _____

7. Explain the difference in this reading compared to task 1, step 11.

_____

_____

_____

8. Remove the jumper wire used to short the bus to ground and reconnect the vehicle battery.   ☐

9. With the engine running, use a jumper wire to short the J1850 bus circuit of the J1962 BOB to battery voltage. What are the observable conditions?

_____

_____

_____

10. Use the scan tool to perform a J1850 module scan and record the results.

_____

_____

11. Use the voltmeter function of the DMM; observe and record the J1850 bus voltage.

_____

12. Turn the ignition switch to the OFF position. Observe and record the J1850 bus voltage.

_____

13. If the voltage changed between step 11 and step 12, what does this indicate?

_____

_____

14. Based on your observations, can the J1850 bus be diagnosed with a DMM?

☐ Yes   ☐ No

Explain your answer.

_____

_____

Task 3

1. A vehicle is brought to the shop with inoperative gauges and several electrical accessories not functioning properly

Can the condition be verified?   ☐ Yes   ☐ No

2. Use the scan tool to perform a J1850 module scan. Are there any modules present?

☐ Yes   ☐ No

3. Disconnect the scan tool. With the ignition switch in the RUN position, use the J1962 BOB and the voltmeter function of the DMM to measure and record the J1850 bus voltage. _____

Is the voltage within normal limits?   ☐ Yes   ☐ No

4. What does the voltmeter reading indicate?

_____

_____

5. Turn the ignition switch to the OFF position, and measure and record the J1850 bus voltage at the BOB.

_____

6. What conclusions, if any, can be made at this time?

_____

_____

_____

7. While the voltmeter is reading the indicated voltage of step 3, monitor the voltage while removing fuses from the power distribution center and junction box one at a time. Which fuses, when removed, caused a change in bus voltage?

_____

8. Use the service information to determine which circuits or systems are implicated by the fuses removed in step 7.

_____

9. At the vehicle, isolate the implicated components one at a time by disconnecting them and observing the voltmeter readings. Does isolating the component correct the problem? ☐ Yes ☐ No

   If yes, identify the faulty component.

   _____

   If no, what is the next step?

   _____

10. If the answer to step 9 was no, perform the tests you indicated that still need to be done. Was the root cause of the fault able to be determined? ☐ Yes ☐ No

   If yes, describe the fault.

   _____

   If no, consult with your instructor.

   _____

Task 4

1. Your instructor will inform you of the customer complaint. Record the problem.

   _____

   _____

2. Can the problem be verified? ☐ Yes ☐ No

3. Use the scan tool to communicate with the module(s) that operate the faulty system. Are communications with the module possible? ☐ Yes ☐ No

   If yes, record any DTCs that are present.

   _____

   _____

   If no, perform a J1850 module scan and record your results.

   _____

   _____

4. What problems are indicated?

   _____

   _____

5. Based on these observations and your understanding of J1850 bus operation, what is the next logical step?

   _____

   _____

6. Based on your answer to step 5, assemble the necessary information needed to diagnose the fault. Record this information.

_____

_____

_____

7. Based on the information gathered, return to the vehicle and determine the cause of the fault. What is the necessary repair action?

_____

_____

_____

Instructor's Response _____

_____

_____

Job Sheet 41

41

Name _____ Date _____

CAN Bus System Diagnosis

Upon completion of this job sheet, you will be able to identify the circuits that make up the CAN bus and determine their operating characteristics.

ASE Correlation

This job sheet is related to the ASE Electrical/Electronic System Certification Exam's content area: *Accessories Diagnosis and Repair;* task: Check for module communication errors using a scan tool.

Tools and Materials

J1962 BOB
Scan tool
DMM
Lab scope
Jumper wires
Service manual
Vehicle with full vehicle CAN bus system

Describe the vehicle being worked on:

Year _____ Make _____ Model _____

VIN _____ Engine type and size _____

Procedure

Task Completed

Task 1

1. Use the service manual to identify the following CAN bus circuit numbers and color codes:

   NOTE: The circuits may be called by other names than those listed here.

   a. CAN B (+) _____

   b. CAN B (−) _____

   c. CAN C (+) _____

   d. CAN C (−) _____

   e. Diagnostic CAN C (+) _____

   f. Diagnostic CAN C (−) _____

2. Identify the cavity numbers the diagnostic CAN bus uses at the DLC.

   a. Diagnostic CAN C (+) _____

   b. Diagnostic CAN C (-) _____

3. Identify the cavity numbers of a connector to a module on the CAN B bus network.

    a. Module _____

      i. CAN B (+) cavity _____

      ii. CAN B (–) cavity _____

4. Identify the cavity numbers for the CAN C bus network at the PCM connector.

    a. CAN B (+) _____

    b. CAN B (–) _____

5. Using the service manual information, determine the baud rate of each of the bus networks.

    a. CAN B (+) _____

    b. CAN B (–) _____

    c. CAN C (+) _____

    d. CAN C (–) _____

    e. Diagnostic CAN C (+) _____

    f. Diagnostic CAN C (–) _____

Task 2

1. Locate the CAN B module from task 1, step 3 on the vehicle. With the ignition switch in the RUN position, use probing tools to connect the voltmeter leads to the CAN B (+) circuit and chassis ground. Record your reading. _____

2. Read the voltage on the CAN B (–) circuit with the ignition switch in the RUN position.

_____

3. Turn the ignition switch to the OFF position, and close all doors so the bus can power down. While the bus is powering down, record the voltages on each circuit.

    a. CAN B (+) _____

    b. CAN B (–) _____

4. Leave the ignition switch in the OFF position, and open a door. Record the bus voltages now.

    a. CAN B (+) _____

    b. CAN B (–) _____

5. Based on your observations, what is required to activate the CAN B bus network?

_____

6. Turn the ignition switch to the OFF position and disconnect the vehicle battery.

7. Measure the resistance between the two CAN B bus circuits. _____

Explain your results. _____

_____

Task 3

1. Connect the J1962 BOB to the DLC, and turn the ignition switch to the RUN position. Use the scan tool to determine which modules on the vehicle are on the CAN B bus and record them.

_____

_____

_____

2. Access the CAN B module connector from task 1, step 3, and use a jumper wire to short one of the CAN B bus circuits to ground. Use the scan tool to determine which modules are active on the CAN B bus network.

_____

_____

_____

3. Using a jumper wire and a probing tool, short both CAN B bus circuits together. Use the scan tool to determine which modules are active on the CAN B bus network.

_____

_____

_____

4. Using a jumper wire and a probing tool, short both CAN B bus circuits to ground. Use the scan tool to determine which modules are active on the CAN B bus network.

_____

_____

_____

5. Use the scan tool and access the central gateway module. Record any fault codes listed.

_____

_____

_____

6. Remove the short from the bus circuit ☐

7. Based on your observations, is this bus network fault tolerant? ☐ Yes ☐ No

Explain your results.

_____

_____

Task 4

1. Use the scan tool to identify the modules that are on the CAN C bus network and record them.

_____

_____

_____

2. Locate the PCM connector identified in task 1, and disconnect it from the PCM. Use a jumper wire and a probing tool to short one of the CAN C bus circuits to chassis ground.

☐

3. Use the scan tool and attempt to establish communications with any of the modules identified in step 1 (except the PCM). Is communication possible?　☐ Yes　☐ No

Explain your results. _____

_____

4. Use a jumper wire and a probing tool to short both of the CAN C bus circuits together. Use the scan tool and attempt to establish communications with any of the modules identified in step 1 (except the PCM). Is communication possible?　☐ Yes　☐ No

Explain your results. _____

_____

5. Based on your observations, is this bus network fault tolerant?　☐ Yes　☐ No

Explain your results.

_____

_____

☐

6. Remove all faults, but leave the PCM connector unplugged.

Task 5

1. Remove the scan tool from the DLC. Use a probing tool and the voltmeter function of the DMM to obtain and record the CAN C bus voltages during the following conditions.

   a. Ignition switch in RUN position.　CAN C (+) _____　CAN C (–) _____

   b. Ignition switch in OFF position.　CAN C (+) _____　CAN C (–) _____

   c. Ignition switch in OFF position and CAN B bus powered down.
   CAN C (+) _____　CAN C (–) _____

   d. Ignition switch in OFF position; open a door.
   CAN C (+) _____　CAN C (–) _____

2. Based on your observations, what is required to activate the CAN B bus network?

_____

☐

3. Turn the ignition switch to the OFF position and disconnect the vehicle battery.

4. Measure the resistance between the two CAN C bus circuits.

_____

Explain your results. _____

_____

5. Reconnect all connectors.

☐

Task 6

1. Connect the scan tool to the J1962 BOB and the DLC. Use a jumper wire to short one of the diagnostic CAN C bus circuits at the BOB to ground. Attempt to communicate with the PCM. Was communication possible?

☐ Yes　　☐ No

2. Remove the short, and move the jumper wire to short the other diagnostic CAN C bus circuit to ground. Attempt to communicate with the PCM. Was communication possible?

☐ Yes　　☐ No

3. Use a jumper wire to short the two diagnostic CAN C bus circuits together. Attempt to communicate with the PCM. Was communication possible?

☐ Yes　　☐ No

4. Based on your observations, what is your conclusion?

_____

_____

_____

5. Turn the ignition switch to the OFF position, and disconnect the vehicle battery.

☐

6. Measure the resistance between the two diagnostic CAN C bus circuits. _____

Explain your results. _____

_____

Instructor's Response _____

_____

_____

_____

Job Sheet 42

42

Name _____ Date _____

LIN Bus System Diagnosis

Upon completion of this job sheet, you will be able to determine normal LIN bus operation. You will also observe faulty bus system operation.

ASE Correlation

This job sheet is related to the ASE Electrical/Electronic System Certification Exam's content area: *Accessories Diagnosis and Repair;* task: Check for module communication errors using a scan tool.

Tools and Materials

Scan tool
DMM
Lab scope
Jumper wires
Service manual
Vehicle with LIN bus system

Describe the vehicle being worked on:

Year _____ Make _____ Model _____

VIN _____ Engine type and size _____

Procedure

Task Completed

1. Use the service manual to identify the function(s) of the LIN bus(es) on the assigned vehicle.

   _____
   _____
   _____

2. Identify the LIN bus cavities of the master module.

   _____
   _____
   _____

3. Identify the LIN bus cavities of the slave module.

_____

_____

_____

4. Does the slave module have any inputs to it from other sources? ☐ Yes ☐ No

If yes, list them.

_____

_____

5. With the ignition switch in the RUN position, connect the scan tool and navigate to the master module. Are any items available in the data display to assist in diagnosing a system failure? ☐ Yes ☐ No

If yes, list them.

_____

_____

6. Press and release any switch inputs that were identified while observing the scan tool. Do the switch states change on the scan tool? ☐ Yes ☐ No

7. Press and hold one of the switch buttons for a minimum of 60 seconds, and then release. Check and record any DTCs.

_____

_____

8. Is the DTC active or stored? _____

a. If active, press and release the same button again. Did it change to stored now?
☐ Yes ☐ No

9. Disconnect the electrical connector to a switch input to the slave module. Check and record any DTCs.

_____

_____

10. If no DTCs were set, why not?

_____

11. Press and release another switch input to the slave module while observing the scan tool. Does the scan tool indicate the switch presses? ☐ Yes ☐ No

12. Explain your results.

_____

_____

13. With the slave module still disconnected, use the voltmeter function of the DMM to measure the voltage on the LIN bus. _____

14. Reconnect the slave module and record the voltage on the LIN bus. _____

15. Are the voltages the same?　　☐ Yes　　☐ No

If no, explain.

_____

16. Use a lab scope to observe the trace of the LIN bus voltages. Describe what is observed.

_____

17. Short the LIN bus circuit to ground by back probing the slave module connector and using a jumper wire connected to a chassis ground. Check and record any DTCs.

_____

_____

18. Caution: Verify circuit identification before performing this step or vehicle damage could occur.

Short the LIN bus circuit to power by back probing the right steering wheel switch connector; use a fused jumper wire connected to vehicle power. Check and record any DTCs.

_____

_____

19. Record your conclusions.

_____

_____

_____

_____

Instructor's Response _____

_____

_____

_____

DIAGNOSTIC CHART 11-1

PROBLEM AREA:	Vehicle communication systems
SYMPTOMS:	Scan tool cannot communicate with a module
POSSIBLE CAUSES:	**1.** Faulty battery feed circuit to module **2.** Faulty ignition feed circuit to module **3.** Poor module ground **4.** Open data bus circuit to module

DIAGNOSTIC CHART 11-2

PROBLEM AREA:	Vehicle communication systems
SYMPTOMS:	Scan tool cannot communicate with any module
POSSIBLE CAUSES:	**1.** Open bus circuit at DLC **2.** Bus circuit shorted to ground **3.** Bus circuit shorted to voltage **4.** Interal failure in a module causing short to power **5.** Internal failure in a module causing short to ground **6.** Open termination resistors **7.** Faulty scan tool

DIAGNOSTIC CHART 11-3

PROBLEM AREA:	Vehicle communication systems
SYMPTOMS:	Total bus failure indicated by default conditions and possible no-start
POSSIBLE CAUSES:	**1.** Open bus circuit at DLC **2.** Bus circuit shorted to ground **3.** Bus circuit shorted to voltage **4.** Internal failure in a module causing short to power **5.** Internal failure in a module causing short to ground **6.** Open termination resistors

Advanced Lighting Systems Diagnosis and Service

Upon completion and review of this chapter, you should be able to:

- ❑ Diagnose computer-controlled concealed headlight systems.
- ❑ Perform a functional test of the automatic headlight system.
- ❑ Diagnose the automatic headlight system.
- ❑ Test the automatic headlight system's photocell.
- ❑ Replace the photocell assembly.

- ❑ Test the automatic headlight system's amplifier.
- ❑ Diagnose Ford's illuminated entry system as an example of control module systems.
- ❑ Diagnose Chrysler's illuminated entry system as an example of BCM-controlled systems.
- ❑ Diagnose fiber optic systems.

Basic Tools

Basic mechanic's tool kit
Service manual

Introduction

Diagnosis of computer-controlled lighting systems is designed to be as easy as possible. The controller may provide trouble codes to assist the technician in diagnosis. Most manufacturers provide a detailed diagnostic chart for the most common symptoms. The most important thing to remember when diagnosing these systems is to follow the diagnostic procedures in order. Do not attempt to get ahead of the chart by assuming the outcome of a test. This will lead to replacement of good parts and lost time.

In this chapter, you will perform selected service samples on the computer-controlled concealed headlight, automatic headlight, and illuminated entry systems. It is out of the scope of this manual to provide service procedures for the different manufacturers that use these systems. Technicians must follow the procedure in the service manual for the vehicle they are diagnosing.

The second section of this chapter covers diagnosis and service of electronic instrument panels.

Computer-Controlled Concealed Headlight Diagnosis

The body control module (BCM) has been utilized by some manufacturers to operate the concealed headlight system. The BCM will receive inputs from the headlight and flash-to-pass switches. The operation of the headlight door motor is controlled by relays that are energized by the BCM.

Customer complaints associated with the concealed headlight system may include:

1. Headlight doors will not open.
2. Headlight doors will not close.
3. Headlight doors will not open for flash-to-pass.
4. Headlight doors do not operate by headlight switch.
5. Headlights do not turn off.

Classroom Manual
Chapter 12, page 343

Special Tools

Scan tool
DMM
Jumper wires
Fender covers
Safety glasses
Service manual

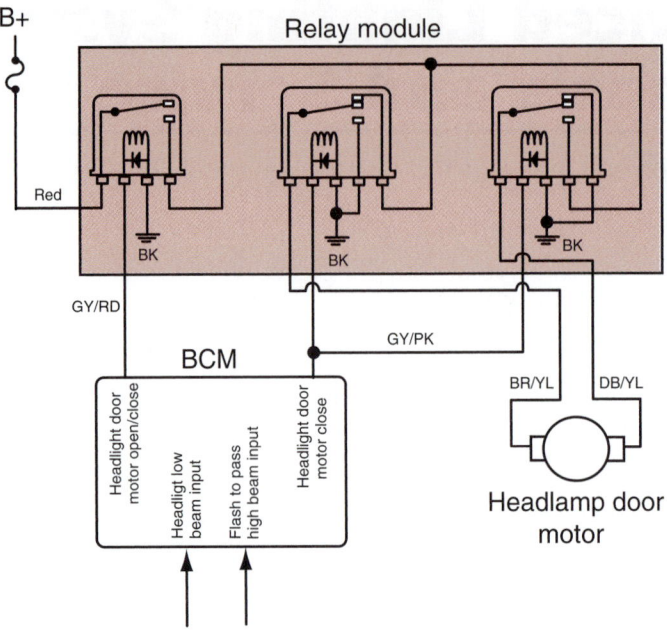

Figure 12-1 Concealed headlight door schematic.

The illustration (Figure 12-1) is a schematic of the concealed headlight system used on some Chrysler vehicles. Confirm the complaint by trying to operate the system. Once the complaint is confirmed, make a complete visual inspection of the system. Look for loose connections, broken wires, damaged components, and so on. If the visual inspection does not reveal an obvious fault, refer to the service manual for the specific tests to be performed as determined by the symptom. Study the schematic of the system you are diagnosing until you understand how it is supposed to operate.

The following is a diagnostic service sample of a customer complaint that the headlight doors will not open. Always refer to the correct service manual procedures for the vehicle you are diagnosing.

Connect the scan tool to the diagnostic connector, then turn the ignition switch to the RUN position. Follow the menus on the scan tool screen to access the BCM. Next, select the "headlight door test" option.

The exact method of getting to the headlight door test may vary depending on the year and scan tool.

The headlight doors should open and close as the scan tool operates the system. If the doors do not operate, go to step 1. If the doors operate, begin at step 4 of the following test procedure.

⚠ **WARNING:** Do not place your hands close to the headlight doors when the test is activated.

1. Check the relay fuse. A blown fuse indicates a possible problem in the circuit to the relay module and door motors. If the fuse is blown, go to step 2. If the fuse is not blown, go to step 3.

2. Disconnect the headlight door relay module (Figure 12-2). Use an ohmmeter to measure the resistance between the connector red wire and ground (Figure 12-3). The resistance should be higher than 5 ohms. If it is not, there is a short to ground in the red wire. If the resistance is above 5 ohms, disconnect the headlight door connector and measure the resistance between the relay module connector DB/YL wire and ground (Figure 12-3).

   ❑ If the resistance is lower than 5 ohms, the DB/YL wire has a short to ground between the relay module and the door motor.

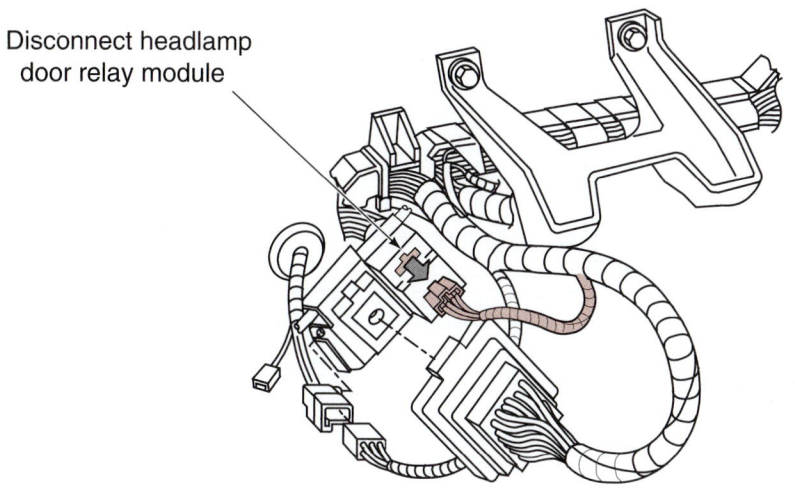

Disconnect headlamp door relay module

Figure 12-2 Headlight door relay module location.

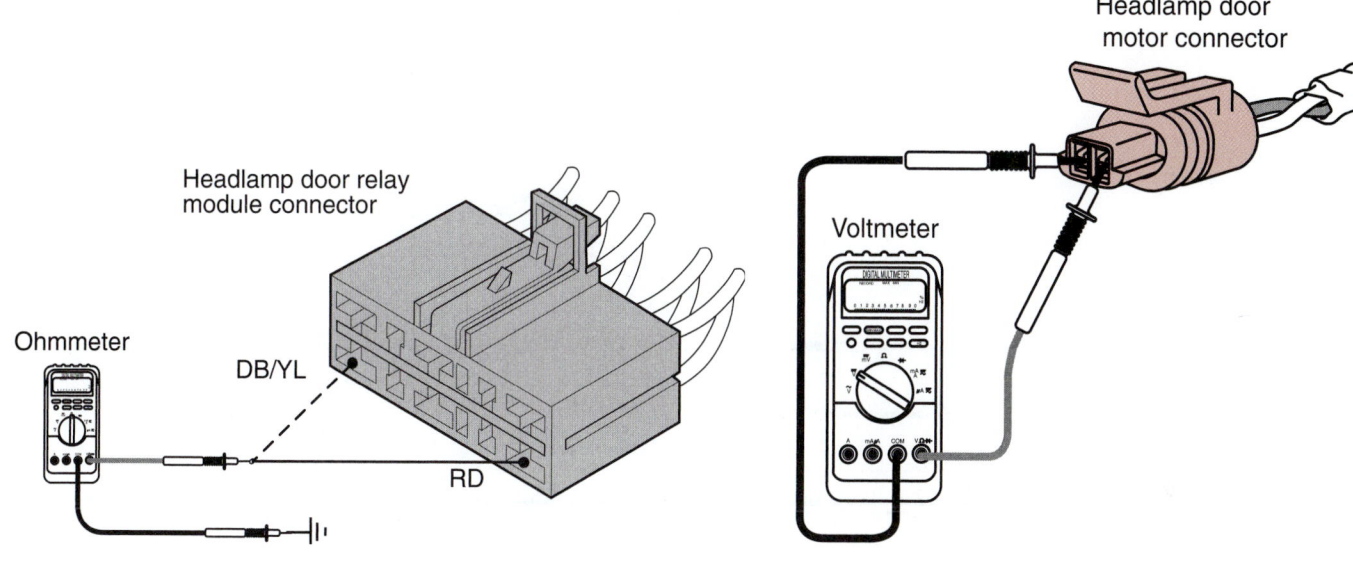

Headlamp door relay module connector

Ohmmeter

DB/YL

RD

Figure 12-3 Door relay module connection test locations.

Headlamp door motor connector

Voltmeter

Figure 12-4 Connect a voltmeter across the door motor connector.

❏ If it is higher than 5 ohms, reconnect the relay module. Use a DMM to test for voltage across the headlight door motor connector (Figure 12-4). Voltage will be present for only a few seconds when the headlight switch is turned on. If the voltage is 10 volts or higher, replace the door motor. If it is less than 10 volts, replace the door relay module.

3. Use a voltmeter to test for applied voltage to the fuse (Figure 12-5). No voltage at this point indicates there is an open or short in the circuit between the battery and the fuse box. If there are more than 10 volts applied to the fuse box, disconnect the headlight door motor connector. Reinstall the fuse and use a DMM to measure the voltage across the connector terminals when the headlight switch is turned on. The voltage will be present for only a few seconds.

❏ If the voltage is higher than 10 volts and a good ground is confirmed, replace the motor.

❏ If the voltage is less than 10 volts, turn off the headlights and back probe the BR/YL wire at the relay module using the voltmeter (Figure 12-6). Voltage should be present for a few seconds when the headlight switch is turned on.

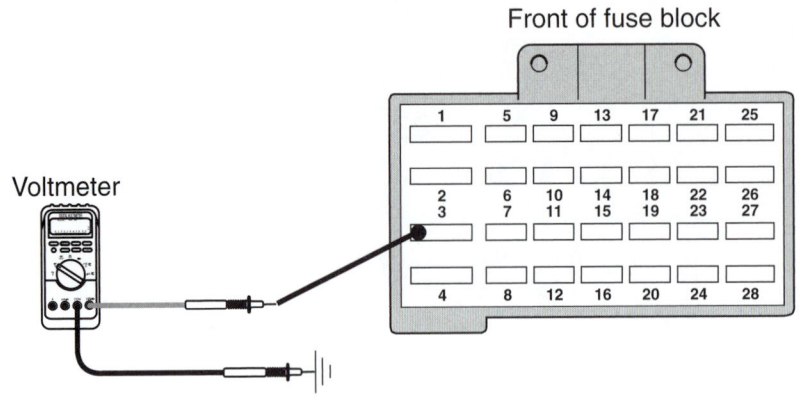

Figure 12-5 Using a voltmeter to check for voltage to the fuse.

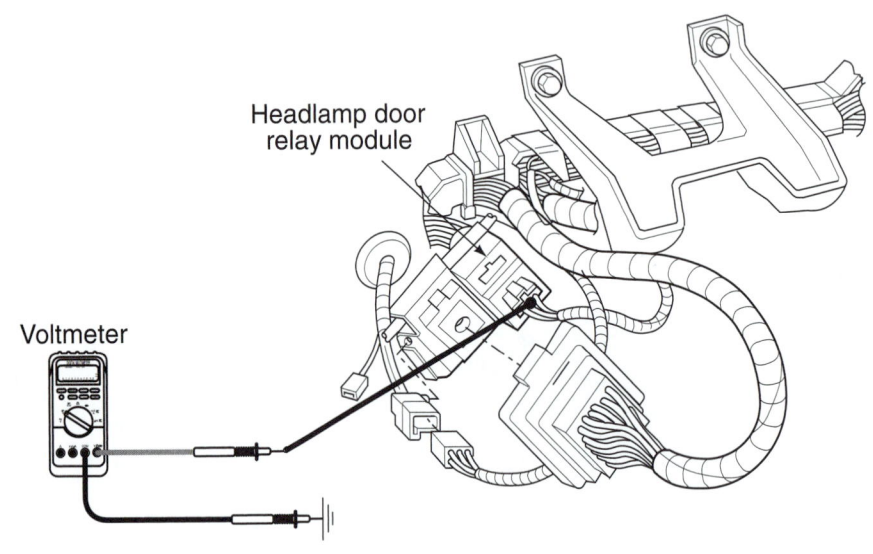

Figure 12-6 Back probing the door relay module for voltage.

❏ If the voltage is higher than 10 volts, there is an open in the BR/YL wire between the relay module and the motor.

❏ If the voltage is less than 10 volts, back probe the RD wire at the relay module connection.

   ❏ If the voltage is lower than 10 volts, there is an open in the RD wire between the relay module and the relay fuse.

   ❏ If the voltage is over 10 volts, turn the ignition switch off and use an ohmmeter to test the resistance at the relay module connector BK wire (Figure 12-7). A resistance reading higher than 5 ohms means there is an open in the BK wire from the connector to ground. If it is lower than 5 ohms, connect a jumper wire from the battery positive terminal to the relay module GY/RD wire. The doors should open. If not, replace the relay module. If the doors open, use a jumper wire to the body controller cavity 24 and ground (Figure 12-8). If the doors open, replace the BCM. If they do not open, there is an open or short in the circuit between the BCM and relay module.

4. If the doors operated properly when the scan tool performed the headlight door test, use the scan tool to access the switch test mode and select "headlight switch" from the menu.

5. While observing the scan tool, turn on the headlight switch. If the scan tool indicates the circuit is closed, replace the BCM.

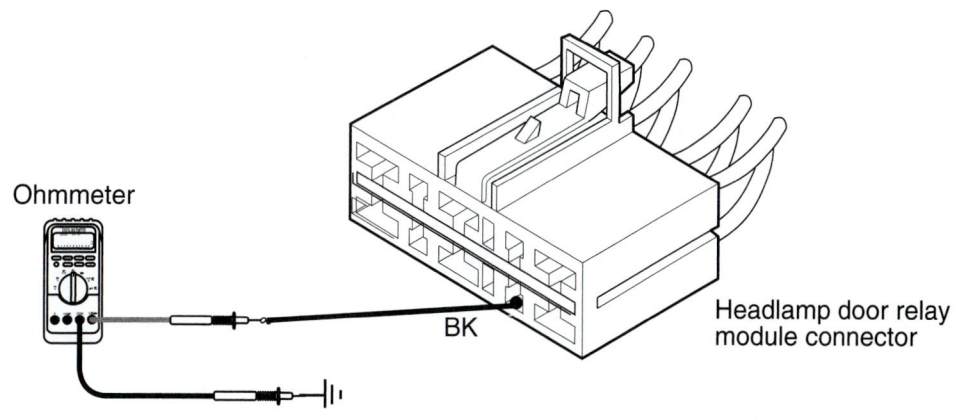

Ohmmeter

BK

Headlamp door relay module connector

Figure 12-7 Using an ohmmeter to test the resistance in the circuit to the door relay module connector.

6. If the scan tool displays that the circuit is open, disconnect the left POD switch connector (Figure 12-9). Use a jumper wire connected between the OR/WT wire terminal and ground.

7. Observe the scan tool. If it reads that the circuit is open, there is an open in the OR/WT wire between the POD and the BCM terminal 18. If a closed circuit is indicated, use an ohmmeter to measure the resistance to ground at the POD switch connector BK/OR wire.

❏ If the resistance is lower than 5 ohms, replace the POD switch.

❏ A reading higher than 5 ohms indicates an open in the circuit to ground.

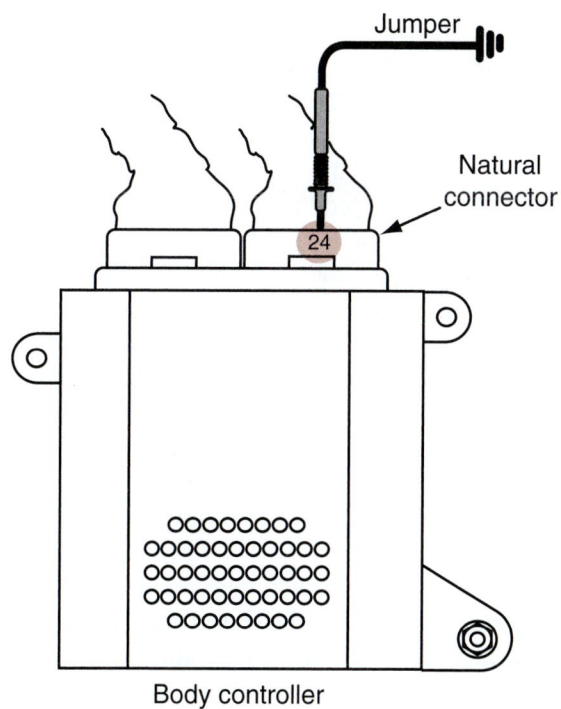

Jumper

Natural connector

24

Body controller

Figure 12-8 Using a jumper wire to ground circuit cavity 24 to the BCM.

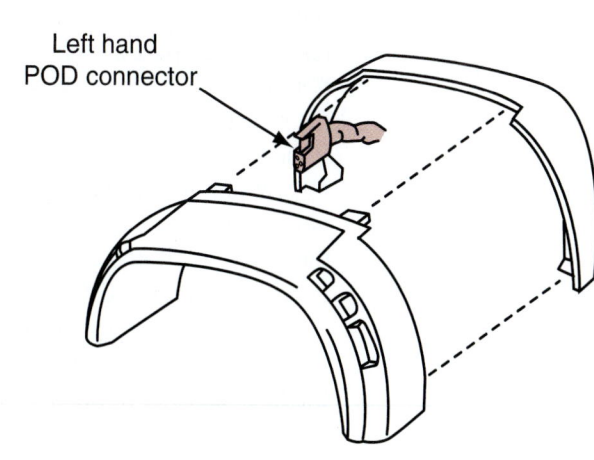

Left hand POD connector

Figure 12-9 Left POD connector location.

The preceding procedure provided an example of testing a concealed headlight system. Regardless of the vehicle being worked on, the diagnosis is broken down into sections. The system uses inputs, a control module, and the actuators. By performing the door open actuator test, you can determine if the fault is isolated to the control module or the motor circuit. If the actuation test passes, then check the inputs that are used.

Classroom Manual
Chapter 12, page 349

Special Tools

Scan tool

DMM

Fused jumper wires

Test light

Testing Computer-Controlled Headlight Systems

A basic computer-controlled headlight system will use the headlight switch as an input to the BCM, which then activates the required relays to perform the requested function. The system may have separate relays for the park lights, fog lights, and headlights. The circuit operation for each of these relays is usually performed by low side drivers to complete the path to ground for the relay coil. The following will focus on testing the headlight system, although diagnostics for the park and fog light systems are similar.

Using Figure 12-10 as a schematic of the system, we will go through the steps to diagnose a customer complaint that the headlights do not come on. Always refer to the proper wiring diagram for the vehicle you are working on. As with any electrical system problem, begin by confirming that the system is malfunctioning. Also, note any other systems that are not working properly. In the system shown, use the multifunction switch to activate the flash-to-pass feature. This will test the operation of the high-beam lights and relay.

Check any associated fuses and repair as needed. Next, conduct a visual inspection and repair any faults found. If the cause of the problem is not located, substitute the low-beam relay with a known good relay of the same type and test operation. If the system works now, the relay was at fault and the repair is complete. If the system still does not work, begin testing the rest of the system.

If a scan tool is available, check to see if any DTCs have been set. If so, use the diagnostic chart for the DTC to locate the problem. If there are no DTCs (not all systems will set codes), use the scan tool to perform an activation test of the headlights. If the headlights work during the activation test, the problem is in the input side of the system. Use the scan tool to monitor the headlight switch while it is placed in all switch positions. Record the voltages for each position and compare to specifications. In this example, if 5 volts is displayed in all positions, there is an open in the circuit. The open could be in the wire between the BCM and the switch, in the switch itself, or in the ground circuit. Test each of these components with an ohmmeter. The switch is a resistive multiplex switch that should have a different resistance value for each switch position. With the switch disconnected, attach the ohmmeter across terminals 1 and 2 of the switch. Place the switch in each position and record the results. Check the results with the service information. If the values are out of range, replace the switch.

When testing the input signal with the scan tool, if the voltage displayed is 0 volts, this would indicate that the circuit is shorted to ground or a faulty BCM. Test the circuit for a short. If there are no shorts, check the voltage input and ground connections of the BCM. If these are good, replace the BCM.

If the headlights did not come on when the scan tool activated the system, the problem is in the relay control circuit, the BCM, the high-current circuit from the battery through the relay to the headlights, or in the headlight ground. To determine which circuit is at fault, remove the low-beam headlight relay. Use a voltmeter or test light to confirm battery voltage is present at pins 30 and 86 of the junction block terminals. If voltage is not present, there is an open or short between the junction block and the battery. If voltage is present at both pins, use a fused jumper wire to connect pin 30 to pin 87. If the lights come on now, the problem is in the control side circuit of the relay (the relay itself was tested earlier when it was substituted).

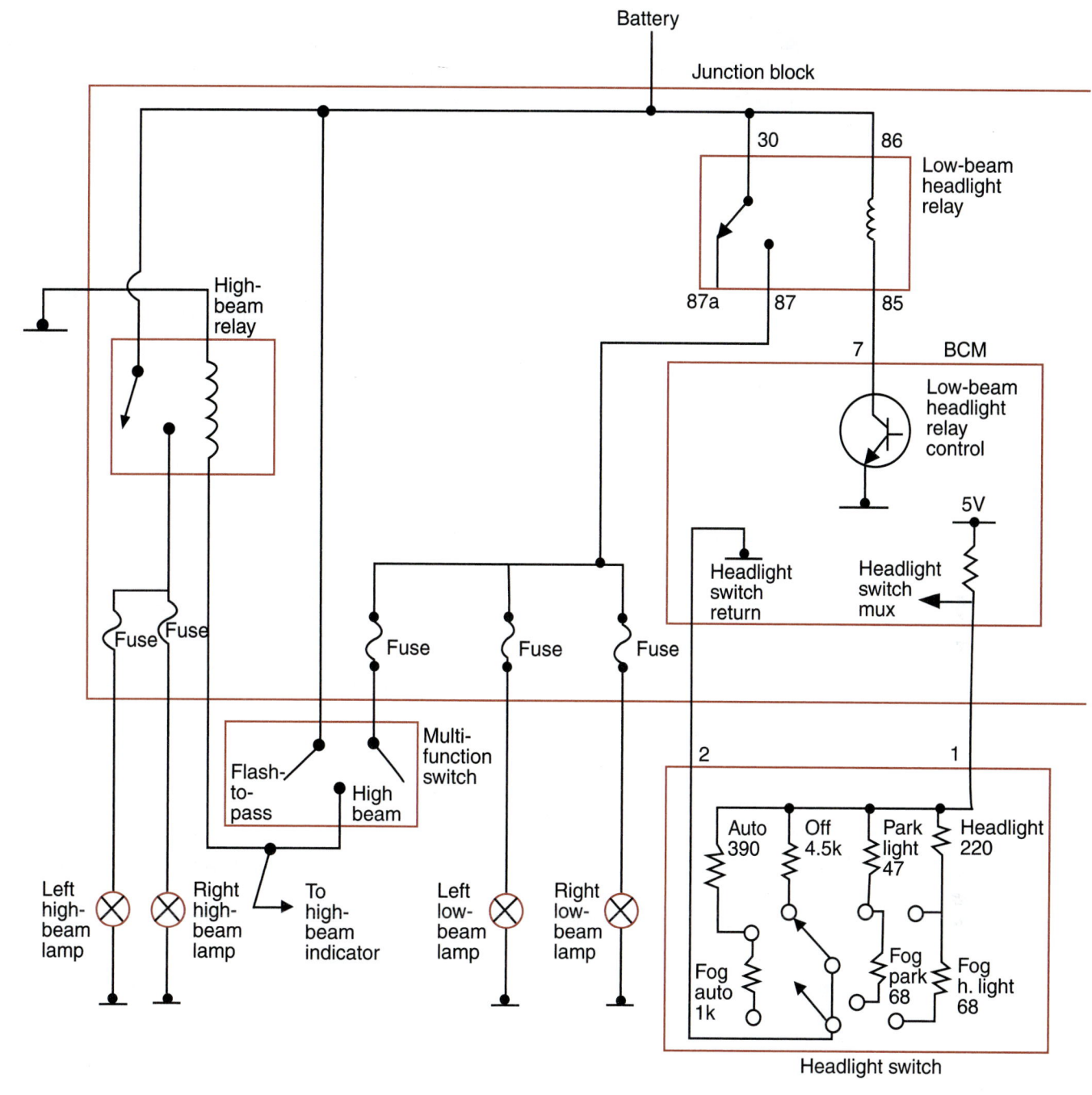

Figure 12-10 Schematic of a computer-controlled headlight system.

To test the control side circuit, connect a test light between pins 86 and 85 of the junction block terminals (relay removed). Place the headlight switch into the headlamp position and observe the test light. Since the BCM should now complete the path to ground, the test light should illuminate. If the test light fails to turn on, go to pin 7 of the BCM and back probe a jumper wire into the connector. Touch the other end of the fused jumper wire to a good ground. If the test light comes on, check the ground connections of the BCM. If the grounds are good, replace the BCM. If the test light still does not come on, there is an open in the circuit between pin 85 of the junction block and the BCM. In this case, the BCM attaches to the back of the junction block and the circuit wire is internal to the junction block. If this circuit is bad, the junction block will need to be replaced.

SERVICE TIP: If a scan tool is not available, the input can be tested along with the control side of the relay by connecting a test light across terminal 85 and 86. If the test light comes on when the switch is placed in the headlamp position, the BCM is receiving the input and carrying out the command.

If the lights do not come on when the fused jumper wire is connected across pins 30 and 87 of the junction block, the problem is between the relay and the headlights. Connect a fused jumper wire from the battery positive post to the feed into one of the headlight's connectors. If the headlight illuminates, check the circuit from the junction block to the headlight. If the circuit is good, the problem is probably at the splice in the junction block and the junction block will need to be replaced. If the headlight still does not turn on, check the ground connections. If the grounds are good, the bulb is burned out. It is unlikely that both headlight bulbs will burn out at the same time unless there are other problems. If both bulbs are burned out, check the charging system for over voltage output. Also, use a lab scope connected to the battery positive post and ground to look for voltage spikes above the charging system's target voltage value. Voltage spikes may indicate that the battery connections are not clean (remember, the battery acts as a buffer), the battery is bad, or there is resistance on the charging system's sense circuit.

Automatic Headlight System Diagnosis

Classroom Manual
Chapter 12, page 351

Special Tools

Bright flashlight

The automatic on/off with time delay has two functions: to turn on the headlights automatically when ambient light decreases to a predetermined level and to allow the headlights to remain on for a certain amount of time after the vehicle has been turned off (providing light for the passengers exiting the vehicle). This system can be used in combination with automatic dimming systems.

The common components of the automatic on/off with time delay include:

1. Photocell and amplifier.
2. Power relay.
3. Timer control that is a potentiometer incorporated into the headlight switch. The timer control unit controls the automatic operation of the system and the length of time the headlights stay on after the ignition switch is turned off.

In a typical system, a photocell is located inside the vehicle's dash to sense outside light. In most systems, the headlight switch must be in the OFF or AUTO position to activate the automatic mode.

If the headlights do not turn on, check the regular headlight system first before condemning the automatic headlight system. If the headlights do not illuminate when the automatic system is turned off, the problem is in the basic circuit.

To perform a "functional test" of the automatic headlight system, turn the headlight switch into the appropriate position to activate the automatic headlight function. Cover the photocell and start the engine. Some systems may be activated with the ignition switch in the RUN position; however, many systems are designed to delay turning on the headlights until an engine run signal is received. The headlights should turn on within 30 seconds. Remove the cover you placed over the photocell and shine a bright light onto it. The lights should turn off after 10 seconds but within 60 seconds. Cover the photocell again. When the lights turn on, wait 15 seconds. Then turn off the ignition switch. The headlights should turn off after the selected amount of time delay.

Once it is determined that the fault is within the automatic headlight system, make a few quick checks of the system. Inspect the photocell lens for obstructions. Check all connections from the headlight switch, as well as all fuses used in the system.

The following are some of the most common complaints that result from problems in the automatic headlight:

1. Lights turning on and off at wrong ambient light levels.
2. Lights that do not turn on in darkness.
3. Lights not turning off in bright light.
4. Lights not staying on for an adjustable time after the ignition switch is turned off.
5. Lights that do not turn off after the ignition switch is turned off.

These problems can be caused by faults in the headlight switch, ignition switch, amplifier, or photocell. To locate the problem, perform the following test series. As a service sampler of automatic headlight system diagnosis, refer to the schematic (Figure 12-11). The following test connections will refer to those this system uses. Use the correct service manual for the system you are diagnosing.

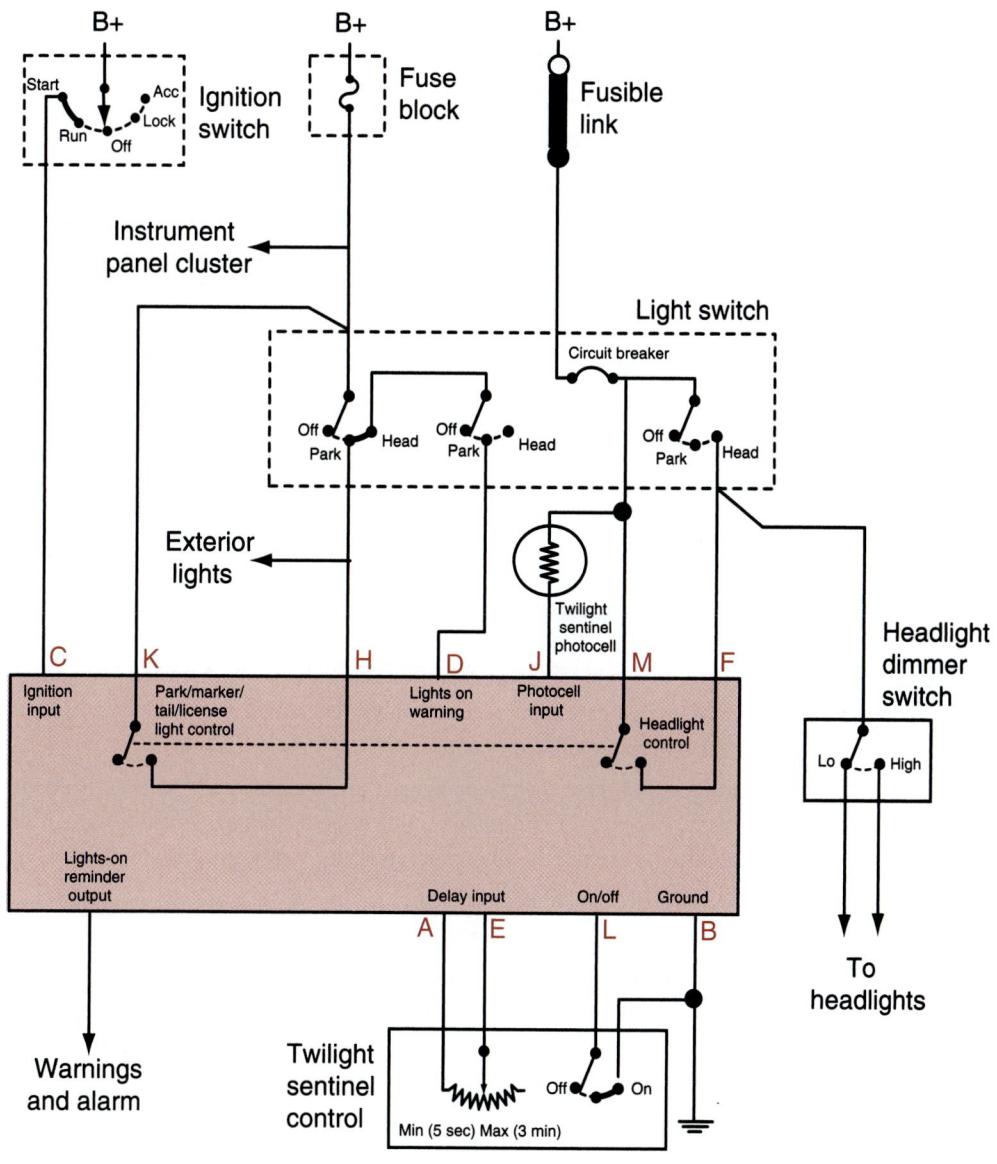

Figure 12-11 Twilight Sentinel schematic.

Classroom Manual
Chapter 12, page 352

Special Tools

DMM

Scan tester

Special Tools

Ohmmeter

Fused jumper wire

Test light

Fender covers

Safety glasses

Service manual

Photocell Test

To test the photocell, use a scan tool or a DMM to measure the voltage of the photocell signal circuit to the BCM. With the engine running, observe the voltage readings. First, cover the photocell's lens with a piece of cardboard. The voltage should read 0 volts. Next, shine a flashlight onto the photocell's lens. The voltage should go up to 5 volts. If the voltages do not change, check the circuit, and if there are no problems, replace the photocell.

Another method of testing the photocell is to unplug the photocell connector and start the engine. Place the headlight switch in the correct position to turn on the automatic headlight feature. If the headlights turn on within 60 seconds, replace the photocell. If the lights do not turn on, test the amplifier (if equipped).

Photocell Replacement. Photo Sequence 20 illustrates the procedure for replacing the photocell.

Amplifier Test

1. Disconnect the wire connector to the amplifier.
2. Turn the headlight switch to the OFF position.
3. Disconnect the negative battery terminal.
4. Turn the control switch to the ON position.
5. Use an ohmmeter to measure the resistance between the wire connector terminal L and ground. There should be 0 ohms of resistance. If there is more than 0 ohms of resistance, check the wire from amplifier terminal L to control switch terminal C for opens. Also, check the circuit from amplifier terminal B for opens. If the ohmmeter indicated 0 ohms of resistance, continue testing.
6. Turn the control switch to the OFF position.
7. The ohmmeter should read infinite when connected between the L terminal and ground. If there is low resistance, check the circuit for a short. If the ohmmeter indicated infinite resistance, continue testing.
8. Connect the ohmmeter between terminals H and D. There should be 0 ohms of resistance. If not, check the two circuits for an open. If the ohmmeter indicated 0 ohms of resistance, continue testing.
9. Place the control switch in the middle of MIN and MAX settings.
10. Connect the ohmmeter between terminals A and E. There should be between 500 and 250,000 ohms of resistance. If the resistance value is not within this range, check the circuits between the amplifier and the control switch for opens.

    ❑ If the circuits are good, observe the ohmmeter as the control switch is moved from the MIN position to the MAX position. The resistance value should change smoothly and consistently from one position to the next. If the ohmmeter indicates a resistance value out of limits or an erratic reading, replace the control switch.
11. If all of the resistance values tested are correct, continue testing.
12. Turn off all switches and reconnect the amplifier wire connector. Reconnect the battery negative terminal.
13. Turn the ignition switch to the RUN position. Turn the headlight switch to the OFF position, and turn the control switch to ON.
14. Connect a fused jumper wire between amplifier terminals J and M. The headlights should go off within 60 seconds. If the headlights go off within 60 seconds, but they do not go off normally in bright light, check the circuit for opens. If the circuit is good, replace the photocell.
15. Turn off all switches.

Use a 20-ampere
fuse in the jumper.

Photo Sequence 20
Typical Procedure for Replacing a Photocell Assembly

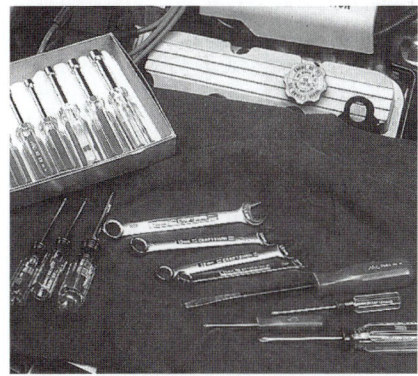

P20-1 Tools required to replace the photocell assembly include nut driver set, thin flat blade screwdriver, phillips screwdriver set, battery terminal pullers, battery pliers, box end wrenches, safety glasses, seat covers, and fender covers.

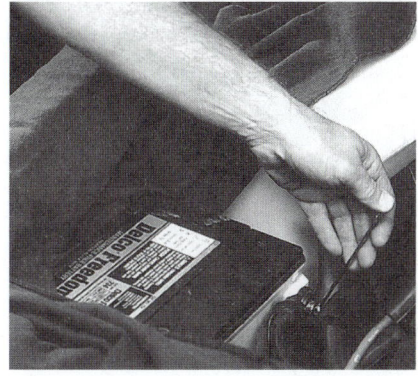

P20-2 Place the fender covers over the vehicle's fenders and disconnect the battery negative terminal.

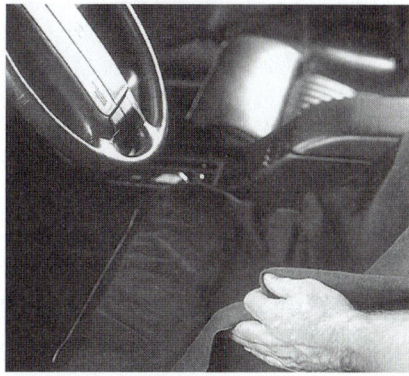

P20-3 Protect the seats by placing the seat covers over them.

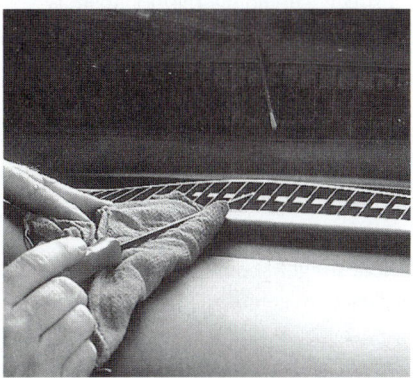

P20-4 Using a clean shop rag to protect the dash pad, insert a long flat blade screwdriver into the defroster and pry it up and out.

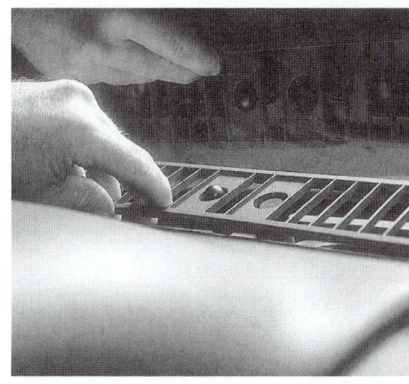

P20-5 Carefully pry out the vent outlets from the panel. Carefully work your way down the length of the grill, pulling it free from its seat.

P20-6 Gently work the grill up and out to gain access to the photocell socket.

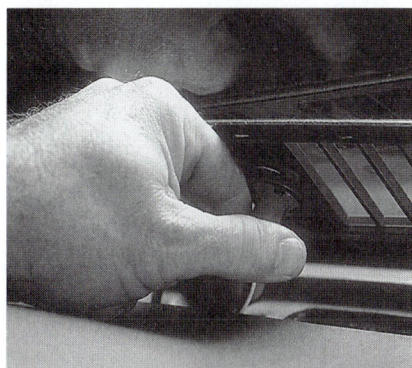

P20-7 Twist the socket to free it from the grill.

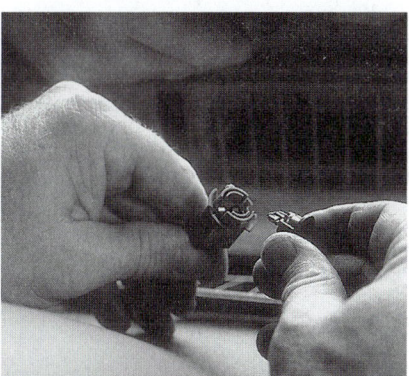

P20-8 Gently work the photocell free from its socket.

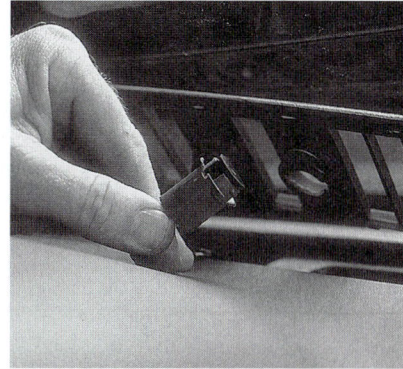

P20-9 Replace the photocell and reinstall its socket into the grill.

Typical Procedure for Replacing a Photocell Assembly (continued)

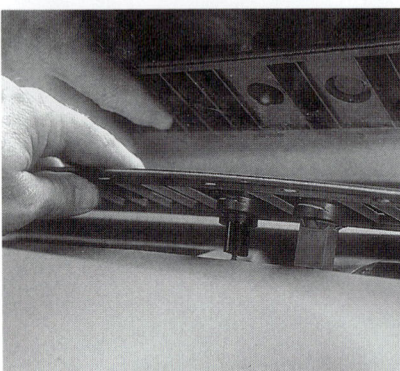

P20-10 Position the back side of the grill into place.

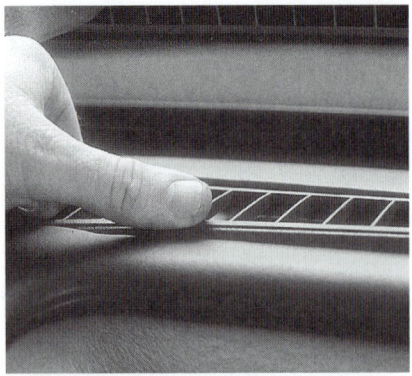

P20-11 Work the leading edge of the grill back into place using your thumb.

16. Disconnect the wire connector to the amplifier.

17. Turn the ignition switch to RUN and turn the headlight switch to OFF.

18. Connect a test light between terminal M and ground. The test light should light. If not, there is a short or open in the circuit between the battery and the amplifier or in the circuit between the amplifier and the photocell. If the test light illuminates, continue testing.

19. Connect the test light between terminals M and B. The test light should light. If not, check the ground circuit for an open. If the test light illuminates, continue testing.

20. Connect the test light between terminal K and ground. The test light should light. If not, check the fuse and the circuit for an open. If the test light illuminates, continue testing.

21. Connect the test light between terminal C and ground. The test light should light. If not, check the circuit for an open. If this circuit is good, check the ignition switch. If the test light illuminates, continue testing.

22. Turn the ignition switch to OFF.

23. With the test light connected as in step 21, the test light should turn off. If not, replace the ignition switch. If the test light turns off, continue testing.

24. Connect the test light between terminal H and ground. If the test light is illuminated, replace the light switch. If the test light is off, continue testing.

25. Connect the test light between terminal F and ground. If the test light is illuminated, replace the light switch. If the test light remains off, continue testing.

26. With the test light connected as in step 25, place the headlight switch in the HEAD position. If the test light does not turn on, check the circuit and the headlight switch for opens. If the test light lights, and all other tests have the correct results, replace the amplifier.

> **CAUTION:** Skipping any of the tests will result in replacement of the amplifier, even if it is good.

Special Tools

Photocell resistance assembly

Resistance Assembly Test

The resistance assembly test is performed when the lights turn on and off at the wrong ambient light levels. Use the illustration (Figure 12-12) to construct a **photocell resistance assembly,** which replaces the photocell to produce predictable results.

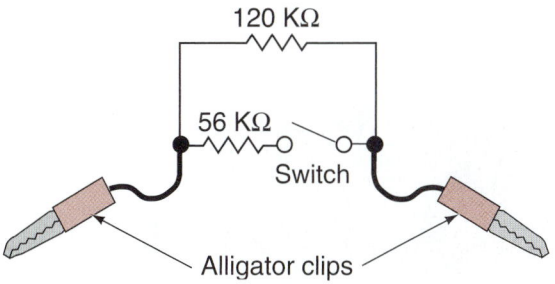

Figure 12-12 Photocell resistance assembly.

Replace the photocell with the resistance assembly. Turn the resistance assembly switch to the OFF (open) position and place the ignition switch in the RUN position. If the lights do not come on within 60 seconds, replace the amplifier.

If the lights turn on, wait 30 seconds and turn the resistance assembly switch on. If the lights turn off within 60 seconds, replace the photocell. If the lights do not turn off within 60 seconds, replace the amplifier.

SmartBeam™ Diagnostics

SmartBeam™ is one of the systems that a manufacturer may use to control automatic high-beam operation. This system is presented as an example of the diagnostic procedures used to determine and repair faults.

Customer concerns that can be related to the operation of SmartBeam™ will usually be in one of two categories: either the system is totally inoperative or it is not performing properly.

 SERVICE TIP: Because the SmartBeam™ requires the function of several vehicle modules, diagnosis of the system may require verifying the correct operation of modules other than the automatic high-beam module (AHBM) itself.

When diagnosing a complaint of SmartBeam™ inoperative, it is important that the following is done:

- ❏ Verify that the headlights work properly on both high and low beam when operated manually.
- ❏ Verify the power and ground circuits of the AHBM.
- ❏ Verify that the system status indicator LED in the mirror (Figure 12-13) is on steady. If the LED is flashing, then a fault has been detected.

Classroom Manual
Chapter 12, page 357

Special Tools

Scan tool
Calibration target
Grease pencil
Tape measure

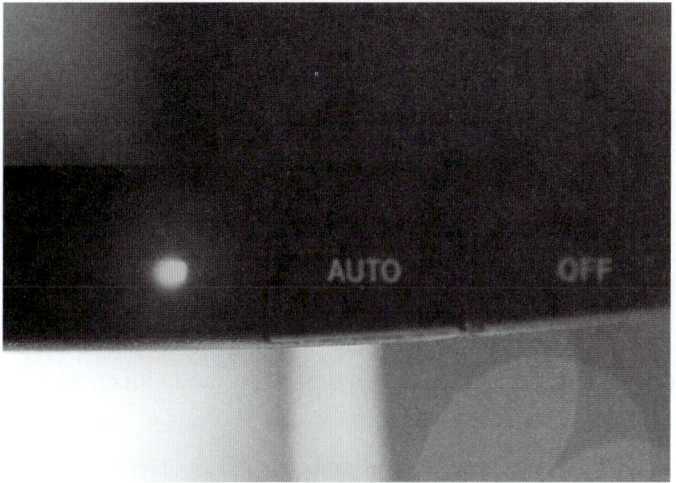

Figure 12-13 The LED can be used to indicate auto high-beam faults.

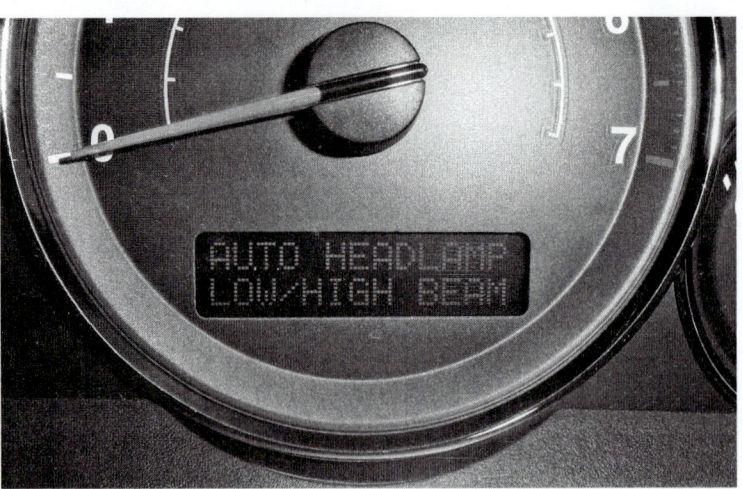

Figure 12-14 For the auto high-beam function to operate, it must be activated.

❏ Verify the automatic high-beam/low-beam function has been enabled (Figure 12-14).

❏ Verify that the headlight switch is set to the AUTO position.

❏ Verify that the headlight beam select switch is in the low-beam position.

Complaints related to poor system performance are usually associated with sensitivity of the system. If the system is oversensitive, the customer may complain that the high beams come on too late and go off too early. If the system is undersensitive, then the high beams will come on too early and go off too late. These types of sensitivity problems can be caused by:

❏ Camera not properly aimed.

❏ Loose camera mounting or improperly positioned mirror button.

❏ Obstruction in front of the camera.

❏ Improper headlight aiming.

❏ Vehicle overloading resulting in the rear of the vehicle sitting lower than the front.

Whenever possible, system operation can be verified on stationary vehicles by using the scan tool to manually activate the system.

SERVICE TIP: Remember that the SmartBeam™ system controls the high-beam portion of the headlamp system. The SmartBeam™ system varies the high-beam head lamp illumination intensity from low-beam level headlamp illumination intensity to full high-beam headlamp illumination intensity, and any level of headlamp illumination intensity in between.

CUSTOMER CARE: Inform the customer that the camera must have an unobstructed view from the front of the windshield for the system to perform properly. Hanging items from the mirror, or placing toll road transponders in front of the camera will result in poor performance of the system.

The SmartBeam™ automatic high-beam system has a **demonstration mode** that may be used to assist in diagnostics. The demonstration mode allows the function of the automatic high beams and high-beam indicator to be demonstrated while the vehicle is stationary and under any ambient lighting conditions. To initiate the demonstration mode function:

1. Begin with the ignition switch in the OFF position.

2. Depress and hold the AUTO button on the inside mirror (refer to Figure 12-13).

3. While continuing to depress the AUTO button, turn the ignition switch to the RUN position.

4. Continue to hold the AUTO button depressed until the demonstration mode begins as indicated by the high beams ramping up in intensity.

5. Release the button.

The system will complete three cycles of ramping up the headlamp high beams to full intensity and then ramping them down. During this time, the high-beam indicator should also come on and go off with the high beams. The high beams will cycle three times; then the system will return to normal operation.

The LED in the rearview mirror is also used to assist in diagnostics. Usually this LED is on steady to indicate that the auto dimming function of the electrochromic mirror is on. A flashing LED indicates the system has detected a problem. The LED can flash at different frequencies and at different sequences to indicate a problem. For example, if the LED is continuously flashing at a rate of 1 flash per second this indicates that the mirror is in need of being calibrated. To correct this condition, the camera calibration procedure must be performed.

If the LED is continuously flashing at a rate of 2 flashes per second, this indicates that the system failed its last attempt to calibrate. To correct this condition, the camera calibration procedure must be performed.

The last possible LED indication is a series of flashes when the ignition switch is first placed in the RUN position, followed by the LED staying on steady. The flashing LED can indicate a hardware failure that may require the AHBM replacement.

Camera Calibration Test and Adjustment. It is critical that the camera's field of view is maintained to specifications. If the camera's field of view is no longer within specifications, the performance of the system is seriously degraded. For proper operation, the camera's field of view must be maintained within 2° of the vehicle centerline and within 10° of horizontal. If the camera is aligned within these specifications, the AHBM can adjust and fine-tune the alignment-based on sensed lighting inputs while driving.

The camera calibration procedure must be performed any time the inside rearview mirror is replaced with a new unit, the rearview mirror mounting button has been replaced, or the windshield has been replaced. The calibration procedure ensures that the field of vision for the camera is aimed at the proper path ahead of the vehicle. If a new camera is installed, it is shipped with the calibration mode initiated. Once the camera is installed and connected, the LED will flash continuously at a rate of once per second while the ignition switch is in the RUN position. Before attempting to calibrate the camera, the following should be performed:

❏ Clean the windshield glass in front of the camera lens.

❏ Check for proper mounting of the mirror assembly and that the set screw that secures the assembly to the button is properly torqued at 15 in. lbs. (1.7 Nm).

❏ Repair or replace any faulty, worn, or damaged suspension components.

❏ Verify proper tire inflation pressures.

❏ Verify that there is no load in the vehicle, except for the driver.

❏ The fuel tank should be full. Add 6.5 pounds (2.94 kilograms) of weight over the fuel tank for each gallon of missing fuel.

To calibrate the camera, the centerline of the vehicle must be established and marked. A special alignment target is also required. The target consists of a black, square field containing three red LEDs positioned in a diagonal pattern (Figure 12-15). The target is placed a specified distance in front of the vehicle. Once the proper height of the target is established to match the height of the camera lens, then the target must be positioned within ½° of centerline of the vehicle. When in calibration mode, the AHBM is programmed to look for the red LED light pattern of the target from the camera. When the target is properly placed, the AHBM will identify it and the center of

Figure 12-15 Calibration target.

the target will become the new center of the field of view. The flashing green LED changes to a steady ON state when calibration is successful. Photo Sequence 21 illustrates the typical procedure for setting up and calibrating the camera.

If the camera cannot locate the target's pattern, the calibration fails. The LED will blink at a 2-Hz rate, indicating that the calibration procedure has failed and that a fault has been set. This may be caused by the target being incorrectly placed or the camera being improperly aligned.

If the camera fails to calibrate, a DTC will be set indicating that the performance is off in any of the four directions. To correct this condition, you will have to verify that the mounting of the camera and the position of the target were correct. If they were correct, then adjustment of camera will be necessary.

Each DTC for alignment has the description of the direction the camera should be moved to make a correction. For example, if the DTC is "High-Beam Camera Alignment Performance Bottom," the camera is locating the target but it is at the bottom of the field of view. To correct this, the camera needs to be tilted downward.

Camera alignment is done by the placement of a spacer between the camera mount and the camera housing (Figure 12-16). The adjustment spacer consists of four stepped shims connected by an integral plastic tree (Figure 12-17).

Figure 12-16 An adjustment shim is used to align the camera.

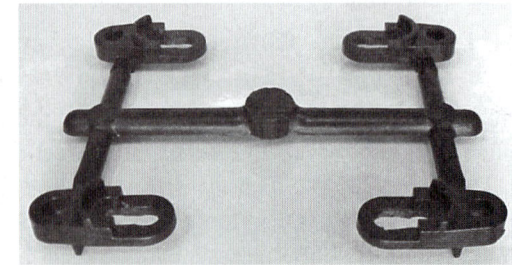

Figure 12-17 The adjustment shim.

Photo Sequence 21
Automatic High Beam Camera Calibration

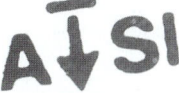

P21-1 Locate the "ASI" mark on the windshield.

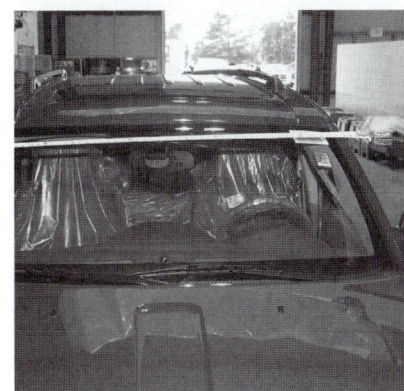

P21-2 Using the "ASI" marks on both sides of the windshield, measure across windshield using the shaft of the tint band arrow as the reference point.

P21-3 Divide the measurement in half and mark that dimension near the lower edge of the tint band using a grease pencil.

P21-4 Measure across the upper edge of the blackout area at the bottom of the windshield using the inside corner of the intersection between the side blackout area and the lower blackout area as the reference point.

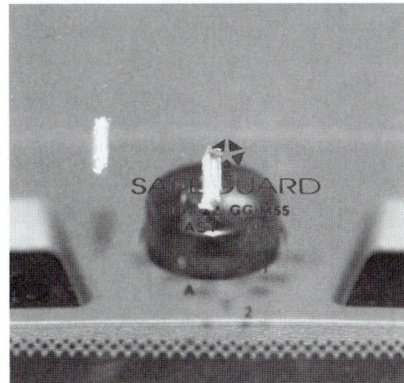

P21-5 Divide the lower measurement in half and mark that dimension near the upper edge of the blackout area using a grease pencil.

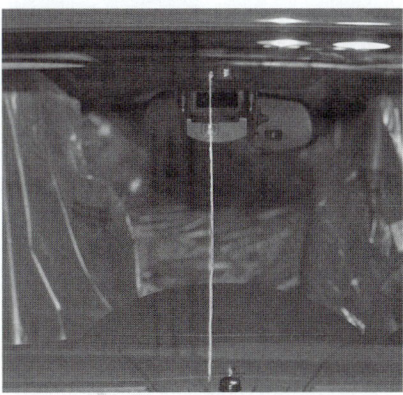

P21-6 To locate the centerline of the camera lens, measure and mark the glass ⅞ inch (21 mm) toward the passenger side of the windshield from the upper and lower glass centerline marks and draw a vertical line connecting these two marks.

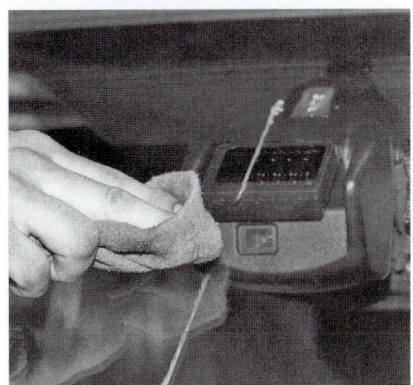

P21-7 Remove the grease pencil mark in front of the lens.

P21-8 To locate the camera centerline on the rear glass, measure across the upper portion of the rear glass using the vertical edges of the body opening as the reference points.

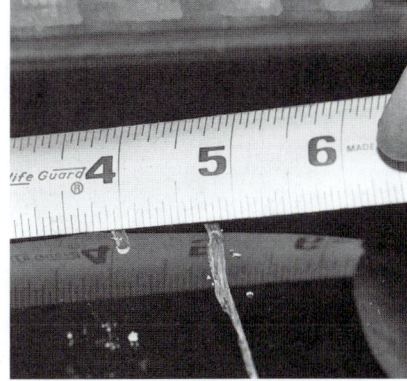

P21-9 Divide the upper measurement in half and mark that dimension on the glass using a grease pencil.

Automatic High Beam Camera Calibration (continued)

P21-10 Measure across the rear glass at the bottom using the vertical edges of the body opening as the reference points.

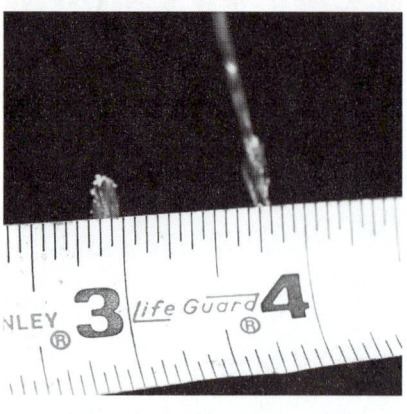

P21-11 Divide the lower measurement in half and mark that dimension on the glass using a grease pencil.

P21-12 Measure and mark ⅞ inch (21 mm) toward the passenger side of the rear glass on both the upper and lower references and draw a vertical line connecting these two marks.

P21-13 Locate the calibration target 50 inches (127 cm) in front of the vehicle measuring from the foremost center of the front fascia.

P21-14 While sighting through the V-notch in the upper edge of the calibration target, move the target left or right to align the target to the camera centerline marks on both the windshield and the rear glass.

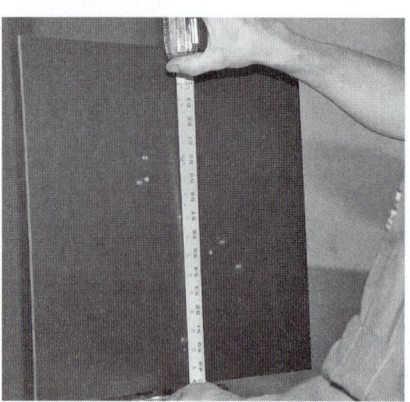

P21-15 Adjust the tripod so that the center LED on the target 57 inches (145 cm) from the floor.

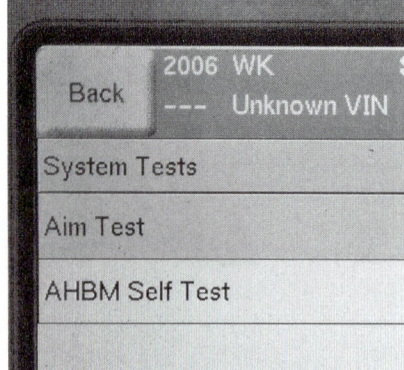

P21-16 Use the scan tool to enter the SmartBeam™ unit into calibration mode. When the calibration mode is entered, the LED in the mirror assembly will flash at a rate of once per second.

P21-17 Turn on the LEDs in the calibration target.

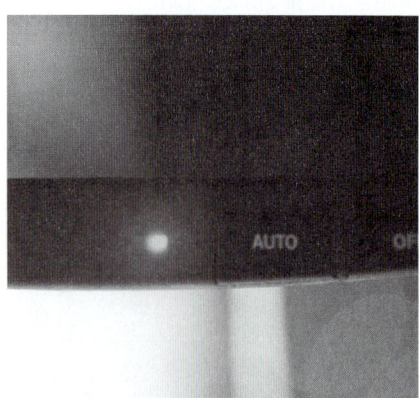

P21-18 The LED in the mirror assembly should continue to flash for five to ten seconds, then will stop flashing to indicate that it has completed calibration.

Determine what shim movement will be required to correct the alignment based on the DTCs retrieved (Figure 12-18). To adjust the camera alignment, move the inside rearview mirror head downward so access to the two screws securing the rear cover can be obtained. Remove the two screws, and then adjust the mirror head up toward the headliner to its uppermost position. Unsnap and pull the upper edge of the rear cover away from the housing far enough to disengage the tabs at the lower edge of the cover; then remove the rear cover.

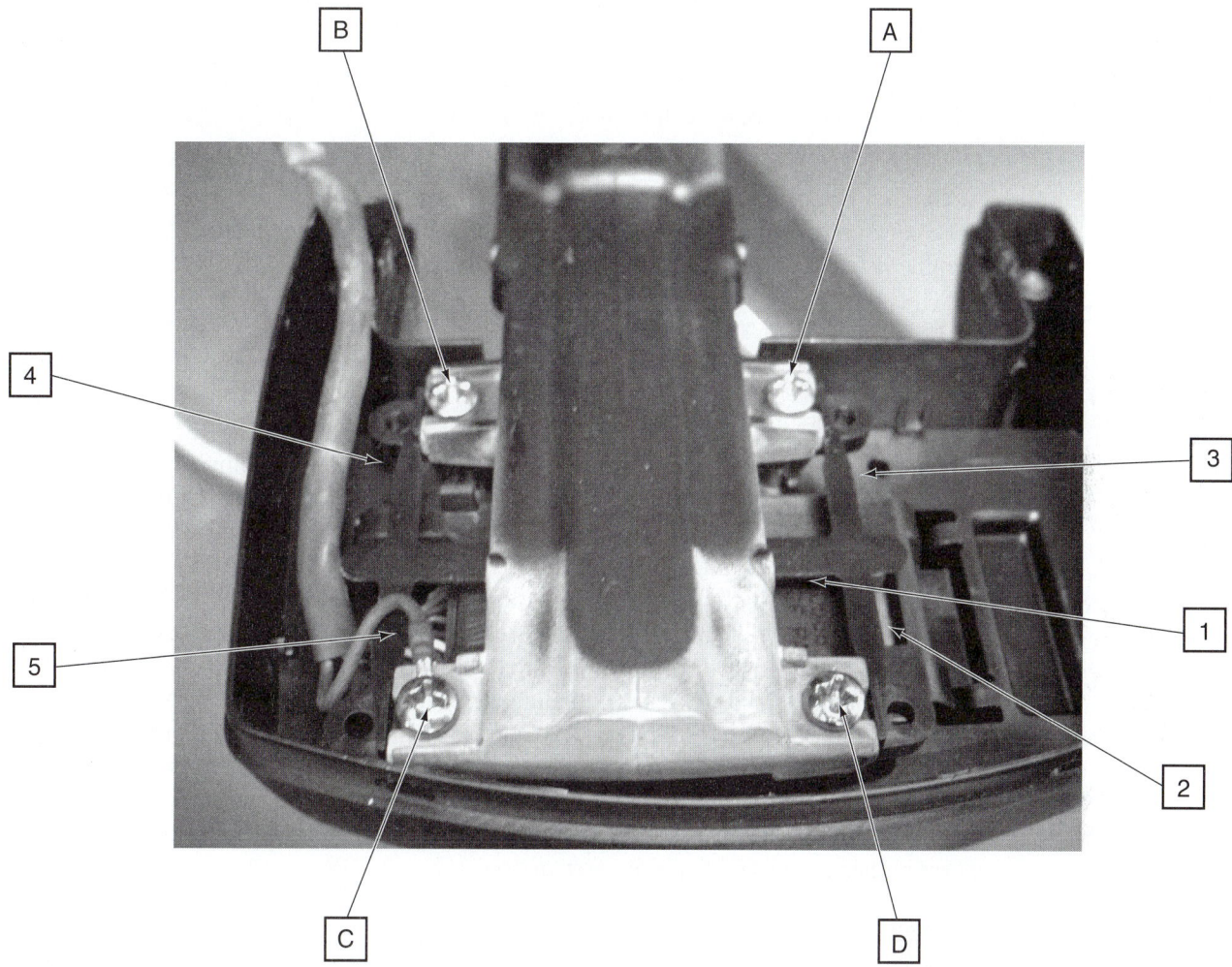

Camera Adjustment Table		
Fault	Tree Cut Location	Shim + Screw Location
Right	1	A + D
Left	1	B + C
Top	3 + 4	A + B
Bottom	2 + 5	C + D

Figure 12-18 Adjustment chart and locations.

With the mounting bracket now accessible, carefully cut the plastic tree for the shims at the appropriate location(s) to allow movement of the shims determined earlier. With the shims separated from the tree, loosen the attaching screws one-half turn. The screw holes in the shims are slotted to allow sliding of the shims. Slide the shim beneath each of the loosened screws to its most outboard position. Once the shims are relocated into the desired position, the attaching screws are then tightened to 7 in. lbs. (0.8 Nm). As the screws are tightened, the camera is pulled into the new alignment position. The total amount of correction is 2 to 2.5°. Only two shims can be moved for correction. The shims moved must be either:

❑ Both top shims, to tilt the camera upward.

❑ Both bottom shims, to tilt the camera downward.

❑ Both left shims, to tilt the camera to the left.

❑ Both right shims, to tilt the camera to the right.

Reinstall the mirror assembly rear cover. Use the scan tool to erase any DTCs, and perform the calibration procedure again. If the correct shims were moved and the necessary correction was within the range of adjustment, calibration should be successful.

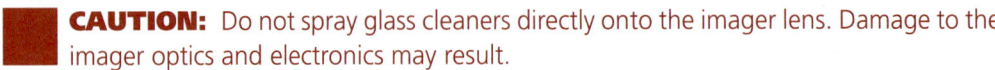 **SERVICE TIP:** If more than one high-beam camera alignment performance fault DTC was retrieved, then one screw and one shim will be common to both of the faults. Make the correction at this location.

Camera Optics Test. If the automatic high-beam system is performing poorly, but the LED does not flash, the camera's optics may be dirty or obstructed. The camera's optics can be tested by entering the demonstration mode. The **optics test** will confirm that the camera can recognize ambient light through the lens and the windshield. To perform the optics test, first initiate the automatic high-beam demonstration mode. While observing the LED, obstruct the view of the imager by placing a piece of cardboard between the camera lens and the windshield. The LED should turn off each time the optics are obstructed by the cardboard. Once the cardboard is removed, the LED should illuminate again.

Failure of the LED to respond to these inputs indicates that the imager optics is obstructed. Clean the lens and the windshield glass and/or remove any obstructions from the windshield in front of the imager. To clean the imager lens, spray a small amount of glass cleaner onto a soft cloth and gently wipe the lens clean. After cleaning the lens and the windshield, repeat the test to confirm proper automatic high-beam system operation.

CAUTION: Do not spray glass cleaners directly onto the imager lens. Damage to the imager optics and electronics may result.

Daytime Running Lamps

Classroom Manual
Chapter 12, page 362

All late-model Canadian vehicles and many newer domestic vehicles are equipped with daytime running lamps. The daytime running lamp (DRL) system includes a solid-state control module assembly, a relay, and an ambient light sensor assembly (Figure 12-19). The system lights the low-beam headlights at a reduced intensity when the ignition switch is in the RUN position during daylight. The daytime running lamp system is designed to light the low-beam headlamps at full intensity when low-light conditions exist.

As the intensity of the light reaching the ambient light sensor increases, the electrical resistance of the sensor assembly decreases. When the DRL control module assembly senses the low resistance, the module allows voltage to be applied to the DRL diode assembly and then to the appropriate headlamps. Because of the voltage drop across the diode assembly, the headlamps are on with a low intensity.

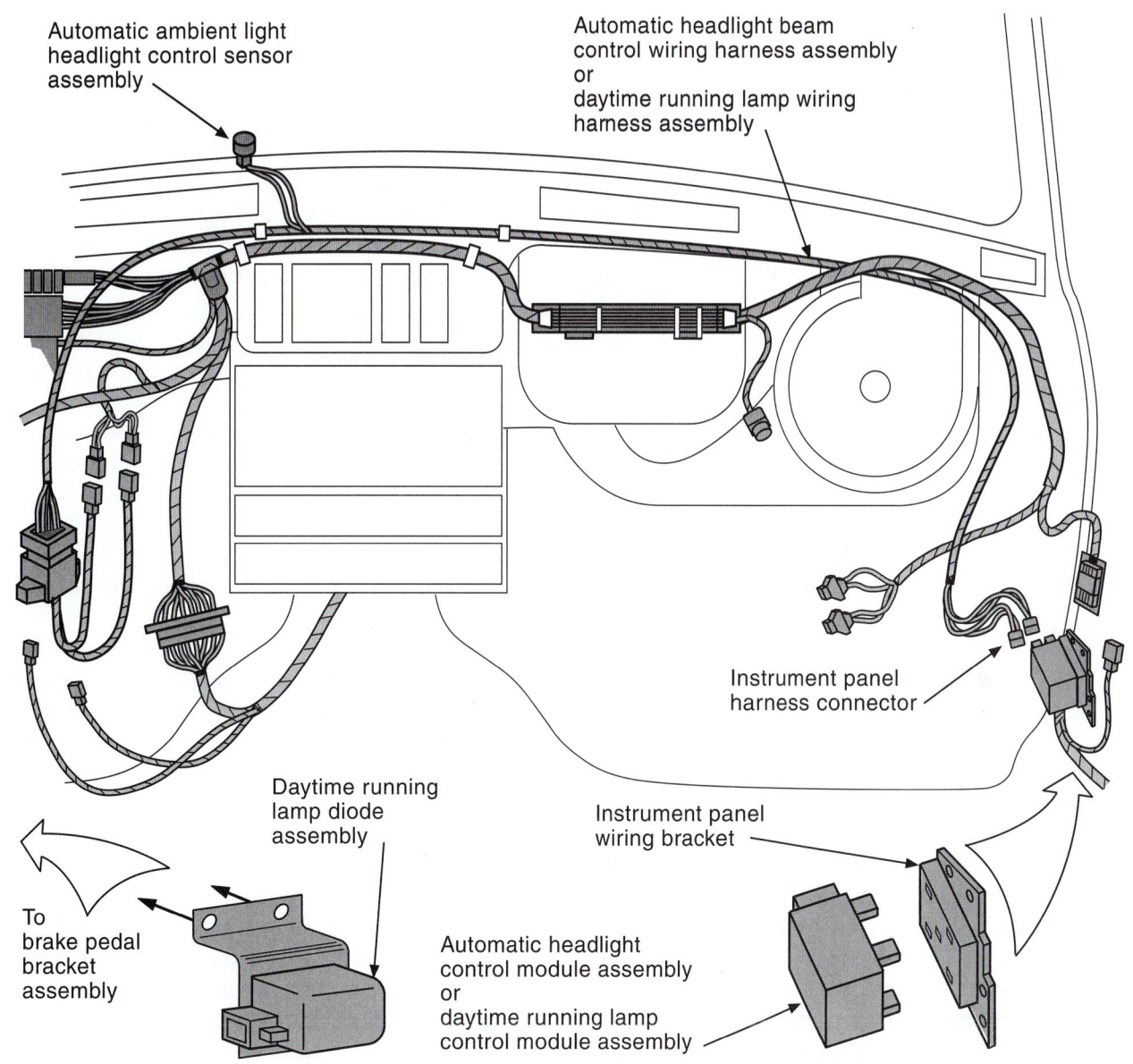

Labels on figure:
- Automatic ambient light headlight control sensor assembly
- Automatic headlight beam control wiring harness assembly or daytime running lamp wiring harness assembly
- Instrument panel harness connector
- Daytime running lamp diode assembly
- To brake pedal bracket assembly
- Instrument panel wiring bracket
- Automatic headlight control module assembly or daytime running lamp control module assembly

Figure 12-19 Headlamp automatic control module, daytime running lamp control module, and lamp diode assemblies.

As the intensity of the light reaching the ambient light sensor decreases, the electrical resistance of the sensor increases. When the DRL module assembly senses high resistance in the sensor, the module closes an internal relay, which allows the headlamps to illuminate with full intensity.

Some manufacturers use a DRL module that receives inputs from either the headlight switch or the BCM to determine switch position. If the switch is in the OFF or AUTO position (and ambient light is high enough that the headlights do not need to be on), the DRL module will send a PWM signal to the headlight beam. This turns the headlight on, but at reduced illumination levels.

Some vehicles will use the parking lamps for the DRL function. In this case, the headlight position is used to determine if the headlights are on. If they are not, the BCM or DRL module will turn on the park lamp relay.

Diagnostics of either type of system is performed by using a scan tool or DMM. As with any system, determine proper power and ground circuit operation. Also, remember that the parking brake switch is used as an input. If the switch should fail and indicates to the DRL module that the parking brakes are applied, the DRL function is disabled.

Illuminated Entry System Diagnosis

The diagnostic procedures for testing the illuminated entry system depend upon the manufacturer. Always refer to the service manual for the vehicle you are working on. Always perform a visual inspection before performing any tests. First check the fuse. Then check to make sure all connections are tight and clean. Inspect the ground wires for good connections. Check all visible wires for fraying or damaged insulation, especially where they go through body parts. Make sure all doors are closed properly and the headlight switch is in the detent position.

The following are typical procedures for diagnosing the illuminated entry system.

Systems that are activated by lifting the outside door handle use a contact switch that momentarily closes to complete the ground circuit of the **illuminated entry actuator** module. The module activates the interior lights for 25 seconds or until the ignition switch is placed in the RUN or ACC position.

A logic circuit is included in the module to prevent battery drain if the door handle is held up for longer than 25 seconds. The system will operate as normal until the 25 seconds have elapsed, and then the module will turn off the lights. The lights will remain off and cannot be reactivated until the handle is returned to the released position.

This type of system has four main components (Figure 12-20). The door lock cylinder uses an LED to provide the illumination of the cylinder. The lens of the LED is built into the cylinder.

On some systems, terminal 4 is used. On these systems, jumping between terminals 6 and 8 will illuminate the courtesy lights. Jumping between terminals 4 and 8 will light the door lock cylinders only. If they do not, trace circuit 464.

CAUTION: The normal operating voltage for the LED is 3 volts. The circuit uses a dropping resistor to protect the LED. Do not apply 12 volts to the LED circuit ahead of the resistor. If the resistor is bypassed, the LED will be destroyed. When applying voltage to test the LED, apply it only to the connector terminals.

To test the system, disconnect the actuator harness from the actuator. Refer to Figure 12-20 and connect the test light between terminal 8 and ground. The test light should illuminate with the ignition switch in the OFF or RUN position. If the test light fails to come on, trace the circuit back to the fuse box to locate the problem.

Connect the test light between terminal 7 and ground. The test light should not glow with the ignition switch in the OFF position. When the ignition switch is turned to the RUN or ACC position, the test light should come on. If the test light does not turn on and off as the ignition switch is turned, trace the circuit to the fuse box and the ignition switch to locate the problem.

Connect a jumper wire between connector terminals 6 and 8. Make sure all doors are closed. The courtesy lights and door lock cylinders should be illuminated. If the lights did not operate, trace the circuit from terminal 26 to the LEDs and ground to locate the problem.

Connect an ohmmeter between connector terminal 2 and chassis ground. The ohmmeter should indicate an infinite reading. However, a minimum of 10,000 ohms is acceptable. Lift up on each of the outside door handles to close the latch switch. Hold the handle up while observing the ohmmeter. The ohmmeter should indicate a resistance value of 50 ohms maximum. If either of the ohmmeter readings are out of specifications, trace the circuit to the latch switches. Also, test the latch switches for correct operation.

With the test light connected between connector terminals 1 and 8, the light should be on. If the test light fails to come on, trace the circuit from terminal 1 to ground.

If the preceding tests did not indicate any problems, the actuator module assembly is faulty. The module must be replaced.

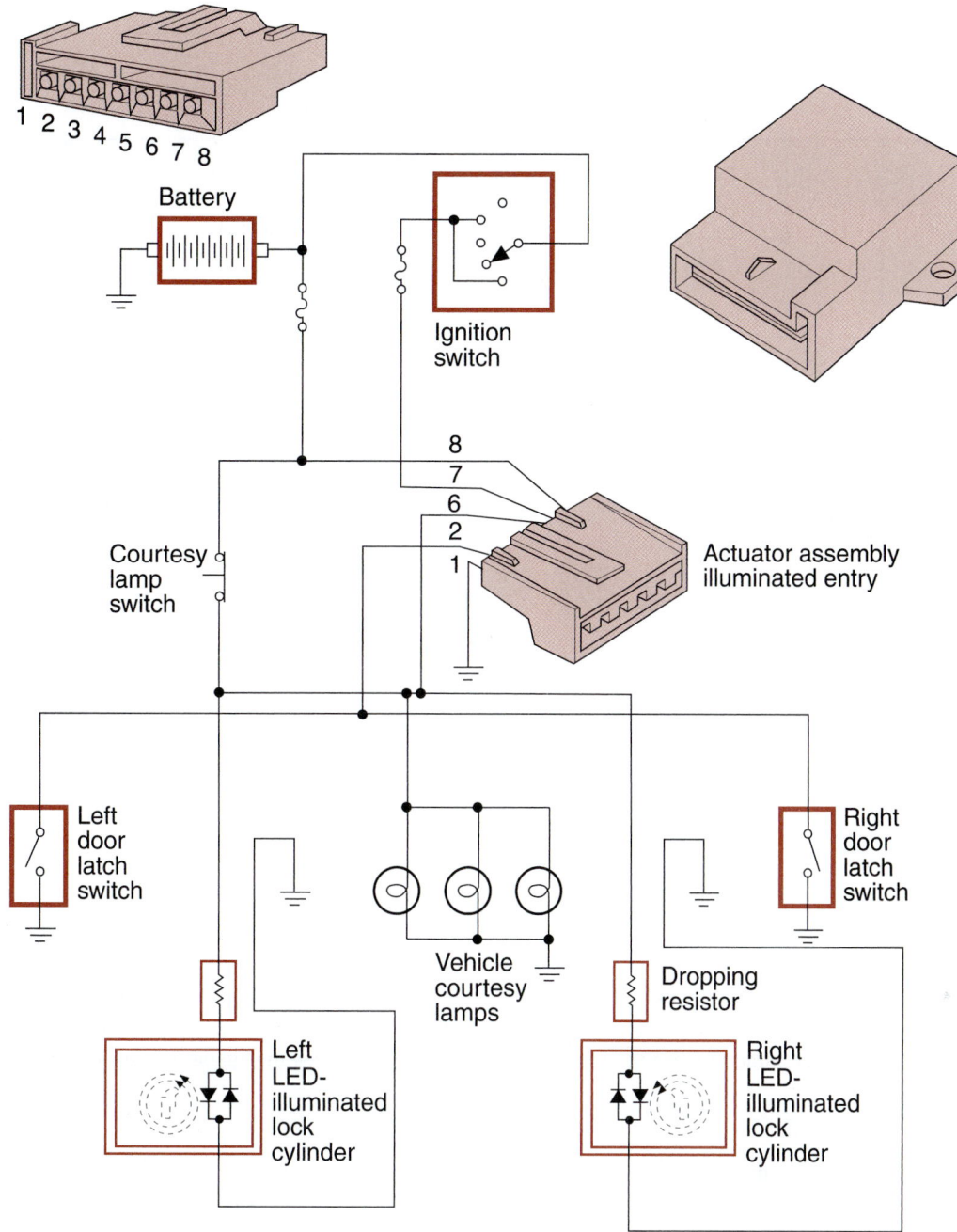

Figure 12-20 Illuminated entry system components and schematic.

(Figure 12-21) illustrates an early BCM-controlled illuminated entry system. Follow these steps to diagnose the system:

1. Move the dimmer control to the center position.

2. Open the driver's-side door to activate the courtesy lights. If none of the courtesy lights turn on, continue testing. If only one bulb is inoperative, check its circuit and the bulb.

3. Lower the driver-side window and close all doors. Manually lock the driver's door. Wait 30 seconds with the ignition switch off.

Special Tools

Scan tool

Service manual

Fused jumper wire

DMM

Test light

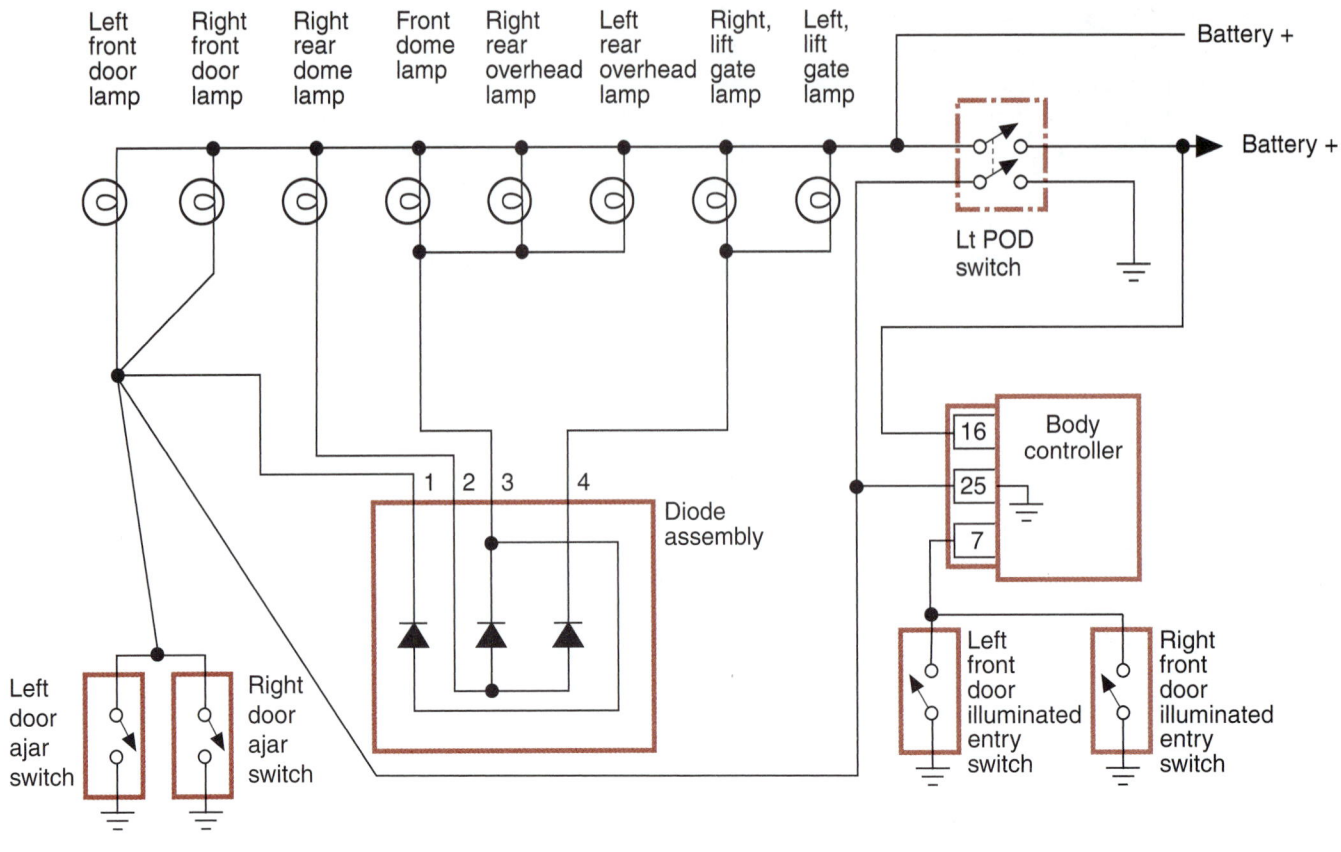

Figure 12-21 Schematic of an illuminated entry system.

4. Activate the illuminated system by lifting the driver's door handle. If the lights come on, repeat the test for the right side door. If the system does not operate when the right side door handle switch is closed, refer to the service manual for the circuits to be tested. The procedure will be the same as when the left door is inoperative. However, the circuit designations are different.

5. If the lights do not turn on when the door handles are lifted, connect the scan tool to the diagnostic connector. Maneuver through the menu screens to locate the door handle switch state.

6. With the ignition switch on, observe the scan tool display while lifting the door handle. The display should indicate the switch closed when the handle was lifted. If the switch operated correctly, go to step 7. If the display did not indicate proper switch operation, connect a jumper wire from the controller terminal identified in the service manual to ground. This would be terminal 7 at the BCM in Figure 12-21. Observe the scan tool. If it indicates the circuit is closed, follow the service manual procedure for testing the door switches. If the display indicates the circuit is open, replace the BCM.

CAUTION: The terminal to jump from the controller is different between years and models. Be sure to refer to the service manual for the correct terminals.

7. Open the driver's-side door. If the courtesy lights do not turn on, go to step 8. Close all doors and jumper the terminal of the BCM identified in the service manual (terminal 25 in Figure 12-21) to ground. If the lights do not turn on, there is an open circuit between the BCM and the lamps. If the lights turned on when the BCM was jumped, replace the BCM.

8. If the lights did not come on in step 7 when the door was opened, gain access to the driver's side door ajar switch harness and disconnect it. Use a jumper wire to jump across the two terminals of the connector on the harness side. This should complete the circuit to ground and the lights come on. If the lights do not turn on, check both circuits for an open. If the lights come on when the connector is jumped, replace the switch. Test the door ajar switch for the passenger side in the same manner. Use a jumper wire to jump across the two switch connector terminals on the harness.

Some BCM-controlled illuminated entry systems do not use a door handle switch. The system is activated when the doors are opened or the remote keyless entry system is used to unlock the doors. The system uses the same lamp driver as the courtesy lamps. Check for operation of the system by opening each door, one at a time. Usually the three passenger door ajar switches are connected in parallel with the driver's door on its own circuit (Figure 12-22). If the system does not activate when the doors are opened, use the remote keyless entry fob and press the UNLOCK button. If the system works now, the problem is in the door ajar input circuits or the BCM. If using the fob does not activate the system, the fault is probably in the BCM or the control circuits to the lamps. However, it is possible that all inputs are missing. To confirm this, use the scan tool to activate the system. If the lamps illuminate now, each input will need to be tested.

> ☑ **SERVICE TIP:** You can also use the headlight switch dimmer function to test the lamps. Turn the switch to the DOME position. If the light come on now, the output circuit from the BCM is working fine and the problem is in the inputs.

To test the inputs to the BCM, use a scan tool to monitor the door ajar switches. The switches should change state as the doors are opened and closed. The scan tool will display the state of the switches, not the door. The switches are open when the door is closed, and the switches are

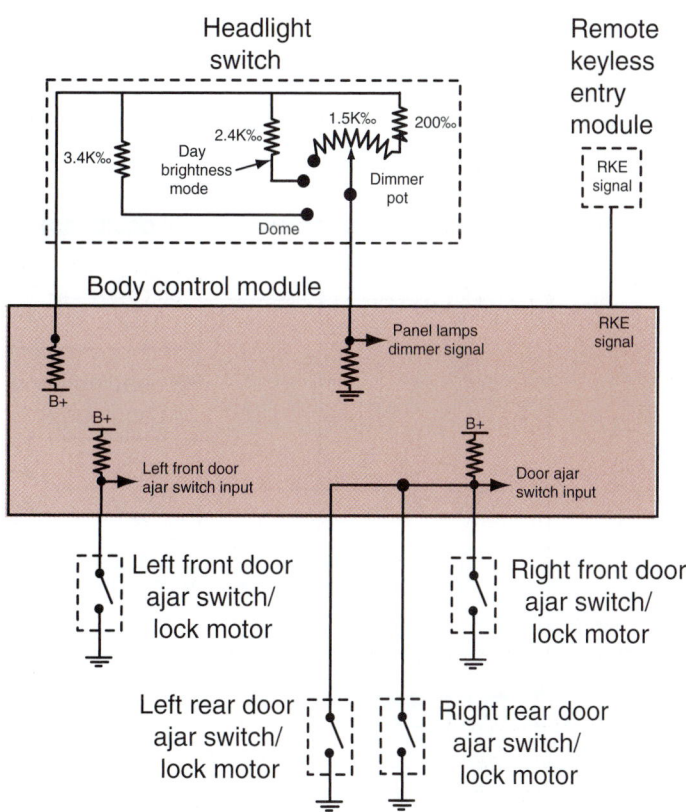

Figure 12-22 The inputs used for the illuminated entry system. Note the passenger door ajar switches are in parallel.

closed when the door is open. If no change of state is seen, test the door ajar signal circuit by disconnecting the door ajar switch. Connect a jumper wire across the connector on the BCM side. If the illuminated entry system activates now (or the scan tool displays the switch as closed), there is a faulty switch. If this still does not activate the system, test for battery voltage to the switch connector. If battery voltage is not present, test the circuit back to the BCM by back probing for voltage at the BCM connector. If voltage is present now, there is an open in the wire between the BCM and the switch. If voltage is not present, check for good battery feed to the BCM and proper ground connections. If these are good, replace the BCM. If battery voltage is present at the switch connector when tested above, the fault is in the ground circuit for the switch.

If the scan tool indicates proper input from the door ajar switches and the activation test failed to illuminate the lamps, then the output side of the system must be tested (Figure 12-23). Locate the proper wire into the BCM connector for the courtesy lamp control. Using a back-probing tool, jump this terminal to a good ground. If the lamp turns on now, check for proper battery feed and grounds to the BCM. If these are good, replace the BCM. If the lamps do not turn on, use the manual switches to turn on each light. If they do not turn on, trace the circuit from the battery to the common connection or splice.

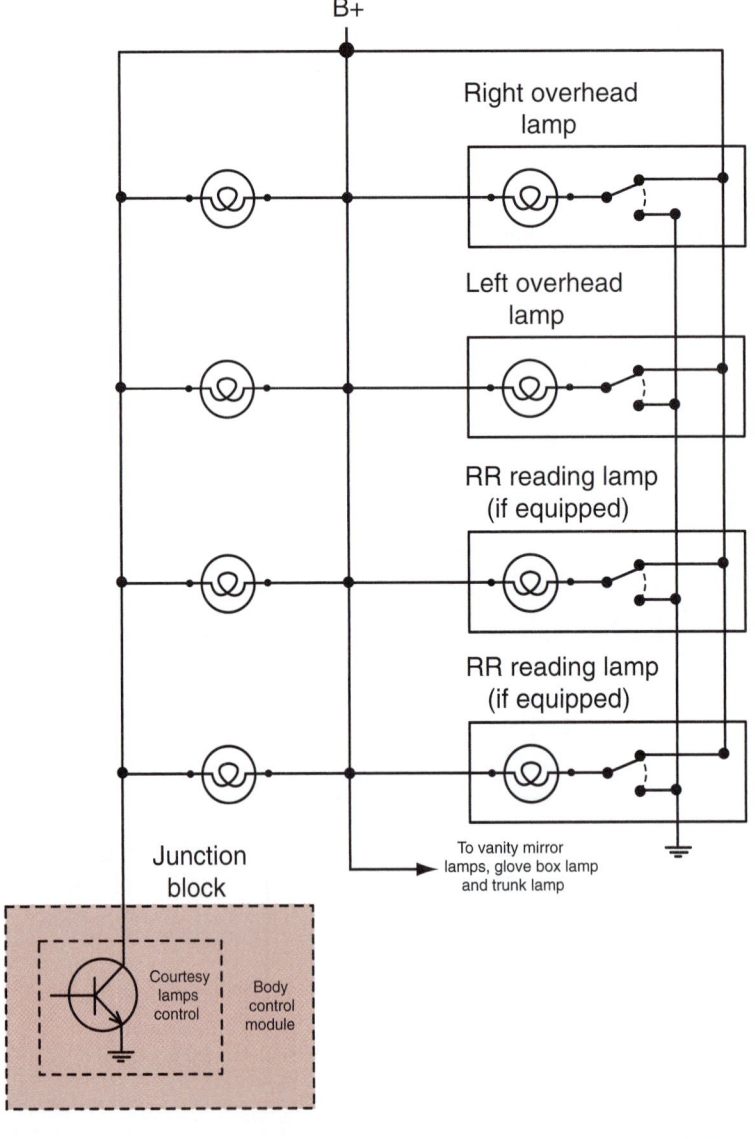

Figure 12-23 Output control of the illuminated entry system.

Fiber Optics Diagnosis

Classroom Manual
Chapter 12, page 367

The fiber optic system uses plastic strands to transmit light from the source to the object to be illuminated. The strands of plastic are sheathed by a polymer that insulates the light rays as they travel within the strands. The light rays travel through the strands by means of internal reflections. Fiber optics can be used to provide light in areas where bulbs would be inaccessible for service.

If the fiber optics do not illuminate, most likely the light source has failed. Check to see that the bulb is illuminating. If the bulb turns on, check that the fiber optic lead is connected to the light source and to the lens. If these are good, the only other cause is that the cable is cut. It will need to be replaced.

CASE STUDY

A customer brings a vehicle into the shop because of an intermittent problem with the headlight delay feature. This person has taken the vehicle to other shops and spent several dollars in repair bills, but the problem has not been corrected. Using a systematic diagnostic approach and following the tests outlined in the service manual leads the technician to test the potentiometer in the control switch. The resistance value is within specifications. However, while moving the potentiometer from the MIN to the MAX position, the ohmmeter reading is erratic in one portion. This area is the usual setting selected by the driver. The technician calls the customer and receives approval to replace the control switch. The new switch cures the problem. The technician opens the old switch and finds that carbon from electrical arcing has built up in the problem area of the potentiometer.

Terms to Know

Demonstration mode

Illuminated entry actuator

Optics test

Photocell resistance assembly

ASE-Style Review Questions

1. The results of a functional test on the automatic headlight system are being discussed.
   Technician A says when the photocell is covered and the engine is running, the headlights should turn on within 30 seconds.
   Technician B says when a bright light is shone onto the photocell, the lights should turn off after the selected amount of time delay.
   Who is correct?
   A. A only
   B. B only
   C. Both A and B
   D. Neither A nor B

2. The results of the photocell resistance test are being discussed.
   Technician A says if the lights do not turn on within 60 seconds when the resistance assembly switch is in the OFF (open) position and the ignition switch is in the RUN position, the photocell should be replaced.
   Technician B says if the lights turn off within 60 seconds after the resistance assembly switch is turned on, the photocell should be replaced.
   Who is correct?
   A. A only
   B. B only
   C. Both A and B
   D. Neither A nor B

3. The automatic high-beam function does not operate properly. The LED in the mirror assembly is flashing continuously at a 1-Hz frequency. This can indicate:
  A. The mirror assembly is in need of being calibrated.
  B. The system failed its last attempt to calibrate.
  C. A hardware failure has occurred.
  D. None of the above.

4. An indicator that uses fiber optics is not functioning. This can be caused by:
  A. A bent fiber-optics cable.
  B. A faulty light source.
  C. Electromagnetic interference.
  D. None of the above.

5. The most likely cause of the automatic headlights failing to activate in low ambient light conditions is:
  A. A faulty ignition switch.
  B. Camera angle alignment out of specifications.
  C. Burned-out headlight elements.
  D. Faulty headlight switch.

6. *Technician A* says problems with the automatic headlight system can be the fault of the headlight switch.
  Technician B says a bad ignition switch may cause the lights to not come on.
  Who is correct?
  A. A only   **C.** Both A and B
  B. B only   **D.** Neither A nor B

7. The fiber-optic indicator is not operating.
  Technician A says the cable can be cut.
  Technician B says the light source may not be operating.
  Who is correct?
  A. A only   **C.** Both A and B
  B. B only   **D.** Neither A nor B

8. The photocell resistance assembly is being discussed.
  Technician A says it is a technician-made test tool consisting of resistors and a switch.
  Technician B says it is a known good replacement photocell.
  Who is correct?
  A. A only   **C.** Both A and B
  B. B only   **D.** Neither A nor B

9. The headlights work in the manual position but do not turn on in the AUTO position. What is the most likely cause?
  A. A bad headlight ground.
  B. A faulty photocell assembly.
  C. A faulty headlight relay.
  D. A bad headlight relay ground connection.

10. *Technician A* says the photocell signal voltage should be 5 volts with the lens covered.
  Technician B says the signal voltage should be 0 volts with the a flashlight shining on the lens.
  Who is correct?
  A. A only   **C.** Both A and B
  B. B only   **D.** Neither A nor B

ASE Challenge Questions

1. An inoperative BCM-controlled automatic headlamp four-door system is being discussed; the doors will open but will not close.
Technician A says that the headlight door motor may be faulty.
Technician B says that the headlamp switch may be faulty.
Who is correct?
   - **A.** A only
   - **B.** B only
   - **C.** Both A and B
   - **D.** Neither A nor B

2. The high-beam headlamps of a vehicle equipped with a Twilight Sentinel automatic headlamp system are inoperative; the low beams are working correctly. A voltmeter connected across the high-beam contacts of the dimmer switch indicate 0.15 volt when the high beams are "on."
Technician A says that the circuit from the Sentinel amplifier to the dimmer switch has excessive resistance.
Technician B says that the dimmer switch is faulty.
Who is correct?
   - **A.** A only
   - **B.** B only
   - **C.** Both A and B
   - **D.** Neither A nor B

3. A DaimlerChrysler vehicle equipped with a BCM-controlled illuminated entry system is being discussed. The system is working fine except for the fact that three interior bulbs (of a total of eight) are inoperative. A voltmeter connected across the terminals of the inoperative bulbs indicates 12.6 volts when the system is in operation.
Technician A says that the BCM is working correctly.
Technician B says that the diode assembly may be faulty.
Who is correct?
   - **A.** A only
   - **B.** B only
   - **C.** Both A and B
   - **D.** Neither A nor B

4. The lamp outage module is illuminating the lamp-out warning light, but the lamp operates.
Technician A says installation of the wrong light bulb can cause this.
Technician B says a corroded connection can cause this.
Who is correct?
   - **A.** A only
   - **B.** B only
   - **C.** Both A and B
   - **D.** Neither A nor B

5. The headlights work normally, except they do not turn on in the automatic mode. What is the LEAST likely cause?
   - **A.** Faulty photocell.
   - **B.** Faulty amplifier.
   - **C.** Faulty headlight switch.
   - **D.** Faulty headlight relay.

Job Sheet 43

43

Name _____ Date _____

Testing an Automatic Headlight System

Upon completion of this job sheet, you should be able to diagnose an automatic headlight system and test the individual components of the system.

ASE Correlation

This job sheet is related to the ASE Electrical/Electronic System Certification Exam's content area: *Headlights, Parking Lights, Taillights, Dash Lights, and Courtesy Lights;* task: Inspect, test, and repair or replace headlight and dimmer switches, relays, control units, sensors, sockets, connectors, and wires of headlight circuits.

Tools and Materials

A vehicle with an automatic headlight system
Wiring diagram for the chosen vehicle
Component locator for the chosen vehicle
Service manual for the chosen vehicle
A fused jumper wire
A DMM

Describe the vehicle being worked on:

Year _____ Make _____ Model _____

VIN _____ Engine type and size _____

Procedure

Task Completed

1. Locate the photocell. Then disconnect the connector to the photocell. ☐
   Turn the ignition switch and the automatic headlamp control to ON. ☐
   Turn the headlight switch off. ☐
   If the lights come on within 60 seconds and the automatic headlights didn't work before, the photocell must be bad. If the lights still don't come on, test the amplifier unit. Describe what happened.

   _____

   _____

   _____

   _____

   _____

   _____

☐
☐
☐
☐

2. Turn the ignition off. Disconnect the negative cable from the battery.
   Locate the amplifier assembly and disconnect the electrical connector to it.
   Turn the headlight switch off.
   Turn the automatic headlight control to its ON position.
   Locate the resistance checks of the amplifier circuit in the service manual. Briefly
   outline those procedures.

   _____

   _____

   _____

   _____

3. Follow the previously described procedures and list the results.

   _____

   _____

   _____

   _____

4. Turn the automatic headlight control to its ON position.
   Locate the resistance checks of the amplifier circuit in the service manual.
   Briefly outline those procedures.

   _____

   _____

   _____

   _____

5. Follow the procedures in step 4 and list the results.

   _____

   _____

   _____

   _____

6. List any additional diagnostic steps that the manufacturer recommends in the case
   where the previously described tests did not identify the problem.

   _____

   _____

   _____

   _____

   _____

   _____

   _____

Instructor's Response _____

_____

_____

Job Sheet 44

44

Name _____ Date _____

Testing the BCM-Controlled Headlight System

Upon completion of this job sheet, you should be able to test the computer-controlled headlight system and determine needed repairs.

ASE Correlation

This job sheet is related to the ASE Electrical/Electronic System Certification Exam's content area: *Headlights, Parking lights, Taillights, Dash Lights, and Courtesy Lights;* tasks: Inspect, test, and repair or replace headlight and dimmer switches, relays, control units, sensors, sockets, connectors, and wires of headlight circuits.

Tools and Materials

A vehicle equipped with BCM-controlled headlights
Scan tool
DVOM

Test light
Fused jumper wires
Wiring diagram for the vehicle

Describe the vehicle being worked on:

Year _____ Make _____ Model _____

VIN _____ Engine type and size _____

Procedure

Task Completed

1. Test the operation of the headlights. Describe the symptoms.

_____

_____

2. Are there any other related symptoms?   ☐ Yes   ☐ No
If yes, describe the symptom(s):

_____

_____

3. Describe the basic operation of the system you are working on.

_____

_____

_____

4. Check any associated fuses and conduct a visual inspection of the system. List any problems found.

_____

_____

Repair any faults found and test system operation. Did this fix the problem?
☐ Yes   ☐ No

5. If used, substitute the headlamp relay with a known good relay of the same type and test operation. Test the system operation. Did this fix the problem?   ☐ Yes   ☐ No

6. Connect the scan tool and record any DTCs.

_____

_____

7. Use the scan tool to perform an activation test of the headlights. Did the headlamps come on during the activation?   ☐ Yes   ☐ No

What do you know about the problem so far?

_____

_____

8. Use the scan tool to monitor the headlight switch as it is placed in all switch positions. Record the voltages for each position and compare to specifications.

_____

_____

9. What do you know about the system so far?

_____

_____

10. Remove the headlight relay and use a voltmeter or test light to confirm battery voltage is present at pins 30 and 86 of the junction block terminals. Is voltage present at these terminals?   ☐ Yes   ☐ No

If NO, what would you test next?

_____

_____

11. If voltage is present at both terminals in step 10, use a fused jumper wire to connect pin 30 to pin 87. Did the lights come on?   ☐ Yes   ☐ No

What do you know about the circuit now?

_____

_____

12. Connect a test light between pins 86 and 85 of the junction block terminals (relay removed). Place the headlight switch into the headlamp position and observe the test light. What does this test check?

_____

_____

13. Based on your results for step 12, what do you know about the problem?

_____

_____

14. What other checks need to be performed to pinpoint the fault?

_____

_____

15. Perform the tests you listed and record your findings and recommendations.

Task Completed

_____ 485 __

_____

Instructor's Response _____

_____

_____

_____

Job Sheet 45

Name _____ Date _____

Diagnosing Automatic High-Beam Systems

Upon completion of this job sheet, you should be able to diagnose the cause of poor automatic high-beam system operation and determine needed repairs.

ASE Correlation

This job sheet is related to the ASE Electrical/Electronic System Certification Exam's content area: *Headlights, Parking Lights, Taillights, Dash Lights, and Courtesy Lights;* task: Diagnose, inspect, test, and repair dimmer control units and determine necessary action.

Tools and Materials

Vehicle with SmartBeam™
Calibration target
Grease pencil
Scan tool
Tape measure
Piece of cardboard
Service manual

Describe the vehicle being worked on:

Year _____ Make _____ Model _____

VIN _____ Engine type and size _____

Procedure

For this job sheet task, you will identify the operational characteristics of the SmartBeam™ automatic high-beam system and calibrate the camera.

Task Completed

1. Turn on the ignition. Observe the SmartBeam™ camera status LED and determine if the camera is calibrated. Is the SmartBeam™ camera calibrated? ☐ Yes ☐ No

   Note: If the camera is not calibrated, it must be calibrated prior to entering demonstration mode.

2. Turn off the ignition. Press and hold the AUTO button on the rearview mirror while turning the ignition switch to the RUN position. Continue to depress the AUTO button until the LED blinks and the high beams turn on, then release the button. Describe what happens.

   _____

   _____

   _____

3. Repeat the steps to enter the demonstration mode again. While the lamps are illuminated during the demonstration mode, slide a piece of cardboard between the camera lens and the windshield. Describe the results.

_____

_____

4. Use the service manual information and determine the alignment specification of the camera. Record the specifications.

_____

_____

5. Connect the scan tool and navigate to the "automatic high-beam module (AHBM)." Locate the camera aim test function and activate the test. Describe the status of the LED.

_____

6. What does this indicate?

_____

7. Turn off the ignition, and use a tape measure and service manual procedures to determine the centerline of the camera lens; mark the windshield and rear glass with a grease pencil.

8. Using a tape measure, determine a location in front of the vehicle that is the specified distance from the vehicle. What is the specified distance? _____

9. Locate the calibration target in front of the vehicle, and align it with the marks on the windshield.

10. Using a tape measure, determine the distance the target's center LED should be from the floor. Record this distance. _____

11. Adjust the calibration target to the required height.

12. Turn on the power supply to the calibration target, and then place the ignition switch in the RUN position. Describe the results.

_____

_____

13. Use the scan tool to determine and record the calibration status. _____

14. Check for DTCs and record.

_____

_____

_____

15. Clear any DTCs.

Instructor's Response _____

_____

_____

DIAGNOSTIC CHART 12-1

PROBLEM AREA:	Automatic headlight system
SYMPTOMS:	Headlights fail to turn on in automatic mode, headlights work in manual mode
POSSIBLE CAUSES:	**1.** Open input circuit from switch to control module **2.** Faulty module **3.** Faulty switch

DIAGNOSTIC CHART 12-2

PROBLEM AREA:	Automatic headlight system
SYMPTOMS:	Headlights turn on in daytime when switch is in AUTO mode
POSSIBLE CAUSES:	**1.** Faulty photocell **2.** Open photocell circuit **3.** Shorted photocell circuit **4.** Faulty control module **5.** Bus communications error **6.** Immobilizer system inoperative

Instrumentation and Warning Lamp System Diagnosis and Repair

Upon completion and review of this chapter, you should be able to:

Basic Tools

Basic mechanic's tool set

Service manual

❏ Remove and replace the instrument cluster.

❏ Remove and replace the printed circuit.

❏ Diagnose and repair causes for erratic and inaccurate speedometer readings.

❏ Diagnose and repair causes for tachometer malfunctions.

❏ Diagnose and repair faulty gauge circuits.

❏ Diagnose and repair the cause of multiple gauge failure.

❏ Diagnose sender units, including thermistors, piezoresistive, and mechanical variable styles.

❏ Diagnose and repair warning light circuits.

❏ Diagnose and repair the cause of multiple warning light failures.

❏ Enter diagnostic mode and retrieve trouble codes from BCM-controlled electronic instrument clusters.

❏ Perform self-diagnostic tests on the electronic instrumentation system.

❏ Determine faults as indicated by the self-test.

❏ Diagnose computer-driven speedometer and odometer instrumentation malfunctions.

❏ Test magnetic pickup speed sensors.

❏ Test optical-type speed sensors.

❏ Determine the cause of constant low gauge readings in computer-controlled instrumentation systems.

❏ Diagnose and locate the cause for constant high gauge readings in computer-controlled instrumentation systems.

Introduction

The instrument panel gauges and warning lamps monitor the various vehicle operating systems and provide information to the driver about their operation. Most problems in the gauges or warning lamps are usually caused by an open circuit in the wiring or the printed circuit; improper gauge calibration; loose connections; excessive resistance; or defective bulbs, gauges, or sending units.

In this chapter you will learn how to diagnose the gauge, lamp, and sending unit of the various styles of conventional instrument panels. You will learn how to remove the instrument panel to replace the printed circuit, and how to repair the speedometer cable core.

This chapter will also introduce you to common service procedures used to diagnose and repair computer-driven instrumentation systems. These systems include the speedometer, odometer, fuel, and engine instrumentation. The computer-driven instrument cluster uses a microprocessor to process information from various sensors and to control the gauge display. Depending on the manufacturer, the microprocessor can be a separate computer that receives direct information from the sensors and makes the calculations, or the body control module (BCM) is used to perform all functions. The computer may control a digital or a quartz swing needle instrument cluster.

It is out of the scope of any textbook to cover service procedures for every type of instrument cluster. To illustrate, in one model year alone, Chrysler offered two different electronic clusters (Huntsville and Motorola). The Huntsville cluster was available in four different variations, one with a message center, one with a trip computer, and two options that offered tachometers. In addition, the system could also have a twenty-four voice alert function. In the same year, Ford offered five different electronic clusters, and each division of General Motors offered its own electronic instrument cluster. Add to this the many different types of import vehicles that use their own systems. This chapter will familiarize you with general procedures; however, it is important to remember that each system uses its own diagnostic procedures.

Usually the technician will be required to isolate the faulty component and replace it. Most instrument panel components are not repaired or serviced in the shop but sent back to the manufacturer or specialty shop for rebuilding.

Special Tools

Fender covers

Safety glasses

Battery terminal
pliers

Instrument Panel and Printed Circuit Removal

Many times it may be necessary to remove the instrument panel to replace defective gauges, lamps, or printed circuits. Before removing the instrument panel, always disconnect the battery negative cable. Consult the service manual for the procedure for the vehicle you are working on. The following is a common method of removing the instrument cluster and printed circuit:

1. Place fender covers on the fenders and disconnect the battery negative cable.
2. Remove the retaining screws to the steering column cover. Then remove the cover.
3. Remove the finish panel retaining screws. On some models it may be necessary to remove the radio knobs.
4. Remove the finish panel.
5. Remove the retaining bolts that hold the cluster to the dash.
6. Reach behind the instrument panel and disconnect the speedometer cable.
7. Gently pull the cluster away from the dash.
8. Disconnect the cluster feed plug from the printed circuit receptacle. Be careful not to damage the printed circuit.
9. Remove the IVR and all illumination and indicator lamp sockets (Figure 13-1).
10. Remove the charging system warning lamp resistor if applicable.
11. Remove all printed circuit attaching nuts and remove the printed circuit.

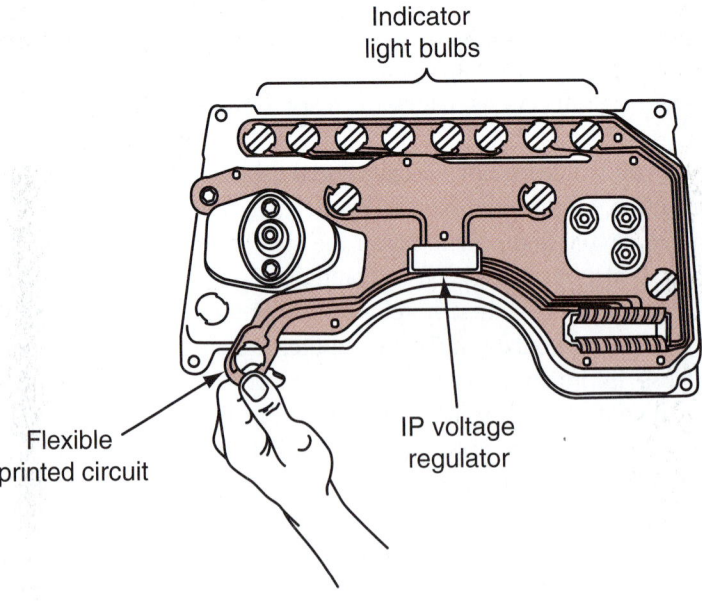

Figure 13-1 Instrument panel printed circuit board.

BCM Diagnostics

The BCM may be capable of running diagnostic checks of the electronic instrument cluster to determine if a fault is present. If the values received from monitored functions are outside of programmed parameters, a DLC is set. This code can be retrieved by the technician to aid in troubleshooting. Depending upon the vehicle, code retrieval is done through the ECC, IPC, jumping terminals in the DLC, or by a scan tool.

To retrieve diagnostic codes in most General Motors vehicles with an electronic climate control (ECC) display, turn the ignition switch to the RUN position and simultaneously press the OFF and WARMER buttons on the climate control panel. ECM codes will be displayed first, followed by BCM codes. The system will display codes twice. The first pass are all codes in memory. Codes that are in the first set but not in the second are history codes. All codes displayed during the second pass are current codes.

After the trouble codes have been retrieved from memory, refer to the proper diagnostic chart to isolate the fault. It is possible for a problem to exist that does not set a trouble code. In these instances, use the symptom or troubleshooting chart in the service manual to locate the fault.

Some manufacturers provide a means of overriding the instrument cluster display and change the parameters to allow for testing. By changing the parameters in this test mode, the gauge will change its indicated reading. If the gauge changes it readings correctly, the fault is in the control module.

Classroom Manual
Chapter 13,
pages 384, 387

Special Tools

Scan tool

Self-Diagnostics

Most instrument panel display modules have a diagnostic mode within their programming. The diagnostic mode allows the module to isolate any faults within the instrument panel cluster. In most systems, if the module is not able to complete its self-diagnostic test, the fault is within the module and it must be replaced. Successful completion of the self-diagnostic test indicates the problem is not the module. The following are examples of self-diagnostic procedures.

Diagnosis of a Typical Electronic Instrument Cluster

> **CAUTION:** VFD displays are easily damaged by physical shock. When handling EICs, do not drop or jar them.

> **CAUTION:** When servicing EICs, follow all service precautions related to static discharge in the vehicle manufacturer's service manual to avoid EIC damage.

All electronic instrument clusters (EICs) are sensitive to static electricity damage and EIC cartons usually have a static electricity warning label. When servicing EICs:

1. Do not open the EIC carton until you are ready to install the component.
2. Ground the carton to a known good ground before opening the package.
3. Always touch a known good ground before handling the component.
4. Do not touch EIC terminals with your fingers.
5. Follow all service precautions and procedures in the vehicle manufacturer's service manual.

Prove-Out Display. Most EICs have a prove-out display each time the ignition switch is turned on. During this display, all the EIC segments are illuminated and then turned off

Figure 13-2 All EIC segments are illuminated during the prove-out display.

momentarily (Figure 13-2). The EIC returns to normal display after the prove-out. If the EIC is not illuminated during the prove-out display, check the power supply and grounds to the EIC. If these are good, replace the EIC.

If some of the segments do not illuminate during the prove-out display, the EIC is defective and must be replaced. During the prove-out mode, the turn signal and high-beam indicators are not illuminated. Other indicator lights remain on when the EIC display is turned off momentarily in the prove-out mode. After the prove-out mode is completed, the indicator lights go out shortly after the EIC returns to normal display.

Function Diagnostic Mode. The diagnostic procedure for EICs varies depending on the vehicle make and model year. Always follow the diagnostic procedures in the vehicle manufacturer's service manual. Some EICs have a function diagnostic mode that provides diagnostic information in the display readings if certain defects occur in the system. For example, if the coolant temperature sender has a shorted circuit, the two top and bottom bars are illuminated in the temperature gauge and the ISO symbol is extinguished (Figure 13-3). If the engine coolant never reaches normal operating temperature or the coolant temperature sender circuit has an open circuit, the bottom bar in the temperature gauge is illuminated with the ISO symbol.

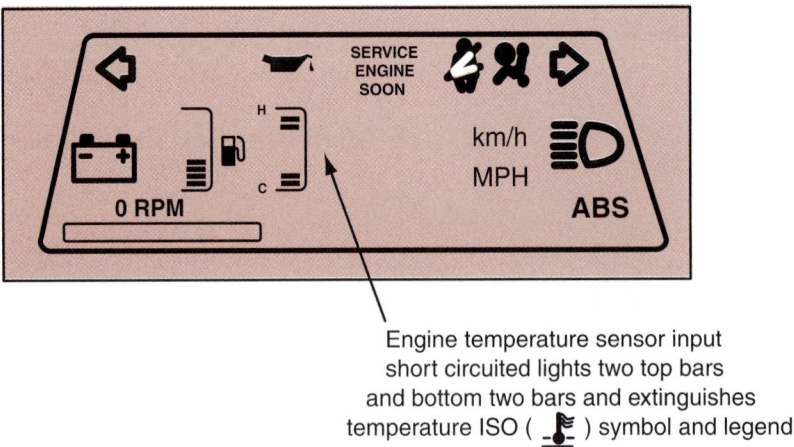

Engine temperature sensor input
short circuited lights two top bars
and bottom two bars and extinguishes
temperature ISO ( ) symbol and legend

Figure 13-3 EIC function diagnostic mode.

If the fuel gauge sender develops a short or open circuit, the two top and bottom bars in the fuel gauge are illuminated and the ISO symbol is not illuminated. A shorted fuel gauge sender causes CS to be displayed in the fuel remaining or distance to empty displays. If the fuel gauge sender has an open circuit, CO is displayed in the fuel remaining and distance to empty displays. When the function diagnostic mode indicates short or open circuits in the inputs, the cause of the problem must be located by performing voltmeter and ohmmeter tests in the circuit with the indicated problem. These voltmeter and ohmmeter tests are included in the vehicle manufacturer's service information.

When the word ERROR appears in the odometer display, the EIC computer cannot read valid odometer information from the nonvolatile memory chip.

Special Test Mode. Most EICs have a special test mode to determine if the display is working properly. To enter the special test mode, press the E/M and SELECT buttons simultaneously and turn the ignition switch from the OFF to the RUN position. When this action is complete, a number appears in the speedometer display and two numbers are illuminated in the odometer display. The gauges and message center displays are not illuminated. If any of the numbers are flashing in the speedometer or odometer displays, the EIC is defective and must be replaced.

Diagnosis of a Typical Import Electronic Instrument Cluster

The display check tests for an open circuit in each display segment and shorts between segments. Press and hold trip reset switch A and turn the ignition switch from OFF to RUN to initiate the display check (Figure 13-4). After this action is taken, all the display segments should illuminate, one after the other. If any segment is not illuminated, the EIC must be replaced.

The preprogrammed signal check tests for defects in various displays. To complete the preprogrammed signal check:

1. Disconnect the negative battery cable. If the vehicle is equipped with an air bag, wait the specified time recommended by the vehicle manufacturer.

2. Remove the EIC power unit.

3. Remove the retaining nuts on the EIC switches; then remove the EIC switches.

4. Remove cluster lid; then remove the EIC assembly.

5. Connect the special self-checking wiring harness to the EIC terminals (Figure 13-5).

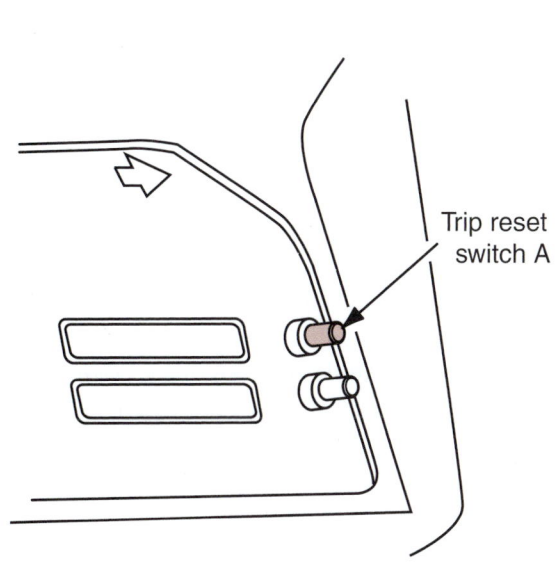

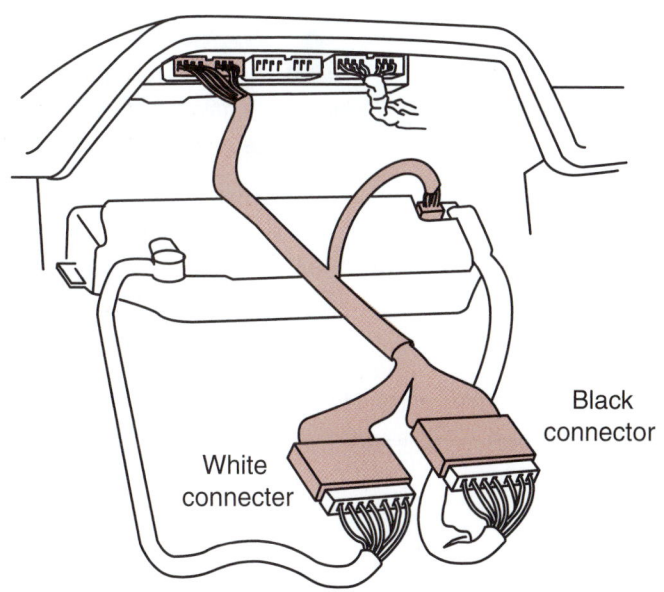

Figure 13-4 EIC reset button.

Figure 13-5 Special self-checking wiring harness connected to EIC terminal.

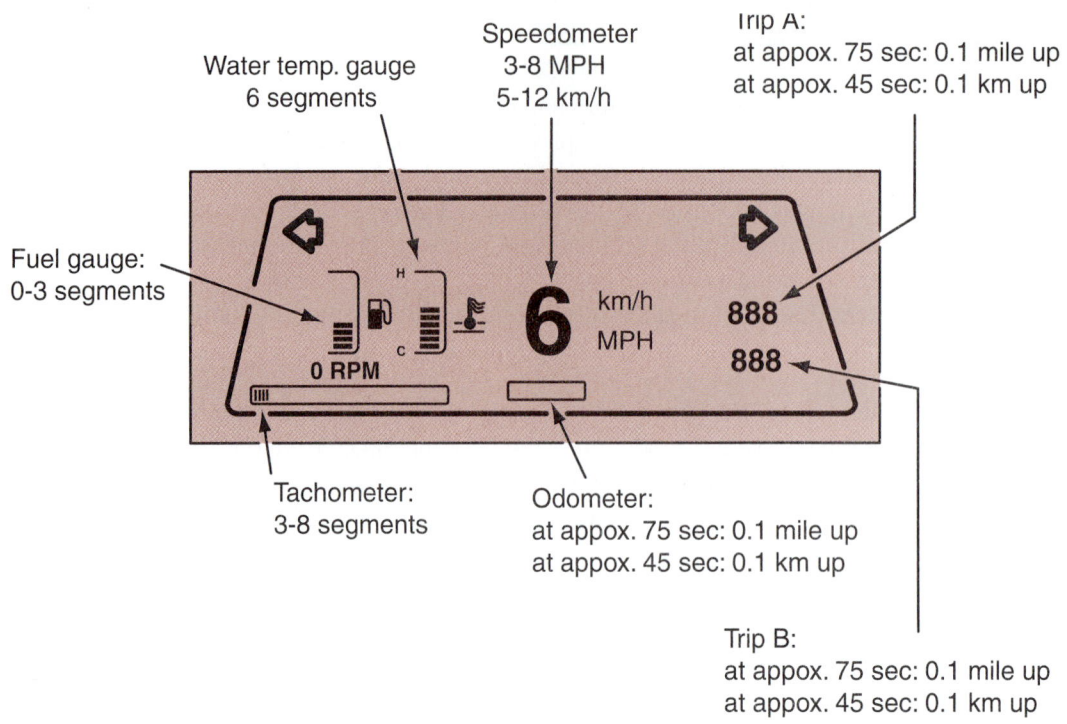

Figure 13-6 Display changes during the preprogrammed signal check.

6. Connect the negative battery cable, turn on the ignition switch, and observe the EIC displays. Each display should change to a specific reading (Figure 13-6). If each display changes as specified by the vehicle manufacturer, the EIC is satisfactory. When some of the displays do not change as specified, voltmeter and ohmmeter tests are required to locate the exact cause of the problem.

After completing the test procedure, turn off the ignition switch and disconnect the negative battery cable. If the vehicle is equipped with an air bag, wait the length of time specified by the vehicle manufacturer. Disconnect the special self-checking wiring harness and connect all EIC connectors securely. Complete installation of the EIC, cluster lid, and switches.

WARNING: If the odometer has been repaired or replaced and it cannot indicate the same mileage as before it was removed, in most areas the law requires that an odometer mileage label must be attached to the left-front door frame. Failure to comply with this procedure could lead to court action.

A defective power unit may cause the EIC displays to be inoperative. The power unit supplies different voltages to various EIC displays. Therefore, it is possible for a defective power unit to cause the failure of specific EIC displays to illuminate.

To perform the power unit test, begin by removing the power unit and leaving the wiring harness connected to the unit. With the ignition switch in the RUN position, test the voltage at the power unit terminals. Each power unit terminal should have the voltage specified by the vehicle manufacturer (Figure 13-7). The power unit ground wire is connected from terminal 9 to ground. With the ignition switch in the OFF position, connect a pair of ohmmeter leads from power unit terminal 9 to ground. If the meter reading is above 0.5 Ω, repair the ground wire. If the power unit does not have the specified voltage at some of the terminals, replace the unit.

A defective speed sensor may show an inoperative or erratic speedometer reading. To test the speed sensor, begin by removing the cluster lid to gain access. Connect a pair of voltmeter leads to terminals 11 and 1 on the EIC with the wiring harness connected (Figure 13-8). Turn the ignition switch to the RUN position and check the voltmeter reading. If the voltage is zero,

Power unit terminals	Voltmeter leads $(+)$	Voltmeter leads $(-)$	Voltage (V)	Remarks
	2		Approx. 12	
	3	9	Approx. 0	Check when no display appears
	5		Approx. 22	
	6		Approx. 26	
		7	Approx. 23	
	9	13	Approx. 14	For speedometer, fuel information, tachometer
		14		
		15	Approx. 19	For temp., trip
		16		

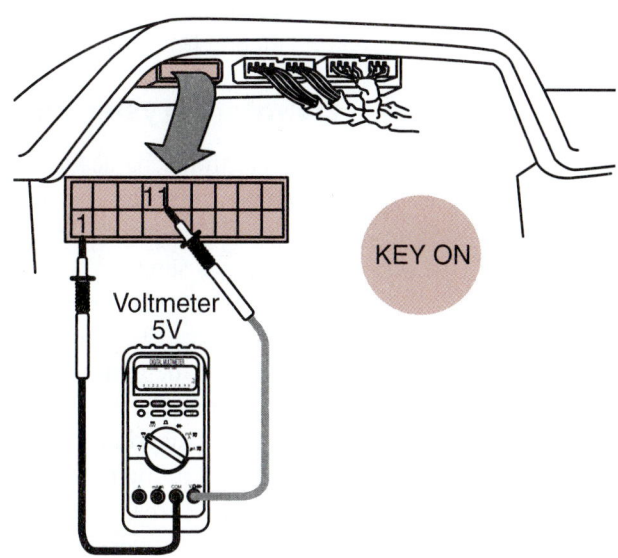

KEY ON

Voltmeter 5V

Figure 13-7 Voltage at various power unit terminals.

Figure 13-8 Voltmeter connections to EIC terminals.

check the power unit. If voltage is present, turn off the ignition switch and disconnect the speedometer cable from the speed sensor. Remove the wiring harness connector containing terminals 1 and 12 from the EIC and connect an analog voltmeter's leads to terminals 1 and 12. Use a small screwdriver to slowly rotate the speed sensor (Figure 13-9). If the voltmeter pointer does not deflect, the speed sensor or the connecting wires are defective. Connect the voltmeter leads directly to the speed sensor terminals and rotate the speed sensor again. If the voltmeter pointer deflects, repair the wires between the speed sensor and EIC terminals 1 and 12. If the voltmeter pointer does not deflect, replace the speed sensor.

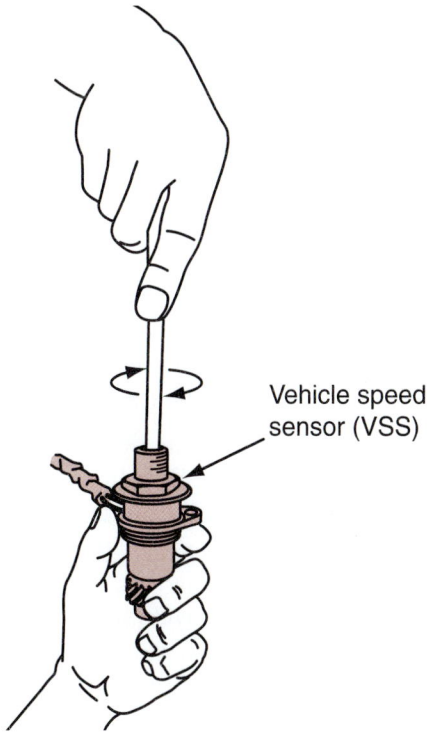

Vehicle speed sensor (VSS)

Figure 13-9 Testing the speed sensor.

SERVICE TIP: While testing the speed sensor, slowly turn the sensor with a small screwdriver. The sensor produces 24 signals per revolution, which are difficult to read during fast sensor rotation.

DaimlerChrysler Electromechanical Cluster Self-Test

Many electromechanical clusters (MIC) can be tested with or without a scan tool. Usually a self-test of the cluster can be performed. This test does not check any of the inputs—only the cluster. As an example of this system, a 1999 Chrysler mini-van is used. However, diagnostics of most MICs is similar.

To enter self-diagnostics, the ignition key must be in the LOCK position. Push and hold the TRIP and RESET buttons on the cluster at the same time. While continuing to hold these buttons, turn the ignition switch to the RUN position. Note that the cluster will illuminate in the UNLOCK position but will not activate self-diagnosis. Continue to hold the two buttons until CODE is displayed in the odometer. Release the buttons. If there are any fault codes, they will be displayed in the odometer. A code 999 means there are no faults. If fault codes are present, use the correct diagnostic manual to diagnose the system.

After the codes are displayed, the cluster will go through a series of tests as follows:

Check 0. Tests all of the VF display segments in the odometer and PRND3L. All segments should be illuminated.

Check 1. Tests the operation of all gauges. The gauge swing needles will move to programmed values.

Check 2. Illuminates each odometer VF segment individually.

Check 3. Tests the PRND3L display.

Check 4. Illuminates all of the warning lamps that are controlled by the MIC.

Observe operation of the MIC during each test. If any of these tests fail proper operation, the MIC must be replaced.

An additional feature of this cluster is that the technician can calibrate the gauges. This requires the use of DaimlerChrysler's DRBIII scan tool. The following steps provide a guide to recalibrating the speedometer. All gauges are calibrated in the same manner.

1. Plug the DRB III cable into the DLC. The ignition switch does not need to be in the RUN position. However, the bus must be active. Opening a door will awake the BCM.
2. Once the DRB III powers up and displays the MAIN MENU, select option number 1.
3. From the SELECT SYSTEMS menu, select BODY.
4. Select ELECTRO/MECH CLUSTER from the BODY menu.
5. Select MISCELLANEOUS.
6. Select CALIBRATE GAUGES.
7. The DRB III will ask if the cluster has a tachometer. Answer with the YES or NO key.
8. The DRB III will ask if the vehicle is a diesel. Answer with the YES or NO key.
9. The DRB III will ask if the cluster units are in MPH. Answer with the YES or NO key.
10. Place the ignition switch into the UNLOCK position.

This will place the DRB III in the mode to run gauge calibration. The first gauge to be calibrated will be the speedometer. The DRB III screen will display that it is sending a signal to the MIC to set the mph at 0. If the needle is not aligned with the 0, then use the up or down arrow keys to move the needle until it is aligned. Once the gauge is calibrated to 0, press the enter key and the DRB III will move to the next calibration unit. The next unit is 20 mph. Follow the same procedure to align the needle with the 20-mph mark on the cluster. After each calibration, press the enter key to move to the next unit. The other calibration units are 55 and 75 mph.

The tachometer, fuel, and temperature gauges are calibrated in the same manner. Once all of the gauges are calibrated, the DRB III will instruct the MIC to write the new values to memory.

Oil gauge	H ≡ N o r m L ≡	Oil pressure sensor input short circuited light top 2 bars and bottom 2 bars and extinguishes oil can ISO symbol	H N o r m L	Low oil pressure warning or oil pressure sensor input open circuited lights bottom bar and flashes ISO symbol
Temp gauge	H ≡ N o r m C ≡	Engine temperature sensor input short circuited lights top 2 bars and bottom 2 bars and extinguishes ISO symbol	H N o r m C	Cold engine temperature indication or engine temperature sensor input open circuited lights bottom bar and ISO symbol
Fuel gauge	H ≡ N o r m L ≡	Fuel level sender input short circuited or open lights top 2 and bottom 2 bars and extinguishes ISO symbol	CO CS	Fuel level sender input short or open circuited displays "CS" (short) or "CO" (open) in driver information center
Odometer		55 ERROR		Odometer malfunction displays ERROR in odometer display
Fuel computer		FFS		Fuel flow signal short or open circuited displays FFS in driver information center

Figure 13-10 Gauge readout indicates the nature of the fault when the system is in the diagnostic mode.

Ford Electronic Cluster Self-Diagnostic Test

The electronic cluster is capable of indicating a fault and providing an explanation of the cause. Use the illustration (Figure 13-10) as a guide to the function of the gauges when in diagnostic mode. Use the gauge display to determine the nature of the fault. Then refer to the service manual for diagnostic charts to locate the problem.

Speedometer Diagnosis and Repair

Speedometer complaints range from chattering noises when cold, to inaccurate readings, to not operating at all. Diagnosis and repair of the speedometer depends on the type, conventional or electronic.

Conventional Speedometer

In instrument clusters that use conventional speedometers, often the problem of noise, erratic, or inaccurate readings can be corrected by lubricating the cable with an approved lubricant. If the cable is dry, it will bind as it attempts to rotate in the housing. However, the cause of the noise must be isolated since just applying lubricant may only stop the noise temporarily. It is a good practice to remove the cable core to clean and inspect it before adding lubricant. If the cable is well lubricated and the problem is still present, check the condition of the speedometer drive and drive gear. If the cable and the gears are not the problem, the speedometer head may be faulty and need to be replaced.

Classroom Manual
Chapter 13, page 383

Special Tools

Fender covers
Safety glasses
Battery terminal pliers
Terminal pliers

The convention speedometer uses **eddy currents** (small induced currents) instead of a direct mechanical connection from the cable to the speedometer head. If the cause of the noise is not in the cable, the bushings in the head may be worn. This would allow the cup and magnet to come in contact and result in noisy operation and inaccurate readings. In this case the speedometer head will have to be replaced.

CAUTION: It is possible to short out wires while reaching behind the instrument panel. Disconnect the battery negative cable before removing the speedometer cable assembly.

To remove the cable core, disconnect the speedometer cable assembly from the back of the speedometer head. For most vehicles, this is done by reaching behind the instrument panel and pressing down on the flat surface of the plastic quick connect. On some vehicles, it may be necessary to remove the instrument panel to gain access to the speedometer cable. With the cable assembly disconnected from the speedometer, visually inspect the cable for kinks or other damage. Raise the drive wheels from the ground and start the engine. Place the transmission in gear and allow the drive wheels to rotate at engine idle. Observe the cable rotation inside of the housing; it should be smooth and constant. Shut off the engine; be sure to apply the brakes to stop transmission movement before attempting to return the shift lever to the PARK position.

CAUTION: Do not allow the drive wheels to rotate faster than 50 mph. Since only one wheel will rotate, one of the differential side gears is remaining stationary as the pinion gears "walk" around it. Excessive speed may result in differential damage.

It may be possible to remove the core by pulling it out of the housing. If the core cannot be removed in this way, disconnect the speedometer retainer from the transaxle (Figure 13-11). Pull the core out of the speed sensor.

CAUTION: If the cable attaches to a speed control sensor, do not attempt to remove spring retainer clip with the speedometer in the sensor.

With the cable core removed from the housing, clean it with solvent and wipe it dry. Place the core on a flat surface and stretch it out straight. Roll it back and forth while looking for signs of kinks or other damage. If the core is damaged, it must be replaced.

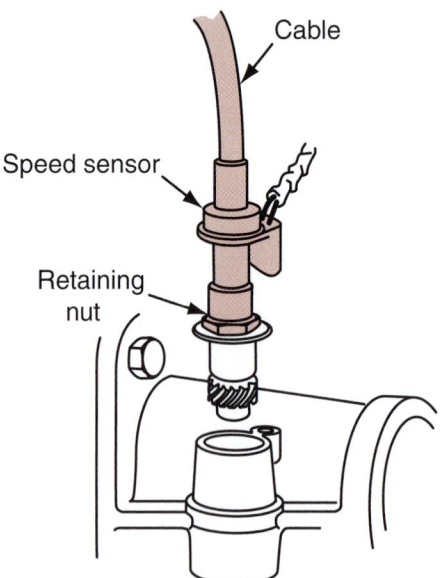

Figure 13-11 Speedometer connection at the transaxle.

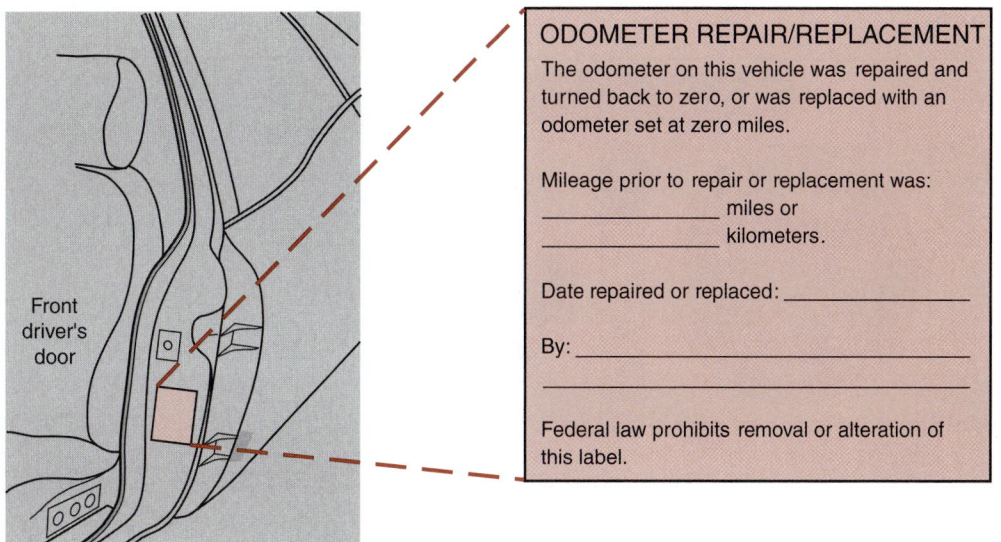

Figure 13-12 An odometer repair label.

When installing the cable to the speedometer head, apply a small amount of approved lubricant to the drive hole. Check that the speedometer cable takes virtually no change of direction for at least 5 inches (127 mm) from the speedometer head.

CAUTION: Changing tire size and differential gear ratios from original equipment specifications will result in speedometer inaccuracy. In some states it is illegal to calibrate the speedometer unless it is performed by a shop that is qualified to perform this task.

CUSTOMER CARE: The speedometer cable should be lubricated every 10,000 miles. This practice will reduce speedometer cable problems that will cause noisy, erratic, or inaccurate readings.

If the speedometer assembly must be replaced, usually a new **odometer** is included with the assembly. The odometer is a mechanical counter in the speedometer unit that indicates total miles driven by the vehicle. Be sure to follow the manufacturer's procedures for setting the odometer to the correct reading.

CAUTION: Federal and state laws prohibit the tampering with the correct mileage as indicated on the odometer. If the odometer must be replaced, it must be set to the reading of the original odometer, or a door sticker must be installed indicating the reading of the odometer when it was replaced (Figure 13-12).

Electronic Speedometers and Odometers

SERVICE TIP: If the electronic speedometer uses a cable-driven sensor, the cable may be serviced as described in the previous section.

Diagnosis of the sending units and input circuits to the gauges of electromechanical instrument clusters typically follow normal testing procedures. The following is an example of diagnosing the **quartz swing needle** electronic speedometer and odometer gauges of an electromechanical

Classroom Manual
Chapter 13, page 389

Classroom Manual
Chapter 13,
pages 382, 387, 389

Special Tools

Scanner

DMM

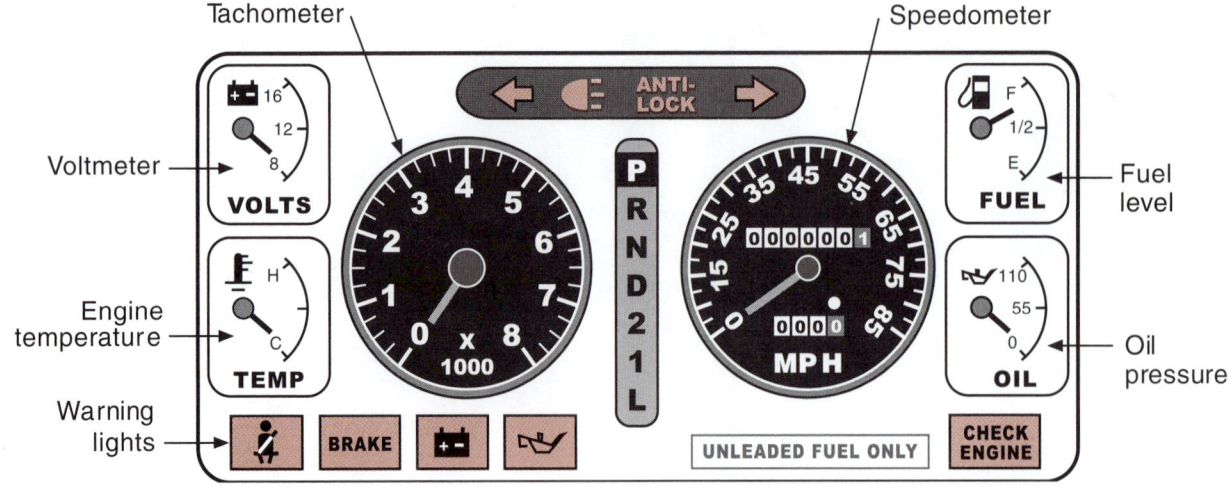

Figure 13-13 Quartz swing needle speedometer used with conventional gauges.

cluster (Figure 13-13). Computer-driven quartz swing needle displays are similar in design to the air-core electromagnetic gauges used in conventional analog instrument panels.

On most systems, the odometer and speedometer receive their input from the vehicle speed sensor. If there is a fault with the speed sensor, other systems (such as cruise control) will also be affected. When test driving the vehicle, attempt to activate the cruise control system to determine if it is operating properly. If the cruise control system fails, the problem can be the speed sensor, its circuit, BCM, or the ECM.

Some systems may provide for replacement of the stepper motor or odometer chip separate of the cluster. Always refer to the service manual for the vehicle being diagnosed.

Generally, if the BCM and/or cluster module pass their self-diagnostic tests, the fault will be in the speed sensor circuit. Common test procedures for the speed sensor are presented later in this chapter.

If the speedometer is not operating, but the odometer works properly, the fault is in the instrument cluster. Likewise, if the speedometer operates but the odometer fails, the fault is in the cluster. In either case the cluster must be replaced. If the speedometer and/or odometer are inaccurate, or both do not operate, check the following items:

1. If the system uses an optical vehicle speed sensor, check the speedometer cable for kinks, twists, or other defects that will cause an inaccurate reading.

2. See if there are shorts in the wiring circuit of the speed sensor.

3. Make sure of proper gear ratio and tire size. Both of these items will affect correct speedometer and odometer operation. Changing tire size from that intended by the manufacturer has the same effect as changing gear ratios in the differential.

If the speedometer and/or odometer display illuminates but remains at zero (or any other digit)—or operates erratically or intermittently—check the connector at the speed sensor for proper installation and corrosion. Next, check the wiring circuit for any shorts or opens. If these tests do not isolate the problem, the speed sensor should be tested. Testing of the speed sensor will depend on the type of sensor used.

Magnetic Pickup Speed Sensor Testing. Disconnect the wire connector at the vehicle speed sensor. With the ignition switch placed in the RUN position, use a jumper wire to make and break the connection between the two wires (Figure 13-14). This should cause the speedometer display to change. Change the rate of speed at which you make and break the connection and the display should indicate the changes in speed. The faster you make and break the connection, the higher the speedometer reading.

Perform this test after the cluster has passed the self-diagnostic test.

Classroom Manual
Chapter 13,
pages 383, 388

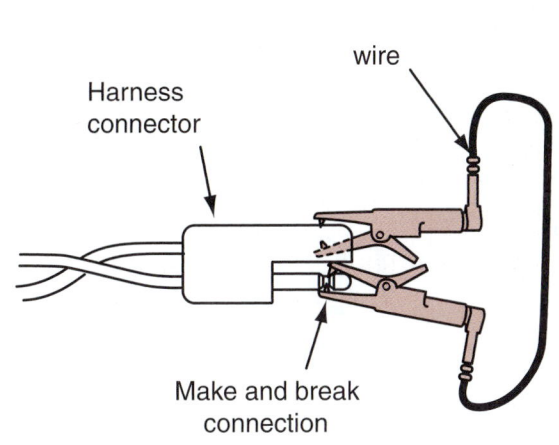

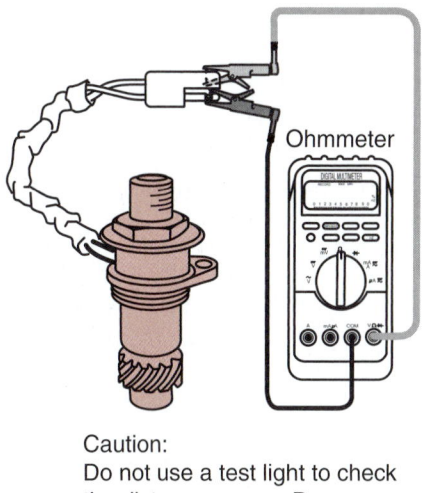

Caution:
Do not use a test light to check
the distance sensor. Damage
to the sensor may result.

Figure 13-14 Testing the magnetic pickup speed sensor circuit. Making and breaking the connection should produce a reading in the speedometer window.

Figure 13-15 Using an ohmmeter to test the speed sensor.

If there is no change in the speedometer display, check for opens and shorts in the sensor circuit. If the cluster passed its self-test (and you did not skip any steps) this is the only area in which the fault can be located.

If the speedometer changed speeds, the problem is in the speed sensor. To test the speed sensor, remove it from the transaxle. Connect an ohmmeter to the connector terminals of the sensor and select the lowest scale (Figure 13-15). Rotate the sensor gear while observing the ohmmeter. Distinct pulses should be detected on the ohmmeter. Compare the number of pulses per revolution with specifications. Also, compare the resistance value with specifications. If the number of pulses and resistance values are within specifications, the sensor is good.

> **CAUTION:** Do not use a test light to test the sensor. Damage to the unit may result.

Optical Speed Sensor Testing. Disconnect the speedometer cable at the transaxle and rotate the cable in its housing as fast as possible. If the speedometer display operates properly, check the speedometer pinion and drive gear for damage.

> **SERVICE TIP:** A reversible, variable speed drill can be used to rotate the speedometer cable if you are sure there are no twists or kinks in the cable.

If there is no speedometer operation, check for a broken speedometer cable. This can usually be determined by feeling for resistance while turning the cable by hand. Little resistance indicates a broken cable. Excessive resistance indicates a damaged cable or sensor head.

If the cable is good, the problem is in the sensor or in the wiring between the sensor and the speedometer. Follow the manufacturer's procedure for removing the instrument cluster. Connect a DMM to read the pulsed speed signal from the sensor (Figure 13-16). Rotate the speedometer cable while observing the voltmeter. Compare the pulses per revolution and pulse output values with specifications. Replace the sensor if the values are not within specifications.

Hall-Effect Switches. In Chapter 10, the operation of the Hall-effect switch was covered. As discussed, using a lab scope will provide a fast and accurate test of the switch. Refer to Chapter 10 for the test procedures of this type of speed sensor.

Special Tools

Jumper wires

Ohmmeter

Lab scope

Use this test to determine erratic speedometer operation.

Classroom Manual
Chapter 13, page 387

Special Tools

DMM

Lab scope

Classroom Manual
Chapter 13, page 388

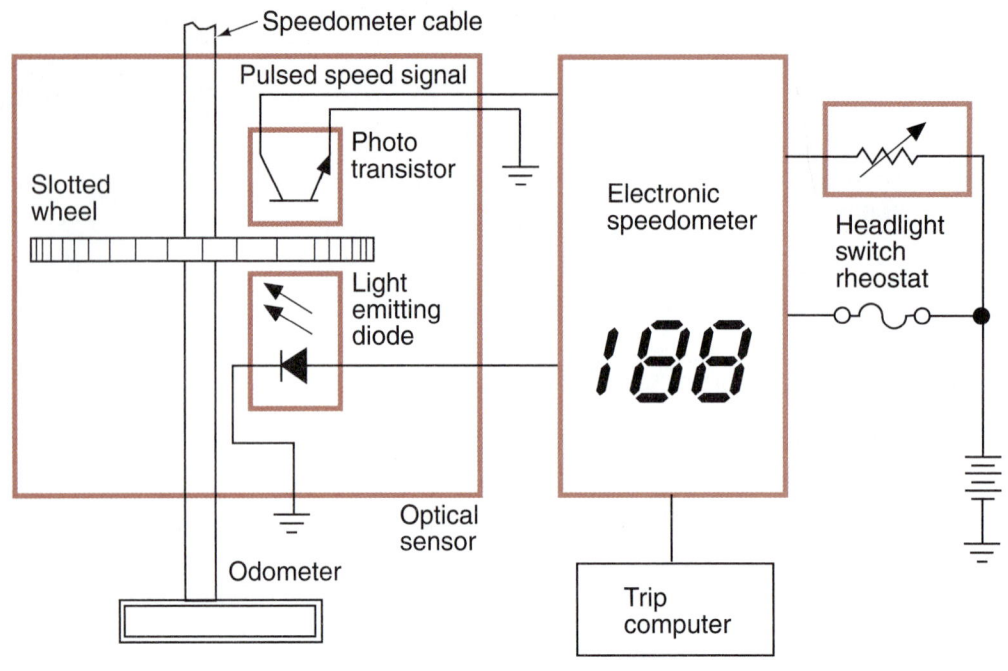

Figure 13-16 Optical speed sensor circuit.

● **CUSTOMER CARE:** Often inaccurate speedometers are caused by improper tire sizes being used on the vehicle. The larger the tire diameter over original equipment, the slower the speedometer will read. Not all manufacturers provide for entering different tire sizes into the computer, and aftermarket "black boxes" may be illegal to install, depending on state and local ordinances. In most cases, it is in the customer's best interest to keep original equipment size tires on the vehicle.

Tachometer

The **tachometer** is a gauge instrument used to display the speed of the engine in revolutions per minute (rpm). Most electrically operated tachometers receive their reference pulses from the ignition system (Figure 13-17). Figure 13-18 illustrates a troubleshooting chart to use as a guide to diagnosing the electrically operated tachometer. If the tachometer is faulty, it must be replaced; there is no servicing the meter itself.

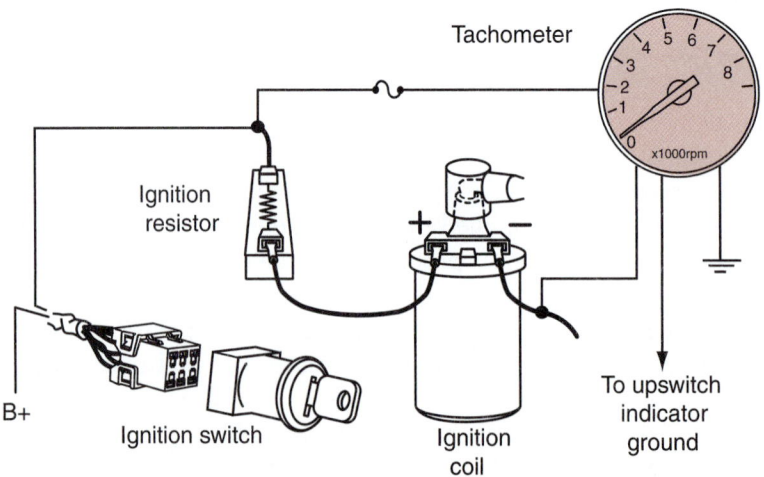

Figure 13-17 Typical wiring diagram of tachometer circuit.

TACHOMETER INOPERATIVE, ERRATIC, WRONG OPERATION

Test step	Result	Action to take
1. **Check OPERATION** Check tachometer operation	~~OK~~ OK	Go to 2. Test complete. Check for intermittent operation.
2. **Check fuse** Check tachometer fuse	~~OK~~ OK	Replace fuse and determine cause. Go to 3.
3. **Check wiring** Check connections and wiring in engine compartment and at instrument cluster	~~OK~~ OK	Repair connections or wiring. Go to 4.
4. **Check resistance and voltage** Disconnect battery. Remove instrument cluster and make resistance and voltage checks at lower wire harness connector as follows: (1) Check pin 5 resistance to ground. Should read 1Ω or less. (2) Check pin 17 resistance to negative terminal of ignition coil. Should read 15Ω or less. (3) Connect battery. Turn ignition switch to RUN position. Check for +12V at pin 14. Turn ignition off and disconnect battery. 	~~OK~~ OK	Repair wiring for open or high resistance. Go to 5.
5. **Check fasteners** Check for loose fasteners on rear of instrument cluster, or damaged printed circuit	~~OK~~ OK	Tighten fasteners / Replace printed circuit Replace tachometer

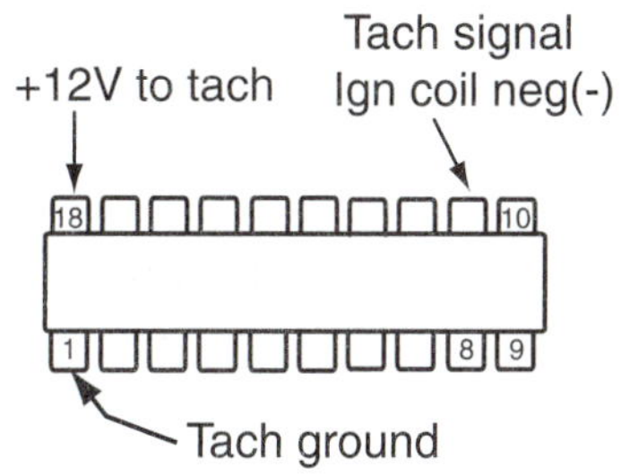

Figure 13-18 Troubleshooting chart for testing the tachometer.

Computer-driven tachometers receive their signals from the crankshaft position sensor (CKP). If the tachometer is not functioning, but the engine starts, check the circuit from the sensor to the instrument cluster. If the instrument cluster receives the signal from the data bus, use the scan tool to determine if the signal is being sent correctly. Usually the CKP sensor is monitored by the PCM then sent on the data bus. If the message is sent to the BCM, then to the instrument cluster, check the input message to both modules. If the bus message is being sent properly, the problem is the gauge or the instrument cluster. Perform the self-diagnostic test, or scan tool activation test, for the tachometer. If the tachometer fails to operate during this test, replace the gauge or cluster. Some manufacturers do not allow for individual gauge replacement; in such cases, the entire instrument cluster must be replaced.

Classroom Manual

Chapter 13, page 378

Conventional Instrument Cluster Guage Diagnosis

A **gauge** is a device that displays the measurement of a monitored system by the use of a needle or pointer that moves along a calibrated scale. The sender unit is the sensor for the gauge. It is a variable resistor that changes resistance values with changing monitored conditions.

> **CAUTION:** These gauges are called analog because they use needle movement to indicate current levels. However, many modern instrument panels use computer-driven analog gauges that operate under different principles. It is important that the technician follow the manufacturer's procedures for testing the gauges or gauge damage will result.

The instrument voltage regulator (IVR) provides a constant voltage to the gauge regardless of the voltage output of the charging system. The gauge is called an electromechanical device because it is operated electrically, but its movement is mechanical.

With the exception of the voltmeter and ammeter, all electromechanical gauges (whether **bimetallic** or **electromagnetic gauges**) use a variable resistance sending unit. Bimetallic gauges (or thermoelectric gauges) are simple dial and needle indicators that transform the heating effect of electricity into mechanical movement. Electromagnetic gauges produce needle movement by magnetic forces instead of heat. The types of tests performed will depend on the nature of the problem and if the system uses an IVR.

Single Gauge Failure

Special Tools

12-volt test light

10-ohm resistor

73-ohm resistor

Fender covers

Safety glasses

If the gauge system does not use an IVR, check the gauge for proper operation as follows:

1. Check the fuse panel for any blown fuses. The gauge that is not operating may share a fuse with some other circuit that is separate from the other gauges.
2. Disconnect the wire connector from the sending unit of the malfunctioning gauge.
3. Check the terminal connectors for signs of corrosion or damage.
4. Use a test light to confirm that voltage is present to the connector with the ignition switch in the RUN position. If the test light does not illuminate, check the circuit back to the gauge and battery.
5. Connect a 10-ohm resistor in series with the lead from the gauge to the sending unit. Connect the lead to ground (Figure 13-19).
6. With the ignition switch in the RUN position, watch the gauge. Depending on the gauge design, the needle should indicate either high or low on the scale. Check the service manual for the correct results.
7. Remove the 10-ohm resistor and replace with a 73-ohm resistor between the sensor lead and ground. Repeat step 6.
8. If the test results are in the acceptable range, the sending unit is faulty.
9. If the gauge did not operate properly in steps 5 and 7, check the wiring to the gauge. If the wiring is good, replace the gauge.

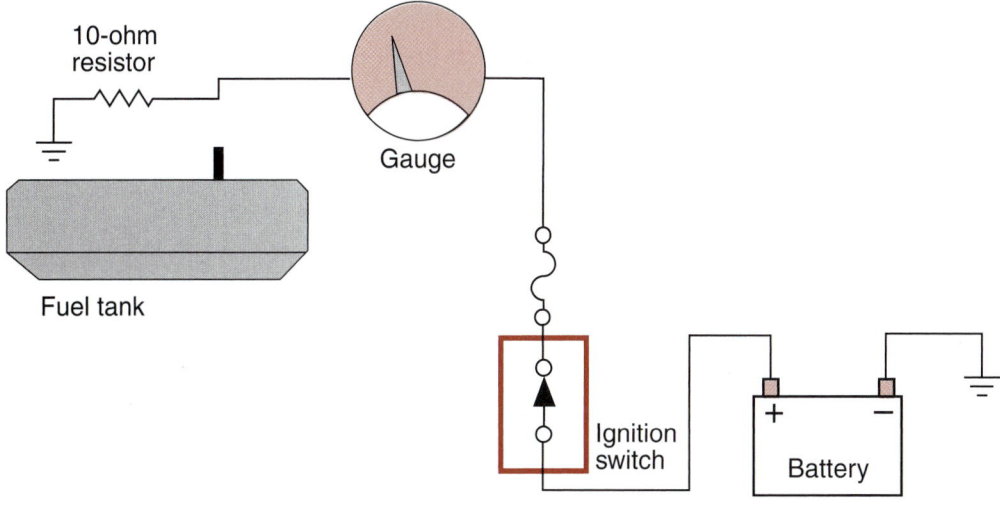

Figure 13-19 Testing gauge operation by putting the gauge lead to ground through a resistor. The resistor is used to protect the circuit.

CAUTION: It is important to know that the sending unit operation is different for electromagnetic gauges than bimetallic gauges. In most bimetallic gauges, the gauge reads high if resistance is low. With many electromagnetic gauges, the gauge reads low if sending unit resistance is low.

CAUTION: Grounding the sender terminal lead directly may damage the gauge. Use a resistor to protect the circuit.

If the gauge circuits use an IVR, follow steps 1 through 4 as described. The test light should flicker on and off. If it did not illuminate, reconnect the sending unit lead and check for voltage at the sender unit side of the gauge (Figure 13-20). If there is voltage at this point,

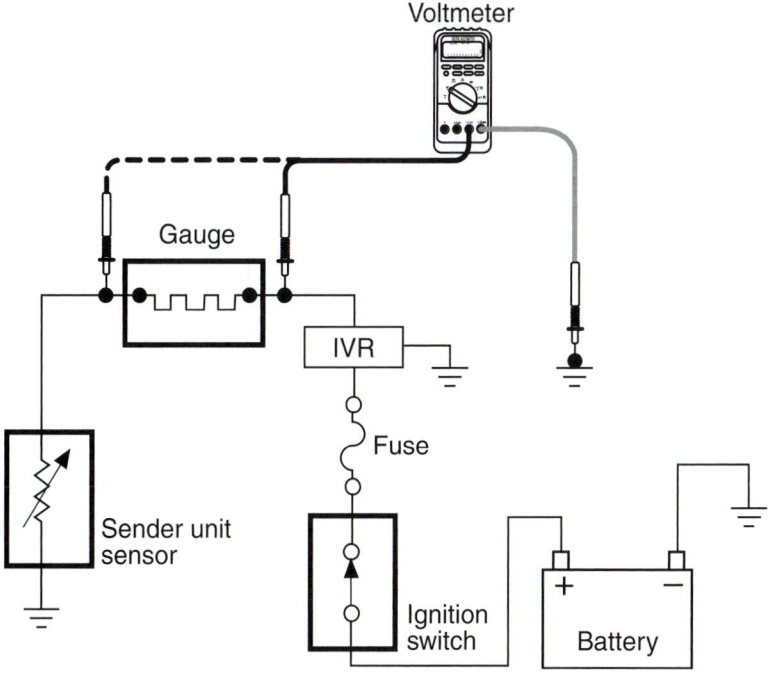

Figure 13-20 Checking for regulated voltage on the sender unit side of the gauge.

repair the circuit between the gauge and the sending unit. If voltage is not shown, test for voltage at the battery side of the gauge. If voltage is present at this point, the gauge is defective and must be replaced. If no voltage is present at this terminal, continue to check the circuit between the battery and the gauge.

If the IVR was working properly and voltage was present to the sender unit, follow steps 5 through 9.

> ✓ **SERVICE TIP:** It is common for the fuel gauge sender unit to get corrosion on the ground wire connection. Before replacing the sending unit, clean the ground connections and test for proper operation.

Special Tools

12-volt test light

10-ohm resistor

73-ohm resistor

Fender covers

Safety glasses

Multiple Gauge Failure

If all gauges fail to operate properly, begin by checking the circuit fuse. Test for voltage to the fuse. If voltage is not present at this point, the fault is between the fuse and the battery. Remember, most systems supply battery voltage to the instrument panel gauges through the ignition switch. If voltage is present at the fuse, then continue through the circuit by testing for voltage at the last common circuit point (Figure 13-21). If voltage is not present at this point, work toward the fuse to find the fault. Keep in mind that this common connection point may be on the printed circuit board. If this is the case, test for voltage at the instrument panel connector first.

If the system uses an IVR, use a voltmeter to test for regulated voltage at a common point to the gauges (Figure 13-22). If the voltage is out of specifications, check the ground circuit of the IVR. If that is good, replace the IVR. If there is no voltage present at the common point, check for voltage on the battery side of the IVR. If voltage is present at this point, then replace the IVR. If battery voltage is not present on the battery side of the IVR, the problem is in the circuit between the fuse and the IVR.

If regulated voltage is within specifications, test the printed circuit from the IVR to the gauges. If there is an open in the printed circuit, replace the board.

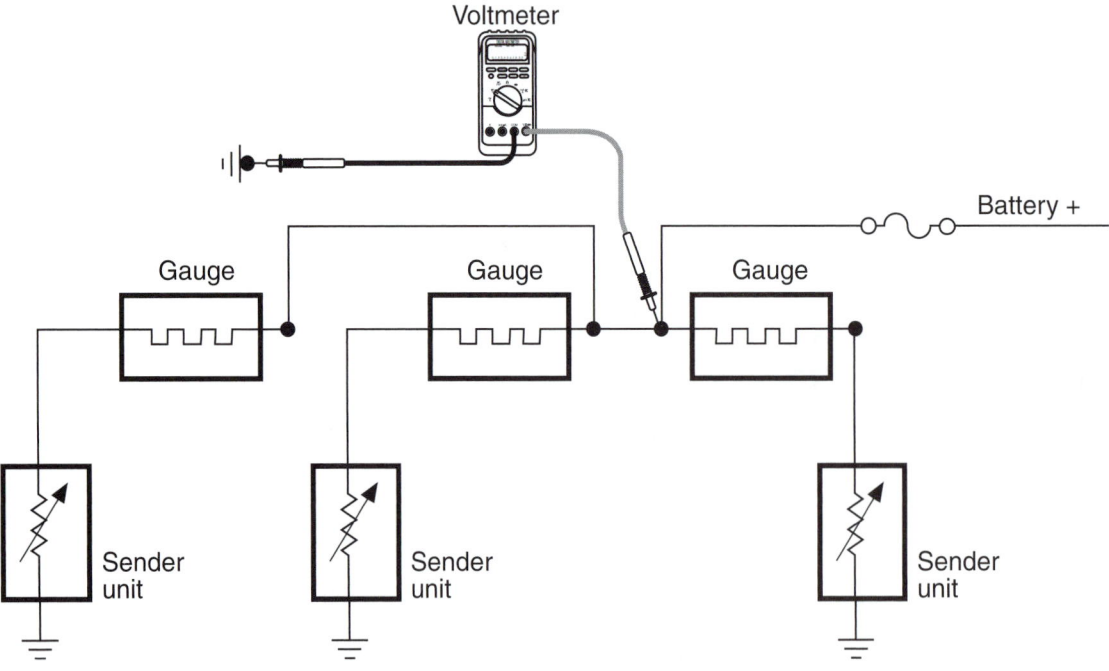

Figure 13-21 Check the last common connection in the circuit for voltage.

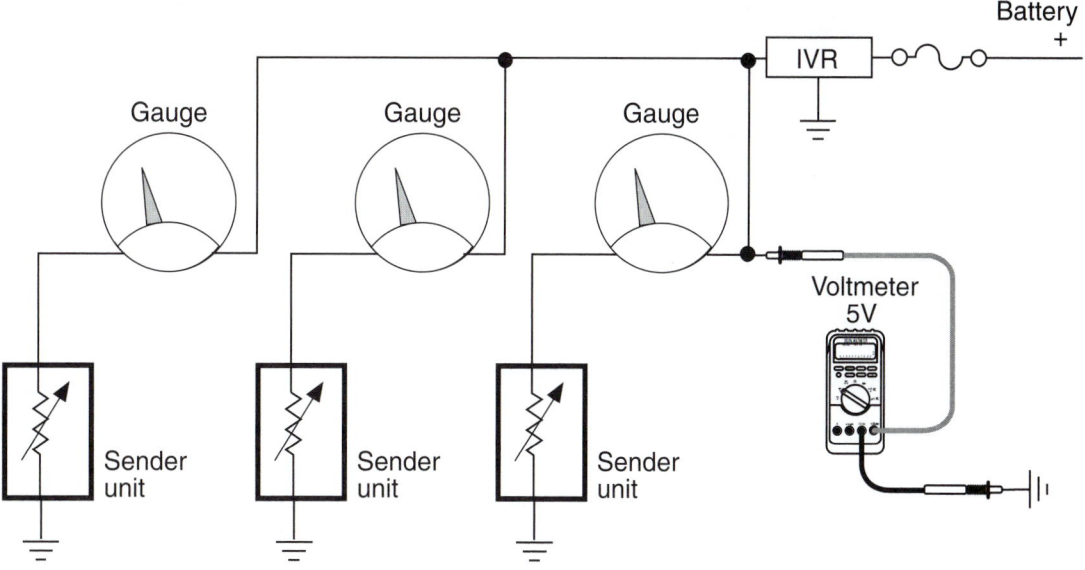

Figure 13-22 Testing for correct IVR operation.

SERVICE TIP: It is unlikely that all of the gauges would fail at the same time. If the diagnostic tests indicate that the gauges are defective, bench test the gauges before replacing them. Use an ohmmeter to check the resistance. Most electromagnetic gauges should read between 10 and 14 ohms. On systems that do not use an IVR and all of the gauges are defective, check the charging system for excessive output.

Electronic Gauge Diagnosis

The following are guidelines for testing the individual gauges for proper operation. These tests are to be conducted after the self-diagnostic test indicates there is no problem with the cluster or the module. Test procedures will vary between manufacturers. To properly troubleshoot the gauges, you will need the manufacturer's diagnostic procedure, specifications, and circuit diagram.

Classroom Manual
Chapter 13,
page 382

Special Tools

Jumper wires

Gauge Reads Low Constantly

A gauge that constantly reads low when the ignition switch is in the RUN position indicates an open in the gauge circuit. To locate the open, follow these steps:

1. Disconnect the wire harness from the sending unit.
2. Connect a jumper wire between the wire circuit from the gauge and ground.
3. Turn the ignition switch to the RUN position. The gauge should indicate maximum.

SERVICE TIP: Although some service manual procedures do not require it, it is a good practice to connect a 10-ohm resistor into the jumper wire when performing these tests. This prevents a nonresistive short to ground, yet does not noticeably affect gauge operation.

CAUTION: Most electronic instrument cluster fuel gauges operate with sensors that decrease resistance values as the fuel level decreases. Jumping the wire to ground would indicate a low reading in these systems. Before faulting the sending unit, refer to the service manual for correct test results for the vehicle you are diagnosing. Do not leave the ignition switch in the RUN position for longer than 30 seconds. This is all the time required to test gauge operation.

CAUTION: Some fuel gauge systems determine fuel level by sending a reference voltage from the module to the sending unit, then measuring the input voltage back from the sending unit. Do not directly ground this system. Use a voltmeter to back probe the connector terminals for the proper voltage values.

If the gauge reads high, check the sending unit ground connection. If the ground is good, the sending unit is faulty and must be replaced.

If the gauge continues to read low, follow the circuit diagram for the vehicle being serviced to test for opens in the wire from the sending unit. If the circuit is good, test the control module following recommended diagnostic procedures.

Gauge Reads High Constantly

A gauge that reads high when the ignition switch is placed in the RUN position indicates there is a short to ground in the circuit. To test the circuit, disconnect the wire harness at the gauge sending unit. Place the ignition switch in the RUN position while observing the gauge. If the gauge reads low, the sending unit is faulty and needs replacement.

CAUTION: Most electronic instrument cluster fuel gauges operate with sensors that decrease resistance values as the fuel level decreases. Opening the wire would indicate a high reading in these systems. Before faulting the sending unit, refer to the service manual for correct test results for the vehicle you are diagnosing.

If the gauge continues to read high, use the circuit diagram to test for shorts to ground in the circuit from the sending unit. If the circuit is good, test the control module following recommended diagnostic procedures.

Inaccurate Gauge Readings

Inaccurate gauge readings are usually caused by faulty sending units. To test the operation of the gauge, you will need the manufacturer's specifications concerning resistance values as they relate to gauge readings. Gauge testers are available to test the units as different resistance values are changed.

SERVICE TIP: If a gauge tester is not available, you can substitute a rheostat of correct resistance range or place different resistors into a jumper wire. Connect the rheostat or resistor between the sending unit wire and ground. For example, if the gauge is designed to read high at zero ohms resistance and low at 90 ohms, placing 45 ohms of resistance in the circuit should produce a reading of midpoint.

Other reasons for inaccurate gauge readings include poor connections, **resistive shorts,** and poor grounds. Resistive shorts are shorts to ground that pass through a form of resistance first. Also, look for damage around the sending unit. For example, a damaged fuel tank can result in inaccurate gauge readings.

Classroom Manual
Chapter 13,
pages 385, 392

The **thermistor** is a sensor that is sensitive to temperature changes, and is often used as a coolant temperature sensor.

Gauge Sending Units

There are three types of sending units associated with electromechanical gauges: a **thermistor,** a piezoresistive sensor, and a mechanical variable resistor. Most of these can be tested before replacement to confirm the fault.

The fuel level sending unit can be tested in or out of the tank. If it is tested in the tank, add and remove fuel to change the level. The easiest method is to bench test the unit. Photo Sequence 22 illustrates how to bench test a mechanical variable resistor sending unit.

Photo Sequence 22
Bench Testing the Fuel Level Sender Unit

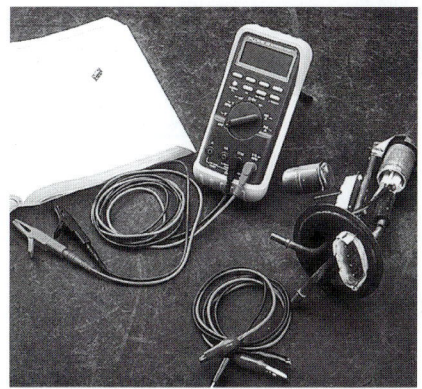

P22-1 Tools required to perform this task: DMM, jumper wires, and service manual.

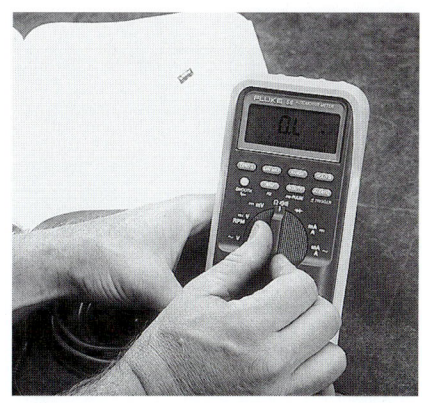

P22-2 Select the ohmmeter function of the DMM.

P22-3 Connect the negative test lead of the DMM to the ground terminal of the sender unit.

P22-4 Connect the positive test lead to the variable resistor terminal.

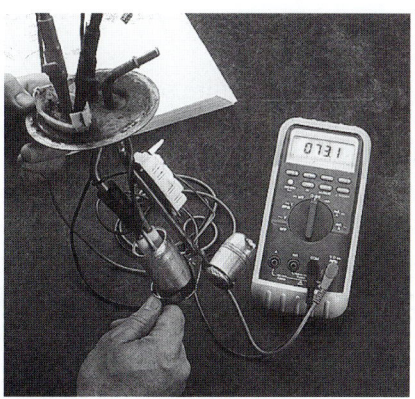

P22-5 Holding the sender unit in its normal position, place the float rod against the empty stop.

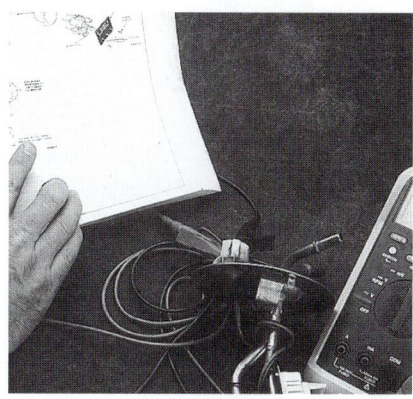

P22-6 Read the ohmmeter and check the results with specifications.

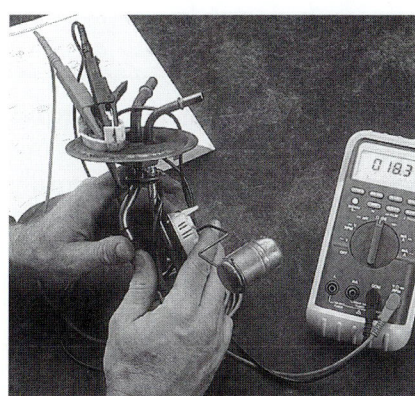

P22-7 Slowly move the float toward the full stop while observing the ohmmeter. The resistance change should be smooth and consistent.

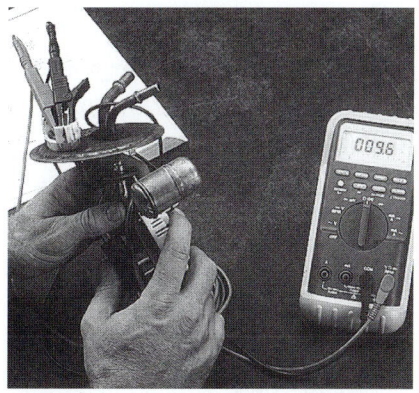

P22-8 Check the resistance value while holding the float against the full stop. Check the results with specifications.

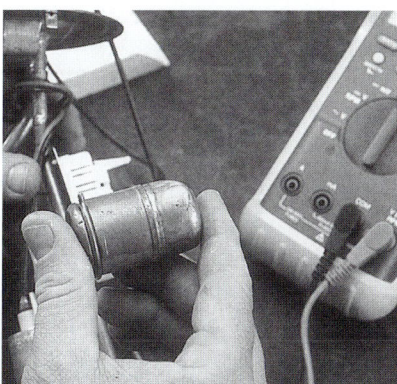

P22-9 Check that the float is not filled with fuel, distorted, or loose.

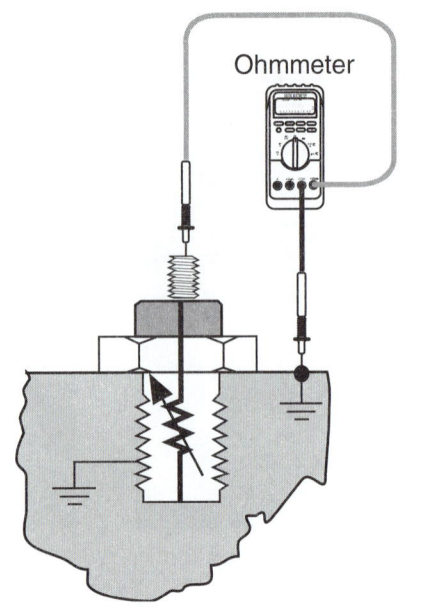

Figure 13-23 Testing a thermistor with an ohmmeter.

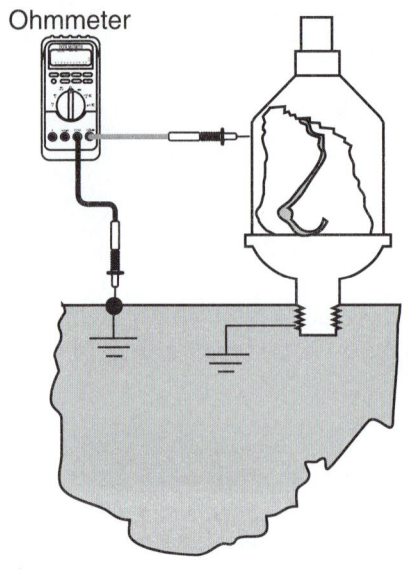

Figure 13-24 Using an ohmmeter to test a piezoresistive sensor.

Special Tools

DMM

12-volt test light

Fender covers

Safety glasses

Special Tools

DMM

Shop oil pressure
gauge

A **piezoresistive
sensor** is sensitive
to pressure changes.
The most common
use of this type of
sensor is to measure
the engine oil
pressure.

To test the coolant temperature sensing unit, use an ohmmeter to measure the resistance between the terminal lead and ground (Figure 13-23). The resistance value of the variable resistor should change in proportion to coolant temperature. Check the test results with manufacturer specifications.

To test a **piezoresistive sensor** sending unit used for oil pressure gauges, connect the ohmmeter to the sending unit terminal and ground (Figure 13-24). Check the resistance with the engine off and compare to specifications. Start the engine and allow it to idle. Check the resistance value and compare to specifications. Before replacing the sending unit, connect a shop oil pressure gauge to confirm that the engine is producing adequate oil pressure.

CAUTION: If the engine is making knocking or other noises, do not test the sending unit first. Immediately connect a shop oil pressure gauge and check for oil pressure. Do not increase the engine speed over idle unless instructed to do so in the service manual.

Warning Lamps

Classroom Manual
Chapter 13, page 403

A warning light may be used to warn of low oil pressure, high coolant temperature, a defective charging system, or a brake failure. Unlike gauge sending units, the sending units for a warning light are nothing more than simple switches. The style of switch can be either normally open or normally closed, depending on the monitored system.

SERVICE TIP: With the bulbs used for most warning light circuits, it is hard to determine whether or not the filament is good. When a test procedure requires that a bulb be checked, it is usually easier to replace the bulb with a known good one.

Special Tools

Jumper wires

scan tool

It is not likely that all of the warning lights would fail at the same time. Check the fuse if all of the lights are not operating properly. Next, check for voltage at the last common connection. If voltage is not present, then trace the circuit back toward the battery. If voltage is present at the common connection, test each circuit branch in the same manner as described here for individual lamps.

To test a faulty warning lamp on a system with a normally open switch (sending unit), turn the ignition switch to the START position. The **prove-out circuit** should light the warning lamp. A prove-out circuit completes the warning light circuit to ground through the ignition switch when it is in the START position. The warning light will be on during engine cranking to indicate to the driver that the bulb is working properly. If the light does not come on during the prove-out, disconnect the sender switch lead (Figure 13-25). Use a jumper wire to connect the sender switch lead to ground. With the ignition switch in the RUN position, the warning lamp should light. If the lamp is illuminated, test the prove-out circuit for an open. If the light does not come on, either the bulb is burned out or the wiring is damaged. Use a test light to confirm voltage is present at the sensor terminal connector. If there is voltage, the bulb is probably bad. If voltage is not present to the sending unit, the bulb may be burned out. At this point, the instrument cluster will need to be removed. With the cluster removed, check for battery voltage to the panel connector. If voltage is present, substitute a known good bulb and test again.

If the system uses a normally closed switch, test in the same manner. However, there will not be a separate prove-out circuit.

> **CAUTION:** Check the manufacturer's service manual to confirm the location of the sensor switch you are testing. For example, there may be a coolant temperature switch for the warning light and a coolant temperature sensor for the engine computer. Grounding the computer terminal lead may result in damage to the computer. Usually a warning light sensor switch will have one lead. The computer sensor will have two to four leads and is contained in a weather-pack connector.

If the customer states that the warning light stays on, test in the following manner: Disconnect the lead to the sender switch. The light should go out with the ignition switch in the RUN position. If it does not, there is a short to ground in the wiring from the sender switch to the lamp. If the light goes out, replace the sender switch.

Computer-driven instrument cluster warning lamps are diagnosed using a scan tool. If the customer states that the lamp does not operate, the scan tool can be used to command activation of the lamp. If the lamp does not come on when commanded, the problem is a faulty lamp, circuit board, or instrument cluster module. If the lamp does light when commanded, the fault is in the signal to the instrument cluster.

If the customer states that the oil pressure light does not go out after the engine is started, use a shop oil pressure gauge to confirm adequate oil pressure.

Classroom Manual
Chapter 13, page 405

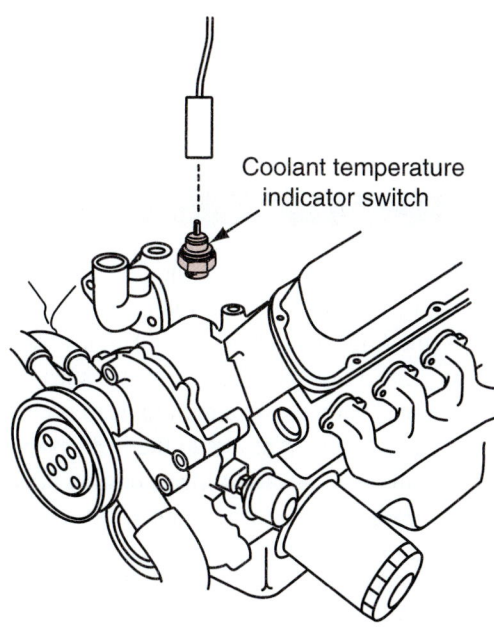

Figure 13-25 Coolant temperature sensor switch and lead.

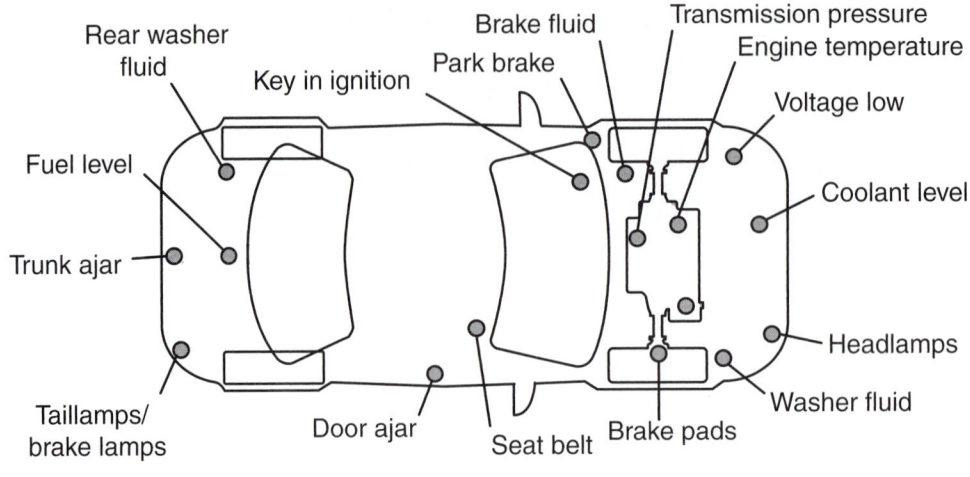

Figure 13-26 Complex information centers use many inputs to monitor the vehicle's subsystems.

If the customer states that the lamp is on all of the time, this indicates a problem in the monitored system. Use the scan tool to check for DTC and use the service manual to diagnose the fault. If there are no problems in the monitored circuit, then you will need to diagnose the module that receives the lamp's sensor input. If there is no problem found in this module, the instrument cluster module will need to be diagnosed. Follow the service manual procedures for all diagnostic tests.

Classroom Manual
Chapter 13, page 400

Trip computer systems may have several different names. The most common are: Traveler, Vehicle Information Center, and Drive Information Centers.

Trip Computers

Simple trip computers use inputs from the speed sensor and the fuel gauge to perform their functions. Like most electronic instrument clusters, trip computers will usually have a self-diagnostic test procedure. Follow the manufacturer's procedure for initiating this test. If the trip computer passes the self-test, check the fuel gauge and speed sensor inputs.

Complex trip computers may receive inputs from several different areas of the vehicle (Figure 13-26). These systems will require the use of specific manufacturer diagnostic charts, specifications, and procedures. Using the skills you have acquired, you should be able to perform these tests competently.

CASE STUDY

A customer brings his vehicle into the shop because the fuel level gauge does not operate. It remains on empty all of the time, regardless of how full the tank is. The technician checks the fuse box for any blown fuses and finds that all are good. He disconnects the lead wire to the fuel tank sending unit and probes for voltage. The test light comes on when he places the ignition switch in the ON position. When the sending unit lead is connected to ground through a 10-ohm resistor, the gauge needle moves to the FULL position. Upon checking with the service manual, the technician determines that this is normal operation. Before draining and removing the fuel tank, he uses a jumper wire to jump from the ground terminal of the sending unit to a known good ground on the frame. When he reconnects the lead wire and places the ignition switch in the RUN position, the gauge works properly. The technician cleans the ground connections for the sending unit and returns the vehicle to the customer.

Terms to Know

Bimetallic gauges

Eddy currents

Electromagnetic gauges

Gauge

Odometer

Piezoresistive sensor

Prove-out circuit

Quartz swing needle

Resistive shorts

Tachometer

Thermistor

ASE-Style Review Questions

1. In a conventional instrument cluster, all of the gauges are inoperable. What is the LEAST likely cause of this?
   A. Faulty IVR.
   B. Blown fuse.
   C. Shorted sending unit.
   D. Faulty printed circuit.

2. All of the following statements concerning gauge sending units are true EXECPT:
   A. A gauge can use a switch as a sensor.
   B. A gauge can use a thermistor as a sensor.
   C. A gauge can use a piezoresistive sensor.
   D. A gauge can use a variable resistor as a sensor.

3. The oil pressure warning light will not turn off with the engine running. A shop gauge confirms good oil pressure. What is the most likely cause?
   A. Faulty IVR.
   B. Damaged printed circuit board.
   C. Open in the wire to the oil pressure switch.
   D. Short to ground in the wire to the oil pressure switch.

4. When testing the IVR-regulated voltage, the indicated reading was 2.5 volts over specifications.
   Technician A says that the IVR ground connection may be faulty.
   Technician B says that the alternator output may be too high.
   Who is correct?
   A. A only
   B. B only
   C. Both A and B
   D. Neither A nor B

5. *Technician A* says to use a voltmeter to test a thermistor sensor unit.
   Technician B says that the thermistor sensor measures pressure.
   Who is correct?
   A. A only
   B. B only
   C. Both A and B
   D. Neither A nor B

6. *Technician A* says when the ignition switch is in the START position the proving circuit should light the warning lamps.
   Technician B says if the sender switch lead is grounded, and the ignition switch is in the RUN position, the warning lamp should come on.
   Who is correct?
   A. A only
   B. B only
   C. Both A and B
   D. Neither A nor B

7. A digital speedometer constantly reads 0 mph.
   Technician A says the speed sensor may be faulty.
   Technician B says the throttle position sensor may have an open.
   Who is correct?
   A. A only
   B. B only
   C. Both A and B
   D. Neither A nor B

8. All gauges read low in a computer-controlled instrument cluster.
   Technician A says the connector to the cluster may be loose.
   Technician B says the cluster module may be at fault.
   Who is correct?
   A. A only
   B. B only
   C. Both A and B
   D. Neither A nor B

9. An electronically controlled instrument cluster is being diagnosed for no speedometer operation. Which of the following statements is most correct?
  A. If the BCM and/or cluster module passes its self-test, the fault is probably in the vehicle speed sensor circuit.
  B. If the speedometer does not work but the odometer does, then the problem is probably in the vehicle speed sensor circuit.
  C. If the speedometer fails to move during the self-test, the problem is probably in the vehicle speed sensor circuit.
  D. All of the above.

10. Ford diagnostic mode displays are being discussed.
  Technician A says a CO displayed in the fuel gauge window indicates that the fuel level is near empty.
  Technician B says a CS indicates an open in the sender circuit.
  Who is correct?
  A. A only
  B. B only
  C. Both A and B
  D. Neither A nor B

ASE Challenge Questions

1. The odometer of a vehicle equipped with a cable-driven speedometer is inoperative; however, the speedometer is working correctly.
Technician A says that the speedometer cable may be faulty.
Technician B says that the speedometer drive gear at the transmission may be stripped.
Who is correct?
  A. A only
  B. B only
  C. Both A and B
  D. Neither A nor B

2. The fuel gauge of a multiple bimetallic-type gauge instrument cluster is reading higher than normal.
Technician A says that the wire leading to the gauge sender unit may be open.
Technician B says that the resistance of the sender unit may be lower than normal.
Who is correct?
  A. A only
  B. B only
  C. Both A and B
  D. Neither A nor B

3. The water temperature warning light of a vehicle is on whenever the ignition key is in the ON position; when the engine is started, the light remains illuminated.
Technician A says that the sending unit wire may be shorted to ground.
Technician B says that the water temperature switch may be electrically open.
Who is correct?
  A. A only
  B. B only
  C. Both A and B
  D. Neither A nor B

4. The diagnosis of a digital instrument cluster that is missing two segments in its display is being discussed.
Technician A says that the cluster has excessive ground circuit resistance.
Technician B says that the cluster should be replaced.
Who is correct?
  A. A only
  B. B only
  C. Both A and B
  D. Neither A nor B

5. A vehicle equipped with an electronic instrument cluster has a "service engine" lamp illuminated; a speed sensor code (indicating the computer does not "see" vehicle speed) has been generated by the computer. The speedometer is working correctly; however, a scan tool that is connected to the vehicle's on-board computer indicates 0 mph whenever the vehicle is in motion.
Technician A says that the vehicle speed sensor may be faulty.
Technician B says that the on-board computer may be faulty.
Who is correct?
  A. A only
  B. B only
  C. Both A and B
  D. Neither A nor B

Job Sheet 47

47

Name _____ Date _____

Checking a Fuel Gauge

Upon completion of this job sheet, you should be able to diagnose an inaccurate or inoperative fuel gauge.

ASE Correlation

This job sheet is related to the ASE Electrical/Electronic System Certification Exam's content area: *Gauges, Warning Devices, and Drive Information Systems Diagnosis and Repair;* task: Inspect, test, and replace gauges and gauge sending units, connectors, wires, and printed circuit boards of gauge circuits.

Tools and Materials

A vehicle   A DMM
Service manual for the chosen vehicle   Miscellaneous resistors
Wiring diagram for the chosen vehicle   A vehicle hoist

Describe the vehicle being worked on:

Year _____ Make _____ Model _____

VIN _____ Engine type and size _____

Procedure

Task Completed

1. From the procedures listed in the service manual, describe the procedures for testing the fuel gauge and fuel gauge sending unit on this vehicle.

   _____
   _____
   _____
   _____
   _____
   _____

2. Locate the fuel gauge circuit in the wiring diagram and draw the circuit below.

3. Follow the procedure to simulate a full fuel tank. Describe what happened when you followed this procedure; include in your answer your conclusions from this test.

_____

_____

_____

_____

_____

4. Follow the procedure to simulate an empty fuel tank. Describe what happened when you followed this procedure; include in your answer your conclusions from this test.

_____

_____

_____

_____

_____

Instructor's Response _____

_____

_____

Job Sheet 48

48

Name _____ _____ Date _____

Testing an Oil Pressure Warning Light Circuit

Upon completion of this job sheet, you should be able to test the oil pressure warning circuit and determine needed repairs.

ASE Correlation

This job sheet is related to the ASE Electrical/Electronic System Certification Exam's content area: *Gauges, Warning Devices, and Drive Information Systems Diagnosis and Repair*; task: Diagnose the cause of constant, intermittent, or no operation of warning light/drive information systems.

Tools and Materials

Vehicle with conventional analog instrument cluster with oil pressure warning lamp.

Describe the vehicle being worked on:

Year _____ Make _____ Model _____

VIN _____ Engine type and size _____

Procedure

Task Completed

1. Describe the normal conditions in which the oil pressure warning lamp should be on.

   _____

   _____

   _____

2. At what pressure should the oil pressure warning lamp turn off? _____

3. Referring to the service information and wiring diagrams, is the oil pressure warning lamp sending unit a normally open or normally closed switch?

   _____

4. Draw the oil pressure warning lamp circuit below.

5. List all possible causes for an oil pressure warning lamp that does not turn on.

_____

_____

_____

_____

_____

_____

6. Unplug the sending unit connector and test for voltage with the ignition key in the RUN position with the engine off. Is voltage present? _____

7. If voltage is not present, what will your next step be? _____

_____

_____

8. If voltage was not present in step 6, test the circuit. What were your results and recommendations?

_____

_____

_____

_____

9. If voltage was present in step 6, describe the method to be used to test the sending unit.

_____

_____

_____

_____

_____

10. Perform the test described in step 9 and record your results and recommendations.

_____

_____

_____

_____

_____

Instructor's Response _____

_____

_____

Job Sheet 49

49

Name _____ Date _____

Testing the Electronic Instrument Cluster

Upon completion of this job sheet, you should be able to test the electronic instrument cluster using a scan tool or stand alone diagnostic routines and determine needed repairs.

ASE Correlation

This job sheet is related to the ASE Electrical/Electronic System Certification Exam's content area: *Gauges, Warning Devices, and Driver Information Systems Diagnosis and Repair;* tasks: Diagnose the cause(s) of intermittent, high, low, or no readings on electronic digital instrument clusters.

Tools and Materials

A vehicle equipped with electronic instrument cluster
Scan tool
Service manual

Describe the vehicle being worked on:

Year _____ Make _____ Model _____

VIN _____ Engine type and size _____

Procedure

Task Completed

1. Test the operation of the gauges. Describe the symptoms.

   _____

   _____

2. Are there any other related symptoms? ☐ Yes ☐ No
   If yes, describe the symptom:

   _____

   _____

3. Refer to the proper service information to determine if a self-diagnostic routine can be performed. If so, describe how to enter the diagnostics.

   _____

   _____

   _____

   _____

4. Is the self-diagnostic routine capable of displaying fault codes? ☐ Yes ☐ No

5. Perform the procedure listed in step 3 and record the results and any DTCs.

   _____

   _____

6. Use a scan tool and access the electronic instrument cluster. Does the scan tool indicate that DTCs are present? ☐ Yes ☐ No

If YES, record the DTCs:

_____

_____

7. Based on the tests so far, is the fault in the instrument cluster or in the sensor circuits?

☐ CLUSTER ☐ SENSOR

8. Based on the results so far, what tests need to be performed to find the cause of the fault?

_____

_____

9. Perform the tests listed in step 8. What is your determination and recommendation?

_____

_____

Instructor's Response _____

_____

_____

_____

DIAGNOSTIC CHART 13-1

PROBLEM AREA:	Instrument cluster gauges
SYMPTOMS:	One or all gauges fluctuate from low or high to normal readings.
POSSIBLE CAUSES:	**1.** Poor ground connection. **2.** Excessive resistance. **3.** Poor connections. **4.** Faulty sending unit. **5.** Defective printed circuit.

DIAGNOSTIC CHART 13-2

PROBLEM AREA:	Instrument cluster gauges
SYMPTOMS:	One or all gauges read high.
POSSIBLE CAUSES:	**1.** Faulty instrument voltage regulator. **2.** Shorted printed circuit. **3.** Faulty sending unit. **4.** Short to ground in sending unit circuit. **5.** Faulty gauge. **6.** Poor sending unit ground connection.

DIAGNOSTIC CHART 13-3

PROBLEM AREA:	Instrument cluster gauges
SYMPTOMS:	One or all gauges read low.
POSSIBLE CAUSES:	**1.** Faulty instrument voltage regulator. **2.** Poor sending unit ground connection. **3.** Improper bulb application. **4.** Improper connections. **5.** Faulty back-up switch contacts.

DIAGNOSTIC CHART 13-4

PROBLEM AREA:	Instrument cluster gauges
SYMPTOMS:	One or all gauges fail to read.
POSSIBLE CAUSES:	**1.** Blown fuse. **2.** Open in the printed circuit. **3.** Faulty gauge. **4.** Poor common ground connection. **5.** Open in the sending unit circuit. **6.** Faulty sending unit. **7.** Faulty instrument voltage regulator.

DIAGNOSTIC CHART 13-5

PROBLEM AREA:	Instrument cluster gauges
SYMPTOMS:	All gauges fail to operate properly.
POSSIBLE CAUSES:	**1.** Faulty instrument voltage regulator. **2.** Poor electrical connection to cluster. **3.** Defective printed circuit board.

DIAGNOSTIC CHART 13-6

PROBLEM AREA:	Instrument cluster gauges
SYMPTOMS:	Inaccurate or no gauge reading.
POSSIBLE CAUSES:	**1.** Faulty gauge. **2.** Faulty sending unit. **3.** Open sending unit circuit. **4.** Shorted sending unit circuit.

DIAGNOSTIC CHART 13-7

PROBLEM AREA:	Warning light operation
SYMPTOMS:	Warning light remains on all of the time.
POSSIBLE CAUSES:	**1.** Sending unit circuit grounded. **2.** Faulty sending unit switch.

DIAGNOSTIC CHART 13-8

PROBLEM AREA:	Warning light operation
SYMPTOMS:	Warning light fails to operate on an intermittent basis.
POSSIBLE CAUSES:	**1.** Loose sending unit circuit connections. **2.** Faulty sending unit.

DIAGNOSTIC CHART 13-9

PROBLEM AREA:	Warning light operation
SYMPTOMS:	One or all warning lights fail to operate.
POSSIBLE CAUSES:	**1.** Blown fuse. **2.** Burned-out bulb. **3.** Open in the circuit. **4.** Defective sending unit switches.

DIAGNOSTIC CHART 13-10

PROBLEM AREA:	Electronic instrument panel
SYMPTOMS:	Digital display does not light.
POSSIBLE CAUSES:	**1.** Blown fuse. **2.** Inoperative power and ground circuit. **3.** Faulty instrument panel.

DIAGNOSTIC CHART 13-11

PROBLEM AREA:	Electronic instrument panel
SYMPTOMS:	Speedometer reads wrong speed.
POSSIBLE CAUSES:	**1.** Faulty speedometer. **2.** Wrong gear on vehicle speed sensor (VSS). **3.** Wrong tire size.

DIAGNOSTIC CHART 13-12

PROBLEM AREA:	Electronic instrument panel
SYMPTOMS:	Fuel gauge displays top and bottom two bars.
POSSIBLE CAUSES:	**1.** Open or short in circuit.

DIAGNOSTIC CHART 13-13

PROBLEM AREA:	Electronic instrument panel
SYMPTOMS:	Fuel computer displays CS or CO.
POSSIBLE CAUSES:	**1.** Open or short in fuel gauge sender. **2.** Inoperative instrument panel.

DIAGNOSTIC CHART 13-14

PROBLEM AREA:	Electronic instrument panel
SYMPTOMS:	Odometer displays error.
POSSIBLE CAUSES:	**1.** Inoperative odometer memory module in instrument panel.

DIAGNOSTIC CHART 13-15

PROBLEM AREA:	Electronic instrument panel
SYMPTOMS:	Fuel gauge display is erratic.
POSSIBLE CAUSES:	**1.** Sticky or inoperative fuel gauge sender. **2.** Fault in circuit. **3.** Inoperative fuel gauge.

DIAGNOSTIC CHART 13-16

PROBLEM AREA:	Electronic instrument panel
SYMPTOMS:	Fuel gauge will not display FULL or EMPTY.
POSSIBLE CAUSES:	**1.** Sticky or inoperative fuel gauge sender.

DIAGNOSTIC CHART 13-17

PROBLEM AREA:	Electronic instrument panel
SYMPTOMS:	Fuel economy function of message center is erratic or inoperative.
POSSIBLE CAUSES:	**1.** Inoperative fuel flow signal. **2.** Faulty wiring. **3.** Inoperative instrument panel.

DIAGNOSTIC CHART 13-18

PROBLEM AREA:	Electronic instrument panel
SYMPTOMS:	Extra or missing display segments.
POSSIBLE CAUSES:	**1.** Inoperative instrument panel.

DIAGNOSTIC CHART 13-19

PROBLEM AREA:	Electronic instrument panel
SYMPTOMS:	Speedometer always reads zero.
POSSIBLE CAUSES:	**1.** Faulty wiring. **2.** Inoperative instrument panel.

DIAGNOSTIC CHART 13-20

PROBLEM AREA:	Electronic instrument panel
SYMPTOMS:	Temperature gauge displays top and bottom two bars.
POSSIBLE CAUSES:	**1.** Short in circuit. **2.** Inoperative coolant temperature sender. **3.** Inoperative instrument panel.

Diagnosis of Electronic Chassis and Accessory Systems

Upon completion and review of this chapter, you should be able to:

- ❏ Diagnose common ATC systems.
- ❏ Test the aspirator assembly for proper operation.
- ❏ Test the sensors used on ATC systems.
- ❏ Enter self-diagnostic tests of common ATC systems.
- ❏ Perform self-diagnostic procedures on electronic cruise control systems.
- ❏ Diagnose causes for no, intermittent, and erratic cruise control operation.
- ❏ Test the cruise control servo assembly for proper operation.
- ❏ Replace the servo assembly and adjust actuator cable.
- ❏ Replace the cruise control switch assembly.
- ❏ Determine the cause of poor, intermittent, and no memory seat operation.
- ❏ Diagnose climate-controlled seats.
- ❏ Determine the circuit fault causing a malfunction in sunroof operation.
- ❏ Diagnose vehicle alarm systems.
- ❏ Perform tests on the antitheft controller to determine proper operation.

- ❏ Use self-diagnostic tests on alarm systems that provide this feature.
- ❏ Diagnose the immobilizer system.
- ❏ Diagnose automatic door lock systems for poor, intermittent, or no operation.
- ❏ Diagnose keyless entry systems.
- ❏ Determine the causes of heated windshield malfunctions.
- ❏ Utilize diagnostic tests (quick and pinpoint) on the AXODE transmission to diagnose problems.
- ❏ Enter diagnostic procedures used for both separate and shared control module speed-sensitive steering systems.
- ❏ Enter self-diagnostic mode of common Bosch and Teves ABS systems.
- ❏ Perform ABS system functional tests.
- ❏ Replace the ABS pump and motor assembly.
- ❏ Diagnose the tire pressure monitoring system.
- ❏ Diagnose vehicle audio and video entertainment systems.

Basic Tools

Basic mechanic's tool set

Service manual

_____

Introduction

In this chapter, you will learn how to properly and safely service automatic temperature control systems and electronic cruise controls. Also, you will learn how to service electronic accessories designed to increase passenger comfort, provide ease of operation, and increase passenger safety. These systems include memory seats, sunroof, automatic door locks, keyless entry, heated windshields, antitheft systems, computer-controlled transmissions, electronically controlled power steering systems, antilock brake systems (ABS), and vehicle radio/stereo audio systems.

Most of these systems are additions to existing systems. For example, the memory seat feature is an addition to the conventional power seat system. As vehicles and accessories become more sophisticated, these "luxury" features will become more commonplace. Today's technician is expected to accurately and quickly diagnose malfunctions in these systems.

Classroom Manual

Chapter 14, page 409

Automatic Temperature Control System Diagnosis

Diagnostics of electronic automatic temperature control (ATC) systems depend on system design. There are two basic system designs: those that use their own microprocessor and those that incorporate the controls of the system into the BCM. When diagnosing the system, remember that the ATC system is a control system and faults within the evaporator and heater systems will have an effect on its operation.

You have learned how to test motors and electromagnetic components in previous chapters. Use these same procedures to check the motor and compressor clutch circuits of the ATC system. You will also need the circuit schematic and specifications for the vehicle you are diagnosing.

The air delivery control systems of ATC systems differ among manufacturers and require specific diagnostic procedures. Most systems can be tested with specific tests designed to troubleshoot that particular system.

ATC systems, even those of different model lines made by the same manufacturer, will require specific troubleshooting procedures found in that model's service manual.

Component Testing

Regardless of the system design, there are some components that have a common function and operate the same. These components are usually tested using common test procedures. These will be discussed first.

Classroom Manual

Chapter 14, page 410

Aspirator Assembly. A paper test is a quick method of testing the aspirator assembly to verify that it is providing enough air flow to the in-vehicle sensor. Set the controls for high blower speed while in the heat mode of operation. Use a piece of paper large enough to cover the aspirator air inlet (Figure 14-1). The suction of the aspirator assembly should be sufficient to hold the paper against the inlet grill. If the paper is not held in place, refer to the aspirator system diagnostic chart (Figure 14-2).

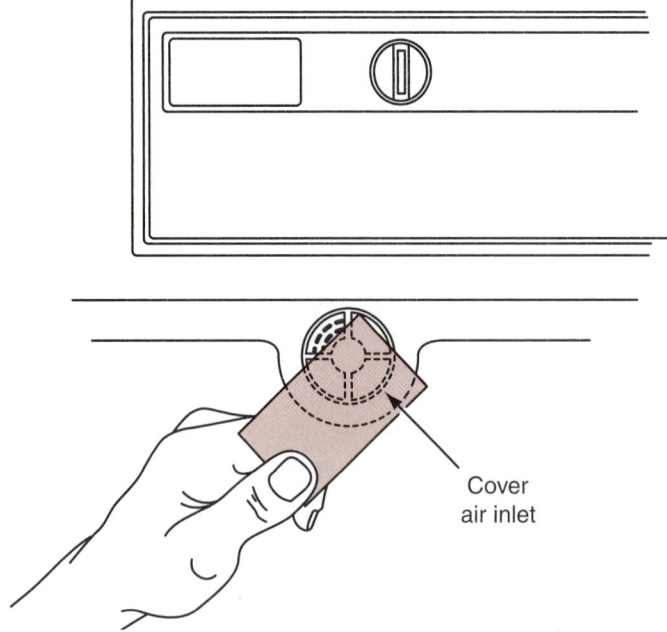

Cover
air inlet

Figure 14-1 When the technician performs the aspirator paper test, the vacuum should hold the paper over the grill.

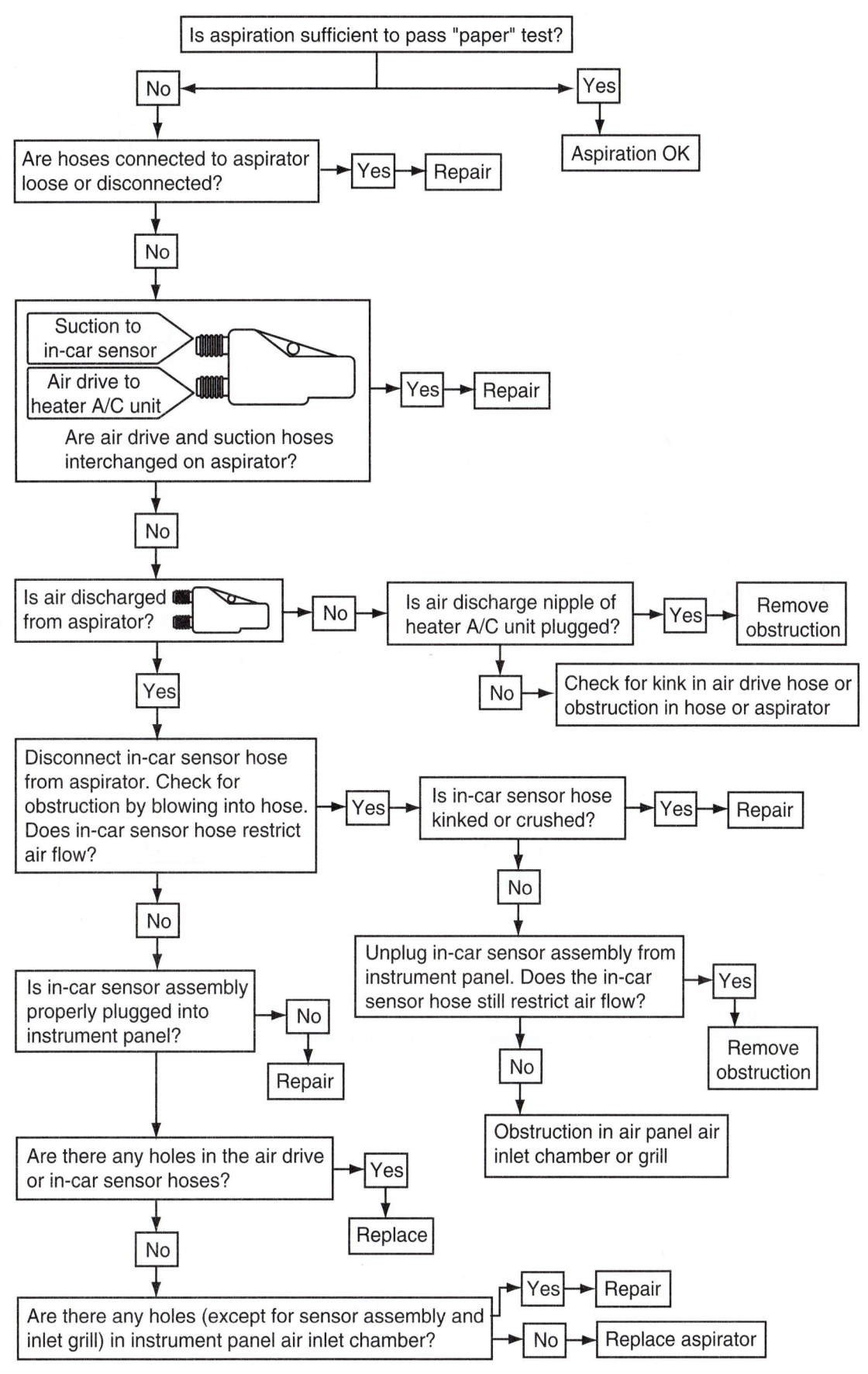

Figure 14-2 Aspirator diagnostic test chart.

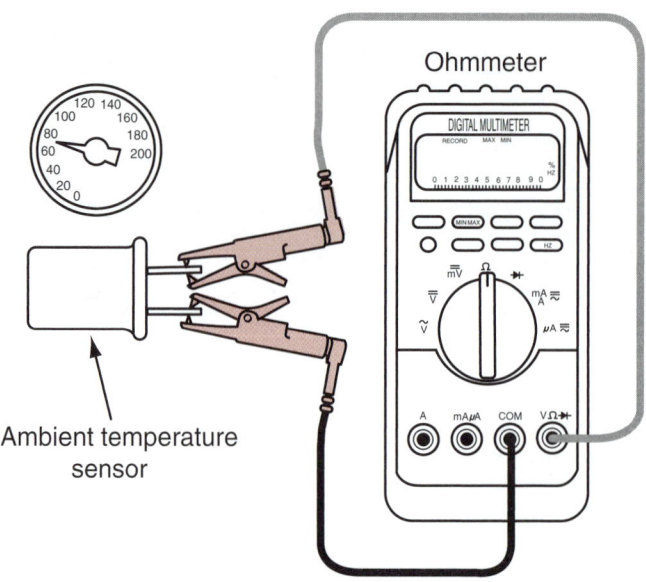

Figure 14-3 Test the ambient temperature sensor at room temperature and avoid touching the sensor during testing. Also, do NOT connect the ohmmeter for longer than 5 seconds.

Special Tools

Ohmmeter

Ambient Temperature Sensor. To test the ambient sensor, remove it from its socket and measure its resistance when the temperature is between 70°F and 80°F (21°C and 27°C). Because current flow through the sensor and body heat will affect the readings, do not hold the sensor in your hand or leave the ohmmeter connected for longer than 5 seconds (Figure 14-3). The resistance through the sensor at room temperature should be compared with specifications. If the resistance levels fall out of the specification range, replace the ambient sensor.

Special Tools

Test thermometer
  with a long stem
Ohmmeter

In-vehicle Sensor. To test the in-vehicle sensor, disconnect its electrical connector. Do not disconnect the aspirator tubes or remove the sensor from the panel. Place the test thermometer stem into the air inlet grill near the in-vehicle sensor.

With the blower motor speed set on the second notch, operate the blower while quickly measuring the resistance of the sensor. The resistance of the in-vehicle sensor should be compared with specifications based on the temperature. Replace the sensor if the resistance is not within specification.

Classroom Manual
Chapter 14, page 413

Servomotor. The **servomotor** is a vacuum or electric motor used to control the position of the mode and blend air doors. Perform the servomotor test only after it has been determined that the aspirator assembly is functioning correctly. To test the servomotor, it may need to be removed from the vehicle. Follow the service manual procedures for removing the motor. Use the following steps to test the motor:

Special Tools

Test light
Jumper wire
Ambient sensor
Door operating
  crank from a
  manual A/C unit
12-volt battery

1. Use a blend door crank from a manual A/C unit to check for free door movement. If the door binds, attempt to locate the cause and repair the unit. If the cause is in the motor, then the servo will have to be replaced.

2. Connect the ambient sensor across the two servo motor terminals on the left side of the connector.

3. Connect a test light between the right side terminal and battery positive and the jumper wire from battery negative to the middle terminal.

4. When the connections are made, the motor should run in the clockwise direction while the test light glows dimly.

5. When the motor reaches its internal stop, the test light should glow bright.

6. Operate the motor in the counterclockwise direction by removing the ambient sensor.

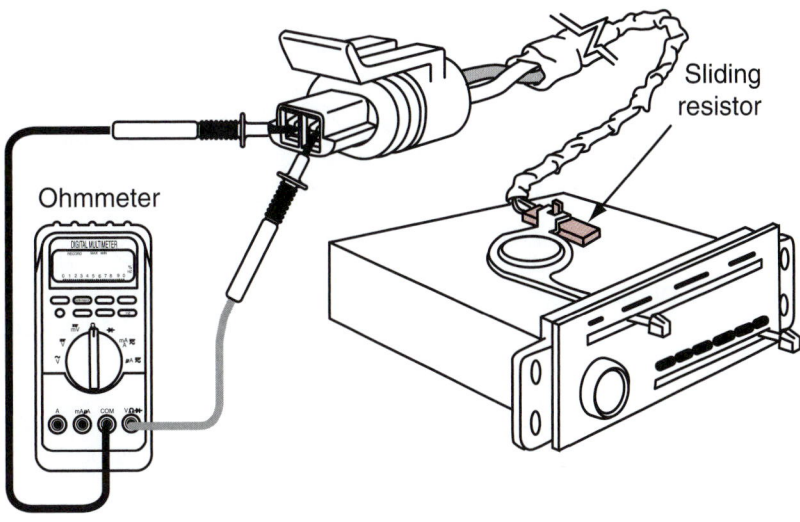

Figure 14-4 Testing the control assembly sliding resistor. Resistor values should be within specifications and should change smoothly.

While operating the motor, check for smooth operation and observe the test light. If the motor encounters a brief jam, the test light illumination level will increase. A test light that flickers while the motor is operating indicates the motor is not moving in a smooth fashion. If the cause cannot be repaired, the servo will need to be replaced. For the motor to be considered acceptable and placed back into service, it must move to the full clockwise and full counterclockwise positions.

Control Assembly Resistor. The **control assembly** provides for driver input into the automatic temperature control microprocessor. The control assembly is also referred to as the control panel. The control assembly resistor can be tested in or out of the vehicle. Disconnect the electrical connector to the sliding resistor and connect the ohmmeter test leads across the terminals (Figure 14-4). Set the comfort control lever to the 65 selection. Check the resistance value in this position with specifications.

Slowly move the comfort control lever toward the right while observing the ohmmeter. Note the ohmmeter reading when the lever is at the 75 setting. Also, the ohmmeter should indicate a smooth increase in resistance. Next, move the lever to the 85 setting. The resistance value should increase smoothly and be compared to specifications.

If the resistance values are not within specifications or the increase in resistance is not smooth, the control assembly must be replaced.

Separate Microprocessor-Controlled Systems

Most ATC systems that use a separate microprocessor have it contained in the control assembly (Figure 14-5). A majority of these systems provide a means of self-diagnostics and has a method of retrieving trouble codes. The following are system examples.

DaimlerChrysler ATC System Troubleshooting. Early DaimlerChrysler ATC systems can be commanded to perform a self-diagnostic check and then display fault codes. This system is not capable of storing any history codes. Therefore, the problem must be present when the self-diagnostic route is run. The following is an example of running the self-diagnostics on DaimlerChrysler ATC systems. However, this is provided to demonstrate how the process and system has evolved through the years. Always refer to the service information for the year and model of vehicle being diagnosed.

Special Tools

Ohmmeter

Classroom Manual
Chapter 14, page 409

Classroom Manual
Chapter 14, page 415

Special Tools

Scan tool

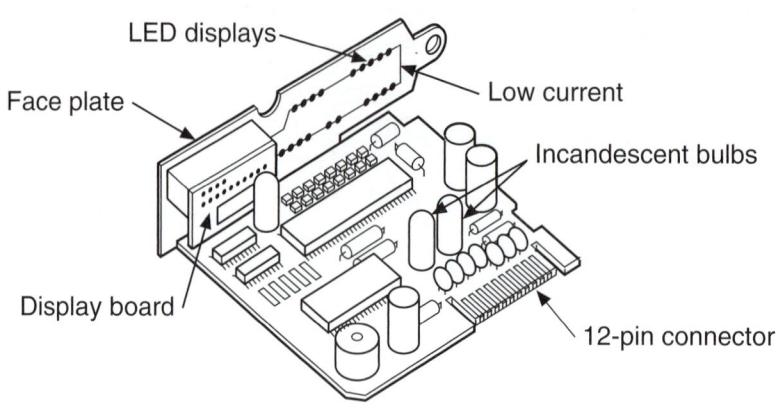

LED displays

Face plate

Low current

Incandescent bulbs

Display board

12-pin connector

Figure 14-5 Most ATC systems use a separate microprocessor that is located in the control assembly.

Before entering self-diagnostics, start the vehicle and allow it to reach normal operating temperature. Make sure all exterior lights are off. Then press the PANEL button. If the display illuminates, the self-diagnostic mode can be entered. If the display does not illuminate, check the fuses and circuits to the control assembly. If the circuits are good, replace the ATC computer.

The self-diagnostic mode is entered by pressing the BI-LEVEL, FLOOR, and DEFROST buttons at the same time (Figure 14-6). If no trouble codes are present, the self-test program will be completed within 90 seconds and display a 75.

During the process of running the self-diagnostic tests, the technician must make four observations that the computer is not able to make by itself. When the test is first initiated, all of the display symbols and indicators should illuminate. The blower motor should operate at its highest speed. Also, air should flow through the panel outlets, and the air temperature should become hot then cycle to cold. The diagnostic flow chart will direct the technician to the cor-

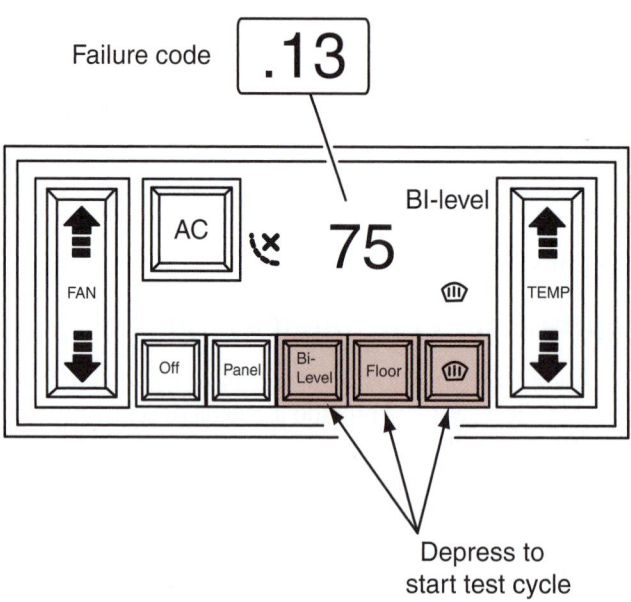

Figure 14-6 Use the panel buttons to enter diagnostics.

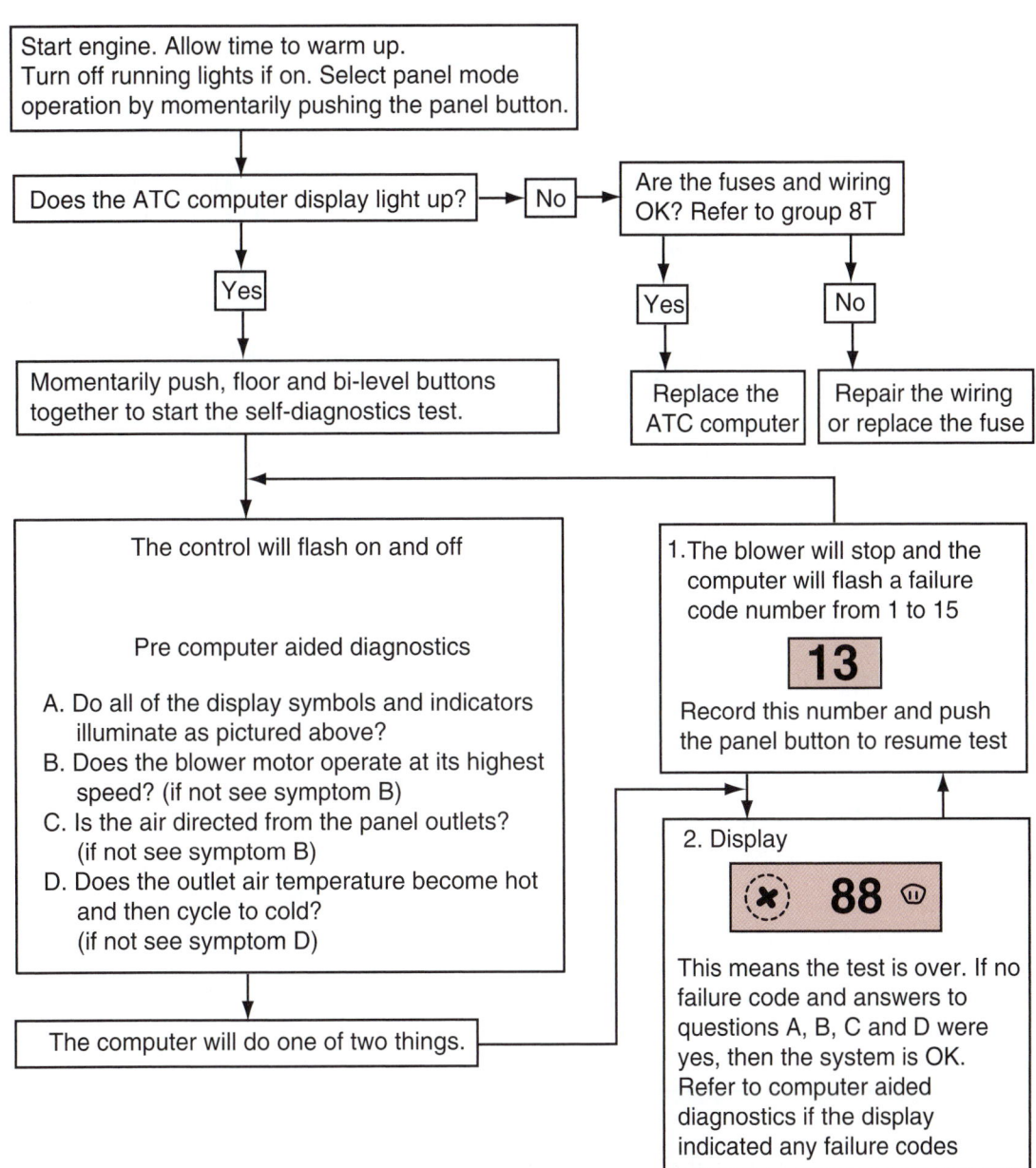

Figure 14-7 ATC diagnostic flow chart for DaimlerChrysler Motors system. While the self-test is being performed, the technician must answer four questions.

rect test to perform if any of the previous functions fail (Figure 14-7). The proper procedures for an observed failure are found in the table (Figure 14-8).

If a fault is detected in the system, a trouble code will be flashed on the display panel. To resume the test, record the trouble code and then press the PANEL button. Refer to the service manual to diagnose the trouble codes received.

Newer ATC systems used by DaimlerChrysler vary depending on vehicle model. Some systems only allow the retrieval of DTCs and other data through the use of a scan tool. Some systems do not support scan tool diagnostics and use only stand alone diagnostics. Other systems support the use of either a scan tool or stand-alone diagnostics. A study of performing self-diagnostic routines on the dual climate control system used on the Jeep® Grand Cherokee is used to illustrate the diagnostic capabilities of these systems.

"NO" ANSWERS	PROBABLE CAUSE	PROCEDURE
A	1) Control	a Replace the Control Module
B	1) Wiring problem 2) Power/Vacuum Module	a) CAUTION: STAY CLEAR OF THE BLOWER MOTOR WHEEL. POWER/VACUUM HEAT SINK IS HOT (12 VOLTS), DO NOT RUN THE POWER/VACUUM MODULE EXCESSIVELY (10 MINUTES) WITH THE UNIT REMOVED FROM THE A/C HEATER HOUSING. b) Check to see if connections are made at the blower motor and at Power/Vacuum Module. c) Did the diagnostic test give an error code of 8 or 12? If yes, refer to Fault Code Page. If no error code then check 30 AMP fuse for blower motor (fuse #4). d) Disconnect blower motor, check voltage at connector (green is +, black is –). Reading should be 3 to 12 volts for 1 to 8 bar segment on the display. If correct then problem is the motor. e) If blower voltage is not correct, then measure volts at green wire to vehicle ground (not blower ground) should be 12 volts (key on). If the voltage is OK, then replace the P/V Module.
C	1) Vacuum Leakage 2) Power/Vacuum Leakage	a) Service, if any codes are found. b) Check all connectors. c) Disconnect 7 port vacuum connector and connect it to a "manual control" and test each mode. Test Check Valve selecting the Panel Mode and disconnecting engine vacuum to see if mode changes quickly. d) Try a new Power/Vacuum Mod.
D	1) Refrigeration System 2) Heater System 3) Blend-Air Door	a) Complete diagnostics test, refer to Fault Code Page if error occurs. b) If a temperature difference of at least 40 Fahrenheit degrees is felt during the diagnostic test, then the Blend-Air door is engaged in the Servo Motor Actuator. If temp difference is less than 40 degrees, then a possible problem is the Blend-Air Door operation. c) Check heater system. 85 temp setting is full heat and 65 is full cool. d) Check refrigeration system. NOTE: Panel, A/C and holding down the bottom of the TEMP button for 4 seconds once 65 is obtained will cause Max A/C.

Figure 14-8 If the technician answers "NO" to any of the questions asked during the self-test, this diagnostic chart is used to isolate the fault.

The automatic zone control (AZC) system performs self-diagnostic routines during normal system operation. During operation, the AZC module continuously monitors various parameters. If a fault is detected, both a current and history fault is set into memory. If the cause of the fault is intermittent, the current code will self-erase but the history fault will be retained. The fault codes can be retrieved either through the use of the scan tool or by reading the display on the front panel of the control head.

The control head is capable of three different types of self-diagnostic routines:

1. Fault code tests.

2. Input circuit tests.

3. Actuator circuit tests.

Self-diagnostics are entered by performing a series of steps using the control head buttons and knobs. To enter self-diagnostics, turn the ignition switch to the RUN position. Depress and hold both the A/C and RECIRC buttons (Figure 14-9). While holding these two buttons down, rotate the left temperature control knob clockwise one detent. As you continue to hold the A/C and RECIRC buttons, the display will illuminate all segments of the control head. All segments should illuminate for as long as you hold the two buttons. If a segment fails to illuminate during this test, replace the control head.

After you have confirmed that all display segments work, release the A/C and RECIRC buttons. The display will clear for a few seconds, then display the fault codes. If a zero is displayed, no fault codes were found. If there are any fault codes, they will be displayed in acceding numeric order. Each code will be displayed for one second until the cycle of codes are complete, then start over again. The codes will continue to be displayed until the left temperature control knob is rotated one detent clockwise or until the ignition switch is turned off.

Fault codes are erased by pressing either the A/C or the RECIRC buttons for three seconds while the system is in the display faults mode. The faults are cleared when two bars appear on the display. Only history codes can be cleared. Current fault codes can be cleared by finding and correcting the cause of the fault.

From the fault code test portion of self-diagnostics, the select test mode may be entered. To enter this mode, turn the left temperature knob until the desired test appears in the display. You will need to reference the proper service information to determine the test number. Once the test number is displayed, press the A/C button to activate the test. For example, if you desire to test the current mode position, rotate the left temperature knob until "21" is displayed.

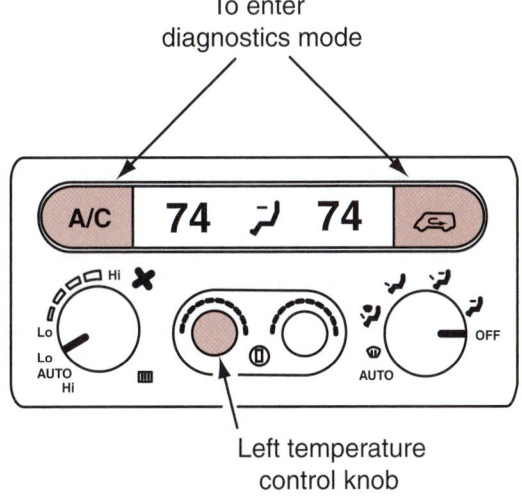

Figure 14-9 The AZC control head buttons and knobs are used to enter self-diagnostics.

Press the A/C button and the display will show a number. This number is the commutator count of the mode door actuator. Placing the mode door in the DEFROST position should show a commutator count of about 3150, a MIX mode of about 2585, a FLOOR mode of about 2000, a BI-LEVEL of about 1500, and a PANEL position of about 1050.

To exit diagnostics, press the A/C and the RECIRC buttons together.

Ford ATC System Troubleshooting. To correctly diagnose Ford's ATC system, you will need the exact system description as well as the exact procedures for trouble code retrieval. Ford uses different versions of ATC systems that have different diagnostic capabilities. The following is a typical service example of performing the self-test. However, this is not the same procedure used on all Ford systems.

1. Turn the ignition switch to the RUN position.
2. Place the temperature selector to the 90 setting and select the OFF mode.
3. Wait 40 seconds while observing the display panel. If the VFD display begins to flash, there is a malfunction in the blend actuator circuit, the actuator, or the control assembly. If the LED light begins to flash, this indicates there is a malfunction in one of the other actuator circuits, the actuators, or the control assembly.
4. If no flashing of displays occurs, place the temperature selection to 60 and select the DEF mode.
5. Wait 40 seconds while observing the VFD and LED displays. If there are no malfunctions in the actuator drive or feedback circuits, the displays will not flash.
6. Regardless of whether or not flashing displays were indicated, continue with self-diagnostics. Press the OFF and DEFROST buttons at the same time.
7. Within 2 seconds, press the AUTO button.

Once you have entered self-diagnostics, if an 88 is displayed, there are no trouble codes present. If there are any trouble codes retrieved, they will be displayed in sequence until the COOLER button is pressed. Always exit self-test mode by pressing the COOLER button before turning the ignition switch to the OFF position. When service repairs have been performed on the system, rerun the self-test to confirm that all faults have been corrected.

GM ETCC System Troubleshooting. General Motors uses several different versions of the microprocessor-controlled electronic touch climate control (ETCC) system. Depending on the GM division and system design, the door controls can be by either vacuum or electric servo motors. Methods of entering diagnostics also vary between divisions and models. For this reason, you will need to have the correct service manual for the system you are servicing in order to perform correct diagnostic procedures. Knowledge of one ETCC system type is no guarantee that you will be able to service other ETCC systems without the use of the service manual. This is true even between models of the same GM division.

Even though the methods and procedures are different between the various system designs, you are able to rely on your knowledge of electrical and electronic component diagnosis to follow the service manual steps and find the cause of the fault. Once a trouble code is retrieved, refer to the diagnostic chart for directions in performing the tests required to isolate the fault.

Many GM ATC systems can be checked for proper operation by using a functional chart (Figure 14-10). In addition, troubleshooting charts that correspond with fault symptoms are a great help to the technician (Figure 14-11).

Nissan ATC Diagnostics. The ATC system used on the Nissan Maxima provides an example of diagnostics used on some import vehicles. The self-diagnostic mode for this system is entered by starting the engine and pressing the OFF button in the A/C control head for 5 seconds. The OFF button must be pressed within 10 seconds after the engine is started and the fresh vent lever must be in the OFF position. The self-diagnostic mode may be canceled by pressing the AUTO button or turning off the ignition switch.

ATC FUNCTIONAL TEST

BEGIN ALL SYSTEM DIAGNOSIS BY CONDUCTING THE FUNCTIONAL TEST IN THE ORDER PRESENTED. IF THE ANSWER TO ANY SYSTEM CHECK QUESTION IS "NO", THEN REFER TO THE SPECIFIC TROUBLE TREE FOR DIAGNOSIS OF THE MALFUNCTION.

PRELIMINARY CHECKS:
1. CHECK FUSES
2. WARM ENGINE
3. OBSERVE LED ABOVE EACH BUTTON DURING SYSTEM CHECKS.

SYSTEM CHECKS	TROUBLE TREE
1. DOES THE CONTROL HEAD DISPLAY IN ALL CONTROL SETTINGS?	1
2. DO COOLER AND WARMER PUSHBUTTONS OPERATE IN ALL CONTROL SETTINGS?	1
3. SET AT 60° A. DOES THE BLOWER OPERATE IN LO, AUTO, AND HI SETTINGS?	2A
B. IS THERE LO BLOWER?	2B
C. IS THERE HI BLOWER?	2C
D. IS THERE AIR FROM A/C OUTLETS IN ECON, LO, AUTO, AND HI SETTINGS?	3
E. DOES COMP CLUTCH ENGAGE IN LO, AUTO, AND HI SETTINGS?	4
F. DOES COMP CLUTCH DISENGAGE IN OFF AND ECON SETTINGS?	5
G. IS THERE COLD AIR FROM A/C OUTLETS? IF NOT: 1) IS THERE HEAT ONLY?	6
2) IS THERE INSUFFICIENT COOLING?	7
H. DOES RECIRC DOOR OPEN FULLY? WAIT UP TO 2 MINUTES FOR SYSTEM TO CHANGE.	8
4. SET TEMPERATURE TO 90° AND CONTROL TO AUTO. A. IS THERE SUFFICIENT HEAT FROM HEATER OUTLETS?	9
5. SET TEMPERATURE TO 85° AND CONTROL TO LO AND AUTO. WAIT UP TO 2 MINUTES FOR SYSTEM TO CHANGE. A. IS THERE WARM OR HOT AIR AT HEATER OUTLET?	10
6. DOES FRONT DEFROSTER OPERATE?	11
7. DOES REAR DEFROSTER OPERATE?	12
8. DOES REAR DEFROSTER TURN OFF?	13

Figure 14-10 GM ATC function test.

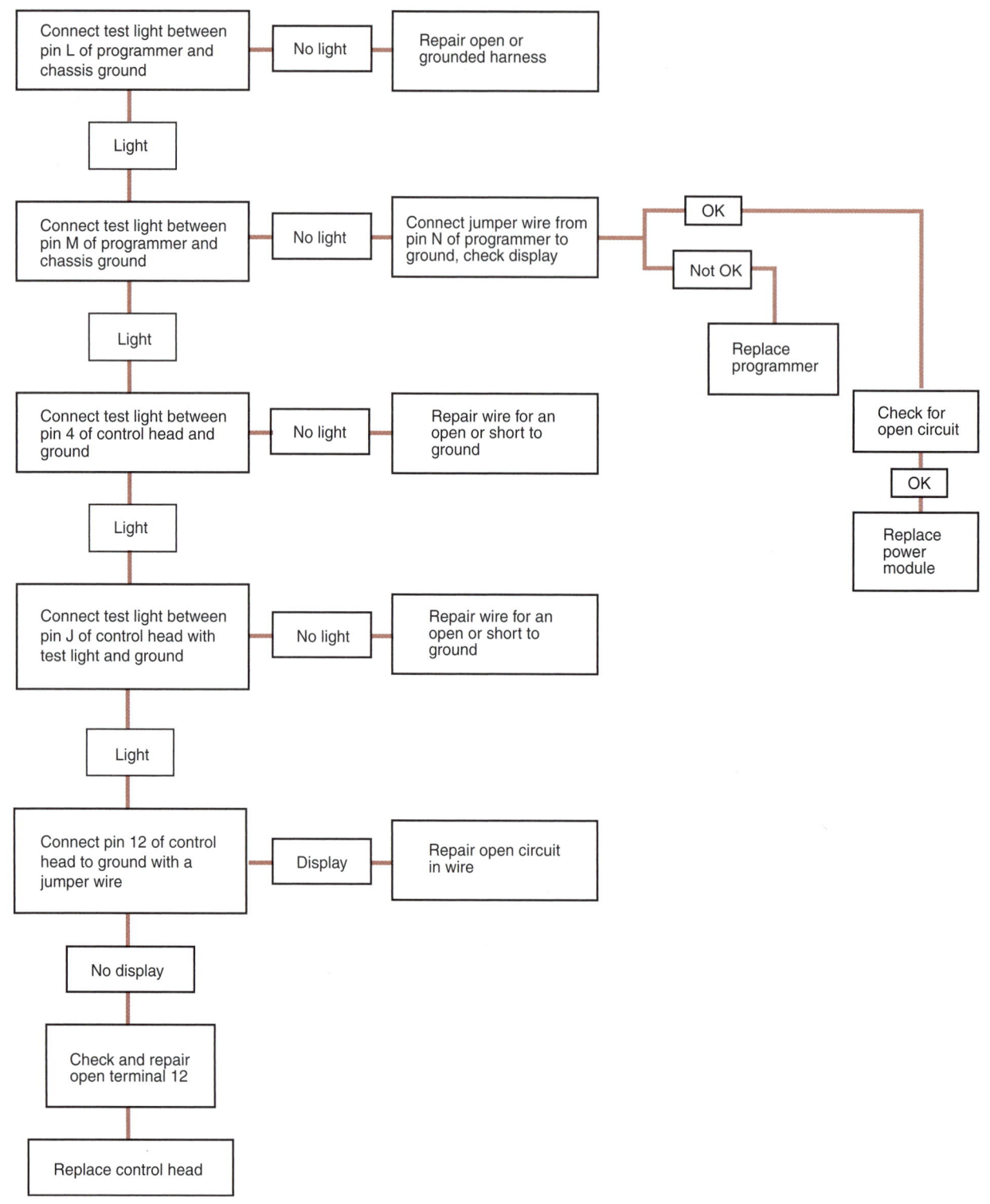

Preliminary Steps:
1. Turn headlamps and twilight sentinel OFF.
2. Start engine.
3. Check for blown fuses.

Figure 14-11 GM ATC troubleshooting chart.

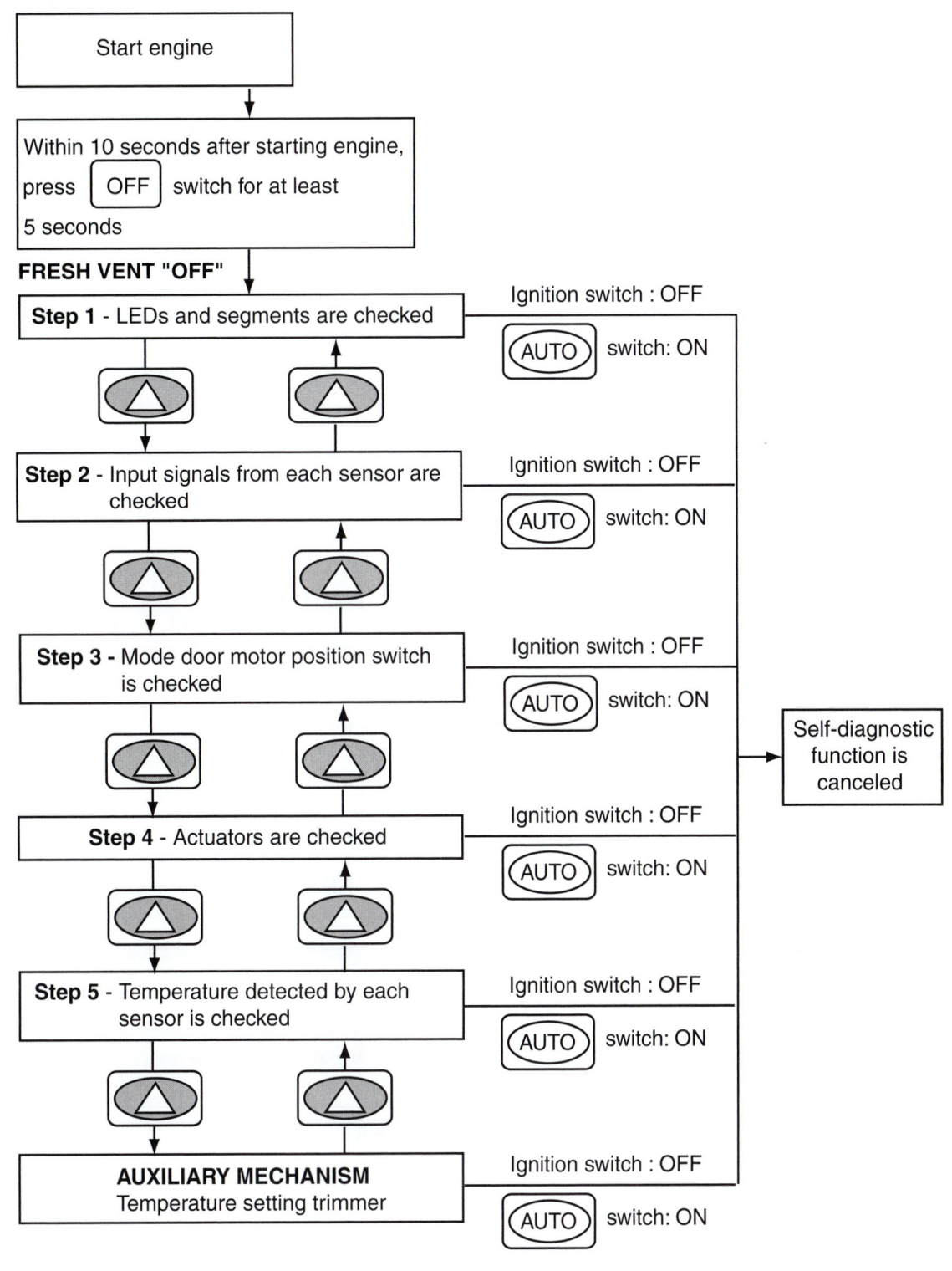

Figure 14-12 Self-diagnostic procedure with five steps.

The self-diagnostic tests are completed in five steps. The UP arrow for temperature setting on the A/C control head is pressed to move to the next step. When the DOWN arrow for temperature setting is pressed, the diagnostic system returns to the previous step (Figure 14-12).

Step 1. All of the LEDs and display segments are illuminated. If some segments or LEDs are not illuminated, these components are defective and the control head must be replaced.

Step 2. When the UP temperature control arrow is pressed to move to step 2, the input sensors are tested by the A/C computer. If there are no defects in the input sensors, the control head will display "20." If any of the input sensors are defective, a code for that sensor is displayed (Figure 14-13).

Step 3. When the UP temperature arrow is pressed to move to step 3, the mode door position switches are tested. If all of these position switches are in satisfactory condition, the display will show "30." A defect in one of the position switches results in a code being displayed ranging from 31 to 36 (Figure 14-14).

Step 4. If the UP temperature arrow is pressed to move to step 4, the A/C control head displays "41" and the A/C computer positions the mode door in the vent position. The intake door is in the recirculation (REC) position and the air mix door is in the full cold position. The defrost (DEF) button is pressed to move to the next mode, which displays "42" in the A/C control head. There are six modes in step 4 and each mode is represented by a number in the A/C control head. These numbers range from 41 to 46. The DEF button is used to select the next mode. In each one, the A/C computer commands specific door positions, blower motor voltage, and compressor clutch operation (Figure 14-15). Door operation may be checked by the air discharge from the various ducts.

Code No.	Malfunctioning sensor (includes circuit)
21	Ambient sensor
22	In-vehicle sensor
24	Intake sensor
25	Sunload sensor #1
26	P.B.R.

Figure 14-13 In step 2, a defective input sensor causes a code to be displayed.

Code No.	Malfunctioning mode door position switch (includes circuit)
31	Vent
32	Bi-level
33	Bi-level
34	Foot / defrost 1
35	Foot / defrost 2
36	Defrost

Figure 14-14 In step 3, a defective mode door switch causes a code to be displayed.

ACTUATOR TEST PATTERN

Code No.	Mode door	Intake door	Air mix door	Blower motor	Com-pressor
41	Vent	REC	Full cold	4-5V	On
42	Bi-level	REC	Full cold	9-11V	On
43	Bi-level	20% FRE	Full hot	7-9V	ON
44	F /D 1	FRE	Full hot	7-9V	OFF
45	F / D 2	FRE	Full hot	7-9V	Off
46	Defrost	FRE	Full hot	10-12V	ON

Figure 14-15 Step 4 contains six modes. In each mode the computer commands a specific door position, blower motor voltage, and compressor clutch operation.

Step 5. When the UP temperature arrow is pressed to select step 5, the temperature detected by each input sensor is displayed. After this mode is entered, a "5" is displayed in the A/C control head. If the DEF button is pressed, the temperature sensed by the ambient sensor is displayed in the A/C control head. Pressing the DEF button a second time displays the temperature detected by the in-vehicle sensor. If the DEF button is pressed again, the temperature detected by the intake sensor is displayed (Figure 14-16). This sensor is mounted near the evaporator outlet. When the temperature displayed varies significantly from the actual temperature, the sensor and connecting wires should be tested with an ohmmeter and a voltmeter.

At the end of step 5, the blower speed button may be pressed to enter the auxiliary mode. Next, the temperature on the display may be adjusted so it is the same as the in-vehicle temperature felt by the driver. After the auxiliary mode is entered, press the up or down temperature control buttons until the A/C control head displays the same temperature as inside the vehicle.

BCM ATC-Controlled Systems

Because the BCM-controlled ATC system incorporates many different microprocessors within its system, diagnostics can be very complex (Figure 14-17). Faults that seem to be unrelated to the ATC system may cause the system to malfunction. Since they were first introduced in 1986, BCM-controlled ATC systems have become increasingly popular on many vehicles. Each model year brings forth revisions and improvements in the system that also require different diagnostic procedures.

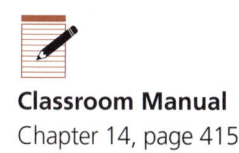

Classroom Manual
Chapter 14, page 415

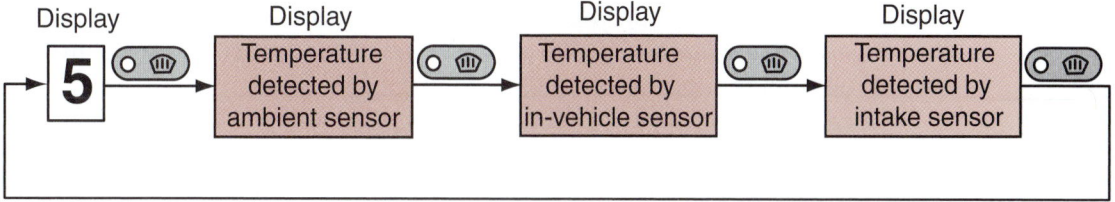

Figure 14-16 Step 5 displays the temperature detected by each sensor.

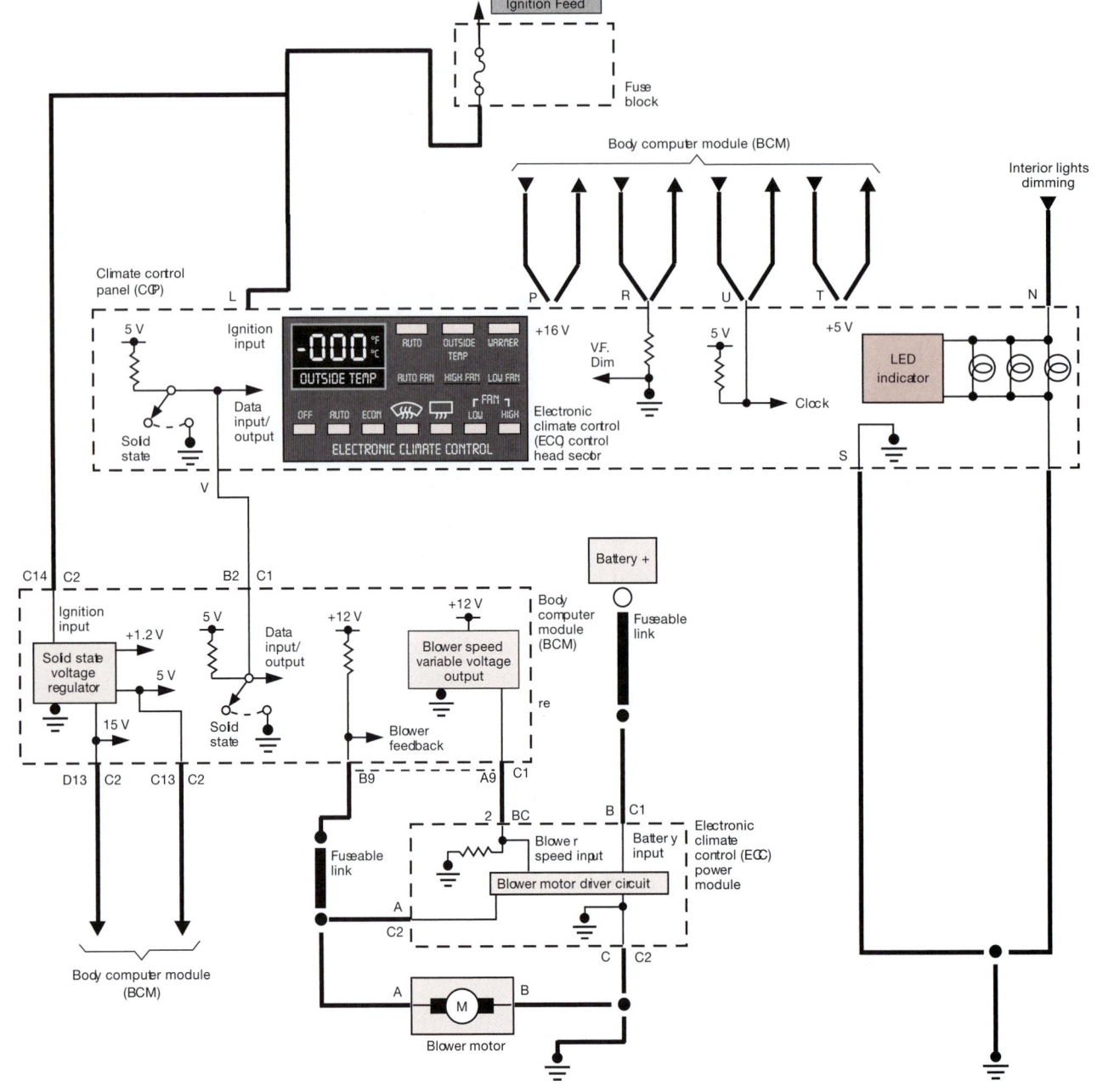

Figure 14-17 The BCM-controlled ATC system has several modules that use multiplexing to share information.

It would be impossible to describe the diagnostic procedures required to service the many systems in use. For this reason, technicians must have the correct service manual for the systems they are diagnosing. There are several methods used to retrieve trouble codes. Be sure to follow the correct procedure.

Once the codes have been retrieved, refer to the correct diagnostic chart. This pinpoint test will lead to the fault in a logical manner. After all repairs to the system are complete, follow the service manual procedure for erasing codes and for resetting the system. Rerun the diagnostic test to confirm that the system is operating properly.

Self-diagnostics. Many BCM computer-controlled air conditioning systems have self-diagnostic capabilities. These self-diagnostic tests vary depending on the vehicle make and model year. Always follow the instructions in the vehicle manufacturer's service manual. On some vehicles, such as the Chrysler LH and LHS cars, the self-diagnostic mode for the automatic temperature control (ATC) system is entered by pressing the FLOOR, MIX, and DEFROST buttons simultaneously, with the engine running and the vehicle not moving (Figure 14-18).

When the diagnostic mode is entered, the digital display in the A/C control head begins blinking. The display continues blinking until self-diagnostic tests are complete. While the display is blinking, the door actuator motors are calibrated to the unit on which they are installed. Any DTCs in the body computer memory are displayed. When one DTC is displayed, the PANEL button may be pressed to scroll through other DTCs. DTC numbers range from 23 to 36 (Figure 14-19). If there are no DTCs in the body computer memory, the system returns to normal

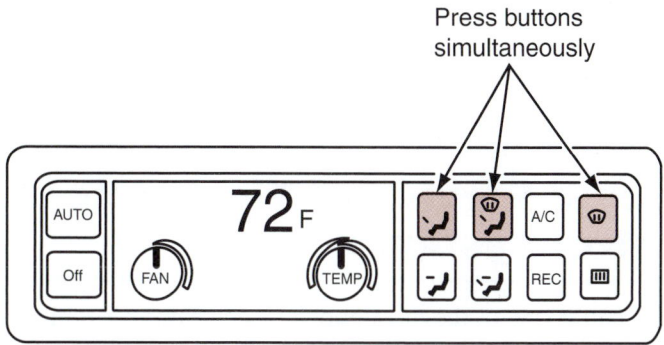

Figure 14-18 The self-diagnostic mode is entered by pressing the FLOOR, MIX, and DEFROST buttons simultaneously with the engine running and the vehicle not moving.

DIAGNOSTIC TROUBLE CODE CHART

Code	Description
23	ATC blend door feedback failure
24	ATC mode door feedback failure
25	Ambient sensor
26	ATC in-car sensor
27	Sun sensor failure
31	ATC recirculation door stall failure
32	ATC blend door stall failure
33	ATC mode door stall failure
34	Engine temperature message not received
35	Evaporator sensor failure
36	ATC head communication failure

Figure 14-19 Diagnostic trouble codes related to the ATC system.

operation as indicated by the temperature display. When DTCs are displayed, voltmeter or ohmmeter tests are usually required to locate the exact cause of the problem. DTCs may be erased with a scan tester or by disconnecting battery voltage from the body computer for 10 minutes.

Special Tools

Scan tool

Scan Tester Diagnosis. The following is a typical procedure for using a scan tool to diagnose the ATC system. The scan tester diagnosis of computer-controlled air conditioning systems varies depending on the vehicle make and model year and by the type of scan tool used. Always use the scan tester diagnostic procedures in the vehicle manufacturer's service manual or the scan tester manufacturer's manual. The proper module for the vehicle make, model year, and air conditioning diagnosis must be installed in the scan tester. The scan tester must be connected to the data link connector.

When the initial menu appears on the scan tester, select the menu item to enter the climate control system and select the desired function.

> **CAUTION:** Do not disconnect the wiring connector from any computer-controlled air conditioning system component with the ignition switch on. This action may result in damage to the air conditioning computer.

Diagnosis and Service of Electronic Cruise Control Systems

Classroom Manual

Chapter 14, page 424

Cruise control systems are also called speed control by some manufacturers.

The cruise control system is one of the most popular electronic accessories installed on today's vehicles. During open-road driving, it will maintain a constant vehicle speed without the continued effort of the driver. This reduces driver fatigue and increases fuel economy.

Problems with the system can vary from no operation, to intermittent operation, to not disengaging. To diagnose these system complaints, today's technicians must be able to rely on their knowledge and diagnostic capabilities. Most of the system is tested using familiar diagnostic procedures. Build on this knowledge and ability to diagnose cruise control problems. Use system schematics, troubleshooting diagnostics, and switch continuity charts to assist in isolating the cause of the fault.

> **WARNING:** When servicing and testing the cruise control system, you will be working close to the air bag and the antilock brake systems. The service manual will instruct you when to disarm and/or depressurize these systems. Failure to follow these procedures can result in injury and additional costly repairs to the vehicle.

Self-Diagnostics

Most vehicle manufacturers have incorporated self-diagnostics into their cruise control systems. This allows some means of retrieving trouble codes to assist the technician in locating system faults. Here are some common methods of retrieving diagnostic codes.

On any vehicle, perform a visual inspection of the system. Check the vacuum hoses for disconnects, pinches, loose connections, and so forth. Inspect all wiring for tight, clean connections. Also, look for good insulation and proper wire routing. Check the fuses for opens and replace as needed. Check and adjust linkage cables or chains, if needed. Some manufacturers will require additional preliminary checks before entering diagnostics. In addition, perform a road test (or simulated road test) in compliance with the service manual to confirm the complaint.

General Motors. General Motors has four different types of electronic cruise control systems. The common procedure for performing self-diagnostics and retrieving codes for system types

1, 3, and 4 follows. Refer to the service manual to determine which system applies to the vehicle you are working on. Type 2 systems do not provide fault code retrieval.

Type 1 Diagnostics. Brougham models with a type 1 system do not provide for trouble code retrieval. Instead a special tool called a "Quick Checker" is used to test the system. To enter self-diagnostics on other vehicles equipped with this system, place the ignition switch in the RUN position and press the OFF and WARMER buttons on the climate control panel (CCP) simultaneously. All of the panel segments should light. If not, the affected panel must be replaced before continuing.

All PCM trouble codes will be displayed, followed by all BCM codes. Engine controller codes are prefixed with an "E" and BCM codes are prefixed with an "F." On Eldorado and Seville models, "Current" or "History" will accompany the codes. On DeVille and Fleetwood models, trouble codes that are displayed on the first pass but not on the second represent history codes. History codes are intermittent faults that have occurred in the past but are not present in the most recent self-test.

Type 3 Diagnostics. Code accessing can be done through the use of a scanner, flashing the "Service Engine Soon" light, or through the ECC panel. The method used depends on the model of vehicle being serviced.

To flash the service light, ground terminal B of the data link connector with the ignition switch in the RUN position (Figure 14-20). Any stored memory codes will be flashed by the light. Also, all PCM-controlled relays and solenoids will be energized.

The indicator light will flash a code 12 (one flash followed by two more flashes) to indicate the system is operating properly. If there are any trouble codes, they are each displayed three times. At the end of the trouble codes, a code 12 will be flashed to indicate code resequencing has begun. If a scan tool is used to retrieve the codes, connect it to the DLC and follow the instructions to retrieve the codes.

Toronado, Reatta, and Riviera models provide for diagnostics through the ECC or CRT panel. Trouble codes are accessed by placing the ignition switch in the RUN position, then pressing the OFF and WARMER buttons on the ECC/CRT panel at the same time. On models equipped with CRT displays, depress the OFF hardkey and the WARM softkey. On Reatta and Riviera models, the WARM button is identified as TEMP-UP.

All ECM trouble codes will be displayed, followed by all BCM codes. On Eldorado and Seville models, "Current" or "History" will accompany the codes. On DeVille and Fleetwood models, trouble codes displayed on the first pass but not on the second, represent history codes.

After the diagnostic service mode is selected, any trouble codes in memory will be displayed. Engine controller codes are prefixed with an "E," BCM codes are prefixed with a "B," instrument panel cluster (IPC) codes are prefixed with an "I," and supplemental restraint system (SRS) codes are prefixed with an "R".

TERMINAL IDENTIFICATION

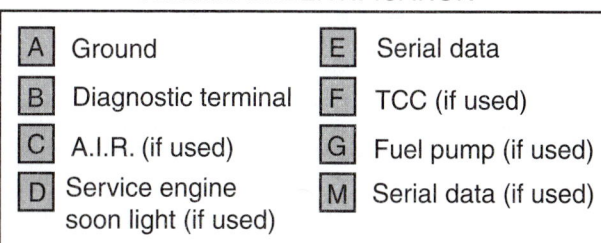

A	Ground	E	Serial data
B	Diagnostic terminal	F	TCC (if used)
C	A.I.R. (if used)	G	Fuel pump (if used)
D	Service engine soon light (if used)	M	Serial data (if used)

Figure 14-20 Data link connector terminal identification.

Observed Symptom	Code Type	Results	Components To Be Diagnosed
• Cruise control switch indicator blinks five times or • Cruise control does not set or • Cruise control inoperative.	Type B	11	• Actuator • Module
		21	• Speed Sensor • Module
		23	• Speedometer • Actuator • Speed Sensor • Vacuum Supply • Vacuum Switch • Module
		31	• Engage Switch • Module
		33	• Engage Switch • Module
	Type A Code 5	No Code	• Speed Sensor • Module
		OK	• Cruise Control Switch • Engage Switch • Stoplamp Switch • Clutch or Neutral Start Switch • Parking Brake Switch • Speedometer • Actuator • Module • Vacuum Hose
Set speed deviated.	Type A Code 3	OK	• Speedometer • Speed Sensor • Vacuum Supply • Vacuum Switch • Actuator • Cruise Control Module
		No Code	• Speed Sensor
Vehicle speed fluctuates when "SET/COAST" switch is activated.			• Speed Sensor • Speedometer • Actuator • Module
Set speed does not cancel when the brake is applied.	Type A Code 4	OK	• Actuator • Stoplamp Switch • Module
		No Code	• Stoplamp Switch • Module
Set speed does not cancel when the parking brake is applied.	Type A Code 4	OK	• Actuator • Module
		No Code	• Parking Brake Switch • Module

Figure 14-21 General Motors Type 4 troubleshooting chart.

Type 4 Diagnostics. Use the chart (Figure 14-21) to determine the type of code to read for the symptom. If it instructs you to read a type A code, refer to the other chart (Figure 14-22) and follow this procedure:

 WARNING: The number 4 code is checked with the drive wheels lifted from the floor and the engine idling.

Observed Symptom	Code Type	Result	Components To Be Diagnosed
Set speed does not cancel when shifted to "N" range (A/T) or clutch pedal is depressed (M/T).	Type A Code 4	OK	• Actuator • Module
		No Code	• Neutral Start Switch • Clutch Switch • Module
Vehicle speed does not decrease when "SET/COAST" is selected.	Type A Code 1	OK	• Actuator • Speedometer • Module
		No Code	• Engage Switch • Module
Vehicle speed does not increase when "RESUME/ACCEL" is selected.	Type A Code 2	OK	• Actuator • Speedometer • Cruise Control Module
		No Code	• Engage Switch • Module
Vehicle speed does not return to set speed when "RESUME/ACCEL" is selected.	Type A Code 2	OK	• Actuator • Speedometer • Module
		No Code	• Engage Switch • Module
Acceleration is sluggish when "RESUME/ACCEL" is selected.	Type A Code 3	OK	• Speedometer • Vacuum Supply • Actuator • Module
		No Code	• Vacuum Switch • Vacuum Supply

Figure 14-21 *(continued)*

1. Turn the ignition switch to the RUN position.
2. Turn the SET/COAST switch on.
3. Push the main switch on.
4. Turn the SET/COAST switch off.
5. Perform the condition requirements listed in the chart.

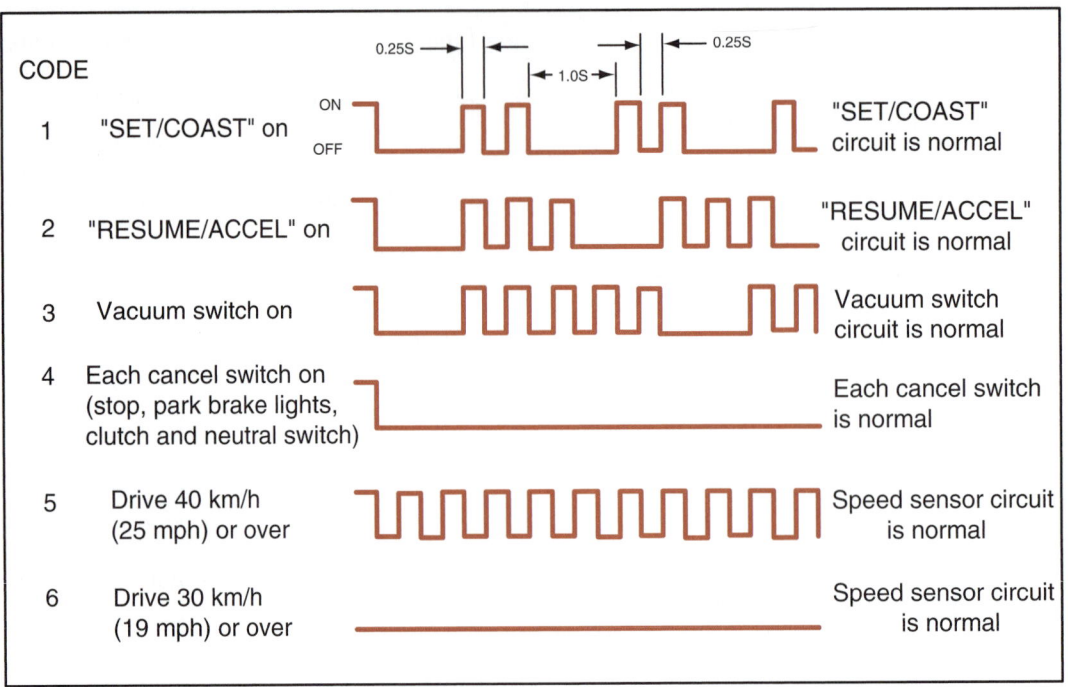

Figure 14-22 Type A code chart.

The diagnostic code is read on the main switch indicator by counting the flashes.

If instructed to read type B codes, refer to the proper chart (Figure 14-23) and follow this procedure:

1. Road test the vehicle.

2. If the system cancels because of a malfunction in the actuator or speed sensor, the indicator light will blink 5 times. If this occurs, do not turn off the ignition switch or the control switch. Inspect the system with the switches on. Turning off the switches will erase the codes from memory.

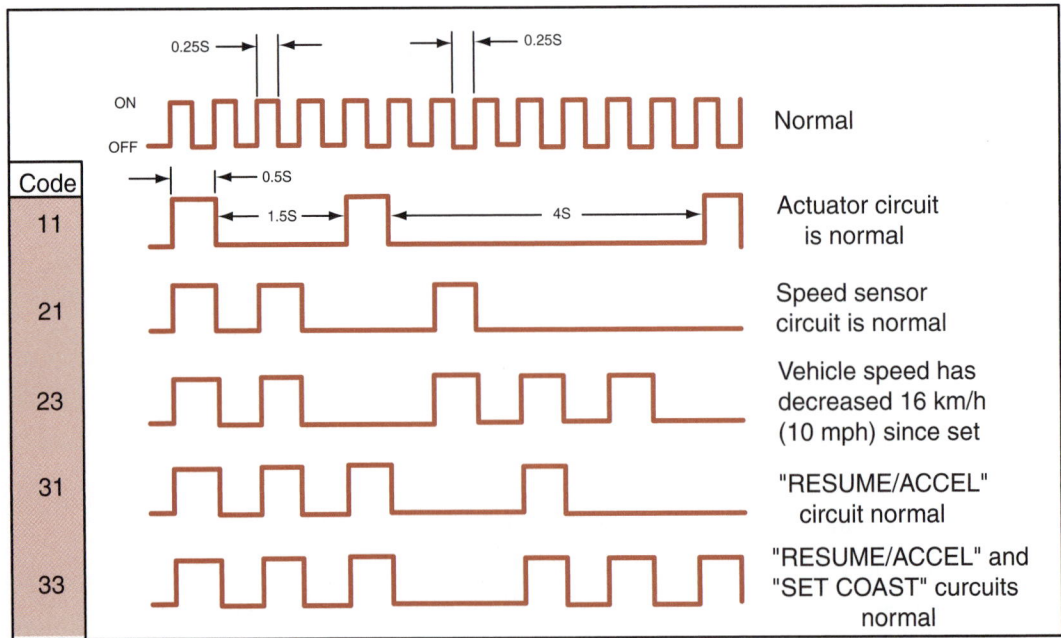

Figure 14-23 Type B code chart.

3. While driving at a speed less than 10 mph, press the SET/COAST button three times within two seconds. Any codes will be displayed through the indicator light.

If no codes are displayed, refer to the service manual to perform diagnostic tests.

Ford IVSC System Diagnostics. Ford's integrated vehicle speed control (IVSC) system has self-test capabilities that are contained within the **KOEO** and **KOER** routine of the ECA (Figure 14-24). KOEO stands for Key On, Engine Off. It is a static test of the IVSC inputs and outputs. KOER stands for Key On, Engine Running. It is a dynamic check of the engine in operation. Testing of the IVSC system is broken down into two divisions: quick tests and pinpoint test.

The quick test will check the operation and function of all system components except the vehicle speed sensor. The quick test is performed first. Then, if any failure codes are displayed, the **pinpoint test** is performed. This is a specific component test service. If there is a complaint with the cruise control system, and the quick test does not indicate any faults, test the speed sensor.

Special Tools

Scan tool

Analog voltmeter

Jumper wires

Fender covers

Service manual

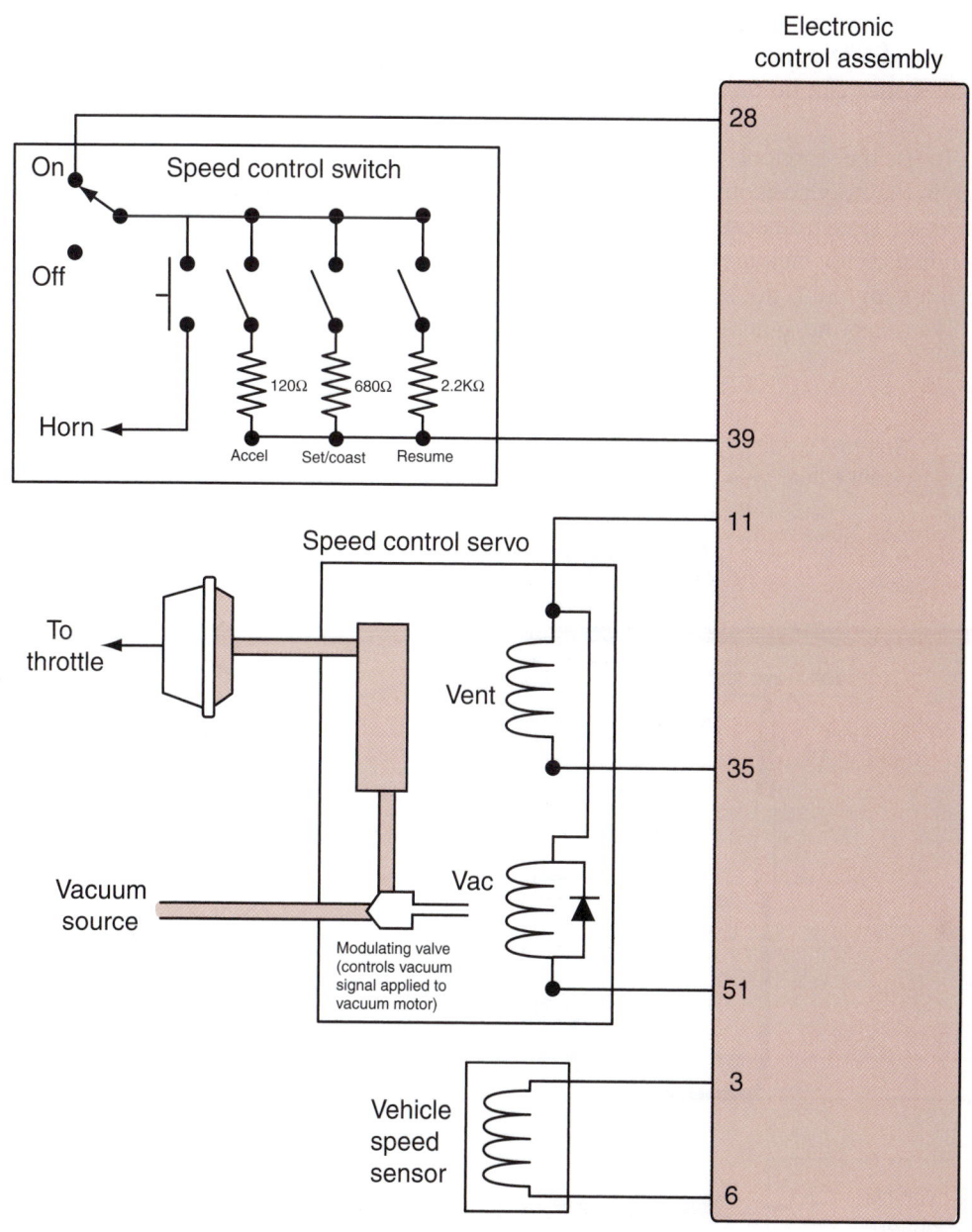

Figure 14-24 The IVSC system is an integrated system where the functions of the amplifier are included in the ECA.

The processor stores the self-test program within its memory. When this test is activated, the processor initiates a function test of the IVSC system to verify that the sensors and actuators are connected and operating properly. The quick test will detect faults that are present at the time of the test. It will not store history codes.

The quick test can be performed with a scan tool or an analog voltmeter. To use an analog voltmeter, place the ignition switch in the OFF position. Connect a jumper wire from the **self-test input (STI)** terminal to pin 2 (single return) of the self-test connector (Figure 14-25). Set the analog voltmeter on the DC 15-volt scale. Then connect the positive voltmeter lead to battery positive and the negative lead to pin 4 (self-test output) of the self-test connector.

Follow the troubleshooting procedures in the service manual. Perform the KOEO, KOER, and intermittent (wiggle) test procedures.

DaimlerChrysler. DaimlerChrysler uses four different types of electronic cruise control systems. The type 3 system does not provide trouble code diagnostics.

Type 1 and 4 Diagnostics. These cruise control types use the PCM to control the system. Trouble codes can be retrieved using a scan tool or by reading the flashes from the "Check Engine" light.

To use the scan tool, connect the tester to the engine diagnostic connector with the ignition switch in the RUN position. Select the correct year of the vehicle you are servicing. When asked to select the system, choose "Engine." Next, select "Read Fault Data" and read the fault messages. Depending on the fault message received, go to the correct test chart to locate the fault.

If a scan tool is not available, trouble codes can be retrieved through the "Check Engine" light by cycling the ignition from OFF to RUN three times within 5 seconds.

After the system has been serviced, perform the quick test to verify proper operation.

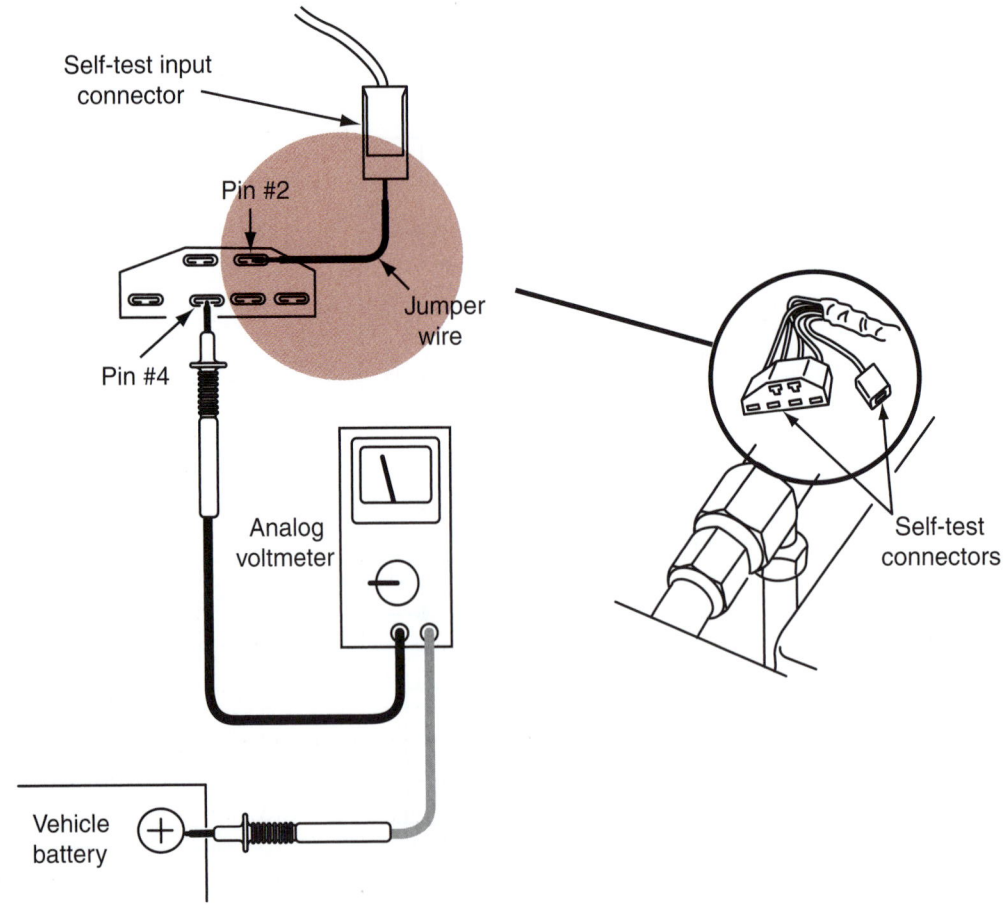

Figure 14-25 Voltmeter connections to the self-test connector and STI.

If fault code 34 is displayed, perform the speed control system test located in the service manual. This series of tests will check the electrical condition of the system, including switches and the servo.

⚠️ **WARNING:** Before conducting the system test, disarm the air bag system as described in the service manual. Failure to disarm the air bag may result in deployment and personal injury. After all service is complete, rearm the air bag according to the service manual procedures.

Special Tool

Scan tool

If the indicator light flashes a code 15, the distance sensor needs to be tested. A code 77 indicates that a speed control relay test must be performed.

Type 2 Diagnostics. This cruise control type is found on DaimlerChrysler vehicles that are produced by Mitsubishi Motors. This system uses a separate cruise control module. There are up to six trouble codes that can be displayed. One of these codes indicates normal operation. To access the codes, connect a voltmeter between the ground and auto-cruise control terminals of the diagnostic connector (Figure 14-26). The connector is located on the lower left side of the instrument panel. With the ignition switch located in the RUN position, read the needle sweeps to determine the code. Once the code is determined, refer to the diagnostic display pattern chart for the vehicle being diagnosed (Figure 14-27). This chart will refer you to the correct check chart to locate the fault.

Special Tool

Analog voltmeter

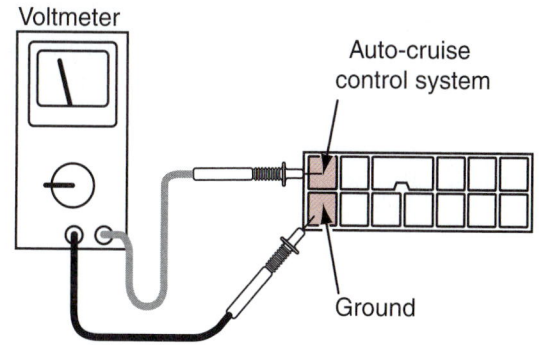

Figure 14-26 Connect the voltmeter between the two identified terminals.

Code number	Display patterns (output codes) (Use an analog voltmeter)	Probable cause	Check chart number
1		Vacuum pump assembly drive output system out of order	5
2		Vehicle speed signal system out of order	4
3		Control switch out of order (When "SET" or "RESUME" switch are kept ON state continuously for more than 60 seconds)	2
4		Control unit out of order	–
5		Throttle position sensor or idle switch out of order	9

Figure 14-27 Trouble code display pattern chart.

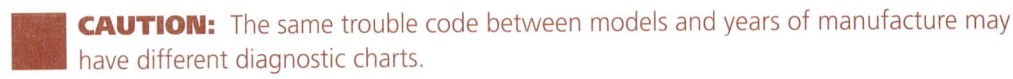

CAUTION: The same trouble code between models and years of manufacture may have different diagnostic charts.

Diagnosing Systems without Trouble Codes

Systems that do not provide for trouble code diagnostics require the technician to perform a series of diagnostic tests. The test performed will depend on the symptom. The following sections discuss areas of generic troubleshooting procedures for all types of systems.

Simulated Road Test. The simulated road test will allow the technician to perform a road test without leaving the shop. Before performing this test, connect the shop's ventilation system to the vehicle's exhaust pipe. Lift the drive wheels from the floor and place jack stands under the vehicle. If the vehicle is equipped with CV joint shafts, place the jack stands under the lower control arms so the shafts are in their normal drive position. If the vehicle is rear-wheel drive with a solid axle, place the jack stands under the axle.

WARNING: Block the wheels that are to remain on the ground. The wheels must remain blocked throughout the test.

Start the engine and place the transmission into drive. Turn the speed control switch into the ON position. Accelerate and hold the speed at 35 mph (56 kmh). Press and release the SET ACCEL button. Maintain a slight foot pressure on the accelerator. The speed should be maintained at 35 mph (56 kmh) for a short period of time, then gradually start to surge. The engine surge is caused by operating the system while there is no load on the engine and is normal.

CAUTION: During this test, it is possible the engine will overspeed. If the system should appear to go out of control, the technician must be ready to turn it off. This can be done by turning off the ignition or turning off the speed control switch.

Press the OFF button and the engine should decelerate to an idle speed. Stop the drive wheels by lightly applying the brakes. Press the ON button and accelerate to 35 mph (56 kmh). Press and hold the SET/ACCEL button and gradually remove your foot from the accelerator pedal. The engine rpm should begin to increase. Continue to hold the SET ACCEL button until the indicated speed reaches 50 mph (80.4 kmh); then release the button. Vehicle speed should remain at 50 mph (80.4 kmh) for a short period of time; then the engine will start to surge.

CAUTION: Do not exceed 50 mph or damage to the differential assembly may result.

CAUTION: Do not attempt to place the transmission back into PARK at any time during the test without first stopping the drive wheels with the brakes. Doing so may result in damage to the transmission.

Press the COAST button and hold it. The engine rpm should return to idle speed. Allow the indicated speed to slow to 35 mph (56 kmh) without applying the brakes. When the speed is returned to 35 mph (56 kmh), release the COAST button. The speed should be held at 35 mph (56 kmh) for a short period of time; then the engine will begin to surge.

Tap the brake pedal, which will cause the speed control system to shut off and engine speed to return to idle. Set the indicated speed to 50 mph (80.4 kmh); then use the brakes to slow to 35 mph (56 kmh). Maintain 35 mph (56 kmh) using the accelerator. Depress the RESUME button and the speed should climb to 50 mph (80.4 kmh).

Diagnosing No Operation

Special Tools

DMM
Vacuum pump

The first step in a verified no-operation complaint is to check all fuses. Next, visually inspect the system for any obvious problems. If the visual inspection does not pinpoint the problem, perform the following steps:

1. Apply the brake pedal to observe proper brake light operation. If the brake lights do not operate, check the switch and circuit.
2. If the vehicle is equipped with a manual transmission, check to assure that the clutch deactivator switch is operating properly. Use an ohmmeter or voltmeter to test its operation.
3. Check for proper operation of the actuator lever and throttle linkage.
4. Disconnect the vacuum hose between the check valve and the servo (on the servo side of the check valve). Apply 18 inches of vacuum to the open end of the hose to test the check valve. It should hold the vacuum. If not, replace the check valve.
5. Check the vacuum dump valve for proper operation.
6. Test control switches and circuits following the procedure already learned. Use the circuit diagram and switch continuity charts to aid in testing.
7. Test servo operation.
8. Test speed sensor operation.
9. If all tests indicate proper operation, yet the system is not operational, replace the amplifier (controller).

Diagnosing Continuously Changing Speeds

If the vehicle speed changes up and down while the cruise control is on, use the following steps to locate the problem:

1. Check the actuator linkage for smooth operation.
2. Check the speedometer for proper routing and to make sure there are no kinks in the cable.
3. Test the servo.
4. Check the speed sensor.
5. Check the operation of the vacuum dump valve.
6. Check all electrical connections.
7. If none of these tests locate the fault, replace the amplifier (controller).

Diagnosing Intermittent Operation

Special Tools

Vacuum Gauge
Ohmmeter

Intermittent operation is usually caused by loose electrical or vacuum connections. If a visual inspection fails to locate the fault, test drive the vehicle and identify when the intermittent problem occurs. If the problem occurs during normal cruising, begin at step 1. If the problem occurs when operating the control buttons, or when the steering wheel is rotated, begin with step 3.

1. Connect the vacuum gauge to the hose entering the servo. There should be at least 2.5 in. Hg (8.5 kPa) of vacuum.
2. Test the servo assembly.
3. Use the service manual's switch continuity chart and system schematic to test switch operation. Turn the steering wheel through its full range while testing the switches. For example, using the Ford system shown (Figure 14-28), this test would be conducted by disconnecting the connector at the amplifier and connecting an ohmmeter between the terminal for circuit 151 and ground (with the ignition switch off). While rotating the steering wheel throughout its full range, make the following checks:

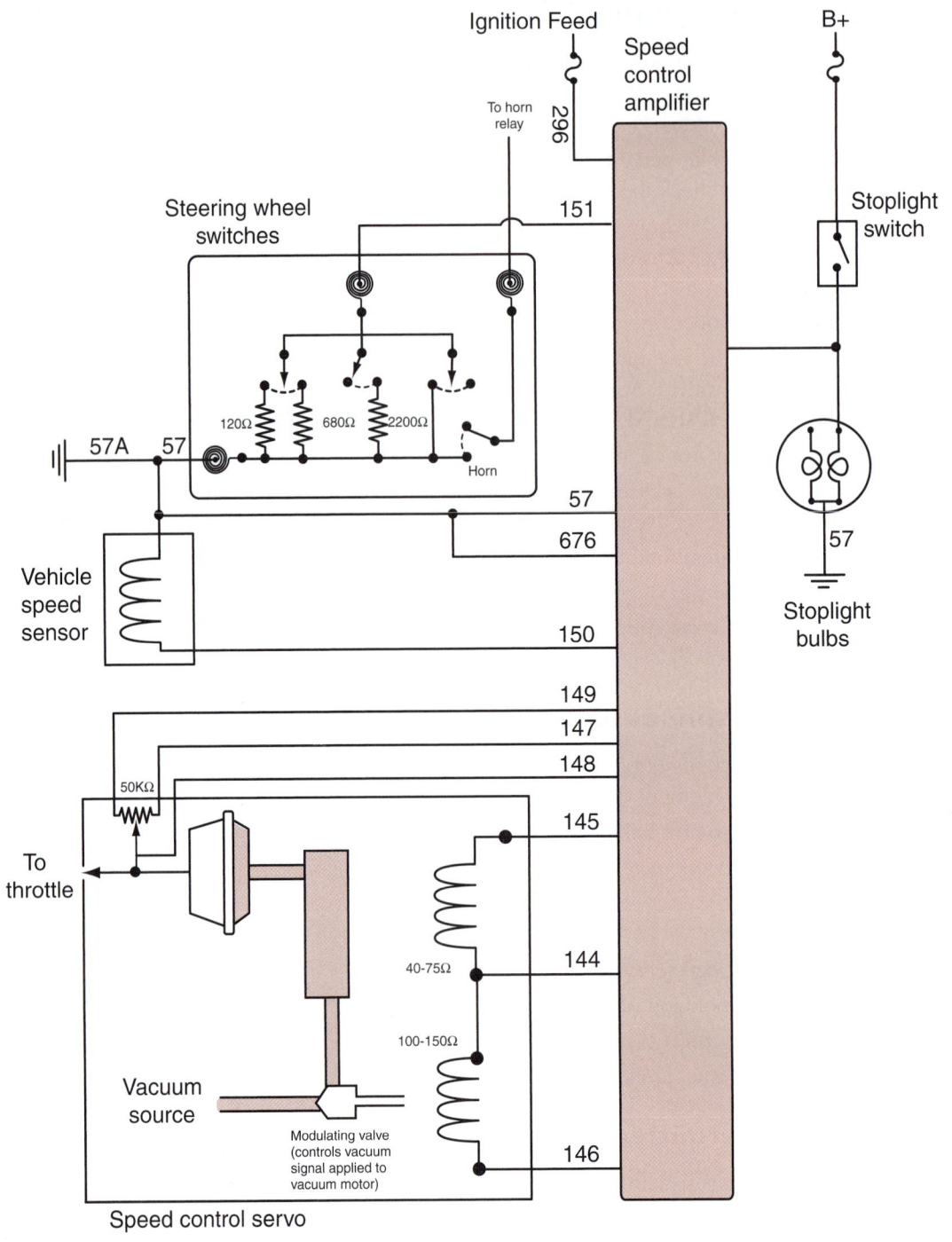

Figure 14-28 Speed control electrical schematic.

- ❏ Depress the OFF button; the ohmmeter reading should read between zero and 1 ohm.
- ❏ Depress the SET/ACCEL button and check for a reading between 646 and 714 ohms.
- ❏ Depress the COAST button and the ohmmeter should read between 126 and 114 ohms.
- ❏ When the RESUME button is depressed, the reading should be between 2,310 and 2,090 ohms.

If the resistance values fluctuate while the steering wheel is being turned, the most likely cause is contamination on the slip rings. Remove the steering wheel and clean the brushes. Apply a light coat of lubricant to the brushes using an approved lubricant. If the resistance values are above specifications, check the switches and ground circuit.

If the preceeding tests (or the road test) fail to identify the fault, conduct a simulated road test while wiggling the electrical and vacuum connections.

Component Testing

Testing of the safety switches and circuits is performed using normal testing procedures you have already learned. Testing of the servo assembly, dump valve, and speed sensor is included to familiarize you with these procedures.

Servo Assembly. Actuator tests vary depending on design. Some manufacturers use vacuum servos and others use stepper motors. Be sure to follow the service manual procedures for the vehicle you are diagnosing. The **servo** controls the position of the throttle by receiving a controlled amount of vacuum. The following servo assembly test is a common test for Ford's cruise control system. Use the schematic (Figure 14-28) to perform the following test.

Disconnect the eight-pin connector to the amplifier. Connect an ohmmeter between circuits 144 and 145. The resistance value should be between 40 and 125 ohms. Move the lead from circuit 145 to circuit 146. The resistance value should read between 60 and 190 ohms. If the resistance levels are out of specifications, check and repair the wiring between the amplifier and the servo. If the wiring is good, replace the servo.

If the resistance values are within specifications, leave the amplifier disconnected and start the engine. Jump 12 volts to circuit 144 and jump circuit 146 to ground. Momentarily jump circuit 145 to ground. The servo actuator arm should pull in and the engine speed should increase.

WARNING: Be sure to have the transmission in PARK or NEUTRAL. Block the wheels and set the parking brake before performing the servo test.

CAUTION: Be ready to abort the test by turning off the ignition, if engine rpm should rise to a level where internal damage may result.

Remove the jumper to ground on circuit 146. The servo should release, and engine speed should return to idle. The servo must be replaced if it does not operate as described.

CAUTION: Do not short the jumper wires from circuit 144 to circuits 145 or 146. Damage to the amplifier will result if the amplifier is connected while this is done.

Dump Valve. The **dump valve** is a safety switch that releases vacuum to the servo when the brake pedal is pressed. A dump valve that is stuck open or leaks will cause a no-operation or erratic-operation complaint. Failure of the dump valve not to release vacuum may not, by itself, be noticed by the driver. It is part of a fail-safe system. If the dump valve does not release, the electrical switch signal is also used to disengage the cruise control system when the brakes are applied. It is good practice to test the dump valve any time the vehicle is in the shop for cruise control service.

To test the dump valve, disconnect the vacuum hose from the servo assembly to the dump valve. Connect a hand vacuum pump to the hose and apply vacuum to the dump valve. If vacuum cannot be applied, either the hose or the dump valve is defective.

Classroom Manual
Chapter 14, page 425

Special Tools

DMM

Jumper wires

Special Tools

Hand vacuum pump

If the valve holds vacuum, press the brake pedal. The vacuum should be released. If not, adjust the dump valve according to the service manual procedures. If the dump valve fails to release vacuum when the brake is applied and it is properly adjusted, it must be replaced.

Special Tools

Ohmmeter

Speed Sensor. Disconnect the six pin connector from the amplifier (Figure 14-28). Connect an ohmmeter between circuits 150 and 57A. The resistance should be approximately 200 ohms. If the resistance value is less than 200 ohms, check for a short in the circuits between the amplifier and the speed sensor. If there is no problem in the wiring, the coil in the sensor is shorted.

If the resistance value is infinite, there is an open in the wires or in the sensor coil.

To test the sensor separate of the wiring harness, disconnect the wire connector from the sensor and connect the ohmmeter between the two terminals. This test should be used after testing at the amplifier connector to determine if there is a fault in the entire circuit.

Component Replacement

The two most common components to be replaced in the cruise control system are the servo and the switches. The following section covers replacement of these units.

Servo Assembly Replacement. Follow Photo Sequence 23 to replace the servo assembly.

Reverse the procedure to install the servo assembly. To adjust the actuator cable, leave the cable adjusting clip off and pull the cable until all slack is removed. Maintain light pressure on the cable and install the adjusting clip. The clip must snap into place.

Switch Replacement. Switch removal differs depending on location. If the switch is a part of the multiple switch assembly on the turn signal stock, refer to the service manual section for removing this switch. The following is a common method of switch replacement for switches contained in the steering wheel.

WARNING: Follow the service manual procedure for disarming the air bag system before performing this task. Failure to disarm the air bag system may result in accidental deployment and personal injury.

With the air bag system properly disarmed, remove the air bag module. Disconnect the electrical connections to the switch assembly. Remove the screws that attach the switch assembly to the steering wheel. Then remove the switch.

To install the new switch assembly, position it into the steering wheel pad cover and attach the retaining screws. If the horn connectors had to be disconnected, attach them to the pad cover. Reinstall the air bag module and rearm the air bag system.

Memory Seat Diagnosis

Classroom Manual
Chapter 14, page 430

Special Tools

Test light
DMM
Scan tool
Service manual

If the seat motors fail to operate under any condition, test the motors and switches as outlined in Chapter 10. This section relates only to that portion of the system that operates the memory function.

Most modern memory seat systems provide diagnostics by use of a scan tool. Although not all systems store fault codes, the scan tool can usually be used to check inputs and perform actuator tests. If the seat operates properly when the power seat switches are used, but not the memory feature, the problem is either the inputs or the memory seat control module. Use the scan tool to check the inputs from all switches. Also, test the input from the keyless entry system if it is tied into the memory seats. If the scan tool indicates that the inputs are operating properly, check the connector from the control module to the motors. Also, check the powers and ground for the module. If these are good, replace the module.

Photo Sequence 23
Typical Procedure for Replacing the Cruise Control Servo Assembly

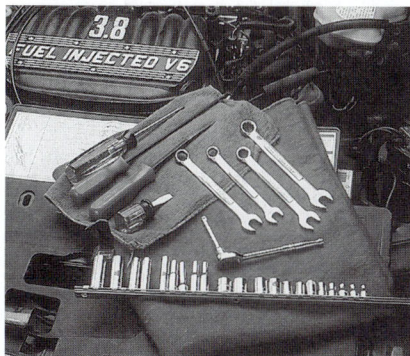

P23-1 Tools required to replace the servo assembly: fender covers, screwdriver set, combination wrench set, ratchet and socket set.

P23-2 Remove the retaining screws attaching the speed control actuator cable to the accelerator cable bracket and intake manifold support bracket.

P23-3 Disconnect the cable from the brackets.

P23-4 Disconnect the speed control cable from the accelerator cable.

P23-5 Disconnect the electrical connection to the servo assembly.

P23-6 Remove the two retaining bolts that attach the servo assembly bracket to the shock tower.

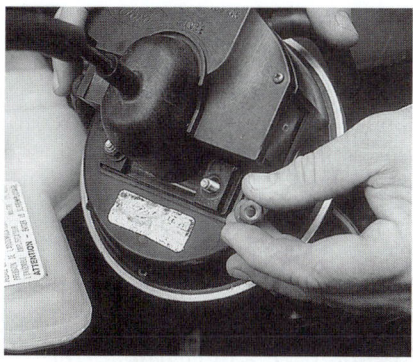

P23-7 Remove the two bolts that attach the servo assembly to the bracket.

P23-8 Remove the servo and cable assembly.

P23-9 Remove the two cable cover to servo assembly retaining bolts and pull off the cover.

Typical Procedure for Replacing the Cruise Control Servo Assembly (continued)

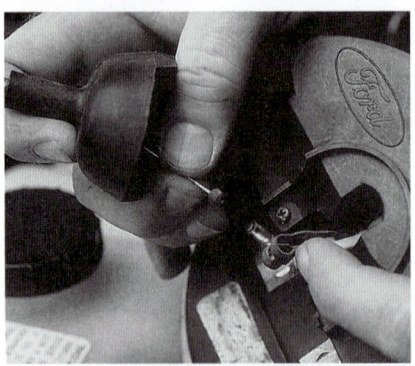

P23-10 Remove the cable from the servo assembly.

If the memory seat system does not support scan tool diagnostics, the problem must be isolated using basic electrical diagnostics. The following is a typical example of testing the system. You will need to reference the proper service information for the vehicle being diagnosed in order to determine what voltage values are used and which circuits to test.

Using the illustration (Figure 14-29) of a memory seat circuit, this system would be diagnosed as follows: All tests are performed at memory seat module connectors C1 and C3. The connectors are disconnected from the module to perform the tests. Place the ignition switch in the RUN position with the gear selector in the PARK position.

With the test light connected between C1 connector terminal B and ground, the lamp should illuminate. If the light does not come on, check for a circuit fault in the battery feed circuit. Connect the test light between terminals A and B at the C1 connector. If the test light fails to illuminate, check the ground circuit for an open. Move the test light between circuit 39 and ground. The light should turn on. If not, there is an open in circuit 39.

Connect the test light between terminal D of connector C3 and ground. If the light does not turn on, there is a problem in the neutral safety switch circuit. Check the adjustment of the neutral safety switch and circuits 75 and 275 for opens. With the test light connected between terminal B of the C3 connector and ground, the test light should remain off. An illuminated light indicates the left-seat switch assembly must be replaced.

Leave the test light connected between terminal B of the C3 connector and ground. Place the memory select switch in position 1. If the test light does not illuminate, check circuit 615 for an open. If the wire is good, use an ohmmeter to test the memory select switch for an open. Release the memory select switch and press the EXIT button. The test light should light. If the light fails to illuminate, there is a fault in the left-seat switch assembly. It therefore must be replaced.

Move the probe of the test light to terminal A of the C3 connector and press the exit button. If the light fails to illuminate, the problem is in circuit 616 or in the exit switch. With the test light still connected to the A terminal of the C3 connector, release the EXIT button. The test light should turn off. If the test light remains illuminated, replace the left-seat switch assembly.

Continue to leave the test light connected to terminal A. Place the memory select switch in the number 2 position. The test light should light. If not, replace the left-seat switch assembly.

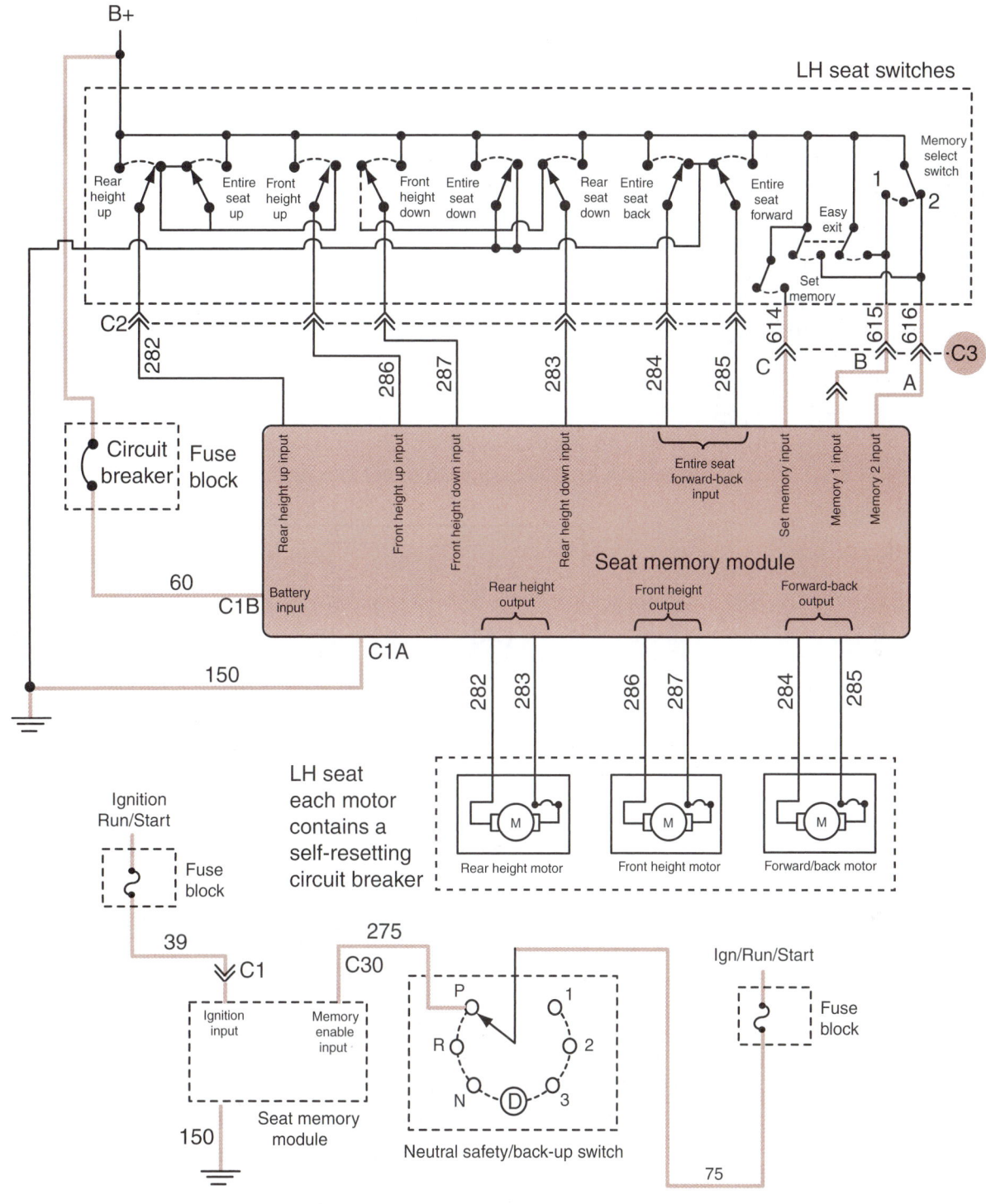

Figure 14-29 To diagnose the memory seat feature, a circuit schematic is required.

With the test light connected between C3 connector terminal C and ground and the memory select switch released, the test light should be off. If it remains on, the seat switch assembly is defective. Press the set memory switch. The test light should illuminate. If not, check the set memory switch and circuit 614 for an open.

If all the test results were correct, the fault is in the control module. This module must be replaced.

Climate-Controlled Seat Diagnosis and Service

Most climate-controlled seat systems can be diagnosed using a scan tool. The system will record DTCs for out-of-range voltage values that may indicate electrical shorts or opens. The system may also diagnose the input switches. Most climate-controlled seat systems rely on several modules to perform their functions. The data bus network is used to transfer the information between these modules. Be sure to include testing of the data bus when diagnosing the system. The following are typical diagnostic procedures for seats that are heated by a grid in the cushion and seat back (Figure 14-30). Note that this system uses a cushion and seat back grid that is wired in parallel. If the seat cushion warms but the seat back does not (or vice versa), the module and ground are not the fault. The fault would be in the shunt circuit to the inoperative grid or the affected grid itself. Some systems use a series connection between the seat cushion and seat back grids.

SERVICE TIP: Some manufacturers use a low-voltage cutoff feature that turns off the heated seat system if battery voltage drops below a specified value. Be certain to check the vehicle's electrical system for proper voltage anytime the power seat system appears inoperative, especially for intermittent operation complaints.

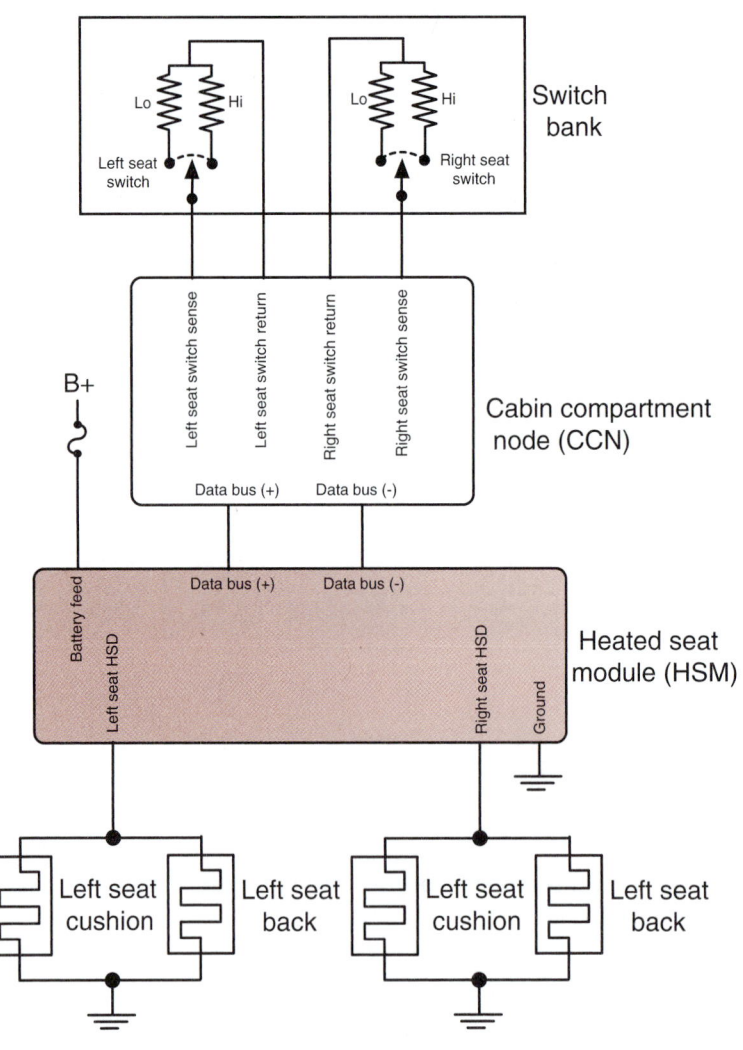

Figure 14-30 Heated seat system schematic.

Since this system uses HSDs to operate the heating elements, they will deactivate the system if a fault is detected. For example, if an open was detected in the circuit prior to any of the elements or on the common ground so that no current was flowing, the HSD will turn off the system. If the technician was to check for voltage from the heated seat module (HSM) at this time, he would read 0.0 volts. This may fool the technician into thinking the HSM is defective when in actuality it is not. For this reason, it is best to test for continuity through the heating element circuits using an ohmmeter.

To test the grids, access the wire harness connectors for the heating elements. Usually this connection is under the seat. With the seat element connector disconnect, measure the resistance between the circuits leading in and out of the inoperative heated seat element. Compare the reading to specifications. The grid wires of each element are wired in parallel so the measured resistance must fall within the specification range listed. A higher-than-specified reading indicates one or more grid wires are open, and a lower-than-specified reading indicates the grid wires are shorted.

> **SERVICE TIP:** When measuring the resistance of the heating element, move the heating element by putting pressure against the cushion. This will help to determine if an intermittent open or short in the element is present.

If the heating element is determined to be faulty, it can usually be replaced without having to replace the entire cushion. Some systems require the faulty grid be removed from underneath the seat trim cover and a new unit be attached. Other systems leave the old grid in place and apply the element directly on top of the old element. In this case, the seat trim cover is removed and the wires that lead from the inoperative heating element are cut off flush with the edge of the heating element. The new element is installed by peeling off the adhesive backing and sticking the element directly on top of the original heating element. It is important to take care not to fold or crease the element assembly while attaching it. If this occurs, premature failure of the new element may result.

If one or both seats fail to operate, diagnose the input switches for proper operation and transmission of data over the bus. Many systems do not wire the switches directly to the HSM. Instead they are hardwired to a close module and bussed over. Study the wiring diagram and service information for the system to determine what modules are involved. Diagnoses of the switches will be done through that module by use of the scan tool. A DMM can be used to check the integrity of the circuit between the switch and the module. Also, an ohmmeter can be used to test the switches.

If switch inputs are being received by the HSM and the elements are not faulty, the module is suspect. Most systems use one control module for both front seats. If one seat is operating but the other is not, then power and ground circuits are not the fault. However, if both seats fail to operate, check powers and grounds before faulting the module.

Electronic Sunroof Diagnosis

Troubleshooting the causes of slow, intermittent, or no sunroof operation is a relatively simple procedure. Unlike many systems, the sunroof operation is not usually integrated with other systems. Because the system stands alone, diagnostics are generally performed in the same manner as testing any other motor-driven accessory. The following is the diagnostic procedure used to troubleshoot the electronic sunroof system shown (Figure 14-31). In addition, refer to the diagnostic chart at the end of this chapter to determine the causes of other system malfunctions.

Slow sunroof operation may be caused by excessive resistance in the circuit or motor. Excessive resistance can be determined by performing a voltage drop test. Obtain the correct schematic for the vehicle being diagnosed and follow through the circuit to locate the cause of the excessive resistance. Resistance can also occur inside the motor as a result of brush wear, bushing wear, and corroded connections.

Classroom Manual
Chapter 14, page 435

Special Tools

Test light
DMM
Ammeter
Service manual

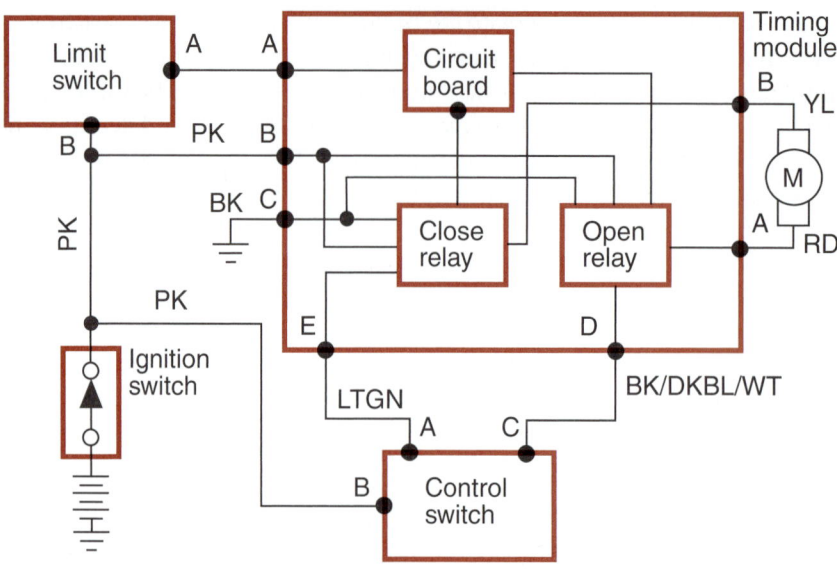

Figure 14-31 Electronic sunroof schematic.

Shorted armature or field coils can also result in slow motor operation. An ammeter can be used to test the current draw of the circuit to determine motor condition. If testing the motor and its circuits does not locate the cause of slow operation, then the problem is mechanical. Check the drive cable and tracks for any signs of wear or damage.

Intermittent problems may be the result of loose or corroded connections. To locate the cause of an intermittent fault, operate the system while wiggling the wires. This will assist in isolating the location of the poor connection. Some systems also use a circuit breaker to protect the motor. It may be overheating and tripping prematurely, or there may be resistance to window movement in the rails. The circuit breaker will trip, then cool down and reset. Check that the glass is able to move easily in the rails. In addition, many intermittent problems are caused by a faulty control switch. Operate the control switch several times while performing the circuit test. Replace the switch if it fails at any time during the test.

Follow the procedures listed in your service manual to locate an intermittent or no operation fault within the circuit. Perform the usual visual inspections of the circuit before continuing. Be sure to check for proper system grounds.

Antitheft System Troubleshooting

Classroom Manual
Chapter 14, page 440

As with many electrical systems, manufacturers take many approaches in designing their antitheft system. Most of the testing of relays, switches, and circuits require only basic electrical troubleshooting capabilities. Use the troubleshooting chart as a guide in locating the fault. Refer to the service manual for the correct procedure of arming the system you are diagnosing.

Self-Diagnostic Systems

Special Tools

Test light

DMM

Scan tool

Service manual

Some antitheft systems offer self-diagnostic capabilities. Follow the service manual procedures for the proper method of entering diagnostics for the vehicle you are working on. The following is a typical example of entering diagnostics.

Some vehicle theft security systems enter the diagnostic mode when the ignition switch is cycled three times from the OFF to accessory position. When the vehicle theft security system is in the diagnostic mode, the horn should sound twice and the parklamps and taillamps should flash. If the horn does not sound or the lights do not flash, voltmeter and ohmmeter tests are required to locate the cause of the problem.

The scan tester may be used to diagnose many vehicle theft alarm systems. Follow the scan tester manufacturer's recommended procedure to enter the vehicle theft alarm system diagnostic mode. When this diagnostic mode is entered, the horn may sound twice to indicate the trunk lock cylinder is in the proper position. When the key is placed in the ignition switch, the parklamps and taillamps should begin to flash.

The following procedures should cause the horn to sound once if the system is operating normally:

1. Activating the power door locks to the LOCKED and UNLOCKED positions.

2. Using the key to lock and unlock each front door.

3. Turning on the ignition switch.

When the ignition switch is turned on in procedure 3, the diagnostic mode is exited.

 CUSTOMER CARE: Always check the indicator lights in a customer's vehicle. These lights may be indicating a dangerous situation, but the customer may not have noticed them. For example, the vehicle theft security system set light may not be flashing when the normal system arming procedure is followed. This indicates an inoperative security system, and someone could break into the car without triggering the alarms. The customer has paid a considerable amount of money to have this system on the car. Therefore, it should be working. If this defect is brought to the customer's attention, he will probably have you repair the system and will appreciate your interest in the vehicle.

Alarm Sounds for No Apparent Reason

Mechanical and corrosion factors on the cylinder tamper switches can cause the system to activate for no apparent reason. If the customer complains of this condition, check the lock cylinder for looseness. Any looseness of the cylinder can cause the switch to activate the alarm.

Other causes of alarm system activation include loose, corroded, or improperly adjusted jamb switches. The switches should be adjusted to assure they remain in the OFF position when the doors are fully closed. The switch is adjusted by a nut located at the base of the switch.

Controller Test

To test the controller used in the illustration (Figure 14-32), disconnect the harness from the controller and the harness to the relay. Connect the test light between the N terminal and ground. The test light should illuminate to indicate voltage to the horns and controller.

CAUTION: Failure to disconnect both harnesses will lead to false test indications.

Move the test light probe to terminal M. The light should turn on only when the electrical door lock switch is moved to the UNLOCK position. Next, connect the test light between terminal B and a 12-volt source. The light should light only if the doors are locked. The light should go out if any doors are unlocked.

Probe terminal K for voltage. The test light should illuminate only when the ignition switch is placed in the RUN position. Check for a blown fuse or an open circuit if it does not light.

To test whether the cylinders are operating properly and have not been tampered with, connect the test light between terminal J and a 12-volt source. The test light should light only if a door is open. If it glows with the doors closed, inspect the lock cylinders for damage.

With the test light connected between terminal H and a 12-volt source, the test light should be on only when the outside door key is turned to the unlock position. Move the test light between terminal G and a good ground. The light should illuminate when the electric door lock switch is operated. This indicates there is electrical power to the switch. The test light should light in the LOCK position and go out in the UNLOCK position.

This system is also referred to as "Vehicle Security System (VSS)" and "Vehicle Theft Security System (VTSS)."

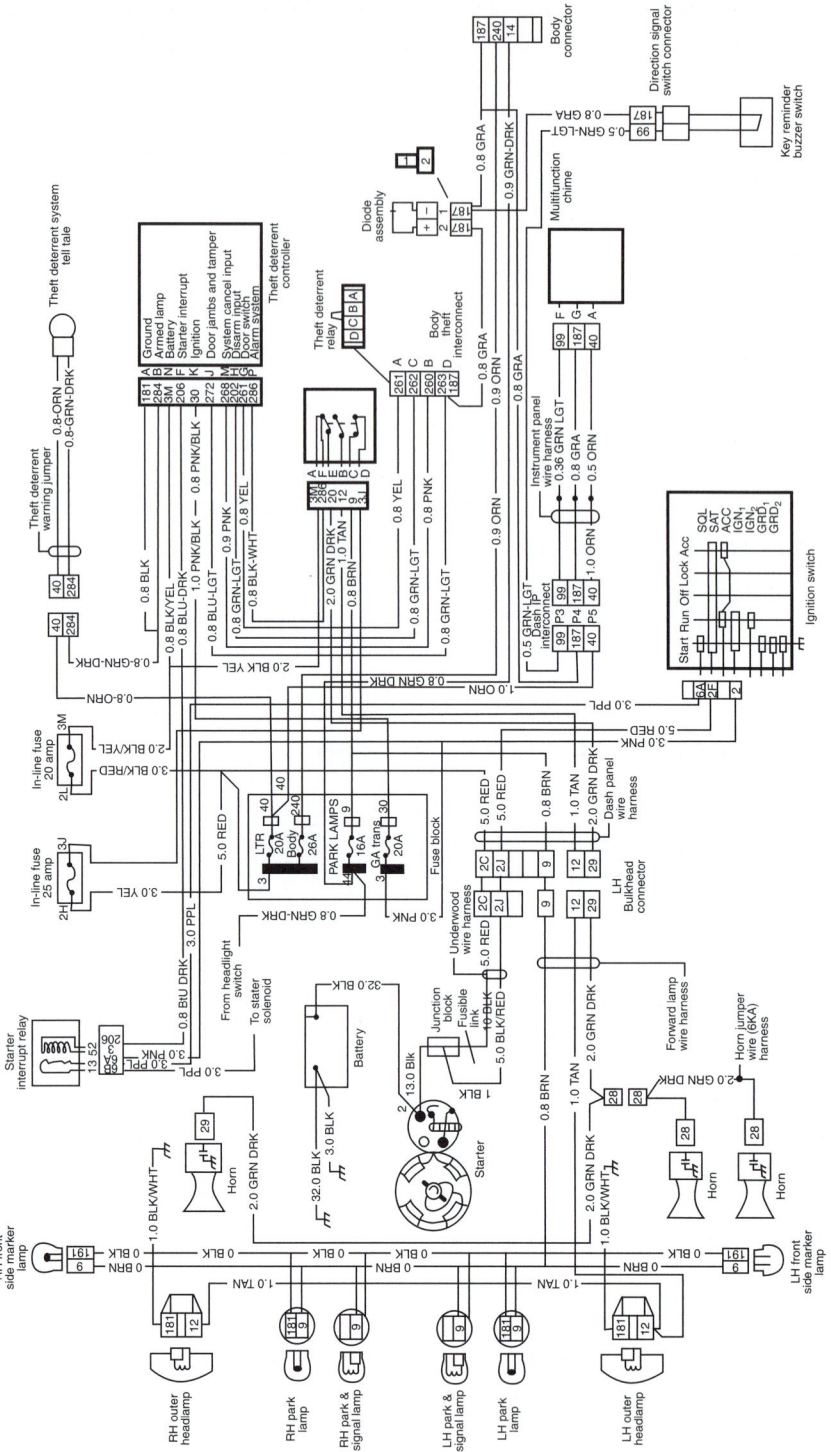

Figure 14-32 Circuit schematic of antitheft system.

Reconnect the relay harness. When the test light is connected between the F terminal of the controller connector and ground, the horn should sound and the lights should turn on. This indicates that the relay coil is functioning.

Next, turn the ignition switch to the RUN position with the test light connected between the E terminal of the controller connector and ground. Use a voltmeter to measure voltage to the starter. There should be zero volts. This indicates the starter interrupt relay is opening to prevent engine starting.

SERVICE TIP: In some instances, it may be easier to attempt starting the engine than to check for voltage at the starter. If the relay is working properly, the engine will not start.

Connect a jumper wire between terminal D and a good ground while observing the security warning light. The warning light should be on. Connect a test light between terminal A of the controller connector and a 12-volt power supply. The test light should light. If not, there is a problem in the ground circuit.

Relay Test

Faulty relays are a leading cause of antitheft system malfunction. Testing the relay is a simple matter of using a jumper wire to bypass the relay. If the circuits operate with the relay bypassed, but not with the relay connected, the relay is probably at fault. However, do not replace the relay until you have tested for the proper amount of applied voltage to the relay and for proper ground switching of the controller. Follow Photo Sequence 24 as a guide for relay testing.

Special Tools

Jumper wires

Service manual

Immobilizer System Service

A common service to the immobilizer system is the addition of keys. Usually the system can record several different key codes. However, the manufacturer may only provide two keys at the time of vehicle purchase. If additional keys are purchased, they will require programming into the system prior to use. Most manufacturers provide a method the customer can follow to program their new keys. This option is not available on all systems. Canadian systems do not allow for customer programming of keys. The following is one method that is used for customer programming of the keys.

In order to perform the customer-programming method, two valid keys must be used to enter the programming routine. The immobilizer module needs to see two unique key IDs before it will allow programming. One key cannot be used two times. Also, the key that is being programmed to the vehicle must be a blank key. If it has been programmed to another vehicle, it cannot be programmed again. If the sequence is not followed, the immobilizer will abort the programming process.

The customer-programming method is as follows:

1. Insert one of the two valid Sentry Keys into the ignition switch and turn the ignition switch to the RUN position. Leave the ignition switch in the RUN position for at least 3 seconds, but not more than 15 seconds.

2. Cycle the ignition switch to the OFF position and remove it.

3. Within 15 seconds, insert the second valid Sentry Key into the ignition switch and turn the ignition switch to the RUN position. Leave the ignition switch in the RUN position while observing the immobilizer warning lamp.

4. In about 10 seconds after the completion of step 3, the security indicator in the instrument cluster will start to flash and a single chime will sound. This indicates that the immobilizer system is in programming mode.

5. Within 60 seconds, turn the ignition switch to the OFF position and remove the key.

6. Insert the blank key into the ignition switch and turn the ignition switch to the RUN position.

7. In about 10 seconds, an audible chime will sound and the security indicator will stay illuminated solid for 3 seconds then turn off. This indicates that the new key has been successfully programmed.

8. Cycle the ignition switch to the OFF position before attempting to start the engine.

Classroom Manual
Chapter 14, page 443

Special Tools

Scan tool

DMM

Photo Sequence 24
Typical Procedure for Testing the Antitheft Relay

P24-1 The tools required to perform this task are a set of jumper wires and the proper service manual.

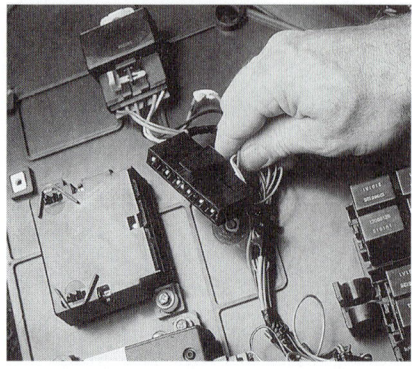

P24-2 Disconnect the relay wiring harness.

P24-3 Connect the jumper wire between terminals A and B. The horns should sound.

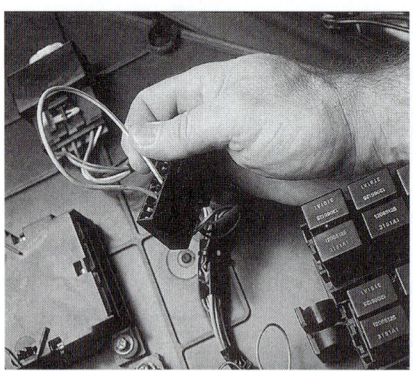

P24-4 Move the jumper wire to connect terminals D and C.

P24-5 The park, tail, and side marker lights should illuminate.

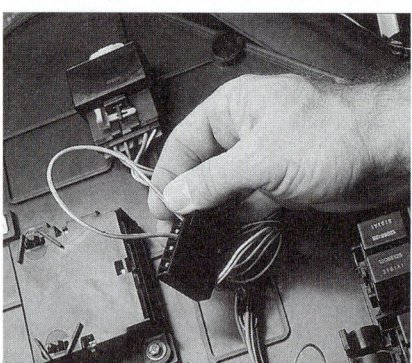

P24-6 Connect the jumper wire between terminals D and E.

P24-7 The low-beam lights should turn on.

P24-8 Reconnect the relay harness.

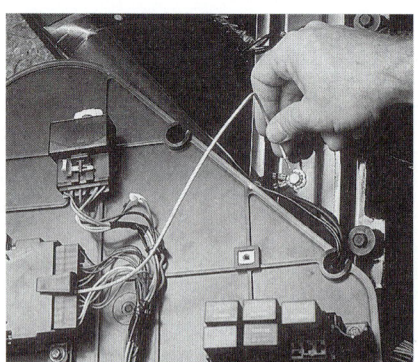

P24-9 Connect the jumper wire from terminal F to a good ground.

Typical Procedure for Testing the Antitheft Relay (continued)

P24-10 Listen for the relay to click. Then the light and horn circuits should be activated.

If additional keys are to be programmed, the entire procedure must be performed for each key. The system automatically exits programming mode after a blank key is programmed.

If the vehicle owner lost a key and needs to program a new one, he may not be able to do so unless he still has two valid keys. In this case, the technician will need to use the secure method of programming the new key. This method requires the use of a diagnostic scan tool. Also, a unique PIN code that is programmed into the immobilizer module will need to be obtained. This is a secure code that will need to be obtained from the vehicle owner, from the original vehicle invoice, or from the vehicle manufacturer.

Problems with the immobilizer system can result in engine no-start conditions. Since the system uses many hardwired components and circuits, these can be diagnosed and tested using normal diagnostic tools and procedures. However, to reliably diagnose the electronic message inputs used to provide the electronic features of the immobilizer system requires the use of a diagnostic scan tool.

If the diagnostics leads to replacement of the immobilizer module or the PCM, specific procedures may need to be followed. Usually the replacement module will require initialization. This will require the use of a scan tool and access to the PIN. This process will transfer the required data between modules so the system will be operative and the existing keys can still be used. After the initialization procedure is performed, the keys may need to be programmed to the new immobilizer module. However, since the secret code data matches, the same keys can be reused.

Automatic Door Lock System Troubleshooting

Some systems offer self-diagnostics through the body computer. The service manual will provide the steps required to enter diagnostics on these vehicles. The following is an example of locating the fault in vehicles that do not provide this feature when the door locks work but they do not lock or unlock automatically. As with any electrical diagnosing, you will need the circuit diagram for the system you are working on. The following steps relate to the system shown (Figure14-33):

1. Locate the controller and back probe for voltage at the power input terminal D with the ignition switch in the RUN position. If there is no voltage present, there is an open in circuit 39.

2. Back probe for voltage between terminals A and D. If there is no voltage, check for an open in circuit 150.

3. Make sure the courtesy lights are off and all doors are closed. With the gear selector in the PARK position, turn the ignition switch to the RUN position.

4. Connect a test light between controller terminal B and a good ground. If the neutral safety switch circuit is operating properly, the test light will light.

Classroom Manual
Chapter 14, page 444

Special Tools

DMM

Scan tool

Test light

Service manual

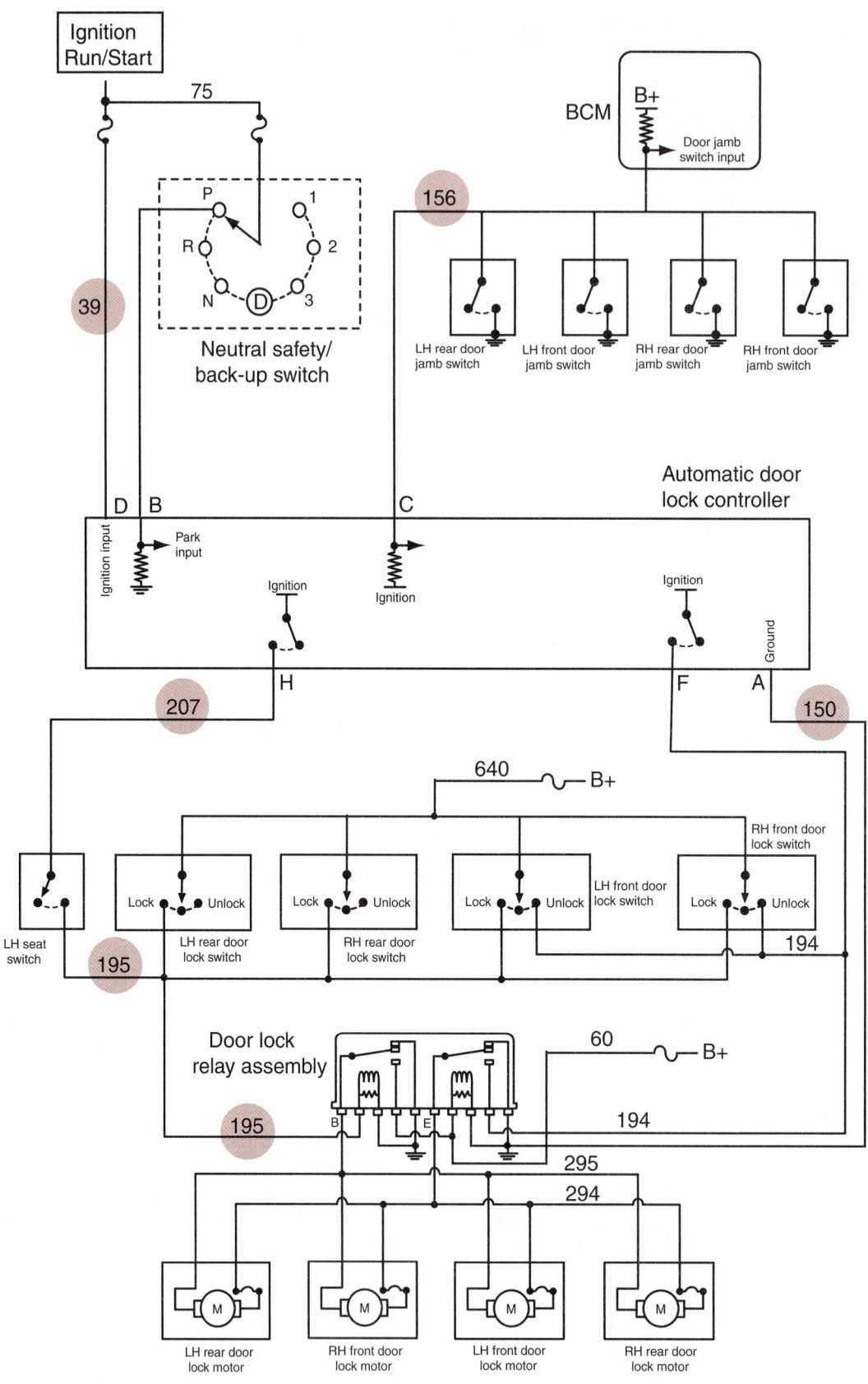

Figure 14-33 Automatic door lock schematic.

5. With the test light connected as in step 4, move the gear selector to any other position. The test light should go out. If the light does not go out, check the neutral safety switch. It may be out of adjustment or faulty.

6. Leave the gear selector as in step 5 and connect the test light between terminals C and D. The test light should not illuminate. If it does, check circuit 156 and the light switch and doorjamb switches.

7. Return the gear selector to the PARK position. Connect the test light between terminal H and ground.

8. Observe the test light while the gear selector is moved from PARK to REVERSE. The test light should flash once. If not, replace the controller.

If the circuits passed all tests, check circuits 207 and 195, and the left switch assembly for opens.

Remote Keyless Entry Diagnosis

Many new vehicles are equipped with a remote keyless entry system that is used to lock and unlock the doors, turn on the interior lights, and release the trunk latch. A small receiver is installed in the vehicle. The transmitter assembly is a hand-held item attached to the key ring (Figure 14-34). It has three buttons that control the functions of the system.

The system operates at a fixed radio frequency. If the unit does not work from a normal distance, check for two conditions: weak batteries in the remote transmitter or a strong radio transmitter close by (radio station, airport transmitter, etc.).

If the system has other problems, make sure the door locks, trunk latch, and interior lamps work normally when manually activated. If these systems check out fine, detailed diagnosis of the remote system is necessary. Follow the manufacturers' recommendations for doing this.

The transponder operation can be tested using an RF signal meter (Figure 14-35). By operating the transponder while the RF signal meter is on, the strength of the signal will be displayed by the LEDs. If the transponder fails to indicate a signal, test the transponder batteries. Usually two 3-volt batteries are used. Also be sure the batteries are installed correctly since they are polarity sensitive.

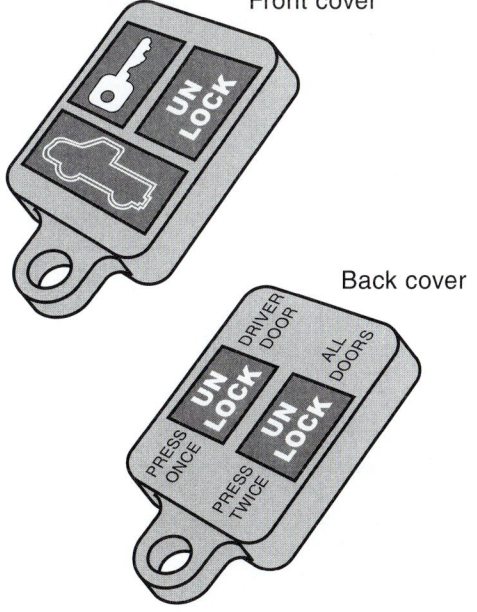

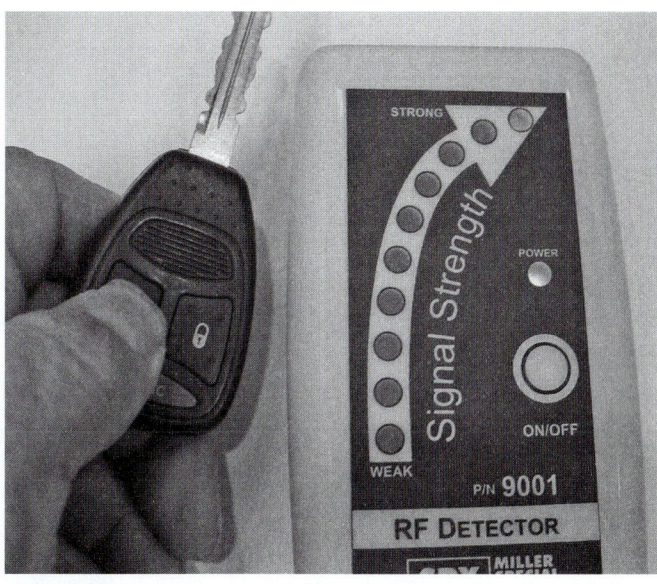

Figure 14-34 Typical door lock control transmitter assembly.

Figure 14-35 RF signal meter indicates the strength of the signal.

If the problem is in the basic electric door lock system, refer to Chapter 11 for diagnostic procedures. This section relates to only that portion of the system that controls automatic operation.

Classroom Manual Chapter 14, page 452

This system is usually called by its acronym RKE, pronounced "Rickie."

Electronically Heated Windshield Service

Special Tools

DMM
Fused jumper wire
Jumper wire

The two basic styles of heated windshields require different approaches to diagnostics. However, in either system, check the operation of the alternator, alternator belt condition, and the windshield for damage before beginning service. Also, the system will not operate if the battery state of charge is low.

During the course of diagnosing the heated windshield system, it will become necessary to override the temperature sensor. The Ford system will not activate unless the interior temperatures are less than 40°F (4.4°C). To override the system, connect a jumper wire between the black test lead pigtail and ground. The test lead is usually located in the engine compartment close to the wiper motor.

For the GM system to turn on, inside vehicle temperatures must be below 65°F (18.3°C). To override the internal thermistor, ground terminal C of the data link connector (Figure 14-36).

CAUTION: Perform the diagnostic test as fast as possible. Prolonged operation of the system at temperatures above 65°F (18.3°C) may cause permanent optical damage to the windshield.

The following is a service sample of the test procedures used to diagnose the GM-style system.

If the customer complains that the system does not turn on, verify this by starting the engine and pressing the activation switch. Observe the LED in the switch. The test procedure is determined by the attitude of the LED.

If the LED comes on for longer than 1/2 second but goes off again within 3 seconds, use the schematic (Figure 14-37) and follow these steps.

1. Measure the voltage at the data link connector terminal C. If it is within 2 volts of battery voltage, check for a short to battery voltage in the yellow wire. If the wire is good, replace the control module. If the voltage is not within 2 volts of battery voltage, continue testing.

2. Ground data link connector terminal C and start the engine. Press the activation switch to turn on the system. If the LED lights and the windshield heats, replace the control module. The internal thermistor is bad. However, it is not serviced separately from the control module.

CAUTION: Check the surface temperature of the windshield during testing. Turn off the system immediately if it gets too hot.

3. Turn the ignition switch to the OFF position. Then return it to the RUN position. Measure the voltage from terminal B6 of the control module. If the measured voltage is less than 11.2 volts, there is an open or short in circuit 2.

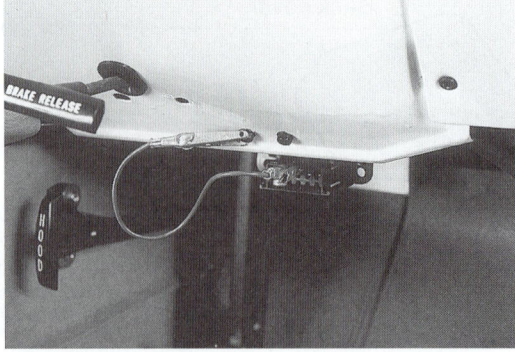

Figure 14-36 Jump the C terminal of the DLC to ground to bypass the thermistor.

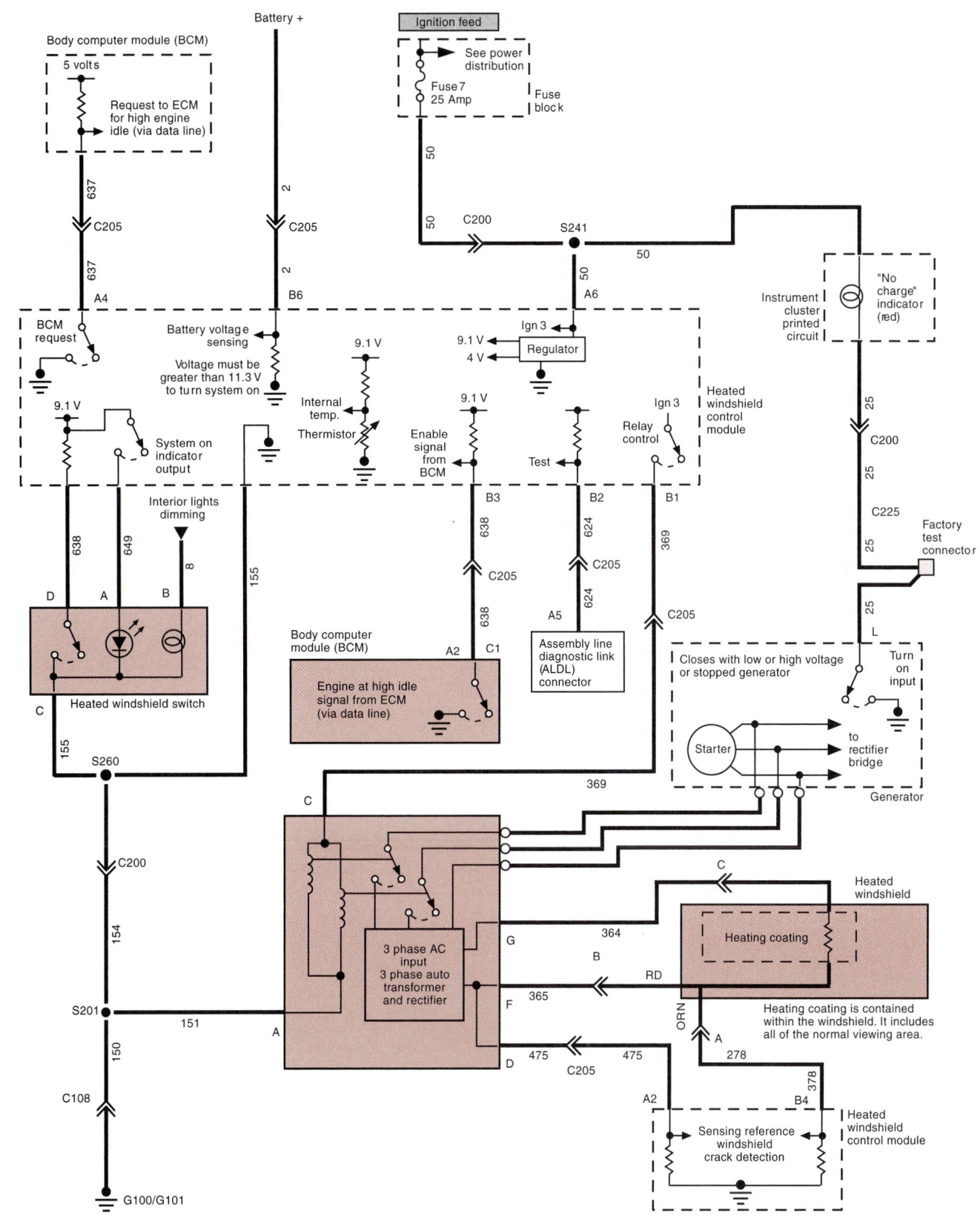

Figure 14-37 General Motors' heated windshield system.

4. With the ignition switch still in the RUN position, measure voltage at terminal A6. If battery voltage is not present, there is an open or short in circuit 50.

5. Measure the voltage between terminals A6 and A8 of the control module. If battery voltage is not present, measure the resistance between circuit 155 and ground. It must be less than 0.5 ohm.

6. With the ignition switch in the RUN position, measure voltage between terminals A6 and A3 while repeatedly pressing the activation switch. Zero volts should be indicated when the switch is pressed and 9.1 volts when it is released. Check circuit 648 for an open if the measured voltage is different than these values. If circuit 648 is good, check the continuity of the activation switch when it is in the released position. Also, check for a good ground connection at terminal C of the switch.

7. Turn the ignition switch to the OFF position and disconnect the windshield connector. Measure the resistance of the windshield at the connector terminals. The resistance between terminals A and B should be less than 10 ohms. It should be less than 6 ohms between B and C. If the measured resistance values are different, inspect the connector. If the connector is good, replace the windshield.

> **CAUTION:** Measure the resistance on the windshield side of the connector. Measuring resistance on the controller side of the connector may damage the controller.

8. Leave the windshield disconnected and measure the resistance of the windshield between each terminal of the connector to ground. All terminals should indicate 10,000 ohms or greater. If less than 10,000 ohms, check for shorts to ground between the windshield and the body.

9. Check circuit 475 for continuity between terminals D and A2 of the control module. Also check for continuity of circuit 378 between the windshield harness side connector terminal A and module terminal B4. If there is not continuity in either one of the circuits, repair the opens. If circuits 475 and 378 are good, replace the control module.

When confirming the complaint, if the LED remains on but the windshield does not heat, test the system as follows:

1. Start the engine and activate the system by pressing the switch.

2. Measure the three-phase voltage at the three posts on the back of the generator. Do not disconnect the connector at this time. Measure the voltage by back probing the connector. The voltmeter must be on a scale higher than 20 volts AC.

3. Refer to the diagram (Figure 14-38) and measure the voltage as follows:
X to Y
X to Z
Y to Z

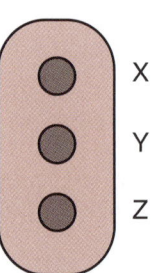

Figure 14-38 Generator connector terminal identification.

4. In all cases, the voltage should be between 9 and 14 volts. If the voltage is within specifications, go to step 6. If the voltage is not within these limits, turn the ignition switch to the OFF position and disconnect the three wires from the generator. Repeat the test again. Replace the generator if the voltage is still not within the limits. If the voltage is between 9 and 14 volts, check the wires from the generator to the power module.

5. Disconnect the windshield connector and measure the resistance between terminals B and C. If the resistance is less than 3 ohms, replace the power module. If the resistance is more than 3 ohms, replace the windshield.

6. Place the ignition switch in the OFF position and disconnect the control module.

7. Connect a fused jumper wire between pin B1 of the control module connector and battery positive.

8. Start the engine and back probe between pins B and C of the connector with a voltmeter. Use 100-volt or greater DC scale. If the voltage is between 50 and 85 volts, replace the control module.

9. With the seven pin connector to the control module disconnected, check to see if circuit 369 has continuity between the control module and terminal B1. Also, check that power module terminal C is free of shorts and grounds.

10. Check that circuit 151 has continuity between power module terminal A and ground.

11. If either circuit 369 or 151 have an open, repair as necessary.

12. If both circuits are good, measure the resistance between terminals G and F of the power module connector. If the resistance is less than 6 ohms, replace the power module. If the resistance is greater than 6 ohms, there is an open in circuit 364 or 365.

When confirming the customer complaint, if the LED illuminates for about 3 seconds and then turns off, test as follows:

CAUTION: It is normal for the LED to turn on and then off after 3 seconds if the ignition switch is in the RUN position but the engine is not started. Be sure to verify the complaint with the engine running.

1. Start the engine and allow it to warm (engine off high idle).

2. Activate the system. If internal temperatures are above 65°F (18.3°C), jump DLC terminal C to ground.

3. The engine idle speed should increase within a few seconds of activating the system. Go to step 4 if the engine does not increase speed. Go to step 7 if the engine speed increases.

4. Turn the ignition switch off and disconnect the control module.

5. Start the engine. Ground terminal A4 of the module connector with a fused jumper wire.

6. If the engine speed increases, there is a problem with the BCM or ECM. Follow the service manual procedures for diagnosing these units. If the engine speed does not increase, replace the control module.

7. With the engine running, back probe terminal B3 of the control module connector with a voltmeter.

8. Activate the system while observing the voltmeter. The voltage should drop from battery voltage to less than 4 volts in approximately 3 seconds. If the voltage fails to decrease, check the wire between A2 of the BCM and B3 of the control module. If the wire is good, the BCM or ECM is defective.

9. If the voltage drops according to specifications, replace the control module.

Speed-Sensitive Steering Diagnostic

Classroom Manual
Chapter 14, page 462

Special Tools

EVO tester
Scan tool

The procedure for diagnosing electronic variable orifice (EVO) steering varies depending on system design. Many vehicles that also have the automatic ride control option incorporate the operation of both systems into one computer.

To diagnose the Ford EVO system with a separate controller, you must use an EVO tester (Figure 14-39) hooked up to the EVO diagnostic connector (Figure 14-40). An EVO tester can be fabricated (Figure 14-41). The connector side is constructed of connecter E9SB-14489-EA, two terminals E8UB-14474-BA, and two 18-inch lengths of 16-gauge wire. The lamp side of the tester is constructed of socket E2DB-13728-A, two terminals D4AB-14490-D, and a 194 bulb. Follow the procedures listed in the service manual for checking output actuator tests, steering wheel rotation sensor, and vehicle speed sensor with the EVO tester.

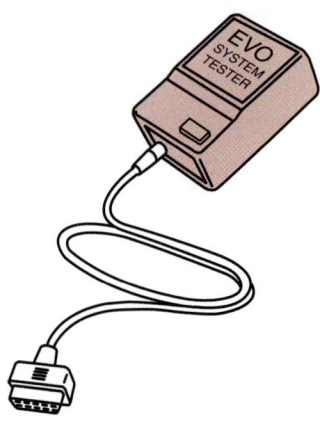

Figure 14-39 EVO system tester.

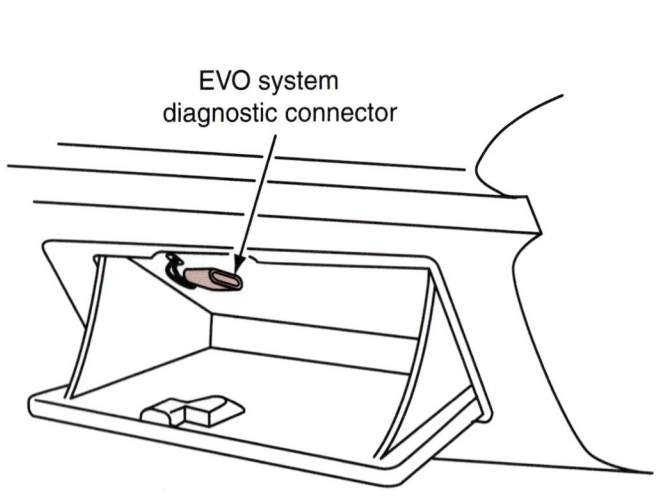

Figure 14-40 EVO diagnostic connector location.

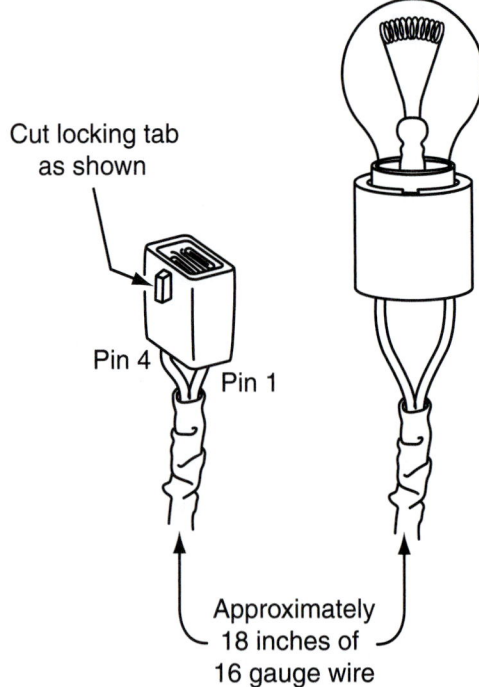

Figure 14-41 Fabricating a service diagnostic lamp.

Systems that incorporate the steering and automatic ride control (ARC) module into the same shared controller, use a "firm ride" lamp to indicate a failure in either system. The lamp will illuminate whenever the system is in the firm ride mode. If there is a trouble code stored in the controller, the code will flash on the lamp when the ignition switch is placed in the RUN position. A SUPERSTAR II tester (or equivalent) can be used to perform a quick test.

Antilock Brake Service

Classroom Manual
Chapter 14, page 466

Antilock brakes (ABS) have become very popular on today's vehicles. There are many different systems used and each has its own service safety requirements. Become familiar with the warnings and cautions associated with the system you are working on by studying the service manual before performing any service.

Certain components of the ABS system are not intended to be serviced individually. Do not attempt to remove or disconnect these components. Only those components with approved removal and installation procedures in the manufacturer's service manual should be serviced.

Some operations require that the tubes, hoses, and fittings be disconnected. Some ABS systems use hydraulic pressures as high as 2,800 psi (193 bar) and an **accumulator** to store this pressurized fluid. An accumulator is a gas-filled pressure chamber that provides hydraulic pressure for ABS operation. Before disconnecting any lines or fittings, the accumulator must be fully depressurized. The following is a common method of depressurizing the ABS system. However, follow the service manual procedures for the vehicle you are working on.

1. Place the ignition switch in the OFF position.
2. Pump the brake pedal between 25 and 50 times.
3. There should be noticeable change in pedal feel when the accumulator is discharged.

Troubleshooting

Troubleshooting of any ABS system begins with a good prediagnosis inspection. Often the cause of the fault can be found in this step. Failure to perform this step can lead to time-consuming activities. Use the illustration (Figure 14-42) as a guide in performing the inspection. The visual inspection involves checking the easily accessible components. It should include the following:

1. Check the master cylinder fluid level.
2. Inspect all brake hoses, lines, and fittings for signs of damage, deterioration, and leakage.
3. Inspect the wheel speed sensors for broken teeth and misadjusted gap.
4. Inspect all electrical connections for signs of corrosion, damage, fraying, and disconnection.

Also, remember that normal braking components that fail may cause the ABS system to shut down. Do not be too quick to condemn the ABS system.

After the visual inspection is completed, test drive the vehicle to evaluate brake system performance. Accelerate to a speed of about 20 mph (32 kmh) and, then bring the vehicle to a stop using normal braking procedures. Look for any signs of swerving or improper operation. Next, return the vehicle speed to approximately 25 mph (40 kmh) and apply hard braking pressure. You should feel the pedal pulsate if the ABS is working properly.

⚠ WARNING: Test the feel of the brake pedal as well as the operation of the brake system before driving the vehicle. It is important for you to be confident that the brake system is working sufficiently to stop the vehicle during the test drive.

ITEM	INSPECT FOR	CORRECTIVE ACTION
BRAKE FLUID RESERVOIR HYDRAULIC UNIT	– LOW FLUID LEVEL – EXTERNAL LEAKS	– FILL RESERVOIR – REPAIR LEAKS AS REQUIRED
PARKING BRAKE	FULL RELEASE	– RELEASE PARKING BRAKE – ADJUST CABLE OR VACUUM RELEASE VALVE IF REQUIRED
BATTERY	ADEQUATE CHARGE ("GREEN EYE")	– CHARGE OR REPLACE BATTERY AS REQUIRED – SERVICE CHARGING SYSTEM AS REQUIRED

FUSES
• ELECTRONIC BRAKE FUSE
  – 5 AMP
  – LOCATED IN FUSE PANEL
    (FUSE #5)
• MAIN RELAY FUSE
  – 30 AMP WITH RED WIRE
  – LOCATED IN FUSE HOLDER
    ON RELAY BRACKET
• PUMP MOTOR FUSE
  – 30 AMP WITH YELLOW WIRE
  – LOCATED IN FUSE HOLDER
    ON RELAY BRACKET
	BLOWN FUSE
	
	BLOWN FUSE
	
	BLOWN FUSE
		– REPLACE FUSE AND VERIFY OPERATION
		
		– REPLACE FUSE AND VERIFY OPERATION
		
		– REPLACE FUSE AND VERIFY OPERATION

CONNECTORS
• MAIN RELAY
  – CONNECTOR WITH 5 WIRES
    ATTACHED
• PUMP MOTOR RELAY
  – CONNECTOR WITH 4 WIRES
    ATTACHED
• PRESSURE SWITCH
• PUMP MOTOR
• MAIN VALVE
• VALVE BLOCK
• FLUID LEVEL SENSOR
  – 3-WIRE CONNECTOR
  – 2-WIRE CONNECTOR
• ELECTRONIC BRAKE CONTROL
  MODULE (EBCM)
	– PROPER ENGAGEMENT OF CONNECTOR
	
	– LOOSE WIRES IN CONNECTOR
		– PROPERLY ENGAGE CONNECTORS
		– REPAIR LOOSE WIRES

GROUNDS
• BODY GROUNDS
  – LOCATED ON STUD AT LEFT
    FENDER RAIL
• HYDRAULIC UNIT GROUND
  – ON HYDRAULIC UNIT
• ELECTRONIC BRAKE GROUND
  – ON GENERATOR BRACKET
	– LOOSE CONNECTIONS
	– BROKEN EYELETS
	– CORROSION
		– TIGHTEN
		– REPAIR WIRE OR EYELET
		– CLEAN CONTACT SURFACES

Figure 14-42 Prediagnostic inspection chart. Perform this check as the first step in diagnosing the ABS system.

Observe the attitude of the brake pedal during the test drive. You can determine the probable cause of a wide variety of brake system problems through interpretation of brake pedal feel.

Self-Diagnosis

All electronic ABS systems provide some form of self-diagnostics. Some systems require the use of special testers and scan tools in order to enter the self-diagnostic mode. Others will flash the code on the BRAKE or ABS warning lamp. All systems also run a functional test when the ignition switch is placed in the RUN position. If any malfunction is detected during this test, the ABS warning light is illuminated and the system is shut down. Whenever the ABS is shut down, the system reverts to normal brake operation.

Lamp Sequence Check. The **lamp sequence check** is used to determine problems by observing the operation of the warning lights under different conditions. The following is a typical method of using the lamp sequence check for those applications requiring this diagnostic approach.

1. While observing the BRAKE and ABS warning lights, turn the ignition switch to the RUN position. The BRAKE light should not turn on; the ABS light should illuminate for approximately 6 seconds and then turn off.
2. Place the ignition switch in the START position while observing both warning lights. Both lights should turn on.
3. When the engine starts, the brake light should turn off immediately. The ABS light will illuminate for approximately 6 seconds.
4. Test drive the vehicle at a speed above 20 mph (32 kmh) for a few minutes. During the drive, both test lights should remain off. If any one of the warning lights turns on, take note of the condition that may have caused it.
5. Stop the vehicle using normal braking pressures. Both lights should remain off during the stopping procedure.
6. Place the gear selector into PARK and observe the warning lights. They should both be off.

Functional Test. The **functional test** will allow the technician to retrieve trouble codes set and recorded by the computer. If the service manual directs you to perform a functional test instead of a lamp sequence check, follow the typical procedure listed and refer to the illustration (Figure 14-43).

The trouble code will determine the troubleshooting chart to be used to isolate the fault. Codes cannot be erased until all codes have been retrieved, all faults have been corrected, and the vehicle has been driven at speeds in excess of 18 mph (29 kmh). After service work is performed on the ABS system, repeat the previous test procedure to confirm that all codes have been erased.

Component Testing

Monitoring input and output signals is possible by using a scan tool or special ABS tester. These testers will allow the technician to see the inputs from the wheel speed sensors, switches used in the system, and the activation of output solenoids. Usually, some form of activation tests are available to allow the technician to check the outputs for proper operation. Some systems will even allow the technician to simulate an ABS stop once vehicle speed is below 5 mph (8 kmh) and the brakes are applied.

If the accumulator is discharged, both warning lights will illuminate until it is recharged by the pump motor (about 30 seconds). If this is encountered while performing step 1, allow the accumulator to recharge and the lights to turn off. Turn the ignition switch to the OFF position and wait 15 seconds before restarting the test.

Special Tools

Scan tool
Lab scope
DMM

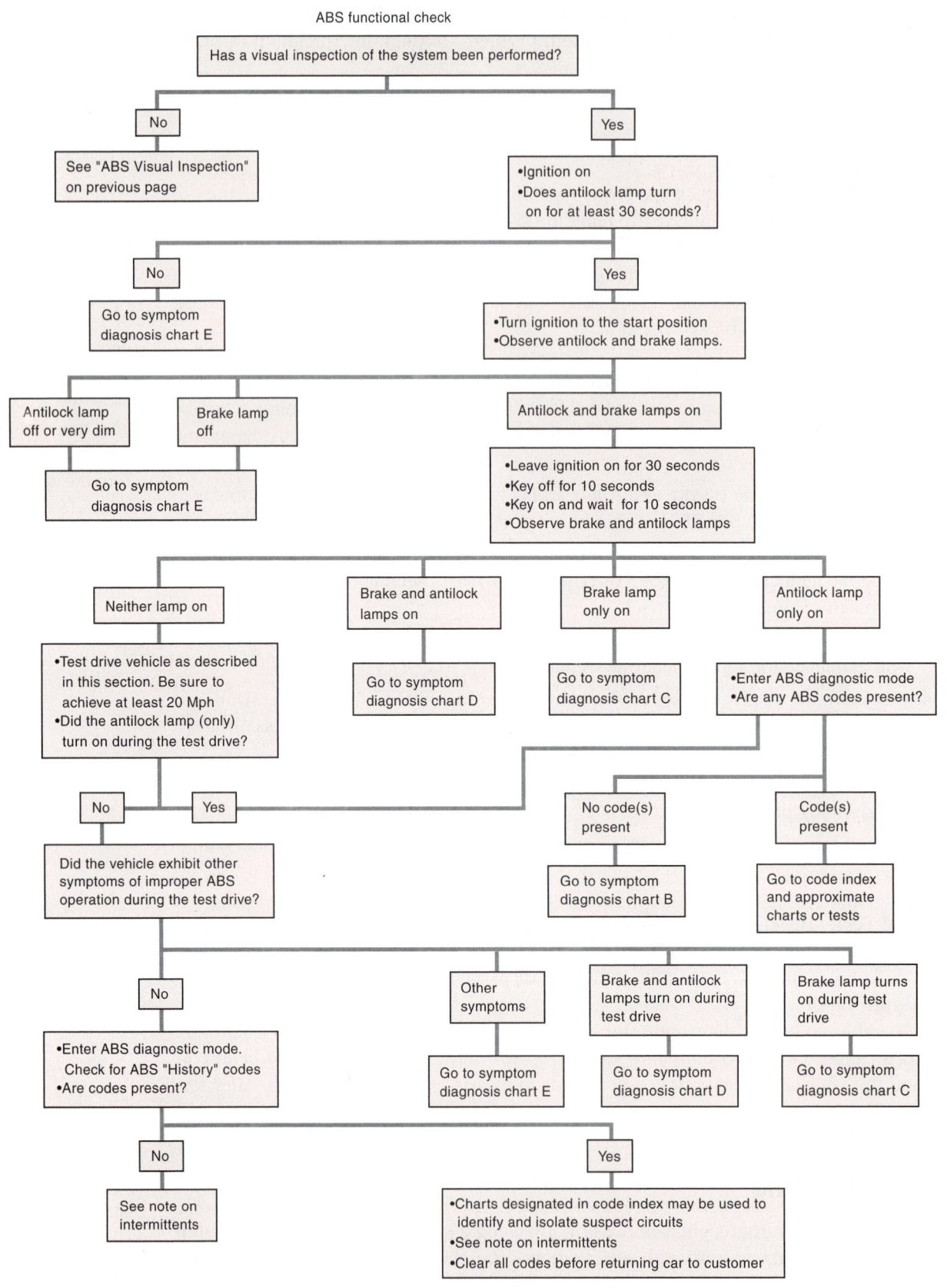

Figure 14-43 Function test flow chart of the GM Teves ABS system.

ABS functional check

Has a visual inspection of the system been performed?

No → See "ABS Visual Inspection" on previous page

Yes →
- Ignition on
- Does antilock lamp turn on for at least 30 seconds?

No → Go to symptom diagnosis chart E

Yes →
- Turn ignition to the start position
- Observe antilock and brake lamps.

Antilock lamp off or very dim / Brake lamp off → Go to symptom diagnosis chart E

Antilock and brake lamps on →
- Leave ignition on for 30 seconds
- Key off for 10 seconds
- Key on and wait for 10 seconds
- Observe brake and antilock lamps

Neither lamp on →
- Test drive vehicle as described in this section. Be sure to achieve at least 20 Mph
- Did the antilock lamp (only) turn on during the test drive?

Brake and antilock lamps on → Go to symptom diagnosis chart D

Brake lamp only on → Go to symptom diagnosis chart C

Antilock lamp only on →
- Enter ABS diagnostic mode
- Are any ABS codes present?

No code(s) present → Go to symptom diagnosis chart B

Code(s) present → Go to code index and approximate charts or tests

No / Yes →

Did the vehicle exhibit other symptoms of improper ABS operation during the test drive?

No →
- Enter ABS diagnostic mode. Check for ABS "History" codes
- Are codes present?

Other symptoms → Go to symptom diagnosis chart E

Brake and antilock lamps turn on during test drive → Go to symptom diagnosis chart D

Brake lamp turns on during test drive → Go to symptom diagnosis chart C

No → See note on intermittents

Yes →
- Charts designated in code index may be used to identify and isolate suspect circuits
- See note on intermittents
- Clear all codes before returning car to customer

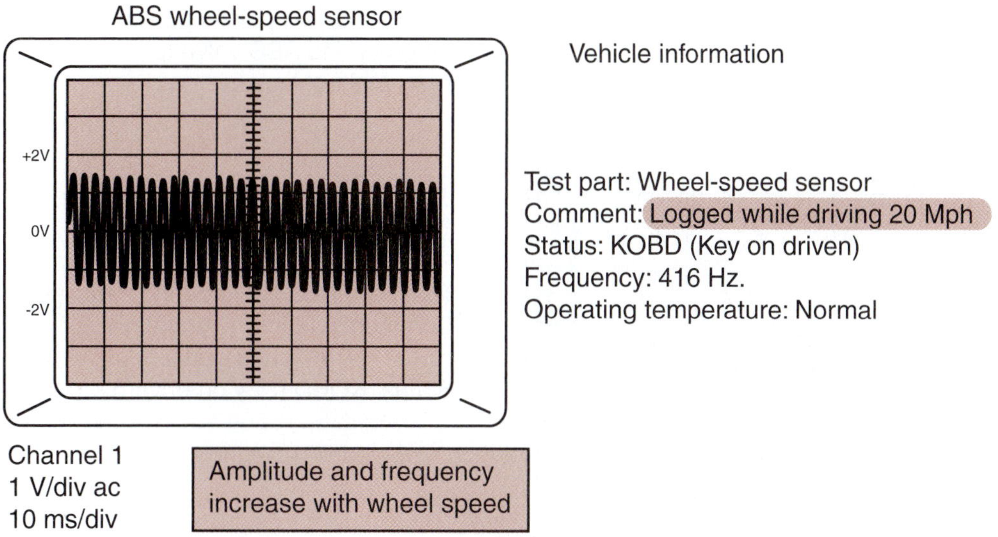

ABS wheel-speed sensor

Vehicle information

Channel 1
1 V/div ac
10 ms/div

Amplitude and frequency
increase with wheel speed

Test part: Wheel-speed sensor
Comment: Logged while driving 20 Mph
Status: KOBD (Key on driven)
Frequency: 416 Hz.
Operating temperature: Normal

Figure 14-44 Wheel speed sensor waveform pattern.

A lab scope is also an excellent instrument for testing the ABS system. The lab scope gives the technician the ability to see all the electrical signals and pulses as they occur. For example, the wheel speed sensor waveform can be used to check for proper signals (Figure 14-44). Defects such as improper gap width, broken tone wheel tooth, excessive resistance, out of round tone wheel, and so forth can be seen on the lab scope trace.

Most ABS wheel speed sensors are PM generator sensors that produce an AC sine wave as the wheel rotates. Recently, many Teves ABS systems were changed to a two-wire magnetoresistive sensor that produces a square wave pattern. The ABS controller sends a 12-volt reference signal on one of the wires while the other wire serves as both sensor signal and ground. The 12 volts power an integrated circuit (IC) in the sensor. The IC supplies the ABS controller with a constant 7 mA. The relationship of the tooth on the tone wheel to the permanent magnet in the sensor signals the IC to enable a second 7-mA supply. As the tone wheel rotates, the teeth shift the magnetic field, which results in a square waveform. When a valley of the tone wheel is aligned with the sensor, a voltage of about 0.90 volts and a current of 7 mA is sent to the controller. When a tooth aligns, the voltage will rise to 1.65 volts and 14 mA. The ABS computer measures the amperage of the digital signal and interprets the wheel speed. This sensor is used because it is more accurate at wheel speeds less than 3 mph (5 kmh).

The lab scope can also be used to monitor the solenoids as they are activated (Figure 14-45). As soon as an ABS stop is initiated, the trace should indicate the pulsing of the solenoid.

Component Replacement

The electrical components of the ABS system are generally very reliable. Common electrical system failures are usually caused by poor or broken connections. Other common faults can be caused by malfunction of the wheel speed sensors, pump and motor assembly, or the valve block assembly. Replacement of most wheel speed sensors is done by first disconnecting the wiring harness to the sensor. The sensor has a retaining bolt that is removed; then the sensor can be slid out of the boss in the knuckle.

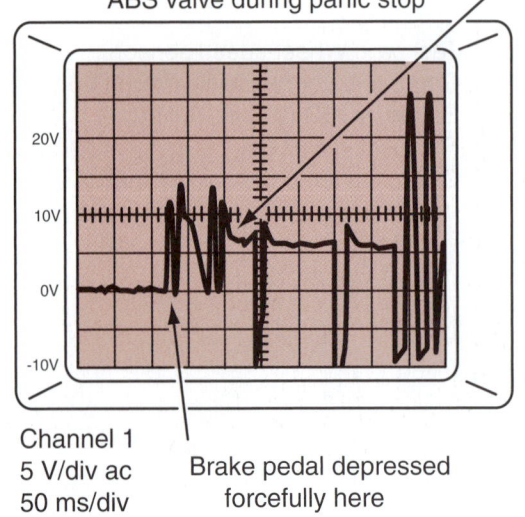

ABS control unit pulses power and ground to ABS valve tp prevent wheel lockup

ABS valve during panic stop

20V
10V
0V
-10V

Vehicle information
Test part: ABS valve during panic stop
Comment: Under full brake pressure in loose gravel
Status: KOBD (Key on driven)

Channel 1
5 V/div ac
50 ms/div

Brake pedal depressed forcefully here

Figure 14-45 As the solenoid valve is activated during an ABS stop, the waveform will depict the electrical activity.

The wheel speed sensor requires a specified gap between itself and the tone wheel in order to produce the proper signal. When installing the wheel speed sensor, this gap must be established (Figure 14-46). Not all vehicles allow for an adjustment. In these vehicles, a component is bent if the gap is not within specifications.

Pump and Motor Assembly Replacement. The service sample illustrated in Photo Sequence 25 is common to the GM Teves ABS system and is similar to other systems. However, it is intended only as an example and a basic guide. Refer to the manufacturer's service manual for the vehicle you are servicing for the exact procedure.

 WARNING: Failure to follow the manufacturer's exact service procedures may result in brake system failure or personal injury.

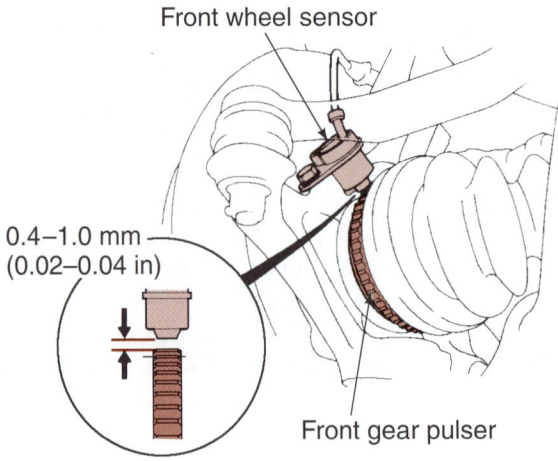

Front wheel sensor

0.4–1.0 mm
(0.02–0.04 in)

Front gear pulser

Figure 14-46. The wheel speed sensor must have the proper air gap in order to provide accurate signals.

Photo Sequence 25
Typical Procedure for Pump and Motor Assembly Replacement

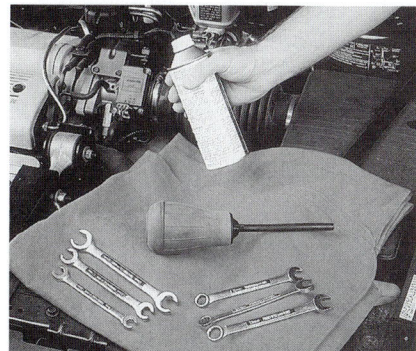

P25-1 To perform this task, you will need the following tools: fender covers, combination wrench set, line (flare) wrench set, syringe, and clean brake fluid.

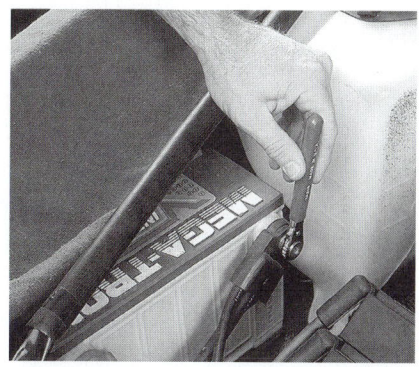

P25-2 Place the fender covers on the vehicle and disconnect the negative battery cable.

P25-3 Depressurize the accumulator by applying and releasing the brake pedal a minimum of 25 times. A noticeable change in pedal height and feel should occur when the accumulator is discharged.

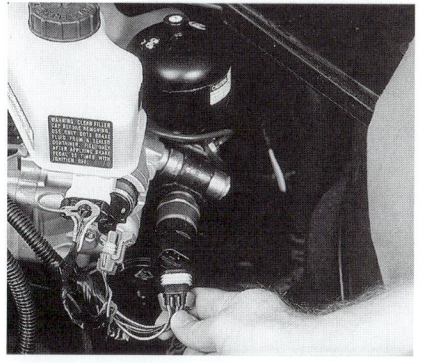

P25-4 Disconnect the electrical connector from the pressure switch and the electric motor.

P25-5 Use a clean syringe to remove the brake fluid from the reservoir.

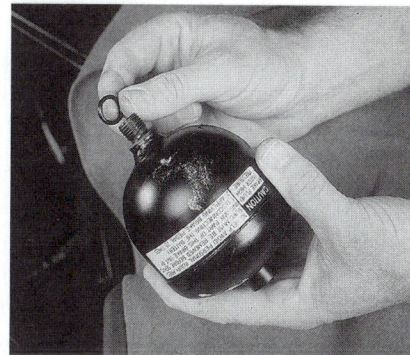

P25-6 Unscrew the accumulator from the hydraulic unit. Remove the O ring from the accumulator.

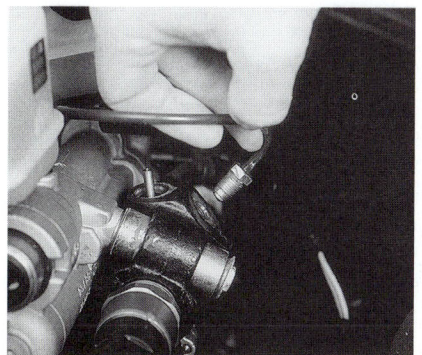

P25-7 Disconnect the high-pressure hose fitting connected to the pump.

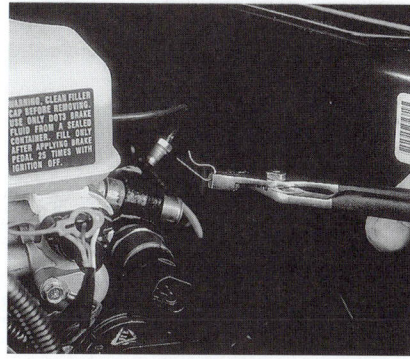

P25-8 Disconnect the wire clip, then pull the return hose fitting out of the pump body.

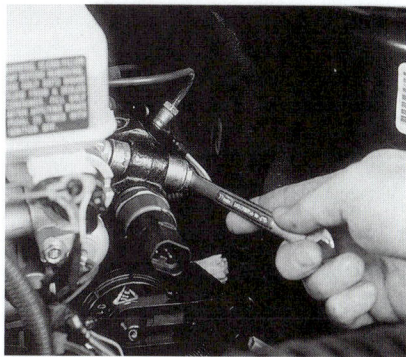

P25-9 Remove the bolt that attaches the pump and motor assembly to the hydraulic unit.

Typical Procedure for Pump and Motor Assembly Replacement (continued)

P25-10 Remove the pump and motor assembly by sliding it off of the locating pin.

To install the replacement pump and motor assembly, reverse the procedure. Use clean brake fluid to lubricate the O rings before installing them. If any of the insulators are damaged or distorted, they must be replaced.

Valve Block Assembly Replacement

CAUTION: This service sample is similar to many systems. However, refer to the manufacturer's service manual for the system you are servicing.

To replace the valve block assembly, first disconnect the negative battery cable and then depressurize the accumulator. Follow the service manual procedure for removing the hydraulic unit from the vehicle.

Remove the three nuts and washers that retain the block to the hydraulic unit (Figure 14-47). Remove the valve block and O rings by sliding the valve block off the studs.

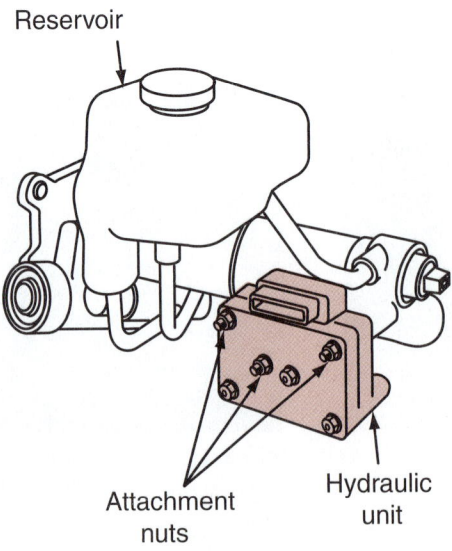

Reservoir

Attachment nuts

Hydraulic unit

Figure 14-47 To remove the valve body assembly from the hydraulic unit, remove the three nuts indicated.

To install the replacement valve block, lubricate the O rings with clean brake fluid and place the block assembly onto the hydraulic unit. Install the three retaining washers and nuts. Then reinstall the hydraulic unit into the vehicle following the service manual procedure. Follow the service manual procedure for bleeding the system.

Audio and Video Entertainment

Classroom Manual
Chapter 14, page 473

Today's audio/video systems integrate several different functions. They may include AM/FM radio, cassette player, CD player, DVD player, MP3 player, hands-free cell phone, and navigational systems. Most are diagnosed using a scan tool since the system will set fault codes when a failure is determined.

Because automotive technicians do not repair radios or system component units, there is a tendency to remove the unit when the customer has described having a particular problem before performing a thorough prediagnosis. In many cases, the units show "NO TROUBLE FOUND" and are sent back to the dealership or shop. Most of the problems could have been solved without taking the radio out of the dash.

Before removing the radio/component, do these simple checks to quickly determine whether the system problems are external:

❏ Test the vehicle's radio system outside, not inside a building. Make sure the hood is down.

❏ Most noise can be located on weak AM stations at the low-frequency end of the tuning band.

❏ Ignition noise on FM usually indicates a problem in the ignition system.

❏ If a test antenna is going to be used, the base must be grounded to the vehicle's body. DON'T HOLD THE MAST.

❏ Ninety percent of radio noise enters by way of the antenna.

❏ Most "rubber" hoses (vacuum, coolant, etc.) are electrically conductive, unless they have a white stripe.

❏ When shielding hoses, wires, the dash, etc., use foil or screening material. Be sure to ground the material.

❏ A weak or fading AM signal is normally caused by an improperly adjusted antenna trimmer (when used).

Special Tool

Scan tool

Diagnostic
    RF sniffer

The technician must determine the exact nature of the problem when performing a diagnosis. Determining whether the problem is intermittent or constant or whether the problem occurs when the vehicle is moving or stationary will help pinpoint the nature of the problem.

The diagnostic chart at the end of this chapter provides a guide to diagnosing radio performance complaints. Remember that the data bus is used for communications between the radio receiver and the amplifier. Be sure to confirm proper operation of this system.

Most complaints associated with the AM/FM radio are about poor reception. This is usually the result of an improper ground connection of the antenna. The base of the antenna provides a path to chassis ground. Resistance in any portion of the vehicle's ground path can affect the overall performance of the audio system. The following conditions should be checked whenever diagnosing a poor radio reception concern:

❏ Loose or corroded battery cable terminals.

❏ High-resistance body and engine grounds.

❏ Loss of antenna and audio system grounds.

When diagnosing the antenna, verify there is continuity between the antenna lead to the tip of the radio connector. Also verify there is no continuity between the antenna lead and chassis ground.

To inspect for the source of radio noise resulting from RFI or EMI, attempt to identify the component that is the source. For example, an ignition system problem can cause radio noise that will be engine speed related. Once the source is identified, the ground path and connections to that component should be checked. All ground connections must be verified prior to replacing any components. Some of the grounds and connections that need to be inspected include:

❏ Radio antenna base ground.

❏ Radio receiver chassis ground.

❏ Generator.

❏ Engine-to-body ground straps.

❏ Electric fuel pump.

❏ Ignition module.

❏ Heater core ground strap.

❏ Wiper motors.

❏ Blower motor.

❏ Exhaust system ground straps.

In addition, spark plugs and spark plug wire routing and condition should be checked.

A "diagnostic RF sniffer" tool can be made from an old piece of antenna lead-in from a mast or power antenna (Figure 14-48). This sniffer can be used, along with the radio, to locate "hot spots" that are generating radio frequency interference (RFI) noise. The noise can be found in wiring harnesses, in the upper part of the dash, or even between the hood and the windshield. When checking for noise on a wire, it is best to hold the sniffer parallel and close to the wire.

If the vehicle is equipped with an amplifier, speaker-related problems are diagnosed through the amplifier. For vehicles that are not equipped with an amplifier, speaker-related problems are diagnosed through the radio. Typical fault conditions that can be detected by the system are speaker circuits shorted to ground, shorted to voltage, shorted together, or open.

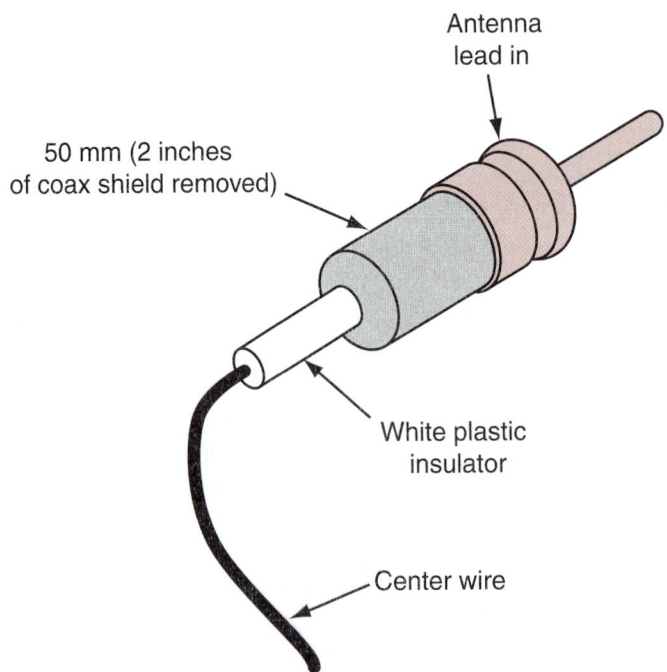

Figure 14-48 RF sniffer tool.

Like the radio, the CD system components generate DTCs for faults that can be retrieved using the scan tool. Use the diagnostic chart at the end of this chapter to assist in identifying common customer complaints for poor CD system performance.

The satellite radio system will also generate DTCs that can be read by the scan tool. The system will perform diagnostic system checks for audio output, presence of the antenna, antenna signal, and current subscription status.

If there is no satellite radio audio out of speakers, first check that the satellite radio subscription is initiated and that it has not expired. Also check to confirm that the satellite antenna is not blocked from the satellite field of view. It is best to have the vehicle in an area that provides an unobstructed line of sight to the entire horizon when diagnosing a no-reception complaint. If the subscription is active and the antenna is not obstructed, then a hardware failure is indicated. This will require repair or replacement of the satellite radio module or the vehicle's radio, amplifier, or wire harness as directed in the service manual diagnostic procedures.

Another concern may be the display of "Updating Channels" on the radio screen. This may indicate a hardware communication failure. Confirm the satellite radio module is communicating on the data bus.

CASE STUDY

The owner of a Continental Mark VII complains the rear of the vehicle raises to an excessive height then returns to normal while driving. The technician follows the service manual procedures for performing pinpoint test A. While in the diagnostic test mode, run tests 6 and 7 are initiated. Run test 6 indicates the left-front solenoid is operating properly. However, run test 7 fails to activate the right-front air spring solenoid. Following the instructions listed in the service manual, the rear solenoids are tested and both pass.

Next, the service manual directs the technician to perform pinpoint test E. It is determined that the right-front solenoid is faulty. The solenoid is replaced and the system operates properly.

The rear of the vehicle would raise to a high height because the front solenoid failed to open to service a front corner leveling request.

CASE STUDY

The owner of a 2005 Charger says the automatic door lock feature fails to operate. The technician checks all accessible connections.

Next, a scan tool is connected to the diagnostic tester and the output test is initiated. The door locks operate properly while in this test mode. The technician then initiates the switch test. The scan tool indicates the driver's-side door ajar switch is constantly reading open. The service manual instructs the technician to disconnect the door ajar switch connector. When disconnected, the scan tool indicates the switch is open. Next, a jumper wire is connected across the harness side connector. The scan tool still indicates the switch is open. According to the service manual, when the jumper wire is connected, the input state should indicate closed.

Voltage from the cabin compartment node (CCN) to the switch connector is tested. The voltage should be 12 volts, but 0 volts is measured. It is found that the wire between the

CCN and the connector is good. According to the service information, the CCN must be replaced since it is not sending out the 12-volt reference signal to the left-front door switch.

The resistance of the terminal lead from the CCN to the switch is measured. Per the service manual specifications, the resistance valve is too high and the CCN requires replacement.

Terms to Know

Accumulator	KOEO	Self-test input (STI)
Control assembly	KOER	Servo
Dump valve	Lamp sequence check	Servomotor
Functional test	Pinpoint test	

ASE-Style Review Questions

1. A customer says there is no airflow from the A/C outlets.
Technician A says the control assembly could be faulty.
Technician B says the ambient temperature sensor is open.
Who is correct?
A. A only
B. B only
C. Both A and B
D. Neither A nor B

2. A customer states the air entering the vehicle is warmer than the selected temperature. All of the following could be the cause EXCEPT:
A. Faulty in-vehicle temperature sensor.
B. Faulty evaporator temperature sensor.
C. Faulty mode door actuator.
D. Faulty blend door actuator.

3. A customer states that her cruise control system is inoperative.
Technician A says the dump valve may be stuck open.
Technician B says the servo may be faulty.
Who is correct?
A. A only
B. B only
C. Both A and B
D. Neither A nor B

4. On a vehicle equipped with an immobilizer system, the engine starts for a few seconds and then dies. What is the LEAST likely cause of this?
A. An invalid key was used.
B. A misadjusted door jam switch.
C. The immobilizer module has not been initialized.
D. Lost communication between immobilizer module and PCM.

5. A customer states that the vehicle alarm will trip when there is no apparent attempt of entry.
Technician A says the fault may be a loose lock cylinder.
Technician B says a misadjusted jamb switch may be the cause.
Who is correct?
A. A only
B. B only
C. Both A and B
D. Neither A nor B

6. A customer says the power seats are not operating.
Technician A says the problem is in the memory seat feature.
Technician B says the switch assembly may be faulty.
Who is correct?
A. A only
B. B only
C. Both A and B
D. Neither A nor B

7. A sunroof is opening slower than normal and appears to be "jerky" in its operation. The most likely cause of this problem is:
A. A shorted armature or field coil in the sunroof motor.
B. A stuck open limit switch.
C. A stuck closed limit switch.
D. A stuck OPEN switch.

8. The engine fails to start.
Technician A says the interrupt relay may be defective.
Technician B says the alarm relay is faulty.
Who is correct?
A. A only
B. B only
C. Both A and B
D. Neither A nor B

9. ABS service safety is being discussed.
Technician A says all components of the system are individually serviceable.
Technician B says do not attempt to remove or disconnect any components that do not have approved procedures in the manufacturer's service manual.
Who is correct?
- **A.** A only
- **B.** B only
- **C.** Both A and B
- **D.** Neither A nor B

10. *Technician A* says weak and fading AM signals could be the result of a bad or ungrounded antenna.
Technician B says that most radio noise enters by way of the radio.
Who is correct?
- **A.** A only
- **B.** B only
- **C.** Both A and B
- **D.** Neither A nor B

ASE Challenge Questions

1. A DaimlerChrysler vehicle equipped with an ATC air conditioning system has an inoperative heater: when the temperature controls are set to maximum heat, only ambient air is discharged from the heater vents. An inspection of the heater case reveals that the servo motor is moving to the full-heat position when maximum heat is selected.
Technician A says that the heater core may be clogged.
Technician B says that the blend-air door or linkage may be faulty.
Who is correct?
- **A.** A only
- **B.** B only
- **C.** Both A and B
- **D.** Neither A nor B

2. A vehicle's cruise control system is inoperative. Which of the following will not cause this problem?
- **A.** Stuck closed brake pedal vacuum switch.
- **B.** Open stoplamp switch.
- **C.** Inoperative speedometer.
- **D.** Open cruise engagement switch.

3. Testing of the alarm relay is being discussed.
Technician A says if the system fails to operate with the relay bypassed, the relay is defective.
Technician B says the relay is generally tested using an ammeter.
Who is correct?
- **A.** A only
- **B.** B only
- **C.** Both A and B
- **D.** Neither A nor B

4. A completely inoperative GM automatic door lock system is being discussed; the door locks will not open or close manually or automatically.
Technician A says that there may be an open in the power feed circuit to the automatic door lock controller.
Technician B says that one or more doorjamb switches may be faulty.
Who is correct?
- **A.** A only
- **B.** B only
- **C.** Both A and B
- **D.** Neither A nor B

5. Radio system problems are being discussed.
Technician A says that on some radios a weak AM signal can be caused by an improperly adjusted antenna trimmer.
Technician B says that it is acceptable to locate radio wires near most rubber hoses.
Who is correct?
- **A.** A only
- **B.** B only
- **C.** Both A and B
- **D.** Neither A nor B

Job Sheet 49

49

Name _____ Date _____

ATC Diagnostics

Upon completion of this job sheet, you should be able to test the electronic automatic temperature control system using stand alone diagnostic routines and determine needed repairs.

ASE Correlation

This job sheet is related to the ASE Heating and Air Conditioning Certification Exam's content area: *Operating Systems and Related Control Diagnosis and Repair, Electrical;* tasks: Diagnose the cause of failures in the electrical control system of heating, ventilating, and A/C system; determine needed repairs.

Tools and Materials

A vehicle equipped with an electronic automatic temperature control system
Service manual

Describe the vehicle being worked on:

Year _____ Make _____ Model _____

VIN _____ Engine type and size _____

Procedure

Task Completed

1. Describe the procedure for entering the self-test.

_____

_____

_____

_____

_____

2. Activate the self-test routine. What indications are provided to confirm that self-diagnostics have been entered?

_____

_____

_____

3. Perform a display panel segment test and record the results.

_____

_____

4. Retrieve DTCs and record.

_____

_____

5. Are the fault codes current or historic codes? _____

6. Perform all diagnostic tests the system is capable of running.

7. Were any problems found? ☐ Yes ☐ No
If yes, use the service information, trace the cause of the fault, and record your conclusions and recommendations.

_____

_____

8. Repair the system and run the self-test again. Was the repair successful?
☐ Yes ☐ No
If no, consult your instructor.

9. Record the procedure for clearing fault codes.

_____

_____

10. Follow the procedure to erase the fault codes. Was the procedure successful?
☐ Yes ☐ No

Instructor's Response _____

_____

_____

Job Sheet 50

50

Name _____ Date _____

Testing the Electronic Cruise Control System

Upon completion of this job sheet, you should be able to test the electronic cruise control system using a scan tool or stand alone diagnostic routines and determine needed repairs.

ASE Correlation

This job sheet is related to the ASE Electrical/Electronic System Certification Exam's content area: *Accessories Diagnosis and Repair;* tasks: Diagnose the cause(s) of unregulated, intermittent, or no operation of cruise control. Inspect, test, and repair switches, relays, and the electronic controller of the cruise control circuits.

Tools and Materials

A vehicle equipped with electronic cruise control
Scan tool
Service manual

Describe the vehicle being worked on:

Year _____ Make _____ Model _____

VIN _____ Engine type and size _____

Procedure

Task Completed

1. Does the system have a stand-alone computer or is the system a part of the PCM functions?
   ☐ STAND-ALONE   ☐ PCM

2. If possible, perform the simulated road test and record your results.

   _____

   _____

3. Are there any other related symptoms?   ☐ Yes ☐ No

   If yes, describe the symptom.

   _____

   _____

4. Refer to the proper service information to determine if a self-diagnostic routine can be performed. If so, describe how to enter diagnostics.

   _____

   _____

   _____

   _____

5. Is the self-diagnostic routine capable of displaying fault codes?   ☐ Yes ☐ No

6. Perform the procedure listed in step 4 and record the results and any DTCs:

_____

_____

7. Use a scan tool and access the electronic cruise control system. Does the scan tool indicate that DTCs are present?   ☐ Yes ☐ No

If yes, record the DTCs.

_____

_____

8. Based on the results so far, what tests need to be performed to find the cause of the fault?

_____

_____

9. Perform the tests listed in step 8. What is your determination and recommendation?

_____

_____

Instructor's Response _____

_____

_____

_____

Job Sheet 51

51

Name _____ Date _____

Testing the Antitheft System Operation

Upon completion of this job sheet, you should be able to test the antitheft system operation through the self-test diagnostic routine and determine needed repairs.

ASE Correlation

This job sheet is related to the ASE Electrical/Electronic System Certification Exam's content area: *Accessories Diagnosis and Repair;* tasks: Diagnose the cause(s) of false, intermittent, or no operation of antitheft systems.

Tools and Materials

A vehicle equipped with an antitheft system
Scan tool
Service manual

Describe the vehicle being worked on:

Year _____ Make _____ Model _____

VIN _____ Engine type and size _____

Procedure

Task Completed

1. Can the system self-diagnostic test be performed stand-alone or through the use of a scan tool?
   ☐ STAND-ALONE   ☐ SCAN TOOL

2. If it can be performed by stand-alone, describe how the process is performed.

   _____
   _____
   _____

3. If a scan tool must be used, list the screen menu selections required to perform the self-test.

   _____
   _____
   _____
   _____

4. Enter the diagnostic self-test and record your results.

   _____
   _____
   _____

5. Based on the results so far, what tests need to be performed to find the cause of the fault?

_____

_____

6. Perform the tests listed in step 5. What is your determination and recommendation?

_____

_____

Instructor's Response _____

_____

_____

Job Sheet 52

52

Name _____ Date _____

Testing an ABS Wheel Speed Sensor

Upon completion and review of this job sheet, you should be able to inspect and test an ABS wheel speed sensor with a graphing multimeter.

ASE Correlation

This job sheet is related to the ASE Brakes Test Content Area: *Antilock Brake System (ABS) Diagnosis and Repair;* task: Diagnose, service, and adjust ABS speed sensors and circuits following manufacturers' recommended procedures (including voltage output, resistance, shorts to voltage/ground, and frequency data).

NATEF CORRELATION

This job sheet is related to NATEF Brake Tasks: *General Brake Systems Diagnosis* and *Antilock Brake and Traction Control Systems;* task: Research applicable vehicle and service information, such as brake system operation, vehicle service history, service precautions, and technical service bulletins; locate and interpret vehicle and major component identification numbers (VIN, vehicle certification labels, calibration decals); test, diagnose, and service ABS speed sensors, toothed ring (tone wheel), and circuits using a graphing multimeter (GMM)/digital storage oscilloscope (DSO) (includes output signal, resistance, shorts to voltage/ground, and frequency data).

Tools and Materials

Service manual
Wiring diagram
Component locator
Graphing multimeter or oscilloscope
Lift or jacks with stands

Describe the vehicle being worked on:

Year _____ Make _____ Model _____

VIN _____ Engine type and size _____

ABS? yes _____ no _____ If yes, type _____

Procedure

	Task Completed
1. If vehicle is FWD, set transaxle to neutral and ignition to ACCESSORY position.	☐
2. Lift the vehicle until one of the front wheel sensors is accessible.	☐
3. Turn wheels to left or right for better access to the sensor.	☐
4. Locate and disconnect the speed sensor.	☐
5. Program the multimeter, if necessary, according to its operator's manual.	☐
6. Connect the multimeter leads to the terminals on the sensor.	☐

Task Completed

 SERVICE TIP: If the multimeter operator can stay clear and keep the leads clear of the wheel, a coworker can rotate the wheel using the engine.

NOTE: If the multimeter has a printer or data storage capabilities, use either or both to print or save the graphs.

7. Rotate the wheel at a constant speed while observing the graph.

   General results _____

   _____

   _____

   _____

8. Speed the wheel faster while observing the graph.

   General results _____

   _____

☐ 9. Stop the wheel(s).

☐ 10. After the wheel(s) have stopped, disconnect the multimeter and reconnect the harness to the sensor.

☐ 11. Lower the vehicle.

☐ 12. Shift the transaxle to park, switch the ignition to off, and set the parking brakes.

13. Record the operational action of this speed sensor and make any recommendation.

   _____

   _____

   _____

   _____

Instructor's Response _____

_____

_____

Job Sheet 53

53

Name _____ Date _____

Programming an Immobilizer Key

Upon completion of this job sheet, you should be able to determine operating characteristics of the immobilizer system and properly program keys.

ASE Correlation

This job sheet is related to the ASE Electrical/Electronic System Certification Exam's content area: *Accessories Diagnosis and Repairs;* task: Diagnose the cause of false, intermittent, or no operation of antitheft system.

Tools and Materials

Vehicle equipped with an immobilizer system
Extra keys
Scan tool
PIN (if required)
Service manual

Describe the vehicle being worked on:

Year _____ Make _____ Model _____

VIN _____ Engine type and size _____

Procedure

Task Completed

Task One

1. According to the service information, does the immobilizer system of the assigned vehicle support customer programming of extra keys? ☐ Yes ☐ No

If yes, describe the procedure.

_____

_____

_____

2. Obtain an extra key from your instructor and attempt to start the vehicle with the new key. Describe the results.

_____

_____

3. Perform the customer-programming method described in step 1 to program a new key. How do you know the key was successfully programmed?

_____

_____

Task Two

☐

1. Connect the scan tool to the DLC and access the immobilizer system.

2. If possible, perform a Module ID and record the following items:

   Version: _____ Part Number: _____ Country Code:_____

3. Does the immobilizer system on the assigned vehicle require the use of a PIN?
   ☐ Yes ☐ No

   If yes, obtain the PIN for the vehicle and record it: _____

4. How do you obtain the PIN?

   _____

   _____

5. If an incorrect PIN is entered or the correct procedure are not followed, are there any actions that must be done to reenter the programming mode? ☐ Yes ☐ No

   If yes, describe what must be done.

   _____

   _____

6. Is there a special procedure that must be followed if the PCM is replaced on the assigned vehicle? ☐ Yes ☐ No

   If yes, describe the procedure.

   _____

   _____

7. Is there a special procedure that must be followed if the immobilizer module is replaced on the assigned vehicle? ☐ Yes ☐ No

   If yes, describe the procedure.

   _____

   _____

8. Do the keys have to be reprogrammed after the SKIM is replaced? ☐ Yes ☐ No

☐

9. Use the scan tool to erase all current ignition keys.

10. Attempt to start the engine and record the results.

    _____

    _____

11. Use the scan tool to read DTCs and record.

    _____

    _____

12. Use the scan tool and access the function to program ignition keys and follow the instructions on the scan tool or in the service manual. Does the vehicle start?
    ☐ Yes ☐ No

☐

13. Clear any DTCs that were set.

Task Three

1. Wrap one of the programmed keys with aluminum foil and attempt to start the engine. Record your results.

_____

_____

2. Use the scan tool to read any DTCs in the system and record.

_____

_____

3. According to the service manual, what could cause the DTCs to set?

_____

_____

Instructor's Response _____

_____

_____

_____

DIAGNOSTIC CHART 14-1

PROBLEM AREA:	Automatic temperature control system operation
SYMPTOMS:	ATC system fails to maintain set temperature.
POSSIBLE CAUSES:	1. Faulty control head. 2. Faulty blend door actuator. 3. Faulty ambient temperature sensor or circuit. 4. Faulty in-car temperature sensor or circuit. 5. Faulty evaporator temperature sensor or circuit. 6. Faulty engine coolant control valve or solenoid. 7. Faulty fan motor or circuit. 8. Faulty control module. 9. High resistance in the control mode power feed circuit. 10. High resistance in control module ground circuit. 11. Problem with base A/C system.

DIAGNOSTIC CHART 14-2

PROBLEM AREA:	Cruise control operation
SYMPTOMS:	Cruise control speed changes over or below set requests.
POSSIBLE CAUSES:	1. Faulty servo. 2. Defective controller. 3. Faulty speed sensor. 4. Throttle linkage adjustment. 5. Faulty amplifier. 6. Faulty dump valve.

DIAGNOSTIC CHART 14-3

PROBLEM AREA:	Cruise control operation
SYMPTOMS:	Cruise control fails to set some of the time.
POSSIBLE CAUSES:	1. Faulty switch. 2. Faulty servo. 3. Defective controller. 4. Poor ground connection. 5. Poor control circuit connections. 6. Faulty relay. 7. Faulty speed sensor. 8. Poor or improper speed sensor circuit. 9. Faulty or misadjusted brake switch.

DIAGNOSTIC CHART 14-4

PROBLEM AREA:	Cruise control operation
SYMPTOMS:	Cruise control fails to set.
POSSIBLE CAUSES:	1. Faulty switch or circuit. 2. Faulty servo or circuit. 3. Defective controller. 4. Poor ground connection. 5. Poor control circuit connections. 6. Faulty relay or circuit. 7. Faulty speed sensor. 8. Faulty speed sensor circuit. 9. Throttle linkage adjustment.

DIAGNOSTIC CHART 14-5

PROBLEM AREA:	Memory seat system operation
SYMPTOMS:	Seat does not move to preset positions; power seat works normally.
POSSIBLE CAUSES:	1. Faulty motor position sensor. 2. Faulty switch. 3. Faulty control module. 4. Open control module power feed circuit. 5. Poor control module ground circuit. 6. Improper park/neutral switch input. 7. Bus communications error.

DIAGNOSTIC CHART 14-6

PROBLEM AREA:	Heated seat system operation
SYMPTOMS:	Seat fails to warm.
POSSIBLE CAUSES:	1. Faulty switch. 2. Open or short in circuit between switch and control module. 3. Open battery feed circuit to control module. 4. Open ignition feed circuit to control module. 5. Poor control module ground circuit. 6. Heated seat grid circuit open. 7. Heated seat grid circuit shorted. 8. Thermistor circuit open. 9. Thermistor circuit shorted. 10. Faulty control module. 11. Bus communications error.

DIAGNOSTIC CHART 14-7

PROBLEM AREA:	Sunroof operation
SYMPTOMS:	Sunroof opens or closes slower than normal.
POSSIBLE CAUSES:	1. Binding track. 2. Excessive circuit resistance. 3. Worn motor. 4. Misadjusted linkage. 5. Trim panel mispositioned. 6. Cable guides mispositioned. 7. Slipping motor clutch.

DIAGNOSTIC CHART 14-8

PROBLEM AREA:	Sunroof operation
SYMPTOMS:	Intermittent sunroof operation
POSSIBLE CAUSES:	1. Worn or defective motor. 2. Faulty controller. 3. Loose ground connection. 4. Poor control circuit connection. 5. Binding linkage and/or tracks.

DIAGNOSTIC CHART 14-9

PROBLEM AREA:	Sunroof operation
SYMPTOMS:	Sunroof fails to move in either direction
POSSIBLE CAUSES:	1. Worn or defective motor. 2. Faulty controller. 3. Loose ground connection. 4. Poor control circuit connection. 5. Binding linkage and/or tracks. 6. Faulty circuit breaker or fuse. 7. Trim panel mispositioned. 8. Cable guides mispositioned. 9. Slipping motor clutch.

DIAGNOSTIC CHART 14-10

PROBLEM AREA:	Antitheft system operation
SYMPTOMS:	System fails to arm or to operate some of the time.
POSSIBLE CAUSES:	**1.** Faulty controller. **2.** Defective switches. **3.** Poor wire connections. **4.** Poor ground connections.

DIAGNOSTIC CHART 14-11

PROBLEM AREA:	Antitheft system operation
SYMPTOMS:	System won't disarm.
POSSIBLE CAUSES:	**1.** Lock cylinder disarm switch loose. **2.** Open in the lock cylinder disarm switch circuit. **3.** Defective lock cylinder disarm switch. **4.** Defective lock cylinder. **5.** Faulty control module power and ground. **6.** Faulty control module.

DIAGNOSTIC CHART 14-12

PROBLEM AREA:	Antitheft system operation
SYMPTOMS:	System trips and sounds alarm by itself.
POSSIBLE CAUSES:	**1.** Doorjamb, hood, or trunk switch loose. **2.** Loose connection in doorjamb switch, hood switch, or trunk switch circuit. **3.** Defective doorjamb, hood, or trunk switch. **4.** Faulty control module.

DIAGNOSTIC CHART 14-13

PROBLEM AREA:	Antitheft system operation
SYMPTOMS:	System won't trip when door is opened.
POSSIBLE CAUSES:	**1.** Doorjamb switch circuit open. **2.** Faulty doorjamb switch. **3.** Faulty control module.

DIAGNOSTIC CHART 14-14

PROBLEM AREA:	Immobilizer system
SYMPTOMS:	Indicator lamp does not illuminate.
POSSIBLE CAUSES:	**1.** Faulty LED. **2.** Blown fuse. **3.** Faulty ground. **4.** Faulty battery feed. **5.** Faulty ignition feed.

DIAGNOSTIC CHART 14-15

PROBLEM AREA:	Immobilizer system
SYMPTOMS:	Indicator lamp flashes during bulb check. Vehicle started and dies.
POSSIBLE CAUSES:	**1.** Invalid key. **2.** Faulty key. **3.** Faulty antenna. **4.** Faulty module.

DIAGNOSTIC CHART 14-16

PROBLEM AREA:	Immobilizer system
SYMPTOMS:	Indicator lamp stays on after bulb check.
POSSIBLE CAUSES:	**1.** Immobilizer fault has been detected. **2.** Immobilizer system inoperative.

DIAGNOSTIC CHART 14-17

PROBLEM AREA:	Keyless entry system operation
SYMPTOMS:	Keyless entry fails to operate some of the time.
POSSIBLE CAUSES:	**1.** Defective key pad. **2.** Open circuit. **3.** Shorted circuit. **4.** Defective controller.

DIAGNOSTIC CHART 14-18

PROBLEM AREA:	Keyless entry system operation
SYMPTOMS:	Keyless entry fails to operate.
POSSIBLE CAUSES:	**1.** Defective key pad. **2.** Open circuit. **3.** Shorted circuit. **4.** Defective controller.

DIAGNOSTIC CHART 14-19

PROBLEM AREA:	Heated windshield
SYMPTOMS:	The heated windshield system does not operate.
POSSIBLE CAUSES:	**1.** No power to heated windshield switch. **2.** No power to heated windshield control module. **3.** No power to heated windshield. **4.** Inoperative AC generator output control relay. **5.** Cracked windshield.

DIAGNOSTIC CHART 14-20

PROBLEM AREA:	Antilock brake system operation
SYMPTOMS:	ABS warning light is on, no ABS operation.
POSSIBLE CAUSES:	**1.** Low brake fluid level. **2.** Faulty wheel speed sensor or circuit. **3.** Faulty accumulator. **4.** Faulty pump motor or circuit. **5.** Defective ABS system relay or circuits. **6.** Faulty brake level sensor or circuit. **7.** Faulty pressure sensor or circuit. **8.** Open or shorted inlet/outlet valve circuit. **9.** Faulty inlet/outlet valve. **10.** Open power feed circuit to control module. **11.** Open ignition feed circuit to control module. **12.** Poor control module ground. **13.** Faulty control module. **14.** Bus communications error.

DIAGNOSTIC CHART 14-21

PROBLEM AREA:	Audio system
SYMPTOMS:	Radio will not turn on.
POSSIBLE CAUSES:	**1.** Open power battery feed circuit. **2.** Open ignition feed circuit. **3.** Poor radio ground circuit. **4.** Loss of bus communications.

DIAGNOSTIC CHART 14-22

PROBLEM AREA:	Audio system
SYMPTOMS:	Radio will not produce sound.
POSSIBLE CAUSES:	1. Open power battery feed circuit. 2. Open ignition feed circuit. 3. Poor ground connection. 4. Defective radio. 5. Open speaker circuit. 6. Defective amplifier. 7. Open power feed to amplifier. 8. Poor amplifier ground circuit. 9. Bus communications error. 10. Stuck MUTE button.

DIAGNOSTIC CHART 14-23

PROBLEM AREA:	Audio system
SYMPTOMS:	No sound in AM or FM mode. CD audio operates normally.
POSSIBLE CAUSES:	1. Faulty antenna connection. 2. Poor antenna ground. 3. Faulty radio.

DIAGNOSTIC CHART 14-24

PROBLEM AREA:	Audio system
SYMPTOMS:	Excessive noise heard in AM audio.
POSSIBLE CAUSES:	1. Faulty antenna connection. 2. Poor antenna ground. 3. Faulty engine to chassis ground.

DIAGNOSTIC CHART 14-25

PROBLEM AREA:	Audio system
SYMPTOMS:	Poor radio reception
POSSIBLE CAUSES:	1. Faulty antenna connection. 2. Poor antenna ground. 3. Faulty radio.

DIAGNOSTIC CHART 14-26

PROBLEM AREA:	Audio CD system
SYMPTOMS:	Radio will not go into CD mode.
POSSIBLE CAUSES:	**1.** CD not inserted into player. **2.** Stuck MODE button. **3.** Faulty player.

DIAGNOSTIC CHART 14-27

PROBLEM AREA:	Audio CD system
SYMPTOMS:	CD will not eject.
POSSIBLE CAUSES:	**1.** CD adhesive label came loose. **2.** Warped CD. **3.** Faulty player.

DIAGNOSTIC CHART 14-28

PROBLEM AREA:	Audio CD system
SYMPTOMS:	CD will not play.
POSSIBLE CAUSES:	**1.** Faulty power feed circuit. **2.** Poor ground connection. **3.** Defective player.

DIAGNOSTIC CHART 14-29

PROBLEM AREA:	Audio CD system
SYMPTOMS:	CD will not insert.
POSSIBLE CAUSES:	**1.** Damaged edges of CD. **2.** CD surface defects. **3.** Wrong disc size. **4.** Faulty player.

DIAGNOSTIC CHART 14-30

PROBLEM AREA:	Audio CD system
SYMPTOMS:	CD skips or jumps tracks.
POSSIBLE CAUSES:	**1.** CD surface defects. **2.** Excessive vehicle vibration. **3.** Loose mounting bolts. **4.** Faulty player.

DIAGNOSTIC CHART 14-31

PROBLEM AREA:	Tire pressure monitoring system operation
SYMPTOMS:	Tire pressure light is on or display shows a low tire.
POSSIBLE CAUSES:	**1.** Low tire pressure. **2.** Faulty tire sensor. **3.** Faulty control module. **4.** Control module power feed circuit open or high resistance. **5.** Control module ground circuit fault. **6.** Bus communications error. **7.** Display module fault.

DIAGNOSTIC CHART 14-32

PROBLEM AREA:	Automatic suspension system operation
SYMPTOMS:	Vehicle is too low and compressor does not operate.
POSSIBLE CAUSES:	**1.** Blown fuse or circuit breaker. **2.** Excessive compressor current draw. **3.** Open compressor control circuit. **4.** Faulty compressor. **5.** Faulty relay or circuits. **6.** Faulty height sensors or circuits. **7.** Bus communications error.

DIAGNOSTIC CHART 14-33

PROBLEM AREA:	Automatic suspension system operation
SYMPTOMS:	Vehicle is too low, compressor does operate.
POSSIBLE CAUSES:	**1.** Air leaks in springs, lines, or compressor. **2.** Plugged or restricted air lines. **3.** Faulty height sensor or circuit. **4.** Faulty compressor. **5.** Faulty solenoid or circuit.

DIAGNOSTIC CHART 14-34

PROBLEM AREA:	Automatic suspension system operation
SYMPTOMS:	Vehicle is too high.
POSSIBLE CAUSES:	**1.** Faulty height sensor or circuit. **2.** Faulty door switch or circuit. **3.** Faulty trunk switch or circuit. **4.** Faulty brake switch or circuit. **5.** Defective exhaust solenoid or circuit. **6.** Faulty spring solenoid or circuit. **7.** Plugged or restricted air lines. **8.** Faulty compressor. **9.** Faulty solenoid or circuit.

Servicing Passive Restraint Systems

Upon completion and review of this chapter, you should be able to:

❏ Diagnose automatic seat belt systems.

❏ Replace the automatic seat belt drive motor assembly.

❏ Replace the drive belt in an automatic seat belt system.

❏ Enter air bag system self-diagnostics.

❏ Use a scan tool to properly retrieve air bag system trouble codes.

❏ Use the flashing warning light to retrieve air bag system trouble codes.

❏ Enter air bag diagnostics through the electronic climate control panel of GM vehicles equipped with digital instrument panels.

❏ Replace the air bag module according to manufacturer's service manual standards.

❏ Replace the clockspring assembly according to manufacturer's service manual standards.

❏ Service the air bag system safely.

❏ Validate the occupant classification system.

Basic Tools

Basic mechanic's tool set

Service manual

Introduction

In this chapter, you will learn how to properly and safely service automatic passive restraint systems. Federal mandates concerning the equipping of these systems has assured that today's technician must be qualified to service them. The safety of the driver and/or passengers depends on the ability of the technician to properly diagnose and repair these systems.

Passive restraint systems can be either automatic seat belts, air bags, or a combination of both. In this chapter, you will learn how to diagnose automatic seat belt systems and replace their main components. You will also learn how to enter self-diagnostics and to retrieve trouble codes in the air bag system. Included in this chapter are procedures for replacing the air bag module and the clockspring. In addition, there are typical procedures for validating the occupant classification system (OCS).

Automatic Seat Belt Service

The **automatic seat belt system** automatically puts the shoulder and/or lap belt around the driver or occupant. The automatic seat belt is operated by DC motors that move the belts by means of carriers on tracks. Even though the components used in the automatic seat belt system vary according to the manufacturer, the basic principles of locating and repairing the cause of a problem are similar. Refer to the service manual to obtain a circuit diagram of the system (Figure 15-1). The circuit schematic, troubleshooting charts, and diagnostic charts will assist you in finding the fault.

In addition, most manufacturers provide a troubleshooting chart for the automatic seat belt system. Troubleshooting the circuits, switches, lamps, and motors as indicated in this chart requires only the skills described in previous chapters. In addition to the troubleshooting chart, an operational logic chart will provide helpful information when testing switches (Figure 15-2). Use a diagnostic chart to test the operation of the system and to determine faults with the module.

Some systems are capable of storing diagnostic trouble codes that can be retrieved by a scan tester. The scan tester is also used to perform a functional test of the system. To perform this test, turn the ignition switch to the RUN position and connect the scan tool to the diagnostic connector. Select the system test and follow the scan tool instructions. If there is a failure in the system, the scan tester will display the code.

Classroom Manual
Chapter 15, page 485

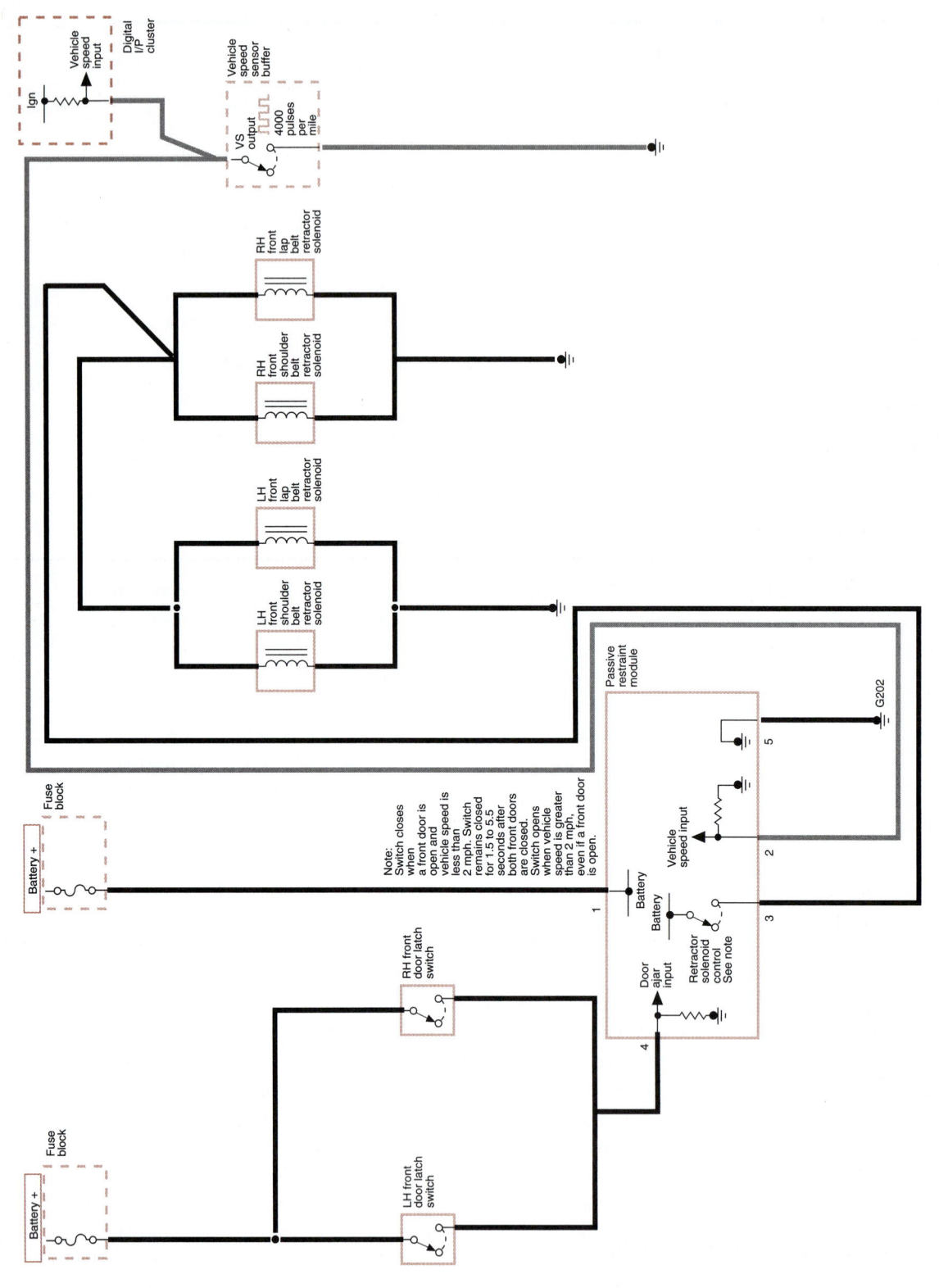

Figure 15-1 The circuit schematic is a vital tool for diagnosing the automatic seat belt system.

PASSIVE SEATBELT SYSTEM   OPERATIONAL LOGIC CHART

Vehicle Condition	Inertia		Door Switch		A Pillar Limit Switch		B Pillar Limit Switch	
	Closed	Open	On	Off	On	Off	On	Off
Ignition switch in RUN position following vehicle entry								
• Door Closed Belt begins to move	X			X		X	X	
• Door Closed Belt travels from A to B-Pillar	X			X	X		X	
• Door Closed Belt located at B-Pillar	X			X	X			X
• Open Door Belt begins to move	X		X		X			X
• Door Opened Belt travels from B to A-Pillar	X		X		X		X	
• Door Opened Belt located at A-Pillar	X		X			X	X	
Ignition switch in OFF position prior to exit								
• Door Closed — Belt located at B-Pillar	X			X	X			X
• Open Door— Belt begins to move	X		X		X			X
• Door Opened— Belt travels from B to A-Pillar	X		X		X		X	
• Door Opened— Belt located at A-Pillar	X		X			X	X	
• Close Door— Belt remains at A-Pillar	X			X		X	X	
Impact—Ignition RUN/Off								
• Ignition in RUN position—Door Opened (during impact) Belt remains at B-Pillar		X	X		X			X
• Ignition Off—Door Open (after impact) Belt travels from B to A-Pillar		X	X			X	X	

Figure 15-2 Switch operational logic chart.

Drive Motor Assembly Replacement

A faulty drive motor can cause slow or no operation of the automatic seat belts. The motor must be replaced if the technician determines that it is the faulty component. A typical procedure for motor assembly replacement is described here.

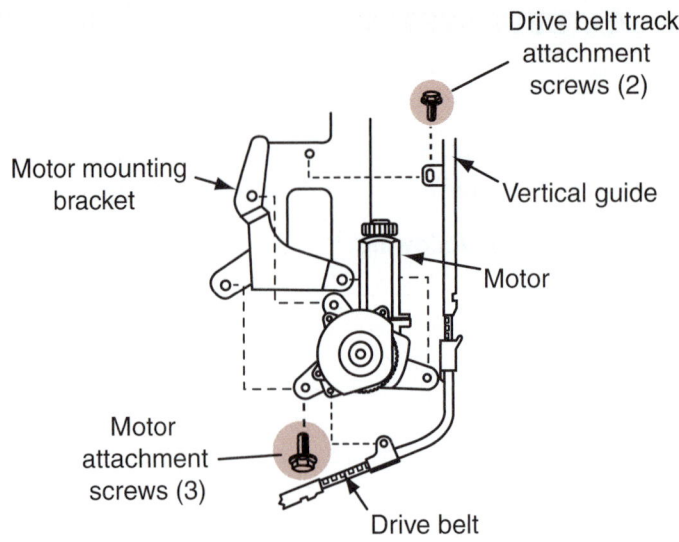

Figure 15-3 Guide and motor attaching screws.

Remove the B-pillar trim to gain access to the drive motor. Then remove the screws that attach the drive belt track to the motor. Next, remove the screw that attaches the vertical guide to the pillar. Be sure to remove only the screw that is next to the motor (Figure 15-3).

Slide the belt track down far enough to disengage the sprockets from the motor drive gear teeth. Disconnect the electrical connectors to the motor. Remove the hex nuts that retain the mounting bracket and remove the motor.

Reverse the order to install the motor.

Drive Belt Replacement

If the drive belt is twisted, kinked, or broken it will cause the seat belt to operate improperly. A binding drive belt may cause the seat belt to not move its full travel or strip the gear teeth on the motor.

To replace a defective drive belt, gain access to the assembly by removing the A- and B-pillar trim covers. Remove the plug button that covers the torx attaching bolt (Figure 15-4). Cycle the shoulder belt to the A-pillar and remove the anchor bolt. Remove the belt anchor cover and the shoulder belt from the belt carrier.

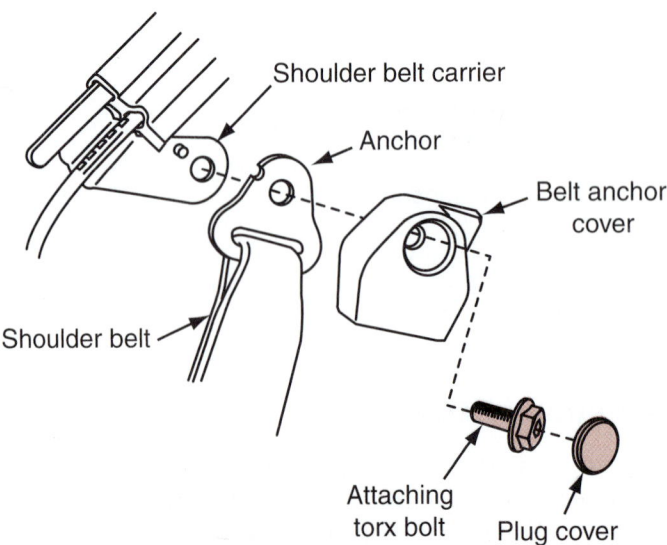

Figure 15-4 Shoulder belt to carrier attachment.

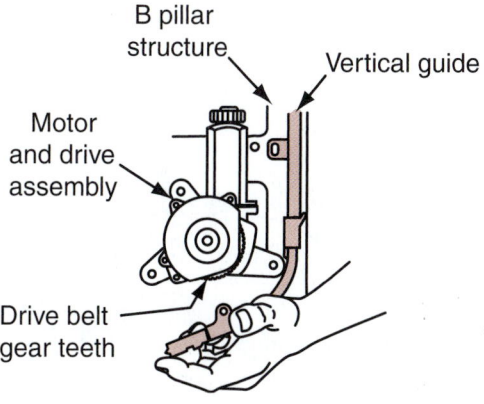

Figure 15-5 Disconnecting the drive belt from the motor.

Next, remove the screws that attach the drive belt track to the drive motor. Then remove the vertical guide retaining screw located next to the motor.

Disconnect the drive belt track from the motor. Allow the belt to pass by the gear teeth without engaging them (Figure 15-5). Remove limit switch A from the retaining bracket (Figure 15-6). Remove the bracket to the A-pillar attaching screw and the two screws that attach the bracket to the track. Slide the shoulder belt carrier forward off the track. The drive belt should slide out of the track with the shoulder belt. Remove the shoulder belt carrier from the drive belt.

Before installing the new drive belt, lubricate the track with approved lubricant. Insert the drive belt into the front of the track at the A-pillar. Install the shoulder belt carrier into the large slot at the front end of the drive belt. Slide the drive belt rearward in the track until it is 1/2 inch (12 mm) in the upper track slot.

Align the sprocket holes in the drive belt with the gear teeth of the motor (Figure 15-5). Install the attaching bolts to secure the vertical guide and track.

Engage the track locator on the A limit switch bracket into the slot in the A-pillar (Figure 15-6). Install the limit switch onto the bracket.

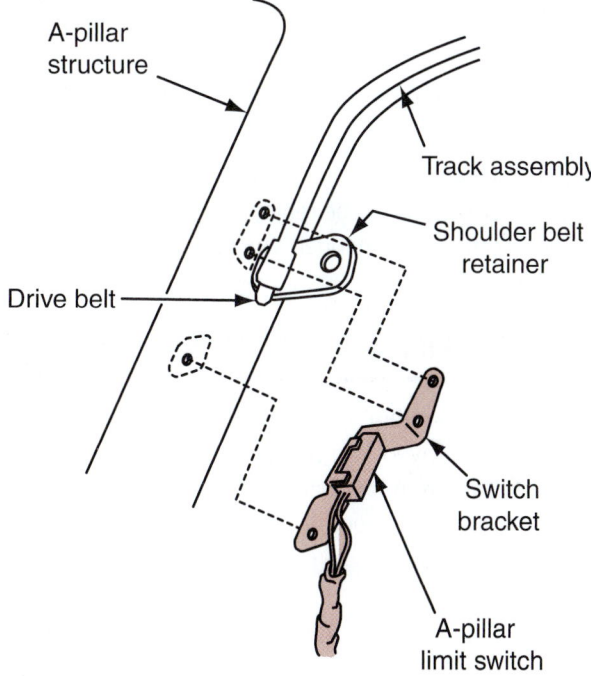

Figure 15-6 Limit switch A retaining bracket attachment to the track assembly.

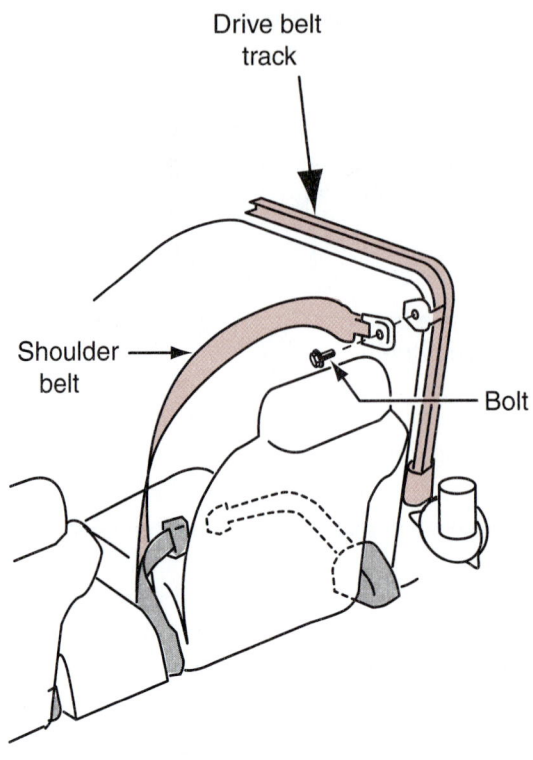

Figure 15-7 Install the shoulder belt to the carrier after checking it for twists and locating the carrier into the B-pillar.

Cycle the belt anchor to the B-pillar. Pull the webbing out of the retractor located in the console (Figure 15-7). Lay the seat belt webbing flat on the seat and check that it has no twists. Also check the seat belt for any fraying or discoloration. If it is frayed or discolored, the seat belt should be replaced. Align the notch in the shoulder belt anchor with the pin on the carrier. Install the anchor cover and bolt. Torque the bolt to specifications and replace the plug button. Replace the A- and B-pillar trim.

 WARNING: The anchor bolt is usually a self-locking bolt. Whenever it is removed, it must be replaced with a new one to prevent breakage in an accident.

CUSTOMER CARE: Usually, if it is properly maintained, the track will last the life of the vehicle. If the track has to be replaced due to wear, remind the customer to bring the vehicle back to you at the recommended service intervals to have the track lubricated.

Air Bag Safety and Service Warnings

Classroom Manual
Chapter 15, page 489

Whenever working on the air bag system, it is important to follow some safety warnings. There are safety concerns with both deployed and live air bag modules. The air bag module is composed of the air bag and inflator assembly that is packaged in a single module. The module is mounted in the center of the steering wheel.

1. Wear safety glasses when servicing the air bag system.

2. Wear safety glasses when handling an air bag module.

3. Always disconnect the battery negative cable, isolate the cable end, and wait for the amount of time specified by the vehicle manufacturer before proceeding with the necessary diagnosis or service. The required amount of time may be as much as 15 minutes. Failure to observe this precaution may cause accidental air bag deployment and personal injury.

4. Handle all sensors with care. Do not strike or jar a sensor in such a manner that deployment may occur.

5. Replacement air bag system parts must have the same part number as the original. Replacement parts of lesser quality or questionable quality must not be used. Improper or inferior components may result in improper air bag deployment and injury to the vehicle occupants.

6. Do not strike or jar a sensor or an air bag system occupant restraint controller (ORC). This may cause air bag deployment or the sensor to become inoperative.

7. Before an air bag system is powered up, all sensors and mounting brackets must be properly mounted and torqued to ensure correct sensor operation. If sensor fasteners do not have the proper torque, improper air bag deployment may result in injury to the vehicle occupants.

8. When carrying a **live module** that has not been deployed, face the trim and bag away from your body.

9. Do not carry the module by its wires or connector.

10. When placing a live module on a bench, face the trim and air bag up.

11. Deployed air bags may have a powdery residue on them. Sodium hydroxide is produced by the deployment reaction and is converted to sodium carbonate when it comes into contact with atmospheric moisture. It is unlikely that sodium hydroxide will still be present. However, wear safety glasses and gloves when handling a deployed air bag. Wash hands after handling.

12. A live air bag module must be deployed before disposal. Because the deployment of an air bag is an explosive process, improper disposal may result in injury and in fines. A deployed air bag should be disposed of according to EPA and manufacturer procedures.

13. Do not use a battery- or AC-powered voltmeter, ohmmeter, or any other type of test equipment not specified in the service manual. Never use a test light to probe for voltage.

14. Never reach across the steering wheel to turn the ignition switch on.

Diagnostic System Check

Before an air bag system is diagnosed, a system check is performed to avoid diagnostic errors. Always consult the manufacturer's specific information because the diagnostic system check may vary depending on the vehicle. The diagnostic system check involves observing the air bag warning light to determine if it is operating normally. A typical diagnostic system check follows:

1. Turn on the ignition switch and observe the air bag warning light. On some General Motors systems, this light should flash 7 to 9 times and then go out. On most other vehicles, the air bag warning light should be illuminated continually for 6 to 8 seconds and then go out. If the air bag warning light does not operate properly, further system diagnosis is necessary.

2. Observe the air bag warning light while cranking the engine. On many General Motors vehicles, this should cause the light to be illuminated continually. Always refer to the vehicle manufacturer's service manual. During engine cranking, if the air bag warning light does not operate as specified by the vehicle manufacturer, a complete system diagnosis is required.

3. Observe the air bag warning light after the engine starts. The light should turn off a few seconds after the engine is started. If the air bag warning light remains off, there are no current DTCs in the air bag system module. If the air bag warning light remains on, obtain the DTCs with a scan tester or flash code method. Not all manufacturers provide for fault code retrieval by flash code methods—these vehicles will require a scan tool.

Classroom Manual
Chapter 15, page 489

Retrieving Fault Codes

Most air bag systems will store fault codes that can be retrieved by either a scan tool or flash codes. In addition, some will allow display of fault codes on the digital panel cluster. After 1996, most manufacturers require the use of a scan tool.

Special Tools

Scan tool

Scan Tool DTC Retrieval

Usually the air bag system will store two types of fault codes, active and stored. Active fault codes will turn the air bag warning lamp on. Stored codes are faults that are intermittent. Some manufacturers will also display how long (in minutes) a code was active. Depending on the manufacturer, the fault codes may be stored in nonvolatile memory. These codes will not be erased if the battery is disconnected or damaged. The only way to erase these codes is by use of the scan tool.

To retrieve fault codes, connect the scan tool to the data link connector (DLC). From 1996 on, this is usually the J1962 connector. On earlier model vehicles, there may be a separate DLC for the air bag system. This connector could be located under the seat, in the glove box, or in a direct connection to the air bag control module. Always refer to the proper service information to determine the DLC location. Turn the ignition switch to the run position.

 WARNING: When turning the ignition switch on, do not reach across the steering wheel. Make sure that your body is away from any air bag modules that may deploy.

Follow the scan tool instruction to request supplement restraint system (SRS) Code Display. Record all stored and active fault codes. Active codes cannot be erased—the cause of the problem must be corrected. Use the proper diagnostic chart to trace the cause of the fault.

Some fault codes require that the ORC be replaced before any further diagnostics can be performed. For example, in some General Motors systems, a displayed code 52 indicates that enough accident information is stored in the EEPROM to fill its memory. In most systems, it takes four simultaneously closed arming and crash sensor events to fill the memory. This code requires that the ORC be replaced. A code 52 cannot be erased nor can any further diagnosis be performed until the ORC is replaced. If a code 71 is set, then an ORC failure is detected. A code 71 requires that the ORC be replaced before any other diagnostic procedures can be performed. A code 71 cannot be erased. If a code 52 or 71 is not displayed and there are other history codes, then the technician will be instructed to go to a diagnostic chart to locate the causes of the intermittent fault. If there are no history codes, diagnose remaining current codes from the lowest to the highest number.

General Motors called its control module as diagnostic energy reserve module (DERM).

Flash Code Retrieval

On vehicles that provide methods to retrieve fault codes by use of the warning lamp, always refer to the proper service information for the correct procedure. The following are typical examples to familiarize you with these methods.

General Motors Flash Code Retrieval. Flash codes allow the technician to retrieve current trouble codes without the use of a scan tool. Flash codes will not provide history codes. However, they will indicate whether there are any in memory. Always consult the proper vehicle manufacturer's service manual before attempting to perform flash code retrieval.

The SRS diagnostic system check, described earlier, must be the first step of any SRS diagnosis. This checks the warning light and ability of the ORC to communicate through the data line.

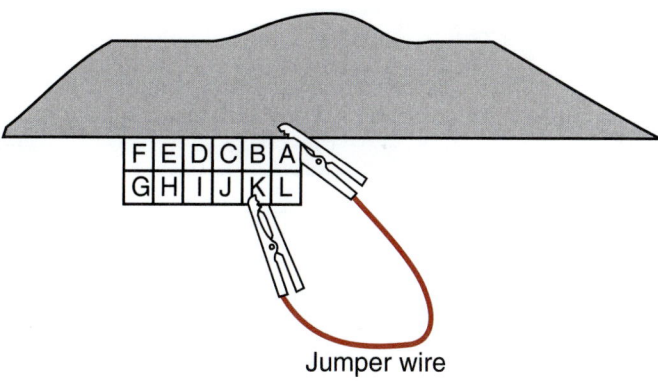

Jumper wire

Figure 15-8 Jump the A terminal to ground to cause the module to enter self-diagnostics and to flash trouble codes.

With the ignition switch in the RUN position, connect a jumper wire from DLC terminal K to terminal A (Figure 15-8). The SRS warning lamp will display trouble codes through a series of flashes. Each displayed code will consist of a number of flashes that represent the first number. This number will be followed by a 1/2-second delay. Then the second number of the code will be flashed. For example, a code 52 would be displayed by five flashes, a 1/2-second pause, followed by two more flashes.

A code 12 will be displayed to indicate that the system is in flash code mode. This is not a fault code. Code 13 will be displayed if there are no history codes stored in memory.

Each code is displayed once, until all codes have been given. The codes will then repeat until the ground is lifted from DLC terminal K.

CAUTION: Be sure to jumper the correct DLC terminals as identified in the service information. Jumping the incorrect terminal to ground may cause controller damage.

Digital Instrument Panel Cluster Display. Late-model Cadillac SRS diagnosis can be done by pressing the OFF and WARMER buttons on the electronic climate control panel (ECC) simultaneously. Press the LOW button four times to enter the SIR system. Then press HIGH to display any recorded codes (Figure 15-9).

Ford Air Bag System Flash Code Diagnosis. On some air bag systems, if a fault exists that could result in unwarranted air bag deployment, the ORC disarms the system. The ORC opens a thermal fuse inside to disarm the system. This fuse is not replaceable.

SERVICE TIP: On some Ford vehicles, the ORC provides an audible tone if the air bag warning light is not operating properly. Under this condition, the ORC provides five sets of five beeps every half hour.

WARNING: On Ford vehicles, the air bag deployment loop is connected to the battery positive terminal even with the ignition switch off. Therefore, air bag deployment is possible with the ignition switch off.

On many Ford vehicles, the air bag warning light begins to flash a DTC when a defect occurs in the air bag system. On many Ford systems, the air bag warning light prioritizes the DTCs and flashes the highest priority DTC if there is more than one fault in the system. When

Special Tools

Jumper wire

Ford calls some of its control modules air bag system diagnostic module (ASDM).

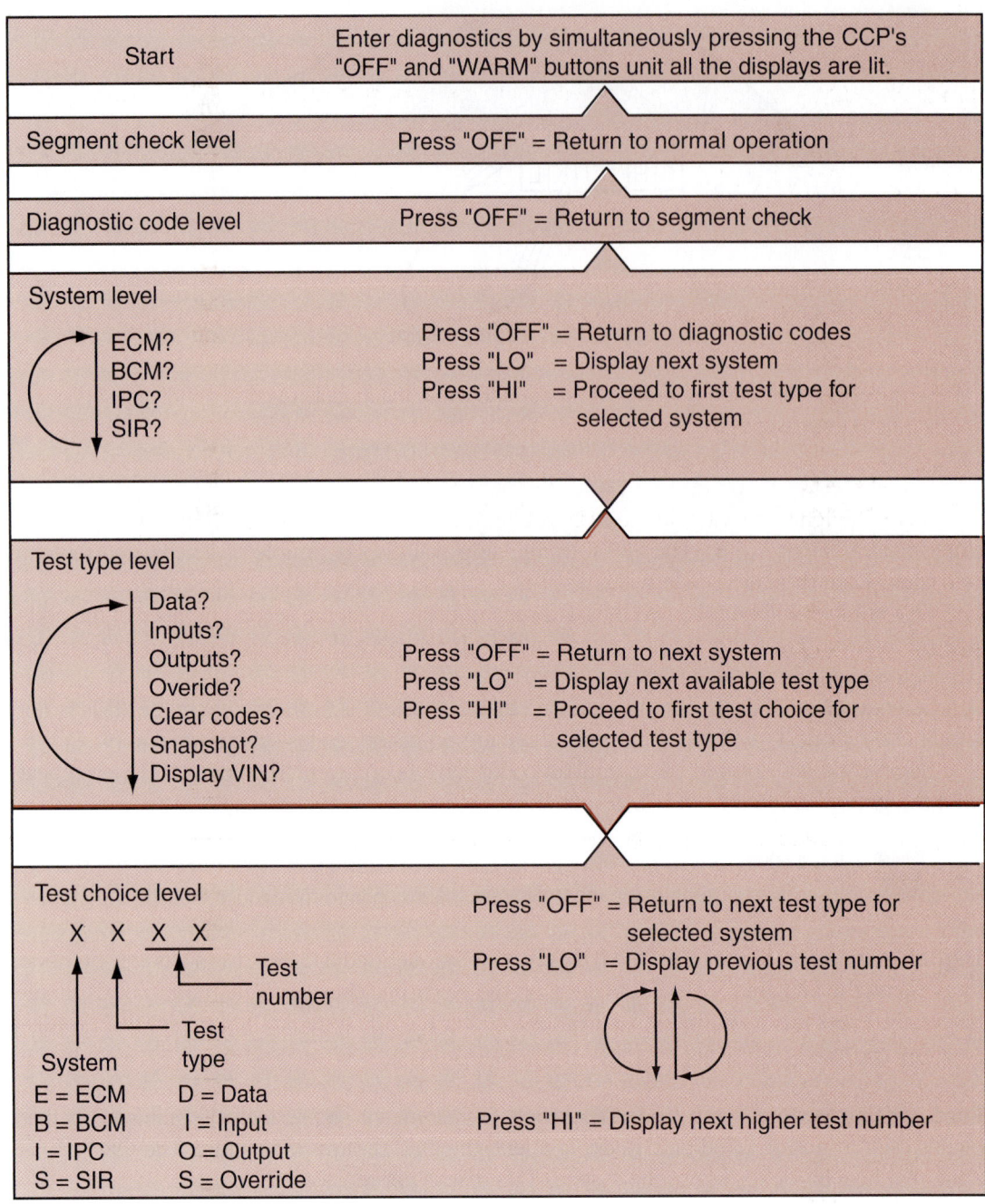

Figure 15-9 On-board diagnostics using the ECC panel.

the fault represented by the flashing air bag warning light is corrected, the light flashes the DTC with the next highest priority—if a second fault exists. Since DTCs vary depending on the model year, the technician must have the DTC list for the model year being diagnosed.

Toyota Air Bag System Flash Code Diagnosis. On Toyota vehicles, air bag flash codes may be obtained by cycling the ignition switch on and off five times. Each time the ignition switch is cycled on or off, the technician must wait 20 seconds. Toyota air bag DTCs may also be obtained by connecting a special jumper wire, supplied by the vehicle manufacturer, between terminals TC and E1 in the DLC2. Before connecting this jumper wire, make sure the ignition switch is in the ACC or ON position. After the ignition switch is in one of these positions, wait 30 seconds. If there are no DTCs in memory, the air bag warning light flashes two times per second. When DTCs are present, the air bag warning light flashes these codes in numerical order. A DTC indicates a fault in a certain area such as a specific air bag sensor. Voltmeter or ohmmeter tests recommended in the vehicle manufacturer's service manual are usually necessary to locate the exact cause of the problem.

✔ **SERVICE TIP:** On Toyota vehicles, if the air bag warning light flashes a DTC that is not on the fault code list for that model year, the control module is defective.

Air Bag System Testing

Classroom Manual
Chapter 15, page 489

Once the DTCs have been recorded, the technician will use the proper diagnostic chart to locate the fault. It is very important to follow *all* of the procedures listed. If the chart calls for the use of an ohmmeter to check a circuit, the technician would have been instructed to disconnect the harness from the air bag module in a prior step. If the technician ignores this step and connects the ohmmeter to the circuit, it is very possible that that air bag(s) will deploy.

Since most systems use shorting bars at the connectors, the technician must remember to lift these up in order to properly diagnose the harness. Most manufacturers provide test connections to be plugged into the harness connectors. These test connectors will lift up the shorting bar and also provide a test location for connecting the DMM.

Honda Air Bag System Voltmeter Diagnosis

⚠ **WARNING:** Use only the vehicle manufacturer's recommended tools and equipment for air bag system service and diagnosis. Failure to observe this precaution may result in unwarranted air bag deployment and personal injury.

⚠ **WARNING:** Do not use battery-powered or AC-powered voltmeters or ohmmeters except those meters specified by the vehicle manufacturer. Failure to observe this precaution may result in unwarranted air bag deployment and personal injury.

⚠ **WARNING:** Do not use nonpowered probe-type test lights or self-powered test lights to diagnose the air bag system. Unwarranted air bag deployment and personal injury may result.

⚠ **WARNING:** Follow the vehicle manufacturer's service and diagnostic procedures. Failure to observe these precautions may cause inaccurate diagnosis, unnecessary repairs, or unwarranted air bag deployment resulting in personal injury.

Honda recommends testing the voltage at specific terminals on the control module, inflator modules, and sensor to diagnose the air bag system. When voltage tests are performed at these terminals, special jumper wires are attached to allow the necessary voltmeter connections without damaging the terminals. Before connecting the special wiring harness, remove and isolate the battery negative cable. Then wait for the time period specified by the vehicle manufacturer.

Wiring harness A is connected to a terminal on the control module. Wiring harness B is connected in series between the large control module wiring connector and the matching terminals on the control module (Figure 15-10). Wiring harness C is connected in series at the inflator module connector and harness D is connected in series at the dash sensor (Figure 15-11). After the special wiring harness is connected, the voltmeter tests provided in the vehicle manufacturer's service manual may be performed to diagnose the system.

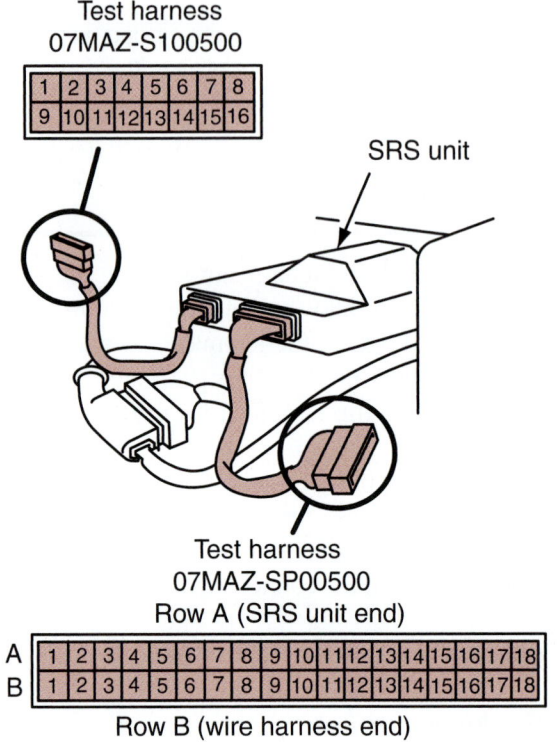

Figure 15-10 Wiring harness A and B connected at the ASDM terminals.

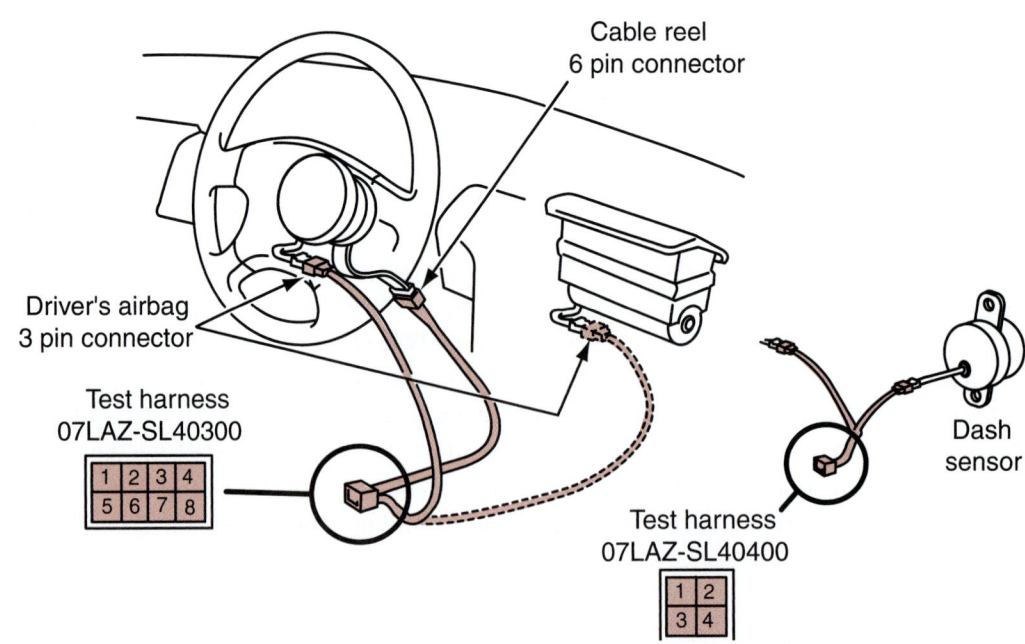

Figure 15-11 Wiring harness C and D connected at the inflator module and dash sensor terminals.

Inspection After an Accident

Any time the vehicle is involved in an accident, even if the air bag was not deployed, all air bag system components should be inspected. The wiring harness must be inspected for damage and repaired or replaced as needed. Any damaged or dented components must also be replaced. Do not attempt to repair any of the sensors or modules. Service is by replacement only.

In the event of deployment, the service manual will provide a list of components that must be replaced. The list of components will vary depending on manufacturer.

Cleanup Procedure After Deployment

If the air bag has been deployed, the residue inside the passenger compartment must be removed before entering the vehicle. Tape the air bag exhaust vents closed to prevent additional powder from escaping. Use the shop vacuum cleaner to remove any powder from the vehicle's interior. Work from the outside to the center of the vehicle. Vacuum the heater and A/C vents. Run the heater fan blower motor on low speed and vacuum any powder that is blown from the plenum.

Component Replacement

None of the sensors, air bag modules, or controllers used in the air bag system are repairable. In addition, seat belt pretensioners are not serviceable. If any of these components are found to be defective, then the component must be replaced.

If wiring repair is required, first refer to the proper service information. Some manufacturers do not recommend wire harness repair and others may have very specific procedures they require.

Module Replacement

WARNING: Before replacing any component of the air bag system, follow the service manual procedure for disarming the system—even if the air bag is deployed.

WARNING: Wear safety glasses when working on the air bag system.

Follow Photo Sequence 26 as a guide to replacement of an air bag module. Always follow the service manual for the vehicle you are working on.

The module bolts must be torqued to the specific value. Usually, new bolts are supplied with the air bag module.

Reverse the procedure to reinstall the module. To rearm the system, connect the yellow two-way electrical connector at the base of the steering wheel and install the CPA. Turn the ignition switch to the RUN position; then replace the SRS fuse and connect the battery. This procedure is done to keep the technician out of the vehicle when the system is powered up.

With the ignition switch in the RUN position, observe the SRS warning light. It should perform its bulb test and then shut off. Perform the SRS diagnostic system check to confirm proper operation.

Classroom Manual
Chapter 15, page 497

Classroom Manual
Chapter 15, page 490

Photo Sequence 26
Typical Procedure for Removing the Air Bag Module

P26-1 Tools required to remove the air bag module: safety glasses, seat covers, screw driver set, torx driver set, battery terminal pullers, battery pliers, assorted wrenches, ratchet and socket set, and service manual.

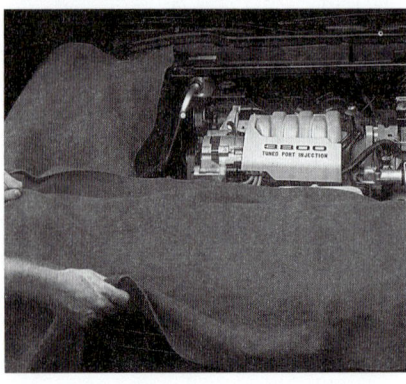

P26-2 Place the seat and fender covers on the vehicle.

P26-3 Place the front wheels in the straight-ahead position and turn the ignition switch to the LOCK position.

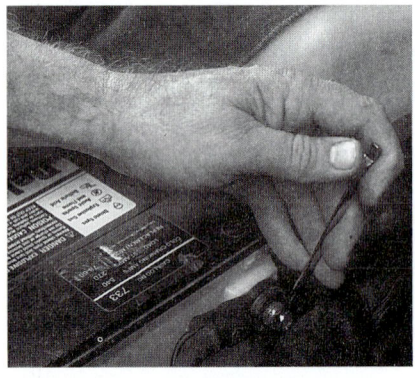

P26-4 Disconnect the negative battery cable.

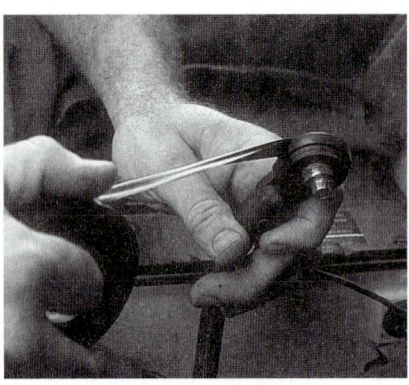

P26-5 Tape the cable terminal to prevent accidental connection with the battery post. Note: A piece of rubber hose can be substituted for the tape.

P26-6 Remove the SIR fuse from the fuse box. Wait 10 minutes to allow the reserve energy to dissipate.

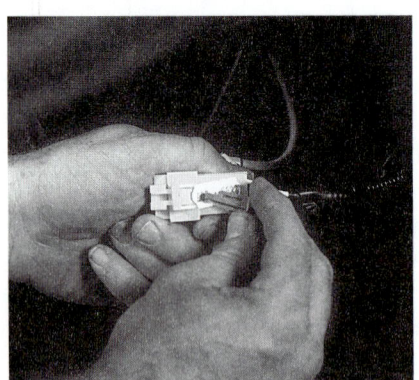

P26-7 Remove the connector position assurance (CPA) from the yellow electrical connector at the base of the steering column.

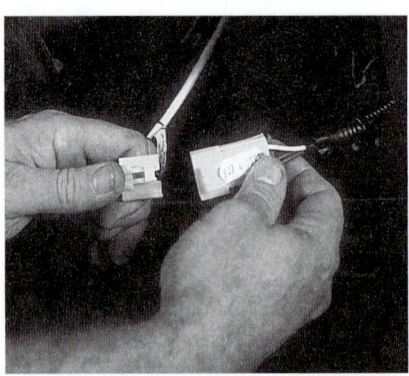

P26-8 Disconnect the yellow two-way electrical connector.

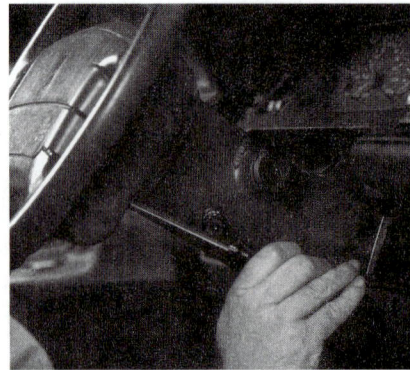

P26-9 Remove the four bolts that secure the module from the rear of the steering wheel.

Typical Procedure for Removing the Air Bag Module (continued)

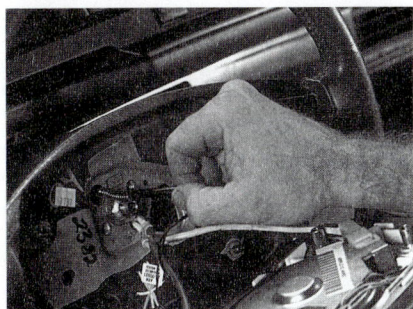

P26-10 Rotate the horn lead 1/4 turn and disconnect it.

P26-11 Disconnect the electrical connectors.

P26-12 Remove the module.

Clockspring Replacement

The clockspring should be inspected any time the vehicle has been involved in an accident, even if the air bag was not deployed. In addition, the heat generated when an air bag is deployed may damage the clockspring. For this reason, it should be replaced whenever the air bag is deployed. Exact procedures vary according to manufacturer. The following is typical for the Ford SRS system.

Classroom Manual
Chapter 15, page 492

 WARNING: Wear safety glasses when servicing the air bag system.

1. Place the front wheels in a straight-ahead position and place the ignition in the LOCK position. Rotate the steering wheel 16 degrees counterclockwise until it locks.
2. Disarm the air bag system.

WARNING: The procedure for disarming the system differs according to the year of the vehicle. Vehicles 1992 and newer require the use of an air bag simulator tool. Do not attempt to service the system if proper disarming is not possible.

3. Remove the nuts that attach the air bag module to the steering wheel.
4. Lift the module from the steering wheel and disconnect the wire connection from the clockspring.
5. Remove and discard the steering wheel attaching bolt.
6. Mark the shaft and steering wheel with index marks for reinstallation.
7. Use a steering wheel puller to remove the steering wheel from the shaft.
8. Remove the upper and lower steering column shrouds.
9. Disconnect the clockspring connector from the steering column harness.
10. Tape the clockspring to prevent rotation of the rotor.
11. Remove the two retaining screws and the clockspring.

Replacement of the clockspring is done in the reverse order. The replacement clockspring will have a locking insert to prevent rotation of the rotor. Do not remove this insert until the clockspring is secured onto the column by the retaining screws. Torque the new steering wheel attaching nut to specifications. When the steering wheel is replaced, rearm the system and perform the verification test.

Classroom Manual
Chapter 15, page 507

Occupant Classification System Service

Diagnosing the electrical functions of the occupant classification system (OCS) is performed using a scan tool. DTCs will be recorded any time a fault is detected in the system. In addition, if a fault is detected, the passenger air bag disable lamp (PADL) will turn on, indicating to the driver that the air bags have been suppressed.

Bladder-type systems produced by Delphi have some unique service requirements. If diagnostics indicate that the occupant classification module (OCM) requires replacement, this is only possible if the original seat is in the vehicle. Since the ORC stores the aging calibrations of the seat, if a new OCM is installed, the ORC will recognize the new module and will continue to use the old seat wear calibrations.

CAUTION: Since the ORC stores the seat-aging calibrations, swapping seats is not a good practice. The ORC will not have the correct calibrations for the replacement seat and the system will not work as intended. This may lead to air bag deployment when an infant is in the seat.

If diagnostics of a bladder-type system lead the technician to replace any of the following items, an **OCS service kit** must be installed (Figure 15-12):

❏ Bladder.

❏ Sensor.

❏ Seat foam.

❏ Cloth seat cover.

The OCS service kit consists of the seat foam, the bladder, the pressure sensor, the occupant classification module (OCM), and the wiring. The service kit is calibrated as an assembly. The service kit OCM has a special identification data bit that is transmitted once it is connected. When the service kit is installed, the ORC will identify the service kit OCM and will clear all calibration data related to the system components.

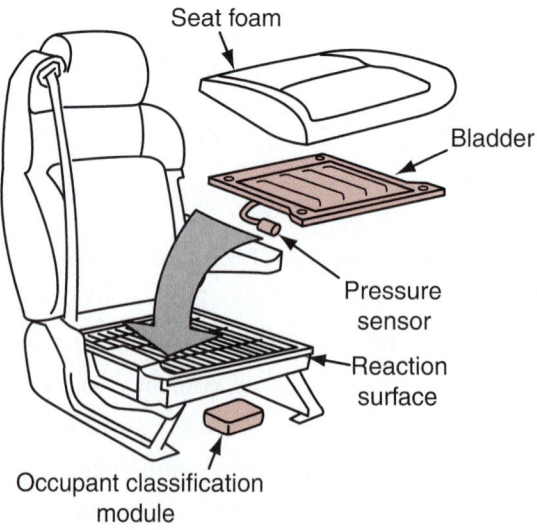

Figure 15-12 The OCS service kit.

The wiring of the service kit is hot glued to the module and the sensor to prevent separation of the components. A tag also may be located on the wiring harness, which identifies it as a service kit (refer to Figure 15-12).

The service kit wiring uses a single connector that mates to the existing vehicle wiring harness. The connection on the vehicle harness that originally connected to the OCM now connects to the service kit harness. The connector on the vehicle harness that originally connected in to the pressure sensor will not be used with the service kit. Tie this wire to the harness to prevent it from getting caught in the seat tracks as the seat is moved.

Strain gauge systems by TRW do not have these service requirements. However, any time a component is replaced on either system, the system must be validated before the vehicle is returned to the owner.

OCS Validation

Whenever the passenger-side front-seat retaining bolts are loosened, the seat is removed from the vehicle, the seat is replaced, the seat trim is replaced, a sensor is replaced, or the OCM is replaced an **OCS validation test** must be performed. Virtually any service that is done to the passenger-side front seat will require the technician to validate OCS. The OCS validation test is done to confirm that the system can properly classify the occupant. This task usually requires the use of special weights (Figure 15-13).

CAUTION: If the seat requires removal from the vehicle, the strain gauges are susceptible to damage. Do not drop the seat or allow anyone to sit in it while it is out of the vehicle.

The special weight set has three parts. The tool approximates federal standards for the weight of occupants:

1. The base weight of the tool weighs 37 lbs. (17 kg) to validate for the classification of a rear-facing infant seat (RFIS) weight.
2. The addition of the 10-pound (4.5-kg) weight to the base validates for the weight classification of a child.

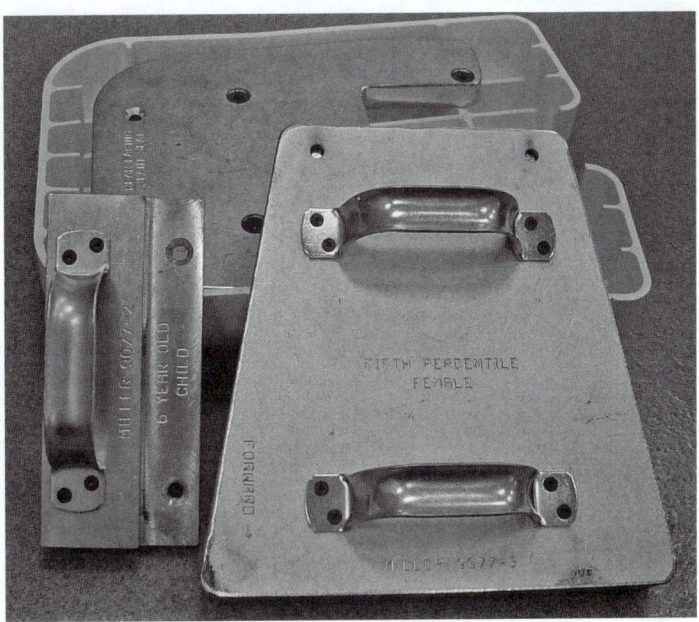

Figure 15-13 Special OCS weights used to validate the system.

3. The addition of the 52-pound (24-kg) weight to the assembly validates for the weight classification of the fifth percentile female.

The total weight of the tool is 99 pounds (45 kg). Weights are added to the base in the proper order and placed in the correct position with the dowel pins. There are differences between the bladder-type and strain gauge–type systems that affect which weights will be used. It is imperative that the correct procedure be followed.

Photo Sequence 27 illustrates a typical procedure for validating the OCS.

Belt Tension Sensor Diagnostic Test

The following is a typical procedure for testing the belt tension sensor (BTS) used on the Delphi bladder OCS. Since this sensor is used to determine if an infant seat is installed in the front passenger-side seat, it is critical that proper diagnostics be followed. Always refer to the service information for the vehicle you are working on.

With the scan tool connected, monitor the output of the BTS. This will usually be listed in counts (Figure 15-14). With the front passenger-side seat unoccupied and the BTS in its static state, the counts should be between 39 and 69. If the counts are outside of this range, the seat belt retractor and the BTS assembly must be replaced.

If the counts are within specified range, use a pull scale (Figure 15-15) to apply a load to the BTS. Connect the hook of the scale through the BTS webbing loop. Grab the pull handle at the top of the spring scale and pull the spring scale straight up, keeping the horizontal bar level

Classroom Manual
Chapter 15, page 510

Special Tools

Scan tool

Pull scale

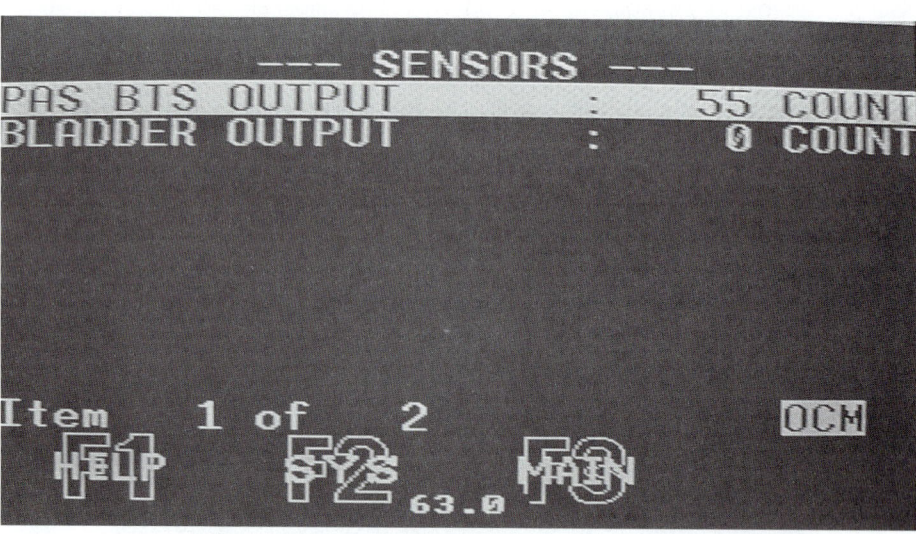

Figure 15-14 Scan tool will show the BTS counts.

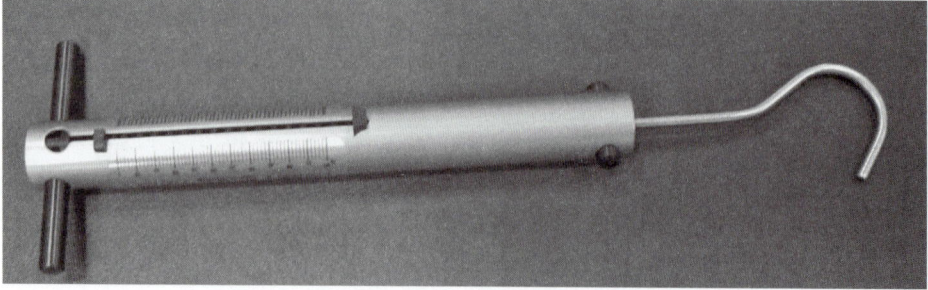

Figure 15-15 Pull scale used to measure pressure against the BTS.

Photo Sequence 27
Occupant Classification Validation

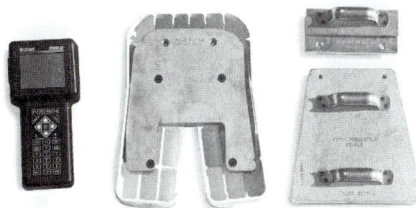

P27-1 Tools required to perform this task include a scan tool and special verification weights.

P27-2 Connect the scan tool and check for active battery voltage, OCM internal failure, or communication DTCs. Correct any of these conditions before continuing.

P27-3 Make sure the seat is empty and is in its full rearward position with the back rest in a normal upright position.

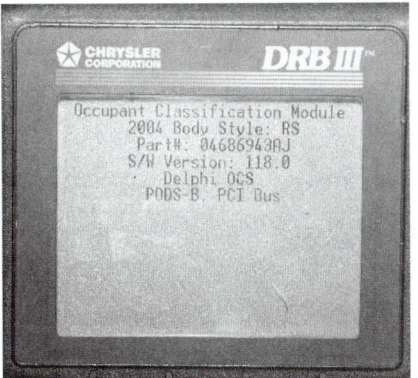

P27-4 Use the scan tool to access the "OCM Verification" screen and follow the instructions to begin the test.

P27-5 Once the test has been started, verify that the PADL is illuminated.

P27-6 When instructed, add the correct weight amount to the seat. Follow the scan tool prompts to complete this phase of the process. The scan tool should confirm the phase was completed successfully.

P27-7 When instructed, add the additional weight and follow the prompts to complete this phase. The scan tool should confirm the phase was completed successfully.

P27-8 Confirm that the PADL is now off.

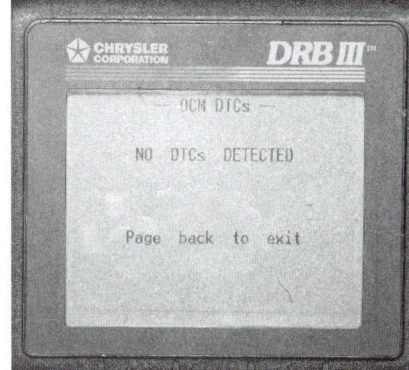

P27-9 Use the scan tool to access DTCs and check for any active codes. If active codes are present, this condition must be repaired and the system verification procedure repeated. Clear any stored codes.

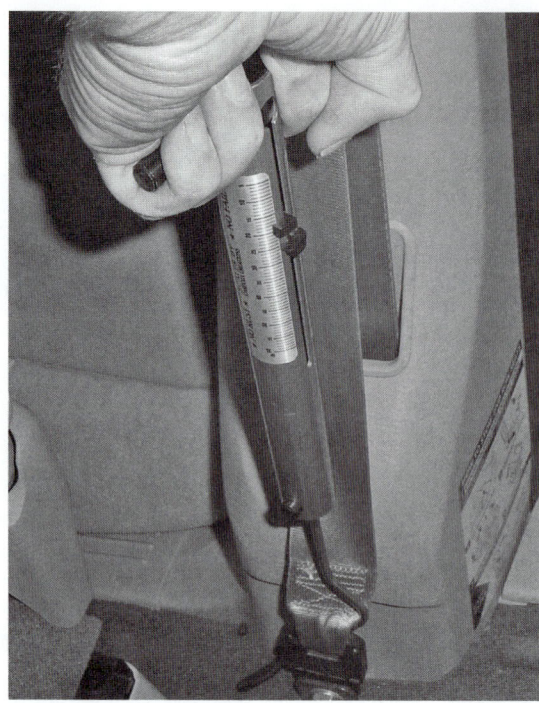

Figure 15-16 Using the pull scale to obtain pressure against the BTS.

with the door sill and the spring scale in line with the BTS (Figure 15-16). Apply and hold 25 pounds (11.5 kg) of pressure on the sensor while monitoring the BTS counts on the scan tool. The counts should increase to between 153 and 204. If the counts are not within this range, replace the passenger seat belt retractor and BTS assembly.

If the counts are within range, release the pressure applied with the pull scale and monitor the BTS counts again. With no load applied to the BTS, the output should return to between 39 and 69 counts within 20 seconds. If the counts are not within this range, replace the passenger seat belt retractor and BTS assembly.

If the above tests all pass, then the BTS is operating as intended. Confirm all electrical connections before returning the vehicle to the customer.

CASE STUDY

A customer brings her vehicle to the repair shop because she has observed that the air bag warning light is always illuminated. The technician confirms the warning lamp operation by performing the diagnostic system check. Next, the technician uses the scan tool to retrieve fault codes. The scan tool displays a current code for an open driver's-side air bag squib circuit. Following the service manual procedure, the technician removes the air bag module and tests the clockspring circuit. Since the clockspring circuit proves to be in good condition, the technician replaces the air bag module. Upon completion of installing the new module, the diagnostic system test is run again to confirm the repair was successful.

The air bag module was replaced due to the process of elimination. Since the clockspring circuit proved to be good, the air bag module had to be faulty. The resistance of the squib in the module was not tested because this can cause the air bag to deploy and the service manual provided several warnings against it.

Terms to Know

Automatic seat belt system

Live module

OCS validation test

OCS service kit

ASE-Style Review Questions

1. The occupant classification system must be validated whenever:
   A. The seat has been removed from the vehicle.
   B. The OCM has been replaced.
   C. A new sensor has been installed.
   D. All of the above.

2. The bladder system service kit is required if any of the following requires replacement EXCEPT:
   A. The bladder.
   B. The BTS.
   C. The pressure sensor.
   D. The cushion.

3. The PADL is off when the seat is empty. This may indicate:
   A. A faulty OCM.
   B. Normal operation.
   C. The BTS is indicating the seat belt is cinched.
   D. All of the above.

4. Replacement of an air bag module is being discussed.
   Technician A says to follow the service manual procedure for disarming the system.
   Technician B says to wear safety glasses.
   Who is correct?
   A. A only
   B. B only
   C. Both A and B
   D. Neither A nor B

5. *Technician A* says the clockspring should be inspected any time the vehicle has been involved in an accident, even if the air bag was not deployed.
   Technician B says the heat that is generated when an air bag is deployed may damage the clockspring.
   Who is correct?
   A. A only
   B. B only
   C. Both A and B
   D. Neither A nor B

6. *Technician A* says air bag residue should be swept from the vehicle's interior using a whisk broom.
   Technician B says whenever a vehicle is involved in an accident the air bag control module must be replaced.
   Who is correct?
   A. A only
   B. B only
   C. Both A and B
   D. Neither A nor B

7. Air bag system service is being discussed.
   Technician A says before an air bag system component is replaced, the negative battery cable should be disconnected and the technician should wait the specified time advised in the service manual.
   Technician B says this waiting period is necessary to dissipate the reserve energy in the air bag system computer.
   Who is correct?
   A. A only
   B. B only
   C. Both A and B
   D. Neither A nor B

8. Air bag sensor service is being discussed.
   Technician A says incorrect torque on air bag sensor fasteners may cause improper air bag deployment.
   Technician B says the arrow on an air bag sensor must face toward the driver's side of the vehicle.
   Who is correct?
   A. A only
   B. B only
   C. Both A and B
   D. Neither A nor B

9. Air bag system diagnosis is being discussed.
   Technician A says on some Ford products, the air bag computer prioritizes faults and flashes the code representing the highest priority fault.
   Technician B says on some air bag systems, the air bag computer disarms the system if a fault occurs that could result in an unwarranted air bag deployment.
   Who is correct?
   A. A only
   B. B only
   C. Both A and B
   D. Neither A nor B

10. Air bag system flash code diagnosis is being discussed.
    Technician A says on Toyota vehicles, if there are no faults in the air bag system, the air bag warning light flashes four times per second in the diagnostic mode.
    Technician B says, on Toyota vehicles, connect a jumper at the DLC to obtain air bag system codes.
    Who is correct?
    A. A only
    B. B only
    C. Both A and B
    D. Neither A nor B

ASE Challenge Questions

1. The motorized seat belt makes a clicking sound in the door pillar during operation.
   Technician A says a bad front limit switch will cause this problem.
   Technician B says the cause is a faulty door ajar switch.
   Who is correct?
   A. A only
   B. B only
   C. Both A and B
   D. Neither A nor B

2. The motorized seat belt fails to move up the door pillar when the door is shut.
   Technician A says the rear limit switch is the fault.
   Technician B says this could be caused by a faulty door switch.
   Who is correct?
   A. A only
   B. B only
   C. Both A and B
   D. Neither A nor B

3. A customer states that while driving the vehicle, the air bag warning lamp illuminates intermittently.
   Technician A says this can be caused by a loose connection to one of the system's sensors.
   Technician B says this may indicate a defect that will set a trouble code.
   Who is correct?
   A. A only
   B. B only
   C. Both A and B
   D. Neither A nor B

4. Passenger-side air bag deactivation is being discussed.
   Technician A says if the vehicle is not equipped with a factory installed ON/OFF switch, disconnect the connector to the air bag module and remove the air bag warning lamp.
   Technician B says the vehicle owner must provide a letter of approval from the NHTSA before a deactivation kit can be installed.
   Who is correct?
   A. A only
   B. B only
   C. Both A and B
   D. Neither A nor B

5. Side-impact air bags are being discussed.
   Technician A says most systems have a control module or sensor located in the B pillar.
   Technician B says the side air bags only deploy when the front air bags deploy.
   Who is correct?
   A. A only
   B. B only
   C. Both A and B
   D. Neither A nor B

Job Sheet 54

54

Name _____ Date _____

Working Safely Around Air Bags

Upon completion of this job sheet, you should be able to work safely around and with air bag systems.

ASE Correlation

This job sheet is related to the ASE Electrical/Electronic System Certification Exam's content area: *Miscellaneous;* task: Inspect, test, and repair or replace the air bag, air bag module, sensors, connectors, and wires of the air bag system circuit(s).

Tools and Materials

A vehicle with air bags
Service manual for the chosen vehicle
Component locator for the chosen vehicle
Safety glasses
A DMM

Describe the vehicle being worked on:

Year _____ Make _____ Model _____

VIN _____ Engine type and size _____

Procedure

Task Completed

1. Locate the information about the air bag system in the service manual.
   How are the critical parts of the system identified in the vehicle?

   _____

   _____

   _____

   _____

   _____

   _____

2. List the main components of the air bag system and describe their locations.

   _____

   _____

   _____

   _____

   _____

   _____

3. There are some very important guidelines to follow when working with and around air bag systems. These are listed below with some key words left out. Read through these and fill in the blanks with the correct words.

   a. Wear _____ _____ when servicing an air bag system and when handling an air bag module.

   b. Wait at least _____ minutes after disconnecting the battery before beginning any service. The reserve _____ module is capable of storing enough energy to deploy the air bag for up to _____ minutes after battery voltage is lost.

   c. Always handle all _____ and other components with extreme care. Never strike or jar a sensor, especially when the battery is connected; this can cause deployment of the air bag.

   d. Never carry an air bag module by its _____ or _____, and, when carrying it, always face the trim and air bag _____ from your body. When placing a module on a bench, always face the trim and air bag.

   e. _____ air bags may have a powdery residue on them. _____ is produced by the deployment reaction and is converted to _____ _____ when it comes in contact with the moisture in the atmosphere. Although it is unlikely that harmful chemicals will still be on the bag, it is wise to wear _____ _____ and _____ when handling a deployed air bag. Wash your hands immediately after handling a deployed air bag.

   f. A live air bag must be _____ before it is disposed of. A deployed air bag should be disposed of in a manner consistent with the _____ and the manufacturer's recommended procedures.

   g. Never use a battery- or AC-powered _____, _____, or any other type of test equipment in the system unless the manufacturer specifically says to. Never probe with a _____ _____ for voltage.

Instructor's Response _____

_____

_____

Job Sheet 55

55

Name _____ Date _____

Diagnosing "Driver Squib Circuit Open" Fault

Upon completion of this job sheet, you should be able to diagnose the cause of a "driver squib circuit open" fault in an air bag system and determine needed repairs.

ASE Correlation

This job sheet is related to the ASE Electrical/Electronic System Certification Exam's content area: *Miscellaneous;* task: Inspect, test, and repair or replace the air bag, air bag module, sensors, connectors, and wires of the air bag system circuit(s).

Tools and Materials

A vehicle equipped with a driver's-side air bag (bugged by instructor prior to performing this task)
Service manual for chosen vehicle
Safety glasses
Battery terminal puller
Jumper wires
Scan tester

Describe the vehicle being worked on:

Year _____ Make _____ Model _____

VIN _____ Engine type and size _____

Procedure

Task Completed

1. Describe the normal conditions in which the air bag warning lamp should operate.

   _____

   _____

   _____

2. Perform the diagnostic system check. How did the air bag warning light respond?

   _____

   _____

3. Follow all safety warnings and cautions listed in the service information and connect the scan tool. Record all fault codes the scan tool displays.

   Active: _____

   _____

   Stored: _____

   _____

4. Is the code for "Driver-Side Squib Circuit" or equivalent active?   ☐ Yes ☐ No
   If no, consult your instructor.

5. List all possible causes that can set this fault code.

   _____

   _____

   _____

   _____

☐ 6. Disconnect the negative battery cable and isolate it.

7. How long does the service information say you must wait before proceeding?

   _____

   _____

☐ 8. Follow the service information instructions to remove the driver's-side air bag module from the steering wheel.

☐ 9. Connect a jumper wire across the upper connector of the clockspring.

☐ 10. With the ignition switch in the RUN position, connect the battery ground cable.

11. Use the scan tool and record the active fault code.

   _____

   _____

12. Does the fault code indicate the circuit is shorted?   ☐ Yes ☐ No

13. If you answered yes to question 12, what is the faulty component?

   _____

   _____

14. If you answered no to question 12, locate the lower clockspring connector. Describe this connector's location.

   _____

   _____

   _____

☐ 15. Disconnect the negative battery cable and isolate it. Wait the recommended amount of time before proceeding.

☐ 16. Disconnect the lower clockspring connector and connect a jumper wire across the control module side of the harness connector.

☐ 17. With the ignition switch in the RUN position, reconnect the battery ground cable.

18. Use the scan tool to retrieve active fault codes and record them.

   _____

19. Does the fault code indicate the circuit is shorted?   ☐ Yes ☐ No

20. If you answered yes to question 19, what is the faulty component?

   _____

21. If you answered no to question 19, locate the connector to the air bag control module. Describe this connector's location.

Task Completed

_____

_____

_____

22. Disconnect the negative battery cable and isolate it. Wait the recommended amount of time before proceeding. ☐

23. Disconnect the air bag control module connector and locate the wires of the driver-side squib circuit. What are the color codes of the wires?

_____

_____

_____

24. Use an ohmmeter to test the wire harness between the air bag control module connector and the lower clockspring connector. Record your results.

_____

_____

25. Based on your results, what have you determined to be the location of the fault?

_____

_____

Instructor's Response _____

_____

_____

_____

Job Sheet 56

56

Name _____ Date _____

Clockspring Replacement and Centering

Upon completion of this job sheet, you should be able to remove, replace, and properly center the clockspring used in an air bag system.

ASE Correlation

This job sheet is related to the ASE Electrical/Electronic System Certification Exam's content area: *Miscellaneous;* task: Inspect, test, and repair or replace the air bag, air bag module, sensors, connectors, and wires of the air bag system circuit(s).

Tools and Materials

A vehicle equipped with a driver's-side air bag
Service manual for chosen vehicle
Safety glasses
Battery terminal puller
Steering wheel puller

Describe the vehicle being worked on:

Year _____ Make _____ Model _____

VIN _____ Engine type and size _____

Procedure

Task Completed

1. According to the service information, in what position must the front wheels be before beginning?

   _____

   _____

2. Place the front wheels in the position described in step 1. ☐

3. Disconnect the battery negative cable and isolate it. ☐

4. According to the service information, how long must you wait before proceeding?

   _____

   _____

5. Remove the air bag module from the steering wheel. ☐

6. Remove the steering wheel attaching bolt or nut. Can this fastener be reused?
   ☐ Yes ☐ No

7. Mark the shaft and steering wheel with index marks for reinstallation. ☐

8. Use a steering wheel puller to remove the steering wheel from the shaft. ☐

9. Remove the upper and lower steering column shrouds. ☐

10. Disconnect the clockspring connector from the steering column harness. ☐

11. Remove the retaining screws (or release the locking tabs) and the clockspring.

12. If the same clockspring is to be reinstalled, describe the procedure for centering the clockspring.

13. In what position must the front wheel and steering column be in to install the clockspring?

14. Install all components.

What is the torque specification for the clockspring fasteners (if used)?

_____

What is the torque specification for the steering wheel fastener?

_____

What is the torque specification for the air bag module fasteners?

_____

Instructor's Response _____

_____

_____

_____

Job Sheet 57

57

Name _____ Date _____

Occupant Classification System Validation Procedure

Upon completion of this job sheet, you should be able to properly perform the OCS validation procedure.

ASE Correlation

This job sheet is related to the ASE Electrical/Electronic System Certification Exams content area: *Accessories Diagnosis and Repair;* task: Diagnose supplemental restraint system (SRS) concerns, determine necessary action.

Tools and Materials

Vehicle with OCS
Scan tool
Validation weight set
Battery charger

Describe the vehicle being worked on:

Year _____ Make _____ Model _____

VIN _____ Engine type and size _____

Procedure

Task Completed

1. Connect a battery charger to the vehicle's battery and set so 13.0 volts is read across the terminals. ☐

2. Is the system on the assigned vehicle a strain gauge or bladder system?
   ☐ Strain Gauge   ☐ Bladder

3. Connect the scan tool to the DLC and access the occupant classification system module. Provide the following information (if available):

   OCM part number _____

   Software version _____

4. Make sure the passenger front seat is empty. What is the state of the PADL?

   _____

5. Sit in the seat and describe the state of the PADL.

   _____

6. Use the scan tool to access the sensor display function. Record the values while remaining in the seat.

   _____

   _____

7. Move out of the seat and record the values.

_____

_____

8. Assure that the seat is empty and navigate the scan tool to the validation function. What instructions are provided on the scan tool screen when the test is started (if any):

_____

_____

_____

_____

9. According to the scan tool, what is the first step that needs to be performed?

_____

_____

10. What is the first weight amount to be placed on the seat? _____

11. During the validation procedure, what is the state of the PADL? _____

12. Complete the validation procedure while listing the required weight amounts.

_____

_____

_____

13. How do you know the procedure was completed?

_____

14. What is the state of the PADL once the procedure is completed? _____

☐ **15.** Clear any DTCs that may have been set.

16. Disconnect a sensor and record any DTCs.

_____

_____

17. Reconnect the sensor. Did the fault go from active to stored? _____

18. What needs to be done to clear the DTC?

_____

☐ **19.** Perform the required task list in step 18.

Instructor's Response _____

_____

_____

DIAGNOSTIC CHART 15-1

PROBLEM AREA:	Automatic seat belt operation
SYMPTOMS:	Belt stalls at A-pillar.
POSSIBLE CAUSES:	**1.** "A" limit switch circuit shorted to ground. **2.** Door switch circuit shorted to ground. **3.** "B" limit switch circuit open. **4.** Obstruction.

DIAGNOSTIC CHART 15-2

PROBLEM AREA:	Automatic seat belt operation
SYMPTOMS:	Belt stalls at B-pillar.
POSSIBLE CAUSES:	**1.** "B" limit switch circuit shorted to ground. **2.** Door switch circuit open. **3.** "A" limit switch circuit open. **4.** Obstruction.

DIAGNOSTIC CHART 15-3

PROBLEM AREA:	Automatic seat belt operation
SYMPTOMS:	Belt will not move.
POSSIBLE CAUSES:	**1.** Motor circuit shorted to ground. **2.** Open motor circuit. **3.** Inertia switch open. **4.** Faulty motor. **5.** Faulty controller. **6.** Obstruction.

DIAGNOSTIC CHART 15-4

PROBLEM AREA:	Automatic seat belt operation
SYMPTOMS:	Belt moves when it is not supposed to.
POSSIBLE CAUSES:	**1.** Faulty doorjamb switches. **2.** Faulty controller.

DIAGNOSTIC CHART 15-5

PROBLEM AREA:	Air bag system operation
SYMPTOMS:	Air bag warning lamp illuminated.
POSSIBLE CAUSES:	**1.** Squib circuit shorted to ground. **2.** Defective clockspring. **3.** Squib circuit open. **4.** Sensor circuit open. **5.** Sensor circuit shorted. **6.** Faulty module. **7.** Poor battery feed circuit to control module. **8.** Poor ignition feed circuit to control module. **9.** Poor control module ground circuit. **10.** Loss of bus communications. **11.** Faulty instrument cluster lamp circuit.

Hybrid and High-Voltage System Service

Upon completion and review of this chapter, you should be able to:

❏ Demonstrate proper safety precautions associated with servicing the hybrid electric vehicle.

❏ Properly remove and install the high-voltage service plug.

❏ Access and interpret DTCs, information codes, freeze frame data, and history data.

❏ Determine the cause of HV battery system failures.

❏ Determine failures of the inverter/converter assembly.

❏ Determine the cause of system main relay failures.

❏ Replace the system main relays.

Basic Tools

Basic mechanic's tool set

Fender covers

Safety glasses

Insulating gloves

Classroom Manual
Chapter 16, page 518

Introduction

This chapter discusses some of the service procedures for a common hybrid system. At the present time, specially trained dealership technicians perform most of the service of the hybrid system. Because of this, the main focus of this chapter will be on safety concerns associated with the high-voltage system.

The hybrid electric vehicle (HEV) system combines the operating characteristics of an internal combustion engine and an electric motor. In addition, the system can use regenerative braking to recover energy that normally would be lost to heat and use it to supplement the power of the engine. The sample HEV used to describe the service procedures in this chapter includes the following components (Figure 16-1):

❏ Hybrid transaxle that integrates the MG1, the MG2, and the planetary gear unit.
❏ Inverter assembly.
❏ HV ECU.
❏ ECM.
❏ HV battery.
❏ Battery ECU.
❏ Service plug.
❏ The system main relay (SMR).
❏ Auxiliary battery.

Safety Precautions

Since the hybrid system can use voltages in excess of 500 volts (both DC and AC), it is vital that the service technician be familiar with, and follow, all safety precautions. Failure to perform the correct procedures can result in electrical shock, battery leakage, or an explosion. The following are the safety procedures that must be followed whenever servicing the HEV's high-voltage systems:

❏ Remove the key from the ignition.
❏ Disconnect the negative (−) terminal of the auxiliary (12-volt) battery. Always disconnect the auxiliary battery prior to removing the high-voltage service plug.
❏ Remove the high-voltage service plug and put it where it cannot be accidentally reinstalled by someone else.

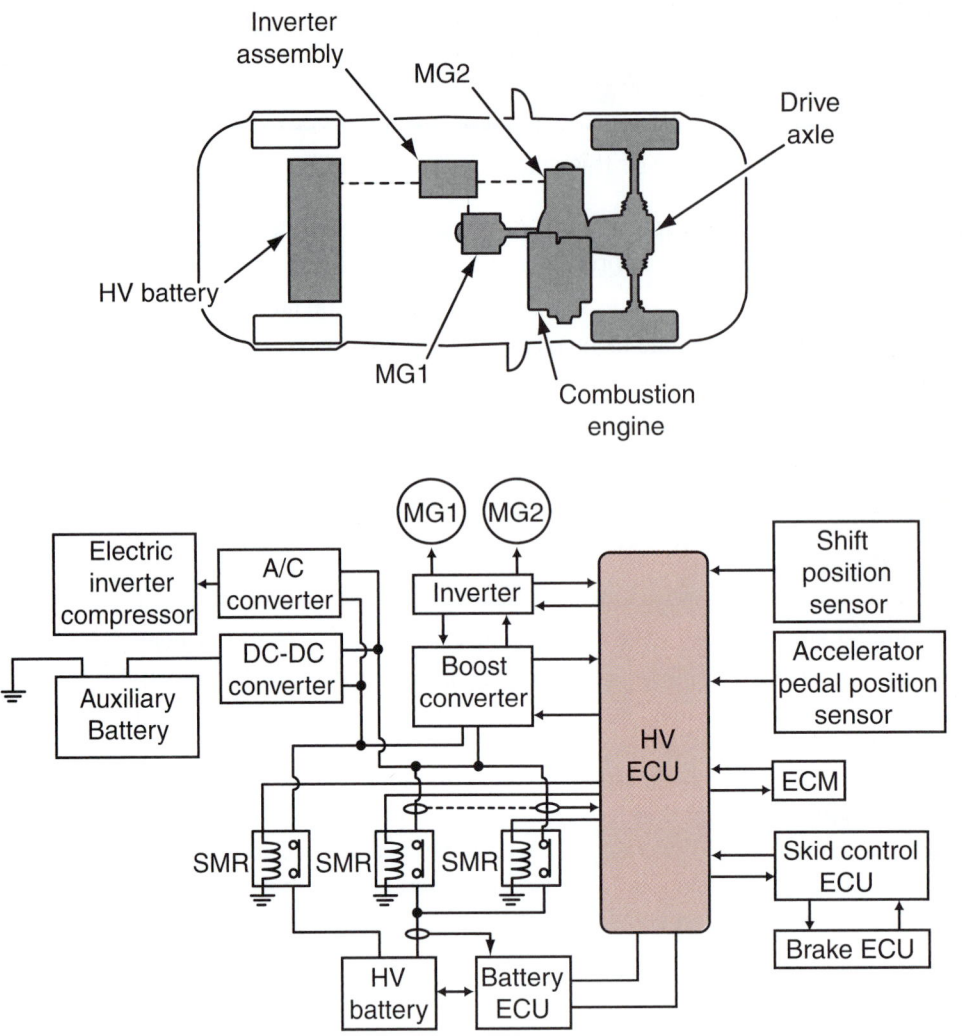

Figure 16-1 Components of an HEV system.

❑ Cover the high-voltage service plug receptacle with insulation tape.

❑ Do not attempt to test or service the system for five minutes after the high-voltage service plug is removed. At least five minutes is required to discharge the high-voltage condenser inside the inverter.

❑ Test the integrity of the insulating gloves prior to use.

❑ Wear high-voltage insulating gloves when disconnecting the service plug.

❑ Never cut the orange high-voltage power cables. The wire harnesses, terminals, and connectors of the high-voltage system are identified by orange. In addition, high-voltage components may have a "High Voltage" caution label attached to them.

❑ Cover the terminals of a disconnected connector with insulation tape.

❑ Never open high-voltage components.

❑ Use a DMM to confirm that high-voltage circuits have 0 V before performing any service procedure.

❑ Use insulated tools when available.

❑ Do not wear metallic objects that may cause electrical shorts.

❑ Follow the service manual diagnostic procedures.

❑ Wear protective safety goggles when inspecting the high-voltage (HV) battery.

❑ Before touching any of the high-voltage system wires or components, wear insulating gloves, make sure the power switch is off, and disconnect the auxiliary battery.

- ❏ Turn the power switch to the OFF position prior to performing a resistance check.
- ❏ Turn the power switch to the OFF position prior to disconnecting or reconnecting any connectors or components.
- ❏ Isolate with insulation tape any high-voltage wires that have been removed.
- ❏ Properly torque the high-voltage terminals.

Insulating Glove Integrity Test

The insulating gloves that the technician wears for protection while servicing the high-voltage system must be tested for integrity before use. If there is a leak in the gloves, high-voltage electricity can travel through the hole to the technician's body. To test a glove, blow air into it and then fold it at the base to seal the air inside. Slowly roll the base of the glove toward the fingers. If the glove holds pressure, its insulating properties are intact. If any leaks are detected, discard the glove.

High-Voltage Service Plug

The HEV is equipped with a **high-voltage service plug** that disconnects the HV battery from the system. Usually, this plug is located near the battery (Figure 16-2). Prior to disconnecting the high-voltage service plug, the vehicle must be turned off and the negative terminal of the auxiliary battery must be disconnected. Once the high-voltage service plug is removed, the high-voltage circuit is shut off at the intermediate position of the HV battery.

The high-voltage service plug assembly contains a safety interlock reed switch. The reed switch is opened when the clip on the high-voltage service plug is lifted. The open reed switch turns off power to the service main relay (SMR). The main fuse for the high-voltage circuit is inside the high-voltage service plug assembly.

However, never assume that the high-voltage circuits are off. The removal of the high-voltage service plug does not disable the individual high-voltage batteries. Use a DMM to verify that 0 volts are in the system before beginning service. When testing the circuit for voltage, set the voltmeter to the 400 VDC scale.

 SERVICE TIP: DTCs will be erased once the batteries are disconnected. Prior to disconnecting the system, be sure to check and record and DTCs.

 CAUTION: Once the high-voltage service plug is removed, do not operate the power switch. Doing so may damage the hybrid vehicle control ECU.

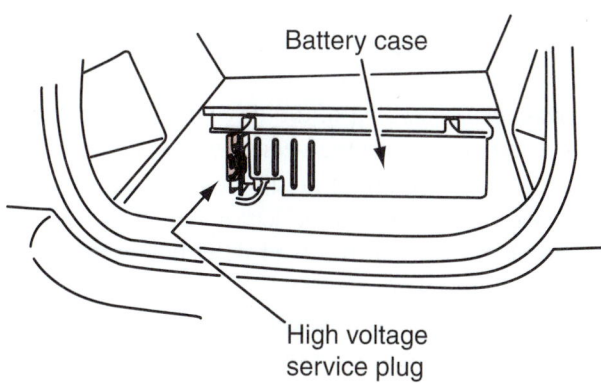

Battery case

High voltage
service plug

Figure 16-2 The high-voltage service plug is usually located near the HV battery.

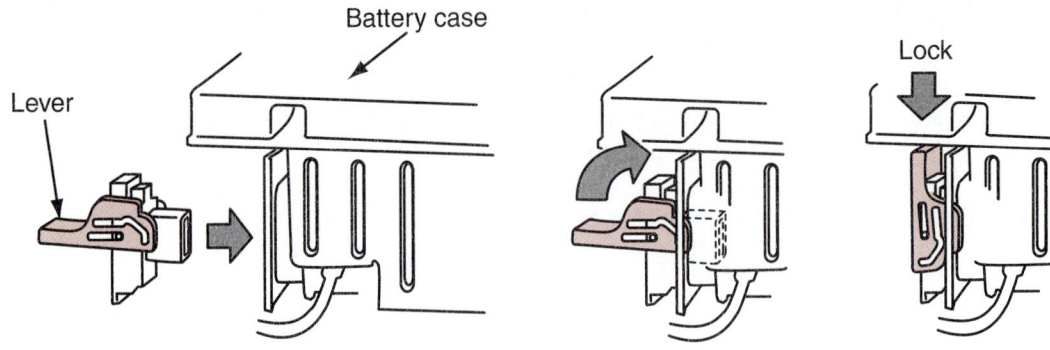

Figure 16-3 To install the service plug, make sure the lever is down and then fully locked once installed.

After the high-voltage service plug is removed, a minimum of five minutes must pass before beginning service on the system. This is required to discharge the high voltage from the condenser in the inverter circuit.

To install the high-voltage service plug, make sure the lever is locked in the DOWN position (Figure 16-3). Slide the plug into the receptacle, and lock it in place by lifting the lever upward. Once it is locked in place, it closes the reed switch.

Self-Diagnosis Capabilities

Classroom Manual
Chapter 16, page 519

Special Tools

Scan tool

The DTCs associated with the HV system may include both SAE codes and manufacturer codes. SAE codes must be set as prescribed by the SAE, while manufacturer codes can be set by a manufacturer.

The **high-voltage electronic control unit (HV ECU)** controls the two motor generators (MG1 and MG2) and the engine based on torque demand. Also, these units are controlled based on regenerative brake control and the HV battery's state of charge (SOC). It is the responsibility of the HV ECU to provide reliable circuit shutdown in the event of a malfunction. The HV ECU uses three relays housed in the SMR assembly to connect and disconnect the high-voltage circuit. If a malfunction is detected, the HV ECU will use the relays to control the system based on programmed instructions stored in its memory. If the system is determined to be malfunctioning, the HV ECU will illuminate the master warning lamp in the instrument cluster. In addition, it may illuminate the HV system warning, the HV battery warning, or the discharge warning lamps.

DTCs are set when the fault occurs. To access the DTCs, a scan tool with the proper interface module (if needed) is connected to the data link connector (DLC). The scan tool will also provide information codes, freeze frame data, and history data.

Information codes are additional codes that provide more information and freeze frame data concerning the DTCs. Information codes along with the DTC indicate more precisely the location of the fault. These codes are accessed using the scan tool while in the HV ECU system screen (Figure 16-4).

The **freeze frame data** is a recording of the driving condition when the malfunction occurred. This is useful for determining how the vehicle was operating at the time and for locating any input or output values that are out of range.

History data information can be useful for determining if a customer's concern is actually a problem with the system. It provides a means of determining if the vehicle owner's driving habits may be the cause of the problem. The data will display information such as if the gear shift lever was moved before the vehicle was ready, if the transmission was shifted into park while the vehicle was moving, if the accelerator pedal was depressed while in the NEUTRAL position, and so on. Use the proper service manual for information on using this data.

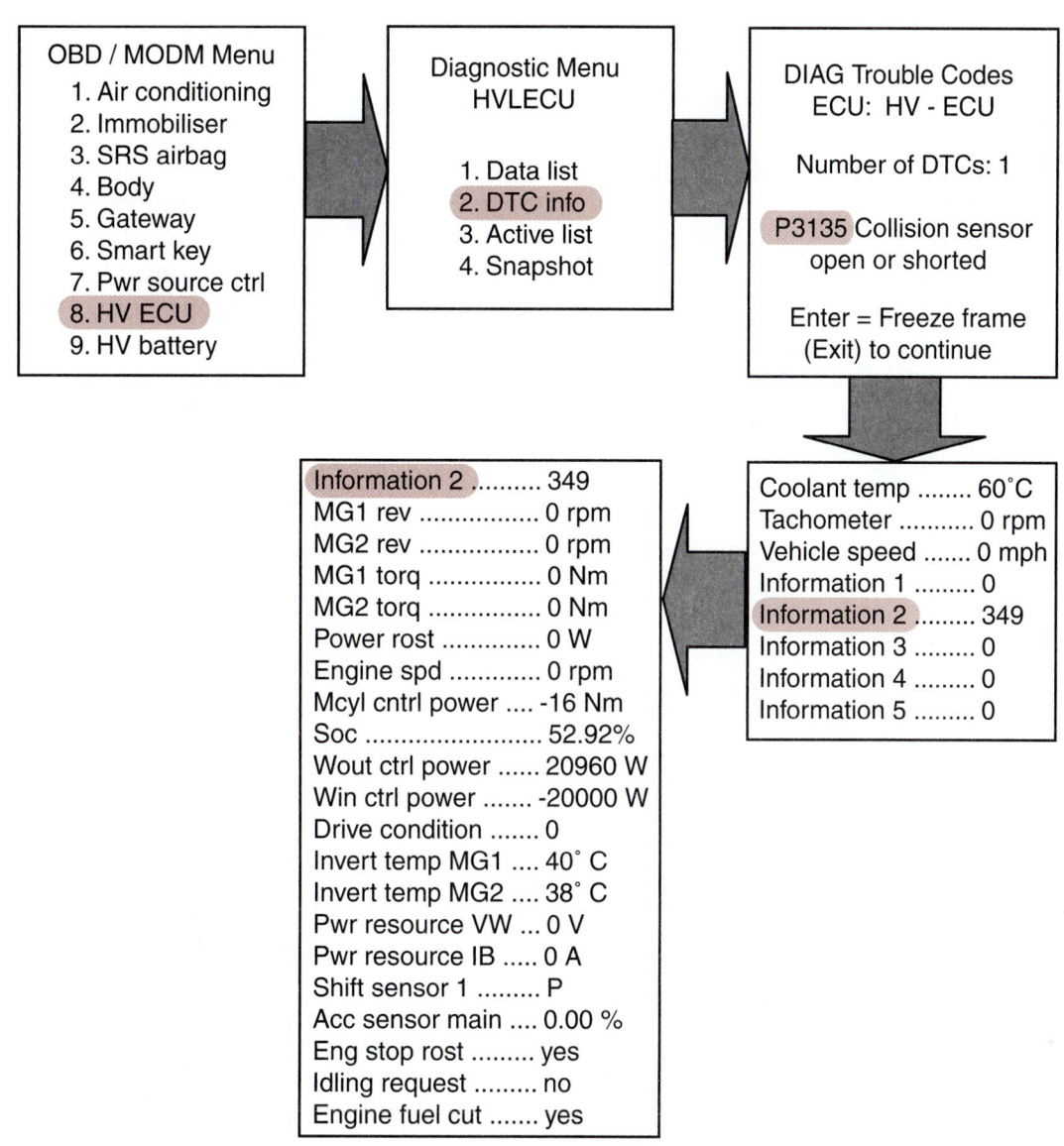

OBD / MODM Menu
1. Air conditioning
2. Immobiliser
3. SRS airbag
4. Body
5. Gateway
6. Smart key
7. Pwr source ctrl
8. HV ECU
9. HV battery

Diagnostic Menu
HVLECU

1. Data list
2. DTC info
3. Active list
4. Snapshot

DIAG Trouble Codes
ECU: HV - ECU

Number of DTCs: 1

P3135 Collision sensor
open or shorted

Enter = Freeze frame
(Exit) to continue

Coolant temp .... 60°C
Tachometer .... 0 rpm
Vehicle speed .... 0 mph
Information 1 0
Information 2 349
Information 3 0
Information 4 0
Information 5 0

Information 2 349
MG1 rev .... 0 rpm
MG2 rev .... 0 rpm
MG1 torq .... 0 Nm
MG2 torq .... 0 Nm
Power rost .... 0 W
Engine spd .... 0 rpm
Mcyl cntrl power .... -16 Nm
Soc 52.92%
Wout ctrl power ...... 20960 W
Win ctrl power .... -20000 W
Drive condition 0
Invert temp MG1 .... 40° C
Invert temp MG2 .... 38° C
Pwr resource VW ... 0 V
Pwr resource IB ..... 0 A
Shift sensor 1 .... P
Acc sensor main .... 0.00 %
Eng stop rost .... yes
Idling request .... no
Engine fuel cut .... yes

Figure 16-4 Accessing information codes.

CUSTOMER CARE: Since the hybrid vehicle is still new on the market, the chances are the vehicle owner has never driven one before. Take the time to explain the proper operation of the vehicle to prevent any misunderstandings of the system function.

In addition, the scan tool may provide activation tests of the HV ECU. Different tests are available. Using the service manual, determine what test is required for the specific test. The following are typical mode functions:

❑ Mode 1. This mode runs the engine continuously in the PARK position and to check HV ECU operation. It can also be used to disable traction control so the speedometer test can be performed.
❑ Mode 2. Cancels the traction control that is affected when the rotational difference between the front and rear wheels is excessive.
❑ Inverter stop. Keeps the inverter power transistor on to determine if there is an internal leak in the inverter or the HV control ECU.

❏ Cranking request. Activates the motor generator continuously to crank the engine in order to measure the compression.

High-Voltage Battery Service

Classroom Manual
Chapter 16, page 520

A **battery ECU** monitors the condition of the HV battery assembly. The battery ECU determines the SOC of the HV battery by monitoring voltage, current, and temperature. The battery ECU collects data and transmits it to the HV ECU to be used for proper charge and discharge control.

The battery ECU also controls the operation of the battery blower motor to maintain proper HV battery temperature.

The HV battery stores power generated by MG1 and recovered by MG2 during regenerative braking (Figure 16-5). The HV battery must also supply power to the electric motor when the vehicle is first started from a stop or when additional power is needed. A typical HV battery uses several nickel-metal-hydride modules and can provide over 270 volts (Figure 16-6)

When the vehicle is moving, the HV battery is subjected to repetitive charge and discharge cycles. The HV battery is discharged by MG2 during acceleration mode and then is recharged

Special Tools

DMM

Scan tool

MG1 functions as the control element for the planetary gear set. It recharges the HV battery and supplies electrical power to drive MG2. MG1 also functions as the starter for the engine. MG2 is used for power at low speeds and for supplemental power when needed at higher speeds.

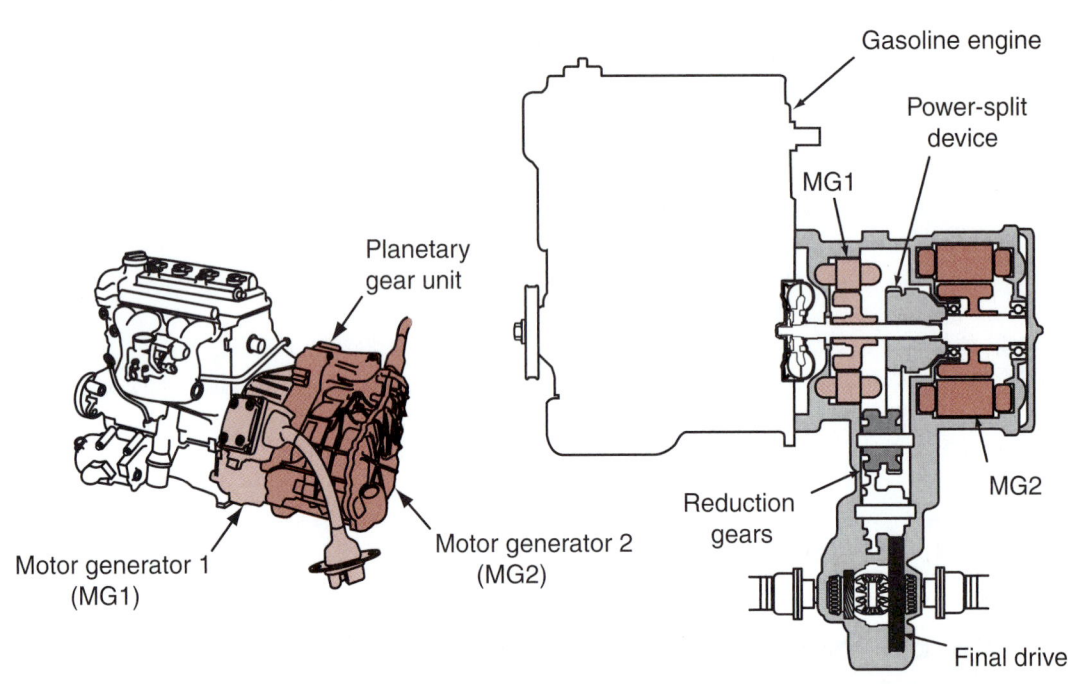

Figure 16-5 Layout of the generating HEV transaxle.

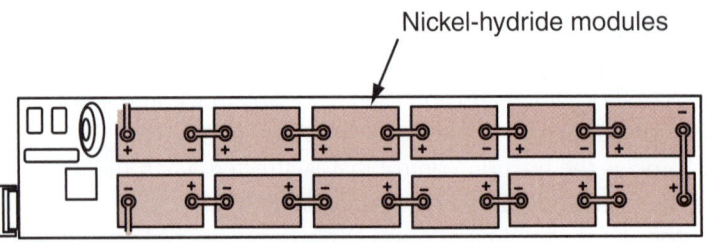

Figure 16-6 The HV battery.

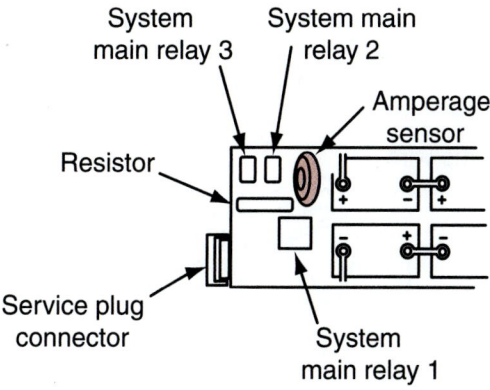

Figure 16-7 The battery ECU and amperage sensor.

by regenerative braking. An amperage sensor (Figure 16-7) is used so the battery ECU can transmit requests to the HV ECU to maintain the SOC of the HV battery. The battery ECU attempts to keep the SOS at 60 percent. The battery ECU also monitors delta SOC to determine if it is capable of maintaining acceptable levels of charge. The normal, low-to-high SOC delta is 20 percent.

If the battery ECU sends abnormal messages to the HV ECU, the HV ECU illuminates the warning light and enters fail-safe control. DTCs and informational codes are set along with freeze frame data. Fail-safe can result in the battery ECU restricting or stopping the charging and discharging of the HV battery.

If there is a leak in the high-voltage system insulation that may seriously harm a person, the system will enter fail-safe control and set DTCs. This will occur if the battery ECU determines the insulation resistance of the power cable is 100k ohms or less.

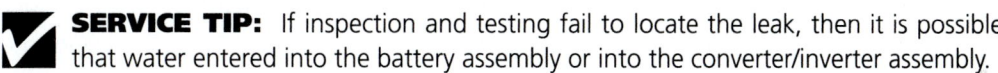

SERVICE TIP: If inspection and testing fail to locate the leak, then it is possible that water entered into the battery assembly or into the converter/inverter assembly.

Whenever an HV battery malfunction occurs, use the scan tool to view the "HV Battery Data List." This provides all HV battery system information.

High-Voltage Battery Charging

If the SOC of the HV battery is too low to allow the engine to run, the HV battery will need to be recharged. This requires the use of a special high-voltage battery charger (Figure 16-8). In addition, most manufacturers will only allow specially trained people to recharge the battery. Some manufacturers will not even supply the charging equipment to the dealer; a representative of the company performs the task of recharging the HV battery.

HV battery recharging must be performed outside. The correct cable is connected between the vehicle and the charger (Figure 16-9). When using the charger, the immediate area must be secured and marked with warning tape. It will require about three hours to recharge the battery to an SOC of about 50 percent.

Special Tools

High-voltage battery charger

Warning tape

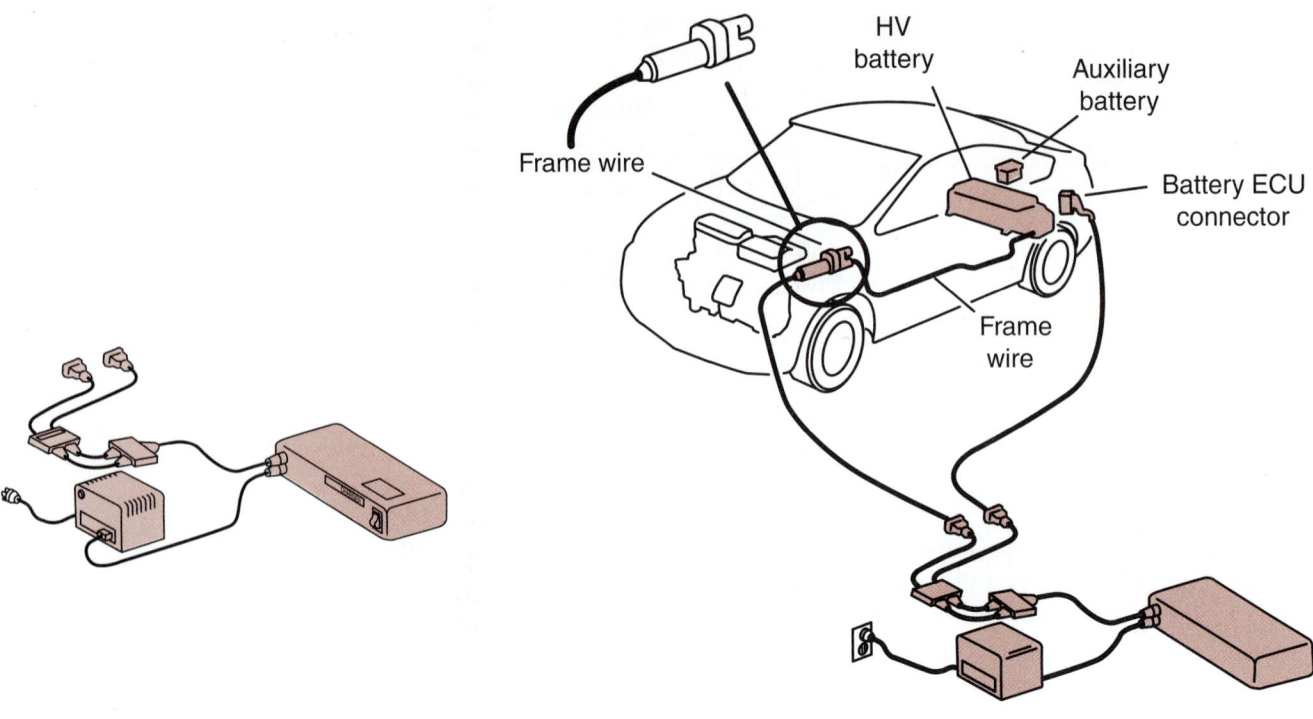

Figure 16-8 HV battery charger.

Figure 16-9 HV battery charger connection.

Classroom Manual
Chapter 16, page 522

Special Tools

Scan tool

Inverter/Converter Assembly

The **inverter** (Figure 16-10) controls the current flow between MG1, MG2, and the HV battery. The inverter converts HV battery DC voltage into three-phase alternating current for MG1 and MG2. It also converts (rectifies) high-voltage AC from MG1 and MG2 to DC voltage to charge the HV battery. The HV ECU controls the activation of the power transistors to perform these functions (Figure 16-11)

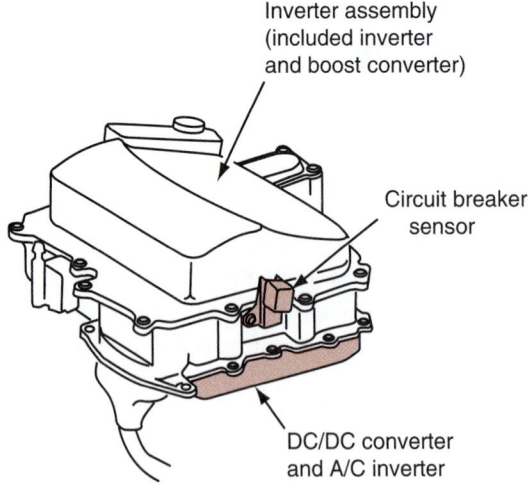

Figure 16-10 The inverter.

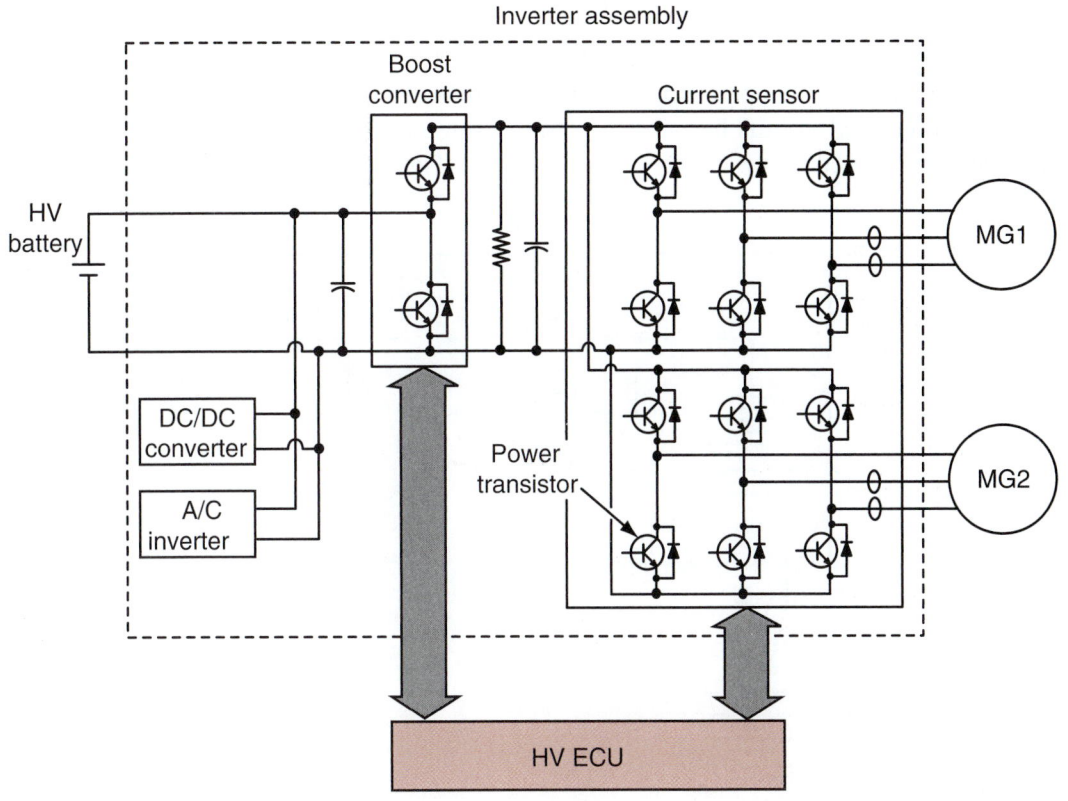

Figure 16-11 Inverter assembly internal electrical circuit.

The **converter** is a DC/DC transformer (Figure 16-12). It converts the voltage from 270 volts DC to 14 volts DC to recharge the auxiliary battery and to power 12-volt electrical components. If the DC/DC converter should malfunction, the auxiliary battery voltage will drop until it is no longer possible to drive the vehicle. The HV ECU monitors operation of the DC/DC converter and will illuminate the warning lamp and set a DTC if a failure is determined.

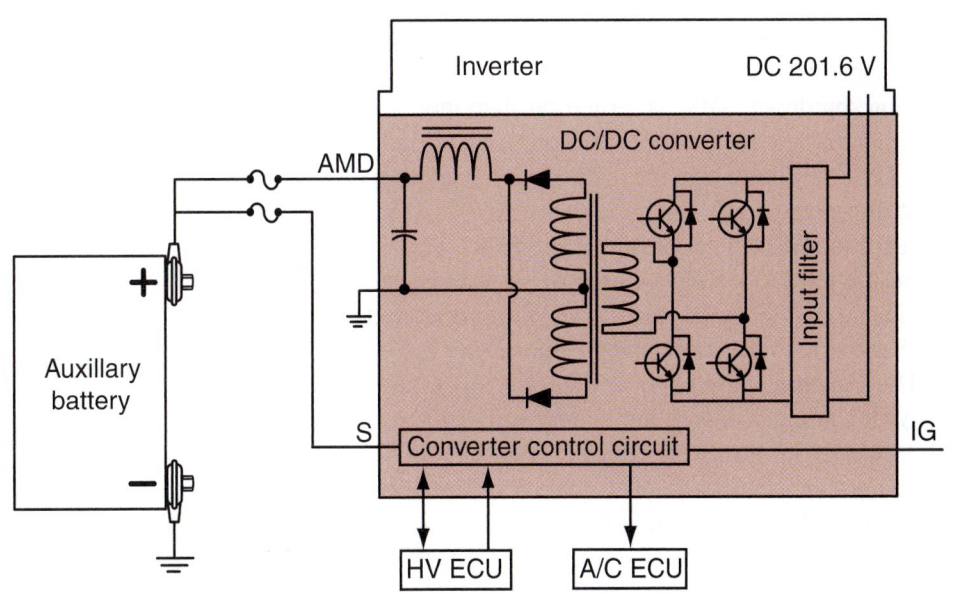

Figure 16-12 The internal circuit of the DC/DC converter.

System Main Relay

The HV ECU controls the operation of the **system main relay (SMR)** to connect and disconnect the power source of the high-voltage system. Three relays are used. One is for the negative side, while the other two are for the positive side (Figure 16-13).

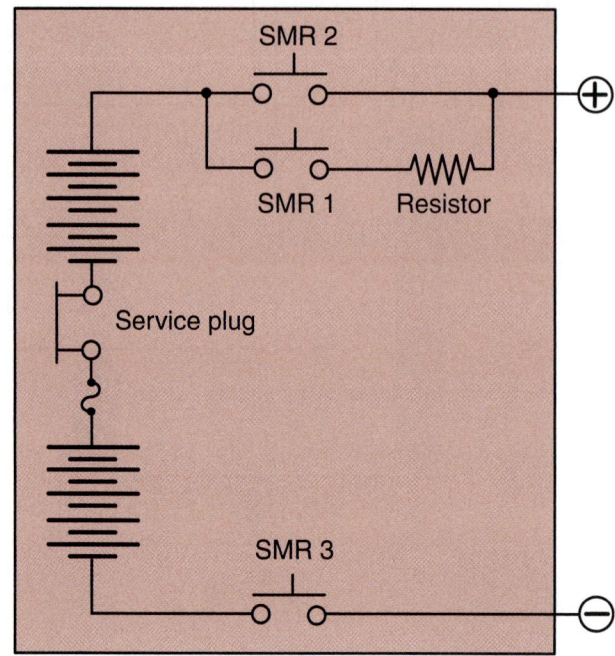

Figure 16-13 Schematic of the SMR.

Initially SMR1 and SMR3 are turned on to energize the circuit for the HV battery. This provides the needed current for the motor generator to start the engine. Then SMR2 is energized and SMR1 is turned off. This makes the current from the generator flow through a resistor, thus controlling the amount of current flow. This protects the circuit from excessive initial current from the generator. Finally, SMR2 is turned on and SMR1 is turned off to allow free flow of current in the circuit.

During shutdown, SMR2 is turned off first, then SMR3. This provides a means for the HV ECU to verify that the relays have been properly turned off.

The HV ECU checks that the system main relay is operating normally. If a fault is detected, DTCs and information data will be stored. The following is a typical procedure to be used if diagnosis leads to replacement of one of the SMRs:

1. Secure the proper service manual, tested insulating gloves, and insulting tape.
2. Gain access to the auxiliary battery and disconnect the negative terminal.
3. Wearing the insulated gloves, remove the high-voltage service plug.
4. Use insulating tape to cover the terminals of the plug receptacle.
5. After gaining access to the battery carrier panel, remove the junction terminal. Be sure to wear your insulating gloves.
6. Remove the grandwire, SMR2 cover, and disconnect the relay connector to remove SMR2.
7. Remove SMR3 by disconnecting the main battery cable, disconnecting the connector, and removing the fasteners.
8. Disconnect the connector and the ground terminal for SMR1.
9. Remove the fasteners and remove SMR1.

An owner of a 2005 Prius has brought his vehicle into the shop. He claims that the master warning light comes on every once in a while but goes out after a couple of starts. The technician retrieves the DTCs from the HV ECU and finds a "Shift before ready" code. Upon investigation of the information code, the technician also refers to history data. Here it is determined that the cause of the fault was that the customer was not waiting for the ready light to stop flashing before placing the transmission into drive. The warning light would go out after three starts if the same condition did not reoccur. The technician took the time to go over the proper startup sequence with the vehicle owner so future problems could be avoided.

Terms to Know

Battery ECU	High-voltage electronic control unit (HV ECU)	Information codes
Converter	High-voltage service plug	Inverter
Freeze frame data	History data	System main relay (SMR)

ASE-Style Review Questions

1. All of the following statements concerning hybrid high-voltage system safety is true EXCEPT:
   A. Disconnect the motor generators prior to turning the ignition off.
   B. Disconnect the negative (–) terminal of the auxiliary battery before removing the service plug.
   C. Do not attempt to test or service the system for five minutes after the high-voltage service plug is removed.
   D. Turn the power switch to the OFF position prior to performing a resistance check.

2. *Technician A* says HEV batteries can provide over 270 volts.
   Technician B says the HEV high-voltage from the MG1 and MG2 to the inverter/converter can be more than 500 volts.
   Who is correct?
   A. A only
   B. B only
   C. Both A and B
   D. Neither A nor B

3. When working on the high-voltage system, which of the following should be done?
   A. Always place the high-voltage service plug where someone will not accidentally reinstall it.
   B. Before servicing, use a voltmeter set on 400 VDC to determine if the high-voltage system voltage is at 0 volts.
   C. Test the integrity of the insulating gloves prior to use.
   D. All of the above.

4. *Technician A* says the main system relay should be removed before disconnecting the service plug.
   Technician B says the high-voltage components are usually identified with a warning label.
   Who is correct?
   A. A only
   B. B only
   C. Both A and B
   D. Neither A nor B

5. To test the integrity of the insulating gloves:
   A. Fill the gloves with water to see if there is a leak.
   B. Fill the gloves with air and submerge in water to see if air bubbles arise from any leaks.
   C. Shine a flashlight into the glove and see if light escapes.
   D. None of the above.

6. The high-voltage service plug:
   A. Disconnects the inverter/converter from the motor generators.
   B. Disconnects the auxiliary battery from the HV battery.
   C. Disconnects the HV battery from the system.
   D. Provides a connection for the battery charger.

7. *Technician A* says once the service plug is disconnected, there is no high voltage in the vehicle systems.
   Technician B says prior to disconnecting the high-voltage service plug, the vehicle must be turned off and the negative terminal of the auxiliary battery must be disconnected.
   Who is correct?
   A. A only
   B. B only
   C. Both A and B
   D. Neither A nor B

8. *Technician A* says the high-voltage electronic control unit (HV ECU) can shut down the high-voltage system if a fault is detected.
   Technician B says if the auxiliary battery voltage goes low, the HV ECU will direct regenerative braking energy to the auxiliary battery.
   Who is correct?
   A. A only
   B. B only
   C. Both A and B
   D. Neither A nor B

9. *Technician A* says the HV battery is charged with a conventional flooded battery charger set at 3.5 amps.
   Technician B says the HV battery can only be charged if it is removed from the vehicle.
   Who is correct?
   A. A only
   B. B only
   C. Both A and B
   D. Neither A nor B

10. *Technician A* says that information codes provide more specific indications of the fault location.
    Technician B says the freeze frame data is a recording of the driving condition when the malfunction occurred.
    Who is correct?
    A. A only
    B. B only
    C. Both A and B
    D. Neither A nor B

ASE Challenge Questions

1. *Technician A* says if an intermittent fault code is being set in the high-voltage system, water may be entering a connector.
   Technician B says to locate the cause of current leakage through the high-voltage cable insulation, spray the cable with water.
   Who is correct?
   A. A only
   B. B only
   C. Both A and B
   D. Neither A nor B

2. *Technician A* says if the vehicle fails to start, the system main relay may have a fault.
   Technician B says if the vehicle fails to start, the HV battery may be too low.
   Who is correct?
   A. A only
   B. B only
   C. Both A and B
   D. Neither A nor B

3. *Technician A* says a faulty reed switch in the service plug may cause the vehicle to not start.
   Technician B says if a fault is set that the HV battery is too hot, the water cooler pump may have failed.
   Who is correct?
   A. A only
   B. B only
   C. Both A and B
   D. Neither A nor B

4. *Technician A* says the engine is started by the auxiliary battery if the HV SOC is below 15 percent.
   Technician B says if the engine fails to start, the inverter may have malfunctioned.
   Who is correct?
   A. A only
   B. B only
   C. Both A and B
   D. Neither A nor B

5. *Technician A* say if the vehicle operates slowly from a stop, the engine may require a tune-up.
   Technician B says the HEV system does not provide a method of performing a compression test on the engine.
   Who is correct?
   A. A only
   B. B only
   C. Both A and B
   D. Neither A nor B

Job Sheet 58

58

Name _____ Date _____

Hybrid Safety

Upon completion of this job sheet, you should be familiar with the critical safety procedures involved in servicing a high-voltage hybrid system.

ASE Correlation

This job sheet is related to the ASE Electrical/Electronic System Certification Exam's content area: *General Electrical System Diagnosis;* task: Identify location of hybrid vehicle high-voltage circuit disconnect and safety procedures.

This job sheet also is related to the ASE Electrical/Electronic System Certification Exam's content area: *Battery Diagnosis and Service;* task: Identify high-voltage circuits of electric or hybrid electric vehicle and related safety precautions.

Tools and Materials

HEV
Service manual
Insulating gloves
Eye protection

Describe the vehicle being worked on:

Year _____ Make _____ Model _____

VIN _____ Engine type and size _____

Procedure

Task Completed

1. Use the service manual information and determine the location on the vehicle for the high-voltage service plug.

   _____

   _____

2. What must be done prior to disconnecting the service plug?

   _____

   _____

3. How long must you wait after the plug is disconnected before servicing the system?

   _____

   _____

☐

4. Access the 12-V auxiliary battery and remove the negative terminal.

5. Test the insulating gloves for leaks. Are the gloves safe to use?   ☐ Yes   ☐ No

If no, inform your instructor.

☐

6. Put on the insulating gloves and eye protection.

7. Remove the service plug. What device(s) are integrated into the service plug assembly?

_____

_____

☐

8. Reinstall the service plug.

☐

9. Review your observations with your instructor.

Instructor's Response _____

_____

_____

_____

Job Sheet 59

59

Name _____ Date _____

Hybrid System DTCs

Upon completion of this job sheet, you should be able to diagnose hybrid faults by retrieving DTCs, information codes, and data values.

ASE Correlation

This job sheet is related to the ASE Electrical/Electronic System Certification Exam's content area: *Accessories Diagnosis and Repairs;* task: Diagnose body electronic systems using a scan tool and determine necessary action.

Tools and Materials

HEV
Scan tool
Service manual

Describe the vehicle being worked on:

Year _____ Make _____ Model _____

VIN _____ Engine type and size _____

Procedure

Task Completed

1. With the vehicle started, observe if any warning lamps are illuminated. If so, which ones?

   _____

   _____

2. Connect the scan tool and check for DTCs in the PCM and the HV ECU. Record any DTCs displayed.

   _____

   _____

3. Access and record any information codes that are displayed.

   _____

   _____

4. Use the service manual to determine what system component is affected. Record your results.

   _____

   _____

5. Access the "HV ECU Data List" screen and record any values that are out of range.

_____

_____

6. Based on your results, what would be the next logical approach to take?

_____

_____

Instructor's Response _____

_____

_____

_____

DIAGNOSTIC CHART 16-1

PROBLEM AREA:	HV system
SYMPTOMS:	Engine will not start. Warning lamp illumination.
POSSIBLE CAUSES:	**1.** Auxiliary battery low. **2.** HV battery low SOC. **3.** Faulty motor generator. **4.** Faulty HV ECU. **5.** Poor electrical connections in HV cable. **6.** Service plug not installed.

DIAGNOSTIC CHART 16-2

PROBLEM AREA:	HV battery
SYMPTOMS:	Low SOC. Warning lamp illumination.
POSSIBLE CAUSES:	**1.** Poor electrical connections. **2.** Faulty motor generator or circuits. **3.** Faulty HV ECU or circuits. **4.** Faulty battery ECU or circuits. **5.** Faulty amperage sensor or circuits. **6.** Faulty HV battery.

APPENDIX A

ASE PRACTICE EXAMINATION

1. The current draw of a window motor is being measured.
   Technician A says the ammeter can be connected on the power supply side of the motor.
   Technician B says the ammeter can be connected on the ground side of the motor.
   Who is correct?
   A. A only
   B. B only
   C. Both A and B
   D. Neither A nor B

2. The digital readout of an auto-ranging DVOM that is in the "volts" position displays "13.7." Which of the following statements is true about this measurement?
   A. The actual value is 13.7 V.
   B. The actual value is 13.7 mV.
   C. The actual value is 137 mV.
   D. More information is needed in order to determine the actual value.

3. A voltmeter that is connected across the input and output terminals of an instrument cluster illumination lamp rheostat indicates 12.6 volts with the switch in the maximum brightness position and the engine off. Which of the following statements is true?
   A. The voltage available at the lamps will be about 12.6 volts.
   B. The voltage available at the lamps will be 0.0 volts.
   C. The rheostat is operating correctly.
   D. More information is needed in order to determine whether the lamps will operate correctly.

4. An analog voltmeter is indicating 0.45 volt. Which of the following represents this measurement?
   A. 0.450 mV.
   B. 450 mV.
   C. 4.5 mV.
   D. 0.045 V.

5. The following information about a fuel injector control signal has been gathered using a lab scope: Frequency is 10 Hz and pulse width is 5 mS.
   Technician A says this means that the injector is being turned on and off 10 times per second and that the length of time the injector is open during each "on" pulse is 5 mS.
   Technician B says this means that the injector is being turned on and off 10 times per second and that the length of time the injector is closed during each "off" cycle is 95 mS.
   Who is correct?
   A. A only
   B. B only
   C. Both A and B
   D. Neither A nor B

6. Oscilloscope testing is being discussed.
   Technician A says the preferred method to use when observing the output of a potentiometer is to select an external trigger.
   Technician B says, when analyzing the output of a low-voltage computer input sensor, a high trigger level should be selected.
   Who is correct?
   A. A only
   B. B only
   C. Both A and B
   D. Neither A nor B

7. The left-rear and right-rear taillights and the left-rear brake light of a vehicle turn on dimly whenever the brake pedal is depressed; however, the right-rear brake light operates at the correct brightness.
   Technician A says the left-rear taillight and brake light may have a poor ground connection.
   Technician B says the brakelight switch may have excessive resistance.
   Who is correct?
   A. A only
   B. B only
   C. Both A and B
   D. Neither A nor B

8. The horn of a vehicle equipped with a horn relay sounds weak and distorted whenever it is applied. Which of the following is the least likely cause of this problem?
   A. High resistance in the relay load circuit.
   B. High resistance in the horn ground circuit.
   C. Excessive voltage drop between the relay load contact and the horn.
   D. Excessive voltage drop across the relay coil winding.

9. The circuit breaker that protects an electric window circuit blows whenever an attempt is made to lower the window.
Technician A says the internal resistance of the motor is too high.
Technician B says the window regulator may be sticking.
Who is correct?
A. A only   **C.** Both A and B
B. B only   **D.** Neither A nor B

10. The relay coil resistance of an inoperative electric antenna circuit is 0.5 K ohms.
Technician A says this could prevent the proper operation of the antenna motor.
Technician B says this could result in the antenna rising but not retracting.
Who is correct?
A. A only   **C.** Both A and B
B. B only   **D.** Neither A nor B

NOTE: Question 11 refers to the horn circuit on page 685.

11. The horn circuit wiring diagram is being discussed.
Technician A says the wire that provides the ground path for the high-note horn is an 18-gauge black wire that is part of circuit Z1.
Technician B says the wire that provides power to the horn relay coil is a violet wire that is connected to terminal 87 of the relay.
Who is correct?
A. A only   **C.** Both A and B
B. B only   **D.** Neither A nor B

NOTE: Question 12 refers to the rear wiper/washer on page 686.

12. The wiring diagram is being discussed.
Technician A says the wire that activates the rear wiper motor module originates at terminal G of the rear window/wiper switch.
Technician B says there are only three wires connected to the accessory switch panel of this vehicle.
Who is correct?
A. A only   **C.** Both A and B
B. B only   **D.** Neither A nor B

13. Domestic and import wire sizes are being discussed.
Technician A says a 14-gauge wire is larger than a 16-gauge wire.
Technician B says a 0.8 wire is smaller than a 1.0 wire.
Who is correct?
A. A only   **C.** Both A and B
B. B only   **D.** Neither A nor B

14. What is the expected total current flow through a 12-volt circuit with two 12-ohm resistors connected in parallel?
A. 0.5 ampere   **C.** 2 amperes
B. 1 ampere   **D.** 32 amperes

15. A hydrometer that is being used to measure the specific gravity of a battery indicates 1.240.
Technician A says if the ambient temperature is 70°F, the corrected specific gravity reading will be 1.236.
Technician B says if the battery temperature is 60°F, the corrected specific gravity reading will be 1.232.
Who is correct?
A. A only   **C.** Both A and B
B. B only   **D.** Neither A nor B

16. The specific gravity of each of the cells of a battery is as follows: 1.220, 1.245, 1.190, 1.205, 1.210, and 1.215. Which of the following procedures should be performed?
A. Charge the battery and then retest the specific gravity.
B. Perform a battery capacity test.
C. Perform a three-minute charge test.
D. Replace the battery.

17. A battery has an open circuit voltage of 12.1 volts.
Technician A says a battery capacity test should now be performed.
Technician B says the battery should now be charged.
Who is correct?
A. A only   **C.** Both A and B
B. B only   **D.** Neither A nor B

18. Battery test series is being discussed.
Technician A says to perform the battery capacity test first.
Technician B says the state of charge test is always the last test performed.
Who is correct?
- **A.** A only
- **B.** B only
- **C.** Both A and B
- **D.** Neither A nor B

19. All of the following could cause a slow cranking condition except:
- **A.** Overadvanced ignition timing.
- **B.** Shorted neutral safety switch.
- **C.** Misaligned starter mounting.
- **D.** Low battery state of charge.

20. A vehicle owner states that occasionally he is unable to start his car: the engine will start cranking at normal speed and then all of a sudden he will hear a "whee" sound and at that point the engine will stop cranking. After four or five attempts, the engine will finally start.
Technician A says the starter drive gear may be slipping.
Technician B says there may be excessive voltage drop across the starter-solenoid contacts.
Who is correct?
- **A.** A only
- **B.** B only
- **C.** Both A and B
- **D.** Neither A nor B

21. A vehicle with a no-crank condition is being tested. When the starter-solenoid "battery" and "start" terminals are connected with a jumper wire, the starter begins to crank the engine.
Technician A says the starter-solenoid may be faulty.
Technician B says the ignition switch may be faulty.
Who is correct?
- **A.** A only
- **B.** B only
- **C.** Both A and B
- **D.** Neither A nor B

22. A vehicle is being tested for a slow-crank condition. One lead of a voltmeter is connected to the positive battery post and the other lead is connected to the motor terminal of the starter-solenoid. With the ignition key "on," the voltmeter indicates 12 volts; when the engine is cranked, the voltmeter indicates 0.2 volt.
Technician A says the positive side of the starter-solenoid load circuit is OK.
Technician B says the negative side of the starter circuit may have excessive voltage drop.
Who is correct?
- **A.** A only
- **B.** B only
- **C.** Both A and B
- **D.** Neither A nor B

23. A technician is performing an output test on an AC generator (alternator) rated at 100 amperes with an internal regulator. During the test, it produces 30 amperes. The cause of the low output may be:
- **A.** Shorted diode.
- **B.** Open sense circuit to the regulator.
- **C.** Worn brushes.
- **D.** Faulty capacitors.

24. A charging system voltage output test reveals the following information:
1. Base voltage: 12.6 volts.
2. At 1,500 rpm: 13.3 volts.
3. At 2,000 rpm with loads on: 13.0 volts.
Technician A says the alternator output voltage is lower than normal.
Technician B says there may be excessive resistance on the insulated side of the charging circuit.
Who is correct?
- **A.** A only
- **B.** B only
- **C.** Both A and B
- **D.** Neither A nor B

25. Charging system voltage and amperage on a vehicle equipped with an external voltage regulator is 12.1 volts and 0 amps at 1,500 rpm with all loads applied. When the alternator is full-fielded under the same conditions, the values rise to 13.8 volts and 95 amps. All of the following statements concerning these tests are correct except:
- **A.** The alternator may have an open field circuit.
- **B.** The voltage regulator's voltage limiter contacts may have excessive resistance.
- **C.** The voltage limiter may have an open coil winding.
- **D.** The alternator is capable of charging correctly.

26. Alternator diode testing is being discussed.
Technician A says an AC voltmeter can be used to check for faulty diodes.
Technician B says an ammeter can be used to check for faulty diodes.
Who is correct?
A. A only
B. B only
C. Both A and B
D. Neither A nor B

27. *Technician A* says that a short to ground in the PCI bus network circuit will result in total bus network failure.
Technician B says an open PCI bus circuit wire to the BCM will result in total bus network failure.
Who is correct?
A. A only
B. B only
C. Both A and B
D. Neither A nor B

28. *Technician A* says that the normal at rest voltage for the PCI bus is 0 volt.
Technician B says that the normal active voltage on the PCI bus is 12 volts.
Who is correct?
A. A only
B. B only
C. Both A and B
D. Neither A nor B

29. *Technician A* says that the location of an open CAN B bus circuit can be located by shorting the other circuit to ground while observing the module communications on the scan tool.
Technician B says if the fault code indicated that a CAN C circuit is shorted, this would cause a total failure of the CAN C bus.
Who is correct?
A. A only
B. B only
C. Both A and B
D. Neither A nor B

30. *Technician A* says the normal voltmeter reading on the CAN B(+) circuit with the ignition key in the RUN position is between 280 and 920mv.
Technician B says normal CAN C bus termination is 60 ohms.
Who is correct?
A. A only
B. B only
C. Both A and B
D. Neither A nor B

31. How much total resistance is in a 12-volt circuit that is drawing 4 amperes?
A. 0.333 ohm
B. 3 ohms
C. 4.8 ohms
D. 48 ohms

32. A replacement halogen bulb for a composite lamp was replaced but it lasted for a very short period of time; all of the other bulbs on the car are working fine. Which of the following could account for the premature failure of the bulb?
A. Excessive charging system voltage.
B. Fingerprints on the bulb.
C. Excessive voltage drop in the power feed circuit to the bulb.
D. Excessive resistance in the ground circuit of the bulb.

33. The turn signals of a vehicle are inoperative. The green indicator bulbs that are supposed to flash when the turn signal switch is moved to either the left- or right-turn position do not turn on at all.
Technician A says the turn signal flasher contacts could have fused closed.
Technician B says the circuit from the turn signal flasher to the turn signal switch may be open.
Who is correct?
A. A only
B. B only
C. Both A and B
D. Neither A nor B

NOTE: Question 34 refers to the headlight switch on page 687.

34. Referring to the wiring diagram: the lighting circuit of this vehicle has the wire that is connected to the headlamp dimmer switch shorted to ground. Which of the following statements concerning this problem is true?
A. The 10-amp fuse will blow when the headlights are turned on.
B. The 15-amp fuse will blow immediately.
C. The 10-amp fuse will blow immediately.
D. The circuit breaker will open the circuit when the headlights are turned on.

NOTE: Question 35 refers to the multifunction switch on page 688.

35. Which of the following statements about the circuit is true?
A. The retract relay-up coil power feed comes from terminal 106 of the multifunction switch.
B. The power feed to the multifunction switch comes from the theft warning main control amp.
C. The retract relay-up load circuit power feed comes from terminal 111 of the multifunction switch.
D. Terminal 106 of the multifunction switch is connected to ground.

36. *Technician A* says a voltmeter connected to the input wire of an IVR should indicate a fluctuating voltage.

Technician B says an open IVR on a single-gauge system could prevent the proper operation of the coolant temperature warning light.

Who is correct?

A. A only
C. Both A and B
B. B only
D. Neither A nor B

37. The fuel gauge of a vehicle equipped with an electromagnetic gauge cluster indicates a full tank, regardless of the actual amount of fuel in the tank. Which of the following could be the cause of this problem?

A. Open IVR.
B. Open fuel gauge sender unit.
C. An open circuit between the gauge and the sending unit.
D. A grounded circuit between the fuel gauge and the fuel gauge sender unit.

NOTE: Question 38 refers to the charging system wiring diagram on page 689.

38. Referring to the wiring diagram: the charge indicator light of this vehicle does not come on when the ignition switch is in the ON position and the engine is turned off. Which of the following statements concerning this problem is true?

A. The 10-ohm resistor inside the voltage regulator may be open.
B. The field relay contacts may be stuck closed.
C. The rotor winding may be shorted.
D. The 50-ohm resistor may be open.

NOTE: Question 39 refers to the two-speed wiper circuit on page 690.

39. The low-speed position of the windshield wiper system is inoperative; the high-speed position is working fine.

Technician A says circuit 58 may be open.
Technician B says circuit 63 may be open.

Who is correct?

A. A only
C. Both A and B
B. B only
D. Neither A nor B

NOTE: Question 40 refers to the blower motor circuit on page 691.

40. The medium-hi speed of the blower motor circuit is inoperative; the rest of the blower speeds are fine.
Technician A says circuit 752 may be open.
Technician B says the middle resistor in the blower motor resistor assembly may be open.

Who is correct?

A. A only
C. Both A and B
B. B only
D. Neither A nor B

NOTE: Question 41 refers to the power window circuit on page 691.

41. The power window motor in the wiring diagram is completely inoperative. With the master window switch placed in the DOWN position, the following voltages are measured at each terminal:

Terminal #	Voltage
1	12 V
2	0 V
3	12 V
4	0 V
5	12 V
6	12 V
7	0 V
8	12 V
9	0 V
10	12 V
11	0 V

Which of the following statements represents the cause of this problem?

A. The master switch is faulty.
B. The window switch is faulty.
C. The motor is faulty.
D. There is a poor ground in the circuit.

NOTE: Question 42 refers to the body computer on page 692.

42. The "liftgate ajar" lamp is remaining on even after the liftgate is closed. Voltage at BCM terminal J2-4 is 0.0 volts.
Technician A says that the BCM may be faulty.
Technician B says that the liftgate switch may need to be replaced.

Who is correct?

A. A only
C. Both A and B
B. B only
D. Neither A nor B

43. A DaimlerChrysler multiplex system is being discussed.

Technician A says a voltmeter connected to the "Bus +" of any multiplexed component should indicate 12 volts with the ignition switch in the ON position.

Technician B says a 12-volt test light can be used to test for power at various multiplexed components. Who is correct?

A. A only   **C.** Both A and B

B. B only   **D.** Neither A nor B

44. Sweep testing a potentiometer with a voltmeter and an ohmmeter is being discussed.

Referring to a voltmeter, *Technician A* says that the positive lead should be connected to the "feedback" terminal and that the negative lead should be connected to the "ground" terminal.

Referring to an ohmmeter, *Technician B* says that the positive lead should be connected to the "feedback" terminal and that the negative lead should be connected to the "reference" terminal. Who is correct?

A. A only   **C.** Both A and B

B. B only   **D.** Neither A nor B

NOTE: Question 45 refers to the diagnostic chart on page 693.

45. The diagnostic chart is being used to troubleshoot a GM vehicle that has a current diagnostic trouble code 24 stored in memory. The ECC display panel is indicating the correct vehicle speed. Which of the following represents the probable cause of the trouble code?

A. Open circuit 437.

B. Faulty speedometer gear.

C. Faulty vehicle speed sensor.

D. Possible ECM problem.

NOTE: Question 46 refers to the Ford illuminated entry system on page 694.

46. The illuminated entry system is inoperative; neither the right nor the left outer door handle will activate the system. The courtesy lights and each lock cylinder LED will operate when the courtesy lamp switch is turned on. Which of the following represents the possible cause of this problem?

A. Open in circuit 54.

B. Short to ground in circuit 465.

C. Open in circuit 57.

D. Short to ground in circuit 54.

NOTE: Question 47 refers to the instrument cluster schematic on page 695.

47. The electronic instrument cluster is totally inoperative; none of the display segments will illuminate.

Technician A says there may be an open in circuit 389.

Technician B says circuit 151 may be open. Who is correct?

A. A only   **C.** Both A and B

B. B only   **D.** Neither A nor B

48. The ambient temperature sensor of a vehicle equipped with an SATC A/C system is being tested. The resistance of the sensor is about 1 K ohms.

Technician A says the sensor is out of range and needs to be replaced.

Technician B says the temperature of the sensor needs to be known before it can be accurately tested. Who is correct?

A. A only   **C.** Both A and B

B. B only   **D.** Neither A nor B

49. Air bag module replacement is being discussed.

Technician A says the negative battery cable should be disconnected at the beginning of the repair.

Technician B says the reserve energy should be allowed to dissipate after the SIR fuse is removed. Who is correct?

A. A only   **C.** Both A and B

B. B only   **D.** Neither A nor B

NOTE: Question 50 refers to the GM electronic sunroof circuit on page 696.

50. The electric sunroof of the vehicle will close but not open. All of the following could be the cause of this problem except:

A. Open wire between the "open" relay and the motor.

B. Faulty "open" relay load circuit contacts.

C. Faulty control switch.

D. Open wire between the control switch and the "open" relay.

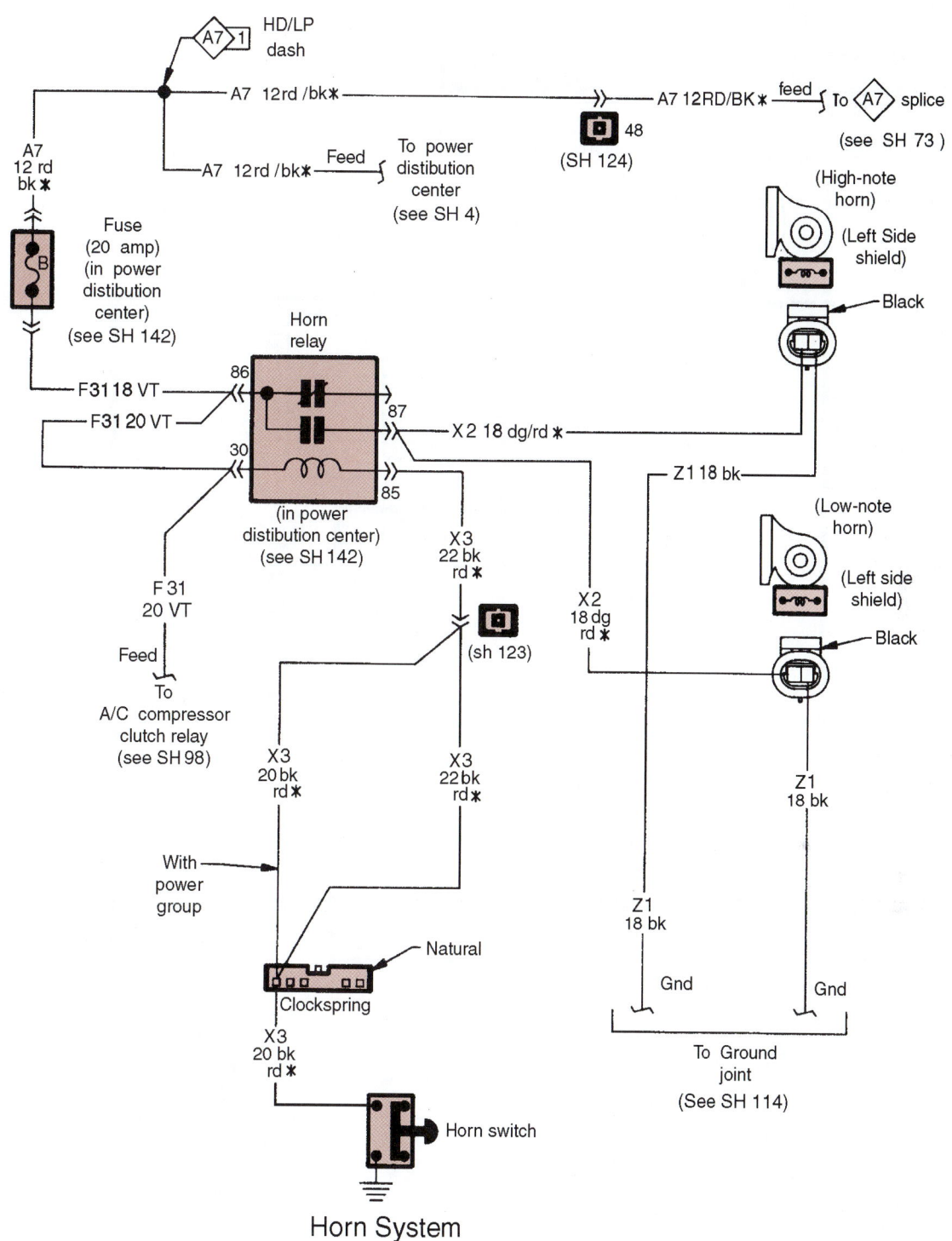

Horn System

Rear Wiper/Washer

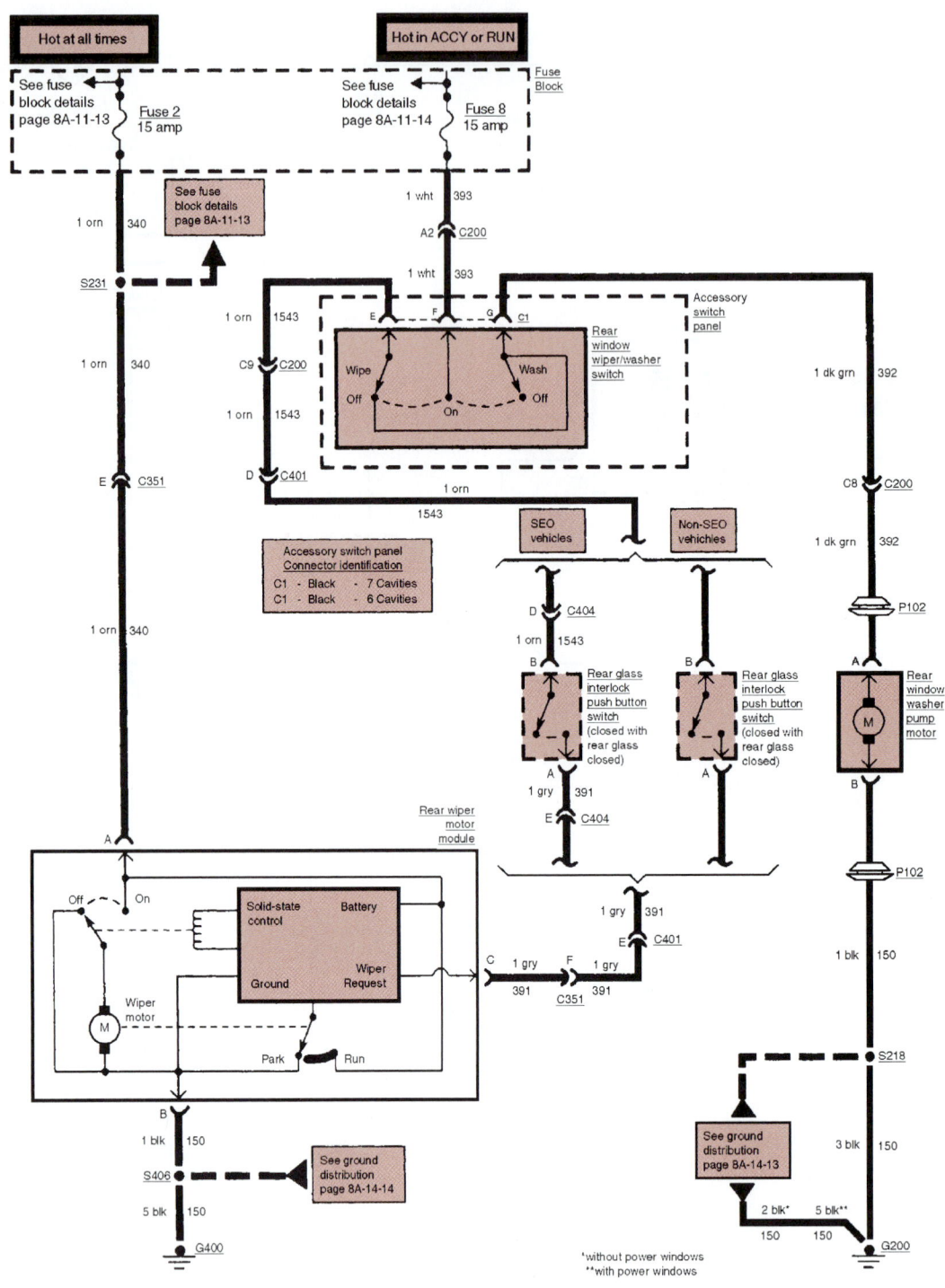

Question 12 Rear wiper/washer circuit.

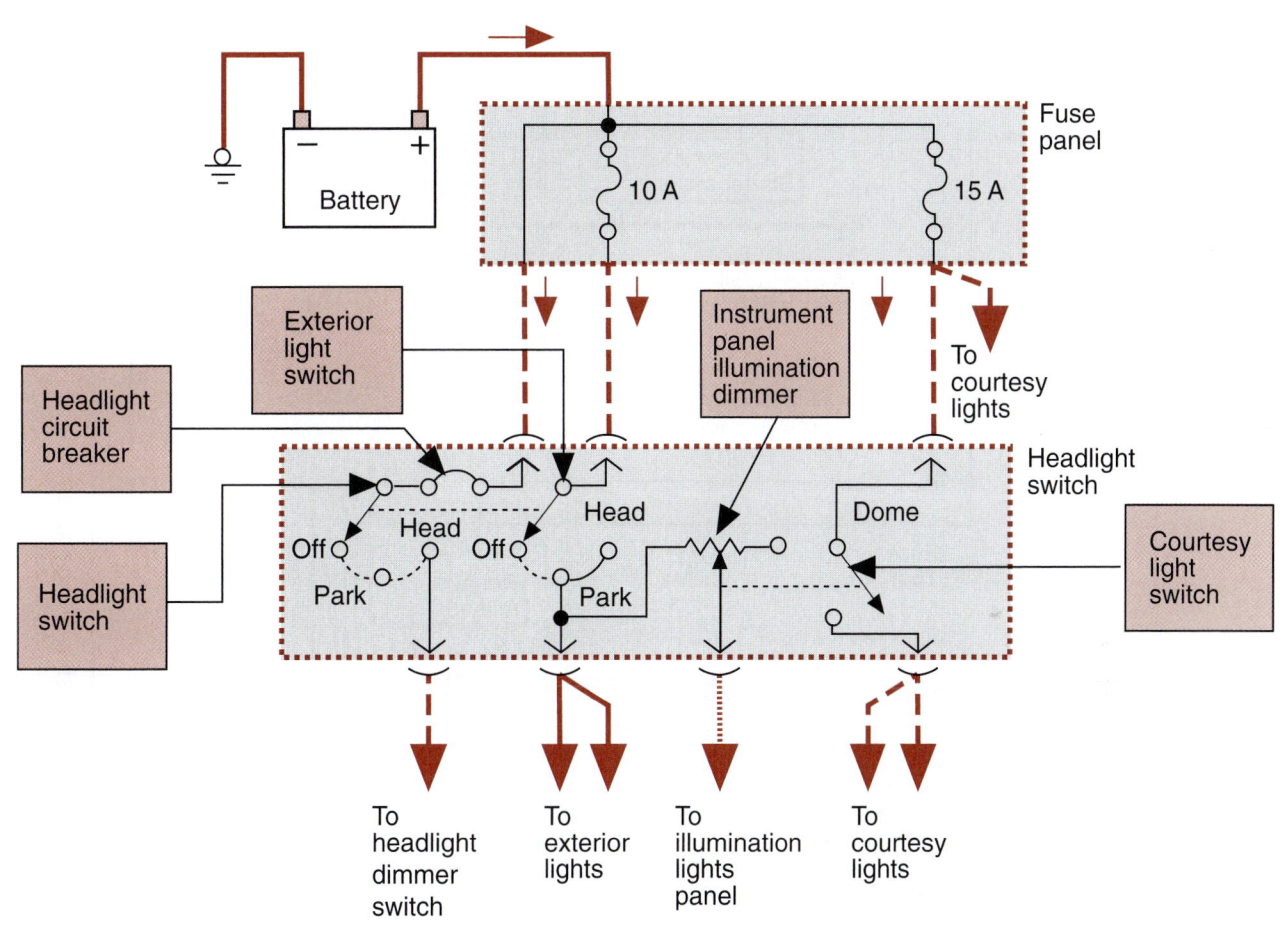

Question 34 Headlight switch circuit.

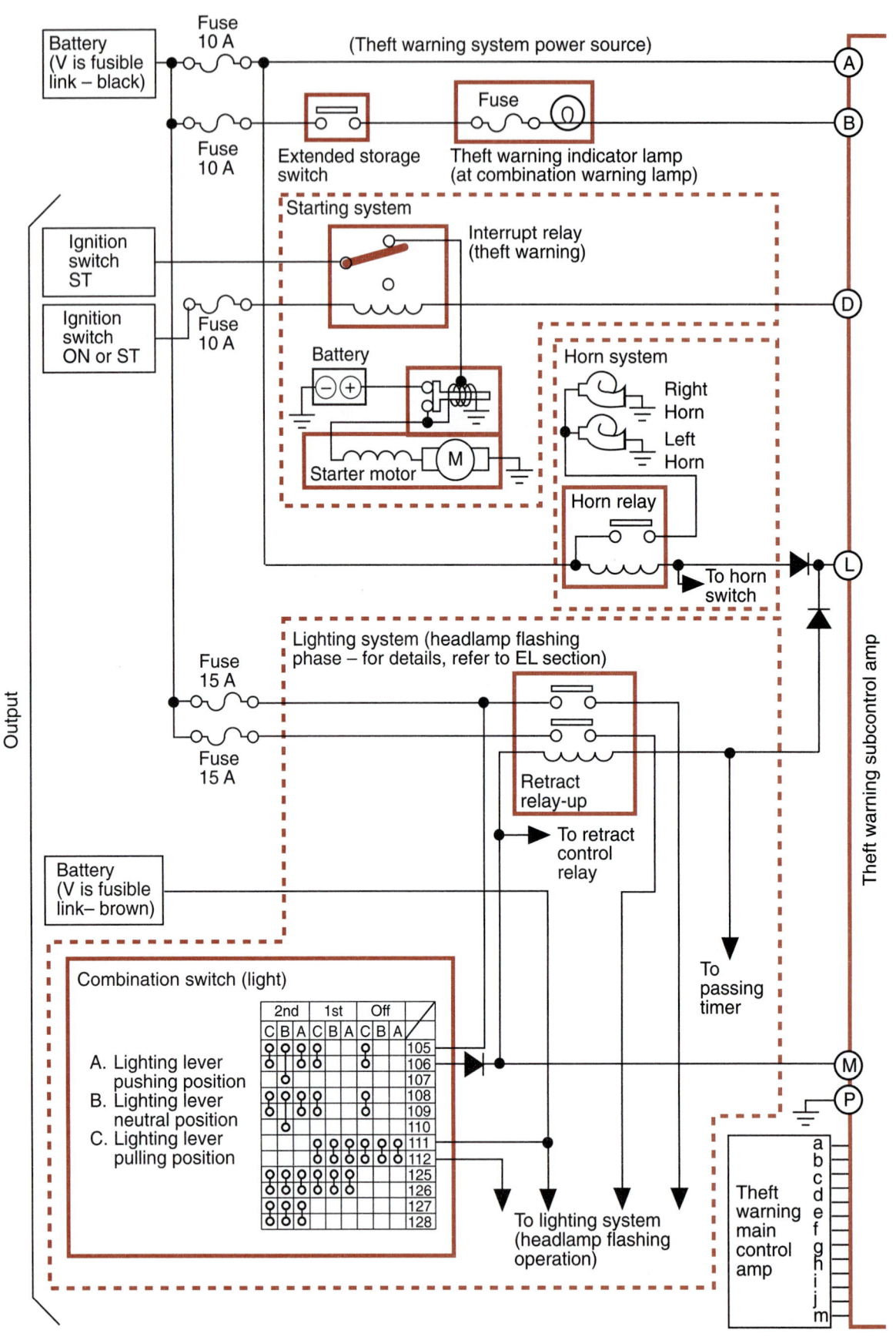

Question 35 Multifunction switch schematic.

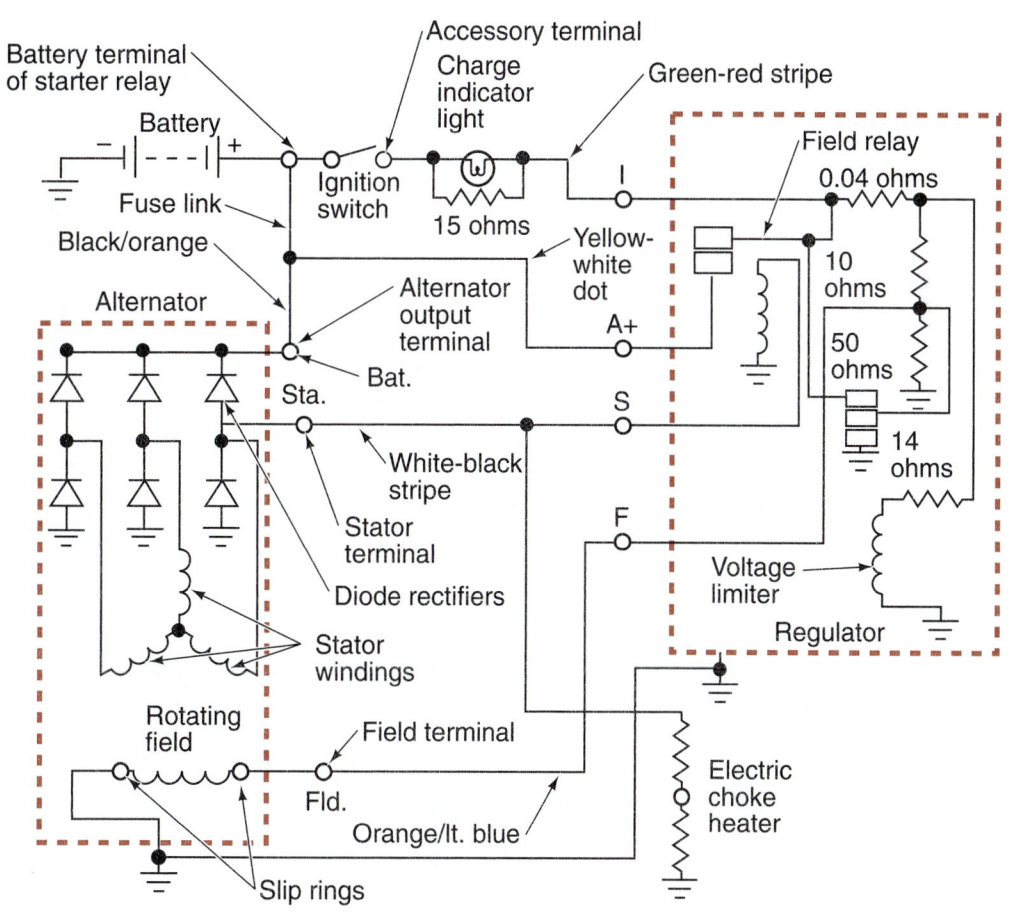

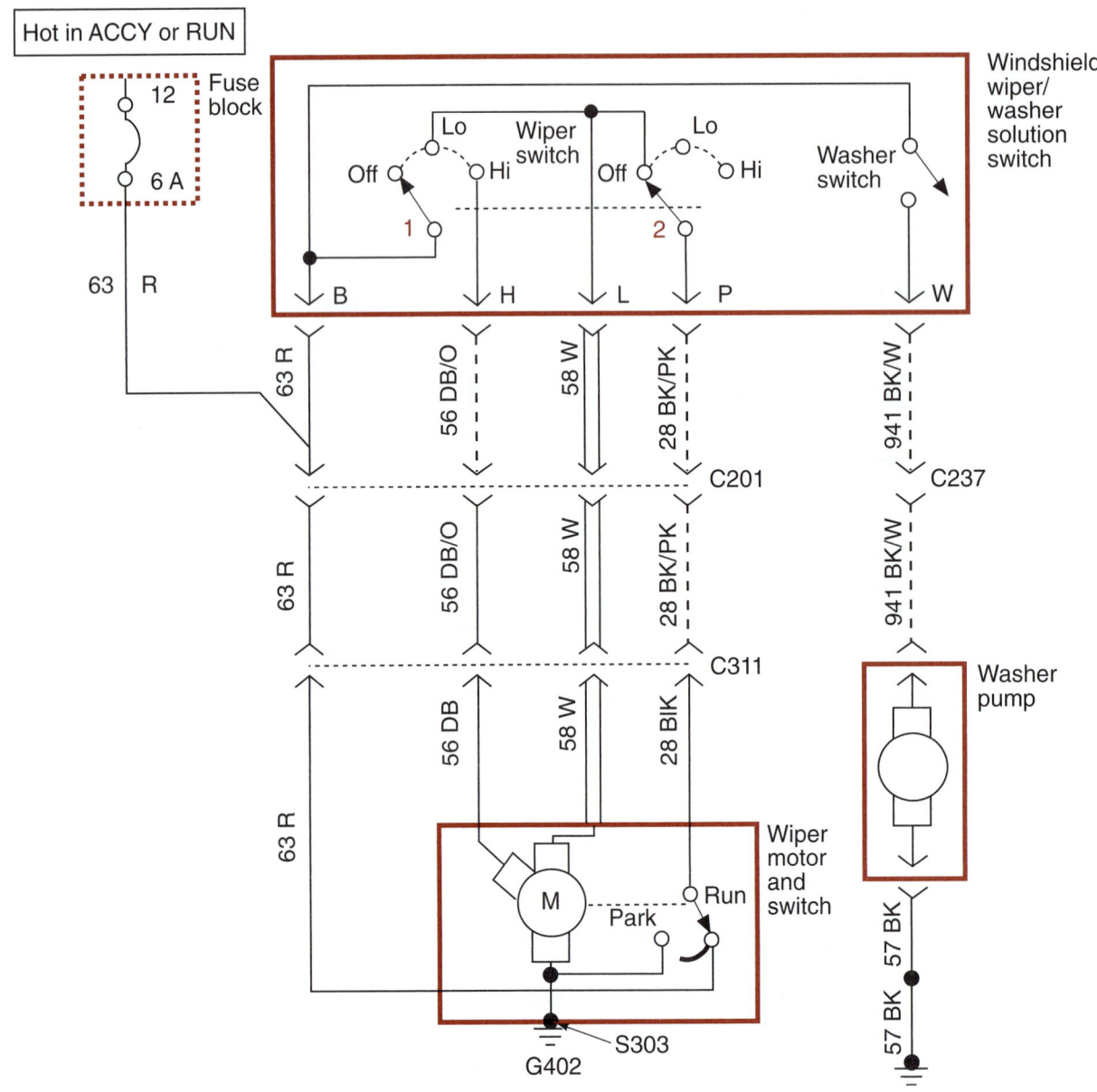

Question 39 Two-speed wiper circuit.

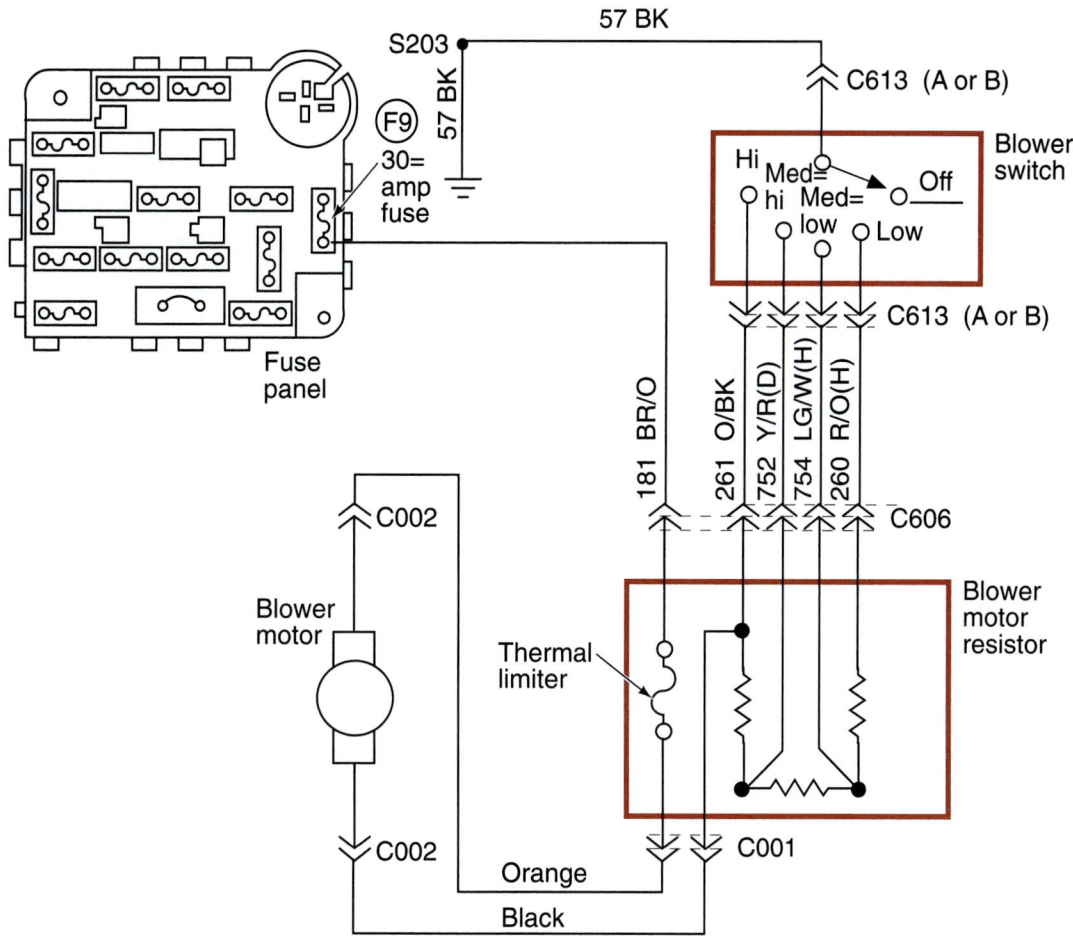

Question 40 Blower motor circuit.

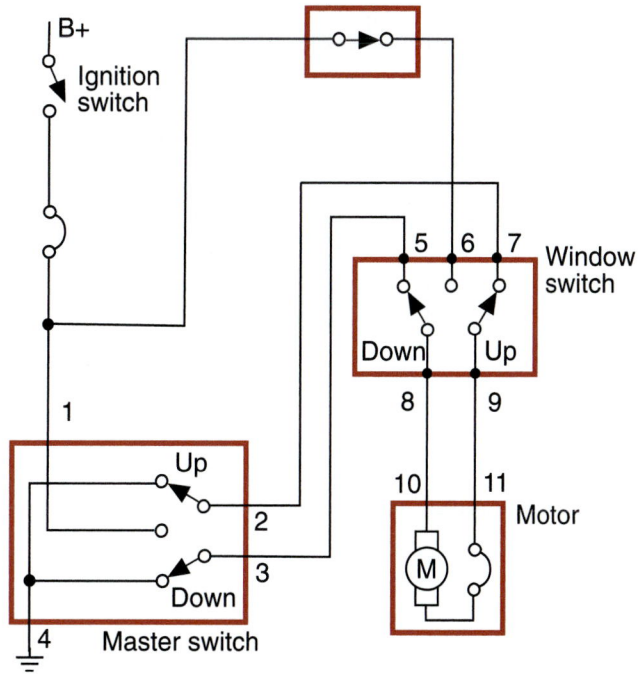

Question 41 Power window circuit.

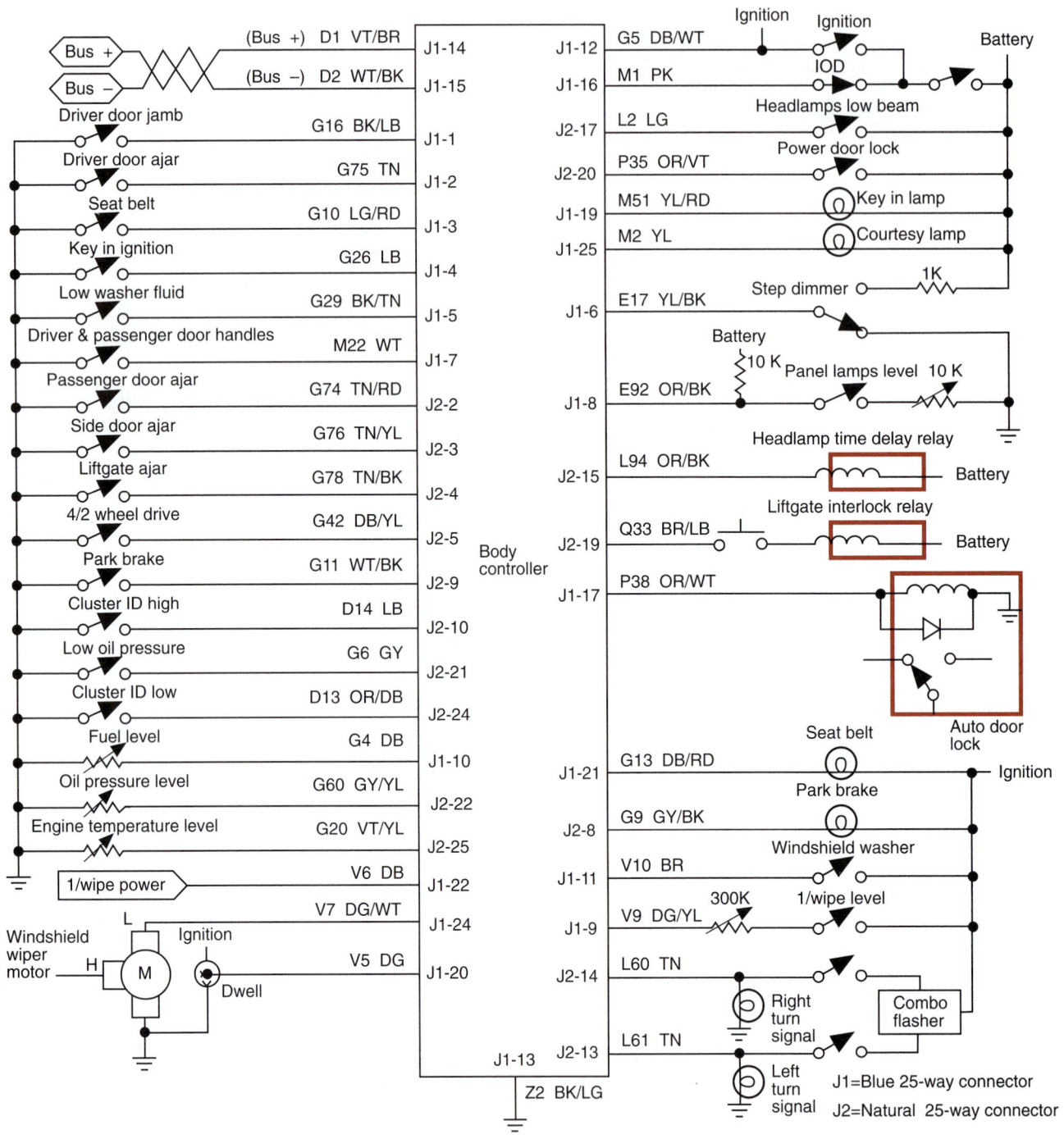

Question 42 Body computer inputs and outputs.

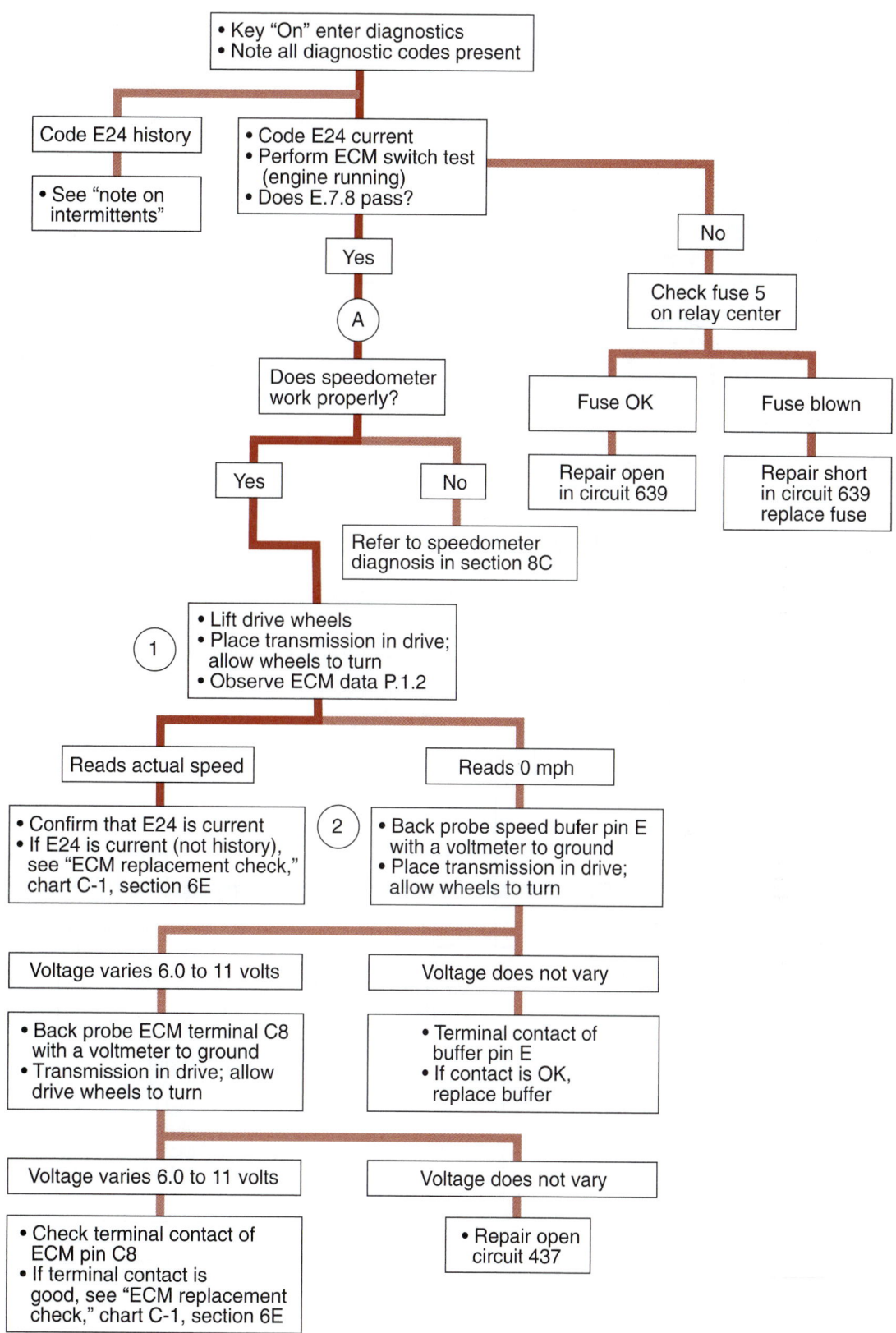

When all diagnoses and repairs are completed, clear codes and verify operation.

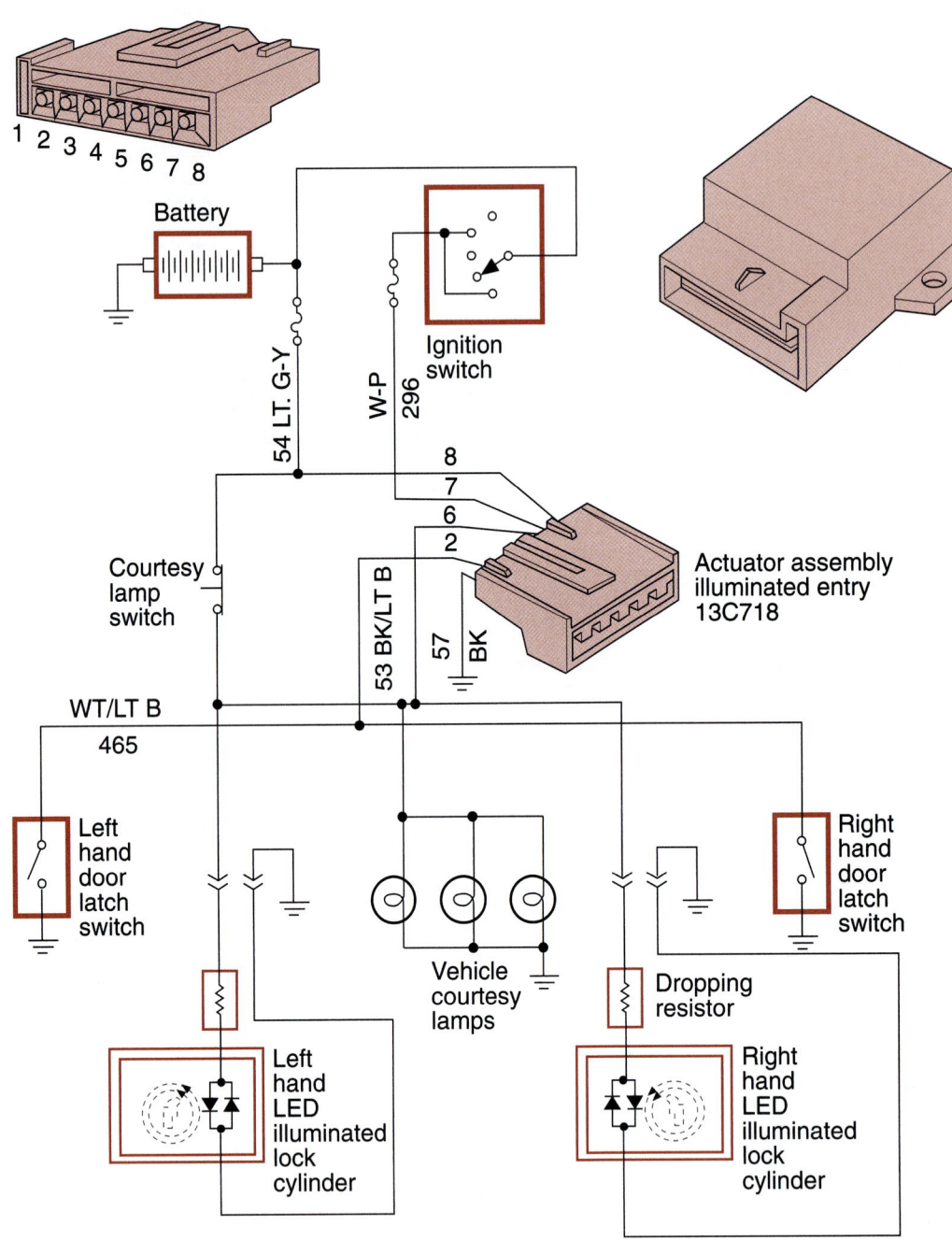

Question 46 Ford Illuminated entry system diagram.

678

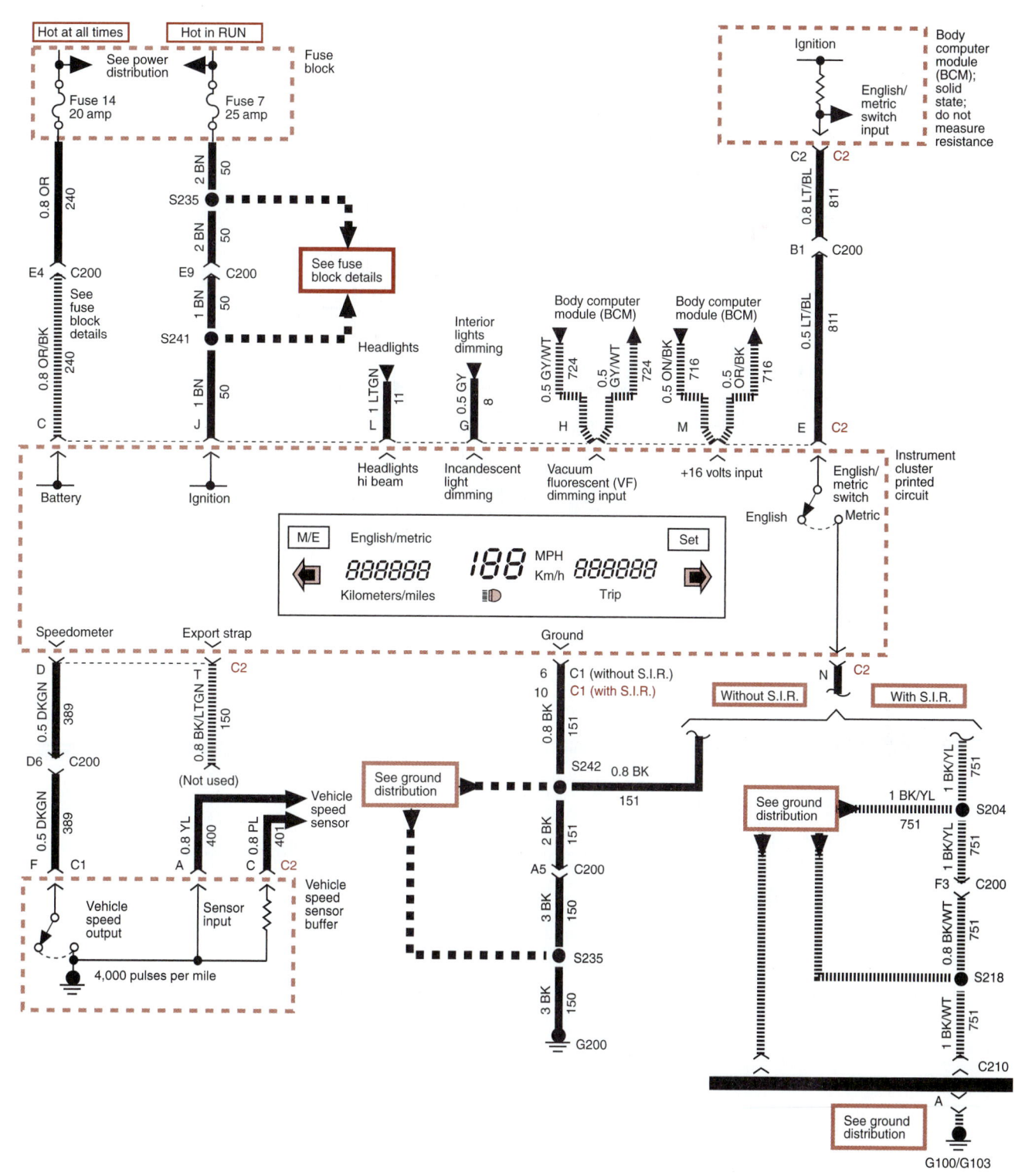

Question 47 Instrument cluster schematic.

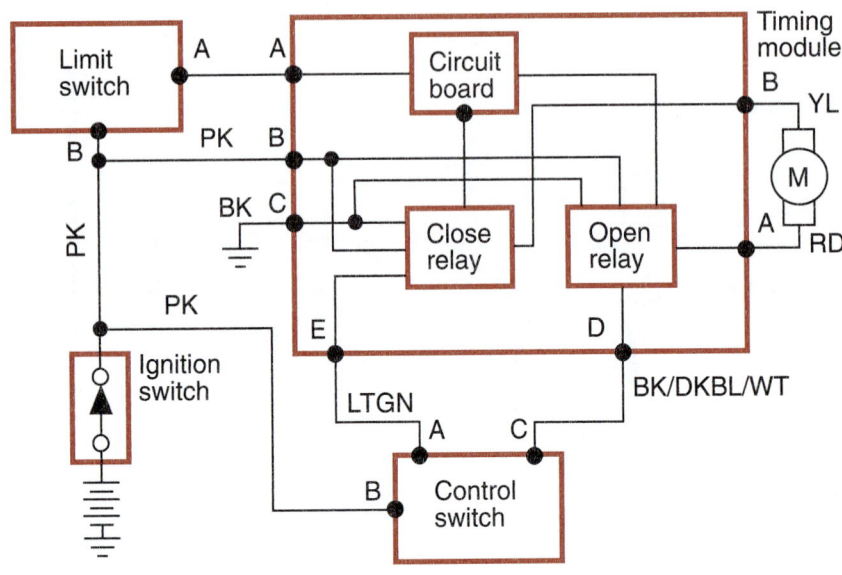

Question 50 GM electronic sunroof diagram.

APPENDIX B

Metric Conversions

	to convert these	to these,	multiply by:
TEMPERATURE	Centigrade degrees	Fahrenheit degrees	1.8 then + 32
	Fahrenheit degrees	Centigrade degrees	0.556 after − 32
LENGTH	Millimeters	Inches	0.03937
	Inches	Millimeters	25.4
	Meters	Feet	3.28084
	Feet	Meters	0.3048
	Kilometers	Miles	0.62137
	Miles	Kilometers	1.60935
AREA	Square centimeters	Square inches	0.155
	Square inches	Square centimeters	6.45159
VOLUME	Cubic centimeters	Cubic inches	0.06103
	Cubic inches	Cubic centimeters	16.38703
	Cubic centimeters	Liters	0.001
	Liters	Cubic centimeters	1,000
	Liters	Cubic inches	61.025
	Cubic inches	Liters	0.01639
	Liters	Quarts	1.05672
	Quarts	Liters	0.94633
	Liters	Pints	2.11344
	Pints	Liters	0.47317
	Liters	Ounces	33.81497
	Ounces	Liters	0.02957
WEIGHT	Grams	Ounces	0.03527
	Ounces	Grams	28.34953
	Kilograms	Pounds	2.20462
	Pounds	Kilograms	0.45359
WORK	Centimeter kilograms	Inch pounds	0.8676
	Inch pounds	Centimeter kilograms	1.15262
	Meter kilograms	Foot pounds	7.23301
	Foot pounds	Newton meters	1.3558
PRESSURE	Kilograms/sq. cm	Pounds/sq. inch	14.22334
	Pounds/sq. inch	Kilograms/sq. cm	0.07031
	Bar	Pounds/sq. inch	14.504
	Pounds/sq. inch	Bar	0.06895

APPENDIX C

Electrical and Electronic Special Tool Suppliers

Association of Automotive Aftermarket Distributors

Allen Test Equipment
Kalamazoo, MI

Alltest, Inc., Div. of Triplett Corp.
Palatine, IL

Baum Tools Unlimited, Inc.
Longboat Key, FL

Big A Auto Parts, APS Inc.
Houston, TX

Carquest Corp.
Tarrytown, NY

Fluke Corp.
Everett, WA

KD Tools, Danaher Tool Group
Lancaster, PA

Kent-Moore, Div. of SPX Corp.
Warren, MI

Mac Tools
Washington Courthouse, OH

Matco Tools
Stow, OH

NAPA Hand/Service Tools
Lancaster, PA

OTC, Div. of SPX Corp.
Owatonna, MN

Parts Plus
Memphis, TN

Snap-On Tools Corp.
Kenosha, WI

Sun Test Equipment, Div. of Snap-On Tools
Kenosha, WI

APPENDIX D

J1930 Terminology List

NOTE: Certain Ford component names have been changed in this Service Manual to conform to Society of Automotive Engineers (SAE) directive J1930.

SAE J1930 standardizes automotive component names for all vehicle manufacturers.

New Term	New Acronyms/ Abbreviations	Old Acronyms/Term
Accelerator pedal	AP	– Accelerator
Air cleaner	ACL	– Air cleaner
Air conditioning	A/C	– A/C – Air conditioning
Barometric pressure	BARO	– BP – Barometric pressure
Battery positive voltage	B+	– BATT+ – Battery positive
Camshaft position	CMP	– Camshaft sensor
Carburetor	CARB	– CARB – Carburetor
Charge air cooler	CAC	– After cooler – Intercooler
Closed loop	CL	– EEC
Closed throttle position	CTP	– CTP – Closed throttle position
Clutch pedal position	CPP	– CES – CIS – Clutch engage switch – Clutch interlock switch
Continuous fuel injection	CFI	– Continuous fuel injection
Continuous trap oxidizer	CTOX	– CTO
Crankshaft position	CKP	– CPS – VRS – Variable reluctance sensor
Data link connector	DLC	– Self-Test connector
Diagnostic test mode	DTM	– Self-Test mode
Diagnostic trouble code	DTC	– Self Test code
Distributor ignition	DI	– CBD – DS – TFI – Closed bowl distributor – Duraspark ignition – Thick film ignition
Early fuel evaporation	EFE	– EFE – Early fuel evaporation
Electrically erasable programmable read only memory	EEPROM	– E2PROM

New Term	New Acronyms/ Abbreviations	Old Acronyms/Term
Electronic ignition	EI	– DIS – EDIS – Distributorless ignition system – Electronic distributorless ignition system
Engine control module	ECM	– ECM – Engine control module
Engine coolant level	ECL	– Engine coolant level
Engine coolant temperature	ECT	– ECT – Engine coolant temperature
Engine speed	RPM	– RPM – Revolutions per minute
Erasable programmable read only memory	EPROM	– EPROM – Erasable programmable read only memory
Evaporative emission	EVAP	– EVP sensor – EVR solenoid
Exhaust gas recirculation	EGR	– EGR – Exhaust gas recirculation
Fan control	FC	– EDF – Electro-Drive fan
Flash electrically erasable programmable read only memory	FEEPROM	– FEEPROM – Flash electrically erasable programmable read only memory
Flash erasable programmable read only memory	FEPROM	– FEPROM – Flash erasable programmable read only memory
Flexible fuel	FF	– FCS – FFS – FFV – Fuel compensation sensor – Flex fuel sensor
Fourth gear	4GR	– Fourth gear
Fuel pump	FP	– FP – Fuel pump
Generator	GEN	– ALT – Alternator
Ground	GND	– GND – Ground
Heated oxygen sensor	HO$_2$S	– HEGO – Heated exhaust gas oxygen sensor
Idle air control	IAC	– IAC – Idle air bypass control
Idle speed control	ISC	– Idle speed control
Ignition control module	ICM	– DIS module – EDIS module – TFI module
Indirect fuel injection	IFI	– IDFI – Indirect fuel injection
Inertia fuel shutoff	IFS	– Inertia switch

New Term	New Acronyms/ Abbreviations	Old Acronyms/ Term
Intake air temperature	IAT	– ACT – Air charge temperature
Knock sensor	KS	– KS – Knock sensor
Malfunction indicator lamp	MIL	– CEL – "CHECK ENGINE" light – "SERVICE ENGINE SOON" light
Manifold absolute pressure	MAP	– MAP – Manifold absolute pressure
Manifold differential pressure	MDP	– MDP – Manifold differential pressure
Manifold surface temperature	MST	– MST – Manifold surface temperature
Manifold vacuum zone	MVZ	– MVZ – manifold vacuum zone
Mass air flow	MAF	– MAF – Mass air flow
Mixture control	MC	– Mixture control
Multiport fuel injection	MFI	– EFI – Electronic fuel injection
Nonvolatile random access memory	NVRAM	– NVM – Non-volatile memory
On-board diagnostic	OBD	– Self-test – On-board diagnostic
Open loop	OL	– OL – Open loop
Oxidation catalytic converter	OC	– COC – Conventional oxidation catalyst
Oxygen sensor	O_2S	– EGO
PARK/NEUTRAL position	PNP	– NDS – NGS – TSN – Neutral drive switch – Neutral gear switch – Transmission select neutral
Periodic trap oxidizer	PTOX	– PTOX – Periodic trap oxidizer
Power steering pressure	PSP	– PSPS – Power steering pressure switch
Powertrain control module	PCM	– ECA – ECM – ECU – EEC processor – Engine control assembly – Engine control module – Engine control unit
Programmable read only memory	PROM	– PROM – Programmable read only memory

New Term	New Acronyms/Abbreviations	Old Acronyms/Term
Pulsed secondary air injection	PAIR	– MPA – PA – Thermactor II – Managed pulse air – Pulse air
Random access memory	RAM	– RAM – Random access memory
Read only memory	ROM	– ROM – Read only memory
Relay module	RM	– RM – Relay module RM
Scan tool	ST	– GST – NGS – Generic scan tool – New generation STAR tester – Enhanced scan tool OBD II ST
Secondary air injection	AIR	– AM – CT – MTA – Air management – Conventional thermactor – Managed thermactor air – Thermactor
Sequential multiport fuel injection	SFI	– SEFI – Sequential electronic fuel injection
Service reminder indicator	SRI	– SRI – Service reminder indicator
Smoke puff limiter	SPL	– SPL – Smoke puff limiter
Supercharger	SC	– SC – Supercharger
Supercharger bypass	SCB	– SCB – Supercharger bypass
System readiness test	SRT	– –
Thermal vacuum valve	TVV	– Thermal vacuum switch
Third gear	3GR	– Third gear
Three-way catalytic converter	TWC	– TWC – Three-way catalytic converter
Three-way + oxidation catalytic converter	TWC+OC	– TWC & COC – Dual bed – Three-way catalyst and conventional oxidation catalyst
Throttle body	TB	– TB – Throttle body
Throttle body fuel injection	TBI	– CFI – Central fuel injection – EFI
Throttle position	TP	– TP – Throttle position

New Term	New Acronyms/ Abbreviations	Old Acronyms/Term
Torque converter clutch	TCC	– CCC – CCO – MCCC – Converter clutch control – Converter clutch override – Modulated converter clutch control
Transmission control module	TCM	– 4EAT module
Transmission range	TR	– PRNDL
Turbocharger	TC	– TC – Turbocharger
Vehicle speed sensor	VSS	– VSS – Vehicle speed sensor
Voltage regulator	VR	– VR – Voltage regulator
Volume air flow	VAF	– VAF – Volume air flow
Warm-up oxidation catalytic converter	WU-OC	– WV-OC – Warm-up oxidation catalytic converter
Warm-up three-way catalytic converter	WU-TWC	– WU-TWC – Warm-up three-way catalytic converter
Wide-open throttle	WOT	– Full throttle – WOT – Wide-open throttle

GLOSSARY

Accumulator   A gas-filled pressure chamber that provides hydraulic pressure for ABS operation.
Acumulador   Cámara de presión que contiene gas y que proporciona gran presión hidráulica para una función ABS.

A circuit   A generator circuit that uses an external grounded field circuit. The regulator is on the ground side of the field coil.
Circuito A   Circuito regulador del generador que utiliza un circuito inductor externo puesto a tierra. En el circuito A, el regulador se encuentra en el lado a tierra de la bobina inductora.

Actuators   Devices that perform the actual work commanded by the computer. They can be in the form of a motor, relay, switch, or solenoid.
Accionadores   Dispositivos que realizan el trabajo efectivo que ordena la computadora. Dichos dispositivos pueden ser un motor, un relé, un conmutador o un solenoide.

Air bag module   Composed of the air bag and inflator assembly that is packaged into a single module.
Unidad del Airbag   Formada por el conjunto del Airbag y el inflador. Este conjunto se empaqueta en una sola unidad.

Air bag system   A supplemental restraint that will deploy a bag out of the steering wheel or passenger-side dash panel to provide additional protection against head and face injuries during an accident.
Sistema de Airbag   Resguardo complementario que expulsa una bolsa del volante o del panel de instrumentos del lado del pasajero para proveer protección adicional contra lesiones a la cabeza y a la cara en caso de un accidente.

Ambient temperature   The temperature of the outside air.
Temperatura ambiente   Temperatura del aire ambiente.

Ambient temperature sensor   Thermistor used to measure the temperature of the air entering the vehicle.
Sensor de temperatura ambiente   Termistor utilizado para medir la temperatura del aire que entra al vehículo.

Ammeter   A test meter used to measure current draw.
Amperímetro   Instrumento de prueba utilizado para medir la intensidad de una corriente.

Amperes   *See* current.
Amperios   *Véase* corriente.

Analog   A voltage signal that is infinitely variable or can be changed within a given range.
Señal analógica   Señal continua y variable que debe traducirse a valores numéricos discontinuos para poder ser tratada por una computadora.

Analog signal   Varying voltage with infinite values within a defined range.
Señal análoga   Voltaje variable con infinidad de valores dentro de un límite definido de velocidad.

Antilock brakes (ABS)   A brake system that automatically pulsates the brakes to prevent wheel lock-up under panic stop and poor traction conditions.
Frenos antibloqueo   Sistema de frenos que pulsa los frenos automáticamente para impedir el bloqueo de las ruedas en casos de emergencia y de tracción pobre.

Antitheft device   A device or system that prevents illegal entry or driving of a vehicle. Most are designed to deter entry.
Dispositivo a prueba de hurto   Un dispositivo o sistema quepreviene la entrada o conducción ilícita de un vehículo. Lamayoría se diseñan para detener la entrada.

A-pillar   The pillar in front of the driver or passenger that supports the windshield.
Soporte A   Soporte enfrente del conductor o del pasajero que sostiene el parabrisas.

Arming sensor   A device that places an alarm system into "ready" to detect an illegal entry.
Sensor de armado   Un dispositivo que pone "listo" un sistema dealarma para detectar una entrada ilícita.

Aspirator   Tubular device that uses a venturi effect to draw air from the passenger compartment over the in-car sensor. Some manufacturers use a suction motor to draw the air over the sensor.
Aspirador   Dispositivo tubular que utiliza un efecto venturi para extraer aire del compartimiento del pasajero sobre el sensor dentro del vehículo. Algunos fabricantes utilizan un motor de succion para extraer el aire sobre el sensor.

Audio system   The sound system for a vehicle; can include radio cassette player, CD player, amplifier, and speakers.
Sistema de audio   El sistema de sonido de un vehículo; puedeincluir el radio, el tocacaset, el toca discos compactos, el amplificador, y las bocinas.

Automatic door locks   A system that automatically locks all doors through the activation of one switch.
Cerraduras automáticas de puerta   Un sistema que cierra todas laspuertas automaticamente al activar un solo conmutador.

Automatic seat belt system   Automatically puts the shoulder and/or lap belt around the driver or occupant. The automatic seat belt is operated by DC motors that move the belts by means of carriers on tracks.
Sistema automático de correas de asiento   Funciona automáticamente, poniendo el cinturón de seguridad sobre el hombro y el pecho del chofer y el pasajero del automóvil. La correa automática del asiento trabaja por medio de motores DC, o sea corriente directa, que da movimiento a las correas transportadoras.

Automatic traction control   A system that prevents slippage of one of the drive wheels. This is done by applying the brake at that wheel and/or decreasing the engine's power output.
Control automático de tracción   Un sistema que previene el patinaje de una de las ruedas de mando. Esto se efectúa aplicando el freno en esa rueda y/o disminuyendo la salida de potencia del motor.

Average responding   A method used to read AC voltage.
Respuesta media   Un método que se emplea para leer la tensión decorriente alterna.

Back probe   A term used to mean that a test is being performed on the circuit while the connector is still connected to the component. The test probes are inserted into the back of the wire connector.
Sonda exploradora de retorno   Término utilizado para expresar que se está llevando a cabo una prueba del circuito mientras el conectador sigue conectado al componente. Las sondas de prueba se insertan a la parte posterior del conectador de corriente.

BAT   Terminal of a generator, starter, or solenoid that has direct battery feed connected to it.
Forma de aproximación del balancín   Terminal de un generador, encendedor o solenoide conectado directamente a la batería.

Battery ECU   Used to monitor the condition of the HV battery assembly in an HEV. The battery ECU determines the SOC of the HV battery by monitoring voltage, current, and temperature.
UCE de la batería   Se utiliza para monitoreard del ensamblado de la batería del VH en un VHE. La UCE de la batería determina el EDC de la batería del VH al monitorear el voltaje, la corriente y la temperatura.

Battery holddowns   Brackets that secure the battery to the chassis of the vehicle.
Portabatería   Los sostenes que fijan la batería al chasis del vehículo.

Battery leakage test   Used to determine if current is discharging across the top of the battery case.
Prueba de pérdida de corriente de la batería   Prueba utilizada para determinar si se está descargando corriente a través de la parte superior de la caja de la batería.

Battery terminal test   Checks for poor electrical connections between the battery cables and terminals. Use a voltmeter to measure voltage drop across the cables and terminals.
Prueba del borne de la batería   Verifica si existen conexiones eléctricas pobres entre los cables y los bornes de la batería. Utiliza un voltímetro para medir caídas de tensión entre los cables y los bornes.

B circuit   A generator regulator circuit that is internally grounded. In the B circuit, the voltage regulator controls the power side of the field circuit.
Circuito B   Circuito regulador del generador puesto internamente a tierra. En el circuito B, el regulador de tensión controla el lado de potencia del circuito inductor.

B-codes DTCs that are assigned to the vehicle's body systems and control modules.

Códigos B Instrucciones de transmisión digital (DTC) que se asignan a los sistemas de la carrocería del vehículo y a los módulos de control.

Bench test A term used to indicate that the unit is to be removed from the vehicle and tested.

Prueba de banco Término utilizado para indicar que la unidad será removida del vehículo para ser examinada.

Bendix drive A type of starter drive that uses the inertia of the spinning starter motor armature to engage the drive gear to the gears of the flywheel. This type starter drive was used on early models of vehicles and is rarely seen today.

Acoplamiento Bendix Un tipo del acoplamiento del motor de arranque que usa la inercia de la armadura del motor de arranque giratorio para endentar el engrenaje de mando con los engrenajes del volante. Este tipo de acoplamiento del motor de arranque se usaba en los modelos vehículos antiguos y se ven raramente.

Bezel The retaining trim around a component.

Bisel El resto del decorado alrededor de un componente.

Bimetallic gauges Simple dial and needle indicators that transform the heating effect of electricity into mechanical movement.

Calibradores bimetálicos Un cuadrante simple y agujas indicadoras que transforman el efecto del calor de la electricidad en un movimiento mecánico.

Binary numbers Strings of zeroes and ones, or on's and off's, which represent a numeric value.

Números binarios Franjas que indican ceros (0) y unos (1), de prendido/apagado, representando el valor numérico.

B-pillar The pillar located over the shoulder of the driver or passenger.

Soporte B Soporte ubicado sobre el hombro del conductor o del pasajero.

Breakout box Allows the technician to test circuits, sensors, and actuators by providing test points.

Caja de interruptores Le permite al técnico probar los circuitos, los monitores y actuadores al indicar los puntos de prueba.

Brushes Electrically conductive sliding contacts, usually made of copper and carbon.

Escobillas Contactos deslizantes de conducción eléctrica, por lo general hechos de cobre y de carbono.

Bucking coil One of the coils in a three-coil gauge. It produces a magnetic field that bucks or opposes the low reading coil.

Bobina compensadora Una de las bobinas de un calibre de tres bobinas. Produce un campo magnético que es contrario o en oposición a la bobina de baja lectura.

Buffer A buffer cleans up a voltage signal. These are used with PM generator sensors to change the AC voltage to a digitalized signal.

Separador Un separador aguza una señal del tensión. Estos se usan con los sensores generadores PM para cambiar la tensión de corriente alterna a una señal digitalizado.

Bulkhead connector A large connector that is used when many wires pass through the bulkhead or firewall.

Conectador del tabique Un conectador que se usa al pasar muchos alambres por el tabique o mamparo de encendios.

Bus bar A common electrical connection to which all of the fuses in the fuse box are attached. The bus bar is connected to battery voltage.

Barra colectora Conexión eléctrica común a la que se conectan todos los fusibles de la caja de fusibles. La barra colectora se conecta a la tensión de la batería.

Capacity test The part of the battery test series that checks the battery's ability to perform when loaded.

Prueba de capacidad Parte de la serie de prueba de la batería que verifica la capacidad de funcionamiento de la batería cuando está cargada.

Captured signal A signal stored in a DSO's memory.

Señal de captura La señal restauradora se mantiene en la memoria indicada con las letras DSO.

Carbon monoxide An odorless, colorless, and toxic gas that is produced as a result of combustion.

Monóxido de carbono Gas inodoro, incoloro y tóxico producido como resultado de la combustión.

Carbon tracking A condition where paths of carbon will allow current to flow to points that are not intended. This condition is most commonly found inside distributor caps.

Rastreo de carbón Una condición en la cual las trayectorias del carbón permiten fluir el corriente a los puntos no indicados. Esta condición se encuentra comunmente dentro de las tapas del distribuidor.

Cartridge fuses *See* maxi-fuse.

Fusibles cartucho *Véase* maxifusible.

Cathode ray tube Similar to a television picture tube. It contains a cathode that emits electrons and an anode that attracts them. The screen of the tube will glow at the points that are hit by the electrons.

Tubo de rayos catódicos Parecidos a un tubo de pantalla de televisor. Contiene un cátodo que emite los electrones y un ánodo que los atrae. La pantalla del tubo iluminará en los puntos en donde pegan los electrones.

Caustic Chemicals that have the ability to destroy or eat through something and that are extremely corrosive.

Cáusticos Químicos que tienen la habilidad de destruir o carcomer algo, y que son extremadamente corrosivos.

Cell element The assembly of a positive and negative plate in a battery.

Elemento de pila La asamblea de una placa positiva y negativa en una batería.

Charge To pass an electric current through the battery in an opposite direction than during discharge.

Cargar Pasar una corriente eléctrica por la batería en una dirección opuesta a la usada durante la descarga.

Charge rate The speed at which the battery can safely be recharged at a set amperage.

Indicador de carga eléctrica La velocidad a la cual la batería puede ser recargada seguramente a un amperaje establecido.

Charging system requirement test Diagnostic test used to determine the total electrical demand of the vehicle's electrical system.

Prueba del requisito del sistema de carga Prueba diagnóstica utilizada para determinar la exigencia eléctrica total del sistema eléctrico del vehículo.

CHMSL The abbreviation for center high mounted stop light, often referred to as the third brake light.

CHMSL La abreviación para el faro de parada montada alto en el centro que suele referirse como el faro de freno tercero.

Circuit The path of electron flow consisting of the voltage source, conductors, load component, and return path to the voltage source.

Circuito Trayectoria del flujo de electrones, compuesto de la fuente de tensión, los conductores, el componente de carga y la trayectoria de regreso a la fuente de tensión.

Clamping diode A diode that is connected in parallel with a coil to prevent voltage spikes from the coil from reaching other components in the circuit.

Diodo de bloqueo Un diodo que se conecta en paralelo con una bobina para prevenir que los impulsos de tensión lleguen a otros componentes en el circuito.

Clock spring Maintains a continuous electrical contact between the wiring harness and the air bag module.

Muelle de reloj Mantiene un contacto eléctrico continuo entre el cableado preformado y la unidad del Airbag.

Closed circuit A circuit that has no breaks in the path and allows current to flow.

Circuito cerrado Circuito de trayectoria ininterrumpida que permite un flujo continuo de corriente.

Cold cranking amps (CCA) Rating indicates the battery's ability to deliver a specified amount of current to start an engine at low ambient temperatures.

Amperios de arranque en frío Tasa indicativa de la capacidad de la batería para producir una cantidad específica de corriente para arrancar un motor a bajas temperaturas ambiente.

Color codes Used to assist in tracing the wires. In most color codes, the first group of letters designates the base color of the insulation and the second group of letters indicates the color of the tracer.

Códigos de colores Utilizados para facilitar la identificación de los alambres. Típicamente, el primer alfabeto representa el color base del aislamiento y el segundo representa el color del indicador.

Common connector A connector that is shared by more than one circuit and/or component.

Conector común Un conector que se comparte entre más de un circuito y/o componente.

Commutator A series of conducting segments located around one end of the armature.

Conmutador Serie de segmentos conductores ubicados alrededor de un extremo de la armadura.

Component locator Service manual used to find where a component is installed in the vehicle. The component locator uses both drawings and text to lead the technician to the desired component.

Manual para indicar los elementos componentes Manual de servico utilizado para localizar dónde se ha instalado un componente en el vehículo. En dicho manual figuran dibujos y texto para guiar al mecánico al componente deseado.

Composite bulb A headlight assembly that has a replaceable bulb in its housing.

Bombilla compuesta Una asamblea de faros cuyo cárter tiene una bombilla reemplazable.

Compound motor A motor that has the characteristics of a series-wound and a shunt-wound motor.

Motor compuesta Un motor que tiene las características de un motor exitado en serie y uno en derivación.

Computer An electronic device that stores and processes data and is capable of operating other devices.

Computadora Dispositivo electrónico que almacena y procesa datos y que es capaz de ordenar a otros dispositivos.

Conductance A measurement of the battery's plate surface that is available for chemical reaction, determining how much power the battery can supply.

Conductancia Medida de la superficie de la placa de la batería que está lista para la reacción química, determinando así cuánta potencia puede suplir la batería.

Conductor A material in which electrons flow or move easily.

Conductor Una material in la cual los electrones circulen o se mueven fácilmente.

Continuity Refers to the circuit being continuous with no opens.

Continuidad Se refiere al circuito ininterrumpido, sin aberturas.

Control assembly Provides for driver input into the automatic temperature control microprocessor. The control assembly is also referred to as the control panel.

Montaje de control Un microprocesador automático que le da al Chofer la información necesaria sobre el control de temperatura. El montaje de control es llamado también panel de control.

Converter A DC/DC transformer that converts the voltage from 270 volts DC to 14 volts DC to recharge the auxiliary battery and to power 12-volt electrical components on an HEV.

Convertidor Transformador CD-CD que convierte el voltaje de 270 voltios a 14 voltios de CD para recargar la batería auxiliar y para darles potencia a los componentes eléctricos de 12 voltios en un VHE.

Corona effect A condition where high voltage leaks through a wire's insulation and produces a light or illumination; worn insulation on spark plug wires causes this.

Efecto corona Una condición en la cual la alta tensión se escapa por la insulación del alambre y produce una luz o una iluminación; esto se causa por la insulación desgastada en los alambres de las bujías.

Cowl The top portion of the front of the automobile body that supports the windshield and dashboard.

Capucha Esta es la parte principal de la carrocería en el frente del automóvil y es la que sostiene el parabrisas y el tablero de instrumentos.

Crash sensor Normally open electrical switch designed to close when subjected to a predetermined amount of jolting or impact.

Sensor de impacto Un conmutador normalmente abierto diseñado a cerrarse al someterse a un sacudo de una fuerza predeterminada o un impacto.

Crimping The process of bending, or deforming by pinching, a connector so that the wire connection is securely held in place.

Engarzado Proceso a través del cual se curva o deforma un conectador mediante un pellizco para que la conexión de alambre se mantenga firme en su lugar.

Crimping tool Has different areas to perform several functions. This single tool will cut the wire, strip the insulation, and crimp the connector.

Herramienta prensadora Tiene diferentes áreas para poder ofrecer varias funciones. Esta simple herramienta cortará el cable, aislará y prensará el conectador.

Crocus cloth Used to polish metals. While polishing, it removes very little metal.

Paño de color azafrán Se usa para pulir los metales. Su acción es suave y al pulir remueve pequeñas partículas de metal.

Cross-fire The undesired firing of a spark plug that results from the firing of another spark plug. This is caused by electromagnetic induction.

Encendido transversal El encendido no deseable de una bujía que resulta del encendido de otra bujía. Esto se causa por la inducción electromagnética.

CRT The common acronym for a cathode ray tube.

CRT La sigla común de un tubo de rayos catódicos.

Curb height The height of the vehicle when it has no passengers or loads, and normal fluid levels and tire pressure.

Altura del contén La altura del vehículo cuando no lleva pasajeros ni cargas, y los niveles de los fluidos y de la presión de las llantas son normales.

Current The aggregate flow of electrons through a wire. One ampere represents the movement of 6.25 billion billion electrons (or one coulomb) past one point in a conductor in one second.

Corriente Flujo combinado de electrones a través de un alambre. Un amperio representa el movimiento de 6,25 mil millones de mil millones de electrones (o un colombio) que sobrepasa un punto en un conductor en un segundo.

Current draw test Diagnostic test used to measure the amount of current that the starter draws when actuated. It determines the electrical and mechanical condition of the starting system.

Prueba de la intensidad de una corriente Prueba diagnóstica utilizada para medir la cantidad de corriente que el arrancador tira cuando es accionado. Determina las condiciones eléctricas y mecánicas del sistema de arranque.

Current output testing Diagnostic test used to determine the maximum output of the AC generator.

Prueba de la salida de una corriente Prueba diagnóstica utilizada para determinar la salida máxima del generador de corriente alterna.

Cycle One set of changes in a signal that repeats itself several times.

Ciclo Una serie de cambios en una señal que se repite varias veces.

Darlington A special type of configuration usually consisting of two transistors fabricated on the same chip or mounted in the same package.

Darlington Tipo especial de configuración que generalmente consiste en 2 transistores fabricados en el mismo chip o montados en el mismo paquete.

d'Arsonval gauge A gauge design that uses the interaction of a permanent magnet and an electromagnet, and the total field effect to cause needle movement.

Calibrador d'Arsonval Calibrador diseñado para utilizar la interacción de un imán permanente y de un electroimán, y el efecto inductor total para generar el movimiento de la aguja.

D'arsonval movement A small coil of wire mounted in the center of a permanent horseshoe-type magnet. A pointer or needle is mounted to the coil.

Movimiento D'arsonval Se refiere a una pequeña bobina de cable montada en el centro de un permanente imán diseñado como la herradura de caballo. El puntero o la aguja está montada sobre la bobina.

Deep cycling Discharging the battery completely before recharging it.

Operacion cíclica completa La descarga completa de la batería previo al recargo.

Demonstration mode SmartBeam™ function that allows operation of the automatic high beams and high-beam indicator while the vehicle is stationary and under any ambient lighting conditions.

Mando de demostración Función del rayo inteligente de marca registrada que permite la operación de rayos altos automáticos y del indicador de rayo alto cuando el vehículo está parado y bajo cualesquiera condiciones de iluminación ambientales.

DERM Designed to provide an energy reserve of 36 volts to assure deployment for a few seconds when vehicle voltage is low or lost. The DERM also maintains constant diagnostic monitoring of the electrical system. It will store a code if a fault is found and provide driver notification by illuminating the warning light.

DERM El significado de este término se refiere al diseño que mantiene una reserva de energía de 36 voltios la que asegura, por unos pocos segundos, la salida de la corriente cuando el voltaje del vehículo es bajo o se ha perdido. El DERM mantiene también un constante diagnóstico monitor del sistema eléctrico. Guarda también una clave en caso de que se encuentre alguna falla y le da al chofer una alerta iluminando las señales.

Diagnostic module Part of an electronic control system that provides self-diagnostics and/or a testing interface.

Módulo de diagnóstico Parte de un sistema controlado electronicamente que provee autodiagnóstico y/o una interfase de pruebas.

Diagnostic trouble codes (DTCs) Fault codes that represent a circuit failure in a monitored system.

Códigos de destello Códigos de fallas de diagnóstico (CFD o DTC) que se muestran por medio de los destellos de una lámpara o diodo luminiscente.

Diaphragm A thin, flexible, circular plate that is held around its outer edge by the horn housing, allowing the middle to flex.

Diafragma Es una fina placa circular flexible que es sostenida alrededor de su borde externo por el cuerno del embrague, permitiendo que el centro se doble.

Digital A voltage signal is either on-off, yes-no, or high-low.

Digital Una señal de tensión está Encendida-Apagada, es Sí-No o Alta-Baja.

Digital multimeter (DMM) Displays values using liquid crystal displays instead of a swinging needle. They are basically computers that determine the measured value and display it for the technician.

Multímetro digital Exhibe valores usando cristal líquido para desplegar, en cambio de una aguja giratoria. Son básicamente computadores que determinan el valor medido y lo muestran al técnico.

Digital signal A voltage value that has two states. The states can be on/off or high/low.

Señal digital El valor del voltaje que tiene dos ventajas. Una señal muestra prendido/apagado y la otra alta/baja.

Dimmer switch A switch in the headlight circuit that provides the means for the driver to select either high-beam or low-beam operation, and to switch between the two. The dimmer switch is connected in series within the headlight circuit and controls the current path for high and low beams.

Conmutador reductor Conmutador en el circuito para faros delanteros que le permite al conductor elegir la luz larga o la luz corta, y conmutar entre las dos. El conmutador reductor se conecta en serie dentro del circuito para faros delanteros y controla la trayectoria de la corriente para la luz larga y la luz corta.

Diode An electrical one-way check valve that will allow current to flow in one direction only.

Diodo Válvula eléctrica de retención, de una vía, que permite que la corriente fluya en una sola dirección.

Diode rectifier bridge A series of diodes that are used to provide a reasonably constant DC voltage to the vehicle's electrical system and battery.

Puente rectificador de diodo Serie de diodos utilizados para proveerles una tensión de corriente continua bastante constante al sistema eléctrico y a la batería del vehículo.

Diode/stator test Performed to determine the condition of the diodes. This is done to eliminate the possibility of damage being done to new components.

Prueba del estator diodo Funciona para determinar la condición de los diodos. Esto se hace para eliminar la posibilidad de daños que puedan suceder en los nuevos componentes.

Diode trio Used by some manufacturers to rectify the stator of an AC generator current so that it can be used to create the magnetic field in the field coil of the rotor.

Trío de diodos Utilizado por algunos fabricantes para rectificar el estátor de la corriente de un generador de corriente alterna y poder así utilizarlo para crear el campo magnético en la bobina inductora del rotor.

Direct drive A situation where the drive power is the same as the power exerted by the device that is driven.

Transmisión directa Una situación en la cual el poder de mando es lo mismo que la potencia empleada por el dispositivo arrastrado.

Discriminating sensors Part of the air bag circuitry; these sensors are calibrated to close with speed changes that are great enough to warrant air bag deployment. These sensors are also referred to as crash sensors.

Sensores discriminadores Una parte del conjunto de circuitos de Airbag; estos sensores se calibran para cerrar con los cambios de la velocidad que son bastante severas para justificar el despliegue del Airbag. Estos sensores también se llaman los sensores de impacto.

Display pattern A pattern of an oscilloscope in which the cylinders of an engine appear one after another according to the firing order.

Patrón visualizador Un patrón en un osciloscopio en el cual los cilindros de un motor aparecen uno tras otro según el orden del encendido.

DRB-III A diagnostic read-out box designed by DaimlerChrysler to gain access to the diagnostic functions of the controller.

DRB-III Esta iniciales se refieren a la caja de lectura de diagnósticos diseñada por Daimler Chrysler para tener acceso a los diagnósticos y funciones del controler.

Drive coil A hollowed field coil used in a positive-engagement starter to attract the movable pole shoe of the starter.

Bobina de excitación Una bobina inductora hueca empleada en un encendedor de acoplamiento directo para atraer la pieza polar móvil del encendedor.

DSO A common acronym for a digital storage oscilloscope.

DSO Una sigla común del osciloscopio de almacenamiento digital.

Dump valve A safety switch that releases vacuum to the servo when the brake pedal is pressed.

Válvula de descargue Un interruptor de seguridad que facilita el vacío al servo cuando se presiona el pedal del freno.

Duty cycle The percentage of on time to total cycle time.

Ciclo de trabajo Porcentaje del trabajo efectivo a tiempo total del ciclo.

Eddy currents Small induced currents.

Corriente de Foucault Pequeñas corrientes inducidas.

Electrical load The working device of the circuit.

Carga eléctrica Dispositivo de trabajo del circuito.

Electrochemical The chemical action of two dissimilar materials in a chemical solution.

Electroquímico Acción química de dos materiales distintos en una solución química.

Electrolysis The producing of chemical changes by passing electrical current through an electrolyte.

Electrólisis La producción de los cambios químicos al pasar un corriente eléctrico por un electrolito.

Electrolyte A solution of 64 percent water and 36 percent sulfuric acid.

Electrolito Solucion de un 64 percent de agua y un 36 percent de ácido sulfúrico.

Electromagnetic gauge Gauge that produces needle movement by magnetic forces.

Calibrador electromagnético Calibrador que genera el movimiento de la aguja mediante fuerzas magnéticas.

Electromagnetic induction The production of voltage and current within a conductor as a result of relative motion within a magnetic field.

Inducción electromagnética Producción de tension y de corriente dentro de un conductor como resultado del movimiento relativo dentro de un campo magnético.

Electromagnetic interference (EMI) An undesirable creation of electromagnetism whenever current is switched on and off.

Interferencia electromagnética Fenómeno de electromagnetismo no deseable que resulta cuando se conecta y se desconecta la corriente.

Electromagnetism A form of magnetism that occurs when current flows through a conductor.

Electromagnetismo Forma de magnetismo que ocurre cuando la corriente fluye a través de un conductor.

Electromechanical A device that uses electricity and magnetism to cause a mechanical action.

Electromecánico Un dispositivo que causa una acción mecánica por medio de la electricidad y el magnetismo.

Electromotive force (EMF) *See* voltage.

Fuerza electromotriz *Véase* tensión.

Electronic level controller (ELC) The computer that controls the automatic suspension system.

Controler de nivel electrónico Este es el computador que controla el sistema automático de suspensión.

Electrostatic discharge (ESD) straps Ground your body to prevent static discharges that may damage electronic components.

Correas electroestáticas de descargue Estas aislan su cuerpo para prevenir descargas estáticas que puedan dañar los componentes electrónicos.

EMI Electro-magnetic interference.

EMI La interferencia electromagnética.

Equivalent series load (equivalent resistance) The total resistance of a parallel circuit. It is equivalent to the resistance of a single load in series with the voltage source.

Carga en serie equivalente (resistencia equivalente) Resistencia total de un circuito en paralelo, equivalente a la resistencia de una sola carga en serie con la fuente de tensión.

Excitation current Current that magnetically excites the field circuit of the ac generator.

Corriente de excitación Corriente que excita magnéticamente al circuito inductor del generador de corriente alterna.

Face shields Clear plastic shields that protect the entire face.

Careta de soldador Caretas de plástico claro que protegen toda la cara.

Failsoft Computer substitution of a fixed input value if a sensor circuit should fail. This provides for system operation, but at a limited function. Also referred to as the "Limp-In" mode.

Falla activa Sustitución por la computadora de un valor fijo de entrada en caso de que ocurra una falla en el circuito de un sensor. Esto asegura el funcionamiento del sistema, pero a una capacidad limitada.

Fast charging Battery charging using a high amperage for a short period of time.

Carga rápida Carga de la batería que utiliza un amperaje máximo por un corto espacio de tiempo.

Feedback 1. Data concerning the effects of the computer's commands are fed back to the computer as an input signal. Used to determine if the desired result has been achieved. 2. A condition that can occur when electricity seeks a path of lower resistance, but the alternate path operates another component than that intended. Feedback can be classified as a short.

Realimentación 1. Datos referentes a los efectos de las órdenes de la computadora se suministran a la misma como señal de entrada. La realimentación se utiliza para determinar si se ha logrado el resultado deseado. 2. Condición que puede ocurrir cuando la electricidad busca una trayectoria de menos resistencia, pero la trayectoria alterna opera otro componente que aquel deseado. La realimentación puede clasificarse como un cortocircuito.

Fiber optics A medium of transmitting for the transmission of light through polymethylmethacrylate plastic that keeps the light rays parallel even if there are extreme bends in the plastic.

Transmisión por fibra óptica Técnica de transmisión de luz por medio de un plástico de polimetacrilato de metilo que mantiene los rayos de luz paralelos aunque el plástico esté sumamente torcido.

Field current draw test Diagnostic test that determines if there is current available to the field windings.

Prueba de la intensidad de una corriente inductora Prueba diagnóstica que determina si se está generando corriente a los devanados inductores.

Field relay The relay that controls the amount of current going to the field windings of a generator. This is the main output control unit for a charging system.

Relé inductor El relé que controla la cantidad del corriente a los devanados inductores de un generador. Es la unedad principal de potencia de salida de un sistema de carga.

Fire extinguisher A portable apparatus that contains chemicals, water, foam, or special gas that can be discharged to extinguish a small fire.

Extinctor de incendios Aparato portátil que contiene elementos químicos, agua, espuma o gas especial que pueden descargarse para extinguir un incendio pequeño.

Firing line That section of a scope waveform that appears as a tall spike and represents the amount of voltage necessary to overcome the resistance in the secondary circuit.

Linea de activación Esa sección de una forma de onda de un osciloscopio que aparece como un punto de impulso y representa la cantidad del tensión que se requiere para sobrepasar la resistencia en el circuito secundario.

Flammable A substance that supports combustion.

Inflamable Sustancia que promueve la combustión.

Flash To remove the existing programming and overwrite it with new software.

Limpiar Así se remueve el programa existente y lo reemplaza con programas nuevos.

Flash codes Diagnostic trouble codes (DTCs) that are displayed by flashing a lamp or LED.

Códigos intermitentes Códigos de fallas de diagnóstico (CFD) que se muestran en una lámpara intermitente o LED.

Flat rate A pay system in which a technicians are paid for the amount of work he does. Each job has a flat rate time. Pay is based on that time, regardless of how long it takes to complete the job.

Tarifa bloque Sistema de pago en el que se le paga al técnico por cada hora de trabajo realizado. Cada trabajo tiene un tiempo de tarifa bloque. El pago se basa en ese tiempo sin importar cuánto tiempo se lleve para terminarlo.

Floor jack A portable hydraulic tool used to raise and lower a vehicle.

Gato de pie Herramienta hidráulica portátil utilizada para levantar y bajar un vehículo.

Forward-bias A positive voltage that is applied to the P-type material and negative voltage to the N-type material of a semiconductor.

Polarización directa Tensión positiva aplicada al material P y tensión negativa aplicada al material N de un semiconductor.

Free speed test Diagnostic test that determines the free rotational speed of the armature. This test is also referred to as the no-load test.

Prueba de velocidad libre Prueba diagnóstica que determina la velocidad giratoria libre de la armadura. A dicha prueba se le llama prueba sin carga.

Freeze frame data A recording of the driving condition when the malfunction occurred. This is useful for determining how the vehicle was operating at the time and for locating any input or output values that are out of range.

Datos de trama fija Archivo de las condiciones de manejo cuando sucede una falla. Es conveniente para determinar cómo operaba el vehículo en ese momento y para localizar cualesquier valores de entrada y salida que estén fuera de banda.

Frequency The number of complete oscillations that occur during a specific time, measured in hertz.

Frecuencia El número de oscilaciones completas que ocurren durante un tiempo específico, medidas en Hertz.

Full field Field windings that are constantly energized with full battery current. Full fielding will produce maximum AC generator output.

Campo completo Devanados inductores que se excitan constantemente con corriente total de la batería. EL campo completo producirá la salida máxima de un generador de corriente alterna.

Full field test Diagnostic test used to isolate if the detected problem lies in the AC generator or the regulator.

Prueba de campo completo Prueba diagnóstica utilizada para determinar si el problema descubierto se encuentra en el generador de corriente alterna o en el regulador.

Functional test Checks the operation of the system as the technician observes the different results.

Prueba funcional Verifica la operación del sistema mientras el técnico observa los diferentes resultados.

Fuse A replaceable circuit protection device that will melt should the current passing through it exceed its rating.

Fusible Dispositivo reemplazable de protección del circuito que se fundirá si la corriente que fluye por el mismo excede su valor determinado.

Fuse box A term used that indicates the central location of the fuses contained in a single holding fixture.

Caja de fusibles Término utilizado para indicar la ubicación central de los fusibles contenidos en un solo elemento permanente.

Fusible link A wire made of meltable material with a special heat-resistant insulation. When there is an overload in the circuit, the link melts and opens the circuit.

Cartucho de fusible Alambre hecho de material fusible con aislamiento especial resistente al calor. Cuando ocurre una sobrecarga en el circuito, el cartucho se funde y abre el circuito.

Gassing The conversion of a battery's electrolyte into hydrogen and oxygen gas.

Burbujeo La conversión del electrolito de una batería al gas de hidrógeno y oxígeno.

Gauge 1. A device that displays the measurement of a monitored system by the use of a needle or pointer that moves along a calibrated scale. 2. The number that is assigned to a wire to indicate its size. The larger the number the smaller the diameter of the conductor.

Calibrador 1. Dispositivo que muestra la medida de un sistema regulado por medio de una aguja o indicador que se mueve a través de una escala calibrada. 2. El número asignado a un alambre indica su tamaño. Mientras mayor sea el número, más pequeño será el diámetro del conductor.

Gauss gauge   A meter that is sensitive to the magnetic field surrounding a wire conducting current. The gauge needle will fluctuate over the portion of the circuit that has current flowing through it. Once the ground has been passed, the needle will stop fluctuating.

Calibador gauss   Instrumento sensible al campo magnético que rodea un alambre conductor de corriente. La aguja del calibrador se moverá sobre la parte del circuito a través del cual fluye la corriente. Una vez se pasa a tierra, la aguja dejará de moverse.

Glitches   Unwanted voltage spikes that are seen on a voltage trace. These are normally caused by intermittent opens or shorts.

Irregularidades espontáneos   Impulsos de tensión no deseables que se ven en una traza de tensión. Estos se causan normalmente por las aberturas o cortos intermitentes.

Grids   A series of horizontal, ceramic, silver-compounded lines that are baked into the surface of the window.

Rejillas   Una serie de líneas horizontales, de cerámica plateada, combinadas y endurecidas (horneadas) en la superficie de la ventana.

Ground   The common negative connection of the electrical system that is the point of lowest voltage.

Tierra   Conexión negativa común del sistema eléctrico. Es el punto de tensión más baja.

Ground circuit test   A diagnostic test performed to measure the voltage drop in the ground side of the circuit.

Prueba del circuito a tierra   Prueba diagnóstica llevada a cabo para medir la caída de tensión en el lado a tierra del circuito.

Ground side   The portion of the circuit that is from the load component to the negative side of the source.

Lado a tierra   Parte del circuito que va del componente de carga al lado negativo de la fuente.

Grounded circuit   An electrical defect that allows current to return to ground before it has reached the intended load component.

Circuito puesto a tierra   Falla eléctrica que permite el regreso de la corriente a tierra antes de alcanzar el componente de carga deseado.

Growler   Test equipment used to test starter armatures for shorts and grounds. It produces a very strong magnetic field that is capable of inducing a current flow and magnetism in a conductor.

Indicador de cortocircuitos   Equipo de prueba utilizado para localizar cortocircuitos y tierra en armaduras de arranque. Genera un campo magnético sumamente fuerte, capaz de inducir flujo de corriente y magnetismo en un conductor.

Hall-effect switch   A sensor that operates on the principle that if a current is allowed to flow through thin conducting material being exposed to a magnetic field, another voltage is produced.

Conmutador de efecto Hall   Sensor que funciona basado en el principio de que si se permite el flujo de corriente a través de un material conductor delgado que ha sido expuesto a un campo magnético, se produce otra tensión.

Halogen   The term used to identify a group of chemically related nonmetallic elements. These elements include chlorine, fluorine, and iodine.

Halógeno   Término utilizado para identificar un grupo de elementos no metálicos relacionados químicamente. Dichos elementos incluyen el cloro, el flúor y el yodo.

Hand tools   Tools that use only the force generated from the body to operate.

Herramientas manuales   Herramientas que sólo utilizan la fuerza que genera el cuerpo para manejarlas.

Hard codes   Failures that were detected the last time the BCM tested the circuit.

Códigos indicadores de dureza   Fallas que fueron detectadas la última vez que el funcionamiento indicado con las letras BCM probó el circuito.

Hard-shell connector   An electrical connector that has a hard plastic shell that holds the connecting terminals of separate wires.

Conector de casco duro   Conector eléctrico con casco duro de plástico que sostiene separados los bornes conectadores de alambres individuales.

Hazard Communication Standard   The original bases of the Right-To-Know laws.

Normalización de Comunicado sobre Riesgos   Las bases originales de las leyes de derecho de información.

Hazardous materials   Materials that can cause illness, injury, or death or that pollute water, air, or land.

Materiales peligrosos   Materiales que pueden causar enfermedades, daños, o la muerte, o que contaminan el agua o la tierra.

Heat-shrink tubing   A hollow insulation material that shrinks to an airtight fit over a connection when exposed to heat.

Tubería contraída térmicamente   Material aislante hueco que se contrae para acomodarse herméticamente sobre una conexión cuando se encuentra expuesto al calor.

Hertz   A measurement of frequency.

Hertzios   Es una unidad de frecuencia.

High-voltage electronic control unit (HV ECU)   In an HEV, controls the motor generators and the engine based on torque demand.

Unidad de control electrónico de alto voltaje (UCE AV)   En un VHE controla los generadores del motor y el motor basándose en la demanda del par motor.

High-voltage service plug   The HEV is equipped with a high-voltage service plug that disconnects the HV battery from the system.

Bujía de servicio de alto voltaje   El VHE está equipado con una bujía de servicio de alto voltaje que desconecta la batería del VH del sistema.

History data   Data that is stored when a fault is set in an HEV, which can be used to determine if a customer's concern is actually a problem with the system.

Historial de datos   Los datos que se archivan cuando ocurre una falla en un VHE y que pueden usarse para determinar si la preocupación del cliente es en realidad un problema con el sistema.

Hoist   A lift that is used to raise an entire vehicle.

Elevador   Montacargas utilizado para elevar el vehículo en su totalidad.

Hold-in winding   A winding that holds the plunger of a solenoid in place after it moves to engage the starter drive.

Devanado de retención   Un devanado que posiciona el núcleo móvil de un solenoide después de que mueva para accionar el acoplamiento del motor de arranque.

Hydrometer   A test instrument used to check the specific gravity of the electrolyte to determine the battery's state of charge.

Hidrómetro   Instrumento de prueba utilizado para verificar la gravedad específica del electrolito y así determinar el estado de carga de la batería.

Hydrostatic lock   Liquid entering the cylinder and preventing the piston from moving upward.

Cierre hidrostático   La entrada de líquido en el cilindro que impide el movimiento ascendente del pistón.

Igniter   A combustible device that converts electric energy into thermal energy to ignite the inflator propellant in an air bag system.

Ignitor   Un dispositivo combustible que convierte la energía eléctrica a la energía termal para encender el propelente inflador en un sistema Airbag.

Illuminated entry actuator   Contains a printed circuit and a relay.

Actuador iluminado de entradas   Contiene un circuito impreso y un relevador.

Impedance   The combined opposition to current created by the resistance, capacitance, and inductance of a test meter or circuit.

Impedancia   Oposición combinada a la corriente generada por la resistencia, la capacitancia y la inductancia de un instrumento de prueba o de un circuito.

Inductive reactance   The result of current flowing through a conductor and the resultant magnetic field around the conductor that opposes the normal flow of current.

Reactancia inductiva   El resultado de un corriente que circule por un conductor y que resulta en un campo magnético alrededor del conductor que opone el flujo normal del corriente.

Information codes   HEV codes that provide more information and freeze frame data concerning the DTCs. Information codes along with the DTC indicate more precisely the location of the fault.

Códigos de información   Códigos del VHE que proporcionan mayor información y datos de trama fija que concierne a los códigos de diagnóstico de fallas (CDF.) Los códigos de información junto con los CDF indican con más precisión la localización de la falla.

Instrument voltage regulator (IVR)   Provides a constant voltage to the gauge regardless of the voltage output of the charging system.

Instrumento regulador de tensión   Le provee tensión constante al calibrador, sin importar cual sea la salida de tensión del sistema de carga.

Insulated circuit resistance test   A voltage drop test that is used to locate high resistance in the starter circuit.

Prueba de la resistencia de un circuito aislado Prueba de la caída de tensión utilizada para localizar alta resistencia en el circuito de arranque.

Insulated side The portion of the circuit from the positive side of the source to the load component.

Lado aislado Parte del circuito que va del lado positivo de la fuente al componente de carga.

Insulator A material that does not allow electrons to flow easily through it.

Aislador Una material que no permite circular fácilmente los electrones.

Integrated circuit (IC chip) A complex circuit of thousands of transistors, diodes, resistors, capacitors, and other electronic devices that are formed onto a small silicon chip. As many as 30,000 transistors can be placed on a chip that is 1/4 inch (6.35 mm) square.

Circuito integrado (Fragmento CI) Circuito complejo de miles de transistores, diodos, resistores, condensadores, y otros dispositivos electrónicos formados en un fragmento pequeño de silicio. En un fragmento de 1/4 de pulgada (6,35 mm) cuadrada, pueden colocarse hasta 30.000 transistores.

Intermediate section The part of an ignition trace on a scope that displays the coil's action after the firing of the plug.

Sección intermedia La parte de un razgo de encendido en un osiloscopio que manifiesta la acción de la bobina después de encenderse la bujía.

Intermittent codes Those that have occurred in the past but were not present during the last BCM test of the circuit.

Códigos intermitentes Son los que han ocurrido en el pasado pero no estuvieron presentes durante la última prueba del circuito, efectuada de acuerdo con las especificaciones de las letras BCM.

Inverter Controls current flow between the motor generators and the HV battery in an HEV. Converts HV battery DC voltage into three-phase alternating current for the motor generators and also converts high-voltage AC from the motor generators to DC voltage to charge the HV battery.

Inversor Controla el flujo de la corriente entre los generadores del motor y la batería del VH en un VHE. Convierte el voltaje de la batería del VH en corriente alterna de tres fases para los generadores del motor, y también convierte la corriente alterna de alto voltaje de los generadores del motor a voltaje de corriente directa para cargar la batería del VH.

J1962 breakout box (BOB) A special tool that provides a pass through test point that connects in series between the DLC and the scan tool. This provides easy testing of voltages and resistance of any of the DLC circuits without risk of damage to the DLC.

Caja de conexiones J1962 Herramienta especial que proporciona un paso mediante el punto de prueba que se conecta en series entre el circuito del control del enlace de datos (DLC) y el instrumento de exploración. Esto proporciona una prueba fácil de los voltajes y la resistencia en cualquiera de los circuitos del DLC sin riesgo de dañar el DLC.

Jack stands Support devices used to hold the vehicle off the floor after it has been raised by the floor jack.

Soportes de gato Dispositivos de soporte utilizados para sostener el vehículo sobre el suelo después de haber sido levantado con el gato de pie.

Jumper wire A wire used in diagnostics that is made up of a length of wire with a fuse or circuit breaker and has alligator clips on both ends.

Cable conector Una alambre empleado en los diagnósticos que se comprende de un trozo de alambre con un fusible o un interruptor y que tiene una pinza de conexión en ambos lados.

Keyless entry A lock system that allows for locking and unlocking of a vehicle with a touch keypad instead of a key.

Entrada sin llave Un sistema de cerradura que permite cerrar y abrir un vehículo por medio de un teclado en vez de utilizar una llave.

KOEO Key ON, Engine OFF.

KOEO Llave puesta, motor apagado.

KOER Key ON, Engine RUNNING.

KOER Llave puesta, motor en marcha.

kV Kilovolt or 1000 volts.

kV Kilovolito o 1000 voltios.

Lamination The process of constructing something with layers of materials that are firmly connected.

Laminación El proceso de construir algo de capas de materiales unidas con mucha fuerza.

Lamp A device that produces light as a result of current flow through a filament. The filament is enclosed within a glass envelope and is a type of resistance wire that is generally made from tungsten.

Lámpara Dispositivo que produce luz como resultado del flujo de corriente a través de un filamento. El filamento es un tipo de alambre de resistencia hecho por lo general de tungsteno, que es encerrado dentro de una bombilla.

Lamp sequence check Used to determine problems by observing the operation of the warning lights under different conditions.

Lámpara verificadora de secuencias Usada para determinar problemas al observar el funcionamiento de las luces de alerta bajo condiciones diferentes.

Light-emitting diode (LED) A gallium-arsenide diode that converts the energy developed when holes and electrons collide during normal diode operation into light.

Diodo emisor de luz Diodo semiconductor de galio y arseniuro que convierte en luz la energía producida por la colisión de agujeros y electrones durante el funcionamiento normal del diodo.

Limit switch A switch used to open a circuit when a predetermined value is reached. Limit switches are normally responsive to a mechanical movement or temperature changes.

Disyuntor de seguridad Un conmutador que se emplea para abrir un circuito al alcanzar un valor predeterminado. Los disyuntores de seguridad suelen ser responsivos a un movimiento mecánico o a los cambios de temperatura.

Liquid crystal display (LCD) A display that sandwiches electrodes and polarized fluid between layers of glass. When voltage is applied to the electrodes, the light slots of the fluid are rearranged to allow light to pass through.

Visualizador de cristal líquido Visualizador digital que consta de dos láminas de vidrio selladas, entre las cuales se encuentran los electrodos y el fluido polarizado. Cuando se aplica tensión a los electrodos, se rompe la disposición de las moléculas para permitir la formación de carácteres visibles.

Live module A module that has not been deployed.

Módulo activo El módulo que no ha sido desplegado.

Logic probe A test instrument used to detect a pulsing signal.

Sonda lógica Un instrumento de prueba que se emplea para detectar una señal pulsante.

Magnetic pulse generator Sensor that uses the principle of magnetic induction to produce a voltage signal. Magnetic pulse generators are commonly used to send data concerning the speed of the monitored component to the computer.

Generador de impulsos magnéticos Sensor que funciona según el principio de inducción magnética para producir una señal de tensión. Los generadores de impulsos magnéticos se utilizan comúnmente para transmitir datos a la computadora relacionados a la velocidad del componente regulado.

Magnetism An energy form resulting from atoms aligning within certain materials, giving the materials the ability to attract other metals.

Magnetismo Forma de energía que resulta de la alineación de átomos dentro de ciertos materiales y que le da a éstos la capacidad de atraer otros metales.

Material Safety Data Sheets (MSDS) Contain information about each hazardous material in the workplace.

Hojas de datos de seguridad del material (HDSM) Contienen información sobre cada material peligroso en el lugar de trabajo.

Maxi-fuse A circuit protection device that looks similar to blade-type fuses except they are larger and have a higher amperage capacity. Maxi-fuses are used because they are less likely to cause an underhood fire when there is an overload in the circuit. If the fusible link burns in two, it is possible that the "hot" side of the fuse could come into contact with the vehicle frame and the wire could catch on fire.

Maxifusible Dispositivo de protección del circuito parecido a un fusible de tipo de cuchilla, pero más grande y con mayor capacidad de amperaje. Se utilizan maxifusibles porque existen menos probabilidades de que ocasionen un incendio debajo de la capota cuando ocurra una sobrecarga en el circuito. Si el cartucho de fusible se quemase en dos partes, es posible que el lado "cargado" del fusible entre en contacto con el armazón del vehículo y que el alambre se encienda.

Metri-pack connector Special wire connectors used in some computer circuits. They seal the wire terminals from the atmosphere, thereby preventing corrosion and other damage.

Conector metri-pack Los conectores de alambres especiales que se emplean en algunos circuitos de computadoras. Impermealizan los bornes de los alambres, así previniendo la corrosión y otros daños.

Millisecond Equals 1/100th of a second.
Milesegundos Equivale a 1/100 de un segundo.

Molded connector An electrical connector that usually has one to four wires that are molded into a one-piece component.
Conectador moldeado Conectador eléctrico que por lo general tiene hasta un máximo cuatro alambres que se moldean en un componente de una sola pieza.

MSDS Material Safety Data Sheet.
MSDS Hojas de Dato de Seguridad de los Materiales.

Multifunction switch Can have a combination of any of the following switches in a single unit: headlights, turn signal, hazard, dimmer, horn, and flash to pass.
Interruptor de múltiples funciones Este permite lograr una combinación de cualquiera de los siguientes interruptores: faroles, señal de doblaje, peligro, interruptor reducidor de luz, cuerno, señal de pasada a otro carril.

Multimeter A test instrument that measures more than one electrical property.
Multímetro Un instrumento diagnóstico que mide más de una propiedad eléctrica.

Multiplying coil Made of ten wraps of wire. This multiplies the ammeter reading so that the tester's scale can be used to read lower amperage. For example, if the needle is pointing to 25 amperes when using the multiplying coil, the actual reading is 2.50 amperes.
Bobina múltiple Está hecha de diez (10) cables enroscados. Esto multiplica la lectura del amperímetro y así la escala de prueba puede ser usada para ver si el amperaje está bajo. Por ejemplo, si la aguja está indicando 25 amperajes cuando se está usando la bobina múltiple la lectura actual es de 2.50 amperajes.

Neutral safety switch A switch used to prevent the starting of an engine unless the transmission is in PARK or NEUTRAL.
Disyuntor de seguridad en neutral Un conmutador que se emplea para prevenir que arranque un motor al menos de que la transmisión esté en posición PARK o NEUTRAL.

No-crank A term used to mean that when the ignition switch is placed in the START position, the starter does not turn the engine.
Sin arranque Término utilizado para expresar que cuando el botón conmutador de encendido está en la posición START, el arrancador no enciende el motor.

No-crank test Diagnostic test performed to locate any opens in the starter or control circuits.
Prueba sin arranque Prueba diagnóstica llevada a cabo para localizar aberturas en los circuitos de arranque o de mando.

Noise An unwanted voltage signal that rides on a signal. Noise is usually the result of radio frequency interface (RFI) or electromagnetic induction (EMI).
Ruido Una señal indeseada del voltaje que aparece montada en una señal. El ruido es generalmente el resultado de la radio frecuencia de contacto (RFI) o de la inducción electromagnética (EMI).

Normally closed (NC) switch A switch designation denoting that the contacts are closed until acted upon by an outside force.
Conmutador normalmente cerrado Nombre aplicado a un conmutador cuyos contactos permanecerán cerrados hasta que sean accionados por una fuerza exterior.

Normally open (NO) switch A switch designation denoting that the contacts are open until acted upon by an outside force.
Conmutador normalmente abierto Nombre aplicado a un conmutador cuyos contactos permanecerán abiertos hasta que sean accionados por una fuerza exterior.

NTC Negative temperature coefficient.
NTC Las iniciales NTC representan la temperatura coeficiente negativa.

OBD II Stands for on-board diagnostics, second generation.
OBD II Las letras OBD II se refieren a los diagnósticos del tablero, segunda generación.

Occupational safety glasses Eye protection that is designed with special high-impact lens and frames, and side protection.
Gafas de protección para el trabajo Gafas diseñadas con cristales y monturas especiales resistentes y provistas de protección lateral.

OCS service kit An OCS part set that consists of the seat foam, the bladder, the pressure sensor, the occupant classification module (OCM), and the wiring. The service kit is calibrated as an assembly. The service kit OCM has a special identification data bit that is transmitted once it is connected.

Kit de servicio SCO Juego de partes SCO que consiste en la hule-espuma del asiento, el depósito, el sensor de presión, el módulo de clasificación del ocupante (MCO) y el alambrado. El kit de servicio está calibrado como un ensamblado. El kit de servicio SCO tiene un bit de datos de identificación especial que se transmiten cuando se haya conectado.

OCS validation test Special procedure that confirms the OCS can properly classify the occupant. This task usually requires the use of special weights.
Prueba de revalidación del SCO Procedimiento especial que confirma que el SCO puede clasificar apropiadamente al ocupante. Esta tarea generalmente requiere el uso de pesas especiales.

Odometer A mechanical counter in the speedometer unit that indicates total miles accumulated on the vehicle.
Odómetro Aparato mecánico en la unidad del velocímetro con el que se cuentan las millas totales recorridas por el vehículo.

Offset Placed off center. Refers to the number of degrees a timing light or meter should be set to provide accurate ignition timing readings.
Desviación Ubicado fuera de lo central. Se refiere al número de grados que se debe ajustar una luz de temporización o un medidor para proveer las lecturas exactas del tiempo de encendido.

Ohm Unit of measure for resistance. One ohm is the resistance of a conductor such that a constant current of one ampere in it produces a voltage of one volt between its ends.
Ohmio Unidad de resistencia eléctrica. Un ohmio es la resistencia de un conductor si una corriente constante de 1 amperio en el conductor produce una tensión de 1 voltio entre los dos extremos.

Ohmmeter A test meter used to measure resistance and continuity in a circuit.
Ohmiómetro Instrumento de prueba utilizado para medir la resistencia y la continuidad en un circuito.

Ohm's law Defines the relationship between current, voltage, and resistance.
Ley de Ohm Define la relación entre la corriente, la tensión y la resistencia.

Open An electrical term used to indicate that the circuit is not complete or is broken.
Abierto Es un término de electricidad, usado para indicar que el circuito no está completo o está roto.

Open circuit A term used to indicate that current flow is stopped. By opening the circuit, the path for electron flow is broken.
Circuito abierto Término utilizado para indicar que el flujo de corriente ha sido detenido. Al abrirse el circuito, se interrumpe la trayectoria para el flujo de electrones.

Open circuit voltage test Used to determine the battery's state of charge. It is used when a hydrometer is not available or cannot be used.
Prueba de la tensión en un circuito abierto Sirve para determinar el estado de carga de la batería. Esta prueba se lleva a cabo cuando no se dispone de un hidrómetro o cuando el mismo no puede utilizarse.

Optics test Test of the automatic high beam system to test the ability to recognize ambient light through the lens and the windshield.
Prueba óptica Prueba del sistema automático de luces altas para averiguar la habilidad de reconocer las luces ambientales mediante la lente y el parabrisas

Overload Excess current flow in a circuit.
Sobrecarga Flujo de corriente superior a la que tiene asignada un circuito.

Overrunning clutch A clutch assembly on a starter drive used to prevent the engine's flywheel from turning the armature of the starter motor.
Embrague de sobremarcha Una asamblea de embrague en un acoplamiento del motor de arranque que se emplea para prevenir que el volante del motor dé vueltas al armazón del motor de arranque.

Oxygen sensor A voltage generating sensor that measures the amount of oxygen present in an engine's exhaust.
Sensor de oxígeno Un sensor generador de tensión que mide la cantidad del oxígeno presente en el gas de escape de un motor.

Parallel circuit A circuit that provides two or more paths for electricity to flow.
Circuito en paralelo Circuito que provee dos o más trayectorias para que circule la electricidad.

Parasitic drains Constant drains on the battery due to accessories that draw small amounts of current.
Drenaje parásita Los constantes drenajes en la batería son causados debido a que los accesorios atraen pequeñas cantidades de corriente.

Parasitic loads Electrical loads that are still present when the ignition switch is in the OFF position.

Cargas parásitas Cargas eléctricas que todavía se encuentran presentes cuando el botón conmutador de encendido está en la posición OFF.

Park switch Contact points located inside the wiper motor assembly that supply current to the motor after the wiper control switch has been turned to the PARK position. This allows the motor to continue operating until the wipers have reached their PARK position.

Conmutador PARK Puntos de contacto ubicados dentro del conjunto del motor del frotador que le suministran corriente al motor después de que el conmutador para el control de los frotadores haya sido colocado en la posición PARK. Esto permite que el motor continue su funcionamiento hasta que los frotadores hayan alcanzado la posición original.

Passive seat belt system Seat belt operation that automatically puts the shoulder and/or lap belt around the driver or occupant. The automatic seat belt is moved by DC motors that move the belts by means of carriers on tracks.

Sistema pasivo de cinturones de seguridad Función de los cinturones de seguridad que automáticamente coloca el cinturón superior y/o inferior sobre el conductor o pasajero. Motores de corriente continua accionan los cinturones automáticos mediante el uso de portadores en pistas.

Peak The highest voltage value in one cycle of an AC voltage sine wave.

Máxima eficiencia El valor más alto del voltaje de un ciclo de corriente alterna (AC) en un voltaje de onda senoidal.

Peak-to-peak voltage The total voltage measured between the peaks of an AC sine wave.

Voltaje de máximas eficiencias La medida total del voltaje cuando los puntos son de máxima eficiencia en una onda senoidal de corriente alterna (AC).

Photocell A variable resistor that uses light to change resistance.

Fotocélula Resistor variable que utiliza luz para cambiar la resistencia.

Photocell resistance assembly A technician-made test tool that replaces the photocell to produce predictable results.

Montaje de resistencia fotocélula Una herramienta técnica de prueba que remplaza la fotocélula para producir resultados visibles.

Phototransistor A transistor that is sensitive to light.

Fototransistor Transistor sensible a la luz.

Photovoltaic diodes Diodes capable of producing a voltage when exposed to radiant energy.

Diodos fotovoltaicos Diodos capaces de generar una tensión cuando se encuentran expuestos a la energía de radiación.

Pick-up coil The stationary component of the magnetic pulse generator consisting of a weak permanent magnet that is wound around by fine wire. As the timing disc rotates in front of it, the changes of magnetic lines of force generate a small voltage signal in the coil.

Bobina captadora Componente fijo del generador de impulsos magnéticos compuesta de un imán permanente débil devanado con alambre fino. Mientras gira el disco sincronizador enfrente de él, los cambios de las líneas de fuerza magnética generan una pequeña señal de tensión en la bobina.

Piezoresistive sensor A sensor that is sensitive to pressure changes.

Sensor piezoresistivo Sensor susceptible a los cambios de presión.

Pinion gear A small gear; typically refers to the drive gear of a starter drive assembly or the small drive gear in a differential assembly.

Engranaje de piñon Un engranaje pequeño; tipicamente se refiere al engranaje de arranque de una asamblea de motor de arranque o al engranaje de mando pequeño de la asamblea del diferencial.

Pinpoint test A specific component test.

Prueba de precisión Una prueba hecha con un componente específico.

Plate straps Metal connectors used to connect the positive or negative plates in a battery.

Abrazaderas de la placa Los conectores metálicos que sirven para conectar las placas positivas o negativas de una batería.

Plates The basic structure of a battery cell; each cell has at least one positive plate and one negative plate.

Placas La estructura básica de una celula de batería; cada celula tiene al menos una placa positiva y una placa negativa.

PMGR An abbreviation for permanent magnet gear reduction.

PMGR Una abreviación de desmultiplicación del engranaje del imán permanente.

Pneumatic tools Power tools that are powered by compressed air.

Herrimientas neumáticas Herramientas mecánicas accionadas por aire comprimido.

Polarizers Glass sheets that make light waves vibrate in only one direction. This converts light into polarized light.

Polarizadores Las láminas de vidrio que hacen vibrar las ondas de luz en un sólo sentido. Esto convierte la luz en luz polarizada.

Polarizing The process of light polarization or of setting one end of a field as a positive or negative point.

Polarizadora El proceso de polarización de la luz o de establecer un lado de un campo como un punto positivo o negativo.

Positive engagement starter A type of starter that uses the magnetic field strength of a field winding to engage the starter drive into the flywheel.

Acoplamiento de arranque positivo Un tipo de arrancador que utilisa la fuerza del campo magnético del devanado inductor para accionar el acoplamiento del arrancador en el volante.

Potential The ability to do something; typically voltage is referred to as the potential. If you have voltage, you have the potential for electricity.

Potencial La capacidad de efectuar el trabajo; típicamente se refiere a la tensión como el potencial. Si tiene tensión, tiene la potencial para la electricidad.

Potentiometer A variable resistor that acts as a circuit divider, providing accurate voltage drop readings proportional to movement.

Potenciómetro Resistor variable que actúa como un divisor de circuito para obtener lecturas de perdidas de tensión precisas en proporción con el movimiento.

Power formula A formula used to calculate the amount of electrical power a component uses. The formula is P = I x E, where P stands for power (measured in watts), I stands for current, and E stands for voltage.

Formula de potencia Una formula que se emplea para calcular la cantidad de potencia eléctrica utilizada por un componente. La formula es P = I x E, en el que el P quiere decir potencia (medida en wats), I representa el corriente y el E representa la tensión.

Power tools Tools that use forces other than those generated from the body. They can use compressed air, electricity, or hydraulic pressure to generate and multiply force.

Herramientas mecánicas Herramientas que utilizan fuerzas distintas a las generadas por el cuerpo. Dichas fuerzas pueden ser el aire comprimido, la electricidad, o la presión hidráulica para generar y multiplicar la fuerza.

Power train control module (PCM) The computer that controls the engine operation.

Módulo de control de potencia El computador que controla la operación del motor.

Pressure control solenoid A solenoid used to control the pressure of a fluid, commonly found in electronically controlled transmissions.

Solenoide de control de la presión Un solenoide que controla la presión de un fluido, suele encontrarse en las transmisiones controladas electronicamente.

Primary wiring Conductors that carry low voltage and low current. The insulation of primary wires is usually thin.

Hilos primarios Hilos conductores de tensión y corriente bajas. El aislamiento de hilos primarios es normalmente delgado.

Program A set of instructions that the computer must follow to achieve desired results.

Programa Conjunto de instrucciones que la computadora debe seguir para lograr los resultados deseados.

PROM (programmable read only memory) Memory chip that contains specific data that pertains to the exact vehicle that the computer is installed in. This information may be used to inform the CPU of the accessories that are equipped on the vehicle.

PROM (memoria de sólo lectura programable) Fragmento de memoria que contiene datos específicos referentes al vehículo particular en el que se instala la computadora. Esta información puede utilizarse para informar a la UCP sobre los accesorios de los cuales el vehículo está dotado.

Protection device Circuit protector that is designed to "turn off" the system that it protects. This is done by creating an open to prevent a complete circuit.

Dispositivo de protección Protector de circuito diseñado para "desconectar" el sistema al que provee protección. Esto se hace abriendo el circuito para impedir un circuito completo.

Prove-out circuit A function of the ignition switch that completes the warning light circuit to ground through the ignition switch when it is in the START position. The warning light is on during engine cranking to indicate to the driver that the bulb is working properly.

Circuito de prueba Función del boton conmutador de encendido que completa el circuito de la luz de aviso para que se ponga a tierra a través del botón conmutador de encendido cuando éste se encuentra en la posición START. La luz de aviso se encenderá durante el arranque del motor para avisarle al conductor que la bombilla funciona correctamente.

PTC Positive temperature coefficient.
PTC Iniciales identificadoras de la temperatura positiva coeficiente.

Pull to seat A method used to install the terminals into the connector.
Tire el asiento Un método usado para instalar los terminales dentro del conectador.

Pulse width The length of time in milliseconds that an actuator is energized.
Duración de impulsos Espacio de tiempo en milisegundos en el que se excita un accionador.

Pulse width modulation On/off cycling of a component. The period of time for each cycle does not change, only the amount of on time in each cycle changes.
Modulación de duración de impulsos Modulación de impulsos de un componente. El espacio de tiempo de cada ciclo no varía; lo que varía es la cantidad de trabajo efectivo de cada ciclo.

Push to seat A method used to install the terminals into the connector.
Empuje el asiento Un método usado para instalar los terminales dentro del conectador.

Quartz swing needle Displays that are similar in design to the air core electromagnetic gauges used in conventional analog instrument panels.
Aguja de oscilación del cuarzo Mostrarios que son similares en diseño al del núcleo de aire de los calibres electromagnéticos usados convencionalmente en los paneles de instrumentos análogos.

Quick test Isolates the problem area and determines whether the starter motor, solenoid, or control circuit is at fault.
Prueba rápida Aisla el área con problemas y determina si el arrancador del motor, el selenoid, y el circuito de control están fallando.

Radial grid A type of battery grid that has its patterns branching out from a common center.
Rejilla radial Un tipo de rejilla de bateria cuyos diseños extienden de un centro común.

Radio choke Absorbs voltage spikes and prevents static in the vehicle's radio.
Impedancia del radio Absorba los impulsos de la tensión y previene la presencia del estático en el radio del vehículo.

Radio frequency interface (RFI) Produced when electromagnetic radio waves of sufficient amplitude escape from a wire or connector.
Radio frecuencia de contacto (RFI) Se produce cuando las ondas de radio eletromagnéticas y de suficiente amplitud se escapan de un cable o conector.

Raster pattern Stacks the voltage patterns on top of each other, with cylinder number one being at the bottom.
Modelos de acumulación Modelos de voltaje acumulados, uno encima del otro, con un cilindro numérico colocado en el fondo.

Ratio A mathematical relationship between two or more things.
Razón Una relación matemática entre dos cosas o más.

Rectification The converting of AC current to DC current.
Rectificación Proceso a través del cual la corriente alterna es transformada en una corriente continua.

Refractometer A special meter used to measure the specific gravity of a liquid by the refraction of light.
Refractómetro Herramienta especial que usa para medir la gravedad específica de un líquido por la refracción de la luz.

Relative compression testing A test method for determining if a cylinder has lower compression relative to other cylinders by using current draw during cranking.
Prueba de compresión relativa Método de prueba para determinar si un cilindro tiene compresión más baja comparada a los otros cilindros al utilizar llamada de corriente durante la desencoladura.

Relay A device that uses low current to control a high-current circuit. Low current is used to energize the electromagnetic coil, while high current is able to pass over the relay contacts.
Relé Dispositivo que utiliza corriente baja para controlar un circuito de corriente alta. La corriente baja se utiliza para excitar la bobina electromagnética, mientras que la corriente alta puede transmitirse a través de los contactos del relé.

Reserve-capacity rating An indicator, in minutes, of how long the vehicle can be driven with the headlights on, if the charging system should fail. The reserve-capacity rating is determined by the length of time, in minutes, that a fully charged battery can be discharged at 25 amperes before battery cell voltage drops below 1.75 volts per cell.
Clasificación de capacidad en reserva Indicación, en minutos, de cuánto tiempo un vehículo puede continuar siendo conducido, con los faros delanteros encendidos, en caso de que ocurriese una falla en el sistema de carga. La clasificación de capacidad en reserva se determina por el espacio de tiempo, en minutos, en el que una batería completamente cargada puede descargarse a 25 amperios antes de que la tensión del acumulador de la batería disminuya a un nivel inferior de 1,75 amperios por acumulador.

Resource Conservation and Recovery Act (RCRA) Law that makes hazardous waste generators responsible for the waste from the time it becomes a waste material until the proper waste disposal is completed.
Ley de la Conservación y Recuperación de los Recursos Ley que hace responsables a los que generan residuos peligrosos por su desecho desde el momento en que se convierte en material de desecho hasta que se completa su destrucción apropiada de desechos.

Resistance Opposition to current flow.
Resistencia Oposición que presenta un conductor al paso de la corriente eléctrica.

Resistance wire A special type of wire that has some resistance built into it. These typically are rated by ohms per foot.
Alambre de resistencia Un tipo de alambre especial que por diseño tiene algo de resistencia. Estos tipicamente tienen un valor nominal de ohm por pie.

Resistive shorts Shorts to ground that pass through a form of resistance first.
Cortocircuitos resistivos Cortocircuitos a tierra que primero pasan por una forma de resistencia.

Resistor block A series of resistors with different values.
Bloque resistor Serie de resistores que tienen valores diferentes.

Reversed-bias A positive voltage is applied to the N-type material and negative voltage is applied to the P-type material of a semiconductor.
Polarización inversa Tensión positiva aplicada al material N y tensión negativa aplicada al material P de un semiconductor.

RFI Common acronym for radio frequency interference.
RFI Una sigla común de la interferencia de radiofrecuencia.

Rheostat A two-terminal variable resistor used to regulate the strength of an electrical current.
Reostático Un resistor variable de dos terminales usado para regular la potencia de un circuito eléctrico.

Right-to-know laws Laws concerning hazardous materials and wastes that protect every employee in a workplace by requiring employers to provide a safe working place as it relates to hazardous materials. The right-to-know laws state that employees have a right to know when the materials they use at work are hazardous.
Leyes de derecho de información Leyes contra materiales peligrosos y desechos que protegen a cada empleado en un lugar de trabajo al requerir que los patrones proporcionen un lugar de trabajo segura contra los materiales peligrosos. Las leyes de derecho de información declaran que los empleados tienen derecho a saber cuando y qué materiales que usan en el trabajo son peligrosos.

RMS Root-mean-square; a method for measuring AC voltage.
RMS Raíz de la media de los cuadrados; un método para medir la tensión del corriente alterna.

Root mean square (RMS) Meters convert the AC signal to a comparable DC voltage signal.
Corriente efectiva (RMS) Los medidores convierten la señal de corriente alterna (AC) a una señal de voltaje comparable (DC).

Rotor The component of the ac generator that is rotated by the drive belt and creates the rotating magnetic field of the ac generator.
Rotor Parte rotativa del generador de corriente alterna accionada por la correa de transmisión y que produce el campo magnético rotativo del generador de corriente alterna.

Safety goggles Eye protection device that fits against the face and forehead to seal off the eyes from outside elements.
Gafas de seguridad Dispositivo protector que se coloca delante de los ojos para preservarlos de elementos extraños.

Safety stands *See* Jack stands.

Soportes de seguridad *Véase* soportes de gato.

Scanner A diagnostic test tool that is designed to communicate with the vehicle's on-board computer.

Dispositivo de exploración Herramienta de prueba diagnóstica diseñada para comunicarse con la computadora instalada en el vehículo.

Scan tool A microprocessor designed to communicate with the BCM. It will access trouble codes, run system operation, actuator, and sensor tests.

Herramienta analizadora Un microprocesador diseñado para comunicarse usando la información de las iniciales BCM. Así se encuentran los códigos que indican problemas, cómo funciona el sistema de conducción, los actuadores y monitores de prueba.

Sealed-beam headlight A self-contained glass unit that consists of a filament, an inner reflector, and an outer glass lens.

Faro delantero sellado Unidad de vidrio que contiene un filamento, un reflector interior y una lente exterior de vidrio.

Secondary wiring Conductors, such as battery cables and ignition spark plug wires, that are used to carry high voltage or high current. Secondary wires have extra-thick insulation.

Hilos secundarios Conductores, tales como cables de batería e hilos de bujías del encendido, utilizados para transmitir tensión o corriente alta. Los hilos secundarios poseen un aislamiento sumamente grueso.

Self-test input (STI) The single pigtail connector located next to the self-test connector.

Autoprueba de entrada (STI) Simple cable flexible de conección localizado junto al conector de autoprueba.

Semiconductors An element that is neither a conductor nor an insulator. Semiconductors are materials that conduct electric current under certain conditions, yet will not conduct under other conditions.

Semiconductores Elemento que no es ni conductor ni aislante. Los semiconductores son materiales que transmiten corriente eléctrica bajo ciertas circunstancias, pero no la transmiten bajo otras.

Sender unit The sensor for the gauge. It is a variable resistor that changes resistance values with changing monitored conditions.

Unidad emisora Sensor para el calibrador. Es un resistor variable que cambia los valores de resistencia según cambian las condiciones reguladas.

Sensitivity controls A potentiometer that allows the driver to adjust the sensitivity of the automatic dimmer system to surrounding ambient light conditions.

Controles de sensibilidad Un potenciómetro que permite que el conductor ajusta la sensibilidad del sistema de intensidad de iluminación automático a las condiciones de luz ambientales.

Sensor Any device that provides an input to the computer.

Sensor Cualquier dispositivo que le transmite información a la computadora.

Series circuit A circuit that provides a single path for current flow from the electrical source through all the circuit's components, and back to the source.

Circuito en serie Circuito que provee una trayectoria única para el flujo de corriente de la fuente eléctrica a través de todos los componentes del circuito, y de nuevo hacia la fuente.

Series-parallel circuit A circuit that has some loads in series and some in parallel.

Circuito en series paralelas Circuito que tiene unas cargas en serie y otras en paralelo.

Series-wound motor A type of motor that has its field windings connected in series with the armature. This type of motor develops its maximum torque output at the time of initial start. Torque decreases as motor speed increases.

Motor con devanados en serie Un tipo de motor cuyos devanados inductores se conectan en serie con la armadura. Este tipo de motor desarrolla la salida máxima de par de torsión en el momento inicial de ponerse en marcha. El par de torsión disminuye al aumentar la velocidad del motor.

Servo Controls the position of the throttle by receiving a controlled amount of vacuum.

Control servo Este sirve para controlar la posición de la válvula reguladora al recibir una cantidad controlada de vacío.

Servomotor An electrical motor that produces rotation of less than a full turn. A feedback mechanism is used to position itself to the exact degree of rotation required.

Servomotor Motor eléctrico que genera rotación de menos de una revolución completa. Utiliza un mecanismo de realimentación para ubicarse al grado exacto de la rotación requerida.

Short An unwanted electrical path; sometimes this path goes directly to ground.

Corto Una trayectoria eléctrica no deseable; a veces este trayectoria viaja directamente a tierra.

Shorted circuit Allows current to bypass part of the normal path.

Circuito corto Este circuito permite que la corriente pase por una parte del recorrido normal.

Short to ground A condition that allows current to return to ground before it has reached the intended load component.

Corto a la tierra Su función permite que la corriente regrese a la tierra, antes de que haya llegado a un componente intencionalmente cargado.

Shunt-wound motor A type of motor whose field windings are wired in parallel to the armature. This type of motor does not decrease its torque as speed increases.

Motor con devanados en derivación Un tipo de motor cuyos devanados inductores se cablean paralelos a la armadura. Este tipo de motor no disminuya su par de torsión al aumentar la velocidad.

Shutter wheel A metal wheel consisting of a series of alternating windows and vanes. It creates a magnetic shunt that changes the strength of the magnetic field from the permanent magnet of the Hall-effect switch or magnetic pulse generator.

Rueda obturadora Rueda metálica compuesta de una serie de ventanas y aspas alternas. Genera una derivación magnética que cambia la potencia del campo magnético, del imán permanente del conmutador de efecto Hall o del generador de impulsos magnéticos.

Sine wave A waveform that shows voltage changing polarity.

Onda senoidal Una forma de onda que muestra un cambio de polaridad en la tensión.

Single phase voltage The sine wave voltage induced in one conductor of the stator during one revolution of the rotor.

Tensión monofásica La tensión en forma de onda senoidal inducida en un conductor del estator durante una revolución del rotor.

Sinusoidal A waveform that is a true sine wave.

Senoidal Una forma de onda que es una onda senoidal verdadera.

Slow charging Battery charging rate between 3 and 15 amps for a long period of time.

Carga lenta Indice de carga de la batería de entre 3 y 15 amperios por un largo espacio de tiempo.

Slow cranking A term used to mean that the starter drive engages the ring gear, but the engine turns too slowly to start.

Arranque lento Término utilizado para expresar que el mecanismo de transmisión de arranque engrana la corona, pero que el motor se enciende de forma demasiado lenta para arrancar.

Soft codes Codes are those that have occurred in the past, but were not present during the last BCM test of the circuit.

Códigos suaves Códigos que han ocurrido en el pasado, pero que no estaban presentes durante la última prueba BCM del circuito.

Soldering The process of using heat and solder (a mixture of lead and tin) to make a splice or connection.

Soldadura Proceso a través del cual se utiliza calor y soldadura (una mezcla de plomo y de estaño) para hacer un empalme o una conexión.

Solderless connectors Hollow metal tubes that are covered with insulating plastic. They can be butt connectors or terminal ends.

Conectadores sin soldadura Tubos huecos de metal cubiertos de plástico aislante. Pueden ser extremos de conectadores o de bornes.

Solenoid An electromagnetic device that uses movement of a plunger to exert a pulling or holding force.

Solenoide Dispositivo electromagnético que utiliza el movimiento de un pulsador para ejercer una fuerza de arrastre o de retención.

Solenoid circuit resistance test Diagnostic test used to determine the electrical condition of the solenoid and the control circuit of the starting system.

Prueba de la resistencia de un circuito solenoide Prueba diagnóstica utilizada para determinar la condición eléctrica del solenoide y del circuito de mando del sistema de arranque.

Spark tester A special tool used to provide a quick way to check ingnition spark.

Probador de bujía Herramienta especial que se utiliza para proporcionar una manera rápida de revisar la bujía de encendido.

Specific gravity The weight of a given volume of a liquid divided by the weight of an equal volume of water.

Gravedad específica El peso de un volumen dado de líquido dividido por el peso de un volumen igual de agua.

Speedometer An instrument panel gauge that indicates the speed of the vehicle.
Velocímetro Calibrador en el panel de instrumentos que marca la velocidad del vehículo.

Splice The joining of single wire ends or the joining of two or more electrical conductors at a single point.
Empalme La unión de los extremos de un alambre o la unión de dos o más conductores eléctricos en un solo punto.

Splice clip A special connector used along with solder to assure a good connection. The splice clip is different from solderless connectors in that it does not have insulation.
Grapa para empalme Conector especial utilizado junto con la soldadura para garantizar una conexión perfecta. La grapa para empalme se diferencia de los conectadores sin soldadura porque no está provista de aislamiento.

Square waves Identified by having straight vertical sides and a flat top.
Ondas cuadradas Para saber cuáles son, deben tener lados rectos verticales y una punta plana.

Stabilize Removing the surface charge of the battery by placing a large load on the battery for 15 seconds.
Estabilizar Para hacerlo hay que remover la superficie cargada de la batería, colocando en la batería, por 15 segundos, una carga grande,

Starter drive The part of the starter motor that engages the armature to the engine flywheel ring gear.
Transmisión de arranque Parte del motor de arranque que engrana la armadura a la corona del volante de la máquina.

State of charge The condition of a battery's electrolyte and plate materials at any given time.
Estado de carga Condición del electrolito y de los materiales de la placa de una batería en cualquier momento dado.

Stator The stationary coil of the AC generator in which current is produced.
Estátor Bobina fija del generador de corriente alterna donde se genera corriente.

Stator neutral junction The common junction of Wye stator windings.
Unión de estátor neutral La unión común de los devanados de un estátor Y.

Stepped resistor A resistor that has two or more fixed resistor values.
Resistor de secciones escalonadas Resistor que tiene dos o más valores de resistencia fija.

Stepper motor An electrical motor that contains a permanent magnet armature with two or four field coils. Can be used to move the controlled device to whatever location is desired. By applying voltage pulses to selected coils of the motor, the armature will turn a specific number of degrees. When the same voltage pulses are applied to the opposite coils, the armature will rotate the same number of degrees in the opposite direction.
Motor paso a paso Motor eléctrico que contiene una armadura magnética fija con dos o cuatro bobinas inductoras. Puede utilizarse para mover el dispositivo regulado a cualquier lugar deseado. Al aplicárseles impulsos de tensión a ciertas bobinas del motor, la armadura girará un número específico de grados. Cuando estos mismos impulsos de tensión se aplican a las bobinas opuestas, la armadura girará el mismo número de grados en la dirección opuesta.

Sulfation A chemical action within the battery that interferes with the ability of the cells to deliver current and accept a charge.
Sulfatado Acción química dentro de la batería que interfiere con la capacidad de los acumuladores de transmitir corriente y recibir una carga.

System main relay (SMR) Used to connect and disconnect the power source of the high-voltage system in an HEV.
Relé principal del sistema del (SMR o RPS) Se usa para conectar y desconectar la fuente de potencia del sistema de alto voltaje en un VHE.

Tachometer An instrument that measures the speed of the engine in revolutions per minute (rpm).
Tacómetro Instrumento que mide la velocidad del motor en revoluciones por minuto (rpm).

Thermistor A solid-state variable resistor made from a semiconductor material that changes resistance in relation to temperature changes.
Termistor Resistor variable de estado sólido hecho de un material semiconductor que cambia su resistencia en relación con los cambios de temperatura.

Test light Checks for electrical power in a circuit.
Luz de prueba Verifica la potencia eléctrica en un circuito.

Three-coil gauge A gauge design that uses the interaction of three electromagnets and the total field effect upon a permanent magnet to cause needle movement.
Calibrador de tres bobinas Calibrador diseñado para utilizar la interacción de tres electroimanes y el efecto inductor total sobre un imán permanente para producir el movimiento de la aguja.

Three-minute charge test A reasonably accurate method for diagnosing a sulfated battery for use on conventional batteries.
Prueba de carga de tres minutos Método bastante preciso en baterías convencionales para diagnosticar una batería sulfatada.

Timer control A potentiometer that is part of the headlight switch in some systems. It controls the amount of time the headlights stay on after the ignition switch is turned off.
Control temporizador Un potenciómetro que es parte del conmutador de los faros en algunos sistemas. Controla la cantidad del tiempo que quedan prendidos los faros después de apagarse la llave del encendido.

Tin Process of applying solder to the tip of the soldering iron to provide better heating control.
Estaño El proceso de aplicar soldadura en la punta de un hierro soldador para dar mejor control al calor.

Torque converter A hydraulic device found on automatic transmissions. It is responsible for controlling the power flow from the engine to the transmission; works like a clutch to engage and disengage the engine's power to the drive line.
Convertidor de par Un dispositivo hidráulico en las transmisiones automáticas. Se encarga de controlar el flujo de la potencia del motor a la transmisión; funciona como un embrague para embragar y desembragar la potencia del motor con la flecha motríz.

Trimotor A three-armature motor.
Trimotor Es un motor de tres armaduras.

Trouble codes Output of the self-diagnostics program in the form of a numbered code that indicates faulty circuits or components. Trouble codes are two or three digital characters that are displayed in the diagnostic display if the testing and failure requirements are both met.
Códigos indicadores de fallas Datos del programa autodiagnóstico en forma de código numerado que indica los circuitos o los componentes defectuosos. Dichos códigos se componen de dos o tres carácteres digitales que se muestran en el visualizador diagnóstico si se llenan los requisitos de prueba y de falla.

Troubleshooting The diagnostic procedure of locating and identifying the cause of the fault. It is a step-by-step process of elimination by use of cause-and-effect.
Detección de fallas Procedimiento diagnóstico a través del cual se localiza e identifica la falla. Es un proceso de eliminación que se lleva a cabo paso a paso por medio de causa y efecto.

TVRS An abbreviation for television-radio-suppression cable.
TVRS Una abreviación del cable de supresión del televisión y radio.

Two-button diagnostics The buttons on the instrument panel that, when pressed in the right combination, place the module into self-test mode.
Diagnósticos de dos botones Estos son los botones que aparecen en el panel de instrumentos, los cuales al presionarlos en la combinación correcta, ponen en el módulo en estado de autoprueba.

Two-coil gauge A gauge design that uses the interaction of two electromagnets and the total field effect upon an armature to cause needle movement.
Calibrador de dos bobinas Calibrador diseñado para utilizar la interacción de dos electroimanes y el efecto inductor total sobre una armadura para generar el movimiento de la aguja.

U-codes Diagnostic trouble codes assigned to the vehicle communication network.
Códigos en U Códigos de fallo de diagnóstico asignadas a la red de comunicación del vehículo.

Vacuum distribution valve A valve used in vacuum-controlled concealed headlight systems. It controls the direction of vacuum to various vacuum motors or to vent.
Válvula de distribución al vacío Válvula utilizada en el sistema de faros delanteros ocultos controlado al vacío. Regula la dirección del vacío a varios motores al vacío o sirve para dar salida del sistema.

Vacuum fluorescent display (VFD) A display type that uses anode segments coated with phosphor and bombarded with tungsten electrons to cause the segments to glow.

Visualización de fluorescencia al vacío Tipo de visualización que utiliza segmentos ánodos cubiertos de fósforo y bombardeados de electrones de tungsteno para producir la luminiscencia de los segmentos.

Valve body A unit that consists of many valves and hydraulic circuits. This unit is the central control point for gear shifting in an automatic transmission.

Cuerpo de la válvula Una unedad que consiste de muchas válvulas y circuitos hidráulicos. Esta unedad es el punto central de mando para los cambios de velocidad en una transmisión automática.

Variable resistor A resistor that provides for an infinite number of resistance values within a range.

Resistor variable Resistor que provee un número infinito de valores de resistencia dentro de un margen.

Vehicle Identification Number (VIN) A number that is assigned to a vehicle for identification purposes. The identification plate is usually located on the cowl, next to the left upper instrument panel.

Número de identificación del vehículo Número asignado a cada vehículo para fines de identificación. Por lo general, la placa de identificación se ubica en la bóveda, al lado del panel de instrumentos superior de la izquierda.

Vehicle lift points The areas that the manufacturer recommends for safe vehicle lifting. They are the areas that are structurally strong enough to sustain the stress of lifting.

Puntos para elevar el vehículo Áreas específicas que el fabricante recomienda para sujetar el vehículo a fin de lograr una elevación segura. Son las áreas del vehículo con una estructura suficientemente fuerte para sostener la presión de la elevación.

Vehicle module scan A special function of the scan tools that will query all of the modules on the bus to respond and then list those that did reply.

Explorador del módulo del vehículo Función especial de los instrumentos de exploración que cuestionarán a todos los módulos en el bus para que respondan, y luego harán una lista de aquellos que respondieron.

Volatile A substance that vaporizes or explodes easily.

Volátil Sustancia que se evapora o explota con facilidad.

Volatility The tendency for a fluid to evaporate quickly or pass off in the form of a vapor.

Volatilidad La tendencia de un fluido a evaporarse rápidamente o disiparse en forma de vapor.

Volt The unit used to measure the amount of electrical force.

Voltio Unidad práctica de tensión para medir la cantidad de fuerza eléctrica.

Voltage The difference or potential that indicates an excess of electrons at the end of the circuit the farthest from the electromotive force. It is the electrical pressure that causes electrons to move through a circuit. One volt is the amount of pressure required to move one amp of current through one ohm of resistance.

Tensión Diferencia o potencial que indica un exceso de electrones al punto del circuito que se encuentra más alejado de la fuerza electromotriz. La presión eléctrica genera el movimiento de electrones a través de un circuito. Un voltio equivale a la cantidad de presión requerida para mover un amperio de corriente a través de un ohmio de resistencia.

Voltage drop A resistance in the circuit that reduces the electrical pressure available after the resistance. The resistance can be either the load component, the conductors, any connections, or unwanted resistance.

Caída de tensión Resistencia en el circuito que disminuye la presión eléctrica disponible después de la resistencia. La resistencia puede ser el componente de carga, los conductores, cualquier conexión o resistencia no deseada.

Voltage drop test Determines if the battery, regulator, and AC generator are all operating at the same potential.

Prueba de la caída de tensión en el voltaje Determina si la batería, el regulador, y el generador de corriente alterna AC están todos funcionando al mismo potencial.

Voltage limiter Connected through the resistor network of a voltage regulator. It determines whether the field will receive high, low, or no voltage. It controls the field voltage for the required amount of charging.

Limitador de tensión Conectado por el red de resistores de un regulador de tensión. Determina si el campo recibirá alta, baja o ninguna tensión. Controla la tensión de campo durante el tiempo indicado de carga.

Voltage output test Used to make a quick determination about whether or not the charging system is working properly. If the charging system is operating correctly, then check for battery drain.

Prueba de salida del voltaje Se usa para determinar en forma rápida si el sistema de carga está trabajando correctamente. Si el sistema de carga está funcionando correctamente, entonces revíselo todo para saber si hay algún drenaje en la batería.

Voltage regulator Used to control the output voltage of the ac generator, based on charging system demands, by controlling field current.

Regulador de tensión Dispositivo cuya función es mantener la tensión de salida del generador de corriente alterna, de acuerdo a las variaciones en la corriente de carga, controlando la corriente inductora.

Voltmeter A test meter used to read the pressure behind the flow of electrons.

Voltímetro Instrumento de prueba utilizado para medir la presión del flujo de electrones.

Warning light A lamp that is illuminated to warn the driver of a possible problem or hazardous condition.

Luz de aviso Lámpara que se enciende para avisarle al conductor sobre posibles problemas o condiciones peligrosas.

Watt The unit of measure of electrical power, which is the equivalent of horsepower. One horsepower is equal to 746 watts.

Watio Unidad de potencia eléctrica, equivalente a un caballo de vapor. 746 watios equivalen a un caballo de vapor (CV).

Wattage A measure of the total electrical work being performed per unit of time.

Vataje Medida del trabajo eléctrico total realizado por unidad de tiempo.

Waveform The electronic trace that appears on a scope; it represents voltage over time.

Forma de onda La trayectoria electrónica que aparece en un osciloscopio; representa la tensión a través del tiempo.

Weather-pack connector An electrical connector that has rubber seals on the terminal ends and on the covers of the connector half to protect the circuit from corrosion.

Conectador resistente a la intemperie Conectador que tiene sellos de caucho en los extremos de los bornes y en las cubiertas de la parte del conectador para proteger el circuito contra la corrosión.

Wet fouling A condition of a spark plug in which it is wet with oil.

Engrase húmedo Una condición de la bujía en la cual se moja de aceite.

Wiring diagram An electrical schematic that shows a representation of actual electrical or electronic components and the wiring of the vehicle's electrical systems.

Esquema de conexciones Esquema en el que se muestran las conexiones internas de los componentes eléctricos o electronicos reales y las de los sistemas eléctricos del vehículo.

Wiring harness A group of wires enclosed in a conduit and routed to specific areas of the vehicle.

Cableado preformado Conjunto de alambres envueltos en un conducto y dirigidos hacia áreas específicas del vehículo.

Workplace hazardous materials information systems (WHMIS) Canadian equivalent to MSDS sheets.

Sistemas de información acerca de los materiales peligrosos en el área de trabajo (SIMPAT o WHMIS) Equivalente canadiense a las hojas de datos de la seguridad de un material.

Worm gear A type of gear whose teeth wrap around the shaft. The action of the gear is much like that of a threaded bolt or screw.

Engranaje de tornillo sin fin Un tipo de engranaje cuyos dientes se envuelven alrededor del vástago. El movimiento del engranaje es muy parecido a un perno enroscado o una tuerca.

Wye connection A type of stator winding in which one end of the individual windings are connected at a common point. The structure resembles the letter "Y."

Conexión Y Un tipo de devanado estátor en el cual una extremidad de los devanados individuales se conectan en un punto común. La estructura parece la letra "Y."

INDEX

A

ABS. *See* antilock brake systems
accumulator, 575
AC generators
    See also charging systems
    component testing, 271–274
        diodes (rectifier bridges), 273–274
        diode trio, 274
        rotors, 272
        stators, 272–273
    current output testing, 256
    diagnostic chart for, 286
    diode pattern testing, 263–266
    diode/stator test, 263
    disassembly, 268–271
    full field tests, 256, 258–262
    inspection of components, 271
    noises, 251–252
    reassembly, 274–275
    removal and replacement, 266–268
actuator tests, 387–389, 555
AC voltage, 43, 52–53, 113
air bag module, 23
air bag system diagnostic module (ASDM), 619
air bag systems
    cleanup procedure after deployment, 623
    component replacement, 623–626
        clockspring replacement, 625–626
        module replacement, 623–625
    defined, 23
    diagnostic charts for, 643
    inspecting, after accident, 623
    retrieving fault codes, 618–621
    safety and services warnings, 616–617
    safety precautions with, 23–25
    system check of, 617
    system testing, 621–622
air conditioning. *See* automatic temperature control
  systems
air tools. *See* pneumatic tools
alarms. *See* antitheft systems
alcohol use, 3
alternating current (AC) voltage, 43, 52–53, 113
aluminum wire repairs, 140
ambient temperature sensor, 530
ammeters, 33, 34, 45–46
ampere, 33
amplifier test, 460
analog gauges, 506
analog meters, 38–39
analog ohmmeters, 99
analog signals, 114, 115
antennas, 583–584
antilock brake systems (ABS), 575–583
    component replacement, 579–583
        pump and motor assembly, 580–582
        valve block assembly, 582–583
    component testing, 577–579
    diagnostic charts for, 606
    safety, 25

    self-diagnosis, 577
    troubleshooting, 575–577
antitheft systems, 562–565
    alarm sounds for no reason, 563
    circuit schematic of, 564
    controller test, 563–565
    diagnostic charts for, 604
    relay tests, 565–567
    self-diagnostic tests, 562–563
appearance, personal safety and, 2
apprentices, 66
armature ground test, 233
armature short test, 232–233
ASDM. *See* air bag system diagnostic module
ASE certification, 68–69
aspirator assembly, 528–529
ATC systems. *See* automatic temperature control (ATC) systems
audio entertainment systems, 583–585, 606–608
automatic door lock systems, 567–569
automatic headlight systems, 458–471
    amplifier test, 460
    photocell replacement, 461–462
    photocell test, 460
    resistance assembly test, 462–463
    SmartBeam diagnostics, 463–470
automatic ride control (ARC) module, 575
automatic seat belt systems, 611–616
    circuit schematic of, 612
    diagnostic charts for, 643
    drive belt replacement, 614–616
    drive motor assembly replacement, 613–614
automatic suspension systems, 609
automatic temperature control (ATC) systems, 528–544
    BCM ATC-controlled systems, 541–544
    component testing, 528–531
        ambient temperature sensor, 530
        aspirator assembly, 528–529
        control assembly resistor, 531
        in-vehicle sensor, 530
        servomotor, 530–531
    DaimlerChrysler ATC systems, 531–536
    diagnostic chart for, 601
    Ford ATC systems, 536
    GM ETCC systems, 536, 537–538
    microprocessor-controlled systems, 531–541
    Nissan ATC systems, 536, 539–541
automatic zone control (AZC) systems, 535
automotive bulbs, 287
Automotive Service Excellence (ASE) certification, 68–69
average responding meters, 43

B

back injuries, 6
backprobe, 41–42
back-up lights, 332–333
batteries, 171–195
    See also charging systems
    battery drain test, 188–190
    battery inspection, 173–175
    charging the battery, 175–178, 651–652

computerized load-testing equipment, 187
diagnostic charts for, 207
general precautions, 172–173
high-voltage, 650–652
introduction to, 171–172
jumping the battery, 194–195
load testing, 250
maintaining and restoring memory functions, 194
removal and cleaning, 190–193
safety precautions with, 21–22, 23
selection of, 187
test series, 178–187
    batter terminal test, 178–179
    battery leakage test, 179–180
    capacity test, 184–186
    conductance testing, 187
    open circuit voltage test, 182–184
    state of charge test, 180–182
    three-minute charge test, 186–187
BAT terminal, 22
battery acid, 172
battery backups, 60
battery ECU, 650
battery leakage test, 179–180
battery load testers, 187
battery stickers, 173
battery terminal test, 178–179
battery test equipment, 171–172
BCM. *See* body control module
BCM ATC-controlled systems, 541–544
B-codes, 410
bearings, faulty, 252
belt tension sensors (BTSs), 628, 630
bimetallic gauges, 506
bipolar transistors, testing, 101–103
blade fuses, 88, 89
blower motor service, 349–352
    constantly operating blowing fans, 352
    inoperable motor with ground side switches, 351
    inoperable motor with insulated switches, 350–351
body computer system. *See* body control module
body control module (BCM), 375–398
    computer system functions, 375
    computer voltage supply and ground circuits, 378–382
    concealed headlight system and, 451–456
    entering diagnostics, 384–387
    flashing the, 396–397
    of instrument panel, 493
    introduction to, 375–377
    PROM replacement, 395–396
    service precautions, 377–378
    testing actuators, 387–389
    testing sensors, 389–394
    trouble codes, 383, 386
    visual inspection, 383
boots, 2
brake lights, 308, 310–314, 331
brake systems. *See* antilock brake systems (ABS)
breakout boxes, 395
brighter-than-normal lights, 298–299
brush assembly checks, 234–235
BTSs (belt tension sensors), 628, 630
bus bar lead repair, 355

bus systems
    CAN bus diagnostics, 418–422
    class A bus system diagnostics, 413–416
    communication fault codes, 410
    enhanced scan tool diagnostics, 423–424
    failures, 409
    ISO 9141–2 bus system diagnostics, 410–412
    ISO-K bus system diagnostics, 412–413
    J1850 bus system diagnostics, 416–418
    LIN bus diagnostics, 422–423

C

cable connections, for jump starts, 194–195
Cadillac BCM trouble codes, 386
camera calibration procedure, 465–470
capacitators, 381–382
capacity test, 184–186
captured signals, 51
carbon monoxide, 13–14
cartridge fuses, 88
CD systems, 585
ceramic fuses, 88
certification, ASE, 68–69
charge rate, 176–177
charging systems, 249–276
    AC generators
        component testing, 271–274
        disassembly, 268–271
        noises, 251–252
        reassembly, 274–275
        removal and replacement, 266–268
    current output testing, 256, 257
    diagnostic charts, 285–286
    diode pattern testing, 263–266
    diode/stator test, 263
    field current draw test, 254–256
    full field tests, 256, 258–262
    introduction to, 249–251
    precautions, 251
    regulator test, 262–263
    requirement test, 266
    safety, 22–23
    troubleshooting, 252
    voltage drop testing, 253–254
    voltage output testing, 252–253
charging the battery, 175–178, 651–652
chokes, 381–382
Chrysler Collision Detection (CCD) multiplexing system, 413–416
cigarette smoking, 3, 15
circuit breakers, 89–92
circuit diagrams, reading, 147–150
circuit grounding, 43
circuit protection devices, 88–92
circuits, 87
    ground, 220–222, 378–382
    open, 87, 103–106, 131
    prove-out, 513
    sensor return, 379–382
circuit test chart, 44
circuit testing
    with ammeter, 45–46
    for defects, 103–111
    ground circuit tests, 220–222
    insulated circuit resistance test, 220
    for opens, 103–106, 222–224

for shorts, 106–108
for short to ground, 107–108
solenoid circuit resistance testing, 222
CKP. *See* crankshaft position sensor
class A bus system diagnostics, 413–416
climate-controlled seats, 560–561, 602
clockspring replacement, 625–626
clothing, personal safety and, 2
Code of Federal Regulations, 21
color codes, 147
combustible materials, storage of, 15–16
communication, with customers, 68
commutator tests, 234
compensation, 66–67
composite headlights
    aiming, 296–298
    replacement, 290–291
compressed air, safety issues, 8
computer-controlled lighting systems, 451–477
    automatic headlight system diagnosis, 458–471
    concealed headlight diagnosis, 451–456
    daytime running lamps, 470–471
    fiber optic diagnosis, 477
    headlight systems, 456–458
    illuminated entry lights, 472–476
    introduction to, 451
computer-driven instrumentation systems. *See* instrument panels
computerized load-testing equipment, 187
computer services, for service information, 66
computer systems. *See* body control module
concealed headlights, 300, 451–456
conductance testing, 187
conductors of electricity, 2, 35
connector terminals
    repairing, 142–147
        hard-shell connectors, 142–143
        metri-pack connectors, 146
        molded connectors, 142, 143
        weather-pack connectors, 144–145
continuity tester, 37
control assembly resistor, 531
Controller Area Network (CAN) bus systems, 418–422
controller test, 563–565
converters, 653
copper wire repairs, 134–140
courtesy lights, 314–317, 330
crankshaft position sensor (CKP), 505
crimping, 135–136
crocus cloth, 234
cruise control systems, 544–556
    component replacement, 556–558
    component testing, 555–556
        dump valves, 555–556
        servo assembly, 555
        speed sensors, 556
    continuously changing speeds diagnosis, 553
    DaimlerChrysler, 550–551
    diagnosing without trouble codes, 552
    diagnostic charts for, 601–602
    Ford, 549–550
    General Motors, 544–549
    intermittent operation diagnosis, 553–555
    no operation diagnosis, 553
    self-diagnostics, 544–551
    speed control electrical schematic, 554

curb height, 294
current, 87
current draw tests, 217–218
current output testing, 256, 257
customer relations, 67–68
cycles, 46, 53, 115

D

DaimlerChrysler
    AC generators, 261
    ATC systems, 531–536
    cruise control systems, 550–551
    electromechanical clusters (MICs), 498
Darlington transistors, 103
D'Arsonval movement, 38
dash-mounted switches, 301–302, 346
data buses. *See* bus systems
data link connectors (DLC), 409
daytime running lamps, 470–471
DC/DC converter, 653
DC voltage, 53
defogger diagnosis and repair, 352–355
Delco Remy starter, 228–230
diagnostic charts
    AC generators, 286
    batteries, 207
    burned fusible links, 169
    burned or defective connectors, insulation, or conductors, 169
    charging systems, 285–286
    diodes, 286
    electrical accessories, 367–373
    electrical opens, 131
    electronic accessories, 601–609
    headlights, 327–328, 489
    high-resistance problems, 131
    hybrid electric vehicle (HEV) systems, 663
    instrument panels, 523–526
    lighting systems, 327–333, 489
    multiplexing systems, 449
    passive restraint systems, 643
    shorts, 131
    starting systems, 247–248
    voltage regulation, 286
diagnostics
    See also troubleshooting
    approach to, 69
    cause-and-effect approach to, 142
    importance of, 69
diagnostic tools. *See* test equipment
diagnostic trees, 63–64
diagnostic trouble codes (DTCs), 57, 375
diesel fuel, 17
digital instrument panel cluster display, 619
digital multimeters (DMMs), 39
digital signals, 114–116
digital storage oscilloscopes (DSOs), 51, 111–116
    screen, 112–113
    selecting voltage and time base, 113
    trigger and trigger slope, 113–114
    types of voltage signals, 114–116
    user-friendly, 116
digital volt/ohmmeters (DVOMs), 39
    additional functions of, 46–47
    reading, 47–48
dim lights, 298–299

dimmer switches, testing and replacement of, 304–308
diode pattern testing, 263–266
diodes
   diagnostic chart for, 286
   testing, 99–101, 273–274
diode/stator test, 263
DLC. *See* data link connectors
door locks, 358
   automatic, 567–569
   remote keyless entry systems, 569
drain test, 188–190
dress, personal safety and, 2
drive belts, 250–251
   loose, 251
   replacement, 614–616
drug use, 3
DSOs. *See* digital storage oscilloscopes
DTCs. *See* diagnostic trouble codes
dump valves, 555–556
duty cycle, 46, 47, 116
DVOMs. *See* digital volt/ohmmeters

E
eddy currents, 500
EEPROM chips, 396
EIC. *See* electronic instrument cluster
electrical accessories, 335–359
   antilock brake systems (ABS), 575–583
   antitheft systems, 562–565
   audio entertainment systems, 583–585
   automatic temperature control systems, 528–544
   blower motor service, 349–352
   climate-controlled seats, 560–561
   cruise control systems, 544–556
   diagnostic charts for, 367–373, 601–609
   electric defoggers, 352–355
   heated windshields, 570–573
   horn diagnosis, 335–339
   immobilizer systems, 565, 567
   introduction to, 335, 527
   memory seats, 556, 558–559
   power door locks, 358, 567–569
   power seats, 357
   power windows, 355–357
   remote keyless entry systems, 569
   speed-sensitive steering systems, 574–575
   sunroofs, 561–562
   video entertainment systems, 583–585
   windshield washer system service, 348–349
   wiper system service, 339–348
electrical components
   testing and replacing, 92–99
      relays, 94–96
      stepped resistors, 96–97
      switches, 92–94
      variable resistors, 97–98
electrical diagram/schematics, 133
   reading, 147–150
electrical disturbances, 50–51
electrical noise, 46, 50–51
electrical opens. *See* open circuits
electrical requirement, testing for, 266
electrical system safety, 21–25
   air bags, 23–25
   antilock brake systems (ABS), 25

batteries, 21–22, 23
charging system, 22–23
starting system service, 22
electrical systems technician, working as a, 66–69
electrical test equipment. *See* test equipment
electrical troubleshooting and service
   basic, 34–35
   introduction to, 87
   process of, 133
   test equipment for, 35–59
   testing and replacing electrical components, 92–99
   testing circuit protection devices, 88–92
   testing diodes, 99–101
   testing for circuit defects, 103–111
   testing for voltage drop, 109–111
   testing transistors, 101–103
   using DSO, 111–116
electric cooling fans, safety and, 6, 14
electric current, 33
electric defogger diagnosis and service, 352–355
electricity, conductors of, 2, 35
electromagnetic gauges, 506
electromagnetic interference (EMI), 378, 380–381, 423
electromechanical clusters (MICs), self-test, 498
electromotive force (EMF), 33
electronic clusters, self-test, 499
electronic cruise control systems. *See* cruise control systems
electronic gauges, 509–510
electronic instrument cluster (EIC)
   diagnosis of, 493–497
   function diagnostic mode, 494–495
   imports, 495–497
   prove-out display, 493–494
   special test mode, 495
electronic service precautions, 377–378
electronic sunroofs, 561–562
electronic touch climate control (ETCC) systems, 536
electronic variable orifice (EVO) steering, 574–575
electrostatic discharge (ESD), 377–378
electrostatic discharge (ESD) straps, 377
employer-employee relations, 67
entertainment systems, 583–585
entry lights, 472–476
Environmental Protection Agency (EPA), 20
equipment safety, 7–8
eye injuries, 4
eye protection, 3–4
eyewash fountains, 4

F
face shields, 4
fail soft action, 376
fan motor service, 349–352
fast charging, 176
fault codes, 383, 410, 618–621
   See also trouble codes
feedback, 106
fiber optic diagnosis, 477
field coil testing, 232
field current draw test, 254–256
fire extinguishers
   types, 14
   use of, 17–18
fire hazards, 14–18
first-aid kits, 5
flammable fumes, 15

flash a computer, 396–397
flash code retrieval, 618–621
flash codes, 383
flashers, 313–314
flat ray system, 66–67
floor jacks, safe use of, 8–12
floor-mounted switches, 307
floor stands, 9–10
footwear, 2
Ford
    AC generators, 255–256, 259–261
    air bag flash codes, 619–620
    ATC systems, 536
    electronic clusters, 499
    electronic variable orifice (EVO) steering, 574–575
    integrated vehicle speed control (IVSC) systems, 549–550
free speed test, 224–225
free spin test, 224–225
freeze frame data, 648
frequency, 47, 53, 115–116
frequency signal, 47
fuel gauges, 509–510
fuel level senders, 510–512
full fielding, 256
full field test, 256, 258–262
functional test, of ABS systems, 577, 578
fuses, 88–92, 142
    blown, 314, 452
    determining current flows through, 189–190
fusible links, 88–91, 141–142

G

gasoline, as fire hazard, 15–17
gasoline containers, 16–17
gauges
    See also instrument panels
    analog, 506
    bimetallic, 506
    defined, 506
    diagnosis of conventional, 506–509
    electromagnetic, 506
    electronic, 509–510
    failure of multiple, 508–509
    failure of single, 506–508
    fuel, 509–510
    Gauss, 108–109
    sending units, 510–512
Gauss gauge, 108–109
gear reduction starters, 230–231
gel cell batteries, recharging, 177–178
General Motors (GM)
    AC generators, 256, 258–259
    actuators, 387
    BCMs, 386
    cruise control systems, 544–549
    ETCC systems, 536, 537–538
    flash code retrieval, 618–619
    heated windshield system, 571
generator drive belts, 250–251
generators, 22–23
    See also AC generators; charging systems
glitches, 50–51
gloves, 5
grids, 352

grid wire repair, 353–354
ground circuits, 220–222, 378–382
ground wires, 43

H

hair, long, 2
Hall-effect sensors, 392–393, 503
hand protection, 5
hand tools, safety issues, 7
hard codes, 383
hard-shell connectors, 142–143
Hazard Communication Standard, 18
hazard lights, 332
hazardous materials, 18–21
hazardous waste, 20–21
headlight aiming pads, 295
headlights, 288–304
    aiming, 293–298
    automatic headlight system diagnosis, 458–471
    concealed, 300, 451–456
    daytime running lamps, 470–471
    diagnostic charts for, 327–328, 489
    dimmer- or brighter-than normal, 298–299
    dimmer switch testing and replacement, 304–308
    HID lamp and ballast service, 292–293
    replacement, 288–293
    switch testing and replacement, 301–304
    testing computer-controlled, 456–458
    types of, 288
heated seat systems, 560–561, 602
heated windshields, 570–573, 606
heat shrink tubes, 136
heavy objects, lifting, 6
hertz, 47
HID lamp and ballast service, 292–293
high beam camera calibration, 465–470
high-resistance problems, 87, 109–111, 131
high-voltage batteries, 650–652
high-voltage electronic control unit (HV ECU), 648–649
high-voltage service plugs, 647–648
high-voltage systems. *See* hybrid electric vehicle (HEV) systems
history data information, 648
hoist pads, 12–13
hoists, safe use of, 12–13
Honda air bag systems, 621–622
horn diagnosis, 335–339
    continuous sound, 338–339
    diagnostic charts, 367
    no horn operation, 335–338
    poor sound quality, 338
hotline services, 66
hybrid electric vehicle (HEV) systems
    diagnostic charts for, 663
    high-voltage battery service, 650–652
    high-voltage service plugs, 647–648
    introduction to, 645
    inverter/converter assembly, 652–653
    safety precautions with, 645–647
    self-diagnosis capabilities, 648–650
    system main relay (SMR), 654–655
hydrometer
    battery, 175
    testing batteries with, 180–182

hydrostatic lock, 210, 214
Hypalon, 141

I

IAR. *See* integral alternator/regulator (IAR) system
illuminated entry actuator, 472
illuminated entry lights, 316, 472–476
immobilizer systems, 565, 567, 605
impedance multimeters, 377
information codes, 648–649
injury prevention. *See* safety
instrument cluster lights, 317, 329
instrument panels
    BCM diagnostics, 493
    conventional gauge diagnosis, 506–509
    diagnostic charts, 523–526
    electronic gauge diagnosis, 509–510
    gauge sending units, 510–512
    introduction to, 491–492
    removal, 492
    self-diagnostics, 493–499
    speedometers, 499–504
    tachometers, 504–506
    trip computers, 514
    warning lamps, 512–514
instrument voltage regulator (IVR), 506, 508–509
insulated circuit resistance test, 220
insulating gloves, 647
integral alternator/regulator (IAR) system, 255–256
    full field testing, 259–261
integrated vehicle speed control (IVSC) systems, 549–550
interior lights, 314–317
    courtesy lights, 314–317
    instrument cluster and panel lights, 317
intermittent codes, 383
intermittent wiper system diagnosis, 345–346
in-vehicle sensor, 530
inverter/converter assembly, 652–653
ISO 9141-2 bus system diagnostics, 410–412
ISO-K bus system diagnostics, 412–413
IVR. *See* instrument voltage regulator

J

J1850 bus system diagnostics, 416–418
J1962 breakout box (BOB), 409
jack stands, 9–10
jewelry, 2
jumper wires, 35–36, 93, 98
jump starts, 194–195

K

keyless entry systems, 569, 605

L

lab scopes, 49–55, 216–217
    testing ABS systems with, 579
    testing actuators with, 388–389
lamp sequence check, 577
LEDs, testing, 100–101
lifting
    heavy objects, 6
    vehicles, 8–13

lift points, 11–12
lighting circuits, 287–318
    brake lights, 310–314
    dimmer switch testing and replacement, 304–308
    flashers, 313–314
    headlights, 288–304
        aiming, 293–298
        concealed, 300
        dimmer- or brighter-than normal, 298–299
        replacement, 288–293
        switch testing and replacement, 301–304
    interior lights, 314–317
    introduction to, 287–288
    taillight assemblies, 308–314
    turn signals, 310–314
lighting systems
    computer-controlled, 451–477
        automatic headlight system diagnosis, 458–471
        concealed headlight diagnosis, 451–456
        daytime running lamps, 470–471
        headlight systems, 456–458
        illuminated entry lights, 472–476
        introduction to, 451
    diagnostic charts for, 327–333, 489
live air bag modules, 617
load testers/testing, 187, 253
local interconnect network (LIN) bus systems, 422–423
locking tangs, 144, 145
logic probes, 37
long hair, 2

M

magnetic pickup speed sensors, 502–503
magnetic pulse generators, 390–391
material safety data sheets (MSDS), 19–20
maxi-fuses, 89
memory functions, maintaining and restoring, 194
memory keepers, 60
memory seats, 556, 558–559, 602
metri-pack connectors, repairing, 146
mileage readings, 501
molded connectors, 142, 143
MSDS. *See* material safety data sheets
multifunction switches, 304–306
multimeters, 38–39
    additional functions of, 46–47
    ammeters, 45–46
    analog meters, 38–39
    digital, 39
    ohmmeters, 43–44
    reading, 47–48
    trigger level of, 47
    voltmeters, 39–43
multiplexing systems, 409–425
    CAN bus diagnostics, 418–422
    class A bus system diagnostics, 413–416
    communication fault codes, 410
    diagnostic charts for, 449
    enhanced scan tool diagnostics, 423–424
    introduction to, 409
    ISO 9141-2 bus system diagnostics, 410–412
    ISO-K bus system diagnostics, 412–413

J1850 bus system diagnostics, 416–418
    LIN bus diagnostics, 422–423
multiplying coil, 254

N

Nissan ATC systems, 536, 539–541
no-crank problems, 210, 211–212, 214, 222
noises, 46, 50–51
    from AC generators, 251–252
    circuit, 380–381
no-load test, 224–225
normally open (NO) switches, 92

O

occupant classification systems (OCSs), 626–630
Occupational Safety and Health Act, 20
Occupational Safety and Health Administration (OSHA), 18
occupational safety glasses, 3
OCS service kits, 626–627
OCS validation test, 627–628
odometer mileage labels, 496
odometers, 501–504
ohm, 33
ohmmeters, 33, 43–44, 90
    testing brush holders with, 234–235
    testing bus system failures, 414–416
    testing diodes with, 99
    testing field coils with, 232
    testing relays with, 96
    testing stepped resistors with, 96–97
    testing switches with, 93–94
    testing variable resistors with, 97–98
open circuits, 87
    diagnostic chart for, 131
    testing for, 103–106
open circuit test, 222–224
open circuit voltage test, 182–184
optical refractometer, 182
optical speed sensors, 503–504
oscilloscopes, 49–55
    controls, 54–55
    digital storage oscilloscope (DSOs), 111–116
    diode pattern testing with, 263–266
    waveforms, 51–53
overloads, 88
overrunning clutch inspection, 235

P

panel lights, 317
parking lamps, 328, 471
park switch, 342–344
passive restraint systems, 611–631
    air bag systems, 616–626
    automatic seat belt service, 611–616
    diagnostic charts for, 643
    introduction to, 611
    occupant classification systems (OCSs), 626–630
pay systems, 66–67
peak, 113
peak-to-peak voltage, 113
personal safety, 1–6
    alcohol and drugs and, 3
    dress and appearance and, 2
    with electric cooling fans, 6
    eye protection, 3–4

eyewash fountains, 4
first-aid kits, 5
hand protection, 5
lifting and, 6
with rotating equipment, 5
smoking and, 3
when working with batteries, 172–173, 175
photocell resistance assembly, 462–463
photocells, 458
    replacement, 461–462
    testing, 460
piezoresistive sensor, 512
pinpoint tests, 549
pneumatic tools, safety issues with, 7
potentiometers, 98, 389–390
power door lock diagnosis, 358, 372–373
power seat diagnosis, 357, 371
power tools
    inspection of, 8
    safety issues, 7
power window diagnosis, 355–357, 370
pressure monitoring system, 609
printed circuit board, removal, 492
PROM replacement, 395–396
protection devices. *See* circuit protection devices
protective gear
    eye protection, 3–4
    hand protection, 5, 647
prove-out circuits, 513
prove-out displays, 493–494
pull to seat connectors, 146
pulse width, 47, 115, 116
pulse width modulation (PWM), 116
push to sear connectors, 146

Q

quartz swing needle, 501–502
quick testing, 215–226

R

radio frequency interference (RFI), 378, 423
radio systems, 583–585
rags, storage of dirty, 15, 17
rear window defogger, 352–355, 371–372
recharging. *See* charging the battery
rectifier bridge, testing, 273–274
refractometer, 182
regulator test, 262–263
relative compression testing, 218–219
relays
    system main relay (SMR), 654–655
    testing, 94–96, 565, 566–567
remote keyless entry systems, 569
resistance, 87
    See also voltage drop
    infinite, 44
    low, 44
resistance assembly test, 462–463
resistance problems, 87
    sensor return circuits and, 379–382
    testing for, 109–111
resistive shorts, 510
resistor block, 349
Resource Conservation and Recovery Act (RCRA), 21
RF sniffer tool, 585

rheostats, 97–98
right-to-know laws, 18–20
root mean square (RMS), 113
root mean square (RMS) meters, 43
rotating equipment, safety and, 5
rotors, testing, 272

S

safety, 1–26
 electrical systems, 21–25
 fire hazards and prevention, 14–18
 with hazardous materials, 18–21
 with hybrid systems, 645–647
 introduction to, 1
 lifting vehicles, 8–13
 personal, 1–6
 running vehicles in shop, 13–14
 tool and equipment, 7–8
safety goggles, 4
safety stands, safe use of, 8–12
satellite radio systems, 585
scan tester diagnostics, 384–387, 544
scan tools, 56–59
 communication between bus system and, 410–413
 for DTC retrieval, 618
 enhanced diagnostics, 423–424
 flashing the BCM with, 396–397
 sensor testing with, 395
 vehicle module scans, 411
scopes. *See* oscilloscopes
sealed-beam headlights
 aiming, 294–296
 replacement, 288–290
seat belt systems. *See* automatic seat belt systems
self-diagnostic tests
 of ABS systems, 577
 for antitheft systems, 562–563
 for BCM ATC-controlled systems, 543–544
 for cruise control systems, 544–551
 for Daimler-Chrysler ATC systems, 531–536
 of hybrid systems, 648–650
 for Nissan ATC systems, 536, 539–541
self-powered continuity tester, 37
self-test input (STI) terminal, 550
sending units, gauge, 510–512
sensor return circuits, 379–382
sensors, testing, 389–394, 395
service history tracking, 66
service information/manuals, 61–66
servo assembly, 555–558
servomotor, 530–531
shoes, 2
short circuits, 87
 testing for, 106–108
short to ground, 107–108
 checking for, 234
Silicone/GXL, 141
simulated road tests, for cruise control systems, 552
sine waves, 43, 52–53
sinusoidal waves, 52
slow charging, 176
slow cranking, 210, 213, 214
SmartBeam diagnostics, 463–470
 camera calibration test and adjustment, 465–470
 demonstration mode, 464–465

smoking, 3, 15
SMR. *See* system main relay
soldering, 136–140
solderless connectors, crimping, 135–136
solenoid circuit resistance testing, 222
solvents, 17
spade fuses, 88
speaker problems, 584
special tools
 memory keepers, 60
 service information/manuals, 61–66
 static straps, 59–60
 test equipment, 35–59
specifications, 63
specific gravity, measuring, 180–182
speedometer readings, 496–497
speedometers, 498
 conventional, 499–501
 diagnosis and repair, 499–504
 electronic, 501–504
speed-sensitive steering systems, 574–575
speed sensors, 390–391, 496–498
 ABS wheel, 579, 580
 magnetic pickup, 502–503
 optical, 503–504
 testing, 556
splice clips, 136
splices, 135, 139
square waves, 53
stabilize the battery, 182
starter/charging system tester, 214–215
starter motors
 component tests, 232–235
  armature ground test, 233
  armature short test, 232–233
  brush inspection, 234–235
  commutator tests, 234
  field coil tests, 232
  overrunning clutch inspection, 234–235
  disassembly, 227–231
 reassembly, 235–237
 removal, 225–227
starter safety switches, 224
starter systems, 209–238
 components, 209–210
 diagnostic charts for, 211–213, 247–248
 introduction to, 209–210
 precautions, 210
 safety, 22
 starter motor component tests, 232–235
 starter motor disassembly, 227–231
 starter motor removal, 225–227
 starter reassembly, 235–237
 testing, 214–225
  current draw tests, 217–218
  free speed test, 224–225
  ground circuit test, 220–222
  insulated circuit resistance test, 220
  lab scope tests, 216–217
  open circuit test, 222–224
  quick tests, 215–226
  relative compression tests, 218–219
  solenoid circuit resistance test, 222
 troubleshooting, 210–214
starting/charging system testers, 249, 262

state of charge (SOC), 648
state of charge test, 180–182
static charges, 59
static electricity, 377–378
static straps, 59–60
stators, testing, 272–273
steering column module (SCM), 422
steering column-mounted switches, 303–304, 307–308, 348
stepped resistors, testing, 96–97
stop lights. *See* brake lights
sulfated battery, 186
sulfation, 186
sulfuric acid, 190
Sun Electrical Corporation VAT-40, 171
sunroofs, 561–562, 603
switches
   cruise control system, 556
   dash-mounted, 301–302, 346
   dimmer, testing and replacement, 304–308
   floor-mounted, 307
   Hall-effect, 392–393, 503
   headlight, testing and replacement, 301–304
   multifunction, 304, 305–306
   power window, 355–357
   steering column-mounted, 303–304, 307–308, 348
   testing, 92–94
   turn signal, 309, 310–314
   wiper, 345, 346, 348
   wiper park, 342–344
switch operational logic hart, 613
system main relay (SMR), 654–655

T
tachometers, 504–506
taillight assemblies, 308–314, 328
tap splice connector, 136
Technical Service Bulletins (TSBs), 61
temperature sensor, overriding, 570
test equipment, 35–37
   for batteries, 171–172
   computerized load-testing equipment, 187
   digital storage oscilloscope (DSOs), 111–116
   J1962 breakout box (BOB), 409
   jumper wires, 35–36
   logic probes, 37
   multimeters, 38–39, 38–48
   ohmmeters, 43–44
   optical refractometer, 182
   oscilloscopes, 49–55
   potentiometers, 389–390
   RF sniffer tool, 585
   scan tools, 56–59, 384–387
   self-powered continuity tester, 37
   test lights, 36–37
   voltmeters, 39–43
test lights, 36–37, 91
   testing power windows with, 356–357
   testing for shorts with, 108
   testing relays with, 95, 96
   testing switches with, 93
thermistors, 389–390, 510
three-minute charge test, 186–187
tin, 355

tinning, 136
tool safety, 7–8
Toyota, air bag flash codes, 620–621
transistors, testing, 101–103
trickle charging, 176
trigger, 113
trigger slope, 114
trip computers, 514
trouble codes, 133
   for air bag systems, 618–621
   BCM, 383, 386
   B-codes, 410
   for cruise control systems, 545–548, 551
   for hybrid systems, 648
   instrument panel, 493
   related to ATC system, 543
   U-codes, 410
troubleshooting
   See also diagnostic charts; electrical troubleshooting and
     service
   antilock brake systems, 575–577
   defined, 133
   electrical problems, 34–35
   process of, 133
turn signals, 310–314, 331–332
turn signal switches, 309–314
twisted/shield wire repairs, 140–141

U
U-codes, 410

V
valve block assembly, replacement, 582–583
vapors, gasoline, 15
variable resistors, testing, 97–98
VAT-40, 249
vehicle identification number (VIN), 61, 147
vehicle lift points, 11–12
vehicle module scans, 411
vehicle safety
   on lifts and hoists, 8–13
   while running in shop, 13–14
vehicle speed sensors (VSSs), 390–391
ventilation systems, testing, 13–14
video entertainment systems, 583–585
VIN numbers, 61, 147
volatile substances, 17
voltage, 33, 87
   AC, 43, 52–53, 113
   DC, 53
   peak-to-peak, 113
   waveforms, 51–53
voltage drop, 40–43, 92
   measuring, 178, 220–222
   testing, 109–111, 253–254
   of headlight system, 299
voltage output testing, 252–253
voltage regulation, diagnostic chart for, 286
voltage signals, types of, 114–116
voltage supply, diagnosis of computer, 378–382
voltmeters, 33, 34, 39–43, 91
   See also oscilloscopes
   testing batteries with, 178–180

testing diodes with, 99
testing relays with, 94–96
testing stepped resistors with, 97
testing switches with, 93
testing variable resistors with, 98
testing voltage drop with, 109–111
volts, 33

W
wages, 66–67
warning lamps, 491–492, 512–514, 524
   for air bag systems, 618–621
waveforms, 51–53, 114–116
weather-pack connectors, repairing, 144–145
wheels, blocking, 10
wheel speed sensors, 579, 580
whining noises, 252
wiggle test, 51
Williams-Steiger Act, 20
window defogger, 352–355, 371–372
windshields, heated, 570–573
windshield washer system service, 348–349
wiper system service, 339–348

diagnostic charts, 367–370
intermittent wiper system diagnosis, 345–346
no operation in one speed only, 340–341
no wiper operation, 341–342
slow speed, 342
three-speed wiper system schematic, 340
two-speed wiper circuit, 341
wiper motor removal and installation, 346, 347
wiper switch removal and installation, 48, 346
wiper will not park, 342–344
wire repair, 133–141
   aluminum, 140
   copper, 134–140
   twisted/shielded, 140–141
wires, checking for shorts, 44, 46
wiring diagrams, reading, 147–150
workplace hazardous materials information systems (WHMIS), 19–20
workplace relationships, 67

Z
Zener diodes
   testing, 99–100, 101

Human Anatomy & Physiology Laboratory Manual

MAIN VERSION

Tenth Edition

Elaine N. Marieb, R.N., Ph.D.
Holyoke Community College

Susan J. Mitchell, Ph.D.
Onondaga Community College

Lori A. Smith, Ph.D.
American River College

PhysioEx™ Version 9.1 authored by

Peter Z. Zao
North Idaho College

Timothy Stabler, Ph.D.
Indiana University Northwest

Lori A. Smith, Ph.D.
American River College

Andrew Lokuta, Ph.D.
University of Wisconsin–Madison

Edwin Griff, Ph.D.
University of Cincinnati

PEARSON

Boston Columbus Indianapolis New York San Francisco Upper Saddle River
Amsterdam Cape Town Dubai London Madrid Milan Munich Paris Montréal Toronto
Delhi Mexico City São Paulo Sydney Hong Kong Seoul Singapore Taipei Tokyo

Editor-in-Chief: Serina Beauparlant
Senior Acquisitions Editor: Brooke Suchomel
Associate Project Editor: Shannon Cutt
Director of Development: Barbara Yien
Assistant Editor: Lisa Damerel
Senior Managing Editor: Deborah Cogan
Production and Design Manager: Michele Mangelli
Production Supervisor: Janet Vail
Art and Photo Coordinator: David Novak
Photo Researcher: Kristin Piljay
Interior and Cover Designer: tani hasegawa
Copyeditor: Anita Wagner
Compositor: Cenveo® Publisher Services
Media Producers: Aimee Pavy and Daniel Ross
PhysioEx Developer: BinaryLabs, Inc.
Senior Manufacturing Buyer: Stacey Weinberger
Executive Marketing Manager: Lauren Harp
Marketing Manager: Derek Perrigo
Cover photograph credit: Digital Vision/Getty Images

Also Available from Pearson Education for Human Anatomy & Physiology

By M. Hutchinson, J. Mallatt, E. N. Marieb, P. B. Wilhelm
A Brief Atlas of the Human Body, Second Edition (2007)

By W. Kapit and L. M. Elson
The Anatomy Coloring Book, Fourth Edition (2014)

By W. Kapit, R. I. Macey, and E. Meisami
The Physiology Coloring Book, Second Edition (2000)

By E. N. Marieb, S. J. Mitchell, and L. A. Smith
Human Anatomy & Physiology Laboratory Manual, Main Version, Tenth Edition (2014)

Human Anatomy & Physiology Laboratory Manual, Cat Version, Eleventh Edition (2014)

Human Anatomy & Physiology Laboratory Manual, Fetal Pig Version, Eleventh Edition (2014)

Human Anatomy & Physiology Laboratory Manual, Rat Version (2014)

Human Anatomy Laboratory Manual with Cat Dissections, Seventh Edition (2014)

By E. N. Marieb
Anatomy & Physiology Coloring Workbook: A Complete Study Guide, Tenth Edition (2012)

Laboratory Manual for Anatomy & Physiology, Fifth Edition (2014)

By R. Heisler, N. Hebert, J. Chinn, K. Krabbenhoft, O. Malakhova
Practice Anatomy Lab 3.0 DVD (2012)
Practice Anatomy Lab 3.0 Lab Guide (2014)

ISBN 10: 0-321-82751-1 (student edition)
ISBN 13: 978-0-321-82751-7 (student edition)
ISBN 10: 0-321-90153-3 (instructor's review copy)
ISBN 13: 978-0-321-90153-8 (instructor's review copy)

4 5 6 7 8 9 10—WBC—17 16 15 14

PEARSON

www.pearsonhighered.com

Contents

Preface to the Instructor   ix

Getting Started—What to Expect,
The Scientific Method, and Metrics   xiv

THE HUMAN BODY: AN ORIENTATION

EXERCISE

1 The Language of Anatomy   1

1 Locating Body Regions   2
2 Practicing Using Correct Anatomical Terminology   4
3 Observing Sectioned Specimens   6
4 Identifying Organs in the Abdominopelvic Cavity   8
5 Locating Abdominal Surface Regions   8
Group Challenge   The Language of Anatomy   10
Review Sheet   11

EXERCISE

2 Organ Systems Overview   15

1 Observing External Structures   17
2 Examining the Oral Cavity   17
3 Opening the Ventral Body Cavity   17
4 Examining the Ventral Body Cavity   18
5 Examining the Human Torso Model   22
Group Challenge   Odd Organ Out   24
Review Sheet   25

THE MICROSCOPE AND ITS USES

EXERCISE

3 The Microscope   27

1 Identifying the Parts of a Microscope   28
2 Viewing Objects Through the Microscope   29
3 Estimating the Diameter of the Microscope Field   32
4 Perceiving Depth   33
5 Preparing and Observing a Wet Mount   33
Review Sheet   35

THE CELL

EXERCISE

4 The Cell: Anatomy and Division   39

1 Identifying Parts of a Cell   40
2 Identifying Components of a Plasma Membrane   41
3 Locating Organelles   42
4 Examining the Cell Model   44
5 Observing Various Cell Structures   44
6 Identifying the Mitotic Stages   48
7 "Chenille Stick" Mitosis   48
Review Sheet   49

EXERCISE

5 The Cell: Transport Mechanisms and Cell Permeability   53

1 Observing Diffusion of Dye Through Agar Gel   55
2 Observing Diffusion of Dye Through Water   56
3 Investigating Diffusion and Osmosis Through Nonliving Membranes   56
4 Observing Osmometer Results   58
5 Investigating Diffusion and Osmosis Through Living Membranes   58
6 Observing the Process of Filtration   60
7 Observing Phagocytosis   61
Group Challenge   Compare and Contrast Membrane Transport Processes   62
Review Sheet   63

HISTOLOGY: BASIC TISSUES OF THE BODY

EXERCISE

6 Classification of Tissues   67

1 Examining Epithelial Tissue Under the Microscope   74
Group Challenge 1   Identifying Epithelial Tissues   74
2 Examining Connective Tissue Under the Microscope   82
Group Challenge 2   Identifying Connective Tissue   83
3 Examining Nervous Tissue Under the Microscope   83
4 Examining Muscle Tissue Under the Microscope   85
Review Sheet   87

THE INTEGUMENTARY SYSTEM

EXERCISE

7 The Integumentary System   93

1 Locating Structures on a Skin Model   94
2 Identifying Nail Structures   97
3 Comparison of Hairy and Relatively Hair-Free Skin Microscopically   98
4 Differentiating Sebaceous and Sweat Glands Microscopically   100
5 Plotting the Distribution of Sweat Glands   100
6 Taking and Identifying Inked Fingerprints   101
Review Sheet   103

THE SKELETAL SYSTEM

EXERCISE
8 Overview of the Skeleton: Classification and Structure of Bones and Cartilages 107

1 Examining a Long Bone 110
2 Examining the Effects of Heat and Hydrochloric Acid on Bones 111
3 Examining the Microscopic Structure of Compact Bone 114
4 Examination of the Osteogenic Epiphyseal Plate 114

Review Sheet 115

EXERCISE
9 The Axial Skeleton 119

1 Identifying the Bones of the Skull 120
Group Challenge Odd Bone Out 128
2 Palpating Skull Markings 129
3 Examining Spinal Curvatures 131
4 Examining Vertebral Structure 135
5 Examining the Relationship Between Ribs and Vertebrae 136
6 Examining a Fetal Skull 136

Review Sheet 139

EXERCISE
10 The Appendicular Skeleton 147

1 Examining and Identifying Bones of the Appendicular Skeleton 148
2 Palpating the Surface Anatomy of the Pectoral Girdle and Upper Limb 152
3 Observing Pelvic Articulations 153
4 Comparing Male and Female Pelves 153
5 Palpating the Surface Anatomy of the Pelvic Girdle and Lower Limb 157
6 Constructing a Skeleton 158

Review Sheet 159

EXERCISE
11 Articulations and Body Movements 167

1 Identifying Fibrous Joints 169
2 Identifying Cartilaginous Joints 169
3 Examining Synovial Joint Structure 171
4 Demonstrating the Importance of Friction-Reducing Structures 171
5 Demonstrating Movements of Synovial Joints 173
6 Demonstrating Actions at the Hip Joint 176
7 Demonstrating Actions at the Knee Joint 178
8 Demonstrating Actions at the Shoulder Joint 178
9 Examining the Action at the TMJ 178
Group Challenge Articulations: "Simon Says" 179

Review Sheet 181

THE MUSCULAR SYSTEM

EXERCISE
12 Microscopic Anatomy and Organization of Skeletal Muscle 185

1 Examining Skeletal Muscle Cell Anatomy 188
2 Observing the Histological Structure of a Skeletal Muscle 188
3 Studying the Structure of a Neuromuscular Junction 190

Review Sheet 191

EXERCISE
13 Gross Anatomy of the Muscular System 195

1 Identifying Head and Neck Muscles 197
2 Identifying Muscles of the Trunk 197
3 Identifying Muscles of the Upper Limb 211
4 Identifying Muscles of the Lower Limb 216
Group Challenge Name That Muscle 220
5 Review of Human Musculature 222
6 Making a Muscle Painting 222

Review Sheet 225

EXERCISE
14 Skeletal Muscle Physiology: Frogs and Human Subjects 233

1 Observing Muscle Fiber Contraction 234
2 Inducing Contraction in the Frog Gastrocnemius Muscle 236
3 Demonstrating Muscle Fatigue in Humans 241
BIOPAC 4 Electromyography in a Human Subject Using BIOPAC® 241

Review Sheet 249

THE NERVOUS SYSTEM

EXERCISE
15 Histology of Nervous Tissue 253

1 Identifying Parts of a Neuron 257
2 Studying the Microscopic Structure of Selected Neurons 257
3 Examining the Microscopic Structure of a Nerve 260

Review Sheet 261

EXERCISE
16 Neurophysiology of Nerve Impulse: Frog Subjects 265

1 Stimulating the Nerve 268
2 Inhibiting the Nerve 268
3 Visualizing the Compound Action Potential with an Oscilloscope 270

Review Sheet 273

Contents v

EXERCISE

17 **Gross Anatomy of the Brain and Cranial Nerves 275**

1 Identifying External Brain Structures 277

2 Identifying Internal Brain Structures 279

3 Identifying and Testing the Cranial Nerves 285

Group Challenge Odd (Cranial) Nerve Out 292

Review Sheet 293

EXERCISE

18 **Electroencephalography 299**

1 Observing Brain Wave Patterns Using an Oscilloscope or Physiograph 300

BIOPAC 2 Electroencephalography Using BIOPAC® 301

Review Sheet 305

EXERCISE

19 **The Spinal Cord and Spinal Nerves 307**

1 Identifying Structures of the Spinal Cord 308

2 Identifying Spinal Cord Tracts 311

3 Identifying the Major Nerve Plexuses and Peripheral Nerves 318

Group Challenge Fix the Sequence 318

Review Sheet 319

EXERCISE

20 **The Autonomic Nervous System 323**

1 Locating the Sympathetic Trunk 324

2 Comparing Sympathetic and Parasympathetic Effects 326

BIOPAC 3 Exploring the Galvanic Skin Response (Electrodermal Activity) Within a Polygraph Using BIOPAC® 326

Review Sheet 333

EXERCISE

21 **Human Reflex Physiology 335**

1 Initiating Stretch Reflexes 328

2 Initiating the Crossed-Extensor Reflex 339

3 Initiating the Plantar Reflex 339

4 Initiating the Corneal Reflex 340

5 Initiating the Gag Reflex 340

6 Initiating Pupillary Reflexes 340

7 Initiating the Ciliospinal Reflex 341

8 Initiating the Salivary Reflex 341

9 Testing Reaction Time for Intrinsic and Learned Reflexes 342

BIOPAC 10 Measuring Reaction Time Using BIOPAC® 342

Review Sheet 345

EXERCISE

22 **General Sensation 349**

1 Studying the Structure of Selected Sensory Receptors 351

2 Determining the Two-Point Threshold 352

3 Testing Tactile Localization 352

4 Demonstrating Adaptation of Touch Receptors 353

5 Demonstrating Adaptation of Temperature Receptors 353

6 Demonstrating the Phenomenon of Referred Pain 354

Group Challenge Odd Receptor Out 355

Review Sheet 357

EXERCISE

23 **Special Senses: Anatomy of the Visual System 359**

1 Identifying Accessory Eye Structures 360

2 Identifying Internal Structures of the Eye 363

3 Studying the Microscopic Anatomy of the Retina 364

4 Predicting the Effects of Visual Pathway Lesions 364

Review Sheet 367

EXERCISE

24 **Special Senses: Visual Tests and Experiments 371**

1 Demonstrating the Blind Spot 371

2 Determining Near Point of Accommodation 372

3 Testing Visual Acuity 373

4 Testing for Astigmatism 374

5 Testing for Color Blindness 374

6 Testing for Depth Perception 374

7 Demonstrating Reflex Activity of Intrinsic and Extrinsic Eye Muscles 375

8 Conducting an Ophthalmoscopic Examination 376

Review Sheet 379

EXERCISE

25 **Special Senses: Hearing and Equilibrium 383**

1 Identifying Structures of the Ear 384

2 Examining the Ear with an Otoscope (Optional) 385

3 Examining the Microscopic Structure of the Cochlea 386

4 Conducting Laboratory Tests of Hearing 387

5 Audiometry Testing 389

6 Examining the Microscopic Structure of the Crista Ampullaris 390

7 Conducting Laboratory Tests on Equilibrium 391

Review Sheet 393

EXERCISE

26 **Special Senses: Olfaction and Taste 397**

1 Microscopic Examination of the Olfactory Epithelium 399

2 Microscopic Examination of Taste Buds 400

3 Stimulating Taste Buds 400

4 Examining the Combined Effects of Smell, Texture, and Temperature on Taste 400

5 Assessing the Importance of Taste and Olfaction in Odor Identification 401

6 Demonstrating Olfactory Adaptation 401

Review Sheet 403

THE ENDOCRINE SYSTEM

EXERCISE

27 Functional Anatomy of the Endocrine Glands 405

1 Identifying the Endocrine Organs 408

2 Examining the Microscopic Structure of Endocrine Glands 409

Group Challenge Odd Hormone Out 411

Review Sheet 413

EXERCISE

28 Endocrine Wet Labs and Human Metabolism 417

1 Determining the Effect of Pituitary Hormones on the Ovary 418

2 Observing the Effects of Hyperinsulinism 418

Group Challenge Thyroid Hormone Case Studies 420

Review Sheet 421

THE CIRCULATORY SYSTEM

EXERCISE

29 Blood 423

1 Determining the Physical Characteristics of Plasma 425

2 Examining the Formed Elements of Blood Microscopically 426

3 Conducting a Differential WBC Count 429

4 Determining the Hematocrit 430

5 Determining Hemoglobin Concentration 431

6 Determining Coagulation Time 434

7 Typing for ABO and Rh Blood Groups 434

8 Observing Demonstration Slides 435

9 Measuring Plasma Cholesterol Concentration 436

Review Sheet 437

EXERCISE

30 Anatomy of the Heart 443

1 Using the Heart Model to Study Heart Anatomy 446

2 Tracing the Path of Blood Through the Heart 446

3 Using the Heart Model to Study Cardiac Circulation 448

4 Examining Cardiac Muscle Tissue Anatomy 449

Review Sheet 453

EXERCISE

31 Conduction System of the Heart and Electrocardiography 457

1A Recording ECGs Using a Standard ECG Apparatus 460

BIOPAC® 1B Electrocardiography Using BIOPAC® 462

Review Sheet 467

EXERCISE

32 Anatomy of Blood Vessels 469

1 Examining the Microscopic Structure of Arteries and Veins 472

2 Locating Arteries on an Anatomical Chart or Model 476

3 Identifying the Systemic Veins 479

Group Challenge Fix the Blood Trace 481

4 Identifying Vessels of the Pulmonary Circulation 482

5 Tracing the Pathway of Fetal Blood Flow 482

6 Tracing the Hepatic Portal Circulation 484

Review Sheet 485

EXERCISE

33 Human Cardiovascular Physiology: Blood Pressure and Pulse Determinations 491

1 Auscultating Heart Sounds 494

2 Palpating Superficial Pulse Points 495

BIOPAC® 3 Measuring Pulse Using BIOPAC® 495

4 Taking an Apical Pulse 497

5 Using a Sphygmomanometer to Measure Arterial Blood Pressure Indirectly 498

6 Estimating Venous Pressure 499

7 Observing the Effect of Various Factors on Blood Pressure and Heart Rate 500

8 Examining the Effect of Local Chemical and Physical Factors on Skin Color 502

Review Sheet 505

EXERCISE

34 Frog Cardiovascular Physiology 511

1 Investigating the Automaticity and Rhythmicity of Heart Muscle 512

BIOPAC® 2 Recording Baseline Frog Heart Activity 514

3 Investigating the Refractory Period of Cardiac Muscle Using the Physiograph 517

BIOPAC® 4 Assessing Physical and Chemical Modifiers of Heart Rate 517

5 Investigating the Effect of Various Factors on the Microcirculation 519

Review Sheet 521

EXERCISE

35 The Lymphatic System and Immune Response 525

1 Identifying the Organs of the Lymphatic System 527

2 Studying the Microscopic Anatomy of a Lymph Node, the Spleen, and a Tonsil 529

Group Challenge Compare and Contrast Lymphoid Organs and Tissues 530

3 Using the Ouchterlony Technique to Identify Antigens 531

Review Sheet 533

THE RESPIRATORY SYSTEM

EXERCISE
36 Anatomy of the Respiratory System 537

1 Identifying Respiratory System Organs 543
2 Demonstrating Lung Inflation in a Sheep Pluck 543
3 Examining Prepared Slides of Trachea and Lung Tissue 543
Review Sheet 545

EXERCISE
37 Respiratory System Physiology 549

1 Operating the Model Lung 550
2 Auscultating Respiratory Sounds 551
3 Measuring Respiratory Volumes Using Spirometers 554
4 Measuring the FVC and FEV_1 557
BIOPAC 5 Measuring Respiratory Volumes Using BIOPAC® 559
6 Visualizing Respiratory Variations 562
7 Demonstrating the Reaction Between Carbon Dioxide (in Exhaled Air) and Water 565
8 Observing the Operation of Standard Buffers 565
9 Exploring the Operation of the Carbonic Acid-Bicarbonate Buffer System 565
Review Sheet 567

THE DIGESTIVE SYSTEM

EXERCISE
38 Anatomy of the Digestive System 573

1 Identifying Alimentary Canal Organs 575
2 Studying the Histologic Structure of Selected Digestive System Organs 579
3 Observing the Histologic Structure of the Small Intestine 582
4 Examining the Histologic Structure of the Large Intestine 584
5 Identifying Types of Teeth 585
6 Studying Microscopic Tooth Anatomy 585
7 Examining Salivary Gland Tissue 586
8 Examining the Histology of the Liver 587
Review Sheet 589

EXERCISE
39 Digestive System Processes: Chemical and Physical 595

1 Assessing Starch Digestion by Salivary Amylase 596
2 Assessing Protein Digestion by Trypsin 599
3 Demonstrating the Emulsification Action of Bile and Assessing Fat Digestion by Lipase 599

4 Reporting Results and Conclusions 601
Group Challenge Odd Enzyme Out 601
5 Observing Movements and Sounds of the Digestion System 602
6 Viewing Segmental and Peristaltic Movements 603
Review Sheet 605

THE URINARY SYSTEM

EXERCISE
40 Anatomy of the Urinary System 609

1 Identifying Urinary System Organs 611
2 Studying Nephron Structure 615
3 Studying Bladder Structure 616
Group Challenge Urinary System Sequencing 617
Review Sheet 619

EXERCISE
41 Urinalysis 621

1 Analyzing Urine Samples 623
2 Analyzing Urine Sediment Microscopically (Optional) 626
Review Sheet 627

THE REPRODUCTIVE SYSTEM, DEVELOPMENT, AND HEREDITY

EXERCISE
42 Anatomy of the Reproductive System 629

1 Identifying Male Reproductive Organs 630
2 Penis 633
3 Seminal Gland 633
4 Epididymis 633
5 Identifying Female Reproductive Organs 634
6 Wall of the Uterus 636
7 Uterine Tube 636
Group Challenge Reproductive Homologues 638
Review Sheet 639

EXERCISE
43 Physiology of Reproduction: Gametogenesis and the Female Cycles 645

1 Identifying Meiotic Phases and Structures 647
2 Examining Events of Spermatogenesis 647
3 Examining Meiotic Events Microscopically 648
4 Examining Oogenesis in the Ovary 650
5 Comparing and Contrasting Oogenesis and Spermatogenesis 651
6 Observing Histological Changes in the Endometrium During the Menstrual Cycle 651
Review Sheet 653

EXERCISE

44 Survey of Embryonic Development 657

1 Microscopic Study of Sea Urchin Development 658

2 Examining the Stages of Human Development 658

3 Identifying Fetal Structures 661

4 Studying Placental Structure 662

Review Sheet 663

EXERCISE

45 Principles of Heredity 667

1 Working Out Crosses Involving Dominant and Recessive Genes 668

2 Working Out Crosses Involving Incomplete Dominance 669

3 Working Out Crosses Involving Sex-Linked Inheritance 670

4 Exploring Probability 670

5 Using Phenotype to Determine Genotype 671

6 Using Agarose Gel Electrophoresis to Identify Normal Hemoglobin, Sickle Cell Anemia, and Sickle Cell Trait 672

Group Challenge Odd Phenotype Out 674

Review Sheet 675

SURFACE ANATOMY

EXERCISE

46 Surface Anatomy Roundup 679

1 Palpating Landmarks of the Head 680

2 Palpating Landmarks of the Neck 681

3 Palpating Landmarks of the Trunk 683

4 Palpating Landmarks of the Abdomen 686

5 Palpating Landmarks of the Upper Limb 687

6 Palpating Landmarks of the Lower Limb 690

Review Sheet 695

PHYSIOEX™ 9.1   COMPUTER SIMULATIONS

1 Cell Transport Mechanisms and Permeability PEx-3

2 Skeletal Muscle Physiology PEx-17

3 Neurophysiology of Nerve Impulses PEx-35

4 Endocrine System Physiology PEx-59

5 Cardiovascular Dynamics PEx-75

6 Cardiovascular Physiology PEx-93

7 Respiratory System Mechanics PEx-105

8 Chemical and Physical Processes of Digestion PEx-119

9 Renal System Physiology PEx-131

10 Acid-Base Balance PEx-149

11 Blood Analysis PEx-161

12 Serological Testing PEx-177

Appendix   The Metric System   BM-1

Credits   BM-2

Index   BM-5

Preface to the Instructor

The philosophy behind the revision of this manual mirrors that of all earlier editions. It reflects a still developing sensibility for the way teachers teach and students learn, engendered by years of teaching the subject and by listening to the suggestions of other instructors as well as those of students enrolled in multifaceted healthcare programs. *Human Anatomy & Physiology Laboratory Manual, Main Version* was originally developed to facilitate and enrich the laboratory experience for both teachers and students. This edition retains those same goals.

This manual, intended for students in introductory human anatomy and physiology courses, presents a wide range of laboratory experiences for students concentrating in nursing, physical therapy, dental hygiene, pharmacology, respiratory therapy, and health and physical education, as well as biology and premedical programs. It differs from other versions of *Human Anatomy & Physiology Laboratory Manual* in that it does not contain detailed guidelines for dissecting a laboratory animal. The manual's coverage is intentionally broad, allowing it to serve both one- and two-semester courses.

Basic Approach and Features

The generous variety of experiments in this manual provides flexibility that enables instructors to gear their laboratory approach to specific academic programs, or to their own teaching preferences. The manual is still independent of any textbook, so it contains the background discussions and terminology necessary to perform all experiments. Such a self-contained learning aid eliminates the need for students to bring a textbook into the laboratory.

Each of the 46 exercises leads students toward a coherent understanding of the structure and function of the human body. The manual begins with anatomical terminology and an orientation to the body, which together provide the necessary tools for studying the various body systems. The exercises that follow reflect the dual focus of the manual—both anatomical and physiological aspects receive considerable attention. As the various organ systems of the body are introduced, the initial exercises focus on organization, from the cellular to the organ system level. As indicated by the table of contents, the anatomical exercises are usually followed by physiological experiments that familiarize students with various aspects of body functioning and promote the critical understanding that function follows structure. Homeostasis is continually emphasized as a requirement for optimal health. Pathological conditions are viewed as a loss of homeostasis; these discussions can be recognized by the homeostatic imbalance logo within the descriptive material of each exercise. This holistic approach encourages an integrated understanding of the human body.

Features

- The numerous physiological experiments for each organ system range from simple experiments that can be performed without specialized tools to more complex experiments using laboratory equipment, computers, and instrumentation techniques.

- The laboratory Review Sheets following each exercise provide space for recording and interpreting experimental results and require students to label diagrams and answer matching and short-answer questions.

- In addition to the figures, isolated animal organs such as the sheep heart and pig kidney are employed to study anatomy because of their exceptional similarity to human organs.

- All exercises involving body fluids (blood, urine, saliva) incorporate current Centers for Disease Control and Prevention (CDC) guidelines for handling human body fluids. Because it is important that nursing students, in particular, learn how to safely handle bloodstained articles, the human focus has been retained. However, the decision to allow testing of human (student) blood or to use animal blood in the laboratory is left to the discretion of the instructor in accordance with institutional guidelines. The CDC guidelines for handling body fluids are reinforced by the laboratory safety procedures described on the inside front cover of this text, in Exercise 29: Blood, and in the *Instructor's Guide*. You can photocopy the inside front cover and post it in the lab to help students become well versed in laboratory safety.

- Five icons alert students to special features or instructions. These include:

The dissection scissors icon appears at the beginning of activities that entail the dissection of isolated animal organs.

The homeostatic imbalance icon directs the student's attention to conditions representing a loss of homeostasis.

A safety icon notifies students that specific safety precautions must be observed when using certain equipment or conducting particular lab procedures. For example, when working with ether, a hood is to be used; and when handling body fluids such as blood, urine, or saliva, gloves are to be worn.

BIOPAC The BIOPAC icon in the materials list for an exercise clearly identifies use of the BIOPAC Student Lab System and alerts you to the equipment needed. BIOPAC is used in Exercises 14, 18, 20, 21, 31, 33, 34, and 37. The instructions in the lab manual are for use with the BIOPAC MP36/35 and MP45 data acquisition unit. Note there are some exercises that are not compatible with the MP45 data acquisition unit. For those exercises, the MP45 will not be listed in the materials section. The instructions included in the lab manual are for use with BSL software 3.7.5 to 3.7.7 for Windows and BSL software 3.7.4 to 3.7.7 for Mac OS X with some exceptions. Refer to the Materials section in each exercise for the applicable software version. For instructors using the MP36 (or MP35/30) data acquisition unit using BSL software versions earlier than 3.7.5 (for Windows) and 3.7.4 (for Mac OS X), access BIOPAC instructions in MasteringA&P for Exercises 14, 18, 20, 21, 31, and 34.

PEx The PhysioEx icon at the end of the materials list for an exercise directs students to the corresponding PhysioEx computer simulation exercise found in the back of the lab manual.

• Other data acquisition instructions are available in MasteringA&P, including:

PowerLab® Instructions

For Exercises 14, 21, 31, 33, 34, and 37, instructors using PowerLab equipment may print these exercises for student handouts.

iWorx® Instructions

For Exercises 14, 18, 21, 31, 33, 34, and 37, instructors using iWorx equipment in their laboratory may print these exercises for student handouts.

Intelitool® Instructions

Four physiological experiments (Exercises 14i, 21i, 31i, and 37i) using Intelitool equipment are available. Instructors using Intelitool equipment in their laboratory may print these exercises for student handouts.

What's New

In this revision, we have continued to try to respond to reviewers' and users' feedback concerning trends that are having an impact on the anatomy and physiology laboratory experience, most importantly:

• The ongoing search for good pedagogy and effective use of laboratory time

• The need to develop critical thinking skills

• The desire for more frequent quizzing delivered in both print and media formats, and for more computer-based practice exercises

• The need for laboratory simulations

The specific changes implemented to address these trends are described next.

NEW! Extensive Instructor Support Materials

Instructor's Resource DVD (0321884981 / 9780321884985) New for this edition, the Instructor's Resource DVD (IRDVD) organizes all instructor media resources into one convenient location and allows for easy quizzing in the lab. The IRDVD provides both **JPEG and PowerPoint files of all figures and tables** from the manual, with enlarged labels and easy-to-read type for optimal presentation.

IRDVD resources include:

• **Labeled and unlabeled JPEG files of all numbered figures and tables.** An extra set of JPEG files provides unlabeled figures with leader lines for **quick and easy quizzing in the lab.**

• **Customizable PowerPoint files** of all figures, with editable leader lines and labels.

• **A&P Flix™ Animations**

• **New Bone and Dissection videos**

• **Updated Electronic Test Bank** of Pre-Lab and Post-Lab Quizzes

• **PAL 3.0 Instructor Resources:** All images from PAL 3.0 in JPEG and PowerPoint format, with editable labels and leader lines, and embedded links to relevant 3-D anatomy animations and bone rotations. Instructors can quickly and easily create assignments based on the structures they cover in their course. Also included is an index of anatomical structures covered in PAL 3.0.

NEW! MasteringA&P® with Pearson eText

MasteringA&P is an online learning and assessment system proven to help students learn. It helps instructors maximize lab time with customizable, easy-to-assign, automatically graded assessments that motivate students to learn outside of class and arrive prepared for lab. The powerful gradebook provides unique insight into student and class performance, even before the first lab exam. As a result, instructors can spend valuable time where students need it most.

MasteringA&P resources include:

• **NEW! Bone and Dissection Videos review key bones and organ dissections found in the lab manual.** Each video is supported with activities with hints and specific wrong answer feedback to help students preview or review for lab exercises involving dissection and bone identification.

• **NEW! Drag-and-Drop Art Labeling Questions** let students assess their knowledge of terms and structures.

• **Updated! Assignable pre-lab and post-lab quizzes for all 46 exercises** in the lab manual.

• **Assignable quizzes and lab practicals from the PAL™ 3.0 Test Bank.**

• **Assignable pre-lab and post-lab quizzes for PhysioEx™ 9.1.**

• **Access to PAL 3.0 and PhysioEx 9.1** in the Study Area.

• **Instructor Access to IRDVD content.** A 24-month subscription to MasteringA&P is included with each new copy of the lab manual, and provides access directions and an access code.

In addition, MasteringA&P New Design is now available and offers:

• **Seamless integration with Blackboard Learn.** Using a Blackboard Building Block, this integration delivers streamlined access to your customizable content and highly personalized study paths, responsive learning tools, and real-time evaluation and diagnostics within the context of Blackboard Learn.

• **Student registration offers temporary access,** allowing students to access their Mastering course materials from day one, but defer payment for up to 17 days while they are waiting for financial aid.

• **Improved registration experience** provides a single point of access for instructors and students who are teaching and learning with multiple Mastering courses.

- **Simple user interface allows for quick and easy access** to Assignments, eText (also available offline with an iPad® app), and Results, and more options for course customization.
- **Communication Tools, including Email, Chat, Discussion Boards, and ClassLive Whiteboard** can be used to foster collaboration, class participation, and group work.

NEW! Customization Options

An enhanced custom program allows instructors to pick and choose content to tailor the lab manual to their own course. Instructors can now customize the lab manual at the activity level, selecting only those activities they assign. Each activity includes relevant background information, full-color figures, tables, and charts.

For information on creating a custom version of this manual, visit www.pearsonlearningsolutions.com/, or contact your Pearson representative for details.

NEW! Group Challenge Activities

Designed to teach critical thinking skills, these new activities challenge students to find the relationships between anatomical structures and physiological concepts, and to use that information to understand anatomy and physiology at a deeper level.

New Group Challenge Activities include:

Ex. 1: The Language of Anatomy, p. 10

Ex. 2: Odd Organ Out, p. 24

Ex. 5: Compare and Contrast Membrane
  Transport Processes, p. 62

Ex. 6: 1 Identifying Epithelial Tissues, p. 74
  2 Identifying Connective Tissues, p. 83

Ex. 9: Odd Bone Out, p. 128

Ex. 11: Articulations: "Simon Says," p. 179

Ex. 13: Name That Muscle, p. 220

Ex. 17: Odd (Cranial) Nerve Out, p. 292

Ex. 19: Fix the Sequence, p. 318

Ex. 22: Odd Receptor Out, p. 355

Ex. 27: Odd Hormone Out, p. 411

Ex. 28: Thyroid Hormone Case Studies, p. 420

Ex. 32: Fix the Blood Trace, p. 481

Ex. 35: Compare and Contrast Lymphoid
  Organs and Tissues, p. 530

Ex. 39: Odd Enzyme Out, p. 601

Ex. 40: Urinary System Sequencing, p. 617

Ex. 42: Reproductive Homologues, p. 638

Ex. 45: Odd Phenotype Out, p. 674

NEW! Improved Organization and Streamlined Text

This edition features key improvements to the lab manual's organization. Select content has been moved and divided for better efficiency in the lab. Important information from two smaller exercises is now integrated into other appropriate exercises:

serous membranes are now more elaborately discussed in Ex. 1 (Language of Anatomy), and the *fetal skull* is covered in Ex. 12 (Axial Skeleton). We have also divided two exercises that were previously very large: The Spinal Cord and Spinal Nerves (Ex. 19) is now covered separately from The Autonomic Nervous System (Ex. 20). In addition, coverage of *vision* has been divided into two exercises: Anatomy of the Visual System (Ex. 23) and Visual Tests and Experiments (Ex. 24). Please refer to the new detailed *Table of Contents* for changes to the numbering of Ex. 8–24 (the numbering of Ex. 1–7 and Ex. 25–46 remains the same as the previous edition).

Other improvements to the lab manual include carefully edited, more accessible language; a new, user-friendly design featuring checklists that help students track their progress; and updated terminology that reflects the most recent information in *Terminologia Anatomica* and *Terminologia Histologica*.

Student Supplements

PAL Practice Anatomy Lab™ 3.0

Practice Anatomy Lab (PAL) 3.0 is an indispensable virtual anatomy study and practice tool that gives students 24/7 access to the most widely used laboratory specimens including human cadaver, cat, and fetal pig as well as anatomical models and histological images that are used in the laboratory.

PAL 3.0 features:

- **An interactive cadaver** that allows students to peel back layers of the human cadaver and view hundreds of brand-new dissection photographs specifically commissioned for this version.
- **Interactive histology** that allows students to view the same tissue slide at varying magnifications.
- **Quizzes give students more opportunity for practice.** Each time the student takes a quiz or lab practical exam, a new set of questions is generated.
- **Integration of nerves, arteries, and veins** across body systems.
- **Integrated muscle animations** of the origin, insertion, action, and innervations of key muscles.
- **Rotatable bones** help students appreciate the three-dimensionality of bone structures.

PAL 3.0 is available in the Study Area of MasteringA&P. The PAL 3.0 DVD can also be packaged with this lab manual at no additional charge.

PEx NEW! PhysioEx™ 9.1

PhysioEx 9.1 provides easy-to-use laboratory simulations in 12 exercises, containing a total of 63 physiology laboratory activities. 9.1 features input data variability that allows you to change variables and test out various hypotheses for the experiments. It can be used to supplement or substitute for wet labs. PhysioEx allows students to repeat labs as often as they like, perform experiments without harming live animals, and conduct experiments that are difficult to perform because of time, cost, or safety concerns.

PhysioEx 9.1 software features:

- **Input data variability** that allows students to change variables and test various hypotheses for the experiments.

• **New online format with easy step-by-step instructions** puts everything students need to do to complete the lab in one convenient place. Students gather data, analyze results, and check their understanding, all on screen.

• **Stop & Think Questions** and **Predict Questions** help students think about the connection between the activities and the physiological concepts they demonstrate.

• **Greater data variability in the results** reflects more realistically the results that students would encounter in a wet lab experiment.

• **New Pre-lab and Post-lab Quizzes** and short-answer **Review Sheets** are offered to help students prepare for and review each activity.

• **Students can save their Lab Report as a PDF,** which they can print and/or email to their instructor.

• **A Test Bank of assignable pre-lab and post-lab quizzes** for use with TestGen® or its course management system is provided for instructors.

• **Seven videos of lab experiments** demonstrate the actual experiments simulated on-screen, making it easy for students to understand and visualize the context of the simulations. Videos demonstrate the following experiments: Skeletal Muscle, Blood Typing, Cardiovascular Physiology, Use of a Water-Filled Spirometer, Nerve Impulses, BMR Measurement, and Cell Transport.

PhysioEx 9.1 topics include:

• Exercise 1: *Cell Transport Mechanisms and Permeability.* Explores how substances cross the cell membrane. Topics include: simple and facilitated diffusion, osmosis, filtration, and active transport.

• Exercise 2: *Skeletal Muscle Physiology.* Provides insights into the complex physiology of skeletal muscle. Topics include: electrical stimulation, isometric contractions, and isotonic contractions.

• Exercise 3: *Neurophysiology of Nerve Impulses.* Investigates stimuli that elicit action potentials, stimuli that inhibit action potentials, and factors affecting the conduction velocity of an action potential.

• Exercise 4: *Endocrine System Physiology.* Investigates the relationship between hormones and metabolism; the effect of estrogen replacement therapy; the diagnosis of diabetes; and the relationship between the levels of cortisol and adrenocorticotropic hormone and a variety of endocrine disorders.

• Exercise 5: *Cardiovascular Dynamics.* Allows students to perform experiments that would be difficult if not impossible to do in a traditional laboratory. Topics include: vessel resistance and pump (heart) mechanics.

• Exercise 6: *Cardiovascular Physiology.* Examines variables influencing heart activity. Topics include: setting up and recording baseline heart activity, the refractory period of cardiac muscle, and an investigation of factors that affect heart rate and contractility.

• Exercise 7: *Respiratory System Mechanics.* Investigates physical and chemical aspects of pulmonary function. Students collect data simulating normal lung volumes. Other activities examine factors such as airway resistance and the effect of surfactant on lung function.

• Exercise 8: *Chemical and Physical Processes of Digestion.* Examines factors that affect enzyme activity by manipulating (in compressed time) enzymes, reagents, and incubation conditions.

• Exercise 9: *Renal System Physiology.* Simulates the function of a single nephron. Topics include: factors influencing glomerular filtration, the effect of hormones on urine function, and glucose transport maximum.

• Exercise 10: *Acid-Base Balance.* Topics include: respiratory and metabolic acidosis/alkalosis, and renal and respiratory compensation.

• Exercise 11: *Blood Analysis.* Topics include: hematocrit determination, erythrocyte sedimentation rate determination, hemoglobin determination, blood typing, and total cholesterol determination.

• Exercise 12: *Serological Testing.* Investigates antigen-antibody reactions and their role in clinical tests used to diagnose a disease or an infection.

The PhysioEx 9.1 CD-ROM is available in a value package of the lab manual for no additional charge and is available in the Study Area of MasteringA&P.

Also Available

Practice Anatomy Lab 3.0 Lab Guide

without PAL 3.0 DVD (0-321-84025-9)
with PAL 3.0 DVD (0-321-85767-4)
by Ruth Heisler, Nora Hebert, Jett Chinn, Karen Krabbenhoft, Olga Malakhova

Written to accompany PAL™ 3.0, the new *Practice Anatomy Lab 3.0 Lab Guide* contains exercises that direct the student to select images and features in PAL 3.0, and then assesses their understanding with labeling, matching, short answer, and fill-in-the-blank questions. Exercises cover three key lab specimens in PAL 3.0—human cadaver, anatomical models, and histology.

The Anatomy Coloring Book, Fourth Edition (0-321-83201-9)
by Kapit and Elson

For more than 35 years, **The Anatomy Coloring Book** has been the best-selling human anatomy coloring book! A useful tool for anyone with an interest in learning anatomical structures, this concisely written text features precise, extraordinary hand-drawn figures that were crafted especially for easy coloring and interactive study. The **Fourth Edition** features user-friendly two-page spreads with enlarged art, clearer, more concise text descriptions, and new boldface headings that make this classic coloring book accessible to a wider range of learners.

A Brief Atlas of the Human Body, 2nd Edition (0-321-66261-X)
by Hutchinson, Mallatt, Marieb, Wilhelm

This full-color atlas includes 107 bone and 47 soft-tissue photographs with easy-to-read labels, and a comprehensive histology photomicrograph section covering basic tissue and organ systems.

Instructor's Guide (0-321-86170-1)
by Marieb, Mitchell, Smith

This guide accompanies all versions of the *Human Anatomy & Physiology Laboratory Manual* and includes detailed directions for setting up the laboratory, time allotments for each exercise, common problems encountered in the lab, alternative activities, and answers to the pre-lab quizzes, activity questions, and review sheets that appear in the Lab Manual.

Acknowledgments

We wish to thank the following reviewers for their contributions to this edition: *Lynne Anderson, Meridian Community College; Claudie Biggers, Amarillo College; Patty Bostwick Taylor, Florence-Darlington Technical College; Ellen Carson, Florida State College at Jacksonville; Audra Richele Day, South Plains College; Smruti Desai, Lone Star College, CyFair Campus; Mary Dettman, Seminole State College of Florida; Paul Emerick, Monroe Community College; Brian Feige, Mott Community College; Linda Flora, Delaware County Community College; Paul Garcia, Houston Community College; Emily Gardner, Onondaga Community College; Patricia Hampf, Northeastern University; D.J. Hennager, Kirkwood Community College; Betsy Hogan, Trident Technical College; Jennifer Hollander, University of Nevada, Reno; Melinda Kay Hutton, McNeese State University; Alexander Ibe, Weatherford College; Anita Kandula, De Anza College; Kimberly Kerr, Troy University, Montgomery Campus; Beth Ann Kersten, State College of Florida, Manatee-Sarasota; Luis Labiste, Miami Dade College; Abigail Mabe Goosie, Walters State Community College; Jane Marone, University of Illinois at Chicago; Linda Menard, Quincy College; Maria Oehler, Florida State College at Jacksonville; Sidney L. Palmer, Brigham Young University-Idaho; Brandon H. Poe, Springfield Technical Community College; Kevin Ragland, Nashville State Community College; Josephine Rogers, University of Cincinnati; Janice Yoder Smith, Tarrant County College, Northwest Campus; Laura Steele, Ivy Tech Community College, Northeast Campus; Judith Tate, Laney College; Deborah Temperly, Delta College; James Thompson, Austin Peay State University; Liz Torrano, American River College; Heather Walker, Clemson University; John E. Whitlock, Hillsborough Community College, Dale Mabry Campus; Peggie Williamson, Central Texas College; Colleen Winters, Towson University.*

We would like to extend a special thank you to the following authors and contributors of Practice Anatomy Lab 3.0: Ruth Heisler, University of Colorado at Boulder; Nora Hebert, Red Rocks Community College; Jett Chinn, Cañada College and College of Marin; Karen Krabbenhoft, University of Wisconsin-Madison; Olga Malakhova, University of Florida at Gainesville, College of Medicine; Lisa M. J. Lee, The Ohio State University College of Medicine; Larry DeLay, Waubonsee Community College; Patricia Brady Wilhelm, Community College of Rhode Island-Warwick; Leslie C. Hendon, University of Alabama-Birmingham; Samuel Chen, Moraine Valley Community College; Leif Saul, University of Colorado at Boulder; Eric Howell, Red Rocks Community College; Stephen Downing, University of Minnesota Medical School-Duluth; Yvonne Baptiste-Szymanski, Niagara County Community College; Charles Venglarik, Jefferson State Community College; Nina Zanetti, Siena College.

Special thanks to Josephine Rogers of the University of Cincinnati, the original author of the pre-lab quizzes in this manual, for reviewing them to ensure accuracy in this edition.

The excellence of PhysioEx 9.1 reflects the expertise of Peter Zao, Timothy Stabler, Lori Smith, Andrew Lokuta, Greta Peterson, Nina Zanetti, and Edwin Griff. They generated the ideas behind the activities and simulations. Credit also goes to the team at BinaryLabs, Inc., for their expert programming and design.

Continued thanks to colleagues and friends at Pearson who worked with us in the production of this edition, especially Serina Beauparlant, Editor-in-Chief; Gretchen Puttkamer, Acquisitions Editor; and Shannon Cutt, Associate Project Editor. Applause also to Daniel Ross who managed MasteringA&P, and Aimee Pavy for her work on PhysioEx 9.1. Many thanks to Stacey Weinberger for her manufacturing expertise. Finally, our Marketing Manager, Derek Perrigo, has efficiently kept us in touch with the pulse of the marketplace.

Kudos also to Michele Mangelli and her production team, who did their usual great job. Janet Vail, production editor for this project, got the job done in jig time. David Novak acted as art and photo coordinator, and Kristin Piljay conducted photo research. Our fabulous interior and cover designs were created by tani hasegawa. Anita Wagner brought her experience to copyediting the text.

We are grateful to the team at BIOPAC®, especially to Jocelyn Kremer, who was extremely helpful in making sure we had the latest updates and answering all of our questions.

Elaine N. Marieb
Susan J. Mitchell
Lori A. Smith
Anatomy and Physiology
Pearson Education
1301 Sansome Street
San Francisco, CA 94111

Getting Started—What to Expect, The Scientific Method, and Metrics

Two hundred years ago science was largely a plaything of wealthy patrons, but today's world is dominated by science and its technology. Whether or not we believe that such domination is desirable, we all have a responsibility to try to understand the goals and methods of science that have seeded this knowledge and technological explosion.

The biosciences are very special and exciting because they open the doors to an understanding of all the wondrous workings of living things. A course in human anatomy and physiology (a minute subdivision of bioscience) provides such insights in relation to your own body. Although some experience in scientific studies is helpful when beginning a study of anatomy and physiology, perhaps the single most important prerequisite is curiosity.

Gaining an understanding of science is a little like becoming acquainted with another person. Even though a written description can provide a good deal of information about the person, you can never really know another unless there is personal contact. And so it is with science—if you are to know it well, you must deal with it intimately.

The laboratory is the setting for "intimate contact" with science. It is where scientists test their ideas (do research), the essential purpose of which is to provide a basis from which predictions about scientific phenomena can be made. Likewise, it will be the site of your "intimate contact" with the subject of human anatomy and physiology as you are introduced to the methods and instruments used in biological research.

For many students, human anatomy and physiology is taken as an introductory-level course; and their scientific background exists, at best, as a dim memory. If this is your predicament, this prologue may be just what you need to fill in a few gaps and to get you started on the right track before your actual laboratory experiences begin. So—let's get to it!

The Scientific Method

Science would quickly stagnate if new knowledge were not continually derived from and added to it. The approach commonly used by scientists when they investigate various aspects of their respective disciplines is called the **scientific method.** This method is *not* a single rigorous technique that must be followed in a lockstep manner. It is nothing more or less than a logical, practical, and reliable way of approaching and solving problems of every kind—scientific or otherwise—to gain knowledge. It includes five major steps.

Step 1: Observation of Phenomena

The crucial first step involves observation of some phenomenon of interest. In other words, before a scientist can investigate anything, he or she must decide on a *problem* or focus for the investigation. In most college laboratory experiments, the problem or focus has been decided for you. However, to illustrate this important step, we will assume that you want to investigate the true nature of apples, particularly green apples. In such a case you would begin your studies by making a number of different observations concerning apples.

Step 2: Statement of the Hypothesis

Once you have decided on a focus of concern, the next step is to design a significant question to be answered. Such a question is usually posed in the form of a **hypothesis,** an unproven conclusion that attempts to explain some phenomenon. (At its crudest level, a hypothesis can be considered to be a "guess" or an intuitive hunch that tentatively explains some observation.) Generally, scientists do not restrict themselves to a single hypothesis; instead, they usually pose several and then test each one systematically.

We will assume that, to accomplish step 1, you go to the supermarket and randomly select apples from several bins. When you later eat the apples, you find that the green apples are sour, but the red and yellow apples are sweet. From this observation, you might conclude *(hypothesize)* that "green apples are sour." This statement would represent your current understanding of green apples. You might also reasonably predict that if you were to buy more apples, any green ones you buy will be sour. Thus, you would have gone beyond your initial observation that "these" green apples are sour to the prediction that "all" green apples are sour.

Any good hypothesis must meet several criteria. First, *it must be testable.* This characteristic is far more important than its being correct. The test data may or may not support the hypothesis, or new information may require that the hypothesis be modified. Clearly the accuracy of a prediction in any scientific study depends on the accuracy of the initial information on which it is based.

In our example, no great harm will come from an inaccurate prediction—that is, were we to find that some green apples are sweet. However, in some cases human life may depend on the accuracy of the prediction. For that and other reasons: (1) Repeated testing of scientific ideas is important, particularly because scientists working on the same problem do not always agree in their conclusions. (2) Careful observation is essential, even at the very outset of a study, because conclusions drawn from scientific tests are only as accurate as the information on which they are based.

A second criterion is that, even though hypotheses are guesses of a sort, *they must be based on measurable, describable facts. No mysticism can be theorized.* We cannot conjure up, to support our hypothesis, forces that have not been shown to exist. For example, as scientists, we cannot say that the tooth fairy took Johnny's tooth unless we can prove that the tooth fairy exists!

Third, a hypothesis *must not be anthropomorphic.* Human beings tend to anthropomorphize—that is, to relate all experiences to human experience. Whereas we could state that bears instinctively protect their young, it would be anthropomorphic to say that bears love their young, because love is a human emotional response. Thus, the initial hypothesis must be stated without interpretation.

Step 3: Data Collection

Once the initial hypothesis has been stated, scientists plan experiments that will provide data (or evidence) to support or disprove their hypotheses—that is, they *test* their hypotheses. Data are accumulated by making qualitative or quantitative observations of some sort. The observations are often aided by the use of various types of equipment such as cameras, microscopes, stimulators, or various electronic devices that allow chemical and physiological measurements to be taken.

Observations referred to as **qualitative** are those we can make with our senses—that is, by using our vision, hearing, or sense of taste, smell, or touch. For some quick practice in qualitative observation, compare and contrast an orange and an apple. (*Compare* means to emphasize the similarities between two things, whereas *contrast* means that the differences are to be emphasized.)

Whereas the differences between an apple and an orange are obvious, this is not always the case in biological observations. Quite often a scientist tries to detect very subtle differences that cannot be determined by qualitative observations; data must be derived from measurements. Such observations based on precise measurements of one type or another are **quantitative observations**. Examples of quantitative observations include careful measurements of body or organ dimensions such as mass, size, and volume; measurement of volumes of oxygen consumed during metabolic studies; determination of the concentration of glucose in urine; and determination of the differences in blood pressure and pulse under conditions of rest and exercise. An apple and an orange could be compared quantitatively by analyzing the relative amounts of sugar and water in a given volume of fruit flesh, the pigments and vitamins present in the apple skin and orange peel, and so on.

A valuable part of data gathering is the use of experiments to support or disprove a hypothesis. An **experiment** is a procedure designed to describe the factors in a given situation that affect one another (that is, to discover cause and effect) under certain conditions.

Two general rules govern experimentation. The first of these rules is that the experiment(s) should be conducted in such a manner that every **variable** (any factor that might affect the outcome of the experiment) is under the control of the experimenter. The **independent variables** are manipulated by the experimenter. For example, if the goal is to determine the effect of body temperature on breathing rate, the independent variable is body temperature. The effect observed or value measured (in this case breathing rate) is called the **dependent** or **response variable**. Its value "depends" on the value chosen for the independent variable. The ideal way to perform such an experiment is to set up and run a series of tests that are all identical, except for one specific factor that is varied.

One specimen (or group of specimens) is used as the **control** against which all other experimental samples are compared. The importance of the control sample cannot be overemphasized. The control group provides the "normal standard" against which all other samples are compared relative to the dependent variable. Taking our example one step further, if we wanted to investigate the effects of body temperature (the independent variable) on breathing rate (the dependent variable), we could collect data on the breathing rate of individuals with "normal" body temperature (the

implicit control group), and compare these data to breathing-rate measurements obtained from groups of individuals with higher and lower body temperatures.

The second rule governing experimentation is that valid results require that testing be done on large numbers of subjects. It is essential to understand that it is nearly impossible to control all possible variables in biological tests. Indeed, there is a bit of scientific wisdom that mirrors this truth—that is, that laboratory animals, even in the most rigidly controlled and carefully designed experiments, "will do as they damn well please." Thus, stating that the testing of a drug for its painkilling effects was successful after having tested it on only one postoperative patient would be scientific suicide. Large numbers of patients would have to receive the drug and be monitored for a decrease in postoperative pain before such a statement could have any scientific validity. Then, other researchers would have to be able to uphold those conclusions by running similar experiments. *Repeatability* is an important part of the scientific method and is the primary basis for support or rejection of many hypotheses.

During experimentation and observation, data must be carefully recorded. Usually, such initial, or raw, data are recorded in table form. The table should be labeled to show the variables investigated and the results for each sample. At this point, *accurate recording* of observations is the primary concern. Later, these raw data will be reorganized and manipulated to show more explicitly the outcome of the experimentation.

Some of the observations that you will be asked to make in the anatomy and physiology laboratory will require that a drawing be made. Don't panic! The purpose of making drawings (in addition to providing a record) is to force you to observe things very closely. You need not be an artist (most biological drawings are simple outline drawings), but you do need to be neat and as accurate as possible. It is advisable to use a 4H pencil to do your drawings because it is easily erased and doesn't smudge. Before beginning to draw, you should examine your specimen closely, studying it as though you were going to have to draw it from memory. For example, when looking at cells you should ask yourself questions such as "What is their shape—the relationship of length and width? How are they joined together?" Then decide precisely what you are going to show and how large the drawing must be to show the necessary detail. After making the drawing, add labels in the margins and connect them by straight lines (leader lines) to the structures being named.

Step 4: Manipulation and Analysis of Data

The form of the final data varies, depending on the nature of the data collected. Usually, the final data represent information converted from the original measured values (raw data) to some other form. This may mean that averaging or some other statistical treatment must be applied, or it may require conversions from one kind of units to another. In other cases, graphs may be needed to display the data.

Elementary Treatment of Data

Only very elementary statistical treatment of data is required in this manual. For example, you will be expected to understand and/or compute an average (mean), percentages, and a range.

Two of these statistics, the mean and the range, are useful in describing the *typical* case among a large number of samples evaluated. Let us use a simple example. We will assume that the following heart rates (in beats/min) were recorded during an experiment: 64, 70, 82, 94, 85, 75, 72, 78. If you put these numbers in numerical order, the **range** is easily computed, because the range is the difference between the highest and lowest numbers obtained (highest number minus lowest number). The **mean** is obtained by summing the items and dividing the sum by the number of items. What is the range and the mean for the set of numbers just provided?

1. _____

The word *percent* comes from the Latin meaning "for 100"; thus *percent,* indicated by the percent sign, %, means parts per 100 parts. Thus, if we say that 45% of Americans have type O blood, what we are really saying is that among each group of 100 Americans, 45 (45/100) can be expected to have type O blood. Any ratio can be converted to a percent by multiplying by 100 and adding the percent sign.

$$.25 \times 100 = 25\% \qquad 5 \times 100 = 500\%$$

It is very easy to convert any number (including decimals) to a percent. The rule is to move the decimal point two places to the right and add the percent sign. If no decimal point appears, it is *assumed* to be at the end of the number; and zeros are added to fill any empty spaces. Two examples follow:

$$0.25 = 0.25 = 25\%$$
$$5 = 5 = 500\%$$

Change the following to percents:

2. 38 = _____   **4.** 1.6 = _____

3. .75 = _____

Note that although you are being asked here to convert numbers to percents, percents by themselves are meaningless. We always speak in terms of a percentage *of* something.

To change a percent to decimal form, remove the percent sign, and divide by 100. Change the following percents to whole numbers or decimals:

5. 800% = _____   **6.** 0.05% = _____

Making and Reading Line Graphs

For some laboratory experiments you will be required to show your data (or part of them) graphically. Simple line graphs allow relationships within the data to be shown interestingly and allow trends (or patterns) in the data to be demonstrated. An advantage of properly drawn graphs is that they save the reader's time because the essential meaning of a large amount of statistical data can be seen at a glance.

To aid in making accurate graphs, graph paper (or a printed grid in the manual) is used. Line graphs have both horizontal (X) and vertical (Y) axes with scales. Each scale should have uniform intervals—that is, each unit measured on the scale should require the same distance along the scale as any other. Variations from this rule may be misleading and result in false interpretations of the data. By convention, the condition that is manipulated (the independent variable) in the experimental series is plotted on the X-axis (the horizontal axis); and the value that we then measure (the dependent variable) is plotted on the Y-axis (the vertical axis). To plot the data, a dot or a small x is placed at the precise point where the two variables (measured for each sample) meet; and then a line (this is called the **curve**) is drawn to connect the plotted points.

Sometimes, you will see the curve on a line graph extended beyond the last plotted point. This is (supposedly) done to predict "what comes next." When you see this done, be skeptical. The information provided by such a technique is only slightly more accurate than that provided by a crystal ball! When constructing a graph, be sure to label the X-axis and Y-axis and give the graph a legend **(Figure G.1)**.

To read a line graph, pick any point on the line, and match it with the information directly below on the X-axis and with that directly to the left of it on the Y-axis. The figure below (Figure G.1) is a graph that illustrates the relationship between breaths per minute (respiratory rate) and body temperature. Answer the following questions about this graph:

7. What was the respiratory rate at a body temperature of

96°F? _____

8. Between which two body temperature readings was the

increase in breaths per minute greatest? _____

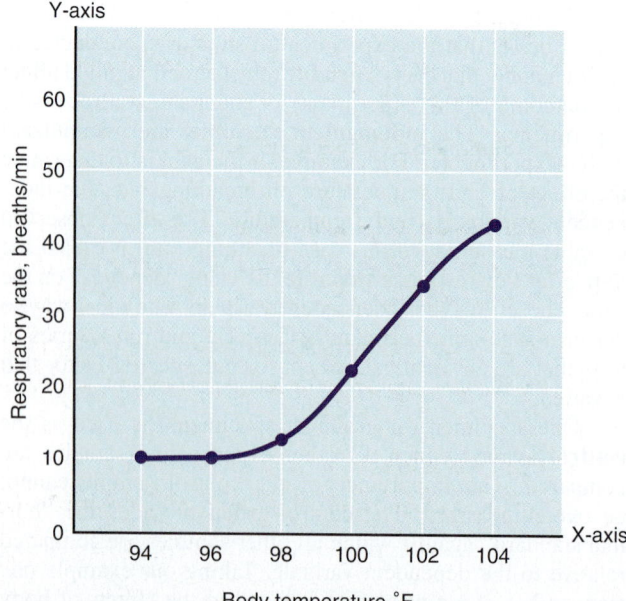

Figure G.1 Example of graphically presented data. Respiratory rate as a function of body temperature.

Step 5: Reporting Conclusions of the Study

Drawings, tables, and graphs alone do not suffice as the final presentation of scientific results. The final step requires that you provide a straightforward description of the conclusions drawn from your results. If possible, your findings should be compared to those of other investigators working on the same

problem. For laboratory investigations conducted by students, these comparative figures are provided by classmates.

It is important to realize that scientific investigations do not always yield the anticipated results. If there are discrepancies between your results and those of others, or what you expected to find based on your class notes or textbook readings, this is the place to try to explain those discrepancies.

Lab Report

Cover Page
• Title of Experiment
• Author's Name
• Course
• Instructor
• Date

Introduction
• Provide background information.
• Describe any relevant observations.
• State hypotheses clearly.

Materials and Methods
• List equipment or supplies needed.
• Provide step-by-step directions for conducting the experiment.

Results
• Present data using a drawing (figure), table, or graph.
• Analyze data.
• Summarize findings briefly.

Discussion and Conclusions
• Conclude whether data gathered support or do not support hypotheses.
• Include relevant information from other sources.
• Explain any uncontrolled variables or unexpected difficulties.
• Make suggestions for further experimentation.

Reference List
• Cite the source of any material used to support this report.

Results are often only as good as the observation techniques used. Depending on the type of experiment conducted, you may need to answer several questions. Did you weigh the specimen carefully enough? Did you balance the scale first? Was the subject's blood pressure actually as high as you recorded it, or did you record it inaccurately? If you did record it accurately, is it possible that the subject was emotionally upset about something, which might have given falsely high data for the variable being investigated? Attempting to explain an unexpected result will often teach you more than you would have learned from anticipated results.

When the experiment produces results that are consistent with the hypothesis, then the hypothesis can be said to have reached a higher level of certainty. The probability that the hypothesis is correct is greater.

A hypothesis that has been validated by many different investigators is called a **theory.** Theories are useful in two important ways. First, they link sets of data; and second, they make predictions that may lead to additional avenues of investigation. (OK, we know this with a high degree of certainty; what's next?)

When a theory has been repeatedly verified and appears to have wide applicability in biology, it may assume the status of a **biological principle.** A principle is a statement that applies with a high degree of probability to a range of events. For example, "Living matter is made of cells or cell products" is a principle stated in many biology texts. It is a sound and useful principle, and will continue to be used as such—unless new findings prove it wrong.

We have been through quite a bit of background concerning the scientific method and what its use entails. Because it is important that you remember the phases of the scientific method, they are summarized here:

1. Observation of some phenomenon

2. Statement of a hypothesis (based on the observations)

3. Collection of data (testing the hypothesis with controlled experiments)

4. Manipulation and analysis of the data

5. Reporting of the conclusions of the study (routinely done by preparing a lab report—see page xvii)

Writing a Lab Report Based on The Scientific Method

A laboratory report is not the same as a scientific paper, but it has some of the same elements and is a formal way to report the results of a scientific experiment. The report should have a cover page that includes the title of the experiment, the author's name, the name of the course, the instructor, and the date. The report should include five separate, clearly marked sections: Introduction, Materials and Methods, Results, Discussion and Conclusions, and References. Use the previous template to guide you through writing a lab report.

Metrics

No matter how highly developed our ability to observe, observations have scientific value only if we can communicate them to others. Without measurement, we would be limited to qualitative description. For precise and repeatable communication of information, the agreed-upon system of measurement used by scientists is the **metric system.**

A major advantage of the metric system is that it is based on units of 10. This allows rapid conversion to workable numbers so that neither very large nor very small figures need be used in calculations. Fractions or multiples of the standard units of length, volume, mass, time, and temperature have been assigned specific names. The metric system **(Table G.1)** shows the commonly used units of the metric system, along with the prefixes used to designate fractions and multiples thereof.

To change from smaller units to larger units, you must *divide* by the appropriate factor of 10 (because there are fewer of the larger units). For example, a milliunit (*milli* = one-thousandth), such as a millimeter, is one step smaller than a centiunit (*centi* = one-hundredth), such as a centimeter. Thus to change milliunits to centiunits, you must divide by 10. On the other hand, when converting from larger units to smaller ones, you must *multiply* by the appropriate factor of 10. A partial scheme for conversions between the metric units is shown on the next page.

Table G.1	Metric System				
A. Commonly used units			**B. Fractions and their multiples**		
Measurement	**Unit**		**Fraction or multiple**	**Prefix**	**Symbol**
Length	Meter (m)		10^6 one million	mega	M
Volume	Liter (L; l with prefix)		10^3 one thousand	kilo	k
Mass	Gram (g)		10^{-1} one-tenth	deci	d
Time*	Second (s)		10^{-2} one-hundredth	centi	c
Temperature	Degree Celsius (°C)		10^{-3} one-thousandth	milli	m
			10^{-6} one-millionth	micro	μ
			10^{-9} one-billionth	nano	n

* The accepted standard for time is the second; and thus hours and minutes are used in scientific, as well as everyday, measurement of time. The only prefixes generally used are those indicating *fractional portions* of seconds—for example, millisecond and microsecond.

$$\text{microunit} \underset{\times 1000}{\overset{\div 1000}{\rightleftharpoons}} \text{milliunit} \underset{\times 10}{\overset{\div 10}{\rightleftharpoons}} \text{centiunit} \underset{\times 100}{\overset{\div 100}{\rightleftharpoons}} \text{unit} \underset{\times 1000}{\overset{\div 1000}{\rightleftharpoons}} \text{kilounit}$$

$$\text{smallest} \rightleftharpoons \text{largest}$$

The objectives of the sections that follow are to provide a brief overview of the most-used measurements in science or health professions and to help you gain some measure of confidence in dealing with them. A listing of the most frequently used conversion factors, for conversions between British and metric system units, is provided in the appendix.

Length Measurements
The metric unit of length is the **meter (m).** Smaller objects are measured in centimeters or millimeters. Subcellular structures are measured in micrometers.

To help you picture these units of length, some equivalents follow:

One meter (m) is slightly longer than one yard (1 m = 39.37 in.).

One centimeter (cm) is approximately the width of a piece of chalk. (Note: There are 2.54 cm in 1 in.)

One millimeter (mm) is approximately the thickness of the wire of a paper clip or of a mark made by a No. 2 pencil lead.

One micrometer (μm) is extremely tiny and can be measured only microscopically.

Make the following conversions between metric units of length:

9. 12 cm = _____ mm

10. 2000 μm = _____ mm

Now, circle the answer that would make the most sense in each of the following statements:

11. A match (in a matchbook) is (0.3, 3, 30) cm long.

12. A standard-size American car is about 4 (mm, cm, m, km) long.

Volume Measurements
The metric unit of volume is the liter. A **liter** (l, or sometimes L, especially without a prefix) is slightly more than a quart (1 L = 1.057 quarts). Liquid volumes measured out for lab experiments are usually measured in milliliters (ml). (The terms *ml* and *cc,* cubic centimeter, are used interchangeably in laboratory and medical settings.)

To help you visualize metric volumes, the equivalents of some common substances follow:

A 12-oz can of soda is a little less than 360 ml.

A fluid ounce is about 30 (it's 29.57) ml (cc).

A teaspoon of vanilla is about 5 ml (cc).

Compute the following:

13. How many 5-ml injections can be prepared from 1 liter

of a medicine? _____

14. A 450-ml volume of alcohol is _____ L.

Mass Measurements
Although many people use the terms *mass* and *weight* interchangeably, this usage is inaccurate. **Mass** is the amount of matter in an object; and an object has a constant mass, regardless of where it is—that is, on earth, or in outer space. However, weight varies with gravitational pull; the greater the gravitational pull, the greater the weight. Thus, our astronauts are said to be weightless when in outer space, but they still have the same mass as they do on earth. (Astronauts are not *really* weightless. It is just that they and their surroundings are being pulled toward the earth at the same speed; and so, in reference to their environment, they appear to float.)

The metric unit of mass is the **gram (g).** Medical dosages are usually prescribed in milligrams (mg) or micrograms (μg); and in the clinical agency, body weight (particularly of infants) is typically specified in kilograms (kg; 1 kg = 2.2 lb).

The following examples are provided to help you become familiar with the masses of some common objects:

Two aspirin tablets have a mass of approximately 1 g.

A nickel has a mass of 5 g.

The mass of an average woman (132 lb) is 60 kg.

Make the following conversions:

15. 300 g = _____ mg = _____ μg

16. 4000 μg = _____ mg = _____ g

17. A nurse must administer to her patient, Mrs. Smith, 5 mg of a drug per kg of body mass. Mrs. Smith weighs 140 lb. How many grams of the drug should the nurse administer to her patient?

_____ g

Temperature Measurements
In the laboratory and in the clinical agency, temperature is measured both in metric units (degrees Celsius, °C) and in British units (degrees Fahrenheit, °F). Thus it helps to be familiar with both temperature scales.

The temperatures of boiling and freezing water can be used to compare the two scales:

The freezing point of water is 0°C and 32°F.

The boiling point of water is 100°C and 212°F.

As you can see, the range from the freezing point to the boiling point of water on the Celsius scale is 100 degrees, whereas the comparable range on the Fahrenheit scale is 180 degrees. Hence, one degree on the Celsius scale represents a greater change in temperature. Normal body temperature is approximately 98.6°F or 37°C.

To convert from the Celsius scale to the Fahrenheit scale, the following equation is used:

$$°C = \frac{5(°F - 32)}{9}$$

To convert from the Fahrenheit scale to the Celsius scale, the following equation is used:

$$°F = (9/5 \, °C) + 32$$

Perform the following temperature conversions:

18. Convert 38°C to °F: _____

19. Convert 158°F to °C:_____

Answers

1. range of 94–64 or 30 beats/min; mean 77.5

2. 3800%

3. 75%

4. 160%

5. 8

6. 0.0005

7. 10 breaths/min

8. interval between 100–102° (went from 22 to 36 breaths/min)

9. 12 cm = 120 mm

10. 2000 μm = 2 mm

11. 3 cm long

12. 4 m long

13. 200

14. 0.45 L

15. 300 g = 3×10^5 mg= 3×10^8 μg

16. 4000 μg = $\underline{4}$ mg = $\underline{4 \times 10^{-3}}$ g (0.004 g)

17. 0.32 g

18. 100.4°F

19. 70°C

The Language of Anatomy

MATERIALS

- ☐ Human torso model (dissectible)
- ☐ Human skeleton
- ☐ Demonstration: sectioned and labeled kidneys [three separate kidneys uncut or cut so that (a) entire, (b) transverse sectional, and (c) longitudinal sectional views are visible]
- ☐ Gelatin-spaghetti molds
- ☐ Scalpel

OBJECTIVES

1. Describe the anatomical position, and explain its importance.
2. Use proper anatomical terminology to describe body regions, orientation and direction, and body planes.
3. Name the body cavities and indicate the important organs in each.
4. Name and describe the serous membranes of the ventral body cavities.
5. Identify the abdominopelvic quadrants and regions on a torso model or image.

PRE-LAB QUIZ

1. Circle True or False. In anatomical position, the body is lying down.
2. Circle the correct underlined term. With regard to surface anatomy, <u>abdominal</u> / <u>axial</u> refers to the structures along the center line of the body.
3. The term *superficial* refers to a structure that is:
   a. attached near the trunk of the body
   b. toward or at the body surface
   c. toward the head
   d. toward the midline
4. The _____ plane runs longitudinally and divides the body into right and left sides.
   a. frontal   c. transverse
   b. sagittal   d. ventral
5. Circle the correct underlined terms. The dorsal body cavity can be divided into the <u>cranial</u> / <u>thoracic</u> cavity, which contains the brain, and the <u>sural</u> / <u>vertebral</u> cavity, which contains the spinal cord.

MasteringA&P® For related exercise study tools, go to the Study Area of MasteringA&P. There you will find:
- Practice Anatomy Lab PAL
- PhysioEx PEx
- A&PFlix A&PFlix
- Practice quizzes, Histology Atlas, eText, Videos, and more!

Most of us are naturally curious about our bodies. This fact is demonstrated by infants, who are fascinated with their own waving hands or their mother's nose. Unlike the infant, however, the student of anatomy must learn to observe and identify the dissectible body structures formally.

A student new to any science is often overwhelmed at first by jargon used in that subject. The study of anatomy is no exception. But without this specialized terminology, confusion is inevitable. For example, what do *over, on top of, superficial to, above,* and *behind* mean in reference to the human body? Anatomists have an accepted set of reference terms that are universally understood. These allow body structures to be located and identified precisely with a minimum of words.

This exercise presents some of the most important anatomical terminology used to describe the body and introduces you to basic concepts of **gross anatomy,** the study of body structures visible to the naked eye.

Anatomical Position

When anatomists or doctors refer to specific areas of the human body, the picture they keep in mind is a universally accepted standard position called the **anatomical position.** It is essential to understand this position because much of the body terminology used in this book refers to this body positioning, regardless of the position the body happens to be in. In the anatomical position the human body is erect, with the feet only slightly apart, head and toes pointed forward, and arms hanging at the sides with palms facing forward **(Figure 1.1a)**.

☐ Assume the anatomical position, and notice that it is not particularly comfortable. The hands are held unnaturally forward rather than hanging with palms toward the thighs.

Check the box when you have completed this task.

Surface Anatomy

Body surfaces provide a wealth of visible landmarks for study. There are two major divisions of the body:

Axial: Relating to head, neck, and trunk, the axis of the body

Appendicular: Relating to limbs and their attachments to the axis

Anterior Body Landmarks

Note the following regions (Figure 1.1a):

Abdominal: Anterior body trunk region inferior to the ribs

Acromial: Point of the shoulder

Antebrachial: Forearm

Antecubital: Anterior surface of the elbow

Axillary: Armpit

Brachial: Arm

Buccal: Cheek

Carpal: Wrist

Cephalic: Head

Cervical: Neck region

Coxal: Hip

Crural: Leg

Digital: Fingers or toes

Femoral: Thigh

Fibular (peroneal): Side of the leg

Frontal: Forehead

Hallux: Great toe

Inguinal: Groin area

Mammary: Breast region

Manus: Hand

Mental: Chin

Nasal: Nose

Oral: Mouth

Orbital: Bony eye socket (orbit)

Palmar: Palm of the hand

Patellar: Anterior knee (kneecap) region

Pedal: Foot

Pelvic: Pelvis region

Pollex: Thumb

Pubic: Genital region

Sternal: Region of the breastbone

Tarsal: Ankle

Thoracic: Chest

Umbilical: Navel

Posterior Body Landmarks

Note the following body surface regions (Figure 1.1b):

Acromial: Point of the shoulder

Brachial: Arm

Calcaneal: Heel of the foot

Cephalic: Head

Dorsum: Back

Femoral: Thigh

Gluteal: Buttocks or rump

Lumbar: Area of the back between the ribs and hips; the loin

Manus: Hand

Occipital: Posterior aspect of the head or base of the skull

Olecranal: Posterior aspect of the elbow

Otic: Ear

Pedal: Foot

Perineal: Region between the anus and external genitalia

Plantar: Sole of the foot

Popliteal: Back of the knee

Sacral: Region between the hips (overlying the sacrum)

Scapular: Scapula or shoulder blade area

Sural: Calf or posterior surface of the leg

Vertebral: Area of the spinal column

ACTIVITY 1

Locating Body Regions

Locate the anterior and posterior body landmarks on yourself, your lab partner, and a human torso model before continuing. ■

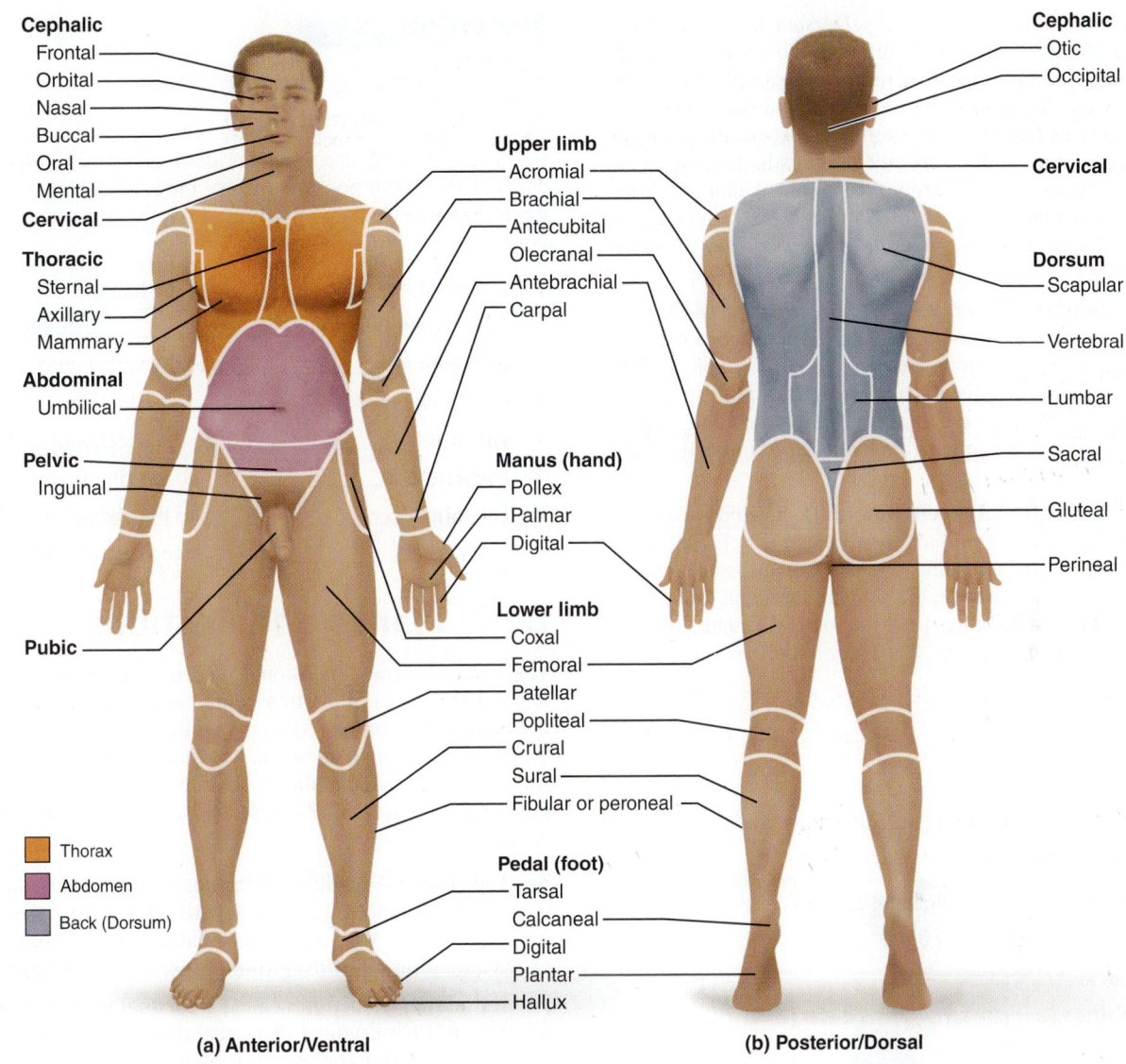

Cephalic
- Frontal
- Orbital
- Nasal
- Buccal
- Oral
- Mental

Cervical

Thoracic
- Sternal
- Axillary
- Mammary

Abdominal
- Umbilical

Pelvic
- Inguinal

Pubic

Upper limb
- Acromial
- Brachial
- Antecubital
- Olecranal
- Antebrachial
- Carpal

Manus (hand)
- Pollex
- Palmar
- Digital

Lower limb
- Coxal
- Femoral
- Patellar
- Popliteal
- Crural
- Sural
- Fibular or peroneal

Pedal (foot)
- Tarsal
- Calcaneal
- Digital
- Plantar
- Hallux

Cephalic
- Otic
- Occipital

Cervical

Dorsum
- Scapular
- Vertebral
- Lumbar
- Sacral
- Gluteal
- Perineal

Thorax
Abdomen
Back (Dorsum)

(a) Anterior/Ventral   **(b) Posterior/Dorsal**

Figure 1.1 Surface anatomy. (a) Anatomical position. **(b)** Heels are raised to illustrate the plantar surface of the foot.

Body Orientation and Direction

Study the terms below (see **Figure 1.2** for a visual aid). Notice that certain terms have a different meaning for a four-legged animal (quadruped) than they do for a human (biped).

Superior/inferior *(above/below):* These terms refer to placement of a structure along the long axis of the body. Superior structures always appear above other structures, and inferior structures are always below other structures. For example, the nose is superior to the mouth, and the abdomen is inferior to the chest.

Anterior/posterior *(front/back):* In humans the most anterior structures are those that are most forward—the face, chest, and abdomen. Posterior structures are those toward the backside of the body. For instance, the spine is posterior to the heart.

Medial/lateral *(toward the midline/away from the midline or median plane):* The sternum (breastbone) is medial to the ribs; the ear is lateral to the nose.

The terms of position just described assume the person is in the anatomical position. The next four term pairs are more absolute. They apply in any body position, and they consistently have the same meaning in all vertebrate animals.

Cephalad (cranial)/caudal *(toward the head/toward the tail):* In humans these terms are used interchangeably with *superior* and *inferior,* but in four-legged animals they are synonymous with *anterior* and *posterior,* respectively.

Dorsal/ventral *(backside/belly side):* These terms are used chiefly in discussing the comparative anatomy of animals, assuming the animal is standing. *Dorsum* is a Latin word meaning "back." Thus, *dorsal* refers to the animal's back or the *back*side of any other structures; for example, the posterior surface of the human leg is its dorsal surface. The term *ventral* derives from the Latin term *venter,* meaning "belly," and always refers to the belly side of animals. In humans the terms *ventral* and *dorsal* are used interchangeably with the terms *anterior* and *posterior,* but in four-legged animals *ventral* and *dorsal* are synonymous with *inferior* and *superior,* respectively.

Proximal/distal *(nearer the trunk or attached end/farther from the trunk or point of attachment):* These terms are used primarily to locate various areas of the body limbs. For example, the fingers are distal to the elbow; the knee is proximal to the toes. However, these terms may also be used to indicate regions (closer to or farther from the head) of internal tubular organs.

Superficial (external)/deep (internal) *(toward or at the body surface/away from the body surface):* These terms locate body organs according to their relative closeness to the body surface. For example, the skin is superficial to the skeletal muscles, and the lungs are deep to the rib cage.

Practicing Using Correct Anatomical Terminology

Before continuing, use a human torso model, a human skeleton, or your own body to specify the relationship between the following structures when the body is in the anatomical position.

1. The wrist is _____ to the hand.
2. The trachea (windpipe) is _____ to the spine.
3. The brain is _____ to the spinal cord.
4. The kidneys are _____ to the liver.
5. The nose is _____ to the cheekbones.
6. The thumb is _____ to the ring finger.
7. The thorax is _____ to the abdomen.
8. The skin is _____ to the skeleton. ▪

Body Planes and Sections

The body is three-dimensional, and in order to observe its internal structures, it is often helpful and necessary to make use of a **section,** or cut. When the section is made through the body wall or through an organ, it is made along an imaginary surface or line called a **plane.** Anatomists commonly refer to three planes **(Figure 1.3)**, or sections, that lie at right angles to one another.

Sagittal plane: A sagittal plane runs longitudinally and divides the body into right and left parts. If it divides the body into equal parts, right down the midline of the body, it is called a **median,** or **midsagittal, plane.**

Frontal plane: Sometimes called a **coronal plane,** the frontal plane is a longitudinal plane that divides the body (or an organ) into anterior and posterior parts.

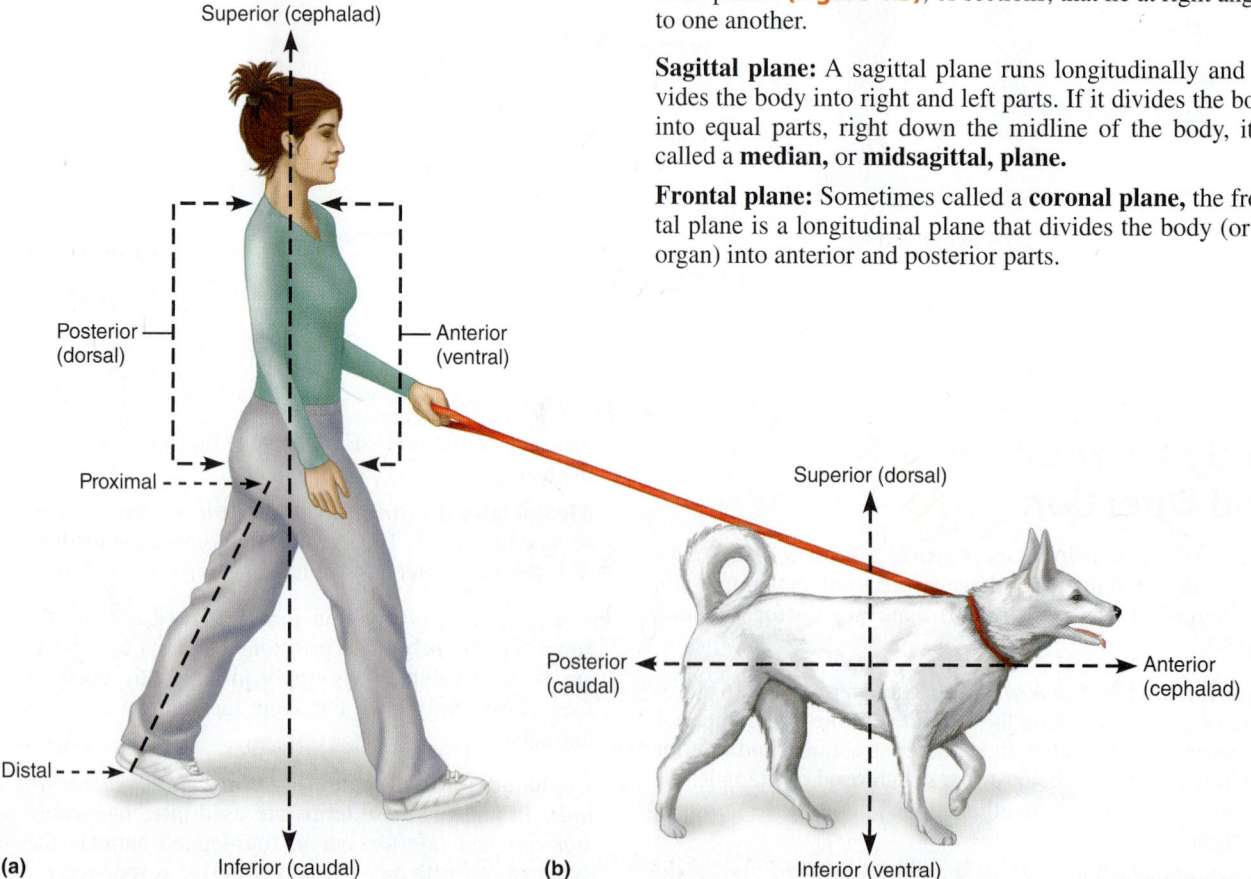

Figure 1.2 Anatomical terminology describing body orientation and direction.
(a) With reference to a human. **(b)** With reference to a four-legged animal.

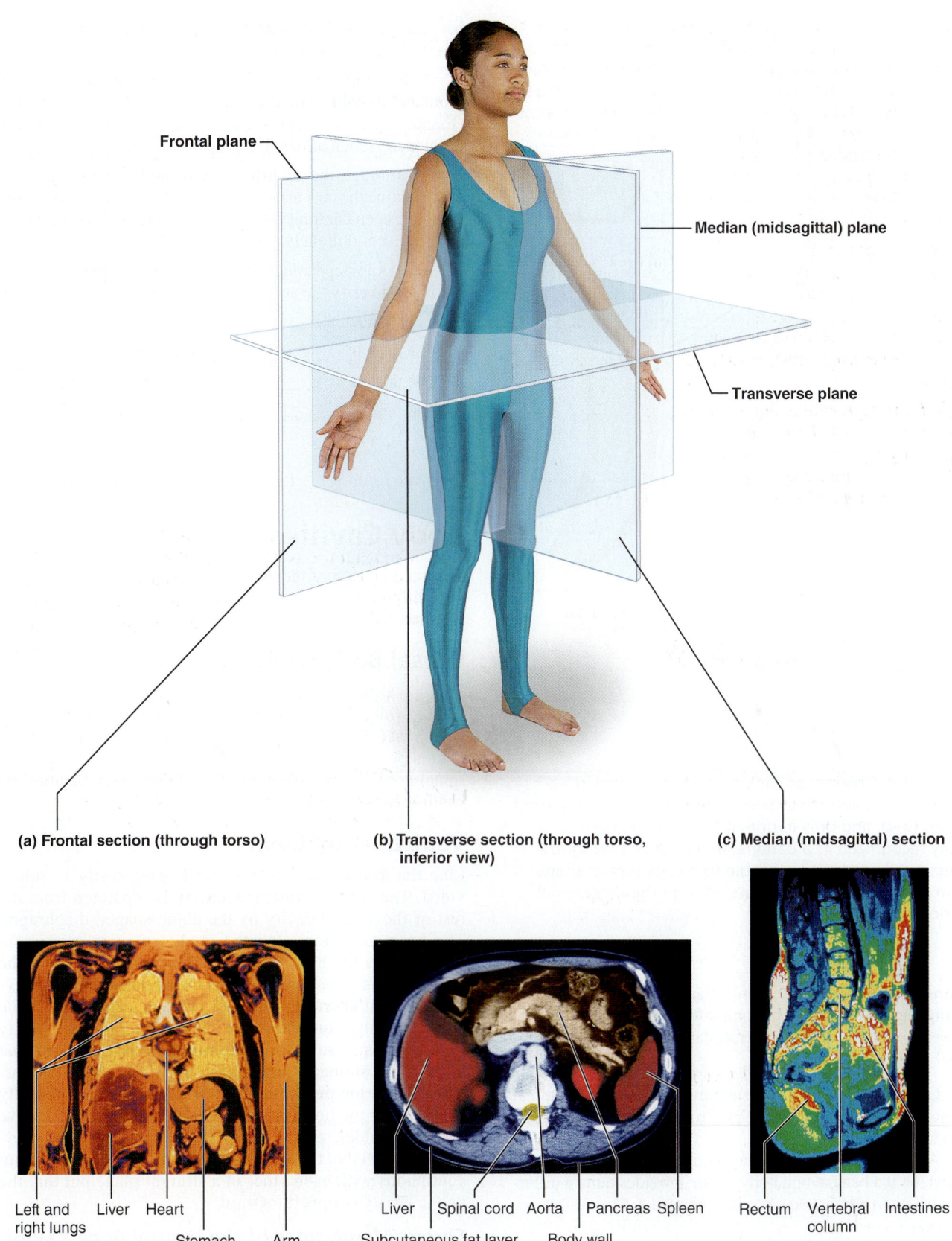

Frontal plane

Median (midsagittal) plane

Transverse plane

(a) Frontal section (through torso)

(b) Transverse section (through torso, inferior view)

(c) Median (midsagittal) section

Left and right lungs   Liver   Heart   Stomach   Arm

Liver   Spinal cord   Aorta   Pancreas   Spleen   Subcutaneous fat layer   Body wall

Rectum   Vertebral column   Intestines

Figure 1.3 Planes of the body with corresponding magnetic resonance imaging (MRI) scans.

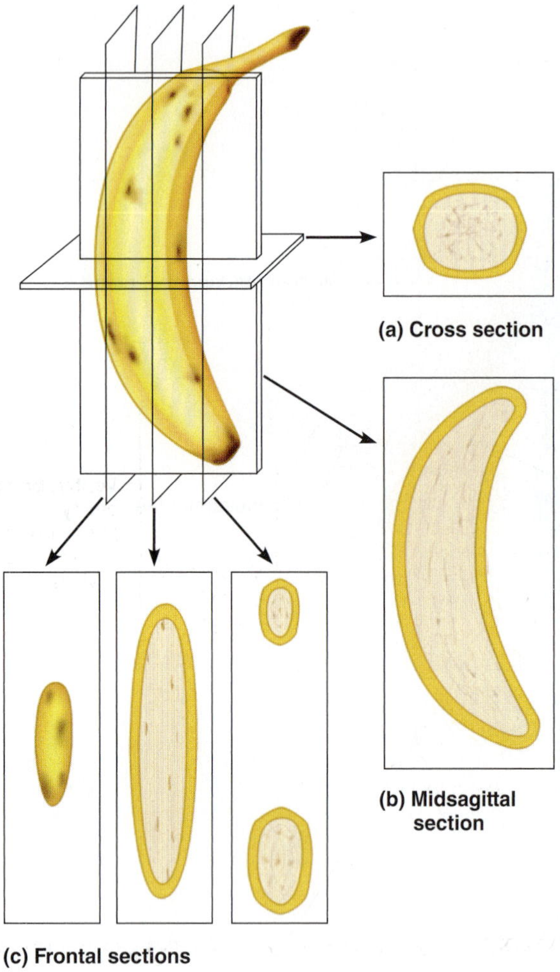

(a) Cross section

(b) Midsagittal section

(c) Frontal sections

Figure 1.4 Objects can look odd when viewed in section. This banana has been sectioned in three different planes **(a–c)**, and only in one of these planes **(b)** is it easily recognized as a banana. If one cannot recognize a sectioned organ, it is possible to reconstruct its shape from a series of successive cuts, as from the three serial sections in **(c)**.

Transverse plane: A transverse plane runs horizontally, dividing the body into superior and inferior parts. When organs are sectioned along the transverse plane, the sections are commonly called **cross sections.**

On microscope slides, the abbreviation for a longitudinal section (sagittal or frontal) is l.s. Cross sections are abbreviated x.s. or c.s.

A sagittal or frontal plane section of any nonspherical object, be it a banana or a body organ, provides quite a different view than a transverse section **(Figure 1.4)**.

ACTIVITY 3

Observing Sectioned Specimens

1. Go to the demonstration area and observe the transversely and longitudinally cut organ specimens (kidneys). Pay close attention to the different structural details in the samples

because you will need to draw these views in the Review Sheet at the end of this exercise.

2. After completing instruction 1, obtain a gelatin-spaghetti mold and a scalpel and bring them to your laboratory bench. (Essentially, this is just cooked spaghetti added to warm gelatin, which is then allowed to gel.)

3. Cut through the gelatin-spaghetti mold along any plane, and examine the cut surfaces. You should see spaghetti strands that have been cut transversely (x.s.), some cut longitudinally, and some cut obliquely.

4. Draw the appearance of each of these spaghetti sections below, and verify the accuracy of your section identifications with your instructor.

Transverse cut   Longitudinal cut   Oblique cut ■■

Body Cavities

The axial portion of the body has two large cavities that provide different degrees of protection to the organs within them **(Figure 1.5)**.

Dorsal Body Cavity

The dorsal body cavity can be subdivided into the **cranial cavity,** which contains the brain within the rigid skull, and the **vertebral** (or **spinal**) **cavity,** within which the delicate spinal cord is protected by the bony vertebral column. Because the spinal cord is a continuation of the brain, these cavities are continuous with each other.

Ventral Body Cavity

Like the dorsal cavity, the ventral body cavity is subdivided. The superior **thoracic cavity** is separated from the rest of the ventral cavity by the dome-shaped diaphragm. The heart and lungs, located in the thoracic cavity, are protected by the bony rib cage. The cavity inferior to the diaphragm is often referred to as the **abdominopelvic cavity.** Although there is no further physical separation of the ventral cavity, some describe the abdominopelvic cavity as two areas: a superior **abdominal cavity,** the area that houses the stomach, intestines, liver, and other organs, and an inferior **pelvic cavity,** the region that is partially enclosed by the bony pelvis and contains the reproductive organs, bladder, and rectum. Notice in the lateral view (Figure 1.5a) that the abdominal and pelvic cavities are not continuous with each other in a straight plane but that the pelvic cavity is tipped forward.

Serous Membranes of the Ventral Body Cavity

The walls of the ventral body cavity and the outer surfaces of the organs it contains are covered with an exceedingly thin, double-layered membrane called the **serosa,** or **serous membrane.** The part of the membrane lining the cavity walls is referred to as the **parietal serosa,** and it is continuous with a similar membrane, the **visceral serosa,** covering the external

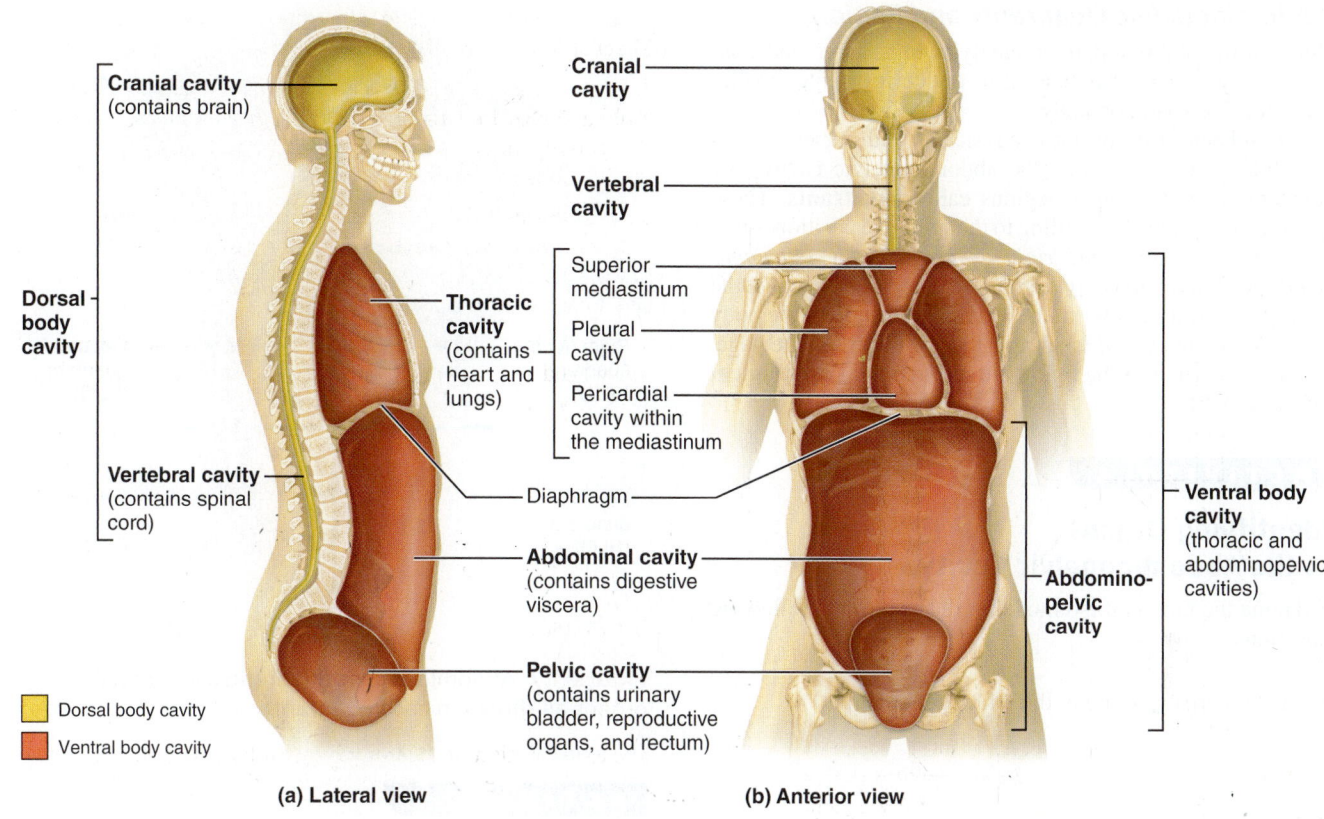

Cranial cavity
(contains brain)

Cranial
cavity

Dorsal
body
cavity

Vertebral
cavity

Superior
mediastinum

Thoracic
cavity
(contains
heart and
lungs)

Pleural
cavity

Pericardial
cavity within
the mediastinum

Vertebral cavity
(contains spinal
cord)

Diaphragm

Abdominal cavity
(contains digestive
viscera)

Ventral body
cavity
(thoracic and
abdominopelvic
cavities)

Abdomino-
pelvic
cavity

Pelvic cavity
(contains urinary
bladder, reproductive
organs, and rectum)

Dorsal body cavity

Ventral body cavity

(a) Lateral view

(b) Anterior view

Figure 1.5 Dorsal and ventral body cavities and their subdivisions.

surface of the organs within the cavity. These membranes produce a thin lubricating fluid that allows the visceral organs to slide over one another or to rub against the body wall with minimal friction. Serous membranes also compartmentalize the various organs so that infection of one organ is prevented from spreading to others.

The specific names of the serous membranes depend on the structures they surround. The serosa lining the abdominal cavity and covering its organs is the **peritoneum,** that enclosing the lungs is the **pleura,** and that around the heart is the **pericardium (Figure 1.6)**.

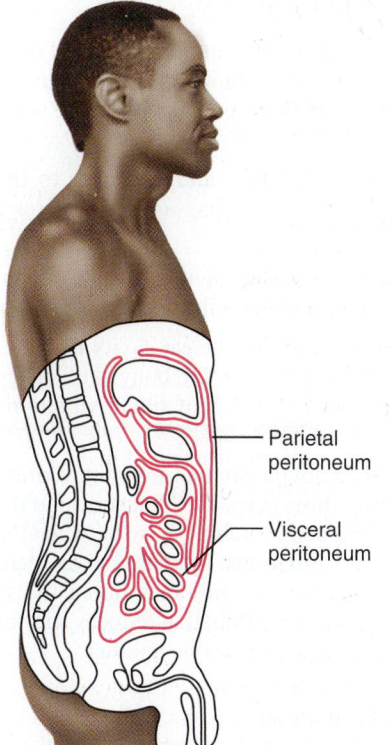

Parietal
peritoneum

Visceral
peritoneum

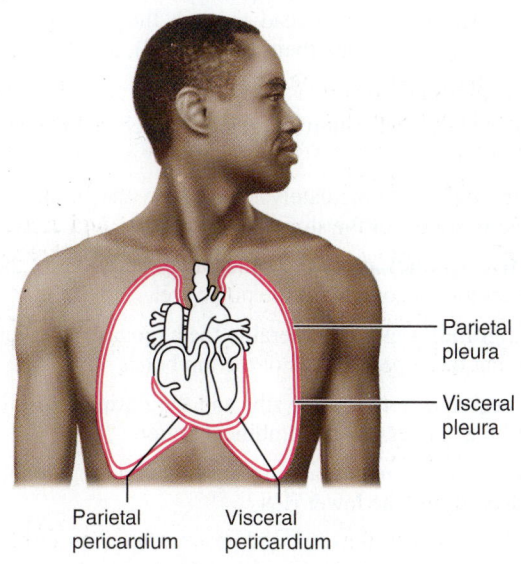

Parietal
pleura

Visceral
pleura

Parietal
pericardium

Visceral
pericardium

Figure 1.6 Serous membranes of the ventral body cavities.

Abdominopelvic Quadrants and Regions

Because the abdominopelvic cavity is quite large and contains many organs, it is helpful to divide it up into smaller areas for discussion or study.

A scheme used by most physicians and nurses divides the abdominal surface and the abdominopelvic cavity into four approximately equal regions called **quadrants**. These quadrants are named according to their relative position—that is, *right upper quadrant, right lower quadrant, left upper quadrant,* and *left lower quadrant* (**Figure 1.7**). Note that the terms left and right refer to the left and right side of the body in the figure, not the left and right side of the art on the page. The left and right of the figure are referred to as **anatomical left and right.**

ACTIVITY 4

Identifying Organs in the Abdominopelvic Cavity

Examine the human torso model to respond to the following questions.

Name two organs found in the left upper quadrant.

_____ and _____

Name two organs found in the right lower quadrant.

_____ and _____

What organ (Figure 1.7) is divided into identical halves by

the median plane? _____ ■

A different scheme commonly used by anatomists divides the abdominal surface and abdominopelvic cavity into nine separate regions by four planes (**Figure 1.8**). Although the names of these nine regions are unfamiliar to you now, with a little patience and study they will become easier to remember. As you read through the descriptions of these nine regions, locate them (Figure 1.8), and note the organs contained in each region.

Umbilical region: The centermost region, which includes the umbilicus (navel)

Epigastric region: Immediately superior to the umbilical region; overlies most of the stomach

Hypogastric (pubic) region: Immediately inferior to the umbilical region; encompasses the pubic area

Iliac, or inguinal, regions: Lateral to the hypogastric region and overlying the superior parts of the hip bones

Lumbar regions: Between the ribs and the flaring portions of the hip bones; lateral to the umbilical region

Hypochondriac regions: Flanking the epigastric region laterally and overlying the lower ribs

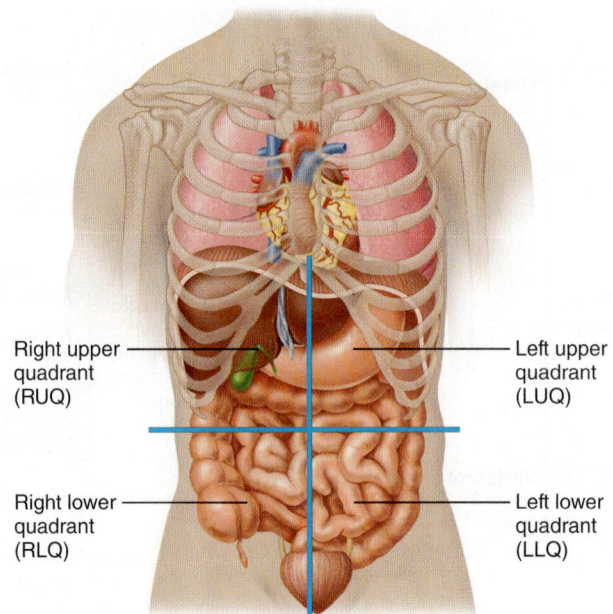

Right upper quadrant (RUQ)

Left upper quadrant (LUQ)

Right lower quadrant (RLQ)

Left lower quadrant (LLQ)

Figure 1.7 Abdominopelvic quadrants. Superficial organs all shown in each quadrant.

ACTIVITY 5

Locating Abdominal Surface Regions

Locate the regions of the abdominal surface on a human torso model and on yourself before continuing. ■

Other Body Cavities

Besides the large, closed body cavities, there are several types of smaller body cavities (**Figure 1.9**). Many of these are in the head, and most open to the body exterior.

Oral cavity: The oral cavity, commonly called the mouth, contains the tongue and teeth. It is continuous with the rest of the digestive tube, which opens to the exterior at the anus.

Nasal cavity: Located within and posterior to the nose, the nasal cavity is part of the passages of the respiratory system.

Orbital cavities: The orbital cavities (orbits) in the skull house the eyes and present them in an anterior position.

Middle ear cavities: Each middle ear cavity lies just medial to an eardrum and is carved into the bony skull. These cavities contain tiny bones that transmit sound vibrations to the hearing receptors in the inner ears.

Synovial cavities: Synovial cavities are joint cavities—they are enclosed within fibrous capsules that surround the freely movable joints of the body, such as those between the vertebrae and the knee and hip joints. Like the serous membranes of the ventral body cavity, membranes lining the synovial cavities secrete a lubricating fluid that reduces friction as the enclosed structures move across one another.

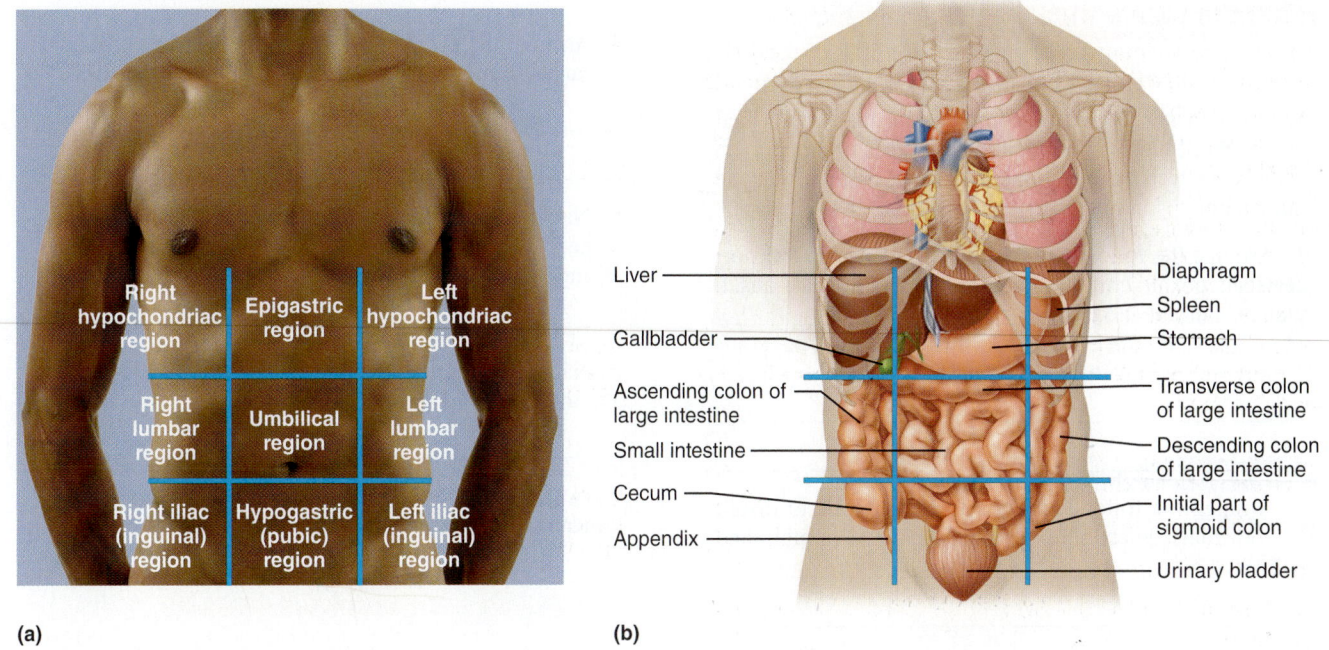

(a)

(b)

Figure 1.8 Abdominopelvic regions. Nine regions delineated by four planes. **(a)** The superior horizontal plane is just inferior to the ribs; the inferior horizontal plane is at the superior aspect of the hip bones. The vertical planes are just medial to the nipples. **(b)** Superficial organs are shown in each region.

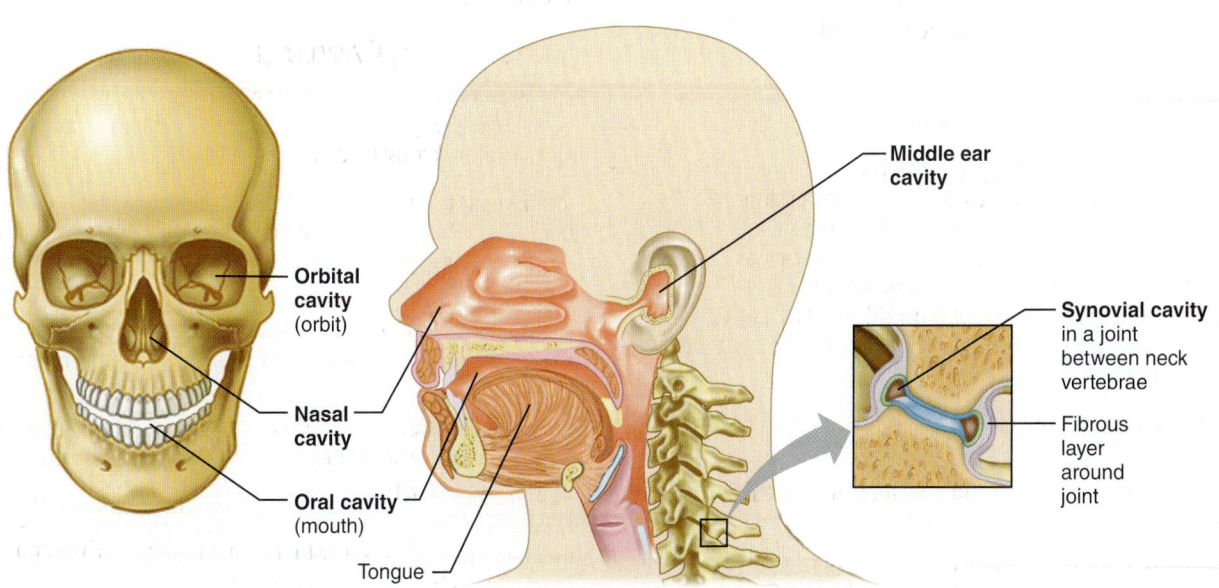

Figure 1.9 Other body cavities. The oral, nasal, orbital, and middle ear cavities are located in the head and open to the body exterior. Synovial cavities are found in joints between many bones such as the vertebrae of the spine, and at the knee, shoulder, and hip.

GROUP CHALLENGE

The Language of Anatomy

Working in small groups, complete the tasks described below. Work together, but don't use a figure or other reference to answer the questions. As usual, assume that the human body is in the anatomical position.

1. Arrange the following terms from superior to inferior: cervical, coxal, crural, femoral, lumbar, mental, nasal, plantar, sternal, and tarsal. _____

_____

_____

_____

2. Arrange the following terms from proximal to distal: antebrachial, antecubital, brachial, carpal, digital, and palmar. _____

_____

3. Arrange the following terms from medial to lateral: acromial, axillary, buccal, otic, pollex, and umbilical.

_____

_____

4. Arrange the following terms from distal to proximal: calcaneal, femoral, hallux, plantar, popliteal, and sural.

_____

_____

5. Name a plane that you could use to section a four-legged chair and still be able to sit in the chair without falling over. _____

_____

6. Name the abdominopelvic region that is both medial and inferior to the right lumbar region.

_____

7. Name the type of inflammation (think "-itis") that is typically accompanied by pain in the lower right quadrant.

_____

Organ Systems Overview

MATERIALS

- ☐ Freshly killed or preserved rat [predissected by instructor as a demonstration or for student dissection (one rat for every two to four students)] or predissected human cadaver
- ☐ Dissection trays
- ☐ Twine or large dissecting pins
- ☐ Scissors
- ☐ Probes
- ☐ Forceps
- ☐ Disposable gloves
- ☐ Human torso model (dissectible)

OBJECTIVES

1. Name the human organ systems and indicate the major functions of each.
2. List several major organs of each system and identify them in a dissected rat, human cadaver or cadaver image, or in a dissectible human torso model.
3. Name the correct organ system for each organ when presented with a list of organs studied in the laboratory.

PRE-LAB QUIZ

1. Name the structural and functional unit of all living things. _____
2. The small intestine is an example of a(n) _____, because it is composed of two or more tissue types that perform a particular function for the body.
   a. epithelial tissue
   b. muscular tissue
   c. organ
   d. organ system
3. The _____ system is responsible for maintaining homeostasis of the body via rapid transmission of electrical signals.
4. The kidneys are part of the _____ system.
5. The thin muscle that separates the thoracic and abdominal cavities is the _____.

MasteringA&P® For related exercise study tools, go to the Study Area of MasteringA&P. There you will find:
- Practice Anatomy Lab PAL
- PhysioEx PEx
- A&PFlix *A&PFlix*
- Practice quizzes, Histology Atlas, eText, Videos, and more!

The basic unit or building block of all living things is the **cell.** Cells fall into four different categories according to their structures and functions. Each of these corresponds to one of the four tissue types: epithelial, muscular, nervous, and connective. A **tissue** is a group of cells that are similar in structure and function. An **organ** is a structure composed of two or more tissue types that performs a specific function for the body. For example, the small intestine, which digests and absorbs nutrients, is made up of all four tissue types.

An **organ system** is a group of organs that act together to perform a particular body function. For example, the organs of the digestive system work together to break down foods and absorb the end products into the bloodstream to provide nutrients and fuel for all the body's cells. In all, there are 11 organ systems (described in **Table 2.1**). The lymphatic system also encompasses a *functional system* called the immune system, which is composed of an army of mobile *cells* that act to protect the body from foreign substances.

Read through this summary of the body's organ systems before beginning your rat dissection or examination of the predissected human cadaver. If a human cadaver is not available, the figures provided in this exercise (Figures 2.3–2.6), will serve as a partial replacement.

Table 2.1	Overview of Organ Systems of the Body	
Organ system	**Major component organs**	**Function**
Integumentary (Skin)	Epidermal and dermal regions; cutaneous sense organs and glands	• Protects deeper organs from mechanical, chemical, and bacterial injury, and drying out • Excretes salts and urea • Aids in regulation of body temperature • Produces vitamin D
Skeletal	Bones, cartilages, tendons, ligaments, and joints	• Body support and protection of internal organs • Provides levers for muscular action • Cavities provide a site for blood cell formation
Muscular	Muscles attached to the skeleton	• Primary function is to contract or shorten; in doing so, skeletal muscles allow locomotion (running, walking, etc.), grasping and manipulation of the environment, and facial expression • Generates heat
Nervous	Brain, spinal cord, nerves, and sensory receptors	• Allows body to detect changes in its internal and external environment and to respond to such information by activating appropriate muscles or glands • Helps maintain homeostasis of the body via rapid transmission of electrical signals
Endocrine	Pituitary, thymus, thyroid, parathyroid, adrenal, and pineal glands; ovaries, testes, and pancreas	• Helps maintain body homeostasis, promotes growth and development; produces chemical messengers called hormones that travel in the blood to exert their effect(s) on various target organs of the body
Cardiovascular	Heart, blood vessels, and blood	• Primarily a transport system that carries blood containing oxygen, carbon dioxide, nutrients, wastes, ions, hormones, and other substances to and from the tissue cells where exchanges are made; blood is propelled through the blood vessels by the pumping action of the heart • Antibodies and other protein molecules in the blood protect the body
Lymphatic/ Immunity	Lymphatic vessels, lymph nodes, spleen, thymus, tonsils, and scattered collections of lymphoid tissue	• Picks up fluid leaked from the blood vessels and returns it to the blood • Cleanses blood of pathogens and other debris • Houses lymphocytes that act via the immune response to protect the body from foreign substances
Respiratory	Nasal passages, pharynx, larynx, trachea, bronchi, and lungs	• Keeps the blood continuously supplied with oxygen while removing carbon dioxide • Contributes to the acid-base balance of the blood via its carbonic acid–bicarbonate buffer system
Digestive	Oral cavity, esophagus, stomach, small and large intestines, and accessory structures including teeth, salivary glands, liver, and pancreas	• Breaks down ingested foods to minute particles, which can be absorbed into the blood for delivery to the body cells • Undigested residue removed from the body as feces
Urinary	Kidneys, ureters, bladder, and urethra	• Rids the body of nitrogen-containing wastes including urea, uric acid, and ammonia, which result from the breakdown of proteins and nucleic acids • Maintains water, electrolyte, and acid-base balance of blood
Reproductive	Male: testes, prostate gland, scrotum, penis, and duct system, which carries sperm to the body exterior	• Provides germ cells called sperm for perpetuation of the species
	Female: ovaries, uterine tubes, uterus, mammary glands, and vagina	• Provides germ cells called eggs; the female uterus houses the developing fetus until birth; mammary glands provide nutrition for the infant

DISSECTION AND IDENTIFICATION: The Organ Systems of the Rat

Many of the external and internal structures of the rat are quite similar in structure and function to those of the human, so a study of the gross anatomy of the rat should help you understand our own physical structure. The following instructions include directions for dissecting and observing a rat. In addition, the descriptions for organ observations (Activity 4, "Examining the Ventral Body Cavity," which begins on page 18) also apply to superficial observations of a previously dissected human cadaver. The general instructions for observing external structures also apply to human cadaver observations. (The photographs in Figures 2.3 to 2.6 will provide visual aids.)

Note that four organ systems (integumentary, skeletal, muscular, and nervous) will not be studied at this time, as they require microscopic study or more detailed dissection. ■

Figure 2.1 Rat dissection: Securing for dissection and the initial incision. (a) Securing the rat to the dissection tray with dissecting pins. **(b)** Using scissors to make the incision on the median line of the abdominal region. **(c)** Completed incision from the pelvic region to the lower jaw. **(d)** Reflection (folding back) of the skin to expose the underlying muscles.

(a)

(b)

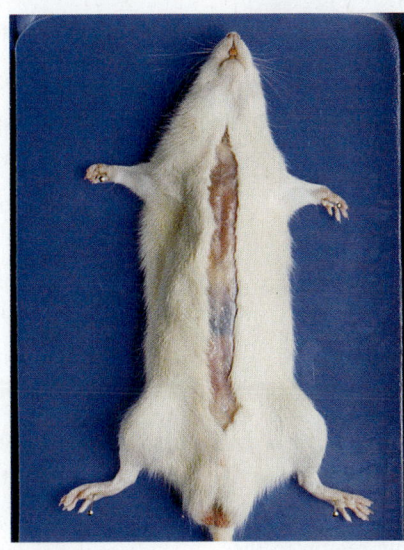

(c)

(d)

ACTIVITY 1

Observing External Structures

1. If your instructor has provided a predissected rat, go to the demonstration area to make your observations. Alternatively, if you and/or members of your group will be dissecting the specimen, obtain a preserved or freshly killed rat, a dissecting tray, dissecting pins or twine, scissors, probe, forceps, and disposable gloves, and bring them to your laboratory bench.

If a predissected human cadaver is available, obtain a probe, forceps, and disposable gloves before going to the demonstration area.

⚠ 2. Don the gloves before beginning your observations. This precaution is particularly important when handling freshly killed animals, which may harbor internal parasites.

3. Observe the major divisions of the body—head, trunk, and extremities. If you are examining a rat, compare these divisions to those of humans. ▪

ACTIVITY 2

Examining the Oral Cavity

Examine the structures of the oral cavity. Identify the teeth and tongue. Observe the extent of the hard palate (the portion underlain by bone) and the soft palate (immediately posterior to the hard palate, with no bony support). Notice that the posterior end of the oral cavity leads into the throat, or pharynx, a passageway used by both the digestive and respiratory systems. ▪

ACTIVITY 3

Opening the Ventral Body Cavity

1. Pin the animal to the wax of the dissecting tray by placing its dorsal side down and securing its extremities to the wax with large dissecting pins **(Figure 2.1a)**.

If the dissecting tray is not waxed, you will need to secure the animal with twine as follows. (Some may prefer

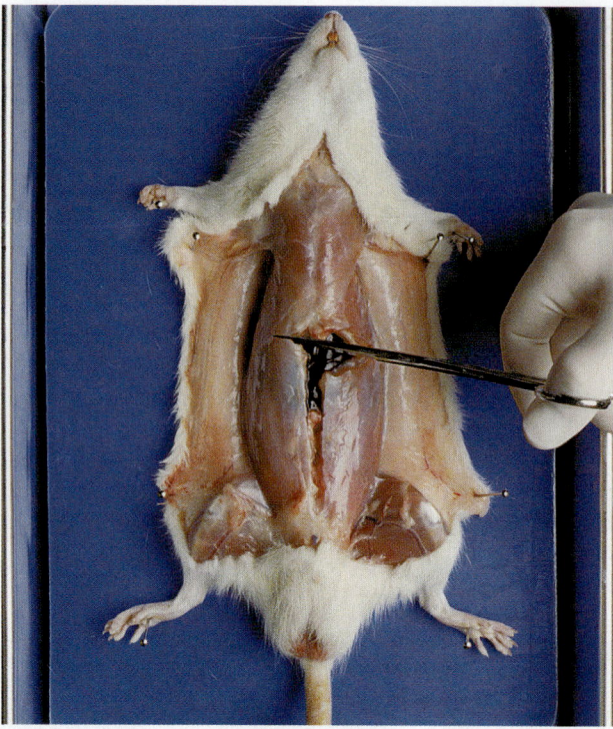

Figure 2.2 Rat dissection: Making lateral cuts at the base of the rib cage.

this method in any case.) Obtain the roll of twine. Make a loop knot around one upper limb, pass the twine under the tray, and secure the opposing limb. Repeat for the lower extremities.

2. Lift the abdominal skin with a forceps, and cut through it with the scissors (Figure 2.1b). Close the scissor blades and insert them flat under the cut skin. Moving in a cephalad direction, open and close the blades to loosen the skin from the underlying connective tissue and muscle. Now, cut the skin along the body midline, from the pubic region to the lower jaw (Figure 2.1c). Finally, make a lateral cut about halfway down the ventral surface of each limb. Complete the job of freeing the skin with the scissor tips, and pin the flaps to the tray (Figure 2.1d). The underlying tissue that is now exposed is the skeletal musculature of the body wall and limbs. It allows voluntary body movement. Notice that the muscles are packaged in sheets of pearly white connective tissue (fascia), which protect the muscles and bind them together.

3.   Carefully cut through the muscles of the abdominal wall in the pubic region, avoiding the underlying organs. Remember, to *dissect* means "to separate"—not mutilate! Now, hold and lift the muscle layer with a forceps and cut through the muscle layer from the pubic region to the bottom of the rib cage. Make two lateral cuts at the base of the rib cage **(Figure 2.2)**. A thin membrane attached to the inferior boundary of the rib cage should be obvious; this is the **diaphragm,** which separates the thoracic and abdominal

cavities. Cut the diaphragm where it attaches to the ventral ribs to loosen the rib cage. Cut through the rib cage on either side. You can now lift the ribs to view the contents of the thoracic cavity. Cut across the flap, at the level of the neck, and remove it. ◼

ACTIVITY 4

Examining the Ventral Body Cavity

1.   Starting with the most superficial structures and working deeper, examine the structures of the thoracic cavity. (Refer to **Figure 2.3,** which shows the superficial organs, as you work.) Choose the appropriate view depending on whether you are examining a rat (a) or a human cadaver (b).

Thymus: An irregular mass of glandular tissue overlying the heart (not illustrated in the human cadaver photograph).

With the probe, push the thymus to the side to view the heart.

Heart: Medial oval structure enclosed within the pericardium (serous membrane sac).

Lungs: Lateral to the heart on either side.

Now observe the throat region to identify the trachea.

Trachea: Tubelike "windpipe" running medially down the throat; part of the respiratory system.

Follow the trachea into the thoracic cavity; notice where it divides into two branches. These are the bronchi.

Bronchi: Two passageways that plunge laterally into the tissue of the two lungs.

To expose the esophagus, push the trachea to one side.

Esophagus: A food chute; the part of the digestive system that transports food from the pharynx (throat) to the stomach.

Diaphragm: A thin muscle attached to the inferior boundary of the rib cage; separates the thoracic and abdominal cavities.

Follow the esophagus through the diaphragm to its junction with the stomach.

Stomach: A curved organ important in food digestion and temporary food storage.

2.   Examine the superficial structures of the abdominopelvic cavity. Lift the **greater omentum,** an extension of the peritoneum that covers the abdominal viscera. Continuing from the stomach, trace the rest of the digestive tract **(Figure 2.4)**.

Small intestine: Connected to the stomach and ending just before the saclike cecum.

Large intestine: A large muscular tube connected to the small intestine and ending at the anus.

Cecum: The initial portion of the large intestine.

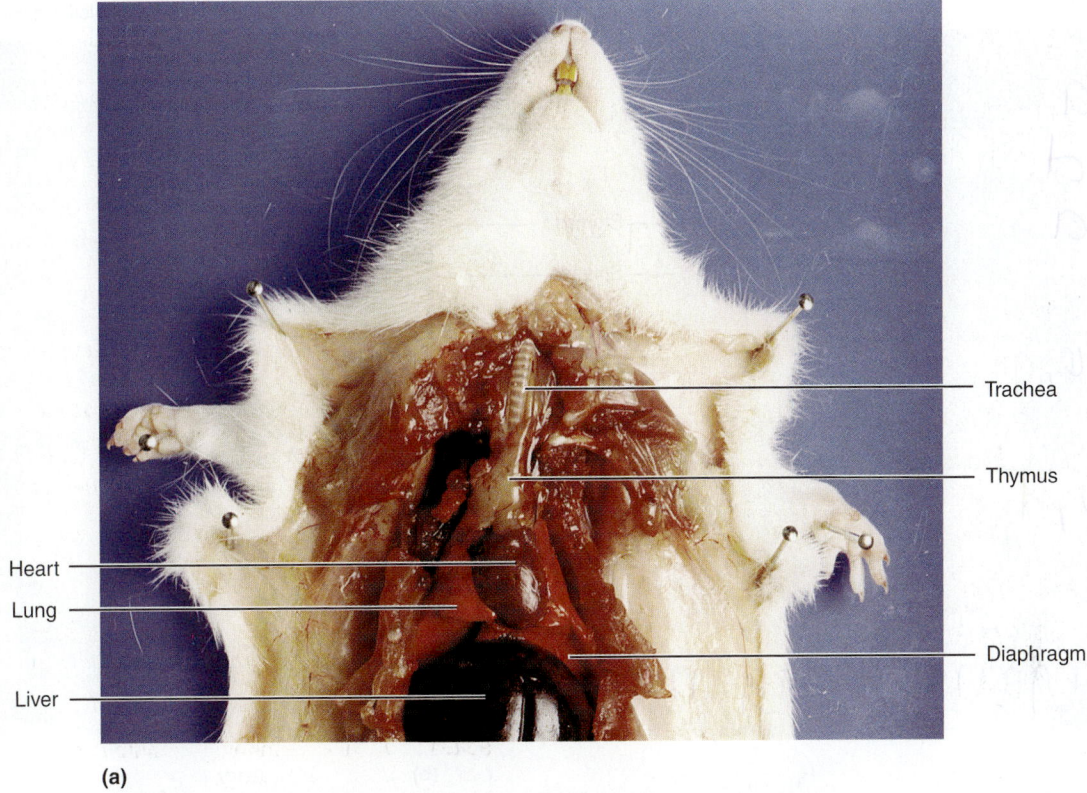

Trachea

Thymus

Heart

Lung

Diaphragm

Liver

(a)

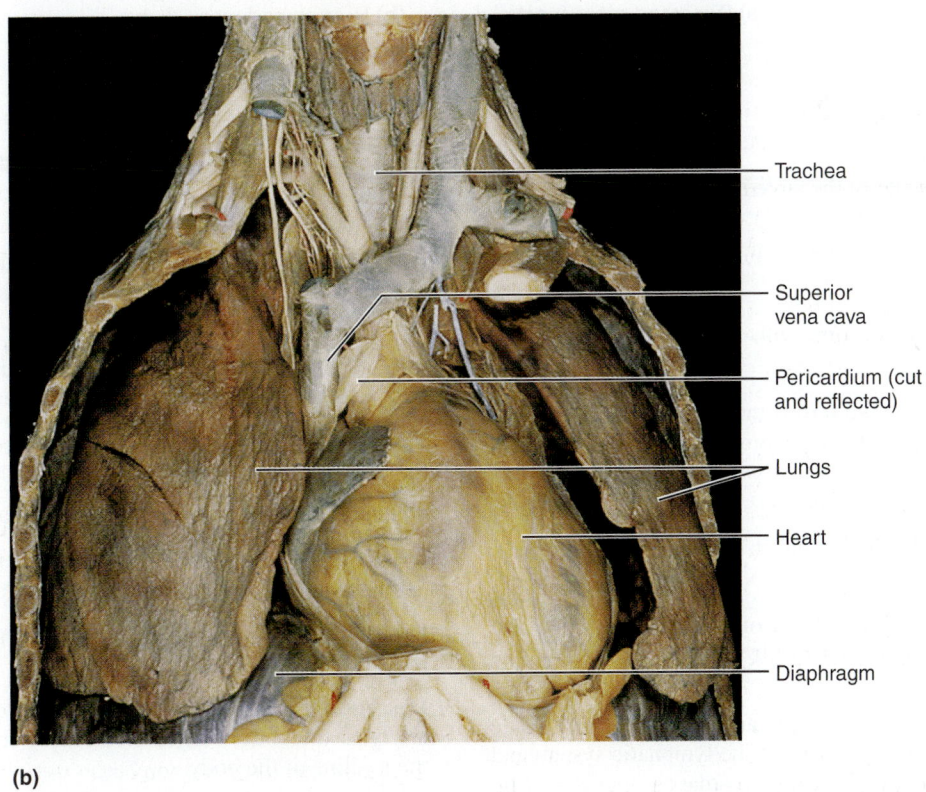

Trachea

Superior
vena cava

Pericardium (cut
and reflected)

Lungs

Heart

Diaphragm

(b)

Figure 2.3 **Superficial organs of the thoracic cavity. (a)** Dissected rat.
(b) Human cadaver.

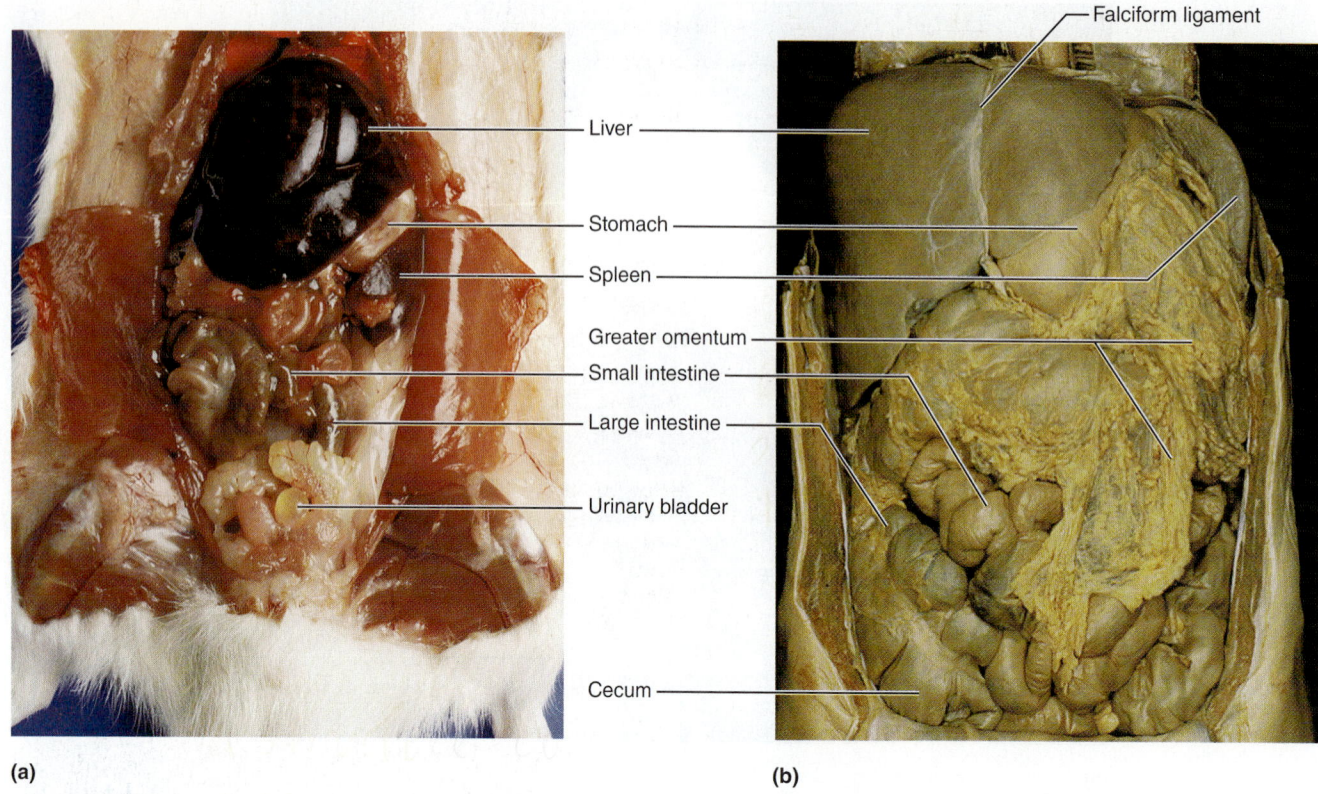

Liver
Stomach
Spleen
Greater omentum
Small intestine
Large intestine
Urinary bladder
Cecum
Falciform ligament

(a)   (b)

Figure 2.4 Abdominal organs. (a) Dissected rat, superficial view. **(b)** Human cadaver, superficial view.

Follow the course of the large intestine to the rectum, which is partially covered by the urinary bladder **(Figure 2.5)**.

Rectum: Terminal part of the large intestine; continuous with the anal canal.

Anus: The opening of the digestive tract (through the anal canal) to the exterior.

Now lift the small intestine with the forceps to view the mesentery.

Mesentery: An apronlike serous membrane; suspends many of the digestive organs in the abdominal cavity. Notice that it is heavily invested with blood vessels and, more likely than not, riddled with large fat deposits.

Locate the remaining abdominal structures.

Pancreas: A diffuse gland; rests dorsal to and in the mesentery between the first portion of the small intestine and the stomach. You will need to lift the stomach to view the pancreas.

Spleen: A dark red organ curving around the left lateral side of the stomach; considered part of the lymphatic system and often called the red blood cell graveyard.

Liver: Large and brownish red; the most superior organ in the abdominal cavity, directly beneath the diaphragm.

3. To locate the deeper structures of the abdominopelvic cavity, move the stomach and the intestines to one side with the probe.

Examine the posterior wall of the abdominal cavity to locate the two kidneys (Figure 2.5).

Kidneys: Bean-shaped organs; retroperitoneal (behind the peritoneum).

Adrenal glands: Large endocrine glands that sit on top of the superior margin of each kidney; considered part of the endocrine system.

Carefully strip away part of the peritoneum with forceps and attempt to follow the course of one of the ureters to the bladder.

Ureter: Tube running from the indented region of a kidney to the urinary bladder.

Urinary bladder: The sac that serves as a reservoir for urine.

4. In the midline of the body cavity lying between the kidneys are the two principal abdominal blood vessels. Identify each.

Inferior vena cava: The large vein that returns blood to the heart from the lower body regions.

Descending aorta: Deep to the inferior vena cava; the largest artery of the body; carries blood away from the heart down the midline of the body.

5. Only a brief examination of reproductive organs will be done. If you are working with a rat, first determine if the animal is a male or female. Observe the ventral body surface beneath the tail. If a saclike scrotum and an opening for the anus are visible, the animal is a male. If three body openings—urethral, vaginal, and anal—are present, it is a female.

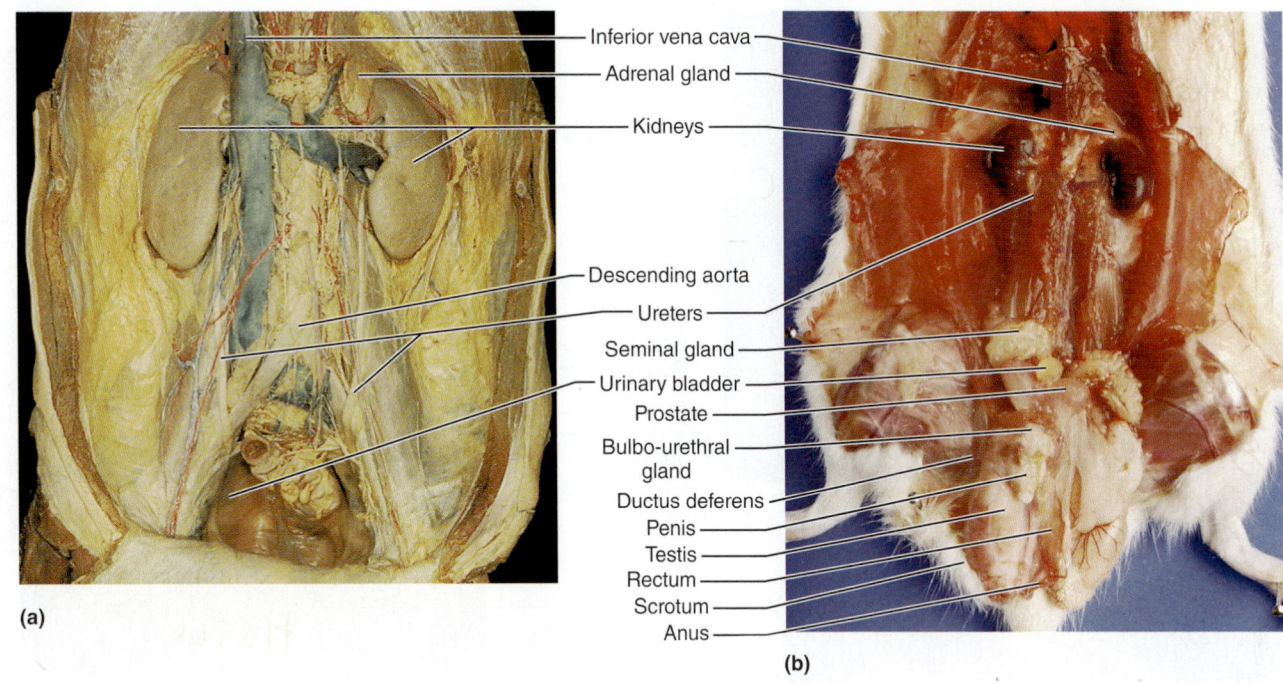

Inferior vena cava
Adrenal gland
Kidneys

Descending aorta
Ureters
Seminal gland
Urinary bladder
Prostate
Bulbo-urethral gland
Ductus deferens
Penis
Testis
Rectum
Scrotum
Anus

(a)

(b)

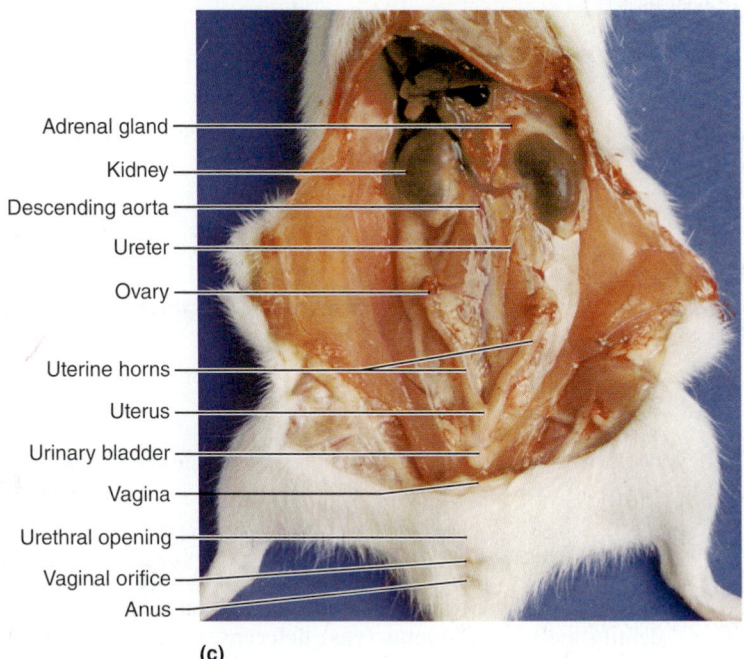

Adrenal gland
Kidney
Descending aorta
Ureter
Ovary

Uterine horns
Uterus
Urinary bladder
Vagina
Urethral opening
Vaginal orifice
Anus

(c)

Figure 2.5 Deep structures of the abdominopelvic cavity. (a) Human cadaver. **(b)** Dissected male rat. (Some reproductive structures also shown.) **(c)** Dissected female rat. (Some reproductive structures also shown.)

Male Animal

Make a shallow incision into the **scrotum.** Loosen and lift out one oval **testis.** Exert a gentle pull on the testis to identify the slender **ductus deferens,** or **vas deferens,** which carries sperm from the testis superiorly into the abdominal cavity and joins with the urethra. The urethra runs through the penis and carries both urine and sperm out of the body. Identify the **penis,** extending from the bladder to the ventral body wall. (Figure 2.5b indicates other glands of the male rat's reproductive system, but they need not be identified at this time.)

Female Animal

Inspect the pelvic cavity to identify the Y-shaped **uterus** lying against the dorsal body wall and superior to the bladder (Figure 2.5c). Follow one of the uterine horns superiorly to identify an **ovary,** a small oval structure at the end of the uterine horn. (The rat uterus is quite different from the uterus of a human female, which is a single-chambered organ about the size and shape of a pear.) The inferior undivided part of the rat uterus is continuous with the **vagina,** which leads to the body exterior. Identify the **vaginal orifice** (external vaginal opening).

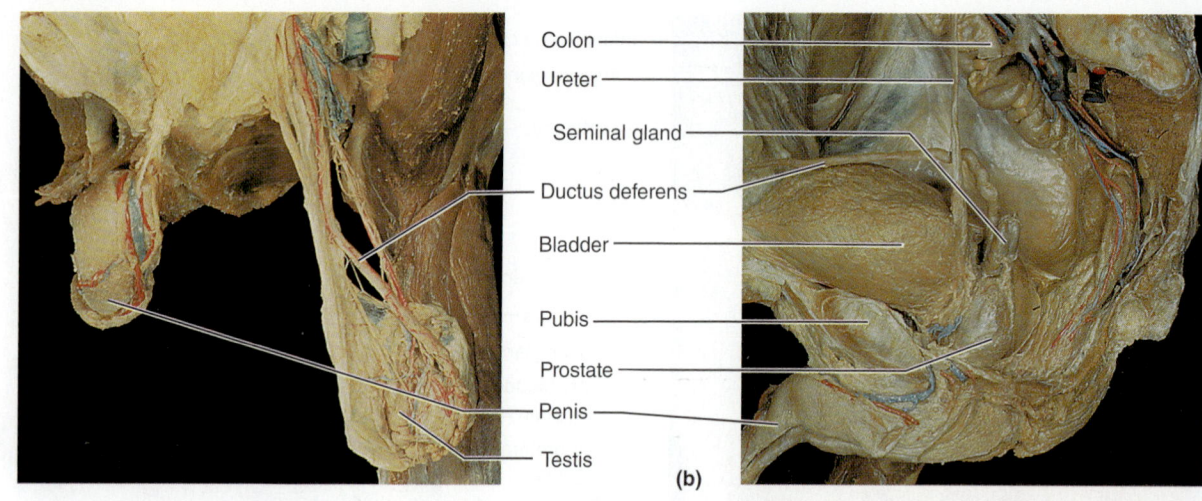

Colon
Ureter
Seminal gland
Ductus deferens
Bladder
Pubis
Prostate
Penis
Testis

(a)   (b)

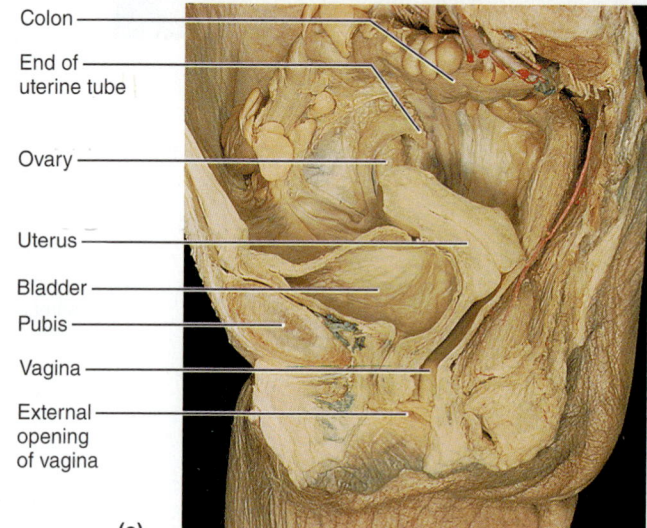

Colon
End of uterine tube
Ovary
Uterus
Bladder
Pubis
Vagina
External opening of vagina

(c)

Figure 2.6 Human reproductive organs. (a) Male external genitalia. **(b)** Sagittal section of the male pelvis. **(c)** Sagittal section of the female pelvis.

If you are working with a human cadaver, proceed as indicated next.

Male Cadaver

Make a shallow incision into the **scrotum (Figure 2.6a).** Loosen and lift out the oval **testis.** Exert a gentle pull on the testis to identify the slender **ductus (vas) deferens,** which carries sperm from the testis superiorly into the abdominal cavity and joins with the urethra (Figure 2.6b). The urethra runs through the penis and carries both urine and sperm out of the body. Identify the **penis,** extending from the bladder to the ventral body wall.

Female Cadaver

Inspect the pelvic cavity to identify the pear-shaped **uterus** lying against the dorsal body wall and superior to the bladder. Follow one of the **uterine tubes** superiorly to identify an **ovary,** a small oval structure at the end of the uterine tube (Figure 2.6c). The inferior part of the uterus is continuous

with the **vagina,** which leads to the body exterior. Identify the **vaginal orifice** (external vaginal opening).

6. When you have finished your observations, rewrap or store the dissection animal or cadaver according to your instructor's directions. Wash the dissecting tools and equipment with laboratory detergent. Dispose of the gloves. Then wash and dry your hands before continuing with the examination of the human torso model. ■

ACTIVITY 5

Examining the Human Torso Model

1. Examine a human torso model to identify the organs listed. Some model organs will have to be removed to see the deeper organs. If a torso model is not available, the photograph of the human torso model **(Figure 2.7)** may be used for this part of the exercise.

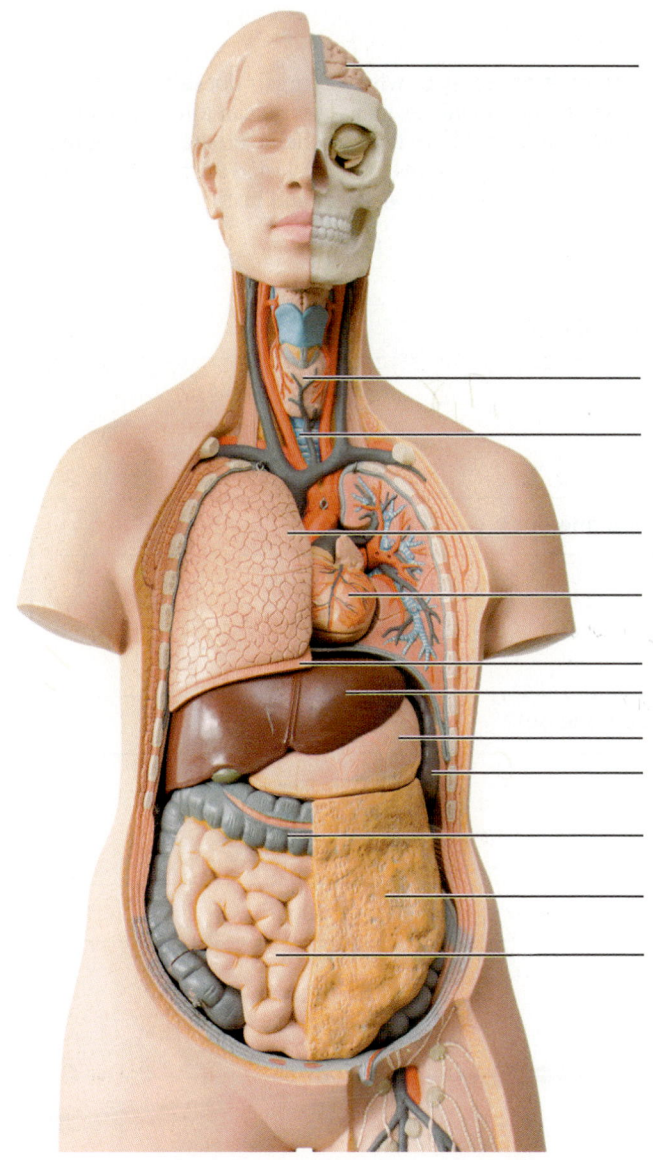

Figure 2.7 **Human torso model.**

Adrenal gland   Lungs
Aortic arch   Mesentery
Brain   Pancreas
Bronchi   Rectum
Descending aorta   Small intestine
Diaphragm   Spinal cord
Esophagus   Spleen
Greater omentum   Stomach
Heart   Thyroid gland
Inferior vena cava   Trachea
Kidneys   Ureters
Large intestine   Urinary bladder
Liver

Abdominopelvic cavity_____

_____

_____

4.   Now, assign each of the organs to one of the organ system categories listed below.

Digestive: _____

_____

Urinary: _____

Cardiovascular: _____

_____

Endocrine: _____

Reproductive: _____

Respiratory: _____

Lymphatic/Immunity: _____

Nervous: _____

2. Using the terms at the right of the torso model photo (Figure 2.7), label each organ supplied with a leader line in the photo.

3.   Place each of the listed organs in the correct body cavity or cavities. For organs found in the abdominopelvic cavity, also indicate which quadrant they occupy.

Dorsal body cavity_____

Thoracic cavity_____

_____

_____

_____

_____

GROUP CHALLENGE

Odd Organ Out

Each of the following sets contains four organs. One of the listed organs in each case does not share a characteristic that the other three do. Circle the organ that doesn't belong with the others and explain why it is singled out. What characteristic is it missing? Sometimes there may be multiple reasons why the organ doesn't belong with the others. Include as many as you can think of but make sure it does not have the key characteristic. Use the overview of organ systems (Table 2.1) and the pictures in your lab manual to help you select and justify your answer.

1. Which is the "odd organ"?	Why is it the odd one out?
Stomach Teeth Small intestine Oral cavity	
2. Which is the "odd organ"?	Why is it the odd one out?
Thyroid gland Thymus Spleen Lymph nodes	
3. Which is the "odd organ"?	Why is it the odd one out?
Ovaries Prostate gland Uterus Uterine tubes	
4. Which is the "odd organ"?	Why is it the odd one out?
Stomach Small intestine Esophagus Large intestine	

The Microscope

MATERIALS

- ☐ Compound microscope
- ☐ Millimeter ruler
- ☐ Prepared slides of the letter e or newsprint
- ☐ Immersion oil
- ☐ Lens paper
- ☐ Prepared slide of grid ruled in millimeters
- ☐ Prepared slide of three crossed colored threads
- ☐ Clean microscope slide and coverslip
- ☐ Toothpicks (flat-tipped)
- ☐ Physiological saline in a dropper bottle
- ☐ Iodine or dilute methylene blue stain in a dropper bottle
- ☐ Filter paper or paper towels
- ☐ Beaker containing fresh 10% household bleach solution for wet mount disposal
- ☐ Disposable autoclave bag
- ☐ Prepared slide of cheek epithelial cells

Note to the Instructor: The slides and coverslips used for viewing cheek cells are to be soaked for 2 hours (or longer) in 10% bleach solution and then drained. The slides and disposable autoclave bag containing coverslips, lens paper, and used toothpicks are to be autoclaved for 15 min at 121°C and 15 pounds pressure to ensure sterility. After autoclaving, the disposable autoclave bag may be discarded in any disposal facility, and the slides and glassware washed with laboratory detergent and prepared for use. These instructions apply as well to any bloodstained glassware or disposable items used in other experimental procedures.

MasteringA&P® For related exercise study tools, go to the Study Area of MasteringA&P. There you will find:
- Practice Anatomy Lab PAL
- PhysioEx PEx
- A&PFlix A&PFlix
- Practice quizzes, Histology Atlas, eText, Videos, and more!

OBJECTIVES

1. Identify the parts of the microscope and list the function of each.
2. Describe and demonstrate the proper techniques for care of the microscope.
3. Demonstrate proper focusing technique.
4. Define *total magnification, resolution, parfocal, field, depth of field* and *working distance.*
5. Measure the field size for one objective lens, calculate it for all the other objective lenses, and estimate the size of objects in each field.
6. Discuss the general relationships between magnification, working distance, and field size.

PRE-LAB QUIZ

1. The microscope slide rests on the _____ while being viewed.
   a. base   c. iris
   b. condenser   d. stage
2. Your lab microscope is *parfocal*. This means that:
   a. The specimen is clearly in focus at this depth.
   b. The slide should be almost in focus when changing to higher magnifications.
   c. You can easily discriminate two close objects as separate.
3. If the ocular lens magnifies a specimen 10✕, and the objective lens used magnifies the specimen 35✕, what is the total magnification being used to observe the specimen? _____
4. How do you clean the lenses of your microscope?
   a. with a paper towel
   b. with soap and water
   c. with special lens paper and cleaner
5. Circle True or False. You should always begin observation of specimens with the oil immersion lens.

With the invention of the microscope, biologists gained a valuable tool to observe and study structures like cells that are too small to be seen by the unaided eye. The information gained helped in establishing many of the theories basic to the understanding of biological sciences. This exercise will familiarize you with the workhorse of microscopes—the compound microscope—and provide you with the necessary instructions for its proper use.

Care and Structure of the Compound Microscope

The **compound microscope** is a precision instrument and should always be handled with care. *At all times you must observe the following rules for its transport, cleaning, use, and storage:*

• When transporting the microscope, hold it in an upright position with one hand on its arm and the other supporting its base. Avoid swinging the instrument during its transport and jarring the instrument when setting it down.

• Use only special grit-free lens paper to clean the lenses. Use a circular motion to wipe the lenses, and clean all lenses before and after use.

• Always begin the focusing process with the lowest-power objective lens in position, changing to the higher-power lenses as necessary.

• Use the coarse adjustment knob only with the lowest-power lens.

• Always use a coverslip with wet mount preparations.

• Before putting the microscope in the storage cabinet, remove the slide from the stage, rotate the lowest-power objective lens into position, wrap the cord neatly around the base, and replace the dust cover or return the microscope to the appropriate storage area.

• Never remove any parts from the microscope; inform your instructor of any mechanical problems that arise.

A C T I V I T Y   1

Identifying the Parts of a Microscope

1. Using the proper transport technique, obtain a microscope and bring it to the laboratory bench.

• Record the number of your microscope in the **Summary Chart** (page 31).

Compare your microscope with the photograph **(Figure 3.1)** and identify the following microscope parts:

Base: Supports the microscope. (*Note:* Some microscopes are provided with an inclination joint, which allows the instrument to be tilted backward for viewing dry preparations.)

Substage light or **mirror:** Located in the base. In microscopes with a substage light source, the light passes directly upward through the microscope: light controls are located on the microscope base. If a mirror is used, light must be reflected from a separate free-standing lamp.

Stage: The platform the slide rests on while being viewed. The stage has a hole in it to permit light to pass through both it and the specimen. Some microscopes have a stage equipped with *spring clips;* others have a clamp-type *mechanical stage* (as shown in Figure 3.1). Both hold the slide in position for viewing; in addition, the mechanical stage has two adjustable knobs that control precise movement of the specimen.

Condenser: Small substage lens that concentrates the light on the specimen. The condenser may have a *rack and pinion knob* that raises and lowers the condenser to vary light delivery. Generally, the best position for the condenser is close to the inferior surface of the stage.

Iris diaphragm lever: Arm attached to the base of the condenser that regulates the amount of light passing through the condenser. The iris diaphragm permits the best possible contrast when viewing the specimen.

Coarse adjustment knob: Used to focus on the specimen.

Fine adjustment knob: Used for precise focusing once coarse focusing has been completed.

Head or **body tube:** Supports the objective lens system, which is mounted on a movable nosepiece, and the ocular lens or lenses.

Arm: Vertical portion of the microscope connecting the base and head.

Ocular (or *eyepiece*): Depending on the microscope, there are one or two lenses at the superior end of the head or body tube. Observations are made through the ocular(s). An ocular lens has a magnification of 10×; it increases the apparent size of the object by ten times or ten diameters. If your microscope has a **pointer** to indicate a specific area of the viewed specimen, it is attached to one ocular and can be positioned by rotating the ocular lens.

Nosepiece: Rotating mechanism at the base of the head. Generally carries three or four objective lenses and permits sequential positioning of these lenses over the light beam passing through the hole in the stage. Use the nosepiece to change the objective lenses. Do not directly grab the lenses.

Objective lenses: Adjustable lens system that permits the use of a **scanning lens,** a **low-power lens,** a **high-power lens,** or an **oil immersion lens.** The objective lenses have different magnifying and resolving powers.

2. Examine the objective lenses carefully; note their relative lengths and the numbers inscribed on their sides. On many microscopes, the scanning lens, with a magnification between 4× and 5×, is the shortest lens. If there is no scanning lens, the low-power objective lens is the shortest and typically has a magnification of 10×. The high-power objective lens is of intermediate length and has a magnification range from 40× to 50×, depending on the microscope. The oil immersion objective lens is usually the longest of the objective lenses and has a magnifying power of 95× to 100×. Some microscopes lack the oil immersion lens.

• Record the magnification of each objective lens of your microscope in the first row of the Summary Chart (page 31). Also, cross out the column relating to a lens that your microscope does not have. Plan on using the same microscope for all microscopic studies.

3. Rotate the lowest-power objective lens until it clicks into position, and turn the coarse adjustment knob about 180 degrees. Notice how far the stage (or objective lens) travels during this adjustment. Move the fine adjustment knob 180 degrees, noting again the distance that the stage (or the objective lens) moves. ■

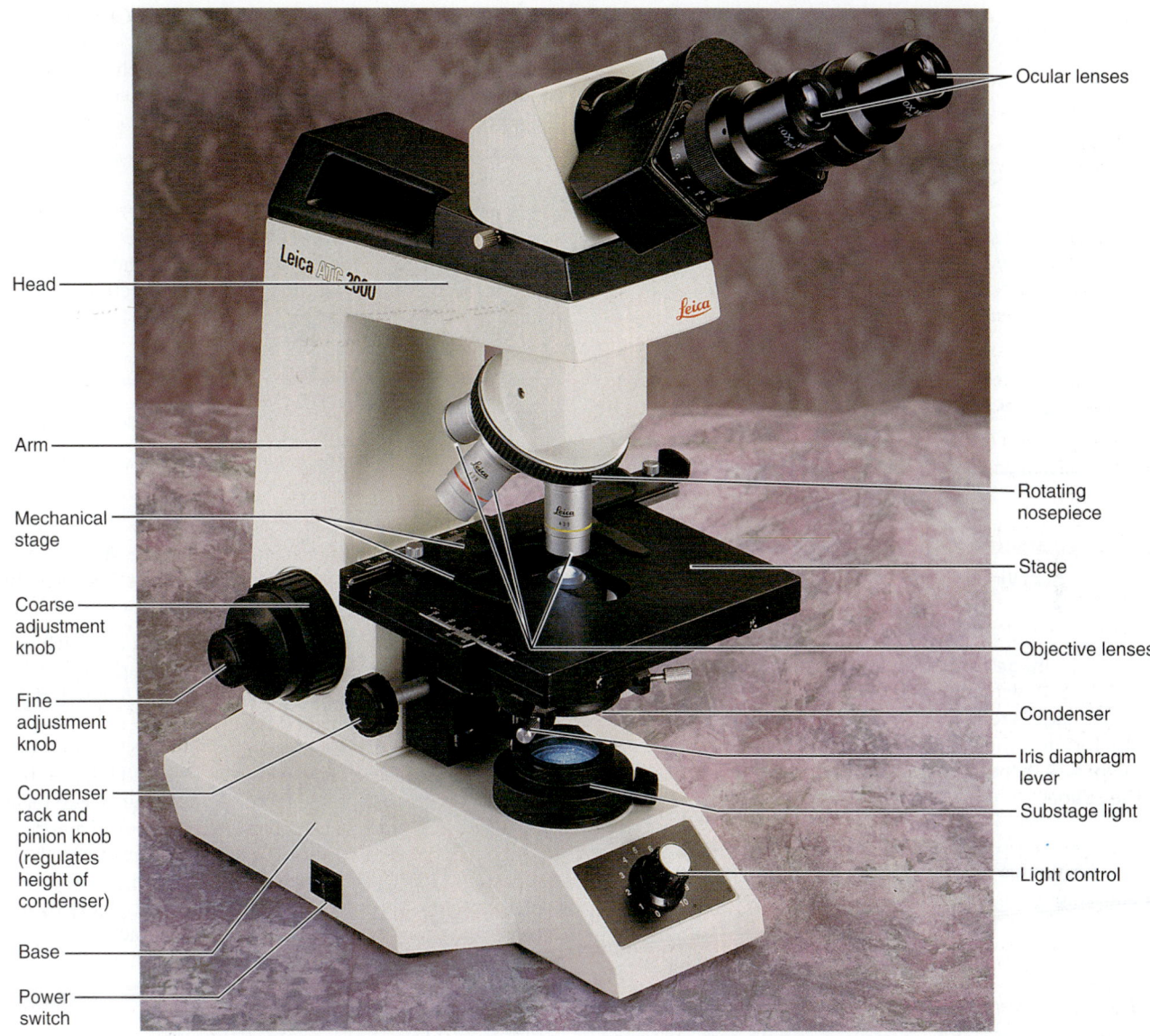

Head

Arm

Mechanical stage

Coarse adjustment knob

Fine adjustment knob

Condenser rack and pinion knob (regulates height of condenser)

Base

Power switch

Ocular lenses

Rotating nosepiece

Stage

Objective lenses

Condenser

Iris diaphragm lever

Substage light

Light control

Figure 3.1 **Compound microscope and its parts.**

Magnification and Resolution

The microscope is an instrument of magnification. In the compound microscope, magnification is achieved through the interplay of two lenses—the ocular lens and the objective lens. The objective lens magnifies the specimen to produce a **real image** that is projected to the ocular. This real image is magnified by the ocular lens to produce the **virtual image** seen by your eye **(Figure 3.2)**.

The **total magnification** (TM) of any specimen being viewed is equal to the power of the ocular lens multiplied by the power of the objective lens used. For example, if the ocular lens magnifies $10\times$ and the objective lens being used magnifies $45\times$, the total magnification is $450\times$ (or 10×45).

• Determine the total magnification you may achieve with each of the objectives on your microscope, and record the figures on the third row of the Summary Chart.

The compound light microscope has certain limitations. Although the level of magnification is almost limitless, the **resolution** (or resolving power), that is, the ability to discriminate two close objects as separate, is not. The human eye can resolve objects about 100 µm apart, but the compound microscope has a resolution of 0.2 µm under ideal conditions. Objects closer than 0.2 µm are seen as a single fused image.

Resolving power is determined by the amount and physical properties of the visible light that enters the microscope. In general, the more light delivered to the objective lens, the greater the resolution. The size of the objective lens aperture (opening) decreases with increasing magnification, allowing less light to enter the objective. Thus, you will probably find it necessary to increase the light intensity at the higher magnifications.

ACTIVITY 2

Viewing Objects Through the Microscope

1. Obtain a millimeter ruler, a prepared slide of the letter *e* or newsprint, a dropper bottle of immersion oil, and some lens paper. Adjust the condenser to its highest position and switch

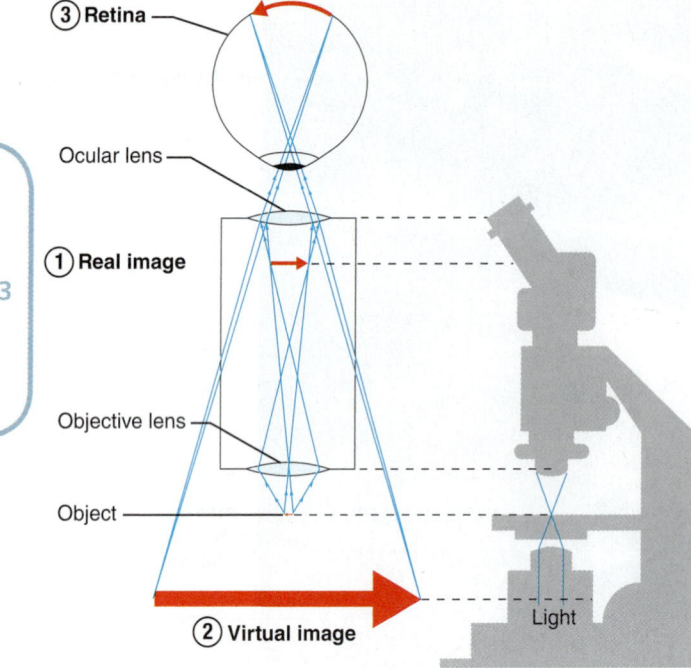

Figure 3.2 Image formation in light microscopy.
Step ① The objective lens magnifies the object, forming the real image. **Step** ② The ocular lens magnifies the real image, forming the virtual image. **Step** ③ The virtual image passes through the lens of the eye and is focused on the retina.

on the light source of your microscope. If the light source is not built into the base, use the curved surface of the mirror to reflect the light up into the microscope.

2.   Secure the slide on the stage so that you can read the slide label and the letter *e* is centered over the light beam passing through the stage. If you are using a microscope with spring clips, make sure the slide is secured at both ends. If your microscope has a mechanical stage, open the jaws of its slide holder by using the control lever, typically located at the rear left corner of the mechanical stage. Insert the slide squarely within the confines of the slide holder. Check that the slide is resting on the stage, not on the mechanical stage frame, before releasing the control lever.

3.   With your lowest-power (scanning or low-power) objective lens in position over the stage, use the coarse adjustment knob to bring the objective lens and stage as close together as possible.

4.   Look through the ocular lens and adjust the light for comfort using the iris diaphragm. Now use the coarse adjustment knob to focus slowly away from the *e* until it is as clearly focused as possible. Complete the focusing with the fine adjustment knob.

5.   Sketch the letter *e* in the circle on the Summary Chart (page 31) just as it appears in the **field**—the area you see through the microscope.

How far is the bottom of the objective lens from the specimen? In other words, what is the **working distance**? Use a millimeter ruler to make this measurement.

Record the working distance in the Summary Chart.

How has the apparent orientation of the *e* changed top to bottom, right to left, and so on?

_____

_____

6.   Move the slide slowly away from you on the stage as you view it through the ocular lens. In what direction does the image move?

_____

Move the slide to the left. In what direction does the image move?

_____

At first this change in orientation may confuse you, but with practice you will learn to move the slide in the desired direction with no problem.

7.   Today most good laboratory microscopes are **parfocal;** that is, the slide should be in focus (or nearly so) at the higher magnifications once you have properly focused. *Without touching the focusing knobs,* increase the magnification by rotating the next higher magnification lens into position over the stage. Make sure it clicks into position. Using the fine adjustment only, sharpen the focus. If you are unable to focus with a new lens, your microscope is not parfocal. Do not try to force the lens into position. Consult your instructor. Note the decrease in working distance. As you can see, focusing with the coarse adjustment knob could drive the objective lens through the slide, breaking the slide and possibly damaging the lens. Sketch the letter *e* in the Summary Chart. What new details become clear?

_____

_____

As best you can, measure the distance between the objective and the slide.

Record the working distance in the Summary Chart.

Is the image larger or smaller? _____

Approximately how much of the letter *e* is visible now?

_____

Is the field larger or smaller? _____

Why is it necessary to center your object (or the portion of the slide you wish to view) before changing to a higher power?

_____

Summary Chart for Microscope # _____			
	Scanning	**Low power**	**High power**	**Oil immersion**
Magnification of objective lens	_____ ×	_____ ×	_____ ×	_____ ×
Magnification of ocular lens	_____10_____ ×	_____10_____ ×	_____10_____ ×	_____10_____ ×
Total magnification	_____ ×	_____ ×	_____ ×	_____ ×
Working distance	_____ mm	_____ mm	_____ mm	_____ mm
Detail observed Letter *e*	◯	◯	◯	◯
Field size (diameter)	____ mm ____ µm	____ mm ____ µm	____ mm ____ µm	____ mm ____ µm

3

Move the iris diaphragm lever while observing the field. What happens?

_____

Is it better to increase *or* decrease the light when changing to a higher magnification?

_____ Why? _____

_____

8. If you have just been using the low-power objective, repeat the steps given in direction 7 using the high-power objective lens. What new details become clear?

_____

_____

Record the working distance in the Summary Chart.

9. Without touching the focusing knob, rotate the high-power lens out of position so that the area of the slide over the opening in the stage is unobstructed. Place a drop of immersion oil over the *e* on the slide and rotate the oil immersion lens into position. Set the condenser at its highest point (closest to the stage), and open the diaphragm fully. Adjust the fine focus and fine-tune the light for the best possible resolution.

Note: If for some reason the specimen does not come into view after adjusting the fine focus, do not go back to the 40× lens to recenter. You do not want oil from the oil immersion lens to cloud the 40× lens. Turn the revolving nosepiece in the other direction to the low-power lens and recenter and refocus the object. Then move the immersion lens back into position, again avoiding the 40× lens. Sketch the letter *e* in the Summary Chart, What new details become clear?

_____

_____

Is the field again decreased in size? _____

As best you can, estimate the working distance, and record it in the Summary Chart. Is the working distance less *or* greater than it was when the high-power lens was focused?

_____

Compare your observations on the relative working distances of the objective lenses with the illustration **(Figure 3.3)**. Explain why it is desirable to begin the focusing process at the lowest power.

_____

_____

_____

10. Rotate the oil immersion lens slightly to the side and remove the slide. Clean the oil immersion lens carefully with lens paper, and then clean the slide in the same manner with a fresh piece of lens paper. ■

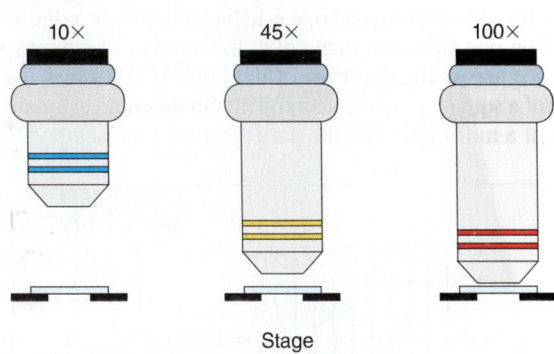

Stage

Figure 3.3 Relative working distances of the 10×, 45×, and 100× objectives.

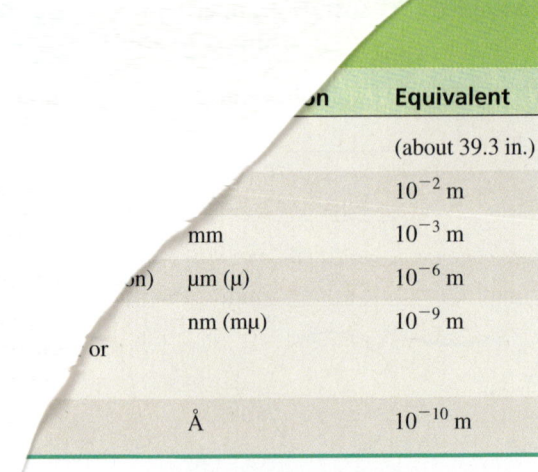

		on	Equivalent
			(about 39.3 in.)
			10^{-2} m
	mm		10^{-3} m
	on)	μm (μ)	10^{-6} m
	nm (mμ)		10^{-9} m
or			
		Å	10^{-10} m

he "Getting Started" exercise [page xiv] for tips on metric
ons.)

he Microscope Field

y this time you should know that the size of the microscope
ield decreases with increasing magnification. For future microscope work, it will be useful to determine the diameter
of each of the microscope fields. This information will allow you to make a fairly accurate estimate of the size of
the objects you view in any field. For example, if you have
calculated the field diameter to be 4 mm and the object being
observed extends across half this diameter, you can estimate
that the length of the object is approximately 2 mm.

Microscopic specimens are usually measured in micrometers and millimeters, both units of the metric system.
(You can get an idea of the relationship and meaning of these
units from **Table 3.1**. A more detailed treatment appears in
the appendix.)

ACTIVITY 3

Estimating the Diameter of the Microscope Field

1. Obtain a grid slide, which is a slide prepared with graph
paper ruled in millimeters. Each of the squares in the grid is
1 mm on each side. Use your lowest-power objective to bring
the grid lines into focus.

2. Move the slide so that one grid line touches the edge of the
field on one side, and then count the number of squares you
can see across the diameter of the field. If you can see only
part of a square, as in the accompanying diagram, estimate the
part of a millimeter that the partial square represents.

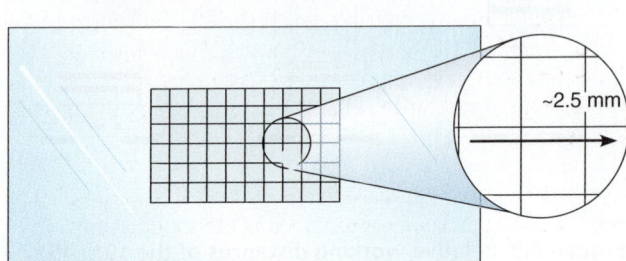

~2.5 mm

Record this figure in the appropriate space marked "field size"
on the Summary Chart (page 31). (If you have been using
the scanning lens, repeat the procedure with the low-power
objective lens.)

Complete the chart by computing the approximate
diameter of the high-power and oil immersion fields.
The general formula for calculating the unknown field
diameter is:

Diameter of field A × total magnification of field A =
diameter of field B × total magnification of field B

where A represents the known or measured field and B represents the unknown field. This can be simplified to

Diameter of field B =

$$\frac{\text{diameter of field } A \times \text{total magnification of field } A}{\text{total magnification of field } B}$$

For example, if the diameter of the low-power field (field A) is
2 mm and the total magnification is 50×, you would compute
the diameter of the high-power field (field B) with a total
magnification of 100× as follows:

Field diameter B = (2 mm × 50)/100
Field diameter B = 1 mm

3. Estimate the length (longest dimension) of the following
microscopic objects. *Base your calculations on the field sizes
you have determined for your microscope.*

Object seen in low-power field:

approximate length:

_____ mm

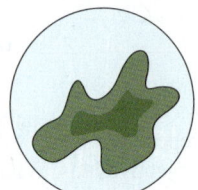

Object seen in high-power field:

approximate length:

_____ mm

or_____ μm

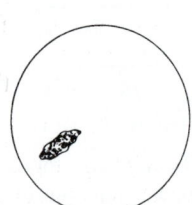

Object seen in oil immersion field:

approximate length:

_____ μm

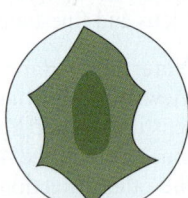

4. If an object viewed with the oil immersion lens looked as it does in the field depicted below, could you determine its approximate size from this view?

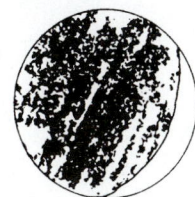

If not, then how could you determine it? _____

_____

_____

_____ ▪

Perceiving Depth

Any microscopic specimen has depth as well as length and width; it is rare indeed to view a tissue slide with just one layer of cells. Normally you can see two or three cell thicknesses. Therefore, it is important to learn how to determine relative depth with your microscope. In microscope work the **depth of field** (the thickness of the plane that is clearly in focus) is greater at lower magnifications. As magnification increases, depth of field decreases.

ACTIVITY 4

Perceiving Depth

1. Obtain a slide with colored crossed threads. Focusing at low magnification, locate the point where the three threads cross each other.

2. Use the iris diaphragm lever to greatly reduce the light, thus increasing the contrast. Focus down with the coarse adjustment until the threads are out of focus, then slowly focus upward again, noting which thread comes into clear focus first. (You will see two or even all three threads, so you must be very careful in determining which one first comes into clear focus.) Observe: As you rotate the adjustment knob forward (away from you), does the stage rise or fall? If the stage rises, then the first clearly focused thread is the top one; the last clearly focused thread is the bottom one.

If the stage descends, how is the order affected? _____

_____

Record your observations, relative to which color of thread is uppermost, middle, or lowest:

Top thread _____

Middle thread _____

Bottom thread _____ ▪

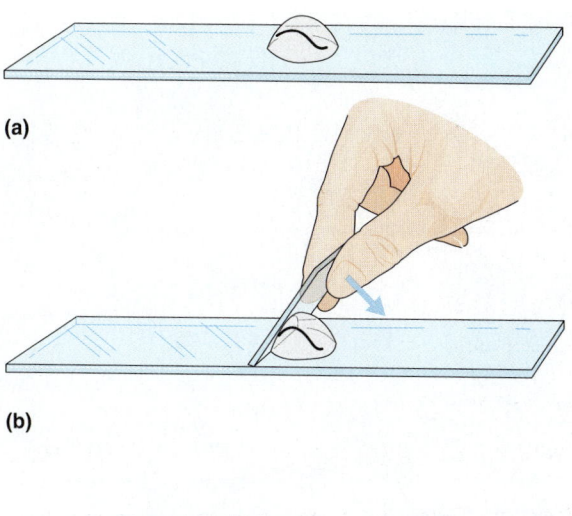

(a)

(b)

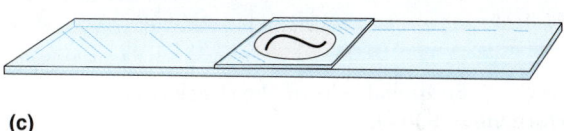

(c)

Figure 3.4 Procedure for preparation of a wet mount. (a) The object is placed in a drop of water (or saline) on a clean slide, **(b)** a coverslip is held at a 45° angle with the fingertips, and **(c)** it is lowered carefully over the water and the object.

Viewing Cells Under the Microscope

There are various ways to prepare cells for viewing under a microscope. Cells and tissues can look very different with different stains and preparation techniques. One method of preparation is to mix the cells in physiological saline (called a wet mount) and stain them with methylene blue stain.

If you are not instructed to prepare your own wet mount, obtain a prepared slide of epithelial cells to make the observations in step 10 of Activity 5.

ACTIVITY 5

Preparing and Observing a Wet Mount

1. Obtain the following: a clean microscope slide and coverslip, two flat-tipped toothpicks, a dropper bottle of physiological saline, a dropper bottle of iodine or methylene blue stain, and filter paper (or paper towels). Handle only your own slides throughout the procedure.

2. Place a drop of physiological saline in the center of the slide. Using the flat end of the toothpick, *gently* scrape the inner lining of your cheek. Transfer your cheek scrapings to the slide by agitating the end of the toothpick in the drop of saline **(Figure 3.4a)**.

 Immediately discard the used toothpick in the disposable autoclave bag provided at the supplies area.

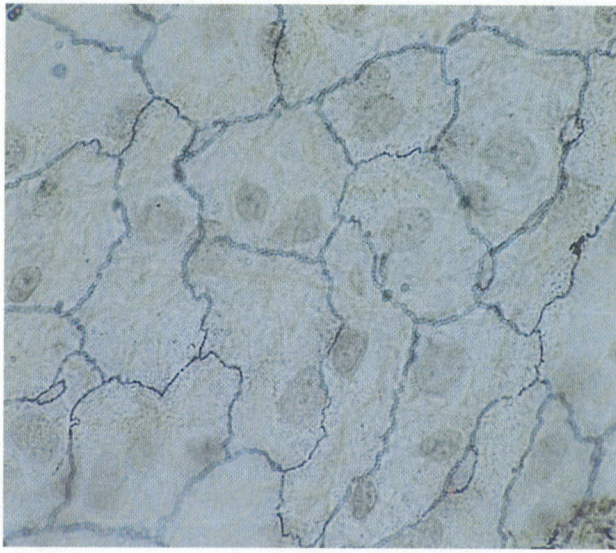

Figure 3.5 Epithelial cells of the cheek cavity (surface view, 600×).

3.   Add a tiny drop of the iodine or methylene blue stain to the preparation. (These epithelial cells are nearly transparent and thus difficult to see without the stain, which colors the nuclei of the cells and makes them look much darker than the cytoplasm.) Stir again.

 Immediately discard the used toothpick in the disposable autoclave bag provided at the supplies area.

4.   Hold the coverslip with your fingertips so that its bottom edge touches one side of the fluid drop (Figure 3.4b), then *carefully* lower the coverslip onto the preparation (Figure 3.4c). *Do not just drop the coverslip,* or you will trap large air bubbles under it, which will obscure the cells. *A coverslip should always be used with a wet mount* to prevent soiling the lens if you should misfocus.

5.   Examine your preparation carefully. The coverslip should be tight against the slide. If there is excess fluid around its edges, you will need to remove it. Obtain a piece of filter paper, fold it in half, and use the folded edge to absorb the excess fluid. You may use a twist of paper towel as an alternative.

 Before continuing, discard the filter paper or paper towel in the disposable autoclave bag.

6.   Place the slide on the stage, and locate the cells at the lowest power. You will probably want to dim the light with the iris diaphragm to provide more contrast for viewing the lightly stained cells. Furthermore, a wet mount will dry out quickly in bright light because a bright light source is hot.

7.   Cheek epithelial cells are very thin, six-sided cells. In the cheek, they provide a smooth, tilelike lining **(Figure 3.5)**. Move to high power to examine the cells more closely.

8.   Make a sketch of the epithelial cells that you observe.

Use information on your Summary Chart (page 31) to estimate the diameter of cheek epithelial cells.

_____ µm

Why do *your* cheek cells look different than those in the photomicrograph (Figure 3.5)? (Hint: What did you have to *do* to your cheek to obtain them?)

_____

_____

 9.   When you complete your observations of the wet mount, dispose of your wet mount preparation in the beaker of bleach solution, and put the coverslips in an autoclave bag.

10.   Obtain a prepared slide of cheek epithelial cells, and view them under the microscope.

Estimate the diameter of one of these cheek epithelial cells using information from the Summary Chart (page 31).

_____ µm

Why are these cells more similar to those in the photograph (Figure 3.5) and easier to measure than those of the wet mount?

_____

_____

_____

11.   Before leaving the laboratory, make sure all other materials are properly discarded or returned to the appropriate laboratory station. Clean the microscope lenses and put the dust cover on the microscope before you return it to the storage cabinet. ■

The Microscope

Care and Structure of the Compound Microscope

1. Label all indicated parts of the microscope.

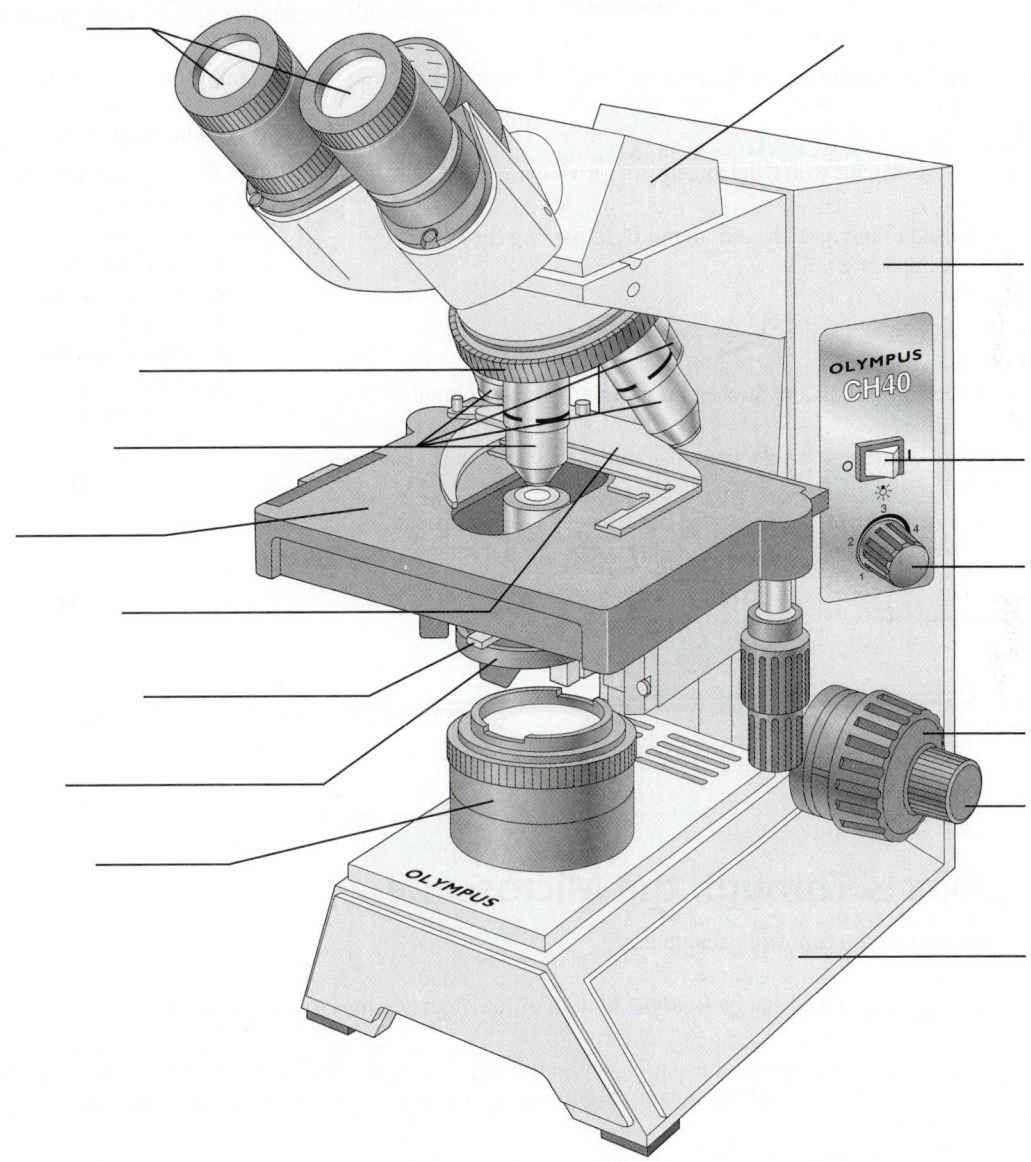

2. Explain the proper technique for transporting the microscope.

_____

3. The following statements are true or false. If true, write *T* on the answer blank. If false, correct the statement by writing on the blank the proper word or phrase to replace the one that is underlined.

_____ 1. The microscope lens may be cleaned <u>with any soft tissue</u>.

_____ 2. The microscope should be stored with the <u>oil immersion</u> lens in position over the stage.

_____ 3. When beginning to focus, use the <u>lowest-power</u> lens.

_____ 4. When focusing, always focus <u>toward</u> the specimen.

_____ 5. A coverslip should always be used <u>with wet mounts and the high-power and oil lenses</u>.

4. Match the microscope structures in column B with the statements in column A that identify or describe them.

Column A

_____ 1. platform on which the slide rests for viewing

_____ 2. used to increase the amount of light passing through the specimen

_____ 3. secure(s) the slide to the stage

_____ 4. delivers a concentrated beam of light to the specimen

_____ 5. used for precise focusing once initial focusing has been done

_____ 6. carries the objective lenses; rotates so that the different objective lenses can be brought into position over the specimen

Column B

a. coarse adjustment knob
b. condenser
c. fine adjustment knob
d. iris diaphragm
e. mechanical stage
f. nosepiece
g. objective lenses
h. ocular
i. spring clips
j. stage

5. Define the following terms.

virtual image: _____

_____

resolution: _____

Viewing Objects Through the Microscope

6. Complete, or respond to, the following statements:

_____ 1. The distance from the bottom of the objective lens to the specimen is called the _____.

_____ 2. Assume there is an object on the left side of the field that you want to bring to the center (that is, toward the apparent right). In what direction would you move your slide? _____

_____ 3. The area of the specimen seen when looking through the microscope is the _____.

_____ 4. If a microscope has a 10× ocular and the total magnification at a particular time is 950×, the objective lens in use at that time is _____ ×.

_____ 5. Why should the light be dimmed when looking at living (nearly transparent) cells?

_____ 6. If, after focusing in low power, only the fine adjustment need be used to focus the specimen at the higher powers, the microscope is said to be _____.

_____ 7. If, when using a 10× ocular and a 15× objective, the field size is 1.5 mm, the approximate field size with a 30× objective is _____ mm.

_____ 8. If the size of the high-power field is 1.2 mm, an object that occupies approximately a third of that field has an estimated diameter of _____ mm.

7. You have been asked to prepare a slide with the letter *k* on it (as shown below). In the circle below, draw the *k* as seen in the low-power field.

k

8. Figure out the magnification of fields 1 and 3, and the field size of 2. (*Hint:* Use your ruler.) Note that the numbers for the field sizes below are too large to represent the typical compound microscope lens system, but the relationships depicted are accurate.

5 mm	_____ mm	0.5 mm
1. →O←	2. →O←	3. →o←
_____ ×	100 ×	_____ ×

9. Say you are observing an object in the low-power field. When you switch to high-power, it is no longer in your field of view.

Why might this occur? _____

_____

What should be done initially to prevent this from happening? _____

_____

10. Do the following factors increase or decrease as one moves to higher magnifications with the microscope?

resolution: _____   amount of light needed: _____

working distance: _____   depth of field: _____

11. A student has the high-dry lens in position and appears to be intently observing the specimen. The instructor, noting a working distance of about 1 cm, knows the student isn't actually seeing the specimen.

How so? _____

12. Describe the proper procedure for preparing a wet mount.

_____

_____

_____

13. Indicate the probable cause of the following situations arising during use of a microscope.

a.   Only half of the field is illuminated: _____

_____

b.   Field does not change as mechanical stage is moved: _____

_____

The Cell: Anatomy and Division

MATERIALS

- ☐ Three-dimensional model of the "composite" animal cell or laboratory chart of cell anatomy
- ☐ Compound microscope
- ☐ Prepared slides of simple squamous epithelium, teased smooth muscle (l.s.), human blood cell smear, and sperm
- ☐ Animation/video of mitosis
- ☐ Three-dimensional models of mitotic stages
- ☐ Prepared slides of whitefish blastulas
- ☐ Chenille sticks (pipe cleaners), two different colors cut into 3-inch pieces, 8 pieces per group

Note to the Instructor: See directions for handling wet mount preparations and disposable supplies (page 34, Exercise 3). For suggestions on the animation/video of mitosis, see the Instructor Guide.

MasteringA&P® For related exercise study tools, go to the Study Area of MasteringA&P. There you will find:

- Practice Anatomy Lab PAL
- PhysioEx PEx
- A&PFlix A&PFlix
- Practice quizzes, Histology Atlas, eText, Videos, and more!

OBJECTIVES

1. Define *cell, organelle,* and *inclusion.*
2. Identify on a cell model or diagram the following cellular regions and list the major function of each: nucleus, cytoplasm, and plasma membrane.
3. Identify the cytoplasmic organelles and discuss their structure and function.
4. Compare and contrast specialized cells with the concept of the "generalized cell."
5. Define *interphase, mitosis,* and *cytokinesis.*
6. List the stages of mitosis and describe the key events of each stage.
7. Identify the mitotic phases on slides or appropriate diagrams.
8. Explain the importance of mitotic cell division and describe its product.

PRE-LAB QUIZ

1. Define *cell.* _____

   _____

2. When a cell is not dividing, the DNA is loosely spread throughout the nucleus in a threadlike form called:
   - a. chromatin
   - b. chromosomes
   - c. cytosol
   - d. ribosomes
3. The plasma membrane not only provides a protective boundary for the cell but also determines which substances enter or exit the cell. We call this characteristic:
   - a. diffusion
   - b. membrane potential
   - c. osmosis
   - d. selective permeability
4. Proteins are assembled on these organelles.

   _____

5. Because these organelles are responsible for providing most of the ATP needed by the cell, they are often referred to as the "powerhouses" of the cell. They are the:
   - a. centrioles
   - b. lysosomes
   - c. mitochondria
   - d. ribosomes
6. Circle the correct underlined term. During <u>cytokinesis</u> / <u>interphase</u> the cell grows and performs its usual activities.
7. Circle True or False. The end product of mitosis is four genetically identical daughter nuclei.
8. How many stages of mitosis are there? _____
9. DNA replication occurs during:
   - a. cytokinesis
   - b. interphase
   - c. metaphase
   - d. prophase
10. Circle True or False. All animal cells have a cell wall.

The **cell,** the structural and functional unit of all living things, is a complex entity. The cells of the human body are highly diverse, and their differences in size, shape, and internal composition reflect their specific roles in the body. Still, cells do have many common anatomical features, and all cells must carry out certain functions to sustain life. For example, all cells can maintain their boundaries, metabolize, digest nutrients and dispose of wastes, grow and reproduce, move, and respond to a stimulus. This exercise focuses on structural similarities found in many cells and illustrated by a "composite," or "generalized," cell **(Figure 4.1a)** and considers only the function of cell reproduction (cell division). (Most other cell functions are considered in detail in later exercises. Exercise 5 explores transport mechanisms, the means by which substances cross a cell's external membrane.)

Anatomy of the Composite Cell

In general, all animal cells have three major regions, or parts, that can readily be identified with a light microscope: the **nucleus,** the **plasma membrane,** and the **cytoplasm.** The nucleus is typically a round or oval structure near the center of the cell. It is surrounded by cytoplasm, which in turn is enclosed by the plasma membrane. Since the invention of the electron microscope, even smaller cell structures—organelles—have been identified. See the diagram (Figure 4.1a) representing the fine structure of the composite cell. An electron micrograph (Figure 4.1b) reveals the cellular structure, particularly of the nucleus.

Nucleus

The nucleus contains the genetic material, DNA, sections of which are called genes. Often described as the control center of the cell, the nucleus is necessary for cell reproduction. A cell that has lost or ejected its nucleus is programmed to die.

When the cell is not dividing, the genetic material is loosely dispersed throughout the nucleus in a threadlike form called **chromatin.** When the cell is in the process of dividing to form daughter cells, the chromatin coils and condenses, forming dense, darkly staining rodlike bodies called **chromosomes**—much in the way a stretched spring becomes shorter and thicker when it is released. Carefully note the appearance of the nucleus—it is somewhat nondescript when a cell is healthy. A dark nucleus and clumped chromatin indicate that the cell is dying and undergoing degeneration.

The nucleus also contains one or more small round bodies, called **nucleoli,** composed primarily of proteins and ribonucleic acid (RNA). The nucleoli are assembly sites for ribosomal particles that are particularly abundant in the cytoplasm. Ribosomes are the actual protein-synthesizing "factories."

The nucleus is bound by a double-layered porous membrane, the **nuclear envelope.** The nuclear envelope is similar in composition to other cellular membranes, but it is distinguished by its large **nuclear pores.** They are spanned by protein complexes that regulate what passes through and permit easy passage of protein and RNA molecules.

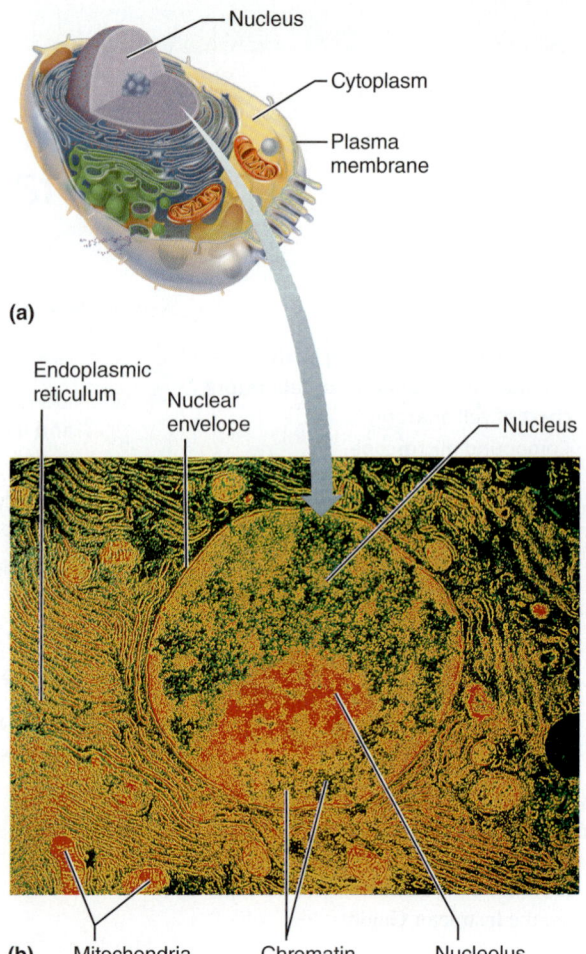

(a)

(b) Mitochondria   Chromatin   Nucleolus

Endoplasmic reticulum   Nuclear envelope   Nucleus

Nucleus
Cytoplasm
Plasma membrane

Figure 4.1 Anatomy of the composite animal cell.
(a) Diagram. **(b)** Transmission electron micrograph (5000×).

ACTIVITY 1

Identifying Parts of a Cell

As able, identify the nuclear envelope, chromatin, nucleolus, and the nuclear pores (see Figure 4.1a and b and Figure 4.3.) ◼

Plasma Membrane

The **plasma membrane** separates cell contents from the surrounding environment. Its main structural building blocks are phospholipids (fats) and globular protein molecules. Some of the externally facing proteins and lipids have sugar (carbohydrate) side chains attached to them that are important in cellular interactions **(Figure 4.2)**. Described by the fluid mosaic model, the membrane is a bilayer of phospholipid molecules in which the protein molecules float. Occasional cholesterol molecules dispersed in the bilayer help stabilize it.

Besides providing a protective barrier for the cell, the plasma membrane plays an active role in determining which substances may enter or leave the cell and in what quantity. Because of its molecular composition, the plasma membrane is selective about what passes through it. It allows

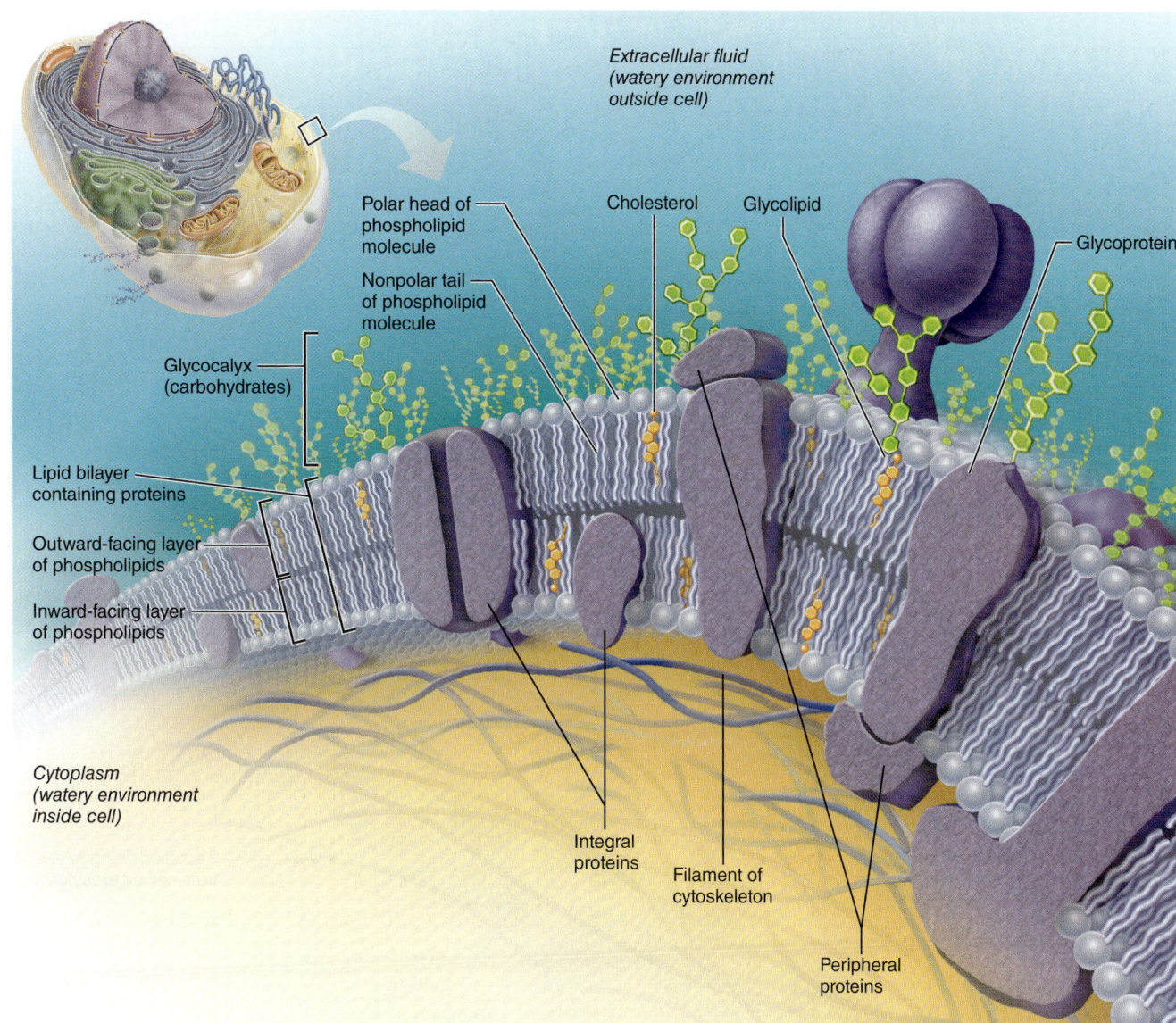

Figure 4.2 Structural details of the plasma membrane.

nutrients to enter the cell but keeps out undesirable substances. By the same token, valuable cell proteins and other substances are kept within the cell, and excreta, or wastes, pass to the exterior. This property is known as **selective permeability.** Transport through the plasma membrane occurs in two basic ways. In *active transport,* the cell must provide energy in the form of adenosine triphosphate, or ATP, to power the transport process. In *passive transport,* the transport process is driven by concentration or pressure differences.

Additionally, the plasma membrane maintains a resting potential that is essential to normal functioning of excitable cells, such as neurons and muscle cells, and plays a vital role in cell signaling and cell-to-cell interactions. In some cells the membrane is thrown into minute fingerlike projections or folds called **microvilli (Figure 4.3)**. Microvilli greatly increase the surface area of the cell available for absorption or passage of materials and for the binding of signaling molecules.

ACTIVITY 2

Identifying Components of a Plasma Membrane

Identify the phospholipid and protein portions of the plasma membrane in the figure (Figure 4.2). Also locate the sugar (*glyco* = carbohydrate) side chains and cholesterol molecules. Identify the microvilli in the generalized cell diagram (Figure 4.3). ■

Cytoplasm and Organelles

The cytoplasm consists of the cell contents between the nucleus and plasma membrane. It is the major site of most activities carried out by the cell. Suspended in the **cytosol,** the fluid cytoplasmic material, are many small structures called **organelles** (literally, "small organs"). The organelles are the metabolic machinery of the cell, and they are highly organized to carry out specific functions for the cell as a

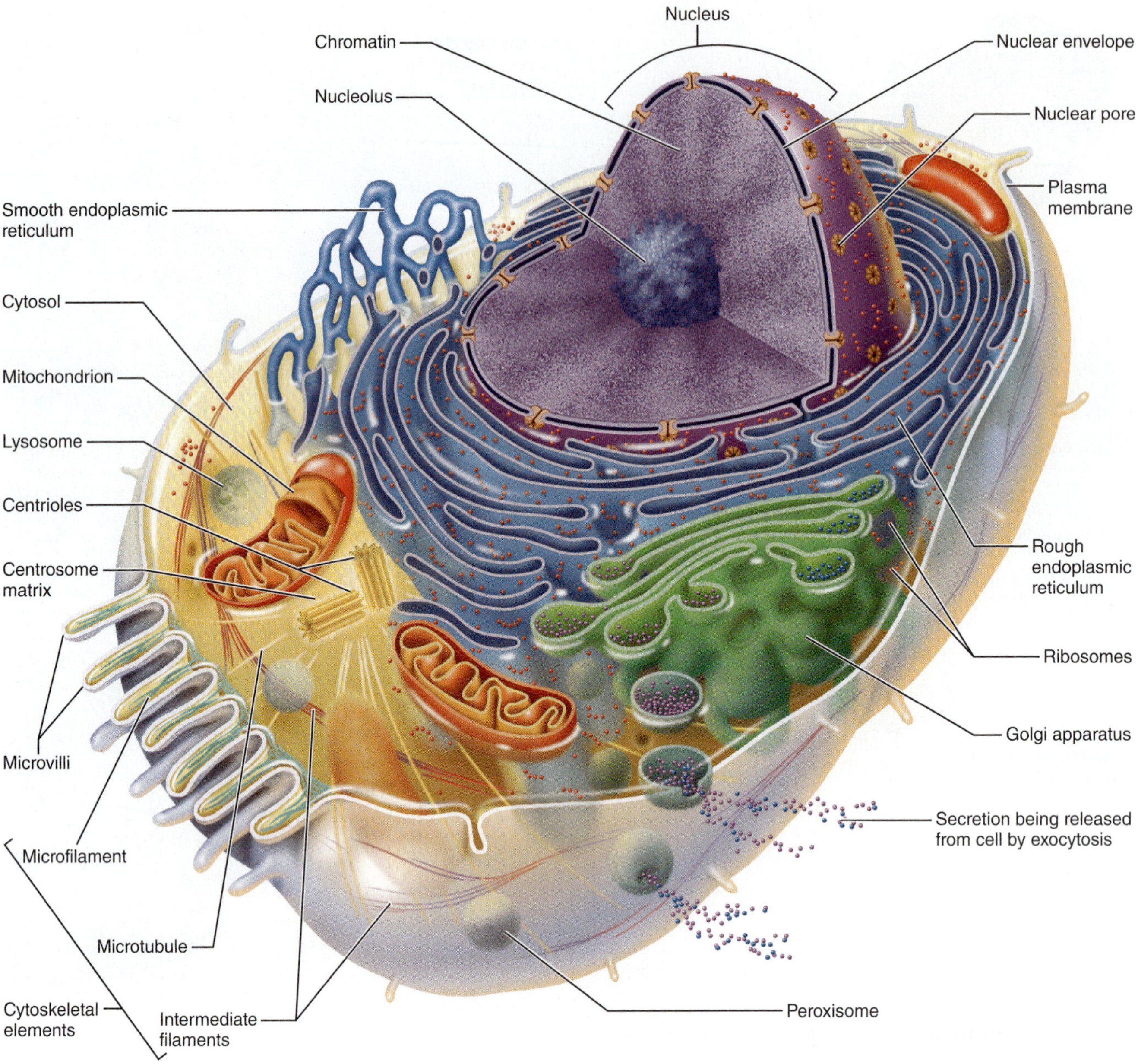

Figure 4.3 Structure of the generalized cell. No cell is exactly like this one, but this composite illustrates features common to many human cells. Not all organelles are drawn to the same scale in this illustration.

whole. The organelles include the ribosomes, endoplasmic reticulum, Golgi apparatus, lysosomes, peroxisomes, mitochondria, cytoskeletal elements, and centrioles.

ACTIVITY 3

Locating Organelles

Each organelle type is described in the following list (and summarized in **Table 4.1**). Read through the list and table and then, as best you can, locate the organelles in the illustrations (Figure 4.1b and 4.3). ■

• **Ribosomes** are densely staining, roughly spherical bodies composed of RNA and protein. They are the actual sites

of protein synthesis. They are seen floating free in the cytoplasm or attached to a membranous structure. When they are attached, the whole ribosome-membrane complex is called the *rough endoplasmic reticulum.*

• The **endoplasmic reticulum (ER)** is a highly folded system of membranous tubules and cisterns (sacs) that extends throughout the cytoplasm. The ER is continuous with the nuclear envelope, forming a system of channels for the transport of cellular substances (primarily proteins) from one part of the cell to another. The ER exists in two forms, **rough ER** and **smooth ER.** A particular cell may have both or only one, depending on its specific functions. The rough ER is studded with ribosomes. Its cisterns modify and store the newly formed proteins and dispatch them to other areas of the cell. The external face of the rough ER is involved in phospholipid

Table 4.1	**Summary of Structure and Function of Cytoplasmic Organelles**
Organelle	**Location and function**
Ribosomes 	Tiny spherical bodies composed of RNA and protein; floating free or attached to a membranous structure (the rough ER) in the cytoplasm. Actual sites of protein synthesis.
Endoplasmic reticulum (ER) 	Membranous system of tubules that extends throughout the cytoplasm; two varieties: rough and smooth. Rough ER is studded with ribosomes; tubules of the rough ER provide an area for storage and transport of the proteins made on the ribosomes to other cell areas; external face synthesizes phospholipids and cholesterol. Smooth ER, which has no function in protein synthesis, is a site of steroid and lipid synthesis, lipid metabolism, and drug detoxification.
Golgi apparatus 	Stack of flattened sacs with bulbous ends and associated small vesicles; found close to the nucleus. Plays a role in packaging proteins or other substances for export from the cell or incorporation into the plasma membrane and in packaging lysosomal enzymes.
Lysosomes 	Various-sized membranous sacs containing digestive enzymes including acid hydrolases; function to digest worn-out cell organelles and foreign substances that enter the cell. Have the capacity of total cell destruction if ruptured.
Peroxisomes	Small lysosome-like membranous sacs containing oxidase enzymes that detoxify alcohol, hydrogen peroxide, and other harmful chemicals.
Mitochondria 	Generally rod-shaped bodies with a double-membrane wall; inner membrane is thrown into folds, or cristae; contain enzymes that oxidize foodstuffs to produce cellular energy (ATP); often referred to as "powerhouses of the cell."
Centrioles 	Paired, cylindrical bodies lie at right angles to each other, close to the nucleus. As part of the centrosome, they direct the formation of the mitotic spindle during cell division; form the bases of cilia and flagella.
Cytoskeletal elements: microfilaments, intermediate filaments, and microtubules 	Provide cellular support; function in intracellular transport. Microfilaments are formed largely of actin, a contractile protein, and thus are important in cell mobility, particularly in muscle cells. Intermediate filaments are stable elements composed of a variety of proteins and resist mechanical forces acting on cells. Microtubules form the internal structure of the centrioles and help determine cell shape.

and cholesterol synthesis. The amount of rough ER is closely correlated with the amount of protein a cell manufactures and is especially abundant in cells that make protein products for export—for example, the pancreas cells that produce digestive enzymes destined for the small intestine.

The smooth ER does not participate in protein synthesis but is present in conspicuous amounts in cells that produce steroid-based hormones—for example, the interstitial endocrine cells of the testes, which produce testosterone. Smooth ER is also abundant in cells that are active in lipid metabolism and drug detoxification activities—liver cells, for instance.

• The **Golgi apparatus** is a stack of flattened sacs with bulbous ends and associated membranous vesicles that is generally found close to the nucleus. Within its cisterns, the proteins delivered to it by transport vesicles from the rough ER are modified, segregated, and packaged into membranous

vesicles that ultimately (1) are incorporated into the plasma membrane, (2) become secretory vesicles that release their contents from the cell, or (3) become lysosomes.

• **Lysosomes,** which appear in various sizes, are membrane-bound sacs containing an array of powerful digestive enzymes. A product of the packaging activities of the Golgi apparatus, the lysosomes contain *acid hydrolases,* enzymes capable of digesting worn-out cell structures and foreign substances that enter the cell via vesicle formation through phagocytosis or endocytosis (see Exercise 5). Because they have the capacity of total cell destruction, the lysosomes are often referred to as the "suicide sacs" of the cell.

• **Peroxisomes,** like lysosomes, are enzyme-containing sacs. However, their *oxidases* have a different task. Using oxygen, they detoxify a number of harmful substances, most importantly free radicals. Peroxisomes are particularly abundant in kidney and liver cells, cells that are actively involved in detoxification.

- **Mitochondria** are generally rod-shaped bodies with a double-membrane wall; the inner membrane is thrown into folds, or *cristae*. Oxidative enzymes on or within the mitochondria catalyze the reactions of the Krebs cycle and the electron transport chain (collectively called aerobic cellular respiration), in which end products of food digestion are broken down to produce energy. The released energy is captured in the bonds of ATP molecules, which are then transported out of the mitochondria to provide a ready energy supply to power the cell. Every living cell requires a constant supply of ATP for its many activities. Because the mitochondria provide the bulk of this ATP, they are referred to as the "powerhouses" of the cell.

- **Cytoskeletal elements** ramify throughout the cytoplasm, forming an internal scaffolding called the *cytoskeleton* that supports and moves substances within the cell. The **microtubules** are slender tubules formed of proteins called *tubulins*. Most microtubules radiate from a region of cytoplasm near the nucleus called the *centrosome,* and they have the ability to aggregate and then disaggregate spontaneously. Microtubules organize the cytoskeleton and form the spindle during cell division. They also transport substances down the length of elongated cells (such as neurons), suspend organelles, and help maintain cell shape by providing rigidity to the soft cellular substance. The stable **intermediate filaments** are proteinaceous cytoskeletal elements that act as internal guy wires to resist mechanical (pulling) forces acting on cells. **Microfilaments,** ribbon or cordlike elements, are formed of contractile proteins, primarily *actin*. Because of their ability to shorten and then relax to assume a more elongated form, these are important in cell mobility and are very conspicuous in muscle cells that are specialized to contract. A cross-linked network of microfilaments called the *terminal web* braces and strengthens the internal face of the plasma membrane.

The cytoskeletal structures are changeable and microscopic. With the exception of the microtubules of the mitotic spindle, which are very obvious during cell division (see pages 46–47), and the microfilaments of skeletal muscle cells, they are rarely seen, even in electron micrographs. (Note that they are not depicted in Figure 4.1b). However, special stains can reveal the plentiful supply of these important structures.

- The paired **centrioles** lie close to the nucleus within the centrosome in cells capable of reproducing themselves. They are rod-shaped bodies that lie at right angles to each other. Internally each centriole is composed of nine triplets of microtubules. During cell division, the centrosome complex that contains the centrioles directs the formation of the mitotic spindle. Centrioles also form the cell projections called cilia and flagella, and in that role are called basal bodies.

The cell cytoplasm contains various other substances and structures, including stored foods (glycogen granules and lipid droplets), pigment granules, crystals of various types, water vacuoles, and ingested foreign materials. However, these are not part of the active metabolic machinery of the cell and are therefore called **inclusions.**

ACTIVITY 4

Examining the Cell Model

Once you have located all of these structures in the art (Figures 4.1b and 4.3), examine the cell model (or cell chart) to repeat and reinforce your identifications. ■

Differences and Similarities in Cell Structure

ACTIVITY 5

Observing Various Cell Structures

1. Obtain a compound microscope and prepared slides of simple squamous epithelium, smooth muscle cells (teased), human blood, and sperm.

2. Observe each slide under the microscope, carefully noting similarities and differences in the cells. See photomicrographs for simple squamous epithelium (Figure 3.5 in Exercise 3) and teased smooth muscle (Figure 6.7c in Exercise 6). The oil immersion lens will be needed to observe blood and sperm. Distinguish the limits of the individual cells, and notice the shape and position of the nucleus in each case. When you look at the human blood smear, direct your attention to the red blood cells, the pink-stained cells that are most numerous. The color photomicrographs illustrating a blood smear (Figure 29.3 in Exercise 29) and sperm (Figure 43.3 in Exercise 43) may be helpful in this cell structure study. Sketch your observations in the circles provided (page 45).

3. Measure the length or diameter of each cell, and record below the appropriate sketch.

4. How do these four cell types differ in shape and size?

_____

_____

_____

_____

How might cell shape affect cell function?

_____

_____

_____

_____

_____

_____

Which cells have visible projections? _____

_____

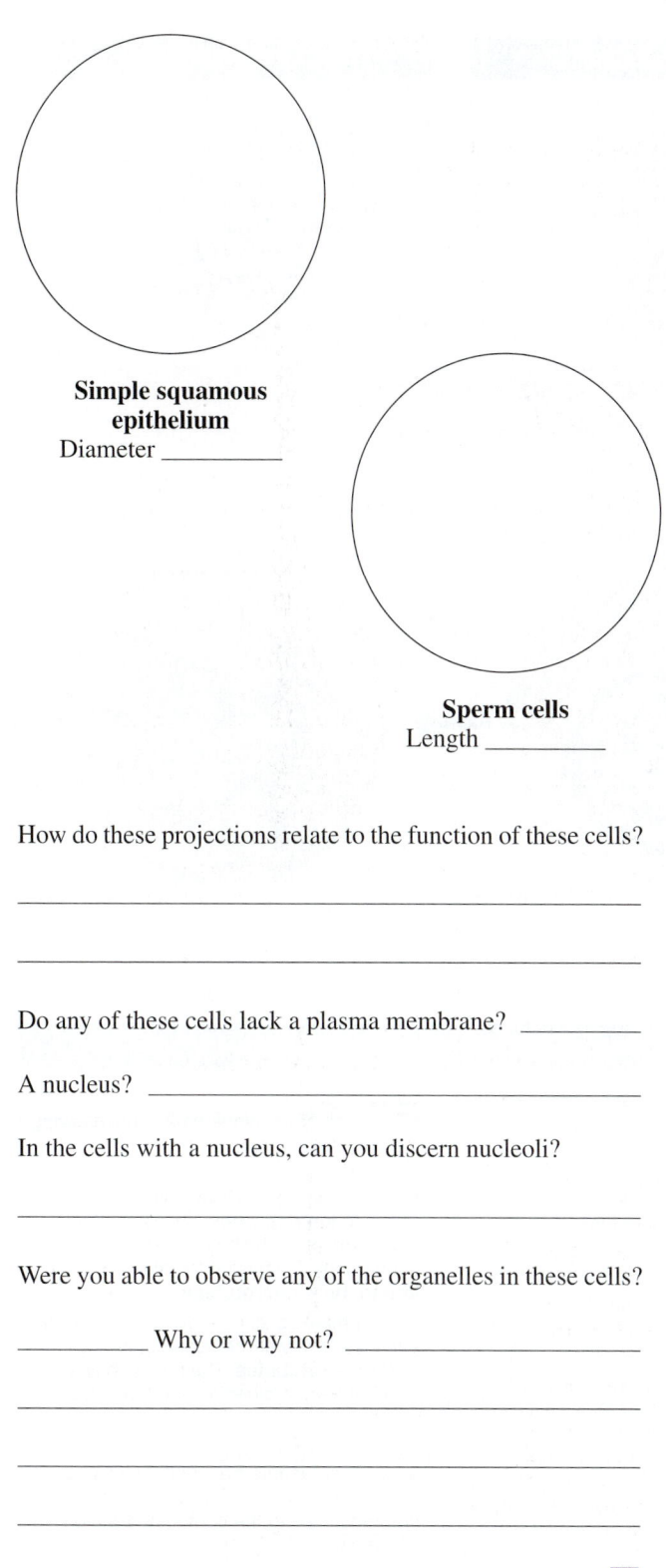

**Simple squamous
epithelium**
Diameter _____

Sperm cells
Length _____

**Human
redblood cells**
Diameter _____

**Teased smooth
muscle cells**
Length _____

How do these projections relate to the function of these cells?

_____

_____

Do any of these cells lack a plasma membrane? _____

A nucleus? _____

In the cells with a nucleus, can you discern nucleoli?

_____

Were you able to observe any of the organelles in these cells?

_____ Why or why not? _____

_____

_____

_____

_____

Cell Division: Mitosis and Cytokinesis

A cell's *life cycle* is the series of changes it goes through from the time it is formed until it reproduces itself. It consists of two stages—**interphase,** the longer period during which the cell grows and carries out its usual activities (Figure 4.4a), and **cell division,** when the cell reproduces itself by dividing. In an interphase cell about to divide, the genetic material (DNA)

is copied exactly. Once this important event has occurred, cell division ensues.

Cell division in all cells other than bacteria consists of two events called mitosis and cytokinesis. **Mitosis** is the division of the copied DNA of the mother cell to two daughter cells. **Cytokinesis** is the division of the cytoplasm, which begins when mitosis is nearly complete. Although mitosis is usually accompanied by cytokinesis, in some instances cytoplasmic division does not occur, leading to the formation of binucleate or multinucleate cells. This is relatively common in the human liver.

The product of **mitosis** is two daughter nuclei that are genetically identical to the mother nucleus. This distinguishes mitosis from **meiosis,** a specialized type of nuclear division (covered in Exercise 43) that occurs only in the reproductive organs (testes or ovaries). Meiosis, which yields four daughter nuclei that differ genetically in composition from the mother nucleus, is used only for the production of gametes (eggs and sperm) for sexual reproduction. The function of cell division, including mitosis and cytokinesis in the body, is to increase the number of cells for growth and repair while maintaining their genetic heritage.

The phases of mitosis include **prophase, metaphase, anaphase,** and **telophase.** (The detailed events of interphase, mitosis, and cytokinesis are described and illustrated in **Figure 4.4.**)

Mitosis is essentially the same in all animal cells, but depending on the tissue, it takes from 5 minutes to several hours to complete. In most cells, centriole replication occurs during interphase of the next cell cycle.

At the end of cell division, two daughter cells exist— each with a smaller cytoplasmic mass than the mother cell but genetically identical to it. The daughter cells grow and carry out the normal spectrum of metabolic processes until it is their turn to divide.

(Text continues on page 48.)

4

Interphase	Early Prophase	Late Prophase

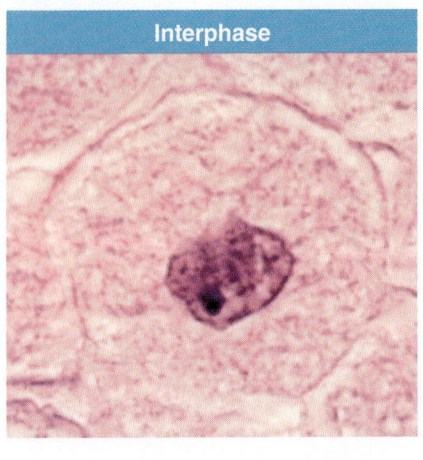
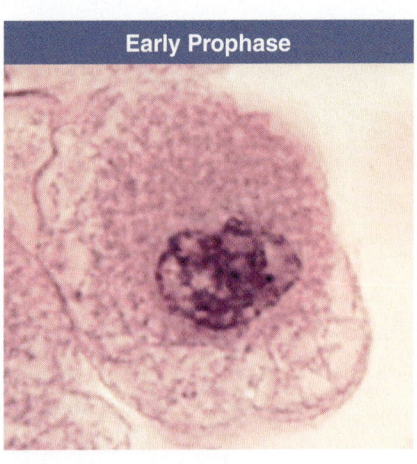
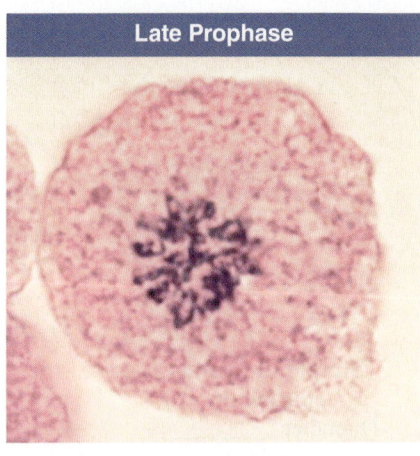

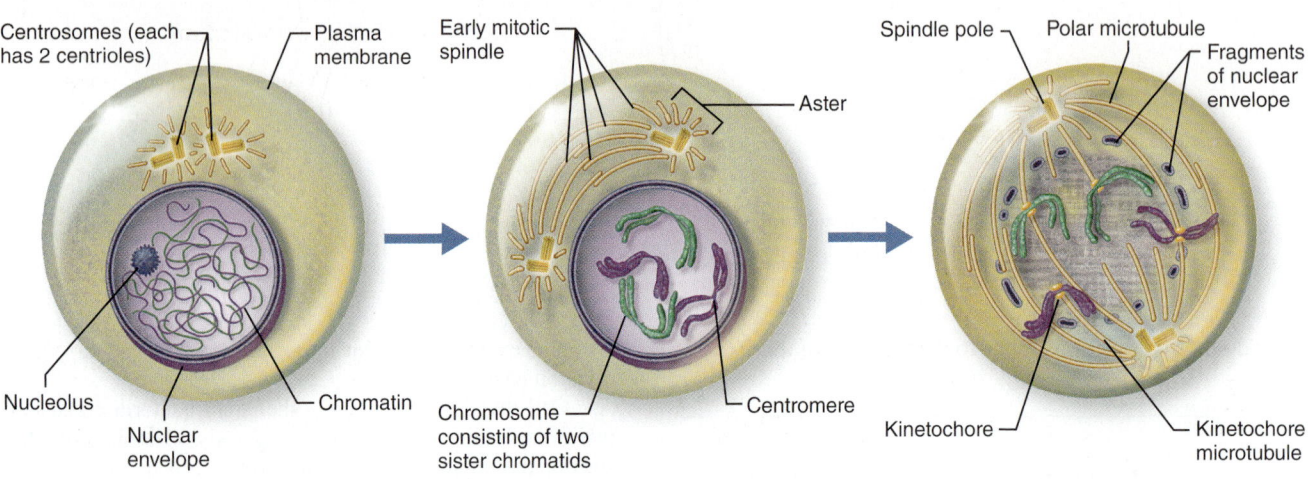

Interphase:
- Centrosomes (each has 2 centrioles)
- Plasma membrane
- Nucleolus
- Nuclear envelope
- Chromatin

Early Prophase:
- Early mitotic spindle
- Aster
- Chromosome consisting of two sister chromatids
- Centromere

Late Prophase:
- Spindle pole
- Polar microtubule
- Fragments of nuclear envelope
- Kinetochore
- Kinetochore microtubule

Interphase	Prophase—first phase of mitosis

Interphase

Interphase is the period of a cell's life when it carries out its normal metabolic activities and grows. Interphase is not part of mitosis.

• During interphase, the DNA-containing material is in the form of chromatin. The nuclear envelope and one or more nucleoli are intact and visible.

• There are three distinct periods of interphase:
  G_1: The centrioles begin replicating.
  S: DNA is replicated.
  G_2: Final preparations for mitosis are completed and centrioles finish replicating.

Early Prophase
• The chromatin condenses, forming barlike chromosomes.

• Each duplicated chromosome consists of two identical threads, called **sister chromatids**, held together at the **centromere**. (Later when the chromatids separate, each will be a new chromosome.)

• As the chromosomes appear, the nucleoli disappear, and the two centrosomes separate from one another.

• The centrosomes act as focal points for growth of a microtubule assembly called the **mitotic spindle**. As the microtubules lengthen, they propel the centrosomes toward opposite ends (poles) of the cell.

• Microtubule arrays called **asters** ("stars") extend from the centrosome matrix.

Late Prophase
• The nuclear envelope breaks up, allowing the spindle to interact with the chromosomes.

• Some of the growing spindle microtubules attach to **kinetochores**, special protein structures at each chromosome's centromere. Such microtubules are called **kinetochore microtubules**.

• The remaining spindle microtubules (not attached to any chromosomes) are called **polar microtubules**. The microtubules slide past each other, forcing the poles apart.

• The kinetochore microtubules pull on each chromosome from both poles in a tug-of-war that ultimately draws the chromosomes to the center, or equator, of the cell.

Figure 4.4 The interphase cell and the events of cell division. The cells shown are from an early embryo of a whitefish. Photomicrographs are above; corresponding diagrams are below. (Micrographs approximately 1600×.)

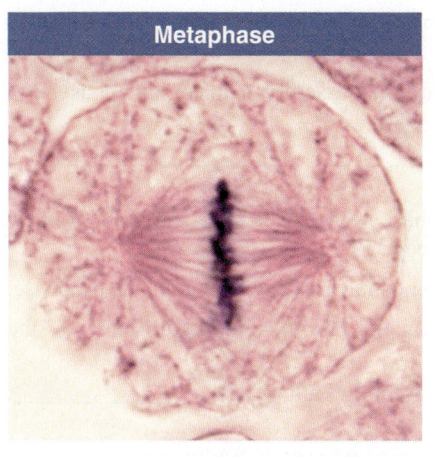
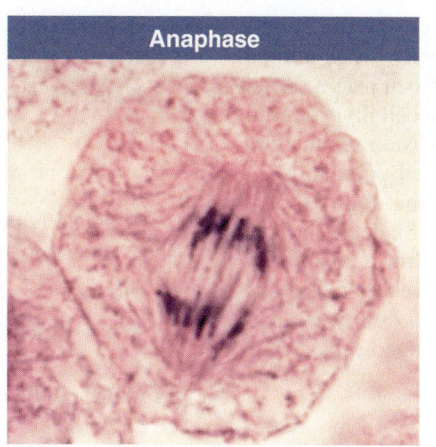
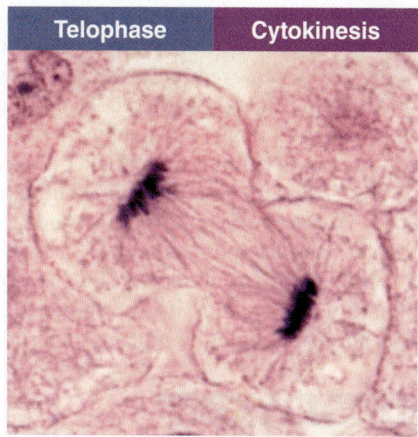

| Metaphase | Anaphase | Telophase | Cytokinesis |

4

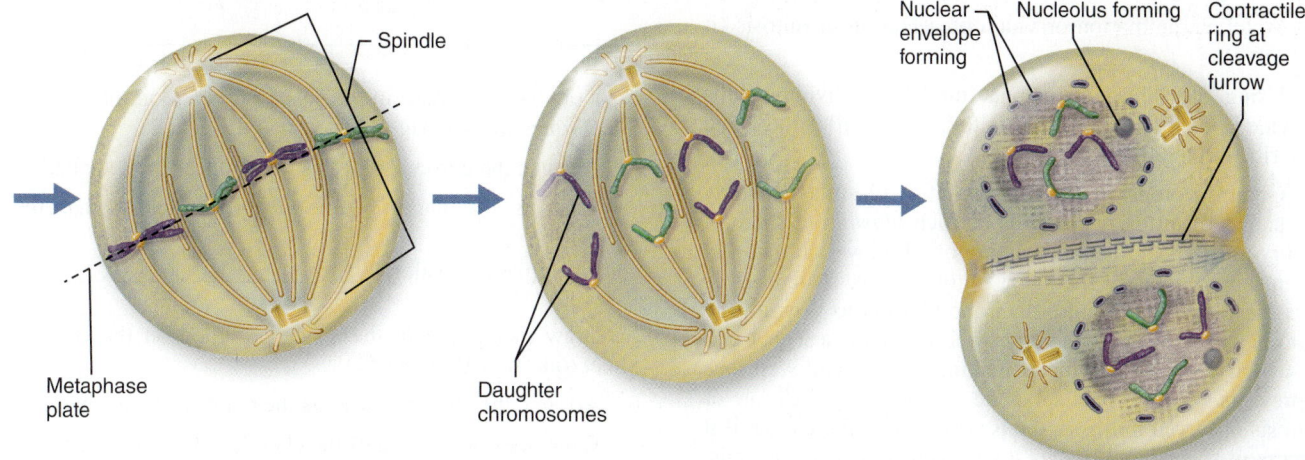

Spindle

Metaphase plate

Daughter chromosomes

Nuclear envelope forming

Nucleolus forming

Contractile ring at cleavage furrow

Metaphase—second phase of mitosis

• The two centrosomes are at opposite poles of the cell.

• The chromosomes cluster at the midline of the cell, with their centromeres precisely aligned at the **equator** of the spindle. This imaginary plane midway between the poles is called the **metaphase plate**.

• Enzymes act to separate the chromatids from each other.

Anaphase—third phase of mitosis

The shortest phase of mitosis, anaphase begins abruptly as the centromeres of the chromosomes split simultaneously. Each chromatid now becomes a chromosome in its own right.

• The kinetochore microtubules, moved along by motor proteins in the kinetochores, gradually pull each chromosome toward the pole it faces.

• At the same time, the polar microtubules slide past each other, lengthen, and push the two poles of the cell apart.

• The moving chromosomes look V shaped. The centromeres lead the way, and the chromosomal "arms" dangle behind them.

• Moving and separating the chromosomes is helped by the fact that the chromosomes are short, compact bodies. Diffuse threads of chromatin would trail, tangle, and break, resulting in imprecise "parceling out" to the daughter cells.

Telophase—final phase of mitosis

Telophase begins as soon as chromosomal movement stops. This final phase is like prophase in reverse.

• The identical sets of chromosomes at the opposite poles of the cell uncoil and resume their threadlike chromatin form.

• A new nuclear envelope forms around each chromatin mass, nucleoli reappear within the nuclei, and the spindle breaks down and disappears.

• Mitosis is now ended. The cell, for just a brief period, is binucleate (has two nuclei) and each new nucleus is identical to the original mother nucleus.

Cytokinesis—division of cytoplasm

Cytokinesis begins during late anaphase and continues through and beyond telophase. A contractile ring of actin microfilaments forms the **cleavage furrow** and pinches the cell apart.

Figure 4.4 *(continued)*

Cell division is extremely important during the body's growth period. Most cells divide until puberty, when normal body size is achieved and overall body growth ceases. After this time in life, only certain cells carry out cell division routinely—for example, cells subjected to abrasion (epithelium of the skin and lining of the gut). Other cell populations—such as liver cells—stop dividing but retain this ability should some of them be removed or damaged. Skeletal muscle, cardiac muscle, and most mature neurons almost completely lose this ability to divide and thus are severely handicapped by injury. Throughout life, the body retains its ability to repair cuts and wounds and to replace some of its aged cells.

ACTIVITY 6

Identifying the Mitotic Stages

1. Watch an animation or video presentation of mitosis (if available).

2. Using the three-dimensional models of dividing cells provided, identify each of the mitotic phases illustrated and described in the figure (Figure 4.4).

3. Obtain a prepared slide of whitefish blastulas to study the stages of mitosis. The cells of each *blastula* (a stage of embryonic development consisting of a hollow ball of cells) are at approximately the same mitotic stage, so it may be necessary to observe more than one blastula to view all the mitotic stages. A good analogy for a blastula is a soccer ball in which each leather piece making up the ball's surface represents an embryonic cell. The exceptionally high rate of mitosis observed in this tissue is typical of embryos, but if it occurs in specialized tissues it can indicate cancerous cells, which also have an extraordinarily high mitotic rate. Examine the slide carefully, identifying the four mitotic phases and the process of cytokinesis. Compare your observations with the photomicrographs (Figure 4.4), and verify your identifications with your instructor. ■

ACTIVITY 7

"Chenille Stick" Mitosis

1. Obtain a total of eight 3-inch pieces of chenille, four of one color and four of another color (e.g., four green and four purple).

2. Assemble the chenille sticks into a total of four chromosomes (each with two sister chromatids) by twisting two sticks of the same color together at the center with a single twist.

What does the twist at the center represent? _____

_____

3. Arrange the chromosomes as they appear in early prophase.

Name the structure that assembles during this phase.

_____

Draw early prophase in the space provided in the Review Sheet (question 10, page 51).

4. Arrange the chromosomes as they appear in late prophase.

What structure on the chromosome centromere do the

growing spindle microtubules attach to? _____.

What structure is now present as fragments? _____

_____

Draw late prophase in the space provided on the Review Sheet (question 10, page 51).

5. Arrange the chromosomes as they appear in metaphase.

What is the name of the imaginary plane that the

chromosomes align along? _____.

Draw metaphase in the space provided on the Review Sheet (question 10, page 51).

6. Arrange the chromosomes as they appear in anaphase.

What does untwisting of the chenille sticks represent?

_____

_____

Each sister chromatid has now become a _____.

Draw anaphase in the space provided on the Review Sheet (question 10, page 51).

7. Arrange the chromosomes as they appear in telophase.

Briefly list four reasons why telophase is like the reverse of prophase.

_____

_____

_____

Draw telophase in the space provided on the Review Sheet (question 10, page 51).

The Cell: Anatomy and Division

Anatomy of the Composite Cell

1. Define the following terms:

organelle: _____

_____

cell: _____

2. Although cells have differences that reflect their specific functions in the body, what functions do they have in common?

_____

3. Identify the following cell parts:

_____   1. external boundary of cell; regulates flow of materials into and out of the cell; site of cell signaling

_____   2. contains digestive enzymes of many varieties; "suicide sac" of the cell

_____   3. scattered throughout the cell; major site of ATP synthesis

_____   4. slender extensions of the plasma membrane that increase its surface area

_____   5. stored glycogen granules, crystals, pigments, and so on

_____   6. membranous system consisting of flattened sacs and vesicles; packages proteins for export

_____   7. control center of the cell; necessary for cell division and cell life

_____   8. two rod-shaped bodies near the nucleus; associated with the formation of the mitotic spindle

_____   9. dense, darkly staining nuclear body; packaging site for ribosomes

_____   10. contractile elements of the cytoskeleton

_____   11. membranous system; involved in intracellular transport of proteins and synthesis of membrane lipids

_____   12. attached to membrane systems or scattered in the cytoplasm; site of protein synthesis

_____   13. threadlike structures in the nucleus; contain genetic material (DNA)

_____   14. site of free radical detoxification

4. In the following diagram, label all parts provided with a leader line.

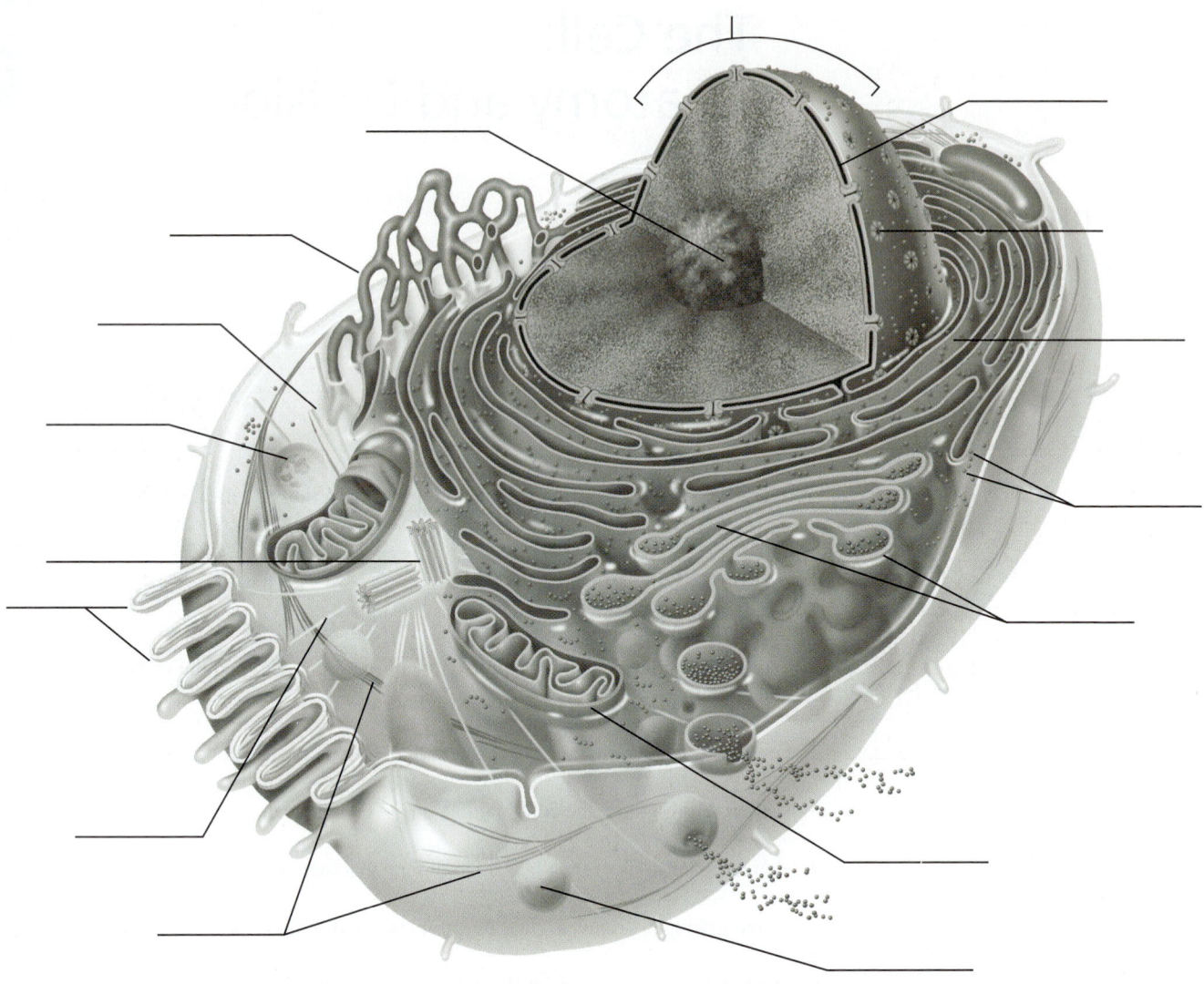

Differences and Similarities in Cell Structure

5. For each of the following cell types, list (a) *one* important structural characteristic observed in the laboratory, and (b) the function that the structure complements or ensures.

squamous epithelium   a. _____

                        b. _____

sperm   a. _____

                        b. _____

smooth muscle   a. _____

                        b. _____

red blood cells   a. _____

                        b. _____

6. What is the significance of the red blood cell being anucleate (without a nucleus)? _____

_____

Did it ever have a nucleus? (Use an appropriate reference.) _____ If so, when? _____

7. Of the four cells observed microscopically (squamous epithelial cells, red blood cells, smooth muscle cells, and sperm),

which has the smallest diameter? _____ Which is longest? _____

Cell Division: Mitosis and Cytokinesis

8. Identify the three phases of mitosis in the following photomicrographs.

a. _____   b. _____   c. _____

9. What is the importance of mitotic cell division? _____

_____

10. Draw the phases of mitosis for a cell that contains four chromosomes as its diploid or 2*n* number.

11. Complete or respond to the following statements:

Division of the __1__ is referred to as mitosis. Cytokinesis is division of the __2__. The major structural difference between chromatin and chromosomes is that the latter are __3__. Chromosomes attach to the spindle fibers by undivided structures called __4__. If a cell undergoes mitosis but not cytokinesis, the product is __5__. The structure that acts as a scaffolding for chromosomal attachment and movement is called the __6__. __7__ is the period of cell life when the cell is not involved in division. Two cell populations in the body that do not routinely undergo cell division are __8__ and __9__.

1. _____

2. _____

3. _____

4. _____

5. _____

6. _____

7. _____

8. _____

9. _____

12. Using the key, categorize each of the events described below according to the phase in which it occurs.

Key: a. anaphase   b. interphase   c. metaphase   d. prophase   e. telophase

_____ 1. Chromatin coils and condenses, forming chromosomes.

_____ 2. The chromosomes are V shaped.

_____ 3. The nuclear envelope re-forms.

_____ 4. Chromosomes stop moving toward the poles.

_____ 5. Chromosomes line up in the center of the cell.

_____ 6. The nuclear envelope fragments.

_____ 7. The mitotic spindle forms.

_____ 8. DNA synthesis occurs.

_____ 9. Centrioles replicate.

_____ 10. Chromosomes first appear to be duplex structures.

_____ 11. Chromosomal centromeres are attached to the kinetochore fibers.

_____ 12. Cleavage furrow forms.

_____ and _____ 13. The nuclear envelope(s) is absent.

13. What is the physical advantage of the chromatin coiling and condensing to form short chromosomes at the onset of mitosis?

_____

_____

The Cell: Transport Mechanisms and Cell Permeability

MATERIALS

Passive Processes

Diffusion of Dye Through Agar Gel

☐ Petri dish containing 12 ml of 1.5% agar-agar
☐ Millimeter-ruled graph paper
☐ Wax marking pencil
☐ 3.5% methylene blue solution (approximately 0.1 M) in dropper bottles
☐ 1.6% potassium permanganate solution (approximately 0.1 M) in dropper bottles
☐ Medicine dropper

Diffusion and Osmosis Through Nonliving Membranes

☐ Four dialysis sacs or small Hefty® sandwich bags
☐ Small funnel
☐ 25-ml graduated cylinder
☐ Wax marking pencil
☐ Fine twine or dialysis tubing clamps
☐ 250-ml beakers
☐ Distilled water
☐ 40% glucose solution
☐ 10% sodium chloride (NaCl) solution
☐ 40% sucrose solution colored with Congo red dye
☐ Laboratory balance
☐ Paper towels

(Text continues on next page.)

MasteringA&P® For related exercise study tools, go to the Study Area of MasteringA&P. There you will find:
• Practice Anatomy Lab PAL
• PhysioEx PEx
• A&PFlix A&PFlix
• Practice quizzes, Histology Atlas, eText, Videos, and more!

OBJECTIVES

1. Define *selective permeability* and explain the difference between active and passive transport processes.
2. Define *diffusion* and explain how simple diffusion and facilitated diffusion differ.
3. Define *osmosis*, and explain the difference between isotonic, hypotonic, and hypertonic solutions.
4. Define *filtration* and discuss where it occurs in the body.
5. Define *vesicular transport*, and describe phagocytosis, pinocytosis, receptor-mediated endocytosis, and exocytosis.
6. List the processes that account for the movement of substances across the plasma membrane and indicate the driving force for each.
7. Name one substance that uses each membrane transport process.
8. Determine which way substances will move passively through a selectively permeable membrane when given appropriate information about their concentration gradients.

PRE-LAB QUIZ

1. Circle the correct underlined term. A passive process, <u>diffusion</u> / <u>osmosis</u> is the movement of solute molecules from an area of greater concentration to an area of lesser concentration.
2. A solution surrounding a cell is *hypertonic* if:
   a. it contains fewer nonpenetrating solute particles than the interior of the cell.
   b. it contains more nonpenetrating solute particles than the interior of the cell.
   c. it contains the same amount of nonpenetrating solute particles as the interior of the cell.
3. Which of the following would require an input of energy?
   a. diffusion
   b. filtration
   c. osmosis
   d. vesicular transport
4. Circle the correct underlined term. In <u>pinocytosis</u> / <u>phagocytosis</u>, parts of the plasma membrane and cytoplasm extend and engulf a relatively large or solid material.
5. Circle the correct underlined term. In <u>active</u> / <u>passive</u> processes, the cell provides energy in the form of ATP to power the transport process.

(Materials list continued.)

- ☐ Hot plate and large beaker for hot water bath
- ☐ Benedict's solution in dropper bottle
- ☐ Silver nitrate (AgNO₃) in dropper bottle
- ☐ Test tubes in rack, test tube holder

Experiment 1

- ☐ Deshelled eggs
- ☐ 400-ml beakers
- ☐ Wax marking pencil
- ☐ Distilled water
- ☐ 30% sucrose solution
- ☐ Laboratory balance
- ☐ Paper towels
- ☐ Graph paper
- ☐ Weigh boat

Experiment 2

- ☐ Clean microscope slides and coverslips
- ☐ Medicine dropper
- ☐ Compound microscope
- ☐ Vials of animal (mammalian) blood obtained from a biological supply house or veterinarian—at option of instructor

- ☐ Freshly prepared physiological (mammalian) saline solution in dropper bottle
- ☐ 5% sodium chloride solution in dropper bottle
- ☐ Distilled water
- ☐ Filter paper
- ☐ Disposable gloves
- ☐ Basin and wash bottles containing 10% household bleach solution
- ☐ Disposable autoclave bag
- ☐ Paper towels

Diffusion Demonstrations

1. Diffusion of a dye through water

Prepared the morning of the laboratory session with setup time noted. Potassium permanganate crystals are placed in a 1000-ml graduated cylinder, and distilled water is added slowly and with as little turbulence as possible to fill to the 1000-ml mark.

2. Osmometer

Just before the laboratory begins, the broad end of a thistle tube is closed with a selectively permeable dialysis membrane, and the tube is secured to a ring stand. Molasses is added to approximately 5 cm above the thistle tube bulb, and the bulb is immersed in a beaker of distilled water. At the beginning of the lab session, the level of the molasses in the tube is marked with a wax pencil.

Filtration

- ☐ Ring stand, ring, clamp
- ☐ Filter paper, funnel
- ☐ Solution containing a mixture of uncooked starch, powdered charcoal, and copper sulfate (CuSO₄)
- ☐ 10-ml graduated cylinder
- ☐ 100-ml beaker
- ☐ Lugol's iodine in a dropper bottle

Active Processes

- ☐ Video showing phagocytosis (if available)
- ☐ Video viewing system

Note to the Instructor: See directions for handling wet mount preparations and disposable supplies (page 33, Exercise 3.)

PEx PhysioEx™ 9.1 Computer Simulation Ex.1 on p. PEx-3.

B ecause of its molecular composition, the plasma membrane is selective about what passes through it. It allows nutrients to enter the cell but keeps out undesirable substances. By the same token, valuable cell proteins and other substances are kept within the cell, and excreta or wastes pass to the exterior. This property is known as **selective,** or **differential, permeability.** Transport through the plasma membrane occurs in two basic ways. In **passive processes,** concentration or pressure differences drive the movement. In **active processes,** the cell provides energy (ATP) to power the transport process.

Passive Processes

The two important passive processes of membrane transport are *diffusion* and *filtration*. Diffusion is an important transport process for every cell in the body. By contrast, filtration usually occurs only across capillary walls.

Molecules possess **kinetic energy** and are in constant motion. As molecules move about randomly at high speeds, they collide and ricochet off one another, changing direction with each collision **(Figure 5.1)**. The driving force for diffusion is kinetic energy of the molecules themselves, and the speed of diffusion depends on molecular size and temperature. Smaller molecules move faster, and molecules move faster as temperature increases.

Diffusion

When a **concentration gradient** (difference in concentration) exists, the net effect of this random molecular movement

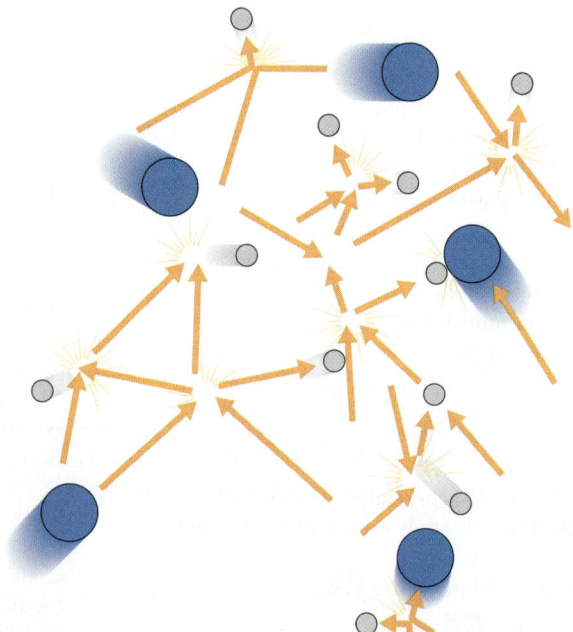

Figure 5.1 Random movement and numerous collisions cause molecules to become evenly distributed. The small spheres represent water molecules; the large spheres represent glucose molecules.

is that the molecules eventually become evenly distributed throughout the environment. **Diffusion** is the movement of molecules from a region of their higher concentration to a region of their lower concentration.

There are many examples of diffusion in nonliving systems. For example, if a bottle of ether was uncorked at the front of the laboratory, very shortly thereafter you would be nodding as the ether molecules became distributed throughout the room. The ability to smell a friend's cologne shortly after he or she has entered the room is another example.

The diffusion of particles into and out of cells is modified by the plasma membrane, which constitutes a physical barrier. In general, molecules diffuse passively through the plasma membrane if they can dissolve in the lipid portion of the membrane, as CO_2 and O_2 can. The unassisted diffusion of solutes (dissolved substances) through a selectively permeable membrane is called **simple diffusion.**

Certain molecules, for example glucose, are transported across the plasma membrane with the assistance of a protein carrier molecule. The glucose binds to the carrier and is ferried across the membrane. Small ions cross the membrane by moving through water-filled protein channels. In both cases, the substances move by a passive transport process called **facilitated diffusion.** As with simple diffusion, the substances move from an area of higher concentration to one of lower concentration, that is, down their concentration gradients.

Osmosis

The flow of water across a selectively permeable membrane is called **osmosis.** During osmosis, water moves down its concentration gradient. The concentration of water is inversely related to the concentration of solutes. If the solutes can diffuse across the membrane, both water and solutes will move down their concentration gradients through the membrane. If the particles in solution are nonpenetrating solutes (prevented from crossing the membrane), water alone will move by osmosis and in doing so will cause changes in the volume of the compartments on either side of the membrane.

Diffusion of Dye Through Agar Gel and Water

The relationship between molecular weight and the rate of diffusion can be examined easily by observing the diffusion of two different types of dye molecules through an agar gel. The dyes used in this experiment are methylene blue, which has a molecular weight of 320 and is deep blue in color, and potassium permanganate, a purple dye with a molecular weight of 158. Although the agar gel appears quite solid, it is primarily (98.5%) water and allows free movement of the dye molecules through it.

ACTIVITY 1

Observing Diffusion of Dye Through Agar Gel

1. Work with members of your group to formulate a hypothesis about the rates of diffusion of methylene blue and potassium permanganate through the agar gel. Justify your hypothesis.

2. Obtain a petri dish containing agar gel, a piece of millimeter-ruled graph paper, a wax marking pencil, dropper

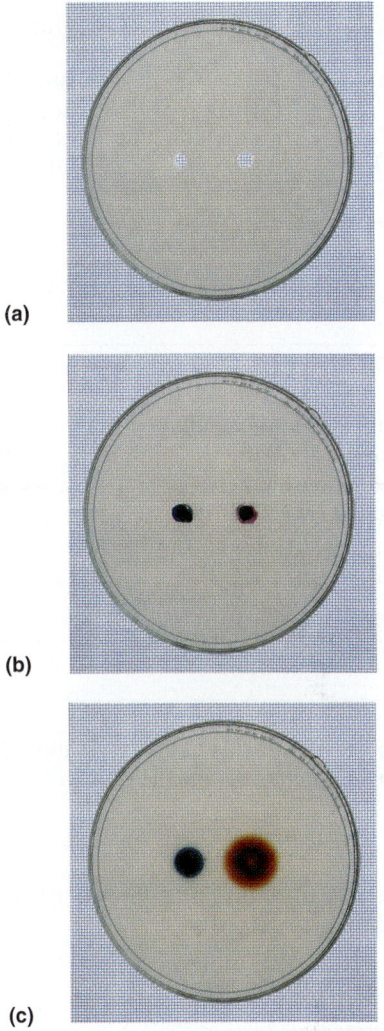

(a)

(b)

(c)

Figure 5.2 Comparing diffusion rates. Agar-plated petri dish as it appears after the diffusion of 0.1 *M* methylene blue placed in one well and 0.1 *M* potassium permanganate placed in another.

bottles of methylene blue and potassium permanganate, and a medicine dropper.

3. Using the wax marking pencil, draw a line on the bottom of the petri dish dividing it into two sections. Place the petri dish on the ruled graph paper.

4. Create a well in the center of each section using the medicine dropper. To do this, squeeze the bulb of the medicine dropper, and push it down into the agar. Release the bulb as you slowly pull the dropper vertically out of the agar. This should remove an agar plug, leaving a well in the agar. (See **Figure 5.2a.**)

5. Carefully fill one well with the methylene blue solution and the other well with the potassium permanganate solution.

Record the time. _____

6. At 15-minute intervals, measure the distance the dye has diffused from each well. Continue these observations for 1 hour, and record the results in the **Activity 1 chart.**

Activity 1: Dye Diffusion Results

Time (min)	Diffusion of methylene blue (mm)	Diffusion of potassium permanganate (mm)
15		
30		
45		
60		

Which dye diffused more rapidly? _____

What is the relationship between molecular weight and rate of molecular movement (diffusion)?

_____

Why did the dye molecules move? _____

_____

Compute the rate of diffusion of the potassium permanganate molecules in millimeters per minute (mm/min) and record.

_____ mm/min

Compute the rate of diffusion of the methylene blue molecules in mm/min and record.

_____ mm/min

7.   Prepare a lab report for these experiments. (See Getting Started, page xiv.) ▬

   Make a mental note to yourself to go to demonstration area 1 at the end of the laboratory session to observe the extent of diffusion of the potassium permanganate dye through water. At that time, follow the directions given next.

ACTIVITY 2

Observing Diffusion of Dye Through Water

1.   Go to diffusion demonstration area 1, and observe the cylinder containing dye crystals and water set up at the beginning of the lab.

2.   Measure the number of millimeters the dye has diffused from the bottom of the graduated cylinder and record.

_____ mm

3.   Record the time the demonstration was set up and the time of your observation. Then compute the rate of the dye's diffusion through water and record below.

Time of setup _____

Time of observation _____

Rate of diffusion _____ mm/min

4.   Does the potassium permanganate dye diffuse more rapidly through water or the agar gel? Explain your answer.

_____

_____ ▬

ACTIVITY 3

Investigating Diffusion and Osmosis Through Nonliving Membranes

The following experiment provides information on the movement of water and solutes through selectively permeable membranes called dialysis sacs. Dialysis sacs have pores of a particular size. The selectivity of living membranes depends on more than just pore size, but using the dialysis sacs will allow you to examine selectivity due to this factor.

1.   Read through the experiments in this activity, and develop a hypothesis for each part.

2.   Obtain four dialysis sacs, a small funnel, a 25-ml graduated cylinder, a wax marking pencil, fine twine or dialysis tubing clamps, and four beakers (250 ml). Number the beakers 1 to 4 with the wax marking pencil, and half fill all of them with distilled water except beaker 2, to which you should add 40% glucose solution.

3.   Prepare the dialysis sacs one at a time. Using the funnel, half fill each with 20 ml of the specified liquid (see below). Press out the air, fold over the open end of the sac, and tie it securely with fine twine or clamp it. Before proceeding to the next sac, rinse it under the tap, and quickly and carefully blot the sac dry by rolling it on a paper towel. Weigh it with a laboratory balance. Record the weight in the **Activity 3 data chart** (page 57), and then drop the sac into the corresponding beaker. Be sure the sac is completely covered by the beaker solution, adding more solution if necessary.

* Sac 1: 40% glucose solution
* Sac 2: 40% glucose solution
* Sac 3: 10% NaCl solution
* Sac 4: Congo red dye in 40% sucrose solution

Allow sacs to remain undisturbed in the beakers for 1 hour. Use this time to continue with other experiments.

4.   After an hour, boil a beaker of water on the hot plate. Obtain the supplies you will need to determine your experimental results: dropper bottles of Benedict's solution and silver nitrate solution, a test tube rack, four test tubes, and a test tube holder.

5.   Quickly and gently blot sac 1 dry and weigh it. (**Note:** Do not squeeze the sac during the blotting process.) Record the weight in the data chart.

Was there any change in weight? _____

| | | | | | Tests— | Tests— |
Beaker	Contents of sac	Initial weight	Final weight	Weight change	beaker fluid	sac fluid
Beaker 1 ½ filled with distilled water	Sac 1, 20 ml of 40% glucose solution				Benedict's test:	Benedict's test:
Beaker 2 ½ filled with 40% glucose solution	Sac 2, 20 ml of 40% glucose solution					
Beaker 3 ½ filled with distilled water	Sac 3, 20 ml of 10% NaCl solution				AgNO₃ test:	
Beaker 4 ½ filled with distilled water	Sac 4, 20 ml of 40% sucrose solution containing Congo red dye				Benedict's test:	

Activity 3: Experimental Data on Diffusion and Osmosis Through Nonliving Membranes

5

Conclusions: _____
_____
_____

Place 5 ml of Benedict's solution in each of two test tubes. Put 4 ml of the beaker fluid into one test tube and 4 ml of the sac fluid into the other. Mark the tubes for identification and then place them in a beaker containing boiling water. Boil 2 minutes. Cool slowly. If a green, yellow, or rusty red precipitate forms, the test is positive, meaning that glucose is present. If the solution remains the original blue color, the test is negative. Record results in the data chart.

Was glucose still present in the sac? _____

Was glucose present in the beaker? _____

Conclusions: _____
_____
_____

6. Blot gently and weigh sac 2. Record the weight in the data chart.

Was there an *increase* or *decrease* in weight? _____

With 40% glucose in the sac and 40% glucose in the beaker, would you expect to see any net movement of water (osmosis) or of glucose molecules (simple diffusion)?

_____ Why or why not? _____
_____

7. Blot gently and weigh sac 3. Record the weight in the data chart.

Was there any change in weight? _____

Conclusions: _____
_____

Take a 5-ml sample of beaker 3 solution and put it in a clean test tube. Add a drop of silver nitrate (AgNO₃). The appearance of a white precipitate or cloudiness indicates the presence of silver chloride (AgCl), which is formed by the reaction of AgNO₃ with NaCl (sodium chloride). Record results in the data chart.

Results: _____

Conclusions: _____

8. Blot gently and weigh sac 4. Record the weight in the data chart.

Was there any change in weight? _____

Did the beaker water turn pink? _____

Conclusions: _____
_____

Take a 1-ml sample of beaker 4 solution and put the test tube in boiling water in a hot water bath. Add 5 drops of Benedict's solution to the tube and boil for 5 minutes. The presence of glucose (one of the hydrolysis products of sucrose) in the test tube is indicated by the presence of a green, yellow, or rusty colored precipitate.

Did sucrose diffuse from the sac into the water in the small beaker? _____

5

Conclusions: _____

_____

9. In which of the test situations did net osmosis occur?

_____

In which of the test situations did net simple diffusion occur?

_____

What conclusions can you make about the relative size of glucose, sucrose, Congo red dye, NaCl, and water molecules?

_____

_____

_____

With what cell structure can the dialysis sac be compared?

_____

10. Prepare a lab report for the experiment. (See Getting Started, page xiv.) Be sure to include in your discussion the answers to the questions proposed in this activity. ■

ACTIVITY 4

Observing Osmometer Results

Before leaving the laboratory, observe demonstration 2, the *osmometer demonstration* set up before the laboratory session to follow the movement of water through a membrane (osmosis). Measure the distance the water column has moved during the laboratory period and record below. (The position of the meniscus [the surface of the water column] in the thistle tube at the beginning of the laboratory period is marked with wax pencil.)

Distance the meniscus has moved: _____ mm

Did net osmosis occur? Why or why not?

_____

_____

_____ ■

ACTIVITY 5

Investigating Diffusion and Osmosis Through Living Membranes

To examine permeability properties of plasma membranes, conduct the following experiments. As you read through the experiments in this activity, develop a hypothesis for each part.

Experiment 1

1. Obtain two deshelled eggs and two 400-ml beakers. Note that the relative concentration of solutes in deshelled eggs is about 14%. Number the beakers 1 and 2 with the wax marking pencil. Half fill beaker 1 with distilled water and beaker 2 with 30% sucrose.

2. Carefully blot each egg by rolling it gently on a paper towel. Place a weigh boat on a laboratory balance and tare the balance (that is, make sure the scale reads 0.0 with the weigh boat on the scale). Weigh egg 1 in the weigh boat, record the initial weight in the **Activity 5 data chart,** and gently place it into beaker 1. Repeat for egg 2, placing it in beaker 2.

3. After 20 minutes, remove egg 1 and gently blot it and weigh it. Record the weight, and replace it into beaker 1. Repeat for egg 2, placing it into beaker 2. Repeat this procedure at 40 minutes and 60 minutes.

4. Calculate the change in weight of each egg at each time period, and enter that number in the data chart. Also calculate the percent change in weight for each time period and enter that number in the data chart.

How has the weight of each egg changed?

Egg 1 _____

Egg 2 _____

Make a graph of your data by plotting the percent change in weight for each egg versus time.

How has the appearance of each egg changed?

Egg 1 _____

Egg 2 _____

A solution surrounding a cell is **hypertonic** if it contains more nonpenetrating solute particles than the interior of the cell. Water moves from the interior of the cell into a surrounding

Activity 5: Experiment 1 Data from Diffusion and Osmosis Through Living Membranes						
Time	Egg 1 (in distilled H$_2$O)	Weight change	% Change	Egg 2 (in 30% sucrose)	Weight change	% Change
Initial weight (g)		—	—		—	—
20 min.						
40 min.						
60 min.						

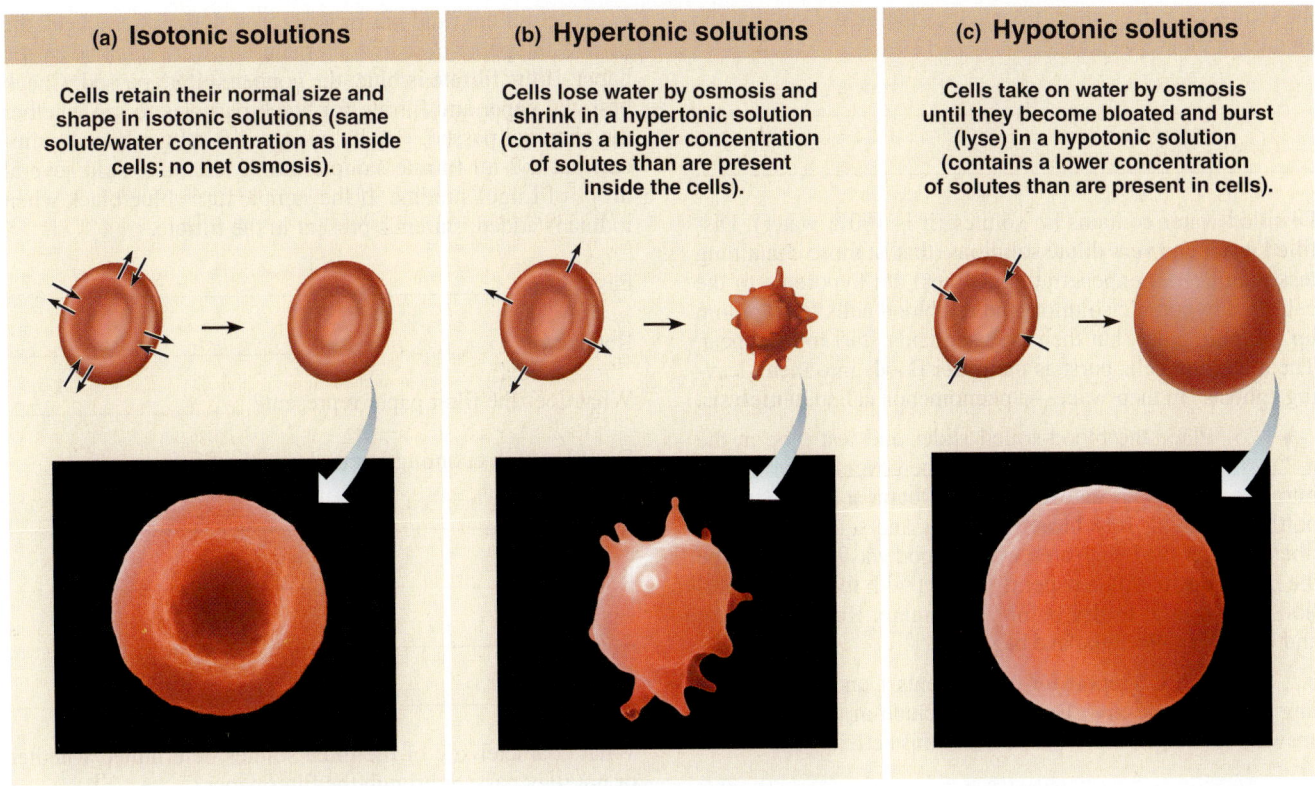

(a) Isotonic solutions

Cells retain their normal size and shape in isotonic solutions (same solute/water concentration as inside cells; no net osmosis).

(b) Hypertonic solutions

Cells lose water by osmosis and shrink in a hypertonic solution (contains a higher concentration of solutes than are present inside the cells).

(c) Hypotonic solutions

Cells take on water by osmosis until they become bloated and burst (lyse) in a hypotonic solution (contains a lower concentration of solutes than are present in cells).

Figure 5.3 Influence of isotonic, hypertonic, and hypotonic solutions on red blood cells.

hypertonic solution by osmosis. A solution surrounding a cell is **hypotonic** if it contains fewer nonpenetrating solute particles than the interior of the cell. Water moves from a hypotonic solution into the cell by osmosis. In both cases, water moved down its concentration gradient. Indicate in your conclusions whether distilled water was a hypotonic or hypertonic solution and whether 30% sucrose was hypotonic or hypertonic.

Conclusions: _____

_____

_____

Experiment 2

Now you will conduct a microscopic study of red blood cells suspended in solutions of varying tonicities. The objective is to determine if these solutions have any effect on cell shape by promoting net osmosis.

1. The following supplies should be available at your laboratory bench to conduct this experimental series: two clean slides and coverslips, a vial of animal blood, a medicine dropper, physiological saline, 5% sodium chloride solution, distilled water, filter paper, and disposable gloves.

 Wear disposable gloves at all times when handling blood (steps 2–5).

2. Place a very small drop of physiological saline on a slide. Using the medicine dropper, add a small drop of animal blood

to the saline on the slide. Tilt the slide to mix, cover with a coverslip, and immediately examine the preparation under the high-power lens. Notice that the red blood cells retain their normal smooth disclike shape (see **Figure 5.3a**). This is because the physiological saline is **isotonic** to the cells. That is, it contains a concentration of nonpenetrating solutes (e.g., proteins and some ions) equal to that in the cells (same solute/water concentration). Consequently, the cells neither gain nor lose water by osmosis. Set this slide aside.

3. Prepare another wet mount of animal blood, but this time use 5% sodium chloride (saline) solution as the suspending medium. Carefully observe the red blood cells under high power. What is happening to the normally smooth disc shape of the red blood cells?

_____

_____

_____

This crinkling-up process, called **crenation,** is due to the fact that the 5% sodium chloride solution is hypertonic to the cytosol of the red blood cell. Under these circumstances, water leaves the cells by osmosis. Compare your observations to the figure above (Figure 5.3b).

4. Add a drop of distilled water to the edge of the coverslip. Fold a piece of filter paper in half and place its folded edge at the opposite edge of the coverslip; it will absorb the saline solution and draw the distilled water across the cells. Watch

the red blood cells as they float across the field. Describe the change in their appearance.

_____

_____

Distilled water contains *no* solutes (it is 100% water). Distilled water and *very* dilute solutions (that is, those containing less than 0.9% nonpenetrating solutes) are hypotonic to the cell. In a hypotonic solution, the red blood cells first "plump up" (Figure 5.3c), but then they suddenly start to disappear. The red blood cells burst as the water floods into them, leaving "ghosts" in their wake—a phenomenon called **hemolysis.**

⚠ 5. Place the blood-soiled slides and test tube in the bleach-containing basin. Put the coverslips you used into the disposable autoclave bag. Obtain a wash (squirt) bottle containing 10% bleach solution, and squirt the bleach liberally over the bench area where blood was handled. Wipe the bench down with a paper towel wet with the bleach solution and allow it to dry before continuing. Remove gloves, and discard in the autoclave bag.

6. Prepare a lab report for experiments 1 and 2. (See Getting Started, page xiv.) Be sure to include in the discussion answers to the questions proposed in this activity. ▬

Filtration

Filtration is a passive process in which water and solutes are forced through a membrane by hydrostatic (fluid) pressure. For example, fluids and solutes filter out of the capillaries in the kidneys and into the kidney tubules because the blood pressure in the capillaries is greater than the fluid pressure in the tubules. Filtration is not selective. The amount of filtrate (fluids and solutes) formed depends almost entirely on the pressure gradient (difference in pressure on the two sides of the membrane) and on the size of the membrane pores.

ACTIVITY 6

Observing the Process of Filtration

1. Obtain the following equipment: a ring stand, ring, and ring clamp; a funnel; a piece of filter paper; a beaker; a 10-ml graduated cylinder; a solution containing uncooked starch, powdered charcoal, and copper sulfate; and a dropper bottle of Lugol's iodine. Attach the ring to the ring stand with the clamp.

2. Fold the filter paper in half twice, open it into a cone, and place it in a funnel. Place the funnel in the ring of the ring stand and place a beaker under the funnel. Shake the starch solution, and fill the funnel with it to just below the top of the filter paper. When the steady stream of filtrate changes to countable filtrate drops, count the number of drops formed in 10 seconds and record.

_____ drops

When the funnel is half empty, again count the number of drops formed in 10 seconds and record the count.

_____ drops

3. After all the fluid has passed through the filter, check the filtrate and paper to see which materials were retained by the paper. If the filtrate is blue, the copper sulfate passed. Check both the paper and filtrate for black particles to see whether the charcoal passed. Finally, using a 10-ml graduated cylinder, put a 2-ml filtrate sample into a test tube. Add several drops of Lugol's iodine. If the sample turns blue/black when iodine is added, starch is present in the filtrate.

Passed: _____

Retained: _____

What does the filter paper represent? _____

During which counting interval was the filtration rate

greatest? _____

Explain: _____

_____

_____

What characteristic of the three solutes determined whether or not they passed through the filter paper?

_____ ▬

Active Processes

Whenever a cell uses the bond energy of ATP to move substances across its boundaries, the process is an *active process*. Substances moved by active means are generally unable to pass by diffusion. They may not be lipid soluble; they may be too large to pass through the membrane channels; or they may have to move against rather than with a concentration gradient. There are two types of active processes: *active transport* and *vesicular transport*.

Active Transport

Like carrier-mediated facilitated diffusion, **active transport** requires carrier proteins that combine specifically with the transported substance. Active transport may be primary, driven directly by hydrolysis of ATP, or secondary, driven indirectly by energy stored in ionic gradients. In most cases the substances move against concentration or electrochemical gradients or both. Some of the substances that are moved into the cells by such carriers are amino acids and some sugars. Both solutes are insoluble in lipid and too large to pass through membrane channels but are necessary for cell life. Sodium ions (Na^+) are ejected from cells by active transport. Active transport is difficult to study in an A&P laboratory and will not be considered further here.

Vesicular Transport

In **vesicular transport,** fluids containing large particles and macromolecules are transported across cellular membranes inside membranous sacs called *vesicles*. Like active

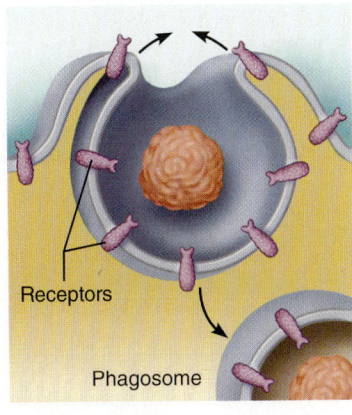

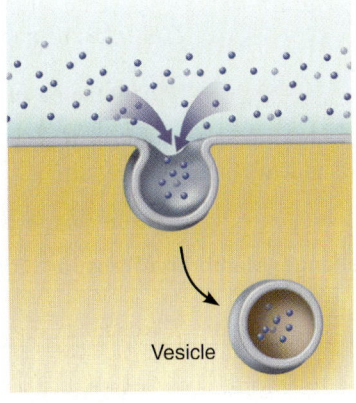

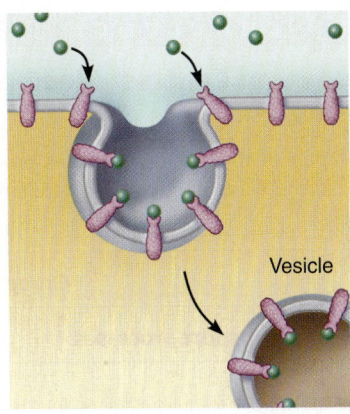

(a) Phagocytosis   **(b) Pinocytosis**   **(c) Receptor-mediated endocytosis**

Figure 5.4 **Three types of endocytosis. (a)** In phagocytosis, cellular extensions flow around the external particle and enclose it within a phagosome. **(b)** In pinocytosis, fluid and dissolved solutes enter the cell in a tiny vesicle. **(c)** In receptor-mediated endocytosis, specific substances attach to cell-surface receptors and enter the cell in protein-coated vesicles.

transport, vesicular transport moves substances into the cell (**endocytosis**) and out of the cell (**exocytosis**). Vesicular transport can combine endocytosis and exocytosis by moving substances into, across, and out of cells as well as moving substances from one area or membranous organelle to another. Vesicular transport requires energy, usually in the form of ATP, and all forms of vesicular transport involve protein-coated vesicles to some extent.

There are three types of endocytosis: phagocytosis, pinocytosis, and receptor-mediated endocytosis. In **phagocytosis** ("cell eating"), the cell engulfs some relatively large or solid material such as a clump of bacteria, cell debris, or inanimate particles **(Figure 5.4a)**. When a particle binds to receptors on the cell's surface, cytoplasmic extensions called pseudopods form and flow around the particle. This produces an endocytotic vesicle called a *phagosome*. In most cases, the phagosome then fuses with a lysosome and its contents are digested. Indigestible contents are ejected from the cell by exocytosis. In the human body, only macrophages and certain other white blood cells perform phagocytosis. These cells help protect the body from disease-causing microorganisms and cancer cells.

In **pinocytosis** ("cell drinking"), also called **fluid-phase endocytosis,** the cell "gulps" a drop of extracellular fluid containing dissolved molecules (Figure 5.4b). Since no receptors are involved, the process is nonspecific. Unlike phagocytosis, pinocytosis is a routine activity of most cells, affording them a way of sampling the extracellular fluid. It is particularly important in cells that absorb nutrients, such as cells that line the intestines.

The main mechanism for *specific* endocytosis of most macromolecules is **receptor-mediated endocytosis** (Figure 5.4c). The receptors for this process are plasma membrane proteins that bind only certain substances. This exquisitely selective mechanism allows cells to concentrate material that is present only in small amounts in the extracellular fluid. The ingested vesicle may fuse with a lysosome that either digests or releases its contents, or it may be transported across the cell to release its contents by exocytosis. The latter case is common in endothelial cells lining blood vessels because it provides a quick means to get substances from blood to extracellular fluid. Substances taken up by receptor-mediated endocytosis include enzymes, insulin and some other hormones, cholesterol (attached to a transport protein), and iron. Unfortunately, flu viruses, diphtheria, and cholera toxins also enter cells by this route.

Exocytosis is a vesicular transport process that ejects substances from the cell into the extracellular fluid. The substance to be removed from the cell is first enclosed in a protein-coated vesicle called a *secretory vesicle*. In most cases the vesicle migrates to the plasma membrane, fuses with it, and then ruptures, spilling its contents out of the cell. Exocytosis is used for hormone secretion, neurotransmitter release, mucus secretion, and ejection of wastes.

ACTIVITY 7

Observing Phagocytosis

Go to the video viewing area and watch the video demonstration of phagocytosis (if available). ■

Note: If you have not already done so, complete Activity 2 ("Observing Diffusion of Dye Through Water," page 56), and Activity 4 ("Observing Osmometer Results," page 58).

Compare and Contrast Membrane Transport Processes

For each pair of membrane transport processes listed in the Group Challenge chart, describe ways in which they are similar and ways in which they differ.

Group Challenge: Membrane Transport Comparison		
Membrane transport processes	**Similarities**	**Differences**
Simple diffusion Osmosis		
Simple diffusion Facilitated diffusion		
Active transport Facilitated diffusion		
Filtration Osmosis		
Pinocytosis Receptor-mediated endocytosis		

Name _____

Lab Time/Date _____

E X E R C I S E

5

The Cell: Transport
Mechanisms and
Permeability

R E V I E W   S H E E T

Choose all answers that apply to questions 1 and 2, and place their letters on the response blanks to the right.

1. Molecular motion _____.
   a. reflects the kinetic energy of molecules
   b. reflects the potential energy of molecules
   c. is ordered and predictable
   d. is random and erratic

2. Velocity of molecular movement _____.
   a. is higher in larger molecules
   b. is lower in larger molecules
   c. increases with increasing temperature
   d. decreases with increasing temperature
   e. reflects kinetic energy

3. Summarize the results of Activity 3, diffusion and osmosis through nonliving membranes, below. List and explain your observations relative to tests used to identify diffusing substances, and changes in sac weight observed.

   Sac 1 containing 40% glucose, suspended in distilled water

   _____

   _____

   Sac 2 containing 40% glucose, suspended in 40% glucose

   _____

   _____

   Sac 3 containing 10% NaCl, suspended in distilled water

   _____

   _____

   Sac 4 containing 40% sucrose and Congo red dye, suspended in distilled water

   _____

   _____

4. What single characteristic of the selectively permeable membranes *used in the laboratory* determines the substances that can

   pass through them? _____

   In addition to this characteristic, what other factors influence the passage of substances through living membranes?

   _____

   _____

5. A semipermeable sac containing 4% NaCl, 9% glucose, and 10% albumin is suspended in a solution with the following composition: 10% NaCl, 10% glucose, and 40% albumin. Assume that the sac is permeable to all substances except albumin. State whether each of the following will (a) move into the sac, (b) move out of the sac, or (c) not move.

glucose: _____   albumin: _____

water: _____   NaCl: _____

6. Summarize the results of Activity 5, Experiment 1 (diffusion and osmosis through living membranes—the egg), below. List and explain your observations.

Egg 1 in distilled water: _____

_____

_____

Egg 2 in 30% sucrose: _____

_____

_____

7. The diagrams below represent three microscope fields containing red blood cells. Arrows show the direction of net osmosis.

Which field contains a hypertonic solution? _____ The cells in this field are said to be _____. Which field

contains an isotonic bathing solution? _____ Which field contains a hypotonic solution? _____ What is

happening to the cells in this field? _____

_____

(a)   (b)   (c)

8. Assume you are conducting the experiment illustrated in the next figure. Both hydrochloric acid (HCl) with a molecular weight of about 36.5 and ammonium hydroxide (NH_4OH) with a molecular weight of 35 are volatile and easily enter the gaseous state. When they meet, the following reaction will occur:

$$HCl + NH_4OH \rightarrow H_2O + NH_4Cl$$

Ammonium chloride (NH_4Cl) will be deposited on the glass tubing as a smoky precipitate where the two gases meet. Predict which gas will diffuse more quickly and indicate to which end of the tube the smoky precipitate will be closer.

a. The faster-diffusing gas is _____.

b. The precipitate forms closer to the _____ end.

Rubber stopper   Cotton wad with HCl   Cotton wad with NH_4OH

Support

9. What determines whether a transport process is active or passive?

_____

10. Characterize membrane transport as fully as possible by choosing all the phrases that apply and inserting their letters on the answer blanks.

Passive processes: _____ Active processes: _____

   a.   account for the movement of fats and respiratory gases through the plasma membrane
   b.   explain solute pumping, phagocytosis, and pinocytosis
   c.   include osmosis, simple diffusion, and filtration
   d.   may occur against concentration and/or electrical gradients
   e.   use hydrostatic pressure or molecular energy as the driving force
   f.   move ions, amino acids, and some sugars across the plasma membrane

11. For the osmometer demonstration (Activity 4), explain why the level of the water column rose during the laboratory session.

_____

_____

12. Define the following terms.

selective permeability: _____

_____

diffusion: _____

_____

simple diffusion: _____

_____

facilitated diffusion: _____

_____

osmosis: _____

_____

filtration: _____

_____

vesicular transport: _____

_____

endocytosis: _____

_____

exocytosis: _____

_____

Classification of Tissues

MATERIALS

- ☐ Compound microscope
- ☐ Immersion oil
- ☐ Prepared slides of simple squamous, simple cuboidal, simple columnar, stratified squamous (nonkeratinized), stratified cuboidal, stratified columnar, pseudostratified ciliated columnar, and transitional epithelium
- ☐ Prepared slides of mesenchyme; of adipose, areolar, reticular, and dense (both regular and irregular connective tissues); of hyaline and elastic cartilage; of fibrocartilage; of bone (x.s.); and of blood
- ☐ Prepared slide of nervous tissue (spinal cord smear)
- ☐ Prepared slides of skeletal, cardiac, and smooth muscle (l.s.)
- ☐ Envelopes containing index cards with color photomicrographs of tissues

MasteringA&P® For related exercise study tools, go to the Study Area of MasteringA&P. There you will find:
- Practice Anatomy Lab PAL
- PhysioEx PEx
- A&PFlix A&PFlix
- Practice quizzes, Histology Atlas, eText, Videos, and more!

OBJECTIVES

1. Name the four primary tissue types in the human body and state a general function of each.
2. Name the major subcategories of the primary tissue types and identify the tissues of each subcategory microscopically or in an appropriate image.
3. State the locations of the various tissues in the body.
4. List the general function and structural characteristics of each of the tissues studied.

PRE-LAB QUIZ

1. Groups of cells that are anatomically similar and share a function are called:
   a. organ systems
   b. organisms
   c. organs
   d. tissues
2. How many primary tissue types are found in the human body?
   _____
3. Circle True or False. Endocrine and exocrine glands are classified as epithelium because they usually develop from epithelial membranes.
4. Epithelial tissues can be classified according to cell shape. _____ epithelial cells are scalelike and flattened.
   a. Columnar
   b. Cuboidal
   c. Squamous
   d. Transitional
5. All connective tissue is derived from an embryonic tissue known as:
   a. cartilage
   b. ground substance
   c. mesenchyme
   d. reticular
6. All the following are examples of connective tissue except:
   a. bones
   b. ligaments
   c. neurons
   d. tendons
7. Circle True or False. Blood is a type of connective tissue.
8. Circle the correct underlined term. Of the two major cell types found in nervous tissue, neurons / neuroglial cells are highly specialized to generate and conduct electrical signals.
9. How many basic types of muscle tissue are there? _____
10. This type of muscle tissue is found in the walls of hollow organs. It has no striations, and its cells are spindle shaped. It is:
    a. cardiac muscle
    b. skeletal muscle
    c. smooth muscle

C ells are the building blocks of life and the all-inclusive functional units of unicellular organisms. However, in higher organisms, cells do not usually operate as isolated, independent entities. In humans and other multicellular organisms, cells depend on one another and cooperate to maintain homeostasis in the body.

With a few exceptions, even the most complex animal starts out as a single cell, the fertilized egg, which divides almost endlessly. The trillions of cells that result become specialized for a particular function; some become supportive bone, others the transparent lens of the eye, still others skin cells, and so on. Thus a division of labor exists, with certain groups of cells highly specialized to perform functions that benefit the organism as a whole. Cell specialization carries with it certain hazards, because when a small specific group of cells is indispensable, any inability to function on its part can paralyze or destroy the entire body.

Groups of cells that are similar in structure and function are called **tissues.** The four primary tissue types—epithelium, connective tissue, nervous tissue, and muscle—have distinctive structures, patterns, and functions. The four primary tissues are further divided into subcategories, as described shortly.

To perform specific body functions, the tissues are organized into **organs** such as the heart, kidneys, and lungs. Most organs contain several representatives of the primary tissues, and the arrangement of these tissues determines the organ's structure and function. Thus **histology,** the study of tissues, complements a study of gross anatomy and provides the structural basis for a study of organ physiology.

The main objective of this exercise is to familiarize you with the major similarities and differences of the primary tissues, so that when the tissue composition of an organ is described, you will be able to more easily understand (and perhaps even predict) the organ's major function. Because epithelial tissue and some types of connective tissue will not be considered again, they are emphasized more than muscle, nervous tissue, and bone (a connective tissue) in this exercise.

Epithelial Tissue

Epithelial tissue, or an **epithelium,** is a sheet of cells that covers a body surface or lines a body cavity. It occurs in the body as (1) covering and lining epithelium and (2) glandular epithelium. Covering and lining epithelium forms the outer layer of the skin and lines body cavities that open to the outside. It covers the walls and organs of the closed ventral body cavity. Since glands almost invariably develop from epithelial sheets, glands are also classed as epithelium.

Epithelial functions include protection, absorption, filtration, excretion, secretion, and sensory reception. For example, the epithelium covering the body surface protects against bacterial invasion and chemical damage; that lining the respiratory tract is ciliated to sweep dust and other foreign particles away from the lungs. Epithelium specialized to absorb substances lines the stomach and small intestine. In the kidney tubules, the epithelium

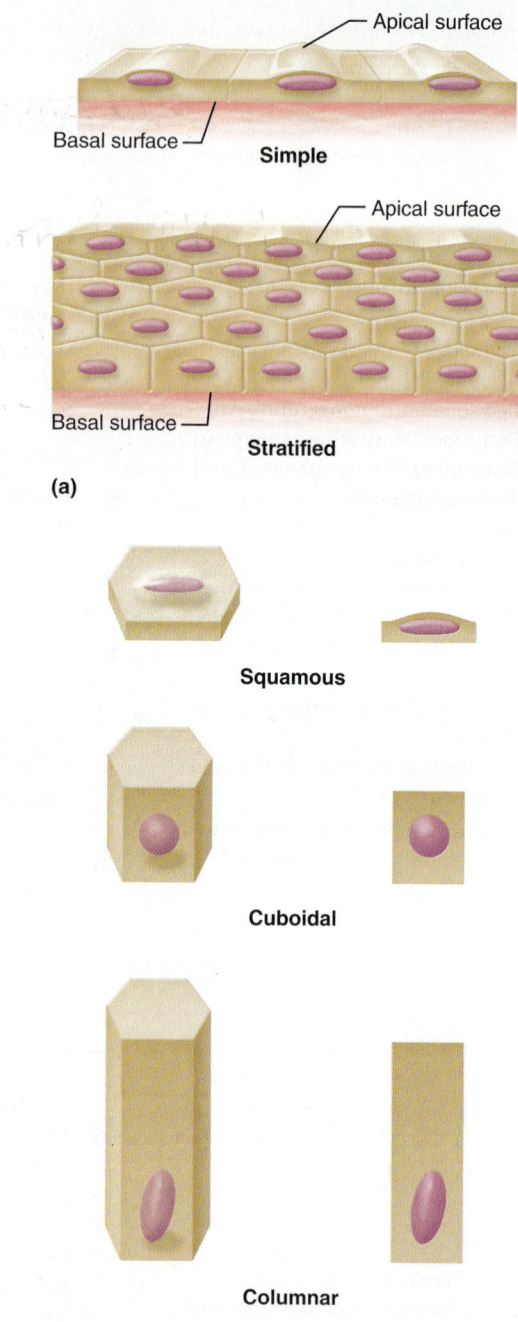

(a)

Simple

Stratified

Squamous

Cuboidal

Columnar

(b)

Figure 6.1 Classification of epithelia. (a) Classification based on number of cell layers. **(b)** Classification based on cell shape. For each category, a whole cell is shown on the left and a longitudinal section is shown on the right.

absorbs, secretes, and filters. Secretion is a specialty of the glands.

The following characteristics distinguish epithelial tissues from other types:

- Polarity. The membranes always have one free surface, called the *apical surface,* and typically that surface is significantly different from the *basal surface.*

- Specialized contacts. Cells fit closely together to form membranes, or sheets of cells, and are bound together by specialized junctions.

- Supported by connective tissue. The cells are attached to and supported by an adhesive **basement membrane,** which is an amorphous material secreted partly by the epithelial cells *(basal lamina)* and connective tissue cells *(reticular lamina)* that lie next to each other.

- Avascular but innervated. Epithelial tissues are supplied by nerves but have no blood supply of their own (are avascular). Instead they depend on diffusion of nutrients from the underlying connective tissue. Glandular epithelia, however, are very vascular.

- Regeneration. If well nourished, epithelial cells can easily divide to regenerate the tissue. This is an important characteristic because many epithelia are subjected to a good deal of friction.

The covering and lining epithelia are classified according to two criteria—arrangement or relative number of layers and cell shape **(Figure 6.1)**. On the basis of arrangement, there are **simple** epithelia, consisting of one layer of cells attached to the basement membrane, and **stratified** epithelia, consisting of two or more layers of cells. The general types based on shape are **squamous** (scalelike), **cuboidal** (cubelike), and **columnar** (column-shaped) epithelial cells. The terms denoting shape and arrangement of the epithelial cells are combined to describe the epithelium fully. *Stratified epithelia are named according to the cells at the apical surface of the epithelial sheet,* not those resting on the basement membrane.

There are, in addition, two less easily categorized types of epithelia. **Pseudostratified epithelium** is actually a simple columnar epithelium (one layer of cells), but because its cells vary in height and the nuclei lie at different levels above the basement membrane, it gives the false appearance of being stratified. This epithelium is often ciliated. **Transitional epithelium** is a rather peculiar stratified squamous epithelium formed of rounded, or "plump," cells with the ability to slide over one another to allow the organ to be stretched. Transitional epithelium is found only in urinary system organs subjected to periodic distension, such as the bladder. The superficial cells are flattened (like true squamous cells) when the organ is distended and rounded when the organ is empty.

Epithelial cells forming glands are highly specialized to remove materials from the blood and to manufacture them into new materials, which they then secrete. There are two types of glands, *endocrine* and *exocrine* **(Figure 6.2)**. **Endocrine glands** lose their surface connection (duct) as they develop; thus they are referred to as ductless glands. They secrete hormones into the extracellular fluid, and from there the hormones enter the blood or the lymphatic vessels that weave through the glands. **Exocrine glands** retain their ducts, and their secretions empty through these ducts either to the body surface or into body cavities. The exocrine glands include the sweat and oil glands, liver, and pancreas. Glands are discussed with the organ systems to which their products are functionally related.

The most common types of epithelia, their characteristic locations in the body, and their functions are described in the accompanying illustrations **(Figure 6.3)**.

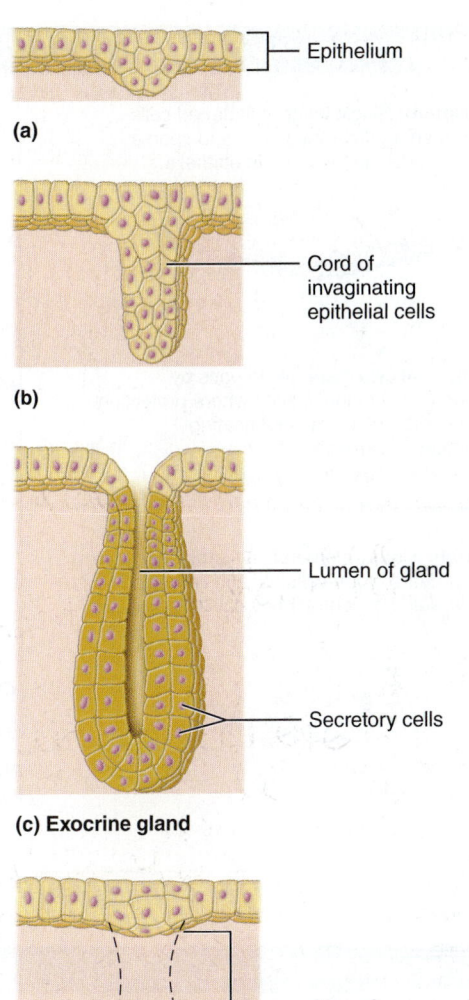

(a)

(b)

(c) Exocrine gland

Epithelium

Cord of invaginating epithelial cells

Lumen of gland

Secretory cells

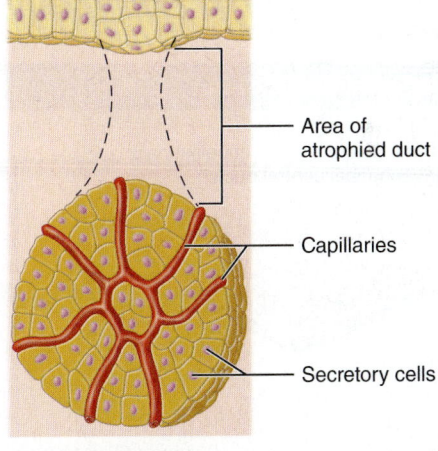

(d) Endocrine gland

Area of atrophied duct

Capillaries

Secretory cells

Figure 6.2 Formation of endocrine and exocrine glands from epithelial sheets. (a) Epithelial cells grow and push into the underlying tissue. **(b)** A cord of epithelial cells forms. **(c)** In an exocrine gland, a lumen (cavity) forms. The inner cells form the duct, the outer cells produce the secretion. **(d)** In a forming endocrine gland, the connecting duct cells atrophy, leaving the secretory cells with no connection to the epithelial surface. However, they do become heavily invested with blood and lymphatic vessels that receive the secretions.

(Text continues on page 74.)

(a) Simple squamous epithelium

Description: Single layer of flattened cells with disc-shaped central nuclei and sparse cytoplasm; the simplest of the epithelia.

Function: Allows materials to pass by diffusion and filtration in sites where protection is not important; secretes lubricating substances in serosae.

Location: Kidney glomeruli; air sacs of lungs; lining of heart, blood vessels, and lymphatic vessels; lining of ventral body cavity (serosae).

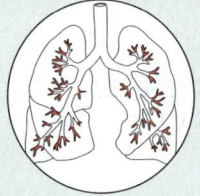

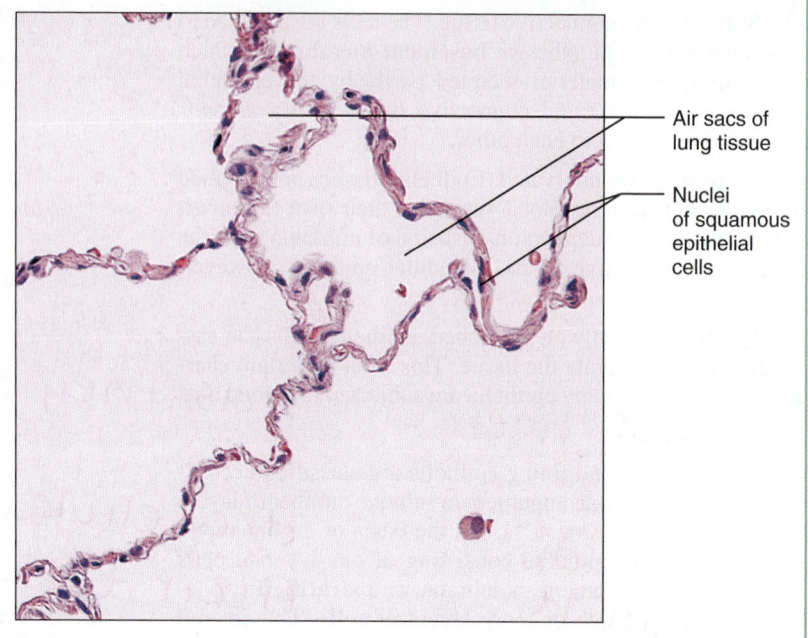

Air sacs of lung tissue

Nuclei of squamous epithelial cells

Photomicrograph: Simple squamous epithelium forming part of the alveolar (air sac) walls (140×).

(b) Simple cuboidal epithelium

Description: Single layer of cubelike cells with large, spherical central nuclei.

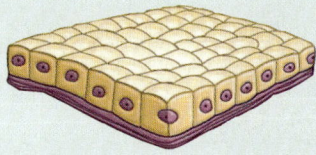

Function: Secretion and absorption.

Location: Kidney tubules; ducts and secretory portions of small glands; ovary surface.

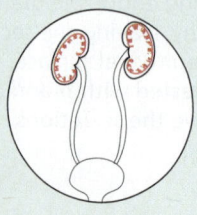

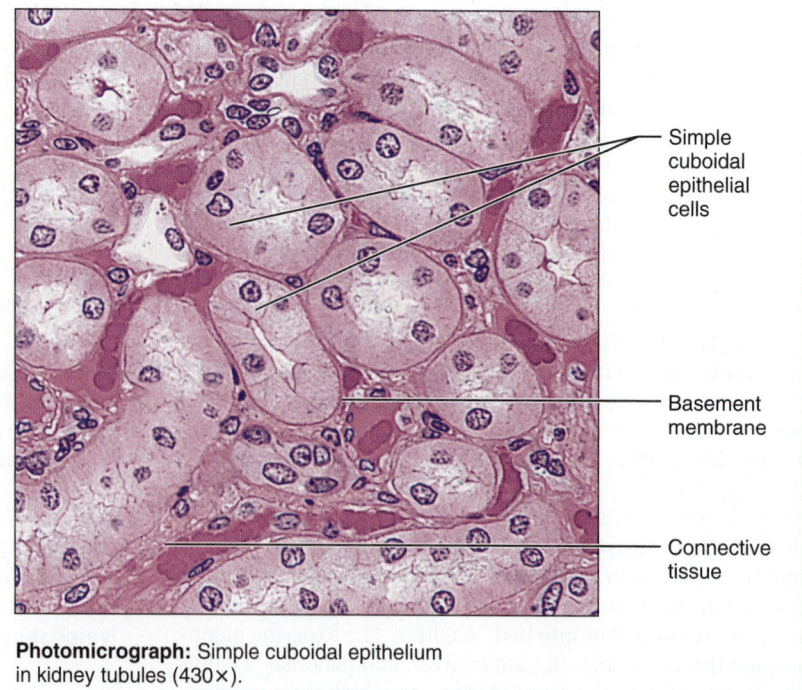

Simple cuboidal epithelial cells

Basement membrane

Connective tissue

Photomicrograph: Simple cuboidal epithelium in kidney tubules (430×).

Figure 6.3 Epithelial tissues. Simple epithelia (**a** and **b**).

(c) Simple columnar epithelium

Description: Single layer of tall cells with *round* to *oval* nuclei; some cells bear cilia; layer may contain mucus-secreting unicellular glands (goblet cells).

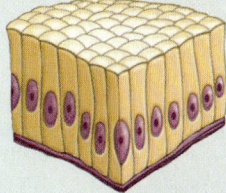

Function: Absorption; secretion of mucus, enzymes, and other substances; ciliated type propels mucus (or reproductive cells) by ciliary action.

Location: Nonciliated type lines most of the digestive tract (stomach to rectum), gallbladder, and excretory ducts of some glands; ciliated variety lines small bronchi, uterine tubes, and some regions of the uterus.

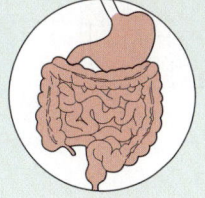

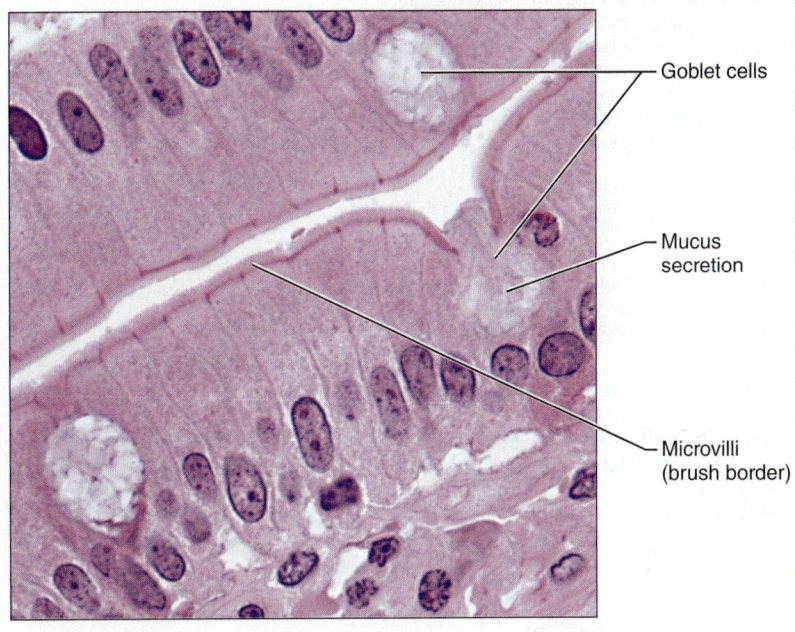

Goblet cells

Mucus secretion

Microvilli (brush border)

Photomicrograph: Simple columnar epithelium containing goblet cells from the small intestine (640×).

(d) Pseudostratified columnar epithelium

Description: Single layer of cells of differing heights, some not reaching the free surface; nuclei seen at different levels; may contain mucus-secreting goblet cells and bear cilia.

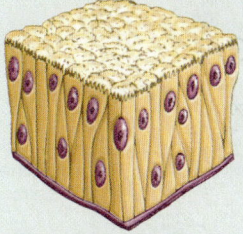

Function: Secretes substances, particularly mucus; propulsion of mucus by ciliary action.

Location: Nonciliated type in male's sperm-carrying ducts and ducts of large glands; ciliated variety lines the trachea, most of the upper respiratory tract.

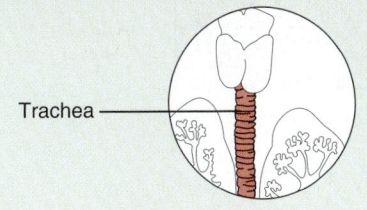

Trachea

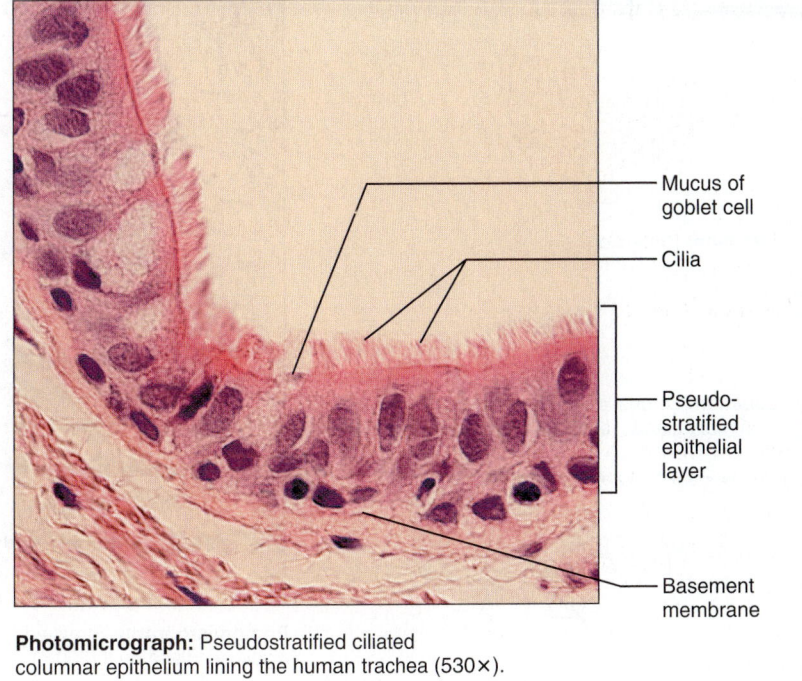

Mucus of goblet cell

Cilia

Pseudo-stratified epithelial layer

Basement membrane

Photomicrograph: Pseudostratified ciliated columnar epithelium lining the human trachea (530×).

6

Figure 6.3 *(continued)* Simple epithelia (**c** and **d**).

(e) Stratified squamous epithelium

Description: Thick membrane composed of several cell layers; basal cells are cuboidal or columnar and metabolically active; surface cells are flattened (squamous); in the keratinized type, the surface cells are full of keratin and dead; basal cells are active in mitosis and produce the cells of the more superficial layers.

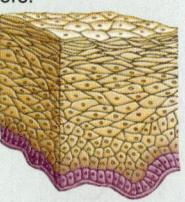

Function: Protects underlying tissues in areas subjected to abrasion.

Location: Nonkeratinized type forms the moist linings of the esophagus, mouth, and vagina; keratinized variety forms the epidermis of the skin, a dry membrane.

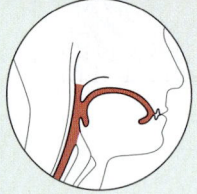

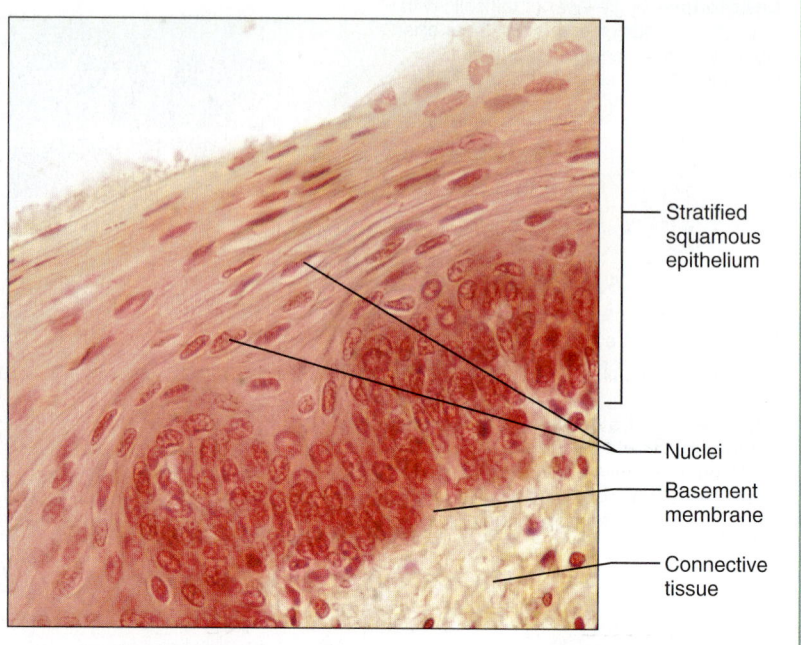

Photomicrograph: Stratified squamous epithelium lining the esophagus (280×).

Stratified squamous epithelium

Nuclei

Basement membrane

Connective tissue

(f) Stratified cuboidal epithelium

Description: Generally two layers of cubelike cells.

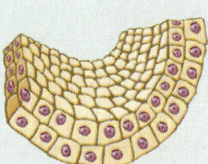

Function: Protection

Location: Largest ducts of sweat glands, mammary glands, and salivary glands.

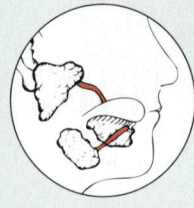

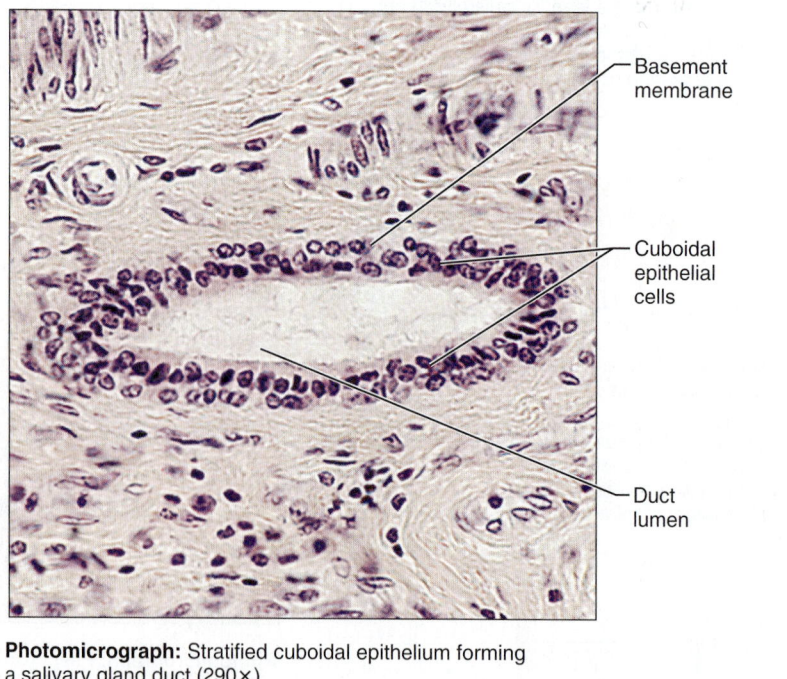

Photomicrograph: Stratified cuboidal epithelium forming a salivary gland duct (290×).

Basement membrane

Cuboidal epithelial cells

Duct lumen

Figure 6.3 *(continued)* **Epithelial tissues.** Stratified epithelia (**e** and **f**).

(g) Stratified columnar epithelium

Description: Several cell layers; basal cells usually cuboidal; superficial cells elongated and columnar.

Function: Protection; secretion.

Location: Rare in the body; small amounts in male urethra and in large ducts of some glands.

Urethra

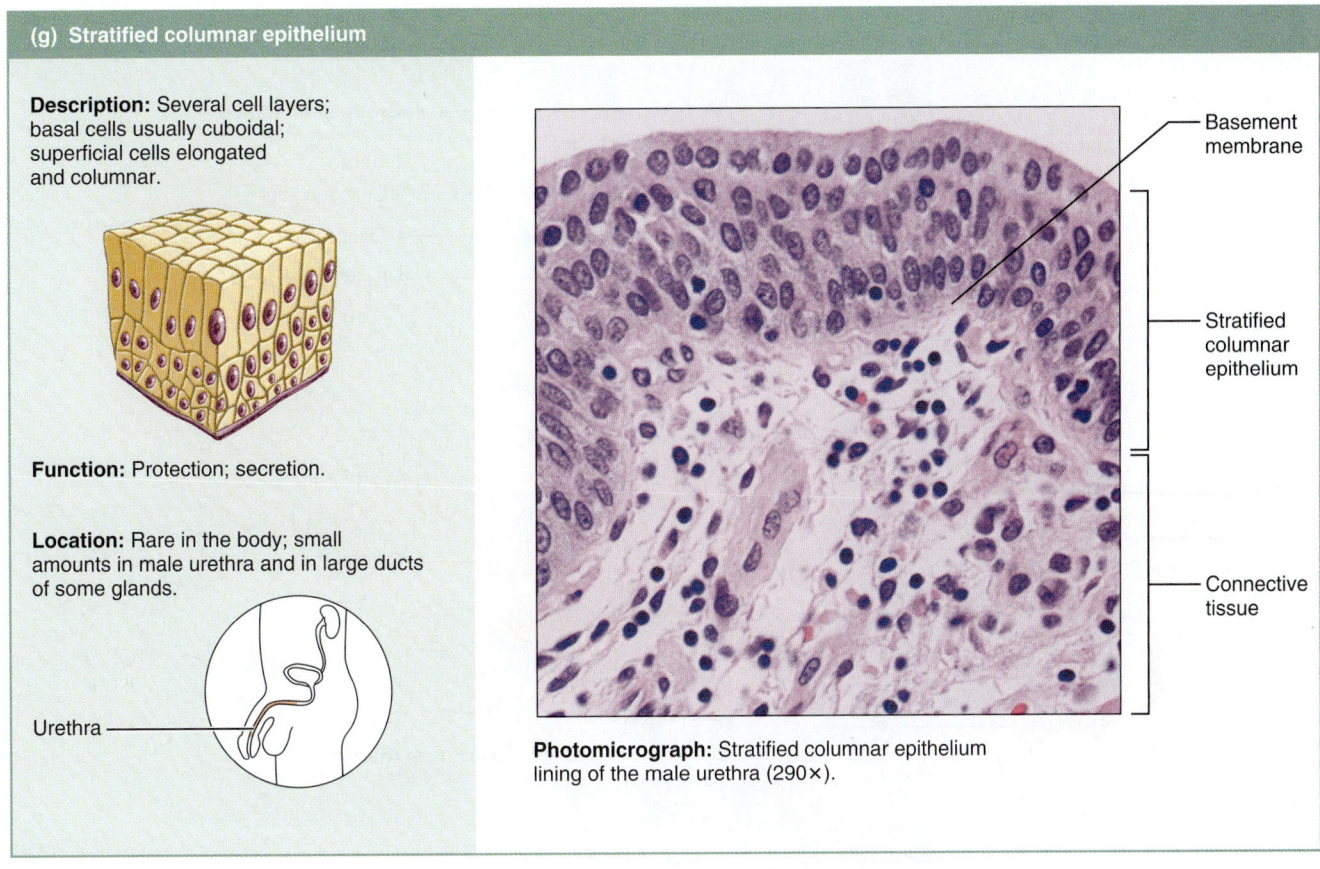

Basement membrane

Stratified columnar epithelium

Connective tissue

Photomicrograph: Stratified columnar epithelium lining of the male urethra (290×).

(h) Transitional epithelium

Description: Resembles both stratified squamous and stratified cuboidal; basal cells cuboidal or columnar; surface cells dome shaped or squamouslike, depending on degree of organ stretch.

Function: Stretches readily and permits distension of urinary organ by contained urine.

Location: Lines the ureters, urinary bladder, and part of the urethra.

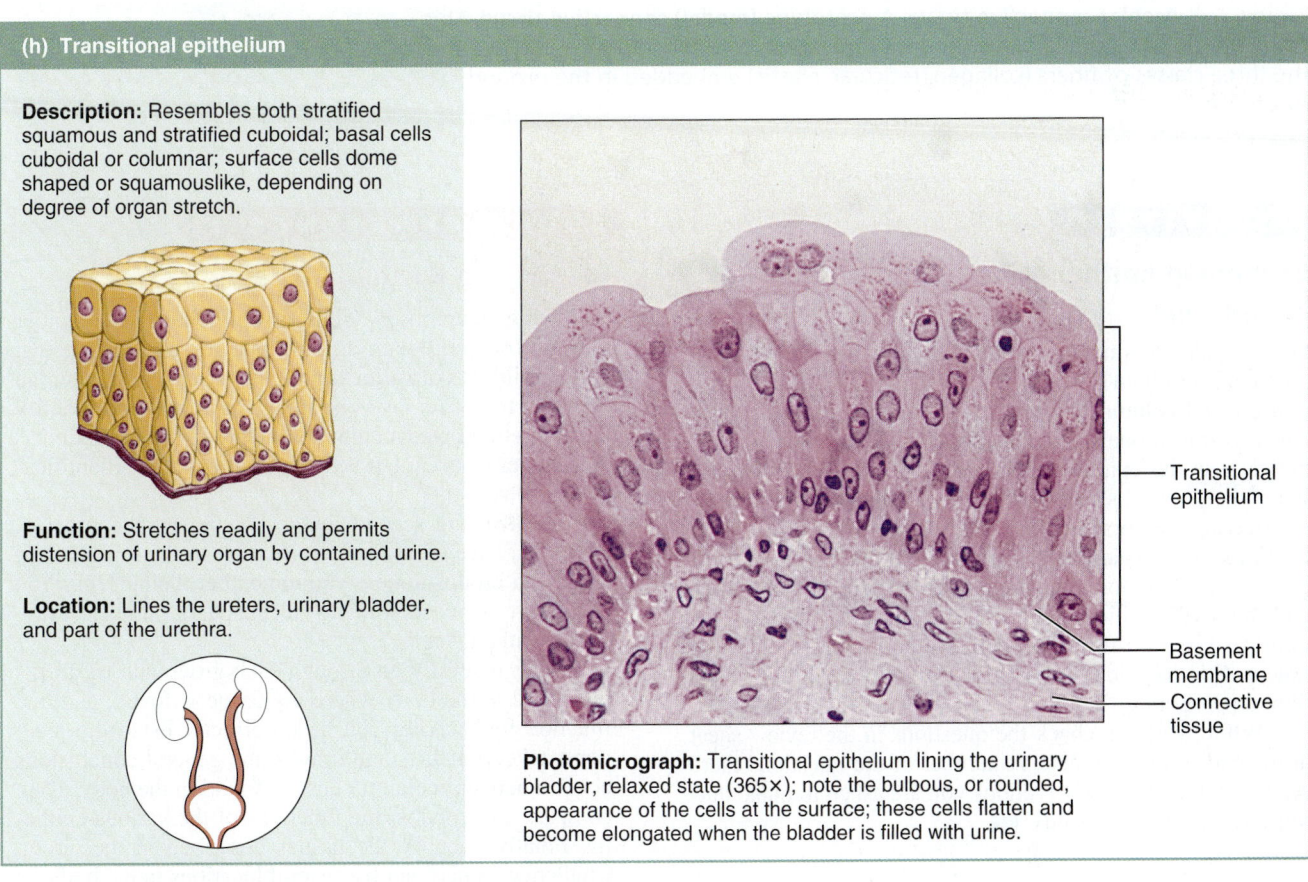

Transitional epithelium

Basement membrane

Connective tissue

Photomicrograph: Transitional epithelium lining the urinary bladder, relaxed state (365×); note the bulbous, or rounded, appearance of the cells at the surface; these cells flatten and become elongated when the bladder is filled with urine.

Figure 6.3 *(continued)* Stratified epithelia (**g** and **h**).

Cell types

Extracellular matrix

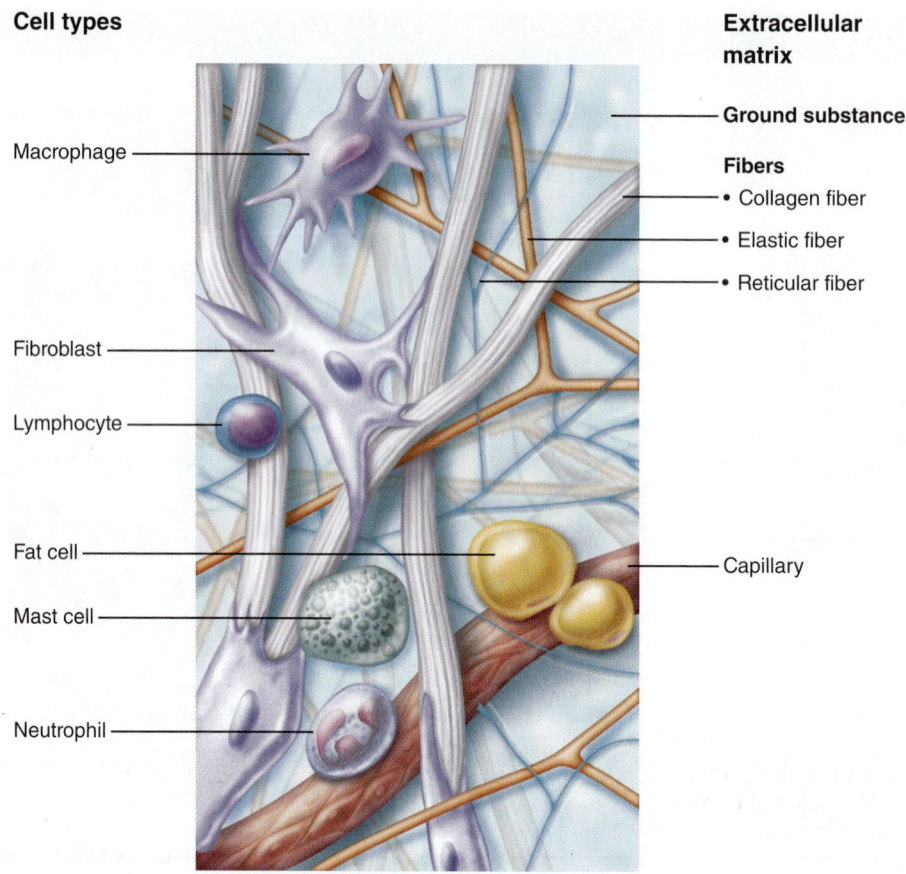

Macrophage

Fibroblast

Lymphocyte

Fat cell

Mast cell

Neutrophil

Ground substance

Fibers
• Collagen fiber
• Elastic fiber
• Reticular fiber

Capillary

Figure 6.4 Areolar connective tissue: A prototype (model) connective tissue. This tissue underlies epithelia and surrounds capillaries. Note the various cell types and the three classes of fibers (collagen, reticular, elastic) embedded in the ground substance.

ACTIVITY 1

Examining Epithelial Tissue Under the Microscope

Obtain slides of simple squamous, simple cuboidal, simple columnar, stratified squamous (nonkeratinized), pseudostratified ciliated columnar, stratified cuboidal, stratified columnar, and transitional epithelia. Examine each carefully, and notice how the epithelial cells fit closely together to form intact sheets of cells, a necessity for a tissue that forms linings or covering membranes. Scan each epithelial type for modifications for specific functions, such as cilia (motile cell projections that help to move substances along the cell surface), and microvilli, which increase the surface area for absorption. Also be alert for goblet cells, which secrete lubricating mucus. Compare your observations with the descriptions and photomicrographs (Figure 6.3.)

While working, check the questions in the review sheet at the end of this exercise. A number of the questions there refer to some of the observations you are asked to make during your microscopic study. ▪

GROUP CHALLENGE 1

Identifying Epithelial Tissues

Following your observations of epithelial tissues under the microscope, obtain an envelope for each group that contains images of various epithelial tissues. With your lab manual closed, remove one image at a time and identify the epithelium. One member of the group shall function as the verifier whose job it is to make sure that the identification is correct.

Remove the second image and repeat the process. After you have correctly identified all of the images, sort them into groups to help you remember them. (*Hint:* You could sort them according to cell shape, or number of layers of epithelial cells.)

Now, carefully go through each group and try to list one place in the body where the tissue is found, and one function for it. After you have correctly listed the locations, take your lists and draw some general conclusions about where epithelial tissues are found in the body. Then compare and contrast the functions of the various epithelia. Finally, identify the tissues described in the Group Challenge 1 chart and list several locations in the body.

Group Challenge 1: Epithelial Tissue IDs		
Magnified appearance	**Tissue type**	**Locations in the body**
• Apical surface has dome-shaped cells (flattened cells may also be mixed in) • Multiple layers of cells are present		
• Cells are mostly columnar • Not all cells reach the apical surface • Nuclei are located at different levels • Cilia are located at the apical surface		
• Apical surface has flattened cells with very little cytoplasm • Cells are not layered		
• Apical surface has square cells with a round nucleus • Cells are not layered		

Connective Tissue

Connective tissue is found in all parts of the body as discrete structures or as part of various body organs. It is the most abundant and widely distributed of the tissue types.

Connective tissues perform a variety of functions, but they primarily protect, support, and bind together other tissues of the body. For example, bones are composed of connective tissue (**bone,** or **osseous tissue**), and they protect and support other body tissues and organs. The ligaments and tendons (**dense connective tissue**) bind the bones together or bind skeletal muscles to bones.

Areolar connective tissue (Figure 6.4) is a soft packaging material that cushions and protects body organs. **Adipose** (fat) tissue provides insulation for the body tissues and a source of stored food. Blood-forming (**hematopoietic**) tissue replenishes the body's supply of red blood cells. Connective tissue also serves a vital function in the repair of all body tissues, since many wounds are repaired by connective tissue in the form of scar tissue.

The characteristics of connective tissue include the following:

• With a few exceptions (cartilages, tendons, and ligaments, which are poorly vascularized), connective tissues have a rich supply of blood vessels.

• Connective tissues are composed of many types of cells.

• There is a great deal of noncellular, nonliving material (matrix) between the cells of connective tissue.

The nonliving material between the cells—the **extracellular matrix**—deserves a bit more explanation because it distinguishes connective tissue from all other tissues. It is produced by the cells and then extruded. The matrix is primarily responsible for the strength associated with connective tissue, but there is variation. At one extreme, adipose tissue is composed mostly of cells. At the opposite extreme, bone and cartilage have few cells and large amounts of matrix.

The matrix has two components—ground substance and fibers. The **ground substance** is composed chiefly of interstitial fluid, cell adhesion proteins, and proteoglycans. Depending on its specific composition, the ground substance may be liquid, semisolid, gel-like, or very hard. When the matrix is firm, as in cartilage and bone, the connective tissue cells reside in cavities in the matrix called *lacunae*. The fibers, which provide support, include **collagen** (white) **fibers, elastic** (yellow) **fibers,** and **reticular** (fine collagen) **fibers.** Of these, the collagen fibers are most abundant.

Generally speaking, the ground substance functions as a molecular sieve, or medium, through which nutrients and other dissolved substances can diffuse between the blood capillaries and the cells. The fibers in the matrix hinder diffusion somewhat and make the ground substance less pliable. The properties of the connective tissue cells and the makeup and arrangement of their matrix elements vary tremendously, accounting for the amazing diversity of this tissue type. Nonetheless, the connective tissues have a common structural plan seen best in *areolar connective tissue* (Figure 6.4), a soft packing tissue that occurs throughout the body. Since all other connective tissues are variations of areolar, it is considered the model or prototype of the connective tissues. Notice that areolar tissue has all three varieties of fibers, but they are sparsely arranged in its transparent gel-like ground substance (Figure 6.4). The cell type that secretes its matrix is the *fibroblast,* but a wide variety of other cells (including phagocytic cells like macrophages and certain white blood cells and mast cells that act in the inflammatory response) are present as well. The more durable connective tissues, such as bone, cartilage, and the dense fibrous varieties, characteristically have a firm ground substance and many more fibers.

There are four main types of adult connective tissue, all of which typically have large amounts of matrix. These are **connective tissue proper** (which includes areolar, adipose, reticular, and dense [fibrous] connective tissues), **cartilage, bone,** and **blood.** All of these derive from an embryonic tissue called *mesenchyme.* The next set of illustrations **(Figure 6.5)** lists the general characteristics, location, and function of some of the connective tissues found in the body.

(Text continues on page 82.)

6

(a) Embryonic connective tissue: Mesenchyme

Description: Embryonic connective tissue; gel-like ground substance containing fibers; star-shaped mesenchymal cells.

Function: Gives rise to all other connective tissue types.

Location: Primarily in embryo.

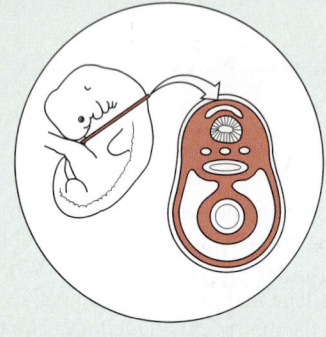

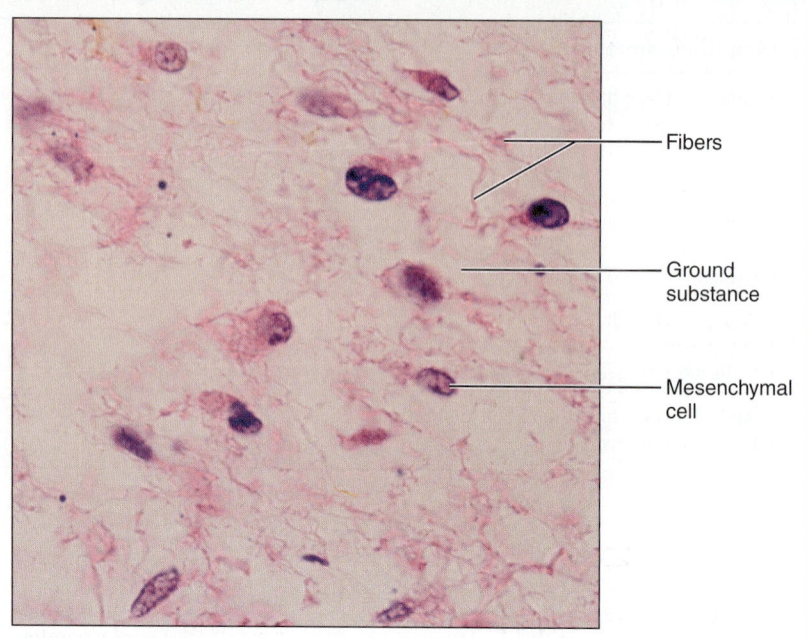

Fibers

Ground substance

Mesenchymal cell

Photomicrograph: Mesenchymal tissue, an embryonic connective tissue (627×); the clear-appearing background is the fluid ground substance of the matrix; notice the fine, sparse fibers.)

(b) Connective tissue proper: loose connective tissue, areolar

Description: Gel-like matrix with all three fiber types; cells: fibroblasts, macrophages, mast cells, and some white blood cells.

Function: Wraps and cushions organs; its macrophages phagocytize bacteria; plays important role in inflammation; holds and conveys tissue fluid.

Location: Widely distributed under epithelia of body, e.g., forms lamina propria of mucous membranes; packages organs; surrounds capillaries.

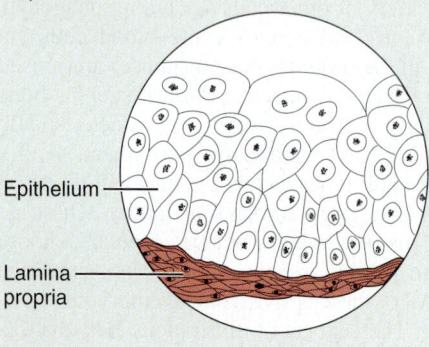

Epithelium

Lamina propria

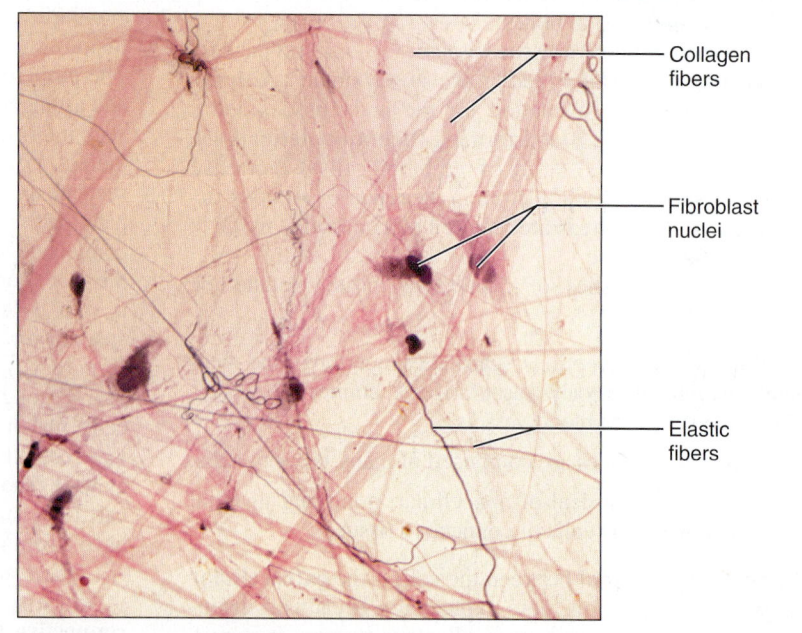

Collagen fibers

Fibroblast nuclei

Elastic fibers

Photomicrograph: Areolar connective tissue, a soft packaging tissue of the body (365×).

Figure 6.5 Connective tissues. Embryonic connective tissue **(a)** and Connective tissue proper **(b)**.

(c) Connective tissue proper: loose connective tissue, adipose

Description: Matrix as in areolar, but very sparse; closely packed adipocytes, or fat cells, have nucleus pushed to the side by large fat droplet.

Function: Provides reserve fuel; insulates against heat loss; supports and protects organs.

Location: Under skin; around kidneys and eyeballs; within abdomen; in breasts.

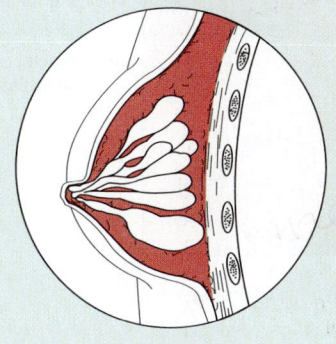

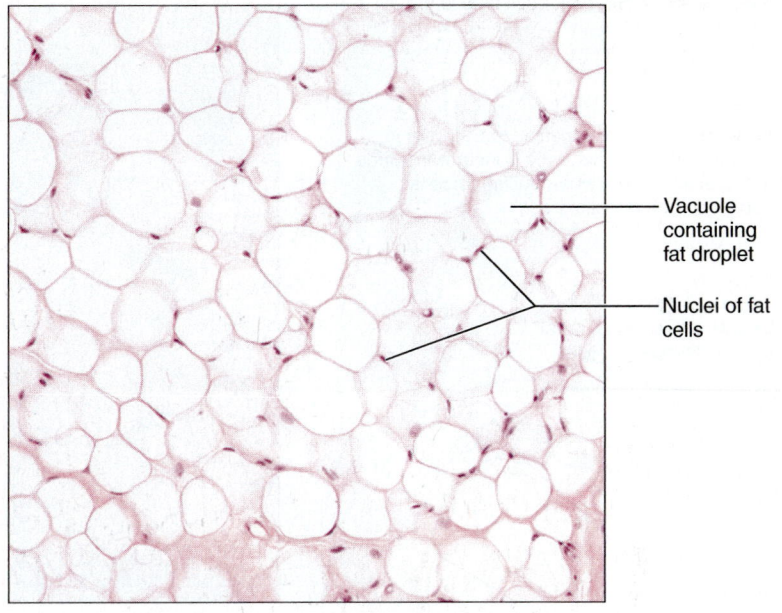

Vacuole containing fat droplet

Nuclei of fat cells

Photomicrograph: Adipose tissue from the subcutaneous layer under the skin (110×).

6

(d) Connective tissue proper: loose connective tissue, reticular

Description: Network of reticular fibers in a typical loose ground substance; reticular cells lie on the network.

Function: Fibers form a soft internal skeleton (stroma) that supports other cell types, including white blood cells, mast cells, and macrophages.

Location: Lymphoid organs (lymph nodes, bone marrow, and spleen).

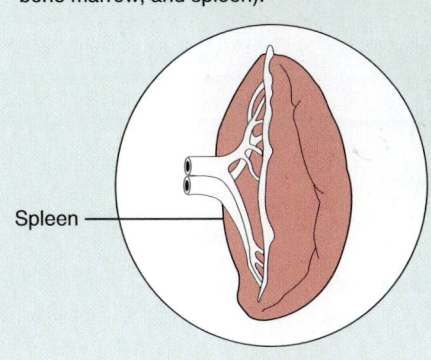

Spleen

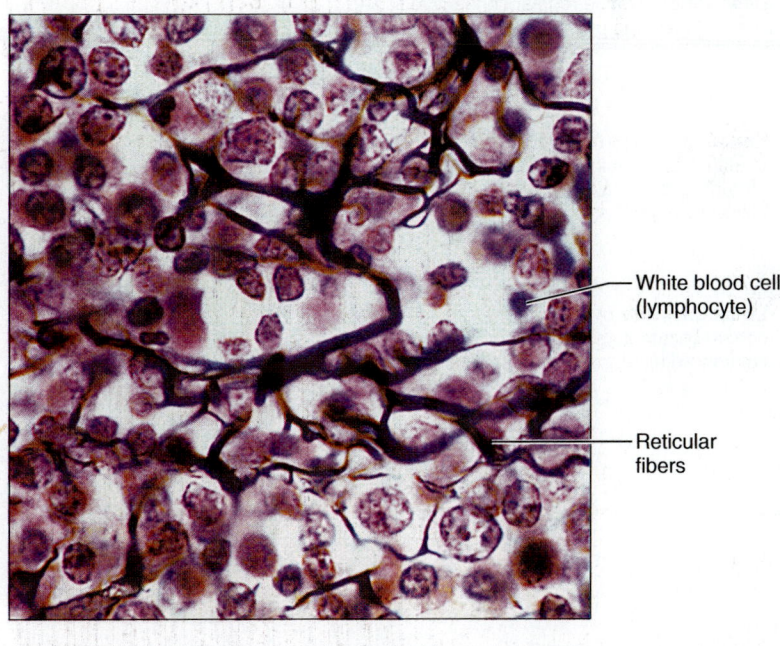

White blood cell (lymphocyte)

Reticular fibers

Photomicrograph: Dark-staining network of reticular connective tissue fibers forming the internal skeleton of the spleen (350×).

Figure 6.5 *(continued)* Connective tissue proper **(c** and **d).**

6

(e) Connective tissue proper: dense connective tissue, dense regular

Description: Primarily parallel collagen fibers; a few elastic fibers; major cell type is the fibroblast.

Function: Attaches muscles to bones or to muscles; attaches bones to bones; withstands great tensile stress when pulling force is applied in one direction.

Location: Tendons, most ligaments, aponeuroses.

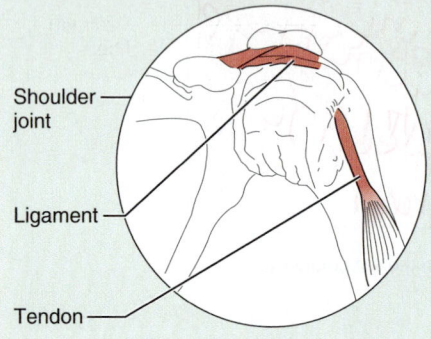

Shoulder joint

Ligament

Tendon

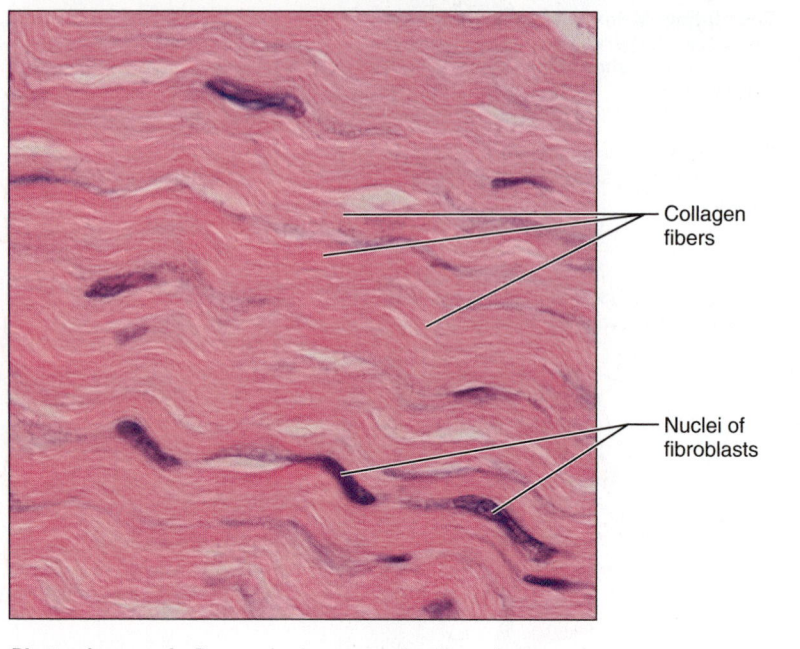

Collagen fibers

Nuclei of fibroblasts

Photomicrograph: Dense regular connective tissue from a tendon (590×).

(f) Connective tissue proper: dense connective tissue, elastic

Description: Dense regular connective tissue containing a high proportion of elastic fibers.

Function: Allows recoil of tissue following stretching; maintains pulsatile flow of blood through arteries; aids passive recoil of lungs following inspiration.

Location: Walls of large arteries; within certain ligaments associated with the vertebral column; within the walls of the bronchial tubes.

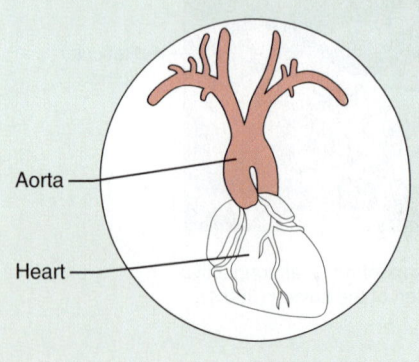

Aorta

Heart

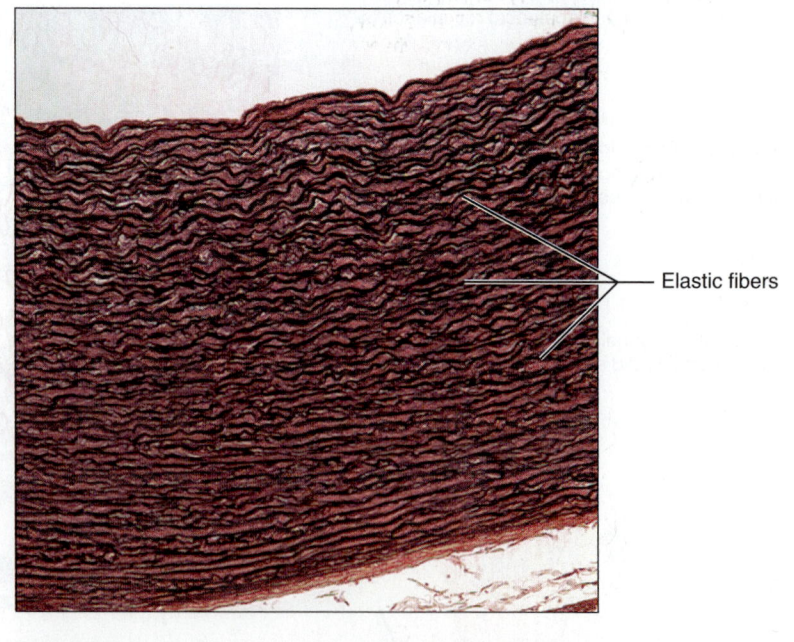

Elastic fibers

Photomicrograph: Elastic connective tissue in the wall of the aorta (250×).

Figure 6.5 *(continued)* **Connective tissues.** Connective tissue proper **(e)** and **(f)**.

(g) Connective tissue proper: dense connective tissue, dense irregular

Description: Primarily irregularly arranged collagen fibers; some elastic fibers; major cell type is the fibroblast.

Function: Able to withstand tension exerted in many directions; provides structural strength.

Location: Fibrous capsules of organs and of joints; dermis of the skin; submucosa of digestive tract.

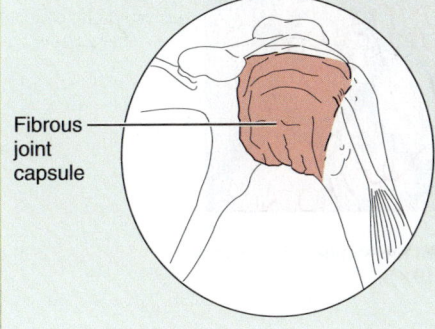

Fibrous joint capsule

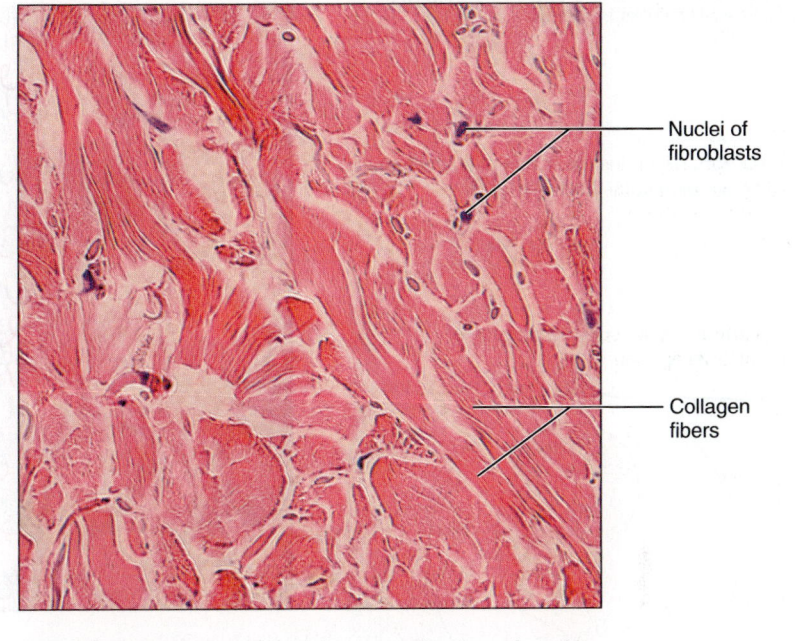

Nuclei of fibroblasts

Collagen fibers

Photomicrograph: Dense irregular connective tissue from the dermis of the skin (210×).

(h) Cartilage: hyaline

Description: Amorphous but firm matrix; collagen fibers form an imperceptible network; chondroblasts produce the matrix and when mature (chondrocytes) lie in lacunae.

Function: Supports and reinforces; serves as resilient cushion; resists compressive stress.

Location: Forms most of the embryonic skeleton; covers the ends of long bones in joint cavities; forms costal cartilages of the ribs; cartilages of the nose, trachea, and larynx.

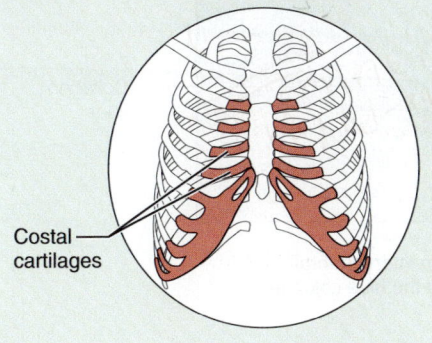

Costal cartilages

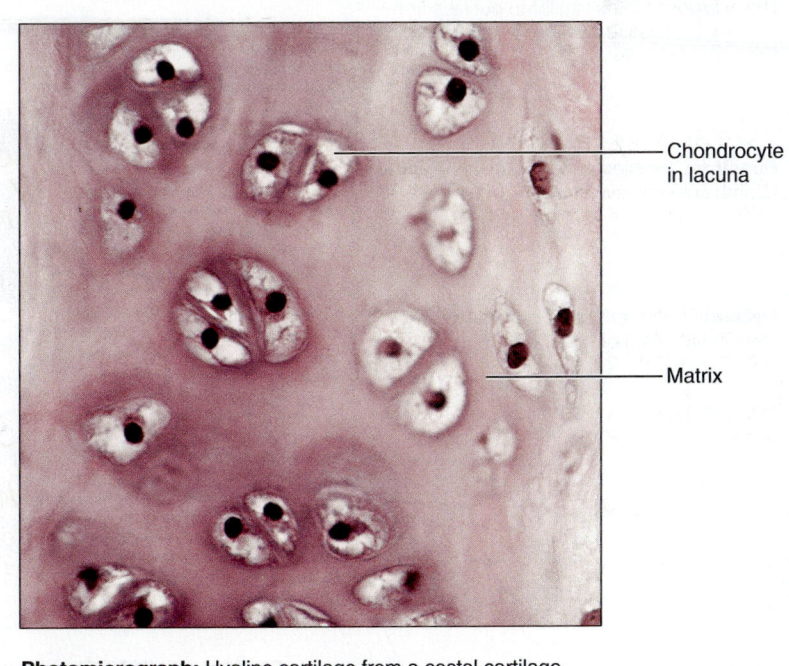

Chondrocyte in lacuna

Matrix

Photomicrograph: Hyaline cartilage from a costal cartilage of a rib (470×).

Figure 6.5 *(continued)* Connective tissue proper **(g)** and Cartilage **(h).**

(i) Cartilage: elastic

Description: Similar to hyaline cartilage, but more elastic fibers in matrix.

Function: Maintains the shape of a structure while allowing great flexibility.

Location: Supports the external ear (auricle); epiglottis.

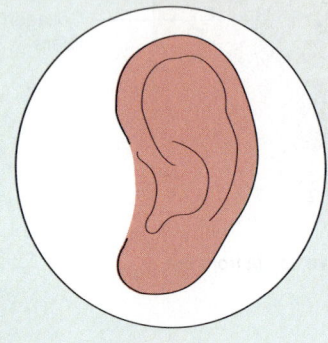

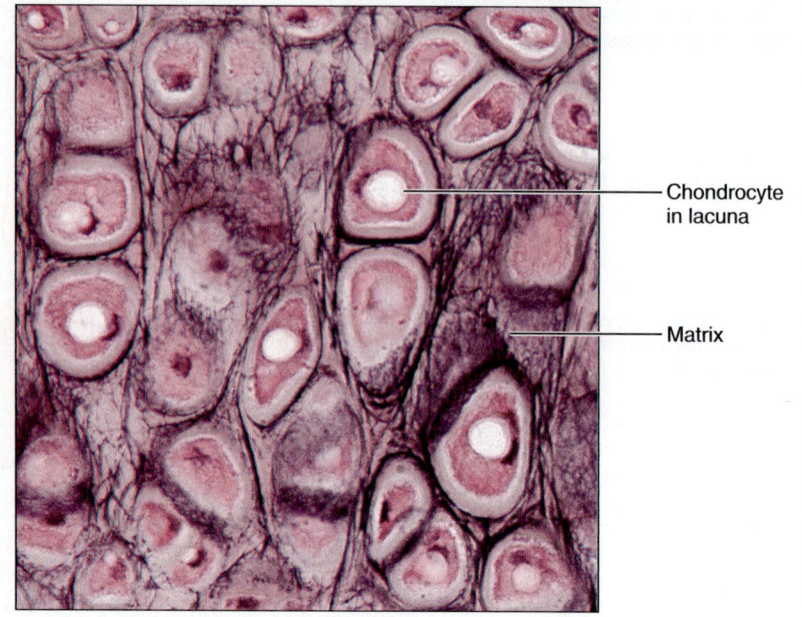

Chondrocyte in lacuna

Matrix

Photomicrograph: Elastic cartilage from the human ear auricle; forms the flexible skeleton of the ear (510×).

(j) Cartilage: fibrocartilage

Description: Matrix similar to but less firm than that in hyaline cartilage; thick collagen fibers predominate.

Function: Tensile strength with the ability to absorb compressive shock.

Location: Intervertebral discs; pubic symphysis; discs of knee joint.

Intervertebral discs

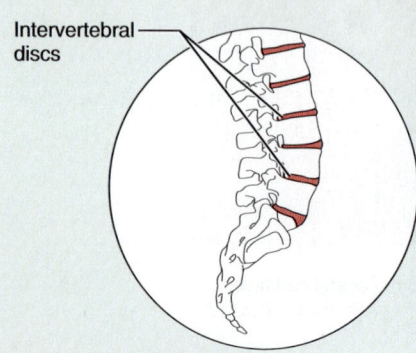

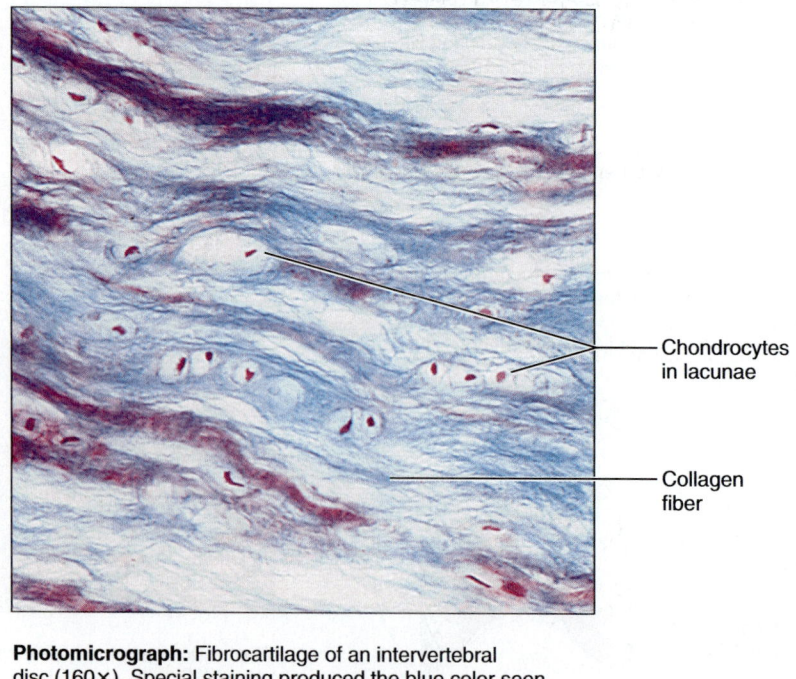

Chondrocytes in lacunae

Collagen fiber

Photomicrograph: Fibrocartilage of an intervertebral disc (160×). Special staining produced the blue color seen.

Figure 6.5 *(continued)* **Connective tissues. Cartilage (i and j).**

(k) Bones (osseous tissue)

Description: Hard, calcified matrix containing many collagen fibers; osteocytes lie in lacunae. Very well vascularized.

Function: Bone supports and protects (by enclosing); provides levers for the muscles to act on; stores calcium and other minerals and fat; marrow inside bones is the site for blood cell formation (hematopoiesis).

Location: Bones

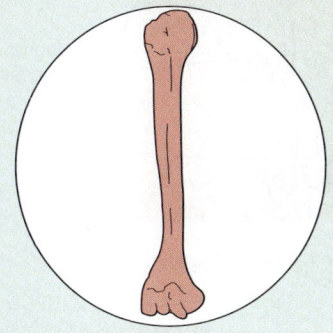

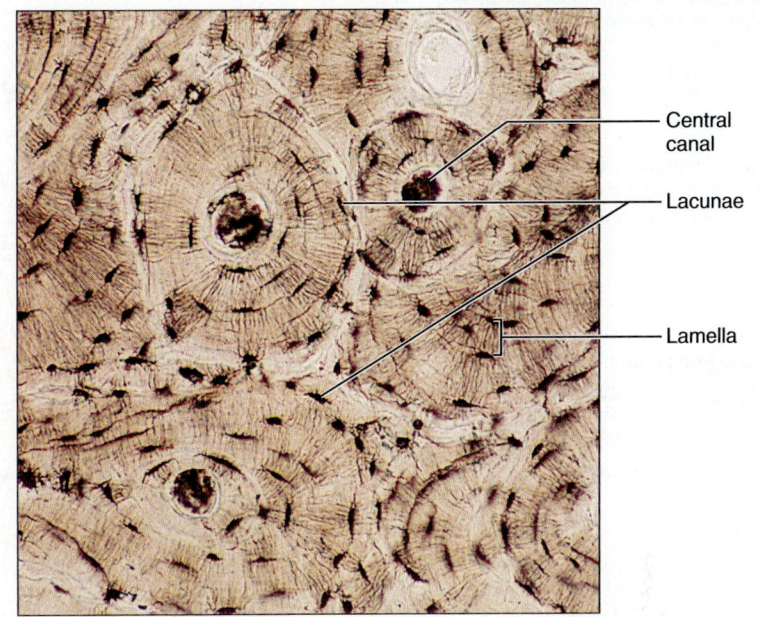

Central canal

Lacunae

Lamella

Photomicrograph: Cross-sectional view of bone (175×).

(l) Blood

Description: Red and white blood cells in a fluid matrix (plasma).

Function: Transport of respiratory gases, nutrients, wastes, and other substances.

Location: Contained within blood vessels.

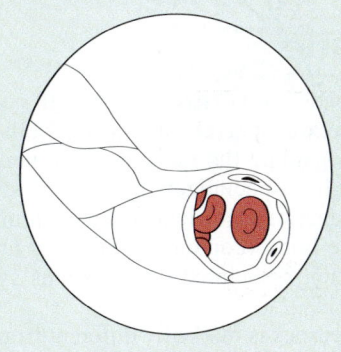

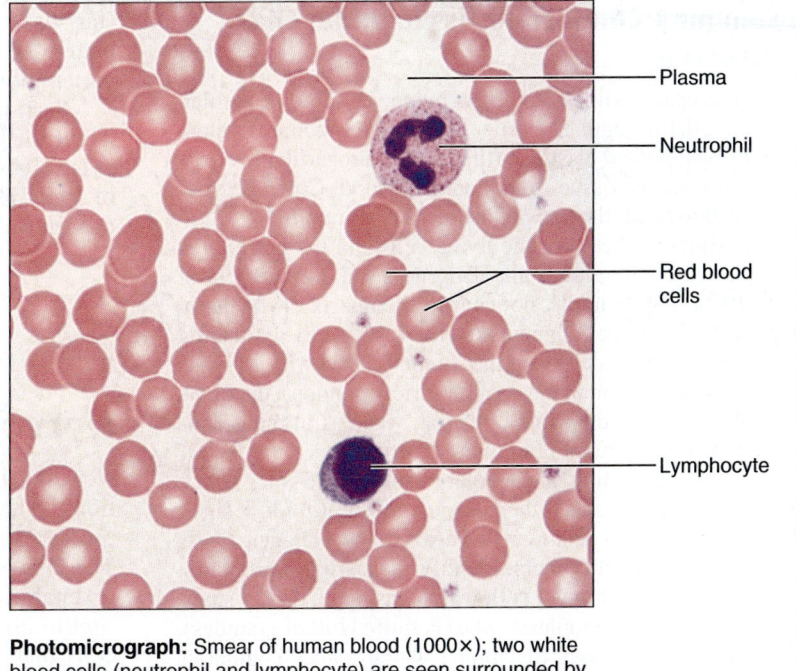

Plasma

Neutrophil

Red blood cells

Lymphocyte

Photomicrograph: Smear of human blood (1000×); two white blood cells (neutrophil and lymphocyte) are seen surrounded by red blood cells.

Figure 6.5 *(continued)* Bone **(k)** and Blood **(l).**

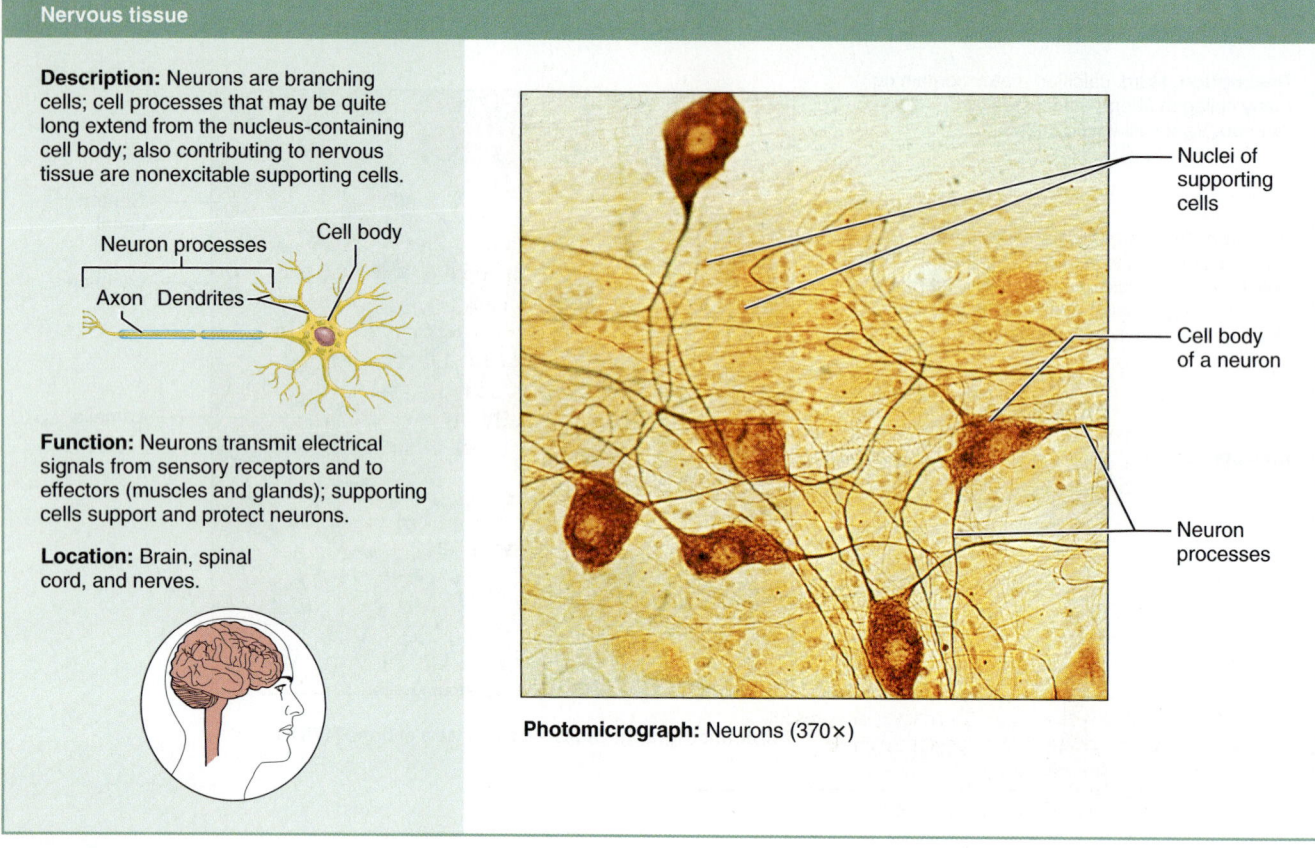

Nervous tissue

Description: Neurons are branching cells; cell processes that may be quite long extend from the nucleus-containing cell body; also contributing to nervous tissue are nonexcitable supporting cells.

Neuron processes
Cell body
Axon   Dendrites

Function: Neurons transmit electrical signals from sensory receptors and to effectors (muscles and glands); supporting cells support and protect neurons.

Location: Brain, spinal cord, and nerves.

Nuclei of supporting cells

Cell body of a neuron

Neuron processes

Photomicrograph: Neurons (370×)

Figure 6.6 Nervous tissue.

ACTIVITY 2

Examining Connective Tissue Under the Microscope

Obtain prepared slides of mesenchyme; of adipose, areolar, reticular, dense regular, elastic, and irregular connective tissue; of hyaline and elastic cartilage and fibrocartilage; of osseous connective tissue (bone); and of blood. Compare your observations with the views illustrated (Figure 6.5).

Distinguish between the living cells and the matrix and pay particular attention to the denseness and arrangement of the matrix. For example, notice how the matrix of the dense regular and irregular connective tissues, respectively making up tendons and the dermis of the skin, is packed with collagen fibers. Note also that in the *regular* variety (tendon), the fibers are all running in the same direction, whereas in the dermis they appear to be running in many directions.

While examining the areolar connective tissue, notice how much empty space there appears to be (*areol* = small empty space), and distinguish between the collagen fibers and the coiled elastic fibers. Identify the starlike fibroblasts. Also, try to locate a **mast cell,** which has large, darkly staining granules in its cytoplasm (*mast* = stuffed full of granules). This cell type releases histamine, which makes capillaries more permeable during inflammatory reactions and allergies and thus is partially responsible for that "runny nose" of some allergies.

In adipose tissue, locate a "signet ring" cell, a fat cell in which the nucleus can be seen pushed to one side by the large, fat-filled vacuole that appears to be a large empty space. Also notice how little matrix there is in adipose (fat) tissue. Distinguish between the living cells and the matrix in the dense fibrous, bone, and hyaline cartilage preparations.

Scan the blood slide at low and then high power to examine the general shape of the red blood cells. Then, switch to the oil immersion lens for a closer look at the various types of white blood cells. How does blood differ from all other connective tissues?

_____

_____

Nervous Tissue

Nervous tissue is made up of two major cell populations. The **neuroglia** are special supporting cells that protect, support, and insulate the more delicate neurons. The **neurons** are highly specialized to receive stimuli (excitability) and to generate electrical signals that may be sent to all parts of the body (conductivity). They are the cells that are most often associated with nervous system functioning.

The structure of neurons is markedly different from that of all other body cells. They have a nucleus-containing cell body, and their cytoplasm is drawn out into long extensions

Identifying Connective Tissue

Following your observations of connective tissues under the microscope, obtain an envelope for each group that contains images of some of the tissues you have studied. With your lab manual closed, remove one image at a time and identify the tissue. One member of the group shall function as the verifier whose job it is to make sure that the identification is correct.

Remove the second image and repeat the process. After you have correctly identified all of the images, sort them into groups according to their primary tissue type and subcategory (if appropriate).

Now, carefully go through each group and try to list one place in the body where the tissue is found, and one function for it. After you have correctly listed locations take your lists and draw some general conclusions about where each primary tissue type is found in the body. Compare and contrast the functions of the primary tissue types, including epithelium. (If you have already completed Group Challenge 1 for epithelial tissues, you need not repeat that work here.)

Next, obtain an envelope from your instructor that contains an image of a section through an organ. Identify all of the tissues that you see in this section, and use it to review the relationship between the location and function of the tissue types that you have studied.

Finally, identify the tissues described in the Group Challenge 2 chart and list several locations in the body.

Group Challenge 2: Connective Tissue IDs		
Magnified appearance	**Tissue type**	**Locations in the body**
• Large, round cells are densely packed • Nucleus is pushed to one side		
• Lacunae (small cavities within the tissue) are present • Lacunae are not arranged in a concentric circle • No visible fibers in the matrix		
• Fibers and cells are loosely packed with visible space between fibers • Fibers overlap but do not form a network		
• Extracellular fibers run parallel to each other • Nuclei of fibroblasts are visible		
• Lacunae are sparsely distributed • Lacunae are not arranged in a concentric circle • Fibers are visible and fairly organized		
• Tapered cells with darkly stained nucleus centrally located • No striations • Cells layered to form a sheet		

(cell processes)—sometimes as long as 1 m (about 3 feet), which allows a single neuron to conduct an electrical signal over relatively long distances. (More detail about the anatomy of the different classes of neurons and neuroglia appears in Exercise 15.)

Examining Nervous Tissue Under the Microscope

Obtain a prepared slide of a spinal cord smear. Locate a neuron and compare it to the photomicrograph **(Figure 6.6)**. Keep the light dim—this will help you see the cellular extensions of the neurons. (See also Figure 15.2 in Exercise 15.) ▪

Muscle Tissue

Muscle tissue (Figure 6.7) is highly specialized to contract and produces most types of body movement. As you might expect, muscle cells tend to be elongated, providing a long axis for contraction. The three basic types of muscle tissue are described briefly here. (Cardiac and skeletal muscles are treated more completely in later exercises.)

Skeletal muscle, the "meat," or flesh, of the body, is attached to the skeleton. It is under voluntary control (consciously controlled), and its contraction moves the limbs and other external body parts. The cells of skeletal muscles are long, cylindrical, and multinucleate (several nuclei per cell), with the nuclei pushed to the periphery of the cells; they have obvious *striations* (stripes).

Cardiac muscle is found only in the heart. As it contracts, the heart acts as a pump, propelling the blood into the blood vessels. Cardiac muscle, like skeletal muscle, has striations, but cardiac cells are branching uninucleate cells that interdigitate (fit together) at junctions called **intercalated discs.** These structural modifications allow the cardiac muscle to act as a unit. Cardiac muscle is under involuntary control, which means that we cannot voluntarily or consciously control the operation of the heart.

(a) Skeletal muscle

Description: Long, cylindrical, multinucleate cells; obvious striations.

Function: Voluntary movement; locomotion; manipulation of the environment; facial expression; voluntary control.

Location: In skeletal muscles attached to bones or occasionally to skin.

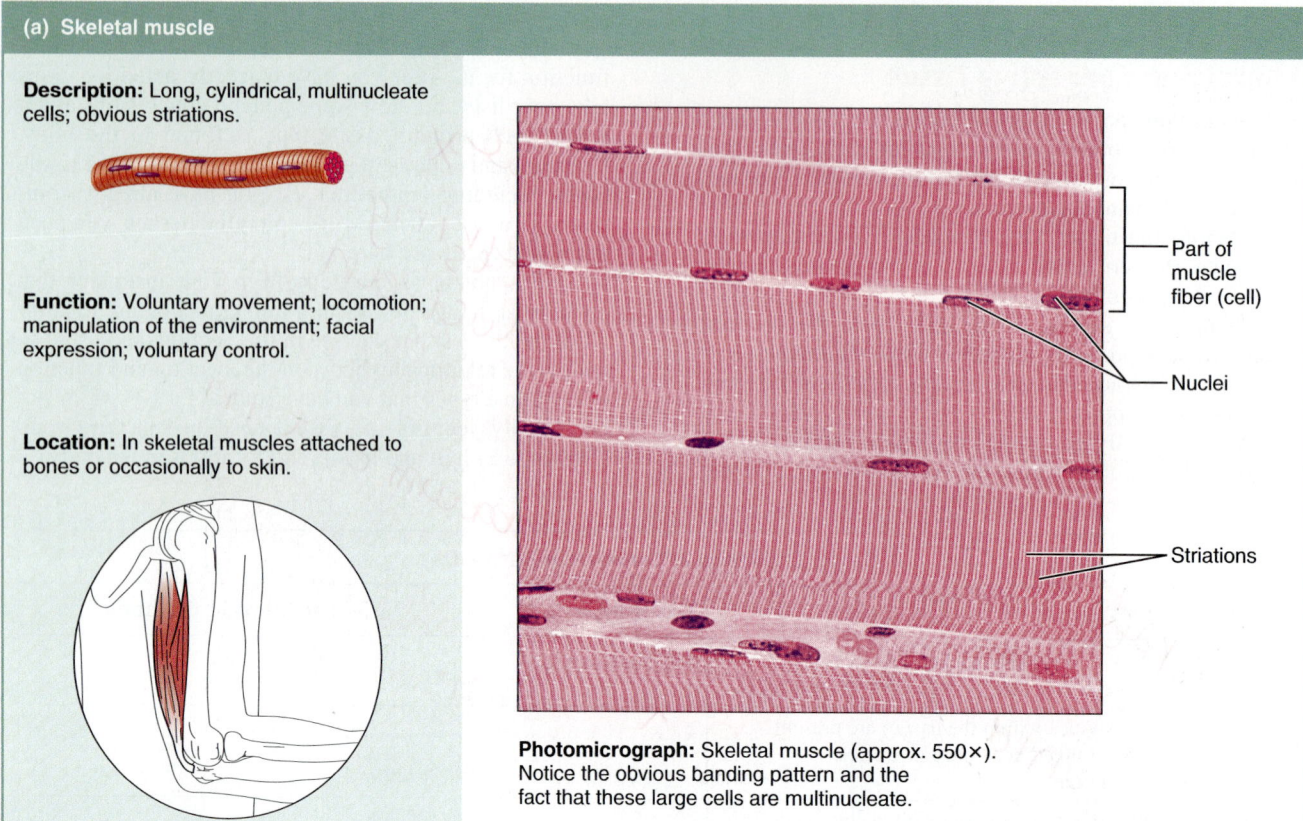

Part of muscle fiber (cell)

Nuclei

Striations

Photomicrograph: Skeletal muscle (approx. 550×). Notice the obvious banding pattern and the fact that these large cells are multinucleate.

(b) Cardiac muscle

Description: Branching, striated, generally uninucleate cells that interdigitate at specialized junctions called intercalated discs.

Function: As it contracts, it propels blood into the circulation; involuntary control.

Location: The walls of the heart.

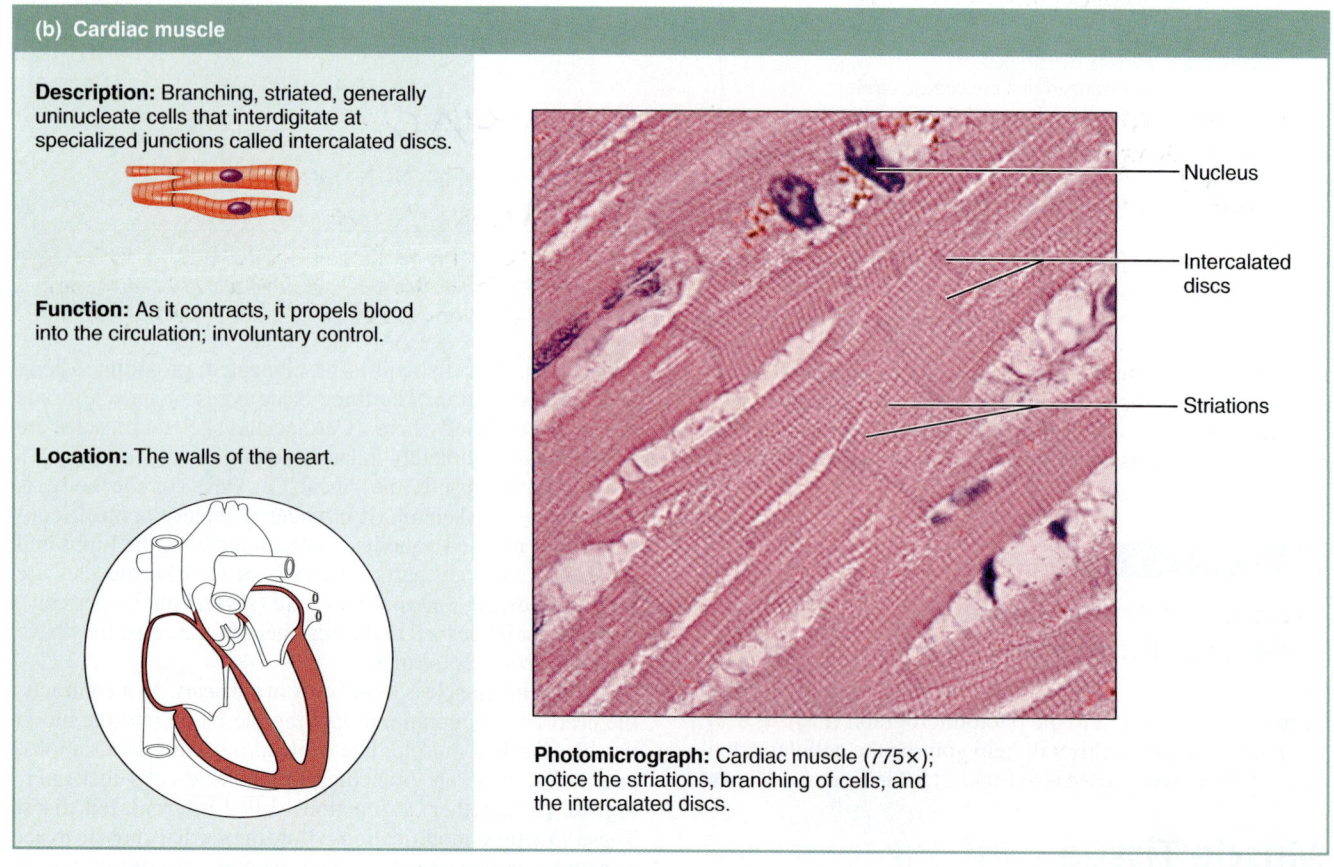

Nucleus

Intercalated discs

Striations

Photomicrograph: Cardiac muscle (775×); notice the striations, branching of cells, and the intercalated discs.

Figure 6.7 Muscle tissues. Skeletal muscle **(a)** and Cardiac muscle **(b).**

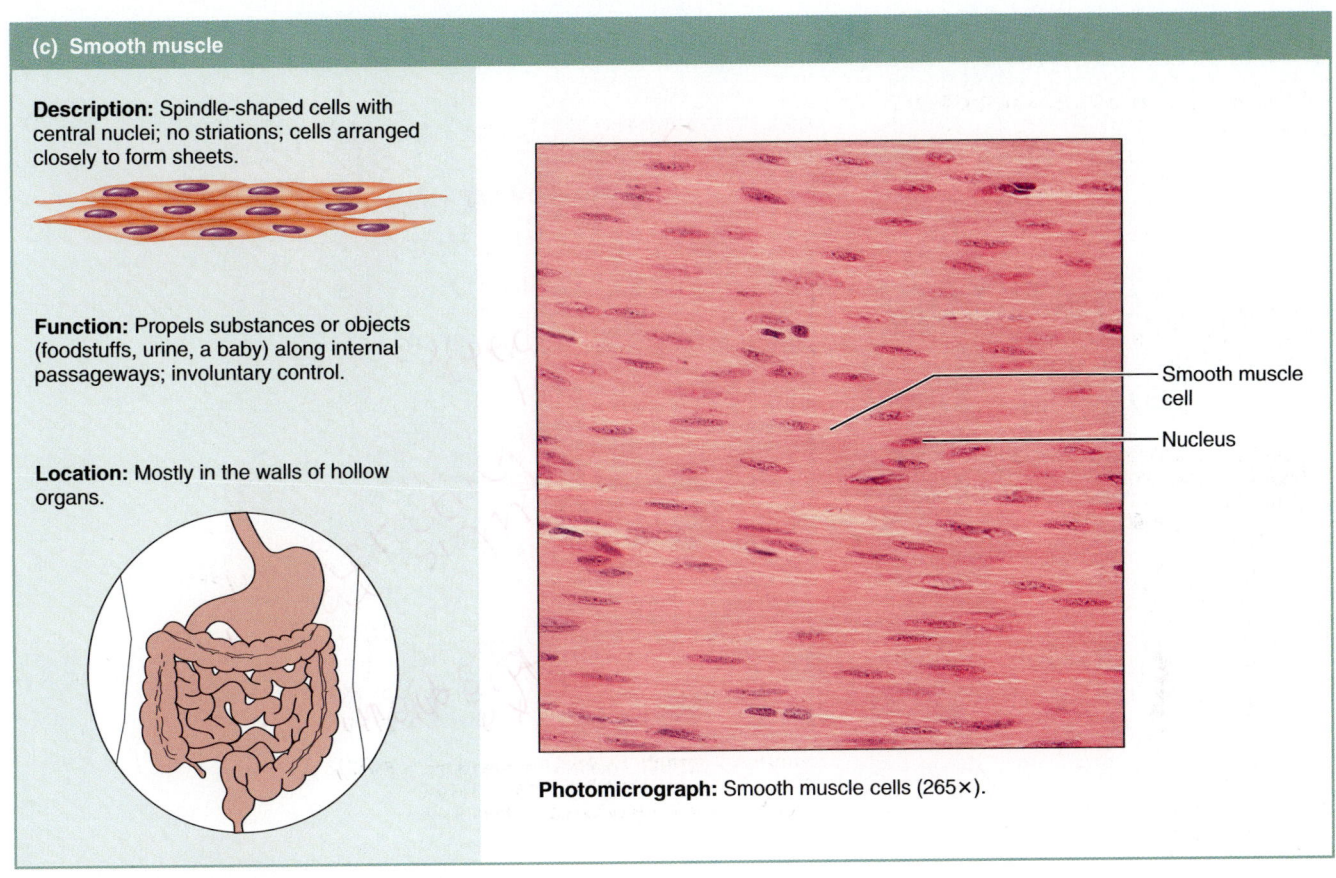

(c) Smooth muscle

Description: Spindle-shaped cells with central nuclei; no striations; cells arranged closely to form sheets.

Function: Propels substances or objects (foodstuffs, urine, a baby) along internal passageways; involuntary control.

Location: Mostly in the walls of hollow organs.

Photomicrograph: Smooth muscle cells (265×).

Smooth muscle cell

Nucleus

Figure 6.7 *(continued)* Smooth muscle **(c)**.

Smooth muscle, or *visceral muscle,* is found mainly in the walls of hollow organs (digestive and urinary tract organs, uterus, blood vessels). Typically it has two layers that run at right angles to each other; consequently its contraction can constrict or dilate the lumen (cavity) of an organ and propel substances along predetermined pathways. Smooth muscle cells are quite different in appearance from those of skeletal or cardiac muscle. No striations are visible, and the uninucleate smooth muscle cells are spindle-shaped. Like cardiac muscle, it is under involuntary control.

ACTIVITY 4

Examining Muscle Tissue Under the Microscope

Obtain and examine prepared slides of skeletal, cardiac, and smooth muscle. Notice their similarities and dissimilarities in your observations and in the illustrations and photomicrographs (Figure 6.7). ■

The Integumentary System

MATERIALS

- ☐ Skin model (three-dimensional, if available)
- ☐ Compound microscope
- ☐ Prepared slide of human scalp
- ☐ Prepared slide of skin of palm or sole
- ☐ Sheet of 20# bond paper ruled to mark off cm^2 areas
- ☐ Scissors
- ☐ Betadine® swabs, or Lugol's iodine and cotton swabs
- ☐ Adhesive tape
- ☐ Disposable gloves
- ☐ Data collection sheet for plotting distribution of sweat glands
- ☐ Porelon® fingerprint pad or portable inking foils
- ☐ Ink cleaner towelettes
- ☐ Index cards (4 in. × 6 in.)
- ☐ Magnifying glasses

MasteringA&P® For related exercise study tools, go to the Study Area of MasteringA&P. There you will find:
- Practice Anatomy Lab PAL
- PhysioEx PEx
- A&PFlix A&PFlix
- Practice quizzes, Histology Atlas, eText, Videos, and more!

OBJECTIVES

1. List several important functions of the skin, or integumentary system.
2. Identify the following skin structures on a model, image, or microscope slide: epidermis, dermis (papillary and reticular layers), hair follicles and hair, sebaceous glands, and sweat glands.
3. Name and describe the layers of the epidermis.
4. List the factors that determine skin color, and describe the function of melanin.
5. Identify the major regions of nails.
6. Describe the distribution and function of hairs, sebaceous glands, and sweat glands.
7. Discuss the difference between eccrine and apocrine sweat glands.
8. Compare and contrast the structure and functions of the epidermis and the dermis.

PRE-LAB QUIZ

1. All the following are functions of the skin except:
   a. excretion of body wastes
   b. insulation
   c. protection from mechanical damage
   d. site of vitamin A synthesis
2. The skin has two distinct regions. The superficial layer is the _____ and the underlying connective tissue is the _____.
3. The most superficial layer of the epidermis is the:
   a. stratum basale   c. stratum granulosum
   b. stratum spinosum   d. stratum corneum
4. Thick skin of the epidermis contains _____ layers.
5. _____ is a yellow-orange pigment found in the stratum corneum and the hypodermis.
   a. Keratin   c. Melanin
   b. Carotene   d. Hemoglobin
6. These cells produce a brown-to-black pigment that colors the skin and protects DNA from ultraviolet radiation damage. The cells are:
   a. dendritic cells   c. melanocytes
   b. keratinocytes   d. tactile cells
7. Circle True or False. Nails originate from the epidermis.
8. The portion of a hair that projects from the scalp surface is known as the:
   a. bulb   c. root
   b. matrix   d. shaft
9. Circle the correct underlined term. The ducts of sebaceous / sweat glands usually empty into a hair follicle but may also open directly on the skin surface.
10. Circle the correct underlined term. Eccrine / Apocrine glands are found primarily in the genital and axillary areas.

The **skin,** or **integument,** is considered an organ system because of its extent and complexity. It is much more than an external body covering; architecturally the skin is a marvel. It is tough yet pliable, a characteristic that enables it to withstand constant insult from outside agents.

The skin has many functions, most concerned with protection. It insulates and cushions the underlying body tissues and protects the entire body from abrasion, exposure to harmful chemicals, temperature extremes, and bacterial invasion. The hardened uppermost layer of the skin prevents water loss from the body surface. The skin's abundant capillary network (under the control of the nervous system) plays an important role in temperature regulation by regulating heat loss from the body surface.

The skin has other functions as well. For example, it acts as a mini-excretory system; urea, salts, and water are lost through the skin pores in sweat. The skin also has important metabolic duties. For example, like liver cells, it carries out some chemical conversions that activate or inactivate certain drugs and hormones, and it is the site of vitamin D synthesis for the body. Finally, the sense organs for touch, pressure, pain, and temperature are located here.

Basic Structure of the Skin

The skin has two distinct regions—the superficial *epidermis* composed of epithelium and an underlying connective tissue, the *dermis* **(Figure 7.1)**. These layers are firmly "cemented" together along a wavy border. But friction, such as the rubbing of a poorly fitting shoe, may cause them to separate, resulting in a blister. Immediately deep to the dermis is the **hypodermis,** or **superficial fascia,** which is not considered part of the skin. It consists primarily of adipose tissue. The main skin areas and structures are described below.

ACTIVITY 1

Locating Structures on a Skin Model

As you read, locate the following structures in the diagram (Figure 7.1) and on a skin model. ▬

Epidermis

Structurally, the avascular epidermis is a keratinized stratified squamous epithelium consisting of four distinct cell types and four or five distinct layers.

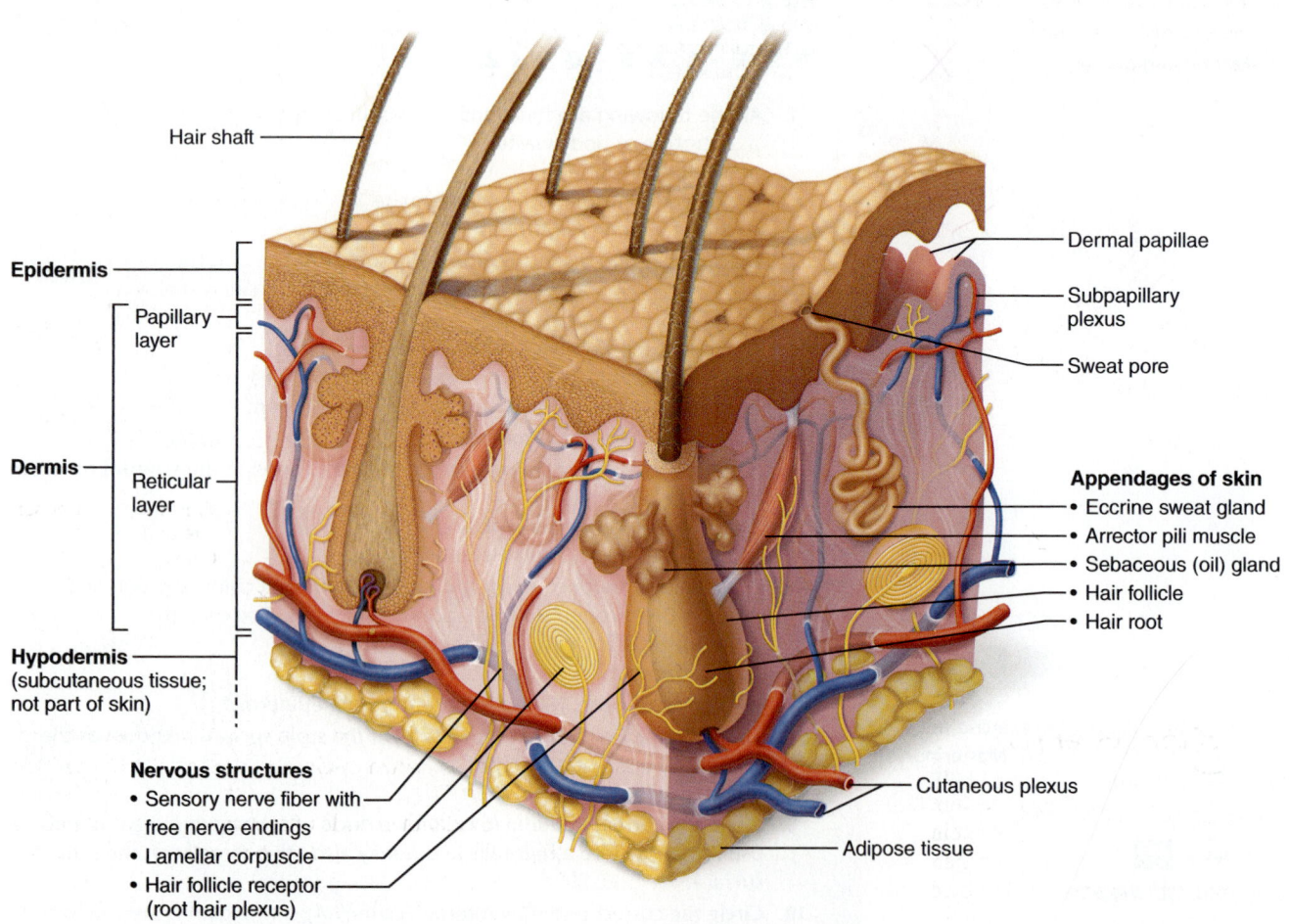

Figure 7.1 Skin structure. Three-dimensional view of the skin and the underlying hypodermis. The epidermis and dermis have been pulled apart at the right corner to reveal the dermal papillae.

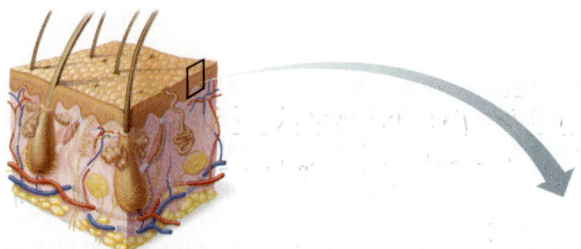

Figure 7.2 The main structural features in epidermis of thin skin.
(a) Photomicrograph depicting the four major epidermal layers
(430×). **(b)** Diagram showing the layers and relative distribution of
the different cell types. Keratinocytes (orange), melanocytes (gray),
dendritic cells (purple), and tactile (Merkel) cells (blue). A sensory
nerve ending (yellow) extending from the dermis is associated with
a tactile cell, forming a tactile disc (touch receptor). Notice that the
keratinocytes are joined by numerous desmosomes. The stratum
lucidum, present in thick skin, is not illustrated here.

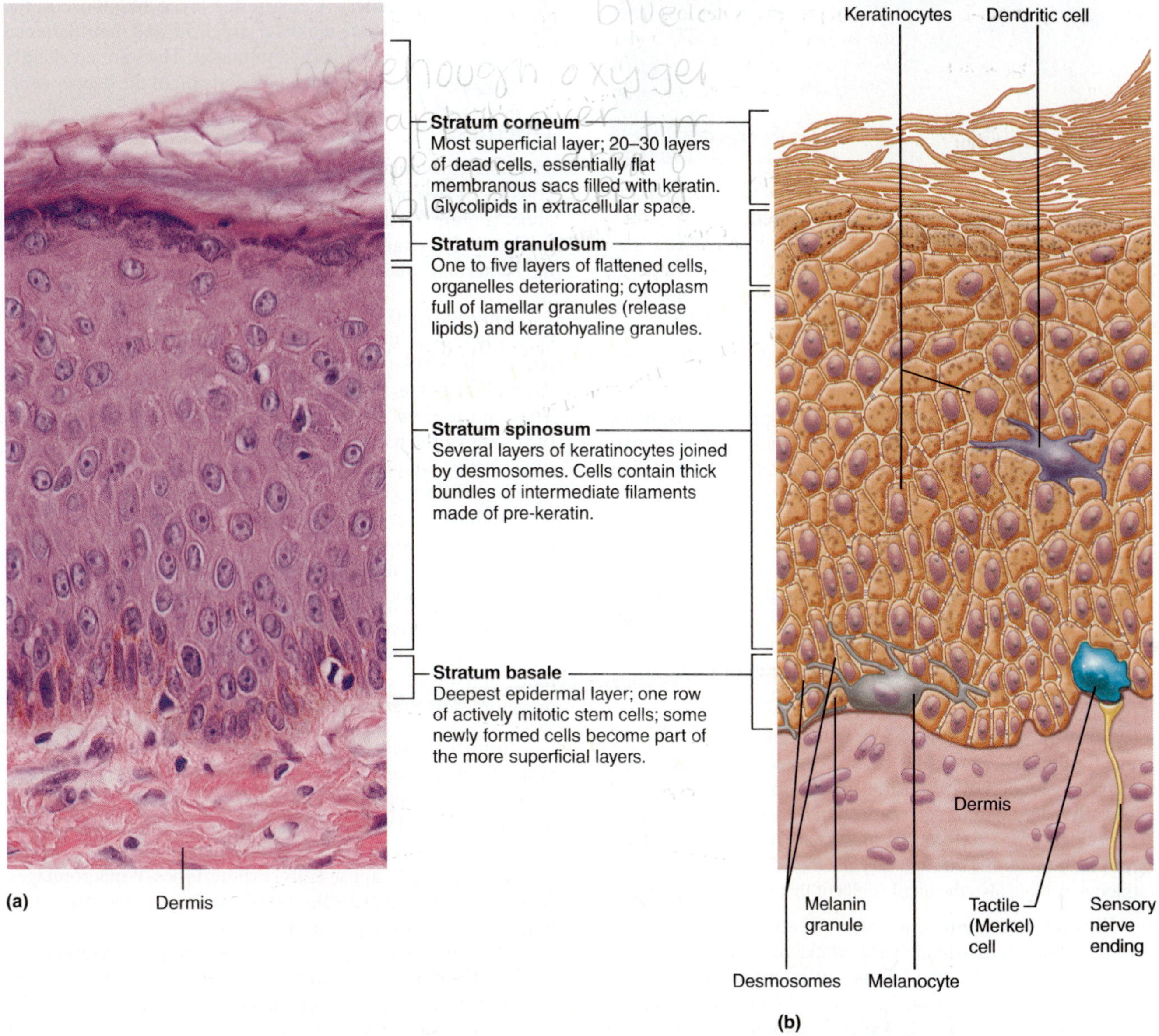

Keratinocytes   Dendritic cell

Stratum corneum
Most superficial layer; 20–30 layers
of dead cells, essentially flat
membranous sacs filled with keratin.
Glycolipids in extracellular space.

Stratum granulosum
One to five layers of flattened cells,
organelles deteriorating; cytoplasm
full of lamellar granules (release
lipids) and keratohyaline granules.

Stratum spinosum
Several layers of keratinocytes joined
by desmosomes. Cells contain thick
bundles of intermediate filaments
made of pre-keratin.

Stratum basale
Deepest epidermal layer; one row
of actively mitotic stem cells; some
newly formed cells become part of
the more superficial layers.

Dermis

Melanin
granule

Tactile
(Merkel)
cell

Sensory
nerve
ending

Desmosomes   Melanocyte

(a)   Dermis

(b)

Cells of the Epidermis

•   **Keratinocytes** (literally, keratin cells): The most abundant epidermal cells, their main function is to produce keratin
fibrils. **Keratin** is a fibrous protein that gives the epidermis
its durability and protective capabilities. Keratinocytes are
tightly connected to each other by desmosomes.

    Far less numerous are the following types of epidermal
cells **(Figure 7.2)**:

•   **Melanocytes:** Spidery black cells that produce the brown-
to-black pigment called **melanin.** The skin tans because
melanin production increases when the skin is exposed to
sunlight. The melanin provides a protective pigment umbrella
over the nuclei of the cells in the deeper epidermal layers, thus
shielding their genetic material (DNA) from the damaging
effects of ultraviolet radiation. A concentration of melanin in
one spot is called a *freckle.*

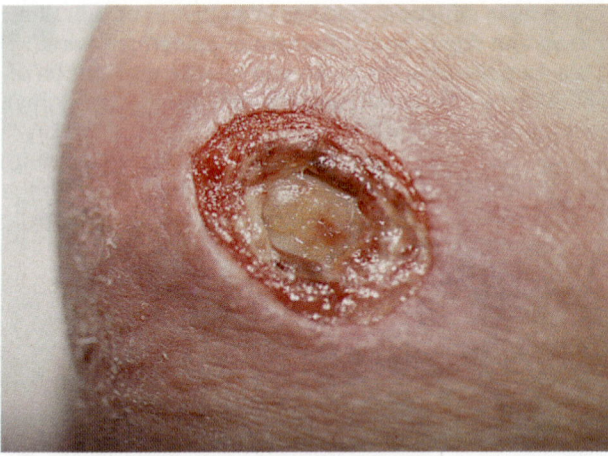

Figure 7.3 Photograph of a deep (stage III) decubitus ulcer.

7

- **Dendritic cells:** Also called *Langerhans cells,* these cells play a role in immunity.
- **Tactile (Merkel) cells:** Occasional spiky hemispheres that, in combination with sensory nerve endings, form sensitive touch receptors called *tactile* or *Merkel discs* located at the epidermal-dermal junction.

Layers of the Epidermis

The epidermis consists of four layers in thin skin, which covers most of the body. Thick skin, found on the palms of the hands and soles of the feet, contains an additional layer, the stratum lucidum. From deep to superficial, the layers of the epidermis are the stratum basale, stratum spinosum, stratum granulosum, stratum lucidum, and stratum corneum (Figure 7.2).

- **Stratum basale** (basal layer): A single row of cells immediately adjacent to the dermis. Its cells are constantly undergoing mitotic cell division to produce millions of new cells daily, hence its alternate name *stratum germinativum.* From 10% to 25% of the cells in this stratum are melanocytes, which thread their processes through this and the adjacent layers of keratinocytes. Note also the tactile cells of this layer (Figure 7.2).
- **Stratum spinosum** (spiny layer): A stratum consisting of several cell layers immediately superficial to the basal layer. Its cells contain thick weblike bundles of intermediate filaments made of a pre-keratin protein. The stratum spinosum cells appear spiky (hence their name) because as the skin tissue is prepared for histological examination, they shrink but their desmosomes hold tight. Cells divide fairly rapidly in this layer, but less so than in the stratum basale. Cells in the basal and spiny layers are the only ones to receive adequate nourishment via diffusion of nutrients from the dermis. So as their daughter cells are pushed upward and away from the source of nutrition, they gradually die. Dendritic cells may occur in the spiny layer (Figure 7.2).
- **Stratum granulosum** (granular layer): A thin layer named for the abundant granules its cells contain. These granules are of two types: (1) *lamellar granules,* which contain a waterproofing glycolipid that is secreted into the extracellular space; and (2) *keratohyaline granules,* which combine with the intermediate filaments in the more superficial layers to form the keratin fibrils. At the upper border of this layer, the cells are beginning to die.

- **Stratum lucidum** (clear layer): A very thin translucent band of flattened dead keratinocytes with indistinct boundaries. It is not present in regions of thin skin.
- **Stratum corneum** (horny layer): This outermost epidermal layer consists of some 20 to 30 cell layers (fewer layers are present in thin skin), and accounts for the bulk of the epidermal thickness. Cells in this layer, like those in the stratum lucidum (where it exists), are dead, and their flattened scalelike remnants are fully keratinized. They are constantly rubbing off and being replaced by division of the deeper cells.

Dermis

The dense irregular connective tissue making up the dermis consists of two principal regions—the papillary and reticular areas (Figure 7.1). Like the epidermis, the dermis varies in thickness. For example, it is particularly thick on the palms of the hands and soles of the feet and is quite thin on the eyelids.

- **Papillary layer:** The more superficial dermal region composed of areolar connective tissue. It is very uneven and has fingerlike projections from its superior surface, the **dermal papillae,** which attach it to the epidermis above. These projections lie on top of the larger dermal ridges. In the palms of the hands and soles of the feet, they produce the *fingerprints,* unique patterns of *epidermal ridges* that remain unchanged throughout life. Abundant capillary networks in the papillary layer furnish nutrients for the epidermal layers and allow heat to radiate to the skin surface. The pain (free nerve endings) and touch receptors (*tactile corpuscles* in hairless skin) are also found here.
- **Reticular layer:** The deepest skin layer. It is composed of dense irregular connective tissue and contains many arteries and veins, sweat and sebaceous glands, and pressure receptors *(lamellar corpuscles).*

Both the papillary and reticular layers are heavily invested with collagenic and elastic fibers. The elastic fibers give skin its exceptional elasticity in youth. In old age, the number of elastic fibers decreases, and the subcutaneous layer loses fat, which leads to wrinkling and inelasticity of the skin. Fibroblasts, adipose cells, various types of macrophages (which are important in the body's defense), and other cell types are found throughout the dermis.

The abundant dermal blood supply allows the skin to play a role in the regulation of body temperature. When body temperature is high, the arterioles serving the skin dilate, and the capillary network of the dermis becomes engorged with the heated blood. Thus body heat is allowed to radiate from the skin surface. If the environment is cool and body heat must be conserved, the arterioles constrict so that blood bypasses the dermal capillary networks temporarily.

Any restriction of the normal blood supply to the skin results in cell death and, if severe enough, skin ulcers **(Figure 7.3). Bedsores (decubitus ulcers)** occur in bedridden patients who are not turned regularly enough. The weight of the body puts pressure on the skin, especially over bony projections (hips, heels, etc.), which leads to restriction of the blood supply and tissue death. ✚

The dermis is also richly provided with lymphatic vessels and nerve fibers. Many of the nerve endings bear highly specialized receptor organs that, when stimulated by environmental changes, transmit messages to the central nervous system for interpretation. Some of these receptors—free nerve endings (pain receptors), a lamellar corpuscle, and a hair follicle receptor (also called a *root hair plexus*)—are shown in the diagram of skin structure (Figure 7.1). (These receptors are discussed in depth in Exercise 22.)

Skin Color

Skin color is a result of the relative amount of melanin in skin, the relative amount of carotene in skin, and the degree of oxygenation of the blood. People who produce large amounts of melanin have brown-toned skin. In light-skinned people, who have less melanin pigment, the dermal blood supply flushes through the rather transparent cell layers above, giving the skin a rosy glow. *Carotene* is a yellow-orange pigment present primarily in the stratum corneum and in the adipose tissue of the hypodermis. Its presence is most noticeable when large amounts of carotene-rich foods (carrots, for instance) are eaten.

Skin color may be an important diagnostic tool. For example, flushed skin may indicate hypertension, fever, or embarrassment, whereas pale skin is typically seen in anemic individuals. When the blood is inadequately oxygenated, as during asphyxiation and serious lung disease, both the blood and the skin take on a bluish or cyanotic cast. **Jaundice,** in which the tissues become yellowed, is almost always diagnostic for liver disease, whereas a bronzing of the skin hints that a person's adrenal cortex is hypoactive **(Addison's disease).** ✚

Accessory Organs of the Skin

The accessory organs of the skin—cutaneous glands, hair, and nails—are all derivatives of the epidermis, but they reside in the dermis. They originate from the stratum basale and grow downward into the deeper skin regions.

Nails

Nails are hornlike derivatives of the epidermis **(Figure 7.4)**. Their named parts are:

- **Body:** The visible attached portion.

- **Free edge:** The portion of the nail that grows out away from the body.

- **Hyponychium:** The region beneath the free edge of the nail.

- **Root:** The part that is embedded in the skin and adheres to an epithelial nail bed.

- **Nail folds:** Skin folds that overlap the borders of the nail.

- **Eponychium:** The thick proximal nail fold commonly called the cuticle.

- **Nail bed:** Extension of the stratum basale beneath the nail.

- **Nail matrix:** The thickened proximal part of the nail bed containing germinal cells responsible for nail growth. As the matrix produces the nail cells, they become heavily keratinized and die. Thus nails, like hairs, are mostly nonliving material.

- **Lunule:** The proximal region of the thickened nail matrix, which appears as a white crescent. Everywhere else, nails

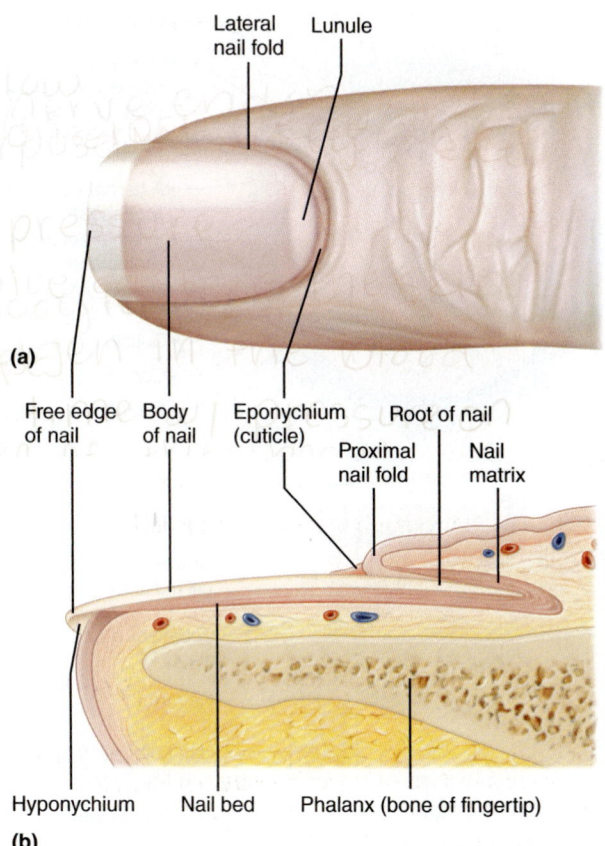

(a)

(b)

Figure 7.4 Structure of a nail. (a) Surface view of the distal part of a finger showing nail parts. The nail matrix that forms the nail lies beneath the lunule; the epidermis of the nail bed underlies the nail. **(b)** Sagittal section of the fingertip.

are transparent and nearly colorless, but they appear pink because of the blood supply in the underlying dermis. When someone is cyanotic because of a lack of oxygen in the blood, the nail beds take on a blue cast.

ACTIVITY 2

Identifying Nail Structures

Identify the parts of a nail (as shown in Figure 7.4) on yourself or your lab partner. ■

Hairs and Associated Structures

Hairs, enclosed in hair follicles, are found all over the entire body surface, except for thick-skinned areas (the palms of the hands and the soles of the feet), parts of the external genitalia, the nipples, and the lips.

- **Hair:** Structure consisting of a medulla, a central region surrounded first by the *cortex* and then by a protective *cuticle* **(Figure 7.5)**. Abrasion of the cuticle results in split ends. Hair color depends on the amount and kind of melanin pigment within the hair cortex. The portion of the hair enclosed within the follicle is called the **root;** that portion projecting from the scalp surface is called the **shaft.** The **hair bulb** is a collection of well-nourished germinal epithelial cells at the

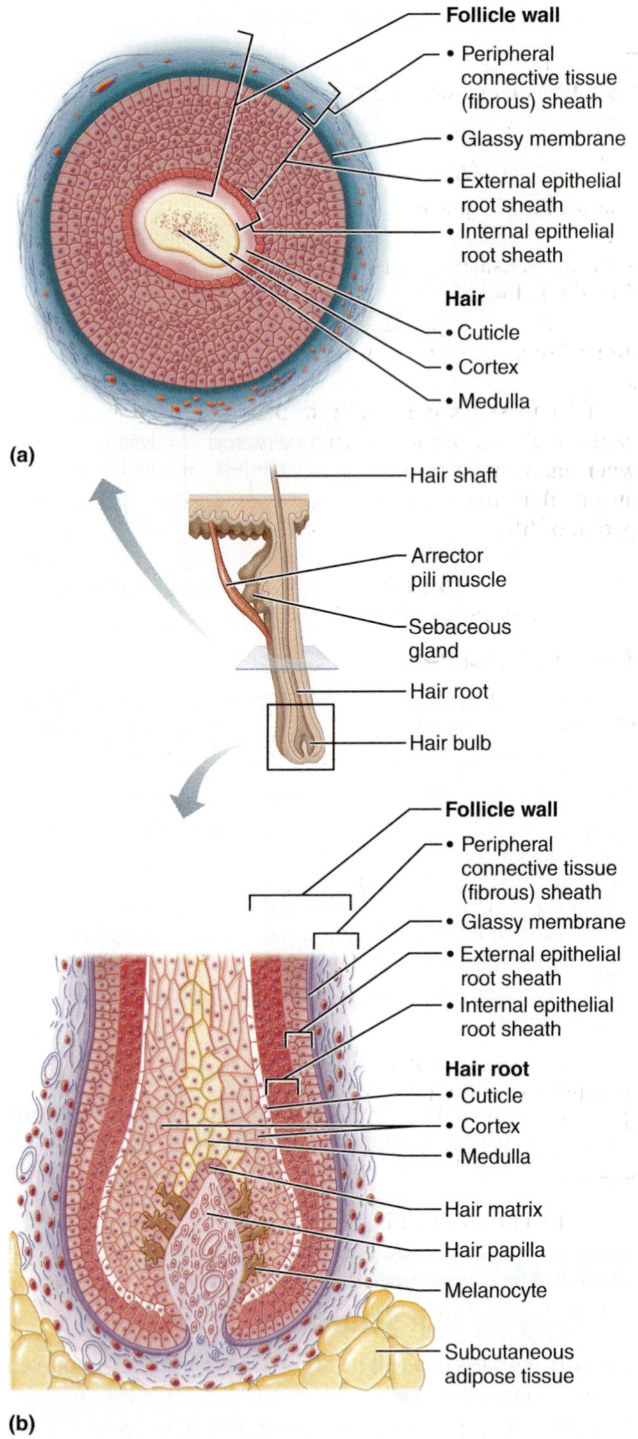

Follicle wall
- Peripheral connective tissue (fibrous) sheath
- Glassy membrane
- External epithelial root sheath
- Internal epithelial root sheath

Hair
- Cuticle
- Cortex
- Medulla

(a)

Hair shaft

Arrector pili muscle

Sebaceous gland

Hair root

Hair bulb

Follicle wall
- Peripheral connective tissue (fibrous) sheath
- Glassy membrane
- External epithelial root sheath
- Internal epithelial root sheath

Hair root
- Cuticle
- Cortex
- Medulla

Hair matrix

Hair papilla

Melanocyte

Subcutaneous adipose tissue

(b)

Figure 7.5 Structure of a hair and hair follicle. (a) Diagram of a cross section of a hair within its follicle. **(b)** Diagram of a longitudinal view of the expanded hair bulb of the follicle, which encloses the matrix, the actively dividing epithelial cells that produce the hair.

basal end of the follicle. As the daughter cells are pushed farther away from the growing region, they die and become keratinized; thus the bulk of the hair shaft, like the bulk of the epidermis, is dead material.

- **Follicle:** A structure formed from both epidermal and dermal cells (Figure 7.5). Its inner epithelial root sheath, with two parts (internal and external), is enclosed by a thickened basement membrane, the glassy membrane, and a peripheral connective tissue (or fibrous) sheath, which is essentially dermal tissue. A small nipple of dermal tissue protrudes into the hair bulb from the peripheral connective tissue sheath and provides nutrition to the growing hair. It is called the **papilla.**

- **Arrector pili muscle:** Small bands of smooth muscle cells connect each hair follicle to the papillary layer of the dermis (Figures 7.1 and 7.5). When these muscles contract (during cold or fright), the slanted hair follicle is pulled upright, dimpling the skin surface with goose bumps. This phenomenon is especially dramatic in a scared cat, whose fur actually stands on end to increase its apparent size. The activity of the arrector pili muscles also puts pressure on the sebaceous glands surrounding the follicle, causing a small amount of sebum to be released.

ACTIVITY 3

Comparison of Hairy and Relatively Hair-Free Skin Microscopically

While thick skin has no hair follicles or sebaceous (oil) glands, thin skin typical of most of the body has both. The scalp, of course, has the highest density of hair follicles.

1. Obtain a prepared slide of the human scalp, and study it carefully under the microscope. Compare your tissue slide to the photomicrograph **(Figure 7.6a)**, and identify as many as possible of the diagrammed structures (Figure 7.1).

How is this stratified squamous epithelium different from that observed in the esophagus (Exercise 6)?

_____

_____

How do these differences relate to the functions of these two similar epithelia?

_____

_____

_____

2. Obtain a prepared slide of hairless skin of the palm or sole (Figure 7.6b). Compare the slide to the previous photomicrograph (Figure 7.6a). In what ways does the thick skin of the palm or sole differ from the thin skin of the scalp?

_____

_____

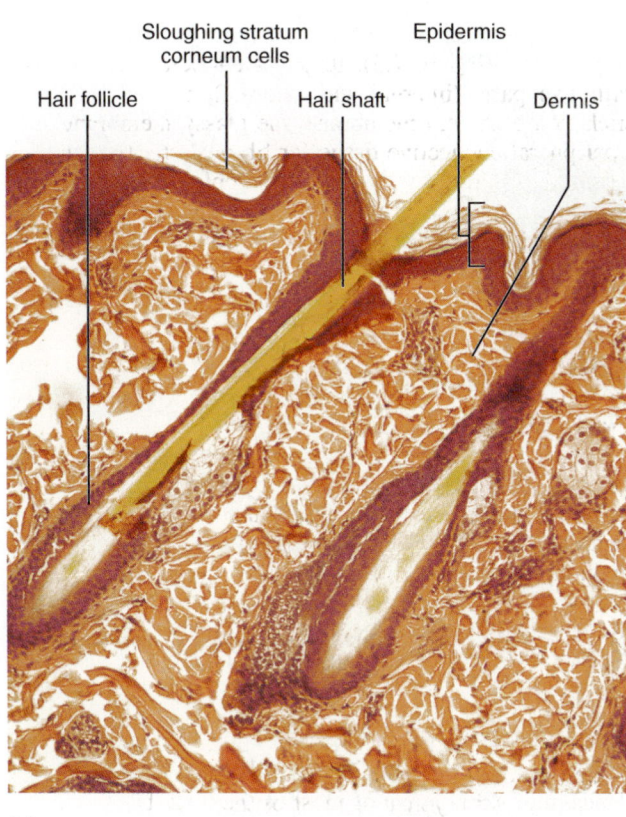

Sloughing stratum corneum cells

Hair follicle

Epidermis

Hair shaft

Dermis

(a)

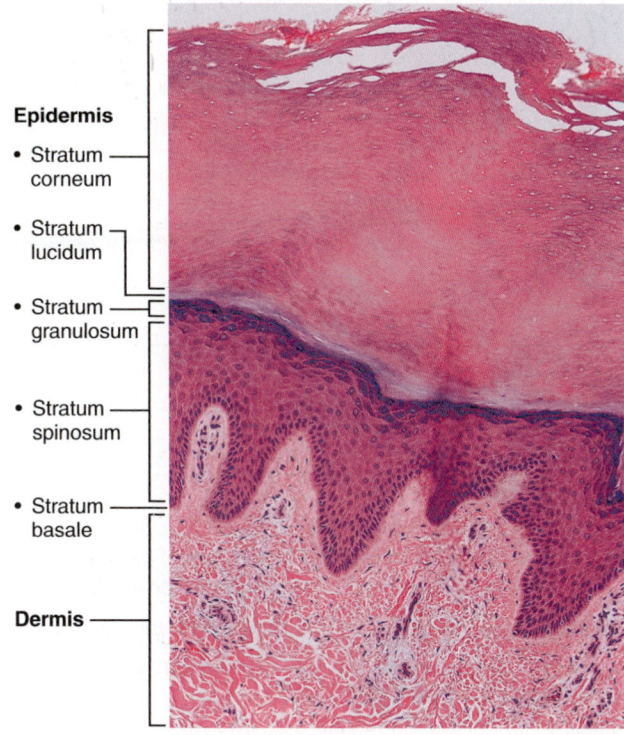

Epidermis
- Stratum corneum
- Stratum lucidum
- Stratum granulosum
- Stratum spinosum
- Stratum basale

Dermis

(b)

Figure 7.6 **Photomicrographs of skin. (a)** Thin skin with hairs (120×). **(b)** Thick hairless skin (75×).

Cutaneous Glands

The cutaneous glands fall primarily into two categories: the sebaceous glands and the sweat glands (Figure 7.1 and **Figure 7.7)**.

Sebaceous (Oil) Glands

The sebaceous glands are found nearly all over the skin, except for the palms of the hands and the soles of the feet. Their ducts usually empty into a hair follicle, but some open directly on the skin surface.

Sebum is the product of sebaceous glands. It is a mixture of oily substances and fragmented cells that acts as a lubricant to keep the skin soft and moist (a natural skin cream) and keeps the hair from becoming brittle. The sebaceous glands become particularly active during puberty when more male hormones (androgens) begin to be produced; thus the skin tends to become oilier during this period of life.

Blackheads are accumulations of dried sebum, bacteria, and melanin from epithelial cells in the oil duct. **Acne** is an active infection of the sebaceous glands. **+**

Sweat (Sudoriferous) Glands

These exocrine glands are widely distributed all over the skin. Outlets for the glands are epithelial openings called *pores*. Sweat glands are categorized by the composition of their secretions.

- **Eccrine glands:** Also called **merocrine sweat glands,** these glands are distributed all over the body. They produce clear perspiration consisting primarily of water, salts (mostly NaCl), and urea. Eccrine sweat glands, under the control of the nervous system, are an important part of the body's heat-regulating apparatus. They secrete perspiration when the external temperature or body temperature is high. When this water-based substance evaporates, it carries excess body heat with it. Thus evaporation of greater amounts of perspiration provides an efficient means of dissipating body heat when the capillary cooling system is not sufficient or is unable to maintain body temperature homeostasis.

- **Apocrine glands:** Found predominantly in the axillary and genital areas, these glands secrete the basic components of eccrine sweat plus proteins and fat-rich substances. Apocrine sweat is an excellent nutrient medium for the microorganisms typically found on the skin. This sweat is odorless, but when bacteria break down its organic components, it begins to smell unpleasant. The function of apocrine glands is not known, but since their activity increases during sexual foreplay and the glands enlarge and recede with the phases of a woman's menstrual cycle, they may be the human equivalent of the sexual scent glands of other animals.

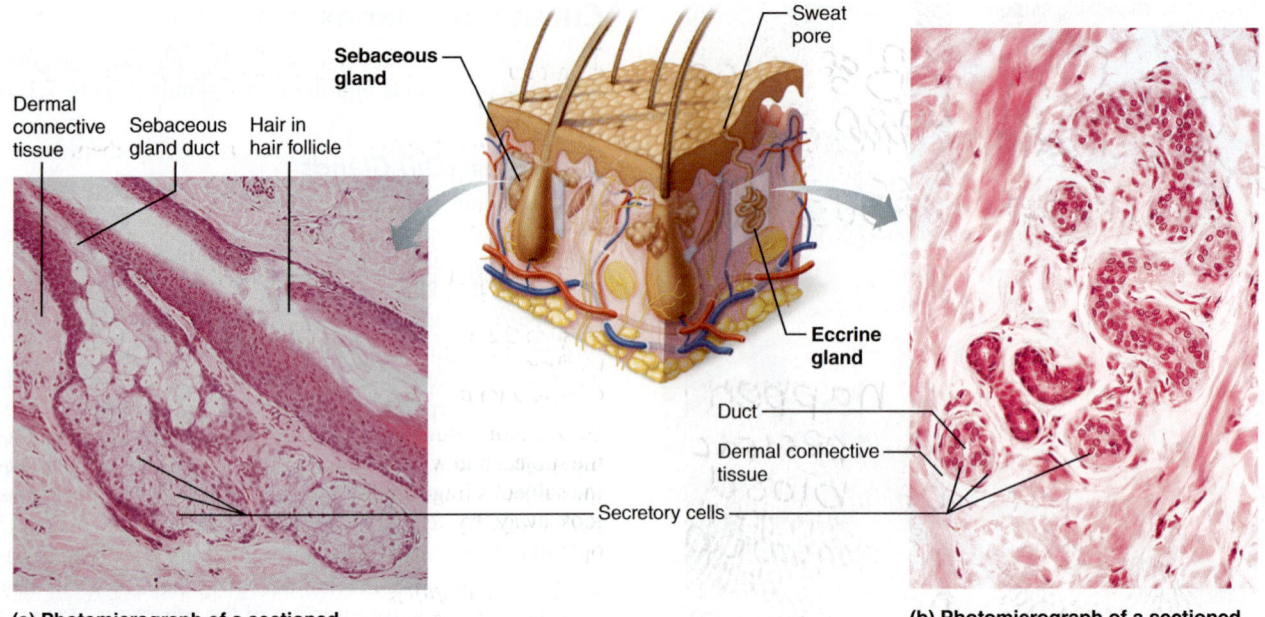

(a) Photomicrograph of a sectioned sebaceous gland (100×)

(b) Photomicrograph of a sectioned eccrine gland (145×)

Figure 7.7 Cutaneous glands.

ACTIVITY 4

Differentiating Sebaceous and Sweat Glands Microscopically

Using the slide *thin skin with hairs,* and the photomicrographs of cutaneous glands (Figure 7.7) as a guide, identify sebaceous and eccrine sweat glands. What characteristics relating to location or gland structure allow you to differentiate these glands?

_____

_____

_____

ACTIVITY 5

Plotting the Distribution of Sweat Glands

1. Form a hypothesis about the relative distribution of sweat glands on the palm and forearm. Justify your hypothesis.

2. The bond paper for this simple experiment has been pre-ruled in cm^2—put on disposable gloves and cut along the lines to obtain the required squares. You will need two squares of bond paper (each 1 cm × 1 cm), adhesive tape, and a Betadine (iodine) swab *or* Lugol's iodine and a cotton-tipped swab.

3. Paint an area of the medial aspect of your left palm (avoid the crease lines) and a region of your left forearm with the iodine solution, and allow it to dry thoroughly. The painted area in each case should be slightly larger than the paper squares to be used.

4. Have your lab partner *securely* tape a square of bond paper over each iodine-painted area, and leave them in place for 20 minutes. (If it is very warm in the laboratory while this

test is being conducted, good results may be obtained within 10 to 15 minutes.)

5. After 20 minutes, remove the paper squares, and count the number of blue-black dots on each square. The presence of a blue-black dot on the paper indicates an active sweat gland. The iodine in the pore is dissolved in the sweat and reacts chemically with the starch in the bond paper to produce the blue-black color. You have produced "sweat maps" for the two skin areas.

6. Which skin area tested has the greater density of sweat glands?

_____

_____

_____

7. Tape your results (bond paper squares) to a data collection sheet labeled "palm" and "forearm" at the front of the lab. Be sure to put your paper squares in the correct columns on the data sheet.

8. Once all the data have been collected, review the class results.

9. Prepare a lab report for the experiment. (See Getting Started, page xiv.) ■

Dermography: Fingerprinting

As noted previously, each of us has a unique genetically determined set of fingerprints. Because of the usefulness of fingerprinting for identifying and apprehending criminals, most people associate this craft solely with criminal investigations. However, civil fingerprints are invaluable in quickly identifying amnesia victims, missing persons, and unknown deceased such as those killed in major disasters.

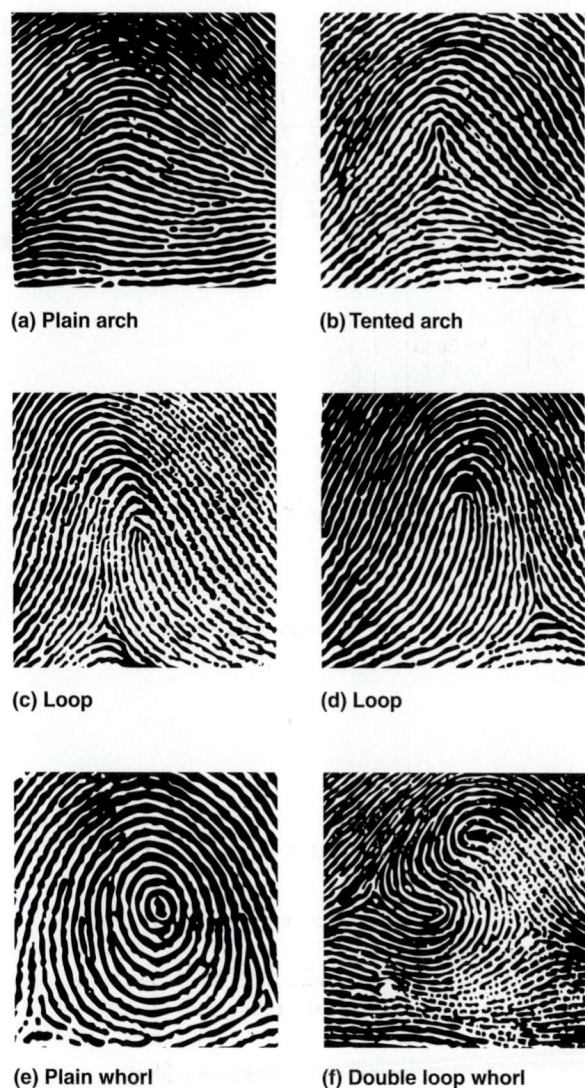

(a) Plain arch   **(b) Tented arch**

(c) Loop   **(d) Loop**

(e) Plain whorl   **(f) Double loop whorl**

Figure 7.8 **Main types of fingerprint patterns. (a–b)** Arches. **(c–d)** Loops. **(e–f)** Whorls.

The friction ridges responsible for fingerprints appear in several patterns, which are clearest when the fingertips are inked and then pressed against white paper. Impressions are also made when perspiration or any foreign material such as blood, dirt, or grease adheres to the ridges and the fingers are then pressed against a smooth, nonabsorbent surface. The three most common patterns are *arches, loops,* and *whorls* **(Figure 7.8)**. The *pattern area* in loops and whorls is the only area of the print used in identification, and it is delineated by the *type lines*—specifically the two innermost ridges that start parallel, diverge, and/or surround or tend to surround the pattern area.

ACTIVITY 6

Taking and Identifying Inked Fingerprints

For this activity, you will be working as a group with your lab partners. Though the equipment for professional fingerprinting is fairly basic, consisting of a glass or metal inking plate, printer's ink (a heavy black paste), ink roller, and standard

8 in. × 8 in. cards, you will be using supplies that are even easier to handle. Each student will prepare two index cards, each bearing his or her thumbprint and index fingerprint of the right hand.

1. Obtain the following supplies and bring them to your bench: two 4 in. × 6 in. index cards per student, Porelon fingerprint pad or portable inking foils, ink cleaner towelettes, and a magnifying glass.

2. The subject should wash and dry the hands. Open the ink pad or peel back the covering over the ink foil, and position it close to the edge of the laboratory bench. The subject should position himself or herself at arm's length from the bench edge and inking object.

3. A second student, called the *operator,* stands to the left of the subject and with two hands holds and directs movement of the subject's fingertip. During this process, the subject should look away, try to relax, and refrain from trying to help the operator.

4. The thumbprint is to be placed on the left side of the index card, the index fingerprint on the right. The operator should position the subject's right thumb or index finger on the side of the bulb of the finger in such a way that the area to be inked spans the distance from the fingertip to just beyond the first joint, and then roll the finger lightly across the inked surface until its bulb faces in the opposite direction. To prevent smearing, the thumb is rolled away from the body midline (from left to right as the subject sees it; see **Figure 7.9**) and the index finger is rolled toward the body midline (from right to left). The same ink foil can be reused for all the students at the bench; the ink pad is good for thousands of prints. Repeat the procedure (still using the subject's right hand) on the second index card.

5. If the prints are too light, too dark, or smeary, repeat the procedure.

6. While subsequent members are making clear prints of their thumb and index finger, those who have completed that activity should clean their inked fingers with a towelette and attempt to classify their own prints as arches, loops, or whorls. Use the magnifying glass as necessary to see ridge details.

7. When all members at a bench have completed the above steps, they are to write their names on the backs of their index

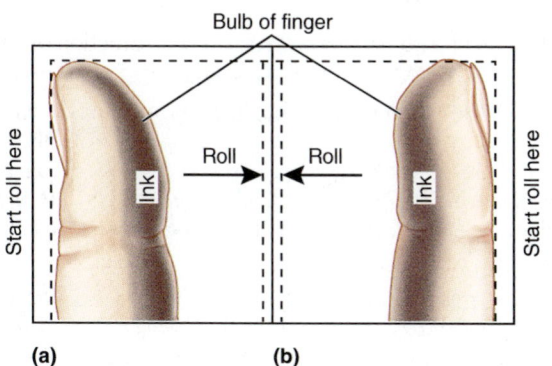

Figure 7.9 **Fingerprinting.** Method of inking and printing **(a)** the thumb and **(b)** the index finger of the right hand.

cards, then combine their cards and shuffle them before transferring them to the bench opposite for classification of pattern and identification of prints made by the same individuals.

How difficult was it to classify the prints into one of the three categories given?

_____

Why do you think this is so?

_____

Was it easy or difficult to identify the prints made by the same individual?

_____

Why do you think this was so?

_____

Overview of the Skeleton: Classification and Structure of Bones and Cartilages

MATERIALS

- ☐ Long bone sawed longitudinally (beef bone from a slaughterhouse, if possible, or prepared laboratory specimen)
- ☐ Disposable gloves
- ☐ Long bone soaked in 10% hydrochloric acid (HCl) (or vinegar) until flexible
- ☐ Long bone baked at 250°F for more than 2 hours
- ☐ Compound microscope
- ☐ Prepared slide of ground bone (x.s.)
- ☐ Three-dimensional model of microscopic structure of compact bone
- ☐ Prepared slide of a developing long bone undergoing endochondral ossification
- ☐ Articulated skeleton

OBJECTIVES

1. Name the two tissue types that form the skeleton.
2. List the functions of the skeletal system.
3. Locate and identify the three major types of skeletal cartilages.
4. Name the four main groups of bones based on shape.
5. Identify surface bone markings and list their functions.
6. Identify the major anatomical areas on a longitudinally cut long bone or on an appropriate image.
7. Explain the role of inorganic salts and organic matrix in providing flexibility and hardness to bone.
8. Locate and identify the major parts of an osteon microscopically, or on a histological model or appropriate image of compact bone.

PRE-LAB QUIZ

1. All the following are functions of the skeleton except:
   a. attachment for muscles
   b. production of melanin
   c. site of red blood cell formation
   d. storage of lipids
2. Circle the correct underlined term. The <u>axial</u> / <u>appendicular</u> skeleton consists of bones that surround the body's center of gravity.
3. The type of cartilage that has the greatest strength and is found in the knee joint and intervertebral discs is
   a. elastic   b. fibrocartilage   c. hyaline
4. Circle the correct underlined term. <u>Compact</u> / <u>Spongy</u> bone looks smooth and homogeneous.
5. _____ bones are generally thin and have a layer of spongy bone between two layers of compact bone.
   a. Flat   b. Irregular   c. Long   d. Short
6. The femur is an example of a(n) _____ bone.
   a. flat   b. irregular   c. long   d. short
7. Circle the correct underlined term. The shaft of a long bone is known as the <u>epiphysis</u> / <u>diaphysis</u>.
8. The structural unit of compact bone is the
   a. osteon   b. canaliculius   c. lacuna
9. Circle True or False. Embryonic skeletons consist primarily of elastic cartilage, which is gradually replaced by bone during development and growth.
10. Circle True or False. Cartilage has a covering made of dense connective tissue called a periosteum.

MasteringA&P® For related exercise study tools, go to the Study Area of MasteringA&P. There you will find:
- Practice Anatomy Lab **PAL**
- PhysioEx **PEx**
- A&PFlix **A&PFlix**
- Practice quizzes, Histology Atlas, eText, Videos, and more!

The **skeleton,** the body's framework, is constructed of two of the most supportive tissues found in the human body—cartilage and bone. In embryos, the skeleton is predominantly made up of hyaline cartilage, but in the adult, most of the cartilage is replaced by more rigid bone. Cartilage persists only in such isolated areas as the external ear, bridge of the nose, larynx, trachea, joints, and parts of the rib cage (see Figure 8.2).

Besides supporting and protecting the body as an internal framework, the skeleton provides a system of levers with which the skeletal muscles work to move the body. In addition, the bones store lipids and many minerals (most importantly calcium). Finally, the red marrow cavities of bones provide a site for hematopoiesis (blood cell formation).

The skeleton is made up of bones that are connected at *joints,* or *articulations.* The skeleton is subdivided into two

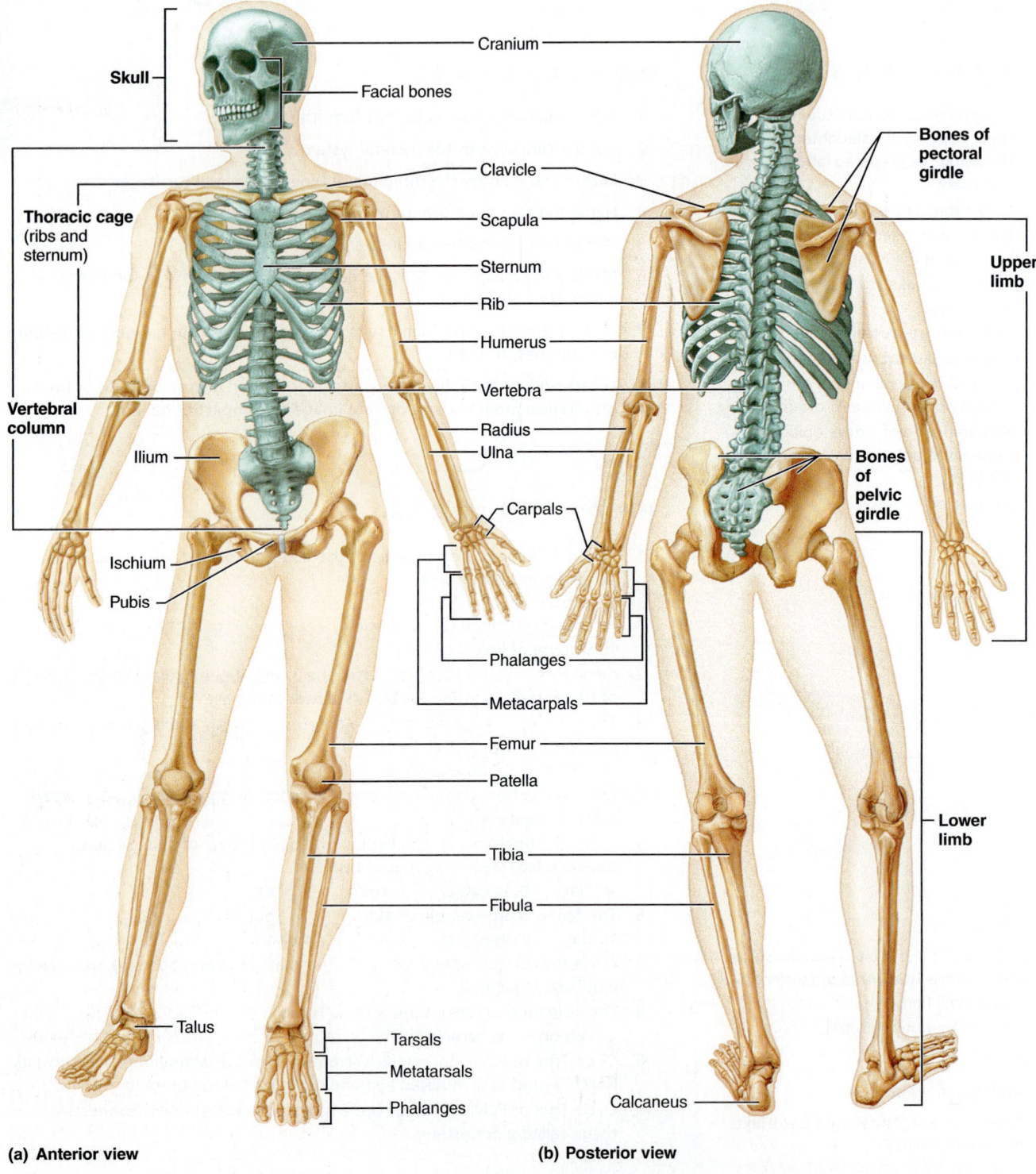

(a) Anterior view

(b) Posterior view

Figure 8.1 The human skeleton. The bones of the axial skeleton are colored green to distinguish them from the bones of the appendicular skeleton.

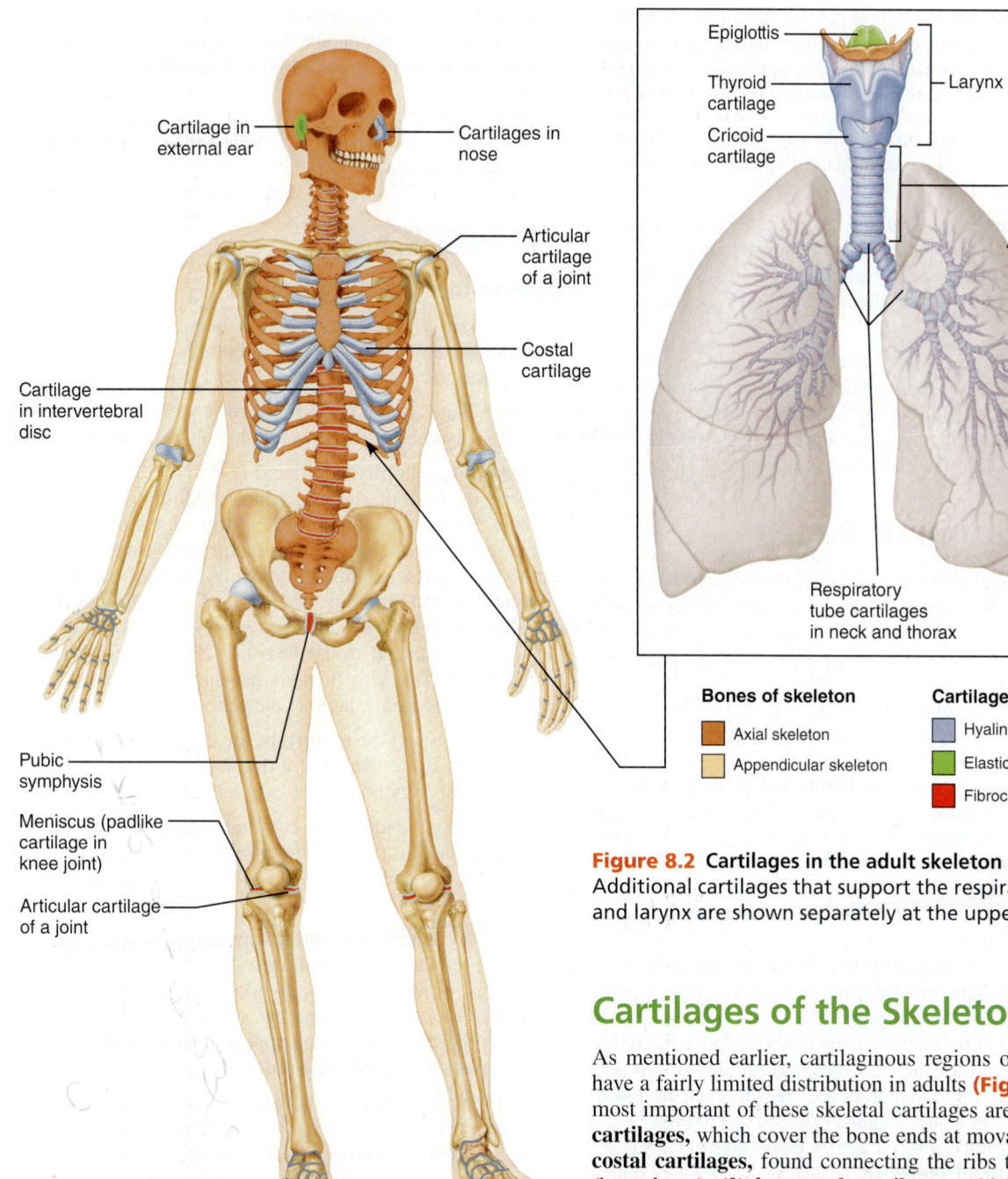

Figure 8.2 Cartilages in the adult skeleton and body. Additional cartilages that support the respiratory tubes and larynx are shown separately at the upper right.

Cartilages of the Skeleton

As mentioned earlier, cartilaginous regions of the skeleton have a fairly limited distribution in adults **(Figure 8.2)**. The most important of these skeletal cartilages are (1) **articular cartilages,** which cover the bone ends at movable joints; (2) **costal cartilages,** found connecting the ribs to the sternum (breastbone); (3) **laryngeal cartilages,** which largely construct the larynx (voice box); (4) **tracheal** and **bronchial cartilages,** which reinforce other passageways of the respiratory system; (5) **nasal cartilages,** which support the external nose; (6) **intervertebral discs,** which separate and cushion bones of the spine (vertebrae); and (7) the cartilage supporting the external ear.

The skeletal cartilages consist of some variety of *cartilage tissue,* which typically consists primarily of water and is fairly resilient. Cartilage tissues are also distinguished by the fact that they contain no nerves and very few blood vessels. Like bones, each cartilage is surrounded by a covering of dense connective tissue, called a *perichondrium* (rather than a periosteum), which acts to resist distortion of the cartilage when it is subjected to pressure, and plays a role in cartilage growth and repair.

divisions: the **axial skeleton** (those bones that lie around the body's center of gravity) and the **appendicular skeleton** (bones of the limbs, or appendages) **(Figure 8.1)**.

Before beginning your study of the skeleton, imagine for a moment that your bones have turned to putty. What if you were running when this change took place? Now imagine your bones forming a continuous metal framework within your body, somewhat like a network of plumbing pipes. What problems could you envision with this arrangement? These images should help you understand how well the skeletal system provides support and protection, as well as facilitating movement.

The skeletal cartilages have representatives from each of the three cartilage tissue types—hyaline, elastic, and fibrocartilage. (Since you have already studied cartilage tissues in Exercise 6, we will only briefly discuss that information here.)

- **Hyaline cartilage** provides sturdy support with some resilience or "give." Most skeletal cartilages are composed of hyaline cartilage (Figure 8.2).

- **Elastic cartilage** is much more flexible than hyaline cartilage, and it tolerates repeated bending better. Only the cartilages of the external ear and the epiglottis (which flops over and covers the larynx when we swallow) are elastic cartilage.

- **Fibrocartilage** consists of rows of chondrocytes alternating with rows of thick collagen fibers. Fibrocartilage, which has great tensile strength and can withstand heavy compression, is used to construct the intervertebral discs and the cartilages within the knee joint (see Figure 8.2).

Classification of Bones

The 206 bones of the adult skeleton are composed of two basic kinds of osseous tissue that differ in their texture. **Compact bone** looks smooth and homogeneous; **spongy** (or *cancellous*) **bone** is composed of small *trabeculae* (bars) of bone and lots of open space.

Bones may be classified further on the basis of their relative gross anatomy into four groups: long, short, flat, and irregular bones.

Long bones, such as the femur and phalanges (bones of the fingers) (Figure 8.1), are much longer than they are wide, generally consisting of a shaft with heads at either end. Long bones are composed predominantly of compact bone. **Short bones** are typically cube shaped, and they contain more spongy bone than compact bone. The tarsals and carpals are examples (see Figure 8.1).

Flat bones are generally thin, with two waferlike layers of compact bone sandwiching a layer of spongy bone between them. Although the name "flat bone" implies a structure that is level or horizontal, many flat bones are curved (for example, the bones of the skull). Bones that do not fall into one of the preceding categories are classified as **irregular bones.** The vertebrae are irregular bones (see Figure 8.1).

Some anatomists also recognize two other subcategories of bones. **Sesamoid bones** are special types of short bones formed in tendons. The patellas (kneecaps) are sesamoid bones. **Sutural bones** are tiny bones between cranial bones. Except for the patellas, the sesamoid and sutural bones are not included in the bone count of 206 because they vary in number and location in different individuals.

Bone Markings

Even a casual observation of the bones will reveal that bone surfaces are not featureless smooth areas but are scarred with an array of bumps, holes, and ridges. These **bone markings** reveal where bones form joints with other bones, where muscles, tendons, and ligaments were attached, and where blood vessels and nerves passed. Bone markings fall into two categories: projections, or processes that grow out from the bone and serve as sites of muscle attachment or help form joints;

and depressions or cavities, indentations or openings in the bone that often serve as conduits for nerves and blood vessels. (The bone markings are summarized in **Table 8.1** on page 112.)

Gross Anatomy of the Typical Long Bone

ACTIVITY 1

Examining a Long Bone

1. Obtain a long bone that has been sawed along its longitudinal axis. If a cleaned dry bone is provided, no special preparations need be made.

 Note: If the bone supplied is a fresh beef bone, don disposable gloves before beginning your observations.

Identify the **diaphysis** or shaft (**Figure 8.3** may help). Observe its smooth surface, which is composed of compact bone. If you are using a fresh specimen, carefully pull away the **periosteum,** or fibrous membrane covering, to view the bone surface. Notice that many fibers of the periosteum penetrate into the bone. These fibers are called **perforating (Sharpey's) fibers.** Blood vessels and nerves travel through the periosteum and invade the bone. *Osteoblasts* (bone-forming cells) and *osteoclasts* (bone-destroying cells) are found on the inner, or osteogenic, layer of the periosteum.

2. Now inspect the **epiphysis,** the end of the long bone. Notice that it is composed of a thin layer of compact bone that encloses spongy bone.

3. Identify the **articular cartilage,** which covers the epiphyseal surface in place of the periosteum. The glassy hyaline cartilage provides a smooth surface to minimize friction at joints.

4. If the animal was still young and growing, you will be able to see the **epiphyseal plate,** a thin area of hyaline cartilage that provides for longitudinal growth of the bone during youth. Once the long bone has stopped growing, these areas are replaced with bone and appear as thin, barely discernible remnants—the **epiphyseal lines.**

5. In an adult animal, the central cavity of the shaft *(medullary cavity)* is essentially a storage region for adipose tissue, or **yellow marrow.** In the infant, this area is involved in forming blood cells, and so **red marrow** is found in the marrow cavities. In adult bones, the red marrow is confined to the interior of the epiphyses, where it occupies the spaces between the trabeculae of spongy bone.

6. If you are examining a fresh bone, look carefully to see if you can distinguish the delicate **endosteum** lining the shaft. The endosteum also covers the trabeculae of spongy bone and lines the canals of compact bone. Like the periosteum, the endosteum contains both osteoblasts and osteoclasts. As the bone grows in diameter on its external surface, it is constantly being broken down on its inner surface. Thus the thickness of the compact bone layer composing the shaft remains relatively constant.

 7. If you have been working with a fresh bone specimen, return it to the appropriate area and properly dispose of your gloves, as designated by your instructor. Wash your hands before continuing to the microscope study. ■

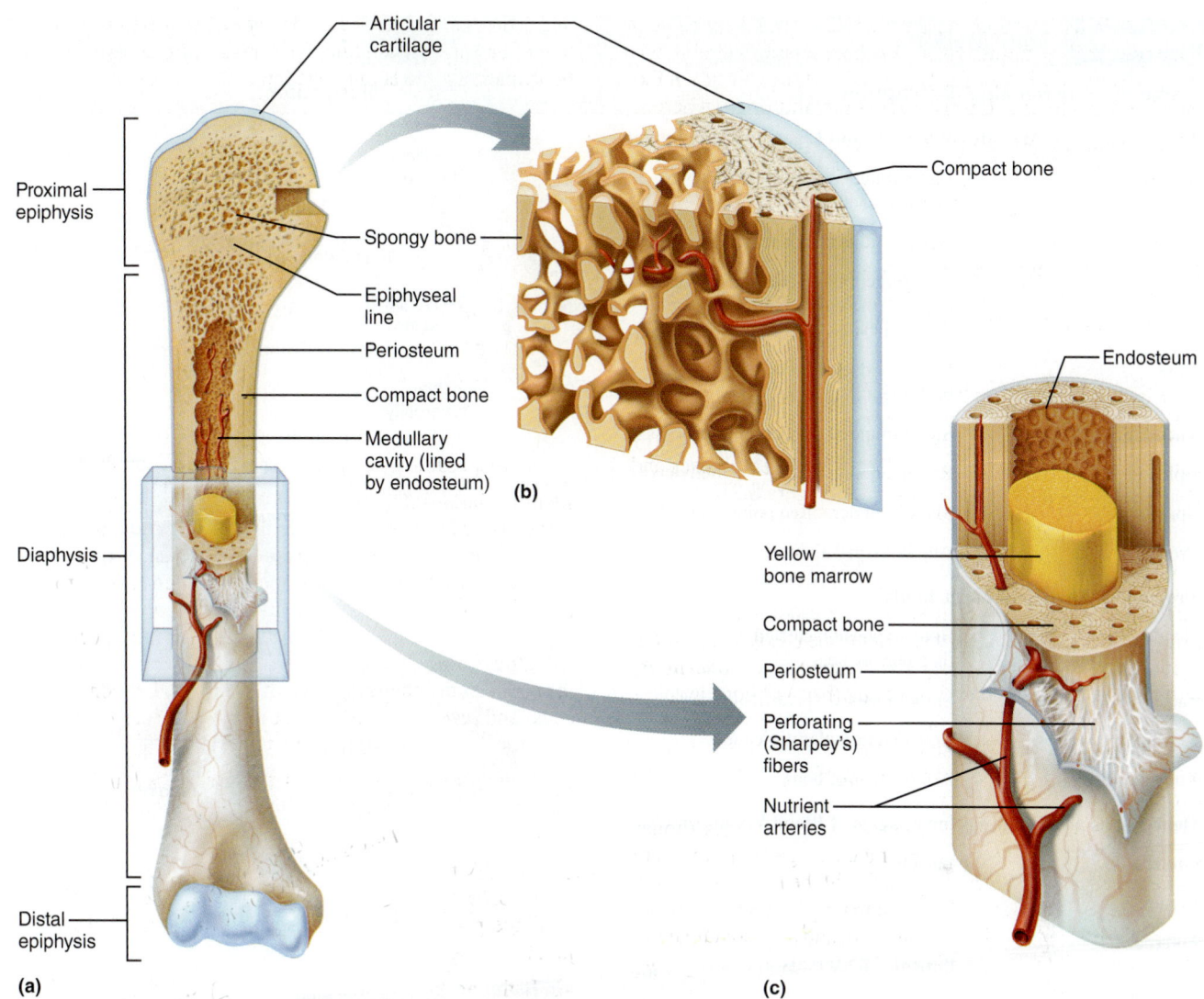

Figure 8.3 The structure of a long bone (humerus of the arm). (a) Anterior view with longitudinal section cut away at the proximal end. **(b)** Pie-shaped, three-dimensional view of spongy bone and compact bone of the epiphysis. **(c)** Cross section of diaphysis (shaft). Note that the external surface of the diaphysis is covered by a periosteum, but the articular surface of the epiphysis is covered with hyaline cartilage.

Longitudinal bone growth at epiphyseal plates (growth plates) follows a predictable sequence and provides a reliable indicator of the age of children exhibiting normal growth. In cases in which problems of long-bone growth are suspected (for example, pituitary dwarfism), X rays are taken to view the width of the growth plates. An abnormally thin epiphyseal plate indicates growth retardation. ✚

Chemical Composition of Bone

Bone is one of the hardest materials in the body. Although relatively light, bone has a remarkable ability to resist tension and shear forces that continually act on it. An engineer would tell you that a cylinder (like a long bone) is one of the strongest structures for its mass. Thus nature has given us an extremely strong, exceptionally simple and flexible supporting system without sacrificing mobility.

The hardness of bone is due to the inorganic calcium salts deposited in its ground substance. Its flexibility comes from the organic elements of the matrix, particularly the collagen fibers.

ACTIVITY 2

Examining the Effects of Heat and Hydrochloric Acid on Bones

Obtain a bone sample that has been soaked in hydrochloric acid (HCl) (or in vinegar) and one that has been baked. Heating removes the organic part of bone, while acid dissolves out the minerals. Do the treated bones retain the structure of untreated specimens?

Table 8.1 Bone Markings

Name of bone marking	Description	Illustration
Projections That Are Sites of Muscle and Ligament Attachment		
Tuberosity	Large rounded projection; may be roughened	
Crest	Narrow ridge of bone; usually prominent	
Trochanter	Very large, blunt, irregularly shaped process (the only examples are on the femur)	
Line	Narrow ridge of bone; less prominent than a crest	
Tubercle	Small rounded projection or process	
Epicondyle	Raised area on or above a condyle	
Spine	Sharp, slender, often pointed projection	
Process	Any bony prominence	
Projections That Help Form Joints		
Head	Bony expansion carried on a narrow neck	
Facet	Smooth, nearly flat articular surface	
Condyle	Rounded articular projection	
Ramus	Armlike bar of bone	
Depressions and Openings for Passage of Blood Vessels and Nerves		
Groove	Furrow	
Fissure	Narrow, slitlike opening	
Foramen	Round or oval opening through a bone	
Notch	Indentation at the edge of a structure	
Others		
Meatus	Canal-like passageway	
Sinus	Bone cavity, filled with air and lined with mucous membrane	
Fossa	Shallow basinlike depression in a bone, often serving as an articular surface	

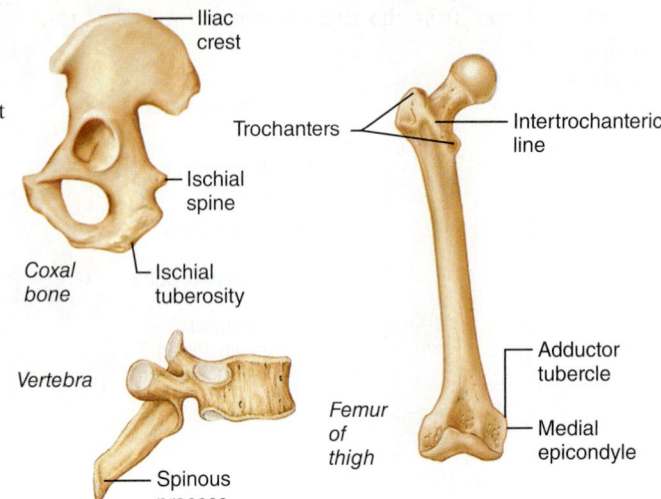

Iliac crest — Trochanters — Intertrochanteric line — Ischial spine — Adductor tubercle — Medial epicondyle — Ischial tuberosity — *Coxal bone* — *Femur of thigh*

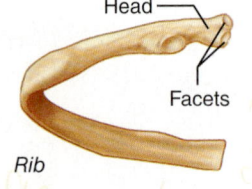

Vertebra — Spinous process

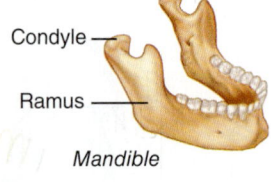

Head — Facets — *Rib* — Condyle — Ramus — *Mandible*

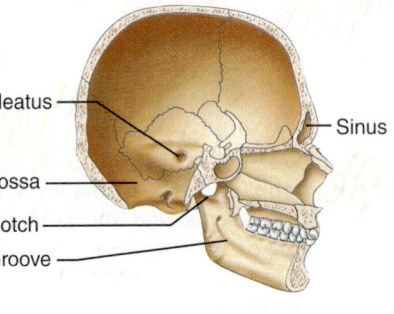

Meatus — Sinus — Fossa — Notch — Groove

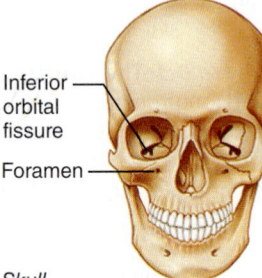

Inferior orbital fissure — Foramen — *Skull*

Gently apply pressure to each bone sample. What happens to the heated bone?

_____

_____

What happens to the bone treated with acid?

_____

What does the acid appear to remove from the bone?

_____

What does baking appear to do to the bone?

_____

_____

In rickets, the bones are not properly calcified. Which of the demonstration specimens would more closely resemble the bones of a child with rickets?

_____

Microscopic Structure of Compact Bone

As you have seen, spongy bone has a spiky, open-work appearance, resulting from the arrangement of the **trabeculae** that compose it, whereas compact bone appears to be dense and homogeneous. However, microscopic examination of compact bone reveals that it is riddled with passageways carrying blood vessels, nerves, and lymphatic vessels that provide the living bone cells with needed substances and a way to eliminate wastes. Indeed, bone histology is much easier to

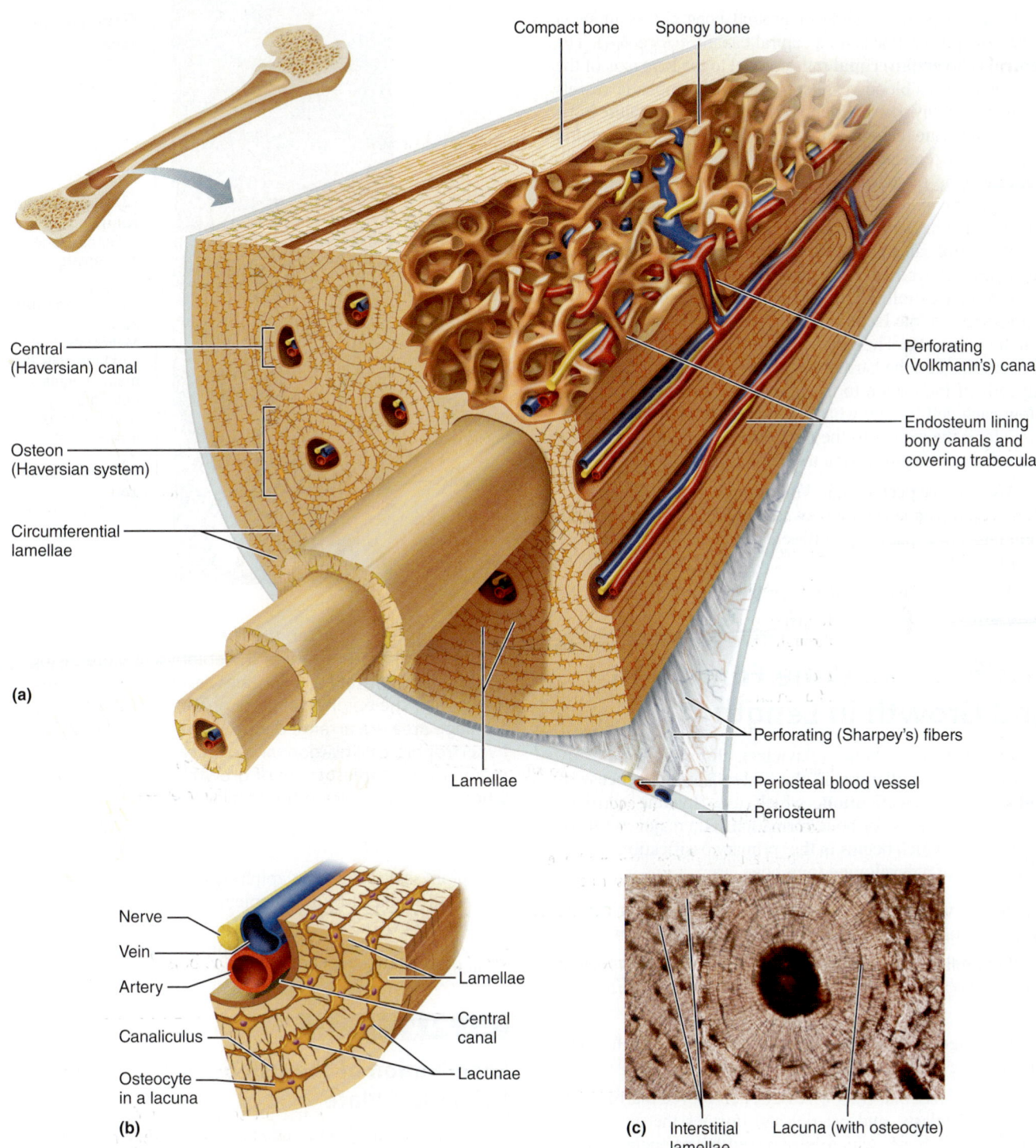

Figure 8.4 Microscopic structure of compact bone. (a) Diagrammatic view of a pie-shaped segment of compact bone, illustrating its structural units (osteons). **(b)** Higher-magnification view of a portion of one osteon. Note the position of osteocytes in lacunae. **(c)** Photomicrograph of a cross-sectional view of an osteon (320×).

understand when you recognize that bone tissue is organized around its blood supply.

ACTIVITY 3

Examining the Microscopic Structure of Compact Bone

1. Obtain a prepared slide of ground bone and examine it under low power. Focus on a central canal **(Figure 8.4)**. The **central (Haversian) canal** runs parallel to the long axis of the bone and carries blood vessels, nerves, and lymphatic vessels through the bony matrix. Identify the **osteocytes** (mature bone cells) in **lacunae** (chambers), which are arranged in concentric circles called **concentric lamellae** around the central canal. Because bone remodeling is going on all the time, you will also see some *interstitial lamellae,* remnants of *circumferential lamellae* that have been broken down (Figure 8.4c).

    A central canal and all the concentric lamellae surrounding it are referred to as an **osteon,** or **Haversian system.** Also identify **canaliculi,** tiny canals radiating outward from a central canal to the lacunae of the first lamella and then from lamella to lamella. The canaliculi form a dense transportation network through the hard bone matrix, connecting all the living cells of the osteon to the nutrient supply. The canaliculi allow each cell to take what it needs for nourishment and to pass along the excess to the next osteocyte. You may need a higher-power magnification to see the fine canaliculi.

2. Also note the **perforating (Volkmann's) canals** (Figure 8.4). These canals run at right angles to the shaft and complete the communication pathway between the bone interior and its external surface.

3. If a model of bone histology is available, identify the same structures on the model. ▄▄

Ossification: Bone Formation and Growth in Length

Except for the collarbones (clavicles), all bones of the body inferior to the skull form in the embryo by the process of **endochondral ossification,** which uses hyaline cartilage "bones" as patterns for bone formation. The major events of this process, which begins in the (primary ossification) center of the shaft of a developing long bone, are:

•   Blood vessels invade the perichondrium covering the hyaline cartilage model and convert it to a periosteum.

•   Osteoblasts at the inner surface of the periosteum secrete bone matrix around the hyaline cartilage model, forming a bone collar.

•   Cartilage in the shaft center calcifies and then hollows out, forming an internal cavity.

•   A *periosteal bud* (blood vessels, nerves, red marrow elements, osteoblasts, and osteoclasts) invades the cavity and forms spongy bone, which is removed by osteoclasts, producing the medullary cavity. This process proceeds in both directions from the *primary ossification center.*

    As bones grow longer, the medullary cavity gets larger and larger. Chondroblasts lay down new cartilage matrix on

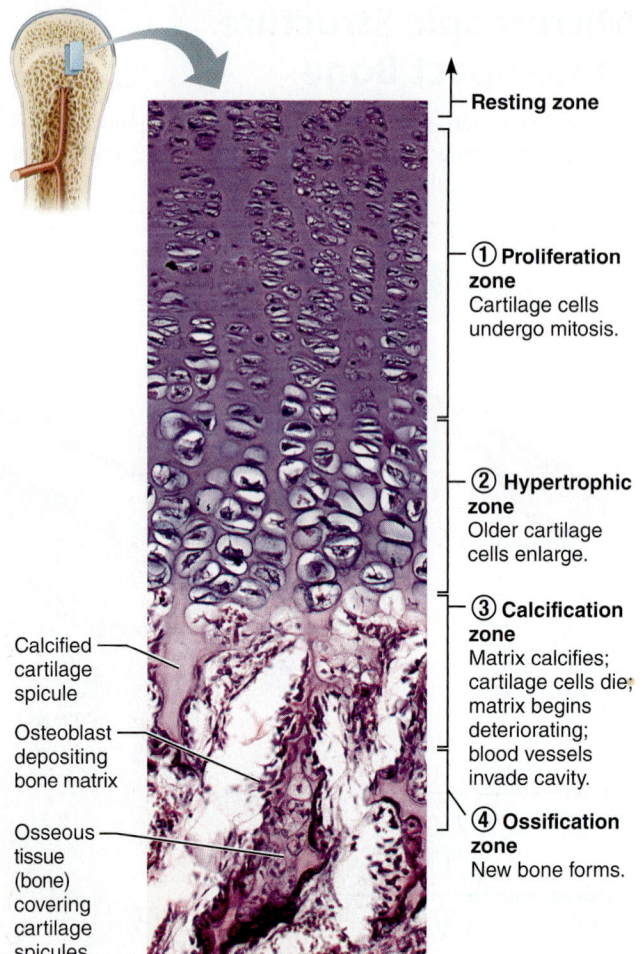

Figure 8.5 Growth in length of a long bone occurs at the epiphyseal plate. The side of the epiphyseal plate facing the epiphysis (distal face) contains resting cartilage cells. The cells of the epiphyseal plate proximal to the resting cartilage area are arranged in four zones—proliferation, hypertrophic, calcification, and ossification—from the region of the earliest stage of growth ① to the region where bone is replacing the cartilage ④ (125×).

the epiphyseal face of the epiphyseal plate, and it is eroded away and replaced by bony spicules on the side facing the medullary cavity **(Figure 8.5)**. This process continues until late adolescence when the entire epiphyseal plate is replaced by bone.

ACTIVITY 4

Examination of the Osteogenic Epiphyseal Plate

Obtain a slide depicting endochondral ossification (cartilage bone formation) and bring it to your bench to examine under the microscope. Identify the proliferation, hypertrophic, calcification, and ossification zones of the epiphyseal plate (Figure 8.5). Then, also identify the area of resting cartilage cells distal to the growth zone, some hypertrophied chondrocytes, bony spicules, the periosteal bone collar, and the medullary cavity. ▄▄

The Axial Skeleton

MATERIALS

- ☐ Intact skull and Beauchene skull
- ☐ X rays of individuals with scoliosis, lordosis, and kyphosis (if available)
- ☐ Articulated skeleton, articulated vertebral column, removable intervertebral discs
- ☐ Isolated cervical, thoracic, and lumbar vertebrae, sacrum, and coccyx
- ☐ Isolated fetal skull

OBJECTIVES

1. Name the three parts of the axial skeleton.
2. Identify the bones of the axial skeleton, either by examining isolated bones or by pointing them out on an articulated skeleton or skull, and name the important bone markings on each.
3. Name and describe the different types of vertebrae.
4. Discuss the importance of intervertebral discs and spinal curvatures.
5. Identify three abnormal spinal curvatures.
6. List the components of the thoracic cage.
7. Identify the bones of the fetal skull by examining an articulated skull or image.
8. Define *fontanelle* and discuss the function and fate of fontanelles in the fetus.
9. Discuss important differences between the fetal and adult skulls.

PRE-LAB QUIZ

1. The axial skeleton can be divided into the skull, the vertebral column, and the _____ .
   a. thoracic cage   c. hip bones
   b. femur   d. humerus
2. Eight bones make up the _____ , which encloses and protects the brain.
   a. cranium   b. face   c. skull
3. How many bones of the skull are considered facial bones? _____
4. Circle the correct underlined term. The lower jawbone, or <u>maxilla</u> / <u>mandible</u>, articulates with the temporal bones in the only freely movable joints in the skull.
5. Circle the correct underlined term. The <u>body</u> / <u>spinous process</u> of a typical vertebra forms the rounded, central portion that faces anteriorly in the human vertebral column.
6. The seven bones of the neck are called _____ vertebrae.
   a. cervical   b. lumbar   c. spinal   d. thoracic
7. The _____ vertebrae articulate with the corresponding ribs.
   a. cervical   b. lumbar   c. spinal   d. thoracic
8. The _____ , commonly referred to as the breastbone, is a flat bone formed by the fusion of three bones: the manubrium, the body, and the xiphoid process.
   a. coccyx   b. sacrum   c. sternum
9. Circle True or False. The first seven pairs of ribs are called floating ribs because they have only indirect cartilage attachments to the sternum.
10. A fontanelle _____ .
    a. is found only in the fetal skull
    b. is a fibrous membrane
    c. allows for compression of the skull during birth
    d. all of the above

MasteringA&P® For related exercise study tools, go to the Study Area of MasteringA&P. There you will find:
- Practice Anatomy Lab PAL
- PhysioEx PEx
- A&PFlix A&PFlix
- Practice quizzes, Histology Atlas, eText, Videos, and more!

The **axial skeleton** (the green portion of Figure 8.1 on page 108) can be divided into three parts: the skull, the vertebral column, and the thoracic cage.

The Skull

The **skull** is composed of two sets of bones. Those of the **cranium** enclose and protect the fragile brain tissue. The **facial bones** support the eyes and position them anteriorly. They also provide attachment sites for facial muscles, which make it possible for us to present our feelings to the world. All but one of the bones of the skull are joined by interlocking joints called *sutures*. The mandible, or lower jawbone, is attached to the rest of the skull by a freely movable joint.

ACTIVITY 1

Identifying the Bones of the Skull

The bones of the skull **(Figures 9.1–9.8)** are described below. As you read through this material, identify each bone on an intact and/or Beauchene skull (see Figure 9.6c).

Note: Important bone markings are listed beneath the bones on which they appear, and a color-coded dot before each bone name corresponds to the bone color in the figures. ■

The Cranium

The cranium may be divided into two major areas for study—the **cranial vault** or **calvaria,** forming the superior, lateral, and posterior walls of the skull, and the **cranial base,** forming the skull bottom. Internally, the cranial base has three distinct depressions: the **anterior, middle,** and **posterior cranial fossae** (see Figure 9.3). The brain sits in these fossae, completely enclosed by the cranial vault.

Eight bones construct the cranium. *With the exception of two paired bones (the parietals and the temporals), all are single bones.* Sometimes the six ossicles of the middle ear, part of the hearing apparatus, are also considered part of the cranium. (The ossicles are described in Exercise 25, Special Senses: Hearing and Equilibrium.)

● *Frontal Bone* (Figures 9.1, 9.3, 9.6, and 9.8) Anterior portion of cranium; forms the forehead, superior part of the orbit, and floor of anterior cranial fossa.

Supraorbital foramen (notch): Opening above each orbit allowing blood vessels and nerves to pass.

Glabella: Smooth area between the eyes.

● *Parietal Bone* (Figures 9.1 and 9.6) Posterolateral to the frontal bone, forming sides of cranium.

Sagittal suture: Midline articulation point of the two parietal bones.

Coronal suture: Point of articulation of parietals with frontal bone.

● *Temporal Bone* (Figures 9.1, 9.2, 9.3, and 9.6) Inferior to parietal bone on lateral skull. The temporals can be divided into three major parts: the **squamous part** borders the parietals; the **tympanic part** surrounds the external ear opening; and the **petrous part** forms the lateral portion of the skull base and contains the mastoid process.

Important markings associated with the flaring squamous part (Figures 9.1 and 9.2) include:

Squamous suture: Point of articulation of the temporal bone with the parietal bone.

Zygomatic process: A bridgelike projection joining the zygomatic bone (cheekbone) anteriorly. Together these two bones form the *zygomatic arch.*

Mandibular fossa: Rounded depression on the inferior surface of the zygomatic process (anterior to the ear); forms the socket for the condylar process of the mandible, where the mandible (lower jaw) joins the cranium.

Tympanic part markings (Figures 9.1 and 9.2) include:

External acoustic meatus: Canal leading to eardrum and middle ear.

Styloid (*stylo* = stake, pointed object) **process:** Needlelike projection inferior to external acoustic meatus; attachment point for muscles and ligaments of the neck. This process is often broken off demonstration skulls.

The petrous part (Figures 9.2 and 9.3), which helps form the middle and posterior cranial fossae, contains the labyrinth (holding the organs of hearing and balance). It exhibits several obvious foramina with important functions and includes:

Jugular foramen: Opening medial to the styloid process through which the internal jugular vein and cranial nerves IX, X, and XI pass.

Carotid canal: Opening medial to the styloid process through which the internal carotid artery passes into the cranial cavity.

Internal acoustic meatus: Opening on posterior aspect (petrous part) of temporal bone allowing passage of cranial nerves VII and VIII (Figure 9.3).

Foramen lacerum: A jagged opening between the petrous temporal bone and the sphenoid providing passage for a number of small nerves and for the internal carotid artery to enter the middle cranial fossa (after it passes through part of the temporal bone).

Stylomastoid foramen: Tiny opening between the mastoid and styloid processes through which cranial nerve VII leaves the cranium.

Mastoid process: Rough projection inferior and posterior to external acoustic meatus; attachment site for muscles.

The mastoid process, full of air cavities and so close to the middle ear—a trouble spot for infections—often becomes infected too, a condition referred to as **mastoiditis.** Because the mastoid area is separated from the brain by only a thin layer of bone, an ear infection that has spread to the mastoid process can inflame the brain coverings, or the meninges. The latter condition is known as **meningitis.** ✚

● *Occipital Bone* (Figures 9.1, 9.2, 9.3, and 9.6). Most posterior bone of cranium—forms floor and back wall. Joins sphenoid bone anteriorly via its narrow basilar part.

Lambdoid suture: Site of articulation of occipital bone and parietal bones.

Foramen magnum: Large opening in base of occipital, which allows the spinal cord to join with the brain.

(Text continues on page 125.)

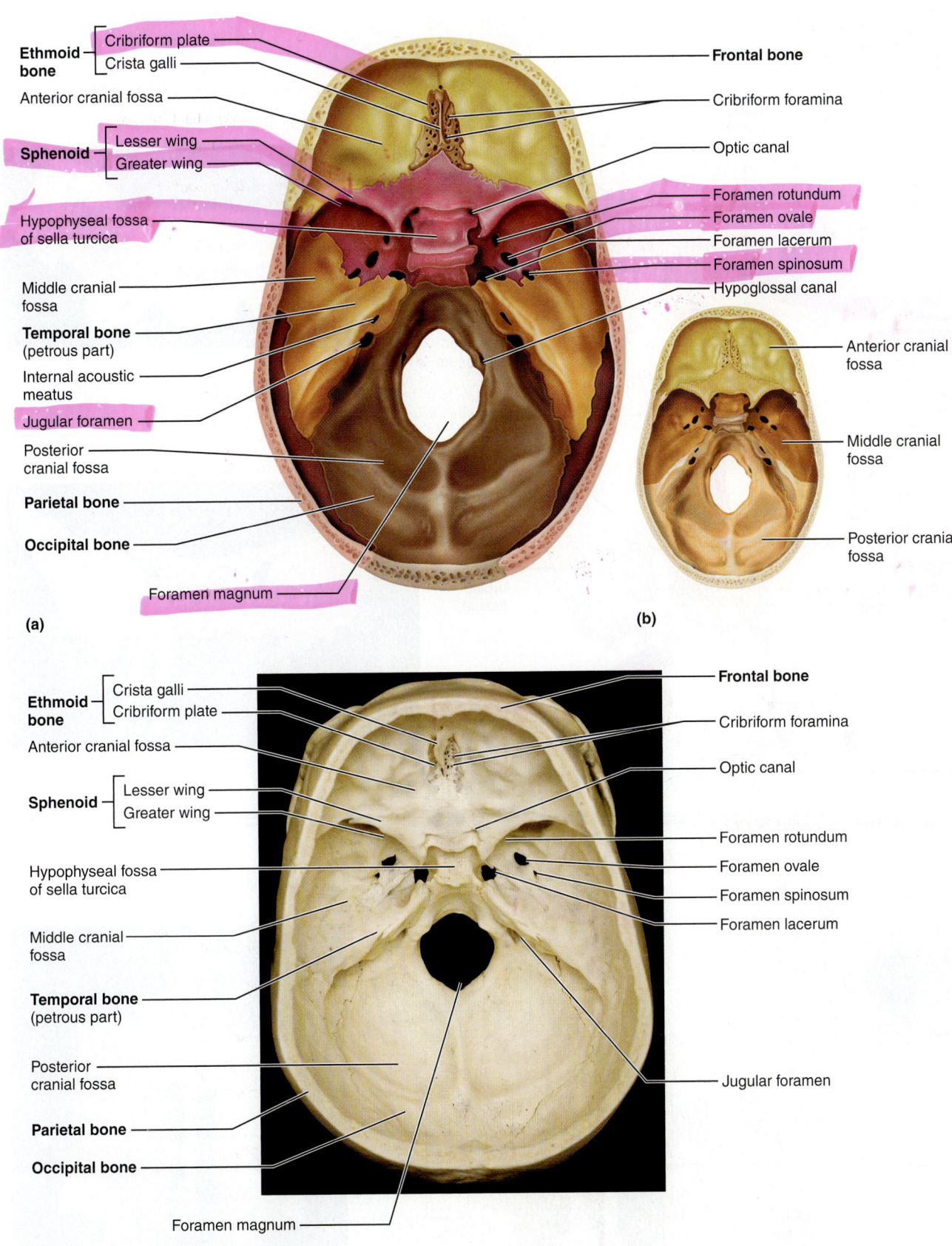

Ethmoid bone
— Cribriform plate
— Crista galli

Anterior cranial fossa

Sphenoid
— Lesser wing
— Greater wing

Hypophyseal fossa of sella turcica

Middle cranial fossa

Temporal bone (petrous part)

Internal acoustic meatus

Jugular foramen

Posterior cranial fossa

Parietal bone

Occipital bone

Foramen magnum

Frontal bone

Cribriform foramina

Optic canal

Foramen rotundum
Foramen ovale
Foramen lacerum
Foramen spinosum
Hypoglossal canal

(a)

Anterior cranial fossa

Middle cranial fossa

Posterior cranial fossa

(b)

Ethmoid bone
— Crista galli
— Cribriform plate

Anterior cranial fossa

Sphenoid
— Lesser wing
— Greater wing

Hypophyseal fossa of sella turcica

Middle cranial fossa

Temporal bone (petrous part)

Posterior cranial fossa

Parietal bone

Occipital bone

Foramen magnum

Frontal bone

Cribriform foramina

Optic canal

Foramen rotundum
Foramen ovale
Foramen spinosum
Foramen lacerum

Jugular foramen

(c)

Figure 9.3 **Internal anatomy of the inferior portion of the skull. (a)** Superior view of the base of the cranial cavity, calvaria removed. **(b)** Schematic view of the cranial base showing the extent of its major fossae. **(c)** Photograph of superior view of the base of the cranial cavity, calvaria removed.

9

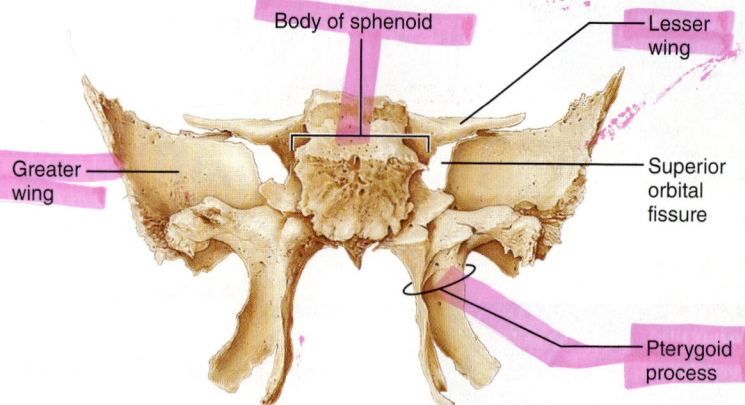

(a) Superior view

Optic canal

Lesser wing

Greater wing

Superior orbital fissure

Hypophyseal fossa of sella turcica

Foramen rotundum

Foramen ovale

Foramen spinosum

Body of sphenoid

9

(b) Posterior view

Body of sphenoid

Lesser wing

Greater wing

Superior orbital fissure

Pterygoid process

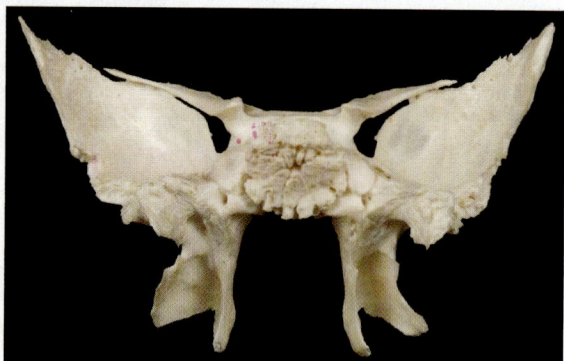

Figure 9.4 **The sphenoid bone.**

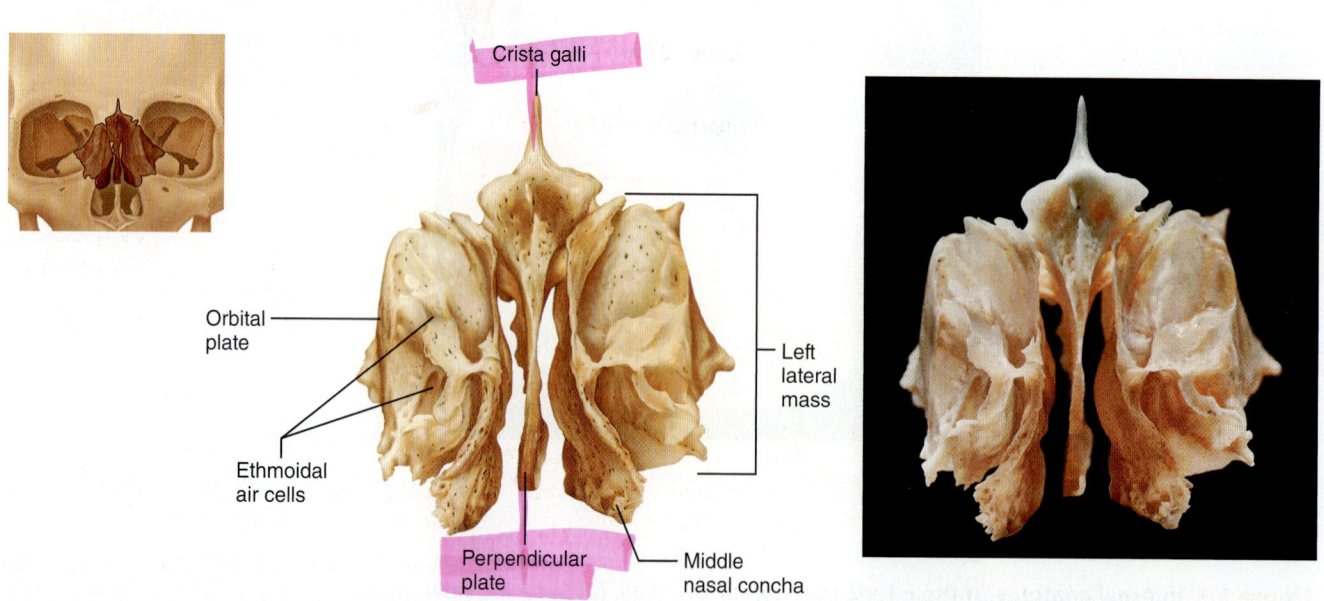

Crista galli

Orbital plate

Left lateral mass

Ethmoidal air cells

Perpendicular plate

Middle nasal concha

Figure 9.5 **The ethmoid bone.** Anterior view. The superior nasal conchae are located posteriorly and are therefore not visible in the anterior view.

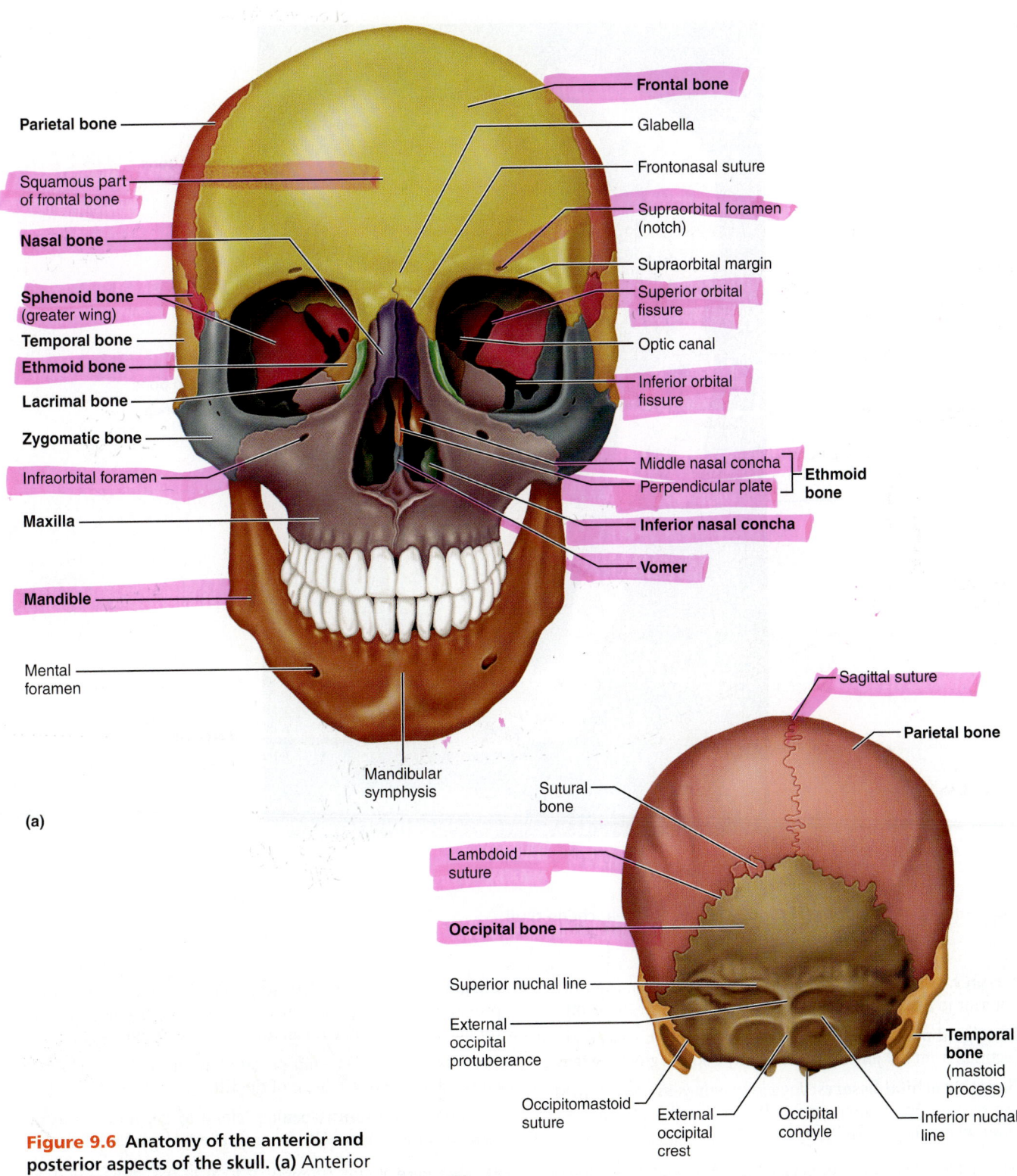

Figure 9.6 **Anatomy of the anterior and posterior aspects of the skull. (a)** Anterior aspect. **(b)** Posterior aspect.

Occipital condyles: Rounded projections lateral to the foramen magnum that articulate with the first cervical vertebra (atlas).

Hypoglossal canal: Opening medial and superior to the occipital condyle through which the hypoglossal nerve (cranial nerve XII) passes.

External occipital crest and protuberance: Midline prominences posterior to the foramen magnum.

● *Sphenoid Bone* (Figures 9.1–9.4, 9.6, and 9.8) Bat-shaped bone forming the anterior plateau of the middle cranial fossa across the width of the skull. The sphenoid bone is the keystone of the cranium because it articulates with all other cranial bones.

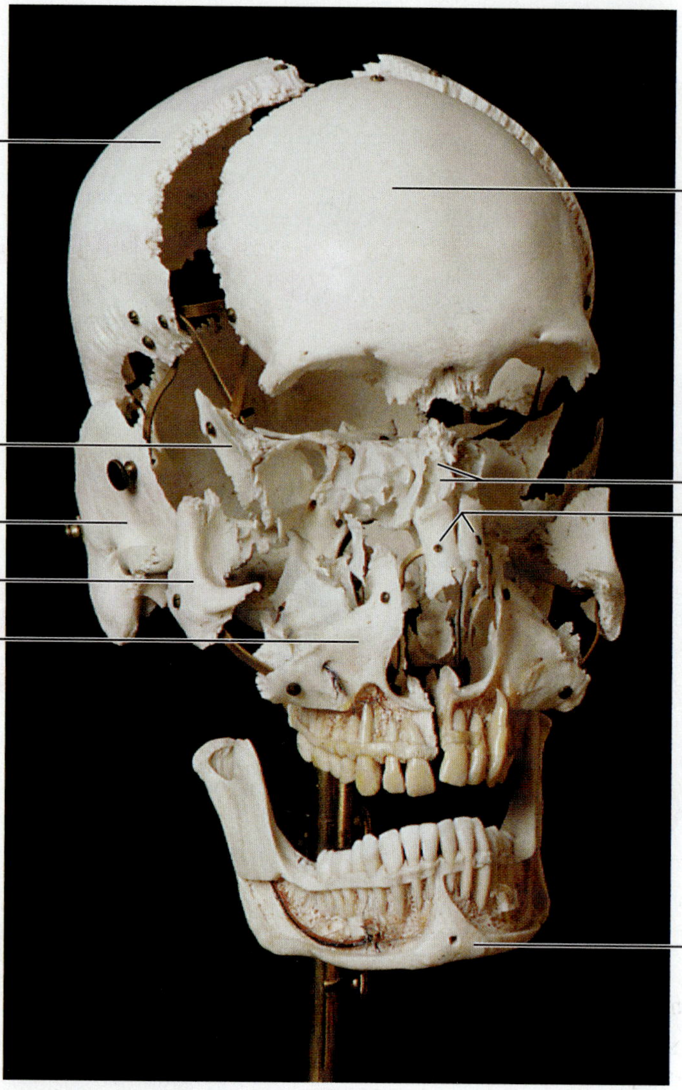

(c)

Figure 9.6 *(continued)* **(c)** Frontal view of the Beauchene skull.

Greater wings: Portions of the sphenoid seen exteriorly anterior to the temporal and forming a part of the eye orbits.

Pterygoid processes: Inferiorly directed trough-shaped projections from the junction of the body and the greater wings.

Superior orbital fissures: Jagged openings in orbits providing passage for cranial nerves III, IV, V, and VI to enter the orbit where they serve the eye.

The sphenoid bone can be seen in its entire width if the top of the cranium (calvaria) is removed (Figure 9.3).

Sella turcica (Turk's saddle): A saddle-shaped region in the sphenoid midline. The seat of this saddle, called the **hypophyseal fossa,** surrounds the pituitary gland (hypophysis).

Lesser wings: Bat-shaped portions of the sphenoid anterior to the sella turcica.

Optic canals: Openings in the bases of the lesser wings through which the optic nerves (cranial nerve II) enter the orbits to serve the eyes.

Foramen rotundum: Opening lateral to the sella turcica providing passage for a branch of the fifth cranial nerve. (This foramen is not visible on an inferior view of the skull.)

Foramen ovale: Opening posterior to the sella turcica that allows passage of a branch of the fifth cranial nerve.

Foramen spinosum: Opening lateral to the foramen ovale through which the middle meningeal artery passes.

• *Ethmoid Bone* (Figures 9.1, 9.3, 9.5, 9.6, and 9.8) Irregularly shaped bone anterior to the sphenoid. Forms the roof of the nasal cavity, upper nasal septum, and part of the medial orbit walls.

Crista galli (cock's comb): Vertical projection providing a point of attachment for the dura mater, helping to secure the brain within the skull.

Cribriform plates: Bony plates lateral to the crista galli through which olfactory fibers (cranial nerve I) pass to the brain from the nasal mucosa through the cribriform foramina.

Figure 9.7 Detailed anatomy of the mandible and maxilla.

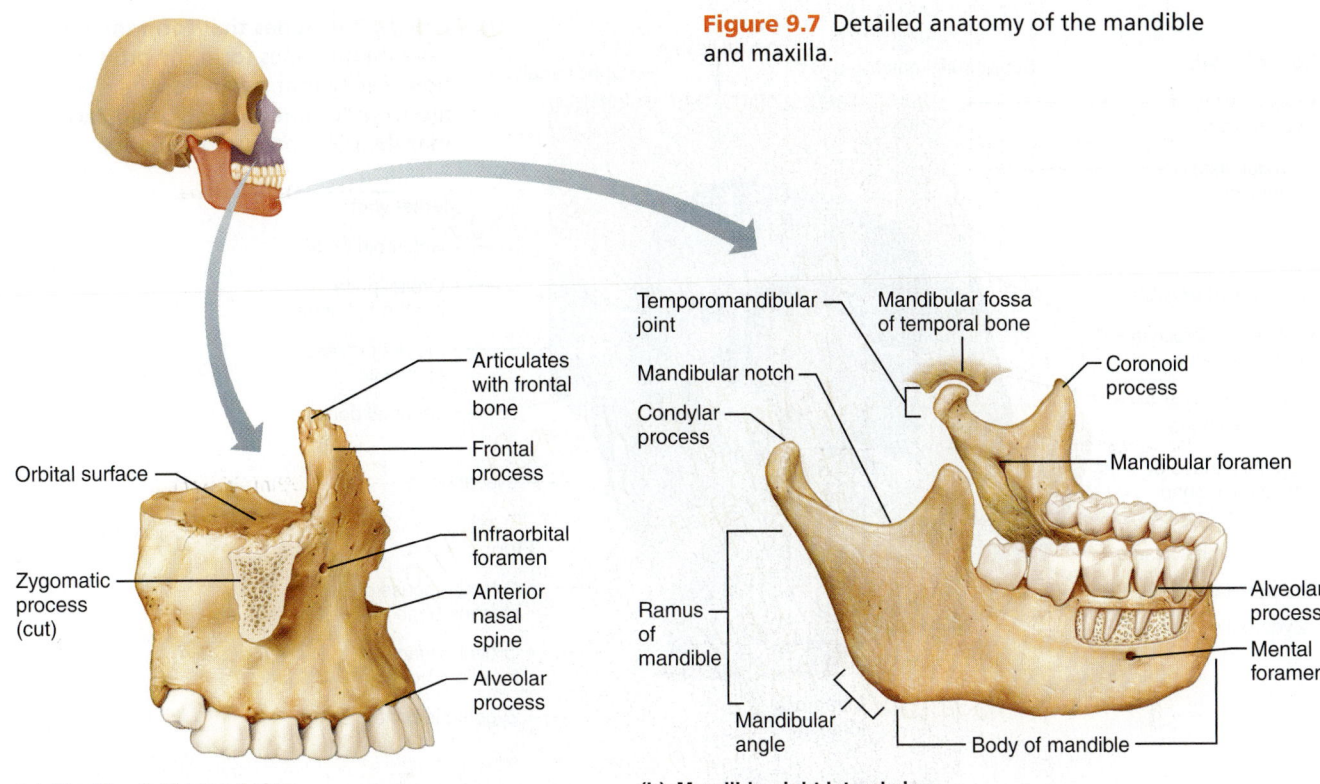

(a) Maxilla, right lateral view

(b) Mandible, right lateral view

Together the cribriform plates and the midline crista galli form the *horizontal plate* of the ethmoid bone.

Perpendicular plate: Inferior projection of the ethmoid that forms the superior part of the nasal septum.

Lateral masses: Irregularly shaped and thin-walled bony regions flanking the perpendicular plate laterally. Their lateral surfaces *(orbital plates)* shape part of the medial orbit wall.

Superior and middle nasal conchae (turbinates): Thin, delicately coiled plates of bone extending medially from the lateral masses of the ethmoid into the nasal cavity. The conchae make air flow through the nasal cavity more efficient and greatly increase the surface area of the mucosa that covers them, thus increasing the mucosa's ability to warm and humidify incoming air.

Facial Bones

Of the 14 bones composing the face, 12 are paired. *Only the mandible and vomer are single bones.* An additional bone, the hyoid bone, although not a facial bone, is considered here because of its location.

●   *Mandible* (Figures 9.1, 9.6, and 9.7) The lower jawbone, which articulates with the temporal bones in the only freely movable joints of the skull.

Mandibular body: Horizontal portion; forms the chin.

Mandibular ramus: Vertical extension of the body on either side.

Condylar process: Articulation point of the mandible with the mandibular fossa of the temporal bone.

Coronoid process: Jutting anterior portion of the ramus; site of muscle attachment.

Mandibular angle: Posterior point at which ramus meets the body.

Mental foramen: Prominent opening on the body (lateral to the midline) that transmits the mental blood vessels and nerve to the lower jaw.

Mandibular foramen: Open the lower jaw of the skull to identify this prominent foramen on the medial aspect of the mandibular ramus. This foramen permits passage of the nerve involved with tooth sensation (mandibular branch of cranial nerve V) and is the site where the dentist injects Novocain to prevent pain while working on the lower teeth.

Alveolar process: Superior margin of mandible; contains sockets in which the teeth lie.

Mandibular symphysis: Anterior median depression indicating point of mandibular fusion.

●   *Maxillae* (Figures 9.1, 9.2, 9.6, and 9.7) Two bones fused in a median suture; form the upper jawbone and part of the orbits. All facial bones, except the mandible, join the maxillae. Thus they are the main, or keystone, bones of the face.

Alveolar process: Inferior margin containing sockets in which teeth lie.

Palatine processes: Form the anterior hard palate; meet medially in the intermaxillary suture.

Infraorbital foramen: Opening under the orbit carrying the infraorbital nerves and blood vessels to the nasal region.

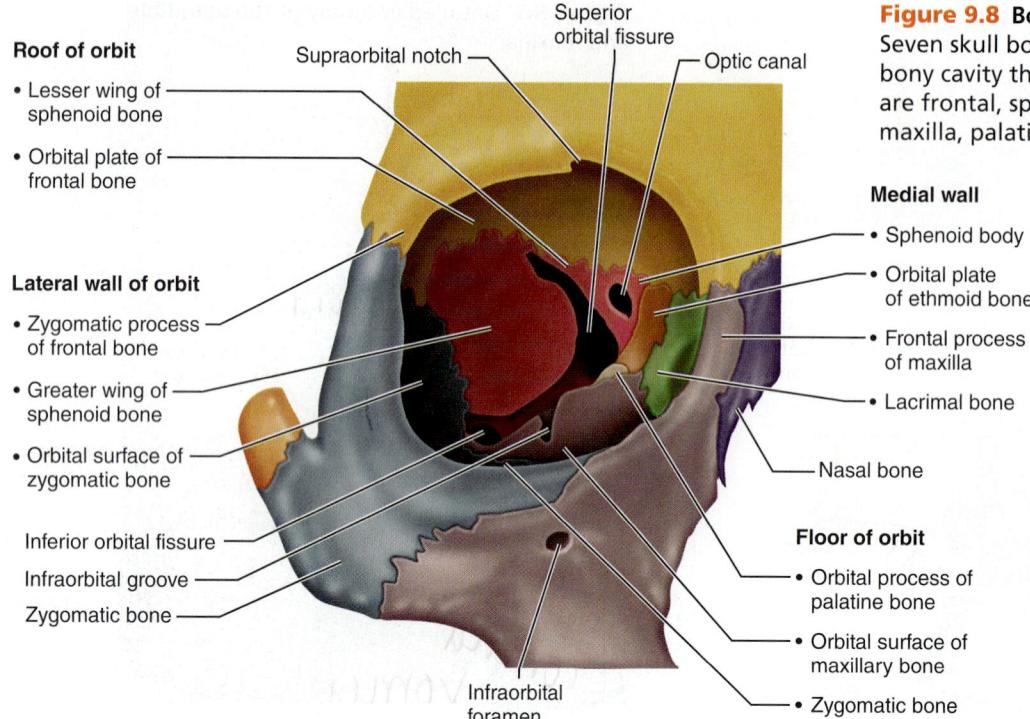

Figure 9.8 Bones that form the orbit. Seven skull bones form the orbit, the bony cavity that surrounds the eye. They are frontal, sphenoid, ethmoid, lacrimal, maxilla, palatine, and zygomatic.

Roof of orbit
- Lesser wing of sphenoid bone
- Orbital plate of frontal bone

Supraorbital notch
Superior orbital fissure
Optic canal

Lateral wall of orbit
- Zygomatic process of frontal bone
- Greater wing of sphenoid bone
- Orbital surface of zygomatic bone

Inferior orbital fissure
Infraorbital groove
Zygomatic bone
Infraorbital foramen

Medial wall
- Sphenoid body
- Orbital plate of ethmoid bone
- Frontal process of maxilla
- Lacrimal bone

Nasal bone

Floor of orbit
- Orbital process of palatine bone
- Orbital surface of maxillary bone
- Zygomatic bone

Incisive fossa: Large bilateral opening located posterior to the central incisor tooth of the maxilla and piercing the hard palate; transmits the nasopalatine arteries and blood vessels.

- **_Lacrimal Bone_** (Figures 9.1 and 9.6a) Fingernail-sized bones forming a part of the medial orbit walls between the maxilla and the ethmoid. Each lacrimal bone is pierced by an opening, the **lacrimal fossa,** which serves as a passageway for tears (*lacrima* = tear).

- **_Palatine Bone_** (Figure 9.2 and Figure 9.8) Paired bones posterior to the palatine processes; form posterior hard palate and part of the orbit; meet medially at the median palatine suture.

- **_Zygomatic Bone_** (Figures 9.1, 9.2, 9.6, and 9.8) Lateral to the maxilla; forms the portion of the face commonly called the cheekbone, and forms part of the lateral orbit. Its three processes are named for the bones with which they articulate.

- **_Nasal Bone_** (Figures 9.1 and 9.6) Small rectangular bones forming the bridge of the nose.

- **_Vomer_** (Figures 9.2 and 9.6) Blade-shaped bone (*vomer* = plow) in median plane of nasal cavity that forms the posterior and inferior nasal septum.

- **_Inferior Nasal Conchae (Turbinates)_** (Figure 9.6a) Thin curved bones protruding medially from the lateral walls of the nasal cavity; serve the same purpose as the turbinate portions of the ethmoid bone.

GROUP CHALLENGE

Odd Bone Out

Each of the following sets contains four bones. One of the listed bones does not share a characteristic that the other three do. Circle the bone that doesn't belong with the others and explain why it is singled out. What characteristic

is it missing? Sometimes there may be multiple reasons why the bone doesn't belong with the others. Include as many as you can think of but make sure it does not have the key characteristic. Use an articulated skull, disarticulated skull bones, and the pictures in your lab manual to help you select and justify your answer.

1. Which is the "odd bone"?	Why is it the odd one out?
Zygomatic bone	
Maxilla	
Vomer	
Nasal bone	
2. Which is the "odd bone"?	**Why is it the odd one out?**
Parietal bone	
Sphenoid bone	
Frontal bone	
Occipital bone	
3. Which is the "odd bone"?	**Why is it the odd one out?**
Lacrimal bone	
Nasal bone	
Zygomatic bone	
Maxilla	

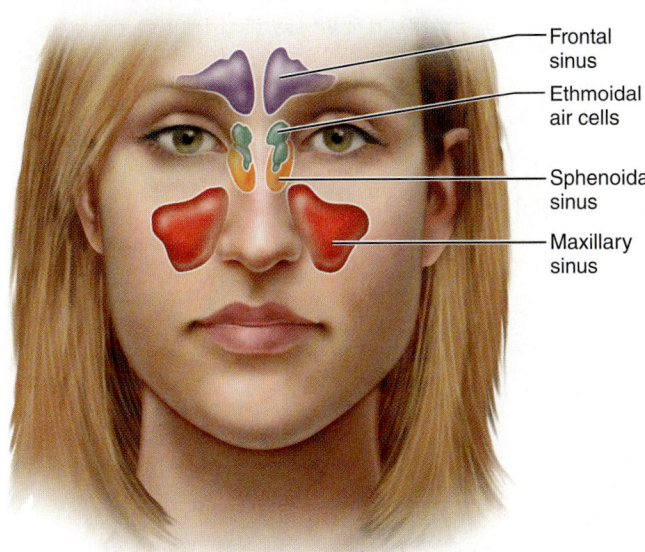

(a)

(b)

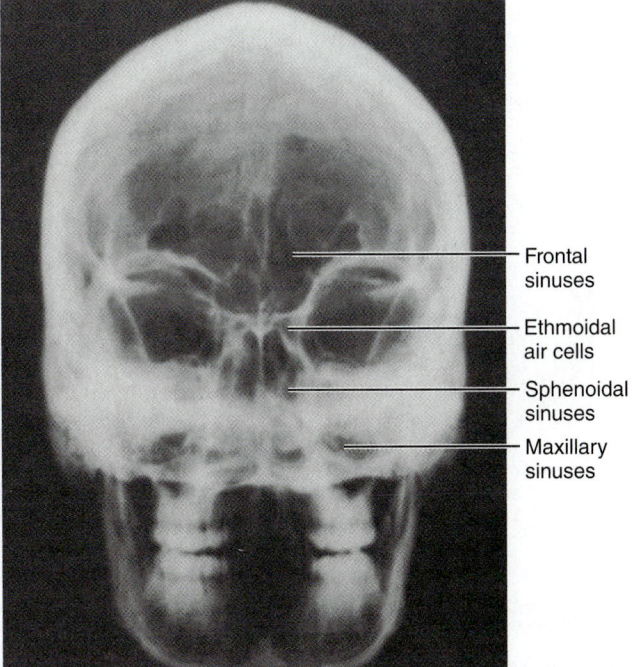

(c)

Figure 9.9 Paranasal sinuses. (a) Anterior aspect. **(b)** Medial aspect. **(c)** Skull X ray showing the paranasal sinuses, anterior view.

Paranasal Sinuses

Four skull bones—maxillary, sphenoid, ethmoid, and frontal—contain sinuses (mucosa-lined air cavities) that lead into the nasal passages (see Figure 9.5 and **Figure 9.9**). These paranasal sinuses lighten the facial bones and may act as resonance chambers for speech. The maxillary sinus is the largest of the sinuses found in the skull.

Sinusitis, or inflammation of the sinuses, sometimes occurs as a result of an allergy or bacterial invasion of the sinus cavities. In such cases, some of the connecting passageways between the sinuses and nasal passages may become blocked with thick mucus or infectious material. Then, as the air in the sinus cavities is absorbed, a partial vacuum forms. The result is a sinus headache localized over the inflamed sinus area. Severe sinus infections may require surgical drainage to relieve this painful condition. ✚

Hyoid Bone

Not really considered or counted as a skull bone, the hyoid bone is located in the throat above the larynx. It serves as a point of attachment for many tongue and neck muscles. It does not articulate with any other bone and is thus unique. It is horseshoe shaped with a body and two pairs of **horns,** or **cornua (Figure 9.10)**.

ACTIVITY 2

Palpating Skull Markings

Palpate the following areas on yourself. Place a check mark in the boxes as you locate the skull markings. Seek assistance from your instructor for any markings that you are unable to locate.

☐   Zygomatic bone and arch. (The most prominent part of your cheek is your zygomatic bone. Follow the posterior course of the zygomatic arch to its junction with your temporal bone.)

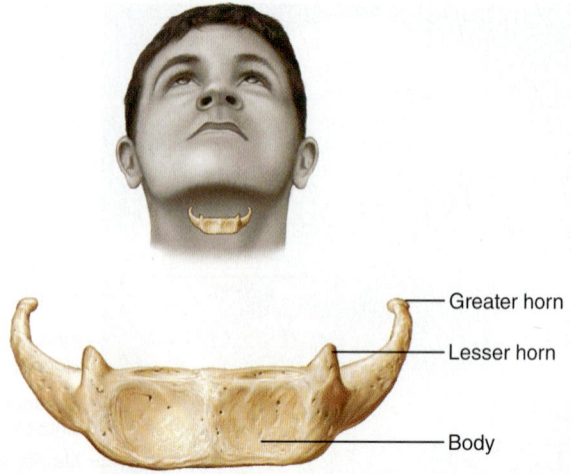

Figure 9.10 Hyoid bone.

☐ Mastoid process (the rough area behind your ear).

☐ Temporomandibular joints. (Open and close your jaws to locate these.)

☐ Greater wing of sphenoid. (Find the indentation posterior to the orbit and superior to the zygomatic arch on your lateral skull.)

☐ Supraorbital foramen. (Apply firm pressure along the superior orbital margin to find the indentation resulting from this foramen.)

☐ Infraorbital foramen. (Apply firm pressure just inferior to the inferomedial border of the orbit to locate this large foramen.)

☐ Mandibular angle (most inferior and posterior aspect of the mandible).

☐ Mandibular symphysis (midline of chin).

☐ Nasal bones. (Run your index finger and thumb along opposite sides of the bridge of your nose until they "slip" medially at the inferior end of the nasal bones.)

☐ External occipital protuberance. (This midline projection is easily felt by running your fingers up the furrow at the back of your neck to the skull.)

☐ Hyoid bone. (Place a thumb and index finger beneath the chin just anterior to the mandibular angles, and squeeze gently. Exert pressure with the thumb, and feel the horn of the hyoid with the index finger.) ▬

The Vertebral Column

The **vertebral column,** extending from the skull to the pelvis, forms the body's major axial support. Additionally, it surrounds and protects the delicate spinal cord while allowing the spinal nerves to emerge from the cord via openings between adjacent vertebrae. The term *vertebral column* might suggest a rigid supporting rod, but this is far from the truth. The vertebral column consists of 24 single bones called **vertebrae** and two composite, or fused, bones (the sacrum and coccyx) that are connected in such a way as to provide a flexible curved structure **(Figure 9.11)**. Of the 24 single

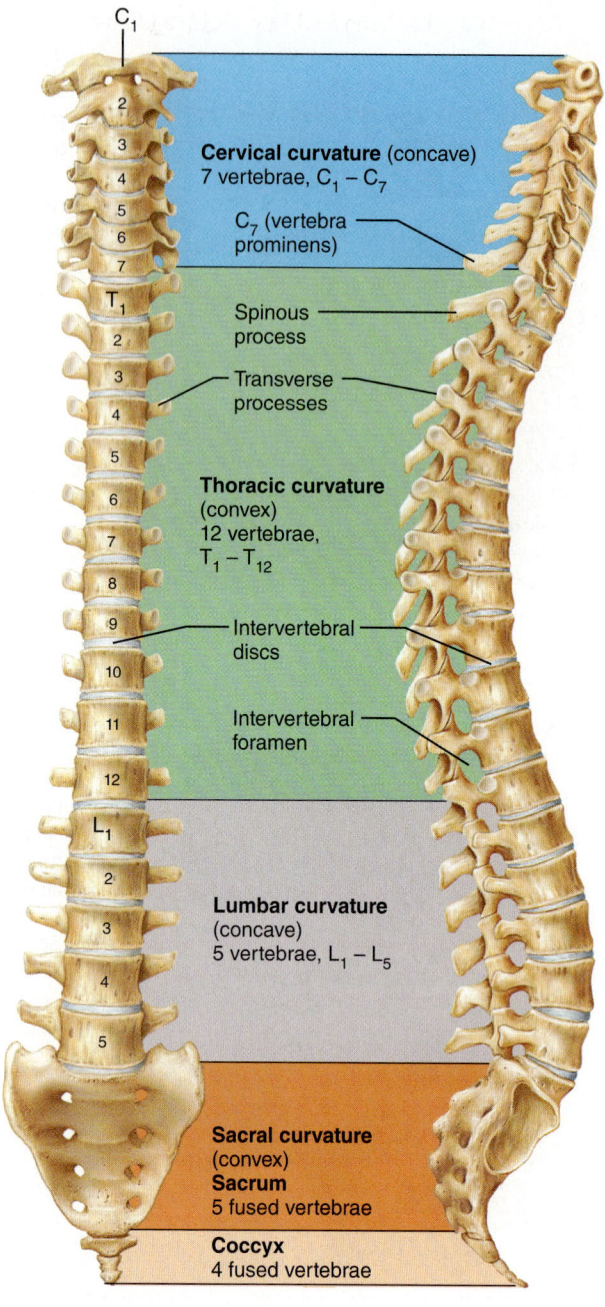

Anterior view   *Right lateral view*

Figure 9.11 The vertebral column. Notice the curvatures in the lateral view. (The terms *convex* and *concave* refer to the curvature of the posterior aspect of the vertebral column.)

vertebrae, the seven bones of the neck are called *cervical vertebrae;* the next 12 are *thoracic vertebrae;* and the 5 supporting the lower back are *lumbar vertebrae.* Remembering common mealtimes for breakfast, lunch, and dinner (7 A.M., 12 noon, and 5 P.M.) may help you to remember the number of bones in each region.

The vertebrae are separated by pads of fibrocartilage, **intervertebral discs,** that cushion the vertebrae and absorb shocks. Each disc has two major regions, a central gelatinous *nucleus pulposus* that behaves like a fluid, and an outer ring of encircling collagen fibers called the *anulus fibrosus* that stabilizes the disc and contains the pulposus.

As a person ages, the water content of the discs decreases (as it does in other tissues throughout the body), and the discs become thinner and less compressible. This situation, along with other degenerative changes such as weakening of the ligaments and tendons of the vertebral column, predisposes older people to a ruptured disc, called a **herniated disc.** In a herniated disc, the anulus fibrosus commonly ruptures and the nucleus pulposus protrudes (herniates) through it. This event typically compresses adjacent nerves, causing pain. ✚

The presence of the discs and the curvatures create a springlike construction of the vertebral column that prevents shock to the head in walking and running and provides flexibility to the body trunk. The thoracic and sacral curvatures of the spine are referred to as *primary curvatures* because they are present and well developed at birth. Later the *secondary curvatures* are formed. The cervical curvature becomes prominent when the baby begins to hold its head up independently, and the lumbar curvature develops when the baby begins to walk.

ACTIVITY 3

Examining Spinal Curvatures

1. Observe the normal curvature of the vertebral column in the articulated vertebral column or laboratory skeleton (compare it to Figure 9.11). Note the differences between normal curvature and three abnormal spinal curvatures seen in the figure—*scoliosis, kyphosis,* and *lordosis* **(Figure 9.12)**. These abnormalities may result from disease or poor posture. Also examine X rays, if they are available, showing these same conditions in a living patient.

2. Then, using the articulated vertebral column (or an articulated skeleton), examine the freedom of movement between two lumbar vertebrae separated by an intervertebral disc.

When the fibrous disc is properly positioned, are the spinal cord or peripheral nerves impaired in any way?

_____

Remove the disc and put the two vertebrae back together. What happens to the nerve?

_____

What would happen to the spinal nerves in areas of malpositioned or "slipped" discs?

_____ ▬

Structure of a Typical Vertebra

Although they differ in size and specific features, all vertebrae have some features in common **(Figure 9.13)**.

Body (or **centrum**): Rounded central portion of the vertebra, which faces anteriorly in the human vertebral column.

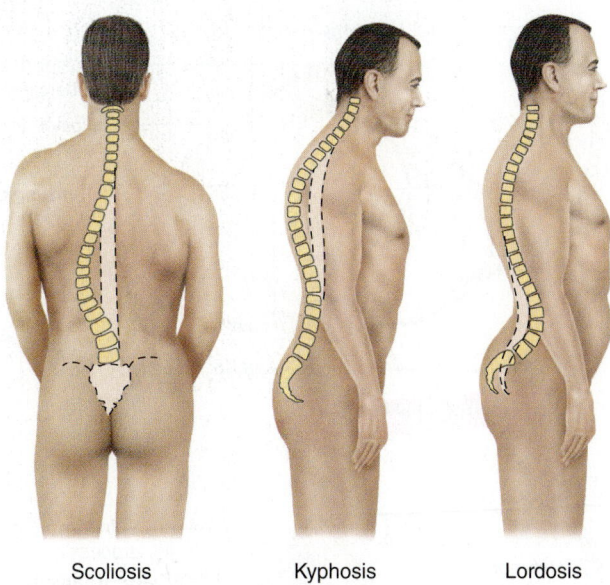

Figure 9.12 Abnormal spinal curvatures

Scoliosis   Kyphosis   Lordosis

Vertebral arch: Composed of pedicles, laminae, and a spinous process, it represents the junction of all posterior extensions from the vertebral body.

Vertebral (spinal) foramen: Opening enclosed by the body and vertebral arch; a passageway for the spinal cord.

Transverse processes: Two lateral projections from the vertebral arch.

Spinous process: Single medial and posterior projection from the vertebral arch.

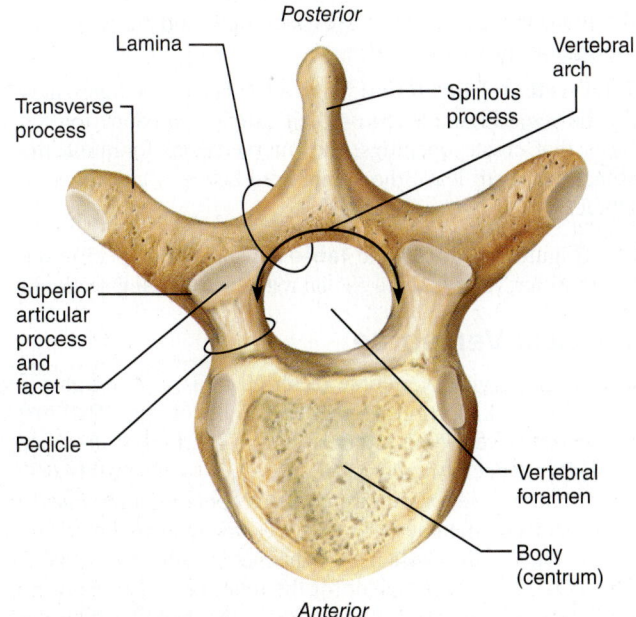

Posterior

Lamina

Transverse process

Superior articular process and facet

Pedicle

Vertebral arch

Spinous process

Vertebral foramen

Body (centrum)

Anterior

Figure 9.13 A typical vertebra, superior view. Inferior articulating surfaces not shown.

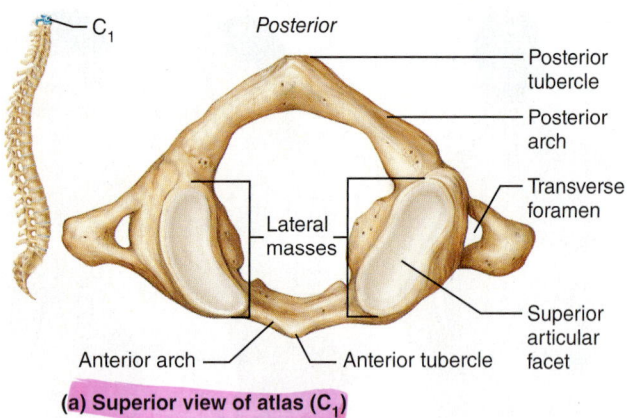

(a) Superior view of atlas (C₁)

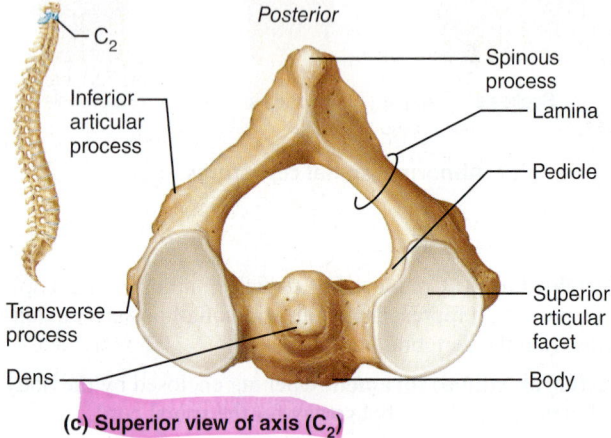

(c) Superior view of axis (C₂)

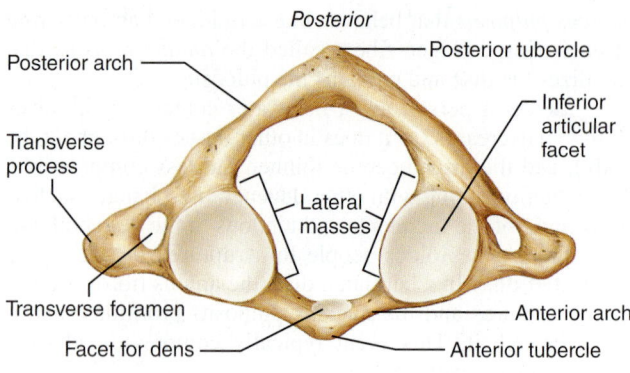

(b) Inferior view of atlas (C₁)

Figure 9.14 The first and second cervical vertebrae.

The more typical cervical vertebrae (C₃ through C₇) are distinguished from the thoracic and lumbar vertebrae by several features (see Table 9.1 and **Figure 9.15**). They are the smallest, lightest vertebrae, and the vertebral foramen is triangular. The spinous process is short and often bifurcated (divided into two branches). The spinous process of C₇ is not branched, however, and is substantially longer than that of the other cervical vertebrae. Because the spinous process of C₇ is visible through the skin, it is called the *vertebra prominens* (Figure 9.11) and is used as a landmark for counting the vertebrae. Transverse processes of the cervical vertebrae are wide, and they contain foramina through which the vertebral arteries pass superiorly on their way to the brain. Any time you see these foramina in a vertebra, you can be sure that it is a cervical vertebra.

☐ Palpate your vertebra prominens. Place a check mark in the box when you locate the structure.

Thoracic Vertebrae

The 12 thoracic vertebrae (referred to as T₁ through T₁₂) may be recognized by the following structural characteristics. They have a larger body than the cervical vertebrae (see Figure 9.15). The body is somewhat heart shaped, with two small articulating surfaces, or **costal facets,** on each side (one superior, the other inferior) close to the origin of the vertebral arch. Sometimes referred to as *costal demifacets* because of their small size, these facets articulate with the heads of the corresponding ribs. The vertebral foramen is oval or round, and the spinous process is long, with a sharp downward hook. The closer the thoracic vertebra is to the lumbar region, the less sharp and shorter the spinous process. Articular facets on the transverse processes articulate with the tubercles of the ribs. Besides forming the thoracic part of the spine, these vertebrae form the posterior aspect of the bony thorax. Indeed, they are the only vertebrae that articulate with the ribs.

Lumbar Vertebrae

The five lumbar vertebrae (L₁ through L₅) have massive block-like bodies and short, thick, hatchet-shaped spinous processes extending directly backward (see Table 9.1 and Figure 9.15). The superior articular facets face posteromedially; the inferior ones

Superior and inferior articular processes: Paired projections lateral to the vertebral foramen that enable articulation with adjacent vertebrae. The superior articular processes typically face toward the spinous process (posteriorly), whereas the inferior articular processes face (anteriorly) away from the spinous process.

Intervertebral foramina: The right and left pedicles have notches (see Figure 9.15) on their inferior and superior surfaces that create openings, the intervertebral foramina, for spinal nerves to leave the spinal cord between adjacent vertebrae.

(Figures 9.14–9.16 and **Table 9.1** show how specific vertebrae differ; refer to them as you read the following sections.)

Cervical Vertebrae

The seven cervical vertebrae (referred to as C₁ through C₇) form the neck portion of the vertebral column. The first two cervical vertebrae (atlas and axis) are highly modified to perform special functions **(Figure 9.14)**. The **atlas** (C₁) lacks a body, and its lateral processes contain large concave depressions on their superior surfaces that receive the occipital condyles of the skull. This joint enables you to nod "yes." The **axis** (C₂) acts as a pivot for the rotation of the atlas (and skull) above. It bears a large vertical process, the **dens,** or **odontoid process,** that serves as the pivot point. The articulation between C₁ and C₂ allows you to rotate your head from side to side to indicate "no."

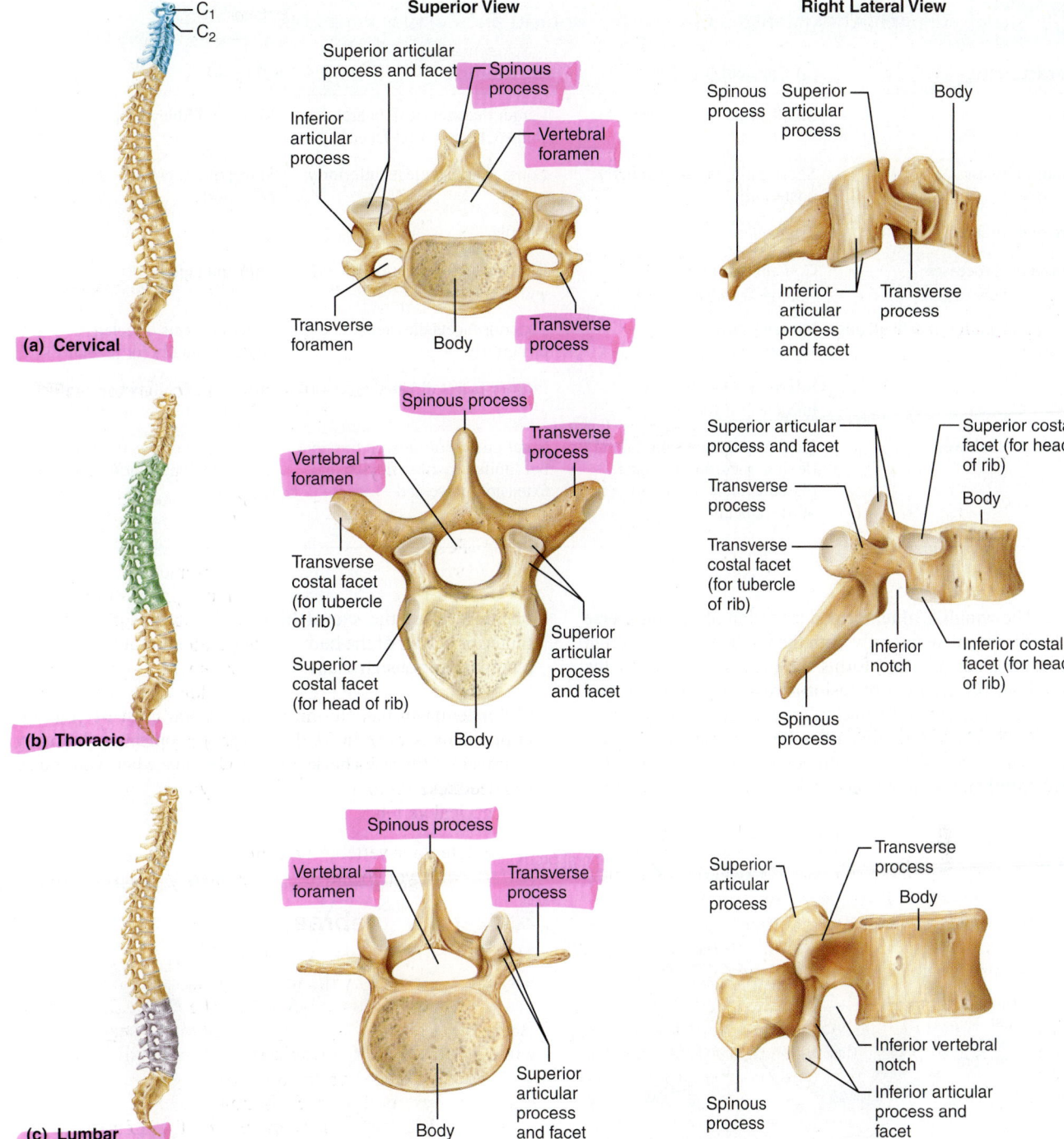

Figure 9.15 Superior and right lateral views of typical vertebrae.

are directed anterolaterally. These structural features reduce the mobility of the lumbar region of the spine. Since most stress on the vertebral column occurs in the lumbar region, these are also the sturdiest of the vertebrae.

The spinal cord ends at the superior edge of L_2, but the outer covering of the cord, filled with cerebrospinal fluid, extends an appreciable distance beyond. Thus a *lumbar puncture* (for examination of the cerebrospinal fluid) or the administration of "saddle block" anesthesia for childbirth is normally done between L_3 and L_4 or L_4 and L_5, where there is little or no chance of injuring the delicate spinal cord.

The Sacrum

The **sacrum (Figure 9.16)** is a composite bone formed from the fusion of five vertebrae. Superiorly it articulates with L_5, and inferiorly it connects with the coccyx. The **median sacral crest** is a remnant of the spinous processes of the fused ver-

Table 9.1	Regional Characteristics of Cervical, Thoracic, and Lumbar Vertebrae		
Characteristic	(a) Cervical (C_3–C_7)	(b) Thoracic	(c) Lumbar
Body	Small, wide side to side	Larger than cervical; heart shaped; bears costal facets	Massive; kidney shaped
Spinous process	Short; bifid; projects directly posteriorly	Long; sharp; projects inferiorly	Short; blunt; projects directly posteriorly
Vertebral foramen	Triangular	Circular	Triangular
Transverse processes	Contain foramina	Bear facets for ribs (except T_{11} and T_{12})	Thin and tapered
Superior and inferior articulating processes	Superior facets directed superoposteriorly	Superior facets directed posteriorly	Superior facets directed posteromedially (or medially)
	Inferior facets directed inferoanteriorly	Inferior facets directed anteriorly	Inferior facets directed anterolaterally (or laterally)
Movements allowed	Flexion and extension; lateral flexion; rotation; the spine region with the greatest range of movement	Rotation; lateral flexion possible but limited by ribs; flexion and extension prevented	Flexion and extension; some lateral flexion; rotation prevented

tebrae. The winglike **alae,** formed by fusion of the transverse processes, articulate laterally with the hip bones. The sacrum is concave anteriorly and forms the posterior border of the pelvis. Four ridges (lines of fusion) cross the anterior part of the sacrum, and **sacral foramina** are located at either end of these ridges. These foramina allow blood vessels and nerves to pass. The vertebral canal continues inside the sacrum as the **sacral canal** and terminates near the coccyx via an enlarged opening called the **sacral hiatus.** The **sacral promontory** (anterior border of the body of S_1) is an important anatomical landmark for obstetricians.

☐   Attempt to palpate the median sacral crest of your sacrum. (This is more easily done by thin people and obviously in privacy.) Place a check mark in the box when you locate the structure.

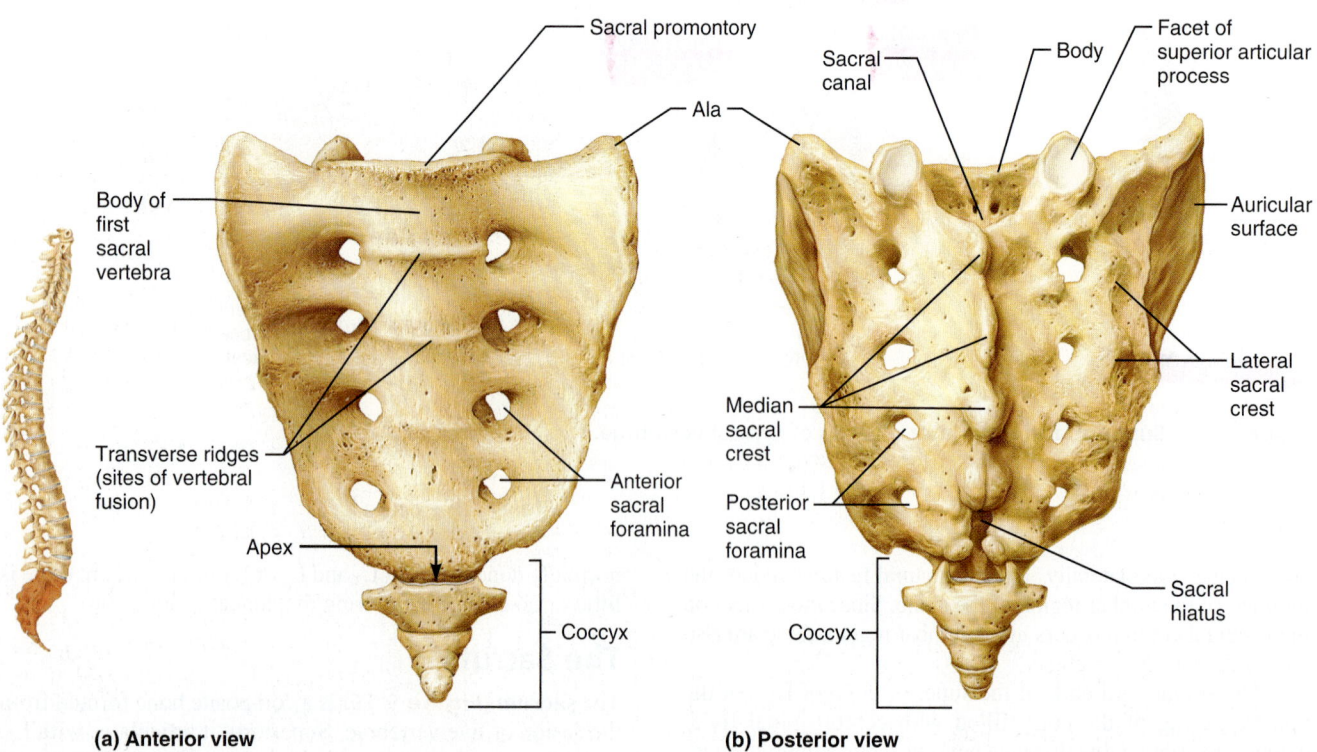

(a) Anterior view   (b) Posterior view

Figure 9.16 Sacrum and coccyx.

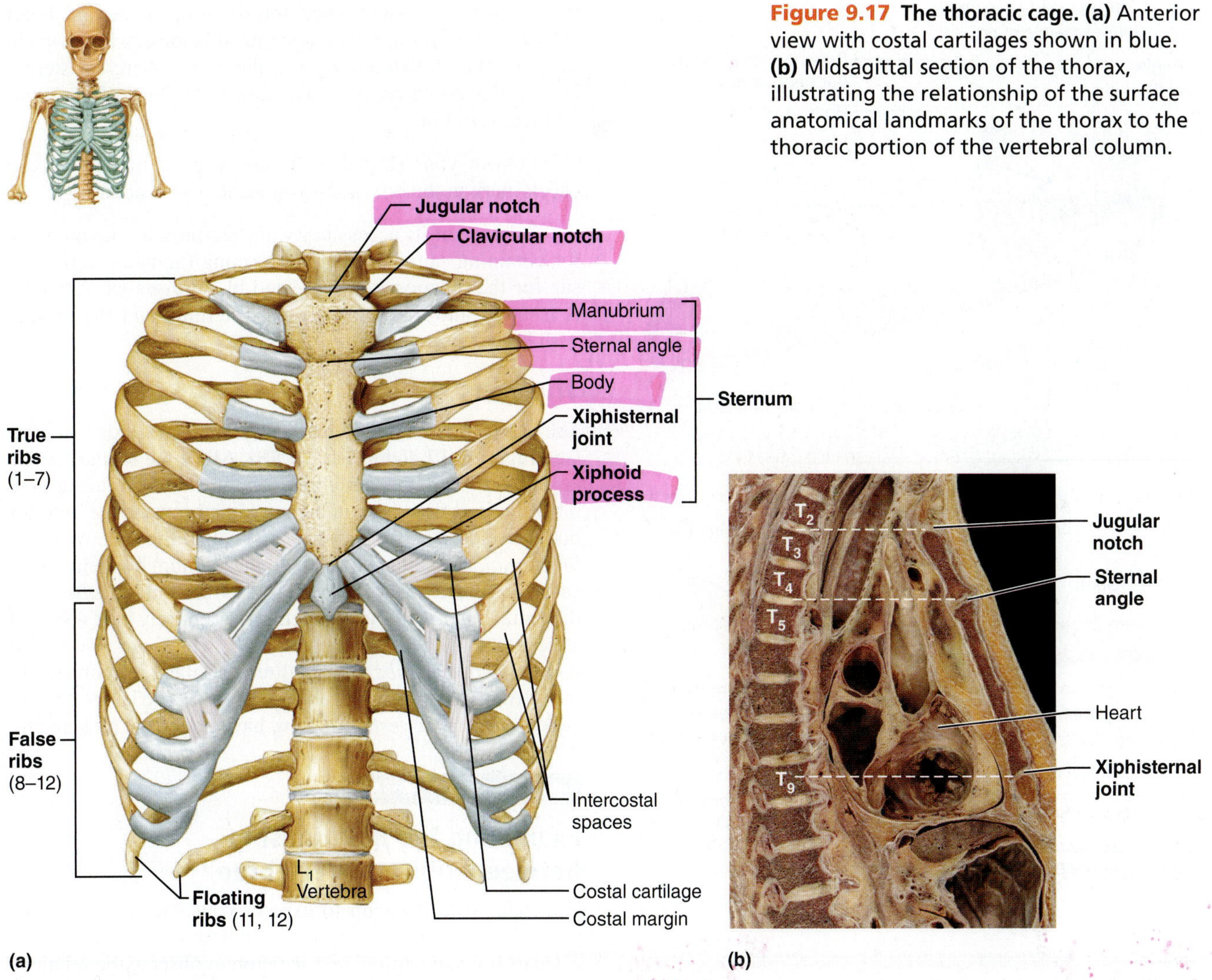

Figure 9.17 The thoracic cage. (a) Anterior view with costal cartilages shown in blue. **(b)** Midsagittal section of the thorax, illustrating the relationship of the surface anatomical landmarks of the thorax to the thoracic portion of the vertebral column.

The Coccyx

The **coccyx** (see Figure 9.16) is formed from the fusion of three to five small irregularly shaped vertebrae. It is literally the human tailbone, a vestige of the tail that other vertebrates have. The coccyx is attached to the sacrum by ligaments.

ACTIVITY 4

Examining Vertebral Structure

Obtain examples of each type of vertebra and examine them carefully, comparing them to the figures and table (Figures 9.14, 9.15, 9.16, and Table 9.1) and to each other. ■

The Thoracic Cage

The **thoracic cage** consists of the bony thorax, which is composed of the sternum, ribs, and thoracic vertebrae, plus the costal cartilages **(Figure 9.17)**. Its cone-shaped cagelike structure protects the organs of the thoracic cavity including the critically important heart and lungs.

The Sternum

The **sternum** (breastbone), a typical flat bone, is a result of the fusion of three bones—the manubrium, body, and xiphoid process. It is attached to the first seven pairs of ribs. The superiormost **manubrium** looks like the knot of a tie; it articulates with the clavicle (collarbone) laterally. The **body** forms the bulk of the sternum. The **xiphoid process** constructs the inferior end of the sternum and lies at the level of the fifth intercostal space. Although it is made of hyaline cartilage in children, it is usually ossified in adults.

In some people, the xiphoid process projects dorsally. This may present a problem because physical trauma to the chest can push such a xiphoid into the underlying heart or liver, causing massive hemorrhage. ✚

The sternum has three important bony landmarks—the jugular notch, the sternal angle, and the xiphisternal joint. The **jugular notch** (concave upper border of the manubrium) can be palpated easily; generally it is at the level of the third thoracic vertebra. The **sternal angle** is a result of the manubrium and body meeting at a slight angle to each other, so that a transverse ridge is formed at the level of the second ribs. It provides a handy reference point for counting ribs to locate

(a)

Transverse costal facet
(for tubercle of rib)

Angle
of rib

Superior costal facet
(for head of rib)

Body of vertebra

Head of rib

Intervertebral disc

Neck of rib

Tubercle of rib

Shaft   Sternum

Cross
section
of rib

Costal groove   Costal cartilage

(b)

Articular facet
on tubercle of rib

Spinous process

Shaft

Ligaments

Neck of rib

Head of rib

Superior costal facet
(for head of rib)

Transverse
costal facet
(for tubercle
of rib)

Body of
thoracic
vertebra

(c)

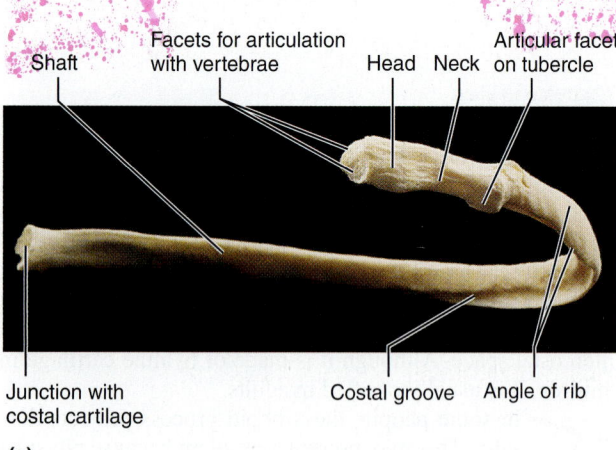

Shaft

Facets for articulation
with vertebrae

Head   Neck

Articular facet
on tubercle

Junction with
costal cartilage

Costal groove   Angle of rib

Figure 9.18 Structure of a typical true rib and its articulations. (a) Vertebral and sternal articulations of a typical true rib. **(b)** Superior view of the articulation between a rib and a thoracic vertebra, with costovertebral ligaments. **(c)** Right rib 6, posterior view.

the second intercostal space for listening to certain heart valves, and is an important anatomical landmark for thoracic surgery. The **xiphisternal joint,** the point where the sternal body and xiphoid process fuse, lies at the level of the ninth thoracic vertebra.

☐ Palpate your sternal angle and jugular notch. Place a check mark in the box when you locate the structures.

Because of its accessibility, the sternum is a favored site for obtaining samples of blood-forming (hematopoietic) tissue for the diagnosis of suspected blood diseases. A needle is inserted into the marrow of the sternum and the sample withdrawn (sternal puncture).

The Ribs

The 12 pairs of **ribs** form the walls of the thoracic cage (see Figure 9.17 and **Figure 9.18**). All of the ribs articulate posteriorly with the vertebral column via their heads and tubercles and then curve downward and toward the anterior body surface. The first seven pairs, called the *true,* or *vertebrosternal, ribs,* attach directly to the sternum by their "own" costal cartilages. The next five pairs are called *false ribs;* they attach indirectly to the sternum or entirely lack a sternal attachment. Of these, rib pairs 8–10, which are also called *vertebrochondral ribs,* have indirect cartilage attachments to the sternum via the costal cartilage of rib 7. The last two pairs, called *floating,* or *vertebral, ribs,* have no sternal attachment.

ACTIVITY 5

Examining the Relationship Between Ribs and Vertebrae

First take a deep breath to expand your chest. Notice how your ribs seem to move outward and how your sternum rises. Then examine an articulated skeleton to observe the relationship between the ribs and the vertebrae.

(Refer to Activity 3, Palpating Landmarks of the Trunk, section on The Thorax: Bones, steps 1 and 3 in Exercise 46, Surface Anatomy Roundup.) ∎

The Fetal Skull

One of the most obvious differences between fetal and adult skeletons is the huge size of the fetal skull relative to the rest of the skeleton. Skull bones are incompletely formed at birth and connected by fibrous membranes called **fontanelles.** The fontanelles allow the fetal skull to be compressed slightly during birth and also allow for brain growth during late fetal life. They ossify (become bone) as the infant ages, completing the process by the time the child is 1½ to 2 years old.

ACTIVITY 6

Examining a Fetal Skull

1. Obtain a fetal skull and study it carefully.

- Does it have the same bones as the adult skull?
- How does the size of the fetal face relate to the cranium?
- How does this compare to what is seen in the adult?

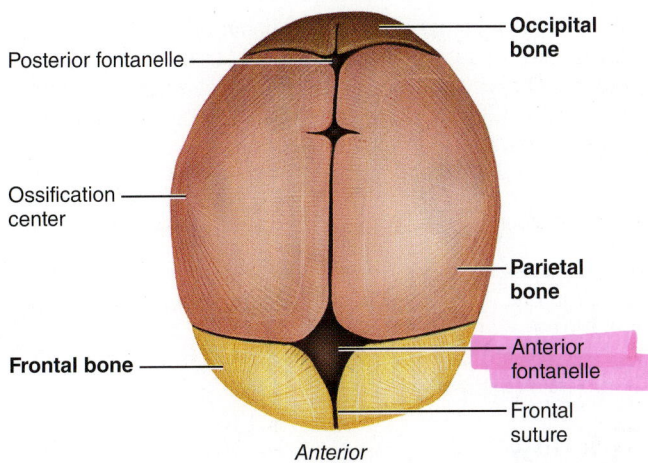

(a) Superior view

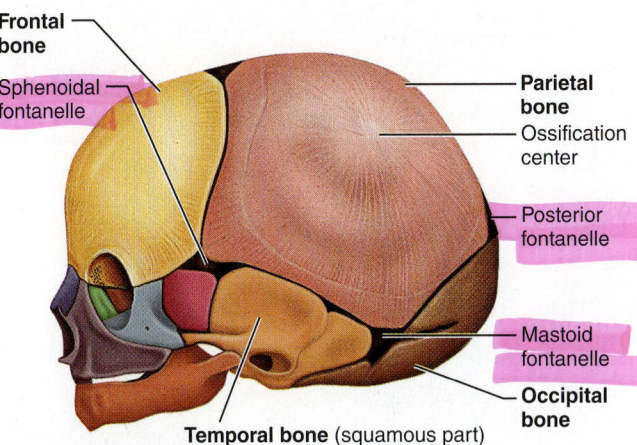

(b) Left lateral view

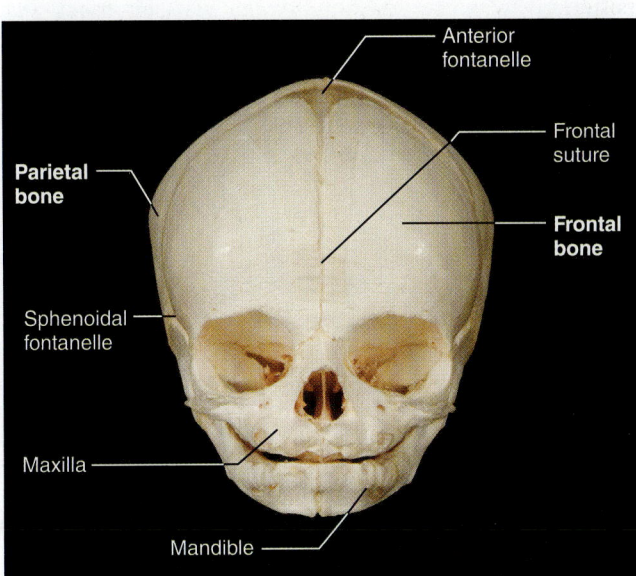

(c) Anterior view

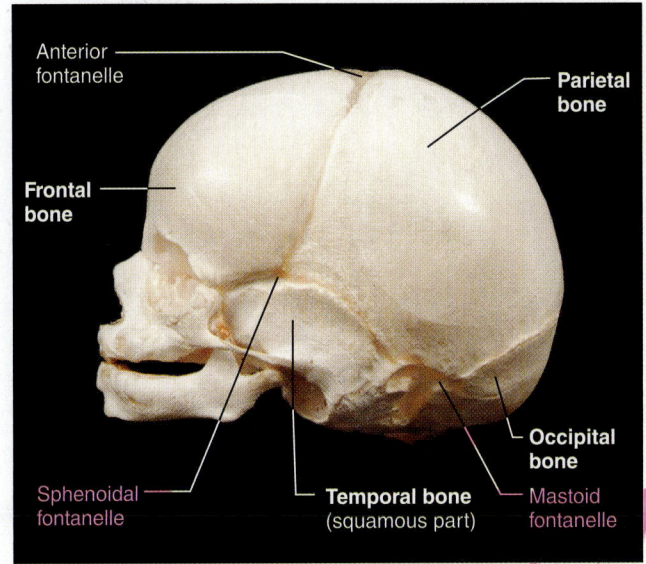

(d) Left lateral view

Figure 9.19 Skull of a newborn.

2. Locate the following fontanelles on the fetal skull (refer to **Figure 9.19**): *anterior (or frontal) fontanelle, mastoid fontanelle, sphenoidal fontanelle,* and *posterior (or occipital) fontanelle.*

3. Notice that some of the cranial bones have conical protrusions. These are **ossification (growth) centers.** Notice also that the frontal bone is still in two parts, and the temporal bone is incompletely ossified, little more than a ring of bone.

4. Before completing this study, check the questions on the review sheet at the end of this exercise to ensure that you have made all of the necessary observations. ▪

The Appendicular Skeleton

OBJECTIVES

1. Identify the bones of the pectoral and pelvic girdles and their attached limbs by examining isolated bones or an articulated skeleton, and name the important bone markings on each.

2. Describe the differences between a male and a female pelvis and explain the importance of these differences.

3. Compare the features of the human pectoral and pelvic girdles, and discuss how their structures relate to their specialized functions.

4. Arrange unmarked, disarticulated bones in their proper places to form an entire skeleton.

PRE-LAB QUIZ

1. The _____ skeleton is made up of 126 bones of the limbs and girdles.
2. Circle the correct underlined term. The <u>pectoral</u> / <u>pelvic</u> girdle attaches the upper limb to the axial skeleton.
3. The _____, on the posterior thorax, are roughly triangular in shape. They have no direct attachment to the axial skeleton but are held in place by trunk muscles.
4. The arm consists of one long bone, the _____.
   a. femur   c. tibia
   b. humerus   d. ulna
5. The hand consists of three groups of bones. The carpals make up the wrist. The _____ make up the palm, and the phalanges make up the fingers.
6. You are studying a pelvis that is wide and shallow. The acetabula are small and far apart. The pubic arch/angle is rounded and greater than 90°. It appears to be tilted forward, with a wide, short sacrum. Is this a male or a female pelvis? _____
7. The strongest, heaviest bone of the body is in the thigh. It is the
   a. femur
   b. fibula
   c. tibia
8. The _____, or "knee cap," is a sesamoid bone that is found within the quadriceps tendon.
9. Circle True or False. The fingers of the hand and the toes of the foot—with the exception of the great toe and the thumb—each have three phalanges.
10. Each foot has a total of _____ bones.

The **appendicular skeleton** (the gold-colored portion of Figure 8.1) is composed of the 126 bones of the appendages and the pectoral and pelvic girdles, which attach the limbs to the axial skeleton. Although the upper and lower limbs differ in their functions and mobility, they have the same fundamental plan, with each limb made up of three major segments connected together by freely movable joints.

10

ACTIVITY 1

Examining and Identifying Bones of the Appendicular Skeleton

Carefully examine each of the bones described in this exercise and identify the characteristic bone markings of each. The markings aid in determining whether a bone is the right or left member of its pair; for example, the glenoid cavity is on the lateral aspect of the scapula and the spine is on its posterior aspect. *This is a very important instruction because you will be constructing your own skeleton to finish this laboratory exercise.* Additionally, when corresponding X rays are available, compare the actual bone specimen to its X-ray image. ■■

Bones of the Pectoral Girdle and Upper Limb

The Pectoral (Shoulder) Girdle

The paired **pectoral,** or **shoulder, girdles (Figure 10.1)** each consist of two bones—the anterior clavicle and the posterior scapula. The shoulder girdles attach the upper limbs to the axial skeleton and provide attachment points for many trunk and neck muscles.

The **clavicle,** or collarbone, is a slender doubly curved bone—convex forward on its medial two-thirds and concave laterally. Its *sternal* (medial) *end,* which attaches to the sternal manubrium, is rounded or triangular in cross section. The sternal end projects above the manubrium and can be felt and usually seen forming the lateral walls of the *jugular notch* (see Figure 9.17, page 135). The *acromial* (lateral) *end* of the clavicle is flattened where it articulates with the scapula to form part of the shoulder joint. On its posteroinferior surface is the prominent **conoid tubercle (Figure 10.2b)**. This projection anchors a ligament and provides a handy landmark for determining whether a given clavicle is from the right or left side of the body. The clavicle serves as an anterior brace, or strut, to hold the arm away from the top of the thorax.

The **scapulae** (Figure 10.2c–e), or shoulder blades, are generally triangular and are commonly called the "wings" of humans. Each scapula has a flattened body and two important processes—the **acromion** (the enlarged, roughened end of the spine of the scapula) and the beaklike **coracoid process** (*corac* = crow, raven). The acromion connects with the clavicle. The coracoid process points anteriorly over the tip of the shoulder joint and serves as an attachment point for some of the upper limb muscles. The **suprascapular notch** at the base of the coracoid process allows nerves to pass. The scapula has no direct attachment to the axial skeleton but is loosely held in place by trunk muscles.

The scapula has three angles: superior, inferior, and lateral. The inferior angle provides a landmark for auscultating (listening to) lung sounds. The **glenoid cavity,** a shallow socket that receives the head of the arm bone (humerus), is located in the blunted lateral angle. The scapula also has three named borders: superior, medial (vertebral), and lateral (axillary). Several shallow depressions (fossae) appear on both sides of the scapula and are named according to location; there are the anterior *subscapular fossa* and the posterior *infraspinous* and *supraspinous fossae.*

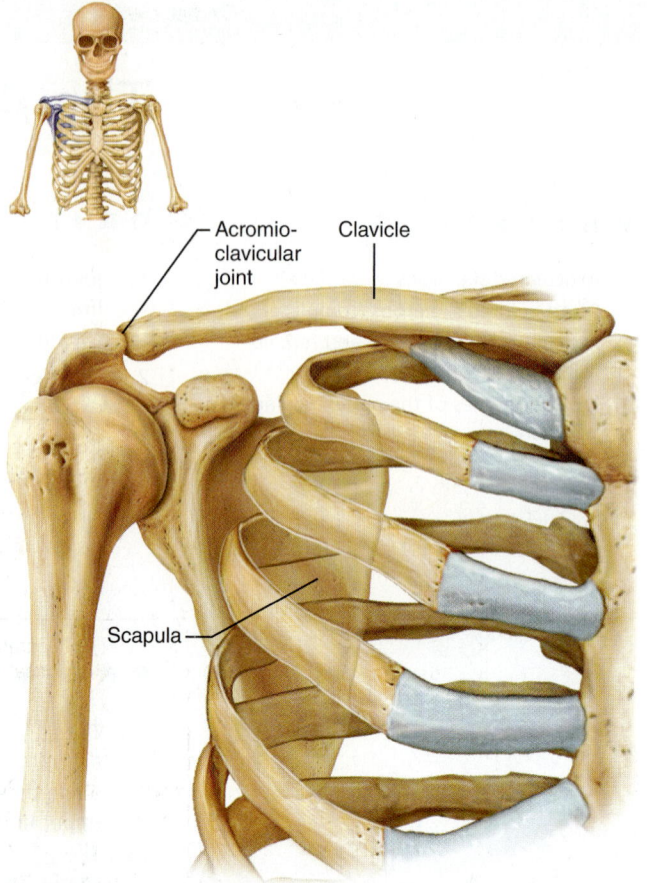

Figure 10.1 Articulated bones of the pectoral (shoulder) girdle. The right pectoral girdle is articulated to show the relationship of the girdle to the bones of the thorax and arm.

The shoulder girdle is exceptionally light and allows the upper limb a degree of mobility not seen anywhere else in the body. This is due to the following factors:

- The sternoclavicular joints are the *only* site of attachment of the shoulder girdles to the axial skeleton.

- The relative looseness of the scapular attachment allows it to slide back and forth against the thorax with muscular activity.

- The glenoid cavity is shallow and does little to stabilize the shoulder joint.

However, this exceptional flexibility exacts a price: the arm bone (humerus) is very susceptible to dislocation, and fracture of the clavicle disables the entire upper limb.

The Arm

The arm or brachium **(Figure 10.3)** contains a single bone—the **humerus,** a typical long bone. Proximally its rounded *head* fits into the shallow glenoid cavity of the scapula. The head is separated from the shaft by the *anatomical neck* and the more constricted *surgical neck,* which is a common site of fracture. Opposite the head are two prominences, the **greater** and **lesser tubercles** (from lateral to medial aspect),

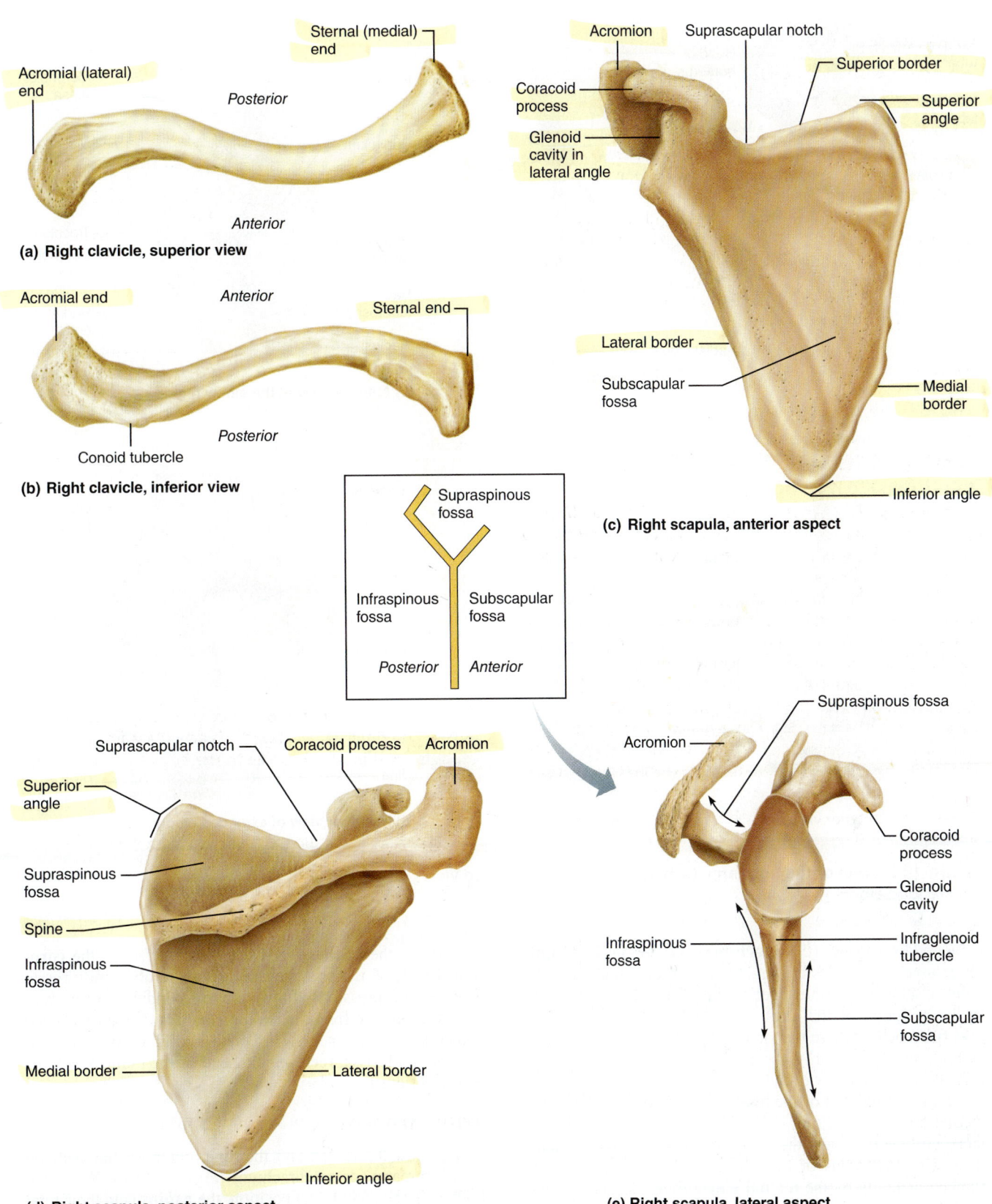

(a) **Right clavicle, superior view**

(b) **Right clavicle, inferior view**

(c) **Right scapula, anterior aspect**

10

(d) **Right scapula, posterior aspect**

(e) **Right scapula, lateral aspect**

Figure 10.2 Individual bones of the pectoral (shoulder) girdle. View **(e)** is accompanied by a schematic representation of its orientation.

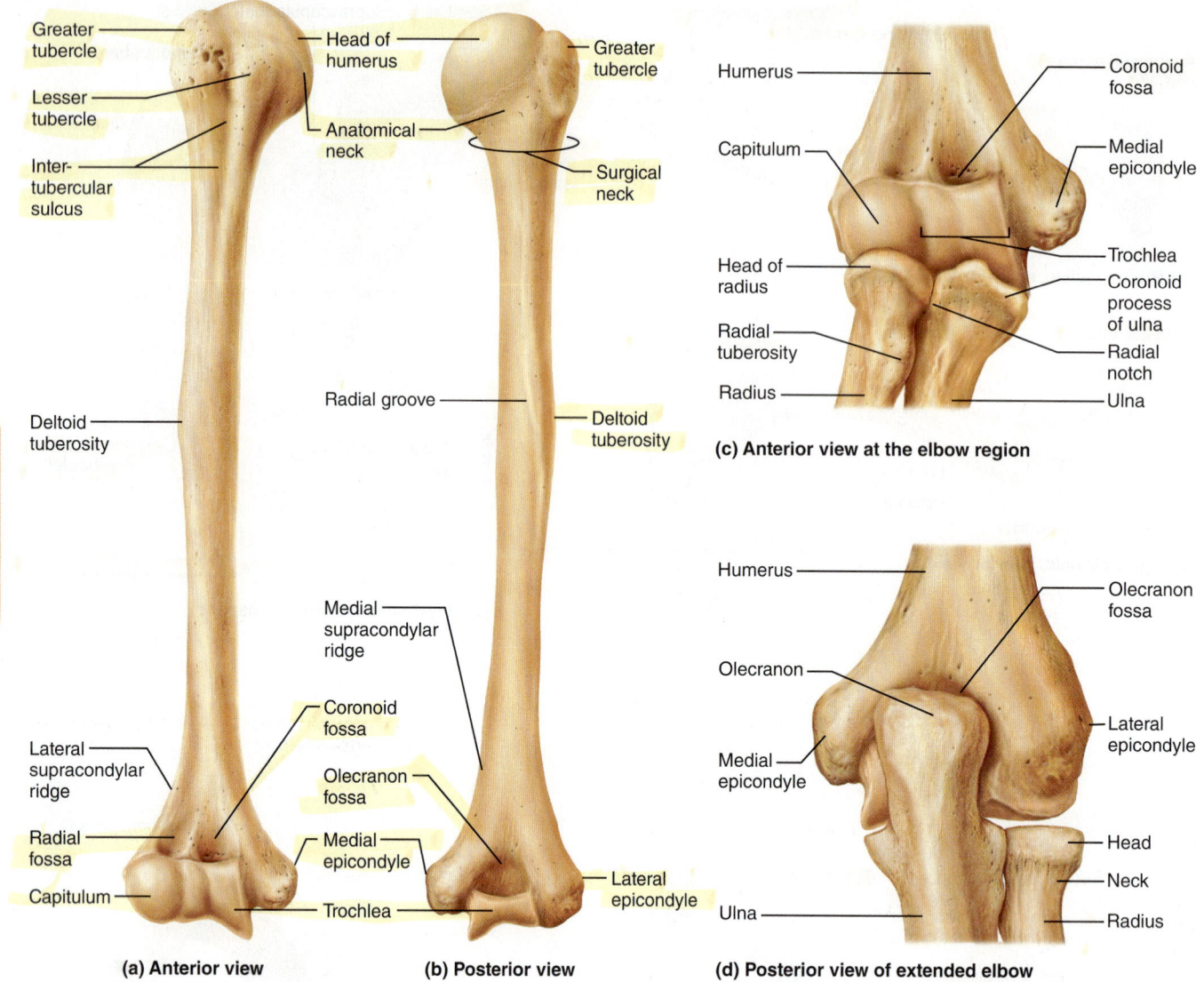

(a) Anterior view

(b) Posterior view

(c) Anterior view at the elbow region

(d) Posterior view of extended elbow

Figure 10.3 Bone of the right arm. (a, b) Humerus. **(c, d)** Detailed views of extended elbow.

separated by a groove (the **intertubercular sulcus,** or **bicipital groove**) that guides the tendon of the biceps muscle to its point of attachment (the superior rim of the glenoid cavity). In the midpoint of the shaft is a roughened area, the **deltoid tuberosity**, where the large fleshy shoulder muscle, the deltoid, attaches. Nearby, the **radial groove** runs obliquely, indicating the pathway of the radial nerve.

At the distal end of the humerus are two condyles—the medial **trochlea** (looking rather like a spool), which articulates with the ulna, and the lateral **capitulum,** which articulates with the radius of the forearm. This condyle pair is flanked medially by the **medial epicondyle** and laterally by the **lateral epicondyle.**

The medial epicondyle is commonly called the "funny bone." The ulnar nerve runs in a groove beneath the medial epicondyle, and when this region is sharply bumped, a temporary, but excruciatingly painful, tingling sensation often occurs. This event is called "hitting the funny bone," a strange expression, because it is certainly *not* funny!

Above the trochlea on the anterior surface is the **coronoid fossa;** on the posterior surface is the **olecranon fossa.** These two depressions allow the corresponding processes of the ulna to move freely when the elbow is flexed (bent) and extended (straightened). The small **radial fossa,** lateral to the coronoid fossa, receives the head of the radius when the elbow is flexed.

The Forearm

Two bones, the radius and the ulna, compose the skeleton of the forearm, or antebrachium **(Figure 10.4)**. When the body is in the anatomical position, the **radius** is in the lateral position in the forearm, and the radius and ulna are parallel. Proximally, the disc-shaped head of the radius articulates with the capitulum of the humerus. Just below the head, on the medial aspect of the shaft, is a prominence called the **radial tuberosity,** the point of attachment for the tendon of the biceps muscle of the arm. Distally, the small **ulnar notch** reveals where the radius articulates with the end of the ulna.

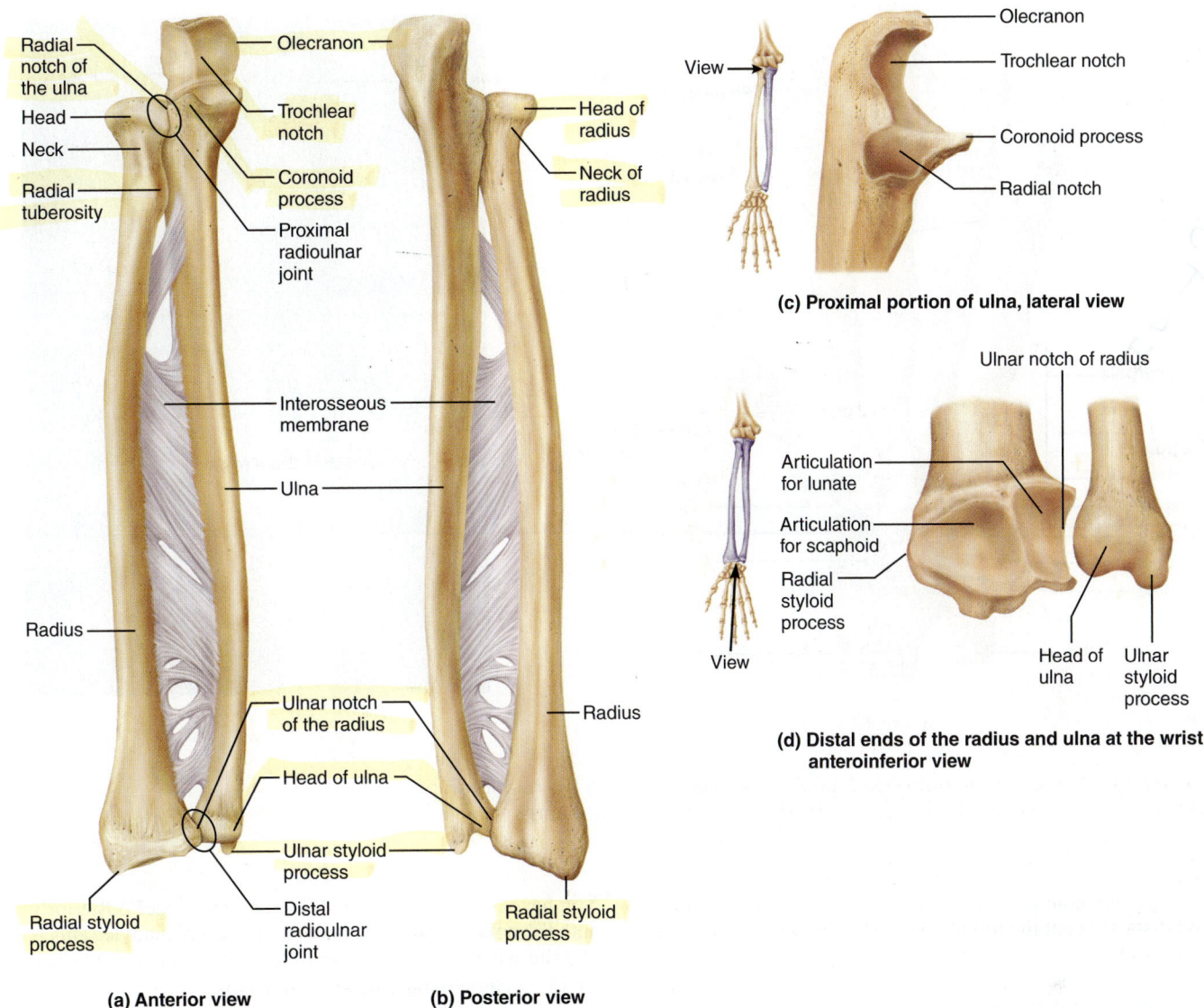

(c) **Proximal portion of ulna, lateral view**

(d) **Distal ends of the radius and ulna at the wrist, anteroinferior view**

(a) **Anterior view**

(b) **Posterior view**

Figure 10.4 Bones of the right forearm. (a, b) Radius and ulna in anterior and posterior views. **(c, d)** Structural details of the articular surfaces between the radius and ulna, and between the radius and bones of the wrist.

The **ulna** is the medial bone of the forearm. Its proximal end bears the anterior **coronoid process** and the posterior **olecranon,** which are separated by the **trochlear notch.** Together these processes grip the trochlea of the humerus in a plierslike joint. The small **radial notch** on the lateral side of the coronoid process articulates with the head of the radius. The slimmer distal end, the ulnar **head,** bears the small medial **ulnar styloid process,** which serves as a point of attachment for the ligaments of the wrist.

The Hand

The skeleton of the hand, or manus **(Figure 10.5)**, includes three groups of bones, those of the carpus (wrist), the metacarpals (bones of the palm), and the phalanges (bones of the fingers).

The wrist is the proximal portion of the hand. It is referred to anatomically as the **carpus;** the eight bones composing it are the **carpals.** (So you actually wear your wristwatch over the distal part of your forearm.) The carpals are arranged in two irregular rows of four bones each (illustrated in Figure 10.5). In the proximal row (lateral to medial) are the *scaphoid, lunate, triquetrum,* and *pisiform bones;* the scaphoid and lunate articulate with the distal end of the radius. In the distal row are the *trapezium, trapezoid, capitate,* and *hamate.* The carpals are bound closely together by ligaments, which restrict movements between them.

The **metacarpals,** numbered I to V from the thumb side of the hand toward the little finger, radiate out from the wrist like spokes to form the palm of the hand. The *bases* of the metacarpals articulate with the carpals of the wrist; their more bulbous *heads* articulate with the phalanges of the fingers distally. When the fist is clenched, the heads of the metacarpals become prominent as the knuckles.

Like the bones of the palm, the fingers are numbered from I to V, beginning from the thumb (*pollex*) side of the hand. The 14 bones of the fingers, or digits, are miniature long bones, called **phalanges** (singular: *phalanx*) as noted above.

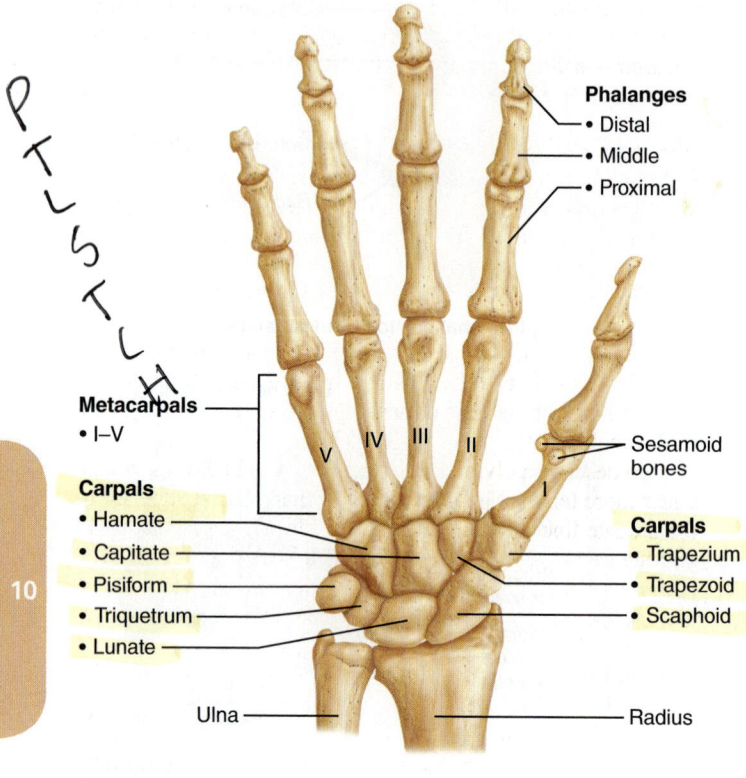

Phalanges
- • Distal
- • Middle
- • Proximal

Metacarpals
- • I–V

Sesamoid bones

Carpals
- • Hamate
- • Capitate
- • Pisiform
- • Triquetrum
- • Lunate

Carpals
- • Trapezium
- • Trapezoid
- • Scaphoid

Ulna   Radius

(a)

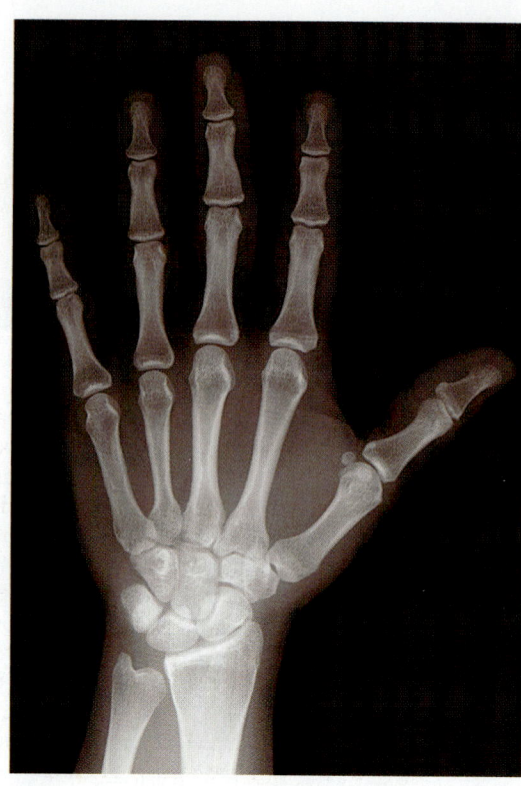

(b)

Figure 10.5 Bones of the right hand. (a) Anterior view showing the relationships of the carpals, metacarpals, and phalanges. **(b)** X ray of the right hand in the anterior view.

Each finger contains three phalanges (proximal, middle, and distal) except the thumb, which has only two (proximal and distal).

ACTIVITY 2

Palpating the Surface Anatomy of the Pectoral Girdle and the Upper Limb

Before continuing on to study the bones of the pelvic girdle, take the time to identify the following bone markings on the skin surface of the upper limb. It is usually preferable to palpate the bone markings on your lab partner since many of these markings can only be seen from the dorsal aspect. Place a check mark in the boxes as you locate the bone markings. Seek assistance from your instructor for any markings that you are unable to locate.

☐ Clavicle: Palpate the clavicle along its entire length from sternum to shoulder.

☐ Acromioclavicular joint: The high point of the shoulder, which represents the junction point between the clavicle and the acromion of the scapular spine.

☐ Spine of the scapula: Extend your arm at the shoulder so that your scapula moves posteriorly. As you do this, your scapular spine will be seen as a winglike protrusion on your dorsal thorax and can be easily palpated by your lab partner.

☐ Lateral epicondyle of the humerus: The inferiormost projection at the lateral aspect of the distal humerus. After

you have located the epicondyle, run your finger posteriorly into the hollow immediately dorsal to the epicondyle. This is the site where the extensor muscles of the hand are attached and is a common site of the excruciating pain of tennis elbow, a condition in which those muscles and their tendons are abused physically.

☐ Medial epicondyle of the humerus: Feel this medial projection at the distal end of the humerus.

☐ Olecranon of the ulna: Work your elbow—flexing and extending—as you palpate its dorsal aspect to feel the olecranon of the ulna moving into and out of the olecranon fossa on the dorsal aspect of the humerus.

☐ Ulnar styloid process: With the hand in the anatomical position, feel out this small inferior projection on the medial aspect of the distal end of the ulna.

☐ Radial styloid process: Find this projection at the distal end of the radius (lateral aspect). It is most easily located by moving the hand medially at the wrist. Once you have palpated the radial styloid process, move your fingers just medially onto the anterior wrist. Press firmly and then let up slightly on the pressure. You should be able to feel your pulse at this pressure point, which lies over the radial artery (radial pulse).

☐ Pisiform: Just distal to the ulnar styloid process feel the rounded pealike pisiform bone.

☐ Metacarpophalangeal joints (knuckles): Clench your fist and find the first set of flexed-joint protrusions beyond the wrist—these are your metacarpophalangeal joints. ■

Bones of the Pelvic Girdle and Lower Limb

The Pelvic (Hip) Girdle

As with the bones of the pectoral girdle and upper limb, pay particular attention to bone markings needed to identify right and left bones.

The **pelvic girdle,** or **hip girdle (Figure 10.6),** is formed by the two **coxal** (*coxa* = hip) **bones** (also called the **ossa coxae,** or hip bones) and the sacrum. The deep structure formed by the hip bones, sacrum, and coccyx is called the **pelvis** or *bony pelvis*. In contrast to the bones of the shoulder girdle, those of the pelvic girdle are heavy and massive, and they attach securely to the axial skeleton. The sockets for the heads of the femurs (thigh bones) are deep and heavily reinforced by ligaments to ensure a stable, strong limb attachment. The ability to bear weight is more important here than mobility and flexibility. The combined weight of the upper body rests on the pelvic girdle (specifically, where the hip bones meet the sacrum).

Each coxal bone is a result of the fusion of three bones— the ilium, ischium, and pubis—which are distinguishable in the young child. The **ilium,** a large flaring bone, forms the major portion of the coxal bone. It connects posteriorly, via its **auricular surface,** with the sacrum at the **sacroiliac joint.** The superior margin of the iliac bone, the **iliac crest,** is rough; when you rest your hands on your hips, you are palpating your iliac crests. The iliac crest terminates anteriorly in the **anterior superior spine** and posteriorly in the **posterior superior spine.** Two inferior spines are located below these. The shallow **iliac fossa** marks its internal surface, and a prominent ridge, the **arcuate line,** outlines the pelvic inlet, or pelvic brim.

The **ischium** is the "sit-down" bone, forming the most inferior and posterior portion of the coxal bone. The most outstanding marking on the ischium is the rough **ischial tuberosity,** which receives the weight of the body when sitting. The **ischial spine,** superior to the ischial tuberosity, is an important anatomical landmark of the pelvic cavity. (See Comparison of the Male and Female Pelves, Table 10.1). The obvious **lesser** and **greater sciatic notches** allow nerves and blood vessels to pass to and from the thigh. The sciatic nerve passes through the latter.

The **pubis** is the most anterior portion of the coxal bone. Fusion of the **rami** of the pubis anteriorly and the ischium posteriorly forms a bar of bone enclosing the **obturator foramen,** through which blood vessels and nerves run from the pelvic cavity into the thigh. The pubis of each hip bone meets anteriorly at the **pubic crest** to form a cartilaginous joint called the **pubic symphysis.** At the lateral end of the pubic crest is the *pubic tubercle* (see Figure 10.6c) to which the important *inguinal ligament* attaches.

The ilium, ischium, and pubis fuse at the deep hemispherical socket called the **acetabulum** (literally, "wine cup"), which receives the head of the thigh bone.

ACTIVITY 3

Observing Pelvic Articulations

Before continuing with the bones of the lower limbs, take the time to examine an articulated pelvis. Notice how each coxal bone articulates with the sacrum posteriorly and how the two coxal bones join at the pubic symphysis. The sacroiliac joint is a common site of lower back problems because of the pressure it must bear. ■

Comparison of the Male and Female Pelves

Although bones of males are usually larger, heavier, and have more prominent bone markings, the male and female skeletons are very similar. The exception to this generalization is pelvic structure.

The female pelvis reflects modifications for childbearing—it is wider, shallower, lighter, and rounder than that of the male. Not only must her pelvis support the increasing size of a fetus, but it must also be large enough to allow the infant's head (its largest dimension) to descend through the birth canal at birth.

To describe pelvic sex differences, we need to introduce a few more terms. The **false pelvis** is that portion superior to the arcuate line; it is bounded by the alae of the ilia laterally and the sacral promontory and lumbar vertebrae posteriorly. Although the false pelvis supports the abdominal viscera, it does not restrict childbirth in any way. The **true pelvis** is the region inferior to the arcuate line that is almost entirely surrounded by bone. Its posterior boundary is formed by the sacrum. The ilia, ischia, and pubic bones define its limits laterally and anteriorly.

The dimensions of the true pelvis, particularly its inlet and outlet, are critical if delivery of a baby is to be uncomplicated. These dimensions are carefully measured by the obstetrician. The **pelvic inlet,** or **pelvic brim,** is the opening delineated by the sacral promontory posteriorly and the arcuate lines of the ilia anterolaterally. It is the superiormost margin of the true pelvis. Its widest dimension is from left to right, that is, along the frontal plane. The **pelvic outlet** is the inferior margin of the true pelvis. It is bounded anteriorly by the pubic arch, laterally by the ischia, and posteriorly by the sacrum and coccyx. Since both the coccyx and the ischial spines protrude into the outlet opening, a sharply angled coccyx or large, sharp ischial spines can dramatically narrow the outlet. The largest dimension of the outlet is the anterior-posterior diameter.

ACTIVITY 4

Comparing Male and Female Pelves

Examine male and female pelves for the following differences:

• The female inlet is larger and more circular.

• The female pelvis as a whole is shallower, and the bones are lighter and thinner.

• The female sacrum is broader and less curved, and the pubic arch is more rounded.

• The female acetabula are smaller and farther apart, and the ilia flare more laterally.

• The female ischial spines are shorter, farther apart, and everted, thus enlarging the pelvic outlet. ■

(The major differences between the male and female pelves are summarized in **Table 10.1**).

(Text continues on page 156.)

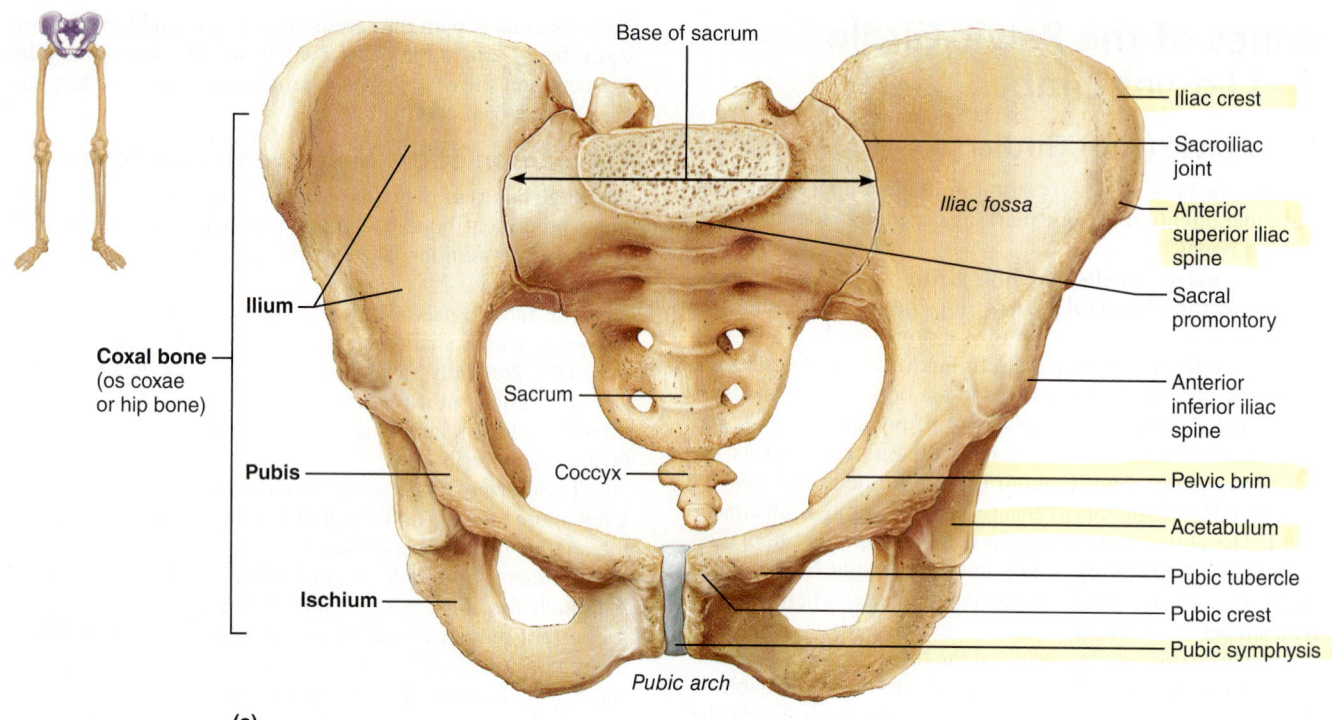

(a)

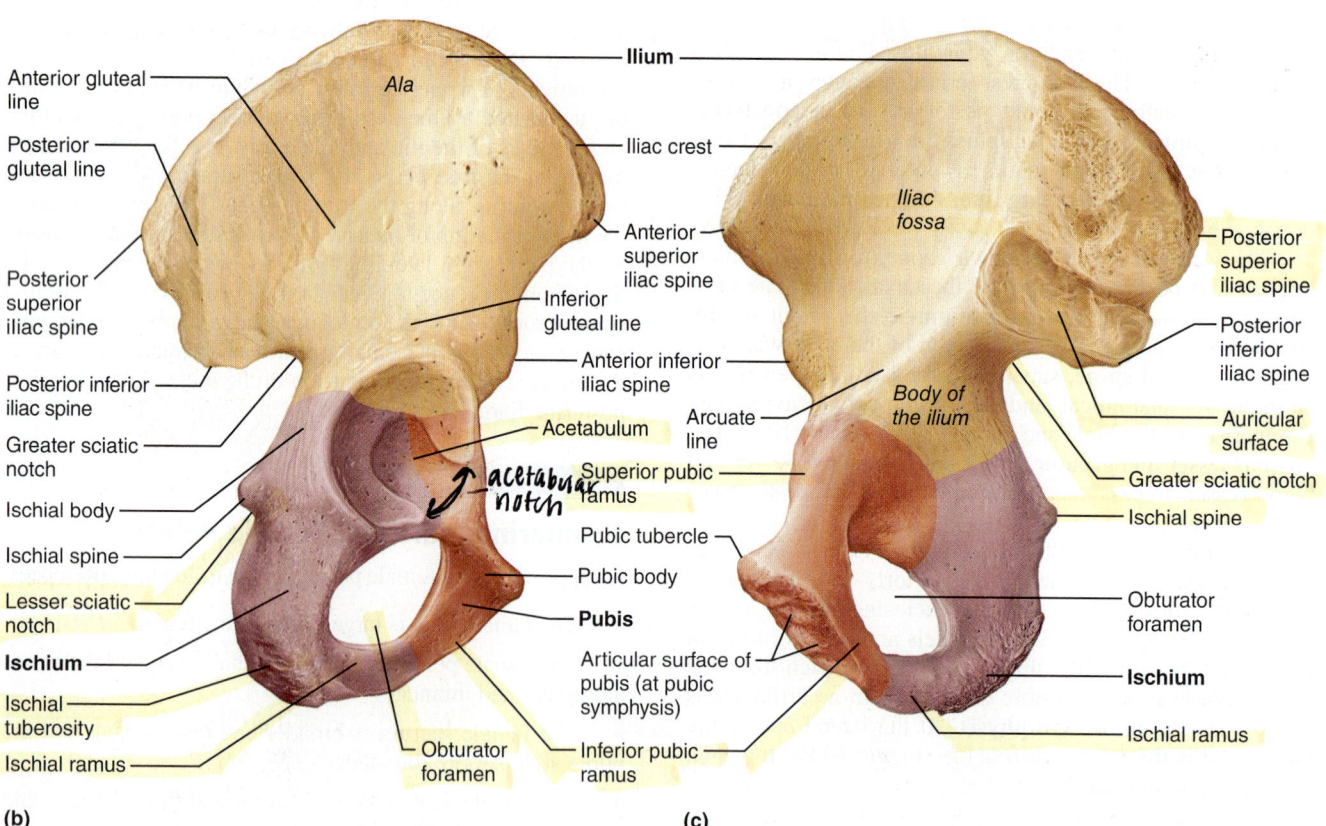

(b)

(c)

Figure 10.6 Bones of the pelvic girdle. (a) Articulated bony pelvis, showing the two hip bones (coxal bones), which together with the sacrum comprise the pelvic girdle, and the coccyx. **(b)** Right hip bone, lateral view, showing the point of fusion of the ilium, ischium, and pubis. **(c)** Right hip bone, medial view.

Table 10.1   Comparison of the Male and Female Pelves

Characteristic	Female	Male
General structure and functional modifications	Tilted forward; adapted for childbearing; true pelvis defines the birth canal; cavity of the true pelvis is broad, shallow, and has a greater capacity	Tilted less far forward; adapted for support of a male's heavier build and stronger muscles; cavity of the true pelvis is narrow and deep
Bone thickness	Less; bones lighter, thinner, and smoother	Greater; bones heavier and thicker, and markings are more prominent
Acetabula	Smaller; farther apart	Larger; closer
Pubic angle/arch	Broader angle (80°–90°); more rounded	Angle is more acute (50°–60°)
Anterior view		

Pelvic brim — Pelvic brim

Pubic arch — Pubic arch

	Female	Male
Sacrum	Wider; shorter; sacrum is less curved	Narrow; longer; sacral promontory more ventral
Coccyx	More movable; straighter; projects inferiorly	Less movable; curves and projects anteriorly
Left lateral view		

	Female	Male
Pelvic inlet (brim)	Wider; oval from side to side	Narrow; basically heart shaped
Pelvic outlet	Wider; ischial spines shorter, farther apart, and everted	Narrower; ischial spines longer, sharper, and point more medially
Posteroinferior view		

Pelvic outlet

10

10

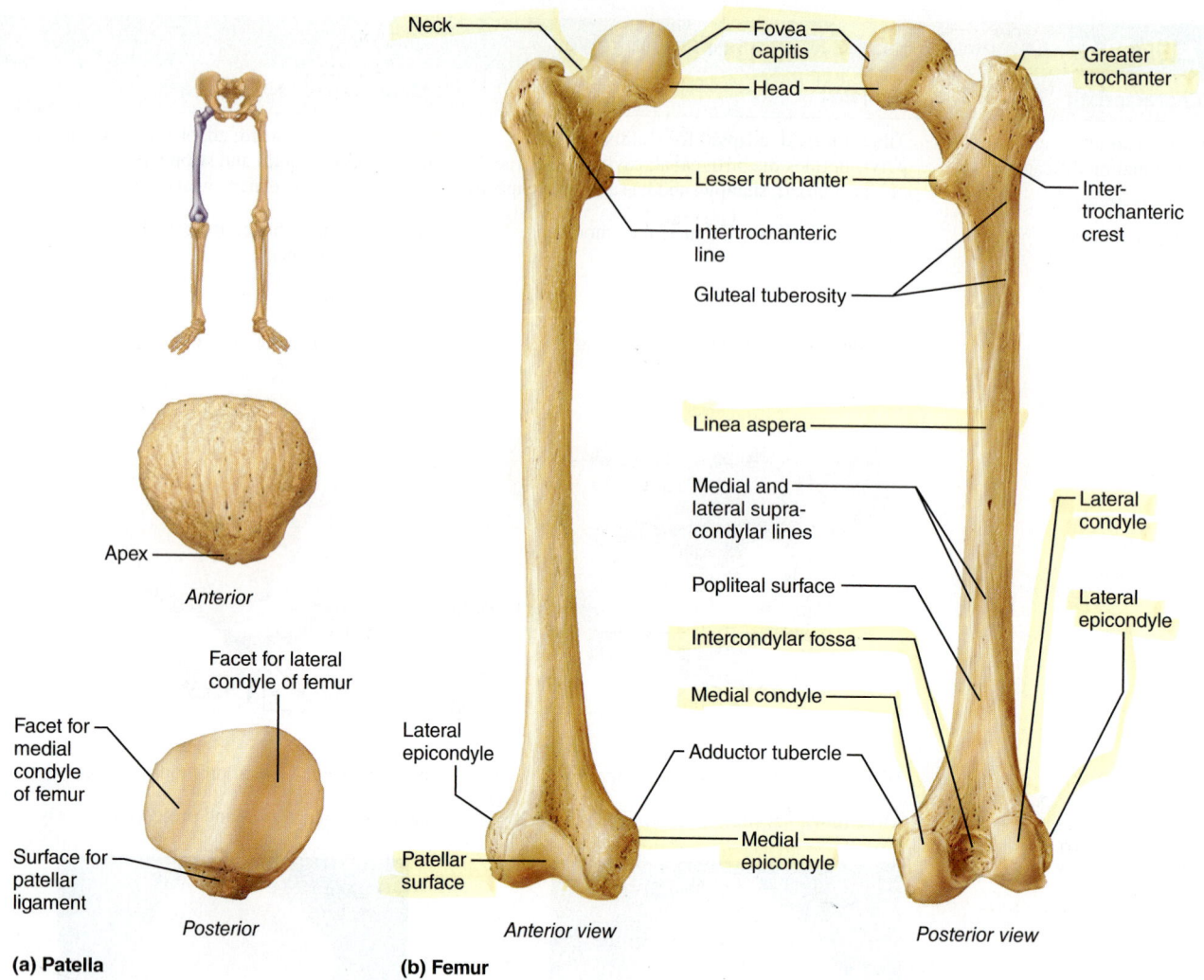

Figure 10.7 Bones of the right knee and thigh.

The Thigh

The **femur,** or thigh bone **(Figure 10.7b)**, is the only bone of the thigh. It is the heaviest, strongest bone in the body. The ball-like head of the femur articulates with the hip bone via the deep, secure socket of the acetabulum. Obvious in the femur's head is a small central pit called the **fovea capitis** ("pit of the head"), from which a small ligament runs to the acetabulum. The head of the femur is carried on a short, constricted *neck,* which angles laterally to join the shaft. The neck is the weakest part of the femur and is a common fracture site (an injury called a broken hip), particularly in the elderly. At the junction of the shaft and neck are the **greater** and **lesser trochanters** separated posteriorly by the **intertrochanteric crest** and anteriorly by the **intertrochanteric line.** The trochanters and trochanteric crest, as well as the **gluteal tuberosity** and the **linea aspera** located on the shaft, are sites of muscle attachment.

The femur inclines medially as it runs downward to the leg bones; this brings the knees in line with the body's center of gravity, or maximum weight. The medial course of the femur is more noticeable in females because of the wider female pelvis.

Distally, the femur terminates in the **lateral** and **medial condyles,** which articulate with the tibia below, and the

patellar surface, which forms a joint with the patella (kneecap) anteriorly. The **lateral** and **medial epicondyles,** just superior to the condyles, are separated by the **intercondylar fossa,** and superior to that on the shaft is the smooth **popliteal surface.** On the superior part of the medial epicondyle is a bump, the **adductor tubercle,** to which the large adductor magnus muscle attaches.

The **patella** (Figure 10.7a) is a triangular sesamoid bone enclosed in the (quadriceps) tendon that secures the anterior thigh muscles to the tibia. It guards the knee joint anteriorly and improves the leverage of the thigh muscles acting across the knee joint.

The Leg

Two bones, the tibia and the fibula, form the skeleton of the leg **(Figure 10.8)**. The **tibia,** or *shinbone,* is the larger and more medial of the two leg bones. At the proximal end, the **medial** and **lateral condyles** (separated by the **intercondylar eminence**) receive the distal end of the femur to form the knee joint. The **tibial tuberosity,** a roughened protrusion on the anterior tibial surface (just below the condyles), is the site of attachment of the patellar (kneecap) ligament. Small facets on the superior and inferior surface of the lateral condyle of the tibia articulate with the fibula. Distally, a process called the

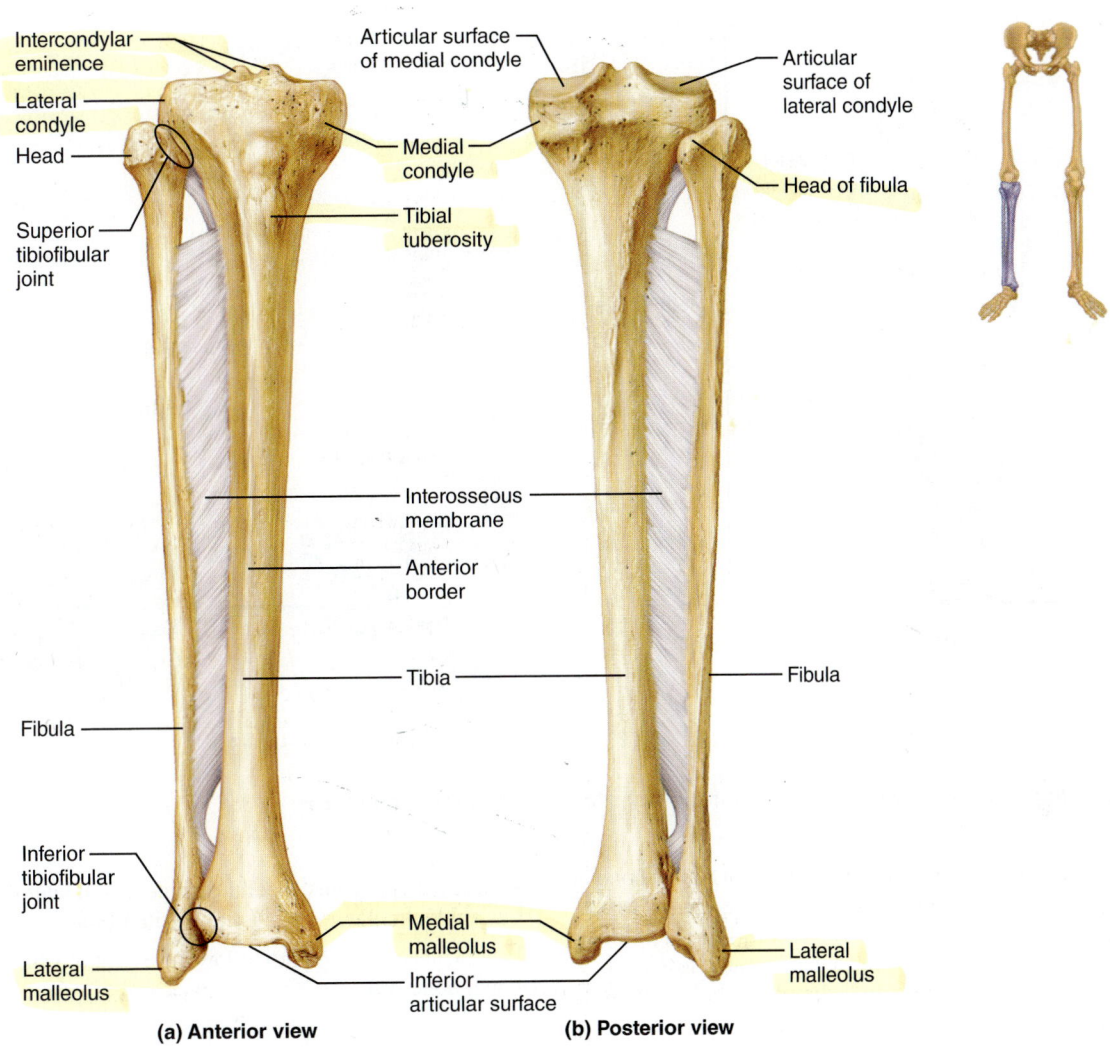

Figure 10.8 Bones of the right leg. Tibia and fibula, anterior and posterior views.

medial malleolus forms the inner (medial) bulge of the ankle. Lateral to this process, the inferior articular surface of the tibia articulates with the talus bone of the foot. The anterior surface of the tibia bears a sharpened ridge that is relatively unprotected by muscles. This so-called **anterior border** is easily felt beneath the skin.

The **fibula,** which lies parallel to the tibia, takes no part in forming the knee joint. Its proximal head articulates with the lateral condyle of the tibia. The fibula is thin and sticklike with a sharp anterior crest. It terminates distally in the **lateral malleolus,** which forms the outer part, or lateral bulge, of the ankle.

The Foot

The bones of the foot include the 7 **tarsal** bones, 5 **metatarsals,** which form the instep, and 14 **phalanges,** which form the toes **(Figure 10.9)**. Body weight is concentrated on the two largest tarsals, which form the posterior aspect of the foot. These are the *calcaneus* (heel bone) and the *talus,* which lies between the tibia and the calcaneus. (The other tarsals are named and identified in Figure 10.9). The metatarsals are numbered I through V, medial to lateral. Like the fingers of

the hand, each toe has three phalanges except the great toe, which has two.

The bones in the foot are arranged to produce three strong arches—two longitudinal arches (medial and lateral) and one transverse arch (Figure 10.9b). Ligaments, binding the foot bones together, and tendons of the foot muscles hold the bones firmly in the arched position but still allow a certain degree of give. Weakened arches are referred to as fallen arches or flat feet.

ACTIVITY 5

Palpating the Surface Anatomy of the Pelvic Girdle and Lower Limb

Locate and palpate the following bone markings on yourself and/or your lab partner. Place a check mark in the boxes as you locate the bone markings. Seek assistance from your instructor for any markings that you are unable to locate.

☐ Iliac crest and anterior superior iliac spine: Rest your hands on your hips—they will be overlying the iliac crests. Trace the crest as far posteriorly as you can and then follow it

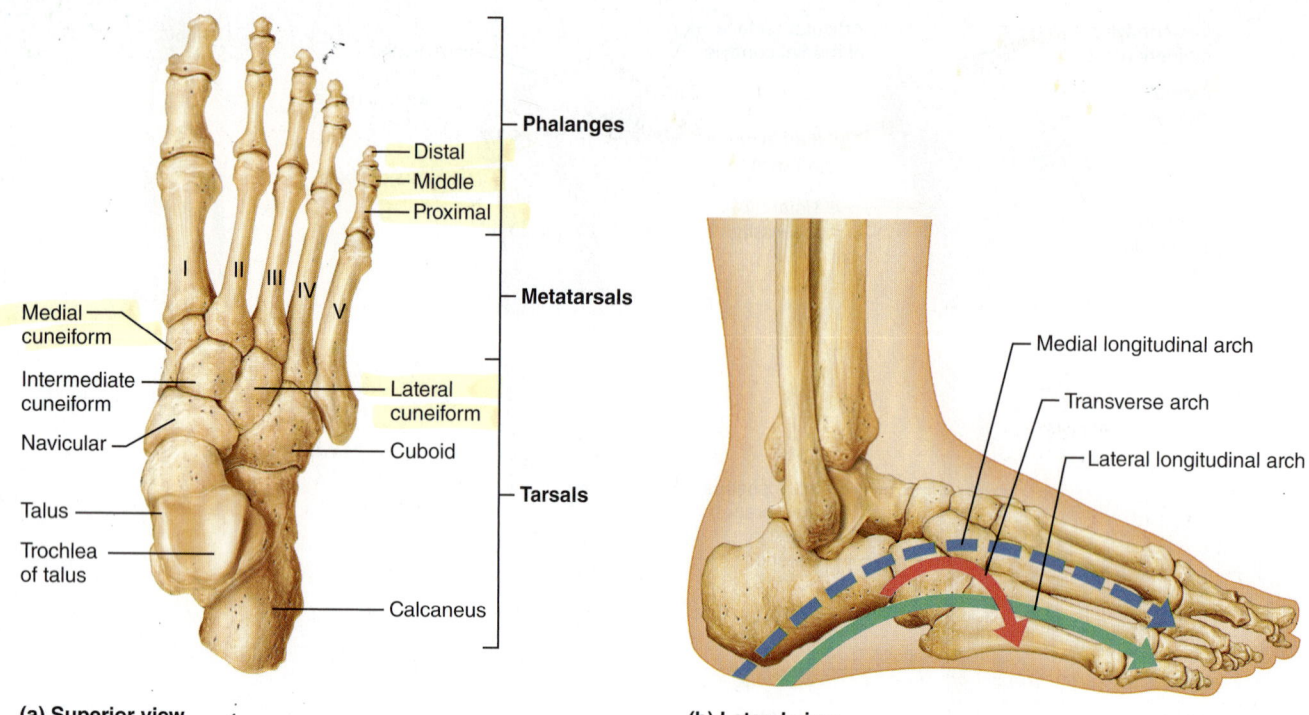

Figure 10.9 Bones of the right foot. Arches of the right foot are diagrammed in **(b)**.

anteriorly to the anterior superior iliac spine. This latter bone marking is easily felt in almost everyone and is clearly visible through the skin (and perhaps the clothing) of very slim people. (The posterior superior iliac spine is much less obvious and is usually indicated only by a dimple in the overlying skin. Check it out in the mirror tonight.)

☐   Greater trochanter of the femur: This is easier to locate in females than in males because of the wider female pelvis; also it is more likely to be clothed by bulky muscles in males. Try to locate it on yourself as the most lateral point of the proximal femur. It typically lies about 6 to 8 inches below the iliac crest.

☐   Patella and tibial tuberosity: Feel your kneecap and palpate the ligaments attached to its borders. Follow the inferior patellar ligament to the tibial tuberosity.

☐   Medial and lateral condyles of the femur and tibia: As you move from the patella inferiorly on the medial (and then the lateral) knee surface, you will feel first the femoral and then the tibial condyle.

☐   Medial malleolus: Feel the medial protrusion of your ankle, the medial malleolus of the distal tibia.

☐   Lateral malleolus: Feel the bulge of the lateral aspect of your ankle, the lateral malleolus of the fibula.

☐   Calcaneus: Attempt to follow the extent of your calcaneus, or heel bone. ■

ACTIVITY 6

Constructing a Skeleton

1. When you finish examining yourself and the disarticulated bones of the appendicular skeleton, work with your lab partner to arrange the disarticulated bones on the laboratory bench in their proper relative positions to form an entire skeleton. Careful observation of bone markings should help you distinguish between right and left members of bone pairs.

2. When you believe that you have accomplished this task correctly, ask the instructor to check your arrangement to ensure that it is correct. If it is not, go to the articulated skeleton and check your bone arrangements. Also review the descriptions of the bone markings as necessary to correct your bone arrangement. ■

Articulations and Body Movements

MATERIALS

- ☐ Skull
- ☐ Articulated skeleton
- ☐ X ray of a child's bone showing the cartilaginous growth plate (if available)
- ☐ Anatomical chart of joint types (if available)
- ☐ Diarthrotic joint (fresh or preserved), preferably a beef knee joint sectioned sagittally (Alternatively, pig's feet with phalanges sectioned frontally could be used)
- ☐ Disposable gloves
- ☐ Water balloons and clamps
- ☐ Functional models of hip, knee, and shoulder joints (if available)
- ☐ X rays of normal and arthritic joints (if available)

MasteringA&P® For related exercise study tools, go to the Study Area of MasteringA&P. There you will find:
- Practice Anatomy Lab **PAL**
- PhysioEx **PEx**
- A&PFlix **A&PFlix**
- Practice quizzes, Histology Atlas, eText, Videos, and more!

OBJECTIVES

1. Name and describe the three functional categories of joints.
2. Name and describe the three structural categories of joints, and discuss how their structure is related to mobility.
3. Identify the types of synovial joints; indicate whether they are nonaxial, uniaxial, biaxial, or multiaxial, and describe the movements made by each.
4. Define *origin* and *insertion* of muscles.
5. Demonstrate or describe the various body movements.
6. Compare and contrast the structure and function of the shoulder and hip joints.
7. Describe the structure and function of the knee and temporomandibular joints.

PRE-LAB QUIZ

1. Name one of the two functions of an articulation, or joint. _____
   _____

2. The functional classification of joints is based on
   a. a joint cavity
   b. amount of connective tissue
   c. amount of movement allowed by the joint

3. Structural classification of joints includes fibrous, cartilaginous, and _____, which have a fluid-filled cavity between articulating bones.

4. Circle the correct underlined term. Sutures, which have their irregular edges of bone joined by short fibers of connective tissue, are an example of <u>fibrous</u> / <u>cartilaginous</u> joints.

5. Circle True or False. All synovial joints are diarthroses, or freely movable joints.

6. Circle the correct underlined term. Every muscle of the body is attached to a bone or other connective tissue structure at two points. The <u>origin</u> / <u>insertion</u> is the more movable attachment.

7. The hip joint is an example of a _____ synovial joint.
   a. ball-and-socket   c. pivot
   b. hinge   d. plane

8. Movement of a limb *away* from the midline or median plane of the body in the frontal plane is known as
   a. abduction   c. extension
   b. eversion   d. rotation

9. Circle the correct underlined term. This type of movement is common in ball-and-socket joints and can be described as the movement of a bone around its longitudinal axis. It is <u>rotation</u> / <u>flexion</u>.

10. Circle True or False. The knee joint is the most freely movable joint in the body.

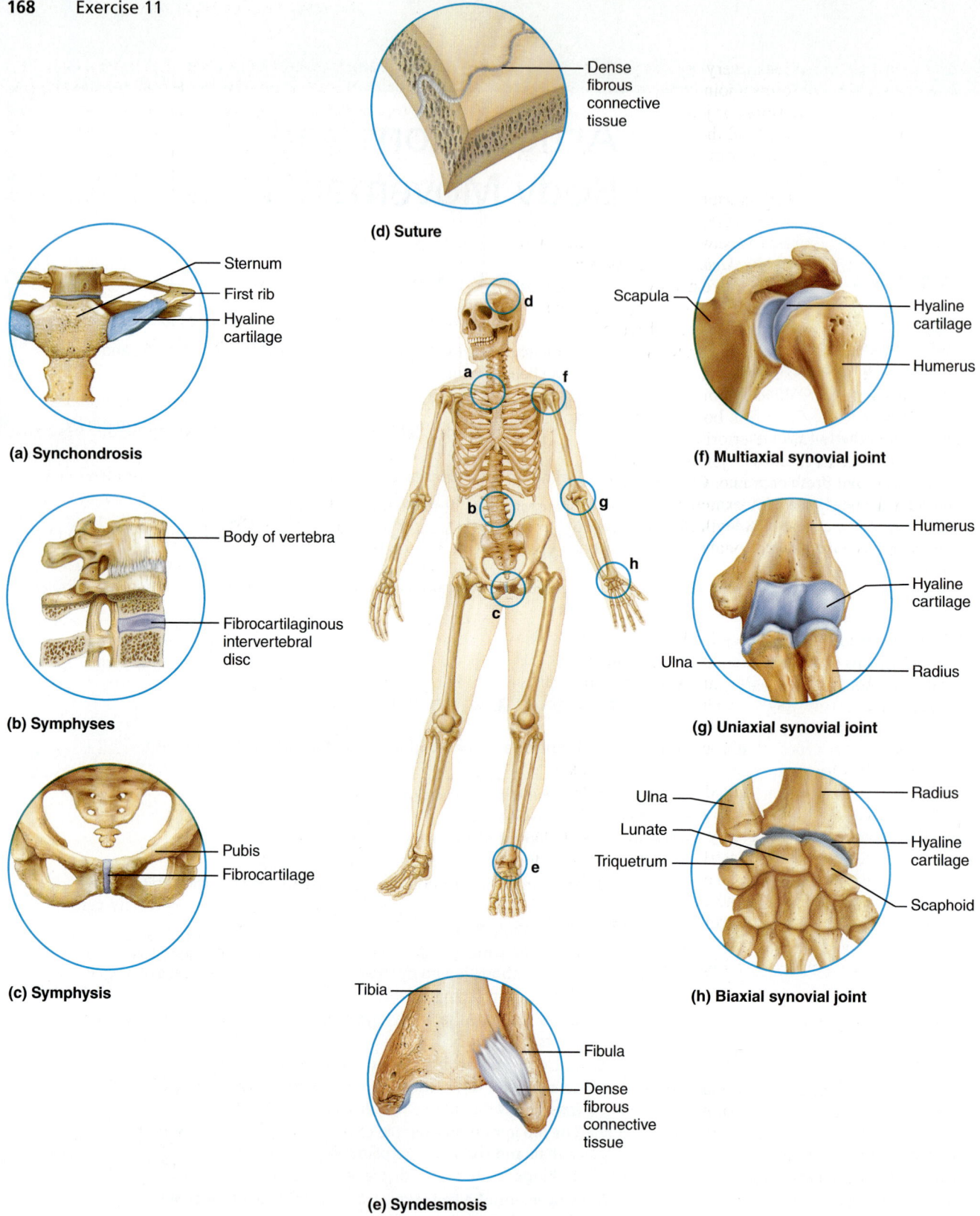

Figure 11.1 Types of joints. Joints to the left of the skeleton are cartilaginous joints; joints above and below the skeleton are fibrous joints; joints to the right of the skeleton are synovial joints. **(a)** Joint between costal cartilage of rib 1 and the sternum. **(b)** Intervertebral discs of fibrocartilage connecting adjacent vertebrae. **(c)** Fibrocartilaginous pubic symphysis connecting the pubic bones anteriorly. **(d)** Dense fibrous connective tissue connecting interlocking skull bones. **(e)** Ligament of dense fibrous connective tissue connecting the inferior ends of the tibia and fibula. **(f)** Shoulder joint. **(g)** Elbow joint. **(h)** Wrist joint.

With rare exceptions, every bone in the body is connected to, or forms a joint with, at least one other bone. **Articulations,** or joints, perform two functions for the body. They (1) hold the bones together and (2) allow the rigid skeletal system some flexibility so that gross body movements can occur.

Joints may be classified structurally or functionally. The structural classification is based on the presence of connective tissue fiber, cartilage, or a joint cavity between the articulating bones. Structurally, there are *fibrous, cartilaginous, and synovial joints.*

The functional classification focuses on the amount of movement allowed at the joint. On this basis, there are **synarthroses,** or immovable joints; **amphiarthroses,** or slightly movable joints; and **diarthroses,** or freely movable joints. Freely movable joints predominate in the limbs, whereas immovable and slightly movable joints are largely restricted to the axial skeleton, where firm bony attachments and protection of enclosed organs are priorities.

As a general rule, fibrous joints are immovable, and synovial joints are freely movable. Cartilaginous joints offer both rigid and slightly movable examples. Since the structural categories are more clear-cut, we will use the structural classification here and indicate functional properties as appropriate.

Fibrous Joints

In **fibrous joints,** the bones are joined by fibrous tissue. No joint cavity is present. The amount of movement allowed depends on the length of the fibers uniting the bones. Although some fibrous joints are slightly movable, most are synarthrotic and permit virtually no movement.

The two major types of fibrous joints are sutures and syndesmoses. In **sutures (Figure 11.1d)** the irregular edges of the bones interlock and are united by very short connective tissue fibers, as in most joints of the skull. In **syndesmoses** the articulating bones are connected by short ligaments of dense fibrous tissue; the bones do not interlock. The joint at the inferior end of the tibia and fibula is an example of a syndesmosis (Figure 11.1e). Although this syndesmosis allows some give, it is classed functionally as a synarthrosis. Not illustrated here is a **gomphosis,** in which a tooth is secured in a bony socket by the periodontal ligament (see Figure 38.12).

ACTIVITY 1

Identifying Fibrous Joints

Examine a human skull. Notice that adjacent bone surfaces do not actually touch but are separated by fibrous connective tissue. Also examine a skeleton and anatomical chart of joint types and the table of joints (**Table 11.1**, pages 170–171) for examples of fibrous joints. ■

Cartilaginous Joints

In **cartilaginous joints,** the articulating bone ends are connected by a plate or pad of cartilage. No joint cavity is present. The two major types of cartilaginous joints are synchondroses and symphyses. Although there is variation, most cartilaginous joints are *slightly movable* (amphiarthrotic) functionally. In **symphyses** (*symphysis* = a growing together), the bones are connected by a broad, flat disc of fibrocartilage. The intervertebral joints between adjacent vertebral bodies and the pubic symphysis of the pelvis are symphyses (see Figure 11.1b and c). In **synchondroses** the bony

portions are united by hyaline cartilage. The articulation of the costal cartilage of the first rib with the sternum (Figure 11.1a) is a synchondrosis, but perhaps the best examples of synchondroses are the epiphyseal plates seen in the long bones of growing children. View an X ray of the cartilaginous growth plate (epiphyseal disc) of a child's bone if one is available. The epiphyseal plates are flexible during childhood, but eventually they are totally ossified.

ACTIVITY 2

Identifying Cartilaginous Joints

Identify the cartilaginous joints on a human skeleton, the table of joints (Table 11.1), and an anatomical chart of joint types. ■

Synovial Joints

Synovial joints are those in which the articulating bone ends are separated by a joint cavity containing synovial fluid (see Figure 11.1f–h). All synovial joints are diarthroses, or freely movable joints. Their mobility varies: some synovial joints permit only small gliding movements, and others can move in several planes. Most joints in the body are synovial joints. All synovial joints have the following structural characteristics **(Figure 11.2)**:

Joint surfaces are enclosed by a two-layered *articular capsule* (a sleeve of connective tissue), creating a joint cavity.

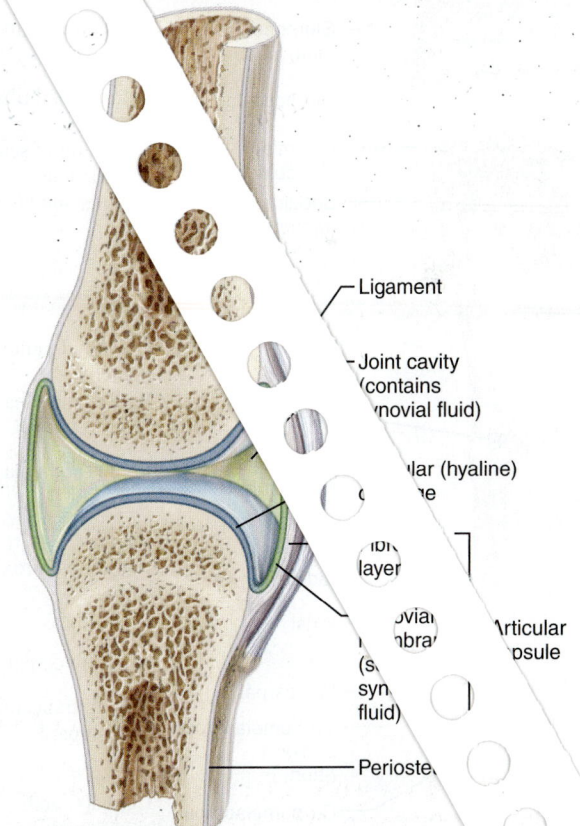

- Ligament
- Joint cavity (contains synovial fluid)
- Articular (hyaline) cartilage
- Fibrous layer
- Synovial membrane (secretes synovial fluid)
- Articular capsule
- Periosteum

Figure 11.2 General structure of a synovial joint. The articulating bone ends are covered with articular cartilage, and enclosed within an articular capsule that is typically reinforced by ligaments externally. Internally the fibrous layer is lined with a smooth synovial membrane that secretes synovial fluid.

Table 11.1		Structural and Functional Characteristics of Body Joints		

Illustration	Joint	Articulating bones	Structural type*	Functional type; movements allowed
	Skull	Cranial and facial bones	Fibrous; suture	Synarthrotic; no movement
	Temporo-mandibular	Temporal bone of skull and mandible	Synovial; modified hinge† (contains articular disc)	Diarthrotic; gliding and uniaxial rotation; slight lateral movement, elevation, depression, protraction, and retraction of mandible
	Atlanto-occipital	Occipital bone of skull and atlas	Synovial; condylar	Diarthrotic; biaxial; flexion, extension, lateral flexion, circumduction of head on neck
	Atlantoaxial	Atlas (C₁) and axis (C₂)	Synovial; pivot	Diarthrotic; uniaxial; rotation of the head
	Intervertebral	Between adjacent vertebral bodies	Cartilaginous; symphysis	Amphiarthrotic; slight movement
	Intervertebral	Between articular processes	Synovial; plane	Diarthrotic; gliding
	Costovertebral	Vertebrae (transverse processes or bodies) and ribs	Synovial; plane	Diarthrotic; gliding of ribs
	Sternoclavicular	Sternum and clavicle	Synovial; shallow saddle (contains articular disc)	Diarthrotic; multiaxial (allows clavicle to move in all axes)
	Sternocostal (first)	Sternum and rib 1	Cartilaginous; synchondrosis	Synarthrotic; no movement
	Sternocostal	Sternum and ribs 2–7	Synovial; double plane	Diarthrotic; gliding
	Acromio-clavicular	Acromion of scapula and clavicle	Synovial; plane (contains articular disc)	Diarthrotic; gliding and rotation of scapula on clavicle
	Shoulder (glenohumeral)	Scapula and humerus	Synovial; ball and socket	Diarthrotic; multiaxial; flexion, extension, abduction, adduction, circumduction, rotation of humerus
	Elbow	Ulna (and radius) with humerus	Synovial; hinge	Diarthrotic; uniaxial; flexion, extension of forearm
		Radius and ulna	Synovial; pivot	Diarthrotic; uniaxial; pivot (head of radius rotates in radial notch of ulna)
	Proximal radioulnar	Radius and ulna	Synovial; pivot (contains articular disc)	Diarthrotic; uniaxial; rotation of radius around long axis of forearm to allow pronation and supination
	Distal radioulnar	Radius and proximal carpals	Synovial; condylar	Diarthrotic; biaxial; flexion, extension, abduction, adduction, circumduction of hand
	Wrist	Adjacent carpals	Synovial; plane	Diarthrotic; gliding
	Intercarpal	Carpal (trapezium) and metacarpal I	Synovial; saddle	Diarthrotic; biaxial; flexion, extension, abduction, adduction, circumduction, opposition of metacarpal I
	Carpometacarpal of digit 1 (thumb)	Carpal(s) and metacarpal(s)	Synovial; plane	Diarthrotic; gliding of metacarpals
	Carpometacarpal of digits 2–5	Metacarpal and proximal phalanx	Synovial; condylar	Diarthrotic; biaxial; flexion, extension, abduction, adduction, circumduction of fingers
	Metacarpo-phalangeal (knuckle)	Adjacent phalanges	Synovial; hinge	Diarthrotic; uniaxial; flexion, extension of fingers
	Interphalangeal (finger)			

Table 11.1   *(continued)*

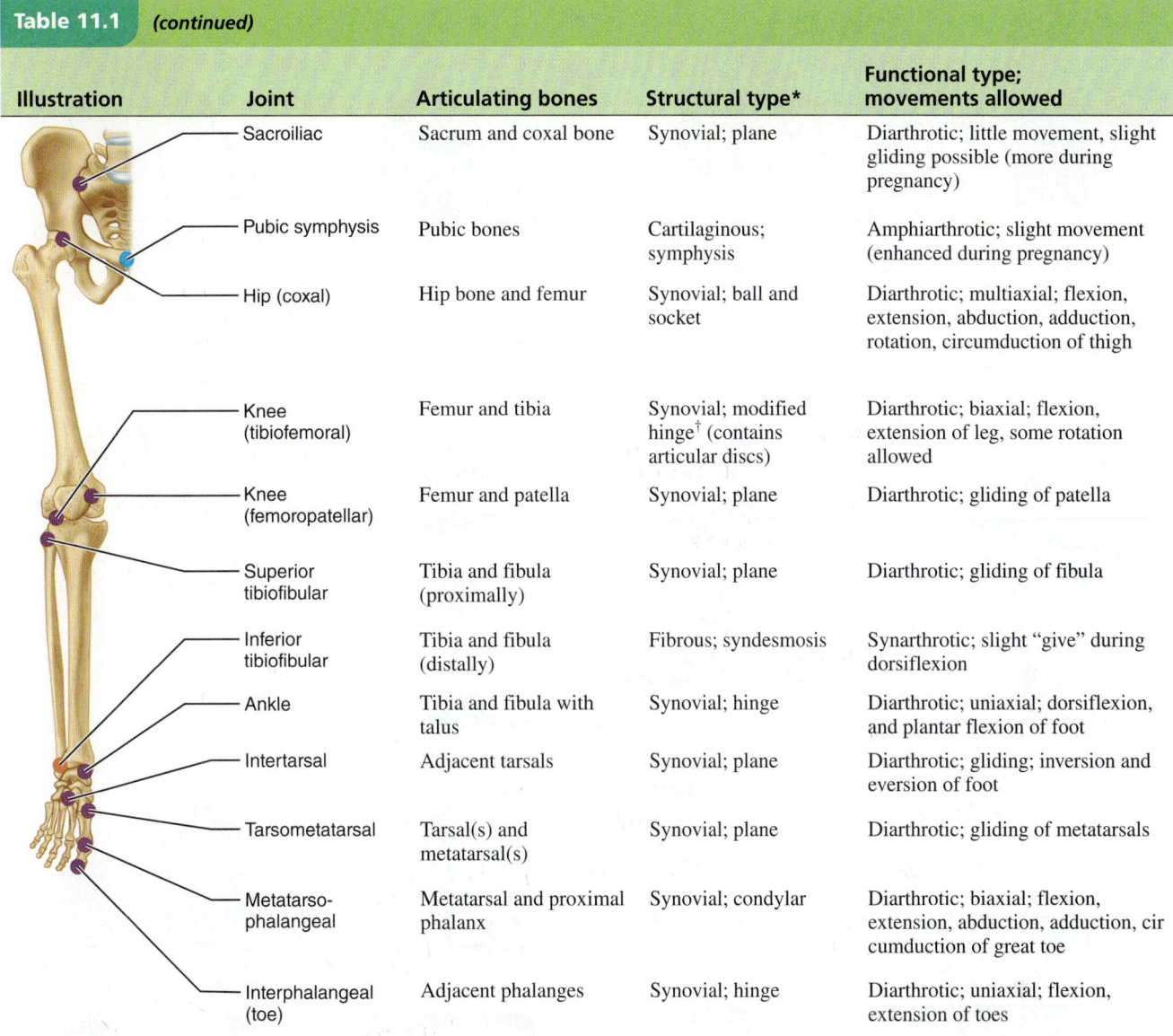

Illustration	Joint	Articulating bones	Structural type*	Functional type; movements allowed
	Sacroiliac	Sacrum and coxal bone	Synovial; plane	Diarthrotic; little movement, slight gliding possible (more during pregnancy)
	Pubic symphysis	Pubic bones	Cartilaginous; symphysis	Amphiarthrotic; slight movement (enhanced during pregnancy)
	Hip (coxal)	Hip bone and femur	Synovial; ball and socket	Diarthrotic; multiaxial; flexion, extension, abduction, adduction, rotation, circumduction of thigh
	Knee (tibiofemoral)	Femur and tibia	Synovial; modified hinge† (contains articular discs)	Diarthrotic; biaxial; flexion, extension of leg, some rotation allowed
	Knee (femoropatellar)	Femur and patella	Synovial; plane	Diarthrotic; gliding of patella
	Superior tibiofibular	Tibia and fibula (proximally)	Synovial; plane	Diarthrotic; gliding of fibula
	Inferior tibiofibular	Tibia and fibula (distally)	Fibrous; syndesmosis	Synarthrotic; slight "give" during dorsiflexion
	Ankle	Tibia and fibula with talus	Synovial; hinge	Diarthrotic; uniaxial; dorsiflexion, and plantar flexion of foot
	Intertarsal	Adjacent tarsals	Synovial; plane	Diarthrotic; gliding; inversion and eversion of foot
	Tarsometatarsal	Tarsal(s) and metatarsal(s)	Synovial; plane	Diarthrotic; gliding of metatarsals
	Metatarso-phalangeal	Metatarsal and proximal phalanx	Synovial; condylar	Diarthrotic; biaxial; flexion, extension, abduction, adduction, circumduction of great toe
	Interphalangeal (toe)	Adjacent phalanges	Synovial; hinge	Diarthrotic; uniaxial; flexion, extension of toes

***Fibrous joint** indicated by orange circles; **cartilaginous joints,** by blue circles; **synovial joints,** by purple circles.
†These modified hinge joints are structurally bicondylar.

• The inner layer is a smooth connective tissue membrane, called the *synovial membrane,* which produces a lubricating fluid (synovial fluid) that reduces friction. The outer layer, or *fibrous layer,* is dense irregular connective tissue.

• *Articular* (hyaline) *cartilage* covers the surfaces of the bones forming the joint.

• The articular capsule is typically reinforced with ligaments and may contain *bursae* (fluid-filled sacs that reduce friction where tendons cross bone).

• Fibrocartilage pads *(articular discs)* may be present within the capsule.

ACTIVITY 3

Examining Synovial Joint Structure

Examine a beef or pig joint to identify the general structural features of diarthrotic joints as listed above.

⚠ If the joint is freshly obtained from the slaughterhouse and you will be handling it, don disposable gloves before beginning your observations. ▬

ACTIVITY 4

Demonstrating the Importance of Friction-Reducing Structures

1. Obtain a small water balloon and clamp. Partially fill the balloon with water (it should still be flaccid), and clamp it closed.

2. Position the balloon atop one of your fists and press down on its top surface with the other fist. Push on the balloon until your two fists touch and move your fists back and forth over one another. Assess the amount of friction generated.

11

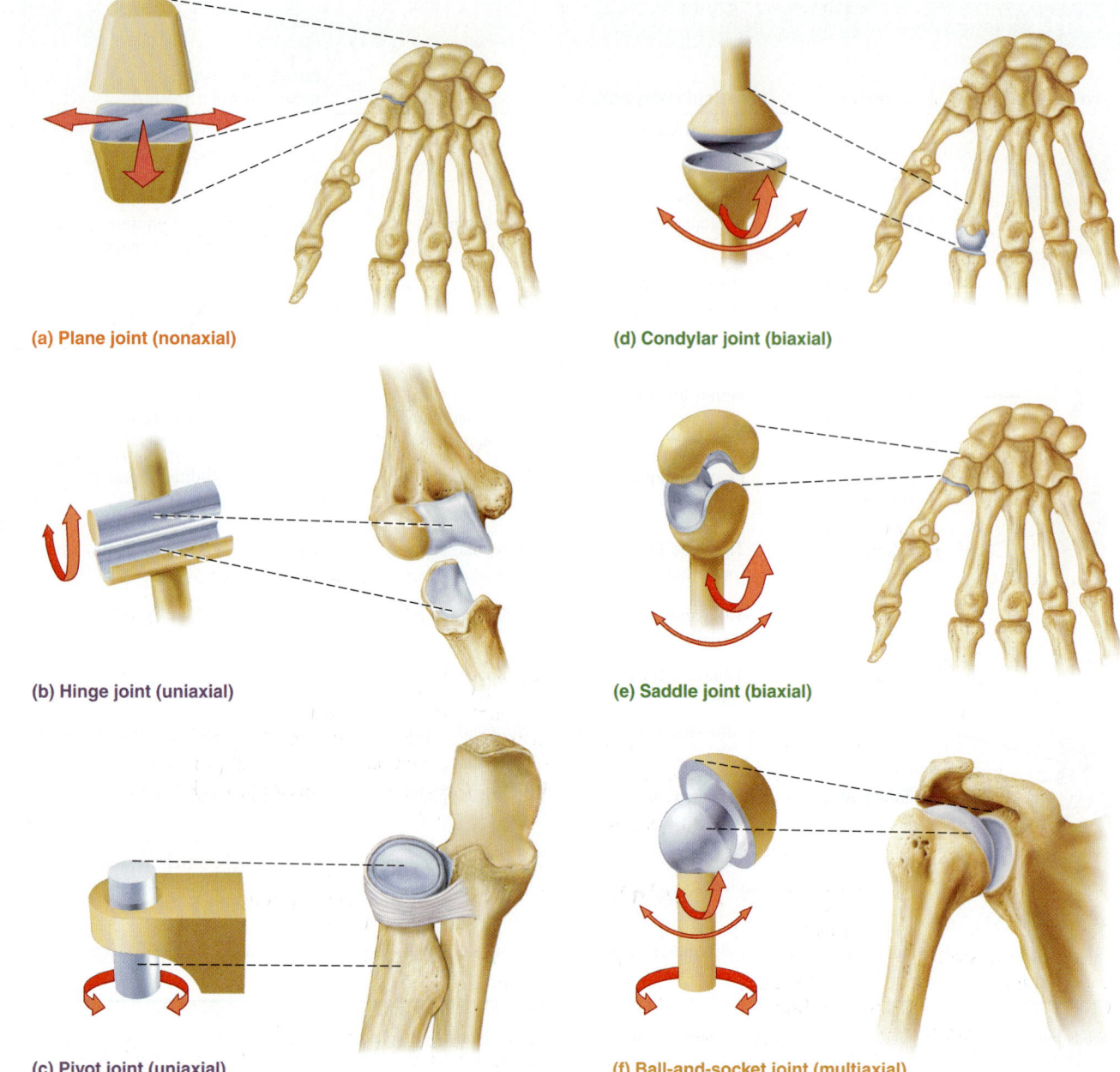

(a) Plane joint (nonaxial)

(b) Hinge joint (uniaxial)

(c) Pivot joint (uniaxial)

(d) Condylar joint (biaxial)

(e) Saddle joint (biaxial)

(f) Ball-and-socket joint (multiaxial)

Figure 11.3 Types of synovial joints. Dashed lines indicate the articulating bones.
(a) Intercarpal joint. **(b)** Elbow. **(c)** Proximal radioulnar joint. **(d)** Metacarpophalangeal joint. **(e)** Carpometacarpal joint of the thumb. **(f)** Shoulder.

3. Unclamp the balloon and add more water. The goal is to get just enough water in the balloon so that your fists cannot come into contact with one another, but instead remain separated by a thin water layer when pressure is applied to the balloon.

4. Repeat the movements in step 2 to assess the amount of friction generated.

How does the presence of a sac containing fluid influence the amount of friction generated?

_____

What anatomical structure(s) does the water-containing balloon mimic?

_____

What anatomical structures might be represented by your fists?

_____

Types of Synovial Joints

The many types of synovial joints can be subdivided according to their function and structure. The shapes of the articular surfaces determine the types of movements that can occur at the joint, and they also determine the structural classification of the joints **(Figure 11.3)**:

• Plane (Nonaxial): Articulating surfaces are flat or slightly curved. These surfaces allow only gliding movements as the surfaces slide past one another. Examples include intercarpal joints, intertarsal joints, and joints between vertebral articular surfaces.

• Hinge (Uniaxial): The rounded or cylindrical process of one bone fits into the concave surface of another bone, allowing movement in one plane, usually flexion and extension. Examples include the elbow and interphalangeal joints.

• Pivot (Uniaxial): The rounded surface of one bone articulates with a shallow depression or foramen in another bone, permitting rotational movement in one plane. Examples include the proximal radioulnar joint and the atlantoaxial joint (between atlas and axis—C_1 and C_2).

• Condylar (Biaxial): The oval condyle of one bone fits into an ellipsoidal depression in another bone to allow movement in two planes, usually flexion/extension and abduction/adduction. Examples include the wrist and metacarpophalangeal (knuckle) joints.

• Saddle (Biaxial): Articulating surfaces are saddle shaped; one surface is convex, and the other is concave. This type of joint permits movement in two planes, flexion/extension and abduction/adduction. Examples include the carpometacarpal joints of the thumbs.

• Ball-and-socket (Multiaxial): The ball-shaped head of one bone fits into a cuplike depression of another bone. These joints permit flexion/extension, abduction/adduction, and rotation, which combine to allow movement in many planes. Examples include the shoulder and hip joints.

Movements Allowed by Synovial Joints

Every muscle of the body is attached to bone (or other connective tissue structures) at two points—the **origin** (the stationary, immovable, or less movable attachment) and the **insertion** (the movable attachment). Body movement occurs when muscles contract across diarthrotic synovial joints **(Figure 11.4)**. When the muscle contracts and its fibers shorten, the insertion moves toward the origin. The type of movement depends on the construction of the joint and on the placement of the muscle relative to the joint. The most common types of body movements are described below (and illustrated in **Figure 11.5**).

ACTIVITY 5

Demonstrating Movements of Synovial Joints

Attempt to demonstrate each movement as you read through the following material:

Flexion (Figure 11.5a–c): A movement, generally in the sagittal plane, that decreases the angle of the joint and reduces the distance between the two bones. Flexion is typical of hinge joints (bending the knee or elbow) but is also common at ball-and-socket joints (bending forward at the hip).

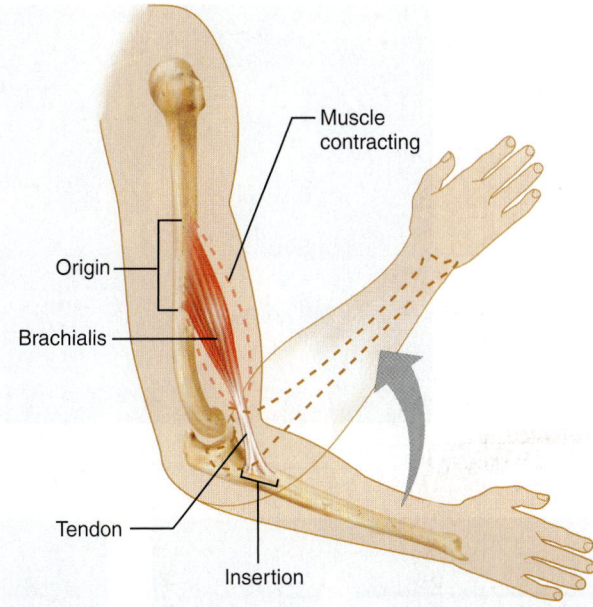

Figure 11.4 Muscle attachments (origin and insertion). When a skeletal muscle contracts, its insertion moves toward its origin.

Extension (Figure 11.5a–c): A movement that increases the angle of a joint and the distance between two bones or parts of the body (straightening the knee or elbow); the opposite of flexion. If extension proceeds beyond anatomical position (bends the trunk backward), it is termed *hyperextension.*

Abduction (Figure 11.5d): Movement of a limb away from the midline or median plane of the body, generally on the frontal plane, or the fanning movement of fingers or toes when they are spread apart.

Adduction (Figure 11.5d): Movement of a limb toward the midline of the body or drawing the fingers or toes together; the opposite of abduction.

Rotation (Figure 11.5e): Movement of a bone around its longitudinal axis without lateral or medial displacement. Rotation, a common movement of ball-and-socket joints, also describes the movement of the atlas around the dens of the axis.

Circumduction (Figure 11.5d): A combination of flexion, extension, abduction, and adduction commonly observed in ball-and-socket joints like the shoulder. The proximal end of the limb remains stationary, and the distal end moves in a circle. The limb as a whole outlines a cone. Condylar and saddle joints also allow circumduction.

Pronation (Figure 11.5f): Movement of the palm of the hand from an anterior or upward-facing position to a posterior or downward-facing position. The distal end of the radius moves across the ulna so that the bones form an X.

Supination (Figure 11.5f): Movement of the palm from a posterior position to an anterior position (the anatomical

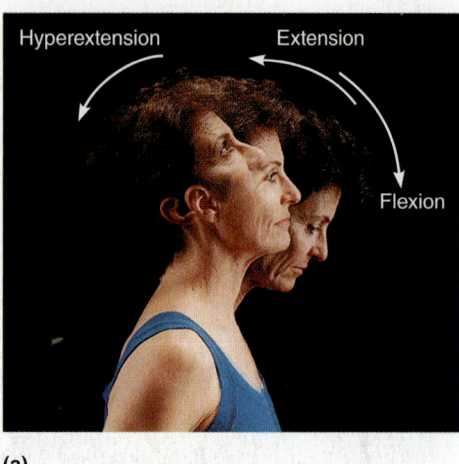

(a)

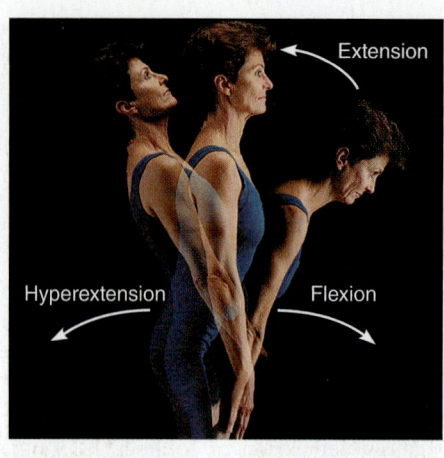

(b)

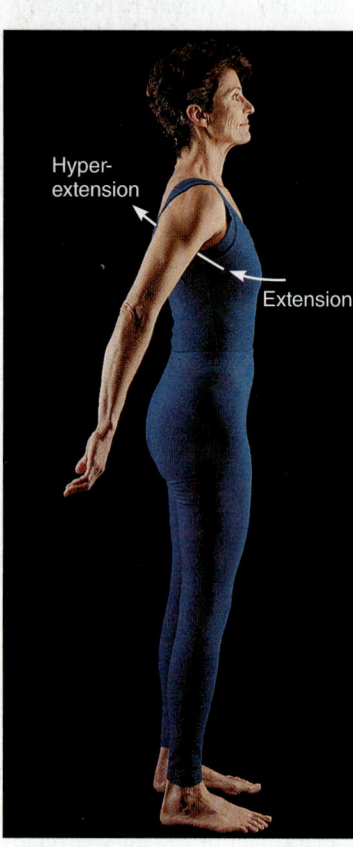

(c)

Figure 11.5 **Movements occurring at synovial joints of the body. (a)** Flexion, extension, and hyperextension of the neck. **(b)** Flexion, extension and hyperextension of the vertebral column. **(c)** Flexion and extension of the knee and shoulder, and hyperextension of the shoulder.

position); the opposite of pronation. During supination, the radius and ulna are parallel.

The last four terms refer to movements of the foot:

Dorsiflexion (Figure 11.5g): A movement of the ankle joint that lifts the foot so that its superior surface approaches the shin.

Plantar flexion (Figure 11.5g): A movement of the ankle joint in which the foot is flexed downward as if standing on one's toes or pointing the toes.

Inversion (Figure 11.5h): A movement that turns the sole of the foot medially.

Eversion (Figure 11.5h): A movement that turns the sole of the foot laterally; the opposite of inversion. ■

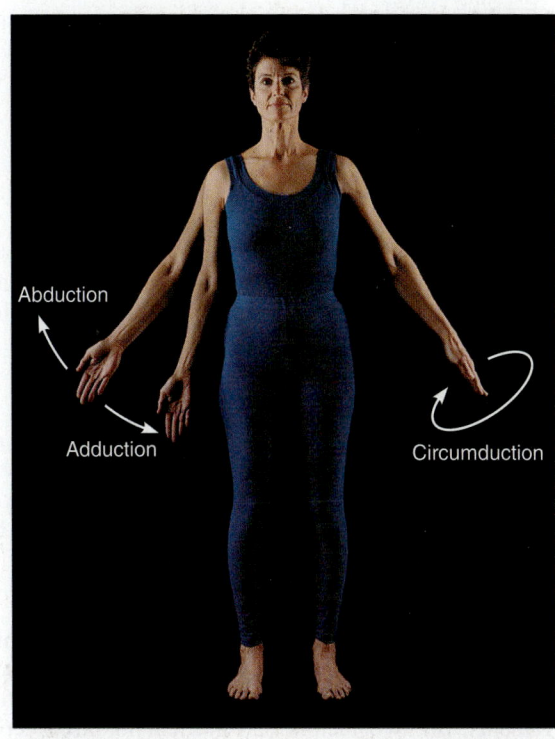

(d)

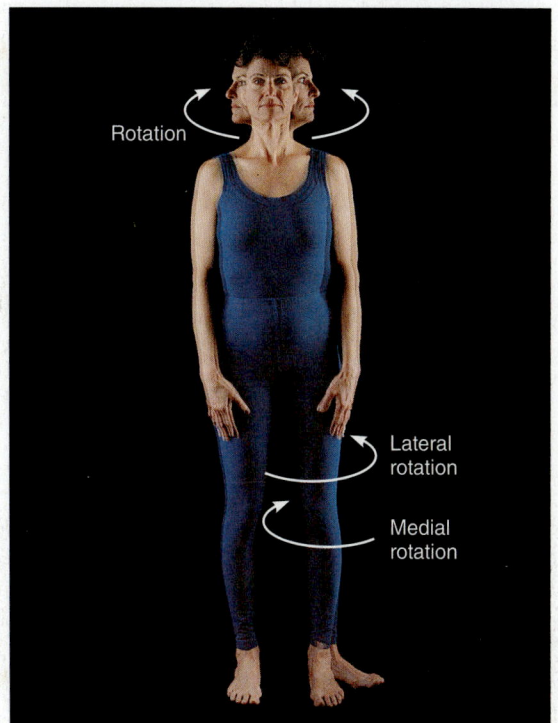

(e)

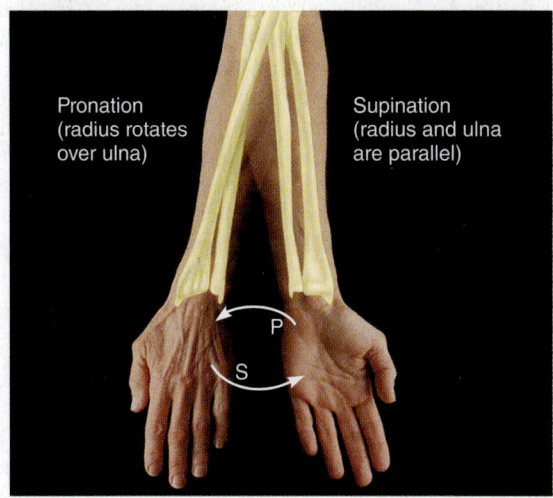

(f)

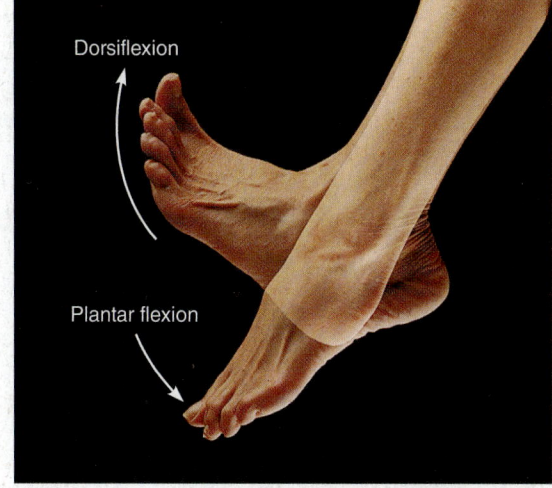

(g)

Figure 11.5 *(continued)* **(d)** Abduction, adduction, and circumduction of the upper limb. **(e)** Rotation of the head and lower limb. **(f)** Pronation and supination of the forearm. **(g)** Dorsiflexion and plantar flexion of the foot. **(h)** Inversion and eversion of the foot.

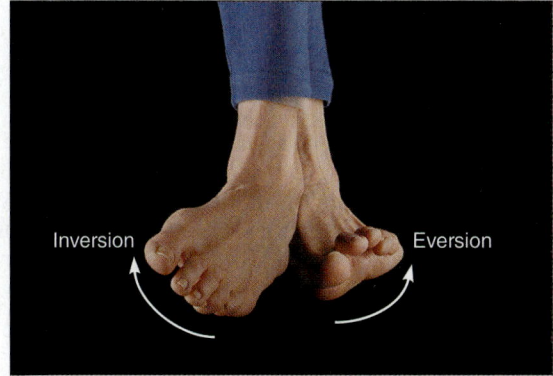

(h)

11

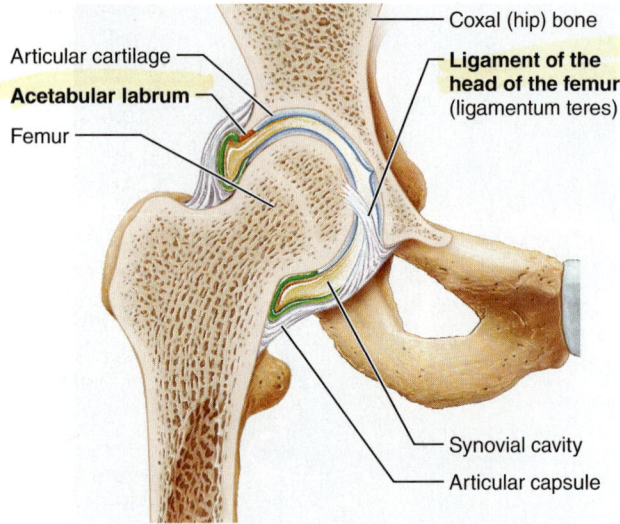

Coxal (hip) bone
Articular cartilage
Acetabular labrum
Femur
Ligament of the head of the femur (ligamentum teres)
Synovial cavity
Articular capsule

(a)

11

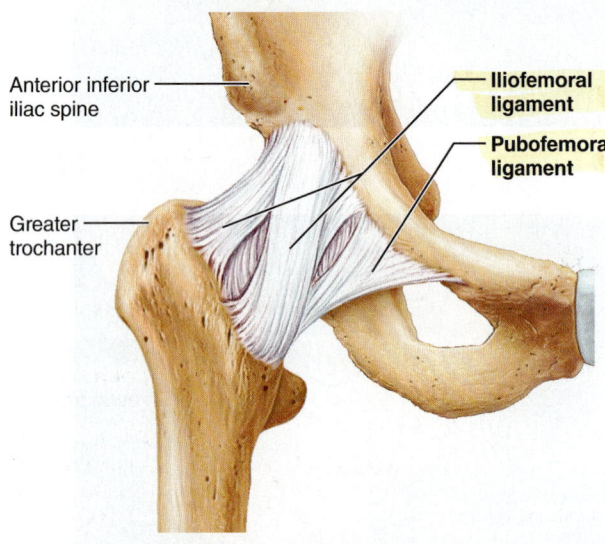

Anterior inferior iliac spine
Greater trochanter
Iliofemoral ligament
Pubofemoral ligament

(b)

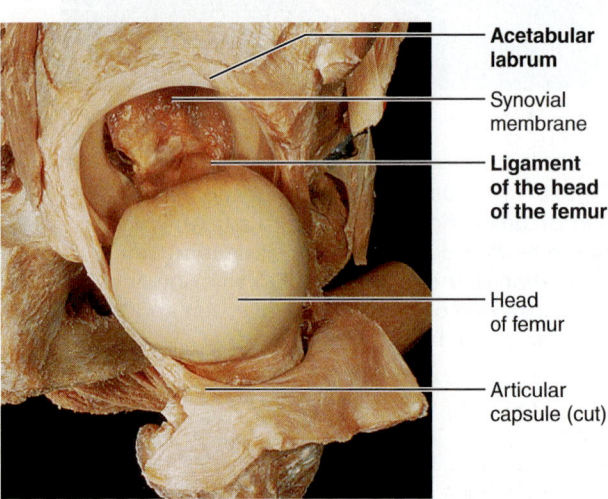

Acetabular labrum
Synovial membrane
Ligament of the head of the femur
Head of femur
Articular capsule (cut)

(c)

Figure 11.6 Hip joint relationships. (a) Frontal section through the right hip joint. **(b)** Anterior superficial view of the right hip joint. **(c)** Photograph of the interior of the hip joint, lateral view.

Selected Synovial Joints

Now you will have the opportunity to compare and contrast the structure of the hip and knee joints and to investigate the structure and movements of the temporomandibular joint and shoulder joint.

The Hip and Knee Joints

Both of these joints are large weight-bearing joints of the lower limb, but they differ substantially in their security. Read through the brief descriptive material below, and look at the questions in the review sheet at the end of this exercise before beginning your comparison.

The Hip Joint

The hip joint is a ball-and-socket joint, so movements can occur in all possible planes. However, its movements are definitely limited by its deep socket and strong reinforcing ligaments, the two factors that account for its exceptional stability **(Figure 11.6)**.

The deeply cupped acetabulum that receives the head of the femur is enhanced by a circular rim of fibrocartilage called the **acetabular labrum.** Because the diameter of the labrum is smaller than that of the femur's head, dislocations of the hip are rare. A short ligament, the **ligament of the head of the femur** (*ligamentum teres*) runs from the pitlike **fovea capitis** on the femur head to the acetabulum where it helps to secure the femur. Several strong ligaments, including the **iliofemoral** and **pubofemoral** anteriorly and the **ischiofemoral** that spirals posteriorly (not shown), are arranged so that they "screw" the femur head into the socket when a person stands upright.

> **ACTIVITY 6**

Demonstrating Actions at the Hip Joint

If a functional hip joint model is available, identify the joint parts and manipulate it to demonstrate the following movements: flexion, extension, abduction, and medial and lateral rotation that can occur at this joint.

Reread the information on what movements the associated ligaments restrict, and verify that information during your joint manipulations. ∎

The Knee Joint

The knee is the largest and most complex joint in the body. Three joints in one **(Figure 11.7)**, it allows extension, flexion, and a little rotation. The **tibiofemoral joint,** actually a duplex joint between the femoral condyles above and the **menisci** (semilunar cartilages) of the tibia below, is functionally a hinge joint, a very unstable one made slightly more secure by the menisci (Figure 11.7b and d). Some rotation occurs when the knee is partly flexed, but during extension, the menisci and ligaments counteract rotation and side-to-side movements. The other joint is the **femoropatellar joint,** the intermediate joint anteriorly (Figure 11.7a and c).

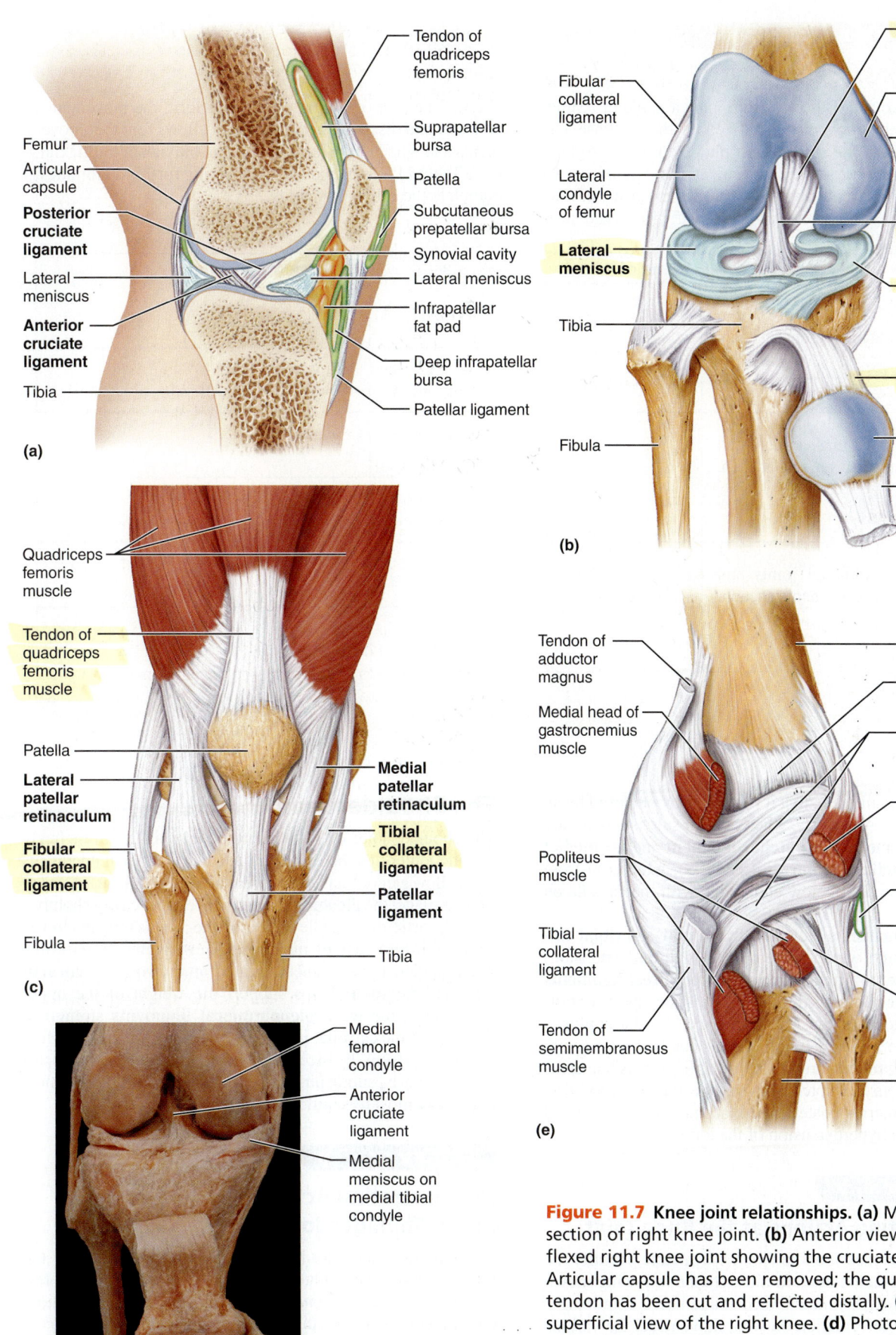

Figure 11.7 Knee joint relationships. (a) Midsagittal section of right knee joint. **(b)** Anterior view of slightly flexed right knee joint showing the cruciate ligaments. Articular capsule has been removed; the quadriceps tendon has been cut and reflected distally. **(c)** Anterior superficial view of the right knee. **(d)** Photograph of an opened knee joint corresponds to view in **(b)**. **(e)** Posterior superficial view of the ligaments clothing the knee joint.

(a) labels:
Femur
Articular capsule
Posterior cruciate ligament
Lateral meniscus
Anterior cruciate ligament
Tibia
Tendon of quadriceps femoris
Suprapatellar bursa
Patella
Subcutaneous prepatellar bursa
Synovial cavity
Lateral meniscus
Infrapatellar fat pad
Deep infrapatellar bursa
Patellar ligament

(b) labels:
Fibular collateral ligament
Lateral condyle of femur
Lateral meniscus
Tibia
Fibula
Posterior cruciate ligament
Medial condyle
Tibial collateral ligament
Anterior cruciate ligament
Medial meniscus
Patellar ligament
Patella
Quadriceps tendon

(c) labels:
Quadriceps femoris muscle
Tendon of quadriceps femoris muscle
Patella
Lateral patellar retinaculum
Fibular collateral ligament
Fibula
Medial patellar retinaculum
Tibial collateral ligament
Patellar ligament
Tibia

(d) labels:
Medial femoral condyle
Anterior cruciate ligament
Medial meniscus on medial tibial condyle
Patella

(e) labels:
Tendon of adductor magnus
Medial head of gastrocnemius muscle
Popliteus muscle
Tibial collateral ligament
Tendon of semimembranosus muscle
Femur
Articular capsule
Oblique popliteal ligament
Lateral head of gastrocnemius muscle
Bursa
Fibular collateral ligament
Arcuate popliteal ligament
Tibia

11

11

Articulations: "Simon Says"

Working in groups of three or four, play a game of "Simon Says" using the movements defined in the exercise (see pages 173–175). One student will play the role of "Simon" while the others perform the movement. For example, when "Simon" says, "Simon says, perform flexion at the elbow," the remaining students would flex their arm. Take turns playing the role of Simon. As you perform the movements, consider and discuss whether the joint allows for other movements and whether the joint is uniaxial, biaxial, or multiaxial. (Use Table 11.1 as a guide.) After playing for 15–20 minutes, complete the following tables.

	Name of joint	Movements allowed
1. List two uniaxial joints and describe the movements at each.		
2. List two biaxial joints and describe the movements at each.		
3. List two multiaxial joints and describe the movements at each.		

The knee is unique in that it is only partly enclosed by an articular capsule. Anteriorly, where the capsule is absent, are three broad ligaments, the **patellar ligament** and the **medial** and **lateral patellar retinacula** (retainers), which run from the patella to the tibia below and merge with the capsule on either side.

Capsular ligaments including the **fibular** and **tibial collateral ligaments** (which prevent rotation during extension) and the **oblique popliteal** and **arcuate popliteal ligaments** are crucial in reinforcing the knee. The knees have a built-in locking device that must be "unlocked" by the popliteus muscles (Figure 11.7e) before the knees can be flexed again. The **cruciate ligaments** are intracapsular ligaments that cross (*cruci* = cross) in the notch between the femoral condyles. They prevent anterior-posterior displacement of the joint and overflexion and hyperextension of the joint.

ACTIVITY 7

Demonstrating Actions at the Knee Joint

If a functional model of a knee joint is available, identify the joint parts and manipulate it to illustrate the following movements: flexion, extension, and medial and lateral rotation.

Reread the information on what movements the various associated ligaments restrict, and verify that information during your joint manipulations. ■

The Shoulder Joint

The shoulder joint or **glenohumeral joint** is the most freely moving joint of the body. The rounded head of the humerus fits the shallow glenoid cavity of the scapula **(Figure 11.8)**. A rim of fibrocartilage, the **glenoid labrum,** deepens the cavity slightly.

The articular capsule enclosing the joint is thin and loose, contributing to ease of movement. Few ligaments reinforce the shoulder, most of them located anteriorly. The **coracohumeral ligament** helps support the weight of the upper limb, and three weak **glenohumeral ligaments** strengthen the front of the capsule. In some people they are absent. Muscle tendons from the biceps brachii and **rotator cuff** muscles (subscapularis, supraspinatus, infraspinatus, and teres minor) contribute most to shoulder stability.

ACTIVITY 8

Demonstrating Actions at the Shoulder Joint

If a functional shoulder joint model is available, identify the joint parts and manipulate the model to demonstrate the following movements: flexion, extension, abduction, adduction, circumduction, and medial and lateral rotation.

Note where the joint is weakest and verify the most common direction of a dislocated humerus. ■

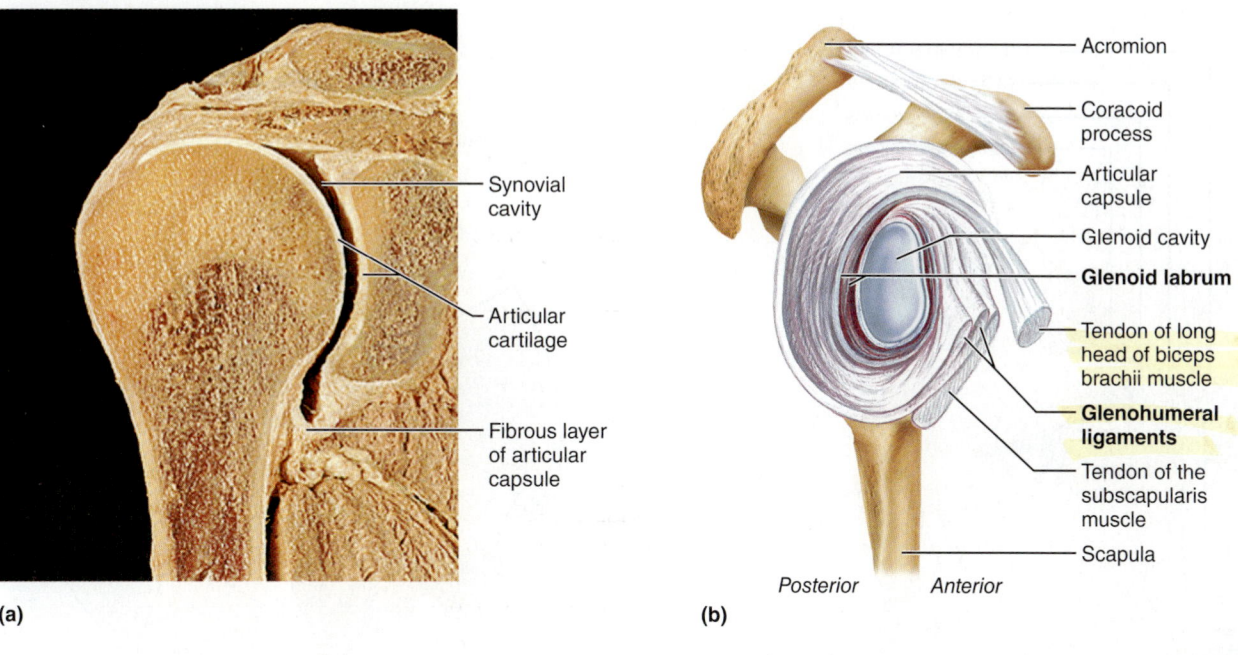

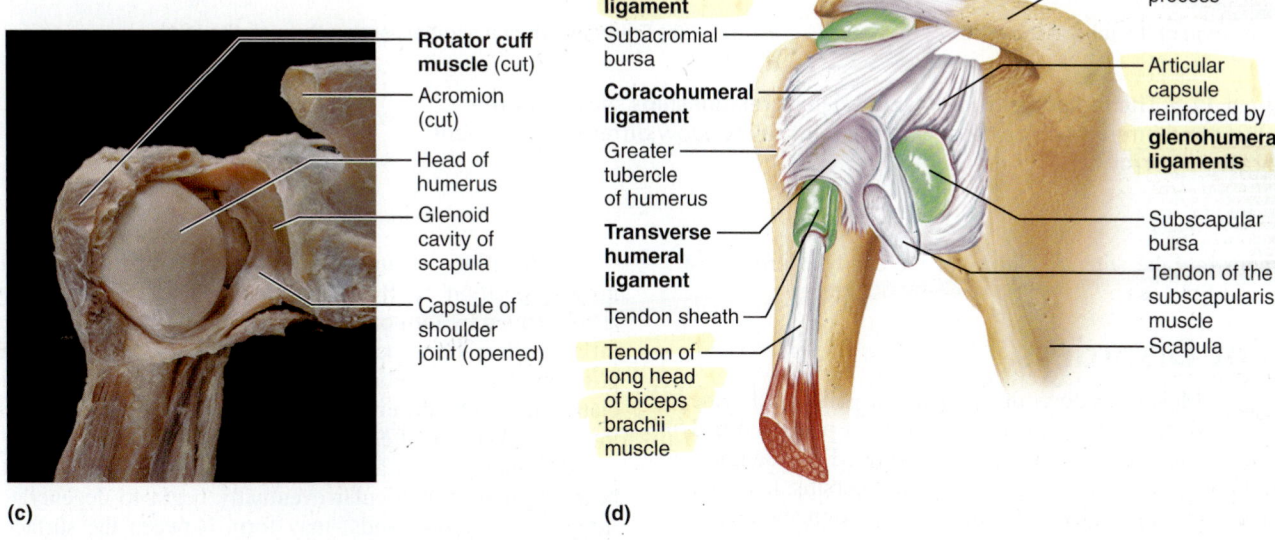

Figure 11.8 Shoulder joint relationships. (a) Frontal section through the shoulder. **(b)** Right shoulder joint, cut open and viewed from the lateral aspect; humerus removed. **(c)** Photograph of the interior of the shoulder joint, anterior view. **(d)** Anterior superficial view of the right shoulder.

The Temporomandibular Joint

The **temporomandibular joint (TMJ)** lies just anterior to the ear **(Figure 11.9)**, where the egg-shaped condylar process of the mandible articulates with the inferior surface of the squamous region of the temporal bone. The temporal bone joint surface has a complicated shape: posteriorly is the **mandibular fossa** and anteriorly is a bony knob called the **articular tubercle.** The joint's articular capsule, though strengthened by the **lateral ligament,** is slack; an articular disc divides the joint cavity into superior and inferior compartments. Typically, the condylar process–mandibular fossa connection allows the familiar hingelike movements

of elevating and depressing the mandible to open and close the mouth. However, when the mouth is opened wide, the condylar process glides anteriorly and is braced against the dense bone of the articular tubercle so that the mandible is not forced superiorly when we bite hard foods.

ACTIVITY 9

Examining the Action at the TMJ

While placing your fingers over the area just anterior to the ear, open and close your mouth to feel the hinge action at the TMJ. Then, keeping your fingers on the TMJ, yawn to

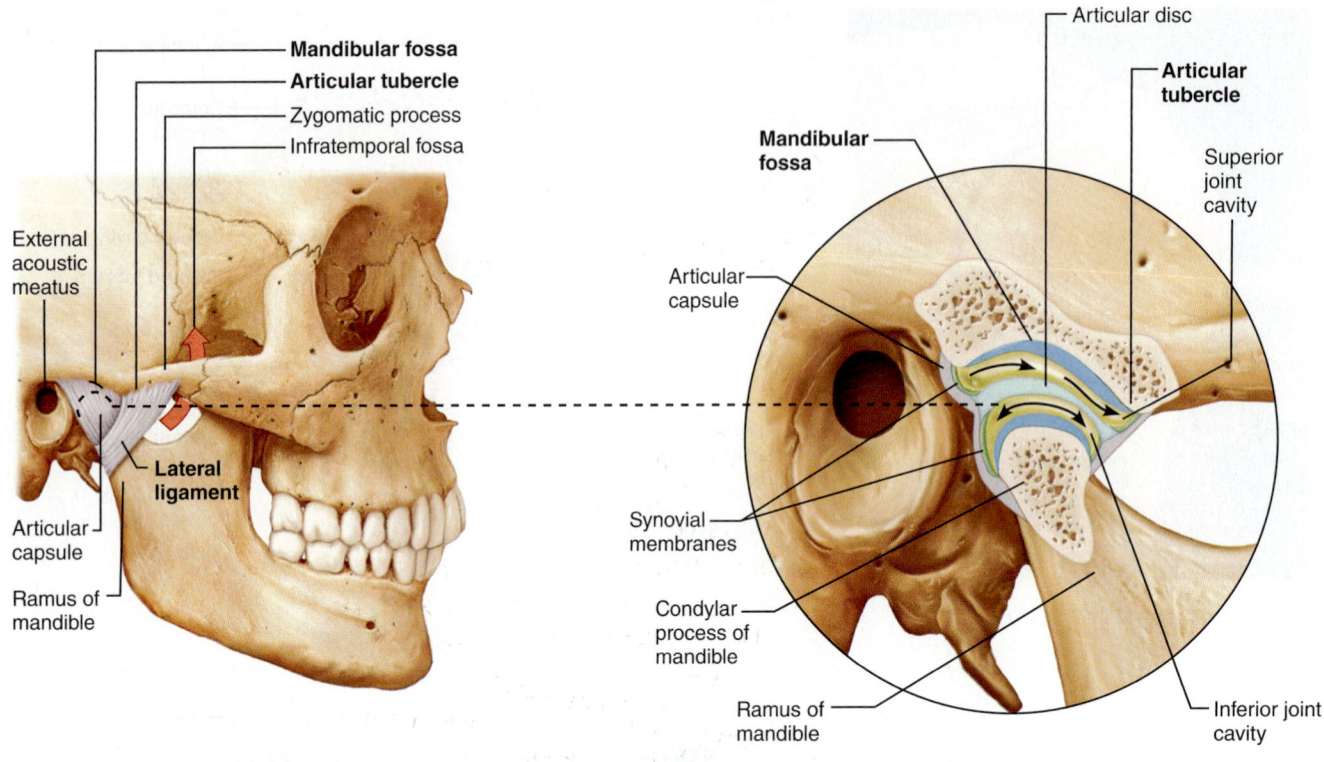

Figure 11.9 **The temporomandibular (jaw) joint relationships.** Note that the superior and inferior compartments of the joint cavity allow different movements indicated by arrows.

demonstrate the anterior gliding of the condylar process of the mandible. ▬

Joint Disorders

Most of us don't think about our joints until something goes wrong with them. Joint pains and malfunctions are caused by a variety of things. For example, a hard blow to the knee can cause a painful bursitis, known as "water on the knee," due to damage to, or inflammation of, the patellar bursa. Slippage of a fibrocartilage pad or the tearing of a ligament may result in a painful condition that persists over a long period, since these poorly vascularized structures heal so slowly.

Sprains and dislocations are other types of joint problems. In a **sprain,** the ligaments reinforcing a joint are damaged by overstretching or are torn away from the bony attachment. Since both ligaments and tendons are cords of dense connective tissue with a poor blood supply, sprains heal slowly and are quite painful.

Dislocations occur when bones are forced out of their normal position in the joint cavity. They are normally accompanied by torn or stressed ligaments and considerable inflammation. The process of returning the bone to its proper position, called reduction, should be done only by a physician. Attempts by the untrained person to "snap the bone back into its socket" are often more harmful than helpful.

Advancing years also take their toll on joints. Weight-bearing joints in particular eventually begin to degenerate. *Adhesions* (fibrous bands) may form between the surfaces where bones join, and extraneous bone tissue *(spurs)* may grow along the joint edges. Such degenerative changes lead to the complaint so often heard from the elderly: "My joints are getting so stiff. . . ."

• If possible, compare an X ray of an arthritic joint to one of a normal joint. ✚

Microscopic Anatomy and Organization of Skeletal Muscle

MATERIALS

- ☐ Three-dimensional model of skeletal muscle cells (if available)
- ☐ Forceps
- ☐ Dissecting needles
- ☐ Clean microscope slides and coverslips
- ☐ 0.9% saline solution in dropper bottles
- ☐ Chicken breast or thigh muscle (freshly obtained from the meat market)
- ☐ Compound microscope
- ☐ Prepared slides of skeletal muscle (l.s. and x.s. views) and skeletal muscle showing neuromuscular junctions
- ☐ Three-dimensional model of skeletal muscle showing neuromuscular junction (if available)

MasteringA&P® For related exercise study tools, go to the Study Area of MasteringA&P. There you will find:

- Practice Anatomy Lab PAL
- PhysioEx PEx
- A&PFlix *A&PFlix*
- Practice quizzes, Histology Atlas, eText, Videos, and more!

OBJECTIVES

1. Define *fiber, myofibril,* and *myofilament* and describe the structural relationship between them.
2. Describe thick (myosin) and thin (actin) filaments and their relation to the sarcomere.
3. Discuss the structure and location of T tubules and terminal cisterns.
4. Define *endomysium, perimysium,* and *epimysium* and relate them to muscle fibers, fascicles, and entire muscles.
5. Define *tendon* and *aponeurosis* and describe the difference between them.
6. Describe the structure of skeletal muscle from gross to microscopic levels.
7. Explain the connection between motor neurons and skeletal muscle and discuss the structure and function of the neuromuscular junction.

PRE-LAB QUIZ

1. Which is *not* true of skeletal muscle?
   a. It enables you to manipulate your environment.
   b. It influences the body's contours and shape.
   c. It is one of the major components of hollow organs.
   d. It provides a means of locomotion.
2. Circle the correct underlined term. Because the cells of skeletal muscle are relatively large and cylindrical in shape, they are also known as <u>fibers</u> / <u>tubules</u>.
3. Circle True or False. Skeletal muscle cells have more than one nucleus.
4. The two contractile proteins that make up the myofilaments of skeletal muscle are_____ and _____.
5. Each muscle cell is surrounded by thin connective tissue called the
   a. aponeuroses   c. endomysium
   b. epimysium   d. perimysium
6. A cordlike structure that connects a muscle to another muscle or bone is
   a. a fascicle
   b. a tendon
   c. deep fascia
7. The junction between an axon and a muscle fiber is called a _____.
8. Circle True or False. The neuron and muscle fiber membranes do not actually touch but are separated by a fluid-filled gap.
9. Circle the correct underlined term. The contractile unit of muscle is the <u>sarcolemma</u> / <u>sarcomere</u>.
10. Circle True or False. Larger, more powerful muscles have relatively less connective tissue than smaller muscles.

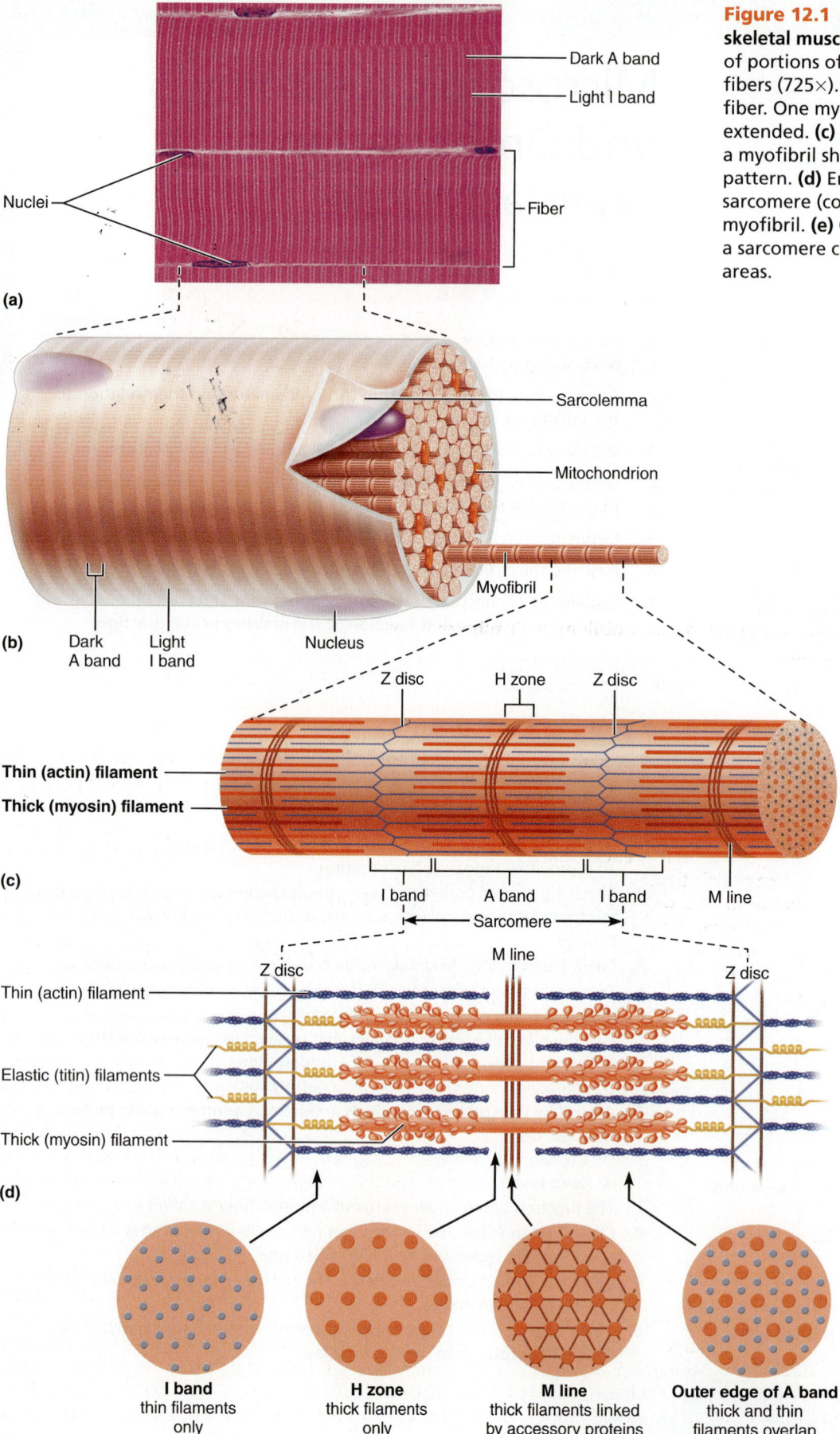

Figure 12.1 Microscopic anatomy of skeletal muscle. (a) Photomicrograph of portions of two isolated muscle fibers (725×). **(b)** Part of a muscle fiber. One myofibril has been extended. **(c)** Enlarged view of a myofibril showing its banding pattern. **(d)** Enlarged view of one sarcomere (contractile unit) of a myofibril. **(e)** Cross-sectional view of a sarcomere cut through in different areas.

(a)

Dark A band
Light I band
Nuclei
Fiber

(b)

Sarcolemma
Mitochondrion
Myofibril
Dark A band
Light I band
Nucleus

(c)

Z disc   H zone   Z disc
Thin (actin) filament
Thick (myosin) filament
I band   A band   I band   M line
Sarcomere

(d)

M line
Z disc   Z disc
Thin (actin) filament
Elastic (titin) filaments
Thick (myosin) filament

(e)

I band
thin filaments only

H zone
thick filaments only

M line
thick filaments linked by accessory proteins

Outer edge of A band
thick and thin filaments overlap

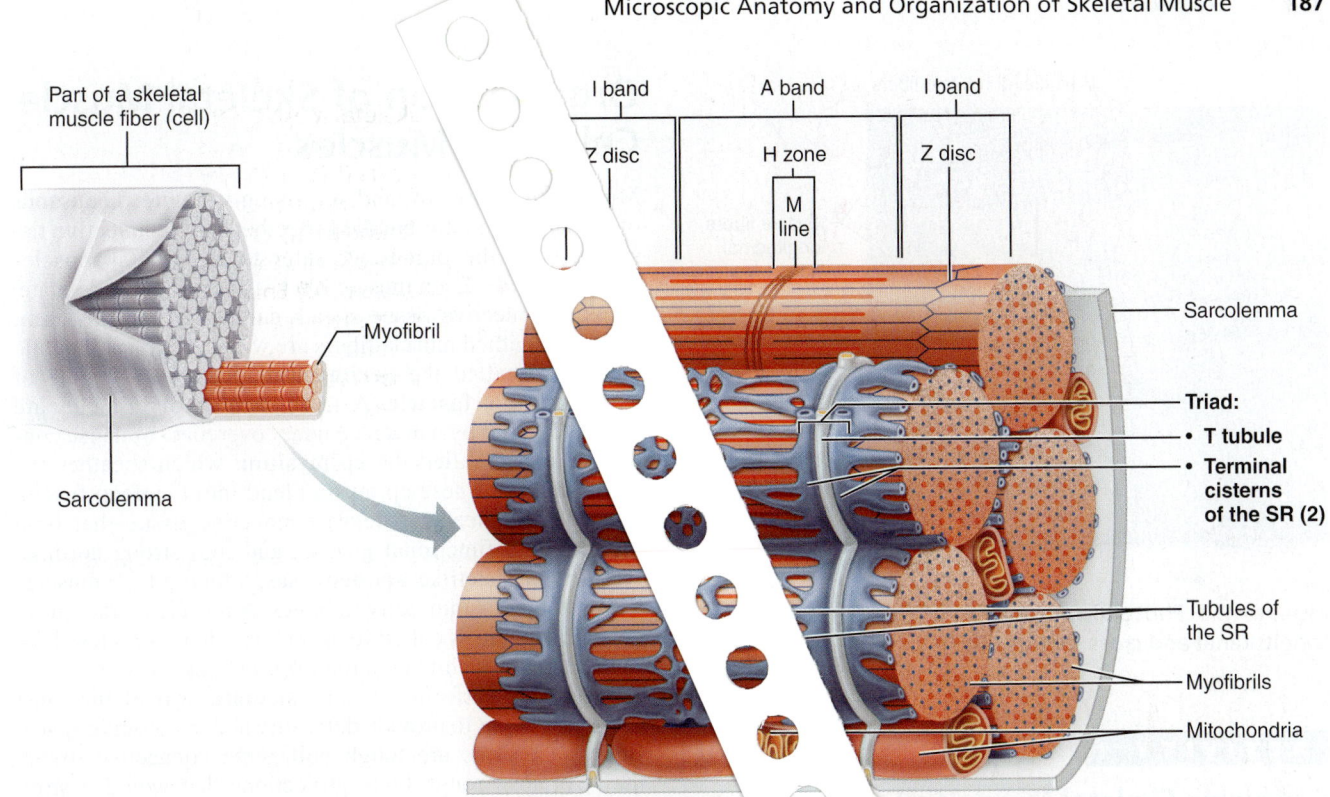

I band   A band   I band

Z disc   H zone   Z disc

M line

Part of a skeletal muscle fiber (cell)

Myofibril

Sarcolemma

Sarcolemma

Triad:
• **T tubule**
• **Terminal cisterns of the SR (2)**

Tubules of the SR

Myofibrils

Mitochondria

Figure 12.2 Relationship of the sarcoplasmic reticulum and T tubules to the myofibrils of skeletal muscle.

M ost of the muscle tissue in the body is **skeletal muscle,** which attaches to the skeleton or associated connective tissue. Skeletal muscle shapes the body and gives you the ability to move—to walk, run, jump, and dance; to draw, paint, and play a musical instrument; and to smile and frown. The remaining muscle tissue of the body consists of smooth muscle that forms the walls of hollow organs and cardiac muscle that forms the walls of the heart. Smooth and cardiac muscle move materials within the body. For example, smooth muscle moves digesting food through the gastrointestinal system, and urine from the kidneys to the exterior of the body. Cardiac muscle moves blood through the blood vessels.

Each of the three muscle types has a structure and function uniquely suited to its task in the body. Our focus here is to investigate the structure of skeletal muscle.

Skeletal muscle is also known as *voluntary muscle* because it can be consciously controlled, and as striated muscle because it appears to be striped. As you might guess from both of these alternative names, skeletal muscle has some special characteristics. Thus an investigation of skeletal muscle begins at the cellular level.

The Cells of Skeletal Muscle

Skeletal muscle is made up of relatively large, long cylindrical cells, sometimes called **fibers.** These cells range from 10 to 100 μm in diameter and some are up to 30 cm long.

Since hundreds of embryonic cells fuse to produce each muscle cell, the cells **(Figure 12.1a and b)** are multinucleate; multiple oval nuclei can be seen just beneath the plasma membrane (called the *sarcolemma* in these cells). The nuclei are pushed peripherally by the longitudinally arranged **myofibrils,** which nearly fill the sarcoplasm. Alternating light (I) and dark (A) bands along the length of the perfectly aligned myofibrils give the muscle fiber as a whole its striped appearance.

Electron microscope studies have revealed that the myofibrils are made up of even smaller threadlike structures called **myofilaments (Figure 12.1c).** The myofilaments are composed largely of two varieties of contractile proteins—**actin** and **myosin**—which slide past each other during muscle activity to bring about shortening, or contraction, of the muscle cells. It is the highly specialized arrangement of the myofilaments within the myofibrils that is responsible for the banding pattern in skeletal muscle. The actual contractile units of muscle, called **sarcomeres,** extend from the middle of one I band (its Z disc) to the middle of the next, along the length of the myofibrils (Figure 12.1c and d). Cross sections of the sarcomere in areas where **thick** and **thin filaments** overlap show that each thick filament is surrounded by six thin filaments; each thin filament is enclosed by three thick filaments (Figure 12.1e).

At each junction of the A and I bands, the sarcolemma indents into the muscle cell, forming a **transverse tubule (T tubule).** These tubules run deep into the muscle cell between cross channels, or **terminal cisterns,** of elaborate smooth endoplasmic reticulum called the **sarcoplasmic reticulum (SR) (Figure 12.2).** Regions where the SR terminal cisterns border a T tubule on each side are called **triads.**

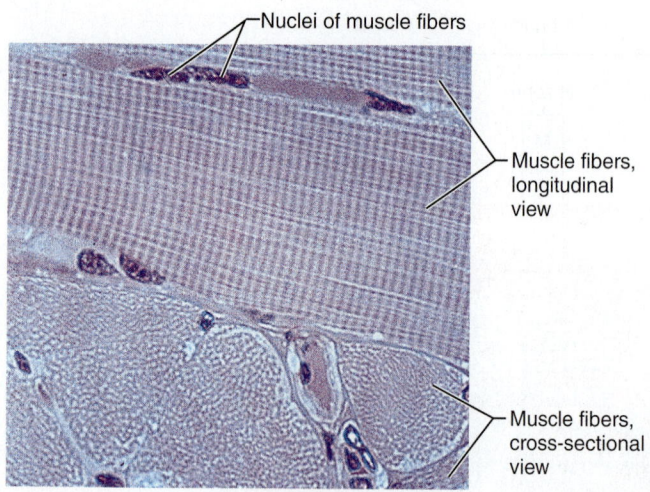

Nuclei of muscle fibers

Muscle fibers, longitudinal view

Muscle fibers, cross-sectional view

Figure 12.3 Photomicrograph of muscle fibers, longitudinal and cross sections (800×).

12

Examining Skeletal Muscle Cell Anatomy

1. Look at the three-dimensional model of skeletal muscle cells, noting the relative shape and size of the cells. Identify the nuclei, myofibrils, and light and dark bands.

2. Obtain forceps, two dissecting needles, slide and coverslip, and a dropper bottle of saline solution. With forceps, remove a very small piece of muscle (about 1 mm diameter) from a fresh chicken breast or thigh. Place the tissue on a clean microscope slide, and add a drop of the saline solution.

3. Pull the muscle fibers apart (tease them) with the dissecting needles until you have a fluffy-looking mass of tissue. Cover the teased tissue with a coverslip, and observe under the high-power lens of a compound microscope. Look for the banding pattern by examining muscle fibers isolated at the edge of the tissue mass. Regulate the light carefully to obtain the highest possible contrast.

4. Now compare your observations with the photomicrograph **(Figure 12.3)** and with what can be seen in professionally prepared muscle tissue. Obtain a slide of skeletal muscle (longitudinal section), and view it under high power. From your observations, draw a small section of a muscle fiber in the space provided below. Label the nuclei, sarcolemma, and A and I bands.

What structural details become apparent with the prepared slide?

_____

_____

Organization of Skeletal Muscle Cells into Muscles

Muscle fibers are soft and surprisingly fragile. Thousands of muscle fibers are bundled together with connective tissue to form the organs we refer to as skeletal muscles **(Figure 12.4)**. Each muscle fiber is enclosed in a delicate, areolar connective tissue sheath called the **endomysium.** Several sheathed muscle fibers are wrapped by a collagenic membrane called the **perimysium,** forming a bundle of fibers called a **fascicle.** A large number of fascicles are bound together by a much coarser "overcoat" of dense connective tissue called the **epimysium,** which sheathes the entire muscle. These epimysia blend into the **deep fascia,** still coarser sheets of dense connective tissue that bind muscles into functional groups, and into strong cordlike **tendons** or sheetlike **aponeuroses,** which attach muscles to each other or indirectly to bones. A muscle's more movable attachment is called its *insertion* whereas its fixed (or immovable) attachment is the *origin* (Exercise 11).

Tendons perform several functions, two of the most important being to provide durability and to conserve space. Because tendons are tough collagenic connective tissue, they can span rough bony projections that would destroy the more delicate muscle tissues. Because of their relatively small size, more tendons than fleshy muscles can pass over a joint.

In addition to supporting and binding the muscle fibers, and providing strength to the muscle as a whole, the connective tissue wrappings provide a route for the entry and exit of nerves and blood vessels that serve the muscle fibers. The larger, more powerful muscles have relatively more connective tissue than muscles involved in fine or delicate movements.

As we age, the mass of the muscle fibers decreases, and the amount of connective tissue increases; thus the skeletal muscles gradually become more sinewy, or "stringier."✚

Observing the Histological Structure of a Skeletal Muscle

Identify the muscle fibers, their peripherally located nuclei, and their connective tissue wrappings—the endomysium, perimysium, and epimysium, if visible (use Figure 12.4 as a reference). ■

The Neuromuscular Junction

The voluntary skeletal muscle cells must be stimulated by motor neurons via nerve impulses. The junction between an axon of a motor neuron and a muscle cell is called a **neuromuscular,** or **myoneural, junction (Figure 12.5)**.

Each axon of the motor neuron usually divides into many branches called *terminal branches* as it approaches the muscle. Each of these branches ends in an axon terminal that participates in forming a neuromuscular junction with a single muscle cell. Thus a single neuron may stimulate many muscle fibers. Together, a neuron and all the muscle fibers it stimulates make up the functional structure called the

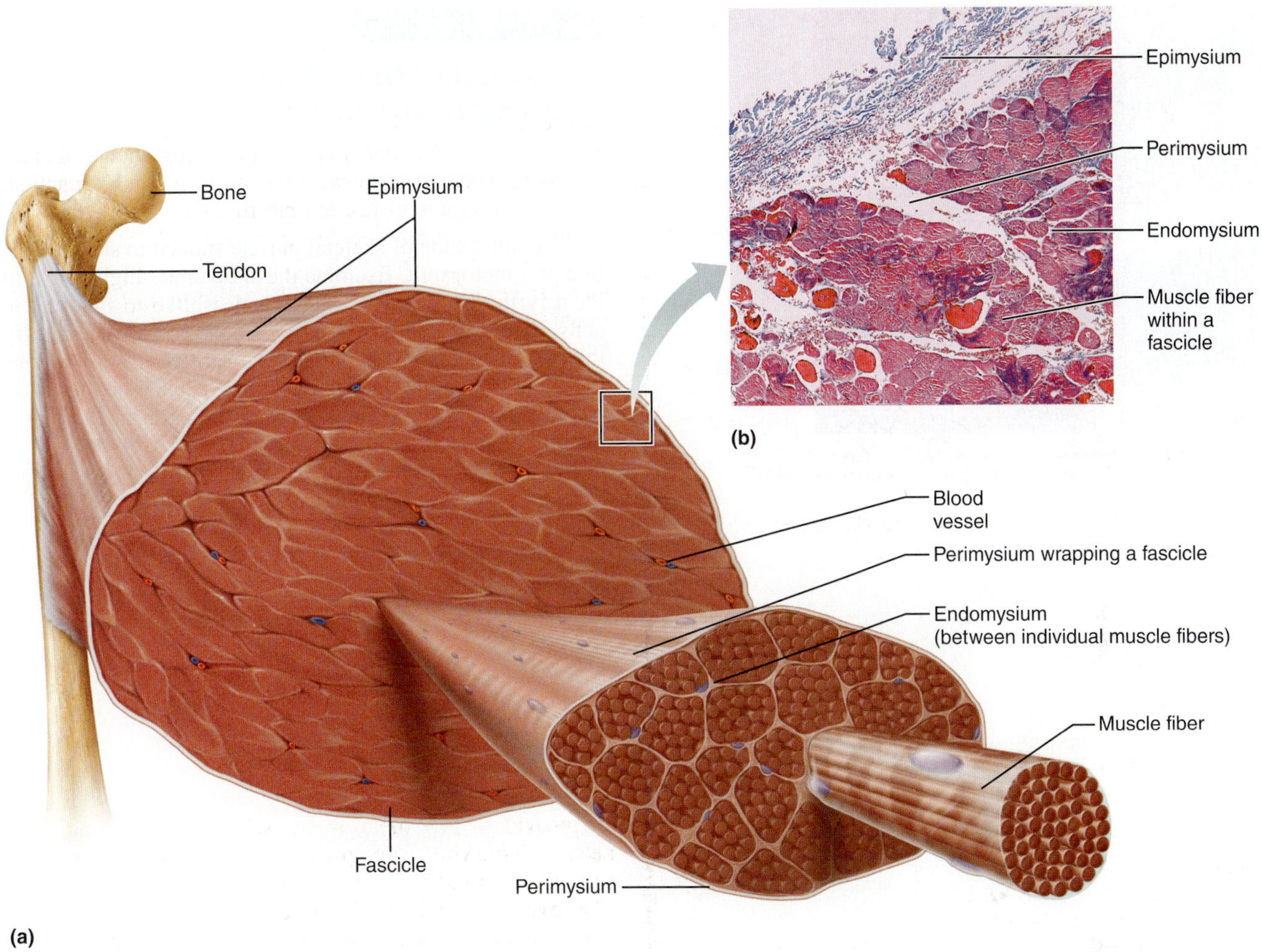

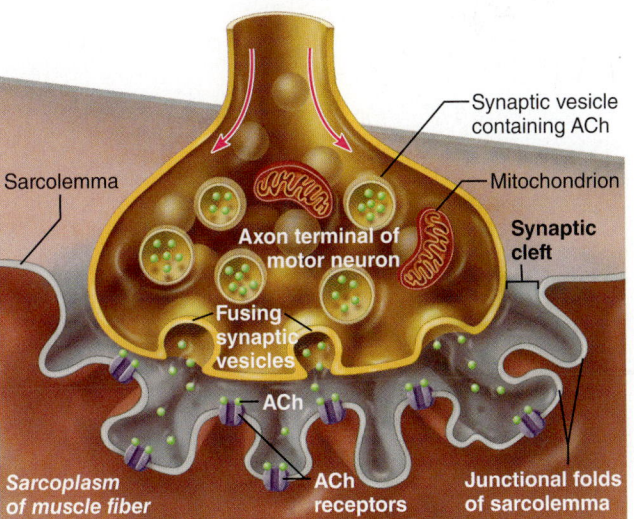

Epimysium

Perimysium

Endomysium

Muscle fiber within a fascicle

(b)

Bone

Epimysium

Tendon

Blood vessel

Perimysium wrapping a fascicle

Endomysium (between individual muscle fibers)

Muscle fiber

Fascicle

Perimysium

(a)

12

Figure 12.4 Connective tissue coverings of skeletal muscle. (a) Diagrammatic view. **(b)** Photomicrograph of a cross section of skeletal muscle (90×).

Synaptic vesicle containing ACh

Sarcolemma

Mitochondrion

Axon terminal of motor neuron

Synaptic cleft

Fusing synaptic vesicles

ACh

Sarcoplasm of muscle fiber

ACh receptors

Junctional folds of sarcolemma

Figure 12.5 The neuromuscular junction. Red arrows indicate arrival of the nerve impulse (action potential), which ultimately causes vesicles to release ACh. The ACh receptor is part of the ion channel that opens briefly, causing depolarization of the sarcolemma.

motor unit. (Part of a motor unit, showing two neuromuscular junctions, is shown in **Figure 12.6.**) The neuron and muscle fiber membranes, close as they are, do not actually touch. They are separated by a small fluid-filled gap called the **synaptic cleft** (see Figure 12.5).

Within the axon terminals are many mitochondria and vesicles containing a neurotransmitter chemical called acetylcholine (ACh). When a nerve impulse reaches the axon terminal, some of these vesicles release their contents into the synaptic cleft. The ACh rapidly diffuses across the junction and combines with the receptors on the sarcolemma. When receptors bind ACh, a change in the permeability of the sarcolemma occurs. Channels that allow both sodium (Na^+) and potassium (K^+) ions to pass open briefly. Because more Na^+ diffuses into the muscle fiber than K^+ diffuses out, depolarization of the sarcolemma and subsequent contraction of the muscle fiber occurs.

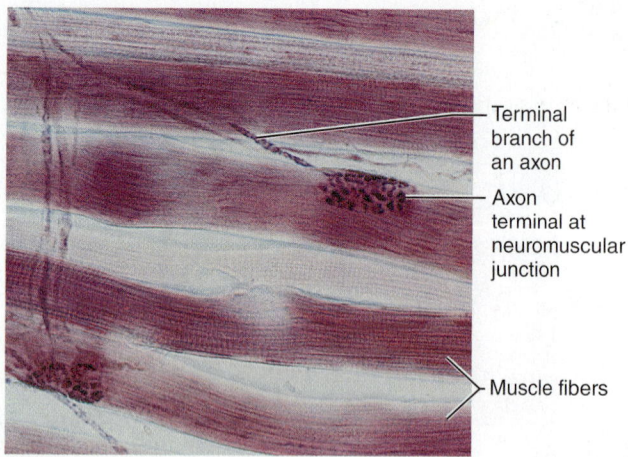

Terminal branch of an axon

Axon terminal at neuromuscular junction

Muscle fibers

Figure 12.6 Photomicrograph of neuromuscular junctions (750×).

12

Studying the Structure of a Neuromuscular Junction

1. If possible, examine a three-dimensional model of skeletal muscle cells that illustrates the neuromuscular junction. Identify the structures just described.

2. Obtain a slide of skeletal muscle stained to show a portion of a motor unit. Examine the slide under high power to identify the axon fibers extending leashlike to the muscle cells. Follow one of the axon fibers to its terminus to identify the oval-shaped axon terminal. Compare your observations to the photomicrograph (Figure 12.6). Sketch a small section in the space provided below. Label the axon of the motor neuron, its terminal branches, and muscle fibers. ■

Gross Anatomy of the Muscular System

MATERIALS

☐ Human torso model or large anatomical chart showing human musculature
☐ Human cadaver for demonstration (if available)
☐ Disposable gloves
☐ *Human Musculature* video
☐ Tubes of body (or face) paint
☐ 1″ wide artist's brushes

✂ For instructions on animal dissections, see the dissection exercises (starting on page 697) in the cat and fetal pig editions of this manual.

OBJECTIVES

1. Define *prime mover (agonist), antagonist, synergist,* and *fixator.*
2. List the criteria used in naming skeletal muscles.
3. Identify the major muscles of the human body on a torso model, a human cadaver, lab chart, or image, and state the action of each.
4. Name muscle origins and insertions as required by the instructor.
5. Explain how muscle actions are related to their location.
6. List antagonists for the major prime movers.

PRE-LAB QUIZ

1. A prime mover or _____ produces a particular type of movement.
   a. agonist   c. fixator
   b. antagonist   d. synergist
2. Skeletal muscles are named on the basis of many criteria. Name one.
   _____
3. Circle True or False. Muscles of facial expression differ from most skeletal muscles because they usually do not insert into a bone.
4. The _____ musculature includes muscles that move the vertebral column and muscles that move the ribs.
   a. head and neck   b. lower limb   c. trunk
5. Muscles that act on the _____ cause movement at the hip, knee, and foot joints.
   a. lower limb   b. trunk   c. upper limb
6. This two-headed muscle bulges when the forearm is flexed. It is the most familiar muscle of the anterior humerus. It is the
   a. biceps brachii   c. extensor digitorum
   b. flexor carpii radialis   d. triceps brachii
7. These abdominal muscles are responsible for giving me my "six-pack." They also stabilize my pelvis when walking. They are the _____ muscles.
   a. internal intercostal   c. quadriceps
   b. rectus abdominis   d. triceps femoris
8. Circle the correct underlined term. This lower limb muscle, which attaches to the calcaneus via the calcaneal tendon and plantar flexes the foot when the knee is extended, is the <u>tibialis anterior</u> / <u>gastrocnemius</u>.
9. The _____ is the largest and most superficial of the gluteal muscles.
   a. gluteus internus   c. gluteus maximus
   b. gluteus medius   d. gluteus minimus
10. Circle True or False. The biceps femoris is located in the anterior compartment of the thigh.

MasteringA&P® For related exercise study tools, go to the Study Area of MasteringA&P. There you will find:
• Practice Anatomy Lab PAL
• PhysioEx PEx
• A&PFlix *A&PFlix*
• Practice quizzes, Histology Atlas, eText, Videos, and more!

Skeletal muscles cause movement. Among the movements are smiling, frowning, speaking, singing, breathing, dancing, running, and playing a musical instrument. Most often, purposeful movements require the coordinated action of several skeletal muscles.

Classification of Skeletal Muscles

Types of Muscles

Muscles that are most responsible for producing a particular movement are called **prime movers,** or **agonists.** Muscles that oppose or reverse a movement are called **antagonists.** When a prime mover is active, the fibers of the antagonist are stretched and in the relaxed state. The antagonist can also regulate the prime mover by providing some resistance, to prevent overshoot or to stop its action. Antagonists can be prime movers in their own right. For example, the biceps muscle of the arm (a prime mover of elbow flexion) is antagonized by the triceps (a prime mover of elbow extension).

Synergists aid the action of agonists either by assisting with the same movement or by reducing undesirable or unnecessary movement. Contraction of a muscle crossing two or more joints would cause movement at all joints spanned if the synergists were not there to stabilize them. For example, you can make a fist without bending your wrist only because synergist muscles stabilize the wrist joint and allow the prime mover to exert its force at the finger joints.

Fixators, or fixation muscles, are specialized synergists. They immobilize the origin of a prime mover so that all the tension is exerted at the insertion. Muscles that help maintain posture are fixators; so too are muscles of the back that stabilize or "fix" the scapula during arm movements.

Naming Skeletal Muscles

Remembering the names of the skeletal muscles is a monumental task, but certain clues help. Muscles are named on the basis of the following criteria:

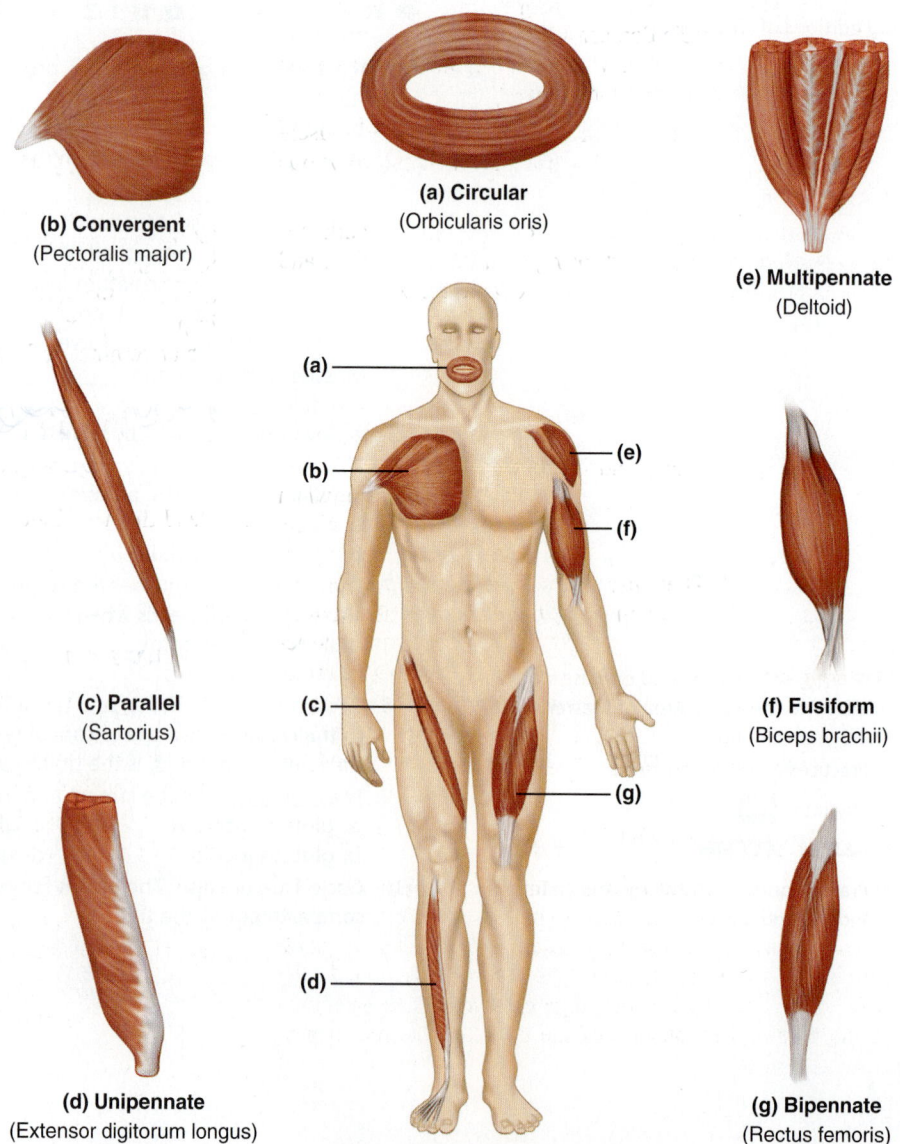

Figure 13.1 Patterns of fascicle arrangement in muscles.

(b) Convergent (Pectoralis major)

(a) Circular (Orbicularis oris)

(e) Multipennate (Deltoid)

(c) Parallel (Sartorius)

(f) Fusiform (Biceps brachii)

(d) Unipennate (Extensor digitorum longus)

(g) Bipennate (Rectus femoris)

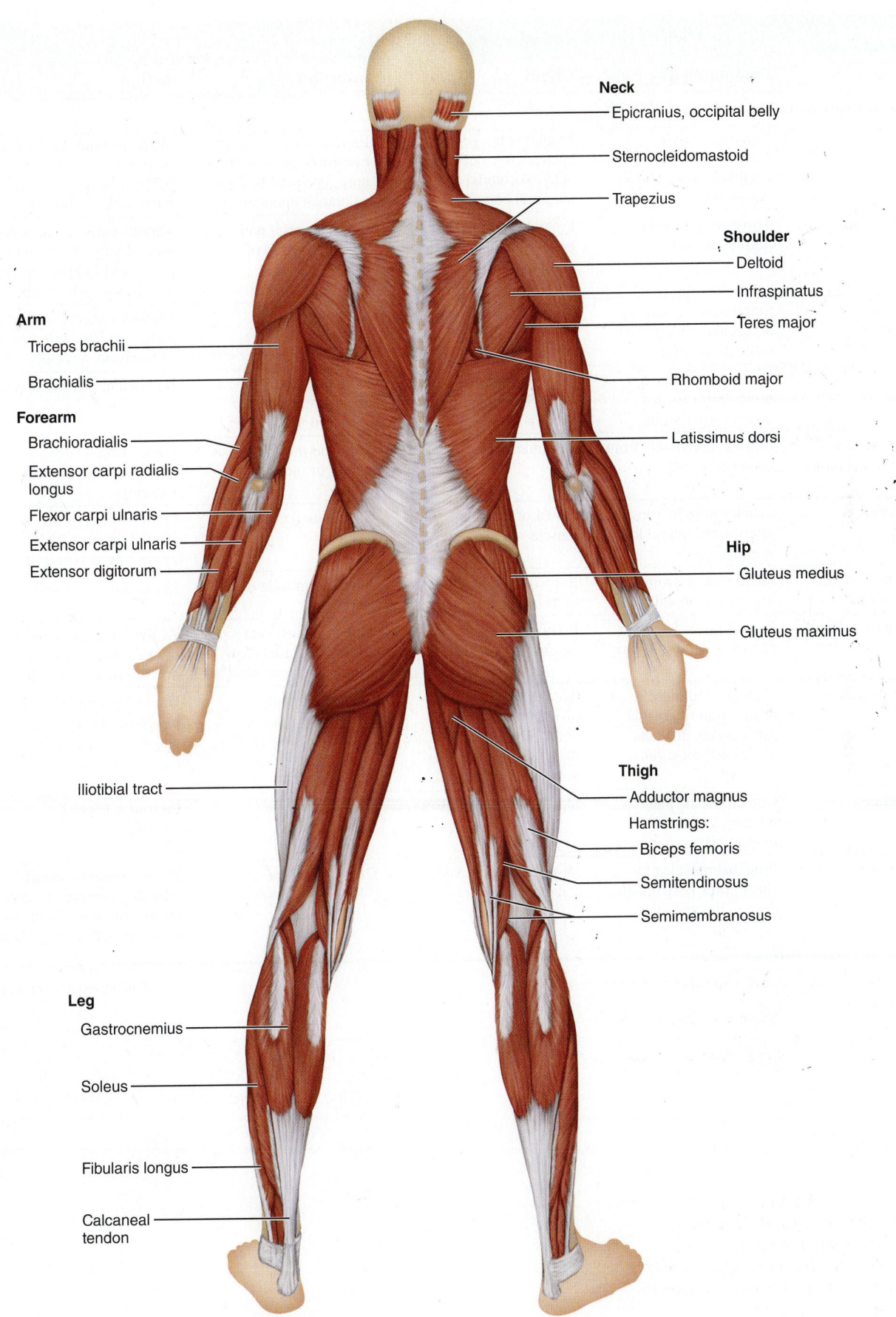

Neck
— Epicranius, occipital belly
— Sternocleidomastoid
— Trapezius

Shoulder
— Deltoid
— Infraspinatus
— Teres major
— Rhomboid major
— Latissimus dorsi

Arm
Triceps brachii
Brachialis

Forearm
Brachioradialis
Extensor carpi radialis longus
Flexor carpi ulnaris
Extensor carpi ulnaris
Extensor digitorum

Hip
Gluteus medius
Gluteus maximus

Thigh
Adductor magnus
Hamstrings:
Biceps femoris
Semitendinosus
Semimembranosus

Iliotibial tract

Leg
Gastrocnemius
Soleus
Fibularis longus
Calcaneal tendon

13

Figure 13.3 Posterior view of superficial muscles of the body.

Table 13.1   Major Muscles of Human Head (see Figure 13.4)

Muscle	Comments	Origin	Insertion	Action
Facial Expression (Figure 13.4a)				
Epicranius— frontal and occipital bellies	Bipartite muscle consisting of frontal and occipital parts, which covers dome of skull	Frontal belly—epicranial aponeurosis; occipital belly—occipital and temporal bones	Frontal belly—skin of eyebrows and root of nose; occipital belly— epicranial aponeurosis	With aponeurosis fixed, frontal belly raises eyebrows; occipital belly fixes aponeurosis and pulls scalp posteriorly
Orbicularis oculi	Tripartite sphincter muscle of eyelids	Frontal and maxillary bones and ligaments around orbit	Encircles orbit and inserts in tissue of eyelid	Various parts can be activated individually; closes eyes, produces blinking, squinting, and draws eyebrows inferiorly
Corrugator supercilii	Small muscle; activity associated with that of orbicularis oculi	Arch of frontal bone above nasal bone	Skin of eyebrow	Draws eyebrows medially and inferiorly; wrinkles skin of forehead vertically
Levator labii superioris	Thin muscle between orbicularis oris and inferior eye margin	Zygomatic bone and infraorbital margin of maxilla	Skin and muscle of upper lip and border of nostril	Raises and furrows upper lip; opens lips
Zygomaticus— major and minor	Extends diagonally from corner of mouth to cheekbone	Zygomatic bone	Skin and muscle at corner of mouth	Raises lateral corners of mouth upward (smiling muscle)
Risorius	Slender muscle; runs inferior and lateral to zygomaticus	Fascia of masseter muscle	Skin at angle of mouth	Draws corner of lip laterally; tenses lip; zygomaticus synergist
Depressor labii inferioris	Small muscle from lower lip to mandible	Body of mandible lateral to its midline	Skin and muscle of lower lip	Draws lower lip inferiorly
Depressor anguli oris	Small muscle lateral to depressor labii inferioris	Body of mandible below incisors	Skin and muscle at angle of mouth below insertion of zygomaticus	Zygomaticus antagonist; draws corners of mouth downward and laterally
Orbicularis oris	Multilayered muscle of lips with fibers that run in many different directions; most run circularly	Arises indirectly from maxilla and mandible; fibers blended with fibers of other muscles associated with lips	Encircles mouth; inserts into muscle and skin at angles of mouth	Closes lips; purses and protrudes lips (kissing and whistling muscle)
Mentalis	One of muscle pair forming V-shaped muscle mass on chin	Mandible below incisors	Skin of chin	Protrudes lower lip; wrinkles chin
Buccinator	Principal muscle of cheek; runs horizontally, deep to the masseter	Molar region of maxilla and mandible	Orbicularis oris	Draws corner of mouth laterally; compresses cheek (as in whistling); holds food between teeth during chewing

(Table continues on page 202.)

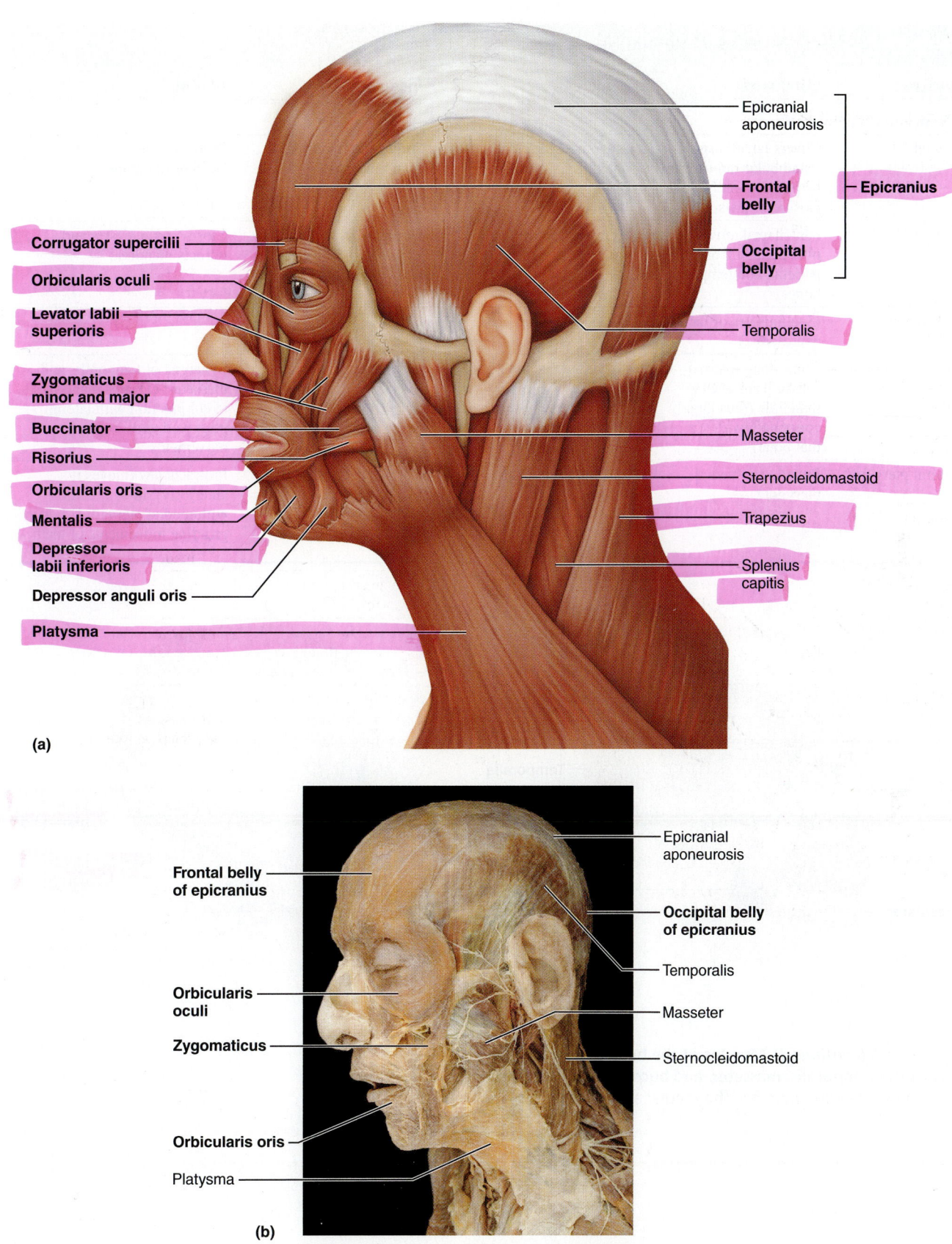

Epicranial
aponeurosis

**Frontal
belly**

**Occipital
belly**

Epicranius

Temporalis

Corrugator supercilii

Orbicularis oculi

**Levator labii
superioris**

**Zygomaticus
minor and major**

Buccinator

Risorius

Orbicularis oris

Mentalis

**Depressor
labii inferioris**

Depressor anguli oris

Platysma

Masseter

Sternocleidomastoid

Trapezius

Splenius
capitis

(a)

13

Epicranial
aponeurosis

**Frontal belly
of epicranius**

**Occipital belly
of epicranius**

Temporalis

Masseter

Sternocleidomastoid

**Orbicularis
oculi**

Zygomaticus

Orbicularis oris

Platysma

(b)

Figure 13.4 Muscles of the head (left lateral view). (a) Superficial muscles.
(b) Photo of superficial structures of head and neck.

Table 13.1   Major Muscles of Human Head (continued)

Muscle	Comments	Origin	Insertion	Action
Mastication (Figure 13.4c, d)				
Masseter	Covers lateral aspect of mandibular ramus; can be palpated on forcible closure of jaws	Zygomatic arch and maxilla	Angle and ramus of mandible	Prime mover of jaw closure; elevates mandible
Temporalis	Fan-shaped muscle lying over parts of frontal, parietal, and temporal bones	Temporal fossa	Coronoid process of mandible	Closes jaw; elevates and retracts mandible
Buccinator	(See muscles of facial expression.)			
Medial pterygoid	Runs along internal (medial) surface of mandible (thus largely concealed by that bone)	Sphenoid, palatine, and maxillary bones	Medial surface of mandible, near its angle	Synergist of temporalis and masseter; elevates mandible; in conjunction with lateral pterygoid, aids in grinding movements
Lateral pterygoid	Superior to medial pterygoid	Greater wing of sphenoid bone	Condylar process of mandible	Protracts jaw (moves it anteriorly); in conjunction with medial pterygoid, aids in grinding movements of teeth

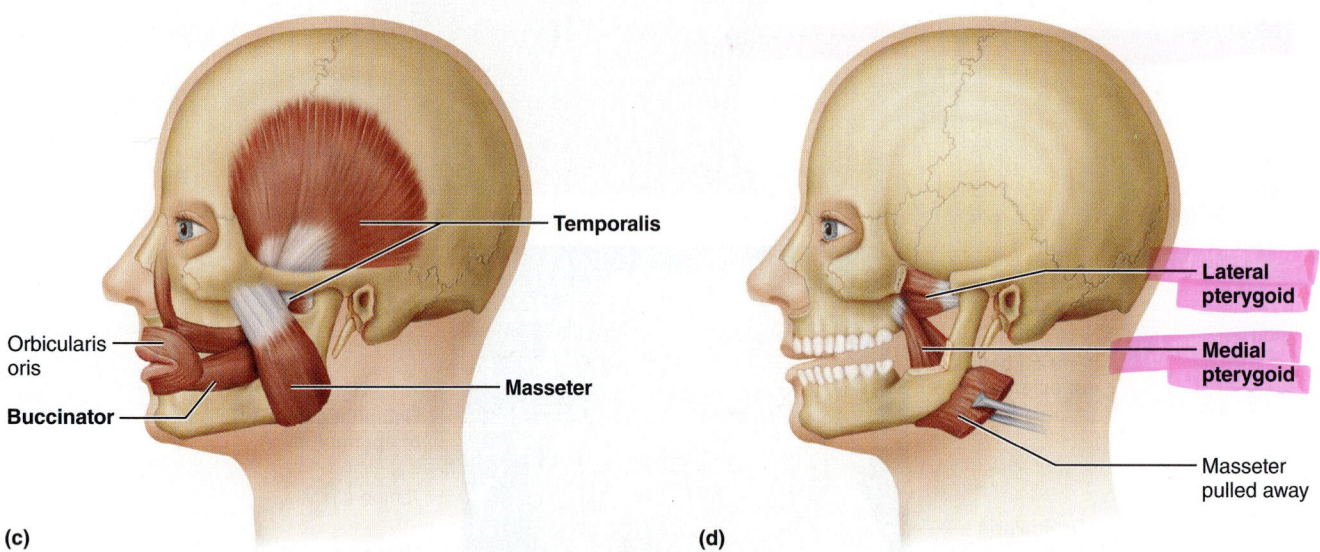

(c)

(d)

Figure 13.4 *(continued)* **Muscles of the head: mastication. (c)** Lateral view of the temporalis, masseter, and buccinator muscles. **(d)** Lateral view of the deep chewing muscles, the medial and lateral pterygoid muscles.

Table 13.2	**Anterolateral Muscles of Human Neck (see Figure 13.5)**			
Muscle	**Comments**	**Origin**	**Insertion**	**Action**
Superficial				
Platysma (see Figure 13.4a)	Unpaired muscle: thin, sheetlike superficial neck muscle, not strictly a head muscle but plays role in facial expression	Fascia of chest (over pectoral muscles) and deltoid	Lower margin of mandible, skin, and muscle at corner of mouth	Tenses skin of neck; depresses mandible; pulls lower lip back and down (i.e., produces downward sag of the mouth)
Sternocleidomastoid	Two-headed muscle located deep to platysma on anterolateral surface of neck; fleshy parts on either side indicate limits of anterior and posterior triangles of neck	Manubrium of sternum and medial portion of clavicle	Mastoid process of temporal bone and superior nuchal line of occipital bone	Simultaneous contraction of both muscles of pair causes flexion of neck forward, generally against resistance (as when lying on the back); acting independently, rotate head toward shoulder on opposite side
Scalenes—anterior, middle, and posterior (see Figure 13.5c)	Located more on lateral than anterior neck; deep to platysma and sternocleidomastoid	Transverse processes of cervical vertebrae	Anterolaterally on ribs 1–2	Flex and slightly rotate neck; elevate ribs 1–2 (aid in inspiration)

(Table continues on page 204.)

13

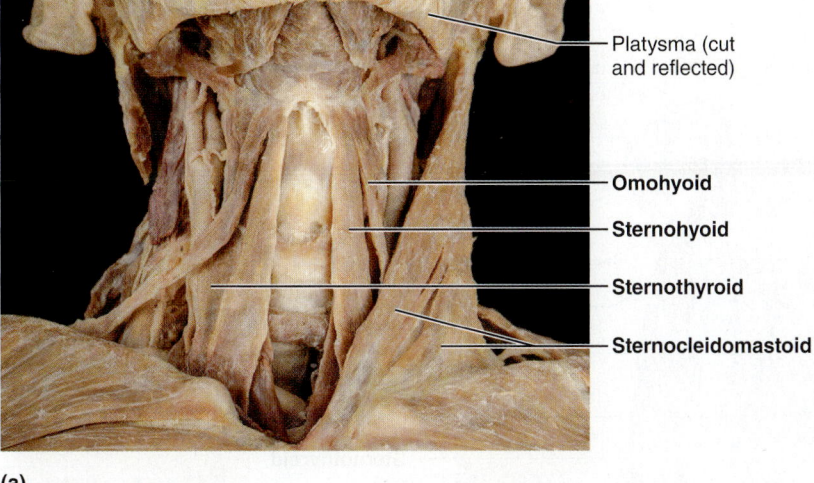

Platysma (cut and reflected)

Omohyoid

Sternohyoid

Sternothyroid

Sternocleidomastoid

(a)

Figure 13.5 Muscles of the anterolateral neck and throat. (a) Photo of the anterior and lateral regions of the neck.

Table 13.2	Anterolateral Muscles of Human Neck *(continued)*			
Muscle	**Comments**	**Origin**	**Insertion**	**Action**
Deep (Figure 13.5a, b)				
Digastric	Consists of two bellies united by an intermediate tendon; assumes a V-shaped configuration under chin	Lower margin of mandible (anterior belly) and mastoid process (posterior belly)	By a connective tissue loop to hyoid bone	Acting in concert, elevate hyoid bone; open mouth and depress mandible
Stylohyoid	Slender muscle parallels posterior border of digastric; below angle of jaw	Styloid process of temporal	Hyoid bone	Elevates and retracts hyoid bone
Mylohyoid	Just deep to digastric; forms floor of mouth	Medial surface of mandible	Hyoid bone and median raphe	Elevates hyoid bone and base of tongue during swallowing
Sternohyoid	Runs most medially along neck; straplike	Manubrium and medial end of clavicle	Lower margin of body of hyoid bone	Acting with sternothyroid and omohyoid, depresses larynx and hyoid bone if mandible is fixed; may also flex skull
Sternothyroid	Lateral and deep to sternohyoid	Posterior surface of manubrium	Thyroid cartilage of larynx	(See Sternohyoid above)
Omohyoid	Straplike with two bellies; lateral to sternohyoid	Superior surface of scapula	Hyoid bone; inferior border	(See Sternohyoid above)
Thyrohyoid	Appears as a superior continuation of sternothyroid muscle	Thyroid cartilage	Hyoid bone	Depresses hyoid bone; elevates larynx if hyoid is fixed

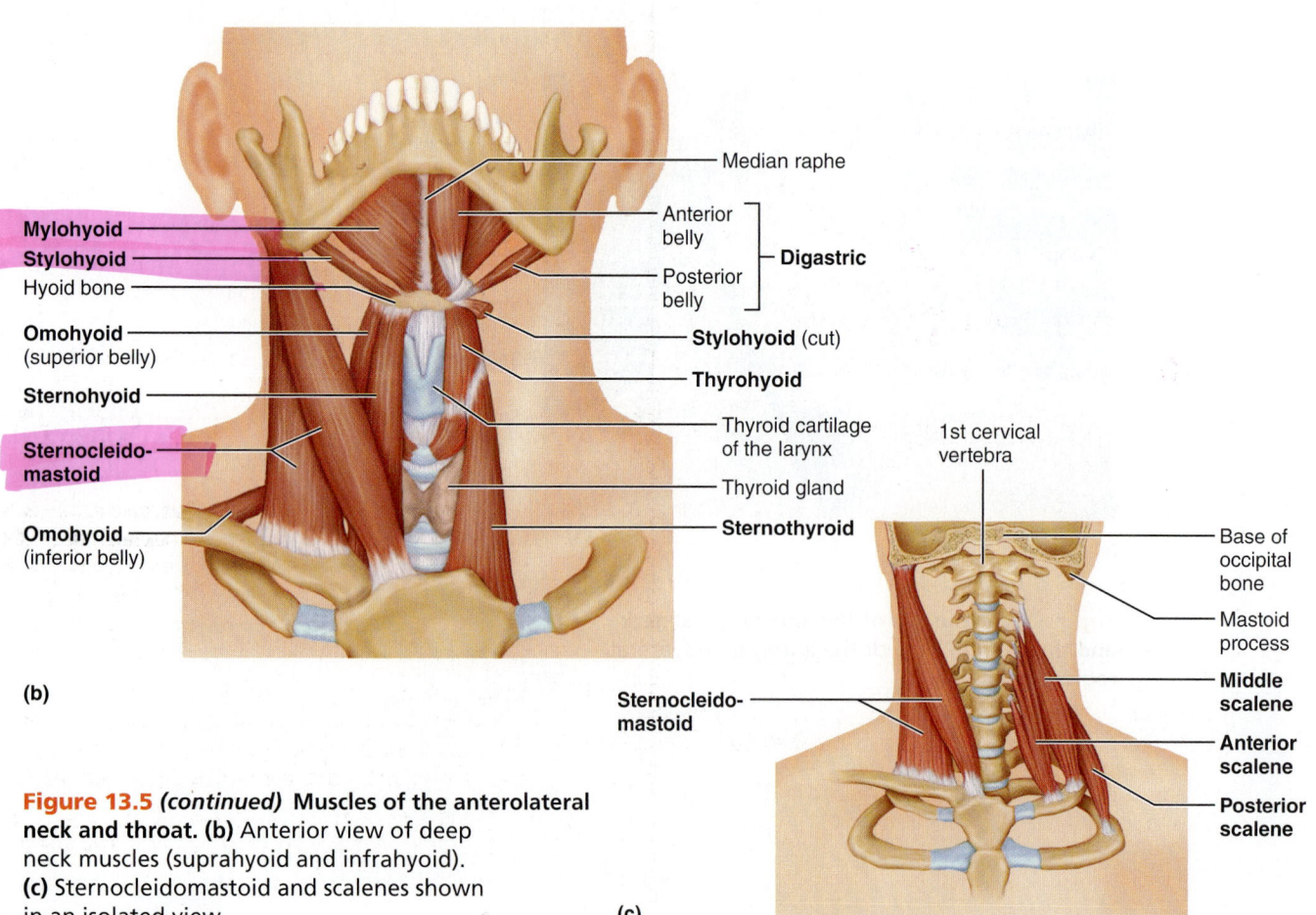

Figure 13.5 *(continued)* **Muscles of the anterolateral neck and throat. (b)** Anterior view of deep neck muscles (suprahyoid and infrahyoid). **(c)** Sternocleidomastoid and scalenes shown in an isolated view.

Table 13.3	Anterior Muscles of Human Thorax, Shoulder, and Abdominal Wall (see Figures 13.6, 13.7, and 13.8)			
Muscle	**Comments**	**Origin**	**Insertion**	**Action**
Thorax and Shoulder, Superficial (Figure 13.6)				
Pectoralis major	Large fan-shaped muscle covering upper portion of chest	Clavicle, sternum, cartilage of ribs 1–6 (or 7), and aponeurosis of external oblique muscle	Fibers converge to insert by short tendon into intertubercular sulcus of humerus	Prime mover of arm flexion; adducts, medially rotates arm; with arm fixed, pulls chest upward (thus also acts in forced inspiration)
Serratus anterior	Fan-shaped muscle deep to scapula; beneath and inferior to pectoral muscles on lateral rib cage	Lateral aspect of ribs 1–8 (or 9)	Vertebral border of anterior surface of scapula	Prime mover to protract and hold scapula against chest wall; rotates scapula, causing inferior angle to move laterally and upward; essential to raising arm; fixes scapula for arm abduction
Deltoid (see also Figure 13.9)	Fleshy triangular muscle forming shoulder muscle mass; intramuscular injection site	Lateral ⅓ of clavicle; acromion and spine of scapula	Deltoid tuberosity of humerus	Acting as a whole, prime mover of arm abduction; when only specific fibers are active, can aid in flexion, extension, and rotation of humerus

(Table continues on page 206.)

13

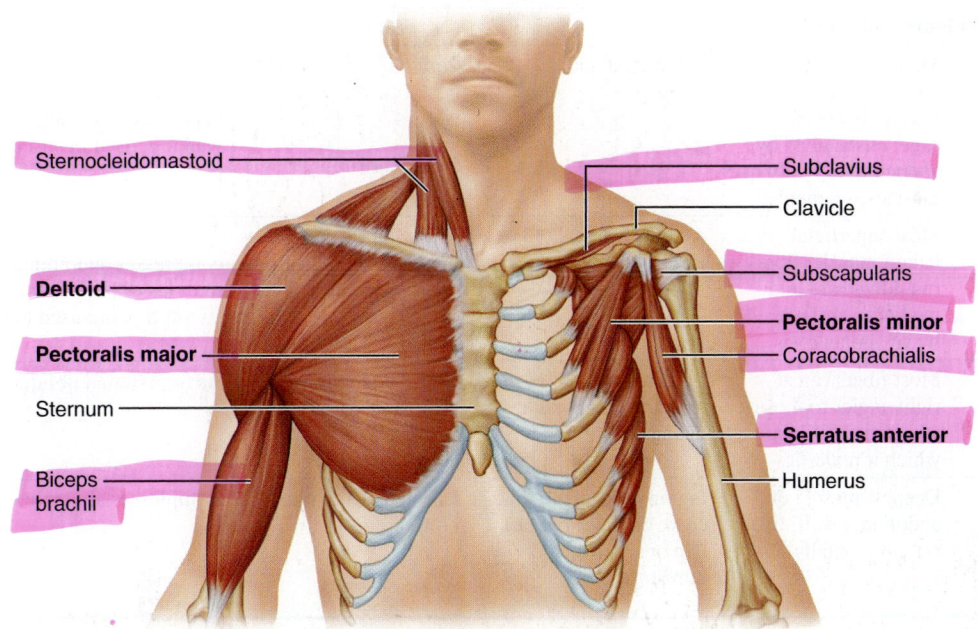

Figure 13.6 **Muscles of the thorax and shoulder acting on the scapula and arm (anterior view).** The superficial muscles, which effect arm movements, are shown on the left side of the figure. These muscles have been removed on the right side of the figure to show the muscles that stabilize or move the pectoral girdle.

Table 13.3	Anterior Muscles of Human Thorax, Shoulder, and Abdominal Wall *(continued)*			
Muscle	**Comments**	**Origin**	**Insertion**	**Action**
Thorax and Shoulder, Superficial *(continued)*				
Pectoralis minor	Flat, thin muscle directly beneath and obscured by pectoralis major	Anterior surface of ribs 3–5, near their costal cartilages	Coracoid process of scapula	With ribs fixed, draws scapula forward and inferiorly; with scapula fixed, draws rib cage superiorly
Thorax, Deep: Muscles of Respiration (Figure 13.7)				
External intercostals	11 pairs lie between ribs; fibers run obliquely downward and forward toward sternum	Inferior border of rib above (not shown in figure)	Superior border of rib below	Pull ribs toward one another to elevate rib cage; aid in inspiration
Internal intercostals	11 pairs lie between ribs; fibers run deep and at right angles to those of external intercostals	Superior border of rib below	Inferior border of rib above (not shown in figure)	Draw ribs together to depress rib cage; aid in forced expiration; antagonistic to external intercostals
Diaphragm	Broad muscle; forms floor of thoracic cavity; dome-shaped in relaxed state; fibers converge from margins of thoracic cage toward a central tendon	Inferior border of rib and sternum, costal cartilages of last six ribs and lumbar vertebrae	Central tendon	Prime mover of inspiration flattens on contraction, increasing vertical dimensions of thorax; increases intra-abdominal pressure
Abdominal Wall (Figure 13.8a and b)				
Rectus abdominis	Medial superficial muscle, extends from pubis to rib cage; ensheathed by aponeuroses of oblique muscles; segmented	Pubic crest and symphysis	Xiphoid process and costal cartilages of ribs 5–7	Flexes and rotates vertebral column; increases abdominal pressure; fixes and depresses ribs; stabilizes pelvis during walking; used in sit-ups and curls
External oblique	Most superficial lateral muscle; fibers run downward and medially; ensheathed by an aponeurosis	Anterior surface of last eight ribs	Linea alba,* pubic crest and tubercles, and iliac crest	See rectus abdominis, above; compresses abdominal wall; also aids muscles of back in trunk rotation and lateral flexion; used in oblique curls
Internal oblique	Most fibers run at right angles to those of external oblique, which it underlies	Lumbar fascia, iliac crest, and inguinal ligament	Linea alba, pubic crest, and costal cartilages of last three ribs	As for external oblique
Transversus abdominis	Deepest muscle of abdominal wall; fibers run horizontally	Inguinal ligament, iliac crest, cartilages of last five or six ribs, and lumbar fascia	Linea alba and pubic crest	Compresses abdominal contents

*The linea alba (white line) is a narrow, tendinous sheath that runs along the middle of the abdomen from the sternum to the pubic symphysis. It is formed by the fusion of the aponeurosis of the external oblique and transversus muscles.

Figure 13.7 Deep muscles of the thorax: muscles of respiration. (a) The external intercostals (inspiratory muscles) are shown on the left and the internal intercostals (expiratory muscles) are shown on the right. These two muscle layers run obliquely and at right angles to each other. **(b)** Inferior view of the diaphragm, the prime mover of inspiration. Notice that its muscle fibers converge toward a central tendon, an arrangement that causes the diaphragm to flatten and move inferiorly as it contracts. The diaphragm and its tendon are pierced by the great vessels (aorta and inferior vena cava) and the esophagus.

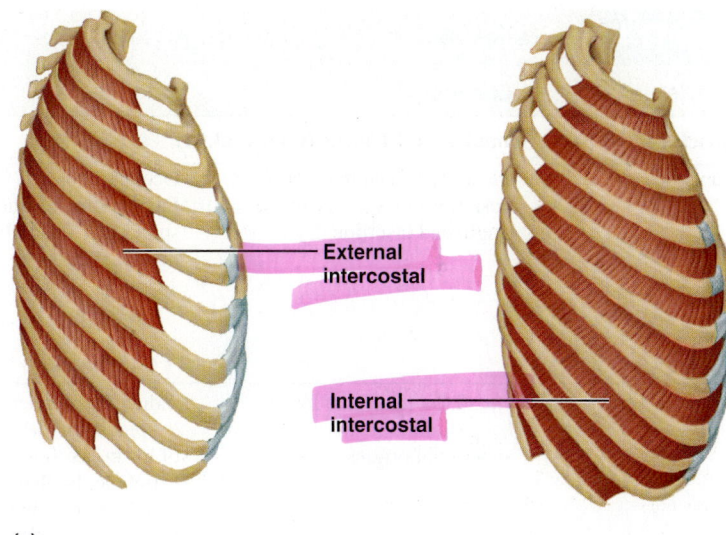

External intercostal

Internal intercostal

(a)

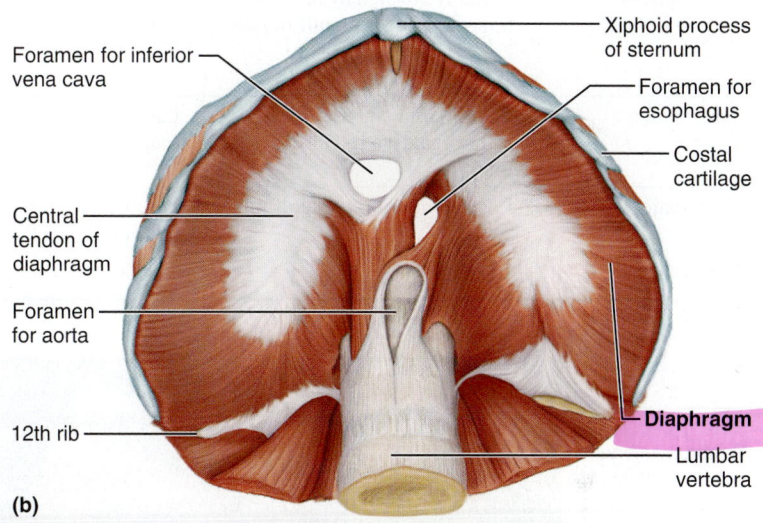

Foramen for inferior vena cava

Xiphoid process of sternum

Foramen for esophagus

Costal cartilage

Central tendon of diaphragm

Foramen for aorta

12th rib

Diaphragm

Lumbar vertebra

(b)

13

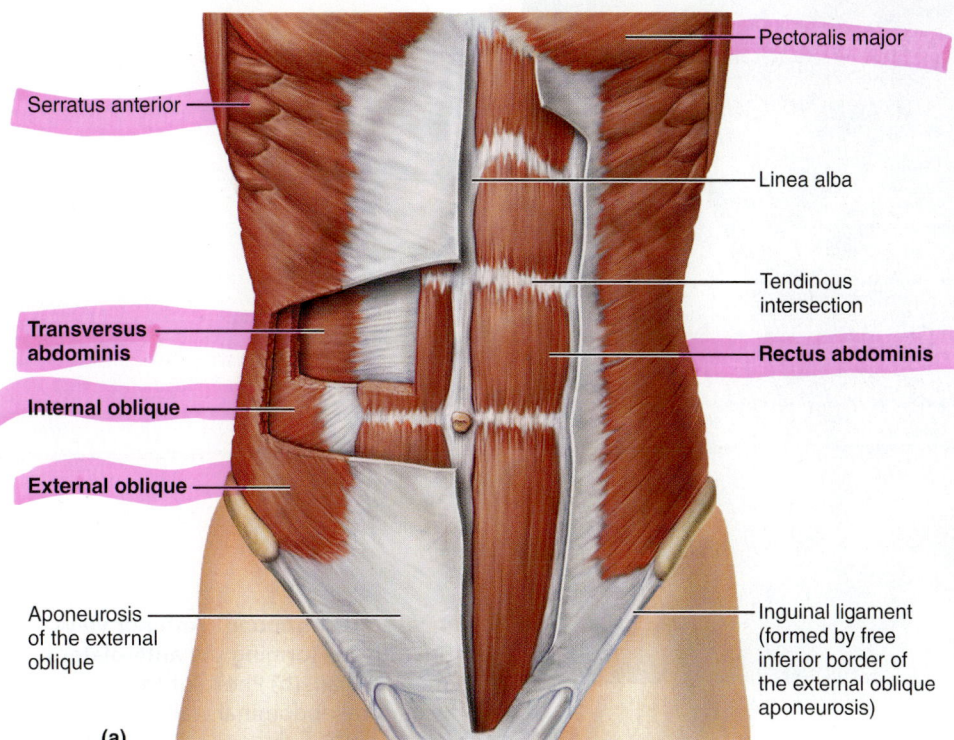

Pectoralis major

Serratus anterior

Linea alba

Tendinous intersection

Transversus abdominis

Rectus abdominis

Internal oblique

External oblique

Aponeurosis of the external oblique

Inguinal ligament (formed by free inferior border of the external oblique aponeurosis)

(a)

Figure 13.8 Anterior view of the muscles forming the anterolateral abdominal wall. (a) The superficial muscles have been partially cut away on the left side of the diagram to reveal the deeper internal oblique and transversus abdominis muscles.

Table 13.4	Posterior Muscles of Human Trunk (see Figure 13.9)			
Muscle	**Comments**	**Origin**	**Insertion**	**Action**
Muscles of the Neck, Shoulder, and Thorax (Figure 13.9a)				
Trapezius	Most superficial muscle of posterior thorax; very broad origin and insertion	Occipital bone; ligamentum nuchae; spines of C_7 and all thoracic vertebrae	Acromion and spinous process of scapula; lateral third of clavicle	Extends head; raises, rotates, and retracts (adducts) scapula and stabilizes it; superior fibers elevate scapula (as in shrugging the shoulders); inferior fibers depress it
Latissimus dorsi	Broad flat muscle of lower back (lumbar region); extensive superficial origins	Indirect attachment to spinous processes of lower six thoracic vertebrae, lumbar vertebrae, last three to four ribs, and iliac crest	Floor of intertubercular sulcus of humerus	Prime mover of arm extension; adducts and medially rotates arm; brings arm down in power stroke, as in striking a blow
Infraspinatus	Partially covered by deltoid and trapezius; a rotator cuff muscle	Infraspinous fossa of scapula	Greater tubercle of humerus	Lateral rotation of humerus; helps hold head of humerus in glenoid cavity; stabilizes shoulder
Teres minor	Small muscle inferior to infraspinatus; a rotator cuff muscle	Lateral margin of scapula	Greater tubercle of humerus	As for infraspinatus
Teres major	Located inferiorly to teres minor	Posterior surface at inferior angle of scapula	Intertubercular sulcus of humerus	Extends, medially rotates, and adducts humerus; synergist of latissimus dorsi

13

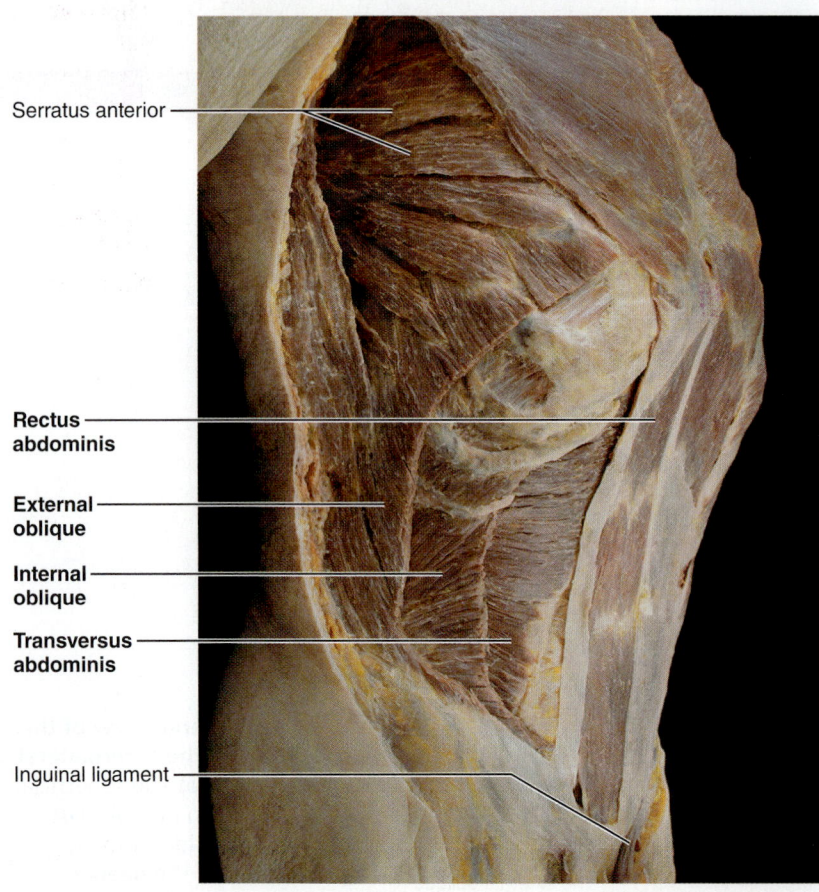

Serratus anterior

Rectus abdominis

External oblique

Internal oblique

Transversus abdominis

Inguinal ligament

(b)

Figure 13.8 *(continued)* **Anterior view of the muscles forming the anterolateral abdominal wall. (b)** Photo of the anterolateral abdominal wall.

Table 13.4	*(continued)*			
Muscle	**Comments**	**Origin**	**Insertion**	**Action**
Supraspinatus	Obscured by trapezius; a rotator cuff muscle	Supraspinous fossa of scapula	Greater tubercle of humerus	Initiates abduction of humerus; stabilizes shoulder joint
Levator scapulae	Located at back and side of neck, deep to trapezius	Transverse processes of C_1–C_4	Medial border of scapula superior to spine	Elevates and adducts scapula; with fixed scapula, laterally flexes neck to the same side
Rhomboids—major and minor	Beneath trapezius and inferior to levator scapulae; rhomboid minor is the more superior muscle	Spinous processes of C_7 and T_1–T_5	Medial border of scapula	Pull scapula medially (retraction); stabilize scapula; rotate glenoid cavity downward
Muscles Associated with the Vertebral Column (Figure 13.9b)				
Semispinalis	Deep composite muscle of the back—thoracis, cervicis, and capitis portions	Transverse processes of C_7–T_{12}	Occipital bone and spinous processes of cervical vertebrae and T_1–T_4	Acting together, extend head and vertebral column; acting independently (right vs. left) causes rotation toward the opposite side

13

(Table continues on page 210.)

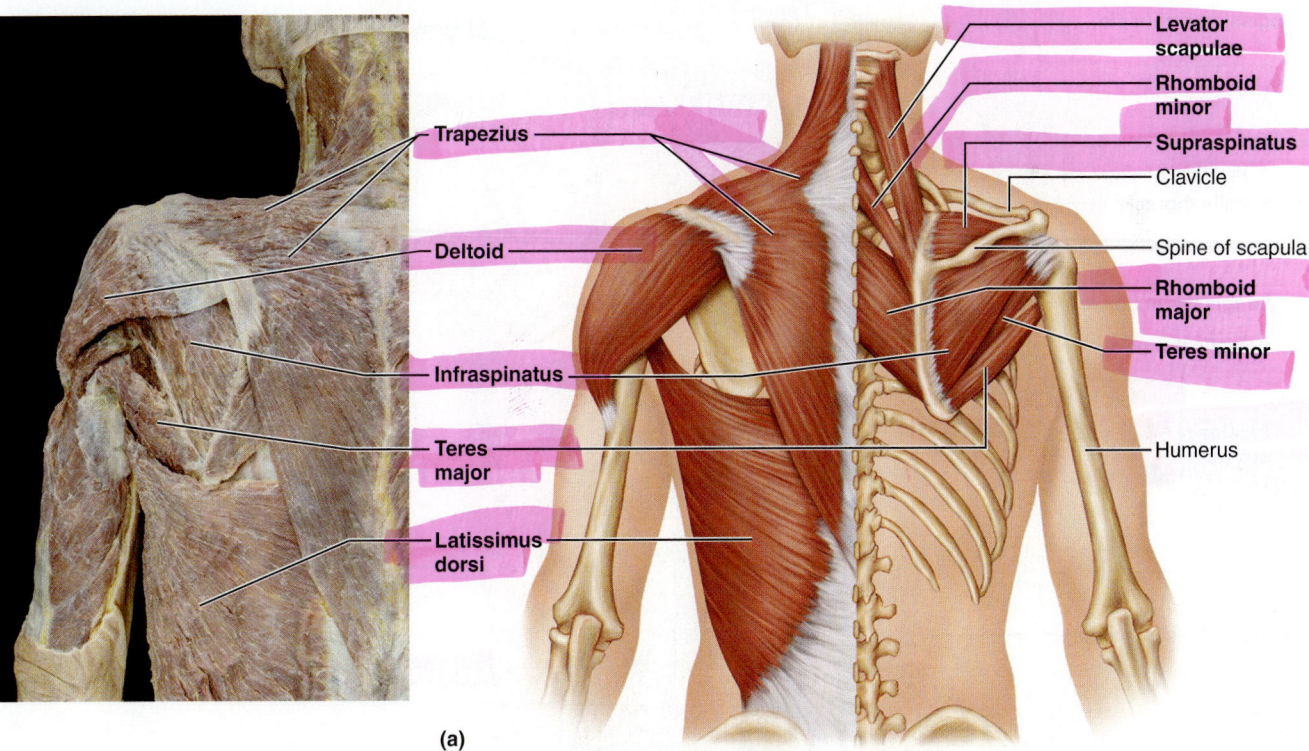

(a)

Figure 13.9 Muscles of the neck, shoulder, and thorax (posterior view). (a) The superficial muscles of the back are shown for the left side of the body, with a corresponding photograph. The superficial muscles are removed on the right side of the illustration to reveal the deeper muscles acting on the scapula and the rotator cuff muscles that help to stabilize the shoulder joint.

Table 13.4	Posterior Muscles of Human Trunk *(continued)*			
Muscle	**Comments**	**Origin**	**Insertion**	**Action**
Muscles Associated with the Vertebral Column *(continued)*				
Erector spinae	A long tripartite muscle composed of iliocostalis (lateral), longissimus, and spinalis (medial) muscle columns; superficial to semispinalis muscles; extends from pelvis to head	Iliac crest, transverse processes of lumbar, thoracic, and cervical vertebrae, and/or ribs 3–6 depending on specific part	Ribs and transverse processes of vertebrae about six segments above origin; longissimus also inserts into mastoid process	Extend and bend the vertebral column laterally; fibers of the longissimus also extend head

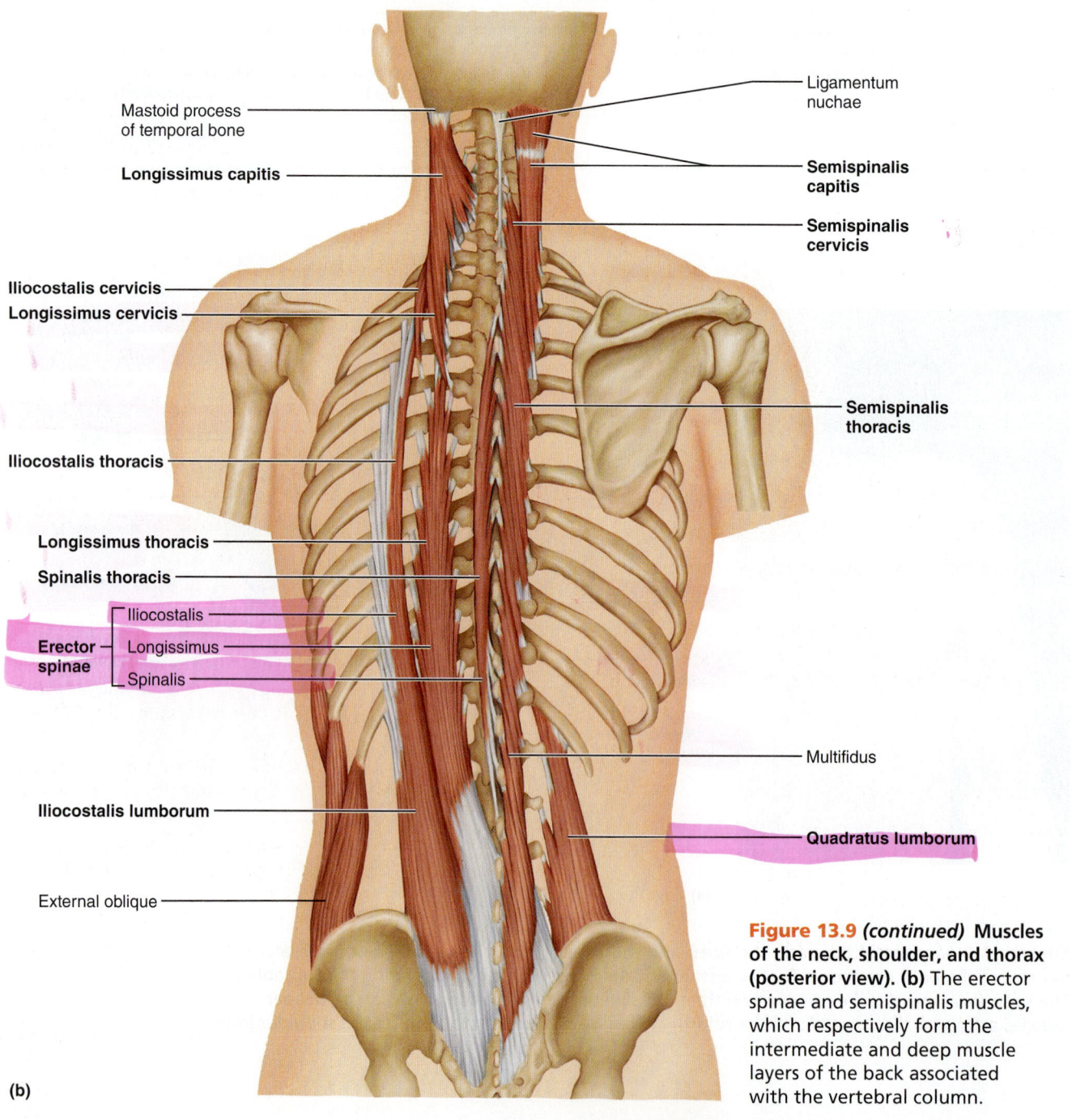

(b)

Figure 13.9 *(continued)* **Muscles of the neck, shoulder, and thorax (posterior view). (b)** The erector spinae and semispinalis muscles, which respectively form the intermediate and deep muscle layers of the back associated with the vertebral column.

Table 13.4	(continued)			
Muscle	**Comments**	**Origin**	**Insertion**	**Action**
Splenius (see Figure 13.9c)	Superficial muscle (capitis and cervicis parts) extending from upper thoracic region to skull	Ligamentum nuchae and spinous processes of C_7–T_6	Mastoid process, occipital bone, and transverse processes of C_2–C_4	As a group, extend or hyperextend head; when only one side is active, head is rotated and bent toward the same side
Quadratus lumborum	Forms greater portion of posterior abdominal wall	Iliac crest and lumbar fascia	Inferior border of rib 12; transverse processes of lumbar vertebrae	Each flexes vertebral column laterally; together extend the lumbar spine and fix rib 12; maintains upright posture

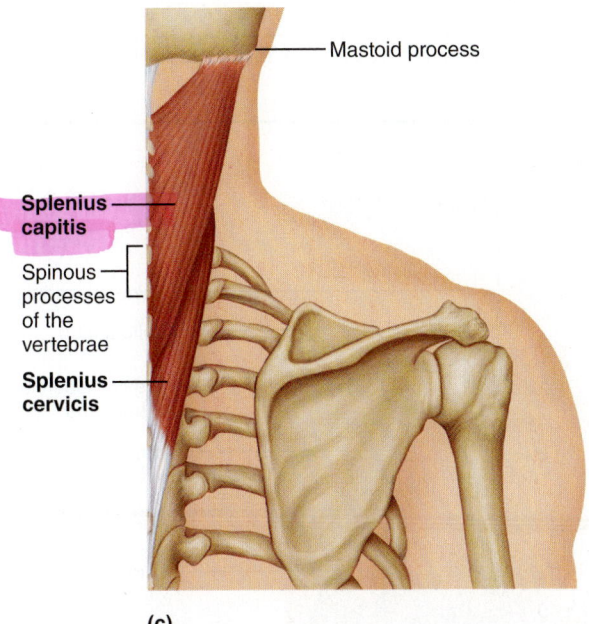

Mastoid process

Splenius capitis

Spinous processes of the vertebrae

Splenius cervicis

(c)

Figure 13.9 *(continued)* **Muscles of the neck, shoulder, and thorax (posterior view). (c)** Deep (splenius) muscles of the posterior neck. Superficial muscles have been removed.

Muscles of the Upper Limb

The muscles that act on the upper limb fall into three groups: those that move the arm, those causing movement at the elbow, and those moving the wrist and hand.

The muscles that cross the shoulder joint to insert on the humerus and move the arm (subscapularis, supraspinatus and infraspinatus, deltoid, and so on) are primarily trunk muscles that originate on the axial skeleton or shoulder girdle. These muscles are included with the trunk muscles.

The second group of muscles, which cross the elbow joint and move the forearm, consists of muscles forming the musculature of the humerus. These muscles arise mainly from the humerus and insert in forearm bones. They are responsible for flexion, extension, pronation, and supination.

The third group forms the musculature of the forearm. For the most part, these muscles insert on the digits and produce movements at the wrist and fingers.

ACTIVITY 3

Identifying Muscles of the Upper Limb

Study the origins, insertions, and actions of muscles that move the forearm and identify them in the figure (**Table 13.5** and **Figure 13.10**).

Do the same for muscles acting on the wrist and hand (**Table 13.6** and **Figure 13.11**). They are more easily identified if their insertion tendons are located first.

Then see if you can identify the upper limb muscles on a torso model, anatomical chart, or cadaver. Complete this portion of the exercise with palpation demonstrations as outlined next.

Demonstrating Operations of Upper Limb Muscles

1. To observe the *biceps brachii,* attempt to flex your forearm (hand supinated) against resistance. The insertion tendon of this biceps muscle can also be felt in the lateral aspect of the antecubital fossa (where it runs toward the radius to attach).

2. If you acutely flex your elbow and then try to extend it against resistance, you can demonstrate the action of your *triceps brachii.*

3. Strongly flex your wrist and make a fist. Palpate your contracting wrist flexor muscles (which originate from the medial epicondyle of the humerus) and their insertion tendons, which can be easily felt at the anterior aspect of the wrist.

4. Flare your fingers to identify the tendons of the *extensor digitorum* muscle on the dorsum of your hand. ■

(Text continues on page 216.)

13

Table 13.5	Muscles of Human Humerus That Act on the Forearm (see Figure 13.10)			
Muscle	**Comments**	**Origin**	**Insertion**	**Action**
Triceps brachii	Sole, large fleshy muscle of posterior humerus; three-headed origin	Long head—inferior margin of glenoid cavity; lateral head—posterior humerus; medial head—distal radial groove on posterior humerus	Olecranon of ulna	Powerful forearm extensor; antagonist of forearm flexors (brachialis and biceps brachii)
Anconeus	Short triangular muscle blended with triceps	Lateral epicondyle of humerus	Lateral aspect of olecranon of ulna	Abducts ulna during forearm pronation; extends elbow
Biceps brachii	Most familiar muscle of anterior humerus because this two-headed muscle bulges when forearm is flexed	Short head: coracoid process; long head: supraglenoid tubercle and lip of glenoid cavity; tendon of long head runs in intertubercular sulcus and within capsule of shoulder joint	Radial tuberosity	Flexion (powerful) of elbow and supination of forearm; "it turns the corkscrew and pulls the cork"; weak arm flexor
Brachioradialis	Superficial muscle of lateral forearm; forms lateral boundary of antecubital fossa	Lateral ridge at distal end of humerus	Base of radial styloid process	Synergist in forearm flexion
Brachialis	Immediately deep to biceps brachii	Distal portion of anterior humerus	Coronoid process of ulna	A major flexor of forearm

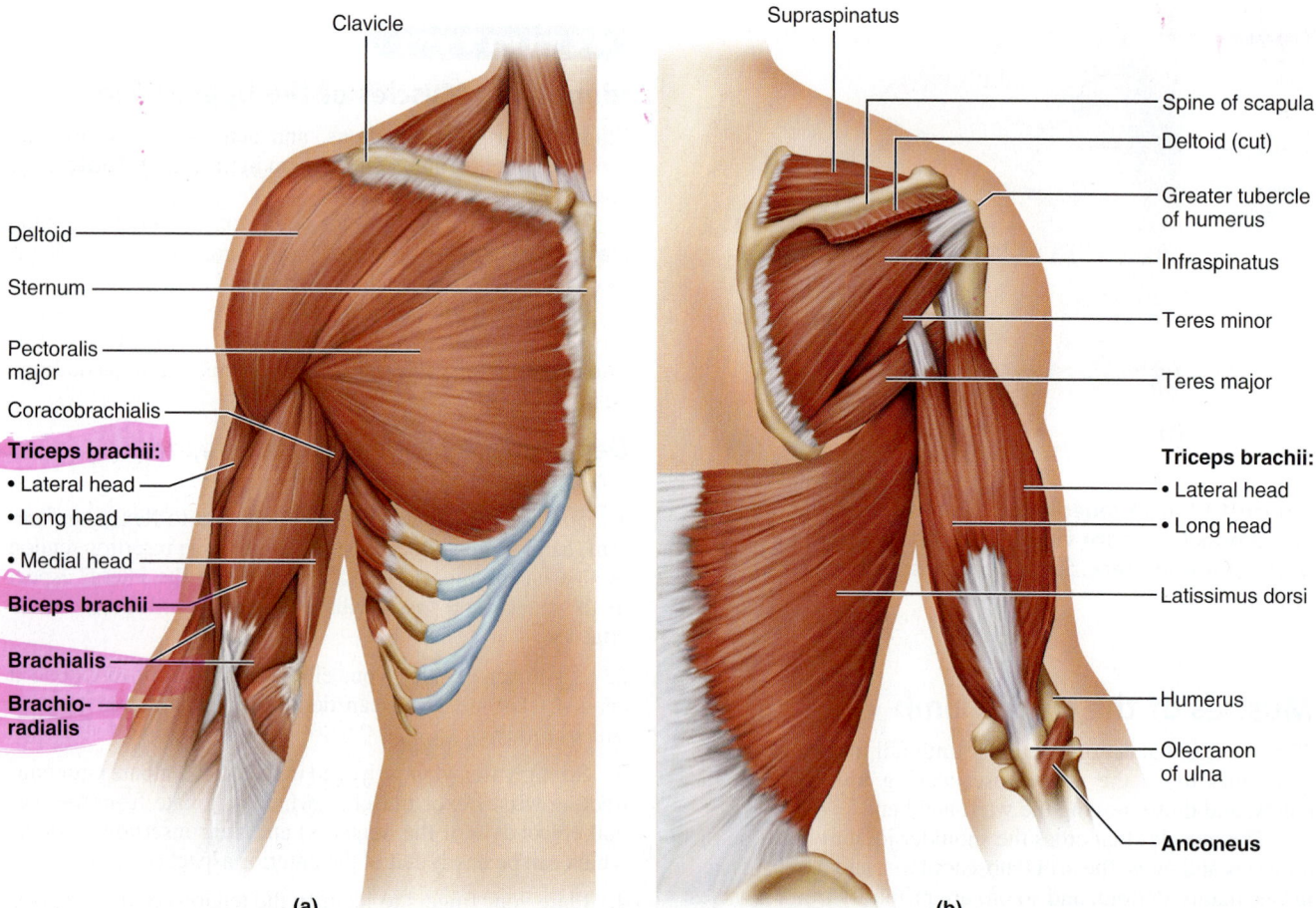

(a)

(b)

Figure 13.10 Muscles causing movements of the arm and forearm. (a) Superficial muscles of the anterior thorax, shoulder, and arm, anterior view. **(b)** Posterior aspect of the arm showing the lateral and long heads of the triceps brachii muscle.

Table 13.6	Muscles of Human Forearm That Act on Hand and Fingers (see Figure 13.11)				
Muscle	**Comments**	**Origin**	**Insertion**	**Action**
Anterior Compartment, Superficial (Figure 13.11a, b, c)				
Pronator teres	Seen in a superficial view between proximal margins of brachioradialis and flexor carpi radialis	Medial epicondyle of humerus and coronoid process of ulna	Midshaft of radius	Acts synergistically with pronator quadratus to pronate forearm; weak elbow flexor
Flexor carpi radialis	Superficial; runs diagonally across forearm	Medial epicondyle of humerus	Base of metacarpals II and III	Powerful flexor of wrist; abducts hand
Palmaris longus	Small fleshy muscle with a long tendon; medial to flexor carpi radialis	Medial epicondyle of humerus	Palmar aponeurosis; skin and fascia of palm	Flexes wrist (weak); tenses skin and fascia of palm

(Table continues on page 214.)

13

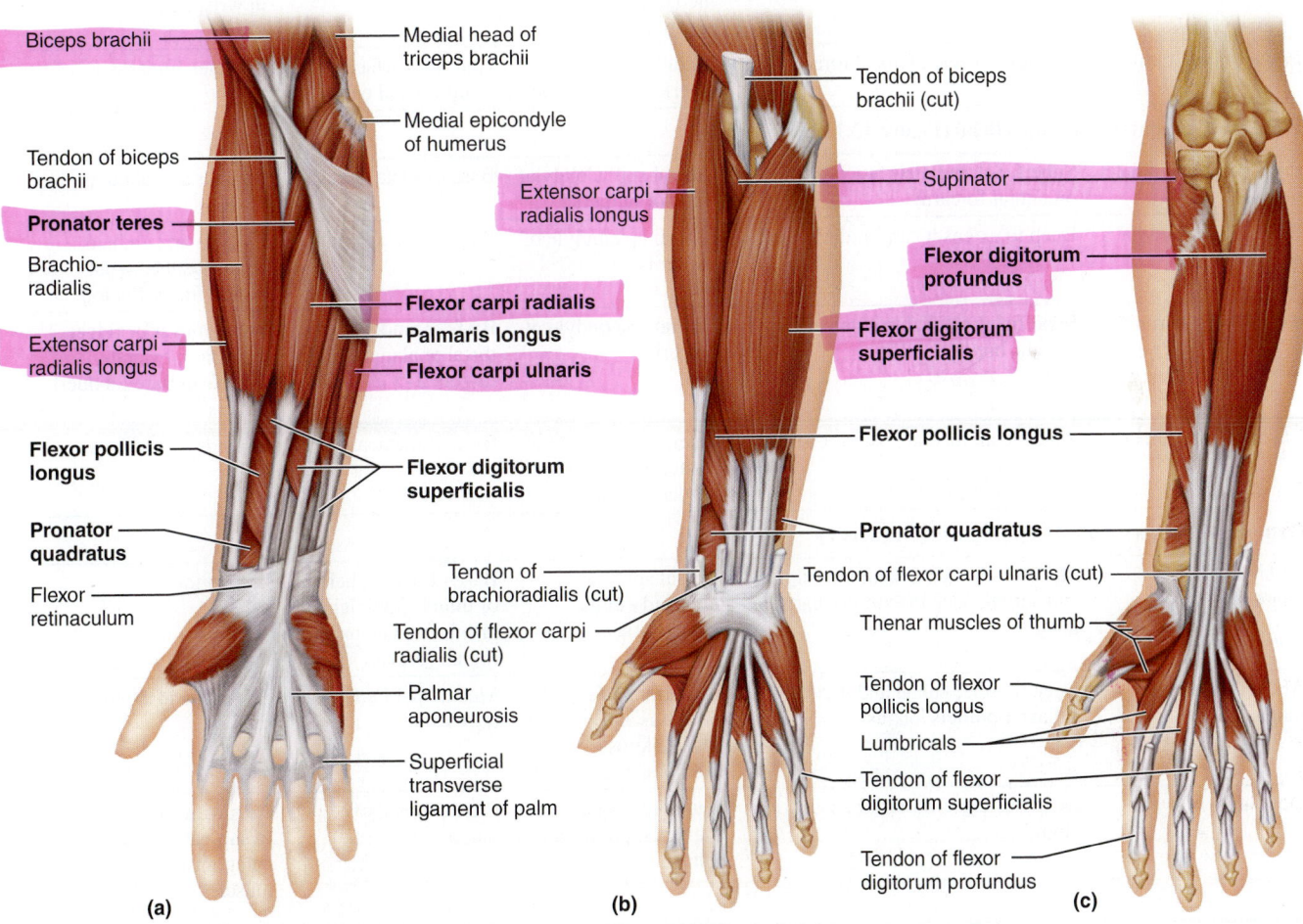

(a)   (b)   (c)

Figure 13.11 Muscles of the forearm and wrist. (a) Superficial anterior view of right forearm and hand. **(b)** The brachioradialis, flexors carpi radialis and ulnaris, and palmaris longus muscles have been removed to reveal the position of the somewhat deeper flexor digitorum superficialis. **(c)** Deep muscles of the anterior compartment. Superficial muscles have been removed. (*Note:* The thenar muscles of the thumb and the lumbricals that help move the fingers are illustrated here but are not described in Table 13.6.)

Table 13.6   Muscles of Human Forearm That Act on Hand and Fingers (continued)

Muscle	Comments	Origin	Insertion	Action
Anterior Compartment, Superficial (continued)				
Flexor carpi ulnaris	Superficial; medial to palmaris longus	Medial epicondyle of humerus and olecranon and posterior surface of ulna	Base of metacarpal; pisiform and hamate bones	Powerful flexor of wrist; adducts hand
Flexor digitorum superficialis	Deeper muscle (deep to muscles named above); visible at distal end of forearm	Medial epicondyle of humerus, coronoid process of ulna, and shaft of radius	Middle phalanges of fingers 2–5	Flexes wrist and middle phalanges of fingers 2–5
Anterior Compartment, Deep (Figure 13.11a, b, c)				
Flexor pollicis longus	Deep muscle of anterior forearm; distal to and paralleling lower margin of flexor digitorum superficialis	Anterior surface of radius, and interosseous membrane	Distal phalanx of thumb	Flexes thumb (*pollex* is Latin for "thumb")
Flexor digitorum profundus	Deep muscle; overlain entirely by flexor digitorum superficialis	Anteromedial surface of ulna, interosseous membrane, and coronoid process	Distal phalanges of fingers 2–5	Sole muscle that flexes distal phalanges; assists in wrist flexion
Pronator quadratus	Deepest muscle of distal forearm	Distal portion of anterior ulnar surface	Anterior surface of radius, distal end	Pronates forearm
Posterior Compartment, Superficial (Figure 13.11d, e, f)				
Extensor carpi radialis longus	Superficial; parallels brachioradialis on lateral forearm	Lateral supracondylar ridge of humerus	Base of metacarpal II	Extends and abducts wrist
Extensor carpi radialis brevis	Deep to extensor carpi radialis longus	Lateral epicondyle of humerus	Base of metacarpal III	Extends and abducts wrist; steadies wrist during finger flexion
Extensor digitorum	Superficial; medial to extensor carpi radialis brevis	Lateral epicondyle of humerus	By four tendons into distal phalanges of fingers 2–5	Prime mover of finger extension; extends wrist; can flare (abduct) fingers
Extensor carpi ulnaris	Superficial; medial posterior forearm	Lateral epicondyle of humerus; posterior border of ulna	Base of metacarpal V	Extends and adducts wrist
Posterior Compartment, Deep (Figure 13.11d, e, f)				
Extensor pollicis longus and brevis	Muscle pair with a common origin and action; deep to extensor carpi ulnaris	Dorsal shaft of ulna and radius, interosseous membrane	Base of distal phalanx of thumb (longus) and proximal phalanx of thumb (brevis)	Extend thumb
Abductor pollicis longus	Deep muscle; lateral and parallel to extensor pollicis longus	Posterior surface of radius and ulna; interosseous membrane	Metacarpal I and trapezium	Abducts and extends thumb
Supinator	Deep muscle at posterior aspect of elbow	Lateral epicondyle of humerus; proximal ulna	Proximal end of radius	Acts with biceps brachii to supinate forearm; antagonist of pronator muscles

13

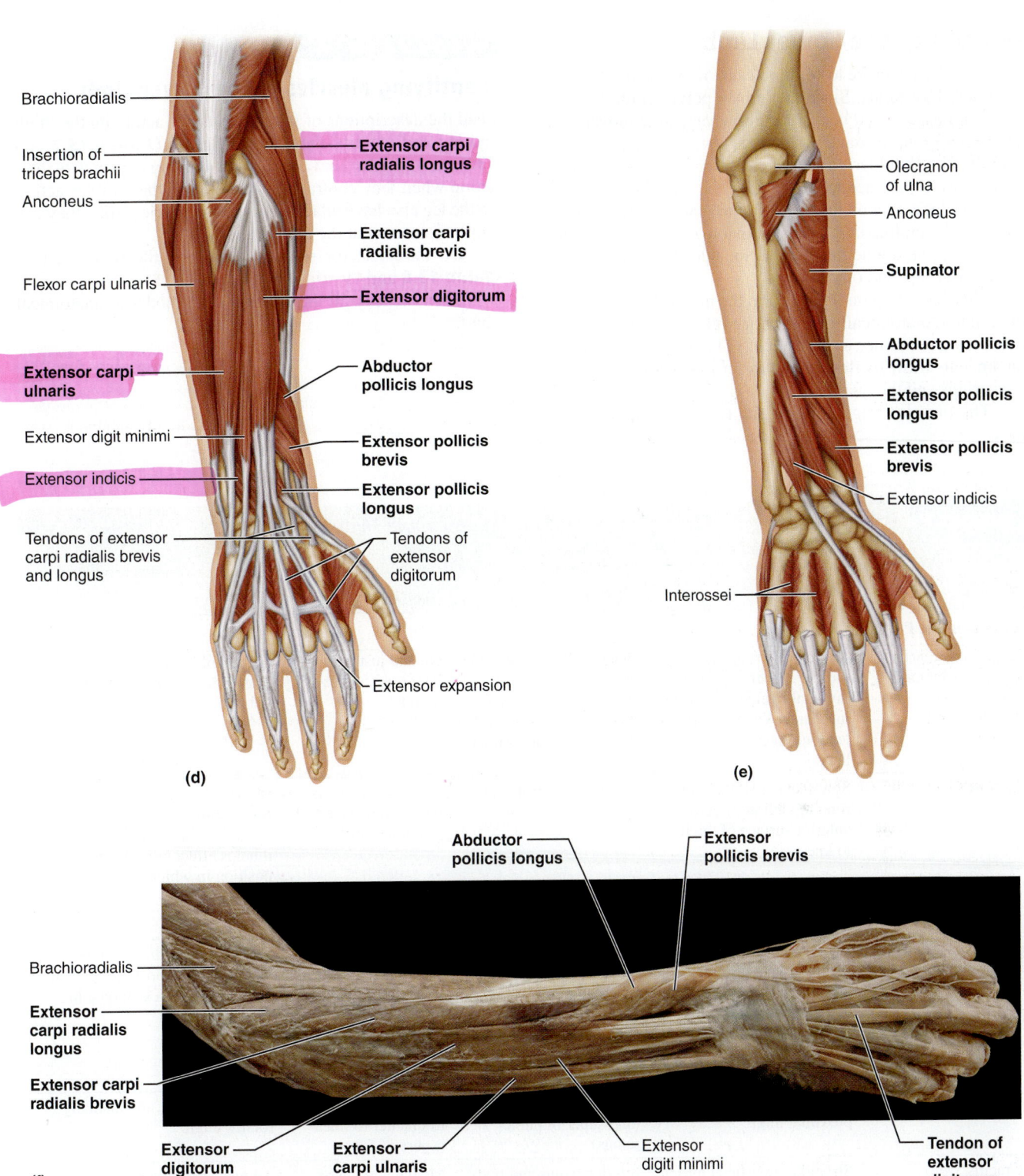

Figure 13.11 *(continued)* **Muscles of the forearm and wrist. (d)** Superficial muscles, posterior view. **(e)** Deep posterior muscles; superficial muscles have been removed. The interossei, the deepest layer of instrinsic hand muscles, are also illustrated. **(f)** Photo of posterior muscles of the right forearm.

13

Muscles of the Lower Limb

Muscles that act on the lower limb cause movement at the hip, knee, and foot joints. Since the human pelvic girdle is composed of heavy, fused bones that allow very little movement, no special group of muscles is necessary to stabilize it. This is unlike the shoulder girdle, where several muscles (mainly trunk muscles) are needed to stabilize the scapulae.

Muscles acting on the thigh (femur) cause various movements at the multiaxial hip joint (flexion, extension, rotation, abduction, and adduction). These include the iliopsoas, the adductor group, and others.

Muscles acting on the leg form the major musculature of the thigh. (Anatomically the term *leg* refers only to that portion between the knee and the ankle.) The thigh muscles cross the knee to allow its flexion and extension. They include the hamstrings and the quadriceps.

The muscles originating on the leg act on the foot and toes.

ACTIVITY 4

Identifying Muscles of the Lower Limb

Read the descriptions of specific muscles acting on the thigh and leg and identify them in the figures (Tables 13.7 and 13.8 and Figures 13.12 and 13.13), trying to visualize their action when they contract. Since some of the muscles acting on the leg also have attachments on the pelvic girdle, they can cause movement at the hip joint.

Do the same for muscles acting on the foot and toes (Table 13.9 and Figures 13.14 and 13.15).

Then identify all the muscles on a model or anatomical chart.

(Text continues on page 222.)

Table 13.7	Muscles Acting on Human Thigh and Leg, Anterior and Medial Aspects (see Figure 13.12)			
Muscle	**Comments**	**Origin**	**Insertion**	**Action**
Origin on the Pelvis				
Iliopsoas—iliacus and psoas major	Two closely related muscles; fibers pass under inguinal ligament to insert into femur via a common tendon; iliacus is more lateral	Iliacus—iliac fossa and crest, lateral sacrum; psoas major—transverse processes, bodies, and discs of T_{12} and lumbar vertebrae	On and just below lesser trochanter of femur	Flex trunk on thigh; flex thigh; lateral flexion of vertebral column (psoas)
Sartorius	Straplike superficial muscle running obliquely across anterior surface of thigh to knee	Anterior superior iliac spine	By an aponeurosis into medial aspect of proximal tibia	Flexes, abducts, and laterally rotates thigh; flexes knee; known as "tailor's muscle" because it helps effect cross-legged position in which tailors are often depicted
Medial Compartment				
Adductors—magnus, longus, and brevis	Large muscle mass forming medial aspect of thigh; arise from front of pelvis and insert at various levels on femur	Magnus—ischial and pubic rami and ischial tuberosity; longus—pubis near pubic symphysis; brevis—body and inferior pubic ramus	Magnus—linea aspera and adductor tubercle of femur; longus and brevis—linea aspera	Adduct and medially rotate and flex thigh; posterior part of magnus is also a synergist in thigh extension
Pectineus	Overlies adductor brevis on proximal thigh	Pectineal line of pubis (and superior pubic ramus)	Inferior from lesser trochanter to linea aspera of femur	Adducts, flexes, and medially rotates thigh
Gracilis	Straplike superficial muscle of medial thigh	Inferior ramus and body of pubis	Medial surface of tibia just inferior to medial condyle	Adducts thigh; flexes and medially rotates leg, especially during walking

(Table continues on page 218.)

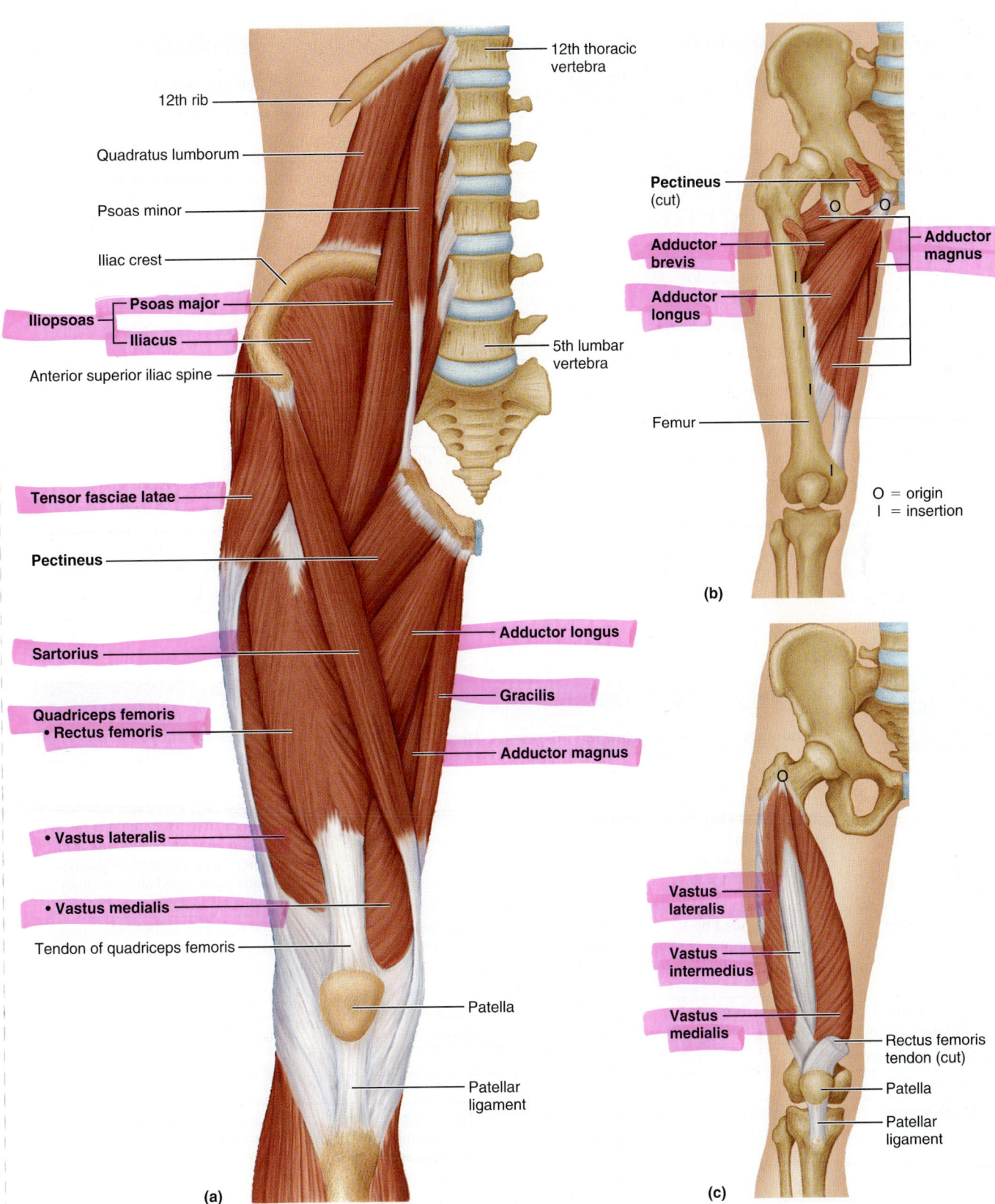

Figure 13.12 Anterior and medial muscles promoting movements of the thigh and leg. (a) Anterior view of the deep muscles of the pelvis and superficial muscles of the right thigh. **(b)** Adductor muscles of the medial compartment of the thigh. **(c)** The vastus muscles (isolated) of the quadriceps group.

13

Table 13.7	Muscles Acting on Human Thigh and Leg, Anterior and Medial Aspects *(continued)*			
Muscle	**Comments**	**Origin**	**Insertion**	**Action**
Anterior Compartment				
Quadriceps femoris*				
Rectus femoris	Superficial muscle of thigh; runs straight down thigh; only muscle of group to cross hip joint	Anterior inferior iliac spine and superior margin of acetabulum	Tibial tuberosity and patella	Extends knee and flexes thigh at hip
Vastus lateralis	Forms lateral aspect of thigh; intramuscular injection site	Greater trochanter, intertrochanteric line, and linea aspera	Tibial tuberosity and patella	Extends and stabilizes knee
Vastus medialis	Forms inferomedial aspect of thigh	Linea aspera and intertrochanteric line	Tibial tuberosity and patella	Extends knee; stabilizes patella
Vastus intermedius	Obscured by rectus femoris; lies between vastus lateralis and vastus medialis on anterior thigh	Anterior and lateral surface of femur	Tibial tuberosity and patella	Extends knee
Tensor fasciae latae	Enclosed between fascia layers of thigh	Anterior aspect of iliac crest and anterior superior iliac spine	Iliotibial tract (lateral portion of fascia lata)	Flexes, abducts, and medially rotates thigh; steadies trunk

*The quadriceps form the flesh of the anterior thigh and have a common insertion in the tibial tuberosity via the patellar tendon. They are powerful leg extensors, enabling humans to kick a football, for example.

Table 13.8	Muscles Acting on Human Thigh and Leg, Posterior Aspect (see Figure 13.13)			
Muscle	**Comments**	**Origin**	**Insertion**	**Action**
Origin on the Pelvis				
Gluteus maximus	Largest and most superficial of gluteal muscles (which form buttock mass); intramuscular injection site	Dorsal ilium, sacrum, and coccyx	Gluteal tuberosity of femur and iliotibial tract*	Complex, powerful thigh extensor (most effective when thigh is flexed, as in climbing stairs—but not as in walking); antagonist of iliopsoas; laterally rotates and abducts thigh
Gluteus medius	Partially covered by gluteus maximus; intramuscular injection site	Upper lateral surface of ilium	Greater trochanter of femur	Abducts and medially rotates thigh; steadies pelvis during walking
Gluteus minimus (not shown in figure)	Smallest and deepest gluteal muscle	External inferior surface of ilium	Greater trochanter of femur	Abducts and medially rotates thigh; steadies pelvis
Posterior Compartment				
Hamstrings†				
Biceps femoris	Most lateral muscle of group; arises from two heads	Ischial tuberosity (long head); linea aspera and distal femur (short head)	Tendon passes laterally to insert into head of fibula and lateral condyle of tibia	Extends thigh; laterally rotates leg; flexes knee

(Table continues on page 219.)

Table 13.8	(continued)			
Muscle	**Comments**	**Origin**	**Insertion**	**Action**
Semitendinosus	Medial to biceps femoris	Ischial tuberosity	Medial aspect of upper tibial shaft	Extends thigh; flexes knee; medially rotates leg
Semimembranosus	Deep to semitendinosus	Ischial tuberosity	Medial condyle of tibia; lateral condyle of femur	Extends thigh; flexes knee; medially rotates leg

* The iliotibial tract, a thickened lateral portion of the fascia lata, ensheathes all the muscles of the thigh. It extends as a tendinous band from the iliac crest to the knee.

†The hamstrings are the fleshy muscles of the posterior thigh. The name comes from the butchers' practice of using the tendons of these muscles to hang hams for smoking. As a group, they are strong extensors of the hip; they counteract the powerful quadriceps by stabilizing the knee joint when standing.

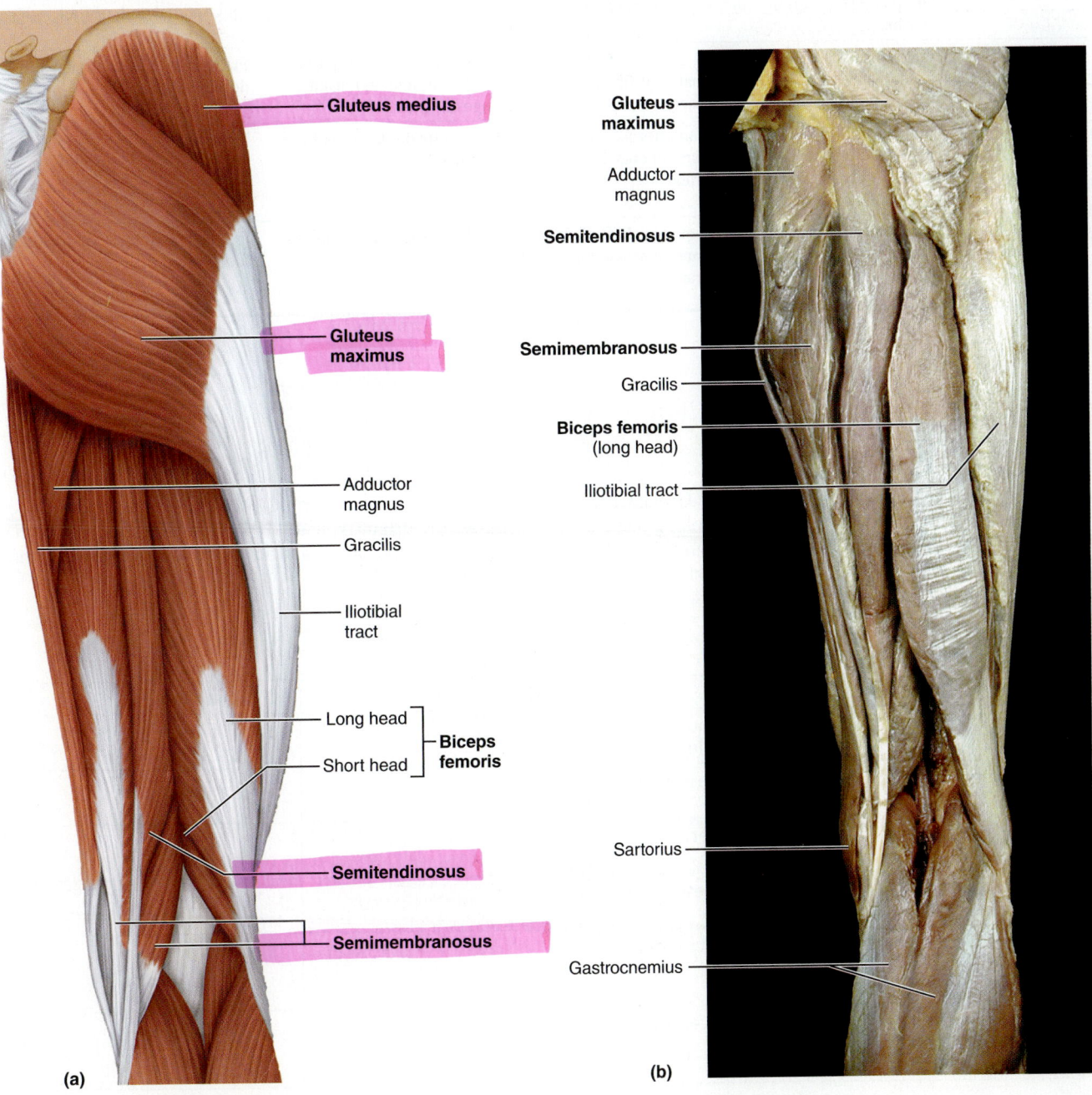

Figure 13.13 Muscles of the posterior aspect of the right hip and thigh.
(a) Superficial view showing the gluteus muscles of the buttock and hamstring muscles of the thigh. (b) Photo of muscles of the posterior thigh.

13

Table 13.9	Muscles Acting on Human Foot and Ankle (see Figures 13.14 and 13.15)			
Muscle	**Comments**	**Origin**	**Insertion**	**Action**
Lateral Compartment (Figure 13.14a, b and Figure 13.15b)				
Fibularis (peroneus) longus	Superficial lateral muscle; overlies fibula	Head and upper portion of fibula	By long tendon under foot to metatarsal I and medial cuneiform	Plantar flexes and everts foot; helps keep foot flat on ground
Fibularis (peroneus) brevis	Smaller muscle; deep to fibularis longus	Distal portion of fibula shaft	By tendon running behind lateral malleolus to insert on proximal end of metatarsal V	Plantar flexes and everts foot, as part of fibularis group
Anterior Compartment (Figure 13.14a, b)				
Tibialis anterior	Superficial muscle of anterior leg; parallels sharp anterior margin of tibia	Lateral condyle and upper $\frac{2}{3}$ of tibia; interosseous membrane	By tendon into inferior surface of first cuneiform and metatarsal I	Prime mover of dorsiflexion; inverts foot; supports longitudinal arch of foot
Extensor digitorum longus	Anterolateral surface of leg; lateral to tibialis anterior	Lateral condyle of tibia; proximal $\frac{3}{4}$ of fibula; interosseous membrane	Tendon divides into four parts; inserts into middle and distal phalanges of toes 2–5	Prime mover of toe extension; dorsiflexes foot
Fibularis (peroneus) tertius	Small muscle; often fused to distal part of extensor digitorum longus	Distal anterior surface of fibula and interosseous membrane	Tendon inserts on dorsum of metatarsal V	Dorsiflexes and everts foot
Extensor hallucis longus	Deep to extensor digitorum longus and tibialis anterior	Anteromedial shaft of fibula and interosseous membrane	Tendon inserts on distal phalanx of great toe	Extends great toe; dorsiflexes foot

(Table continues on page 222.)

GROUP CHALLENGE

Name That Muscle

Work in groups of three or four to fill out the Group Challenge chart for muscle IDs. Refrain from looking back at the tables. Use the "brain power" of your group and the appropriate muscle models. To assist in this task, recall that when a muscle contracts, the muscle's insertion moves toward the muscle's origin. Also, in the muscles of the limbs, the origin typically lies proximal to the insertion. Sometimes the origin and insertion are even part of the muscle's name!

Group Challenge: Muscle IDs

Origin	Insertion	Muscle	Primary action
Zygomatic arch and maxilla	Angle and ramus of the mandible		
Anterior surface of ribs 3–5	Coracoid process of the scapula		
Inferior border of rib above	Superior border of rib below		
Distal portion of anterior humerus	Coronoid process of the ulna		
Anterior inferior iliac spine and superior margin of acetabulum	Tibial tuberosity and patella		
By two heads from medial and lateral condyles of femur	Calcaneus via calcaneal tendon		

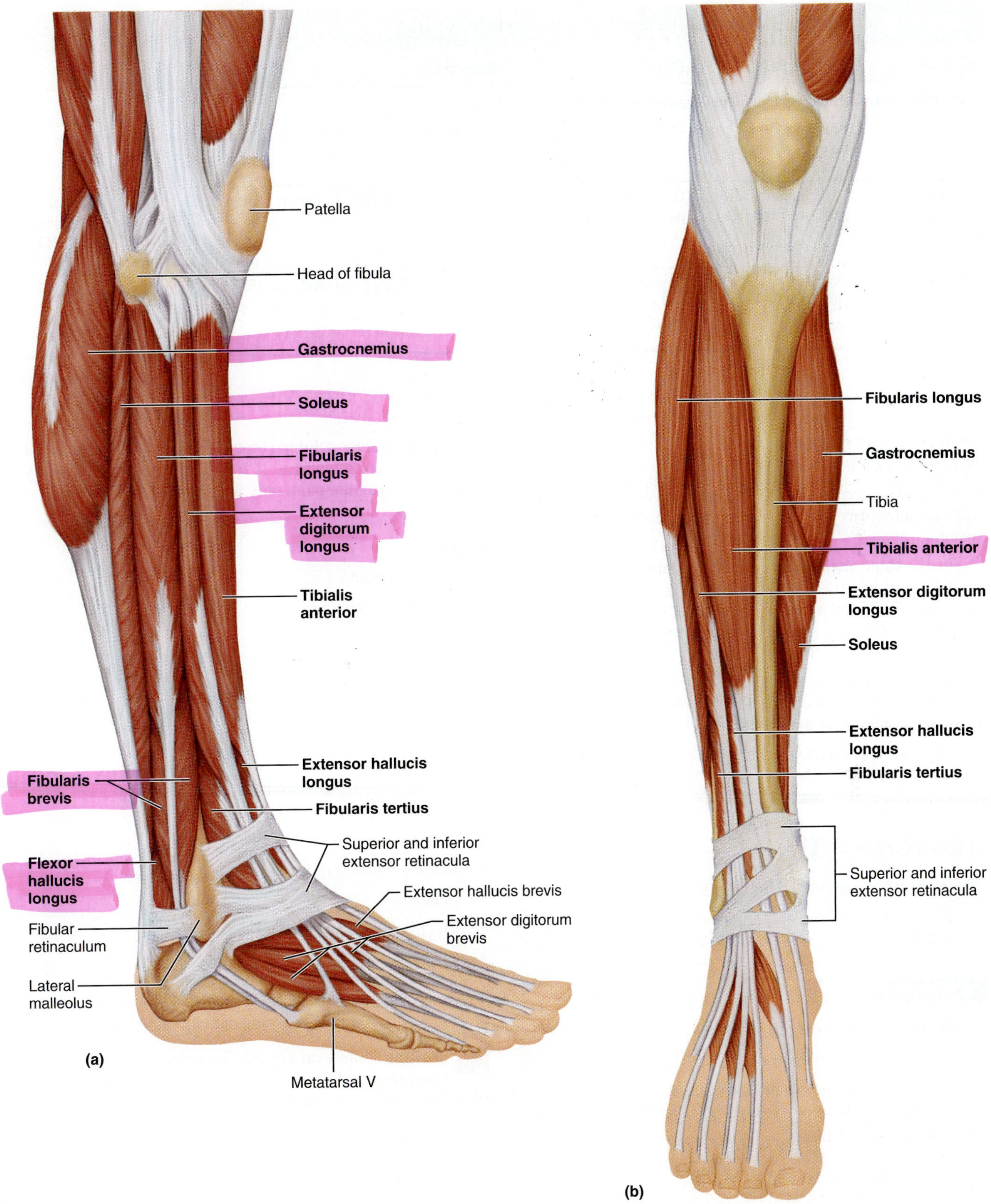

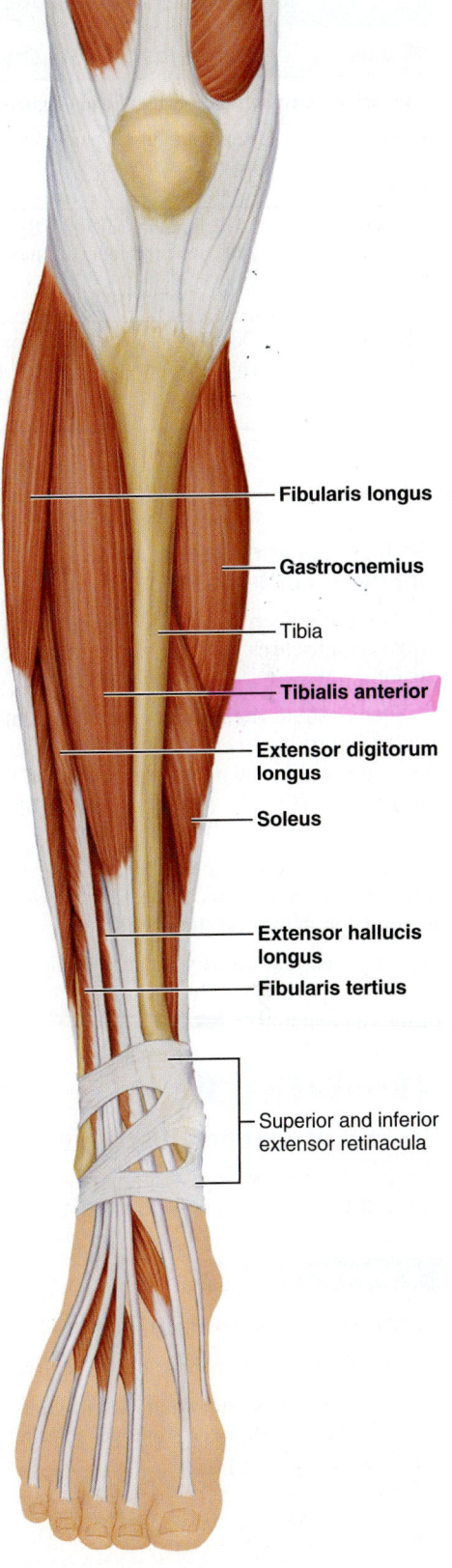

Patella

Head of fibula

Gastrocnemius

Soleus

Fibularis longus

Extensor digitorum longus

Tibialis anterior

Extensor hallucis longus

Fibularis tertius

Superior and inferior extensor retinacula

Extensor hallucis brevis

Extensor digitorum brevis

Fibularis brevis

Flexor hallucis longus

Fibular retinaculum

Lateral malleolus

Metatarsal V

(a)

Fibularis longus

Gastrocnemius

Tibia

Tibialis anterior

Extensor digitorum longus

Soleus

Extensor hallucis longus

Fibularis tertius

Superior and inferior extensor retinacula

(b)

13

Figure 13.14 Muscles of the anterolateral aspect of the right leg.
(a) Superficial view of lateral aspect of the leg, illustrating the positioning of the lateral compartment muscles (fibularis longus and brevis) relative to anterior and posterior leg muscles. **(b)** Superficial view of anterior leg muscles.

Table 13.9   Muscles Acting on Human Foot and Ankle (continued)

Muscle	Comments	Origin	Insertion	Action
Posterior Compartment, Superficial (Figure 13.15a; also Figure 13.14)				
Triceps surae	Refers to muscle pair below that shapes posterior calf		Via common tendon (calcaneal) into calcaneus of the heel	Plantar flex foot
Gastrocnemius	Superficial muscle of pair; two prominent bellies	By two heads from medial and lateral condyles of femur	Calcaneus via calcaneal tendon	Plantar flexes foot when knee is extended; crosses knee joint; thus can flex knee (when foot is dorsiflexed)
Soleus	Deep to gastrocnemius	Proximal portion of tiba and fibula; interosseous membrane	Calcaneus via calcaneal tendon	Plantar flexion; is an important muscle for locomotion

(Table continues on page 224.)

Demonstrating Operations of Lower Limb Muscles

1. Go into a deep knee bend and palpate your own *gluteus maximus* muscle as you extend your hip to resume the upright posture.

2. Demonstrate the contraction of the anterior *quadriceps femoris* by trying to extend your knee against resistance. Do this while seated and note how the patellar tendon reacts. The *biceps femoris* of the posterior thigh comes into play when you flex your knee against resistance.

3. Now stand on your toes. Have your partner palpate the lateral and medial heads of the *gastrocnemius* and follow it to its insertion in the calcaneal tendon.

4. Dorsiflex and invert your foot while palpating your *tibialis anterior* muscle (which parallels the sharp anterior crest of the tibia laterally). ■

ACTIVITY 5

Review of Human Musculature

Review the muscles by watching the *Human Musculature* video. ■

ACTIVITY 6

Making a Muscle Painting

1. Choose a male student to be "muscle painted."

2. Obtain brushes and water-based paints from the supply area while the "volunteer" removes his shirt and rolls up his pant legs (if necessary).

3. Using different colored paints, identify the muscles listed below by painting his skin. If a muscle covers a large body area, you may opt to paint only its borders.

- biceps brachii
- deltoid
- erector spinae
- pectoralis major
- rectus femoris
- tibialis anterior
- triceps brachii
- vastus lateralis
- biceps femoris
- extensor carpi radialis longus
- latissimus dorsi
- rectus abdominis
- sternocleidomastoid
- trapezius
- triceps surae
- vastus medialis

4. Check your "human painting" with your instructor before cleaning your bench and leaving the laboratory. ■

For instructions on animal dissections, see the dissection exercises (starting on page 697) in the cat and fetal pig editions of this manual.

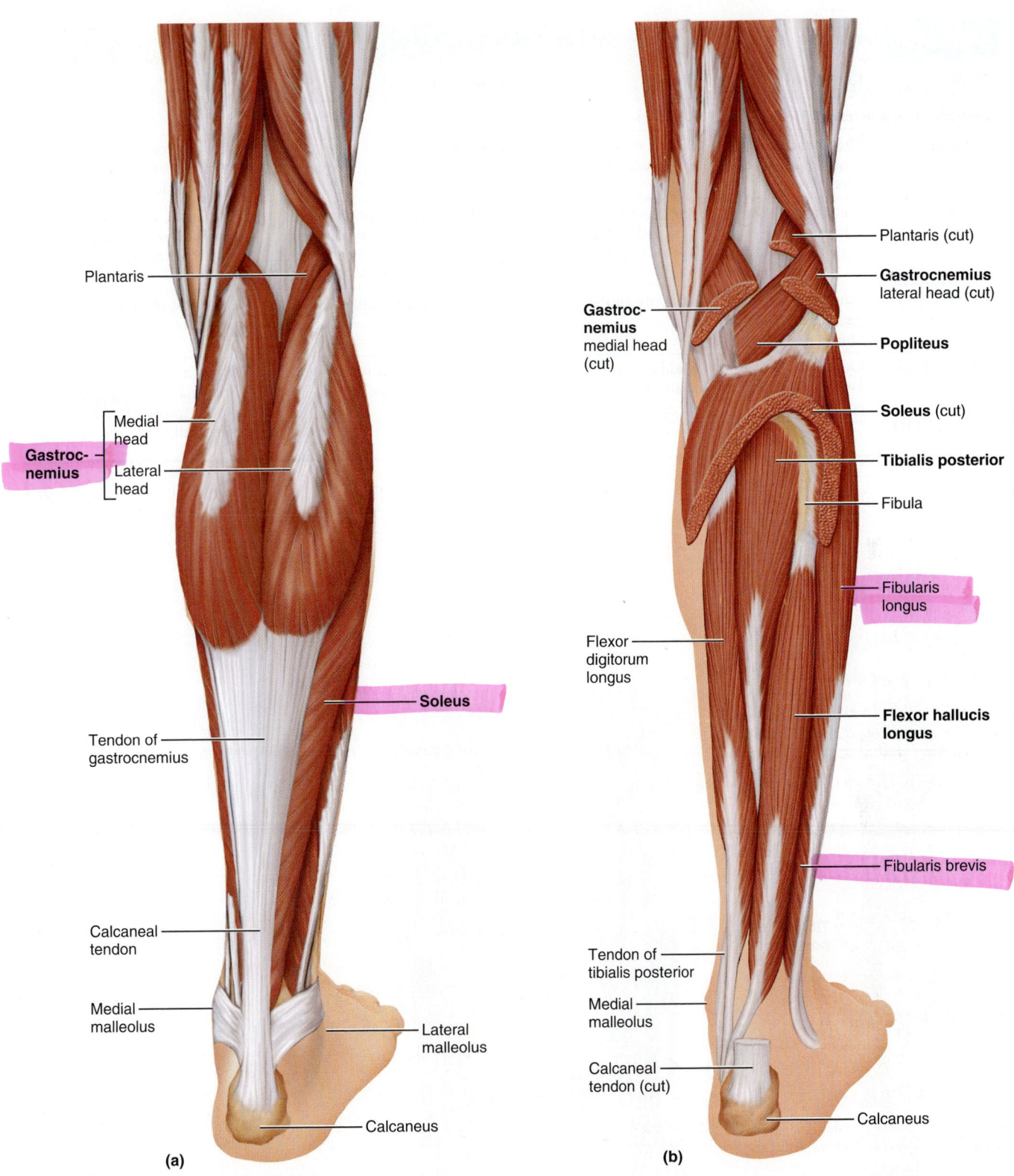

(a)

(b)

13

Figure 13.15 **Muscles of the posterior aspect of the right leg. (a)** Superficial view of the posterior leg. **(b)** The triceps surae has been removed to show the deep muscles of the posterior compartment.

Table 13.9	Muscles Acting on Human Foot and Ankle *(continued)*			
Muscle	**Comments**	**Origin**	**Insertion**	**Action**
Posterior Compartment, Deep (Figure 13.15b–e)				
Popliteus	Thin muscle at posterior aspect of knee	Lateral condyle of femur and lateral meniscus	Proximal tibia	Flexes and rotates leg medially to "unlock" extended knee when knee flexion begins
Tibialis posterior	Thick muscle deep to soleus	Superior portion of tibia and fibula and interosseous membrane	Tendon passes obliquely behind medial malleolus and under arch of foot; inserts into several tarsals and metatarsals II–IV	Prime mover of foot inversion; plantar flexes foot; stabilizes longitudinal arch of foot
Flexor digitorum longus	Runs medial to and partially overlies tibialis posterior	Posterior surface of tibia	Distal phalanges of toes 2–5	Flexes toes; plantar flexes and inverts foot
Flexor hallucis longus (see also Figure 13.14a)	Lies lateral to inferior aspect of tibialis posterior	Middle portion of fibula shaft; interosseous membrane	Tendon runs under foot to distal phalanx of great toe	Flexes great toe (*hallux* = great toe); plantar flexes and inverts foot; the "push-off muscle" during walking

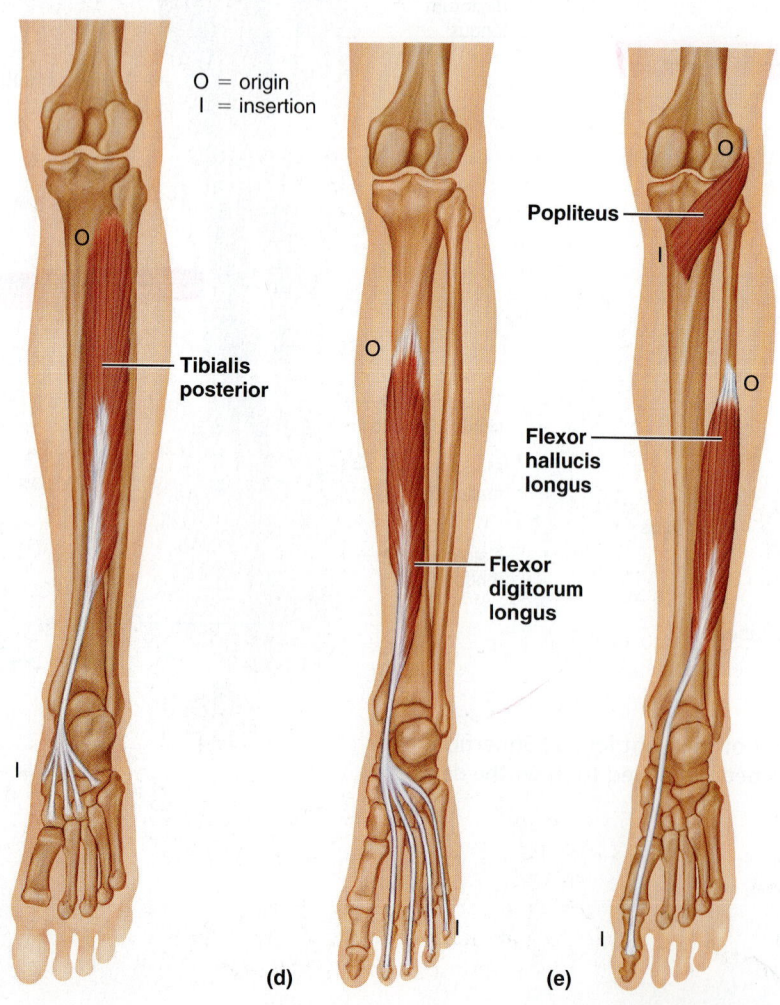

O = origin
I = insertion

Popliteus

Tibialis posterior

Flexor digitorum longus

Flexor hallucis longus

(c)   (d)   (e)

Figure 13.15 *(continued)* **Muscles of the posterior aspect of the right leg.** (c–e) Individual muscles are shown in isolation so that their origins and insertions may be visualized.

13

Skeletal Muscle Physiology: Frogs and Human Subjects

MATERIALS

- ☐ ATP muscle kits (glycerinated rabbit psoas muscle;* ATP and salt solutions obtainable from Carolina Biological Supply)
- ☐ Petri dishes
- ☐ Microscope slides
- ☐ Coverslips
- ☐ Millimeter ruler
- ☐ Compound microscope
- ☐ Stereomicroscope
- ☐ Pointed glass probes (teasing needles)
- ☐ Small beakers (50 ml)
- ☐ Distilled water
- ☐ Glass-marking pencil
- ☐ Textbooks or other heavy books
- ☐ Watch or timer
- ☐ Ringer's solution (frog)
- ☐ Scissors
- ☐ Metal needle probes
- ☐ Medicine dropper
- ☐ Cotton thread
- ☐ Forceps
- ☐ Disposable gloves
- ☐ Glass or porcelain plate
- ☐ Pithed bullfrog†
- ☐ Physiograph or BIOPAC® equipment:‡ Physiograph apparatus, physiograph paper and ink, force transducer, pin electrodes, stimulator, stimulator output extension cable, transducer stand and cable, straight pins, frog board

  BIOPAC® BIOPAC® BSL System for Windows with BSL software version 3.7.5 to 3.7.7, or BSL System for Mac OS X with BSL software version 3.7.4 to 3.7.7, MP 36/35 or MP45

(Text continues on next page.)

MasteringA&P® For related exercise study tools, go to the Study Area of MasteringA&P. There you will find:
- Practice Anatomy Lab **PAL**
- PhysioEx **PEx**
- A&PFlix **A&PFlix**
- Practice quizzes, Histology Atlas, eText, Videos, and more!

OBJECTIVES

1. Define the terms *resting membrane potential, depolarization, repolarization, action potential,* and *absolute and relative refractory periods* and explain the physiological basis of each.
2. Observe muscle contraction microscopically and describe the role of ATP and various ions in muscle contraction.
3. Define *muscle twitch* and describe its three phases.
4. Differentiate between subthreshold stimulus, threshold stimulus, and maximal stimulus.
5. Define *tetanus* in skeletal muscle, and explain how it comes about.
6. Define *muscle fatigue* and describe the effects of load on muscle fatigue.
7. Define *motor unit,* and relate recruitment of motor units and temporal summation to production of a graded contraction.
8. Demonstrate how a physiograph or computer with data acquisition unit can be used to record skeletal muscle activity.
9. Explain the significance of recordings obtained during experimentation.

PRE-LAB QUIZ

1. Circle the correct underlined term. The potential difference, or voltage, across the plasma membrane is the result of the difference in membrane permeability to anions / cations, most importantly Na^+ and K^+.
2. The _____ wave follows the depolarization wave across the sarcolemma.
   a. hyperpolarization
   b. refraction
   c. repolarization
3. Circle True or False. The voltage at which the first noticeable contractile response is achieved is called the threshold stimulus.
4. Circle True or False. A single muscle is made up of many motor units, and the gradual activation of these motor units results in a graded contraction of the whole muscle.
5. A sustained, smooth, muscle contraction that is a result of high-frequency stimulation is
   a. tetanus   b. tonus   c. twitch

The contraction of skeletal and cardiac muscle fibers can be considered in terms of three events—electrical excitation of the muscle cell, excitation contraction coupling, and shortening of the muscle cell due to sliding of the myofilaments within it.

At rest, all cells maintain a potential difference, or voltage, across their plasma membrane; the inner face of the membrane is approximately −60 to −90 millivolts (mV) compared with the cell exterior. This potential difference is a

(Materials list continued.)

data acquisition unit, PC or Mac computer, BIOPAC Student Lab electrode lead set, hand dynamometer, headphones, metric tape measure, disposable vinyl electrodes, and conduction gel.

Instructors using the MP36 (or MP35/30) data acquisition unit with BSL software versions earlier than 3.7.5 (for Windows) and 3.7.4 (for Mac OS X) will need slightly different channel settings and collection strategies. Instructions for using the older data acquisition unit can be found on MasteringA&P.

Notes to the Instructor: *At the beginning of the lab, the muscle bundle should be removed from the test tube and cut into approximately 2-cm lengths. Both the cut muscle segments and the entubed glycerol should be put into a petri dish. One muscle *segment* is sufficient for every two to four students making observations.

†Bullfrogs to be pithed by lab instructor as needed for student experimentation. (If instructor prefers that students pith their own specimens, an instructional sheet on

that procedure suitable for copying for student handouts is provided in the Instructor's Guide.)

‡Additionally, instructions for Activity 3 using a kymograph can be found in the Instructor Guide. Instructions for using PowerLab® equipment can be found on MasteringA&P.

PEx PhysioEx™ 9.1 Computer Simulation Ex.2 on p. PEx-17.

---

result of differences in membrane permeability to cations, most importantly sodium (Na^+) and potassium (K^+) ions. Intracellular potassium concentration is much greater than its extracellular concentration, and intracellular sodium concentration is considerably less than its extracellular concentration. Hence, steep concentration gradients across the membrane exist for both cations. Because the plasma membrane is more permeable to K^+ than to Na^+, the cell's **resting membrane potential** is more negative inside than outside. The resting membrane potential is of particular interest in excitable cells, like muscle cells and neurons, because changes in that voltage underlie their ability to do work (to contract and/or issue electrical signals).

Action Potential

When a muscle cell is stimulated, the sarcolemma becomes temporarily more permeable to Na^+, which enters the cell. This sudden influx of Na^+ alters the membrane potential. That is, the cell interior becomes less negatively charged at that point, an event called **depolarization.** When depolarization reaches a certain level and the sarcolemma momentarily changes its polarity, a depolarization wave travels along the sarcolemma. Even as the influx of Na^+ occurs, the sarcolemma becomes less permeable to Na^+ and more permeable to K^+. Consequently, K^+ ions move out of the cell, restoring the resting membrane potential (but not the original ionic conditions), an event called **repolarization.** The repolarization wave follows the depolarization wave across the sarcolemma. This rapid depolarization and repolarization of the membrane that is propagated along the entire membrane from the point of stimulation is called the **action potential.**

The **absolute refractory period** is the period of time when Na^+ permeability of the sarcolemma is rapidly changing and maximal, and the following period when Na^+ permeability becomes restricted. During this period there is no possibility of generating another action potential. As Na^+ permeability is gradually restored to resting levels during repolarization, an especially strong stimulus to the muscle cell may provoke another action potential. This period of time is the **relative refractory period.** Repolarization restores the muscle cell's normal excitability. If the muscle cell is stimulated to contract rapidly again and again, the changes in Na^+ and K^+ concentrations near the membrane begin to reduce its ability to respond. The sodium-potassium pump,

which actively transports K^+ into the cell and Na^+ out of the cell, must be "revved up" (become more active) to reestablish the ionic concentrations of the resting state.

Contraction

Propagation of the action potential along the sarcolemma causes the release of calcium ions (Ca^{2+}) from storage in the sarcoplasmic reticulum within the muscle cell. When the calcium ions bind to the regulatory protein troponin on the actin myofilaments, they act as an ionic trigger that initiates contraction, and the actin and myosin filaments slide past each other. Once the action potential ends, the calcium ions are almost immediately transported back into the sarcoplasmic reticulum. Instantly the muscle cell relaxes.

The events of the contraction process can most simply be summarized as follows: Muscle cell contraction is initiated by generation and transmission of an action potential along the sarcolemma. This electrical event is coupled to the sliding of the myofilaments—contraction—by the release of calcium ions (Ca^{2+}). Keep in mind this sequence of events as you conduct the experiments.

ACTIVITY 1

Observing Muscle Fiber Contraction

In this simple observational experiment, you will have the opportunity to review your understanding of muscle cell anatomy and to watch fibers respond to the presence of ATP and/or a solution of K^+ and magnesium ions (Mg^{2+}).

This experiment uses preparations of glycerinated muscle. The glycerination process denatures troponin and tropomyosin. Consequently, calcium, so critical for contraction in vivo, is not necessary here. The role of magnesium and potassium salts as cofactors in the contraction process is not well understood, but magnesium and potassium salts seem to be required for ATPase activity in this system.

1. Talk with other members of your lab group to develop a hypothesis about requirements for muscle fiber contraction for this experiment. The hypothesis should have three parts: (1) salts only, (2) ATP only, and (3) salts and ATP.

2. Obtain the following materials from the supply area: two glass teasing needles; six glass microscope slides and six

coverslips; millimeter ruler; dropper bottles containing the following solutions: (a) 0.25% ATP in triply distilled water, (b) 0.25% ATP plus 0.05 M KCl plus 0.001 M MgCl$_2$ in distilled water, and (c) 0.05 M KCl plus 0.001 M MgCl$_2$ in distilled water; a petri dish; a beaker of distilled water; a glass-marking pencil; and a small portion of a previously cut muscle bundle segment. While you are at the supply area, place the muscle fibers in the petri dish and pour a small amount of glycerol (the fluid in the supply petri dish) over your muscle cells. Also obtain both a compound and a stereo-microscope and bring them to your laboratory bench.

3. Using clean fine glass needles, tease the muscle segment to separate its fibers. The objective is to isolate *single* muscle cells or fibers for observation. Be patient and work carefully so that the fibers do not get torn during this isolation procedure.

4. Transfer one or more of the fibers (or the thinnest strands you have obtained) onto a clean microscope slide with a glass needle, and cover it with a coverslip. Examine the fiber under the compound microscope at low- and then high-power magnifications to observe the striations and the smoothness of the fibers when they are in the relaxed state.

5. Clean three microscope slides well and rinse in distilled water. Label the slides A, B, and C.

6. Transfer three or four fibers to microscope slide A with a glass needle. Using the needle as a prod, carefully position the fibers so that they are parallel to one another and as straight as possible. Place this slide under a *stereomicroscope* and measure the length of each fiber by holding a millimeter ruler adjacent to it. Alternatively, you can rest the microscope slide *on* the millimeter ruler to make your length determinations. Record the data on the **Activity 1 chart.**

7. Flood the fibers (situated under the stereomicroscope) with several drops of the solution containing ATP, K$^+$, and Mg^{2+}. Watch the reaction of the fibers after adding the solution. After 30 seconds (or slightly longer), remeasure each fiber and record the observed lengths on the chart. Also, observe the fibers to see if any width changes have occurred. Calculate the degree (or percentage) of contraction by using the simple formula below, and record this data on the chart.

$$\begin{array}{ccc} \text{Initial} & \text{contracted} & \text{degree of} \\ \text{length (mm)} & - \text{length (mm)} & = \text{contraction (mm)} \end{array}$$

then:

$$\frac{\text{Degree of contraction (mm)}}{\text{initial length (mm)}} \times 100 = \underline{\hspace{1cm}} \% \text{ contraction}$$

8. Carefully transfer one of the contracted fibers to a clean, unmarked microscope slide, cover with a coverslip, and observe with the compound microscope. Mentally compare your initial observations with the view you are observing now. What differences do you see? (Be specific.)

_____

_____

_____

What zones (or bands) have disappeared?

_____

9. Repeat steps 6 through 8 twice more, using clean slides and fresh muscle cells. On slide B use the solution of ATP in distilled water (no salts). Then, on slide C use the solution

Activity 1: Observations of Muscle Fiber Contraction				
Salts and ATP, slide A	**Muscle fiber 1**	**Muscle fiber 2**	**Muscle fiber 3**	**Average**
Initial length (mm)				
Contracted length (mm)				
% Contraction				
ATP only, slide B				
Initial length (mm)				
Contracted length (mm)				
% Contraction				
Salts only, slide C				
Initial length (mm)				
Contracted length (mm)				
% Contraction				

14

containing only salts (no ATP) for the third series. Record data on the Activity 1 chart.

10. Collect the data from all the groups in your laboratory and use these data to prepare a lab report. (See Getting Started, page xiv.) Include in your discussion the following questions:

What degree of contraction was observed when ATP was applied in the absence of K^+ and Mg^{2+}?

_____

What degree of contraction was observed when the muscle fibers were flooded with a solution containing K^+ and Mg^{2+}, and lacking ATP?

_____

What conclusions can you draw about the importance of ATP, K^+, and Mg^{2+} to the contractile process?

_____

_____

Can you draw exactly the same conclusions from the data provided by each group? List some variables that might have been introduced into the procedure and that might account for any differences.

_____

_____

Inducing Contraction in the Frog Gastrocnemius Muscle

Physiologists have learned a great deal about the way muscles function by isolating muscles from laboratory animals and then stimulating these muscles to observe their responses. Various stimuli—electrical shock, temperature changes, extremes of pH, certain chemicals—elicit muscle activity, but laboratory experiments of this type typically use electrical shock. This is because it is easier to control the onset and cessation of electrical shock, as well as the strength of the stimulus.

Various types of apparatus are used to record muscle contraction. All include a way to mark time intervals, a way to indicate exactly when the stimulus was applied, and a way to measure the magnitude of the contractile response. Instructions are provided here for setting up a physiograph apparatus **(Figure 14.1)**. Specific instructions for use of recording apparatus during recording will be provided by your instructor.

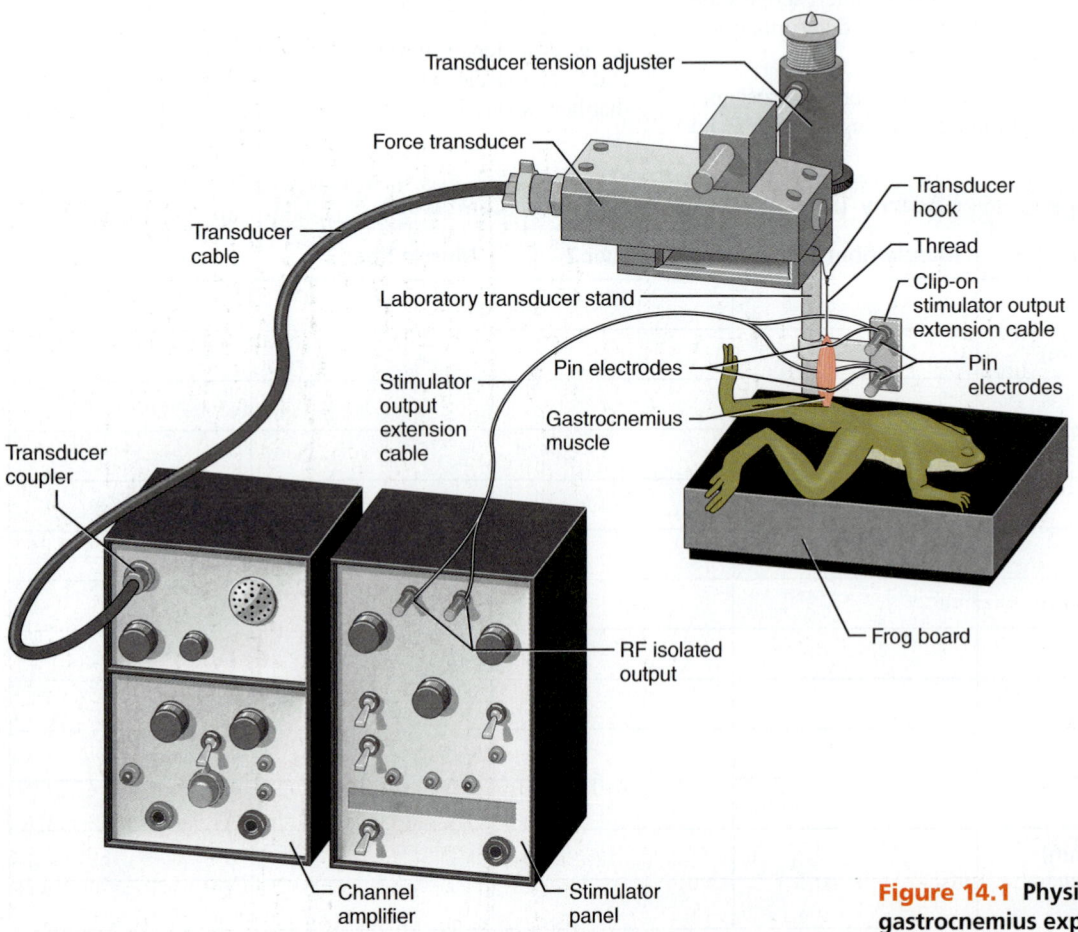

Figure 14.1 Physiograph setup for frog gastrocnemius experiments.

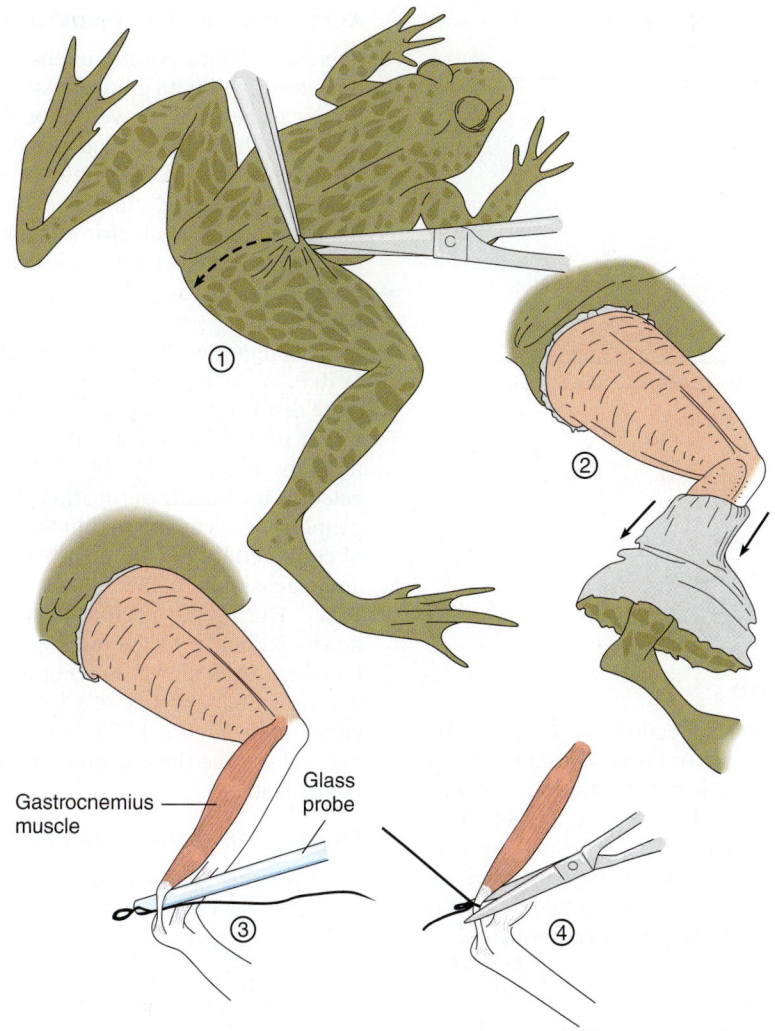

Gastrocnemius muscle

Glass probe

Figure 14.2 **Preparation of the frog gastrocnemius muscle.** Numbers indicate the sequence of steps for preparing the muscle.

Preparing a Muscle for Experimentation

The preparatory work that precedes the recording of muscle activity tends to be quite time-consuming. If you work in teams of two or three, the work can be divided. While one of you is setting up the recording apparatus, one or two students can dissect the frog leg **(Figure 14.2)**. Experimentation should begin as soon as the dissection is completed.

Materials

Channel amplifier and transducer cable

Stimulator panel and stimulator output extension cable

Force transducer

Transducer tension adjuster

Transducer stand

Two pin electrodes

Frog board and straight pins

Prepared frog (gastrocnemius muscle freed and calcaneal tendon ligated)

Frog Ringer's solution

Procedure

1. Connect transducer to tranducer stand and attach frog board to stand.

2. Attach tranducer cable to transducer and to input connection on channel amplifier.

3. Attach stimulator output extension cable to output on stimulator panel (red to red, black to black).

4. Using clip at opposite end of extension cable, attach cable to bottom of transducer stand adjacent to frog board.

5. Attach two pin electrodes securely to electrodes on clip.

6. Place knee of prepared frog in clip-on frog board and secure by inserting a straight pin through tissues of frog. Keep frog muscle moistened with Ringer's solution.

7. Attach thread from the calcaneal tendon of frog to transducer spring hook.

8. Adjust position of tranducer on stand to produce a constant tension on thread attached to muscle (taut but not tight). Gastrocnemius muscle should hang vertically directly below hook.

9. Insert free ends of pin electrodes into the muscle, one at proximal end and the other at distal end.

14

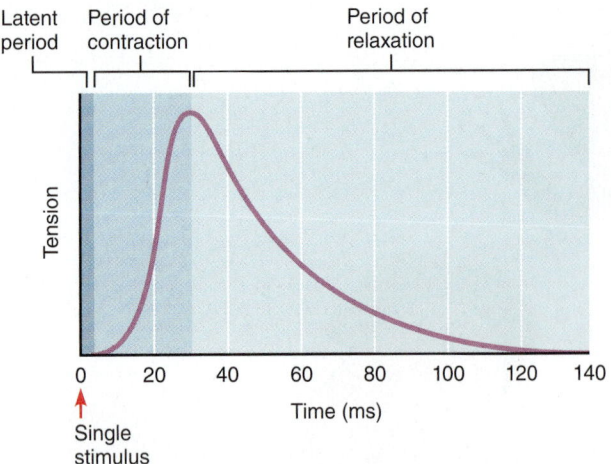

Figure 14.3 **Tracing of a muscle twitch.**

14

✂ DISSECTION:
Frog Hind Limb

1. Before beginning the frog dissection, have the following supplies ready at your laboratory bench: a small beaker containing 20 to 30 ml of frog Ringer's solution, scissors, a metal needle probe, a glass probe with a pointed tip, a medicine dropper, cotton thread, forceps, a glass or porcelain plate, and disposable gloves. While these supplies are being gathered, one member of your team should notify the instructor that you are ready to begin experimentation, so that a frog can be prepared (pithed). Preparation of a frog in this manner renders it unable to feel pain and prevents reflex movements (like hopping) that would interfere with the experiments.

⚠ 2. All students who will be handling the frog should don disposable gloves. Obtain a pithed frog and place it ventral surface down on the glass plate. Make an incision into the skin approximately midthigh (Figure 14.2), and then continue the cut completely around the thigh. Grasp the skin with the forceps and strip it from the leg and hindfoot. The skin adheres more at the joints, but a careful, persistent pulling motion—somewhat like pulling off a nylon stocking—will enable you to remove it in one piece. _From this point on, the exposed muscle tissue should be kept moistened with the Ringer's solution_ to prevent spontaneous twitches.

3. Identify the gastrocnemius muscle (the fleshy muscle of the posterior calf) and the calcaneal tendon that secures it to the heel.

4. Slip a glass probe under the gastrocnemius muscle and run it along the entire length and under the calcaneal tendon to free them from the underlying tissues.

5. Cut a piece of thread about 10 inches long and use the glass probe to slide the thread under the calcaneal tendon. Knot the thread firmly around the tendon and then sever the tendon distal to the thread. Alternatively, you can bend a common pin into a Z shape and insert the pin securely into the tendon. The thread is then attached to the opposite end of the pin. Once the tendon has been tied or pinned, the frog is ready for experimentation (see Figure 14.2). ■

Recording Muscle Activity

Skeletal muscles consist of thousands of muscle cells and react to stimuli with graded responses. Thus muscle contractions can be weak or vigorous, depending on the requirements of the task. Graded responses (different degrees of shortening) of a skeletal muscle depend on the number of muscle cells being stimulated. In the intact organism, the number of motor units firing at any one time determines how many muscle cells will be stimulated. In this laboratory, the frequency and strength of an electrical current determines the response.

A single contraction of skeletal muscle is called a **muscle twitch.** A tracing of a muscle twitch **(Figure 14.3)** shows three distinct phases: latent, contraction, and relaxation. The **latent period** is the interval from stimulus application until the muscle begins to shorten. Although no activity is indicated on the tracing during this phase, excitation-contraction coupling is occurring within the muscle. During the **period of contraction,** the muscle fibers shorten; the tracing shows an increasingly higher needle deflection and the tracing peaks. During the **period of relaxation,** represented by a downward curve of the tracing, the muscle fibers relax and lengthen. On a slowly moving recording surface, the single muscle twitch appears as a spike (rather than a bell-shaped curve, as in Figure 14.3), but on a rapidly moving recording surface, the three distinct phases just described become recognizable.

Determining the Threshold Stimulus

1. Assuming that you have already set up the recording apparatus, set the time marker to deliver one pulse per second and set the paper speed at a slow rate, approximately 0.1 cm per second.

2. Set the duration control on the stimulator between 7 and 15 milliseconds (msec), multiplier ×1. Set the voltage control at 0 V, multiplier ×1. Turn the sensitivity control knob of the stimulator fully clockwise (lowest value, greatest sensitivity).

3. Administer single stimuli to the muscle at 1-minute intervals, beginning with 0.1 V and increasing each successive stimulus by 0.1 V until a contraction is obtained (shown by a spike on the paper).

At what voltage did contraction occur? _____ V

The voltage at which the first perceptible contractile response is obtained is called the **threshold stimulus.** All stimuli applied prior to this point are termed **subthreshold stimuli,** because at those voltages no response was elicited.

4. Stop the recording and mark the record to indicate the threshold stimulus, voltage, and time. _Do not remove the record from the recording surface;_ continue with the next experiment. _Remember: keep the muscle preparation moistened with Ringer's solution at all times._

Observing Graded Muscle Response to Increased Stimulus Intensity

1. Follow the previous setup instructions, but set the voltage control at the threshold voltage (as determined in the first experiment).

2. Deliver single stimuli at 1-minute intervals. Initially increase the voltage between shocks by 0.5 V; then increase the voltage by 1 to 2 V between shocks as the experiment continues, until contraction height increases no further. Stop the recording apparatus.

What voltage produced the highest spike (and thus the maximal strength of contraction)? _____ V

This voltage, called the **maximal stimulus** (for *your* muscle specimen), is the weakest stimulus at which all muscle cells are being stimulated.

3. Mark the record *maximal stimulus*. Record the maximal stimulus voltage and the time you completed the experiment.

4. What is happening to the muscle as the voltage is increased?

_____

_____

What is another name for this phenomenon? (Use an appropriate reference if necessary.)

_____

5. Explain why the strength of contraction does not increase once the maximal stimulus is reached.

_____

_____

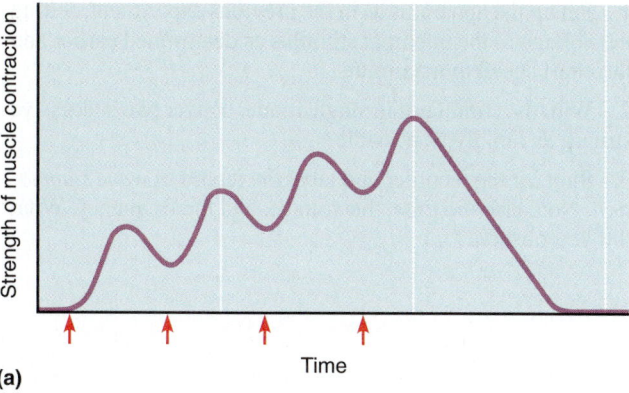

(a)

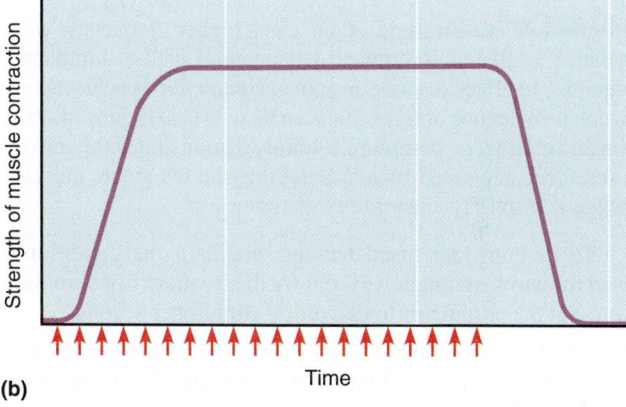

(b)

Figure 14.4 Muscle response to stimulation. Arrows represent stimuli. **(a)** Wave summation at low-frequency stimulation. **(b)** Fused tetanus occurs as stimulation rate is increased.

Timing the Muscle Twitch

1. Follow the previous setup directions, but set the voltage for the maximal stimulus (as determined in the preceding experiment) and set the paper advance or recording speed at maximum. Record the paper speed setting:

_____ mm/sec

2. Determine the time required for the paper to advance 1 mm by using the formula

$$\frac{1 \text{ mm}}{\text{mm/sec (paper speed)}}$$

(Thus, if your paper speed is 25 mm/sec, each mm on the chart equals 0.04 sec.) Record the computed value:

1 mm = _____ sec

3. Deliver single stimuli at 1-minute intervals to obtain several "twitch" curves. Stop the recording.

4. Determine the duration of the latent, contraction, and relaxation phases of the twitches and record the data below.

Duration of latent period: _____ sec

Duration of period of contraction: _____ sec

Duration of period of relaxation: _____ sec

5. Label the record to indicate the point of stimulus, the beginning of contraction, the end of contraction, and the end of relaxation.

6. Allow the muscle to rest (but keep it moistened with Ringer's solution) before continuing with the next experiment.

Observing Graded Muscle Response to Increased Stimulus Frequency

Muscles subjected to frequent stimulation, without a chance to relax, exhibit two kinds of responses—wave summation and tetanus—depending on the level of stimulus frequency **(Figure 14.4)**.

Wave Summation

If a muscle is stimulated with a rapid series of stimuli of the same intensity before it has had a chance to relax completely, the response to the second and subsequent stimuli will be greater than to the first stimulus (see Figure 14.4a). This phenomenon, called **wave summation,** or **temporal summation,** occurs because the muscle is already in a partially contracted state when subsequent stimuli are delivered.

14

1. Set up the apparatus as in the previous experiment, setting the voltage to the maximal stimulus as determined earlier and the chart speed to maximum.

2. With the stimulator in single mode, deliver two successive stimuli as rapidly as possible.

3. Shut off the recorder and label the record as *wave summation.* Note also the time, the voltage, and the frequency. What did you observe?

_____

_____

Tetanus

Stimulation of a muscle at an even higher frequency will produce a "fusion" (complete tetanization) of the summated twitches. In effect, a single sustained contraction is achieved in which no evidence of relaxation can be seen (see Figure 14.4b). **Fused tetanus,** or **complete tetanus,** demonstrates the maximum force generated by a skeletal muscle; the single muscle twitch is primarily a laboratory phenomenon.

1. To demonstrate fused tetanus, maintain the conditions used for wave summation except for the frequency of stimulation. Set the stimulator to deliver 60 stimuli per second.

2. As soon as you obtain a single smooth, sustained contraction (with no evidence of relaxation), discontinue stimulation and shut off the recorder.

3. Label the tracing with the conditions of experimentation, the time, and the area of *fused* or *complete tetanus.*

Inducing Muscle Fatigue

Muscle fatigue is a reversible physiological condition in which a muscle is unable to contract even though it is being stimulated. Fatigue can occur with short-duration maximal contraction or long-duration submaximal contraction. Although the phenomenon of muscle fatigue is not completely understood, several factors appear to contribute to it. Most affect excitation-contraction coupling. One theory involves the buildup of inorganic phosphate (P_i) from ATP and creatine phosphate breakdown, which may block calcium release from the sarcoplasmic reticulum (SR). Another theory suggests that potassium accumulation in the T tubules may block calcium release from the SR and alter the membrane potential of the muscle fiber. Lactic acid buildup, long implicated as a cause of fatigue, does not appear to play a role.

1. To demonstrate muscle fatigue, set up an experiment like the tetanus experiment but continue stimulation until the muscle completely relaxes and the contraction curve returns to the base line.

2. Measure the time interval between the beginning of complete tetanus and the beginning of fatigue (when the tracing begins its downward curve). Mark the record appropriately.

3. Determine the time required for complete fatigue to occur (the time interval from the beginning of fatigue until the return of the curve to the base line). Mark the record appropriately.

4. Allow the muscle to rest (keeping it moistened with Ringer's solution) for 10 minutes, and then repeat the experiment.

What was the effect of the rest period on the fatigued muscle?

_____

_____

What might be the physiological basis for this reaction?

_____

_____

Determining the Effect of Load on Skeletal Muscle

When the fibers of a skeletal muscle are slightly stretched by a weight or tension, the muscle responds by contracting more forcibly and thus is capable of doing more work. When the actin and myosin barely overlap, sliding can occur along nearly the entire length of the actin filaments. If the load is increased beyond the optimum, the latent period becomes longer, contractile force decreases, and relaxation (fatigue) occurs more quickly. With excessive stretching, the muscle is unable to develop any active tension and no contraction occurs. Since the filaments no longer overlap at all with this degree of stretching, the sliding force cannot be generated.

If your equipment allows you to add more weights to the muscle specimen or to increase the tension on the muscle, perform the following experiment to determine the effect of loading on skeletal muscle and to develop a work curve for the frog's gastrocnemius muscle.

1. Set the stimulator to deliver the maximal voltage as previously determined.

2. Stimulate the unweighted muscle with single shocks at 1– to 2– second intervals to achieve three or four muscle twitches.

3. Stop the recording apparatus and add 10 g of weight or tension to the muscle. Restart and advance the recording about 1 cm, and then stimulate again to obtain three or four spikes.

4. Repeat the previous step seven more times, increasing the weight by 10 g each time until the total load on the muscle is 80 g or the muscle fails to respond. If the calcaneal tendon tears, the weight will drop, which ends the trial. In such cases, you will need to prepare another frog's leg to continue the experiments and the maximal stimulus will have to be determined for the new muscle preparation.

5. When these "loading" experiments are completed, discontinue recording. Mark the curves on the record to indicate the load (in grams).

6. Measure the height of contraction (in millimeters) for each sequence of twitches obtained with each load, and insert this information on the **Activity 2 chart** (page 241).

7. Compute the work done by the muscle for each twitch (load) sequence.

Weight of load (g) $\times$ distance load lifted (mm) = work done

Enter these calculations into the chart in the column labeled Work done, Trial 1.

8. Allow the muscle to rest for 5 minutes. Then conduct a second trial in the same manner (i.e., repeat steps 2 through

7). Record this second set of measurements and calculations in the columns labeled Trial 2. Be sure to keep the muscle well moistened with Ringer's solution during the resting interval.

9. Using two different colors, plot a line graph of work done against the weight on the grid accompanying the chart for each trial. Label each plot appropriately.

10. Dismantle all apparatus and prepare the equipment for storage. Dispose of the frog remains in the appropriate container. Discard the gloves as instructed and wash and dry your hands.

11. Inspect your records of the experiments and make sure each is fully labeled with the experimental conditions, the date, and the names of those who conducted the experiments. For future reference, attach a tracing (or a copy of the tracing) for each experiment to this page. ■■

Activity 2: Results for Effect of Load on Skeletal Muscle

Load (g)	Distance load lifted (mm)		Work done	
	Trial 1	Trial 2	Trial 1	Trial 2
0				
10				
20				
20				
30				
40				
50				
60				
70				
80				

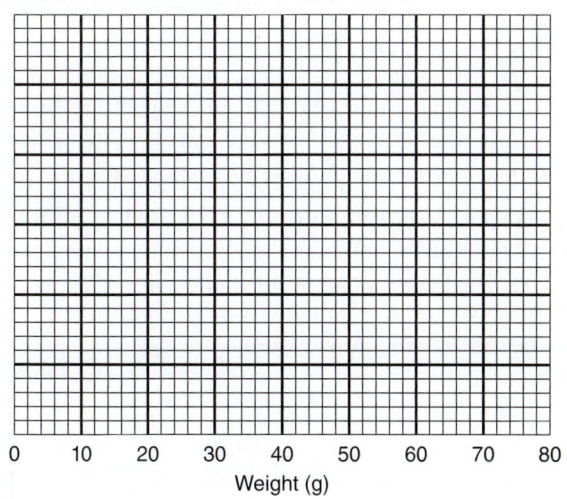

Weight (g)

ACTIVITY 3

Demonstrating Muscle Fatigue in Humans

1. Work in small groups. In each group select a subject, a timer, and a recorder.

2. Obtain a copy of the laboratory manual and a copy of the textbook. Weigh each book separately, and then record the weight of each in the **Activity 3 chart** in step 6.

3. The subject is to extend an upper limb straight out in front of him or her, holding the position until the arm shakes or the muscles begin to ache. Record the time to fatigue on the chart.

4. Allow the subject to rest for several minutes. Now ask the subject to hold the laboratory manual while keeping the arm and forearm in the same position as in step 3 above. Record the time to fatigue on the chart.

5. Allow the subject to rest again for several minutes. Now ask the subject to hold the textbook while keeping the upper limb in the same position as in steps 3 and 4 above. Record the time to fatigue on the chart.

6. Each person in the group should take a turn as the subject, and all data should be recorded in the chart below.

Activity 3: Results for Human Muscle Fatigue

Load	Weight of object	Time elapsed until fatigue		
		Subject 1	Subject 2	Subject 3
Appendage	N/A			
Lab Manual				
Textbook				

7. What can you conclude about the effect of load on muscle fatigue? Explain.

_____

_____

_____ ■■

ACTIVITY 4

Electromyography in a Human Subject Using BIOPAC®

Part 1: Temporal and Multiple Motor Unit Summation

This activity is an introduction to a procedure known as **electromyography,** the recording of skin-surface voltage that indicates underlying skeletal muscle contraction. The actual visible recording of the resulting voltage waveforms is called an **electromyogram** (EMG).

A single skeletal muscle consists of numerous elongated *skeletal muscle cells,* also called *skeletal muscle fibers* (Exercise 12). These muscle cells are excited by *motor neurons* of the central nervous system whose *axons* terminate at the muscle. An axon of a motor neuron branches profusely at the muscle. Each branch produces multiple **axon terminals,** each of which innervates a single fiber. The number of muscle cells controlled by a single motor neuron can vary greatly, from five (for fine control needed in the hand) to 500 (for gross control, such as in the buttocks). The most important organizational concept in the physiology of muscle contraction is the **motor unit,** a single motor neuron and all of the cells within a muscle that it activates (see Figure 12.6 page 190). Understanding gross muscular contraction depends upon realizing that a single muscle consists of multiple motor units, and that the gradual and coordinated activation of these motor units results in **graded contraction** of the whole muscle.

14

14

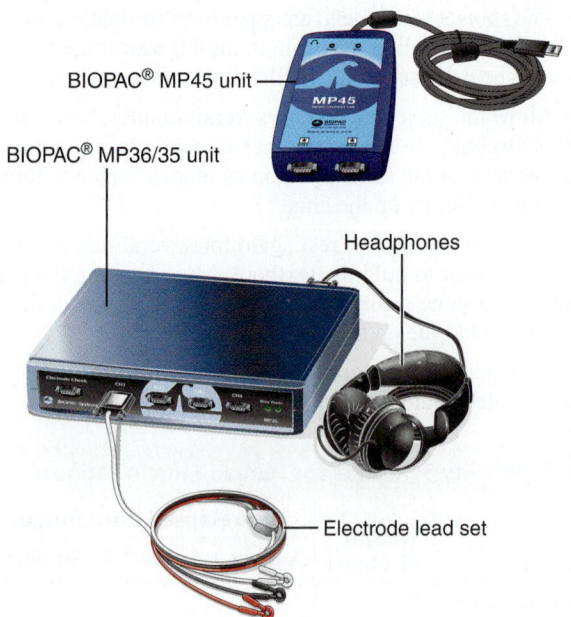

Figure 14.5 Setting up the BIOPAC® equipment to observe recruitment and temporal summation. Plug the headphones into the back of the MP36/35 data acquisition unit or into the top of the MP45 unit, and the electrode lead set into Channel 1. Electrode leads and headphones are shown connected to the MP36/35 unit.

The nervous system controls muscle contraction by two mechanisms:

- **Recruitment (multiple motor unit summation):** the gradual activation of more and more motor units

- **Temporal (wave) summation:** an increase in the *frequency* of nerve impulses for each active motor unit

Thus, increasing the force of contraction of a muscle arises from gradually increasing the number of motor units being activated and increasing the frequency of nerve impulses delivered by those active motor units.

A final phenomenon, which is hardly noticeable except when performing electromyography, is **tonus,** a constant state of slight excitation of a muscle while it is in the relaxed state. Even while "at rest," a small number of motor units to a skeletal muscle remain slightly active to prepare the muscle for possible contraction.

Setting Up the Equipment

1. Connect the BIOPAC® unit to the computer, then turn the computer **ON.**

2. Make sure the BIOPAC® unit is **OFF.**

3. Plug in the equipment (as shown in **Figure 14.5**).

- Electrode lead set—CH 1

- Headphones—back of MP36/35 unit or top of MP45 unit

4. Turn the BIOPAC® unit **ON.**

5. Attach three electrodes to the subject's dominant forearm (as shown in **Figure 14.6**) and attach the electrode leads according to the colors indicated.

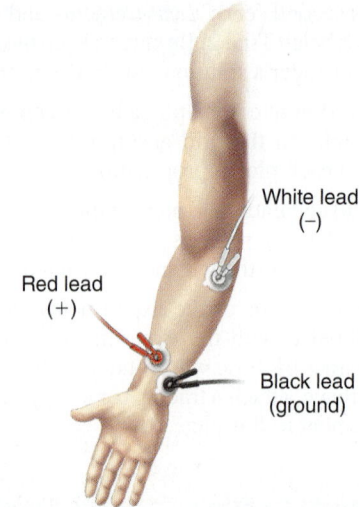

Figure 14.6 Placement of electrodes and the appropriate attachment of electrode leads by color.

6. Start the BIOPAC® Student Lab program by double-clicking the icon on the desktop or by following your instructor's guidance.

7. Select lesson **L01-EMG-1** from the menu and click **OK.**

8. Type in a filename that will save this subject's data on the computer hard drive. You may want to use the subject's last name followed by EMG-1 (for example, SmithEMG-1). Then click **OK.**

Calibrating the Equipment

1. With the subject in a still position, click **Calibrate.** This initiates the process by which the computer automatically establishes parameters to record the data properly for the subject.

2. After you click **OK,** have the subject wait for 2 seconds, clench the fist tightly for 2 seconds, then release the fist and relax. The computer then automatically stops the recording.

3. Observe the recording of the calibration data, which should look like the waveform in the example **(Figure 14.7)**.

- If the data look very different, click **Redo Calibration** and repeat the steps above.

- If the data look similar, proceed to the next section.

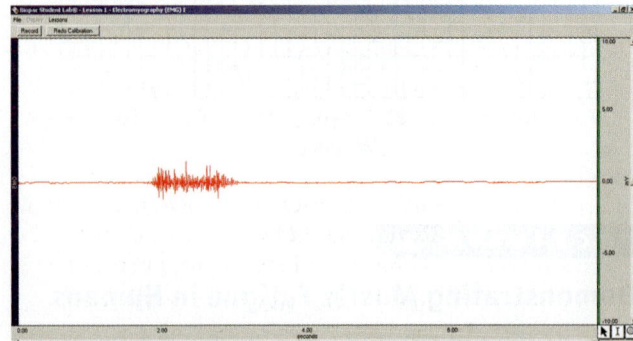

Figure 14.7 Example of waveform during the calibration procedure.

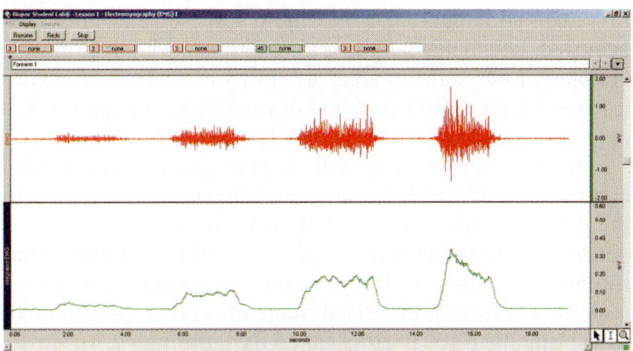

Figure 14.8 Example of waveforms during the recording of data. Note the increased signal strength with the increasing force of the clench.

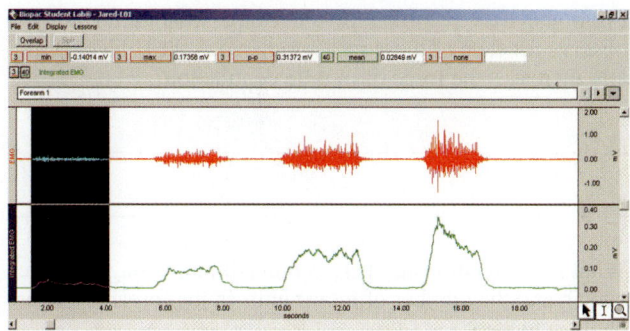

Figure 14.9 Using the I-beam cursor to highlight a cluster of data for analysis.

Recording the Data

1. Tell the subject that the recording will be of a series of four fist clenches, with the following instructions: First clench the fist softly for 2 seconds, then relax for 2 seconds, then clench harder for 2 seconds, and relax for 2 seconds, then clench even harder for 2 seconds, and relax for 2 seconds, and finally clench with maximum strength, then relax. The result should be a series of four clenches of increasing intensity.

When the subject is ready to do this, click **Record;** then click **Suspend** when the subject is finished.

2. Observe the recording of the data, which should look like the waveforms in the example **(Figure 14.8)**.

- If the data look very different, click **Redo** and repeat the steps above.

- If the data look similar, click **STOP.** Click **YES** in response to the question, "Are you finished with both forearm recordings?"

Optional: Anyone can use the headphones to listen to an "auditory version" of the electrical activity of contraction by clicking **Listen** and having the subject clench and relax. Note that the frequency of the auditory signal corresponds with the frequency of action potentials stimulating the muscles. The signal will continue to run until you click **STOP.**

3. Click **Done,** and then remove all electrodes from the forearm.

- If you wish to record from another subject, choose the **Record from another subject** option and return to step 5 under Setting Up the Equipment.

- If you are finished recording, choose **Analyze current data file** and click **OK.** Proceed to Data Analysis, step 2.

Data Analysis

1. If you are just starting the BIOPAC® program to perform data analysis, enter **Review Saved Data** mode and choose the file with the subject's EMG data (for example, SmithEMG-1).

2. Observe the **Raw EMG** recording and computer-calculated **Integrated EMG.** The raw EMG is the actual recording of the voltage (in mV) at each instant in time, while the integrated EMG reflects the absolute intensity of the voltage from baseline at each instant in time.

3. To analyze the data, set up the first four pairs of channel/measurement boxes at the top of the screen by selecting the following channels and measurement types from the drop-down menus.

Channel	Measurement	Data
CH 1	min	raw EMG
CH 1	max	raw EMG
CH 1	p-p	raw EMG
CH 40	mean	integrated EMG

4. Use the arrow cursor and click the I-beam cursor box at the lower right of the screen to activate the "area selection" function. Using the activated I-beam cursor, highlight the first EMG cluster, representing the first fist-clenching **(Figure 14.9)**.

5. Notice that the computer automatically calculates the **min, max, p-p,** and **mean** for the selected area. These measurements, calculated from the data by the computer, represent the following:

min: Displays the *minimum* value in the selected area

max: Displays the *maximum* value in the selected area

p-p (peak-to-peak): Measures the difference in value between the highest and lowest values in the selected area

mean: Displays the average value in the selected area

6. Write down the data for clench 1 in the chart in step 7 (round to the nearest 0.01 mV).

7. Using the I-beam cursor, highlight the clusters for clenches 2, 3, and 4, and record the data in the following chart.

EMG Cluster Results			
	Min	Max	P-P	Mean
Clench 1				
Clench 2				
Clench 3				
Clench 4				

From the data recorded in the chart, what trend do you observe for each of these measurements as the subject gradually increases the force of muscle contraction?

_____

_____

What is the relationship between maximum voltage for each clench and the number of motor units in the forearm that are being activated?

_____

_____

Part 2: Force Measurement and Fatigue

In this set of activities you will be observing graded muscle contractions and fatigue in a subject. **Graded muscle contractions,** which represent increasing levels of force generated by a muscle, depend upon: (1) the gradual activation of more motor units, and (2) increasing the frequency of motor neuron action potentials for each active motor unit. This permits a range of forces to be generated by any given muscle or group of muscles, all the way up to the maximum force.

For example, the biceps muscle will have more active motor units and exert more force when lifting a 10-kg object than when lifting a 2-kg object. In addition, the motor neuron of each active motor unit will increase the frequency of action potentials delivered to the motor units, resulting in tetanus. When all of the motor units of a muscle are activated and

in a state of tetanus, the maximum force of that muscle is achieved. Recall that fatigue is a condition in which the muscle gradually loses some or all of its ability to contract after contracting for an extended period of time. Recent experimental evidence suggests that this is mostly due to ionic imbalances.

In this exercise, you will observe and measure graded contractions of the fist, and then observe fatigue, in both the dominant and nondominant arms. To measure the force generated during fist contraction, you will use a **hand dynamometer** (*dynamo* = force; *meter* = measure). The visual recording of force is called a **dynagram,** and the procedure of measuring the force itself is called **dynamometry.**

You will first record data from the subject's **dominant arm** (forearm 1) indicated by his or her "handedness," then repeat the procedures on the subject's **nondominant arm** (forearm 2) for comparison.

Setting Up the Equipment

1. Connect the BIOPAC® unit to the computer and turn the computer **ON.**

2. Make sure the BIOPAC® unit is **OFF.**

3. Plug in the equipment (as shown in **Figure 14.10**).

- Electrode lead set—CH 1

- Hand dynamometer—CH 2

- Headphones—back of MP36/35 unit or top of MP45 unit

4. Turn the BIOPAC® unit **ON.**

5. Attach three electrodes to the subject's *dominant* forearm (forearm 1; see the attachments in **Figure 14.11**), and attach the electrode leads according to the colors indicated.

6. Start the BIOPAC® Student Lab program on the computer by double-clicking the icon on the desktop or by following your instructor's guidance.

7. Select lesson **L02-EMG-2** from the menu and click **OK.**

8. Type in a filename that will save this subject's data on the computer hard drive. You may want to use subject's last name followed by EMG-2 (for example, SmithEMG-2). Then click **OK.**

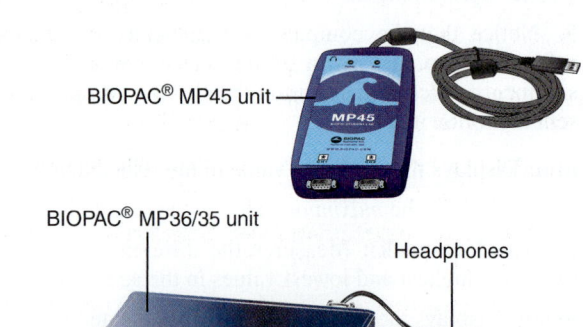
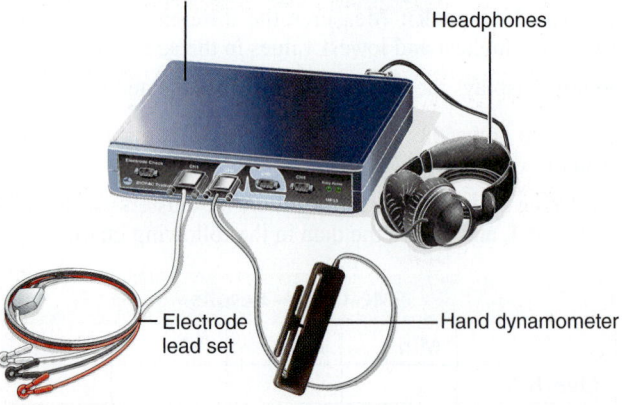

BIOPAC® MP45 unit

BIOPAC® MP36/35 unit

Headphones

Electrode lead set

Hand dynamometer

Figure 14.10 Setting up the BIOPAC® equipment to observe graded muscle contractions and muscle fatigue. Plug the headphones into the back of the MP36/35 data acquisition unit or into the top of the MP45 unit, the electrode lead set into Channel 1, and the hand dynamometer into Channel 2. Electrode leads and dynamometer are shown connected to the MP36/35 unit.

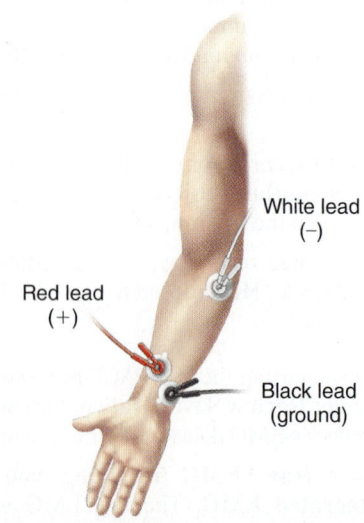

White lead (−)

Red lead (+)

Black lead (ground)

Figure 14.11 Placement of electrodes and the appropriate attachment of electrode leads by color.

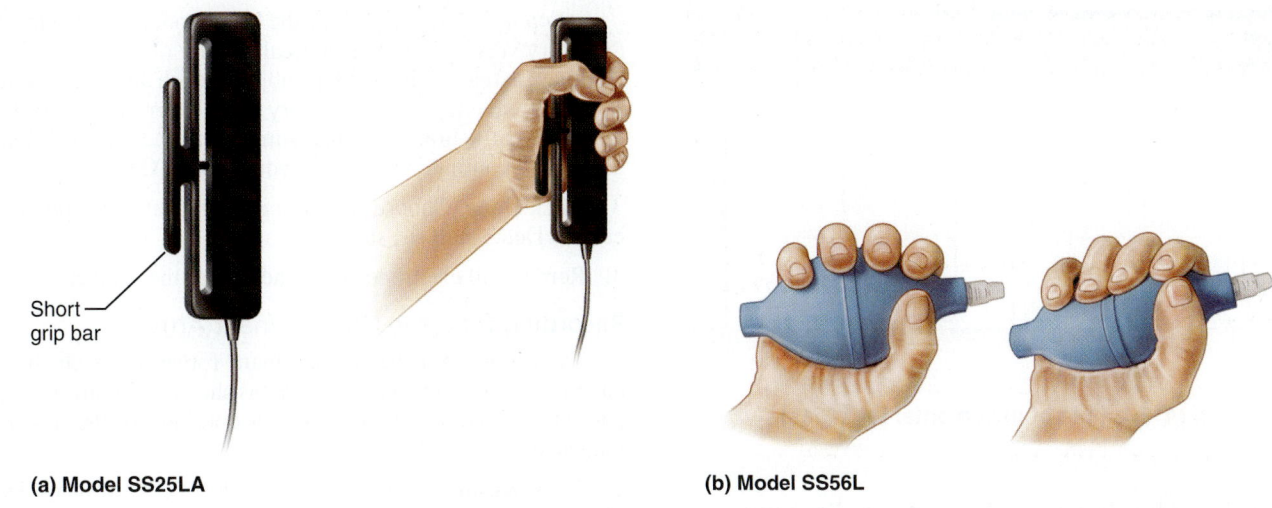

(a) Model SS25LA

Short grip bar

(b) Model SS56L

Figure 14.12 Proper grasp of the hand dynamometer using either model SS25LA or model SS56L.

14

Calibrating the Equipment

1. With the hand dynamometer at rest on the table, click **Calibrate.** This initiates the process by which the computer automatically establishes parameters to record the data properly for the subject.

2. A pop-up window prompts the subject to remove any grip force. This is to ensure that the dynamometer has been at rest on the table and that no force is being applied. When this is so, click **OK.**

3. As instructed by the pop-up window, the subject is to grasp the hand dynamometer with the dominant hand. With model SS25LA, grasp the dynamometer with the palm of the hand against the short grip bar (as shown in **Figure 14.12a**). Hold model SS25LA vertically. With model SS56L, wrap the hand around the bulb (as shown in Figure 14.12b). Do not curl the fingers into the bulb. (The older SS25L hand dynamometer may also be used with any of the data acquisition units.) Then, click **OK.** The instructions that follow apply to this model of dynamometer.

4. Tell the subject that the instructions will be to wait 2 seconds, then squeeze the hand dynamometer as hard as possible for 2 seconds, and then relax. The computer will automatically stop the calibration.

5. When the subject is ready to proceed, click **OK** and follow the instructions in step 4, which are also in the pop-up window. The calibration will stop automatically.

6. Observe the recording of the calibration data, which should look like the waveforms in the calibration example **(Figure 14.13)**.

- If the data look very different, click **Redo Calibration** and repeat the steps above.

- If the data look similar, proceed to the next section.

Recording Incremental Force Data for the Forearm

1. Using the force data from the calibration procedure, estimate the **maximum force** that the subject generated (kg).

2. Divide that maximum force by four. In the following steps, the subject will gradually increase the force in approximately

these increments. For example, if the maximum force generated was 20 kg, the increment will be 20/4 = 5 kg. The subject will grip at 5 kg, then 10 kg, then 15 kg, and then 20 kg. The subject should watch the tracing on the computer screen and compare it to the scale on the right to determine each target force. Click **Continue**.

3. After you click **Record,** have the subject wait 2 seconds, clench at the first force increment 2 seconds (for example, 5 kg), then relax 2 seconds, clench at the second force increment 2 seconds (10 kg), then relax 2 seconds, clench at the third force increment 2 seconds (15 kg), then relax 2 seconds, then clench with the maximum force 2 seconds (20 kg), and then relax. When the subject relaxes after the maximum clench, click **Suspend** to stop the recording.

4. Observe the recording of the data, which should look similar to the data in the example **(Figure 14.14)**.

- If the data look very different, click **Redo** and repeat the steps above.

- If the data look similar, click **Continue** and proceed to observation and recording of muscle fatigue.

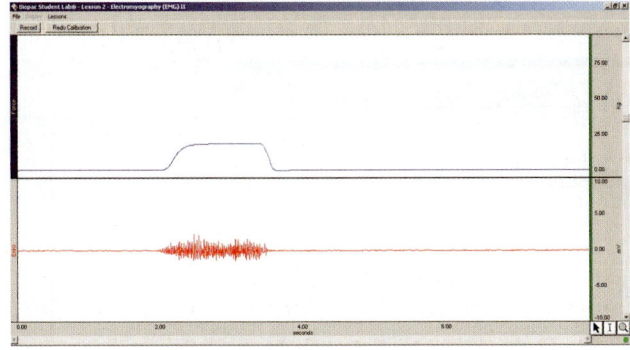

Figure 14.13 Example of calibration data. Force is measured in kilograms or kgf/m^2 at the top and electromyography is measured in millivolts at the bottom.

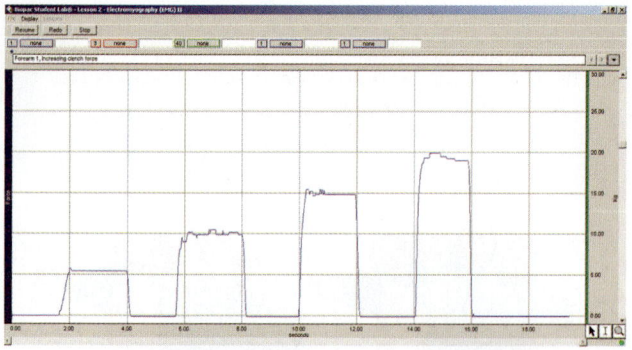

Figure 14.14 Example of incremental force data.

Recording Muscle Fatigue Data for the Forearm

Continuing from the end of the incremental force recording, the subject will next record muscle fatigue.

1. After you click **Resume,** the recording will continue from where it stopped and the subject will clench the dynamometer with maximum force for as long as possible. A "marker" will appear at the top of the data, denoting the beginning of this recording segment. When the subject's clench force falls below 50% of the maximum (for example, below 10 kg for a subject with 20 kg maximum force), click **Suspend.** The subject should not watch the screen during this procedure; those helping can inform the subject when it is time to relax.

2. Observe the recording of the data, which should look similar to the data in the muscle fatigue data example **(Figure 14.15)**.

- If the data look very different, click **Redo** and repeat the steps above.

- If the data look similar, and you want to record from the nondominant arm, click **Continue** and proceed to Recording from the Nondominant Arm.

- If the data look similar, and you do not want to record from the nondominant arm or you have just finished recording the nondominant arm, click **STOP.** A dialog box comes up asking if you are sure you want to stop recording. Click **NO** to return to the **Resume** or **Stop** options, providing one last chance to redo the fatigue recording. Click **YES** to end the recording session and automatically save your data.

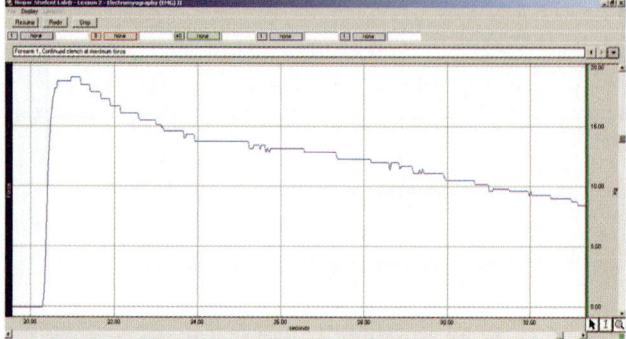

Figure 14.15 Example of muscle fatigue data.

Optional: Anyone can use the headphones to listen to an "auditory version" of the electrical activity of contraction by clicking **Listen** and having the subject clench and relax. Note that the frequency of the auditory signal corresponds to the frequency of action potentials stimulating the muscles. The signal will continue to run until you click **STOP.**

3. Click **Done**. Choose **Analyze current data file** and proceed to Data Analysis, step 2.

4. Remove all electrodes from the arm of the subject.

Recording from the Nondominant Arm

1. To record from the nondominant forearm, attach three electrodes to the subject's forearm (as shown in Figure 14.11) and attach the electrode leads according to the colors indicated.

2. Click **Resume.** Repeat the Clench-Release-Wait cycles with increasing clench force as performed with the dominant arm.

3. Observe the recording of the data, which should look similar to the data in the muscle fatigue data example (Figure 14.15). If the data look very different, click **Redo** and repeat.

4. End the session by repeating the steps for muscle fatigue. Click **Continue** and repeat steps 1 and 2 of the muscle fatigue section, recording the nondominant arm.

Data Analysis

1. In **Review Saved Data** mode, select the file that is to be analyzed (for example, SmithEMG-2-1-L02).

2. Observe the recordings of the clench **Force** (kg or kgf/m^2), **Raw EMG** (mV), and computer-calculated **Integrated EMG** (mV). The force is the actual measurement of the strength of clench in kilograms at each instant in time. The raw EMG is the actual recording of the voltage (in mV) at each instant in time, and the integrated EMG indicates the absolute intensity of the voltage from baseline at each instant in time.

3. To analyze the data, set up the first three pairs of channel/measurement boxes at the top of the screen. Select the following channels and measurement types:

Channel	Measurement	Data
CH 41	mean	clench force
CH 40	mean	integrated EMG

4. Use the arrow cursor and click the I-beam cursor box on the lower right side of the screen to activate the "area selection" function. Using the activated I-beam cursor, highlight the "plateau phase" of the first clench cluster. The plateau should be a relatively flat force in the middle of the cluster **(Figure 14.16)**.

5. Observe that the computer automatically calculates the **p-p** and **mean** values for the selected area. These measurements, calculated from the data by the computer, represent the following:

p-p (peak-to-peak): Measures the difference in value between the highest and lowest values in the selected area

mean: Displays the average value in the selected area

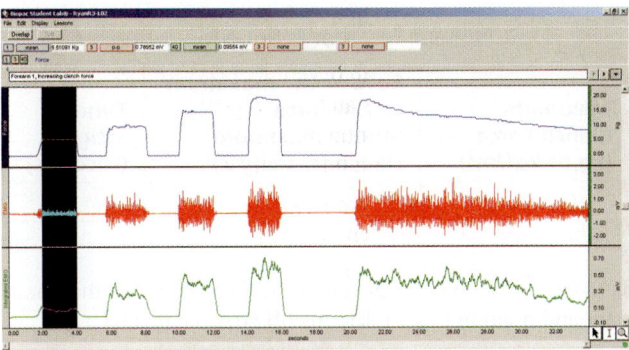

Figure 14.16 Highlighting the "plateau" of clench cluster 1.

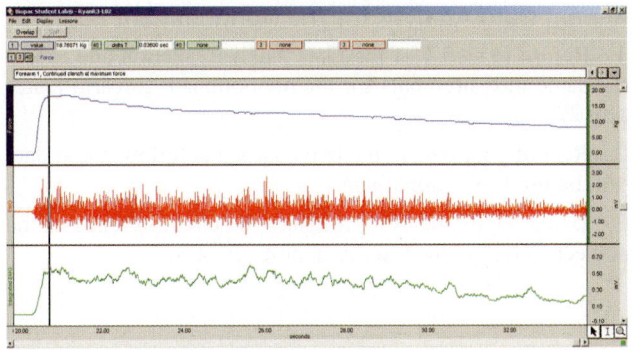

Figure 14.17 Selection of single point of maximum clench.

6. In the chart in step 7, record the data for clench 1 (for example, 5-kg clench) to the nearest 0.01.

7. Using the I-beam cursor, highlight the clusters for the subsequent clenches and record the data in the chart.

Dominant Forearm Clench Increments		
	Force at plateau Mean (kg or kgf/m²)	Integrated EMG Mean (mV-sec)
Clench 1		
Clench 2		
Clench 3		
Clench 4		

8. Scroll along the bottom of the data page to the segment that includes the recording of muscle fatigue (it should begin after the "marker" that appears at the top of the data).

9. Change the channel/measurement boxes so that the first two selected appear as follows (the third should be set to "none"):

Channel	Measurement	data
CH 41	value	force
CH 40	delta T	integrated EMG

value: Measures the highest value in the selected area (Force measured in kg with SS25LA or kgf/m² with the SS56L.)

delta T: Measures the time elapsed in the selected area (Time measured in seconds.)

10. Use the arrow cursor and click on the I-beam cursor box on the lower right side of the screen to activate the "area selection" function.

11. Using the activated I-beam cursor, select just the single point of maximum clench strength at the start of this data segment (as shown in **Figure 14.17**).

12. In the following chart, note the maximum force measurement for this point.

Dominant Forearm Fatigue Measurement		
Maximum clench force (kg or kgf/m²)	50% of the maximum clench force (divide maximum clench force by 2)	Time to fatigue (seconds)

13. Calculate the value of 50% of the maximum clench force, and record this in the data chart.

14. Using a metric tape measure, measure the circumference of the subject's dominant forearm at its greatest diameter:

_____ cm

15. Measure the amount of time that elapsed between the initial maximum force and the point at which the subject fatigued to 50% of this level. Using the activated I-beam cursor, highlight the area from the point of 50% clench force back to the point of maximal clench force, as shown in the example **(Figure 14.18)**.

16. Note the time it took the subject to reach this point of fatigue (CH 40 delta T) and record this data in the Dominant Forearm Fatigue Measurement chart.

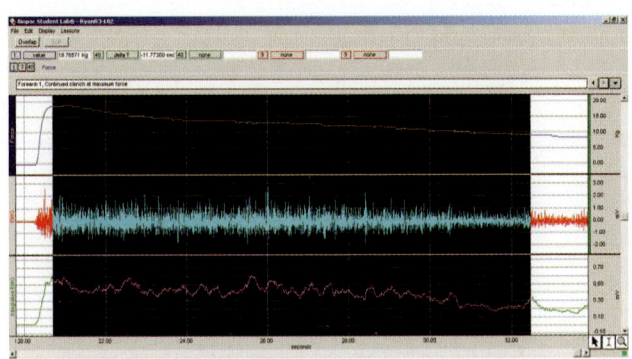

Figure 14.18 Highlighting to measure elapsed time to 50% of the maximum clench force.

Repeat Data Analysis for the Nondominant Forearm

1. Return to step 1 of the Data Analysis section and repeat the same measurements for the nondominant forearm (forearm 2).

2. Record your data in the two charts that follow; these data will be used for comparison.

3. Using a tape measure, measure the circumference of the subject's nondominant forearm at its greatest:

_____ cm

4. When finished, exit the program by going to the **File** menu at the top of the screen and clicking **Quit.**

Is there a difference in maximal force that was generated between the dominant and nondominant forearms? If so, how much?

_____

Calculate the percentage difference in force between the dominant maximal force and nondominant maximal force.

_____

_____

Nondominant Forearm Clench Increments		
	Force at plateau Mean (kg or kgf/m^2)	Integrated EMG Mean (mV-sec)
Clench 1		
Clench 2		
Clench 3		
Clench 4		

Nondominant Forearm Fatigue Measurement		
Maximum Clench Force (kg or kgf/m^2)	50% of the maximum clench force (divide maximum clench force by 2)	Time to fatigue (seconds)

Is there a difference in the circumference between the dominant and nondominant forearms? If so, how much?

_____

If there is a difference in circumference, is this difference likely to be due to a difference in the *number* of muscle fibers in each forearm or in the *diameter* of each muscle fiber in the forearms? Explain. Use an appropriate reference if needed.

_____

_____

Compare the time to fatigue between the two forearms.

_____

_____

Skeletal Muscle Physiology: Frogs and Human Subjects

Muscle Activity

1. The following group of incomplete statements begins with a muscle cell in the resting state just before stimulation. Complete each statement by choosing the correct response from the key items.

Key: a. Na^+ diffuses out of the cell
 b. K^+ diffuses out of the cell
 c. Na^+ diffuses into the cell
 d. K^+ diffuses into the cell
 e. inside the cell
 f. outside the cell

 g. relative ionic concentrations on the two sides of the membrane
 h. electrical conditions
 i. activation of the sodium-potassium pump, which moves K^+ into the cell and Na^+ out of the cell
 j. activation of the sodium-potassium pump, which moves Na^+ into the cell and K^+ out of the cell

There is a greater concentration of Na^+ _____; there is a greater concentration of K^+ _____.

When the stimulus is delivered, the permeability of the membrane at that point is changed; and _____, initiating

the depolarization of the membrane. Almost as soon as the depolarization wave has begun, a repolarization wave follows it

across the membrane. This occurs as _____. Repolarization restores the _____ of the resting cell

membrane. The _____ is (are) reestablished by _____.

2. Number the following statements in the proper sequence to describe the contraction mechanism in a skeletal muscle cell. Number 1 has already been designated.

_____ 1 _____ Depolarization occurs, and the action potential is generated.

_____ The muscle cell relaxes and lengthens.

_____ The calcium ion concentrations at the myofilaments increase; the myofilaments slide past one another, and the cell shortens.

_____ The action potential, carried deep into the cell by the T tubules, triggers the release of calcium ions from the sarcoplasmic reticulum.

_____ The concentration of the calcium ions at the myofilaments decreases as they are actively transported into the sarcoplasmic reticulum.

3. Refer to your observations of muscle fiber contraction in Activity 1 to answer the following questions.

a. Did your data support your hypothesis? _____

b. *Explain* your observations fully. _____

_____

_____

c. Draw a relaxed and a contracted sarcomere below.

Relaxed   **Contracted**

Induction of Contraction in the Frog Gastrocnemius Muscle

4. Why is it important to destroy the brain and spinal cord of a frog before conducting physiological experiments on muscle

contraction? _____

_____

5. What kind of stimulus (electrical or chemical) travels from the motor neuron to skeletal muscle? _____

What kind of stimulus (electrical or chemical) travels from the axon terminal to the sarcolemma? _____

6. Give the name and duration of each of the three phases of the muscle twitch, and describe what is happening during each phase.

a. _____, _____ msec, _____

_____

b. _____, _____ msec, _____

c. _____, _____ msec, _____

7. Use the items in the key to identify the conditions described.

Key:

a. maximal stimulus
b. multiple motor unit summation
c. subthreshold stimulus

d. tetanus
e. threshold stimulus
f. wave summation

_____ 1. sustained contraction without any evidence of relaxation

_____ 2. stimulus that results in no perceptible contraction

_____ 3. stimulus at which the muscle first contracts perceptibly

_____ 4. increasingly stronger contractions owing to stimulation at a rapid rate

_____ 5. increasingly stronger contractions owing to increased stimulus strength

_____ 6. weakest stimulus at which all muscle cells in the muscle are contracting

8. Complete the following statements by writing the appropriate words on the corresponding numbered blanks at the right.

When a weak but smooth muscle contraction is desired, a few motor units are stimulated at a __1__ rate. Within limits, as the load on a muscle is increased, the muscle contracts __2__ (more/less) strongly.

1. _____

2. _____

9. During the frog experiment on muscle fatigue, how did the muscle contraction pattern change as the muscle began to fatigue?

_____

_____

How long was stimulation continued before fatigue was apparent? _____

If the sciatic nerve that stimulates the living frog's gastrocnemius muscle had been left attached to the muscle and the stimulus had been applied to the nerve rather than the muscle, would fatigue have become apparent sooner, later, or at the same time?

_____

10. What will happen to a muscle in the body when its nerve supply is destroyed or badly damaged?

_____

_____

11. Explain the relationship between the load on a muscle and its strength of contraction. _____

_____

_____

12. The skeletal muscles are maintained in a slightly stretched condition for optimal contraction. How is this accomplished?

_____

Why does stretching a muscle beyond its optimal length reduce its ability to contract? (Include an explanation of the events

at the level of the myofilaments.) _____

_____

_____

13. If the length but not the tension of a muscle is changed, the contraction is called an isotonic contraction. In an isometric contraction, the tension is increased but the muscle does not shorten. Which type of contraction did you observe most often

during the laboratory experiments? _____

Electromyography in a Human Subject Using BIOPAC®

14. If you were a physical therapist applying a constant voltage to the forearm, what might you observe if you gradually increased the *frequency* of stimulatory impulses, keeping the voltage constant each time?

_____

_____

15. Describe what is meant by the term *motor unit recruitment.* _____

_____

16. Describe the physiological processes occurring in the muscle cells that account for the gradual onset of muscle fatigue.

_____

_____

_____

17. Most subjects use their dominant forearm far more than their nondominant forearm. What does this indicate about degree of activation of motor units and these factors: muscle fiber diameter, maximum muscle fiber force, and time to muscle fatigue? (You may need to use your textbook for help with this one.)

_____

_____

_____

_____

18. Define *dynamometry.* _____

_____

19. How might dynamometry be used to assess patients in a clinical setting? _____

_____

_____

Histology of Nervous Tissue

MATERIALS

- ☐ Model of a "typical" neuron (if available)
- ☐ Compound microscope
- ☐ Immersion oil
- ☐ Prepared slides of an ox spinal cord smear and teased myelinated nerve fibers
- ☐ Prepared slides of Purkinje cells (cerebellum), pyramidal cells (cerebrum), and a dorsal root ganglion
- ☐ Prepared slide of a nerve (x.s.)

OBJECTIVES

1. Discuss the functional differences between neurons and neuroglia.
2. List six types of neuroglia and indicate where each is found in the nervous system.
3. Identify the important anatomical features of a neuron on an appropriate image.
4. List the functions of dendrites, axons, and axon terminals.
5. Explain how a nerve impulse is transmitted from one neuron to another.
6. State the function of myelin sheaths and explain how Schwann cells myelinate axons in the peripheral nervous system.
7. Classify neurons structurally and functionally.
8. Differentiate between a nerve and a tract, and between a ganglion and a CNS nucleus.
9. Define nerve.
10. Identify endoneurium, perineurium, and epineurium microscopically or in an appropriate image and cite their functions.

PRE-LAB QUIZ

1. Circle the correct underlined term. Nervous tissue is made up of <u>two</u> / <u>three</u> main cell types.
2. Neuroglia of the peripheral nervous system include
   a. ependymal cells and satellite cells
   b. oligodendrocytes and astrocytes
   c. satellite cells and Schwann cells
3. _____ are the functional units of nervous tissue.
4. These branching neuron processes serve as receptive regions and transmit electrical signals toward the cell body. They are:
   a. axons   c. dendrites
   b. collaterals   d. neuroglia
5. Circle True or False. Axons are the neuron processes that generate and conduct nerve impulses.
6. Most axons are covered with a fatty material called _____, which insulates the fibers and increases the speed of neurotransmission.
7. Circle the correct underlined term. Neuron fibers (axons) running through the central nervous system form <u>tracts</u> / <u>nerves</u> of white matter.
8. Neurons can be classified according to structure. _____ neurons have many processes that issue from the cell body.
   a. Bipolar   b. Multipolar   c. Unipolar
9. Circle the correct underlined term. Neurons can be classified according to function. <u>Afferent</u> / <u>Efferent</u> or motor neurons carry electrical signals from the central nervous system primarily to muscles or glands.
10. Within a nerve, each axon is surrounded by a covering called the:
    a. endoneurium   b. epineurium   c. perineurium

MasteringA&P® For related exercise study tools, go to the Study Area of MasteringA&P. There you will find:

- Practice Anatomy Lab PAL
- PhysioEx PEx
- A&PFlix A&PFlix
- Practice quizzes, Histology Atlas, eText, Videos, and more!

The nervous system is the master integrating and coordinating system, continuously monitoring and processing sensory information both from the external environment and from within the body. Every thought, action, and sensation is a reflection of its activity. Like a computer, it processes and integrates new "inputs" with information previously fed into it to produce an appropriate response. However, no computer can possibly compare in complexity and scope to the human nervous system.

Two primary divisions make up the nervous system: the central nervous system, or CNS, consisting of the brain and spinal cord, and the peripheral nervous system, or PNS, which includes all the nervous elements located outside the central nervous system. PNS structures include nerves, sensory receptors, and some clusters of nerve cells.

Despite its complexity, nervous tissue is made up of just two principal cell types; neurons and neuroglia.

Neuroglia

The **neuroglia** ("nerve glue") or **glial cells** of the CNS include *astrocytes, oligodendrocytes, microglial cells,* and *ependymal cells* **(Figure 15.1)**. The neuroglia found in the PNS include *Schwann cells,* also called neurolemmocytes, and *satellite cells.*

Neuroglia serve the needs of the delicate neurons by bracing and protecting them. In addition, they act as phagocytes (microglial cells), myelinate the cytoplasmic extensions of the neurons (oligodendrocytes and Schwann cells), play a role in capillary-neuron exchanges, and control the chemical environment around neurons (astrocytes). Although neuroglia resemble neurons in some ways (they have fibrous cellular extensions), they are not capable of generating and transmitting nerve impulses, a capability that is highly developed in neurons. Our focus in this exercise is the highly excitable neurons.

Neurons

Neurons, or nerve cells, are the basic functional units of nervous tissue. They are highly specialized to transmit messages from one part of the body to another in the form of nerve impulses. Although neurons differ structurally, they have many identifiable features in common **(Figure 15.2a and b)**. All have a **cell body** from which slender processes extend. The cell body is both the *biosynthetic center* of the neuron and part of its *receptive region.* Neuron cell bodies make up the gray matter of the CNS, and form clusters there that are called **nuclei.** In the PNS, clusters of neuron cell bodies are called **ganglia.**

The neuron cell body contains a large round nucleus surrounded by cytoplasm. Two prominent structures are found in the cytoplasm: One is cytoskeletal elements called **neurofibrils**, which provide support for the cell and a means

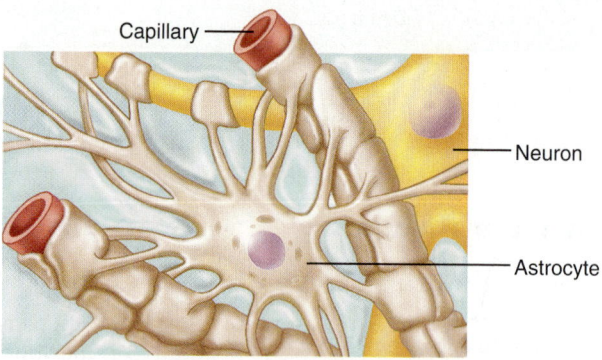

Capillary — Neuron — Astrocyte

(a) Astrocytes are the most abundant CNS neuroglia.

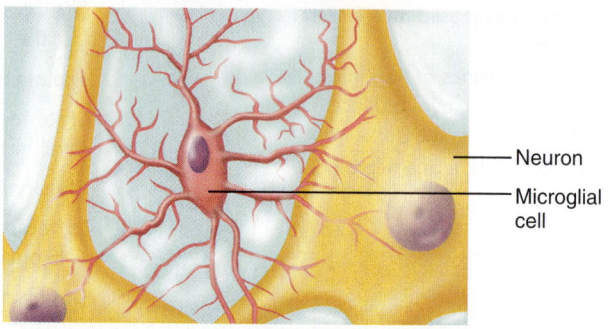

Neuron — Microglial cell

(b) Microglial cells are defensive cells in the CNS.

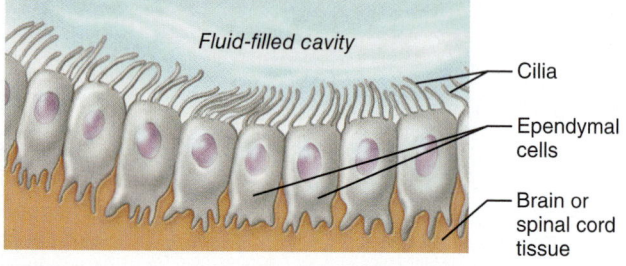

Fluid-filled cavity — Cilia — Ependymal cells — Brain or spinal cord tissue

(c) Ependymal cells line cerebrospinal fluid–filled cavities.

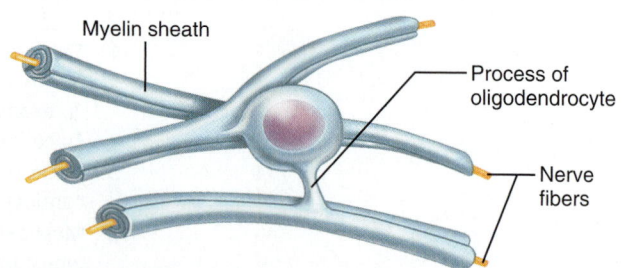

Myelin sheath — Process of oligodendrocyte — Nerve fibers

(d) Oligodendrocytes have processes that form myelin sheaths around CNS nerve fibers.

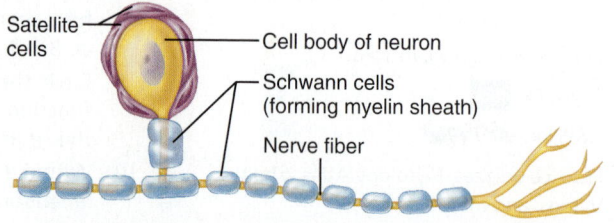

Satellite cells — Cell body of neuron — Schwann cells (forming myelin sheath) — Nerve fiber

(e) Satellite cells and Schwann cells (which form myelin) surround neurons in the PNS.

Figure 15.1 Neuroglia. (a–d) The four types of neuroglia in the central nervous system. **(e)** Neuroglia of the peripheral nervous system.

15

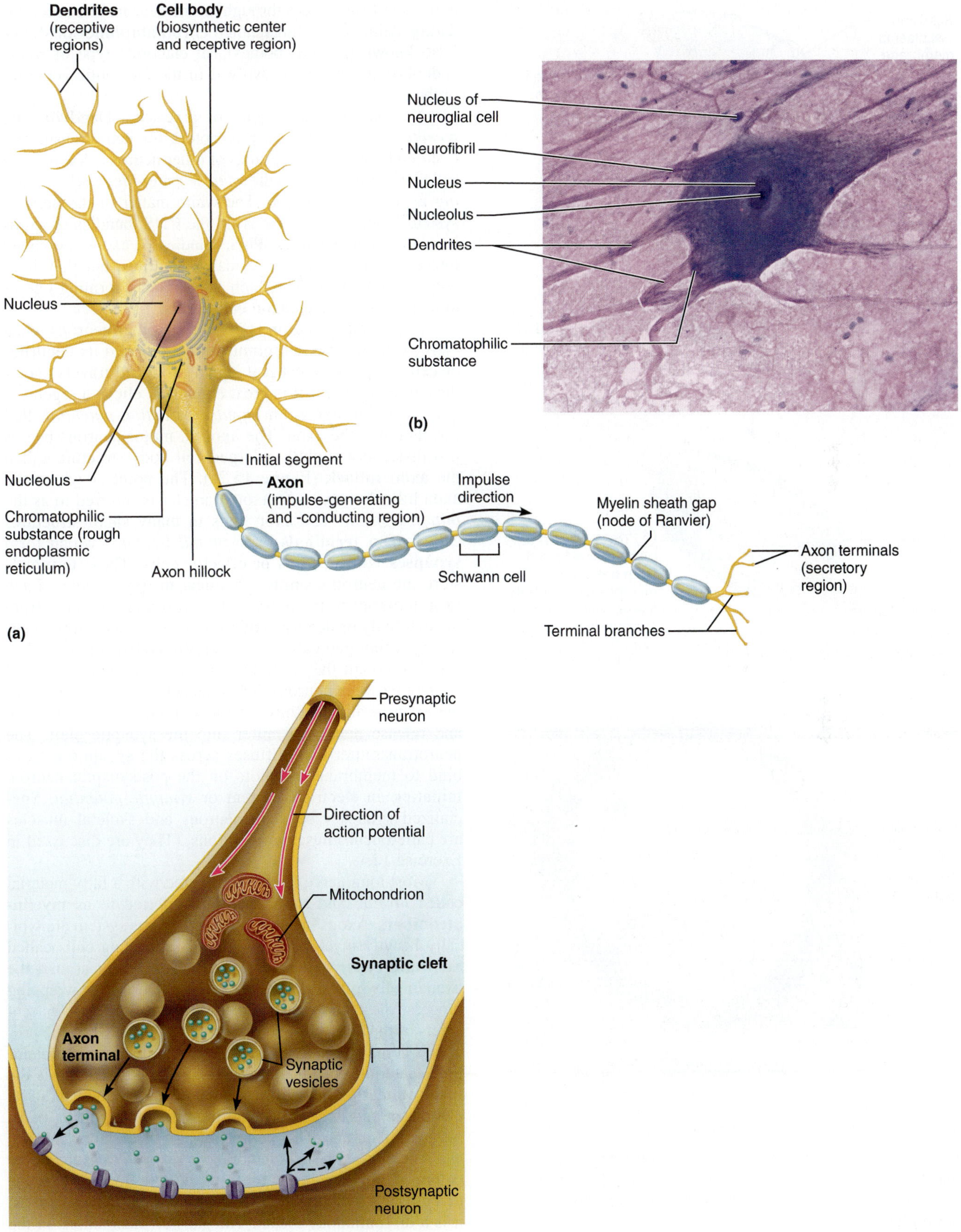

Dendrites
(receptive regions)

Cell body
(biosynthetic center and receptive region)

Nucleus of neuroglial cell

Neurofibril

Nucleus

Nucleolus

Dendrites

Chromatophilic substance

Nucleus

Nucleolus

Chromatophilic substance (rough endoplasmic reticulum)

Axon hillock

Initial segment

Axon
(impulse-generating and -conducting region)

Impulse direction

Myelin sheath gap (node of Ranvier)

Schwann cell

Axon terminals (secretory region)

Terminal branches

(b)

(a)

15

Presynaptic neuron

Direction of action potential

Mitochondrion

Synaptic cleft

Axon terminal

Synaptic vesicles

Postsynaptic neuron

(c)

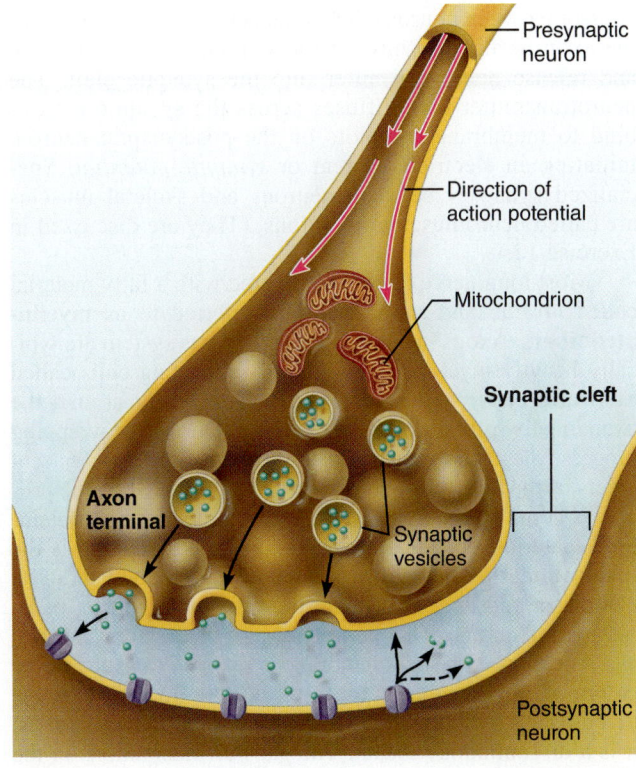

Figure 15.2 Structure of a typical motor neuron. (a) Diagram.
(b) Photomicrograph (450×). **(c)** Enlarged diagram of a synapse.

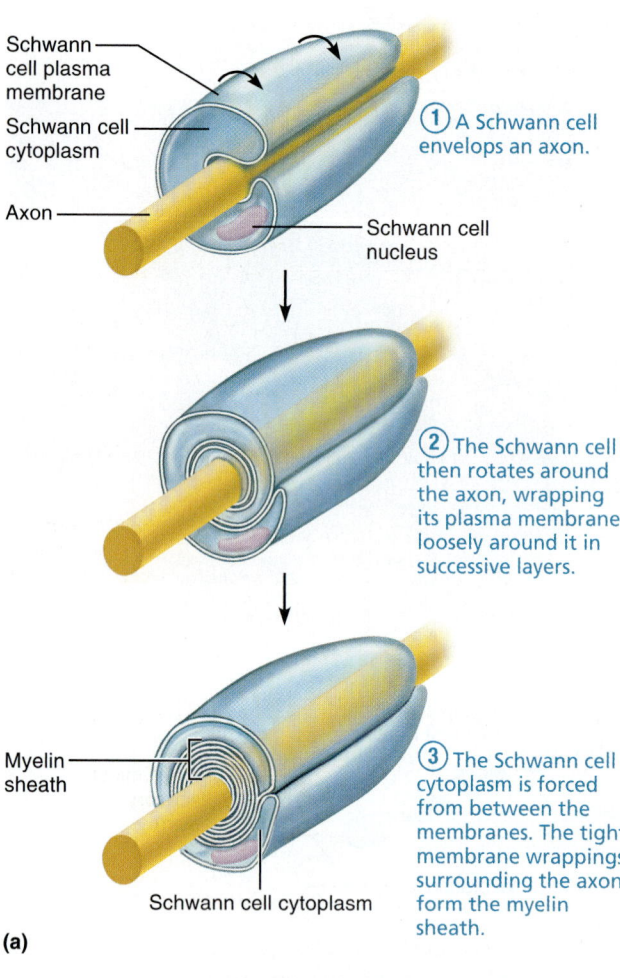

① A Schwann cell envelops an axon.

Schwann cell plasma membrane

Schwann cell cytoplasm

Axon

Schwann cell nucleus

② The Schwann cell then rotates around the axon, wrapping its plasma membrane loosely around it in successive layers.

Myelin sheath

③ The Schwann cell cytoplasm is forced from between the membranes. The tight membrane wrappings surrounding the axon form the myelin sheath.

Schwann cell cytoplasm

(a)

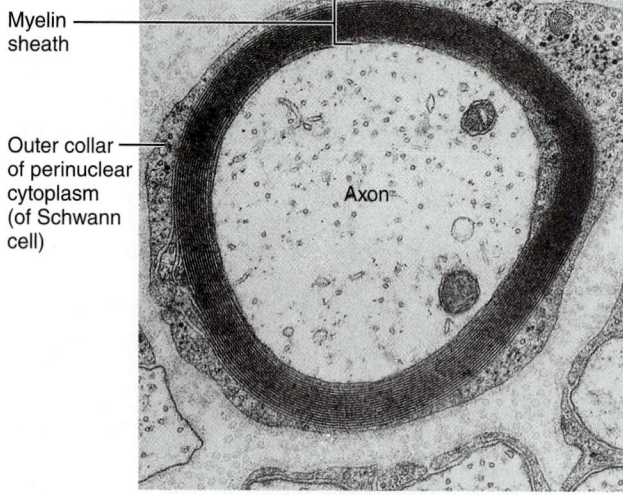

Myelin sheath

Outer collar of perinuclear cytoplasm (of Schwann cell)

Axon

(b)

Figure 15.3 Myelination of a nerve fiber (axon) by Schwann cells. (a) Nerve fiber myelination. **(b)** Electron micrograph of cross section through a myelinated axon (7500×).

to transport substances throughout the neuron. The second is darkly staining structures called **chromatophilic substance** (also known as Nissl bodies), an elaborate type of rough endoplasmic reticulum involved in the metabolic activities of the cell.

Neurons have two types of processes. **Dendrites** are *receptive regions* that bear receptors for neurotransmitters released by the axon terminals of other neurons. **Axons,** also called *nerve fibers,* form the *impulse generating and conducting region* of the neuron. The white matter of the nervous system is made up of axons. In the CNS, bundles of axons are called **tracts;** in the PNS, bundles of axons are called **nerves.** Neurons may have many dendrites, but they have only a single axon. The axon may branch, forming one or more processes called **axon collaterals.**

In general, a neuron is excited by other neurons when their axons release neurotransmitters close to its dendrites or cell body. The electrical signal produced travels across the cell body and if it is great enough, it elicits a regenerative electrical signal, an *impulse* or *action potential,* that travels down the axon. The axon in motor neurons begins just distal to a slightly enlarged cell body structure called the **axon hillock** (Figure 15.2a). The point at which the axon hillock narrows to axon diameter is referred to as the *initial segment.* The axon ends in many small structures called **axon terminals,** or *terminal boutons,* which form **synapses** with neurons or effector cells. These terminals store the neurotransmitter chemical in tiny vesicles. Each axon terminal of the presynaptic neuron is separated from the cell body or dendrites of the next, or postsynaptic, neuron by a tiny gap called the **synaptic cleft** (Figure 15.2c). Thus, although they are close, there is no actual physical contact between neurons. When an action potential reaches the axon terminals, some of the *synaptic vesicles* rupture and release neurotransmitter into the synaptic cleft. The neurotransmitter then diffuses across the synaptic cleft to bind to membrane receptors on the postsynaptic neuron, initiating an electrical current or *synaptic potential.* Specialized synapses between neurons and skeletal muscles are called neuromuscular junctions. (They are discussed in Exercise 12.)

Most long nerve fibers are covered with a fatty material called *myelin,* and such fibers are referred to as **myelinated fibers.** Axons in the peripheral nervous system are typically heavily myelinated by special supporting cells called **Schwann cells,** which wrap themselves tightly around the axon in jelly roll fashion **(Figure 15.3).** During the wrapping process, the cytoplasm is squeezed from between adjacent layers of the Schwann cell membranes, so that when the process is completed a tight core of plasma membrane (protein-lipid material) encompasses the axon. This wrapping is the **myelin sheath.** The Schwann cell nucleus and the bulk of its cytoplasm end up just beneath the outermost portion of its plasma membrane. This peripheral part of the Schwann cell and its exposed plasma membrane is referred to as the **outer collar of perinuclear cytoplasm** (Figure 15.3). Since the myelin sheath is formed by many individual Schwann cells, it is a discontinuous sheath. The gaps or indentations in the sheath are called **myelin sheath gaps** or **nodes of Ranvier** (see Figure 15.2a).

Within the CNS, myelination is accomplished by neuroglia called **oligodendrocytes** (see Figure 15.1d). Because of its chemical composition, myelin electrically insulates the fibers and greatly increases the transmission speed of nerve impulses.

ACTIVITY 1

Identifying Parts of a Neuron

1. Study the illustration of a typical motor neuron (Figure 15.2), noting the structural details described above, and then identify these structures on a neuron model.

2. Obtain a prepared slide of the ox spinal cord smear, which has large, easily identifiable neurons. Study one representative neuron under oil immersion and identify the cell body; the nucleus; the large, prominent "owl's eye" nucleolus; and the granular chromatophilic substance. If possible, distinguish the axon from the many dendrites.

   Sketch the cell in the space provided below, and label the important anatomical details you have observed. (Compare your sketch to Figure 15.2b.)

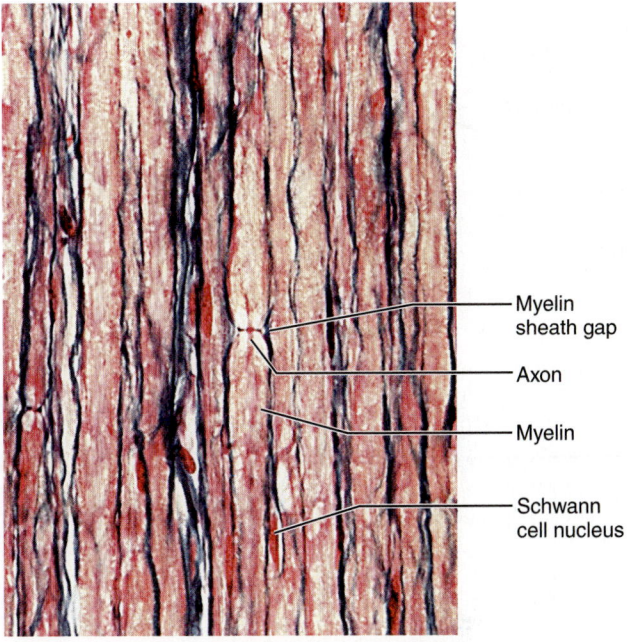

Myelin sheath gap

Axon

Myelin

Schwann cell nucleus

Figure 15.4 **Photomicrograph of a small portion of a peripheral nerve in longitudinal section (400×).**

3. Obtain a prepared slide of teased myelinated nerve fibers. Identify the following (use **Figure 15.4** as a guide): myelin sheath gaps, axon, Schwann cell nuclei, and myelin sheath.

   Sketch a portion of a myelinated nerve fiber in the space provided below, illustrating a myelin sheath gap. Label the axon, myelin sheath, myelin sheath gap, and the outer collar of perinuclear cytoplasm.

Neuron Classification

Neurons may be classified on the basis of structure or of function.

Classification by Structure

Structurally, neurons may be differentiated by the number of processes attached to the cell body **(Figure 15.5a)**. In **unipolar neurons,** one very short process, which divides into *peripheral* and *central processes,* extends from the cell body. Functionally, only the most distal parts of the peripheral process act as receptive endings; the rest acts as an axon along with the central process. Unipolar neurons are more accurately called **pseudounipolar neurons** because they are derived from bipolar neurons. Nearly all neurons that conduct impulses toward the CNS are unipolar.

   Bipolar neurons have two processes attached to the cell body. This neuron type is quite rare, typically found only as part of the receptor apparatus of the eye, ear, and olfactory mucosa.

   Many processes issue from the cell body of **multipolar neurons,** all classified as dendrites except for a single axon. Most neurons in the brain and spinal cord and those whose axons carry impulses away from the CNS fall into this last category.

Do the gaps seem to occur at consistent intervals, or are they

irregularly distributed? _____

Explain the functional significance of this finding: _____

_____

_____ ▬

ACTIVITY 2

Studying the Microscopic Structure of Selected Neurons

Obtain prepared slides of pyramidal cells of the cerebral cortex, Purkinje cells of the cerebellar cortex, and a dorsal root ganglion. As you observe them under the microscope, try to pick

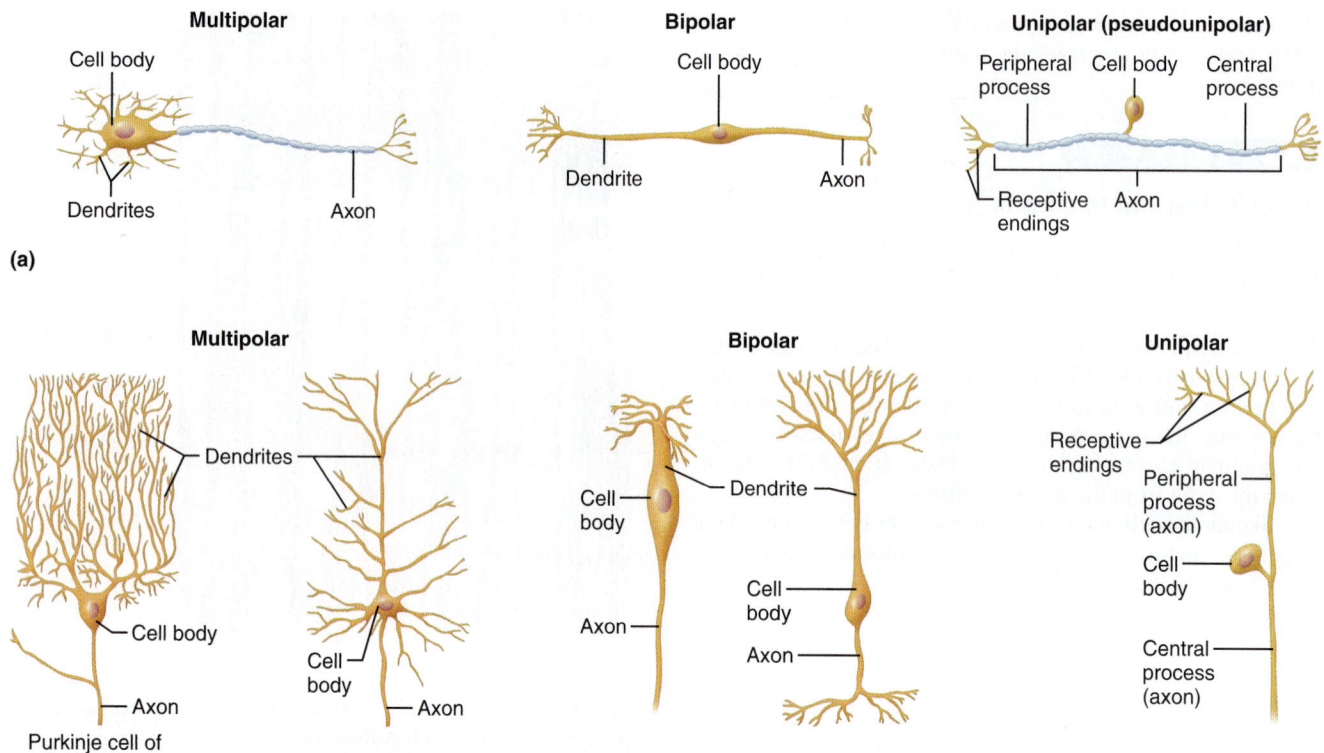

(a)

(b)

Figure 15.5 Classification of neurons according to structure. (a) Classification of neurons based on structure (number of processes extending from the cell body). **(b)** Structural variations within the classes.

out the anatomical details (compare the cells to Figure 15.5b and **Figure 15.6**). Notice that the neurons of the cerebral and cerebellar tissues (both brain tissues) are extensively branched; in contrast, the neurons of the dorsal root ganglion are more rounded. The many small nuclei visible surrounding the neurons are those of bordering neuroglia.

Which of these neuron types would be classified as multipolar neurons?

_____

Which as unipolar? _____

Classification by Function

In general, neurons carrying impulses from sensory receptors in the internal organs (viscera), the skin, skeletal muscles, joints, or special sensory organs are termed **sensory, or afferent, neurons (Figure 15.7)**. The receptive endings of sensory neurons are often equipped with specialized receptors that are stimulated by specific changes in their immediate environment. (The structure and function of these receptors are considered separately in Exercise 22, General Sensation.) The cell bodies of sensory neurons are always found in a ganglion outside the CNS, and these neurons are typically unipolar.

Neurons carrying impulses from the CNS to the viscera and/or body muscles and glands are termed **motor, or efferent, neurons.** Motor neurons are most often multipolar, and their cell bodies are almost always located in the CNS.

The third functional category of neurons is **interneurons** or *association neurons,* which are situated between and contribute to pathways that connect sensory and motor neurons. Their cell bodies are always located within the CNS, and they are multipolar neurons structurally.

Structure of a Nerve

A nerve is a bundle of axons found in the PNS. Wrapped in connective tissue coverings, nerves extend to and/or from the CNS and visceral organs or structures of the body periphery such as skeletal muscles, glands, and skin.

Like neurons, nerves are classified according to the direction in which they transmit impulses. **Sensory (afferent) nerves** conduct impulses only toward the CNS. A few of the cranial nerves are pure sensory nerves. **Motor (efferent) nerves** carry impulses only away from the CNS. The ventral roots of the spinal cord are motor nerves. Nerves carrying both sensory (afferent) and motor (efferent) fibers are called **mixed nerves;** most nerves of the body, including all spinal nerves, are mixed nerves.

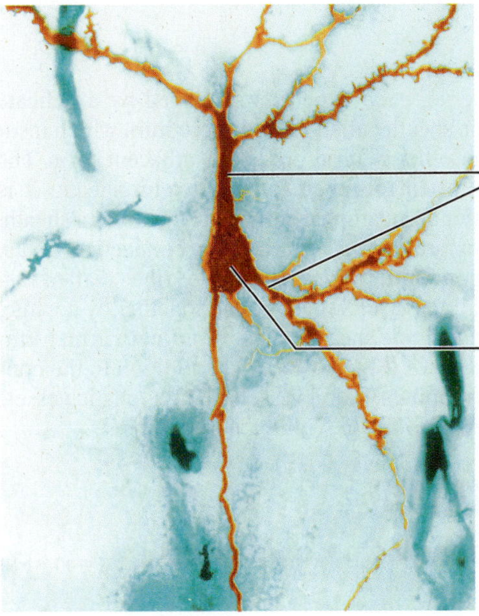

Dendrites

Cell body

(a)

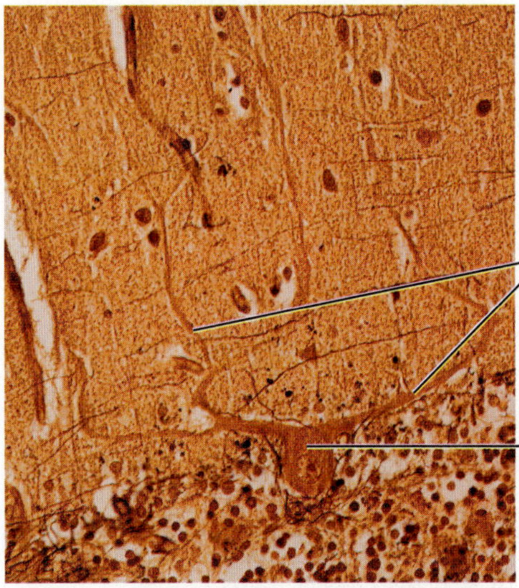

Dendrites

Cell body

(b)

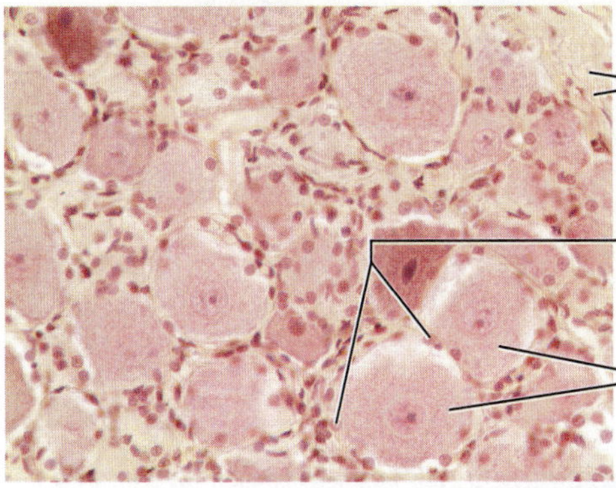

Nerve
fibers

Satellite
cells

Cell bodies

(c)

Figure 15.6 Photomicrographs of neurons.
(a) Pyramidal neuron from the cerebral cortex (600×).
(b) Purkinje cell from the cerebellar cortex (600×).
(c) Dorsal root ganglion cells (245×).

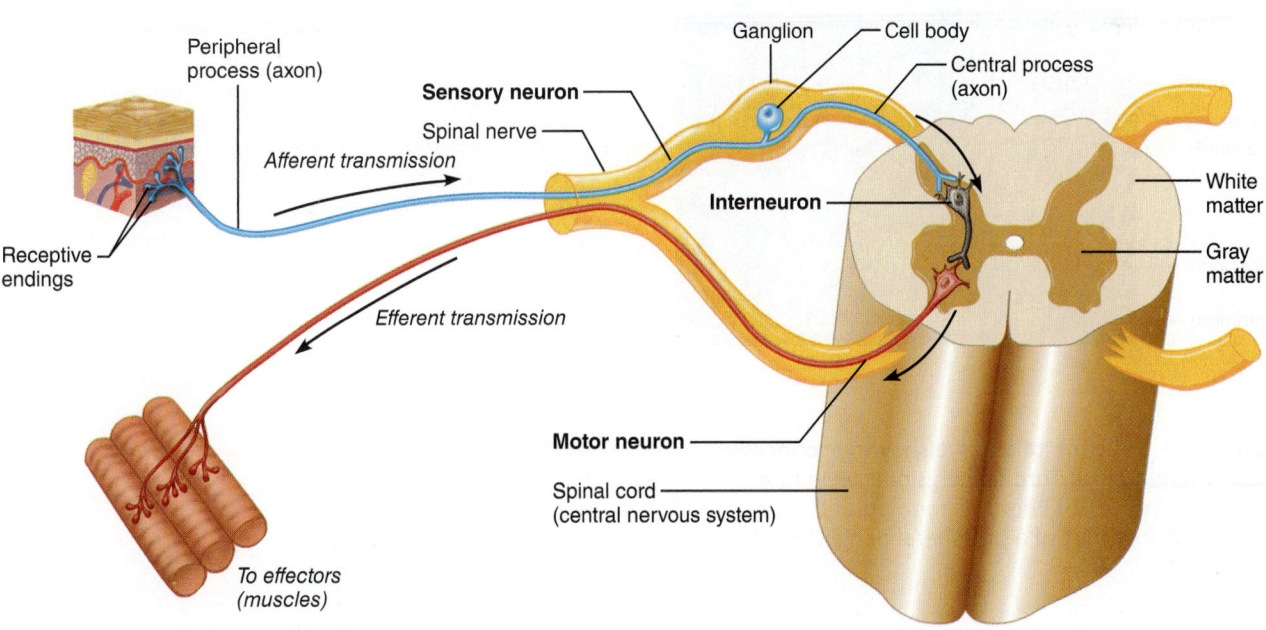

Figure 15.7 Classification of neurons on the basis of function. Sensory (afferent)
neurons conduct impulses from the body's sensory receptors to the central nervous
system; most are unipolar neurons with their nerve cell bodies in ganglia in the
peripheral nervous system (PNS). Motor (efferent) neurons transmit impulses
from the CNS to effectors (muscles). Interneurons (association neurons) complete
the communication line between sensory and motor neurons. They are typically
multipolar, and their cell bodies reside in the CNS.

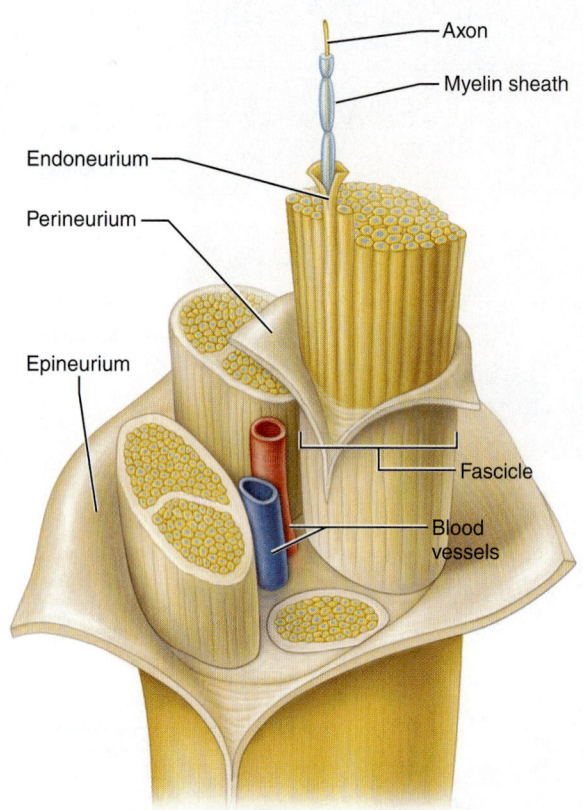

(a)

Axon

Myelin sheath

Endoneurium

Perineurium

Epineurium

Fascicle

Blood vessels

Within a nerve, each axon is surrounded by a delicate connective tissue sheath called an **endoneurium,** which insulates it from the other neuron processes adjacent to it. The endoneurium is often mistaken for the myelin sheath; it is instead an additional sheath that surrounds the myelin sheath. Groups of axons are bound by a coarser connective tissue, called the **perineurium,** to form bundles of fibers called **fascicles.** Finally, all the fascicles are bound together by a white, fibrous connective tissue sheath called the **epineurium,** forming the cordlike nerve **(Figure 15.8)**. In addition to the connective tissue wrappings, blood vessels and lymphatic vessels serving the fibers also travel within a nerve.

ACTIVITY 3

Examining the Microscopic Structure of a Nerve

Use the compound microscope to examine a prepared cross section of a peripheral nerve. Using the photomicrograph (Figure 15.8b) as an aid, identify axons, myelin sheaths, fascicles, and endoneurium, perineurium, and epineurium sheaths. If desired, sketch the nerve in the space below. ■

Myelin sheath

Nonmyelinated axon

Endoneurium

Heavily myelinated axons

Perineurium

Epineurium

(b)

Figure 15.8 Structure of a nerve showing connective tissue wrappings. (a) Three-dimensional view of a portion of a nerve. **(b)** Photomicrograph of a cross-sectional view of part of a peripheral nerve (510×).

Histology of Nervous Tissue

1. The basic functional unit of the nervous system is the neuron. What is the major function of this cell type?

_____

2. Name four types of neuroglia in the CNS, and list a function for each of these cells. (You will need to consult your text-book for this.)

Types		**Functions**
a. _____		a. _____
		_____
b. _____		b. _____
		_____
c. _____		c. _____
		_____
d. _____		d. _____
		_____

Name the PNS neuroglial cell that forms myelin. _____

Name the PNS neuroglial cell that surrounds dorsal root ganglion neurons. _____

3. Match each description with a term from the key.

Key: a. afferent neuron   e. interneuron   i. nuclei
b. central nervous system   f. neuroglia   j. peripheral nervous system
c. efferent neuron   g. neurotransmitters   k. synapse
d. ganglion   h. nerve   l. tract

_____ 1. the brain and spinal cord collectively

_____ 2. specialized supporting cells in the CNS

_____ 3. junction or point of close contact between neurons

_____ 4. a bundle of axons inside the CNS

_____ 5. neuron serving as part of the conduction pathway between sensory and motor neurons

_____ 6. ganglia and spinal and cranial nerves

_____ 7. collection of nerve cell bodies found outside the CNS

_____ 8. neuron that conducts impulses away from the CNS to muscles and glands

_____ 9. neuron that conducts impulses toward the CNS from the body periphery

_____ 10. chemicals released by neurons that stimulate or inhibit other neurons or effectors

Neuron Anatomy

4. Match the following anatomical terms (column B) with the appropriate description or function (column A).

Column A

_____ 1. region of the cell body from which the axon originates

_____ 2. secretes neurotransmitters

_____ 3. receptive region of a neuron

_____ 4. insulates the nerve fibers

_____ 5. site of the nucleus and most important metabolic area

_____ 6. involved in the transport of substances within the neuron

_____ 7. essentially rough endoplasmic reticulum, important metabolically

_____ 8. impulse generator and transmitter

Column B

a. axon

b. axon terminal

c. axon hillock

d. chromatophilic substance

e. dendrite

f. myelin sheath

g. neurofibril

h. neuronal cell body

5. Draw a "typical" multipolar neuron in the space below. Include and label the following structures on your diagram: cell body, nucleus, nucleolus, chromatophilic substance, dendrites, axon, axon collateral branch, myelin sheath, myelin sheath gaps, axon terminals, and neurofibrils.

6. What substance is found in synaptic vesicles of the axon terminal?_____

What role does this substance play in neurotransmission? _____

_____

_____

7. What anatomical characteristic determines whether a particular neuron is classified as unipolar, bipolar, or multipolar?

_____

Make a simple line drawing of each type here.

Unipolar neuron   **Bipolar neuron**   **Multipolar neuron**

8. Correctly identify the sensory (afferent) neuron, interneuron (association neuron), and motor (efferent) neuron in the figure below.

Which of these neuron types is/are unipolar? _____

Which is/are most likely multipolar? _____

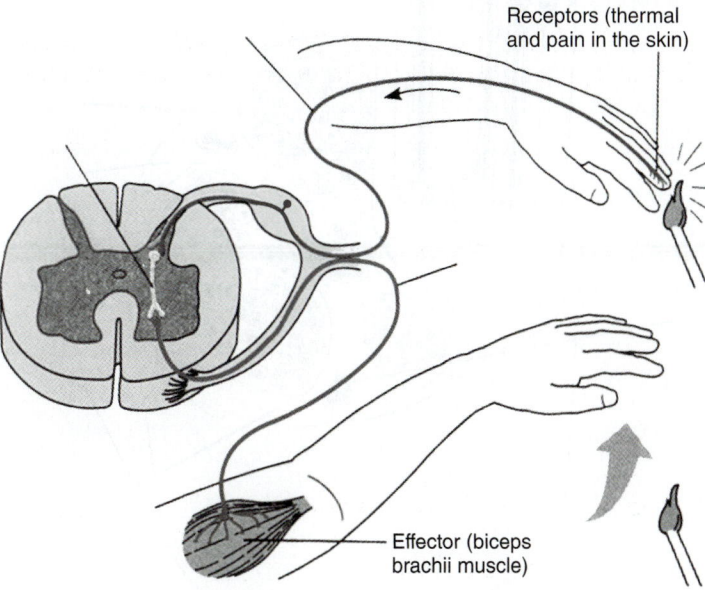

Receptors (thermal and pain in the skin)

Effector (biceps brachii muscle)

9. Describe how the Schwann cells form the myelin sheath and the outer collar of perinuclear cytoplasm encasing the nerve fibers.

_____

_____

Structure of a Nerve

10. What is a nerve? _____

_____

11. State the location of each of the following connective tissue coverings.

endoneurium: _____

perineurium: _____

epineurium: _____

12. What is the function of the connective tissue wrappings found in a nerve? _____

_____

13. Define *mixed nerve.* _____

_____

14. Identify all indicated parts of the nerve section.

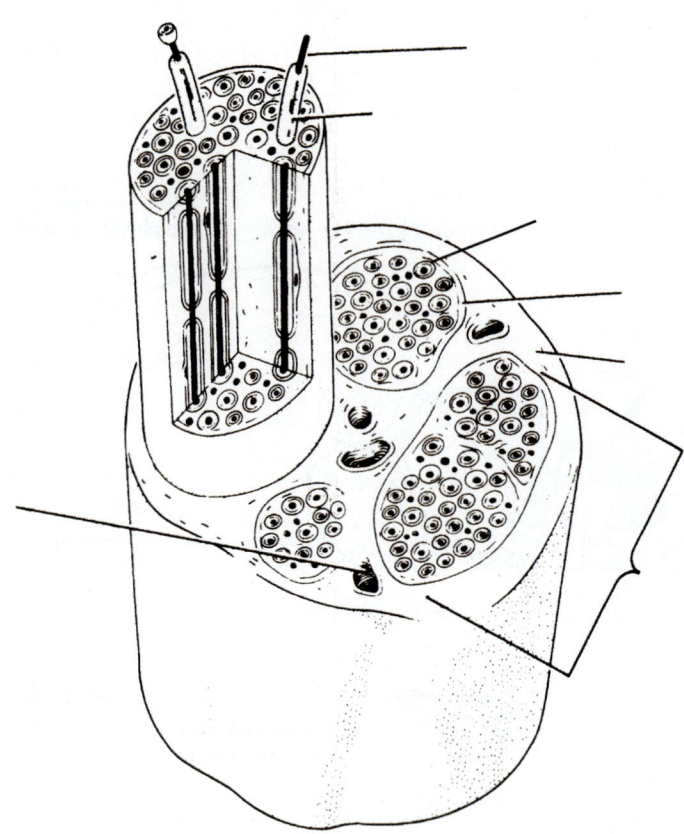

Neurophysiology of Nerve Impulses: Frog Subjects

MATERIALS

- [ ] *Rana pipiens**
- [ ] Dissecting instruments and tray
- [ ] Disposable gloves
- [ ] Ringer's solution (frog) in dropper bottles, some at room temperature and some in an ice bath
- [ ] Thread
- [ ] Glass rods or probes
- [ ] Glass plates or slides
- [ ] Ring stand and clamp
- [ ] Stimulator; platinum electrodes
- [ ] Forceps
- [ ] Filter paper
- [ ] 0.01% hydrochloric acid (HCl) solution
- [ ] Sodium chloride (NaCl) crystals
- [ ] Heat-resistant mitts
- [ ] Bunsen burner
- [ ] Safety goggles
- [ ] Absorbent cotton
- [ ] Ether
- [ ] Pipettes
- [ ] 1-cc syringe with small-gauge needle
- [ ] 0.5% tubocurarine solution
- [ ] Frog board
- [ ] Oscilloscope
- [ ] Nerve chamber

*Instructor to provide freshly pithed frogs (*Rana pipiens*) for student experimentation.

PEx PhysioEx™ 9.1 Computer Simulation Ex.3 on p. PEx-35.

MasteringA&P® For related exercise study tools, go to the Study Area of MasteringA&P. There you will find:

- Practice Anatomy Lab **PAL**
- PhysioEx **PEx**
- A&PFlix **A&PFlix**
- Practice quizzes, Histology Atlas, eText, Videos, and more!

OBJECTIVES

1. Describe the resting membrane potential in neurons.
2. Define *depolarization, repolarization, action potential,* and *relative refractory period* and *absolute refractory period*.
3. Describe the events that lead to the generation and conduction of an action potential.
4. Explain briefly how a nerve impulse is transmitted from one neuron to another, and how a neurotransmitter may be either excitatory or inhibitory to the recipient cell.
5. Define *compound action potential* and discuss how it differs from an action potential in a single neuron.
6. Describe the preparation used to examine contraction of the frog gastrocnemius muscle.
7. List various substances and factors that can stimulate neurons.
8. State the site of action of the blocking agents ether and curare.
9. Describe the experimental setup used to record compound action potentials in the frog sciatic nerve.

PRE-LAB QUIZ

1. Circle the correct underlined term. <u>Excitability</u> / <u>Conductivity</u> is the ability to transmit nerve impulses to other neurons.
2. When a neuron is stimulated, the membrane becomes more permeable to Na^+ ions, which diffuse into the cell and cause
   a. depolarization
   b. hyperpolarization
   c. repolarization
3. As an action potential progresses, the permeability to Na^+ decreases and the permeability to this ion increases:
   a. Ca^{2+}
   b. K^+
   c. Na^+
4. The period of time when the neuron is totally insensitive to further stimulation and cannot generate another action potential is
   a. absolute refractory period
   b. membrane potential
   c. repolarization
   d. threshold
5. What muscle and nerve will you need to isolate to study the physiology of nerve fibers?
   a. gastrocnemius and sciatic
   b. sartorius and femoral
   c. triceps brachii and radial

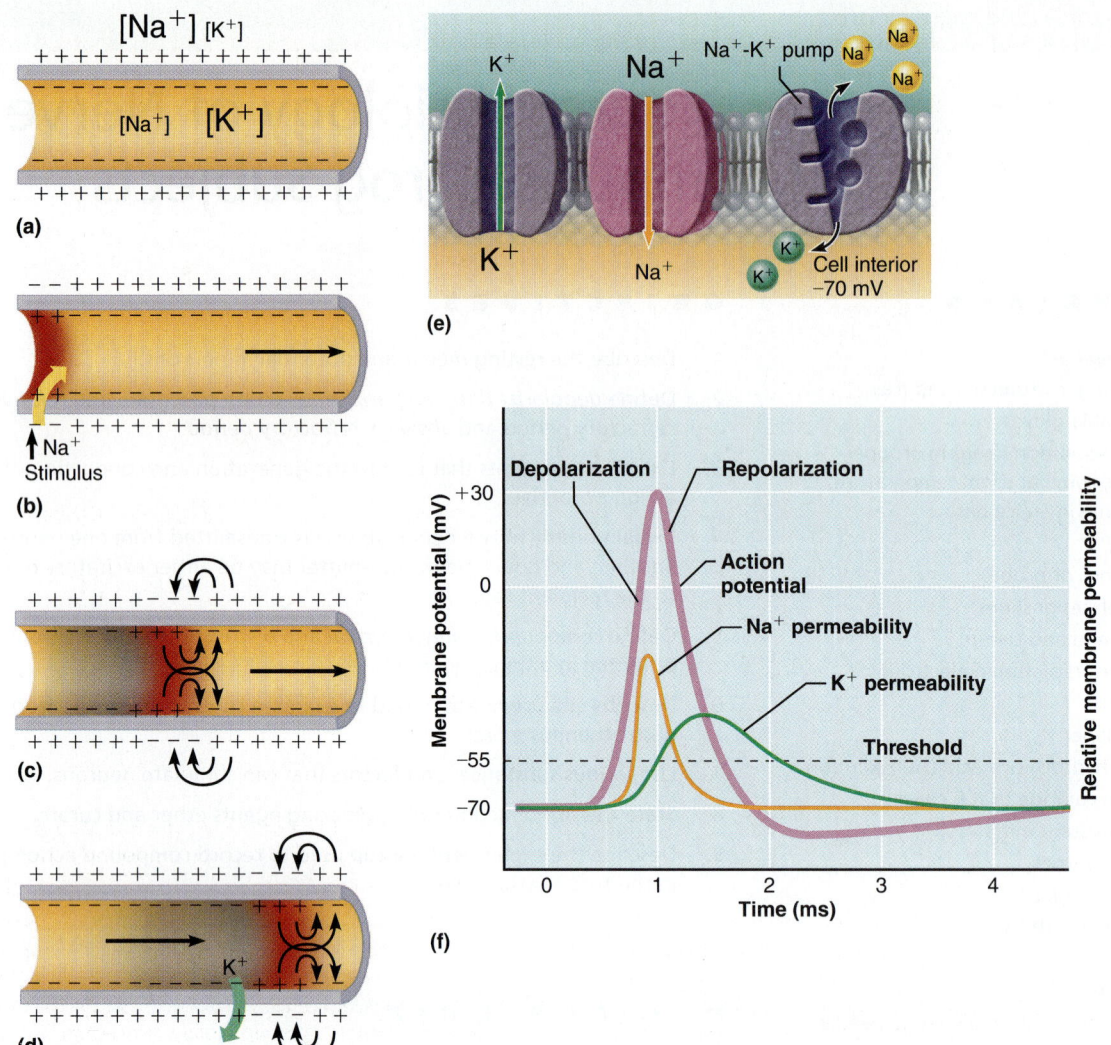

Figure 16.1 The action potential.
(a) Resting membrane potential (RMP). There is an excess of positive ions at the external cell surface, with Na⁺ the predominant extracellular fluid ion and K⁺ the predominant intracellular ion. The plasma membrane has a low permeability to Na⁺. **(b)** Depolarization—reversal of the RMP. Application of a stimulus changes the membrane permeability, and Na⁺ ions are allowed to diffuse rapidly into the cell. **(c)** Generation of the action potential or nerve impulse. If the stimulus is of adequate intensity, the depolarization wave spreads rapidly along the entire length of the membrane. **(d)** Repolarization— reestablishment of the RMP. The negative charge on the internal plasma membrane surface and the positive charge on its external surface are reestablished by diffusion of K⁺ ions out of the cell, proceeding in the same direction as in depolarization. **(e)** In the resting state, Na⁺ ions leak into the cell and K⁺ ions leak out. The RMP is maintained by the active sodium-potassium pump. **(f)** The action potential is caused by permeability changes in the plasma membrane.

Neurons are **excitable;** they respond to stimuli by producing an elecrical signal. Excited neurons communicate—they transmit electrical signals to neurons, muscles, glands, and other tissues of the body, a property called **conductivity.** In a resting neuron, the interior of the cell membrane is slightly more negatively charged than the exterior surface **(Figure 16.1)**. The difference in electrical charge produces a **resting membrane potential** across the membrane that is measured in millivolts. As in most cells, the predominant intracellular cation is K⁺; Na⁺ is the predominant cation in the extracellular fluid. In a resting neuron, Na⁺ leaks into the cell and K⁺ leaks out. The resting membrane potential is maintained by the sodium-potassium pump, which transports Na⁺ back out of the cell and K⁺ back into the cell.

The Action Potential

When a neuron receives an excitatory stimulus, the membrane becomes more permeable to sodium ions, and Na⁺ diffuses down its electrochemical gradient into the cell. As a result, the interior of the membrane becomes less negative (Figure 16.1b), an event called **depolarization.** If the stimulus is great enough to depolarize the initial segment of the axon to **threshold,** an **action potential** is generated. The initial segment of the axon in multipolar neurons is at the axon hillock of the cell body. In peripheral sensory neurons, the initial segment is just proximal to the sensory receptor, far from the cell body located in the dorsal root ganglion.

When the threshold voltage is reached, the membrane permeability to Na^+ increases rapidly (Figure 16.1f). As the neuron depolarizes, the polarity of the membrane reverses: the interior surface now becomes more positive than the exterior (Figure 16.1c). As the membrane permeability to Na^+ falls, the permeability to K^+ increases, and K^+ diffuses down its electrochemical gradient and out of the cell (Figure 16.1d). Once again the interior of the membrane becomes more negative than the exterior. This event is called **repolarization.** As you can see, the action potential is a brief reversal of the neuron's membrane potential.

The period of time when Na^+ permeability is rapidly changing and maximal, and the period immediately following when Na^+ permeability becomes restricted, together correspond to a time when the neuron is insensitive to further stimulation and cannot generate another action potential. This period is called the **absolute refractory period.** As Na^+ permeability is gradually restored to resting levels during repolarization, an especially strong stimulus to the neuron may provoke another action potential. This period of time is the **relative refractory period.** Restoration of the resting membrane potential restores the neuron's normal excitability.

Once generated, the action potential propagates along the entire length of the axon. It is never partially transmitted. Furthermore, it retains a constant amplitude and duration; the action potential is not small when a stimulus is small and large when a stimulus is large. Since the action potential of a given neuron is always the same, it is said to be an all-or-none response. When an action potential reaches the axon terminals, it causes neurotransmitter to be released. The neurotransmitter may be excitatory or inhibitory to the next cell in the transmission chain, depending on the receptor types on that cell. (The experiments in this exercise consider only excitatory neurotransmitters.)

The concentration of Na^+ and K^+ both inside and outside an active neuron change very little during a single action potential. Even when many action potentials are generated in a given area, the sodium-potassium pumps maintain the concentration differences across the membrane that are needed for the normal function of these excitable cells (Figure 16.1e).

Physiology of Nerves

The sciatic nerve is a bundle of axons that vary in diameter. An electrical signal recorded from a nerve represents the summed electrical activity of all the axons in the nerve. This summed activity is called a **compound action potential.** Unlike an action potential in a single axon, the compound action potential varies in shape according to which axons are producing action potentials. When a nerve is stimulated by external electrodes, as in our experiments, the largest axons reach threshold first and generate action potentials. Higher-intensity stimuli are required to produce action potentials in smaller axons.

In this laboratory session, you will investigate the functioning of a nerve by subjecting the sciatic nerve of a frog to

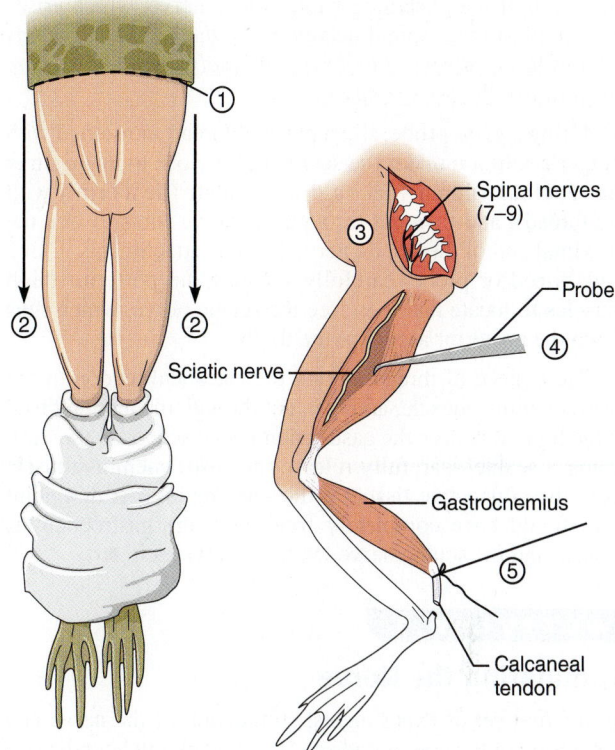

Figure 16.2 **Removal of the sciatic nerve and gastrocnemius muscle. (1)** Cut through the frog's skin around the circumference of the trunk. **(2)** Pull the skin down over the trunk and legs. **(3)** Make a longitudinal cut through the abdominal musculature and expose the roots of the sciatic nerve (arising from spinal nerves 7–9). Ligate the nerve and cut the roots proximal to the ligature. **(4)** Use a glass probe to expose the sciatic nerve beneath the posterior thigh muscles. **(5)** Ligate the calcaneal tendon and cut it free distal to the ligature. Release the gastrocnemius muscle from the connective tissue of the knee region.

various types of stimuli and blocking agents. Work in groups of two to four to lighten the workload.

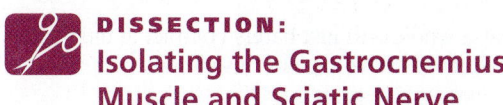 **DISSECTION:**
Isolating the Gastrocnemius Muscle and Sciatic Nerve

⚠ 1.   Don gloves to protect yourself from any parasites the frogs might have. Obtain a pithed frog from your instructor, and bring it to your laboratory bench. Also obtain dissecting instruments, a tray, thread, two glass rods or probes, and frog Ringer's solution at room temperature from the supply area.

2.   Prepare the sciatic nerve as illustrated **(Figure 16.2)**. Place the pithed frog on the dissecting tray, dorsal side down. Make a cut through the skin around the circumference of the frog approximately halfway down the trunk, and then pull the skin down over the muscles of the legs. Open the abdominal cavity and push the abdominal organs to one side to expose

the origin of the glistening white sciatic nerve, which arises from the last three spinal nerves. *Once the sciatic nerve has been exposed, it should be kept continually moist with room temperature Ringer's solution.*

3. Using a glass probe, slip a piece of thread moistened with Ringer's solution under the sciatic nerve close to its origin at the vertebral column. Make a single ligature (tie it firmly with the thread), and then cut through the nerve roots to free the proximal end of the sciatic nerve from its attachments. Using a glass rod or probe, carefully separate the posterior thigh muscles to locate and then free the sciatic nerve, which runs down the posterior aspect of the thigh.

4. Tie a piece of thread around the calcaneal tendon of the gastrocnemius muscle, and then cut through the tendon distal to the ligature to free the gastrocnemius muscle from the heel. Using a scalpel, carefully release the gastrocnemius muscle from the connective tissue in the knee region. At this point you should have completely freed both the gastrocnemius muscle and the sciatic nerve, which innervates it. ■

ACTIVITY 1

Stimulating the Nerve

In this first set of experiments, stimulation of the nerve and generation of the compound action potential will be indicated by contraction of the gastrocnemius muscle. Because you will make no mechanical recording (unless your instructor asks you to), you must keep complete and accurate records of all experimental procedures and results.

1. Obtain a glass slide or plate, ring stand and clamp, stimulator, electrodes, salt (NaCl), forceps, filter paper, 0.01% hydrochloric acid (HCl) solution, Bunsen burner, and heat-resistant mitts. With glass rods, transfer the isolated muscle-nerve preparation to a glass plate or slide, and then attach the slide to a ring stand with a clamp. Allow the end of the sciatic nerve to hang over the free edge of the glass slide, so that it is easily accessible for stimulation. *Remember to keep the nerve moist at all times.*

2. You are now ready to investigate the response of the sciatic nerve to various stimuli, beginning with electrical stimulation. Using the stimulator and platinum electrodes, stimulate the sciatic nerve with single shocks, gradually increasing the intensity of the stimulus until the threshold stimulus is determined.

The muscle as a whole will just barely contract at the threshold stimulus. Record the voltage of this stimulus:

Threshold stimulus: _____ V

Continue to increase the voltage until you find the point beyond which no further increase occurs in the strength of muscle contraction—that is, the point at which the maximal contraction of the muscle is obtained. Record this voltage below.

Maximal stimulus: _____ V

Delivering multiple or repeated shocks to the sciatic nerve causes volleys of impulses in the nerve. Shock the nerve with multiple stimuli. Observe the response of the muscle. How does this response compare with the response to the single electrical shocks?

_____

_____

3. To investigate mechanical stimulation, pinch the free end of the nerve by firmly pressing it between two glass rods or by pinching it with forceps. What is the result?

_____

4. Chemical stimulation can be tested by applying a small piece of filter paper saturated with HCl solution to the free end of the nerve. What is the result?

_____

Drop a few grains of salt (NaCl) on the free end of the nerve. What is the result?

_____

5. Now test thermal stimulation. Wearing the heat-resistant mitts, heat a glass rod for a few moments over a Bunsen burner. Then touch the rod to the free end of the nerve. What is the result?

_____

What do these muscle reactions say about the excitability and conductivity of neurons?

_____

_____ ■

Most neurons within the body are stimulated to the greatest degree by a particular stimulus (in many cases, a chemical neurotransmitter), but a variety of other stimuli may trigger nerve impulses, as seen in the experimental series just conducted. Generally, no matter what type of stimulus is present, if the affected part responds by becoming activated, it will always react in the same way. Familiar examples are the well-known phenomenon of "seeing stars" when you receive a blow to the head or press on your eyeball (try it), both of which trigger impulses in your optic nerves.

ACTIVITY 2

Inhibiting the Nerve

Numerous physical factors and chemical agents can impair the ability of nerve fibers to function. For example, deep pressure and cold temperature both block nerve impulse transmission by preventing the local blood supply from reaching the nerve fibers. Local anesthetics, alcohol, and numerous other chemicals are also very effective at blocking nerve transmission. Ether, one such chemical blocking agent, will be investigated first.

⚠ Since ether is extremely volatile and explosive, perform this experiment in a vented hood. *Don safety goggles before beginning this procedure.*

1. Obtain another glass slide or plate, absorbent cotton, ether, and a pipette. Clamp the new glass slide to the ring stand slightly below the first slide of the apparatus setup for the previous experiment. With glass rods, gently position the sciatic nerve on this second slide, allowing a small portion of the nerve's distal end to extend over the edge. Place a piece of absorbent cotton soaked with ether under the midsection of the nerve on the slide, prodding it into position with a glass rod. Using a voltage slightly above the threshold stimulus, stimulate the distal end of the nerve at 2-minute intervals until the muscle fails to respond. (If the cotton dries before this, re-wet it with ether using a pipette.) How long did it take for anesthesia to occur?

_____ sec

2. Once anesthesia has occurred, stimulate the nerve beyond the anesthetized area, between the ether-soaked pad and the muscle. What is the result?

_____

3. Remove the ether-soaked pad and flush the nerve fibers with Ringer's solution. Again stimulate the nerve at its distal end at 2-minute intervals. How long does it take for recovery?

_____

Does ether exert its blocking effect on the nerve *or* on the muscle cells?

_____ Explain your reasoning.

_____

_____

   If sufficient frogs are available and time allows, you may do the following classic experiment. In the 1800s Claude Bernard described an investigation into the effect of curare on nerve-muscle interaction. *Curare* was used by some South American Indian tribes to tip their arrows. Victims struck with these arrows were paralyzed, but the paralysis was not accompanied by loss of sensation.

1. Prepare another frog as described in steps 1 through 3 of the dissection instructions. However, in this case position the frog ventral side down on a frog board. In exposing the sciatic nerve, take care not to damage the blood vessels in the thigh region, as the success of the experiment depends on maintaining the blood supply to the muscles of the leg.

2. Expose and gently tie the left sciatic nerve so that it can be lifted away from the muscles of the leg for stimulation. Slip another length of thread under the nerve, and then tie the thread tightly around the thigh muscles to cut off circulation to the leg. The sciatic nerve should be above the thread and *not in* the ligated tissue. Expose and ligate the sciatic nerve of the *right* leg in the same manner, but this time do *not* ligate the thigh muscles.

⚠ 3. *Take great care in handling tubocurarine, because it is extremely poisonous.* Do not get any on your skin. Get a syringe and needle, and a vial of 0.5% tubocurarine. Obtain 1 cc of the tubocurarine by injecting 1 cc of air into the vial through the rubber membrane, and then drawing up 1 cc of the chemical into the syringe. Slowly and carefully inject 1 cc of the tubocurarine into the dorsal lymph sac of the frog. The dorsal lymph sacs are located dorsally at the level of the scapulae, so introduce the needle of the syringe just beneath the skin between the scapulae and toward one side of the spinal column.

4. Wait 15 minutes after injection of the tubocurarine to allow it to be distributed throughout the body in the blood and lymphatic stream. Then electrically stimulate the left sciatic nerve. Be careful not to touch any of the other tissues with the electrode. Gradually increasing the voltage, deliver single shocks until the threshold stimulus is determined for this specimen.

Threshold stimulus: _____ V

Now stimulate the right sciatic nerve with the same voltage intensity. Is there any difference in the reaction of the two muscles?

_____ If so, explain. _____

_____

_____

If you did not find any difference, wait an additional 10 to 15 minutes and restimulate both sciatic nerves.

What is the result? _____

_____

5. To determine the site at which tubocurarine acts, directly stimulate each gastrocnemius muscle. What is the result?

_____

_____

_____

Explain the difference between the responses of the right and left sciatic nerves.

_____

_____

_____

16

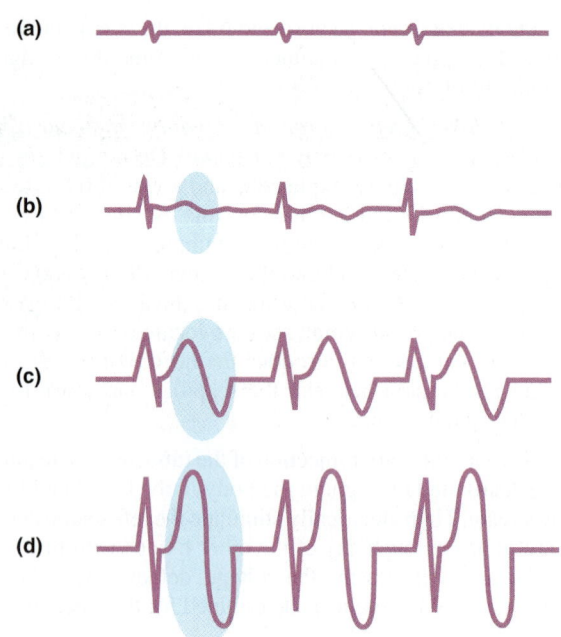

Figure 16.3 Recordings of the compound action potential from the sciatic nerve. The first compound action potential in each scan is circled. **(a)** Stimulus artifacts only. **(b–d)** Increasing stimulus strengths reveal the graded nature of the compound action potential.

Explain the results when the muscles were stimulated directly.

_____

_____

_____

At what site does tubocurarine (or curare) act?

_____

The Oscilloscope: An Experimental Tool

In this exercise, a frog's sciatic nerve will be electrically stimulated, and the compound action potential generated will be observed on the oscilloscope. The **oscilloscope** is an instrument that visually displays the rapid changes in voltage that occur during an action potential. The dissected nerve will be placed in contact with two pairs of electrodes—*stimulating* and *recording*. The stimulating electrodes will be used to deliver a pulse of electricity to a point on the sciatic nerve. At another point on the nerve, a pair of recording electrodes connected to the oscilloscope will deliver the signal to the oscilloscope, and the electrical pulse will be visible on the screen as a vertical deflection, or a *stimulus artifact* **(Figure 16.3)**. As the nerve is stimulated with increasingly higher voltage, the stimulus artifact increases in amplitude as well. When the stimulus voltage reaches a high enough level, a compound action potential will be generated by the nerve, and a *second* vertical deflection will appear on the screen, approximately 2 milliseconds after the stimulus artifact (Figure 16.3 b–d). This second deflection reports the potential difference that represents the compound action potential.

ACTIVITY 3

Visualizing the Compound Action Potential with an Oscilloscope

1. Obtain a nerve chamber, an oscilloscope, a stimulator, frog Ringer's solution (room temperature), a dissecting needle, and glass probes. Set up the experimental apparatus as illustrated **(Figure 16.4)**. Connect the two stimulating electrodes to the output terminals of the stimulator and the two recording electrodes to the oscilloscope.

2. Obtain another pithed frog, and prepare one of its sciatic nerves for experimentation as indicated in steps 1 through 3 of the dissection instructions (page 267). While working, be careful not to touch the nerve with your fingers, and do not allow the nerve to touch the frog's skin.

3. When you have freed the sciatic nerve to the knee region with the glass probe, slip another thread length beneath that end of the nerve and make a ligature. Cut the nerve distal to this tied thread and then carefully lift the cut nerve away from the thigh of the frog by holding the threads at the nerve's proximal and distal ends. Place the nerve in the nerve chamber so that it rests across all four electrodes—the two stimulating and two recording electrodes (see Figure 16.4). Flush the nerve with room temperature frog Ringer's solution.

4. Adjust the horizontal sweep according to the instructions given in the manual or by your instructor, and set the stimulator duration, frequency, and amplitude to their lowest settings.

5. Begin to stimulate the nerve with single stimuli, slowly increasing the voltage until a threshold stimulus is achieved. The compound action potential will appear as a small rounded "hump" immediately following the stimulus artifact. Record the voltage of the threshold stimulus:

Threshold stimulus: _____ V

6. Flush the nerve with the Ringer's solution and continue to increase the voltage, watching as the vertical deflections produced by the compound action potential become diphasic (show both upward and downward vertical deflections). Record the voltage at which the compound action potential reaches its maximal amplitude; this is the maximal stimulus.

Maximal stimulus: _____ V

7. Set the stimulus voltage at a level just slightly lower than the maximal stimulus and gradually increase the frequency of stimulation. What is the effect on the amplitude of the compound action potential?

_____

16

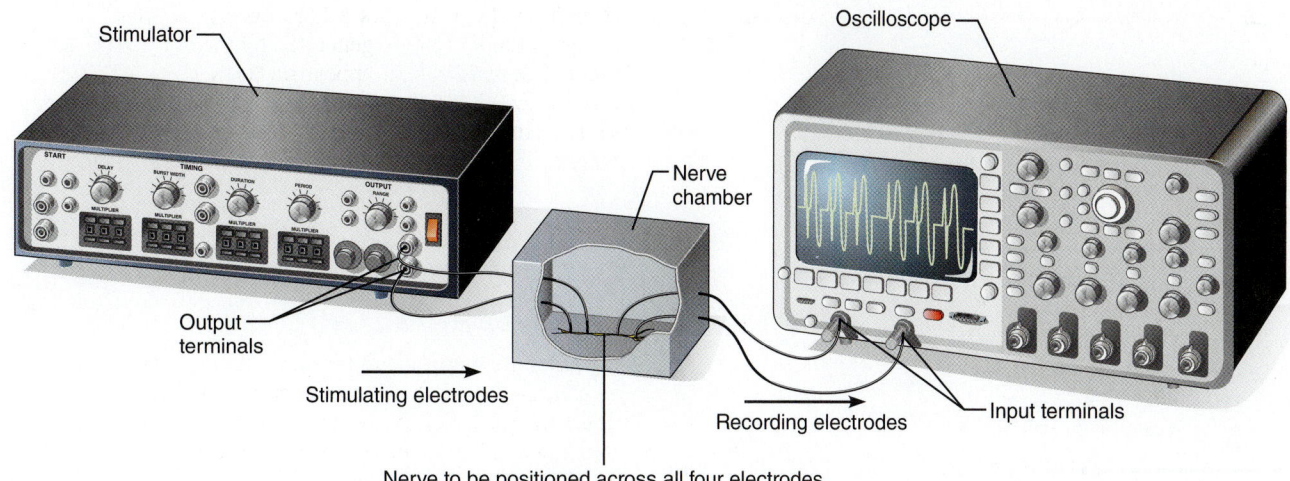

Figure 16.4 **Setup for oscilloscope visualization of action potentials in a nerve.**

8. Flush the nerve with room temperature Ringer's solution once again, and allow the nerve to sit for a few minutes while you obtain a bottle of Ringer's solution from the ice bath. Repeat steps 5 and 6 while your partner continues to flush the nerve preparation with the cold saline. Record the threshold and maximal stimuli, and watch the oscilloscope pattern carefully to detect any differences in the velocity or speed of conduction from what was seen previously.

Threshold stimulus: _____ V

Maximal stimulus: _____ V

9. Flush the nerve preparation with room temperature Ringer's solution again and then gently lift the nerve by its attached threads. Then turn the nerve around so that the end formerly resting on the stimulating electrodes now rests on the recording electrodes and vice versa. Stimulate the nerve. Is the impulse conducted in the opposite direction?

_____

10. Dispose of the frog remains and gloves in the appropriate containers, clean the lab bench and equipment, and return your equipment to the proper supply area. ▪▬

16

Neurophysiology of Nerve Impulses: Frog Subjects

The Action Potential

1. Match the terms in column B to the appropriate definition in column A.

Column A

_____ 1. period of depolarization of the neuron membrane during which it cannot respond to a second stimulus

_____ 2. reversal of the resting potential due to an influx of sodium ions

_____ 3. period during which potassium ions diffuse out of the neuron because of a change in membrane permeability

_____ 4. period of repolarization when only a strong stimulus will elicit an action potential

_____ 5. mechanism in which ATP is used to move sodium out of the cell and potassium into the cell; restores the resting membrane voltage and intracellular ionic concentrations

Column B

a. absolute refractory period

b. action potential

c. depolarization

d. relative refractory period

e. repolarization

f. sodium-potassium pump

2. Define the term *depolarization*. _____

_____

How does an action potential differ from simple depolarization? _____

_____

3. Would a substance that decreases membrane permeability to sodium increase or decrease the probability of generating an action potential? Why?

_____

_____

4. The diagram here represents a section of an axon. Complete the figure by illustrating an area of resting membrane potential, an area of depolarization, and local current flow. Indicate the direction of the depolarization wave.

$[Na^+]$ $[K^+]$

$[Na^+]$ $[K^+]$

Physiology of Nerves Stimulating and Inhibiting the Nerve

5. Respond appropriately to each question posed below. Insert your responses in the corresponding numbered blanks to the right.

1–3.   Name three types of stimuli that resulted in action potential generation in the sciatic nerve of the frog.

4.   Which of the stimuli resulted in the most effective nerve stimulation?

5.   Which of the stimuli employed in that experiment might represent types of stimuli to which nerves in the human body are subjected?

6.   What is the usual mode of stimulus transfer in neuron-to-neuron interactions?

7.   Since the action potentials themselves were not visualized with an oscilloscope during this initial set of experiments, how did you recognize that impulses were being transmitted?

1. _____

2. _____

3. _____

4. _____

5. _____

6. _____

7. _____

6. How did the site of action of ether and tubocurarine differ? _____

_____

In the curare experiment, why was one of the frog's legs ligated? _____

_____

Visualizing the Compound Action Potential with an Oscilloscope

7. Explain why the amplitude of the compound action potential recorded from the frog sciatic nerve increased when the voltage

of the stimulus was increased above the threshold value. _____

_____

8. What was the effect of cold temperature (flooding the nerve with iced Ringer's solution) on the functioning of the sciatic

nerve tested? _____

_____

9. When the nerve was reversed in position, was the impulse conducted in the opposite direction? _____

How can this result be reconciled with the concept of one-way conduction in neurons? _____

_____

_____

_____

Gross Anatomy of the Brain and Cranial Nerves

MATERIALS

- ☐ Human brain model (dissectible)
- ☐ Preserved human brain (if available)
- ☐ Three-dimensional model of ventricles
- ☐ Coronally sectioned human brain slice (if available)
- ☐ Materials as needed for cranial nerve testing (see Table 17.1): aromatic oils (e.g., vanilla and cloves); eye chart; ophthalmoscope; penlight; safety pin; blunt probe (hot and cold); cotton; solutions of sugar, salt, vinegar, and quinine; ammonia; tuning fork; and tongue depressor
- ☐ Preserved sheep brain (meninges and cranial nerves intact)
- ☐ Dissecting instruments and tray
- ☐ Disposable gloves

OBJECTIVES

1. List the elements of the central and peripheral divisions of the nervous system.
2. Discuss the difference between the sensory and motor portions of the nervous system and name the two divisions of the motor portion.
3. Recognize the terms that describe the development of the human brain and discuss the relationships between the terms.
4. As directed by your instructor, identify the bold terms associated with the cerebral hemispheres, diencephalon, brain stem, and cerebellum on a dissected human brain, brain model, or appropriate image, and state their functions.
5. State the difference between gyri, fissures, and sulci.
6. Describe the composition of gray matter and white matter in the nervous system.
7. Name and describe the three meninges that cover the brain, state their functions, and locate the falx cerebri, falx cerebelli, and tentorium cerebelli.
8. Discuss the formation, circulation, and drainage of cerebrospinal fluid.
9. Identify the cranial nerves by number and name on a model or image, stating the origin and function of each.
10. Identify at least four pertinent anatomical differences between the human and sheep brain.

PRE-LAB QUIZ

1. Circle the correct underlined term. The <u>central nervous system</u> / <u>peripheral nervous system</u> consists of the brain and spinal cord.
2. Circle the correct underlined term. The most superior portion of the brain is the <u>cerebral hemispheres</u> / <u>brain stem.</u>
3. Circle True or False. Deep grooves within the cerebral hemispheres are known as gyri.
4. On the ventral surface of the brain, you can observe the optic nerves and chiasma, the pituitary gland, and the mammillary bodies. These externally visible structures form the floor of the
   - a. brain stem
   - b. diencephalon
   - c. frontal lobe
   - d. occipital lobe
5. Circle the correct underlined term. The inferior region of the brain stem, the <u>medulla oblongata</u> / <u>cerebellum</u> houses many vital autonomic centers involved in the control of heart rate, respiratory rhythm, and blood pressure.
6. Directly under the occipital lobes of the cerebrum is a large cauliflower-like structure known as the _____.
   - a. brain stem
   - b. cerebellum
   - c. diencephalon

(Text continues on next page.)

MasteringA&P® For related exercise study tools, go to the Study Area of MasteringA&P. There you will find:
- Practice Anatomy Lab PAL
- PhysioEx PEx
- A&PFlix *A&PFlix*
- Practice quizzes, Histology Atlas, eText, Videos, and more!

7. Circle the correct underlined term. The outer cortex of the brain contains the cell bodies of cerebral neurons and is known as <u>white matter</u> / <u>gray matter.</u>

8. The brain and spinal cord are covered and protected by three connective tissue layers called
   a. lobes
   b. meninges
   c. sulci
   d. ventricles

9. Circle True or False. Cerebrospinal fluid is produced by the frontal lobe of the cerebrum and is unlike any other body fluid.

10. How many pairs of cranial nerves are there? _____

When viewed alongside all nature's animals, humans are indeed unique, and the key to their uniqueness is found in the brain. Each of us is a composite reflection of our brain's experience. If all past sensory input could mysteriously and suddenly be "erased," we would be unable to walk, talk, or communicate in any manner. Spontaneous movement would occur, as in a fetus, but no voluntary integrated function of any type would be possible. Clearly we would cease to be the same individuals.

Because of the complexity of the nervous system, its anatomical structures are usually considered in terms of two principal divisions: the central nervous system and the peripheral nervous system. The **central nervous system (CNS)** consists of the brain and spinal cord, which primarily interpret incoming sensory information and issue instructions based on that information and on past experience. The **peripheral nervous system (PNS)** consists of the cranial and spinal nerves, ganglia, and sensory receptors. These structures serve as communication lines as they carry impulses—from the sensory receptors to the CNS and from the CNS to the appropriate glands, muscles, or other effector organs.

The PNS has two major subdivisions: the **sensory portion,** which consists of nerve fibers that conduct impulses toward the CNS, and the **motor portion,** which contains nerve fibers that conduct impulses away from the CNS. The motor portion, in turn, consists of the **somatic division** (sometimes called the *voluntary system*), which controls the skeletal muscles, and the **autonomic nervous system (ANS),** which controls smooth and cardiac muscles and glands. The ANS is often referred to as the *involuntary nervous system*. Its sympathetic and parasympathetic branches play a major role in maintaining homeostasis.

In this exercise the brain (CNS) and cranial nerves (PNS) will be studied because of their close anatomical relationship.

The Human Brain

During embryonic development of all vertebrates, the CNS first makes its appearance as a simple tubelike structure, the **neural tube,** that extends down the dorsal median plane. By the fourth week, the human brain begins to form as an expansion of the anterior or rostral end of the neural tube (the end toward the head). Shortly thereafter, constrictions appear, dividing the developing brain into three major regions—**forebrain, midbrain,** and **hindbrain (Figure 17.1)**. The remainder of the neural tube becomes the spinal cord.

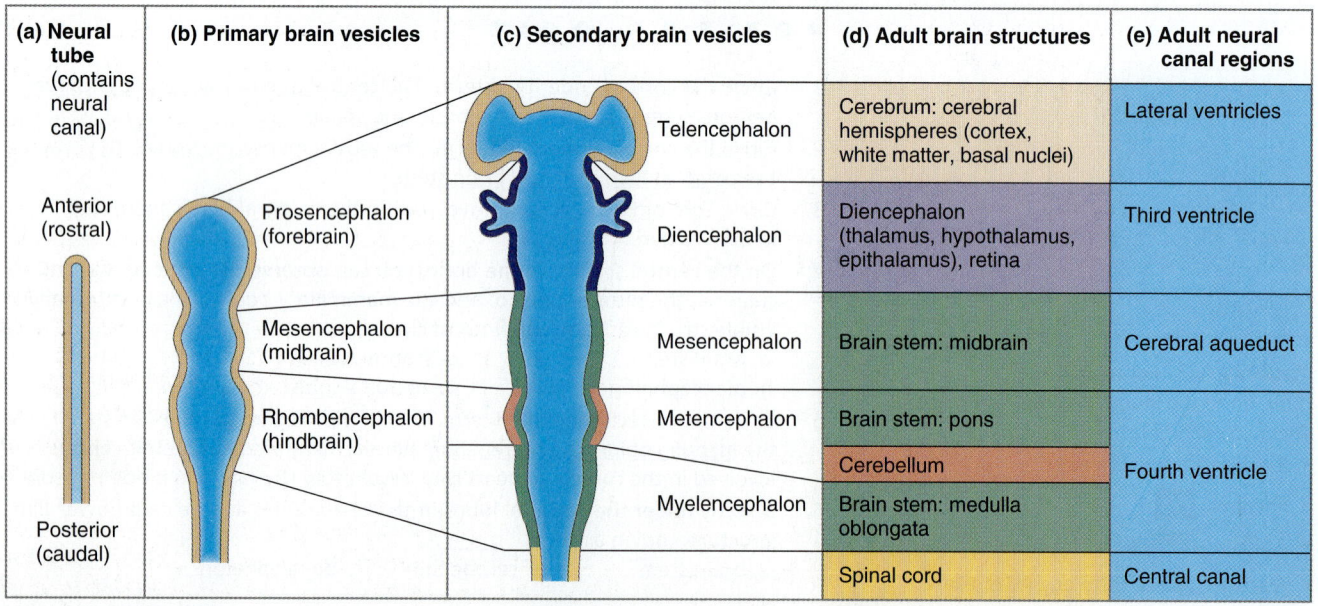

(a) Neural tube (contains neural canal)	(b) Primary brain vesicles	(c) Secondary brain vesicles	(d) Adult brain structures	(e) Adult neural canal regions
Anterior (rostral)	Prosencephalon (forebrain)	Telencephalon	Cerebrum: cerebral hemispheres (cortex, white matter, basal nuclei)	Lateral ventricles
		Diencephalon	Diencephalon (thalamus, hypothalamus, epithalamus), retina	Third ventricle
	Mesencephalon (midbrain)	Mesencephalon	Brain stem: midbrain	Cerebral aqueduct
	Rhombencephalon (hindbrain)	Metencephalon	Brain stem: pons	Fourth ventricle
			Cerebellum	
Posterior (caudal)		Myelencephalon	Brain stem: medulla oblongata	
			Spinal cord	Central canal

Figure 17.1 Embryonic development of the human brain. (a) The neural tube subdivides into **(b)** the primary brain vesicles, which subsequently form **(c)** the secondary brain vesicles, which differentiate into **(d)** the adult brain structures. **(e)** The adult structures derived from the neural canal.

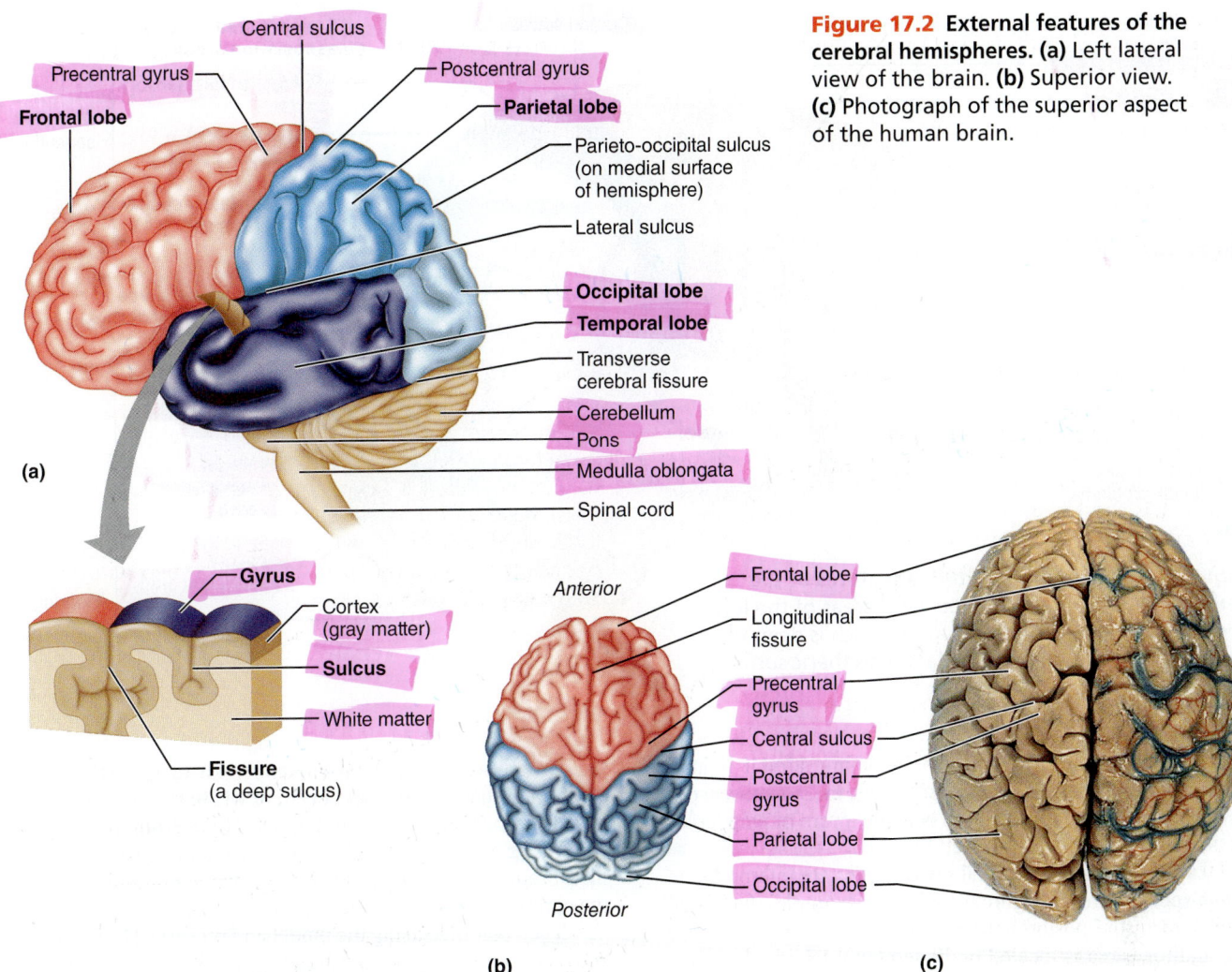

Figure 17.2 External features of the cerebral hemispheres. (a) Left lateral view of the brain. **(b)** Superior view. **(c)** Photograph of the superior aspect of the human brain.

During fetal development, two anterior outpocketings extend from the forebrain and grow rapidly to form the cerebral hemispheres. The skull imposes space restrictions that force the cerebral hemispheres to grow posteriorly and inferiorly, and they finally end up enveloping and obscuring the rest of the forebrain and most midbrain structures. Somewhat later in development, the dorsal hindbrain also enlarges to produce the cerebellum. The central canal of the neural tube, which remains continuous throughout the brain and cord, enlarges in four regions of the brain, forming chambers called **ventricles** (see Figure 17.8a and b, page 284).

(see Figure 17.8a and b, page 284).

ACTIVITY 1

Identifying External Brain Structures

Identify external brain structures using the figures cited. Also use a model of the human brain and other learning aids as they are mentioned.

Generally, the brain is studied in terms of four major regions: the cerebral hemispheres, diencephalon, brain stem, and cerebellum. It's useful to be aware of the relationship between these four anatomical regions and the structures of the forebrain, midbrain, and hindbrain (Figure 17.1).

Cerebral Hemispheres

The **cerebral hemispheres** are the most superior portion of the brain **(Figure 17.2)**. Their entire surface is thrown into elevated ridges of tissue called **gyri** that are separated by shallow grooves called **sulci** or deeper grooves called **fissures.** Many of the fissures and gyri are important anatomical landmarks.

The cerebral hemispheres are divided by a single deep fissure, the **longitudinal fissure.** The **central sulcus** divides the **frontal lobe** from the **parietal lobe,** and the **lateral sulcus** separates the **temporal lobe** from the parietal lobe. The **parieto-occipital sulcus** on the medial surface of each hemisphere divides the **occipital lobe** from the parietal lobe. It is not visible externally. A fifth lobe of each cerebral hemisphere, the **insula,** is buried deep within the lateral sulcus, and is covered by portions of the temporal, parietal, and frontal lobes. Notice that most cerebral hemisphere lobes are named for the cranial bones that lie over them.

Some important functional areas of the cerebral hemispheres have also been located (Figure 17.2d). The **primary somatosensory cortex** is located in the **postcentral gyrus** of the parietal lobe. Impulses traveling from the body's sensory receptors (such as those for pressure, pain, and temperature) are localized in this area of the brain. ("This information is from my big toe.") Immediately posterior to the primary somatosensory area is the **somatosensory association**

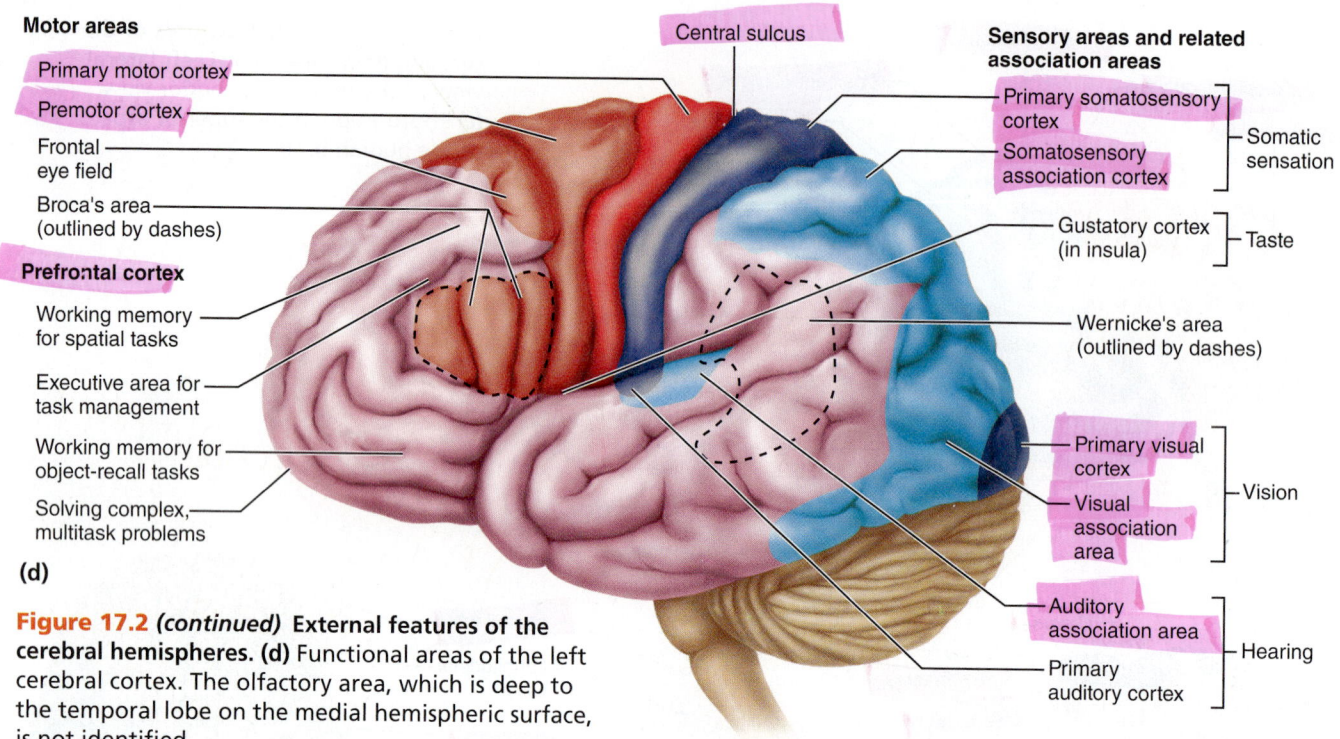

Motor areas

Primary motor cortex

Premotor cortex

Frontal eye field

Broca's area (outlined by dashes)

Prefrontal cortex

Working memory for spatial tasks

Executive area for task management

Working memory for object-recall tasks

Solving complex, multitask problems

Central sulcus

Sensory areas and related association areas

Primary somatosensory cortex

Somatosensory association cortex ⎤ Somatic sensation

Gustatory cortex (in insula) ⎤ Taste

Wernicke's area (outlined by dashes)

Primary visual cortex

Visual association area ⎤ Vision

Auditory association area

Primary auditory cortex ⎤ Hearing

(d)

Figure 17.2 (continued) External features of the cerebral hemispheres. **(d)** Functional areas of the left cerebral cortex. The olfactory area, which is deep to the temporal lobe on the medial hemispheric surface, is not identified.

17

cortex, in which the meaning of incoming stimuli is analyzed. ("Ouch! I have a *pain* there.") Thus, the somatosensory association cortex allows you to become aware of pain, coldness, a light touch, and the like.

Impulses from the special sense organs are interpreted in other specific areas (Figure 17.2d). For example, the visual areas are in the posterior portion of the occipital lobe, and the auditory area is located in the temporal lobe in the gyrus bordering the lateral sulcus. The olfactory area is deep within the temporal lobe along its medial surface, in a region called the **uncus** (see Figure 17.4a).

The **primary motor cortex,** which is responsible for conscious or voluntary movement of the skeletal muscles, is located in the **precentral gyrus** of the frontal lobe. A specialized motor speech area called **Broca's area** is found at the base of the precentral gyrus just above the lateral sulcus. Damage to this area (which is located in only one cerebral hemisphere, usually the left) reduces or eliminates the ability to articulate words. Many areas involved in intellect, complex reasoning, and personality lie in the anterior portions of the frontal lobes, in a region called the **prefrontal cortex.**

A rather poorly defined region at the junction of the parietal and temporal lobes is **Wernicke's area,** an area in which unfamiliar words are sounded out. Like Broca's area, Wernicke's area is located in one cerebral hemisphere only, typically the left.

Although there are many similar functional areas in both cerebral hemispheres, each hemisphere is also a "specialist" in certain ways. For example, the left hemisphere is the "language brain" in most of us, because it houses centers associated with language skills and speech. The right hemisphere is more concerned with abstract, conceptual, or spatial processes—skills associated with artistic or creative pursuits.

The cell bodies of cerebral neurons involved in these functions are found only in the outermost gray matter of the cerebrum, the **cerebral cortex.** Most of the balance of cerebral tissue—the deeper **cerebral white matter**—is composed of fiber tracts carrying impulses to or from the cortex.

Using a model of the human brain (and a preserved human brain, if available), identify the areas and structures of the cerebral hemispheres described above.

Then continue using the model and preserved brain along with the figures as you read about other structures.

Diencephalon

The **diencephalon** is embryologically part of the forebrain, along with the cerebral hemispheres.

Turn the brain model so the ventral surface of the brain can be viewed. Starting superiorly (and using **Figure 17.3** as a guide), identify the externally visible structures that mark the position of the floor of the diencephalon. These are the **olfactory bulbs** (synapse point of cranial nerve I) and **tracts, optic nerves** (cranial nerve II), **optic chiasma** (where the fibers of the optic nerves partially cross over), **optic tracts, pituitary gland,** and **mammillary bodies.**

Brain Stem

Continue inferiorly to identify the **brain stem** structures—the **cerebral peduncles** (fiber tracts in the **midbrain** connecting the pons below with cerebrum above), the pons, and the medulla oblongata. *Pons* means "bridge," and the **pons** consists primarily of motor and sensory fiber tracts connecting the brain with lower CNS centers. The lowest brain stem region, the **medulla oblongata,** is also composed primarily of fiber tracts. You can see the **decussation of pyramids,** a crossover point for the major motor tracts (pyramidal tracts) descending from the motor areas of the cerebrum to the cord, on the medulla's surface. The medulla also houses many vital autonomic centers involved in the control of heart rate, respiratory rhythm, and blood pressure as well as involuntary centers involved in vomiting, swallowing, and so on.

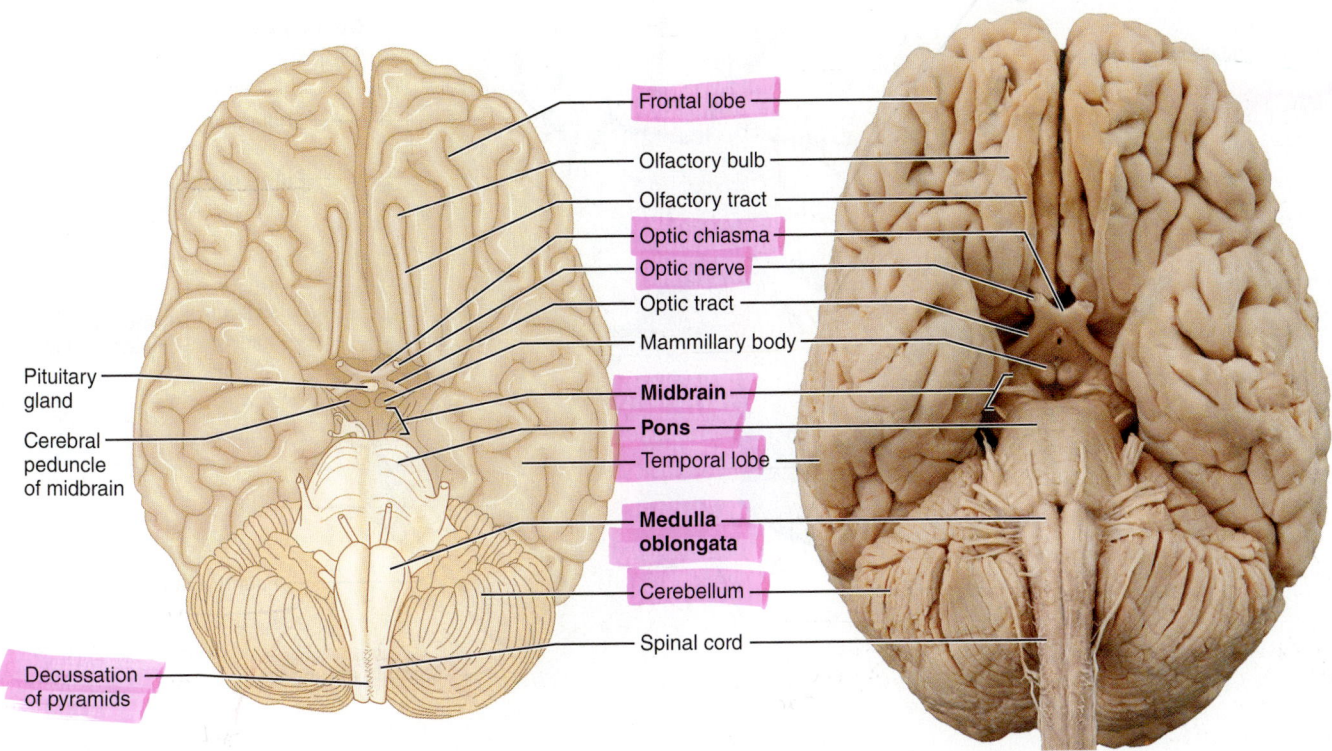

Frontal lobe

Olfactory bulb

Olfactory tract

Optic chiasma

Optic nerve

Optic tract

Mammillary body

Midbrain

Pons

Temporal lobe

Medulla oblongata

Cerebellum

Spinal cord

Pituitary gland

Cerebral peduncle of midbrain

Decussation of pyramids

Figure 17.3 **Ventral (inferior) aspect of the human brain, showing the three regions of the brain stem.** Only a small portion of the midbrain can be seen; the rest is surrounded by other brain regions.

Cerebellum

1. Turn the brain model so you can see the dorsal aspect. Identify the large cauliflower-like **cerebellum,** which projects dorsally from under the occipital lobes of the cerebrum. Notice that, like the cerebrum, the cerebellum has two major hemispheres and a convoluted surface (see Figure 17.6). It also has an outer cortex made up of gray matter with an inner region of white matter.

2. Remove the cerebellum to view the **corpora quadrigemina** (Figure 17.4), located on the posterior aspect of the midbrain, a brain stem structure. The two superior prominences are the **superior colliculi** (visual reflex centers); the two smaller inferior prominences are the **inferior colliculi** (auditory reflex centers). ■

ACTIVITY 2

Identifying Internal Brain Structures

The deeper structures of the brain have also been well mapped. Like the external structures, these can be studied in terms of the four major regions. As the internal brain areas are described, identify them on the figures cited. Also, use the brain model as indicated to help you in this study.

Cerebral Hemispheres

1. Take the brain model apart so you can see a median sagittal view of the internal brain structures **(Figure 17.4).** Observe the model closely to see the extent of the outer cortex (gray matter), which contains the cell bodies of cerebral neurons. The pyramidal cells of the cerebral motor cortex

(studied in Exercise 15 and Figure 15.5) are representative of the neurons seen in the precentral gyrus.

2. Observe the deeper area of white matter, which is composed of fiber tracts. The fiber tracts found in the cerebral hemisphere white matter are called *association tracts* if they connect two portions of the same hemisphere, *projection tracts* if they run between the cerebral cortex and lower brain structures or spinal cord, and *commissures* if they run from one hemisphere to another. Observe the large **corpus callosum,** the major commissure connecting the cerebral hemispheres. The corpus callosum arches above the structures of the diencephalon and roofs over the lateral ventricles. Notice also the **fornix,** a bandlike fiber tract concerned with olfaction as well as limbic system functions, and the membranous **septum pellucidum,** which separates the lateral ventricles of the cerebral hemispheres.

3. In addition to the gray matter of the cerebral cortex, there are several clusters of neuron cell bodies called **nuclei** buried deep within the white matter of the cerebral hemispheres. One important group of cerebral nuclei, called the **basal nuclei** or **basal ganglia,*** flank the lateral and third ventricles. You can see these nuclei if you have a dissectible model or a coronally or cross-sectioned human brain slice. (Otherwise, **Figure 17.5** will suffice.)

The basal nuclei, part of the *indirect pathway,* are involved in regulating voluntary motor activities. The most important of them are the arching, comma-shaped **caudate**

---

*The historical term for these nuclei, *basal ganglia,* is misleading because ganglia are PNS structures. Although technically not the correct anatomical term, "basal ganglia" is included here because it is widely used in clinical settings.

17

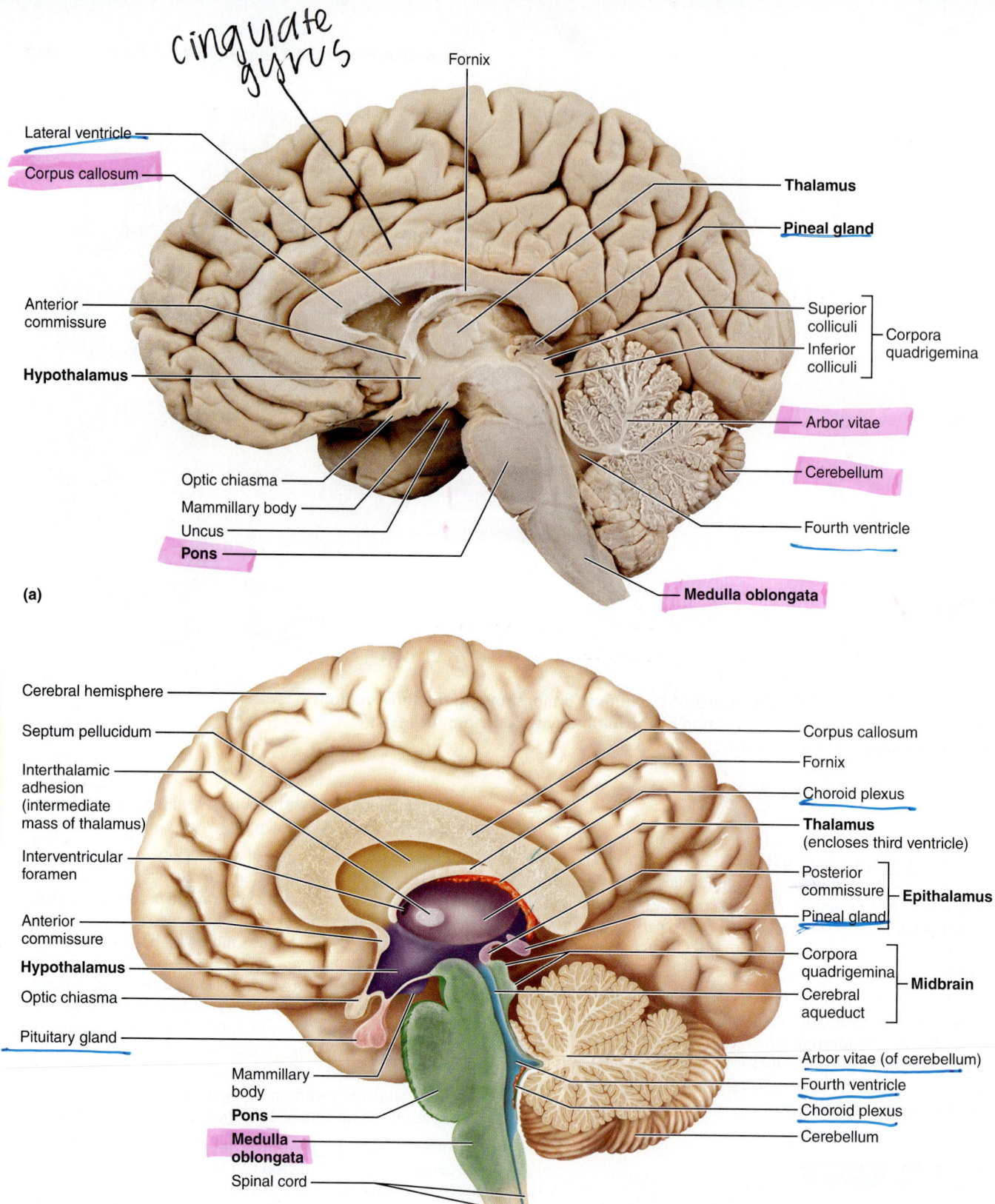

cinguiate gyrus

(a)

- Lateral ventricle
- Corpus callosum
- Anterior commissure
- **Hypothalamus**
- Optic chiasma
- Mammillary body
- Uncus
- **Pons**
- Fornix
- **Thalamus**
- **Pineal gland**
- Superior colliculi
- Inferior colliculi
- Corpora quadrigemina
- Arbor vitae
- Cerebellum
- Fourth ventricle
- **Medulla oblongata**

(b)

- Cerebral hemisphere
- Septum pellucidum
- Interthalamic adhesion (intermediate mass of thalamus)
- Interventricular foramen
- Anterior commissure
- **Hypothalamus**
- Optic chiasma
- Pituitary gland
- Mammillary body
- **Pons**
- **Medulla oblongata**
- Spinal cord
- Corpus callosum
- Fornix
- Choroid plexus
- **Thalamus** (encloses third ventricle)
- Posterior commissure
- Pineal gland
- **Epithalamus**
- Corpora quadrigemina
- Cerebral aqueduct
- **Midbrain**
- Arbor vitae (of cerebellum)
- Fourth ventricle
- Choroid plexus
- Cerebellum

Figure 17.4 Diencephalon and brain stem structures as seen in a sagittal section of the brain. (a) Photograph. **(b)** Diagram.

17

nucleus, the **putamen,** and the **globus pallidus.** The closely associated *amygdaloid body* (located at the tip of the caudate nucleus) is part of the *limbic system.*

The **corona radiata,** a spray of projection fibers coursing down from the precentral (motor) gyrus, combines with sensory fibers traveling to the sensory cortex to form a broad band of fibrous material called the **internal capsule.** The internal capsule passes between the diencephalon and the basal nuclei and through parts of the basal nuclei, giving them a striped appearance. This is why the caudate nucleus and the

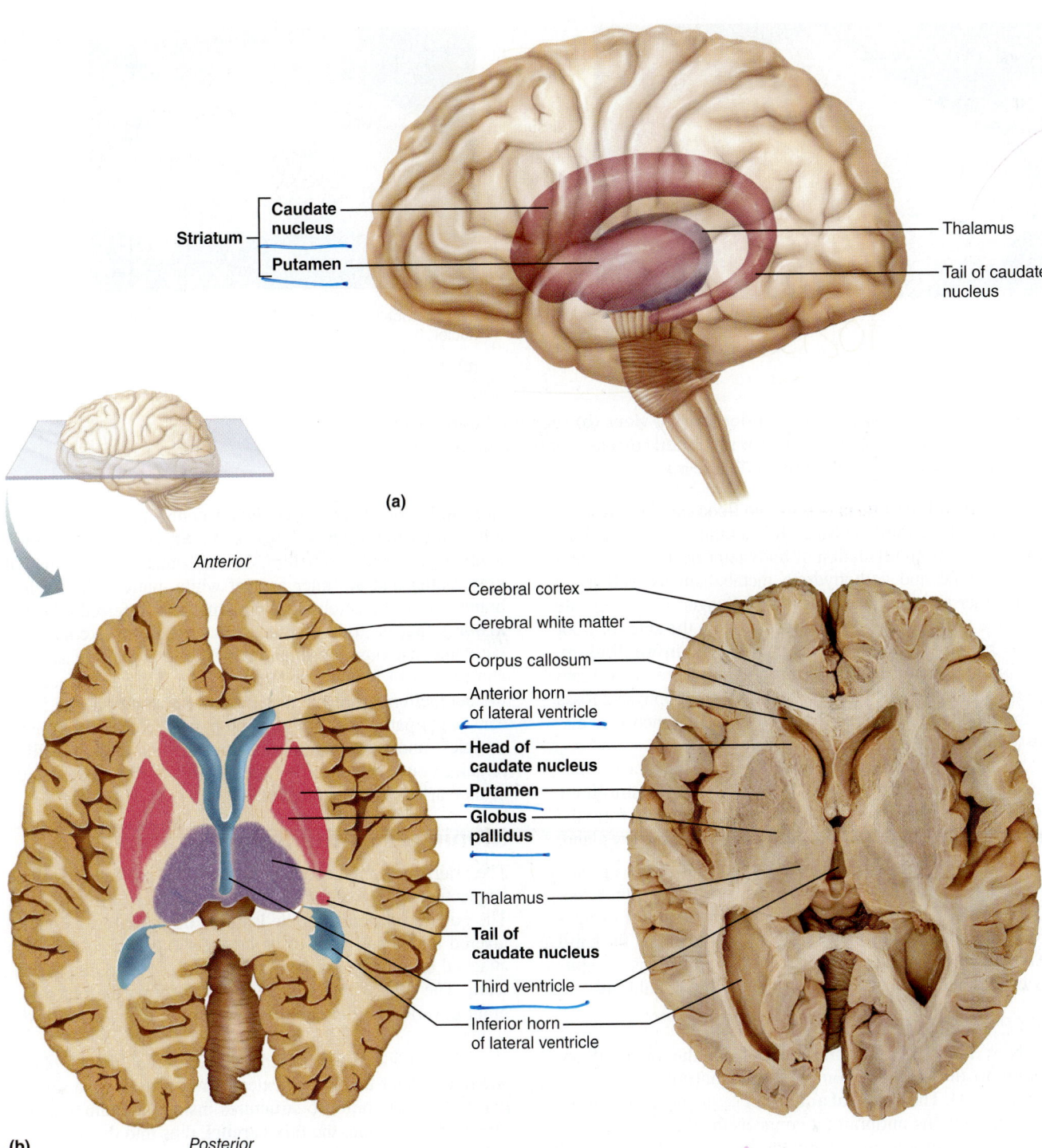

Figure 17.5 **Basal nuclei. (a)** Three-dimensional view of the basal nuclei showing their positions within the cerebrum. **(b)** A transverse section of the cerebrum and diencephalon showing the relationship of the basal nuclei to the thalamus and the lateral and third ventricles.

putamen are sometimes referred to collectively as the **striatum,** or "striped body" (Figure 17.5a).

4. Examine the relationship of the lateral ventricles and corpus callosum to the diencephalon structures; that is, thalamus and third ventricle—from the cross-sectional viewpoint (see Figure 17.5b).

Diencephalon

1. The major internal structures of the diencephalon are the thalamus, hypothalamus, and epithalamus (see Figure 17.4).

The **thalamus** consists of two large lobes of gray matter that laterally enclose the shallow third ventricle of the brain. A slender stalk of thalamic tissue, the **interthalamic adhesion,** or **intermediate mass,** connects the two thalamic lobes and bridges the ventricle. The thalamus is a major integrating and relay station for sensory impulses passing upward to the cortical sensory areas for localization and interpretation. Locate also the **interventricular foramen,** a tiny opening connecting the third ventricle with the lateral ventricle on the same side.

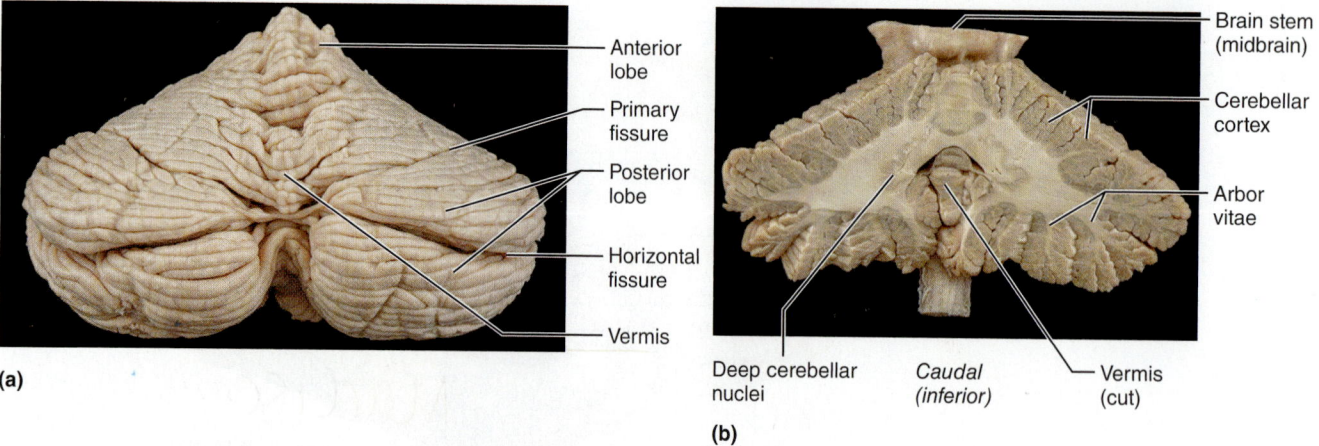

Figure 17.6 Cerebellum. (a) Posterior (dorsal) view. **(b)** Sectioned to reveal the cerebellar cortex. (The cerebellum is sectioned coronally and the brain stem is sectioned transversely in this posterior view.)

2. The **hypothalamus** makes up the floor and the inferolateral walls of the third ventricle. It is an important autonomic center involved in regulation of body temperature, water balance, and fat and carbohydrate metabolism as well as in many other activities and drives (sex, hunger, thirst). Locate again the pituitary gland, which hangs from the anterior floor of the hypothalamus by a slender stalk, the **infundibulum.** The pituitary gland is usually not present in preserved brain specimens. In life, the pituitary rests in the hypophyseal fossa of the sella turcica of the sphenoid bone. (Its function is discussed in Exercise 27.)

Anterior to the pituitary, identify the optic chiasma portion of the optic pathway to the brain. The **mammillary bodies,** relay stations for olfaction, bulge exteriorly from the floor of the hypothalamus just posterior to the pituitary gland.

3. The **epithalamus** forms the roof of the third ventricle and is the most dorsal portion of the diencephalon. Important structures in the epithalamus are the **pineal gland** (a neuroendocrine structure), and the **choroid plexus** of the third ventricle. The choroid plexuses, knotlike collections of capillaries within each ventricle, form the cerebrospinal fluid.

Brain Stem

1. Now trace the short midbrain from the mammillary bodies to the rounded pons below. (Continue to refer to Figure 17.4). The **cerebral aqueduct** is a slender canal traveling through the midbrain; it connects the third ventricle to the fourth ventricle in the hindbrain below. The cerebral peduncles and the rounded corpora quadrigemina make up the midbrain tissue anterior and posterior (respectively) to the cerebral aqueduct.

2. Locate the hindbrain structures. Trace the rounded pons to the medulla oblongata below, and identify the fourth ventricle posterior to these structures. Attempt to identify the single median aperture and the two lateral apertures, three openings found in the walls of the fourth ventricle. These apertures serve as passageways for cerebrospinal fluid to circulate into the subarachnoid space from the fourth ventricle.

Cerebellum

Examine the cerebellum. Notice that it is composed of two lateral hemispheres, each with three lobes (*anterior, poste-*

rior, and a deep *flocculonodular*) connected by a midline lobe called the **vermis (Figure 17.6)**. As in the cerebral hemispheres, the cerebellum has an outer cortical area of gray matter and an inner area of white matter. The treelike branching of the cerebellar white matter is referred to as the **arbor vitae,** or "tree of life." The cerebellum is concerned with unconscious coordination of skeletal muscle activity and control of balance and equilibrium. Fibers converge on the cerebellum from the equilibrium apparatus of the inner ear, visual pathways, proprioceptors of tendons and skeletal muscles, and from many other areas. Thus the cerebellum remains constantly aware of the position and state of tension of the various body parts. ■

Meninges of the Brain

The brain and spinal cord are covered and protected by three connective tissue membranes called **meninges (Figure 17.7)**. The outermost meninx is the leathery **dura mater,** a double-layered membrane. One of its layers (the *periosteal layer*) is attached to the inner surface of the skull, forming the periosteum. The other (the *meningeal layer*) forms the outermost brain covering and is continuous with the dura mater of the spinal cord.

The dural layers are fused together except in three places where the inner membrane extends inward to form a septum that secures the brain to structures inside the cranial cavity. One such extension, the **falx cerebri,** dips into the longitudinal fissure between the cerebral hemispheres to attach to the crista galli of the ethmoid bone of the skull (Figure 17.7a). The cavity created at this point is the large **superior sagittal sinus,** which collects blood draining from the brain tissue. The **falx cerebelli,** separating the two cerebellar hemispheres, and the **tentorium cerebelli,** separating the cerebrum from the cerebellum below, are two other important inward folds of the inner dural membrane.

The middle meninx, the weblike **arachnoid mater,** underlies the dura mater and is partially separated from it by the **subdural space.** Threadlike projections bridge the **subarachnoid space** to attach the arachnoid to the innermost meninx, the **pia mater.** The delicate pia mater is highly vascular and clings tenaciously to the surface of the brain, following its convolutions.

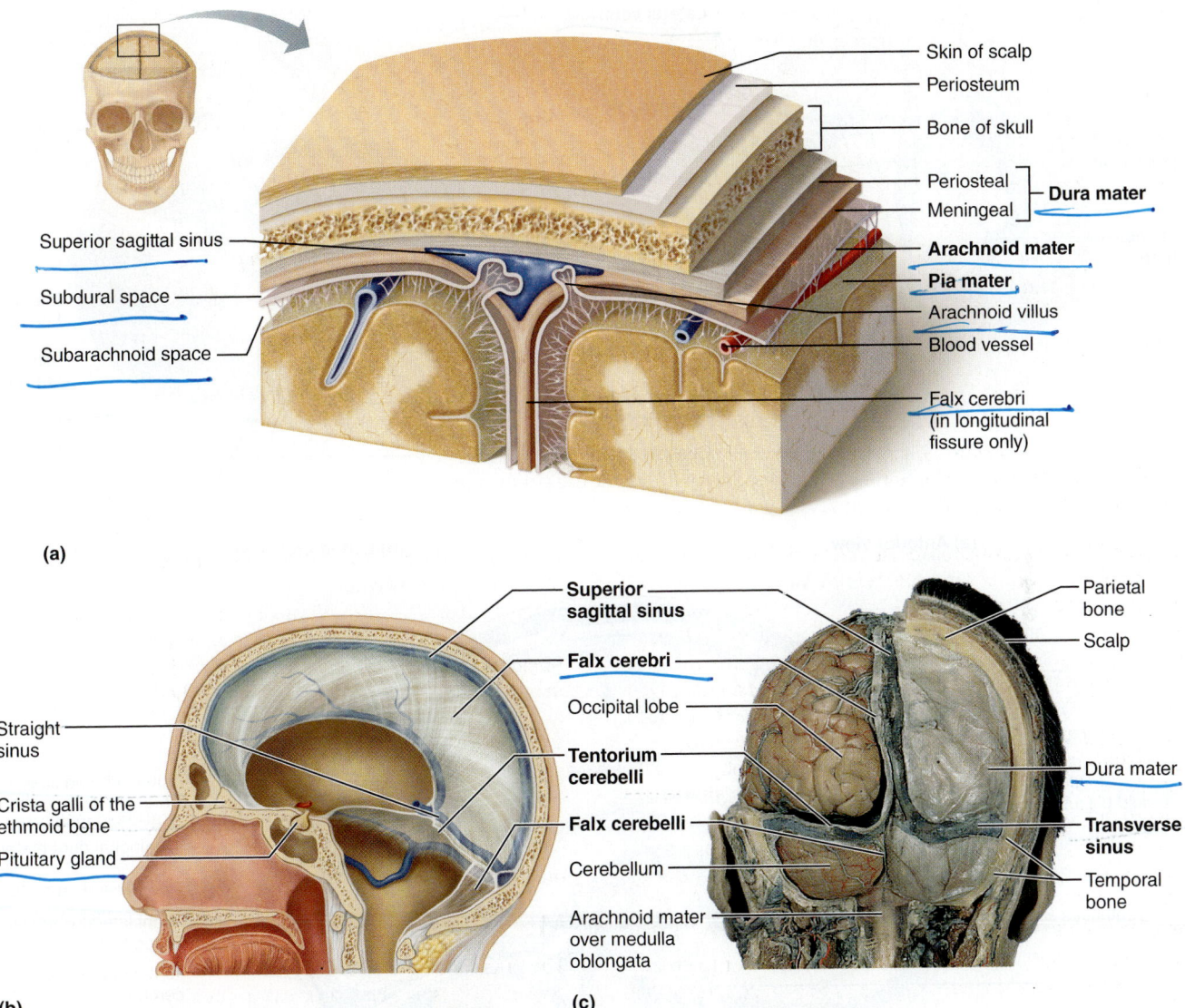

Figure 17.7 Meninges of the brain. (a) Three-dimensional frontal section showing the relationship of the dura mater, arachnoid mater, and pia mater. The meningeal dura forms the falx cerebri fold, which extends into the longitudinal fissure and attaches the brain to the ethmoid bone of the skull. The superior sagittal sinus is enclosed by the dural membranes superiorly. Arachnoid villi, which return cerebrospinal fluid to the dural sinus, are also shown. **(b)** Midsagittal view showing the position of the dural folds: the falx cerebri, tentorium cerebelli, and falx cerebelli. **(c)** Posterior view of the brain in place, surrounded by the dura mater. Sinuses between periosteal and meningeal dura contain venous blood.

In life, the subarachnoid space is filled with cerebrospinal fluid. Specialized projections of the arachnoid tissue called **arachnoid villi** protrude through the dura mater. These villi allow the cerebrospinal fluid to drain back into the venous circulation via the superior sagittal sinus and other dural sinuses.

Meningitis, inflammation of the meninges, is a serious threat to the brain because of the intimate association between the brain and meninges. Should infection spread to the neural tissue of the brain itself, life-threatening **encephalitis** may occur. Meningitis is often diagnosed by taking a sample of cerebrospinal fluid from the subarachnoid space. ✚

Cerebrospinal Fluid

The cerebrospinal fluid (CSF), much like plasma in composition, is continually formed by the **choroid plexuses,** small capillary knots hanging from the roof of the ventricles of the brain. The cerebrospinal fluid in and around the brain forms a watery cushion that protects the delicate brain tissue against blows to the head.

Within the brain, the cerebrospinal fluid circulates from the two lateral ventricles (in the cerebral hemispheres) into the third ventricle via the **interventricular foramina,** and then through the cerebral aqueduct of the midbrain into the fourth ventricle in the hindbrain **(Figure 17.8)**. CSF enters

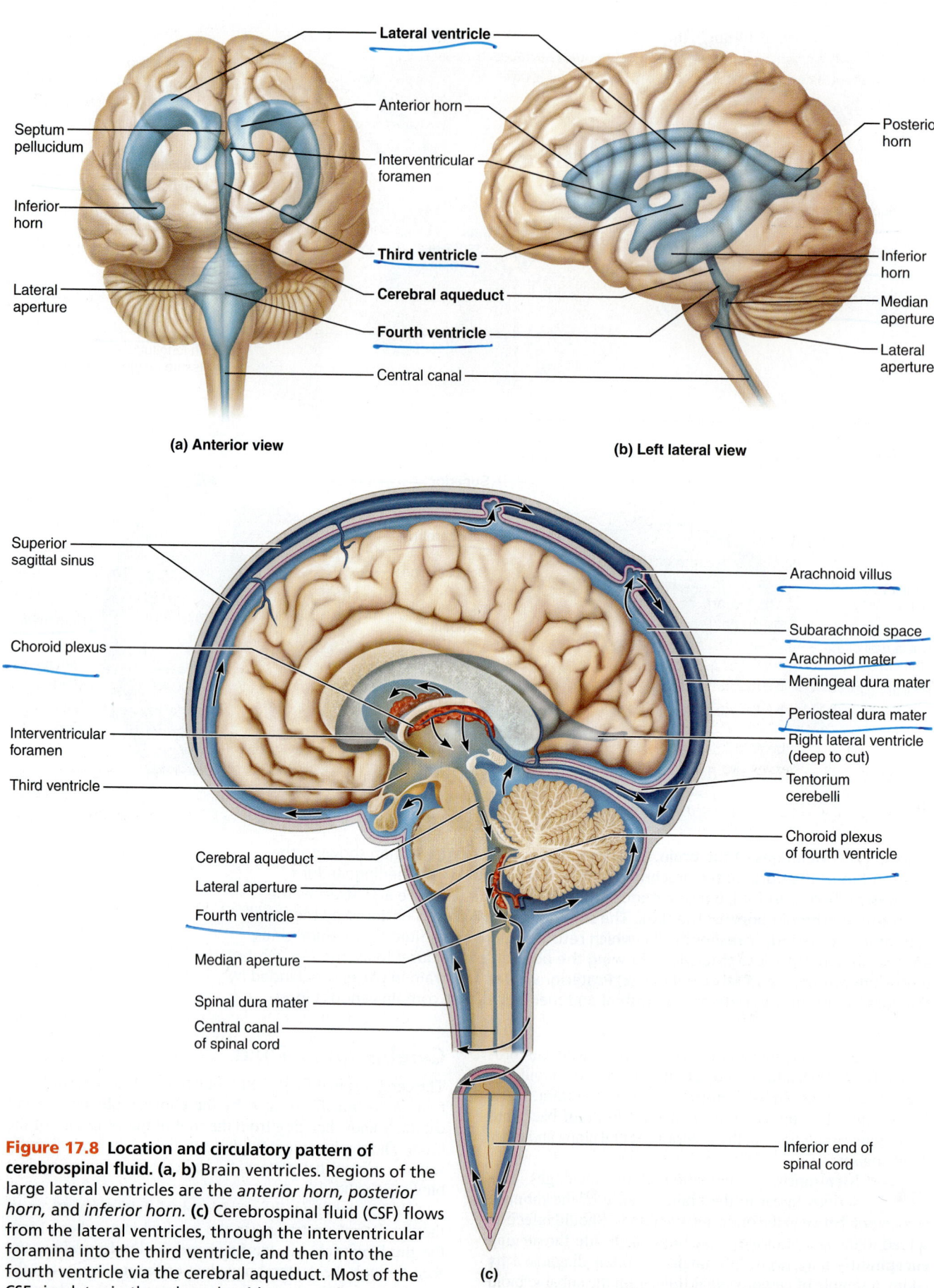

(a) Anterior view

(b) Left lateral view

(c)

Figure 17.8 Location and circulatory pattern of cerebrospinal fluid. (a, b) Brain ventricles. Regions of the large lateral ventricles are the *anterior horn, posterior horn,* and *inferior horn.* **(c)** Cerebrospinal fluid (CSF) flows from the lateral ventricles, through the interventricular foramina into the third ventricle, and then into the fourth ventricle via the cerebral aqueduct. Most of the CSF circulates in the subarachnoid space and returns to the blood through arachnoid villi.

the subarachnoid space through the three foramina in the walls of the fourth ventricle. There it bathes the outer surfaces of the brain and spinal cord. The fluid returns to the blood in the dural sinuses via the arachnoid villi.

Ordinarily, cerebrospinal fluid forms and drains at a constant rate. However, under certain conditions—for example, obstructed drainage or circulation resulting from tumors or anatomical deviations—cerebrospinal fluid accumulates and exerts increasing pressure on the brain which, uncorrected, causes neurological damage in adults. In infants, **hydrocephalus** (literally, "water on the brain") is indicated by a gradually enlarging head. The infant's skull is still flexible and contains fontanelles, so it can expand to accommodate the increasing size of the brain. ✚

Cranial Nerves

The **cranial nerves** are part of the peripheral nervous system and not part of the brain proper, but they are most appropriately identified while studying brain anatomy. The 12 pairs of cranial nerves primarily serve the head and neck. Only one pair, the vagus nerves, extends into the thoracic and abdominal cavities. All but the first two pairs (olfactory and optic nerves) arise from the brain stem and pass through foramina in the base of the skull to reach their destination.

The cranial nerves are numbered consecutively, and in most cases their names reflect the major structures they control. The cranial nerves are described by name, number (Roman numeral), origin, course, and function in the list **(Table 17.1)**. This information should be committed to memory. A mnemonic device that might be helpful for remembering the cranial nerves in order is "*On* *o*ccasion, *o*ur *t*rusty *t*ruck *a*cts *f*unny—*v*ery *g*ood *v*ehicle *a*ny*h*ow." The first letter of each word and the "a" and "h" of the final word "anyhow" will remind you of the first letter of the cranial nerve name.

Most cranial nerves are mixed nerves (containing both motor and sensory fibers). But close scrutiny of the list (Table 17.1) will reveal that three pairs of cranial nerves (optic, olfactory, and vestibulocochlear) are purely sensory in function.

Recall that the cell bodies of neurons are always located within the central nervous system (cortex or nuclei) or in specialized collections of cell bodies (ganglia) outside the CNS. Neuron cell bodies of the sensory cranial nerves are located in ganglia; those of the mixed cranial nerves are found both within the brain and in peripheral ganglia.

ACTIVITY 3

Identifying and Testing the Cranial Nerves

1. Observe the ventral surface of the brain model to identify the cranial nerves. (**Figure 17.9** may also aid you in this study.) Notice that the first (olfactory) cranial nerves are not visible on the model because they consist only of short axons that run from the nasal mucosa through the cribriform plate of the ethmoid bone. (However, the synapse points of the first cranial nerves, the *olfactory bulbs,* are visible on the model.)

2. Testing cranial nerves, is an important part of any neurological examination. (See the last column of Table 17.1 for techniques you can use for such tests.) Conduct tests of cranial nerve function following directions given in the "testing" column of the table. The results may help you understand

cranial nerve function, especially as it pertains to some aspects of brain function.

3. Several cranial nerve ganglia are named in the **Activity 3 chart.** *Using your textbook or an appropriate reference,* fill in the chart by naming the cranial nerve the ganglion is associated with and stating its location. ▬

Activity 3: Cranial Nerve Ganglia		
Cranial nerve ganglion	**Cranial nerve**	**Site of ganglion**
Trigeminal		
Geniculate		
Inferior		
Superior		
Spiral		
Vestibular		

DISSECTION:
The Sheep Brain

The sheep brain is enough like the human brain to warrant comparison. Obtain a sheep brain, disposable gloves, dissecting tray, and instruments, and bring them to your laboratory bench.

1. Don disposable gloves. If the dura mater is present, remove it as described here. Place the intact sheep brain ventral surface down on the dissecting pan, and observe the dura mater. Feel its consistency and note its toughness. Cut through the dura mater along the line of the longitudinal fissure (which separates the cerebral hemispheres) to enter the superior sagittal sinus. Gently force the cerebral hemispheres apart laterally to expose the corpus callosum deep to the longitudinal fissure.

2. Carefully remove the dura mater and examine the superior surface of the brain. Notice that its surface, like that of the human brain, is thrown into convolutions (fissures and gyri). Locate the arachnoid mater, which appears on the brain surface as a delicate "cottony" material spanning the fissures. In contrast, the innermost meninx, the pia mater, closely follows the cerebral contours.

3. Before beginning the dissection, turn your sheep brain so that you are viewing its left lateral aspect. Compare the various areas of the sheep brain (cerebrum, brain stem, cerebellum) to the photo of the human brain **(Figure 17.10)**. Relatively speaking, which of these structures is obviously much larger in the human brain?

Ventral Structures

Turn the brain so that its ventral surface is uppermost. (**Figure 17.11a** and **b** shows the important features of the ventral surface of the brain.)

1. Look for the clublike olfactory bulbs anteriorly, on the inferior surface of the frontal lobes of the cerebral hemispheres. Axons of olfactory neurons run from the nasal mucosa through

17

Table 17.1   The Cranial Nerves (see Figure 17.9)

Number and name	Origin and course	Function*	Testing
I. Olfactory	Fibers arise from olfactory epithelium and run through cribriform plate of ethmoid bone to synapse in olfactory bulbs.	Purely sensory—carries afferent impulses associated with sense of smell.	Person is asked to sniff aromatic substances, such as oil of cloves and vanilla, and to identify each.
II. Optic	Fibers arise from retina of eye to form the optic nerve and pass through optic canal of orbit. Fibers partially cross over at the optic chiasma and continue on to the thalamus as the optic tracts. Final fibers of this pathway travel from the thalamus to the visual cortex as the optic radiation.	Purely sensory—carries afferent impulses associated with vision.	Vision and visual field are determined with eye chart and by testing the point at which the person first sees an object (finger) moving into the visual field. Fundus of eye viewed with ophthalmoscope to detect papilledema (swelling of optic disc, or point at which optic nerve leaves the eye) and to observe blood vessels.
III. Oculomotor	Fibers emerge from dorsal midbrain and course ventrally to enter the orbit. They exit from skull via superior orbital fissure.	Primarily motor—somatic motor fibers to inferior oblique and superior, inferior, and medial rectus muscles, which direct eyeball, and to levator palpebrae muscles of the superior eyelid; parasympathetic fibers to iris and smooth muscle controlling lens shape (reflex responses to varying light intensity and focusing of eye for near vision).	Pupils are examined for size, shape, and equality. Pupillary reflex is tested with penlight (pupils should constrict when illuminated). Convergence for near vision is tested, as is subject's ability to follow objects with the eyes.
IV. Trochlear	Fibers emerge from midbrain and exit from skull via superior orbital fissure.	Primarily motor—provides somatic motor fibers to superior oblique muscle that moves the eyeball.	Tested in common with cranial nerve III.
V. Trigeminal	Fibers run from face to pons and form three divisions: mandibular division fibers pass through foramen ovale in sphenoid bone, maxillary division fibers pass via foramen rotundum in sphenoid bone, and ophthalmic division fibers pass through superior orbital fissure of eye socket.	Mixed—major sensory nerve of face; conducts sensory impulses from skin of face and anterior scalp, from mucosae of mouth and nose, and from surface of eyes; mandibular division also contains motor fibers that innervate muscles of mastication and muscles of floor of mouth.	Sensations of pain, touch, and temperature are tested with safety pin and hot and cold objects. Corneal reflex tested with wisp of cotton. Motor branch assessed by asking person to clench the teeth, open mouth against resistance, and move jaw side to side.
VI. Abducens	Fibers leave inferior pons and exit from skull via superior orbital fissure to run to eye.	Carries somatic motor fibers to lateral rectus muscle that moves the eyeball.	Tested in common with cranial nerve III.
VII. Facial	Fibers leave pons and travel through temporal bone via internal acoustic meatus, exiting via stylomastoid foramen to reach the face.	Mixed—supplies somatic motor fibers to muscles of facial expression and parasympathetic motor fibers to lacrimal and salivary glands; carries sensory fibers from taste receptors of anterior portion of tongue.	Anterior two-thirds of tongue is tested for ability to taste sweet (sugar), salty, sour (vinegar), and bitter (quinine) substances. Symmetry of face is checked. Subject is asked to close eyes, smile, whistle, and so on. Tearing is assessed with ammonia fumes.

17

Table 17.1	(continued)		
Number and name	**Origin and course**	**Function***	**Testing**
VIII. Vestibulocochlear	Fibers run from inner-ear equilibrium and hearing apparatus, housed in temporal bone, through internal acoustic meatus to enter pons.	Purely sensory—vestibular branch transmits impulses associated with sense of equilibrium from vestibular apparatus and semicircular canals; cochlear branch transmits impulses associated with hearing from cochlea.	Hearing is checked by air and bone conduction using tuning fork.
IX. Glossopharyngeal	Fibers emerge from medulla and leave skull via jugular foramen to run to throat.	Mixed—somatic motor fibers serve pharyngeal muscles, and parasympathetic motor fibers serve salivary glands; sensory fibers carry impulses from pharynx, tonsils, posterior tongue (taste buds), and from chemoreceptors and pressure receptors of carotid artery.	A tongue depressor is used to check the position of the uvula. Gag and swallowing reflexes are checked. Subject is asked to speak and cough. Posterior third of tongue may be tested for taste.
X. Vagus	Fibers emerge from medulla and pass through jugular foramen and descend through neck region into thorax and abdomen.	Mixed—fibers carry somatic motor impulses to pharynx and larynx and sensory fibers from same structures; very large portion is composed of parasympathetic motor fibers, which supply heart and smooth muscles of abdominal visceral organs; transmits sensory impulses from viscera.	As for cranial nerve IX (IX and X are tested in common, since they both innervate muscles of throat and mouth).
XI. Accessory	Fibers arise from the superior aspect of spinal cord, enter the skull, and then travel through jugular foramen to reach muscles of neck and back.	Mixed (but primarily motor in function)—provides somatic motor fibers to sternocleidomastoid and trapezius muscles and to muscles of soft palate, pharynx, and larynx (spinal and medullary fibers respectively).	Sternocleidomastoid and trapezius muscles are checked for strength by asking person to rotate head and shrug shoulders against resistance.
XII. Hypoglossal	Fibers arise from medulla and exit from skull via hypoglossal canal to travel to tongue.	Mixed (but primarily motor in function)—carries somatic motor fibers to muscles of tongue.	Person is asked to protrude and retract tongue. Any deviations in position are noted.

*Does not include sensory impulses from proprioceptors.

the perforated cribriform plate of the ethmoid bone to synapse with the olfactory bulbs.

How does the size of these olfactory bulbs compare with those of humans?

_____

Is the sense of smell more important as a protective and a food-getting sense in sheep or in humans?

_____

2. The optic nerve (II) carries sensory impulses from the retina of the eye. Thus this cranial nerve is involved in the sense of vision. Identify the optic nerves, optic chiasma, and optic tracts.

3. Posterior to the optic chiasma, two structures protrude from the ventral aspect of the hypothalamus—the infundibulum (stalk of the pituitary gland) immediately posterior to the optic chiasma and the mammillary body. Notice that the sheep's mammillary body is a single rounded eminence. In humans it is a double structure.

4. Identify the cerebral peduncles on the ventral aspect of the midbrain, just posterior to the mammillary body of the hypothalamus. The cerebral peduncles are fiber tracts connecting the cerebrum and medulla oblongata. Identify the large oculomotor nerves (III), which arise from the ventral midbrain surface, and the tiny trochlear nerves (IV), which can be seen at the junction of the midbrain and pons. Both of

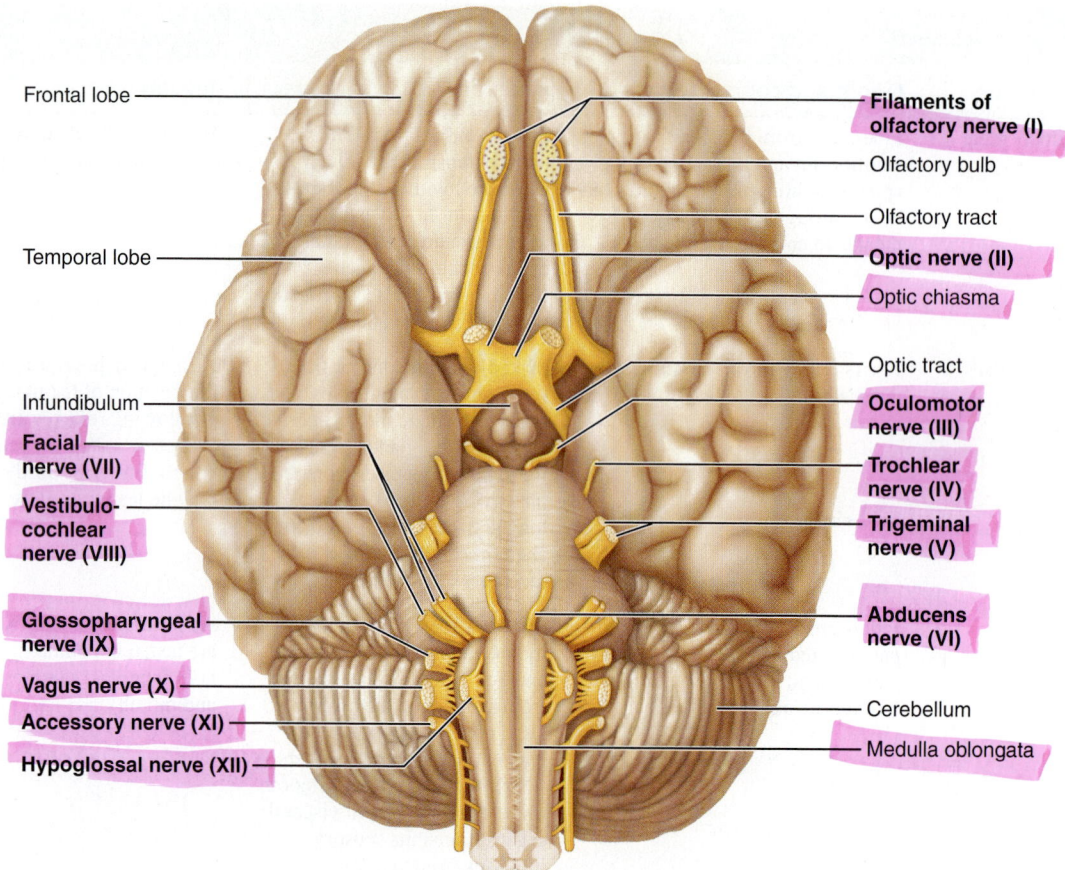

Frontal lobe

Filaments of olfactory nerve (I)

Olfactory bulb

Olfactory tract

Temporal lobe

Optic nerve (II)

Optic chiasma

Optic tract

Infundibulum

Oculomotor nerve (III)

Facial nerve (VII)

Trochlear nerve (IV)

Vestibulo-cochlear nerve (VIII)

Trigeminal nerve (V)

Glossopharyngeal nerve (IX)

Abducens nerve (VI)

Vagus nerve (X)

Accessory nerve (XI)

Cerebellum

Hypoglossal nerve (XII)

Medulla oblongata

Figure 17.9 Ventral aspect of the human brain, showing the cranial nerves. (See also Figure 17.3.)

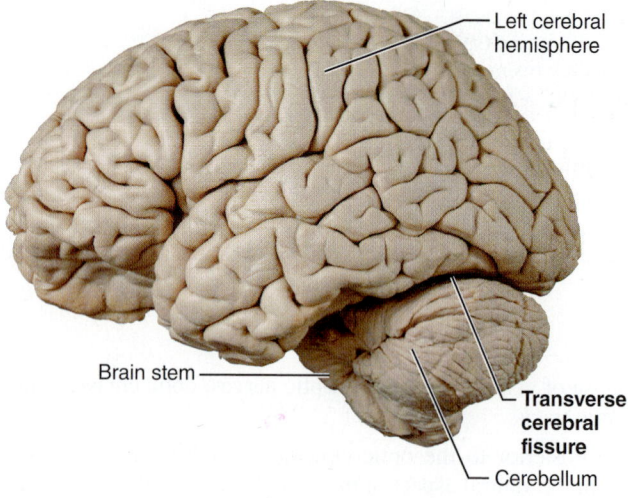

Left cerebral hemisphere

Brain stem

Transverse cerebral fissure

Cerebellum

Figure 17.10 Photograph of lateral aspect of the human brain.

these cranial nerves provide motor fibers to extrinsic muscles of the eyeball.

5. Move posteriorly from the midbrain to identify first the pons and then the medulla oblongata, both hindbrain structures composed primarily of ascending and descending fiber tracts.

6. Return to the junction of the pons and midbrain, and proceed posteriorly to identify the following cranial nerves, all arising from the pons. Check them off as you locate them.

☐ Trigeminal nerves (V), which are involved in chewing and sensations of the head and face.

☐ Abducens nerves (VI), which abduct the eye (and thus work in conjunction with cranial nerves III and IV)

☐ Facial nerves (VII), large nerves involved in taste sensation, gland function (salivary and lacrimal glands), and facial expression.

7. Continue posteriorly to identify and check off:

☐ Vestibulocochlear nerves (VIII), purely sensory nerves that are involved with hearing and equilibrium.

☐ Glossopharyngeal nerves (IX), which contain motor fibers innervating throat structures and sensory fibers transmitting taste stimuli (in conjunction with cranial nerve VII).

☐ Vagus nerves (X), often called "wanderers," which serve many organs of the head, thorax, and abdominal cavity.

☐ Accessory nerves (XI), which serve muscles of the neck, larynx, and shoulder; actually arise from the spinal cord (C_1 through C_5) and travel superiorly to enter the skull before running to the muscles that they serve.

☐ Hypoglossal nerves (XII), which stimulate tongue and neck muscles.

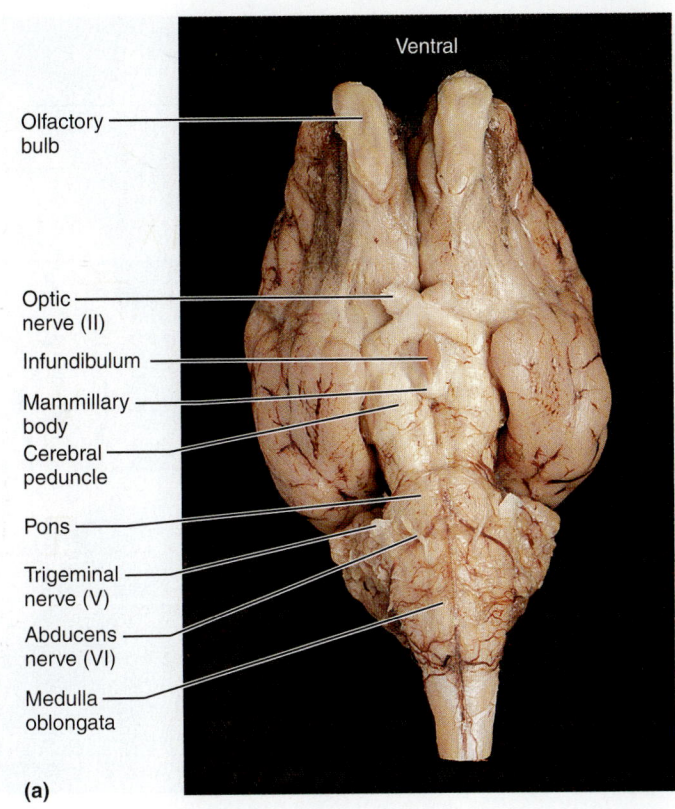

Ventral

Olfactory bulb

Optic nerve (II)

Infundibulum

Mammillary body

Cerebral peduncle

Pons

Trigeminal nerve (V)

Abducens nerve (VI)

Medulla oblongata

(a)

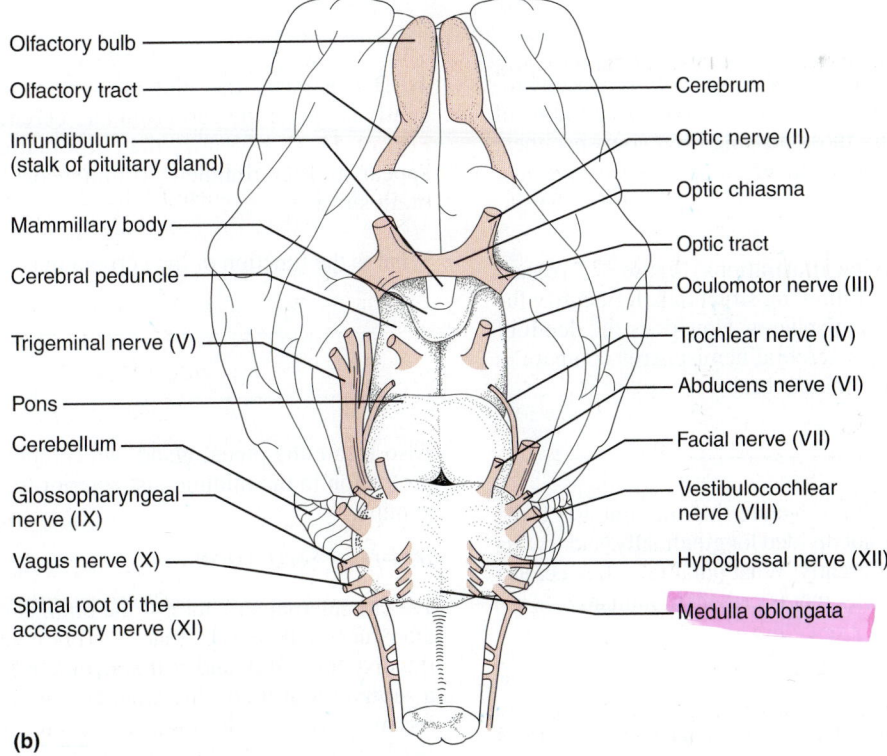

Olfactory bulb

Olfactory tract

Infundibulum (stalk of pituitary gland)

Mammillary body

Cerebral peduncle

Trigeminal nerve (V)

Pons

Cerebellum

Glossopharyngeal nerve (IX)

Vagus nerve (X)

Spinal root of the accessory nerve (XI)

Cerebrum

Optic nerve (II)

Optic chiasma

Optic tract

Oculomotor nerve (III)

Trochlear nerve (IV)

Abducens nerve (VI)

Facial nerve (VII)

Vestibulocochlear nerve (VIII)

Hypoglossal nerve (XII)

Medulla oblongata

(b)

Figure 17.11 Intact sheep brain. (a) Photograph of ventral view. **(b)** Diagrammatic ventral view.

17

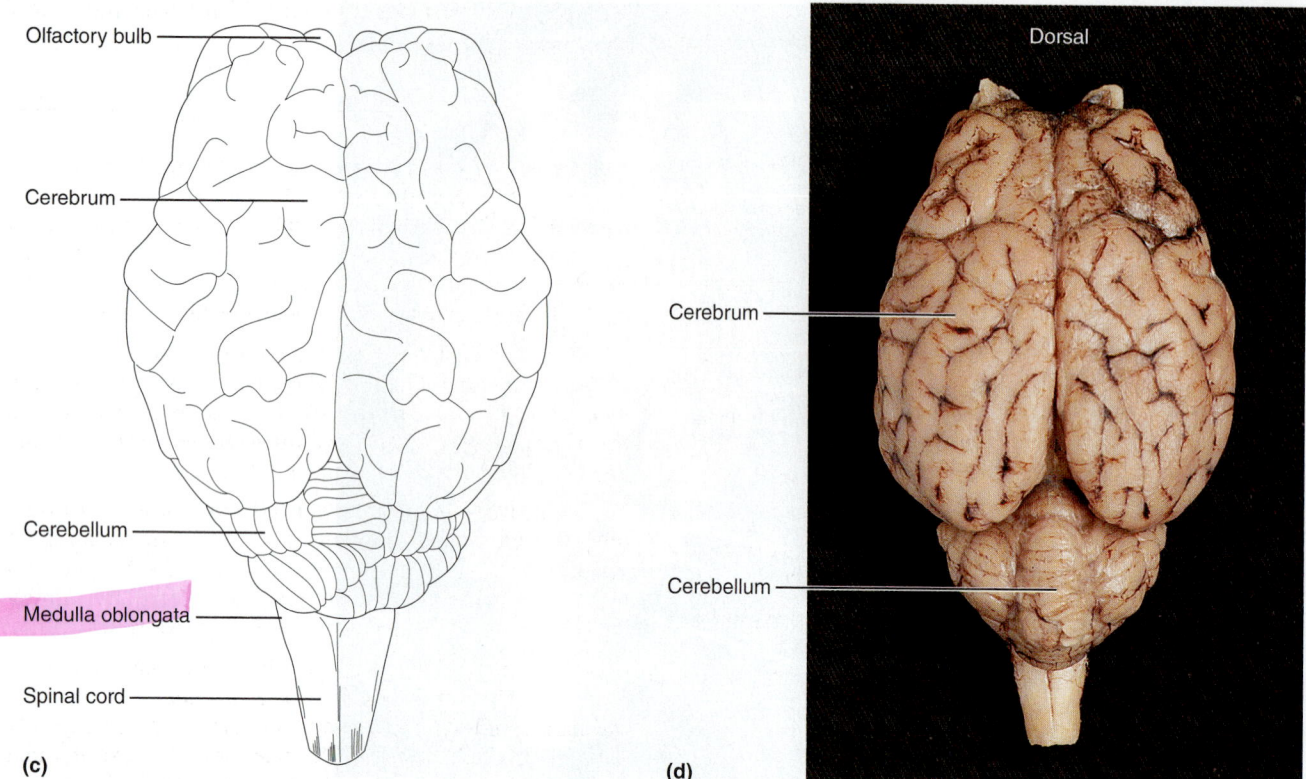

Olfactory bulb

Cerebrum

Cerebellum

Medulla oblongata

Spinal cord

(c)

Dorsal

Cerebrum

Cerebellum

(d)

Figure 17.11 *(continued)* **Intact sheep brain. (c, d)** Diagram and photograph of the dorsal view, respectively.

It is likely that some of the cranial nerves will have been broken off during brain removal. If so, observe sheep brains of other students to identify those missing from your specimen, using your check marks as a guide.

Dorsal Structures

1. Refer to the dorsal view illustrations (Figure 17.11c) as a guide in identifying the following structures. Reidentify the now exposed cerebral hemispheres. How does the depth of the fissures in the sheep's cerebral hemispheres compare to that of the fissures in the human brain?

_____

2. Examine the cerebellum. Notice that, in contrast to the human cerebellum, it is not divided longitudinally, and that its fissures are oriented differently. What dural falx (falx cerebri or falx cerebelli) is missing that is present in humans?

_____

3. Locate the three pairs of cerebellar peduncles, fiber tracts that connect the cerebellum to other brain structures, by lifting the cerebellum dorsally away from the brain stem. The most posterior pair, the inferior cerebellar peduncles, connect the cerebellum to the medulla. The middle cerebellar peduncles attach the cerebellum to the pons, and the superior cerebellar peduncles run from the cerebellum to the midbrain.

4. To expose the dorsal surface of the midbrain, gently separate the cerebrum and cerebellum (as shown in **Figure 17.12**.) Identify the corpora quadrigemina, which appear as four rounded prominences on the dorsal midbrain surface.

What is the function of the corpora quadrigemina?

_____

_____

Also locate the pineal gland, which appears as a small oval protrusion in the midline just anterior to the corpora quadrigemina.

Internal Structures

1. The internal structure of the brain can be examined only after further dissection. Place the brain ventral side down on the dissecting tray and make a cut completely through it in a superior to inferior direction. Cut through the longitudinal fissure, corpus callosum, and midline of the cerebellum. (Refer to **Figure 17.13** as you work.)

2. A thin nervous tissue membrane immediately ventral to the corpus callosum that separates the lateral ventricles is the septum pellucidum. If it is still intact, pierce this membrane and probe the lateral ventricle cavity. The fiber tract ventral to the septum pellucidum and anterior to the third ventricle is the fornix.

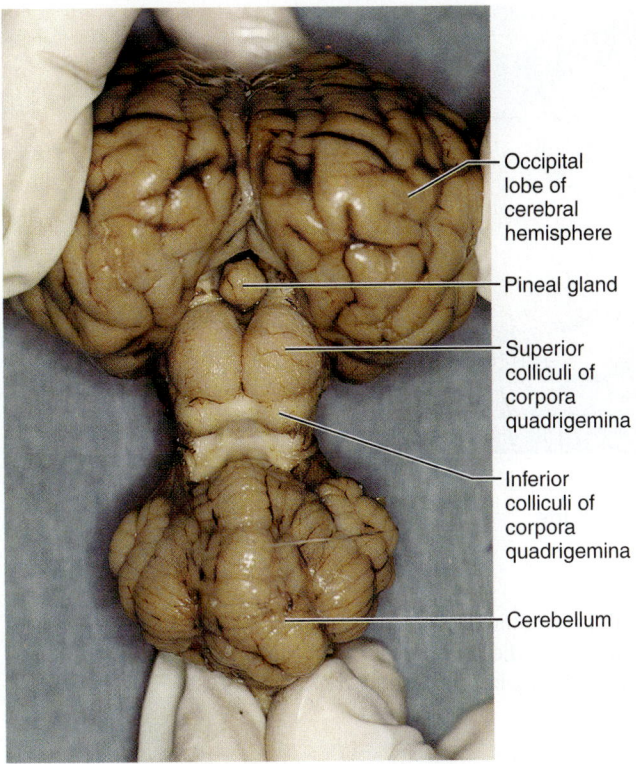

Figure 17.12 Means of exposing the dorsal midbrain structures of the sheep brain.

How does the size of the fornix in this brain compare with the size of the human fornix?

Why do you suppose this is so? (Hint: What is the function of this band of fibers?)

3. Identify the thalamus, which forms the walls of the third ventricle and is located posterior and ventral to the fornix. The intermediate mass spanning the ventricular cavity appears as an oval protrusion of the thalamic wall. Anterior to the intermediate mass, locate the interventricular foramen, a canal connecting the lateral ventricle on the same side with the third ventricle.

4. The hypothalamus forms the floor of the third ventricle. Identify the optic chiasma, infundibulum, and mammillary body on its exterior surface. You can see the pineal gland at the superoposterior end of the third ventricle, just beneath the junction of the corpus callosum and fornix.

5. Locate the midbrain by identifying the corpora quadrigemina that form its dorsal roof. Follow the cerebral aqueduct (the narrow canal connecting the third and fourth ventricles) through the midbrain tissue to the fourth ventricle. Identify the cerebral peduncles, which form its anterior walls.

6. Identify the pons and medulla oblongata, which lie anterior to the fourth ventricle. The medulla continues into the spinal cord without any obvious anatomical change, but the point at which the fourth ventricle narrows to a small canal is generally accepted as the beginning of the spinal cord.

7. Identify the cerebellum posterior to the fourth ventricle. Notice its internal treelike arrangement of white matter, the arbor vitae.

8. If time allows, obtain another sheep brain and section it along the coronal plane so that the cut passes through the infundibulum. Compare your specimen with the photograph of a coronal section **(Figure 17.14)**, and attempt to identify all the structures shown in the figure.

17

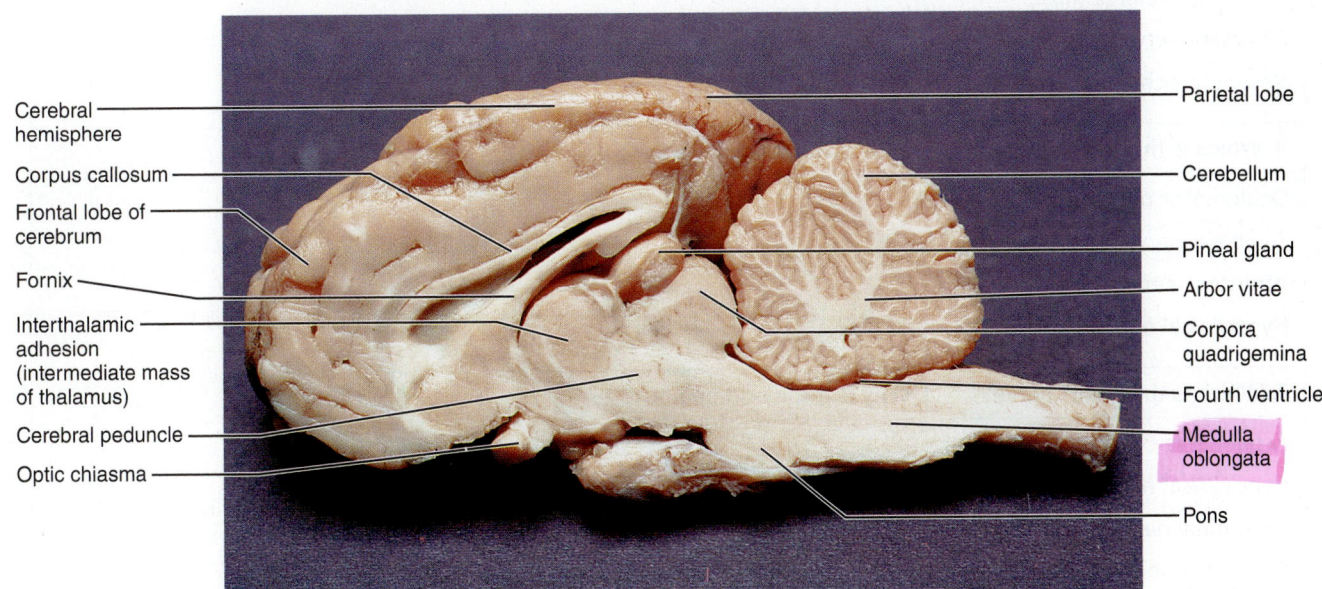

Figure 17.13 Photograph of sagittal section of the sheep brain showing internal structures.

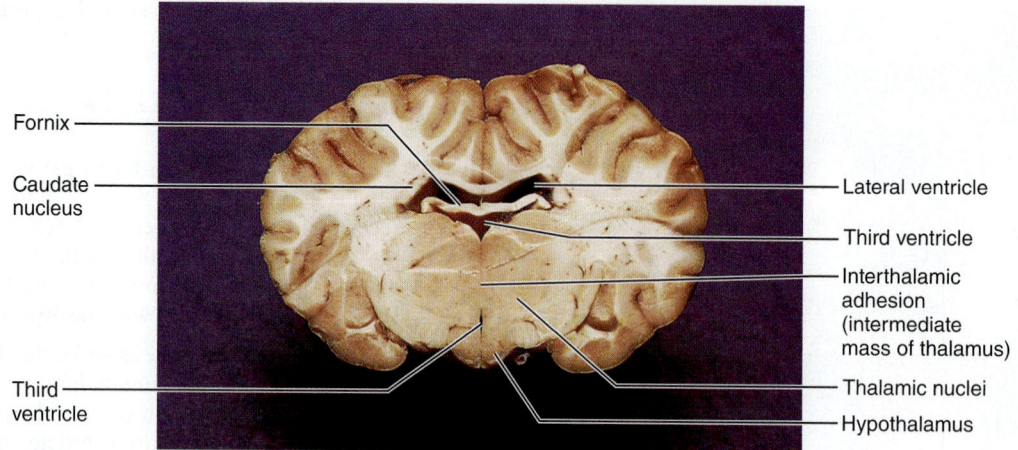

Figure 17.14 Coronal section of a sheep brain. Major structures include the thalamus, hypothalamus, and lateral and third ventricles.

9. Check with your instructor to determine if a small portion of the spinal cord from your brain specimen should be saved for spinal cord studies (Exercise 19.) Otherwise, dispose of all the organic debris in the appropriate laboratory containers and clean the laboratory bench, the dissection instruments, and the tray before leaving the laboratory. ■

GROUP CHALLENGE

Odd (Cranial) Nerve Out

The following boxes each contain four cranial nerves. One of the listed nerves does not share a characteristic with the other three. Circle the cranial nerve that doesn't belong with the others and explain why it is singled out.

What characteristic is it missing? Sometimes there may be multiple reasons why the cranial nerve doesn't belong with the others.

1. Which is the "odd" nerve?	Why is it the odd one out?
Optic nerve (II)	
Oculomotor nerve (III)	
Olfactory nerve (I)	
Vestibulocochlear nerve (VIII)	

2. Which is the "odd" nerve?	Why is it the odd one out?
Oculomotor nerve (III)	
Trochlear nerve (IV)	
Abducens nerve (VI)	
Hypoglossal nerve (XII)	

3. Which is the "odd" nerve?	Why is it the odd one out?
Facial nerve (VII)	
Hypoglossal nerve (XII)	
Trigeminal nerve (V)	
Glossopharyngeal nerve (IX)	

Electroencephalography

MATERIALS

☐ Oscilloscope and EEG lead-selector box or physiograph and high-gain preamplifier
☐ Cot (if available) or pillow
☐ Electrode gel
☐ EEG electrodes and leads
☐ Collodion gel or long elastic EEG straps

🔺 **BIOPAC®** BIOPAC® BSL System for Windows with BSL software version 3.7.5 to 3.7.7, or BSL System for Mac OS X with BSL software version 3.7.4 to 3.7.7, MP36/35 or MP45 data acquisition unit, PC or Mac computer, electrode lead set, disposable vinyl electrodes, Lycra® swim cap (such as Speedo® brand) or supportive wrap (such as 3M Coban™ Self-adhering Support Wrap) to press electrodes against head for improved contact, and a cot or lab bench and pillow.

Instructors using the MP36 (or MP35/30) data acquisition unit with BSL software versions earlier than 3.7.5 (for Windows) and 3.7.4 (for Mac OS X) will need slightly different channel settings and collection strategies. Instructions for using the older data acquisition unit can be found on MasteringA&P.

OBJECTIVES

1. Define *electroencephalogram (EEG)*, and discuss its clinical significance.
2. Describe or recognize typical tracings of alpha, beta, theta, and delta brain waves, and indicate the conditions when each is most likely to occur.
3. Indicate the source of brain waves.
4. Define *alpha block*.
5. Monitor the EEG in a human subject.
6. Describe the effect of a sudden sound, mental concentration, and respiratory alkalosis on the EEG.

PRE-LAB QUIZ

1. What does an electroencephalogram (EEG) measure?
   a. electrical activity of the brain
   b. electrical activity of the heart
   c. emotions
   d. physical activity of the subject
2. Circle the correct underlined term. <u>Alpha waves</u> / <u>Beta waves</u> are typical of the attentive or awake state.
3. Circle True or False. Brain waves can change with age, sensory stimuli, and the chemical state of the body.
4. Where will you place the indifferent (ground) electrode on your subject?
   a. the earlobe
   b. the forehead
   c. over the occipital lobe
   d. over the temporal bone
5. During today's activity, students will instruct subjects to *hyperventilate*. What should the subjects do?
   a. breathe in a normal manner
   b. breathe rapidly
   c. breathe very slowly
   d. hold their breath until they almost pass out

MasteringA&P® For related exercise study tools, go to the Study Area of MasteringA&P. There you will find:
• Practice Anatomy Lab **PAL**
• PhysioEx **PEx**
• A&PFlix **A&PFlix**
• Practice quizzes, Histology Atlas, eText, Videos, and more!

As curious humans we are particularly interested in how the brain thinks, reasons, learns, remembers, and controls consciousness. As students we have learned that the brain accomplishes its tasks through electrical activities of neurons. The remarkable noninvasive technologies of twenty-first-century neuroscience have advanced our understanding of brain functions, as has the long-used technique of electroencephalography—the recording of electrical activity from the surface portions of the brain. It is incredible that the sophisticated equipment used to record an electroencephalogram (EEG) is commonly available in undergraduate laboratories. As you use it, you begin to explore the complex higher functions of the human brain.

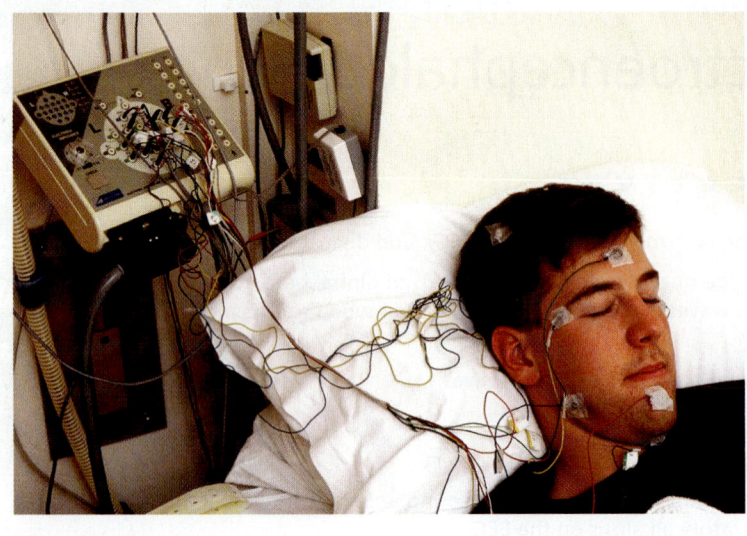

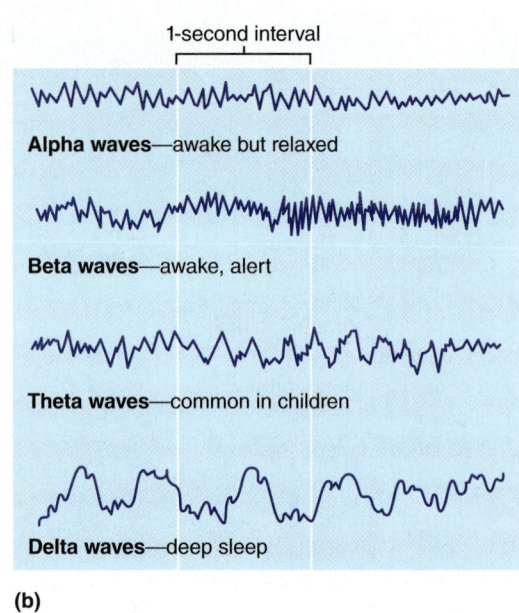

1-second interval

Alpha waves—awake but relaxed

Beta waves—awake, alert

Theta waves—common in children

Delta waves—deep sleep

(a) **(b)**

Figure 18.1 Electroencephalography and brain waves. (a) Scalp electrodes are positioned on the patient to record brain waves. **(b)** Typical EEGs.

Brain Wave Patterns and the Electroencephalogram

The **electroencephalogram (EEG),** a record of the electrical activity of the brain, can be obtained through electrodes placed at various points on the skin or scalp of the head. This electrical activity, which is recorded as waves **(Figure 18.1)**, represents the summed synaptic activity of many neurons.

Certain characteristics of brain waves are known. They have a frequency of 1 to 30 hertz (Hz) or cycles per second, a dominant rhythm of 10 Hz, and an average amplitude (voltage) of 20 to 100 microvolts (μV). They vary in frequency in different brain areas, occipital waves having a lower frequency than those associated with the frontal and parietal lobes.

The first of the brain waves to be described by scientists were the alpha waves (or alpha rhythm). **Alpha waves** have an average frequency range of 8 to 13 Hz and are produced when the individual is in a relaxed state with the eyes closed. **Alpha block,** suppression of the alpha rhythm, occurs if the eyes are opened or if the individual begins to concentrate on some mental problem or visual stimulus. Under these conditions, the waves decrease in amplitude but increase in frequency. Under conditions of fright or excitement, the frequency increases still more.

Beta waves, closely related to alpha waves, are faster (14 to 30 Hz) and have a lower amplitude. They are typical of the attentive or alert state.

Very large (high-amplitude) waves with a frequency of 4 Hz or less that are seen in deep sleep are **delta waves. Theta waves** are large, abnormally contoured waves with a frequency of 4 to 7 Hz. Although theta waves are normal in children, they are abnormal in awake adults.

Brain waves change with age, sensory stimuli, brain pathology, and the chemical state of the body. Glucose deprivation, oxygen poisoning, and sedatives all interfere with the rhythmic activity of brain output by disturbing the metabolism of neurons. Sleeping individuals and patients in a coma have EEGs that are slower (lower frequency) than the alpha rhythm of normal adults. Fright, epileptic seizures, and various types of drug intoxication can be associated with comparatively faster cortical activity. As these examples show, impairment of cortical function is indicated by neuronal activity that is either too fast or too slow; unconsciousness occurs at both extremes of the frequency range.

Because spontaneous brain waves are always present, even during unconsciousness and coma, the absence of brain waves (a "flat" EEG) is taken as clinical evidence of death. The EEG is used clinically to diagnose and localize many types of brain lesions, including epileptic foci, infections, abscesses, and tumors. ✚

ACTIVITY 1

Observing Brain Wave Patterns Using an Oscilloscope or Physiograph

If one electrode (the *active electrode*) is placed over a particular cortical area and another (the *indifferent electrode*) is placed over an inactive part of the head, such as the earlobe, all of the activity of the cortex underlying the active electrode will, theoretically, be recorded. The inactive area provides a zero reference point, or a baseline, and the EEG represents the difference between "activities" occurring under the two electrodes.

1. Connect the EEG lead-selector box to the oscilloscope preamplifier, or connect the high-gain preamplifier to the physiograph channel amplifier. Adjust the horizontal sweep and sensitivity according to the directions given in the instrument manual or by your instructor.

2. Prepare the subject. The subject should lie undisturbed on a cot or on the lab bench with eyes closed in a quiet, dimly lit area. (Someone who is able to relax easily makes a

good subject.) Apply a small amount of electrode gel to the subject's forehead above the left eye and on the left earlobe. Press an electrode to each prepared area and secure each by (1) applying a film of collodion gel to the electrode surface and the adjacent skin or (2) using a long elastic EEG strap (knot tied at the back of the head). If collodion gel is used, allow it to dry before you continue.

3. Connect the active frontal lead (forehead) to the EEG lead-selector box outlet marked "L Frontal." Connect the lead from the indifferent electrode (earlobe) to the ground outlet (or to the appropriate input terminal on the high-gain preamplifier).

4. Turn the oscilloscope or physiograph on, and observe the EEG pattern of the relaxed subject for a period of 5 minutes. If the subject is truly relaxed, you should see a typical alpha-wave pattern. (If the subject is unable to relax and the alpha-wave pattern does not appear in this time interval, test another subject.) Since the electrical activity of muscles interferes with EEG recordings, discourage all muscle movement during the monitoring period. If 60-cycle "noise" (appearing as fast, regular, low-amplitude waves superimposed on the more irregular brain waves) is present in your record because of the presence of other electronic equipment, consult your instructor to eliminate it.

5. Abruptly and loudly clap your hands. The subject's eyes should open, and alpha block should occur. Observe the immediate brain wave pattern. How do the frequency and amplitude of the brain waves change?

_____

_____

Would you characterize this as beta rhythm? _____

Why? _____

_____

6. Allow the subject about 5 minutes to achieve complete relaxation once again, then ask him or her to compute a number problem that requires concentration (for example, add 3 and 36, subtract 7, multiply by 2, add 50, etc.). Observe the brain wave pattern during the period of mental computation.

Observations: _____

_____

_____

7. Once again allow the subject to relax until alpha rhythm resumes. Then, instruct him or her to hyperventilate for 3 minutes. *Be sure to tell the subject when to stop hyperventilating.* Hyperventilation rapidly flushes carbon dioxide out of the lungs, decreasing carbon dioxide levels in the blood and producing respiratory alkalosis.

Observe the changes in the rhythm and amplitude of the brain waves occurring during the period of hyperventilation.

Observations: _____

_____

_____

8. Think of other stimu... ...ct brain wave patterns. Test your hypothese... ...w... stimuli you tested and what responses you ob... ve...

_____

_____

_____

_____

ACTIVITY 2

Electroencephalography U... ...®

In this activity, the EEG of the subject ... ...l during a relaxed state, first with the eyes clo... ...e eyes open while silently counting to ten, an... ...e eyes closed again.

Setting Up the Equipment

1. Connect the BIOPAC® unit to the co... ... the computer **ON.**

2. Make sure the BIOPAC® unit is **OFF.**

3. Plug in the equipment (as shown in **Figu...**

• Electrode lead set–CH 1

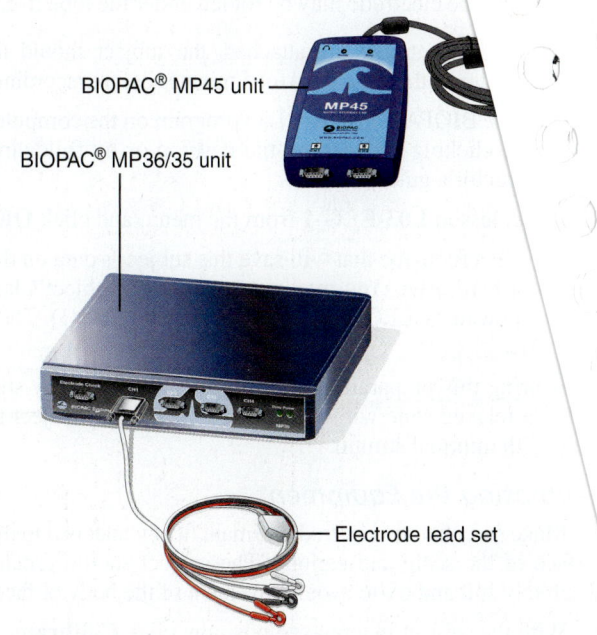

BIOPAC® MP45 unit

BIOPAC® MP36/35 unit

Electrode lead set

Figure 18.2 Setting up the BIOPAC® equipment. Plug the electrode set into Channel 1. Electrode leads are shown connected to the MP36/35 unit.

18

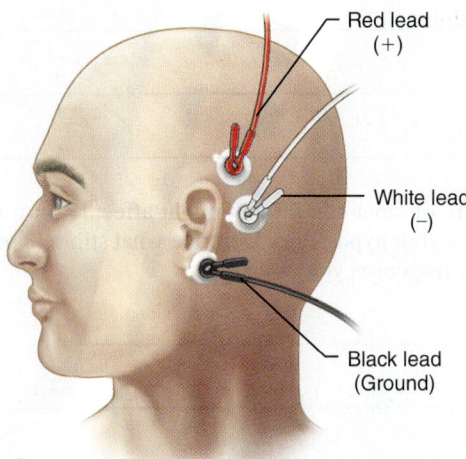

Figure 18.3 Placement of electrodes and the appropriate attachment of electrode leads by color.

4. Turn the BIOPAC® unit **ON.**

5. Attach three electrodes to the subject's scalp and ear (as shown in **Figure 18.3**). Follow these important guidelines to assist in effective electrode placement:

• Select subjects with the easiest access to the scalp.

• Move as much hair out of the way as possible.

• Apply a dab of electrode gel to the spots where the electrodes will be attached.

• Apply pressure to the electrodes for 1 minute to ensure attachment.

• Use a swimcap or supportive wrap to maintain attachment.

• Do not touch the electrodes while recording.

• The earlobe electrode may be folded under the lobe itself.

6. When the electrodes are attached, the subject should lie down and relax with eyes closed for 5 minutes before recording.

7. Start the BIOPAC® Student Lab program on the computer by double-clicking the icon on the desktop or by following your instructor's guidance.

8. Select lesson **L03-EEG-1** from the menu, and click **OK.**

9. Type in a filename that will save this subject's data on the computer hard drive. You may want to use the subject's last name followed by EEG-1 (for example, SmithEEG-1). Then click **OK.**

10. During this preparation, the subject should be very still and in a relaxed state with eyes closed. Allow the subject to relax with minimal stimuli.

Calibrating the Equipment

1. Make sure that the electrodes remain firmly attached to the surface of the scalp and earlobe. The subject should remain absolutely still and try to avoid movement of the body or face.

2. With the subject in a relaxed position, click **Calibrate.**

3. You will be prompted to check electrode attachment one final time. When ready, click **OK;** the computer will record for 8 seconds and stop automatically.

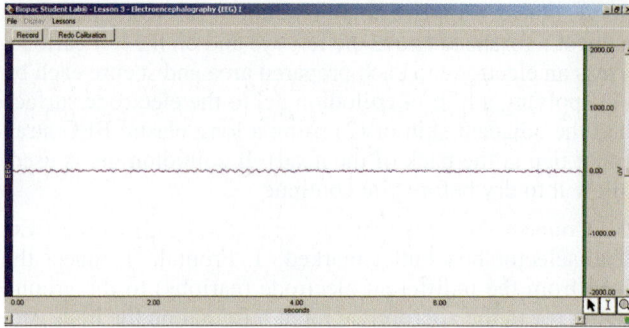

Figure 18.4 Example of calibration data.

4. Observe the recording of the calibration data (it should look like **Figure 18.4** with baseline at zero).

• If the data look very different, click **Redo Calibration** and repeat the steps above.

• If the data look similar, proceed to the next section.

Recording the Data

1. The subject should remain relaxed with eyes closed.

2. After clicking **Record,** the "director" will instruct the subject to keep his or her eyes closed for the first 20 seconds of recording, then open the eyes and *mentally* (not verbally) count to twenty, then close the eyes again and relax for 20 seconds. The director will insert a marker by pressing the **F4** key (PC or Mac) when the command to open eyes is given, and another marker by pressing the **F5** key (PC or Mac) when the subject reaches the count of twenty and closes the eyes. Click **Suspend** 20 seconds after the subject recloses the eyes.

3. Observe the recording of the data (it should look similar to the data in **Figure 18.5**).

• If the subject moved too much during the recording, it is likely that artifact spikes will appear in the data. Remind the subject to be very still.

• Look carefully at the **alpha rhythm** band of data. The intensity of the alpha signal should decrease during the "eyes open" phase of the recording. If the data do not demonstrate this change, make sure that the electrodes are firmly attached.

• If the data show artifact spikes or the alpha signal fails to decrease when the eyes are open, click **Redo.**

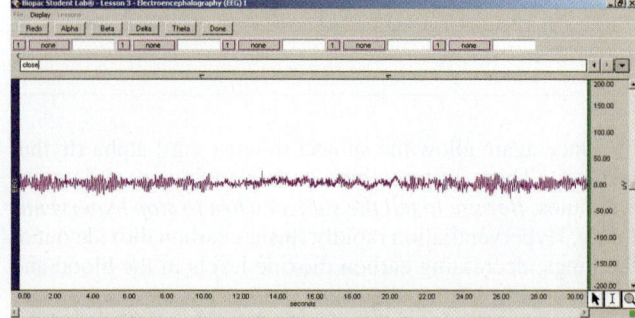

Figure 18.5 Example of EEG data.

- If the data look similar to the example (Figure 18.5), proceed to the next step.

4. When finished, click **Done.** If you are certain you want to stop recording, click **YES.** Remove the electrodes from the subject's scalp.

5. A pop-up window will appear. To record from another subject, select **Record from another subject** and return to step 5 under Setting Up the Equipment. If continuing to the Data Analysis section, select **Analyze current data file** and proceed to step 2.

Data Analysis

1. If just starting the BIOPAC® program to perform data analysis, enter **Review Saved Data** mode and choose the file with the subject's EEG data (for example, SmithEEG-1).

2. Observe the way the channel numbers are designated: CH 40–**alpha;** CH 41–**beta;** CH 42–**delta;** and CH 43–**theta.** CH 1 (raw EEG) is hidden. The software used it to extract and display each frequency band. If you want to see CH 1, hold down the **Ctrl** key (PC) or **Option** key (Mac) while using the cursor to click channel box 1 (the small box with a 1 at the upper left of the screen).

3. To analyze the data, set up the first four pairs of channel/measurement boxes at the top of the screen by selecting the following channels and measurement types from the drop-down menus:

Channel	Measurement	Data
CH 40	stddev	alpha
CH 41	stddev	beta
CH 42	stddev	delta
CH 43	stddev	theta

stddev (standard deviation): This is a statistical calculation that estimates the variability of the data in the area highlighted by the I-beam cursor. This function minimizes the effects of extreme values and electrical artifacts that may unduly influence interpretation of the data.

4. Use the arrow cursor and click the I-beam cursor box at the lower right of the screen to activate the "area selection" function. Using the activated I-beam cursor, highlight the first 20-second segment of EEG data, which represents the subject at rest with eyes closed **(Figure 18.6).**

5. Observe that the computer automatically calculates the stddev for each of the channels of data (alpha, beta, delta, and theta).

6. Record the data for each rhythm in the **Standard Deviations chart,** rounding to the nearest 0.01 µV.

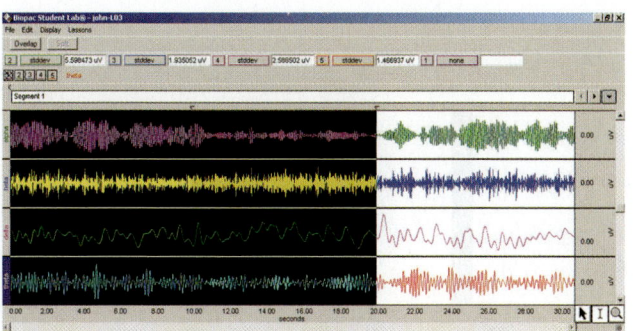

Figure 18.6 Highlighting the first data segment.

7. Repeat steps 4–6 to analyze and record the data for the next two segments of data, with eyes open, and with eyes reclosed. The triangular markers inserted at the top of the data should provide guidance for highlighting.

8. To continue the analysis, change the settings in the first four pairs of channel/measurement boxes. Select the following channels and measurement types:

Channel	Measurement	Data
CH 40	Freq	alpha
CH 41	Freq	beta
CH 42	Freq	delta
CH 43	Freq	theta

Freq (frequency): This gives the frequency in hertz (Hz) of an individual wave that is highlighted by the I-beam cursor.

9. To view an individual wave from among the high-frequency waveforms, you must use the zoom function. To activate the zoom function, use the cursor to click the magnifying glass at the lower-right corner of the screen (near the I-beam cursor box). The cursor will become a magnifying glass.

10. As the analysis begins, CH 40—the alpha data—will be automatically activated. To examine individual waves within the **alpha** data, click that band with the magnifying glass until it is possible to observe the peaks and troughs of individual waves within **Segment 1.**

- To properly view each of the waveforms, you may have to click the **Display** menu and select **Autoscale Waveforms.** This function rescales the data for the rhythm band that is selected.

Standard Deviations (stddev) of Signals in Each Segment				
Rhythm	Channel	Eyes closed Segment 1 Seconds 0–20	Eyes open Segment 2 Seconds 21–40	Eyes reclosed Segment 3 Seconds 41–60
Alpha	CH 40			
Beta	CH 41			
Delta	CH 42			
Theta	CH 43			

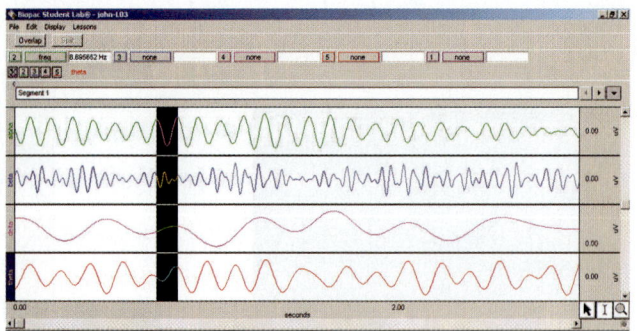

Figure 18.7 Highlighting a single alpha wave from peak to peak.

11. At this time, focus on alpha waves only. Reactivate the I-beam cursor by clicking its box in the lower-right corner. Highlight a *single* alpha wave from peak to peak (as shown in **Figure 18.7**).

12. Read the calculated frequency (in Hz) in the measurement box for CH 40, and record this as the frequency of Wave 1 for alpha rhythm in the **Frequencies of Waves chart.**

13. Use the I-beam cursor to select two more individual **alpha** waves and record their frequencies in the chart.

14. You will now perform the same frequency measurements for three waves in each of the **beta** (CH 41), **delta** (CH 42), and **theta** (CH 43) data sets. Record these measurements in the chart.

15. Calculate the average of the three waves measured for each of the brain rhythms, and record the average in the chart.

16. When finished, answer the following questions and then exit the program by going to the **File** menu at the top of the page and clicking **Quit.**

Look at the waveforms you recorded and carefully examine all three segments of the alpha rhythm record. Is there a difference in electrical activity in this frequency range when the eyes are open versus closed? Describe your observations.

_____

_____

_____

Carefully examine all three segments of the beta rhythm record. Is there a difference in electrical activity in this frequency range when the eyes are open versus closed? Describe what you observe.

_____

_____

_____

This time, compare the intensity (height) of the alpha and beta waveforms throughout all three segments. Does the intensity of one signal appear more varied than the other in the record? Describe your observations.

_____

_____

_____

Examine the data for the delta and theta rhythms. Is there any change in the waveform as the subject changes states? If so, describe the change observed.

_____

_____

_____

The degree of variation in the intensity of the signal was estimated by calculating the standard deviation of the waves in each segment of data. In which time segment (eyes open, eyes closed, or eyes reclosed) is the difference in the standard deviations the greatest?

_____

Frequencies of Waves for Each Rhythm (Hz)					
Rhythm	**Channel**	**Wave 1**	**Wave 2**	**Wave 3**	**Average**
Alpha	CH 40				
Beta	CH 41				
Delta	CH 42				
Theta	CH 43				

Electroencephalography

Brain Wave Patterns and the Electroencephalogram

1. Define *EEG*. _____

2. Identify the type of brain wave pattern described in each statement below.

    _____ below 4 Hz; slow, large waves; normally seen during deep sleep

    _____ rhythm generally apparent when an individual is in a relaxed, nonattentive state with the eyes closed

    _____ correlated to the alert state; usually about 14 to 30 Hz

3. What is meant by the term *alpha block*? _____

   _____

4. List at least four types of brain lesions that may be determined by EEG studies. _____

   _____

5. What is the common result of hypoactivity or hyperactivity of the brain neurons? _____

Observing Brain Wave Patterns

6. How was alpha block demonstrated in the laboratory experiment? _____

   _____

7. What was the effect of mental concentration on the brain wave pattern? _____

   _____

   _____

8. What effect on the brain wave pattern did hyperventilation have? _____

Electroencephalography Using BIOPAC®

9. Observe the average frequency of the waves you measured for each rhythm. Did the calculated average for each fall within the specified range indicated in the introduction to encephalograms?

_____

_____

10. Suggest the possible advantages and disadvantages of using electroencephalography in a clinical setting.

_____

_____

_____

_____

The Spinal Cord and Spinal Nerves

MATERIALS

☐ Spinal cord model (cross section)
☐ Three-dimensional models or laboratory charts of the spinal cord and spinal nerves
☐ Red and blue pencils
☐ Preserved cow spinal cord sections with meninges and nerve roots intact (or spinal cord segment saved from the brain dissection in Exercise 17)
☐ Dissecting instruments and tray
☐ Disposable gloves
☐ Stereomicroscope
☐ Prepared slide of spinal cord (x.s.)
☐ Compound microscope

For instructions on animal dissections, see the dissection exercises (starting on p. 697) in the cat and fetal pig editions of this manual.

OBJECTIVES

1. List two major functions of the spinal cord.
2. Define *conus medullaris, cauda equina,* and *filum terminale.*
3. Name the meningeal coverings of the spinal cord, and state their function.
4. Indicate two major areas where the spinal cord is enlarged, and explain the reasons for the enlargement.
5. Identify important anatomical areas on a model or image of a cross section of the spinal cord, and where applicable name the neuron type found in these areas.
6. Locate on a diagram the fiber tracts in the spinal cord, and state their functions.
7. Note the number of pairs of spinal nerves that arise from the spinal cord, describe their division into groups, and identify the number of pairs in each group.
8. Describe the origin and fiber composition of the spinal nerves, differentiating between roots, the spinal nerve proper, and rami, and discuss the result of transecting these structures.
9. Discuss the distribution of the dorsal and ventral rami of the spinal nerves.
10. Identify the four major nerve plexuses on a model or image, name the major nerves of each plexus, and describe the destination and function of each.

PRE-LAB QUIZ

1. The spinal cord extends from the foramen magnum of the skull to the first or second lumbar vertebra, where it terminates in the
   a. conus medullaris
   b. denticulate ligament
   c. filum terminale
   d. gray matter
2. How many pairs of spinal nerves do humans have?
   a. 10   c. 31
   b. 12   d. 47
3. Circle the correct underlined term. In cross section, the <u>gray</u> / <u>white</u> matter of the spinal cord looks like a butterfly or the letter H.
4. Circle True or False. The cell bodies of sensory neurons are found in an enlarged area of the dorsal root called the gray commissure.
5. Circle the correct underlined term. Fiber tracts conducting impulses to the brain are called ascending or <u>sensory</u> / <u>motor</u> tracts.
6. Circle True or False. Because the spinal nerves arise from fusion of the ventral and dorsal roots of the spinal cord, and contain motor and sensory fibers, all spinal nerves are considered mixed nerves.

MasteringA&P® For related exercise study tools, go to the Study Area of MasteringA&P. There you will find:
• Practice Anatomy Lab PAL
• PhysioEx PEx
• A&PFlix A&PFlix
• Practice quizzes, Histology Atlas, eText, Videos, and more!

(Text continues on next page.)

7. The ventral rami of all spinal nerves except T_2 through T_{12} form complex networks of nerves known as _____.
   - a. fissures
   - b. ganglia
   - c. plexuses
   - d. sulci
8. Severe injuries to the _____ plexus cause weakness or paralysis of the entire upper limb.
   - a. brachial
   - b. cervical
   - c. lumbar
   - d. sacral
9. Circle True or False. The femoral nerve is the largest nerve from the sacral plexus.
10. Circle the correct underlined term. The sciatic nerve divides into the tibial and <u>posterior femoral cutaneous</u> / <u>common fibular</u> nerves.

The cylindrical **spinal cord,** a continuation of the brain stem, is an association and communication center. It plays a major role in spinal reflex activity and provides neural pathways to and from higher nervous centers.

Anatomy of the Spinal Cord

Enclosed within the vertebral canal of the spinal column, the spinal cord extends from the foramen magnum of the skull to the first or second lumbar vertebra, where it terminates in the cone-shaped **conus medullaris (Figure 19.1)**. Like the brain, the cord is cushioned and protected by meninges. The dura mater and arachnoid meningeal coverings extend beyond the conus medullaris, approximately to the level of S_2, and the **filum terminale,** a fibrous extension of the pia mater, extends even farther into the coccygeal canal to attach to the posterior coccyx. **Denticulate ligaments,** saw-toothed shelves of pia mater, secure the spinal cord to the bony wall of the vertebral column all along its length (Figure 19.1c).

The cerebrospinal fluid–filled meninges extend well beyond the end of the spinal cord, providing an excellent site for removing cerebrospinal fluid without endangering the delicate spinal cord. Analysis of the fluid can provide important information about suspected bacterial or viral infections of the spinal cord or meninges. This procedure, called a *lumbar tap,* is usually performed below L_3. Additionally, "saddle block," or caudal anesthesia for childbirth, is normally administered (injected) between L_3 and L_5.

In humans, 31 pairs of spinal nerves arise from the spinal cord and pass through intervertebral foramina to serve the body area at their approximate level of emergence. The cord is about the size of a finger in circumference for most of its length, but there are obvious enlargements in the *cervical* and *lumbar* areas where the nerves serving the upper and lower limbs issue from the cord.

Because the spinal cord does not extend to the end of the vertebral column, the spinal nerves emerging from the inferior end of the cord must travel through the vertebral canal for some distance before exiting at the appropriate intervertebral foramina. This collection of spinal nerves passing through the inferior end of the vertebral canal is called the **cauda equina** (Figure 19.1a and d) because of its similarity to a horse's tail (the literal translation of *cauda equina*).

ACTIVITY 1

Identifying Structures of the Spinal Cord

Obtain a three-dimensional model or laboratory chart of a cross section of a spinal cord and identify its structures as they are described next. ■

Gray Matter

In cross section, the **gray matter** of the spinal cord looks like a butterfly or the letter H **(Figure 19.2)**. The two dorsal projections are called the **dorsal (posterior) horns.** The two ventral projections are the **ventral (anterior) horns.** The tips of the ventral horns are broader and less tapered than those of the dorsal horns. In the thoracic and lumbar regions of the cord, there is also a lateral outpocketing of gray matter on each side referred to as the **lateral horn.** The central area of gray matter connecting the two vertical regions is the **gray commissure,** which surrounds the **central canal** of the cord.

Neurons with specific functions can be localized in the gray matter. The dorsal horns contain interneurons and sensory fibers that enter the cord from the body periphery via the **dorsal root.** The cell bodies of these sensory neurons are found in an enlarged area of the dorsal root called the **dorsal root ganglion.** The ventral horns mainly contain cell bodies of motor neurons of the somatic nervous system, which send their axons out via the **ventral root** of the cord to enter the adjacent spinal nerve. Because they are formed by the fusion of the dorsal and ventral roots, the **spinal nerves** are **mixed nerves** containing both sensory and motor fibers. The lateral horns, where present, contain nerve cell bodies of motor neurons of the autonomic nervous system, sympathetic division. Their axons also leave the cord via the ventral roots, along with those of the motor neurons of the ventral horns.

White Matter

The **white matter** of the spinal cord is nearly bisected by fissures (Figure 19.2). The more open ventral fissure is the **ventral median fissure,** and the dorsal one is the shallow **dorsal median sulcus.** The white matter is composed of myelinated and nonmyelinated fibers—some running to higher centers, some traveling from the brain to the cord, and some conducting impulses from one side of the cord to the other.

19

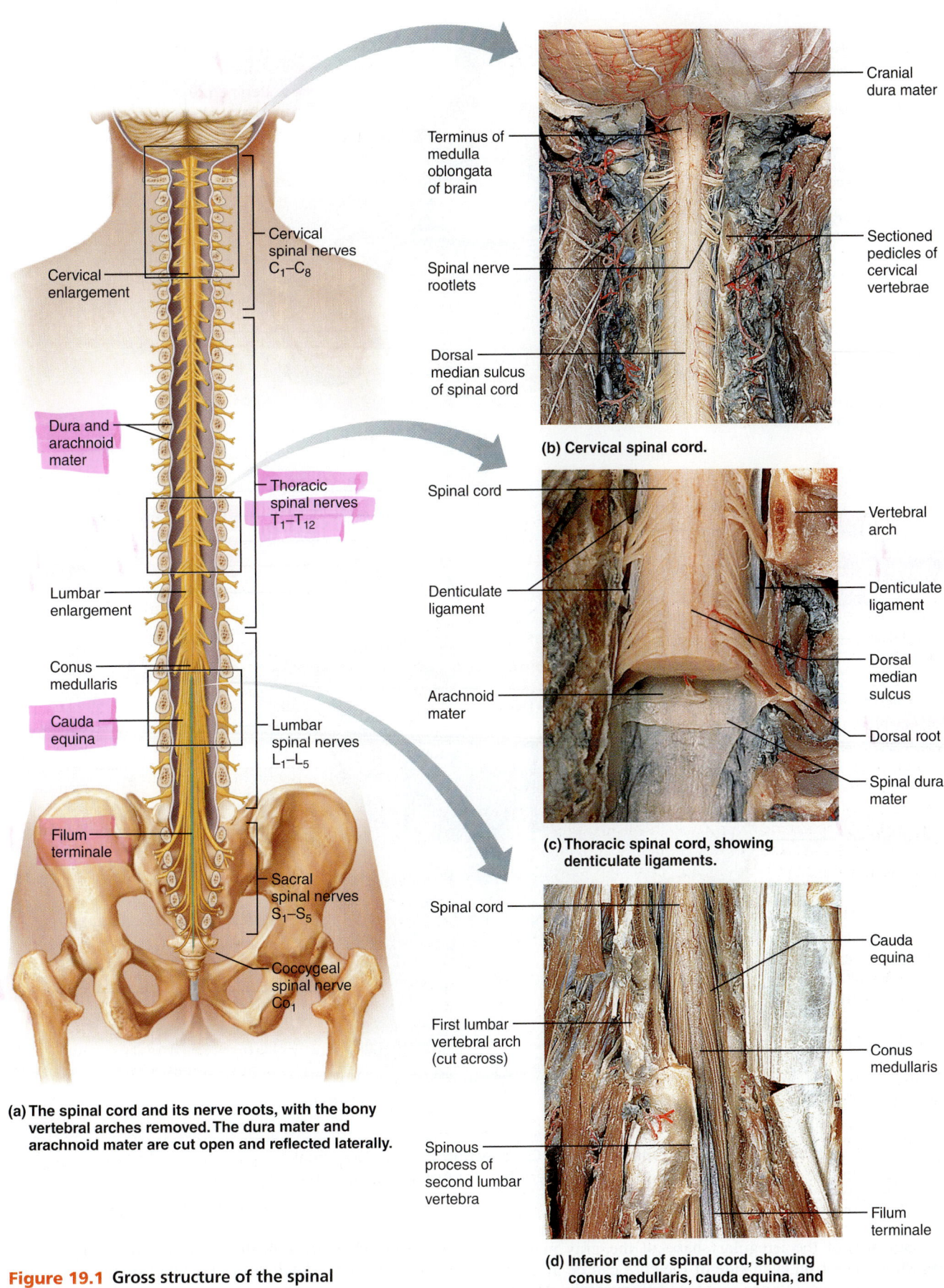

(a) The spinal cord and its nerve roots, with the bony vertebral arches removed. The dura mater and arachnoid mater are cut open and reflected laterally.

Figure 19.1 Gross structure of the spinal cord, dorsal view.

(b) Cervical spinal cord.

(c) Thoracic spinal cord, showing denticulate ligaments.

(d) Inferior end of spinal cord, showing conus medullaris, cauda equina, and filum terminale.

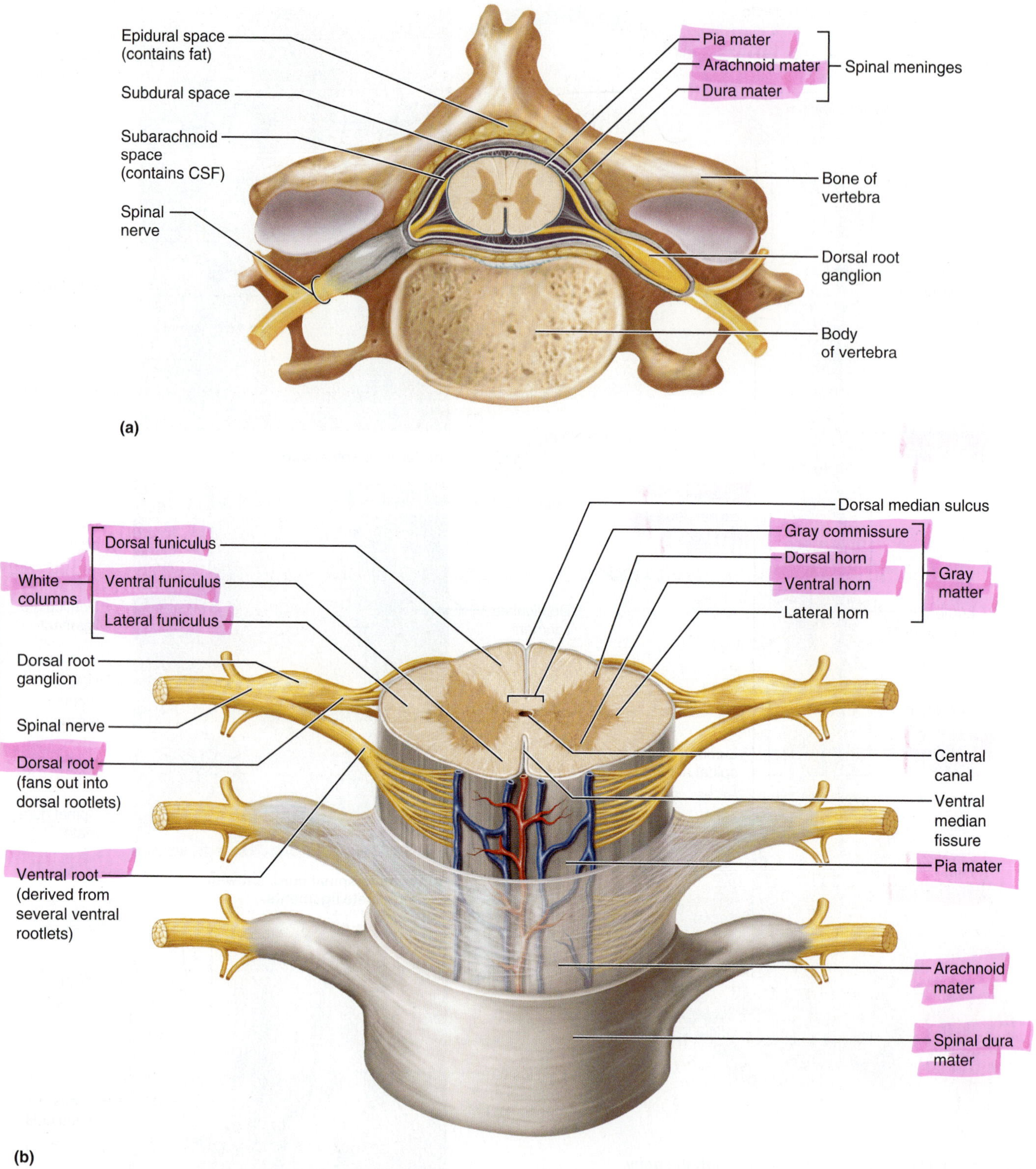

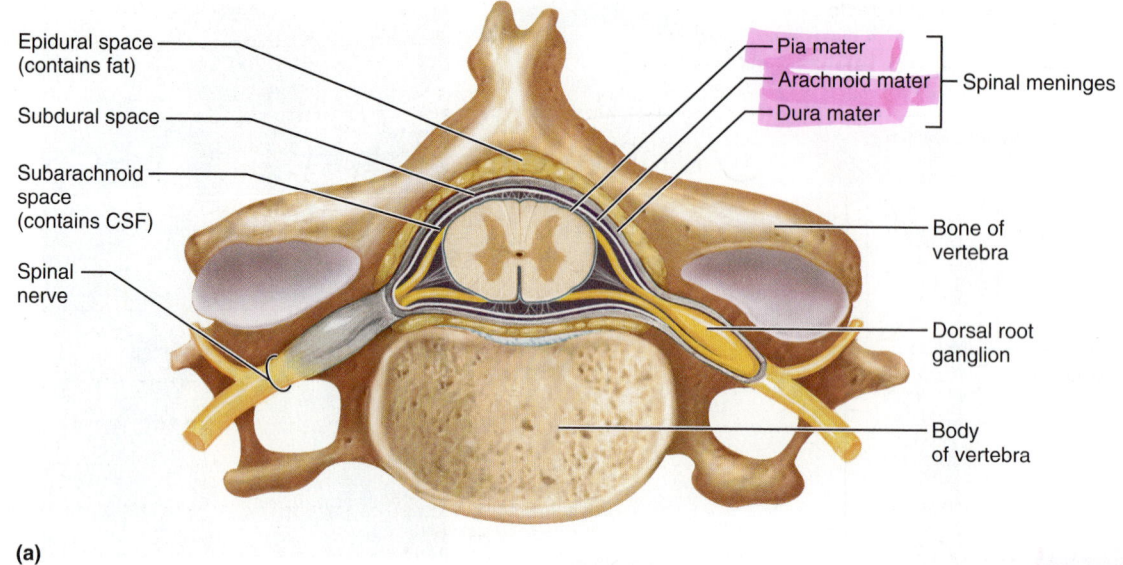

Epidural space
(contains fat)

Subdural space

Subarachnoid
space
(contains CSF)

Spinal
nerve

Pia mater

Arachnoid mater — Spinal meninges

Dura mater

Bone of
vertebra

Dorsal root
ganglion

Body
of vertebra

(a)

Dorsal median sulcus

Gray commissure

Dorsal horn

Ventral horn — Gray
matter

Lateral horn

Dorsal funiculus

White
columns   Ventral funiculus

Lateral funiculus

Dorsal root
ganglion

Spinal nerve

Dorsal root
(fans out into
dorsal rootlets)

Ventral root
(derived from
several ventral
rootlets)

Central
canal

Ventral
median
fissure

Pia mater

Arachnoid
mater

Spinal dura
mater

(b)

Figure 19.2 Anatomy of the human spinal cord. (a) Cross section through the spinal cord illustrating its relationship to the surrounding vertebra. **(b)** Anterior view of the spinal cord and its meningeal coverings.

Because of the irregular shape of the gray matter, the white matter on each side of the cord can be divided into three primary regions or **white columns:** the **dorsal (posterior), lateral,** and **ventral (anterior) funiculi.** Each funiculus contains a number of fiber **tracts** composed of axons with the same origin, terminus, and function. Tracts conducting sensory impulses to the brain are called *ascending* or *sensory tracts;* those carrying impulses from the brain to the skeletal muscles are *descending* or *motor tracts.*

Ascending tracts   Descending tracts

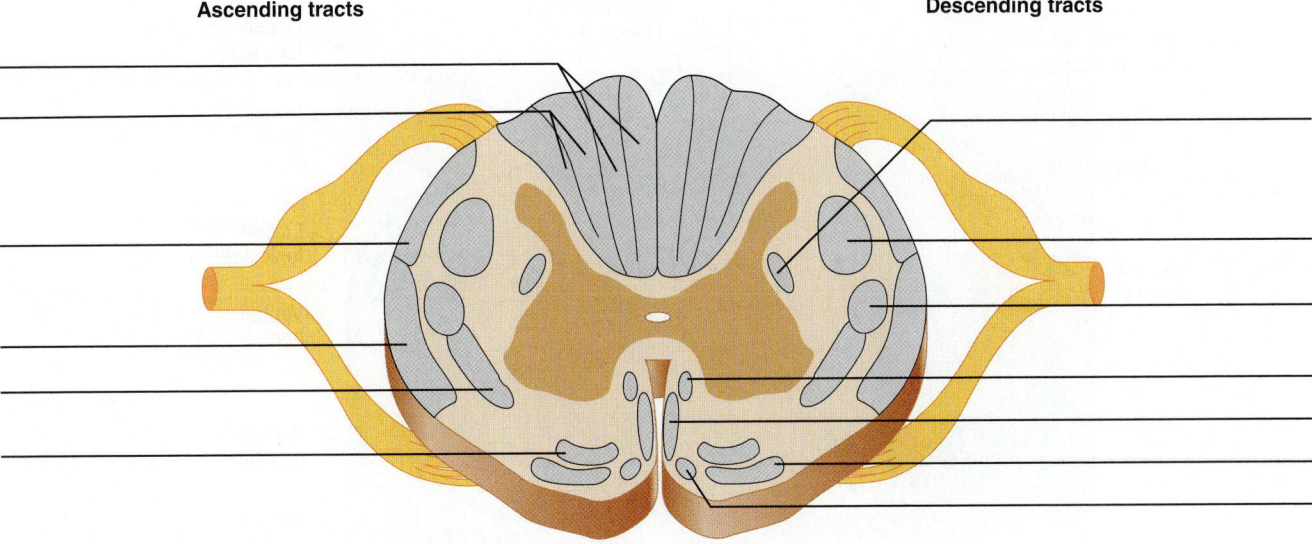

Figure 19.3 **Cross section of the spinal cord showing the relative positioning of its major tracts.**

Because it serves as the transmission pathway between the brain and the body periphery, the spinal cord is extremely important functionally. Even though it is protected by meninges and cerebrospinal fluid in the vertebral canal, it is highly vulnerable to traumatic injuries, such as might occur in an automobile accident.

When the cord is transected (or severely traumatized), both motor and sensory functions are lost in body areas normally served by that region and lower regions of the spinal cord. Injury to certain spinal cord areas may even result in a permanent flaccid paralysis of both legs, called **paraplegia,** or of all four limbs, called **quadriplegia.** ✚

ACTIVITY 2

Identifying Spinal Cord Tracts

With the help of your textbook, label the spinal cord diagram **(Figure 19.3)** with the tract names that follow. Each tract is represented on both sides of the cord, but for clarity, label the motor tracts on the right side of the diagram and the sensory tracts on the left side of the diagram. *Color ascending tracts blue and descending tracts red.* Then fill in the functional importance of each tract beside its name below. As you work, try to be aware of how the naming of the tracts is related to their anatomical distribution.

Dorsal columns

Fasciculus gracilis _____

Fasciculus cuneatus _____

Dorsal spinocerebellar _____

Ventral spinocerebellar _____

Lateral spinothalamic _____

Ventral spinothalamic _____

Lateral corticospinal _____

Ventral corticospinal _____

Rubrospinal _____

Tectospinal _____

Vestibulospinal _____

Medial reticulospinal _____

Lateral reticulospinal _____

DISSECTION:
Spinal Cord

1. Obtain a dissecting tray and instruments, disposable gloves, and a segment of preserved spinal cord (from a cow or saved from the brain specimen used in Exercise 17). Identify the tough outer meninx (dura mater) and the weblike arachnoid mater.

What name is given to the third meninx, and where is it found?

_____

_____

19

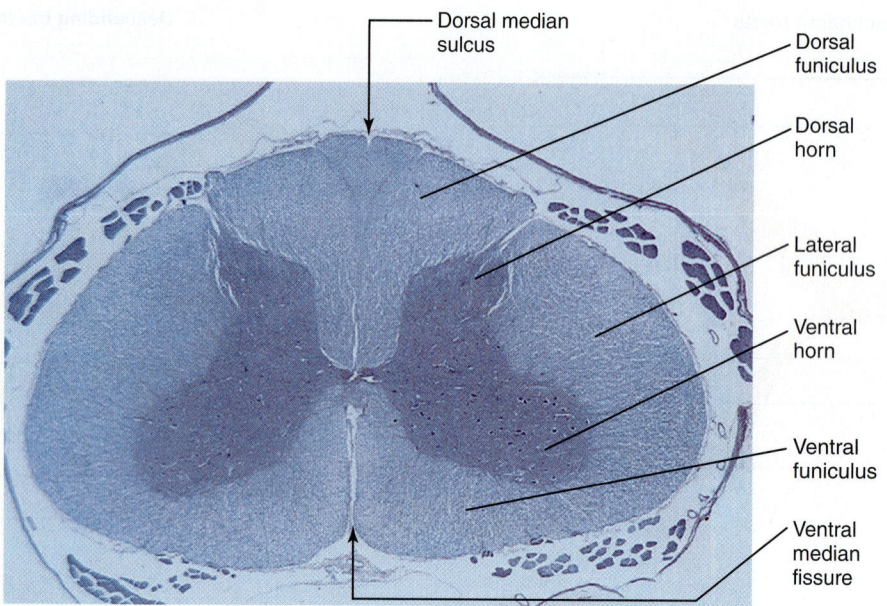

Figure 19.4 Cross section of the spinal cord (10×).

Peel back the dura mater and observe the fibers making up the dorsal and ventral roots. If possible, identify a dorsal root ganglion.

2. Cut a thin cross section of the cord and identify the ventral and dorsal horns of the gray matter with the naked eye or with the aid of a dissecting microscope.

How can you be certain that you are correctly identifying the ventral and dorsal horns?

_____

_____

Also identify the central canal, white matter, ventral median fissure, dorsal median sulcus, and dorsal, ventral, and lateral funiculi.

3. Obtain a prepared slide of the spinal cord (cross section) and a compound microscope. Examine the slide carefully under low power (refer to **Figure 19.4** to identify spinal cord features). Observe the shape of the central canal.

Is it basically circular or oval? _____

Name the neuroglial cell type that lines this canal. _____

Can any neuron cell bodies be seen? _____

If so, where, and what type of neurons would these most

likely be—motor, sensory, or interneuron? _____

_____ ■

Spinal Nerves and Nerve Plexuses

The 31 pairs of human spinal nerves arise from the fusions of the ventral and dorsal roots of the spinal cord (see Figure 19.2a). There are 8 pairs of cervical nerves (C_1–C_8), 12 pairs of thoracic nerves (T_1–T_{12}), 5 pairs of lumbar nerves (L_1–L_5), 5 pairs of sacral nerves (S_1–S_5), and 1 pair of coccygeal nerves (Co_1) **(Figure 19.5a)**. The first pair of spinal nerves leaves the vertebral canal between the base of the occiput and the atlas, but all the rest exit via the intervertebral foramina. The first through seventh pairs of cervical nerves emerge *above* the vertebra for which they are named; C_8 emerges between C_7 and T_1. (Notice that there are 7 cervical vertebrae, but 8 pairs of cervical nerves.) The remaining spinal nerve pairs emerge from the spinal cord area *below* the same-numbered vertebra.

Almost immediately after emerging, each nerve divides into **dorsal** and **ventral rami.** Thus each spinal nerve is only about 1 or 2 cm long. The rami, like the spinal nerves, contain both motor and sensory fibers. The smaller dorsal rami serve the skin and musculature of the posterior body trunk at their approximate level of emergence. The ventral rami of spinal nerves T_2 through T_{12} pass anteriorly as the **intercostal nerves** to supply the muscles of intercostal spaces, and the skin and muscles of the anterior and lateral trunk. The ventral rami of all other spinal nerves form complex networks of nerves called **nerve plexuses.** These plexuses primarily serve the muscles and skin of the limbs. The fibers of the ventral rami unite in the plexuses (with a few rami supplying fibers to more than one plexus). From the plexuses the fibers diverge again to form peripheral nerves, each of which contains fibers from more than one spinal nerve. (The four major nerve plexuses and their chief peripheral nerves are described in Tables 19.1–19.4 and illustrated in Figures 19.6–19.9. Their names and site of origin should be committed to memory). The tiny

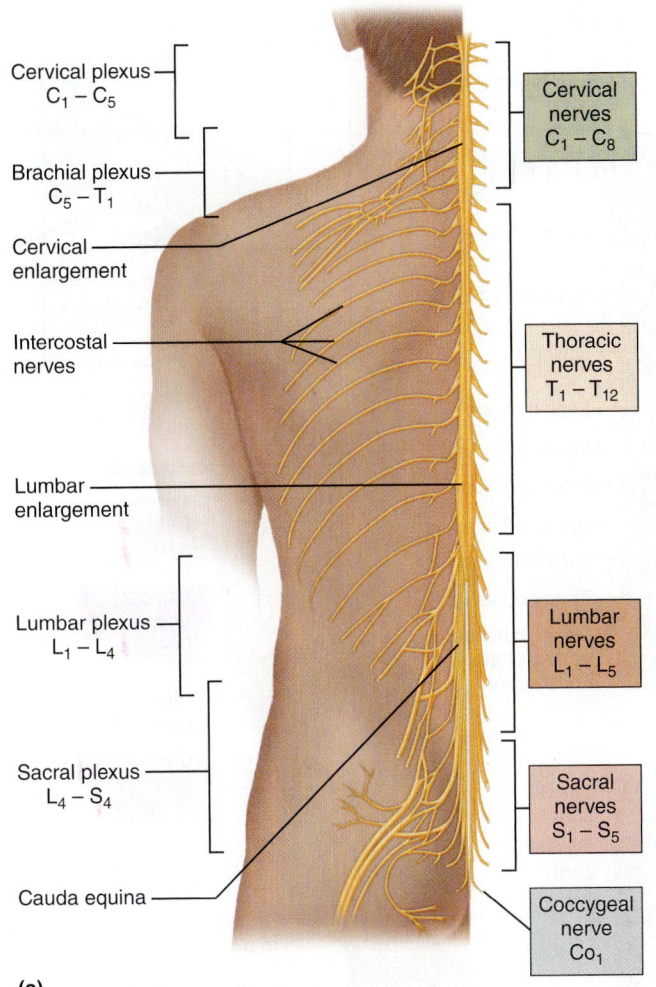

Cervical plexus
$C_1 - C_5$

Brachial plexus
$C_5 - T_1$

Cervical
enlargement

Intercostal
nerves

Lumbar
enlargement

Lumbar plexus
$L_1 - L_4$

Sacral plexus
$L_4 - S_4$

Cauda equina

Cervical
nerves
$C_1 - C_8$

Thoracic
nerves
$T_1 - T_{12}$

Lumbar
nerves
$L_1 - L_5$

Sacral
nerves
$S_1 - S_5$

Coccygeal
nerve
Co_1

(a)

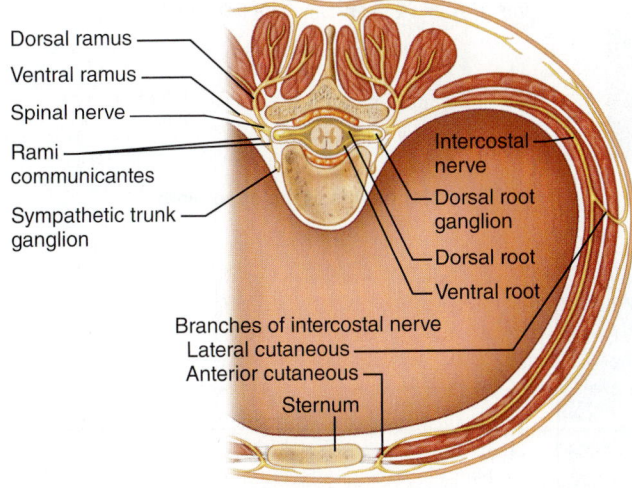

Dorsal ramus

Ventral ramus

Spinal nerve

Rami
communicantes

Sympathetic trunk
ganglion

Intercostal
nerve

Dorsal root
ganglion

Dorsal root

Ventral root

Branches of intercostal nerve
Lateral cutaneous
Anterior cutaneous
Sternum

(b)

Figure 19.5 Human spinal nerves. (a) Spinal nerves are shown at right; ventral rami and the major nerve plexuses are shown at left. **(b)** Relative distribution of the ventral and dorsal rami of a spinal nerve (cross section of thorax).

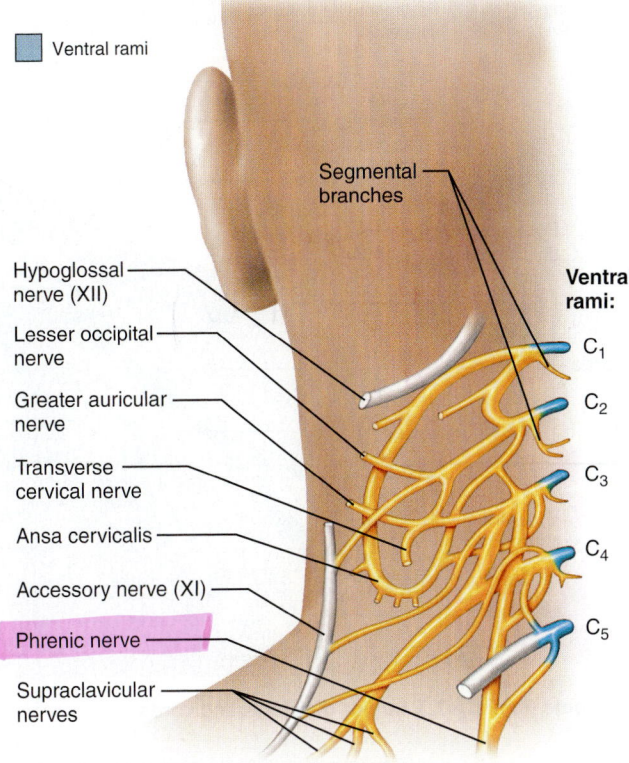

Ventral rami

Segmental
branches

Hypoglossal
nerve (XII)

Lesser occipital
nerve

Greater auricular
nerve

Transverse
cervical nerve

Ansa cervicalis

Accessory nerve (XI)

Phrenic nerve

Supraclavicular
nerves

**Ventral
rami:**

C_1

C_2

C_3

C_4

C_5

Figure 19.6 The cervical plexus. The nerves colored gray connect to the plexus but do not belong to it. (See Table 19.1.)

S_5 and Co_1 spinal nerves contribute to a small plexus that serves part of the pelvic floor.

Cervical Plexus and the Neck

The **cervical plexus** (**Figure 19.6** and **Table 19.1**) arises from the ventral rami of C_1 through C_5 to supply muscles of the shoulder and neck. The major motor branch of this plexus is the **phrenic nerve,** which arises from C_3 through C_4 (plus some fibers from C_5) and passes into the thoracic cavity in front of the first rib to innervate the diaphragm. The primary danger of a broken neck is that the phrenic nerve may be severed, leading to paralysis of the diaphragm and cessation of breathing. A jingle to help you remember the rami (roots) forming the phrenic nerves is "C_3, C_4, C_5 keep the diaphragm alive."

Brachial Plexus and the Upper Limb

The **brachial plexus** is large and complex, arising from the ventral rami of C_5 through C_8 and T_1 (**Table 19.2**). The plexus, after being rearranged consecutively into *trunks, divisions,* and *cords,* finally becomes subdivided into five major *peripheral nerves* (**Figure 19.7**).

The **axillary nerve,** which serves the muscles and skin of the shoulder, has the most limited distribution. The large **radial nerve** passes down the posterolateral surface of the arm and forearm, supplying all the extensor muscles of the arm, forearm, and hand and the skin along its course.

(Text continues on page 316.)

19

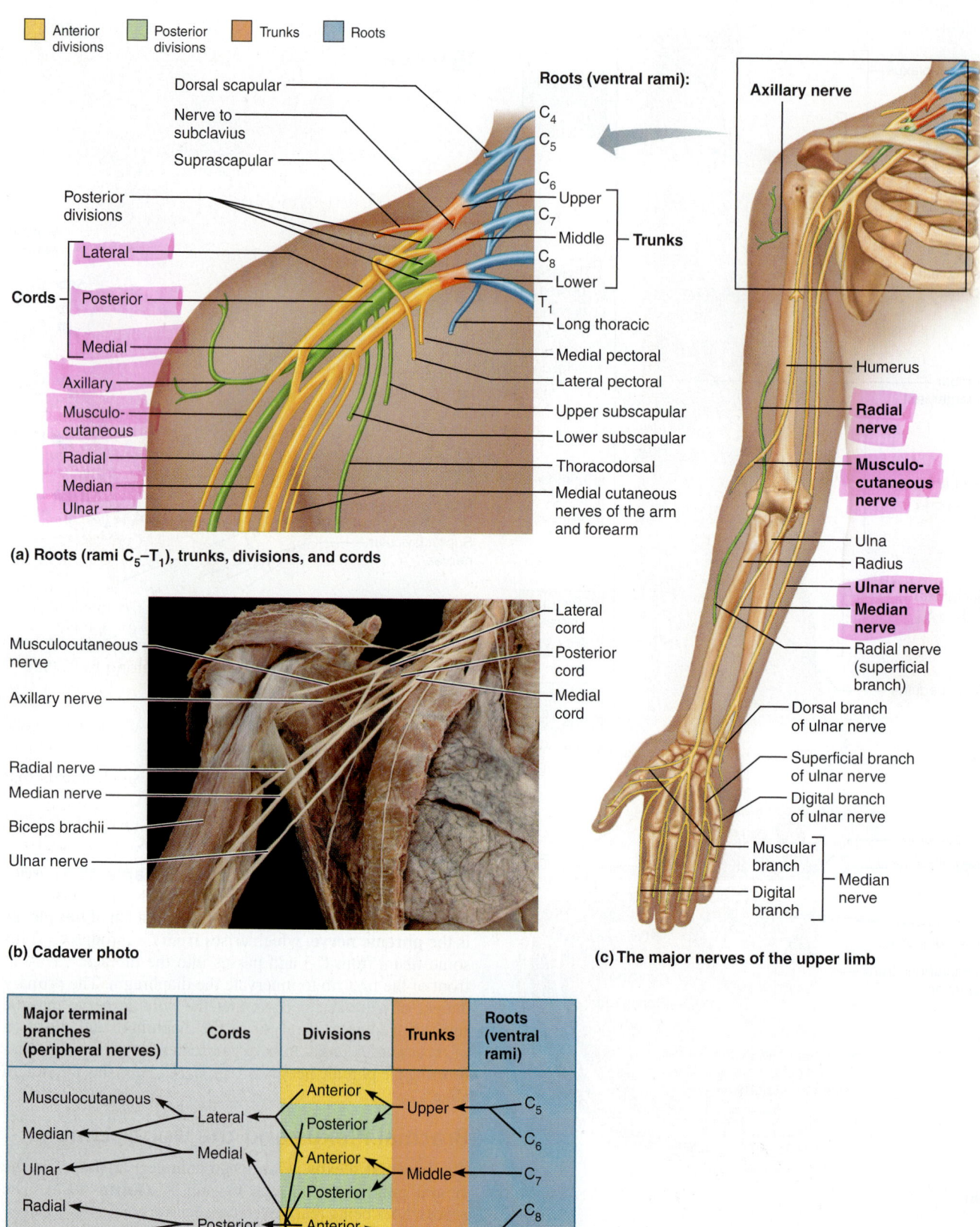

Anterior divisions | Posterior divisions | Trunks | Roots

(a) Roots (rami C₅–T₁), trunks, divisions, and cords

Roots (ventral rami):

C₄
C₅
C₆   Upper
C₇   Middle   Trunks
C₈   Lower
T₁

Dorsal scapular
Nerve to subclavius
Suprascapular
Posterior divisions
Lateral
Posterior   Cords
Medial
Axillary
Musculo-cutaneous
Radial
Median
Ulnar

Long thoracic
Medial pectoral
Lateral pectoral
Upper subscapular
Lower subscapular
Thoracodorsal
Medial cutaneous nerves of the arm and forearm

Axillary nerve

(b) Cadaver photo

Musculocutaneous nerve
Axillary nerve
Radial nerve
Median nerve
Biceps brachii
Ulnar nerve

Lateral cord
Posterior cord
Medial cord

(c) The major nerves of the upper limb

Humerus
Radial nerve
Musculo-cutaneous nerve
Ulna
Radius
Ulnar nerve
Median nerve
Radial nerve (superficial branch)
Dorsal branch of ulnar nerve
Superficial branch of ulnar nerve
Digital branch of ulnar nerve
Muscular branch
Digital branch   Median nerve

(d) Flowchart summarizing relationships within the brachial plexus

Major terminal branches (peripheral nerves)	Cords	Divisions	Trunks	Roots (ventral rami)

Musculocutaneous
Median   Lateral   Anterior   Upper   C₅
Ulnar   Medial   Posterior   C₆
Radial   Posterior   Anterior   Middle   C₇
Axillary   Posterior   C₈
    Anterior   Lower   T₁
    Posterior

Figure 19.7 The brachial plexus. (See Table 19.2.)

19

Table 19.1	Branches of the Cervical Plexus (See Figure 19.6)		
Nerves		**Ventral rami**	**Structures served**
Cutaneous Branches (Superficial)			
Lesser occipital		C_2 (C_3)	Skin on posterolateral aspect of neck
Greater auricular		C_2, C_3	Skin of ear, skin over parotid gland
Transverse cervical		C_2, C_3	Skin on anterior and lateral aspect of neck
Supraclavicular (medial, intermediate, and lateral)		C_3, C_4	Skin of shoulder and clavicular region
Motor Branches (Deep)			
Ansa cervicalis (superior and inferior roots)		C_1–C_3	Infrahyoid muscles of neck (omohyoid, sternohyoid, and sternothyroid)
Segmental and other muscular branches		C_1–C_5	Deep muscles of neck (geniohyoid and thyrohyoid) and portions of scalenes, levator scapulae, trapezius, and sternocleidomastoid muscles
Phrenic		C_3–C_5	Diaphragm (sole motor nerve supply)

Table 19.2	Branches of the Brachial Plexus (See Figure 19.7)	
Nerves	**Cord and ventral rami**	**Structures served**
Axillary	Posterior cord (C_5, C_6)	Muscular branches: deltoid and teres minor muscles Cutaneous branches: some skin of shoulder region
Musculocutaneous	Lateral cord (C_5–C_7)	Muscular branches: flexor muscles in anterior arm (biceps brachii, brachialis, coracobrachialis) Cutaneous branches: skin on anterolateral forearm (extremely variable)
Median	By two branches, one from medial cord (C_8, T_1) and one from the lateral cord (C_5–C_7)	Muscular branches to flexor group of anterior forearm (palmaris longus, flexor carpi radialis, flexor digitorum superficialis, flexor pollicis longus, lateral half of flexor digitorum profundus, and pronator muscles); intrinsic muscles of lateral palm and digital branches to the fingers Cutaneous branches: skin of lateral two-thirds of hand on ventral side and dorsum of fingers 2 and 3
Ulnar	Medial cord (C_8, T_1)	Muscular branches: flexor muscles in anterior forearm (flexor carpi ulnaris and medial half of flexor digitorum profundus); most intrinsic muscles of hand Cutaneous branches: skin of medial third of hand, both anterior and posterior aspects
Radial	Posterior cord (C_5–C_8, T_1)	Muscular branches: posterior muscles of arm and forearm (triceps brachii, anconeus, supinator, brachioradialis, extensors carpi radialis longus and brevis, extensor carpi ulnaris, and several muscles that extend the fingers) Cutaneous branches: skin of posterolateral surface of entire limb (except dorsum of fingers 2 and 3)
Dorsal scapular	Branches of C_5 rami	Rhomboid muscles and levator scapulae
Long thoracic	Branches of C_5–C_7 rami	Serratus anterior muscle
Subscapular	Posterior cord; branches of C_5 and C_6 rami	Teres major and subscapularis muscles
Suprascapular	Upper trunk (C_5, C_6)	Shoulder joint; supraspinatus and infraspinatus muscles
Pectoral (lateral and medial)	Branches of lateral and medial cords (C_5–T_1)	Pectoralis major and minor muscles

19

The radial nerve is often injured in the axillary region by the pressure of a crutch or by hanging one's arm over the back of a chair. The **median nerve** passes down the anteromedial surface of the arm to supply most of the flexor muscles in the forearm and several muscles in the hand (plus the skin of the lateral surface of the palm of the hand).

•   Hyperextend your wrist to identify the long, obvious tendon of your palmaris longus muscle, which crosses the exact midline of the anterior wrist. Your median nerve lies immediately deep to that tendon, and the radial nerve lies just *lateral* to it.

The **musculocutaneous nerve** supplies the arm muscles that flex the forearm and the skin of the lateral surface of the forearm. The **ulnar nerve** travels down the posteromedial surface of the arm. It courses around the medial epicondyle of the humerus to supply the flexor carpi ulnaris, the ulnar head of the flexor digitorum profundus of the forearm, and all intrinsic muscles of the hand not served by the median nerve. It supplies the skin of the medial third of the hand, both the anterior and posterior surfaces. Trauma to the ulnar nerve, which often occurs when the elbow is hit, produces a smarting sensation commonly referred to as "hitting the funny bone."

Severe injuries to the brachial plexus cause weakness or paralysis of the entire upper limb. Such injuries may occur when the upper limb is pulled hard and the plexus is stretched (as when a football tackler yanks the arm of the halfback), and by blows to the shoulder that force the humerus inferiorly (as when a cyclist is pitched headfirst off his motorcycle and grinds his shoulder into the pavement). ✚

Lumbosacral Plexus and the Lower Limb

The **lumbosacral plexus,** which serves the pelvic region of the trunk and the lower limbs, is actually a complex of two plexuses, the lumbar plexus and the sacral plexus (Figures 19.8 and 19.9). These plexuses interweave considerably and many fibers of the lumbar plexus contribute to the sacral plexus.

The Lumbar Plexus

The **lumbar plexus** arises from ventral rami of L_1 through L_4 (and sometimes T_{12}). Its nerves serve the lower abdominopelvic region and the anterior thigh (**Table 19.3** and **Figure 19.8**). The largest nerve of this plexus is the **femoral nerve,** which passes beneath the inguinal ligament to innervate the anterior thigh muscles. The cutaneous

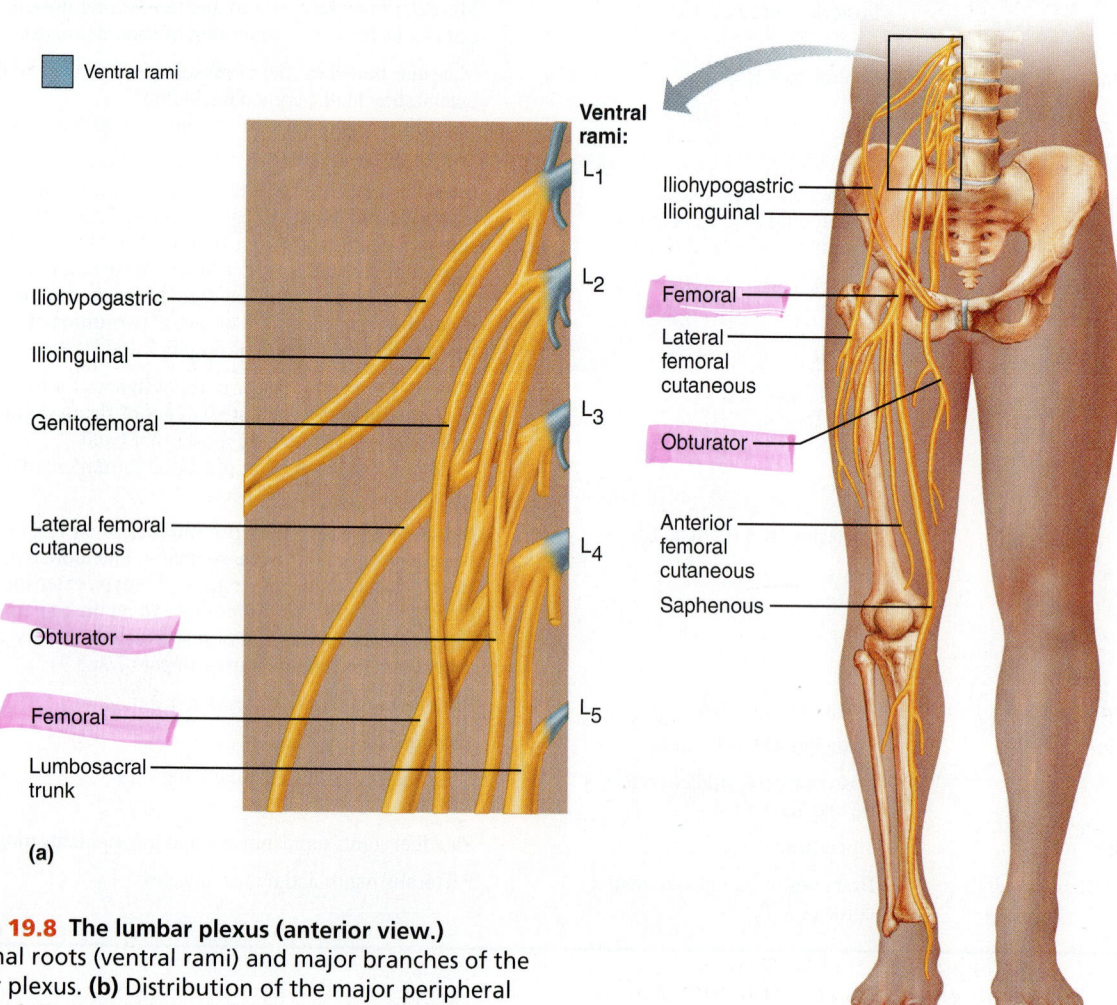

Figure 19.8 The lumbar plexus (anterior view.)
(a) Spinal roots (ventral rami) and major branches of the lumbar plexus. **(b)** Distribution of the major peripheral nerves of the lumbar plexus in the lower limb. (See Table 19.3.)

Table 19.3	Branches of the Lumbar Plexus (See Figure 19.8)	
Nerves	**Ventral rami**	**Structures served**
Femoral	L_2–L_4	Skin of anterior and medial thigh via *anterior femoral cutaneous* branch; skin of medial leg and foot, hip and knee joints via *saphenous* branch; motor to anterior muscles (quadriceps and sartorius) of thigh and to pectineus, iliacus
Obturator	L_2–L_4	Motor to adductor magnus (part), longus, and brevis muscles, gracilis muscle of medial thigh, obturator externus; sensory for skin of medial thigh and for hip and knee joints
Lateral femoral cutaneous	L_2, L_3	Skin of lateral thigh; some sensory branches to peritoneum
Iliohypogastric	L_1	Skin of lower abdomen and hip; muscles of anterolateral abdominal wall (obliques and transversus abdominis)
Ilioinguinal	L_1	Skin of external genitalia and proximal medial aspect of the thigh; inferior abdominal muscles
Genitofemoral	L_1, L_2	Skin of scrotum in males, of labia majora in females, and of anterior thigh inferior to middle portion of inguinal region; cremaster muscle in males

branches of the femoral nerve (median and anterior femoral cutaneous and the saphenous nerves) supply the skin of the anteromedial surface of the entire lower limb.

The Sacral Plexus

Arising from L_4 through S_4, the nerves of the **sacral plexus** supply the buttock, the posterior surface of the thigh, and virtually all sensory and motor fibers of the leg and foot (**Table 19.4** and **Figure 19.9**). The major peripheral nerve of this plexus is the **sciatic nerve,** the largest nerve in the body. The sciatic nerve leaves the pelvis through the greater sciatic notch and travels down the posterior thigh, serving its flexor muscles and skin. In the popliteal region, the sciatic nerve divides into the **common fibular nerve** and the **tibial nerve,**

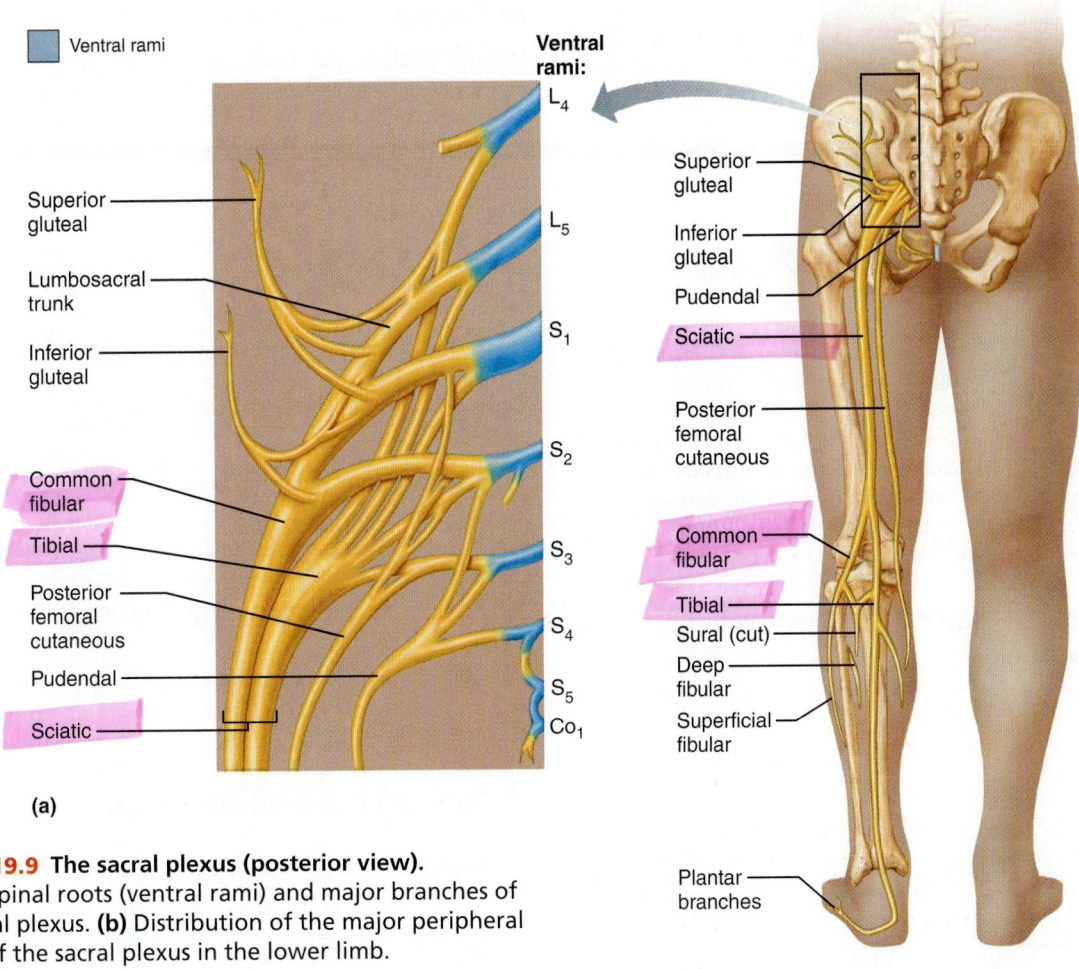

Figure 19.9 The sacral plexus (posterior view).
(a) The spinal roots (ventral rami) and major branches of the sacral plexus. **(b)** Distribution of the major peripheral nerves of the sacral plexus in the lower limb. (See Table 19.4.)

Table 19.4	Branches of the Sacral Plexus (See Figure 19.9)	

Nerves	Ventral rami	Structures served
Sciatic nerve	L_4–S_3	Composed of two nerves (tibial and common fibular) in a common sheath; they diverge just proximal to the knee
• Tibial (including sural, medial and lateral plantar, and medial calcaneal branches)	L_4–S_3	Cutaneous branches: to skin of posterior surface of leg and sole of foot Motor branches: to muscles of back of thigh, leg, and foot (hamstrings [except short head of biceps femoris], posterior part of adductor magnus, triceps surae, tibialis posterior, popliteus, flexor digitorum longus, flexor hallucis longus, and intrinsic muscles of foot)
• Common fibular (superficial and deep branches)	L_4–S_2	Cutaneous branches: to skin of anterior and lateral surface of leg and dorsum of foot Motor branches: to short head of biceps femoris of thigh, fibularis muscles of lateral leg, tibialis anterior, and extensor muscles of toes (extensor hallucis longus, extensors digitorum longus and brevis)
Superior gluteal	L_4–S_1	Motor branches: to gluteus medius and minimus and tensor fasciae latae
Inferior gluteal	L_5–S_2	Motor branches: to gluteus maximus
Posterior femoral cutaneous	S_1–S_3	Skin of buttock, posterior thigh, and popliteal region; length variable; may also innervate part of skin of calf and heel
Pudendal	S_2–S_4	Supplies most of skin and muscles of perineum (region encompassing external genitalia and anus and including clitoris, labia, and vaginal mucosa in females, and scrotum and penis in males); external anal sphincter

which together supply the balance of the leg muscles and skin, both directly and via several branches.

Injury to the proximal part of the sciatic nerve, as might follow a fall or disc herniation, results in a number of lower limb impairments. **Sciatica** (si-at′ĭ-kah), characterized by stabbing pain radiating over the course of the sciatic nerve, is common. When the sciatic nerve is completely severed, the leg is nearly useless. The leg cannot be flexed and the foot drops into plantar flexion (dangles), a condition called **footdrop.** ✛

ACTIVITY 3

Identifying the Major Nerve Plexuses and Peripheral Nerves

Identify each of the four major nerve plexuses and their major nerves (Figures 19.6–19.9) on a large laboratory chart or model. Trace the courses of the nerves and relate those observations to the information provided (Tables 19.1–19.4). ■

GROUP CHALLENGE

Fix the Sequence

Listed below are sets of a plexus, a nerve, and a muscle possibly innervated by the listed nerve. Working in small groups, decide if each set is correct for the sequence of a motor signal or needs to be corrected. If correct, simply write "all correct." If incorrect, suggest a corrected flow. Note that there may be more than one way to correct the sequence. Depend only on each other. Refrain from using a figure or other reference to help with your decision.

1. Cervical plexus, phrenic nerve, diaphragm _____

_____

2. Brachial plexus, ulnar nerve, palmaris longus _____

_____

3. Brachial plexus, radial nerve, triceps brachii _____

_____

4. Cervical plexus, axillary nerve, deltoid _____

_____

5. Lumbar plexus, femoral nerve, gracilis _____

_____

6. Lumbar plexus, sciatic nerve, common fibular nerve, tibialis anterior _____

_____

7. Sacral plexus, superior gluteal nerve, gluteus maximus _____

_____

The Spinal Cord and Spinal Nerves

Anatomy of the Spinal Cord

1. Match each anatomical term in the key to the descriptions given below.

Key: a.   cauda equina   b.   conus medullaris   c.   filum terminale   d.   foramen magnum

_____ 1.   most superior boundary of the spinal cord

_____ 2.   meningeal extension beyond the spinal cord terminus

_____ 3.   spinal cord terminus

_____ 4.   collection of spinal nerves traveling in the vertebral canal below the terminus of the spinal cord

2. Match the key letters on the diagram with the following terms.

_____ 1. arachnoid mater

_____ 2. central canal

_____ 3. dorsal horn

_____ 4. dorsal ramus of spinal nerve

_____ 5. dorsal root ganglion

_____ 6. dorsal root of spinal nerve

_____ 7. dura mater

_____ 8. gray commissure

_____ 9. lateral horn

_____ 10. pia mater

_____ 11. spinal nerve

_____ 12. ventral horn

_____ 13. ventral ramus of spinal nerve

_____ 14. ventral root of spinal nerve

_____ 15. white matter

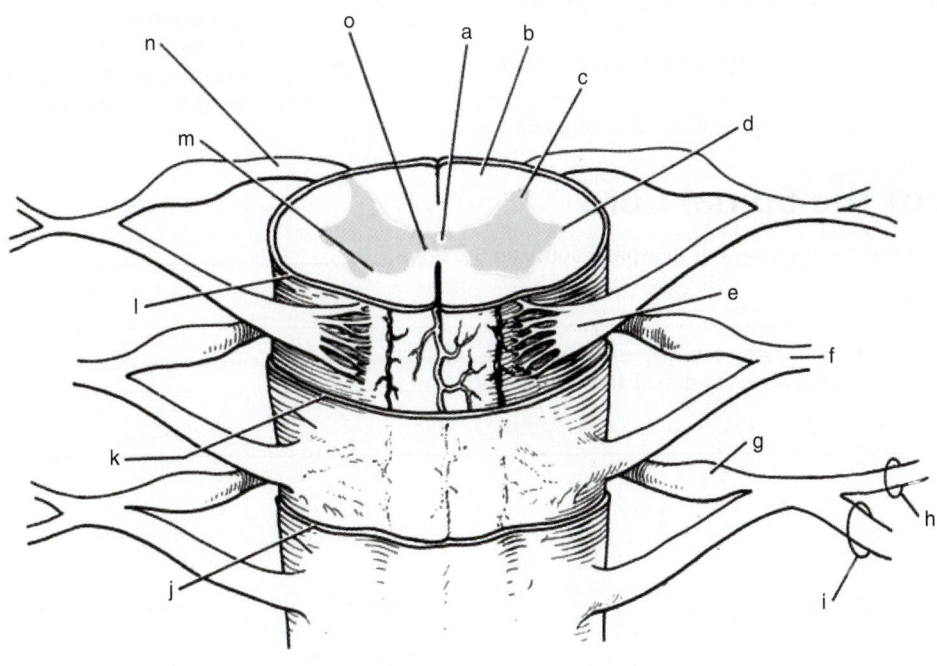

3. Choose the proper answer from the following key to respond to the descriptions relating to spinal cord anatomy. (Some terms are used more than once.)

Key: a.   sensory   b.   motor   c.   both sensory and motor   d.   interneurons

_____ 1.   neuron type found in dorsal horn   _____ 4.   fiber type in ventral root

_____ 2.   neuron type found in ventral horn   _____ 5.   fiber type in dorsal root

_____ 3.   neuron type in dorsal root ganglion   _____ 6.   fiber type in spinal nerve

4. Where in the vertebral column is a lumbar puncture generally done? _____

Why is this the site of choice? _____

_____

5. The spinal cord is enlarged in two regions, the _____ and the _____ regions.

What is the significance of these enlargements? _____

_____

6. How does the position of the gray and white matter differ in the spinal cord and the cerebral hemispheres?

_____

_____

7. From the key, choose the name of the tract that might be damaged when the following conditions are observed. (More than one choice may apply; some terms are used more than once.)

_____ 1.   uncoordinated movement

_____ 2.   lack of voluntary movement

_____ 3.   tremors, jerky movements

_____ 4.   diminished pain perception

_____ 5.   diminished sense of touch

Key: a.   dorsal columns (fasciculus cuneatus and fasciculus gracilis)
         b.   lateral corticospinal tract
         c.   ventral corticospinal tract
         d.   tectospinal tract
         e.   rubrospinal tract
         f.   vestibulospinal tract
         g.   lateral spinothalamic tract
         h.   ventral spinothalamic tract

Dissection of the Spinal Cord

8. Compare and contrast the meninges of the spinal cord and the brain. _____

_____

9. How can you distinguish between the dorsal and ventral horns? _____

_____

Spinal Nerves and Nerve Plexuses

10. In the human, there are 31 pairs of spinal nerves, named according to the region of the vertebral column from which they issue. The spinal nerves are named below. Indicate how they are numbered.

cervical nerves _____   sacral nerves _____

lumbar nerves _____   thoracic nerves _____

11. The ventral rami of spinal nerves C_1 through T_1 and T_{12} through S_4 take part in forming _____ ,

which serve the _____ of the body. The ventral rami of T_2 through T_{12} run

between the ribs to serve the _____. The dorsal rami of the spinal nerves

serve _____.

12. What would happen if the following structures were damaged or transected? (Use the key choices for responses.)

Key: a. loss of motor function   b. loss of sensory function   c. loss of both motor and sensory function

_____ 1. dorsal root of a spinal nerve   _____ 3. ventral ramus of a spinal nerve

_____ 2. ventral root of a spinal nerve

13. Define *plexus.* _____

_____

14. Name the major nerves that serve the following body areas.

_____ 1. head, neck, shoulders (name plexus only)

_____ 2. diaphragm

_____ 3. posterior thigh

_____ 4. leg and foot (name two)

_____ 5. anterior forearm muscles (name two)

_____ 6. arm muscles (name two)

_____ 7. abdominal wall (name plexus only)

_____ 8. anterior thigh

_____ 9. medial side of the hand

The Autonomic Nervous System

MATERIALS

☐ Laboratory chart or three-dimensional model of the sympathetic trunk (chain)

BIOPAC® BIOPAC® BSL System for Windows with BSL software version 3.7.5 to 3.7.7, or BSL System for Mac OS X with BSL software version 3.7.4 to 3.7.7, MP36/35 data acquisition unit, PC or Mac computer, respiratory transducer belt, EDA/GSR finger leads or disposable finger electrodes with EDA pinch leads, electrode lead set, disposable vinyl electrodes, conduction gel, and nine 8½ × 11 inch sheets of paper of different colors (white, black, red, blue, green, yellow, orange, brown, and purple) to be viewed in this sequence.

Instructors using the MP36 (or MP35/30) data acquisition unit with BSL software versions earlier than 3.7.5 (for Windows) and 3.7.4 (for Max OS X) will need slightly different channel settings and collection strategies. Instructions for using the older data acquisition unit can be found on MasteringA&P.

For instructions on animal dissections, see the dissection exercises (starting on p. 697) in the cat and fetal pig editions of this manual.

OBJECTIVES

1. Identify the site of origin and the function of the sympathetic and parasympathetic divisions of the autonomic nervous system.

2. State how the autonomic nervous system differs from the somatic nervous system.

3. Identify the neurotransmitters associated with the sympathetic and parasympathetic fibers.

4. Record and analyze data associated with the galvanic skin response.

PRE-LAB QUIZ

1. The _____ nervous system is the subdivision of the peripheral nervous system that regulates body activities that are generally not under conscious control.
   a. autonomic   c. somatic
   b. cephalic   d. vascular

2. Circle the correct underlined term. The parasympathetic division of the autonomic nervous system is also known as the craniosacral / thoracolumbar division.

3. Circle True or False. Cholinergic fibers release epinephrine.

4. The _____ division of the autonomic nervous system is responsible for the "fight-or-flight" response because it adapts the body for extreme conditions such as exercise.

5. Circle True or False. The galvanic skin response measures an increase in water and electrolytes at the skin surface.

MasteringA&P® For related exercise study tools, go to the Study Area of MasteringA&P. There you will find:

• Practice Anatomy Lab PAL

• PhysioEx PEx

• A&PFlix A&PFlix

• Practice quizzes, Histology Atlas, eText, Videos, and more!

The **autonomic nervous system (ANS)** is the subdivision of the peripheral nervous system (PNS) that regulates body activities that are generally not under conscious control. It is composed of a special group of motor neurons serving smooth muscle, cardiac muscle, and glands. The ANS is also called the *involuntary nervous system,* which reflects its subconscious control.

There is a basic anatomical difference between the motor pathways of the **somatic** (voluntary) **nervous system,** which innervates the skeletal muscles, and those of the autonomic nervous system. In the somatic division, the cell bodies of the motor neurons reside in the brain stem or ventral horns of the spinal cord, and their axons, sheathed in cranial or spinal nerves, extend directly to the skeletal muscles they serve. However, the autonomic nervous system consists of chains of two motor neurons. The first motor neuron of each pair, called the *preganglionic neuron,* resides in the brain stem or lateral horn of the spinal cord. Its axon leaves the central nervous system (CNS) to synapse with the second motor neuron, the *postganglionic neuron,* whose cell body is located in a ganglion outside the CNS. The axon of the postganglionic neuron then extends to the organ it serves.

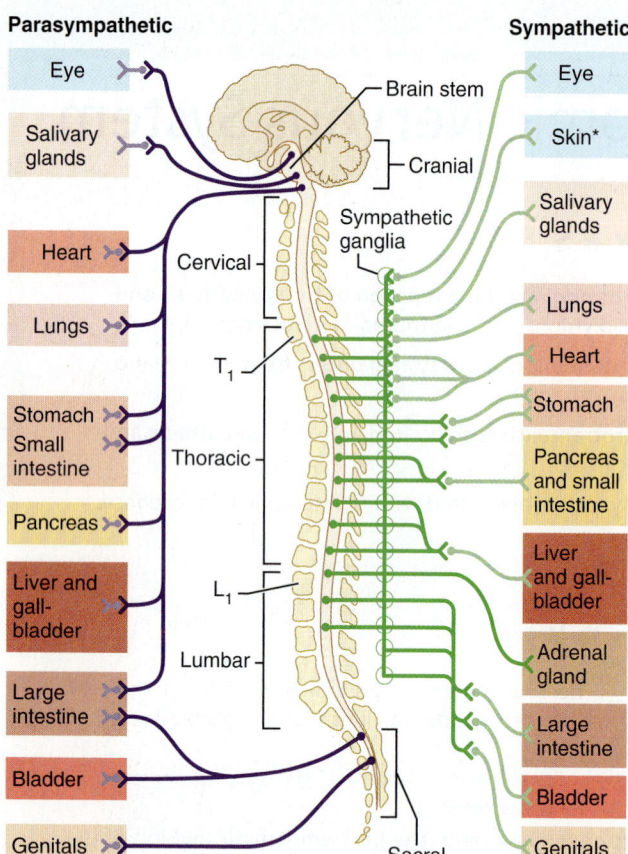

Parasympathetic   **Sympathetic**

Figure 20.1 Overview of the subdivisions of the autonomic nervous system. The parasympathetic and sympathetic divisions differ anatomically in (1) the sites of origin of their nerves, (2) the relative lengths of preganglionic and postganglionic fibers, and (3) the locations of their ganglia. Although sympathetic innervation to the skin(*) is shown only for the cervical area, all nerves to the periphery carry postganglionic sympathetic fibers.

The ANS has two major functional subdivisions **(Figure 20.1)**: the sympathetic and parasympathetic divisions. Both serve most of the same organs but generally cause opposing, or antagonistic, effects.

Parasympathetic (Craniosacral) Division

The preganglionic neurons of the **parasympathetic,** or **craniosacral,** division are located in brain stem nuclei of cranial nerves III, VII, IX, X and in the S_2 through S_4 level of the spinal cord. The axons of preganglionic neurons of the cranial region travel in their respective cranial nerves to the *immediate area* of the head and neck organs to be stimulated. There in a **terminal,** or **intramural** (literally, "within the walls"), **ganglion** they synapse with postganglionic neurons. The postganglionic neuron then sends out a very short postganglionic axon to the organ it serves. In the sacral region, the preganglionic axons leave the ventral roots of the spinal cord and collectively form the **pelvic splanchnic nerves,**

which travel to the pelvic cavity. In the pelvic cavity, the preganglionic axons synapse with the postganglionic neurons in ganglia located on or close to the organs served.

Sympathetic (Thoracolumbar) Division

The preganglionic neurons of the **sympathetic,** or **thoracolumbar,** division are located in the lateral horns of the gray matter of the spinal cord from T_1 through L_2. The preganglionic axons leave the cord via the ventral root with the axons of the somatic motor neurons. They enter the spinal nerve, and then travel briefly in the ventral ramus **(Figure 20.2)**. From the ventral ramus, they pass through a small branch called the **white ramus communicans** to enter a **sympathetic trunk ganglion.** These two trunks or *chains* lie alongside the vertebral column and are also called *paravertebral ganglia.*

Having reached the ganglion, a preganglionic axon may take one of three main courses (Figure 20.2b). First, it may synapse with a postganglionic neuron in the sympathetic trunk at that level. Second, the axon may travel upward or downward through the sympathetic trunk to synapse with a postganglionic neuron at another level. In either of these two instances, the postganglionic axons then reenter the spinal nerve via a **gray ramus communicans** and travel in branches of a dorsal or ventral ramus to innervate skin structures including sweat glands, arrector pili muscles attached to hair follicles, and the smooth muscles of blood vessel walls and thoracic organs. Third, the axon may pass through the ganglion without synapsing and form part of a **splanchnic nerve,** which travels to the viscera to synapse with a postganglionic neuron in a **collateral,** or **prevertebral, ganglion.** The major collateral ganglia—the *celiac, superior mesenteric, inferior mesenteric,* and *inferior hypogastric ganglia*—supply the abdominal and pelvic visceral organs. The postganglionic axon then leaves the ganglion and travels to a nearby visceral organ that it innervates.

ACTIVITY 1

Locating the Sympathetic Trunk

Locate the sympathetic trunk (chain) on the spinal nerve chart or three-dimensional model. ■

Autonomic Functioning

As noted earlier, most body organs served by the autonomic nervous system receive fibers from both the sympathetic and parasympathetic divisions. The only exceptions are the structures of the skin (sweat glands and arrector pili muscles attached to the hair follicles), the adrenal medulla, and essentially all blood vessels except those of the external genitalia, all of which receive sympathetic innervation only. When both divisions serve an organ, they usually have opposite effects. This is because their postganglionic axons release different neurotransmitters. The parasympathetic fibers, called **cholinergic fibers,** release acetylcholine; the sympathetic postganglionic fibers, called **adrenergic fibers,** release norepinephrine. However, there are isolated examples of postganglionic sympathetic fibers, such as those serving sweat glands that release acetylcholine. The preganglionic fibers of both divisions release acetylcholine.

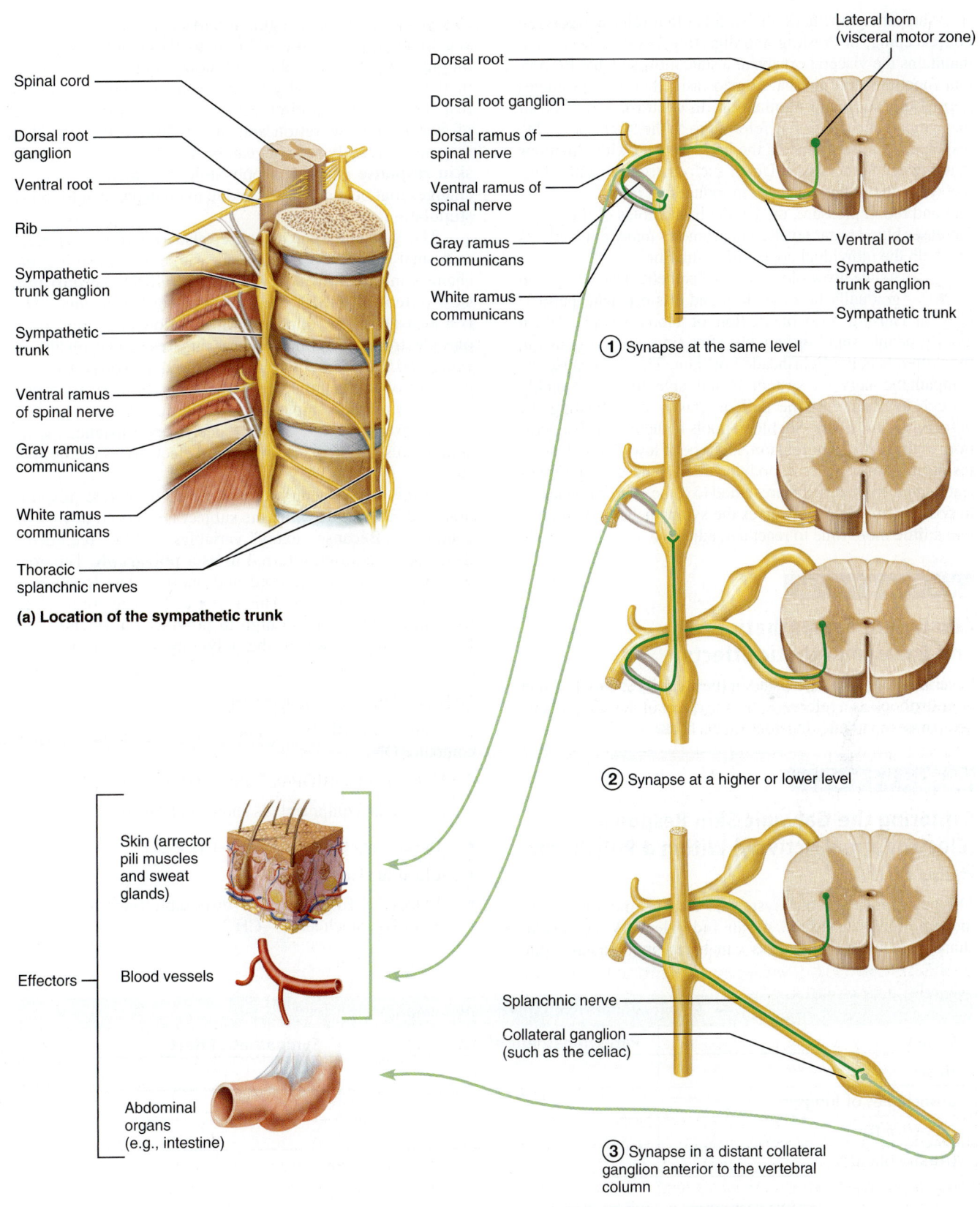

Spinal cord

Dorsal root ganglion

Ventral root

Rib

Sympathetic trunk ganglion

Sympathetic trunk

Ventral ramus of spinal nerve

Gray ramus communicans

White ramus communicans

Thoracic splanchnic nerves

(a) Location of the sympathetic trunk

Lateral horn (visceral motor zone)

Dorsal root

Dorsal root ganglion

Dorsal ramus of spinal nerve

Ventral ramus of spinal nerve

Gray ramus communicans

White ramus communicans

Ventral root

Sympathetic trunk ganglion

Sympathetic trunk

① Synapse at the same level

② Synapse at a higher or lower level

Splanchnic nerve

Collateral ganglion (such as the celiac)

③ Synapse in a distant collateral ganglion anterior to the vertebral column

Effectors

Skin (arrector pili muscles and sweat glands)

Blood vessels

Abdominal organs (e.g., intestine)

(b) Three pathways of sympathetic innervation

20

Figure 20.2 Sympathetic trunks and pathways. (a) Diagram of the right sympathetic trunk in the posterior thorax. **(b)** Synapses between preganglionic and postganglionic sympathetic neurons can occur at three different locations—in a sympathetic trunk ganglion at the same level, in a sympathetic trunk ganglion at a different level, or in a collateral ganglion.

The parasympathetic division is often referred to as the housekeeping, or "resting and digesting," system because it maintains the visceral organs in a state most suitable for normal functions and internal homeostasis; that is, it promotes normal digestion and elimination. In contrast, activation of the sympathetic division is referred to as the "fight-or-flight" response because it readies the body to cope with situations that threaten homeostasis. Under such emergency conditions, the sympathetic nervous system induces an increase in heart rate and blood pressure, dilates the bronchioles of the lungs, increases blood sugar levels, and promotes many other effects that help the individual cope with a stressor.

As we grow older, our sympathetic nervous system gradually becomes less and less efficient, particularly in causing vasoconstriction of blood vessels. When elderly people stand up quickly after sitting or lying down, they often become light-headed or faint. This is because the sympathetic nervous system is not able to react quickly enough to counteract the pull of gravity by activating the vasoconstrictor fibers. So, blood pools in the feet. This condition, **orthostatic hypotension,** is a type of low blood pressure resulting from changes in body position as described. Orthostatic hypotension can be prevented to some degree if changes in position are *slow*. This gives the sympathetic nervous system a little more time to react and adjust. +

ACTIVITY 2

Comparing Sympathetic and Parasympathetic Effects

Several body organs are listed in the **Activity 2 chart.** Using your textbook as a reference, list the effect of the sympathetic and parasympathetic divisions on each. ■

ACTIVITY 3

Exploring the Galvanic Skin Response (Electrodermal Activity) Within a Polygraph Using BIOPAC®

The autonomic nervous system is closely integrated with the emotions, or affect, of an individual. A sad event, sharp pain, or simple stress can bring about measurable changes in autonomic regulation of heart rate, respiration, and blood pressure. In addition to these obvious physiological signs, more subtle autonomic changes can occur in the skin. Specifically, changes in autonomic tone in response to external circumstances can influence the rate of sweat gland secretion and blood flow to the skin that may not be readily seen but can be measured. The **galvanic skin response** is an electrophysiological measurement of changes that occur in the skin due to changes in autonomic stimulation.

The galvanic skin response, also referred to as electrodermal activity (EDA), is measured by recording the changes in **galvanic skin resistance (GSR)** and **galvanic skin potential (GSP).** Resistance, recorded in *ohms* (Ω), is a measure of the opposition to the flow of current from one electrode to another. Increasing resistance results in decreased current. Potential, measured in *volts* (V), is a measure of the amount of charge separation between two points. Increased sympathetic stimulation of sweat glands, in response to change in affect, decreases resistance on the skin because of increased water and electrolytes on the skin surface.

In this experiment you will record heart rate, respiration, and EDA/GSR while the subject is exposed to various conditions. Because "many" variables will be "recorded," this process is often referred to as a **polygraph.** The goal of this exercise is to record and analyze data to observe how this process works. This is not a "lie detector test," as its failure rate is far too high to provide true scientific or legal certainty. However, the polygraph can be used as an investigative tool.

Setting Up the Equipment

1. Connect the BIOPAC® unit to the computer and turn the computer **ON.**

2. Make sure the BIOPAC® unit is **OFF.**

3. Plug in the equipment (as shown in **Figure 20.3**).

- Respiratory transducer belt—CH 1

- Electrode lead set—CH 2

- EDA/GSR finger leads or disposable finger electrodes and EDA pinch leads—CH 3

Activity 2: Parasympathetic and Sympathetic Effects		
Organ	**Parasympathetic effect**	**Sympathetic Effect**
Heart		
Bronchioles of lungs		
Digestive tract		
Urinary bladder		
Iris of the eye		
Blood vessels (most)		
Penis/clitoris		
Sweat glands		
Adrenal medulla		
Pancreas		

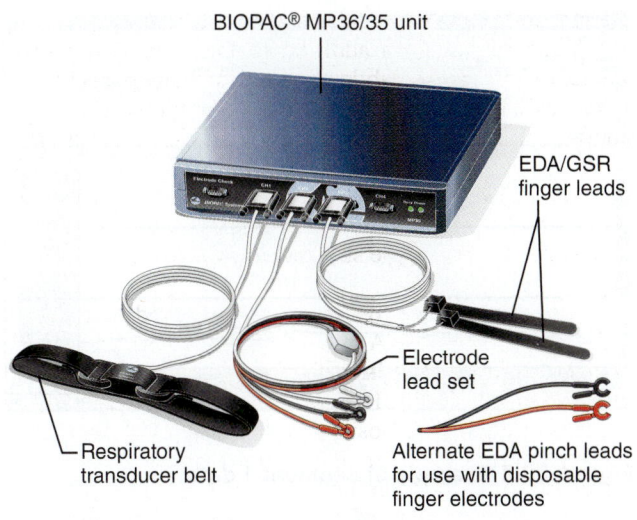

BIOPAC® MP36/35 unit

EDA/GSR finger leads

Electrode lead set

Alternate EDA pinch leads for use with disposable finger electrodes

Respiratory transducer belt

Figure 20.3 Setting up the BIOPAC® equipment. Plug the respiratory transducer belt into Channel 1, the electrode lead set into Channel 2, and the EDA/GSR finger leads into Channel 3.

4. Turn the BIOPAC® unit **ON.**

5. Attach the respiratory transducer belt to the subject (as shown in **Figure 20.4**). It should be fastened so that it is slightly tight even at the point of maximal expiration.

6. To pick up a good EDA/GSR signal, it is important that the subject's hand have enough sweat (as it normally would). *The subject should not have freshly washed or cold hands.*

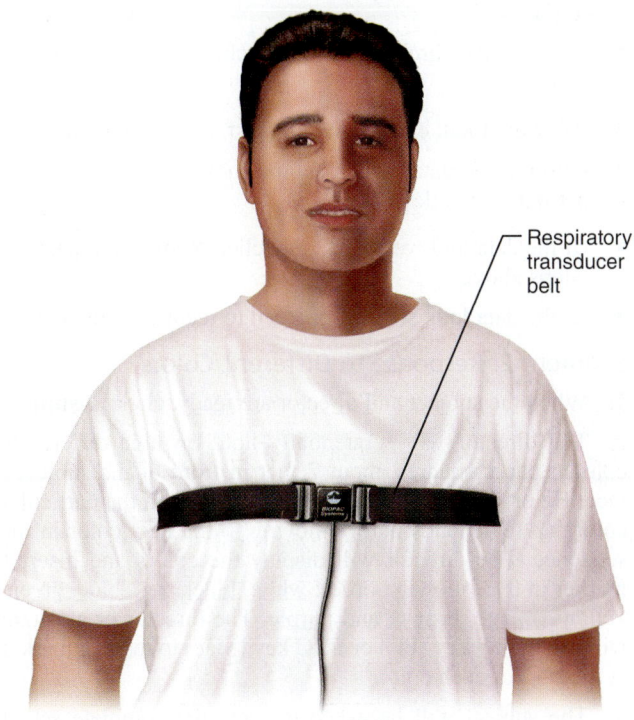

Respiratory transducer belt

Figure 20.4 Proper placement of the respiratory transducer belt around the subject's thorax.

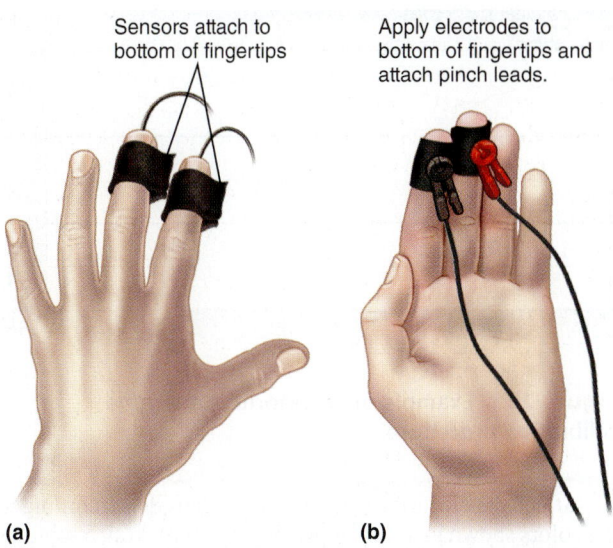

Sensors attach to bottom of fingertips

Apply electrodes to bottom of fingertips and attach pinch leads.

(a)   (b)

Figure 20.5 Placement of the EDA/GSR finger lead sensors or disposable electrodes on the fingers.

Place the electrodes on the middle and index fingers with the sensors on the skin, not the fingernail. They should fit snugly but not be so tight as to cut off circulation. If using EDA/GSR finger lead sensors, fill both cavities of the leads with conduction gel, and attach the sensors to the subject's fingers (as shown in **Figure 20.5a**). If using disposable finger electrodes, apply to subject's fingers and attach pinch leads (as shown in Figure 20.5b). Attach the electrodes at least 5 minutes before recording.

7. In order to record the heart rate, place the electrodes on the subject (as shown in **Figure 20.6**). Place an electrode on the medial surface of each leg, just above the ankle. Place another electrode on the right anterior forearm just above the wrist.

20

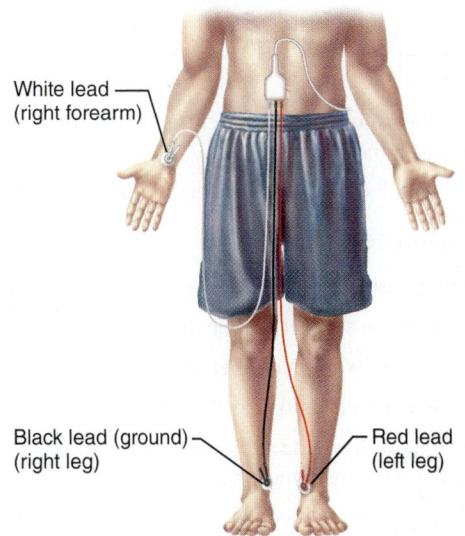

White lead (right forearm)

Black lead (ground) (right leg)

Red lead (left leg)

Figure 20.6 Placement of electrodes and the appropriate attachment of electrode leads by color.

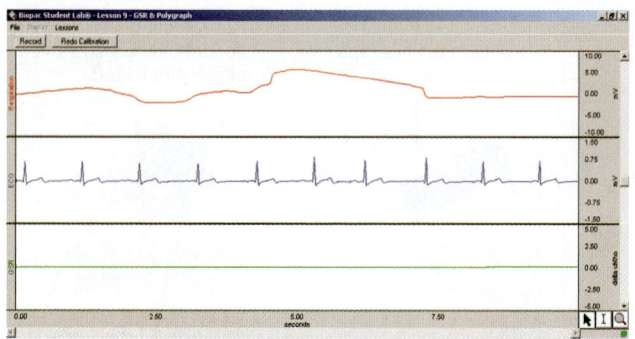

Figure 20.7 **Example of waveforms during the calibration procedure.**

8. Attach the electrode lead set to the electrodes according to the colors shown in the example (Figure 20.6). Wait 5 minutes before starting the calibration procedure.

9. Start the BIOPAC® Student Lab program on the computer by double-clicking the icon on the desktop or by following your instructor's guidance.

10. Select lesson **L09-Poly-1** from the menu and click **OK.**

11. Type in a filename that will save this subject's data on the computer hard drive. You may want to use the subject's last name followed by Poly-1 (for example, SmithPoly-1), then click **OK.**

Calibrating the Equipment

1. Have the subject sit facing the director, but do not allow the subject to see the computer screen. The subject should remain immobile but be relaxed with legs and arms in a comfortable position.

2. When the subject is ready, click **Calibrate** and then click **Yes** and **OK** if prompted. After 3 seconds, the subject will hear a beep and should inhale and exhale deeply for one breath.

3. Wait for the calibration to stop automatically after 10 seconds.

4. Observe the data, which should look similar to that in **(Figure 20.7)**.

- If the data look very different, click **Redo Calibration** and repeat the steps above.
- If the data look similar, proceed to the next section.

Recording the Data

Hints to obtaining the best data:

- Do not let the subject see the data as it is being recorded.
- Conduct the exam in a quiet setting.
- Keep the subject as still as possible.
- Take care to have the subject move the mouth as little as possible when responding to questions.
- Make sure the subject is relaxed at resting heart rate before the exam begins.

The data will be recorded in three segments. The director must read through the directions for the entire segment before

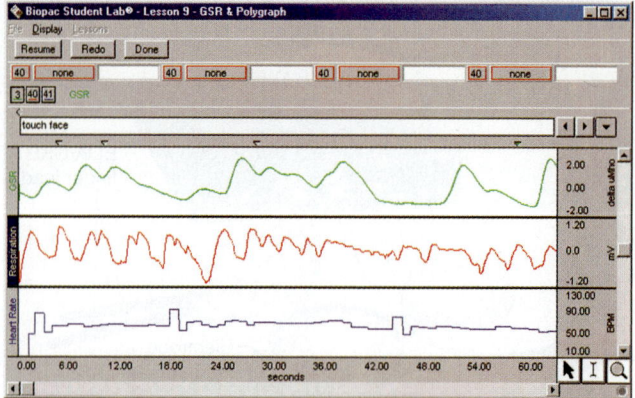

Figure 20.8 **Example of Segment 1 data.**

proceeding so that the subject can be prompted and questioned appropriately.

Segment 1: Baseline Data

1. When the subject and director are ready, click **Record.**

2. After waiting 5 seconds, the director will ask the subject to respond to the following questions and should remind the subject to minimize mouth movements when answering. Use the **F9** key (PC) or **ESC** key (Mac) to insert a marker after each response. Wait about 5 seconds after each answer.

- Quietly state your name.
- Slowly count down from ten to zero.
- Count backward from 30 by odd numbers (29, 27, 25, etc.).
- Finally, the director lightly touches the subject on the cheek.

3. After the final, cheek-touching test, click **Suspend.**

4. Observe the data, which should look similar to the Segment 1 data example **(Figure 20.8)**.

- If the data look very different, click **Redo** and repeat the steps above.
- If the data look similar, proceed to record Segment 2.

Segment 2: Response to Different Colors

1. When the subject and director are ready, click **Resume.**

2. The director will sequentially hold up nine differently colored paper squares about 2 feet in front of the subject's face. He or she will ask the subject to focus on the particular color for 10 seconds before moving to the next color in the sequence. The director will display the colors and insert a marker in the following order: white, black, red, blue, green, yellow, orange, brown, and purple. The director or assistant will use the **F9** key (PC) or **ESC** key (Mac) to insert a marker at the start of each color.

3. The subject will be asked to view the complete set of colors. After the color purple, click **Suspend.**

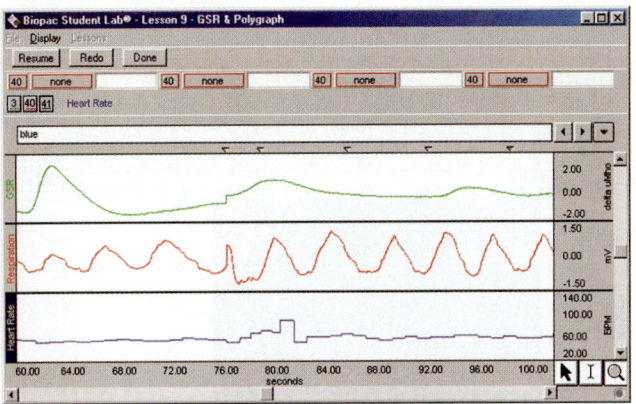

Figure 20.9 Example of Segment 2 data.

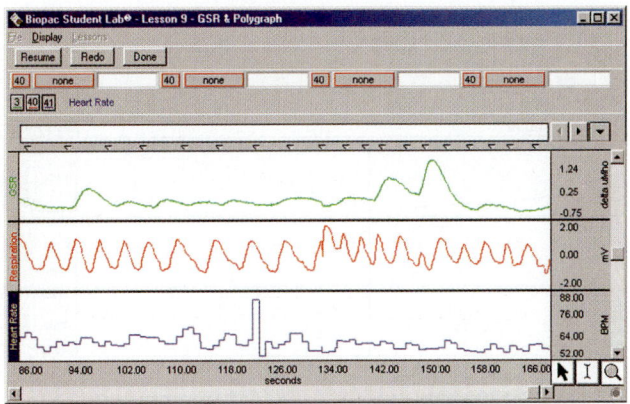

Figure 20.10 Example of Segment 3 data.

4. Observe the data, which should look similar to the Segment 2 data example **(Figure 20.9)**.

- If the data look very different, click **Redo** and repeat the steps above.

- If the data look similar, proceed to record Segment 3.

Segment 3: Response to Different Questions

1. When the subject and director are ready, click **Resume.**

2. The director will ask the subject the ten questions in step 3 and note if the answer is Yes or No. In this segment, the recorder will use the **F9** key (PC) or **ESC** key (Mac) to insert a marker at the end of each question and the end of each answer. The director will circle the Yes or No response of the subject in the "Response" column of the Segment 3 Measurements chart (page 331).

3. The following questions are to be asked and answered either Yes or No:

- Are you currently a student?

- Are your eyes blue?

- Do you have any brothers?

- Did you earn an "A" on the last exam?

- Do you drive a motorcycle?

- Are you less than 25 years old?

- Have you ever traveled to another planet?

- Have aliens from another planet ever visited you?

- Do you watch *Sesame Street*?

- Have you answered all of the preceding questions truthfully?

4. After the last question is answered, click **Suspend.**

5. Observe the data, which should look similar to the Segment 3 data example **(Figure 20.10)**.

- If the data look very different, click **Redo** and repeat the steps above.

- If the data look similar, click **Done.** Click **Yes** if you are finished recording.

6. Without recording, simply ask the subject to respond once again to all of the questions as honestly as possible. The director circles the Yes or No response of the subject in the "Truth" column of the Segment 3 Measurements chart (page 331).

7. Remove all of the sensors and equipment from the subject, and continue to Data Analysis.

Data Analysis

1. If you are just starting the BIOPAC® program to perform data analysis, enter **Review Saved Data** mode and choose the file with the subject's EDA/GSR data (for example, SmithPoly-1). If **Analyze Current Data File** was previously chosen, proceed to analysis.

2. Observe how the channel numbers are designated (as shown in **Figure 20.11**): CH 3—**EDA/GSR;** CH 40—**Respiration;** CH 41—**Heart Rate.**

3. You may need to use the following tools to adjust the data in order to clearly view and analyze the first 5 seconds of the recording.

- Click the magnifying glass in the lower right corner of the screen (near the I-beam box) to activate the **zoom** function. Use the magnifying glass cursor to click on the

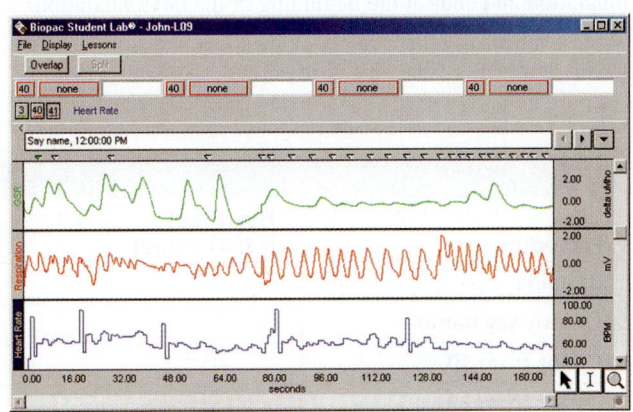

Figure 20.11 Example of polygraph recording with EDA/GSR, respiration, and heart rate.

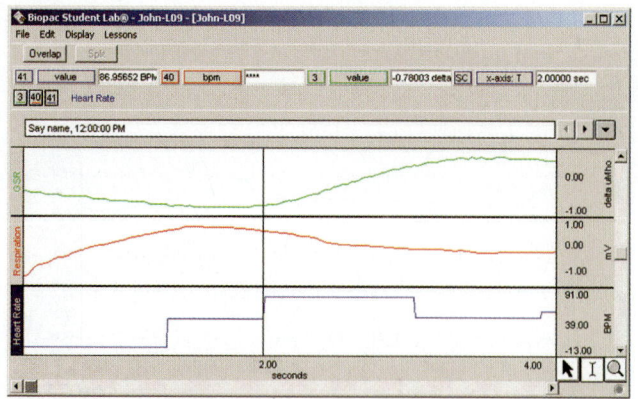

Figure 20.12 Selecting the two-second point for data analysis.

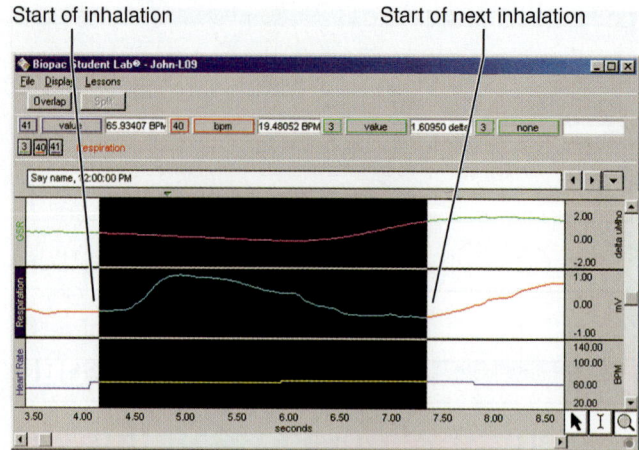

Figure 20.13 Highlighting the waveforms from the start of one inhalation to the start of the next.

very first waveforms until the first 5 seconds of data are represented (see horizontal time scale at the bottom of the screen).

- Select the **Display** menu at the top of the screen and click **Autoscale Waveforms** in the drop-down menu. This function will adjust the data for better viewing.

4. To analyze the data, note the first three pairs of channel/measurement boxes at the top of the screen. (Each box activates a drop-down menu when you click it.) The following channels and measurement types should already be set:

Channel	Measurement	Data
CH 41	value	heart rate
CH 40	value	respiration
CH 3	value	EDA/GSR

Value: Displays the value of the measurement (for example, heart rate or EDA/GSR) at the point in time that is selected.

BPM: In this analysis, the BPM calculates breaths per minute when the area that is highlighted starts at the beginning of one inhalation and ends at the beginning of the next inhalation.

5. Use the arrow cursor and click the I-beam cursor box at the lower right side of the screen to activate the "area selection"

function. Using the activated I-beam cursor, select the 2-second point on the data (as shown in **Figure 20.12**). Record the heart rate and EDA/GSR values for Segment 1 data in the **Segment 1 Measurements chart.** This point represents the resting or baseline data.

6. Using data from the first 5 seconds, use the I-beam cursor tool to highlight an area from the start of one inhalation to the start of the next inhalation (as shown in **Figure 20.13**). The start of an inhalation is indicated by the beginning of the ascension of the waveform. Record this as the baseline respiratory rate in the Segment 1 Measurements chart.

7. Using the markers as guides, scroll along the bottom scroll bar until the data from Segment 1 appears.

8. Analyze all parts of Segment 1. Using the tools described in steps 5 and 6, acquire the measurements for the heart rate, EDA/GSR, and respiration rate soon after each subject response. Use the maximum EDA/GSR value in that time frame as the point of measurement for EDA/GSR and heart rate. Use the beginning of two consecutive inhalations in that same time frame to measure respiration rate. Record these data in the Segment 1 Measurements chart.

9. Repeat these same procedures to measure EDA/GSR, heart rate, and respiration rate for each color in Segment 2. Record these data in the **Segment 2 Measurements chart.**

Segment 1 Measurements			
Procedure	Heart rate [CH 41 value]	Respiratory rate [CH 40 BPM]	EDA/GSR [CH 3 value]
Baseline			
Quietly say name			
Count from 10			
Count from 30			
Face is touched			

Segment 2 Measurements

Color	Heart rate [CH 41 value]	Respiratory rate [CH 40 BPM]	EDA/GSR [CH 3 value]
White			
Black			
Red			
Blue			
Green			
Yellow			
Orange			
Brown			
Purple			

Segment 3 Measurements

Question	Response		Truth		Heart rate [CH 41 value]	Resp. rate [CH 40 BPM]	EDA/GSR [CH 3 value]
Student?	Y	N	Y	N			
Blue eyes?	Y	N	Y	N			
Brothers?	Y	N	Y	N			
Earn "A"?	Y	N	Y	N			
Motorcycle?	Y	N	Y	N			
Under 25?	Y	N	Y	N			
Planet?	Y	N	Y	N			
Aliens?	Y	N	Y	N			
Sesame?	Y	N	Y	N			
Truthful?	Y	N	Y	N			

10. Repeat these same procedures to measure EDA/GSR, heart rate, and respiration rate for responses to each question in Segment 3. Record these data in the **Segment 3 Measurements chart.**

11. Examine EDA/GSR, heart rate, and respiration rate of the baseline data in the Segment 1 Measurements chart.

12. For every condition to which the subject was exposed, write **H** if that value is higher than baseline, write **L** if the value is lower, and write **NC** if there is no significant change. Repeat this analysis for Segments 2 and 3.

Examine the data in the Segment 1 Measurements chart. Is there any noticeable difference between the baseline EDA/GSR, heart rate, and respiration rate after each prompt? Under which prompts is the most significant change noted?

_____

_____

_____

Examine the data in the Segment 2 Measurements chart. Is there any noticeable difference between the baseline EDA/GSR, heart rate, and respiration rate after each color presentation? Under which colors is the most significant change noted?

_____

_____

_____

Examine the data in the Segment 3 Measurements chart. Is there any noticeable difference between the baseline EDA/GSR, heart rate, and respiration rate after each question? After which is the most significant change noted?

_____

_____

_____

20

Speculate as to the reasons why a subject may demonstrate a change in EDA/GSR from baseline under different color conditions.

_____

_____

_____

Speculate as to the reasons why a subject may demonstrate a change in EDA/GSR from baseline when a particular question is asked.

_____

_____

_____

Which branch of the autonomic nervous system is dominant during a galvanic skin response?

_____

20

The Autonomic Nervous System

Parasympathetic and Sympathetic Divisions

1. For the most part, sympathetic and parasympathetic fibers serve the same organs and structures. How can they exert opposite effects? (After all, nerve impulses are nerve impulses—aren't they?)

_____

2. Name three structures that receive sympathetic but not parasympathetic innervation.

_____

3. A pelvic splanchnic nerve contains (circle one):

   a.   preganglionic sympathetic fibers   c.   preganglionic parasympathetic fibers

   b.   postganglionic sympathetic fibers   d.   postganglionic parasympathetic fibers

4. The following chart states a number of conditions. Use a check mark to show which division of the autonomic nervous system is involved in each.

Sympathetic division	Condition	Parasympathetic division
	Postganglionic fibers secrete norepinephrine; adrenergic fibers	
	Postganglionic fibers secrete acetylcholine; cholinergic fibers	
	Long preganglionic axon; short postganglionic axon	
	Short preganglionic axon; long postganglionic axon	
	Arises from cranial and sacral nerves	
	Arises from spinal nerves T_1 through L_3	
	Normally in control	
	"Fight-or-flight" system	
	Has more specific control (Look it up!)	

Galvanic Skin Response (Electrodermal Activity) Within a Polygraph Using BIOPAC®

5. Describe exactly how, from a physiological standpoint, EDA/GSR can be correlated with activity of the autonomic nervous system.

_____

_____

_____

6. Based on this brief and unprofessional exposure to a polygraph, explain why this might not be an exact tool for testing the sincerity and honesty of a subject. Refer to your data to support your conclusions.

_____

_____

_____

Human Reflex Physiology

MATERIALS

- ☐ Reflex hammer
- ☐ Sharp pencils
- ☐ Cot (if available)
- ☐ Absorbent cotton (sterile)
- ☐ Tongue depressor
- ☐ Metric ruler
- ☐ Flashlight
- ☐ 100- or 250-ml beaker
- ☐ 10- or 25-ml graduated cylinder
- ☐ Lemon juice in dropper bottle
- ☐ Wide-range pH paper
- ☐ Large laboratory bucket containing freshly prepared 10% household bleach solution for saliva-soiled glassware
- ☐ Disposable autoclave bag
- ☐ Wash bottle containing 10% bleach solution
- ☐ Reaction time ruler (if available)

BIOPAC BIOPAC® BSL System for Windows with BSL software version 3.7.5 to 3.7.7, or BSL System for Mac OS X with BSL software version 3.7.4 to 3.7.7, MP36/35 data acquisition unit, PC or Mac computer, hand switch, and headphones.

Instructors using the MP36 (or MP35/30) data acquisition unit with BSL software versions earlier than 3.7.5 (for Windows) and 3.7.4 (for Mac OS X) will need slightly different channel settings and collection strategies. Instructions for using the older data acquisition unit can be found on MasteringA&P.

Note: Instructions for using PowerLab® equipment can be found on MasteringA&P.

MasteringA&P® For related exercise study tools, go to the Study Area of MasteringA&P. There you will find:

- Practice Anatomy Lab PAL
- PhysioEx PEx
- A&PFlix A&PFlix
- Practice quizzes, Histology Atlas, eText, Videos, and more!

OBJECTIVES

1. Define *reflex* and *reflex arc*.
2. Describe the differences between autonomic and somatic reflexes.
3. Explain why reflex testing is an important part of every physical examination.
4. Name, identify, and describe the function of each element of a reflex arc.
5. Describe and discuss several types of reflex activities as observed in the laboratory; indicate the functional or clinical importance of each; and categorize each as a somatic or autonomic reflex action.
6. Explain why cord-mediated reflexes are generally much faster than those involving input from the higher brain centers.
7. Investigate differences in reaction time between intrinsic and learned reflexes.

PRE-LAB QUIZ

1. Define *reflex*. _____
   _____

2. Circle the correct underlined term. <u>Autonomic</u> / <u>Somatic</u> reflexes include all those reflexes that involve stimulation of skeletal muscles.

3. In a reflex arc, the _____ transmits afferent impulses to the central nervous system.
   a. integration center
   b. motor neuron
   c. receptor
   d. sensory neuron

4. Circle True or False. Most reflexes are simple, two-neuron, monosynaptic reflex arcs.

5. Stretch reflexes are initiated by tapping a _____, which stretches the associated muscle.
   a. bone
   b. muscle
   c. tendon or ligament

6. An example of an autonomic reflex that you will be studying in today's lab is the _____ reflex.
   a. crossed-extensor   c. plantar
   b. gag   d. salivary

7. Circle True or False. A reflex that occurs on the same side of the body that was stimulated is an ipsilateral response.

8. Name one of the pupillary reflexes you will be examining today. _____

9. Circle the correct underlined term. The effectors of the salivary reflex are <u>muscles</u> / <u>glands.</u>

10. Circle True or False. Learned reflexes involve far fewer neural pathways and fewer types of higher intellectual activities than intrinsic reflexes, which shortens their response time.

Reflexes are rapid, predictable, involuntary motor responses to stimuli; they are mediated over neural pathways called reflex arcs. Many of the body's control systems are reflexes, which can be either inborn or learned. *Inborn* or *intrinsic reflexes* are wired into our nervous system and are unlearned. *Learned* or *acquired reflexes* result from practice or repetition.

Another way to categorize reflexes is into one of two large groups: autonomic reflexes and somatic reflexes. **Autonomic** (or visceral) **reflexes** are mediated through the autonomic nervous system, and we are not usually aware of them. These reflexes activate smooth muscles, cardiac muscle, and the glands of the body, and they regulate body functions such as digestion, elimination, blood pressure, salivation, and sweating. **Somatic reflexes** include all those reflexes that involve stimulation of skeletal muscles by the somatic division of the nervous system. An example of such a reflex is the rapid withdrawal of a hand from a hot object.

Reflex testing is an important diagnostic tool for assessing the condition of the nervous system. Distorted, exaggerated, or absent reflex responses may indicate degeneration or pathology of portions of the nervous system, often before other signs are apparent.

If the spinal cord is damaged, the easily performed reflex tests can help pinpoint the area (level) of spinal cord injury. Motor nerves above the injured area may be unaffected, whereas those at or below the lesion site may be unable to participate in normal reflex activity. ✚

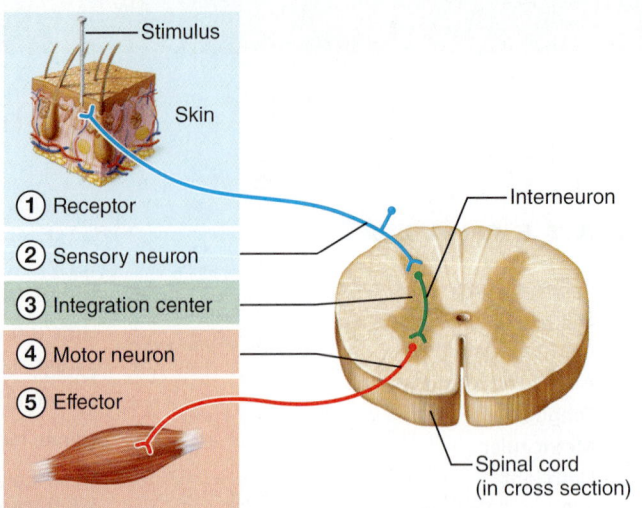

Figure 21.1 The five basic components of reflex arcs. The reflex illustrated is polysynaptic.

Components of a Reflex Arc

Reflex arcs have five basic components **(Figure 21.1)**:

1. The *receptor* is the site of stimulus action.
2. The *sensory neuron* transmits afferent impulses to the CNS.

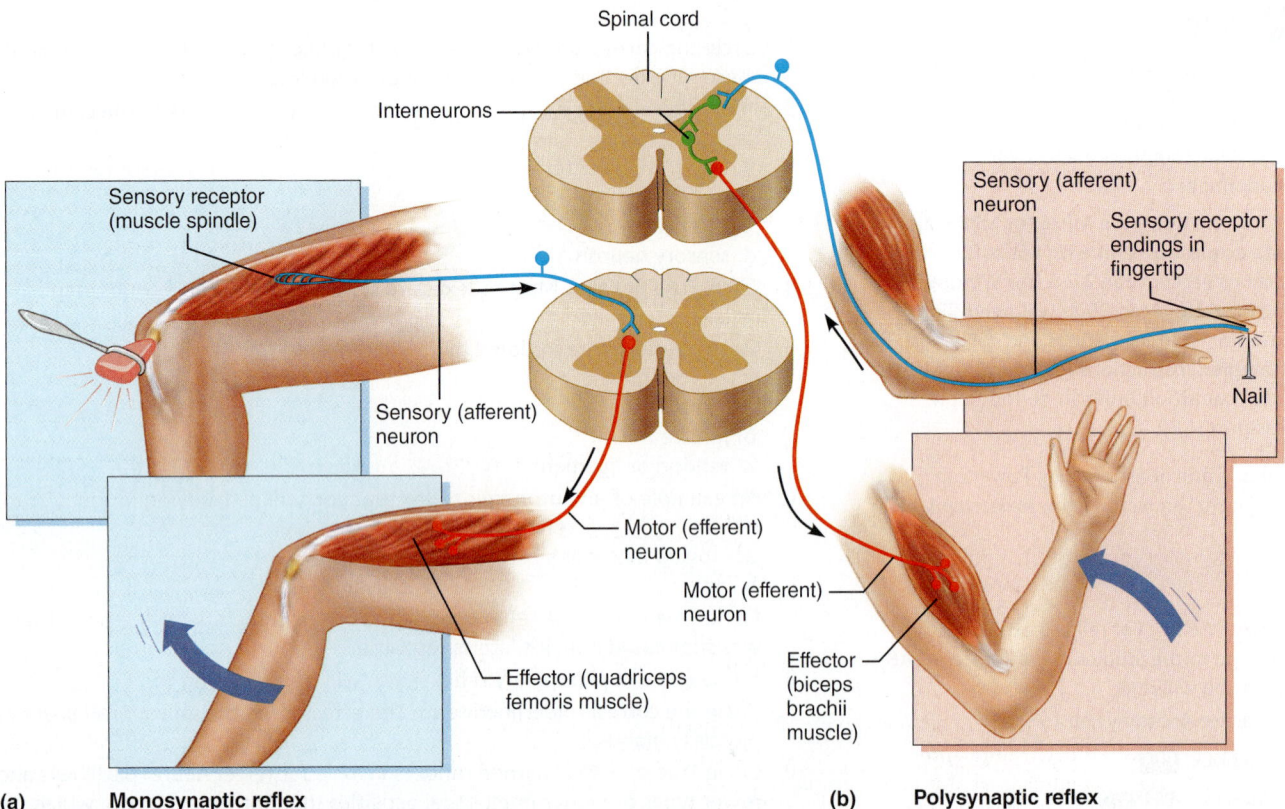

(a)   **Monosynaptic reflex**

(b)   **Polysynaptic reflex**

Figure 21.2 Monosynaptic and polysynaptic reflex arcs. The integration center is in the spinal cord, and in each example the receptor and effector are in the same limb. **(a)** The patellar reflex, a two-neuron monosynaptic reflex. **(b)** A flexor reflex, an example of a polysynaptic reflex.

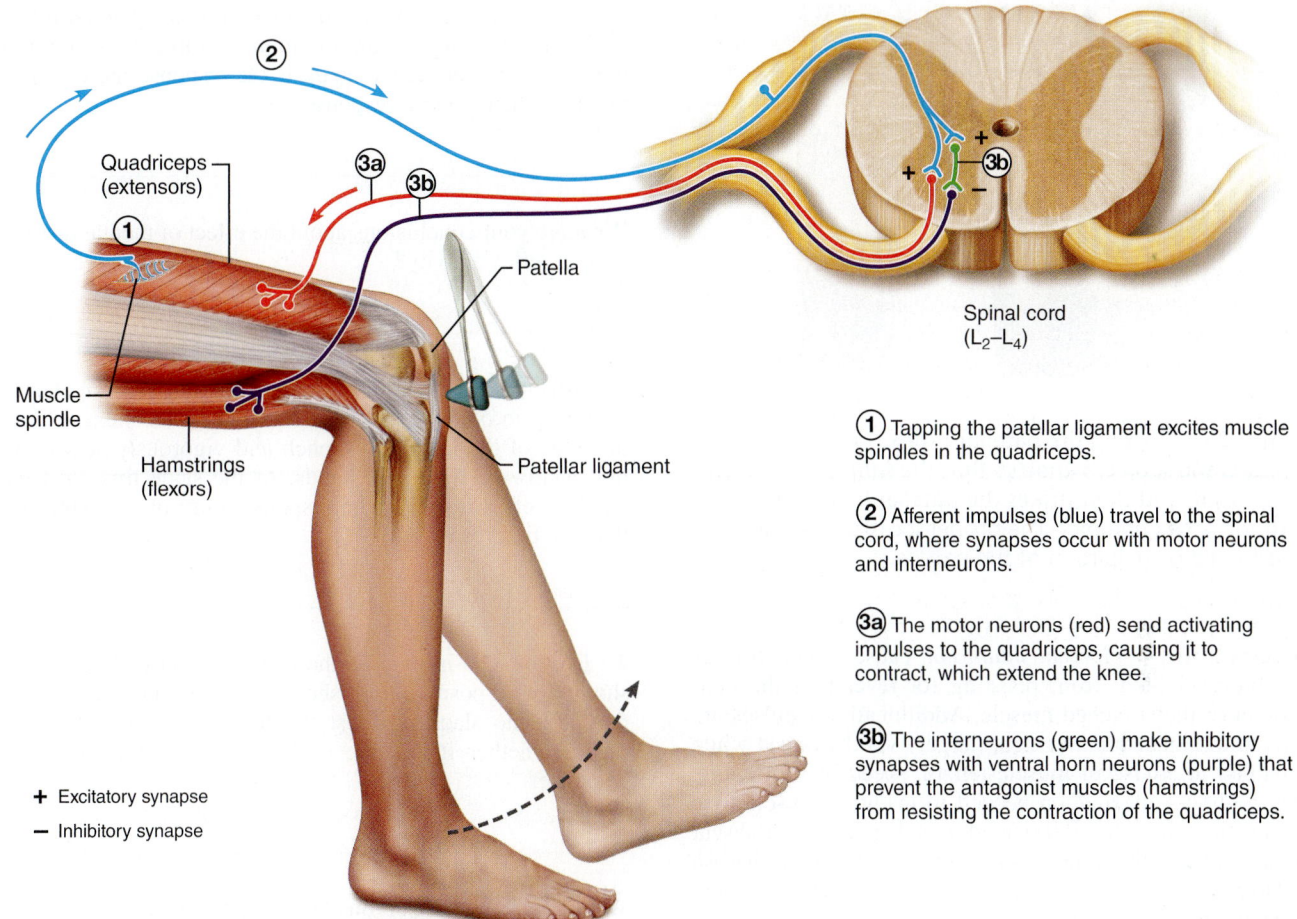

① Tapping the patellar ligament excites muscle spindles in the quadriceps.

② Afferent impulses (blue) travel to the spinal cord, where synapses occur with motor neurons and interneurons.

③a The motor neurons (red) send activating impulses to the quadriceps, causing it to contract, which extend the knee.

③b The interneurons (green) make inhibitory synapses with ventral horn neurons (purple) that prevent the antagonist muscles (hamstrings) from resisting the contraction of the quadriceps.

Figure 21.3 The patellar (knee-jerk) reflex—a specific example of a stretch reflex.

3. The *integration center* consists of one or more neurons in the CNS.

4. The *motor neuron* conducts efferent impulses from the integration center to an effector organ.

5. The *effector,* a muscle fiber or a gland cell, responds to efferent impulses by contracting or secreting, respectively.

The simple patellar or knee-jerk reflex **(Figure 21.2a)** is an example of a simple, two-neuron, *monosynaptic* (literally, "one synapse") reflex arc. It will be demonstrated in the laboratory. However, most reflexes are more complex and *polysynaptic,* involving the participation of one or more interneurons in the reflex arc pathway. An example of a polysynaptic reflex is the flexor reflex (Figure 21.2b). Since delay or inhibition of the reflex may occur at the synapses, the more synapses encountered in a reflex pathway, the more time is required for the response.

Reflexes of many types may be considered programmed into the neural anatomy. Many *spinal reflexes,* reflexes that are initiated and completed at the spinal cord level, occur without the involvement of higher brain centers. Generally these reflexes are present in animals whose brains have been destroyed, as long as the spinal cord is functional. Conversely, other reflexes require the involvement of the brain, since many different inputs must be evaluated before the appropriate reflex is determined. Superficial cord reflexes and

pupillary responses to light are in this category. In addition, although many spinal reflexes do not require the involvement of higher centers, the brain is "advised" of spinal cord reflex activity and may alter it by facilitating or inhibiting the reflexes.

Somatic Reflexes

There are several types of somatic reflexes, including several that you will be eliciting during this laboratory session—the stretch, crossed-extensor, superficial cord, corneal, and gag reflexes. Some require only spinal cord activity; others require brain involvement as well.

Spinal Reflexes

Stretch Reflexes

Stretch reflexes are important postural reflexes, normally acting to maintain posture, balance, and locomotion. Stretch reflexes are initiated by tapping a tendon or ligament, which stretches the muscle to which the tendon is attached **(Figure 21.3)**. This stimulates the muscle spindles and causes reflex contraction of the stretched muscle or muscles. Branches of the afferent fibers from the muscle spindles also synapse with interneurons controlling the antagonist muscles. The inhibition of those interneurons and the antagonist

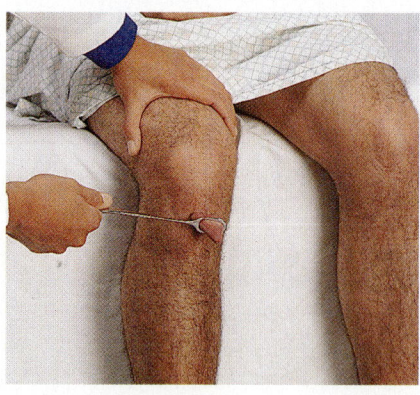

Figure 21.4 Testing the patellar reflex. The examiner supports the subject's knee so that the subject's muscles are relaxed, and then strikes the patellar ligament with the reflex hammer. The proper location may be ascertained by palpation of the patella.

muscles, called *reciprocal inhibition,* causes them to relax and prevents them from resisting (or reversing) the contraction of the stretched muscle. Additionally, impulses are relayed to higher brain centers (largely via the dorsal white columns) to advise of muscle length, speed of shortening, and the like—information needed to maintain muscle tone and posture. Stretch reflexes tend to be hypoactive or absent in cases of peripheral nerve damage or ventral horn disease and hyperactive in corticospinal tract lesions. They are absent in deep sedation and coma.

ACTIVITY 1

Initiating Stretch Reflexes

1. Test the **patellar** or **knee-jerk reflex** by seating a subject on the laboratory bench with legs hanging free (or with knees crossed). Tap the patellar ligament sharply with the reflex hammer just below the knee between the patella and the tibial tuberosity (as shown in **Figure 21.4**). The knee-jerk response assesses the L_2–L_4 level of the spinal cord. Test both knees and record your observations. (Sometimes a reflex can be altered by your actions. If you encounter difficulty, consult your instructor for helpful hints.)

_____

_____

Which muscles contracted? _____

_____

What nerve is carrying the afferent and efferent impulses?

_____

2. Test the effect of mental distraction on the patellar reflex by having the subject add a column of three-digit numbers while you test the reflex again. Is the response more *or* less vigorous than the first response?

_____

What are your conclusions about the effect of mental distraction on reflex activity?

_____

3. Now test the effect of muscular activity occurring simultaneously in other areas of the body. Have the subject clasp the edge of the laboratory bench and vigorously attempt to pull it upward with both hands. At the same time, test the patellar reflex again. Is the response more or less vigorous than the first response?

_____

4. Fatigue also influences the reflex response. The subject should jog in position until she or he is very fatigued (*really fatigued*—no slackers). Test the patellar reflex again, and record whether it is more or less vigorous than the first response.

_____

Would you say that nervous system activity *or* muscle function is responsible for the changes you have just observed?

Explain your reasoning. _____

_____

_____

_____

5. The **calcaneal tendon** or **ankle-jerk reflex** assesses the first two sacral segments of the spinal cord. With your shoe removed and your foot dorsiflexed slightly to increase the tension of the gastrocnemius muscle, have your partner sharply tap your calcaneal tendon with the broad side of the reflex hammer **(Figure 21.5)**.

What is the result? _____

_____

During walking, what is the action of the gastrocnemius at the ankle?

_____

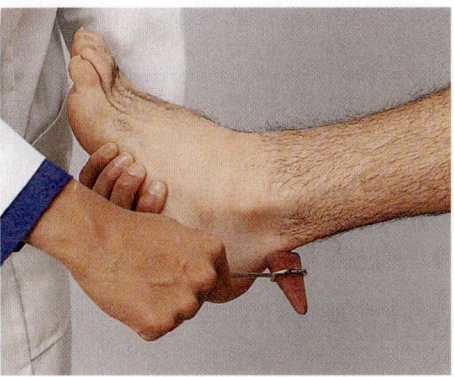

Figure 21.5 Testing the calcaneal tendon reflex. The examiner slightly dorsiflexes the subject's ankle by supporting the foot lightly in the hand, and then taps the calcaneal tendon just above the ankle.

Crossed-Extensor Reflex

The **crossed-extensor reflex** is more complex than the stretch reflex. It consists of a flexor, or withdrawal, reflex followed by extension of the opposite limb.

This reflex is quite obvious when, for example, a stranger suddenly and strongly grips one's arm. The immediate response is to withdraw the clutched arm and push the intruder away with the other arm. The reflex is more difficult to demonstrate in a laboratory because it is anticipated, and under these conditions the extensor part of the reflex may be inhibited.

ACTIVITY 2

Initiating the Crossed-Extensor Reflex

The subject should sit with eyes closed and with the dorsum of one hand resting on the laboratory bench. Obtain a sharp pencil, and suddenly prick the subject's index finger. What are the results?

_____

_____

_____

Did the extensor part of this reflex occur simultaneously or more slowly than the other reflexes you have observed?

_____

What are the reasons for this? _____

_____

_____

The reflexes that have been demonstrated so far—the stretch and crossed-extensor reflexes—are examples of reflexes in which the reflex pathway is initiated and completed at the spinal cord level.

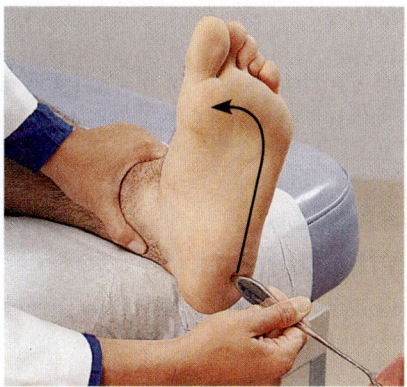

Figure 21.6 Testing the plantar reflex. Using a moderately sharp object, the examiner strokes the lateral border of the subject's sole, starting at the heel and continuing toward the great toe across the ball of the foot.

Superficial Cord Reflexes

The **superficial cord reflexes** (abdominal, cremaster, and plantar reflexes) result from pain and temperature changes. They are initiated by stimulation of receptors in the skin and mucosae. The superficial cord reflexes depend *both* on functional upper-motor pathways and on the cord-level reflex arc. Since only the plantar reflex can be tested conveniently in a laboratory setting, we will use this as our example.

The **plantar reflex,** an important neurological test, is elicited by stimulating the cutaneous receptors in the sole of the foot. In adults, stimulation of these receptors causes the toes to flex and move closer together. Damage to the corticospinal tract, however, produces *Babinski's sign,* an abnormal response in which the toes flare and the great toe moves in an upward direction. In newborn infants, it is normal to see Babinski's sign due to incomplete myelination of the nervous system.

ACTIVITY 3

Initiating the Plantar Reflex

Have the subject remove a shoe and lie on the cot or laboratory bench with knees slightly bent and thighs rotated so that the posterolateral side of the foot rests on the cot. Alternatively, the subject may sit up and rest the lateral surface of the foot on a chair. Draw the handle of the reflex hammer firmly along the lateral side of the exposed sole from the heel to the base of the great toe **(Figure 21.6)**.

What is the response? _____

_____

Is this a normal plantar reflex or a Babinski's sign?

_____

Cranial Nerve Reflex Tests

In these experiments, you will be working with your lab partner to illustrate two somatic reflexes mediated by cranial nerves.

21

Corneal Reflex

The **corneal reflex** is mediated through the trigeminal nerve (cranial nerve V). The absence of this reflex is an ominous sign because it often indicates damage to the brain stem resulting from compression of the brain or other trauma.

ACTIVITY 4

Initiating the Corneal Reflex

Stand to one side of the subject; the subject should look away from you toward the opposite wall. Wait a few seconds and then quickly, *but gently,* touch the subject's cornea (on the side toward you) with a wisp of absorbent cotton. What reflexive reaction occurs when something touches the cornea?

_____

What is the function of this reflex?

_____

_____

_____ ▬

Gag Reflex

The **gag reflex** tests the somatic motor responses of cranial nerves IX and X. When the oral mucosa on the side of the uvula is stroked, each side of the mucosa should rise, and the amount of elevation should be equal. The uvula is the fleshy tab hanging from the roof of the mouth just above the root of the tongue.

ACTIVITY 5

Initiating the Gag Reflex

For this experiment, select a subject who does not have a queasy stomach, because regurgitation is a possibility. Gently stroke the oral mucosa on each side of the subject's uvula with a tongue depressor. What happens?

_____

_____

⚠ Discard the used tongue depressor in the disposable autoclave bag before continuing. *Do not* lay it on the laboratory bench at any time. ▬

Autonomic Reflexes

The autonomic reflexes include the pupillary, ciliospinal, and salivary reflexes, as well as a multitude of other reflexes. Work with your partner to demonstrate the four autonomic reflexes described next.

Pupillary Reflexes

There are several types of pupillary reflexes. The **pupillary light reflex** and the **consensual reflex** will be examined here. In both of these pupillary reflexes, the retina of the eye is the receptor, the optic nerve (cranial nerve II) contains the afferent fibers, the oculomotor nerve (cranial nerve III) is responsible for conducting efferent impulses to the eye, and the smooth muscle of the iris is the effector. Many central nervous system centers are involved in the integration of these responses. Absence of normal pupillary reflexes is generally a late indication of severe trauma or deterioration of the vital brain stem tissue due to metabolic imbalance.

ACTIVITY 6

Initiating Pupillary Reflexes

1. Conduct the reflex testing in an area where the lighting is relatively dim. Before beginning, obtain a metric ruler and a flashlight. Measure and record the size of the subject's pupils as best you can.

Right pupil: _____ mm   Left pupil: _____ mm

2. Stand to the left of the subject to conduct the testing. The subject should shield his or her right eye by holding a hand vertically between the eye and the right side of the nose.

3. Shine a flashlight into the subject's left eye. What is the pupillary response?

_____

Measure the size of the left pupil: _____ mm

4. Without moving the flashlight, observe the right pupil. Has the same type of change (called a *consensual response*) occurred in the right eye?

_____

Measure the size of the right pupil: _____ mm

The consensual response, or any reflex observed on one side of the body when the other side has been stimulated, is called a **contralateral response.** The pupillary light response, or any reflex occurring on the same side stimulated, is referred to as an **ipsilateral response.**

What does the occurrence of a contralateral response indicate about the pathways involved?

_____

_____

Was the sympathetic *or* the parasympathetic division of the autonomic nervous system active during the testing of these reflexes?

_____

What is the function of these pupillary responses?

_____

_____

_____ ■

Ciliospinal Reflex

The **ciliospinal reflex** is another example of reflex activity in which pupillary responses can be observed. This response may initially seem a little bizarre, especially in view of the consensual reflex just demonstrated.

ACTIVITY 7

Initiating the Ciliospinal Reflex

1. While observing the subject's eyes, gently stroke the skin (or just the hairs) on the left side of the back of the subject's neck, close to the hairline.

What is the reaction of the left pupil? _____

The reaction of the right pupil? _____

2. If you see no reaction, repeat the test using a gentle pinch in the same area.
   The response you should have noted—pupillary dilation— is consistent with the pupillary changes occurring when the sympathetic nervous system is stimulated. Such a response may also be elicited in a single pupil when more impulses from the sympathetic nervous system reach it for any reason. For example, when the left side of the subject's neck was stimulated, sympathetic impulses to the left iris increased, resulting in the ipsilateral reaction of the left pupil.
   On the basis of your observations, would you say that the sympathetic innervation of the two irises is closely integrated?

_____ Why or why not? _____

_____

_____ ■

Salivary Reflex

Unlike the other reflexes, in which the effectors were smooth or skeletal muscles, the effectors of the **salivary reflex** are glands. The salivary glands secrete varying amounts of saliva in response to reflex activation.

ACTIVITY 8

Initiating the Salivary Reflex

1. Obtain a small beaker, a graduated cylinder, lemon juice, and wide-range pH paper. After refraining from swallowing for 2 minutes, the subject is to expectorate (spit) the accumulated saliva into a small beaker. Using the graduated

cylinder, measure the volume of the expectorated saliva and determine its pH.

Volume: _____ cc   pH: _____

2. Now place 2 or 3 drops of lemon juice on the subject's tongue. Allow the lemon juice to mix with the saliva for 5 to 10 seconds, and then determine the pH of the subject's saliva by touching a piece of pH paper to the tip of the tongue.

pH: _____

As before, the subject is to refrain from swallowing for 2 minutes. After the 2 minutes is up, again collect and measure the volume of the saliva and determine its pH.

Volume: _____ cc   pH: _____

3. How does the volume of saliva collected after the application of the lemon juice compare with the volume of the first saliva sample?

_____

How does the final saliva pH reading compare to the initial reading?

_____

How does the final saliva pH reading compare to that obtained 10 seconds after the application of lemon juice?

_____

What division of the autonomic nervous system mediates the reflex release of saliva?

_____

⚠ Dispose of the saliva-containing beakers and the graduated cylinders in the laboratory bucket that contains bleach and put the used pH paper into the disposable autoclave bag. Wash the bench down with 10% bleach solution before continuing. ■

Reaction Time of Intrinsic and Learned Reflexes

The time required for reaction to a stimulus depends on many factors—sensitivity of the receptors, velocity of nerve conduction, the number of neurons and synapses involved, and the speed of effector activation, to name just a few. There is no clear-cut distinction between intrinsic and learned reflexes, as most reflex actions are subject to modification by learning or conscious effort. In general, however, if the response involves a simple reflex arc, the response time is short. Learned reflexes involve a far larger number of neural pathways and many types of higher intellectual activities, including choice and decision making, which lengthens the response time.

There are various ways of testing reaction time of reflexes. The tests range from simple to ultrasophisticated. The following activities provide an opportunity to demonstrate the major time difference between simple and learned reflexes and to measure response time under various conditions.

ACTIVITY 9

Testing Reaction Time for Intrinsic and Learned Reflexes

1. Using a reflex hammer, elicit the patellar reflex in your partner. Note the relative reaction time needed for this intrinsic reflex to occur.

2. Now test the reaction time for learned reflexes. The subject should hold a hand out, with the thumb and index finger extended. Hold a metric ruler so that its end is exactly 3 cm above the subject's outstretched hand. The ruler should be in the vertical position with the numbers reading from the bottom up. When the ruler is dropped, the subject should be able to grasp it between thumb and index finger as it passes, without having to change position. Have the subject catch the ruler five times, varying the time between trials. The relative speed of reaction can be determined by reading the number on the ruler at the point of the subject's fingertips. [*] (Thus if the number at the fingertips is 15 cm, the subject was unable to catch the ruler until 18 cm of length had passed through his or her fingers; 15 cm of ruler length plus 3 cm to account for the distance of the ruler above the hand.) [†] Record the number of centimeters that pass through the subject's fingertips (or the number of seconds required for reaction) for each trial:

Trial 1: _____ cm   Trial 4: _____ cm
_____ sec   _____ sec
Trial 2: _____ cm   Trial 5: _____ cm
_____ sec   _____ sec
Trial 3: _____ cm
_____ sec

3. Perform the test again, but this time say a simple word each time you release the ruler. Designate a specific word as a signal for the subject to catch the ruler. On all other words, the subject is to allow the ruler to pass through his fingers. Trials in which the subject erroneously catches the ruler are to be disregarded. Record the distance the ruler travels (or the number of seconds required for reaction) in five *successful* trials:

Trial 1: _____ cm   Trial 2: _____ cm
_____ sec   _____ sec

*Distance *(d)* can be converted to time *(t)* using the simple formula:

$$d \text{ (in cm)} = (1/2)(980 \text{ cm/sec}^2)t^2$$

$$t^2 = (d/490 \text{ cm/sec}^2)$$

$$t = \sqrt{(d/(490 \text{ cm/sec}^2))}$$

†An alternative would be to use a reaction time ruler, which converts distance to time (seconds).

Trial 3: _____ cm   Trial 5: _____ cm
_____ sec   _____ sec
Trial 4: _____ cm
_____ sec

Did the addition of a specific word to the stimulus increase or decrease the reaction time?

_____

4. Perform the testing once again to investigate the subject's reaction to word association. As you drop the ruler, say a word—for example, *hot*. The subject is to respond with a word he or she associates with the stimulus word—for example, *cold*—catching the ruler while responding. If unable to make a word association, the subject must allow the ruler to pass through his or her fingers. Record the distance the ruler travels (or the number of seconds required for reaction) in five successful trials, as well as the number of times the ruler is not caught by the subject.

Trial 1: _____ cm   Trial 4: _____ cm
_____ sec   _____ sec
Trial 2: _____ cm   Trial 5: _____ cm
_____ sec   _____ sec
Trial 3: _____ cm
_____ sec

Number of times the subject did not catch the ruler:

_____

You should have noticed quite a large variation in reaction time in this series of trials. Why is this so?

_____

_____

_____

_____

ACTIVITY 10

Measuring Reaction Time Using BIOPAC®
Setting Up the Equipment

1. Connect the BIOPAC® unit to the computer and turn the computer **ON.**

2. Make sure the BIOPAC® unit is **OFF.**

3. Plug in the equipment (as shown in **Figure 21.7**).

• Hand switch—CH1

• Headphones—back of MP36/35 unit

4. Turn the BIOPAC® unit **ON.**

5. Start the BIOPAC® Student Lab program on the computer by double-clicking the icon on the desktop or by following your instructor's guidance.

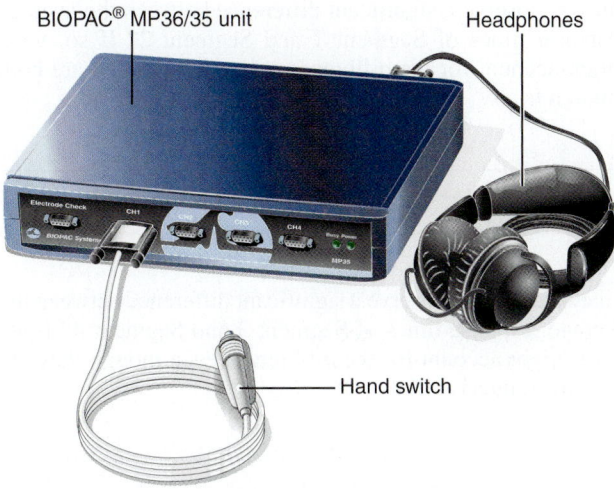

BIOPAC® MP36/35 unit   Headphones

Hand switch

Figure 21.7 Setting up the BIOPAC® equipment. Plug the headphones into the back of the MP36/35 data acquisition unit and the hand switch into Channel 1. Hand switch and headphones are shown connected to the MP36/35 unit.

6. Select lesson **L11-React-1** from the menu and click **OK.**

7. Type in a filename that will save this subject's data on the computer hard drive. You may want to use the subject's last name followed by React-1 (for example, SmithReact-1), then click **OK.**

Calibrating the Equipment

1. Seat the subject comfortably so that he or she cannot see the computer screen and keyboard.

2. Put the headphones on the subject and give the subject the hand switch to hold.

3. Tell the subject that he or she is to push the hand switch button when a "click" is heard.

4. Click **Calibrate,** and then click **OK** when the subject is ready.

5. Observe the recording of the calibration data, which should look like the waveforms in the calibration example **(Figure 21.8)**.

- If the data look very different, click **Redo Calibration** and repeat the steps above.

- If the data look similar, proceed to the next section.

Recording the Data

In this experiment, you will record four different segments of data. In Segments 1 and 2, the subject will respond to random click stimuli. In Segments 3 and 4, the subject will respond to click stimuli at fixed intervals (about 4 seconds). The director will click **Record** to initiate the Segment 1 recording, and **Resume** to initiate Segments 2, 3, and 4. The subject should focus only on responding to the sound.

Segment 1: Random Trial 1

1. Each time a sound is heard, the subject should respond by pressing the button on the hand switch as quickly as possible.

2. When the subject is ready, the director should click **Record** to begin the stimulus-response sequence. The recording will stop automatically after ten clicks.

- A triangular marker will be inserted above the data each time a "click" stimulus occurs.

- An upward-pointing "pulse" will be inserted each time the subject responds to the stimulus.

3. Observe the recording of the data, which should look similar to the data-recording example **(Figure 21.9)**.

- If the data look very different, click **Redo** and repeat the steps above.

- If the data look similar, move on to recording the next segment.

Segment 2: Random Trial 2

1. Each time a sound is heard, the subject should respond by pressing the button on the hand switch as quickly as possible.

2. When the subject is ready, the director should click **Resume** to begin the stimulus-response sequence. The recording will stop automatically after ten clicks.

3. Observe the recording of the data, which should again look similar to the data-recording example (Figure 21.9).

- If the data look very different, click **Redo** and repeat the steps above.

- If the data look similar, move on to recording the next segment.

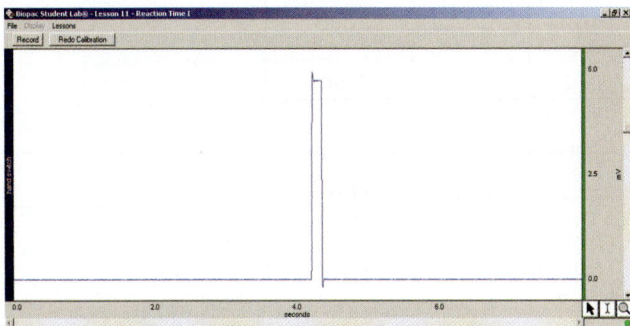

Figure 21.8 Example of waveforms during the calibration procedure.

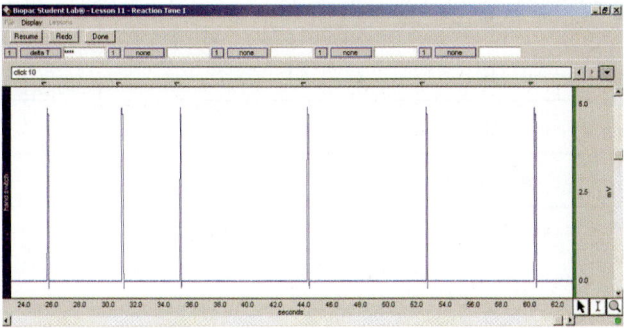

Figure 21.9 Example of waveforms during the recording of data.

Segment 3: Fixed Interval Trial 3

1. Repeat the steps for Segment 2 above.

Segment 4: Fixed Interval Trial 4

1. Repeat the steps for Segment 2 above.

2. If the data after this final segment are fine, click **Done.** A pop-up window will appear; to record from another subject select **Record from another subject,** and return to step 7 under Setting Up the Equipment. If continuing to the Data Analysis section, select **Analyze current data file** and proceed to step 2 in the Data Analysis section.

Data Analysis

1. If just starting the BIOPAC® program to perform data analysis, enter **Review Saved Data** mode and choose the file with the subject's reaction data (for example, SmithReact-1).

2. Observe that all ten reaction times are automatically calculated for each segment and are placed in the journal at the bottom of the computer screen.

3. Write the ten reaction times for each segment in the chart **Reaction Times.**

4. Delete the highest and lowest values of each segment, then calculate and record the average for the remaining eight data points.

5. When finished, exit the program by going to the **File** menu at the top of the page and clicking **Quit.**

Do you observe a significant difference between the average response times of Segment 1 and Segment 2? If so, what might account for the difference, even though they are both random trials?

_____

_____

Likewise, do you observe a significant difference between the average response times of Segment 3 and Segment 4? If so, what might account for the difference, even though they are both fixed interval trials?

_____

_____

Optional Activity with BIOPAC® Reaction Time Measurement

To expand the experiment, choose another variable to test. Response to visual cues may be tested, or you may have the subject change the hand used when clicking the hand switch button. Design the experiment, conduct the test, then record and analyze the data as described above. ■

	Reaction Times (seconds)			
	Random		**Fixed Interval**	
Stimulus #	**Segment 1**	**Segment 2**	**Segment 3**	**Segment 4**
1				
2				
3				
4				
5				
6				
7				
8				
9				
10				
Average				

21

Human Reflex Physiology

Name _____

Lab Time/Date _____

The Reflex Arc

1. Define *reflex.* _____

_____

2. Name five essential components of a reflex arc: _____, _____,

_____, _____, and _____

3. In general, what is the importance of reflex testing in a routine physical examination? _____

_____

_____

Somatic and Autonomic Reflexes

4. Use the key terms to complete the statements given below. (Some terms are used more than once.)

Key: a. abdominal reflex   d. corneal reflex   g. patellar reflex
         b. calcaneal tendon reflex   e. crossed-extensor reflex   h. plantar reflex
         c. ciliospinal reflex   f. gag reflex   i. pupillary light reflex

Reflexes classified as somatic reflexes include a _____, _____, _____, _____, _____, _____, and _____.

Of these, the stretch reflexes are _____ and _____, and the superficial cord reflexes are _____ and _____.

Reflexes classified as autonomic reflexes include _____ and _____.

5. Name two spinal cord–mediated reflexes. _____ and _____

_____

Name two somatic reflexes in which the higher brain centers participate. _____

and _____

6. Can the stretch reflex be elicited in a pithed animal (that is, an animal in which the brain has been destroyed)? _____

Explain your answer. _____

_____

7. Trace the reflex arc, naming efferent and afferent nerves, receptors, effectors, and integration centers, for the two reflexes listed. (Hint: Remember which nerve innervates the anterior thigh, and which nerve innervates the posterior thigh.)

patellar reflex: _____

_____

_____

calcaneal tendon reflex: _____

_____

_____

8. Three factors that influence the speed and effectiveness of reflex arcs were investigated in conjunction with patellar reflex testing—mental distraction, effect of simultaneous muscle activity in another body area, and fatigue.

Which of these factors increases the excitatory level of the spinal cord? _____

Which factor decreases the excitatory level of the muscles? _____

When the subject was concentrating on an arithmetic problem, did the change noted in the patellar reflex indicate that brain

activity is necessary for the patellar reflex or only that it may modify it? _____

_____

9. Name the division of the autonomic nervous system responsible for each of the reflexes listed.

ciliospinal reflex: _____   salivary reflex: _____

pupillary light reflex: _____

10. The pupillary light reflex, the crossed-extensor reflex, and the corneal reflex illustrate the purposeful nature of reflex activity. Describe the protective aspect of each.

pupillary light reflex: _____

corneal reflex: _____

crossed-extensor reflex: _____

_____

_____

_____

11. Was the pupillary consensual response contralateral or ipsilateral? _____

Why would such a response be of significant value in this particular reflex? _____

_____

12. Differentiate between the types of activities accomplished by somatic and autonomic reflexes. _____

_____

_____

_____

13. Several types of reflex activity were not investigated in this exercise. The most important of these are autonomic reflexes, which are difficult to illustrate in a laboratory situation. To rectify this omission, complete the following chart, using references as necessary.

Reflex	Organ involved	Receptors stimulated	Action
Micturition (urination)			
Defecation			
Carotid sinus			

Reaction Time of Intrinsic and Learned Reflexes

14. How do intrinsic and learned reflexes differ? _____

_____

15. Name at least three factors that may modify reaction time to a stimulus. _____

_____

16. In general, how did the response time for the learned activity performed in the laboratory compare to that for the simple

patellar reflex? _____

17. Did the response time without verbal stimuli decrease with practice? _____ Explain the reason for this.

_____

18. Explain, in detail, why response time increased when the subject had to react to a word stimulus.

_____

_____

19. When measuring reaction time in the BIOPAC® activity, was there a difference in reaction time when the stimulus was predictable versus unpredictable? Explain your answer.

_____

_____

General Sensation

MATERIALS

☐ Compound microscope
☐ Immersion oil
☐ Prepared slides (longitudinal sections) of lamellar corpuscles, tactile corpuscles, tendon organs, and muscle spindles
☐ Calipers or esthesiometer
☐ Small metric rulers
☐ Fine-point, felt-tipped markers (black, red, and blue)
☐ Large beaker of ice water; chipped ice
☐ Hot water bath set at 45°C; laboratory thermometer
☐ Towel
☐ Four coins (nickels or quarters)
☐ Three large finger bowls or 1000-ml beakers

MasteringA&P® For related exercise study tools, go to the Study Area of MasteringA&P. There you will find:
• Practice Anatomy Lab PAL
• PhysioEx PEx
• A&PFlix A&PFlix
• Practice quizzes, Histology Atlas, eText, Videos, and more!

OBJECTIVES

1. List the stimuli that activate general sensory receptors.
2. Define *exteroceptor, interoceptor,* and *proprioceptor.*
3. Recognize and describe the various types of general sensory receptors as studied in the laboratory, and list the function and locations of each.
4. Explain the tactile two-point discrimination test, and state its anatomical basis.
5. Define *tactile localization,* and describe how this ability varies in different areas of the body.
6. Define *adaptation,* and describe how this phenomenon can be demonstrated.
7. Discuss *negative afterimages* as they are related to temperature receptors.
8. Define *referred pain,* give an example for it, and define *projection.*

PRE-LAB QUIZ

1. Name one of the special senses. _____
2. Sensory receptors can be classified according to their source of stimulus. _____ are found close to the body surface and react to stimuli in the external environment.
   a. Exteroceptors
   b. Interoceptors
   c. Proprioceptors
   d. Visceroceptors
3. Circle True or False. General sensory receptors are widely distributed throughout the body and respond to, among other things, touch, pain, stretch, and changes in position.
4. Tactile corpuscles respond to light touch. Where would you expect to find tactile corpuscles?
   a. deep within the dermal layer of hairy skin
   b. in the dermal papillae of hairless skin
   c. in the hypodermis of hairless skin
   d. in the uppermost portion of the epidermis
5. Lamellar corpuscles respond to
   a. deep pressure and vibrations
   b. light touch
   c. pain and temperature
6. Circle True or False. A map of the sensory receptors for touch, heat, cold, and pain shows that they are not evenly distributed throughout the body.
7. Circle the correct underlined term. Two-point threshold / Tactile localization is the ability to determine where on the body the skin has been touched.
8. When a stimulus is applied for a prolonged period, the rate of receptor discharge slows, and conscious awareness of the stimulus declines. This phenomenon is known as
   a. accommodation
   b. adaptation
   c. adjustment
   d. discernment
9. Circle True or False. Pain is always perceived in the same area of the body that is receiving the stimulus.

(Text continues on next page.)

10. You will test referred pain in this activity by immersing the subject's
    a. face in ice water to test the cranial nerve response
    b. elbow in ice water to test the ulnar nerve response
    c. hand in ice water to test the axillary nerve response
    d. leg in ice water to test the sciatic nerve response

People are very responsive to *stimuli,* which are changes within a person's environment. Hold a sizzling steak before them and their mouths water. Flash your high beams in their eyes on the highway and they cuss. Tickle them and they giggle. These and many other stimuli continually assault us.

The body's **sensory receptors** react to stimuli. The tiny sensory receptors of the **general senses** react to touch, pressure, pain, heat, cold, stretch, vibration, and changes in position and are distributed throughout the body. In contrast to these widely distributed *general sensory receptors,* the receptors of the special senses are large, complex *sense organs* or small, localized groups of receptors. The **special senses** include vision, hearing, equilibrium, smell, and taste.

Sensory receptors may be classified by the type of stimulus they detect (for example touch, pain, or temperature), their structure (free nerve endings or complex encapsulated structures), or their body location. **Exteroceptors** react to stimuli in the external environment, and typically they are found close to the body surface. Exteroceptors include the simple cutaneous receptors in the skin and the highly specialized receptor structures of the special senses (the vision apparatus of the eye, and the hearing and equilibrium receptors of the ear, for example). **Interoceptors** or *visceroceptors* respond to stimuli arising within the body. Interoceptors are found in the internal visceral organs and include stretch receptors (in walls of hollow organs), chemoreceptors, and others. **Proprioceptors,** like interoceptors, respond to internal stimuli but are restricted to skeletal muscles, tendons, joints, ligaments, and connective tissue coverings of bones and muscles. They provide information about body movements and position by monitoring the degree of stretch of those structures.

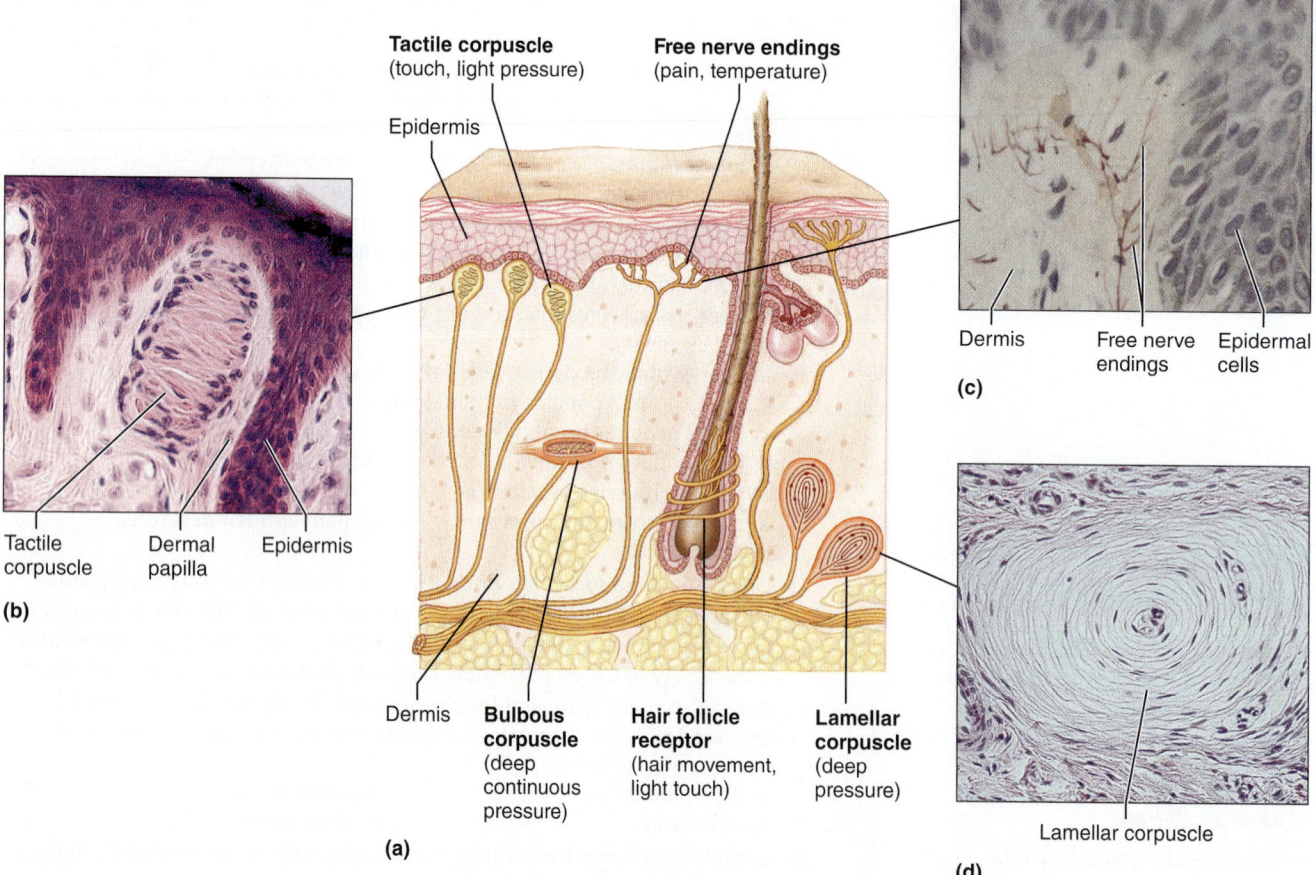

Tactile corpuscle
(touch, light pressure)

Free nerve endings
(pain, temperature)

Epidermis

Dermis   Free nerve endings   Epidermal cells

(c)

Tactile corpuscle   Dermal papilla   Epidermis

(b)

Dermis   **Bulbous corpuscle**
(deep continuous pressure)

Hair follicle receptor
(hair movement, light touch)

Lamellar corpuscle
(deep pressure)

(a)

Lamellar corpuscle

(d)

Figure 22.1 Examples of cutaneous receptors. Drawing **(a)** and photomicrographs **(b–d)**. **(a)** Free nerve endings, hair follicle receptor, tactile corpuscles, lamellar corpuscles, and bulbous corpuscle. Tactile (Merkel) discs are not illustrated. **(b)** Tactile corpuscle in a dermal papilla (300×). **(c)** Free nerve endings at dermal-epidermal junction (330×). **(d)** Cross section of a lamellar corpuscle in the dermis (220×).

22

The receptors of the special sense organs are complex and deserve considerable study. (The special senses are covered separately in Exercises 23–26). Only the anatomically simpler **general sensory receptors**—cutaneous receptors and proprioceptors—will be studied in this exercise.

Structure of General Sensory Receptors

You cannot become aware of changes in the environment unless your sensory neurons and their receptors are operating properly. Sensory receptors are either modified dendritic endings or specialized cells associated with the dendrites that are sensitive to specific environmental stimuli. They react to such stimuli by initiating a nerve impulse. Several histologically distinct types of general sensory receptors have been identified in the skin. (Their structures are depicted in **Figure 22.1**.)

Many references link receptor types to specific stimuli; however, one type of receptor can respond to several kinds of stimuli. Likewise, several different types of receptors can respond to similar stimuli. Certainly, intense stimulation of any of them is always interpreted as pain.

The least specialized of the cutaneous receptors are the **nonencapsulated (free) nerve endings** of sensory neurons (Figure 22.1c), which respond chiefly to pain and temperature. The pain receptors are widespread in the skin and make up a sizable portion of the visceral interoceptors. Certain free nerve endings associate with specific epidermal cells to form **tactile (Merkel) discs,** or entwine in hair follicles to form **hair follicle receptors.** Both tactile discs and hair follicle receptors function as light touch receptors.

The other cutaneous receptors are a bit more complex, and the nerve endings are **encapsulated** by connective tissue

capsules. **Tactile corpuscles** respond to light touch. They are located in the dermal papillae of hairless (glabrous) skin only (Figure 22.1b). **Bulbous corpuscles** appear to respond to deep pressure and stretch stimuli. **Lamellar corpuscles** are anatomically more distinctive than bulbous corpuscles and lie deepest in the dermis (Figure 22.1d). Lamellar corpuscles respond only when deep pressure is first applied. They are best suited to monitor high-frequency vibrations.

ACTIVITY 1

Studying the Structure of Selected Sensory Receptors

1. Obtain a compound microscope and histologic slides of lamellar and tactile corpuscles. Locate, under low power, a tactile corpuscle in the dermal layer of the skin. As mentioned above, these are usually found in the dermal papillae. Then switch to the oil immersion lens for a detailed study. Notice that the free nerve fibers within the capsule are aligned parallel to the skin surface. Compare your observations to the photomicrograph of a tactile corpuscle (Figure 22.1b).

2. Next observe a lamellar corpuscle located much deeper in the dermis. Try to identify the slender naked nerve ending in the center of the receptor and the heavy capsule of connective tissue surrounding it (which looks rather like an onion cut lengthwise). Also, notice how much larger the lamellar corpuscles are than the tactile corpuscles. Compare your observations to the photomicrograph of a lamellar corpuscle (Figure 22.1d).

3. Obtain slides of muscle spindles and tendon organs, the two major types of proprioceptors **(Figure 22.2)**. In the slide of **muscle spindles,** note that minute extensions

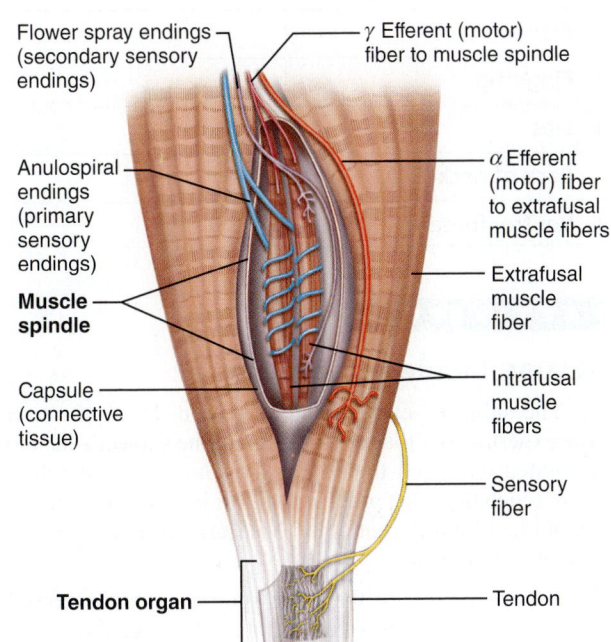

(a)

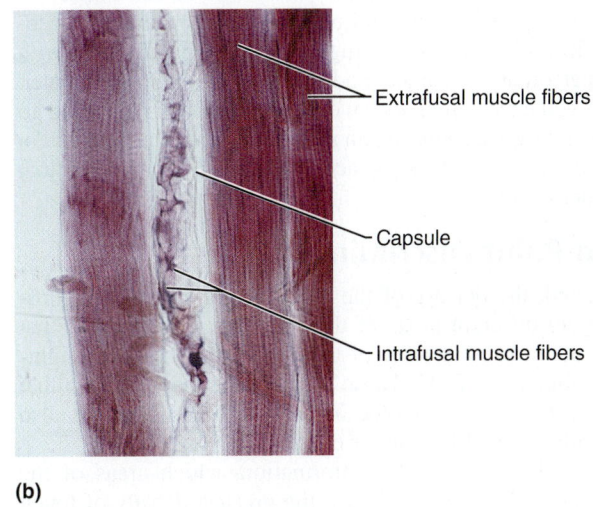

(b)

Figure 22.2 Proprioceptors. (a) Diagram of a muscle spindle and tendon organ. Myelin has been omitted from all nerve fibers for clarity. **(b)** Photomicrograph of a muscle spindle (80×).

of the nerve endings of the sensory neurons coil around specialized slender skeletal muscle cells called **intrafusal fibers.** The **tendon organs** are composed of nerve endings that ramify through the tendon tissue close to the attachment between muscle and tendon. Stretching of muscles or tendons excites these receptors, which then transmit impulses that ultimately reach the cerebellum for interpretation. Compare your observations to the proprioceptor art (Figure 22.2). ▄

Receptor Physiology

Sensory receptors act as **transducers,** changing environmental *stimuli* into nerve impulses that are relayed to the CNS. *Sensation* (awareness of the stimulus) and *perception* (interpretation of the meaning of the stimulus) occur in the brain. Nerve impulses from cutaneous receptors are relayed to the primary somatosensory cortex, where stimuli from different body regions form a body map. Therefore, each location on the body is represented by a specific cortical area. It is this cortical organization that allows us to know exactly where a sensation comes from. Further interpretation of the sensory information occurs in the somatosensory association cortex.

Four qualities of cutaneous sensations have traditionally been recognized: tactile (touch), heat, cold, and pain. Mapping these sensations on the skin has revealed that the sensory receptors for these qualities are not distributed uniformly. Instead, they have discrete locations and are characterized by clustering at certain points—**punctate distribution.**

The simple pain receptors, extremely important in protecting the body, are the most numerous. Touch receptors cluster where greater sensitivity is desirable, as on the hands and face. It may be surprising to learn that rather large areas of the skin are quite insensitive to touch because of a relative lack of touch receptors.

There are several simple experiments you can conduct to investigate the location and physiology of cutaneous receptors. In each of the following activities, work in pairs with one person as the subject and the other as the experimenter. After you have completed an experiment, switch roles and go through the procedures again so that all class members obtain individual results. Keep an accurate account of each test that you perform.

Two-Point Discrimination Test

As noted, the density of the touch receptors varies significantly in different areas of the body. In general, areas that have the greatest density of tactile receptors have a heightened ability to "feel." These areas correspond to areas that receive the greatest motor innervation; thus they are also typically areas of fine motor control.

On the basis of this information, which areas of the body do you *predict* will have the greatest density of touch receptors?

ACTIVITY 2

Determining the Two-Point Threshold

1. Using calipers or an esthesiometer and a metric ruler, test the ability of the subject to differentiate two distinct

sensations when the skin is touched simultaneously at two points. Beginning with the face, start with the caliper arms completely together. Gradually increase the distance between the arms, testing the subject's skin after each adjustment. Continue with this testing procedure until the subject reports that *two points* of contact can be felt. This measurement, the smallest distance at which two points of contact can be felt, is the **two-point threshold.**

2. Repeat this procedure on the back and palm of the hand, fingertips, lips, back of the neck, and ventral forearm. Record your results in the chart **Determining Two-Point Threshold.**

3. Which area has the smallest two-point threshold?

_____ ▄

Tactile Localization

Tactile localization is the ability to determine which portion of the skin has been touched. The tactile receptor field of the body periphery has a corresponding "touch" field in the brain's primary somatosensory cortex. Some body areas are well represented with touch receptors, allowing tactile stimuli to be localized with great accuracy, but touch-receptor density in other body areas allows only crude discrimination.

Determining Two-Point Threshold	
Body area tested	Two-point threshold (mm)
Face	
Back of hand	
Palm of hand	
Fingertip	
Lips	
Back of neck	
Ventral forearm	

ACTIVITY 3

Testing Tactile Localization

1. The subject's eyes should be closed during the testing. The experimenter touches the palm of the subject's hand with a pointed black felt-tipped marker. The subject should then try to touch the exact point with his or her own marker, which should be of a different color. Measure the error of localization in millimeters.

2. Repeat the test in the same spot twice more, recording the error of localization for each test. Average the results of the three determinations, and record it in the chart **Testing Tactile Localization.**

Testing Tactile Localization	
Body area tested	**Average error (mm)**
Palm of hand	
Fingertip	
Ventral forearm	
Back of hand	
Back of neck	

Does the ability to localize the stimulus improve the second

time? _____ The third time? _____

Explain. _____

_____

3. Repeat the preceding procedure on a fingertip, the ventral forearm, the back of a hand, and the back of the neck. Record the averaged results in the chart above.

4. Which area has the smallest error of localization?

_____

Adaptation of Sensory Receptors

The number of impulses transmitted by sensory receptors often changes both with the intensity of the stimulus and with the length of time the stimulus is applied. In many cases, when a stimulus is applied for a prolonged period, the rate of receptor discharge slows and conscious awareness of the stimulus declines or is lost until some type of stimulus change occurs. This phenomenon is referred to as **adaptation.** The touch receptors adapt particularly rapidly, which is highly desirable. Who, for instance, would want to be continually aware of the pressure of clothing on their skin? The simple experiments to be conducted next allow you to investigate the phenomenon of adaptation.

ACTIVITY 4

Demonstrating Adaptation of Touch Receptors

1. The subject's eyes should be closed. Obtain four coins. Place one coin on the anterior surface of the subject's forearm, and determine how long the sensation persists for the subject. Duration of the sensation:

_____ sec

2. Repeat the test, placing the coin at a different forearm location. How long does the sensation persist at the second location?

_____ sec

3.   After awareness of the sensation has been lost at the second site, stack three more coins atop the first one.

Does the pressure sensation return? _____

If so, for how long is the subject aware of the pressure in this instance?

_____ sec

Are the same receptors being stimulated when the four coins,

rather than the one coin, are used? _____

Explain._____

_____

4. To further illustrate the adaptation of touch receptors—in this case, the hair follicle receptors—gently and slowly bend one hair shaft with a pen or pencil until it springs back (away from the pencil) to its original position. Is the tactile sensation greater when the hair is being slowly bent or when it springs back?

_____

Why is the adaptation of the touch receptors in the hair follicles particularly important to a woman who wears her hair in a ponytail? If the answer is not immediately apparent, consider the opposite phenomenon: what would happen, in terms of sensory input from her hair follicles, if these receptors did not exhibit adaptation?

_____

_____

ACTIVITY 5

Demonstrating Adaptation of Temperature Receptors

Adaptation of the temperature receptors can be tested using some very unsophisticated methods.

1. Obtain three large finger bowls or 1000-ml beakers and fill the first with 45°C water. Have the subject immerse her or his left hand in the water and report the sensation. Keep the left hand immersed for 1 minute and then also immerse the right hand in the same bowl.

What is the sensation of the left hand when it is first immersed?

_____

What is the sensation of the left hand after 1 minute as compared to the sensation in the right hand just immersed?

_____

Had adaptation occurred in the left hand? _____

2. Rinse both hands in tap water, dry them, and wait 5 minutes before conducting the next test. Just before beginning the test, refill the finger bowl with fresh 45°C water, fill

a second with ice water, and fill a third with water at room temperature.

3. Place the *left* hand in the ice water and the *right* hand in the 45°C water. What is the sensation in each hand after 2 minutes as compared to the sensation perceived when the hands were first immersed?

_____

_____

Which hand seemed to adapt more quickly?

_____

4. After reporting these observations, the subject should then place both hands simultaneously into the finger bowl containing the water at room temperature. Record the sensa-

tion in the left hand: _____

The right hand:_____

The sensations that the subject experiences when both hands were put into room-temperature water are called **negative afterimages.** They are explained by the fact that sensations of heat and cold depend on the speed of heat loss or gain by the skin and differences in the temperature gradient. ▬

Referred Pain

Experiments on pain receptor localization and adaptation are commonly conducted in the laboratory. However, there are certain problems with such experiments. Pain receptors are densely distributed in the skin, and they adapt very little, if at all. This lack of adaptability is due to the protective function of the receptors. The sensation of pain often indicates tissue damage or trauma to body structures. Thus no attempt will be made in this exercise to localize the pain receptors or to prove their nonadaptability, since both would cause needless discomfort to those of you acting as subjects and would not add any additional insight.

However, the phenomenon of referred pain is easily demonstrated in the laboratory, and such experiments provide information that may be useful in explaining common examples of this phenomenon. **Referred pain** is a sensory experience in which pain is perceived as arising in one area of the body when in fact another, often quite remote area, is receiving the painful stimulus. Thus the pain is said to be "referred" to a different area. The phenomenon of **projection,** the process by which the brain refers sensations to their *usual* point of stimulation, provides the simplest explanation of such experiences. Many of us have experienced referred pain as a radiating pain in the forehead, sometimes referred to as "brain freeze," after quickly swallowing an ice-cold drink. Referred pain is important in many types of clinical diagnosis because damage to many visceral organs results in this phenomenon. For example, inadequate oxygenation of the heart muscle often results in pain being referred to the chest wall and left shoulder *(angina pectoris)*, and the reflux of gastric juice into the esophagus causes a sensation of intense discomfort in the thorax referred to as *heartburn*.

ACTIVITY 6

Demonstrating the Phenomenon of Referred Pain

Immerse the subject's elbow in a finger bowl containing ice water. In the chart **Demonstrating Referred Pain,** record the quality (such as discomfort, tingling, or pain) and the quality progression of the sensations he or she reports for the intervals indicated. The elbow should be removed from ice water after the 2-minute reading. The last recording is to occur 3 minutes after removal of the subject's elbow from the ice water.

Also record the location of the perceived sensations. The ulnar nerve, which serves the medial third of the hand, is involved in the phenomenon of referred pain experienced during this test. How does the localization of this referred pain correspond to the areas served by the ulnar nerve?

_____ ▬

Demonstrating Referred Pain		
Time of observation	Quality of sensation	Localization of sensation
On immersion		
After 1 min		
After 2 min		
3 min after removal		

GROUP CHALLENGE

Odd Receptor Out

Each group below contains four receptors. One of the listed receptors does not share a characteristic with the other three. Circle the receptor that doesn't belong with the others and explain why it is singled out. What characteristic is it missing? Remember to consider both structural and functional similarities. Sometimes there may be multiple reasons why the receptor doesn't belong with the others.

1. Which is the "odd receptor"?	Why it is the odd one out?
Bulbous corpuscle Lamellar corpuscle Tendon organ Tactile corpuscle	
2. Which is the "odd receptor"?	**Why it is the odd one out?**
Tendon organ Muscle spindle Hair follicle receptor Free nerve endings	

22

General Sensation

Structure of General Sensory Receptors

1. Differentiate between interoceptors and exteroceptors relative to location and stimulus source.

interoceptor: _____

exteroceptor: _____

2. A number of activities and sensations are listed in the chart below. For each, check whether the receptors would be exteroceptors or interoceptors; and then name the specific receptor types. (Because visceral receptors were not described in detail in this exercise, you need only indicate that the receptor is a visceral receptor if it falls into that category.)

Activity or sensation	Exteroceptor	Interoceptor	Specific receptor type
Backing into a sun-heated iron railing			
Someone steps on your foot			
Reading a book			
Leaning on your elbows			
Doing sit-ups			
The "too full" sensation			
Seasickness			

Receptor Physiology

3. Explain how the sensory receptors act as transducers. _____

_____

4. Define *stimulus*. _____

5. What was demonstrated by the two-point discrimination test? _____

_____

How well did your results correspond to your predictions? _____

_____

What is the relationship between the accuracy of the subject's tactile localization and the results of the two-point discrimi-

nation test? _____

6. Define *punctate distribution.*_____

_____

7. Several questions regarding general sensation are posed below. Answer each by placing your response in the appropriately numbered blanks to the right.

    1. Which cutaneous receptors are the most numerous?   1. _____

    2–3. Which two body areas tested were most sensitive to touch?   2-3. _____

    4–5. Which two body areas tested were least sensitive to touch?   4-5. _____

    6–8. Where would referred pain appear if the following organs were receiving painful stimuli: (6) gallbladder, (7) kidneys, and (8) appendix? (Use your textbook if necessary.)   6. _____

            7. _____

    9. Where was referred pain felt when the elbow was immersed in ice water during the laboratory experiment?   8. _____

            9. _____

    10. What region of the cerebrum interprets the kind and intensity of stimuli that cause cutaneous sensations?   10. _____

8. Define *adaptation of sensory receptors.*_____

_____

9. Why is it advantageous to have pain receptors that are sensitive to all vigorous stimuli, whether heat, cold, or pressure?

_____

Why is the nonadaptability of pain receptors important?_____

_____

10. Imagine yourself without any cutaneous sense organs. Why might this be very dangerous?_____

_____

_____

11. Define *referred pain.*_____

_____

What is the probable explanation for referred pain? (Consult your textbook or an appropriate reference if necessary.)

_____

_____

Special Senses: Anatomy of the Visual System

MATERIALS

- ☐ Chart of eye anatomy
- ☐ Dissectible eye model
- ☐ Prepared slide of longitudinal section of an eye showing retinal layers
- ☐ Compound microscope
- ☐ Preserved cow or sheep eye
- ☐ Dissecting instruments and tray
- ☐ Disposable gloves

OBJECTIVES

1. Identify the external, internal, and accessory anatomical structures of the eye on a model or appropriate image and list the function(s) of each; identify the structural components that are present in a preserved sheep or cow eye (if available).

2. Define *conjunctivitis, cataract,* and *glaucoma.*

3. Describe the cellular makeup of the retina.

4. Explain the difference between rods and cones with respect to visual perception and retinal localization.

5. Trace the visual pathway to the primary visual cortex, and indicate the effects of damage to various parts of this pathway.

PRE-LAB QUIZ

1. Name the mucous membrane that lines the internal surface of the eyelids and continues over the anterior surface of the eyeball.

   _____

2. How many extrinsic eye muscles are attached to the exterior surface of each eyeball?
   - a. three
   - b. four
   - c. five
   - d. six

3. The wall of the eye has three layers. The outermost fibrous layer is made up of the opaque white sclera and the transparent _____.
   - a. choroid
   - b. ciliary gland
   - c. cornea
   - d. lacrima

4. Circle the correct underlined term. The <u>aqueous humor</u> / <u>vitreous humor</u> is a clear, watery fluid that helps to maintain the intraocular pressure of the eye and provides nutrients for the avascular lens and cornea.

5. Circle True or False. At the optic chiasma, the fibers from the medial side of each eye cross over to the opposite side.

MasteringA&P® For related exercise study tools, go to the Study Area of MasteringA&P. There you will find:
- Practice Anatomy Lab PAL
- PhysioEx PEx
- A&PFlix *A&PFlix*
- Practice quizzes, Histology Atlas, eText, Videos, and more!

Anatomy of the Eye

External Anatomy and Accessory Structures

The adult human eye is a sphere measuring about 2.5 cm (1 inch) in diameter. Only about one-sixth of the eye's anterior surface is observable **(Figure 23.1)**; the remainder is enclosed and protected by a cushion of fat and the walls of the bony orbit.

The **lacrimal apparatus** consists of the lacrimal gland, lacrimal canaliculi, lacrimal sac, and the nasolacrimal duct. The **lacrimal glands** are situated superior

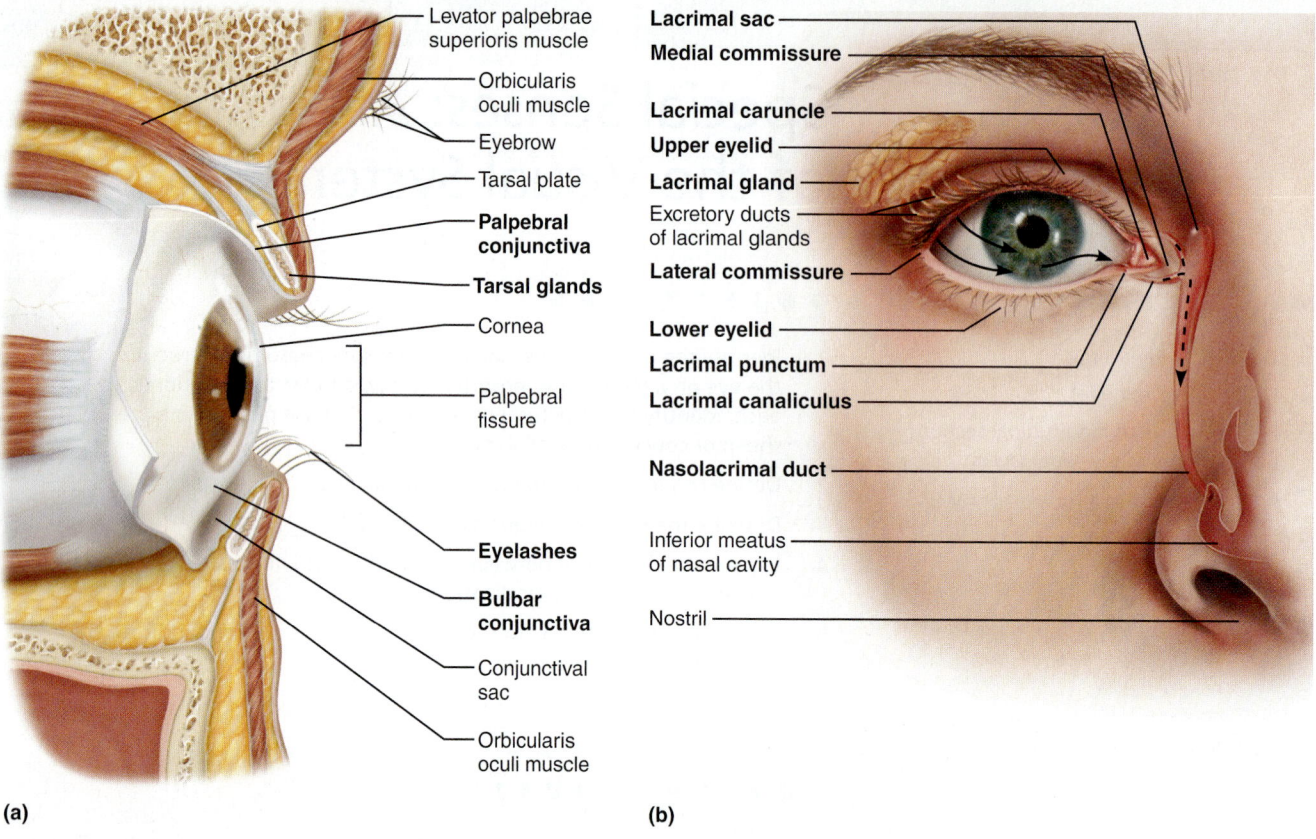

Levator palpebrae
superioris muscle

Orbicularis
oculi muscle

Eyebrow

Tarsal plate

**Palpebral
conjunctiva**

Tarsal glands

Cornea

Palpebral
fissure

Eyelashes

**Bulbar
conjunctiva**

Conjunctival
sac

Orbicularis
oculi muscle

(a)

Lacrimal sac

Medial commissure

Lacrimal caruncle

Upper eyelid

Lacrimal gland

Excretory ducts
of lacrimal glands

Lateral commissure

Lower eyelid

Lacrimal punctum

Lacrimal canaliculus

Nasolacrimal duct

Inferior meatus
of nasal cavity

Nostril

(b)

Figure 23.1 External anatomy of the eye and accessory structures. (a) Lateral view; some structures shown in sagittal section. **(b)** Anterior view with lacrimal apparatus.

to the lateral aspect of each eye. They continually release a dilute salt solution (tears) that flows onto the anterior surface of the eyeball through several small ducts. The tears flush across the eyeball and through the **lacrimal puncta,** the tiny openings of the **lacrimal canaliculi** medially, then into the **lacrimal sac,** and finally into the **nasolacrimal duct,** which empties into the nasal cavity. The lacrimal secretion also contains **lysozyme,** an antibacterial enzyme. Because it constantly flushes the eyeball, the lacrimal fluid cleanses and protects the eye surface as it moistens and lubricates it. As we age, our eyes tend to become dry due to decreased lacrimation, and thus are more vulnerable to bacterial invasion and irritation.

The anterior surface of each eye is protected by the **eyelids** or **palpebrae** (Figure 23.1). The medial and lateral junctions of the upper and lower eyelids are referred to as the **medial** and **lateral commissures** *(canthi),* respectively. The **lacrimal caruncle,** a fleshy raised area at the medial commissure, produces a whitish oily secretion. A mucous membrane, the **conjunctiva,** lines the internal surface of the eyelids (as the *palpebral conjunctiva*) and continues over the anterior surface of the eyeball to its junction with the corneal epithelium (as the *bulbar conjunctiva*). The conjunctiva secretes mucus, which aids in lubricating the eyeball. Inflammation of the conjunctiva, often accompanied by redness of the eye, is called **conjunctivitis.**

Projecting from the border of each eyelid is a row of short hairs, the **eyelashes.** The **ciliary glands,** modified sweat glands, lie between the eyelash hair follicles and help

lubricate the eyeball. Small sebaceous glands associated with the hair follicles and the larger **tarsal glands,** located posterior to the eyelashes, secrete an oily substance. An inflammation of one of the ciliary glands or a small oil gland is called a **sty.**

Six **extrinsic eye muscles** attached to the exterior surface of each eyeball control eye movement and make it possible for the eye to follow a moving object. (The names and positioning of these extrinsic muscles are noted in **Figure 23.2**). Their actions are given in the chart (Figure 23.2c).

ACTIVITY 1

Identifying Accessory Eye Structures

Using a chart of eye anatomy or the art of the extrinsic eye muscles (Figure 23.1), observe the eyes of another student, and identify as many of the accessory structures as possible. Ask the student to look to the left. Which extrinsic eye muscles are responsible for this action?

Right eye: _____

Left eye: _____

Internal Anatomy of the Eye

Anatomically, the wall of the eye is constructed of three layers **(Figure 23.3)**. The outermost **fibrous layer** is a protective layer composed of dense avascular connective tissue.

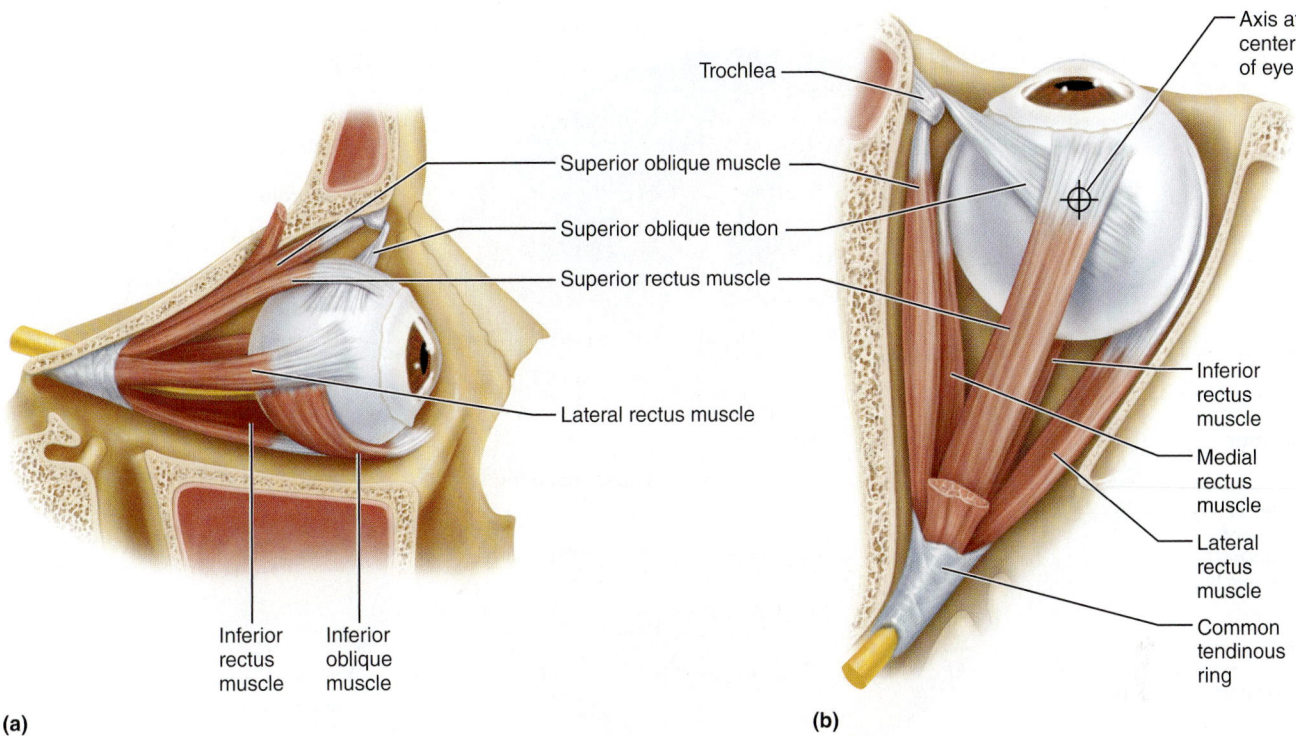

Figure 23.2 Extrinsic muscles of the eye. (a) Lateral view of the right eye.
(b) Superior view of the right eye. **(c)** Summary of actions of the extrinsic eye muscles
and cranial nerves that control them.

Muscle	Action	Controlling cranial nerve
Lateral rectus	Moves eye laterally	VI (abducens)
Medial rectus	Moves eye medially	III (oculomotor)
Superior rectus	Elevates eye and turns it medially	III (oculomotor)
Inferior rectus	Depresses eye and turns it medially	III (oculomotor)
Inferior oblique	Elevates eye and turns it laterally	III (oculomotor)
Superior oblique	Depresses eye and turns it laterally	IV (trochlear)

It has two obviously different regions: The opaque white **sclera** forms the bulk of the fibrous layer and is observable anteriorly as the "white of the eye." Its anteriormost portion is modified structurally to form the transparent **cornea,** through which light enters the eye.

The middle layer is the **vascular layer,** also called the *uvea.* Its posteriormost part, the **choroid,** is a blood-rich nutritive region containing a dark pigment that prevents light scattering within the eye. Anteriorly, the choroid is modified to form the **ciliary body,** which is chiefly composed of *ciliary muscles,* which are smooth muscles important in controlling lens shape, and *ciliary processes.* The ciliary processes secrete aqueous humor. The most anterior part of the vascular layer is the pigmented **iris.** The iris is incomplete, resulting in a rounded opening, the **pupil,** through which light passes.

The iris is composed of circularly and radially arranged smooth muscle fibers and acts as a reflexively activated diaphragm to regulate the amount of light entering the eye. In close vision and bright light, the sphincter pupillae (circular muscles) of the iris contract, and the pupil constricts. In distant vision and in dim light, the dilator pupillae (radial muscles) contract, enlarging (dilating) the pupil and allowing more light to enter the eye.

Together the sphincter pupillae and dilator pupillae muscles of the iris and the ciliary muscles are the *intrinsic muscles* of the eye, controlled by the autonomic nervous system.

The innermost **sensory layer** of the eye is the delicate, two-layered **retina** (Figure 23.3 and **Figure 23.4**). The outer **pigmented layer** abuts the choroid and extends anteriorly to cover the ciliary body and the posterior side of the iris. The pigment cells, like those of the choroid, absorb light and prevent it from scattering in the eye. They also participate in photoreceptor cell renewal by acting as phagocytes, and they store vitamin A needed by the photoreceptor cells. The transparent inner **neural layer** extends anteriorly only to the ciliary body. It contains the photoreceptors, *rods* and *cones,* which begin the chain of electrical events that ultimately result in the

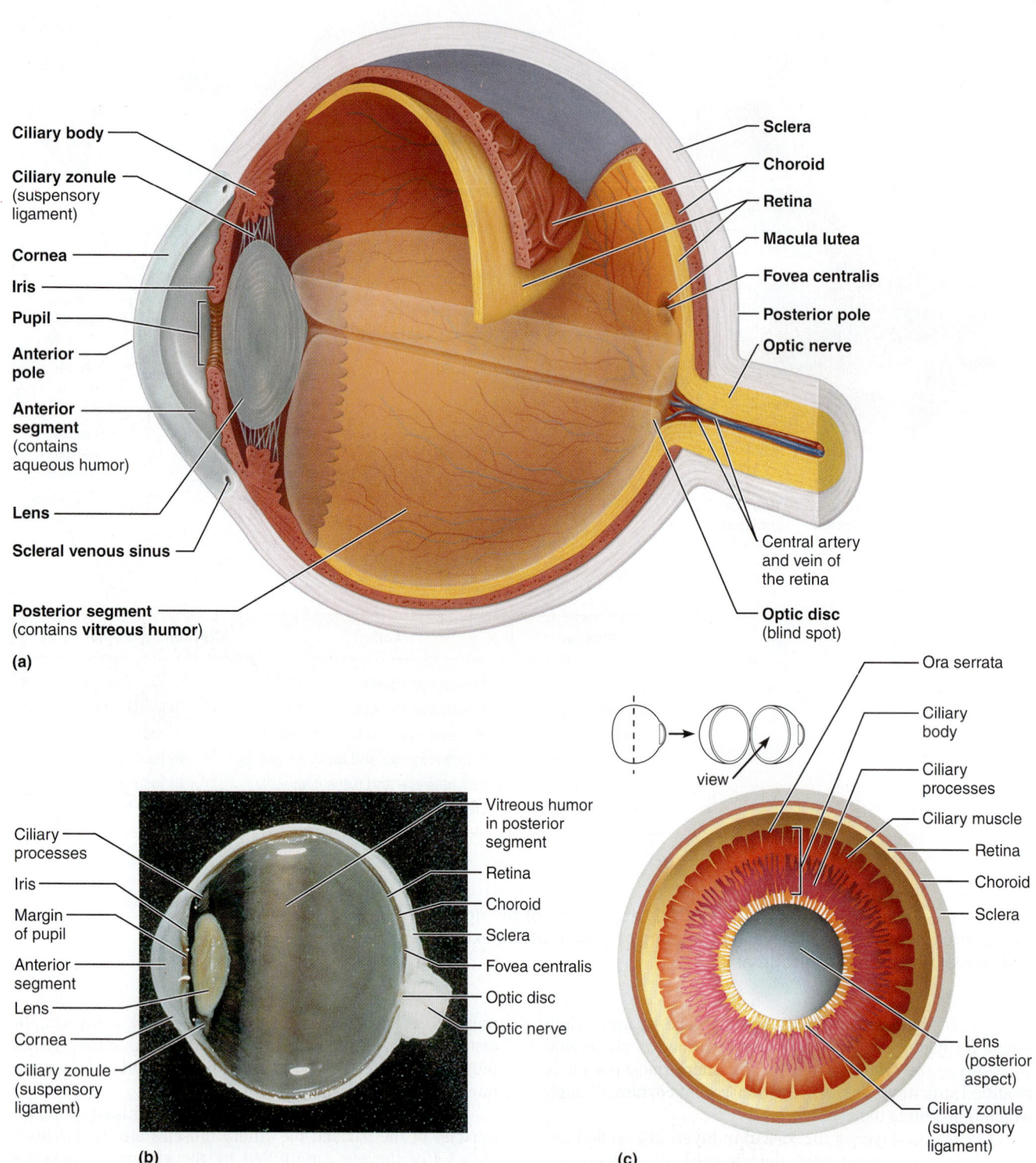

Figure 23.3 Internal anatomy of the eye. (a) Diagram of sagittal section of the eye. The vitreous humor is illustrated only in the bottom half of the eyeball. **(b)** Photograph of the human eye. **(c)** Posterior view of anterior half of the eye.

transduction of light energy into nerve impulses that are transmitted to the primary visual cortex of the brain. Vision is the result. The photoreceptor cells are distributed over the entire neural retina, except where the optic nerve leaves the eyeball. This site is called the **optic disc,** or *blind spot,* and is located in a weak spot in the **fundus** (posterior wall). Lateral to each

blind spot, and directly posterior to the lens, is an area called the **macula lutea** ("yellow spot"), an area of high cone density. In its center is the **fovea centralis,** a tiny pit about 0.4 mm in diameter, which contains only cones and is the area of greatest visual acuity. Focusing for discriminative vision occurs in the fovea centralis.

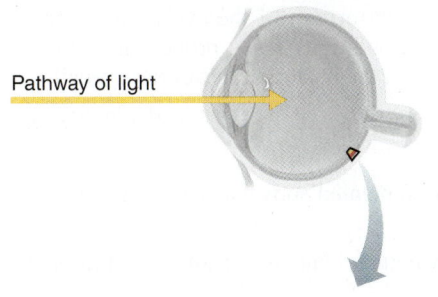

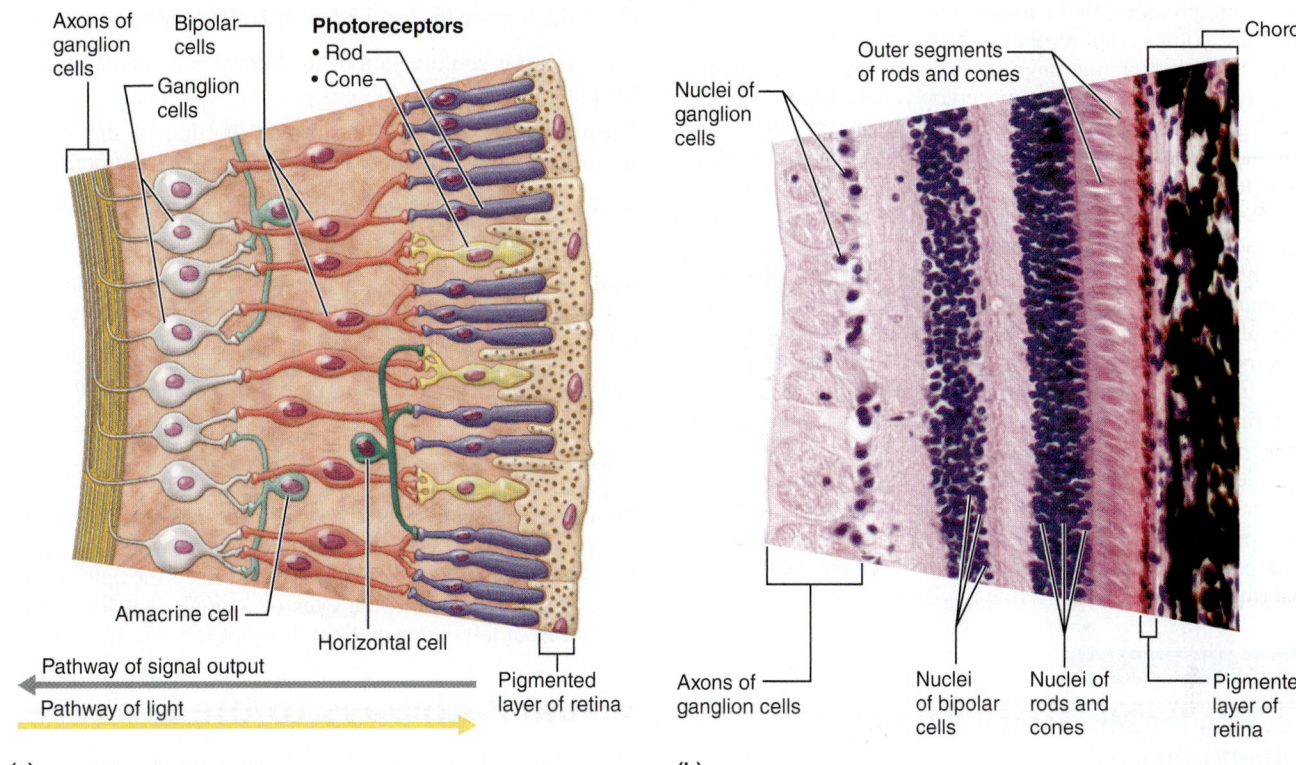

Figure 23.4 Microscopic anatomy of the retina. (a) Diagram of cells of the neural retina. Note the pathway of light through the retina. Neural signals (output of the retina) flow in the opposite direction. **(b)** Photomicrograph of the retina (140×).

Light entering the eye is focused on the retina by the **lens,** a flexible crystalline structure held vertically in the eye's interior by the **ciliary zonule** *(suspensory ligament)* attached to the ciliary body. Activity of the ciliary muscle, which accounts for the bulk of ciliary body tissue, changes lens thickness to allow light to be properly focused on the retina.

In the elderly the lens becomes increasingly hard and opaque. **Cataracts,** which often result from this process, cause vision to become hazy or entirely obstructed. ✚

The lens divides the eye into two segments: the **anterior segment** anterior to the lens, which contains a clear watery fluid called the **aqueous humor,** and the **posterior segment** behind the lens, filled with a gel-like substance, the **vitreous humor,** or **vitreous body.** The anterior segment is further divided into **anterior** and **posterior chambers,** located before and after the iris, respectively. The aqueous humor is continually formed by the capillaries of the **ciliary processes** of the ciliary body. It helps to maintain the intraocular

pressure of the eye and provides nutrients for the avascular lens and cornea. The aqueous humor is reabsorbed into the **scleral venous sinus.** The vitreous humor provides the major internal reinforcement of the posterior part of the eyeball, and helps to keep the retina pressed firmly against the wall of the eyeball. It is formed *only* before birth.

Anything that interferes with drainage of the aqueous fluid increases intraocular pressure. When intraocular pressure reaches dangerously high levels, the retina and optic nerve are compressed, resulting in pain and possible blindness, a condition called **glaucoma.** ✚

ACTIVITY 2

Identifying Internal Structures of the Eye

Obtain a dissectible eye model and identify its internal structures described above. (As you work, also refer to Figure 23.3.) ▪

Microscopic Anatomy of the Retina

Cells of the retina include the pigment cells of the outer pigmented layer and the inner photoreceptors and neurons, which are in contact with the vitreous humor (see Figure 23.4). The inner neural layer is composed of three major populations of cells. These are, from outer to inner aspect, the **photoreceptors,** the **bipolar cells,** and the **ganglion cells.**

The **rods** are the specialized receptors for dim light. Visual interpretation of their activity is in gray tones. The **cones** are color receptors that permit high levels of visual acuity, but they function only under conditions of high light intensity; thus, for example, no color vision is possible in moonlight. The fovea contains only cones, the macula contains mostly cones, and from the edge of the macula to the retina periphery, cone density declines gradually. By contrast, rods are most numerous in the periphery, and their density decreases as the macula is approached.

Light must pass through the ganglion cell layer and the bipolar cell layer to reach and excite the rods and cones. As a result of a light stimulus, the photoreceptors undergo changes in their membrane potential that influence the bipolar cells. These in turn stimulate the ganglion cells, whose axons leave the retina in the tight bundle of fibers known as the **optic nerve** (Figure 23.3). The retinal layer is thickest where the optic nerve attaches to the eyeball because an increasing number of ganglion cell axons converge at this point. It thins as it approaches the ciliary body. In addition to these three major cell types, the retina also contains horizontal cells and amacrine cells, which play a role in visual processing.

ACTIVITY 3

Studying the Microscopic Anatomy of the Retina

Use a compound microscope to examine a histologic slide of a longitudinal section of the eye. Identify the retinal layers by comparing your view to the photomicrograph (Figure 23.4b). ▬

DISSECTION:
The Cow (Sheep) Eye

1. Obtain a preserved cow or sheep eye, dissecting instruments, and a dissecting tray. Don disposable gloves.

2. Examine the external surface of the eye, noting the thick cushion of adipose tissue. Identify the optic nerve (cranial nerve II) as it leaves the eyeball, the remnants of the extrinsic eye muscles, the conjunctiva, the sclera, and the cornea. The normally transparent cornea is opalescent or opaque if the eye has been preserved. (Refer to **Figure 23.5** as you work.)

3. Trim away most of the fat and connective tissue, but leave the optic nerve intact. Holding the eye with the cornea facing downward, carefully make an incision with a sharp scalpel into the sclera about 6 mm (¼ inch) above the cornea. (The sclera of the preserved eyeball is *very* tough, so you will have to apply substantial pressure to penetrate it.) Using scissors, complete the incision around the circumference of the eyeball paralleling the corneal edge.

4. Carefully lift the anterior part of the eyeball away from the posterior portion. Conditions being proper, the vitreous body should remain with the posterior part of the eyeball.

5. Examine the anterior part of the eye, and identify the following structures:

Ciliary body: Black pigmented body that appears to be a halo encircling the lens.

Lens: Biconvex structure that is opaque in preserved specimens.

Carefully remove the lens and identify the adjacent structures:

Iris: Anterior continuation of the ciliary body penetrated by the pupil.

Cornea: More convex anteriormost portion of the sclera; normally transparent but cloudy in preserved specimens.

6. Examine the posterior portion of the eyeball. Carefully remove the vitreous humor, and identify the following structures:

Retina: The neural layer of the retina appears as a delicate tan, probably crumpled membrane that separates easily from the pigmented choroid.

Note its point of attachment. What is this point called?

_____

Pigmented choroid coat: Appears iridescent in the cow or sheep eye owing to a special reflecting surface called the **tapetum lucidum.** This specialized surface reflects the light within the eye and is found in the eyes of animals that live under conditions of low-intensity light. It is not found in humans. ■

Visual Pathways to the Brain

The axons of the ganglion cells of the retina converge at the posterior aspect of the eyeball and exit from the eye as the optic nerve. At the **optic chiasma,** the fibers from the medial side of each eye cross over to the opposite side **(Figure 23.6)**. The fiber tracts thus formed are called the **optic tracts.** Each optic tract contains fibers from the lateral side of the eye on the same side and from the medial side of the opposite eye.

The optic tract fibers synapse with neurons in the **lateral geniculate body** of the thalamus, whose axons form the **optic radiation,** terminating in the **primary visual cortex** in the occipital lobe of the brain. Here they synapse with the cortical neurons, and visual interpretation occurs.

ACTIVITY 4

Predicting the Effects of Visual Pathway Lesions

After examining the visual pathway diagram (Figure 23.6a), determine what effects lesions in the following areas would have on vision:

In the right optic nerve: _____

_____

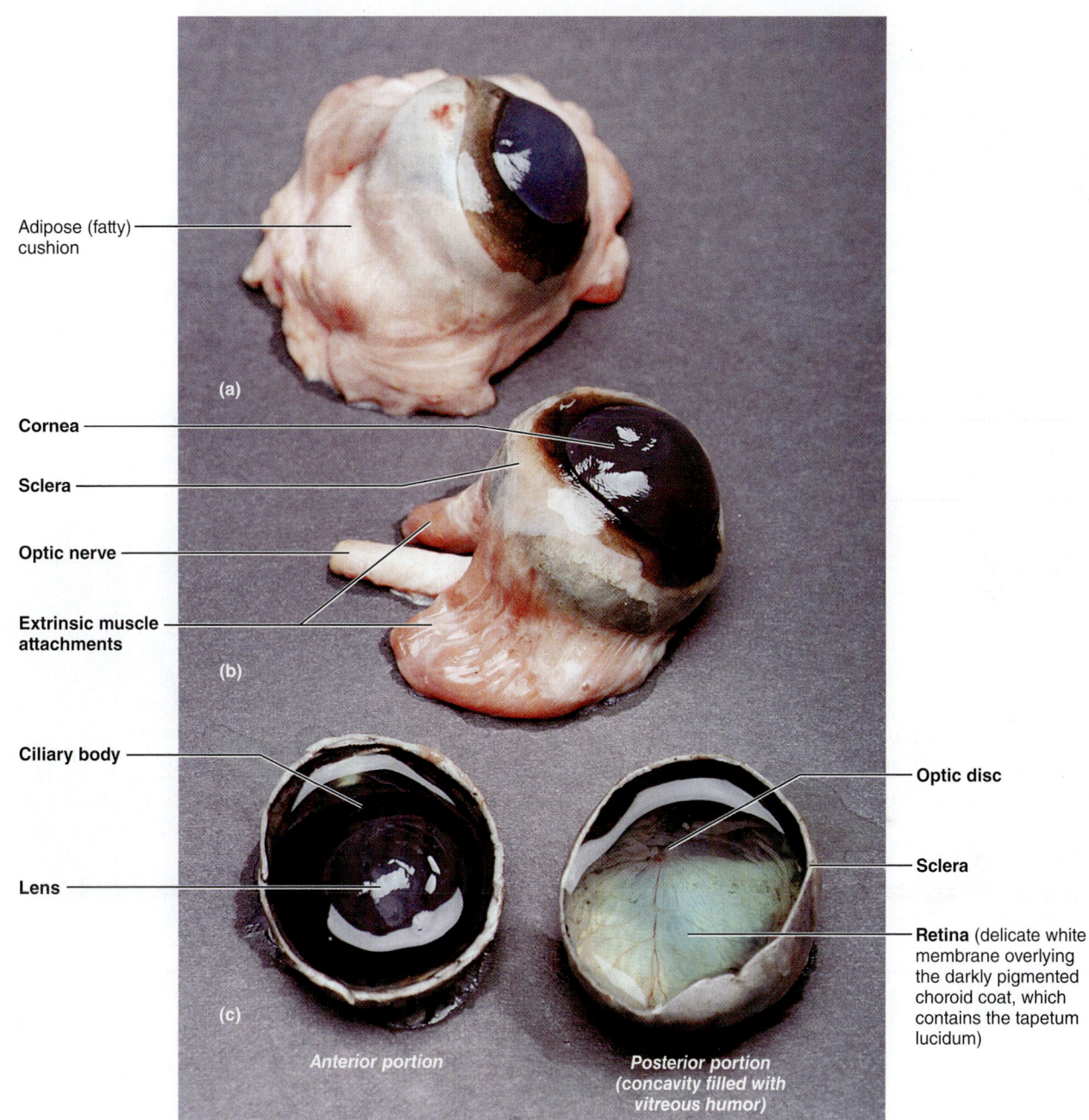

Adipose (fatty) cushion

(a)

Cornea

Sclera

Optic nerve

Extrinsic muscle attachments

(b)

Ciliary body

Lens

(c)

Anterior portion

Posterior portion (concavity filled with vitreous humor)

Optic disc

Sclera

Retina (delicate white membrane overlying the darkly pigmented choroid coat, which contains the tapetum lucidum)

Figure 23.5 Anatomy of the cow eye. (a) Cow eye (entire) removed from orbit (notice the large amount of fat cushioning the eyeball). **(b)** Cow eye (entire) with fat removed to show the extrinsic muscle attachments and optic nerve. **(c)** Cow eye cut along the frontal plane to reveal internal structures.

Through the optic chiasma: _____   In the right cerebral cortex (visual area): _____

_____   _____

In the left optic tract: _____

_____

23

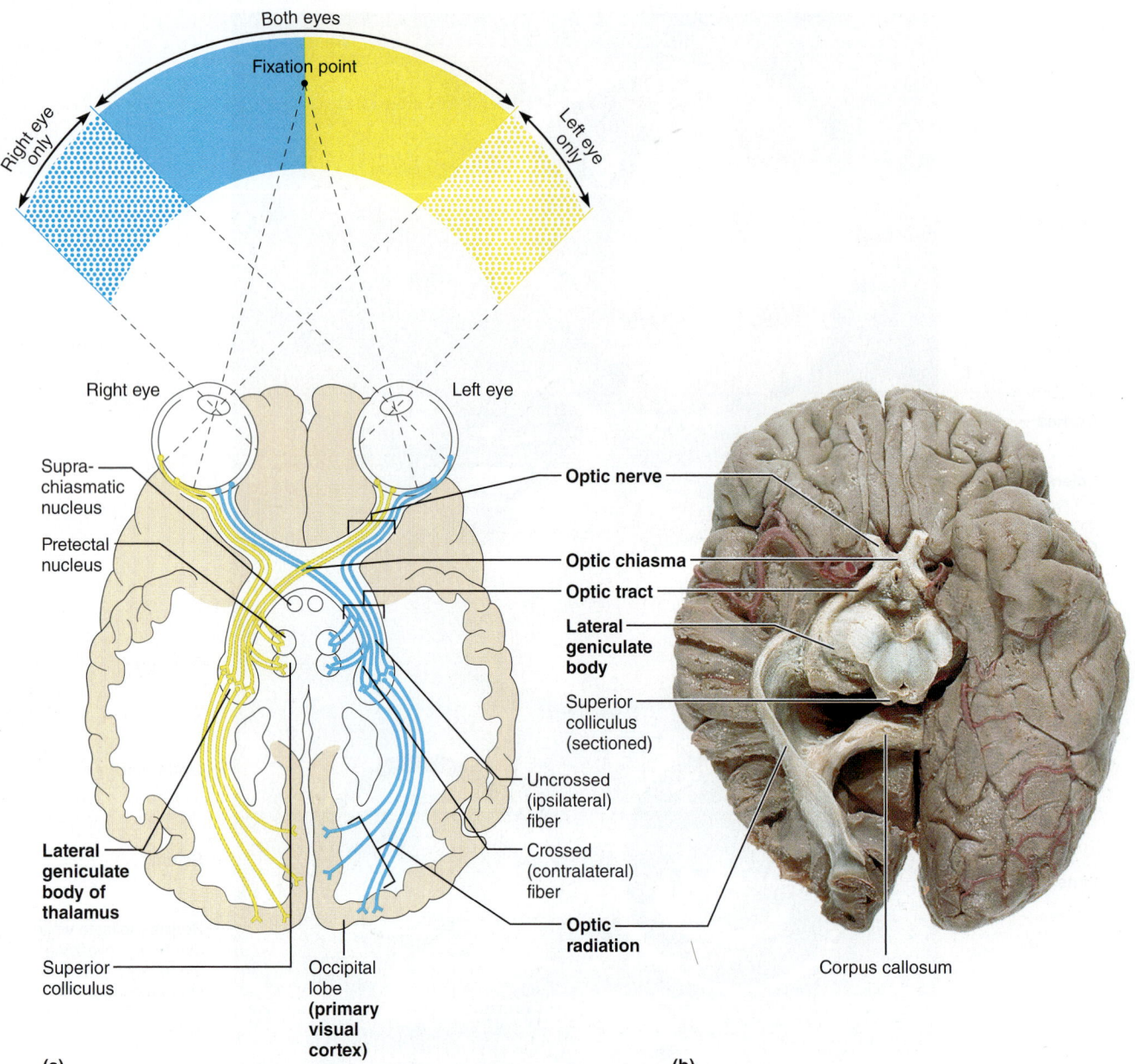

Figure 23.6 Inferior views of the visual pathway to the brain. (a) Diagram. Note that fibers from the lateral portion of each retinal field do not cross at the optic chiasma. **(b)** Photograph. Right side is dissected to reveal internal structures.

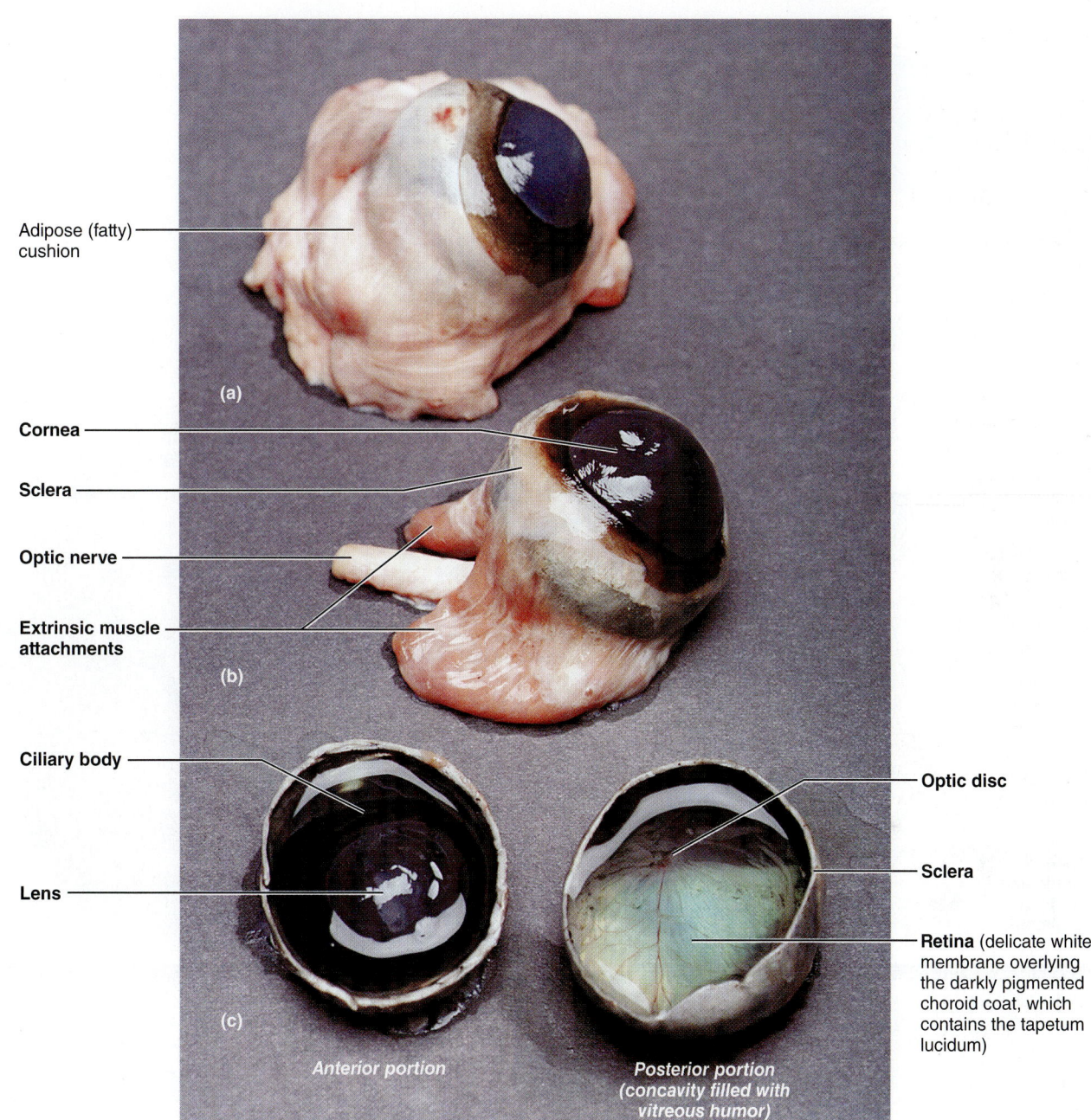

Figure 23.5 Anatomy of the cow eye. (a) Cow eye (entire) removed from orbit (notice the large amount of fat cushioning the eyeball). **(b)** Cow eye (entire) with fat removed to show the extrinsic muscle attachments and optic nerve. **(c)** Cow eye cut along the frontal plane to reveal internal structures.

Through the optic chiasma: _____

_____

In the left optic tract: _____

_____

In the right cerebral cortex (visual area): _____

_____

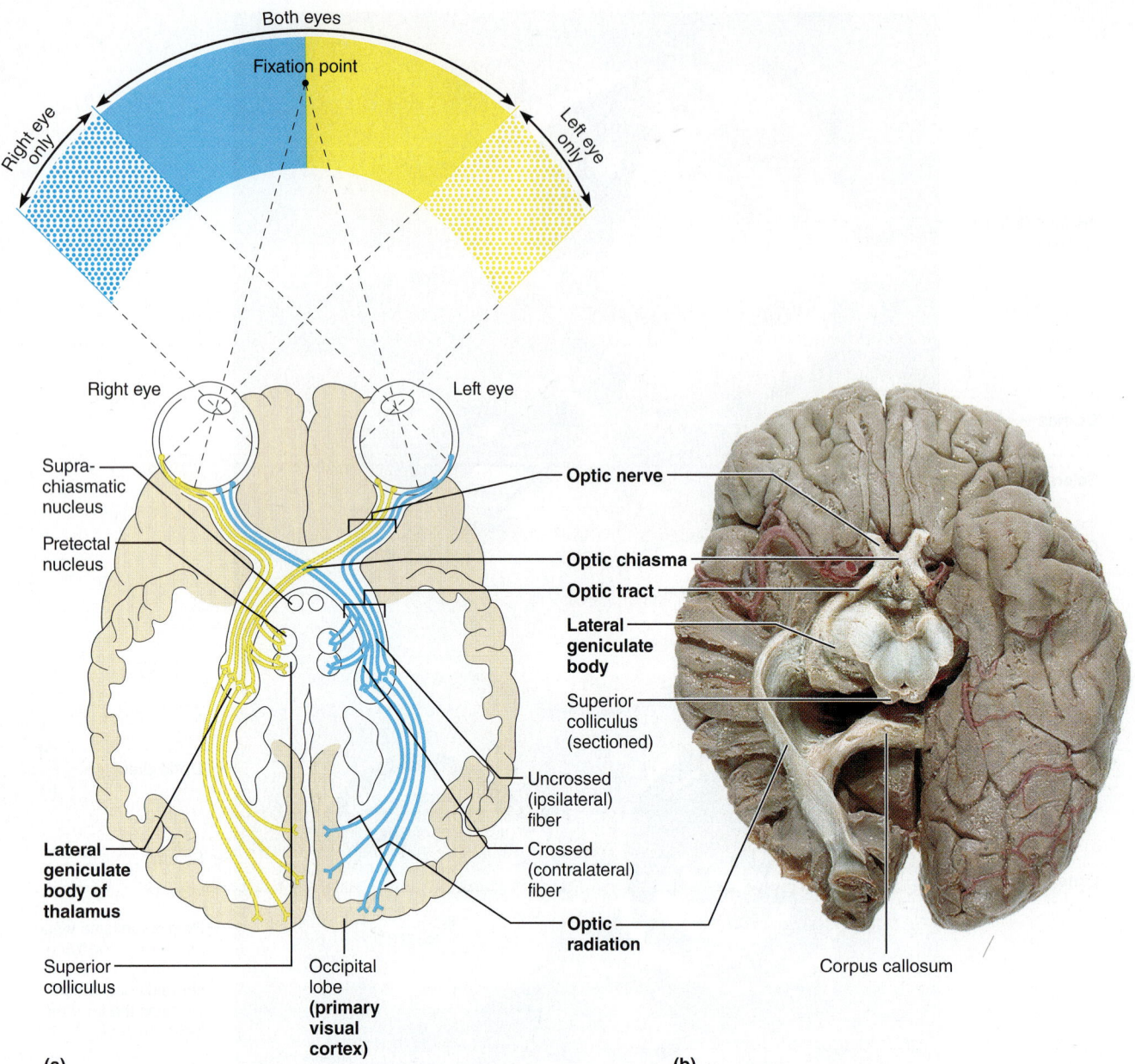

Figure 23.6 Inferior views of the visual pathway to the brain. (a) Diagram. Note that fibers from the lateral portion of each retinal field do not cross at the optic chiasma. **(b)** Photograph. Right side is dissected to reveal internal structures.

Special Senses: Visual Tests and Experiments

MATERIALS

- ☐ Metric ruler; meter stick
- ☐ Common straight pins
- ☐ Snellen eye chart, floor marked with chalk or masking tape to indicate 20-ft distance from posted Snellen chart
- ☐ Ishihara's color plates
- ☐ Two pencils
- ☐ Test tubes large enough to accommodate a pencil
- ☐ Laboratory lamp or penlight
- ☐ Ophthalmoscope (if available)

OBJECTIVES

1. Discuss the mechanism of image formation on the retina.
2. Define the following terms: *accommodation, astigmatism, emmetropic, hyperopia, myopia, refraction,* and *presbyopia,* and describe several simple visual tests to which the terms apply.
3. Discuss the benefits of binocular vision.
4. Define *convergence* and discuss the importance of the pupillary and convergence reflexes.
5. State the importance of an ophthalmoscopic examination.

PRE-LAB QUIZ

1. Circle the correct underlined term. Photoreceptors are distributed over the entire neural retina, except where the optic nerve leaves the eyeball. This site is called the macula lutea / optic disc.
2. Circle True or False. People with difficulty seeing objects at a distance are said to have myopia.
3. A condition that results in the loss of elasticity of the lens and difficulty focusing on a close object is called
   a. myopia   c. hyperopia
   b. presbyopia   d. astigmatism
4. Photoreceptors of the eye include rods and cones. Which one is responsible for interpreting color; which can function only under conditions of high light intensity?

   _____

5. Circle the correct underlined term. Extrinsic / Intrinsic eye muscles are controlled by the autonomic nervous system.

The Optic Disc

In this exercise, you will perform several visual tests and experiments focusing on the physiology of vision. The first test involves demonstrating the blind spot (optic disc), the site where the optic nerve exits the eyeball.

MasteringA&P® For related exercise study tools, go to the Study Area of MasteringA&P. There you will find:
- Practice Anatomy Lab PAL
- PhysioEx PEx
- A&PFlix *A&PFlix*
- Practice quizzes, Histology Atlas, eText, Videos, and more!

ACTIVITY 1

Demonstrating the Blind Spot

1. Hold the figure for the blind spot test **(Figure 24.1)** about 46 cm (18 inches) from your eyes. Close your left eye, and focus your right eye on the X, which should be positioned so that it is directly in line with your right eye. Move the figure slowly toward your face, keeping your right eye focused on the X. When the dot focuses on the blind spot, which lacks photoreceptors, it will disappear.

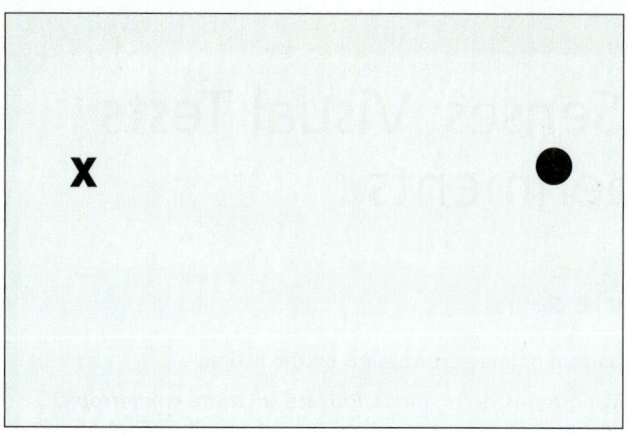

Figure 24.1 Blind spot test figure.

2. Have your laboratory partner record in metric units the distance at which this occurs. The dot will reappear as the figure is moved closer. Distance at which the dot disappears:

Right eye _____

Repeat the test for the left eye, this time closing the right eye and focusing the left eye on the dot. Record the distance at which the X disappears:

Left eye _____ ▬

Refraction, Visual Acuity, and Astigmatism

When light rays pass from one medium to another, their velocity, or speed of transmission, changes, and the rays are bent, or **refracted.** Thus the light rays in the visual field are refracted as they encounter the cornea, lens, and vitreous humor of the eye.

The refractive index (bending power) of the cornea and vitreous humor are constant. But the lens's refractive index can be varied by changing the lens's shape—that is, by making it more or less convex so that the light is properly converged and focused on the retina. The greater the lens convexity, or bulge, the more the light will be bent and the stronger the lens. Conversely, the less the lens convexity (the flatter it is), the less it bends the light.

In general, light from a distant source (over 6 m, or 20 feet) approaches the eye as parallel rays, and no change in lens convexity is necessary for it to focus properly on the retina. However, light from a close source tends to diverge, and the convexity of the lens must increase to make close vision possible. To achieve this, the ciliary muscle contracts, decreasing the tension on the ciliary zonule attached to the lens and allowing the elastic lens to "round up." Thus, a lens capable of bringing a *close* object into sharp focus is stronger (more convex) than a lens focusing on a more distant object. The ability of the eye to focus differentially for objects of near vision (less than 6 m, or 20 feet) is called **accommodation.** It should be noted that the image formed on the retina as a result of the refractory activity of the lens **(Figure 24.2)** is a

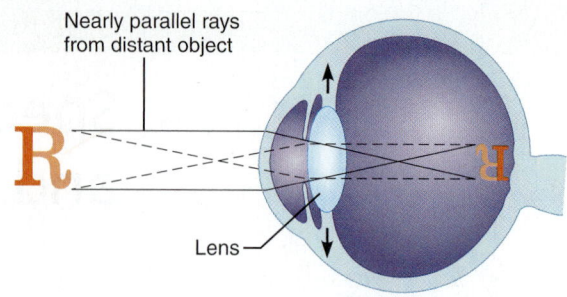

Figure 24.2 Refraction and real images. The refraction of light in the eye produces a real image (reversed, inverted, and reduced) on the retina.

real image (reversed from left to right, inverted, and smaller than the object).

The normal, or **emmetropic, eye** is able to accommodate properly **(Figure 24.3a)**. However, visual problems may result (1) from lenses that are too strong or too "lazy" (over-converging and underconverging, respectively), (2) from structural problems such as an eyeball that is too long or too short to provide for proper focusing by the lens, or (3) from a cornea or lens with improper curvatures.

Individuals in whom the image normally focuses in front of the retina are said to have **myopia,** or near-sightedness (Figure 24.3b); they can see close objects without difficulty, but distant objects are blurred or seen indistinctly. Correction requires a concave lens, which causes the light reaching the eye to diverge.

If the image focuses behind the retina, the individual is said to have **hyperopia,** or farsightedness. Such persons have no problems with distant vision but need glasses with convex lenses to augment the converging power of the lens for close vision (Figure 24.3c).

Irregularities in the curvatures of the lens and/or the cornea lead to a blurred vision problem called **astigmatism.** Cylindrically ground lenses, which compensate for inequalities in the curvatures of the refracting surfaces, are prescribed to correct the condition. ✚

Near-Point Accommodation

The elasticity of the lens decreases dramatically with age, resulting in difficulty in focusing for near or close vision. This condition is called **presbyopia**—literally, old vision. Lens elasticity can be tested by measuring the **near point of accommodation.** The near point of vision is about 10 cm from the eye in young adults. It is closer in children and farther in old age.

ACTIVITY 2

Determining Near Point of Accommodation

To determine your near point of accommodation, hold a common straight pin at arm's length in front of one eye. (If desired, the text in the lab manual can be used rather than a pin.) Slowly move the pin toward that eye until the pin image becomes distorted. Have your lab partner use a metric ruler to measure the distance in centimeters from your eye to the pin at this point, and record the distance below. Repeat the procedure for the other eye.

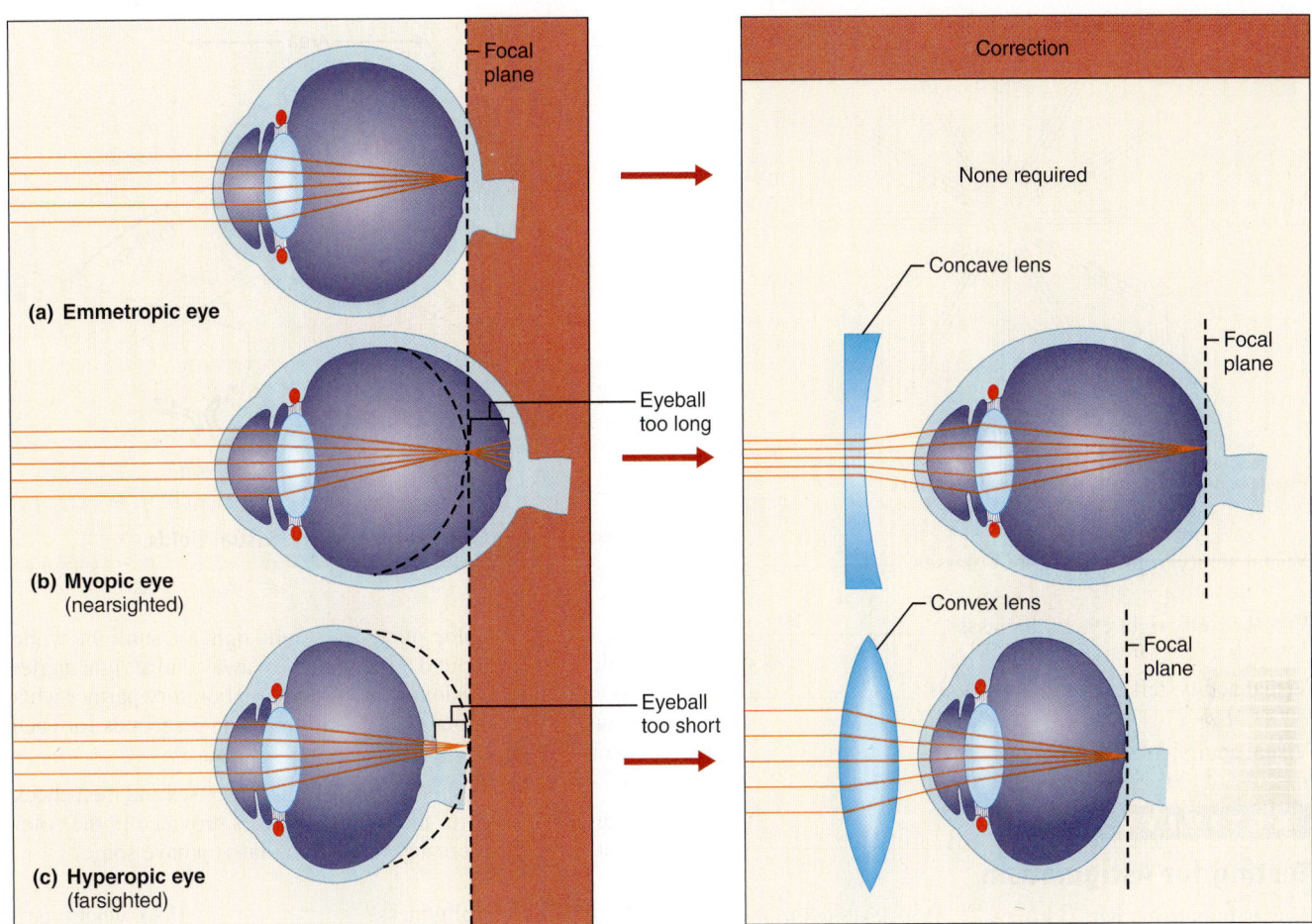

Figure 24.3 Problems of refraction. (a) In the emmetropic (normal) eye, light from both near and far objects is focused properly on the retina. **(b)** In a myopic eye, light from distant objects is brought to a focal point before reaching the retina. It then diverges. Applying a concave lens focuses objects properly on the retina. **(c)** In the hyperopic eye, light from a near object is brought to a focal point behind the retina. Applying a convex lens focuses objects properly on the retina. The refractory effect of the cornea is ignored here.

Near point for right eye: _____

Near point for left eye: _____ ▬

Visual Acuity

Visual acuity, or sharpness of vision, is generally tested with a Snellen eye chart, which consists of letters of various sizes printed on a white card. This test is based on the fact that letters of a certain size can be seen clearly by eyes with normal vision at a specific distance. The distance at which the normal, or emmetropic, eye can read a line of letters is printed at the end of that line.

ACTIVITY 3

Testing Visual Acuity

1. Have your partner stand 6 m (20 feet) from the posted Snellen eye chart and cover one eye with a card or hand. As your partner reads each consecutive line aloud, check

for accuracy. If this individual wears glasses, give the test twice—first with glasses off and then with glasses on. *Do not remove contact lenses, but note that they were in place during the test.*

2. Record the number of the line with the smallest-sized letters read. If it is 20/20, the person's vision for that eye is normal. If it is 20/40, or any ratio with a value less than one, he or she has less than the normal visual acuity. (Such an individual is myopic.) If the visual acuity is 20/15, vision is better than normal, because this person can stand at 6 m (20 feet) from the chart and read letters that are discernible by the normal eye only at 4.5 m (15 feet). Give your partner the number of the line corresponding to the smallest letters read, to record in step 4.

3. Repeat the process for the other eye.

4. Have your partner test and record your visual acuity. If you wear glasses, the test results *without* glasses should be recorded first.

24

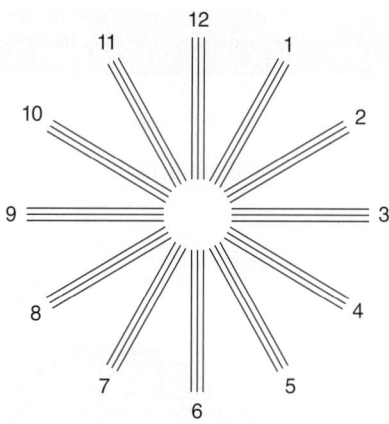

Figure 24.4 Astigmatism testing chart.

Visual acuity, right eye without glasses: _____

Visual acuity, right eye with glasses: _____

Visual acuity, left eye without glasses: _____

Visual acuity, left eye with glasses: _____ ■

ACTIVITY 4

Testing for Astigmatism

The astigmatism chart **(Figure 24.4)** is designed to test for defects in the refracting surface of the lens and/or cornea.

View the chart first with one eye and then with the other, focusing on the center of the chart. If all the radiating lines appear equally dark and distinct, there is no distortion of your refracting surfaces. If some of the lines are blurred or appear less dark than others, at least some degree of astigmatism is present.

Is astigmatism present in your left eye? _____

Right eye? _____ ■

Color Blindness

Ishihara's color plates are designed to test for deficiencies in the cones or color photoreceptor cells. There are three cone types, each containing a different light-absorbing pigment. One type primarily absorbs the red wavelengths of the visible light spectrum, another the blue wavelengths, and a third the green wavelengths. Nerve impulses reaching the brain from these different photoreceptor types are then interpreted (seen) as red, blue, and green, respectively. Interpretation of the intermediate colors of the visible light spectrum is a result of overlapping input from more than one cone type.

ACTIVITY 5

Testing for Color Blindness

1. Find the interpretation table that accompanies the Ishihara color plates, and prepare a sheet to record data for the test. Note which plates are patterns rather than numbers.

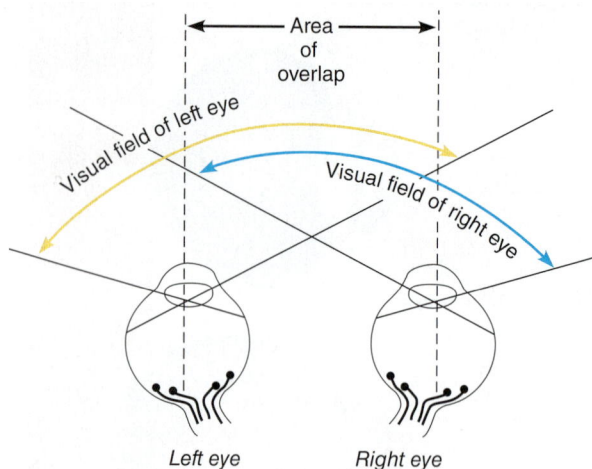

Figure 24.5 Overlapping of the visual fields.

2. View the color plates in bright light or sunlight while holding them about 0.8 m (30 inches) away and at right angles to your line of vision. Report to your laboratory partner what you see in each plate. Take no more than 3 seconds for each decision.

3. Your partner should record your responses and then check their accuracy with the correct answers provided in the color plate book. Is there any indication that you have some

degree of color blindness? _____ If so, what type?

_____

_____

Repeat the procedure to test your partner's color vision. ■

Binocular Vision

Humans, cats, predatory birds, and most primates are endowed with *binocular vision*. Their visual fields, each about 170 degrees, overlap to a considerable extent, and each eye sees a slightly different view **(Figure 24.5)**. The primary visual cortex fuses the slightly different images, providing **depth perception** (or **three-dimensional vision**). This provides an accurate means of locating objects in space.

In contrast, the eyes of rabbits, pigeons, and many other animals are on the sides of their head. Such animals see in two different directions and thus have a panoramic field of view and *panoramic vision*. A mnemonic device to keep these straight is "Eyes in the front—likes to hunt; eyes to the side—likes to hide."

ACTIVITY 6

Testing for Depth Perception

1. To demonstrate that a slightly different view is seen by each eye, perform the following simple experiment.

Close your left eye. Hold a pencil at arm's length directly in front of your right eye. Position another pencil directly beneath it and then move the lower pencil about half the distance toward you. As you move the lower pencil, make sure it remains in the *same plane* as the stationary pencil, so that the two pencils continually form a straight line. Then, without moving the pencils, close your right eye and open your left eye. Notice that with only the right eye open, the moving pencil stays in the same plane as the fixed pencil, but that when viewed with the left eye, the moving pencil is displaced laterally away from the plane of the fixed pencil.

2. To demonstrate the importance of two-eyed binocular vision for depth perception, perform this second simple experiment.

Have your laboratory partner hold a test tube erect about arm's length in front of you. With both eyes open, quickly insert a pencil into the test tube. Remove the pencil, bring it back close to your body, close one eye, and quickly and without hesitation insert the pencil into the test tube. *(Do not feel for the test tube with the pencil!)* Repeat with the other eye closed.

Was it as easy to dunk the pencil with one eye closed as with both eyes open?

_____   ▄

Eye Reflexes

Both intrinsic (internal) and extrinsic (external) muscles are necessary for proper eye functioning. The *intrinsic muscles,* controlled by the autonomic nervous system, are those of the ciliary body (which alters the lens curvature in focusing) and the sphincter pupillae and dilator pupillae muscles of the iris (which control pupillary size and thus regulate the amount of light entering the eye). The *extrinsic muscles* are the rectus and oblique muscles, which are attached to the eyeball exterior (see Figure 23.2). These muscles control eye movement and make it possible to keep moving objects focused on the fovea centralis. They are also responsible for **convergence,** or medial eye movements, which is essential for near vision. When convergence occurs, both eyes are directed toward the near object viewed. The extrinsic eye muscles are controlled by the somatic nervous system.

ACTIVITY 7

Demonstrating Reflex Activity of Intrinsic and Extrinsic Eye Muscles

Involuntary activity of both the intrinsic and extrinsic muscle types is brought about by reflex actions that can be observed in the following experiments.

Photopupillary Reflex

Sudden illumination of the retina by a bright light causes the pupil to constrict reflexively in direct proportion to the light intensity. This protective response prevents damage to the delicate photoreceptor cells.

Obtain a laboratory lamp or penlight. Have your laboratory partner sit with eyes closed and hands over his or her eyes. Turn on the light and position it so that it shines on the subject's right hand. After 1 minute, ask your partner to uncover and open the right eye. Quickly observe the pupil of that eye. What happens to the pupil?

_____

Shut off the light and ask your partner to uncover and open the opposite eye. What are your observations of the pupil?

_____

_____

Accommodation Pupillary Reflex

Have your partner gaze for approximately 1 minute at a distant object in the lab—*not* toward the windows or another light source. Observe your partner's pupils. Then hold some printed material 15 to 25 cm (6 to 10 inches) from his or her face, and direct him or her to focus on it.

How does pupil size change as your partner focuses on the printed material?

_____

_____

Explain the value of this reflex. _____

_____

_____

_____

Convergence Reflex

Repeat the previous experiment, this time using a pen or pencil as the close object to be focused on. Note the position of your partner's eyeballs while he or she gazes at the distant object, and then at the close object. Do they change position as the object of focus is changed?

_____ In what way? _____

_____

_____

_____

Explain the importance of the convergence reflex.

_____

_____

_____   ▄

Ophthalmoscopic Examination of the Eye (Optional)

The ophthalmoscope is an instrument used to examine the *fundus,* or eyeball interior, to determine visually the condition of the retina, optic disc, and internal blood vessels. Certain pathological conditions such as diabetes mellitus, arteriosclerosis, and degenerative changes of the optic nerve and retina can be detected by such an examination. The ophthalmoscope

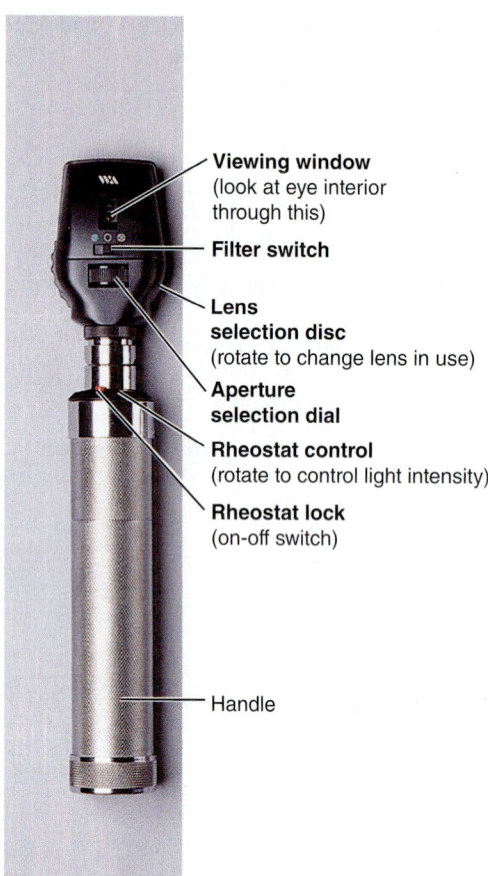

Viewing window
(look at eye interior through this)

Filter switch

Lens selection disc
(rotate to change lens in use)

Aperture selection dial

Rheostat control
(rotate to control light intensity)

Rheostat lock
(on-off switch)

Handle

(a)

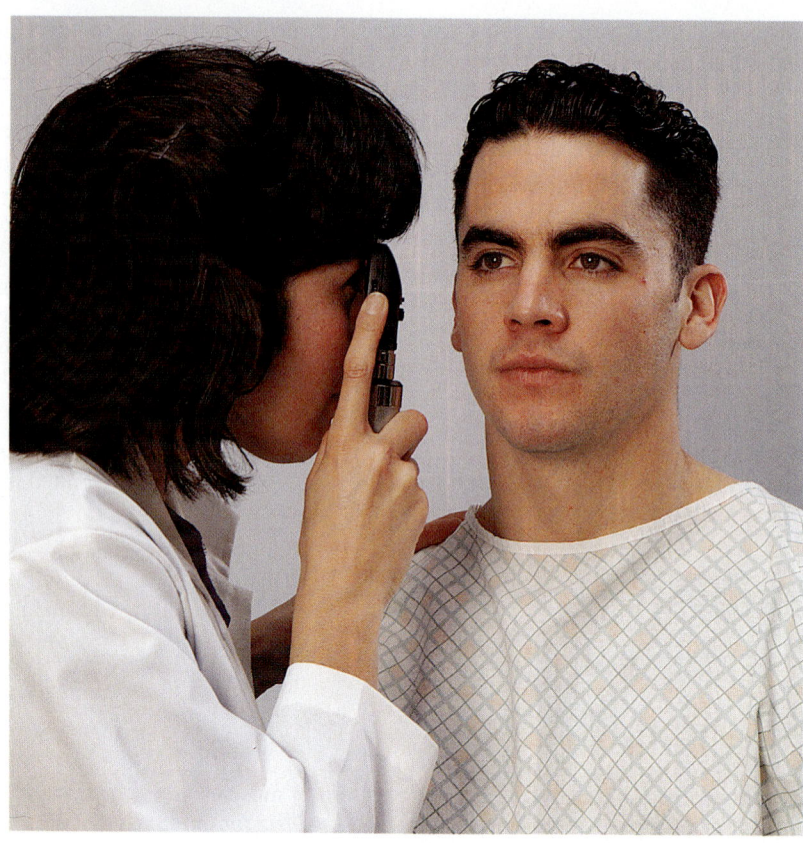

(b)

Figure 24.6 Structure and use of an ophthalmoscope. (a) Structure of an ophthalmoscope. **(b)** Proper position for beginning to examine the right eye with an ophthalmoscope.

consists of a set of lenses mounted on a rotating disc (the **lens selection disc**), a light source regulated by a **rheostat control,** and a mirror that reflects the light so that the eye interior can be illuminated **(Figure 24.6a)**.

The lens selection disc is positioned in a small slit in the mirror, and the examiner views the eye interior through this slit, appropriately called the **viewing window.** The focal length of each lens is indicated in diopters preceded by a plus (+) sign if the lens is convex and by a negative (−) sign if the lens is concave. When the zero (0) is seen in the **diopter window,** on the examiner side of the instrument, there is no lens positioned in the slit. The depth of focus for viewing the eye interior is changed by changing the lens.

The light is turned on by depressing the red **rheostat lock button** and then rotating the rheostat control in the clockwise direction. The **aperture selection dial** on the front of the instrument allows the nature of the light beam to be altered. The **filter switch,** also on the front, allows the choice of a green, unfiltered, or polarized light beam. Generally, green light allows for clearest viewing of the blood vessels in the eye interior and is most comfortable for the subject.

Once you have examined the ophthalmoscope and have become familiar with it, you are ready to conduct an eye examination.

ACTIVITY 8

Conducting an Ophthalmoscopic Examination

1. Conduct the examination in a dimly lit or darkened room with the subject comfortably seated and gazing straight ahead. To examine the right eye, sit face-to-face with the subject, hold the instrument in your right hand, and use your right eye to view the eye interior (Figure 24.6b). You may want to steady yourself by resting your left hand on the subject's shoulder. To view the left eye, use your left eye, hold the instrument in your left hand, and steady yourself with your right hand.

2. Begin the examination with the 0 (no lens) in position. Grasp the instrument so that the lens disc may be rotated with the index finger. Holding the ophthalmoscope about 15 cm (6 inches) from the subject's eye, direct the light into the pupil at a slight angle—through the pupil edge rather than directly through its center. You will see a red circular area that is the illuminated eye interior.

3. Move in as close as possible to the subject's cornea (to within 5 cm, or 2 inches) as you continue to observe the area.

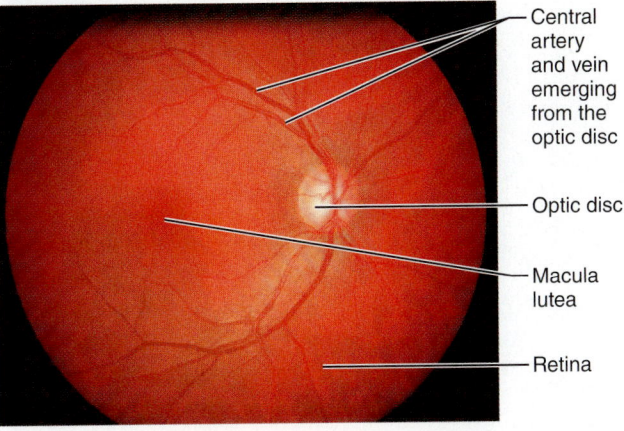

Central
artery
and vein
emerging
from the
optic disc

Optic disc

Macula
lutea

Retina

Figure 24.7 **Fundus (posterior wall) of right retina.**

Steady your instrument-holding hand on the subject's cheek if necessary. If both your eye and that of the subject are normal, the fundus can be viewed clearly without further adjustment of the ophthalmoscope. If the fundus cannot be focused, slowly rotate the lens disc counterclockwise until the fundus can be clearly seen. When the ophthalmoscope is correctly set, the fundus of the right eye should appear as in the photograph **(Figure 24.7)**. (**Note:** If a positive [convex] lens is required and your eyes are normal, the subject has hyperopia. If a negative [concave] lens is necessary to view the fundus and your eyes are normal, the subject is myopic.)

When the examination is proceeding correctly, the subject can often see images of retinal vessels in his own eye that appear rather like cracked glass. If you are unable to achieve a sharp focus or to see the optic disc, move medially or laterally and begin again.

4. Examine the optic disc for color, elevation, and sharpness of outline, and observe the blood vessels radiating from near its center. Locate the macula, lateral to the optic disc. It is a darker area in which blood vessels are absent, and the fovea appears to be a slightly lighter area in its center. The macula is most easily seen when the subject looks directly into the light of the ophthalmoscope.

 Do not examine the macula for longer than 1 second at a time.

5. When you have finished examining your partner's retina, shut off the ophthalmoscope. Change places with your partner (become the subject) and repeat steps 1–4. ▪

24

Special Senses: Visual Tests and Experiments

The Optic Disc, Refraction, Visual Acuity, and Astigmatism

1. Explain why vision is lost when light hits the blind spot. _____

2. Match the terms in column B with the descriptions in column A.

Column A

_____ 1. light bending

_____ 2. ability to focus for close (less than 20 feet) vision

_____ 3. normal vision

_____ 4. inability to focus well on close objects (farsightedness)

_____ 5. nearsightedness

_____ 6. blurred vision due to unequal curvatures of the lens or cornea

_____ 7. medial movement of the eyes during focusing on close objects

Column B

a. accommodation

b. astigmatism

c. convergence

d. emmetropia

e. hyperopia

f. myopia

g. refraction

3. Complete the following statements:

In farsightedness, the light is focused __1__ the retina. The lens required to treat myopia is a __2__ lens. The "near point" increases with age because the __3__ of the lens decreases as we get older. A convex lens, like that of the eye, produces an image that is upside down and reversed from left to right. Such an image is called a __4__ image.

1. _____

2. _____

3. _____

4. _____

4. Use terms from the key to complete the statements concerning near and distance vision. (Some choices will be used more than once.)

Key: a. contracted   b. decreased   c. increased   d. relaxed   e. taut

During distance vision, the ciliary muscle is _____, the ciliary zonule is _____, the convexity of the lens

is _____, and light refraction is _____. During close vision, the ciliary muscle is _____, the ciliary zonule is

_____, lens convexity is _____, and light refraction is _____.

5. Using your Snellen eye test results, answer the following questions.

Is your visual acuity normal, less than normal, or better than normal? _____

Explain your answer. _____

_____

Explain why each eye is tested separately when using the Snellen eye chart. _____

_____

Explain 20/40 vision. _____

_____

Explain 20/10 vision. _____

_____

6. Define *astigmatism*. _____

_____

How can it be corrected? _____

7. Define *presbyopia*. _____

_____

What causes it? _____

Color Blindness

8. To which wavelengths of light do the three cone types of the retina respond maximally?

_____, _____, and _____

9. How can you explain the fact that we see a great range of colors even though only three cone types exist?

_____

_____

Binocular Vision

10. Explain the difference between binocular and panoramic vision. _____

_____

_____

What is the advantage of binocular vision? _____

What factor(s) are responsible for binocular vision? _____

_____

Eye Reflexes

11. In the experiment on the convergence reflex, what happened to the position of the eyeballs as the object was moved closer

to the subject's eyes? _____

What extrinsic eye muscles control the movement of the eyes during this reflex? _____

What is the value of this reflex? _____

_____

12. In the experiment on the photopupillary reflex, what happened to the pupil of the eye exposed to light?

_____ What happened to the pupil of the nonilluminated eye? _____

Explanation? _____

_____

Ophthalmoscopic Examination

13. Why is the ophthalmoscopic examination an important diagnostic tool? _____

_____

14. Many college students struggling through mountainous reading assignments are told that they need glasses for "eyestrain." Why is it more of a strain on the extrinsic and intrinsic eye muscles to look at close objects than at far objects?

_____

_____

Special Senses: Hearing and Equilibrium

MATERIALS

- ☐ Three-dimensional dissectible ear model and/or chart of ear anatomy
- ☐ Otoscope (if available)
- ☐ Disposable otoscope tips (if available) and autoclave bag
- ☐ Alcohol swabs
- ☐ Compound microscope
- ☐ Prepared slides of the cochlea of the ear
- ☐ Absorbent cotton
- ☐ Pocket watch or clock that ticks
- ☐ Metric ruler
- ☐ Tuning forks (range of frequencies)
- ☐ Rubber mallet
- ☐ Audiometer and earphones
- ☐ Red and blue pencils
- ☐ Demonstration: Microscope focused on a slide of a crista ampullaris receptor of a semicircular canal
- ☐ Three coins of different sizes
- ☐ Rotating chair or stool
- ☐ Blackboard and chalk or whiteboard and markers

MasteringA&P® For related exercise study tools, go to the Study Area of MasteringA&P. There you will find:

- Practice Anatomy Lab **PAL**
- PhysioEx **PEx**
- A&PFlix **A&PFlix**
- Practice quizzes, Histology Atlas, eText, Videos, and more!

OBJECTIVES

1. Identify the anatomical structures of the external, middle, and internal ear on a model or appropriate diagram, and explain their functions.
2. Describe the anatomy of the organ of hearing (spiral organ in the cochlea), and explain its function in sound reception.
3. Discuss how one is able to localize the source of sounds.
4. Define *sensorineural deafness* and *conduction deafness* and relate these conditions to the Weber and Rinne tests.
5. Describe the anatomy of the organs of equilibrium in the internal ear (cristae ampullares and maculae), and explain their relative function in maintaining equilibrium.
6. State the locations and functions of endolymph and perilymph.
7. Discuss the effects of acceleration on the semicircular canals.
8. Define *nystagmus* and relate this event to the balance and Barany tests.
9. State the purpose of the Romberg test.
10. Explain the role of vision in maintaining equilibrium.

PRE-LAB QUIZ

1. Circle the correct underlined term. The ear is divided into three / four major areas.
2. The external ear is composed primarily of the _____ and the external acoustic meatus.
   - a. auricle
   - b. cochlea
   - c. eardrum
   - d. stapes
3. Circle the correct underlined term. Sound waves that enter the external acoustic meatus eventually encounter the tympanic membrane / oval window, which then vibrates at the same frequency as the sound waves hitting it.
4. Three small bones found within the middle ear are the malleus, incus, and _____.
   - a. auricle
   - b. cochlea
   - c. eardrum
   - d. stapes
5. The snail-like _____, found in the internal ear, contains sensory receptors for hearing.
   - a. cochlea
   - b. lobule
   - c. semicircular canals
   - d. vestibule
6. Circle the correct underlined term. Today you will use an ophthalmoscope / otoscope to examine the ear.
7. The _____ test is used for comparing bone and air-conduction hearing.
   - a. Barany
   - b. Rinne
   - c. Weber

(Text continues on next page.)

8. The equilibrium apparatus of the ear, the vestibular apparatus, is found in the
   a. external ear
   b. internal ear
   c. middle ear
9. Circle the correct underlined terms. The crista ampullaris / macula located in the semicircular duct / vestibule is essential for detecting static equilibrium.
10. Nystagmus is
    a. ability to hear only high-frequency tones
    b. ability to hear only low-frequency tones
    c. involuntary trailing of eyes in one direction, then rapid movement in the other
    d. sensation of dizziness

The ear is a complex structure containing sensory receptors for hearing and equilibrium. The ear is divided into three major areas: the *external ear,* the *middle ear,* and the *internal ear* **(Figure 25.1)**. The external and middle ear structures serve the needs of the sense of hearing *only,* whereas internal ear structures function both in equilibrium and hearing reception.

Anatomy of the Ear

Gross Anatomy

ACTIVITY 1

Identifying Structures of the Ear

25

Obtain a dissectible ear model or chart of ear anatomy and identify the structures described below. (Refer to Figure 25.1 as you work.)

The **external (outer) ear** is composed primarily of the auricle and the external acoustic meatus. The **auricle,** or **pinna,*** is the skin-covered cartilaginous structure encircling the auditory canal opening. In many animals, it collects and directs sound waves into the external auditory canal. In humans this function of the pinna is largely lost. The portion of the pinna lying inferior to the external auditory canal is the **lobule.**

The **external acoustic meatus,** or **external auditory canal,*** is a short, narrow (about 2.5 cm long by 0.6 cm wide) chamber carved into the temporal bone. In its skin-lined walls are wax-secreting glands called **ceruminous glands.** Sound waves that enter the external auditory meatus eventually encounter the **tympanic membrane,** or **eardrum,** which vibrates at exactly the same frequency as the sound wave(s) hitting it. The membranous eardrum separates the external from the middle ear.

The **middle ear** is essentially a small chamber—the **tympanic cavity**—found within the temporal bone. The

*Although the preferred anatomical terms for *pinna* and *external auditory canal* are *auricle* and *external acoustic meatus,* "pinna" and "external auditory canal" are heard often in clinical situations and will continue to be used here.

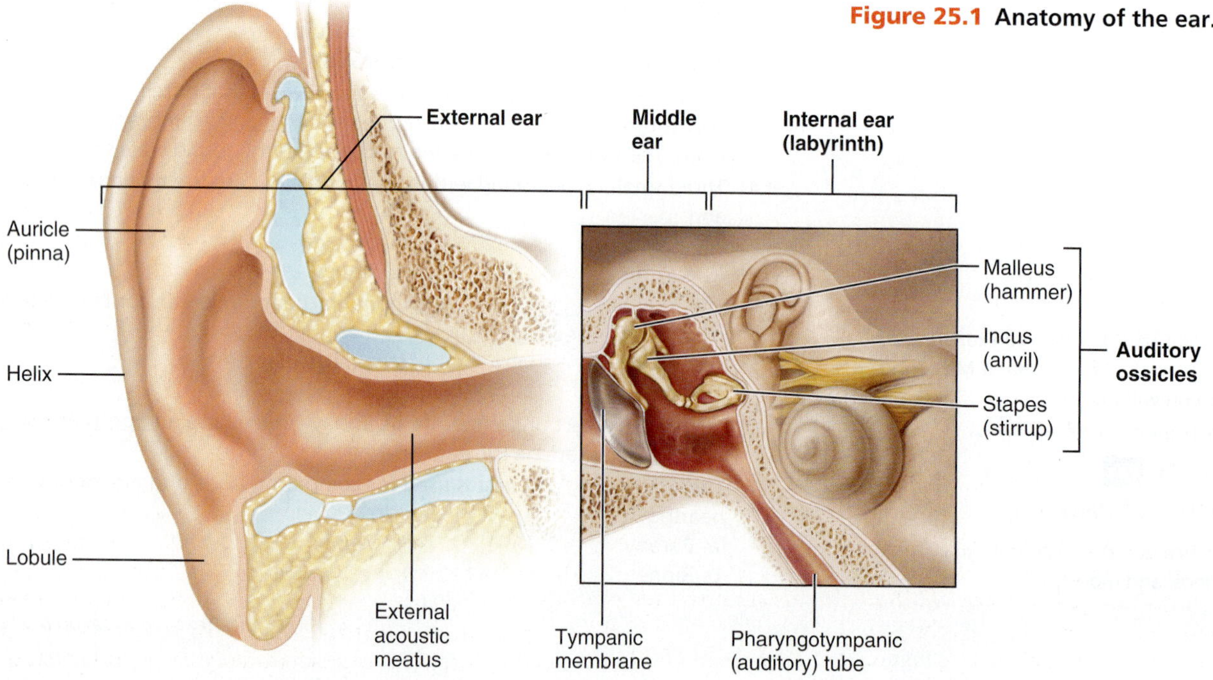

Figure 25.1 Anatomy of the ear.

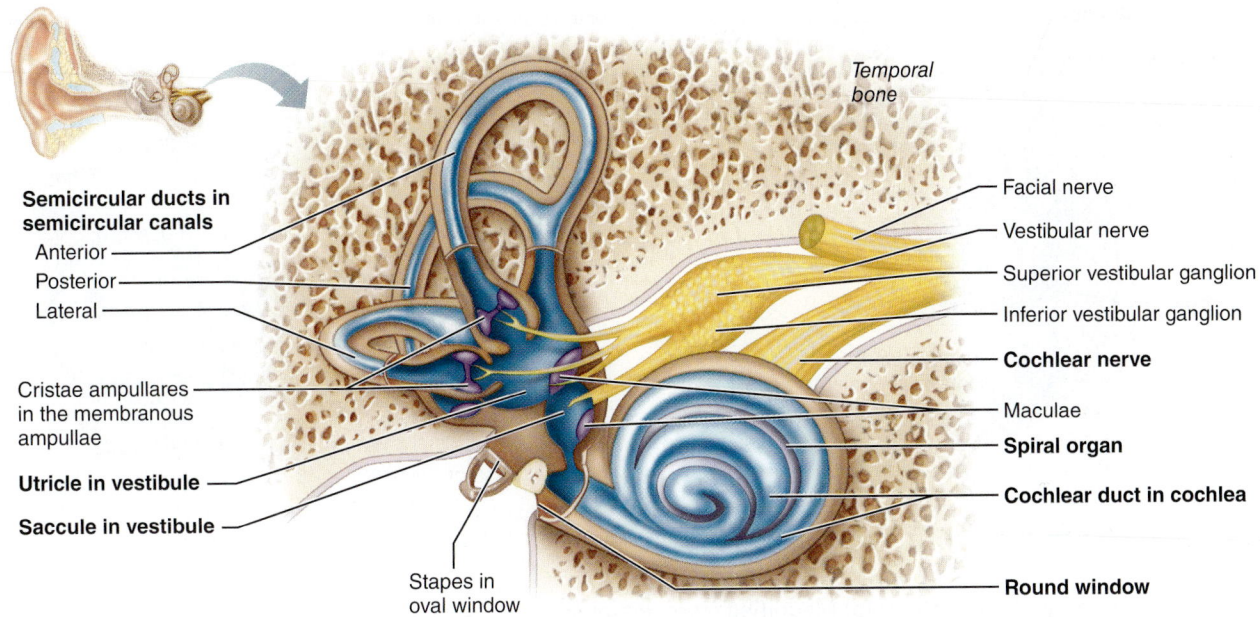

Semicircular ducts in
semicircular canals
Anterior
Posterior
Lateral

Cristae ampullares
in the membranous
ampullae

Utricle in vestibule

Saccule in vestibule

Stapes in
oval window

Temporal bone

Facial nerve
Vestibular nerve
Superior vestibular ganglion
Inferior vestibular ganglion
Cochlear nerve
Maculae
Spiral organ
Cochlear duct in cochlea
Round window

Figure 25.2 **Internal ear.** Right membranous labyrinth (blue) shown within the bony labyrinth (tan). The locations of sensory organs for hearing and equilibrium are shown in purple.

cavity is spanned by three small bones, collectively called the **auditory ossicles** (**malleus, incus,** and **stapes**), which articulate to form a lever system that amplifies and transmits the vibratory motion of the eardrum to the fluids of the inner ear via the **oval window.** The ossicles are often referred to by their common names: hammer, anvil, and stirrup, respectively.

Connecting the middle ear chamber with the nasopharynx is the **pharyngotympanic (auditory) tube** (formerly known as the eustachian tube). Normally this tube is flattened and closed, but swallowing or yawning can cause it to open temporarily to equalize the pressure of the middle ear cavity with external air pressure. This is an important function. The eardrum does not vibrate properly unless the pressure on both of its surfaces is the same.

Because the mucosal membranes of the middle ear cavity and nasopharynx are continuous through the pharyngotympanic tube, **otitis media,** or inflammation of the middle ear, is a fairly common condition, especially among youngsters prone to sore throats. In cases where large amounts of fluid or pus accumulate in the middle ear cavity, an emergency myringotomy (lancing of the eardrum) may be necessary to relieve the pressure. Frequently, tiny ventilating tubes are put in during the procedure. ✚

The **internal ear** consists of a system of bony and rather tortuous chambers called the **bony labyrinth,** which is filled with an aqueous fluid called **perilymph (Figure 25.2).** Suspended in the perilymph is the **membranous labyrinth,** a system that mostly follows the contours of the bony labyrinth. The membranous labyrinth is filled with a more viscous fluid called **endolymph.** The three subdivisions of the bony labyrinth are the cochlea, the vestibule, and the semicircular canals, with the vestibule situated between the cochlea and semicircular canals. The **vestibule** and the **semicircular canals** are involved with equilibrium.

The snail-like **cochlea** (see Figure 25.2 and **Figure 25.3**) contains the sensory receptors for hearing. The membranous

cochlear duct is a soft wormlike tube about 3.8 cm long. It winds through the full two and three-quarter turns of the cochlea and separates the perilymph-containing cochlear cavity into upper and lower chambers, the **scala vestibuli** and **scala tympani.** The scala vestibuli terminates at the oval window, which "seats" the foot plate of the stirrup located laterally in the tympanic cavity. The scala tympani is bounded by a membranous area called the **round window.** The cochlear duct is the middle **scala media.** It is filled with endolymph and supports the **spiral organ,** which contains the receptors for hearing—the sensory hair cells and nerve endings of the **cochlear nerve,** a division of the vestibulocochlear nerve (VIII).

ACTIVITY 2

Examining the Ear with an Otoscope (Optional)

1. Obtain an otoscope and two alcohol swabs. Inspect your partner's external auditory canal and then select the largest—*diameter* (not length!) speculum that will fit comfortably into his or her ear to permit full visibility. Clean the speculum thoroughly with an alcohol swab, and then attach the speculum to the battery-containing otoscope handle. Before beginning, check that the otoscope light beam is strong. If not, obtain another otoscope or new batteries. Some otoscopes come with disposable tips. Be sure to use a new tip for each ear examined. Dispose of these tips in an autoclave bag after use.

2. When you are ready to begin the examination, hold the lighted otoscope securely between your thumb and forefinger (like a pencil), and rest the little finger of the otoscope-holding hand against your partner's head. This maneuver forms a brace that allows the speculum to move as your partner moves and prevents the speculum from penetrating too deeply into the external auditory canal during unexpected movements.

25

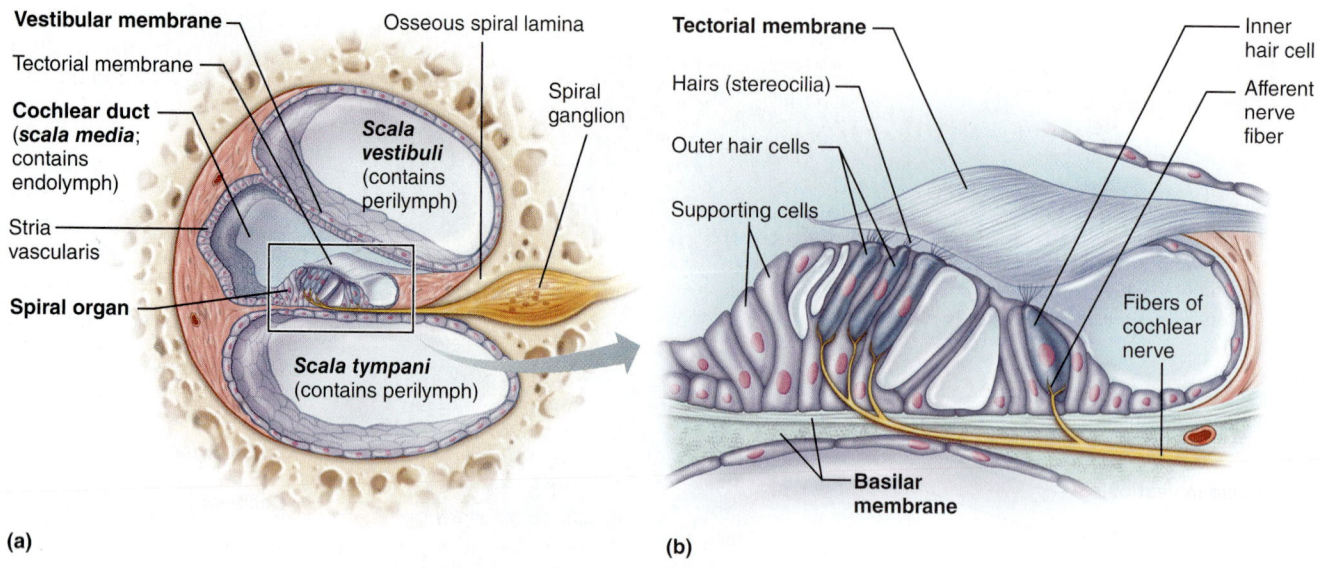

(a)   **(b)**

Figure 25.3 Anatomy of the cochlea. (a) Magnified cross-sectional view of one turn of the cochlea, showing the relationship of the three scalae. The scalae vestibuli and tympani contain perilymph; the cochlear duct (scala media) contains endolymph. **(b)** Detailed structure of the spiral organ.

3. Grasp the ear pinna firmly and pull it up, back, and slightly laterally. If your partner experiences pain or discomfort when the pinna is manipulated, an inflammation or infection of the external ear may be present. If this occurs, do not attempt to examine the ear canal.

4. Carefully insert the speculum of the otoscope into the external auditory canal in a downward and forward direction only far enough to permit examination of the tympanic membrane, or eardrum. Note its shape, color, and vascular network. The healthy tympanic membrane is pearly white. During the examination, notice if there is any discharge or redness in the external auditory canal and identify earwax.

5. After the examination, thoroughly clean the speculum with the second alcohol swab before returning the otoscope to the supply area. ■

Microscopic Anatomy of the Spiral Organ and the Mechanism of Hearing

In the spiral organ, the auditory receptors are hair cells that rest on the **basilar membrane,** which forms the floor of the cochlear duct (Figure 25.3). Their "hairs" are stereocilia that project into a gelatinous membrane, the **tectorial membrane,** that overlies them. The roof of the cochlear duct is called the **vestibular membrane.**

ACTIVITY 3

Examining the Microscopic Structure of the Cochlea

Obtain a compound microscope and a prepared microscope slide of the cochlea and identify the areas shown in the photomicrograph **(Figure 25.4).** ■

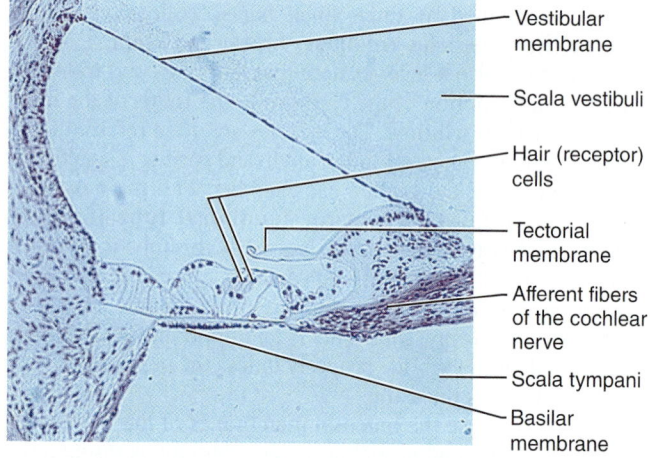

Figure 25.4 Histological image of the spiral organ (100×)

The mechanism of hearing begins as sound waves pass through the external auditory canal and through the middle ear into the internal ear, where the vibration eventually reaches the spiral organ, which contains the receptors for hearing.

Vibration of the stirrup at the oval window initiates traveling pressure waves in the perilymph that cause maximal displacements of the basilar membrane where they peak and stimulate the hair cells of the spiral organ in that region. Since the area at which the traveling waves peak is a high-pressure area, the vestibular membrane is compressed at this point and, in turn, compresses the endolymph and the basilar membrane of the cochlear duct. The resulting pressure on the perilymph in the scala tympani causes the membrane of the round window to bulge outward into the middle ear chamber, thus acting as a relief valve for the compressional wave. High-frequency waves (high-pitch sounds) peak close to the oval window and low-frequency waves (low-pitched sounds)

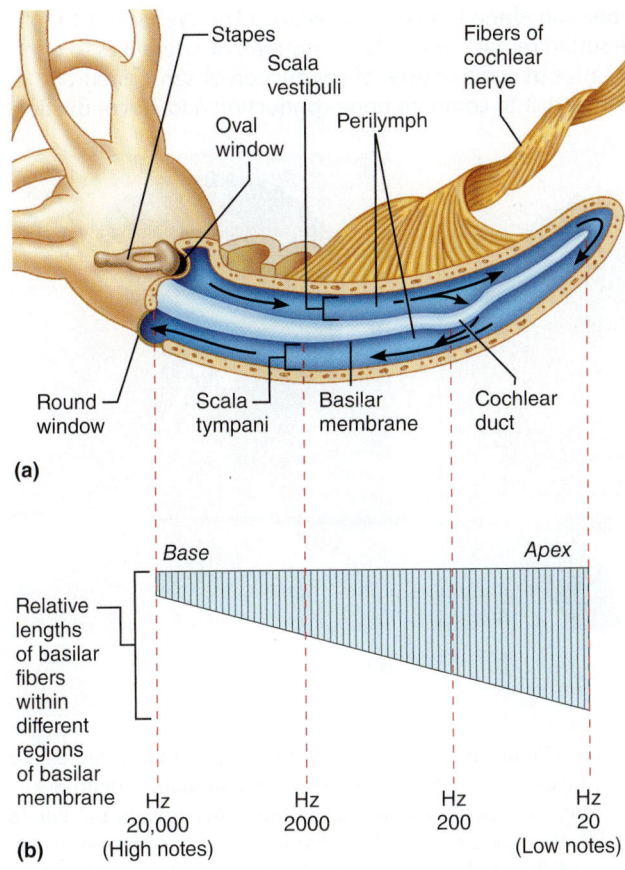

(a)

(b)

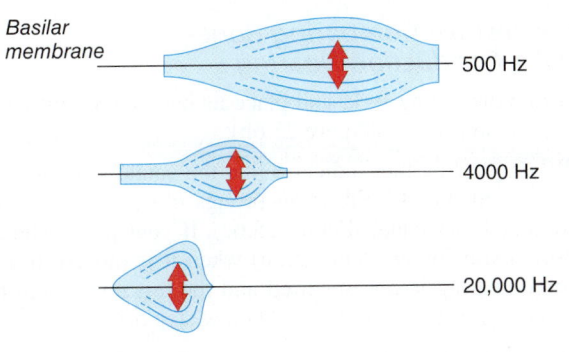

(c)

Figure 25.5 Resonance of the basilar membrane. The cochlea is depicted as if it has been uncoiled. **(a)** Fluid movement in the cochlea following the stirrup thrust at the oval window. The compressional wave thus created causes the round window to bulge into the middle ear. Pressure waves set up vibrations in the basilar membrane. **(b)** Fibers span the basilar membrane. The length of the fibers "tunes" specific regions to vibrate at specific frequencies. **(c)** Different frequencies of pressure waves in the cochlea stimulate particular hair cells and neurons.

peak farther up the basilar membrane near the apex of the cochlea. The mechanism of sound reception by the spiral organ is complex. Hair cells at any given spot on the basilar membrane are stimulated by sounds of a specific frequency

and amplitude. Once stimulated, they depolarize and begin the chain of nervous impulses that travel along the cochlear nerve to the auditory centers of the temporal lobe cortex. This series of events results in the phenomenon we call hearing **(Figure 25.5)**.

Sensorineural deafness results from damage to neural structures anywhere from the cochlear hair cells through neurons of the auditory cortex. **Presbycusis** is a type of sensorineural deafness that occurs commonly in people by the time they are in their sixties. It results from a gradual deterioration and atrophy of the spiral organ and leads to a loss in the ability to hear high tones and speech sounds. Because many elderly people refuse to accept their hearing loss and resist using hearing aids, they begin to rely more and more on their vision for clues as to what is going on around them and may be accused of ignoring people.

Although presbycusis is considered to be a disability of old age, it is becoming much more common in younger people as our world grows noisier. Prolonged or excessive noise tears the cilia from hair cells, and the damage is progressive and cumulative. Each assault causes a bit more damage. Music played and listened to at deafening levels definitely contributes to the deterioration of hearing receptors. ✚

ACTIVITY 4

Conducting Laboratory Tests of Hearing

Perform the following hearing tests in a quiet area. Test both the right and left ears.

Acuity Test

Have your lab partner pack one ear with cotton and sit quietly with eyes closed. Obtain a ticking clock or pocket watch and hold it very close to his or her *unpacked* ear. Then slowly move it away from the ear until your partner signals that the ticking is no longer audible. Record the distance in centimeters at which ticking is inaudible and then remove the cotton from the packed ear.

Right ear: _____ Left ear: _____

Is the threshold of audibility sharp or indefinite?

_____

Sound Localization

Ask your partner to close both eyes. Hold the pocket watch at an audible distance (about 15 cm) from his or her ear, and move it to various locations (front, back, sides, and above his or her head). Have your partner locate the position by pointing in each instance. Can the sound be localized equally well at all positions?

_____

If not, at what position(s) was the sound less easily located?

_____

_____

25

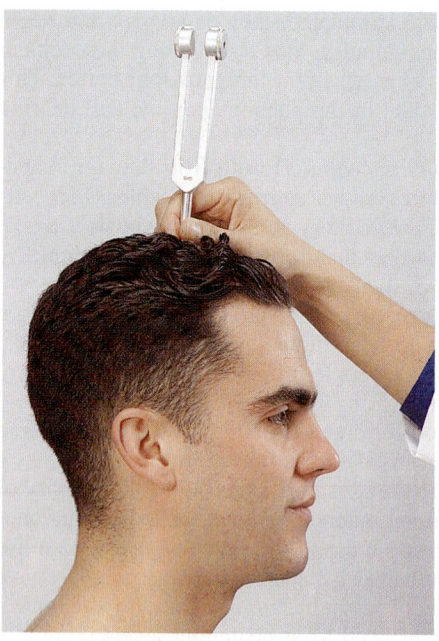

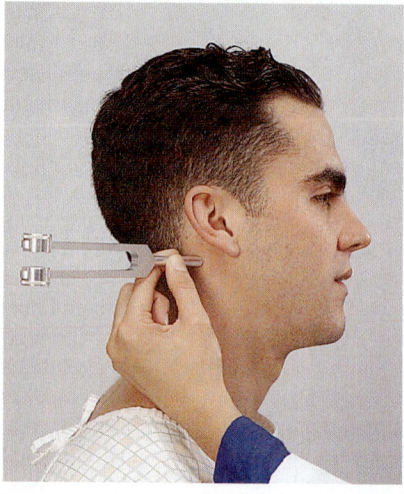

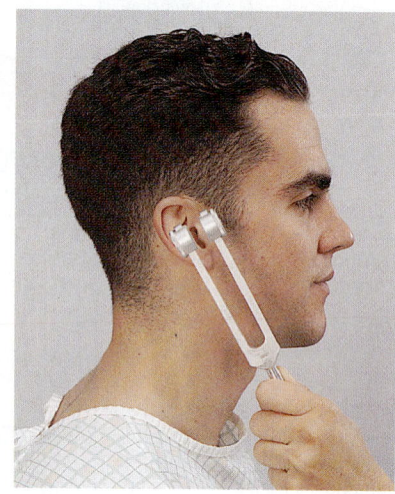

Figure 25.6 The Weber and Rinne tuning fork tests. (a) The Weber test to evaluate whether the sound remains centralized (normal) or lateralizes to one side or the other (indicative of some degree of conduction or sensorineural deafness). **(b, c)** The Rinne test to compare bone conduction and air conduction.

(a)

(b)

(c)

The ability to localize the source of a sound depends on two factors—the difference in the loudness of the sound reaching each ear and the time of arrival of the sound at each ear. How does this information help to explain your findings?

_____

_____

Frequency Range of Hearing

Obtain three tuning forks: one with a low frequency (75 to 100 Hz [cps]), one with a frequency of approximately 1000 Hz, and one with a frequency of 4000 to 5000 Hz. Strike the lowest-frequency fork on the heel of your hand or with a rubber mallet, and hold it close to your partner's ear. Repeat with the other two forks.

Which fork was heard most clearly and comfortably?

_____ Hz

Which was heard least well? _____ Hz

Weber Test to Determine Conduction and Sensorineural Deafness

Strike a tuning fork and place the handle of the tuning fork medially on your partner's head **(Figure 25.6a)**. Is the tone equally loud in both ears, or is it louder in one ear?

_____

If it is equally loud in both ears, you have equal hearing or equal loss of hearing in both ears. If sensorineural deafness is present in one ear, the tone will be heard in the unaffected ear but not in the ear with sensorineural deafness.

    Conduction deafness occurs when something prevents sound waves from reaching the fluids of the internal ear. Compacted earwax, a perforated eardrum, inflammation of

the middle ear (otitis media), and damage to the ossicles are all causes of conduction deafness. If conduction deafness is present, the sound will be heard more strongly in the ear in which there is a hearing loss due to sound conduction by the bone of the skull. Conduction deafness can be simulated by plugging one ear with cotton.

Rinne Test for Comparing Bone- and Air-Conduction Hearing

1. Strike the tuning fork, and place its handle on your partner's mastoid process (Figure 25.6b).

2. When your partner indicates that the sound is no longer audible, hold the still-vibrating prongs close to his or her external auditory canal (Figure 25.6c). If your partner hears the fork again (by air conduction) when it is moved to that position, hearing is not impaired and the test result is to be recorded as positive (+). (Record below step 5.)

3. Repeat the test on the same ear, but this time test air-conduction hearing first.

4. After the tone is no longer heard by air conduction, hold the handle of the tuning fork on the bony mastoid process. If the subject hears the tone again by bone conduction after hearing by air conduction is lost, there is some conduction deafness and the result is recorded as negative (−).

5. Repeat the sequence for the opposite ear.

Right ear: _____ Left ear: _____

Does the subject hear better by bone or by air conduction?

_____

Audiometry

When the simple tuning fork tests reveal a problem in hearing, audiometer testing is usually prescribed to determine the precise nature of the hearing deficit. An *audiometer* is

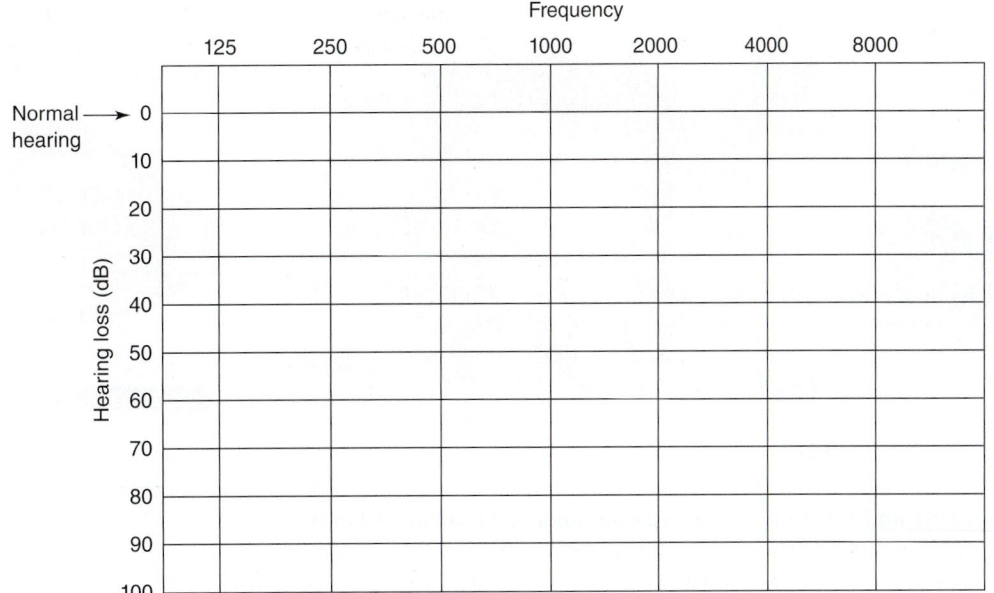

an instrument (specifically, an electronic oscillator with earphones) used to determine hearing acuity by exposing each ear to sound stimuli of differing *frequencies* and *intensities*. The hearing range of human beings during youth is from 20 to 20,000 Hz, but hearing acuity declines with age, with reception for the high-frequency sounds lost first. Though this loss represents a major problem for some people, such as musicians, most of us tend to be fairly unconcerned until we begin to have problems hearing sounds in the range of 125 to 8000 Hz, the normal frequency range of speech.

The basic procedure of audiometry is to initially deliver tones of different frequencies to one ear of the subject at an intensity of 0 decibels (dB). (Zero decibels is not the complete absence of sound, but rather the softest sound intensity that can be heard by a person of normal hearing at each frequency.) If the subject cannot hear a particular frequency stimulus of 0 dB, the hearing threshold level control is adjusted until the subject reports that he or she can hear the tone. The number of decibels of intensity required above 0 dB is recorded as the hearing loss. For example, if the subject cannot hear a particular frequency tone until it is delivered at 30 dB intensity, then he or she has a hearing loss of 30 dB for that frequency.

ACTIVITY 5

Audiometry Testing

1. Obtain an audiometer and earphones, and a red and a blue pencil. Before beginning the tests, examine the audiometer to identify the two tone controls: one to regulate frequency and a second to regulate the intensity (loudness) of the sound stimulus. Identify the two output control switches that regulate the delivery of sound to one ear or the other (*red* to the right ear, *blue* to the left ear). Also find the *hearing threshold level control,* which is calibrated to deliver a basal tone of 0 dB to the subject's ears.

2. Place the earphones on the subject's head so that the red cord or ear-cushion is over the right ear and the blue cord or

ear-cushion is over the left ear. Instruct the subject to raise one hand when he or she hears a tone.

3. Set the frequency control at 125 Hz and the intensity control at 0 dB. Press the red output switch to deliver a tone to the subject's right ear. If the subject does not respond, raise the sound intensity slowly by rotating the hearing level control counterclockwise until the subject reports (by raising a hand) that a tone is heard. Repeat this procedure for frequencies of 250, 500, 1000, 2000, 4000, and 8000.

4. Record the results in the grid (above) for frequency versus hearing loss by marking a small red circle on the grid at each frequency-dB junction at which a tone was heard. Then connect the circles with a red line to produce a hearing acuity graph for the right ear.

5. Repeat steps 3 and 4 for the left (blue) ear, and record the results with blue circles and connecting lines on the grid. ■

Microscopic Anatomy of the Equilibrium Apparatus and Mechanisms of Equilibrium

The equilibrium receptors of the internal ear are collectively called the **vestibular apparatus,** and are found in the vestibule and semicircular canals of the bony labyrinth. Their chambers are filled with perilymph, in which membranous labyrinth structures are suspended. The vestibule contains the saclike **utricle** and **saccule,** and the semicircular chambers contain **membranous semicircular ducts.** Like the cochlear duct, these membranes are filled with endolymph and contain receptor cells that are activated by the bending of their cilia.

Semicircular Canals

The semicircular canals monitor angular movements of the head. This process is called **dynamic equilibrium.** The canals are 1.2 cm in circumference and are oriented in three planes—horizontal, frontal, and sagittal. At the base of each semicircular duct is an enlarged region, the **ampulla,** which communicates with the utricle of the vestibule. Within

25

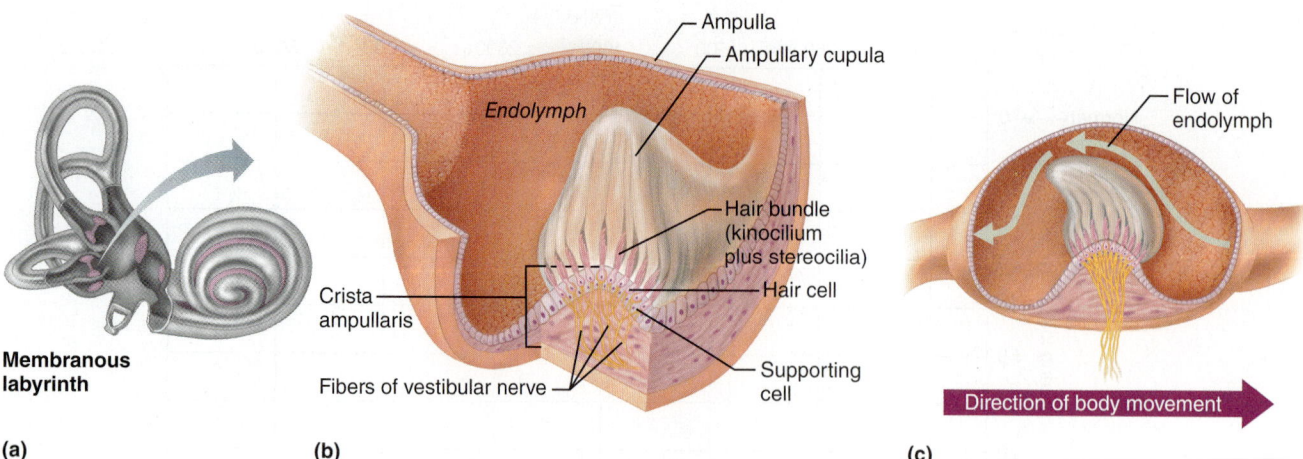

Figure 25.7 Structure and function of the crista ampullaris. (a) Arranged in the three spatial planes, the semicircular ducts in the semicircular canals each have a swelling called an ampulla at their base. **(b)** Each ampulla contains a crista ampullaris, a receptor that is essentially a cluster of hair cells with hairs projecting into a gelatinous cap called the ampullary cupula. **(c)** Movement of the cupula during angular acceleration of the head.

each ampulla is a receptor region called a **crista ampullaris,** which consists of a tuft of hair cells covered with a gelatinous cap, or **ampullary cupula (Figure 25.7).**

The cristae respond to changes in the velocity of rotational head movements. During acceleration, as when you begin to twirl around, the endolymph in the canal lags behind the head movement due to inertia pushing the ampullary cupula—like a swinging door—in the opposite direction. The head movement depolarizes the hair cells, and results in enhanced impulse transmission in the vestibular division of the eighth cranial nerve to the brain (Figure 25.7c). If the body continues to rotate at a constant rate, the endolymph eventually comes to rest and moves at the same speed as the body. The ampullary cupula returns to its upright position, hair cells are no longer stimulated, and you lose the sensation of spinning. When rotational movement stops suddenly, the endolymph keeps on going in the direction of head movement. This pushes the ampullary cupula in the *same* direction as the previous head movement and hyperpolarizes the hair cells, resulting in fewer impulses being transmitted to the brain. This tells the brain that you have stopped moving and accounts for the reversed motion sensation you feel when you stop twirling suddenly.

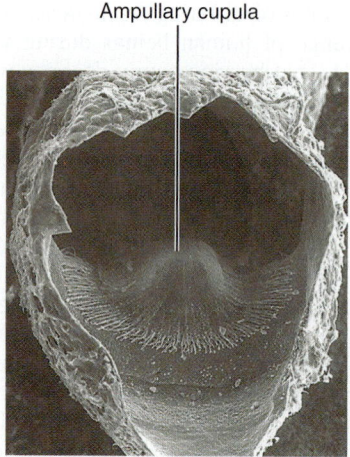

Figure 25.8 Scanning electron micrograph of a crista ampullaris (14×).

ACTIVITY 6

Examining the Microscopic Structure of the Crista Ampullaris

Go to the demonstration area and examine the slide of a crista ampullaris. Identify the areas depicted in the photomicrograph **(Figure 25.8)** and labeled diagram (Figure 25.7b). ■

Maculae

Maculae in the vestibule contain another set of **hair cells,** receptors that in this case monitor head position and acceleration in a straight line. This monitoring process is called **static equilibrium.** The maculae respond to gravitational pull, thus providing information on which way is up or down as well as changes in linear speed. They are located on the walls of the saccule and utricle. The hair cells in each macula are embedded in the **otolith membrane,** a gelatinous material containing small grains of calcium carbonate called **otolith.** When the head moves, the otoliths move in response to variations in gravitational pull. As they deflect different hair cells, they trigger hyperpolarization or depolarization of the hair cells and modify the rate of impulse transmission along the vestibular nerve **(Figure 25.9).**

Although the receptors of the semicircular canals and the vestibule are responsible for dynamic and static equilibrium respectively, they rarely act independently. Complex interaction of many of the receptors is the rule. Processing is also complex and involves the brain stem and cerebellum as well as input from proprioceptors and the eyes.

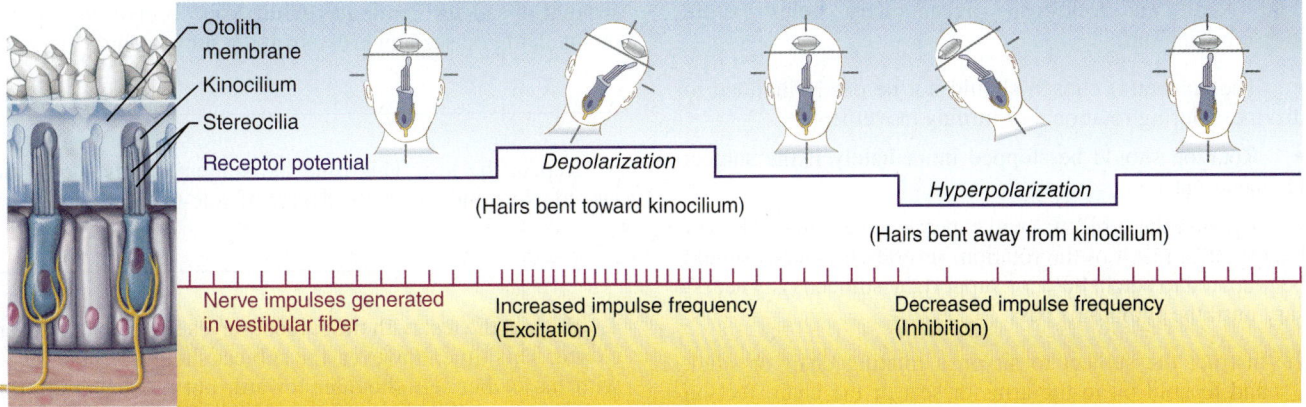

Figure 25.9 The effect of gravitational pull on a macula receptor in the utricle. When movement of the otolith membrane bends the hair cells in the direction of the kinocilium, the hair cells depolarize, exciting the nerve fibers, which generates action potentials more rapidly. When the hairs are bent in the direction away from the kinocilium, the hair cells become hyperpolarized, inhibiting the nerve fibers and decreasing the action potential rate (i.e., below the resting rate of discharge).

ACTIVITY 7

Conducting Laboratory Tests on Equilibrium

The function of the semicircular canals and vestibule are not routinely tested in the laboratory, but the following simple tests illustrate normal equilibrium apparatus function as well as some of the complex processing interactions.

In the first balance test and the Barany test, you will look for **nystagmus,** which is the involuntary rolling of the eyes in any direction or the trailing of the eyes slowly in one direction, followed by their rapid movement in the opposite direction. During rotation, the slow drift of the eyes is related to the backflow of endolymph in the semicircular canals. The rapid movement represents CNS compensation to find a new fixation point. Nystagmus is normal during and after rotation; abnormal otherwise. The direction of nystagmus is that of its quick phase on acceleration.

Nystagmus is often accompanied by **vertigo**—is a sensation of dizziness and rotational movement when such movement is not occurring or has ceased.

Balance Tests

1. Have your partner walk a straight line, placing one foot directly in front of the other.

Is he or she able to walk without undue wobbling from side to

side? _____

Did he or she experience any dizziness? _____

The ability to walk with balance and without dizziness, unless subject to rotational forces, indicates normal function of the equilibrium apparatus.

Was nystagmus present? _____

2. Place three coins of different sizes on the floor. Ask your lab partner to pick up the coins, and carefully observe his or her muscle activity and coordination.

Did your lab partner have any difficulty locating and picking

up the coins? _____

Describe your observations and your lab partner's observations during the test.

_____

_____

What kinds of interactions involving balance and coordination must occur for a person to move fluidly during this test?

_____

_____

3. If a person has a depressed nervous system, mental concentration may result in a loss of balance. Ask your lab partner to stand up and count backward from ten as rapidly as possible.

Did your lab partner lose balance? _____

Barany Test (Induction of Nystagmus and Vertigo)

This experiment evaluates the semicircular canals and should be conducted as a group effort to protect the test subject(s) from possible injury.

Read the following precautionary notes before beginning:

• The subject(s) chosen should not be easily inclined to dizziness during rotational or turning movements.

• Rotation should be stopped immediately if the subject feels nauseated.

• Because the subject(s) will experience vertigo and loss of balance as a result of the rotation, several classmates should be prepared to catch, hold, or support the subject(s) as necessary until the symptoms pass.

1. Instruct the subject to sit on a rotating chair or stool, and to hold on to the arms or seat of the chair, feet on stool rungs. The subject's head should be tilted forward approximately 30 degrees (almost touching the chest). The horizontal (lateral) semicircular canal is stimulated when the head is in this position. The subject's eyes are to remain *open* during the test.

2. Four classmates should position themselves so that the subject is surrounded on all sides. The classmate posterior to the subject will rotate the chair.

3. Rotate the chair to the subject's right approximately 10 revolutions in 10 seconds, then suddenly stop the rotation.

4. Immediately note the direction of the subject's resultant nystagmus; and ask him or her to describe the feelings of movement, indicating speed and direction sensation. Record this information below.

_____

_____

_____

If the semicircular canals are operating normally, the subject will experience a sensation that the stool is still rotating immediately after it has stopped and *will* demonstrate nystagmus.

When the subject is rotated to the right, the ampullary cupula will be bent to the left, causing nystagmus during rotation in which the eyes initially move slowly to the left and then quickly to the right. Nystagmus will continue until the ampullary cupula has returned to its initial position. Then, when rotation is stopped abruptly, the ampullary cupula will be bent to the right, producing nystagmus with its slow phase to the right and its rapid phase to the left. In many subjects, this will be accompanied by a feeling of vertigo and a tendency to fall to the right.

Romberg Test

The Romberg test determines the integrity of the dorsal white column of the spinal cord, which transmits impulses to the brain from the proprioceptors involved with posture.

1. Have your partner stand with his or her back to the blackboard or whiteboard.

2. Draw one line parallel to each side of your partner's body. He or she should stand erect, with eyes open and staring straight ahead for 2 minutes while you observe any movements. Did you see any gross swaying movements?

_____

3. Repeat the test. This time the subject's eyes should be closed. Note and record the degree of side-to-side movement.

_____

4. Repeat the test with the subject's eyes first open and then closed. This time, however, the subject should be positioned with his or her left shoulder toward, but not touching, the board so that you may observe and record the degree of front-to-back swaying.

_____

Do you think the equilibrium apparatus of the internal ear was operating equally well in all these tests?

_____

The proprioceptors? _____

Why was the observed degree of swaying greater when the eyes were closed?

_____

_____

What conclusions can you draw regarding the factors necessary for maintaining body equilibrium and balance?

_____

_____

Role of Vision in Maintaining Equilibrium

To further demonstrate the role of vision in maintaining equilibrium, perform the following experiment. (Ask your lab partner to record observations and act as a "spotter.") Stand erect, with your eyes open. Raise your left foot approximately 30 cm off the floor, and hold it there for 1 minute.

Record the observations: _____

_____

_____

Rest for 1 or 2 minutes; and then repeat the experiment with the same foot raised but with your eyes closed. Record the observations:

_____

_____

Special Senses: Olfaction and Taste

MATERIALS

- ☐ Prepared slides: nasal olfactory epithelium (l.s.); the tongue showing taste buds (x.s.)
- ☐ Compound microscope
- ☐ Small mirror
- ☐ Paper towels
- ☐ Packets of granulated sugar
- ☐ Disposable autoclave bag
- ☐ Paper plates
- ☐ Equal-size food cubes of cheese, apple, raw potato, dried prunes, banana, raw carrot, and hard-cooked egg white (These prepared foods should be in an opaque container; a foil-lined egg carton would work well.)
- ☐ Toothpicks
- ☐ Disposable gloves
- ☐ Cotton-tipped swabs
- ☐ Paper cups
- ☐ Flask of distilled or tap water
- ☐ Prepared vials of oil of cloves, oil of peppermint, and oil of wintergreen or corresponding flavors found in the condiment section of a supermarket
- ☐ Chipped ice
- ☐ Five numbered vials containing common household substances with strong odors (herbs, spices, etc.)
- ☐ Nose clips
- ☐ Absorbent cotton

MasteringA&P® For related exercise study tools, go to the Study Area of MasteringA&P. There you will find:
- Practice Anatomy Lab PAL
- PhysioEx PEx
- A&PFlix A&PFlix
- Practice quizzes, Histology Atlas, eText, Videos, and more!

OBJECTIVES

1. State the location and cellular composition of the olfactory epithelium.
2. Describe the structure of olfactory sensory neurons and state their function.
3. Discuss the locations and cellular composition of taste buds.
4. Describe the structure of gustatory epithelial cells and state their function.
5. Identify the cranial nerves that carry the sensations of olfaction and taste.
6. Name five basic qualities of taste sensation, and list the chemical substances that elicit them.
7. Explain the interdependence between the senses of smell and taste.
8. Name two factors other than olfaction that influence taste appreciation of foods.
9. Define *olfactory adaptation*.

PRE-LAB QUIZ

1. Circle True or False. Receptors for olfaction and taste are classified as chemoreceptors because they respond to dissolved chemicals.
2. The organ of smell is the _____, located in the roof of the nasal cavity.
   a. nares   c. olfactory epithelium
   b. nostrils   d. olfactory nerve
3. Circle the correct underlined term. Olfactory receptors are <u>bipolar</u> / <u>unipolar</u> sensory neurons whose olfactory cilia extend outward from the epithelium.
4. Most taste buds are located in _____, peglike projections of the tongue mucosa.
   a. cilia   c. papillae
   b. concha   d. supporting cells
5. Circle the correct underlined term. Vallate papillae are arranged in a V formation on the <u>anterior</u> / <u>posterior</u> surface of the tongue.
6. Circle the correct underlined term. Most taste buds are made of <u>two</u> / <u>three</u> types of modified epithelial cells.
7. There are five basic taste sensations. Name one. _____
8. Circle True or False. Taste buds typically respond optimally to one of the five basic taste sensations.
9. Circle True or False. Texture, temperature, and smell have little or no effect on the sensation of taste.
10. You will use absorbent cotton and oil of wintergreen, peppermint, or cloves to test for olfactory
    a. accommodation   c. identification
    b. adaptation   d. recognition

The receptors for olfaction and taste are classified as **chemoreceptors** because they respond to chemicals in solution. Although five relatively specific types of taste receptors have been identified, the olfactory receptors are considered sensitive to a much wider range of chemical sensations. The sense of smell is the least understood of the special senses.

Location and Anatomy of the Olfactory Receptors

The **olfactory epithelium** is the organ of smell. It occupies an area of about 5 cm^2 in the roof of the nasal cavity **(Figure 26.1a)**. Since the air entering the human nasal cavity must make a hairpin turn to enter the respiratory passages below, the nasal epithelium is in a rather poor position for performing its function. This is why sniffing, which brings more air into contact with the receptors, increases your ability to detect odors.

The specialized receptor cells in the olfactory epithelium are **olfactory sensory neurons.** They are surrounded by epithelial **supporting cells.** The bipolar neurons have **olfactory cilia** that extend outward from the epithelium. Axons emerging from their basal ends penetrate the cribriform plate of the ethmoid bone and proceed as the *olfactory nerve filaments* (cranial nerve I) to synapse in the olfactory bulbs lying on either side of the crista galli of the ethmoid bone. Impulses from neurons of the olfactory bulbs are then conveyed to the olfactory portion of the cortex without synapsing in the thalamus.

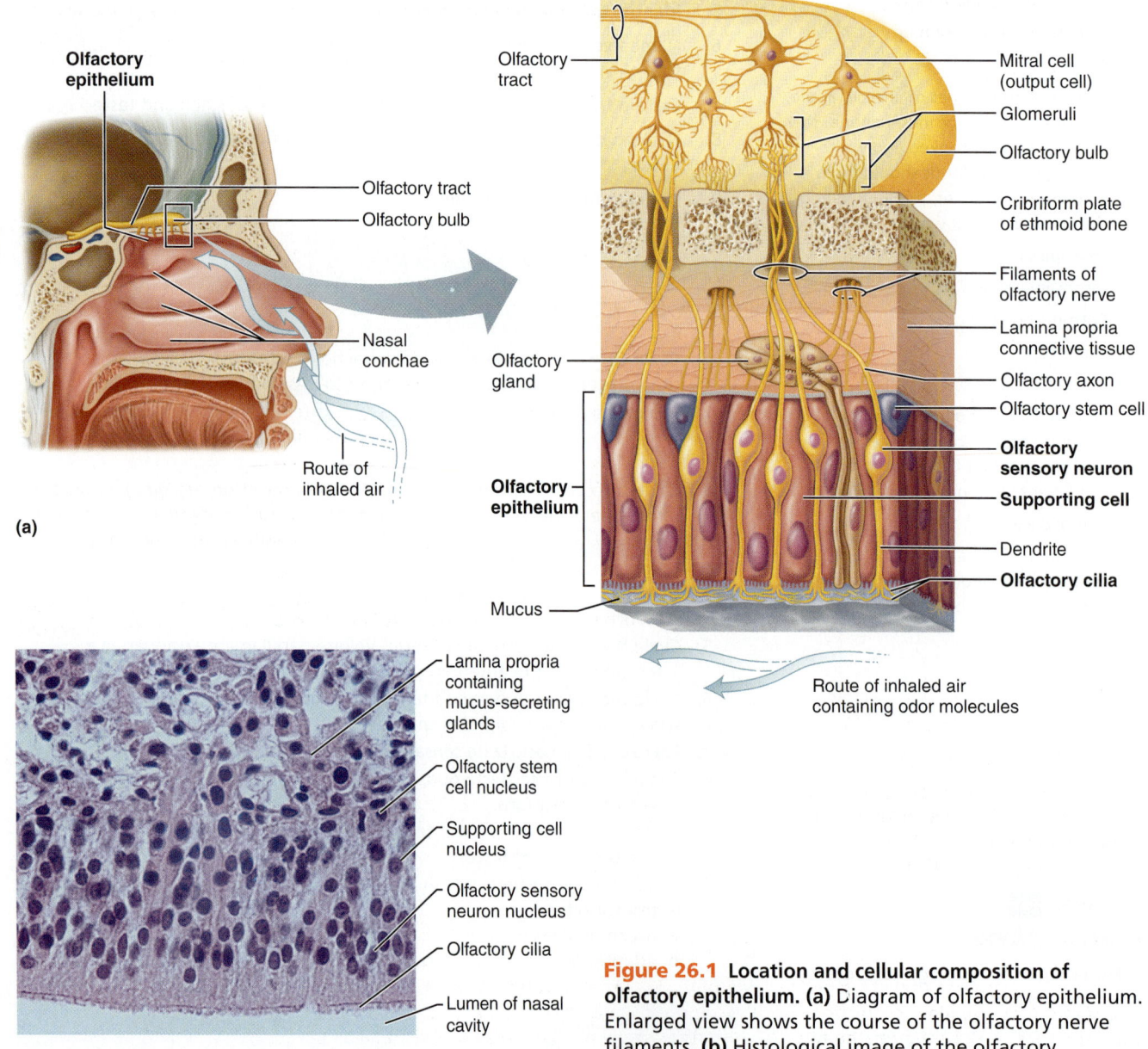

(a)

(b)

Figure 26.1 Location and cellular composition of olfactory epithelium. (a) Diagram of olfactory epithelium. Enlarged view shows the course of the olfactory nerve filaments. **(b)** Histological image of the olfactory epithelium (275×).

Microscopic Examination of the Olfactory Epithelium

Obtain a longitudinal section of olfactory epithelium. Examine it closely using a compound microscope, comparing it to the photomicrograph (Figure 26.1b). ▬

Location and Anatomy of Taste Buds

The **taste buds,** containing specific receptors for the sense of taste, are widely but not uniformly distributed in the oral cavity. Most are located in **papillae,** peglike projections of the mucosa, on the dorsal surface of the tongue (as described next). A few are found on the soft palate, epiglottis, pharynx, and inner surface of the cheeks.

Taste buds are located primarily on the sides of the large **vallate papillae** (arranged in a V formation on the posterior surface of the tongue); in the side walls of the **foliate papillae;** and on the tops of the more numerous, mushroom-shaped **fungiform papillae (Figure 26.2)**.

- Use a mirror to examine your tongue. Which of the various papillae types can you pick out? _____

_____

Each taste bud consists largely of a globular arrangement of two types of modified epithelial cells: the **gustatory epithelial cells,** which are the actual receptor cells for taste, and basal epithelial cells. Several nerve fibers enter each taste bud and supply sensory nerve endings to each of the gustatory epithelial cells. The long microvilli of the receptor cells penetrate the epithelial surface through an opening called the **taste pore.** When these microvilli, called **gustatory hairs,** contact specific chemicals in the solution, the receptor cells depolarize. The afferent fibers from the taste buds to the somatosensory cortex in the postcentral gyrus of the brain are carried in three cranial nerves: the *facial nerve (VII)* serves the anterior two-thirds of the tongue; the *glossopharyngeal nerve (IX)* serves the posterior third of the tongue; and the *vagus nerve (X)* carries a few fibers from the pharyngeal region.

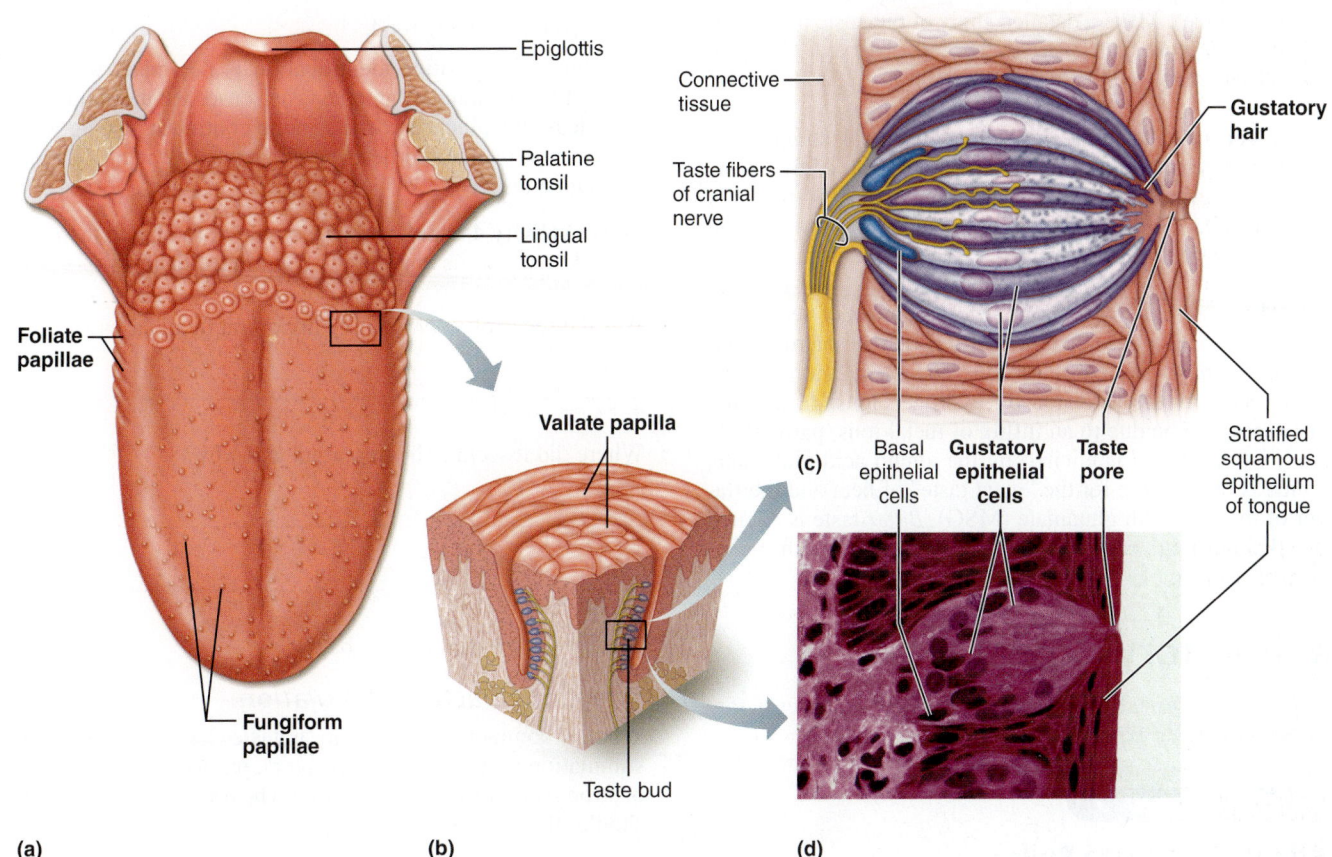

(a)   (b)   (c)   (d)

Figure 26.2 Location and structure of taste buds. (a) Taste buds on the tongue are associated with papillae, projections of the tongue mucosa. **(b)** A sectioned vallate papilla shows the position of the taste buds in its lateral walls. **(c)** An enlarged view of a taste bud. **(d)** Photomicrograph of a taste bud (445×).

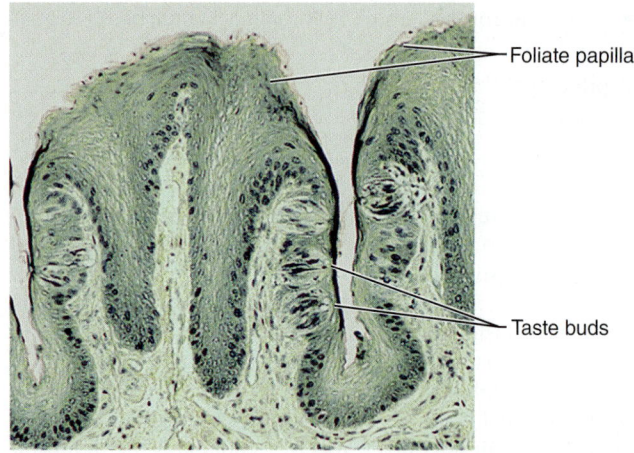

Figure 26.3 Taste buds on the lateral aspects of foliate papillae of the tongue (140×).

ACTIVITY 2

Microscopic Examination of Taste Buds

Obtain a microscope and a prepared slide of a tongue cross section. Locate the taste buds on the tongue papillae (use Figure 26.2b as guide). Make a detailed study of one taste bud. Identify the taste pore and gustatory hairs if observed. Compare your observations to the photomicrograph **(Figure 26.3)**. ▬

When taste is tested with pure chemical compounds, most taste sensations can be grouped into one of five basic qualities—sweet, sour, bitter, salty, or umami (oo-mom′ ē; "delicious"). Although all taste buds are believed to respond in some degree to all five classes of chemical stimuli, each type responds optimally to only one.

The *sweet* receptors respond to a number of seemingly unrelated compounds such as sugars (fructose, sucrose, glucose), saccharine, some lead salts, and some amino acids. *Sour* receptors are activated by hydrogen ions (H^+). *Salty* taste seems to be due to an influx of metal ions, particularly Na^+, while *umami* is elicited by the amino acid glutamate, which is responsible for the "meat taste" of beef and the flavor of monosodium glutamate (MSG). *Bitter* taste is elicited by alkaloids (e.g., caffeine and quinine) and other substances such as aspirin.

Laboratory Experiments

 Notify instructor of any food or scent allergies before beginning experiments.

ACTIVITY 3

Stimulating Taste Buds

1. Obtain several paper towels, a sugar packet, and a disposable autoclave bag and bring them to your bench.

2. With a paper towel, dry the dorsal surface of your tongue.

 Immediately dispose of the paper towel in the autoclave bag.

3. Tear off a corner of the sugar packet and shake a few sugar crystals on your dried tongue. Do *not* close your mouth.

Time how long it takes to taste the sugar. _____ sec

Why couldn't you taste the sugar immediately?

_____ ▬

ACTIVITY 4

Examining the Combined Effects of Smell, Texture, and Temperature on Taste

Effects of Smell and Texture

1. Ask the subject to sit with eyes closed and to pinch his or her nostrils shut.

2. Using a paper plate, obtain samples of the food items provided by your laboratory instructor. At no time should the subject be allowed to see the foods being tested. Wear disposable gloves and use toothpicks to handle food.

3. For each test, place a cube of food in the subject's mouth and ask him or her to identify the food by using the following sequence of activities:

- First, manipulate the food with the tongue.

- Second, chew the food.

- Third, if a positive identification is not made with the first two techniques and the taste sense, ask the subject to release the pinched nostrils and to continue chewing with the nostrils open to determine if a positive identification can be made.

In the **Activity 4 chart,** record the type of food, and then put a check mark in the appropriate column for the result.

Was the sense of smell equally important in all cases?

_____

Where did it seem to be important and why?

_____

_____

Discard gloves in autoclave bag.

Effect of Olfactory Stimulation

What is commonly referred to as taste depends heavily on stimulation of the olfactory receptors, particularly in the case of strongly odoriferous substances. The following experiment should illustrate this fact.

1. Obtain vials of oil of wintergreen, peppermint, and cloves, paper cup, flask of water, paper towels, and some fresh cotton-tipped swabs. Ask the subject to sit so that he or she cannot see which vial is being used, and to dry the tongue and close the nostrils.

Activity 4: Identification by Texture and Smell				
Food tested	Texture only	Chewing with nostrils pinched	Chewing with nostrils open	Identification not made

2. Use a cotton swab to apply a drop of one of the oils to the subject's tongue. Can he or she distinguish the flavor?

_____

 Put the used swab in the autoclave bag. *Do not redip the swab into the oil.*

3. Have the subject open the nostrils, and record the change in sensation he or she reports.

_____

4. Have the subject rinse the mouth well and dry the tongue.

5. Prepare two swabs, each with one of the two remaining oils.

6. Hold one swab under the subject's open nostrils, while touching the second swab to the tongue.

Record the reported sensations. _____

_____

_____

 7. Dispose of the used swabs and paper towels in the autoclave bag before continuing.

Which sense, taste or smell, appears to be more important in the proper identification of a strongly flavored volatile substance?

_____

Effect of Temperature

In addition to the effect that olfaction and food texture have in determining our taste sensations, the temperature of foods also helps determine if the food is appreciated or even tasted. To illustrate this, have your partner hold some chipped ice on the tongue for approximately a minute and then close his

or her eyes. Immediately place any of the foods previously identified in his or her mouth and ask for an identification.

Results? _____

_____

_____

ACTIVITY 5

Assessing the Importance of Taste and Olfaction in Odor Identification

1. Go to the designated testing area. Close your nostrils with a nose clip, and breathe through your mouth. Breathing through your mouth only, attempt to identify the odors of common substances in the numbered vials at the testing area. Do not look at the substance in the container. Record your responses on the chart above.

2. Remove the nose clips, and repeat the tests using your nose to sniff the odors. Record your responses in the **Activity 5 chart.**

3. Record any other observations you make as you conduct the tests.

4. Which method gave the best identification results?

_____

What can you conclude about the effectiveness of the senses of taste and olfaction in identifying odors?

_____

ACTIVITY 6

Demonstrating Olfactory Adaptation

Obtain some absorbent cotton and two of the following oils (oil of wintergreen, peppermint, or cloves). Place several drops of oil on the absorbent cotton. Press one nostril shut.

26

Activity 5: Identification by Mouth and Nasal Inhalation			
Vial number	Identification with nose clips	Identification without nose clips	Other observations
1			
2			
3			
4			
5			

Hold the cotton under the open nostril and exhale through the mouth. Record the time required for the odor to disappear (for olfactory adaptation to occur).

_____ sec

Repeat the procedure with the other nostril.

_____ sec

Immediately test another oil with the nostril that has just experienced olfactory adaptation. What are the results?

_____

What conclusions can you draw? _____

_____

Special Senses: Olfaction and Taste

Location and Anatomy of the Olfactory Receptors

1. Describe the location and cellular composition of the olfactory epithelium. _____

   _____

   _____

2. How and why does sniffing increase your ability to detect an odor? _____

   _____

Location and Anatomy of Taste Buds

3. Name five sites where receptors for taste are found, and circle the predominant site.

   _____ , _____ , _____ ,

   _____ , and _____

4. Describe the cellular makeup and arrangement of a taste bud. (Use a diagram, if helpful.) _____

   _____

Laboratory Experiments

5. Taste and smell receptors are both classified as _____ , because they both

   respond to _____

6. Why is it impossible to taste substances with a dry tongue? _____

   _____

7. The basic taste sensations are mediated by specific chemical substances or groups. Name them for the following taste modalities.

   salt: _____   sour: _____   umami: _____

   bitter: _____   sweet: _____

8. Name three factors that influence our appreciation of foods. Substantiate each choice with an example from the laboratory experience.

1. _____ Substantiation: _____

_____

2. _____ Substantiation: _____

_____

3. _____ Substantiation: _____

_____

Which of the factors chosen is most important? _____ Substantiate your choice with an example from

everyday life. _____

_____

Expand on your explanation and choices by explaining why a cold, greasy hamburger is unappetizing to most people.

_____

_____

9. How palatable is food when you have a cold? _____ Explain your answer. _____

_____

10. In your opinion, is olfactory adaptation desirable? _____ Explain your answer.

_____

_____

Functional Anatomy of the Endocrine Glands

MATERIALS

☐ Human torso model

☐ Anatomical chart of the human endocrine system

☐ Compound microscope

☐ Prepared slides of the anterior pituitary and pancreas (with differential staining), posterior pituitary, thyroid gland, parathyroid glands, and adrenal gland

 For instructions on animal dissections, see the dissection exercises (starting on page 697) in the cat and fetal pig editions of this manual.

OBJECTIVES

1. Identify the major endocrine glands and tissues of the body when provided with an appropriate image.

2. List the hormones produced by the endocrine glands and discuss the general function of each.

3. Explain how hormones contribute to body homeostasis by giving appropriate examples of hormonal actions.

4. Discuss some mechanisms that stimulate release of hormones from endocrine glands.

5. Describe the structural and functional relationship between the hypothalamus and the pituitary gland.

6. Cite a major pathological consequence of hypersecretion and hyposecretion of several of the hormones studied.

7. Correctly identify the histological structure of the thyroid, parathyroid, pancreas, anterior and posterior pituitary, adrenal cortex, and adrenal medulla by microscopic inspection or in an image.

8. Name and point out the specialized hormone-secreting cells in the above tissues.

PRE-LAB QUIZ

1. Define *hormone*. _____

2. Circle the correct underlined term. An <u>endocrine</u> / <u>exocrine</u> gland is a ductless gland that empties its hormone into the extracellular fluid, from which it enters the blood.

3. The pituitary gland, also known as the _____, is located in the sella turcica of the sphenoid bone.
   a. hypophysis
   b. hypothalamus
   c. thalamus

4. Circle True or False. The anterior pituitary gland is also referred to as the master endocrine gland because it controls the activity of many other endocrine glands.

5. The _____ gland is composed of two lobes and located in the throat, just inferior to the larynx.
   a. pancreas   c. thymus
   b. posterior pituitary   d. thyroid

6. The pancreas produces two hormones that are responsible for regulating blood sugar levels. Name the hormone that increases blood glucose levels.

   _____

7. Circle True or False. The gonads are considered to be both endocrine and exocrine glands.

MasteringA&P® For related exercise study tools, go to the Study Area of MasteringA&P. There you will find:
- Practice Anatomy Lab PAL
- PhysioEx PEx
- A&PFlix A&PFlix
- Practice quizzes, Histology Atlas, eText, Videos, and more!

(Text continues on next page.)

8. This gland is rather large in an infant, begins to atrophy at puberty, and is relatively inconspicuous by old age. It produces hormones that direct the maturation of T cells. It is the _____ gland.
   a. pineal
   b. testes
   c. thymus
   d. thyroid

9. Circle the correct underlined term. <u>Pancreatic islets</u> / <u>Acinar cells</u> form the endocrine portion of the pancreas.

10. The outer cortex of the adrenal gland is divided into three areas. Which one produces aldosterone?
    a. zona fasciculata
    b. zona glomerulosa
    c. zona reticularis

The **endocrine system** is the second major control system of the body. Acting with the nervous system, it helps coordinate and integrate the activity of the body. The nervous system uses electrochemical impulses to bring about rapid control, whereas the more slowly acting endocrine system uses chemical messengers, or **hormones,** which ultimately enter the blood to be transported throughout the body.

The term *hormone* comes from a Greek word meaning "to arouse." The body's hormones, which are steroids or amino acid–based molecules, arouse the body's tissues and cells by stimulating changes in their metabolic activity. These changes lead to growth and development and to the physiological homeostasis of many body systems. Although all hormones are bloodborne, a given hormone affects only the biochemical activity of a specific organ or organs. Organs that respond to a particular hormone are referred to as the **target organs** of that hormone. The ability of the target tissue to respond depends on the ability of the hormone to bind with specific receptors occurring on the cells' plasma membrane or within the cells.

Although the function of some hormone-producing glands (the anterior pituitary, thyroid, adrenals, parathyroids) is purely endocrine, the function of others (the pancreas and gonads) is mixed—both endocrine and exocrine. Both types of glands are derived from epithelium, but the endocrine glands release their hormones directly into the extracellular fluid, from which the hormones enter blood or lymph. The exocrine glands release their products at the body's surface or upon an epithelial membrane via ducts. In addition, there are hormone-producing cells in the heart, the gastrointestinal tract, kidney, skin, adipose tissue, skeleton, and placenta, organs whose functions are primarily nonendocrine. Only the major endocrine organs, plus the pineal gland and the thymus, are considered here.

Gross Anatomy and Basic Function of the Endocrine Glands

Pituitary Gland (Hypophysis)

The **pituitary gland,** or **hypophysis,** is located in the sella turcica of the sphenoid bone. It consists largely of two functional *lobes,* the **adenohypophysis,** or **anterior pituitary,** and the **neurohypophysis,** consisting of the **posterior pituitary** and the **infundibulum**—the stalk that attaches the pituitary gland to the hypothalamus (**Figure 27.1**).

Anterior Pituitary Hormones

The anterior pituitary produces and secretes a number of hormones, four of which are **tropic hormones.** The target organ of a tropic hormone is another endocrine gland, which secretes its hormone in response to stimulation. Hormones from these target glands exert their effects on other body organs and tissues.

Because the anterior pituitary controls the activity of many other endocrine glands, it has been called the *master endocrine gland.* However, because *releasing* or *inhibiting hormones* from neurons of the ventral hypothalamus control anterior pituitary cells, the hypothalamus has superseded the anterior pituitary as the major controller of endocrine glands.

The anterior pituitary *tropic hormones* include:

- **Gonadotropins–follicle-stimulating hormone (FSH)** and **luteinizing hormone (LH)**—regulate gamete production and hormonal activity of the gonads (ovaries and testes). (The precise roles of the gonadotropins are described in Exercise 43 along with other considerations of reproductive system physiology.)

- **Adrenocorticotropic hormone (ACTH)** regulates the endocrine activity of the adrenal cortex.

- **Thyroid-stimulating hormone (TSH),** or **thyrotropin,** influences the growth and activity of the thyroid gland.

The two other important hormones produced by the anterior pituitary are not directly involved in the regulation of other endocrine glands of the body. They are:

- **Growth hormone (GH)** is a general metabolic hormone that plays an important role in determining body size. It affects many tissues of the body; however, its major effects are exerted on the growth of muscle and the long bones of the body.

  Hyposecretion results in pituitary dwarfism in children. Hypersecretion causes gigantism in children and **acromegaly** (overgrowth of bones in hands, feet, and face) in adults. ✚

- **Prolactin (PRL)** stimulates milk production by the breasts. The role of prolactin in males is not well understood.

The ventral hypothalamic hormones control production and secretion of the tropic hormones, GH, and PRL. The hypothalamic hormones reach the cells of the anterior pituitary

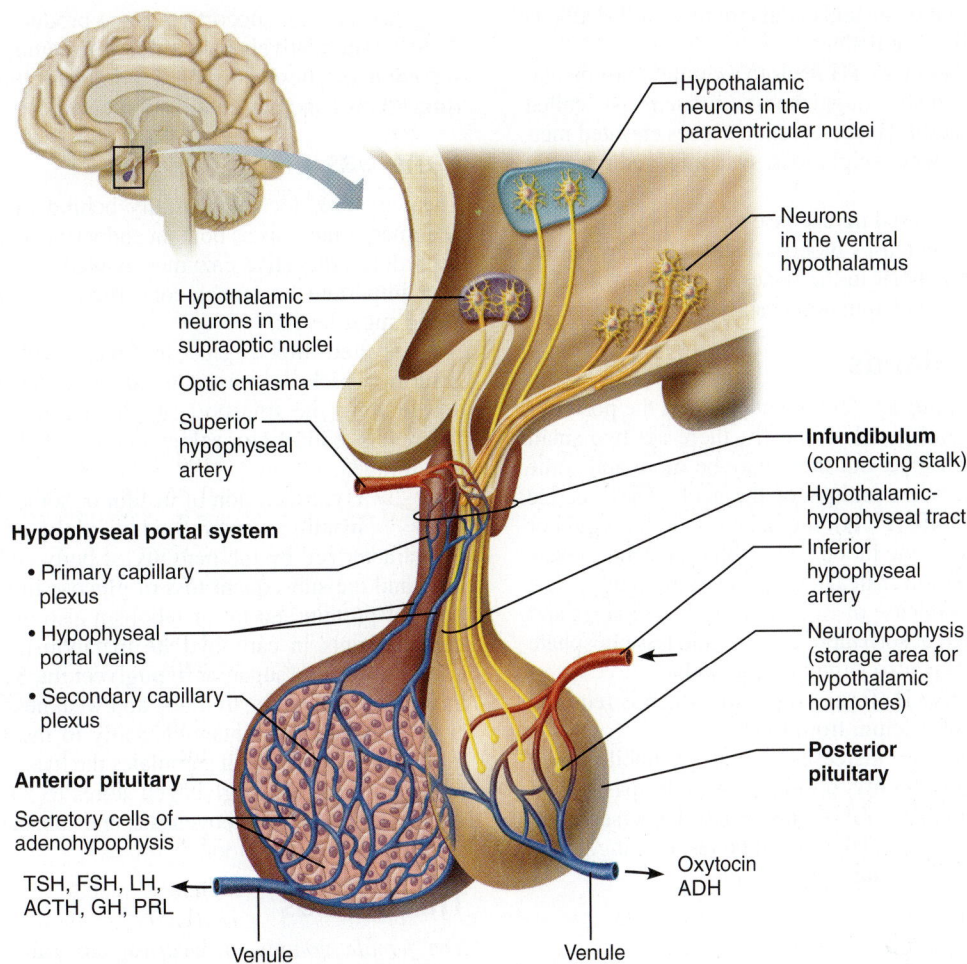

Figure 27.1 Hypothalamus and pituitary gland. Neural and vascular relationships between the hypothalamus and the anterior and posterior lobes of the pituitary are depicted.

through the **hypophyseal portal system** (Figure 27.1), a complex vascular arrangement of two capillary beds that are connected by the hypophyseal portal veins.

Posterior Pituitary Hormones

The posterior pituitary is not an endocrine gland because it does not synthesize the hormones it releases. Instead, it acts as a storage area for two *neurohormones* transported to it via the axons of neurons in the paraventricular and supraoptic nuclei of the hypothalamus. The hormones are released in response to nerve impulses from these neurons. The first of these hormones is **oxytocin,** which stimulates powerful uterine contractions during birth and also causes milk ejection in the lactating mother. The second, **antidiuretic hormone (ADH),** causes the tubules of the kidneys to reabsorb more water from the urinary filtrate, thereby reducing urine output and conserving body water.

Hyposecretion of ADH results in dehydration from excessive urine output, a condition called **diabetes insipidus.** Individuals with this condition experience an insatiable thirst. Hypersecretion results in edema, headache, and disorientation. ✚

Pineal Gland

The *pineal gland* is a small cone-shaped gland located in the roof of the third ventricle of the brain. Its major endocrine product is **melatonin,** which exhibits a diurnal (daily) cycle. It peaks at night, making us drowsy, and is lowest around noon.

The endocrine role of the pineal gland in humans is still controversial, but it is known to play a role in mating and migratory behavior of other animals. In humans, melatonin appears to exert some inhibitory effect on the reproductive system that prevents precocious sexual maturation. Changing levels of melatonin may also affect biological rhythms associated with body temperature, sleep, and appetite.

Thyroid Gland

The *thyroid gland* is composed of two lobes joined by a central mass, or isthmus. It is located in the throat, just inferior to the larynx. It produces two major hormones, thyroid hormone and calcitonin.

Thyroid hormone (TH) is actually two physiologically active hormones known as T_4 (**thyroxine**) and T_3 (**triiodothyronine**). Because its primary function is to control the rate of body metabolism and cellular oxidation, TH affects virtually every cell in the body.

Hyposecretion of TH leads to a condition of mental and physical sluggishness, which is called **myxedema** in the adult. Hypersecretion causes elevated metabolic rate, nervousness, weight loss, sweating, and irregular heartbeat. ✚

Calcitonin is released in response to high blood calcium levels. Although it decreases blood calcium levels by stimulating calcium salt deposit in the bones, it is not involved in day-to-day control of calcium homeostasis.

Parathyroid Glands

The *parathyroid glands* are found embedded in the posterior surface of the thyroid gland. Typically, there are two small oval glands on each lobe, but there may be more and some may be located in other regions of the neck. They secrete **parathyroid hormone (PTH),** the most important regulator of calcium balance of the blood. When blood calcium levels decrease below a certain critical level, the parathyroids release PTH, which causes release of calcium from bone matrix and prods the kidney to reabsorb more calcium and less phosphate from the filtrate. PTH also stimulates the kidneys to convert vitamin D to its active D_3 form, *calcitriol*, which is required for the absorption of calcium from food.

Hyposecretion increases neural excitability and may lead to **tetany,** prolonged muscle spasms that can result in respiratory paralysis and death. Hypersecretion of PTH results in loss of calcium from bones, causing deformation, softening, and spontaneous fractures. ✚

Thymus

The *thymus* is a bilobed gland situated in the superior thorax, posterior to the sternum and anterior to the heart and lungs. Conspicuous in the infant, it begins to atrophy at puberty, and by old age it is relatively inconspicuous. The thymus produces several different families of hormones including **thymulin, thymosins,** and **thymopoietins.** These hormones are thought to be involved in the development of T lymphocytes and the immune response. Their role is poorly understood; they appear to act locally as paracrines.

Adrenal Glands

The two *adrenal,* or *suprarenal, glands* are located atop or close to the kidneys. Anatomically, the **adrenal medulla** develops from neural crest tissue, and it is directly controlled by the sympathetic nervous system. The medullary cells respond to this stimulation by releasing a hormone mix of **epinephrine** (80%) and **norepinephrine** (20%), which act with the sympathetic nervous system to elicit the fight-or-flight response to stressors.

The **adrenal cortex** produces three major groups of steroid hormones, collectively called **corticosteroids.** The **mineralocorticoids,** chiefly *aldosterone,* regulate water and electrolyte balance in the extracellular fluids, mainly by regulating sodium ion reabsorption by kidney tubules. The **glucocorticoids** include (*cortisol* [*hydrocortisone*], *cortisone,* and *corticosterone*), but only cortisol is secreted in significant amounts in humans. It enables the body to resist long-term stressors, primarily by increasing blood glucose levels. The **gonadocorticoids,** or **sex hormones,** produced by the adrenal cortex are chiefly *androgens* (male sex hormones), but some *estrogens* (female sex hormones) are also formed.

The gonadocorticoids are produced throughout life in relatively insignificant amounts; however, hypersecretion of these hormones produces abnormal hairiness **(hirsutism),** and masculinization occurs. ✚

Pancreas

The *pancreas,* located partially behind the stomach in the abdomen, functions as both an endocrine and exocrine gland. It produces digestive enzymes as well as insulin and glucagon, important hormones concerned with the regulation of blood sugar levels.

Elevated blood glucose levels stimulate release of **insulin,** which decreases blood sugar levels, primarily by accelerating the transport of glucose into the body cells, where it is oxidized for energy or converted to glycogen or fat for storage.

Hyposecretion of insulin or some deficiency in the insulin receptors leads to **diabetes mellitus,** which is characterized by the inability of body cells to utilize glucose and the subsequent loss of glucose in the urine. Alterations of protein and fat metabolism also occur secondary to derangements in carbohydrate metabolism. Hypersecretion causes low blood sugar, or **hypoglycemia.** Symptoms include anxiety, nervousness, tremors, and weakness. ✚

Glucagon acts antagonistically to insulin. When blood glucose levels are low, it stimulates the liver, its primary target organ, to break down glycogen stores to glucose, to synthesize glucose by gluconeogenesis, and subsequently to release the glucose into the blood.

The Gonads

The *female gonads,* or *ovaries,* are paired, almond-sized organs located in the pelvic cavity. In addition to producing the female sex cells (ova), the ovaries produce two steroid hormone groups, the estrogens and progesterone. The endocrine and exocrine functions of the ovaries do not begin until the onset of puberty. The **estrogens** are responsible for the development of the secondary sex characteristics of the female at puberty (primarily maturation of the reproductive organs and development of the breasts) and act with progesterone to bring about cyclic changes of the uterine lining that occur during the menstrual cycle. The estrogens also help prepare the mammary glands for lactation. During pregnancy progesterone maintains the uterine musculature in a quiescent state and helps to prepare the breast tissue for lactation.

The paired oval *testes* of the male are suspended in a pouchlike sac, the scrotum, outside the pelvic cavity. In addition to the male sex cells (sperm), the testes produce the male sex hormone, **testosterone.** Testosterone promotes the maturation of the reproductive system accessory structures, brings about the development of the male secondary sex characteristics, and is responsible for sexual drive, or libido. Both the endocrine and exocrine functions of the testes begin at puberty. (For a more detailed discussion of the function and histology of the ovaries and testes, see Exercises 42 and 43.)

ACTIVITY 1

Identifying the Endocrine Organs

Locate the endocrine organs on the figure of the body **(Figure 27.2)**. Also locate these organs on the anatomical charts or torso model. ■

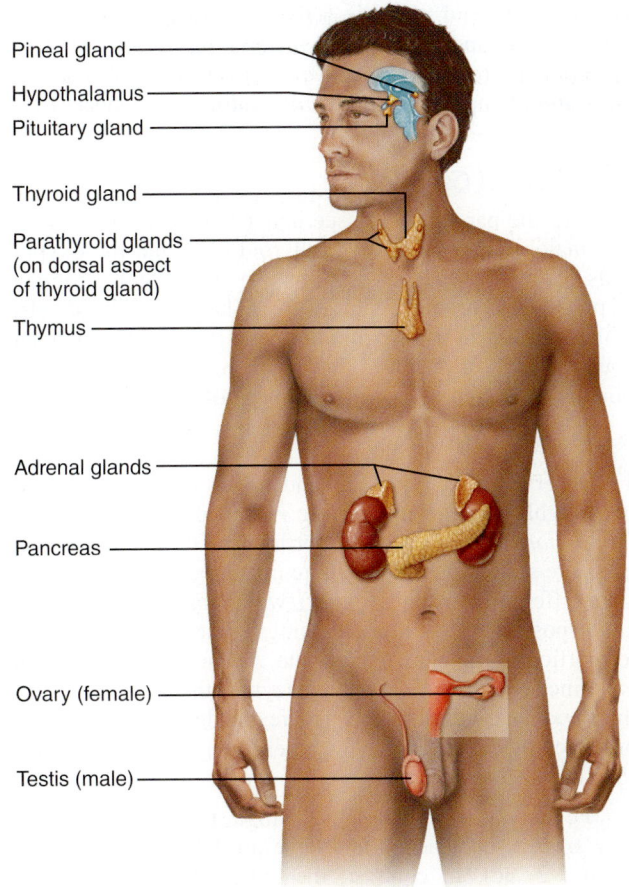

Pineal gland
Hypothalamus
Pituitary gland
Thyroid gland
Parathyroid glands (on dorsal aspect of thyroid gland)
Thymus
Adrenal glands
Pancreas
Ovary (female)
Testis (male)

Figure 27.2 Human endocrine organs.

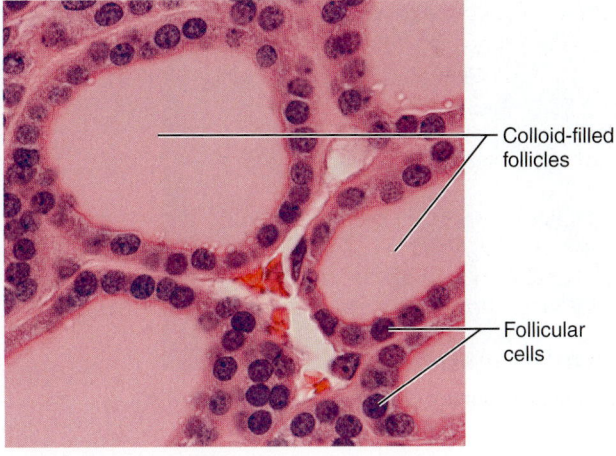

Colloid-filled follicles
Follicular cells

(a) Thyroid gland (360×)

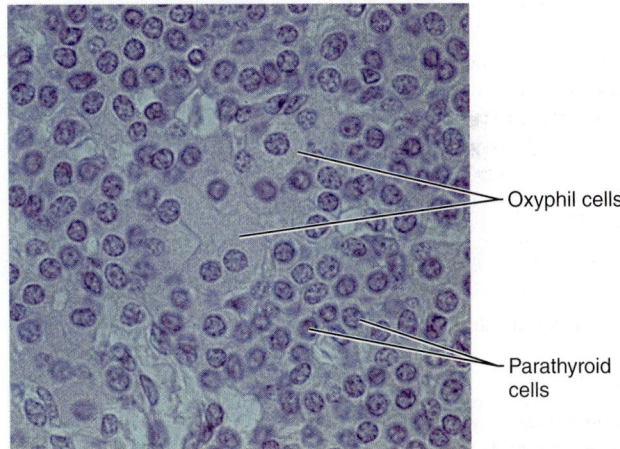

Oxyphil cells
Parathyroid cells

(b) Parathyroid gland (375×)

27

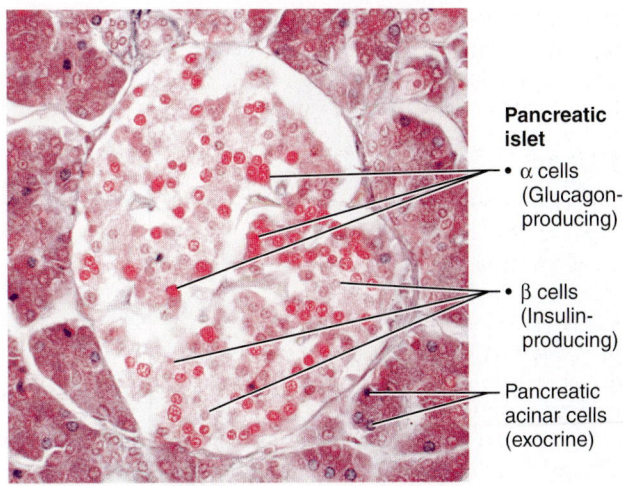

Pancreatic islet
• α cells (Glucagon-producing)
• β cells (Insulin-producing)
Pancreatic acinar cells (exocrine)

(c) Pancreatic islet (205×)

Figure 27.3 Microscopic anatomy of selected endocrine organs.

Microscopic Anatomy of Selected Endocrine Glands

ACTIVITY 2

Examining the Microscopic Structure of Endocrine Glands

Obtain a microscope and one of each slide on the materials list. We will study only organs in which it is possible to identify the endocrine-producing cells. Compare your observations with the histology images **(Figure 27.3a–f)**.

Thyroid Gland

1. Scan the thyroid under low power, noting the **follicles,** spherical sacs containing a pink-stained material *(colloid).* Stored T_3 and T_4 are attached to the protein colloidal material stored in the follicles as **thyroglobulin** and are released gradually to the blood. Compare the tissue viewed to the photomicrograph of thyroid tissue (Figure 27.3a).

2. Observe the tissue under high power. Notice that the walls of the follicles are formed by simple cuboidal or squamous epithelial cells that synthesize the follicular products. The **parafollicular,** or **C, cells** you see between the follicles are responsible for calcitonin production.

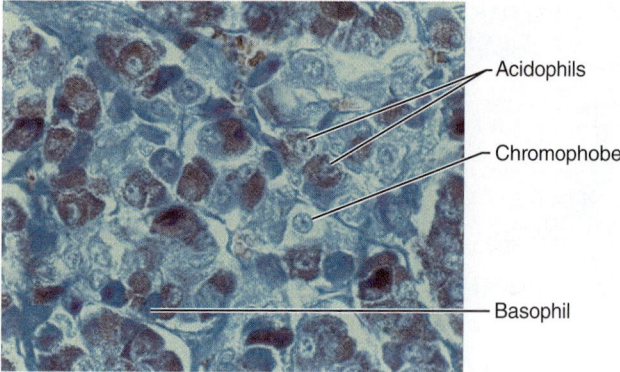

(d) Anterior pituitary (380×)

Acidophils
Chromophobe
Basophil

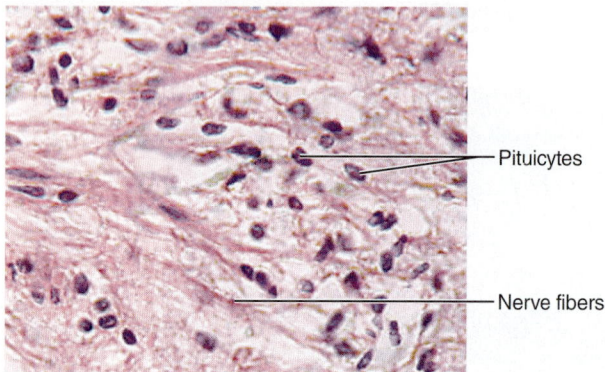

(e) Posterior pituitary (345×)

Pituicytes
Nerve fibers

27

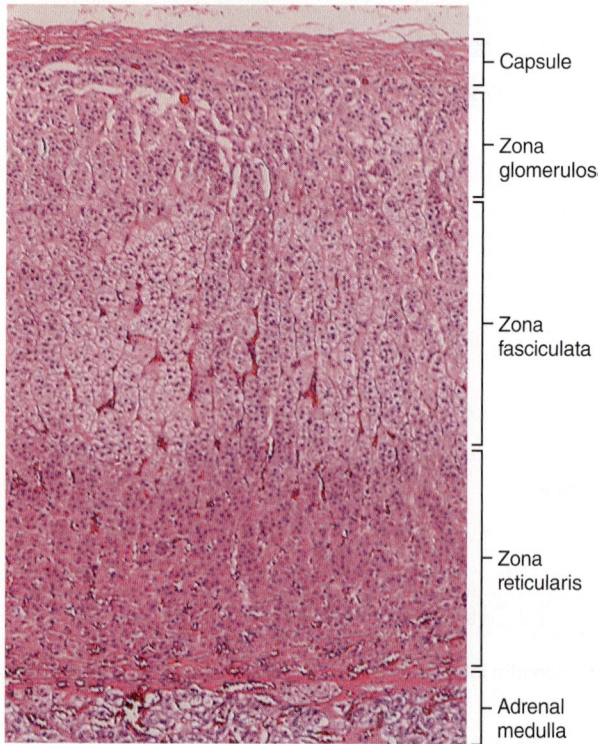

(f) Adrenal gland (60×)

Capsule
Zona glomerulosa
Zona fasciculata
Zona reticularis
Adrenal medulla

Figure 27.3 *(continued)* **Microscopic anatomy of selected endocrine organs.**

When the thyroid gland is actively secreting, the follicles appear small, and the colloidal material has a ruffled border. When the thyroid is hypoactive or inactive, the follicles are large and plump, and the follicular epithelium appears to be squamouslike.

Parathyroid Glands

Observe the parathyroid tissue under low power to view its two major cell types, the parathyroid cells and the oxyphil cells. Compare your observations to the photomicrograph of parathyroid tissue (Figure 27.3b). The **parathyroid cells,** which synthesize parathyroid hormone (PTH), are small and abundant, and arranged in thick branching cords. The function of the scattered, much larger **oxyphil cells** is unknown.

Pancreas

1. Observe pancreas tissue under low power to identify the roughly circular **pancreatic islets** (also called *islets of Langerhans*), the endocrine portions of the pancreas. The islets are scattered amid the more numerous **acinar cells** and stain differently (usually lighter), which makes their identification possible. The deeper-staining acinar cells form the major portion of the pancreatic tissue. Acinar cells produce the exocrine secretion of digestive enzymes that is released into the duodenum through the pancreatic duct. Alkaline fluid produced by duct cells accompanies the hydrolytic enzymes. (See Figure 27.3c.)

2. Focus on islet cells under high power. Notice that they are densely packed and have no definite arrangement (Figure 27.3c). In contrast, the cuboidal acinar cells are arranged around secretory ducts. If special stains are used, it will be possible to distinguish the **alpha (α) cells,** which tend to cluster at the periphery of the islets and produce glucagon, from the **beta (β) cells,** which synthesize insulin. With these specific stains, the beta cells are larger and stain gray-blue, and the alpha cells are smaller and appear bright pink.

Pituitary Gland

1. Observe the general structure of the pituitary gland under low power to differentiate between the glandular anterior pituitary and the neural posterior pituitary.

2. Using the high-power lens, focus on the nests of cells of the anterior pituitary. When differential stains are used it is possible to identify the specialized cell types that secrete the specific hormones. Using the anterior pituitary photomicrograph (Figure 27.3d) as a guide, locate the reddish brown–stained **acidophil cells,** which produce growth hormone and prolactin, and the **basophil cells,** whose deep-blue granules are responsible for the production of the tropic hormones (TSH, ACTH, FSH, and LH). **Chromophobes,** the third cellular population, do not take up the stain and appear rather dull and colorless. The role of the chromophobes is controversial, but they apparently are not directly involved in hormone production.

3. Switch your focus to the posterior pituitary where two hormones (oxytocin and ADH) synthesized by hypothalamic neurons are stored. Observe the axons of hypothalamic neurons that compose most of this portion of the pituitary. Also note the neuroglia, or **pituicytes** (Figure 27.3e).

Adrenal Gland

1. Hold the slide of the adrenal gland up to the light to distinguish the outer cortex and inner medulla areas. Then scan the cortex under low power to distinguish the differences in cell appearance and arrangement in the three cortical areas. Refer to the adrenal gland photomicrograph (Figure 27.3f) as you work. In the outermost **zona glomerulosa,** where most mineralocorticoid production occurs, the tightly packed cells are arranged in spherical clusters. The deeper intermediate **zona fasciculata** produces glucocorticoids. This is the thickest part of the cortex. Its cells are arranged in parallel cords. The innermost cortical zone, the **zona reticularis** produces sex hormones and some glucocorticoids. The cells here stain intensely and form a branching network.

2. Switch to higher power to view the large, lightly stained cells of the adrenal medulla, which produce epinephrine and norepinephrine. Notice their clumped arrangement. ▬

GROUP CHALLENGE

Odd Hormone Out

Each box below contains four hormones. One of the listed hormones does not share a characteristic that the other three do. Circle the hormone that doesn't belong with the others and explain why it is singled out. What characteristic is it missing? Sometimes there may be multiple reasons why the hormone doesn't belong with the others.

1. Which is the "odd hormone"?	Why is it the odd one out?
ACTH oxytocin LH FSH	
2. Which is the "odd hormone"?	**Why is it the odd one out?**
aldosterone cortisol epinephrine ADH	
3. Which is the "odd hormone"?	**Why is it the odd one out?**
PTH testosterone LH FSH	
4. Which is the "odd hormone"?	**Why is it the odd one out?**
insulin cortisol calcitonin glucagon	

27

Functional Anatomy
of the Endocrine Glands

Gross Anatomy and Basic Function of the Endocrine Glands

1. Both the endocrine and nervous systems are major regulating systems of the body; however, the nervous system has been compared to an airmail delivery system and the endocrine system to the Pony Express. Briefly explain this comparison.

_____

_____

_____

2. Define *hormone.* _____

_____

3. Chemically, hormones belong chiefly to two molecular groups, the _____

and the _____.

4. Define *target organ.* _____

_____

5. If hormones travel in the bloodstream, why don't all tissues respond to all hormones? _____

_____

6. Identify the endocrine organ described by each of the following statements.

_____ 1. located in the throat; bilobed gland connected by an isthmus

_____ 2. found atop the kidney

_____ 3. a mixed gland, located close to the stomach and small intestine

_____ 4. paired glands suspended in the scrotum

_____ 5. ride "horseback" on the thyroid gland

_____ 6. found in the pelvic cavity of the female, concerned with ova and female hormone production

_____ 7. found in the upper thorax overlying the heart; large during youth

_____ 8. found in the roof of the third ventricle

7. The table below lists the functions of many of the hormones you have studied. From the keys below, fill in the hormones responsible for each function, and the endocrine glands that produce each hormone. Glands may be used more than once.

Hormones Key:

ACTH	FSH	prolactin
ADH	glucagon	PTH
aldosterone	insulin	T_3/T_4
cortisol	LH	testosterone
epinephrine	oxytocin	TSH
estrogens	progesterone	

Glands Key:

adrenal cortex	parathyroid glands
adrenal medulla	posterior pituitary
anterior pituitary	testes
hypothalamus	thyroid gland
ovaries	
pancreas	

Function	Hormone(s)	Gland(s)
Regulate the function of another endocrine gland	1.	
	2.	
	3.	
	4.	
Maintenance of salt and water balance in the extracellular fluid	1.	
	2.	
Directly involved in milk production and ejection	1.	
	2.	
Controls the rate of body metabolism and cellular oxidation	1.	
Regulates blood calcium levels	1.	
Regulate blood glucose levels; produced by the same "mixed" gland	1.	
	2.	
Released in response to stressors	1.	
	2.	
Drive development of secondary sex characteristics in males	1.	
Directly responsible for regulation of the menstrual cycle	1.	
	2.	

8. Although the pituitary gland is often referred to as the master gland of the body, the hypothalamus exerts control over the pituitary gland. How does the hypothalamus control both anterior and posterior pituitary functioning?

_____

_____

9. Indicate whether the release of the hormones listed below is stimulated by (A) another hormone; (B) the nervous system (neurotransmitters, or neurosecretions); or (C) humoral factors (the concentration of specific nonhormonal substances in the blood or extracellular fluid). (Use your textbook as necessary.)

_____ 1. ACTH   _____ 4. insulin   _____ 7. T_4/T_3

_____ 2. calcitonin   _____ 5. norepinephrine   _____ 8. testosterone

_____ 3. estrogens   _____ 6. parathyroid hormone   _____ 9. TSH, FSH

10. Name the hormone(s) produced in *inadequate* amounts that directly result in the following conditions.

_____ 1. tetany

_____ 2. excessive diuresis without high blood glucose levels

_____ 3. loss of glucose in the urine

_____ 4. abnormally small stature, normal proportions

_____ 5. low BMR, mental and physical sluggishness

11. Name the hormone(s) produced in *excessive* amounts that directly result in the following conditions.

_____ 1. large hands and feet in the adult, large facial bones

_____ 2. nervousness, irregular pulse rate, sweating

_____ 3. demineralization of bones, spontaneous fractures

Microscopic Anatomy of Selected Endocrine Glands

12. Choose a response from the key below to name the hormone(s) produced by the cell types listed.

Key: a. calcitonin   d. glucocorticoids   g. PTH
       b. GH, prolactin   e. insulin   h. T_4/T_3
       c. glucagon   f. mineralocorticoids   i. TSH, ACTH, FSH, LH

_____ 1. parafollicular cells of the thyroid   _____ 6. zona fasciculata cells

_____ 2. follicular cells of the thyroid   _____ 7. zona glomerulosa cells

_____ 3. beta cells of the pancreatic islets   _____ 8. parathyroid cells

_____ 4. alpha cells of the pancreatic islets   _____ 9. acidophil cells of the anterior pituitary

_____ 5. basophil cells of the anterior pituitary

13. Six diagrams of the microscopic structures of the endocrine glands are presented here. Identify each and _name all structures indicated by a leader line or bracket._

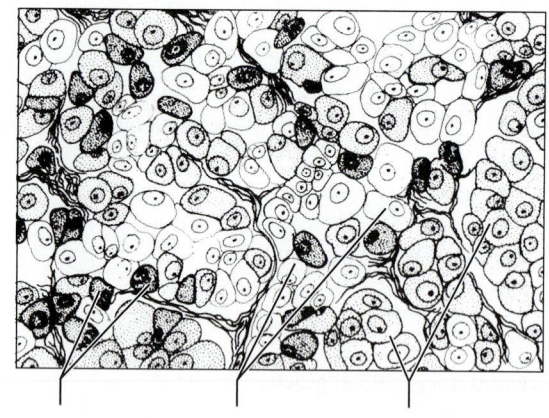

(a) _____

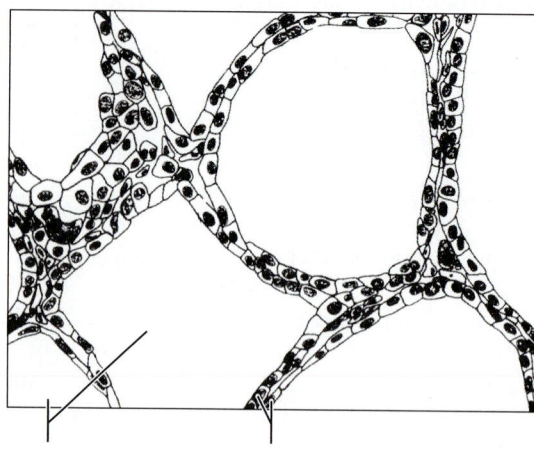

(d) _____

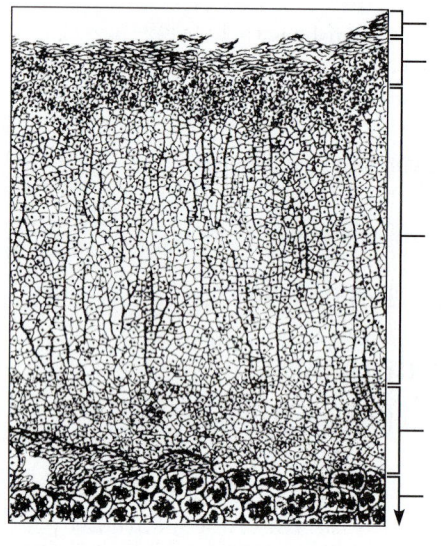

(b) _____

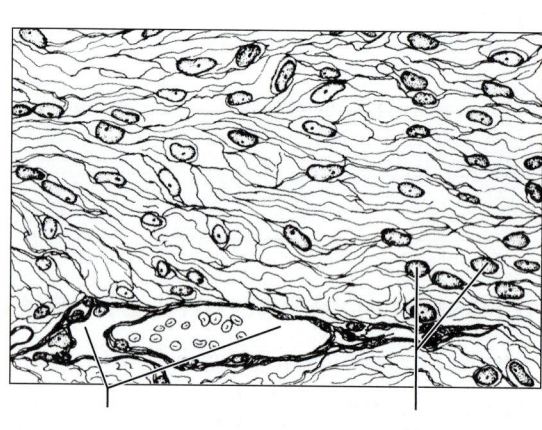

(e) _____

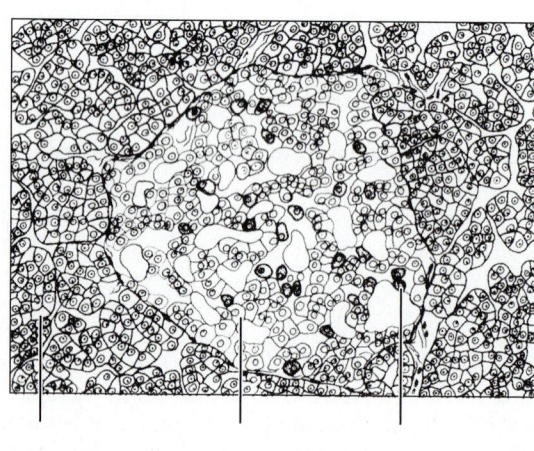

(c) _____

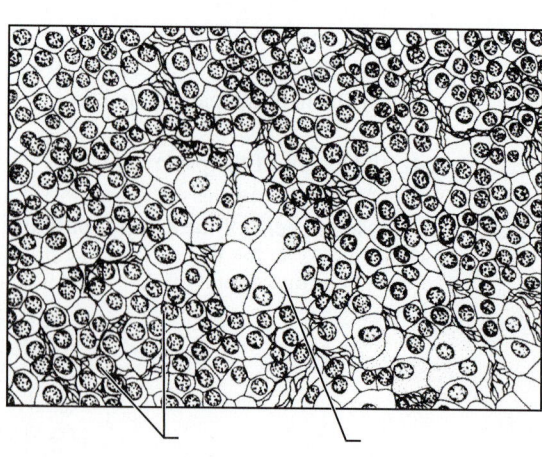

(f) _____

Endocrine Wet Labs and Human Metabolism

MATERIALS

Activity 1: Pituitary hormone and ovary*
- ☐ Female frogs (*Rana pipiens*)
- ☐ Disposable gloves
- ☐ Battery jars
- ☐ Syringe (2-ml capacity)
- ☐ 20- to 25-gauge needle
- ☐ Frog pituitary extract
- ☐ Physiological saline
- ☐ Spring or pond water
- ☐ Wax marking pencils

Activity 2: Hyperinsulinism*
- ☐ 500- or 600-ml beakers
- ☐ 20% glucose solution
- ☐ Commercial insulin solution (400 international units [IU] per 100 ml of H_2O)
- ☐ Finger bowls
- ☐ Small (4–5 cm, or 1½–2 in.) freshwater fish (guppy, bluegill, or sunfish–listed in order of preference)
- ☐ Wax marking pencils

The Selected Actions of Hormones and Other Chemical Messengers video (available to qualified adopters from Pearson Education) may be used in lieu of student participation in Activities 1 and 2.

PEx PhysioEx™ 9.1 Computer Simulation Ex. 4 on p. PEx-59.

MasteringA&P® For related exercise study tools, go to the Study Area of MasteringA&P. There you will find:
- Practice Anatomy Lab **PAL**
- PhysioEx **PEx**
- A&PFlix **A&PFlix**
- Practice quizzes, Histology Atlas, eText, Videos, and more!

OBJECTIVES

1. Describe the effects of pituitary extract in the frog and indicate which hormone(s) is/are responsible for these effects.
2. Describe the symptoms of hyperinsulinism in the fish and explain how these symptoms were reversed.
3. Define *metabolism.*
4. State the functions of thyroid hormone in the body.
5. Explain how negative feedback mechanisms regulate thyroid hormone secretion.
6. Describe and explain the various pathologies associated with hypothyroidism and hyperthyroidism.

PRE-LAB QUIZ

1. Circle True or False. Gonadotropins are produced by the anterior pituitary gland.
2. Circle the correct underlined term. Many people with diabetes mellitus need injections of <u>insulin</u> / <u>glucagon</u> to maintain homeostasis.
3. Circle the correct underlined term. <u>Catabolism</u> / <u>Anabolism</u> is the process by which substances are broken down into simpler compounds.
4. _____ is the single most important hormone responsible for influencing the rate of cellular metabolism and body heat production.
   a. Calcitonin   c. Insulin
   b. Estrogen   d. Thyroid hormone
5. Basal metabolic rate (BMR) is
   a. decreased in individuals with hyperthyroidism
   b. increased in individuals with hyperthyroidism
   c. increased in obese individuals

The endocrine system exerts many complex and interrelated effects on the body as a whole, as well as on specific organs and tissues. Most scientific knowledge about this system is recent, and new information is constantly being presented. Many experiments on the endocrine system require relatively large laboratory animals; are time-consuming (requiring days to weeks of observation); and often involve technically difficult surgical procedures to remove the glands or parts of them, all of which makes it difficult to conduct more general types of laboratory experiments. Nevertheless, the two technically unsophisticated experiments presented here should illustrate how dramatically hormones affect body functioning. (Also, students may perform simulated endocrine wet labs in PhysioEx Exercise 4.)

Endocrine Experiments: Gonadotropins and Insulin

ACTIVITY 1

Determining the Effect of Pituitary Hormones on the Ovary

The anterior pituitary gonadotropic hormones—follicle-stimulating hormone (FSH) and luteinizing hormone (LH)—regulate the ovarian cycles of the female (see Exercise 43). Although amphibians normally ovulate seasonally, many can be stimulated to ovulate "on demand" by injecting an extract of pituitary hormones. In the following experiment, you will need to inject the frog the day before the lab session or return to check results the day after the scheduled lab session.

⚠ 1. Don disposable gloves, and obtain two frogs. Place them in separate battery jars to bring them to your laboratory bench. Also bring back a syringe and needle, a wax marking pencil, pond or spring water, and containers of pituitary extract and physiological saline.

2. Before beginning, examine each frog for the presence of eggs. Hold the frog firmly with one hand and exert pressure on its abdomen toward the cloaca (in the direction of the legs). If ovulation has occurred, any eggs present in the oviduct will be forced out and will appear at the cloacal opening. If no eggs are present, continue with step 3.

If eggs are expressed, return the animal to your instructor and obtain another frog for experimentation. Repeat the procedure for determining if eggs are present until two frogs that lack eggs have been obtained.

3. Aspirate 1 to 2 ml of the pituitary extract into a syringe. Inject the extract subcutaneously into the anterior abdominal (peritoneal) cavity of the frog you have selected to be the experimental animal. To inject into the peritoneal cavity, hold the frog with its ventral surface superiorly. Insert the needle through the skin and muscles of the abdominal wall in the lower quarter of the abdomen. Do not insert the needle far enough to damage any of the vital organs. With a wax marker, label its large battery jar "experimental," and place the frog in it. Add a small amount of pond water to the battery jar before continuing.

4. Aspirate 1 to 2 ml of physiological saline into a syringe and inject it into the peritoneal cavity of the second frog—this will be the control animal. (Make sure you inject the same volume of fluid into both frogs.) Place this frog into the second battery jar, marked "control." Allow the animals to remain undisturbed for 24 hours.

5. After 24 hours, again check each frog for the presence of eggs in the cloaca. (See step 2.) If no eggs are present, make arrangements with your laboratory instructor to return to the lab on the next day (at 48 hours after injection) to check your frogs for the presence of eggs.

6. Return the frogs to the terrarium before leaving or continuing with the lab.

In which of the prepared frogs was ovulation induced?

_____

Specifically, what hormone in the pituitary extract causes ovulation to occur?

_____ ▬

ACTIVITY 2

Observing the Effects of Hyperinsulinism

Many people with diabetes mellitus need injections of insulin to maintain normal blood glucose levels. Adequate amounts of blood glucose are essential for proper functioning of the nervous system; thus, the administration of insulin must be carefully controlled. If blood glucose levels fall precipitously, the patient will go into insulin shock.

A small fish will be used to demonstrate the effects of hyperinsulinism. Since the action of insulin on the fish parallels that in the human, this experiment should provide valid information concerning its administration to humans.

1. Prepare two finger bowls. Using a wax marking pencil, mark one A and the other B. To finger bowl A, add 100 ml of the commercial insulin solution. To finger bowl B, add 200 ml of 20% glucose solution.

2. Place a small fish in finger bowl A and observe its actions carefully as the insulin diffuses into its bloodstream through the capillary circulation of its gills.

Approximately how long did it take for the fish to become comatose?

_____

What types of activity did you observe in the fish before it became comatose?

_____

_____

_____

_____

3. When the fish is comatose, carefully transfer it to finger bowl B and observe its actions. What happens to the fish after it is transferred?

_____

_____

_____

Approximately how long did it take for this recovery?

_____

4. After all observations have been made and recorded, carefully return the fish to the aquarium. ▬

Human Metabolism and Thyroid Hormones

Metabolism is a broad term referring to all chemical reactions that are necessary to maintain life. It involves both *catabolism,* enzymatically controlled processes in which substances are broken down to simpler substances, and *anabolism,* processes in which larger molecules or structures are built from smaller ones. Most catabolic reactions in the body are accompanied by a net release of energy. Some of the liberated energy is captured to make ATP, the energy-rich molecule used by body cells to energize all their activities; the balance is lost in the form of thermal energy or heat. Maintaining body temperature is linked to the heat-liberating aspects of metabolism.

Various foodstuffs make different contributions to the process of metabolism. For example, carbohydrates, particularly glucose, are generally broken down or oxidized to make ATP, whereas fats are utilized to form cell membranes and myelin sheaths, and to insulate the body with a fatty cushion. Fats are used secondarily for producing ATP, particularly when the diet is inadequate in carbohydrates. Proteins and amino acids tend to be conserved by body cells, and understandably so, since most structural elements of the body are built with proteins.

Thyroid hormone (TH, collectively T_3 and T_4), produced by the thyroid gland, is the single most important hormone influencing an individual's basal metabolic rate (BMR) and body heat production. Basal metabolic rate, often called the "energy cost of living," is the energy needed to perform essential activity such as breathing and maintaining organ function. The level of thyroid hormone produced directly affects BMR, the more thyroid hormone produced, the higher the BMR. In addition, thyroid hormone regulates growth and development and is especially important for the maturation and normal function of the nervous system.

The tropic hormone thyroid-stimulating hormone (TSH), produced by the anterior pituitary, controls the secretory activity of the thyroid gland. The hypothalamic hormone thyrotropin-releasing hormone (TRH) stimulates the release of TSH from cells of the anterior pituitary gland. Rising levels of thyroid hormone act on both the anterior pituitary and the hypothalamus to inhibit secretion of TSH. (**Figure 28.1** illustrates the feedback loop that regulates thyroid hormone secretion.)

A **goiter** is an enlargement of the thyroid gland. Both *hypothyroidism* and *hyperthyroidism* can result in production of a goiter. In either case, the goiter is a result of excessive stimulation of the thyroid gland.

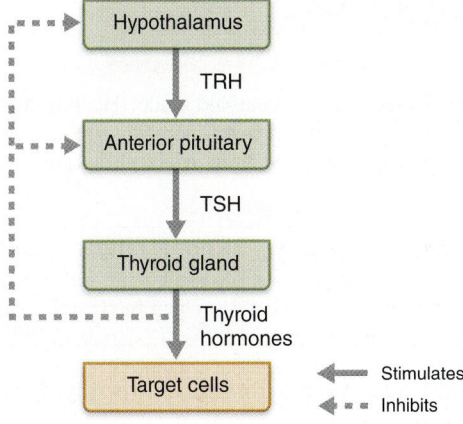

Figure 28.1 Regulation of thyroid hormone secretion.

Hypothyroidism, also called **myxedema,** produces symptoms including low metabolic rate; feeling chilled; constipation; thick, dry skin and puffy skin ("bags") beneath the eyes; edema; lethargy; and mental sluggishness. A goiter occurs when hypothyroidism is caused by (1) primary failure of the thyroid gland or (2) an iodine-deficient diet that prevents the thyroid gland from producing TH. In both cases, the low levels of TH remove the inhibition for secretion of TSH, and its levels rise. When hypothyroidism is secondary to hypothalamic or anterior pituitary failure, TRH and/or TSH levels fall, and no goiter is observed.

Symptoms of *hyperthyroidism* include elevated metabolism; sweating; a rapid, more forceful heartbeat; nervousness; weight loss; difficulty concentrating; and changes in skin texture. The most common cause of hyperthyroidism is Graves' disease. Protrusion of the eyeballs sometimes occurs in patients with Graves' disease and is a unique symptom of this type of hyperthyroidism. Graves' disease is an autoimmune disorder in which the body makes abnormal antibodies that mimic the action of TSH on follicular cells of the thyroid. Despite low levels of TSH, the thyroid is being powerfully stimulated and produces a large goiter. Hyperthyroidism can also arise secondary to excess hypothalamic or anterior pituitary secretion. In this case, TSH levels are high and a goiter occurs. A hypersecreting thyroid tumor also causes hyperthyroidism. TSH levels are low when such a tumor is present, and there is no goiter. ✚

Use the information above to answer the questions associated with the case studies in the following Group Challenge.

28

GROUP CHALLENGE

Thyroid Hormone Case Studies

Case 1: Marty is a 24-year-old male. He has noticed a bulge on his neck that has been increasing in size over the past few months. His physician orders a blood test with the following results:

Component	Results	Normal range	Units
TSH	<0.1	0.1–5.5	µIU/ml
Free T$_4$	5.3	0.8–1.7	ng/dL

Does Marty have hypothyroidism or hyperthyroidism?

_____

Name and briefly describe the most likely cause of his

thyroid disorder. _____

_____

What other signs and symptoms might Marty be

experiencing? _____

_____

Case 2: Heather is a 60-year-old female. She complains of swelling in her limbs and fatigue. Her physician orders a blood test with the following results:

Component	Results	Normal range	Units
TSH	5.7	0.1–5.5	µIU/ml
Free T$_4$	0.5	0.8–1.7	ng/dL

Does Heather have hypothyroidism or hyperthyroidism?

_____

Name and briefly describe the most likely cause(s) of her

thyroid disorder. _____

_____

Does Heather have a goiter? _____

What other signs and symptoms might Heather be

experiencing? _____

_____

Endocrine Wet Labs and Human Metabolism

Determining the Effect of Pituitary Hormones on the Ovary

1. In the experiment on the effects of pituitary hormones, two anterior pituitary hormones caused ovulation to occur in the experimental animal. Which of these actually triggered ovulation or egg expulsion?

   _____ The normal function of the second hormone involved, _____,

   is to _____.

2. Why was a second frog injected with saline? _____

Observing the Effects of Hyperinsulinism

3. Briefly explain what was happening within the fish's system when the fish was immersed in the insulin solution.

   _____

   _____

4. What is the mechanism of the recovery process observed? _____

   _____

5. What would you do to help a friend who had inadvertently taken an overdose of insulin? _____

   _____ Why? _____

   _____

Human Metabolism and Thyroid Hormones

6. Use an appropriate reference to indicate which of the following would be associated with increased or decreased BMR. Indicate increase by ↑ and decrease by ↓.

   increased exercise _____   aging _____   infection/fever _____

   small/slight stature _____   obesity _____   sex (♂ or ♀) _____

7. What are some possible treatments for myxedema? (Use your textbook or another appropriate reference.)

_____

_____

8. What are some possible treatments for Graves' disease? (Use your textbook or another appropriate reference.)

_____

_____

Blood

MATERIALS

General supply area:*

☐ Disposable gloves
☐ Safety glasses (student-provided)
☐ Bucket or large beaker containing 10% household bleach solution for slide and glassware disposal
☐ Spray bottles containing 10% bleach solution
☐ Autoclave bag
☐ Designated lancet (sharps) disposal container
☐ Plasma (obtained from an animal hospital or prepared by centrifuging animal [for example, cattle or sheep] blood obtained from a biological supply house)
☐ Test tubes and test tube racks
☐ Wide-range pH paper
☐ Stained smears of human blood from a biological supply house or, if desired by the instructor, heparinized animal blood obtained from a biological supply house or an animal hospital (for example, dog blood), or EDTA-treated red cells (reference cells[†]) with blood type labels obscured (available from Immucor, Inc.)

***Note to the Instructor:** See directions for handling of soiled glassware and disposable items (page 424).

[†]The blood in these kits (each containing four blood cell types—A1, A2, B, and O—individually supplied in 10-ml vials) is used to calibrate cell counters and other automated clinical laboratory equipment. This blood has been carefully screened and can be safely used by students for blood typing and determining hematocrits. It is not usable for hemoglobin determinations or coagulation studies.

(Text continues on next page.)

MasteringA&P® For related exercise study tools, go to the Study Area of MasteringA&P. There you will find:

• Practice Anatomy Lab PAL
• PhysioEx PEx
• A&PFlix A&PFlix
• Practice quizzes, Histology Atlas, eText, Videos, and more!

OBJECTIVES

1. Name the two major components of blood, and state their average percentages in whole blood.
2. Describe the composition and functional importance of plasma.
3. Define *formed elements* and list the cell types composing them, state their relative percentages, and describe their major functions.
4. Identify erythrocytes, basophils, eosinophils, monocytes, lymphocytes, and neutrophils when provided with a microscopic preparation or appropriate image.
5. Provide the normal values for a total white blood cell count and a total red blood cell count, and state the importance of these tests.
6. Conduct the following blood tests in the laboratory, and state their norms and the importance of each: differential white blood cell count, hematocrit, hemoglobin determination, clotting time, and plasma cholesterol concentration.
7. Define *leukocytosis, leukopenia, leukemia, polycythemia,* and *anemia* and cite a possible cause for each condition.
8. Perform an ABO and Rh blood typing test in the laboratory and discuss the reason for transfusion reactions resulting from the administration of mismatched blood.

PRE-LAB QUIZ

1. Circle True or False. There are no special precautions that I need to observe when performing today's lab.
2. Three types of formed elements found in blood include erythrocytes, leukocytes, and _____.
   a. electrolytes   b. fibers   c. platelets   d. sodium salts
3. Circle the correct underlined term. Mature <u>erythrocytes</u> / <u>leukocytes</u> are the most numerous blood cells and do not have a nucleus.
4. The least numerous but largest of all agranulocytes is the
   a. basophil   b. lymphocyte   c. monocyte   d. neutrophil
5. _____ are the leukocytes responsible for releasing histamine and other mediators of inflammation.
   a. Basophils   b. Eosinophils   c. Monocytes   d. Neutrophils
6. _____ are essential for blood clotting.
7. Circle the correct underlined term. When determining the <u>hematocrit</u> / <u>hemoglobin</u>, you will centrifuge whole blood in order to allow the formed elements to sink to the bottom of the sample.
8. Circle the correct underlined term. The normal hematocrit value for <u>females</u> / <u>males</u> is generally higher than that of the opposite sex.
9. Circle the correct underlined term. Blood typing is based on the presence of proteins known as <u>antigens</u> / <u>antibodies</u> on the outer surface of the red blood cell plasma membrane.
10. Circle True or False. If an individual is transfused with the wrong blood type, the recipient's antibodies react with the donor's antigens, eventually clumping and hemolyzing the donated RBCs.

(Materials list continued.)

☐ Clean microscope slides
☐ Glass stirring rods
☐ Wright's stain in a dropper bottle
☐ Distilled water in a dropper bottle
☐ Sterile lancets
☐ Absorbent cotton balls
☐ Alcohol swabs (wipes)
☐ Paper towels
☐ Compound microscope
☐ Immersion oil
☐ Assorted slides of white blood count pathologies labeled "Unknown Sample _____"
☐ Timer

Because many blood tests are to be conducted in this exercise, it is advisable to set up a number of appropriately labeled supply areas for the various tests, as designated below. Some needed supplies are located in the general supply area.

Note: Artificial blood prepared by Ward's Natural Science can be used for differential counts, hematocrit, and blood typing.

Activity 4: Hematocrit

☐ Heparinized capillary tubes
☐ Microhematocrit centrifuge and reading gauge (if the reading gauge is not available, a millimeter ruler may be used)
☐ Capillary tube sealer or modeling clay

Activity 5: Hemoglobin determination

☐ Hemoglobinometer, hemolysis applicator, and lens paper; or Tallquist hemoglobin scale and test paper

Activity 6: Coagulation time

☐ Capillary tubes (nonheparinized)
☐ Fine triangular file

Activity 7: Blood typing

☐ Blood typing sera (anti-A, anti-B, and anti-Rh [anti-D])
☐ Rh typing box

☐ Wax marking pencil
☐ Toothpicks
☐ Medicine dropper
☐ Blood test cards or microscope slides

Activity 8: Demonstration

☐ Microscopes set up with prepared slides demonstrating the following bone (or bone marrow) conditions: macrocytic hypochromic anemia, microcytic hypochromic anemia, sickle cell anemia, lymphocytic leukemia (chronic), and eosinophilia

Activity 9: Cholesterol measurement

☐ Cholesterol test cards and color scale

PEx PhysioEx™ 9.1 Computer Simulation Ex. 11 on p. PEx-161.

In this exercise you will study plasma and formed elements of blood and conduct various hematologic tests. These tests are useful diagnostic tools for the physician because blood composition (number and types of blood cells, and chemical composition) reflects the status of many body functions and malfunctions.

⚠️ **ALERT: Special precautions when handling blood.** This exercise provides information on blood from several sources: human, animal, human treated, and artificial blood. The decision to use animal blood for testing or to have students test their own blood will be made by the instructor in accordance with the educational goals of the student group. For example, for students in the nursing or laboratory technician curricula, learning how to safely handle human blood or other human wastes is essential. Whenever blood is being handled, special attention must be paid to safety precautions. Instructors who opt to use human blood are responsible for its safe handling. Precautions should be used regardless of the source of the blood. This will both teach good technique and ensure the safety of the students.

Follow exactly the safety precautions listed below.

1. Wear safety gloves at all times. Discard appropriately.

2. Wear safety glasses throughout the exercise.

3. Handle only your own, freshly drawn (human) blood.

4. Be sure you understand the instructions and have all supplies on hand before you begin any part of the exercise.

5. Do not reuse supplies and equipment once they have been exposed to blood.

6. Keep the lab area clean. Do not let anything that has come in contact with blood touch surfaces or other individuals in the lab. Pay attention to the location of any supplies and equipment that come into contact with blood.

7. Dispose of lancets immediately after use in a designated disposal container. Do not put them down on the lab bench, even temporarily.

8. Dispose of all used cotton balls, alcohol swabs, blotting paper, and so forth in autoclave bags and place all soiled glassware in containers of 10% bleach solution.

9. Wipe down the lab bench with 10% bleach solution when you finish.

Composition of Blood

Circulating blood is a rather viscous substance that varies from bright scarlet to a dull brick red, depending on the amount of oxygen it is carrying. The average volume of blood in the body is about 5–6 L in adult males and 4–5 L in adult females.

Blood is classified as a type of connective tissue because it consists of a nonliving fluid matrix (the **plasma**) in which living cells (**formed elements**) are suspended. The fibers typical of a connective tissue matrix become visible in blood only when clotting occurs. They then appear as fibrin threads, which form the structural basis for clot formation.

More than 100 different substances are dissolved or suspended in plasma **(Figure 29.1)**, which is over 90% water. These include nutrients, gases, hormones, various wastes and metabolites, many types of proteins, and electrolytes. The composition of plasma varies continuously as cells remove or add substances to the blood.

Three types of formed elements are present in blood **(Table 29.1)**. Most numerous are **erythrocytes,** or **red blood cells (RBCs),** which are literally sacs of hemoglobin molecules that transport the bulk of the oxygen carried in the blood (and a small percentage of the carbon dioxide). **Leukocytes,** or **white blood cells (WBCs),** are part of the

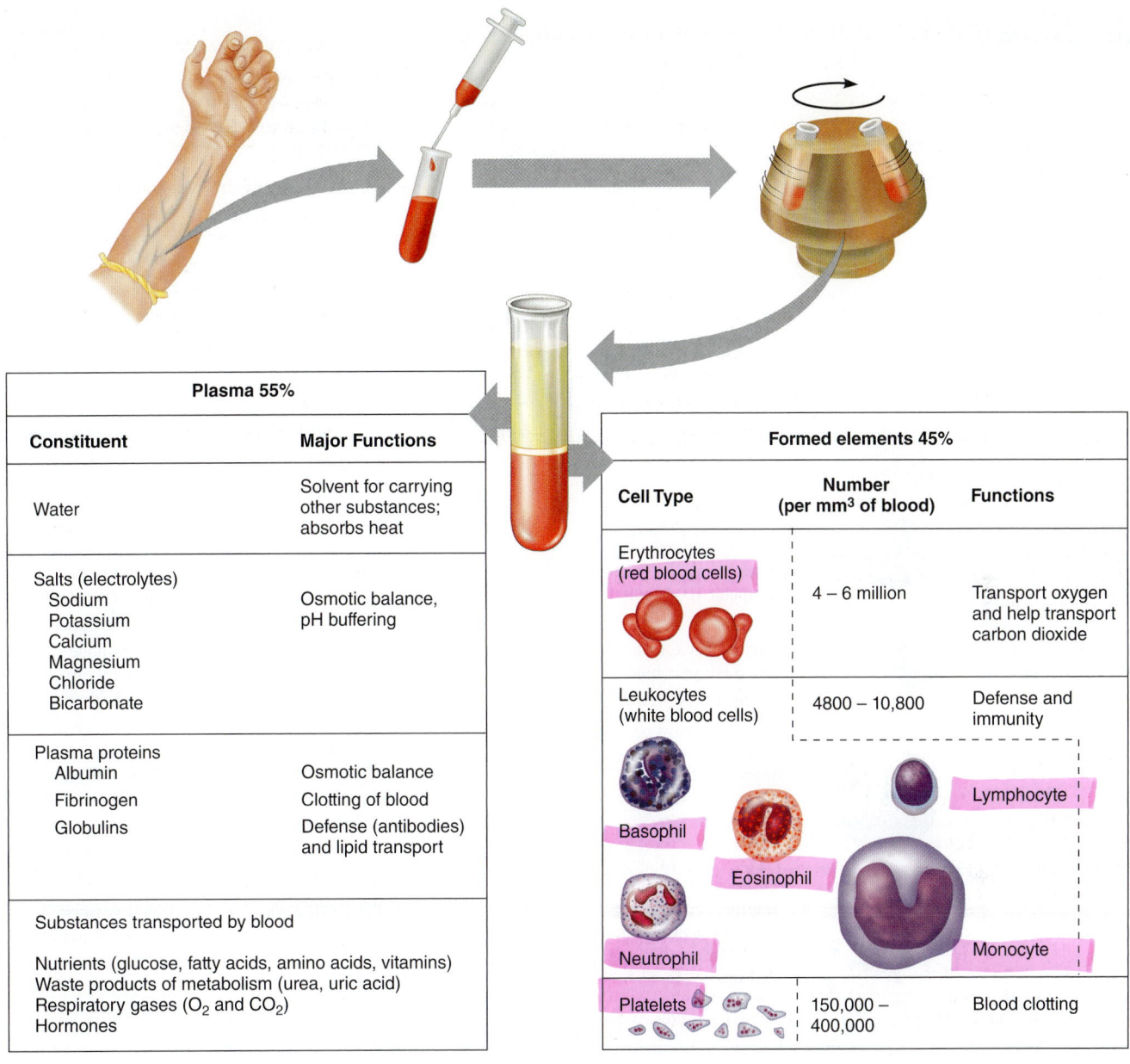

Figure 29.1 **The composition of blood.** Note that leukocytes and platelets are found in the band between plasma (above) and erythrocytes (below).

body's nonspecific defenses and the immune system, and **platelets** function in hemostasis (blood clot formation); together they make up <1% of whole blood. Formed elements normally constitute 45% of whole blood; plasma accounts for the remaining 55%.

Determining the Physical Characteristics of Plasma

Go to the general supply area and carefully pour a few milliliters of plasma into a test tube. Also obtain some wide-range pH paper, and then return to your laboratory bench to make the following simple observations.

pH of Plasma

Test the pH of the plasma with wide-range pH paper. Record

the pH observed. _____

Color and Clarity of Plasma

Hold the test tube up to a source of natural light. Note and record its color and degree of transparency. Is it clear, translucent, or opaque?

Color _____

Degree of transparency _____

29

| Table 29.1 | Summary of Formed Elements of the Blood |

Cell type	Illustration	Description*	Cells/mm3 (µl) of blood	Duration of development (D) and life span (LS)	Function
Erythrocytes (red blood cells, RBCs)		Biconcave, anucleate disc; salmon-colored; diameter 7–8 µm	4–6 million	D: about 15 days LS: 100–120 days	Transport oxygen and carbon dioxide
Leukocytes (white blood cells, WBCs)		Spherical, nucleated cells	4800–10,800		
Granulocytes Neutrophil		Nucleus multilobed; inconspicuous cytoplasmic granules; diameter 10–12 µm	3000–7000	D: about 14 days LS: 6 hours to a few days	Phagocytize bacteria
Eosinophil		Nucleus bilobed; red cytoplasmic granules; diameter 10–14 µm	100–400	D: about 14 days LS: about 5 days	Kill parasitic worms; complex role in allergy and asthma
Basophil		Nucleus lobed; large blue-purple cytoplasmic granules; diameter 10–14 µm	20–50	D: 1–7 days LS: a few hours to a few days	Release histamine and other mediators of inflammation; contain heparin, an anticoagulant
Agranulocytes Lymphocyte		Nucleus spherical or indented; pale blue cytoplasm; diameter 5–17 µm	1500–3000	D: days to weeks LS: hours to years	Mount immune response by direct cell attack or via antibodies
Monocyte		Nucleus U- or kidney-shaped; gray-blue cytoplasm; diameter 14–24 µm	100–700	D: 2–3 days LS: months	Phagocytosis; develop into macrophages in tissues
Platelets		Discoid cytoplasmic fragments containing granules; stain deep purple; diameter 2–4 µm	150,000–400,000	D: 4–5 days LS: 5–10 days	Seal small tears in blood vessels; instrumental in blood clotting

*Appearance when stained with Wright's stain.

Consistency

While wearing gloves, dip your finger and thumb into plasma and then press them firmly together for a few seconds. Gently pull them apart. How would you describe the consistency of plasma (slippery, watery, sticky, granular)? Record your observations.

_____ ▬

ACTIVITY 2

Examining the Formed Elements of Blood Microscopically

In this section, you will observe blood cells on an already prepared (purchased) blood slide or on a slide prepared from your own blood or blood provided by your instructor.

• Those using the purchased blood slide are to obtain a slide and begin their observations at step 6.

• Those testing blood provided by a biological supply source or an animal hospital are to obtain a tube of the supplied blood, disposable gloves, and the supplies listed in step 1, except for the lancets and alcohol swabs. After donning gloves, those students will go to step 3b to begin their observations.

• If you are examining your own blood, you will perform all the steps described below *except* step 3b.

1. Obtain two glass slides, a glass stirring rod, dropper bottles of Wright's stain and distilled water, two or three lancets, cotton balls, and alcohol swabs. Bring this equipment to the laboratory bench. Clean the slides thoroughly and dry them.

2. Open the alcohol swab packet and scrub your third or fourth finger with the swab. (Because the pricked finger may be a little sore later, it is better to prepare a finger on the nondominant hand.) Circumduct your hand (swing it in a cone-shaped path) for 10 to 15 seconds. This will dry the alcohol and cause your fingers to become engorged with

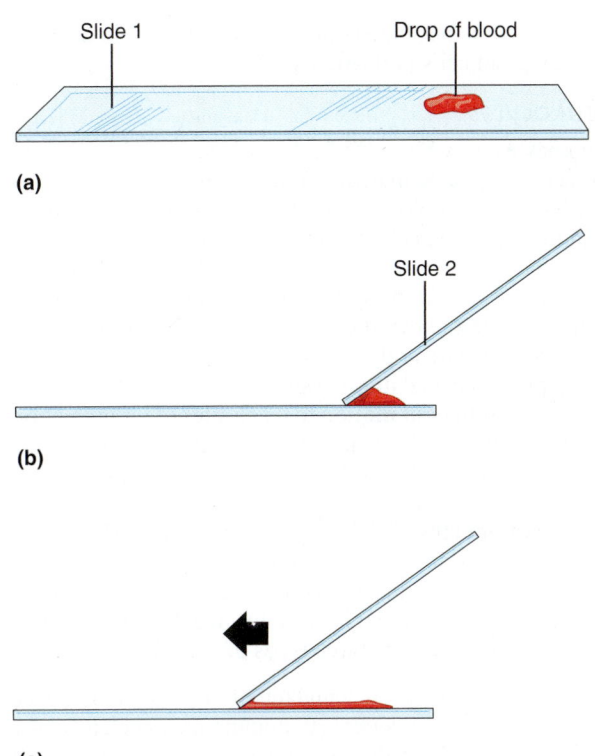

(a)

(b)

(c)

Figure 29.2 Procedure for making a blood smear.
(a) Place a drop of blood on slide 1 approximately
½ inch from one end. **(b)** Hold slide 2 at a 30° to 40° angle
to slide 1 (it should touch the drop of blood) and allow
blood to spread along entire bottom edge of angled
slide. **(c)** Smoothly advance slide 2 to end of slide 1 (blood
should run out before reaching the end of slide 1). Then
lift slide 2 away from slide 1 and place it on a paper
towel.

blood. Then, open the lancet packet and grasp the lancet by
its blunt end. Quickly jab the pointed end into the prepared
finger to produce a free flow of blood. It is *not* a good idea
to squeeze or "milk" the finger, as this forces out tissue fluid
as well as blood. If the blood is not flowing freely, another
puncture should be made.

⚠️ _Under no circumstances is a lancet to be used for more
than one puncture._ Dispose of the lancets in the designated disposal container immediately after use.

3a. With a cotton ball, wipe away the first drop of blood;
then allow another large drop of blood to form. Touch the
blood to one of the cleaned slides approximately 1.3 cm, or
½ inch, from the end. Then quickly (to prevent clotting) use
the second slide to form a blood smear **(Figure 29.2)**. When
properly prepared, the blood smear is uniformly thin. If the
blood smear appears streaked, the blood probably began to
clot or coagulate before the smear was made, and another
slide should be prepared. Continue at step 4.

3b. Dip a glass rod in the blood provided, and transfer a generous drop of blood to the end of a cleaned microscope slide.
For the time being, lay the glass rod on a paper towel on the
bench. Then, as described in step 3a (Figure 29.2), use the
second slide to make your blood smear.

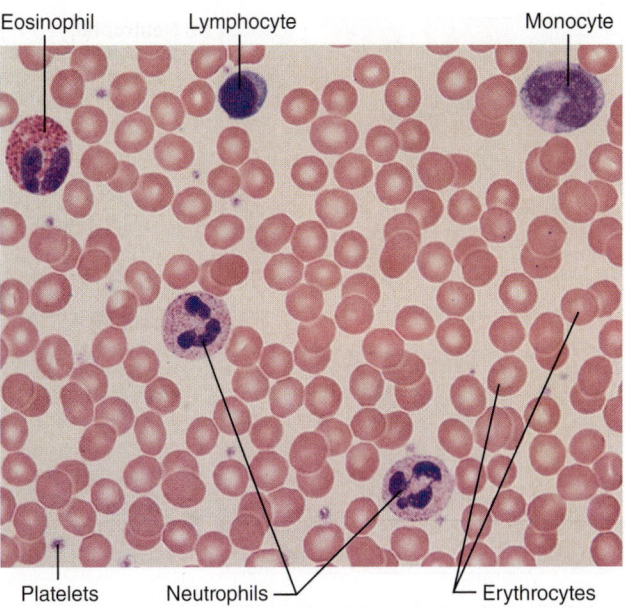

**Figure 29.3 Photomicrograph of a human blood smear
stained with Wright's stain (765×).**

4. Dry the slide by waving it in the air. When it is completely
dry, it will look dull. Place it on a paper towel, and flood it
with Wright's stain. Count the number of drops of stain used.
Allow the stain to remain on the slide for 3 to 4 minutes, and
then flood the slide with an equal number of drops of distilled
water. Allow the water and Wright's stain mixture to remain
on the slide for 4 or 5 minutes or until a metallic green film or
scum is apparent on the fluid surface. Blow on the slide gently
every minute or so to keep the water and stain mixed during
this interval.

5. Rinse the slide with a stream of distilled water. Then
flood it with distilled water, and allow it to lie flat until the
slide becomes translucent and takes on a pink cast. Then
stand the slide on its long edge on the paper towel, and
allow it to dry completely. Once the slide is dry, you can
begin your observations.

6. Obtain a microscope and scan the slide under low power
to find the area where the blood smear is the thinnest.
After scanning the slide in low power to find the areas with
the largest numbers of nucleated WBCs, read the following
descriptions of cell types, and find each one in the art illustrating blood cell types (in Figure 29.1 and Table 29.1). (The
formed elements are also shown in **Figure 29.3** and Figure
29.4.) Then, switch to the oil immersion lens, and observe the
slide carefully to identify each cell type.

7. Set your prepared slide aside for use in Activity 3.

Erythrocytes

Erythrocytes, or red blood cells, which average 7.5 μm in
diameter, vary in color from a salmon red color to pale pink,
depending on the effectiveness of the stain. They have a distinctive biconcave disc shape and appear paler in the center
than at the edge (see Figure 29.3).

   As you observe the slide, notice that the red blood cells
are by far the most numerous blood cells seen in the field.
Their number averages 4.5 million to 5.5 million cells per
cubic millimeter of blood (for women and men, respectively).

29

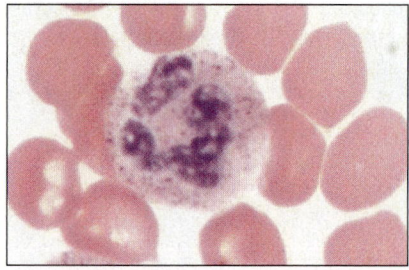

(a) Neutrophil; multilobed nucleus

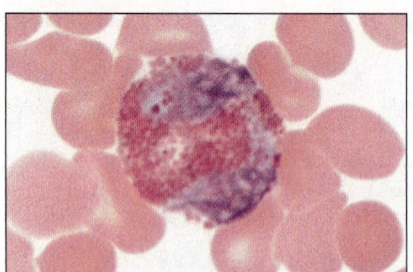

(b) Eosinophil; bilobed nucleus, red cytoplasmic granules

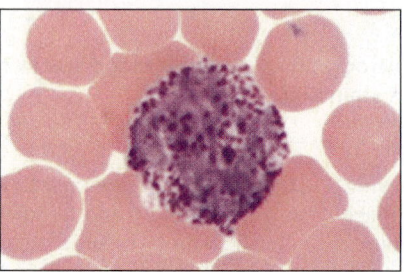

(c) Basophil; bilobed nucleus, purplish-black cytoplasmic granules

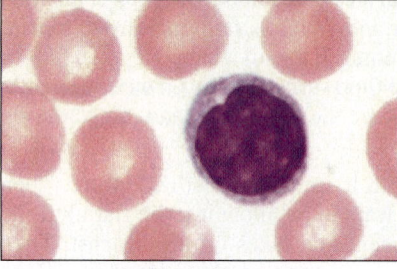

(d) Small lymphocyte; large spherical nucleus

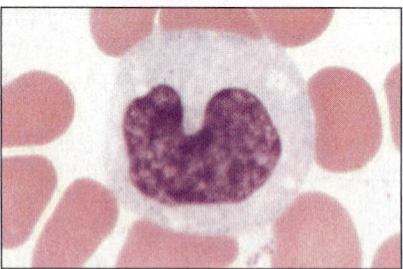

(e) Monocyte; kidney-shaped nucleus

Figure 29.4 Leukocytes. In each case the leukocytes are surrounded by erythrocytes (1330×, Wright's stain).

Red blood cells differ from the other blood cells because they are anucleate (lacking a nucleus) when mature and circulating in the blood. As a result, they are unable to reproduce or repair damage and have a limited life span of 100 to 120 days, after which they begin to fragment and are destroyed, mainly in the spleen.

In various anemias, the red blood cells may appear pale (an indication of decreased hemoglobin con-tent) or may be nucleated (an indication that the bone marrow is turning out cells prematurely). ✚

Leukocytes

Leukocytes, or white blood cells, are nucleated cells that are formed in the bone marrow from the same stem cells (*hemocytoblast*) as red blood cells. They are much less numerous than the red blood cells, averaging from 4800 to 10,800 cells per cubic millimeter. Basically, white blood cells are protective, pathogen-destroying cells that are transported to all parts of the body in the blood or lymph. Important to their protective function is their ability to move in and out of blood vessels, a process called **diapedesis**, and to wander through body tissues by **amoeboid motion** to reach sites of inflammation or tissue destruction. They are classified into two major groups, depending on whether or not they contain conspicuous granules in their cytoplasm.

Granulocytes make up the first group. The granules in their cytoplasm stain differentially with Wright's stain, and they have peculiarly lobed nuclei, which often consist of expanded nuclear regions connected by thin strands of nucleoplasm. There are three types of granulocytes:

Neutrophil: The most abundant of the white blood cells (50% to 70% of the leukocyte population); nucleus consists of 3 to 6 lobes and the pale lilac cytoplasm contains fine cytoplasmic granules, which are generally indistinguishable and take up both the acidic (red) and basic (blue) dyes (*neutrophil* = neutral loving) **(Figure 29.4a)**; functions as an active phagocyte. The number of neutrophils increases exponentially during acute infections.

Eosinophil: Represents 2% to 4% of the leukocyte population; nucleus is generally figure-8 or bilobed in shape; contains large cytoplasmic granules (elaborate lysosomes) that stain red-orange with the acid dyes in Wright's stain (see Figure 29.4b). Eosinophils are about the size of neutrophils and play a role in counterattacking parasitic worms. Eosinophils have complex roles in many other diseases, especially in allergy and asthma.

Basophil: Least abundant leukocyte type representing less than 1% of the population; large U- or S-shaped nucleus with two or more indentations. Cytoplasm contains coarse, sparse granules that are stained deep purple by the basic dyes in Wright's stain (see Figure 29.4c). The granules contain several chemicals, including histamine, a vasodilator that is discharged on exposure to antigens and helps mediate the inflammatory response. Basophils are about the size of neutrophils.

The second group, **agranulocytes,** or **agranular leukocytes,** contains no *visible* cytoplasmic granules. Although found in the bloodstream, they are much more abundant in lymphoid tissues. Their nuclei tend to be closer to the norm, that is, spherical, oval, or kidney-shaped. Specific characteristics of the two types of agranulocytes are listed below.

Lymphocyte: The smallest of the leukocytes, approximately the size of a red blood cell (see Figure 29.4d). The nucleus stains dark blue to purple, is generally spherical or slightly indented, and accounts for most of the cell mass. Sparse cytoplasm appears as a thin blue rim around the nucleus. Concerned with immunological responses in the body; one population, the *B lymphocytes,* gives rise to *plasma cells* that produce antibodies released to blood. The second population,

T lymphocytes, plays a regulatory role and destroys grafts, tumors, and virus-infected cells. Represents 25% or more of the WBC population.

Monocyte: The largest of the leukocytes; approximately twice the size of red blood cells (see Figure 29.4e). Represents 3% to 8% of the leukocyte population. Dark blue nucleus is generally kidney-shaped; abundant cytoplasm stains gray-blue. Once in the tissues, monocytes convert to macrophages, which are active phagocytes (the "long-term cleanup team"). They increase dramatically in number during chronic infections such as tuberculosis.

Students are often asked to list the leukocytes in order from the most abundant to the least abundant. The following silly phrase may help you with this task: *N*ever *l*et *m*onkeys *e*at *b*ananas (neutrophils, lymphocytes, monocytes, eosinophils, basophils).

Platelets

Platelets are cell fragments of large multinucleate cells (**megakaryocytes**) formed in the bone marrow. They appear as darkly staining, irregularly shaped bodies interspersed among the blood cells (see Figure 29.3). The normal platelet count in blood ranges from 150,000 to 400,000 per cubic millimeter. Platelets are instrumental in the clotting process that occurs in plasma when blood vessels are ruptured.

After you have identified these cell types on your slide, observe charts and three-dimensional models of blood cells if these are available. Do not dispose of your slide, as you will use it later for the differential white blood cell count. ▬

Hematologic Tests

When someone enters a hospital as a patient, several hematologic tests are routinely done to determine general level of health as well as the presence of pathologic conditions. You will be conducting the most common of these tests in this exercise.

⚠ Materials such as cotton balls, lancets, and alcohol swabs are used in nearly all of the following diagnostic tests. These supplies are at the general supply area and should be properly disposed of (glassware to the bleach bucket, lancets in a designated disposal container, and disposable items to the autoclave bag) immediately after use.

Other necessary supplies and equipment are at specific supply areas marked according to the test with which they are used. Since nearly all of the tests require a finger stab, if you will be using your own blood it might be wise to quickly read through the tests to determine in which instances more than one preparation can be done from the same finger stab. A little planning will save you the discomfort of a multiple-punctured finger.

An alternative to using blood obtained from the finger stab technique is using heparinized blood samples supplied by your instructor. The purpose of using heparinized tubes is to prevent the blood from clotting. Thus blood collected and stored in such tubes will be suitable for all tests except coagulation time testing.

Total White and Red Blood Cell Counts

A **total WBC count** or **total RBC count** determines the total number of that cell type per unit volume of blood. Total WBC and RBC counts are a routine part of any physical exam. Most clinical agencies use computers to conduct these counts. Since the hand counting technique typically done in college labs is rather outdated, total RBC and WBC counts will not be done here, but the importance of such counts (both normal and abnormal values) is briefly described below.

Total White Blood Cell Count

Since white blood cells are an important part of the body's defense system, it is essential to note any abnormalities in them.

Leukocytosis, an abnormally high WBC count, may indicate bacterial or viral infection, metabolic disease, hemorrhage, or poisoning by drugs or chemicals. A decrease in the white cell number below 4000/mm^3 (**leukopenia**) may indicate typhoid fever, measles, infectious hepatitis or cirrhosis, tuberculosis, or excessive antibiotic or X-ray therapy. A person with leukopenia lacks the usual protective mechanisms. **Leukemia,** a malignant disorder of the lymphoid tissues characterized by uncontrolled proliferation of abnormal WBCs accompanied by a reduction in the number of RBCs and platelets, is detectable not only by a total WBC count but also by a differential WBC count. ✚

Total Red Blood Cell Count

Since RBCs are absolutely necessary for oxygen transport, a doctor typically investigates any excessive change in their number immediately.

An increase in the number of RBCs (**polycythemia**) may result from bone marrow cancer or from living at high altitudes where less oxygen is available. A decrease in the number of RBCs results in anemia. The term **anemia** simply indicates a decreased oxygen-carrying capacity of blood that may result from a decrease in RBC number or size or a decreased hemoglobin content of the RBCs. A decrease in RBCs may result suddenly from hemorrhage or more gradually from conditions that destroy RBCs or hinder RBC production. ✚

Differential White Blood Cell Count

To make a **differential white blood cell count,** 100 WBCs are counted and classified according to type. Such a count is routine in a physical examination and in diagnosing illness, since any abnormality or significant elevation in percentages of WBC types may indicate a problem or the source of pathology.

ACTIVITY 3

Conducting a Differential WBC Count

1. Use the slide prepared for the identification of the blood cells in Activity 2. Begin at the edge of the smear and move the slide in a systematic manner on the microscope stage—either up and down or from side to side (as indicated in **Figure 29.5**).

2. Record each type of white blood cell you observe by making a count in the first blank column of the **Activity 3 chart** (for example, 𝍤 𝍠 = 7 cells) until you have observed and recorded a total of 100 WBCs. Using the following equation, compute the percentage of each WBC type counted, and

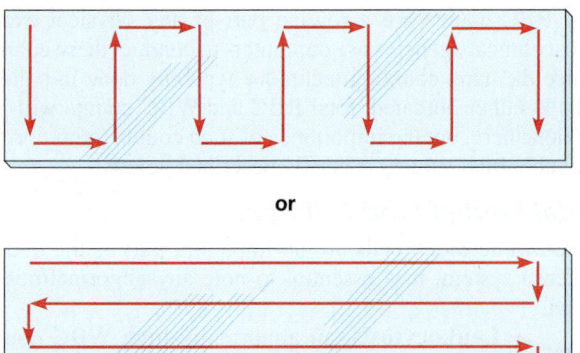

Figure 29.5 Alternative methods of moving the slide for a differential WBC count.

record the percentages on the Hematologic Test Data Sheet on the last page of the exercise, preceding the Review Sheet.

$$\text{Percent (\%)} = \frac{\text{\# observed}}{\text{Total \# counted (100)}} \times 100$$

3. Select a slide marked "Unknown sample," record the slide number, and use the count chart below to conduct a differential count. Record the percentages on the data sheet (page 436).

How does the differential count from the unknown sample slide compare to the normal percentages given for each type in the earlier Leukocytes text section (pages 428–429)?

_____

_____

Activity 3: Count of 100 WBCs		
	Number observed	
Cell type	Student blood smear	Unknown sample # ____
Neutrophils		
Eosinophils		
Basophils		
Lymphocytes		
Monocytes		

Using the text and other references, try to determine the blood pathology on the unknown slide. Defend your answer.

_____

_____

4. How does your differential white blood cell count compare to the percentages given in the earlier Leukocytes text section?

_____

Hematocrit

The **hematocrit,** or **packed cell volume (PCV),** is routinely determined when anemia is suspected. Centrifuging whole blood spins the formed elements to the bottom of the tube, with plasma forming the top layer (see Figure 29.1). Since the blood cell population is primarily RBCs, the PCV is generally considered equivalent to the RBC volume, and this is the only value reported. However, the relative percentage of WBCs can be differentiated, and both WBC and plasma volume will be reported here. Normal hematocrit values for the male and female, respectively, are 47.0 ± 7 and 42.0 ± 5.

ACTIVITY 4

Determining the Hematocrit

The hematocrit is determined by the micromethod, so only a drop of blood is needed. If possible (and the centrifuge allows), all members of the class should prepare their capillary tubes at the same time so the centrifuge can be run only once.

1. Obtain two heparinized capillary tubes, capillary tube sealer or modeling clay, a lancet, alcohol swabs, and some cotton balls.

2. If you are using your own blood, cleanse a finger, and allow the blood to flow freely. Wipe away the first few drops and, holding the red-line-marked end of the capillary tube to the blood drop, allow the tube to fill at least three-fourths full by capillary action **(Figure 29.6a)**. If the blood is not flowing freely, the end of the capillary tube will not be completely submerged in the blood during filling, air will enter, and you will have to prepare another sample.

   If you are using instructor-provided blood, simply immerse the red-marked end of the capillary tube in the blood sample and fill it three-quarters full as just described.

3. Plug the blood-containing end by pressing it into the capillary tube sealer or clay (Figure 29.6b). Prepare a second tube in the same manner.

4. Place the prepared tubes opposite one another in the radial grooves of the microhematocrit centrifuge with the sealed ends abutting the rubber gasket at the centrifuge periphery (Figure 29.6c). This loading procedure balances the centrifuge and prevents blood from spraying everywhere by centrifugal force. *Make a note of the numbers of the grooves your tubes are in.* When all the tubes have been loaded, make sure the centrifuge is properly balanced, and secure the centrifuge cover. Turn the centrifuge on, and set the timer for 4 or 5 minutes.

5. Determine the percentage of RBCs, WBCs, and plasma by using the microhematocrit reader. The RBCs are the bottom layer, the plasma is the top layer, and the WBCs are the buff-colored layer between the two. If the reader is not available, use a millimeter ruler to measure the length of the

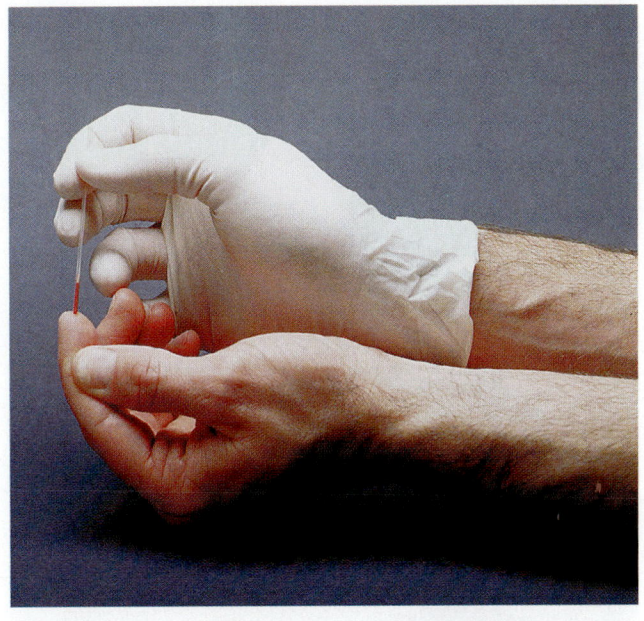

(a)

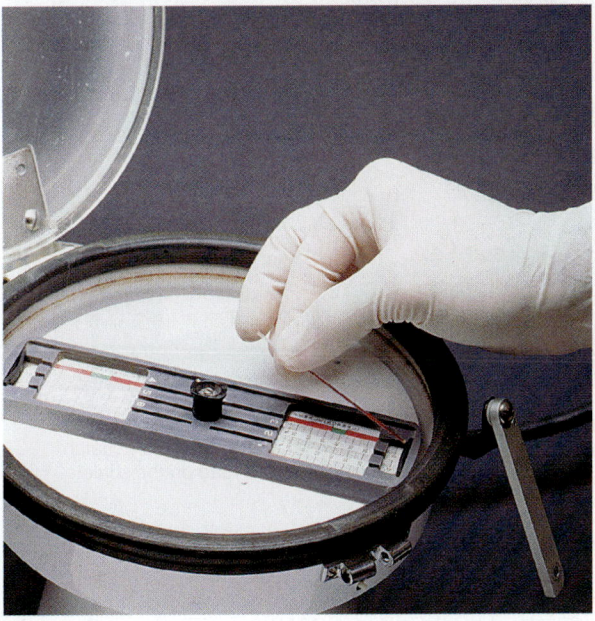

(c)

Figure 29.6 Steps in a hematocrit determination.
(a) Load a heparinized capillary tube with blood.
(b) Plug the blood-containing end of the tube with
clay. **(c)** Place the tube in a microhematocrit centrifuge.
(Centrifuge must be balanced.)

filled capillary tube occupied by each element, and compute
its percentage by using the following formula:

$$\frac{\text{Height of the column composed of the element (mm)}}{\text{Height of the original column of whole blood (mm)}} \times 100$$

Record your calculations below and on the data sheet
(page 436).

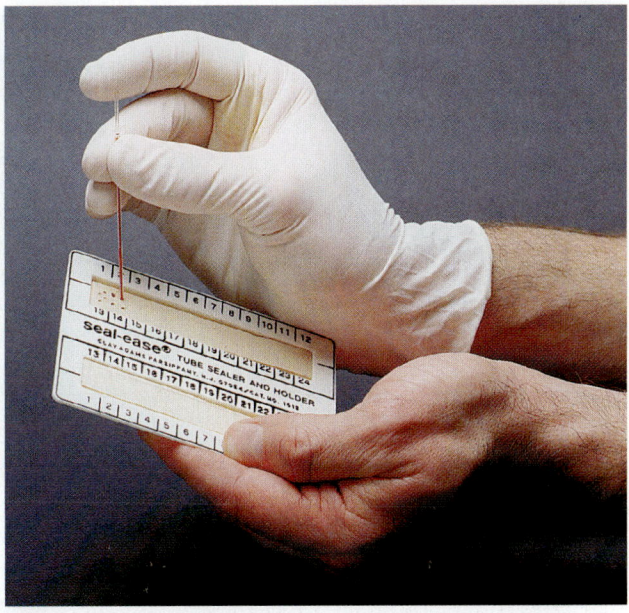

(b)

% RBC _____ % WBC _____ % plasma _____

Usually WBCs constitute 1% of the total blood volume. How
do your blood values compare to this figure and to the normal
percentages for RBCs and plasma? (See Figure 29.1.)

_____

As a rule, a hematocrit is considered a more accurate test than
the total RBC count for determining the RBC composition of
the blood. A hematocrit within the normal range generally
indicates a normal RBC number, whereas an abnormally high
or low hematocrit is cause for concern. ■■

Hemoglobin Concentration

As noted earlier, a person can be anemic even with a normal RBC count. Since hemoglobin (Hb) is the RBC protein
responsible for oxygen transport, perhaps the most accurate
way of measuring the oxygen-carrying capacity of the blood
is to determine its hemoglobin content. Oxygen, which combines reversibly with the heme (iron-containing portion) of
the hemoglobin molecule, is picked up by the blood cells in
the lungs and unloaded in the tissues. Thus, the more hemoglobin molecules the RBCs contain, the more oxygen they
will be able to transport. Normal blood contains 12 to 18 g
of hemoglobin per 100 ml of blood. Hemoglobin content in
men is slightly higher (13 to 18 g) than in women (12 to 16 g).

ACTIVITY 5

Determining Hemoglobin Concentration

Several techniques have been developed to estimate the
hemoglobin content of blood, ranging from the old, rather
inaccurate Tallquist method to expensive colorimeters,
which are precisely calibrated and yield highly accurate
results. Directions for both the Tallquist method and a
hemoglobinometer are provided here.

29

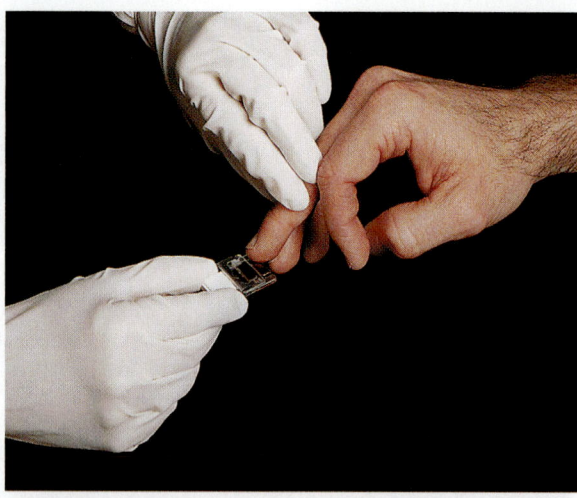

(a) A drop of blood is added to the moat plate of the blood chamber. The blood must flow freely.

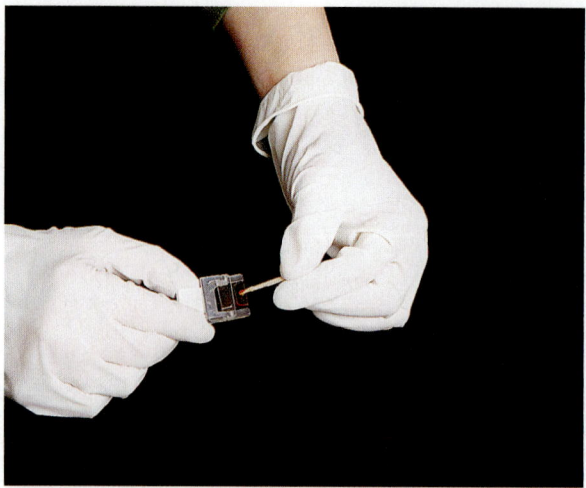

(b) The blood sample is hemolyzed with a wooden hemolysis applicator. Complete hemolysis requires 35 to 45 seconds.

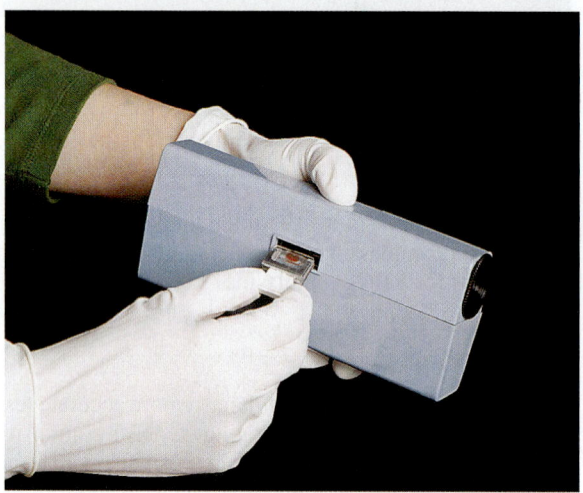

(c) The charged blood chamber is inserted into the slot on the side of the hemoglobinometer.

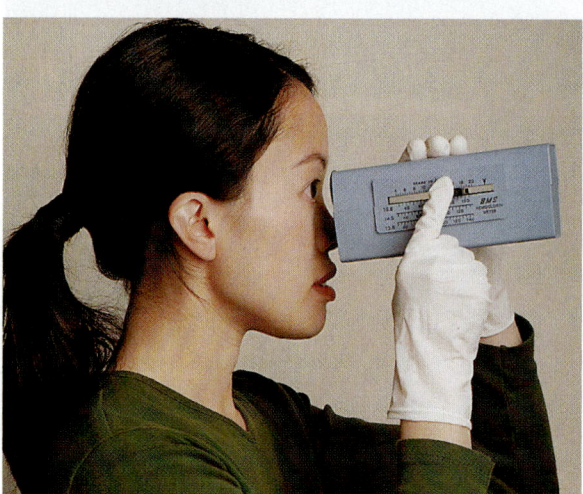

(d) The colors of the green split screen are found by moving the slide with the right index finger. When the two colors match in density, the grams/100 ml and % Hb are read on the scale.

Figure 29.7 Hemoglobin determination using a hemoglobinometer.

Tallquist Method

1. Obtain a Tallquist hemoglobin scale, test paper, lancets, alcohol swabs, and cotton balls.

2. Use instructor-provided blood or prepare the finger as previously described. (For best results, make sure the alcohol evaporates before puncturing your finger.) Place one good-sized drop of blood on the special absorbent paper provided with the color scale. The blood stain should be larger than the holes on the color scale.

3. As soon as the blood has dried and loses its glossy appearance, match its color, under natural light, with the color standards by moving the specimen under the comparison scale so that the blood stain appears at all the various apertures. (The blood should not be allowed to dry to a brown color, as this will result in an inaccurate reading.) Because the colors on the scale represent 1% variations in hemoglobin content, it may be necessary to estimate the percentage if the color of your blood sample is intermediate between two color standards.

4. On the data sheet (page 436) record your results as the percentage of hemoglobin concentration and as grams per 100 ml of blood.

Hemoglobinometer Determination

1. Obtain a hemoglobinometer, hemolysis applicator, alcohol swab, and lens paper, and bring them to your bench. Test the hemoglobinometer light source to make sure it is working; if not, request new batteries before proceeding and test it again.

2. Remove the blood chamber from the slot in the side of the hemoglobinometer and disassemble the blood chamber by separating the glass plates from the metal clip. Notice as you do this that the larger glass plate has an H-shaped depression cut into it that acts as a moat to hold the blood, whereas the smaller glass piece is flat and serves as a coverslip.

3. Clean the glass plates with an alcohol swab, and then wipe them dry with lens paper. Hold the plates by their sides to prevent smearing during the wiping process.

4. Reassemble the blood chamber (remember: larger glass piece on the bottom with the moat up), but leave the moat plate about halfway out to provide adequate exposed surface to charge it with blood.

5. Obtain a drop of blood (from the provided sample or from your fingertip as before), and place it on the depressed area of the moat plate that is closest to you **(Figure 29.7a)**.

6. Using the wooden hemolysis applicator, stir or agitate the blood to rupture (lyse) the RBCs (Figure 29.7b). This usually takes 35 to 45 seconds. Hemolysis is complete when the blood appears transparent rather than cloudy.

7. Push the blood-containing glass plate all the way into the metal clip and then firmly insert the charged blood chamber back into the slot on the side of the instrument (Figure 29.7c).

8. Hold the hemoglobinometer in your left hand with your left thumb resting on the light switch located on the underside of the instrument. Look into the eyepiece and notice that there is a green area divided into two halves (a split field).

9. With the index finger of your right hand, slowly move the slide on the right side of the hemoglobinometer back and forth until the two halves of the green field match (Figure 29.7d).

10. Note and record on the data sheet (page 436) the grams of Hb (hemoglobin)/100 ml of blood indicated on the uppermost scale by the index mark on the slide. Also record % Hb, indicated by one of the lower scales.

11. Disassemble the blood chamber once again, and carefully place its parts (glass plates and clip) into a bleach-containing beaker.

Generally speaking, the relationship between the PCV and grams of hemoglobin per 100 ml of blood is 3:1—for example, a PCV of 36 with 12 g of Hb per 100 ml of blood is a ratio of 3:1. How do your values compare?

Record on the data sheet (page 436) the value obtained from your data. ■

Bleeding Time

Normally a sharp prick of the finger or earlobe results in bleeding that lasts from 2 to 7 minutes (Ivy method) or 0 to 5 minutes (Duke method), although other factors such as altitude affect the time. How long the bleeding lasts is referred to as **bleeding time** and tests the ability of platelets to stop bleeding in capillaries and small vessels. Absence of some clotting factors may affect bleeding time, but prolonged bleeding time is most often associated with deficient or abnormal platelets.

Coagulation Time

Blood clotting, or **coagulation,** is a protective mechanism that minimizes blood loss when blood vessels are ruptured. This process requires the interaction of many substances normally present in the plasma (clotting factors, or procoagulants) as well as some released by platelets and injured tissues. Basically hemostasis proceeds as follows **(Figure 29.8a)**: The injured tissues and platelets release **tissue factor (TF)** and **PF_3** respectively, which trigger the clotting mechanism, or cascade. Tissue factor and PF_3

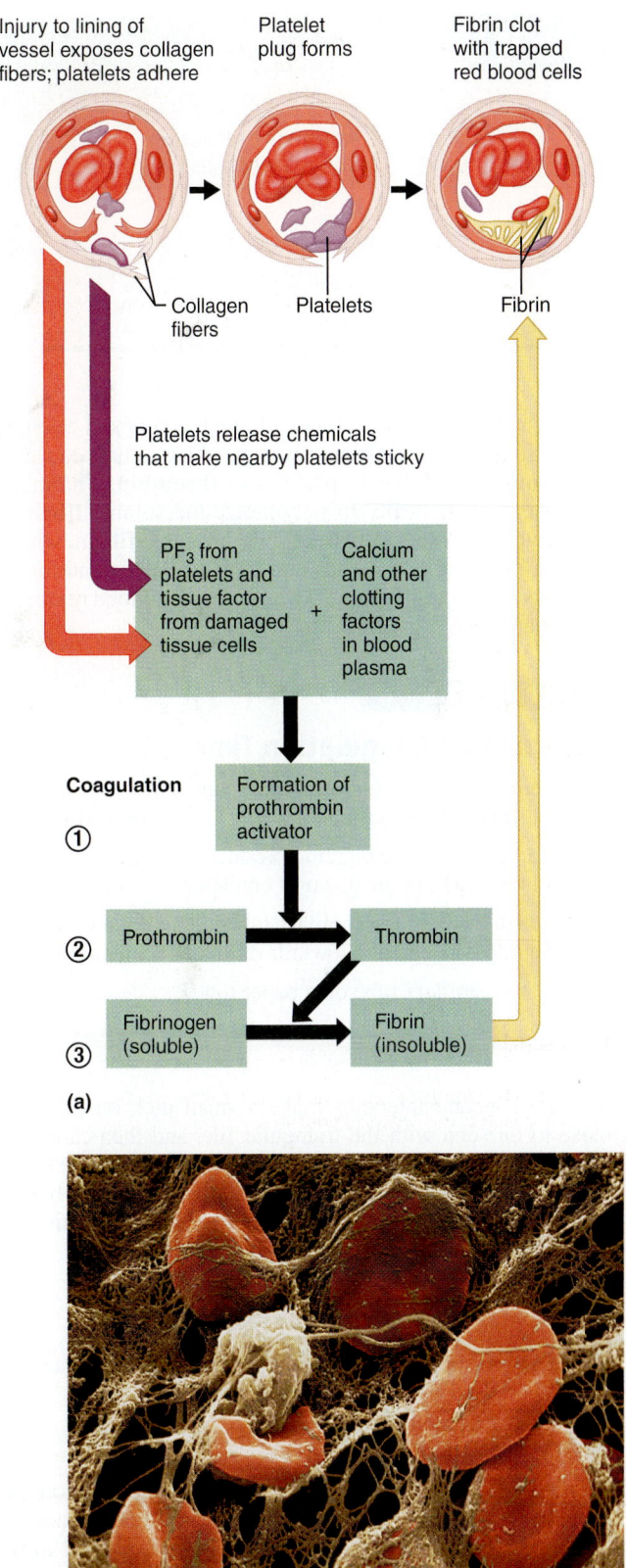

Figure 29.8 Events of hemostasis and blood clotting.
(a) Simple schematic of events. Steps numbered 1–3 represent the major events of coagulation.
(b) Photomicrograph of RBCs trapped in a fibrin mesh (2700×).

29

Table 29.2	ABO Blood Typing		% of U.S. population		
ABO blood type	Antigens present on RBC membranes	Antibodies present in plasma	White	Black	Asian
A	A	Anti-B	40	27	28
B	B	Anti-A	11	20	27
AB	A and B	None	4	4	5
O	Neither	Anti-A and anti-B	45	49	40

interact with other blood protein clotting factors and calcium ions to form **prothrombin activator,** which in turn converts **prothrombin** (present in plasma) to **thrombin.** Thrombin then acts enzymatically to polymerize the soluble **fibrinogen** proteins (present in plasma) into insoluble **fibrin,** which forms a meshwork of strands that traps the RBCs and forms the basis of the clot (Figure 29.8b). Normally, blood removed from the body clots within 2 to 6 minutes.

ACTIVITY 6

Determining Coagulation Time

1. Obtain a *nonheparinized* capillary tube, a timer (or watch), a lancet, cotton balls, a triangular file, and alcohol swabs.

2. Clean and prick the finger to produce a free flow of blood. Discard the lancet in the disposal container.

3. Place one end of the capillary tube in the blood drop, and hold the opposite end at a lower level to collect the sample.

4. Lay the capillary tube on a paper towel.

Record the time. _____

5. At 30-second intervals, make a small nick on the tube close to one end with the triangular file, and then carefully break the tube. Slowly separate the ends to see if a gel-like thread of fibrin spans the gap. When this occurs, record below and on the data sheet (page 436) the time for coagulation to occur. Are your results within the normal time range?

_____

6. Put used supplies in the autoclave bag and broken capillary tubes into the sharps container. ■

Blood Typing

Blood typing is a system of blood classification based on the presence of specific glycoproteins on the outer surface of the RBC plasma membrane. Such proteins are called **antigens,** or **agglutinogens,** and are genetically determined. In many cases, these antigens are accompanied by plasma proteins, called **antibodies** or **agglutinins,** that react with RBCs bearing different antigens, causing them to be clumped, agglutinated, and eventually hemolyzed. It is because of this phenomenon that a person's blood must be carefully typed before a whole blood or packed cell transfusion.

Several blood typing systems exist, based on the various possible antigens, but the factors routinely typed for are antigens of the ABO and Rh blood groups which are most commonly involved in transfusion reactions. Other blood factors, such as Kell, Lewis, M, and N, are not routinely typed for unless the individual will require multiple transfusions. (The basis of the ABO typing is shown in **Table 29.2.**)

Individuals whose red blood cells carry the Rh antigen are Rh positive (approximately 85% of the U.S. population); those lacking the antigen are Rh negative. Unlike ABO blood groups, the blood of neither Rh-positive (Rh^+) nor Rh-negative (Rh^-) individuals carries preformed anti-Rh antibodies. This is understandable in the case of the Rh-positive individual. However, Rh-negative persons who receive transfusions of Rh-positive blood become sensitized by the Rh antigens of the donor RBCs, and their systems begin to produce anti-Rh antibodies. On subsequent exposures to Rh-positive blood, typical transfusion reactions occur, resulting in the clumping and hemolysis of the donor blood cells.

Although the blood of dogs and other mammals does react with some of the human agglutinins (present in the antisera), the reaction is not as pronounced and varies with the animal blood used. For this reason, the most accurate and predictable blood typing results are obtained with human blood. The artificial blood kit does not use any body fluids and produces results similar to but not identical to results for human blood.

ACTIVITY 7

Typing for ABO and Rh Blood Groups

Blood may be typed on glass slides or using blood test cards. Each method is described in this activity.

Typing Blood Using Glass Slides

1. Obtain two clean microscope slides, a wax marking pencil, anti-A, anti-B, and anti-Rh typing sera, toothpicks, lancets, alcohol swabs, medicine dropper, and the Rh typing box.

2. Divide slide 1 into halves with the wax marking pencil. Label the lower left-hand corner "anti-A" and the lower right-hand corner "anti-B." Mark the bottom of slide 2 "anti-Rh."

3. Place one drop of anti-A serum on the *left* side of slide 1. Place one drop of anti-B serum on the *right* side of slide 1. Place one drop of anti-Rh serum in the center of slide 2.

4. If you are using your own blood, cleanse your finger with an alcohol swab, pierce the finger with a lancet, and wipe away the first drop of blood. Obtain 3 drops of freely flowing blood, placing one drop on each side of slide 1 and a drop on slide 2. Immediately dispose of the lancet in a designated disposal container.

Blood being tested   **Serum**

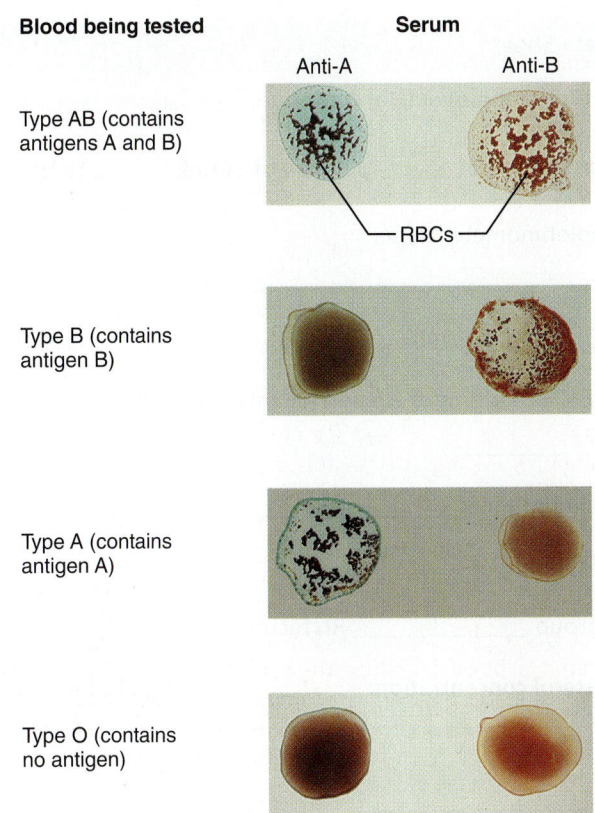

Figure 29.9 **Blood typing of ABO blood types.** When serum containing anti-A or anti-B antibodies (agglutinins) is added to a blood sample, agglutination will occur between the antibody and the corresponding antigen (agglutinogen A or B). As illustrated, agglutination occurs with both sera in blood group AB, with anti-B serum in blood group B, with anti-A serum in blood group A, and with neither serum in blood group O.

If using instructor-provided animal blood or EDTA-treated red cells, use a medicine dropper to place one drop of blood on each side of slide 1 and a drop of blood on slide 2.

⚠️ 5. Quickly mix each blood-antiserum sample with a *fresh* toothpick. Then dispose of the toothpicks and used alcohol swab in the autoclave bag.

6. Place slide 2 on the Rh typing box and rock gently back and forth. (A slightly higher temperature is required for precise Rh typing than for ABO typing.)

7. After 2 minutes, observe all three blood samples for evidence of clumping. The agglutination that occurs in the positive test for the Rh factor is very fine and difficult to perceive; thus if there is any question, observe the slide under the microscope. Record your observations in the **Activity 7: Blood Typing chart.**

8. Interpret your ABO results (see the examples of each type) in **Figure 29.9**. If clumping was observed on slide 2, you are Rh positive. If not, you are Rh negative.

9. Record your blood type on the data sheet (page 436).

10. Put the used slides in the bleach-containing bucket at the general supply area; put disposable supplies in the autoclave bag.

Activity 7: Blood Typing		
Result	Observed (+)	Not observed (−)
Presence of clumping with anti-A		
Presence of clumping with anti-B		
Presence of clumping with anti-Rh		

Using Blood Typing Cards

1. Obtain a blood typing card marked A, B, and Rh, dropper bottles of anti-A serum, anti-B serum, and anti-Rh serum, toothpicks, lancets, and alcohol swabs.

2. Place a drop of anti-A serum in the spot marked anti-A, place a drop of anti-B serum on the spot marked anti-B, and place a drop of anti-Rh serum on the spot marked anti-Rh (or anti-D).

3. Carefully add a drop of blood to each of the spots marked "Blood" on the card. If you are using your own blood, refer to step 4 in the Activity 7 section Typing Blood Using Glass Slides. Immediately discard the lancet in the designated disposal container.

4. Using a new toothpick for each test, mix the blood sample with the antibody. Dispose of the toothpicks appropriately.

5. Gently rock the card to allow the blood and antibodies to mix.

6. After 2 minutes, observe the card for evidence of clumping. The Rh clumping is very fine and may be difficult to observe. Record your observations in the Activity 7: Blood Typing chart. (Use Figure 29.9 to interpret your results.)

7. Record your blood type on the data sheet (page 436), and discard the card in an autoclave bag. ■

ACTIVITY 8

Observing Demonstration Slides

Before continuing on to the cholesterol determination, take the time to look at the slides of *macrocytic hypochromic anemia, microcytic hypochromic anemia, sickle cell anemia, lymphocytic leukemia* (chronic), and *eosinophilia* that have been put on demonstration by your instructor. Record your observations in the appropriate section of the Review Sheet. You can refer to your notes, the text, and other references later to respond to questions about the blood pathologies represented on the slides. ■

Cholesterol Concentration in Plasma

Atherosclerosis is the disease process in which the body's blood vessels become increasingly occluded, or blocked, by plaques. By narrowing the arteries, the plaques can contribute to hypertensive heart disease. They also serve as starting points for the formation of blood clots (thrombi), which may break away and block smaller vessels farther

29

Hematologic Test Data Sheet				
Differential WBC count:			**Hemoglobin (Hb) content:**	

Differential WBC count:

WBC	Student blood smear	Unknown sample # ____
% neutrophils	_____	_____
% eosinophils	_____	_____
% basophils	_____	_____
% lymphocytes	_____	_____
% monocytes	_____	_____

Hematocrit (PCV):

RBC _____ % of blood volume

WBC _____ % of blood volume   not generally reported

Plasma _____ % of blood

Hemoglobin (Hb) content:

Tallquist method: _____ g/100 ml of blood; _____ % Hb

Hemoglobinometer (type:

_____)

_____ g/100 ml of blood; _____ %Hb

Ratio (PCV to grams of Hb per 100 ml of blood):

_____

Coagulation time: _____

Blood typing:

ABO group _____ Rh factor _____

Cholesterol concentration: _____ mg/dl of blood

downstream in the circulatory pathway and cause heart attacks or strokes.

Ever since medical clinicians discovered that cholesterol is a major component of the smooth muscle plaques formed during atherosclerosis, it has had a bad press. Today, virtually no physical examination of an adult is considered complete until cholesterol levels are assessed along with other lifestyle risk factors. A normal value for plasma cholesterol in adults ranges from 130 to 200 mg per 100 ml of plasma; you will use blood to make such a determination.

Although the total plasma cholesterol concentration is valuable information, it may be misleading, particularly if a person's high-density lipoprotein (HDL) level is high and low-density lipoprotein (LDL) level is relatively low. Cholesterol, being water insoluble, is transported in the blood complexed to lipoproteins. In general, cholesterol bound into HDLs is destined to be degraded by the liver and then eliminated from the body, whereas that forming part of the LDLs is "traveling" to the body's tissue cells. When LDL levels are excessive, cholesterol is deposited in the blood vessel walls; hence, LDLs are considered to carry the "bad" cholesterol.

ACTIVITY 9

Measuring Plasma Cholesterol Concentration

1. Go to the appropriate supply area, and obtain a cholesterol test card and color scale, lancet, and alcohol swab.

2. Clean your fingertip with the alcohol swab, allow it to dry, then prick it with a lancet. Place a drop of blood on the test area of the card. Put the lancet in the designated disposal container.

3. After 3 minutes, remove the blood sample strip from the card and discard in the autoclave bag.

4. Analyze the underlying test spot, using the included color scale. Record the cholesterol level below and on the **Hematologic Test Data Sheet.**

Cholesterol level _____ mg/dl

⚠ 5. Before leaving the laboratory, use the spray bottle of bleach solution and saturate a paper towel to thoroughly wash down your laboratory bench. ▪

Blood being tested

Serum

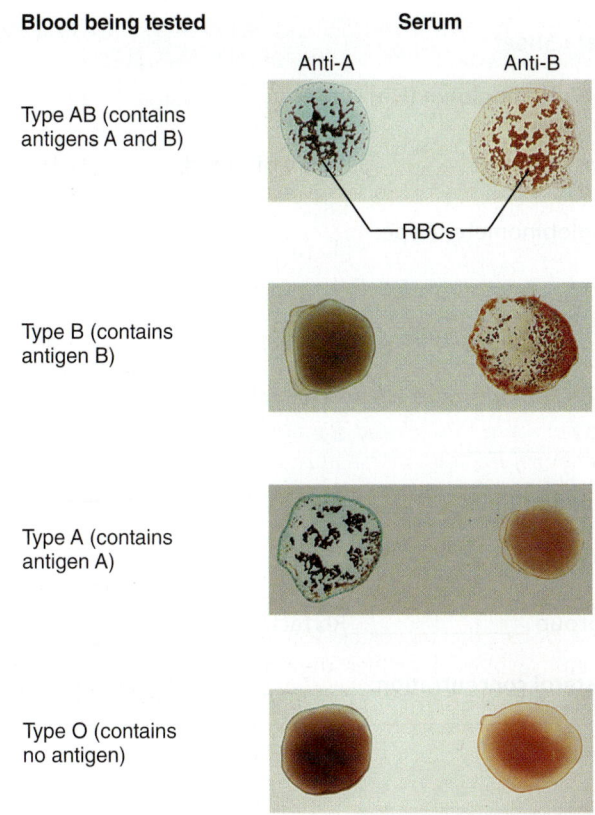

Anti-A   Anti-B

Type AB (contains antigens A and B)

RBCs

Type B (contains antigen B)

Type A (contains antigen A)

Type O (contains no antigen)

Figure 29.9 Blood typing of ABO blood types. When serum containing anti-A or anti-B antibodies (agglutinins) is added to a blood sample, agglutination will occur between the antibody and the corresponding antigen (agglutinogen A or B). As illustrated, agglutination occurs with both sera in blood group AB, with anti-B serum in blood group B, with anti-A serum in blood group A, and with neither serum in blood group O.

If using instructor-provided animal blood or EDTA-treated red cells, use a medicine dropper to place one drop of blood on each side of slide 1 and a drop of blood on slide 2.

⚠ 5. Quickly mix each blood-antiserum sample with a *fresh* toothpick. Then dispose of the toothpicks and used alcohol swab in the autoclave bag.

6. Place slide 2 on the Rh typing box and rock gently back and forth. (A slightly higher temperature is required for precise Rh typing than for ABO typing.)

7. After 2 minutes, observe all three blood samples for evidence of clumping. The agglutination that occurs in the positive test for the Rh factor is very fine and difficult to perceive; thus if there is any question, observe the slide under the microscope. Record your observations in the **Activity 7: Blood Typing chart.**

8. Interpret your ABO results (see the examples of each type) in **Figure 29.9**. If clumping was observed on slide 2, you are Rh positive. If not, you are Rh negative.

9. Record your blood type on the data sheet (page 436).

10. Put the used slides in the bleach-containing bucket at the general supply area; put disposable supplies in the autoclave bag.

Activity 7: Blood Typing		
Result	Observed (+)	Not observed (−)
Presence of clumping with anti-A		
Presence of clumping with anti-B		
Presence of clumping with anti-Rh		

Using Blood Typing Cards

1. Obtain a blood typing card marked A, B, and Rh, dropper bottles of anti-A serum, anti-B serum, and anti-Rh serum, toothpicks, lancets, and alcohol swabs.

2. Place a drop of anti-A serum in the spot marked anti-A, place a drop of anti-B serum on the spot marked anti-B, and place a drop of anti-Rh serum on the spot marked anti-Rh (or anti-D).

3. Carefully add a drop of blood to each of the spots marked "Blood" on the card. If you are using your own blood, refer to step 4 in the Activity 7 section Typing Blood Using Glass Slides. Immediately discard the lancet in the designated disposal container.

4. Using a new toothpick for each test, mix the blood sample with the antibody. Dispose of the toothpicks appropriately.

5. Gently rock the card to allow the blood and antibodies to mix.

6. After 2 minutes, observe the card for evidence of clumping. The Rh clumping is very fine and may be difficult to observe. Record your observations in the Activity 7: Blood Typing chart. (Use Figure 29.9 to interpret your results.)

7. Record your blood type on the data sheet (page 436), and discard the card in an autoclave bag. ■

ACTIVITY 8

Observing Demonstration Slides

Before continuing on to the cholesterol determination, take the time to look at the slides of *macrocytic hypochromic anemia, microcytic hypochromic anemia, sickle cell anemia, lymphocytic leukemia* (chronic), and *eosinophilia* that have been put on demonstration by your instructor. Record your observations in the appropriate section of the Review Sheet. You can refer to your notes, the text, and other references later to respond to questions about the blood pathologies represented on the slides. ■

Cholesterol Concentration in Plasma

Atherosclerosis is the disease process in which the body's blood vessels become increasingly occluded, or blocked, by plaques. By narrowing the arteries, the plaques can contribute to hypertensive heart disease. They also serve as starting points for the formation of blood clots (thrombi), which may break away and block smaller vessels farther

29

Hematologic Test Data Sheet

Differential WBC count:

WBC	Student blood smear	Unknown sample # ___
% neutrophils	_____	_____
% eosinophils	_____	_____
% basophils	_____	_____
% lymphocytes	_____	_____
% monocytes	_____	_____

Hematocrit (PCV):

RBC _____ % of blood volume

WBC _____ % of blood volume   not generally reported

Plasma _____ % of blood

Hemoglobin (Hb) content:

Tallquist method: _____ g/100 ml of blood; _____ % Hb

Hemoglobinometer (type:

_____)

_____ g/100 ml of blood; _____ %Hb

Ratio (PCV to grams of Hb per 100 ml of blood):

_____

Coagulation time: _____

Blood typing:

ABO group _____ Rh factor _____

Cholesterol concentration: _____ mg/dl of blood

---

downstream in the circulatory pathway and cause heart attacks or strokes.

Ever since medical clinicians discovered that cholesterol is a major component of the smooth muscle plaques formed during atherosclerosis, it has had a bad press. Today, virtually no physical examination of an adult is considered complete until cholesterol levels are assessed along with other lifestyle risk factors. A normal value for plasma cholesterol in adults ranges from 130 to 200 mg per 100 ml of plasma; you will use blood to make such a determination.

Although the total plasma cholesterol concentration is valuable information, it may be misleading, particularly if a person's high-density lipoprotein (HDL) level is high and low-density lipoprotein (LDL) level is relatively low. Cholesterol, being water insoluble, is transported in the blood complexed to lipoproteins. In general, cholesterol bound into HDLs is destined to be degraded by the liver and then eliminated from the body, whereas that forming part of the LDLs is "traveling" to the body's tissue cells. When LDL levels are excessive, cholesterol is deposited in the blood vessel walls; hence, LDLs are considered to carry the "bad" cholesterol.

ACTIVITY 9

Measuring Plasma Cholesterol Concentration

1. Go to the appropriate supply area, and obtain a cholesterol test card and color scale, lancet, and alcohol swab.

2. Clean your fingertip with the alcohol swab, allow it to dry, then prick it with a lancet. Place a drop of blood on the test area of the card. Put the lancet in the designated disposal container.

3. After 3 minutes, remove the blood sample strip from the card and discard in the autoclave bag.

4. Analyze the underlying test spot, using the included color scale. Record the cholesterol level below and on the **Hematologic Test Data Sheet.**

Cholesterol level _____ mg/dl

5. Before leaving the laboratory, use the spray bottle of bleach solution and saturate a paper towel to thoroughly wash down your laboratory bench. ▪

29

Anatomy of the Heart

MATERIALS

- [ ] X ray of the human thorax for observation of the position of the heart in situ; X-ray viewing box
- [ ] Three-dimensional heart model and torso model or laboratory chart showing heart anatomy
- [ ] Red and blue pencils
- [ ] Highlighter
- [ ] Three-dimensional models of cardiac and skeletal muscle
- [ ] Compound microscope
- [ ] Prepared slides of cardiac muscle (l.s.)
- [ ] Preserved or fresh sheep hearts, pericardial sacs intact (if possible)
- [ ] Dissecting instruments and tray
- [ ] Pointed glass rods or blunt probes
- [ ] Small plastic metric rulers
- [ ] Disposable gloves
- [ ] Container for disposal of organic debris
- [ ] Laboratory detergent
- [ ] Spray bottle with 10% household bleach solution

✀ For instructions on animal dissections, see the dissection exercises (starting on p. 697) in the cat and fetal pig editions of this manual.

MasteringA&P® For related exercise study tools, go to the Study Area of MasteringA&P. There you will find:
- Practice Anatomy Lab PAL
- PhysioEx PEx
- A&PFlix A&PFlix
- Practice quizzes, Histology Atlas, eText, Videos, and more!

OBJECTIVES

1. Describe the location of the heart.
2. Name and describe the covering and lining tissues of the heart.
3. Name and locate the major anatomical areas and structures of the heart when provided with an appropriate model, image, or dissected sheep heart, and describe the function of each.
4. Explain how the atrioventricular and semilunar valves operate.
5. Distinguish between blood vessels carrying oxygen-rich blood and those carrying carbon dioxide–rich blood and describe the system used to color code them in images.
6. Explain why the heart is called a double pump, and compare the pulmonary and systemic circuits.
7. Trace the pathway of blood through the heart.
8. Trace the functional blood supply of the heart and name the associated blood vessels.
9. Describe the histology of cardiac muscle, and state the importance of its intercalated discs and the spiral arrangement of its cells.

PRE-LAB QUIZ

1. The heart is enclosed in a double-walled sac called the
   a. apex   b. mediastinum   c. pericardium   d. thorax
2. The heart is divided into _____ chambers.
   a. two   b. three   c. four   d. five
3. What is the name of the two receiving chambers of the heart?
   _____
4. The left ventricle discharges blood into the _____, from which all systemic arteries of the body diverge to supply the body tissues.
   a. aorta   c. pulmonary vein
   b. pulmonary artery   d. vena cava
5. Circle True or False. Blood flows through the heart in one direction—from the atria to the ventricles.
6. Circle the correct underlined term. The right atrioventricular valve, or tricuspid valve / mitral valve, prevents backflow into the right atrium when the right ventricle is contracting.
7. Circle the correct underlined term. The heart serves as a double pump. The right / left side serves as the pulmonary circulation pump, shunting carbon dioxide–rich blood to the lungs.
8. The blood vessels that supply blood to the heart itself are the
   a. aortas   c. coronary arteries
   b. carotid arteries   d. pulmonary trunks
9. Two microscopic features of cardiac cells that help distinguish them from other types of muscle cells are branching of the cells and
   a. intercalated discs   c. sarcolemma
   b. myosin fibers   d. striations
10. Circle the correct underlined term. In the heart, the left / right ventricle has thicker walls and a basically circular cavity shape.

he major function of the **cardiovascular system** is transportation. Using blood as the transport vehicle, the system carries oxygen, digested foods, cell wastes, electrolytes, and many other substances vital to the body's homeostasis to and from the body cells. The system's propulsive force is the contracting heart, which can be compared to a muscular pump equipped with one-way valves. As the heart contracts, it forces blood into a closed system of large and small plumbing tubes (blood vessels) within which the blood is confined and circulated. This exercise deals with the structure of the heart, or circulatory pump. (The anatomy of the blood vessels is considered separately in Exercise 32.)

Gross Anatomy of the Human Heart

The **heart,** a cone-shaped organ approximately the size of a fist, is located within the mediastinum, or medial cavity, of the thorax. It is flanked laterally by the lungs, posteriorly by the vertebral column, and anteriorly by the sternum **(Figure 30.1)**. Its more pointed **apex** extends slightly to the left and rests on the diaphragm, approximately at the level of the fifth intercostal space. Its broader **base,** from which the great vessels emerge, lies beneath the second rib and points toward the right shoulder. In situ, the right ventricle of the heart forms most of its anterior surface.

The apical pulse may be heard in the 5th intercostal space at the point of maximal intensity (PMI).

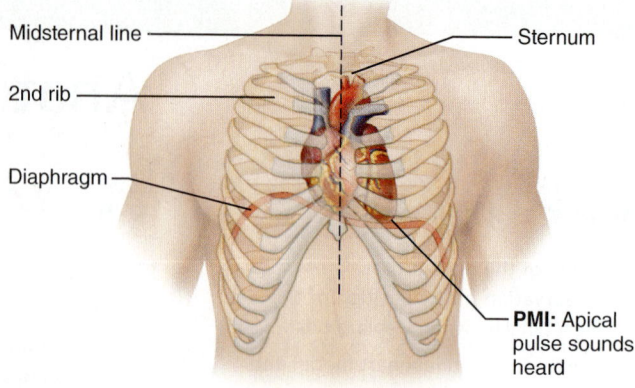

Figure 30.1 Location of the heart in the thorax. PMI is the point of maximal intensity where the apical pulse is heard.

• If an X ray of a human thorax is available, verify the relationships described above (otherwise, Figure 30.1 should suffice).

The gross anatomy figure **(Figure 30.2)** shows three views of the heart—external anterior and posterior views and

Figure 30.2 Gross anatomy of the human heart. (a) External anterior view.

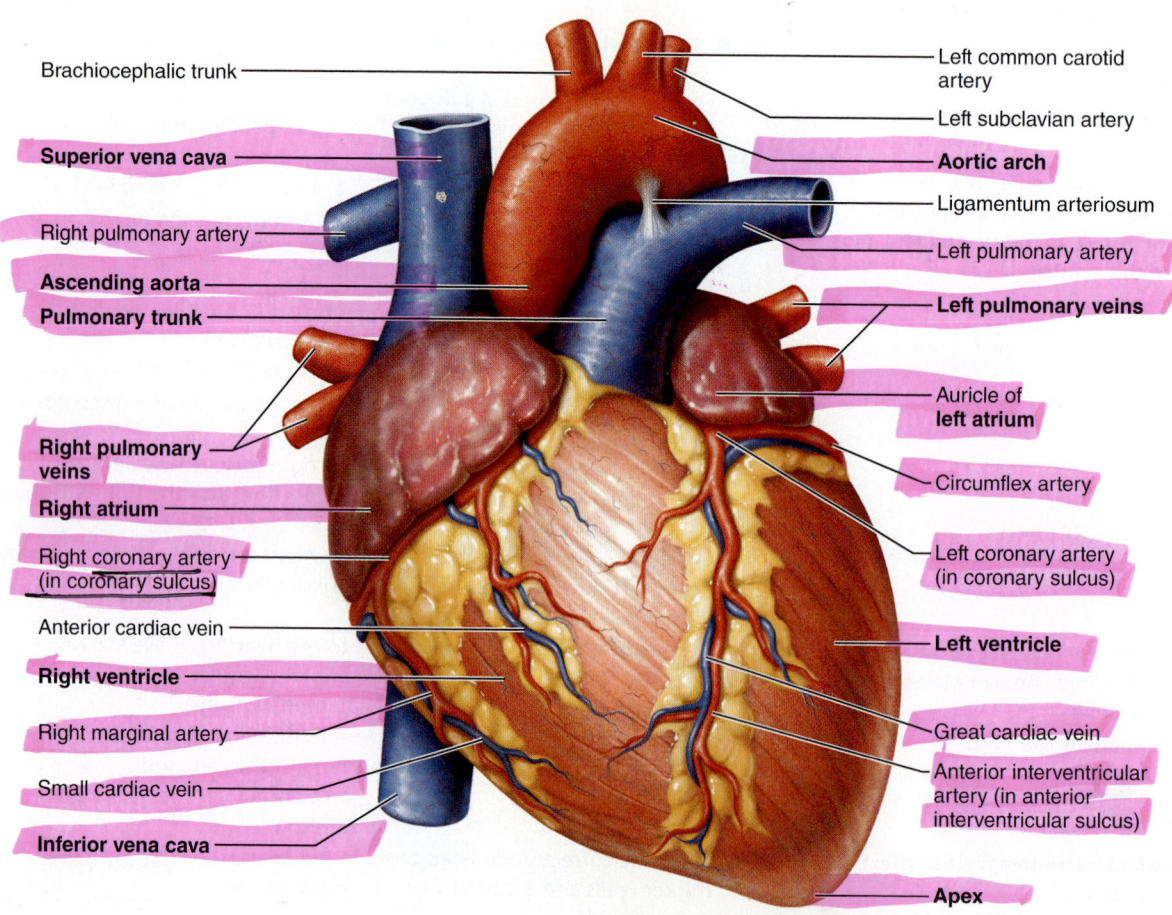

(a)

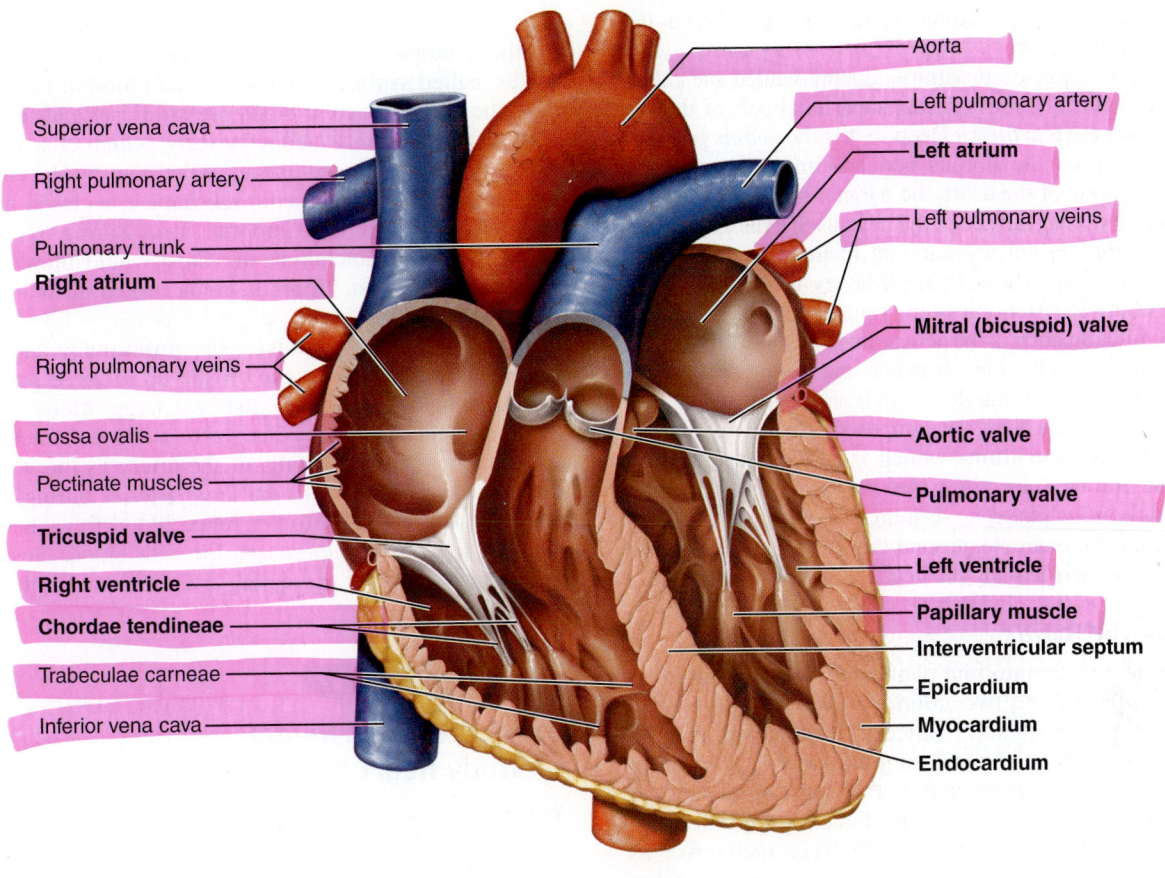

Aorta

Superior vena cava

Right pulmonary artery

Pulmonary trunk

Right atrium

Right pulmonary veins

Fossa ovalis

Pectinate muscles

Tricuspid valve

Right ventricle

Chordae tendineae

Trabeculae carneae

Inferior vena cava

Left pulmonary artery

Left atrium

Left pulmonary veins

Mitral (bicuspid) valve

Aortic valve

Pulmonary valve

Left ventricle

Papillary muscle

Interventricular septum

Epicardium

Myocardium

Endocardium

(b)

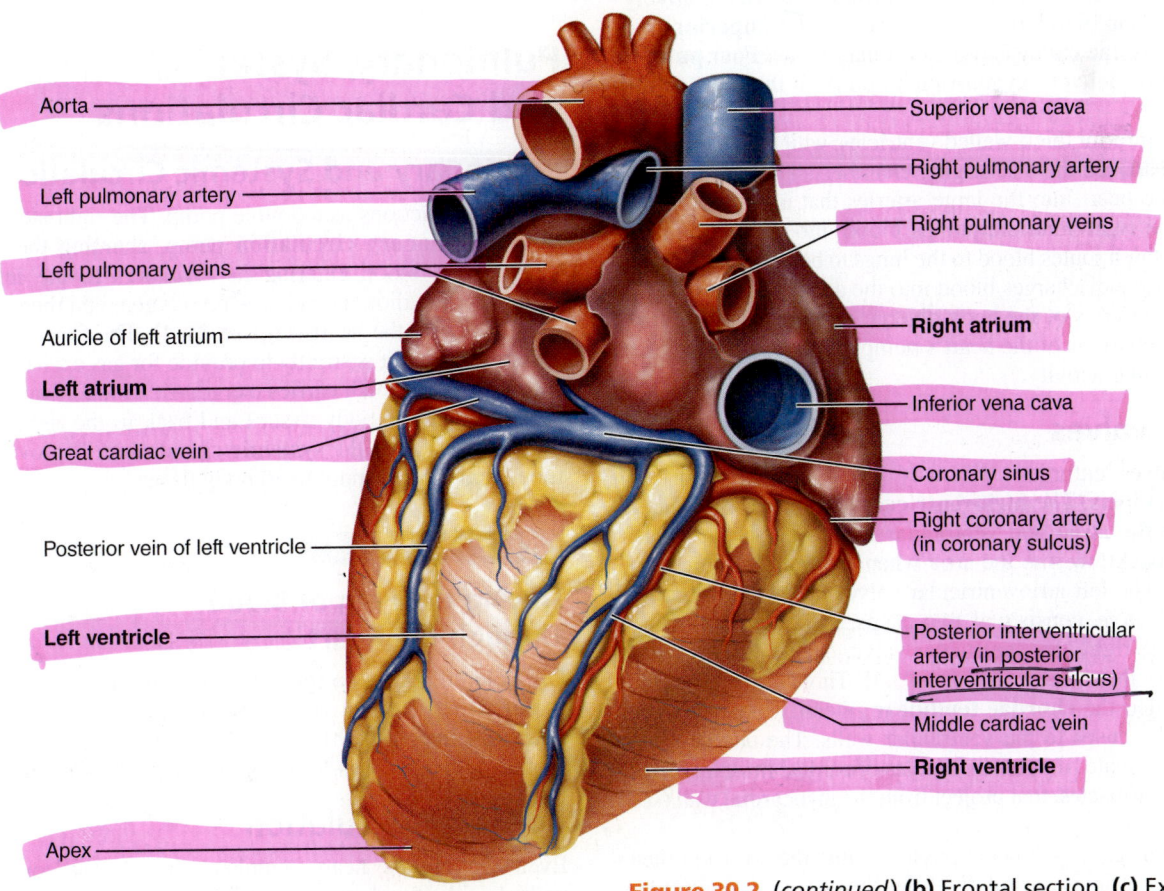

Aorta

Left pulmonary artery

Left pulmonary veins

Auricle of left atrium

Left atrium

Great cardiac vein

Posterior vein of left ventricle

Left ventricle

Apex

Superior vena cava

Right pulmonary artery

Right pulmonary veins

Right atrium

Inferior vena cava

Coronary sinus

Right coronary artery (in coronary sulcus)

Posterior interventricular artery (in posterior interventricular sulcus)

Middle cardiac vein

Right ventricle

(c)

Figure 30.2 (*continued*) **(b)** Frontal section. **(c)** Exterior posterior view.

30

a frontal section. As its anatomical areas are described in the text, consult the figure.

The heart is enclosed within a double-walled sac called the pericardium. The loose-fitting superficial part of the sac is the **fibrous pericardium.** Deep to it is the *serous pericardium,* which lines the fibrous pericardium as the **parietal layer.** At the base of the heart, the parietal layer reflects back to cover the external surface of the heart as the **visceral layer,** or **epicardium.** The epicardium is an integral part of the heart wall. Serous fluid produced by these layers allows the heart to beat in a relatively frictionless environment.

Inflammation of the pericardium, **pericarditis,** causes painful adhesions between the serous pericardial layers. These adhesions interfere with heart movements. ✚

The walls of the heart are composed primarily of cardiac muscle—the **myocardium**—which is reinforced internally by a dense fibrous connective tissue network. This network—the fibrous *cardiac skeleton*—is more elaborate and thicker in certain areas, for example, around the valves and at the base of the great vessels leaving the heart.

Heart Chambers

The heart is divided into four chambers: two superior **atria** (singular: *atrium*) and two inferior **ventricles,** each lined by thin serous endothelium called the **endocardium.** The septum that divides the heart longitudinally is referred to as the **interatrial** or **interventricular septum,** depending on which chambers it partitions. Functionally, the atria are receiving chambers and are relatively ineffective as pumps. Blood flows into the atria under low pressure from the veins of the body. The right atrium receives relatively oxygen-poor blood from the body via the **superior** and **inferior venae cavae** and the coronary sinus. Four **pulmonary veins** deliver oxygen-rich blood from the lungs to the left atrium.

The inferior thick-walled ventricles, which form the bulk of the heart, are the discharging chambers. They force blood out of the heart into the large arteries that emerge from its base. The right ventricle pumps blood into the **pulmonary trunk,** which routes blood to the lungs to be oxygenated. The left ventricle discharges blood into the **aorta,** from which all systemic arteries of the body diverge to supply the body tissues. Discussions of the heart's pumping action usually refer to ventricular activity.

Heart Valves

Four valves enforce a one-way blood flow through the heart chambers. The **atrioventricular (AV) valves,** located between the atrial and ventricular chambers on each side, prevent backflow into the atria when the ventricles are contracting. The left atrioventricular valve, called the **mitral** or *bicuspid valve,* consists of two cusps, or flaps, of endocardium. The right atrioventricular valve, called the **tricuspid valve,** has three cusps **(Figure 30.3)**. Tiny white collagenic cords called the **chordae tendineae** (literally, heart strings) anchor the cusps to the ventricular walls. The chordae tendineae originate from small bundles of cardiac muscle, called **papillary muscles,** that project from the myocardial wall (see Figure 30.2b).

When blood is flowing passively into the atria and then into the ventricles during **diastole** (the period of ventricular filling), the AV valve flaps hang limply into the ventricular chambers and then are carried passively toward the atria by the accumulating blood. The contraction of the ventricles, called **systole,** compresses the blood in their chambers; the intraventricular blood pressure rises and causes the valve flaps to be reflected superiorly, which closes the AV valves. The chordae tendineae, pulled taut by the contracting papillary muscles, anchor the flaps in a closed position that prevents backflow into the atria during ventricular contraction. If unanchored, the flaps would blow upward into the atria like an umbrella being turned inside out by a strong wind.

The second set of valves, the **pulmonary** and **aortic (semilunar, SL) valves**, each made up of three pocketlike cusps, guards the bases of the two large arteries leaving the ventricular chambers. The valve cusps are forced open and flatten against the walls of the artery as the ventricles discharge their blood into the large arteries during systole. However, when the ventricles relax, blood flows backward toward the heart and the cusps fill with blood, closing the semilunar valves and preventing arterial blood from reentering the heart.

ACTIVITY 1

Using the Heart Model to Study Heart Anatomy

When you have located in the gross anatomy figure (Figure 30.2) all the structures described above, observe the human heart model and laboratory charts and reidentify the same structures without referring to the figure. ■

Pulmonary, Systemic, and Cardiac Circulations

Pulmonary and Systemic Circulations

The heart functions as a double pump. The right side serves as the **pulmonary circulation** pump, shunting the carbon dioxide–rich blood entering its chambers to the lungs to unload carbon dioxide and pick up oxygen, and then back to the left side of the heart **(Figure 30.4)**. The function of the pulmonary circuit is strictly to provide for gas exchange. The second circuit, which carries oxygen-rich blood from the left heart through the body tissues and back to the right side of the heart, is called the **systemic circulation.** It provides the functional blood supply to all body tissues.

ACTIVITY 2

Tracing the Path of Blood Through the Heart

Use colored pencils to trace the pathway of a red blood cell through the heart by adding arrows to the frontal section diagram (Figure 30.2b). Use red arrows for the oxygen-rich blood and blue arrows for the carbon dioxide–rich blood. ■

Coronary Circulation

Even though the heart chambers are almost continually bathed with blood, this contained blood does not nourish the myocardium. The functional blood supply of the heart

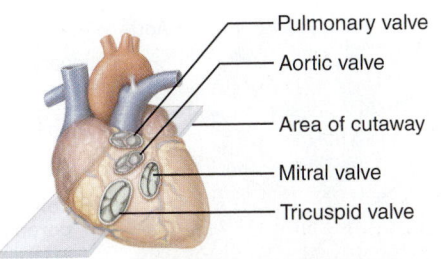

Pulmonary valve
Aortic valve
Area of cutaway
Mitral valve
Tricuspid valve

Figure 30.3 Heart valves. (a) Superior view of the two sets of heart valves (atria removed). **(b)** Photograph of the heart valves, superior view. **(c)** Photograph of the tricuspid valve. This inferior-to-superior view shows the valve as seen from the right ventricle. **(d)** Frontal section of the heart.

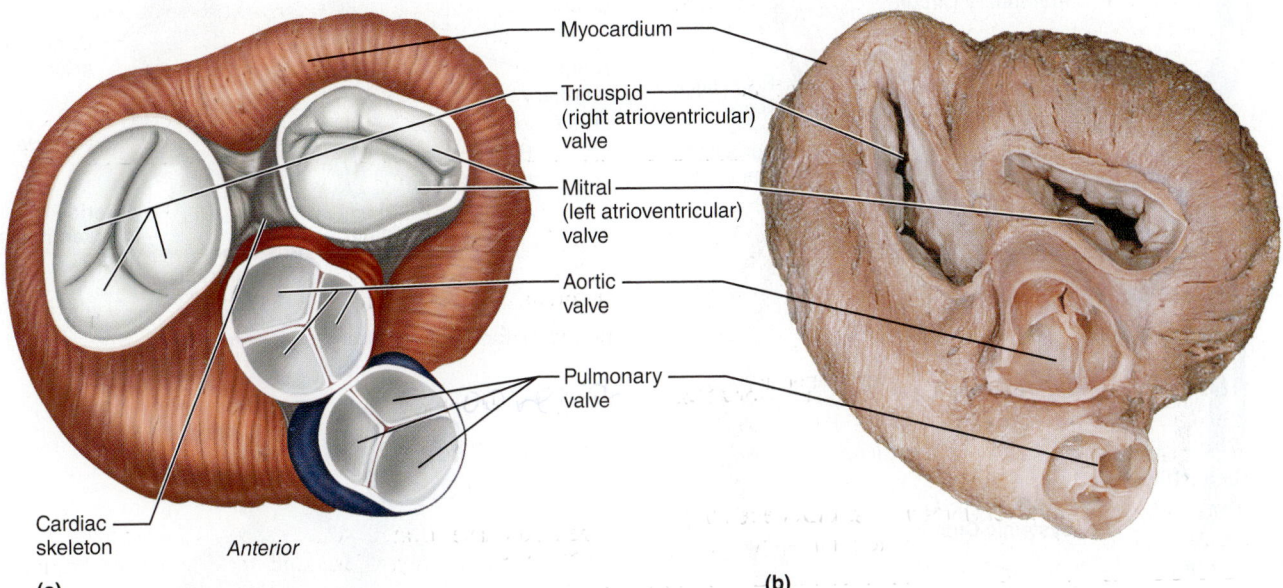

Myocardium

Tricuspid (right atrioventricular) valve

Mitral (left atrioventricular) valve

Aortic valve

Pulmonary valve

Cardiac skeleton   *Anterior*

(a)

(b)

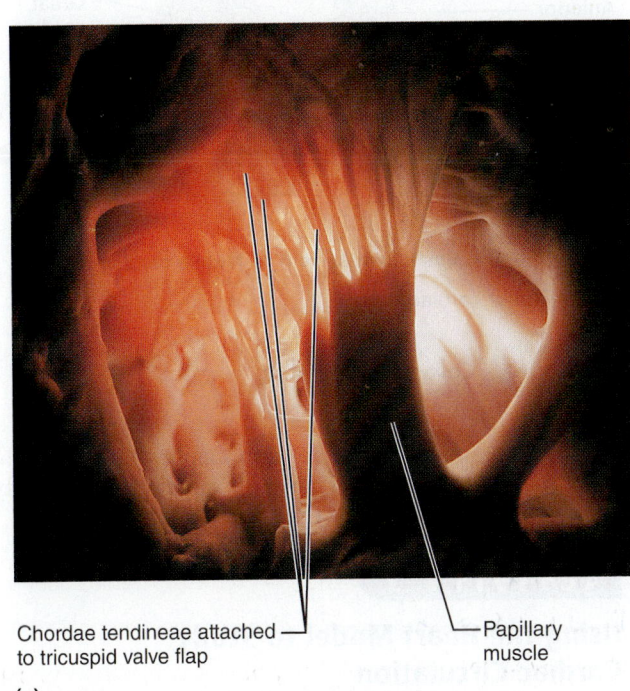

Chordae tendineae attached to tricuspid valve flap

Papillary muscle

(c)

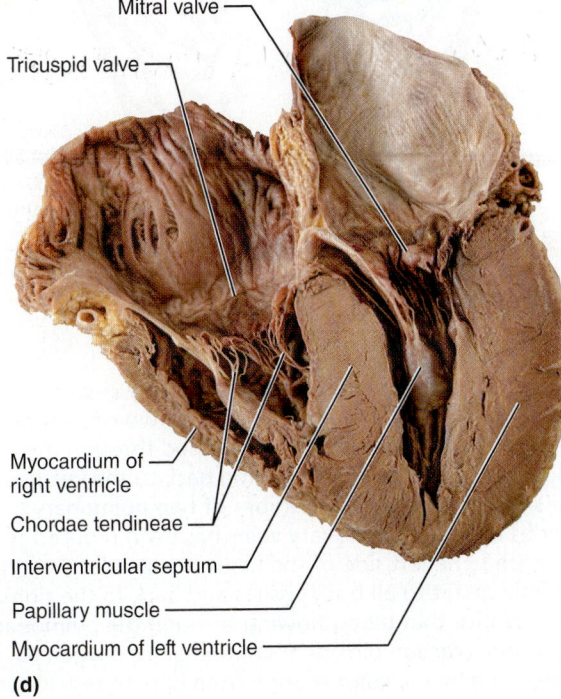

Mitral valve

Tricuspid valve

Myocardium of right ventricle

Chordae tendineae

Interventricular septum

Papillary muscle

Myocardium of left ventricle

(d)

30

is provided by the coronary arteries (see Figure 30.2 and **Figure 30.5**). The **right** and **left coronary arteries** issue from the base of the aorta just above the aortic semilunar valve and encircle the heart in the **coronary sulcus** at the junction of the atria and ventricles. They then ramify over the heart's surface, the right coronary artery supplying the

posterior surface of the ventricles and the lateral aspect of the right side of the heart, largely through its **posterior interventricular** and **right marginal artery** branches. The left coronary artery supplies the anterior ventricular walls and the laterodorsal part of the left side of the heart via its two major branches, the **anterior interventricular artery** (also

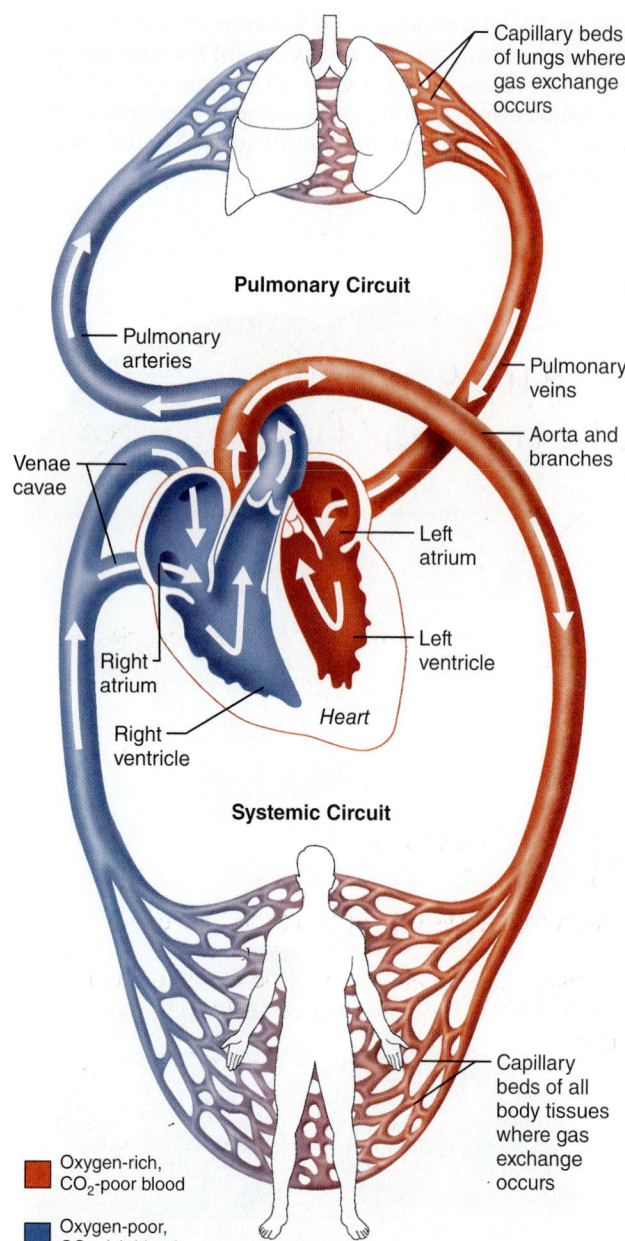

Figure 30.4 The systemic and pulmonary circuits.
The heart is a double pump that serves two circulations. The right side of the heart pumps blood through the pulmonary circuit to the lungs and back to the left heart. (For simplicity, the actual number of two pulmonary arteries and four pulmonary veins has been reduced to one each.) The left side of the heart pumps blood via the systemic circuit to all body tissues and back to the right heart. Notice that blood flowing through the pulmonary circuit loses carbon dioxide (CO_2) and gains oxygen (O_2) as depicted by the color change from blue to red. Blood flowing through the systemic circuit loses oxygen and picks up carbon dioxide (red to blue color change).

called the *left anterior descending artery*) and the **circum-flex artery.** The coronary arteries and their branches are compressed during systole and fill when the heart is relaxed.

The myocardium is largely drained by the **great, middle,** and **small cardiac veins,** which empty into the **coronary**

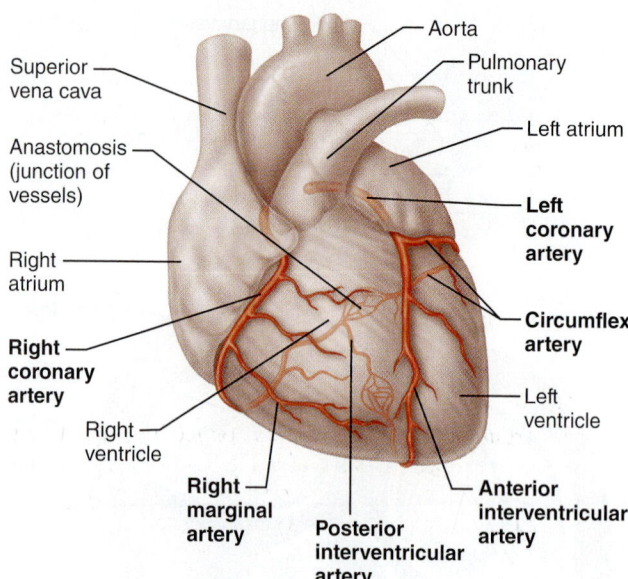

(a) The major coronary arteries

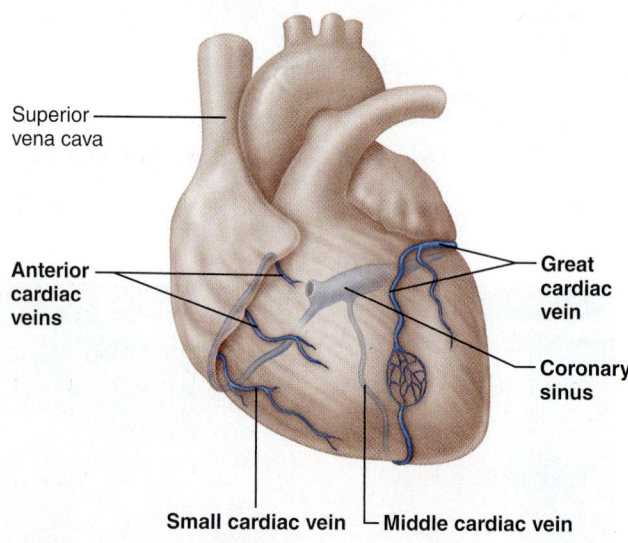

(b) The major cardiac veins

Figure 30.5 Coronary circulation.

sinus. The coronary sinus, in turn, empties into the right atrium. In addition, several **anterior cardiac veins** empty directly into the right atrium (Figure 30.5).

ACTIVITY 3

Using the Heart Model to Study Cardiac Circulation

1. Obtain a highlighter and highlight all the cardiac blood vessels in the external gross anatomy illustrations (Figure 30.2a and c). Note how arteries and veins travel together.

2. On a model of the heart, locate all the cardiac blood vessels shown in the coronary circulation illustration (Figure 30.5). Use your finger to trace the pathway of blood from the right coronary artery to the lateral aspect of the right side of the heart and back to the right atrium. Name the arteries and

veins along the pathway. Trace the pathway of blood from the left coronary artery to the anterior ventricular walls and back to the right atrium. Name the arteries and veins along the pathway. Note that there are multiple different pathways to distribute blood to these parts of the heart. ■

Microscopic Anatomy of Cardiac Muscle

Cardiac muscle is found in only one place—the heart. The heart acts as a vascular pump, propelling blood to all tissues of the body; cardiac muscle is thus very important to life. Cardiac muscle is involuntary, ensuring a constant blood supply.

The cardiac cells, crisscrossed by connective tissue fibers for strength, are arranged in spiral or figure-8-shaped bundles **(Figure 30.6)**. When the heart contracts, its internal chambers become smaller (or are temporarily obliterated), forcing the blood into the large arteries leaving the heart.

ACTIVITY 4

Examining Cardiac Muscle Tissue Anatomy

1. Observe the three-dimensional model of cardiac muscle, examining its branching cells and the areas where the cells interdigitate, the **intercalated discs.** These two structural features provide a continuity to cardiac muscle not seen in other muscle tissues and allow close coordination of heart activity.

2. Compare the model of cardiac muscle to the model of skeletal muscle. Note the similarities and differences between the two kinds of muscle tissue.

3. Obtain and observe a longitudinal section of cardiac muscle under high power. Identify the nucleus, striations,

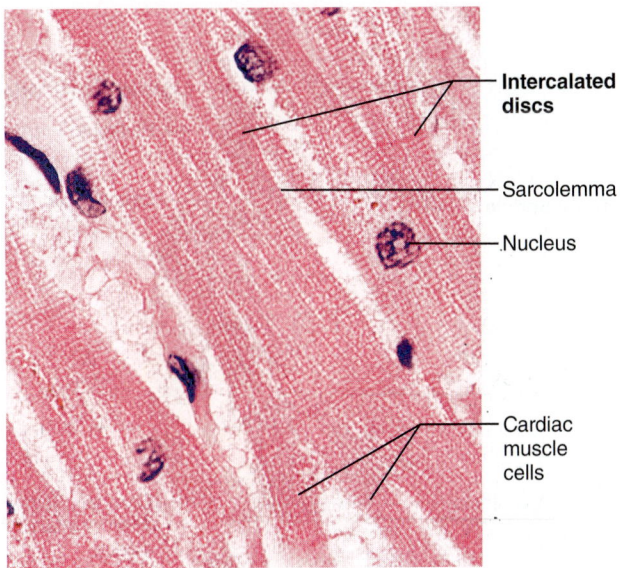

Figure 30.7 Photomicrograph of cardiac muscle (665×).

intercalated discs, and sarcolemma of the individual cells and then compare your observations to the photomicrograph **(Figure 30.7)**. ■

DISSECTION:
The Sheep Heart

Dissection of a sheep heart is valuable because it is similar in size and structure to the human heart. Also, a dissection experience allows you to view structures in a way not possible with models and diagrams. (Refer to **Figure 30.8** as you proceed with the dissection.)

1. Obtain a preserved sheep heart, a dissecting tray, dissecting instruments, a glass probe, a plastic ruler, and gloves. Rinse the sheep heart in cold water to remove excessive preservatives and to flush out any trapped blood clots. Now you are ready to make your observations.

2. Observe the texture of the fibrous pericardium. Also, note its point of attachment to the heart. Where is it attached?

_____

3. If the fibrous pericardial sac is still intact, slit it open and cut it from its attachments. Observe the slippery parietal pericardium that lines the sac and the visceral pericardium (epicardium) that covers the heart wall. Using a sharp scalpel, carefully pull a little of the epicardium away from the myocardium. How do its position, thickness, and apposition to the heart differ from those of the parietal pericardium?

_____

_____

_____

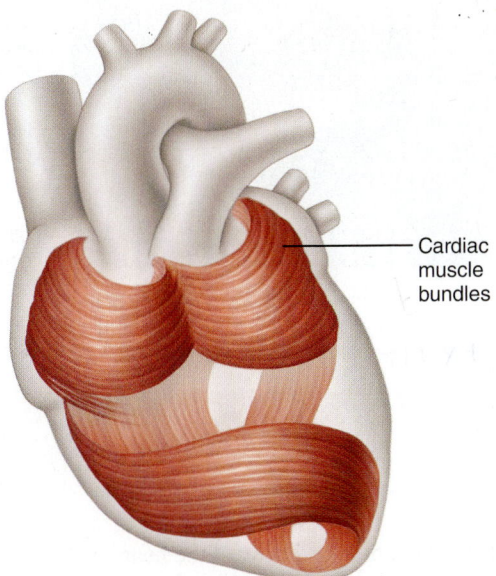

Figure 30.6 The circular and spiral arrangement of cardiac muscle bundles in the myocardium of the heart.

30

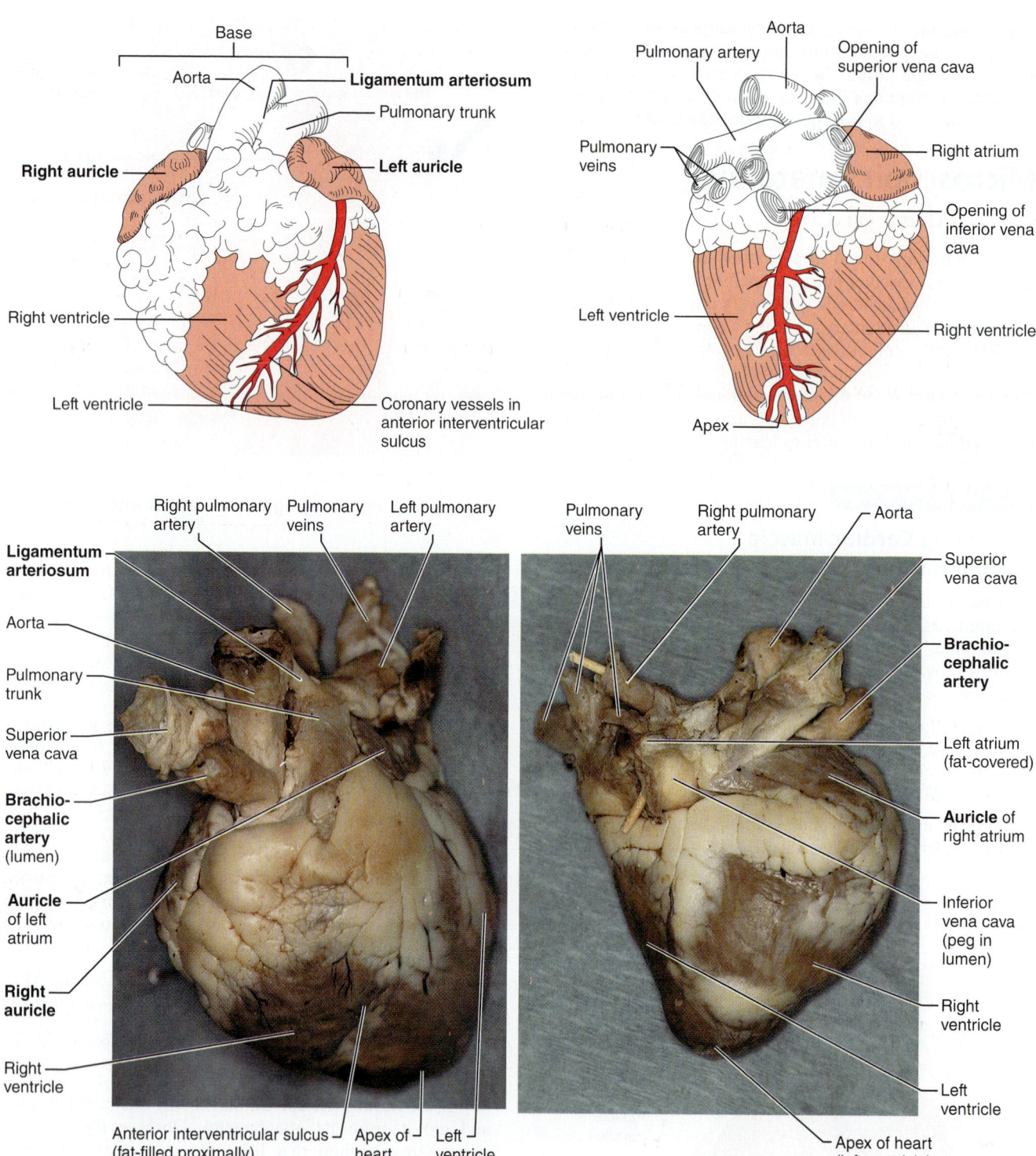

Figure 30.8 Anatomy of the sheep heart. (a) Anterior view. **(b)** Posterior view. Diagrammatic views at top; photographs at bottom.

4.   Examine the external surface of the heart. Notice the accumulation of adipose tissue, which in many cases marks the separation of the chambers and the location of the coronary arteries that nourish the myocardium. Carefully scrape away some of the fat with a scalpel to expose the coronary blood vessels.

5.   Identify the base and apex of the heart, and then identify the two wrinkled **auricles,** earlike flaps of tissue projecting from the atrial chambers. The balance of the heart muscle is ventricular tissue. To identify the left ventricle, compress the ventricular chambers on each side of the longitudinal fissures carrying the coronary blood vessels. The side that feels thicker and more solid is the left ventricle. The right ventricle

feels much thinner and somewhat flabby when compressed. This difference reflects the greater demand placed on the left ventricle, which must pump blood through the much longer systemic circulation, a pathway with much higher resistance than the pulmonary circulation served by the right ventricle. Hold the heart in its anatomical position (Figure 30.8a), with the anterior surface uppermost. In this position the left ventricle composes the entire apex and the left side of the heart.

6. Identify the pulmonary trunk and the aorta extending from the superior aspect of the heart. The pulmonary trunk is more anterior, and you may see its division into the right and left pulmonary arteries if it has not been cut too closely to the heart. The thicker-walled aorta, which branches almost immediately, is located just beneath the pulmonary trunk. The first observable branch of the sheep aorta, the **brachiocephalic artery (trunk),** is identifiable unless the aorta has been cut immediately as it leaves the heart. The brachiocephalic artery splits to form the right carotid and subclavian arteries, which supply the right side of the head and right forelimb, respectively.

Gently pull on the aorta with your gloved fingers or forceps to stretch it. Repeat with the venae cavae.

Which vessel is easier to stretch? _____

How does the elasticity of each vessel relate to its ability to withstand pressure?

_____

Carefully clear away some of the fat between the pulmonary trunk and the aorta to expose the **ligamentum arteriosum,** a cordlike remnant of the **ductus arteriosus.** (In the fetus, the ductus arteriosus allows blood to pass directly from the pulmonary trunk to the aorta, thus bypassing the nonfunctional fetal lungs.)

7. Cut through the wall of the aorta until you see the aortic (semilunar) valve. Identify the two openings to the coronary arteries just above the valve. Insert a probe into one of these holes to see if you can follow the course of a coronary artery across the heart.

8. Turn the heart to view its posterior surface (compare it to the view in Figure 30.8b). Notice that the right and left ventricles appear equal-sized in this view. Try to identify the four thin-walled pulmonary veins entering the left atrium. Identify the superior and inferior venae cavae entering the right atrium. Because of the way the heart is trimmed, the pulmonary veins and superior vena cava may be very short or missing. If possible, compare the approximate diameter of the superior vena cava with the diameter of the aorta.

Which is larger? _____

Which has thicker walls? _____

Why do you suppose these differences exist?

_____

_____

_____

9. Insert a probe into the superior vena cava, through the right atrium, and out the inferior vena cava. Use scissors to cut along the probe so that you can view the interior of the right atrium. Observe the tricuspid valve.

How many flaps does it have? _____

Pour some water into the right atrium and allow it to flow into the ventricle. *Slowly and gently* squeeze the right ventricle to watch the closing action of this valve. (If you squeeze too vigorously, you'll get a face full of water!) Drain the water from the heart before continuing.

10. Return to the pulmonary trunk and cut through its anterior wall until you can see the pulmonary (semilunar) valve **(Figure 30.9)**. Pour some water into the base of the pulmonary trunk to observe the closing action of this valve. How does its action differ from that of the tricuspid valve?

_____

_____

_____

After observing pulmonary valve action, drain the heart once again. Extend the cut through the pulmonary trunk into the right ventricle. Cut down, around, and up through the tricuspid valve to make the cut continuous with the cut across the right atrium (see Figure 30.9).

11. Reflect the cut edges of the superior vena cava, right atrium, and right ventricle to obtain the view seen in the dissection photo (Figure 30.9). Observe the comblike ridges of muscle throughout most of the right atrium. This is called **pectinate muscle** (*pectin* = comb). Identify, on the ventral atrial wall, the large opening of the inferior vena cava and follow it to its external opening with a probe. Notice that the atrial walls in the vicinity of the venae cavae are smooth and lack the roughened appearance (pectinate musculature) of the other regions of the atrial walls. Just below the inferior vena caval opening, identify the opening of the **coronary sinus,** which returns venous blood of the coronary circulation to the right atrium. Nearby, locate an oval depression, the **fossa ovalis,** in the interatrial septum. This depression marks the site of an opening in the fetal heart, the **foramen ovale,** which allows blood to pass from the right to the left atrium, thus bypassing the fetal lungs.

12. Identify the papillary muscles in the right ventricle, and follow their attached chordae tendineae to the flaps of the tricuspid valve. Notice the pitted and ridged appearance **(trabeculae carneae)** of the inner ventricular muscle.

13. Identify the **moderator band** (septomarginal band), a bundle of cardiac muscle fibers connecting the interventricular septum to anterior papillary muscles. It contains a branch of the atrioventricular bundle and helps coordinate contraction of the ventricle.

14. Make a longitudinal incision through the left atrium and continue it into the left ventricle. Notice how much thicker the myocardium of the left ventricle is than that of the right

30

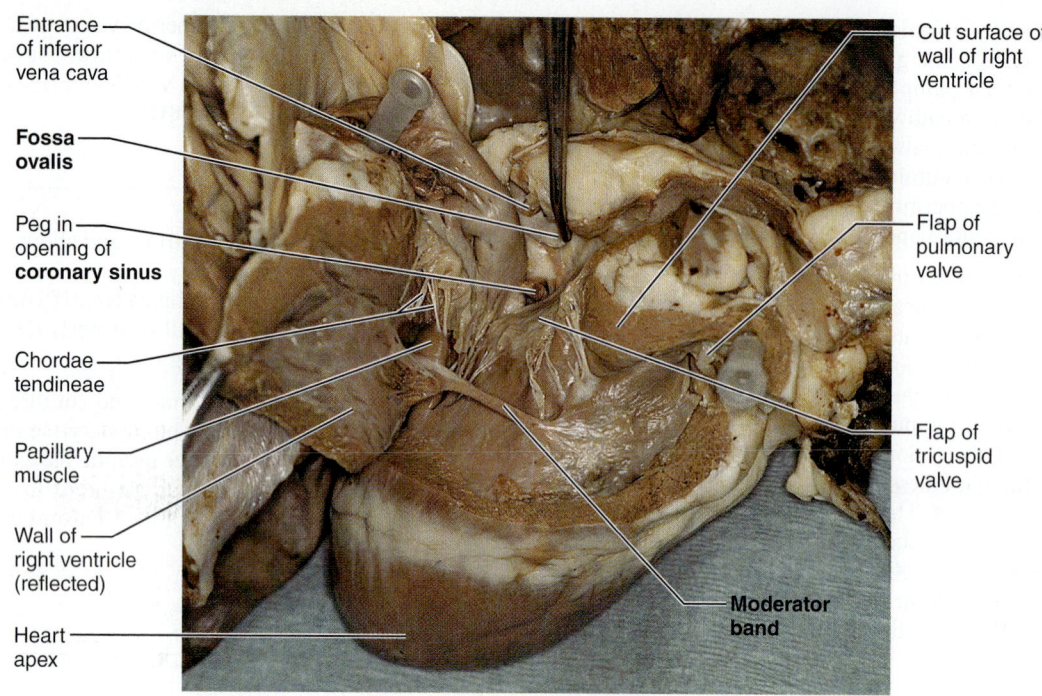

Entrance of inferior vena cava

Fossa ovalis

Peg in opening of **coronary sinus**

Chordae tendineae

Papillary muscle

Wall of right ventricle (reflected)

Heart apex

Cut surface of wall of right ventricle

Flap of pulmonary valve

Flap of tricuspid valve

Moderator band

Figure 30.9 Right side of the sheep heart opened and reflected to reveal internal structures.

ventricle. Measure the thickness of right and left ventricular walls in millimeters and record the numbers.

_____

How do your numbers compare with those of your classmates?

_____

Compare the *shape* of the left ventricular cavity to the shape of the right ventricular cavity. (See **Figure 30.10**.)

_____

_____

Are the papillary muscles and chordae tendineae observed in

the right ventricle also present in the left ventricle? _____

Count the number of cusps in the mitral valve. How does this compare with the number seen in the tricuspid valve?

_____

How do the sheep valves compare with their human counterparts?

_____

15. Reflect the cut edges of the atrial wall, and attempt to locate the entry points of the pulmonary veins into the left atrium.

Follow the pulmonary veins, if present, to the heart exterior with a probe. Notice how thin-walled these vessels are.

16. Dispose of the organic debris in the designated container, clean the dissecting tray and instruments with detergent and water, and wash the lab bench with bleach solution before leaving the laboratory. ▬

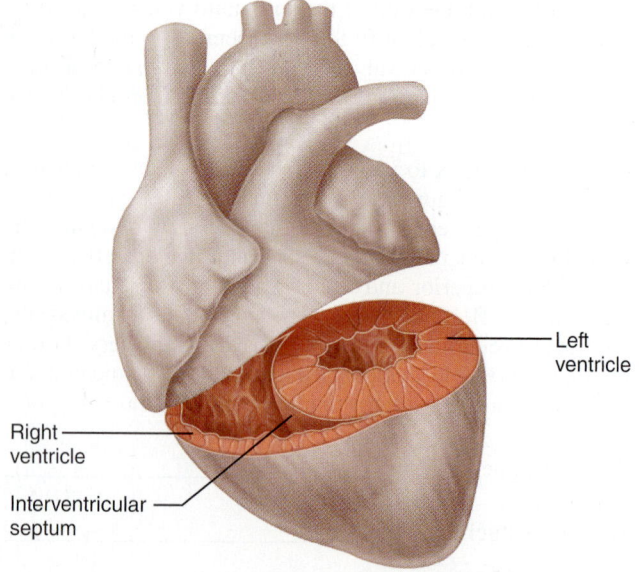

Left ventricle

Right ventricle

Interventricular septum

Figure 30.10 Anatomical differences between the right and left ventricles. The left ventricle has thicker walls, and its cavity is basically circular. By contrast, the right ventricle cavity is crescent-shaped and wraps around the left ventricle.

feels much thinner and somewhat flabby when compressed. This difference reflects the greater demand placed on the left ventricle, which must pump blood through the much longer systemic circulation, a pathway with much higher resistance than the pulmonary circulation served by the right ventricle. Hold the heart in its anatomical position (Figure 30.8a), with the anterior surface uppermost. In this position the left ventricle composes the entire apex and the left side of the heart.

6. Identify the pulmonary trunk and the aorta extending from the superior aspect of the heart. The pulmonary trunk is more anterior, and you may see its division into the right and left pulmonary arteries if it has not been cut too closely to the heart. The thicker-walled aorta, which branches almost immediately, is located just beneath the pulmonary trunk. The first observable branch of the sheep aorta, the **brachiocephalic artery (trunk),** is identifiable unless the aorta has been cut immediately as it leaves the heart. The brachiocephalic artery splits to form the right carotid and subclavian arteries, which supply the right side of the head and right forelimb, respectively.

Gently pull on the aorta with your gloved fingers or forceps to stretch it. Repeat with the venae cavae.

Which vessel is easier to stretch? _____

How does the elasticity of each vessel relate to its ability to withstand pressure?

_____

Carefully clear away some of the fat between the pulmonary trunk and the aorta to expose the **ligamentum arteriosum,** a cordlike remnant of the **ductus arteriosus.** (In the fetus, the ductus arteriosus allows blood to pass directly from the pulmonary trunk to the aorta, thus bypassing the nonfunctional fetal lungs.)

7. Cut through the wall of the aorta until you see the aortic (semilunar) valve. Identify the two openings to the coronary arteries just above the valve. Insert a probe into one of these holes to see if you can follow the course of a coronary artery across the heart.

8. Turn the heart to view its posterior surface (compare it to the view in Figure 30.8b). Notice that the right and left ventricles appear equal-sized in this view. Try to identify the four thin-walled pulmonary veins entering the left atrium. Identify the superior and inferior venae cavae entering the right atrium. Because of the way the heart is trimmed, the pulmonary veins and superior vena cava may be very short or missing. If possible, compare the approximate diameter of the superior vena cava with the diameter of the aorta.

Which is larger? _____

Which has thicker walls? _____

Why do you suppose these differences exist?

_____

_____

_____

9. Insert a probe into the superior vena cava, through the right atrium, and out the inferior vena cava. Use scissors to cut along the probe so that you can view the interior of the right atrium. Observe the tricuspid valve.

How many flaps does it have? _____

Pour some water into the right atrium and allow it to flow into the ventricle. *Slowly and gently* squeeze the right ventricle to watch the closing action of this valve. (If you squeeze too vigorously, you'll get a face full of water!) Drain the water from the heart before continuing.

10. Return to the pulmonary trunk and cut through its anterior wall until you can see the pulmonary (semilunar) valve **(Figure 30.9).** Pour some water into the base of the pulmonary trunk to observe the closing action of this valve. How does its action differ from that of the tricuspid valve?

_____

_____

_____

After observing pulmonary valve action, drain the heart once again. Extend the cut through the pulmonary trunk into the right ventricle. Cut down, around, and up through the tricuspid valve to make the cut continuous with the cut across the right atrium (see Figure 30.9).

11. Reflect the cut edges of the superior vena cava, right atrium, and right ventricle to obtain the view seen in the dissection photo (Figure 30.9). Observe the comblike ridges of muscle throughout most of the right atrium. This is called **pectinate muscle** (*pectin* = comb). Identify, on the ventral atrial wall, the large opening of the inferior vena cava and follow it to its external opening with a probe. Notice that the atrial walls in the vicinity of the venae cavae are smooth and lack the roughened appearance (pectinate musculature) of the other regions of the atrial walls. Just below the inferior vena caval opening, identify the opening of the **coronary sinus,** which returns venous blood of the coronary circulation to the right atrium. Nearby, locate an oval depression, the **fossa ovalis,** in the interatrial septum. This depression marks the site of an opening in the fetal heart, the **foramen ovale,** which allows blood to pass from the right to the left atrium, thus bypassing the fetal lungs.

12. Identify the papillary muscles in the right ventricle, and follow their attached chordae tendineae to the flaps of the tricuspid valve. Notice the pitted and ridged appearance **(trabeculae carneae)** of the inner ventricular muscle.

13. Identify the **moderator band** (septomarginal band), a bundle of cardiac muscle fibers connecting the interventricular septum to anterior papillary muscles. It contains a branch of the atrioventricular bundle and helps coordinate contraction of the ventricle.

14. Make a longitudinal incision through the left atrium and continue it into the left ventricle. Notice how much thicker the myocardium of the left ventricle is than that of the right

30

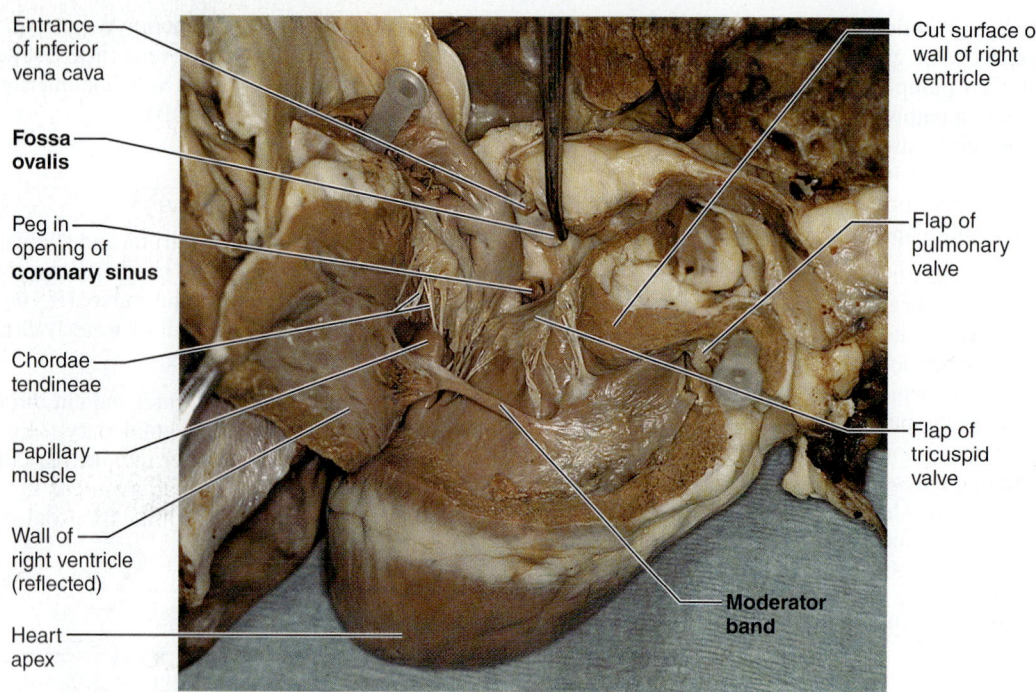

Figure 30.9 **Right side of the sheep heart opened and reflected to reveal internal structures.**

Labels on figure:
- Entrance of inferior vena cava
- **Fossa ovalis**
- Peg in opening of **coronary sinus**
- Chordae tendineae
- Papillary muscle
- Wall of right ventricle (reflected)
- Heart apex
- Cut surface of wall of right ventricle
- Flap of pulmonary valve
- Flap of tricuspid valve
- **Moderator band**

ventricle. Measure the thickness of right and left ventricular walls in millimeters and record the numbers.

_____

How do your numbers compare with those of your classmates?

_____

Compare the *shape* of the left ventricular cavity to the shape of the right ventricular cavity. (See **Figure 30.10**.)

_____

_____

Are the papillary muscles and chordae tendineae observed in

the right ventricle also present in the left ventricle? _____

Count the number of cusps in the mitral valve. How does this compare with the number seen in the tricuspid valve?

_____

How do the sheep valves compare with their human counterparts?

_____

15. Reflect the cut edges of the atrial wall, and attempt to locate the entry points of the pulmonary veins into the left atrium.

Follow the pulmonary veins, if present, to the heart exterior with a probe. Notice how thin-walled these vessels are.

16. Dispose of the organic debris in the designated container, clean the dissecting tray and instruments with detergent and water, and wash the lab bench with bleach solution before leaving the laboratory. ■

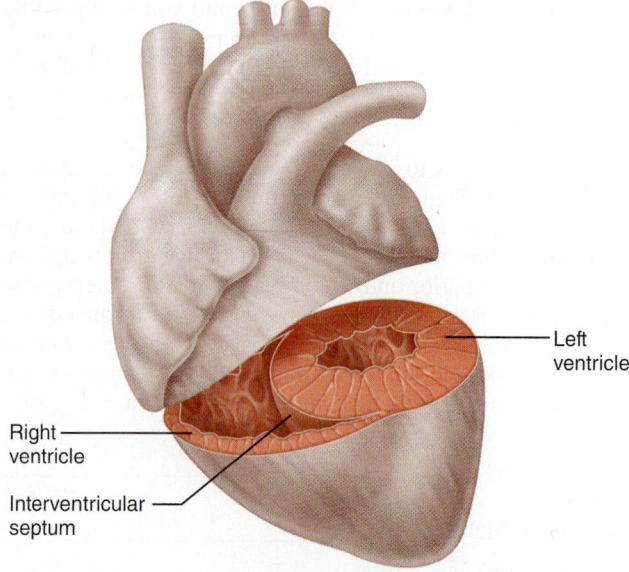

Labels on figure:
- Right ventricle
- Interventricular septum
- Left ventricle

Figure 30.10 **Anatomical differences between the right and left ventricles.** The left ventricle has thicker walls, and its cavity is basically circular. By contrast, the right ventricle cavity is crescent-shaped and wraps around the left ventricle.

Conduction System of the Heart and Electrocardiography

MATERIALS

☐ ECG or BIOPAC® equipment:* ECG recording apparatus, electrode paste, alcohol swabs, rubber straps or disposable electrodes

BIOPAC® BIOPAC® BSL System for Windows with BSL software version 3.7.5 to 3.7.7, or BSL System for Mac OS X with BSL software version 3.7.4 to 3.7.7, MP36/35 or MP45 data acquisition unit, PC or Mac computer, electrode lead set, disposable vinyl electrodes

Instructors using the MP36 (or MP35/30) data acquisition unit with BSL software versions earlier than 3.7.5 (for Windows) and 3.7.4 (for Mac Os X) will need slightly different channel settings and collection strategies. Instructions for using the older data acquisition unit can be found on MasteringA&P.

☐ Cot or lab table; pillow (optional)

☐ Millimeter ruler

*Note: Instructions for using PowerLab® equipment can be found on Mastering A&P.

MasteringA&P® For related exercise study tools, go to the Study Area of MasteringA&P. There you will find:

- Practice Anatomy Lab **PAL**
- PhysioEx **PEx**
- A&PFlix **A&PFlix**
- Practice quizzes, Histology Atlas, eText, Videos, and more!

OBJECTIVES

1. State the function of the intrinsic conduction system of the heart.
2. List and identify the elements of the intrinsic conduction system, and describe how impulses are initiated and conducted through this system and the myocardium.
3. Interpret the ECG in terms of depolarization and repolarization events occurring in the myocardium; and identify the P, QRS, and T waves on an ECG recording using an ECG recorder or BIOPAC®.
4. Define *tachycardia, bradycardia,* and *fibrillation.*
5. Calculate the heart rate, durations of the QRS complex, P-R interval, and Q-T interval from an ECG obtained during the laboratory period, and recognize normal values for the durations of these events.
6. Describe and explain the changes in the ECG observed during experimental conditions such as exercise or breath holding.

PRE-LAB QUIZ

1. Circle True or False. Cardiac muscle cells are electrically connected by gap junctions and behave as a single unit.
2. Because it sets the rate of depolarization for the normal heart, the _____ node is known as the pacemaker of the heart.
   a. atrioventricular   b. Purkinje   c. sinoatrial
3. Circle True or False. Stimulation by the nerves of the autonomic nervous system is essential for cardiac muscle to contract.
4. Today you will create a graphic recording of the electrical changes that occur during a cardiac cycle. This is known as an
   a. electrocardiogram
   b. electroencephalogram
   c. electromyogram
5. Circle the correct underlined term. The typical ECG has three / six normally recognizable deflection waves.
6. In a typical ECG, the _____ wave signals the depolarization of the atria immediately before they contract.
   a. P   c. R
   b. Q   d. T
7. Circle True or False. The repolarization of the atria is usually masked by the large QRS complex.
8. Circle the correct underlined term. A heart rate over 100 beats/minute is known as tachycardia / bradycardia.
9. How many electrodes will you place on your subject for today's activity if you use a standard ECG apparatus?
   a. 3   c. 10
   b. 4   d. 12
10. Circle True or False. ECG can be used to calculate heart rate.

Figure 31.1 The intrinsic conduction system of the heart. Dashed-line arrows indicate transmission of the impulse from the SA node through the atria. Solid yellow arrow indicates transmission of the impulse from the SA node to the AV node via the internodal pathway.

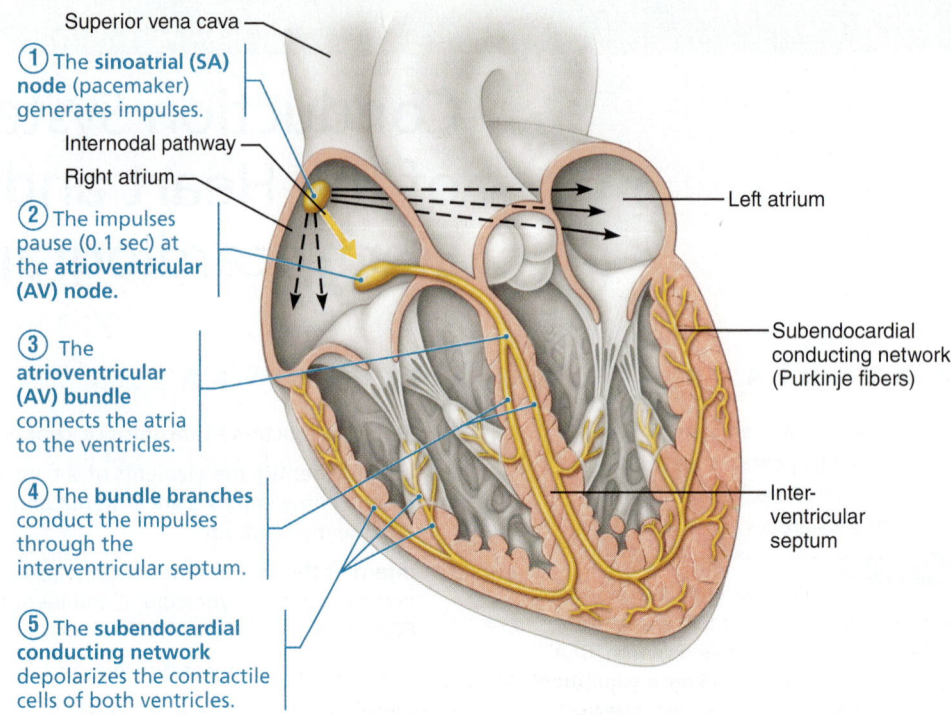

Superior vena cava

① The **sinoatrial (SA) node** (pacemaker) generates impulses.

Internodal pathway

Right atrium

② The impulses pause (0.1 sec) at the **atrioventricular (AV) node.**

③ The **atrioventricular (AV) bundle** connects the atria to the ventricles.

④ The **bundle branches** conduct the impulses through the interventricular septum.

⑤ The **subendocardial conducting network** depolarizes the contractile cells of both ventricles.

Left atrium

Subendocardial conducting network (Purkinje fibers)

Inter-ventricular septum

Heart contraction results from a series of depolarization waves that travel through the heart preliminary to each beat. Because cardiac muscle cells are electrically connected by gap junctions, the entire myocardium behaves like a single unit, a **functional syncytium**.

The Intrinsic Conduction System

The ability of cardiac muscle to beat is intrinsic—it does not depend on impulses from the nervous system to initiate its contraction and will continue to contract rhythmically even if all nerve connections are severed. However, two types of controlling systems exert their effects on heart activity. One of these involves nerves of the autonomic nervous system, which accelerate or decelerate the heartbeat rate depending on which division is activated. The second system is the **intrinsic conduction system** of the heart, consisting of specialized noncontractile myocardial tissue. The intrinsic conduction system ensures that heart muscle depolarizes in an orderly and sequential manner, from atria to ventricles, and that the heart beats as a coordinated unit.

The components of the intrinsic conduction system include the **sinoatrial (SA) node,** located in the right atrium just inferior to the entrance to the superior vena cava; the **atrioventricular (AV) node** in the lower atrial septum at the junction of the atria and ventricles; the **AV bundle (bundle of His)** and right and left **bundle branches,** located in the interventricular septum; and the **subendocardial conducting network.** The subendocardial conducting network, also called **Purkinje fibers,** consists essentially of long strands of barrel-shaped cells called *Purkinje myocytes,* which ramify within the muscle bundles of the ventricular walls. This network is much denser and more elaborate in

the left ventricle because of the larger size of this chamber **(Figure 31.1)**.

The SA node, which has the highest rate of discharge, provides the stimulus for contraction. Because it sets the rate of depolarization for the heart as a whole, the SA node is often referred to as the *pacemaker.* From the SA node, the impulse spreads throughout the atria and to the AV node. This electrical wave is immediately followed by atrial contraction. At the AV node, the impulse is momentarily delayed (approximately 0.1 sec), allowing the atria to complete their contraction. It then passes through the AV bundle, the right and left bundle branches, and the subendocardial conducting network, finally resulting in ventricular contraction. Note that the atria and ventricles are separated from one another by a region of electrically inert connective tissue, so the depolarization wave can be transmitted to the ventricles only via the tract between the AV node and AV bundle. Thus, any damage to the AV node-bundle pathway partially or totally insulates the ventricles from the influence of the SA node. Although cardiac pacemaker cells are found throughout the heart, their rates of spontaneous depolarization differ. The nodal system sets the rate of heart depolarization and synchronizes heart activity.

Electrocardiography

The conduction of impulses through the heart generates electrical currents that eventually spread throughout the body. These impulses can be detected on the body's surface and recorded with an instrument called an *electrocardiograph.* The graphic recording of the electrical changes occurring during the cardiac cycle is called an **electrocardiogram (ECG or EKG) (Figure 31.2)**. For analysis, the ECG is divided into segments and intervals (see **Table 31.1**

31

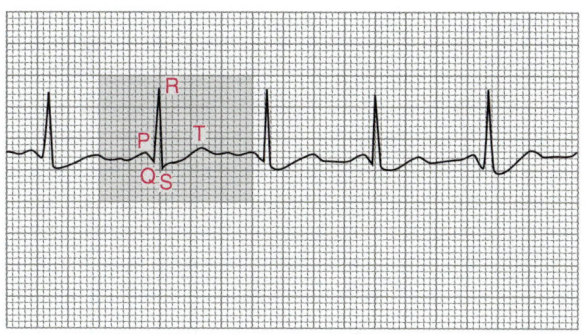

(a)

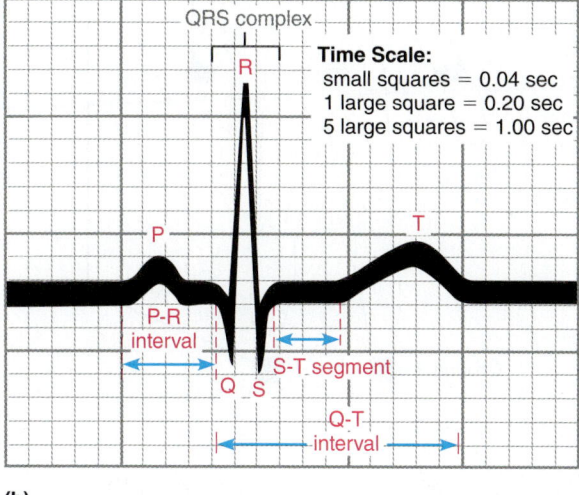

(b)

Figure 31.2 The normal electrocardiogram. (a) Regular sinus rhythm. **(b)** Waves, segments, and intervals of a normal ECG.

for the definitions). The deflection waves of an ECG and sequential excitation of the heart have specific relationships **(Figure 31.3)**.

Table 31.1	Boundaries of Each ECG Component
Feature	**Boundaries**
P wave	Start of P deflection to return to isoelectric line
P-R interval	Start of P deflection to start of Q deflection
QRS complex	Start of Q deflection to S return to isoelectric line
S-T segment	End of S deflection to start of T wave
Q-T interval	Start of Q deflection to end of T wave
T wave	Start of T deflection to return to isoelectric line
T-P segment	End of T wave to start of next P wave
R-R interval	Peak of R wave to peak of next R wave

It is important to understand what an ECG does *and does not* show: First, an ECG is a record of voltage and time—nothing else. Although we can and do infer that muscle contraction follows its excitation, sometimes it does not. Second, an ECG records electrical events occurring in relatively large amounts of muscle tissue (i.e., the bulk of the heart muscle), *not* the activity of nodal tissue which, like muscle contraction, can only be inferred. Nonetheless, abnormalities of the deflection waves and changes in the time intervals of the ECG are useful in detecting myocardial infarcts (heart attacks) or problems with the conduction system of the heart. The P-R interval represents the time between the beginning of atrial depolarization and ventricular depolarization. Thus, it typically includes the period during which the depolarization wave passes to the AV node, atrial systole, and the passage of the excitation wave to the balance of the conducting system. Generally, the P-R interval is about 0.12–0.20 sec. A longer interval may suggest a partial AV heart block caused by damage to the AV node. In total heart block, no impulses are transmitted through the AV node, and the atria and ventricles beat independently of one another—the atria at the SA node rate and the ventricles at their intrinsic rate, which is considerably slower.

31

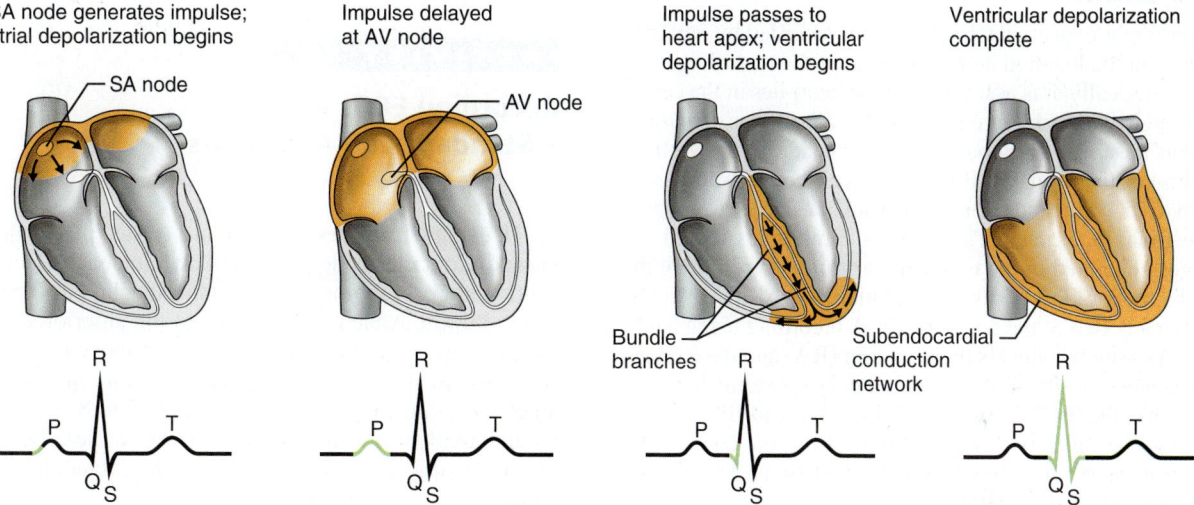

Figure 31.3 The sequence of excitation of the heart related to the deflection waves of an ECG tracing.

The S-T segment is a very important area to examine when evaluating the ECG. Elevation of this segment is characteristic of a myocardial infarct.

A prolonged QRS complex (normally about 0.08 sec) may indicate a right or left bundle branch block in which one ventricle is contracting later than the other. The Q-T interval is the period from the beginning of ventricular depolarization through repolarization and includes the time of ventricular contraction (the S-T segment). With a heart rate of 70 beats/min, this interval is normally 0.32–0.38 sec. As the rate increases, this interval becomes shorter; conversely, when the heart rate drops, the interval is longer. The repolarization of the atria, which occurs during the QRS interval, is generally obscured by the large QRS complex.

A heart rate over 100 beats/min is referred to as **tachycardia;** a rate below 60 beats/min is **bradycardia.** Although neither condition is pathological, prolonged tachycardia may progress to **fibrillation,** a condition of rapid uncoordinated heart contractions which makes the heart useless as a pump. Bradycardia in athletes is a positive finding; that is, it indicates an increased efficiency of cardiac functioning. Because *stroke volume* (the amount of blood ejected by a ventricle with each contraction) increases with physical conditioning, the heart can contract more slowly and still meet circulatory demands.

Twelve standard leads are used to record an ECG for diagnostic purposes. Three of these are bipolar leads that measure the voltage difference between the arms, or an arm and a leg, and nine are unipolar leads. Together the 12 leads provide a fairly comprehensive picture of the electrical activity of the heart.

For this investigation, four electrodes are used **(Figure 31.4)**, and results are obtained from the three *standard limb leads* (also shown in Figure 31.4). Several types of physiographs or ECG recorders are available. Your instructor will provide specific directions on how to set up and use the available apparatus if standard ECG apparatus is used (Activity 1A). Instructions for use of BIOPAC® apparatus (Activity 1B) follow (pages 462–466).

Understanding the Standard Limb Leads

As you might expect, electrical activity recorded by any lead depends on the location and orientation of the recording electrodes. Clinically, it is assumed that the heart lies in the center of a triangle with sides of equal lengths (*Einthoven's triangle*) and that the recording connections are made at the vertices (corners) of that triangle. But in practice, the electrodes connected to each arm and to the left leg are considered to connect to the triangle vertices. The standard limb leads record the voltages generated in the extracellular fluids surrounding the heart by the ion flows occurring simultaneously in many cells between any two of the connections. A recording using lead I (RA-LA), which connects the right arm (RA) and the left arm (LA), is most sensitive to electrical activity spreading horizontally across the heart. Lead II (RA-LL) and lead III (LA-LL) record activity along the vertical axis (from the base of the heart to its apex) but from different orientations. The significance of Einthoven's triangle is that the sum of the voltages of leads I and III equals that in lead II (Einthoven's law). Hence, if the voltages of two of the standard leads are recorded, that of the third lead can be determined mathematically.

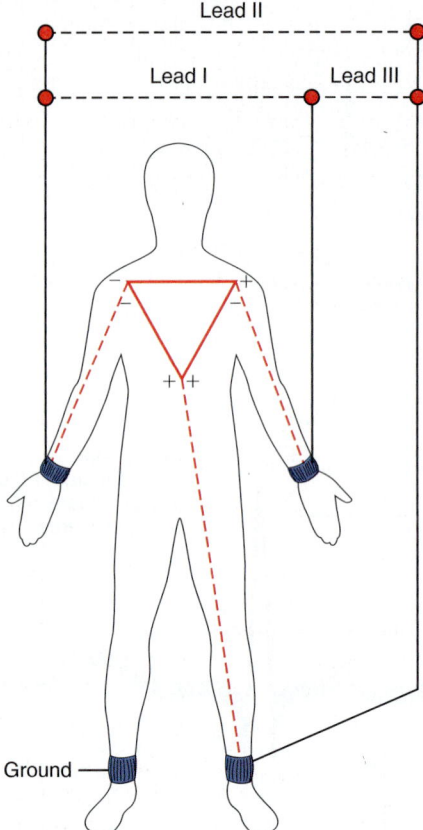

Figure 31.4 ECG recording positions for the standard limb leads.

Once the subject is prepared, the ECG will be recorded first under baseline (resting) conditions and then under conditions of fairly strenuous activity. Finally, recordings will be made while the subject holds his or her breath or carries out deep breathing. The activity recordings and those involving changes in respiratory rate or depth will be compared to the baseline recordings, and you will be asked to determine the reasons for the observed differences in the recordings.

ACTIVITY 1A

Recording ECGs Using a Standard ECG Apparatus
Preparing the Subject

1. If using electrodes that require paste, place electrode paste on four electrode plates and position each electrode as follows after scrubbing the skin at the attachment site with an alcohol swab. Attach an electrode to the anterior surface of each forearm, about 5 to 8 cm (2 to 3 in.) above the wrist, and secure them with rubber straps. In the same manner, attach an electrode to each leg, approximately 5 to 8 cm above the medial malleolus (inner aspect of the ankle). Disposable electrodes may be placed directly on the subject in the same areas.

2. Attach the appropriate tips of the patient cable to the electrodes. The cable leads are marked RA (right arm), LA (left arm), LL (left leg), and RL (right leg, the ground).

Making a Baseline Recording

1. Position the subject comfortably in a supine position on a cot (if available), or sitting relaxed on a laboratory chair.

2. Turn on the power switch and adjust the sensitivity knob to 1. Set the paper speed to 25 mm/sec and the lead selector to the position corresponding to recording from lead I (RA-LA).

3. Set the control knob at the **RUN** position and record the subject's at-rest ECG from lead I for 2 to 3 minutes or until the recording stabilizes. (You will need a tracing long enough to provide each student in your group with a representative segment.) The subject should try to relax and not move unnecessarily, because the skeletal muscle action potentials will also be picked up and recorded.

4. Stop the recording and mark it "lead I."

5. Repeat the recording procedure for leads II (RA-LL) and III (LA-LL).

6. Each student should take a representative segment of one of the lead recordings and label the record with the name of the subject and the lead used. Identify and label the P, QRS, and T waves. The calculations you perform for your recording should be based on the following information: Because the paper speed was 25 mm/sec, each millimeter of paper corresponds to a time interval of 0.04 sec. Thus, if an interval requires 4 mm of paper, its duration is 4 mm × 0.04 sec/mm = 0.16 sec.

7. Compute the heart rate. Obtain a millimeter ruler and measure the distance from the R wave of one QRS complex to the R wave of the next QRS complex. Enter this value into the following equation to find the time for one heartbeat.

_____ mm/beat × 0.04 sec/mm = _____ sec/beat

Now find the beats per minute, or heart rate, by using the figure just computed for seconds per beat in the following equation:

$$\frac{1}{\text{_____ sec/beat}} \times 60 \text{ sec/min} = \text{_____ beats/min}$$

Is the obtained value within normal limits? _____

Measure the QRS complex, and compute its duration.

_____

Measure the Q-T interval, and compute its duration.

_____

Measure the P-R interval, and compute its duration.

_____

Are the computed values within normal limits?

_____

8. At the bottom of this page, attach segments of the ECG recordings from leads I through III. Make sure you indicate the paper speed, lead, and subject's name on each tracing. To the recording on which you based your previous computations, add your calculations for the duration of the QRS complex and the P-R and Q-T intervals above the respective area of tracing. Also record the heart rate on that tracing.

Recording the ECG for Running in Place

1. Make sure the electrodes are securely attached to prevent electrode movement while recording the ECG.

2. Set the paper speed to 25 mm/sec, and prepare to make the recording using lead I.

3. Record the ECG while the subject is running in place for 3 minutes. Then have the subject sit down, but continue to record the ECG for an additional 4 minutes. *Mark the recording* at the end of the 3 minutes of running and at 1 minute after cessation of activity.

4. Stop the recording. Compute the beats/min during the third minute of running, at 1 minute after exercise, and at 4 minutes after exercise. Record below:

_____ beats/min while running in place

_____ beats/min at 1 minute after exercise

_____ beats/min at 4 minutes after exercise

5. Compare this recording with the previous recording from lead I. Which intervals are shorter in the "running" recording?

_____

6. Does the subject's heart rate return to resting level by 4 minutes after exercise?

_____

Recording the ECG During Breath Holding

1. Position the subject comfortably in the sitting position.

2. Using lead I and a paper speed of 25 mm/sec, begin the recording. After approximately 10 seconds, instruct the subject to begin breath holding and mark the record to indicate the onset of the 1-minute breath-holding interval.

3. Stop the recording after 1 minute and remind the subject to breathe. Compute the beats/minute during the 1-minute experimental (breath-holding) period.

Beats/min during breath holding: _____

4. Compare this recording with the lead I recording obtained under resting conditions.

What differences are seen? _____

_____

Attempt to explain the physiological reason for the differences you have seen. (Hint: A good place to start might be to

31

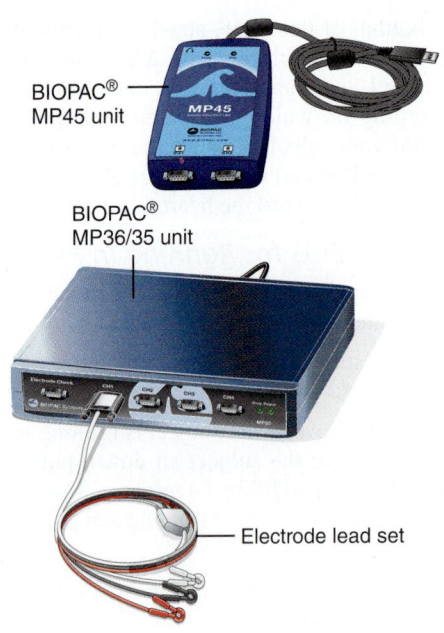

Figure 31.5 Setting up the BIOPAC® unit. Plug the electrode lead set into Channel 1. Leads are shown plugged into the MP36/35 unit.

check hypoventilation or the role of the respiratory system in acid-base balance of the blood.)

_____

_____

_____

_____

ACTIVITY 1B

Electrocardiography Using BIOPAC®

In this activity, you will record the electrical activity of the heart under three different conditions: (1) while the subject is lying down, (2) after the subject sits up and breathes normally, and (3) after the subject has exercised and is breathing deeply.

Since the electrodes are not placed directly over the heart, artifacts can result from the recording of unwanted skeletal muscle activity. In order to obtain a clear ECG, it is important that the subject:

- Remain still during the recording.

- Refrain from laughing or talking during the recording.

- When in the sitting position, keep arms and legs steady and relaxed.

- Remove metal watches and bracelets.

Setting Up the Equipment

1. Connect the BIOPAC® unit to the computer and turn the computer **ON.**

2. Make sure the BIOPAC® unit is **OFF.**

3. Plug in the equipment (as shown in **Figure 31.5**):

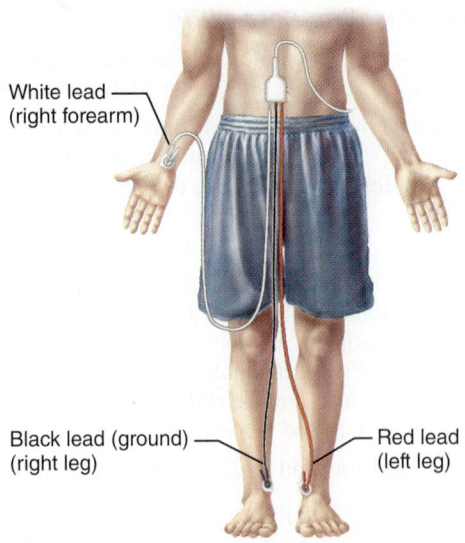

Figure 31.6 Placement of electrodes and the appropriate attachment of electrode leads by color.

- Electrode lead set—CH 1

4. Turn the BIOPAC® unit **ON.**

5. Place the three electrodes on the subject (as shown in **Figure 31.6**), and attach the electrode leads according to the colors indicated. The electrodes should be placed on the medial surface of each leg, 5 to 8 cm (2 to 3 in.) superior to the ankle. The other electrode should be placed on the right anterior forearm 5 to 8 cm above the wrist.

6. The subject should lie down and relax in a comfortable position with eyes closed. A chair or place to sit up should be available nearby.

7. Start the BIOPAC® Student Lab program on the computer by double-clicking the icon on the desktop or by following your instructor's guidance.

8. Select lesson **L05-ECG-1** from the menu, and click **OK.**

9. Type in a filename that will save this subject's data on the computer hard drive. You may want to use the subject's last name followed by ECG-1 (for example, SmithECG-1), then click **OK.**

10. Because we are not recording all available lesson options, click the File Menu, choose **Lesson Preferences,** choose **Heart Rate Data,** and click **OK.** Choose **Do Not Calculate** and click **OK.** Click the file menu and choose **Lesson Preferences** again, select **Lesson Segments** and click **OK,** click the box for **Deep Breathing** to deselect it, and then click **OK.**

Calibrating the Equipment

- Examine the electrodes and the electrode leads to be certain they are properly attached.

1. The subject must remain supine, still, and relaxed. With the subject in a still position, click **Calibrate.** This will initiate the process whereby the computer will automatically establish parameters to record the data.

2. The calibration procedure will stop automatically after 8 seconds.

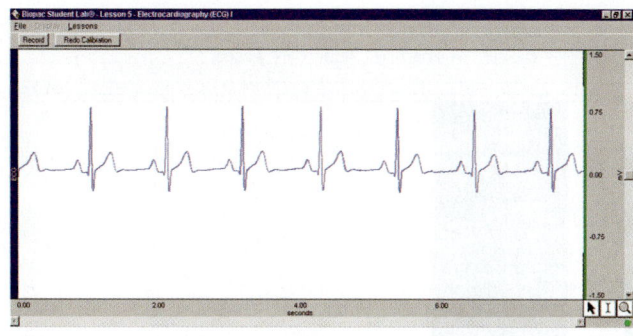

Figure 31.7 **Example of calibration data.**

3. Observe the recording of the calibration data, which should look similar to the data example **(Figure 31.7)**.

- If the data look very different, click **Redo Calibration** and repeat the steps above.

- If the data look similar, proceed to the next section. **Don't** click **Done** until you have completed all 3 segments.

Recording Segment 1: Subject Lying Down

1. To prepare for the recording, remind the subject to remain still and relaxed while lying down.

2. Click **Continue** and when prepared, click **Record** and gather data for 20 seconds. At the end of 20 seconds, click **Suspend.**

3. Observe the data, which should look similar to the data example **(Figure 31.8)**.

- If the data look very different, click **Redo** and repeat the steps above. Be certain to check attachment of the electrodes and leads, and remind the subject not to move, talk, or laugh.

- If the data look similar, move on to the next recording segment.

Recording Segment 2: After Subject Sits Up, with Normal Breathing

1. Tell the subject to be ready to sit up in the designated location. With the exception of breathing, the subject should try to remain motionless after assuming the seated position. *If*

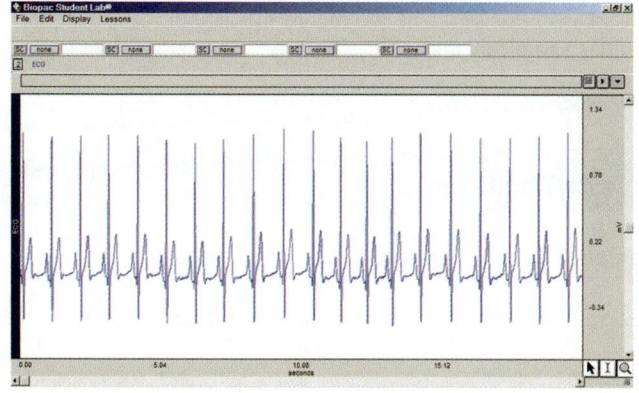

Figure 31.8 **Example of ECG data while the subject is lying down.**

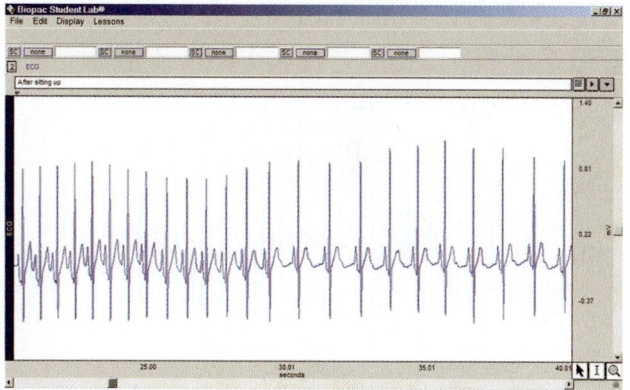

Figure 31.9 **Example of ECG data after the subject sits up and breathes normally.**

the subject moves too much during recording after sitting up, unwanted skeletal muscle artifacts will affect the recording.

2. Click **Continue** and when prepared, instruct the subject to sit up. Immediately after the subject assumes a motionless state, click **Record,** and the data will begin recording.

3. At the end of 20 seconds, click **Suspend** to stop recording.

4. Observe the data, which should look similar to the data example **(Figure 31.9)**.

- If the data look very different, have the subject lie down, then click **Redo.** Be certain to check attachment of the electrodes, then repeat steps 1–4 above. Do not click **Record** until the subject is motionless.

- If the data look similar, move on to the next recording segment.

Recording Segment 3: After Subject Exercises, with Deep Breathing

1. Click **Continue,** but **don't** click **Record** until after the subject has exercised. Remove the electrode pinch connectors from the electrodes on the subject.

2. Have the subject do a brief round of exercise, such as jumping jacks or running in place for 1 minute, in order to elevate the heart rate.

3. As quickly as possible after the exercise, have the subject resume a motionless, seated position and reattach the pinch connectors. Once again, if the subject moves too much during recording, unwanted skeletal muscle artifacts will affect the data. After exercise, the subject is likely to be breathing deeply, but otherwise should remain as still as possible.

4. Immediately after the subject assumes a motionless, seated state, click **Record,** and the data will begin recording. Record the ECG for 60 seconds in order to observe post-exercise recovery.

5. After 60 seconds, click **Suspend** to stop recording.

6. Observe the data, which should look similar to the data example **(Figure 31.10)**.

- If the data look very different, click **Redo** and repeat the steps above. Be certain to check attachment of the electrodes and leads, and remember not to click **Record** until the subject is motionless.

31

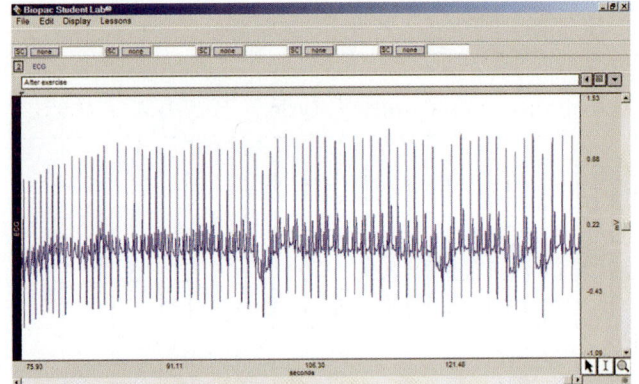

Figure 31.10 Example of ECG data after the subject exercises.

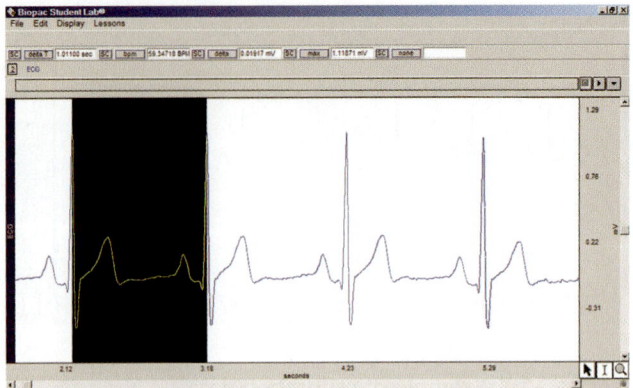

Figure 31.11 Example of highlighting from R wave to R wave.

7. When finished, click **Done** and then **Yes.** Remove the electrodes from the subject.

8. A pop-up window will appear. To record from another subject, select **Record from another subject** and return to step 5 under Setting Up the Equipment. If continuing to the Data Analysis section, select **Analyze current data file** and proceed to step 2 of the Data Analysis section.

Data Analysis

1. If just starting the BIOPAC® program to perform data analysis, enter **Review Saved Data** mode and choose the file with the subject's ECG data (for example, SmithECG-1).

2. Use the following tools to adjust the data in order to clearly view and analyze four consecutive cardiac cycles:

- Click the magnifying glass (near the I-beam cursor box) to activate the **zoom** function. Use the magnifying glass cursor to click on the very first waveforms until there are about 4 seconds of data represented (see horizontal time scale at the bottom of the screen).

- Select the **Display** menu at the top of the screen, and click **Autoscale Waveforms** (or click Ctrl + Y). This function will adjust the data for better viewing.

- Click the **Adjust Baseline** button. Two new buttons will appear; simply click these buttons to move the waveforms **Up** or **Down** so they appear clearly in the center of the display window. Once they are centered, click **Exit.**

3. Note that the first two pairs of channel/measurement boxes at the top of the screen are set to Delta T and bpm.

Channel	Measurement	Data
CH 1	Delta T	ECG
CH 1	bpm	ECG

Analysis of Segment 1: Subject Lying Down

1. Use the arrow cursor and click the I-beam cursor box for the "area selection" function.

2. First measure **Delta T** and **bpm** in Segment 1 (approximately seconds 0–20). Using the I-beam cursor, highlight from the peak of one R wave to the peak of the next R wave, (as shown in **Figure 31.11**).

3. Observe that the computer automatically calculates the **Delta T** and **bpm** for the selected area. These measurements represent the following:

Delta T (difference in time): Computes the elapsed time between the beginning and end of the highlighted area

bpm (beats per minute): Computes the beats per minute when the area from the R wave of one cycle to the R wave of another cycle is highlighted

4. Record these data in the **Segment 1 Samples chart** under R to R Sample 1 (round to the nearest 0.01 second and 0.1 beat per minute).

5. Using the I-beam cursor, highlight two other pairs of R to R areas in this segment and record the data in the same chart under Samples 2 and 3.

Segment 1 Samples for Delta T and bpm					
Measure	Channel	R to R Sample 1	R to R Sample 2	R to R Sample 3	Mean
Delta T	CH 1				
bpm	CH 1				

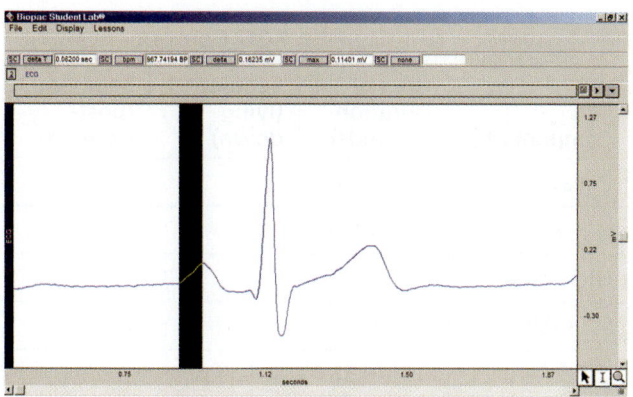

Figure 31.12 **Example of a single ECG waveform with the first part of the P wave highlighted.**

6. Calculate the means of the data in this chart.

7. Next, use the **zoom, Autoscale Waveforms,** and **Adjust Baseline** tools described in step 2 to focus in on one ECG waveform within Segment 1. (See the example in **Figure 31.12**).

8. Once a single ECG waveform is centered for analysis, click the I-beam cursor box to activate the "area selection" function.

9. Using the highlighting function and **Delta T** computation, measure the duration of every component of the ECG waveform. (Refer to Figure 31.2b and Table 31.1 for guidance in highlighting each component.)

10. Highlight each component of one cycle. Observe the elapsed time, and record this data under Cycle 1 in the **Segment 1 Elapsed Time chart.**

11. Scroll along the horizontal axis at the bottom of the data to view and analyze two additional cycles in Segment 1. Record the elapsed time for every component of Cycle 2 and Cycle 3 in the Segment 1 Elapsed Time chart.

12. In the same chart, calculate the means for the three cycles of data and record.

Segment 1 Elapsed Time for ECG Components (seconds)				
Component	Cycle 1	Cycle 2	Cycle 3	Mean
P wave				
P-R interval				
QRS complex				
S-T segment				
Q-T interval				
T wave				
T-P segment				
R-R interval				

Analysis of Segment 2: Subject Sitting Up and Breathing Normally

1. Scroll along the horizontal time bar until you reach the data for Segment 2 (approximately seconds 20–40). A marker with "Seated" should denote the beginning of this data.

2. As in the analysis of Segment 1, use the I-beam tool to highlight and measure the **Delta T** and **bpm** between three different pairs of R waves in this segment, and record the data in the **Segment 2 Samples chart.**

Analysis of Segment 3: After Exercise with Deep Breathing

1. Scroll along the horizontal time bar until you reach the data for Segment 3 (approximately seconds 40–100). A marker with "After exercise" should denote the beginning of this data.

31

Segment 2 Samples for Delta T and bpm					
Measure	Channel	R to R Sample 1	R to R Sample 2	R to R Sample 3	Mean
Delta T	CH 1				
bpm	CH 1				

Segment 3 Samples for Delta T and bpm					
Measure	Channel	R to R Sample 1	R to R Sample 2	R to R Sample 3	Mean
Delta T	CH 1				
bpm	CH 1				

Segment 3 Elapsed Time for ECG Components (seconds)	
Component	Cycle 1
P wave	
P-R interval	
QRS complex	
S-T segment	
Q-T interval	
T wave	
T-P segment	
R-R interval	

Average Duration for ECG Components			
ECG Component	Normal Duration (seconds)	Segment 1 (lying down)	Segment 3 (post-exercise)
P wave	0.07–0.18		
P–R interval	0.12–0.20		
QRS complex	0.06–0.12		
S-T segment	<0.20		
Q-T interval	0.32–0.38		
T wave	0.10–0.25		
T-P segment	0–0.40		
R-R interval	varies		

2. As before, use the I-beam tool to highlight and measure the **Delta T** and **bpm** between three pairs of R waves in this segment, and record the data in the **Segment 3 Samples chart** on the previous page.

3. Using the instructions for steps 7–9 in the section Analysis of Segment 1, highlight and observe the elapsed time for each component of one cycle, and record these data under Cycle 1 in the **Segment 3 Elapsed Time chart.**

4. When finished, **Exit** from the file menu to quit.

Compare the average **Delta T** times and average **bpm** between the data in Segment 1 (lying down) and the data in Segment 3 (after exercise). Which is greater in each case?

_____

_____

What is the relationship between the elapsed time (**Delta T**) between R waves and the heart rate?

_____

What event does the period between R waves correspond to?

_____

Is there a change in heart rate when the subject makes the transition from lying down (Segment 1) to a sitting position (Segment 2)?

_____

Examine the average duration of each of the ECG components in Segment 1 and the data in Segment 3. In the **Average Duration chart,** record the average values for Segment 1 and the data for Segment 3. Draw a circle around those measures that fit within the normal range.

Compare the Q-T intervals in the data while the subject is at rest versus after exercise; this interval corresponds closely to the duration of contraction of the ventricles. Describe and explain any difference.

_____

_____

Compare the duration in the period from the end of each T wave to the next P wave while the subject is at rest versus after exercise. Describe and explain any difference.

_____

_____

A patient presents with a P-R interval three times longer than the normal duration. What might be the cause of this abnormality?

_____

_____

Conduction System of the Heart and Electrocardiography

The Intrinsic Conduction System

1. List the elements of the intrinsic conduction system in order, starting from the SA node.

   SA node → _____ → _____ →

   _____ → _____

   At what structure in the transmission sequence is the impulse temporarily delayed? _____

   Why? _____

   _____

2. Even though cardiac muscle has an inherent ability to beat, the nodal system plays a critical role in heart physiology.

   What is that role? _____

   _____

Electrocardiography

3. Define *ECG*. _____

   _____

4. Draw an ECG wave form representing one heartbeat. Label the P, QRS, and T waves; the P-R interval; the S-T segment, and the Q-T interval.

5. Why does heart rate increase during running? _____

   _____

   _____

6. Describe what happens in the cardiac cycle in the following situations.

1.   immediately before the P wave: _____

2.   during the P wave: _____

3.   immediately after the P wave: _____

4.   during the QRS wave: _____

5.   immediately after the QRS wave (S-T segment): _____

6.   during the T wave: _____

7. Define the following terms.

1.   *tachycardia:* _____

2.   *bradycardia:* _____

3.   *fibrillation:* _____

8. Which would be more serious, atrial or ventricular fibrillation? _____

Why? _____

_____

9. Abnormalities of heart valves can be detected more accurately by auscultation than by electrocardiography. Why is this so?

_____

_____

_____

Anatomy of Blood Vessels

MATERIALS

☐ Compound microscope

☐ Prepared microscope slides showing cross sections of an artery and vein

☐ Anatomical charts of human arteries and veins (or a three-dimensional model of the human circulatory system)

☐ Anatomical charts of the following specialized circulations: pulmonary circulation, hepatic portal circulation, fetal circulation, arterial supply of the brain (or a brain model showing this circulation)

For instructions on animal dissections, see the dissection exercises (starting on page 697) in the cat and fetal pig editions of this manual.

OBJECTIVES

1. Describe the tunics of blood vessel walls, and state the function of each layer.

2. Correlate differences in artery, vein, and capillary structure with the functions of these vessels.

3. Recognize a cross-sectional view of an artery and vein when provided with a microscopic view or appropriate image.

4. List and identify the major arteries arising from the aorta, and indicate the body region supplied by each.

5. Describe the cerebral arterial circle, and discuss its importance in the body.

6. List and identify the major veins draining into the superior and inferior venae cavae, and indicate the body regions drained.

7. Describe these special circulations in the body: pulmonary circulation, hepatic portal system, and fetal circulation, and discuss the important features of each.

PRE-LAB QUIZ

1. Circle the correct underlined term. <u>Arteries</u> / <u>Veins</u> drain tissues and return blood to the heart.

2. Circle True or False. Gas exchange takes place between tissue cells and blood through capillary walls.

3. The _____ is the largest artery of the body.
   a. aorta
   b. carotid artery
   c. femoral artery
   d. subclavian artery

4. Circle the correct underlined term. The largest branch of the abdominal aorta, the <u>renal</u> / <u>superior mesenteric</u> artery, supplies most of the small intestine and the first half of the large intestine.

5. The anterior tibial artery terminates with the _____ artery, which is often palpated in patients with circulatory problems to determine the circulatory efficiency of the lower limb.
   a. dorsalis pedis
   b. external iliac
   c. obturator
   d. tibial

6. Circle the correct underlined term. Veins draining the head and upper extremities empty into the <u>superior</u> / <u>inferior</u> vena cava.

7. Located in the lower limb, the _____ is the longest vein in the body.
   a. external iliac
   b. fibular
   c. great saphenous
   d. internal iliac

8. Circle the correct underlined term. The <u>renal</u> / <u>hepatic</u> veins drain the liver.

9. The function of the _____ is to drain the digestive viscera and carry dissolved nutrients to the liver for processing.
   a. fetal circulation
   b. hepatic portal circulation
   c. pulmonary circulation system

10. Circle the correct underlined term. In the developing fetus, the umbilical <u>artery</u> / <u>vein</u> carries blood rich in nutrients and oxygen to the fetus.

MasteringA&P® For related exercise study tools, go to the Study Area of MasteringA&P. There you will find:

• Practice Anatomy Lab PAL

• PhysioEx PEx

• A&PFlix A&PFlix

• Practice quizzes, Histology Atlas, eText, Videos, and more!

The blood vessels form a closed transport system. As the heart contracts, blood is propelled into the large arteries leaving the heart. It moves into successively smaller arteries and then to the arterioles, which feed the capillary beds in the tissues. Capillary beds are drained by the venules, which in turn empty into veins that ultimately converge on the great veins entering the heart.

Arteries, carrying blood away from the heart, and veins, which drain the tissues and return blood to the heart, function simply as conducting vessels or conduits. Only the tiny capillaries that connect the arterioles and venules and branch throughout the tissues directly serve the needs of the body's cells. It is through the capillary walls that exchanges are made between tissue cells and blood. Respiratory gases, nutrients, and wastes move along diffusion gradients. Thus, oxygen and nutrients diffuse from the blood to the tissue cells, and carbon dioxide and metabolic wastes move from the cells to the blood.

In this exercise you will examine the microscopic structure of blood vessels and identify the major arteries and veins of the systemic circulation and other special circulations.

Microscopic Structure of the Blood Vessels

Except for the microscopic capillaries, the walls of blood vessels are constructed of three coats, or *tunics* (**Figure 32.1**). The **tunica intima,** which lines the lumen of a vessel, is a

Figure 32.1 Generalized structure of arteries, veins, and capillaries. (a) Photomicrograph of a muscular artery and the corresponding vein in cross section (76×). **(b)** Comparison of wall structure of arteries, veins, and capillaries. Note that the tunica media is thick in arteries and thin in veins, while the tunica externa is thin in arteries and relatively thicker in veins. Capillaries have only endothelium and a sparse basal lamina.

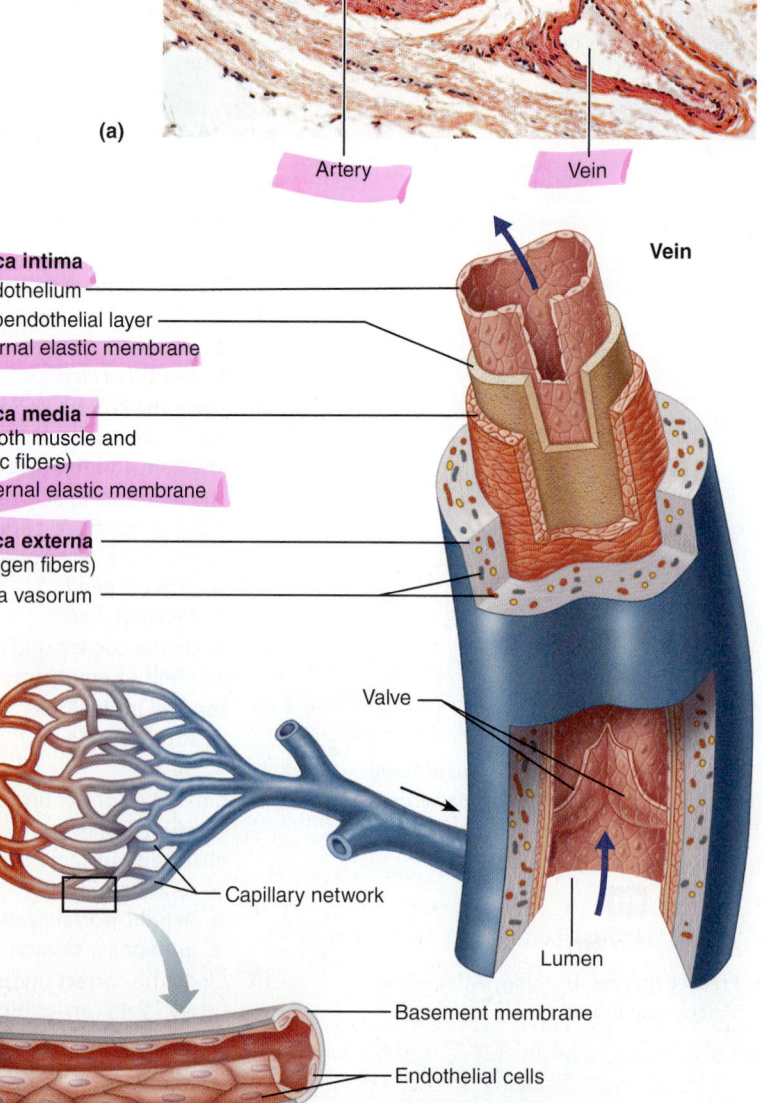

(a)

Artery   Vein

Artery

Tunica intima
• Endothelium
• Subendothelial layer
• Internal elastic membrane

Tunica media
(smooth muscle and elastic fibers)
• External elastic membrane

Tunica externa
(collagen fibers)
• Vasa vasorum

Vein

Lumen

Valve

Capillary network

Lumen

Basement membrane

Capillary

Endothelial cells

(b)

32

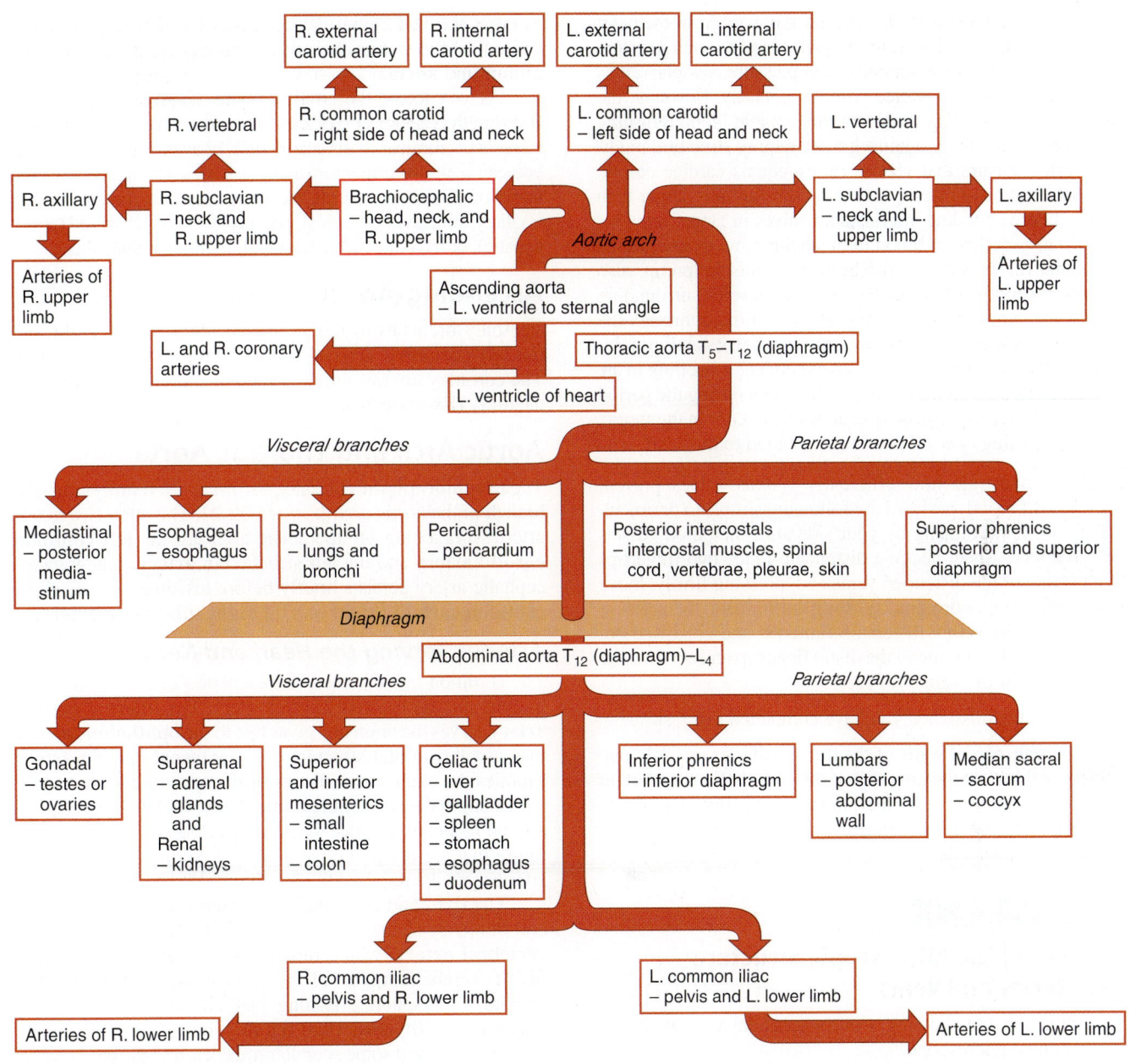

Figure 32.2 Schematic of the systemic arterial circulation. (R. = right, L. = left)

single thin layer of *endothelium* (squamous cells underlain by a scant basal lamina) that is continuous with the endocardium of the heart. Its cells fit closely together, forming an extremely smooth blood vessel lining that helps decrease resistance to blood flow.

The **tunica media** is the more bulky middle coat and is composed primarily of smooth muscle and elastin. The smooth muscle, under the control of the sympathetic nervous system, plays an active role in regulating the diameter of blood vessels, which in turn alters peripheral resistance and blood pressure.

The **tunica externa,** the outermost tunic, is composed of areolar or fibrous connective tissue. Its function is basically supportive and protective. In larger vessels, the tunica externa contains a system of tiny blood vessels, the **vasa vasorum.**

In general, the walls of arteries are thicker than those of veins. The tunica media in particular tends to be much heavier

and contains substantially more smooth muscle and elastic tissue. This anatomical difference reflects a functional difference in the two types of vessels. Arteries, which are closer to the pumping action of the heart, must be able to expand as an increased volume of blood is propelled into them during systole and then recoil passively as the blood flows off into the circulation during diastole. Their walls must be sufficiently strong and resilient to withstand such pressure fluctuations. Since these larger arteries have such large amounts of elastic tissue in their media, they are often referred to as *elastic arteries*. Smaller arteries, further along in the circulatory pathway, are exposed to less extreme pressure fluctuations. They have less elastic tissue but still have substantial amounts of smooth muscle in their media. For this reason, they are called *muscular arteries*. (A schematic of the systemic arteries is provided in **Figure 32.2.**)

By contrast, veins, which are far removed from the heart in the circulatory pathway, are not subjected to such pressure fluctuations and are essentially low-pressure vessels. Thus, veins may be thinner-walled without jeopardy. However, the low-pressure condition itself and the fact that blood returning to the heart often flows against gravity require structural modifications to ensure that venous return equals cardiac output. Thus, the lumens of veins tend to be substantially larger than those of corresponding arteries, and valves in larger veins act to prevent backflow of blood in much the same manner as the semilunar valves of the heart. The skeletal muscle "pump" also promotes venous return; as the skeletal muscles surrounding the veins contract and relax, the blood is milked through the veins toward the heart. Anyone who has been standing relatively still for an extended time has experienced swelling in the ankles, caused by blood pooling in their feet during the period of muscle inactivity. Pressure changes that occur in the thorax during breathing also aid the return of blood to the heart.

☐ To demonstrate how efficiently venous valves prevent backflow of blood, perform the following simple experiment. Allow one hand to hang by your side until the blood vessels on the dorsal aspect become distended. Place two fingertips against one of the distended veins and, pressing firmly, move the superior finger proximally along the vein and then release this finger. The vein will remain flattened and collapsed despite gravity. Then remove the distal fingertip and observe the rapid filling of the vein.

Check the box when you have completed this task.

The transparent walls of the tiny capillaries are only one cell layer thick, consisting of just the endothelium underlain by a basal lamina, that is, the tunica intima. Because of this exceptional thinness, exchanges are easily made between the blood and tissue cells.

ACTIVITY 1

Examining the Microscopic Structure of Arteries and Veins

1. Obtain a slide showing a cross-sectional view of blood vessels and a microscope.

2. Scan the section to identify a thick-walled artery (use Figure 32.1 as a guide). Very often, but not always, an arterial lumen will appear scalloped due to the constriction of its walls by the elastic tissue of the media.

3. Identify a vein. Its lumen may be elongated or irregularly shaped and collapsed, and its walls will be considerably thinner. Notice the difference in the relative amount of elastic fibers in the media of the two vessels. Also, note the thinness of the intima layer, which is composed of flat squamous cells. ■

Major Systemic Arteries of the Body

The **aorta** is the largest artery of the body. Extending upward as the ascending aorta from the left ventricle, it arches posteriorly and to the left (aortic arch) and then courses downward as the descending aorta through the thoracic cavity. Called the **thoracic aorta** from T_5 to T_{12}, it penetrates the diaphragm to enter the abdominal cavity just anterior to the vertebral column. As it enters the abdominal cavity, it becomes the **abdominal aorta.**

As you locate the arteries on the diagram (Figure 32.2) showing the relationship of the aorta and its major branches and on other anatomical charts and models, be aware of ways in which you can make your memorization task easier. In many cases the name of the artery reflects the body region it travels through (axillary, subclavian, brachial, popliteal), the organ served (renal, hepatic), or the bone followed (tibial, femoral, radial, ulnar).

Ascending Aorta

The only branches of the ascending aorta are the **right** and the **left coronary arteries,** which supply the myocardium. The coronary arteries are described in conjunction with heart anatomy (Exercise 30).

Aortic Arch and Thoracic Aorta

The **brachiocephalic** (literally, "arm-head") **trunk** is the first branch of the aortic arch **(Figure 32.3)**. The other two major arteries branching off the aortic arch are the **left common carotid artery** and the **left subclavian artery.** The brachiocephalic artery persists briefly before dividing into the **right common carotid artery** and the **right subclavian artery.**

Arteries Serving the Head and Neck

The common carotid artery on each side divides to form an internal and an external carotid artery. The **internal carotid artery** serves the brain and gives rise to the **ophthalmic artery** that supplies orbital structures. The **external carotid artery** supplies the extracranial tissues of the neck and head, largely via its **superficial temporal, maxillary, facial,** and **occipital** arterial branches. (Notice that several arteries are shown in the figure that are not described here. Ask your instructor which arteries you are required to identify.)

The right and left subclavian arteries each give off several branches to the head and neck. The first of these is the **vertebral artery,** which runs up the posterior neck to supply the cerebellum, part of the brain stem, and the posterior cerebral hemispheres. Issuing just lateral to the vertebral artery are the **thyrocervical trunk,** which mainly serves the thyroid gland and some scapular muscles, and the **costocervical trunk,** which supplies deep neck muscles and some of the upper intercostal muscles. In the armpit, the subclavian artery becomes the axillary artery, which serves the upper limb.

Arteries Serving the Brain

A continuous blood supply to the brain is crucial because oxygen deprivation for even a few minutes causes irreparable damage to the delicate brain tissue. The brain is supplied by two pairs of arteries arising from the region of the aortic arch—the internal carotid arteries and the vertebral arteries. (Figure 32.3b is a diagram of the brain's arterial supply.)

Within the cranium, each internal carotid artery divides into **anterior** and **middle cerebral arteries,** which supply the bulk of the cerebrum. The right and left anterior cerebral arteries are connected by a short shunt called the **anterior communicating artery.** This shunt, along with shunts from each of the middle cerebral arteries, called the **posterior communicating arteries,** contribute to the formation of the **cerebral arterial circle** (circle of Willis), an arterial anastomosis at the base of the brain surrounding the pituitary gland and the optic chiasma.

32

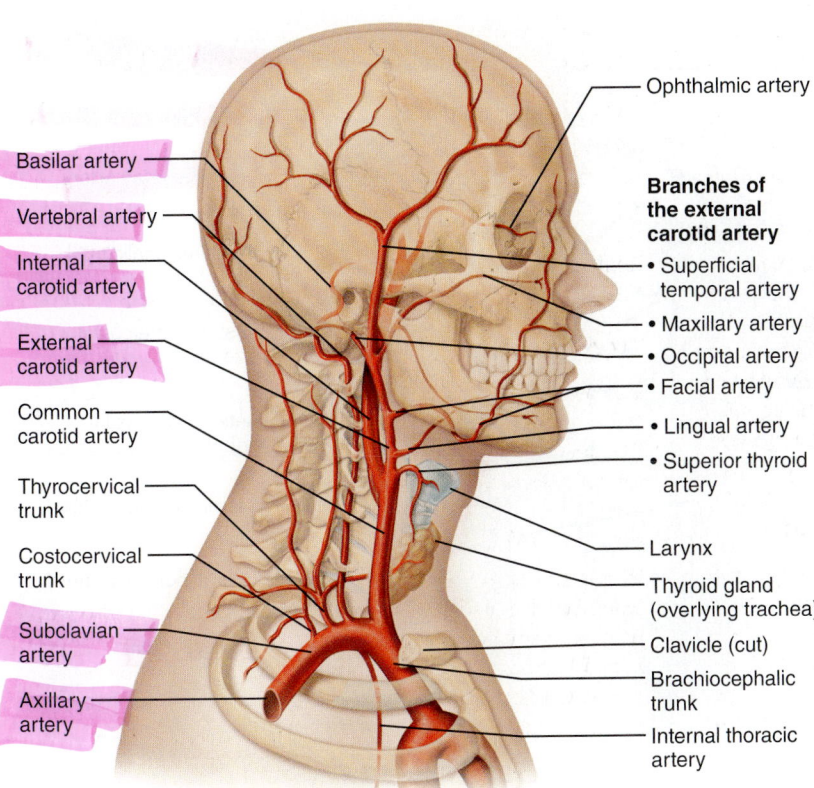

Figure 32.3 Arteries of the head, neck, and brain. (a) Right aspect. **(b)** Drawing of the cerebral arteries. Cerebellum is not shown on the left side of the figure. **(c)** Cerebral arterial circle (circle of Willis) in a human brain.

Ophthalmic artery

Basilar artery

Vertebral artery

Internal carotid artery

External carotid artery

Common carotid artery

Thyrocervical trunk

Costocervical trunk

Subclavian artery

Axillary artery

Branches of the external carotid artery
- Superficial temporal artery
- Maxillary artery
- Occipital artery
- Facial artery
- Lingual artery
- Superior thyroid artery

Larynx

Thyroid gland (overlying trachea)

Clavicle (cut)

Brachiocephalic trunk

Internal thoracic artery

(a)

Anterior

Frontal lobe

Olfactory bulb

Optic chiasma

Middle cerebral artery

Internal carotid artery

Pituitary gland

Temporal lobe

Pons

Occipital lobe

Vertebral artery

Cerebral arterial circle (*circle of Willis*)
- Anterior communicating artery
- Anterior cerebral artery
- Posterior communicating artery
- Posterior cerebral artery

Basilar artery

Cerebellum

Posterior

(b)

(c)

32

The paired **vertebral arteries** diverge from the subclavian arteries and pass superiorly through the foramina of the transverse process of the cervical vertebrae to enter the skull through the foramen magnum. Within the skull, the vertebral arteries unite to form a single **basilar artery,** which continues superiorly along the ventral aspect of the brain stem, giving off branches to the pons, cerebellum, and inner ear. At the base of the cerebrum, the basilar artery divides to form the **posterior cerebral arteries.** These supply portions of the temporal and occipital lobes of the cerebrum and complete the cerebral arterial circle posteriorly.

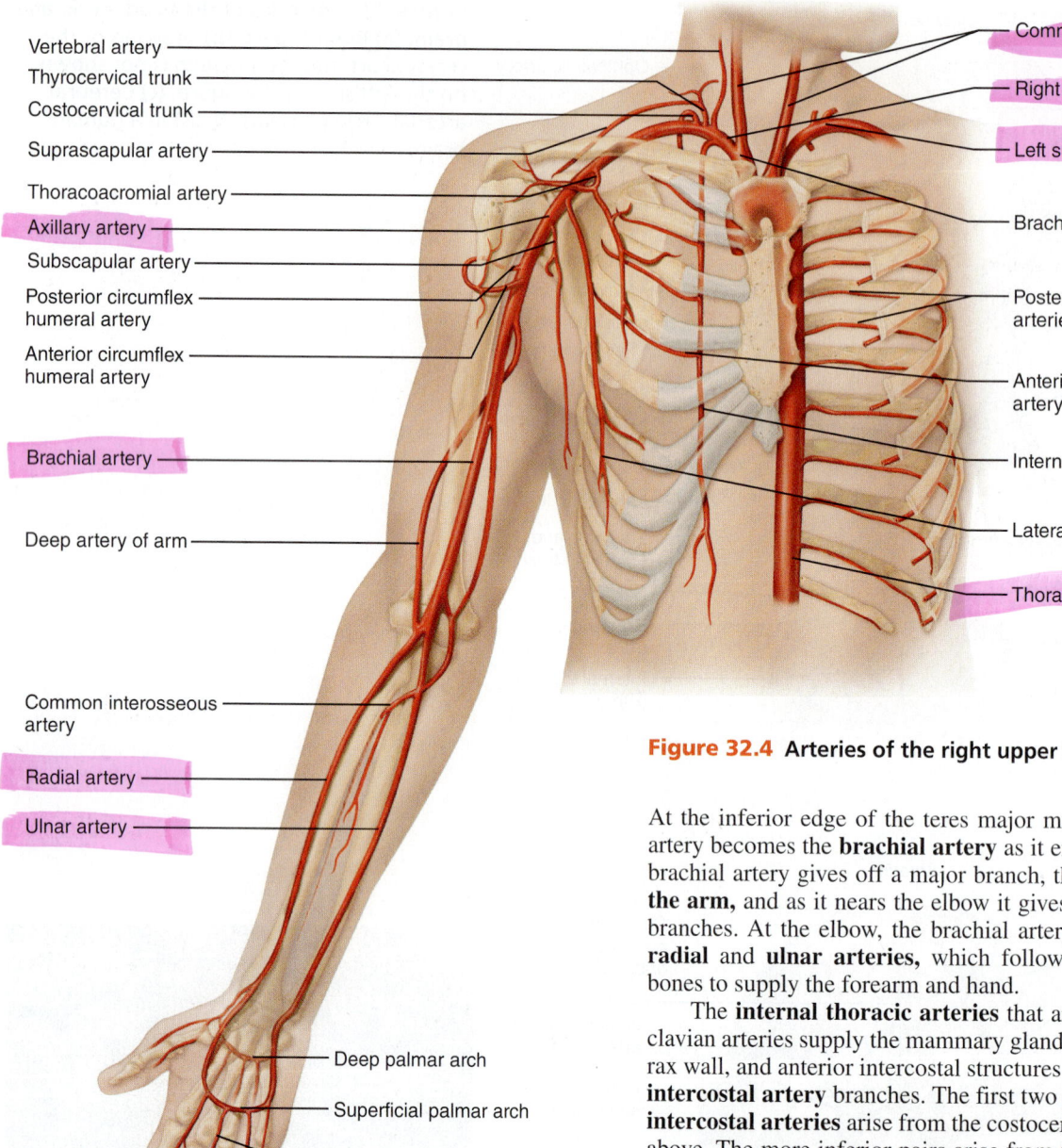

Vertebral artery
Thyrocervical trunk
Costocervical trunk
Suprascapular artery
Thoracoacromial artery
Axillary artery
Subscapular artery
Posterior circumflex humeral artery
Anterior circumflex humeral artery
Brachial artery
Deep artery of arm
Common interosseous artery
Radial artery
Ulnar artery
Deep palmar arch
Superficial palmar arch
Digital arteries

Common carotid arteries
Right subclavian artery
Left subclavian artery
Brachiocephalic trunk
Posterior intercostal arteries
Anterior intercostal artery
Internal thoracic artery
Lateral thoracic artery
Thoracic aorta

Figure 32.4 Arteries of the right upper limb and thorax.

At the inferior edge of the teres major muscle, the axillary artery becomes the **brachial artery** as it enters the arm. The brachial artery gives off a major branch, the **deep artery of the arm,** and as it nears the elbow it gives off several small branches. At the elbow, the brachial artery divides into the **radial** and **ulnar arteries,** which follow the same-named bones to supply the forearm and hand.

The **internal thoracic arteries** that arise from the subclavian arteries supply the mammary glands, most of the thorax wall, and anterior intercostal structures via their **anterior intercostal artery** branches. The first two pairs of **posterior intercostal arteries** arise from the costocervical trunk, noted above. The more inferior pairs arise from the thoracic aorta. (Not shown in Figure 32.4 are the small arteries that serve the diaphragm *[phrenic arteries]*, esophagus *[esophageal arteries]*, bronchi *[bronchial arteries]*, and other structures of the mediastinum *[mediastinal and pericardial arteries]*.)

Abdominal Aorta

Although several small branches of the descending aorta serve the thorax (see the previous section), the more major branches of the descending aorta are those serving the abdominal organs and ultimately the lower limbs **(Figure 32.5)**.

Arteries Serving Abdominal Organs

The **celiac trunk** (Figure 32.5a) is an unpaired artery that subdivides almost immediately into three branches: the **left gastric artery** supplying the stomach, the **splenic artery** supplying the spleen, and the **common hepatic artery,** which runs superiorly and gives off branches to the stomach (**right gastric artery**), duodenum, and pancreas. Where the **gastroduodenal artery** branches off, the common hepatic artery becomes the **hepatic artery proper,** which serves the liver. The **right** and **left gastroepiploic arteries,** branches of the

The uniting of the blood supply of the internal carotid arteries and the vertebral arteries via the cerebral arterial circle is a protective device that theoretically provides an alternative set of pathways for blood to reach the brain tissue in the case of arterial occlusion or impaired blood flow anywhere in the system. In actuality, the communicating arteries are tiny, and in many cases the communicating system is not functional.

Arteries Serving the Thorax and Upper Limbs

As the **axillary artery** runs through the axilla, it gives off several branches to the chest wall and shoulder girdle **(Figure 32.4)**. These include the **thoracoacromial artery** (to shoulder and pectoral region), the **lateral thoracic artery** (lateral chest wall), the **subscapular artery** (to scapula and dorsal thorax), and the **anterior** and **posterior circumflex humeral arteries** (to the shoulder and the deltoid muscle).

32

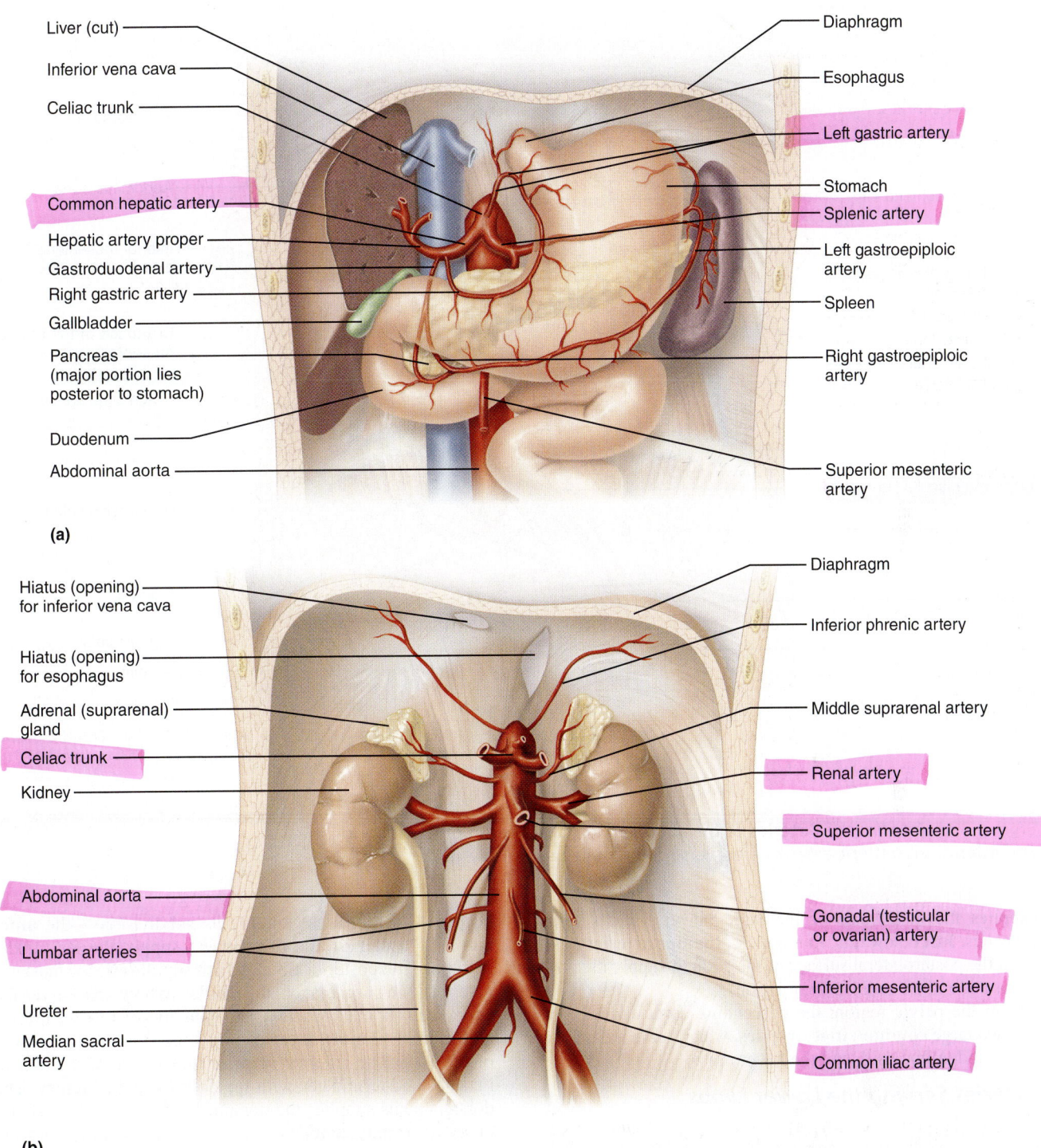

Figure 32.5 Arteries of the abdomen. (a) The celiac trunk and its major branches.
(b) Major branches of the abdominal aorta.

gastroduodenal and splenic arteries respectively, serve the greater curvature of the stomach.

The largest branch of the abdominal aorta, the **superior mesenteric artery** (Figure 32.5b and 32.5c), supplies most of the small intestine (via the intestinal arteries) and the first half of the large intestine (via the ileocolic and colic arteries). Flanking the superior mesenteric artery on the left and right are the **middle suprarenal arteries** serving the adrenal glands that sit atop the kidneys.

The paired **renal arteries** (Figure 32.5b) supply the kidneys, and the **gonadal arteries,** arising from the ventral aortic surface just below the renal arteries, run inferiorly to serve the gonads. They are called **ovarian arteries** in the female and **testicular arteries** in the male. Since these vessels must travel through the inguinal canal to supply the testes, they are considerably longer in the male than the female.

The final major branch of the abdominal aorta is the **inferior mesenteric artery** (Figure 32.5b and 32.5c), which

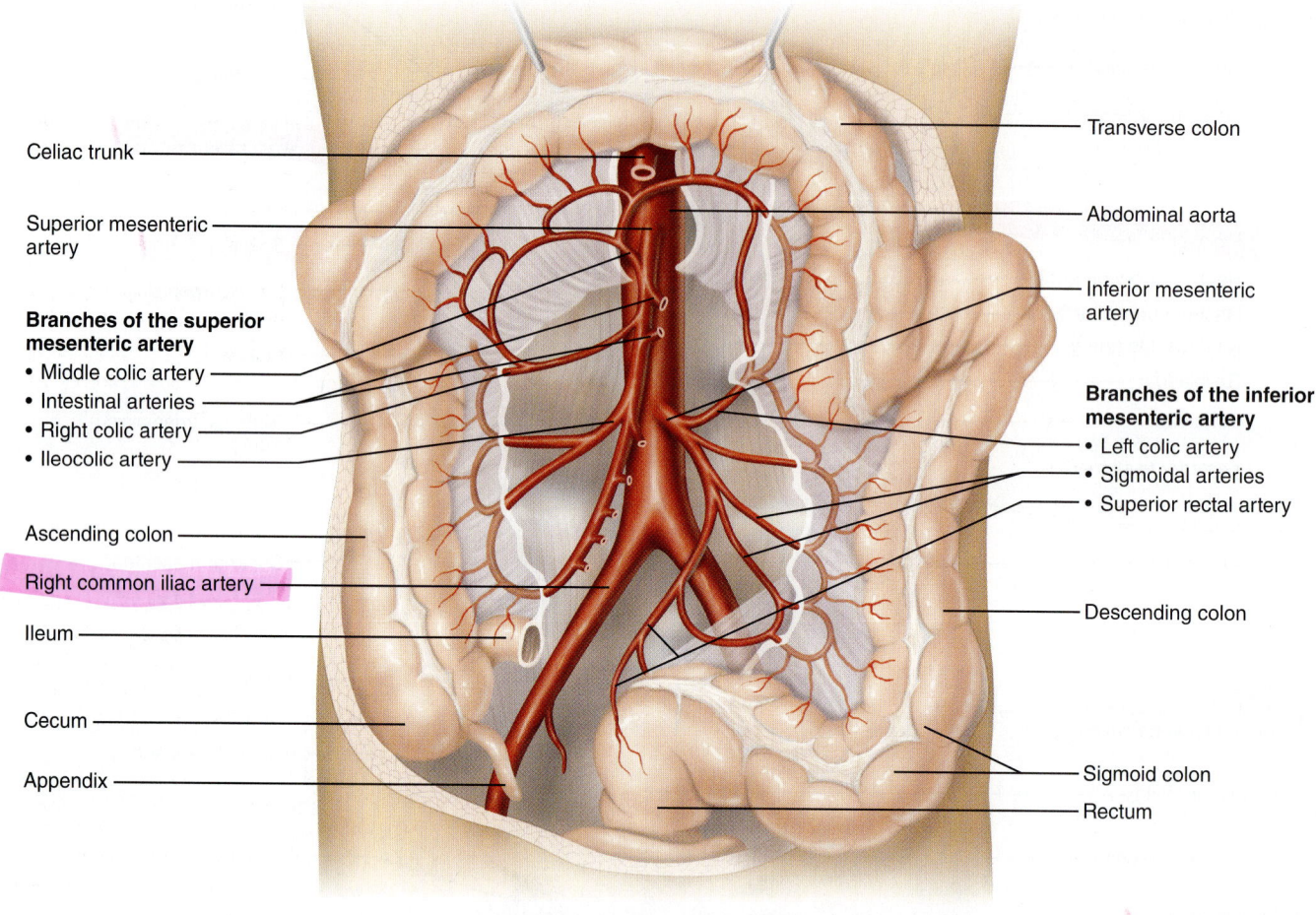

Celiac trunk

Superior mesenteric artery

Branches of the superior mesenteric artery
- Middle colic artery
- Intestinal arteries
- Right colic artery
- Ileocolic artery

Ascending colon

Right common iliac artery

Ileum

Cecum

Appendix

Transverse colon

Abdominal aorta

Inferior mesenteric artery

Branches of the inferior mesenteric artery
- Left colic artery
- Sigmoidal arteries
- Superior rectal artery

Descending colon

Sigmoid colon

Rectum

(c)

Figure 32.5 *(continued)* **Arteries of the abdomen. (c)** Distribution of the superior and inferior mesenteric arteries, transverse colon pulled superiorly.

supplies the distal half of the large intestine via several branches. Just below this, four pairs of **lumbar arteries** arise from the posterolateral surface of the aorta to supply the posterior abdominal wall (lumbar region).

In the pelvic region, the descending aorta divides into the two large **common iliac arteries,** which serve the pelvis, lower abdominal wall, and the lower limbs.

Arteries Serving the Lower Limbs

Each of the common iliac arteries extends for about 5 cm (2 inches) into the pelvis before it divides into the internal and external iliac arteries **(Figure 32.6)**. The **internal iliac artery** supplies the gluteal muscles via the **superior** and **inferior gluteal arteries** and the adductor muscles of the medial thigh via the **obturator artery,** as well as the external genitalia and perineum (via the *internal pudendal artery,* not illustrated).

The **external iliac artery** supplies the anterior abdominal wall and the lower limb. As it continues into the thigh, its name changes to **femoral artery.** Proximal branches of the femoral artery, the **circumflex femoral arteries,** supply the head and neck of the femur and the hamstring muscles. The femoral artery gives off a deep branch, the **deep artery of the thigh** (also called the *deep femoral artery*), which is the main supply to the thigh muscles (hamstrings, quadriceps, and adductors). In the knee region, the femoral artery briefly

becomes the **popliteal artery;** its subdivisions—the **anterior** and **posterior tibial arteries**—supply the leg, ankle, and foot. The posterior tibial, which supplies flexor muscles, gives off one main branch, the **fibular artery,** that serves the lateral calf (fibular muscles), and then divides into the **lateral** and **medial plantar arteries** that supply blood to the sole of the foot. The anterior tibial artery supplies the extensor muscles and terminates with the **dorsalis pedis artery.** The dorsalis pedis supplies the dorsum of the foot and continues on as the **arcuate artery** which issues the **dorsal metatarsal arteries** to the metatarsus of the foot. The dorsalis pedis is often palpated in patients with circulation problems of the leg to determine the circulatory efficiency to the limb as a whole.

☐ Palpate your own dorsalis pedis artery.

Check the box when you have completed this task.

ACTIVITY 2

Locating Arteries on an Anatomical Chart or Model

Now that you have identified the arteries in the illustrations (Figures 32.2–32.6), attempt to locate and name them

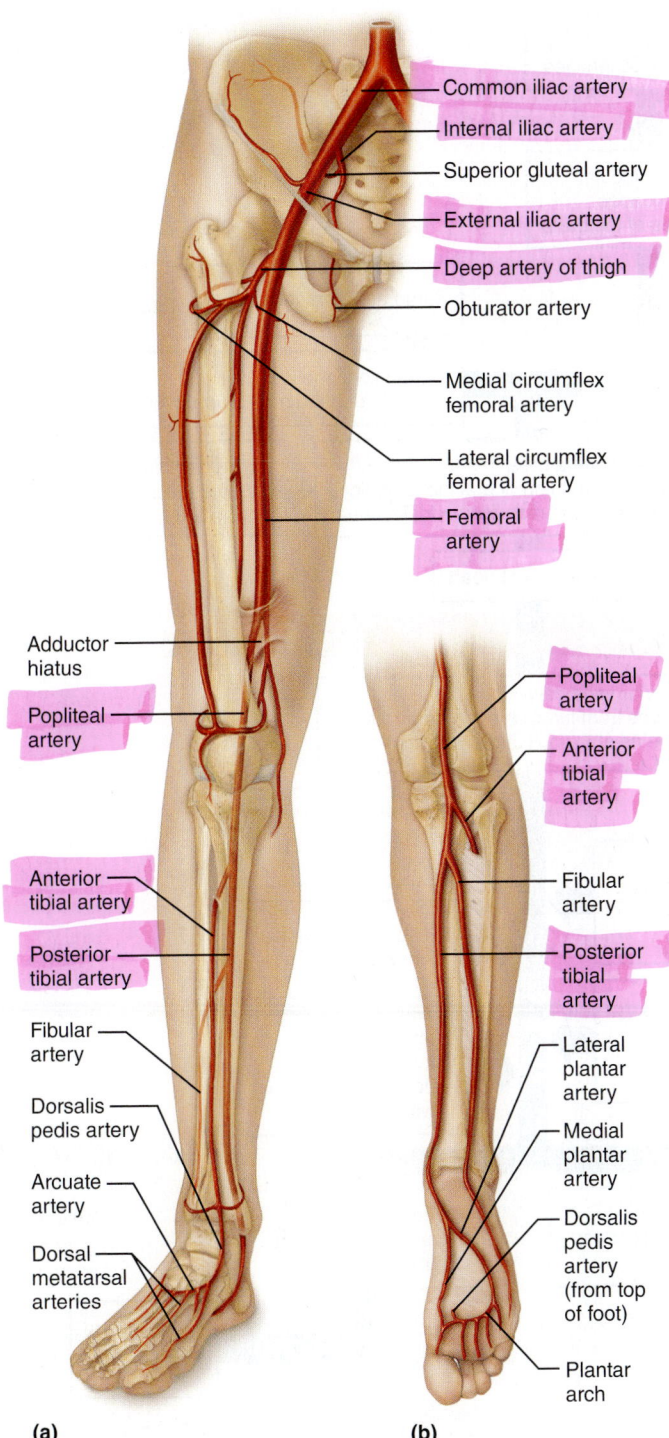

Common iliac artery
Internal iliac artery
Superior gluteal artery
External iliac artery
Deep artery of thigh
Obturator artery
Medial circumflex femoral artery
Lateral circumflex femoral artery
Femoral artery
Adductor hiatus
Popliteal artery
Anterior tibial artery
Posterior tibial artery
Fibular artery
Dorsalis pedis artery
Arcuate artery
Dorsal metatarsal arteries

Popliteal artery
Anterior tibial artery
Fibular artery
Posterior tibial artery
Lateral plantar artery
Medial plantar artery
Dorsalis pedis artery (from top of foot)
Plantar arch

(a)   (b)

Figure 32.6 Arteries of the right pelvis and lower limb.
(a) Anterior view. **(b)** Posterior view.

(without a reference) on a large anatomical chart or three-dimensional model of the vascular system. ▬

Major Systemic Veins of the Body

Arteries are generally located in deep, well-protected body areas. However, many veins follow a more superficial course and are often easily seen and palpated on the body surface. Most deep veins parallel the course of the major arteries, and in many cases the naming of the veins and arteries is identical except for the designation of the vessels as veins. Whereas the major systemic arteries branch off the aorta, the veins tend to converge on the venae cavae, which enter the right atrium of the heart. Veins draining the head and upper extremities empty into the **superior vena cava,** and those draining the lower body empty into the **inferior vena cava.** The schematic **(Figure 32.7)** of the systemic veins and their relationship to the venae cavae will get you started.

Veins Draining into the Inferior Vena Cava

The inferior vena cava, a much longer vessel than the superior vena cava, returns blood to the heart from all body regions below the diaphragm (see Figure 32.7). It begins in the lower abdominal region with the union of the paired **common iliac veins,** which drain venous blood from the legs and pelvis.

Veins of the Lower Limbs

Each common iliac vein is formed by the union of the **internal iliac vein,** draining the pelvis, and the **external iliac vein,** which receives venous blood from the lower limb **(Figure 32.8).** Veins of the leg include the **anterior** and **posterior tibial veins,** which serve the calf and foot. The anterior tibial vein is a superior continuation of the **dorsalis pedis vein** of the foot. The posterior tibial vein is formed by the union of the **medial** and **lateral plantar veins,** and ascends deep in the calf muscles. It receives the **fibular vein** in the calf and then joins with the anterior tibial vein at the knee to produce the **popliteal vein,** which crosses the back of the knee. The popliteal vein becomes the **femoral vein** in the thigh; the femoral vein in turn becomes the external iliac vein in the inguinal region.

The **great saphenous vein,** a superficial vein, is the longest vein in the body. Beginning in common with the **small saphenous vein** from the **dorsal venous arch,** it extends up the medial side of the leg, knee, and thigh to empty into the femoral vein. The small saphenous vein runs along the lateral aspect of the foot and through the calf muscle, which it drains, and then empties into the popliteal vein at the knee (Figure 32.8b).

Veins of the Abdomen

Moving superiorly in the abdominal cavity **(Figure 32.9),** the inferior vena cava receives blood from the posterior abdominal wall via several pairs of **lumbar veins,** and from the right ovary or testis via the **right gonadal vein.** (The **left gonadal [ovarian or testicular] vein** drains into the left renal vein superiorly.) The paired **renal veins** drain the kidneys. Just above the right renal vein, the **right suprarenal vein** (receiving blood from the adrenal gland on the same side) drains into the inferior vena cava, but its partner, the **left suprarenal vein,** empties into the left renal vein inferiorly. The **hepatic veins** drain the liver. The unpaired veins draining the digestive tract organs empty into a special vessel, the hepatic portal

32

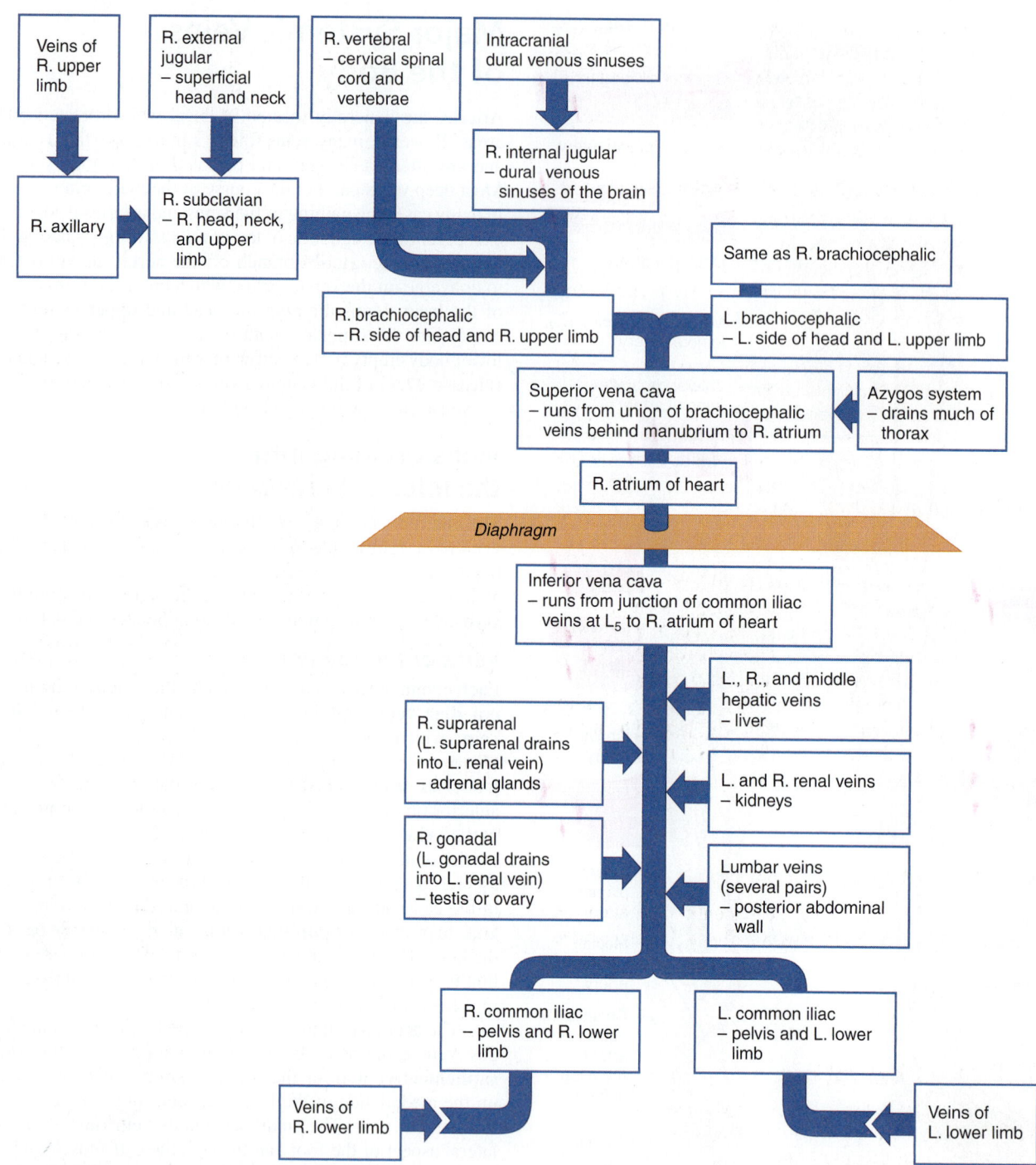

Figure 32.7 **Schematic of systemic venous circulation.**

vein, which carries blood to the liver to be processed before it enters the systemic venous system. (The hepatic portal system is discussed separately on page 484.)

Veins Draining into the Superior Vena Cava

Veins draining into the superior vena cava are named from the superior vena cava distally, *but remember that the flow of blood is in the opposite direction*.

Veins of the Head and Neck

The **right** and **left brachiocephalic veins** drain the head, neck, and upper extremities and unite to form the superior vena cava **(Figure 32.10)**. Notice that although there is only one brachiocephalic artery, there are two brachiocephalic veins.

Branches of the brachiocephalic veins include the internal jugular, vertebral, and subclavian veins. The **internal jugular veins** are large veins that drain the superior sagittal sinus and other **dural sinuses** of the brain. As they run inferiorly, they receive blood from the head and neck via the **superficial**

temporal and **facial veins.** The **vertebral veins** drain the posterior aspect of the head including the cervical vertebrae and spinal cord. The **subclavian veins** receive venous blood from the upper extremity. The **external jugular vein** joins the subclavian vein near its origin to return the venous drainage of the extracranial (superficial) tissues of the head and neck.

Veins of the Upper Limb and Thorax

As the subclavian vein passes through the axilla, it becomes the **axillary vein** and then the **brachial vein** as it courses along the posterior aspect of the humerus **(Figure 32.11)**. The brachial vein is formed by the union of the deep **radial** and **ulnar veins** of the forearm. The superficial venous drainage of the arm includes the **cephalic vein,** which courses along the lateral aspect of the arm and empties into the axillary vein; the **basilic vein,** found on the medial aspect of the arm and entering the brachial vein; and the **median cubital vein,** which runs between the cephalic and basilic veins in the anterior aspect of the elbow (this vein is often the site of choice for removing blood for testing purposes). The **median antebrachial vein** lies between the radial and ulnar veins, and terminates variably by entering the cephalic or basilic vein at the elbow.

The **azygos system** (Figure 32.11) drains the intercostal muscles of the thorax and provides an accessory venous system to drain the abdominal wall. The **azygos vein,** which drains the right side of the thorax, enters the dorsal aspect of the superior vena cava immediately before that vessel enters the right atrium. Also part of the azygos system are the **hemiazygos** (a continuation of the **left ascending lumbar vein** of the abdomen) and the **accessory hemiazygos veins,** which together drain the left side of the thorax and empty into the azygos vein.

ACTIVITY 3

Identifying the Systemic Veins

Identify the important veins of the systemic circulation on the large anatomical chart or model without referring to the figures. ■

Special Circulations

Pulmonary Circulation

The pulmonary circulation (discussed previously in relation to heart anatomy on page 446) differs in many ways from systemic circulation because it does not serve the metabolic needs of the body tissues with which it is associated (in this case, lung tissue). It functions instead to bring the blood into close contact with the alveoli of the lungs to permit gas exchanges that rid the blood of excess carbon dioxide and replenish its supply of vital oxygen. The arteries of the pulmonary circulation are structurally much like veins, and they create a low-pressure bed in the lungs. (If the arterial pressure in the systemic circulation is 120/80, the pressure in the pulmonary artery is likely to be approximately 24/8.) The functional blood supply of the lungs is provided by the **bronchial arteries** (not shown), which diverge from the thoracic portion of the descending aorta.

Pulmonary circulation begins with the large **pulmonary trunk,** which leaves the right ventricle and divides into the **right** and **left pulmonary arteries** about 5 cm (2 inches)

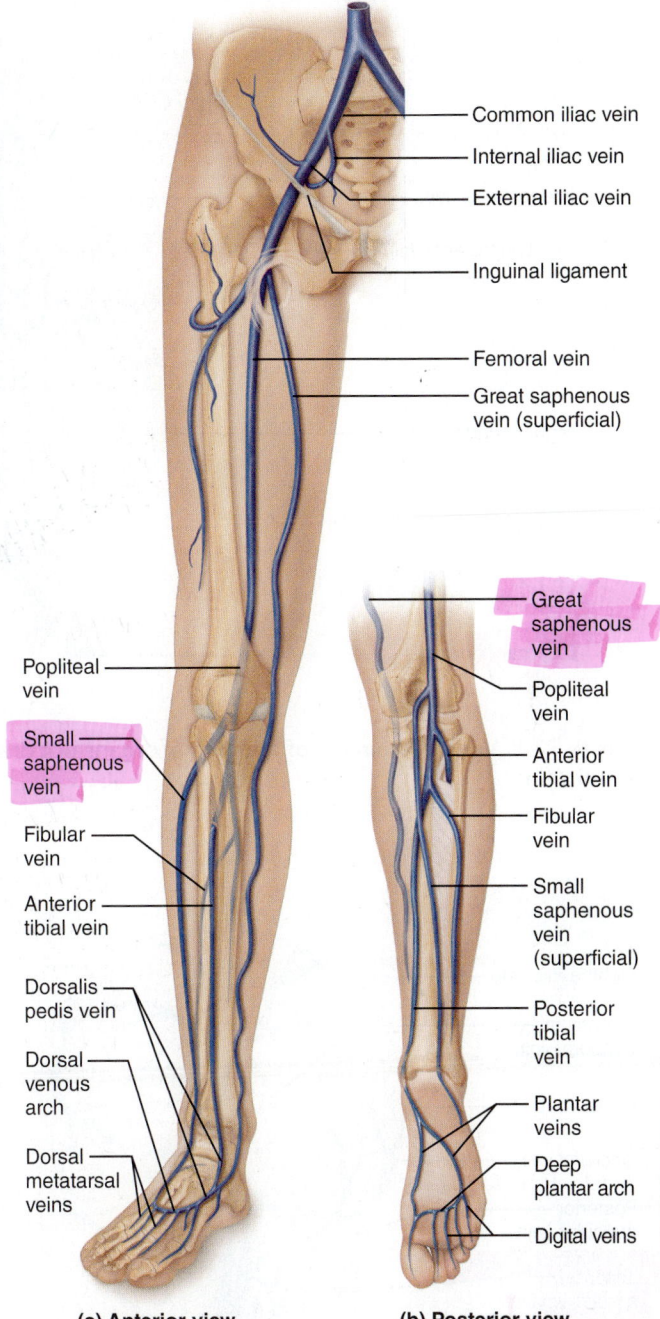

(a) Anterior view **(b) Posterior view**

Figure 32.8 Veins of the right pelvis and lower limb.
(a) Anterior view. **(b)** Posterior view.

above its origin. The right and left pulmonary arteries plunge into the lungs, where they subdivide into **lobar arteries** (three on the right and two on the left). The lobar arteries accompany the main bronchi into the lobes of the lungs and branch extensively within the lungs to form arterioles, which finally terminate in the capillary networks surrounding the alveolar sacs of the lungs. Diffusion of the respiratory gases occurs across the walls of the alveoli and **pulmonary capillaries.** The pulmonary capillary beds are drained by venules, which converge to form sequentially larger veins and finally the four **pulmonary veins** (two leaving each lung), which return the blood to the left atrium of the heart.

32

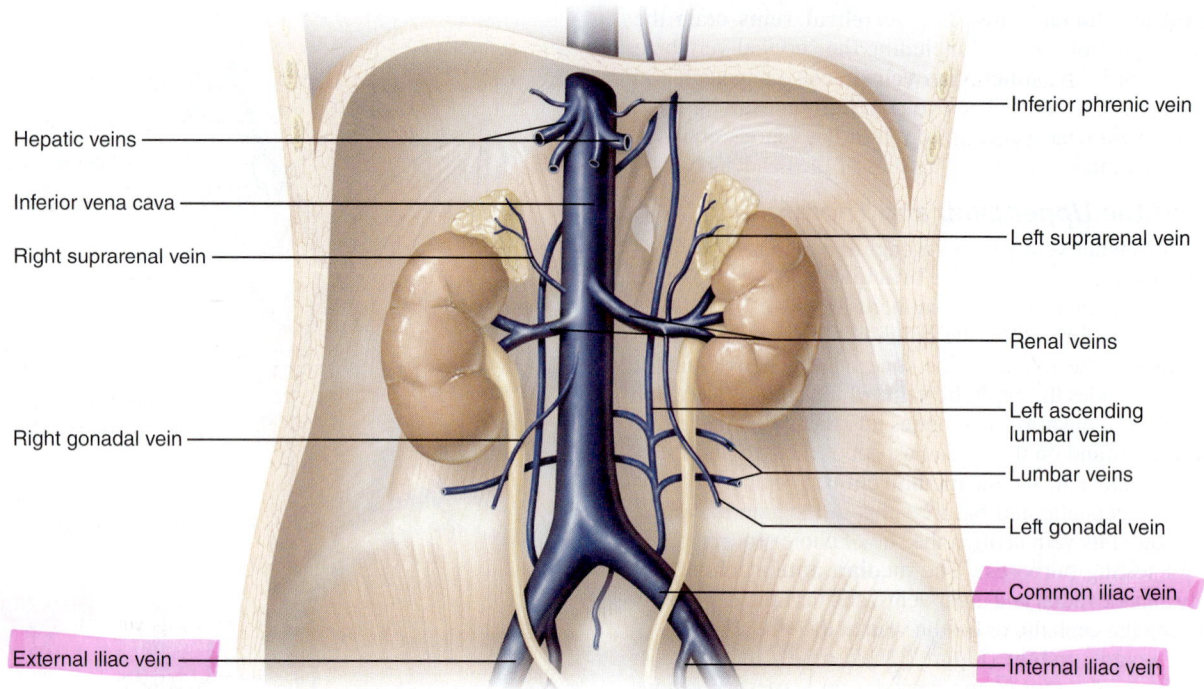

Hepatic veins

Inferior vena cava

Right suprarenal vein

Right gonadal vein

External iliac vein

Inferior phrenic vein

Left suprarenal vein

Renal veins

Left ascending lumbar vein

Lumbar veins

Left gonadal vein

Common iliac vein

Internal iliac vein

Figure 32.9 **Venous drainage of abdominal organs not drained by the hepatic portal vein.**

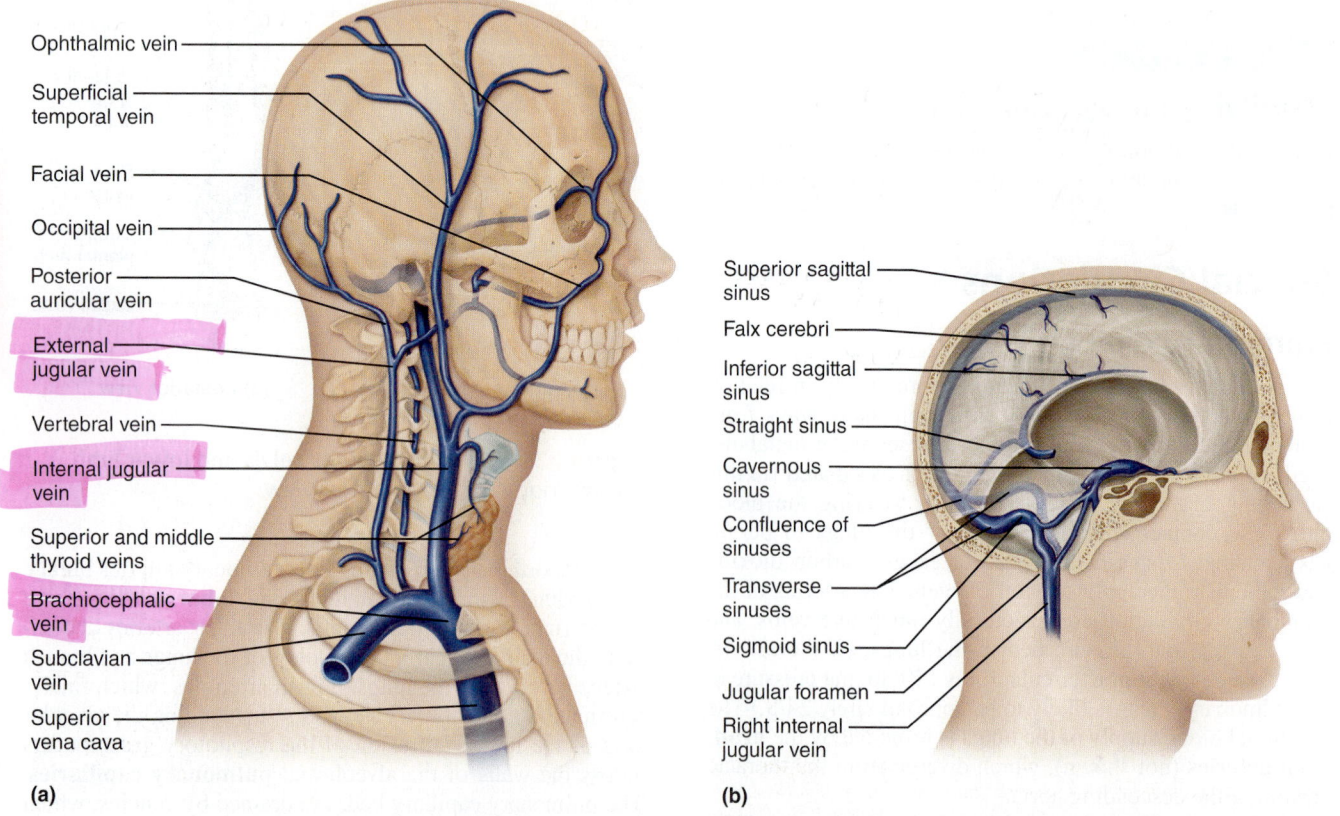

Ophthalmic vein

Superficial temporal vein

Facial vein

Occipital vein

Posterior auricular vein

External jugular vein

Vertebral vein

Internal jugular vein

Superior and middle thyroid veins

Brachiocephalic vein

Subclavian vein

Superior vena cava

Superior sagittal sinus

Falx cerebri

Inferior sagittal sinus

Straight sinus

Cavernous sinus

Confluence of sinuses

Transverse sinuses

Sigmoid sinus

Jugular foramen

Right internal jugular vein

(a)

(b)

Figure 32.10 **Venous drainage of the head, neck, and brain. (a)** Veins of the head and neck, right superficial aspect. **(b)** Dural sinuses of the brain, right aspect.

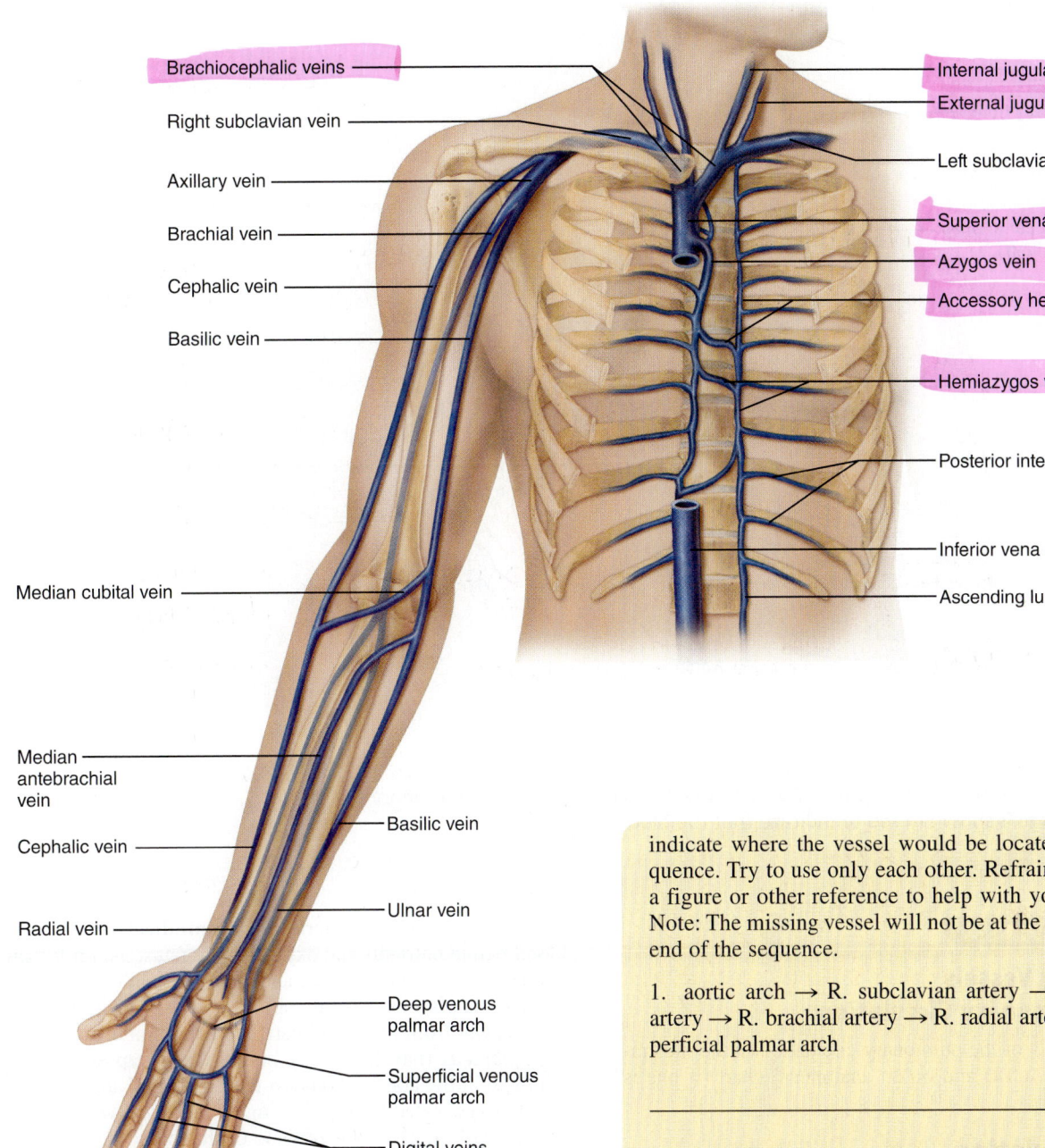

Brachiocephalic veins

Right subclavian vein

Axillary vein

Brachial vein

Cephalic vein

Basilic vein

Internal jugular vein

External jugular vein

Left subclavian vein

Superior vena cava

Azygos vein

Accessory hemiazygos vein

Hemiazygos vein

Posterior intercostals

Inferior vena cava

Ascending lumbar vein

Median cubital vein

Median antebrachial vein

Cephalic vein

Radial vein

Basilic vein

Ulnar vein

Deep venous palmar arch

Superficial venous palmar arch

Digital veins

Figure 32.11 Veins of the thorax and right upper limb. For clarity, the abundant branching and anastomoses of these vessels are not shown.

GROUP CHALLENGE

Fix the Blood Trace

Several artery or vein sequences are listed below. Working in small groups, decide if each sequence of blood vessels is correct or if a blood vessel is missing. If correct, simply write "all correct." If incorrect, list the missing vessel and draw an insertion mark ("v") on the arrow to indicate where the vessel would be located in the sequence. Try to use only each other. Refrain from using a figure or other reference to help with your decision. Note: The missing vessel will not be at the beginning or end of the sequence.

1. aortic arch → R. subclavian artery → R. axillary artery → R. brachial artery → R. radial artery → R. superficial palmar arch

_____

2. abdominal aorta → R. common iliac artery → R. femoral artery → R. popliteal artery → R. anterior tibial artery → R. dorsalis pedis artery

_____

3. ascending aorta → aortic arch → L. common carotid artery → L. internal carotid artery → L. anterior cerebral artery

_____

4. R. median antebrachial vein → R. basilic vein → R. axillary vein → R. brachiocephalic vein → superior vena cava

_____

32

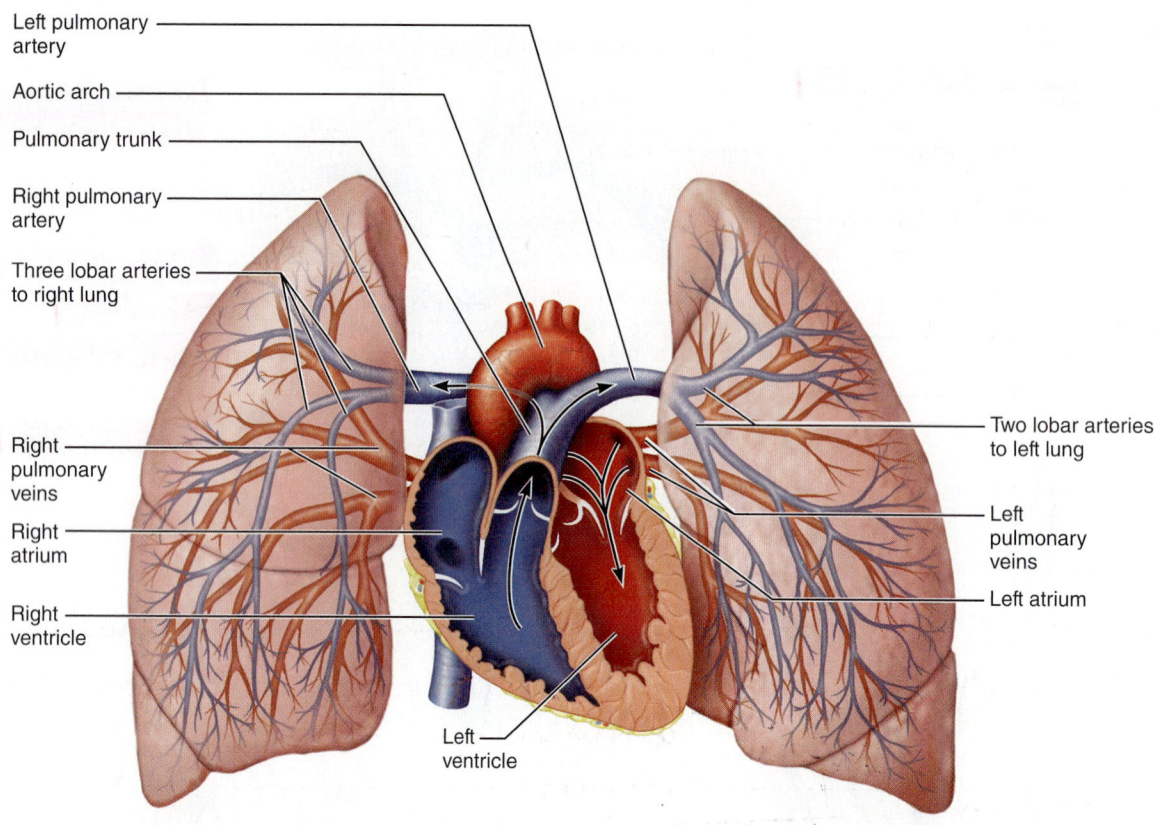

Left pulmonary artery

Aortic arch

Pulmonary trunk

Right pulmonary artery

Three lobar arteries to right lung

Right pulmonary veins

Right atrium

Right ventricle

Left ventricle

Two lobar arteries to left lung

Left pulmonary veins

Left atrium

Figure 32.12 **The pulmonary circulation.** The pulmonary arterial system is shown in blue to indicate that the blood it carries is relatively oxygen-poor. The pulmonary venous drainage is shown in red to indicate that the blood it transports is oxygen-rich.

ACTIVITY 4

Identifying Vessels of the Pulmonary Circulation

Find the vessels of the pulmonary circulation in the illustration **(Figure 32.12)** and on an anatomical chart (if one is available). ▬

Fetal Circulation

In a developing fetus, the lungs and digestive system are not yet functional, and all nutrient, excretory, and gaseous exchanges occur through the placenta **(Figure 32.13a)**. Nutrients and oxygen move across placental barriers from the mother's blood into fetal blood, and carbon dioxide and other metabolic wastes move from the fetal blood supply to the mother's blood.

ACTIVITY 5

Tracing the Pathway of Fetal Blood Flow

Trace the pathway of fetal blood flow using the illustration of fetal structures (Figure 32.13a) and an anatomical chart (if available). Locate all the named vessels. Identify the named remnants of the foramen ovale and fetal vessels (use Figure 32.13b). ▬

Fetal blood travels through the umbilical cord, which contains three blood vessels: one large umbilical vein and two smaller umbilical arteries. The **umbilical vein** carries blood rich in nutrients and oxygen to the fetus; the **umbilical arteries** carry carbon dioxide and waste-laden blood from the fetus to the placenta. The umbilical arteries, which transport blood away from the fetal heart, meet the umbilical vein at the *umbilicus* (navel, or belly button) and wrap around the vein within the cord en route to their placental attachments. Newly oxygenated blood flows in the umbilical vein superiorly toward the fetal heart. Some of this blood perfuses the liver, but the larger proportion is ducted through the relatively nonfunctional liver to the inferior vena cava via a shunt vessel called the **ductus venosus,** which carries the blood to the right atrium of the heart.

Because fetal lungs are nonfunctional and collapsed, two shunting mechanisms ensure that blood almost entirely bypasses the lungs. Much of the blood entering the right atrium is shunted into the left atrium through the **foramen ovale,** a flaplike opening in the interatrial septum. The left ventricle then pumps the blood out the aorta to the systemic circulation. Blood that does enter the right ventricle and is pumped out of the pulmonary trunk encounters a second shunt, the **ductus arteriosus,** a short vessel connecting the pulmonary trunk and the aorta. Because the collapsed lungs present an extremely high-resistance pathway, blood more readily enters the systemic circulation through the ductus arteriosus.

The aorta carries blood to the tissues of the body; this blood ultimately finds its way back to the placenta via the umbilical arteries. The only fetal vessel that carries highly

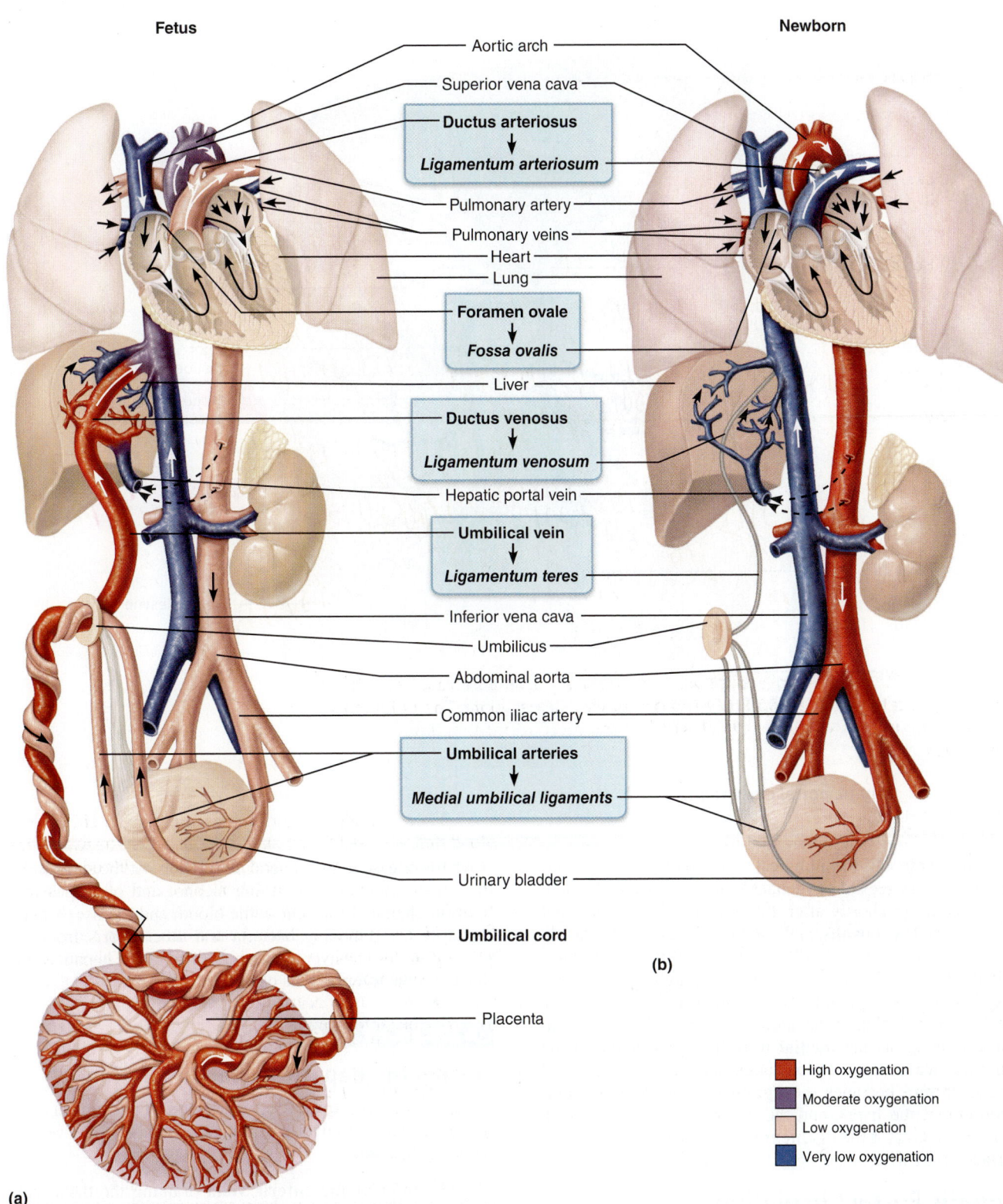

Fetus

Newborn

Aortic arch
Superior vena cava
Ductus arteriosus
↓
Ligamentum arteriosum
Pulmonary artery
Pulmonary veins
Heart
Lung
Foramen ovale
↓
Fossa ovalis
Liver
Ductus venosus
↓
Ligamentum venosum
Hepatic portal vein
Umbilical vein
↓
Ligamentum teres
Inferior vena cava
Umbilicus
Abdominal aorta
Common iliac artery
Umbilical arteries
↓
Medial umbilical ligaments
Urinary bladder
Umbilical cord

(b)

Placenta

■ High oxygenation
■ Moderate oxygenation
■ Low oxygenation
■ Very low oxygenation

(a)

32

Figure 32.13 Circulation in fetus and newborn. Arrows indicate direction of blood flow. Arrows in the blue boxes go from the fetal structure to what it becomes after birth. **(a)** Special adaptations for embryonic and fetal life. The umbilical vein (red) carries oxygen- and nutrient-rich blood from the placenta to the fetus. The umbilical arteries (pink) carry waste-laden blood from the fetus to the placenta. **(b)** Changes in the cardiovascular system at birth. The umbilical vessels are occluded, as are the liver and lung bypasses (ductus venosus and arteriosus, and the foramen ovale).

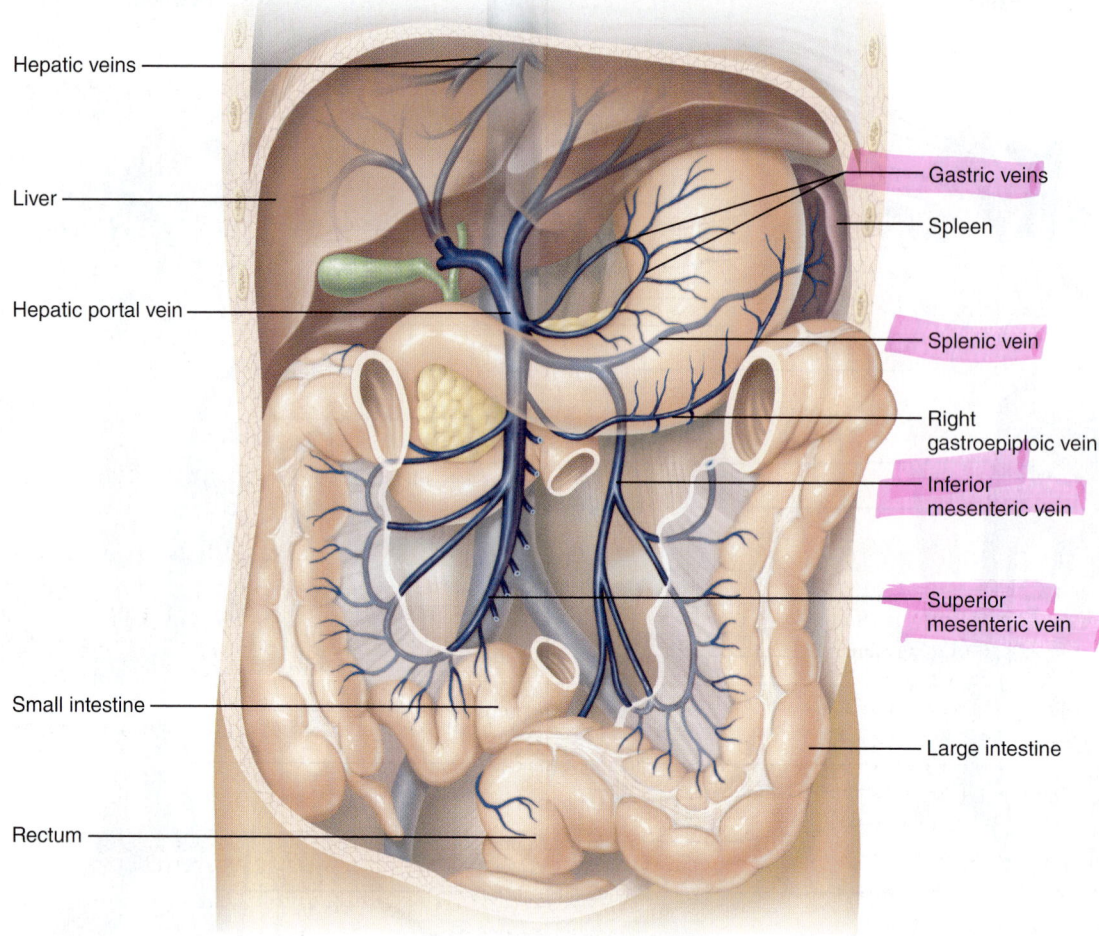

Figure 32.14 Hepatic portal circulation.

oxygenated blood is the umbilical vein. All other vessels contain varying degrees of oxygenated and deoxygenated blood.

At birth, or shortly after, the foramen ovale closes and becomes the **fossa ovalis,** and the ductus arteriosus collapses and is converted to the fibrous **ligamentum arteriosum** (Figure 32.13b). Lack of blood flow through the umbilical vessels leads to their eventual obliteration, and the circulatory pattern becomes that of the adult. Remnants of the umbilical arteries persist as the **medial umbilical ligaments** on the inner surface of the anterior abdominal wall; the occluded umbilical vein becomes the **ligamentum teres** (or **round ligament** of the liver); and the ductus venosus becomes a fibrous band called the **ligamentum venosum** on the inferior surface of the liver.

Hepatic Portal Circulation

Blood vessels of the hepatic portal circulation drain the digestive viscera, spleen, and pancreas and deliver this blood to the liver for processing via the **hepatic portal vein.** If a meal has recently been eaten, the hepatic portal blood will be nutrient-rich. The liver is the key body organ involved in maintaining proper sugar, fatty acid, and amino acid concentrations in the blood, and this system ensures that these substances pass through the liver before entering the systemic circulation. As blood percolates through the liver sinusoids, some of the nutrients are removed to be stored or processed in various

ways for release to the general circulation. At the same time, the hepatocytes are detoxifying alcohol and other possibly harmful chemicals present in the blood, and the liver's macrophages are removing bacteria and other debris from the passing blood. The liver in turn is drained by the hepatic veins that enter the inferior vena cava.

ACTIVITY 6

Tracing the Hepatic Portal Circulation

Locate on the art **(Figure 32.14)**, and on an anatomical chart of the hepatic portal circulation (if available), the vessels named below. ■

The **inferior mesenteric vein,** draining the distal portions of the large intestine, joins the **splenic vein,** which drains the spleen and part of the pancreas and stomach. The splenic vein and the **superior mesenteric vein,** which receives blood from the small intestine and the ascending and transverse colon, unite to form the hepatic portal vein. The **left gastric vein,** which drains the lesser curvature of the stomach, drains directly into the hepatic portal vein.

For instructions on animal dissections, see the dissection exercises (starting on page 697) in the cat and fetal pig editions of this manual.

32

Human Cardiovascular Physiology: Blood Pressure and Pulse Determinations

MATERIALS

☐ Recording of "Interpreting Heart Sounds" (if available on free loan from the local chapters of the American Heart Association) or any of the suitable Internet sites on heart sounds

☐ Stethoscope

☐ Alcohol swabs

☐ Watch (or clock) with second hand

BIOPAC® BIOPAC® BSL System for Windows with BSL software version 3.7.5 to 3.7.7, or BSL System for Mac OS X with BSL software version 3.7.4 to 3.7.7, MP36/35 or MP45 data acquisition unit, PC or Mac computer, BIOPAC® pulse plethysmograph

Instructors using the MP36 (or MP35/30) data acquisition unit with BSL software versions earlier than 3.7.5 (for Windows) and 3.7.4 (for Mac Os X) will need slightly different channel settings and collection strategies. Instructions for using the older data acquisition unit can be found on MasteringA&P.

☐ Sphygmomanometer

☐ Felt marker

☐ Meter stick

☐ Cot (if available)

☐ Step stools (0.4 m [16 in.] and 0.5 m [20 in.] in height)

☐ Small basin suitable for the immersion of one hand

☐ Ice

☐ Laboratory thermometer

PEx PhysioEx™ 9.1 Computer Simulation Ex. 5 on p. PEx-75

Note: Instructions for using PowerLab® equipment can be found on MasteringA&P.

MasteringA&P® For related exercise study tools, go to the Study Area of MasteringA&P. There you will find:

• Practice Anatomy Lab PAL

• PhysioEx PEx

• A&PFlix A&PFlix

• Practice quizzes, Histology Atlas, eText, Videos, and more!

OBJECTIVES

1. Define *systole*, *diastole*, and *cardiac cycle*.

2. Indicate the normal length of the cardiac cycle, the relative pressure changes occurring within the atria and ventricles, the timing of valve closure, and the volume changes occurring in the ventricles during the cycle.

3. Correlate the events of the ECG with the events of the cardiac cycle.

4. Use the stethoscope to auscultate heart sounds, relate heart sounds to cardiac cycle events, and describe the clinical significance of heart murmurs.

5. Demonstrate thoracic locations where the first and second heart sounds are most accurately auscultated.

6. Define *pulse, pulse pressure, pulse deficit, blood pressure, sounds of Korotkoff,* and *MAP*.

7. Determine a subject's apical and radial pulse.

8. Determine a subject's blood pressure with a sphygmomanometer, and relate systolic and diastolic pressures to events of the cardiac cycle.

9. Compare the value of venous pressure to systemic blood pressure, and describe how venous pressure is measured.

10. Investigate the effects of exercise on blood pressure, pulse, and cardiovascular fitness.

11. Indicate factors affecting blood flow and skin color.

PRE-LAB QUIZ

1. Circle the correct underlined term. According to general usage, <u>systole</u> / <u>diastole</u> refers to ventricular relaxation.

2. A graph illustrating the pressure and volume changes during one heartbeat is called the
   a. blood pressure   c. conduction system of the heart
   b. cardiac cycle   d. electrical events of the heartbeat

3. Circle True or False. When ventricular systole begins, intraventricular pressure increases rapidly, closing the atrioventricular (AV) valves.

4. The average heart beats approximately _____ times per minute.
   a. 50   c. 100
   b. 75   d. 125

5. Circle the correct underlined term. Abnormal heart sounds called <u>murmurs</u> / <u>stroke</u> can indicate valve problems.

6. The term _____ refers to the alternating surges of pressure in an artery that occur with each contraction and relaxation of the left ventricle.
   a. diastole   c. pulse
   b. murmur   d. systole

(Text continues on next page.)

7. Circle the correct underlined term. The pulse is most often taken at the lateral aspect of the wrist, above the thumb, by compressing the popliteal / radial artery.

8. What device will you use today to measure your subject's blood pressure?

_____

9. In reporting a blood pressure of 120/90, which number represents the *diastolic* pressure? _____

10. The _____ are characteristic sounds that indicate the resumption of blood flow to the artery being occluded when taking blood pressure.
   a. ectopic heartbeats
   b. heart murmurs
   c. heart rhythms
   d. sounds of Korotkoff

A ny comprehensive study of human cardiovascular physiology takes much more time than a single laboratory period. However, it is possible to conduct investigations of a few phenomena such as pulse, heart sounds, and blood pressure, all of which reflect the heart in action and the function of blood vessels. (The electrocardiogram is studied separately in Exercise 31.) A discussion of the cardiac cycle will provide a basis for understanding and interpreting the various physiological measurements taken.

Cardiac Cycle

In a healthy heart, the two atria contract simultaneously. As they begin to relax, the ventricles contract simultaneously. According to general usage, the terms **systole** and **diastole** refer to events of ventricular contraction and relaxation, respectively. The **cardiac cycle** is equivalent to one complete heartbeat—during which both atria and ventricles contract and then relax. It is marked by a succession of changes in blood volume and pressure within the heart.

The events of the cardiac cycle for the left side of the heart can be graphed in several ways **(Figure 33.1)**. Although pressure changes in the right side are less dramatic than those in the left, the same relationships apply.

We will begin the discussion of the cardiac cycle with the heart in complete relaxation (diastole). At this point, pressure in the heart is very low, blood is flowing passively from the pulmonary and systemic circulations into the atria and on through to the ventricles; the semilunar valves are closed, and the AV valves are open. Shortly, atrial contraction occurs and atrial pressure increases, forcing residual blood into the ventricles. Then ventricular systole begins and intraventricular pressure increases rapidly, closing the AV valves. When ventricular pressure exceeds that of the large arteries leaving the heart, the semilunar valves are forced open, and the blood in the ventricular chambers is expelled through the valves. During this phase, the aortic pressure reaches approximately 120 mm Hg in a healthy young adult. During ventricular systole, the atria relax and their chambers fill with blood, which results in gradually increasing atrial pressure. At the end of ventricular systole, the ventricles relax; the semilunar valves snap shut, preventing backflow, and momentarily, the ventricles are closed chambers. When the aortic (semilunar) valve snaps shut, a momentary increase in the aortic pressure results from the elastic recoil of the aorta after valve closure. This event results in the pressure fluctuation called the *dicrotic notch* (see Figure 33.1a). As the ventricles relax, the pressure

within them begins to drop. When intraventricular pressure is again less than atrial pressure, the AV valves are forced open, and the ventricles again begin to fill with blood. Atrial and aortic pressures decrease, and the ventricles rapidly refill, completing the cycle.

The average heart beats approximately 75 beats per minute, and so the length of the cardiac cycle is about 0.8 second. Of this time period, atrial contraction occupies the first 0.1 second, which is followed by atrial relaxation and ventricular contraction for the next 0.3 second. The remaining 0.4 second is a period of total heart relaxation, the **quiescent period**. When the heart beats at a more rapid pace than normal, the quiescent period decreases.

Notice that two different types of events control the movement of blood through the heart: the alternate contraction and relaxation of the myocardium, and the opening and closing of valves, which is entirely dependent on the pressure changes within the heart chambers.

Study the cardiac cycle illustration (Figure 33.1) carefully to make sure you understand what has been discussed before continuing with the next portion of the exercise.

Heart Sounds

Sounds heard in the cardiovascular system result from turbulent blood flow. Two distinct sounds can be heard during each cardiac cycle. These heart sounds are commonly described by the monosyllables "lub" and "dup"; and the sequence is designated lub-dup, pause, lub-dup, pause, and so on. The first heart sound (lub) is referred to as S_1 and is associated with closure of the AV valves at the beginning of ventricular systole. The second heart sound (dup), called S_2, occurs as the semilunar valves close and corresponds with the end of systole. (Figure 33.1a indicates the correlation of heart sounds with events of the cardiac cycle.)

☐ Listen to the recording "Interpreting Heart Sounds" or another suitable recording so that you may hear both normal and abnormal heart sounds.

Check the box when you have completed the task.

Abnormal heart sounds are called **murmurs** and often indicate valvular problems. In valves that do not close tightly, closure is followed by a swishing sound due to the backflow of blood (regurgitation). Distinct sounds, often described as high-pitched screeching, are associated with the tortuous flow of blood through constricted, or stenosed, valves. +

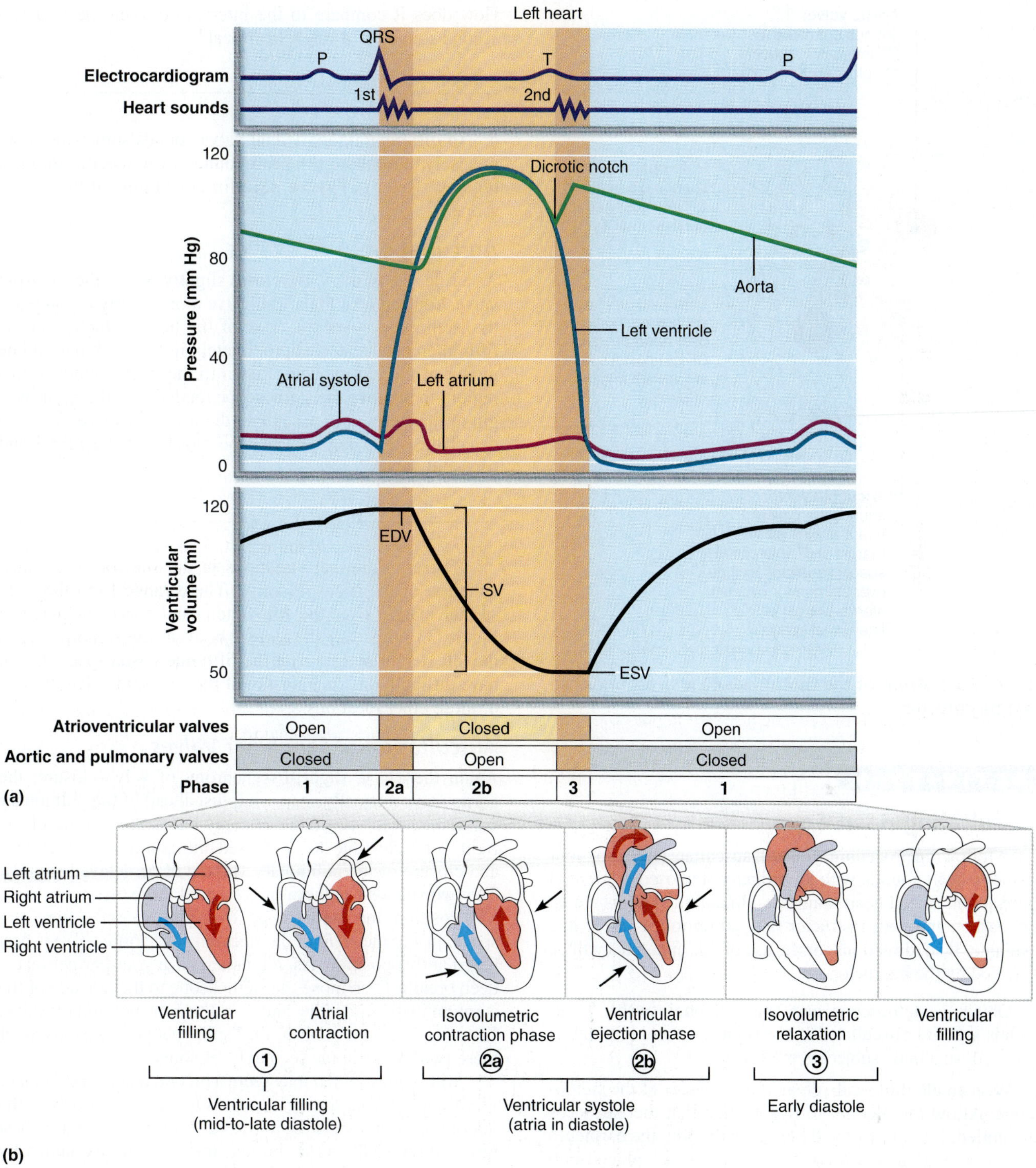

Figure 33.1 Summary of events occurring in the heart during the cardiac cycle.
(a) An ECG tracing is superimposed on the graph (top) so that electrical events can
be related to pressure and volume changes (center) in the left side of the heart.
Pressures are lower in the right side of the heart. Time occurrence of heart sounds is
also indicated. (EDV = end diastolic volume, SV = stroke volume, ESV = end systolic
volume) **(b)** Events of phases 1 through 3 of the cardiac cycle are diagrammed.

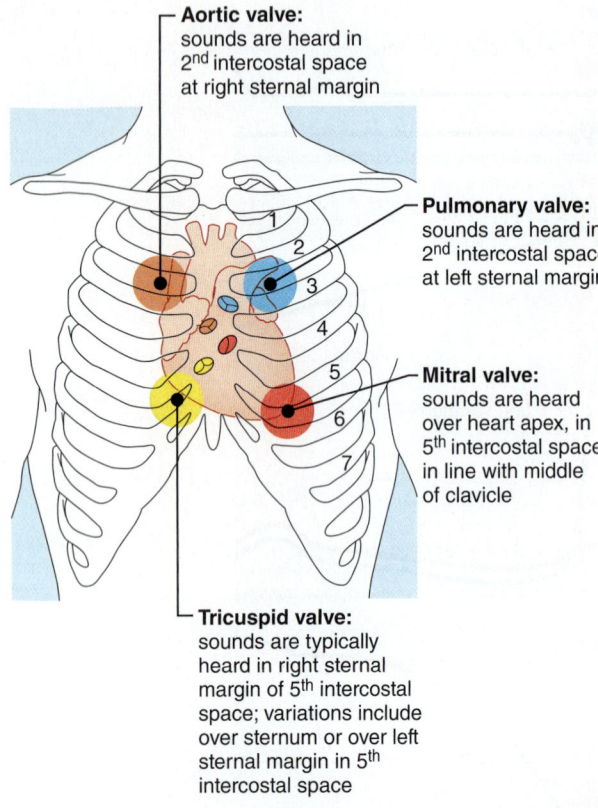

Aortic valve: sounds are heard in 2nd intercostal space at right sternal margin

Pulmonary valve: sounds are heard in 2nd intercostal space at left sternal margin

Mitral valve: sounds are heard over heart apex, in 5th intercostal space in line with middle of clavicle

Tricuspid valve: sounds are typically heard in right sternal margin of 5th intercostal space; variations include over sternum or over left sternal margin in 5th intercostal space

Figure 33.2 Areas of the thorax where heart sounds can best be detected.

ACTIVITY 1

Auscultating Heart Sounds

In the following procedure, you will auscultate (listen to) your partner's heart sounds with an ordinary stethoscope. Several more sophisticated heart-sound amplification systems are on the market, and your instructor may prefer to use one if it is available. If so, directions for the use of this apparatus will be provided by the instructor.

1. Obtain a stethoscope and some alcohol swabs. Heart sounds are best auscultated if the subject's outer clothing is removed, so a male subject is preferable.

2. With an alcohol swab, clean the earpieces of the stethoscope. Allow the alcohol to dry. Notice that the earpieces are angled. For comfort and best auscultation, the earpieces should be angled in a *forward* direction when placed into the ears.

3. Don the stethoscope. Place the diaphragm of the stethoscope on your partner's thorax, just to the sternal side of the left nipple at the fifth intercostal space, and listen carefully for heart sounds. The first sound will be a longer, louder (more booming) sound than the second, which is short and sharp. After listening for a couple of minutes, try to time the pause between the second sound of one heartbeat and the first sound of the subsequent heartbeat.

How long is this interval? _____ sec

How does it compare to the interval between the first and second sounds of a single heartbeat?

_____

4. To differentiate individual valve sounds somewhat more precisely, auscultate the heart sounds over specific thoracic regions. (Refer to **Figure 33.2** for positioning of the stethoscope.)

Auscultation of AV Valves

As a rule, the mitral valve closes slightly before the tricuspid valve. You can hear the mitral valve more clearly if you place the stethoscope over the apex of the heart, which is at the fifth intercostal space, approximately in line with the middle region of the left clavicle. Listen to the heart sounds at this region; then move the stethoscope medially to the right margin of the sternum to auscultate the tricuspid valve. Can you detect the slight lag between the closure of the mitral and tricuspid valves?

_____

There are normal variations in the site for "best" auscultation of the tricuspid valve. These range from the right sternal margin over the fifth intercostal space (depicted in Figure 33.2) to over the sternal body in the same plane, to the left sternal margin over the fifth intercostal space. If you have difficulty hearing closure of the tricuspid valve, try one of these other locations.

Auscultation of Semilunar Valves

Again there is a slight dissynchrony of valve closure; the aortic valve normally snaps shut just ahead of the pulmonary valve. If the subject inhales deeply but gently, filling of the right ventricle (due to decreased intrapulmonary pressure) and closure of the pulmonary valve will be delayed slightly. The two sounds can therefore be heard more distinctly.

Position the stethoscope over the second intercostal space, just to the *right* of the sternum. The aortic valve is best heard at this position. As you listen, have your partner take a deep breath. Then move the stethoscope to the *left* side of the sternum in the same line, and auscultate the pulmonary valve. Listen carefully; try to hear the "split" between the closure of these two valves in the second heart sound.

Although at first it may seem a bit odd that the pulmonary valve issuing from the *right* heart is heard most clearly to the *left* of the sternum and the aortic valve of the left heart is best heard at the right sternal border, this is easily explained by reviewing heart anatomy. Because the heart is twisted, with the right ventricle forming most of the anterior ventricular surface, the pulmonary trunk actually crosses to the left as it issues from the right ventricle. Similarly, the aorta issues from the left ventricle at the left side of the pulmonary trunk before arching up and over that vessel. ■

The Pulse

The term **pulse** refers to the alternating surges of pressure (expansion and then recoil) in an artery that occur with each contraction and relaxation of the left ventricle. This difference between systolic and diastolic pressure is called the **pulse pressure.** (See page 499.) Normally the pulse rate (pressure surges per minute) equals the heart rate (beats per minute), and the pulse averages 70 to 76 beats per minute in the resting state.

Parameters other than pulse rate are also useful clinically. You may also assess the regularity (or rhythmicity) of the pulse, and its amplitude and/or tension—does the blood vessel expand and recoil (sometimes visibly) with the pressure waves? Can you feel it strongly, or is it difficult to detect? Is it regular like the ticking of a clock, or does it seem to skip beats?

ACTIVITY 2

Palpating Superficial Pulse Points

The pulse may be felt easily on any superficial artery when the artery is compressed over a bone or firm tissue. Palpate the following pulse or pressure points on your partner by placing the fingertips of the first two or three fingers of one hand over the artery. It helps to compress the artery firmly as you begin your palpation and then immediately ease up on the pressure slightly. In each case, notice the regularity of the pulse, and assess the degree of tension or amplitude. (**Figure 33.3** illustrates the superficial pulse points to be palpated.) Check off the boxes as you locate each pulse point.

☐ **Superficial temporal artery:** Anterior to the ear, in the temple region.

☐ **Facial artery:** Clench the teeth, and palpate the pulse just anterior to the masseter muscle on the mandible (in line with the corner of the mouth).

☐ **Common carotid artery:** At the side of the neck.

☐ **Brachial artery:** In the cubital fossa, at the point where it bifurcates into the radial and ulnar arteries.

☐ **Radial artery:** At the lateral aspect of the wrist, above the thumb.

☐ **Femoral artery:** In the groin.

☐ **Popliteal artery:** At the back of the knee.

☐ **Posterior tibial artery:** Just above the medial malleolus.

☐ **Dorsalis pedis artery:** On the dorsum of the foot.

Which pulse point had the greatest amplitude?

_____

Which had the least? _____

Can you offer any explanation for this? _____

_____

_____

_____

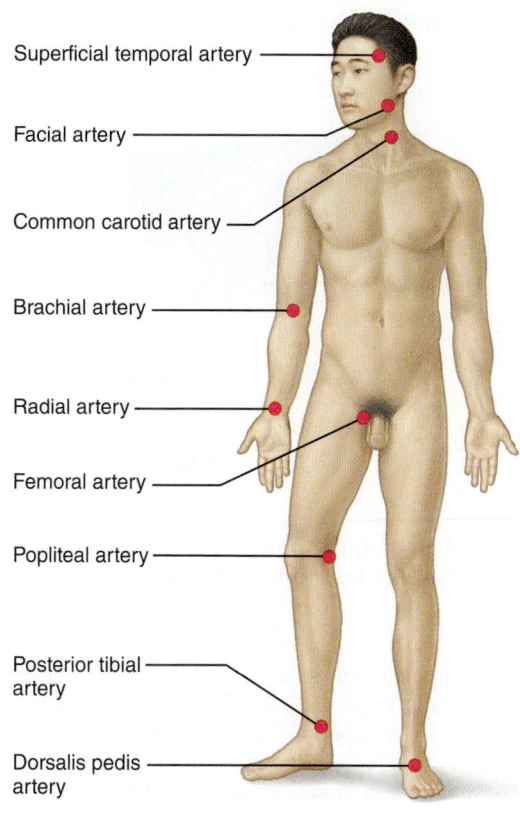

Figure 33.3 Body sites where the pulse is most easily palpated.

Because of its easy accessibility, the pulse is most often taken on the radial artery. With your partner sitting quietly, practice counting the radial pulse for 1 minute. Make three counts and average the results.

count 1 _____count 2 _____

count 3 _____average _____ ■ 33

Due to the elasticity of the arteries, blood pressure decreases and smooths out as blood moves farther away from the heart. A pulse, however, can still be felt in the fingers. A device called a plethysmograph or a piezoelectric pulse transducer can measure this pulse.

ACTIVITY 3

Measuring Pulse Using BIOPAC®
Setting Up the Equipment

1. Connect the BIOPAC® unit to the computer and turn the computer **ON.**

2. Make sure the BIOPAC® unit is **OFF.**

3. Plug in the equipment (as shown in **Figure 33.4**).

- Pulse transducer (plethysmograph)—CH 2

4. Turn the BIOPAC® unit **ON.**

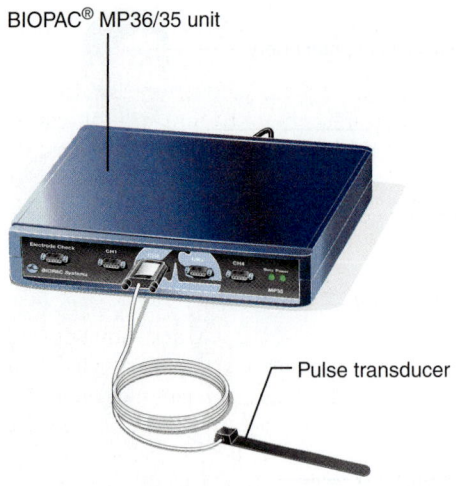

BIOPAC® MP36/35 unit

Pulse transducer

Figure 33.4 Setting up the BIOPAC®equipment.
Plug the pulse tranducer into Channel 2.

5. Wrap the pulse transducer around the tip of the index finger (as shown in **Figure 33.5**). Wrap the Velcro around the finger gently (if wrapped too tight, it will reduce circulation to the finger and obscure the recording). *Do not wiggle the finger or move the plethysmograph cord during recording.*

6. Have the subject sit down with the forearms supported and relaxed.

7. Start the BIOPAC® Student Lab program on the computer by double-clicking the icon on the desktop or by following your instructor's guidance.

8. Select lesson **L07-ECG&P-1** from the menu, and click **OK.**

9. Type in a filename that will save this subject's data on the computer hard drive. You may want to use the subject's last name followed by Pulse (for example, SmithPulse-1), then click **OK.**

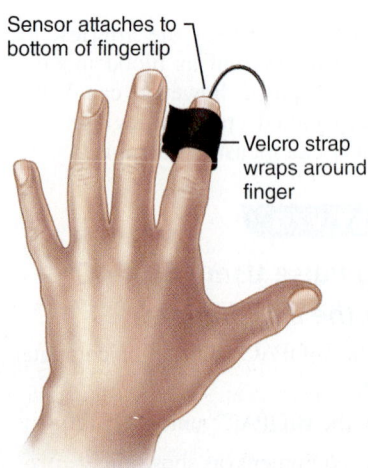

Sensor attaches to bottom of fingertip

Velcro strap wraps around finger

Figure 33.5 Placement of the pulse transducer around the tip of the index finger.

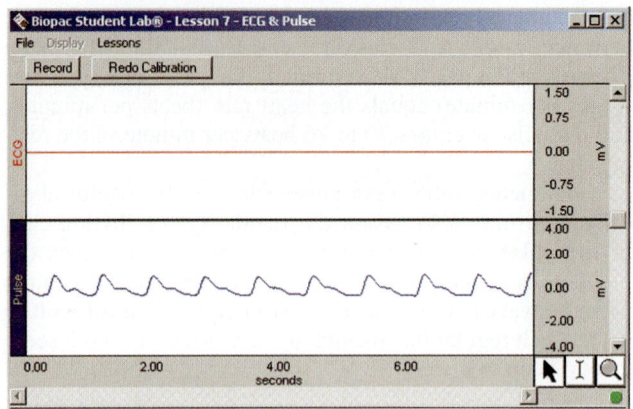

Figure 33.6 Example of waveforms during the calibration procedure.

Calibrating the Equipment

1. When the subject is relaxed, click **Calibrate.** Click **Ignore** when prompted for SS2L on Ch.1. The calibration will stop automatically after 8 seconds.

2. Observe the recording of the calibration data, which should look like the waveforms in the example **(Figure 33.6)**. ECG data is not recorded in this activity.

- If the data look very different, click **Redo Calibration** and repeat the steps above. If there is no signal, you may need to loosen the Velcro on the finger and check all attachments.

- If the data look similar, proceed to the next section.

Recording the Data

1. When the subject is ready, click **Record** to begin recording the pulse. After 30 seconds, click **Suspend.**

2. Observe the data, which should look similar to the pulse data example **(Figure 33.7)**.

- If the data look very different, click **Redo** and repeat the steps above.

- If the data look similar, go to the next step.

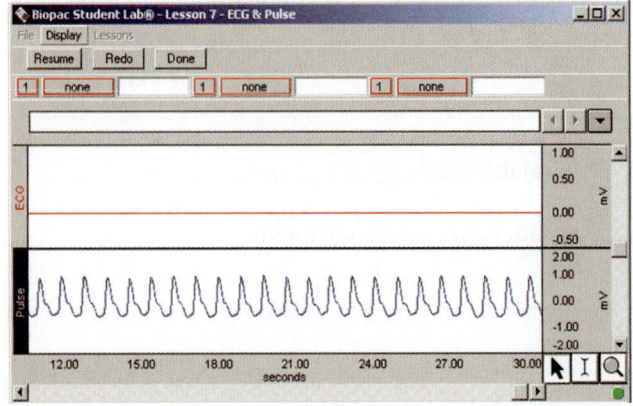

Figure 33.7 Example of waveforms during the recording of data.

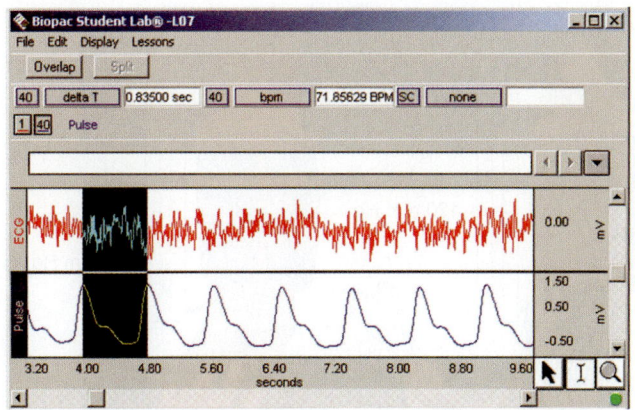

Figure 33.8 Using the I-beam cursor to highlight data for analysis.

3. When you are finished, click **Done** and then click **Yes.** A pop-up window will appear; to record from another subject select **Record from another subject** and return to step 5 under Setting Up the Equipment. If continuing to the Data Analysis section, select **Analyze current data file** and proceed to step 2 in the Data Analysis section.

Data Analysis

1. If just starting the BIOPAC® program to perform data analysis, enter **Review Saved Data** mode and choose the file with the subject's pulse data (for example, SmithPulse-1).

2. Observe that pulse data is in the lower scale.

3. To analyze the data, set up the first channel/measurement box at the top of the screen.

Channel	Measurement	Data
CH 40	Delta T	pulse

Delta T: Measures the time elapsed in the selected area

4. Use the arrow cursor and click the I-beam cursor box on the lower right side of the screen to activate the "area selection" function. Using the activated I-beam cursor, highlight from the peak of one pulse to the peak of the next pulse (as shown in **Figure 33.8**).

Observe the elapsed time between heartbeats and record here

(to the nearest 0.01 second): _____

5. Calculate the beats per minute by inserting the elapsed time into this formula:

$$(1 \text{ beat}/ \underline{\hspace{2cm}} \text{ sec}) \times (60 \text{ sec/min})$$

$$= \underline{\hspace{2cm}} \text{ beats/min}$$

Optional Activity with BIOPAC® Pulse Measurement

To expand the experiment, you can measure the effects of heat and cold on pulse rate. To do this, you can submerge the subject's hand (the hand without the plethysmograph!) in hot and/or ice water for 2 minutes, and then record the pulse. Alternatively, you can investigate change in pulse rate after brief exercise such as jogging in place. ▬

Apical-Radial Pulse

The correlation between the apical and radial pulse rates can be determined by simultaneously counting them. The **apical pulse** (actually the counting of heartbeats) may be slightly faster than the radial because of a slight lag in time as the blood rushes from the heart into the large arteries where it can be palpated. However, any *large* difference between the values observed, referred to as a **pulse deficit,** may indicate cardiac impairment (a weakened heart that is unable to pump blood into the arterial tree to a normal extent), low cardiac output, or abnormal heart rhythms. In the case of atrial fibrillation or ectopic heartbeats, for instance, the second beat may follow the first so quickly that no second pulse is felt even though the apical pulse can still be auscultated. Apical pulse counts are routinely ordered for those with cardiac decompensation.

ACTIVITY 4

Taking an Apical Pulse

With the subject sitting quietly, one student, using a stethoscope, should determine the apical pulse rate while another simultaneously counts the radial pulse rate. The stethoscope should be positioned over the fifth left intercostal space. The person taking the radial pulse should determine the starting point for the count and give the stop-count signal exactly 1 minute later. Record your values below.

apical count _____ beats/min

radial count _____ pulses/min

pulse deficit _____ pulses/min ▬

Blood Pressure Determinations

Blood pressure (BP) is defined as the pressure the blood exerts against any unit area of the blood vessel walls, and it is generally measured in the arteries. Because the heart alternately contracts and relaxes, the resulting rhythmic flow of blood into the arteries causes the blood pressure to rise and fall during each beat. Thus you must take two blood pressure readings: the **systolic pressure,** which is the pressure in the arteries at the peak of ventricular ejection, and the **diastolic pressure,** which reflects the pressure during ventricular relaxation. Blood pressures are reported in millimeters of mercury (mm Hg), with the systolic pressure appearing first; 120/80 translates to 120 over 80, or a systolic pressure of 120 mm Hg and a diastolic pressure of 80 mm Hg. Normal blood pressure varies considerably from one person to another.

33

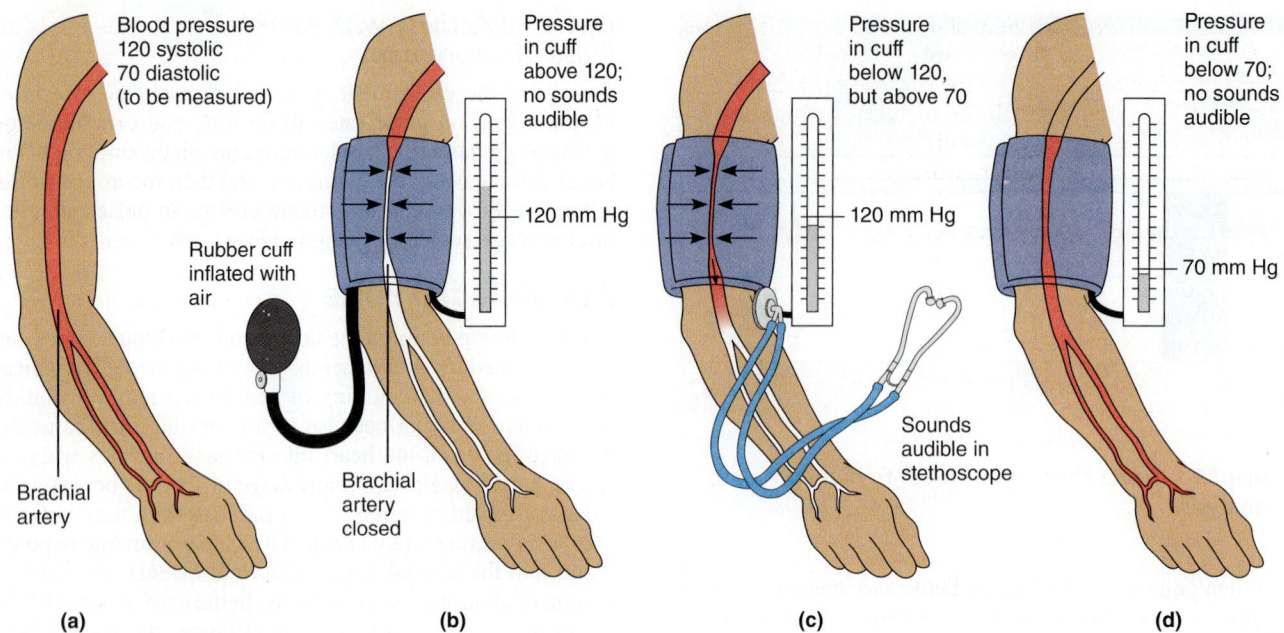

Blood pressure
120 systolic
70 diastolic
(to be measured)

Rubber cuff
inflated with
air

Brachial
artery

Brachial
artery
closed

Pressure
in cuff
above 120;
no sounds
audible

120 mm Hg

Pressure
in cuff
below 120,
but above 70

120 mm Hg

Sounds
audible in
stethoscope

Pressure
in cuff
below 70;
no sounds
audible

70 mm Hg

(a)   (b)   (c)   (d)

Figure 33.9 Procedure for measuring blood pressure. (a) The course of the brachial artery of the arm. Assume a blood pressure of 120/70. **(b)** The blood pressure cuff is wrapped snugly around the arm just above the elbow and inflated until blood flow into the forearm is stopped and no brachial pulse can be felt or heard. **(c)** Pressure in the cuff is gradually reduced while the examiner listens (auscultates) for sounds (of Korotkoff) in the brachial artery with a stethoscope. The pressure, read as the first soft tapping sounds are heard (the first point at which a small amount of blood is spurting through the constricted artery), is recorded as the systolic pressure. **(d)** As the pressure is reduced still further, the sounds become louder and more distinct, but when the artery is no longer restricted and blood flows freely, the sounds can no longer be heard. The pressure at which the sounds disappear is routinely recorded as the diastolic pressure.

In this procedure, you will measure arterial pressure by indirect means and under various conditions. You will investigate and demonstrate factors affecting blood pressure, and the rapidity of blood pressure changes.

ACTIVITY 5

Using a Sphygmomanometer to Measure Arterial Blood Pressure Indirectly

The **sphygmomanometer,** commonly called a *blood pressure cuff,* is an instrument used to obtain blood pressure readings by the auscultatory method **(Figure 33.9)**. It consists of an inflatable cuff with an attached pressure gauge. The cuff is placed around the arm and inflated to a pressure higher than systolic pressure to occlude circulation to the forearm. As cuff pressure is gradually released, the examiner listens with a stethoscope for characteristic sounds called the **sounds of Korotkoff,** which indicate the resumption of blood flow into the forearm. The pressure at which the first soft tapping sounds can be detected is recorded as the systolic pressure. As the pressure is reduced further, blood flow becomes more turbulent, and the sounds become louder. As the pressure is reduced still further, below the diastolic pressure, the artery is no longer compressed; and blood flows freely and without turbulence. At this point, the sounds of Korotkoff can no longer be detected.

The pressure at which the sounds disappear is recorded as the diastolic pressure.

1. Work in pairs to obtain radial artery blood pressure readings. Obtain a felt marker, stethoscope, alcohol swabs, and a sphygmomanometer. Clean the earpieces of the stethoscope with the alcohol swabs, and check the cuff for the presence of trapped air by compressing it against the laboratory table. (A partially inflated cuff will produce erroneous measurements.)

2. The subject should sit in a comfortable position with one arm resting on the laboratory table (approximately at heart level if possible). Wrap the cuff around the subject's arm, just above the elbow, with the inflatable area on the medial arm surface. The cuff may be marked with an arrow; if so, the arrow should be positioned over the brachial artery (Figure 33.9). Secure the cuff by tucking the distal end under the wrapped portion or by bringing the Velcro areas together.

3. Palpate the brachial pulse, and lightly mark its position with a felt pen. Don the stethoscope, and place its diaphragm over the pulse point.

⚠️ *The cuff should not be kept inflated for more than 1 minute.* If you have any trouble obtaining a reading within this time, deflate the cuff, wait 1 or 2 minutes, and try again. (A prolonged interference with BP homeostasis can lead to fainting.)

33

4. Inflate the cuff to approximately 160 mm Hg pressure, and slowly release the pressure valve. Watch the pressure gauge as you listen carefully for the first soft thudding sounds of the blood spurting through the partially occluded artery. Mentally note this pressure (systolic pressure), and continue to release the cuff pressure. You will notice first an increase, then a muffling, of the sound. For the diastolic pressure, note the pressure at which the sound becomes muffled or disappears. Controversy exists over which of the two points should be recorded as the diastolic pressure; so in some cases you may see readings such as 120/80/78, which indicates the systolic pressure followed by the *first* and *second diastolic end points.* The first diastolic end point is the pressure at which the sound muffles; the second is the pressure at which the sound disappears. It makes little difference here which of the two diastolic pressures is recorded, but be consistent. Make two blood pressure determinations, and record your results below.

First trial: **Second trial:**

systolic pressure _____   systolic pressure _____

diastolic pressure _____   diastolic pressure _____

5. Compute the **pulse pressure** for each trial. The pulse pressure is the difference between the systolic and diastolic pressures, and it indicates the amount of blood forced from the heart during systole, or the actual "working" pressure. A narrowed pulse pressure (less than 30 mm Hg) may be a signal of severe aortic stenosis, constrictive pericarditis, or tachycardia. A widened pulse pressure (over 40 mm Hg) is common in hypertensive individuals.

Pulse pressure:

first trial _____   second trial _____

6. Compute the **mean arterial pressure (MAP)** for each trial using the following equation:

$$\text{MAP} = \text{diastolic pressure} + \frac{\text{pulse pressure}}{3}$$

first trial _____   second trial _____ ▪

ACTIVITY 6

Estimating Venous Pressure

It is not possible to measure venous pressure with the sphygmomanometer. The methods available for measuring it produce estimates at best, because venous pressures are so much lower than arterial pressures. The difference in pressure becomes obvious when these vessels are cut. If a vein is cut, the blood flows evenly from the cut. A lacerated artery produces rapid spurts of blood. In this activity, you will estimate venous pressures.

1. Obtain a meter stick, and ask your lab partner to stand with his or her right side toward the blackboard, arms hanging freely at the sides. On the board, mark the approximate level

of the right atrium. (This will be just slightly higher than the point at which you ausculated the apical pulse.)

2. Observe the superficial veins on the dorsum of the right hand as the subject alternately raises and lowers it. Notice the collapsing and filling of the veins as internal pressures change. Have the subject repeat this action until you can determine the point at which the veins have just collapsed. Mark this hand level on the board. Then measure, in millimeters, the distance in the vertical plane from this point to the level of the right atrium (previously marked). Record this value. Distance of right arm from right atrium at point of venous collapse:

_____ mm

3. Compute the venous pressure (P_V), in millimeters of mercury, with the following formula:

$$P_v = \frac{1.056 \text{ (specific gravity of blood)} \times \text{mm (measured)}}{13.6 \text{ (specific gravity of Hg)}}$$

Venous pressure computed: _____ mm Hg

Normal venous pressure varies from approximately 30 to 90 mm Hg. That of the hand ranges between 30 and 40 mm Hg. How does your computed value compare?

_____

4. Because venous walls are so thin, pressure within them is readily affected by external factors such as muscle activity, deep pressure, and pressure changes occurring in the thorax during breathing. The Valsalva maneuver, which increases intrathoracic pressure, is used to demonstrate the effect of thoracic pressure changes on venous pressure.

To perform this maneuver take a deep breath, and then mimic the motions of exhaling forcibly, but without actually exhaling. In reaction to this, the glottis will close; and intrathoracic pressure will increase. (Most of us have performed this maneuver unknowingly in acts of defecation in which there is "straining at stool.") Have the same subject again stand next to the blackboard mark for the level of his or her right atrium. While the subject performs the Valsalva maneuver and raises and lowers one hand, determine the point of venous collapse and mark it on the board. Measure the distance of that mark from the right atrium level and record it below. Then compute the venous pressure and record it.

_____ mm   Venous pressure: _____ mm Hg

How does this value compare with the venous pressure measurement computed for the relaxed state?

_____

Explain: _____

_____

_____ ▪

33

Posture				
	Trial 1		Trial 2	
	BP	Pulse	BP	Pulse
Sitting quietly				
Reclining (after 2 to 3 min)				
Immediately on standing from the reclining position ("at attention" stance)				
After standing for 3 min				

ACTIVITY 7

Observing the Effect of Various Factors on Blood Pressure and Heart Rate

Arterial blood pressure is directly proportional to cardiac output (CO, amount of blood pumped out of the left ventricle per unit time) and peripheral resistance (PR) to blood flow, that is,

$$BP = CO \times PR$$

Peripheral resistance is increased by blood vessel constriction (most importantly the arterioles), by an increase in blood viscosity, and by a loss of elasticity of the arteries (seen in arteriosclerosis). Any factor that increases either the cardiac output or the peripheral resistance causes an almost immediate reflex rise in blood pressure. A close examination of these relationships reveals that many factors—age, weight, time of day, exercise, body position, emotional state, and various drugs, for example—alter blood pressure. The influence of a few of these factors is investigated here.

The following tests are done most efficiently if one student acts as the subject; two are examiners (one taking the radial pulse and the other auscultating the brachial blood pressure); and a fourth student collects and records data. The sphygmomanometer cuff should be left on the subject's arm throughout the experiments (in a deflated state, of course) so that, at the proper times, the blood pressure can be taken quickly. In each case, take the measurements at least twice. For each of the following tests, students should formulate hypotheses, collect data, and write lab reports. (See Getting Started, page xiv.) Conclusions should be shared with the class.

Posture

To monitor circulatory adjustments to changes in position, take blood pressure and pulse measurements under the conditions noted in the **Posture chart** above. Record your results on the chart.

Exercise

Blood pressure and pulse changes during and after exercise provide a good yardstick for measuring one's overall cardiovascular fitness. Although there are more sophisticated and more accurate tests that evaluate fitness according to a specific point system, the *Harvard step test* described here is a quick way to compare the relative fitness level of a group of people.

You will be working in groups of four, duties assigned as indicated above, except that student 4, in addition to recording the data, will act as the timer and call the cadence.

 Any student with a known heart problem should refuse to participate as the subject.

All four students may participate as the subject in turn, if desired, but the bench stepping is to be performed *at least twice* in each group—once with a well-conditioned person acting as the subject, and once with a poorly conditioned subject.

Bench stepping is the following series of movements repeated sequentially:

1. Place one foot on the step.

2. Step up with the other foot so that both feet are on the platform. Straighten the legs and the back.

3. Step down with one foot.

4. Bring the other foot down.

The pace for the stepping will be set by the "timer" (student 4), who will repeat "Up-2-3-4, up-2-3-4" at such a pace that each "up-2-3-4" sequence takes 2 sec (30 cycles/min).

1. Student 4 should obtain the step (0.5 m [20-in.] height for male subject or 0.4 m [16-in.] for a female subject) while baseline measurements are being obtained on the subject.

2. Once the baseline pulse and blood pressure measurements have been recorded on the **Exercise chart** (next page), the subject is to stand quietly at attention for 2 minutes to allow his or her blood pressure to stabilize before beginning to step.

3. The subject is to perform the bench stepping for as long as possible, up to a maximum of 5 minutes, according to the cadence called by the timer. The subject is to be watched for and warned against crouching (posture must remain erect). If he or she is unable to keep the pace for a span of 15 seconds, the test is to be terminated.

4. When the subject is stopped by the pacer for crouching, stops voluntarily because he or she is unable to continue, or has completed 5 minutes of bench stepping, he or she is to sit down. The duration of exercise (in seconds) is to be recorded, and the blood pressure and pulse are to be measured immediately and thereafter at 1-minute intervals for 3 minutes post-exercise.

Duration of exercise: _____ sec

Exercise										
			Interval Following Test							
	Baseline		Immediately		1 min		2 min		3 min	
Harvard step test for 5 min at 30/min	BP	P	BP	P	BP	P	BP	P	BP	P
Well-conditioned individual	___	___	___	___	___	___	___	___	___	___
Poorly conditioned individual	___	___	___	___	___	___	___	___	___	___

5. The subject's *index of physical fitness* is to be calculated using the following formula:

$$\text{Index} = \frac{\text{duration of exercise in seconds} \times 100}{2 \times \text{sum of the three pulse counts in recovery}}$$

Scores are interpreted according to the following scale:

below 55	poor physical condition
55 to 62	low average
63 to 71	average
72 to 79	high average
80 to 89	good
90 and over	excellent

6. Record the test values on the Exercise chart above, and repeat the testing and recording procedure with the second subject.

When did you notice a greater elevation of blood pressure and pulse?

_____

Explain: _____

_____

Was there a sizable difference between the after-exercise values for well-conditioned and poorly conditioned individuals?

_____ Explain: _____

_____

Did the diastolic pressure also increase? _____

Explain: _____

_____

A Noxious Sensory Stimulus (Cold)

Blood pressure can be affected by emotions and pain. This lability of blood pressure will be investigated through use of the **cold pressor test,** in which one hand will be immersed in unpleasantly (even painfully) cold water.

1. Measure the blood pressure and pulse of the subject as he or she sits quietly. Record these as the baseline values on the **Noxious Sensory Stimulus chart** (next page).

2. Obtain a basin and thermometer, fill the basin with ice cubes, and add water. When the temperature of the ice bath has reached 5°C, immerse the subject's other hand (the non-cuffed limb) in the ice water. With the hand still immersed, take blood pressure and pulse readings at 1-minute intervals for a period of 3 minutes, and record the values on the chart.

How did the blood pressure change during cold exposure?

_____

_____

Was there any change in pulse? _____

3. Subtract the respective baseline readings of systolic and diastolic blood pressure from the highest single reading of systolic and diastolic pressure obtained during cold immersion. (For example, if the highest experimental reading is 140/88 and the baseline reading is 120/70, then the differences in blood pressure would be systolic pressure, 20 mm Hg, and diastolic pressure, 18 mm Hg.) These differences are called the index of response. According to their index of response, subjects can be classified as follows:

Hyporeactors (stable blood pressure): Exhibit a rise of diastolic and/or systolic pressure ranging from 0 to 22 mm Hg or a drop in pressures

Hyperreactors (labile blood pressure): Exhibit a rise of 23 mm Hg or more in the diastolic and/or systolic blood pressure

Is the subject tested a hypo- or hyperreactor?

_____

33

A Noxious Sensory Stimulus (Cold)							
Baseline		1 min		2 min		3 min	
BP	P	BP	P	BP	P	BP	P

Skin Color as an Indicator of Local Circulatory Dynamics

Skin color reveals with surprising accuracy the state of the local circulation, and allows inferences concerning the larger blood vessels and the circulation as a whole. The Activity 8 experiments on local circulation illustrate a number of factors that affect blood flow to the tissues.

Clinical expertise often depends upon good observation skills, accurate recording of data, and logical interpretation of the findings. A single example will be given to demonstrate this statement: A massive hemorrhage may be internal and hidden (thus, not obvious) but will still threaten the blood delivery to the brain and other vital organs. One of the earliest compensatory responses of the body to such a threat is constriction of cutaneous blood vessels, which reduces blood flow to the skin and diverts it into the circulatory mainstream to serve other, more vital tissues. As a result, the skin of the face and particularly of the extremities becomes pale, cold, and eventually moist with perspiration. Therefore, pale, cold, clammy skin should immediately lead the careful diagnostician to suspect that the circulation is dangerously inefficient. Other conditions, such as local arterial obstruction and venous congestion, as well as certain pathologies of the heart and lungs, also alter skin texture, color, and circulation in characteristic ways.

ACTIVITY 8

Examining the Effect of Local Chemical and Physical Factors on Skin Color

The local blood supply to the skin (indeed, to any tissue) is influenced by (1) local metabolites, (2) oxygen supply, (3) local temperature, (4) autonomic nervous system impulses, (5) local vascular reflexes, (6) certain hormones, and (7) substances released by injured tissues. A number of these factors are examined in the simple experiments that follow. Each experiment should be conducted by students in groups of three or four. One student will act as the subject; the others will conduct the tests and make and record observations.

Vasodilation and Flushing of the Skin Due to Local Metabolites

1. Obtain a sphygmomanometer (blood pressure cuff) and stethoscope. You will also need a watch or clock with a second hand.

2. The subject should bare both arms by rolling up the sleeves as high as possible and then lay the forearms side by side on the bench top.

3. Observe the general color of the subject's forearm skin, and the normal contour and size of the veins. Notice whether skin color is bilaterally similar. Record your observations:

_____

_____

_____

4. Apply the blood pressure cuff to one arm, and inflate it to 250 mm Hg. Keep it inflated for 1 minute. During this period, repeat the observations made above, and record the results:

_____

_____

5. Release the pressure in the cuff (leaving the deflated cuff in position), and again record the forearm skin color and the condition of the forearm veins. Make this observation immediately after deflation and then again 30 seconds later.

Immediately after deflation: _____

_____

30 sec after deflation: _____

_____

The above observations constitute your baseline information. Now conduct the following tests.

6. Instruct the subject to raise the cuffed arm above his or her head and to clench the fist as tightly as possible. While the hand and forearm muscles are tightly contracted, rapidly inflate the cuff to 240 mm Hg or more. This maneuver partially empties the hand and forearm of blood and stops most blood flow to the hand and forearm. Once the cuff has been inflated, the subject is to relax the fist and return the forearm to the bench top so that it can be compared to the other forearm.

7. Leave the cuff inflated for exactly 1 minute. During this interval, compare the skin color in the "ischemic" (blood-deprived) hand to that of the "normal" (non-cuffed-limb) hand. Quickly release the pressure immediately after 1 minute.

What are the subjective effects (sensations felt by the subject, such as pain, cold, warmth, tingling, weakness) of stopping blood flow to the arm and hand for 1 minute? These sensations are "symptoms" of a change in function.

_____

_____

What are the objective effects (color of skin and condition of veins)?

_____

_____

How long does it take for the subject's ischemic hand to regain its normal color?

_____

Effects of Venous Congestion

1. Again, but with a different subject, observe and record the appearance of the skin and veins on the forearms resting on the bench top. This time, pay particular attention to the color of the fingers, particularly the distal phalanges, and the nail beds. Record this information:

_____

_____

2. Wrap the blood pressure cuff around one of the subject's arms, and inflate it to 40 mm Hg. Maintain this pressure for 5 minutes. Make a record of the subjective and objective findings just before the 5 minutes are up, and then again immediately after release of the pressure at the end of 5 minutes.

Subjective (arm cuffed): _____

_____

Objective (arm cuffed): _____

_____

Subjective (pressure released): _____

_____

Objective (pressure released): _____

_____

3. With still another subject, conduct the following simple experiment: Raise one arm above the head, and let the other hang by the side for 1 minute. After 1 minute, quickly lay both arms on the bench top, and compare their color.

Color of raised arm: _____

Color of dependent arm: _____

From this and the two preceding observations, analyze the factors that determine tint of color (pink or blue) and intensity of skin color (deep pink or blue as opposed to light pink or blue). Record your conclusions.

_____

_____

_____

_____

_____

Collateral Blood Flow

In some diseases, blood flow to an organ through one or more arteries may be completely and irreversibly obstructed. Fortunately, in most cases a given body area is supplied both by one main artery and by anastomosing channels connecting the main artery with one or more neighboring blood vessels. Consequently, an organ may remain viable even though its main arterial supply is occluded, as long as the **collateral vessels** are still functional.

The effectiveness of collateral blood flow in preventing ischemia can be easily demonstrated.

1. Check the subject's hands to be sure they are _warm_ to the touch. If not, choose another subject, or warm the subject's hands in 35°C water for 10 minutes before beginning.

2. Palpate the subject's radial and ulnar arteries approximately 2.5 cm (1 in.) above the wrist flexure, and mark their locations with a felt marker.

3. Instruct the subject to supinate one forearm and to hold it in a partially flexed (about a 30° angle) position, with the elbow resting on the bench top.

4. Face the subject and grasp his or her forearm with both of your hands, the thumb and fingers of one hand compressing the marked radial artery and the thumb and fingers of the other hand compressing the ulnar artery. Maintain the pressure for 5 minutes, noticing the progression of the subject's hand to total ischemia.

5. At the end of 5 minutes, release the pressure abruptly. Record the subject's sensations, as well as the intensity and duration of the flush in the previously occluded hand. (Use the other hand as a baseline for comparison.)

_____

_____

_____

33

6. Allow the subject to relax for 5 minutes; then repeat the maneuver, but this time *compress only the radial artery.* Record your observations.

_____

_____

How do the results of the first test differ from those of the second test with respect to color changes during compression and to the intensity and duration of reactive hyperemia (redness of the skin)?

_____

_____

_____

_____

7. Once again allow the subject to relax for 5 minutes. Then repeat the maneuver, *with only the ulnar artery compressed.* Record your observations:

_____

_____

What can you conclude about the relative sizes of, and hand areas served by, the radial and ulnar arteries?

_____

_____

Effect of Mechanical Stimulation of Blood Vessels of the Skin

With moderate pressure, draw the blunt end of your pen across the skin of a subject's forearm. Wait 3 minutes to observe the effects, and then repeat with firmer pressure.

What changes in skin color do you observe with light-to-moderate pressure?

_____

With heavy pressure? _____

The redness, or *flare,* observed after mechanical stimulation of the skin results from a local inflammatory response promoted by chemical mediators released by injured tissues. These mediators stimulate increased blood flow into the area and leaking of fluid (from the capillaries) into the local tissues. (**Note:** People differ considerably in skin sensitivity. Those most sensitive will show **dermatographism,** a condition in which the direct line of stimulation will swell quite obviously. This excessively swollen area is called a *wheal*.) ◼

33

Human Cardiovascular Physiology: Blood Pressure and Pulse Determinations

Cardiac Cycle

1. Using the grouped sets of terms to the right of the diagram, correctly identify each trace, valve closings and openings, and each time period of the cardiac cycle.

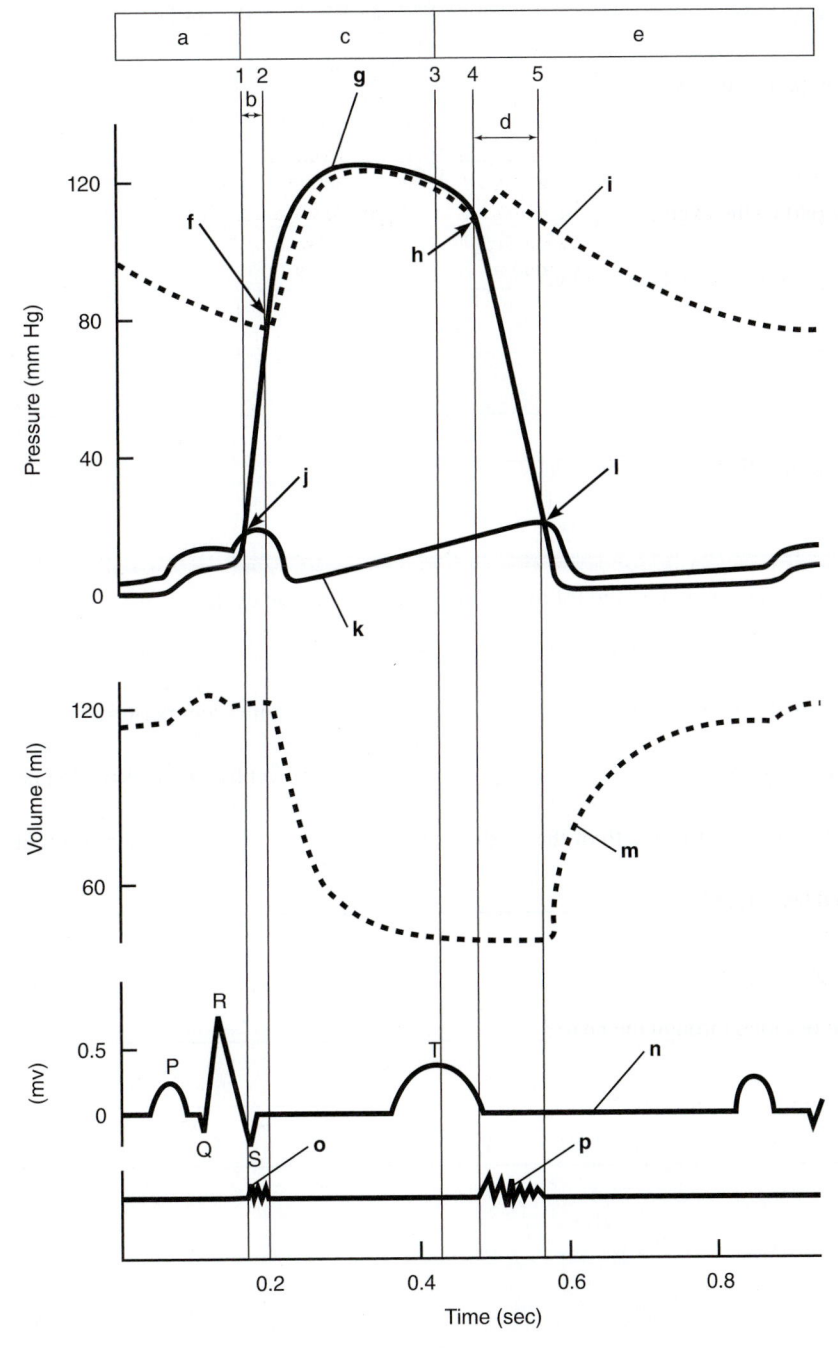

_____ 1. aortic pressure

_____ 2. atrial pressure

_____ 3. ECG

_____ 4. first heart sound

_____ 5. second heart sound

_____ 6. ventricular pressure

_____ 7. ventricular volume

_____ 8. aortic (semilunar) valve closes

_____ 9. aortic (semilunar) valve opens

_____ 10. AV and semilunar valves closed (2 letters)

_____ 11. AV valve closes

_____ 12. AV valve opens

_____ 13. ventricular diastole (2 letters)

_____ 14. ventricular systole

2. Define the following terms.

systole: _____

*diastole:*_____

*cardiac cycle:*_____

3. Answer the following questions concerning events of the cardiac cycle.

When are the AV valves closed? _____

What event within the heart causes the AV valves to open? _____

When are the semilunar valves closed? _____

What event causes the semilunar valves to open? _____

Are both sets of valves closed during any part of the cycle? _____

If so, when? _____

Are both sets of valves open during any part of the cycle? _____

At what point in the cardiac cycle is the pressure in the heart highest? _____

Lowest _____

_____

What event results in the pressure deflection called the dicrotic notch? _____

_____

4. Using the key below, indicate the time interval occupied by the following events of the cardiac cycle.

Key: a. *0.8 sec*   b. *0.4 sec*   c. *0.3 sec*   d. *0.1 sec*

_____ 1. the length of the normal cardiac cycle   _____ 3. the quiescent period

_____ 2. the time interval of atrial systole   _____ 4. the ventricular contraction period

5. If an individual's heart rate is 80 beats/min, what is the length of the cardiac cycle? _____ What portion of the

cardiac cycle decreases with a more rapid heart rate? _____

_____

6. What two factors promote the movement of blood through the heart?_____

_____ and _____

_____

Heart Sounds

7. Complete the following statements.

      The monosyllables describing the heart sounds are __1__.
The first heart sound is a result of closure of the __2__ valves,
whereas the second is a result of closure of the __3__ valves.
The heart chambers that have just been filled when you hear
the first heart sound are the __4__, and the chambers that have
just emptied are the __5__. Immediately after the second heart
sound, both the __6__ and __7__ are filling with blood.

1. _____

2. _____

3. _____

4. _____

5. _____

6. _____

7. _____

8. As you listened to the heart sounds during the laboratory session, what differences in pitch, length, and amplitude (loudness)

of the two sounds did you observe? _____

_____

_____

9. In order to auscultate most accurately, indicate where you would place your stethoscope for the following sounds:

closure of the tricuspid valve: _____

closure of the aortic valve: _____

apical heartbeat: _____

Which valve is heard most clearly when the apical heartbeat is auscultated?_____

10. No one expects you to be a full-fledged physician on such short notice, but on the basis of what you have learned about heart
sounds, how might abnormal sounds be used to diagnose heart problems?

_____

_____

_____

The Pulse

11. Define *pulse*. _____

12. Describe the procedure used to take the pulse. _____

_____

_____

13. Identify the artery palpated at each of the pressure points listed.

at the wrist: _____

in front of the ear: _____

on the dorsum of the foot: _____

at the side of the neck: _____

14. When you were palpating the various pulse or pressure points, which appeared to have the greatest amplitude or tension?

_____ Why do you think this was so? _____

_____

15. Assume someone has been injured in an auto accident and is hemorrhaging badly. What pressure point would you compress to help stop bleeding from each of the following areas?

the thigh: _____   the calf: _____

the forearm: _____   the thumb: _____

16. How could you tell by simple observation whether bleeding is arterial or venous? _____

_____

17. You may sometimes observe a slight difference between the value obtained from an apical pulse (beats/min) and that from an arterial pulse taken elsewhere on the body. What is this difference called?

_____

Blood Pressure Determinations

18. Define *blood pressure.* _____

19. Identify the phase of the cardiac cycle to which each of the following apply.

systolic pressure: _____   diastolic pressure: _____

20. What is the name of the instrument used to compress the artery and record pressures in the auscultatory method of determining

blood pressure? _____

21. What are the sounds of Korotkoff? _____

_____

What causes the systolic sound? _____

_____

What causes the disappearance of the sound? _____

22. Interpret the pressure reading for each of the three numbers listed: 145/85/82. _____

_____

_____

23. Assume the following BP measurement was recorded for an elderly patient with severe arteriosclerosis: 170/110/–. Explain the inability to obtain the third reading.

_____

_____

24. Define *pulse pressure.* _____

Why is this measurement important? _____

_____

25. How do venous pressures compare to arterial pressures? _____

Why? _____

26. What maneuver to increase the thoracic pressure illustrates the effect of external factors on venous pressure? _____

How is it performed? _____

_____

27. What might an abnormal increase in venous pressure indicate? (Think!) _____

_____

Observing the Effect of Various Factors on Blood Pressure and Heart Rate

28. What effect do the following have on blood pressure? (Indicate increase by ↑ and decrease by ↓.)

_____ 1. increased diameter of the arterioles   _____ 4. hemorrhage

_____ 2. increased blood viscosity   _____ 5. arteriosclerosis

_____ 3. increased cardiac output   _____ 6. increased pulse rate

29. In which position (sitting, reclining, or standing) is the blood pressure normally the highest?

_____ The lowest? _____

What immediate changes in blood pressure did you observe when the subject stood up after being in the sitting or reclining

position? _____

_____

What changes in the blood vessels might account for the change? _____

_____

_____

After the subject stood for 3 minutes, what changes in blood pressure were observed? _____

_____

How do you account for this change? _____

30. What was the effect of exercise on blood pressure? _____

On pulse rate? _____ Do you think these effects reflect changes in cardiac output *or* in

peripheral resistance? _____

_____

_____

Why are there normally no significant increases in diastolic pressure after exercise? _____

_____

31. What effects of the following did you observe on blood pressure in the laboratory?

cold temperature: _____

What do you think the effect of heat would be? _____

Why? _____

32. Differentiate between a hypo- and a hyperreactor relative to the cold pressor test. _____

_____

Skin Color as an Indicator of Local Circulatory Dynamics

33. Describe normal skin color and the appearance of the veins in the subject's forearm before any testing was conducted.

_____

34. What changes occurred when the subject emptied the forearm of blood (by raising the arm and making a fist) and the flow

was occluded with the cuff? _____

What changes occurred during venous congestion? _____

_____

35. What is the importance of collateral blood supplies? _____

_____

36. Explain the mechanism by which mechanical stimulation of the skin produced a flare. _____

_____

Frog Cardiovascular Physiology

MATERIALS

- ☐ Dissecting instruments and tray
- ☐ Disposable gloves
- ☐ Petri dishes
- ☐ Medicine dropper
- ☐ Millimeter ruler
- ☐ Disposal container for organic debris
- ☐ Frog Ringer's solutions (at room temperature, 5°C, and 32°C)
- ☐ Frogs*
- ☐ Thread
- ☐ Large rubber bands
- ☐ Fine common pins
- ☐ Frog board
- ☐ Cotton balls
- ☐ Physiograph or BIOPAC® equipment:

Physiograph (polygraph), physiograph paper and ink, force transducer, transducer cable, transducer stand, stimulator output extension cable, electrodes

🔵 **BIOPAC®** BIOPAC® BSL *PRO* software, MP36/35 or MP45 data acquisition unit, PC or Mac computer, BIOPAC® HDW100A tension adjuster (or equivalent), BIOPAC® SS12LA force transducer with S-hook, small hook with thread, and transducer (or ring) stand

*Instructor will double-pith frogs as required for student experimentation.

Note: Instructions for using PowerLab® equipment can be found on MasteringA&P.

(Text continues on next page.)

MasteringA&P® For related exercise study tools, go to the Study Area of MasteringA&P. There you will find:
- Practice Anatomy Lab **PAL**
- PhysioEx **PEx**
- A&PFlix **A&PFlix**
- Practice quizzes, Histology Atlas, eText, Videos, and more!

OBJECTIVES

1. Describe the properties of automaticity and rhythmicity as they apply to cardiac muscle.
2. Discuss the anatomical differences between frog and human hearts.
3. Compare the intrinsic rate of contraction of the "pacemaker" of the frog heart (sinus venosus) to that of the atria and ventricle.
4. Define *extrasystole,* and explain when an extrasystole can occur on a tracing of the contractile activity of the heart.
5. Explain why it is important that cardiac muscle cannot be tetanized.
6. Describe the effects of the following on heart rate: cold, heat, vagal stimulation, pilocarpine, atropine sulfate, epinephrine, digitalis, and potassium, sodium, and calcium ions.
7. Define *ectopic pacemaker, vagal escape,* and *partial* and *total heart block.*
8. Name the blood vessels associated with a capillary bed and describe microcirculation.
9. Identify an arteriole, venule, and capillaries in a frog's web, and cite the differences between relative size of these vessels and the rate of blood flow through them.
10. Discuss the effect of heat, cold, local irritation, and histamine on blood flow in capillaries, and explain how these responses help maintain homeostasis.

PRE-LAB QUIZ

1. Circle True or False. Heart muscle can depolarize spontaneously in the absence of any external stimulation.
2. Spontaneous depolarization-repolarization events occur in a regular and continuous manner in cardiac muscle, a property known as
   a. automaticity
   b. rhythmicity
   c. synchronicity
3. How many chambers does the frog heart have?
   a. two   b. three   c. four
4. Circle True or False. Heart rate can be modified by extrinsic impulses from the autonomic nerves.
5. What is an extrasystole? _____
   _____
6. Which chemical agent will you use to modify the frog heart rate?
   a. caffeine
   b. digitalis
   c. magnesium solution
   d. Ringer's solution

(Text continues on next page.)

(Materials list continued)

Instructors using the MP36 (or MP35/30) data acquisition unit with BSL software versions earlier than 3.7.5 (for Windows) and 3.7.4 (for Mac Os X) will need slightly different channel settings and collection strategies. Instructions for using the older data acquisition unit can be found on MasteringA&P.

☐ Dropper bottles of freshly prepared solutions (using frog Ringer's solution as the solvent) of the following:

2.5% pilocarpine

5% atropine sulfate

1% epinephrine

2% digitalis

2% calcium chloride ($CaCl_2$)

0.7% sodium chloride (NaCl)

5% potassium chloride (KCl)

0.01% histamine

0.01 *N* HCl

☐ Dissecting pins

☐ Paper towels

☐ Compound microscope

PEx PhysioEx™ 9.1 Computer Simulation Ex. 6 on p. PEx-93

7. The _____ nerve carries parasympathetic impulses to the heart.
   a. cardiac   c. phrenic
   b. olfactory   d. vagus

8. Circle the correct underlined term. The phenomenon of <u>vagal escape</u> / <u>heart block</u> occurs when the heart stops momentarily then begins to beat again.

9. Circle True or False. The flow of blood through capillary beds is slow and intermittent.

10. _____ causes extensive vasodilation when applied to the frog web.
    a. Calcium
    b. Epinephrine
    c. Histamine
    d. Ringer's solution

nvestigations of human cardiovascular physiology are very interesting, but many areas obviously do not lend themselves to experimentation. It would be tantamount to murder to inject a human subject with various drugs to observe their effects on heart activity or to expose the human heart in order to study the length of its refractory period. However, this type of investigation can be done on frogs or small laboratory animals and provides valuable data because the physiological mechanisms in these animals are similar, if not identical, to those in humans.

In this exercise, you will conduct the cardiac investigations just mentioned and others. In addition, you will observe the microcirculation in a frog's web and subject it to various chemical and thermal agents to demonstrate their influence on local blood flow.

Special Electrical Properties of Cardiac Muscle: Automaticity and Rhythmicity

Cardiac muscle differs from skeletal muscle both functionally and in its fine structure. Skeletal muscle must be electrically stimulated to contract. In contrast, heart muscle can and does depolarize spontaneously in the absence of external stimulation. This property, called **automaticity,** is due to plasma membranes that have reduced permeability to potassium ions but still allow sodium ions to slowly leak into the cells. This leakage causes the muscle cells to gradually depolarize until the action potential threshold is reached and *fast calcium channels* open, allowing Ca^{2+} entry from the extracellular fluid. Shortly thereafter, contraction occurs. Also, the spontaneous depolarization-repolarization events occur in a regular, continuous manner in cardiac muscle, a property called **rhythmicity.**

In the following experiment, you will observe these properties of cardiac muscle in vitro (that is, removed from the body). Work together in groups of three or four. (The instructor may choose to demonstrate this procedure if time or frogs are at a premium.)

ACTIVITY 1

Investigating the Automaticity and Rhythmicity of Heart Muscle

1. Obtain a dissecting tray and instruments, disposable gloves, two petri dishes, frog Ringer's solution, a metric ruler, and a medicine dropper, and bring them to your laboratory bench.

2. Don the gloves, and then request and obtain a doubly pithed frog from your instructor. Quickly open the thoracic cavity and observe the heart rate in situ (at the site or within the body).

Record the heart rate: _____ beats/min

3. Dissect out the heart and the gastrocnemius muscle of the calf and place the removed organs in separate petri dishes containing frog Ringer's solution. (**Note:** Just in case you don't remember, the procedure for removing the gastrocnemius muscle is provided on page 237 in Exercise 14. The extreme care used in that procedure for the removal of the gastrocnemius muscle need not be exercised here.)

4. Observe the activity of the two organs for a few seconds.

Which is contracting? _____

At what rate? _____ beats/min

Is the contraction rhythmic?) _____

5. Sever the sinus venosus from the heart **(Figure 34.1)**. The **sinus venosus** of the frog's heart corresponds to the SA node of the human heart.

Does the sinus venosus continue to beat? _____

If not, lightly touch it with a probe to stimulate it. Record its rate of contraction.

Rate: _____ beats/min

6. Sever the right atrium from the heart; then remove the left atrium. Does each atrium continue to beat?

_____ Rate: _____ beats/min

Does the ventricle continue to beat? _____

Rate: _____ beats/min

7. Notice that frogs have a single ventricle (Figure 34.1). Fragment the ventricle to determine how small the ventricular fragments must be before the automaticity of ventricular muscle is abolished. Measure these fragments and record their approximate size.

_____ mm × _____ mm × _____ mm

Which portion of the heart exhibited the most marked automaticity?

_____

Which showed the least? _____

8. Properly dispose of the frog and heart fragments in the appropriate container before continuing. ■

Baseline Frog Heart Activity

The heart's effectiveness as a pump is dependent both on intrinsic (within the heart) and extrinsic (external to the heart) controls. In this activity, you will investigate some of these factors.

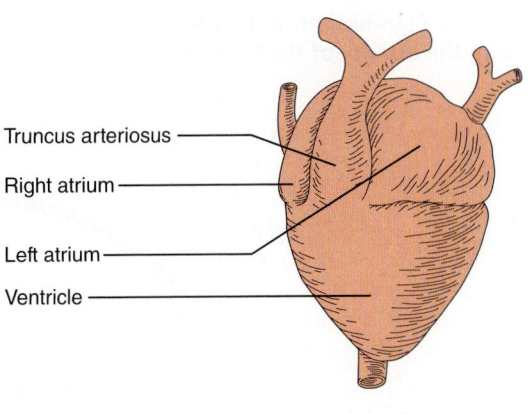

(a)

(b)

(c)

Figure 34.1 Anatomy of the frog heart. (a) Ventral view showing the single truncus arteriosus leaving the undivided ventricle. **(b)** Longitudinal section showing the two atrial and single ventricular chambers. **(c)** Dorsal view showing the sinus venosus (pacemaker).

The nodal system, in which the "pacemaker" imposes its depolarization rate on the rest of the heart, is one intrinsic factor that influences the heart's pumping action. If its impulses fail to reach the ventricles (as in heart block), the ventricles continue to beat but at their own inherent rate, which is much slower than that usually imposed on them. Although heart contraction does not depend on nerve impulses, its rate can be modified by extrinsic impulses reaching it through the autonomic nerves. Additionally, cardiac activity is modified

34

Figure 34.2 Physiograph setup for recording the activity of the frog heart.

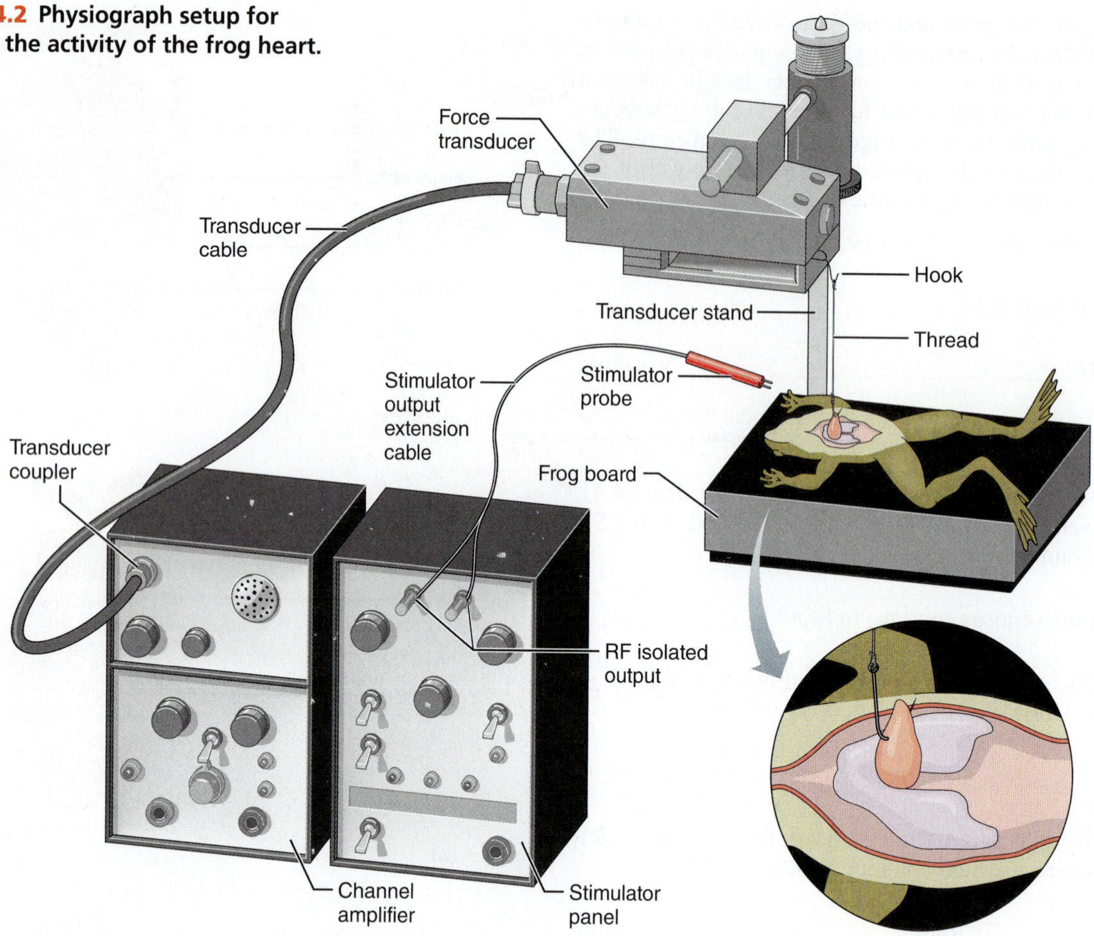

by various chemicals, hormones, ions, and metabolites. The effects of several of these chemical factors are examined in the next experimental series.

The frog heart has two atria and a single, incompletely divided ventricle (see Figure 34.1). The pacemaker is located in the sinus venosus, an enlarged region between the venae cavae and the right atrium. The SA node of mammals may have evolved from the sinus venosus.

ACTIVITY 2

Recording Baseline Frog Heart Activity

To record baseline frog heart activity, work in groups of four—two students handling the equipment setup and two preparing the frog for experimentation. Two sets of instructions are provided for apparatus setup—one for the physiograph **(Figure 34.2)**, the other for BIOPAC® **(Figure 34.3)**. Follow the procedure outlined for the apparatus you will be using.

Apparatus Setups

Physiograph Apparatus Setup

1. Obtain a force transducer, transducer cable, and transducer stand, and bring them to the recording site.

2. Attach the force transducer to the transducer stand (as shown in Figure 34.2).

3. Then attach the transducer cable to the transducer coupler (input) on the channel amplifier of the physiograph and to the force transducer.

4. Attach the stimulator output extension cable to output on the stimulator panel (red to red, black to black).

BIOPAC® Apparatus Setup

1. Connect the BIOPAC® apparatus to the computer and turn the computer **ON.**

2. Make sure the BIOPAC® unit is **OFF.**

3. Set up the equipment (as shown in Figure 34.3).

4. Turn the BIOPAC® unit **ON.**

5. Launch the BIOPAC® BSL *PRO* software by clicking the icon on the desktop or by following your instructor's guidance.

6. Open the Frog Heart template. MP36/35 users: Go to the **File** menu at the top of the screen and choose **Open > Files of Type = Graph Template (GTL) > FrogHeart.gtl.** MP45 users: Click the **BSL *PRO*** tab and navigate to look in the *PRO* lessons. Select **a04.gtl** and click **OK.**

7. Put the tension adjuster (BIOPAC® HDW100A, or equivalent) on the transducer stand, and attach the BIOPAC® SS12LA force transducer with the hook holes pointing down. Level the force transducer both horizontally and vertically.

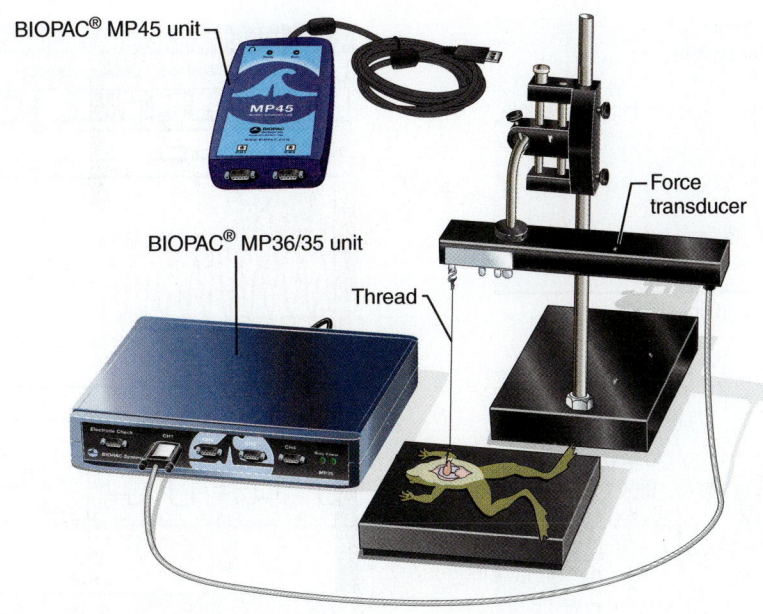

BIOPAC® MP45 unit

BIOPAC® MP36/35 unit

Force transducer

Thread

Figure 34.3 **BIOPAC® setup for recording the activity of the frog heart.** Plug the force transducer into channel 1. Transducer is shown plugged into the MP36/35 unit.

8. Set the tension adjuster to approximately one quarter of its full range. (**Note**: *Do not firmly tighten any of the thumbscrews at this stage.*) Select a force range of 0 to 50 grams for this experiment.

9. Select and attach the small S-hook to the force transducer.

Preparation of the Frog

1. Obtain room-temperature frog Ringer's solution, a medicine dropper, dissecting instruments and tray, disposable gloves, fine common pins (physiograph) or small hook (BIOPAC®), cotton ball, frog board, large rubber bands, and some thread, and bring them to your bench.

 2. Don the gloves, and obtain a doubly pithed frog from your instructor.

3. Make a longitudinal incision through the abdominal and thoracic walls with scissors, and then cut through the sternum to expose the heart.

4. Grasp the pericardial sac with forceps, and cut it open so that the beating heart can be observed.

Is the sequence an atrial-ventricular one? _____

5. Locate the vagus nerve, which runs down the lateral aspect of the neck and parallels the trachea and carotid artery. (In good light, it appears to be striated.) Slip an 18-inch length of thread under the vagus nerve so that it can later be lifted away from the surrounding tissues by the thread. Then place a Ringer's solution–soaked cotton ball over the nerve to keep it moistened until you are ready to stimulate it later in the procedure.

6. Using a medicine dropper, flush the heart with Ringer's solution. *From this point on the heart must be kept continually moistened with room-temperature Ringer's solution unless other solutions are being used for the experimentation.*

7. Attach the frog to the frog board using large rubber bands.

Physiograph Frog Heart Preparation

1. Bend a common pin to a 90° angle, and tie to its head a thread 0.46–0.5 m (18–20 inches) long. Take care not to penetrate the ventricular chamber as you force the pin through the apex of the heart until the apex is well secured in the angle of the pin.

2. Tie the thread from the heart to the hook on the force transducer. Do not pull the thread too tightly. It should be taut enough to lift the heart apex upward, away from the thorax, but should *not* stretch the heart. Adjust the force transducer as necessary. (See Figure 34.2.)

BIOPAC® Frog Heart Preparation

1. Attach a small hook tied with thread to the frog heart, following the instructions in step 1 of the physiograph instructions above to insert it through the apex of the heart. Confirm that the prepared frog is firmly attached to the frog board, positioned below the ring stand with the line running vertically from the frog heart to the transducer.

2. Slide the tension adjuster/force transducer assembly down the ring stand until you can hang the loop loosely from the S-hook, then slide it back up until the line is taut but the heart muscle is not stretched (be careful not to tear the heart).

3. Position the tension adjuster and/or force transducer so that the top is level, with approximately 10 cm (4 inches) of line from the heart to the S-hook. Adjust the assembly so that the thread line runs vertically; for a true reflection of the muscle's contractile force, the muscle must not be pulled at an angle.

4. Use the tension adjuster knob to make the line taut and tighten all thumbscrews to secure positioning of the assembly. Let the setup sit for a minute, then recheck the tension to make sure nothing has slipped or stretched.

5. Data may be distorted if the transducer line is not pulling directly vertical from the frog heart to the S-hook. Once again, align the frog as described above and make sure that the heart is not twisting the thread. If it is twisting, you will need to *carefully* remove the hook and repeat the setup.

34

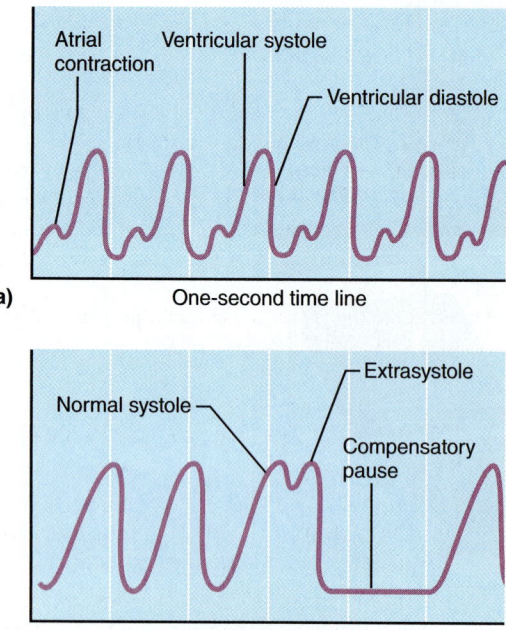

(a) One-second time line

(b) One-second time line

Figure 34.4 Physiograph recording of contractile activity of a frog heart. (a) Normal heartbeat. **(b)** Induction of an extrasystole.

6. BIOPAC® calculates *rate* data, which always trails the actual rate by one cycle. Data collection is a sensitive process and may display artifacts from table movement, heart movement (for instance, from breathing on the heart), and chemicals touching the heart. To get the best data, keep the experimental area stable, clean, and clear of obstructions. Hints for obtaining best data:

- The SS12LA force transducer must be level on the horizontal and vertical planes.

- Set up the tension adjuster and force transducer in positions that minimize their movement when tension is applied. Keep the point of S-hook attachment as close as possible to the ring stand support.

- Position the tension adjuster so that you will not bump the cables or frog board when using the adjustment knob.

- Position and/or tape the force transducer cables where they will not be pulled or bumped easily.

- Make sure the frog board is on a stable surface.

- Make sure the frog is firmly attached to the frog board so it will not rise up when tension is applied.

Making the Baseline Recording

Using the Physiograph

1. Turn the amplifier on and balance the apparatus according to instructions provided by your instructor. Set the paper speed at 0.5 cm/sec. Press the record and paper advance buttons.

2. Set the signal magnet or time marker at 1/sec.

3. Record 12 to 15 normal heartbeats. Be sure you can distinguish atrial and ventricular contractions **(Figure 34.4)**. Then adjust the paper or scroll speed so that the peaks of ventricular contractions are approximately 2 cm apart. (Peaks

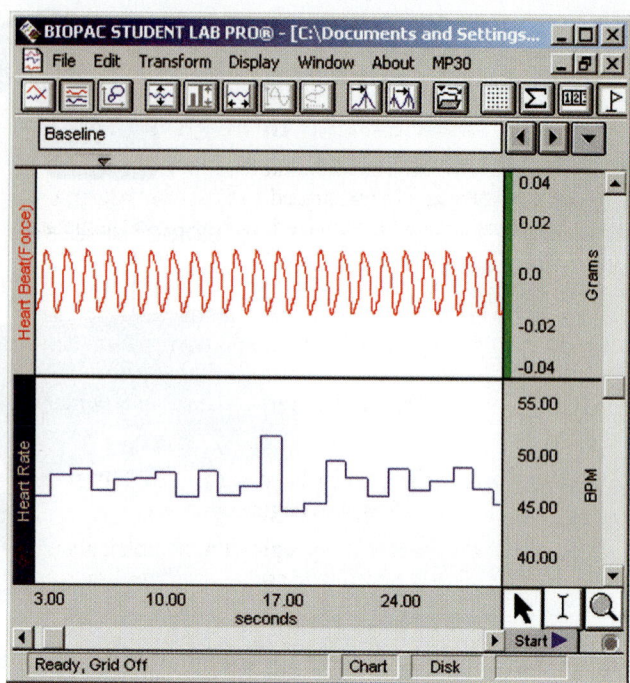

Figure 34.5 Example of baseline frog heart rate data.

indicate systole; troughs indicate diastole.) Pay attention to the relative force of heart contractions while recording.

Using BIOPAC®

1. Click **Start** to begin recording.

2. Observe at least 5 heart rate cycles, then click **Stop** to stop recording. Your data should look like that in the example **(Figure 34.5)**.

3. Choose **Save** from the File menu, and type in a filename to save the recorded data. You may want to save by your team's name followed by FrogHeart-1 (for example, Smith-FrogHeart-1).

Analyzing the Baseline Data

Count the number of ventricular contractions per minute from your physiograph or BIOPAC® data, and record:

_____ beats/min

Compute the A–V interval (period from the beginning of atrial contraction to the beginning of ventricular contraction).

_____ sec

How do the two tracings compare in time?

_____

_____

Mark the atrial and ventricular systoles on the record. *Remember to keep the heart moistened with Ringer's solution.* ■

ACTIVITY 3

Investigating the Refractory Period of Cardiac Muscle Using the Physiograph

Repeated rapid stimuli can cause skeletal muscle to remain in a contracted state (as demonstrated in Exercise 14). In other words, the muscle can be tetanized. This is possible because of the relatively short refractory period of skeletal muscle. In this experiment, you will use the physiograph to investigate the refractory period of cardiac muscle and its response to stimulation.* During the procedure, one student should keep the stimulating electrodes in constant contact with the frog heart ventricle while another student operates the stimulator panel.

1. Using the physiograph, set the stimulator to deliver 20-V shocks of 2-msec duration, and begin recording.

2. Deliver single shocks at the beginning of ventricular contraction, the peak of ventricular contraction, and then later and later in the cardiac cycle.

3. Observe the recording for **extrasystoles,** which are extra beats that show up riding on the ventricular contraction peak. Also note the **compensatory pause,** which allows the heart to get back on schedule after an extrasystole. (See Figure 34.4b.)

During which portion of the cardiac cycle was it possible to induce an extrasystole?

_____

_____

4. Attempt to tetanize the heart by stimulating it at the rate of 20 to 30 impulses per second. What is the result?

_____

_____

Considering the function of the heart, why is it important that heart muscle cannot be tetanized?

_____

_____

ACTIVITY 4

Assessing Physical and Chemical Modifiers of Heart Rate

Now that you have observed normal frog heart activity, you will have an opportunity to investigate the effects of various factors that modify heart activity. In each case, record a few normal heartbeats before introducing the modifying factor. After removing the agent, allow the heart to return to

its normal rate before continuing with the testing. On each record, indicate the point of introduction and removal of the modifying agent.

For each physical agent or solution that is applied:

If using the physiograph, increase the scroll or paper speed so that heartbeats appear as spikes 4 to 5 mm apart.

If using BIOPAC®, after applying each solution, click the **Start** button. When the effect is observed, record the effect for five cycles, then click **Stop.** Choose **Save** from the File menu to save your data.

(Note: Repeat these steps for each physical agent or chemical solution that is being applied.)

Temperature

1. Obtain 5°C and 32°C frog Ringer's solutions and medicine droppers.

2. Bathe the heart with 5°C Ringer's solution, and continue to record until the recording indicates a change in cardiac activity and five cardiac cycles have been recorded.

3. Stop recording, pipette off the cold Ringer's solution (remove the fluid by sucking it into the barrel of a medicine dropper), and flood the heart with room-temperature Ringer's solution.

4. Start recording again to determine the resumption of the normal heart rate. When this has been achieved, flood the heart with 32°C Ringer's solution, and again record five cardiac cycles after a change is noted.

5. Stop the recording, pipette off the warm Ringer's solution, and bathe the heart with room-temperature Ringer's solution once again.

What change occurred with the cold (5°C) Ringer's solution?

_____

What change occurred with the warm (32°C) Ringer's solution?

_____

6. Count the heart rate at the two temperatures, and record the data below.

_____ beats/min at 5°C; _____ beats/min at 32°C

Chemical Agents

Pilocarpine

Flood the heart with a 2.5% solution of pilocarpine. Record until a change in the pattern of the ECG is noticed. Pipette off the excess pilocarpine solution, and proceed immediately to the next test, which uses atropine as the testing solution. What happened when the heart was bathed in the pilocarpine solution?

_____

*BIOPAC® users may investigate the refractory period of cardiac muscle by using PhysioEx Exercise 6.

34

Pilocarpine simulates the effect of parasympathetic (vagal) nerve stimulation by enhancing acetylcholine release; such drugs are called parasympathomimetic drugs.

Is pilocarpine an agonist or an antagonist of acetylcholine?

_____

Atropine Sulfate

Apply a few drops of atropine sulfate to the frog's heart, and observe the recording. If no changes are observed within 2 minutes, apply a few more drops. When you observe a response, pipette off the excess atropine sulfate and flood the heart with room-temperature Ringer's solution. What happens when the atropine sulfate is added?

_____

_____

Atropine is a drug that blocks the effect of the neurotransmitter acetylcholine, which is liberated by the parasympathetic nerve endings. Do your results accurately reflect this effect of atropine?

_____

Is atropine an agonist or an antagonist of acetylcholine?

_____

Epinephrine

Flood the frog heart with epinephrine solution, and continue to record until a change in heart activity is noted.

What are the results? _____

_____

Which division of the autonomic nervous system does its effect mimic?

_____

Digitalis

Pipette off the excess epinephrine solution, and rinse the heart with room-temperature Ringer's solution. Continue recording, and when the heart rate returns to baseline values, bathe it in digitalis solution. What is the effect of digitalis on the heart?

_____

_____

_____

Digitalis is a drug commonly prescribed for heart patients with congestive heart failure. It slows heart rate, providing more time for venous return and decreasing the work of the weakened heart. These effects are thought to be due to inhibition of the sodium-potassium pump and enhancement of Ca^{2+} entry into the myocardial fibers.

Various Ions

To test the effect of various ions on the heart, apply the designated solution until you observe a change in heart rate or in strength of contraction. Pipette off the solution, flush with room-temperature Ringer's solution, and allow the heart to resume its normal rate before continuing. *Do not allow the heart to stop.* If the rate should decrease dramatically, flood the heart with room-temperature Ringer's solution.

Effect of Ca^{2+} (use 2% $CaCl_2$) _____

_____

Effect of Na^+ (use 0.7% NaCl) _____

_____

Effect of K^+ (use 5% KCl) _____

_____

Potassium ion concentration is normally higher within cells than in the extracellular fluid. *Hyperkalemia* decreases the resting potential of plasma membranes, thus decreasing the force of heart contraction. In some cases, the conduction rate of the heart is so depressed that **ectopic pacemakers** (pacemakers appearing erratically and at abnormal sites in the heart muscle) appear in the ventricles, and fibrillation may occur. Was there any evidence of premature beats in the

recording of potassium ion effects? _____

Was arrhythmia produced with any of the ions tested?

_____ If so, which? _____

Vagus Nerve Stimulation

The vagus nerve carries parasympathetic impulses to the heart, which modify heart activity. If you are using the physiograph, you can test this by stimulating the vagus nerve.*

1. Remove the cotton placed over the vagus nerve. Using the previously tied thread, lift the nerve away from the tissues and place the nerve on the stimulating electrodes.

2. Using a duration of 0.5 msec at a voltage of 1 mV, stimulate the nerve at a rate of 50/sec. Continue stimulation until the heart stops momentarily and then begins to beat again (**vagal escape).** If no effect is observed, increase stimulus intensity and try again. If no effect is observed after a substantial increase in stimulus voltage, reexamine your "vagus nerve" to make sure that it is not simply strands of connective tissue.

3. Discontinue stimulation after you observe vagal escape, and flush the heart with room-temperature Ringer's solution

*BIOPAC® users may observe the effects of vagal stimulation by using PhysioEx Exercise 6.

until the normal heart rate resumes. What is the effect of vagal stimulation on heart rate?

The phenomenon of vagal escape demonstrates that many factors are involved in heart regulation and that any deleterious factor (in this case, excessive vagal stimulation) will be overcome, if possible, by other physiological mechanisms such as activation of the sympathetic division of the autonomic nervous system (ANS).

Intrinsic Conduction System Disturbance (Heart Block)

1. Moisten a 25-cm (10-inch) length of thread and make a Stannius ligature (loop the thread around the heart at the junction of the atria and ventricle).

2. If using a physiograph, decrease the scroll or paper speed to achieve intervals of approximately 2 cm between the ventricular contractions, and record a few normal heartbeats.

3. Tighten the ligature in a stepwise manner while observing the atrial and ventricular contraction curves. As heart block occurs, the atria and ventricle will no longer show a 1:1 contraction ratio. Record a few beats each time you observe a different degree of heart block—a 2:1 ratio of atrial to ventricular contractions, 3:1, 4:1, and so on. As long as you can continue to count a whole number ratio between the two chamber types, the heart is in **partial heart block.** When you can no longer count a whole number ratio, the heart is in **total,** or **complete, heart block.**

4. When total heart block occurs, release the ligature to see if the normal A–V rhythm is reestablished. What is the result?

---

5. Attach properly labeled recordings (or copies of the recordings) made during this procedure to the last page of this exercise for future reference.

6. Dispose of the frog remains and gloves in appropriate containers, and dismantle the experimental apparatus before continuing. ▄▄

The Microcirculation and Local Blood Flow

The thin web of a frog's foot provides an excellent opportunity to observe the flow of blood to, from, and within the capillary beds, where the real business of the circulatory system occurs. The flow of blood through a capillary bed is called the **microcirculation.** Arterioles carry blood to the capillary bed; venules carry blood away. Most capillary beds consist of a vascular shunt, called the **metarteriole–thoroughfare channel,** and true capillaries, the actual exchange vessels **(Figure 34.6)**.

 The total cross-sectional area of the capillaries in the body is much greater than that of the veins and arteries combined. Thus, the velocity of flow through the capillary beds is quite slow. Capillary flow is also intermittent, because if all

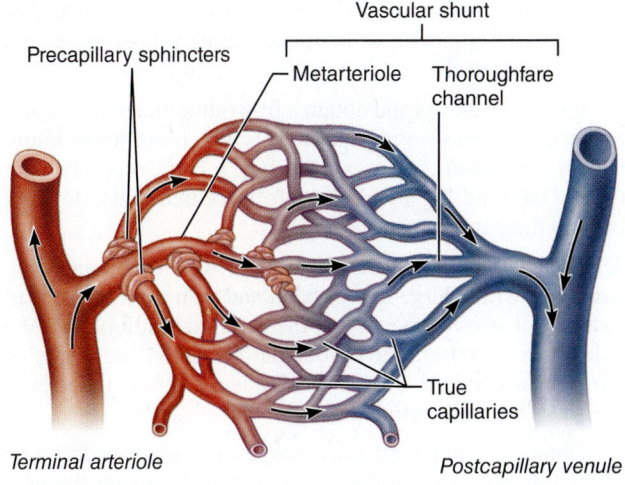

(a) Sphincters open

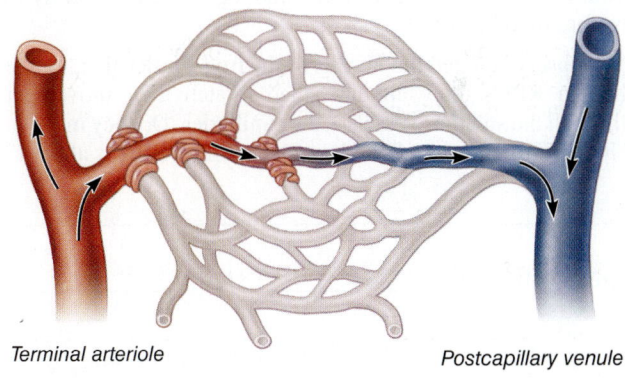

(b) Sphincters closed

Figure 34.6 Anatomy of a capillary bed. The composite metarteriole—thoroughfare channels act as shunts to bypass the true capillaries when precapillary sphincters controlling blood entry into the true capillaries are constricted.

capillary beds were filled with blood at the same time, there would be no blood at all in the large vessels. The flow of blood into the capillary beds is regulated by the activity of muscular *terminal arterioles,* which feed the beds, and by *precapillary sphincters* at entrances to the true capillaries. The amount of blood flowing into the true capillaries of the bed is regulated most importantly by local chemical controls (local concentrations of carbon dioxide, histamine, pH). Thus a capillary bed may be flooded with blood or almost entirely bypassed depending on what is happening within the body or in a particular body region at any one time. You will investigate some of the local controls in the next group of experiments.

ACTIVITY 5

Investigating the Effect of Various Factors on the Microcirculation

1. Obtain a frog board (with a hole at one end), dissecting pins, disposable gloves, frog Ringer's solution (room

temperature, 5°C, and 32°C), 0.01 N HCl, 0.01% histamine solution, 1% epinephrine solution, a large rubber band, and some paper towels.

2. Put on the gloves and obtain a frog (alive and hopping, *not* pithed). Moisten several paper towels with room-temperature Ringer's solution, and wrap the frog's body securely with them. One hind leg should be left unsecured and extending beyond the paper cocoon.

3. Attach the frog to the frog board (or other supporting structure) with a large rubber band and then carefully spread (but do not stretch) the web of the exposed hindfoot over the hole in the support. Have your partners hold the edges of the web firmly for viewing. Alternatively, secure the toes to the board with dissecting pins.

4. Obtain a compound microscope, and observe the web under low power to find a capillary bed. Focus on the vessels in high power. Keep the web moistened with Ringer's solution as you work. If the circulation seems to stop during your observations, massage the hind leg of the frog gently to restore blood flow.

5. Observe the red blood cells of the frog. Notice that, unlike human RBCs, they are nucleated. Watch their movement through the smallest vessels—the capillaries. Do they move in single file, or do they flow through two or three cells abreast?

_____

Are they flexible? _____ Explain. _____

Can you see any white blood cells in the capillaries?

_____ If so, which types? _____

6. Notice the relative speed (velocity) of blood flow through the blood vessels. Differentiate between the arterioles, which feed the capillary bed, and the venules, which drain it. This may be tricky, because images are reversed in the microscope. Thus, the vessel that appears to feed into the capillary bed will actually be draining it. You can distinguish between the vessels, however, if you consider that the flow is more pulsating and turbulent in the arterioles and smoother and steadier in the venules. How does the velocity of flow in the arterioles compare with that in the venules?

_____

In the capillaries? _____

What is the relative difference in the diameter of the arterioles and capillaries?

_____

Temperature

1. To investigate the effect of temperature on blood flow, flood the web with 5°C Ringer's solution two or three times to chill the entire area. Is a change in vessel diameter noticeable?

_____ Which vessels are affected? _____

How? _____

2. Blot the web gently with a paper towel, and then bathe the web with warm (32°C) Ringer's solution. Record your observations.

_____

Inflammation

1. Pipette 0.01 N HCl onto the frog's web. Hydrochloric acid will act as an irritant and cause a localized inflammatory response. Is there an increase or decrease in the blood flow into the capillary bed following the application of HCl?

_____

What purpose do these local changes serve during a localized inflammatory response?

_____

_____

_____

2. Flush the web with room-temperature Ringer's solution and blot.

Histamine

1. Histamine, which is released in large amounts during allergic responses, causes extensive vasodilation. Investigate this effect by adding a few drops of histamine solution to the frog web. What happens?

_____

_____

How does this response compare to that produced by HCl?

_____

2. Blot the web and flood with 32°C Ringer's solution as before. Now add a few drops of 1% epinephrine solution, and observe the web. What are epinephrine's effects on the blood vessels?

_____

Epinephrine is used clinically to reverse the vasodilation seen in severe allergic attacks (such as asthma), which are mediated by histamine and other vasoactive molecules.

3. Return the dropper bottles to the supply area and the frog to the terrarium. Properly clean your work area before leaving the lab. ■

The Lymphatic System and Immune Response

MATERIALS

- [ ] Large anatomical chart of the human lymphatic system
- [ ] Prepared slides of lymph node, spleen, and tonsil
- [ ] Compound microscope
- [ ] Wax marking pencil
- [ ] Petri dish containing simple saline agar
- [ ] Medicine dropper
- [ ] Dropper bottles of red and green food color
- [ ] Dropper bottles of goat antibody to horse serum albumin, goat antibody to bovine serum albumin, goat antibody to swine serum albumin, horse serum albumin diluted to 20% with physiological saline, unknown albumin sample diluted to 20% (prepared from horse, swine, and/or bovine albumin)
- [ ] Colored pencils

✂ For instructions on animal dissections, see the dissection exercises (starting on page 697) in the cat and fetal pig editions of this manual.

PEx PhysioEx™ 9.1 Computer Simulation Ex. 12 on p. PEx-177

OBJECTIVES

1. State the function of the lymphatic system, name its components, and compare its function to that of the blood vascular system.
2. Describe the formation and composition of lymph, and discuss how it is transported through the lymphatic vessels.
3. Relate immunological memory, specificity, and differentiation of self from nonself to immune function.
4. Differentiate between the roles of B cells and T cells in the immune response.
5. Describe the structure and function of lymph nodes, and indicate the location of T cells, B cells, and macrophages in a typical lymph node.
6. Describe the major microanatomical features of the spleen and tonsils.
7. Draw or describe the structure of the antibody monomer, and name the five immunoglobulin subclasses.
8. Differentiate between antigen and antibody.
9. Explain how the Ouchterlony test detects antigens by using the antigen-antibody reaction.

PRE-LAB QUIZ

1. Circle True or False. The lymphatic system protects the body by removing foreign material such as bacteria from the lymphatic stream.
2. Lymph is
   a. excess blood that has escaped from veins
   b. excess tissue fluid that has leaked out of capillaries
   c. excess tissue fluid that has escaped from arteries
3. Circle True or False. Collecting lymphatic vessels have three tunics and are equipped with valves like veins.
4. _____, which serve as filters for the lymphatic system, occur at various points along the lymphatic vessels.
   a. Glands   b. Lymph nodes   c. Valves
5. Circle True or False. The immune response is a systemic response that occurs when the body recognizes a substance as foreign and acts to destroy or neutralize it.
6. Three characteristics of the immune response are the ability to distinguish self from nonself, memory, and
   a. autoimmunity   b. specificity   c. susceptibility
7. Circle the correct underlined term. B cells / T cells differentiate in the thymus.
8. Circle the correct underlined term. T cells mediate humoral / cellular immunity because they destroy cells infected with viruses and certain bacteria and parasites.
9. Circle True or False. Antibodies are produced by plasma cells in response to antigens and are found in all body secretions.
10. Antibodies that have only one structural unit (monomer) consist of _____ protein chains, connected by disulfide bonds.
    a. two   c. four
    b. three   d. six

MasteringA&P® For related exercise study tools, go to the Study Area of MasteringA&P. There you will find:
- Practice Anatomy Lab PAL
- PhysioEx PEx
- A&PFlix A&PFlix
- Practice quizzes, Histology Atlas, eText, Videos, and more!

The overall function of the lymphatic system is twofold: (1) it transports tissue fluid (lymph) to the blood vessels, and (2) it protects the body by removing foreign material such as bacteria from the lymphatic stream and by serving as a site for lymphocyte "policing" of body fluids and lymphocyte multiplication.

The Lymphatic System

The **lymphatic system** consists of a network of lymphatic vessels (lymphatics), lymphoid tissue, lymph nodes, and a number of other lymphoid organs, such as the tonsils, thymus, and spleen. We will focus on the lymphatic vessels and lymph nodes in this section. The white blood cells that are the central actors in body immunity are described later in this exercise.

Distribution and Function of Lymphatic Vessels and Lymph Nodes

As blood circulates through the body, the hydrostatic and osmotic pressures operating at the capillary beds result in fluid outflow at the arterial end of the bed and in its return at the venous end. However, not all of the lost fluid is returned

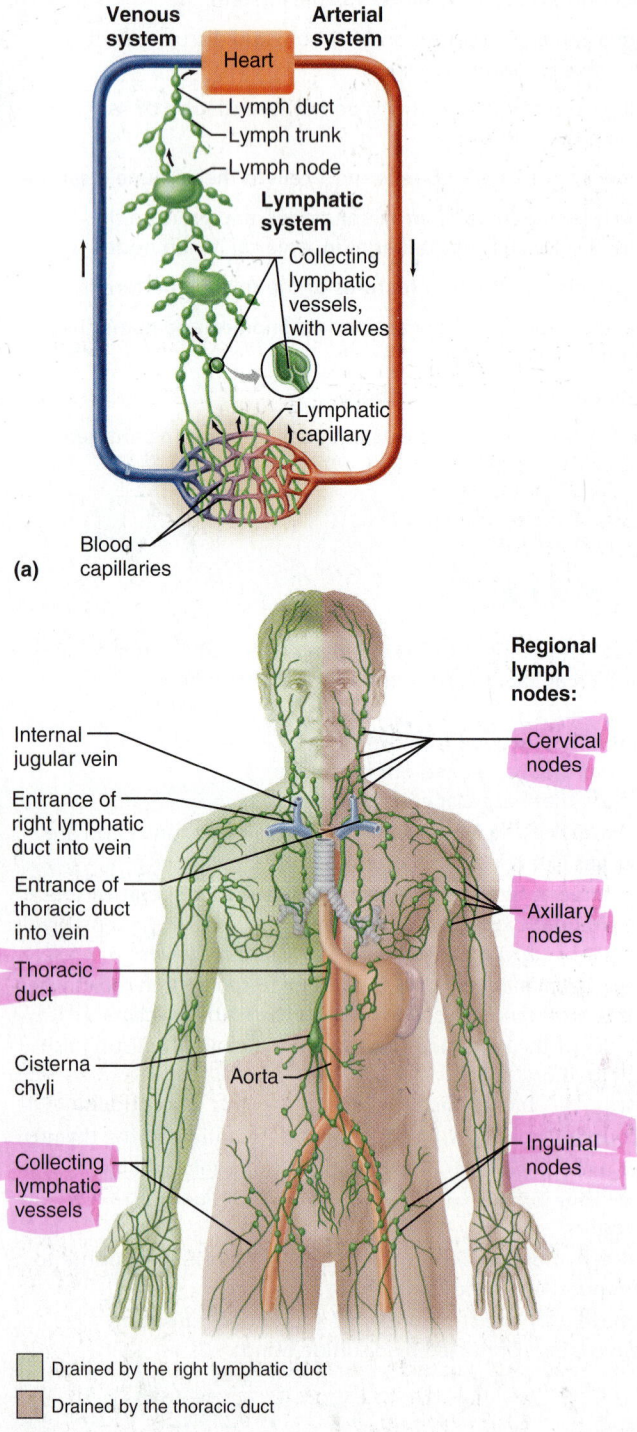

Figure 35.1 Lymphatic system. (a) Simplified scheme of the relationship of lymphatic vessels to blood vessels of the cardiovascular system. **(b)** Distribution of lymphatic vessels and lymph nodes. The green-shaded area represents body area drained by the right lymphatic duct. **(c)** Major veins in the superior thorax showing entry points of the thoracic and right lymphatic ducts. The major lymphatic trunks are also identified.

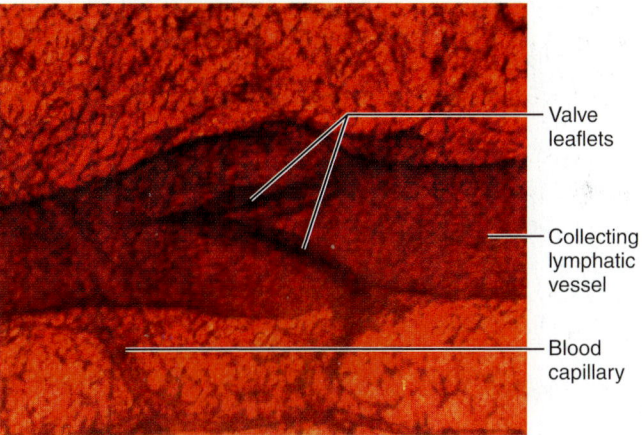

Figure 35.2 Collecting lymphatic vessel (700×).

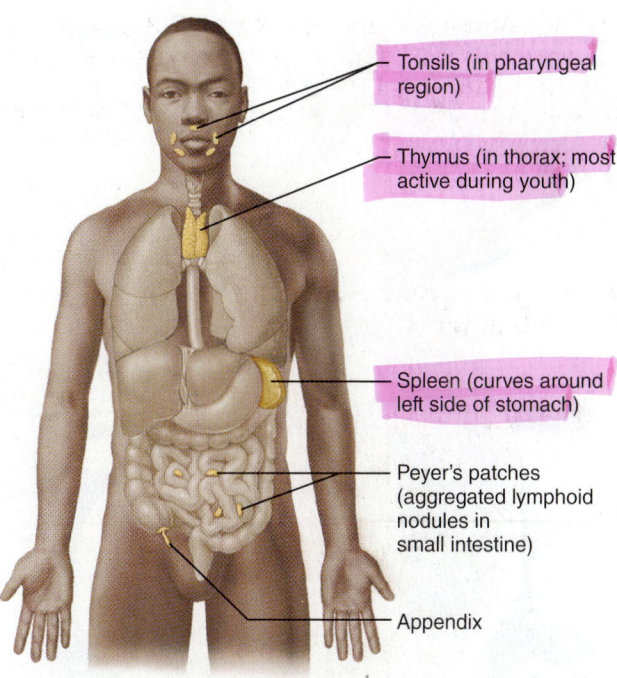

Tonsils (in pharyngeal region)

Thymus (in thorax; most active during youth)

Spleen (curves around left side of stomach)

Peyer's patches (aggregated lymphoid nodules in small intestine)

Appendix

Figure 35.3 Lymphoid organs. Locations of the tonsils, thymus, spleen, appendix, and Peyer's patches.

to the bloodstream by this mechanism, and the fluid that lags behind in the tissue spaces must eventually return to the blood if the vascular system is to operate properly. When fluid is not returned, it accumulates in the tissues, producing a condition called *edema*. It is the microscopic, blind-ended **lymphatic capillaries (Figure 35.1a)**, which branch through nearly all the tissues of the body, that pick up this leaked fluid (primarily water and a small amount of dissolved proteins) and carry it through successively larger vessels—**collecting lymphatic vessels** to **lymphatic trunks**—until the lymph finally returns to the blood vascular system through one of the two large ducts in the thoracic region (Figure 35.1b). The **right lymphatic duct,** present in some but not all individuals, drains lymph from the right upper extremity, head, and thorax delivered by the jugular, subclavian, and bronchomediastinal trunks. In individuals without a right lymphatic duct, those trunks open directly into veins of the neck. The large **thoracic duct** receives lymph from the rest of the body (see Figure 35.1c). In humans, both ducts empty the lymph into the venous circulation at the junction of the internal jugular vein and the subclavian vein, on their respective sides of the body. Notice that the lymphatic system, lacking both a contractile "heart" and arteries, is a one-way system; it carries lymph only toward the heart.

Like veins of the blood vascular system, the collecting lymphatic vessels have three tunics and are equipped with valves **(Figure 35.2)**. However, lymphatics tend to be thinner-walled, to have *more* valves, and to anastomose (form branching networks) more than veins. Since the lymphatic system is a pumpless system, lymph transport depends largely on the milking action of the skeletal muscles and on pressure changes within the thorax that occur during breathing.

As lymph is transported, it filters through bean-shaped **lymph nodes,** which cluster along the lymphatic vessels of the body. There are thousands of lymph nodes, but because they are usually embedded in connective tissue, they are not ordinarily seen. Within the lymph nodes are **macrophages,** phagocytes that destroy bacteria, cancer cells, and other foreign matter in the lymphatic stream, thus rendering many harmful substances or cells harmless before the lymph enters the bloodstream. Particularly large collections of lymph nodes are found in the inguinal, axillary, and cervical regions of the body. Although we are not usually aware of

the filtering and protective nature of the lymph nodes, most of us have experienced "swollen glands" during an active infection. This swelling is a manifestation of the trapping function of the nodes.

Other lymphoid organs—the tonsils, thymus, and spleen **(Figure 35.3)**—resemble the lymph nodes histologically, and house similar cell populations (lymphocytes and macrophages).

ACTIVITY 1

Identifying the Organs of the Lymphatic System

Study the large anatomical chart to observe the general plan of the lymphatic system. Notice the distribution of lymph nodes, various lymphatics, the lymphatic trunks, and the location of the right lymphatic duct and the thoracic duct. Also identify the **cisterna chyli,** the enlarged terminus of the thoracic duct that receives lymph from the digestive viscera. ▬▬

For instructions on animal dissections, see the dissection exercises (starting on page 697) in the cat and fetal pig editions of this manual.

The Immune Response

The **adaptive immune system** is a functional system that recognizes something as foreign and acts to destroy or neutralize it. This response is known as the **immune response.** It is a systemic response and is not restricted to the initial infection site. When operating effectively, the immune response protects us from bacterial and viral infections, bacterial toxins, and cancer. When it fails or malfunctions, the body is quickly devastated by pathogens or assaults by its own immune system.

35

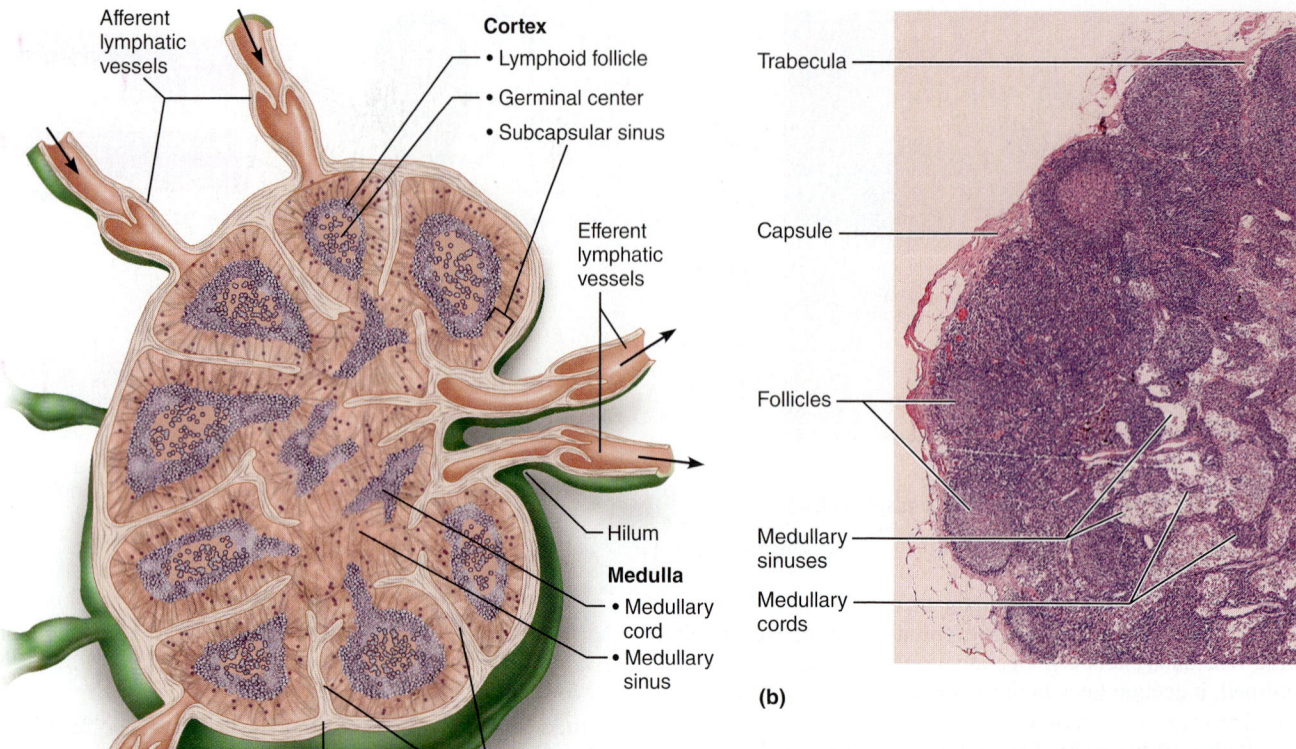

(a)

(b)

Figure 35.4 Structure of lymph node. (a) Longitudinal view of the internal structure of a lymph node and associated lymphatics. Notice that the afferent vessels outnumber the efferent vessels, which slows the rate of lymph flow. The arrows indicate the direction of the lymph flow. **(b)** Photomicrograph of part of a lymph node (20×).

Major Characteristics of the Immune Response

The most important characteristics of the immune response are its (1) **memory,** (2) **specificity,** and (3) **ability to differentiate self from nonself.** Not only does the immune system have a "memory" for previously encountered foreign antigens (the chicken pox virus for example), but this memory is also remarkably accurate and highly specific.

An almost limitless variety of things are *antigens*—that is, anything capable of provoking an immune response and reacting with the products of the response. Nearly all foreign proteins, many polysaccharides, bacteria and their toxins, viruses, mismatched RBCs, cancer cells, and many small molecules (haptens), when linked to our own body proteins, exhibit this capability. The cells that recognize antigens and initiate the immune response are lymphocytes, the second most numerous members of the leukocyte, or white blood cell (WBC), population. Each immunocompetent lymphocyte is virtually monospecific; that is, it has receptors on its surface allowing it to bind with only one or a few very similar antigens.

As a rule, our own proteins are tolerated, a fact that reflects the ability of the immune system to distinguish our own tissues (self) from foreign antigens (nonself). Nevertheless, an inability to recognize self can and does occasionally happen, and our own tissues are attacked by the immune system. This phenomenon is called *autoimmunity.* Autoimmune diseases include multiple sclerosis (MS), myasthenia gravis, Graves' disease, glomerulonephritis, rheumatoid arthritis (RA), and type 1 (insulin-dependent) diabetes mellitus.

Organs, Cells, and Cell Interactions of the Immune Response

The immune system uses as part of its arsenal the **lymphoid organs** and **lymphoid tissues,** including the thymus, lymph nodes, spleen, tonsils, appendix, and bone marrow. Of these, the thymus and bone marrow are considered to be the *primary lymphoid organs.* The others are *secondary lymphoid organs and tissues.*

The stem cells that give rise to the immune system arise in the bone marrow. Their subsequent differentiation into one of the two populations of immunocompetent lymphocytes occurs in the primary lymphoid organs. The **B cells** (B lymphocytes) differentiate in bone marrow, and the **T cells** (T lymphocytes) differentiate in the thymus. While in their "programming organs" the lymphocytes become *immunocompetent,* an event indicated by the appearance of specific cell-surface proteins that enable the lymphocytes to respond (by binding) to a particular antigen.

After differentiation, the B and T cells leave the bone marrow and thymus, respectively; enter the bloodstream; and travel to peripheral (secondary) lymphoid organs, where clonal selection occurs. **Clonal selection** is triggered when an antigen binds to the specific cell-surface receptors of a T or B cell. This event causes the lymphocyte to proliferate rapidly, forming a clone of like cells, all bearing the same antigen-specific receptors. Then, in the presence of certain

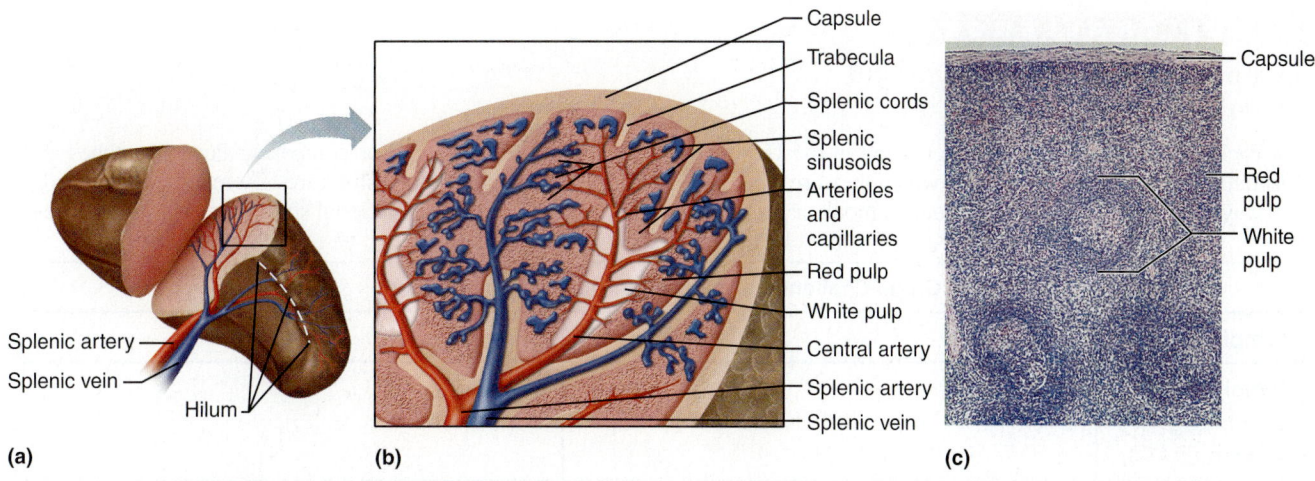

Figure 35.5 **The spleen. (a)** Gross structure. **(b)** Diagram of the histological structure. **(c)** Photomicrograph of spleen tissue showing white and red pulp regions (30×).

regulatory signals, the members of the clone specialize, or differentiate—some forming memory cells and others becoming effector or regulatory cells. Upon subsequent meetings with the same antigen, the immune response proceeds considerably faster because the troops are already mobilized and awaiting further orders, so to speak.

In the case of B cell clones, some become **memory B cells;** the others form antibody-producing **plasma cells**. Because the B cells act indirectly through the antibodies that their progeny release into the bloodstream (or other body fluids), they are said to provide **humoral immunity.** T cell clones are more diverse. Although all T cell clones also contain memory cells, some clones contain *cytotoxic T cells* (effector cells that directly attack virus-infected tissue cells). Others contain *helper T cells,* which help activate the B cells and cytotoxic T cells, and still others contain *regulatory T cells,* which can inhibit the immune response. Because certain T cells act directly to destroy cells infected with viruses, certain bacteria or parasites, and cancer cells, and to reject foreign grafts, T cells are said to mediate **cellular immunity.**

Absence or failure of thymic differentiation of T lymphocytes results in a marked depression of both antibody and cell-mediated immune functions. Additionally, the observation that the thymus naturally shrinks with age has been correlated with the relatively immune-deficient status of elderly individuals. ✚

All lymphoid tissues except the thymus and bone marrow contain both T and B cell–dependent regions.

ACTIVITY 2

Studying the Microscopic Anatomy of a Lymph Node, the Spleen, and a Tonsil

1. Obtain a compound microscope and prepared slides of a lymph node, spleen, and a tonsil. As you examine the lymph node slide, notice the following anatomical features (depicted in **Figure 35.4**). The node is enclosed within a fibrous **capsule,** from which connective tissue septa **(trabeculae)** extend inward to divide the node into several compartments.

Very fine strands of reticular connective tissue issue from the trabeculae, forming the stroma of the gland within which cells are found.

In the outer region of the node, the **cortex,** some of the cells are arranged in globular masses, referred to as germinal centers. The **germinal centers** contain rapidly dividing B cells. The rest of the cortical cells are primarily T cells that circulate continuously, moving from the blood into the node and then exiting from the node in the lymphatic stream.

In the internal portion of the gland, the **medulla,** the cells are arranged in cordlike fashion. Most of the medullary cells are macrophages. Macrophages are important not only for their phagocytic function but also because they play an essential role in "presenting" the antigens to the T cells.

Lymph enters the node through a number of *afferent vessels,* circulates through *lymph sinuses* within the node, and leaves the node through *efferent vessels* at the **hilum.** Since each node has fewer efferent than afferent vessels, the lymph flow stagnates somewhat within the node. This allows time for the generation of an immune response and for the macrophages to remove debris from the lymph before it reenters the blood vascular system.

2. As you observe the slide of the spleen, look for the areas of lymphocytes suspended in reticular fibers, the **white pulp,** clustered around central arteries **(Figure 35.5)**. The remaining tissue in the spleen is the **red pulp,** which is composed of splenic sinusoids and areas of reticular tissue and macrophages called the **splenic cords.** The white pulp, composed primarily of lymphocytes, is responsible for the immune functions of the spleen. Macrophages remove worn-out red blood cells, debris, bacteria, viruses, and toxins from blood flowing through the sinuses of the red pulp.

3. As you examine the tonsil slide, notice the **follicles** containing **germinal centers** surrounded by scattered lymphocytes. The characteristic **tonsillar crypts** (invaginations of the mucosal epithelium) of the tonsils trap bacteria and other foreign material **(Figure 35.6)**. Eventually the bacteria work their way into the lymphoid tissue and are destroyed. ■

35

GROUP CHALLENGE

Compare and Contrast Lymphoid Organs and Tissues

For each pair of lymphoid structures listed in the **Group Challenge chart,** describe ways in which they are similar and ways in which they differ. Use your textbook or another appropriate reference for comparing Peyer's patches and the thymus with other structures. Remember to consider both structural and functional similarities and differences.

Group Challenge: Lymphoid Structure Comparisons		
Lymphoid pair	**Similarities**	**Differences**
Lymph node Spleen		
Lymph node Tonsil		
Peyer's patches Tonsils		
Tonsil Spleen		
Thymus Spleen		

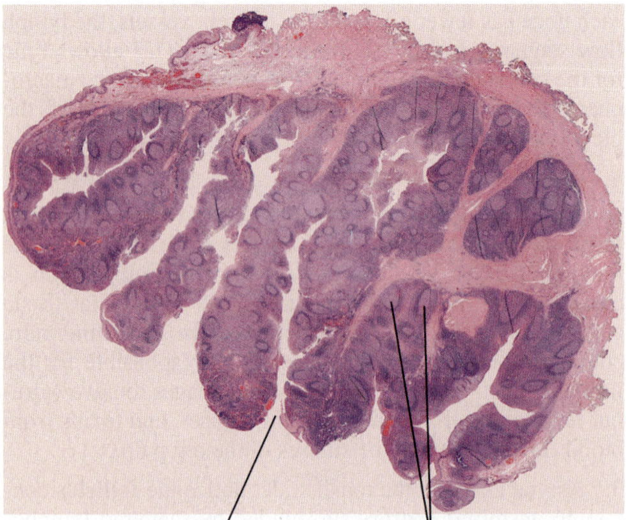

Figure 35.6 Histology of a palatine tonsil. The luminal surface is covered with epithelium that invaginates deeply to form crypts (10×).

Tonsillar crypt

Germinal centers in lymphoid follicles

Antibodies and Tests for Their Presence

Antibodies, or **immunoglobulins (Igs),** produced by sensitized B cells and their plasma cell offspring in response to an antigen, are a heterogeneous group of proteins that make up the general class of plasma proteins called **gamma globulins.** Antibodies are found not only in plasma but also (to greater or lesser extents) in all body secretions. Five major classes of immunoglobulins have been identified: IgM, IgG, IgD, IgA, and IgE. The immunoglobulin classes share a common basic structure but differ functionally and in their localization in the body.

All Igs are composed of one or more structural units called **antibody monomers.** A monomer consists of four protein chains bound together by disulfide bridges **(Figure 35.7)**. Two of the chains are quite large and have a high molecular weight; these are the **heavy chains.** The other two chains are only half as long and have a low molecular weight. These are called **light chains.** The two heavy chains have a *constant (C) region,* in which the amino acid sequence is the same in a class of immunoglobulins, and a *variable (V) region,* which differs in the Igs formed in response to different antigens. The same is true of the two light chains; each has a constant and a variable region.

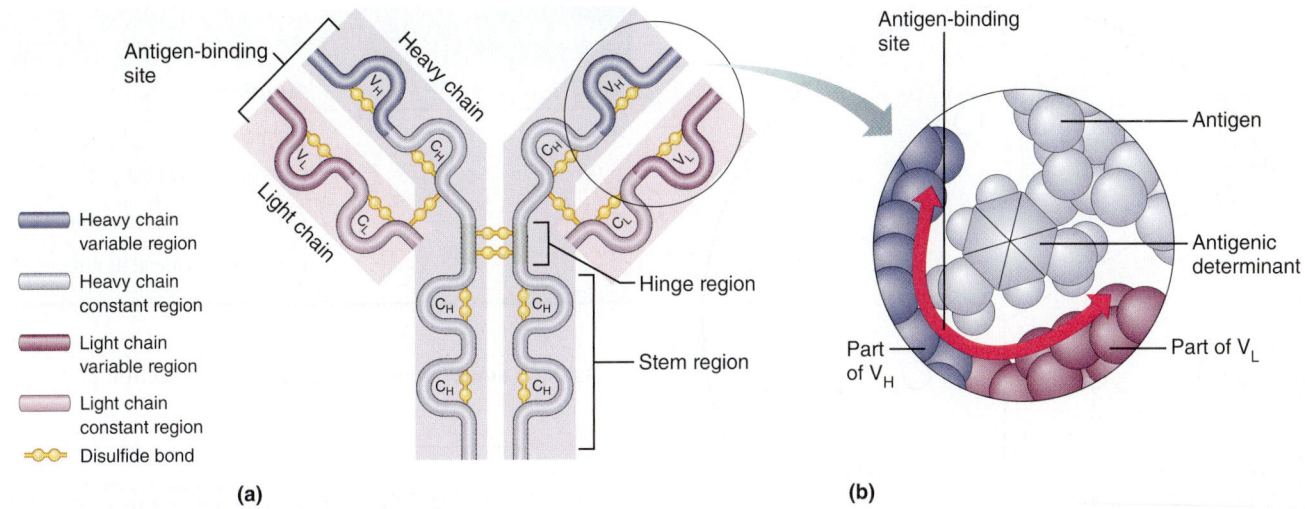

Figure 35.7 Antibody structure. (a) Schematic antibody structure consists of two heavy chains and two light chains connected by disulfide bonds. **(b)** Enlargement of an antigen-binding site of an immunoglobulin.

The intact Ig molecule has a three-dimensional shape that is generally Y shaped. Together, the variable regions of the light and heavy chains in each "arm" construct one **antigen-binding site** uniquely shaped to "fit" a specific *antigenic determinant* (portion) of an antigen. Thus, each Ig monomer bears two identical sites that bind to identical antigenic determinants. Binding of the immunoglobulins to their complementary antigen(s) effectively immobilizes the antigens until they can be phagocytized or lysed by complement fixation.

Although the role of the immune system is to protect the body, symptoms of certain diseases involve excessively high antibody synthesis (as in multiple myeloma, a cancer of the bone marrow and adjacent bony structures) and/or the production of abnormal antibodies.

The antigen-antibody reaction is used diagnostically in a variety of ways. One of the most familiar is blood typing. (See Exercise 29 for instructions on ABO and Rh blood typing.) Pregnancy tests also use the antigen-antibody reaction to test for the presence of human chorionic gonadotropin (hCG), a hormone produced early in pregnancy. Another technique for detecting antigens is the enzyme-linked immunosorbent assay (ELISA). Originally designed to measure antibody titer, ELISA has been modified for use in HIV-1 blood screening. The antigen-antibody test that we will use in this laboratory session is the Ouchterlony technique, which is used mainly for rapid screening of suspected antigens.

Ouchterlony Double-Gel Diffusion, an Immunological Technique

The Ouchterlony double-gel diffusion technique was developed in 1948 to detect the presence of particular antigens in sera or extracts. Antigens and antibodies are placed in wells in a gel and allowed to diffuse toward each other. If an antigen reacts with an antibody, a thin white line called a *precipitin line* forms. In the following activity, the double-gel diffusion technique will be used to identify antigens. Work in groups of no more than three.

Using the Ouchterlony Technique to Identify Antigens

1. Obtain one each of the materials for conducting the Ouchterlony test: petri dish containing saline agar; medicine dropper; wax marking pencil; and dropper bottles of red and green food dye, horse serum albumin, an unknown serum albumin sample, and antibodies to horse, bovine, and swine albumin. Put your initials and the number of the unknown albumin sample used on the bottom of the petri dish near the edge.

2. Use the wax marking pencil and the template **(Figure 35.8)** to divide the dish into three sections, and mark them I, II, and III.

3. Prepare sample wells (again using the template in Figure 35.8). Squeeze the medicine dropper bulb, and gently touch the tip to the surface of the agar. While releasing the bulb, push the tip down through the agar to the bottom of the dish. Lift the dropper vertically; this should leave a straight-walled well in the agar.

4. Repeat step 3 so that section I has two wells and sections II and III have four wells each.

5. To observe diffusion through the gel, nearly fill one well in section I with red dye and the other well with green dye. Be careful not to overfill the wells. Observe periodically for 30 to 45 minutes as the dyes diffuse through the agar. Draw your results as instructed in the Results section.

6. To demonstrate positive and negative results, fill the wells in section II as instructed **(Table 35.1)**. A precipitin line should form only between wells 1 and 2.

7. To test the unknown sample, fill the wells in section III as instructed **(Table 35.2)**.

8. Replace the cover on the petri dish, and incubate at room temperature for at least 16 hours. Make arrangements to observe the agar for precipitin lines after 16 hours. *The lines may begin to fade after 48 hours.* Draw the results as

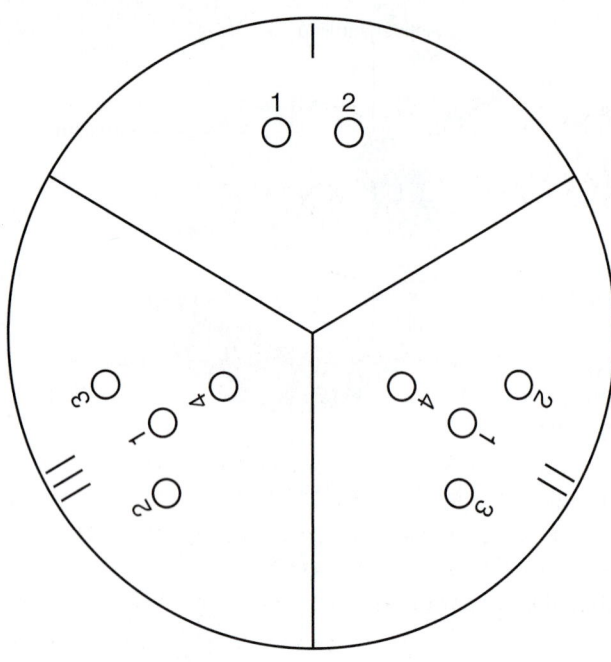

Figure 35.8 Template for well preparation for the Ouchterlony double-gel diffusion experiment.

instructed in the following section, parts 1–3, indicating the location of all precipitin lines that form.

Results

1. For the demonstration of diffusion using dye, draw on the template (Figure 35.8) the appearance of section I of your dish after 30 to 45 minutes. Use colored pencils.

2. Draw section II on the template (Figure 35.8) as it appears after incubation, 16 to 48 hours. Be sure the wells are numbered.

3. Draw section III on the template as it appears after incubation, 16 to 48 hours. Be sure the wells are numbered.

Unknown # _____

4. What evidence for diffusion did you observe in section I?

_____

_____

Table 35.1	Section II
Well	**Solution**
1	Horse serum albumin
2	Goat anti–horse albumin
3	Goat anti–bovine albumin
4	Goat anti–swine albumin

Table 35.2	Section III
Well	**Solution**
1	Unknown # _____
2	Goat anti–horse albumin
3	Goat anti–bovine albumin
4	Goat anti–swine albumin

5. Is there any evidence of a precipitate in section I?

_____

6. Which of the sera functioned as an antigen in section II?

_____

7. Which antibody reacted with the antigen in section II?

_____

How do you know? (Be specific about your observations.)

_____

_____

8. If swine albumin had been placed in well 1, what would you expect to happen? Explain.

_____

_____

9. If chicken albumin had been placed in well 1, what would you expect to happen? Explain.

_____

_____

10. What antigens were present in the unknown solution?

_____

How do you know? (Be specific about your observations.)

_____

_____

Anatomy of the Respiratory System

MATERIALS

- ☐ Resin cast of the respiratory tree (if available)
- ☐ Human torso model
- ☐ Thoracic cavity structures model and/or chart of the respiratory system
- ☐ Larynx model (if available)
- ☐ Preserved inflatable lung preparation (obtained from a biological supply house) or sheep pluck fresh from the slaughterhouse
- ☐ Source of compressed air
- ☐ Dissecting tray
- ☐ Disposable gloves
- ☐ Disposable autoclave bag
- ☐ Prepared slides of the following (if available): trachea (cross section), lung tissue, both normal and pathological specimens (for example, sections taken from lung tissues exhibiting bronchitis, pneumonia, emphysema, or lung cancer)
- ☐ Compound and stereomicroscopes

✂ For instructions on animal dissections, see the dissection exercises (starting on page 697) in the cat and fetal pig, editions of this manual.

MasteringA&P® For related exercise study tools, go to the Study Area of MasteringA&P. There you will find:
- Practice Anatomy Lab PAL
- PhysioEx PEx
- A&PFlix A&PFlix
- Practice quizzes, Histology Atlas, eText, Videos, and more!

OBJECTIVES

1. State the major functions of the respiratory system.
2. Define the following terms: *pulmonary ventilation*, *external respiration*, and *internal respiration*.
3. Identify the major respiratory system structures on models or appropriate images, and describe the function of each.
4. Describe the difference between the conducting and respiratory zones and indicate which is referred to as anatomical dead space.
5. Name the serous membrane that encloses each lung and describe its structure.
6. Demonstrate lung inflation in a fresh sheep pluck or preserved tissue specimen.
7. Recognize the histologic structure of the trachea and lung tissue microscopically or in an image, and describe the functions served by the observed structures.

PRE-LAB QUIZ

1. The major role of the respiratory system is to
   a. dispose of waste products in a solid form
   b. permit the flow of nutrients through the body
   c. supply the body with carbon dioxide and dispose of oxygen
   d. supply the body with oxygen and dispose of carbon dioxide
2. Circle True or False. Four processes—pulmonary ventilation, external respiration, transport of respiratory gases, and internal respiration—must all occur in order for the respiratory system to function fully.
3. The upper respiratory structures include the nose, the larynx, and the
   a. epiglottis   c. pharynx
   b. lungs   d. trachea
4. Circle the correct underlined term. The thyroid cartilage / arytenoid cartilage is the largest and most prominent of the laryngeal cartilages.
5. Circle True or False. The epiglottis forms a lid over the larynx when we swallow food: it closes off the respiratory passageway to incoming food or drink.
6. Air flows from the larynx to the trachea, and then enters the
   a. left and right lungs   c. pharynx
   b. left and right main bronchi   d. segmental bronchi
7. Circle the correct underlined term. The lining of the trachea is pseudostratified ciliated columnar epithelium / transitional epithelium, which propels dust particles, bacteria, and other debris away from the lungs.
8. Circle True or False. All but the smallest branches of the bronchial tree have cartilaginous reinforcements in their walls.
9. _____, tiny balloonlike structures, are composed of a single thin layer of squamous epithelium. They are the main structural and functional units of the lung and the actual sites of gas exchange.
10. Circle the correct underlined term. Fissures divide the lungs into lobes, three on the right and two / three on the left.

Body cells require an abundant and continuous supply of oxygen. As the cells use oxygen, they release carbon dioxide, a waste product that the body must get rid of. These oxygen-using cellular processes, collectively referred to as *cellular respiration,* are more appropriately described along with the topic of cellular metabolism. The major role of the **respiratory system,** our focus in this exercise, is to supply the body with oxygen and dispose of carbon dioxide. To fulfill this role, at least four distinct processes, collectively referred to as **respiration,** must occur:

Pulmonary ventilation: The tidelike movement of air into and out of the lungs so that the gases in the alveoli are continuously changed and refreshed. Also more simply called *ventilation,* or *breathing.*

External respiration: The gas exchange between the blood and the air-filled chambers of the lungs (oxygen loading/carbon dioxide unloading).

Transport of respiratory gases: The transport of respiratory gases between the lungs and tissue cells of the body accomplished by the cardiovascular system, using blood as the transport vehicle.

Internal respiration: Exchange of gases between systemic blood and tissue cells (oxygen unloading and carbon dioxide loading).

Only the first two processes are exclusive to the respiratory system, but all four must occur for the respiratory system to "do its job." Hence, the respiratory and circulatory systems are irreversibly linked. If either system fails, cells begin to die from oxygen starvation and accumulation of carbon dioxide. Uncorrected, this situation soon causes death of the entire organism.

Upper Respiratory System Structures

The upper respiratory system structures—the nose, pharynx, and larynx—are described below (and illustrated in **Figure 36.1**). As you read through the descriptions, identify each structure in the figure.

Air generally passes into the respiratory tract through the **nostrils** or **nares,** and enters the **nasal cavity,** which is divided by the **nasal septum.** It then flows posteriorly over three pairs of lobelike structures, the **inferior, superior,** and **middle nasal conchae,** which increase the air turbulence. As the air passes through the nasal cavity, it is also warmed, moistened, and filtered by the nasal mucosa. The air that flows directly beneath the superior part of the nasal cavity may chemically stimulate the olfactory receptors located in the mucosa of that region. The nasal cavity is surrounded by the **paranasal sinuses** in the frontal, sphenoid, ethmoid, and maxillary bones. These sinuses, named for the bones in which they are located, act as resonance chambers in speech. Their mucosae, like that of the nasal cavity, warm and moisten the incoming air.

The nasal passages are separated from the oral cavity below by a partition composed anteriorly of the **hard palate** and posteriorly by the **soft palate.**

The genetic defect called **cleft palate** results from failure of the palatine bones and/or the palatine processes of the maxillary bones to fuse medially. It causes difficulty in breathing and oral cavity functions such as sucking and, later, chewing and speech. ✚

Of course, air may also enter the body via the mouth. From there it passes through the oral cavity to move into the pharynx posteriorly, where the oral and nasal cavities are joined temporarily.

Commonly called the *throat,* the funnel-shaped **pharynx** connects the nasal and oral cavities to the larynx and esophagus inferiorly. It has three named parts (Figure 36.1):

1. The **nasopharynx** lies posterior to the nasal cavity and is continuous with it via the **posterior nasal aperture.** It lies above the soft palate; hence, it serves only as an air passage. High on its posterior wall is the *pharyngeal tonsil,* masses of lymphoid tissue that help to protect the respiratory passages from invading pathogens. The *pharyngotympanic (auditory) tubes,* which allow middle ear pressure to become equalized to atmospheric pressure, drain into the lateral aspects of the nasopharynx. The *tubal tonsils* surround the openings of these tubes into the nasopharynx.

Because of the continuity of the middle ear and nasopharyngeal mucosae, nasal infections may invade the middle ear cavity and cause **otitis media** (middle ear inflammation), which is difficult to treat. ✚

2. The **oropharynx** is continuous posteriorly with the oral cavity. Since it extends from the soft palate to the epiglottis of the larynx inferiorly, it serves as a common conduit for food and air. In its lateral walls are the *palatine tonsils.* The *lingual tonsil* covers the base of the tongue.

3. The **laryngopharynx,** like the oropharynx, accommodates both ingested food and air. It lies directly posterior to the upright epiglottis and extends to the larynx, where the common pathway divides into the respiratory and digestive channels. From the laryngopharynx, air enters the lower respiratory passageways by passing through the larynx (voice box) and into the trachea below.

The **larynx (Figure 36.2)** consists of nine cartilages. The two most prominent are the large shield-shaped **thyroid cartilage,** whose anterior medial laryngeal prominence is commonly referred to as *Adam's apple,* and the inferiorly located, ring-shaped **cricoid cartilage,** whose widest dimension faces posteriorly. All the laryngeal cartilages are composed of hyaline cartilage except the flaplike **epiglottis,** a flexible elastic cartilage located superior to the opening of the larynx. The epiglottis, sometimes referred to as the "guardian of the airways," forms a lid over the larynx when we swallow. This closes off the respiratory passageways to incoming food or drink, which is routed into the posterior esophagus, or food chute.

☐ Palpate your larynx by placing your hand on the anterior neck surface approximately halfway down its length. Swallow. Can you feel the cartilaginous larynx rising?

Check the box when you have completed the task.

If anything other than air enters the larynx, a cough reflex attempts to expel the substance. Note that this reflex operates only when a person is conscious. Therefore, you should never try to feed or pour liquids down the throat of an unconscious person.

Anatomy of the Respiratory System

MATERIALS

☐ Resin cast of the respiratory tree (if available)

☐ Human torso model

☐ Thoracic cavity structures model and/or chart of the respiratory system

☐ Larynx model (if available)

☐ Preserved inflatable lung preparation (obtained from a biological supply house) or sheep pluck fresh from the slaughterhouse

☐ Source of compressed air

☐ Dissecting tray

☐ Disposable gloves

☐ Disposable autoclave bag

☐ Prepared slides of the following (if available): trachea (cross section), lung tissue, both normal and pathological specimens (for example, sections taken from lung tissues exhibiting bronchitis, pneumonia, emphysema, or lung cancer)

☐ Compound and stereomicroscopes

✂ For instructions on animal dissections, see the dissection exercises (starting on page 697) in the cat and fetal pig, editions of this manual.

MasteringA&P® For related exercise study tools, go to the Study Area of MasteringA&P. There you will find:

• Practice Anatomy Lab PAL

• PhysioEx PEx

• A&PFlix A&PFlix

• Practice quizzes, Histology Atlas, eText, Videos, and more!

OBJECTIVES

1. State the major functions of the respiratory system.

2. Define the following terms: *pulmonary ventilation, external respiration,* and *internal respiration.*

3. Identify the major respiratory system structures on models or appropriate images, and describe the function of each.

4. Describe the difference between the conducting and respiratory zones and indicate which is referred to as anatomical dead space.

5. Name the serous membrane that encloses each lung and describe its structure.

6. Demonstrate lung inflation in a fresh sheep pluck or preserved tissue specimen.

7. Recognize the histologic structure of the trachea and lung tissue microscopically or in an image, and describe the functions served by the observed structures.

PRE-LAB QUIZ

1. The major role of the respiratory system is to
   a. dispose of waste products in a solid form
   b. permit the flow of nutrients through the body
   c. supply the body with carbon dioxide and dispose of oxygen
   d. supply the body with oxygen and dispose of carbon dioxide

2. Circle True or False. Four processes—pulmonary ventilation, external respiration, transport of respiratory gases, and internal respiration—must all occur in order for the respiratory system to function fully.

3. The upper respiratory structures include the nose, the larynx, and the
   a. epiglottis   c. pharynx
   b. lungs   d. trachea

4. Circle the correct underlined term. The <u>thyroid cartilage</u> / <u>arytenoid cartilage</u> is the largest and most prominent of the laryngeal cartilages.

5. Circle True or False. The epiglottis forms a lid over the larynx when we swallow food: it closes off the respiratory passageway to incoming food or drink.

6. Air flows from the larynx to the trachea, and then enters the
   a. left and right lungs   c. pharynx
   b. left and right main bronchi   d. segmental bronchi

7. Circle the correct underlined term. The lining of the trachea is pseudostratified ciliated <u>columnar epithelium</u> / <u>transitional epithelium,</u> which propels dust particles, bacteria, and other debris away from the lungs.

8. Circle True or False. All but the smallest branches of the bronchial tree have cartilaginous reinforcements in their walls.

9. _____, tiny balloonlike structures, are composed of a single thin layer of squamous epithelium. They are the main structural and functional units of the lung and the actual sites of gas exchange.

10. Circle the correct underlined term. Fissures divide the lungs into lobes, three on the right and <u>two</u> / <u>three</u> on the left.

Body cells require an abundant and continuous supply of oxygen. As the cells use oxygen, they release carbon dioxide, a waste product that the body must get rid of. These oxygen-using cellular processes, collectively referred to as *cellular respiration,* are more appropriately described along with the topic of cellular metabolism. The major role of the **respiratory system,** our focus in this exercise, is to supply the body with oxygen and dispose of carbon dioxide. To fulfill this role, at least four distinct processes, collectively referred to as **respiration,** must occur:

Pulmonary ventilation: The tidelike movement of air into and out of the lungs so that the gases in the alveoli are continuously changed and refreshed. Also more simply called *ventilation,* or *breathing.*

External respiration: The gas exchange between the blood and the air-filled chambers of the lungs (oxygen loading/carbon dioxide unloading).

Transport of respiratory gases: The transport of respiratory gases between the lungs and tissue cells of the body accomplished by the cardiovascular system, using blood as the transport vehicle.

Internal respiration: Exchange of gases between systemic blood and tissue cells (oxygen unloading and carbon dioxide loading).

Only the first two processes are exclusive to the respiratory system, but all four must occur for the respiratory system to "do its job." Hence, the respiratory and circulatory systems are irreversibly linked. If either system fails, cells begin to die from oxygen starvation and accumulation of carbon dioxide. Uncorrected, this situation soon causes death of the entire organism.

Upper Respiratory System Structures

The upper respiratory system structures—the nose, pharynx, and larynx—are described below (and illustrated in **Figure 36.1**). As you read through the descriptions, identify each structure in the figure.

Air generally passes into the respiratory tract through the **nostrils** or **nares,** and enters the **nasal cavity,** which is divided by the **nasal septum.** It then flows posteriorly over three pairs of lobelike structures, the **inferior, superior,** and **middle nasal conchae,** which increase the air turbulence. As the air passes through the nasal cavity, it is also warmed, moistened, and filtered by the nasal mucosa. The air that flows directly beneath the superior part of the nasal cavity may chemically stimulate the olfactory receptors located in the mucosa of that region. The nasal cavity is surrounded by the **paranasal sinuses** in the frontal, sphenoid, ethmoid, and maxillary bones. These sinuses, named for the bones in which they are located, act as resonance chambers in speech. Their mucosae, like that of the nasal cavity, warm and moisten the incoming air.

The nasal passages are separated from the oral cavity below by a partition composed anteriorly of the **hard palate** and posteriorly by the **soft palate.**

The genetic defect called **cleft palate** results from failure of the palatine bones and/or the palatine processes of the maxillary bones to fuse medially. It causes difficulty in breathing and oral cavity functions such as sucking and, later, chewing and speech. ✚

Of course, air may also enter the body via the mouth. From there it passes through the oral cavity to move into the pharynx posteriorly, where the oral and nasal cavities are joined temporarily.

Commonly called the *throat,* the funnel-shaped **pharynx** connects the nasal and oral cavities to the larynx and esophagus inferiorly. It has three named parts (Figure 36.1):

1. The **nasopharynx** lies posterior to the nasal cavity and is continuous with it via the **posterior nasal aperture.** It lies above the soft palate; hence, it serves only as an air passage. High on its posterior wall is the *pharyngeal tonsil,* masses of lymphoid tissue that help to protect the respiratory passages from invading pathogens. The *pharyngotympanic (auditory) tubes,* which allow middle ear pressure to become equalized to atmospheric pressure, drain into the lateral aspects of the nasopharynx. The *tubal tonsils* surround the openings of these tubes into the nasopharynx.

Because of the continuity of the middle ear and nasopharyngeal mucosae, nasal infections may invade the middle ear cavity and cause **otitis media** (middle ear inflammation), which is difficult to treat. ✚

2. The **oropharynx** is continuous posteriorly with the oral cavity. Since it extends from the soft palate to the epiglottis of the larynx inferiorly, it serves as a common conduit for food and air. In its lateral walls are the *palatine tonsils.* The *lingual tonsil* covers the base of the tongue.

3. The **laryngopharynx,** like the oropharynx, accommodates both ingested food and air. It lies directly posterior to the upright epiglottis and extends to the larynx, where the common pathway divides into the respiratory and digestive channels. From the laryngopharynx, air enters the lower respiratory passageways by passing through the larynx (voice box) and into the trachea below.

The **larynx (Figure 36.2)** consists of nine cartilages. The two most prominent are the large shield-shaped **thyroid cartilage,** whose anterior medial laryngeal prominence is commonly referred to as *Adam's apple,* and the inferiorly located, ring-shaped **cricoid cartilage,** whose widest dimension faces posteriorly. All the laryngeal cartilages are composed of hyaline cartilage except the flaplike **epiglottis,** a flexible elastic cartilage located superior to the opening of the larynx. The epiglottis, sometimes referred to as the "guardian of the airways," forms a lid over the larynx when we swallow. This closes off the respiratory passageways to incoming food or drink, which is routed into the posterior esophagus, or food chute.

☐ Palpate your larynx by placing your hand on the anterior neck surface approximately halfway down its length. Swallow. Can you feel the cartilaginous larynx rising?

Check the box when you have completed the task.

If anything other than air enters the larynx, a cough reflex attempts to expel the substance. Note that this reflex operates only when a person is conscious. Therefore, you should never try to feed or pour liquids down the throat of an unconscious person.

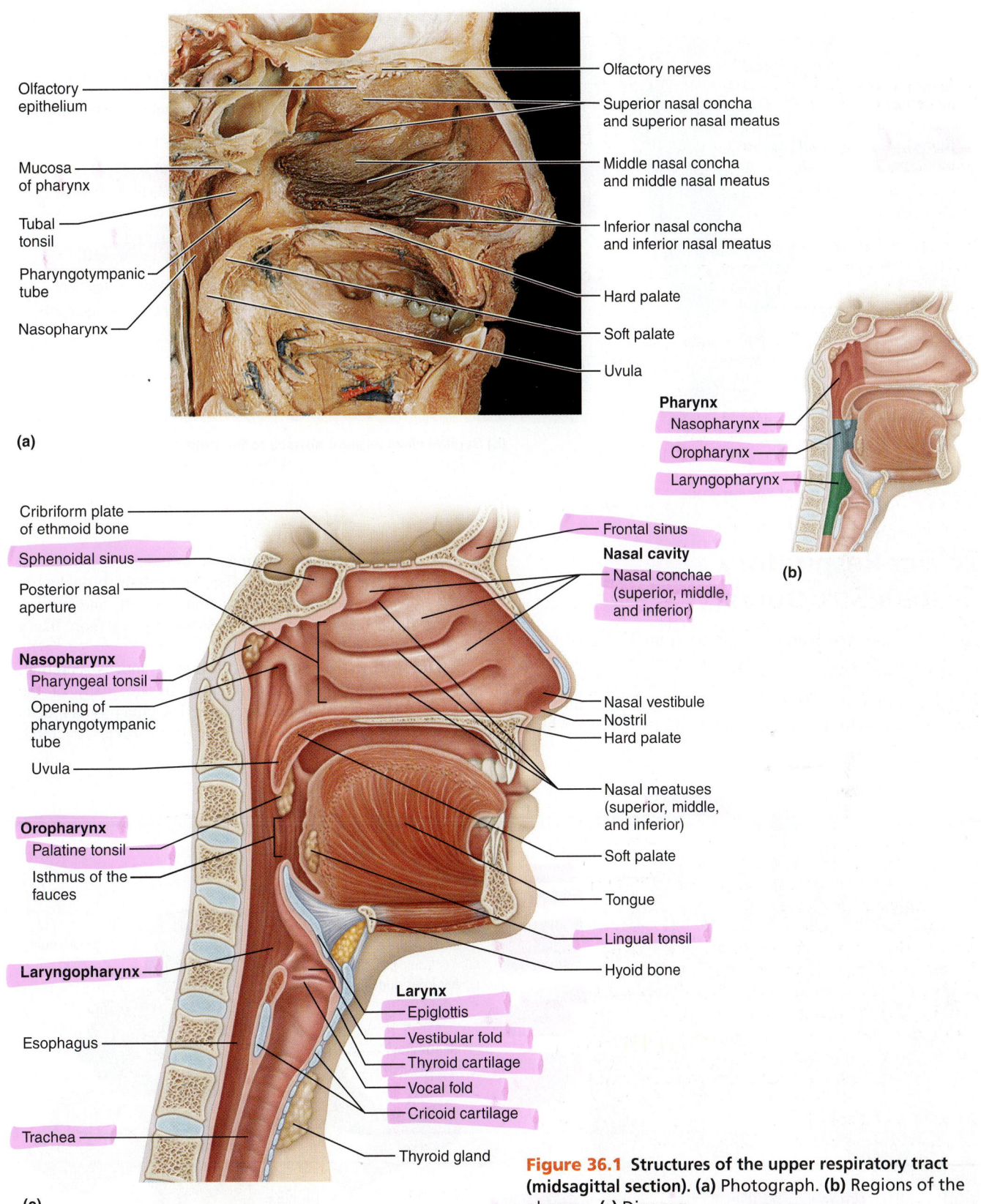

(a)

Olfactory epithelium
Mucosa of pharynx
Tubal tonsil
Pharyngotympanic tube
Nasopharynx
Olfactory nerves
Superior nasal concha and superior nasal meatus
Middle nasal concha and middle nasal meatus
Inferior nasal concha and inferior nasal meatus
Hard palate
Soft palate
Uvula

Pharynx
Nasopharynx
Oropharynx
Laryngopharynx

(b)

Cribriform plate of ethmoid bone
Sphenoidal sinus
Posterior nasal aperture
Nasopharynx
Pharyngeal tonsil
Opening of pharyngotympanic tube
Uvula
Oropharynx
Palatine tonsil
Isthmus of the fauces
Laryngopharynx
Esophagus
Trachea
(c)

Frontal sinus
Nasal cavity
Nasal conchae (superior, middle, and inferior)
Nasal vestibule
Nostril
Hard palate
Nasal meatuses (superior, middle, and inferior)
Soft palate
Tongue
Lingual tonsil
Hyoid bone
Larynx
Epiglottis
Vestibular fold
Thyroid cartilage
Vocal fold
Cricoid cartilage
Thyroid gland

Figure 36.1 Structures of the upper respiratory tract (midsagittal section). (a) Photograph. **(b)** Regions of the pharynx. **(c)** Diagram.

36

The mucous membrane of the larynx is thrown into two pairs of folds—the upper **vestibular folds,** also called the **false vocal cords,** and the lower **vocal folds,** or **true vocal cords,** which vibrate with expelled air for speech. The vocal cords are attached posterolaterally to the small triangular **arytenoid cartilages** by the *vocal ligaments*. The vocal folds and the slitlike passageway between them is called the **glottis.**

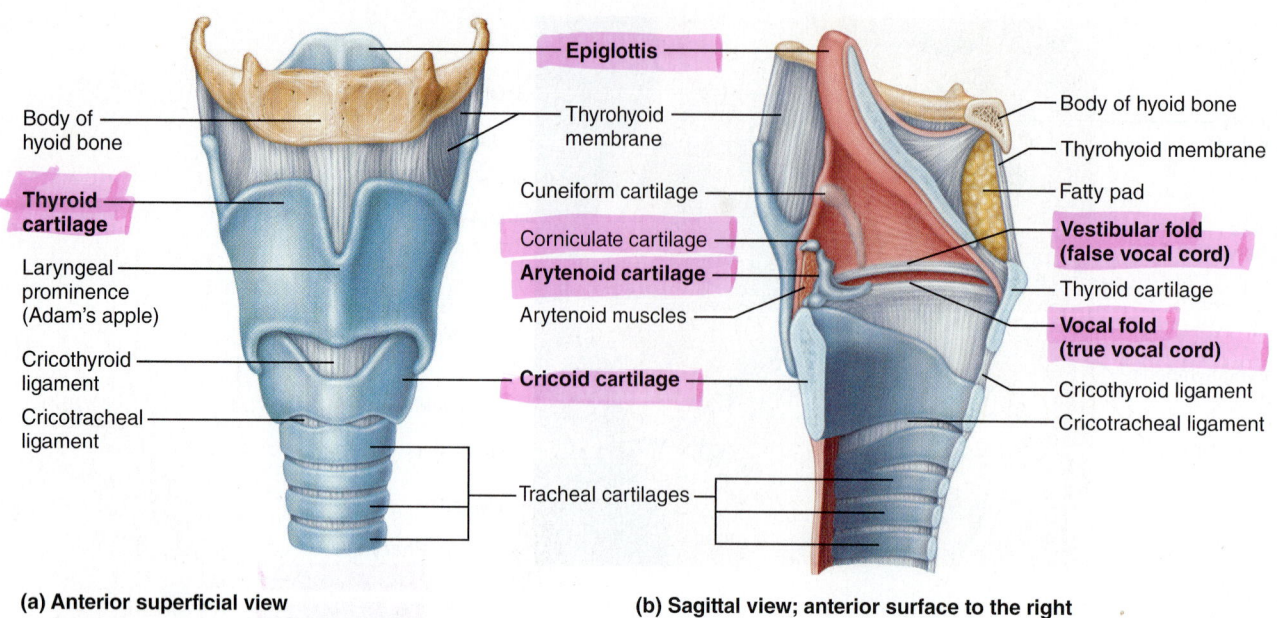

(a) Anterior superficial view

(b) Sagittal view; anterior surface to the right

Figure 36.2 Structure of the larynx.

Lower Respiratory System Structures

Air entering the **trachea,** or windpipe, from the larynx travels down its length (about 11.0 cm or 4 inches) to the level of the *sternal angle* (or the disc between the fourth and fifth thoracic vertebrae). There the passageway divides into the right and left **main (primary) bronchi (Figure 36.3)**, which plunge into their respective lungs at an indented area called the **hilum** (see Figure 36.5c). The right main bronchus is wider, shorter, and more vertical than the left, and foreign objects that enter the respiratory passageways are more likely to become lodged in it.

The trachea is lined with a ciliated, mucus-secreting, pseudostratified columnar epithelium, as are many of the other respiratory system passageways. The cilia propel mucus (produced by goblet cells) laden with dust particles, bacteria,

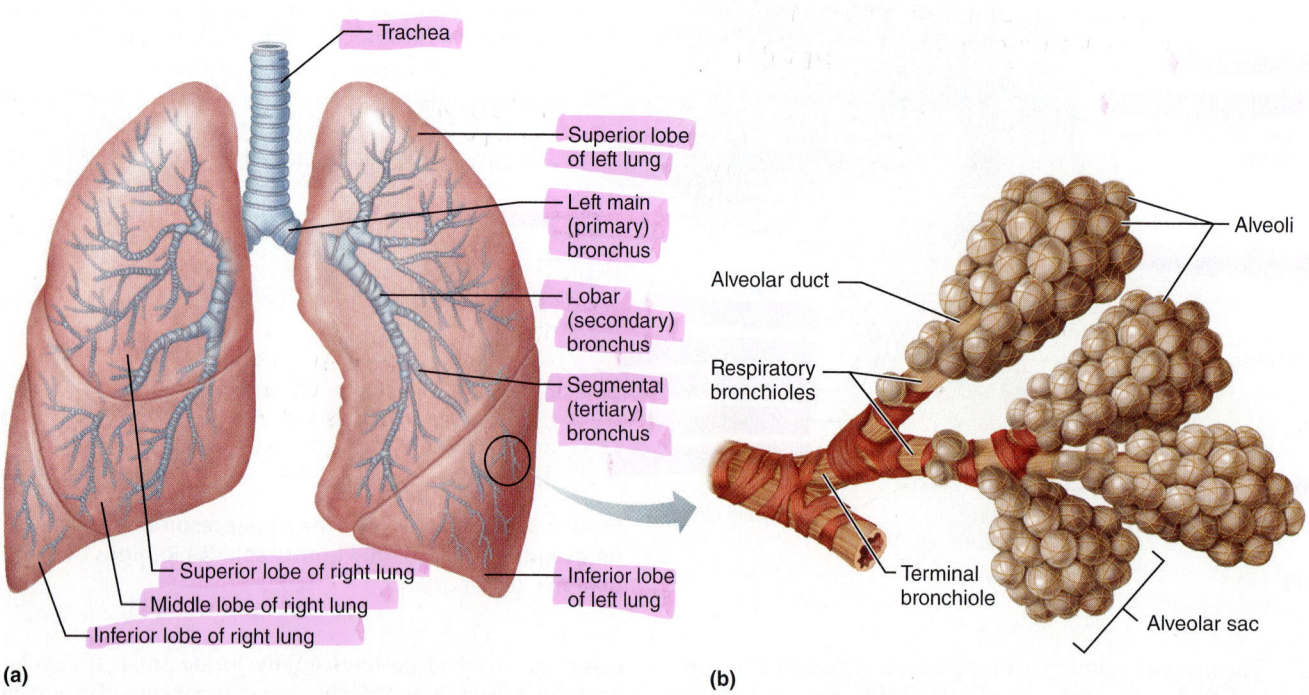

Figure 36.3 Structures of the lower respiratory tract. **(a)** Diagram. **(b)** Enlarged view of alveoli.

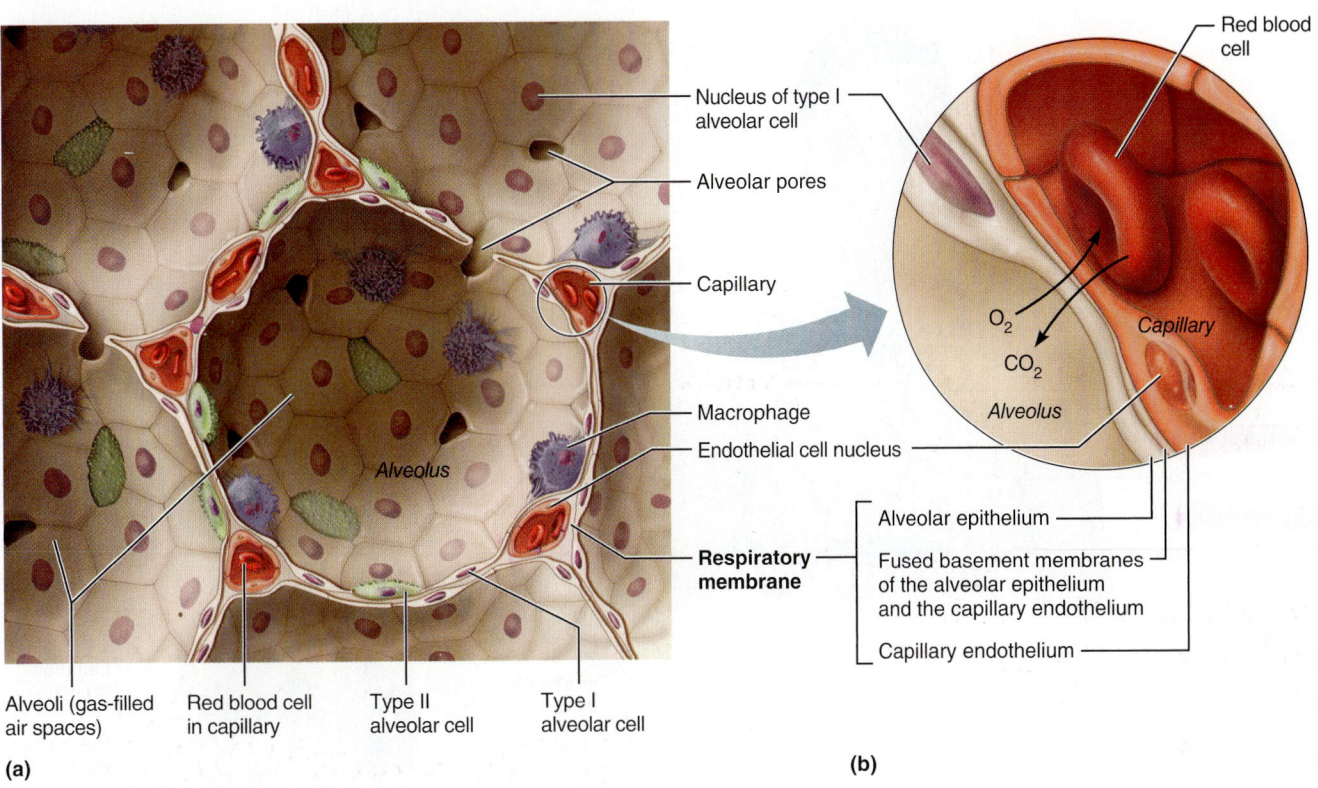

Nucleus of type I alveolar cell
Alveolar pores
Capillary
Macrophage
Endothelial cell nucleus
Respiratory membrane
Alveolus
Red blood cell
O₂
CO₂
Capillary
Alveolar epithelium
Fused basement membranes of the alveolar epithelium and the capillary endothelium
Capillary endothelium

Alveoli (gas-filled air spaces)
Red blood cell in capillary
Type II alveolar cell
Type I alveolar cell
(a)
(b)

Figure 36.4 **Diagram of the relationship between the alveoli and pulmonary capillaries involved in gas exchange. (a)** One alveolus surrounded by capillaries. **(b)** Enlargement of the respiratory membrane.

and other debris away from the lungs and toward the throat, where it can be expectorated or swallowed. The walls of the trachea are reinforced with **C**-shaped cartilaginous rings, the incomplete portion located posteriorly (see Figure 36.6, page 543). These **C**-shaped cartilages serve a double function: The incomplete parts allow the esophagus to expand anteriorly when a large food bolus is swallowed. The solid portions reinforce the trachea walls to maintain its open passageway regardless of the pressure changes that occur during breathing.

The main bronchi further divide into smaller and smaller branches (the secondary, tertiary, on down), finally becoming the **bronchioles,** which have terminal branches called **respiratory bronchioles** (Figure 36.3b). All but the most minute branches have cartilaginous reinforcements in their walls, usually in the form of small plates of hyaline cartilage rather than cartilaginous rings. As the respiratory tubes get smaller and smaller, the relative amount of smooth muscle in their walls increases as the amount of cartilage declines and finally disappears. The complete layer of smooth muscle present in the bronchioles enables them to provide considerable resistance to air flow under certain conditions (asthma, hay fever, etc.). The continuous branching of the respiratory passageways in the lungs is often referred to as the **bronchial tree.** The comparison becomes much more meaningful in a resin cast of the respiratory passages.

• Observe a resin cast of respiratory passages if one is available.

The respiratory bronchioles in turn subdivide into several **alveolar ducts,** which terminate in alveolar sacs that

rather resemble clusters of grapes. **Alveoli,** tiny balloonlike expansions along the alveolar sacs and occasionally found protruding from alveolar ducts and respiratory bronchioles, are composed of a single thin layer of squamous epithelium overlying a wispy basal lamina. The external surfaces of the alveoli are densely spiderwebbed with a network of pulmonary capillaries **(Figure 36.4)**. Together, the alveolar and capillary walls and their fused basement membranes form the **respiratory membrane,** also called the *blood air barrier.*

Because gas exchanges occur by simple diffusion across the respiratory membrane—oxygen passing from the alveolar air to the capillary blood and carbon dioxide leaving the capillary blood to enter the alveolar air—the alveolar sacs, alveolar ducts, and respiratory bronchioles are referred to collectively as **respiratory zone structures.** All other respiratory passageways (from the nasal cavity to the terminal bronchioles) simply serve as access or exit routes to and from these gas exchange chambers and are called **conducting zone structures.** Because the conducting zone structures have no exchange function, they are also referred to as *anatomical dead space.*

The Lungs and Their Pleural Coverings

The paired lungs are soft, spongy organs that occupy the entire thoracic cavity except for the *mediastinum,* which houses the heart, bronchi, esophagus, and other organs **(Figure 36.5)**. Each lung is connected to the mediastinum by a **root** containing its vascular and bronchial attachments. The structures of the root enter (or leave) the lung via a medial indentation called the *hilum.* All structures distal to the primary bronchi

36

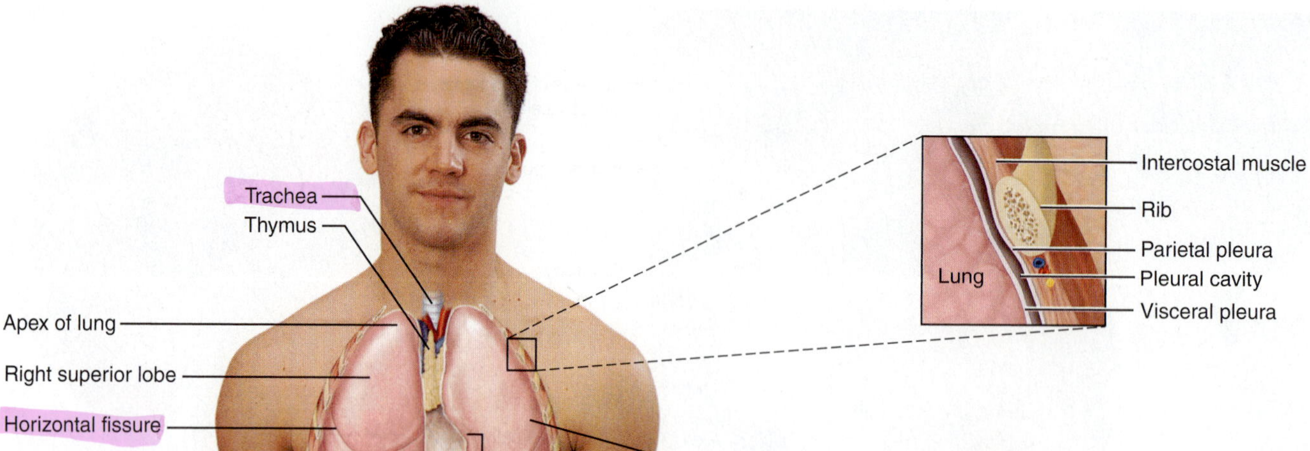

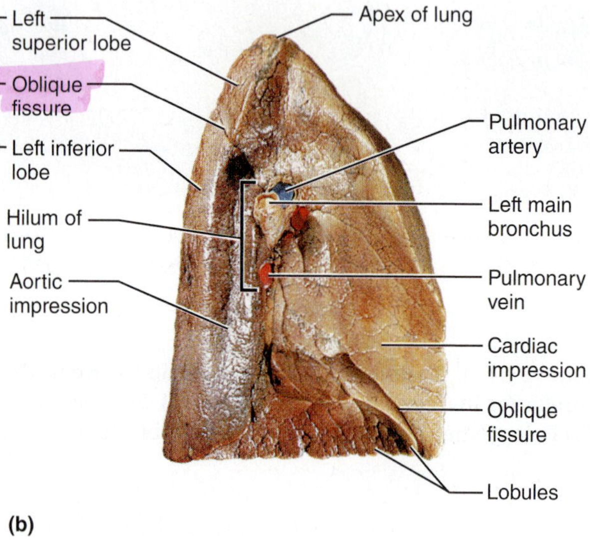

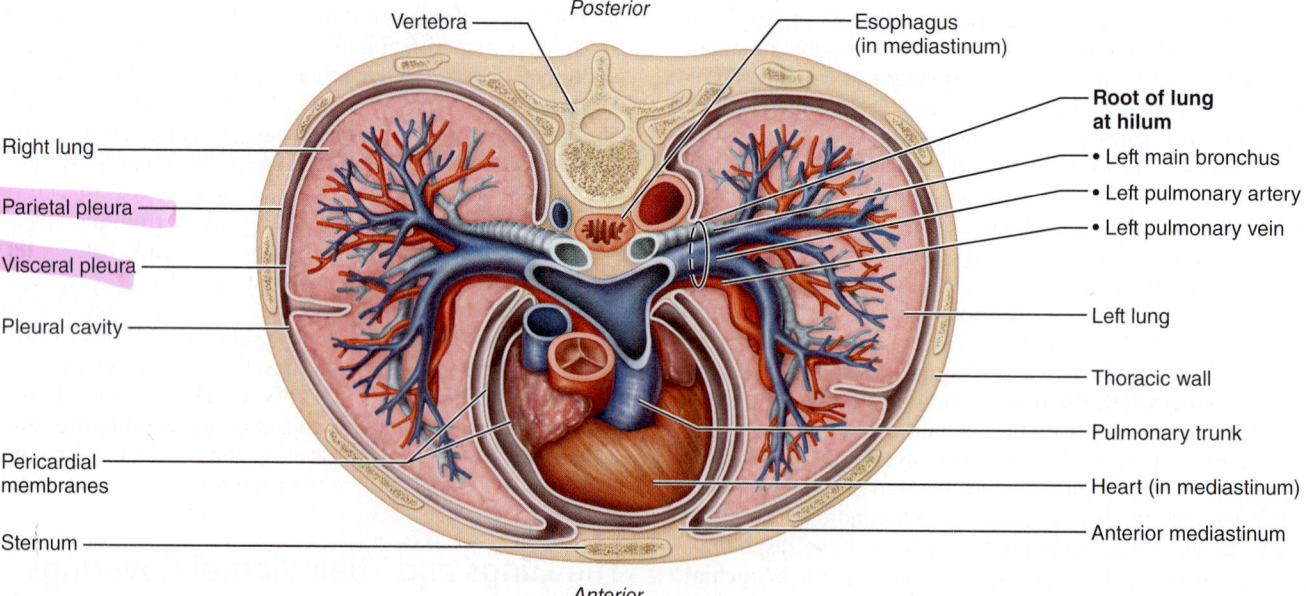

Figure 36.5 Anatomical relationships of organs in the thoracic cavity. (a) Anterior view of the thoracic organs. The lungs flank the central mediastinum. The inset at upper right depicts the pleurae and the pleural cavity. **(b)** Photograph of medial aspect of left lung. **(c)** Transverse section through the superior part of the thorax, showing the lungs and the main organs in the mediastinum.

(a)

(b)

(c)

36

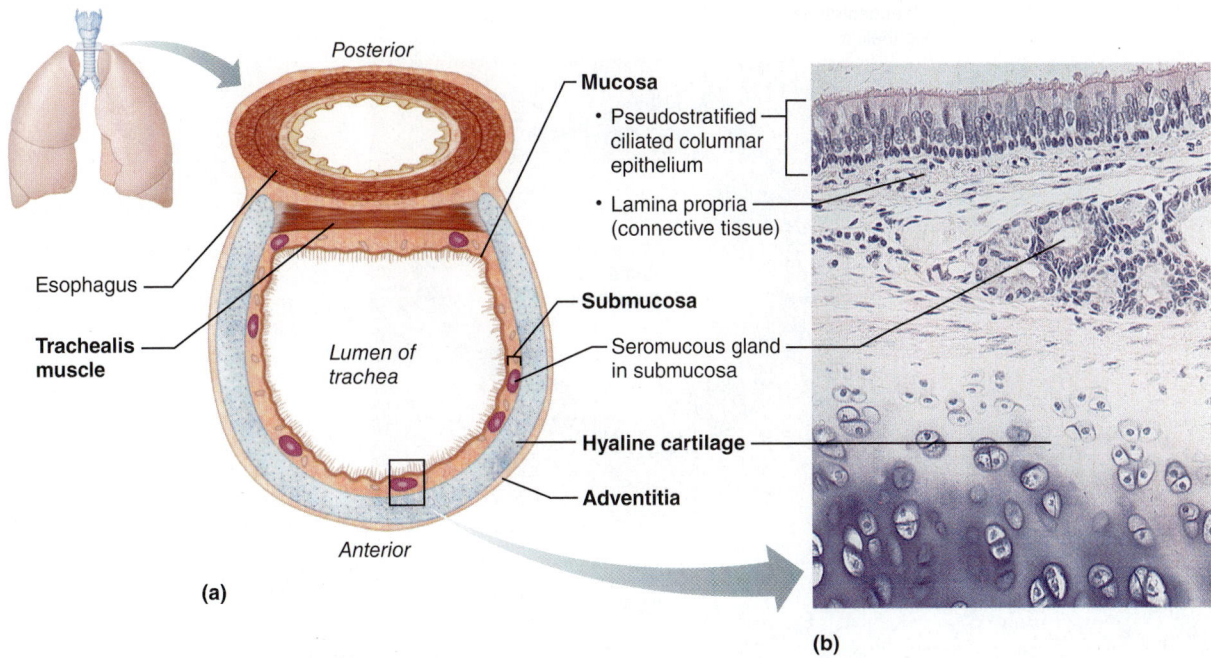

Posterior

Mucosa
• Pseudostratified ciliated columnar epithelium
• Lamina propria (connective tissue)

Esophagus

Trachealis muscle

Lumen of trachea

Submucosa

Seromucous gland in submucosa

Hyaline cartilage

Adventitia

Anterior

(a)

(b)

Figure 36.6 Tissue composition of the tracheal wall. (a) Cross-sectional view of the trachea. **(b)** Photomicrograph of a portion of the tracheal wall (125×).

are found within the lung substance. A lung's **apex,** the narrower superior aspect, lies just deep to the clavicle, and its **base,** the inferior concave surface, rests on the diaphragm. Anterior, lateral, and posterior lung surfaces are in close contact with the ribs and, hence, are collectively called the **costal surface.** The medial surface of the left lung exhibits a concavity called the **cardiac notch,** which accommodates the heart where it extends left from the body midline. Fissures divide the lungs into a number of **lobes**—two in the left lung and three in the right. Other than the respiratory passageways and air spaces that make up the bulk of their volume, the lungs are mostly elastic connective tissue, which allows them to recoil passively during expiration.

Each lung is enclosed in a double-layered sac of serous membrane called the **pleura.** The outer layer, the **parietal pleura,** is attached to the thoracic walls and the **diaphragm;** the inner layer, covering the lung tissue, is the **visceral pleura.** The two pleural layers are separated by the **pleural cavity,** which is more of a potential space than an actual one. The pleural layers produce lubricating serous fluid that causes them to adhere closely to one another, holding the lungs to the thoracic wall and allowing them to move easily against one another during the movements of breathing.

ACTIVITY 1

Identifying Respiratory System Organs

Before proceeding, be sure to locate on the torso model, thoracic cavity structures model, larynx model, or an anatomical chart all the respiratory structures described—both upper and lower respiratory system organs. ■

 For instructions on animal dissections, see the dissection exercises (starting on page 697) in the cat and fetal pig editions of this manual.

ACTIVITY 2

Demonstrating Lung Inflation in a Sheep Pluck

A *sheep pluck* includes the larynx, trachea with attached lungs, the heart and pericardium, and portions of the major blood vessels found in the mediastinum (aorta, pulmonary artery and vein, venae cavae). If a sheep pluck is not available, a good substitute is a preserved inflatable pig lung.

⚠ Don disposable gloves, obtain a dissecting tray and a fresh sheep pluck (or a preserved pluck of another animal), and identify the lower respiratory system organs. Once you have completed your observations, insert a hose from an air compressor (vacuum pump) into the trachea and alternately allow air to flow in and out of the lungs. Notice how the lungs inflate. This observation is educational in a preserved pluck but it is a spectacular sight in a fresh one. Another advantage of using a fresh pluck is that the lung pluck changes color (becomes redder) as hemoglobin in trapped RBCs becomes loaded with oxygen.

⚠ Dispose of the gloves in the autoclave bag immediately after use. ■

ACTIVITY 3

Examining Prepared Slides of Trachea and Lung Tissue

1. Obtain a compound microscope and a slide of a cross section of the tracheal wall. Identify the smooth muscle layer, the hyaline cartilage supporting rings, and the pseudostratified ciliated epithelium (use **Figure 36.6** as a guide). Also try to identify a few goblet cells in the epithelium (see Figure 6.3c and d on page 71).

36

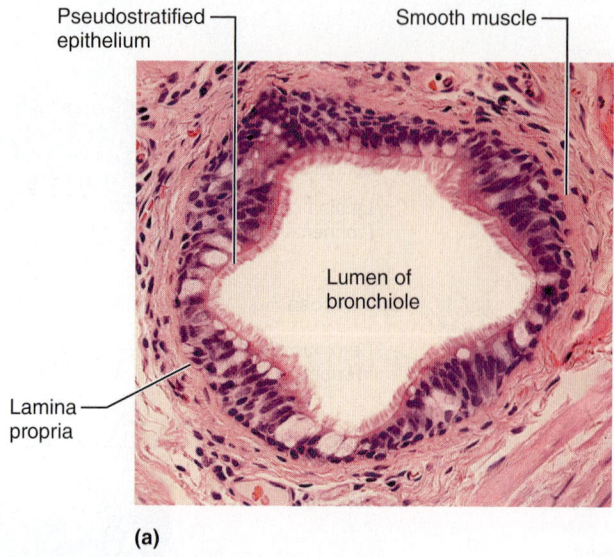

Pseudostratified epithelium

Smooth muscle

Lumen of bronchiole

Lamina propria

(a)

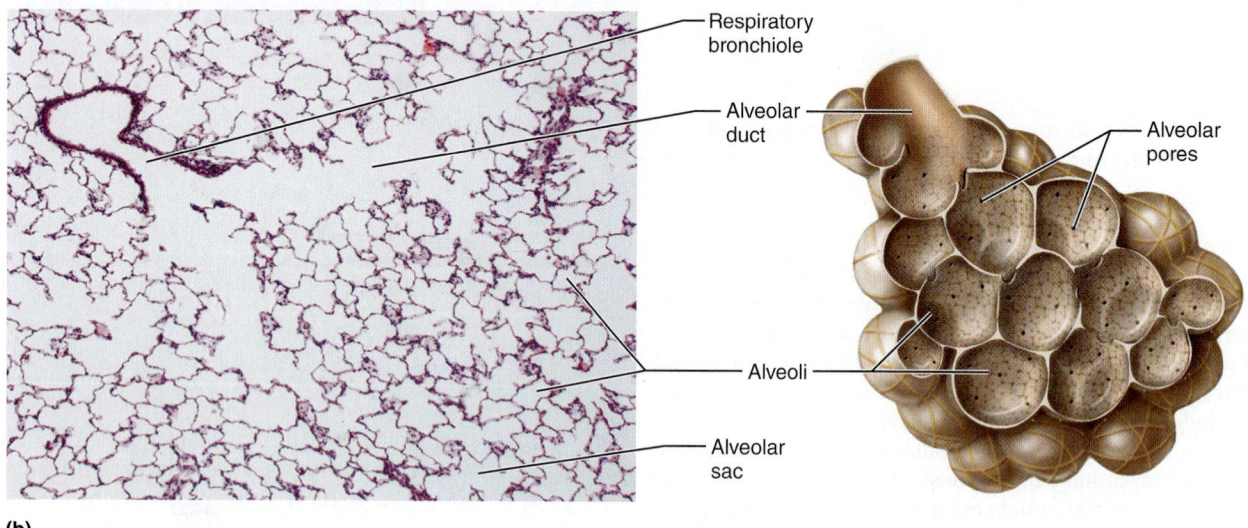

Respiratory bronchiole

Alveolar duct

Alveolar pores

Alveoli

Alveolar sac

(b)

Figure 36.7 Microscopic structure of a bronchiole and alveoli. (a) Photomicrograph of a section of a bronchiole (180×). **(b)** Photomicrograph showing the final divisions of the bronchial tree (50×) and diagram of alveoli.

2. Obtain a slide of lung tissue for examination. The alveolus is the main structural and functional unit of the lung and is the actual site of gas exchange. Identify a bronchiole **(Figure 36.7a)** and the thin squamous epithelium of the alveolar walls (Figure 36.7b).

3. Examine slides of pathological lung tissues, and compare them to the normal lung specimens. Record your observations in the review sheet at the end of this exercise. ■

36

Respiratory System Physiology

MATERIALS

- ☐ Model lung (bell jar demonstrator)
- ☐ Tape measure with centimeter divisions (cloth or plastic)
- ☐ Stethoscope
- ☐ Alcohol swabs
- ☐ Spirometer or BIOPAC® equipment:

Spirometer, disposable cardboard mouthpieces, nose clips, table (on board) for recording class data, disposable autoclave bag, battery jar containing 70% ethanol solution

BIOPAC® BIOPAC® BSL System for Windows with BSL software version 3.7.5 to 3.7.7, or BSL System for Mac OS X with BSL software version 3.7.4 to 3.7.7, MP36/35 or MP45 data acquisition unit, PC or Mac computer, BIOPAC® airflow transducer, BIOPAC® calibration syringe, disposable mouthpiece, nose clip, and bacteriological filter

Instructors using the MP36 (or MP35/30) data acquisition unit with BSL software versions earlier than 3.7.5 (for Windows) and 3.7.4 (for Mac Os X) will need slightly different channel settings and collection strategies. Instructions for using the older data acquisition unit can be found on MasteringA&P.

- ☐ Paper bag
- ☐ 0.05 M NaOH
- ☐ Phenol red in a dropper bottle

(Text continues on next page.)

MasteringA&P® For related exercise study tools, go to the Study Area of MasteringA&P. There you will find:

- Practice Anatomy Lab PAL
- PhysioEx PEx
- A&PFlix A&PFlix
- Practice quizzes, Histology Atlas, eText, Videos, and more!

OBJECTIVES

1. Define the following and provide volume figures if applicable:

   inspiration   inspiratory reserve volume (IRV)

   expiration   minute respiratory volume (MRV)

   tidal volume (TV)   forced vital capacity (FVC)

   vital capacity (VC)   forced expiratory volume (FEV_T)

   expiratory reserve volume (ERV)

2. Explain the role of muscles and volume changes in the mechanical process of breathing.

3. Describe bronchial and vesicular breathing sounds.

4. Demonstrate proper usage of a spirometer or an airflow transducer and associated BIOPAC® equipment.

5. Discuss the relative importance of various mechanical and chemical factors in producing respiratory variations.

6. Explain the importance of the carbonic acid–bicarbonate buffer system in maintaining blood pH.

PRE-LAB QUIZ

1. Circle the correct underlined term. <u>Inspiration</u> / <u>Expiration</u> is the phase of pulmonary ventilation when air passes out of the lungs.

2. Which of the following processes does *not* occur during inspiration?
   a. diaphragm moves to a flattened position
   b. gas pressure inside the lungs is lowered
   c. inspiratory muscles relax
   d. size of thoracic cavity increases

3. Circle True or False. Vesicular breathing sounds are produced by air rushing through the trachea and bronchi.

4. During normal quiet breathing, about _____ ml of air moves into and out of the lungs with each breath.
   a. 250   c. 1000
   b. 500   d. 2000

5. Circle the correct underlined term. <u>Tidal volume</u> / <u>Vital capacity</u> is the maximum amount of air that can be exhaled after a maximal inspiration.

6. Circle True or False. The neural centers that control respiratory rhythm and maintain a rate of 12–18 respirations per minute are located in the medulla and thalamus.

7. Circle the correct underlined term. Changes in pH and oxygen concentrations in the blood are monitored by chemoreceptor regions in the <u>medulla</u> / <u>aortic and carotid bodies</u>.

8. The carbonic acid-bicarbonate buffer system stabilizes arterial blood pH at:
   a. 2.0 ± 1.00   c. 7.4 ± 0.02
   b. 6.2 ± 0.07   d. 9.5 ±1.15

(Text continues on next page.)

(Materials list continued.)

☐ 100-ml beakers
☐ Distilled water
☐ Straws
☐ Concentrated HCl and NaOH in dropper bottles
☐ 250- and 50-ml beakers
☐ Plastic wash bottles containing distilled water
☐ pH meter (standardized with buffer of pH 7)
☐ Buffer solution (pH 7)
☐ Graduated cylinder (100 ml)
☐ Glass stirring rod
☐ Animal plasma
☐ 0.01 *M* HCl in dropper bottles
☐ **PEx** PhysioEx™ 9.1 Computer Simulation Ex. 7 on p. PEx-105

Note: Instructions for using PowerLab® equipment can be found on MasteringA&P.

9. Circle the correct underlined term. <u>Acids</u> / <u>Bases</u> released into the blood by the body cells tend to lower the pH of the blood and cause it to become acidic.

10. Circle True or False. Rate and depth of breathing, hyperventilation, and hypoventilation should have little or no effect on the acid-base balance of blood.

The body's trillions of cells require O_2 and give off CO_2 as a waste the body must get rid of. The **respiratory system** provides the link with the external environment for both taking in O_2 and eliminating CO_2, but it doesn't work alone. The cardiovascular system via its contained blood provides the watery medium for transporting O_2 and CO_2 in the body. Let's look into how the respiratory system carries out its role.

Mechanics of Respiration

Pulmonary ventilation, or **breathing,** consists of two phases: **inspiration,** during which air is taken into the lungs, and **expiration,** during which air passes out of the lungs. As the inspiratory muscles (external intercostals and diaphragm) contract during inspiration, the size of the thoracic cavity increases. The diaphragm moves from its relaxed dome shape to a flattened position, increasing the superoinferior volume. The external intercostals lift the rib cage, increasing the anteroposterior and lateral dimensions **(Figure 37.1)**. Because the lungs adhere to the thoracic walls like flypaper owing to the presence of serous fluid in the pleural cavity, the intrapulmonary volume (volume within the lungs) also increases, lowering the air (gas) pressure inside the lungs. The gases then expand to fill the available space, creating a partial vacuum that causes air to flow into the lungs—constituting the act of inspiration. During expiration, the inspiratory muscles relax, and the natural tendency of the elastic lung tissue to recoil decreases the intrathoracic and intrapulmonary volumes. As the gas molecules within the lungs are forced closer together, the intrapulmonary pressure rises to a point higher than atmospheric pressure. This causes gases to flow out of the lungs to equalize the pressure inside and outside the lungs—the act of expiration.

ACTIVITY 1

Operating the Model Lung

Observe the model lung, which demonstrates the principles involved in gas flows into and out of the lungs. It is a simple apparatus with a bottle "thorax," a rubber membrane "diaphragm," and balloon "lungs."

1. Go to the demonstration area and work the model lung by moving the rubber diaphragm up and down. The balloons will not fully inflate or deflate, but notice the *relative* changes in balloon (lung) size as the volume of the thoracic cavity is alternately increased and decreased.

2. Check the appropriate columns in the chart concerning these observations in the review sheet at the end of this exercise.

3. A pneumothorax is a condition in which air has entered the pleural cavity, as with a puncture wound. Simulate a pneumothorax: Inflate the balloon lungs by pulling down on the diaphragm. Ask your lab partner to let air into the bottle "thorax" by loosening the rubber stopper.

What happens to the balloon lungs?

_____

4. After observing the operation of the model lung, conduct the following tests on your lab partner. Use the tape measure to determine his or her chest

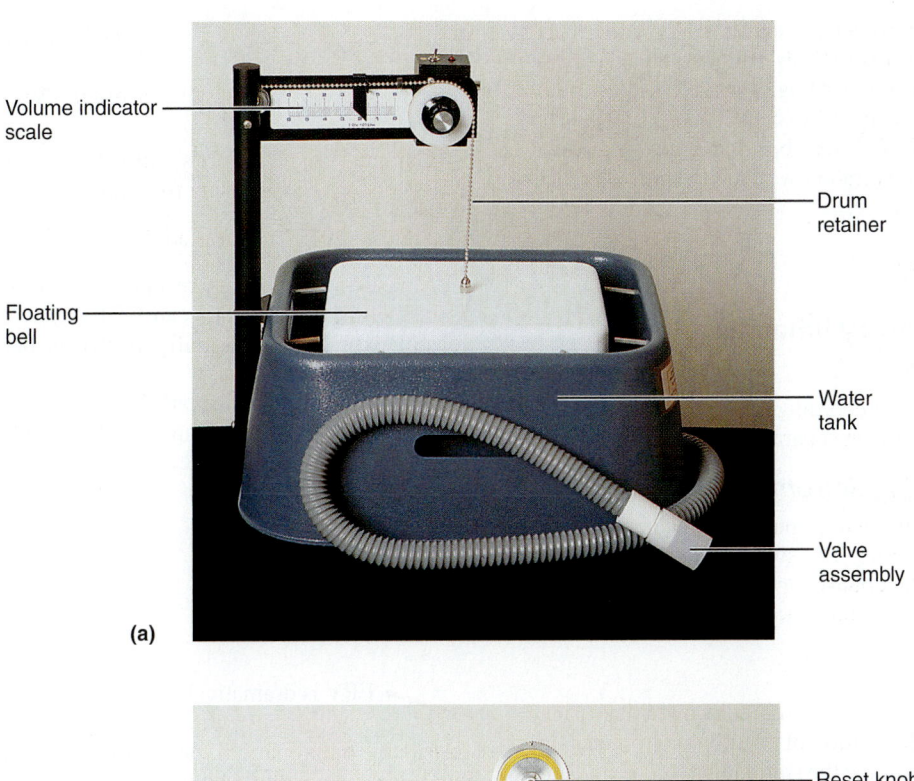

Volume indicator scale
Drum retainer
Floating bell
Water tank
Valve assembly

(a)

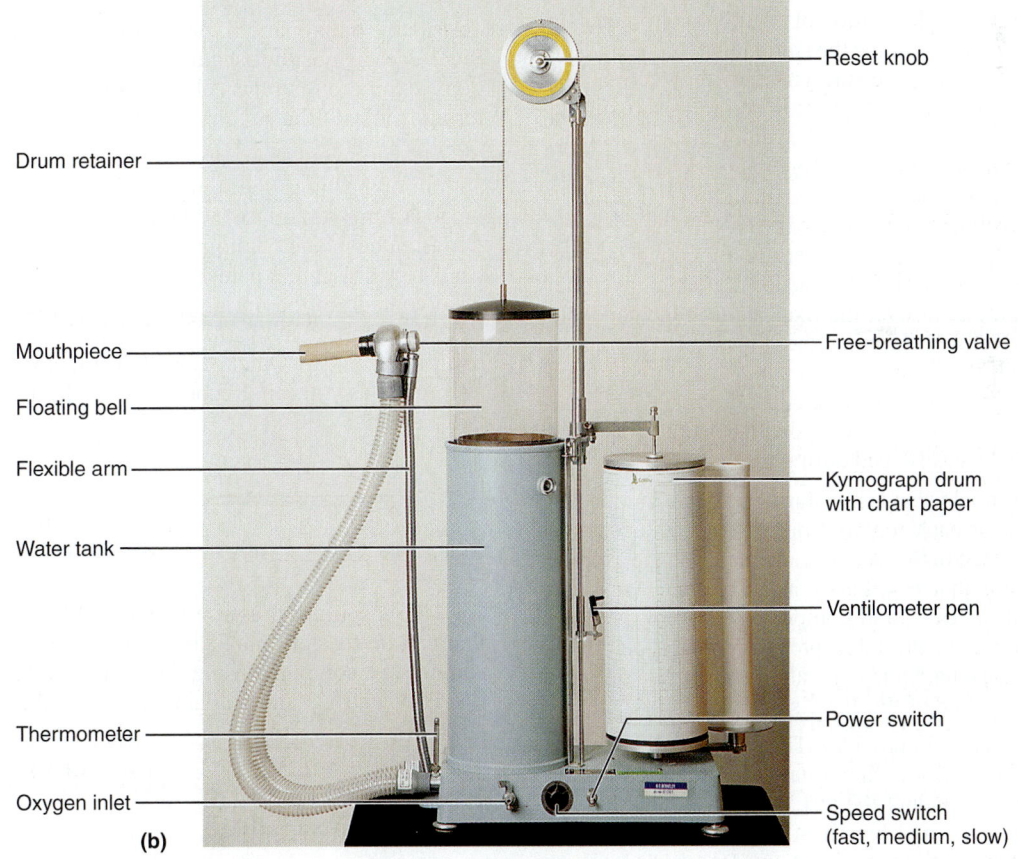

Drum retainer
Reset knob
Mouthpiece
Free-breathing valve
Floating bell
Flexible arm
Kymograph drum with chart paper
Water tank
Ventilometer pen
Thermometer
Power switch
Oxygen inlet
Speed switch (fast, medium, slow)

(b)

Figure 37.4 Wet spirometers. (a) The Phipps and Bird wet spirometer.
(b) The Collins-9L wet recording spirometer.

37

In nonrecording spirometers, an indicator moves as air is *exhaled,* and only expired air volumes can be measured directly. By contrast, recording spirometers allow both inspired and expired gas volumes to be measured.

Alternatively, BIOPAC® may be used with an airflow transducer to measure respiratory volumes. Those instructions appear in Activity 5.

ACTIVITY 3

Measuring Respiratory Volumes Using Spirometers

The steps for using a nonrecording spirometer and a wet recording spirometer are given separately below.

Using a Nonrecording Spirometer

1. Before using the spirometer, count and record the subject's normal respiratory rate. The subject should face away from you as you make the count.

Respirations per minute: _____

Now identify the parts of the spirometer you will be using by comparing it to the illustration of a similar spirometer (in Figure 37.3 or 37.4a). Examine the spirometer volume indicator *before beginning* to make sure you know how to read the scale. Work in pairs, with one person acting as the subject while the other records the data of the volume determinations. *Reset the indicator to zero before beginning each trial.*

Obtain a disposable cardboard mouthpiece. Prior to inserting the cardboard mouthpiece, clean the valve assembly with an alcohol swab. Then insert the mouthpiece in the open end of the valve assembly (attached to the flexible tube) of the wet spirometer or over the fixed stem of the handheld dry spirometer. Before beginning, the subject should practice exhaling through the mouthpiece without exhaling through the nose, or prepare to use the nose clips. If you are using the handheld spirometer, make sure its dial faces upward so that the volumes can be easily read during the tests.

2. The subject should stand erect during testing. Conduct the test three times for each required measurement. Record the data where indicated in this section, and then find the average volume figure for that respiratory measurement. After you have completed the trials and computed the averages, enter the average values on the table prepared on the board for tabulation of class data,* and copy all averaged data onto the review sheet at the end of the exercise.

3. Measuring tidal volume (TV). The TV, or volume of air inhaled and exhaled with each normal respiration, is approximately 500 ml. To conduct the test, inhale a normal breath, and then exhale a normal breath of air into the spirometer mouthpiece. (Do not force the expiration!) Record the volume and repeat the test twice.

*****Note to the Instructor:** The format of class data tabulation can be similar to that shown here. However, it would be interesting to divide the class into smokers and nonsmokers and then compare the mean average VC and ERV for each group. Such a comparison might help to determine if smokers are handicapped in any way. It also might be a good opportunity for an informal discussion of the early warning signs of bronchitis and emphysema, which are primarily smokers' diseases.

trial 1: _____ ml   trial 2: _____ ml

trial 3: _____ ml   average TV: _____ ml

4. Compute the subject's **minute respiratory volume (MRV)** using the following formula:

$$\text{MRV} = \text{TV} \times \text{respirations/min} = _____ \text{ ml/min}$$

5. Measuring expiratory reserve volume (ERV). The ERV is the volume of air that can be forcibly exhaled after a normal expiration. Normally it ranges between 700 and 1200 ml.

Inhale and exhale normally two or three times, then insert the spirometer mouthpiece and exhale forcibly as much of the additional air as you can. Record your results, and repeat the test twice again.

trial 1: _____ ml   trial 2: _____ ml

trial 3: _____ ml   average ERV: _____ ml

ERV is dramatically reduced in conditions in which the elasticity of the lungs is decreased by a chronic obstructive pulmonary disease (COPD) such as **emphysema.** Since energy must be used to *deflate* the lungs in such conditions, expiration is physically exhausting to individuals suffering from COPD. ✚

6. Measuring vital capacity (VC). The VC, or total exchangeable air of the lungs (the sum of TV + IRV + ERV), normally ranges from 3100 ml to 4800 ml.

Breathe in and out normally two or three times, and then bend forward and exhale all the air possible. Then, as you raise yourself to the upright position, inhale as fully as possible. It is important to *strain* to inhale the maximum amount of air that you can. Quickly insert the mouthpiece, and exhale as forcibly as you can. Record your results and repeat the test twice again.

trial 1: _____ ml   trial 2: _____ ml

trial 3: _____ ml   average VC: _____ ml

7. The inspiratory reserve volume (IRV), or volume of air that can be forcibly inhaled following a normal inspiration, can now be computed using the average values obtained for TV, ERV, and VC and plugging them into the equation:

$$\text{IRV} = \text{VC} - (\text{TV} + \text{ERV})$$

Record your average IRV: _____ ml

The normal IRV is substantial, ranging from 1900 to 3100 ml. How does your computed value compare?

_____

Steps 8–10, which provide common directions for use of both nonrecording and recording spirometers, continue (on pages 556–557) after the wet recording spirometer directions.

Using a Recording Spirometer

1. In preparation for recording, familiarize yourself with the spirometer by comparing it to the illustrated equipment (Figure 37.4b).

2. Examine the chart paper, noting that its horizontal lines represent milliliter units. To apply the chart paper to the recording drum, first lift the drum retainer and then remove the kymograph drum. Wrap a sheet of chart paper around the drum, *making sure that the right edge overlaps the left.* Fasten it with tape, and then replace the kymograph drum and lower the drum retainer into its original position in the hole in the top of the drum.

3. Raise and lower the floating bell several times, noting as you do so that the *ventilometer pen* moves up and down on the drum. This pen, which writes in black ink, will be used for recording and should be adjusted so that it records in the approximate middle of the chart paper. This adjustment is made by repositioning the floating bell using the *reset knob* on the metal pulley at the top of the spirometer apparatus. The other pen, the respirometer pen, which records in red ink, will not be used for these tests and should be moved away from the drum's recording surface.

4. Recording your normal respiratory rate. Clean the nose clips with an alcohol swab. While you wait for the alcohol to air dry, count and record your normal respiratory rate.

Respirations per minute: _____

5. Recording tidal volume. After the alcohol has air dried, apply the nose clips to your nose. This will enforce mouth breathing.

Open the *free-breathing valve.* Insert a disposable cardboard mouthpiece into the end (valve assembly) of the breathing tube, and then insert the mouthpiece into your mouth. Practice breathing for several breaths to get used to the apparatus. At this time, you are still breathing room air.

Set the spirometer switch to **SLOW** (32 mm/min). Close the free-breathing valve, and breathe in a normal manner for 2 minutes to record your tidal volume—the amount of air inspired or expired with each normal respiratory cycle. This recording should show a regular pattern of inspiration-expiration spikes and should gradually move upward on the chart paper. (A downward slope indicates that there is an air leak somewhere in the system—most likely at the mouthpiece.) Notice that on an apparatus using a counterweighted pen (such as the Collins-9L Ventilometer shown in Figure 37.4b), inspirations are recorded by upstrokes and expirations are recorded by downstrokes.*

6. Recording vital capacity. To record your vital capacity, take the deepest possible inspiration you can and then exhale to the greatest extent possible—really *push* the air out. (The recording obtained should resemble that shown in Figure 37.5). Repeat the vital capacity measurement twice again. Then turn off the spirometer and remove the chart paper from the kymograph drum.

7. Determine and record your measured, averaged, and corrected respiratory volumes. Because the pressure and

*If a Collins survey spirometer is used, the situation is exactly opposite: Upstrokes are expirations and downstrokes are inspirations.

temperature inside the spirometer are influenced by room temperature and differ from those in the body, all measured values are to be multiplied by a **BTPS** (body temperature, atmospheric pressure, and water saturation) **factor.** At room temperature, the BTPS factor is typically 1.1 or very close to that value. Hence, you will multiply your average measured values by 1.1 to obtain your corrected respiratory volume values. Copy the averaged and corrected values onto the review sheet at the end of this exercise.

- Tidal volume (TV). Select a typical resting tidal breath recording. Subtract the millimeter value of the trough (exhalation) from the millimeter value of the peak (inspiration). Record this value below as *measured TV 1*. Select two other TV tracings to determine the TV values for the TV 2 and TV 3 measurements. Then, determine your average TV and multiply it by 1.1 to obtain the BTPS-corrected average TV value.

measured TV 1: _____ ml   average TV: _____ ml

measured TV 2: _____ ml   corrected average TV:

measured TV 3: _____ ml   _____ ml

Also compute your **minute respiratory volume (MRV)** using the following formula:

$$MRV = TV \times respirations/min = \text{_____} ml/min$$

- Inspiratory capacity (IC). In the first vital capacity recording, find the expiratory trough immediately preceding the maximal inspiratory peak achieved during vital capacity determination. Subtract the milliliter value of that expiration from the value corresponding to the peak of the maximal inspiration that immediately follows. For example, according to our typical recording **(Figure 37.5)**, these values would be

$$6600 - 3650 = 2950 \text{ ml}$$

Record your computed value and the results of the two subsequent tests on the appropriate lines below. Then calculate the measured and corrected inspiratory capacity averages and record.

measured IC 1: _____ ml   average IC: _____ ml

measured IC 2: _____ ml   corrected
                                   average IC: _____ ml
measured IC 3: _____ ml

- Inspiratory reserve volume (IRV). Subtract the corrected average tidal volume from the corrected average for the inspiratory capacity and record below.

$$IRV = corrected \text{ } average \text{ } IC - corrected \text{ } average \text{ } TV$$

corrected average IRV _____ ml

- Expiratory reserve volume (ERV). Subtract the number of milliliters corresponding to the trough of the maximal

37

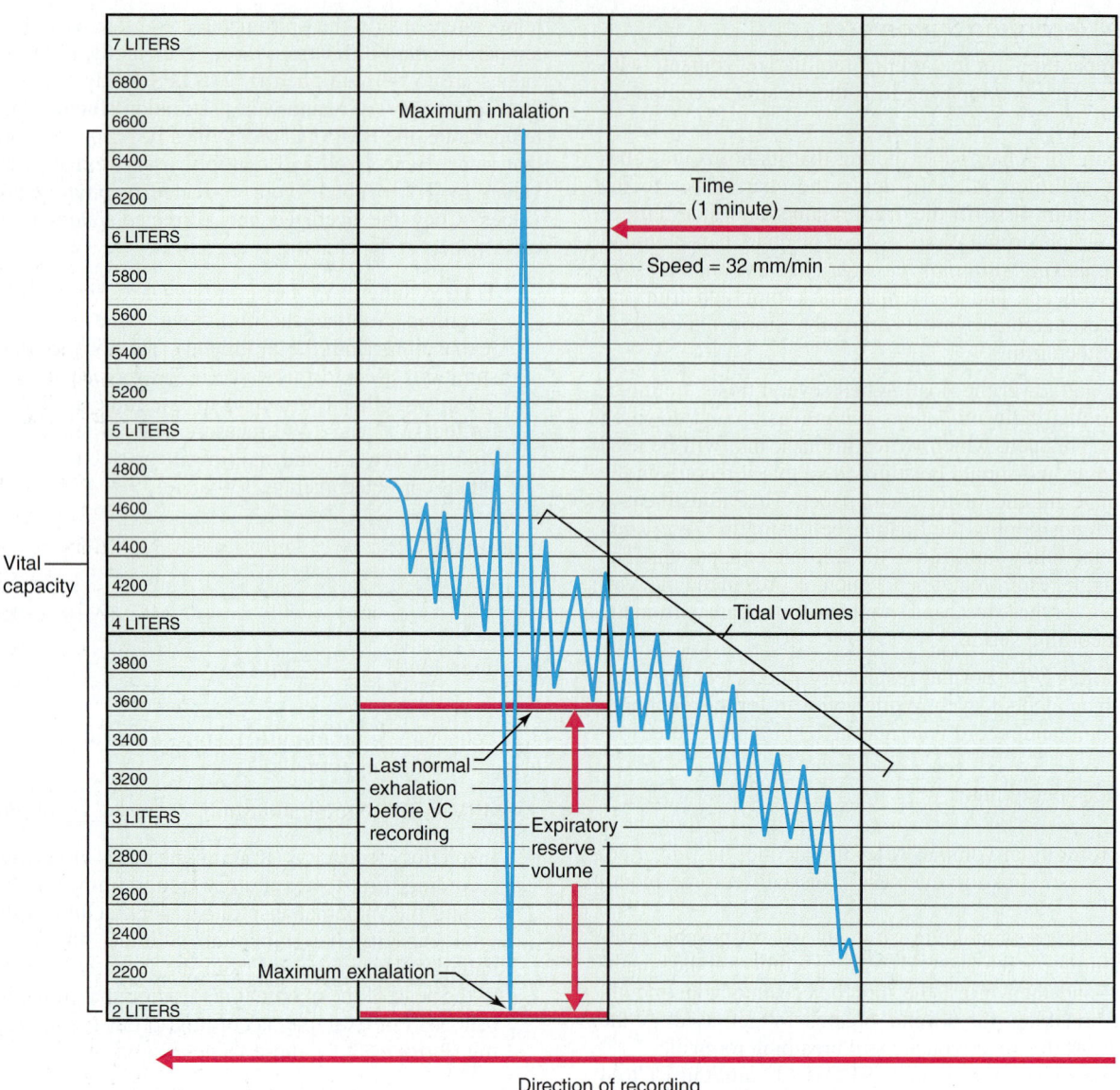

Figure 37.5 A typical spirometry recording of tidal volume, inspiratory capacity, expiratory reserve volume, and vital capacity. At a drum speed of 32 mm/min, each vertical column of the chart represents a time interval of 1 minute. (Note that downstrokes represent exhalations and upstrokes represent inhalations.)

expiration obtained during the vital capacity recording from milliliters corresponding to the last *normal* expiration before the VC maneuver is performed. For example, according to our typical recording (Figure 37.5), these values would be

$$3650 \text{ ml} - 2050 \text{ ml} = 1600 \text{ ml}$$

Record your measured and averaged values (three trials) below.

measured ERV 1: _____ ml   average ERV: _____ ml

measured ERV 2: _____ ml   corrected average ERV:

measured ERV 3: _____ ml   _____ ml

• Vital capacity (VC). Add your corrected values for ERV and IC to obtain the corrected average VC. Record below and on the review sheet at the end of this exercise.

corrected average VC: _____ ml

Now continue with step 8 (below) whether you are following the procedure for the nonrecording or recording spirometer.

8. Figure out how closely your measured average vital capacity volume compares with the *predicted values* for someone your age, sex, and height. Obtain the predicted figure either from the equation on the next page or the appropriate table (see your instructor for the printed table). Notice that you will have to convert your height in inches to centimeters (cm) to find the

corresponding value. This is easily done by multiplying your height in inches by 2.54.

Computed height: _____ cm

Male VC = (0.052) H − (0.022) A − 3.60

Female VC = (0.041) H − (0.018) A − 2.69

Note: (VC) = vital capacity in liters, (H) = height in centimeters, and (A) = age in years.

Predicted VC value (obtained from the equation or appropriate table):

_____ ml

Use the following equation to compute your VC as a percentage of the predicted VC value:

$$\% \text{ of predicted VC } = \frac{\text{average VC}}{\text{predicted value} \times 100}$$

% predicted VC value: _____ %

Carefully examine the idealized tracing (Figure 37.2 on page 552) of the respiratory volumes described and tested in this exercise. How closely do your test results compare to the values in the tracing?

_____

9. Computing residual volume. A respiratory volume that cannot be experimentally demonstrated here is the residual volume (RV). RV is the amount of air remaining in the lungs after a maximal expiratory effort. The presence of residual air (usually about 1200 ml) that cannot be voluntarily flushed from the lungs is important because it allows gas exchange to go on continuously—even between breaths.

   Although the residual volume cannot be measured directly, it can be approximated by using one of the following factors:

For ages 16–34   Factor = 0.250

For ages 35–49   Factor = 0.305

For ages 50–69   Factor = 0.445

Compute your predicted RV using the following equation:

$$RV = VC \times \text{factor}$$

 10. Recording is finished for this subject. Before continuing with the next member of your group:

- Dispose of used cardboard mouthpieces in the autoclave bag.

- Swish the valve assembly (if removable) in the 70% ethanol solution, then rinse with tap water.

- Put a fresh mouthpiece into the valve assembly (or on the stem of the handheld spirometer). Using the procedures outlined above, measure and record the respiratory volumes for all members of your group. ■

Forced Expiratory Volume (FEV$_T$) Measurement

Though not really diagnostic, pulmonary function tests can help the clinician distinguish between obstructive and restrictive pulmonary diseases. (In obstructive disorders, like chronic bronchitis and asthma, airway resistance is increased, whereas in restrictive diseases, such as polio and tuberculosis, total lung capacity declines.) Two highly useful pulmonary function tests used for this purpose are the FVC and the FEV$_T$ **(Figure 37.6)**.

   The **FVC** (forced vital capacity) measures the amount of gas expelled when the subject takes the deepest possible breath and then exhales forcefully and rapidly. This volume is reduced in those with restrictive pulmonary disease. The **FEV$_T$** (forced expiratory volume) involves the same basic testing procedure, but it specifically looks at the percentage of the vital capacity that is exhaled during specific time intervals of the FVC test. FEV$_1$, for instance, is the amount exhaled during the first second. Healthy individuals can expire 75% to 85% of their FVC in the first second. The FEV$_1$ is low in those with obstructive disease.

ACTIVITY 4

Measuring the FVC and FEV$_1$

Directions provided here for the FEV$_T$ determination apply only to the recording spirometer.

1. Prepare to make your recording as described for the recording spirometer, steps 1–5 on page 555.

2. At a signal agreed upon by you and your lab partner, take the deepest inspiration possible and hold it for 1 to 2 seconds. As the inspiratory peak levels off, your partner is to change the drum speed to **FAST** (1920 mm/min) so that the distance between the vertical lines on the chart represents 1 second.

3. Once the drum speed is changed, exhale as much air as you can as rapidly and forcibly as possible.

4. When the tracing plateaus (bottoms out), stop recording and determine your FVC. Subtract the milliliter reading in the expiration trough (the bottom plateau) from the preceding inhalation peak (the top plateau). Record this value.

FVC: _____ ml

5. Prepare to calculate the FEV$_1$. Draw a vertical line intersecting with the spirogram tracing at the precise point that exhalation began. Identify this line as *line 1*. From line 1, measure 32 mm horizontally to the left, and draw a second vertical line. Label this as *line 2*. The distance between the two lines represents 1 second, and the volume exhaled in the first second is read where line 2 intersects the spirogram tracing. Subtract that milliliter value from the milliliter value of the inhalation peak (at the intersection of line 1), to determine the volume of gas expired in the first second. According to the values given in the example (Figure 37.6), that figure would be 3400 ml (6800 ml – 3400 ml). Record your measured value below.

Milliliters of gas expired in second 1:   _____ ml

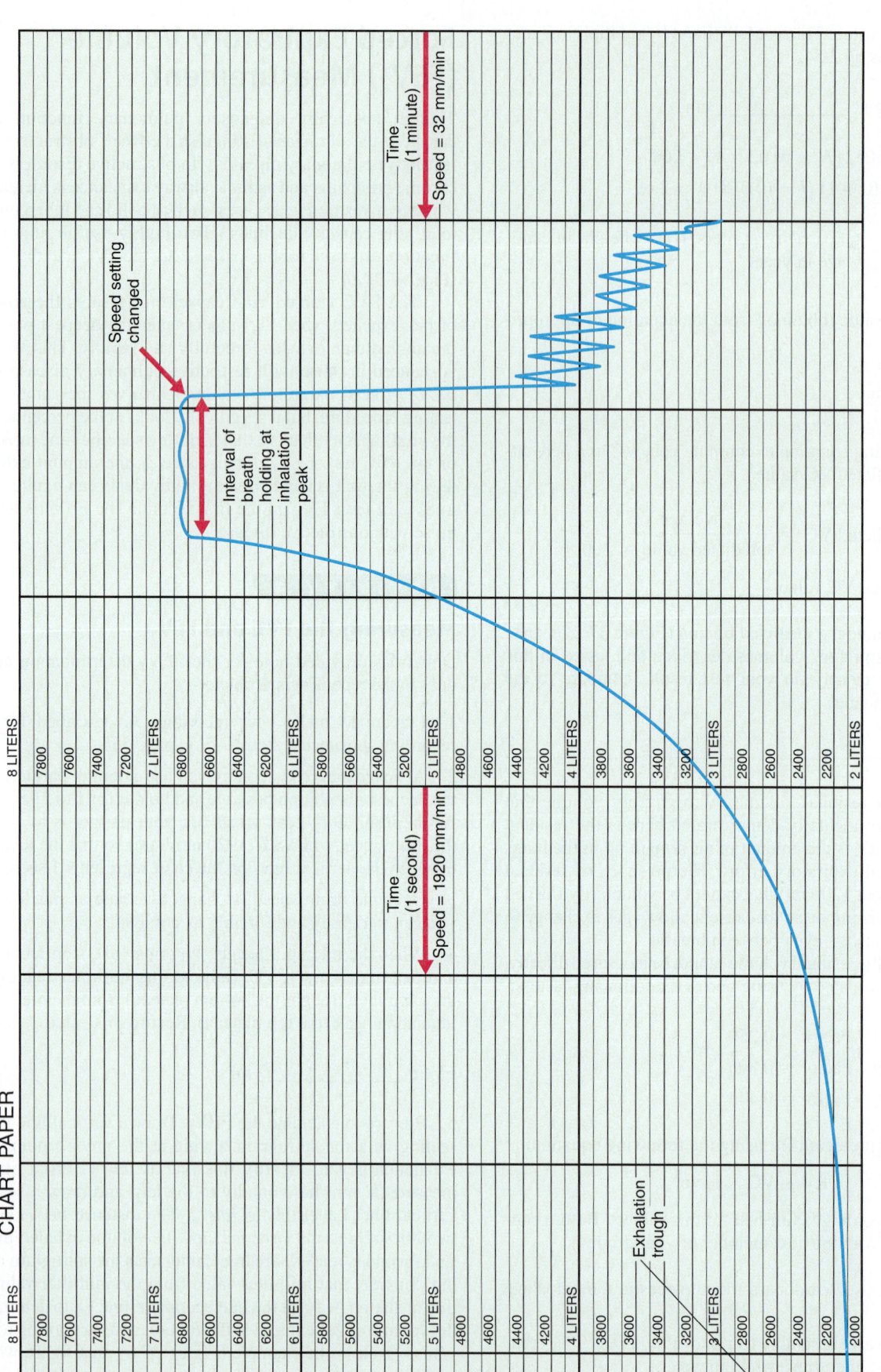

Figure 37.6 A recording of the forced vital capacity (FVC) and forced expiratory volume (FEV) or timed vital capacity test.

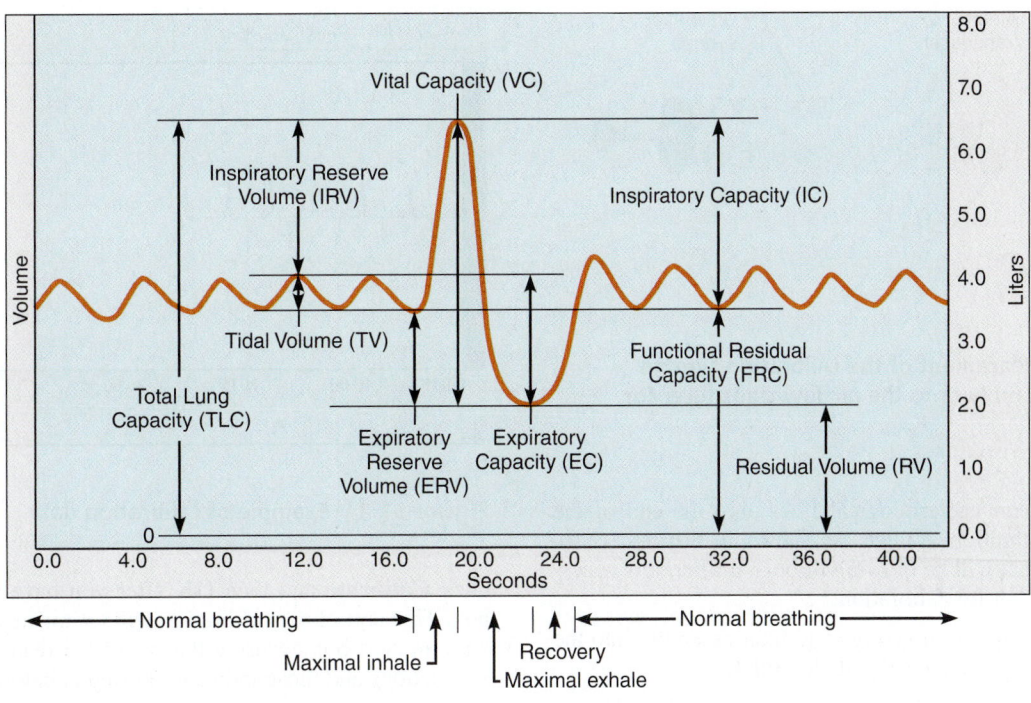

Figure 37.7 **Example of a computer-generated spirogram.**

6. To compute the FEV_1 use the following equation:

$$FEV_1 = \frac{\text{volume expired in second 1}}{\text{FVC volume}} \times 100\%$$

Record your calculated value below and on the review sheet at the end of this exercise.

FEV_1: _____ % of FVC ▬

ACTIVITY 5

Measuring Respiratory Volumes Using BIOPAC®

In this activity, you will measure respiratory volumes using the BIOPAC® airflow transducer. An example of these volumes is demonstrated in the computer-generated spirogram **(Figure 37.7)**. Since it is not possible to measure **residual volume (RV)** using the airflow transducer, assume that it is 1.0 liter for each subject, which is a reasonable estimation. Or enter a volume between 1 and 5 liters via Preferences. It is also important to estimate the **predicted vital capacity** of the subject for comparison to the measured value. A rough estimate of the vital capacity in liters (VC) of a subject can be calculated using the following formulas based on height in centimeters (*H*) and age in years (*A*).

Male VC $= (0.052)H - (0.022)A - 3.60$

Female VC $= (0.041)H - (0.018)A - 2.69$

Because many factors besides height and age influence vital capacity, it should be assumed that measured values up to 20% above or below the calculated predicted value are normal.

Setting Up the Equipment

1. Connect the BIOPAC® unit to the computer and turn the computer **ON.**

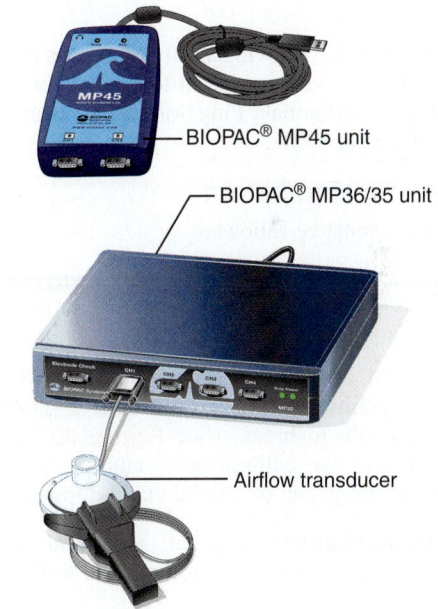

— BIOPAC® MP45 unit

— BIOPAC® MP36/35 unit

— Airflow transducer

Figure 37.8 **Setting up the BIOPAC® equipment.** Plug the airflow transducer into Channel 1. Transducer is shown plugged into the MP36/35 unit.

2. Make sure the BIOPAC® unit is **OFF.**

3. Plug in the equipment (as shown in **Figure 37.8**).

• Airflow transducer—CH 1

4. Turn the BIOPAC® unit **ON.**

37

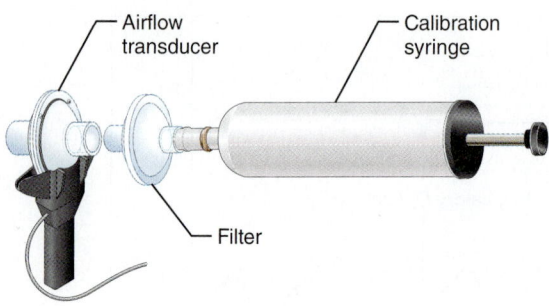

Figure 37.9 Placement of the calibration syringe and filter assembly onto the airflow transducer for calibration.

5. Place a *clean* bacteriological filter onto the end of the BIOPAC® calibration syringe (as shown in **Figure 37.9**). Since the subject will be blowing through a filter, it is necessary to use a filter for calibration.

6. Insert the calibration syringe and filter assembly into the airflow transducer on the side labeled **Inlet.**

7. Start the BIOPAC® Student Lab program on the computer by double-clicking the icon on the desktop or by following your instructor's guidance.

8. Select lesson **L12-Pulmonary Functions-1** from the menu, and click **OK.**

9. Type in a filename that will save this subject's data on the computer hard drive. You may want to use the subject's last name followed by Pulmonary Functions (PF)-1 (for example, SmithPF-1), then click **OK.**

Calibrating the Equipment

Two precautions must be followed:

• The airflow transducer is sensitive to gravity, so it must be held directly parallel to the ground during calibration and recording.

• Do not hold onto the airflow transducer when it is attached to the calibration syringe and filter assembly—the syringe tip is likely to break. (See **Figure 37.10** for the proper handling of the calibration assembly). The size of the calibration syringe can be altered via Preferences.

1. Make sure the plunger is pulled all the way out. While the assembly is held in a steady position parallel to the ground,

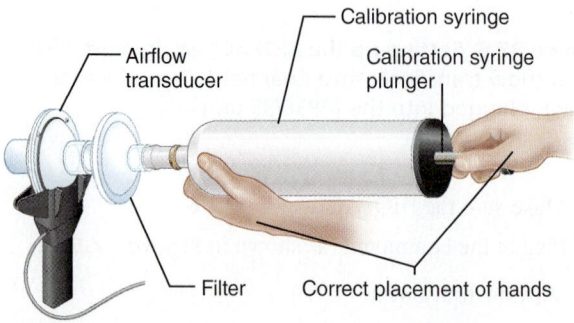

Figure 37.10 Proper handling of the calibration assembly.

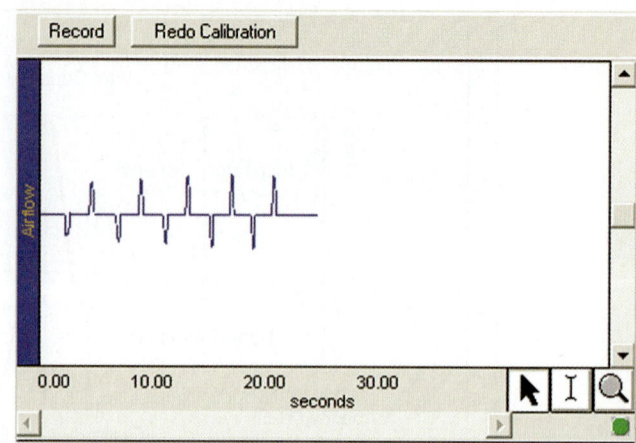

Figure 37.11 Example of calibration data.

click **Calibrate** and then **OK** after you have read the alert box. This part of the calibration will terminate automatically with an alert box ensuring that you have read the on-screen instructions and those indicated in step 2, below.

2. The final part of the calibration involves simulating five breathing cycles using the calibration syringe. A single cycle consists of:

• Pushing the plunger in (taking 1 second for this stroke)

• Waiting for 2 seconds

• Pulling the plunger out (taking 1 second for this stroke)

• Waiting 2 seconds

Remember to hold the airflow transducer directly parallel to the ground during calibration and recording.

3. When ready to perform this second stage of the calibration, click **Yes.** After you have completed five cycles, click **End Calibration.**

4. Observe the data, which should look similar to that in the example **(Figure 37.11)**.

• If the data look very different, click **Redo Calibration** and repeat the steps above.

• If the data look similar, gently remove the calibration syringe, leaving the air filter attached to the transducer. Proceed to the next section.

Recording the Data

Follow these procedures precisely, because the airflow transducer is very sensitive. Hints to obtain the best data:

• Always insert air filter on, and breathe through, the transducer side labeled **Inlet.**

• Keep the airflow transducer upright at all times.

• The subject should not look at the computer screen during the recording of data.

• The subject must keep a nose clip on throughout the experiment.

1. Insert a clean mouthpiece into the air filter that is already attached to the airflow transducer. *Be sure that the filter is attached to the **Inlet** side of the airflow transducer.*

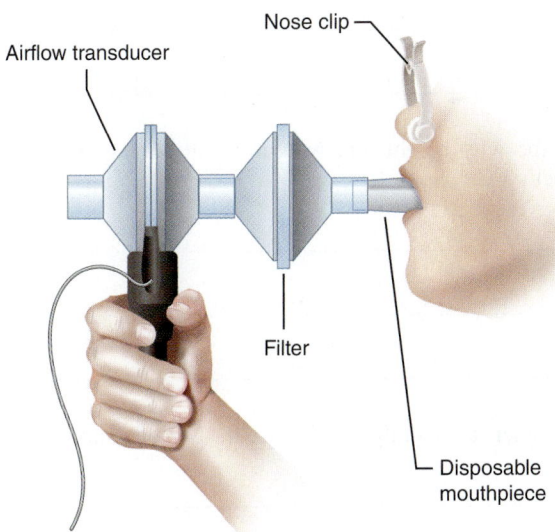

Figure 37.12 Proper equipment setup for recording data.

2. Write the name of the subject on the mouthpiece and air filter. For safety purposes, each subject must use his or her own air filter and mouthpiece.

3. The subject should now place the nose clip on the nose (or hold the nose very tightly with finger pinch), wrap the lips tightly around the mouthpiece, and begin breathing normally through the airflow transducer (as shown in **Figure 37.12**).

4. When prepared, the subject will complete the following unbroken series with nose plugged and lips tightly sealed around the mouthpiece:

• Take five normal breaths (1 breath = inhale + exhale).

• Inhale as much air as possible.

• Exhale as much air as possible.

• Take five normal breaths.

5. When the subject is prepared to proceed, click **Record** on the first normal inhalation and proceed. When the subject finishes the last exhalation at the end of the series, click **Stop.**

6. Observe the data, which should look similar to that in the example **(Figure 37.13)**.

• If the data look very different, click **Redo** and repeat the steps above. Be certain that the lips are sealed around the mouthpiece, the nose is completely plugged, and the transducer is upright.

• If the data look similar, proceed to step 7.

7. When finished, click **Done.** A pop-up window will appear.

• Click **Yes** if you are done and want to stop recording.

• To record from another subject, select **Record from another Subject** and return to step 1 under Recording the Data. You will not need to redo the calibration procedure for the second subject.

• If continuing to the Data Analysis section, select **Analyze current data file** and proceed to step 2 of the Data Analysis section.

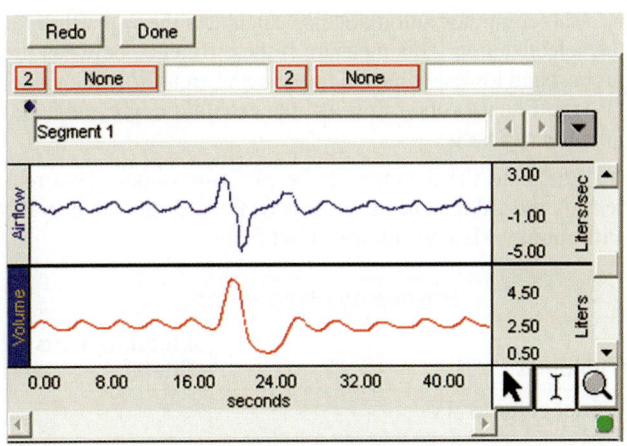

Figure 37.13 Example of pulmonary data.

Data Analysis

1. If just starting the BIOPAC® program to perform data analysis, enter **Review Saved Data** mode and choose the file with the subject's PF data (for example, SmithPF-1).

2. Observe how the channel numbers are designated: CH 1— Airflow; CH 2—Volume.

3. To set up the display for optimal viewing, hide CH 1—Airflow. To do this, hold down the Ctrl key (PC) or Option key (Mac) while using the cursor to click the Channel box 1 (the small box with a 1 at the upper left of the screen).

4. To analyze the data, set up the first pair of channel/measurement boxes at the top of the screen by selecting the following channel and measurement type from the drop-down menu:

Channel	Measurement	Data
CH 2	p-p	volume

5. Take two measures for an averaged TV calculation: Use the arrow cursor and click the I-beam cursor box on the lower right side of the screen to activate the "area selection" function. Using the activated I-beam cursor, highlight the inhalation of cycle 3 (as shown in **Figure 37.14**).

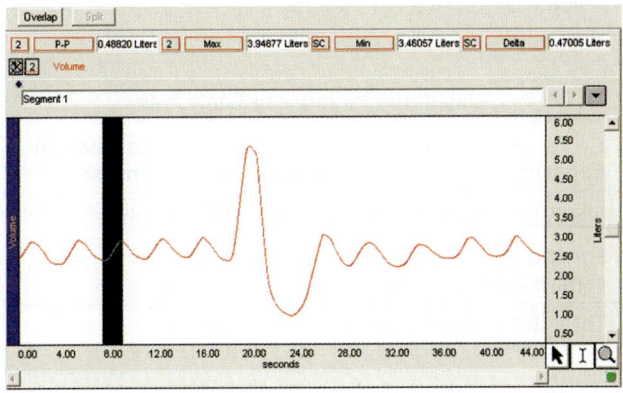

Figure 37.14 Highlighting data for the inhalation of the third breath.

37

6. The computer automatically calculates the **p-p** value for the selected area. This measure is the difference between the highest and lowest values in the selected area. Note the value. Use the I-beam cursor to select the exhalation of cycle 3 and note the **p-p** value.

7. Calculate the average of the two **p-p** values. This represents the **tidal volume** (in liters). Record the value in the **Pulmonary Measurements chart** below:

Pulmonary Measurements

Volumes	Measurements (liters)
Tidal volume (TV)	
Inspiratory reserve volume (IRV)	
Expiratory reserve volume (ERV)	
Vital capacity (VC)	
Residual volume (RV)	1.00 (assumed)

8. Use the I-beam cursor to measure the IRV: Highlight from the peak of maximum inhalation to the peak of the last normal inhalation just before it (see Figure 37.7 for an example of IRV). Observe and record the Δ **(delta)** value in the chart (to the nearest 0.01 liter).

9. Use the I-beam cursor to measure the ERV: Highlight from the trough of maximum exhalation to the trough of the last normal exhalation just before it (see Figure 37.7 for an example of ERV). Observe and record the Δ **(delta)** value in the chart (to the nearest 0.01 liter).

10. Last, use the I-beam cursor to measure the VC: Highlight from the trough of maximum exhalation to the peak of maximum inhalation (see Figure 37.7 for an example of VC). Observe and record the **p-p** value in the chart (to the nearest 0.01 liter).

11. When finished, choose **File menu** and **Quit** to close the program.

Using the measured data, calculate the capacities listed in the **Calculated Pulmonary Capacities chart**.

Use the formula in the introduction of this activity (page 559) to calculate the predicted vital capacity of the subject based on height and age.

Predicted VC: _____ liters

Calculated Pulmonary Capacities

Capacity	Formula	Calculation (liters)
Inspiratory capacity (IC)	= TV + IRV	
Functional residual capacity (FRC)	= ERV + RV	
Total lung capacity (TLC)	= TV + RV + IRV + ERV	

How does the measured vital capacity compare to the predicted vital capacity?

_____

_____

_____

Describe why height and weight might correspond with a subject's VC.

_____

_____

_____

What other factors might influence the VC of a subject?

_____

_____

_____

Factors Influencing Rate and Depth of Respiration

The neural centers that control respiratory rhythm and maintain a rate of 12 to 18 respirations/min are located in the medulla and pons. On occasion, input from the stretch receptors in the lungs (via the vagus nerve to the medulla) modifies the respiratory rate, as in cases of extreme overinflation of the lungs (Hering-Breuer reflex).

Death occurs when medullary centers are completely suppressed, as from an overdose of sleeping pills or gross overindulgence in alcohol, and respiration ceases completely. +

Although the nervous system centers initiate the basic rhythm of breathing, there is no question that physical phenomena such as talking, yawning, coughing, and exercise can modify the rate and depth of respiration. So, too, can chemical factors such as changes in oxygen or carbon dioxide concentrations in the blood or fluctuations in blood pH. This is especially important in initiating breathing in a newborn. The buildup of carbon dioxide in the blood triggers the baby's first breath. Changes in carbon dioxide blood levels seem to act directly on the medulla control centers, whereas changes in pH and oxygen concentrations are monitored by chemoreceptor regions in the aortic and carotid bodies, which in turn send input to the medulla. The experimental sequence in Activity 6 is designed to test the relative importance of various physical and chemical factors in the process of respiration.

ACTIVITY 6

Visualizing Respiratory Variations

In this activity, you will count the respiratory rate of the subject visually by observing the movement of the chest or abdomen.

1. Record quiet breathing for 1 minute with the subject in a sitting position.

Breaths per minute: _____

2. Record the subject's breathing as he or she performs activities from the following list. Record your results on the review sheet at the end of this exercise.

talking	swallowing water
yawning	coughing
laughing	lying down
standing	running in place

doing a math problem
  (concentrating)

3. Without recording, have the subject breathe normally for 2 minutes, then inhale deeply and hold his or her breath for as long as he or she can.

Breath-holding interval: _____ sec

As the subject exhales, record the recovery period (time to return to normal breathing—usually slightly over 1 minute):

Time of recovery period: _____ sec

Did the subject have the urge to inspire *or* expire during

breath holding? _____

Without recording, repeat the above experiment, but this time exhale completely and forcefully *after* taking the deep breath.

Breath-holding interval _____ sec

Time of recovery period _____ sec

Did the subject have the urge to inspire *or* expire? _____

Explain the results. (Hint: The vagus nerve is the sensory nerve of the lungs and plays a role here.)

_____

_____

_____

4. During the next task, a sensation of dizziness may develop. As the carbon dioxide is washed out of the blood by hyperventilation, the blood pH increases, leading to a decrease in blood pressure and reduced cerebral circulation.

⚠ If you have a history of dizzy spells or a heart condition, do not perform this task.
   The subject may experience a lack of desire to breathe after forced breathing is stopped. If the period of breathing cessation—apnea—is extended, cyanosis of the lips may occur.
   Have the subject hyperventilate (breathe deeply and forcefully at the rate of 1 breath/4 sec) for about 30 sec.

Is the respiratory rate after hyperventilation faster *or* slower than during normal quiet breathing?

_____

5. Repeat the hyperventilation step. After hyperventilation, the subject is to hold his or her breath as long as possible.

Breath-holding interval: _____

Can the breath be held for a longer or shorter time after hyperventilating?

_____

6.   Without recording, have the subject breathe into a paper bag for 3 minutes, then record his or her breathing movements.

⚠ *During the bag-breathing exercise, the subject's partner should watch the subject carefully for any untoward reactions.*

Is the breathing rate faster *or* slower than that recorded during normal quiet breathing?

_____

After hyperventilating? _____.

7. Run in place for 2 minutes, and then have your partner determine how long you can hold your breath.

Breath-holding interval: _____ sec

8. To prove that respiration has a marked effect on circulation, conduct the following test. Have your lab partner record the rate and relative force of your radial pulse before you begin.

Rate: _____ beats/min   Relative force: _____

Inspire forcibly. Immediately close your mouth and nose to retain the inhaled air, and then make a forceful and prolonged expiration. Your lab partner should observe and record the condition of the blood vessels of your neck and face, and again immediately palpate the radial pulse.

Observations: _____

Radial pulse: _____ beats/min   Relative force: _____

Explain the changes observed. _____

_____

_____

37

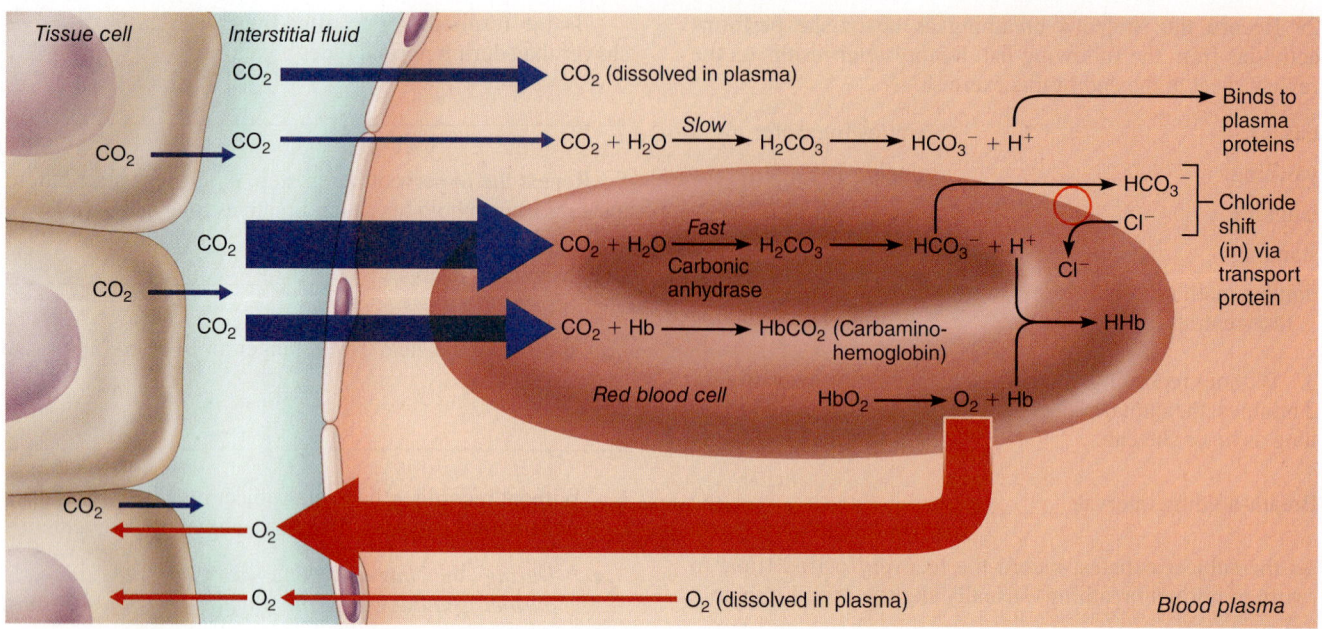

Figure 37.15 Oxygen release and carbon dioxide pickup at the tissues.

Dispose of the paper bag in the autoclave bag. Observation of the test results should enable you to determine which chemical factor, carbon dioxide or oxygen, has the greatest effect on modifying the respiratory rate and depth. ▬

Role of the Respiratory System in Acid-Base Balance of Blood

As you have already learned, pulmonary ventilation is necessary for continuous oxygenation of the blood and removal of carbon dioxide (a waste product of cellular respiration) from the blood. Blood pH must be relatively constant for the cells of the body to function optimally. The carbonic acid–bicarbonate buffer system of the blood is extremely important because it helps stabilize arterial blood pH at 7.4 ± 0.02.

When carbon dioxide diffuses into the blood from the tissue cells, much of it enters the red blood cells, where it combines with water to form carbonic acid **(Figure 37.15)**:

$$H_2O + CO_2 \xrightarrow[\text{enzyme present in RBC}]{\text{carbonic anhydrase}} H_2CO_3$$

Some carbonic acid is also formed in the plasma, but that reaction is very slow because of the lack of the carbonic anhydrase enzyme. Shortly after it forms, carbonic acid dissociates to release bicarbonate (HCO_3^-) and hydrogen ions (H^+). The hydrogen ions that remain in the cells are neutralized, or buffered, when they combine with hemoglobin molecules. If they were not neutralized, the intracellular pH would become very acidic as H^+ ions accumulated. The bicarbonate ions diffuse out of the red blood cells into the plasma, where they become part of the carbonic acid–bicarbonate buffer system.

As HCO_3^- follows its concentration gradient into the plasma, an electrical imbalance develops in the RBCs that draws Cl^- into them from the plasma. This exchange phenomenon is called the *chloride shift*.

Acids (more precisely, H^+) released into the blood by the body cells tend to lower the pH of the blood and to cause it to become acidic. On the other hand, basic substances that enter the blood tend to cause the blood to become more alkaline and the pH to rise. Both of these tendencies are resisted in large part by the carbonic acid–bicarbonate buffer system. If the H^+ concentration in the blood begins to increase, the H^+ ions combine with bicarbonate ions to form carbonic acid (a weak acid that does not tend to dissociate at physiological or acid pH) and are thus removed.

$$H^+ + HCO_3^- \rightarrow H_2CO_3$$

Likewise, as blood H+ concentration drops below what is desirable and blood pH rises, H_2CO_3 dissociates to release bicarbonate ions and H^+ ions to the blood.

$$H_2CO_3 \rightarrow H^+ + HCO_3^-$$

The released H^+ lowers the pH again. The bicarbonate ions, being *weak* bases, are poorly functional under alkaline conditions and have little effect on blood pH unless and until blood pH drops toward acid levels.

In the case of excessively slow or shallow breathing (hypoventilation) or fast deep breathing (hyperventilation), the amount of carbonic acid in the blood can be greatly modified—increasing dramatically during hypoventilation and decreasing substantially during hyperventilation. In either situation, if the buffering ability of the blood is inadequate, respiratory acidosis or alkalosis can result. Therefore, maintaining the normal rate and depth of breathing is important for proper control of blood pH.

Demonstrating the Reaction Between Carbon Dioxide (in Exhaled Air) and Water

1. Fill a beaker with 100 ml of distilled water.

2. Add 5 ml of 0.05 M NaOH and five drops of phenol red. Phenol red is a pH indicator that turns yellow in acidic solutions.

3. Blow through a straw into the solution.

What do you observe?

_____

_____

What chemical reaction is taking place in the beaker?

_____

_____

4. Discard the straw in the autoclave bag. ▬

Observing the Operation of Standard Buffers

1. A **buffer** is a molecule or molecular system that stabilizes the pH of a solution. To observe the action of a buffer system, obtain five 250-ml beakers and a wash bottle containing distilled water. Set up the following experimental samples:

Beaker 1:
(150 ml distilled water)   pH _____

Beaker 2:
(150 ml distilled water and
1 drop concentrated HCl)   pH _____

Beaker 3:
(150 ml distilled water and
1 drop concentrated NaOH)   pH _____

Beaker 4:
(150 ml standard buffer solution
[pH 7] and 1 drop concentrated HCl)   pH _____

Beaker 5:
(150 ml standard buffer solution
[pH 7] and 1 drop concentrated NaOH)   pH _____

2. Using a pH meter standardized with a buffer solution of pH 7, determine the pH of the contents of each beaker and record above. After *each and every* pH recording, the pH meter switch should be turned to **STANDBY,** and the electrodes rinsed thoroughly with a stream of distilled water from the wash bottle.

3. Add 3 more drops of concentrated HCl to beaker 4, stir,

and record the pH:_____

4. Add 3 more drops of concentrated NaOH to beaker 5, stir,

and record the pH:_____

How successful was the buffer solution in resisting pH changes when a strong acid (HCl) or a strong base (NaOH) was added?

_____ ▬

Exploring the Operation of the Carbonic Acid-Bicarbonate Buffer System

To observe the ability of the carbonic acid-bicarbonate buffer system of blood to resist pH changes, perform the following simple experiment.

1. Obtain two small beakers (50 ml), animal plasma, graduated cylinder, glass stirring rod, and a dropper bottle of 0.01 M HCl. Using the pH meter standardized with the buffer solution of pH 7.0, measure the pH of the animal plasma. Use only enough plasma to allow immersion of the electrodes and measure the volume used carefully.

pH of the animal plasma: _____

2. Add 2 drops of the 0.01 M HCl solution to the plasma; stir and measure the pH again.

pH of plasma plus 2 drops of HCl: _____

3. Turn the pH meter switch to **STANDBY,** rinse the electrodes, and then immerse them in a quantity of distilled water (pH 7) exactly equal to the amount of animal plasma used. Measure the pH of the distilled water.

pH of distilled water: _____

4. Add 2 drops of 0.01 M HCl, swirl, and measure the pH again.

pH of distilled water plus the two drops of HCl: _____

Is the plasma a good buffer? _____

What component of the plasma carbonic acid–bicarbonate buffer system was acting to counteract a change in pH when HCl was added?

_____ ▬

Respiratory System Physiology

Mechanics of Respiration

1. For each of the following cases, check the column appropriate to your observations on the operation of the model lung.

Change	Diaphragm pushed up		Diaphragm pulled down	
	Increased	Decreased	Increased	Decreased
In internal volume of the bell jar (thoracic cage)				
In internal pressure				
In the size of the balloons (lungs)				
In direction of air flow	Into lungs	Out of lungs	Into lungs	Out of lungs

2. Base your answers to the following on your observations in question 1.

   Under what internal conditions does air tend to flow into the lungs? _____

   _____

   Under what internal conditions does air tend to flow out of the lungs? Explain why this is so. _____

   _____

3. Activation of the diaphragm and the external intercostal muscles begins the inspiratory process. What effect does contraction of these muscles have on thoracic volume, and how is this accomplished? _____

   _____

   _____

4. What was the approximate increase in diameter of chest circumference during a quiet inspiration? _____ cm

   During forced inspiration? _____ cm

What temporary physiological advantage is created by the substantial increase in chest circumference during forced

inspiration? _____

_____

5. The presence of a partial vacuum between the pleural membranes is integral to normal breathing movements. What would happen if an opening were made into the chest cavity, as with a puncture wound?

_____

What must be done to treat this condition medically? _____

Respiratory Sounds

6. Which of the respiratory sounds is heard during both inspiration and expiration? _____

Which is heard primarily during inspiration? _____

7. Where did you best hear the vesicular respiratory sounds? _____

Respiratory Volumes and Capacities—Spirometry or BIOPAC®

8. Write the respiratory volume term and the normal value that is described by the following statements.

Volume of air present in the lungs after a forceful expiration: _____

Volume of air that can be expired forcibly after a normal expiration: _____

Volume of air that is breathed in and out during a normal respiration: _____

Volume of air that can be inspired forcibly after a normal inspiration: _____

Volume of air corresponding to TV + IRV + ERV: _____

9. For the spirometer activities, record experimental respiratory volumes as determined in the laboratory. (Corrected values and FEV_1 are for the recording spirometer only.)

Average TV: _____ ml   Average ERV: _____ ml

Corrected value for TV: _____ ml   Corrected value for ERV: _____ ml

Average IRV: _____ ml   Average VC: _____ ml

Corrected value for IRV: _____ ml   Corrected value for VC: _____ ml

MRV: _____ ml/min   % predicted VC: _____ %

                                        FEV_1: _____ % FVC

For the BIOPAC® activity, record the following experimental respiratory volumes as determined in the laboratory.

TV: _____ L   IRV: _____ L

ERV: _____ L   VC: _____ L

10. Would your vital capacity measurement differ if you performed the test while standing? _____ While lying down? _____

_____ Explain. _____

_____

11. Which respiratory ailments can respiratory volume tests be used to detect?

_____

_____

12. Using an appropriate reference, complete the chart below.

		O_2	CO_2	N_2
% of composition of air	Inspired			
	Expired			

Factors Influencing Rate and Depth of Respiration

13. Where are the neural control centers of respiratory rhythm? _____ and _____

For questions 14–21, use your Activity 6 data.

14. In your data, what was the rate of quiet breathing?

Initial testing _____ breaths/min

Test performed	Observations (breaths per minute)
Talking	
Yawning	
Laughing	
Standing	
Concentrating	
Swallowing water	
Coughing	
Lying down	
Running in place	

15. Record student data below.

Breath-holding interval after a deep inhalation: _____ sec   length of recovery period: _____ sec

Breath-holding interval after a forceful expiration: _____ sec   length of recovery period: _____ sec

After breathing quietly and taking a deep breath (which you held), was your urge to inspire *or* expire? _____

After exhaling and then holding one's breath, was the desire for inspiration *or* expiration? _____

Explain these results. (Hint: What reflex is involved here?) _____

_____

_____

16. Observations after hyperventilation: _____

_____

17. Breath-holding interval after hyperventilation: _____ sec

Why does hyperventilation produce apnea or a reduced respiratory rate? _____

_____

18. Observations for rebreathing air: _____

_____

Why does rebreathing air produce an increased respiratory rate? _____

_____

19. What was the effect of running in place (exercise) on the duration of breath holding? _____

Explain this effect. _____

_____

20. Record student data from the test illustrating the effect of respiration on circulation.

Radial pulse before beginning test: _____ /min   Radial pulse after testing: _____ /min

Relative pulse force before beginning test: _____   Relative force of radial pulse after testing: _____

Condition of neck and facial veins after testing: _____

Explain these data. _____

_____

21. Do the following factors generally increase (indicate ↑) or decrease (indicate ↓) the respiratory rate and depth?

increase in blood CO_2: _____

increase in blood pH: _____

decrease in blood O_2: _____

decrease in blood pH: _____

Did it appear that CO_2 or O_2 had a more marked effect on modifying the respiratory rate? _____

22. Where are sensory receptors sensitive to changes in blood pressure located? _____

_____

23. Where are sensory receptors sensitive to changes in O_2 levels in the blood located? _____

_____

24. What is the primary factor that initiates breathing in a newborn infant? _____

_____

25. Which, if any, of the measurable respiratory volumes would likely be increased in a person who is cardiovascularly fit, such as a runner or a swimmer?

_____

_____

Which, if any, of the measurable respiratory volumes would likely be decreased in a person who has smoked a lot for over twenty years?

_____

_____

26. Blood CO_2 levels and blood pH are related. When blood CO_2 levels increase, does the pH increase or decrease?

_____ Explain why. _____

_____

Role of the Respiratory System in Acid-Base Balance of Blood

27. Define *buffer*. _____

28. How successful was the laboratory buffer (pH 7) in resisting changes in pH when the acid was added? _____

_____

When the base was added? _____

How successful was the buffer in resisting changes in pH when the additional drops of the acid and base were added to the

original samples? _____

29. What buffer system operates in blood plasma? _____

Which member of the buffer system resists a *drop* in pH? _____ Which resists a *rise* in pH? _____

30. Explain how the carbonic acid–bicarbonate buffer system of the blood operates. _____

_____

_____

_____

31. What happened when the carbon dioxide in exhaled air mixed with water? _____

_____

What role does exhalation of carbon dioxide play in maintaining relatively constant blood pH? _____

_____

_____

Anatomy of the Digestive System

MATERIALS

- ☐ Dissectible torso model
- ☐ Anatomical chart of the human digestive system
- ☐ Prepared slides of the liver and mixed salivary glands; of longitudinal sections of the gastroesophageal junction and a tooth; and of cross sections of the stomach, duodenum, ileum, and large intestine
- ☐ Compound microscope
- ☐ Three-dimensional model of a villus (if available)
- ☐ Jaw model or human skull
- ☐ Three-dimensional model of liver lobules (if available)

✂ For instructions on animal dissections, see the dissection exercises (starting on page 697) in the cat and fetal pig editions of this manual.

OBJECTIVES

1. State the overall function of the digestive system.
2. Describe the general histologic structure of the alimentary canal wall and identify the following structures on an appropriate image of the wall: mucosa, submucosa, muscularis externa, and serosa or adventitia.
3. Identify on a model or image the organs of the alimentary canal, and name their subdivisions if any.
4. Describe the general function of each of the digestive system organs or structures.
5. List and explain the specializations of the structure of the stomach and small intestine that contribute to their functional roles.
6. Name and identify the accessory digestive organs listing a function for each.
7. Describe the anatomy of the generalized tooth, and name the human deciduous and permanent teeth.
8. List the major enzymes or enzyme groups produced by the salivary glands, stomach, small intestine, and pancreas.
9. Recognize microscopically or in an image the histologic structure of the following organs:

small intestine	tooth	liver
salivary glands	stomach	

PRE-LAB QUIZ

1. The digestive system
   a. eliminates undigested food
   b. provides the body with nutrients
   c. provides the body with water
   d. all of the above
2. Circle the correct underlined term. Digestion / Absorption occurs when small molecules pass through epithelial cells into the blood for distribution to the body cells.
3. The _____ abuts the lumen of the alimentary canal and consists of epithelium, lamina propria, and muscularis mucosae.
   a. mucosa
   b. serosa
   c. submucosa
4. Circle the correct underlined term. Approximately 25 cm long, the esophagus / alimentary canal conducts food from the pharynx to the stomach.
5. Wavelike contractions of the digestive tract that propel food along are called
   a. digestion
   b. elimination
   c. ingestion
   d. peristalsis

(Text continues on next page.)

6. The _____ is located on the left side of the abdominal cavity and is hidden by the liver and diaphragm.
   a. gallbladder
   b. large intestine
   c. small intestine
   d. stomach

7. Circle True or False. Nearly all nutrient absorption occurs in the small intestine.

8. Circle the correct underlined term. The <u>ascending colon</u> / <u>descending colon</u> traverses down the left side of the abdominal cavity and becomes the sigmoid colon.

9. A tooth consists of two major regions, the crown and the
   a. dentin
   b. enamel
   c. gingiva
   d. root

10. Located inferior to the diaphragm, the _____ is the largest gland in the body.
   a. gallbladder
   b. liver
   c. pancreas
   d. thymus

The **digestive system** provides the body with the nutrients, water, and electrolytes essential for health. The organs of this system ingest, digest, and absorb food and eliminate the undigested remains as feces.

The digestive system consists of a hollow tube extending from the mouth to the anus, into which various accessory organs or glands empty their secretions **(Figure 38.1)**. Food material within this tube, the *alimentary canal,* is technically outside the body because it has contact only with the cells lining the tract. For ingested food to become available to the body cells, it must first be broken down *physically* (by chewing or churning) and *chemically* (by enzymatic hydrolysis) into its smaller diffusible molecules—a process called **digestion.** The digested end products can then pass through the epithelial cells lining the tract into the blood for distribution to the body cells—a process termed **absorption.** In one sense, the digestive tract can be viewed as a disassembly line, in which food is carried from one stage of its digestive processing to the next by muscular activity, and its nutrients are made available to the cells of the body en route.

The organs of the digestive system are traditionally separated into two major groups: the **alimentary canal,** or **gastrointestinal (GI) tract,** and the **accessory digestive organs.** The alimentary canal is approximately 9 meters long in a cadaver but is considerably shorter in a living person due to muscle tone. It consists of the mouth, pharynx, esophagus, stomach, and small and large intestines. The accessory structures include the teeth, which physically break down foods, and the salivary glands, gallbladder, liver, and pancreas, which secrete their products into the alimentary canal.

General Histological Plan of the Alimentary Canal

From the esophagus to the anal canal, the basic structure of the alimentary canal is similar. So, it makes sense to begin our study by learning the features of this structure. As we study individual parts of the alimentary canal, we will note how this basic plan is modified to provide the unique digestive functions of each subsequent organ.

Essentially the alimentary canal wall has four basic **tunics** (layers). From the lumen outward, these are the *mucosa,* the *submucosa,* the *muscularis externa,* and either a *serosa* or *adventitia* **(Figure 38.2)**. Each of these tunics has a predominant tissue type and a specific function in the digestive process.

Mucosa (mucous membrane): The mucosa is the wet epithelial membrane abutting the alimentary canal lumen. It consists of a surface *epithelium* (in most cases, a simple columnar), a *lamina propria* (areolar connective tissue on which the epithelial layer rests), and a *muscularis mucosae* (a scant layer of smooth muscle fibers that enable local movements of the mucosa). The major functions of the mucosa are secretion (of enzymes, mucus, hormones, etc.), absorption of digested foodstuffs, and protection (against bacterial invasion). A particular mucosal region may be involved in one or all three functions.

Submucosa: The submucosa is moderately dense connective tissue containing blood and lymphatic vessels, scattered lymphoid follicles, and nerve fibers. Its intrinsic nerve supply is called the *submucosal plexus.* Its vessels absorb and transport nutrients, and its abundant elastic fibers help maintain the normal shape of each organ.

Muscularis externa: The muscularis externa, also simply called the *muscularis,* typically is a bilayer of smooth muscle, with the inner layer running circularly and the outer layer running longitudinally. This layer moves the contents of the canal along by segmentation and peristalsis. An important intrinsic nerve plexus, the *myenteric plexus,* associated with this tunic is the major regulator of GI motility.

Serosa: The outermost covering of the intraperitoneal organs is the serosa, also called the *visceral peritoneum.* It consists of mesothelium associated with a thin layer of areolar connective tissue. The serosa reduces friction as the mobile digestive system organs work and slide across one another and the cavity walls. In the esophagus, which is *outside* the abdominopelvic cavity, the serosa is replaced by an **adventitia,** a layer of coarse fibrous connective tissue that binds the organ to surrounding tissues. The adventitia anchors and protects the surrounded organ.

38

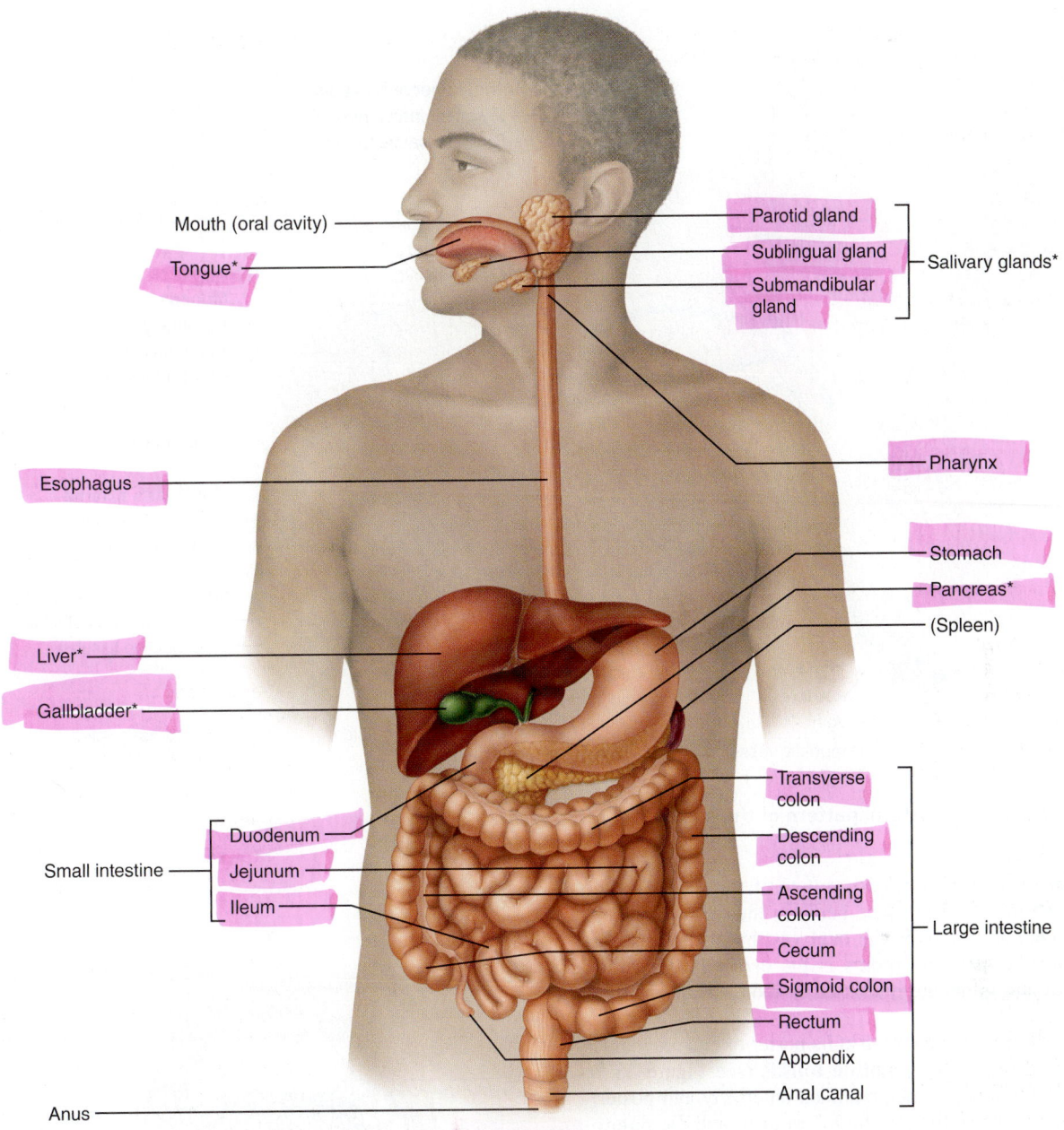

Mouth (oral cavity)

Tongue*

Parotid gland

Sublingual gland

Submandibular gland

Salivary glands*

Pharynx

Esophagus

Stomach

Pancreas*

(Spleen)

Liver*

Gallbladder*

Transverse colon

Duodenum

Descending colon

Small intestine   Jejunum

Ascending colon

Ileum

Cecum

Large intestine

Sigmoid colon

Rectum

Appendix

Anal canal

Anus

Figure 38.1 The human digestive system: alimentary tube and accessory organs. Organs with asterisks are accessory organs. Those without asterisks are alimentary canal organs (except the spleen, a part of the lymphatic system).

Organs of the Alimentary Canal

ACTIVITY 1

Identifying Alimentary Canal Organs

The sequential pathway and fate of food as it passes through the alimentary canal organs are described in the next sections. Identify each structure in the digestive system illustration (Figure 38.1) and on the torso model or anatomical chart of the digestive system as you work. ■

Oral Cavity or Mouth

Food enters the digestive tract through the **oral cavity,** or **mouth (Figure 38.3)**. Within this mucous membrane–lined

cavity are the gums, teeth, tongue, and openings of the ducts of the salivary glands. The **lips (labia)** protect the opening of the chamber anteriorly, the **cheeks** form its lateral walls, and the **palate,** its roof. The anterior portion of the palate is referred to as the **hard palate** because the palatine processes of the maxillae and horizontal plates of the palatine bones underlie it. The posterior **soft palate** is a fibromuscular structure that is unsupported by bone. The **uvula,** a fingerlike projection of the soft palate, extends inferiorly from its posterior margin. The soft palate rises to close off the oral cavity from the nasal and pharyngeal passages during swallowing. The floor of the oral cavity is occupied by the muscular **tongue,** which is largely supported by the *mylohyoid muscle* **(Figure 38.4)** and attaches to the hyoid bone, mandible, styloid processes, and pharynx. A membrane called the **lingual**

38

38

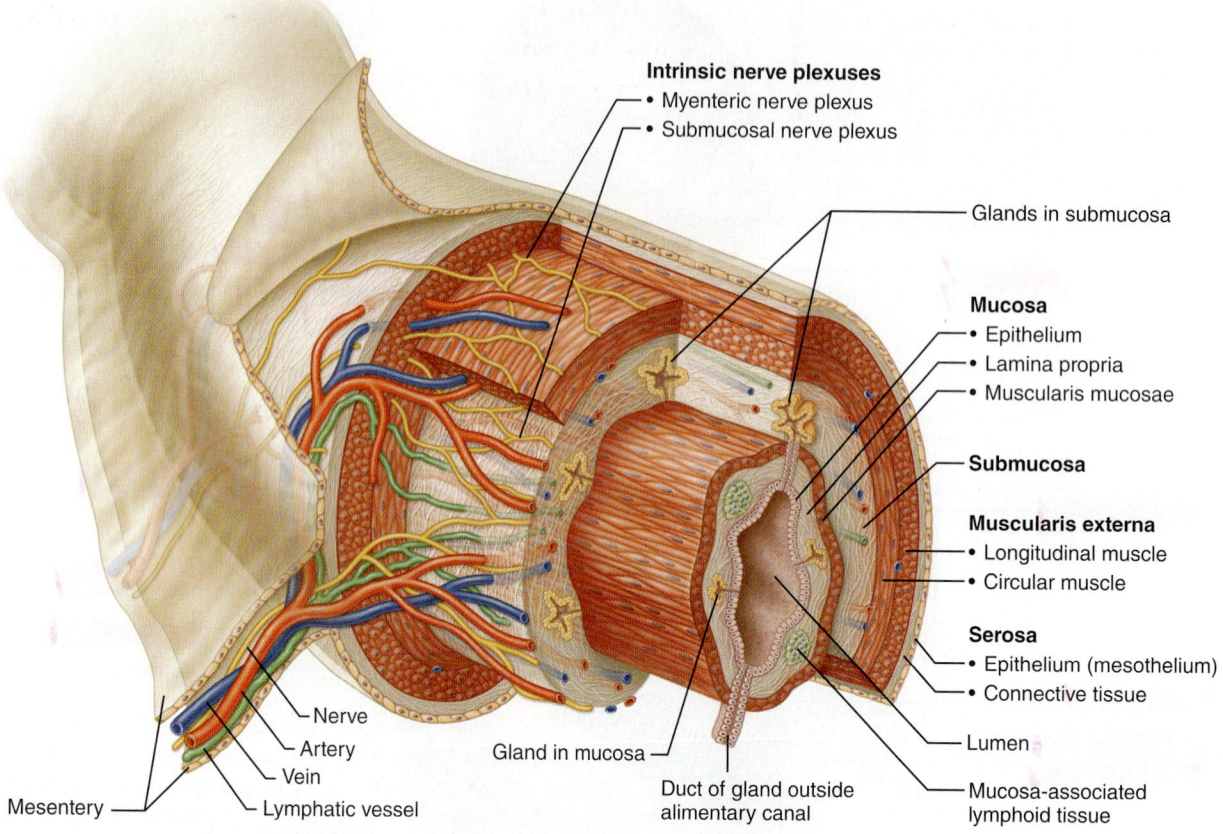

Figure 38.2 Basic structural pattern of the alimentary canal wall.

frenulum secures the inferior midline of the tongue to the floor of the mouth. The space between the lips and cheeks and the teeth and gums is the **oral vestibule;** the area that lies within the teeth and gums is the **oral cavity proper.** (The teeth and gums are discussed in more detail on pages 584–585.)

On each side of the mouth at its posterior end are masses of lymphoid tissue, the **palatine tonsils** (see Figure 38.3). Each lies in a concave area bounded anteriorly and posteriorly by membranes, the **palatoglossal arch** and the **palatopharyngeal arch,** respectively. Another mass of lymphoid tissue, the **lingual tonsil** (see Figure 38.4), covers the base of the tongue, posterior to the oral cavity proper. The tonsils, in common with other lymphoid tissues, are part of the body's defense system.

Very often in young children, the palatine tonsils become inflamed and enlarge, partially blocking the entrance to the pharynx posteriorly and making swallowing difficult and painful. This condition is called **tonsillitis.** ✚

Three pairs of salivary glands duct their secretion, saliva, into the oral cavity. One component of saliva, salivary amylase, begins the digestion of starchy foods within the oral cavity. (The salivary glands are discussed in more detail on page 585.)

As food enters the mouth, it is mixed with saliva and masticated (chewed). The cheeks and lips help hold the food between the teeth during mastication, and the highly mobile tongue manipulates the food during chewing and initiates swallowing. Thus the mechanical and chemical breakdown of food begins before the food has left the oral cavity. The surface of the tongue is covered with papillae, many of which

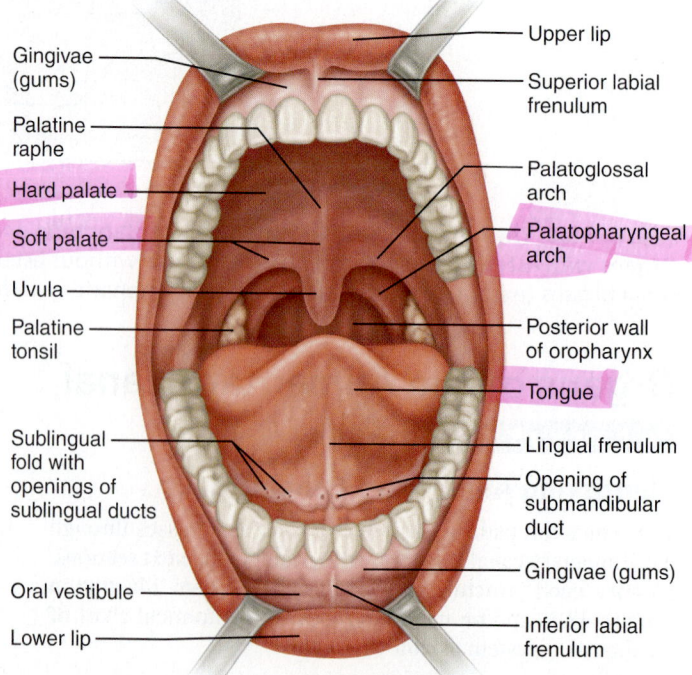

Figure 38.3 Anterior view of the oral cavity.

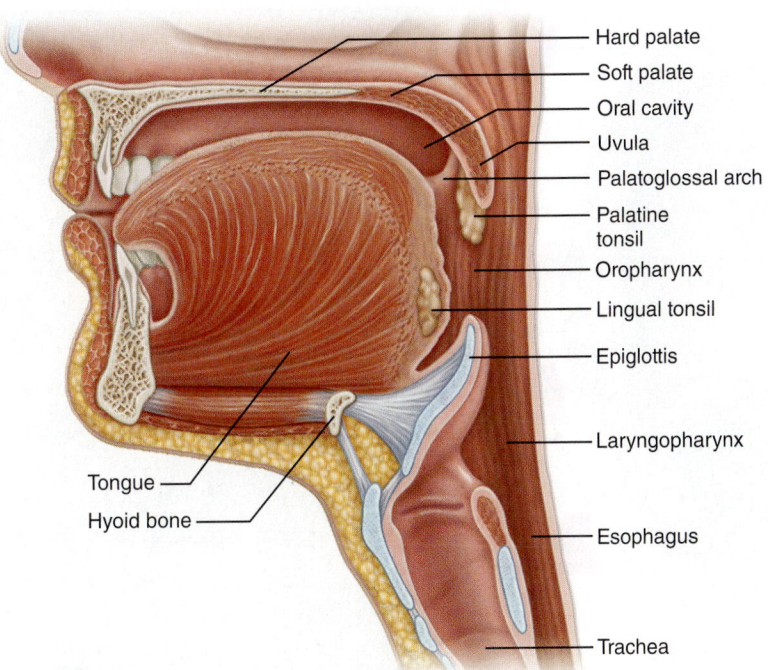

Figure 38.4 **Sagittal view of the head showing oral cavity and pharynx.**

Hard palate
Soft palate
Oral cavity
Uvula
Palatoglossal arch
Palatine tonsil
Oropharynx
Lingual tonsil
Epiglottis
Laryngopharynx
Esophagus
Trachea
Tongue
Hyoid bone

contain taste buds. Taste cells within the taste buds are the receptors for taste sensation. So, in addition to its manipulative function, the tongue permits the enjoyment and appreciation of food.

Pharynx

When the tongue initiates swallowing, the food passes posteriorly into the pharynx, a common passageway for food, fluid, and air (see Figure 38.4). The pharynx is subdivided anatomically into three parts—the **nasopharynx** (behind the nasal cavity), the **oropharynx** (behind the oral cavity extending from the soft palate to the epiglottis), and the **laryngopharynx** (extending from the epiglottis to the base of the larynx).

The walls of the pharynx consist largely of two layers of skeletal muscles: an inner layer of longitudinal muscle and an outer layer of circular constrictor muscles. Together these initiate wavelike contractions that propel the food inferiorly into the esophagus. The mucosa of the oropharynx and laryngopharynx, like that of the oral cavity, contains a friction-resistant stratified squamous epithelium.

Esophagus

The **esophagus,** or gullet, extends from the pharynx through the diaphragm to the gastroesophageal sphincter in the superior aspect of the stomach. Approximately 25 cm long in humans, it is essentially a food passageway that conducts food to the stomach in a wavelike peristaltic motion. The esophagus has no digestive or absorptive function. The walls at its superior end contain skeletal muscle, which is replaced by smooth muscle in the area nearing the stomach. The **gastroesophageal sphincter,** a slight thickening of the smooth muscle layer at the esophagus-stomach junction, controls food passage into the stomach **(Figure 38.5)**.

Stomach

The **stomach** (Figures 38.1 and 38.5) is on the left side of the abdominal cavity and is hidden by the liver and diaphragm.

The stomach is made up of several regions. The **cardial part** or **cardia** is the area surrounding the cardial orifice through which food enters the stomach. The **fundus** is a dome-shaped portion of the stomach found superolaterally to the cardia. The **body** forms the midportion of the stomach, which leads to the funnel-shaped **pyloric part.** The wide superior area of the pyloric part is called the *pyloric antrum;* it narrows to form the *pyloric canal,* which terminates in the *pylorus.* The pylorus is continuous with the small intestine through the **pyloric sphincter** or **valve.**

The concave medial surface of the stomach is called the **lesser curvature;** its convex lateral surface is the **greater curvature.** Extending from these curvatures are two mesenteries, called *omenta.* The **lesser omentum** extends from the liver to the lesser curvature of the stomach. The **greater omentum,** a saclike mesentery, extends from the greater curvature of the stomach, reflects downward over the abdominal contents to cover them in an apronlike fashion, and then blends with the **mesocolon** attaching the transverse colon to the posterior body wall. (Figure 38.7 on page 580 illustrates the omenta as well as the other peritoneal attachments of the abdominal organs.)

The stomach is a temporary storage region for food as well as a site for mechanical and chemical breakdown of food. It contains a third (innermost) *obliquely* oriented layer of smooth muscle in its muscularis externa that allows it to churn, mix, and pummel the food, physically reducing it to smaller fragments. **Gastric glands** of the mucosa secrete hydrochloric acid (HCl) and hydrolytic enzymes—primarily pepsinogen, the inactive form of *pepsin,* which digests protein. The *mucosal glands* also secrete a viscous mucus that helps prevent the stomach itself from being digested by the proteolytic enzymes. Most digestive activity occurs in the pyloric part of the stomach. After the food is processed in the stomach, it resembles a creamy mass called **chyme,** which enters the small intestine through the pyloric sphincter.

38

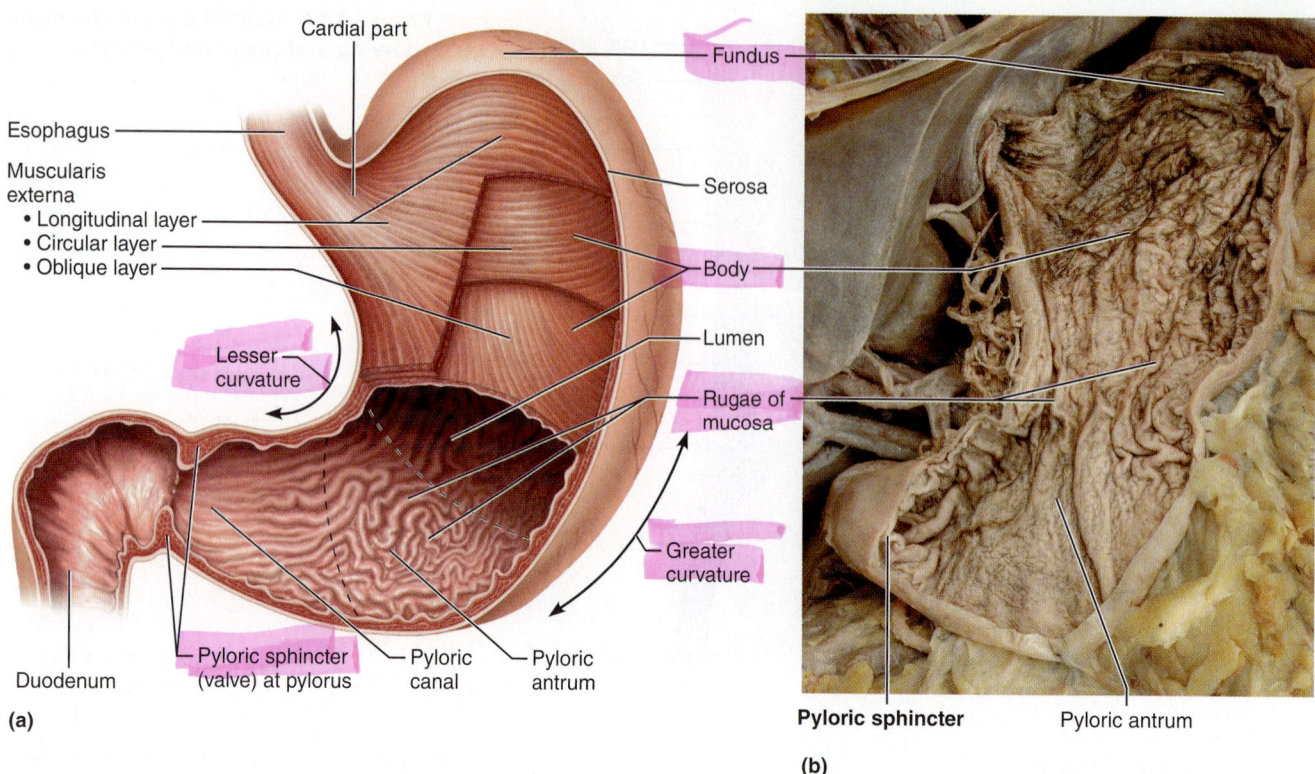

(a)

Pyloric sphincter   Pyloric antrum

(b)

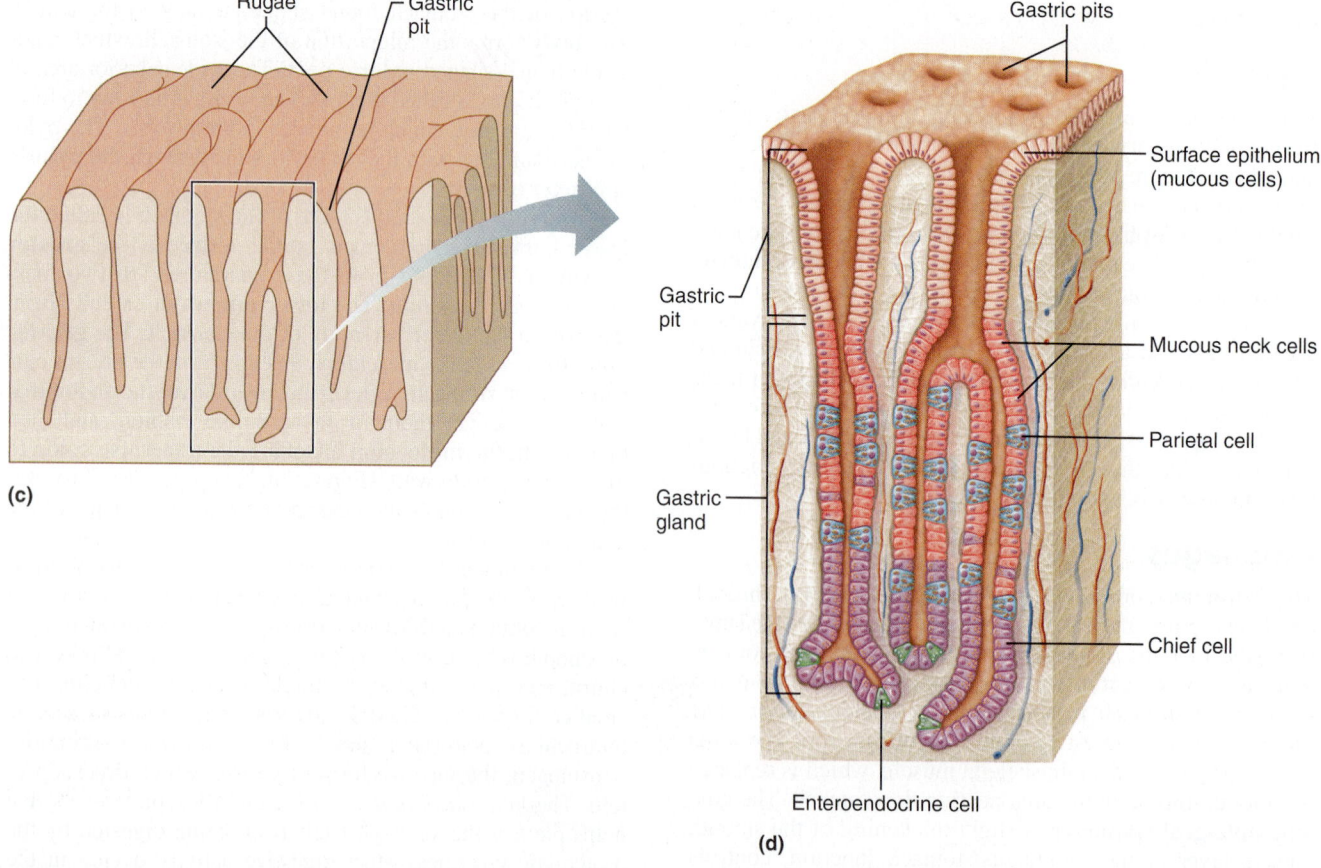

(c)

(d)

Figure 38.5 Anatomy of the stomach. (a) Gross internal and external anatomy. **(b)** Photograph of internal aspect of stomach. **(c, d)** Section of the stomach wall showing rugae and gastric pits.

Studying the Histologic Structure of Selected Digestive System Organs

To prepare for the histologic study you will be conducting now and later in the lab, obtain a microscope and the following slides: salivary glands (submandibular or sublingual); liver; cross sections of the large intestine, duodenum, ileum, and stomach; and longitudinal sections of a tooth and the gastroesophageal junction.

1. **Stomach:** View the stomach slide first. Refer to the photomicrograph **(Figure 38.6a)** as you scan the tissue under low power to locate the muscularis externa; then move to high power to more closely examine this layer. Try to pick out the three smooth muscle layers. How does the extra oblique layer of smooth muscle found in the stomach correlate with the stomach's churning movements?

_____

_____

_____

Identify the gastric glands and the gastric pits (see Figures 38.5 and 38.6b). If the section is taken from the stomach fundus and is appropriately stained, you can identify, in the gastric glands, the blue-staining **chief cells,** which produce pepsinogen, and the red-staining **parietal cells,** which secrete HCl. The enteroendocrine cells that release hormones are indistinguishable. Draw a small section of the stomach wall, and label it appropriately.

2. **Gastroesophageal junction:** Scan the slide under low power to locate the mucosal junction between the end of the esophagus and the beginning of the stomach, the gastroesophageal junction. Compare your observations to the photomicrograph (Figure 38.6c). What is the functional importance of the epithelial differences seen in the two organs?

_____

_____

_____

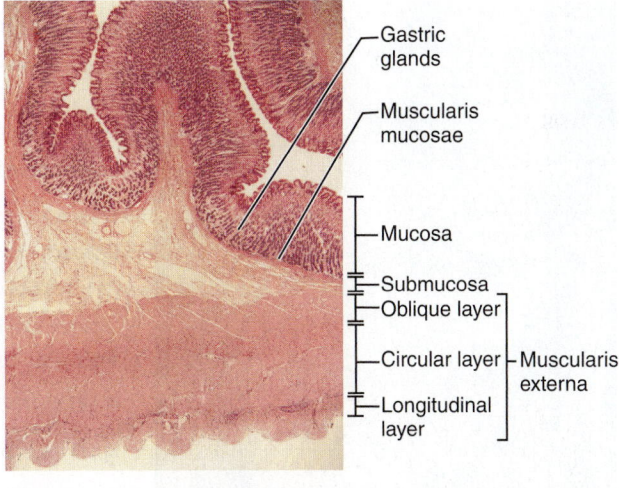

- Gastric glands
- Muscularis mucosae
- Mucosa
- Submucosa
- Oblique layer
- Circular layer — Muscularis externa
- Longitudinal layer

(a)

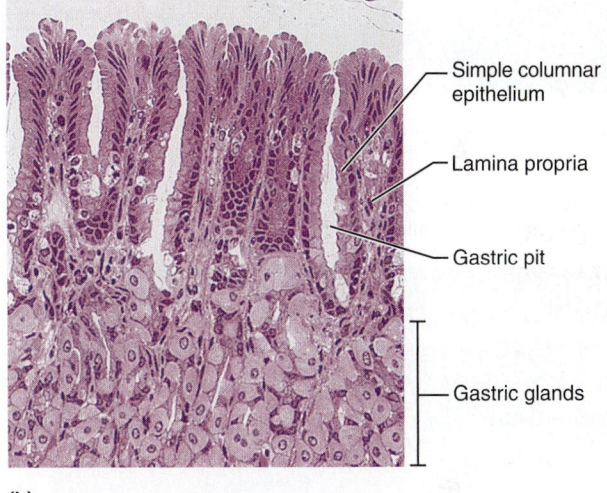

- Simple columnar epithelium
- Lamina propria
- Gastric pit
- Gastric glands

(b)

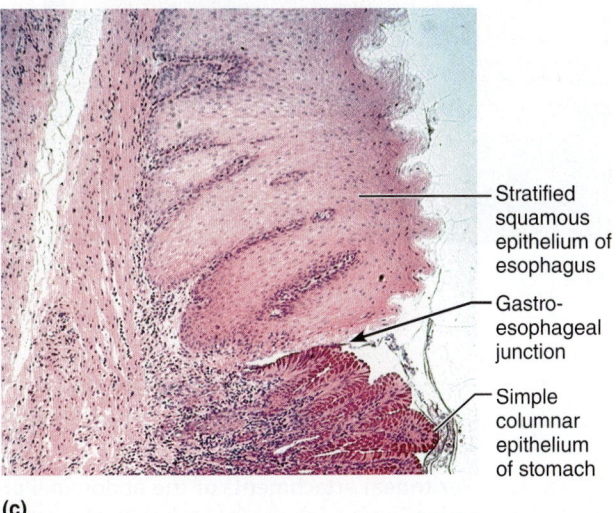

- Stratified squamous epithelium of esophagus
- Gastroesophageal junction
- Simple columnar epithelium of stomach

(c)

Figure 38.6 Histology of selected regions of the stomach and gastroesophageal junction. (a) Stomach wall (12×). **(b)** Gastric pits and glands (130×). **(c)** Gastroesophageal junction, longitudinal section (60×).

38

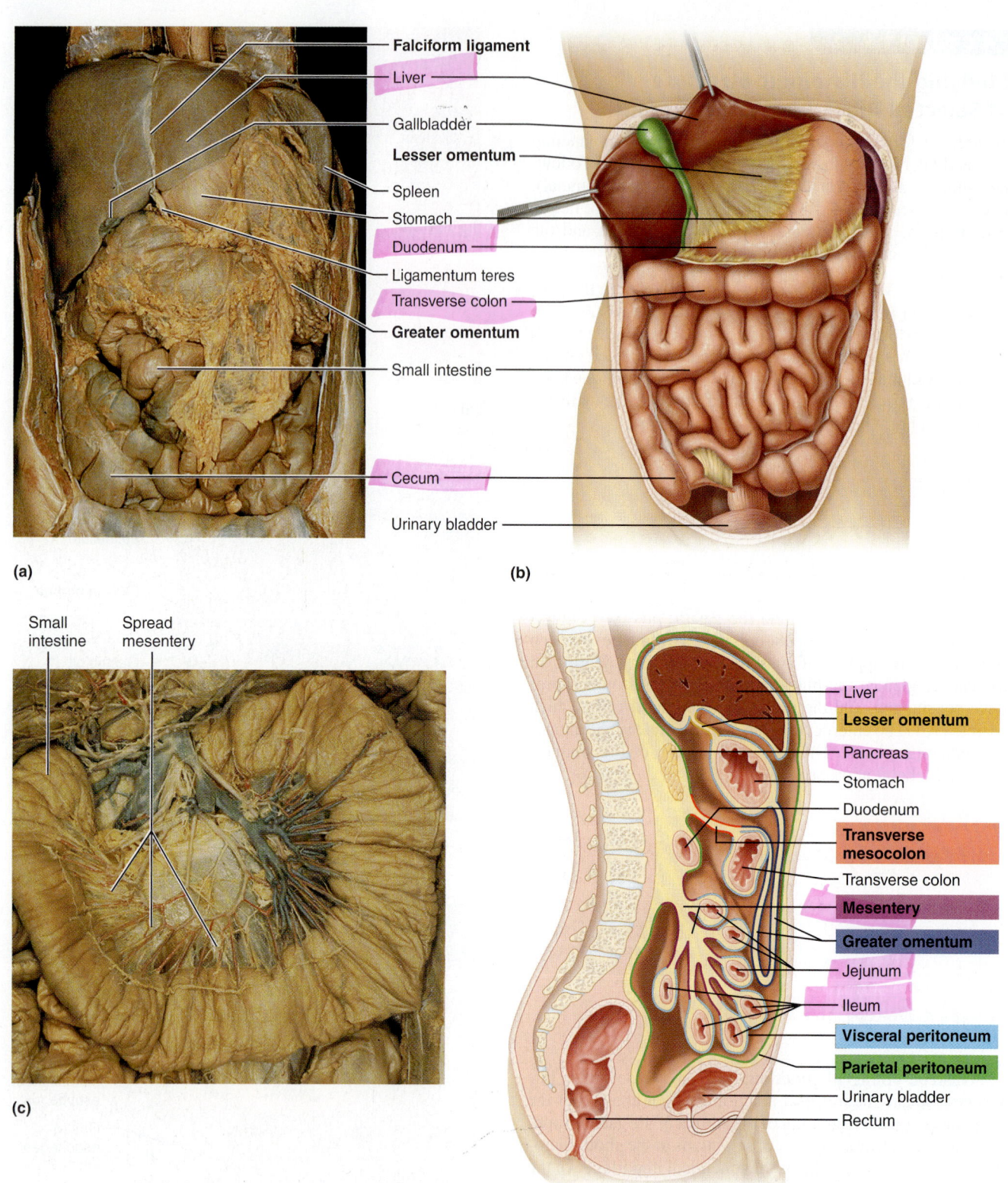

Figure 38.7 Peritoneal attachments of the abdominal organs. Superficial anterior views of abdominal cavity: **(a)** photograph with the greater omentum in place and **(b)** diagram showing greater omentum removed and liver and gallbladder reflected superiorly. **(c)** Mesentery of the small intestine. **(d)** Sagittal view of a male torso. Mesentery labels appear in colored boxes.

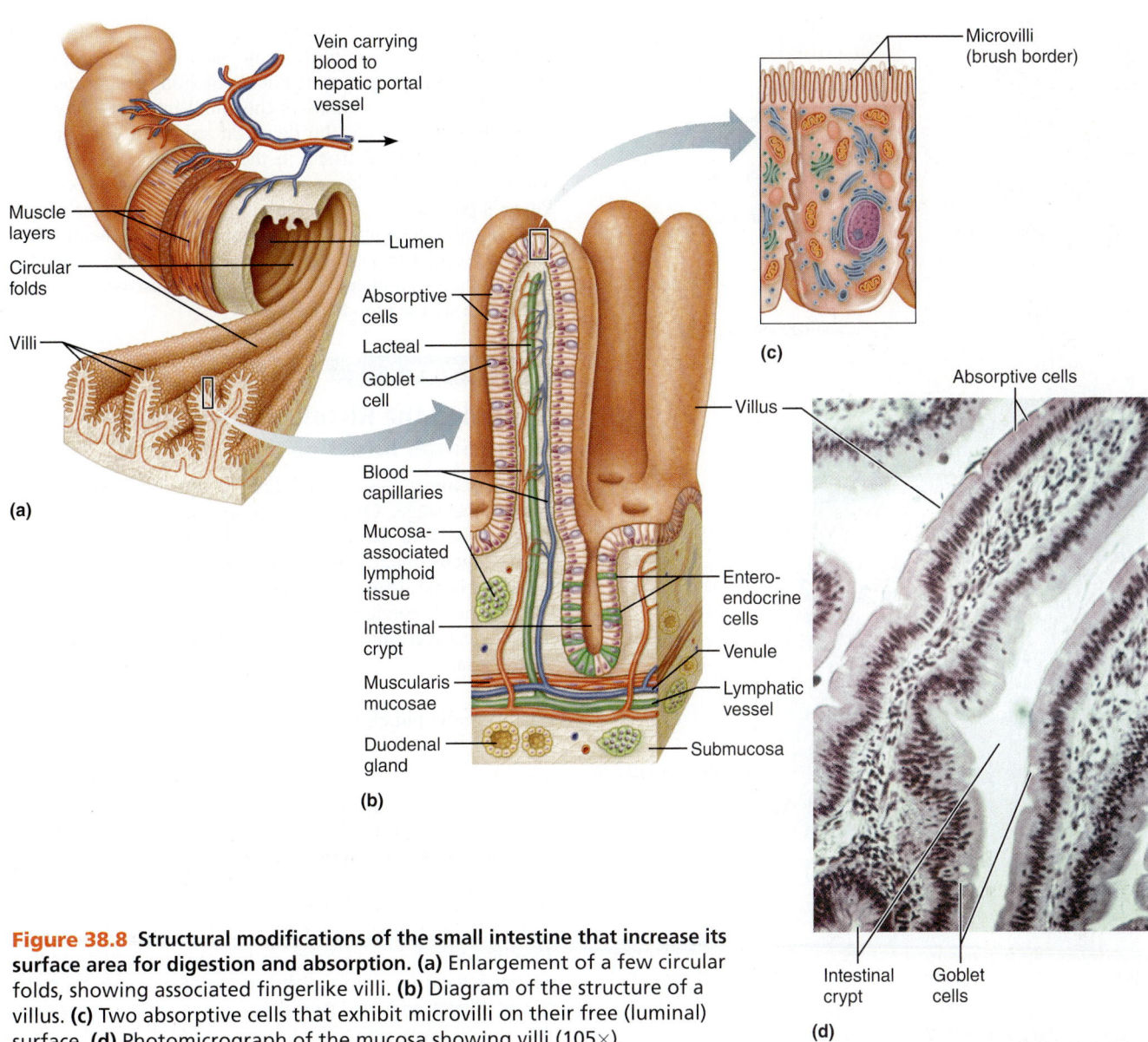

Figure 38.8 Structural modifications of the small intestine that increase its surface area for digestion and absorption. (a) Enlargement of a few circular folds, showing associated fingerlike villi. **(b)** Diagram of the structure of a villus. **(c)** Two absorptive cells that exhibit microvilli on their free (luminal) surface. **(d)** Photomicrograph of the mucosa showing villi (105×).

38

Small Intestine

The **small intestine** is a convoluted tube, 6 to 7 meters (about 20 feet) long in a cadaver but only about 2 m (6 feet) long during life because of its muscle tone. It extends from the pyloric sphincter to the ileocecal valve. The small intestine is suspended by a double layer of peritoneum, the fan-shaped **mesentery,** from the posterior abdominal wall **(Figure 38.7)**, and it lies, framed laterally and superiorly by the large intestine, in the abdominal cavity. The small intestine has three subdivisions (see Figure 38.1): (1) the **duodenum** extends from the pyloric sphincter for about 25 cm (10 inches) and curves around the head of the pancreas; most of the duodenum lies in a retroperitoneal position. (2) The **jejunum,** continuous with the duodenum, extends for 2.5 m (about 8 feet). Most of the jejunum occupies the umbilical region of the abdominal cavity. (3) The **ileum,** the terminal portion of the small intestine, is about 3.6 m (12 feet) long and joins the large intestine at the **ileocecal valve.** It is located inferiorly and somewhat to the right in the abdominal cavity, but its major portion lies in the hypogastric region.

Brush border enzymes, hydrolytic enzymes bound to the microvilli of the columnar epithelial cells, and, more importantly, enzymes produced by the pancreas and ducted into the duodenum largely via the **main pancreatic duct** complete the enzymatic digestion process in the small intestine. Bile (formed in the liver) also enters the duodenum via the **bile duct** in the same area. At the duodenum, the ducts join to form the bulblike **hepatopancreatic ampulla** and empty their products into the duodenal lumen through the **major duodenal papilla,** an orifice controlled by a muscular valve called the **hepatopancreatic sphincter** (see Figure 38.15 on page 587).

Nearly all nutrient absorption occurs in the small intestine, where three structural modifications increase the absorptive surface of the mucosa: the microvilli, villi, and circular folds **(Figure 38.8)**. **Microvilli** are minute projections of the surface plasma membrane of the columnar epithelial lining cells of the mucosa. **Villi** are the fingerlike projections of the mucosa tunic that give it a velvety appearance and texture. The **circular folds** are deep, permanent folds of the mucosa and submucosa layers that force chyme to spiral through the

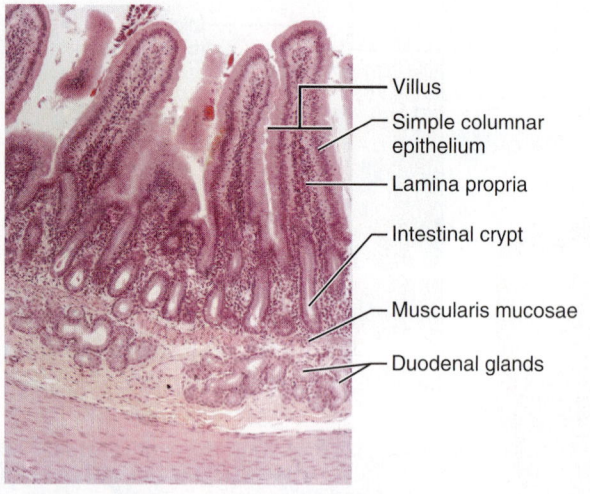

(a)

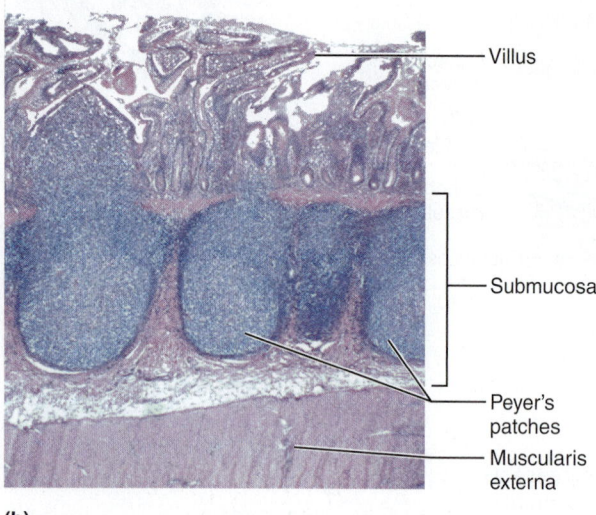

(b)

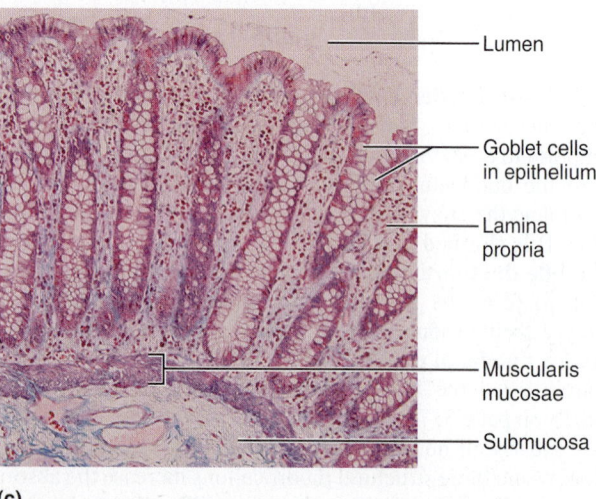

(c)

Figure 38.9 Histology of selected regions of the small and large intestines. Cross-sectional views. **(a)** Duodenum of the small intestine (95×). **(b)** Ileum of the small intestine (20×). **(c)** Large intestine (80×).

intestine, mixing it and slowing its progress. These structural modifications decrease in frequency and size toward the end of the small intestine. Any residue remaining undigested and unabsorbed at the terminus of the small intestine enters the large intestine through the ileocecal valve. In contrast, the amount of lymphoid tissue in the submucosa of the small intestine (especially the aggregated lymphoid nodules called **Peyer's patches, Figure 38.9b**) increases along the length of the small intestine and is very apparent in the ileum. This reflects the fact that the remaining undigested food residue contains large numbers of bacteria that must be prevented from entering the bloodstream.

ACTIVITY 3

Observing the Histologic Structure of the Small Intestine

1. **Duodenum:** Secure the slide of the duodenum to the microscope stage. Observe the tissue under low power to identify the four basic tunics of the intestinal wall—that is, the **mucosa** and its three sublayers, the **submucosa,** the **muscularis externa,** and the **serosa** or *visceral peritoneum.* Consult the photomicrograph (Figure 38.9a) to help you identify the scattered mucus-producing **duodenal glands** in the submucosa.

What type of epithelium do you see here? _____

_____

Examine the large leaflike *villi,* which increase the surface area for absorption. Notice the scattered mucus-producing goblet cells in the epithelium of the villi. Note also the **intestinal crypts** (see also Figure 38.8), invaginated areas of the mucosa between the villi containing the cells that produce intestinal juice, a watery mucus-containing mixture that serves as a carrier fluid for absorption of nutrients from the chyme. Sketch and label a small section of the duodenal wall, showing all layers and villi.

2. **Ileum:** The structure of the ileum resembles that of the duodenum, except that the villi are less elaborate (because most of the absorption has occurred by the time that chyme reaches the ileum). Secure a slide of the ileum to the microscope stage for viewing. Observe the villi, and identify

38

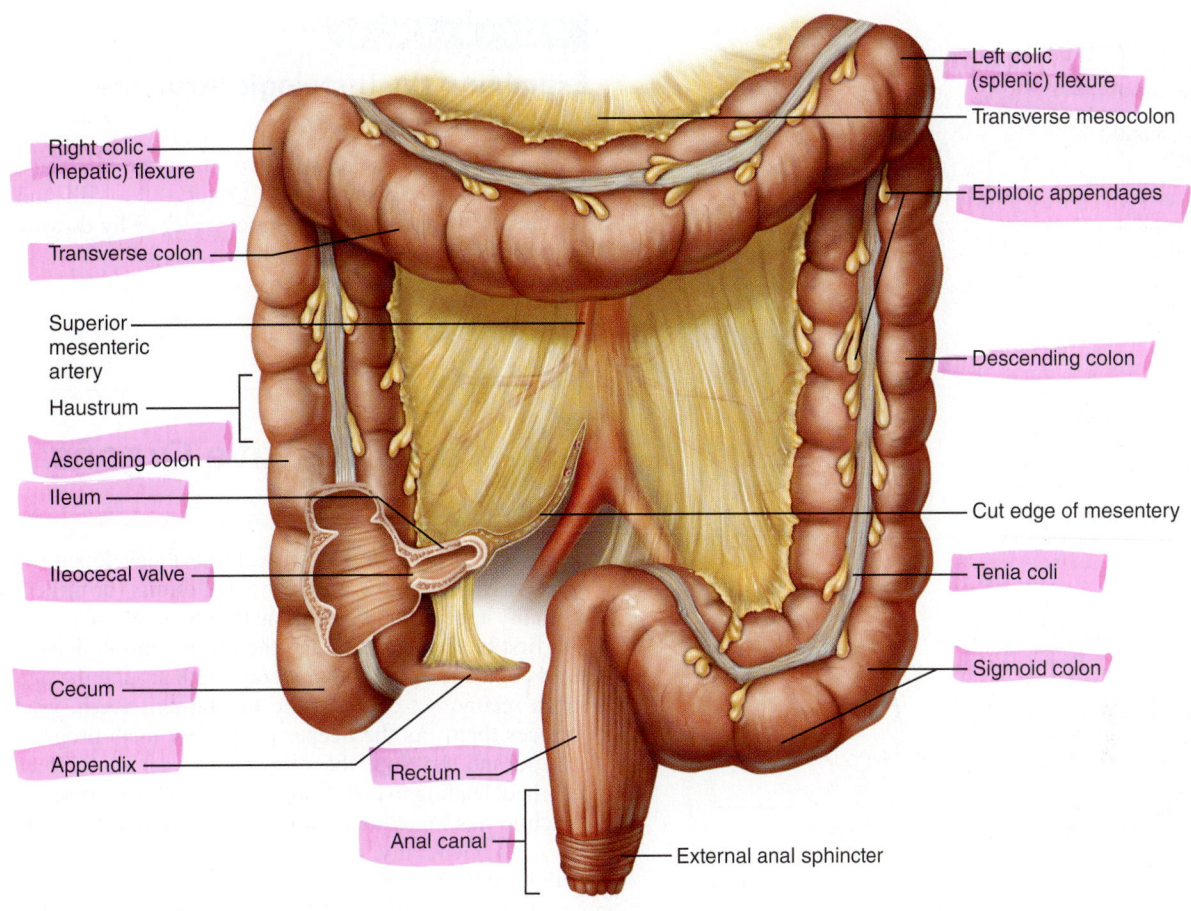

Figure 38.10 The large intestine. (Section of the cecum removed to show the ileocecal valve.)

Labels:
Left colic (splenic) flexure
Transverse mesocolon
Right colic (hepatic) flexure
Epiploic appendages
Transverse colon
Superior mesenteric artery
Descending colon
Haustrum
Ascending colon
Ileum
Cut edge of mesentery
Ileocecal valve
Tenia coli
Cecum
Sigmoid colon
Appendix
Rectum
Anal canal
External anal sphincter

the four layers of the wall and the large, generally spherical Peyer's patches (Figure 38.9b). What tissue type are Peyer's patches?

3. If a villus model is available, identify the following cells or regions before continuing: absorptive epithelium, goblet cells, lamina propria, slips of the muscularis mucosae, capillary bed, and lacteal. If possible, also identify the intestinal crypts. ▪

Large Intestine

The **large intestine (Figure 38.10)** is about 1.5 m (5 feet) long and extends from the ileocecal valve to the anus. It encircles the small intestine on three sides and consists of the following subdivisions: **cecum, appendix, colon, rectum,** and **anal canal.**

The blind wormlike appendix, which hangs from the cecum, is a trouble spot in the large intestine. Since it is generally twisted, it provides an ideal location for

bacteria to accumulate and multiply. Inflammation of the appendix, or appendicitis, is the result. ✚

The colon is divided into several distinct regions. The **ascending colon** travels up the right side of the abdominal cavity and makes a right-angle turn at the **right colic (hepatic) flexure** to cross the abdominal cavity as the **transverse colon.** It then turns at the **left colic (splenic) flexure** and continues down the left side of the abdominal cavity as the **descending colon,** where it takes an **S**-shaped course as the **sigmoid colon.** The sigmoid colon, rectum, and the anal canal lie in the pelvis anterior to the sacrum and thus are not considered abdominal cavity structures. Except for the transverse and sigmoid colons, which are secured to the dorsal body wall by mesocolons (see Figure 38.7), the colon is retroperitoneal.

The anal canal terminates in the **anus,** the opening to the exterior of the body. The anal canal has two sphincters, a voluntary *external anal sphincter* composed of skeletal muscle, and an involuntary *internal anal sphincter* composed of smooth muscle. The sphincters are normally closed except during defecation, when undigested food and bacteria are eliminated from the body as feces.

38

Incisors
Central (6–8 mo)

Lateral (8–10 mo)

Canine (eyetooth)
(16–20 mo)

Molars
First molar
(10–15 mo)

Second molar
(about 2 yr)

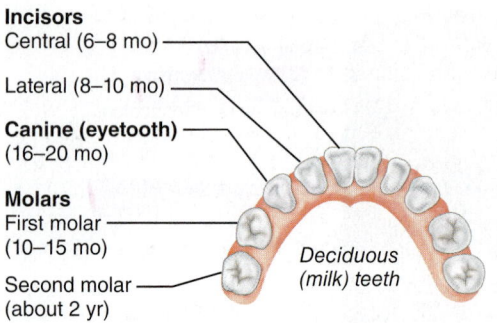

*Deciduous
(milk) teeth*

Incisors
Central (7 yr)

Lateral (8 yr)

Canine (eyetooth)
(11 yr)

**Premolars
(bicuspids)**
First premolar
(11 yr)

Second premolar
(12–13 yr)

Molars
First molar (6–7 yr)

Second molar
(12–13 yr)

Third molar
(wisdom tooth)
(17–25 yr)

*Permanent
teeth*

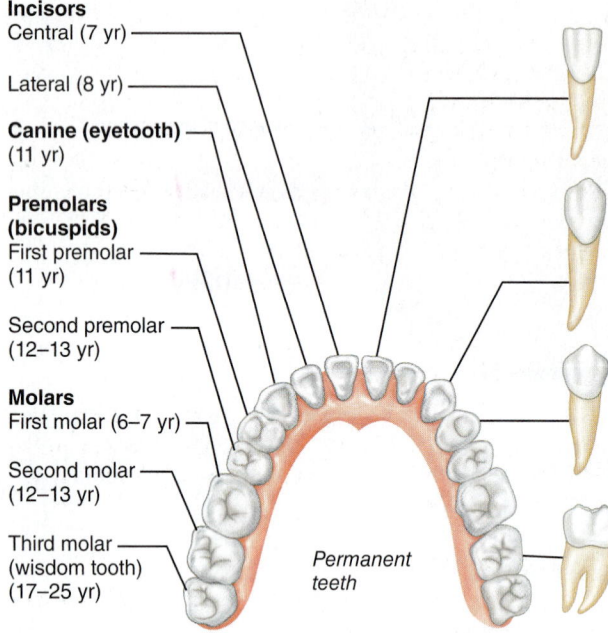

Figure 38.11 Human deciduous teeth and permanent teeth. (Approximate time of teeth eruption shown in parentheses.)

38

In the large intestine, the longitudinal muscle layer of the muscularis externa is reduced to three longitudinal muscle bands called the **teniae coli.** Since these bands are shorter than the rest of the wall of the large intestine, they cause the wall to pucker into small pocketlike sacs called **haustra.** Fat-filled pouches of visceral peritoneum, called *epiploic appendages,* hang from the colon's surface.

The major function of the large intestine is to consolidate and propel the unusable fecal matter toward the anus and eliminate it from the body. While it does that chore, it (1) provides a site where intestinal bacteria manufacture vitamins B and K; and (2) reclaims most of the remaining water from undigested food, thus conserving body water.

Watery stools, or **diarrhea,** result from any condition that rushes undigested food residue through the large intestine before it has had sufficient time to absorb the water. Conversely, when food residue remains in the large intestine for extended periods, excessive water is absorbed and the stool becomes hard and difficult to pass, causing **constipation. ✚**

ACTIVITY 4

Examining the Histologic Structure of the Large Intestine

Large intestine: Secure a slide of the large intestine to the microscope stage for viewing. Observe the villi and note the numerous goblet cells (Figure 38.9c). Why do you think the large intestine produces so much mucus?

_____

_____

Accessory Digestive Organs

Teeth

By the age of 21, two sets of teeth have developed **(Figure 38.11)**. The initial set, called the **deciduous** (or **milk**) **teeth,** normally appears between the ages of 6 months and 2½ years. The first of these to erupt are the lower central incisors. The child begins to shed the deciduous teeth around the age of 6, and a second set of teeth, the **permanent teeth,** gradually replaces them. As the deeper permanent teeth progressively enlarge and develop, the roots of the deciduous teeth are resorbed, leading to their final shedding. During years 6 to 12, the child has mixed dentition—both permanent and deciduous teeth. Generally, by the age of 12, all of the deciduous teeth have been shed.

Teeth are classified as **incisors, canines** *(eye teeth),* **premolars** *(bicuspids),* and **molars.** Teeth names reflect differences in relative structure and function. The incisors are chisel shaped and exert a shearing action used in biting. Canines are cone shaped or fanglike, the latter description being much more applicable to the canines of animals whose teeth are used for the tearing of food. Incisors, canines, and premolars typically have single roots, though the first upper premolars may have two. The lower molars have two roots, but the upper molars usually have three. The premolars have two *cusps* (grinding surfaces); the molars have broad crowns with rounded cusps specialized for the fine grinding of food.

Dentition is described by means of a **dental formula,** which designates the numbers, types, and position of the teeth in one side of the jaw. (Because tooth arrangement is bilaterally symmetrical, it is only necessary to designate one side of the jaw.) The complete dental formula for the deciduous teeth from the medial aspect of each jaw and proceeding posteriorly is as follows:

$$\frac{\text{Upper teeth: 2 incisors, 1 canine, 0 premolars, 2 molars}}{\text{Lower teeth: 2 incisors, 1 canine, 0 premolars, 2 molars}} \times 2$$

This formula is generally abbreviated to read as follows:

$$\frac{2,1,0,2}{2,1,0,2} \times 2 = 20 \text{ (number of deciduous teeth)}$$

The 32 permanent teeth are then described by the following dental formula:

$$\frac{2,1,2,3}{2,1,2,3} \times 2 = 32 \text{ (number of permanent teeth)}$$

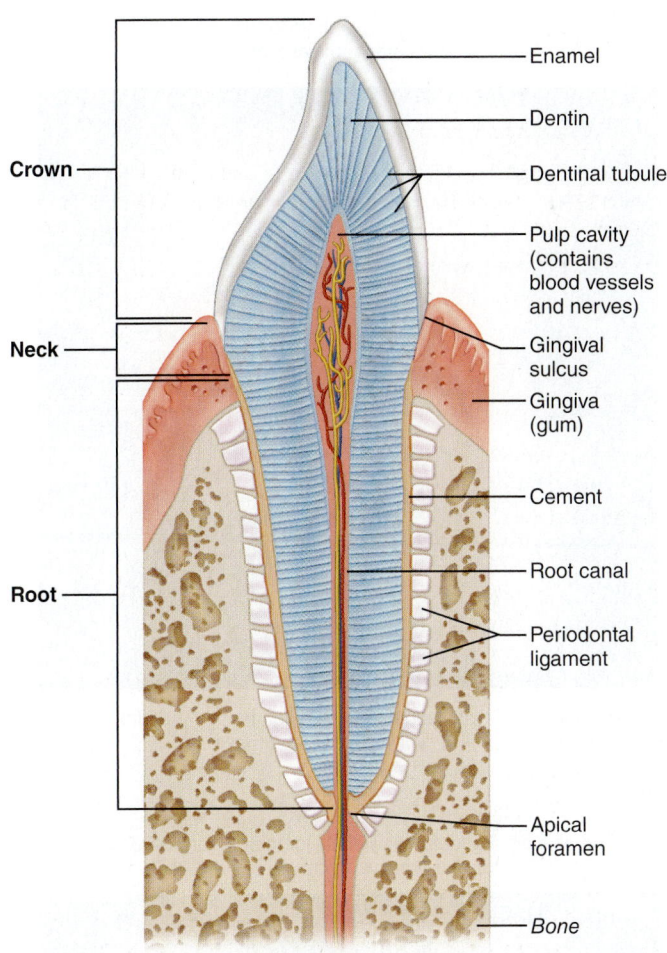

Crown

Neck

Root

Enamel

Dentin

Dentinal tubules

Pulp cavity (contains blood vessels and nerves)

Gingival sulcus

Gingiva (gum)

Cement

Root canal

Periodontal ligament

Apical foramen

Bone

Figure 38.12 **Longitudinal section of human canine tooth within its bony socket (alveolus).**

Although 32 is designated as the normal number of permanent teeth, not everyone develops a full complement. In many people, the third molars, commonly called *wisdom teeth,* never erupt.

ACTIVITY 5

Identifying Types of Teeth

Identify the four types of teeth (incisors, canines, premolars, and molars) on the jaw model or human skull. ◼

A tooth consists of two major regions, the **crown** and the **root.** A longitudinal section made through a tooth shows the following basic anatomical plan **(Figure 38.12)**. The crown is the superior portion of the tooth. The portion of the crown visible above the **gingiva,** or **gum,** is referred to as the *clinical crown.* The entire area covered by **enamel** is called the *anatomical crown.* Enamel is the hardest substance in the body and is fairly brittle. It consists of 95% to 97% inorganic

calcium salts and thus is heavily mineralized. The crevice between the end of the anatomical crown and the upper margin of the gingiva is referred to as the *gingival sulcus.*

That portion of the tooth embedded in the alveolar portion of the jaw is the root, and the root and crown are connected by a slight constriction, the **neck.** The outermost surface of the root is covered by **cement,** which is similar to bone in composition and less brittle than enamel. The cement attaches the tooth to the **periodontal ligament,** which holds the tooth in the tooth socket and exerts a cushioning effect. **Dentin,** which composes the bulk of the tooth, is the bonelike material interior to the enamel and cement.

The **pulp cavity** occupies the central portion of the tooth. **Pulp,** connective tissue liberally supplied with blood vessels, nerves, and lymphatics, occupies this cavity and provides for tooth sensation and supplies nutrients to the tooth tissues. **Odontoblasts,** specialized cells that reside in the outer margins of the pulp cavity, produce the dentin. The pulp cavity extends into distal portions of the root and becomes the **root canal.** An opening at the root apex, the **apical foramen,** provides a route of entry into the tooth for blood vessels, nerves, and other structures from the tissues beneath.

ACTIVITY 6

Studying Microscopic Tooth Anatomy

Observe a slide of a longitudinal section of a tooth, and compare your observations with the structures detailed in the illustration (Figure 38.12). Identify as many of these structures as possible. ◼

Salivary Glands

Three pairs of major **salivary glands** (see Figure 38.1) empty their secretions into the oral cavity.

Parotid glands: Large glands located anterior to the ear and ducting into the mouth over the second upper molar through the parotid duct.

Submandibular glands: Located along the medial aspect of the mandibular body in the floor of the mouth, and ducting under the tongue to the base of the lingual frenulum.

Sublingual glands: Small glands located most anteriorly in the floor of the mouth and emptying under the tongue via several small ducts.

Food in the mouth and mechanical pressure (even chewing rubber bands or wax) stimulate the salivary glands to secrete saliva. Saliva consists primarily of a viscous glycoprotein called *mucin,* which moistens the food and helps to bind it together into a mass called a **bolus,** and a clear serous fluid containing the enzyme *salivary amylase.* Salivary amylase begins the digestion of starch, breaking it down into disaccharides and glucose. Parotid gland secretion is mainly serous; the submandibular is a mixed gland that produces both mucin and serous components; and the sublingual gland produces mostly mucin.

38

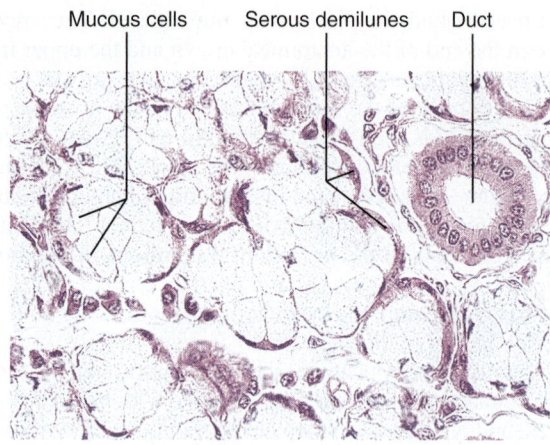

Mucous cells   Serous demilunes   Duct

Figure 38.13 Histology of a mixed salivary gland. Sublingual gland (170×).

Examining Salivary Gland Tissue

Examine salivary gland tissue under low power and then high power to become familiar with the appearance of a glandular tissue. Notice the clustered arrangement of the cells around their ducts. The cells are basically triangular, with their pointed ends facing the duct opening. If possible, differentiate between mucus-producing cells, which look hollow or have a clear cytoplasm, and serous cells, which produce the clear, enzyme-containing fluid and have granules in their cytoplasm. The serous cells often form *demilunes* (caps) around the more central mucous cells. (**Figure 38.13** may be helpful in this task.) ▬

Liver and Gallbladder

The **liver** (see Figure 38.1), the largest gland in the body, is located inferior to the diaphragm, more to the right than the left side of the body. As noted earlier, it hides the stomach from view in a superficial observation of abdominal contents. The human liver has four lobes and is suspended from the diaphragm and anterior abdominal wall by the **falciform ligament (Figure 38.14)**.

The liver is one of the body's most important organs, and it performs many metabolic roles. However, its digestive function is to produce bile, which leaves the liver through the **common hepatic duct** and then enters the duodenum through the **bile duct (Figure 38.15)**. Bile has no enzymatic action but emulsifies fats, breaking up large fat particles into smaller ones, which creates a larger surface area for more efficient lipase activity. Without bile, very little fat digestion or absorption occurs.

When digestive activity is not occurring in the digestive tract, bile backs up into the **cystic duct** and enters the **gallbladder,** a small, green sac on the inferior surface of the liver. Bile is stored there until needed for the digestive process. While in the gallbladder, bile is concentrated by the removal of water and some ions. When fat-rich food enters the duodenum, a hormonal stimulus causes the gallbladder to contract, releasing the stored bile and making it available to the duodenum.

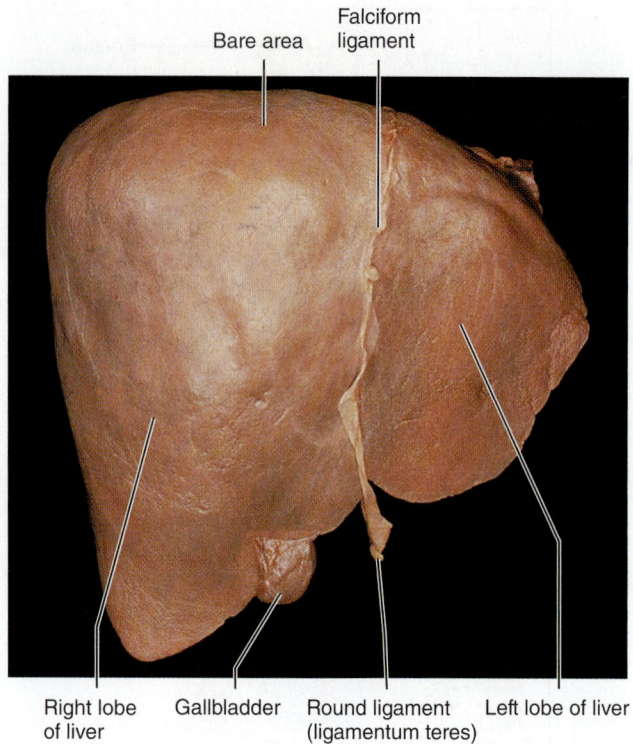

Bare area   Falciform ligament

Right lobe of liver   Gallbladder   Round ligament (ligamentum teres)   Left lobe of liver

(a)

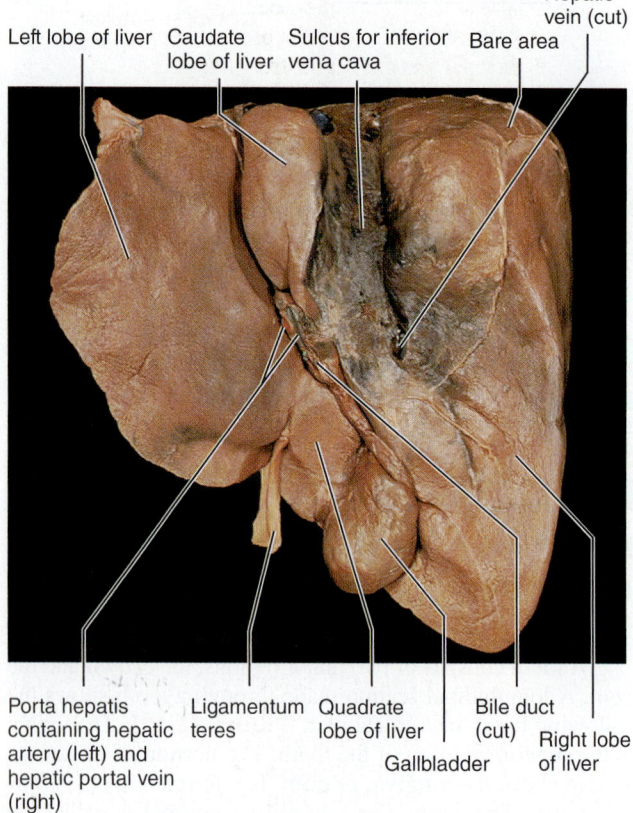

Left lobe of liver   Caudate lobe of liver   Sulcus for inferior vena cava   Bare area   Hepatic vein (cut)

Porta hepatis containing hepatic artery (left) and hepatic portal vein (right)   Ligamentum teres   Quadrate lobe of liver   Bile duct (cut)   Gallbladder   Right lobe of liver

(b)

Figure 38.14 Gross anatomy of the human liver. **(a)** Anterior view. **(b)** Posteroinferior aspect. The four liver lobes are separated by a group of fissures in this view.

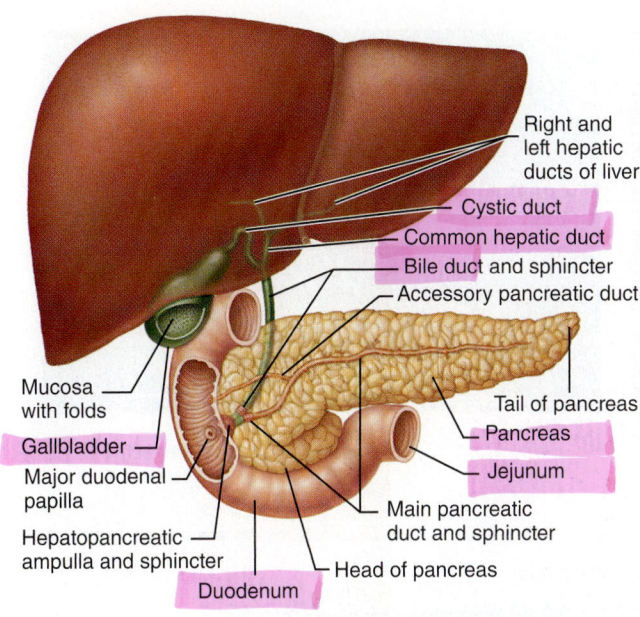

Right and left hepatic ducts of liver

Cystic duct
Common hepatic duct
Bile duct and sphincter
Accessory pancreatic duct

Mucosa with folds
Gallbladder
Major duodenal papilla
Hepatopancreatic ampulla and sphincter

Tail of pancreas
Pancreas
Jejunum
Main pancreatic duct and sphincter
Head of pancreas
Duodenum

Figure 38.15 **Ducts of accessory digestive organs.**

If the common hepatic or bile duct is blocked (for example, by wedged gallstones), bile is prevented from entering the small intestine, accumulates, and eventually backs up into the liver. This exerts pressure on the liver cells, and bile begins to enter the bloodstream. As the bile circulates through the body, the tissues become yellow, or jaundiced.

Blockage of the ducts is just one cause of jaundice. More often it results from actual liver problems such as **hepatitis,** (which is any inflammation of the liver,) or **cirrhosis,** a condition in which the liver is severely damaged and becomes hard and fibrous. Cirrhosis is prevalent in those who drink excessive alcohol for many years. ✚

As demonstrated by its highly organized anatomy, the liver **(Figure 38.16)** is very important in the initial processing of the nutrient-rich blood draining the digestive organs. Its structural and functional units are called **lobules.** Each lobule is a basically hexagonal structure consisting of cordlike arrays of **hepatocytes** or *liver cells,* which radiate outward from a central vein running upward in the longitudinal axis of the lobule. At each of the six corners of the lobule is a **portal triad,** so named because three basic structures are always present there: a *portal arteriole* (a branch of the *hepatic artery,* the functional blood supply of the liver), a *portal venule* (a branch of the *hepatic portal vein* carrying nutrient-rich blood from the digestive viscera), and a *bile duct.* Between the liver cells are blood-filled spaces, or **sinusoids,** through which blood from the hepatic portal vein and hepatic artery percolates. **Stellate macrophages,** special phagocytic cells, also called **hepatic macrophages,** line the sinusoids and remove debris such as bacteria from the blood as it flows past, while the hepatocytes pick up oxygen and nutrients. Much of the glucose transported to the liver from the digestive system

is stored as glycogen in the liver for later use, and amino acids are taken from the blood by the liver cells and utilized to make plasma proteins. The sinusoids empty into the central vein, and the blood ultimately drains from the liver via the *hepatic veins.*

Bile is continuously being made by the hepatocytes. It flows through tiny canals, the **bile canaliculi,** which run between adjacent cells toward the bile duct branches in the triad regions, where the bile eventually leaves the liver. Notice that the directions of blood and bile flow in the liver lobule are exactly opposite.

ACTIVITY 8

Examining the Histology of the Liver

Examine a slide of liver tissue and identify as many as possible of the structural features (see Figure 38.16). Also examine a three-dimensional model of liver lobules if this is available. Reproduce a small pie-shaped section of a liver lobule in the space below. Label the hepatocytes, the stellate macrophages, sinusoids, a portal triad, and a central vein. ▬

Pancreas

The **pancreas** is a soft, triangular gland that extends horizontally across the posterior abdominal wall from the spleen to the duodenum (see Figure 38.1). Like the duodenum, it is a retroperitoneal organ (see Figure 38.7). The pancreas has both an endocrine function producing the hormones insulin and glucagon and an exocrine function. Its exocrine secretion includes many hydrolytic enzymes, and is secreted into the duodenum through the pancreatic duct. Pancreatic juice is very alkaline. Its high concentration of bicarbonate ion (HCO_3^-) neutralizes the acidic chyme entering the duodenum from the stomach, enabling the pancreatic and intestinal enzymes to operate at their optimal pH, which is slightly alkaline. (See Figure 27.3c.)

For instructions on animal dissections, see the dissection exercises (starting on page 697) in the cat and fetal pig editions of this manual.

38

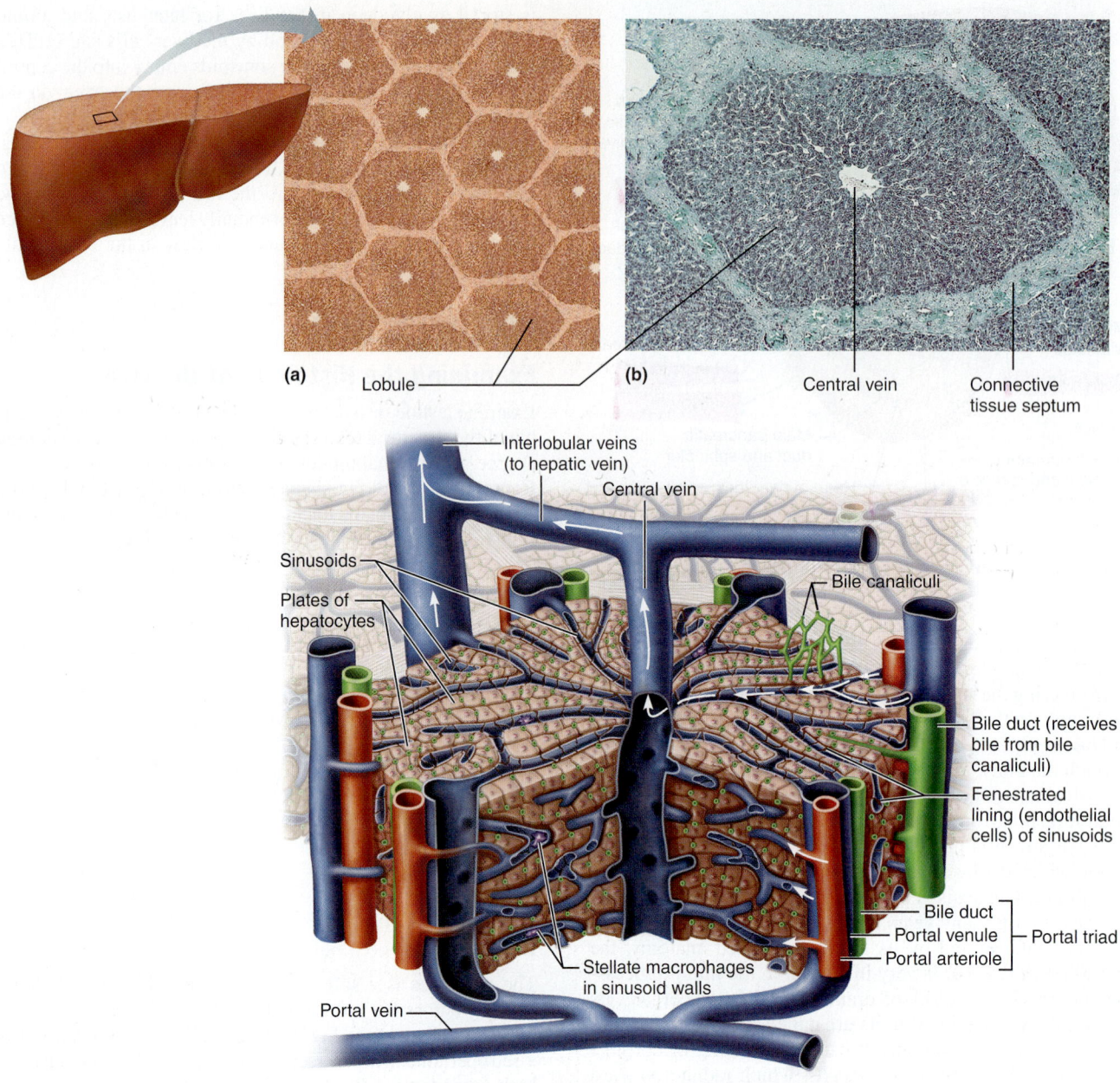

(a) Lobule

(b) Central vein   Connective tissue septum

Interlobular veins (to hepatic vein)
Central vein
Sinusoids
Plates of hepatocytes
Bile canaliculi
Bile duct (receives bile from bile canaliculi)
Fenestrated lining (endothelial cells) of sinusoids
Bile duct
Portal venule   Portal triad
Portal arteriole
Stellate macrophages in sinusoid walls
Portal vein

(c)

Figure 38.16 Microscopic anatomy of the liver. (a) Schematic view of the cut surface of the liver showing the hexagonal nature of its lobules. **(b)** Photomicrograph of one liver lobule (55×). **(c)** Enlarged three-dimensional diagram of one liver lobule. Arrows show direction of blood flow. Bile flows in the opposite direction toward the bile ducts.

38

Digestive System Processes: Chemical and Physical

MATERIALS

Part I: Enzyme Action

General Supply Area

- ☐ Hot plates
- ☐ 250-ml beakers
- ☐ Boiling chips
- ☐ Test tubes and test tube rack
- ☐ Wax markers
- ☐ Water bath set at 37°C (if not available, incubate at room temperature and double the time)
- ☐ Ice water bath
- ☐ Chart on board for recording class results

Activity 1: Starch Digestion

- ☐ Dropper bottle of distilled water
- ☐ Dropper bottles of the following:
  1% alpha-amylase solution*
  1% boiled starch solution, freshly prepared†
  1% maltose solution
  Lugol's iodine solution (IKI)
  Benedict's solution
- ☐ Spot plate

Activity 2: Protein Digestion

- ☐ Dropper bottles of 1% trypsin and 0.01% BAPNA solution

*The alpha-amylase must be a low-maltose preparation for good results.

†Prepare by adding 1 g starch to 100 ml distilled water; boil and cool; add a pinch of salt (NaCl). Prepare fresh daily.

(Text continues on next page.)

MasteringA&P® For related exercise study tools, go to the Study Area of MasteringA&P. There you will find:
- Practice Anatomy Lab **PAL**
- PhysioEx **PEx**
- A&PFlix **A&PFlix**
- Practice quizzes, Histology Atlas, eText, Videos, and more!

OBJECTIVES

1. List the digestive system enzymes involved in the digestion of proteins, fats, and carbohydrates; state their site of origin; and summarize the conditions promoting their optimal functioning.
2. Name the end products of protein, fat, and carbohydrate digestion.
3. Define *enzyme, catalyst, control, substrate,* and *hydrolase.*
4. Describe the different types of enzyme assays, and the appropriate chemical tests to determine if digestion of a particular foodstuff has occurred.
5. Discuss the role of temperature and pH in the regulation of enzyme activity.
6. State the function of bile in the digestive process.
7. Explain why swallowing is both a voluntary and a reflex activity, and discuss the role of the tongue, larynx, and gastroesophageal sphincter in swallowing.
8. Compare and contrast segmentation and peristalsis as mechanisms of mixing and propulsion in digestive tract organs.

PRE-LAB QUIZ

1. Circle the correct underlined term. Enzymes are <u>catalysts</u> / <u>substrates</u> that increase the rate of chemical reactions without becoming a part of the product.
2. Circle True or False. Breakdown products of fats are absorbed by the lymphatic system and are then transported into the systemic circulation by lymph.
3. A(n) _____ is a specimen or standard against which all experimental samples are compared.
   a. assay   b. control   c. substrate   d. trial
4. One enzyme that you will be studying today, produced by the salivary glands and secreted into the mouth, hydrolyzes starch to maltose. It is _____.
5. Circle True or False. When you use iodine to test for starch, a color change to blue-black indicates a positive starch test.
6. If Benedict's test in the starch assay produces a _____ precipitate, then your test will be recorded as positive for maltose.
   a. blue to black   b. green to orange   c. white
7. The enzyme _____, produced by the pancreas, is responsible for breaking down proteins.
   a. amylase   b. kinase   c. lipase   d. trypsin
8. Circle the correct underlined term. The enzyme <u>pancreatic lipase</u> / <u>pepsin</u> hydrolyzes neutral fats to their component monoglycerides and fatty acids.
9. Circle True or False. Both smooth and skeletal muscles are involved in the propulsion of foodstuffs along the alimentray canal.
10. _____ movements are local contractions that mix foodstuffs with digestive juices and increase the rate of absorption.
    a. Deglutition   b. Elimination   c. Peristaltic   d. Segmental

(Materials list continued.)

Activity 3: Bile Action and Fat Digestion

☐ Dropper bottles of 1% pancreatin solution, litmus cream (fresh cream to which powdered litmus is added to achieve a deep blue color), 0.1 *N* HCl, and vegetable oil

☐ Bile salts (sodium taurocholate)

☐ Parafilm® (small squares to cover the test tubes)

Part II: Physical Processes

Activity 5: Observing Digestive Movements

☐ Water pitcher
☐ Paper cups
☐ Stethoscope
☐ Alcohol swab
☐ Disposable autoclave bag
☐ Watch, clock, or timer

Activity 6: Video Viewing

☐ Television and VCR or DVD player for independent viewing of video by student

☐ *Interactive Physiology*®, Digestive System

PEx PhysioEx™ 9.1 Computer Simulation Ex. 8 on p. PEx-119

The food we eat must be processed so that its nutrients can reach the cells of our body. First the food is mechanically broken down into small particles, and then the particles are chemically (enzymatically) digested into the molecules that can be absorbed. Food digestion is a prerequisite to food absorption. (You have already studied mechanisms of passive and active absorption in Exercise 5 and/or PhysioEx Exercise 1. Before proceeding, review that material.)

Digestion of Foodstuffs: Enzymatic Action

Enzymes are large protein molecules produced by body cells. They are biological **catalysts,** meaning that they increase the rate of a chemical reaction without themselves becoming part of the product. The digestive enzymes are hydrolytic enzymes, or **hydrolases.** Their **substrates,** or the molecules on which they act, are organic food molecules which they break down by adding water to the molecular bonds, thus cleaving the bonds between the chemical building blocks, or monomers.

The various hydrolytic enzymes are highly specific in their action. Each enzyme hydrolyzes only one or a small group of substrate molecules, and specific environmental conditions are necessary for it to function optimally. Since digestive enzymes actually function outside the body cells in the digestive tract, their hydrolytic activity can also be studied in a test tube. Such a study provides a convenient laboratory setting for investigating the effect of such changes in environmental conditions on enzymatic activity.

See the flowchart **(Figure 39.1)** of the progressive digestion of carbohydrates, proteins, fats, and nucleic acids. It summarizes the specific enzymes involved, their site of formation, and their site of action. Acquaint yourself with the flowchart before beginning this experiment, and refer to it as necessary during the laboratory session.

General Instructions for Activities 1–3

Work in groups of four, with each group taking responsibility for setting up and conducting one of the following experiments. In each of the digestive procedures being studied and in the amylase assay, you are directed to boil the contents of one or more test tubes. To do this, obtain a 250-ml beaker, boiling chips, and a hot plate from the general supply area. Place a few boiling chips into the beaker, add about 125 ml of

water, and bring to a boil. Place the test tube for each specimen in the water for the number of minutes specified in the directions. You will also be using a 37°C bath and an ice water bath for parts of these experiments. You will need to use your time very efficiently in order to set up and perform each test properly.

Upon completion of the experiments, each group should communicate its results to the rest of the class by recording them in a chart on the board. Each assay contains one or more **controls,** the specimens against which experimental samples are compared. All members of the class should observe the controls as well as the experimental results. Additionally, all members of the class should be able to explain the tests used and the results observed and anticipated for each experiment. Note that water baths and hot plates are at the general supply area.

ACTIVITY 1

Assessing Starch Digestion by Salivary Amylase

1. From the general supply area, obtain a test tube rack, 10 test tubes, and a wax marking pencil. From the Activity 1 supply area, obtain a dropper bottle of distilled water and dropper bottles of maltose, amylase, and starch solutions.

2. In this experiment you will investigate the hydrolysis of starch to maltose by **salivary amylase.** You will need to be able to identify the presence of starch and maltose, the breakdown product of starch, to determine to what extent the enzymatic activity has occurred. Thus controls must be prepared to provide a known standard against which comparisons can be made. Starch decreases and sugar increases as digestion occurs, according to the following formula:

$$\text{Starch} + \text{water} \xrightarrow{\text{amylase}} \text{maltose}$$

Two students should prepare the controls (tubes 1A to 3A) while the other two prepare the experimental samples (tubes 4A to 6A).

• Mark each tube with a wax pencil and load the tubes as indicated in the **Salivary Amylase chart** (page 598), using 3 drops (gtt) of each indicated substance.

• Place all tubes in a rack in the 37°C water bath for approximately 1 hour. Shake the rack gently from time to time to keep the contents evenly mixed.

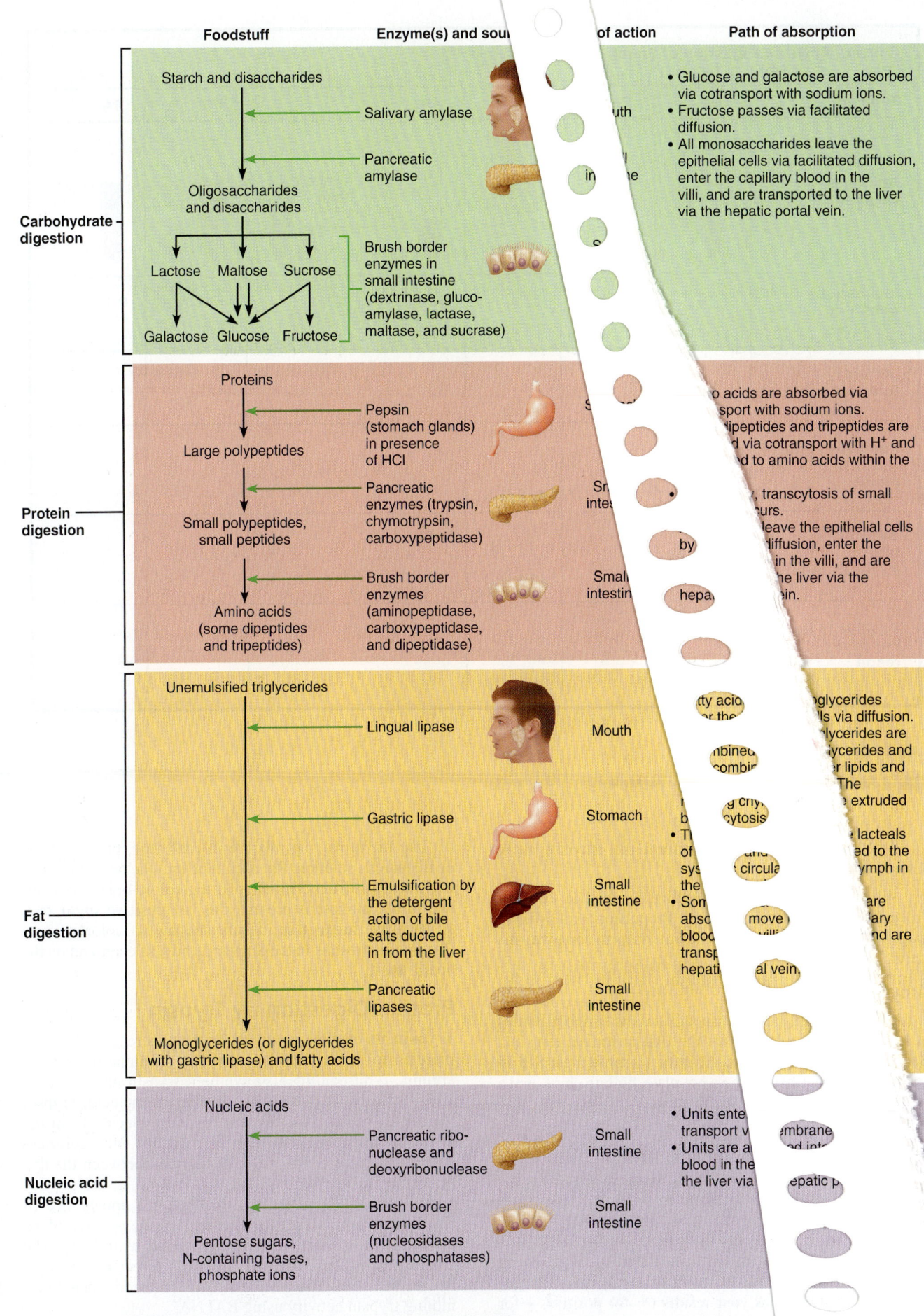

Figure 39.1 Flowchart of digestion and absorption of foodstuffs.

Salivary Amylase Digestion of Starch

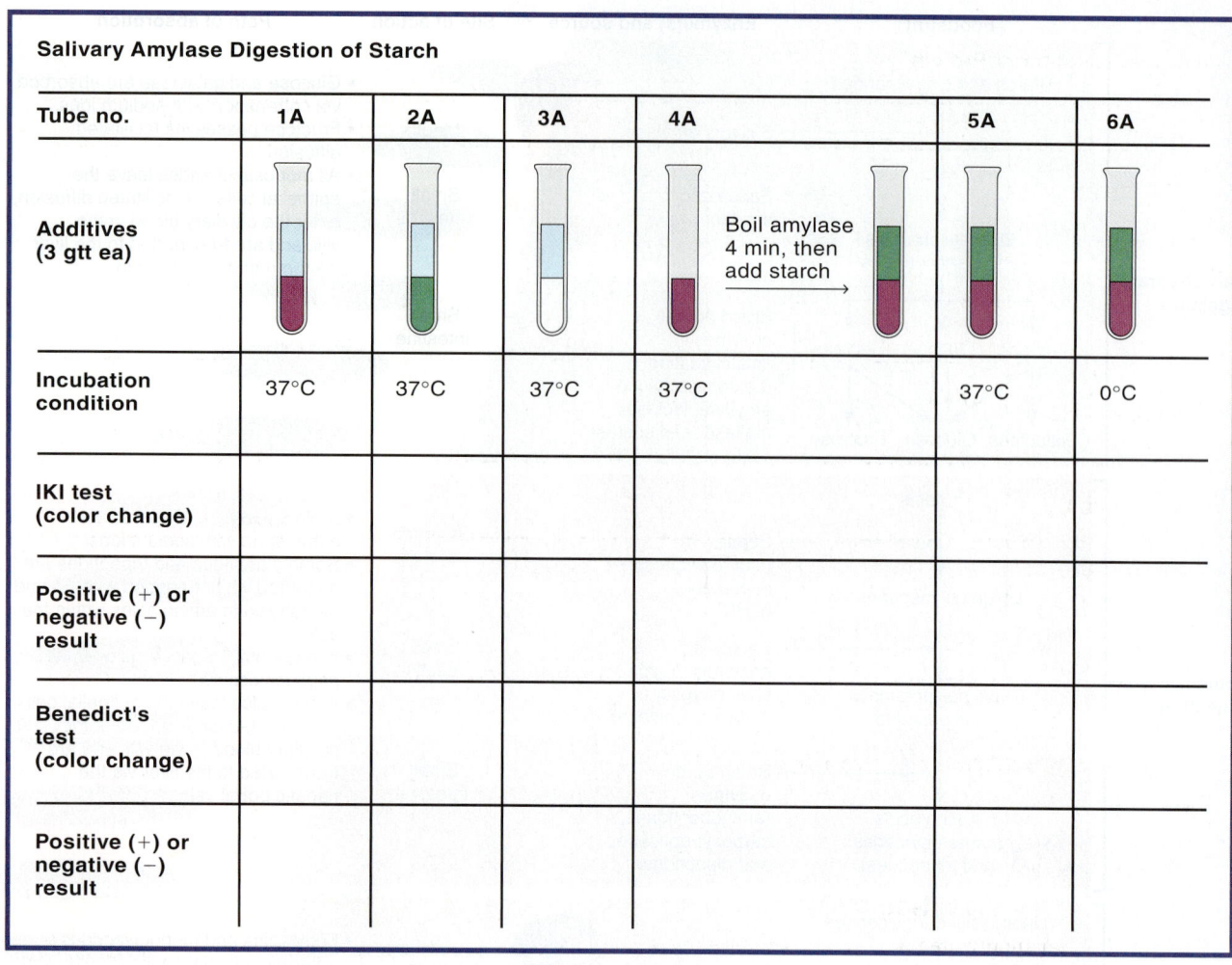

Tube no.	1A	2A	3A	4A		5A	6A
Additives (3 gtt ea)				Boil amylase 4 min, then add starch →			
Incubation condition	37°C	37°C	37°C	37°C		37°C	0°C
IKI test (color change)							
Positive (+) or negative (−) result							
Benedict's test (color change)							
Positive (+) or negative (−) result							

Additive key:

■ = Amylase   ■ = Starch   □ = Maltose   ■ = Water

- At the end of the hour, perform the amylase assay described below.

- While these tubes are incubating, proceed to Physical Processes: Mechanisms of Food Propulsion and Mixing (page 602). Be sure to monitor the time so as to complete this activity as needed.

Amylase Assay

1. After one hour, obtain a spot plate and dropper bottles of Lugol's iodine solution (for the IKI, or iodine, test) and Benedict's solution from the Activity 1 supply area. Set up your boiling water bath using a hot plate, boiling chips, and a 250-ml beaker obtained from the general supply area.

2. While the water is heating, mark six depressions of the spot plate 1A–6A (A for amylase) for sample identification.

3. Pour about a drop of the sample from each of the tubes 1A–6A into the appropriately numbered spot. Into each sample droplet, place a drop of Lugol's iodine (IKI) solution. A blue-black color indicates the presence of starch and is referred to as a **positive starch test.** If starch is not present, the mixture will not turn blue, which is referred to as a **negative starch test.** Record your results (+ for positive, − for negative) in the Salivary Amylase chart and on the board.

4. Into the remaining mixture in each tube, place 3 drops of Benedict's solution. Put each tube into the beaker of boiling water for about 5 minutes. If a green-to-orange precipitate forms, maltose is present; this is a **positive sugar test.** A **negative sugar test** is indicated by no color change. Record your results in the Salivary Amylase chart and on the board. ■

Protein Digestion by Trypsin

Trypsin, an enzyme produced by the pancreas, hydrolyzes proteins to small peptides. BAPNA (*N*-alpha-benzoyl-L-arginine-*p*-nitroanilide) is a synthetic trypsin substrate consisting of a dye covalently bound to an amino acid. Trypsin hydrolysis of BAPNA cleaves the dye molecule from the amino acid, causing the solution to change from colorless to bright yellow. Since the covalent bond between the dye molecule and the amino acid is the same as the peptide bonds that link amino acids together, the appearance of a yellow color indicates the presence and activity of an enzyme that is capable of peptide bond hydrolysis. The color change from clear to yellow is direct evidence of hydrolysis, so additional tests are not required when determining trypsin activity using BAPNA.

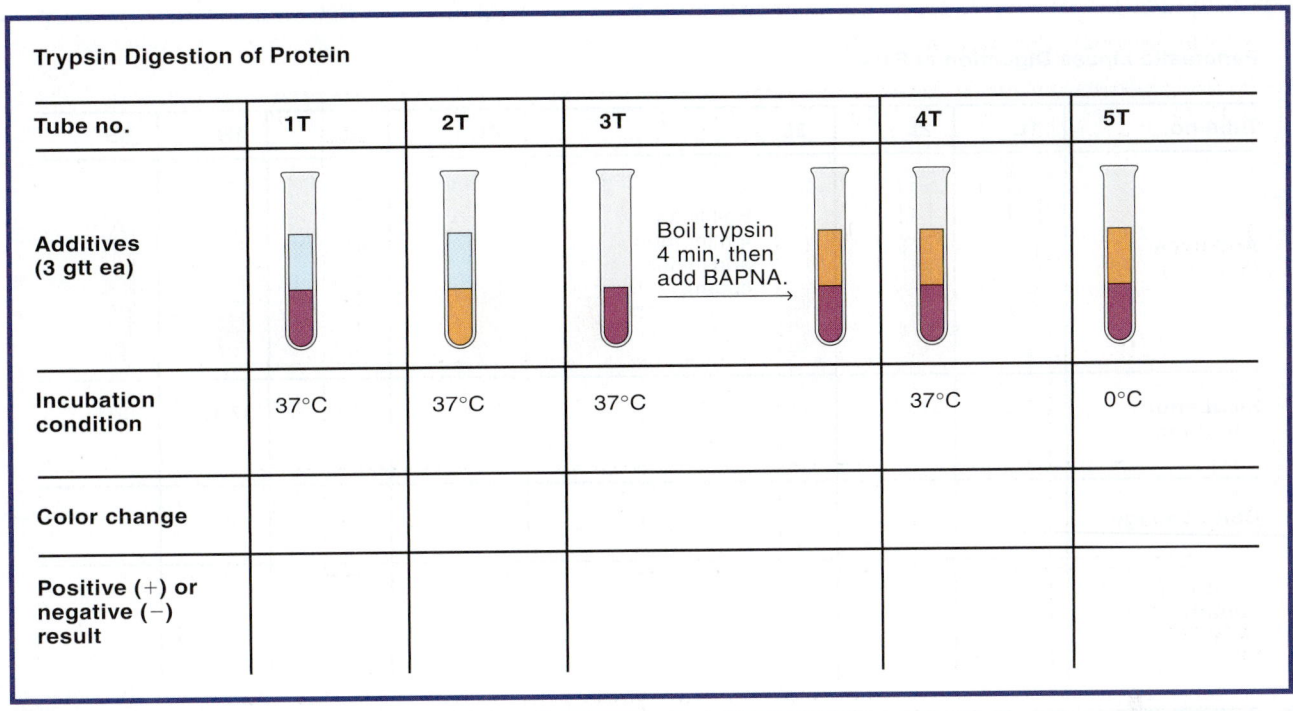

Trypsin Digestion of Protein

Tube no.	1T	2T	3T		4T	5T
Additives (3 gtt ea)			Boil trypsin 4 min, then add BAPNA. →			
Incubation condition	37°C	37°C	37°C		37°C	0°C
Color change						
Positive (+) or negative (−) result						

Additive key:

■ = Trypsin   ■ = BAPNA   ■ = Water

ACTIVITY 2

Assessing Protein Digestion by Trypsin

1. From the general supply area, obtain five test tubes and a test tube rack, and from the Activity 2 supply area get a dropper bottle of trypsin and one of BAPNA and bring them to your bench.

2. Two students should prepare the controls (tubes 1T and 2T) while the other two prepare the experimental samples (tubes 3T to 5T).

• Mark each tube with a wax pencil and load the tubes as indicated in the Trypsin chart, using 3 drops (gtt) of each indicated substance.

• Place all tubes in a rack in the appropriate water bath for approximately 1 hour. Shake the rack occasionally to keep the contents well mixed.

• At the end of the hour, examine the tubes for the results of the trypsin assay (detailed below).

• While these tubes are incubating, proceed to Physical Processes: Mechanisms of Food Propulsion and Mixing (page 602).

Trypsin Assay

Since BAPNA is a synthetic color-producing substrate, the presence of yellow color indicates a **positive hydrolysis test;** the dye molecule has been cleaved from the amino acid. If the sample mixture remains clear, a **negative hydrolysis test** has occurred.

Record the results in the **Trypsin chart** and on the board. ■■

Pancreatic Lipase Digestion of Fats and the Action of Bile

The treatment that fats and oils go through during digestion in the small intestine is a bit more complicated than that of carbohydrates or proteins—pretreatment with bile to physically emulsify the fats is required. Hence, two sets of reactions occur.

First:

$$\text{Fats/oils} \xrightarrow[\text{(emulsification)}]{\text{bile}} \text{minute fat/oil droplets}$$

Then:

$$\text{Fat/oil droplets} \xrightarrow[\text{(digestion)}]{\text{lipase}} \text{monoglycerides and fatty acids}$$

The term **pancreatin** describes the enzymatic product of the pancreas, which includes enzymes that digest proteins, carbohydrates, nucleic acids, and fats. It is used here to investigate the properties of **pancreatic lipase,** which hydrolyzes fats and oils to their component monoglycerides and two fatty acids.

Since fatty acids are organic acids, they acidify solutions, decreasing the pH. An easy way to recognize that digestion is ongoing or completed is to test pH. You will be using a pH indicator called *litmus blue* to follow these changes; it changes from blue to pink as the test tube contents become acidic.

ACTIVITY 3

Demonstrating the Emulsification Action of Bile and Assessing Fat Digestion by Lipase

1. From the general supply area, obtain nine test tubes and a test tube rack, plus one dropper bottle of each of the solutions in the Activity 3 supply area.

39

Pancreatic Lipase Digestion of Fats

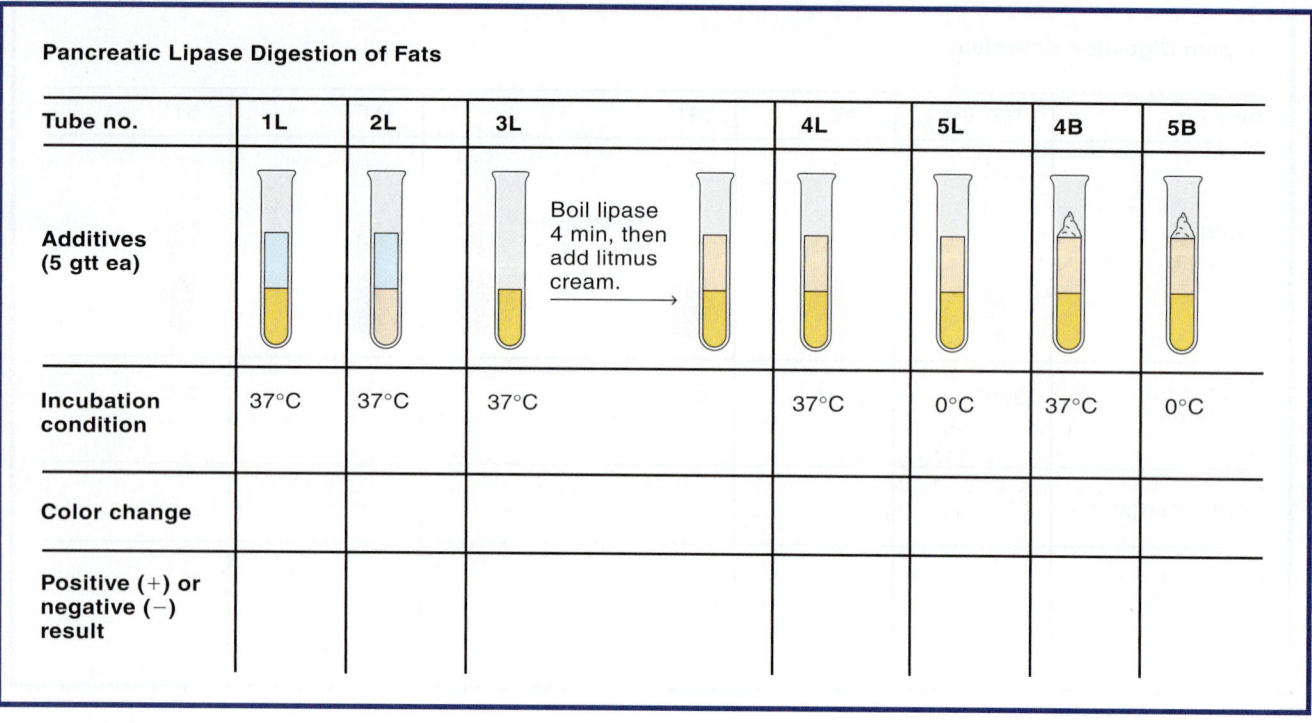

Tube no.	1L	2L	3L		4L	5L	4B	5B
Additives (5 gtt ea)			Boil lipase 4 min, then add litmus cream. →					
Incubation condition	37°C	37°C	37°C		37°C	0°C	37°C	0°C
Color change								
Positive (+) or negative (−) result								

Additive key:

☐ = Lipase   ☐ = Litmus cream   ☐ = Water   △ = Pinch bile salts

2. Although *bile,* a secretory product of the liver, is not an enzyme, it is important to fat digestion because of its emulsifying action. It physically breaks down large fat particles into smaller ones. Emulsified fats provide a larger surface area for enzymatic activity. To demonstrate the action of bile on fats, prepare two test tubes and mark them 1E and 2E (*E* for emulsified fats).

- To tube 1E, add 10 drops of water and 2 drops of vegetable oil.

- To tube 2E, add 10 drops of water, 2 drops of vegetable oil, and a pinch of bile salts.

- Cover each tube with a small square of Parafilm, shake vigorously, and allow the tubes to stand at room temperature.

After 10 to 15 minutes, observe both tubes. If emulsification has not occurred, the oil will be floating on the surface of the water. If emulsification has occurred, the fat droplets will be suspended throughout the water, forming an emulsion.

In which tube has emulsification occurred?_____

3. Two students should prepare the controls (1L and 2L, *L* for lipase), while the other two students in the group set up the experimental samples (3L to 5L, 4B, and 5B, where *B* is for bile) as illustrated in the Pancreatic Lipase chart.

- Mark each tube with a wax pencil and load the tubes using 5 drops (gtt) of each indicated solution.

- Place a pinch of bile salts in tubes 4B and 5B.

- Cover each tube with a small square of Parafilm, and shake to mix the contents of the tube.

- Remove the Parafilm, and place all tubes in a rack in the appropriate water bath for approximately 1 hour. Shake the test tube rack from time to time to keep the contents well mixed.

- At the end of the hour, perform the lipase assay below.

- While these tubes are incubating, proceed to Physical Processes: Mechanisms of Food Propulsion and Mixing (p. 602). Be sure to monitor the time so as to complete this activity when needed.

Lipase Assay

Fresh cream provides the fat substrate for this assay; add litmus powder to it to make litmus cream. The basis of this assay is a pH change that is detected by the litmus powder indicator. Alkaline or neutral solutions containing litmus are blue but will turn reddish in the presence of acid. If digestion occurs, the fatty acids produced will turn the litmus cream from blue to pink. Because the effect of hydrolysis is directly seen, additional assay reagents are not necessary.

1. To prepare a color control, add 0.1 *N* HCl drop by drop to tubes 1L and 2L (covering the tubes with a square of Parafilm after each addition and shaking to mix) until the cream turns pink.

2. Record the color of the tubes in the **Pancreatic Lipase chart** and on the board. ■

ACTIVITY 4

Reporting Results and Conclusions

1. Share your results with the class as directed in the General Instructions (page 596).

2. Suggest additional experiments, and carry out experiments if time permits.

3. Prepare a lab report for the experiments on digestion. (See Getting Started on page xiv.) ■

GROUP CHALLENGE

Odd Enzyme Out

The following boxes each contain four digestive enzymes. One of the listed enzymes does not share a characteristic that the other three do. Circle the enzyme that doesn't belong with the others and explain why it is singled out.

What characteristic is it missing? Sometimes there may be multiple reasons why the enzyme doesn't belong with the others. Include as many as you can think of but make sure it does not have the key characteristic.

1. Which is the "odd enzyme"?	Why is it the odd one out?
Trypsin Carboxypeptidase Pepsin Chymotrypsin	
2. Which is the "odd enzyme"?	**Why is it the odd one out?**
Lactase Pepsin Aminopeptidase Trypsin	
3. Which is the "odd enzyme"?	**Why is it the odd one out?**
Maltase Pancreatic lipase Nucleosidase Dipeptidase	
4. Which is the "odd enzyme"?	**Why is it the odd one out?**
Sucrase Dextrinase Glucoamylase Chymotrypsin	

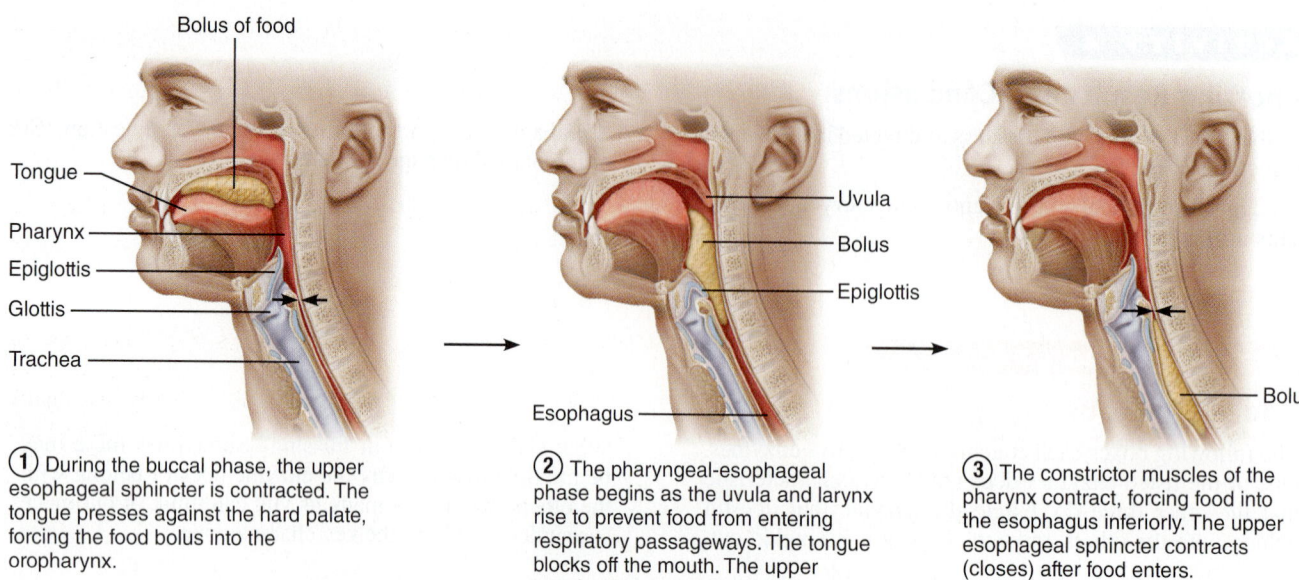

① During the buccal phase, the upper esophageal sphincter is contracted. The tongue presses against the hard palate, forcing the food bolus into the oropharynx.

② The pharyngeal-esophageal phase begins as the uvula and larynx rise to prevent food from entering respiratory passageways. The tongue blocks off the mouth. The upper esophageal sphincter relaxes, allowing food to enter the esophagus.

③ The constrictor muscles of the pharynx contract, forcing food into the esophagus inferiorly. The upper esophageal sphincter contracts (closes) after food enters.

Figure 39.2 Swallowing. The process of swallowing consists of voluntary (buccal) (step ①) and involuntary (pharyngeal-esophageal) phases (steps ② – ③).

Physical Processes: Mechanisms of Food Propulsion and Mixing

Although enzyme activity is a very important part of the overall digestion process, foods must also be processed physically (by chewing and churning) and moved by mechanical means along the tract if digestion and absorption are to be completed. Muscles are involved in producing the movements of foodstuffs along the gastrointestinal tract. Although we tend to think only of smooth muscles when visceral activities are involved, both skeletal and smooth muscles are involved in the physical processes. This fact is amply demonstrated by the simple activities that follow.

Deglutition (Swallowing)

Swallowing, or **deglutition,** which is largely the result of skeletal muscle activity, occurs in two phases: *buccal* (mouth) and *pharyngeal-esophageal*. The buccal phase (**Figure 39.2** step ①) is voluntarily controlled and initiated by the tongue. Once begun, the process continues involuntarily in the pharynx and esophagus through peristalsis, resulting in the delivery of the swallowed contents to the stomach (Figure 39.2 steps ② – ③).

ACTIVITY 5

Observing Movements and Sounds of the Digestive System

1. Obtain a pitcher of water, a stethoscope, a paper cup, an alcohol swab, and an autoclave bag in preparation for making the following observations.

2. While swallowing a mouthful of water, consciously note the movement of your tongue during the process. Record your observations.

_____

_____

_____

3. Repeat the swallowing process while your laboratory partner watches the externally visible movements of your larynx. (This movement is more obvious in a male, since males have a larger Adam's apple.) Record your observations.

_____

_____

What do these movements accomplish? _____

_____

_____

4. Before donning the stethoscope, your lab partner should clean the earpieces with an alcohol swab. Then, he or she should place the diaphragm of the stethoscope over your abdominal wall, approximately 2.5 cm (1 inch) below the xiphoid process and slightly to the left, to listen for sounds as you again take two or three swallows of water. There should be two audible sounds—one when the water splashes against the gastroesophageal sphincter and the

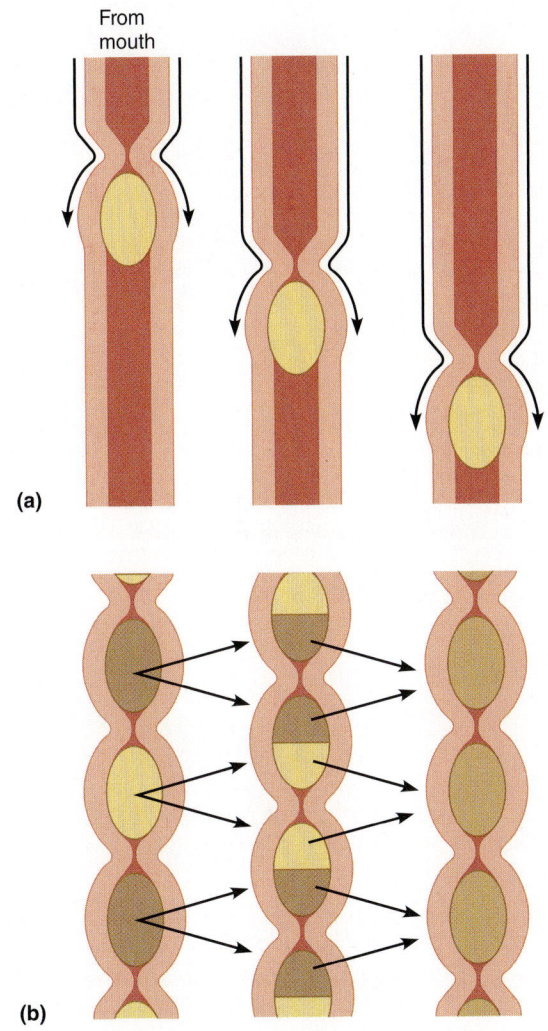

From mouth

(a)

(b)

Figure 39.3 Peristaltic and segmental movements of the digestive tract. (a) Peristalsis: neighboring segments of the intestine alternately contract and relax, moving food along the tract. **(b)** Segmentation: single segments of intestine alternately contract and relax. Because inactive segments exist between active segments, food mixing occurs to a greater degree than food movement. Peristalsis is superimposed on segmentation movements.

second when the peristaltic wave of the esophagus arrives at the sphincter and the sphincter opens, allowing water to gurgle into the stomach. Determine, as accurately as possible, the time interval between these two sounds and record it below.

Interval between arrival of water at the sphincter and the

opening of the sphincter: _____ sec

This interval gives a fair indication of the time it takes for the peristaltic wave to travel down the 25 cm (10 inches) of the esophagus. (Actually the time interval is slightly less than it seems, because pressure causes the sphincter to relax before the peristaltic wave reaches it.)

 Dispose of the used paper cup in the autoclave bag. ■

Segmentation and Peristalsis

Although several types of movements occur in the digestive tract organs, peristalsis and segmentation are most important as mixing and propulsive mechanisms **(Figure 39.3)**.

Peristaltic movements are the major means of propelling food through most of the digestive viscera. Essentially they are waves of contraction followed by waves of relaxation that squeeze foodstuffs through the alimentary canal, and they are superimposed on segmental movements.

Segmental movements are local constrictions of the organ wall that occur rhythmically. They serve mainly to mix the foodstuffs with digestive juices and to increase the rate of absorption by continually moving different portions of the chyme over adjacent regions of the intestinal wall. However, segmentation is also an important means of food propulsion in the small intestine, and slow segmenting movements called haustral contractions are frequently seen in the large intestine.

ACTIVITY 6

Viewing Segmental and Peristaltic Movements

If a video showing some of the propulsive movements is available, go to a viewing station to view it before leaving the laboratory. Alternatively, use the *Interactive Physiology*® module on the Digestive System to observe gut motility. ■

39

Digestive System Processes: Chemical and Physical

Digestion of Foodstuffs: Enzymatic Action

1. Match the following definitions with the proper choices from the key.

Key: a. catalyst   b. control   c. enzyme   d. substrate

_____ 1. substance on which a catalyst works

_____ 2. biologic catalyst; protein in nature

_____ 3. increases the rate of a chemical reaction without becoming part of the product

_____ 4. provides a standard of comparison for test results

2. List the three characteristics of enzymes. _____

_____

_____

3. The enzymes of the digestive system are classified as hydrolases. What does this mean?

_____

_____

4. Fill in the following chart about the various digestive system enzymes encountered in this exercise.

Enzyme	Organ producing it	Site of action	Substrate(s)	Optimal pH
Salivary amylase				
Trypsin				
Lipase (pancreatic)				

5. Name the end products of digestion for the following types of foods.

proteins: _____ carbohydrates: _____

fats: _____ and _____

6. You used several indicators or tests in the laboratory to determine the presence or absence of certain substances. Choose the correct test or indicator from the key to correspond to the condition described below.

Key: a. Lugol's iodine (IKI)   b. Benedict's solution   c. litmus   d. BAPNA

_____ 1. used to test for protein hydrolysis, which was indicated by a yellow color

_____ 2. used to test for the presence of starch, which was indicated by a blue-black color

_____ 3. used to test for the presence of fatty acids, which was evidenced by a color change from blue to pink

_____ 4. used to test for the presence of reducing sugars (maltose, sucrose, glucose) as indicated by a blue to green or orange color change

7. What conclusions can you draw when an experimental sample gives both a positive starch test and a positive maltose test

after incubation? _____

Why was 37°C the optimal incubation temperature? _____

Why did very little, if any, starch digestion occur in test tube 4A? _____

_____

When starch was incubated with amylase at 0°C, did you see any starch digestion? _____

Why or why not? _____

_____

Assume you have made the statement to a group of your peers that amylase is capable of starch hydrolysis to maltose. If you

had not done control tube 1A, what objection to your statement could be raised? _____

_____

What if you had not done tube 2A? _____

8. In the exercise concerning trypsin function, why was an enzyme assay like Benedict's or Lugol's iodine (IKI), which test

for the presence of a reaction product, not necessary? _____

Why was tube 1T necessary? _____

Why was tube 2T necessary? _____

Trypsin is a protease similar to pepsin, the protein-digesting enzyme in the stomach. Would trypsin work well in the

stomach? _____ Why? _____

9. In the procedure concerning pancreatic lipase digestion of fats and the action of bile salts, how did the appearance of tubes

1E and 2E differ? _____

_____

Explain the reason for the difference. _____

_____

Why did the litmus indicator change from blue to pink during fat hydrolysis? _____

_____

Why is bile not considered an enzyme? _____

How did the tubes containing bile compare with those not containing bile? _____

What role does bile play in fat digestion? _____

_____

10. The three-dimensional structure of a functional protein is altered by intense heat or nonphysiological pH even though peptide bonds may not break. Such inactivation is called denaturation, and denatured enzymes are nonfunctional. Explain why.

_____

_____

What specific experimental conditions resulted in denatured enzymes? _____

_____

11. Pancreatic and intestinal enzymes operate optimally at a pH that is slightly alkaline, yet the chyme entering the duodenum from the stomach is very acid. How is the proper pH for the functioning of the pancreatic-intestinal enzymes ensured?

_____

_____

12. Assume you have been chewing a piece of bread for 5 or 6 minutes. How would you expect its taste to change during this

interval? _____

Why? _____

13. Note the mechanism of absorption (passive or active transport) of the following food breakdown products, and indicate by a check mark (✓) whether the absorption would result in their movement into the blood capillaries or the lymphatic capillaries (lacteals).

Substance	Mechanism of absorption	Blood	Lymph
Monosaccharides			
Fatty acids and monoglycerides			
Amino acids			
Water			
Na^+, Cl^-, Ca^{2+}			

14. People on a strict diet to lose weight begin to metabolize stored fats at an accelerated rate. How does this condition affect

blood pH? _____

15. Using a flowchart, trace the pathway of a ham sandwich (ham = protein and fat; bread = starch) from the mouth to the site of absorption of its breakdown products, noting where digestion occurs and what specific enzymes are involved.

16. Some of the digestive organs have groups of secretory cells that liberate hormones into the blood. These exert an effect on the digestive process by acting on other cells or structures and causing them to release digestive enzymes, expel bile, or increase the motility of the digestive tract. For each hormone below, note the organ producing the hormone and its effects on the digestive process. Include the target organs affected.

Hormone	Produced by	Target organ(s) and effects
Secretin		
Gastrin		
Cholecystokinin		

Physical Processes: Mechanisms of Food Propulsion and Mixing

17. Complete the following statements.

Swallowing, or __1__, occurs in two phases—the __2__ and __3__. One of these phases, the __4__ phase, is voluntary. During the voluntary phase, the __5__ is used to push the food into the back of the throat. During swallowing, the __6__ rises to ensure that its passageway is covered by the epiglottis so that the ingested substances don't enter the respiratory passageways. It is possible to swallow water while standing on your head because the water is carried along the esophagus involuntarily by the process of __7__. The pressure exerted by the foodstuffs on the __8__ sphincter causes it to open, allowing the foodstuffs to enter the stomach.

The two major types of propulsive movements that occur in the small intestine are __9__ and __10__. One of these movements, __11__, acts to continually mix the foods and to increase the absorption rate by moving different parts of the chyme mass over the intestinal mucosa, but it has less of a role in moving foods along the digestive tract.

1. _____

2. _____

3. _____

4. _____

5. _____

6. _____

7. _____

8. _____

9. _____

10. _____

11. _____

Anatomy of the Urinary System

MATERIALS

- ☐ Human dissectible torso model, three-dimensional model of the urinary system, and/or anatomical chart of the human urinary system
- ☐ Dissecting instruments and tray
- ☐ Pig or sheep kidney, doubly or triply injected
- ☐ Disposable gloves
- ☐ Three-dimensional models of the cut kidney and of a nephron (if available)
- ☐ Compound microscope
- ☐ Prepared slides of a longitudinal section of kidney and cross sections of the bladder

✂ For instructions on animal dissections, see the dissection exercises (starting on page 697) in the cat and fetal pig editions of this manual.

OBJECTIVES

1. List the functions of the urinary system.
2. Identify, on a model or image, the urinary system organs and state the general function of each.
3. Compare the course and length of the urethra in males and females.
4. Identify these regions of the dissected kidney (longitudinal section): hilum, cortex, medulla, medullary pyramids, major and minor calyces, pelvis, renal columns, and fibrous and perirenal fat capsules.
5. Trace the blood supply of the kidney from the renal artery to the renal vein.
6. Define *nephron,* and describe its anatomy.
7. Define *glomerular filtration, tubular reabsorption,* and *tubular secretion,* and indicate the nephron areas involved in these processes.
8. Define *micturition,* and explain the differences in the control of the internal and external urethral sphincters.
9. Recognize the histologic structure of the kidney and ureter microscopically or in an image.

MasteringA&P® For related exercise study tools, go to the Study Area of MasteringA&P. There you will find:
- Practice Anatomy Lab PAL
- PhysioEx PEx
- A&PFlix A&PFlix
- Practice quizzes, Histology Atlas, eText, Videos, and more!

PRE-LAB QUIZ

1. Circle the correct underlined term. In its excretory role, the urinary system is primarily concerned with the removal of <u>carbon-containing</u> / <u>nitrogenous</u> wastes from the body.
2. The _____ perform(s) the excretory and homeostatic functions of the urinary system.
   - a. kidneys
   - b. ureters
   - c. urinary bladder
   - d. all of the above
3. Circle the correct underlined term. The <u>cortex</u> / <u>medulla</u> of the kidney is segregated into triangular regions with a striped appearance.
4. Circle the correct underlined term. As the renal artery approaches a kidney, it is divided into branches known as the <u>segmental arteries</u> / <u>afferent arterioles.</u>
5. What do we call the anatomical units responsible for the formation of urine? _____
6. This knot of coiled capillaries, found in the kidneys, forms the filtrate. It is the
   - a. arteriole
   - b. glomerulus
   - c. podocyte
   - d. tubule
7. The section of the renal tubule closest to the glomerular capsule is the
   - a. collecting duct
   - b. distal convoluted tubule
   - c. nephron loop
   - d. proximal convoluted tubule
8. Circle the correct underlined term. The <u>afferent</u> / <u>efferent</u> arteriole drains the glomerular capillary bed.
9. Circle True or False. During tubular reabsorption, components of the filtrate move from the bloodstream into the tubule.
10. Circle the correct underlined term. The <u>internal</u> / <u>external</u> urethral sphincter consists of skeletal muscle and is voluntarily controlled.

Metabolism of nutrients by the body produces wastes including carbon dioxide, nitrogenous wastes, and ammonia that must be eliminated from the body if normal function is to continue. Excretory processes involve multiple organ systems, with the **urinary system** primarily responsible for the removal of nitrogenous wastes from the body. In addition to this excretory function, the kidney maintains the electrolyte, acid-base, and fluid balances of the blood and is thus a major, if not *the* major, homeostatic organ of the body.

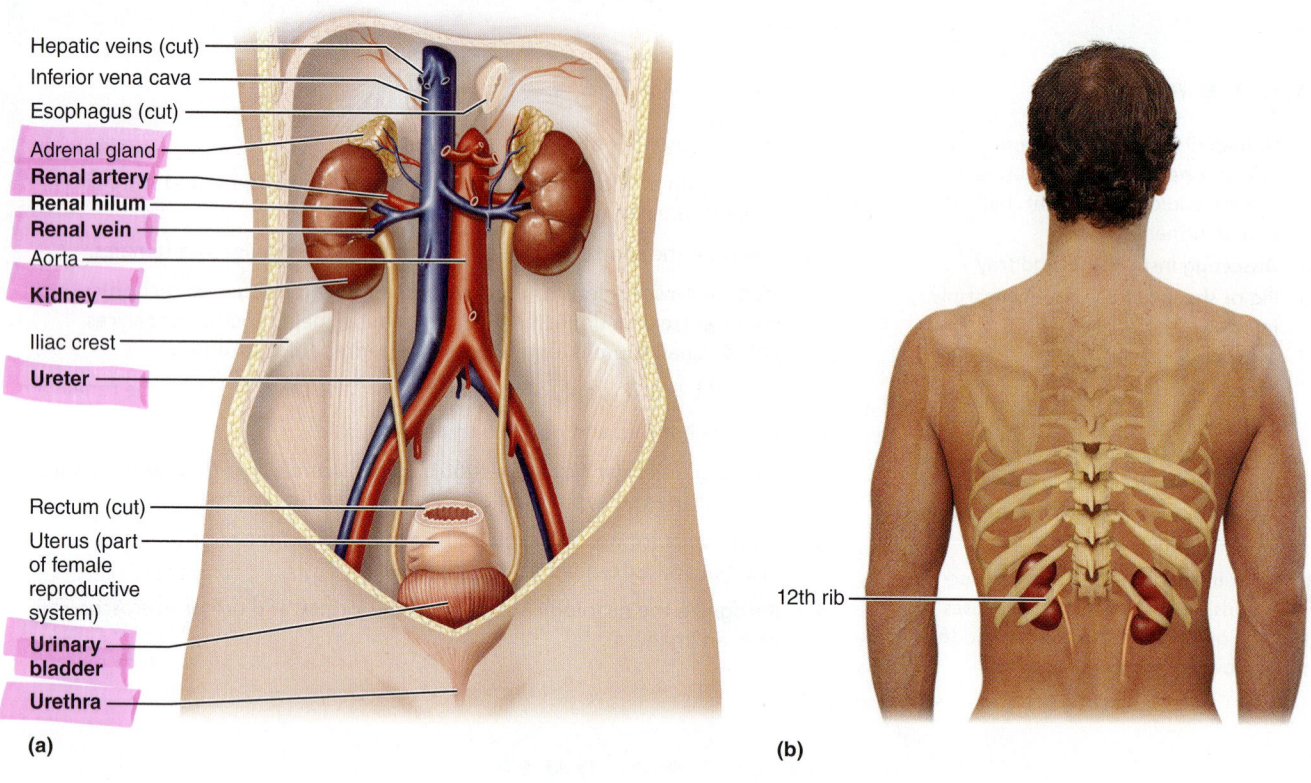

Hepatic veins (cut)
Inferior vena cava
Esophagus (cut)
Adrenal gland
Renal artery
Renal hilum
Renal vein
Aorta
Kidney
Iliac crest
Ureter
Rectum (cut)
Uterus (part of female reproductive system)
Urinary bladder
Urethra

(a)

12th rib

(b)

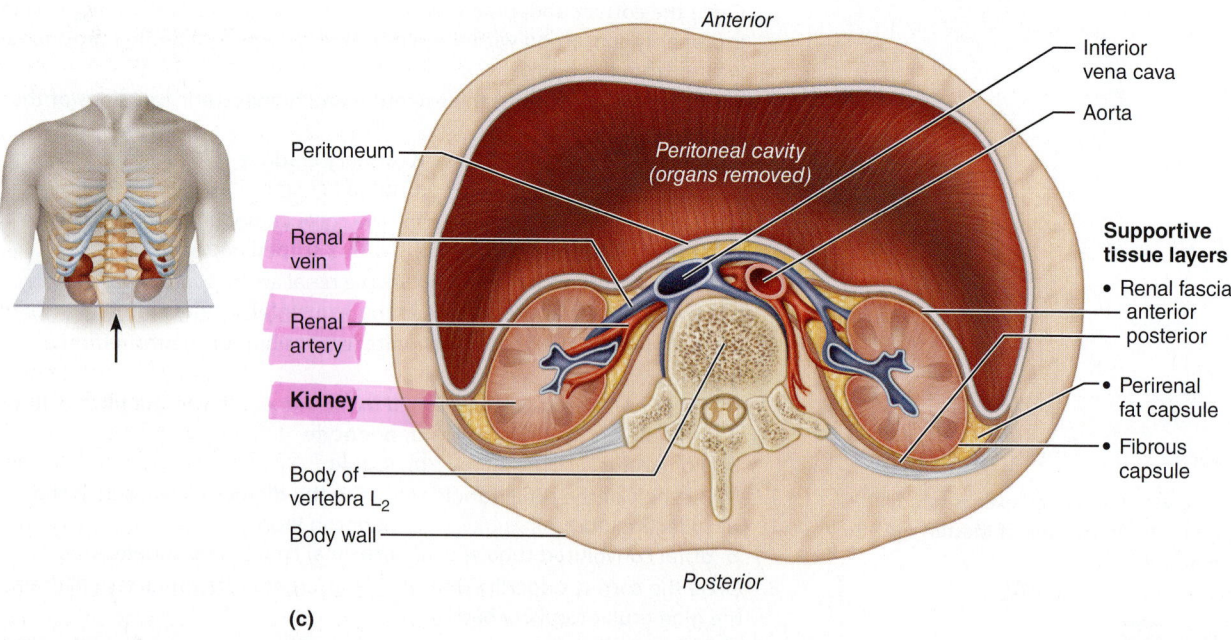

Anterior

Inferior vena cava
Aorta

Peritoneum
Peritoneal cavity (organs removed)

Supportive tissue layers
- Renal fascia
  anterior
  posterior
- Perirenal fat capsule
- Fibrous capsule

Renal vein
Renal artery
Kidney
Body of vertebra L$_2$
Body wall

Posterior

(c)

Figure 40.1 **Organs of the urinary system. (a)** Anterior view of the female urinary organs. Most unrelated abdominal organs have been removed. **(b)** Posterior in situ view showing the position of the kidneys relative to the twelfth ribs. **(c)** Cross section of the abdomen viewed from inferior direction. Note the retroperitoneal position and supportive tissue layers of the kidneys.

40

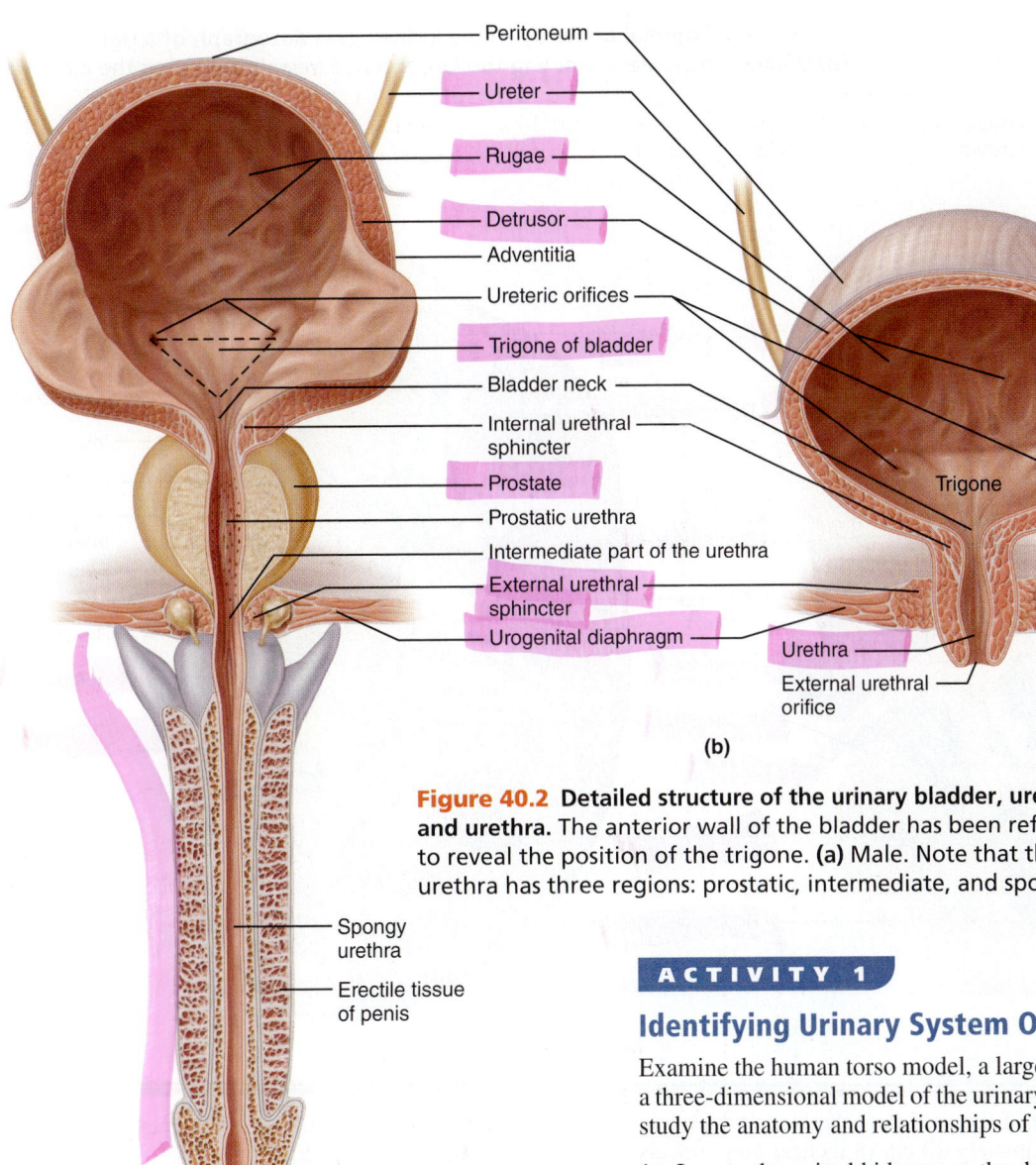

Peritoneum

Ureter

Rugae

Detrusor

Adventitia

Ureteric orifices

Trigone of bladder

Bladder neck

Internal urethral sphincter

Prostate

Prostatic urethra

Intermediate part of the urethra

External urethral sphincter

Urogenital diaphragm

Trigone

Urethra

External urethral orifice

(b)

Spongy urethra

Erectile tissue of penis

External urethral orifice

(a)

Figure 40.2 Detailed structure of the urinary bladder, urethral sphincters, and urethra. The anterior wall of the bladder has been reflected or omitted to reveal the position of the trigone. **(a)** Male. Note that the long male urethra has three regions: prostatic, intermediate, and spongy. **(b)** Female.

To perform its functions, the kidney acts first as a blood filter, and then as a filtrate processor. It allows toxins, metabolic wastes, and excess ions to leave the body in the urine, while retaining needed substances and returning them to the blood. Malfunction of the urinary system, particularly of the kidneys, leads to a failure in homeostasis which, unless corrected, is fatal.

Gross Anatomy of the Human Urinary System

The urinary system **(Figure 40.1)** consists of the paired kidneys and ureters and the single urinary bladder and urethra. The **kidneys** perform the functions described above and manufacture urine in the process. The remaining organs of the system provide temporary storage reservoirs or transportation channels for urine.

ACTIVITY 1

Identifying Urinary System Organs

Examine the human torso model, a large anatomical chart, or a three-dimensional model of the urinary system to locate and study the anatomy and relationships of the urinary organs.

1. Locate the paired kidneys on the dorsal body wall in the superior lumbar region. Notice that they are not positioned at exactly the same level. Because it is crowded by the liver, the right kidney is slightly lower than the left kidney. Three layers of support tissue surround each kidney. Beginning with the innermost layer they are: (1) a transparent *fibrous capsule,* (2) a *perirenal fat capsule,* and (3) the fibrous *renal fascia* that holds the kidneys in place in a retroperitoneal position.

In cases of rapid weight loss or in very thin individuals, the fat capsule may be reduced or meager in amount. Since the kidneys are less securely anchored, they may drop to a more inferior position in the abdominal cavity. This phenomenon is called **ptosis.**

2. Observe the **renal arteries** as they diverge from the descending aorta and plunge into the indented medial region, called the **hilum,** of each kidney. Note also the **renal veins,** which drain the kidneys, and the two **ureters,** which carry urine from the kidneys, moving it by peristalsis to the bladder for temporary storage.

3. Locate the **urinary bladder,** and observe the point of entry of the two ureters into this organ. Also locate the single **urethra,** which drains the bladder. The triangular region of the bladder delineated by the openings of the ureters and the urethra is referred to as the **trigone (Figure 40.2).**

40

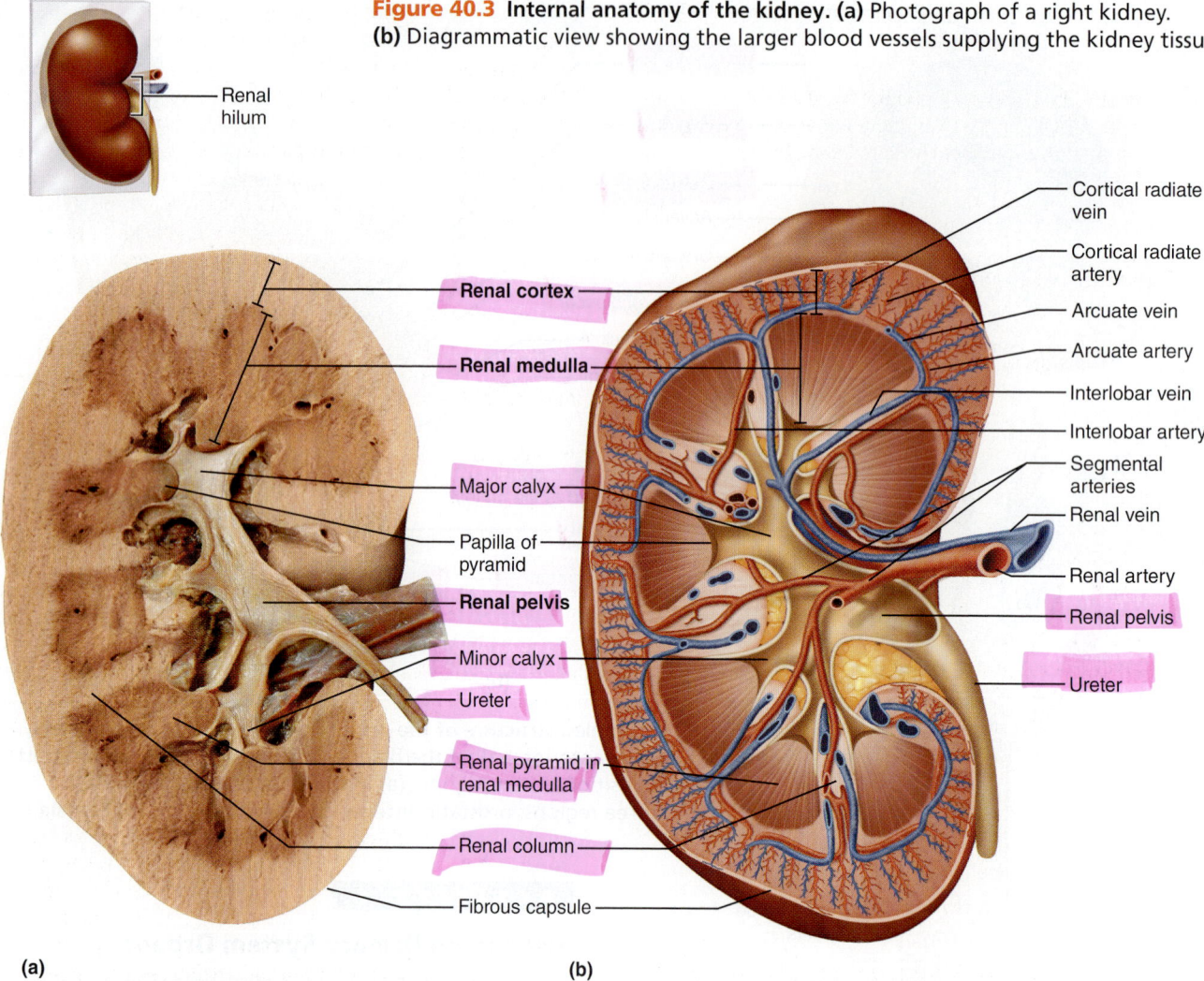

Figure 40.3 **Internal anatomy of the kidney. (a)** Photograph of a right kidney. **(b)** Diagrammatic view showing the larger blood vessels supplying the kidney tissue.

Renal hilum

Renal cortex

Renal medulla

Major calyx

Papilla of pyramid

Renal pelvis

Minor calyx

Ureter

Renal pyramid in renal medulla

Renal column

Fibrous capsule

Cortical radiate vein

Cortical radiate artery

Arcuate vein

Arcuate artery

Interlobar vein

Interlobar artery

Segmental arteries

Renal vein

Renal artery

Renal pelvis

Ureter

(a)   (b)

4. Follow the course of the urethra to the body exterior. In the male, it is approximately 20 cm (8 inches) long, travels the length of the **penis,** and opens at its tip. Its three named regions are the *prostatic, intermediate part,* and *spongy urethrae* (described in more detail in Exercise 42). The male urethra has a dual function: it carries urine to the body exterior, and it provides a passageway for semen ejaculation. Thus in the male, the urethra is part of both the urinary and reproductive systems. In females, the urethra is short, approximately 4 cm (1½ inches) long (see Figure 40.2). There are no common urinary-reproductive pathways in the female, and the female's urethra serves only to transport urine to the body exterior. Its external opening, the **external urethral orifice,** lies anterior to the vaginal opening.

DISSECTION:
Gross Internal Anatomy of the Pig or Sheep Kidney

1. In preparation for dissection, don gloves. Obtain a preserved sheep or pig kidney, dissecting tray, and instruments. Observe the kidney to identify the **fibrous capsule,** a smooth, transparent membrane that adheres tightly to the external aspect of the kidney.

2. Find the ureter, renal vein, and renal artery at the hilum. The renal vein has the thinnest wall and will be collapsed.

The ureter is the largest of these structures and has the thickest wall.

3. Section the kidney in the frontal plane through the longitudinal axis and locate the anatomical areas described below (and depicted in **Figure 40.3**).

Renal cortex: The superficial kidney region, which is lighter in color. If the kidney is doubly injected with latex, you will see a predominance of red and blue latex specks in this region indicating its rich vascular supply.

Renal medulla: Deep to the cortex; a darker, reddish-brown color. The medulla is segregated into triangular regions that have a striped appearance—the **medullary (renal) pyramids.** The base of each pyramid faces toward the cortex. Its more pointed *papilla,* or *apex,* points to the innermost kidney region.

Renal columns: Areas of tissue that are more like the cortex in appearance, the columns dip inward between the pyramids, separating them.

Renal pelvis: Extending inward from the hilum; a relatively flat, basinlike cavity that is continuous with the **ureter,** which exits from the hilum region. Fingerlike extensions of the pelvis should be visible. The larger, or primary, extensions are called the **major calyces** (singular: *calyx*); subdivisions of the major calyces are the **minor calyces.** Notice that the minor calyces terminate in cuplike areas that enclose the apexes of the medullary pyramids and collect urine draining from the pyramidal tips into the pelvis.

4. If the preserved kidney is doubly or triply injected, follow the renal blood supply from the renal artery to the **glomeruli.** The glomeruli appear as little red and blue specks in the cortex region.

Approximately a fourth of the total blood flow of the body is delivered to the kidneys each minute by the large **renal arteries.** As a renal artery approaches the kidney, it breaks up into branches called **segmental arteries,** which enter the hilum. Each segmental artery, in turn, divides into several **interlobar arteries,** which ascend toward the cortex in the renal columns. At the top of the medullary region, these arteries give off arching branches, the **arcuate arteries,** which curve over the bases of the medullary pyramids. Small **cortical radiate arteries** branch off the arcuate arteries and ascend into the cortex, giving off the individual **afferent arterioles,** which lead to the capillary beds associated with the nephrons, the functional units of the kidney. Blood draining from the nephron capillary beds enters the **cortical radiate veins** and then drains through the **arcuate veins** and the **interlobar veins** to finally enter the **renal vein** in the pelvis region. There are no segmental veins.

Dispose of the kidney specimen as your instructor specifies. ■

Functional Microscopic Anatomy of the Kidney and Bladder

Kidney

Each kidney contains over a million **nephrons,** the structural and functional units responsible for filtering the blood and forming urine. (**Figure 40.4** depicts the detailed structure and the relative positioning of the nephrons in the kidney.)

Each nephron consists of two major structures: a *renal corpuscle* and a *renal tubule*. All of the **renal corpuscles** are located in the cortex of the kidney. Each consists of a cuplike structure called the **glomerular** (or **Bowman's**) **capsule** that contains a cluster of capillaries called the **glomerulus.** This special capillary bed resembles a ball of yarn. The capsule has a visceral layer of **podocytes** that have long branching *foot processes*. The foot processes interdigitate and cling to the basement membrane of the glomerulus, forming part of the filtration membrane. Blood in the glomerulus is filtered into the capsule to begin its journey along the renal tubule.

The **renal tubule** exits from the glomerular capsule to form the highly coiled **proximal convoluted tubule (PCT)** that then becomes the descending limb of the hairpin-like **nephron loop.** The ascending limb of the nephron loop leads to the coiled **distal convoluted tubule (DCT)** that empties into a collecting duct. A single layer of epithelial cells resting on a basement membrane forms the walls of both the renal tubule and the collecting duct. The cells of each part of the tubule and collecting duct are specialized to perform a particular function in processing filtrate. The cuboidal cells of the PCT have dense microvilli that greatly increase the surface area exposed to filtrate in the lumen. They are specialized for reabsorption. In contrast, the cuboidal cells of the DCT have far fewer microvilli, reflecting the decreased role of reabsorption in this part of the tubule.

There are two kinds of nephrons, cortical and juxtamedullary (see Figure 40.4). **Cortical nephrons** are most numerous, making up about 85% of nephrons. They are located almost entirely within the renal cortex except for small parts of their nephron loops that dip into the renal medulla. The renal corpuscles of **juxtamedullary nephrons** are located deep in the cortex at the border with the medulla; their long nephron loops penetrate deeply into the medulla. Juxtamedullary nephrons play an important role in concentrating urine.

Filtrate leaving the renal tubule enters a **collecting duct** where its volume and concentration can be modified. Each collecting duct receives filtrate from many nephrons. The ducts travel through the medullary pyramids, giving them a striped appearance, then fuse near the renal pelvis. They deliver urine into the minor calyces via the papillae of the pyramids.

The function of the nephron depends on several unique features of the renal circulation (**Figure 40.5**). The capillary vascular supply consists of two distinct capillary beds, the *glomerulus* and the *peritubular capillary bed*. Vessels leading to and from the glomerulus, the first capillary bed, are both arterioles: the **afferent arteriole** feeds the bed while the **efferent arteriole** drains it. The glomerular capillary bed is unique in the body. It is a high-pressure bed along its entire length. Its high pressure is a result of two major factors: (1) the bed is *fed and drained* by arterioles, and (2) the afferent feeder arteriole is larger in diameter than the efferent arteriole draining the bed. The high hydrostatic pressure created by these two anatomical features forces fluid and blood components smaller than proteins out of the glomerulus into the glomerular capsule. That is, it forms the filtrate that is processed by the nephron tubule.

The **peritubular capillary bed** arises from the efferent arteriole draining the glomerulus. This set of capillaries clings intimately to the renal tubule and empties into the cortical radiate veins that leave the cortex. The peritubular capillaries are *low-pressure,* porous capillaries adapted for absorption rather than filtration and readily take up the solutes and water reabsorbed from the filtrate by the tubule cells. Efferent arterioles that supply juxtaglomerular nephrons tend not to form peritubular capillaries. Instead they form long, straight, highly interconnected vessels called **vasa recta** that run parallel and close to the long nephron loops. The vasa recta is essential for the formation of concentrated urine.

Each nephron also has a **juxtaglomerular complex (JGC)** (**Figure 40.6a**) located where the most distal portion of the ascending limb of the nephron loop touches the afferent arteriole. Helping to form the JGC are (1) *granular cells* (also called *juxtaglomerular [JG] cells*) in the arteriole walls that sense blood pressure in the afferent arteriole, and (2) a group of columnar cells in the nephron loop called the *macula densa* that monitors NaCl concentration in the filtrate. The role of the JGC is to regulate the rate of filtration and systemic blood pressure.

Urine formation is a result of three processes: *filtration, reabsorption,* and *secretion* (see Figure 40.5). **Filtration,** the role of the glomerulus, is largely a passive process in which a portion of the blood passes from the glomerular capillary into the glomerular capsule. This filtrate then enters the proximal convoluted tubule where tubular reabsorption and secretion begin. During **tubular reabsorption,** many of the filtrate components move through the tubule cells and return to the blood in the peritubular capillaries. Some of this reabsorption is passive, such as that of water which passes by osmosis, but the reabsorption of most substances depends on active transport processes and is highly selective. Which substances are reabsorbed at a particular time depends on the composition of the blood and needs of the

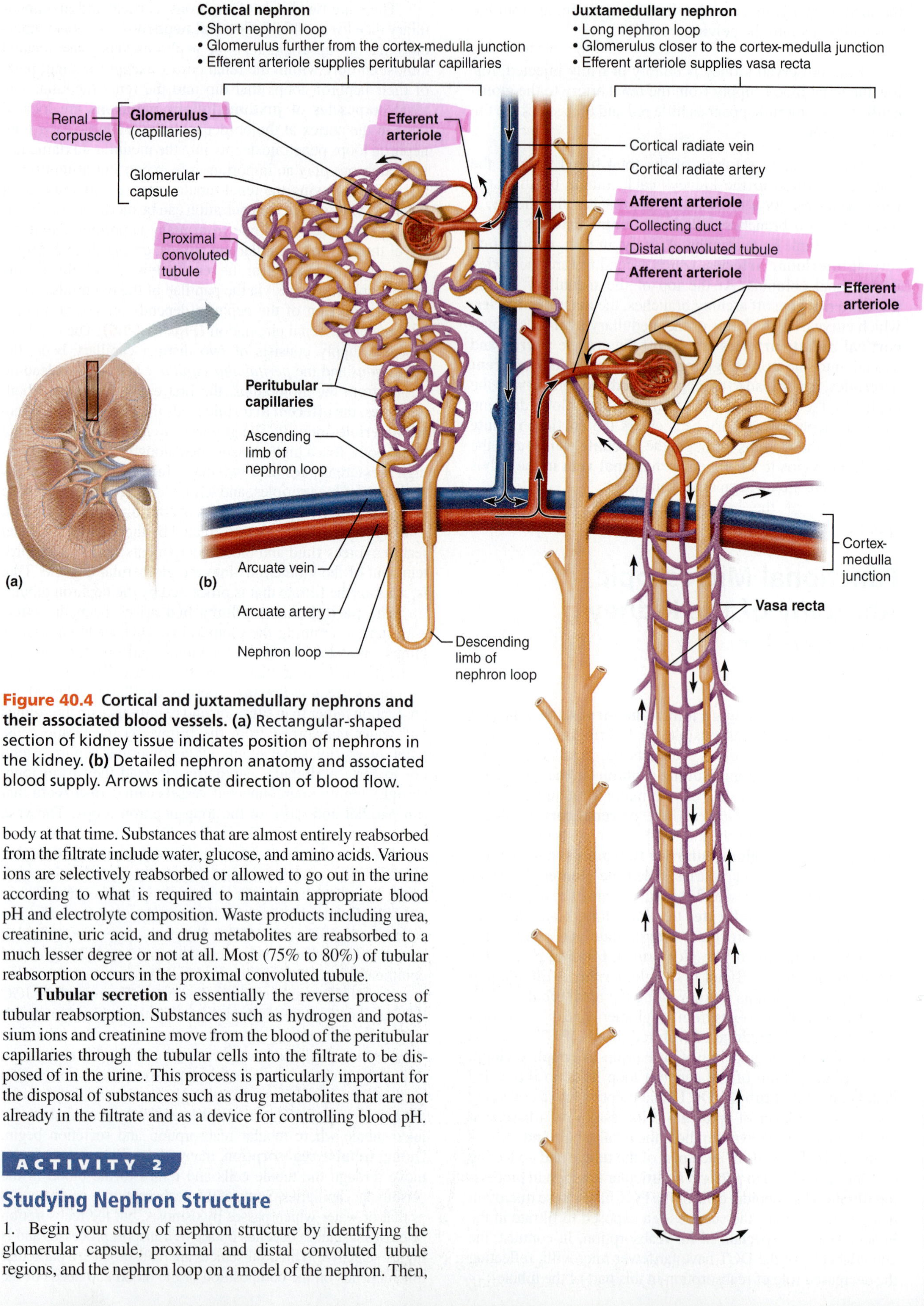

Figure 40.4 Cortical and juxtamedullary nephrons and their associated blood vessels. (a) Rectangular-shaped section of kidney tissue indicates position of nephrons in the kidney. **(b)** Detailed nephron anatomy and associated blood supply. Arrows indicate direction of blood flow.

body at that time. Substances that are almost entirely reabsorbed from the filtrate include water, glucose, and amino acids. Various ions are selectively reabsorbed or allowed to go out in the urine according to what is required to maintain appropriate blood pH and electrolyte composition. Waste products including urea, creatinine, uric acid, and drug metabolites are reabsorbed to a much lesser degree or not at all. Most (75% to 80%) of tubular reabsorption occurs in the proximal convoluted tubule.

Tubular secretion is essentially the reverse process of tubular reabsorption. Substances such as hydrogen and potassium ions and creatinine move from the blood of the peritubular capillaries through the tubular cells into the filtrate to be disposed of in the urine. This process is particularly important for the disposal of substances such as drug metabolites that are not already in the filtrate and as a device for controlling blood pH.

ACTIVITY 2

Studying Nephron Structure

1. Begin your study of nephron structure by identifying the glomerular capsule, proximal and distal convoluted tubule regions, and the nephron loop on a model of the nephron. Then,

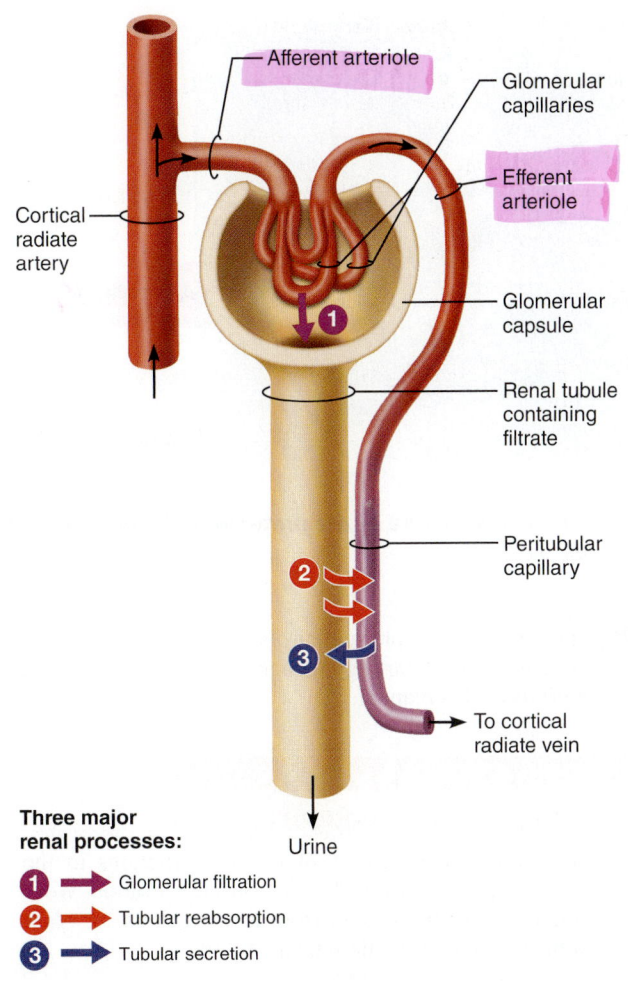

Three major renal processes:

1. ➡ Glomerular filtration
2. ➡ Tubular reabsorption
3. ➡ Tubular secretion

Figure 40.5 A schematic, uncoiled nephron. A kidney actually has millions of nephrons acting in parallel. The three major renal processes by which the kidneys adjust the composition of plasma are **(1)** glomerular filtration, **(2)** tubular reabsorption, and **(3)** tubular secretion. Black arrows show the path of blood flow through the renal microcirculation.

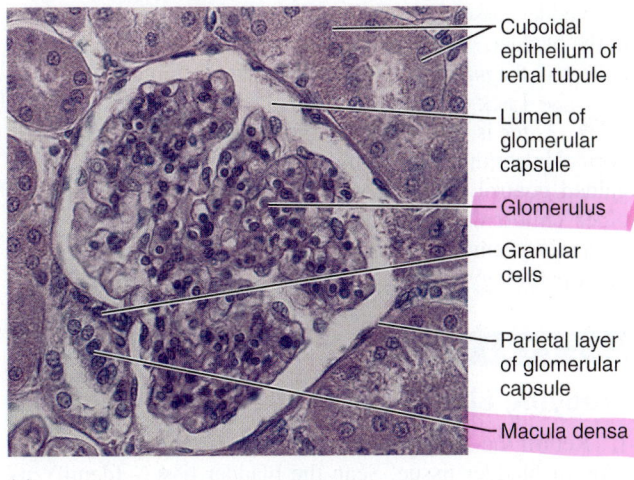

(a)

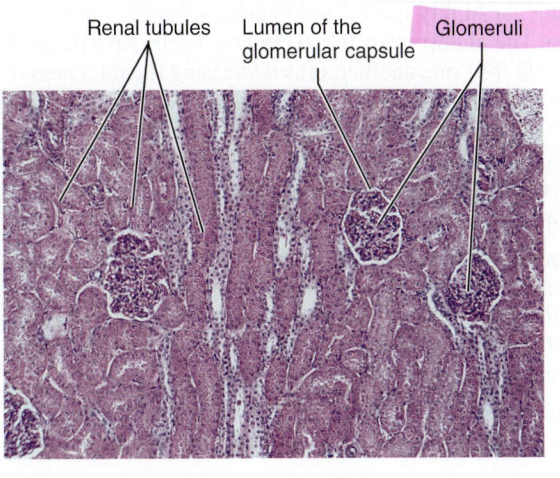

(b)

Figure 40.6 Microscopic structure of kidney tissue.
(a) Detailed structure of the glomerulus (225×).
(b) Low-power view of the renal cortex (67×).

obtain a compound microscope and a prepared slide of kidney tissue to continue with the microscope study of the kidney.

2. Hold the longitudinal section of the kidney up to the light to identify cortical and medullary areas. Then secure the slide on the microscope stage, and scan the slide under low power.

3. Move the slide so that you can see the cortical area. Identify a glomerulus, which appears as a ball of tightly packed material containing many small nuclei (Figure 40.6). It is usually surrounded by a vacant-appearing region corresponding to the space between the visceral and parietal layers of the glomerular capsule that surrounds it.

4. Notice that the renal tubules are cut at various angles. Try to differentiate between the fuzzy cuboidal epithelium of the proximal convoluted tubule, which has dense microvilli, and that of the distal convoluted tubule with sparse microvilli. Also identify the thin-walled nephron loop. ▬

Bladder

Although urine production by the kidney is a continuous process, urine is usually removed from the body when voiding is convenient. In the meantime the **urinary bladder,** which receives urine via the ureters and discharges it via the urethra, stores it temporarily.

Voiding, or **micturition,** is the act of emptying the bladder. Two sphincter muscles (see Figure 40.2), the **internal urethral sphincter** and the **external urethral sphincter,** control the outflow of urine from the bladder. Ordinarily, the bladder continues to collect urine until about 200 ml have accumulated, at which time the stretching of the bladder wall activates stretch receptors. Impulses transmitted to the central nervous system subsequently produce reflex contractions of the bladder wall through parasympathetic nervous system pathways via the pelvic splanchnic nerves. As contractions increase in force and frequency, stored urine is forced past the internal sphincter, which is a smooth muscle involuntary sphincter, into the superior part of the urethra. It is then that a person feels the urge to void. The inferior external sphincter consists of skeletal muscle and is voluntarily controlled. If it is not convenient to void, the opening of this sphincter can be inhibited. Conversely, if the time is convenient, the sphincter may be relaxed and the stored urine flushed from the body. If voiding is inhibited, the reflex contractions of the bladder cease temporarily and urine continues to accumulate in the bladder. After another 200 to 300 ml of urine have been collected, the *micturition reflex* will again be initiated.

Lack of voluntary control over the external sphincter is referred to as **incontinence.** Incontinence is normal in children 2 years old or younger, as they have not yet gained control over the voluntary sphincter. In adults and older children, incontinence generally results from spinal cord injury, emotional problems, bladder irritability, or some other pathology of the urinary tract. ✚

ACTIVITY 3

Studying Bladder Structure

1. Return the kidney slide to the supply area, and obtain a slide of bladder tissue. Scan the bladder tissue. Identify its three layers: mucosa, muscular layer, and fibrous adventitia.

2. Study the highly specialized transitional epithelium of the mucosa. The plump, transitional epithelial cells have the ability to slide over one another, thus decreasing the thickness of the mucosa layer as the bladder fills and stretches to accommodate the increased urine volume. Depending on the degree of stretching of the bladder, the mucosa may be three to eight cell layers thick. (Compare the transitional epithelium of the mucosa to that shown in Figure 6.3h on page 73).

3. Examine the heavy muscular wall (detrusor), which consists of three irregularly arranged muscular layers. The innermost and outermost muscle layers are arranged longitudinally; the middle layer is arranged circularly. Attempt to differentiate the three muscle layers.

4. Draw a small section of the bladder wall, and label all regions or tissue areas.

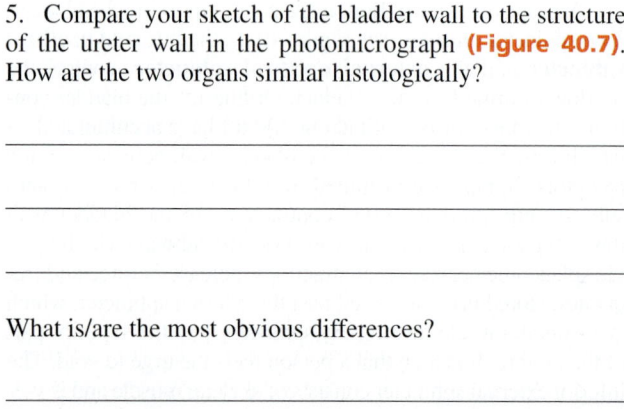

5. Compare your sketch of the bladder wall to the structure of the ureter wall in the photomicrograph **(Figure 40.7)**. How are the two organs similar histologically?

_____

_____

_____

What is/are the most obvious differences?

_____

_____

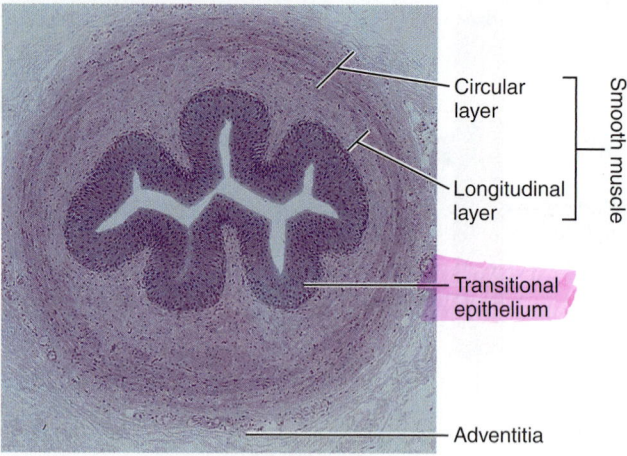

Figure 40.7 Structure of the ureter wall. Cross section of ureter (35×).

For instructions on animal dissections, see the dissection exercises (starting on page 697) in the cat and fetal pig editions of this manual.

GROUP CHALLENGE

Urinary System Sequencing

Arrange the following sets of urinary structures in the correct order for the flow of urine, filtrate, or blood. Work in small groups, but refrain from using a figure or other reference to determine the sequence.

1. renal pelvis, minor calyx, renal papilla, urinary bladder, ureter, major calyx, and urethra

_____

_____

2. distal convoluted tubule, ascending limb of the nephron loop, glomerulus, collecting duct, descending limb of the nephron loop, proximal convoluted tubule, and glomerular capsule

_____

_____

3. segmental artery, afferent arteriole, cortical radiate artery, glomerulus, renal artery, interlobar artery, and arcuate artery _____

_____

4. arcuate vein, inferior vena cava, peritubular capillaries, renal vein, interlobar vein, cortical radiate vein, and efferent arteriole _____

_____

Anatomy of the Urinary System

Gross Anatomy of the Human Urinary System

1. Complete the following statements.

The kidney is referred to as an excretory organ because it excretes __1__ wastes. It is also a major homeostatic organ because it maintains the electrolyte, __2__, and __3__ balance of the blood.

Urine is continuously formed by the __4__ and is routed down the __5__ by the mechanism of __6__ to a storage organ called the __7__. Eventually, the urine is conducted to the body __8__ by the urethra. In the male, the urethra is __9__ centimeters long and transports both urine and __10__. The female urethra is __11__ centimeters long and transports only urine.

Voiding or emptying the bladder is called __12__. Voiding has both voluntary and involuntary components. The voluntary sphincter is the __13__ sphincter. An inability to control this sphincter is referred to as __14__.

1. _____

2. _____

3. _____

4. _____

5. _____

6. _____

7. _____

8. _____

9. _____

10. _____

11. _____

12. _____

13. _____

14. _____

2. What is the function of the fat cushion that surrounds the kidneys in life? _____

_____

3. Define *ptosis.* _____

_____

4. Why is incontinence a normal phenomenon in the child under 1½ to 2 years old? _____

_____

What events may lead to its occurrence in the adult? _____

_____

_____

5. Complete the labeling of the diagram to correctly identify the urinary system organs.

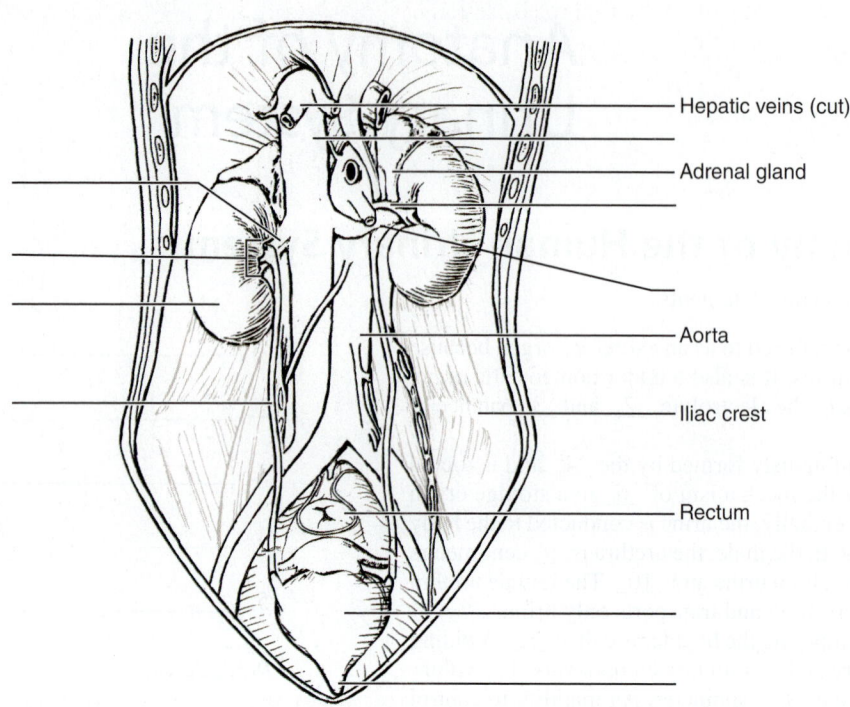

Hepatic veins (cut)

Adrenal gland

Aorta

Iliac crest

Rectum

Gross Internal Anatomy of the Pig or Sheep Kidney

6. Match the appropriate structure in column B to its description in column A. The items in column B may be used more than once.

Column A

_____ 1. smooth membrane, tightly adherent to the kidney surface

_____ 2. portion of the kidney containing mostly collecting ducts

_____ 3. portion of the kidney containing the bulk of the nephron structures

_____ 4. superficial region of kidney tissue

_____ 5. basinlike area of the kidney, continuous with the ureter

_____ 6. a cup-shaped extension of the pelvis that encircles the apex of a pyramid

_____ 7. area of cortical tissue running between the medullary pyramids

Column B

a. cortex

b. fibrous capsule

c. medulla

d. minor calyx

e. renal column

f. renal pelvis

Functional Microscopic Anatomy of the Kidney and Bladder

7. Label the blood vessels and parts of the nephron by selecting the letter for the correct structure from the key below. The items in the key may be used more than once.

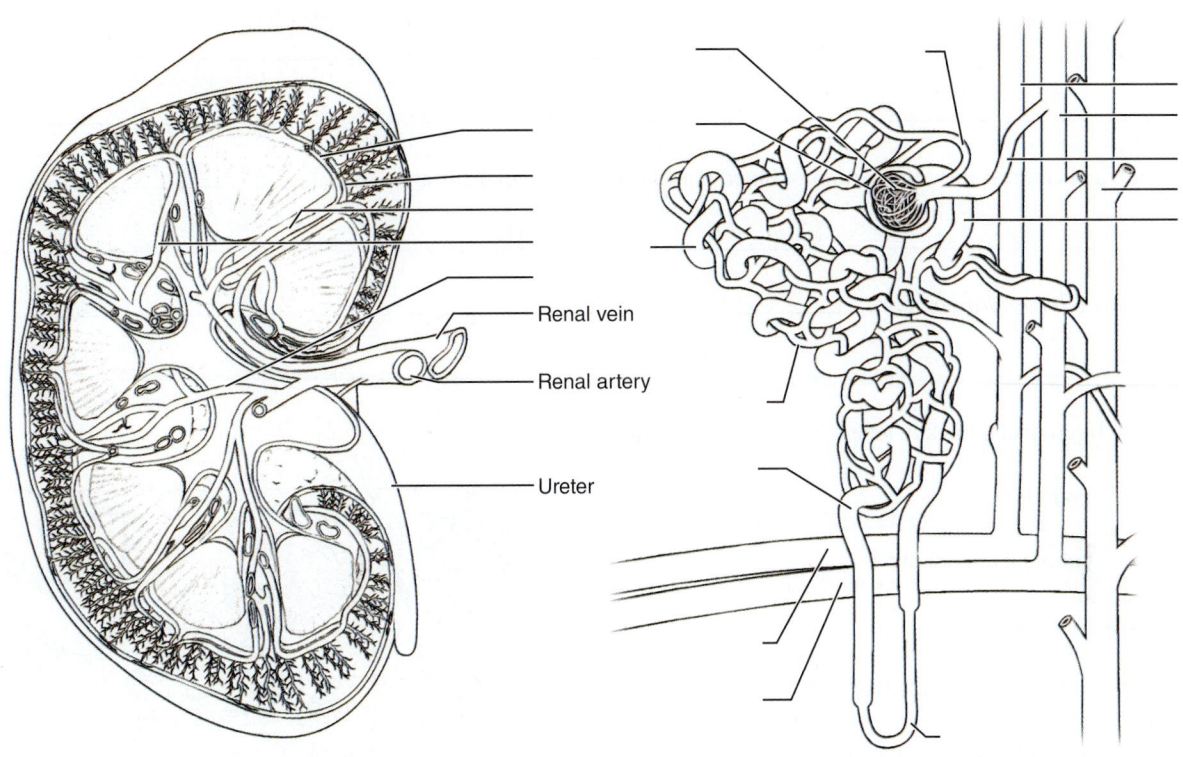

Renal vein

Renal artery

Ureter

Key:

a. afferent arteriole

b. arcuate artery

c. arcuate vein

d. collecting duct

e. cortical radiate artery

f. cortical radiate vein

g. distal convoluted tubule

h. efferent arteriole

i. glomerular capsule

j. glomerulus

k. interlobar artery

l. interlobar vein

m. nephron loop—ascending limb

n. nephron loop— descending limb

o. peritubular capillary

p. proximal convoluted tubule

q. segmental artery

8. For each of the following descriptions of a structure, find the matching name in the question 7 key.

_____ 1. capillary specialized for filtration

_____ 2. capillary specialized for reabsorption

_____ 3. cuplike part of the renal corpuscle

_____ 4. location of macula densa

_____ 5. primary site of tubular reabsorption

_____ 6. receives urine from many nephrons

9. Explain *why* the glomerulus is such a high-pressure capillary bed. _____

_____

How does its high-pressure condition aid its function of filtrate formation? _____

_____

10. What structural modification of certain tubule cells enhances their ability to reabsorb substances from the filtrate?

_____

11. Explain the mechanism of tubular secretion, and explain its importance in the urine-formation process. _____

_____

_____

12. Compare and contrast the composition of blood plasma and glomerular filtrate. _____

_____

13. Define *juxtaglomerular complex.* _____

_____

14. Label the figure using the key letters of the correct terms.

    Key: a.   granular cells

           b.   cuboidal epithelium

           c.   macula densa

           d.   glomerular capsule (parietal layer)

           e.   ascending limb of the nephron loop

15. What is important functionally about the specialized epithelium (transitional epithelium) in the bladder?

_____

_____

Urinalysis

MATERIALS

- ☐ Disposable gloves
- ☐ Student urine samples collected at the beginning of the laboratory or "normal" artificial urine provided by the instructor*
- ☐ Numbered "pathological" urine specimens provided by the instructor*
- ☐ Wide-range pH paper
- ☐ Dipsticks: individual (Clinistix®, Ketostix®, Albustix®, Hemastix®) or combination (Chemstrip® or Multistix®)
- ☐ Urinometer
- ☐ Test tubes, test tube rack, and test tube holders
- ☐ 10-cc graduated cylinders
- ☐ Test reagents for sulfates: 10% barium chloride solution, dilute hydrochloric acid (HCl)
- ☐ Hot plate
- ☐ 500-ml beaker
- ☐ Test reagent for phosphates: dilute nitric acid (HNO_3), dilute ammonium molybdate
- ☐ Glass stirring rod
- ☐ Test reagent for chloride: 3.0% silver nitrate solution ($AgNO_3$), freshly prepared

*Directions for making artificial urine are provided in the Instructor's Guide for this manual.

(Text continues on next page.)

MasteringA&P® For related exercise study tools, go to the Study Area of MasteringA&P. There you will find:

- Practice Anatomy Lab PAL
- PhysioEx PEx
- A&PFlix A&PFlix
- Practice quizzes, Histology Atlas, eText, Videos, and more!

OBJECTIVES

1. List the physical characteristics of urine, and indicate the normal pH and specific gravity ranges.
2. List substances that are normal urinary constituents.
3. Conduct various urinalysis tests and procedures, and use them to determine the substances present in a urine specimen.
4. Define the following urinary conditions:

calculi	*albuminuria*	*hemoglobinuria*
casts	*glycosuria*	*ketonuria*
	hematuria	*pyuria*

5. Discuss the possible causes and implications of conditions listed in objective 4.

PRE-LAB QUIZ

1. Normal urine is usually pale yellow to amber in color, due to the presence of
   a. hemochrome
   b. melanin
   c. urochrome
2. Circle the correct underlined term. The average pH value of urine is 6.0 / 11.0.
3. Circle True or False. Glucose can usually be found in all normal urine.
4. _____, like other blood proteins, is/are too large to pass through the glomerular filtration membrane and is/are normally not found in urine.
   a. Albumin
   b. Chloride
   c. Nitrates
   d. Sulfate
5. Circle the correct underlined term. Hematuria / Ketonuria, the appearance of red blood cells in the urine, almost always indicates pathology of the urinary system.
6. The appearance of bile pigments in the urine, a condition known as _____, can be an indication of liver disease.
   a. albuminuria
   b. bilirubinuria
   c. ketonuria
   d. pyuria
7. Circle the correct underlined term. Albuminuria / Pyuria, the presence of white blood cells or pus in the urine, is consistent with inflammation of the urinary tract.
8. Circle the correct underlined term. Casts / Calculi are hardened cell fragments formed in the distal convoluted tubules and collecting ducts and flushed out of the urinary tract.
9. Circle True or False. When testing the pH of the urine, you will use the same piece of wide-range pH paper for each test.
10. When determining the presence of inorganic constituents such as sulfates, phosphates, and chlorides, you will be looking for the "formation of a precipitate." What is a precipitate? _____ _____

(Materials list continued.)

- ☐ Clean microscope slide and coverslip
- ☐ Compound microscope
- ☐ Test reagent for urea: concentrated nitric acid in dropper bottles
- ☐ Test reagent for glucose: Clinitest® tablets; Clinitest color chart

- ☐ Medicine droppers
- ☐ Timer (watch or clock with a second hand)
- ☐ Ictotest® reagent tablets and test mat
- ☐ Flasks and laboratory buckets containing 10% bleach solution

- ☐ Disposable autoclave bags
- ☐ *Demonstration:* Instructor-prepared specimen of urine sediment set up for microscopic analysis
- **PEx** PhysioEx™ 9.1 Computer Simulation Ex. 9 on p. PEx-131.

Blood composition depends on three major factors: diet, cellular metabolism, and urinary output. In 24 hours, the kidneys' 2 million nephrons filter 150 to 180 liters of blood plasma through their glomeruli into the tubules, where it is selectively processed by tubular reabsorption and secretion. In the same period, urinary output, which contains by-products of metabolism and excess ions, is 1.0 to 1.8 liters. In healthy people, the kidneys can maintain blood constancy despite wide variations in diet and metabolic activity. With certain pathological conditions, urine composition often changes dramatically.

Characteristics of Urine

To be valuable as a diagnostic tool, a urinalysis must be done within 30 minutes after the urine is voided or on refrigerated urine. Freshly voided urine is generally clear and pale yellow to amber in color. This normal yellow color is due to *urochrome,* a pigment metabolite that arises from the body's destruction of hemoglobin and travels to the kidney as bilirubin or bile pigments. As a rule, color variations from pale yellow to deeper amber indicate the relative concentration of solutes to water in the urine. The greater the solute concentration, the deeper the color. Abnormal urine color may be due to certain foods, such as beets, various drugs, bile, or blood.

The odor of freshly voided urine is slightly aromatic, but bacterial action gives it an ammonia-like odor when left standing. Some drugs, vegetables (such as asparagus), and various disease processes (such as diabetes mellitus) alter the characteristic odor of urine. For example, the urine of a person with uncontrolled diabetes mellitus (and elevated levels of ketones) smells fruity or acetone-like.

The pH of urine ranges from 4.5 to 8.0, but its average value, 6.0, is slightly acidic. Diet may markedly influence the pH of the urine. For example, a diet high in protein (meat, eggs, cheese) and whole wheat products increases the acidity of urine. Such foods are called *acid ash foods.* On the other hand, a vegetarian diet contains *alkaline ash foods* that increase the alkalinity of the urine. The naming of foods as acid or alkaline *ash* comes from the fact that if an acid ash food is burned to ash, the pH of the ash is acidic. The ash of alkaline ash foods is alkaline. A bacterial infection of the urinary tract may also cause the urine to become more alkaline.

Specific gravity is the relative weight of a specific volume of liquid compared with an equal volume of distilled water. The specific gravity of distilled water is 1.000, because 1 ml weighs 1 g. Since urine contains dissolved solutes, it weighs more than water, and its customary specific gravity ranges from 1.001 to 1.030. Urine with a specific gravity of 1.001 contains few solutes and is considered very dilute. Dilute urine commonly results when a person drinks excessive amounts of water, uses diuretics,

or suffers from diabetes insipidus or chronic renal failure. Conditions that produce urine with a high specific gravity include limited fluid intake, fever, and kidney inflammation, called *pyelonephritis.* If urine becomes excessively concentrated, some of the substances normally held in solution begin to precipitate or crystallize, forming **kidney stones,** or **renal calculi.**

Water is the largest component of urine, accounting for 95% of its volume. Of the solutes that make up the remaining 5% of urine, urea is the largest component. *Urea* comes from the breakdown of protein and is one of the *nitrogenous wastes. Uric acid*, from the breakdown of nucleic acids, and *creatinine*, a metabolite of creatine phosphate that is present in large amounts in skeletal muscle, make up the remaining nitrogenous wastes.

Normal solute constituents of urine, in order of decreasing concentration, include urea, sodium, potassium, phosphate, and sulfate ions; creatinine, and uric acid. Much smaller but highly variable amounts of calcium, magnesium, and bicarbonate ions are also found in the urine. Abnormally high concentrations of any of these urinary constituents may indicate a pathological condition.

Abnormal Urinary Constituents

Abnormal urinary constituents are substances not normally present in the urine when the body is operating properly. ✚

Glucose

The presence of glucose in the urine, a condition called **glycosuria,** indicates abnormally high blood sugar levels. Normally, blood sugar levels are maintained between 80 and 100 mg/100 ml of blood. At this level all glucose in the filtrate is reabsorbed by the tubular cells and returned to the blood. Glycosuria may result from carbohydrate intake so excessive that normal physiological and hormonal mechanisms cannot clear it from the blood quickly enough. In such cases, the active transport reabsorption mechanisms of the renal tubules for glucose are exceeded—but only temporarily.

Pathological glycosuria occurs in conditions such as uncontrolled diabetes mellitus, in which the body cells are unable to absorb glucose from the blood because the pancreatic islet cells produce inadequate amounts of the hormone insulin, or there is some abnormality of the insulin receptors. Under such circumstances, the body cells increase their metabolism of fats, and the excess and unusable glucose spills out in the urine.

Albumin

Albuminuria, or the presence of albumin in urine, is an abnormal finding. Albumin is the single most abundant blood protein

41

and is very important in maintaining the osmotic pressure of the blood. Albumin, like other blood proteins, is too large to pass through the glomerular filtration membrane. Thus, albuminuria is generally indicative of abnormally increased permeability of the glomerular membrane. Certain nonpathological conditions, such as excessive exertion, pregnancy, or overabundant protein intake, can temporarily increase the membrane permeability, leading to *physiological albuminuria*. Pathological conditions resulting in albuminuria include events that damage the glomerular membrane, such as kidney trauma due to blows, the ingestion of poisons or heavy metals, bacterial toxins, glomerulonephritis, and hypertension.

Ketone Bodies

Ketone bodies including acetoacetic acid, beta-hydroxybutyric acid, and acetone normally appear in the urine in very small amounts. **Ketonuria,** the presence of these intermediate products of fat metabolism in excessive amounts, usually indicates that abnormal metabolic processes are occurring. The result may be *acidosis* and its complications. Ketonuria is an expected finding during starvation, or diets very low in carbohydrates, when inadequate food intake forces the body to use its fat stores. Ketonuria coupled with a finding of glycosuria is generally diagnostic for diabetes mellitus.

Red Blood Cells

Hematuria, the appearance of red blood cells in the urine, almost always indicates pathology of the urinary tract, because erythrocytes are too large to pass through the glomerular pores. Possible causes include irritation of the urinary tract organs by calculi (kidney stones), which produces clinically evident bleeding; infection or tumors of the urinary tract; or physical trauma to the urinary organs. In healthy menstruating females, it may reflect accidental contamination of the urine sample with the menstrual flow.

Hemoglobin

Hemoglobinuria, the presence of hemoglobin in the urine, is a result of the fragmentation, or hemolysis, of red blood cells. As a result, hemoglobin is liberated into the plasma and subsequently appears in the kidney filtrate. Hemoglobinuria indicates various pathological conditions including hemolytic anemias, transfusion reactions, burns, poisonous snake bites, or renal disease.

Nitrites

The presence of urinary nitrites might indicate a bacterial infection, particularly *E. coli* or other gram-negative rods. The presence of nitrites is a valuable finding for early detection of bladder infections.

Bile Pigments

Bilirubinuria, the appearance of bilirubin (bile pigments) in urine, is an abnormal finding and usually indicates liver pathology, such as hepatitis, cirrhosis, or bile duct blockage. Bilirubinuria is signaled by a yellow foam that forms when the urine sample is shaken.

Urobilinogen is produced in the intestine from bilirubin and gives feces a brown color. Some urobilinogen is reabsorbed into the blood and either excreted back into the intestine by the liver or excreted by the kidneys in the urine.

Complete absence of urobilinogen may indicate renal disease or obstruction of bile flow in the liver. Increased levels may indicate hepatitis A, cirrhosis, or biliary disease.

White Blood Cells

Pyuria is the presence of white blood cells or other pus constituents in the urine. It indicates inflammation of the urinary tract.

Casts

Any complete discussion of the varieties and implications of casts is beyond the scope of this exercise. However, because they always represent a pathological condition of the kidney or urinary tract, they should at least be mentioned. **Casts** are hardened cell fragments, usually cylindrical, which are formed in the distal convoluted tubules and collecting ducts and then flushed out of the urinary tract. Hyaline casts are formed from a mucoprotein secreted by tubule cells (Figure 41.1b, page 626). These casts form when the filtrate flow rate is slow, the pH is low, or the salt concentration is high, all conditions which cause protein to denature. Red blood cell casts are typical in glomerulonephritis, as red blood cells leak through the filtration membrane and stick together in the tubules. White blood cell casts form when the kidney is inflamed, which is typically a result of pyelonephritis but sometimes occurs with glomerulonephritis. Degenerated renal tubule cells form granular or waxy casts (Figure 41.1b). Broad waxy casts may indicate end-stage renal disease.

ACTIVITY 1

Analyzing Urine Samples

In this part of the exercise, you will use prepared dipsticks and perform chemical tests to determine the characteristics of normal urine as well as to identify abnormal urinary components. You will investigate two or more urine samples. The first, designated as the *standard urine specimen* in the **Urinalysis Results chart** (page 625), will be either yours or a "normal" sample provided by your instructor. The second will be an unknown urine specimen provided by your instructor. Make the following determinations on both samples, and record your results by circling the appropriate item or description or by adding data to complete the chart. If you have more than one unknown sample, accurately identify each sample by number.

⚠ *Obtain and wear disposable gloves throughout this laboratory session.* Although the instructor-provided urine samples are actually artificial urine (concocted in the laboratory to resemble real urine), you should still observe the techniques of safe handling of body fluids as part of your learning process. When you have completed the laboratory procedures: (1) dispose of the gloves, used pH paper strips, and dipsticks in the autoclave bag; (2) put used glassware in the bleach-containing laboratory bucket; (3) wash the lab bench down with 10% bleach solution.

Determination of the Physical Characteristics of Urine

1. Determine the color, transparency, and odor of your "normal" sample and one of the numbered pathological samples, and circle the appropriate descriptions in the Urinalysis Results chart.

2. Obtain a roll of wide-range pH paper to determine the pH of each sample. Use a fresh piece of paper for each test, and dip the strip into the urine to be tested two or three times before comparing the color obtained with the chart on the dispenser. Record your results in the chart. (If you will be using one of the combination dipsticks—Chemstrip or Multistix—this pH determination can be done later.)

3. To determine specific gravity, obtain a urinometer cylinder and float. Mix the urine well, and fill the urinometer cylinder about two-thirds full with urine.

4. Examine the urinometer float to determine how to read its markings. In most cases, the scale has numbered lines separated by a series of unnumbered lines. The numbered lines give the reading for the first two decimal places. You must determine the third decimal place by reading the lower edge of the meniscus—the curved surface representing the urine-air junction—on the stem of the float.

5. Carefully lower the urinometer float into the urine. Make sure it is floating freely before attempting to take the reading. Record the specific gravity of both samples in the chart. _Do not dispose of this urine if the samples that you have are less than 200 ml in volume_ because you will need to make several more determinations.

Determination of Inorganic Constituents in Urine

Sulfates

Using a 10-cc graduated cylinder, add 5 ml of urine to a test tube, and then add a few drops of dilute hydrochloric acid and 2 ml of 10% barium chloride solution. The appearance of a white precipitate (barium sulfate) indicates the presence of sulfates in the sample. Clean the graduated cylinder and the test tubes well after use. Record your results.

Phosphates

Obtain a hot plate and a 500-ml beaker. To prepare the hot water bath, half fill the beaker with tap water and heat it on the hot plate. Add 5 ml of urine to a test tube, and then add three or four drops of dilute nitric acid and 3 ml of ammonium molybdate. Mix well with a glass stirring rod, and then heat gently in a hot water bath. Formation of a yellow precipitate indicates the presence of phosphates in the sample. Record your results.

Chlorides

Place 5 ml of urine in a test tube, and add several drops of silver nitrate. The appearance of a white precipitate (silver chloride) is a positive test for chlorides. Record your results.

Nitrites

Use a combination dipstick to test for nitrites. Record your results.

Determination of Organic Constituents in Urine

Individual dipsticks or combination dipsticks (Chemstrip or Multistix) may be used for many of the tests in this section. If combination dipsticks are used, be prepared to take the readings on several factors (pH, protein [albumin], glucose, ketones, blood/hemoglobin, leukocytes, urobilinogen, bilirubin, and nitrites) at the same time. Generally speaking, results for all of these tests may be read _during_ the second minute after immersion, but readings taken after 2 minutes have passed should be considered invalid. Pay careful attention to the directions for method and time of immersion and disposal of excess urine from the strip, regardless of the dipstick used. Identify the dipsticks that you use in the chart. If you are testing your own urine and get an unanticipated result, it is helpful to know that most of the combination dipsticks produce false positive or negative results for certain solutes when the subject is taking vitamin C, aspirin, or certain drugs.

Urea

Put two drops of urine on a clean microscope slide and _carefully_ add one drop of concentrated nitric acid to the urine. Slowly warm the mixture on a hot plate until it begins to dry at the edges, but do not allow it to boil or to evaporate to dryness. When the slide has cooled, examine the edges of the preparation under low power to identify the rhombic or hexagonal crystals of urea nitrate, which form when urea and nitric acid react chemically. Keep the light low for best contrast. Record your results.

Glucose

Use a combination dipstick or obtain a vial of Clinistix, and conduct the dipstick test according to the instructions on the vial. Record your results in the Urinalysis Results chart.

Because the Clinitest reagent is routinely used in clinical agencies for glucose determinations in pediatric patients, it is worthwhile to conduct this test as well. Obtain the Clinitest tablets and the associated color chart. You will need a timer (watch or clock with a second hand) for this test. Using a medicine dropper, put 5 drops of urine into a test tube; then rinse the dropper and add 10 drops of water to the tube. Add a Clinitest tablet. Wait 15 seconds and then compare the color obtained to the color chart. Record your results.

Albumin

Use a combination dipstick or obtain the Albustix dipsticks, and conduct the determinations as indicated on the vial. Record your results.

Ketones

Use a combination dipstick or obtain the Ketostix dipsticks. Conduct the determinations as indicated on the vial. Record your results.

Blood/Hemoglobin

Test your urine samples for the presence of hemoglobin by using a Hemastix dipstick or a combination dipstick according to the directions on the vial. Usually a short drying period is required before making the reading, so read the directions carefully. Record your results.

Bilirubin

Using a combination dipstick, determine if there is any bilirubin in your urine samples. Record your results.

Also conduct the Ictotest for the presence of bilirubin. Using a medicine dropper, place one drop of urine in the center of one of the special test mats provided with the Ictotest reagent tablets. Place one of the reagent tablets over the drop of urine, and then add two drops of water directly to the tablet. If the mixture turns purple when you add water, bilirubin is present. Record your results.

Leukocytes

Use a combination dipstick to test for leukocytes. Record your results.

Urobilinogen

Use a combination dipstick to test for urobilinogen. Record your results.

Clean up your area following the procedures described at the beginning of this activity. ■

Urinalysis Results			
Observation or test	Normal values	Standard urine specimen	Unknown specimen (# _____)
Physical Characteristics			
Color	Pale yellow	Yellow: pale medium dark other _____	Yellow: pale medium dark other _____
Transparency	Transparent	Clear Slightly cloudy Cloudy	Clear Slightly cloudy Cloudy
Odor	Characteristic	Describe: _____	Describe: _____
pH	4.5–8.0	_____	_____
Specific gravity	1.001–1.030	_____	_____
Inorganic Components			
Sulfates	Present	Present Absent	Present Absent
Phosphates	Present	Present Absent	Present Absent
Chlorides	Present	Present Absent	Present Absent
Nitrites	Absent	Present Absent	Present Absent
Organic Components			
Urea	Present	Present Absent	Present Absent
Glucose Dipstick: _____	Negative	Record results: _____	Record results: _____
Clinitest	Negative	_____	_____
Albumin Dipstick: _____	Negative	_____	_____
Ketone bodies Dipstick: _____	Negative	_____	
RBCs/hemoglobin Dipstick: _____	Negative	_____	_____
Bilirubin Dipstick:_____	Negative	_____	_____
Ictotest	Negative (no color change)	Negative Positive (purple)	Negative Positive (purple)
Leukocytes	Absent	Present Absent	Present Absent
Urobilinogen	Present	Present Absent	Present Absent

41

ACTIVITY 2

Analyzing Urine Sediment Microscopically (Optional)

If your instructor so indicates, conduct a microscopic analysis of urine sediment in "real" urine. The urine sample to be analyzed microscopically has been centrifuged to spin the more dense urine components to the bottom of a tube, and some of the sediment has been mounted on a slide and stained with Sedi-stain to make the components more visible.

Go to the demonstration microscope to conduct this study. Using the lowest light source possible, examine the slide under low power to determine if any common sediments can be seen **(Figure 41.1)**.

Unorganized sediments: Chemical substances that form crystals or precipitate from solution; for example, calcium oxalates, carbonates, and phosphates; uric acid; ammonium ureates; and cholesterol. Also, if one has been taking antibiotics or certain drugs such as sulfa drugs, these may be detectable in the urine in crystalline form. Normal urine contains very small amounts of crystals, but conditions such as urinary retention or urinary tract infection may cause the appearance of much larger amounts. The high-power lens may be needed to view the various crystals, which tend to be much more minute than the organized cellular sediments.

Organized sediments: Include epithelial cells (rarely of any pathological significance), pus cells (white blood cells), red blood cells, and casts. Urine is normally negative for organized sediments, and the presence of white blood cells, red blood cells, and casts other than trace amounts always indicates kidney pathology. ■

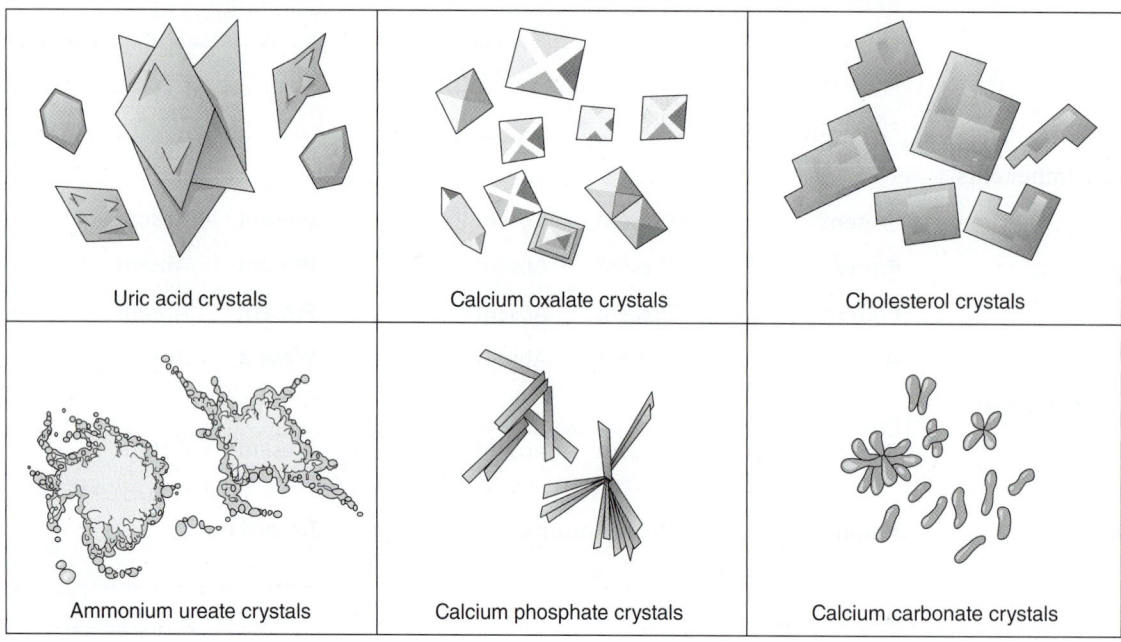

Uric acid crystals

Calcium oxalate crystals

Cholesterol crystals

Ammonium ureate crystals

Calcium phosphate crystals

Calcium carbonate crystals

(a) Unorganized sediments

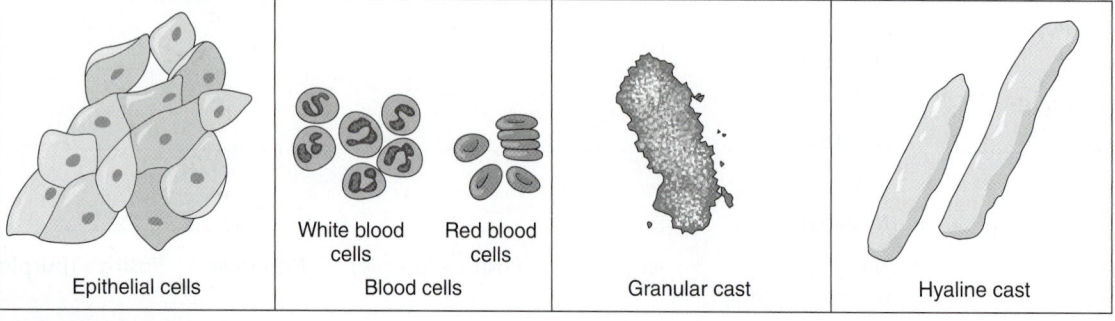

Epithelial cells

White blood cells

Red blood cells

Blood cells

Granular cast

Hyaline cast

(b) Organized sediments

Figure 41.1 Examples of sediments.

Urinalysis

Characteristics of Urine

1. What is the normal volume of urine excreted in a 24-hour period? _____

2. Assuming normal conditions, note whether each of the following substances would be (a) in greater relative concentration in the urine than in the glomerular filtrate, (b) in lesser concentration in the urine than in the glomerular filtrate, or (c) absent from both the urine and the glomerular filtrate. Use an appropriate reference as needed.

_____ 1. water   _____ 6. amino acids   _____ 11. uric acid

_____ 2. phosphate ions _____ 7. glucose   _____ 12. creatinine

_____ 3. sulfate ions   _____ 8. albumin   _____ 13. pus (WBCs)

_____ 4. potassium ions _____ 9. red blood cells   _____ 14. nitrites

_____ 5. sodium ions   _____ 10. urea

3. Explain why urinalysis is a routine part of any good physical examination. _____

_____

4. What substance is responsible for the normal yellow color of urine? _____

5. Which has a greater specific gravity: 1 ml of urine or 1 ml of distilled water? _____ Explain your answer. _____

_____

6. Explain the relationship between the color, specific gravity, and volume of urine. _____

_____

Abnormal Urinary Constituents

7. A microscopic examination of urine may reveal the presence of certain abnormal urinary constituents.

Name three constituents that might be present if a urinary tract infection exists. _____,

_____, and _____

8. How does a urinary tract infection influence urine pH? _____

How does starvation influence urine pH? _____

9. All urine specimens become alkaline and cloudy on standing at room temperature. Explain why. _____

_____

10. Several specific terms have been used to indicate the presence of abnormal urine constituents. Identify each of the abnormalities described below by inserting a term from the key at the right that names the condition.

_____ 1. presence of erythrocytes in the urine

_____ 2. presence of hemoglobin in the urine

_____ 3. presence of glucose in the urine

_____ 4. presence of albumin in the urine

_____ 5. presence of ketone bodies in the urine

_____ 6. presence of pus (white blood cells) in the urine

Key:

a. albuminuria

b. glycosuria

c. hematuria

d. hemoglobinuria

e. ketonuria

f. pyuria

11. What are renal calculi, and what conditions favor their formation? _____

_____

12. Glucose and albumin are both normally absent in the urine, but the reason for their exclusion differs. Explain the reason for

the absence of glucose. _____

Explain the reason for the absence of albumin. _____

13. The presence of abnormal constituents or conditions in urine may be associated with diseases, disorders, or other causes listed in the key. Select and list all conditions associated with each numbered item. Some numbered items will have multiple letters.

_____ 1. low specific gravity

_____ 2. high specific gravity

_____ 3. glucose

_____ 4. albumin

_____ 5. blood cells

_____ 6. hemoglobin

_____ 7. bilirubin

_____ 8. ketone bodies

_____ 9. casts

_____ 10. pus

Key:

a. cystitis (inflammation of the bladder)

b. diabetes insipidus

c. diabetes mellitus

d. eating a 5-lb box of sweets for lunch

e. glomerulonephritis

f. gonorrhea

g. hemolytic anemias

h. hepatitis, cirrhosis of the liver

i. kidney stones

j. pregnancy, exertion

k. pyelonephritis

l. starvation

14. Name the three major nitrogenous wastes found in the urine. _____,

_____, and _____

15. Explain the difference between organized and unorganized sediments. _____

_____

_____

Anatomy of the Reproductive System

MATERIALS

☐ Three-dimensional models or large laboratory charts of the male and female reproductive tracts

☐ Prepared slides of cross sections of the penis, seminal glands, epididymis, uterus showing endometrium (secretory phase), and uterine tube

☐ Compound microscope

✂ For instructions on animal dissections, see the dissection exercises (starting on page 697) in the cat and fetal pig editions of this manual.

OBJECTIVES

1. Discuss the general function of the reproductive system.
2. Identify the structures of the male and female reproductive systems on an appropriate model or image, and list the general function of each.
3. Define *semen,* state its composition, and name the organs involved in its production.
4. Trace the pathway followed by a sperm from its site of formation to the external environment.
5. Define *erection* and *ejaculation*.
6. Define *gonad,* and name the gametes and endocrine products of the testes and ovaries, indicating the cell types or structures responsible for the production of each.
7. Describe the microscopic structure of the penis, seminal glands, epididymis, uterine wall and uterine tube, and relate structure to function.
8. Explain the role of the fimbriae and ciliated epithelium of the uterine tubes in the movement of the "egg" from the ovary to the uterus.
9. Identify the fundus, body, and cervical regions of the uterus.
10. Define *endometrium, myometrium,* and *ovulation*.
11. Describe the anatomy and discuss the reproduction-related function of female mammary glands.

PRE-LAB QUIZ

1. The essential organs of reproduction are the _____, which produce the sex cells.
   a. accessory male glands   c. seminal glands
   b. gonads   d. uterus
2. Circle the correct underlined term. The paired oval testes lie in the scrotum / prostate outside the abdominopelvic cavity, where they are kept slightly cooler than body temperature.
3. After sperm are produced, they enter the first part of the duct system, the _____.
   a. ductus deferens   c. epididymis
   b. ejaculatory duct   d. urethra
4. The prostate, seminal glands, and bulbo-urethral glands produce _____, the liquid medium in which sperm leaves the body.
   a. seminal fluid   c. urine
   b. testosterone   d. water
5. Circle the correct underlined term. The interstitial endocrine cells / seminiferous tubules produce testosterone, the hormonal product of the testis.
6. The endocrine products of the ovaries are estrogen and
   a. luteinizing hormone   c. prolactin
   b. progesterone   d. testosterone

MasteringA&P® For related exercise study tools, go to the Study Area of MasteringA&P. There you will find:
- Practice Anatomy Lab PAL
- PhysioEx PEx
- A&PFlix *A&PFlix*
- Practice quizzes, Histology Atlas, eText, Videos, and more!

(Text continues on next page.)

7. Circle the correct underlined term. The <u>labia majora</u> / <u>clitoris</u> are/is homologous to the penis.

8. The _____ is a pear-shaped organ that houses the embryo or fetus during its development.
   a. bladder   c. uterus
   b. cervix   d. vagina

9. Circle the correct underlined term. The <u>endometrium</u> / <u>myometrium</u>, the thick mucosal lining of the uterus, has a superficial layer that sloughs off periodically.

10. Circle the correct underlined term. A developing egg is ejected from the ovary at the appropriate stage of maturity in an event known as <u>menstruation</u> / <u>ovulation</u>.

Most organ systems of the body function from the time they are formed to sustain the existing individual. However, the **reproductive system** begins its biological function, the perpetuation of the species, at puberty.

The essential organs of reproduction are the **gonads**, the testes and the ovaries, which produce the sex cells or **gametes** and the sex hormones. The reproductive role of the male is to manufacture sperm and to deliver them to the female reproductive tract. The female, in turn, produces eggs. If the time is suitable, the combination of sperm and egg produces a fertilized egg, which is the first cell of a new individual. Once fertilization has occurred, the female uterus provides a nurturing, protective environment in which the embryo, later called the fetus, develops until birth.

Gross Anatomy of the Human Male Reproductive System

The primary reproductive organs of the male are the **testes**, which produce sperm and the male sex hormones. All other reproductive structures are conduits or sources of secretions, which aid in the safe delivery of the sperm to the body exterior or female reproductive tract.

ACTIVITY 1

Identifying Male Reproductive Organs

As the following organs and structures are described, locate them on the illustration **(Figure 42.1)** and then identify them on a three-dimensional model of the male reproductive system or on a large laboratory chart. ■■

Figure 42.1 Reproductive organs of the human male. (a) Sagittal view.

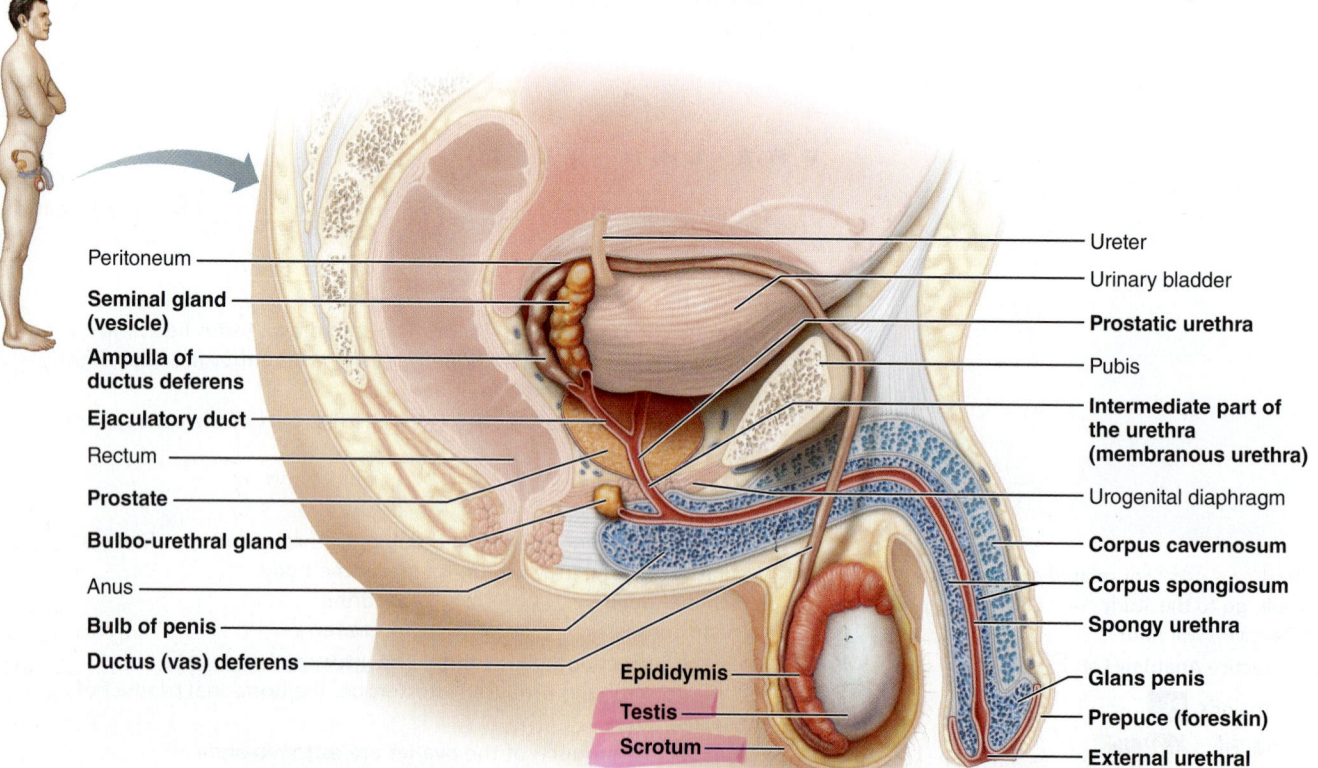

Peritoneum
Seminal gland (vesicle)
Ampulla of ductus deferens
Ejaculatory duct
Rectum
Prostate
Bulbo-urethral gland
Anus
Bulb of penis
Ductus (vas) deferens

Ureter
Urinary bladder
Prostatic urethra
Pubis
Intermediate part of the urethra (membranous urethra)
Urogenital diaphragm
Corpus cavernosum
Corpus spongiosum
Spongy urethra
Glans penis
Prepuce (foreskin)
External urethral orifice

Epididymis
Testis
Scrotum

(a)

42

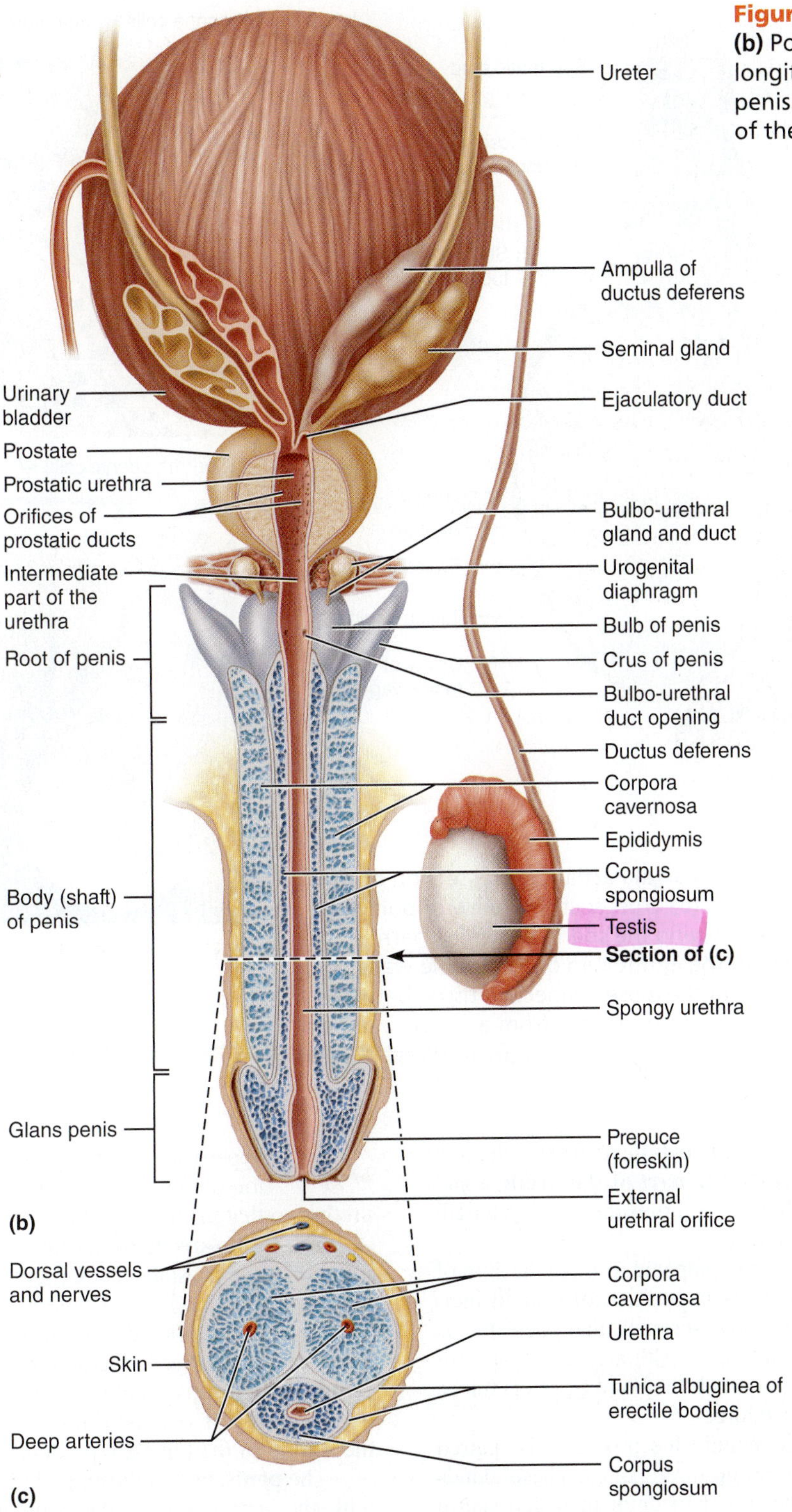

Figure 42.1 *(continued)*
(b) Posterior view showing a longitudinal section of the penis. **(c)** Transverse section of the penis.

Ureter

Ampulla of ductus deferens

Seminal gland

Ejaculatory duct

Urinary bladder

Prostate

Prostatic urethra

Orifices of prostatic ducts

Bulbo-urethral gland and duct

Intermediate part of the urethra

Urogenital diaphragm

Bulb of penis

Root of penis

Crus of penis

Bulbo-urethral duct opening

Ductus deferens

Corpora cavernosa

Epididymis

Corpus spongiosum

Body (shaft) of penis

Testis

Section of (c)

Spongy urethra

Glans penis

Prepuce (foreskin)

External urethral orifice

(b)

Dorsal vessels and nerves

Corpora cavernosa

Urethra

Skin

Tunica albuginea of erectile bodies

Deep arteries

Corpus spongiosum

(c)

The paired oval testes lie in the **scrotum** outside the abdominopelvic cavity. The temperature there (approximately 94°F, or 34°C) is slightly lower than body temperature, a requirement for producing viable sperm.

The accessory structures forming the *duct system* are the epididymis, the ductus deferens, the ejaculatory duct, and the urethra. The **epididymis** is an elongated structure running up the posterolateral aspect of the testis and capping its superior aspect. The epididymis forms the first portion of the duct system and provides a site for immature sperm entering it from the testis to complete their maturation process. The **ductus deferens**, or **vas deferens** (sperm duct), arches superiorly from the epididymis, passes through the inguinal canal into the pelvic cavity, and courses over the superior aspect of the urinary bladder. In life, the ductus deferens is enclosed along with blood vessels and nerves in a connective tissue sheath called the **spermatic cord** (Figure 42.2). The terminus of the ductus deferens enlarges to form the region called the **ampulla**, which empties into the **ejaculatory duct.** During **ejaculation,** contraction of the ejaculatory duct propels the

42

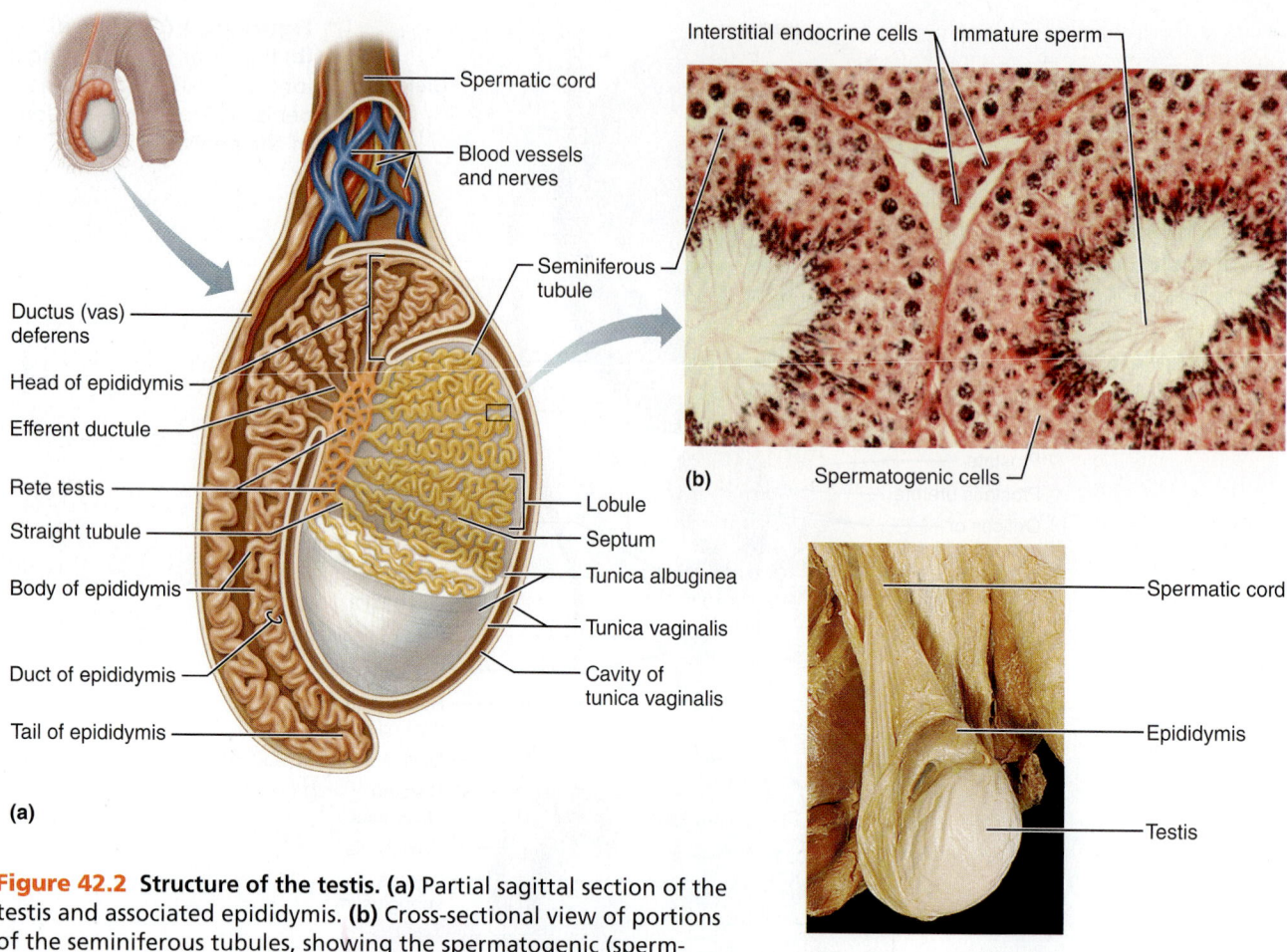

Spermatic cord

Blood vessels and nerves

Ductus (vas) deferens

Head of epididymis

Efferent ductule

Rete testis

Straight tubule

Body of epididymis

Duct of epididymis

Tail of epididymis

Seminiferous tubule

Lobule

Septum

Tunica albuginea

Tunica vaginalis

Cavity of tunica vaginalis

(a)

Interstitial endocrine cells — Immature sperm

(b)

Spermatogenic cells

Spermatic cord

Epididymis

Testis

(c)

Figure 42.2 Structure of the testis. (a) Partial sagittal section of the testis and associated epididymis. **(b)** Cross-sectional view of portions of the seminiferous tubules, showing the spermatogenic (sperm-forming) cells, which make up the epithelium of the tubule walls, and the interstitial endocrine cells in the loose connective tissue between the tubules (200×). **(c)** External view of a testis from a cadaver; same orientation as in (a).

sperm through the prostate to the **prostatic urethra,** which in turn empties into the **intermediate part of the urethra** and then into the **spongy urethra,** which runs through the length of the penis to the body exterior.

The spermatic cord is easily palpated through the skin of the scrotum. When a *vasectomy* is performed, a small incision is made in each side of the scrotum, and each ductus deferens is cut through or cauterized. Although sperm are still produced, they can no longer reach the body exterior; thus a man is sterile after this procedure.

The *accessory glands* include the prostate, the paired seminal glands, and the bulbo-urethral glands. These glands produce **seminal fluid,** the liquid medium in which sperm leave the body. The **seminal glands,** which produce about 70% of seminal fluid, lie at the posterior wall of the urinary bladder close to the terminus of the ductus deferens. They produce a viscous alkaline secretion containing fructose and other substances that nourish the sperm passing through the tract or that promote the fertilizing capability of sperm in some way. The duct of each seminal gland merges with a ductus deferens to form the ejaculatory duct; thus sperm and seminal fluid enter the urethra together.

The **prostate** encircles the urethra just inferior to the bladder. It secretes a slightly acidic, milky fluid into the urethra, which plays a role in activating the sperm.

Hypertrophy of the prostate, a troublesome condition commonly seen in elderly men, constricts the urethra so that urination is difficult.

The **bulbo-urethral glands** are tiny, pea-shaped glands inferior to the prostate. They produce a thick, clear, alkaline mucus that drains into the intermediate part of the urethra. This secretion lubricates the tip of the penis and neutralizes traces of acidic urine in the urethra just prior to ejaculation of **semen,** which consists of sperm plus seminal fluid. The relative alkalinity of seminal fluid also buffers the sperm against the acidity of the female reproductive tract.

The **penis,** part of the external genitalia of the male along with the scrotal sac, is the copulatory organ of the male. Designed to deliver sperm into the female reproductive tract, it consists of a body, or shaft, which terminates in an enlarged tip, the **glans penis** (Figure 42.1a and b). The skin covering the penis is loosely applied, and it reflects downward to form a circular fold of skin, the **prepuce,** or **foreskin,** around the proximal end of the glans. The foreskin may be removed in the surgical procedure called *circumcision.* Internally, the penis consists primarily of three elongated cylinders of erectile tissue, which engorge with blood during sexual excitement. This causes the penis to become rigid and enlarged so that it may more adequately serve as a penetrating device. This event is called **erection.** The paired dorsal cylinders are

42

the **corpora cavernosa.** The single ventral **corpus spongiosum** surrounds the spongy urethra (Figure 42.1c).

Microscopic Anatomy of Selected Male Reproductive Organs

Each **testis** is covered by a dense connective tissue capsule called the **tunica albuginea** (literally, "white tunic"). Extensions of this sheath enter the testis, dividing it into a number of lobes, each of which houses one to four highly coiled **seminiferous tubules,** the sperm-forming factories **(Figure 42.2)**. The seminiferous tubules of each lobe converge to empty the sperm into another set of tubules, the **rete testis,** at the posterior of the testis. Sperm traveling through the rete testis then enter the epididymis, located on the exterior aspect of the testis, as previously described. Lying between the seminiferous tubules and softly padded with connective tissue are the **interstitial endocrine cells,** which produce testosterone, the main hormonal product of the testis. Microscopic study of the testis is not included here (it is in Exercise 43) unless your instructor adds it.

ACTIVITY 2

Penis

Obtain a slide of a cross section of the penis. Scan the tissue under low power to identify the urethra and the cavernous bodies. Compare your observations to the penile cross-section images (Figure 42.1c and **Figure 42.3**). Observe the lumen of the urethra carefully. What type of epithelium do you see?

_____

Explain the function of this type of epithelium.

_____

_____

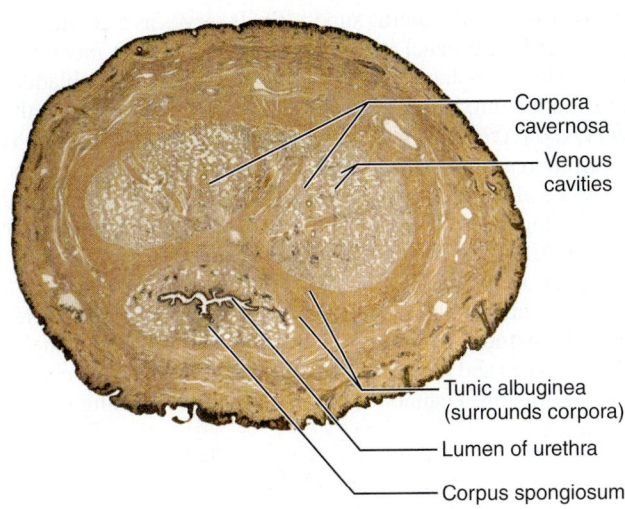

Corpora
cavernosa

Venous
cavities

Tunic albuginea
(surrounds corpora)

Lumen of urethra

Corpus spongiosum

Figure 42.3 Transverse section of the penis (3×).

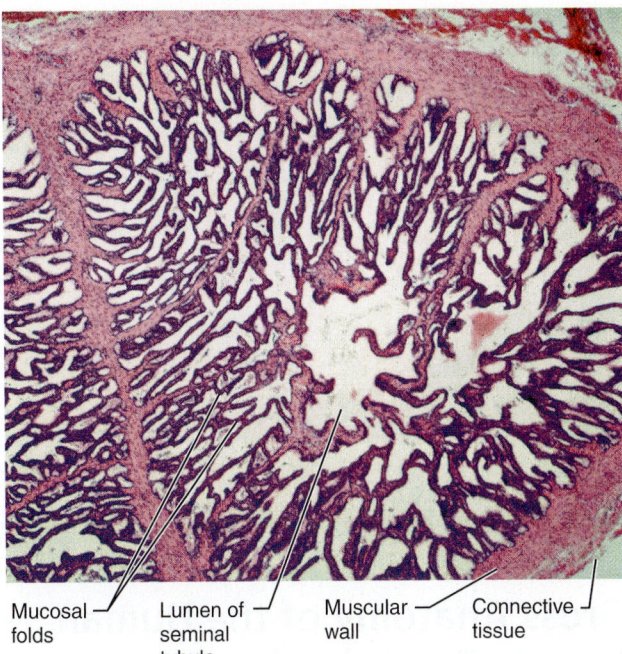

Mucosal folds — Lumen of seminal tubule — Muscular wall — Connective tissue

Figure 42.4 Cross-sectional view of a seminal gland with its elaborate network of mucosal folds. Glandular secretion is seen in the lumen (25×).

ACTIVITY 3

Seminal Gland

Obtain a slide showing a cross-sectional view of the seminal gland. Examine the slide at low magnification to get an overall view of the highly folded mucosa of this gland. Switch to higher magnification, and notice that the folds of the gland protrude into the lumen where they divide further, giving the lumen a honeycomb look **(Figure 42.4)**. Notice that the loose connective tissue lamina propria is underlain by smooth muscle fibers—first a circular layer, and then a longitudinal layer. Identify the glandular secretion in the lumen, a viscous substance that is rich in fructose and prostaglandins. ■

ACTIVITY 4

Epididymis

Obtain a slide of a cross section of the epididymis. Notice the abundant tubule cross sections resulting from the fact that the coiling epididymis tubule has been cut through many times in the specimen **(Figure 42.5)**. Look for sperm in the lumen of the tubule. Examine the composition of the tubule wall carefully. Identify the _stereocilia_ of the pseudostratified columnar epithelial lining. These nonmotile microvilli absorb excess fluid and pass nutrients to the sperm in the lumen. Now identify the smooth muscle layer. What do you think the function of the smooth muscle is?

_____

_____

42

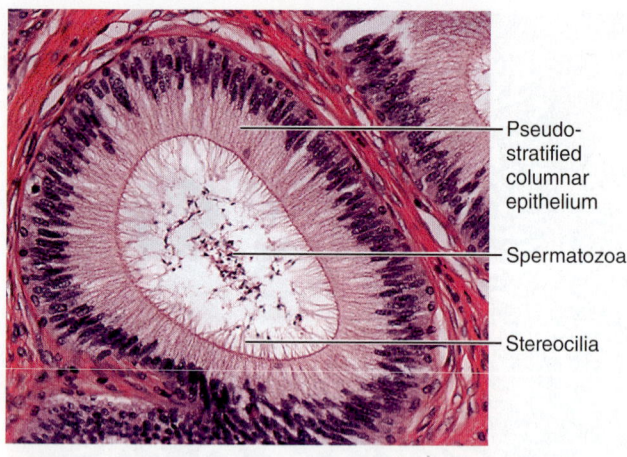

Figure 42.5 Cross section of epididymis (120×).

Gross Anatomy of the Human Female Reproductive System

The **ovaries** are the primary reproductive organs of the female. Like the testes of the male, the ovaries produce gametes (in this case eggs, or ova) and also sex hormones (estrogens and progesterone). The other accessory structures of the female reproductive system transport, house, nurture, or otherwise serve the needs of the reproductive cells and/or the developing fetus.

The reproductive structures of the female are generally considered in terms of internal organs and external organs, or external genitalia.

ACTIVITY 5

Identifying Female Reproductive Organs

As you read the descriptions of these structures, locate them on the illustrations of internal and external female structures **(Figure 42.6** and **Figure 42.7)** and then on the female reproductive system model or large laboratory chart. ◼

External Genitalia

The **external genitalia (vulva)** consist of the mons pubis, the labia majora and minora, the clitoris, the external urethral and vaginal orifices, the hymen, and the greater vestibular glands. The **mons pubis** is a rounded fatty eminence overlying the pubic symphysis. Running inferiorly and posteriorly from the mons pubis are two elongated, pigmented, hair-covered skin folds, the **labia majora,** which are homologous to the scrotum. These enclose two smaller hair-free folds, the **labia minora.** (Terms indicating only one of the two folds in each case are *labium majus* and *minus,* respectively.) The labia minora, in turn, enclose a region called the **vestibule,** which contains many structures—the clitoris, most anteriorly, followed by the external urethral orifice and the vaginal orifice. The diamond-shaped region between the anterior end of the labial folds, the ischial tuberosities laterally, and the anus posteriorly is called the **perineum.**

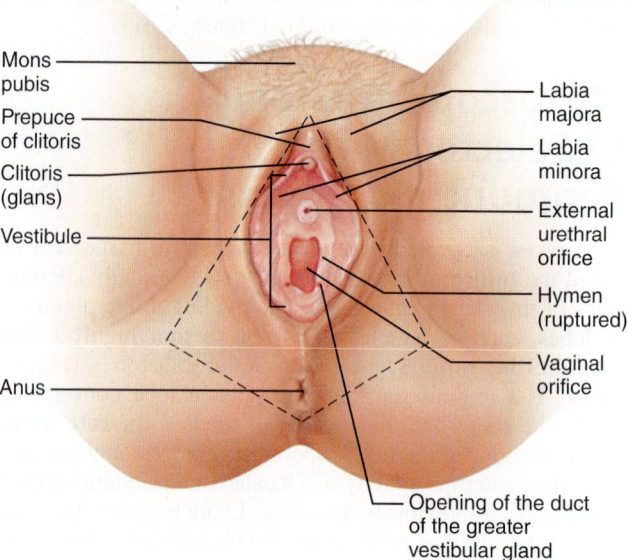

Figure 42.6 **External genitalia of the human female.** The region enclosed by dashed lines is the perineum.

The **clitoris** is a small protruding structure, homologous to the penis. Like its counterpart, it is composed of highly sensitive, erectile tissue. It is hooded by skin folds of the anterior labia minora, referred to as the **prepuce of the clitoris.** The external urethral orifice, which lies posterior to the clitoris, is the outlet for the urinary system and has no reproductive function in the female. The vaginal opening is partially closed by a thin fold of mucous membrane called the **hymen** and is flanked by the pea-sized, mucus-secreting **greater vestibular glands.** These glands (Figures 42.6 and 42.7a) lubricate the distal end of the vagina during coitus and are homologous to the bulbo-urethral glands of males.

Internal Organs

The internal female organs include the vagina, uterus, uterine tubes, ovaries, and the ligaments and supporting structures that suspend these organs in the pelvic cavity (Figure 42.7). The **vagina** extends for approximately 10 cm (4 inches) from the vestibule to the uterus superiorly. It serves as a copulatory organ and birth canal and permits passage of the menstrual flow. The pear-shaped **uterus,** situated between the bladder and the rectum, is a muscular organ with its narrow end, the **cervix,** directed inferiorly. The major portion of the uterus is referred to as the **body;** its superior rounded region above the entrance of the uterine tubes is called the **fundus.** A fertilized egg is implanted in the uterus, which houses the embryo or fetus during its development.

In some cases, the fertilized egg may implant in a uterine tube or even on the abdominal viscera, creating an **ectopic pregnancy.** Such implantations are usually unsuccessful and may even endanger the mother's life because the uterine tubes cannot accommodate the increasing size of the fetus. ✚

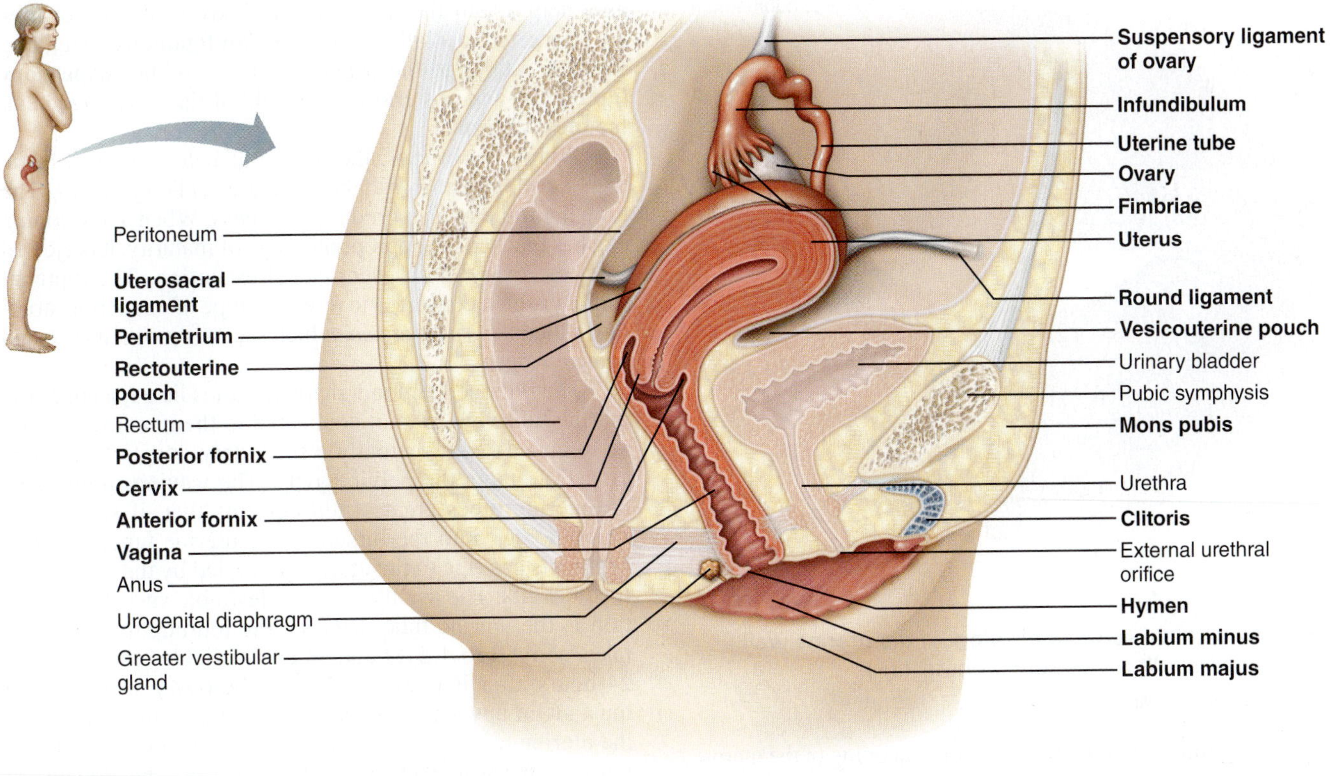

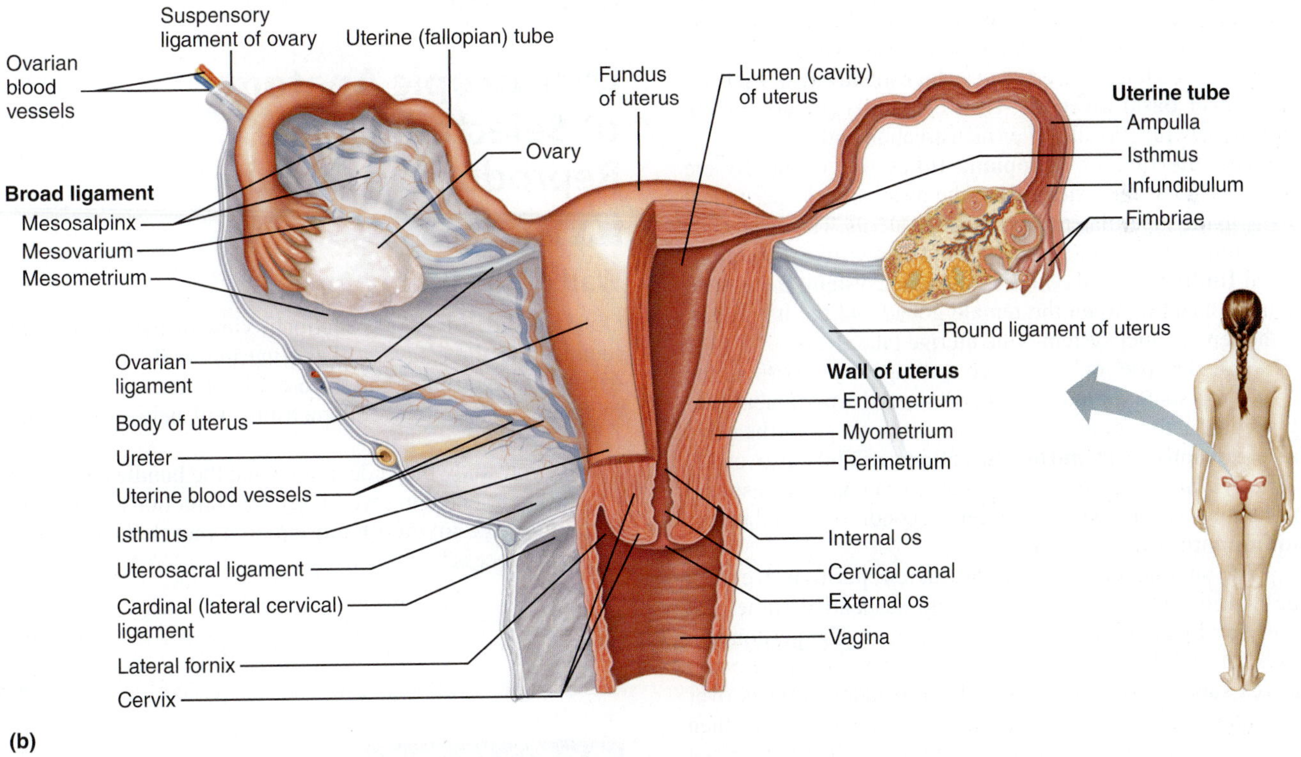

(a)

(b)

Figure 42.7 **Internal reproductive organs of the human female. (a)** Midsagittal section of the human female reproductive system. **(b)** Posterior view. The posterior walls of the vagina, uterus, and uterine tubes, and the broad ligament have been removed on the right side to reveal the shape of the lumen of these organs.

42

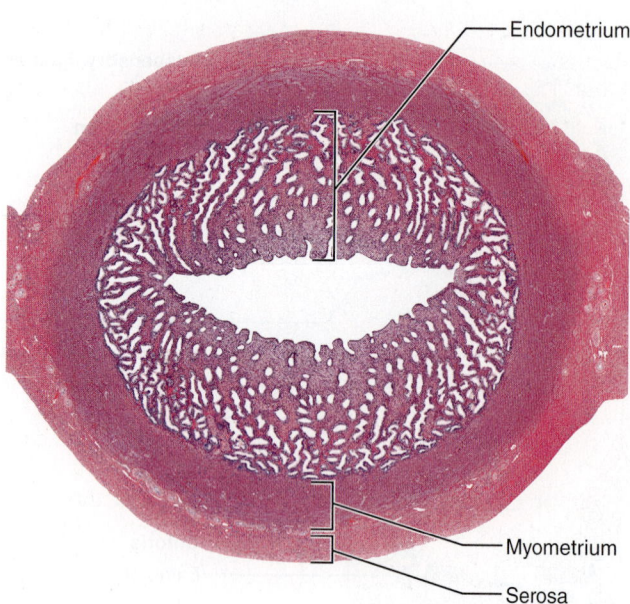

Endometrium

Myometrium

Serosa

Figure 42.8 Cross-sectional view of the uterine wall. The mucosa is in the secretory phase. (3×).

The **endometrium,** the thick mucosal lining of the uterus, has a superficial **functional layer,** or **stratum functionalis,** that sloughs off periodically (about every 28 days) in response to cyclic changes in the levels of ovarian hormones in the woman's blood. This sloughing-off process, which is accompanied by bleeding, is referred to as **menstruation,** or **menses.** The deeper **basal layer,** or **stratum basalis** (Figure 43.6b), forms a new functionalis after menstruation ends.

The **uterine,** or **fallopian, tubes** are about 10 cm (4 inches) long and extend from the ovaries in the peritoneal cavity to the superolateral region of the uterus. The distal ends of the tubes are funnel-shaped and have fingerlike projections called **fimbriae.** Unlike in the male duct system, there is no actual contact between the female gonad and the initial part of the female duct system—the uterine tube.

Because of this open passageway between the female reproductive organs and the peritoneal cavity, reproductive system infections, such as gonorrhea and other **sexually transmitted infections (STIs),** also called *sexually transmitted diseases (STDs),* can cause widespread inflammations of the pelvic viscera, a condition called **pelvic inflammatory disease (PID).** ✚

The internal female organs are all retroperitoneal, except the ovaries. They are supported and suspended somewhat freely by ligamentous folds of peritoneum. The peritoneum takes an undulating course. From the pelvic cavity floor it moves superiorly over the top of the bladder, reflects over the anterior and posterior surfaces of the uterus, and then over the rectum, and up the posterior body wall. The fold that encloses the uterine tubes and uterus and secures them to the lateral body walls is the **broad ligament** (Figure 42.7b). The part of the broad ligament specifically anchoring the uterus is called the **mesometrium** and that anchoring the uterine tubes, the **mesosalpinx.** The **round ligaments,** fibrous cords that run from the uterus to the labia majora, and the **uterosacral ligaments,** which course posteriorly to the sacrum,

also help attach the uterus to the body wall. The ovaries are supported medially by the **ovarian ligament** (extending from the uterus to the ovary), laterally by the **suspensory ligaments,** and posteriorly by a fold of the broad ligament, the **mesovarium.**

Within the ovaries, the female gametes, or eggs, begin their development in saclike structures called *follicles.* The growing follicles also produce *estrogens.* When a developing egg has reached the appropriate stage of maturity, it is ejected from the ovary in an event called **ovulation.** The ruptured follicle is then converted to a second type of endocrine structure called a *corpus luteum,* which secretes progesterone and some estrogens.

The flattened almond-shaped ovaries lie adjacent to the uterine tubes but are not connected to them; consequently, an ovulated "egg," actually a secondary oocyte (see Exercise 43), enters the pelvic cavity. The waving fimbriae of the uterine tubes create fluid currents that, if successful, draw the egg into the lumen of the uterine tube, where it begins its passage to the uterus, propelled by the cilia of the tubule walls. The usual and most desirable site of fertilization is the uterine tube, because the journey to the uterus takes about 3 to 4 days and an egg is viable only for up to 24 hours after it is expelled from the ovary. Thus, sperm must swim upward through the vagina and uterus and into the uterine tubes to reach the egg. This must be an arduous journey, because they must swim against the downward current created by ciliary action—rather like swimming against the tide!

Microscopic Anatomy of Selected Female Reproductive Organs

ACTIVITY 6

Wall of the Uterus

Obtain a slide of a cross-sectional view of the uterine wall. Identify the three layers of the uterine wall—the endometrium, myometrium, and serosa. A photomicrograph that includes the secretory endometrium will help in this study **(Figure 42.8).**

As you study the slide, notice that the bundles of smooth muscle are oriented in several different directions. What is the function of the **myometrium** (smooth muscle layer) during the birth process?

_____

_____ ■

ACTIVITY 7

Uterine Tube

Obtain a slide of a cross-sectional view of a uterine tube for examination. Notice that the mucosal folds nearly fill the tubule lumen **(Figure 42.9).** Then switch to high power to examine the ciliated secretory epithelium. ■

The Mammary Glands

The **mammary glands** exist within the breasts in both sexes, but they normally have a reproduction-related function only in females. Since the function of the mammary glands is to produce milk to nourish the newborn infant, their importance is more closely associated with events that occur when reproduction has already been accomplished. Periodic stimulation by the female sex hormones, especially estrogens, increases the size of the female mammary glands at puberty. During this period, the duct system becomes more elaborate, and fat is deposited—fat deposition is the more important contributor to increased breast size.

The rounded, skin-covered mammary glands lie anterior to the pectoral muscles of the thorax, attached to them by connective tissue. Slightly below the center of each breast is a pigmented area, the **areola,** which surrounds a centrally protruding **nipple (Figure 42.10).**

Internally each mammary gland consists of 15 to 25 **lobes** which radiate around the nipple and are separated by fibrous connective tisse and adipose, or fatty, tissue. Within each lobe are smaller chambers called **lobules,** containing the glandular **alveoli** that produce milk during lactation. The alveoli of each lobule pass the milk into a number of **lactiferous ducts,** which join to form an expanded storage chamber, the **lactiferous sinus,** as they approach the nipple. The sinuses open to the outside at the nipple.

For instructions on animal dissections, see the dissection exercises (starting on page 697) in the cat and fetal pig editions of this manual.

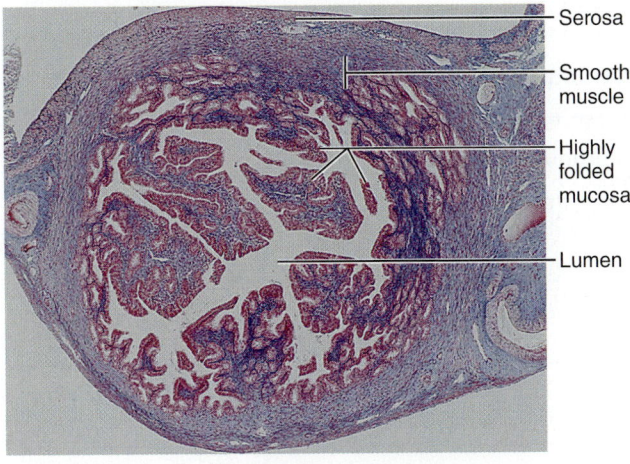

Figure 42.9 Cross-sectional view of the uterine tube (12×).

Serosa

Smooth muscle

Highly folded mucosa

Lumen

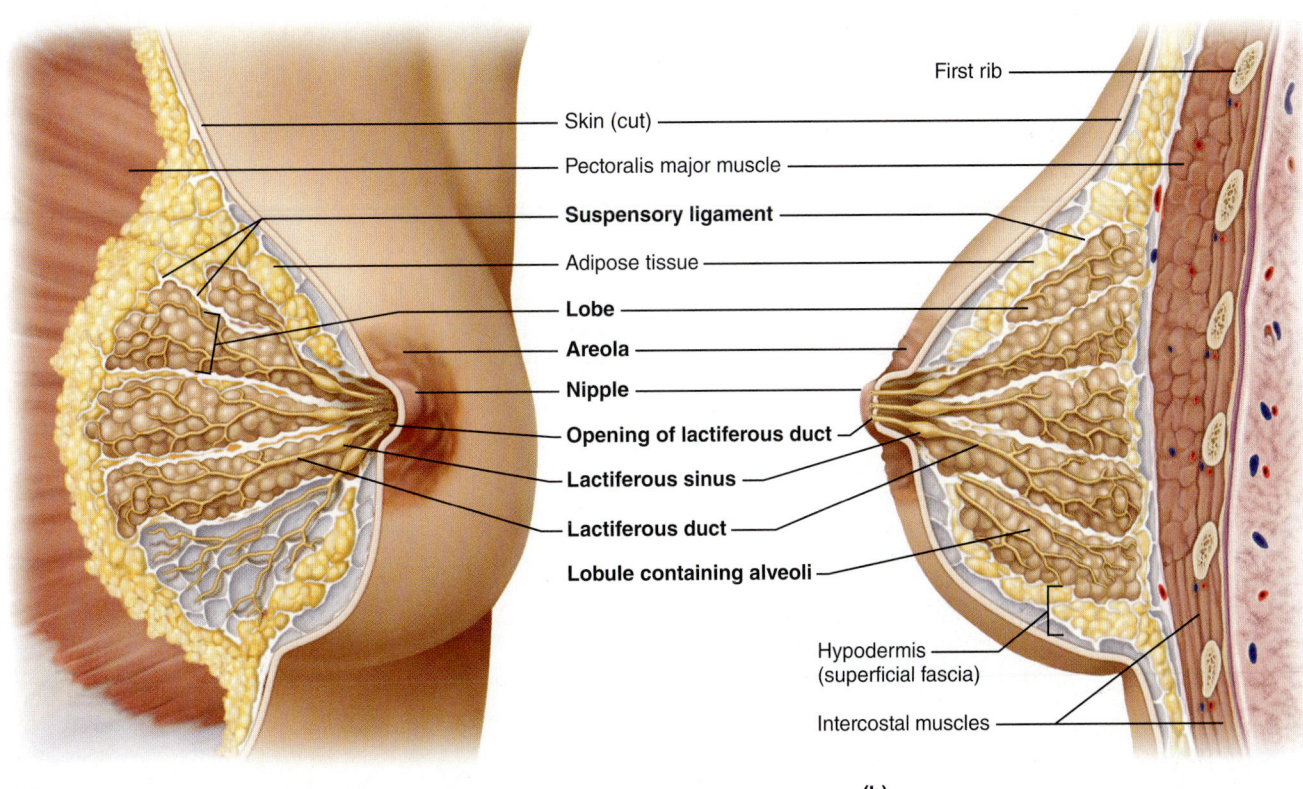

First rib

Skin (cut)

Pectoralis major muscle

Suspensory ligament

Adipose tissue

Lobe

Areola

Nipple

Opening of lactiferous duct

Lactiferous sinus

Lactiferous duct

Lobule containing alveoli

Hypodermis (superficial fascia)

Intercostal muscles

(a)

(b)

Figure 42.10 Anatomy of lactating mammary gland. (a) Anterior view of partially dissected breast. **(b)** Sagittal section of the breast.

42

Reproductive Homologues

The reproductive system has many homologues, organs that are structurally similar in the male and female because they are derived from the same embryonic tissue. Some homologous structures also provide a similar function. Think of as many reproductive homologues as you can and list them in the chart below. Also, describe their similar functions. The first one is done for you.

Group Challenge: Reproductive Homologues

Male homologue	Female homologue	Similar function
testis	ovary	Both produce gametes.

Name _____

Lab Time/Date _____

Anatomy of the Reproductive System

Gross Anatomy of the Human Male Reproductive System

1. List the two principal functions of the testis. _____

   and _____

2. Identify all indicated structures or portions of structures on the diagrammatic view of the male reproductive system below.

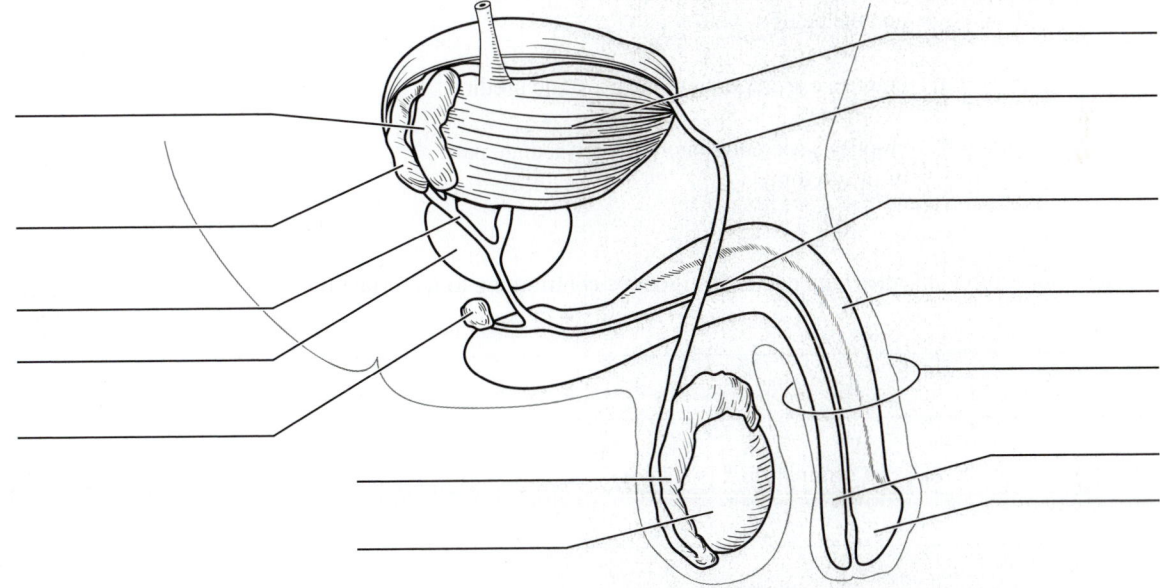

3. A common part of any physical examination of the male is palpation of the prostate. How is this accomplished?

   (Think!) _____

4. How might enlargement of the prostate interfere with urination or the reproductive ability of the male?

   _____

   _____

5. Why are the testes located in the scrotum rather than inside the ventral body cavity? _____

   _____

6. Match the terms in column B to the descriptive statements in column A.

Column A

      _____ 1. copulatory organ/penetrating device

      _____ 2. muscular passageway conveying sperm to the ejaculatory duct; in the spermatic cord

      _____ 3. transports both sperm and urine

      _____ 4. sperm maturation site

      _____ 5. location of the testis in adult males

      _____ 6. loose fold of skin encircling the glans penis

      _____ 7. portion of the urethra between the prostate and the penis

      _____ 8. empties a secretion into the prostatic urethra

      _____ 9. empties a secretion into the intermediate part of the urethra

Column B

a. bulbo-urethral glands

b. ductus (vas) deferens

c. epididymis

d. intermediate part of the urethra

e. penis

f. prepuce

g. prostate

h. prostatic urethra

i. scrotum

j. seminal gland

k. spongy urethra

7. Describe the composition of semen, and name all structures contributing to its formation. _____

_____

_____

8. Of what importance is the fact that seminal fluid is alkaline? _____

_____

9. What structures compose the spermatic cord? _____

_____

Where is it located? _____

10. Using the following terms, trace the pathway of sperm from the testes to the urethra: rete testis, epididymis, seminiferous tubule, ductus deferens.

_____ → _____ → _____ → _____

Gross Anatomy of the Human Female Reproductive System

11. Name the structures composing the external genitalia, or vulva, of the female. _____

_____

_____ _____

12. On the diagram below of a frontal section of a portion of the female reproductive system, identify all indicated structures.

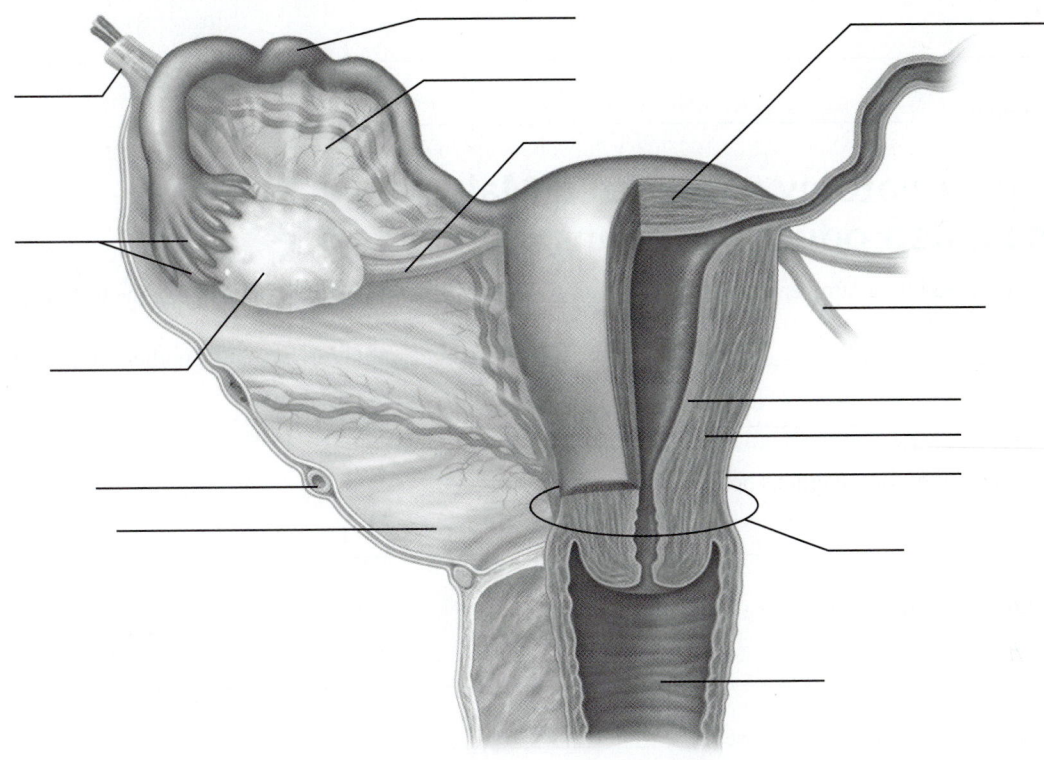

13. Identify the female reproductive system structures described below.

_____   1.   site of fetal development

_____   2.   copulatory canal

_____   3.   egg typically fertilized here

_____   4.   becomes erect during sexual excitement

_____   5.   duct extending from ovaries to the uterus

_____   6.   partially closes the vaginal canal; a membrane

_____   7.   produces oocytes, estrogens, and progesterone

_____   8.   fingerlike ends of the uterine tube

14. Do any sperm enter the pelvic cavity of the female? Why or why not? _____

_____

15. What is an ectopic pregnancy, and how can it happen? _____

_____

_____

16. Put the following vestibular-perineal structures in their proper order from the anterior to the posterior aspect: vaginal orifice, anus, external urethral opening, and clitoris.

Anterior limit: _____ → _____ → _____ → _____

17. Assume a couple has just consummated the sex act and the sperm have been deposited in the vagina. Trace the pathway of the sperm through the female reproductive tract.

_____

18. Define *ovulation.* _____

Microscopic Anatomy of Selected Male and Female Reproductive Organs

19. The testis is divided into a number of lobes by connective tissue. Each of these lobes contains one to four _____

_____, which converge to empty sperm into another set of tubules called the

_____.

20. What is the function of the cavernous bodies seen in the penis? _____

_____

21. Name the three layers of the uterine wall from the inside out.

_____, _____, _____

Which of these is sloughed during menses? _____

Which contracts during childbirth? _____

22. Describe the epithelium found in the uterine tube. _____

_____

23. Describe the arrangement of the layers of smooth muscle in the seminal gland. _____

_____

24. What is the function of the stereocilia exhibited by the epithelial cells of the mucosa of the epididymis? _____

_____

25. On the diagram showing the sagittal section of the human testis, correctly identify all structures provided with leader lines.

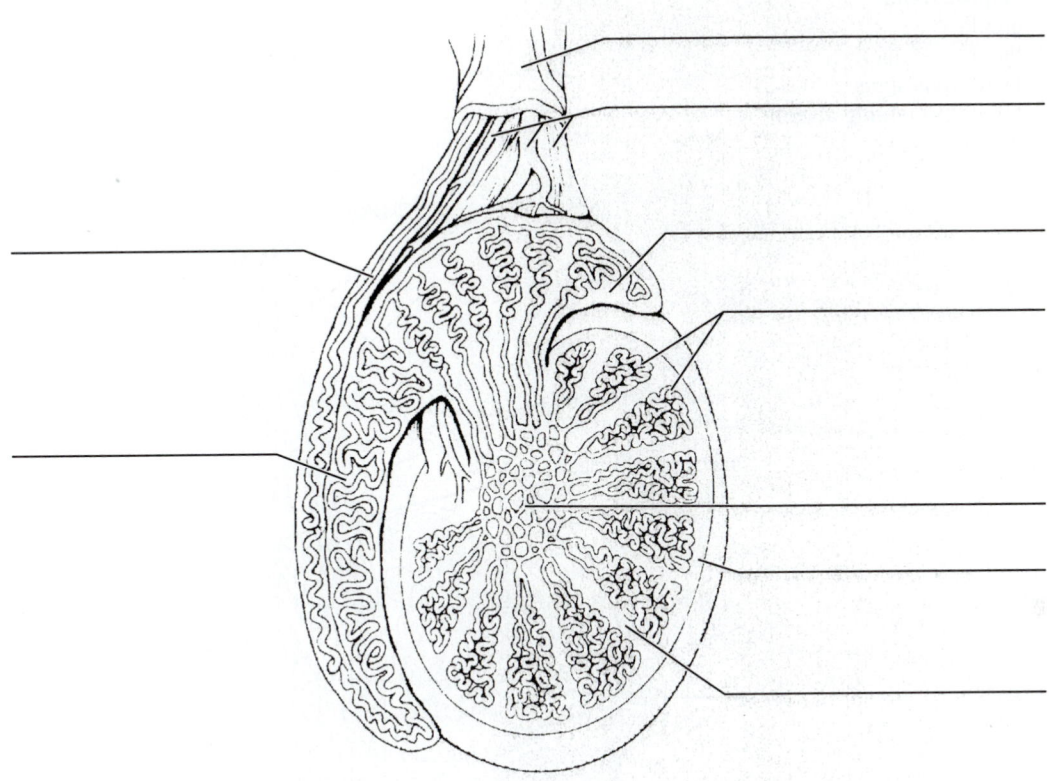

The Mammary Glands

26. Match the key term with the correct description.

_____ glands that produce milk during lactation

_____ subdivision of mammary lobes that contains alveoli

_____ enlarged storage chamber for milk

_____ duct connecting alveoli to the storage chambers

_____ pigmented area surrounding the nipple

_____ releases milk to the outside

Key:

a.   alveoli

b.   areola

c.   lactiferous duct

d.   lactiferous sinus

e.   lobule

f.   nipple

27. Using the key terms, correctly identify breast structures.

Key: a. adipose tissue
b. areola
c. lactiferous duct
d. lactiferous sinus
e. lobule containing alveoli
f. nipple

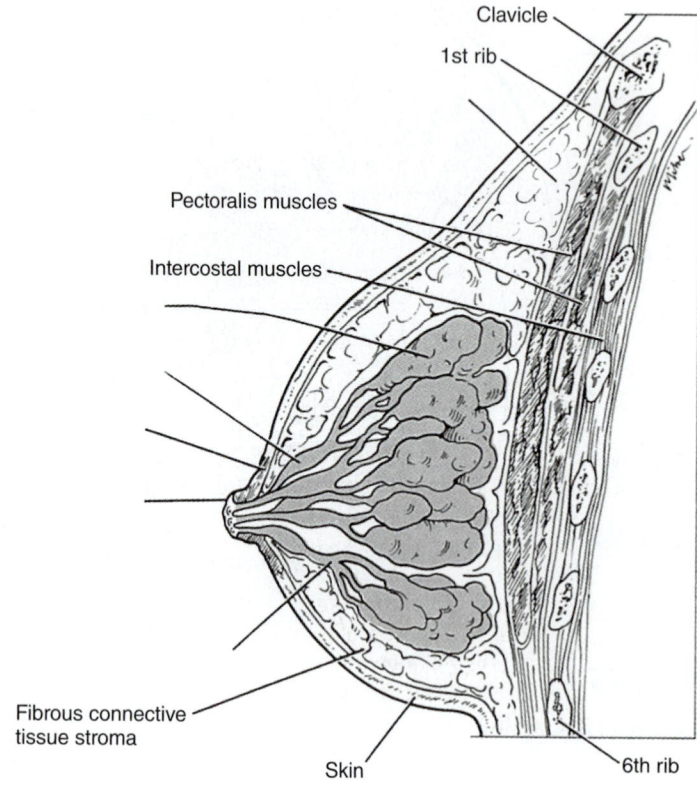

28. Describe the procedure for self-examination of the breasts. (Men are not exempt from breast cancer, you know!)

_____

_____

_____

_____

Physiology of Reproduction: Gametogenesis and the Female Cycles

MATERIALS

- ☐ Three-dimensional models illustrating meiosis, spermatogenesis, and oogenesis
- ☐ Sets of "pop it" beads in two colors with magnetic centromeres, available in Chromosome Simulation Lab Activity from Ward's Natural Science
- ☐ Compound microscope
- ☐ Prepared slides of testis and human sperm
- ☐ *Demonstration:* microscopes set up to demonstrate the following stages of oogenesis in *Ascaris megalocephala:*
  Slide 1: Primary oocyte with fertilization membrane, sperm nucleus, and aligned tetrads apparent
  Slide 2: Formation of the first polar body
  Slide 3: Secondary oocyte with dyads aligned
  Slide 4: Formation of the ovum and second polar body
  Slide 5: Fusion of the male and female pronuclei to form the fertilized egg
- ☐ Prepared slides of ovary and uterine endometrium (showing menstrual, proliferative, and secretory phases)

MasteringA&P® For related exercise study tools, go to the Study Area of MasteringA&P. There you will find:
- Practice Anatomy Lab **PAL**
- PhysioEx **PEx**
- A&PFlix **A&PFlix**
- Practice quizzes, Histology Atlas, eText, Videos, and more!

OBJECTIVES

1. Define *meiosis, gametogenesis, oogenesis, spermatogenesis, synapsis, haploid, zygote,* and *diploid.*
2. Cite similarities and differences between mitosis and meiosis.
3. Describe the stages of spermatogenesis and relate each to the cross-sectional structure of the seminiferous tubule.
4. Define *spermiogenesis* and relate the anatomy of sperm to their function.
5. Describe the effects of FSH and LH on testicular function.
6. Discuss the microscopic structure of the ovary and identify primary, secondary, and vesicular follicles, and the corpus luteum; list the hormones produced by the follicles and the corpus luteum.
7. Relate the stages of oogenesis to follicle development in the ovary.
8. Compare and contrast spermatogenesis and oogenesis.
9. Discuss the effect of FSH and LH on the ovary, and describe the feedback relationship between anterior pituitary gonadotropins and ovarian hormones.
10. List the phases of the menstrual cycle and discuss the hormonal control of each.

PRE-LAB QUIZ

1. Human gametes contain _____ chromosomes.
   a. 13   b. 23   c. 36   d. 46
2. The end product of meiosis is
   a. two diploid daughter cells   c. four diploid daughter cells
   b. two haploid daughter cells   d. four haploid daughter cells
3. Circle the correct underlined terms. A grouping of four chromatids, known as a <u>dyad</u> / <u>tetrad</u>, occurs only during <u>mitosis</u> / <u>meiosis</u>.
4. _____ extend inward from the periphery of the seminiferous tubule and provide nourishment to the spermatids as they begin their transformation into sperm.
   a. Interstitial endocrine cells   c. Sustentocytes
   b. Granulosa cells   d. Follicle cells
5. Circle the correct underlined term. The <u>acrosome</u> / <u>midpiece</u> of the sperm contains enzymes involved in the penetration of the egg.
6. Circle the correct underlined term. Within each ovary, the immature ovum develops in a saclike structure called a <u>corpus</u> / <u>follicle</u>.
7. As the primordial follicle grows and its epithelium changes from squamous to cuboidal cells, it becomes a(n) _____ and begins to produce estrogens.
   a. oocyte   c. primary follicle
   b. oogonia   d. primary ovum

(Text continues on next page.)

8. Circle True or False. A sudden release of luteinizing hormone by the anterior pituitary triggers ovulation.

9. Circle the correct underlined term. The <u>corpus luteum</u> / <u>corpus albicans</u> is a solid glandular structure with a scalloped lumen that develops from a ruptured follicle.

10. The _____ phase of the female cycle occurs from days 1–5 and is signaled by the sloughing off of the thick functional layer of the endometrium.
   a. endometrial
   b. menstrual
   c. proliferative
   d. secretory

H uman beings develop from the union of egg and sperm. Each of these gametes is a unique cell produced either in the ovary or testis. Unlike all other body cells, gametes have only half the normal chromosome number and they are produced by a special type of nuclear division called meiosis.

Meiosis

The normal number of chromosomes in most human body cells is 46, the **diploid** or **2n** chromosomal number. This number is made up of two sets of similar chromosomes, one set of 23 from each parent. Thus each body cell contains 23 pairs of similar chromosomes called **homologous chromosomes** or homologues **(Figure 43.1)**. Each member of a homologous pair contains genes that code for the same traits.

Gametes contain only one member of each homologous pair of chromosomes. Therefore each human gamete contains a total of 23 chromosomes, the **haploid** or **n** chromosomal number. When egg and sperm fuse they form a **zygote** that restores the diploid number of chromosomes. The zygote divides by the process of **mitosis** to produce the multicellular human body. Recall that mitosis is the process by which most body cells divide and it produces two diploid daughter cells each containing 46 chromosomes that are identical to those of the mother cell (see Exercise 4).

Gametogenesis is the process of gamete formation. It involves nuclear division by **meiosis,** which reduces the number of chromosomes by half. Before meiosis begins, the chromosomes in the *mother cells* or stem cells are replicated just as they are before mitosis. The identical copies remain together as *sister chromatids*. They are held together by a centromere, forming a structure called a dyad (Figure 43.1).

Two nuclear divisions, called meiosis I and meiosis II, occur during meiosis. Each has the same phases as mitosis—prophase, metaphase, anaphase, and telophase. Meiosis I is the *reduction division*. During prophase of meiosis I, the homologous chromosomes pair up in a process called **synapsis** that forms little groups of four chromatids, called **tetrads** (Figure 43.1). During synapsis, the free ends of adjacent maternal and paternal chromatids wrap around each other at one or more points, forming **crossovers** or **chiasmata.** Crossovers allow maternal and paternal chromosomes to exchange genetic material. The tetrads align randomly at the metaphase plate so that either the maternal or paternal chromosome may be on a given side of the plate. Then the two homologous chromosomes, each still composed of two sister chromatids, are pulled to opposite ends of the cell. At the end of meiosis I, each haploid daughter cell contains one member of each original homologous pair.

Meiosis II begins immediately without replication of the chromosomes. The dyads align on the metaphase plate and the two sister chromatids are pulled apart, each now becoming a full chromosome. The net result of meiosis is four haploid daughter cells, each containing 23 chromosomes. The events of crossover and the random alignment of tetrads during meiosis I introduce great genetic variability. As a result, it is unlikely that any gamete is exactly like another.

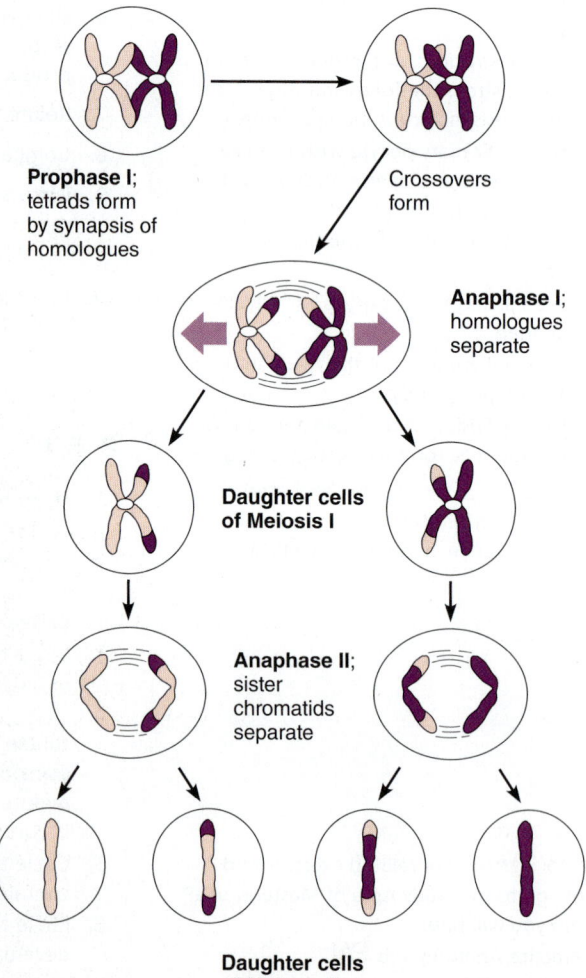

Prophase I; tetrads form by synapsis of homologues

Crossovers form

Anaphase I; homologues separate

Daughter cells of Meiosis I

Anaphase II; sister chromatids separate

Daughter cells of Meiosis II

Figure 43.1 Events of meiosis involving one pair of homologous chromosomes. Male homologue is purple; female homologue is pink.

43

Identifying Meiotic Phases and Structures

1. Obtain a model depicting the events of meiosis, and follow the sequence of events during meiosis I and II. Identify prophase, metaphase, anaphase, and telophase in each. Also identify tetrads and chiasmata during meiosis I and dyads during meiosis II. Note ways in which the daughter cells resulting from meiosis I differ from the mother cell and how the gametes differ from both cell populations. Use the key on the model, your textbook, or an appropriate reference as necessary to aid you in these observations.

2. Using strings of colored "pop it" beads with magnetic centromeres, demonstrate the phases of meiosis, including crossing over, for a cell with a diploid (2n) number of 4. Use one bead color for the male chromosomes and another color for the female chromosomes.

3. Ask your instructor to verify the accuracy of your "creation" before returning the beads to the supply area. ■■

Spermatogenesis

Human sperm production, or **spermatogenesis,** begins at puberty and continues without interruption throughout life. The average male ejaculation contains about a quarter billion sperm. Because only one sperm fertilizes an ovum, it seems that nature has tried to ensure that the perpetuation of the species will not be endangered for lack of sperm.

Spermatogenesis, the process of gametogenesis in males, occurs in the seminiferous tubules of the testes (**Figure 43.2** and Exercise 42). The primitive stem cells or **spermatogonia,** found at the tubule periphery, divide extensively to build up the stem cell line. Before puberty, all divisions are mitotic divisions that produce more spermatogonia. At puberty, however, under the influence of follicle-stimulating hormone (FSH) secreted by the anterior pituitary gland, each mitotic division of a spermatogonium produces one spermatogonium and one **primary spermatocyte,** which is destined to undergo meiosis. As meiosis occurs, the dividing cells approach the lumen of the tubule. Thus the progression of meiotic events can be followed from the tubule periphery to the lumen. It is important to recognize that **spermatids,** haploid cells that are the actual product of meiosis, are not functional gametes. They are nonmotile cells and have too much excess baggage to function well in a reproductive capacity. A subsequent process, called **spermiogenesis,** strips away the extraneous cytoplasm from the spermatid, converting it to a motile, streamlined **sperm.**

Examining Events of Spermatogenesis

1. Obtain a slide of the testis and a microscope. Examine the slide under low power to identify the cross-sectional views of the cut seminiferous tubules. Then rotate the high-power lens into position and observe the wall of one of the cut tubules (Figure 43.2).

2. Scrutinize the cells at the periphery of the tubule. The cells in this area are the spermatogonia. About half of these will form primary spermatocytes, which begin meiosis. These are recognizable by their pale-staining nuclei with centrally

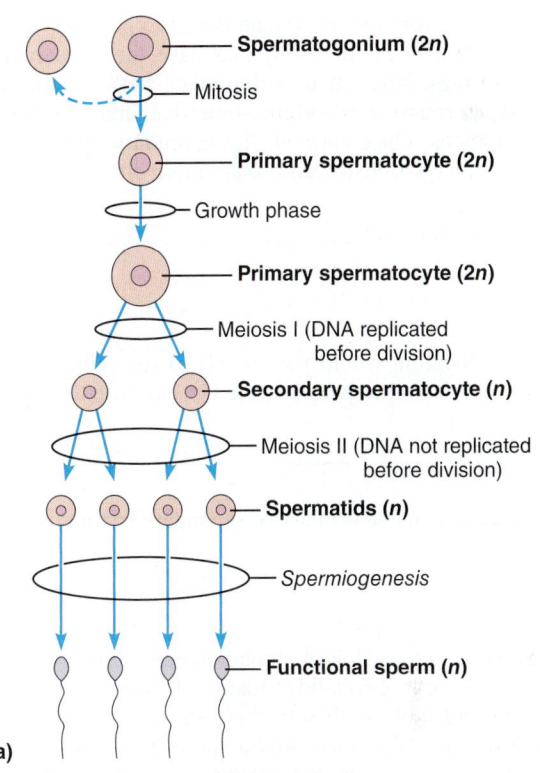

Spermatogonium (2n)

Mitosis

Primary spermatocyte (2n)

Growth phase

Primary spermatocyte (2n)

Meiosis I (DNA replicated before division)

Secondary spermatocyte (n)

Meiosis II (DNA not replicated before division)

Spermatids (n)

Spermiogenesis

Functional sperm (n)

(a)

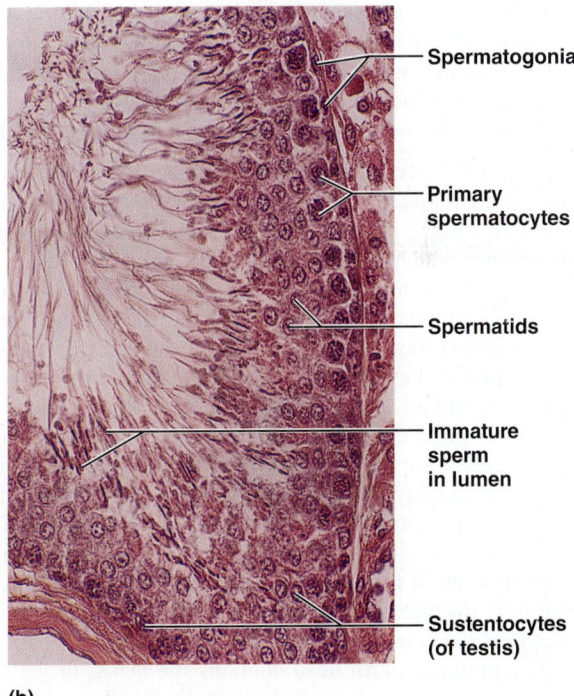

Spermatogonia

Primary spermatocytes

Spermatids

Immature sperm in lumen

Sustentocytes (of testis)

(b)

Figure 43.2 Spermatogenesis. (a) Flowchart of meiotic events and spermiogenesis. **(b)** Micrograph of an active seminiferous tubule (275×).

located nucleoli. The remaining daughter cells resulting from mitotic divisions of spermatogonia stay at the tubule periphery to maintain the germ cell line.

3. Observe the cells in the middle of the tubule wall. There you should see a large number of spermatocytes that are obviously undergoing a nuclear division process. Look for coarse

clumps of chromatin or threadlike chromosomes (visible only during nuclear division) that have the appearance of coiled springs. Attempt to differentiate between the larger primary spermatocytes and the somewhat smaller secondary spermatocytes. Once formed, the secondary spermatocytes quickly undergo division and so are more difficult to find.

Can you see tetrads? _____

Is there evidence of crossover? _____

In which location would you expect to see cells containing tetrads, closer to the spermatogonia or closer to the lumen?

_____

Would these cells be primary or secondary spermatocytes?

_____

4. Examine the cells at the tubule lumen. Identify the small round-nucleated spermatids, many of which may appear lopsided and look as though they are starting to lose their cytoplasm. See if you can find a spermatid embedded in an elongated cell type—a **sustentocyte,** or *Sertoli cell*—which extends inward from the periphery of the tubule. The sustentocytes nourish the spermatids as they begin their transformation into sperm. Also in the adluminal area (area toward the lumen), locate immature sperm, which can be identified by their tails. The sperm develop directly from the spermatids by the loss of extraneous cytoplasm and the development of a propulsive tail.

5. Identify the **interstitial endocrine cells,** also called *Leydig cells,* lying external to and between the seminiferous tubules. LH (luteinizing hormone) prompts these cells to produce testosterone, which acts synergistically with FSH (follicle-stimulating hormone) to stimulate sperm production. Both LH and FSH are named for their effects on the female gonad.

In the next stage of sperm development, spermiogenesis, all the superficial cytoplasm is sloughed off, and the remaining cell organelles are compacted into the three regions of the mature sperm. At the risk of oversimplifying, these anatomical regions are the *head*, the *midpiece*, and the *tail*, which correspond roughly to the activating and genetic region, the metabolic region, and the locomotor region, respectively. The mature sperm is a streamlined cell equipped with an organ of locomotion and a high rate of metabolism that enable it to move long distances quickly to get to the egg. It is a prime example of the correlation of form and function.

The pointed sperm head contains the DNA, or genetic material, of the chromosomes. Essentially it is the nucleus of the spermatid. Anterior to the nucleus is the **acrosome,** which contains enzymes necessary for penetration of the egg.

In the midpiece of the sperm is a centriole which gives rise to the filaments that structure the sperm tail. Wrapped tightly around the centriole are mitochondria that provide the ATP needed for contractile activity of the tail.

The tail is a typical flagellum produced by a centriole. When powered by ATP, the tail propels the sperm.

6. Obtain a prepared slide of human sperm, and view it with the oil immersion lens. Identify the head, acrosome, and tail regions of the sperm **(Figure 43.3)**. Deformed sperm, for

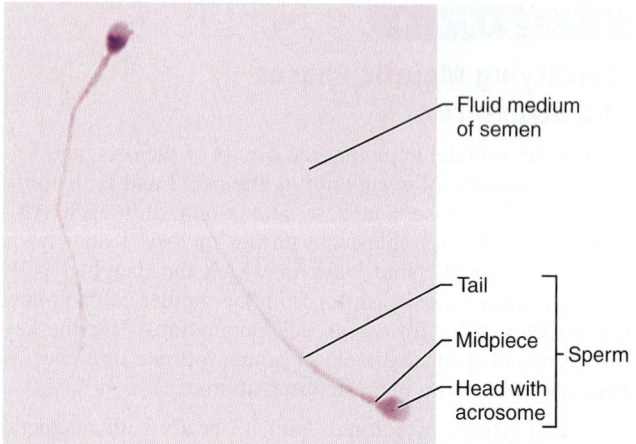

Figure 43.3 **Sperm in semen.** (1000×).

example sperm with multiple heads or tails, are sometimes present in such preparations. Did you observe any?

_____ If so, describe them. _____

_____

7. Examine the model of spermatogenesis to identify the spermatogonia, the primary and secondary spermatocytes, the spermatids, and the functional sperm. ▪

Demonstration of Oogenesis in *Ascaris* (Optional)

Oogenesis (the process of producing an egg) in mammals is difficult to demonstrate. However, the process may be studied rather easily in the transparent eggs of *Ascaris megalocephala*, an invertebrate roundworm parasite found in the intestine of mammals. Since its diploid chromosome number is 4, the chromosomes are easily counted.

ACTIVITY 3

Examining Meiotic Events Microscopically

Go to the demonstration area where the slides are set up, and make the following observations:

1. Scan the first demonstration slide to identify a *primary oocyte*, the cell type that begins the meiotic process. It will have what appears to be a relatively thick cell membrane; this is the *fertilization membrane* that the oocyte produces after sperm penetration. Find and study a primary oocyte that is undergoing meiosis I. Look for a barrel-shaped spindle with two tetrads (two groups of four beadlike chromosomes) in it. Most often the spindle is located at the periphery of the cell. The sperm nucleus may or may not be seen, depending on how the cell was cut.

2. Observe slide 2. Locate a cell in which half of each tetrad (a dyad) is being extruded from the cell surface into a smaller cell called the *first polar body*.

3. On slide 3, attempt to locate a *secondary oocyte* (a daughter cell produced during meiosis I) undergoing meiosis II.

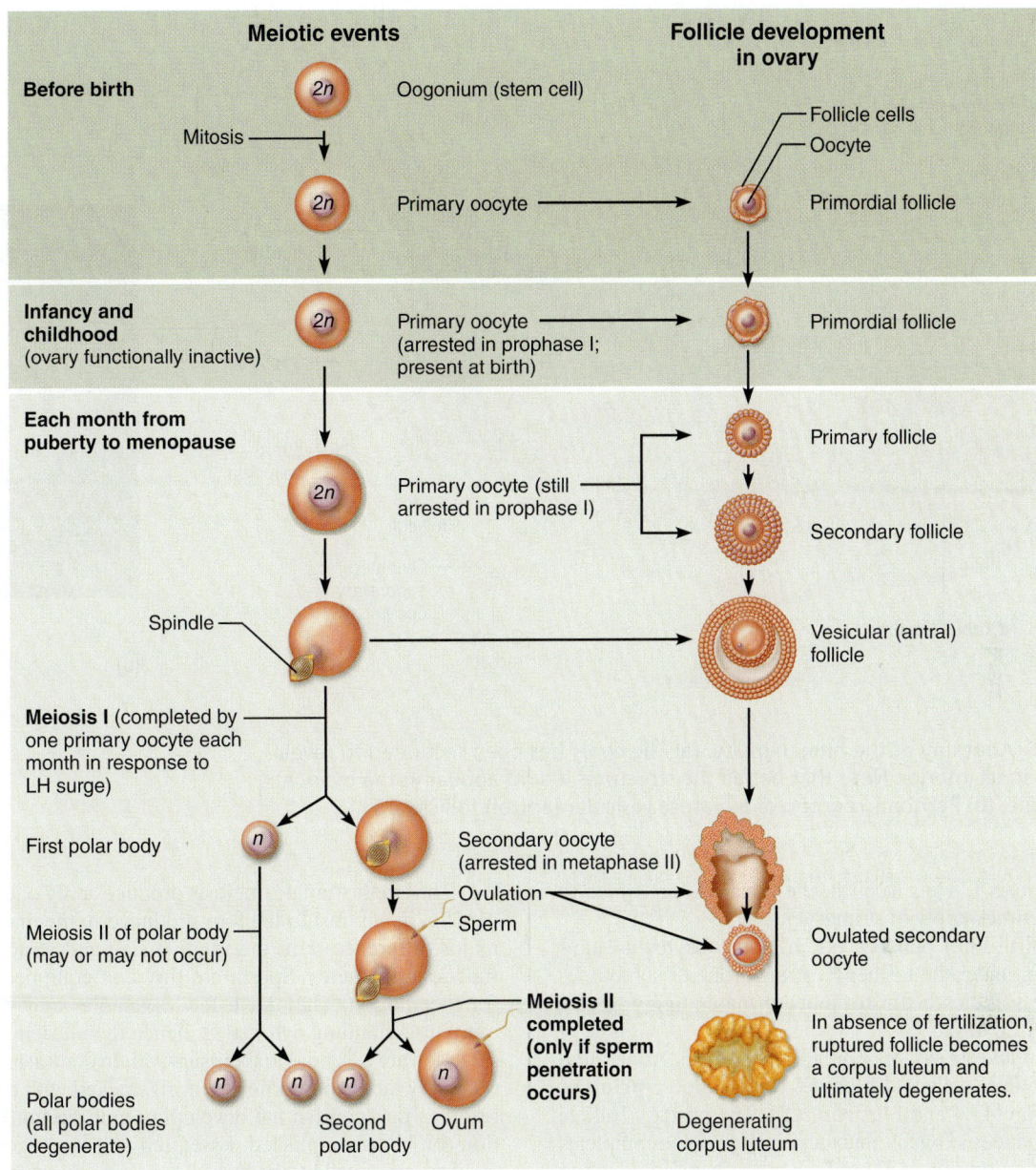

Figure 43.4 Oogenesis. Left, flowchart of meiotic events. Right, correlation with follicular development and ovulation in the ovary.

In this view, you should see two dyads (each with two bead-like chromosomes) on the spindle.

4. On slide 4, locate a cell in which the *second polar body* is being formed. In this case, both it and the ovum will now contain two chromosomes, the haploid number for *Ascaris*.

5. On slide 5, identify a *fertilized egg* or a cell in which the sperm and ovum nuclei (actually *pronuclei*) are fusing to form a single nucleus containing four chromosomes. ■■

Human Oogenesis
and the Ovarian Cycle

Once the adult ovarian cycle is established, gonadotropic hormones produced by the anterior pituitary influence the development of ova in the ovaries and their cyclic production

of female sex hormones. Within an ovary, each immature ovum develops within a saclike structure called a *follicle,* where it is encased by one or more layers of smaller cells. The surrounding cells are called **follicle cells** if one layer is present and **granulosa cells** when more than one layer is present.

The process of **oogenesis,** or female gamete formation, which occurs in the ovary, is similar to spermatogenesis occurring in the testis, but there are some important differences. Oogenesis begins with primitive stem cells called **oogonia,** located in the ovarian cortices of the developing female fetus **(Figure 43.4)**. During fetal development, the oogonia undergo mitosis thousands of times until their number reaches 2 million or more. They then become encapsulated by a single layer of squamouslike follicle cells and form the **primordial follicles** of the ovary. By the time the female child is born, most of her oogonia have increased in size and have become **primary oocytes,** which are in the prophase

43

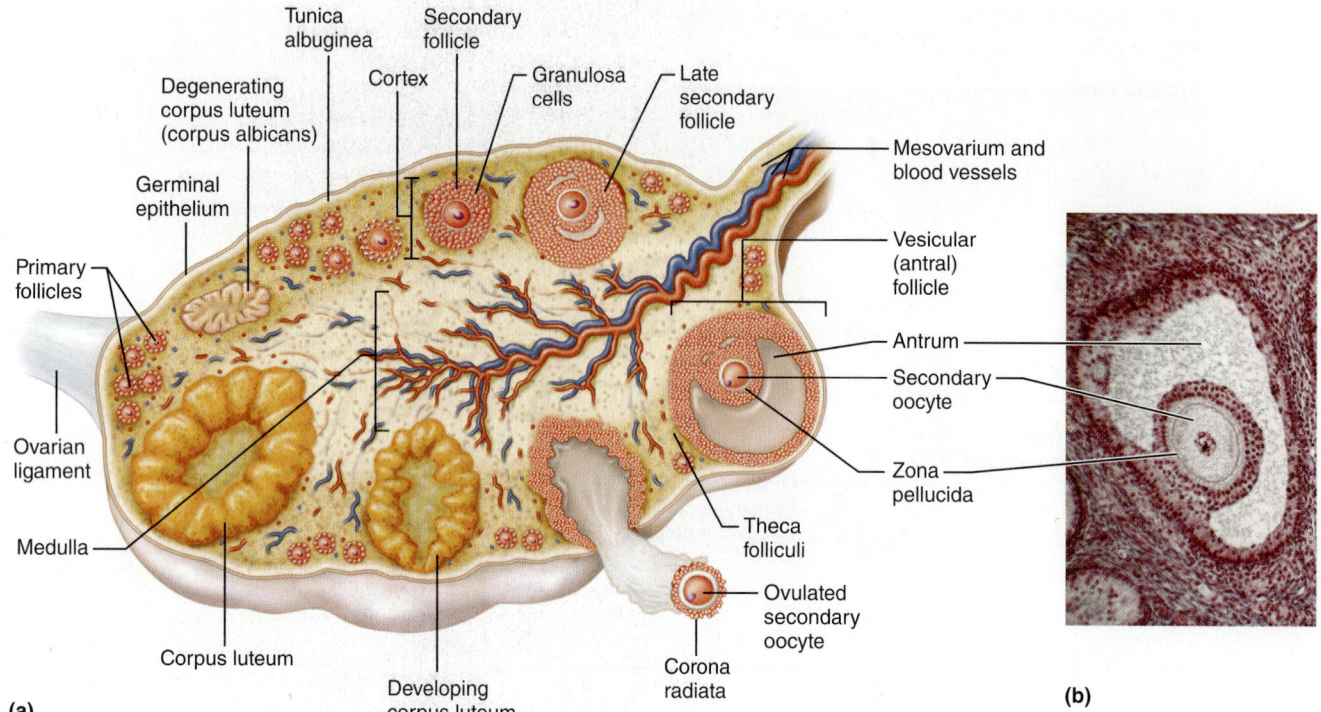

Figure 43.5 Anatomy of the human ovary. (a) The ovary has been sectioned to reveal the follicles in its interior. Note that not all the structures would appear in the ovary at the same time. **(b)** Photomicrograph of a mature vesicular (antral) follicle (75×).

stage of meiosis I. Thus at birth, the female is presumed to have her lifetime supply of primary oocytes.

From birth until puberty, the primary oocytes are quiescent. Then, under the influence of FSH, one or sometimes more of the follicles begin to undergo maturation approximately every 28 days.

As a follicle grows, its epithelium changes from squamous to cuboidal cells and it comes to be called a **primary follicle** (Figure 43.4 and **Figure 43.5**). The primary follicle begins to produce estrogens, and the primary oocyte completes its first maturation division, producing two haploid daughter cells that are very disproportionate in size. One of these is the **secondary oocyte,** which contains nearly all of the cytoplasm in the primary oocyte. The other is the tiny **first polar body.** The first polar body may then complete the second maturation division, producing two more polar bodies. These eventually disintegrate for lack of sustaining cytoplasm.

As the follicle containing the secondary oocyte continues to enlarge, blood levels of estrogens rise. Initially, estrogen exerts a negative feedback influence on the release of gonadotropins by the anterior pituitary. However, approximately in the middle of the 28-day cycle, as the follicle reaches the mature **vesicular (antral) follicle** stage, rising estrogen levels become highly stimulatory and a sudden burstlike release of LH (and, to a lesser extent, FSH) by the anterior pituitary triggers ovulation. The secondary oocyte is extruded and begins its journey along the uterine tube to the uterus. If penetrated en route by a sperm, the secondary oocyte will undergo meiosis II, producing one large **ovum** and a tiny **second polar body.** When meiosis II is complete, the chromosomes of the egg and sperm combine to form the diploid nucleus of the fertilized egg. If sperm penetration does not occur, the secondary oocyte simply disintegrates without ever producing the female gamete.

Thus in the female, meiosis produces only one functional gamete, in contrast to the four produced in the male. Another major difference is in the relative size and structure of the functional gametes. Sperm are tiny and equipped with tails for locomotion. They have few organelles and virtually no nutrient-containing cytoplasm; hence the nutrients contained in semen are essential to their survival. In contrast, the egg is a relatively large nonmotile cell, well stocked with cytoplasmic reserves that nourish the developing embryo until implantation can be accomplished. Essentially all the zygote's organelles are provided by the egg.

Once the secondary oocyte has been expelled from the ovary, LH transforms the ruptured follicle into a **corpus luteum,** which begins producing progesterone and estrogen. Rising blood levels of the two ovarian hormones inhibit FSH release by the anterior pituitary (Figure 43.7). As FSH declines, its stimulatory effect on follicular production of estrogens ends, and estrogen blood levels begin to decline. Since rising estrogen levels triggered LH release by the anterior pituitary, falling estrogen levels result in declining levels of LH in the blood. Corpus luteum secretory function is maintained by high blood levels of LH. Thus as LH blood levels begin to drop toward the end of the 28-day cycle, progesterone production ends and the corpus luteum begins to degenerate and is replaced by scar tissue, the **corpus albicans.**

ACTIVITY 4

Examining Oogenesis in the Ovary

Because many different stages of ovarian development exist within the ovary at any one time, a single microscopic preparation will contain follicles at many different stages of development (Figure 43.5). Obtain a cross section of ovary tissue, and identify the following structures.

Germinal epithelium: Outermost layer of the ovary.

Primary follicle: One or a few layers of cuboidal cells surrounding the larger central developing ovum.

Secondary follicles: Follicles consisting of several layers of granulosa cells surrounding the central developing ovum, and beginning to show evidence of fluid accumulation and **antrum** (central cavity) formation. Follicle development may take more than one cycle.

Vesicular (antral) follicle: At this stage of development, the follicle has a large antrum containing fluid produced by the granulosa cells. The developing secondary oocyte is pushed to one side of the follicle and is surrounded by a capsule of several layers of granulosa cells called the **corona radiata** (radiating crown). When the secondary oocyte is released, it enters the uterine tubes with its corona radiata intact. The connective tissue adjacent to the mature follicle forms a capsule, called the **theca folliculi,** that encloses the follicle.

Corpus luteum: A solid glandular structure or a structure containing a scalloped lumen that develops from the ruptured follicle. ▄▄

Comparing and Contrasting Oogenesis and Spermatogenesis

Examine the model of oogenesis, and compare it with the spermatogenesis model. Note differences in the number, size, and structure of the functional gametes. ▄▄

The Menstrual Cycle

The **uterine cycle,** or **menstrual cycle,** is hormonally controlled by estrogens and progesterone secreted by the ovary. It is normally divided into three stages: menstrual, proliferative, and secretory. Notice how the endometrial changes **(Figure 43.6)** correlate with hormonal and ovarian changes **(Figure 43.7)**.

If fertilization has occurred, the embryo will produce a hormone much like LH, which will maintain the function of the corpus luteum. Otherwise, as the corpus luteum begins to deteriorate, lack of ovarian hormones in the blood causes blood vessels supplying the endometrium to kink and become spastic, setting the stage for menstruation to begin by the 28th day.

Although 28 days is a common length for the menstrual cycle (Figure 43.7d), its length is highly variable, sometimes as short as 21 days or as long as 38.

Observing Histological Changes in the Endometrium During the Menstrual Cycle

Obtain slides showing the menstrual, secretory, and proliferative phases of the uterine endometrium. Observe each carefully, comparing their relative thicknesses and vascularity. As you work, refer to the corresponding photomicrographs (Figure 43.6). ▄▄

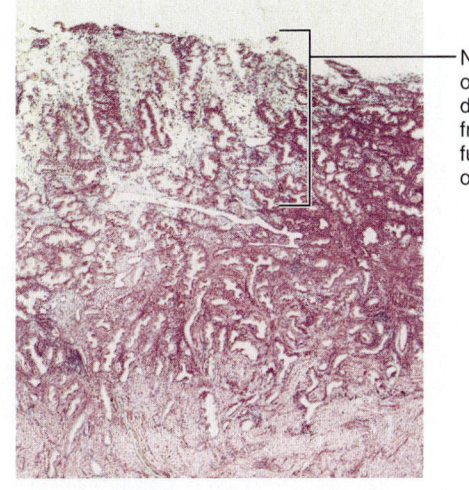

(a)

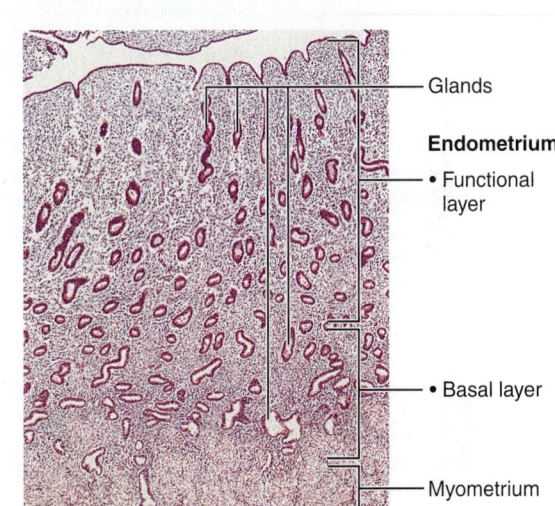

(b)

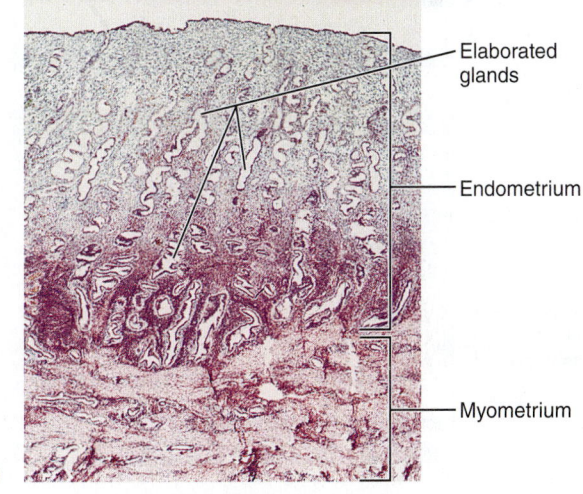

(c)

Figure 43.6 Endometrial changes during the menstrual cycle. (a) Onset of menstruation (8×). **(b)** Early proliferative phase (10×). **(c)** Early secretory phase (8×).

43

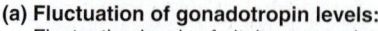

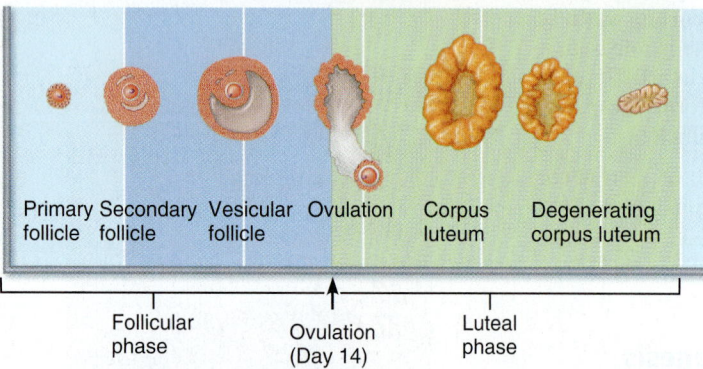

(a) Fluctuation of gonadotropin levels: Fluctuating levels of pituitary gonadotropins (follicle-stimulating hormone and luteinizing hormone) in the blood regulate the events of the ovarian cycle.

(b) Ovarian cycle: Structural changes in the ovarian follicles during the ovarian cycle are correlated with (d) changes in the endometrium of the uterus during the uterine cycle.

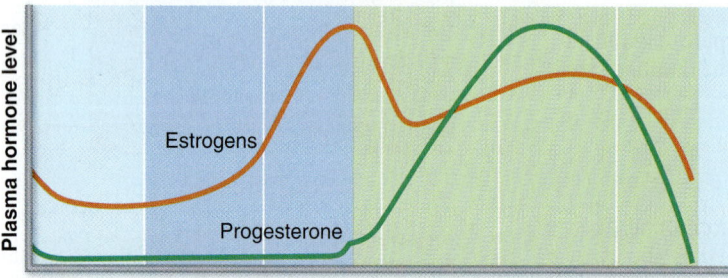

(c) Fluctuation of ovarian hormone levels: Fluctuating levels of ovarian hormones (estrogens and progesterone) cause the endometrial changes of the uterine cycle. The high estrogen levels are also responsible for the LH/FSH surge in (a).

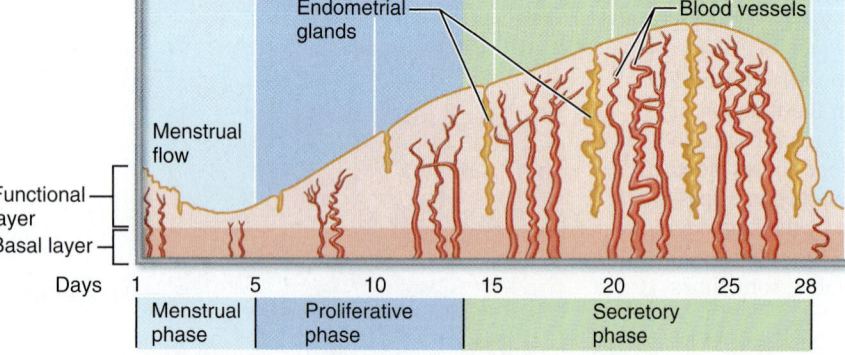

(d) The three phases of the uterine cycle:
- **Menstrual:** The functional layer of the endometrium is shed. Approximately days 1–5.
- **Proliferative:** The functional layer of the endometrium is rebuilt under influence of estrogens. Approximately days 6–14 . Ovulation occurs at the end of this phase.
- **Secretory:** Begins immediately after ovulation under the influence of progesterone. Enrichment of the blood supply and glandular secretion of nutrients prepare the endometrium to receive an embryo.

Figure 43.7 Correlation of anterior pituitary and ovarian hormones with structural changes in the ovary and uterus. The time bar applies to all parts of the figure.

43

Physiology of Reproduction: Gametogenesis and the Female Cycles

Meiosis

1. The following statements refer to events occurring during mitosis and/or meiosis. For each statement, decide if the event occurs in (a) mitosis only, (b) meiosis only, or (c) both mitosis and meiosis.

_____ 1. dyads are visible

_____ 2. tetrads are visible

_____ 3. product is two diploid daughter cells genetically identical to the mother cell

_____ 4. product is four haploid daughter cells quantitatively and qualitatively different from the mother cell

_____ 5. involves the phases prophase, metaphase, anaphase, and telophase

_____ 6. occurs throughout the body

_____ 7. occurs only in the ovaries and testes

_____ 8. provides cells for growth and repair

_____ 9. homologues synapse; crossovers are seen

_____ 10. chromosomes are replicated before the division process begins

_____ 11. provides cells for perpetuation of the species

_____ 12. consists of two consecutive nuclear divisions, without chromosomal replication occurring before the second division

2. Describe the process of synapsis. _____

_____

3. How does crossover introduce variability in the daughter cells? _____

_____

4. Define *homologous chromosomes*. _____

_____

Spermatogenesis

5. The cell types seen in the seminiferous tubules are listed in the key. Match the correct cell type or types with the descriptions given below.

Key: a.   primary spermatocyte   c.   spermatogonium   e.   spermatid
       b.   secondary spermatocyte   d.   sustentocyte   f.   sperm

_____ 1. primitive stem cell   _____ 4. products of meiosis II

_____ 2. haploid   _____ 5. product of spermiogenesis

_____ 3. provides nutrients to   _____ 6. product of meiosis I
                    developing sperm

6. Why are spermatids not considered functional gametes? _____

_____

7. Differentiate between *spermatogenesis* and *spermiogenesis*. _____

_____

_____

8. Draw a sperm below, and identify the acrosome, head, midpiece, and tail. Then beside each label, note the composition and function of each of these sperm structures.

9. The life span of a sperm is very short. What anatomical characteristics might lead you to suspect this even if you don't know

its life span? _____

Oogenesis, the Ovarian Cycle, and the Menstrual Cycle

10. The sequence of events leading to germ cell formation in the female begins during fetal development. By the time the child

is born, all viable oogonia have been converted to _____.

In view of this fact, how does the total germ cell potential of the female compare to that of the male?

_____

11. The female gametes develop in structures called *follicles*. What is a follicle? _____

_____

How are primary and vesicular follicles anatomically different? _____

_____

_____

_____

What is a corpus luteum? _____

_____

12. What is the major hormone produced by the vesicular follicle? _____

By the corpus luteum? _____

13. Use the key to identify the cell type you would expect to find in the following structures. The items in the key may be used once, more than once, or not at all.

Key: a.   oogonium   b.   primary oocyte   c.   secondary oocyte   d.   ovum

_____ 1.   forming part of the primary follicle in the ovary

_____ 2.   in the uterine tube before fertilization

_____ 3.   in the mature vesicular follicle of the ovary

_____ 4.   in the uterine tube shortly after sperm penetration

14. The cellular product of spermatogenesis is four _____; the final product of oogenesis is one

_____ and three _____. What is the function of this unequal cytoplasmic

division seen during oogenesis in the female? _____

_____

_____

What is the fate of the three tiny cells produced during oogenesis? _____

Why? _____

15. The following statements deal with anterior pituitary and ovarian hormones and with hormonal interrelationships. Name the hormone(s) described in each statement.

_____ 1.   produced by primary follicles in the ovary

_____ 2.   ovulation occurs after its burstlike release

_____ and _____ 3.   exert negative feedback on the anterior pituitary relative to FSH secretion

_____ 4.   stimulates LH release by the anterior pituitary

_____ 5.   stimulates the corpus luteum to produce progesterone and estrogen

_____ 6.   maintains the hormonal production of the corpus luteum in a nonpregnant woman

16. Why does the corpus luteum deteriorate toward the end of the ovarian cycle? _____

_____

17. For each statement below dealing with hormonal blood levels during the female ovarian and menstrual cycles, decide whether the condition in column A is usually (a) greater than, (b) less than, or (c) essentially equal to the condition in column B.

Column A   **Column B**

_____ 1.   amount of LH in the blood . . . .   amount of LH in the blood at ovulation
during menstruation

_____ 2.   amount of FSH in the blood . . . .   amount of FSH in the blood on day 20 of the cycle
on day 6 of the cycle

_____ 3.   amount of estrogen in the blood . . . .   amount of estrogen in the blood at ovulation
during menstruation

_____ 4.   amount of progesterone in the blood . . . .   amount of progesterone in the blood on day 23
on day 14

_____ 5.   amount of estrogen in the blood . . . .   amount of progesterone in the blood on day 10
on day 10

18. What uterine tissue undergoes dramatic changes during the menstrual cycle? _____

19. When during the female menstrual cycle would fertilization be unlikely? Explain why. _____

_____

20. Assume that a woman could be an "on demand" ovulator like the rabbit, in which copulation stimulates the hypothalamic-pituitary-gonadal axis and causes LH release, and an oocyte was ovulated and fertilized on day 26 of her 28-day cycle. Why would a successful pregnancy be unlikely at this time?

_____

_____

_____

21. The menstrual cycle depends on events within the female ovary. The stages of the menstrual cycle are listed below. For each, note its approximate time span and the related events in the uterus; and then to the right, record the ovarian events occurring simultaneously. Pay particular attention to hormonal events.

Menstrual cycle stage	Uterine events	Ovarian events
Menstruation		
Proliferative		
Secretory		

Survey of Embryonic Development

MATERIALS

- ☐ Prepared slides of sea urchin development (zygote through larval stages)
- ☐ Compound microscope
- ☐ Three-dimensional human development models or plaques (if available)
- ☐ *Demonstration:* Phases of human development in *Life Before Birth: Normal Fetal Development,* Second Edition (by Marjorie A. England, 1996, Mosby-Wolfe)
- ☐ Disposable gloves
- ☐ *Demonstration:* Pregnant cat, rat, or pig uterus (one per laboratory session) with uterine wall dissected to allow student examination
- ☐ Three-dimensional model of pregnant human torso
- ☐ *Demonstration:* Fresh or formalin-preserved human placenta (obtained from a clinical agency)
- ☐ Prepared slide of placenta tissue

MasteringA&P® For related exercise study tools, go to the Study Area of MasteringA&P. There you will find:
- Practice Anatomy Lab PAL
- PhysioEx PEx
- A&PFlix *A&PFlix*
- Practice quizzes, Histology Atlas, eText, Videos, and more!

OBJECTIVES

1. Define *fertilization* and *zygote.*
2. Define and discuss the function of *cleavage* and *gastrulation.*
3. Differentiate between the blastula and gastrula forms of the sea urchin and human using appropriate models or diagrams.
4. Define *blastocyst.*
5. Identify the following structures of a human blastocyst on an appropriate diagram, and state the function of each.

inner cell mass	chorionic villi	yolk sac
trophoblast	amnion	allantois

6. Describe the process and timing of implantation in the human.
7. Define *decidua basalis* and *decidua capsularis.*
8. Name the three primary germ layers and list the organs or organ systems that arise from each in the human.
9. Describe developmental direction.
10. Describe the gross anatomy and general function of the human placenta.

PRE-LAB QUIZ

1. Circle the correct underlined term. The fertilized egg, or zygote / embryo, appears as a single cell surrounded by a fertilization membrane and a jellylike membrane.
2. The uniting of the egg and sperm nuclei is known as
   - a. embryogenesis
   - b. fertilization
   - c. implantation
   - d. mitosis
3. Circle True or False. Cleavage is a series of mitotic divisions without any intervening growth periods and results in a multicellular embryonic body.
4. Circle the correct underlined term. As a result of gastrulation, a three / four-layered embryo forms, with each layer corresponding to a primary germ layer.
5. The _____ implants in the uterine wall.
   - a. zygote
   - b. morula
   - c. blastocyst
   - d. gastrula
6. The _____ gives rise to the epidermis of the skin and the nervous system.
   - a. ectoderm
   - b. endoderm
   - c. mesoderm
7. By the ninth week of development, the embryo is referred to as a
   - a. blastocyst
   - b. blastomere
   - c. fetus
   - d. gastrula
8. Circle True or False. The placenta is composed solely of embryonic membranes.
9. Circle the correct underlined term. The allantois / amnion encases the young embryonic body in a fluid-filled chamber that acts to protect the developing embryo against trauma.
10. What is the function of the placenta? _____
    _____

Because reproduction is such a familiar event, we tend to lose sight of the wonder of the process. One part of that process, the development of the embryo, is the concern of embryologists who study the changes in structure that occur from the time of fertilization until the time of birth.

Early development in all animals involves three basic types of activities, which are integrated to ensure the formation of a viable offspring: (1) an increase in cell number and subsequent cell growth; (2) cellular specialization; and (3) morphogenesis, the formation of functioning organ systems. This exercise first provides a rather broad overview of the changes in structure that take place during embryonic development in sea urchins. The pattern of changes in this marine animal provides a basis of comparison with developmental events in the human.

Developmental Stages of Sea Urchins and Humans

ACTIVITY 1

Microscopic Study of Sea Urchin Development

1. Obtain a compound microscope and a set of slides depicting embryonic development of the sea urchin.

2. Observe the fertilized egg, or **zygote,** which appears as a single cell immediately surrounded by a jellylike membrane and a fertilization membrane. After an egg is penetrated by a sperm, the egg and the sperm nuclei fuse to form a single nucleus. This process is called **fertilization**. Within 2 to 5 minutes after sperm penetration, a fertilization membrane forms beneath the jelly coat to prevent the entry of additional sperm. Draw the zygote and label the fertilization and jelly membranes.

Zygote

3. Observe the cleavage stages. Once fertilization has occurred, the zygote begins to divide, forming a mass of successively smaller and smaller cells, called **blastomeres.** This series of mitotic divisions without intervening growth periods is referred to as **cleavage,** and it results in a multicellular embryonic body. The cleavage stage of embryonic development provides a large number of building blocks (cells) with which to fashion the forming body. (If this is a little difficult to understand, consider trying to build a structure with a huge block of granite rather than with small bricks.)

As the division process continues, a solid ball of cells forms. At the 32-cell stage, it is called the **morula,** and the embryo resembles a raspberry in form. Then the cell mass hollows out to become the embryonic form called the **blastula,** which is a ball of cells surrounding a central cavity. The blastula is the final product of cleavage.

Identify and sketch the blastula stage of cleavage—a ball of cells with an apparently lighter center due to the presence of the central cavity.

Blastula

4. Identify the **early gastrula** form, which follows the blastula in the developmental sequence. The gastrula looks as if one end of the blastula has been indented or pushed into the central cavity, forming a two-layered embryo. In time, a third layer of cells appears between the initial two cell layers. Thus, as a result of **gastrulation,** a three-layered embryo forms, each layer corresponding to a **primary germ layer** from which all body tissues develop. The innermost layer, the **endoderm,** and the middle layer, the **mesoderm,** form the internal organs; the outermost layer, the **ectoderm,** forms the surface tissues of the body.

Draw a gastrula below. Label the ectoderm and endoderm. If you can see the third layer of cells, the mesoderm, budding off between the other two layers, label that also.

Gastrula

5. Gastrulation in the sea urchin is followed by the appearance of the free-swimming larval form, in which the three germ layers have differentiated into the various tissues and organs of the animal's body.

The larvae exist for a few days in the unattached form and then settle to the ocean bottom to attach and develop into the sessile adult form. If time allows, observe the larval form on the prepared slides. ■

ACTIVITY 2

Examining the Stages of Human Development

Use the images in this exercise (**Figure 44.1** and **Figure 44.2**) and models or plaques of human development to identify the various stages of development and respond to the questions posed below.

1. Observe the fertilized egg, or zygote, which appears as a single cell immediately surrounded by a jellylike *zona pellucida* and then a crown of granulosa cells (the *corona radiata*).

2. Next, observe the cleavage stages.

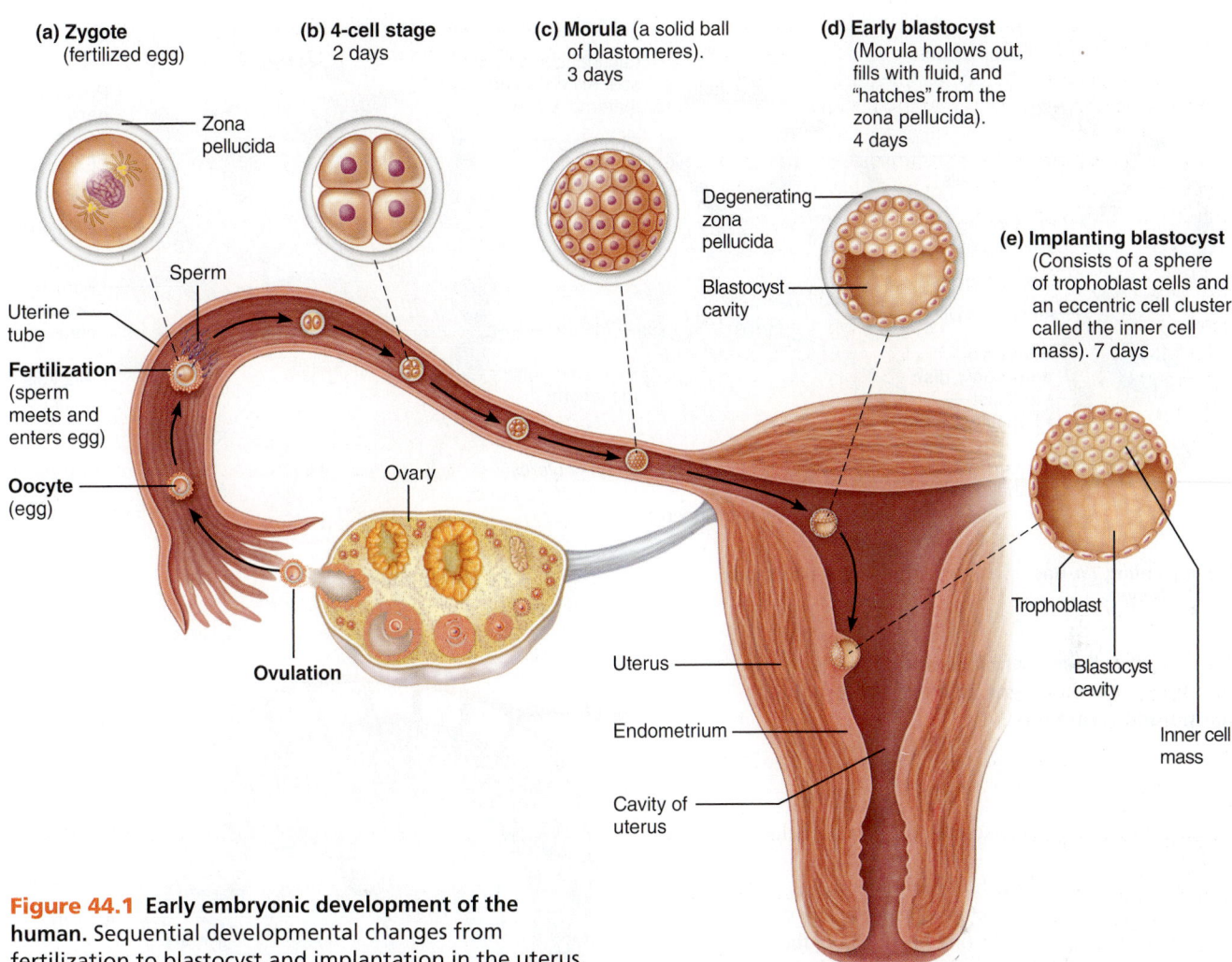

(a) Zygote
(fertilized egg)

Zona pellucida

Sperm

Uterine tube

Fertilization
(sperm meets and enters egg)

Oocyte
(egg)

Ovulation

Ovary

(b) 4-cell stage
2 days

(c) Morula (a solid ball of blastomeres).
3 days

(d) Early blastocyst
(Morula hollows out, fills with fluid, and "hatches" from the zona pellucida).
4 days

Degenerating zona pellucida

Blastocyst cavity

(e) Implanting blastocyst
(Consists of a sphere of trophoblast cells and an eccentric cell cluster called the inner cell mass). 7 days

Trophoblast

Blastocyst cavity

Inner cell mass

Uterus

Endometrium

Cavity of uterus

Figure 44.1 Early embryonic development of the human. Sequential developmental changes from fertilization to blastocyst and implantation in the uterus.

Is the human cleavage process similar to that in the sea urchin?

_____

Why do you suppose this is so? _____

_____

Observe the blastula, the final product of cleavage, which is called the **blastocyst** in the human. Unlike the sea urchin, only a portion of the blastocyst cells in the human contributes to the formation of the embryonic body—those seen at one side of the blastocyst (Figure 44.1e) forming the **inner cell mass (ICM).** The rest of the blastocyst—that enclosing the central cavity and overriding the ICM—is referred to as the **trophoblast.** The trophoblast becomes an extraembryonic membrane called the **chorion,** which forms the fetal portion of the **placenta.** The ICM becomes the **embryonic disc,** which forms the embryo proper.

3. Observe the *implanting* blastocyst shown on the model or in the figures. By approximately the seventh day after ovulation, a developing human embryo (blastocyst) is floating free in the uterine cavity. About that time, it adheres to the uterine wall over the ICM area, and implantation begins. The trophoblast cells secrete enzymes that erode the uterine mucosa at the point of attachment to reach the vascular supply in the submucosa. By the fourteenth day after ovulation, implantation is completed and the uterine mucosa has grown over the burrowed-in embryo. The portion of the uterine wall beneath the ICM, destined to take part in placenta formation, is called the **decidua basalis** and that surrounding the rest of the blastocyst is called the **decidua capsularis.** Identify these regions.

By the time implantation has been completed, embryonic development has progressed to the **gastrula stage,** and the three primary germ layers are present and are beginning to differentiate (Figure 44.2). Within the next 6 weeks, virtually all of the body organ systems will have been laid down at least in rudimentary form by the germ layers. The outermost layer, *ectoderm*, gives rise to the epidermis of the skin and the nervous system. The deepest layer, the *endoderm*, forms the mucosa of the digestive and respiratory tracts and associated glands. *Mesoderm*, the middle layer, forms virtually everything lying between the two (skeleton, walls of the digestive

44

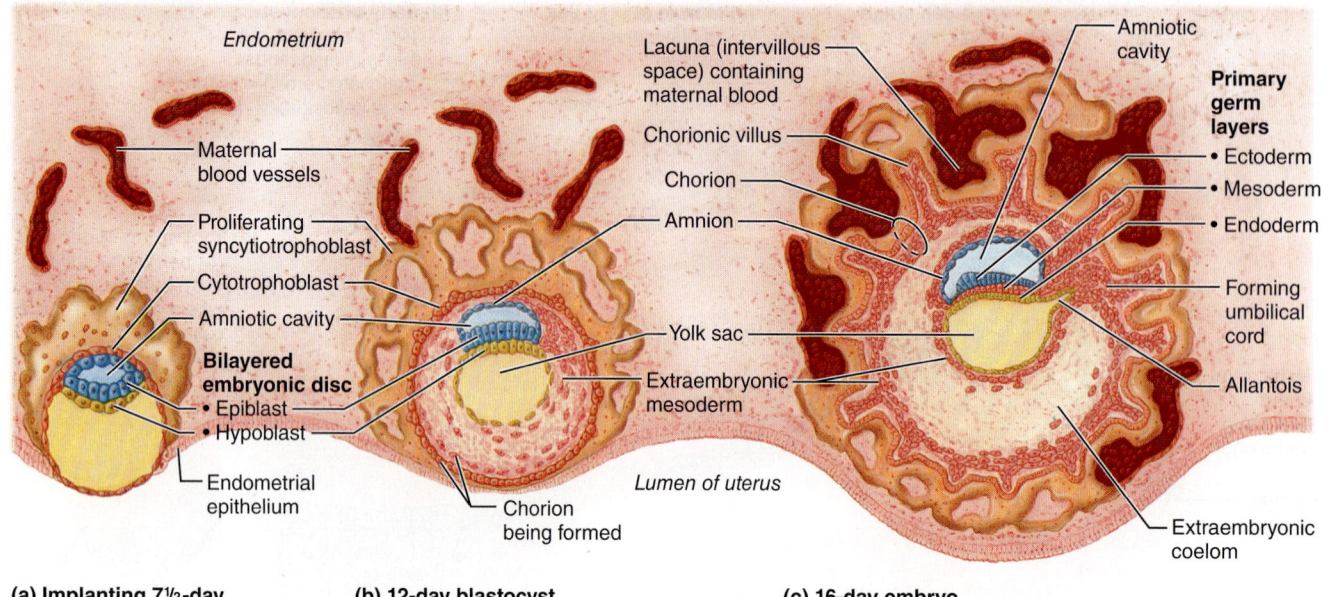

(a) Implanting 7½-day blastocyst.

(b) 12-day blastocyst.

(c) 16-day embryo.

Figure 44.2 Events of placentation, early embryonic development, and extraembryonic membrane formation.

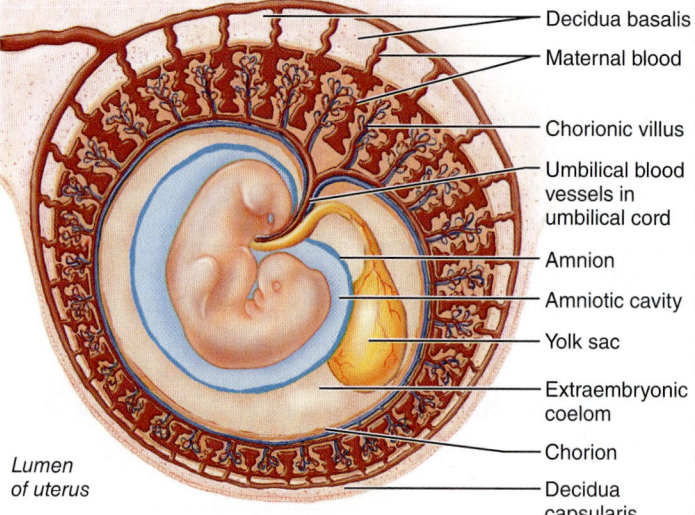

(d) 4½-week embryo.

organs, urinary system, skeletal muscles, circulatory system, and others).

All the groundwork has been completed by the eighth week. By the ninth week of development, the embryo is referred to as a **fetus,** and from this point on, the major activities are growth and tissue and organ specialization.

4. Again observe the blastocyst to follow the formation of the embryonic membranes and the placenta (Figure 44.2). Notice the villus extensions of the trophoblast. By the time implantation is complete, the trophoblast has differentiated into the chorion, and its large elaborate villi are lying in the blood-filled sinusoids in the uterine tissue. This composite of uterine tissue and **chorionic villi** is called the **placenta** (Figure 44.3), and all exchanges to and from the embryo occur through the chorionic membranes.

Three embryonic membranes (originating in the ICM) have also formed by this time—the amnion, the allantois, and the yolk sac (Figure 44.2). Attempt to identify each.

- The **amnion** encases the young embryonic body in a fluid-filled chamber that protects the embryo against mechanical trauma and temperature extremes and prevents adhesions during rapid embryonic growth.

- The **yolk sac** in humans has lost its original function, which was to pass nutrients to the embryo after digesting the yolk mass. The placenta has taken over that task; also, the human egg has very little yolk. However, the yolk sac is not totally useless. The embryo's first blood cells originate here, and the primordial germ cells migrate from it into the embryo's body to seed the gonadal tissue.

- The **allantois,** which protrudes from the posterior end of the yolk sac, is also largely redundant in humans because of the placenta. In birds and reptiles, it is a repository for

embryonic wastes. In humans, it is the structural basis on which the mesoderm migrates to form the **umbilical cord,** which attaches the embryo to the placenta.

5. Go to the demonstration area to view the photographic series, *Life Before Birth: Normal Fetal Development*. These photographs illustrate human development in a way you will long remember. After viewing them, respond to the following questions.

In your own words, what do the chorionic villi look like?

_____

_____

_____

44

What organs or organ systems appear *very* early in embryonic development?

_____

_____

Does development occur in a rostral to caudal (head to toe) direction, or vice versa?

_____

Does development occur in a distal to proximal direction, or vice versa?

_____

Does spontaneous movement occur in utero? _____

How does the mother recognize this? _____

_____

The very young embryo has been described as resembling "an astronaut suspended and floating in space." Do you think this definition is appropriate?

_____ Why or why not? _____

_____

_____

What is vernix caseosa? _____

_____

_____

What is lanugo? _____

_____

In Utero Development

ACTIVITY 3

Identifying Fetal Structures

1. Put on disposable gloves. Go to the appropriate demonstration area and observe the fetuses in the Y-shaped animal (cat, rat, or pig) uterus. Identify the following fetal or fetal-related structures:

- **Placenta** (a composite structure formed from the uterine mucosa and the fetal chorion).

Describe its appearance. _____

_____

_____

_____

- **Umbilical cord.** Describe its relationship to the placenta and fetus.

_____

_____

_____

- **Amniotic sac.** Identify the transparent amnion surrounding a fetus. Open one amniotic sac and note the amount, color, and consistency of the fluid.

_____

_____

_____

Remove a fetus and observe the degree of development of the head, body, and extremities. Is the skin thick or thin?

_____

What is the basis for your response? _____

_____

_____

2. Observe the model of a pregnant human torso. Identify the placenta. How does it differ in shape from the animal placenta observed?

_____

_____

_____

Identify the umbilical cord. In what region of the uterus does implantation usually occur, as indicated by the position of the placenta?

_____

_____

What might be the consequence if it occurred lower?

_____

_____

_____

Why would a feet-first position (breech presentation) be less desirable than the positioning of the model?

_____

_____

_____

Gross and Microscopic Anatomy of the Placenta

The placenta is a remarkable temporary organ. Composed of maternal and fetal tissues, it is responsible for providing nutrients and oxygen to the embryo and fetus while removing carbon dioxide and metabolic wastes.

ACTIVITY 4

Studying Placental Structure

1. Notice that the human placenta on display has two very different-appearing surfaces—one smooth and the other spongy, roughened, and torn-looking.

Which is the fetal side? _____

What is the basis for your conclusion? _____

_____

_____

Identify the umbilical cord. Within the cord, identify the umbilical vein and two umbilical arteries. What is the function of the umbilical vein?

_____

_____

The umbilical arteries? _____

_____

2. Obtain a microscope slide of placental tissue. Observe the tissue carefully and identify the *intervillous spaces* (lacunae), which are blood-filled in life **(Figure 44.3)**. Identify the villi, and notice their rich vascular supply. Draw a small representative diagram of your observations below and label it appropriately. ▬

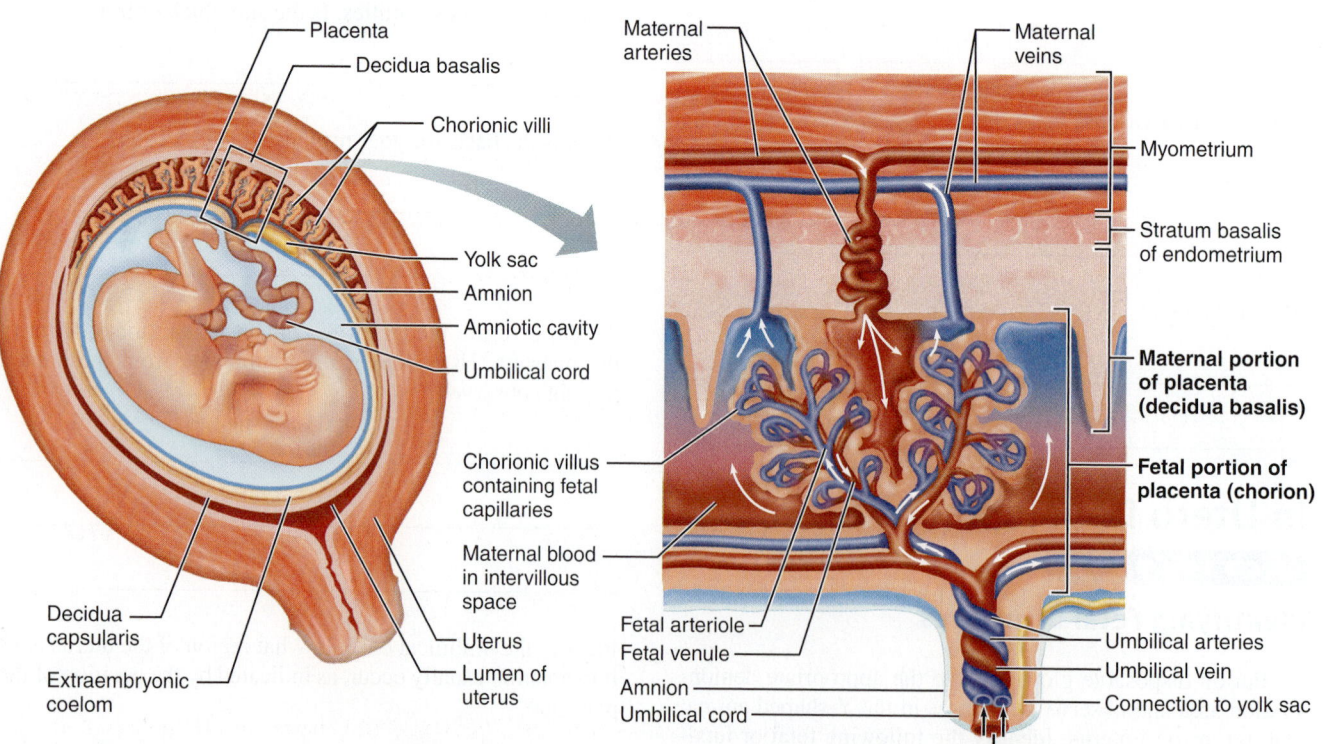

Figure 44.3 Diagrammatic representation of the structure of the placenta for a 13-week fetus.

44

Survey of Embryonic Development

Developmental Stages of Sea Urchins and Humans

1. Define *zygote*. _____

2. Describe how you were able to tell by observation when a sea urchin egg was fertilized. _____

_____

3. Use the key choices to identify the embryonic stage or process described below.

Key: a. blastocyst (blastula in sea urchins)   c. fertilization   e. morula
      b. cleavage   d. gastrulation   f. zygote

_____ 1. process of male and female pronuclei fusion

_____ 2. solid ball of embryonic cells

_____ 3. process of rapid mitotic cell division without intervening growth periods

_____ 4. cell resulting from combination of egg and sperm

_____ 5. process involving cell rearrangements to form the three primary germ layers

_____ 6. embryonic stage in which the embryo consists of a hollow ball of cells

4. What is the importance of cleavage in embryonic development? _____

_____

How is cleavage different from mitotic cell division, which occurs later in life? _____

_____

_____

5. The cells of the human blastocyst have various fates. Which blastocyst derivatives have the following fates?

_____ 1. forms the embryo proper

_____ 2. becomes the chorion and cooperates with uterine tissues to form the placenta

_____ 3. produces the amnion, yolk sac, and allantois

_____ 4. produces the primordial germ cells

_____ 5. an embryonic membrane that provides the structural basis for the umbilical cord

6. Using the letters on the diagram, correctly identify each of the following maternal or embryonic structures.

_____ amnion   _____ chorionic villi   _____ decidua capsularis   _____ forming umbilical cord

_____ chorion   _____ decidua basalis   _____ ectoderm   _____ mesoderm

_____ endoderm   _____ uterine cavity

7. Explain the process and importance of gastrulation. _____

_____

_____

8. What is the function of the amnion and the amniotic fluid? _____

_____

_____

9. Describe the process of implantation, noting the role of the trophoblast cells. _____

_____

_____

10. How many days after fertilization is implantation generally completed? _____ What event in the female menstrual cycle

ordinarily occurs just about this time if implantation does not occur? _____

_____

11. What name is given to the part of the uterine wall directly under the implanting embryo? _____

That surrounding the rest of the embryonic structure? _____

12. Using an appropriate reference, find out what *decidua* means and state the definition. _____

_____

How is this terminology applicable to the deciduas of pregnancy? _____

_____

13. Referring to the illustrations and text of *Life Before Birth: Normal Fetal Development,* answer the following:

Which two organ systems are extensively developed in the *very young* embryo?

_____ and _____

Describe the direction of development by circling the correct descriptions below:

proximal-distal   distal-proximal   caudal-rostral   rostral-caudal

Does body control during infancy develop in the same directions? Think! Can an infant pick up a common pin (pincer grasp) or wave his arms earlier? Is arm-hand or leg-foot control achieved earlier?

_____

14. Note whether each of the following organs or organ systems develops from the (a) ectoderm, (b) endoderm, or (c) mesoderm. Use an appropriate reference as necessary.

_____ 1. skeletal muscle   _____ 4. respiratory mucosa   _____ 7. nervous system

_____ 2. skeleton   _____ 5. circulatory system   _____ 8. serosa membrane

_____ 3. lining of gut   _____ 6. epidermis of skin   _____ 9. liver, pancreas

In Utero Development

15. Make the following comparisons between a human and the dissected structures of a pregnant animal.

Comparison object	Human	Dissected animal
Shape of the placenta		
Shape of the uterus		

16. Where in the human uterus do implantation and placentation ordinarily occur? _____

_____

17. Describe the function(s) of the placenta. _____

_____

What embryonic membranes has the placenta more or less "put out of business"? _____

_____

18. When does the human embryo come to be called a fetus? _____

19. What is the usual and most desirable fetal position in utero? _____

Why is this the most desirable position? _____

_____

_____

_____

Gross and Microscopic Anatomy of the Placenta

20. Describe fully the gross structure of the human placenta as observed in the laboratory. _____

_____

_____

21. What is the tissue origin of the placenta: fetal, maternal, or both? _____

22. What placental barriers must be crossed to exchange materials? _____

_____

Principles of Heredity

MATERIALS

- ☐ Pennies (for coin tossing)
- ☐ PTC (phenylthiocarbamide) taste strips
- ☐ Sodium benzoate taste strips
- ☐ Chart drawn on board for tabulation of class results of human phenotype/genotype determinations

Activity 5: Blood typing

- ☐ Anti-A and Anti-B sera
- ☐ Clean microscope slides
- ☐ Toothpicks
- ☐ Wax pencils
- ☐ Sterile lancets
- ☐ Alcohol swabs
- ☐ Beaker containing 10% bleach solution
- ☐ Disposable autoclave bag

Activity 6: Hemoglobin phenotyping

- ☐ Electrophoresis equipment and power supply
- ☐ 1.2% agarose gels
- ☐ 1X TBE (Tris-Borate/EDTA) buffer pH 8.4
- ☐ Micropipette or variable automicropipette with tips
- ☐ Marking pen
- ☐ Safety goggles (student-provided)
- ☐ Metric ruler
- ☐ Plastic baggies
- ☐ Disposable gloves
- ☐ Coomassie blue protein stain solution
- ☐ Coomassie blue de-stain solution
- ☐ Distilled water
- ☐ Staining tray
- ☐ 100-ml graduated cylinder
- ☐ Hemoglobin samples dissolved in TBE solubilizing buffer with bromophenol blue: HbA, labeled A; HbS, labeled S; HbA + HbS, labeled AS; and unknown samples of each

MasteringA&P® For related exercise study tools, go to the Study Area of MasteringA&P. There you will find:

- Practice Anatomy Lab **PAL**
- PhysioEx **PEx**
- A&PFlix **A&PFlix**
- Practice quizzes, Histology Atlas, eText, Videos, and more!

OBJECTIVES

1. Define *allele, heterozygous, homozygous, dominance, recessiveness, genotype, phenotype,* and *incomplete dominance.*
2. Work simple genetics problems using a Punnett square.
3. State the basic laws of probability.
4. Observe selected human phenotypes, and determine their genotype basis.
5. Separate variants of hemoglobin using agarose gel electrophoresis.

PRE-LAB QUIZ

1. Circle the correct underlined term. Genes that code for the same genetic trait are <u>alleles</u> / <u>sister chromatids</u>.
2. Circle True or False. A heterozygous individual will have two of the same alleles in a chromosome pair.
3. The allele with less potency, which is present but not expressed, is the _____ allele.
   a. dominant
   b. genotypic
   c. homozygous
   d. recessive
4. Circle the correct underlined term. The physical appearance of an individual's genetic makeup, the characteristics that we can see, is that person's <u>genotype</u> / <u>phenotype</u>.
5. A _____ is used to demonstrate the genotypes of potential offspring that might result from mating.
   a. karyotype
   b. phenotype cross
   c. Punnett square
6. A condition known as _____ can result when heterozygous individuals exhibit a phenotype intermediate between homozygous individuals, and both alleles are expressed in the offspring.
   a. incomplete dominance
   b. sex-linked inheritance
   c. total dominance
   d. total heterozygosity
7. Circle the correct underlined term. Possession of the <u>X</u> / <u>Y</u> chromosome determines maleness.
8. Circle True or False. The Y chromosome is only about one-third the size of the X chromosome, and it lacks many of the genes that are found on the X chromosome.
9. Circle True or False. Males are more likely to inherit hemophilia because it is a result of receiving the sex-linked gene from the father.
10. Circle True or False. The inheritance of ABO blood type involves three possible alleles.

The field of genetics is bristling with excitement. Complex gene-splicing techniques have allowed researchers to precisely isolate genes coding for specific proteins and then to use those genes to harvest large amounts of specific proteins and even to cure some dreaded human diseases. At present, growth hormone, insulin, erythropoietin, and interferon produced by these genetic engineering techniques are available for clinical use, and the list is growing daily.

Understanding the genetics involved in such studies requires arduous training. However, anyone can gain a basic understanding and appreciation of how genes regulate our various traits (dimples and hair color, for example). This exercise provides a "genetics sampler," a relatively simple introduction to the principles of heredity.

Introduction to the Language of Genetics

In humans all cells, except eggs and sperm, contain 46 chromosomes, that is, the diploid number. This number is established when fertilization occurs and the egg and sperm fuse, combining 23 chromosomes—the haploid number—each is carrying. The diploid chromosomal number is maintained throughout life in nearly all cells of the body by the process of mitosis. The diploid chromosomal number actually represents two complete (or nearly complete) sets of genetic instructions—one from the mother and the other from the father—or 23 pairs of *homologous chromosomes*.

Genes coding for the same traits on each pair of homologous chromosomes are called **alleles.** The alleles may be identical or different in their influence. For example, the pair of alleles coding for hairline shape on your forehead may specify either straight across or widow's peak. When both alleles in a homologous chromosome pair have the same expression, the individual is **homozygous** for that trait. When the alleles differ in their expression, the individual is **heterozygous** for the given trait; and often only one of the alleles, called the **dominant gene,** will exert its effects. The allele with less potency, the **recessive gene,** will be present but suppressed. Whereas dominant genes, or alleles, exert their effects in both homozygous and heterozygous conditions, as a rule recessive alleles *must* be present in double dose (homozygosity) to exert their influence.

An individual's actual genetic makeup, that is, whether he is homozygous or heterozygous for the various alleles, is called his **genotype.** The expression of the genotype, for example, the presence of a widow's peak or not (Figure 45.2) is referred to as a **phenotype.**

The complete story of heredity is much more complex than just outlined, and in actuality the expression of many traits (for example, eye color) is determined by the interaction of many allele pairs. However, our emphasis here will be to investigate only the less complex aspects of genetics.

Dominant-Recessive Inheritance

One of the best ways to master the terminology and learn the principles of heredity is to work out the solutions to some genetic crosses in much the same way Gregor Mendel did in his classic experiments on pea plants. Mendel, an Austrian

monk of the mid-1800s, found evidence in these experiments that each gamete contributes just one allele to each pair in the zygote.

To work out the various simple monohybrid (one pair of alleles) crosses in this exercise, you will be given the genotype of the parents. You will then determine the possible genotypes of their offspring by using a grid called the *Punnett square,* and you will record the percentages of both genotype and phenotype. To illustrate the procedure, an example of one of Mendel's pea plant crosses is outlined next.

Alleles: *T* (determines *tallness;* dominant)

t (determines *dwarfism;* recessive)

Genotypes of parents: *TT* (♂) × *tt* (♀)

Phenotypes of parents: Tall × dwarf

To use the Punnett, or checkerboard, square, write the alleles (actually gametes) of one parent across the top and the gametes of the other parent down the left side. Then combine the gametes across and down to achieve all possible combinations, as follows:

Results: Genotypes 100% *Tt* (all heterozygous)
Phenotypes 100% tall (because *T*, which determines tallness, is dominant and all contain the *T* allele)

ACTIVITY 1

Working Out Crosses Involving Dominant and Recessive Genes

For each of the following crosses, draw your own Punnett square and use the technique outlined above to determine the genotypes and phenotypes of the offspring.

1. Genotypes of parents: *Tt* (♂) × *tt* (♀)

% of each genotype: _____

% of each phenotype: _____% tall, _____% dwarf

2. Genotypes of parents: *Tt* (♂) × *Tt* (♀)

% of each genotype: _____

% of each phenotype: _____% tall, _____% dwarf

3. Genotypes of parents: *TT* (♂) × *Tt* (♀)

% of each genotype: _____

% of each phenotype: _____% tall, _____% dwarf ■

Incomplete Dominance

The concepts of dominance and recessiveness are somewhat arbitrary and artificial in some instances because so-called dominant genes may be expressed differently in homozygous and heterozygous individuals. This produces a condition called **incomplete dominance,** or *intermediate inheritance.* In such cases, both alleles express themselves in the offspring. The crosses are worked out in the same manner as indicated previously, but heterozygous offspring exhibit a phenotype intermediate between that of the homozygous individuals. Some examples follow.

ACTIVITY 2

Working Out Crosses Involving Incomplete Dominance

1. The inheritance of flower color in snapdragons illustrates the principle of incomplete dominance. The genotype *RR* is expressed as a red flower, *Rr* yields pink flowers, and *rr* produces white flowers. Work out the following crosses to determine the expected phenotypes and both genotypic and phenotypic percentages.

a. Genotypes of parents: *RR* × *rr*

% of each genotype: _____

% of each phenotype: _____

b. Genotypes of parents: *Rr* × *rr*

% of each genotype: _____

% of each phenotype: _____

_____

c. Genotypes of parents: *Rr* × *Rr*

% of each genotype: _____

% of each phenotype: _____

_____

2. In humans, the inheritance of sickle cell anemia/trait is determined by a single pair of alleles that exhibit incomplete dominance. Individuals homozygous for the sickling gene *(s)* have *sickle cell anemia.* In double dose *(ss),* the sickling gene causes production of a very abnormal hemoglobin, which crystallizes and becomes sharp and spiky under conditions of oxygen deficit. This, in turn, leads to clumping and hemolysis of red blood cells in the circulation, which causes a great deal of pain and can be fatal. Heterozygous individuals *(Ss)* have the *sickle cell trait;* they make both normal and sickling hemoglobin. Usually these individuals are healthy, but prolonged decreases in blood oxygen levels can lead to a sickle cell crisis. Individuals with the genotype *SS* form normal hemoglobin. Work out the following crosses:

a. Parental genotypes: *SS* × *ss*

% of each genotype: _____

% of each phenotype: _____

b. Parental genotypes: *Ss* × *Ss*

% of each genotype: _____

% of each phenotype: _____

_____

c. Parental genotypes: *ss* × *Ss*

% of each genotype: _____

% of each phenotype: _____ ■

Sex-Linked Inheritance

A cell's chromosomes can be stained, photographed, and digitally rearranged to produce an image called a *karyotype,* which shows the complete human diploid chromosomal complement displayed in homologous pairs **(Figure 45.1)**.

Of the 23 pairs of homologous chromosomes, 22 pairs are referred to as **autosomes.** The autosomes guide the expression of most body traits. The 23rd pair, the **sex chromosomes,** determine the sex of an individual, that is, whether an individual will be male or female. Normal females

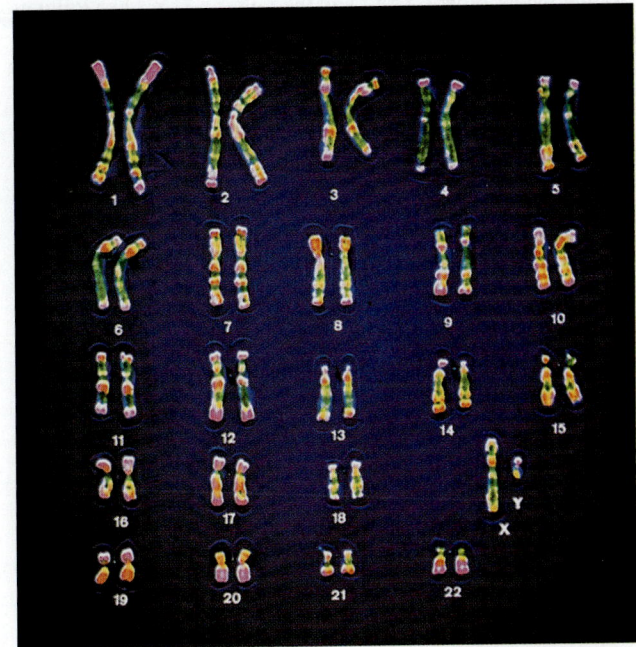

Figure 45.1 Karyotype (chromosomal complement) of human male. Each pair of homologous chromosomes is numbered except the sex chromosomes, which are identified by their letters, X and Y.

45

possess two sex chromosomes that look alike, the X chromosomes. Males possess two dissimilar sex chromosomes, referred to as X and Y. Possession of the Y chromosome determines maleness. The Y sex chromosome is only about a third the size of the X sex chromosome, and it lacks many of the genes that are found on the X.

Inherited traits determined by genes on the sex chromosomes are said to be *sex-linked,* and genes present *only* on the X sex chromosome are said to be *X-linked.* Some examples of X-linked genes include those that determine normal color vision (or, conversely, color blindness), and normal clotting ability (as opposed to hemophilia). The alleles that determine color blindness and hemophilia are recessive alleles. In females, *both* X chromosomes must carry the recessive alleles for a woman to express either of these conditions, and thus they tend to be infrequently seen. However, should a male receive an X-linked recessive allele for these conditions, he will exhibit the recessive phenotype because his Y chromosome does not contain alleles for that gene.

The critical point to understand about X-linked inheritance is the *absence* of male to male (that is, father to son) transmission of sex-linked genes. The X of the father *will* pass to each of his daughters but to none of his sons. Males always inherit sex-linked conditions from their mothers via the X chromosome.

ACTIVITY 3

Working Out Crosses Involving Sex-Linked Inheritance

1. A heterozygous woman carrying the recessive gene for color blindness marries a man who is color-blind. Assume the dominant gene is X^C (allele for normal color vision) and the recessive gene is X^c (determines color blindness). The mother's genotype is $X^C X^c$ and the father's $X^c Y$. Do a Punnett square to determine the answers to the following questions.

According to the laws of probability, what percentage of all their children will be color-blind?

_____ %

What is the percentage of color-blind individuals by sex?

_____% males; _____% females

What percentage of all children will be carriers? _____%

What is the sex of the carriers? _____

2. A heterozygous woman carrying the recessive gene for hemophilia marries a man who is not a hemophiliac. Assume the dominant gene is X^H and the recessive gene is X^h. The woman's genotype is $X^H X^h$ and her husband's genotype is $X^H Y$. What is the potential percentage and sex of their offspring who will be hemophiliacs?

_____% males; _____% females

What percentage can be expected to lack the allele for hemophilia?

_____%

What is the anticipated sex and percentage of individuals who will be carriers for hemophilia?

_____%; _____sex ▬

Probability

Parceling out of chromosomes to gametes during meiosis and the combination of egg and sperm are random events. Hence, the possibility that certain genomes will arise and be expressed is based on the laws of probability. The randomness of gene recombination from each parent determines individual uniqueness and explains why siblings, however similar, never have totally corresponding traits (unless, of course, they are identical twins). The Punnett square method that you have been using to work out the genetics problems actually provides information on the *probability* of the appearance of certain genotypes considering all possible events. Probability *(P)* is defined as:

$$P = \frac{\text{number of specific events or cases}}{\text{total number of events or cases}}$$

If an event is certain to happen, its probability is 1. If it happens one out of every two times, its probability is ½; if one out of four times, its probability is ¼, and so on.

When figuring the probability of separate events occurring together (or consecutively), the probability of each event must be multiplied together to get the final probability figure. For example, the probability of a penny coming up "heads" in each toss is ½ (because it has two sides—heads and tails). But the probability of a tossed penny coming up heads four times in a row is ½ × ½ × ½ × ½ = $\frac{1}{16}$.

ACTIVITY 4

Exploring Probability

1. Obtain two pennies and perform the following simple experiment to explore the laws of probability.

a. Toss one penny into the air ten times, and record the number of heads (H) and tails (T) observed.

_____ heads _____ tails

Probability: _____/10 tails; _____/10 heads

b. Now simultaneously toss two pennies into the air for 24 tosses, and record the results of each toss below. In each case, report the probability in the lowest fractional terms.

# of HH: _____   Probability: _____

# of HT: _____   Probability: _____

# of TT: _____   Probability: _____

Does the first toss have any influence on the second?

_____

Does the third toss have any influence on the fourth?

_____

c. Do a Punnett square using HT for the "alleles" of one "parental" coin and HT for the "alleles" of the other.

Probability of HH: _____

Probability of HT: _____

Probability of TT: _____

How closely do your coin-tossing results correlate with the percentages obtained from the Punnett square results?

_____

_____

2. Determine the probability of having a boy or girl offspring for each conception.

Parental genotypes: XY × XX

Probability of males: _____%

Probability of females: _____%

3. Dad wants a baseball team! What are the chances of his having nine sons in a row?

_____ (Sorry, Dad! Let the girls play.) ▬

Genetic Determination of Selected Human Characteristics

Most human traits are determined by multiple alleles or the interaction of several gene pairs. However, a few visible human traits or phenotypes can be traced to a single gene pair. It is some of these that will be investigated here.

ACTIVITY 5

Using Phenotype to Determine Genotype

For each of the characteristics described here, determine (as best you can) both your own phenotype and genotype, and record this information on the **Record of Human Genotypes/Phenotypes** chart. Since it is impossible to know if you are homozygous or heterozygous for a nondetrimental trait when you exhibit its dominant expression, you are to record your genotype as *A*—(or *B*—, and so on, depending on the letter used to indicate the alleles) in such cases. As you will see, all the traits examined here are nonharmful characteristics.

Record of Human Genotypes/Phenotypes		
Characteristic	Phenotype	Genotype
PTC taste *(P,p)*		
Sodium benzoate taste *(S,s)*		
Sex *(X,Y)*		
Dimples *(D,d)*		
Widow's peak *(W,w)*		
Proximal finger hair *(H,h)*		
Freckles *(F,f)*		
Blaze *(B,b)*		
ABO blood type *(I^A, I^B, i)*		

Generally speaking, in humans dominant gene disorders are presumed to be heterozygous (one dominant and one recessive allele), because having two such defective mutated alleles is usually not compatible with life. On the other hand, if you exhibit the recessive trait, you are homozygous for the recessive allele and should record it accordingly as *aa* (*bb*, *cc*, and so on). When you have completed your observations, also record your data in the chart on the board for tabulation of class results.

PTC taste: Obtain a PTC taste strip. PTC, or phenylthiocarbamide, is a harmless chemical that some people can taste and others find tasteless. Chew the strip. If it tastes slightly bitter, you are a "taster" and possess the dominant gene *(P)* for this trait. If you cannot taste anything, you are a nontaster and are homozygous recessive *(pp)* for the trait. Approximately 70% of the people in the United States are tasters.

Sodium benzoate taste: Obtain a sodium benzoate taste strip and chew it. A different pair of alleles (from that determining PTC taste) determines the ability to taste sodium benzoate. If you can taste it, you have at least one of the dominant alleles *(S)*. If not, you are homozygous recessive *(ss)* for the trait. Also record whether sodium benzoate tastes salty, bitter, or sweet to you (if a taster). Even though PTC and sodium benzoate taste are inherited independently, they interact to determine a person's taste sensations. Individuals who find PTC bitter and sodium benzoate salty tend to like sauerkraut, buttermilk, spinach, and other slightly bitter or salty foods.

Sex: The genotype XX determines the female phenotype, whereas XY determines the male phenotype.

Dimpled cheeks: The presence of dimples in one or both cheeks is due to a dominant gene *(D)*. Absence of dimples indicates the homozygous recessive condition *(dd)*.

Widow's peak: A distinct downward V-shaped hairline at the middle of the forehead is referred to as a widow's peak. It is determined by a dominant allele *(W)*, whereas the straight or continuous forehead hairline is determined by the homozygous recessive condition *(ww)* **(Figure 45.2)**.

Proximal finger hair: Critically examine the dorsum of the proximal phalanx of fingers 3 and 4. If no hair is obvious, you are recessive *(hh)* for this condition. If hair is seen, you have the dominant gene *(H)* for this trait (which, however, is determined by multigene inheritance) (Figure 45.2).

Freckles: Freckles are the result of a dominant gene. Use *F* as the dominant allele and *f* as the recessive allele (Figure 45.2).

Dominant traits

Recessive traits

Widow's peak

Straight hairline

Finger hair

No finger hair

Freckles

No freckles

Figure 45.2 Selected examples of human phenotypes.

Blaze: A lock of hair different in color from the rest of scalp hair is called a blaze; it is determined by a dominant gene. Use *B* for the dominant gene and *b* for the recessive gene.

Blood type: Some genes exhibit more than two allele forms, leading to a phenomenon called **multiple-allele inheritance.** Inheritance of the ABO blood type is based on the existence of three alleles designated as I^A, I^B, and i. Both I^A and I^B are dominant over i, but neither is dominant over the other. The I^A and I^B alleles are *codominant.* Thus the possession of I^A and I^B will yield type AB blood, whereas the possession of the I^A and i alleles will yield type A blood, and so on (as explained in Exercise 29). The four ABO blood groups or phenotypes are A, B, AB, and O (their correlation to genotype is indicated in the Blood Groups table **(Table 45.1)**.

Table 45.1	**Blood Groups**
ABO blood group (phenotype)	**Genotype**
A	$I^A I^A$ or $I^A i$
B	$I^B I^B$ or $I^B i$
AB	$I^A I^B$
O	ii

If you have previously typed your blood, record your phenotype and genotype in the chart (page 671). If not, type your blood following your instructor's instructions (see Exercise 29, pages 434–435), and then enter your results in the table.

⚠ Dispose of any blood-soiled supplies by placing the glassware in the bleach-containing beaker and all other items in the autoclave bag.

Once class data have been tabulated, scrutinize the results. Is there a single trait that is expressed in an identical manner by all members of the class?

_____

Because all human beings have 23 pairs of homologues and each pair segregates independently at meiosis, the number of possible combinations at segregation is more than 8 million! On the basis of this information, what would you guess are the chances of any two individuals in the class having identical phenotypes for all 14 traits investigated?

_____

Hemoglobin Phenotype Identification Using Agarose Gel Electrophoresis

Agarose gel electrophoresis separates molecules based on charge. In the appropriate buffer with an alkaline pH, hemoglobin molecules will move toward the anode of the apparatus at different speeds, based on the number of negative charges on the molecules. Agarose gel provides a medium for travel and slows the migration down a bit.

Sickle cell anemia and sickle cell trait are discussed in Activity 2 of this exercise (and slides are observed in Exercise 29). The beta chains of hemoglobin S (HbS) contain a base substitution where a valine replaces glutamic acid. As a result of the substitution, HbS has fewer negative charges than the predominant form of adult hemoglobin (HbA) and can be separated from HbA using agarose gel electrophoresis.

ACTIVITY 6

Using Agarose Gel Electrophoresis to Identify Normal Hemoglobin, Sickle Cell Anemia, and Sickle Cell Trait

1. You will need an electrophoresis unit and power supply **(Figure 45.3)**, a 1.2% agarose gel with eight wells, 1X TBE buffer, micropipettes or a variable automatic micropipette (2–20 μl) with tips, samples of hemoglobin (marked A, AS, S, and unknown #_____) dissolved in TBE solubilizing buffer containing bromophenol blue, a marking pen, safety goggles, metric ruler, plastic baggie, and disposable gloves.

2. Record the number of your unknown sample. _____

3. Place the agarose gel into the electrophoresis unit.

4. Using a micropipette, carefully add 15 μl of hemoglobin sample A to wells number 1 and 5, AS to wells 2 and 6, S to wells 3 and 7, and the unknown samples to wells 4 and 8.

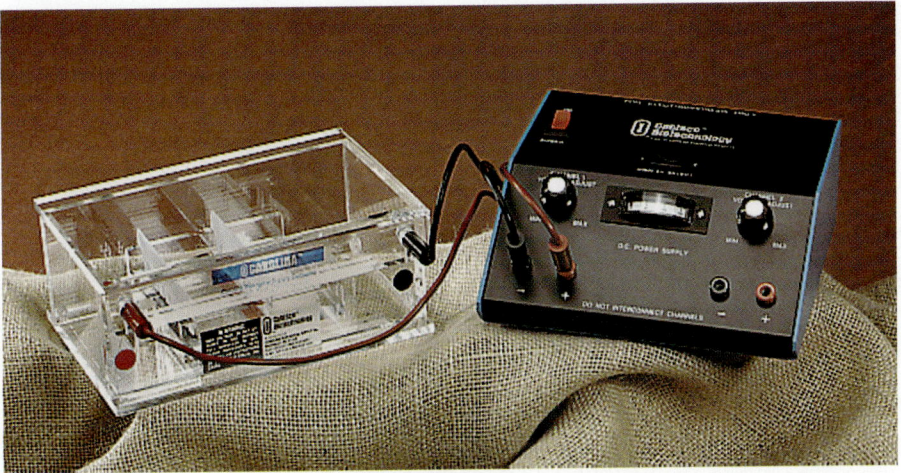

Figure 45.3 Agarose gel electrophoresis equipment and power supply.

5. Slowly add electrophoresis buffer until the gels are covered with about 0.25 cm of buffer.

⚠ 6. The electrophoresis unit runs with high voltage. Do not attempt to open it while the power supply is attached. Close and lock the electrophoresis unit, and connect the unit to the power source, red to red and black to black.

7. Run the unit about 50 minutes at 120 volts until the bromophenol blue is about 0.25 cm from the anode.

8. Turn the power supply **OFF**. Disconnect the cables.

9. Open the electrophoresis unit, carefully remove the gel, and slide the gel into a plastic baggie. You should be able to see the hemoglobin bands on the gel. Mark each of the bands on the gel with the marking pen.

Alternatively the gels may be stained with Coomassie blue. Obtain a flask of Coomassie blue stain, a flask of de-staining solution, a flask of distilled water, a staining tray, and a 100-ml graduated cylinder, and do the following:

a. Carefully remove the gel from the plastic plate, and place the gel into a staining dish.

b. Add about 30 ml of stain (enough stain to cover the gel), and be sure the agarose is not stuck to the dish.

c. Allow the gel to remain in the stain for at least an hour (more time might be necessary), then remove the stain and pour into an appropriate waste container. Rinse the gel and dish with distilled water.

d. Add about 100 ml of de-staining solution. Change the solution after a day. If the background stain has been reduced enough to see the bands, place the staining dish over a light source, and observe the bands. If the stain is still too dark, repeat the de-staining process until the bands can be observed.

e. To store the gels, refrigerate in a baggie with a small amount of de-stain solution or dry on a glass plate.

10. Draw the banding patterns for samples 1 through 8 in the figure for question 16 in the Review Sheet (page 678). Based on the banding patterns of the known samples, what are the genotypes of your unknown samples? Record on the Review Sheet chart.

11. Rinse the electrophoresis unit in distilled or de-ionized water, and clean the glass plates with soap and water. ▬

45

Odd Phenotype Out

The following boxes each contain four phenotypes. One of the listed traits in each case does not belong with the other three for some reason. Circle the trait that doesn't belong with the others and explain how it is different from the others.

1. Which is the "odd phenotype"?	Why is it the odd one out?
Freckles Widow's peak Sickle cell trait Sodium benzoate "taster"	
2. Which is the "odd phenotype"?	**Why is it the odd one out?**
Blaze Type AB blood PTC "taster" Widow's peak	
3. Which is the "odd phenotype"?	**Why is it the odd one out?**
No proximal finger hair Straight hairline Color blindness Sodium benzoate "nontaster"	
4. Which is the "odd phenotype"?	**Why is it the odd one out?**
Dimpled cheek Straight hairline Freckles Blaze	

45

Principles of Heredity

Introduction to the Language of Genetics

1. Match the key choices with the definitions given below.

Key: a. alleles   d. genotype   g. phenotype
      b. autosomes   e. heterozygous   h. recessive
      c. dominant   f. homozygous   i. sex chromosomes

_____ 1. actual genetic makeup

_____ 2. chromosomes determining maleness/femaleness

_____ 3. situation in which an individual has identical alleles for a particular trait

_____ 4. genes not expressed unless they are present in homozygous condition

_____ 5. expression of a genetic trait

_____ 6. situation in which an individual has different alleles making up his genotype for a particular trait

_____ 7. genes for the same trait that may have different expressions

_____ 8. chromosomes regulating most body characteristics

_____ 9. the more potent gene allele; masks the expression of the less potent allele

Dominant-Recessive Inheritance

2. In humans, farsightedness is inherited by possession of a dominant allele *(A)*. If a man who is homozygous for normal vision *(aa)* marries a woman who is heterozygous for farsightedness *(Aa),* what proportion of their children would be expected to be farsighted?

_____%

3. A metabolic disorder called phenylketonuria (PKU) is due to an abnormal recessive gene *(p)*. Only homozygous recessive individuals exhibit this disorder. What percentage of the offspring will be anticipated to have PKU if the parents are *Pp* and *pp*?

_____%

4. A man obtained 32 spotted and 10 solid-color rabbits from a mating of two spotted rabbits.

Which trait is dominant? _____ Recessive? _____

If the dominant allele is *S,* what is the probable genotype of the rabbit parents? _____ × _____

5. Assume that the allele controlling brown eyes *(B)* is dominant over that controlling blue eyes *(b)* in human beings. (In actuality, eye color in humans is an example of polygenic inheritance, which is much more complex than this.) A blue-eyed man marries a brown-eyed woman, and they have six children, all brown-eyed. What is the most likely genotype of the father?

_____ Of the mother? _____ If the seventh child had *blue* eyes, what could you conclude about the parents' genotypes?

_____

_____

Incomplete Dominance

6. Tail length on a bobcat is controlled by incomplete dominance. The alleles are *T* for normal tail length and *t* for tail-less.

What name could/would you give to the tails of heterozygous *(Tt)* cats? _____

How would their tail length compare with that of *TT* or *tt* bobcats? _____

7. If curly-haired individuals are genotypically *CC*, straight-haired individuals are *cc*, and wavy-haired individuals are heterozygotes *(Cc),* what percentage of the various phenotypes would be anticipated from a cross between a *CC* woman and a *cc* man?

_____% curly   _____% wavy   _____% straight

Sex-Linked Inheritance

8. What does it mean when someone says a particular characteristic is sex-linked? _____

_____

9. You are a male, and you have been told that hemophilia "runs in your genes." Whose ancestors, your mother's or your

father's, should you investigate? _____ Why? _____

_____

_____

10. An $X^C X^c$ female marries an $X^C Y$ man. Do a Punnett square for this match.

What is the probability of producing a color-blind son? _____

A color-blind daughter? _____

A daughter who is a carrier for the color-blind allele? _____

11. Why are consanguineous marriages (marriages between blood relatives) prohibited in most cultures?

_____

_____

Probability

12. What is the probability of having three daughters in a row? _____

13. A man and a woman, each of seemingly normal intellect, marry. Although neither is aware of the fact, each is a heterozygote

for the allele for mental retardation. Is the allele for mental retardation dominant or recessive? _____

What are the chances of their having one mentally retarded child? _____

What are the chances that all their children (they plan a family of four) will be mentally retarded? _____

Genetic Determination of Selected Human Characteristics

14. Look back at your data to complete this section. For each of the situations described here, determine if an offspring with the characteristics noted is possible with the parental genotypes listed. Check (✓) the appropriate column.

Parental genotypes	Phenotype of child	Possibility	
		Yes	No
$Ff \times ff$	Freckles		
$DD \times dd$	Dimples		
$HH \times Hh$	Proximal finger hair		
$I^A i \times I^B i$	Type O blood		
$I^A I^B \times ii$	Type B blood		

15. You have dimples, and you would like to know if you are homozygous or heterozygous for this trait. You have six brothers and sisters. By observing your siblings, how could you tell, with some degree of certainty, that you are a heterozygote?

_____

_____

_____

Using Agarose Gel Electrophoresis to Identify Hemoglobin Phenotypes

16. Draw the banding patterns you obtained on the figure below.

Sample	Well	Banding pattern
1. A	1. ☐	
2. AS	2. ☐	
3. S	3. ☐	
4. Unknown	4. ☐	
5. A	5. ☐	
6. AS	6. ☐	
7. S	7. ☐	
8. Unknown	8. ☐	

17. What is the genotype of sickle cell anemia? _____ Sickle cell trait? _____

18. Why does sickle cell hemoglobin behave differently from normal hemoglobin during agarose gel electrophoresis?

_____

_____

Surface Anatomy Roundup

MATERIALS

- ☐ Articulated skeletons
- ☐ Three-dimensional models or charts of the skeletal muscles of the body
- ☐ Hand mirror
- ☐ Stethoscope
- ☐ Alcohol swabs
- ☐ Washable markers

OBJECTIVES

1. Define *surface anatomy* and explain why it is an important field of study; define *palpation*.

2. Describe and palpate the major surface features of the cranium, face, and neck.

3. Describe the easily palpated bony and muscular landmarks of the back, and locate the vertebral spines on the living body.

4. List the bony surface landmarks of the thoracic cage, explain how they relate to the major soft organs of the thorax, and explain how to find the second to eleventh ribs.

5. Name and palpate the important surface features on the anterior abdominal wall, and explain how to palpate a full bladder.

6. Define and explain the following: *linea alba, umbilical hernia,* examination for an inguinal hernia, *linea semilunaris,* and *McBurney's point.*

7. Locate and palpate the main surface features of the upper limb.

8. Explain the significance of the cubital fossa, pulse points in the distal forearm, and the anatomical snuff box.

9. Describe and palpate the surface landmarks of the lower limb.

10. Explain exactly where to administer an injection in the gluteal region and in the other major sites of intramuscular injection.

PRE-LAB QUIZ

1. Why is it useful to study surface anatomy?
   a. You can easily locate deep muscle insertions.
   b. You can relate external surface landmarks to the location of internal organs.
   c. You can study cadavers more easily.
   d. You really can't learn that much by studying surface anatomy; it's a gimmick.

2. Circle the correct underlined term. <u>Palpation</u> / <u>Dissection</u> allows you to feel internal structures through the skin.

3. The epicranial aponeurosis binds to the subcutaneous tissue of the cranium to form the
   a. mastoid process   c. true scalp
   b. occipital protuberance   d. xiphoid process

4. The _____ is the most prominent neck muscle and also the neck's most important landmark.
   a. buccinator   c. masseter
   b. epicranius   d. sternocleidomastoid

5. The three boundaries of the _____ are the trapezius medially, the latissimus dorsi inferiorly, and the scapula laterally.
   a. torso triangle   c. triangle of back muscles
   b. triangle of ausculation   d. triangle of McBurney

6. Circle True or False. The lungs do not fill the inferior region of the pleural cavity.

(Text continues on next page.)

MasteringA&P® For related exercise study tools, go to the Study Area of MasteringA&P. There you will find:

- Practice Anatomy Lab PAL
- PhysioEx PEx
- A&PFlix *A&PFlix*
- Practice quizzes, Histology Atlas, eText, Videos, and more!

7. Circle True or False. With the exception of a full bladder, most internal pelvic organs are not easily palpated through the skin of the body surface.

8. On the dorsum of your hand is a grouping of superficial veins known as the _____, which provides a site for drawing blood and inserting intravenous catheters.
   a. anatomical snuff box
   b. dorsal venous network
   c. radial and ulnar veins
   d. palmar arches

9. Circle True or False. To avoid harming major nerves and blood vessels, clinicians who administer intramuscular injections in the gluteal region of adults use the gluteus medius muscle.

10. The large femoral artery and vein descend vertically through the _____, formed by the border of the inguinal ligament, the medial border of the adductor longus muscle, and the medial border of the sartorius muscle.
    a. femoral triangle
    b. lateral condyle
    c. medial condyle
    d. quadriceps

S urface anatomy is a valuable branch of anatomical and medical science. True to its name, **surface anatomy** does indeed study the *external surface* of the body, but more importantly, it also studies *internal* organs as they relate to external surface landmarks and as they are seen and felt through the skin. Feeling internal structures through the skin with the fingers is called **palpation** (literally, "touching").

Surface anatomy is living anatomy, better studied in live people than in cadavers. It can provide a great deal of information about the living skeleton (almost all bones can be palpated) and about the muscles and blood vessels that lie near the body surface. Furthermore, a skilled examiner can learn a good deal about the heart, lungs, and other deep organs by performing a surface assessment. Thus, surface anatomy serves as the basis of the standard physical examination. For those planning a career in the health sciences or physical education, a study of surface anatomy will show you where to take pulses, where to insert tubes and needles, where to locate broken bones and inflamed muscles, and where to listen for the sounds of the lungs, heart, and intestines.

We will take a regional approach to surface anatomy, exploring the head first and proceeding to the trunk and the limbs. You will be observing and palpating your own body as you work through the exercise, because your body is the best learning tool of all. To aid your exploration of living anatomy, skeletons and muscle models or charts are provided around the lab so that you can review the bones and muscles you will encounter. For skin sites you are asked to mark that you cannot reach on your own body, it probably would be best to choose a male student as a subject.

ACTIVITY 1

Palpating Landmarks of the Head

The head (**Figure 46.1** and **Figure 46.2**) is divided into the cranium and the face.

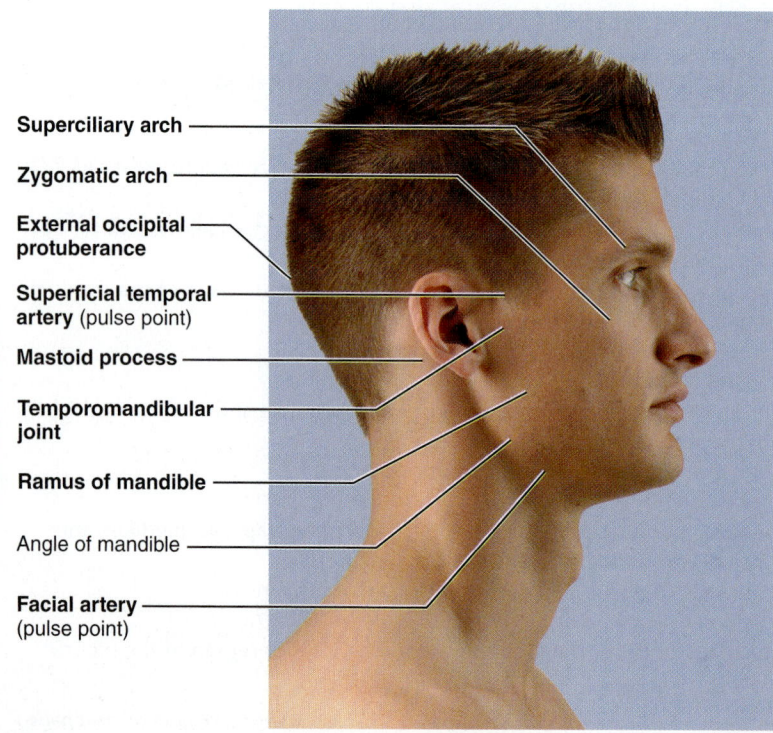

Superciliary arch
Zygomatic arch
External occipital protuberance
Superficial temporal artery (pulse point)
Mastoid process
Temporomandibular joint
Ramus of mandible
Angle of mandible
Facial artery (pulse point)

(a)

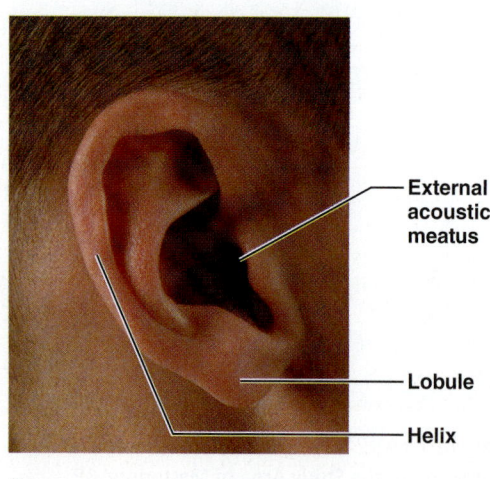

External acoustic meatus
Lobule
Helix

(b)

Figure 46.1 Surface anatomy of the head. (a) Lateral aspect. **(b)** Close-up of an auricle.

46

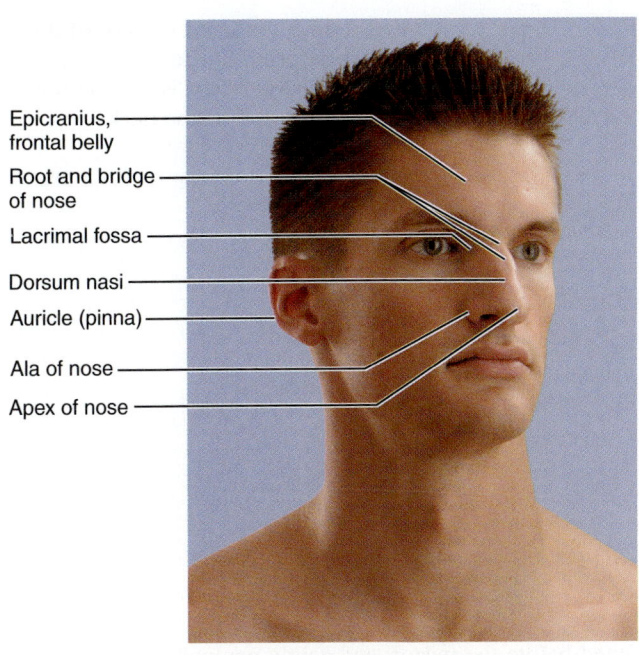

Epicranius, frontal belly

Root and bridge of nose

Lacrimal fossa

Dorsum nasi

Auricle (pinna)

Ala of nose

Apex of nose

Figure 46.2 Surface structures of the face.

Cranium

1. Run your fingers over the superior surface of your head. Notice that the underlying cranial bones lie very near the surface. Proceed to your forehead and palpate the **super-ciliary arches** (brow ridges) directly superior to your orbits (Figure 46.1).

2. Move your hand to the posterior surface of your skull, where you can feel the knoblike **external occipital protuber-ance.** Run your finger directly laterally from this projection to feel the ridgelike *superior nuchal line* on the occipital bone. This line, which marks the superior extent of the muscles of the posterior neck, serves as the boundary between the head and the neck. Now feel the prominent **mastoid process** on each side of the cranium just posterior to your ear.

3. The **frontal belly** of the epicranius (Figure 46.2) inserts superiorly onto the broad aponeurosis called the *epicranial aponeurosis* (Table 13.1, page 200) that covers the superior surface of the cranium. This aponeurosis binds tightly to the overlying subcutaneous tissue and skin to form the true **scalp.** Push on your scalp, and confirm that it slides freely over the underlying cranial bones. Because the scalp is only loosely bound to the skull, people can easily be "scalped" (in indus-trial accidents, for example). The scalp is richly vascularized by a large number of arteries running through its subcutane-ous tissue. Most arteries of the body constrict and close after they are cut or torn, but those in the scalp are unable to do so because they are held open by the dense connective tissue surrounding them.

What do these facts suggest about the amount of bleeding that accompanies scalp wounds?

_____

_____

Face

The surface of the face is divided into many different regions, including the *orbital, nasal, oral* (mouth), and *auricular* (ear) areas.

1. Trace a finger around the entire margin of the bony orbit. The **lacrimal fossa,** which contains the tear-gathering lacri-mal sac, may be felt on the medial side of the eye socket.

2. Touch the most superior part of your nose, its **root,** which lies between the eyebrows (Figure 46.2). Just inferior to this, between your eyes, is the **bridge** of the nose formed by the nasal bones. Continue your finger's progress inferiorly along the nose's anterior margin, the **dorsum nasi,** to its tip, the **apex.** Place one finger in a nostril and another finger on the flared winglike **ala** that defines the nostril's lateral border.

3. Grasp your **auricle,** the shell-like part of the external ear that surrounds the opening of the **external acoustic meatus** (Figure 46.1). Now trace the ear's outer rim, or **helix,** to the **lobule** (earlobe) inferiorly. The lobule is easily pierced, and since it is not highly sensitive to pain, it provides a conve-nient place to hang an earring or obtain a drop of blood for clinical blood analysis. Next, place a finger on your temple just anterior to the auricle. There, you will be able to feel the pulsations of the **superficial temporal artery,** which ascends to supply the scalp (Figure 46.1).

4. Run your hand anteriorly from your ear toward the orbit, and feel the **zygomatic arch** just deep to the skin. This bony arch is easily broken by blows to the face. Next, place your fingers on the skin of your face, and feel it bunch and stretch as you contort your face into smiles, frowns, and grimaces. You are now monitoring the action of several of the subcuta-neous **muscles of facial expression** (Table 13.1, page 200).

5. On your lower jaw, palpate the parts of the bony **mandible:** its anterior body and its posterior ascending **ramus.** Press on the skin over the mandibular ramus, and feel the **masseter muscle** bulge when you clench your teeth. Palpate the anterior border of the masseter, and trace it to the mandible's inferior margin. At this point, you will be able to detect the pulse of your **facial artery** (Figure 46.1). Finally, to feel the **temporomandibular joint,** place a finger directly anterior to the external acoustic meatus of your ear, and open and close your mouth several times. The bony structure you feel moving is the *condylar process of the mandible.* ■

ACTIVITY 2

Palpating Landmarks of the Neck
Bony Landmarks

1. Run your fingers inferiorly along the back of your neck, in the posterior midline, to feel the *spinous processes* of the cervical vertebrae. The spine of C_7, the *vertebra prominens,* is especially prominent.

2. Now, beginning at your chin, run a finger inferiorly along the anterior midline of your neck **(Figure 46.3)**. The first hard structure you encounter will be the U-shaped **hyoid bone,** which lies in the angle between the floor of the mouth and the vertical part of the neck. Directly inferior to this, you will feel the **laryngeal prominence** (Adam's apple) of the thyroid cartilage. Just inferior to the laryngeal prominence, your finger will sink into a soft depression (formed by the

46

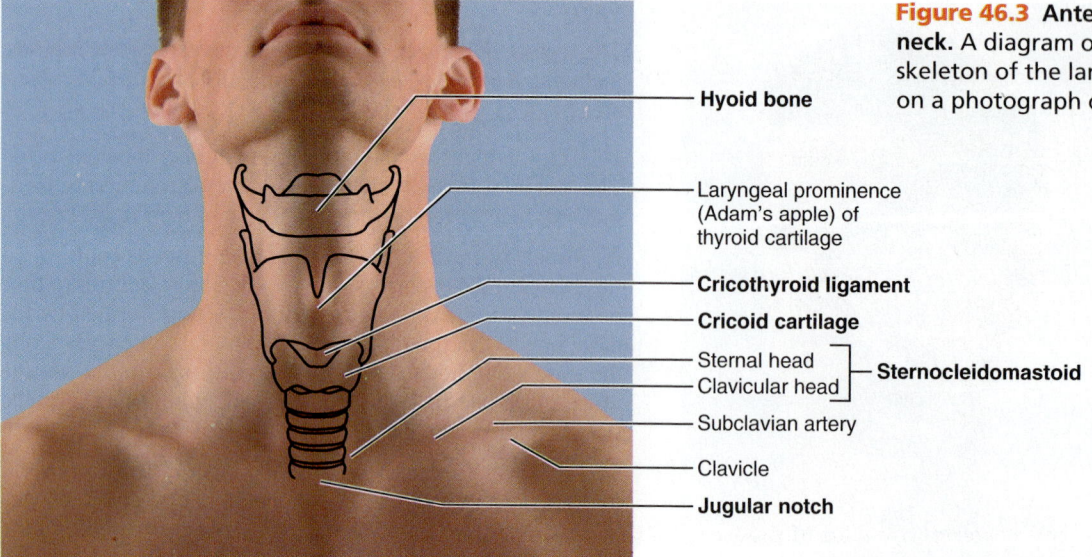

Hyoid bone

Laryngeal prominence (Adam's apple) of thyroid cartilage

Cricothyroid ligament

Cricoid cartilage

Sternal head
Clavicular head — Sternocleidomastoid

Subclavian artery

Clavicle

Jugular notch

cricothyroid ligament) before proceeding onto the rounded surface of the **cricoid cartilage.** Now swallow several times, and feel the whole larynx move up and down.

3. Continue inferiorly to the trachea. Attempt to palpate the *isthmus of the thyroid gland,* which feels like a spongy cushion over the second to fourth tracheal rings (Figure 46.3). Then, try to palpate the two soft lateral *lobes* of your thyroid gland along the sides of the trachea.

4. Move your finger all the way inferiorly to the root of the neck, and rest it in the **jugular notch,** the depression in the superior part of the sternum between the two clavicles. By pushing deeply at this point, you can feel the cartilage rings of the trachea.

Muscles

The **sternocleidomastoid** is the most prominent muscle in the neck and the neck's most important surface landmark. You can best see and feel it when you turn your head to the side.

Obtain a hand mirror, hold it in front of your face, and turn your head sharply from right to left several times. You will be able to see both heads of this muscle, the **sternal head** medially and the **clavicular head** laterally (Figure 46.3). Several important structures lie beside or beneath the sternocleidomastoid:

• The *cervical lymph nodes* lie both superficial and deep to this muscle. (Swollen cervical nodes provide evidence of infections or cancer of the head and neck.)

• The *common carotid artery* and *internal jugular vein* lie just deep to the sternocleidomastoid, a relatively superficial location that exposes these vessels to danger in slashing wounds to the neck.

• Just lateral to the inferior part of the sternocleidomastoid is the large **subclavian artery** on its way to supply the upper limb. By pushing on the subclavian artery at this point, one can stop the bleeding from a wound anywhere in the associated limb.

• Just anterior to the sternocleidomastoid, superior to the level of your larynx, you can feel a carotid pulse—the pulsations of the **external carotid artery (Figure 46.4).**

• The *external jugular vein* descends vertically, just superficial to the sternocleidomastoid and deep to the skin (Figure 46.4b). To make this vein "appear" on your neck,

stand before the mirror, and gently compress the skin superior to your clavicle with your fingers.

Triangles of the Neck

The sternocleidomastoid muscles divide each side of the neck into the posterior and anterior triangles (Figure 46.4a).

1. The **posterior triangle** is defined by the sternocleidomastoid anteriorly, the trapezius posteriorly, and the clavicle inferiorly. Palpate the borders of the posterior triangle.

The **anterior triangle** is defined by the inferior margin of the mandible superiorly, the midline of the neck anteriorly, and the sternocleidomastoid posteriorly.

2. The contents of these two triangles include nerves, glands, blood vessels, and small muscles (Figure 46.4b). The posterior triangle contains the **accessory nerve** (cranial nerve XI), most of the **cervical plexus,** and the **phrenic nerve.** In the inferior part of the triangle are the **external jugular vein,** the trunks of the **brachial plexus,** and the **subclavian artery.** These structures are relatively superficial and are easily cut or injured by wounds to the neck.

In the neck's anterior triangle, important structures include the **submandibular gland,** the **suprahyoid** and **infrahyoid muscles,** and parts of the **carotid arteries** and **jugular veins** that lie superior to the sternocleidomastoid.

• Palpate your carotid pulse.

A wound to the posterior triangle of the neck can lead to long-term loss of sensation in the skin of the neck and shoulder, as well as partial paralysis of the sternocleidomastoid and trapezius muscles. Explain these effects. ✚

_____

_____ ▬

ACTIVITY 3

Palpating Landmarks of the Trunk

The trunk of the body consists of the thorax, abdomen, pelvis, and perineum. The *back* includes parts of all of these regions, but for convenience it is treated separately.

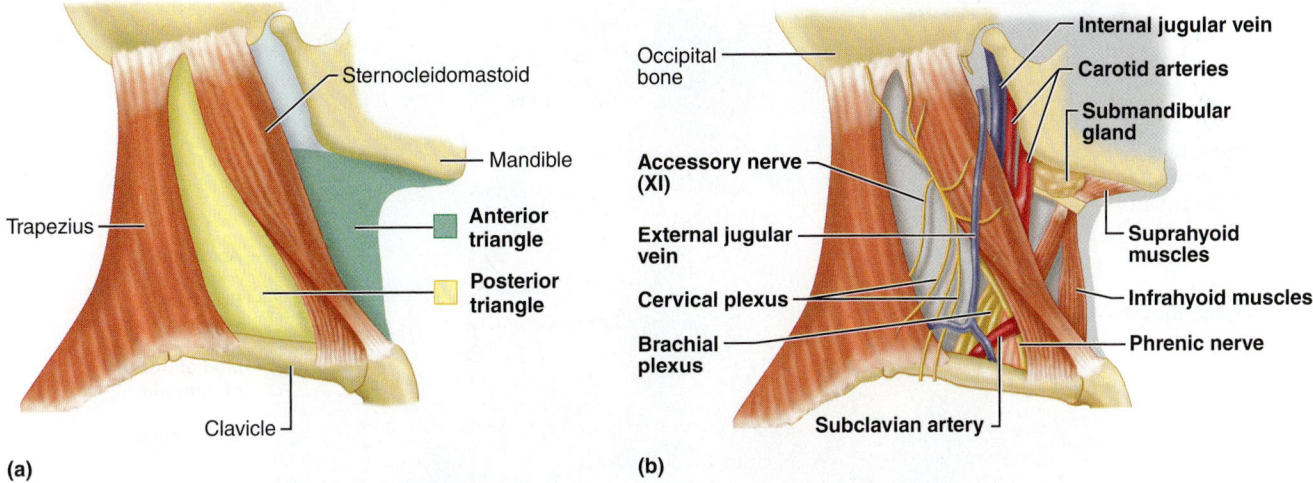

Figure 46.4 Anterior and posterior triangles of the neck. (a) Boundaries of the triangles. **(b)** Some contents of the triangles.

The Back

Bones

1. The vertical groove in the center of the back is called the **posterior median furrow (Figure 46.5).** The *spinous processes* of the vertebrae are visible in the furrow when the spinal column is flexed.

- Palpate a few of these processes on your partner's back (C_7 and T_1 are the most prominent and the easiest to find).

- Also palpate the posterior parts of some ribs, as well as the prominent **spine of the scapula** and the scapula's long **medial border.**

   The scapula lies superficial to ribs 2 to 7; its **inferior angle** is at the level of the spinous process of vertebra T_7. The medial end of the scapular spine lies opposite the T_3 spinous process.

2. Now feel the **iliac crests** (superior margins of the iliac bones) in your own lower back. You can find these crests effortlessly by resting your hands on your hips. Locate the most superior point of each crest, a point that lies roughly halfway between the posterior median furrow and the lateral side of the body (Figure 46.5). A horizontal line through these two superior points, the **supracristal line,** intersects L_4, providing a simple way to locate that vertebra. The ability to locate L_4 is essential for performing a *lumbar puncture,* a procedure in which the clinician inserts a needle into the vertebral canal of the spinal column directly superior or inferior to L_4 and withdraws cerebrospinal fluid.

3. The *sacrum* is easy to palpate just superior to the cleft in the buttocks. You can feel the *coccyx* in the extreme inferior part of that cleft, just posterior to the anus.

Muscles

The largest superficial muscles of the back are the **trapezius** superiorly and **latissimus dorsi** inferiorly (Figure 46.5). Furthermore, the deeper **erector spinae** muscles are very evident in the lower back, flanking the vertebral column like thick vertical cords.

1. Shrug your shoulders to feel the trapezius contracting just deep to the skin.

2. Feel your partner's erector spinae muscles contract and bulge as he straightens his spine from a slightly bent-over position.

   The superficial muscles of the back fail to cover a small area of the rib cage called the **triangle of auscultation** (Figure 46.5). This triangle lies just medial to the inferior part of the scapula. Its three boundaries are formed by the trapezius medially, the latissimus dorsi inferiorly, and the scapula laterally. The physician places a stethoscope over the skin of this triangle to listen for lung sounds (*auscultation* = listening). To hear the lungs clearly, the doctor first asks the patient to fold the arms together in front of the chest and then flex the trunk.

   What do you think is the precise reason for having the patient take this action?

_____

_____

3. Have your partner assume the position just described. After cleaning the earpieces with an alcohol swab, use the stethoscope to auscultate the lung sounds. Compare the clarity of the lung sounds heard over the triangle of auscultation to that over other areas of the back.

The Thorax

Bones

1. Start exploring the anterior surface of your partner's bony *thoracic cage* (**Figure 46.6** and **Figure 46.7**) by defining the extent of the *sternum.* Use a finger to trace the sternum's triangular *manubrium* inferior to the jugular notch, its flat *body,* and the tongue-shaped **xiphoid process.** Now palpate the ridgelike **sternal angle,** where the manubrium meets the body of the sternum. Locating the sternal angle is important because it directs you to the second ribs (which attach to it). Once you find the second rib, you can count down to identify every other rib in the thorax (except the first and sometimes the twelfth rib, which lie too deep to be palpated). The sternal angle is a highly reliable landmark—it is easy to locate, even in overweight people.

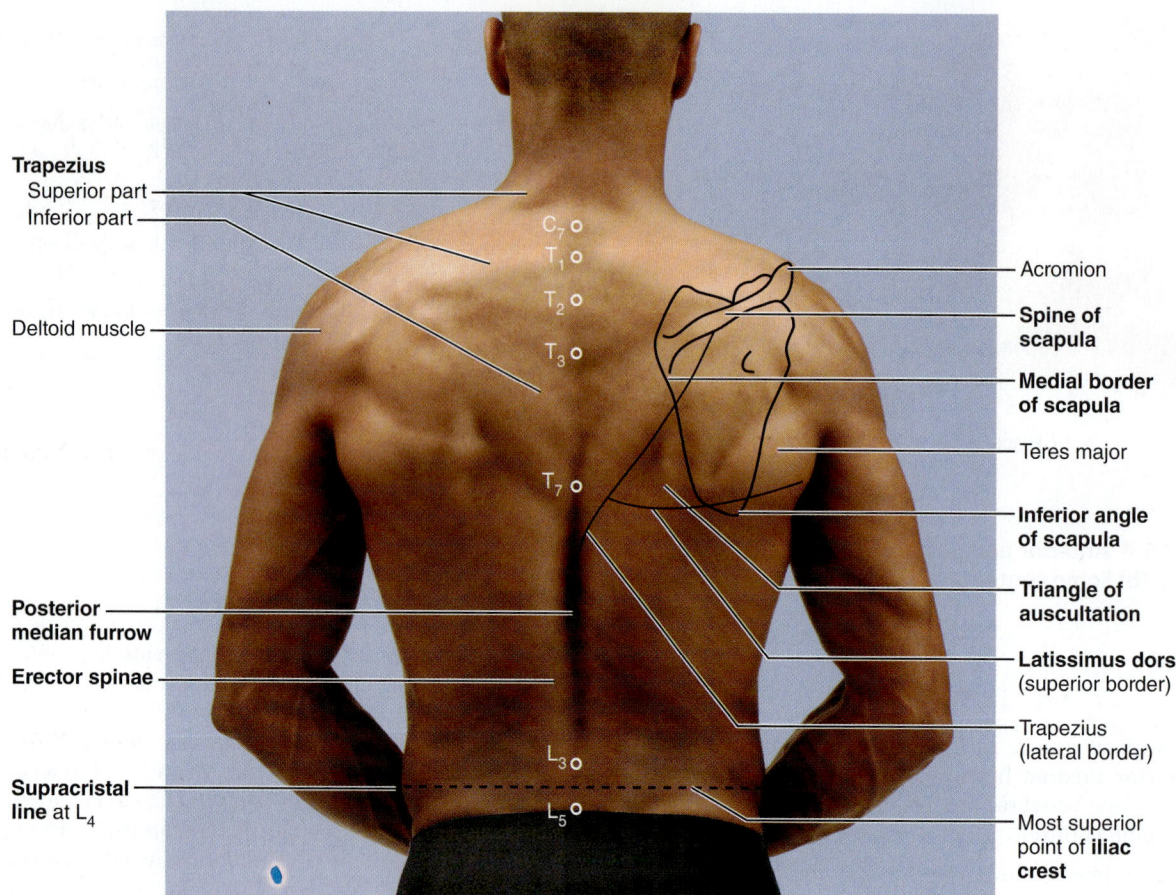

Figure 46.5 Surface anatomy of the back.

2. By locating the individual ribs, you can mentally "draw" a series of horizontal lines of "latitude" that you can use to map and locate the underlying visceral organs of the thoracic cavity. Such mapping also requires lines of "longitude," so let us construct some vertical lines on the wall of your partner's trunk. As he lifts an arm straight up in the air, extend a line inferiorly from the center of the axilla onto his lateral thoracic wall. This is the **midaxillary line** (Figure 46.6a). Now estimate the midpoint of his **clavicle,** and run a vertical line inferiorly from that point toward the groin. This is the **midclavicular line,** and it will pass about 1 cm medial to the nipple.

3. Next, feel along the V-shaped inferior edge of the rib cage, the **costal margin.** At the **infrasternal angle,** the superior angle of the costal margin, lies the **xiphisternal joint.** The heart lies on the diaphragm deep to the xiphisternal joint.

4. The thoracic cage provides many valuable landmarks for locating the vital organs of the thoracic and abdominal cavities. On the anterior thoracic wall, ribs 2–6 define the superior-to-inferior extent of the female breast, and the fourth intercostal space indicates the location of the **nipple** in men, children, and small-breasted women. The right costal margin runs across the anterior surface of the liver and gallbladder. Surgeons must be aware of the inferior margin of the *pleural cavities* because if they accidentally cut into one of these cavities, a lung collapses. The inferior pleural margin lies adjacent to vertebra T_{12} near the posterior midline (Figure 46.6b)

and runs horizontally across the back to reach rib 10 at the midaxillary line. From there, the pleural margin ascends to rib 8 in the midclavicular line (Figure 46.6a) and to the level of the xiphisternal joint near the anterior midline. The *lungs* do not fill the inferior region of the pleural cavity. Instead, their inferior borders run at a level that is two ribs superior to the pleural margin, until they meet that margin near the xiphisternal joint.

5. Let's review the relationship of the *heart* to the thoracic cage. The superior right corner of the heart lies at the junction of the third rib and the sternum; the superior left corner lies at the second rib, near the sternum; the inferior left corner lies in the fifth intercostal space in the midclavicular line; and the inferior right corner lies at the sternal border of the sixth rib. You may wish to outline the heart on your chest or that of your lab partner by connecting the four corner points with a washable marker.

Muscles

The main superficial muscles of the anterior thoracic wall are the **pectoralis major** and the anterior slips of the **serratus anterior** (Figure 46.7).

• Palpate these two muscles on your chest. They both contract during push-ups, and you can confirm this by pushing yourself up from your desk with one arm while palpating the muscles with your opposite hand. ■

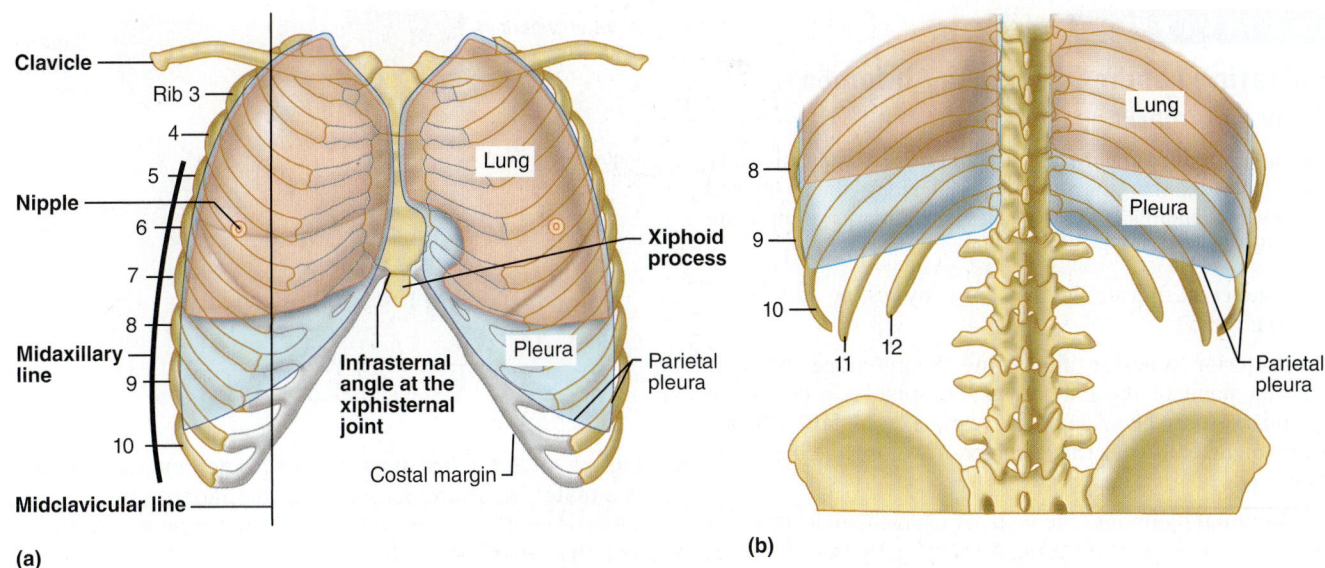

(a)

(b)

Figure 46.6 The bony rib cage as it relates to the underlying lungs and pleural cavities. Both the pleural cavities (blue) and the lungs (pink) are outlined. **(a)** Anterior view. **(b)** Posterior view.

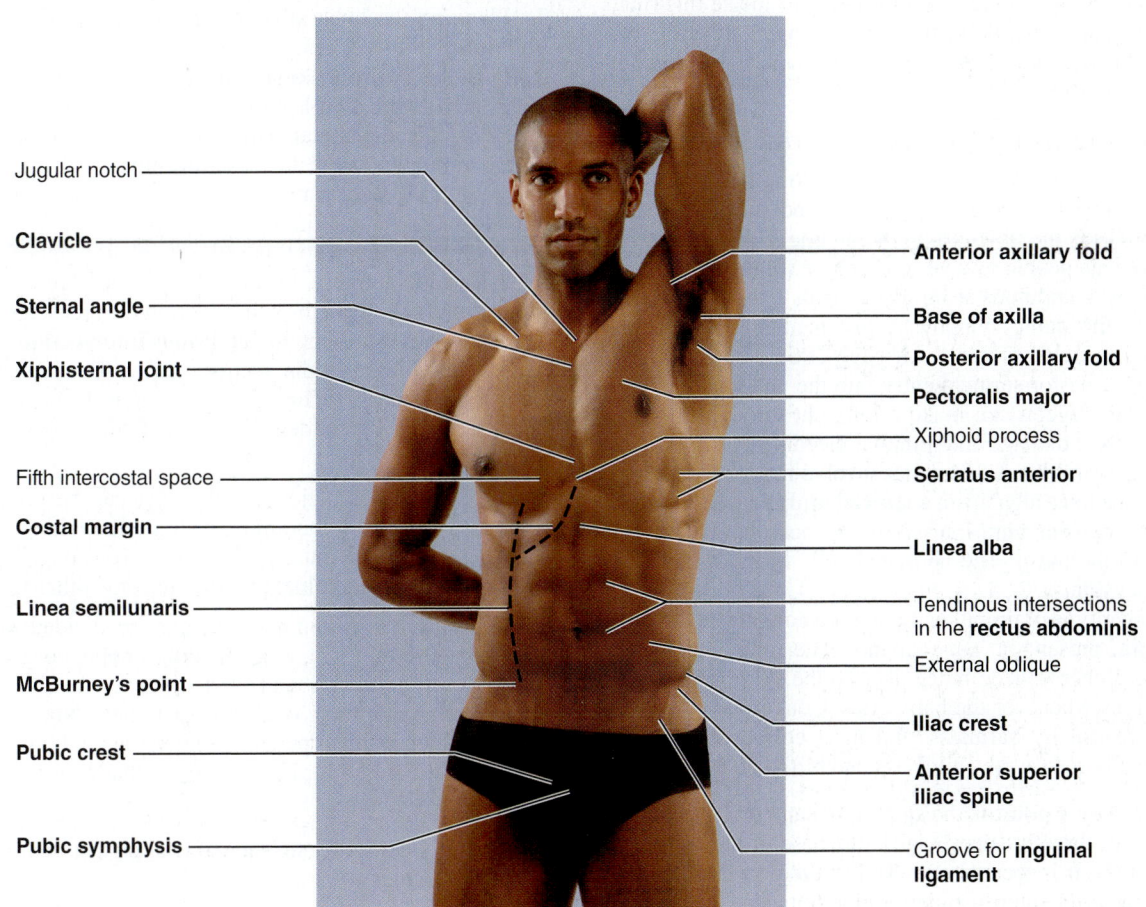

Figure 46.7 The anterior thorax and abdomen.

46

ACTIVITY 4

Palpating Landmarks of the Abdomen
Bony Landmarks

The anterior abdominal wall (Figure 46.7) extends inferiorly from the costal margin to an inferior boundary that is defined by several landmarks. Palpate these landmarks as they are described below.

1. **Iliac crest.** Locate the iliac crests by resting your hands on your hips.

2. **Anterior superior iliac spine.** Representing the most anterior point of the iliac crest, this spine is a prominent landmark. It can be palpated in everyone, even those who are overweight. Run your fingers anteriorly along the iliac crest to its end.

3. **Inguinal ligament.** The inguinal ligament, indicated by a groove on the skin of the groin, runs medially from the anterior superior iliac spine to the pubic tubercle of the pubis.

4. **Pubic crest.** You will have to press deeply to feel this crest on the pubis near the median **pubic symphysis.** The **pubic tubercle,** the most lateral point of the pubic crest, is easier to palpate, but you will still have to push deeply.

Inguinal hernias occur immediately superior to the inguinal ligament and may exit from a medial opening called the **superficial inguinal ring.** To locate this ring, one would palpate the pubic tubercle. An inguinal hernia in a male can be detected by pushing into the superficial inguinal ring **(Figure 46.8). +**

Muscles and Other Surface Features

The central landmark of the anterior abdominal wall is the *umbilicus* (navel). Running superiorly and inferiorly from the umbilicus is the **linea alba** ("white line"), represented in the skin of lean people by a vertical groove (Figure 46.7). The linea alba is a tendinous seam that extends from the xiphoid process to the pubic symphysis, just medial to the rectus abdominis muscles (Table 13.3, page 206). The linea alba is a favored site for surgical entry into the abdominal cavity because the surgeon can make a long cut through this line with no muscle damage and minimal bleeding.

Several kinds of hernias involve the umbilicus and the linea alba. In an **acquired umbilical hernia,** the linea alba weakens until intestinal coils push through it just superior to the navel. The herniated coils form a bulge just deep to the skin.

Another type of umbilical hernia is a **congenital umbilical hernia,** present in some infants: The umbilical hernia is seen as a cherry-sized bulge deep to the skin of the navel that enlarges whenever the baby cries. Congenital umbilical hernias are usually harmless, and most correct themselves automatically before the child's second birthday. **+**

1. **McBurney's point** is the spot on the anterior abdominal skin that lies directly superficial to the base of the appendix (Figure 46.7). It is located one-third of the way along a line between the right anterior superior iliac spine and the umbilicus. Try to find it on your body.

McBurney's point is the most common site of incision in appendectomies, and it is often the place where the pain of

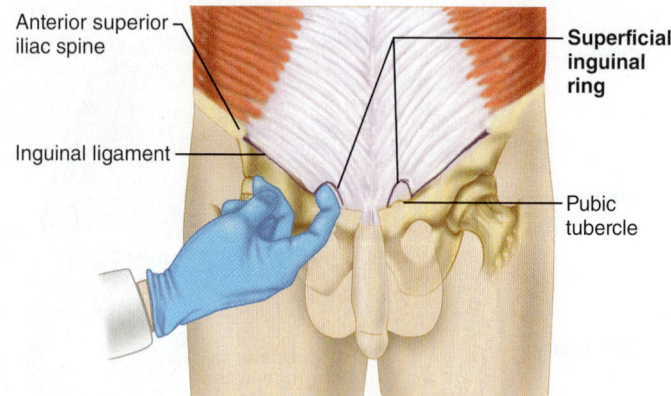

Figure 46.8 Clinical examination for an inguinal hernia in a male. The examiner palpates the patient's pubic tubercle, pushes superiorly to invaginate the scrotal skin into the superficial inguinal ring, and asks the patient to cough. If an inguinal hernia exists, it will push inferiorly and touch the examiner's fingertip.

appendicitis is experienced most acutely. Pain at McBurney's point after the pressure is removed (rebound tenderness) can indicate appendicitis. This is not a *precise* method of diagnosis, however.

2. Flanking the linea alba are the vertical straplike **rectus abdominis** muscles (Figure 46.7). Feel these muscles contract just deep to your skin as you do a bent-knee sit-up (or as you bend forward after leaning back in your chair). In the skin of lean people, the lateral margin of each rectus muscle makes a groove known as the **linea semilunaris** (half-moon line). On your right side, estimate where your linea semilunaris crosses the costal margin of the rib cage. The *gallbladder* lies just deep to this spot, so this is the standard point of incision for gallbladder surgery. In muscular people, three horizontal grooves can be seen in the skin covering the rectus abdominis. These grooves represent the **tendinous intersections,** fibrous bands that subdivide the rectus muscle. Because of these subdivisions, each rectus abdominis muscle presents four distinct bulges. Try to identify these insertions on yourself or your partner.

3. The only other major muscles that can be seen or felt through the anterior abdominal wall are the lateral **external obliques.** Feel these muscles contract as you cough, strain, or raise your intra-abdominal pressure in some other way.

4. The anterior abdominal wall can be divided into four quadrants (Figure 1.7). A clinician listening to a patient's **bowel sounds** places the stethoscope over each of the four abdominal quadrants, one after another. Normal bowel sounds, which result as peristalsis moves air and fluid through the intestine, are high-pitched gurgles that occur every 5 to 15 seconds.

- Use the stethoscope to listen to your own or your partner's bowel sounds.

Abnormal bowel sounds can indicate intestinal disorders. Absence of bowel sounds indicates a halt in intestinal activity, which follows long-term obstruction of the

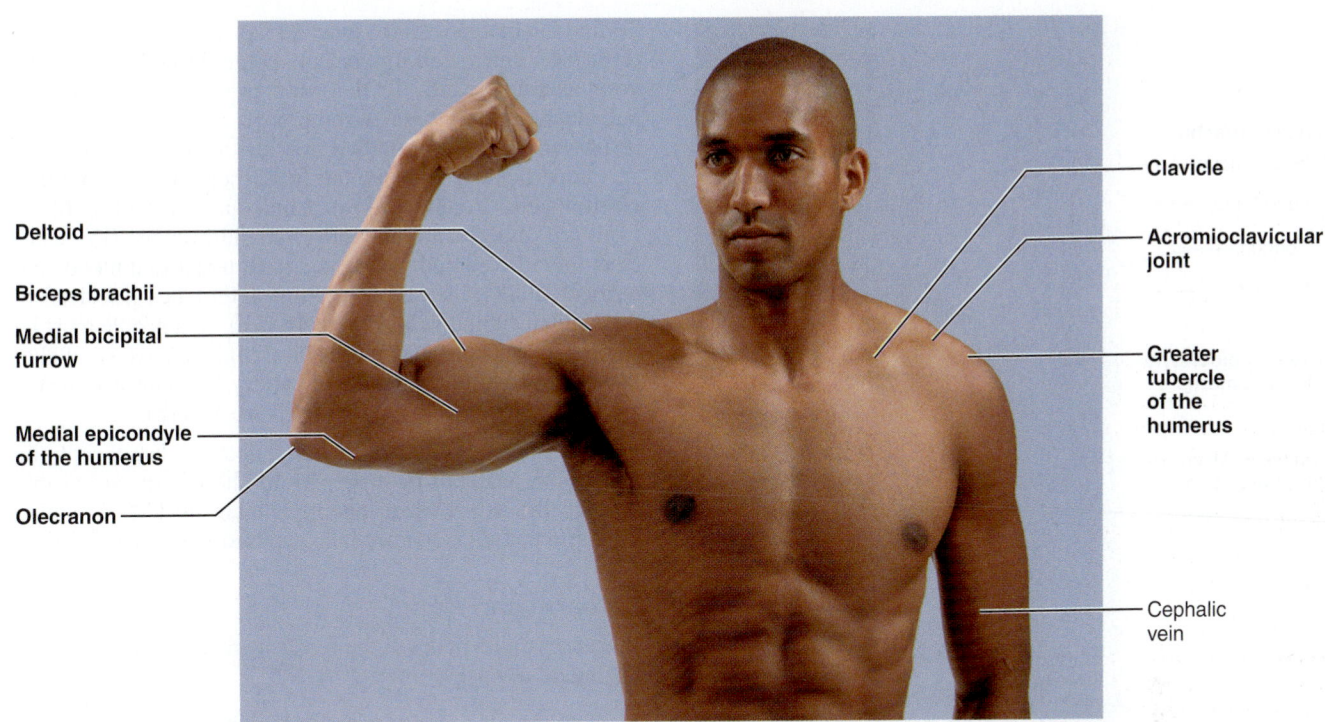

Deltoid

Biceps brachii

Medial bicipital furrow

Medial epicondyle of the humerus

Olecranon

Clavicle

Acromioclavicular joint

Greater tubercle of the humerus

Cephalic vein

Figure 46.9 Shoulder and arm.

intestine, surgical handling of the intestine, peritonitis, or other conditions. Loud tinkling or splashing sounds, by contrast, indicate an increase in intestinal activity. Such loud sounds may accompany gastroenteritis (inflammation and upset of the GI tract) or a partly obstructed intestine. ✚

The Pelvis and Perineum

The bony surface features of the *pelvis* are considered with the bony landmarks of the abdomen (page 686) and the gluteal region (page 690). Most *internal* pelvic organs are not palpable through the skin of the body surface. A full *bladder,* however, becomes firm and can be felt through the abdominal wall just superior to the pubic symphysis. A bladder that can be palpated more than a few centimeters above this symphysis is retaining urine and dangerously full, and it should be drained by catheterization. ◼

ACTIVITY 5

Palpating Landmarks of the Upper Limb

Axilla

The **base of the axilla** is the groove in which the underarm hair grows (Figure 46.7). Deep to this base lie the axillary *lymph nodes* (which swell and can be palpated in breast cancer), the large *axillary vessels* serving the upper limb, and much of the brachial plexus. The base of the axilla forms a "valley" between two thick, rounded ridges, the **axillary folds.** Just anterior to the base, clutch your **anterior axillary fold,** formed by the pectoralis major muscle. Then grasp your **posterior axillary fold.** This fold is formed by the latissimus dorsi and teres major muscles of the back as they course toward their insertions on the humerus.

Shoulder

1. Again locate the prominent spine of the scapula posteriorly (Figure 46.5). Follow the spine to its lateral end, the flattened **acromion** on the shoulder's summit. Then, palpate the **clavicle** anteriorly, tracing this bone from the sternum to the shoulder (Figure 46.9). Notice the clavicle's curved shape.

2. Now locate the junction between the clavicle and the acromion on the superolateral surface of your shoulder, at the **acromioclavicular joint.** To find this joint, thrust your arm anteriorly repeatedly until you can palpate the precise point of pivoting action.

3. Next, place your fingers on the **greater tubercle** of the humerus. This is the most lateral bony landmark on the superior surface of the shoulder. It is covered by the thick **deltoid muscle,** which forms the rounded superior part of the shoulder. Intramuscular injections are often given into the deltoid, about 5 cm (2 inches) inferior to the greater tubercle (Figure 46.17a, page 691).

Arm

Remember, according to anatomists, the arm runs only from the shoulder to the elbow, and not beyond.

1. In the arm, palpate the humerus along its entire length, especially along its medial and lateral sides.

2. Feel the **biceps brachii** muscle contract on your anterior arm when you flex your forearm against resistance. The medial boundary of the biceps is represented by the **medial bicipital furrow (Figure 46.9).** This groove contains the large *brachial artery,* and by pressing on it with your

46

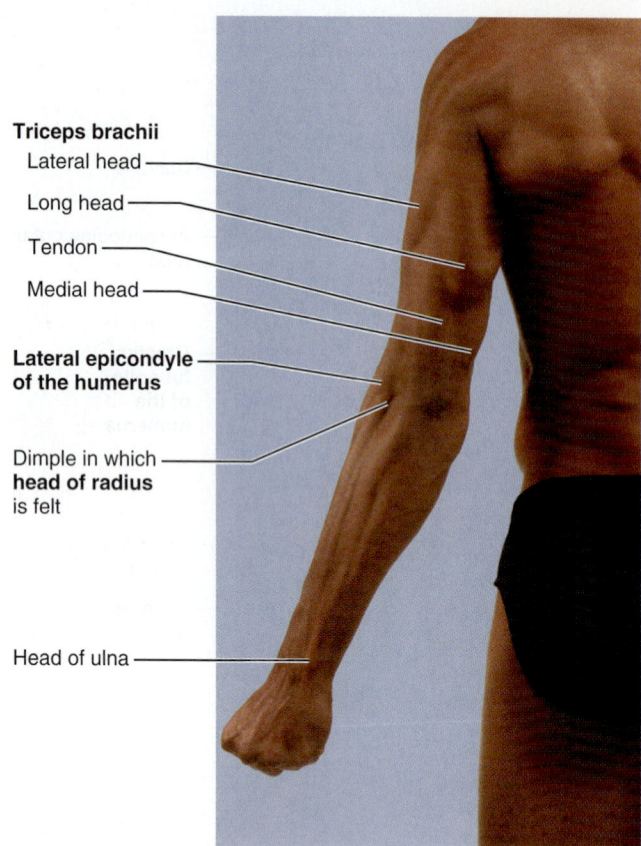

Triceps brachii
   Lateral head
   Long head
   Tendon
   Medial head

**Lateral epicondyle
of the humerus**

Dimple in which
head of radius
is felt

Head of ulna

**Figure 46.10 Surface anatomy of the upper limb,
posterior view.**

fingertips you can feel your *brachial pulse*. Recall that the brachial artery is the artery routinely used in measuring blood pressure with a sphygmomanometer.

3.   All three heads of the **triceps brachii** muscle (lateral, long, and medial) are visible through the skin of a muscular person **(Figure 46.10)**.

Elbow Region

1.   In the distal part of your arm, near the elbow, palpate the two projections of the humerus, the **lateral** and **medial epicondyles** (Figures 46.9 and 46.10). Midway between the epicondyles, on the posterior side, feel the **olecranon,** which forms the point of the elbow.

2.   Confirm that the two epicondyles and the olecranon all lie in the same horizontal line when the elbow is extended. If these three bony processes do not line up, the elbow is dislocated.

3.   Now feel along the posterior surface of the medial epicondyle. You are palpating your ulnar nerve.

4.   On the anterior surface of the elbow is a triangular depression called the **cubital fossa (Figure 46.11)**. The triangle's superior *base* is formed by a horizontal line between the humeral epicondyles; its two inferior sides are defined by the **brachioradialis** and **pronator teres** muscles (Figure 46.11b). Try to define these boundaries on your own limb.

To find the brachioradialis muscle, flex your forearm against resistance, and watch this muscle bulge through the skin of your lateral forearm. To feel your pronator teres contract, palpate the cubital fossa as you pronate your forearm against resistance. (Have your partner provide the resistance.)

   Superficially, the cubital fossa contains the **median cubital vein** (Figure 46.11a). Clinicians often draw blood from this superficial vein and insert intravenous (IV) catheters into it to administer medications, transfused blood, and nutrient fluids. The large **brachial artery** lies just deep to the median cubital vein (Figure 46.11b), so a needle must be inserted into the vein from a shallow angle (almost parallel to the skin) to avoid puncturing the artery. Tendons and nerves are also found deep in the fossa (Figure 46.11b).

5.   The median cubital vein interconnects the larger **cephalic** and **basilic veins** of the upper limb. These veins are visible through the skin of lean people (Figure 46.11a). Examine your arm to see if your cephalic and basilic veins are visible.

Forearm and Hand

The two parallel bones of the forearm are the medial *ulna* and the lateral *radius*.

1.   Feel the ulna along its entire length as a sharp ridge on the posterior forearm (confirm that this ridge runs inferiorly from the olecranon). As for the radius, you can feel its distal half, but most of its proximal half is covered by muscle. You can, however, feel the rotating **head** of the radius. To do this, extend your forearm, and note that a dimple forms on the posterior lateral surface of the elbow region (Figure 46.10). Press three fingers into this dimple, and rotate your free hand as if you were turning a doorknob. You will feel the head of the radius rotate as you perform this action.

2.   Both the radius and ulna have a knoblike **styloid process** at their distal ends. Palpate these processes at the wrist **(Figure 46.12)**. Do not confuse the ulnar styloid process with the conspicuous **head of the ulna,** from which the styloid process stems. Confirm that the radial styloid process lies about 1 cm (0.4 inch) distal to that of the ulna.

   Colles' fracture of the wrist is an impacted fracture in which the distal end of the radius is pushed proximally into the shaft of the radius. This sometimes occurs when someone falls on outstretched hands, and it most often happens to elderly women with osteoporosis. Colles' fracture bends the wrist into curves that resemble those on a fork. ✚

Can you deduce how physicians use palpation to diagnose a Colles' fracture?

_____

_____

3.   Next, feel the major groups of muscles within your forearm. Flex your hand and fingers against resistance, and feel the anterior *flexor muscles* contract. Then extend your hand at the wrist, and feel the tightening of the posterior *extensor muscles.*

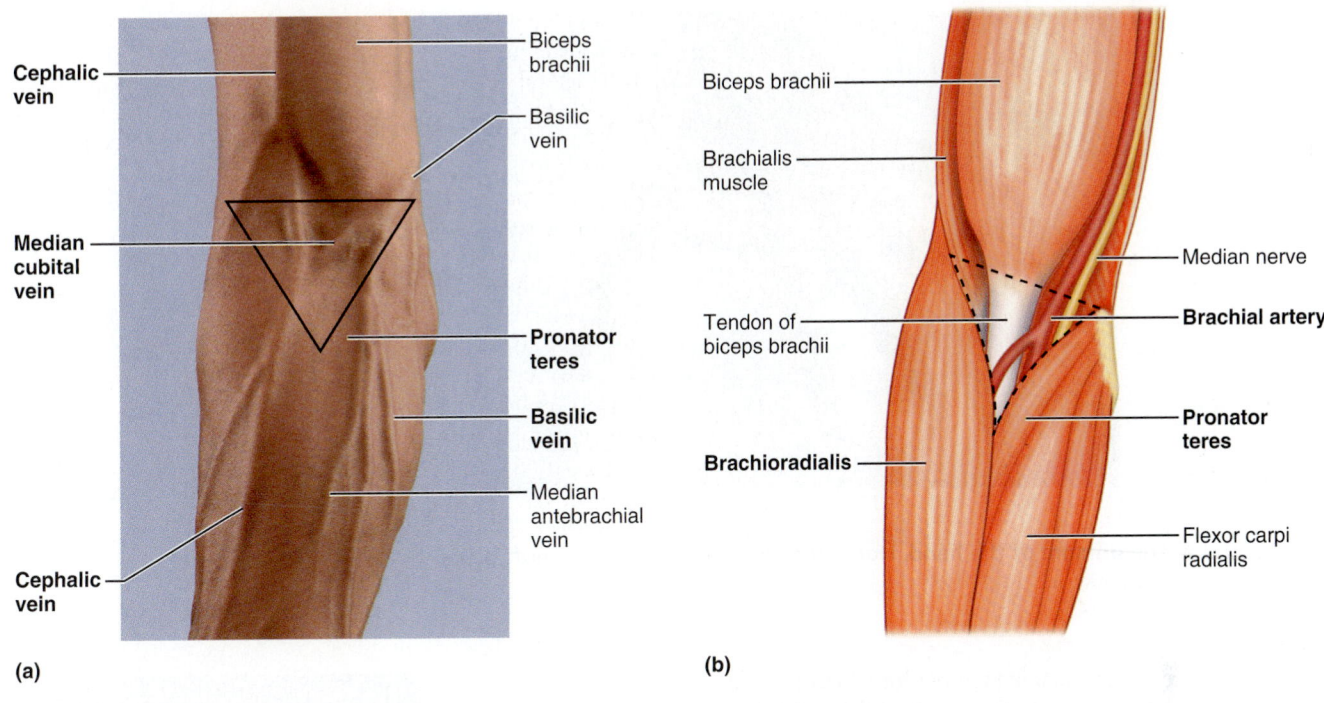

(a)

(b)

Figure 46.11 The cubital fossa on the anterior surface of the right elbow (outlined by the triangle). (a) Photograph. **(b)** Diagram of deeper structures in the fossa.

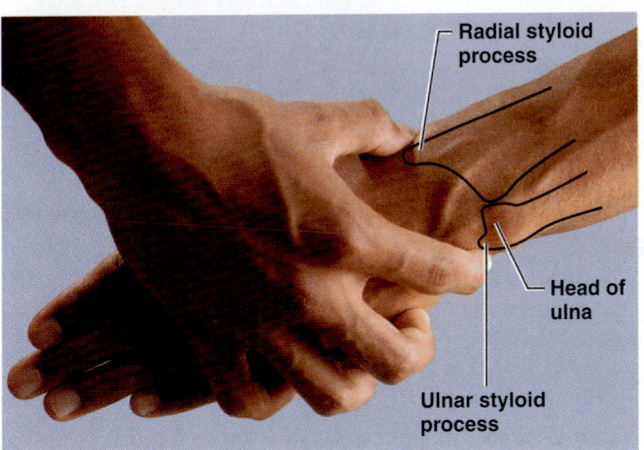

Figure 46.12 A way to locate the ulnar and radial styloid processes. The right hand is palpating the left hand in this picture. Note that the head of the ulna is not the same as the ulnar styloid process. The radial styloid process lies about 1 cm distal to the ulnar styloid process.

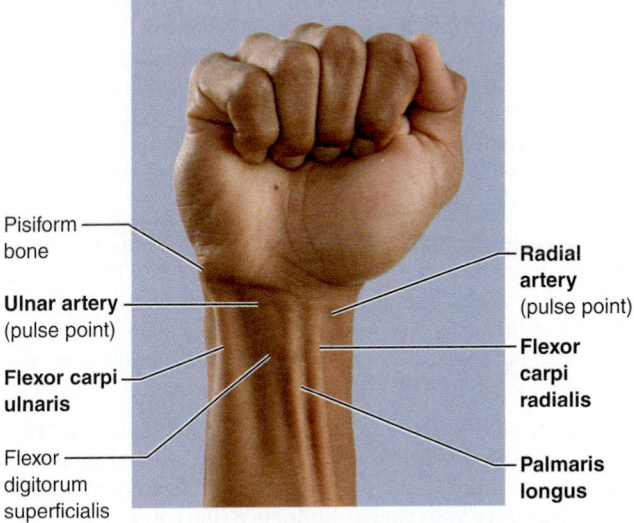

Figure 46.13 The anterior surface of the distal forearm and fist. The tendons of the flexor muscles guide the clinician to several sites for pulse taking.

46

4. Near the wrist, the anterior surface of the forearm reveals many significant features **(Figure 46.13)**. Flex your fist against resistance; the tendons of the main wrist flexors will bulge the skin of the distal forearm. The tendons of the **flexor carpi radialis** and **palmaris longus** muscles are most obvious. The palmaris longus, however, is absent from at least one arm in 30% of all people, so your forearm may exhibit just one prominent tendon instead of two. The **radial artery** lies just lateral to (on the thumb side of) the flexor carpi radialis tendon, where the pulse is easily detected (Figure 46.13). Feel your radial pulse here. The *median nerve,* which innervates the thumb, lies deep to the palmaris longus tendon. Finally, the **ulnar artery** lies on the medial side of the forearm, just lateral to the tendon of the **flexor carpi ulnaris.** Locate and feel your ulnar arterial pulse (Figure 46.13).

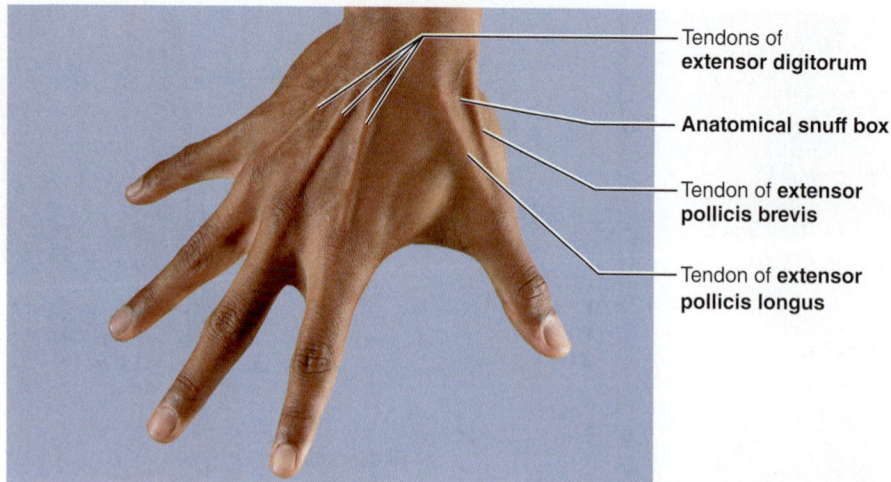

Figure 46.14 The dorsum of the hand. Note especially the anatomical snuff box and dorsal venous network.

5. Extend your thumb and point it posteriorly to form a triangular depression in the base of the thumb on the back of your hand. This is the **anatomical snuff box (Figure 46.14)**. Its two elevated borders are defined by the tendons of the thumb extensor muscles, **extensor pollicis brevis** and **extensor pollicis longus.** The radial artery runs within the snuff box, so this is another site for taking a radial pulse. The main bone on the floor of the snuff box is the scaphoid bone of the wrist, but the radial styloid process is also present here. If displaced by a bone fracture, the radial styloid process will be felt outside of the snuff box rather than within it. The "snuff box" took its name from the fact that people once put snuff (tobacco for sniffing) in this hollow before lifting it up to the nose.

6. On the dorsum of your hand, observe the superficial veins just deep to the skin. This is the **dorsal venous network,** which drains superiorly into the cephalic vein. This venous network provides a site for drawing blood and inserting intravenous catheters and is preferred over the median cubital vein for these purposes. Next, extend your hand and fingers, and observe the tendons of the **extensor digitorum** muscle.

7. The anterior surface of the hand also contains some features of interest **(Figure 46.15)**. These features include the *epidermal ridges* (fingerprints) and many **flexion creases** in the skin. Grasp your **thenar eminence** (the bulge on the palm that contains the thumb muscles) and your **hypothenar eminence** (the bulge on the medial palm that contains muscles that move the little finger). ■

ACTIVITY 6

Palpating Landmarks of the Lower Limb

Gluteal Region

Dominating the gluteal region are the two *prominences* (cheeks) of the buttocks **(Figure 46.16)**. These are formed by subcutaneous fat and by the thick **gluteus maximus** muscles. The midline groove between the two prominences is called the **natal cleft** (*natal* = rump) or **gluteal cleft.** The inferior margin of

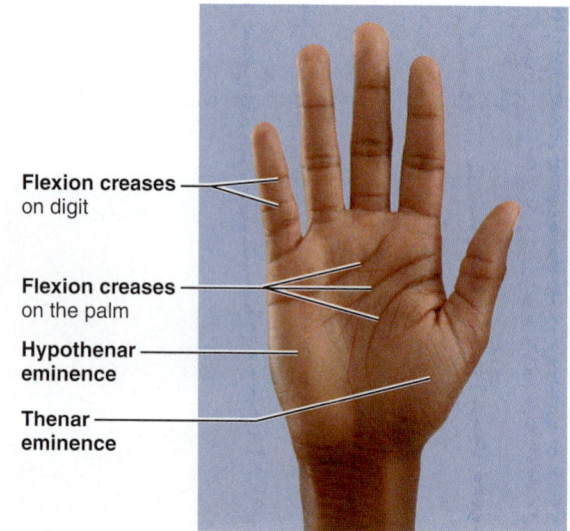

Figure 46.15 The palmar surface of the hand.

each prominence is the horizontal **gluteal fold,** which roughly corresponds to the inferior margin of the gluteus maximus.

1. Try to palpate your **ischial tuberosity** just above the medial side of each gluteal fold (it will be easier to feel if you sit down or flex your thigh first). The ischial tuberosities are the robust inferior parts of the ischial bones, and they support the body's weight during sitting.

2. Next, palpate the **greater trochanter** of the femur on the lateral side of your hip. This trochanter lies just anterior to a hollow and about 10 cm (one hand's breadth, or 4 inches) inferior to the iliac crest. To confirm that you have found the greater trochanter, alternately flex and extend your thigh. Because this trochanter is the most superior point on the lateral femur, it moves with the femur as you perform this movement.

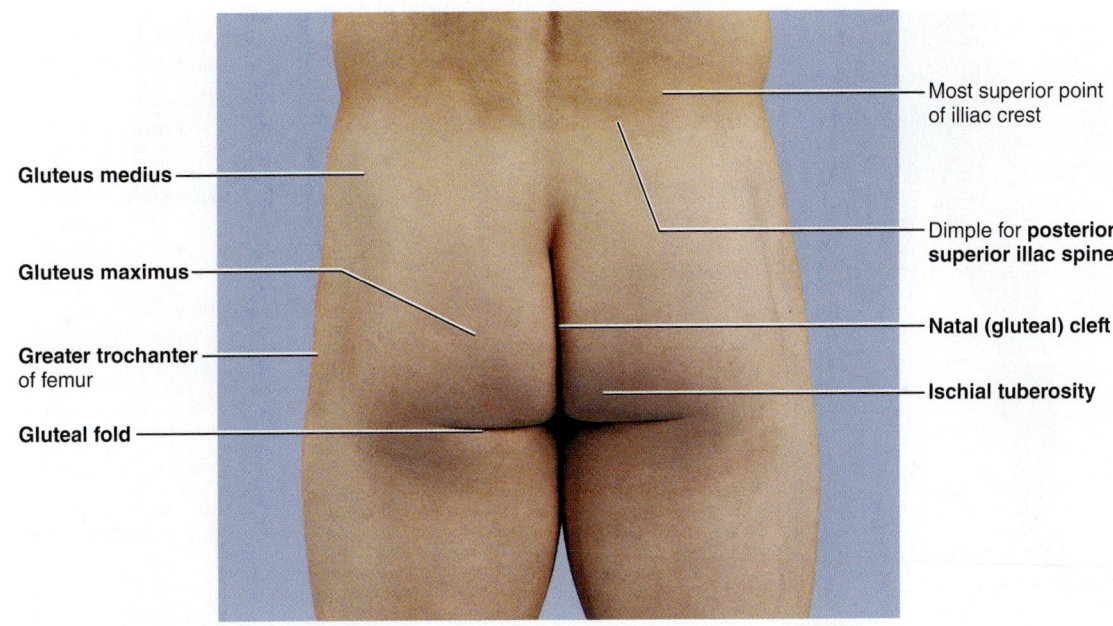

Gluteus medius

Gluteus maximus

Greater trochanter of femur

Gluteal fold

Most superior point of illiac crest

Dimple for **posterior superior illac spine**

Natal (gluteal) cleft

Ischial tuberosity

Figure 46.16 The gluteal region. The region extends from the iliac crests superiorly to the gluteal folds inferiorly. Therefore, it includes more than just the prominences of the buttock.

3. To palpate the sharp **posterior superior iliac spine,** locate your iliac crests again, and trace each to its most posterior point. You may have difficulty feeling this spine, but it is indicated by a distinct dimple in the skin that is easy to find. This dimple lies two to three finger breadths lateral to the midline of the back. The dimple also indicates the position of the *sacroiliac joint,* where the hip bone attaches to the sacrum of the spinal column. You can check *your* "dimples" out in the privacy of your home.

The gluteal region is a major site for administering intramuscular injections. When giving such injections, extreme care must be taken to avoid piercing the major nerve that lies just deep to the gluteus maximus muscle.

This thick *sciatic nerve* innervates much of the lower limb. Furthermore, the needle must avoid the gluteal nerves and gluteal blood vessels, which also lie deep to the gluteus maximus.

To avoid harming these structures, the injections are most often applied to the **gluteus** *medius* (not maximus) muscle superior to the cheeks of the buttocks, in a safe area called the **ventral gluteal site (Figure 46.17b)**. To locate this site, mentally draw a line laterally from the posterior superior iliac spine (dimple) to the greater trochanter; the injection would be given 5 cm (2 inches) superior to the midpoint of that line. Another safe way to locate the ventral gluteal site is to approach the lateral side of the patient's left hip with your

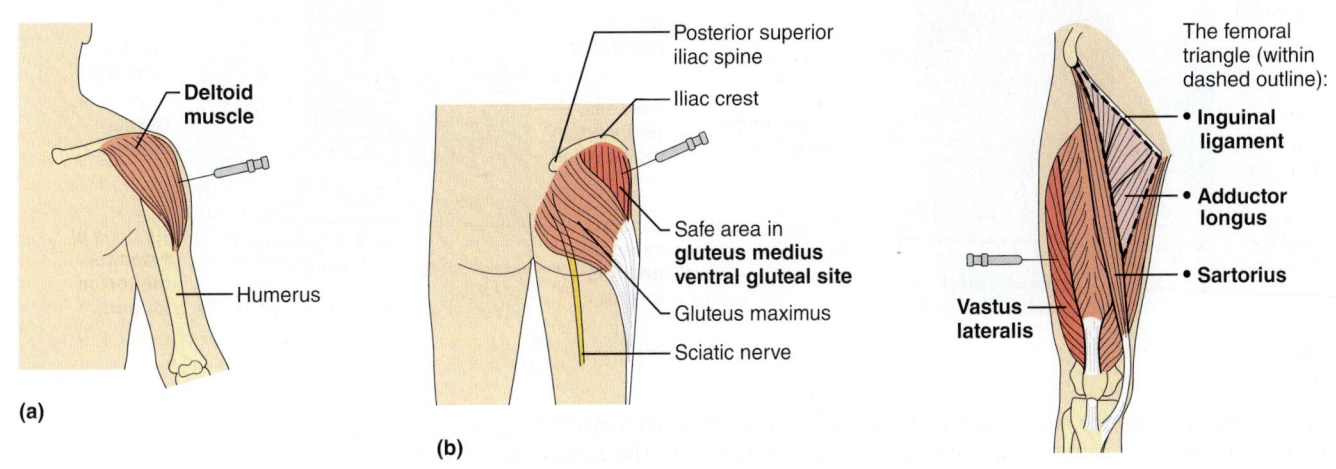

Deltoid muscle

Humerus

(a)

Posterior superior iliac spine

Iliac crest

Safe area in gluteus medius ventral gluteal site

Gluteus maximus

Sciatic nerve

(b)

The femoral triangle (within dashed outline):

• **Inguinal ligament**

• **Adductor longus**

• **Sartorius**

Vastus lateralis

(c)

46

Figure 46.17 Three major sites of intramuscular injections. (a) Deltoid muscle of the arm. **(b)** Ventral gluteal site (gluteus medius). **(c)** Vastus lateralis in the lateral thigh. The femoral triangle is also shown.

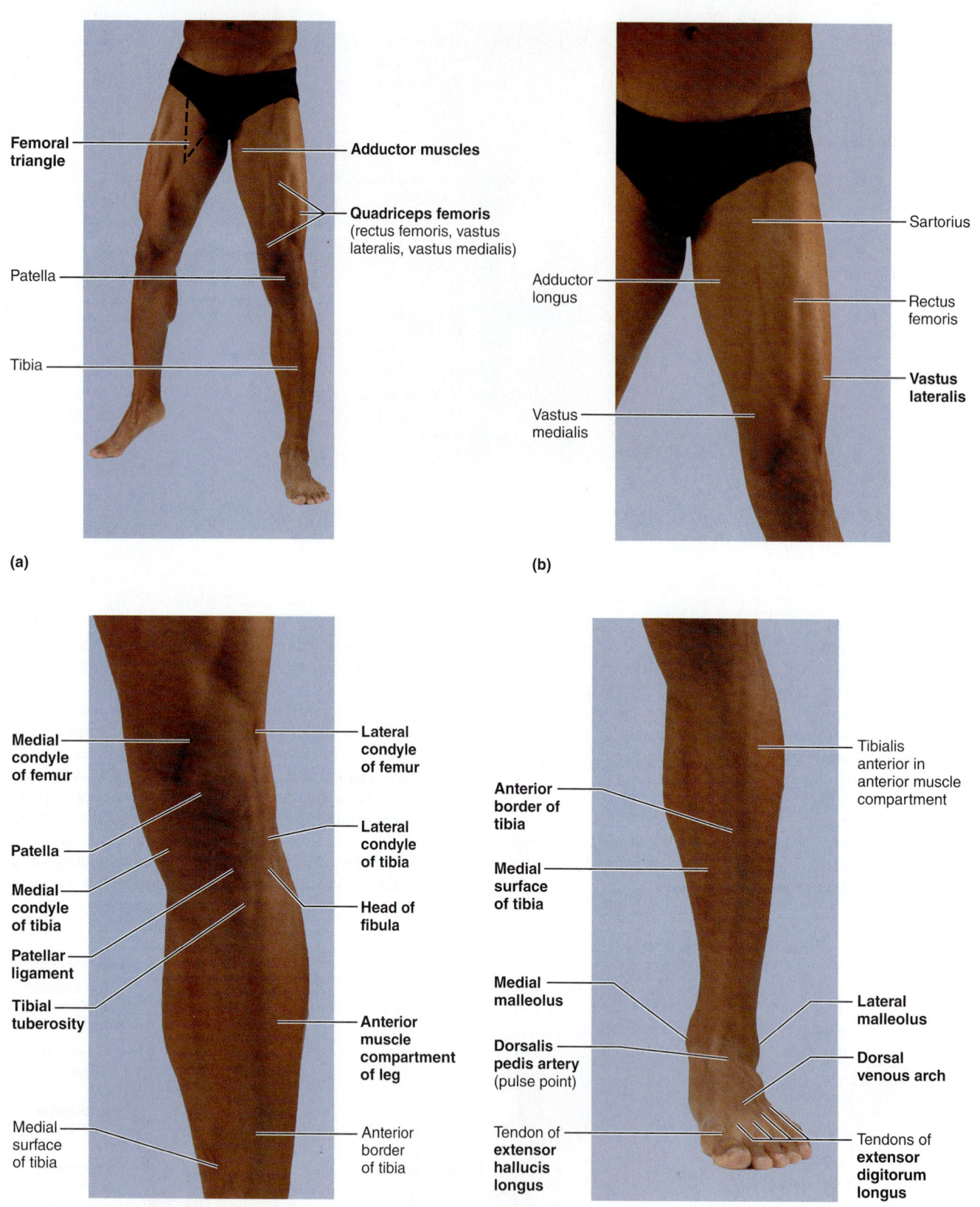

Figure 46.18 Anterior surface of the lower limb. (a) Both limbs, with the right limb revealing its medial aspect. The femoral triangle is outlined on the right limb. **(b)** Enlarged view of the left thigh. **(c)** The left knee region. **(d)** The dorsum of the left foot.

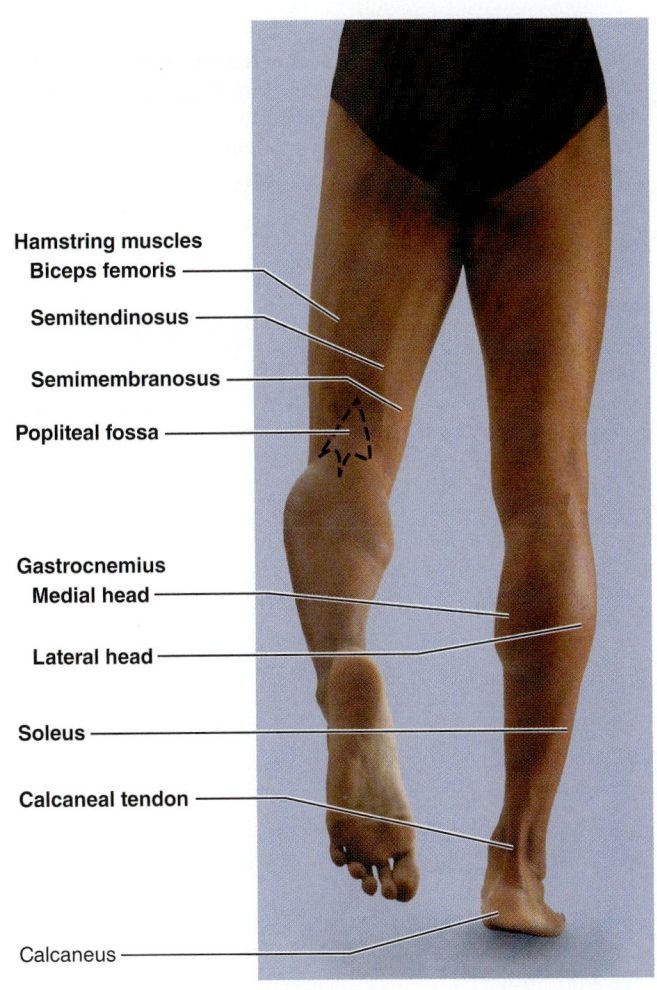

Hamstring muscles
Biceps femoris
Semitendinosus
Semimembranosus
Popliteal fossa

Gastrocnemius
Medial head
Lateral head

Soleus

Calcaneal tendon

Calcaneus

Figure 46.19 Posterior surface of the lower limb. Notice the diamond-shaped popliteal fossa posterior to the knee.

extended right hand (or the right hip with your left hand). Then, place your thumb on the anterior superior iliac spine and your index finger as far posteriorly on the iliac crest as it can reach. The heel of your hand comes to lie on the greater trochanter, and the needle is inserted in the angle of the V formed between your thumb and index finger about 4 cm (1.5 inches) inferior to the iliac crest.

   Gluteal injections are not given to small children because their "safe area" is too small to locate with certainty and because the gluteal muscles are thin at this age. Instead, infants and toddlers receive intramuscular shots in the prominent **vastus lateralis** muscle of the thigh.

Thigh

Much of the femur is clothed by thick muscles, so the thigh has few palpable bony landmarks (**Figure 46.18** and **Figure 46.19**).

1. Distally, feel the **medial** and **lateral condyles of the femur** and the **patella** anterior to the condyles (Figure 46.18c and a).

2. Next, palpate your three groups of thigh muscles—the **quadriceps femoris muscles** anteriorly, the **adductor muscles** medially, and the **hamstrings** posteriorly (Figures 46.18a and b and 46.19). The **vastus lateralis,** the lateral muscle of the quadriceps group, is a site for intramuscular injections. Such injections are administered about halfway down the length of this muscle (Figure 46.17c).

3. The anterosuperior surface of the thigh exhibits a three-sided depression called the **femoral triangle** (Figure 46.18a). As shown in Figure 46.17c, the superior border of this triangle is formed by the **inguinal ligament,** and its two inferior borders are defined by the **sartorius** and **adductor longus** muscles. The large *femoral artery* and *vein* descend vertically through the center of the femoral triangle. To feel the pulse of your femoral artery, press inward just inferior to your midinguinal point (halfway between the anterior superior iliac spine and the pubic tubercle). Be sure to push hard, because the artery lies somewhat deep. By pressing very hard on this point, one can stop the bleeding from a hemorrhage in the lower limb. The femoral triangle also contains most of the *inguinal lymph nodes,* which are easily palpated if swollen.

Leg and Foot

1. Locate your patella again, then follow the thick **patellar ligament** inferiorly from the patella to its insertion on the superior tibia (Figure 46.18c). Here you can feel a rough projection, the **tibial tuberosity.** Continue running your fingers inferiorly along the tibia's sharp **anterior border** and its flat **medial surface**—bony landmarks that lie very near the surface throughout their length.

2. Now, return to the superior part of your leg, and palpate the expanded **lateral** and **medial condyles of the tibia** just inferior to the knee. You can distinguish the tibial condyles from the femoral condyles because you can feel the tibial condyles move with the tibia during knee flexion. Feel the bulbous **head of the fibula** in the superolateral region of the leg (Figure 46.18c). Try to feel the *common fibular nerve* where it wraps around the fibula's *neck* just inferior to its head. This nerve, which serves the anterior leg and foot, is often bumped against the bone here and damaged.

3. In the most distal part of the leg, feel the **lateral malleolus** of the fibula as the lateral prominence of the ankle (Figure 46.18d). Notice that this lies slightly inferior to the **medial malleolus** of the tibia, which forms the ankle's medial prominence. Place your finger just posterior to the medial malleolus to feel the pulse of your *posterior tibial artery.*

4. On the posterior aspect of the knee is a diamond-shaped hollow called the **popliteal fossa** (Figure 46.19). Palpate the large muscles that define the four borders of this fossa: The **biceps femoris** forming the superolateral border, the **semitendinosus** and **semimembranosus** defining the superomedial border, and the two heads of the **gastrocnemius** forming the inferior border. The main vessels to the leg, the *popliteal artery* and *vein,* lie deep within this fossa. To feel a popliteal pulse, flex your leg at the knee and push your fingers firmly into the popliteal fossa. If a physician is unable to feel a patient's popliteal pulse, the femoral artery may be narrowed by atherosclerosis.

46

5. Observe the dorsum (superior surface) of your foot. You may see the superficial **dorsal venous arch** overlying the proximal part of the metatarsal bones (Figure 46.18d). This arch gives rise to both saphenous veins (the main superficial veins of the lower limb). Visible in lean people, the *great saphenous vein* ascends along the medial side of the entire limb (Figure 32.8, page 479). The *small saphenous vein* ascends through the center of the calf.

As you extend your toes, observe the tendons of the **extensor digitorum longus** and **extensor hallucis longus** muscles on the dorsum of the foot. Finally, place a finger on the extreme proximal part of the space between the first and second metatarsal bones. Here you should be able to feel the pulse of the **dorsalis pedis artery.** ◼

Surface Anatomy Roundup

_____ 1. A blow to the cheek is most likely to break what superficial bone or bone part? (a) superciliary arches, (b) mastoid process, (c) zygomatic arch, (d) ramus of the mandible

_____ 2. Rebound tenderness (a) occurs in appendicitis, (b) is whiplash of the neck, (c) is a sore foot from playing basketball, (d) occurs when the larynx falls back into place after swallowing.

_____ 3. The anatomical snuff box (a) is in the nose, (b) contains the radial styloid process, (c) is defined by tendons of the flexor carpi radialis and palmaris longus, (d) cannot really hold snuff.

_____ 4. Some landmarks on the body surface can be seen or felt, but others are abstractions that you must construct by drawing imaginary lines. Which of the following pairs of structures is abstract and invisible? (a) umbilicus and costal margin, (b) anterior superior iliac spine and natal cleft, (c) linea alba and linea semilunaris, (d) McBurney's point and midaxillary line, (e) lacrimal fossa and sternocleidomastoid

_____ 5. Many pelvic organs can be palpated by placing a finger in the rectum or the vagina, but only one pelvic organ is readily palpated through the skin. This is the (a) nonpregnant uterus, (b) prostate, (c) full bladder, (d) ovaries, (e) rectum.

_____ 6. A muscle that contributes to the posterior axillary fold is the (a) pectoralis major, (b) latissimus dorsi, (c) trapezius, (d) infraspinatus, (e) pectoralis minor, (f) a and e.

_____ 7. Which of the following is not a pulse point? (a) anatomical snuff box, (b) inferior margin of mandible anterior to masseter muscle, (c) center of distal forearm at palmaris longus tendon, (d) medial bicipital furrow on arm, (e) dorsum of foot between the first two metatarsals

_____ 8. Which pair of ribs inserts on the sternum at the sternal angle? (a) first, (b) second, (c) third, (d) fourth, (e) fifth

_____ 9. The inferior angle of the scapula is at the same level as the spinous process of which vertebra? (a) C_5, (b) C_7, (c) T_3, (d) T_7, (e) L_4

_____ 10. An important bony landmark that can be recognized by a distinct dimple in the skin is the (a) posterior superior iliac spine, (b) ulnar styloid process, (c) shaft of the radius, (d) acromion.

_____ 11. A nurse missed a patient's median cubital vein while trying to withdraw blood and then inserted the needle far too deeply into the cubital fossa. This error could cause any of the following problems, except this one: (a) paralysis of the ulnar nerve, (b) paralysis of the median nerve, (c) bruising the insertion tendon of the biceps brachii muscle, (d) blood spurting from the brachial artery.

_____ 12. Which of these organs is almost impossible to study with surface anatomy techniques? (a) heart, (b) lungs, (c) brain, (d) nose

_____ 13. A preferred site for inserting an intravenous medication line into a blood vessel is the (a) medial bicipital furrow on arm, (b) external carotid artery, (c) dorsal venous network of hand, (d) popliteal fossa.

_____ 14. One listens for bowel sounds with a stethoscope placed (a) on the four quadrants of the abdominal wall; (b) in the triangle of auscultation; (c) in the right and left midaxillary line, just superior to the iliac crests; (d) inside the patient's bowels (intestines), on the tip of an endoscope.

_____ 15. A stab wound in the posterior triangle of the neck could damage any of the following structures except the (a) accessory nerve, (b) phrenic nerve, (c) external jugular vein, (d) external carotid artery.

PhysioEx™ 9.1

PhysioEx™ 9.1 by
Peter Zao, North Idaho College
Timothy Stabler, Indiana University Northwest
Lori Smith, American River College
Andrew Lokuta, University of Wisconsin–Madison
Edwin Griff, University of Cincinnati

Exercise 1 Cell Transport Mechanisms and Permeability
Exercise 2 Skeletal Muscle Physiology
Exercise 3 Neurophysiology of Nerve Impulses
Exercise 4 Endocrine System Physiology
Exercise 5 Cardiovascular Dynamics
Exercise 6 Cardiovascular Physiology
Exercise 7 Respiratory System Mechanics
Exercise 8 Chemical and Physical Processes of Digestion
Exercise 9 Renal System Physiology
Exercise 10 Acid-Base Balance
Exercise 11 Blood Analysis
Exercise 12 Serological Testing

Cell Transport Mechanisms and Permeability

PRE-LAB QUIZ

1. Circle the correct underlined term: In <u>active transport</u> / <u>passive transport</u> processes, the cell must provide energy in the form of ATP to power the process.

2. The movement of particles from an area of greater concentration to an area of lesser concentration is:
   a. diffusion
   b. osmosis
   c. active transport
   d. kinetic energy

3. All of the following are true of active transport *except:*
   a. ATP is used to power active transport.
   b. Solutes are moving with their concentration gradient.
   c. It uses a membrane-bound carrier protein.
   d. It can only occur in certain animals.

4. Circle the correct underlined term: In Exercise 1, the dialysis tubing will mimic the <u>nucleus</u> / <u>plasma membrane</u> of a cell.

5. Circle the correct underlined term: The larger the <u>molecular weight</u> / <u>concentration</u> of a compound, the larger the pore size required for passive transport of that compound.

Exercise Overview

The molecular composition of the plasma membrane allows it to be selective about what passes through it. It allows nutrients and appropriate amounts of ions to enter the cell and keeps out undesirable substances. For that reason, we say the plasma membrane is **selectively permeable.** Valuable cell proteins and other substances are kept within the cell, and metabolic wastes pass to the exterior.

Transport through the plasma membrane occurs in two basic ways: either passively or actively. In **passive processes,** the transport process is driven by concentration or pressure differences *(gradients)* between the interior and exterior of the cell. In **active processes,** the cell provides energy (ATP) to power the transport.

Two key passive processes of membrane transport are **diffusion** and **filtration.** Diffusion is an important transport process for every cell in the body. **Simple diffusion** occurs without the assistance of membrane proteins, and **facilitated diffusion** requires a membrane-bound carrier protein that assists in the transport.

In both simple and facilitated diffusion, the substance being transported moves *with* (or *along* or *down*) the *concentration gradient* of the solute (from a region of its higher concentration to a region of its lower concentration). The process does not require energy from the cell. Instead, energy in the form of **kinetic energy** comes from the constant motion of the molecules. The movement of solutes continues until the solutes are evenly dispersed throughout the solution. At this point, the solution has reached **equilibrium.**

A special type of diffusion across a membrane is **osmosis.** In osmosis, water moves with its concentration gradient, from a higher concentration of water to a lower concentration of water. It moves in response to a higher concentration of solutes on the other side of a membrane.

In the body, the other key passive process, **filtration,** usually occurs only across capillary walls. Filtration depends upon a *pressure gradient* as its driving force. It is not a selective process. It is dependent upon the size of the pores in the filter.

The two key active processes (recall that active processes require energy) are **active transport** and **vesicular transport.** Like facilitated diffusion, active transport uses a membrane-bound carrier protein. Active transport differs from facilitated diffusion because the solutes move *against* their concentration gradient and because ATP is used to power the transport. Vesicular transport includes phagocytosis, endocytosis, pinocytosis, and exocytosis. These processes are not covered in this exercise. The activities in this exercise will explore the cell transport mechanisms individually.

Simulating Dialysis (Simple Diffusion)

OBJECTIVES

1. To understand that diffusion is a passive process dependent upon a solute concentration gradient.
2. To understand the relationship between molecular weight and molecular size.
3. To understand how solute concentration affects the rate of diffusion.
4. To understand how molecular weight affects the rate of diffusion.

Introduction

Recall that all molecules possess *kinetic energy* and are in constant motion. As molecules move about randomly at high speeds, they collide and bounce off one another, changing direction with each collision. For a given temperature, all matter has about the same average kinetic energy. Smaller molecules tend to move faster than larger molecules because kinetic energy is directly related to both mass and velocity ($KE = \frac{1}{2}\ mv^2$).

When a **concentration gradient** (difference in concentration) exists, the net effect of this random molecular movement is that the molecules eventually become evenly distributed throughout the environment—in other words, diffusion occurs. **Diffusion** is the movement of molecules from a region of their higher concentration to a region of their lower concentration. The driving force behind diffusion is the kinetic energy of the molecules themselves.

The diffusion of particles into and out of cells is modified by the plasma membrane, which is a physical barrier. In general, molecules diffuse passively through the plasma membrane if they are small enough to pass through its pores (and are aided by an electrical and/or concentration gradient) or if they can dissolve in the lipid portion of the membrane (as in the case of CO_2 and O_2). A membrane is called *selectively permeable, differentially permeable,* or *semipermeable* if it allows some solute particles (molecules) to pass but not others.

The diffusion of *solute particles* dissolved in water through a selectively permeable membrane is called **simple diffusion.** The diffusion of *water* through a differentially permeable membrane is called **osmosis.** Both simple diffusion and osmosis involve movement of a substance from an area of its higher concentration to an area of its lower concentration, that is, *with* (or *along* or *down)* its concentration gradient.

This activity provides information on the passage of water and solutes through selectively permeable membranes. You can apply what you learn to the study of transport mechanisms in living, membrane-bounded cells. The dialysis membranes used each have a different *molecular weight cutoff (MWCO),* indicated by the number below it. You can think of MWCO in terms of pore size: the larger the MWCO number, the larger the pores in the membrane. The molecular weight of a solute is the number of grams per mole, where a mole is the constant Avogadro's number 6.02×10^{23} molecules/mole. The larger the molecular weight, the larger the mass of the molecule. The term molecular mass is sometimes used instead of molecular weight.

> **EQUIPMENT USED** The following equipment will be depicted on-screen: left and right beakers—used for diffusion of solutes; dialysis membranes with various molecular weight cutoffs (MWCOs).

Experiment Instructions

Go to the home page in the PhysioEx software and click **Exercise 1: Cell Transport Mechanisms and Permeability.** Click **Activity 1: Simulating Dialysis (Simple Diffusion),** and take the online **Pre-lab Quiz** for Activity 1.

After you take the online Pre-lab Quiz, click the **Experiment** tab and begin the experiment. The experiment instructions are reprinted here for your reference. The opening screen for the experiment is shown below.

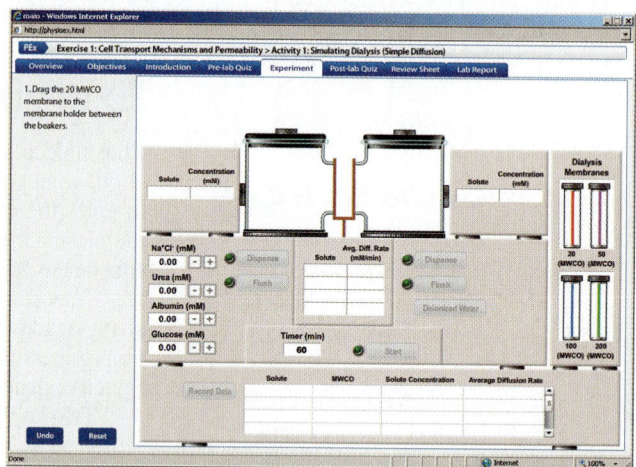

1. Drag the 20 MWCO membrane to the membrane holder between the beakers.

2. Increase the Na^+Cl^- concentration to be dispensed to the left beaker to 9.00 m*M* by clicking the + button beside the Na^+Cl^- display. Click **Dispense** to fill the left beaker with 9.00 m*M* Na^+Cl^- solution.

3. Note that the concentration of Na^+Cl^- in the left beaker is displayed in the concentration window to the left of the beaker. Click **Deionized Water** and then click **Dispense** to fill the right beaker with deionized water.

4. After you start the run, the barrier between the beakers will descend, allowing the solutions in each beaker to have access to the dialysis membrane separating them. You will be

able to determine the amount of solute that passes through the membrane by observing the concentration display to the side of each beaker. A level above zero in Na^+Cl^- concentration in the right beaker indicates that Na^+ and Cl^- ions are diffusing from the left beaker into the right beaker through the selectively permeable dialysis membrane. Note that the timer is set to 60 minutes. The simulation compresses the 60-minute time period into 10 seconds of real time. Click **Start** to start the run and watch the concentration display to the side of each beaker for any activity.

5. Click **Record Data** to display your results in the grid (and record your results in Chart 1).

CHART 1	Dialysis Results (average diffusion rate in mM/min)			
	Membrane MWCO			
Solute	**20**	**50**	**100**	**200**
Na^+Cl^-				
Urea				
Albumin				
Glucose				

? PREDICT Question 1
The molecular weight of urea is 60.07. Do you think urea will diffuse through the 20 MWCO membrane?

_____

6. Click **Flush** beneath each of the beakers to prepare for the next run.

7. Increase the urea concentration to be dispensed to the left beaker to 9.00 mM by clicking the + button beside the urea display. Click **Dispense** to fill the left beaker with 9.00 mM urea solution.

8. Click **Deionized Water** and then click **Dispense** to fill the right beaker with deionized water.

9. Click **Start** to start the run and watch the concentration display to the side of each beaker for any activity.

10. Click **Record Data** to display your results in the grid (and record your results in Chart 1).

11. Click the 20 MWCO membrane in the membrane holder to automatically return it to the membrane cabinet and then click **Flush** beneath each beaker to prepare for the next run.

12. Drag the 50 MWCO membrane to the membrane holder between the beakers. Increase the Na^+Cl^- concentration to be dispensed to the left beaker to 9.00 mM. Click **Dispense** to fill the left beaker with 9.00 mM Na^+Cl^- solution.

13. Click **Deionized Water** and then click **Dispense** to fill the right beaker with deionized water.

14. Click **Start** to start the run and watch the concentration display to the side of each beaker for any activity.

15. Click **Record Data** to display your results in the grid (and record your results in Chart 1).

16. Click **Flush** beneath each of the beakers to prepare for the next run.

17. Increase the Na^+Cl^- concentration to be dispensed to the left beaker to 18.00 mM. Click **Dispense** to fill the left beaker with 18.00 mM Na^+Cl^- solution.

18. Click **Deionized Water** and then click **Dispense** to fill the right beaker with deionized water.

19. Click **Start** to start the run and watch the concentration display to the side of each beaker for any activity.

20. Click **Record Data** to display your results in the grid (and record your results in Chart 1).

21. Click the 50 MWCO membrane in the membrane holder to automatically return it to the membrane cabinet and then click **Flush** beneath each beaker to prepare for the next run.

22. Drag the 100 MWCO membrane to the membrane holder between the beakers. Increase the Na^+Cl^- concentration to be dispensed to the left beaker to 9.00 mM. Click **Dispense** to fill the left beaker with 9.00 mM Na^+Cl^- solution.

23. Click **Deionized Water** and then click **Dispense** to fill the right beaker with deionized water.

24. Click **Start** to start the run and watch the concentration display to the side of each beaker for any activity.

25. Click **Record Data** to display your results in the grid (and record your results in Chart 1).

26. Click **Flush** beneath each of the beakers to prepare for the next run.

27. Increase the urea concentration to be dispensed to the left beaker to 9.00 mM. Click **Dispense** to fill the left beaker with 9.00 mM urea solution.

28. Click **Deionized Water** and then click **Dispense** to fill the right beaker with deionized water.

29. Click **Start** to start the run and watch the concentration display to the side of each beaker for any activity.

30. Click **Record Data** to display your results in the grid (and record your results in Chart 1).

31. Click the 100 MWCO membrane in the membrane holder to automatically return it to the membrane cabinet and then click **Flush** beneath each beaker to prepare for the next run.

? PREDICT Question 2
Recall that glucose is a monosaccharide, albumin is a protein with 607 amino acids, and the average molecular weight of a single amino acid is 135 g/mole. Will glucose or albumin be able to diffuse through the 200 MWCO membrane?

_____

32. Drag the 200 MWCO membrane to the membrane holder between the beakers. Increase the glucose concentration to be dispensed to the left beaker to 9.00 m*M*. Click **Dispense** to fill the left beaker with 9.00 m*M* glucose solution.

33. Click **Deionized Water** and then click **Dispense** to fill the right beaker with deionized water.

34. Click **Start** to start the run and watch the concentration display to the side of each beaker for any activity.

35. Click **Record Data** to display your results in the grid (and record your results in Chart 1).

36. Click **Flush** beneath each of the beakers to prepare for the next run.

37. Increase the albumin concentration to be dispensed to the left beaker to 9.00 m*M*. Click **Dispense** to fill the left beaker with 9.00 m*M* albumin solution.

38. Click **Deionized Water** and then click **Dispense** to fill the right beaker with deionized water.

39. Click **Start** to start the run and watch the concentration display to the side of each beaker for any activity.

40. Click **Record Data** to display your results in the grid (and record your results in Chart 1).

After you complete the experiment, take the online **Post-lab Quiz** for Activity 1.

Activity Questions

1. Did any solutes move through the 20 MWCO membrane? Why or why not?

_____

_____

2. Did Na^+Cl^- move through the 50 MWCO membrane?

_____

_____

3. Describe how the size of a molecule (molecular weight) affects its rate of diffusion.

_____

_____

4. What happened to the rate of diffusion when you increased the Na^+Cl^- solute concentration?

_____

_____

Simulated Facilitated Diffusion

OBJECTIVES

1. To understand that some solutes require a carrier protein to pass through a membrane because of size or solubility limitations.

2. To observe how the concentration of solutes affects the rate of facilitated diffusion.

3. To observe how the number of transport proteins affects the rate of facilitated diffusion.

4. To understand how transport proteins can become saturated.

Introduction

Some molecules are lipid insoluble or too large to pass through pores in the cell's plasma membrane. Instead, they pass through the membrane by a passive transport process called **facilitated diffusion.** For example, sugars, amino acids, and ions are transported by facilitated diffusion. In this form of transport, solutes combine with carrier-protein molecules in the membrane and are then transported *with* (or *along* or *down*) their concentration gradient. The carrier-protein molecules in the membrane might have to change shape slightly to accommodate the solute, but the cell does not have to expend the energy of ATP.

Because facilitated diffusion relies on carrier proteins, solute transport varies with the number of available carrier-protein molecules in the membrane. The carrier proteins can become saturated if too much solute is present and the maximum transport rate is reached. The carrier proteins are embedded in the plasma membrane and act like a shield, protecting the hydrophilic solute from the lipid portions of the membrane.

Facilitated diffusion typically occurs in one direction for a given solute. The greater the concentration difference between one side of the membrane and the other, the greater the rate of facilitated diffusion.

> **EQUIPMENT USED** The following equipment will be depicted on-screen: left and right beakers—used for diffusion of solutes; dialysis membranes with various molecular weight cutoffs (MWCOs); membrane builder—used to build membranes with different numbers of glucose protein carriers.

Experiment Instructions

Go to the home page in the PhysioEx software and click **Exercise 2: Cell Transport Mechanisms and Permeability.** Click **Activity 2: Simulated Facilitated Diffusion,** and take the online **Pre-lab Quiz** for Activity 2.

After you take the online Pre-lab Quiz, click the **Experiment** tab and begin the experiment. The experiment instructions are reprinted here for your reference. The opening screen for the experiment is shown on the following page.

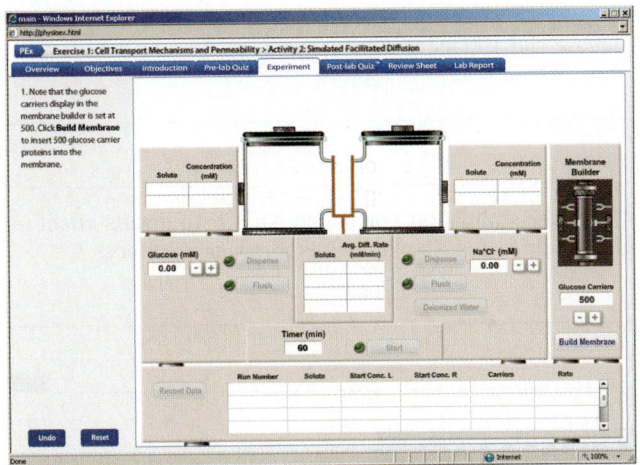

1. Note that the glucose carriers display in the membrane builder is set at 500. Click **Build Membrane** to insert 500 glucose carrier proteins into the membrane.

2. Drag the membrane to the membrane holder between the beakers.

3. Increase the glucose concentration to be dispensed to the left beaker to 2.00 m*M* by clicking the + button beside the glucose display. Click **Dispense** to fill the left beaker with 2.00 m*M* glucose solution.

4. Note that the concentration of glucose in the left beaker is displayed in the concentration window to the left of the beaker. Click **Deionized Water** and then click **Dispense** to fill the right beaker with deionized water.

5. After you start the run, the barrier between the beakers will descend, allowing the solutions in each beaker to have access to the dialysis membrane separating them. You will be able to determine the amount of solute that passes through the membrane by observing the concentration display to the side of each beaker. A level above zero in glucose concentration in the right beaker indicates that glucose is diffusing from the left beaker into the right beaker through the selectively permeable dialysis membrane. Note that the timer is set to 60 minutes. The simulation compresses the 60-minute time period into 10 seconds of real time. Click **Start** to start the run and watch the concentration display to the side of each beaker for any activity.

6. Click **Record Data** to display your results in the grid (and record your results in Chart 2).

CHART 2	Facilitated Diffusion Results (glucose transport rate, m*M*/min)		
	Number of glucose carrier proteins		
Glucose concentration	500	700	100
2 m*M*			
8 m*M*			
10 m*M*			
2 m*M* w/2.00 m*M* Na⁺Cl⁻			

7. Click **Flush** beneath each of the beakers to prepare for the next run.

8. Increase the glucose concentration to be dispensed to the left beaker to 8.00 m*M* by clicking the + button beside the glucose display. Click **Dispense** to fill the left beaker with 8.00 m*M* glucose solution.

9. Click **Deionized Water** and then click **Dispense** to fill the right beaker with deionized water.

10. Click **Start** to start the run and watch the concentration display to the side of each beaker for any activity.

11. Click **Record Data** to display your results in the grid (and record your results in Chart 2).

12. Click the membrane in the membrane holder to automatically return it to the membrane builder and then click **Flush** beneath each beaker to prepare for the next run.

> **? PREDICT Question 1**
> What effect do you think increasing the number of protein carriers will have on the glucose transport rate?
>
> _____

13. Increase the number of glucose carriers to 700 by clicking the + button beneath the glucose carriers display. Click **Build Membrane** to insert 700 glucose carrier proteins into the membrane.

14. Drag the membrane to the membrane holder between the beakers. Increase the glucose concentration to be dispensed to the left beaker to 2.00 m*M*. Click **Dispense** to fill the left beaker with 2.00 m*M* glucose solution.

15. Click **Deionized Water** and then click **Dispense** to fill the right beaker with deionized water.

16. Click **Start** to start the run and watch the concentration display to the side of each beaker for any activity.

17. Click **Record Data** to display your results in the grid (and record your results in Chart 2).

18. Click **Flush** beneath each of the beakers to prepare for the next run.

19. Increase the glucose concentration to be dispensed to the left beaker to 8.00 m*M*. Click **Dispense** to fill the left beaker with 8.00 m*M* glucose solution.

20. Click **Deionized Water** and then click **Dispense** to fill the right beaker with deionized water.

21. Click **Start** to start the run and watch the concentration display to the side of each beaker for any activity.

22. Click **Record Data** to display your results in the grid (and record your results in Chart 2).

23. Click the membrane in the membrane holder to automatically return it to the membrane builder and then click **Flush** beneath each beaker to prepare for the next run.

24. Decrease the number of glucose carriers to 100 by clicking the – button beneath the glucose carriers display. Click **Build Membrane** to insert 100 glucose carrier proteins into the membrane.

25. Drag the membrane to the membrane holder between the beakers. Increase the glucose concentration to be dispensed to the left beaker to 10.00 mM. Click **Dispense** to fill the left beaker with 10.00 mM glucose solution.

26. Click **Deionized Water** and then click **Dispense** to fill the right beaker with deionized water.

27. Click **Start** to start the run and watch the concentration display to the side of each beaker for any activity.

28. Click **Record Data** to display your results in the grid (and record your results in Chart 2).

29. Click the membrane in the membrane holder to automatically return it to the membrane builder and then click **Flush** beneath each beaker to prepare for the next run.

30. Increase the number of glucose carriers to 700. Click **Build Membrane** to insert 700 glucose carrier proteins into the membrane.

> **?** **PREDICT** Question 2
> What effect do you think adding Na$^+$Cl$^-$ will have on the glucose transport rate?
>
> _____

31. Increase the glucose concentration to be dispensed to the left beaker to 2.00 mM. Click **Dispense** to fill the left beaker with 2.00 mM glucose solution.

32. Increase the Na$^+$Cl$^-$ concentration to be dispensed to the right beaker to 2.00 mM. Click **Dispense** to fill the right beaker with 2.00 mM Na$^+$Cl$^-$ solution.

33. Click **Start** to start the run and watch the concentration display to the side of each beaker for any activity.

34. Click **Record Data** to display the results in the grid (and record your results in Chart 2).

After you complete the experiment, take the online **Post-lab Quiz** for Activity 2.

Activity Questions

1. Are the solutes moving with or against their concentration gradient in facilitated diffusion?

_____

_____

_____

2. What happened to the rate of facilitated diffusion when the number of carrier proteins was increased?

_____

_____

3. Explain why equilibrium was not reached with 10 mM glucose and 100 membrane carriers.

_____

_____

4. In the simulation you added Na$^+$Cl$^-$ to test its effect on glucose diffusion. Explain why there was no effect.

_____

_____   ▬

A C T I V I T Y   3

Simulating Osmotic Pressure

OBJECTIVES

1. To explain how osmosis is a special type of diffusion.
2. To understand that osmosis is a passive process that depends upon the concentration gradient of water.
3. To explain how tonicity of a solution relates to changes in cell volume.
4. To understand conditions that affect osmotic pressure.

Introduction

A special form of diffusion, called **osmosis,** is the diffusion of water through a selectively permeable membrane. (A membrane is called *selectively permeable, differentially permeable,* or *semipermeable* if it allows some molecules to pass but not others.) Because water can pass through the pores of most membranes, it can move from one side of a membrane to the other relatively freely. Osmosis takes place whenever there is a difference in water concentration between the two sides of a membrane.

If we place distilled water on both sides of a membrane, *net* movement of water does not occur. Remember, however, that water molecules would still move between the two sides of the membrane. In such a situation, we would say that there is no *net* osmosis.

The concentration of water in a solution depends on the number of solute particles present. For this reason, increasing the solute concentration coincides with decreasing the water concentration. Because water moves down its concentration gradient (from an area of its higher concentration to an area of its lower concentration), it always moves *toward* the solution with the highest concentration of solutes. Similarly, solutes also move down their concentration gradients.

If we position a *fully* permeable membrane (permeable to solutes and water) between two solutions of differing concentrations, then all substances—solutes and water—diffuse freely, and an equilibrium will be reached between the two sides of the membrane. However, if we use a selectively permeable membrane that is impermeable to the solutes, then we have established a condition where water moves but solutes do not. Consequently, water moves toward the more concentrated solution, resulting in a *volume increase* on that side of the membrane.

By applying this concept to a closed system where volumes cannot change, we can predict that the *pressure* in the more concentrated solution will rise. The force that would need to be applied to oppose the osmosis in a closed system is the **osmotic pressure.** Osmotic pressure is measured in *millimeters of mercury (mm Hg).* In general, the more impermeable the solutes, the higher the osmotic pressure.

Osmotic changes can affect the volume of a cell when it is placed in various solutions. The concept of **tonicity** refers to the way a solution affects the volume of a cell. The tonicity of a solution tells us whether or not a cell will shrink or swell. If the concentration of impermeable solutes is the *same* inside and outside of the cell, the solution is **isotonic.** If there is a *higher* concentration of impermeable solutes *outside* the cell than in the cell's interior, the solution is **hypertonic.** Because the net movement of water would be out of the cell, the cell would *shrink* in a hypertonic solution. Conversely, if the concentration of impermeable solutes is *lower* outside of the cell than in the cell's interior, then the solution is **hypotonic.** The net movement of water would be into the cell, and the cell would *swell* and possibly burst.

> **EQUIPMENT USED** The following equipment will be depicted on-screen: left and right beakers—used for diffusion of solutes; dialysis membranes with various molecular weight cutoffs (MWCOs).

Experiment Instructions

Go to the home page in the PhysioEx software and click **Exercise 1: Cell Transport Mechanisms and Permeability.** Click **Activity 3: Simulating Osmotic Pressure,** and take the online **Pre-lab Quiz** for Activity 3.

After you take the online Pre-lab Quiz, click the **Experiment** tab and begin the experiment. The experiment instructions are reprinted here for your reference. The opening screen for the experiment is shown below.

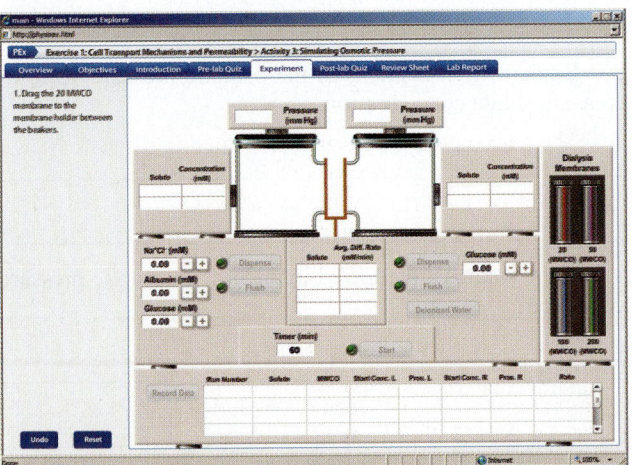

1. Drag the 20 MWCO membrane to the membrane holder between the beakers.

2. Increase the Na$^+$Cl$^-$ concentration to be dispensed to the left beaker to 5.00 m*M* by clicking the + button beside the Na$^+$Cl$^-$ display. Click **Dispense** to fill the left beaker with 5.00 m*M* Na$^+$Cl$^-$ solution.

3. Note that the concentration of Na$^+$Cl$^-$ in the left beaker is displayed in the concentration window to the left of the beaker. Click **Deionized Water** and then click **Dispense** to fill the right beaker with deionized water.

4. After you start the run, the barrier between the beakers will descend, allowing the solutions in each beaker to have access to the dialysis membrane separating them. You can observe the changes in pressure in the two beakers by watching the pressure display above each beaker. You will also be able to determine the amount of solute that passes through the membrane by observing the concentration display to the side of each beaker. A level above zero in Na$^+$Cl$^-$ concentration in the right beaker indicates that Na$^+$ and Cl$^-$ ions are diffusing from the left beaker into the right beaker through the selectively permeable dialysis membrane. Note that the timer is set to 60 minutes. The simulation compresses the 60-minute time period into 10 seconds of real time. Click **Start** to start the run and watch the pressure display above each beaker for any activity.

5. Click **Record Data** to display your results in the grid (and record your results in Chart 3).

CHART 3	Osmosis Results		
Solute	Membrane (MWCO)	Pressure on left (mm Hg)	Diffusion rate (m*M*/min)
Na$^+$Cl$^-$			
Na$^+$Cl$^-$			
Na$^+$Cl$^-$			
Glucose			
Glucose			
Glucose			
Albumin w/glucose			

6. Click **Flush** beneath each of the beakers to prepare for the next run.

7. Increase the Na$^+$Cl$^-$ concentration to be dispensed to the left beaker to 10.00 m*M* by clicking the + button beside the Na$^+$Cl$^-$ display. Click **Dispense** to fill the left beaker with 10.00 m*M* Na$^+$Cl$^-$ solution.

8. Click **Deionized Water** and then click **Dispense** to fill the right beaker with deionized water.

> **? PREDICT Question 1**
> What effect do you think increasing the Na$^+$Cl$^-$ concentration will have?
>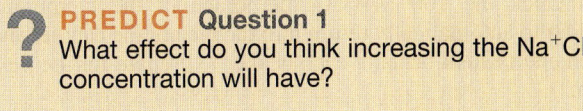

9. Click **Start** to start the run and watch the pressure display above each beaker for any activity.

10. Click **Record Data** to display your results in the grid (and record your results in Chart 3).

11. Click the 20 MWCO membrane in the membrane holder to automatically return it to the membrane cabinet and then click **Flush** beneath each beaker to prepare for the next run.

12. Drag the 50 MWCO membrane to the membrane holder between the beakers. Increase the Na^+Cl^- concentration to be dispensed to the left beaker to 10.00 m*M*. Click **Dispense** to fill the left beaker with 10.00 m*M* Na^+Cl^- solution.

13. Click **Deionized Water** and then click **Dispense** to fill the right beaker with deionized water.

14. Click **Start** to start the run and watch the pressure display above each beaker for any activity.

15. Click **Record Data** to display your results in the grid (and record your results in Chart 3).

16. Click the 50 MWCO membrane in the membrane holder to automatically return it to the membrane cabinet and then click **Flush** beneath each beaker to prepare for the next run.

17. Drag the 100 MWCO membrane to the membrane holder between the beakers. Increase the glucose concentration to be dispensed to the left beaker to 8.00 m*M* by clicking the + button beside the glucose display beneath the left beaker. Click **Dispense** to fill the left beaker with 8.00 m*M* glucose solution.

18. Click **Deionized Water** and then click **Dispense** to fill the right beaker with deionized water.

19. Click **Start** to start the run and watch the pressure display above each beaker for any activity.

20. Click **Record Data** to display your results in the grid (and record your results in Chart 3).

21. Click **Flush** beneath each of the beakers to prepare for the next run.

22. Increase the glucose concentration to be dispensed to the left beaker to 8.00 m*M*. Click **Dispense** to fill the left beaker with 8.00 m*M* glucose solution.

23. Increase the glucose concentration to be dispensed to the right beaker to 8.00 m*M* by clicking the + button beside the glucose display beneath the right beaker. Click **Dispense** to fill the right beaker with 8.00 m*M* glucose solution.

24. Click **Start** to start the run and watch the pressure display above each beaker for any activity.

25. Click **Record Data** to display your results in the grid (and record your results in Chart 3).

26. Click the 100 MWCO membrane in the membrane holder to automatically return it to the membrane cabinet and then click **Flush** beneath each beaker to prepare for the next run.

27. Drag the 200 MWCO membrane to the membrane holder between the beakers. Increase the glucose concentration to be dispensed to the left beaker to 8.00 m*M*. Click **Dispense** to fill the left beaker with 8.00 m*M* glucose solution.

28. Click **Deionized Water** and then click **Dispense** to fill the right beaker with deionized water.

29. Click **Start** to start the run and watch the pressure display above each beaker for any activity.

30. Click **Record Data** to display your results in the grid (and record your results in Chart 3).

31. Click **Flush** beneath each of the beakers to prepare for the next run.

32. Increase the albumin concentration to be dispensed to the left beaker to 9.00 m*M*. Click **Dispense** to fill the left beaker with 9.00 m*M* albumin solution.

33. Increase the glucose concentration to be dispensed to the right beaker to 10.00 m*M*. Click **Dispense** to fill the right beaker with 10.00 m*M* glucose solution.

> **? PREDICT Question 2**
> What do you think will be the pressure result of the current experimental conditions?
> _____

34. Click **Start** to start the run and watch the pressure display above each beaker for any activity.

35. Click **Record Data** to display your results in the grid (and record your results in Chart 3).

After you complete the experiment, take the online **Post-lab Quiz** for Activity 3.

Activity Questions

1. Which membrane resulted in the greatest pressure with Na^+Cl^- as the solute? Why?

_____

_____

2. Explain what happens to the osmotic pressure with increasing solute concentration.

_____

_____

3. If the solutes are allowed to diffuse, is osmotic pressure generated?

_____

_____

4. If the solute concentrations are equal, is osmotic pressure generated? Why or why not?

_____

_____

Simulating Filtration

OBJECTIVES

1. To understand that filtration is a passive process dependent upon a pressure gradient.
2. To understand that filtration is not a selective process.
3. To explain that the size of the membrane pores will determine what passes through.
4. To explain the effect that increasing the hydrostatic pressure has on the filtration rate and how this correlates to events in the body.
5. To understand the relationship between molecular weight and molecular size.

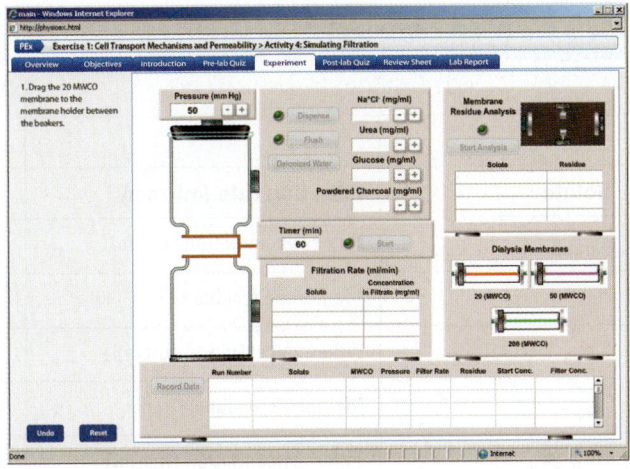

Introduction

Filtration is the process by which water and solutes pass through a membrane (such as a dialysis membrane) from an area of higher hydrostatic (fluid) pressure into an area of lower hydrostatic pressure. Like diffusion, filtration is a passive process. For example, fluids and solutes filter out of the capillaries in the kidneys into the kidney tubules because blood pressure in the capillaries is greater than the fluid pressure in the tubules. So, if blood pressure increases, the rate of filtration increases.

Filtration is not a selective process. The amount of *filtrate*—the fluids and solutes that pass through the membrane—depends almost entirely on the *pressure gradient* (the difference in pressure between the solutions on the two sides of the membrane) and on the *size* of the *membrane pores*. Solutes that are too large to pass through are retained by the capillaries. These solutes usually include blood cells and proteins. Ions and smaller molecules, such as glucose and urea, can pass through.

In this activity the pore size is measured as a *molecular weight cutoff (MWCO),* which is indicated by the number below the filtration membrane. You can think of MWCO in terms of pore size: the larger the MWCO number, the larger the pores in the filtration membrane. The molecular weight of a solute is the number of grams per mole, where a mole is the constant Avogadro's number 6.02×10^{23} molecules/mole. You will also analyze the filtration membrane for the presence or absence of solutes that might be left sticking to the membrane.

> **EQUIPMENT USED** The following equipment will be depicted on-screen: top and bottom beakers—used for filtration of solutes; dialysis membranes with various molecular weight cutoffs (MWCOs); membrane residue analysis station—used to analyze the filtration membrane.

Experiment Instructions

Go to the home page in the PhysioEx software and click **Exercise 1: Cell Transport Mechanisms and Permeability.** Click **Activity 4: Simulating Filtration** and take the online **Pre-lab Quiz** for Activity 4.

After you take the online Pre-lab Quiz, click the **Experiment** tab and begin the experiment. The experiment instructions are reprinted here for your reference. The opening screen for the experiment is shown above.

1. Drag the 20 MWCO membrane to the membrane holder between the beakers.

2. Increase the concentration of Na$^+$Cl$^-$, urea, glucose, and powdered charcoal to be dispensed to 5.00 mg/ml by clicking the + button beside the display for each solute. Click **Dispense** to fill the top beaker.

3. After you start the run, the membrane holder below the top beaker retracts, and the solution will filter through the membrane into the beaker below. You will be able to determine whether solute particles are moving through the filtration membrane by observing the concentration displays beside the bottom beaker. A rise in detected solute concentration indicates that the solute particles are moving through the filtration membrane. Note that the pressure is set at 50 mm Hg and the timer is set to 60 minutes. The simulation compresses the 60-minute time period into 10 seconds of real time. Click **Start** to start the run and watch the concentration displays beside the bottom beaker for any activity.

4. Drag the 20 MWCO membrane to the holder in the membrane residue analysis unit. Click **Start Analysis** to begin analysis (and cleaning) of the membrane.

5. Click **Record Data** to display your results in the grid (and record your results in Chart 4).

6. Click the 20 MWCO membrane in the membrane holder to automatically return it to the membrane cabinet and then click **Flush** to prepare for the next run.

> 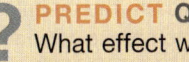 **PREDICT** Question 1
> What effect will increasing the pore size of the filter have on the filtration rate?

7. Drag the 50 MWCO membrane to the membrane holder between the beakers. With the concentration of Na$^+$Cl$^-$, urea, glucose, and powdered charcoal still set to 5.00 mg/ml, click **Dispense** to fill the top beaker.

8. Click **Start** to start the run and watch the concentration displays beside the bottom beaker for any activity.

CHART 4	Filtration Results	Membrane (MWCO)			
		20	50	200	200
Solute	**Filtration rate (ml/min)**				
Na^+Cl^-	Filter concentration (mg/ml)				
	Membrane residue				
Urea	Filter concentration (mg/ml)				
	Membrane residue				
Glucose	Filter concentration (mg/ml)				
	Membrane residue				
Powdered charcoal	Filter concentration (mg/ml)				
	Membrane residue				

9. Drag the 50 MWCO membrane to the holder in the membrane residue analysis unit. Click **Start Analysis** to begin analysis (and cleaning) of the membrane.

10. Click **Record Data** to display your results in the grid (and record your results in Chart 4).

11. Click the 50 MWCO membrane in the membrane holder to automatically return it to the membrane cabinet and then click **Flush** to prepare for the next run.

12. Drag the 200 MWCO membrane to the membrane holder between the beakers. With the concentration of Na^+Cl^-, urea, glucose, and powdered charcoal still set to 5.00 mg/ml, click **Dispense** to fill the top beaker.

13. Click **Start** to start the run and watch the concentration displays beside the bottom beaker for any activity.

14. Drag the 200 MWCO membrane to the holder in the membrane residue analysis unit. Click **Start Analysis** to begin analysis (and cleaning) of the membrane.

15. Click **Record Data** to display your results in the grid (and record your results in Chart 4).

16. Click the 200 MWCO membrane in the membrane holder to automatically return it to the membrane cabinet and then click **Flush** to prepare for the next run.

? PREDICT Question 2
What will happen if you increase the pressure above the beaker (the driving pressure)?

17. Increase the pressure to 100 mm Hg by clicking on the + button beside the pressure display above the top beaker.

18. Drag the 200 MWCO membrane to the membrane holder between the beakers. With the concentration of Na^+Cl^-, urea, glucose, and powdered charcoal still set to 5.00 mg/ml, click **Dispense** to fill the top beaker.

19. Click **Start** to start the run and watch the concentration displays beside the bottom beaker for any activity.

20. Drag the 200 MWCO membrane to the holder in the membrane residue analysis unit. Click **Start Analysis** to begin analysis (and cleaning) of the membrane.

21. Click **Record Data** to display your results in the grid (and record your results in Chart 4).

After you complete the Experiment, take the online **Post-lab Quiz** for Activity 4.

Activity Questions

1. Explain your results with the 20 MWCO filter. Why weren't any of the solutes present in the filtrate?

_____

_____

2. Describe two variables that affected the rate of filtration in your experiments.

_____

_____

3. Explain how you can increase the filtration rate through living membranes.

_____

_____

4. Judging from the filtration results, indicate which solute has the largest molecular weight.

_____

Simulating Active Transport

OBJECTIVES

1. To understand that active transport requires cellular energy in the form of ATP.

2. To explain how the balance of sodium and potassium is maintained by the Na^+-K^+ pump, which moves both ions against their concentration gradients.

3. To understand coupled transport and be able to explain how the movement of sodium and potassium is independent of other solutes, such as glucose.

Introduction

Whenever a cell uses cellular energy (ATP) to move substances across its membrane, the process is an *active transport process*. Substances moved across cell membranes by an active transport process are generally unable to pass by diffusion. There are several reasons why a substance might not be able to pass through a membrane by diffusion: it might be too large to pass through the membrane pores, it might not be lipid soluble, or it might have to move *against*, rather than with, a concentration gradient.

In one type of active transport, substances move across the membrane by combining with a carrier-protein molecule. This kind of process resembles an enzyme-substrate interaction. ATP hydrolysis provides the driving force, and, in many cases, the substances move *against* concentration gradients or electrochemical gradients or both. The carrier proteins are commonly called **solute pumps.** Substances that are moved into cells by solute pumps include amino acids and some sugars. Both of these kinds of solutes are necessary for the life of the cell, but they are lipid insoluble and too large to pass through membrane pores.

In contrast, sodium ions (Na^+) are ejected from the cells by active transport. There is more Na^+ outside the cell than inside the cell, so Na^+ tends to remain in the cell unless actively transported out. In the body, the most common type of solute pump is the Na^+-K^+ (sodium-potassium) pump, which moves Na^+ and K^+ in opposite directions across cellular membranes. Three Na^+ ions are ejected from the cell for every two K^+ ions entering the cell. Note that there is more K^+ inside the cell than outside the cell, so K^+ tends to remain outside the cell unless actively transported in.

Membrane carrier proteins that move more than one substance, such as the Na^+-K^+ pump, participate in *coupled transport*. If the solutes move in the same direction, the carrier is a *symporter*. If the solutes move in opposite directions, the carrier is an *antiporter*. A carrier that transports only a single solute is a *uniporter*.

> **EQUIPMENT USED** The following equipment will be depicted on-screen: Simulated cell inside a large beaker.

Experiment Instructions

Go to the home page in the PhysioEx software and click **Exercise 1: Cell Transport Mechanisms and Permeability.** Click **Activity 5: Simulating Active Transport** and take the online **Pre-lab Quiz** for Activity 5.

After you take the online Pre-lab Quiz, click the **Experiment** tab and begin the experiment. The experiment instructions are reprinted here for your reference. The opening screen for the experiment is shown below.

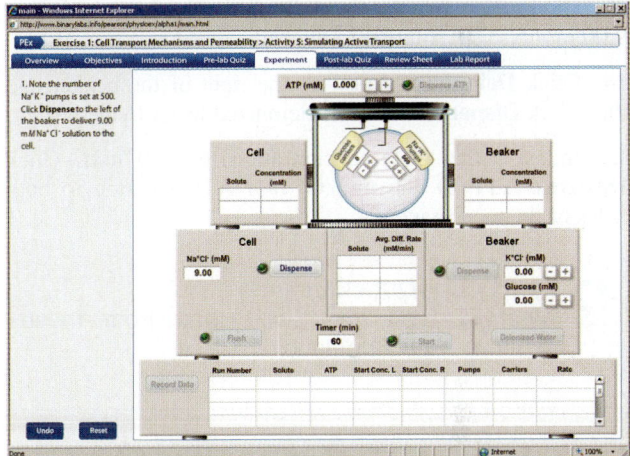

1. Note the number of Na^+-K^+ pumps is set at 500. Click **Dispense** to the left of the beaker to deliver 9.00 mM Na^+Cl^- solution to the cell.

2. Increase the K^+Cl^- concentration to be delivered to the beaker to 6.00 mM by clicking the + button beside the K^+Cl^- display. Click **Dispense** to the right of the beaker to deliver 6.00 mM K^+Cl^- solution to the beaker.

3. Increase the ATP concentration to 1.00 mM by clicking the + button beside the ATP display above the beaker. Click **Dispense ATP** to deliver 1.00 mM ATP solution to both sides of the membrane.

4. After you start the run, the solutes will move across the cell membrane, simulating active transport. You will be able to determine the amount of solute that is transported across the membrane by observing the concentration displays on both sides of the beaker (the display on the left shows the concentrations inside the cell and the display on the right shows the concentrations inside the beaker). Note that the timer is set to 60 minutes. The simulation compresses the 60-minute time period into 10 seconds of real time. Click **Start** to start the run and watch the concentration displays on both sides of the beaker for any activity.

5. Click **Record Data** to display your results in the grid.

6. Click **Flush** to reset the beaker and simulated cell.

7. Click **Dispense** to the left of the beaker to deliver 9.00 mM Na$^+$Cl$^-$ solution to the cell.

8. Increase the K$^+$Cl$^-$ concentration to be delivered to the beaker to 6.00 mM by clicking the + button beside the K$^+$Cl$^-$ display. Click **Dispense** to the right of the beaker to deliver 6.00 mM K$^+$Cl$^-$ solution to the beaker.

9. Increase the ATP concentration to 3.00 mM by clicking the + button beside the ATP display above the beaker. Click **Dispense ATP** to deliver 3.00 mM ATP solution to both sides of the membrane.

10. Click **Start** to start the run and watch the concentration displays on both sides of the beaker for any activity.

11. Click **Record Data** to display your results in the grid.

12. Click **Flush** to reset the beaker and simulated cell.

13. Click **Dispense** to the left of the beaker to deliver 9.00 mM Na$^+$Cl$^-$ solution to the cell.

14. Click **Deionized Water** to the right of the beaker and then click **Dispense** to deliver deionized water to the beaker.

15. Increase the ATP concentration to 3.00 mM. Click **Dispense ATP** to deliver 3.00 mM ATP solution to both sides of the membrane.

> **? PREDICT Question 1**
> What do you think will result from these experimental conditions?
> _____

16. Click **Start** to start the run and watch the concentration displays on both sides of the beaker for any activity.

17. Click **Record Data** to display your results in the grid.

18. Click **Flush** to reset the beaker and simulated cell.

19. Increase the number of Na$^+$-K$^+$ pumps to 800 by clicking the + button beneath the Na$^+$-K$^+$ pump display. Click **Dispense** to the left of the beaker to deliver 9.00 mM Na$^+$Cl$^-$ solution to the cell.

20. Increase the K$^+$Cl$^-$ concentration to be delivered to the beaker to 6.00 mM. Click **Dispense** to the right of the beaker to deliver 6.00 mM K$^+$Cl$^-$ solution to the beaker.

21. Increase the ATP concentration to 3.00 mM. Click **Dispense ATP** to deliver 3.00 mM ATP solution to both sides of the membrane.

22. Click **Start** to start the run and watch the concentration displays on both sides of the beaker for any activity.

23. Click **Record Data** to display your results in the grid.

24. Click **Flush** to reset the beaker and simulated cell.

25. With the number of Na$^+$-K$^+$ pumps still set to 800, increase the number of glucose carriers to 400 by clicking the + button beneath the glucose carriers display. Click **Dispense** to the left of the beaker to deliver 9.00 mM Na$^+$Cl$^-$ solution to the cell.

> **? PREDICT Question 2**
> Do you think the addition of glucose carriers will affect the transport of sodium or potassium?
> _____

26. Increase the K$^+$Cl$^-$ concentration to be delivered to the beaker to 6.00 mM. Increase the glucose concentration to be delivered to 10.00 mM. Click **Dispense** to the right of the beaker to deliver 6.00 mM K$^+$Cl$^-$ and 10.00 mM glucose solution to the beaker.

27. Increase the ATP concentration to 3.00 mM. Click **Dispense ATP** to deliver 3.00 mM ATP solution to both sides of the membrane.

28. Click **Start** to start the run and watch the concentration displays on both sides of the beaker for any activity.

29. Click **Record Data** to display your results in the grid.

After you complete the experiment, take the online **Post-lab Quiz** for Activity 5.

Activity Questions

1. In the initial trial the number of Na$^+$-K$^+$ pumps is set to 500, the Na$^+$Cl$^-$ concentration is set to 9.00 mM, the K$^+$Cl$^-$ concentration is set to 6.00 mM, and the ATP concentration is set to 1.00 mM. Explain what happened and why. What would happen if no ATP had been dispensed?

_____

_____

2. Why was there no transport when you dispensed only Na$^+$Cl$^-$, even though ATP was present?

_____

_____

3. What happens to the rate of transport of Na$^+$ and K$^+$ when you increase the number of Na$^+$-K$^+$ pumps?

_____

_____

4. Explain why the Na$^+$ and K$^+$ transports were unaffected by the addition of glucose.

_____

_____

NAME _____

LAB TIME/DATE _____

Cell Transport Mechanisms and Permeability

ACTIVITY 1 **Simulating Dialysis (Simple Diffusion)**

1. Describe two variables that affect the rate of diffusion. _____

    _____

2. Why do you think the urea was not able to diffuse through the 20 MWCO membrane? How well did the results compare

    with your prediction? _____

3. Describe the results of the attempts to diffuse glucose and albumin through the 200 MWCO membrane. How well did the

    results compare with your prediction? _____

    _____

4. Put the following in order from smallest to largest molecular weight: glucose, sodium chloride, albumin, and urea. _____

    _____

ACTIVITY 2 **Simulated Facilitated Diffusion**

1. Explain one way in which facilitated diffusion is the same as simple diffusion and one way in which it differs. _____

    _____

2. The larger value obtained when more glucose carriers were present corresponds to an increase in the rate of glucose transport.

    Explain why the rate increased. How well did the results compare with your prediction? _____

    _____

3. Explain your prediction for the effect Na^+Cl^- might have on glucose transport. In other words, explain why you picked the

    choice that you did. How well did the results compare with your prediction? _____

ACTIVITY 3 **Simulating Osmotic Pressure**

1. Explain the effect that increasing the Na^+Cl^- concentration had on osmotic pressure and why it has this effect. How well did

    the results compare with your prediction? _____

    _____

2. Describe one way in which osmosis is similar to simple diffusion and one way in which it is different. _____

   _____

3. Solutes are sometimes measured in milliosmoles. Explain the statement, "Water chases milliosmoles." _____

   _____

4. The conditions were 9 mM albumin in the left beaker and 10 mM glucose in the right beaker with the 200 MWCO membrane

   in place. Explain the results. How well did the results compare with your prediction? _____

   _____

ACTIVITY 4   Simulating Filtration

1. Explain in your own words why increasing the pore size increased the filtration rate. Use an analogy to support your state-

   ment. How well did the results compare with your prediction? _____

   _____

2. Which solute did not appear in the filtrate using any of the membranes? Explain why. _____

   _____

3. Why did increasing the pressure increase the filtration rate but not the concentration of solutes? How well did the results

   compare with your prediction? _____

   _____

ACTIVITY 5   Simulating Active Transport

1. Describe the significance of using 9 mM sodium chloride inside the cell and 6 mM potassium chloride outside the cell,

   instead of other concentration ratios. _____

   _____

2. Explain why there was no sodium transport even though ATP was present. How well did the results compare with your

   prediction? _____

3. Explain why the addition of glucose carriers had no effect on sodium or potassium transport. How well did the results

   compare with your prediction? _____

4. Do you think glucose is being actively transported or transported by facilitated diffusion in this experiment? Explain your

   answer. _____

   _____

Skeletal Muscle Physiology

Exercise Overview

Humans make voluntary decisions to walk, talk, stand up, and sit down. Skeletal muscles, which are usually attached to the skeleton, make these actions possible (view Figure 2.1). Skeletal muscles characteristically span two joints and attach to the skeleton via **tendons,** which attach to the periosteum of a bone. Skeletal muscles are composed of hundreds to thousands of individual cells called **muscle fibers,** which produce **muscle tension** (also referred to as **muscle force**). Skeletal muscles are remarkable machines. They provide us with the manual dexterity to create magnificent works of art and can generate the brute force needed to lift a 45-kilogram sack of concrete.

When a skeletal muscle is isolated from an experimental animal and mounted on a **force transducer,** you can generate **muscle contractions** with controlled **electrical stimulation.** Importantly, the contractions of this isolated muscle are known to mimic those of working muscles in the body. That is, in vitro experiments reproduce in vivo functions. Therefore, the activities you perform in this exercise will give you valuable insight into skeletal muscle physiology.

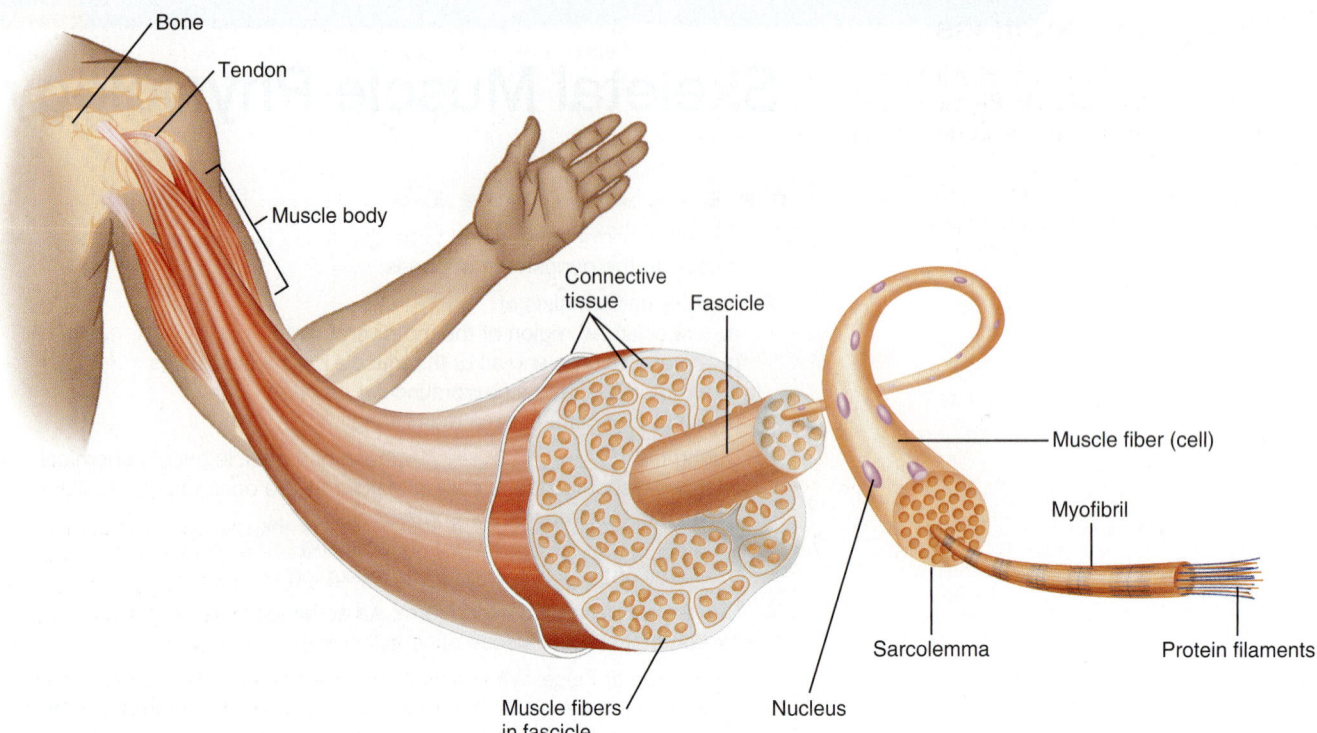

FIGURE 2.1   Structure of a skeletal muscle.

The Muscle Twitch and the Latent Period

OBJECTIVES

1. To understand the terms *excitation-contraction coupling, electrical stimulus, muscle twitch, latent period, contraction phase,* and *relaxation phase.*

2. To initiate muscle twitches with electrical stimuli of varying intensity.

3. To identify and measure the duration of the latent period.

Introduction

A **motor unit** consists of a **motor neuron** and all of the **muscle fibers** it innervates. The motor neuron and a muscle fiber intersect at the **neuromuscular junction.** Specifically, the neuromuscular junction is the location where the axon terminal of the neuron meets a specialized region of the muscle fiber's plasma membrane. This specialized region is called the **motor end plate.**

An action potential in a motor neuron triggers the release of acetylcholine from its terminal. Acetylcholine then diffuses onto the muscle fiber's plasma membrane (or **sarcolemma**) and binds to receptors in the motor end plate, initiating a change in ion permeability that results in a *graded depolarization* of the muscle plasma membrane (the end-plate potential). The events that occur at the neuromuscular junction lead to the **end-plate potential.** The end-plate potential triggers a series of events that results in the contraction of a muscle cell. This entire process is called **excitation-contraction coupling.**

You will be simulating excitation-contraction coupling in this and subsequent activities, but you will be using electrical pulses, rather than acetylcholine, to trigger action potentials. The pulses will be administered by an electrical stimulator that can be set for the precise voltage, frequency, and duration of shock desired. When applied to a muscle that has been surgically removed from an animal, a single electrical stimulus will result in a **muscle twitch**—the mechanical response to a single action potential. A muscle twitch has three phases: the *latent period,* the *contraction phase,* and the *relaxation phase.*

1. The **latent period** is the period of time that elapses between the generation of an action potential in a muscle cell and the start of muscle contraction. Although no force is generated during the latent period, chemical changes (including the release of calcium from the sarcoplasmic reticulum) occur intracellularly in preparation for contraction.

2. The **contraction phase** starts at the end of the latent period and ends when muscle tension peaks.

3. The **relaxation phase** is the period of time from peak tension until the end of the muscle contraction

EQUIPMENT USED   The following equipment will be depicted on-screen: intact, viable skeletal muscle dissected off the leg of a frog; electrical stimulator—delivers the desired amount and duration of stimulating voltage to the muscle via electrodes resting on the muscle; mounting stand—includes a force transducer to measure the amount of force, or tension, developed by the muscle; oscilloscope—displays the stimulated muscle twitch and the amount of active, passive, and total force developed by the muscle.

Experiment Instructions

Go to the home page in the PhysioEx software and click **Exercise 2: Skeletal Muscle Physiology.** Click **Activity 1: The Muscle Twitch and the Latent Period,** and take the online **Pre-lab Quiz** for Activity 1.

After you take the online Pre-lab Quiz, click the **Experiment** tab and begin the experiment. The experiment instructions are reprinted here for your reference. The opening screen for the experiment is shown below.

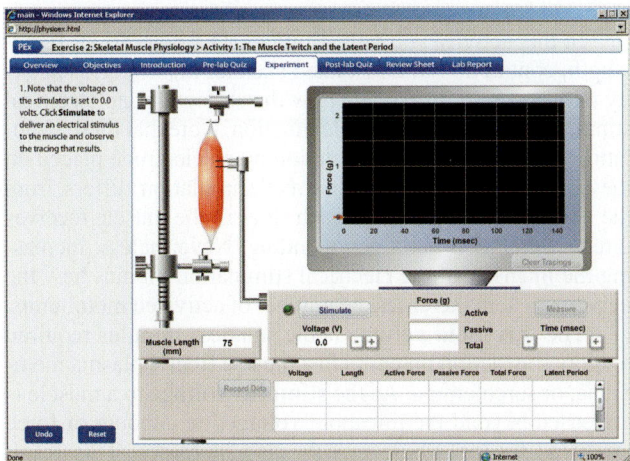

1. Note that the voltage on the stimulator is set to 0.0 volts. Click **Stimulate** to deliver an electrical stimulus to the muscle and observe the tracing that results.

2. The tracing on the oscilloscope indicates active muscle force. Note whether any muscle force developed with the voltage set to zero. Click **Record Data** to display your results in the grid (and record your results in Chart 1).

CHART 1	Latent Period Results	
Voltage	**Active force (g)**	**Latent period (msec)**

3. Increase the voltage to 3.0 volts by clicking the + button beside the voltage display.

4. Click **Stimulate** and observe the tracing that results.

5. Note the muscle force that developed. Click **Record Data** to display your results in the grid (and record your results in Chart 1).

6. Click **Clear Tracings** to remove the tracings from the oscilloscope.

7. Increase the voltage to 4.0 volts by clicking the + button beside the voltage display.

8. Click **Stimulate** and observe the tracing that results. Note that the trace starts at the left side of the screen and stays flat for a short period of time. Remember that the X-axis displays elapsed time in milliseconds. Also note how the force during the twitch also changes.

9. Click **Measure** on the stimulator. A thin, vertical yellow line appears at the far left side of the oscilloscope screen. To measure the length of the latent period, you measure the time between the application of the stimulus and the beginning of the first observable response (here, an increase in force). Click the + button beside the time display. You will see the vertical yellow line start to move across the screen. Watch what happens in the time (msec) display as the line moves across the screen. Keep clicking the + button until the yellow line reaches the point in the tracing where the graph stops being a flat line and begins to rise (this is the point at which muscle tension starts to develop). If the yellow line moves past the desired point, click the – button to move it backward.

When the yellow line is positioned correctly, click **Record Data** to display the latent period in the grid (and record your results in Chart 1).

10. Click **Clear Tracings** to remove the tracings from the oscilloscope.

> **? PREDICT Question 1**
> Will changes to the stimulus voltage alter the duration of the latent period? Explain.

11. You will now gradually increase the voltage to observe how changes to the stimulus voltage alter the duration of the latent period.

- Increase the voltage by 2.0 volts.

- Click **Stimulate** and observe the tracing that results.

- Click **Measure** on the stimulator and then click the + button until the yellow line reaches the point in the tracing where the graph stops being a flat line and begins to rise.

- Click **Record Data** (and record your results in Chart 1).

Repeat this step until you reach 10.0 volts.

After you complete the experiment, take the online **Post-lab Quiz** for Activity 1.

Activity Questions

1. Draw a graph that depicts a single skeletal muscle twitch, placing time on the X-axis and force on the Y-axis. Label the phases of this muscle twitch and describe what is happening in the muscle during each phase.

2. During the latent period of a skeletal muscle twitch, there is an apparent lack of muscle activity. Describe the electrical and chemical changes that occur in the muscle during this period.

_____

_____

_____

ACTIVITY 2

The Effect of Stimulus Voltage on Skeletal Muscle Contraction

OBJECTIVES

1. To understand the terms *motor neuron, muscle twitch, motor unit, recruitment, stimulus voltage, threshold stimulus,* and *maximal stimulus.*

2. To understand how motor unit recruitment can increase the tension a whole muscle develops.

3. To identify a threshold stimulus voltage.

4. To observe the effect of increases in stimulus voltage on a whole muscle.

5. To understand how increasing stimulus voltage to an isolated muscle in an experiment mimics motor unit recruitment in the body.

Introduction

A skeletal muscle produces **tension** (also known as **muscle force**) when nervous or electrical stimulation is applied. The force generated by a whole muscle reflects the number of active **motor units** at a given moment. A strong muscle contraction implies that many motor units are activated, with each unit developing its maximal tension, or force. A weak muscle contraction implies that fewer motor units are activated, but each motor unit still develops its maximal tension. By increasing the number of active motor units, we can produce a steady increase in muscle force, a process called **motor unit recruitment.**

Regardless of the number of **motor units** activated, a single stimulated contraction of whole skeletal muscle is called

a **muscle twitch.** A tracing of a muscle twitch is divided into three phases: the latent period, the contraction phase, and the relaxation phase. The latent period is a short period between the time of muscle stimulation and the beginning of a muscle response. Although no force is generated during this interval, chemical changes occur intracellularly in preparation for contraction (including the release of calcium from the sarcoplasmic reticulum). During the contraction phase, the myofilaments utilize the cross-bridge cycle and the muscle develops tension. Relaxation takes place when the contraction has ended and the muscle returns to its normal resting state and length.

In this activity you will stimulate an isometric, or fixed-length, contraction of an isolated skeletal muscle. This activity allows you to investigate how the strength of an electrical stimulus affects whole-muscle function. Note that these simulations involve indirect stimulation by an electrode placed on the surface of the muscle. Indirect stimulation differs from the situation in vivo, where each fiber in the muscle receives direct stimulation via a nerve ending. Nevertheless, increasing the intensity of the electrical stimulation mimics how the nervous system increases the number of activated motor units.

The **threshold voltage** is the smallest stimulus required to induce an action potential in a muscle fiber's plasma membrane, or sarcolemma. As the **stimulus voltage** to a muscle is increased beyond the threshold voltage, the amount of force produced by the whole muscle also increases. This result occurs because, as more voltage is delivered to the whole muscle, more muscle fibers are activated and, thus, the total force produced by the muscle increases. Maximal tension in the whole muscle occurs when all the muscle fibers have been activated by a sufficiently strong stimulus (referred to as the **maximal voltage**). Stimulation with voltages greater than the maximal voltage will not increase the force of contraction. This experiment is analogous to, and accurately mimics, muscle activity in vivo, where the recruitment of additional motor units increases the total muscle force produced. This phenomenon is called *motor unit recruitment.*

> **EQUIPMENT USED** The following equipment will be depicted on-screen: intact, viable skeletal muscle dissected off the leg of a frog; electrical stimulator—delivers the desired amount and duration of stimulating voltage to the muscle via electrodes resting on the muscle; mounting stand—includes a force transducer to measure the amount of force, or tension, developed by the muscle; oscilloscope—displays the stimulated muscle twitch and the amount of active, passive, and total force developed by the muscle.

Experiment Instructions

Go to the home page in the PhysioEx software and click **Exercise 2: Skeletal Muscle Physiology.** Click **Activity 2: The Effect of Stimulus Voltage on Skeletal Muscle Contraction,** and take the online **Pre-lab Quiz** for Activity 2.

After you take the online Pre-lab Quiz, click the **Experiment** tab and begin the experiment. The experiment instructions are reprinted here for your reference. The opening screen for the experiment is shown on the following page.

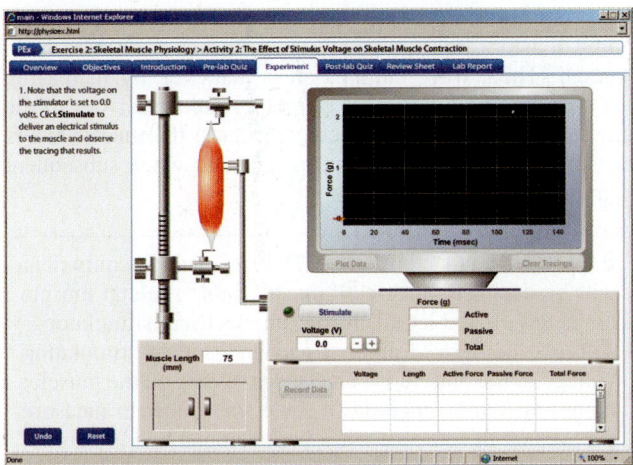

CHART 2	Effect of Stimulus Voltage on Skeletal Muscle Contraction
Voltage	Active force (g)

1. Note that the voltage on the stimulator is set to 0.0 volts. Click **Stimulate** to deliver an electrical stimulus to the muscle and observe the tracing that results.

2. Note the active force display and then click **Record Data** to display your results in the grid (and record your results in Chart 2).

3. Increase the voltage to 0.2 volts by clicking the + button beside the voltage display. Click **Stimulate** to deliver an electrical stimulus to the muscle and observe the tracing that results.

4. Note the active force display and then click **Record Data** to display your results in the grid (and record your results in Chart 2).

5. You will now gradually increase the voltage and stimulate the muscle to determine the minimum voltage required to generate active force.

   • Increase the voltage by 0.1 volts and then click **Stimulate.**

   • If no active force is generated, increase the voltage by 0.1 volts and stimulate the muscle again. When active force is generated, click **Record Data** to display your results in the grid (and record your results in Chart 2).

6. Enter the threshold voltage for this experiment in the field below and then click **Submit** to record your answer in the lab report. _____ volts

7. Click **Clear Tracings** to clear the tracings on the oscilloscope.

? PREDICT Question 1
As the stimulus voltage is increased from 1.0 volt up to 10 volts, what will happen to the amount of active force generated with each stimulus?

8. Increase the voltage on the stimulator to 1.0 volt and then click **Stimulate.**

9. Note the active force display and then click **Record Data** to display your results in the grid (and record your results in Chart 2).

10. You will now gradually increase the voltage and stimulate the muscle to determine the maximal voltage.

   • Increase the voltage by 0.5 volts.

   • Click **Stimulate** and observe the tracing that results.

   • Note the active force display and then click **Record Data** to display your results in the grid (and record your results in Chart 2).

   Repeat this step until you reach 10.0 volts.

11. Click **Plot Data** to view a summary of your data on a plotted grid. Click **Submit** to record your plot in the lab report.

12. Enter the maximal voltage for this experiment in the field below and then click **Submit** to record your answer in the lab report. _____ volts

After you complete the experiment, take the online **Post-lab Quiz** for Activity 2.

Activity Questions

1. For a single skeletal muscle twitch, explain the effect of increasing stimulus voltage.

_____

_____

2. How is this effect achieved in vivo?

_____

_____ ▬

The Effect of Stimulus Frequency on Skeletal Muscle Contraction

OBJECTIVES

1. To understand the terms *stimulus frequency, wave summation,* and *treppe.*
2. To observe the effect of an increasing stimulus frequency on the force developed by an isolated skeletal muscle.
3. To understand how increasing stimulus frequency to an isolated skeletal muscle induces the summation of twitch force.

Introduction

As demonstrated in Activity 2, increasing the stimulus voltage to an isolated skeletal muscle (up to a maximal value) results in an increase of force produced by the whole muscle. This experimental result is analogous to motor unit recruitment in the body. Importantly, this result relies on being able to increase the single stimulus intensity in the experiment. You will now explore another way to increase the force produced by an isolated skeletal muscle.

When a muscle first contracts, the force it is able to produce is less than the force it is able to produce with subsequent stimulations within a relatively short time span. **Treppe** is the progressive increase in force generated when a muscle is stimulated in succession, such that muscle twitches follow one another closely, with each successive twitch peaking slightly higher than the one before. This step-like increase in force is why treppe is also known as the staircase effect. For the first few twitches, each successive twitch produces slightly more force than the previous twitch as long as the muscle is allowed to fully relax between stimuli and the stimuli are delivered relatively close together.

When a skeletal muscle is stimulated repeatedly, such that the stimuli arrive one after another within a short period of time, muscle twitches can overlap with each other and result in a stronger muscle contraction than a stand-alone twitch. This phenomenon is known as wave summation.

Wave summation occurs when muscle fibers that are developing tension are stimulated again before the fibers have relaxed. Thus, wave summation is achieved by increasing the **stimulus frequency,** or rate of stimulus delivery to the muscle. Wave summation occurs because the muscle fibers are already in a partially contracted state when subsequent stimuli are delivered.

> **EQUIPMENT USED** The following equipment will be depicted on-screen: intact, viable skeletal muscle dissected off the leg of a frog; an electrical stimulator—delivers the desired amount and duration of stimulating voltage to the muscle via electrodes resting on the muscle; mounting stand—includes a force transducer to measure the amount of force, or tension, developed by the muscle; oscilloscope—displays the stimulated muscle twitch and the amount of active, passive, and total force developed by the muscle.

Experiment Instructions

Go to the home page in the PhysioEx software and click **Exercise 2: Skeletal Muscle Physiology.** Click **Activity 3: The Effect of Stimulus Frequency on Skeletal Muscle Contraction,** and take the online **Pre-lab Quiz** for Activity 3.

After you take the online Pre-lab Quiz, click the **Experiment** tab and begin the experiment. The experiment instructions are reprinted here for your reference. The opening screen for the experiment is shown below.

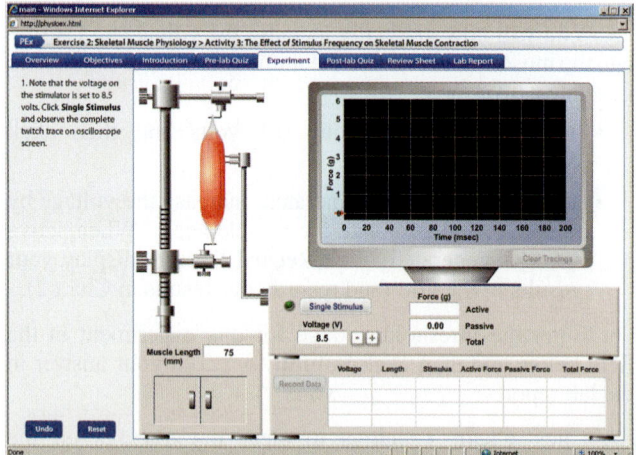

1. Note that the voltage on the stimulator is set to 8.5 volts. Click **Single Stimulus** and observe the tracing that results on the oscilloscope.

2. Note the active force display and then click **Record Data** to display your results in the grid (and record your results in Chart 3).

3. Click **Single Stimulus** and allow the trace to rise and completely fall. *Immediately after* the trace has returned to baseline, click **Single Stimulus** again.

CHART 3	Effect of Stimulus Frequency on Skeletal Muscle Contraction	
Voltage	**Stimulus**	**Active force (g)**

4. Note the active force for the second muscle twitch and click **Record Data** to display your results in the grid (and record your results in Chart 3).

5. You should have observed an increase in active force generated by the muscle with the immediate second stimulus. This increase demonstrates the phenomenon of treppe. Click **Clear Tracings** to clear the tracings on the oscilloscope.

6. You will now investigate the process of wave summation. Click **Single Stimulus** and watch the trace rise and begin to fall. *Before* the trace falls completely back to the baseline, click **Single Stimulus** again. (You can simply click **Single Stimulus** twice in quick succession in order to achieve this.)

7. Note the active force for the second muscle twitch and click **Record Data** to display your results in the grid (and record your results in Chart 3).

> **? PREDICT Question 1**
> As the stimulus frequency increases, what will happen to the muscle force generated with each successive stimulus? Will there be a limit to this response?
>
> _____
>
> _____
>
> _____

8. Now stimulate the muscle at a higher frequency by clicking **Single Stimulus** four times in rapid succession.

9. Note the active force display and then click **Record Data** to display your results in the grid (and record your results in Chart 3).

10. Click **Clear Tracings** to clear the tracings on the oscilloscope.

> **? PREDICT Question 2**
> In order to produce sustained muscle contractions with an active force value of 5.2 grams, do you think you need to increase the stimulus voltage?

11. Increase the voltage to 10.0 volts by clicking the + button beside the voltage display. After setting the voltage, click **Single Stimulus** four times in rapid succession.

12. Note the active force display and then click **Record Data** to display your results in the grid (and record your results in Chart 3).

13. Click **Clear Tracings** to clear the tracings on the oscilloscope.

14. Return the voltage to 8.5 volts by clicking the – button beside the voltage display. After setting the voltage, click **Single Stimulus** as many times as you can in rapid succession. Note the active force display. If you did not achieve an active force of 5.2 grams, click **Clear Tracings** and then click **Single Stimulus** even more rapidly. Repeat this step until you achieve an active force of 5.2 grams.

When you achieve an active force of 5.2 grams, click **Record Data** to display your results in the grid (and record your results in Chart 3).

After you complete the experiment, take the online **Post-lab Quiz** for Activity 3.

Activity Questions

1. Why is treppe also known as the staircase effect?

_____

_____

2. What changes are thought to occur in the skeletal muscle to allow treppe to be observed?

_____

_____

3. How does the frequency of stimulation affect the amount of force generated by a skeletal muscle?

_____

_____

4. Explain how wave summation is achieved in vivo.

_____

_____

ACTIVITY 4

Tetanus in Isolated Skeletal Muscle

OBJECTIVES

1. To understand the terms *stimulus frequency, unfused tetanus, fused tetanus,* and *maximal tetanic tension.*
2. To observe the effect of an increasing stimulus frequency on an isolated skeletal muscle.
3. To understand how increasing the stimulus frequency to an isolated skeletal muscle leads to unfused or fused tetanus.

Introduction

As demonstrated in Activity 3, increasing the **stimulus frequency** to an isolated skeletal muscle results in an increase in force produced by the whole muscle. Specifically, you observed that, if electrical stimuli are applied to a skeletal muscle in quick succession, the overlapping twitches generated more force with each successive stimulus. However, if stimuli continue to be applied frequently to a muscle over a prolonged period of time, the maximum possible muscle force from each stimulus will eventually reach a plateau—a state known as **unfused tetanus.** If stimuli are then applied with even greater frequency, the twitches will begin to fuse so that the peaks and valleys of each twitch become indistinguishable from one another—this state is known as **complete (fused) tetanus.** When the stimulus frequency reaches a value beyond which no further increases in force are generated by the muscle, the muscle has reached its **maximal tetanic tension.**

> **EQUIPMENT USED** The following equipment will be depicted on-screen: intact, viable skeletal muscle dissected off the leg of a frog; electrical stimulator—delivers the desired amount and duration of stimulating voltage to the muscle via electrodes resting on the muscle; mounting stand—includes a force transducer to measure the amount of force, or tension, developed by the muscle; oscilloscope—displays the stimulated muscle twitch and the amount of active, passive, and total force developed by the muscle.

Experiment Instructions

Go to the home page in the PhysioEx software and click **Exercise 2: Skeletal Muscle Physiology.** Click **Activity 4: Tetanus in Isolated Skeletal Muscle** and take the online **Pre-lab Quiz** for Activity 4.

After you take the online Pre-lab Quiz, click the **Experiment** tab and begin the experiment. The experiment instructions are reprinted here for your reference. The opening screen for the experiment is shown above.

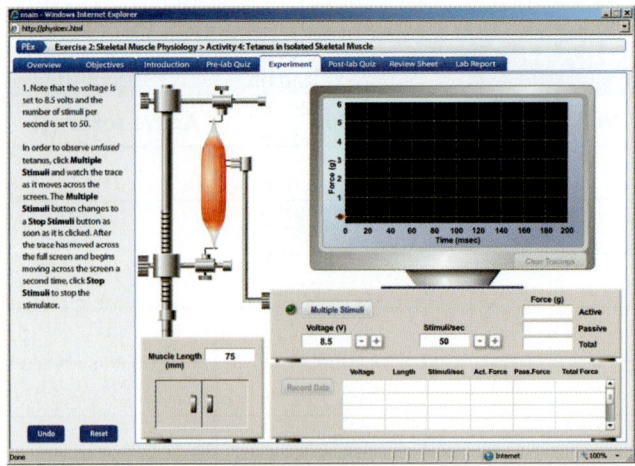

1. Note that the voltage is set to 8.5 volts and the number of stimuli per second is set to 50. To observe *unfused* tetanus, click **Multiple Stimuli** and watch the trace as it moves across the screen. The **Multiple Stimuli** button changes to a **Stop Stimuli** button after it is clicked. After the trace has moved across the full screen and begins moving across the screen a second time, click **Stop Stimuli** to stop the stimulator.

2. Click **Record Data** to display your results in the grid (and record your results in Chart 4).

CHART 4	Tetanus in Isolated Skeletal Muscle
Stimuli/second	**Active force (g)**

? PREDICT Question 1
As the stimulus frequency increases further, what will happen to the muscle tension and twitch appearance with each successive stimulus? Will there be a limit to this response?

_____

_____

3. In order to observe *fused* tetanus, increase the stimuli/sec setting to 130 by clicking the + button beside the stimuli/sec display. Click **Multiple Stimuli** and observe the resulting trace. After the trace has moved across the full screen and begins moving across the screen a second time, click **Stop Stimuli.**

4. Note the fused tetanus and click **Record Data** to display your results in the grid (and record your results in Chart 4).

5. Click **Clear Tracings** to clear the oscilloscope screen.

6. Increase the stimuli/sec setting to 140 by clicking the + button beside the stimuli/sec display. Click **Multiple Stimuli** and observe the resulting trace. After the trace has moved across the full screen and begins moving across the screen a second time, click **Stop Stimuli.**

7. Note the fused tetanus and click **Record Data** to display your results in the grid (and record your results in Chart 4).

8. Click **Clear Tracings** to clear the oscilloscope screen.

9. You will now observe the effect of incremental increases in the number of stimuli per second above 140 stimuli per second.

- Increase the stimuli/sec setting by 2.

- Click **Multiple Stimuli** and observe the resulting trace. After the trace has moved across the full screen and begins moving across the screen a second time, click **Stop Stimuli.**

- Click **Record Data** to display your results in the grid (and record your results in Chart 4).

- Click **Clear Tracings** to clear the oscilloscope screen.

    Repeat this step until you reach 150 stimuli per second.

After you complete the experiment, take the online **Post-lab Quiz** for Activity 4.

Activity Questions

1. Explain what you think is being summated in the skeletal muscle to allow a high stimulus frequency to induce a smooth, continuous skeletal muscle contraction.

_____

_____

2. Why do many toddlers receive a tetanus shot (and then subsequent booster shots, as needed, later in life)? How does the condition known as "lockjaw" relate to tetanus shots?

_____

_____

ACTIVITY 5

Fatigue in Isolated Skeletal Muscle

OBJECTIVES

1. To understand the terms *stimulus frequency, complete (fused) tetanus, fatigue,* and *rest period.*

2. To observe the development of skeletal muscle fatigue.

3. To understand how the length of intervening rest periods determines the onset of fatigue.

Introduction

As demonstrated in Activities 3 and 4, increasing the stimulus frequency to an isolated skeletal muscle induces an increase of force produced by the whole muscle. Specifically, if voltage stimuli are applied to a muscle frequently in quick succession, the skeletal muscle generates more force with each successive stimulus.

However, if stimuli continue to be applied frequently to a muscle over a prolonged period of time, the maximum force of each twitch eventually reaches a plateau—a state known as *unfused tetanus.* If stimuli are then applied with even greater frequency, the twitches begin to fuse so that the peaks and valleys of each twitch become indistinguishable from one another—this state is known as **complete (fused) tetanus.** When the **stimulus frequency** reaches a value beyond which no further increase in force is generated by the muscle, the muscle has reached its **maximal tetanic tension.**

In this activity you will observe the phenomena of skeletal muscle *fatigue.* Fatigue refers to a decline in a skeletal muscle's ability to maintain a constant level of force, or tension, after prolonged, repetitive stimulation. You will also demonstrate how intervening **rest periods** alter the onset of fatigue in skeletal muscle. The causes of fatigue are still being investigated and multiple molecular events are thought to be involved, though the accumulations of lactic acid, ADP, and P_i in muscles are thought to be the major factors causing fatigue in the case of high-intensity exercise.

Common definitions for **fatigue** are:

- The failure of a muscle fiber to produce tension because of previous contractile activity.

- A decline in the muscle's ability to maintain a constant force of contraction after prolonged, repetitive stimulation.

> **EQUIPMENT USED** The following equipment will be depicted on-screen: intact, viable skeletal muscle dissected off the leg of a frog; electrical stimulator—delivers the desired amount and duration of stimulating voltage to the muscle via electrodes resting on the muscle; mounting stand—includes a force transducer to measure the amount of force, or tension, developed by the muscle; oscilloscope—displays the stimulated muscle twitch and the amount of active, passive, and total force developed by the muscle.

Experiment Instructions

Go to the home page in the PhysioEx software and click **Exercise 2: Skeletal Muscle Physiology.** Click **Activity 5: Fatigue in Isolated Skeletal Muscle,** and take the online **Pre-lab Quiz** for Activity 5.

After you take the online Pre-lab Quiz, click the **Experiment** tab and begin the experiment. The experiment instructions are reprinted here for your reference. The opening screen for the experiment is shown on the following page.

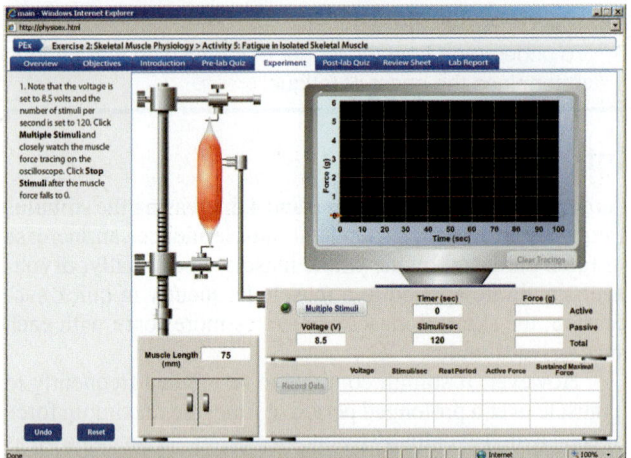

1. Note that the voltage is set to 8.5 volts and the number of stimuli per second is set to 120. Click **Multiple Stimuli** and closely watch the muscle force tracing on the oscilloscope. Click **Stop Stimuli** after the muscle force falls to 0.

2. Click **Record Data** to display your results in the grid (and record your results in Chart 5).

CHART 5	Fatigue Results	
Rest period (sec)	Active force (g)	Sustained maximal force (sec)

3. Click **Clear Tracings** to clear the oscilloscope screen.

> **? PREDICT Question 1**
> If the stimulator is briefly turned off for defined periods of time, what will happen to the length of time that the muscle is able to sustain maximal developed tension when the stimulator is turned on again?
>
> _____
>
> _____

4. To demonstrate the onset of fatigue after a variable rest period, you will be clicking the **Multiple Stimuli** button on and off three times. Read through the steps below before proceeding. Watch the timer closely to help you determine when to turn the stimulator back on.

- Click **Multiple Stimuli.**
- After the muscle force falls to 0, click **Stop Stimuli** to turn off the stimulator.
- Wait 10 seconds, then click **Multiple Stimuli** to turn the stimulator back on.
- Click **Stop Stimuli** after the muscle force falls to 0.

- Wait 20 seconds, then click **Multiple Stimuli** to turn the stimulator back on.
- Click **Stop Stimuli** after the muscle force falls to 0.

5. Click **Record Data** to display your results in the grid (and record your results in Chart 5).

After you complete the experiment, take the online **Post-lab Quiz** for Activity 5.

Activity Questions

1. What proposed mechanisms most likely explain why fatigue develops?

_____

_____

2. What would you recommend to an interested friend as the best ways to delay the onset of fatigue?

_____

_____

The Skeletal Muscle Length-Tension Relationship

OBJECTIVES

1. To understand the terms *isometric contraction, active force, passive force, total force,* and *length-tension relationship.*

2. To understand the effect that resting muscle length has on tension development when the muscle is maximally stimulated in an isometric experiment.

3. To explain the molecular basis of the skeletal muscle length-tension relationship.

Introduction

Skeletal muscle contractions are either isometric or isotonic. When a muscle attempts to move a load that is equal to the force generated by the muscle, the muscle contracts isometrically. During an **isometric** contraction, the muscle stays at a fixed length (*isometric* means "same length"). An example of isometric muscle contraction is when you stand in a doorway and push on the doorframe. The load that you are attempting to move (the doorframe) can easily equal the force generated by your muscles, so your muscles do not shorten even though they are actively contracting.

Isometric contractions are accomplished experimentally by keeping both ends of the muscle in a fixed position while electrically stimulating the muscle. Resting length (the length of the muscle before stimulation) is an important factor in determining the amount of force that a muscle can develop when stimulated. **Passive force** is generated by stretching the muscle and results from the elastic recoil of the tissue itself. This passive force is largely caused by the protein titin, which acts as a molecular bungee cord. **Active force** is generated when myosin thick filaments bind to actin thin filaments, thus engaging the cross bridge cycle and ATP hydrolysis. Think of the skeletal muscle as having two force properties: it exerts passive force when it is stretched (like a rubber band exerts passive force) and active force when it is stimulated. **Total force** is the sum of passive and active forces.

This activity allows you to set and hold constant the length of the isolated skeletal muscle and subsequently stimulate it with individual maximal voltage stimuli. A graph relating the three forces generated and the fixed length of the muscle will be automatically plotted after you stimulate the muscle. In muscle physiology this graph is known as the **isometric length-tension relationship.** The results of this simulation can be applied to human muscles to understand how optimum resting length will result in maximum force production.

To understand why muscle tissue behaves as it does, you must understand tension at the cellular level. If you have difficulty understanding the results of this activity, review the sliding filament model of muscle contraction. Think of the length-tension relationship in terms of those sarcomeres that are too short, those that are too long, and those that have the ideal amount of thick and thin filament overlap.

> **EQUIPMENT USED** The following equipment will be depicted on-screen: intact, viable skeletal muscle dissected off the leg of a frog; electrical stimulator—delivers the desired amount and duration of stimulating voltage to the muscle via electrodes resting on the muscle; mounting stand—includes (1) a force transducer to measure the amount of force, or tension, developed by the muscle and (2) a gearing system that allows the hook through the muscle's lower tendon to be moved up or down, thus altering the fixed length of the muscle; oscilloscope—displays the stimulated muscle twitch and the amount of active, passive, and total force developed by the muscle.

Experiment Instructions

Go to the home page in the PhysioEx software and click **Exercise 2, Skeletal Muscle Physiology.** Click **Activity 6, The Skeletal Muscle Length-Tension Relationship,** and take the online **Pre-lab Quiz** for Activity 6.

After you take the online Pre-lab Quiz, click the **Experiment** tab and begin the experiment. The experiment instructions are reprinted here for your reference. The opening screen for the experiment is shown below.

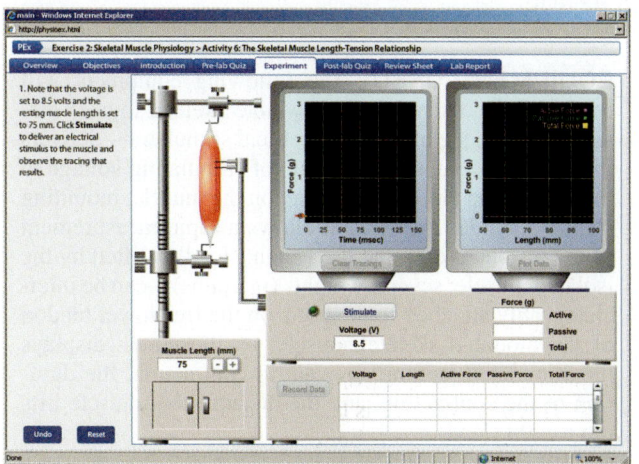

1. Note that the voltage is set to 8.5 volts and the resting muscle length is set to 75 mm. Click **Stimulate** to deliver an electrical stimulus to the muscle and observe the tracing that results.

2. You should see a single muscle twitch tracing on the left oscilloscope display and three data points (representing active, passive, and total force generated during this twitch) plotted on the right display. The yellow box represents the total force, the red dot contained within the yellow box represents the active force, and the green square represents the passive force. Click **Record Data** to display your results in the grid (and record your results in Chart 6).

CHART 6	Skeletal Muscle Length-Tension Relationship		
Length (mm)	Active force (g)	Passive force (g)	Total force (g)

> **? PREDICT Question 1**
> As the resting length of the muscle is changed, what will happen to the amount of total force the muscle generates during the stimulated twitch?
> _____
> _____

3. You will now gradually shorten the muscle to determine the effect of muscle length on active, passive, and total force.

- Shorten the muscle by 5 mm by clicking the − button beside the muscle length display.

- Click **Stimulate** to deliver an electrical stimulus to the muscle and note the values of the total, active, and passive forces relative to those observed at the original 75 mm.

- Click **Record Data** to display your results in the grid (and record your results in Chart 6).

Repeat these steps until you reach a muscle length of 50 mm.

4. Click **Clear Tracings** to clear the left oscilloscope display.

5. Lengthen the muscle to 80 mm by clicking the + button beside the muscle length display. Click **Stimulate** to deliver an electrical stimulus to the muscle and note the values of the total, active, and passive forces relative to those observed at the original 75 mm.

6. Click **Record Data** to display your results in the grid (and record your results in Chart 6).

7. You will now gradually lengthen the muscle to determine the effect of muscle length on active, passive, and total force.

- Lengthen the muscle by 10 mm by clicking the + button beside the muscle length display.

- Click **Stimulate** to deliver an electrical stimulus to the muscle and note the values of the total, active, and passive forces relative to those observed at the original 75 mm.

- Click **Record Data** to display your results in the grid(and record your results in Chart 6).

Repeat these steps until you reach a muscle length of 100 mm.

8. Click **Plot Data** to view a summary of your data on a plotted grid. Click **Submit** to record your plot in the lab report.

After you complete the experiment, take the online **Post-lab Quiz** for Activity 6.

Activity Questions

1. Explain what happens in the skeletal muscle sarcomere to result in the changes in active, passive, and total force when the resting muscle length is changed.

_____

_____

2. Explain the dip in the total force curve as the muscle was stretched to longer lengths. (Hint: Keep in mind that you are measuring the sum of active and passive forces.)

_____

_____ ▬

Isotonic Contractions and the Load-Velocity Relationship

OBJECTIVES

1. To understand the terms *isotonic concentric contraction, load, latent period, shortening velocity,* and *load-velocity relationship.*
2. To understand the effect that increasing load (that is, weight) has on an isolated skeletal muscle when the muscle is stimulated in an isotonic contraction experiment.
3. To understand the load-velocity relationship in isolated skeletal muscle.

Introduction

Skeletal muscle contractions can be described as either isometric or isotonic. When a muscle attempts to move an object (the **load**) that is equal in weight to the force generated by the muscle, the muscle is observed to contract isometrically. In an isometric contraction, the muscle stays at a fixed length (*isometric* means "same length").

During an **isotonic contraction,** the skeletal muscle length changes and, thus, the load moves a measurable distance. If the muscle length shortens as the load moves, the contraction is called an **isotonic** *concentric* **contraction.** An isotonic concentric contraction occurs when a muscle generates a force greater than the load attached to the muscle's end. In this type of contraction, there is a **latent period** during which there is a rise in muscle tension but no observable movement of the weight. After the muscle tension exceeds the weight of the load, an isotonic concentric contraction can begin. Thus, the latent period gets longer as the weight of the load gets larger. When the building muscle force exceeds the load, the muscle shortens and the weight moves. Eventually, the force of the muscle contraction will decrease as the muscle twitch begins the relaxation phase, and the load will therefore start to return to its original position.

An isotonic twitch is not an all-or-nothing event. If the load is increased, the muscle must generate more force to move it and the latent period will therefore get longer because it will take more time for the necessary force to be generated by the muscle. The speed of the contraction (muscle **shortening velocity**) also depends on the load that the muscle is attempting to move. Maximal shortening velocity is attained with minimal load attached to the muscle. Conversely, the heavier the load, the slower the muscle twitch. You can think of lifting an object from the floor as an example. A light object can be lifted quickly (high velocity), whereas a heavier object will be lifted with a slower velocity for a shorter duration.

In an isotonic muscle contraction experiment, one end of the muscle remains free (unlike in an isometric contraction experiment, where both ends of the muscle are held in a fixed position). Different weights (loads) can then be attached to the free end of the isolated muscle, while the other end is held in a fixed position by the force transducer. If the weight (the load) is less than the tension generated by the whole muscle, then the muscle will be able to lift it with a measurable distance, velocity, and duration. In this activity, you will change the weight (load) that the muscle will try to move as it shortens.

EQUIPMENT USED The following equipment will be depicted on-screen: intact, viable skeletal muscle dissected off the leg of a frog; electrical stimulator—delivers the desired amount and duration of stimulating voltage to the muscle via electrodes resting on the muscle; mounting stand—includes a ruler that allows a rapid measurement of the distance (cm) that the weight (load) is lifted by the isolated muscle; several weights (in grams)—can be interchangeably attached to the hook on the free lower tendon of the mounted skeletal muscle; oscilloscope—displays the stimulated isotonic concentric contraction, the duration of the contraction, and the distance that muscle lifts the weight (load).

Experiment Instructions

Go to the home page in the PhysioEx software and click **Exercise 2: Skeletal Muscle Physiology.** Click **Activity 7: Isotonic Contractions and the Load-Velocity Relationship,** and take the online **Pre-lab Quiz** for Activity 7.

After you take the online Pre-lab Quiz, click the **Experiment** tab and begin the experiment. The experiment instructions are reprinted here for your reference. The opening screen for the experiment is shown below.

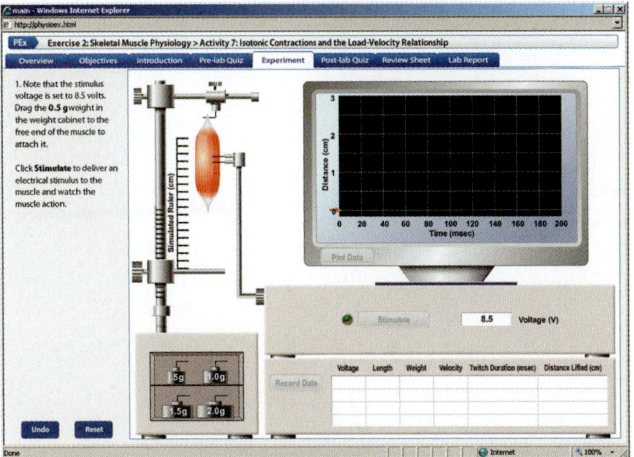

1. Note that the stimulus voltage is set to 8.5 volts. Drag the 0.5-g weight in the weight cabinet to the free end of the muscle to attach it. Click **Stimulate** to deliver an electrical stimulus to the muscle and watch the muscle action.

2. Observe that, as the muscle shortens in length, it lifts the weight off the platform. The muscle then lengthens as it relaxes and lowers the weight back down to the platform. Click **Stimulate** again and try to watch both the muscle and the oscilloscope screen at the same time.

3. Click **Record Data** to display your results in the grid (and record your results in Chart 7).

CHART 7	Isotonic Contraction Results		
Weight (g)	Velocity (cm/sec)	Twitch duration (msec)	Distance lifted (cm)

? PREDICT Question 1
As the load on the muscle *increases*, what will happen to the latent period, the shortening velocity, the distance that the weight moved, and the contraction duration?

_____

_____

4. Remove the 0.5-g weight by dragging it back to the weight cabinet. Drag the 1.0-g weight to the free end of the muscle to attach it. Click **Stimulate** and observe the muscle and the oscilloscope screen.

5. Click **Record Data** to display your results in the grid (and record your results in Chart 7).

6. Remove the 1.0-g weight by dragging it back to the weight cabinet. Drag the 1.5-g weight to the free end of the muscle to attach it. Click **Stimulate** and observe the muscle and the oscilloscope screen.

7. Click **Record Data** to display your results in the grid (and record your results in Chart 7).

8. Remove the 1.5-g weight by dragging it back to the weight cabinet. Drag the 2.0-g weight to the free end of the muscle to attach it. Click **Stimulate** and observe the muscle and the oscilloscope screen.

9. Click **Record Data** to display your results in the grid (and record your results in Chart 7).

10. Click **Plot Data** to generate a muscle load-velocity relationship. Watch the display carefully as the program animates the development of a load-velocity relationship for the data you have collected. Click **Submit** to record your plot in the lab report.

After you complete the experiment, take the online **Post-lab Quiz** for Activity 7.

Activity Questions

1. Explain the relationship between the load attached to a skeletal muscle and the initial velocity of skeletal muscle shortening.

_____

_____

_____

2. Explain why it will take you longer to perform ten repetitions lifting a 20-pound weight than it would to perform the same number of repetitions with a 5-pound weight.

_____

_____

_____

NAME _____

LAB TIME/DATE _____

Skeletal Muscle Physiology

ACTIVITY 1   **The Muscle Twitch and the Latent Period**

1. Define the terms *skeletal muscle fiber*, *motor unit*, *skeletal muscle twitch*, *electrical stimulus*, and *latent period*. _____

   _____

   _____

   _____

   _____

2. What is the role of acetylcholine in a skeletal muscle contraction? _____

   _____

3. Describe the process of excitation-contraction coupling in skeletal muscle fibers. _____

   _____

   _____

4. Describe the three phases of a skeletal muscle twitch. _____

   _____

   _____

   _____

5. Does the duration of the latent period change with different stimulus voltages? How well did the results compare with your

   prediction? _____

   _____

6. At the threshold stimulus, do sodium ions start to move into or out of the cell to bring about the membrane depolarization?

   _____

   _____

ACTIVITY 2 The Effect of Stimulus Voltage on Skeletal Muscle Contraction

1. Describe the effect of increasing stimulus voltage on isolated skeletal muscle. Specifically, what happened to the muscle

   force generated with stronger electrical stimulations and why did this change occur? How well did the results compare with

   your prediction? _____

   _____

2. How is this change in whole-muscle force achieved in vivo? _____

   _____

3. What happened in the isolated skeletal muscle when the maximal voltage was applied? _____

   _____

ACTIVITY 3 The Effect of Stimulus Frequency on Skeletal Muscle Contraction

1. What is the difference between stimulus intensity and stimulus frequency? _____

2. In this experiment you observed the effect of stimulating the isolated skeletal muscle multiple times in a short period with

   complete relaxation between the stimuli. Describe the force of contraction with each subsequent stimulus. Are these results

   called treppe or wave summation? _____

   _____

3. How did the frequency of stimulation affect the amount of force generated by the isolated skeletal muscle when the fre-

   quency of stimulation was increased such that the muscle twitches did not fully relax between subsequent stimuli? Are these

   results called treppe or wave summation? How well did the results compare with your prediction? _____

   _____

   _____

4. To achieve an active force of 5.2 g, did you have to increase the stimulus voltage above 8.5 volts? If not, how did you

   achieve an active force of 5.2 g? How well did the results compare with your prediction? _____

   _____

   _____

5. Compare and contrast frequency-dependent wave summation with motor unit recruitment (previously observed by increas-

   ing the stimulus voltage). How are they similar? How was each achieved in the experiment? Explain how each is achieved

   in vivo. _____

   _____

ACTIVITY 4 Tetanus in Isolated Skeletal Muscle

1. Describe how increasing the stimulus frequency affected the force developed by the isolated whole skeletal muscle in this activity. How well did the results compare with your prediction? _____

   _____

2. Indicate what type of force was developed by the isolated skeletal muscle in this activity at the following stimulus frequencies: at 50 stimuli/sec, at 140 stimuli/sec, and above 146 stimuli/sec. _____

   _____

3. Beyond what stimulus frequency is there no further increase in the peak force? What is the muscle tension called at this frequency? _____

ACTIVITY 5 Fatigue in Isolated Skeletal Muscle

1. When a skeletal muscle fatigues, what happens to the contractile force over time? _____

   _____

2. What are some proposed causes of skeletal muscle fatigue? _____

   _____

3. Turning the stimulator off allows a small measure of muscle recovery. Thus, the muscle will produce more force for a longer time period if the stimulator is briefly turned off than if the stimuli were allowed to continue without interruption. Explain why this might occur. How well did the results compare with your prediction? _____

   _____

   _____

   _____

4. List a few ways that humans could delay the onset of fatigue when they are vigorously using their skeletal muscles. _____

   _____

ACTIVITY 6 The Skeletal Muscle Length-Tension Relationship

1. What happens to the amount of total force the muscle generates during the stimulated twitch? How well did the results compare with your prediction? _____

   _____

   _____

2. What is the key variable in an isometric contraction of a skeletal muscle? _____

3. Based on the unique arrangement of myosin and actin in skeletal muscle sarcomeres, explain why active force varies with changes in the muscle's resting length. _____

_____

_____

4. What skeletal muscle lengths generated passive force? (Provide a range.) _____

5. If you were curling a 7-kg dumbbell, when would your bicep muscles be contracting isometrically? _____

ACTIVITY 7   Isotonic Contractions and the Load-Velocity Relationship

1. If you were using your bicep muscles to curl a 7-kg dumbbell, when would your muscles be contracting isotonically?

_____

2. Explain why the latent period became longer as the load became heavier in the experiment. How well did the results compare with your prediction?_____

_____

_____

3. Explain why the shortening velocity became slower as the load became heavier in this experiment. How well did the results compare with your prediction? _____

_____

_____

4. Describe how the shortening distance changed as the load became heavier in this experiment. How well did the results compare with your prediction? _____

_____

_____

5. Explain why it would take you longer to perform 10 repetitions lifting a 10-kg weight than it would to perform the same number of repetitions with a 5-kg weight. _____

_____

6. Describe what would happen in the following experiment: A 2.5-g weight is attached to the end of the isolated whole skeletal muscle used in these experiments. Simultaneously, the muscle is maximally stimulated by 8.5 volts and the platform supporting the weight is removed. Will the muscle generate force? Will the muscle change length? What is the name for this type of contraction? _____

_____

PhysioEx 9.1

Neurophysiology of Nerve Impulses

PRE-LAB QUIZ

1. Circle the correct underlined term: Neurons / Neuroglial cells are capable of generating an electrical signal.

2. The _____ is the potential difference between the inside of a cell and the outside of a cell across the membrane.
   a. conductance
   b. resting membrane potential
   c. permeable potential
   d. active membrane potential

3. If the response to a stimulus is a change from a negative potential to a less negative potential, the change is called:
   a. repolarization
   b. stimulus recovery
   c. depolarization
   d. stimulus response

4. What type of channels open when the membrane depolarizes?
   a. voltage-gated potassium channels
   b. ligand-gated sodium channels
   c. voltage-gated sodium channels
   d. ligand-gated potassium channels

5. The region where the neurotransmitter is released from one neuron and binds to a receptor on a target cell is the:
   a. chemical synapse
   b. receptor potential
   c. postsynaptic synapse
   d. postsynaptic potential

Exercise Overview

The nervous system contains two general types of cells: **neurons** and neuroglia (or glial cells). This exercise focuses on neurons. Neurons respond to their local environment by generating an electrical signal. For example, sensory neurons in the nose generate a signal (called a **receptor potential**) when odor molecules interact with receptor proteins on the membrane of these olfactory sensory neurons. Thus, sensory neurons can respond directly to sensory stimuli. The receptor potential can trigger another electrical signal (called an **action potential**), which travels along the membrane of the sensory neuron's axon to the brain—you could say that the action potential is conducted to the brain.

The action potential causes the release of **chemical neurotransmitters** onto neurons in olfactory regions of the brain. These chemical neurotransmitters bind to receptor proteins on the membrane of these brain **interneurons.** In general, interneurons respond to chemical neurotransmitters released by other neurons. In the nose the odor molecules are sensed by sensory neurons. In the brain the odor is perceived by the activity of interneurons responding to neurotransmitters. Any resulting action or behavior is caused by the subsequent activity of **motor neurons,** which can stimulate muscles to contract (see Exercise 2).

In general each neuron has three functional regions for signal transmission: a receiving region, a conducting region, and an output region, or secretory region. Sensory neurons often have a receptive ending specialized to detect a specific sensory stimulus, such as odor, light, sound, or touch. The **cell body** and **dendrites** of interneurons receive stimulation by neurotransmitters at structures called **chemical synapses** and produce **synaptic potentials.** The conducting

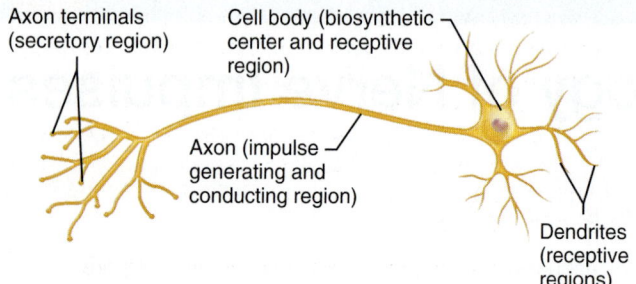

Axon terminals (secretory region)

Cell body (biosynthetic center and receptive region)

Axon (impulse generating and conducting region)

Dendrites (receptive regions)

FIGURE 3.1 A neuron with functional areas identified.

region is usually an **axon,** which ends in an output region (the axon terminal) where neurotransmitter is released (view Figure 3.1).

Although the neuron is a single cell surrounded by a continuous plasma membrane, each region contains distinct membrane proteins that provide the basis for the functional differences. Thus, the receiving end has receptor proteins and proteins that generate the receptor potential, the conducting region has proteins that generate and conduct action potentials, and the output region has proteins to package and release neurotransmitters. Membrane proteins are found throughout the neuronal membrane—many of these proteins transport ions (see Exercise 1).

The signals generated and conducted by neurons are electrical. In ordinary household devices, electric current is carried by electrons. In biological systems, currents are carried by positively or negatively charged **ions.** Like charges repel each other and opposite charges attract. In general, ions cannot easily pass through the lipid bilayer of the plasma membrane and must pass through **ion channels** formed by integral membrane proteins. Some channels are usually open (leak channels) and others are gated, meaning that the channel can be in an open or closed configuration. Channels can also be selective for which ions are allowed to pass. For example, sodium channels are mostly permeable to sodium ions when open, and potassium channels are mostly permeable to potassium ions when open. The term **conductance** is often used to describe **permeability.** In general, ions will flow through an open channel from a region of higher concentration to a region of lower concentration (see Exercise 1). In this exercise you will explore some of these characteristics applied to neurons.

Although it is possible to measure the ionic currents through the membrane (even the currents passing through single ion channels), it is more common to measure the potential difference, or voltage, across the membrane. This membrane voltage is usually called the **membrane potential,** and the units are **millivolts (mV).** One can think of the membrane as a battery, a device that separates and stores charge. A typical household battery has a positive and a negative pole so that when it is connected, for example through a lightbulb in a flashlight, current flows through the bulb. Similarly, the plasma membrane can store charge and has a relatively positive side and a relatively negative side. Thus, the membrane is said to be **polarized.** When these two sides (intracellular and extracellular) are connected through open ion channels, current in the form of ions can flow in or out across the membrane and thus change the membrane voltage.

The Resting Membrane Potential

OBJECTIVES

1. To define the term *resting membrane potential*.
2. To measure the resting membrane potential in different parts of a neuron.
3. To determine how the resting membrane potential depends on the concentrations of potassium and sodium.
4. To understand the ion conductances/ion channels involved in the resting membrane potential.

Introduction

The receptor potential, synaptic potentials, and action potentials are important signals in the nervous system. These potentials refer to changes in the membrane potential from its resting level. In this activity you will explore the nature of the resting potential. The **resting membrane potential** is really a potential difference between the inside of the cell (intracellular) and the outside of the cell (extracellular) across the membrane. It is a steady-state condition that depends on the resting permeability of the membrane to ions and on the intracellular and extracellular concentrations of those ions to which the membrane is permeable.

For many neurons, Na^+ and K^+ are the most important ions, and the concentrations of these ions are established by transport proteins, such as the Na^+-K^+ pump, so that the intracellular Na^+ concentration is low and the intracellular K^+ concentration is high. Inside a typical cell, the concentration of K^+ is ~150 mM and the concentration of Na^+ is ~5 mM. Outside a typical cell, the concentration of K^+ is ~5 mM and the concentration of Na^+ is ~150 mM. If the membrane is permeable to a particular ion, that ion will diffuse down its concentration gradient from a region of higher concentration to a region of lower concentration. In the generation of the resting membrane potential, K^+ ions diffuse out across the membrane, leaving behind a net negative charge—large anions that cannot cross the membrane.

The membrane potential can be measured with an amplifier. In the experiment the extracellular solution is connected to a ground (literally, the earth) which is defined as 0 mV. To record the voltage across the membrane, a microelectrode is inserted through the membrane without significantly damaging it. Typically, the microelectrode is made by pulling a thin glass pipette to a fine hollow point and filling the pulled pipette with a salt solution. The salt solution conducts electricity like a wire, and the glass insulates it. Only the tip of the microelectrode is inserted through the membrane, and the filled tip of the microelectrode makes electrical contact with the intracellular solution. A wire connects the microelectrode to the input of the amplifier so that the amplifier records the membrane potential, the voltage across the membrane between the intracellular and grounded extracellular solutions.

The membrane potential and the various signals can be observed on an oscilloscope. An electron beam is pulled up or down according to the voltage as it sweeps across a phosphorescent screen. Voltages below 0 mV are negative and voltages above 0 mV are positive. For this first activity, the time of the sweep is set for 1 second per division, and the sensitivity is set to 10 mV per division; a division is the distance between gridlines on the oscilloscope.

> **EQUIPMENT USED**   The following equipment will be depicted on-screen: neuron (in vitro)—a large, dissociated (or cultured) neuron; three extracellular solutions—control, high potassium, and low sodium; microelectrode—a probe with a very small tip that can impale a single neuron (In an actual wet lab, a microelectrode manipulator is used to position the microelectrode. For simplicity, the microelectrode manipulator will not be depicted in this activity.); microelectrode manipulator controller—controls movement of the manipulator; microelectrode amplifier—used to measure the voltage between the microelectrode and a reference; oscilloscope—used to observe voltage changes.

Experiment Instructions

Go to the home page in the PhysioEx software, and click **Exercise 3: Neurophysiology of Nerve Impulses.** Click **Activity 1: The Resting Membrane Potential,** and take the online **Pre-lab Quiz** for Activity 1.

After you take the online Pre-lab Quiz, click the **Experiment** tab and begin the experiment. The experiment instructions are reprinted here for your reference. The opening screen for the experiment is shown below.

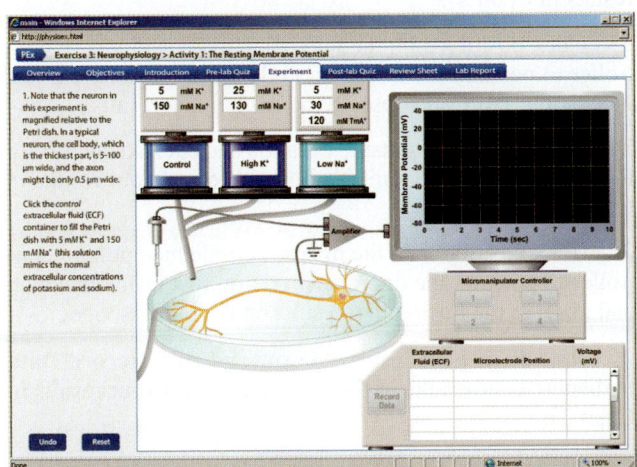

1. Note that the neuron in this experiment is magnified relative to the petri dish. In a typical neuron, the cell body, which is the thickest part, is 5–100 µm wide, and the axon might be only 0.5 µm wide.

Click the *control* **extracellular fluid (ECF)** container to fill the petri dish with 5 mM K$^+$ and 150 mM Na$^+$ (this solution mimics the normal extracellular concentrations of potassium and sodium).

2. Note that a reference electrode is already positioned in the petri dish. This reference electrode is connected to ground through the amplifier.

Click position **1** on the microelectrode manipulator controller to position the microelectrode tip in the solution, just outside the cell body, and observe the tracing that results on the oscilloscope.

3. Note the oscilloscope tracing of the voltage outside the cell body and click **Record Data** to display your results in the grid (and record your results in Chart 1).

4. Click position **2** on the microelectrode manipulator controller to position the microelectrode tip just inside the cell body and observe the tracing that results.

5. Note the oscilloscope tracing of the voltage inside the cell body and click **Record Data** to display your results in the grid (and record your results in Chart 1). This is the resting membrane potential; that is, the potential difference between intracellular and extracellular membrane voltages. By convention, the extracellular resting membrane voltage is taken to be 0 mV.

6. Click position **3** on the microelectrode manipulator controller to position the microelectrode tip in the solution, just outside the axon, and observe the tracing that results.

7. Note the oscilloscope tracing of the voltage outside the axon and click **Record Data** to display your results in the grid (and record your results in Chart 1).

CHART 1 Resting Membrane Potential		
Extracellular fluid (ECF)	**Microelectrode position**	**Voltage (mV)**

8. Click position **4** on the microelectrode manipulator controller to position the microelectrode tip just inside the axon and observe the tracing that results.

9. Note the oscilloscope tracing of the voltage inside the axon and click **Record Data** to display your results in the grid (and record your results in Chart 1).

> **?** **PREDICT Question 1**
> Predict what will happen to the resting membrane potential if the extracellular K^+ concentration is increased.
>
> _____
>
> _____

10. You will now change the concentrations of the ions in the extracellular fluid to determine which ions contribute most to the separation of charge across the membrane. The extracellular potassium concentration is normally low, so you will first increase the extracellular potassium concentration.

In the high K^+ ECF solution the K^+ concentration has been increased fivefold, from 5 to 25 mM. To keep the number of positive charges in the extracellular solution constant, the Na^+ concentration has been reduced by 20 mM, from 150 to 130 mM. As you will see, this relatively small decrease in Na^+ will not by itself change the membrane potential. Note that in this activity, the generation of the action potential (which is covered in Activities 3–9) is blocked with a toxin. Click the **high K^+ ECF** container to change the solution in the petri dish to 25 mM K^+ and 130 mM Na^+.

11. Note the voltage inside the axon and click **Record Data** to display your results in the grid (and record your results in Chart 1).

12. Click position **3** on the microelectrode manipulator controller to position the microelectrode tip in the solution, just outside the axon, and observe the tracing that results.

13. Note the voltage outside the axon and click **Record Data** to display your results in the grid (and record your results in Chart 1).

14. Click position **1** on the microelectrode manipulator controller to position the microelectrode tip in the solution, just outside the cell body, and observe the tracing that results.

15. Note the voltage outside the cell body and click **Record Data** to display your results in the grid (and record your results in Chart 1).

16. Click position **2** on the microelectrode manipulator controller to position the microelectrode tip just inside the cell body and observe the tracing that results on the oscilloscope.

17. Note the voltage inside the cell body and click **Record Data** to display your results in the grid (and record your results in Chart 1).

18. Click the *control* ECF container to change back to the normal K^+ concentration and note the change in voltage inside the cell body.

19. You will now decrease the extracellular Na^+ concentration (the extracellular Na^+ concentration is normally high).

The extracellular sodium concentration in the low Na^+ solution has been decreased fivefold, from 150 mM to 30 mM. To keep the number of positive charges constant in the extracellular solution, the Na^+ has been replaced by the same amount of a large monovalent cation. Note that the extracellular Na^+ concentration, even in the low Na^+ ECF, is higher than the intracellular Na^+ concentration. Click the **low Na^+ ECF** container to change the solution in the petri dish to 5 mM K^+ and 30 mM Na^+.

20. Note the voltage inside the cell body and click **Record Data** to display your results in the grid (and record your results in Chart 1).

21. Click position **1** on the microelectrode manipulator controller to position the microelectrode tip in the solution, just outside the cell body, and observe the tracing that results.

22. Note the voltage outside the cell body and click **Record Data** to display your results in the grid (and record your results in Chart 1).

23. Click position **3** on the microelectrode manipulator controller to position the microelectrode tip in the solution, just outside the axon, and observe the tracing that results.

24. Note the voltage outside the axon and click **Record Data** to display your results in the grid (and record your results in Chart 1).

25. Click position **4** on the microelectrode manipulator controller to position the microelectrode tip just inside the axon and observe the tracing that results on the oscilloscope.

26. Note the voltage inside the axon and click **Record Data** to display your results in the grid (and record your results in Chart 1).

After you complete the experiment, take the online **Post-lab Quiz** for Activity 1.

Activity Questions

1. Explain why the resting membrane potential had the same value in the cell body and in the axon.

_____

_____

2. Describe what would happen to a resting membrane potential if the sodium-potassium transport pump was blocked.

_____

_____

3. Describe what would happen to a resting membrane potential if the concentration of large intracellular anions that are unable to cross the membrane is experimentally increased.

_____

_____

Receptor Potential

OBJECTIVES

1. To define the terms *sensory receptor, receptor potential, sensory transduction, stimulus modality,* and *depolarization.*
2. To determine the *adequate stimulus* for different sensory receptors.
3. To demonstrate that the receptor potential amplitude increases with stimulus intensity.

Introduction

The receiving end of a sensory neuron, the **sensory receptor,** has receptor proteins (as well as other membrane proteins) that can generate a signal called the **receptor potential** when the sensory neuron is stimulated by an appropriate, adequate stimulus. In this activity you will use the same recording instruments and microelectrode that you used in Activity 1. However, in this activity, you will record from the sensory receptor of three different sensory neurons and examine how these neurons respond to sensory stimuli of different modalities.

The sensory region will be shown disconnected from the rest of the neuron so that you can record the receptor potential in isolation. Similar results can sometimes be obtained by treating a whole neuron with chemicals that block the responses generated by the axon. The molecules localized to the sensory receptor ending are able to generate a receptor potential when an adequate stimulus is applied. The energy in the stimulus (for example, chemical, physical, or heat) is changed into an electrical response that involves the opening or closing of membrane ion channels. The general process that produces this change is called **sensory transduction,** which occurs at the receptor ending of the sensory neuron. Sensory transduction can be thought of as a type of signal transduction where the signal is the sensory stimulus.

You will observe that, with an appropriate stimulus, the amplitude of the receptor potential increases with stimulus intensity. Such a response is an example of a potential that is graded with stimulus intensity. These responses are sometimes referred to as *graded potentials,* or *local potentials.* Thus, the receptor potential is a graded, or local, potential. If the response (receptor potential) is a change in membrane potential from the negative resting potential to a less negative level, the membrane becomes less polarized and the change is called **depolarization.**

EQUIPMENT USED The following equipment will be depicted on-screen: three sensory receptors—Pacinian (lamellar) corpuscle, olfactory receptor, and free nerve ending; microelectrode—a probe with a very small tip that can impale a single neuron (In an actual wet lab, a microelectrode manipulator is used to position the microelectrodes. For simplicity, the microelectrode manipulator will not be depicted in this activity.); microelectrode amplifier—used to measure the voltage between the microelectrode and a reference; stimulator—used to select the stimulus modality (pressure, chemical, heat, or light) and intensity (low, moderate, or high); oscilloscope—used to observe voltage changes.

Experiment Instructions

Go to the home page in the PhysioEx software, and click **Exercise 3: Neurophysiology of Nerve Impulses.** Click **Activity 2: Receptor Potential,** and take the online **Pre-lab Quiz** for Activity 2.

After you take the online Pre-lab Quiz, click the **Experiment** tab and begin the experiment. The experiment instructions are reprinted here for your reference. The opening screen for the experiment is shown below.

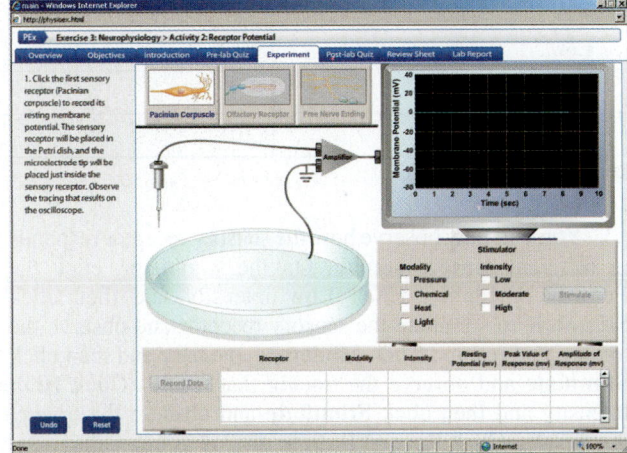

1. Note that the timescale on the oscilloscope has been changed from 1 second per division to 10 milliseconds per division, so that you can observe the responses recorded in the sensory receptors more clearly. Click the first sensory receptor (Pacinian corpuscle) to record its resting membrane potential. The sensory receptor will be placed in the petri dish, and the microelectrode tip will be placed just inside the sensory receptor. Observe the tracing that results on the oscilloscope.

2. Note the voltage inside the sensory receptor and click **Record Data** to display your results in the grid (and record your results in Chart 2).

> **? PREDICT Question 1**
> The adequate stimulus for a Pacinian corpuscle is pressure or vibration on the skin. For a Pacinian corpuscle, which modality will induce a receptor potential of the largest amplitude?
>
> _____
>
> _____

CHART 2	Receptor Potential		
	Receptor potential (mV)		
Stimulus modality	Pacinian (lamellar) corpuscle	Olfactory receptor	Free nerve ending
None			
Pressure			
Low			
Moderate			
High			
Chemical			
Low			
Moderate			
High			
Heat			
Low			
Moderate			
High			
Light			
Low			
Moderate			
High			

3. You will now observe how the sensory receptor responds to different sensory stimuli. On the stimulator, click the **Pressure** modality. Click **Low** intensity and then click **Stimulate** to stimulate the sensory receptor and observe the tracing that results. Click **Moderate** intensity and then click **Stimulate** and observe the tracing that results. Click **High** intensity and then click **Stimulate** and observe the tracing that results. Click **Record Data** to display your results in the grid (and record your results in Chart 2).

4. On the stimulator, click the **Chemical** (odor) modality. Click **Low** intensity and then click **Stimulate** to stimulate the sensory receptor and observe the tracing that results. Click **Moderate** intensity and then click **Stimulate** and observe the tracing that results. Click **High** intensity and then click **Stimulate** and observe the tracing that results. Click **Record Data** to display your results in the grid (and record your results in Chart 2).

5. On the stimulator, click the **Heat** modality. Click **Low** intensity and then click **Stimulate** to stimulate the sensory receptor and observe the tracing that results. Click **Moderate** intensity and then click **Stimulate** and observe the tracing that results. Click **High** intensity and then click **Stimulate** and observe the tracing that results. Click **Record Data** to display your results in the grid (and record your results in Chart 2).

6. On the stimulator, click the **Light** modality. Click **Low** intensity and then click **Stimulate** to stimulate the sensory receptor and observe the tracing that results. Click **Moderate** intensity and then click **Stimulate** and observe the tracing that results. Click **High** intensity and then click **Stimulate** and

observe the tracing that results. Click **Record Data** to display your results in the grid (and record your results in Chart 2).

> **? PREDICT Question 2**
> The adequate stimuli for olfactory receptors are chemicals, typically odorant molecules. For an olfactory receptor, which modality will induce a receptor potential of the largest amplitude?
>
> _____

7–12. Repeat steps 1–6 with the next sensory receptor: olfactory receptor.

13–18. Repeat steps 1–6 with the next sensory receptor: free nerve ending.

After you complete the experiment, take the online **Post-lab Quiz** for Activity 2.

Activity Questions

1. Are graded receptor potentials always depolarizing? Do graded receptor potentials always make it easier to induce action potentials?

_____

_____

2. Based on the definition of membrane depolarization in this activity, define membrane *hyperpolarization.*

_____

_____

3. What do you think is the adequate stimulus for sensory receptors in the ear? Can you think of a stimulus that would inappropriately activate the sensory receptors in the ear if the stimulus had enough intensity?

_____

_____ ▬

The Action Potential: Threshold

OBJECTIVES

1. To define the terms *action potential, nerve, axon hillock, trigger zone,* and *threshold.*
2. To predict how an increase in extracellular K^+ could trigger an action potential.

Introduction

In this activity you will explore changes in potential that occur in the axon. Axons are long, thin structures that conduct a signal called the **action potential.** A **nerve** is a bundle of axons.

Axons are typically studied in a nerve chamber. In this activity the axon will be draped over wires that make electrical contact with the axon and can therefore record the electrical activity in the axon. Because the axon is so thin, it is very difficult to insert an electrode across the membrane into the axon. However, some of the charge (ions) that crosses the membrane to generate the action potential can be recorded from outside the membrane (extracellular recording), as you will do in this activity. The molecular mechanisms underlying the action potential were explored more than 50 years ago with intracellular recording using the giant axons of the squid, which are about 1 millimeter in diameter.

In this activity the axon will be artificially disconnected from the cell body and dendrites. In a typical multipolar neuron (view Figure 3.1 in the Exercise Overview), the axon extends from the cell body at a region called the **axon hillock.** In a myelinated axon, this first region is called the initial segment. An action potential is usually initiated at the junction of the axon hillock and the initial segment; therefore, this region is also referred to as the **trigger zone.**

You will use an electrical stimulator to explore the properties of the action potential. Current passes from the stimulator to one of the stimulation wires, then across the axon, and then back to the stimulator through a second wire. This current will depolarize the axon. Normally, in a sensory neuron, the depolarizing receptor potential spreads passively to the axon hillock and produces the depolarization needed to evoke the action potential. Once an action potential is generated, it is regenerated down the membrane of the axon. In other words, the action potential is **propagated,** or *conducted,* down the axon (see Activity 6).

You will now generate an action potential at one end of the axon by stimulating it electrically and record the action potential that is propagated down the axon. The extracellular action potential that you record is similar to one that would be recorded across the membrane with an intracellular microelectrode, but much smaller. For simplicity, only one axon is depicted in this activity.

> **EQUIPMENT USED** The following equipment will be depicted on-screen: nerve chamber; axon; oscilloscope—used to observe timing of stimuli and voltage changes in the axon; stimulator—used to set the stimulus voltage and to deliver pulses that depolarize the axon; stimulation wires (S); recording electrodes (wires R1 and R2)—used to record voltage changes in the axon. (The first set of recording electrodes, R1, is 2 centimeters from the stimulation wires, and the second set of recording electrodes, R2, is 2 centimeters from R1.)

Experiment Instructions

Go to the home page in the PhysioEx software, and click **Exercise 3: Neurophysiology of Nerve Impulses.** Click **Activity 3: The Action Potential: Threshold,** and take the online **Pre-lab Quiz** for Activity 3.

After you take the online Pre-lab Quiz, click the **Experiment** tab and begin the experiment. The experiment instructions are reprinted here for your reference. The opening screen for the experiment is shown below.

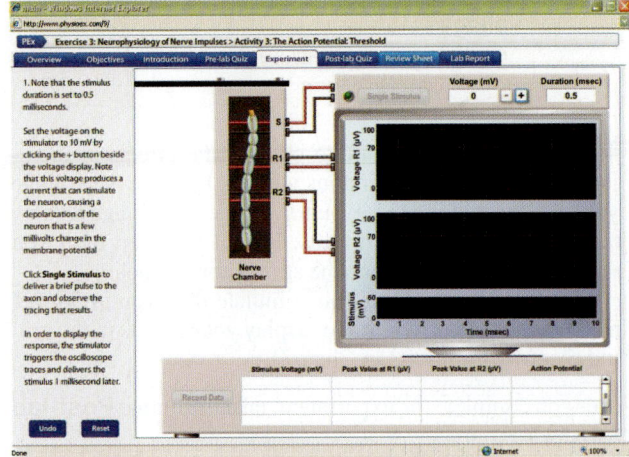

1. Note that the stimulus duration is set to 0.5 milliseconds. Set the voltage on the stimulator to 10 mV by clicking the + button beside the voltage display. Note that this voltage produces a current that can stimulate the neuron, causing a depolarization of the neuron that is a change of a few millivolts in the membrane potential.

Click **Single Stimulus** to deliver a brief pulse to the axon and observe the tracing that results. In order to display the response, the stimulator triggers the oscilloscope traces and delivers the stimulus 1 millisecond later.

2. Note that the recording electrodes R1 and R2 record the extracellular voltage, rather than the actual membrane potential. The 10 mV depolarization at the site of stimulation only occurs locally at that site and is not recorded farther down the axon. At this initial stimulus voltage, there was no action potential. Click **Record Data** to display your results in the grid (and record your results in Chart 3).

CHART 3	Threshold		
Stimulus voltage (mV)	Peak value at R1 (μV)	Peak value at R2 (μV)	Action potential

3. You will increase the stimulus voltage until you observe an action potential at recording electrode 1 (R1). Increase the voltage by 10 mV by clicking the **+** button beside the voltage display and then click **Single Stimulus.** The voltage at which you first observe an action potential is the **threshold voltage.** Note that the action potential recorded extracellularly is quite small. Intracellularly, the membrane potential would change from −70 mV to about +30 mV. Click **Record Data** to display your results in the grid (and record your results in Chart 3).

> **? PREDICT Question 1**
> How will the action potential at R1 (or R2) change as you continue to increase the stimulus voltage?

4. You will now continue to observe the effects of incremental increases of the stimulus voltage. Increase the voltage by 10 mV by clicking the **+** button beside the voltage display and then click **Single Stimulus.** Repeat this step until you reach the maximum voltage the stimulator can deliver.

Repeat this step until you stimulate the axon at 50 mV and then click **Record Data** to display your results in the grid (and record your results in Chart 3).

After you complete the experiment, take the online **Post-lab Quiz** for Activity 3.

Activity Questions

1. Explain why the threshold voltage is not always the same value (between axons and within an axon).

2. Describe how the action potential is regenerated by local ion flux at each location on the axon.

3. Why doesn't the peak value of the action potential increase with stronger stimuli?

The Action Potential: Importance of Voltage-Gated Na$^+$ Channels

OBJECTIVES

1. To define the term *voltage-gated channel.*
2. To describe the effect of tetrodotoxin on the voltage-gated Na$^+$ channel.
3. To describe the effect of lidocaine on the voltage-gated Na$^+$ channel.
4. To examine the effects of tetrodotoxin and lidocaine on the action potential.
5. To predict the effect of lidocaine on pain perception and to predict the site of action in the sensory neurons (nociceptors) that sense pain.

Introduction

The action potential (as seen in Activity 3) is generated when voltage-gated sodium channels open in sufficient numbers. **Voltage-gated sodium channels** open when the membrane depolarizes. Each sodium channel that opens allows Na$^+$ ions to diffuse into the cell down their electrochemical gradient. When enough sodium channels open so that the amount of sodium ions that enters via these voltage-gated channels overcomes the leak of potassium ions (recall that the potassium leak via passive channels establishes and maintains the negative resting membrane potential), threshold for the action potential is reached, and an action potential is generated.

In this activity you will observe what happens when these voltage-gated sodium channels are blocked with chemicals. One such chemical is tetrodotoxin (TTX), a toxin found in puffer fish, which is extremely poisonous. Another such chemical is lidocaine, which is typically used to block pain in dentistry and minor surgery.

> **EQUIPMENT USED** The following equipment will be depicted on-screen: nerve chamber; axon; oscilloscope—used to observe timing of stimuli and voltage changes in the axon; stimulator—used to set the stimulus voltage and the interval between stimuli and to deliver pulses that depolarize the axon; stimulation wires (S); recording electrodes (wires R1 and R2)—used to record voltage changes in the axon (The first set of recording electrodes, R1, is 2 centimeters from the stimulation wires, and the second set of recording electrodes, R2, is 2 centimeters from R1.); tetrodotoxin (TTX); lidocaine.

Experiment Instructions

Go to the home page in the PhysioEx software and click **Exercise 3: Neurophysiology of Nerve Impulses.** Click **Activity 4: The Action Potential: Importance of**

Voltage-Gated Na⁺ Channels, and take the online **Pre-lab Quiz** for Activity 4.

After you take the online Pre-lab Quiz, click the **Experiment** tab and begin the experiment. The experiment instructions are reprinted here for your reference. The opening screen for the experiment is shown below.

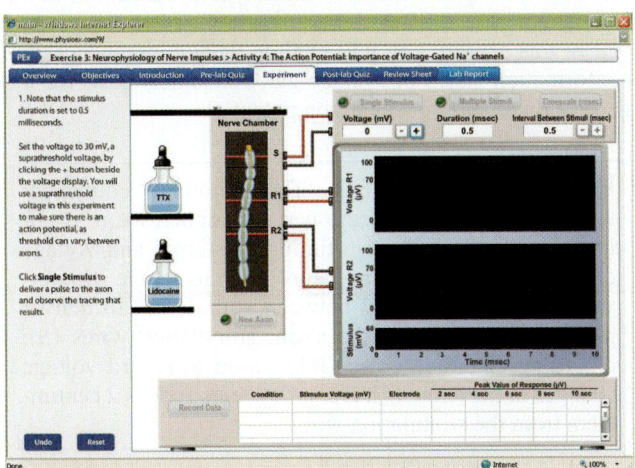

1. Note that the stimulus duration is set to 0.5 milliseconds. Set the voltage to 30 mV, a suprathreshold voltage, by clicking the **+** button beside the voltage display. You will use a suprathreshold voltage in this experiment to make sure there is an action potential, as threshold can vary between axons. Click **Single Stimulus** to deliver a pulse to the axon and observe the tracing that results.

2. Enter the peak value of the response at R1 and R2 in the field below and then click **Submit** to record your answer in the lab report. _____ μV

3. Click **Timescale** on the stimulator to change the timescale on the oscilloscope from milliseconds to seconds.

4. You will now deliver successive stimuli separated by 2.0-second intervals to observe what the control action potentials look like at this timescale. Set the interval between stimuli to 2.0 seconds by clicking the **+** button beside the "Interval between Stimuli" display. Click **Multiple Stimuli** to deliver pulses to the axon every 2 seconds. The stimuli will be stopped after 10 seconds.

5. Note the peak values of the responses at R1 and R2 and click **Record Data** to display your results in the grid (and record your results in Chart 4).

> **? PREDICT Question 1**
> If you apply TTX between recording electrodes R1 and R2, what effect will the TTX have on the action potentials at R1 and R2?
>
> _____
>
> _____

6. Drag the dropper cap of the TTX bottle to the axon between recording electrodes R1 and R2 to apply a drop of TTX to the axon.

7. Click **Multiple Stimuli** to deliver pulses to the axon every 2 seconds. The stimuli will be stopped after 10 seconds.

8. Note the peak values of the responses at R1 and R2 and click **Record Data** to display your results in the grid (and record your results in Chart 4).

9. Click **New Axon** to select a new axon. TTX is irreversible and there is no known antidote for TTX poisoning.

> **? PREDICT Question 2**
> If you apply lidocaine between recording electrodes R1 and R2, what effect will the lidocaine have on the action potentials at R1 and R2?
>
> _____
>
> _____

10. Drag the dropper cap of the lidocaine bottle to the axon between recording electrodes R1 and R2 to apply a drop of lidocaine to the axon.

11. Set the interval between stimuli to 2.0 seconds by clicking the **+** button beside the "Interval between Stimuli" display. Click **Multiple Stimuli** to deliver pulses to the axon every 2 seconds. The stimuli will be stopped after 10 seconds.

CHART 4	Effects of Tetrodotoxin and Lidocaine						
			Peak value of response (µV)				
Condition	Stimulus voltage (mV)	Electrodes	2 sec	4 sec	6 sec	8 sec	10 sec

12. Note the peak values of the responses at R1 and R2. For simplicity, this experiment was performed on a single axon, where the action potential is an "all-or-none" event. If you had treated a bundle of axons (a nerve), each with a slightly different threshold and sensitivity to the drugs, you would likely see the peak values of the action potentials decrease more gradually as more and more axons were blocked. Click **Record Data** to display your results in the grid (and record your results in Chart 4).

After you complete the experiment, take the online **Post-lab Quiz** for Activity 4.

Activity Questions

1.   If depolarizing membrane potentials open voltage-gated sodium channels, what closes them?

_____

2.   Why must a sushi chef go through years of training to prepare puffer fish for human consumption?

_____

3.   For action potential generation and propagation, are there any other cation channels that could substitute for the voltage-gated sodium channels if the sodium channels were blocked?

_____

A C T I V I T Y   5

The Action Potential: Measuring Its Absolute and Relative Refractory Periods

OBJECTIVES

1. To define *inactivation* as it applies to a voltage-gated sodium channel.

2. To define the *absolute refractory period* and *relative refractory period* of an action potential.

3. To define the relationship between stimulus frequency and the generation of action potentials.

Introduction

Voltage-gated sodium channels in the plasma membrane of an excitable cell open when the membrane depolarizes. About 1–2 milliseconds later, these same channels inactivate, meaning they no longer allow sodium to go through the channel. These inactivated channels cannot be reopened by depolarization for an additional period of time (usually many milliseconds). Thus, during this time, fewer sodium channels can be opened. There are also voltage-gated potassium channels that open during the action potential. These potassium channels open more slowly. They contribute to the repolarization of the action potential from its peak, as more potassium flows out through this second type of potassium channel (recall

there are also passive potassium channels that let potassium leak out, and these leak channels are always open). The flux through extra voltage-gated potassium channels opposes the depolarization of the membrane to threshold, and it also causes the membrane potential to become transiently more negative than the resting potential at the end of an action potential. This phase is called after-hyperpolarization, or the undershoot.

In this activity you will explore what consequences the conformation states of voltage-gated channels have for the generation of subsequent action potentials.

> **EQUIPMENT USED** The following equipment will be depicted on-screen: nerve chamber; axon; oscilloscope—used to observe timing of stimuli and voltage changes in the axon; stimulator—used to set the stimulus voltage and the interval between stimuli and to deliver pulses that depolarize the axon; stimulation wires (S); recording electrode (wires R1)—used to record voltage changes in the axon. (The recording electrode is 2 centimeters from the stimulation wires.)

Experiment Instructions

Go to the home page in the PhysioEx software and click **Exercise 3: Neurophysiology of Nerve Impulses.** Click **Activity 5: The Action Potential: Measuring Its Absolute and Relative Refractory Periods,** and take the online **Pre-lab Quiz** for Activity 5.

After you take the online Pre-lab Quiz, click the **Experiment** tab and begin the experiment. The experiment instructions are reprinted here for your reference. The opening screen for the experiment is shown below.

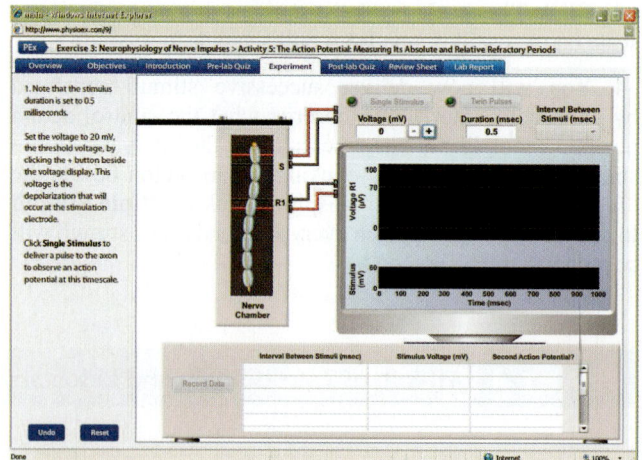

1. Note that the stimulus duration is set to 0.5 milliseconds. Set the voltage to 20 mV, the threshold voltage, by clicking the + button beside the voltage display. This voltage is the depolarization that will occur at the stimulation electrode. Click **Single Stimulus** to deliver a pulse to observe an action potential at this timescale.

2. You will now deliver two successive stimuli separated by 250 milliseconds. Set the interval between stimuli to 250 milliseconds by selecting 250 in the "Interval between Stimuli"

pull-down menu. Click **Twin Pulses** to deliver two pulses to the axon and observe the tracing that results. Click **Record Data** to display your results in the grid (and record your results in Chart 5).

CHART 5	Absolute and Relative Refractory Periods	
Interval between stimuli (msec)	Stimulus voltage (mV)	Second action potential?

3. Decrease the interval between stimuli to 125 milliseconds by selecting 125 in the "Interval between Stimuli" pull-down menu. Click **Twin Pulses** to deliver two pulses to the axon and observe the tracing that results. Click **Record Data** to display your results in the grid (and record your results in Chart 5).

4. Decrease the interval between stimuli to 60 milliseconds by selecting 60 in the "Interval between Stimuli" pull-down menu. Click **Twin Pulses** to deliver two pulses to the axon and observe the tracing that results.
Note that, at this stimulus interval, the second stimulus did not generate an action potential. Click **Record Data** to display your results in the grid (and record your results in Chart 5).

5. A second action potential can be generated at this stimulus interval, but the stimulus intensity must be increased. This interval is part of the relative refractory period, the time after an action potential when a second action potential can be generated if the stimulus intensity is increased.
Increase the stimulus intensity by 5 mV by clicking the + button beside the voltage display and then click **Twin Pulses** to deliver two pulses to the axon. Repeat this step until you generate a second action potential. After you generate a second action potential, click **Record Data** to display your results in the grid (and record your results in Chart 5).

? PREDICT Question 1
If you further decrease the interval between the stimuli, will the threshold for the second action potential change?

_____

_____

6. You will now decrease the interval until the second action potential fails again. (So that you can clearly observe two action potentials at the shorter interval between stimuli, the timescale on the oscilloscope has been set to 10 msec per division.) Decrease the interval between stimuli by 50% and then click **Twin Pulses** to deliver two pulses to the axon. When the second action potential fails, click **Record Data** to display your results in the grid (and record your results in Chart 5).

7. You will now increase the stimulus intensity until a second action potential is generated again. Increase the stimulus intensity by 5 mV by clicking the + button beside the voltage display and then click **Twin Pulses** to deliver two pulses to the axon. Repeat this step until you generate a second action potential. After you generate a second action potential, click **Record Data** to display your results in the grid (and record your results in Chart 5).

8. You will now determine the interval between stimuli at which a second action potential cannot be generated, no matter how intense the stimulus. Increase the stimulus intensity to 60 mV (the highest voltage on the stimulator). Decrease the interval between stimuli by 50% and then click **Twin Pulses** to deliver two pulses to the axon. Repeat this step until the second action potential fails.
The interval at which the second action potential fails is the **absolute refractory period**, the time after an action potential when the neuron cannot fire a second action potential, no matter how intense the stimulus. Click **Record Data** to display your results in the grid (and record your results in Chart 5).

After you complete the experiment, take the online **Post-lab Quiz** for Activity 5.

Activity Questions

1. Explain how the absolute refractory period ensures directionality of action potential propagation.

_____

_____

2. Some tissues (for example, cardiac muscle) have long absolute refractory periods. Why would this be beneficial?

_____

_____

3. What do you think is the benefit of a relative refractory period in an axon of a sensory neuron?

_____

_____

The Action Potential: Coding for Stimulus Intensity

OBJECTIVES

1. To observe the response of axons to longer periods of stimulation.
2. To examine the relationship between stimulus intensity and the frequency of action potentials.

Introduction

As seen in Activity 3, the action potential has a constant amplitude, regardless of the stimulus intensity—it is an "all-or-none" event. As seen in Activity 5, the absolute refractory period is the time after an action potential when the neuron cannot fire a second action potential, no matter how intense the stimulus, and the relative refractory period is the time after an action potential when a second action potential can be generated if the stimulus intensity is increased.

In this activity you will use these concepts to begin to explore how the axon codes the stimulus intensity as *frequency,* the number of events (in this case, action potentials) per unit time. To demonstrate this phenomenon you will use longer periods of stimulation that are more representative of real-life stimuli. For example, when you encounter an odor, the odor is normally present for seconds (or longer), unlike the very brief stimuli used in Activities 3–5. These longer stimuli allow the axon of the neuron to generate additional action potentials as soon as it has recovered from the first. As seen in Activity 5, the length of this recovery period changes depending on the stimulus intensity. For example, at threshold, a second action potential can occur only after the axon has recovered from the absolute refractory period and the entire relative refractory period.

We will not consider the phenomenon of adaptation, which is a decrease in the response amplitude that often occurs with prolonged stimuli. For example, with most odors, after many seconds, you no longer smell the odor, even though it is still present. This decrease in response is due to adaptation.

EQUIPMENT USED The following equipment will be depicted on-screen: nerve chamber; axon; oscilloscope—used to observe timing of stimuli and voltage changes in the axon; stimulator—used to set the voltage and duration of stimuli and to deliver pulses that depolarize the axon; stimulation wires (S); recording electrode (wires R1)—used to record voltage changes in the axon. (The recording electrode is 2 centimeters from the stimulation wires.)

Experiment Instructions

Go to the home page in the PhysioEx software and click **Exercise 3: Neurophysiology of Nerve Impulses.** Click **Activity 6: The Action Potential: Coding for Stimulus Intensity,** and take the online **Pre-lab Quiz** for Activity 6.

After you take the online Pre-lab Quiz, click the **Experiment** tab and begin the experiment. The experiment instructions are reprinted here for your reference. The opening screen for the experiment is shown below.

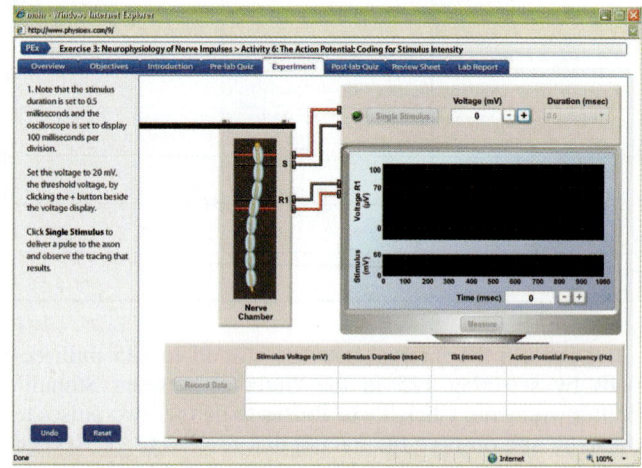

1. Note that the stimulus duration is set to 0.5 milliseconds and the oscilloscope is set to display 100 milliseconds per division. Set the voltage to 20 mV, the threshold voltage, by clicking the **+** button beside the voltage display. Click **Single Stimulus** to deliver a pulse to the axon and observe the tracing that results.

2. Note how the action potential looks at this timescale and click **Record Data** to display your results in the grid (and record your results in Chart 6).

CHART 6	Frequency of Action Potentials		
Stimulus voltage (mV)	Stimulus duration (msec)	ISI (msec)	Action potential frequency (Hz)

3. Increase the stimulus duration to 500 milliseconds by selecting 500 from the duration pull-down menu. Click **Single Stimulus** to deliver a pulse to the axon and observe the tracing that results. The stimulus is delivered after a delay of 100 milliseconds so that you can easily see the timing of the stimulus.

4. At the site of stimulation, the stimulus keeps the membrane of the axon at threshold for a long time, but this depolarization does not spread to the recording electrode. After one action potential has been generated and the axon has fully recovered from its absolute and relative refractory periods, the stimulus is still present to generate another action potential.

Measure the time (in milliseconds) between action potentials. This interval should be a bit longer than the relative refractory period (measured in Activity 5). Click **Measure** to help determine the time between action potentials. A thin, vertical yellow line appears at the far left side of the oscilloscope screen. You can move the line in 10-millisecond increments by clicking the + and − buttons beside the time display, which shows the time at the line. Click **Submit** to display your answer in the data table (and record your results in Chart 6).

5. The interval between action potentials is sometimes called the interspike interval (ISI). Action potentials are sometimes referred to as spikes because of their rapid time course. From the ISI, you can calculate the action potential frequency. The frequency is the reciprocal of the interval and is usually expressed in hertz (Hz), which is events (action potentials) per second. From the ISI you entered, calculate the frequency of action potentials with a prolonged (500 msec) threshold stimulus intensity. Frequency = 1/ISI. Click **Submit** to display your answer in the data table (and record your results in Chart 6).

6. A stimulus intensity of 30 mV was able to generate a second action potential toward the end of the relative refractory period in Activity 5. With this stronger stimulus, the second action potential can occur after a shorter time. Increase the stimulus intensity to 30 mV by clicking the + button beside the voltage display. Click **Single Stimulus** to deliver this stronger stimulus and observe the tracing that results.

7. Click **Submit** to display your answer in the data table (and record your results in Chart 6). Click **Measure** to help determine the time between action potentials. A thin, vertical yellow line appears at the far left side of the oscilloscope screen. You can move the line in 10-millisecond increments by clicking the + and − buttons beside the time display, which shows the time at the line.

8. From the ISI you entered, calculate the frequency of action potentials with a prolonged (500 msec) 30-mV stimulus intensity. Frequency = 1/ISI. Click **Submit** to display your answer in the data table (and record your results in Chart 6).

9. A stimulus intensity of 45 mV was able to generate a second action potential in the middle of the relative refractory period in Activity 5. With this even stronger stimulus, the second action potential can occur after an even shorter time. Increase the stimulus intensity to 45 mV.

? **PREDICT Question 1**
What effect will the increased stimulus intensity have on the frequency of action potentials?

10. Click **Single Stimulus** to deliver the stronger, 45-mV stimulus and observe the tracing that results.

11. Click **Submit** to display your answer in the data table (and record your results in Chart 6). Click **Measure** to help determine the time between action potentials. A thin, vertical yellow line appears at the far left side of the oscilloscope screen. You can move the line in 10-millisecond increments by clicking the + and − buttons beside the time display, which shows the time at the line.

12. From the ISI you entered, calculate the frequency of action potentials with a prolonged (500 msec) 45-mV stimulus intensity. Frequency = 1/ISI. Click **Submit** to display your answer in the data table (and record your results in Chart 6).

After you complete the experiment, take the online **Post-lab Quiz** for Activity 6.

Activity Questions

1. Compare the action potential frequency in a temperature-sensitive sensory neuron exposed to warm water and then hot water.

2. When a long-duration stimulus is applied, what two determinants of an action potential refractory period are being overcome?

3. Suggest several ways to pharmacologically overcome a neuron's refractory period and thereby increase the action potential frequency.

ACTIVITY 7

The Action Potential: Conduction Velocity

OBJECTIVES

1. To define and measure *conduction velocity* for an action potential.

2. To examine the effect of myelination on conduction velocity.

3. To examine the effect of axon diameter on conduction velocity.

Introduction

Once generated, the action potential is propagated, or conducted, down the axon. In other words, all-or-none action potentials are regenerated along the entire length of the axon. This propagation ensures that the amplitude of the action potential does not diminish as it is conducted along the axon. In some cases, such as the sensory neuron traveling from your toe to the spinal cord, the axon can be quite long (in this case, up to 1 meter). Propagation/conduction occurs because there are voltage-gated sodium and potassium channels located along the axon and because the large depolarization that constitutes the action potential (once generated at the trigger zone) easily brings the next region of the axon to threshold. The **conduction velocity** can be easily calculated by knowing both the distance the action potential travels and the amount of time it takes. Velocity has the units of distance per time, typically meters/second. An experimental stimulus artifact (see Activity 3) provides a convenient marker of the stimulus time because it travels very quickly (for our purposes, instantaneously) along the axon.

Several parameters influence the conduction velocity in an axon, including the axon diameter and the amount of myelination. **Myelination** refers to a special wrapping of the membrane from glial cells (or neuroglia) around the axon. In the central nervous system, oligodendrocytes are the glia that wrap around the axon. In the peripheral nervous system, the Schwann cells are the glia that wrap around the axon. Many glial cells along the axon contribute a myelin sheath, and the myelin sheaths are separated by gaps called nodes of Ranvier.

In this activity you will compare the conduction velocities of three axons: (1) a large-diameter, heavily myelinated axon, often called an A fiber (the terms axon and fiber are synonymous), (2) a medium-diameter, lightly myelinated axon (called the B fiber), and (3) a thin, unmyelinated fiber (called the C fiber). Examples of these axon types in the body include the axon of the sensory Pacinian corpuscle (an A fiber), the axon of both the olfactory sensory neuron and a free nerve ending (C fibers), and a visceral sensory fiber (a B fiber).

EQUIPMENT USED The following equipment will be depicted on-screen: nerve chamber; three axons— A fiber, B fiber, and C fiber; oscilloscope—used to observe timing of stimuli and voltage changes in the axon; stimulator—used to set the stimulus voltage and to deliver pulses that depolarize the axon; stimulation wires (S); recording electrodes (wires R1 and R2)—used to record voltage changes in the axon. (The first set of recording electrodes, R1, is 2 centimeters from the stimulation wires, and the second set of recording electrodes, R2, is 2 centimeters from R1.)

Experiment Instructions

Go to the home page in the PhysioEx software and click **Exercise 3: Neurophysiology of Nerve Impulses.** Click **Activity 7: The Action Potential: Conduction Velocity,** and take the online **Pre-lab Quiz** for Activity 7.

After you take the online Pre-lab Quiz, click the **Experiment** tab and begin the experiment. The experiment instructions are reprinted here for your reference. The opening screen for the experiment is shown below.

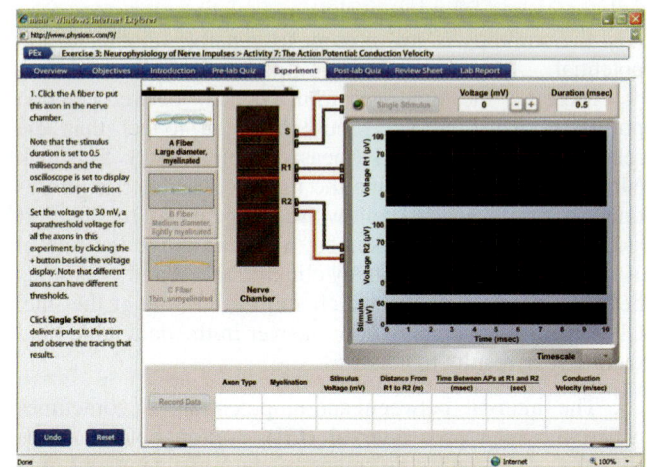

1. Click the A fiber to put this axon in the nerve chamber. Note that the stimulus duration is set to 0.5 milliseconds and the oscilloscope is set to display 1 millisecond per division.

Set the voltage to 30 mV, a suprathreshold voltage for all the axons in this experiment, by clicking the **+** button beside the voltage display. Note that different axons can have different thresholds. Click **Single Stimulus** to deliver a pulse to the axon and observe the tracing that results.

2. Click **Record Data** to display your results in the grid (and record your results in Chart 7).

3. Note the difference in time between the action potential recorded at R1 and the action potential recorded at R2. The distance between these sets of recording electrodes is 10 centimeters (0.1 m). Convert the time from milliseconds to seconds and then click **Submit** to display your results in the grid (and record your results in Chart 7).

4. Calculate the conduction velocity in meters/second by dividing the distance between R1 and R2 (0.1 m) by the time it took for the action potential to travel from R1 to R2. Click

CHART 7	Conduction Velocity					
Axon type	Myelination	Stimulus voltage (mV)	Distance from R1 to R2 (m)	Time between action potentials at R1 and R2 (msec)	(sec)	Conduction velocity (m/sec)

Submit to display your results in the grid (and record your results in Chart 7).

? **PREDICT** Question 1

How will the conduction velocity in the B fiber compare with that in the A fiber?

_____

_____

5. Click the **B fiber** to put this axon in the nerve chamber. Set the timescale on the oscilloscope to 10 milliseconds per division by selecting 10 in the timescale pull-down menu. Click **Single Stimulus** to deliver a pulse to the axon and observe the tracing that results.

6–8. Repeat steps 2–4 with the B fiber (and record your results in Chart 7).

? **PREDICT** Question 2

How will the conduction velocity in the C fiber compare with that in the B fiber?

_____

_____

9. Click the **C fiber** to put this axon in the nerve chamber. Set the timescale on the oscilloscope to 50 milliseconds per division by selecting 50 in the timescale pull-down menu. Click **Single Stimulus** to deliver a pulse to the axon and observe the tracing that results.

10–12. Repeat steps 2–4 with the C fiber (and record your results in Chart 7).

After you complete the experiment, take the online **Post-lab Quiz** for Activity 7.

Activity Questions

1. The squid utilizes a very large-diameter, unmyelinated axon to execute a rapid escape response when it perceives danger. How is this possible, given that the axon is unmyelinated?

_____

_____

2. When you burn your finger on a hot stove, you feel sharp, immediate pain, which later becomes slow, throbbing pain. These two types of pain are carried by different pain axons. Speculate on the axonal diameter and extent of myelination of these axons.

_____

_____

3. Why do humans possess a mixture of axons, some large-diameter, heavily myelinated axons and some small-diameter, relatively unmyelinated axons?

_____

_____

ACTIVITY 8

Chemical Synaptic Transmission and Neurotransmitter Release

OBJECTIVES

1. To define *neurotransmitter, chemical synapse, synaptic vesicle,* and *postsynaptic potential.*
2. To determine the role of calcium ions in neurotransmitter release.

Introduction

A major function of the nervous system is communication. The axon conducts the action potential from one place to another. Often, the axon has branches so that the action potential is conducted to several places at about the same time. At the end of each branch, there is a region called the axon terminal that is specialized to release packets of chemical neurotransmitters from small (~30-nm diameter) intracellular membrane-bound vesicles, called **synaptic vesicles. Neurotransmitters** are extracellular signal molecules that act on local targets as paracrine agents, on the neuron releasing the chemical as autocrine agents, and sometimes as hormones (endocrine agents) that reach their target(s) via the circulation. These chemicals are released by exocytosis and diffuse across a small extracellular space (called the synaptic gap, or synaptic cleft) to the target (most often the receiving end of another neuron or a muscle or gland). The neurotransmitter molecules often bind to membrane receptor proteins on the target, setting in motion a sequence of molecular events that can open or close membrane ion channels and cause the membrane potential in the target cell to change. This region where the neurotransmitter is released from one neuron and binds to a receptor on a target cell is called a **chemical synapse,** and the change in membrane potential of the target is called a synaptic potential, or **postsynaptic potential.**

In this activity you will explore some of the steps in neurotransmitter release from the axon terminal. Exocytosis of synaptic vesicles is normally triggered by an increase in calcium ions in the axon terminal. The calcium enters from outside the cell through membrane calcium channels that are opened by the depolarization of the action potential. The axon terminal has been greatly magnified in this activity so that you can visualize the release of neurotransmitter. Different from the other activities in this exercise, however, this procedure of directly seeing neurotransmitter release is not easily done in the lab; rather, neurotransmitter is usually detected by the postsynaptic potentials it triggers or by collecting and analyzing chemicals at the synapse after robust stimulation of the neurons.

EQUIPMENT USED The following equipment will be depicted on-screen: neuron (in vitro)—a large, dissociated (or cultured) neuron with magnified axon terminals; four extracellular solutions—control Ca^{2+}, no Ca^{2+}, low Ca^{2+}, and Mg^{2+}.

Experiment Instructions

Go to the home page in the PhysioEx software and click **Exercise 3: Neurophysiology of Nerve Impulses.** Click **Activity 8: Chemical Synaptic Transmission and Neurotransmitter Release,** and take the online **Pre-lab Quiz** for Activity 8.

After you take the online Pre-lab Quiz, click the **Experiment** tab and begin the experiment. The experiment instructions are reprinted here for your reference. The opening screen for the experiment is shown below.

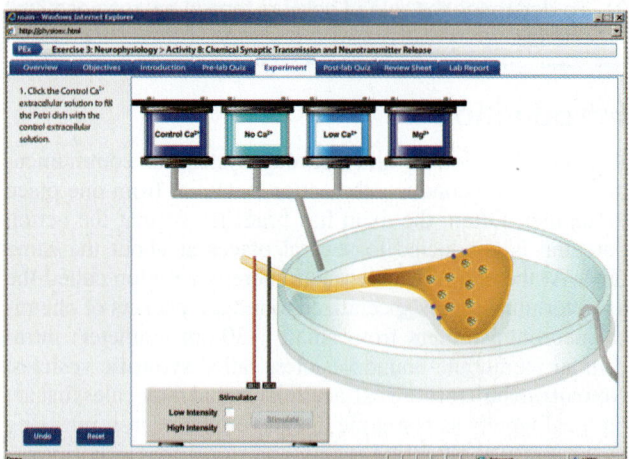

1. Click the **control Ca^{2+}** extracellular solution to fill the petri dish with the control extracellular solution.

2. Click **Low Intensity** on the stimulator and then click **Stimulate** to stimulate the neuron (axon) with a threshold stimulus that generates a low frequency of action potentials. Observe the release of neurotransmitter.

3. Click **High Intensity** on the stimulator and then click **Stimulate** to stimulate the neuron with a longer, more intense stimulus to generate a burst of action potentials. Observe the release of neurotransmitter.

> **? PREDICT Question 1**
> You have just observed that each action potential in a burst can trigger additional neurotransmitter release. If calcium ions are removed from the extracellular solution, what will happen to neurotransmitter release at the nerve terminal?
>
> _____
>
> _____

4–6. Repeat steps 1–3 with the *no Ca^{2+}* extracellular solution.

> **? PREDICT Question 2**
> What will happen to the amount of neurotransmitter release when low amounts of calcium are added back to the extracellular solution?
>
> _____
>
> _____

7–9. Repeat steps 1–3 with the *low Ca^{2+}* extracellular solution.

> **? PREDICT Question 3**
> What will happen to neurotransmitter release when magnesium is added to the extracellular solution?
>
> _____
>
> _____

10–12. Repeat steps 1–3 with the *Mg^{2+}* extracellular solution.

After you complete the experiment, take the online **Post-lab Quiz** for Activity 8.

Activity Questions

1. If you added more sodium to the extracellular solution, could the sodium substitute for the missing calcium?

_____

_____

2. How does botulinum toxin block synaptic transmission? Why is it used for cosmetic procedures?

_____

_____

_____

ACTIVITY 9

The Action Potential: Putting It All Together

OBJECTIVES

1. To identify the functional areas (for example, the sensory ending, axon, and postsynaptic membrane) of a two-neuron circuit.

2. To predict and test the responses in each functional area to a very weak, subthreshold stimulus.

3. To predict and test the responses in each functional area to a moderate stimulus.

4. To predict and test the responses in each functional area to an intense stimulus.

Introduction

In the nervous system, sensory neurons respond to adequate sensory stimuli, generating action potentials in the axon if the stimulus is strong enough to reach threshold (the action potential is an "all-or-nothing" event). Via chemical synapses, these sensory neurons communicate with interneurons that process the information. Interneurons also communicate with motor neurons that stimulate muscles and glands, again, usually via chemical synapses.

After performing Activities 1–8, you should have a better understanding of how neurons function by generating changes from their resting membrane potential. If threshold is reached, an action potential is generated and propagated. If the stimulus is more intense, then action potentials are generated at a higher frequency, causing the release of more neurotransmitter at the next synapse. At an excitatory synapse the chemical neurotransmitter binds to receptors at the receiving end of the next cell (usually the cell body or dendrites of an interneuron), causing ion channels to open, resulting in a depolarization toward threshold for an action potential in the interneuron's axon. This depolarizing synaptic potential (called an excitatory postsynaptic potential) is graded in amplitude, depending on the amount of neurotransmitter and the number of channels that open. In the axon, the amplitude of this synaptic potential is coded as the frequency of action potentials. Neurotransmitters can also cause inhibition, which will not be covered in this activity.

In this activity you will stimulate a sensory neuron, predict the response of that cell and its target, and then test those predictions.

EQUIPMENT USED   The following equipment will be depicted on-screen: neuron (in vitro)—a large, dissociated (or cultured) neuron; interneuron (in vitro)—a large, dissociated (or cultured) interneuron; microelectrodes— small probes with very small tips that can impale a single neuron (In an actual wet lab, a microelectrode manipulator is used to position the microelectrodes. For simplicity, the microelectrode manipulator will not be depicted in this activity.); microelectrode amplifier—used to measure the voltage between the microelectrodes and a reference; oscilloscope—used to observe the changes in voltage across the membrane of the neuron and interneuron; stimulator—used to set the stimulus intensity (low or high) and to deliver pulses to the neuron.

Experiment Instructions

Go to the home page in the PhysioEx software and click **Exercise 3: Neurophysiology of Nerve Impulses.** Click **Activity 9: The Action Potential: Putting It All Together,** and take the online **Pre-lab Quiz** for Activity 9.

After you take the online Pre-lab Quiz, click the **Experiment** tab and begin the experiment. The experiment instructions are reprinted here for your reference. The opening screen for the experiment is shown above.

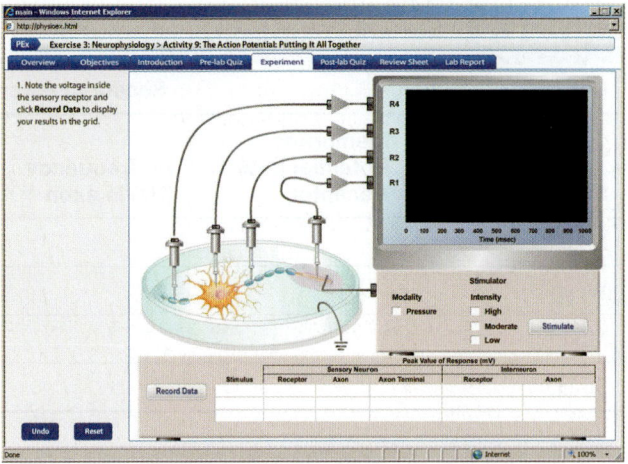

1. Note the membrane potential at the sensory receptor and the receiving end of the interneuron and click **Record Data** to display your results in the grid (and record your results in Chart 9).

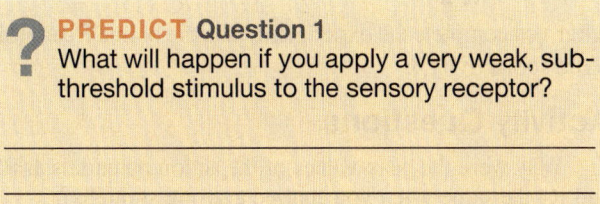

? PREDICT Question 1
What will happen if you apply a very weak, sub-threshold stimulus to the sensory receptor?

2. Click **Very Weak** intensity on the stimulator and then click **Stimulate** to stimulate the receiving end of the sensory neuron and observe the tracing that results.

3. Click **Record Data** to display your results in the grid (and record your results in Chart 9). The stimulus lasts 500 msec.

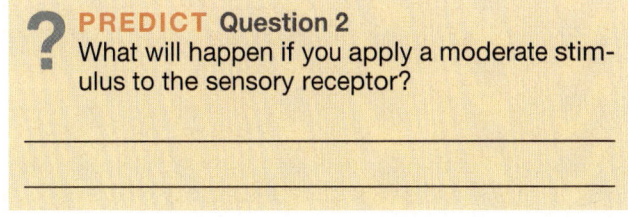

? PREDICT Question 2
What will happen if you apply a moderate stimulus to the sensory receptor?

4. Click **Moderate** intensity on the stimulator and then click **Stimulate** to stimulate the sensory receptor and observe the tracing that results.

5. Click **Record Data** to display your results in the grid (and record your results in Chart 9).

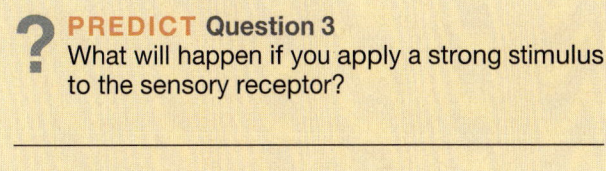

? PREDICT Question 3
What will happen if you apply a strong stimulus to the sensory receptor?

CHART 9	Putting It All Together					
	Sensory neuron			Interneuron	
Stimulus	Membrane Potential (mV) Receptor	AP frequency (Hz) in axon	Vesicles released from axon terminal	Membrane potential (mV) receiving end	AP frequency (Hz) in axon
None					
Weak					
Moderate					
Strong					

6. Click **Strong** intensity on the stimulator and then click **Stimulate** to stimulate the sensory receptor and observe the tracing that results.

7. Click **Record Data** to display your results in the grid (and record your results in Chart 9).

After you complete the experiment, take the online **Post-lab Quiz** for Activity 9.

Activity Questions

1. Why were the peak values of the action potentials at R2 and R4 the same when you applied a strong stimulus?

_____

_____

2. If the axons were unmyelinated, would the peak value of the action potential at R4 change relative to that at R2?

_____

_____

NAME _____

LAB TIME/DATE _____

Neurophysiology of Nerve Impulses

ACTIVITY 1   The Resting Membrane Potential

1. Explain why increasing extracellular K^+ reduces the net diffusion of K^+ out of the neuron through the K^+ leak channels.

   _____

2. Explain why increasing extracellular K^+ causes the membrane potential to change to a less negative value. How well did the

   results compare with your prediction? _____

   _____

   _____

3. Explain why a change in extracellular Na^+ did not alter the membrane potential in the resting neuron. _____

   _____

4. Discuss the relative permeability of the membrane to Na^+ and K^+ in a resting neuron. _____

   _____

   _____

5. Discuss how a change in Na^+ or K^+ conductance would affect the resting membrane potential. _____

   _____

ACTIVITY 2   Receptor Potential

1. Sensory neurons have a resting potential based on the efflux of potassium ions (as demonstrated in Activity 1). What passive

   channels are likely found in the membrane of the olfactory receptor, in the membrane of the Pacinian corpuscle, and in the

   membrane of the free nerve ending? _____

   _____

   _____

2. What is meant by the term *graded potential*? _____

   _____

   _____

3. Identify which of the stimulus modalities induced the largest amplitude receptor potential in the Pacinian corpuscle. How

   well did the results compare with your prediction? _____

   _____

4. Identify which of the stimulus modalities induced the largest-amplitude receptor potential in the olfactory receptors. How

   well did the results compare with your prediction? _____

   _____

5. The olfactory receptor also contains a membrane protein that recognizes isoamyl acetate and, via several other molecules,

   transduces the odor stimulus into a receptor potential. Does the Pacinian corpuscle likely have this isoamyl acetate receptor

   protein? Does the free nerve ending likely have this isoamyl acetate receptor protein? _____

   _____

6. What type of sensory neuron would likely respond to a green light? _____

   _____

   _____

ACTIVITY 3 **The Action Potential: Threshold**

1. Define the term *threshold* as it applies to an action potential. _____

   _____

   _____

2. What change in membrane potential (depolarization or hyperpolarization) triggers an action potential? _____

   _____

3. How did the action potential at R1 (or R2) change as you increased the stimulus voltage above the threshold voltage? How

   well did the results compare with your prediction? _____

   _____

4. An action potential is an "all-or-nothing" event. Explain what is meant by this phrase. _____

   _____

   _____

5. What part of a neuron was investigated in this activity? _____

   _____

ACTIVITY 4 The Action Potential: Importance of Voltage-Gated Na⁺ Channels

1. What does TTX do to voltage-gated Na⁺ channels? _____

_____

_____

2. What does lidocaine do to voltage-gated Na⁺ channels? How does the effect of lidocaine differ from the effect of TTX?

_____

_____

3. A nerve is a bundle of axons, and some nerves are less sensitive to lidocaine. If a nerve, rather than an axon, had been used in

the lidocaine experiment, the responses recorded at R1 and R2 would be the sum of all the action potentials (called a compound

action potential). Would the response at R2 after lidocaine application necessarily be zero? Why or why not? _____

_____

_____

4. Why are fewer action potentials recorded at R2 when TTX is applied between R1 and R2? How well did the results compare

with your prediction?_____

_____

_____

5. Why are fewer action potentials recorded at R2 when lidocaine is applied between R1 and R2? How well did the results

compare with your prediction? _____

_____

_____

6. Pain-sensitive neurons (called nociceptors) conduct action potentials from the skin or teeth to sites in the brain involved in

pain perception. Where should a dentist inject the lidocaine to block pain perception? _____

_____

_____

ACTIVITY 5 The Action Potential: Measuring Its Absolute and Relative Refractory Periods

1. Define *inactivation* as it applies to a voltage-gated sodium channel. _____

_____

_____

2. Define the *absolute refractory period*. _____

_____

_____

3. How did the threshold for the second action potential change as you further decreased the interval between the stimuli?

How well did the results compare with your prediction? _____

_____

4. Why is it harder to generate a second action potential during the relative refractory period? _____

_____

_____

A C T I V I T Y 6 **The Action Potential: Coding for Stimulus Intensity**

1. Why are multiple action potentials generated in response to a long stimulus that is above threshold? _____

_____

_____

2. Why does the frequency of action potentials increase when the stimulus intensity increases? How well did the results

compare with your prediction? _____

_____

3. How does threshold change during the relative refractory period? _____

_____

_____

4. What is the relationship between the interspike interval and the frequency of action potentials? _____

_____

_____

A C T I V I T Y 7 **The Action Potential: Conduction Velocity**

1. How did the conduction velocity in the B fiber compare with that in the A fiber? How well did the results compare with your

prediction? _____

_____

_____

2. How did the conduction velocity in the C fiber compare with that in the B fiber? How well did the results compare with your

   prediction? _____

   _____

   _____

3. What is the effect of axon diameter on conduction velocity? _____

   _____

   _____

4. What is the effect of the amount of myelination on conduction velocity? _____

   _____

   _____

5. Why did the time between the stimulation and the action potential at R1 differ for each axon? _____

   _____

6. Why did you need to change the timescale on the oscilloscope for each axon? _____

   _____

   _____

ACTIVITY 8   **Chemical Synaptic Transmission and Neurotransmitter Release**

1. When the stimulus intensity is increased, what changes: the number of synaptic vesicles released or the amount of

   neurotransmitter per vesicle? _____

   _____

   _____

2. What happened to the amount of neurotransmitter release when you switched from the control extracellular fluid to the

   extracellular fluid with no Ca^{2+}? How well did the results compare with your prediction? _____

   _____

   _____

3. What happened to the amount of neurotransmitter release when you switched from the extracellular fluid with no Ca^{2+} to

   the extracellular fluid with low Ca^{2+}? How well did the results compare with your prediction? _____

   _____

   _____

4. How did neurotransmitter release in the Mg^{2+} extracellular fluid compare to that in the control extracellular fluid? How well did the result compare with your prediction? _____

_____

_____

5. How does Mg^{2+} block the effect of extracellular calcium on neurotransmitter release? _____

_____

_____

ACTIVITY 9 **The Action Potential: Putting It All Together**

1. Why is the resting membrane potential the same value in both the sensory neuron and the interneuron? _____

_____

_____

2. Describe what happened when you applied a very weak stimulus to the sensory receptor. How well did the results compare with your prediction? _____

_____

_____

3. Describe what happened when you applied a moderate stimulus to the sensory receptor. How well did the results compare with your prediction? _____

_____

_____

4. Identify the type of membrane potential (graded receptor potential or action potential) that occurred at R1, R2, R3, and R4 when you applied a moderate stimulus. (View the response to the stimulus.) _____

_____

_____

5. Describe what happened when you applied a strong stimulus to the sensory receptor. How well did the results compare with your prediction? _____

_____

_____

Endocrine System Physiology

P R E - L A B   Q U I Z

1. Define *metabolism.* _____

2. Circle the correct underlined term: Hormones are chemicals secreted from <u>endocrine</u> / <u>exocrine</u> glands.

3. The most important hormone for maintaining metabolism and body heat is:
   a. steroid hormone
   b. thyroxine
   c. thyroid-stimulating hormone
   d. adrenaline

4. Circle the correct underlined term: <u>Negative feedback mechanisms</u> / <u>Positive feedback mechanisms</u> ensure that if the body needs a hormone it will be produced until there is too much of it.

5. After menopause, the ovaries will stop producing and secreting:
   a. progesterone
   b. follicle-stimulating hormone
   c. estrogen
   d. androgen

6. _____ is the hormone produced by the beta cells of the pancreas that allows our cells to absorb glucose from the bloodstream.
   a. Insulin   c. Cortisol
   b. Glucagon   d. Mellitol

7. Circle True or False: Cortisol is a hormone secreted by the adrenal medulla.

8. Circle the correct underlined term: Diabetes mellitus <u>type 1</u> / <u>type 2</u> results when the body is able to produce enough insulin but fails to respond to it.

Exercise Overview

In the human body the **endocrine system** (in addition to the nervous system) coordinates and integrates the functions of different physiological systems (view Figure 4.1). Thus, the endocrine system plays a critical role in maintaining **homeostasis.** This role begins with chemicals, called **hormones,** secreted from ductless **endocrine glands,** which are tissues that have an epithelial origin. Endocrine glands secrete hormones into the extracellular fluid compartments. More specifically, the blood usually carries hormones (sometimes attached to specific plasma proteins) to their **target cells.** Target cells can be very close to, or very far from, the source of the hormone.

Hormones bind to high-affinity **receptors** located on the target cell's surface, in its cytosol, or in its nucleus. These hormone receptors have remarkable **sensitivity,** as the hormone concentration in the blood can range from 10^{-9} to 10^{-12} molar! A hormone-receptor complex forms and can then exert a **biological action** through signal-transduction cascades and alteration of gene transcription at the target cell. The physiological response to hormones can vary from seconds to hours to days, depending on the chemical nature of the hormone and its receptor location in the target cell.

The chemical structure of the hormone is important in determining how it will interact with target cells. *Peptide* and *catecholamine hormones* are fast-acting hormones that attach to a plasma-membrane receptor and cause a second-messenger cascade in the cytoplasm of the target cell. For example, a chemical called cAMP (cyclic adenosine monophosphate) is synthesized from a molecule of ATP. The

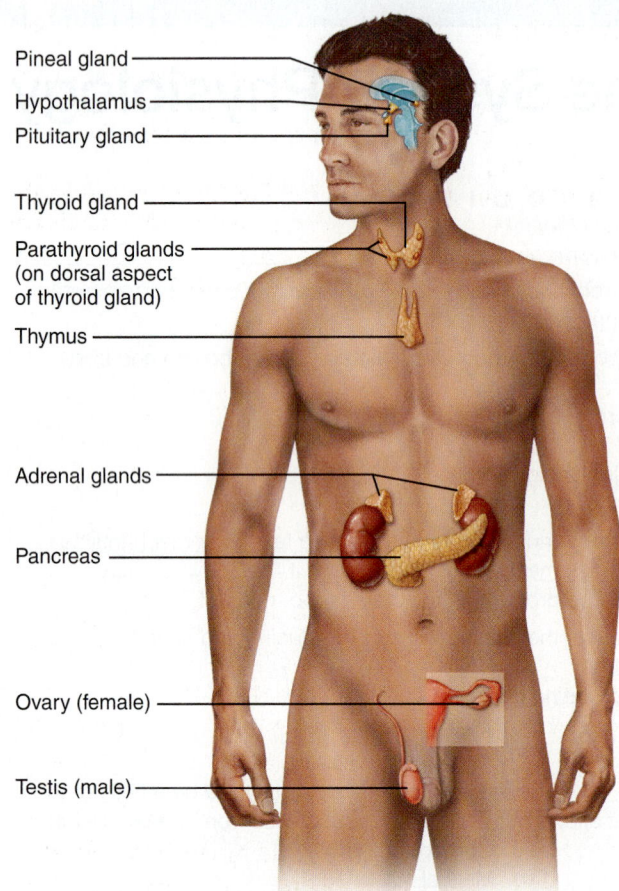

Pineal gland

Hypothalamus

Pituitary gland

Thyroid gland

Parathyroid glands
(on dorsal aspect
of thyroid gland)

Thymus

Adrenal glands

Pancreas

Ovary (female)

Testis (male)

FIGURE 4.1 Selected endocrine organs of the body.

synthesis of this chemical makes the cell more metabolically active and, therefore, more able to respond to a stimulus.

Steroid hormones and *thyroxine* (thyroid hormone) are slow-acting hormones that enter the target cell and interact with the nucleus to affect the transcription of various proteins that the cell can synthesize. The hormones enter the nucleus and attach at specific points on the DNA. Each attachment causes the production of a specific mRNA, which is then moved to the cytoplasm, where ribosomes can translate the mRNA into a protein.

Keep in mind that the organs of the endocrine system do not function independently. The activities of one endocrine gland are often coordinated with the activities of other glands. No one system functions independently of any other system. For this reason, we will be stressing feedback mechanisms and how we can use them to predict, explain, and understand hormone effects.

Given the powerful influence that hormones have on homeostasis, **negative feedback mechanisms** are important in regulating hormone secretion, synthesis, and effectiveness at target cells. Negative feedback ensures that if the body needs a particular hormone, that hormone will be produced until there is too much of it. When there is too much of the hormone, its release will be inhibited.

Rarely, the body regulates hormones via a *positive feedback mechanism.* The release of *oxytocin* from the posterior pituitary is one of these rare instances. Oxytocin is a hormone that causes the muscle layer of the uterus, called the *myometrium,* to contract during childbirth. This contraction of the myometrium causes additional oxytocin to be released,

allowing stronger contractions. Unlike what happens in negative feedback mechanisms, the increase in circulating levels of oxytocin does not inhibit oxytocin secretion.

Many experimental methods can be used to study the functions of an endocrine gland. These methods include removing the gland from an animal and then injecting, implanting, or feeding glandular extracts into a normal animal or an animal deprived of the gland being studied. In this exercise you will use these methods to gain a deeper understanding of the *function* and *regulation* of some of the endocrine glands.

ACTIVITY 1

Metabolism and Thyroid Hormone

OBJECTIVES

1. To understand the terms *basal metabolic rate (BMR), thyroid-stimulating hormone (TSH), thyroxine, goiter, hypothyroidism, hyperthyroidism, thyroidectomized,* and *hypophysectomized.*

2. To observe how negative feedback mechanisms regulate hormone release.

3. To understand thyroxine's role in maintaining the basal metabolic rate.

4. To understand the effect of TSH on the basal metabolic rate.

5. To understand the role of the hypothalamus in regulating the secretion of thyroxine and TSH.

Introduction

Metabolism is the broad range of biochemical reactions occurring in the body. Metabolism includes *anabolism* and *catabolism.* Anabolism is the building up of small molecules into larger, more complex molecules via enzymatic reactions. Energy is stored in the chemical bonds formed when larger, more complex molecules are formed.

Catabolism is the breakdown of large, complex molecules into smaller molecules via enzymatic reactions. The breaking of chemical bonds in catabolism releases energy that the cell can use to perform various activities, such as forming ATP. The cell does not use all the energy released by bond breaking. Much of the energy is released as heat to maintain a fixed body temperature, especially in humans. Humans are *homeothermic* organisms that need to maintain a fixed body temperature to maintain the activity of the various metabolic pathways in the body.

The most important hormone for maintaining metabolism and body heat is **thyroxine** (thyroid hormone), also known as *tetraiodothyronine,* or T_4. Thyroxine is secreted by the thyroid gland, located in the neck.

The production of thyroxine is controlled by the pituitary gland, or hypophysis, which secretes **thyroid-stimulating hormone (TSH).** The blood carries TSH to its target tissue, the thyroid gland. TSH causes the thyroid gland to increase in size and secrete thyroxine into the general circulation. If TSH levels are too high, the thyroid gland enlarges. The resulting glandular swelling in the neck is called a **goiter.**

The **hypothalamus** in the brain is also a vital participant in thyroxine and TSH production. It is a primary endocrine gland that secretes several hormones that affect the pituitary gland, or hypophysis, which is also located in the brain.

Thyrotropin-releasing hormone (TRH) is directly linked to thyroxine and TSH secretion. TRH from the hypothalamus stimulates the anterior pituitary to produce TSH, which then stimulates the thyroid to produce thyroxine.

These events are part of a classic negative feedback mechanism. When circulation levels of thyroxine are low, the hypothalamus secretes more TRH to stimulate the pituitary gland to secrete more TSH. The increase in TSH further stimulates the secretion of thyroxine from the thyroid gland. The increased levels of thyroxine will then influence the hypothalamus to reduce its production of TRH.

TRH travels from the hypothalamus to the pituitary gland via the **hypothalamic-pituitary portal system.** This specialized arrangement of blood vessels consists of a single **portal vein** that connects two capillary beds. The hypothalamic-pituitary portal system transports many other hormones from the hypothalamus to the pituitary gland. The hypothalamus primarily secretes *tropic* hormones, which stimulate the secretion of other hormones. TRH is an example of a tropic hormone because it stimulates the release of TSH from the pituitary gland. TSH itself is also an example of a tropic hormone because it stimulates production of thyroxine.

In this activity you will investigate the effects of thyroxine and TSH on a rat's metabolic rate. The metabolic rate will be indicated by the amount of oxygen the rat consumes per time per body mass. You will perform four experiments on three rats: a normal rat, a thyroidectomized rat (a rat whose thyroid gland has been surgically removed), and a hypophysectomized rat (a rat whose pituitary gland has been surgically removed). You will determine (1) the rat's basal metabolic rate, (2) its metabolic rate after it has been injected with thyroxine, (3) its metabolic rate after it has been injected with TSH, and (4) its metabolic rate after it has been injected with propylthiouracil, a drug that inhibits the production of thyroxine.

EQUIPMENT USED The following equipment will be depicted on-screen: three refillable syringes—used to inject the rats with propylthiouracil (a drug that inhibits the production of thyroxine by blocking the incorporation of iodine into the hormone precursor molecule), thyroid-stimulating hormone (TSH), and thyroxine; airtight, glass animal chamber—provides an isolated, sealed system in which to measure the amount of oxygen consumed by the rat in a specified amount of time (Opening the clamp on the left tube allows outside air into the chamber, and closing the clamp will create a closed, airtight system. The T-connector on the right tube allows you to connect the chamber to the manometer or to connect the fluid-filled manometer to the syringe filled with air.); soda lime (found at the bottom of the glass chamber)—absorbs the carbon dioxide given off by the rat; manometer—U-shaped tube containing fluid (As the rat consumes oxygen in the isolated, sealed system, this fluid will rise in the left side of the U-shaped tube and fall in the right side of the tube.); syringe—used to inject air into the tube and thus measure the amount of air that is needed to return the fluid columns in the manometer to their original levels; animal scale—used to measure body weight; three white rats—a *normal* rat, a *thyroidectomized* (Tx) rat (a rat whose thyroid gland has been surgically removed), and a *hypophysectomized* (Hypox) rat (a rat whose pituitary gland has been surgically removed).

Experiment Instructions

Go to the home page in the PhysioEx software and click **Exercise 4: Endocrine System Physiology.** Click **Activity 1: Metabolism and Thyroid Hormone,** and take the online **Pre-lab Quiz** for Activity 1.

After you take the online Pre-lab Quiz, click the **Experiment** tab and begin the experiment. The experiment instructions are reprinted here for your reference. The opening screen for the experiment is shown below.

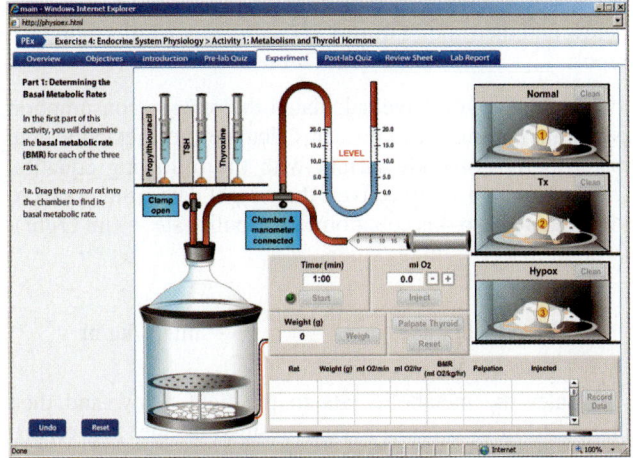

Part 1: Determining the Basal Metabolic Rates

In the first part of this activity, you will determine the basal metabolic rate (BMR) for each of the three rats.

1a. Drag the *normal* rat into the chamber to find its BMR.

1b. Click **Weigh** to determine the rat's weight.

1c. Click the clamp on the left tube (top of the chamber) to close it. This will prevent any outside air from entering the chamber and ensure that the only oxygen the rat is breathing is the oxygen inside the closed system.

1d. Note that the timer is set to one minute. Click **Start** beneath the timer to measure the amount of oxygen consumed by the rat in one minute in the sealed chamber. Note what happens to the water levels in the manometer as time progresses.

1e. Click the T-connector knob to connect the manometer and syringe.

1f. Click the clamp on the left tube (top of the chamber) to open it so the rat can breathe outside air.

1g. Observe the difference between the level in the left and right arms of the manometer. Estimate the volume of O_2 that you will need to inject to make the levels equal by counting the divisions on both sides. This volume is equivalent to the amount of oxygen that the rat consumed during the minute in the sealed chamber. Click the + button under the ml O_2 display until you reach the estimated volume. Then click **Inject** and watch what happens to the fluid in the two arms. When the volume levels are equalized, the word "Level" will appear and stay on the screen.

- If you have not injected enough oxygen, the word "Level" will not appear. Click the + to increase the volume and then click **Inject** again.

- If you have injected too much oxygen, the word "Level" will flash and then disappear. Click the button to decrease the volume and then click **Inject** again. Click **Record Data** when the levels are equalized.

1h. Calculate the oxygen consumption per hour for this rat using the following equation:

$$\frac{\text{ml O}_2 \text{ consumed}}{1 \text{ minute}} \times \frac{60 \text{ minutes}}{1 \text{ hr}} = \text{ml O}_2/\text{hr}$$

Enter the oxygen consumption per hour in the field below and then click **Submit** to record your results in the lab report. _____ ml O$_2$/hr

1i. Now that you have calculated the oxygen consumption per hour for this rat, you can calculate the metabolic rate per kilogram of body weight with the following equation (note that you need to convert the weight data from grams to kilograms to use this equation): Metabolic rate = (ml O$_2$/hr)/ (weight in kg) = ml O$_2$/kg/hr.

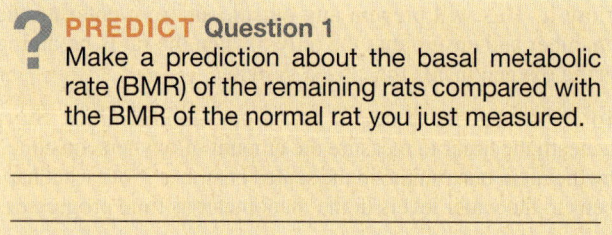

$$\text{Metabolic rate} = \frac{\text{ml O}_2/\text{hr}}{\text{weight in kg}} = \text{ml O}_2/\text{kg/hr}$$

Enter the metabolic rate in the field below and then click **Submit** to record your results in the lab report. _____ ml O$_2$/kg/hr

1j. Click **Palpate Thyroid** to manually check the size of the thyroid and, thus, whether a goiter is present. After reviewing the findings, click **Submit** to record your results in the lab report.

1k. Drag the rat from the chamber back to its cage and then click **Restore** (beneath **Palpate Thyroid**) to restore the apparatus to its initial state.

> **? PREDICT Question 1**
> Make a prediction about the basal metabolic rate (BMR) of the remaining rats compared with the BMR of the normal rat you just measured.
>
> _____
>
> _____

2a.–2k. Repeat steps 1a–1k for the *thyroidectomized (Tx)* rat.

3a.–3k. Repeat steps 1a–1k for the *hypophysectomized (Hypox)* rat.

> **? PREDICT Question 2**
> What do you think will happen to the metabolic rates of the rats after you inject them with thyroxine?
>
> _____
>
> _____

Part 2: Determining the Effect of Thyroxine on Metabolic Rate

In this part of the activity, you will investigate the effects of thyroxine injections on the metabolic rates of all three rats.

4a. Drag the syringe filled with *thyroxine* to the *normal* rat's hindquarters. Release the mouse button to inject thyroxine into the rat. (In this experiment, the effects of the injection are immediate. In a wet lab, you would have to inject the rats daily with thyroxine for 1–2 weeks).

4b. In this part of the activity, the rat's weight, the amount of oxygen consumed by the rat in one minute, the rat's oxygen consumption per hour, the rat's metabolic rate, and the result of the thyroid palpation will be generated automatically after you drag the rat into the chamber.

Drag the injected rat into the chamber and note the results (and record your results in Chart 1).

4c. Drag the rat from the chamber back to its cage and then click **Clean** to clear all traces of thyroxine from the rat and clean the syringe. (In this experiment, the thyroxine is removed instantly. In a wet lab, clearance would take weeks or require that a different rat be used.)

5a.–5c. Repeat steps 4a–4c with the *thyroidectomized (Tx)* rat (and record your results in Chart 1).

6a.–6c. Repeat steps 4a–4c with the *hypophysectomized (Hypox)* rat (and record your results in Chart 1).

> **? PREDICT Question 3**
> What do you think will happen to the metabolic rates of the rats after you inject them with TSH?
>
> _____
>
> _____

Part 3: Determining the Effect of TSH on Metabolic Rate

In this part of the activity, you will investigate the effects of TSH injections on the metabolic rates of all three rats.

7a. Drag the syringe filled with *TSH* to the *normal* rat's hindquarters. Release the mouse button to inject TSH into the rat. (In this experiment, the effects of the injection are immediate. In a wet lab, you would have to inject the rats daily with TSH for 1–2 weeks.)

7b. In this part of the activity, the rat's weight, the amount of oxygen consumed by the rat in one minute, the rat's oxygen consumption per hour, the rat's metabolic rate, and the result of the thyroid palpation will be generated automatically after you drag the rat into the chamber.

Drag the injected rat into the chamber and note the results (and record your results in Chart 1).

7c. Drag the rat from the chamber back to its cage and then click **Clean** to clear all traces of TSH from the rat and clean the syringe. (In this experiment, the TSH is removed instantly. In a wet lab, clearance would take weeks or require that a different rat be used.)

8a.–8c. Repeat steps 7a–7c with the *thyroidectomized (Tx)* rat (and record your results in Chart 1).

9a.–9c. Repeat steps 7a–7c with the *hypophysectomized (Hypox)* rat (and record your results in Chart 1).

CHART 1	Effects of Hormones on Metabolic Rate		
	Normal rat	**Thyroidectomized rat**	**Hypophysectomized rat**
Baseline			
Weight	_____ grams	_____ grams	_____ grams
ml O$_2$ used in 1 minute	_____ ml	_____ ml	_____ ml
ml O$_2$ used per hour	_____ ml	_____ ml	_____ ml
Metabolic rate	_____ ml O$_2$/kg/hr	_____ ml O$_2$/kg/hr	_____ ml O$_2$/kg/hr
Palpation results	_____	_____	_____
With thyroxine			
Weight	_____ grams	_____ grams	_____ grams
ml O$_2$ used in 1 minute	_____ ml	_____ ml	_____ ml
ml O$_2$ used per hour	_____ ml	_____ ml	_____ ml
Metabolic rate	_____ ml O$_2$/kg/hr	_____ ml O$_2$/kg/hr	_____ ml O$_2$/kg/hr
Palpation results	_____	_____	_____
With TSH			
Weight	_____ grams	_____ grams	_____ grams
ml O$_2$ used in 1 minute	_____ ml	_____ ml	_____ ml
ml O$_2$ used per hour	_____ ml	_____ ml	_____ ml
Metabolic rate	_____ ml O$_2$/kg/hr	_____ ml O$_2$/kg/hr	_____ ml O$_2$/kg/hr
Palpation results	_____	_____	_____
With propylthiouracil			
Weight	_____ grams	_____ grams	_____ grams
ml O$_2$ used in 1 minute	_____ ml	_____ ml	_____ ml
ml O$_2$ used per hour	_____ ml	_____ ml	_____ ml
Metabolic rate	_____ ml O$_2$/kg/hr	_____ ml O$_2$/kg/hr	_____ ml O$_2$/kg/hr
Palpation results	_____	_____	_____

? PREDICT Question 4
Propylthiouracil (PTU) is a drug that inhibits the production of thyroxine by blocking the attachment of iodine to tyrosine residues in the follicle cells of the thyroid gland (iodinated tyrosines are linked together to form thyroxine). What do you think will happen to the metabolic rates of the rats after you inject them with PTU?

_____

_____

Part 4: Determining the Effect of Propylthiouracil on Metabolic Rate

In this part of the activity, you will investigate the effects of propylthiouracil injections on the metabolic rates of all three rats.

10a. Drag the syringe filled with *propylthiouracil* to the *normal* rat's hindquarters. Release the mouse button to inject propylthiouracil into the rat. (In this experiment, the effects of the injection are immediate. In a wet lab, you would have to inject the rats daily with propylthiouracil for 1–2 weeks).

10b. In this part of the activity, the rat's weight, the amount of oxygen consumed by the rat in one minute, the rat's oxygen consumption per hour, the rat's metabolic rate, and the result of the thyroid palpation will be generated automatically after you drag the rat into the chamber.

Drag the injected rat into the chamber and note the results (and record your results in Chart 1).

10c. Drag the rat from the chamber back to its cage and then click **Clean** to clear all traces of propylthiouracil from the rat and clean the syringe. (In this experiment, the propylthiouracil is removed instantly. In a wet lab, clearance would take weeks or require that a different rat be used.)

11a.–11c. Repeat steps 10a–10c with the *thyroidectomized (Tx)* rat (and record your results in Chart 1).

12a.–12c. Repeat steps 10a–10c with the *hypophysectomized (Hypox)* rat (and record your results in Chart 1).

After you complete the experiment, take the online **Post-lab Quiz** for Activity 1.

Activity Questions

1. Using a water-filled manometer, you observed the amount of oxygen consumed by rats in a sealed chamber. What happened to the carbon dioxide the rat produced while in the sealed chamber?

2. What would happen to the fluid levels of the manometer (and, thus, the results of the metabolism experiment) if the rats in the sealed chamber were engaged in physical activity (such as running in a wheel)?

3. Describe the role of the hypothalamus in the production of thyroxine.

4. What does it mean if a hormone is a *tropic* hormone?

5. How could you treat a thyroidectomized rat so that it functions like a "normal" rat? How would you verify that your treatments were safe and effective?

6. What is the role of the hypothalamus in the production of thyroid-stimulating hormone (TSH)?

7. How does thyrotropin-releasing hormone (TRH) travel from the hypothalamus to the pituitary gland?

8. Why didn't the administration of TSH have any effect on the metabolic rate of the thyroidectomized rat?

9. Why didn't the administration of propylthiouracil have any effect on the metabolic rate of either the thyroidectomized rat or the hypophysectomized rat?

10. Propylthiouracil inhibits the production of thyroxine by blocking the attachment of iodine to the amino acid tyrosine. What naturally occurring problem in some parts of the world does this drug mimic?

ACTIVITY 2

Plasma Glucose, Insulin, and Diabetes Mellitus

OBJECTIVES

1. To understand the use of the terms *insulin, type 1 diabetes mellitus, type 2 diabetes mellitus,* and *glucose standard curve.*

2. To understand how fasting plasma glucose levels are used to diagnose diabetes mellitus.

3. To understand the assay that is used to measure plasma glucose.

Introduction

Insulin is a hormone produced by the beta cells of the endocrine portion of the pancreas. This hormone is vital to the regulation of **plasma glucose** levels, or "blood sugar," because the hormone enables our cells to absorb glucose from the bloodstream. Glucose absorbed from the blood is either used as fuel for metabolism or stored as glycogen (also known as animal starch), which is most notable in liver and muscle cells. About 75% of glucose consumed during a meal is stored as glycogen. As humans do not feed continuously (we are considered "discontinuous feeders"), the production of glycogen from a meal ensures that a supply of glucose will be available for several hours after a meal.

Furthermore, the body has to maintain a certain level of plasma glucose to continuously serve nerve cells because these cell types use only glucose for metabolic fuel. When glucose levels in the plasma fall below a certain value, the alpha cells of the pancreas are stimulated to release the hormone **glucagon.** Glucagon stimulates the breakdown of stored glycogen into glucose, which is then released back into the blood.

When the pancreas does not produce enough insulin, **type 1 diabetes mellitus** results. When the pancreas produces sufficient insulin but the body fails to respond to it, **type 2 diabetes mellitus** results. In either case, glucose remains in the bloodstream, and the body's cells are unable to take it up to serve as the primary fuel for metabolism. The kidneys then filter the excess glucose out of the plasma. Because the reabsorption of filtered glucose involves a finite number of transporters in kidney tubule cells, some of the excess glucose is not reabsorbed into the circulation. Instead, it passes out of the body in urine (hence *sweet urine,* as the name **diabetes mellitus** suggests).

The inability of body cells to take up glucose from the blood also results in skeletal muscle cells undergoing protein catabolism to free up amino acids to be used in forming glucose in the liver. This action puts the body into a negative nitrogen balance from the resulting protein depletion and tissue wasting. Other associated problems include poor wound healing and poor resistance to infections.

This activity is divided into two parts. In Part 1, you will generate a **glucose standard curve,** which will be explained in the experiment. In Part 2, you will use the glucose standard curve to measure the fasting plasma glucose levels from several patients to diagnose the presence or absence of diabetes mellitus. A patient with FPG values greater than or equal to 126 mg/dl in two FPG tests is diagnosed with diabetes. FPG values between 110 and 126 mg/dl indicate impairment or borderline impairment of insulin-mediated glucose uptake by cells. FPG values less than 110 mg/dl are considered normal.

EQUIPMENT USED   The following equipment will be depicted on-screen: deionized water—used to adjust the volume so that it is the same for each reaction; glucose standard; enzyme color reagent; barium hydroxide; heparin; blood samples from five patients; test tubes—used as reaction vessels for the various tests; test tube incubation unit—used to incubate, mix, and centrifuge the samples; spectrophotometer—used to measure the amount of light absorbed or transmitted by a pigmented solution.

Experiment Instructions

Go to the home page in the PhysioEx software and click **Exercise 4: Endocrine System Physiology.** Click **Activity 2: Plasma Glucose, Insulin, and Diabetes Mellitus,** and take the online **Pre-lab Quiz** for Activity 2.

After you take the online Pre-lab Quiz, click the **Experiment** tab and begin the experiment. The experiment instructions are reprinted here for your reference. The opening screen for the experiment is shown below.

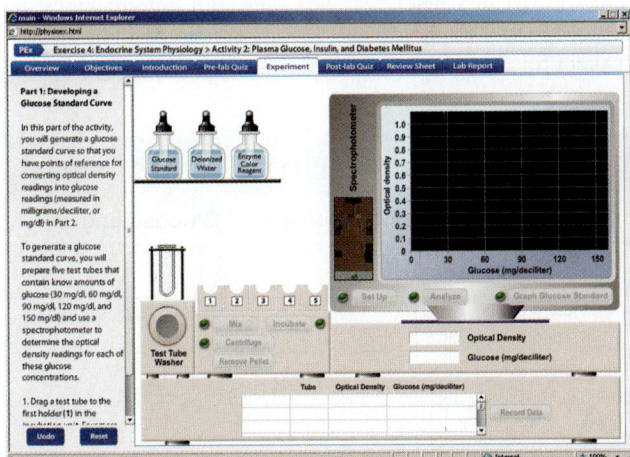

Part 1: Developing a Glucose Standard Curve

In this part of the activity, you will generate a glucose standard curve so that you have points of reference for converting optical density readings into glucose readings (measured in milligrams/deciliter, or mg/dl) in Part 2.

To generate a glucose standard curve, you will prepare five test tubes that contain known amounts of glucose (30 mg/dl, 60 mg/dl, 90 mg/dl, 120 mg/dl, and 150 mg/dl) and use a spectrophotometer to determine the optical density readings for each of these glucose concentrations.

1. Drag a test tube to the first holder (**1**) in the incubation unit. Four more test tubes will automatically be placed in the incubation unit.

2. Drag the dropper cap of the glucose standard bottle to the first tube in the incubation unit to dispense one drop of glucose standard solution into the tube. The dropper will automatically move across and dispense glucose standard to the remaining tubes. Note that each tube receives one additional drop of glucose standard (tube 2 receives 2 drops, tube 3 receives 3 drops, tube 4 receives 4 drops, and tube 5 receives 5 drops).

3. Drag the dropper cap of the deionized water bottle to the first tube in the incubation unit to dispense four drops of deionized water into the tube. The dropper will automatically move across and dispense deionized water to the remaining tubes. Note that each tube receives one less drop of deionized water (tube 2 receives 3 drops, tube 3 receives 2 drops, tube 4 receives 1 drop, and tube 5 does not receive any drops).

4. Click **Mix** to mix the contents of the tubes.

5. Click **Centrifuge** to centrifuge the contents of the tubes. After the centrifugation process, the tubes will automatically rise.

6. Click **Remove Pellet** to remove any pellets formed during the centrifugation process. Pellets can contain reagent precipitates and debris from the laboratory environment.

7. Drag the dropper cap of the enzyme color reagent bottle to the first tube in the incubation unit to dispense five drops of enzyme color reagent into each tube.

8. Click **Incubate** to incubate the contents of the tubes. The incubation unit will gently agitate the test tube rack, evenly mixing the contents of all test tubes throughout the incubation.

9. Click **Set Up** on the spectrophotometer to warm up the instrument and get it ready for your sample readings.

10. Drag tube 1 to the spectrophotometer.

11. Click **Analyze** to analyze the sample. A data point will appear on the monitor to show the optical density and the glucose concentration of the sample. These values will also appear in the optical density and glucose displays.

12. Click **Record Data** to display your results in the grid (and record your results in Chart 2.1). The tube will automatically be placed in the test tube washer.

CHART 2.1	Glucose Standard Curve Results	
Tube	**Optical density**	**Glucose (mg/dl)**
1		
2		
3		
4		
5		

13. You will now analyze the samples in the remaining tubes.

- Drag the next tube into the spectrophotometer.

- Click **Analyze** to analyze the sample. A data point will appear on the monitor to show the optical density and the glucose concentration of the sample. These values will also appear in the optical density and glucose displays.

- Click **Record Data** to display your results in the grid (and record your results in Chart 2.1). The tube will automatically be placed in the test tube washer.

   Repeat this step until you analyze all five tubes.

14. Click **Graph Glucose Standard** to generate the glucose standard curve on the monitor. You will use this graph in Part 2.

? PREDICT Question 1
How would you measure the amount of plasma glucose in a patient sample?

_____

_____

Part 2: Measure Fasting Plasma Glucose Levels

In this part of the activity, you will use the glucose standard curve you generated in Part 1 to measure the fasting plasma glucose levels from five patients to diagnose the presence or absence of diabetes mellitus. Note the addition of two reagent bottles (barium hydroxide and heparin) and blood samples from the five patients. To undergo the fasting plasma glucose (FPG) test, patients must fast for a minimum of 8 hours prior to the blood draw.

   A patient with FPG values greater than or equal to 126 mg/dl in two FPG tests is diagnosed with diabetes. FPG values between 110 and 126 mg/dl indicate impairment or borderline impairment of insulin-mediated glucose uptake by cells. FPG values less than 110 mg/dl are considered normal.

15. Drag a test tube to the first holder (**1**) in the incubation unit. Four more test tubes will automatically be placed in the incubation unit.

16. Drag the dropper cap of the first patient blood sample to the first tube in the incubation unit to dispense three drops of the sample. Three drops from each sample will automatically be dispensed into a separate tube.

17. Drag the dropper cap of the deionized water bottle to the first tube in the incubation unit to dispense five drops of deionized water into each tube.

18. Barium hydroxide dissolves and thus clears both proteins and cell membranes (so that clear glucose readings can be obtained). Drag the dropper cap of the barium hydroxide bottle to the first tube in the incubation unit to dispense five drops of barium hydroxide into each tube.

19. Drag the dropper cap of the heparin bottle to the first tube in the incubation unit to dispense a drop of heparin into each tube. Heparin prevents blood clots, which would interfere with clear glucose readings.

20. Click **Mix** to mix the contents of the tubes.

21. Click **Centrifuge** to centrifuge the contents of the tubes. After the centrifugation process, the tubes will automatically rise.

22. Click **Remove Pellet** to remove any pellets formed during the centrifugation process. Pellets can contain reagent precipitates and debris from the laboratory environment.

23. Drag the dropper cap of the enzyme color reagent bottle to the first tube in the incubation unit to dispense five drops of enzyme color reagent into each tube.

24. Click **Incubate** to incubate the contents of the tubes. The incubation unit will gently agitate the test tube rack, evenly mixing the contents of all test tubes throughout the incubation.

25. Click **Set Up** on the spectrophotometer to warm up the instrument and get it ready for your sample readings.

26. Click **Graph Glucose Standard** to display the glucose standard curve you generated in Part 1 on the monitor.

27. Drag tube 1 to the spectrophotometer.

28. Click **Analyze** to analyze the sample. A horizontal line will appear on the monitor to show the optical density of the sample. The optical density will also appear in the optical density display.

29. Drag the movable ruler (the vertical red line on the right side of the monitor) to the intersection of the horizontal yellow line (the optical density of the sample) and the glucose standard curve. Note the change in the glucose display as you move the line. The glucose concentration where the lines intersect is the fasting plasma glucose for this patient. Click **Record Data** to display your results in the grid (and record your results in Chart 2.2). The tube will automatically be placed in the test tube washer, and the monitor will be cleared (except for the glucose standard curve).

CHART 2.2	Fasting Plasma Glucose Results	
Sample	Optical density	Glucose (mg/dl)
1		
2		
3		
4		
5		

30. You will now analyze the samples in the remaining tubes.

- Drag the next tube into the spectrophotometer.

- Click **Analyze** to analyze the sample. A data point will appear on the monitor to show the optical density and the glucose concentration of the sample. These values will also appear in the optical density and glucose displays.

- Click **Record Data** to display your results in the grid. The tube will automatically be placed in the test tube washer (and record your results in Chart 2.2).

  Repeat this step until you analyze all five tubes.

After you complete the experiment, take the online **Post-lab Quiz** for Activity 2.

Activity Questions

1. How would you know if your glucose standard curve was aberrant and thus inappropriate for patient diagnostics?

_____

_____

2. What are potential sources of variability when generating a glucose standard curve?

_____

_____

3. What recommendations would you make to a patient with fasting plasma glucose levels in the impaired/borderline-impaired range who was in the impaired/borderline-impaired range for the oral glucose tolerance test?

_____

_____

4. The amount of corn syrup in the American diet has been described as alarmingly high (especially in the foods that children eat). In the context of this activity, predict the likely trends in the fasting plasma glucose levels of our children as they mature.

_____

_____

ACTIVITY 3

Hormone Replacement Therapy

OBJECTIVES

1. To understand the terms *hormone replacement therapy, follicle-stimulating hormone (FSH), estrogen, calcitonin, osteoporosis, ovariectomized,* and *T score.*
2. To understand how estrogen levels affect bone density.
3. To understand the potential benefits of hormone replacement therapy.

Introduction

Follicle-stimulating hormone (FSH) is an anterior pituitary peptide hormone that stimulates ovarian follicle growth. Developing ovarian follicles then produce and secrete a steroid hormone called **estrogen** into the plasma. Estrogen has numerous effects on the female body and homeostasis, including the stimulation of bone growth and protection against **osteoporosis** (a reduction in the quantity of bone characterized by decreased bone mass and increased susceptibility to fractures).

After menopause, the ovaries stop producing and secreting estrogen. One of the effects and potential health problems of menopause is a loss of bone density that can result in osteoporosis and bone fractures. For this reason, post-menopausal treatments to prevent osteoporosis often include hormone replacement therapy. Estrogen can be administered to increase bone density. Calcitonin (secreted by C cells in the thyroid gland) is another peptide hormone that can be administered to counteract the development of osteoporosis. Calcitonin inhibits osteoclast activity and stimulates calcium uptake and deposition in long bones.

In this activity you will use three **ovariectomized** rats that are no longer producing estrogen because their ovaries have been surgically removed. A **T score** is a quantitative measurement of the mineral content of bone, used as an indicator of the structural strength of the bone and as a screen for osteoporosis. The three rats were chosen because each has a baseline T score of 2.61, indicating osteoporosis. T scores are interpreted as follows: normal = +1 to −0.99; osteopenia (bone thinning) = −1.0 to −2.49; osteoporosis = −2.5 and below.

You will administer either estrogen therapy or calcitonin therapy to these rats, representing two types of **hormone replacement therapy.** The third rat will serve as an untreated control and receive daily injections of saline. The vertebral bone density (VBD) of each rat will be measured with dual X-ray absorptiometry (DXA) to obtain its T score after treatment.

EQUIPMENT USED The following equipment will be depicted on-screen: three ovariectomized rats (Note that if this were an actual wet lab, the ovariectomies would have been performed on the rats a month before the experiment to ensure that no residual hormones remained in the rats' systems.); saline; estrogen; calcitonin; reusable syringe—used to inject the rats; anesthesia—used to immobilize the rats for the X-ray scanning; dual X-ray absorptiometry bone-density scanner (DXA)—used to measure vertebral bone density of the rats.

Experiment Instructions

Go to the home page in the PhysioEx software and click **Exercise 4: Endocrine System Physiology.** Click **Activity 3: Hormone Replacement Therapy,** and take the online **Pre-lab Quiz** for Activity 3.

After you take the online Pre-lab Quiz, click the **Experiment** tab and begin the experiment. The experiment instructions are reprinted here for your reference. The opening screen for the experiment is shown on the following page.

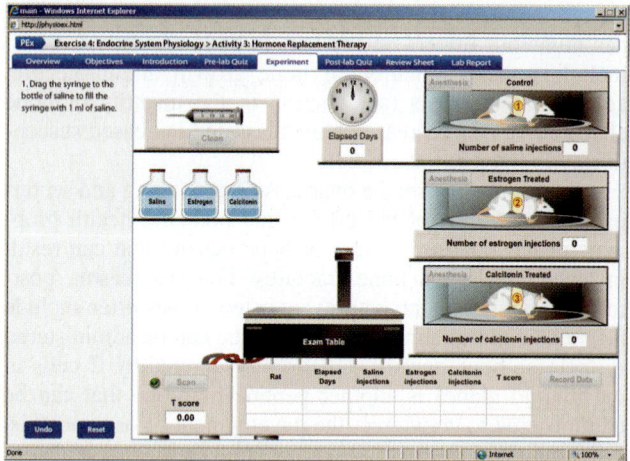

1. Drag the syringe to the bottle of saline to fill the syringe with 1 ml of saline.

2. Drag the syringe to the *control* rat, placing the tip of the needle in the rat's lower abdominal area. Injections into this area are considered *intraperitoneal* and will quickly be circulated by the abdominal blood vessels.

3. Click **Clean** beneath the syringe holder to clean the syringe of all residues.

4. Drag the syringe to the bottle of estrogen to fill the syringe with 1 ml of estrogen.

5. Drag the syringe to the *estrogen-treated* rat, placing the tip of the needle in the rat's lower abdominal area.

6. Click **Clean** beneath the syringe holder to clean the syringe of all residues.

7. Drag the syringe to the bottle of calcitonin to fill the syringe with 1 ml of calcitonin.

8. Drag the syringe to the *calcitonin-treated* rat, placing the tip of the needle in the rat's lower abdominal area.

9. Click **Clean** beneath the syringe holder to clean the syringe of all residues.

10. Click the clock face to advance one day (24 hours).

11. Each rat must receive seven injections over the course of seven days (one injection per day). The remaining injections will be automated. Click the clock face to repeat the series of injections until you have injected each of the rats seven times.

> **?** **PREDICT Question 1**
> What effect will the saline injections have on the control rat's vertebral bone density?

> **?** **PREDICT Question 2**
> What effect will the estrogen injections have on the estrogen-treated rat's vertebral bone density?

> **?** **PREDICT Question 3**
> What effect will the calcitonin injections have on the calcitonin-treated rat's vertebral bone density?

12. Click **Anesthesia** above the *control* rat's cage to immobilize the control rat with a gaseous anesthetic for X-ray scanning.

13. Drag the anesthetized rat to the exam table for X-ray scanning.

14. Click **Scan** to activate the scanner. The T score will appear in the T score display. Click **Record Data** to record your results in the grid (and record your results in Chart 3). The control rat will be automatically returned to its cage.

CHART 3	Hormone Replacement Therapy Results
Rat	**T score**

15. You will now obtain the T scores for the remaining rats. Perform these steps to obtain the T score for the *estrogen-treated* rat, then repeat these steps to obtain the T score for the *calcitonin-treated* rat.

- Click **Anesthesia** above the rat's cage to immobilize the rat with a gaseous anesthetic for X-ray scanning.

- Drag the anesthetized rat to the exam table for X-ray scanning.

- Click **Scan** to activate the scanner. The T score will appear in the T score display.

- Click **Record Data** to record your results in the grid (and record your results in Chart 3). The rat will be automatically returned to its cage.

After you complete the experiment, take the online **Post-lab Quiz** for Activity 3.

Activity Questions

1. Recently, hormone replacement therapy has been prominent in the popular press. Describe a hormone replacement therapy that you have seen in the news, and highlight its benefits, its potential risks, the reasons to continue and the reasons to discontinue its use.

2. In hormone replacement therapy, how is the hormone dose determined by the prescribing physician?

_____

_____ ▬

Measuring Cortisol and Adrenocorticotropic Hormone

OBJECTIVES

1. To understand the terms *cortisol, adrenocorticotropic hormone (ACTH), corticotropin-releasing hormone (CRH), Cushing's syndrome, iatrogenic, Cushing's disease,* and *Addison's disease.*

2. To understand how CRH controls ACTH secretion and ACTH controls cortisol secretion.

3. To understand how negative feedback mechanisms influence the levels of tropic CRH and ACTH.

4. To measure the blood levels of cortisol and ACTH in five patients and correlate these readings with symptoms and diagnoses.

5. To distinguish between Cushing's syndrome and Cushing's disease.

Introduction

Cortisol, a hormone secreted by the *adrenal cortex,* is important in the body's response to many kinds of stress. Cortisol release is stimulated by **adrenocorticotropic hormone (ACTH),** a tropic hormone released by the anterior pituitary. A *tropic* hormone stimulates the secretion of another hormone. ACTH release, in turn, is stimulated by **corticotropin-releasing hormone (CRH),** a tropic hormone from the hypothalamus. Increased levels of cortisol negatively feed back to inhibit the release of both ACTH and CRH.

Increased cortisol in the blood, or *hypercortisolism,* is referred to as **Cushing's syndrome** if the increase is caused by an adrenal gland tumor. Cushing's syndrome can also be **iatrogenic** (that is, physician induced). For example, physician-induced Cushing's syndrome can occur when glucocorticoid hormones, such as prednisone, are administered to treat rheumatoid arthritis, asthma, or lupus. Cushing's syndrome is often referred to as "steroid diabetes" because it results in hyperglycemia. In contrast, **Cushing's disease** is hypercortisolism caused by an anterior pituitary tumor. People with Cushing's disease exhibit increased levels of ACTH and cortisol.

Decreased cortisol in the blood, or *hypocortisolism,* can occur because of adrenal insufficiency. In primary adrenal insufficiency, also known as **Addison's disease,** the low cortisol is directly caused by gradual destruction of the adrenal cortex and ACTH levels are typically elevated as a compensatory effect. Secondary adrenal insufficiency also results in low levels of cortisol, usually caused by damage to the anterior pituitary. Therefore, the levels of ACTH are also low in secondary adrenal insufficiency.

As you can see, a variety of endocrine disorders can be related to both high and low levels of cortisol and ACTH. Table 4.1 summarizes these endocrine disorders.

TABLE 4.1	**Cortisol and ACTH Disorders**	
	Cortisol level	ACTH level
Cushing's syndrome (primary hypercortisolism)	High	Low
Iatrogenic Cushing's syndrome	High	Low
Cushing's disease (secondary hypercortisolism)	High	High
Addison's disease (primary adrenal insufficiency)	Low	High
Secondary adrenal insufficiency (hypopituitarism)	Low	Low

EQUIPMENT USED The following equipment will be depicted on-screen: plasma samples from five patients; HPLC (high-performance liquid chromatography) column—used to quantitatively measure the amount of cortisol and ACTH in the patient samples; HPLC detector—provides the hormone concentration in the patient sample; reusable syringe—used to inject the patient samples into the HPLC injection port; HPLC injection port—used to inject the patient samples into the HPLC column.

Experiment Instructions

Go to the home page in the PhysioEx software and click **Exercise 4: Endocrine System Physiology.** Click **Activity 4: Measuring Cortisol and Adrenocorticotropic Hormone,** and take the online **Pre-lab Quiz** for Activity 4.

After you take the online Pre-lab Quiz, click the **Experiment** tab and begin the experiment. The experiment instructions are reprinted here for your reference. The opening screen for the experiment is shown below.

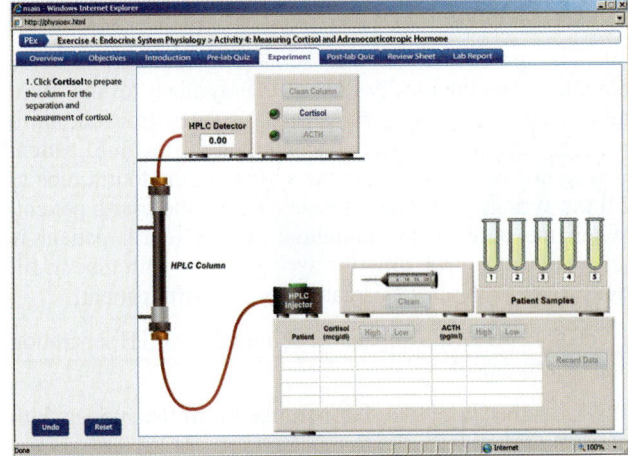

1. Click **Cortisol** to prepare the column for the separation and measurement of cortisol.

2. Drag the syringe to the first tube to fill the syringe with plasma isolated from the first patient.

3. Drag the syringe to the HPLC injector. The sample will enter the tubing and flow through the column. The cortisol concentration in the patient sample will appear in the HPLC detector display.

4. Click **Record Data** to display your results in the grid (and record your results in Chart 4).

CHART 4	Measurement of Cortisol			
Patient	Cortisol (mcg/dl)	Cortisol level	ACTH (pg/ml)	ACTH level
1				
2				
3				
4				
5				

5. Click **Clean** beneath the syringe to prepare it for the next sample. Click **Clean Column** to remove residual cortisol from the column.

6. Drag the syringe to the second tube to fill the syringe with plasma isolated from the second patient.

7. Drag the syringe to the HPLC injector. The sample will enter the tubing and flow through the column. The cortisol concentration in the patient sample will appear in the HPLC detector display.

8. Click **Record Data** to display your results in the grid (and record your results in Chart 4).

9. Click **Clean** beneath the syringe to prepare it for the next sample. Click **Clean Column** to remove residual cortisol from the column.

10. The procedure for the remaining samples will be completed automatically. Drag the syringe to the third tube to fill the syringe with plasma isolated from the third patient. When the cortisol concentration for the third patient is recorded in the grid, drag the syringe to the fourth tube to fill the syringe with plasma isolated from the fourth patient. When the cortisol concentration for the fourth patient is recorded in the grid, drag the syringe to the fifth tube to fill the syringe with plasma isolated from the fifth patient.

11. Click **ACTH** to prepare the column for ACTH separation and measurement.

12. Drag the syringe to the first tube to fill the syringe with plasma isolated from the first patient.

13. Drag the syringe to the HPLC injector. The sample will enter the tubing and flow through the column. The ACTH concentration in the patient sample will appear in the HPLC detector display.

14. Click **Record Data** to display your results in the grid.

15. Click **Clean** beneath the syringe to prepare it for the next sample. Click **Clean Column** to remove residual ACTH from the column.

16. Drag the syringe to the second tube to fill the syringe with plasma isolated from the second patient.

17. Drag the syringe to the HPLC injector. The sample will enter the tubing and flow through the column. The ACTH concentration in the patient sample will appear in the HPLC detector display.

18. Click **Record Data** to display your results in the grid (and record your results in Chart 4).

19. Click **Clean** beneath the syringe to prepare it for the next sample. Click **Clean Column** to remove residual ACTH from the column.

20. The procedure for the remaining samples will be completed automatically. Drag the syringe to the third tube to fill the syringe with plasma isolated from the third patient. When the ACTH concentration for the third patient is recorded in the grid, drag the syringe to the fourth tube to fill the syringe with plasma isolated from the fourth patient. When the ACTH concentration for the fourth patient is recorded in the grid, drag the syringe to the fifth tube to fill the syringe with plasma isolated from the fifth patient.

21. Indicate whether the cortisol and ACTH concentrations (levels) for each patient are high or low using the breakpoints shown in Table 4.2. Click the row of the patient and then click **High** or **Low** next to cortisol and ACTH.

TABLE 4.2	Abnormal Morning Cortisol and ACTH Levels	
ACTH level	High	Low
Cortisol	$\geq$23 mcg/dl	<5 mcg/dl
ACTH	$\geq$80 pg/ml	<20 pg/ml

Note: 1 mcg = 1 μg = 1 microgram

After you complete the experiment, take the online **Post-lab Quiz** for Activity 4.

Activity Questions

1. Discuss the benefits and drawbacks of giving glucocorticoids to young children that have significant allergy-induced asthma.

_____

_____

2. Explain the difference between Cushing's syndrome and Cushing's disease.

_____

_____

Endocrine System Physiology

NAME _____

LAB TIME/DATE _____

ACTIVITY 1 **Metabolism and Thyroid Hormone**

Part 1

1. Which rat had the fastest basal metabolic rate (BMR)? _____

   _____

2. Why did the metabolic rates differ between the normal rat and the surgically altered rats? How well did the results compare

   with your prediction? _____

   _____

3. If an animal has been thyroidectomized, what hormone(s) would be missing in its blood? _____

   _____

4. If an animal has been hypophysectomized, what effect would you expect to see in the hormone levels in its body? _____

   _____

   _____

Part 2

5. What was the effect of thyroxine injections on the normal rat's BMR? _____

   _____

   _____

6. What was the effect of thyroxine injections on the thyroidectomized rat's BMR? How does the BMR in this case compare

   with the normal rat's BMR? Was the dose of thyroxine in the syringe too large, too small, or just right? _____

   _____

   _____

7. What was the effect of thyroxine injections on the hypophysectomized rat's BMR? How does the BMR in this case compare

with the normal rat's BMR? Was the dose of thyroxine in the syringe too large, too small, or just right? _____

_____

_____

Part 3

8. What was the effect of thyroid-stimulating hormone (TSH) injections on the normal rat's BMR? _____

_____

_____

9. What was the effect of TSH injections on the thyroidectomized rat's BMR? How does the BMR in this case compare with

the normal rat's BMR? Why was this effect observed? _____

_____

10. What was the effect of TSH injections on the hypophysectomized rat's BMR? How does the BMR in this case compare with

the normal rat's BMR? Was the dose of TSH in the syringe too large, too small, or just right? _____

_____

_____

Part 4

11. What was the effect of propylthiouracil (PTU) injections on the normal rat's BMR? Why did this rat develop a palpable goiter?

_____

_____

12. What was the effect of PTU injections on the thyroidectomized rat's BMR? How does the BMR in this case compare with

the normal rat's BMR? Why was this effect observed? _____

_____

13. What was the effect of PTU injections on the hypophysectomized rat's BMR? How does the BMR in this case compare with

the normal rat's BMR? Why was this effect observed? _____

_____

_____

ACTIVITY 2 Plasma Glucose, Insulin, and Diabetes Mellitus

1. What is a glucose standard curve, and why did you need to obtain one for this experiment? Did you correctly predict how you would measure the amount of plasma glucose in a patient sample using the glucose standard curve? _____

   _____

   _____

2. Which patient(s) had glucose reading(s) in the diabetic range? Can you say with certainty whether each of these patients has type 1 or type 2 diabetes? Why or why not? _____

   _____

   _____

3. Describe the diagnosis for patient 3, who was also pregnant at the time of this assay. _____

   _____

   _____

4. Which patient(s) had normal glucose reading(s)? _____

   _____

5. What are some lifestyle choices these patients with normal plasma glucose readings might recommend to the borderline impaired patients? _____

   _____

   _____

ACTIVITY 3 Hormone Replacement Therapy

1. Why were ovariectomized rats used in this experiment? How does the fact that the rats are ovariectomized explain their baseline T scores? _____

   _____

   _____

2. What effect did the administration of saline injections have on the control rat? How well did the results compare with your prediction? _____

   _____

   _____

3. What effect did the administration of estrogen injections have on the estrogen-treated rat? How well did the results compare with your prediction? _____

_____

_____

4. What effect did the administration of calcitonin injections have on the calcitonin-treated rat? How well did the results compare with your prediction? _____

_____

_____

5. What are some health risks that postmenopausal women must consider when contemplating estrogen hormone replacement therapy? _____

_____

_____

ACTIVITY 4 **Measuring Cortisol and Adrenocorticotropic Hormone**

1. Which patient would most likely be diagnosed with Cushing's disease? Why? _____

_____

_____

2. Which two patients have hormone levels characteristic of Cushing's syndrome? _____

_____

_____

3. Patient 2 is being treated for rheumatoid arthritis with prednisone. How does this information change the diagnosis? _____

_____

_____

4. Which patient would most likely be diagnosed with Addison's disease? Why? _____

_____

_____

Cardiovascular Dynamics

Exercise Overview

The cardiovascular system is composed of a pump—the heart—and blood vessels that distribute blood containing oxygen and nutrients to every cell of the body. The principles governing blood flow are the same physical laws that apply to the flow of liquid through a system of pipes. For example, one very basic law in fluid mechanics is that the flow rate of a liquid through a pipe is directly proportional to the difference between the pressures at the two ends of the pipe (the **pressure gradient**) and inversely proportional to the pipe's **resistance** (a measure of the degree to which the pipe hinders, or resists, the flow of the liquid).

$$\text{Flow} = \text{pressure gradient/resistance} = \Delta P/R$$

This basic law also applies to blood flow. The "liquid" is blood, and the "pipes" are blood vessels. The pressure gradient is the difference between the pressure in arteries and the pressure in veins that results when blood is pumped into arteries. Blood flow rate is directly proportional to the pressure gradient and inversely proportional to resistance.

Blood flow is the amount of blood moving through a body area or the entire cardiovascular system in a given amount of time. Total blood flow is proportional to **cardiac output** (the amount of blood the heart is able to pump per minute). Blood flow to specific body areas can vary dramatically in a given time period. Organs differ in their requirements from moment to moment, and blood vessels have different-sized diameters in their lumen (opening) to regulate local blood flow to various areas in response to the tissues' immediate needs. Consequently, blood flow can increase to some areas and decrease to other areas at the same time.

Resistance is a measure of the degree to which the blood vessel hinders, or resists, the flow of blood. The main factors that affect resistance are (1) blood vessel *radius,* (2) blood vessel *length,* and (3) blood *viscosity.*

Radius

The smaller the blood vessel radius, the greater the resistance, because of frictional drag between the blood and the vessel walls. Contraction of smooth muscle of the blood vessel, or **vasoconstriction,** results in a decrease in the blood vessel radius. Lipid deposits can also cause the radius of an artery to decrease, preventing blood from reaching the coronary tissue, which frequently leads to a heart attack. Alternately, relaxation of smooth muscle of the blood vessel, or **vasodilation,** causes an increase in the blood vessel radius. Blood vessel radius is the single most important factor in determining blood flow resistance.

Length

The longer the vessel length, the greater the resistance—again, because of friction between the blood and vessel walls. The length of a person's blood vessels change only as a person grows. Otherwise, the length generally remains constant.

Viscosity

Viscosity is blood "thickness," determined primarily by **hematocrit**—the fractional contribution of red blood cells to total blood volume. The higher the hematocrit, the greater the viscosity. Under most physiological conditions, hematocrit does not vary much and blood viscosity remains more or less constant.

The Effect of Blood Pressure and Vessel Resistance on Blood Flow

Blood flow is directly proportional to blood pressure because the pressure difference (ΔP) between the two ends of a vessel is the driving force for blood flow. Peripheral resistance is the friction that opposes blood flow through a blood vessel. This relationship is represented in the following equation:

$$\text{Blood flow (ml/min)} = \frac{\Delta P}{\text{peripheral resistance}}$$

Three factors that contribute to peripheral resistance are blood viscosity (η), blood vessel length (L), and the radius of

the blood vessel (r). These relationships are expressed in the following equation:

$$\text{Peripheral resistance} = \frac{8L\eta}{\pi r^4}$$

From this equation you can see that the viscosity of the blood and the length of the blood vessel are directly proportional to peripheral resistance. The peripheral resistance is inversely proportional to the fourth power of the vessel radius. If you combine the two equations, you get the following result:

$$\text{Blood flow (ml/min)} = \frac{\Delta P \pi r^4}{8L\eta}$$

From this combination you can see that blood flow is directly proportional to the fourth power of vessel radius, which means that small changes in vessel radius result in dramatic changes in blood flow.

ACTIVITY 1

Studying the Effect of Blood Vessel Radius on Blood Flow Rate

OBJECTIVES

1. To understand how blood vessel radius affects blood flow rate.

2. To understand how vessel radius is changed in the body.

3. To understand how to interpret a graph of blood vessel radius versus blood flow rate.

Introduction

Controlling **blood vessel radius** (one-half of the diameter) is the principal method of controlling blood flow. Controlling blood vessel radius is accomplished by contracting or relaxing the smooth muscle within the blood vessel walls (vasoconstriction or vasodilation).

To understand why radius has such a pronounced effect on blood flow, consider the physical relationship between blood and the vessel wall. Blood in direct contact with the vessel wall flows relatively slowly because of the friction, or drag, between the blood and the lining of the vessel. In contrast, blood in the center of the vessel flows more freely because it is not rubbing against the vessel wall. The free-flowing blood in the middle of the vessel is called the **laminar flow.** Now picture a fully constricted (small-radius) vessel and a fully dilated (large-radius) vessel. In the fully constricted vessel, proportionately more blood is in contact with the vessel wall and there is less laminar flow, significantly impeding the rate of blood flow in the fully constricted vessel relative to that in the fully dilated vessel.

In this activity you will study the effect of blood vessel radius on blood flow. The experiment includes two glass beakers and a tube connecting them. Imagine that the left beaker is your heart, the tube is an artery, and the right beaker is a destination in your body, such as another organ.

EQUIPMENT USED The following equipment will be depicted on-screen: left beaker—simulates blood flowing from the heart; flow tube between the left and right beaker—simulates an artery; right beaker—simulates another organ (for example, the biceps brachii muscle).

Experiment Instructions

Go to the home page in the PhysioEx software and click **Exercise 5: Cardiovascular Dynamics.** Click **Activity 1: Studying the Effect of Blood Vessel Radius on Blood Flow Rate,** and take the online **Pre-lab Quiz** for Activity 1.

After you take the online Pre-lab Quiz, click the **Experiment** tab and begin the experiment. The experiment instructions are reprinted here for your reference. The opening screen for the experiment is shown below.

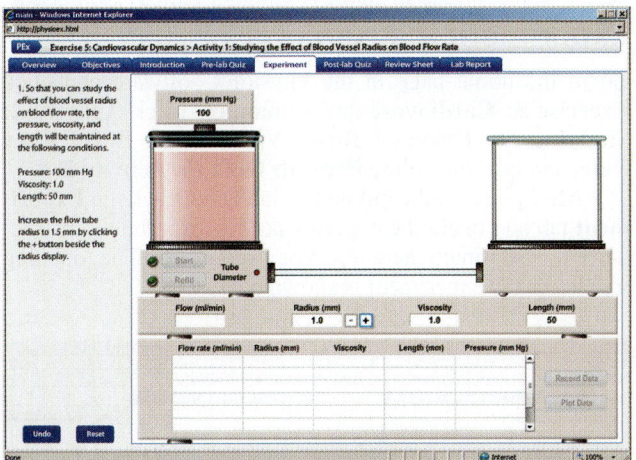

1. So that you can study the effect of blood vessel radius on blood flow rate, the pressure, viscosity, and length will be maintained at the following conditions:

Pressure: 100 mm Hg

Viscosity: 1.0

Length: 50 mm

Increase the flow tube radius to 1.5 mm by clicking the + button beside the radius display.

2. Click **Start** and then watch the fluid move into the right beaker. (Fluid moves slowly under some conditions—be patient!) Pressure propels fluid from the left beaker to the right beaker through the flow tube. The flow rate is shown in the flow rate display after the left beaker has finished draining.

3. Click **Record Data** to display your results in the grid (and record your results in Chart 1).

4. Click **Refill** to replenish the left beaker.

? PREDICT Question 1
What do you think will happen to the flow rate if the radius is increased by 0.5 mm?

CHART 1	Effect of Blood Vessel Radius on Blood Flow Rate	
Flow (ml/min)	**Radius (mm)**

5. Increase the flow tube radius to 2.0 mm by clicking the + button beside the radius display. Click **Start** and watch the fluid move into the right beaker.

6. Click **Record Data** to display your results in the grid (and record your results in Chart 1).

7. Click **Refill** to replenish the left beaker.

8. You will now observe the effect of incremental increases in flow tube radius.

- Increase the flow tube radius by 0.5 mm.

- Click **Start** and then watch the fluid move into the right beaker.

- Click **Record Data** to display your results in the grid (and record your results in Chart 1).

- Click **Refill** to replenish the left beaker.

  Repeat this step until you reach a flow tube radius of 5.0 mm.

? PREDICT Question 2
Do you think a graph plotted with radius on the X-axis and flow rate on the Y-axis will be linear (a straight line)?

9. Click **Plot Data** to view a summary of your data on a plotted grid. Radius will be displayed on the X-axis and flow rate will be displayed on the Y-axis. Click **Submit** to record your plot in the lab report.

After you complete the experiment, take the online **Post-lab Quiz** for Activity 1.

Activity Questions

1. Describe the relationship between vessel radius and blood flow rate.

2. In this activity you altered the radius of the flow tube by clicking the + and − buttons. Explain how and why the radius of blood vessels is altered in the human body.

_____

_____

3. Describe the appearance of your plot of blood vessel radius versus blood flow rate and relate the plot to the relationship between these two variables.

_____

_____

4. Describe an advantage of slower blood velocity in some areas of the body, for example, in the capillaries of our fingers.

_____

_____

Studying the Effect of Blood Viscosity on Blood Flow Rate

OBJECTIVES

1. To understand how blood viscosity affects blood flow rate.

2. To list the components in the blood that contribute to blood viscosity.

3. To explain conditions that might lead to viscosity changes in the blood.

4. To understand how to interpret a graph of viscosity versus blood flow.

Introduction

Viscosity is the thickness, or "stickiness," of a fluid. The more viscous a fluid, the more resistance to flow. Therefore, the flow rate will be slower for a more viscous solution. For example, consider how much more slowly maple syrup pours out of a container than milk does.

The viscosity of blood is due to the presence of plasma proteins and formed elements, which include white blood cells (leukocytes), red blood cells (erythrocytes), and platelets (thrombocytes). Formed elements and plasma proteins in the blood slide past one another, increasing the resistance to flow. With a viscosity of 3–5, blood is much more viscous than water (usually given a viscosity value of 1).

A body in homeostatic balance has a relatively stable blood consistency. Nevertheless, it is useful to examine the effects of blood viscosity on blood flow to predict what might occur in the human cardiovascular system when homeostatic imbalances occur. Factors such as dehydration and altered blood cell numbers do alter blood viscosity. For example,

polycythemia is a condition in which excess red blood cells are present, and certain types of anemia result in fewer red blood cells. Increasing the number of red blood cells increases blood viscosity, and decreasing the number of red blood cells decreases blood viscosity.

In this activity you will examine the effects of blood viscosity on blood flow rate. The experiment includes two glass beakers and a tube connecting them. Imagine that the left beaker is your heart, the tube is an artery, and the right beaker is a destination in your body, such as another organ.

> **EQUIPMENT USED** The following equipment will be depicted on-screen: left beaker—simulates blood flowing from the heart; flow tube between the left and right beaker—simulates an artery; right beaker—simulates another organ (for example, the biceps brachii muscle).

Experiment Instructions

Go to the home page in the PhysioEx software and click **Exercise 5: Cardiovascular Dynamics.** Click **Activity 2: Studying the Effect of Blood Viscosity on Blood Flow Rate,** and take the online **Pre-lab Quiz** for Activity 2.

After you take the online Pre-lab Quiz, click the **Experiment** tab and begin the experiment. The experiment instructions are reprinted here for your reference. The opening screen for the experiment is shown below.

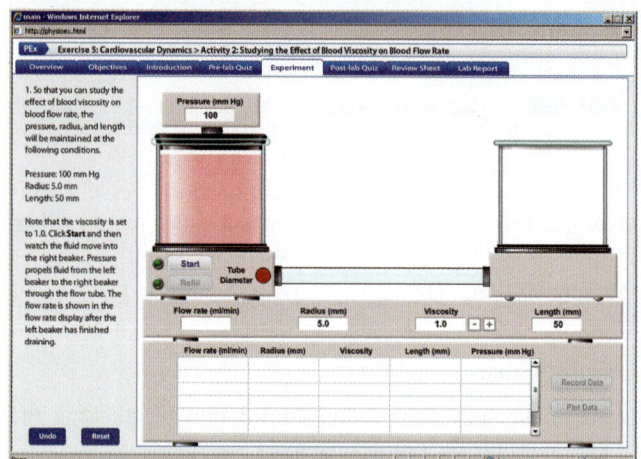

1. So that you can study the effect of blood viscosity on blood flow rate, the pressure, radius, and length will be maintained at the following conditions:

Pressure: 100 mm Hg

Radius: 5.0 mm

Length: 50 mm

Note that the viscosity is set to 1.0. Click **Start** and then watch the fluid move into the right beaker. Pressure propels fluid from the left beaker to the right beaker through the flow tube. The flow rate is shown in the flow rate display after the left beaker has finished draining.

2. Click **Record Data** to display your results in the grid (and record your results in Chart 2).

CHART 2	Effect of Blood Viscosity on Blood Flow Rate	
Flow (ml/min)	**Viscosity**	

3. Click **Refill** to replenish the left beaker.

> **? PREDICT Question 1**
> What effect do you think increasing the viscosity will have on the fluid flow rate?

4. Increase the fluid viscosity to 2.0 by clicking the + button beside the viscosity display. Click **Start** and then watch the fluid move into the right beaker.

5. Click **Record Data** to display your results in the grid (and record your results in Chart 2).

6. Click **Refill** to replenish the left beaker.

7. You will now observe the effect of incremental increases in viscosity.

- Increase the viscosity by 1.0.
- Click **Start** and then watch the fluid move into the right beaker.
- Click **Record Data** to display your results in the grid (and record your results in Chart 2).
- Click **Refill** to replenish the left beaker.

   Repeat this step until you reach a viscosity of 8.0.

8. Click **Plot Data** to view a summary of your data on a plotted grid. Viscosity will be displayed on the X-axis and flow rate will be displayed on the Y-axis. Click **Submit** to record your plot in the lab report.

After you complete the experiment, take the online **Post-lab Quiz** for Activity 2.

Activity Questions

1. Describe the effect on blood flow rate when blood viscosity was increased.

2. Explain why the relationship between viscosity and blood flow rate is inversely proportional.

3. What might happen to blood flow if you increased the number of blood cells?

ACTIVITY 3

Studying the Effect of Blood Vessel Length on Blood Flow Rate

OBJECTIVES

1. To understand how blood vessel length affects blood flow rate.
2. To explain conditions that can lead to blood vessel length changes in the body.
3. To compare the effect of blood vessel length changes with the effect of blood vessel radius changes on blood flow rate.

Introduction

Blood vessel lengths increase as we grow to maturity. The longer the vessel, the greater the resistance to blood flow through the blood vessel because there is a larger surface area in contact with the blood cells. Therefore, when blood vessel length increases, friction increases. Our blood vessel lengths stay fairly constant in adulthood, unless we gain or lose weight. If we gain weight, blood vessel lengths can increase, and if we lose weight, blood vessel lengths can decrease.

   In this activity you will study the physical relationship between blood vessel length and blood flow. Specifically, you will study how blood flow changes in blood vessels of constant radius but different lengths. The experiment includes two glass beakers and a tube connecting them. Imagine that the left beaker is your heart, the tube is an artery, and the right beaker is a destination in your body, such as another organ.

> **EQUIPMENT USED** The following equipment will be depicted on-screen: left beaker—simulates blood flowing from the heart; flow tube between the left and right beaker—simulates an artery; right beaker—simulates another organ (for example, the biceps brachii muscle).

Experiment Instructions

Go to the home page in the PhysioEx software and click **Exercise 5: Cardiovascular Dynamics.** Click **Activity 3: Studying the Effect of Blood Vessel Length on Blood Flow Rate,** and take the online **Pre-lab Quiz** for Activity 3.

   After you take the online Pre-lab Quiz, click the **Experiment** tab and begin the experiment. The experiment instructions are reprinted here for your reference. The opening screen for the experiment is shown on the following page.

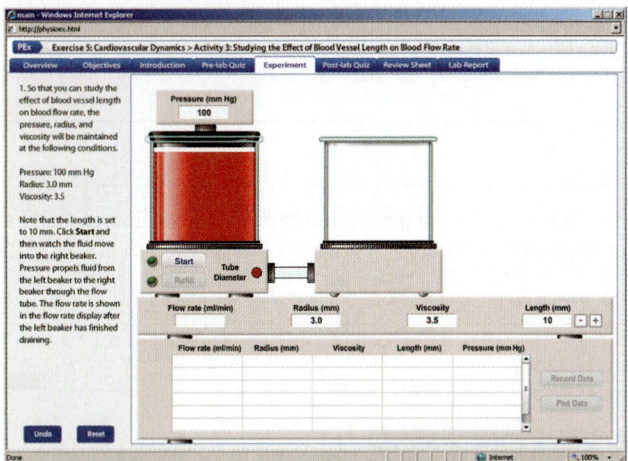

1. So that you can study the effect of blood vessel length on blood flow rate, the pressure, radius, and viscosity will be maintained at the following conditions:

Pressure: 100 mm Hg

Radius: 3.0 mm

Viscosity: 3.5

Note that the length is set to 10 mm. Click **Start** and then watch the fluid move into the right beaker. Pressure propels fluid from the left beaker to the right beaker through the flow tube. The flow rate is shown in the flow rate display after the left beaker has finished draining.

2. Click **Record Data** to display your results in the grid (and record your results in Chart 3).

CHART 3	Effect of Blood Vessel Length on Blood Flow Rate	
Flow (ml/min)	**Flow Tube length (mm)**	

3. Click **Refill** to replenish the left beaker.

? PREDICT Question 1
What effect do you think increasing the flow tube length will have on the fluid flow rate?

4. Increase the flow tube length to 15 mm by clicking the + button beside the length display. Click **Start** and then watch the fluid move into the right beaker.

5. Click **Record Data** to display your results in the grid (and record your results in Chart 3).

6. Click **Refill** to replenish the left beaker.

7. You will now observe the effect of incremental increases in flow tube length.

- Increase the flow tube length by 5 mm.
- Click **Start** and then watch the fluid move into the right beaker.
- Click **Record Data** to display your results in the grid (and record your results in Chart 3).
- Click **Refill** to replenish the left beaker.

Repeat this step until you reach a flow tube length of 40 mm.

8. Click **Plot Data** to view a summary of your data on a plotted grid. Length will be displayed on the X-axis and flow rate will be displayed on the Y-axis. Click **Submit** to record your plot in the lab report.

After you complete the experiment, take the online **Post-lab Quiz** for Activity 3.

Activity Questions

1. Is the relationship between blood vessel length and blood flow rate directly proportional or inversely proportional? Why?

_____

_____

2. Which of the following can vary in size more quickly: blood vessel diameter or blood vessel length?

_____

_____

3. Describe what happens to resistance when blood vessel length increases.

_____

_____

ACTIVITY 4

Studying the Effect of Blood Pressure on Blood Flow Rate

OBJECTIVES

1. To understand how blood pressure affects blood flow rate.

2. To understand what structure produces blood pressure in the human body.

3. To compare the plot generated for pressure versus blood flow to those generated for radius, viscosity, and length.

Introduction

The pressure difference between the two ends of a blood vessel is the driving force behind blood flow. This pressure difference is referred to as a pressure gradient. In the cardiovascular system, the force of contraction of the heart provides the initial pressure and vascular resistance contributes to the pressure gradient. If the heart changes its force of contraction, the blood vessels need to be able to respond to the change in force. Large arteries close to the heart have more elastic tissue in their tunics in order to accommodate these changes.

In this activity you will look at the effect of pressure changes on blood flow (recall from the blood flow equation that a change in blood flow is directly proportional to the pressure gradient). The experiment includes two glass beakers and a tube connecting them. Imagine that the left beaker is your heart, the tube is an artery, and the right beaker is a destination in your body, such as another organ.

> **EQUIPMENT USED** The following equipment will be depicted on-screen: left beaker—simulates blood flowing from the heart; flow tube between the left and right beaker—simulates an artery; right beaker—simulates another organ (for example, the biceps brachii muscle).

Experiment Instructions

Go to the home page in the PhysioEx software and click **Exercise 5: Cardiovascular Dynamics.** Click **Activity 4: Studying the Effect of Blood Pressure on Blood Flow Rate,** and take the online **Pre-lab Quiz** for Activity 4.

After you take the online Pre-lab Quiz, click the **Experiment** tab and begin the experiment. The experiment instructions are reprinted here for your reference. The opening screen for the experiment is shown below.

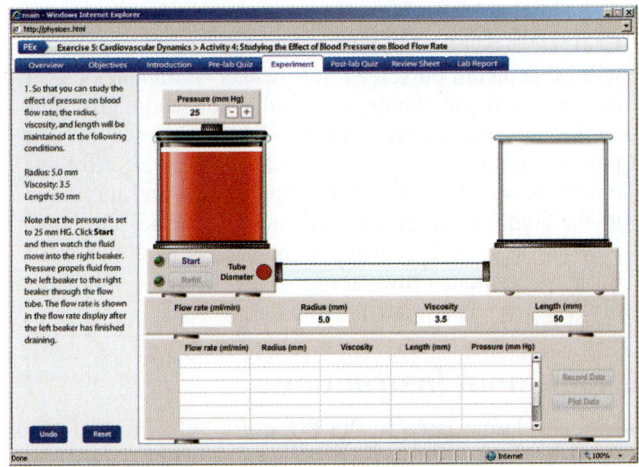

1. So that you can study the effect of pressure on blood flow rate, the radius, viscosity, and length will be maintained at the following conditions:

Radius: 5.0 mm

Viscosity: 3.5

Length: 50 mm

Note that the pressure is set to 25 mm Hg. Click **Start** and then watch the fluid move into the right beaker. Pressure pro-

pels fluid from the left beaker to the right beaker through the flow tube. The flow rate is shown in the flow rate display after the left beaker has finished draining.

2. Click **Record Data** to record your results in the grid (and record your results in Chart 4).

CHART 4	Effect of Blood Pressure on Blood Flow Rate
Flow (ml/min)	**Pressure (mm Hg)**

3. Click **Refill** to replenish the left beaker.

> **? PREDICT Question 1**
> What effect do you think increasing the pressure will have on the fluid flow rate?

4. Increase the pressure to 50 mm Hg by clicking the + button beside the pressure display. Click **Start** and then watch the fluid move into the right beaker.

5. Click **Record Data** to record your results in the grid (and record your results in Chart 4).

6. Click **Refill** to replenish the left beaker.

7. You will now observe the effect of incremental increases in pressure.

- Increase the pressure by 25 mm Hg.

- Click **Start** and then watch the fluid move into the right beaker.

- Click **Record Data** to display your results in the grid (and record your results in Chart 4).

- Click **Refill** to replenish the left beaker.

Repeat this step until you reach a pressure of 200 mm Hg.

> **? PREDICT Question 2**
> Do you think a graph plotted with pressure on the X-axis and flow rate on the Y-axis will be linear (a straight line)?

8. Click **Plot Data** to view a summary of your data on a plotted grid. Pressure will be displayed on the X-axis and flow rate will be displayed on the Y-axis. Click **Submit** to record your plot in the lab report.

After you complete the experiment, take the online **Post-lab Quiz** for Activity 4.

Activity Questions

1. How does increasing the driving pressure affect the blood flow rate?

_____

_____

2. Is the relationship between blood pressure and blood flow rate directly proportional or inversely proportional? Why?

_____

_____

3. How does the cardiovascular system increase pressure?

_____

_____

4. Although changing blood pressure can be used to alter the blood flow rate, this approach causes problems if it continues indefinitely. Explain why.

_____

_____ ▬▬

Studying the Effect of Blood Vessel Radius on Pump Activity

OBJECTIVES

1. To understand the terms *systole* and *diastole*.
2. To predict how a change in blood vessel radius will affect flow rate.
3. To predict how a change in blood vessel radius will affect heart rate.
4. To observe the compensatory mechanisms for maintaining blood pressure.

Introduction

In the human body, the heart beats approximately 70 strokes each minute. Each heartbeat consists of a filling interval, when blood moves into the chambers of the heart, and an ejection period, when blood is actively pumped into the aorta and the pulmonary trunk.

The pumping activity of the heart can be described in terms of the phases of the cardiac cycle. Heart chambers fill during **diastole** (relaxation of the heart) and pump blood out during **systole** (contraction of the heart). As you can imagine, the length of time the heart is relaxed is one factor that determines the amount of blood within the heart at the end of the filling interval. Up to a point, increasing ventricular filling time results in a corresponding increase in ventricular volume. The volume in the ventricles at the end of diastole, just before cardiac contraction, is called the **end diastolic volume,** or **EDV.** The volume ejected by a single ventricular contraction is the **stroke volume,** and the volume remaining in the ventricle after contraction is the **end systolic volume,** or **ESV.**

The human heart is a complex, four-chambered organ consisting of two individual pumps (the right and left sides). The right side of the heart pumps blood through the lungs into the left side of the heart. The left side of the heart, in turn, delivers blood to the systems of the body. Blood then returns to the right side of the heart to complete the circuit.

Recall that cardiac output (**CO**) is equal to blood flow. To determine CO, you multiply heart rate (HR) by stroke volume (SV): CO = HR × SV. From the equation for flow (flow = $\Delta P/R$), you can determine the equation for blood pressure: ΔP = flow × R. Substituting CO in the equation for flow, you get: ΔP = HR × SV × R.

Therefore, to maintain blood pressure, the cardiovascular system can alter heart rate, stroke volume, or resistance. For example, if resistance decreases, heart rate can increase to maintain the pressure difference.

In this activity you will explore the operation of a simple, one-chambered pump and apply the physical concepts in the experiment to the operation of either of the two pumps of the human heart. The stroke volume and the difference in pressure will remain constant. You will explore the effect that a change in resistance has on heart rate and the compensatory mechanisms that the cardiovascular system uses to maintain blood pressure.

> **EQUIPMENT USED** The following equipment will be depicted on-screen: left beaker—simulates blood coming from the lungs; flow tube connecting the left beaker and the pump—simulates the pulmonary veins; pump—simulates the left ventricle (the valve to the left of the pump simulates the bicuspid valve, and the valve to the right of the pump simulates the aortic semilunar valve); flow tube connecting the pump and the right beaker—simulates the aorta; right beaker—simulates blood going to the systemic circuit.

Experiment Instructions

Go to the home page in the PhysioEx software and click **Exercise 5: Cardiovascular Dynamics.** Click **Activity 5: Compensation: Studying the Effect of Blood Vessel Radius on Pump Activity** and take the online **Pre-lab Quiz** for Activity 5.

After you take the online Pre-lab Quiz, click the **Experiment** tab and begin the experiment. The experiment instructions are reprinted here for your reference. The opening screen for the experiment is shown on the following page.

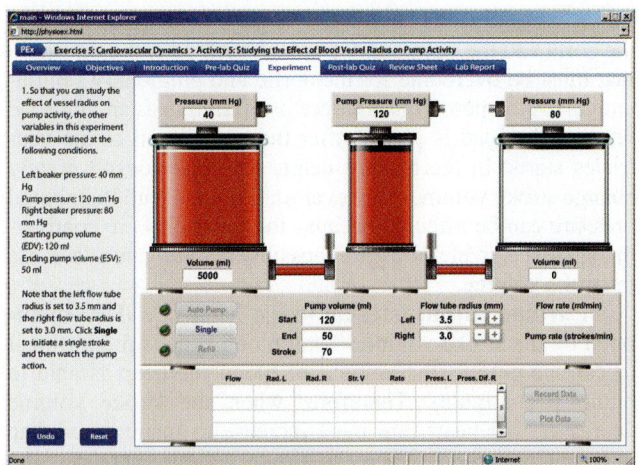

1. So that you can study the effect of vessel radius on pump activity, the other variables in this experiment will be maintained at the following conditions:

Left beaker pressure: 40 mm Hg

Pump pressure: 120 mm Hg

Right beaker pressure: 80 mm Hg

Starting pump volume (EDV): 120 ml

Ending pump volume (ESV): 50 ml

Note that the left flow tube radius is set to 3.5 mm and the right flow tube radius is set to 3.0 mm. Click **Single** to initiate a single stroke and then watch the pump action.

2. Click **Auto Pump** to initiate 10 strokes and then watch the pump action. The flow rate is shown in the flow rate display and the pump rate is shown in the pump rate display after the left beaker has finished draining.

3. Click **Record Data** to display your results in the grid (and record your results in Chart 5).

CHART 5	Effect of Blood Vessel Radius on Pump Activity	
Flow rate (ml/min)	Right radius (mm)	Pump rate (strokes/min)

4. Click **Refill** to replenish the left beaker.

? PREDICT Question 1
If you increase the flow tube radius, what will happen to the pump rate to maintain constant pressure?

5. Increase the right flow tube radius to 3.5 mm by clicking the + button beside the right flow tube radius display. Click **Auto Pump** to initiate 10 strokes and then watch the pump action.

6. Click **Record Data** to display your results in the grid (and record your results in Chart 5).

7. Click **Refill** to replenish the left beaker.

8. You will now observe the effect of incremental increases in the right flow tube radius.

- Increase the right flow tube radius by 0.5 mm.

- Click **Auto Pump** to initiate 10 strokes and then watch the pump action.

- Click **Record Data** to display your results in the grid (and record your results in Chart 5).

- Click **Refill** to replenish the left beaker.

Repeat this step until you reach a right flow tube radius of 5.0 mm.

9. Click **Plot Data** to view a summary of your data on a plotted grid. Right flow tube radius will be displayed on the X-axis and flow rate will be displayed on the Y-axis. Click **Submit** to record your plot in the lab report.

After you complete the experiment, take the online **Post-lab Quiz** for Activity 5.

Activity Questions

1. Describe the position of the pump during diastole.

_____

_____

2. Describe the position of the pump during systole.

_____

_____

3. Describe what happened to the flow rate when the blood vessel radius was increased.

_____

_____

4. Explain what happened to the resistance and the pump rate to maintain pressure when the radius was increased.

_____

_____

Studying the Effect of Stroke Volume on Pump Activity

OBJECTIVES

1. To understand the effect a change in venous return has on stroke volume.
2. To explain how stroke volume is changed in the heart.
3. To explain the Frank-Starling law of the heart.
4. To define *preload, contractility,* and *afterload.*
5. To distinguish between intrinsic and extrinsic control of contractility of the heart.
6. To explore how heart rate and stroke volume contribute to cardiac output and blood flow.

Introduction

In a normal individual, 60% of the blood contained within the heart is ejected from the heart during ventricular systole, leaving 40% of the blood behind. The blood ejected by the heart—the **stroke volume**—is the difference between the **end diastolic volume (EDV),** the volume in the ventricles at the end of diastole, just before cardiac contraction, and **end systolic volume (ESV),** the volume remaining in the ventricle after contraction. That is, stroke volume = EDV – ESV. Many factors affect stroke volume, the most important of which include *preload, contractility,* and *afterload.* We will look at these defining factors and how they relate to stroke volume.

The Frank-Starling law of the heart states that, when more than the normal volume of blood is returned to the heart by the venous system, the heart muscle will be stretched, resulting in a more forceful contraction of the ventricles. This, in turn, will cause more than normal blood to be ejected by the heart, raising the stroke volume. The degree to which the ventricles are stretched by the end diastolic volume (EDV) is referred to as the **preload.** Thus, the preload results from the amount of ventricular filling between strokes, or the magnitude of the EDV. Ventricular filling could increase when the heart rate is slow because there will be more time for the ventricles to fill. Exercise increases venous return and, therefore, EDV. Factors such as severe blood loss and dehydration decrease venous return and EDV.

The **contractility** of the heart refers to strength of the cardiac muscle contraction (usually the ventricles) and its ability to generate force. A number of extrinsic mechanisms, including the sympathetic nervous system and hormones, control the force of cardiac muscle contraction, but they are not the focus of this activity. The focus of this activity will be the intrinsic controls of contractility (those that reside entirely within the heart). When the end diastolic volume increases, the cardiac muscle fibers of the ventricles stretch and lengthen. As the length of the cardiac sarcomere increases, so does the force of contraction. Cardiac muscle, like skeletal muscle, demonstrates a **length-tension relationship.** At rest, cardiac muscles are at a less than optimum overlap length for maximum tension production in the healthy heart. Therefore, when the heart experiences an increase in stretch with an increase in venous return and, therefore, EDV, it can respond by increasing the force of contraction, yielding a corresponding increase in stroke volume.

Afterload is the back pressure generated by the blood in the aorta and the pulmonary trunk. Afterload is the threshold that must be overcome for the aortic and pulmonary semilunar valves to open. This pressure is referred to as an *after*load because the load is placed after the contraction of the ventricles starts. In the healthy heart, afterload doesn't greatly change stroke volume. However, individuals with high blood pressure can be affected because the ventricles are contracting against a greater pressure, possibly resulting in a decrease in stroke volume.

Cardiac output is equal to the heart rate (HR) multiplied by the stroke volume. Total blood flow is proportional to cardiac output (the amount of blood the heart is able to pump per minute). Therefore, when the stroke volume decreases, the heart rate must increase to maintain cardiac output. Conversely, when the stroke volume increases, the heart rate must decrease to maintain cardiac output.

Even though our simple pump in this experiment does not work exactly like the human heart, you can apply the concepts presented to basic cardiac function. In this activity you will examine how the activity of the pump is affected by changing the starting (EDV) and ending volumes (ESV).

> **EQUIPMENT USED** The following equipment will be depicted on-screen: left beaker—simulates blood coming from the lungs; flow tube connecting the left beaker and the pump—simulates the pulmonary veins; pump—simulates the left ventricle (the valve to the left of the pump simulates the bicuspid valve, and the valve to the right of the pump simulates the aortic semilunar valve); flow tube connecting the pump and the right beaker—simulates the aorta; right beaker—simulates blood going to the systemic circuit.

Experiment Instructions

Go to the home page in the PhysioEx software and click **Exercise 5: Cardiovascular Dynamics.** Click **Activity 6: Studying the Effect of Stroke Volume on Pump Activity** and take the online **Pre-lab Quiz** for Activity 6.

After you take the online Pre-lab Quiz, click the **Experiment** tab and begin the experiment. The experiment instructions are reprinted here for your reference. The opening screen for the experiment is shown below.

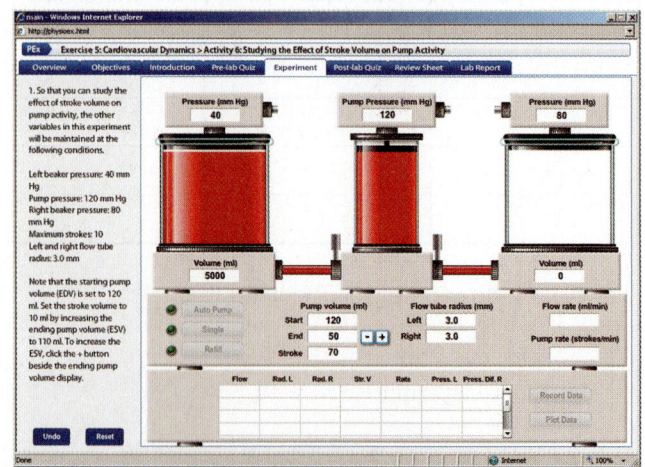

1. So that you can study the effect of stroke volume on pump activity, the other variables in this experiment will be maintained at the following conditions:

Left beaker pressure: 40 mm Hg

Pump pressure: 120 mm Hg

Right beaker pressure: 80 mm Hg

Maximum strokes: 10

Left and right flow tube radius: 3.0 mm

Note that the starting pump volume (EDV) is set to 120 ml. Set the stroke volume to 10 ml by increasing the ending pump volume (ESV) to 110 ml. To increase the ESV, click the + button beside the ending pump volume display.

2. Click **Auto Pump** to initiate 10 strokes and then watch the pump action. The flow rate is shown in the flow rate display and the pump rate is shown in the pump rate display after the left beaker has finished draining.

3. Click **Record Data** to display your results in the grid (and record your results in Chart 6).

CHART 6	Effect of Stroke Volume on Pump Activity	
Flow rate (ml/min)	Stroke volume (ml)	Pump rate (strokes/min)

4. Click **Refill** to replenish the left beaker.

> **? PREDICT Question 1**
> If the pump rate is analogous to the heart rate, what do you think will happen to the rate when you increase the stroke volume?

5. Increase the stroke volume to 20 ml by decreasing the ESV. To decrease the ending pump volume, click the − button beside the ending pump volume display.

6. Click **Auto Pump** to initiate 10 strokes and then watch the pump action.

7. Click **Record Data** to display your results in the grid (and record your results in Chart 6).

8. Click **Refill** to replenish the left beaker.

9. You will now observe the effect of incremental increases in the stroke volume.

- Increase the stroke volume by 10 ml by decreasing the ending pump volume (ESV).

- Click **Auto Pump** to initiate 10 strokes and then watch the pump action.

- Click **Record Data** to display your results in the grid (and record your results in Chart 6).

- Click **Replenish** to refill the left beaker.

  Repeat this step until you reach a stroke volume of 60 ml.

10. Increase the stroke volume by 20 ml by decreasing the ending pump volume (ESV). Click **Auto Pump** to initiate 10 strokes and then watch the pump action.

11. Click **Record Data** to display your results in the grid (and record your results in Chart 6).

12. Click **Refill** to replenish the left beaker.

13. Increase the stroke volume by 20 ml by decreasing the ending pump volume (ESV). Click **Auto Pump** to initiate 10 strokes and then watch the pump action.

14. Click **Record Data** to display your results in the grid (and record your results in Chart 6).

15. Click **Plot Data** to view a summary of your data on a plotted grid. Stroke volume will be displayed on the X-axis and flow rate will be displayed on the Y-axis. Click **Submit** to record your plot in the lab report.

After you complete the Experiment, take the online **Post-lab Quiz** for Activity 6.

Activity Questions

1. Describe how the heart responds to an increase in end diastolic volume (include the terms *preload* and *contractility* in your explanation).

_____

_____

2. Explain what happened to the pump rate when the stroke volume increased. Why?

_____

_____

3. Judging from the simulation results, explain why an athlete's resting heart rate might be lower than that of an average person.

_____

_____

ACTIVITY 7

Compensation in Pathological Cardiovascular Conditions

OBJECTIVES

1. To understand how aortic stenosis affects flow of blood through the heart.
2. To explain ways in which the cardiovascular system might compensate for changes in peripheral resistance.
3. To understand how the heart compensates for changes in afterload.
4. To explain how valves affect the flow of blood through the heart.

Introduction

If a blood vessel is compromised, your cardiovascular system can compensate to some degree. Aortic valve stenosis is a condition where there is a partial blockage of the aortic semilunar valve, increasing resistance to blood flow and left ventricular **afterload.** Therefore, the pressure that must be reached to open the aortic valve increases. The heart could compensate for a change in afterload by increasing contractility, the force of contraction. Increasing contractility will increase cardiac output by increasing stroke volume. To increase contractility, the myocardium becomes thicker. Athletes similarly improve their hearts through cardiovascular conditioning. That is, the thickness of the myocardium increases in diseased hearts with aortic valve stenosis and in athletes' hearts (though the chamber volume increases in athletes' hearts and decreases in diseased hearts).

Valves are important in the heart because they ensure that blood flows in one direction through the heart. The valves in the activity will ensure that blood moves in a single direction. Because the right flow tube represents the aorta (which is actually on the left side of the heart), decreasing the right flow tube radius simulates stenosis, or narrowing of the aortic valve.

Plaques in the arteries, known as **atherosclerosis,** can similarly cause an increase in resistance. An increase in peripheral resistance results in a decreased flow rate. Atherosclerosis is a type of **arteriosclerosis** in which the arteries have lost their elasticity. Atherosclerosis is one of the conditions that leads to heart disease.

In this activity you will test three different compensation mechanisms and predict which mechanism will make the best improvement in flow rate. The three mechanisms include (1) increasing the left flow tube radius (that is, increasing preload), (2) increasing the pump's pressure (that is, increasing contractility), and (3) decreasing the pressure in the right beaker (that is, decreasing afterload).

> **EQUIPMENT USED** The following equipment will be depicted on-screen: left beaker—simulates blood coming from the lungs; flow tube connecting the left beaker and the pump—simulates the pulmonary veins; pump—simulates the left ventricle (the valve to the left of the pump simulates the bicuspid valve, and the valve to the right of the pump simulates the aortic semilunar valve); flow tube connecting the pump and the right beaker—simulates the aorta; right beaker—simulates blood going to the systemic circuit.

Experiment Instructions

Go to the home page in the PhysioEx software and click **Exercise 5: Cardiovascular Dynamics.** Click **Activity 7: Compensation in Pathological Cardiovascular Conditions** and take the online **Pre-lab Quiz** for Activity 7.

After you take the online Pre-lab Quiz, click the **Experiment** tab and begin the experiment. The experiment instructions are reprinted here for your reference. The opening screen for the experiment is shown below.

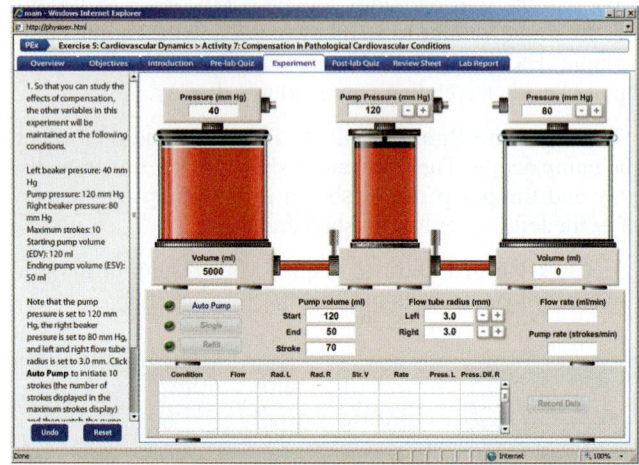

1. So that you can study the effects of compensation, the other variables in this experiment will be maintained at the following conditions:

Left beaker pressure: 40 mm Hg

Maximum strokes: 10

Starting pump volume (EDV): 120 ml

Ending pump volume (ESV): 50 ml

Note that the pump pressure is set to 120 mm Hg, the right beaker pressure is set to 80 mm Hg, and left and right flow tube radius is set to 3.0 mm. Click **Auto Pump** to initiate 10 strokes (the number of strokes displayed in the maximum strokes display) and then watch the pump action. The flow rate is shown in the flow rate display and the pump rate is shown in the pump rate display after the left beaker has finished draining.

2. Click **Record Data** to display your results in the grid (and record your results in Chart 7). This will be your baseline, or "normal," data point for flow rate.

3. Click **Refill** to replenish the left beaker.

4. Decrease the right flow tube radius to 2.5 mm by clicking the − button beside the right flow tube radius display. Click **Auto Pump** to initiate 10 strokes and then watch the pump action.

5. Click **Record Data** to display your results in the grid (and record your results in Chart 7).

6. Click **Refill** to replenish the left beaker.

CHART 7	Compensation Results						
Condition		Flow rate (ml/min)	Left radius (mm)	Right radius (mm)	Pump rate (strokes/min)	Pump pressure (mm Hg)	Right beaker pressure (mm Hg)

? **PREDICT Question 1**
You will now test three mechanisms to compensate for the decrease in flow rate caused by the decreased flow tube radius. Which mechanism do you think will have the greatest compensatory effect?

_____

7. Increase the left flow tube radius to 3.5 mm by clicking the + button beside the left flow tube radius display. Click **Auto Pump** to initiate 10 strokes and then watch the pump action.

8. Click **Record Data** to display your results in the grid (and record your results in Chart 7).

9. Click **Refill** to replenish the left beaker.

10. Increase the left flow tube radius to 4.0 mm. Click **Auto Pump** to initiate 10 strokes and then watch the pump action.

11. Click **Record Data** to display your results in the grid (and record your results in Chart 7).

12. Click **Refill** to replenish the left beaker.

13. Increase the left flow tube radius to 4.5 mm. Click **Auto Pump** to initiate 10 strokes and then watch the pump action.

14. Click **Record Data** to display your results in the grid (and record your results in Chart 7).

15. Click **Refill** to replenish the left beaker.

16. Decrease the left flow tube radius to 3.0 mm by clicking the − button beside the left flow tube radius display and increase the pump pressure to 130 mm Hg by clicking the

+ button beside the pump pressure display. Click **Auto Pump** to initiate 10 strokes and then watch the pump action.

17. Click **Record Data** to display your results in the grid (and record your results in Chart 7).

18. Click **Refill** to replenish the left beaker.

19. Increase the pump pressure to 140 mm Hg by clicking the + button beside the pump pressure display. Click **Auto Pump** to initiate 10 strokes and then watch the pump action.

20. Click **Record Data** to display your results in the grid (and record your results in Chart 7).

21. Click **Refill** to replenish the left beaker.

22. Increase the pump pressure to 150 mm Hg. Click **Auto Pump** to initiate 10 strokes and then watch the pump action.

23. Click **Record Data** to display your results in the grid (and record your results in Chart 7).

24. Click **Refill** to replenish the left beaker.

25. Decrease the pump pressure to 120 mm Hg by clicking the − button beside the pump pressure display and decrease the right (destination) beaker pressure to 70 mm Hg by clicking the − button beside the right beaker pressure display. Click **Auto Pump** to initiate 10 strokes and then watch the pump action.

26. Click **Record Data** to display your results in the grid (and record your results in Chart 7).

27. Click **Refill** to replenish the left beaker.

28. Decrease the right (destination) beaker pressure to 60 mm Hg by clicking the − button beside the right beaker pressure display. Click **Auto Pump** to initiate 10 strokes and then watch the pump action.

29. Click **Record Data** to display your results in the grid (and record your results in Chart 7).

30. Click **Refill** to replenish the left beaker.

31. Decrease the right (destination) beaker pressure to 50 mm Hg. Click **Auto Pump** to initiate 10 strokes and then watch the pump action.

32. Click **Record Data** to display your results in the grid (and record your results in Chart 7).

33. Click **Refill** to replenish the left beaker.

> **?** **PREDICT** Question 2
> What do you think will happen if the pump pressure and the beaker pressure are the same?

34. Increase the right (destination) beaker pressure to 120 mm Hg by clicking the + button beside the right beaker pressure display. Click **Auto Pump** to initiate 10 strokes and then watch the pump action.

After you complete the experiment, take the online **Post-lab Quiz** for Activity 7.

Activity Questions

1. Explain why a thicker myocardium is seen in both the athlete's heart and the diseased heart.

_____

_____

2. Describe what the term *afterload* means.

_____

_____

3. Explain which mechanism in the simulation had the greatest compensatory effect.

_____

_____

4. Describe the mechanism used in the human heart to compensate for aortic stenosis.

_____

_____

Cardiovascular Dynamics

NAME _____

LAB TIME/DATE _____

ACTIVITY 1   Studying the Effect of Blood Vessel Radius on Blood Flow Rate

1. Explain how the body establishes a pressure gradient for fluid flow. _____

_____

2. Explain the effect that the flow tube radius change had on flow rate. How well did the results compare with your prediction?

_____

_____

3. Describe the effect that radius changes have on the laminar flow of a fluid. _____

_____

4. Why do you think the plot was not linear? (Hint: Look at the relationship of the variables in the equation.) How well did the

results compare with your prediction? _____

_____

ACTIVITY 2   Studying the Effect of Blood Viscosity on Blood Flow Rate

1. Describe the components in the blood that affect viscosity. _____

_____

2. Explain the effect that the viscosity change had on flow rate. How well did the results compare with your prediction?

_____

_____

3. Describe the graph of flow versus viscosity. _____

_____

4. Discuss the effect that polycythemia would have on viscosity and on blood flow. _____

_____

_____

ACTIVITY 3 Studying the Effect of Blood Vessel Length on Blood Flow Rate

1. Which is more likely to occur, a change in blood vessel radius or a change in blood vessel length? Explain why.

    _____

    _____

2. Explain the effect that the change in blood vessel length had on flow rate. How well did the results compare with your

    prediction? _____

    _____

3. Explain why you think blood vessel radius can have a larger effect on the body than changes in blood vessel length (use the

    blood flow equation). _____

    _____

4. Describe the effect that obesity would have on blood flow and why. _____

    _____

ACTIVITY 4 Studying the Effect of Blood Pressure on Blood Flow Rate

1. Explain the effect that pressure changes had on flow rate. How well did the results compare with your prediction?

    _____

2. How does the plot differ from the plots for tube radius, viscosity, and tube length? How well did the results compare with

    your prediction? _____

    _____

3. Explain why pressure changes are not the best way to control blood flow. _____

    _____

4. Use your data to calculate the increase in flow rate in ml/min/mm Hg. _____

    _____

ACTIVITY 5 Studying the Effect of Blood Vessel Radius on Pump Activity

1. Explain the effect of increasing the right flow tube radius on the flow rate, resistance, and pump rate. _____

    _____

2. Describe what the left and right beakers in the experiment correspond to in the human heart. _____

_____

3. Briefly describe how the human heart could compensate for flow rate changes to maintain blood pressure. _____

_____

ACTIVITY 6 **Studying the Effect of Stroke Volume on Pump Activity**

1. Describe the Frank-Starling law in the heart. _____

_____

2. Explain what happened to the pump rate when you increased the stroke volume. Why do you think this occurred? How well

did the results compare with your prediction? _____

_____

3. Describe how the heart alters stroke volume. _____

_____

4. Describe the intrinsic factors that control stroke volume. _____

_____

ACTIVITY 7 **Compensation in Pathological Cardiovascular Conditions**

1. Explain how the heart could compensate for changes in peripheral resistance. _____

_____

2. Which mechanism had the greatest compensatory effect? How well did the results compare with your prediction? _____

_____

3. Explain what happened when the pump pressure and the beaker pressure were the same. How well did the results compare

with your prediction? _____

_____

4. Explain whether it would be better to adjust heart rate or blood vessel diameter to achieve blood flow changes at a local level

(for example, in just the digestive system). _____

_____

Cardiovascular Physiology

Exercise Overview

Cardiac muscle and some types of smooth muscle contract spontaneously, without any external stimuli. Skeletal muscle is unique in that it requires depolarizing signals from the nervous system to contract. The heart's ability to trigger its own contractions is called **autorhythmicity.**

If you isolate cardiac pacemaker muscle cells, place them into cell culture, and observe them under a microscope, you can see the cells contract. Autorhythmicity occurs because the plasma membrane in cardiac pacemaker muscle cells has reduced permeability to potassium ions but still allows sodium and calcium ions to slowly leak into the cells. This leakage causes the muscle cells to slowly depolarize until the action potential threshold is reached and L-type calcium channels open, allowing Ca^{2+} entry from the extracellular fluid. Shortly thereafter, contraction of the remaining cardiac muscle occurs prior

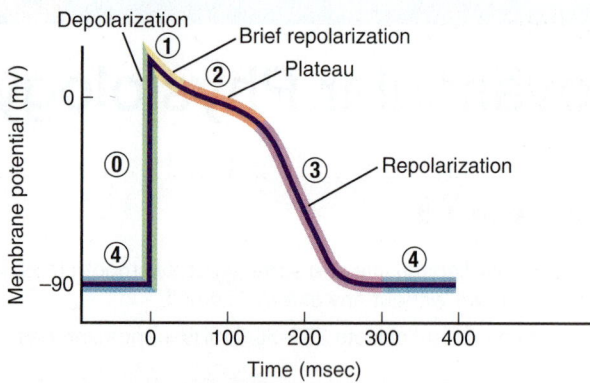

FIGURE 6.1 The cardiac action potential.

to potassium-dependent repolarization. The spontaneous depolarization-repolarization events occur in a regular and continuous manner in cardiac pacemaker muscle cells, leading to **cardiac action potentials** in the majority of cardiac muscle.

There are five main phases of membrane polarization in a cardiac action potential (view Figure 6.1).

- **Phase 0** is similar to depolarization in the neuronal action potential. Depolarization causes voltage-gated sodium channels in the cell membrane to open, increasing the flow of sodium ions into the cell and increasing the membrane potential.

- In **phase 1,** the open sodium channels begin to inactivate, decreasing the flow of sodium ions into the cell and causing the membrane potential to fall slightly. At the same time, voltage-gated potassium channels close and voltage-gated calcium channels open. The subsequent decrease in the flow of potassium out of the cell and increase in the flow of calcium into the cell act to depolarize the membrane and curb the fall in membrane potential caused by the inactivation of sodium channels.

- In **phase 2,** known as the **plateau phase,** the membrane remains in a depolarized state. Potassium channels stay closed, and long-lasting (L-type) calcium channels stay open. This plateau lasts about 0.2 seconds, or 200 milliseconds.

- In **phase 3,** the membrane potential gradually falls to more negative values when a second set of potassium channels that began opening in phases 1 and 2 allows significant amounts of potassium to flow out of the cell. The falling membrane potential causes calcium channels to close, reducing the flow of calcium into the cell and repolarizing the membrane until the resting potential is reached.

- In **phase 4,** the resting membrane potential is again established in cardiac muscle cells and is maintained until the next depolarization arrives from neighboring cardiac pacemaker cells.

The total cardiac action potential lasts 250–300 milliseconds.

Investigating the Refractory Period of Cardiac Muscle

OBJECTIVES

1. To observe the autorhythmicity of the heart.
2. To understand the phases of the cardiac action potential.
3. To induce extrasystoles and observe them on the oscilloscope tracing of contractile activity in the isolated, intact frog heart.
4. To relate the presence or absence of wave summation and tetanus in cardiac muscle to the refractory period of the cardiac action potential.

Introduction

Recall that **wave summation** occurs when a skeletal muscle is stimulated with such frequency that muscle twitches overlap and result in a stronger contraction than a single muscle twitch. When the stimulations are frequent enough, the muscle reaches a state of fused tetanus, during which the individual muscle twitches cannot be distinguished. Tetanus occurs in skeletal muscle because skeletal muscle has a relatively short **absolute refractory period** (a period during which action potentials cannot be generated no matter how strong the stimulus).

Unlike skeletal muscle, cardiac muscle has a relatively long refractory period and is thus incapable of wave summation. In fact, cardiac muscle is incapable of reacting to *any* stimulus before approximately the middle of phase 3, and will not respond to a normal cardiac stimulus before phase 4. The period of time between the beginning of the cardiac action potential and the approximate middle of phase 3 is the **absolute refractory period.** The period of time between the absolute refractory period and phase 4 is the **relative refractory period.** The total refractory period of cardiac muscle is 200–250 milliseconds—almost as long as the contraction of the cardiac muscle.

In this activity you will use external stimulation to better understand the refractory period of cardiac muscle. You will use a frog heart, which is anatomically similar to the human heart. The frog heart has two atria and a single, incompletely divided ventricle.

> **EQUIPMENT USED** The following equipment will be depicted on-screen: oscilloscope display—displays the contractile activity from the frog heart; electrical stimulator—used to apply electrical shocks to the frog heart; electrode holder—locks electrodes in place for stimulation; external stimulation electrode; apparatus for sustaining an isolated frog heart—includes 23°C Ringer's solution; frog heart.

Experiment Instructions

Go to the home page in the PhysioEx software and click **Exercise 6: Cardiovascular Physiology.** Click **Activity 1: Investigating the Refractory Period of Cardiac Muscle,** and take the online **Pre-lab Quiz** for Activity 1.

After you take the online Pre-lab Quiz, click the **Experiment** tab and begin the experiment. The experiment

instructions are reprinted here for your reference. The opening screen for the experiment is shown below.

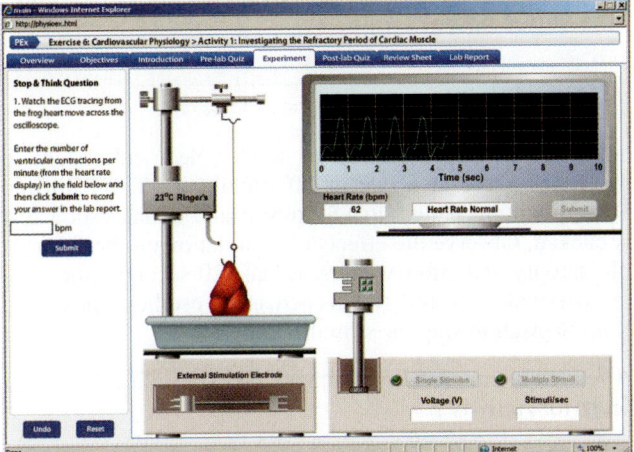

1. Watch the contractile activity from the frog heart on the oscilloscope. Enter the number of ventricular contractions per minute (from the heart rate display) in the field below and then click **Submit** to record your answer in the lab report.

_____ beats/min

2. Drag the external stimulation electrode to the electrode holder to the right of the frog heart. The electrode will touch the ventricular muscle tissue.

> **? PREDICT Question 1**
> When you increase the frequency of the stimulation, what do you think will happen to the amplitude (height) of the ventricular systole wave?

3. Deliver single shocks in succession by clicking **Single Stimulus** rapidly. You might need to practice to acquire the correct technique. You should see a "doublet," or double peak, which contains an **extrasystole,** or extra contraction of the ventricle, and then a compensatory pause, which allows the heart to get back on schedule after the extrasystole. When you see a doublet, click **Submit** to record the tracing in the lab report.

> **? PREDICT Question 2**
> If you deliver multiple stimuli (20 stimuli per second) to the heart, what do you think will happen?

4. Click **Multiple Stimuli** to deliver electrical shocks to the heart at a rate of 20 stimuli/sec. The **Multiple Stimuli** button changes to a **Stop Stimuli** button as soon as it is clicked. Observe the effects of stimulation on the contractile activity and, after a few seconds, click **Stop Stimuli** to stop the stimuli.

After you complete the experiment, take the online **Post-lab Quiz** for Activity 1.

Activity Questions

1. Describe how the frog heart and human heart differ anatomically.

_____

_____

2. What does an extrasystole correspond to? How did you induce an extrasystole?

_____

_____

3. Explain why it is important that wave summation and tetanus do not occur in the cardiac muscle.

_____

_____

ACTIVITY 2

Examining the Effect of Vagus Nerve Stimulation

OBJECTIVES

1. To understand the role that the sympathetic and parasympathetic nervous systems have on heart activity.
2. To explain the consequences of vagal stimulation and vagal escape.
3. To explain the functionality of the sinoatrial node.

Introduction

The autonomic nervous system has two branches: the **sympathetic** nervous system ("fight or flight") and **parasympathetic** nervous system ("resting and digesting"). At rest both the sympathetic and parasympathetic nervous systems are working but the parasympathetic branch is more active. The sympathetic nervous system becomes more active when needed, for example, during exercise and when confronting danger.

Both the parasympathetic and sympathetic nervous systems supply nerve impulses to the heart. Stimulation of the sympathetic nervous system increases the rate and force of contraction of the heart. Stimulation of the parasympathetic nervous system decreases the heart rate without directly changing the force of contraction. The vagus nerve (cranial nerve X) carries the signal to the heart. If stimulation of the vagus nerve (vagal stimulation) is excessive, the heart will stop beating. After a short time, the ventricles will begin to beat again. The resumption of the heartbeat is referred to as **vagal escape** and can be the result of sympathetic reflexes or initiation of a rhythm by the Purkinje fibers.

The **sinoatrial node (SA node)** is a cluster of autorhythmic cardiac cells found in the right atrial wall in the human heart. The SA node has the fastest rate of spontaneous depolarization, and, for that reason, it determines the heart rate and is therefore referred to as the heart's **"pacemaker."** In the absence of parasympathetic stimulation, sympathetic stimulation, and hormonal controls, the SA node generates action potentials 100 times per minute.

> **EQUIPMENT USED** The following equipment will be depicted on-screen: oscilloscope display—displays the contractile activity from the frog heart; electrical stimulator—used to apply electrical shocks to the frog heart; electrode holder—locks electrodes in place for stimulation; vagus nerve stimulation electrode; apparatus for sustaining an isolated, intact frog heart—includes 23°C Ringer's solution; frog heart with vagus nerve (thin, white strand to the right).

Experiment Instructions

Go to the home page in the PhysioEx software and click **Exercise 6: Cardiovascular Physiology.** Click **Activity 2: Examining the Effect of Vagus Nerve Stimulation,** and take the online **Pre-lab Quiz** for Activity 2.

After you take the online Pre-lab Quiz, click the **Experiment** tab and begin the experiment. The experiment instructions are reprinted here for your reference. The opening screen for the experiment is shown below.

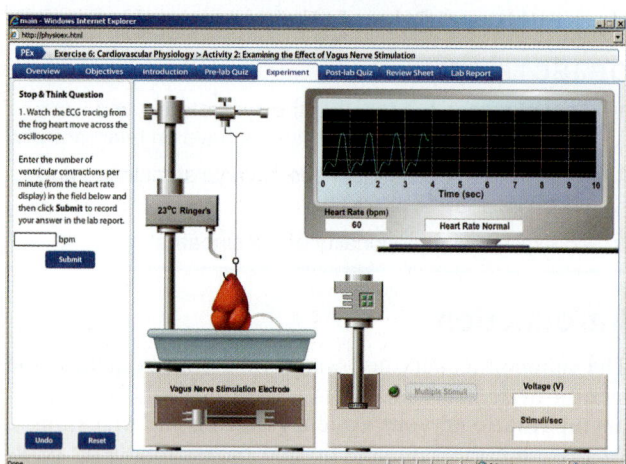

1. Watch the contractile activity from the frog heart on the oscilloscope. Enter the number of ventricular contractions per minute (from the heart rate display) in the field below and then click **Submit** to record your answer in the lab report.

_____ beats/min

2. Drag the vagus nerve stimulation electrode to the electrode holder to the right of the heart. Note that, when the electrode locks in place, the vagus nerve is draped over the electrode. Stimuli will go directly to the vagus nerve and indirectly to the heart.

3. Enter the number of ventricular contractions per minute (from the heart rate display) in the field below and then click **Submit** to record your answer in the lab report.

_____ beats/min

> **? PREDICT Question 1**
> What do you think will happen if you apply multiple stimuli to the heart by indirectly stimulating the vagus nerve?

4. Click **Multiple Stimuli** to deliver electrical shocks to the vagus nerve at a rate of 50 stimuli/sec. The **Multiple Stimuli** button changes to a **Stop Stimuli** button as soon as it is clicked. Observe the effects of stimulation on the contractile activity and, after waiting at least 20 seconds (the tracing will make two full sweeps across the oscilloscope), click **Stop Stimuli** to stop the stimuli.

After you complete the experiment, take the online **Post-lab Quiz** for Activity 2.

Activity Questions

1. Describe how stimulation of the vagus nerves affects the heart rate.

_____

_____

2. How does the sympathetic nervous system affect heart rate and the force of contraction?

_____

_____

3. Describe the mechanism of vagal escape.

_____

_____

4. What would happen to the heart rate if the vagus nerve were cut?

_____

_____

ACTIVITY 3

Examining the Effect of Temperature on Heart Rate

OBJECTIVES

1. To define the terms *hyperthermia* and *hypothermia*.
2. To contrast the terms *homeothermic* and *poikilothermic*.
3. To understand the effect that temperature has on the frog heart.
4. To understand the effect that temperature could have on the human heart.

Introduction

Humans are **homeothermic,** which means that the human body maintains an internal body temperature within the 35.8–38.2°C range even though the external temperature is changing. When the external temperature is elevated, the hypothalamus is signaled to activate heat-releasing mechanisms, such as sweating and vasodilation, to maintain the body's internal temperature. During extreme external temperature conditions, the body might not be able to maintain homeostasis and either **hyperthermia** (elevated body temperature) or **hypothermia** (low body temperature) could result. In contrast, the frog is a **poikilothermic** animal. Its internal body temperature changes depending on the temperature of its external environment because it lacks internal homeostatic regulatory mechanisms.

Ringer's solution, also known as Ringer's irrigation, consists of essential electrolytes (chloride, sodium, potassium, calcium, and magnesium) in a physiological solution and is required to keep the isolated, intact heart viable. In this activity you will explore the effect of temperature on heart rate using a Ringer's solution incubated at different temperatures.

EQUIPMENT USED The following equipment will be depicted on-screen: oscilloscope display—displays the contractile activity from the frog heart; electrical stimulator—used to apply electrical shocks to the frog heart; electrode holder—locks electrodes in place for stimulation; external stimulation electrode; apparatus for sustaining an isolated, intact frog heart—includes 5°C, 23°C, and 32°C Ringer's solution; frog heart.

Experiment Instructions

Go to the home page in the PhysioEx software and click **Exercise 6: Cardiovascular Physiology.** Click **Activity 3: Examining the Effect of Temperature on Heart Rate,** and take the online **Pre-lab Quiz** for Activity 3.

After you take the online Pre-lab Quiz, click the **Experiment** tab and begin the experiment. The experiment instructions are reprinted here for your reference. The opening screen for the experiment is shown below.

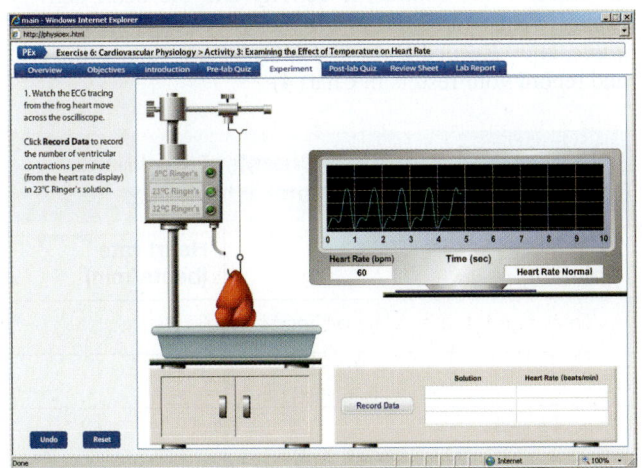

1. Watch the contractile activity from the frog heart on the oscilloscope. Click **Record Data** to record the number of ventricular contractions per minute (from the heart rate display) in 23°C Ringer's solution.

? PREDICT Question 1
What effect will decreasing the temperature of the Ringer's solution have on the heart rate of the frog?

---

2. Click **5°C Ringer's** to observe the effects of lowering the temperature.

3. When the heart activity display reads *Heart Rate Stable*, click **Record Data** to display your results in the grid (and record your results in Chart 3).

CHART 3	Effect of Temperature on Heart Rate
Solution	**Heart rate (beats/min)**

4. Click **23°C Ringer's** to bathe the heart and return it to room temperature. When the heart activity display reads *Heart Rate Normal*, you can proceed.

? PREDICT Question 2
What effect will increasing the temperature of the Ringer's solution have on the heart rate of the frog?

---

5. Click **32°C Ringer's** to observe the effects of increasing the temperature.

6. When the heart activity display reads *Heart Rate Stable*, click **Record Data** to display your results in the grid (and record your results in Chart 3).

After you complete the experiment, take the online **Post-lab Quiz** for Activity 3.

Activity Questions

1. Explain the importance of Ringer's solution (essential electrolytes in physiological saline) in maintaining the autorhythmicity of the heart.

2.   Describe the effect of lower temperature on heart rate.

_____

_____

3.   Explain the effect that fever would have on heart rate. Explain why.

_____

_____ — ▬

A C T I V I T Y   4

Examining the Effects of Chemical Modifiers on Heart Rate

OBJECTIVES

1. To distinguish between cholinergic and adrenergic modifiers of heart rate.

2. To define agonist and antagonist modifiers of heart rate.

3. To observe the effects of epinephrine, pilocarpine, atropine, and digitalis on heart rate.

4. To relate chemical modifiers of the heart rate to sympathetic and parasympathetic activation.

Introduction

Although the heart does not need external stimulation to beat, it can be affected by extrinsic controls, most notably the autonomic nervous system. The sympathetic nervous system is activated in times of "fight or flight," and sympathetic nerve fibers release **norepinephrine** (also known as **noradrenaline**) and **epinephrine** (also known as **adrenaline**) at their cardiac synapses.

Norepinephrine and epinephrine increase the frequency of action potentials by binding to β_1 adrenergic receptors embedded in the plasma membrane of **sinoatrial (SA) node** (pacemaker) cells. Working through a cAMP second-messenger mechanism, binding of the ligand opens sodium and calcium channels, increasing the rate of depolarization and shortening the period of repolarization, thus increasing the heart rate.

The parasympathetic nervous system, our "resting and digesting branch," usually dominates, and parasympathetic nerve fibers release **acetylcholine** at their cardiac synapses. Acetylcholine decreases the frequency of action potentials by binding to muscarinic cholinergic receptors embedded in the plasma membrane of the SA node cells. Acetylcholine indirectly opens potassium channels and closes calcium and sodium channels, decreasing the rate of depolarization and, thus, decreasing heart rate.

Chemical modifiers that inhibit, mimic, or enhance the action of acetylcholine in the body are labeled **cholinergic.** Chemical modifiers that inhibit, mimic, or enhance the action of epinephrine in the body are **adrenergic.** If the modifier works in the same fashion as the neurotransmitter (acetylcholine or norepinephrine), it is an **agonist.** If the modifier works in opposition to the neurotransmitter, it is an **antagonist.** In this activity you will explore the effects of pilocarpine, atropine, epinephrine, and digitalis on heart rate.

> **EQUIPMENT USED** The following equipment will be depicted on-screen: oscilloscope display— displays the contractile activity; apparatus for sustaining an isolated intact frog heart—includes 23°C Ringer's solution; pilocarpine; atropine; epinephrine; digitalis; frog heart.

Experiment Instructions

Go to the home page in the PhysioEx software and click **Exercise 6: Cardiovascular Physiology.** Click **Activity 4: Examining the Effects of Chemical Modifiers on Heart Rate,** and take the online **Pre-lab Quiz** for Activity 4.

After you take the online Pre-lab Quiz, click the **Experiment** tab and begin the experiment. The experiment instructions are reprinted here for your reference. The opening screen for the experiment is shown below.

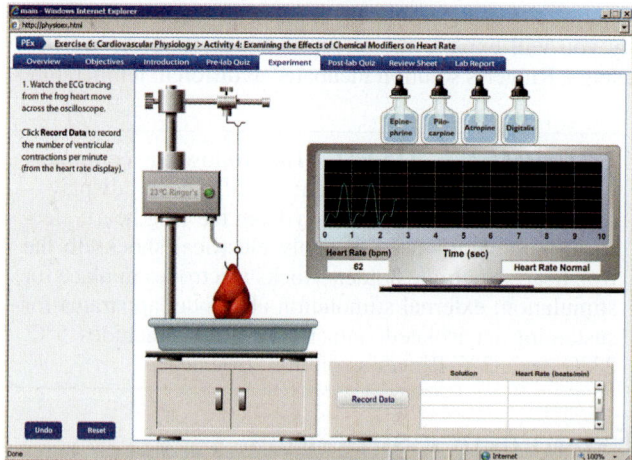

1.   Watch the contractile activity from the frog heart on the oscilloscope. Click **Record Data** to record the number of ventricular contractions per minute (from the heart rate display) and record your results in Chart 4.

2.   Drag the dropper cap of the epinephrine bottle to the frog heart to release epinephrine onto the heart.

3.   Observe the contractile activity and the heart activity display. When the heart activity display reads *Heart Rate Stable,* click **Record Data** to display your results in the grid (and record your results in Chart 4).

CHART 4	Effects of Chemical Modifiers on Heart Rate
Solution	**Heart rate (beats/min)**

4. Click **23°C Ringer's** (room temperature) to bathe the heart and flush out the epinephrine. When the heart activity display reads *Heart Rate Normal,* you can proceed.

> **? PREDICT Question 1**
> Pilocarpine is a cholinergic drug, an acetylcholine agonist. Predict the effect that pilocarpine will have on heart rate.
>
> _____

5. Drag the dropper cap of the pilocarpine bottle to the frog heart to release pilocarpine onto the heart.

6. Observe the contractile activity and the heart activity display. When the heart activity display reads *Heart Rate Stable,* click **Record Data** to display your results in the grid (and record your results in Chart 4).

7. Click **23°C Ringer's** (room temperature) to bathe the heart and flush out the pilocarpine. When the heart activity display reads *Heart Rate Normal,* you can proceed.

> **? PREDICT Question 2**
> Atropine is another cholinergic drug, an acetylcholine antagonist. Predict the effect that atropine will have on heart rate.
>
> _____

8. Drag the dropper cap of the atropine bottle to the frog heart to release atropine onto the heart.

9. Observe the contractile activity and the heart activity display. When the heart activity display reads *Heart Rate Stable,* click **Record Data** to display your results in the grid (and record your results in Chart 4).

10. Click **23°C Ringer's** (room temperature) to bathe the heart and flush out the atropine. When the heart activity display reads *Heart Rate Normal,* you can proceed.

11. Drag the dropper cap of the digitalis bottle to the frog heart to release digitalis onto the heart.

12. Observe the contractile activity and the heart activity display. When the heart activity display reads *Heart Rate Stable,* click **Record Data** to display your results in the grid (and record your results in Chart 4).

After you complete the experiment, take the online **Post-lab Quiz** for Activity 4.

Activity Questions

1. Define *agonist* and *antagonist*. Clearly distinguish between the two and give examples used in this activity.

_____

_____

2. Describe the effect of epinephrine on heart rate and force of contraction.

_____

_____

3. What is the effect of atropine on heart rate?

_____

_____

4. Describe the effect of digitalis on heart rate and force of contraction.

_____

_____

ACTIVITY 5

Examining the Effects of Various Ions on Heart Rate

OBJECTIVES

1. To understand the movement of ions that occurs during the cardiac action potential.
2. To describe the potential effect of potassium, sodium, and calcium ions on heart rate.
3. To explain how calcium channel blockers might be used pharmaceutically to treat heart patients.
4. To define the terms *inotropic* and *chronotropic.*

Introduction

In cardiac muscle cells, action potentials are caused by changes in permeability to ions due to the opening and closing of ion channels. The permeability changes that occur for the cardiac muscle cell involve potassium, sodium, and calcium ions. The concentration of potassium is greater inside the cardiac muscle cell than outside the cell. Sodium and calcium are present in larger quantities outside the cell than inside the cell.

The resting cell membrane favors the movement of potassium more than sodium or calcium. Therefore, the resting membrane potential of cardiac cells is determined mainly by the ratio of extracellular and intracellular concentrations of potassium. View Table 6.1 for a summary of the phases of the cardiac action potential and ion movement during each phase.

Calcium channel blockers are used to treat high blood pressure and abnormal heart rates. They block the movement of calcium through its channels throughout all phases of the cardiac action potentials. Consequently, because less calcium gets through, both the rate of depolarization and the force of the contraction are reduced. Modifiers that affect heart rate are **chronotropic,** and modifiers that affect the force of contraction are **inotropic.** Modifiers that lower heart rate are negative chronotropic, and modifiers that increase heart rate are positive chronotropic. The same adjectives describe inotropic modifiers. Therefore, negative inotropic drugs decrease the force of contraction of the heart and positive inotropic drugs increase the force of contraction of the heart.

TABLE 6.1	
Phase of cardiac action potential	**Ion movement**
Phase 0 (rapid depolarization)	Sodium moves in
Phase 1 (small repolarization)	Sodium movement decreases
Phase 2 (plateau)	Potassium movement out decreases Calcium moves in
Phase 3 (repolarization)	Potassium moves out Calcium movement decreases
Phase 4 (resting potential)	Potassium moves out Little sodium or calcium moves in

EQUIPMENT USED The following equipment will be depicted on-screen: oscilloscope display—displays the contractile activity from the frog heart; apparatus for sustaining frog heart—includes 23°C Ringer's solution; calcium ions; sodium ions; potassium ions; frog heart.

Experiment Instructions

Go to the home page in the PhysioEx software and click **Exercise 6: Cardiovascular Physiology.** Click **Activity 5: Examining the Effects of Various Ions on Heart Rate,** and take the online **Pre-lab Quiz** for Activity 5.

After you take the online Pre-lab Quiz, click the **Experiment** tab and begin the experiment. The experiment instructions are reprinted here for your reference. The opening screen for the experiment is shown below.

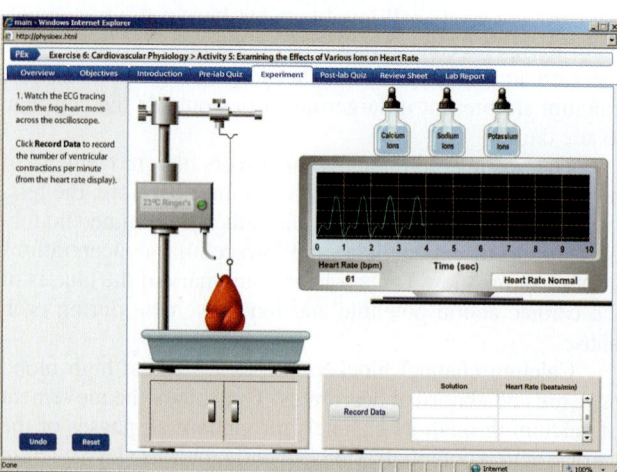

1. Watch the contractile activity from the frog heart move on the oscilloscope. Click **Record Data** to record the number of ventricular contractions per minute (from the heart rate display).

? PREDICT Question 1
Because calcium channel blockers are negative chronotropic and negative inotropic, what effect do you think increasing the concentration of calcium will have on heart rate?

2. Drag the dropper cap of the calcium ions bottle to the frog heart to release calcium ions onto the heart. Note the change in heart rate after you drop the calcium ions onto the heart.

3. When the heart activity display reads *Heart Rate Stable,* click **Record Data** to display your results in the grid (and record your results in Chart 5).

CHART 5	Effects of Various Ions on Heart Rate
Solution	**Heart rate (beats/min)**

4. **Click 23°C Ringer's** (room temperature) to bathe the heart and flush out the calcium. When the heart activity display reads *Heart Rate Normal,* you can proceed.

5. Drag the dropper cap of the sodium ions bottle to the frog heart to release sodium ions onto the heart. Note the immediate change in the heart rate and the change in heart rate over time after you drop the sodium ions onto the heart.

6. After waiting at least 20 seconds (the tracing will make two full sweeps across the oscilloscope), click **Record Data** to display your results in the grid (and record your results in Chart 5).

7. Click **23°C Ringer's** (room temperature) to bathe the heart and flush out the sodium. When the heart activity display reads *Heart Rate Normal,* you can proceed.

? PREDICT Question 2
Excess potassium outside of the cardiac cell decreases the resting potential of the plasma membrane, thus decreasing the force of contraction. What effect (if any) do you think it will *initially* have on heart rate?

8. Drag the dropper cap of the potassium ions bottle to the frog heart to release potassium ions onto the heart. Note the immediate change in heart rate and the change in heart rate over time after you drop the potassium ions onto the heart.

9. After waiting at least 20 seconds (the tracing will make two full sweeps across the oscilloscope), click **Record Data** to display your results in the grid (and record your results in Chart 5).

After you complete the experiment, take the online **Post-lab Quiz** for Activity 5.

Activity Questions

1. Define chronotropic and inotropic effects on the heart.

_____

_____

2. Describe the effect of adding calcium ions to the frog heart.

_____

_____

3. Calcium channel blockers are often used to treat high blood pressure. Explain how their effects would benefit individuals with high blood pressure.

_____

_____

4. Describe the initial effect of adding potassium ions to the frog heart.

_____

_____

NAME _____

LAB TIME/DATE _____

Cardiovascular Physiology

ACTIVITY 1 Investigating the Refractory Period of Cardiac Muscle

1. Explain why the larger waves seen on the oscilloscope represent ventricular contraction.

    _____

    _____

2. Explain why the amplitude of the wave did not change when you increased the frequency of the stimulation. (Hint: Relate

    your response to the refractory period of the cardiac action potential.) How well did the results compare with your prediction?

    _____

    _____

3. Why is it only possible to induce an extrasystole during relaxation? _____

    _____

4. Explain why wave summation and tetanus are not possible in cardiac muscle tissue. How well did the results compare with

    your prediction? _____

    _____

ACTIVITY 2 Examining the Effect of Vagus Nerve Stimulation

1. Explain the effect that extreme vagus nerve stimulation had on the heart. How well did the results compare with your prediction?

    _____

    _____

2. Explain two ways that the heart can overcome excessive vagal stimulation. _____

    _____

3. Describe how the sympathetic and parasympathetic nervous systems work together to regulate heart rate. _____

    _____

4. What do you think would happen to the heart rate if the vagus nerve was cut? _____

    _____

ACTIVITY 3 Examining the Effect of Temperature on Heart Rate

1. Explain the effect that decreasing the temperature had on the frog heart. How do you think the human heart would respond? How well did the results compare with your prediction? _____

    _____

2. Describe why Ringer's solution is required to maintain heart contractions. _____

    _____

3. Explain the effect that increasing the temperature had on the frog heart. How do you think the human heart would respond? How well did the results compare with your prediction? _____

    _____

ACTIVITY 4 Examining the Effects of Chemical Modifiers on Heart Rate

1. Describe the effect that pilocarpine had on the heart and why it had this effect. How well did the results compare with your prediction? _____

    _____

2. Atropine is an acetylcholine antagonist. Does atropine inhibit or enhance the effects of acetylcholine? Describe your results and how they correlate with how the drug works. How well did the results compare with your prediction? _____

    _____

3. Describe the benefits of administering digitalis. _____

    _____

4. Distinguish between cholinergic and adrenergic chemical modifiers. Include examples of each in your discussion. _____

    _____

ACTIVITY 5 Examining the Effects of Various Ions on Heart Rate

1. Describe the effect that increasing the calcium ions had on the heart. How well did the results compare with your prediction?

    _____

2. Describe the effect that increasing the potassium ions initially had on the heart in this activity. Relate this to the resting membrane potential of the cardiac muscle cell. How well did the results compare with your prediction? _____

    _____

3. Describe how calcium channel blockers are used to treat patients and why. _____

    _____

Respiratory System Mechanics

P R E - L A B   Q U I Z

1. Circle the correct underlined term: Blood enriched with <u>carbon dioxide</u> / <u>oxygen</u> returns to the heart from the body tissues.

2. Which of the following occurs during expiration?
   a. external intercostal muscles contract
   b. diaphragm contracts
   c. diaphragm relaxes
   d. abdominal wall muscles contract

3. Circle True or False: When the diaphragm contracts, the volume in the thoracic cavity decreases

4. The volume of air remaining in the lungs after a forceful and complete respiration is known as:
   a. tidal volume
   b. vital capacity
   c. residual volume
   d. reserve volume

5. Circle the correct underlined term: <u>Obstructive</u> / <u>Restrictive</u> lung diseases reduce both respiratory volumes and capacities.

6. Circle True or False: During an acute asthma attack, there is a significant loss of elastic recoil as the disease destroys the alveoli of the lungs.

7. Circle the correct underlined term: <u>Total lung capacity</u> / <u>Vital capacity</u> is the maximum amount of air that can be inspired and then expired after maximal effort.

8. The actual site of gas exchange in the lungs occurs in the:
   a. alveoli
   b. diaphragm
   c. external intercostals
   d. internal intercostals

9. Circle the correct underlined term: A <u>spirometer</u> / <u>inhaler</u> is a device that can measure the volume of air inspired and expired in a specific period of time.

10. Circle True or False: During heavy exercise, both the rate of breathing and the residual volume will increase to their maximum limits.

Exercise Overview

The physiological function of the respiratory system is essential to life. If problems develop in most other physiological systems, we can survive for some time without addressing them. But if a persistent problem develops within the respiratory system (or the circulatory system), death can occur in minutes.

The primary role of the respiratory system is to distribute oxygen to, and remove carbon dioxide from, *all* the cells of the body. The respiratory system works together with the circulatory system to achieve this. **Respiration** includes **ventilation,** or the movement of air into and out of the lungs (breathing), and the transport (via blood) of oxygen and carbon dioxide between the lungs and body cells (view Figure 7.1). The heart pumps deoxygenated blood to pulmonary capillaries, where gas exchange occurs between blood and **alveoli** (air sacs in the lungs), thus oxygenating the blood. The heart then pumps the oxygenated blood to body tissues, where oxygen is used for cell metabolism. At the same time, carbon dioxide (a waste product of metabolism) from body tissues diffuses into the blood. This carbon dioxide–enriched, oxygen-reduced blood then returns to the heart, completing the circuit.

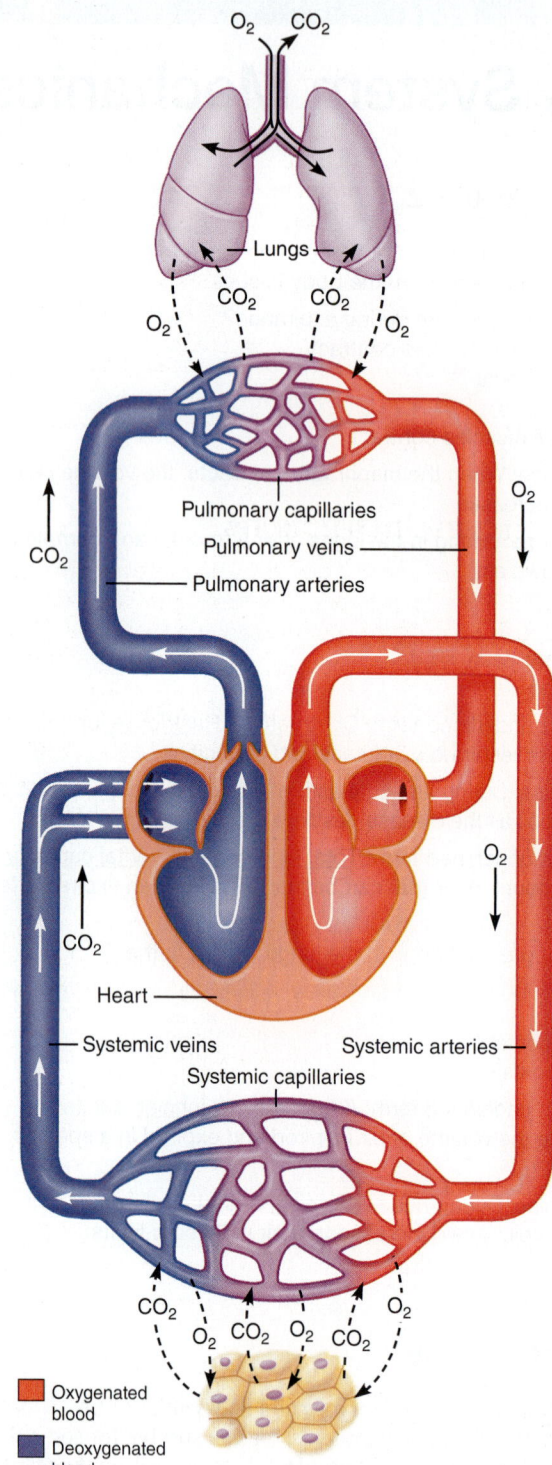

FIGURE 7.1 Relationship between external respiration and internal respiration.

Ventilation is the result of skeletal muscle contraction. When the **diaphragm**—a dome-shaped muscle that divides the thoracic and abdominal cavities—and the **external intercostal muscles** contract, the volume in the thoracic cavity increases. This increase in thoracic volume reduces the pressure in the thoracic cavity, allowing atmospheric gas to enter

the lungs (a process called **inspiration**). When the diaphragm and the external intercostals relax, the pressure in the thoracic cavity increases as the volume decreases, forcing air out of the lungs (a process called **expiration**). Inspiration is considered an *active* process because muscle contraction requires the use of ATP, whereas expiration is usually considered a *passive* process because the muscles relax, rather than contract. When a person is running, however, expiration becomes an active process, resulting from the contraction of **internal intercostal muscles** and **abdominal muscles.** In this case, both inspiration and expiration are considered *active* processes because muscle contraction is needed for both.

The amount of air that flows into and out of the lungs in 1 minute is the pulmonary **minute ventilation,** which is calculated by multiplying the **frequency of breathing** by the volume of each breath (the **tidal volume**). Ventilation must be regulated at all times to maintain oxygen in arterial blood and carbon dioxide in venous blood at their normal levels—that is, at their normal **partial pressures.** The *partial pressure* of a gas is the proportion of pressure that the gas exerts in a mixture. For example, in the atmosphere at sea level, the total pressure is 760 mm Hg. Oxygen makes up 21% of the total atmosphere and, therefore, has a partial pressure (P_{O_2}) of 160 mm Hg (760 mm Hg $\times$ 0.21).

Oxygen and carbon dioxide diffuse down their partial pressure gradients, from high partial pressures to low partial pressures. Oxygen diffuses from the alveoli of the lungs into the blood, where it can dissolve in plasma and attach to hemoglobin, and then diffuses from the blood into the tissues. Carbon dioxide (produced by the metabolic reactions of the tissues) diffuses from the tissues into the blood and then diffuses from the blood into the alveoli for export from the body.

In this exercise you will investigate the basic mechanics and regulation of the respiratory system. The concepts you will explore with a simulated lung will help you understand the operation of the human respiratory system in better detail.

A C T I V I T Y   1

Measuring Respiratory Volumes and Calculating Capacities

OBJECTIVES

1. To understand the use of the terms *ventilation, inspiration, expiration, diaphragm, external intercostals, internal intercostals, abdominal-wall muscles, expiratory reserve volume (ERV), forced vital capacity (FVC), tidal volume (TV), inspiratory reserve volume (IRV), residual volume (RV),* and *forced expiratory volume in one second (FEV $_1$).*

2. To understand the roles of skeletal muscles in the mechanics of breathing.

3. To understand the volume and pressure changes in the thoracic cavity during ventilation of the lungs.

4. To understand the effects of airway radius and, thus, resistance on airflow.

Introduction

The two phases of **ventilation,** or breathing, are (1) **inspiration,** during which air is taken into the lungs, and (2) **expiration,** during which air is expelled from the lungs. Inspiration occurs as the **external intercostal muscles** and the **diaphragm** contract. The diaphragm, normally a dome-shaped muscle, flattens as it moves inferiorly while the external intercostal muscles, situated between the ribs, lift the rib cage. These cooperative actions increase the thoracic volume. Air rushes into the lungs because this increase in thoracic volume creates a partial vacuum.

During quiet expiration, the inspiratory muscles relax, causing the diaphragm to rise superiorly and the chest wall to move inward. Thus, the **thorax** returns to its normal shape because of the elastic properties of the lung and thoracic wall. As in a deflating balloon, the pressure in the lungs rises, forcing air out of the lungs and airways. Although expiration is normally a *passive* process, **abdominal-wall muscles** and the **internal intercostal muscles** can also contract during expiration to force additional air from the lungs. Such forced expiration occurs, for example, when you exercise, blow up a balloon, cough, or sneeze.

Normal, quiet breathing moves about 500 ml (0.5 liter) of air (the **tidal volume**) into and out of the lungs with each breath, but this amount can vary due to a person's size, sex, age, physical condition, and immediate respiratory needs. In this activity you will measure the following respiratory volumes (the values given for the normal adult male and female are approximate).

Tidal volume (TV): Amount of air inspired and then expired with each breath under resting conditions (500 ml)

Inspiratory reserve volume (IRV): Amount of air that can be forcefully inspired after a normal tidal volume inspiration (male, 3100 ml; female, 1900 ml)

Expiratory reserve volume (ERV): Amount of air that can be forcefully expired after a normal tidal volume expiration (male, 1200 ml; female, 700 ml)

Residual volume (RV): Amount of air remaining in the lungs after forceful and complete expiration (male, 1200 ml; female, 1100 ml)

Respiratory capacities are calculated from the respiratory volumes. In this activity you will calculate the following respiratory capacities.

Total lung capacity (TLC): Maximum amount of air contained in lungs after a maximum inspiratory effort: $TLC = TV + IRV + ERV + RV$ (male, 6000 ml; female, 4200 ml)

Vital capacity (VC): Maximum amount of air that can be inspired and then expired with maximal effort: $VC = TV + IRV + ERV$ (male, 4800 ml; female 3100 ml)

You will also perform two pulmonary function tests in this activity.

Forced vital capacity (FVC): Amount of air that can be expelled when the subject takes the deepest possible inspiration and forcefully expires as completely and rapidly as possible

Forced expiratory volume (FEV$_1$): Measures the percentage of the vital capacity that is expired during 1 second of the FVC test (normally 75%–85% of the vital capacity)

EQUIPMENT USED The following equipment will be depicted on-screen: simulated human lungs suspended in a glass bell jar; rubber diaphragm—used to seal the jar and change the volume and, thus, pressure in the jar (As the diaphragm moves inferiorly, the volume in the bell jar increases and the pressure drops slightly, creating a partial vacuum in the bell jar. This partial vacuum causes air to be sucked into the tube at the top of the bell jar and then into the simulated lungs. As the diaphragm moves up, the decreasing volume and rising pressure within the bell jar forces air out of the lungs.); adjustable airflow tube—connects the lungs to the atmosphere; oscilloscope; three different breathing patterns: normal tidal volumes, expiratory reserve volume (ERV), and forced vital capacity (FVC).

Experiment Instructions

Go to the home page in the PhysioEx software and click **Exercise 7: Respiratory System Mechanics.** Click **Activity 1: Measuring Respiratory Volumes and Calculating Capacities,** and take the online **Pre-lab Quiz** for Activity 1.

After you take the online Pre-lab Quiz, click the **Experiment** tab and begin the experiment. The experiment instructions are reprinted here for your reference. The opening screen for the experiment is shown below.

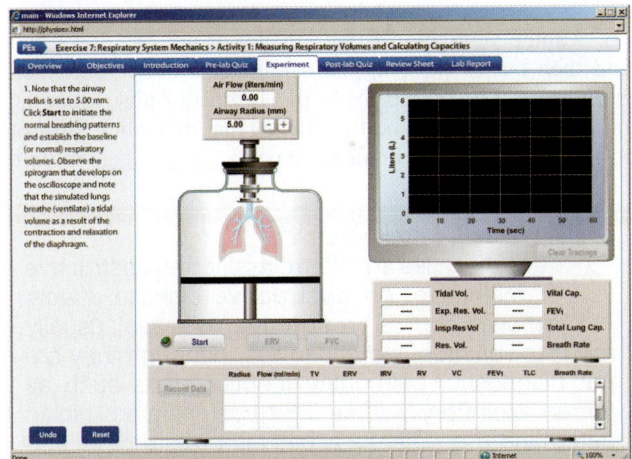

1. Note that the airway radius is set to 5.00 mm. Click **Start** to initiate the normal breathing patterns and establish the baseline (or normal) respiratory volumes. Observe the spirogram that develops on the oscilloscope and note that the simulated lungs breathe (ventilate) a tidal volume as a result of the contraction and relaxation of the diaphragm.

2. Click **Record Data** to display your results in the grid (and record your results in Chart 1).

3. Click **Clear Tracings** to clear the spirogram on the oscilloscope.

4. You will now complete the measurement of respiratory volumes and determine the respiratory capacities. First, click **Start** to initiate the normal breathing pattern. After 10 seconds, click **ERV.** Wait another 10 seconds and then click **FVC** to

CHART 1	Respiratory Volumes and Capacities								
Radius (mm)	Flow (ml/min)	TV (ml)	ERV (ml)	IRV (ml)	RV (ml)	VC (ml)	FEV$_1$ (ml)	TLC (ml)

complete the measurement of respiratory volumes. When you click ERV, the program will simulate forced expiration using the contraction of the internal intercostal muscles and abdominal-wall muscles. When you click FVC, the lungs will first inspire maximally and then expire fully to demonstrate forced vital capacity.

5. Note that, in addition to the tidal volume, the expiratory reserve volume, inspiratory reserve volume, and residual volume were measured. The vital capacity and total lung capacity were calculated from those volumes. Click **Record Data** to display your results in the grid (and record your results in Chart 1).

6. Minute ventilation is the amount of air that flows into and then out of the lungs in a minute. Minute ventilation (ml/min) = TV (ml/breath) × BPM (breaths/min). Enter the minute ventilation in the field below and then click **Submit** to record your answer in the lab report. _____ ml/min

? PREDICT Question 1

Lung diseases are often classified as obstructive or restrictive. An **obstructive** disease affects *airflow,* and a **restrictive** disease usually reduces *volumes and capacities*. Although they are not diagnostic, pulmonary function tests such as forced expiratory volume (FEV$_1$) can help a clinician determine the difference between obstructive and restrictive diseases. Specifically, an FEV$_1$ is the forced volume expired in 1 second.

In obstructive diseases such as chronic bronchitis and asthma, airway radius is decreased. Thus, FEV$_1$ will:

_____ .

7. You will now explore what effect changing the airway radius has on pulmonary function. Decrease the airway radius to 4.50 mm by clicking the – button beneath the airway radius display.

8. Click **Start** to initiate the normal breathing pattern. After 10 seconds, click **ERV.** Wait another 10 seconds and then click **FVC.** The FEV$_1$ will appear in the FEV$_1$ display beneath the oscilloscope.

9. Click **Record Data** to display your results in the grid (and record your results in Chart 1).

10. You will now gradually decrease the airway radius.

- Decrease the airway radius by 0.50 mm by clicking the – button beneath the airway radius display.

- Click **Start** to initiate the normal breathing pattern. After 10 seconds, click **ERV.** Wait another 10 seconds and then click **FVC.** The FEV$_1$ will appear in the FEV$_1$ display beneath the oscilloscope.

- Click **Record Data** to display your results in the grid (and record your results in Chart 1).

Repeat this step until you reach an airway radius of 3.00 mm.

11. A useful way to express FEV$_1$ is as a percentage of the forced vital capacity (FVC). Using the FEV$_1$ and FVC values from the data grid, calculate the FEV$_1$ (%) by dividing the FEV$_1$ volume by the FVC volume (in this case, the VC is equal to the FVC) and multiply by 100%. Enter the FEV$_1$ (%) for an airway radius of 5.0 mm in the field below and then click **Submit** to record your answer in the lab report.

FEV$_1$ (%) for an airway radius of 5.0 (mm): _____

12. Enter the FEV$_1$ (%) for an airway radius of 3.00 mm in the field below and then click **Submit** to record your answer in the lab report.

FEV$_1$ (%) for an airway radius of 3.00 (mm): _____

After you complete the experiment, take the online **Post-lab Quiz** for Activity 1.

Activity Questions

1. When you forcefully exhale your entire expiratory reserve volume, any air remaining in your lungs is called the residual volume (RV). Why is it impossible to further exhale the RV (that is, *where* is this air volume trapped, and *why* is it trapped)?

_____

_____

2. How do you measure a person's RV in a laboratory?

_____

_____

3. Draw a spirogram that depicts a person's volumes and capacities before and during a significant cough.

ACTIVITY 2

Comparative Spirometry

OBJECTIVES

1. To understand the terms *spirometry, spirogram, emphysema, asthma, inhaler, moderate exercise, heavy exercise, tidal volume (TV), expiratory reserve volume (ERV), inspiratory reserve volume (IRV), residual volume (RV), vital capacity (VC), total lung capacity (TLC), forced vital capacity (FVC),* and *forced expiratory volume in one second (FEV$_1$).*

2. To observe and compare spirograms collected from resting, healthy patients to those taken from an emphysema patient.

3. To observe and compare spirograms collected from resting, healthy patients to those taken from a patient suffering an acute asthma attack.

4. To observe and compare the spirogram collected from an asthmatic patient *while* suffering an acute asthma attack to that taken after the patient uses an inhaler for relief.

5. To observe and compare spirograms collected from volunteers engaged in moderate exercise and heavy exercise.

Introduction

In this activity you will explore the changes to normal respiratory volumes and capacities when pathophysiology develops and during aerobic exercise by recruiting volunteers to breathe into a water-filled spirometer. The spirometer is a device that measures the volume of air inspired and expired by the lungs over a specified period of time. Several lung capacities and flow rates can be calculated from this data to assess pulmonary function. With your knowledge of respiratory mechanics, you can predict, document, and explain changes to the volumes and capacities in each state.

Emphysema breathing: With emphysema, there is a significant loss of elastic recoil in the lung tissue. This loss of elastic recoil occurs as the disease destroys the walls of the alveoli. Airway resistance is also increased as the lung tissue

in general becomes more flimsy and exerts less anchoring on the surrounding airways. Thus, the lung becomes overly compliant and expands easily. Conversely, a great effort is required to expire because the lungs can no longer passively recoil and deflate. Each expiration requires a noticeable and exhausting muscular effort, and a person with emphysema expires slowly.

Acute asthma attack breathing: During an acute asthma attack, bronchiole smooth muscle spasms and, thus, the airways become constricted (that is, reduced in diameter). They also become clogged with thick mucus secretions. These changes lead to significantly increased airway resistance.

Underlying these symptoms is an airway inflammatory response brought on by triggers such as allergens (for example, dust and pollen), extreme temperature changes, and even exercise. Like with emphysema, the airways collapse and pinch closed before a forced expiration is completed. Thus, the volumes and peak flow rates are significantly reduced during an asthma attack. Unlike with emphysema, the elastic recoil is not diminished in an acute asthma attack.

When an acute asthma attack occurs, many people seek to relieve symptoms with an inhaler, which atomizes the medication and allows for direct application onto the afflicted airways. Usually, the medication includes a smooth muscle relaxant (for example, a β_2 agonist or an acetylcholine antagonist) that relieves the bronchospasms and induces bronchiole dilation. The medication can also contain an anti-inflammatory agent, such as a corticosteroid, that suppresses the inflammatory response. The use of the inhaler reduces airway resistance.

Breathing during exercise: During *moderate* aerobic exercise, the human body has an increased metabolic demand, which is met, in part, by changes in respiration. Specifically, both the rate of breathing and the tidal volume increase. These two respiratory variables do not increase by the same amount. The increase in the tidal volume is greater than the increase in the rate of breathing. During *heavy* exercise, further changes in respiration are required to meet the extreme metabolic demands of the body. In this case both the rate of breathing and the tidal volume increase to their maximum tolerable limits.

> **EQUIPMENT USED** The following equipment will be depicted on-screen: a classic water-filled spirometer with an attached rotating drum that records the analog spirogram in real time; breathing patterns from a variety of patients: unforced breathing and forced vital capacity for a "normal" patient, a patient with emphysema, and a patient with asthma (during an attack and after using an inhaler); and the breathing patterns from a patient during moderate and heavy exercise.

Experiment Instructions

Go to the home page in the PhysioEx software and click **Exercise 7: Respiratory System Mechanics.** Click **Activity 2: Comparative Spirometry,** and take the online **Pre-lab Quiz** for Activity 2.

After you take the online Pre-lab Quiz, click the **Experiment** tab and begin the experiment. The experiment

instructions are reprinted here for your reference. The opening screen for the experiment is shown below.

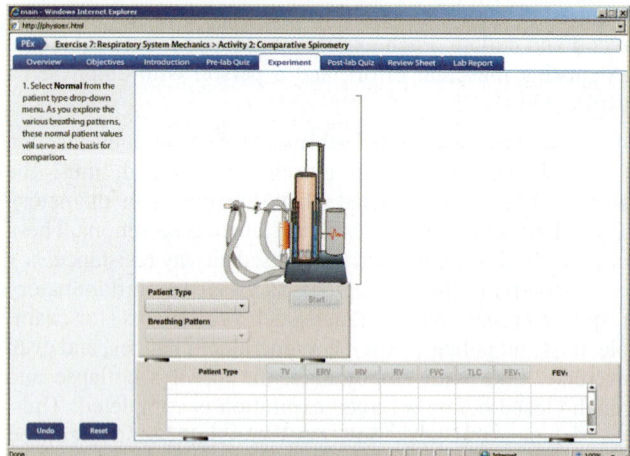

1. Select **Normal** from the patient type drop-down menu. As you explore the various breathing patterns, these normal patient values will serve as the basis for comparison.

2. Select **Unforced Breathing** from the breathing pattern drop-down menu.

3. Click **Start** to record the patient's unforced breathing pattern and watch as the drum starts turning and the spirogram develops on the paper rolling off the drum.

4. Note the volume levels (in milliliters) on the Y-axis of the spirogram. When half the screen is filled with unforced tidal volumes and the spirogram has paused, select **Forced Vital Capacity** from the breathing pattern drop-down menu.

5. Click **Start** to record the patient's forced vital capacity. The spirogram ends as the paper rolls to the right edge of the screen.

6. Click on each of the buttons in the data recorder to measure respiratory volumes and capacities. Start with tidal volume (TV) and work your way to the right. When you measure each volume or capacity, (1) a bracket appears on

the spirogram to indicate where that measurement originates and (2) the value (in milliliters) displays in the grid. After you complete all the measurements, the FEV_1 (%) ratio will automatically be calculated. The FEV_1 (%) = $(FEV_1/FVC) \times 100\%$. Record your results in Chart 2.

> **? PREDICT Question 1**
> With emphysema, there is a significant loss of elastic recoil in the lung tissue and a noticeable, exhausting muscular effort is required for each expiration. Inspiration actually becomes easier because the lung is now overly compliant. What lung values will change (from those of the normal patient) in the spirogram when the patient with emphysema is selected?
> _____

7. Select **Emphysema** from the patient type drop-down menu.

8. Select **Unforced Breathing** from the breathing pattern drop-down menu.

9. Click **Start** to record the patient's unforced breathing pattern and watch as the drum starts turning and the spirogram develops on the paper rolling off the drum.

10. Note the volume levels on the Y-axis of the spirogram. When half the screen is filled with unforced tidal volumes and the spirogram has paused, select **Forced Vital Capacity** from the breathing pattern drop-down menu.

11. Click **Start** to record the patient's forced vital capacity. The spirogram ends as the paper rolls to the right edge of the screen.

12. Click on each of the buttons in the data recorder to measure respiratory volumes and capacities. Start with tidal volume (TV) and work your way to the right. Record your results in Chart 2.

CHART 2	Spirometry Results							
Patient type	TV (ml)	ERV (ml)	IRV (ml)	RV (ml)	FVC (ml)	TLC (ml)	FEV_1 (ml)	FEV_1 (%)
Normal								
Emphysema								
Acute asthma attack								
Plus inhaler								
Moderate exercise								
Heavy exercise								

? PREDICT Question 2
During an acute asthma attack, airway resistance is significantly increased by (1) increased thick mucous secretions and (2) airway smooth muscle spasms. What lung values will change (from those of the normal patient) in the spirogram for a patient suffering an acute asthma attack?

_____

13. Select **Acute Asthma Attack** from the patient type drop-down menu.

14. Select **Unforced Breathing** from the breathing pattern drop-down menu.

15. Click **Start** to record the patient's uforced breathing pattern and watch as the drum starts turning and the spirogram develops on the paper rolling off the drum.

16. Note the volume levels on the Y-axis of the spirogram. When half the screen is filled with unforced tidal volumes and the spirogram has paused, select **Forced Vital Capacity** from the breathing pattern drop-down menu.

17. Click **Start** to record the patient's forced vital capacity. The spirogram ends as the paper rolls to the right edge of the screen.

18. Click on each of the buttons in the data recorder to measure respiratory volumes and capacities. Start with tidal volume (TV) and work your way to the right. Record your results in Chart 2.

? PREDICT Question 3
When an acute asthma attack occurs, many people seek relief from the increased airway resistance by using an inhaler. This device atomizes the medication and induces bronchiole dilation (though it can also contain an anti-inflammatory agent). What lung values will change _back_ to those of the normal patient in the spirogram after the asthma patient uses an inhaler?

_____

19. Select **Plus Inhaler** from the patient type drop-down menu.

20. Select **Unforced Breathing** from the breathing pattern drop-down menu.

21. Click **Start** to record the patient's unforced breathing pattern and watch as the drum starts turning and the spirogram develops on the paper rolling off the drum.

22. Note the volume levels on the Y-axis of the spirogram. When half the screen is filled with unforced tidal volumes and the spirogram has paused, select **Forced Vital Capacity** from the breathing pattern drop-down menu.

23. Click **Start** to record the patient's forced vital capacity. The spirogram ends as the paper rolls to the right edge of the screen.

24. Click on each of the buttons in the data recorder to measure respiratory volumes and capacities. Start with tidal volume (TV) and work your way to the right. Record your results in Chart 2.

? PREDICT Question 4
During moderate aerobic exercise, the human body will change its respiratory cycle in order to meet increased metabolic demands. During heavy exercise, further changes in respiration are required to meet the extreme metabolic demands of the body. Which lung value will change more during moderate exercise, the ERV or the IRV?

_____

25. Select **Moderate Exercise** from the patient type drop-down menu. Note that the selection of a breathing pattern is not applicable because our central nervous system automatically adjusts and maintains the depth and frequency of breathing to meet the increased metabolic demands while we exercise. We do not normally alter this pattern with conscious intervention.

26. Click **Start** to record the patient's breathing pattern and watch as the drum starts turning and the spirogram develops on the paper rolling off the drum.

27. Click on each of the buttons in the data recorder to measure respiratory volumes and capacities. Start with tidal volume (TV) and work your way to the right. _ND_ indicates this measurement or calculation was not done. Record your results in Chart 2.

28. Select **Heavy Exercise** from the patient type drop-down menu.

29. Click **Start** to record the patient's breathing pattern and watch as the drum starts turning and the spirogram develops on the paper rolling off the drum.

30. Click on each of the buttons in the data recorder to measure respiratory volumes and capacities. Start with tidal volume (TV) and work your way to the right. Record your results in Chart 2.

After you complete the experiment, take the online **Post-lab Quiz** for Activity 2.

Activity Questions

1. Why is residual volume (RV) above normal in a patient with emphysema?

_____

_____

2. Why did the asthmatic patient's inhaler medication fail to return all volumes and capacities to normal values right away?

_____

_____

3. Looking at the spirograms generated in this activity, state an easy way to determine whether a person's exercising effort is moderate or heavy.

_____

_____ ▬

Effect of Surfactant and Intrapleural Pressure on Respiration

OBJECTIVES

1. To understand the terms *surfactant, surface tension, intrapleural space, intrapleural pressure, pneumothorax,* and *atelectasis.*

2. To understand the effect of surfactant on surface tension and lung function.

3. To understand how negative intrapleural pressure prevents lung collapse.

Introduction

At any gas-liquid boundary, the molecules of the liquid are attracted more strongly to each other than they are to the gas molecules. This unequal attraction produces tension at the liquid surface, called **surface tension.** Because surface tension resists any force that tends to increase surface area of the gas-liquid boundary, it acts to decrease the size of hollow spaces, such as the alveoli, or microscopic air spaces within the lungs.

If the film lining the air spaces in the lung were pure water, it would be very difficult, if not impossible, to inflate the lungs. However, the aqueous film covering the alveolar surfaces contains **surfactant,** a detergent-like mixture of lipids and proteins that decreases surface tension by reducing the attraction of water molecules to each other. You will explore the importance of surfactant in this activity.

Between breaths, the pressure in the pleural cavity, the **intrapleural pressure,** is less than the pressure in the alveoli. Two forces cause this negative pressure condition: (1) the tendency of the lung to recoil because of its elastic properties and the surface tension of the alveolar fluid and (2) the tendency of the compressed chest wall to recoil and expand outward. These two forces pull the lungs away from the thoracic wall, creating a partial vacuum in the pleural cavity.

Because the pressure in the intrapleural space is lower than atmospheric pressure, any opening created in the pleural membranes equalizes the intrapleural pressure with atmospheric pressure by allowing air to enter the pleural cavity, a condition called **pneumothorax.** A pneumothorax can then lead to lung collapse, a condition called **atelectasis.** In this activity, the **intrapleural space** is the space between the wall of the glass bell jar and the outer wall of the lung it contains.

EQUIPMENT USED The following equipment will be depicted on-screen: simulated human lungs suspended in a glass bell jar; rubber diaphragm—used to seal the jar and change the volume and, thus, pressure in the jar (As the diaphragm moves inferiorly, the volume in the bell jar increases and the pressure drops slightly, creating a partial vacuum in the bell jar. This partial vacuum causes air to be sucked into the tube at the top of the bell jar and then into the simulated lungs. As the diaphragm moves up, the decreasing volume and rising pressure within the bell jar forces air out of the lungs.); valve—allows intrapleural pressure in the left side of the bell jar to equalize with atmospheric pressure; surfactant—amphipathic lipids (dipalmitoylphosphatidylcholine, phosphatidylglycerol, and palmitic acid) and short, synthetic peptides in a mixture that mimics the surfactant found in human lungs (surfactant molecules reduce surface tension in alveoli by adsorbing to the air-water interface, with their hydrophilic parts in the water and their hydrophobic parts facing toward the air); oscilloscope.

Experiment Instructions

Go to the home page in the PhysioEx software and click **Exercise 7: Respiratory System Mechanics.** Click **Activity 3: Effect of Surfactant and Intrapleural Pressure on Respiration,** and take the online **Pre-lab Quiz** for Activity 3.

After you take the online Pre-lab Quiz, click the **Experiment** tab and begin the experiment. The experiment instructions are reprinted here for your reference. The opening screen for the experiment is shown below.

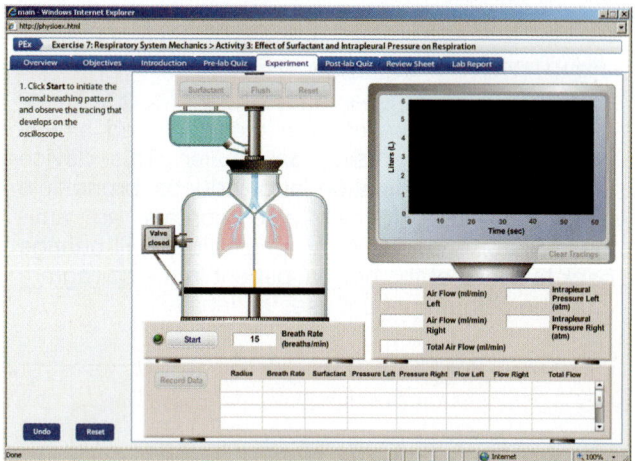

1. Click **Start** to initiate the normal breathing pattern and observe the tracing that develops on the oscilloscope.

2. Click **Record Data** to display your results in the grid (and record your results in Chart 3). This data represents breathing in the absence of surfactant.

3. Click **Surfactant** twice to dispense two aliquots of the synthetic lipids and peptides onto the interior lining of the lungs.

4. Click **Start** to initiate breathing in the presence of surfactant and observe the tracing that develops.

5. Click **Record Data** to display your results in the grid (and record your results in Chart 3).

CHART 3	Effect of Surfactant and Intrapleural Pressure on Respiration				
Surfactant	Intrapleural pressure left (atm)	Intrapleural pressure right (atm)	Airflow left (ml/min)	Airflow right (ml/min)	Total airflow (ml/min)

? PREDICT Question 1
What effect will adding more surfactant have on these lungs?

_____

6. Click **Surfactant** twice to dispense two more aliquots of the synthetic lipids and proteins onto the interior lining of the lungs.

7. Click **Start** to initiate breathing in the presence of additional surfactant and observe the tracing that develops.

8. Click **Record Data** to display your results in the grid (and record your results in Chart 3).

9. Click **Clear Tracings** to clear the tracing on the oscilloscope.

10. Click **Flush** to clear the lungs of surfactant from the previous run.

11. Click **Start** to initiate breathing and observe the tracing that develops. Notice the negative pressure condition displayed below the oscilloscope when the lungs inflate.

12. Click **Record Data** to display your results in the grid (and record your results in Chart 3).

13. Click the valve on the left side of the glass bell jar to open it.

14. Click **Start** to initiate breathing and observe the tracing that develops.

15. Click **Record Data** to display your results in the grid (and record your results in Chart 3).

? PREDICT Question 2
What will happen to the collapsed lung in the left side of the glass bell jar if you close the valve?

16. Click the valve on the left side of the glass bell jar to close it.

17. Click **Start** to initiate breathing and observe the tracing that develops.

18. Click **Record Data** to display your results in the grid (and record your results in Chart 3).

19. Click the **Reset** button above the glass bell jar to draw the air out of the intrapleural space and return the lung to its normal resting condition.

20. Click **Start** to initiate breathing and observe the tracing that develops.

21. Click **Record Data** to display your results in the grid (and record your results in Chart 3).

After you complete the experiment, take the online **Post-lab Quiz** for Activity 3.

Activity Questions

1. Why is normal quiet breathing so difficult for premature infants?

_____

_____

2. Why does a pneumothorax frequently lead to atelectasis?

_____

_____

NAME _____

LAB TIME/DATE _____

Respiratory System Mechanics

ACTIVITY 1   Measuring Respiratory Volumes and Calculating Capacities

1. What would be an example of an everyday respiratory event the ERV button simulates? _____

2. What additional skeletal muscles are utilized in an ERV activity?

   _____

3. What was the FEV_1 (%) at the initial radius of 5.00 mm?

   _____

4. What happened to the FEV_1 (%) as the radius of the airways decreased? How well did the results compare with your prediction?

   _____

5. Explain why the results from the experiment suggest that there is an obstructive, rather than a restrictive, pulmonary problem.

   _____

ACTIVITY 2   Comparative Spirometry

1. What lung values changed (from those of the normal patient) in the spirogram when the patient with emphysema was selected? Why did these values change as they did? How well did the results compare with your prediction?

   _____

   _____

2. Which of these two parameters changed more for the patient with emphysema, the FVC or the FEV_1? _____

3. What lung values changed (from those of the normal patient) in the spirogram when the patient experiencing an acute asthma attack was selected? Why did these values change as they did? How well did the results compare with your prediction?

   _____

   _____

   _____

4. How is having an acute asthma attack similar to having emphysema? How is it different? _____

   _____

5. Describe the effect that the inhaler medication had on the asthmatic patient. Did all the spirogram values return to "normal"? Why do you think some values did not return all the way to normal? How well did the results compare with your prediction?

_____

_____

_____

6. How much of an increase in FEV_1 do you think is required for it to be considered significantly improved by the medication?

_____

7. With moderate aerobic exercise, which changed more from normal breathing, the ERV or the IRV? How well did the results compare with your prediction?

_____

_____

8. Compare the breathing rates during normal breathing, moderate exercise, and heavy exercise. _____

_____

ACTIVITY 3 **Effect of Surfactant and Intrapleural Pressure on Respiration**

1. What effect does the addition of surfactant have on the airflow? How well did the results compare with your prediction?

_____

_____

2. Why does surfactant affect airflow in this manner? _____

_____

3. What effect did opening the valve have on the left lung? Why does this happen?

_____

_____

4. What effect on the collapsed lung in the left side of the glass bell jar did you observe when you closed the valve? How well did the results compare with your prediction?

_____

_____

5. What emergency medical condition does opening the left valve simulate?

_____

6. In the last part of this activity, you clicked the Reset button to draw the air out of the intrapleural space and return the lung to its normal resting condition. What emergency procedure would be used to achieve this result if these were the lungs in a living person?

_____

_____

7. What do you think would happen when the valve is opened if the two lungs were in a single large cavity rather than separate

cavities? _____

_____

Chemical and Physical Processes of Digestion

PRE-LAB QUIZ

1. Circle the correct underlined term: The <u>liver</u> / <u>stomach</u> produces pepsin which, in the presence of hydrochloric acid, digests protein.

2. _____ is a hydrolytic enzyme that breaks starch down to maltose.
   a. Pepsin
   b. Salivary amylase
   c. Pancreatic lipase
   d. Bile

3. A _____ is made in order to compare a known standard to an experimental standard.

4. Circle True or False: A positive IKI test for starch will yield a bright orange to green color.

5. An enzyme has a pocket or _____ , which the substrate or substrates must fit into temporarily in order for catalysis to occur.
   a. reagent
   b. hydrolase
   c. active site
   d. polysaccharide

6. During digestion, the chief cells of the stomach secrete _____ , which is responsible for the digestion of protein.
   a. peptidase
   b. amylase
   c. lipase
   d. pepsin

7. Circle the correct underlined term: <u>Positive</u> / <u>Negative</u> controls are used to determine whether there are any contaminating substances in the reagents used in an experiment.

8. Circle True or False: At room temperature, both fats and oils are liquid and are soluble in water.

9. A solution containing fatty acids formed by lipase activity will have a _____ than a solution without such fatty acid production.
   a. lower pH
   b. higher pH
   c. lower temperature
   d. higher temperature

Exercise Overview

The **digestive system,** also called the gastrointestinal system, consists of the digestive tract (also called the gastrointestinal tract, or GI tract) and accessory glands that secrete enzymes and fluids needed for digestion. The digestive tract includes the mouth, pharynx, esophagus, stomach, small intestine, colon, rectum, and anus. The major functions of the digestive system are to ingest food, to break food down to its simplest components, to extract nutrients from these components for absorption into the body, and to eliminate wastes.

Most of the food we consume cannot be absorbed into our bloodstream without first being broken down into smaller subunits. **Digestion** is the process of

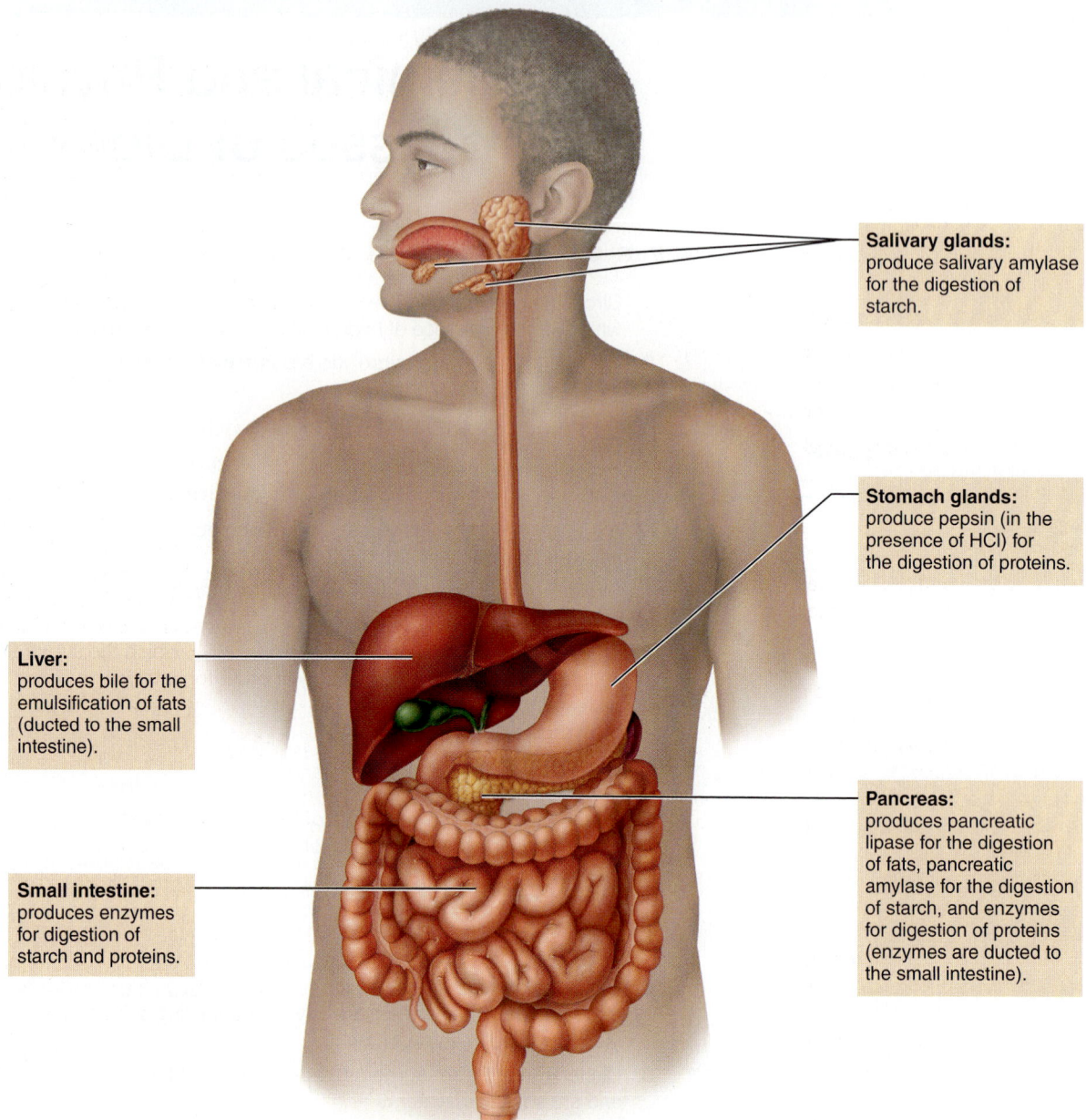

Salivary glands: produce salivary amylase for the digestion of starch.

Stomach glands: produce pepsin (in the presence of HCl) for the digestion of proteins.

Liver: produces bile for the emulsification of fats (ducted to the small intestine).

Pancreas: produces pancreatic lipase for the digestion of fats, pancreatic amylase for the digestion of starch, and enzymes for digestion of proteins (enzymes are ducted to the small intestine).

Small intestine: produces enzymes for digestion of starch and proteins.

FIGURE 8.1 **The human digestive system.** A few sites of chemical digestion and the organs that produce the enzymes of chemical digestion.

breaking down food molecules into smaller molecules with the aid of enzymes in the digestive tract. **Enzymes** are large protein molecules produced by body cells. They are biological catalysts that increase the rate of a chemical reaction without becoming part of the product. The digestive enzymes are hydrolytic enzymes, or **hydrolases,** which break down organic food molecules, or **substrates,** by adding water to the molecular bonds, thus cleaving the bonds between the subunits, or monomers.

A hydrolytic enzyme is highly specific in its action. Each enzyme hydrolyzes one substrate molecule or, at most, a small group of substrate molecules. Specific environmental

conditions are necessary for an enzyme to function optimally. For example, in extreme environments, such as high temperature, an enzyme can unravel, or denature, because of the effect that temperature has on the three-dimensional structure of the protein.

Because digestive enzymes actually function outside the body cells in the digestive tract lumen, their hydrolytic activity can also be studied in vitro in a test tube. Such in vitro studies provide a convenient laboratory environment for investigating the effect of various factors on enzymatic activity. View Figure 8.1 for an overview of chemical digestion sites in the body.

Assessing Starch Digestion by Salivary Amylase

OBJECTIVES

1. To explain how enzyme activity can be assessed with enzyme assays: the IKI assay and the Benedict's assay.
2. To define *enzyme, catalyst, hydrolase, substrate,* and *control.*
3. To understand the specificity of amylase action.
4. To name the end products of carbohydrate digestion.
5. To perform the appropriate chemical tests to determine whether digestion of a particular food has occurred.
6. To discuss the possible effect of temperature and pH on amylase activity.

Introduction

In this activity you will investigate the hydrolysis of starch to maltose by **salivary amylase,** the enzyme produced by the salivary glands and secreted into the mouth. For you to be able to detect whether or not enzymatic action has occurred, you need to be able to identify the presence of the substrate and the product to determine to what extent hydrolysis has occurred. Thus, **controls** must be prepared to provide a known standard against which comparisons can be made. With positive controls, all of the required substances are included and a positive result is expected. Sometimes negative controls are included. With negative controls, a negative result is expected. Negative results with negative controls validate the experiment. Negative controls are used to determine whether there are any contaminating substances in the reagents. So, when a positive result is produced but a negative result is expected, one or more contaminating substances are present to cause the change.

With amylase activity, starch decreases and maltose increases as digestion proceeds according to the following equation.

$$\text{Starch} + \text{water} \xrightarrow{\text{amylase}} \text{maltose}$$

Because the chemical changes that occur as starch is digested to maltose cannot be seen by the naked eye, you need to conduct an **enzyme assay,** the chemical method of detecting the presence of digested substances. You will perform two enzyme assays on each sample. The IKI assay detects the presence of starch, and the Benedict's assay tests for the presence of reducing sugars, such as glucose or maltose, which are the digestion products of starch. Normally a caramel-colored solution, IKI turns blue-black in the presence of starch. Benedict's reagent is a bright blue solution that changes to green to orange to reddish brown with increasing amounts of maltose. It is important to understand that enzyme assays only indicate the presence or absence of substances. It is up to you to analyze the results of the experiments to decide whether enzymatic hydrolysis has occurred.

EQUIPMENT USED The following equipment will be depicted on-screen: amylase—an enzyme that digests starch; starch—a complex carbohydrate substrate; maltose—a disaccharide substrate; pH buffers—solutions used to adjust the pH of the solution; deionized water—used to adjust the volume so that it is the same for each reaction; test tubes—used as reaction vessels for the various tests; incubators—used for temperature treatments (boiling, freezing, and 37°C incubation); IKI—found in the assay cabinet; used to detect the presence of starch; Benedict's reagent—found in the assay cabinet; used to detect the products of starch digestion (this includes the reducing sugars maltose and glucose).

Experiment Instructions

Go to the home page in the PhysioEx software and click **Exercise 8: Chemical and Physical Processes of Digestion.** Click **Activity 1: Assessing Starch Digestion by Salivary Amylase,** and take the online **Pre-lab Quiz** for Activity 1.

After you take the online Pre-lab Quiz, click the **Experiment** tab and begin the experiment. The experiment instructions are reprinted here for your reference. The opening screen for the experiment is shown below.

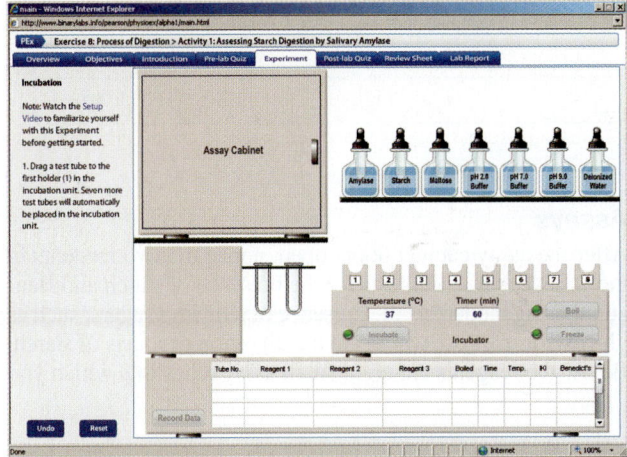

Incubation

1. Drag a test tube to the first holder (**1**) in the incubation unit. Seven more test tubes will automatically be placed in the incubation unit.

2. Add the substances indicated below to tubes 1 through 7.

Tube 1: amylase, starch, pH 7.0 buffer

Tube 2: amylase, starch, pH 7.0 buffer

Tube 3: amylase, starch, pH 7.0 buffer

Tube 4: amylase, deionized water, pH 7.0 buffer

Tube 5: deionized water, starch, pH 7.0 buffer

Tube 6: deionized water, maltose, pH 7.0 buffer

Tube 7: amylase, starch, pH 2.0 buffer

Tube 8: amylase, starch, pH 9.0 buffer

To add a substance to a test tube, drag the dropper cap of the bottle on the solutions shelf to the top of the test tube.

3. Click the number (1) under the first test tube. The tube will descend into the incubation unit. All other tubes should remain in the raised position.

4. Click **Boil** to boil tube 1. After boiling for a few moments, the tube will automatically rise.

5. Click the number (2) under the second test tube. The tube will descend into the incubation unit. All other tubes should remain in the raised position.

6. Click **Freeze** to freeze tube 2. After freezing for a few moments, the tube will automatically rise.

7. Click **Incubate** to start the run. Note that the incubation temperature is set at 37°C and the timer is set at 60 min. The incubation unit will gently agitate the test tube rack, evenly mixing the contents of all test tubes throughout the incubation. The simulation compresses the 60-minute time period into 10 seconds of real time, so what would be a 60-minute incubation in real time will take only 10 seconds in the simulation. When the incubation time elapses, the test tube rack will automatically rise, and the doors to the assay cabinet will open.

> **?** **PREDICT Question 1**
> What effect do you think boiling and freezing will have on the activity of the amylase enzyme?
> _____
> _____

Assays

After the assay cabinet doors open, notice the two reagents in the assay cabinet. IKI tests for the presence of starch and Benedict's reagent detects the presence of reducing sugars, such as glucose or maltose, which are the digestion products of starch. Below the reagents are eight small assay tubes into which you

will dispense a small amount of test solution from the incubated samples in the incubation unit, plus a drop of IKI.

8. Drag the first tube in the incubation unit to the first small assay tube on the left side of the assay cabinet to decant approximately half of the contents in the test tube into the assay tube. The decanting step will automatically repeat for the remaining tubes in the incubation unit.

9. Drag the IKI dropper cap to the first assay tube to dispense a drop of IKI into the assay tube. The dropper will automatically move across and dispense IKI to the remaining tubes.

10. Inspect the tubes for color change. A blue-black color indicates a positive starch test. If starch is not present, the mixture will look like diluted IKI, a negative starch test. Intermediate starch amounts result in a pale-gray color. Click **Record Data** to display your results in the grid (and record your results in Chart 1).

11. Drag the Benedict's reagent dropper cap to the test tube in the first holder (1) in the incubation unit to dispense five drops of Benedict's reagent into the tube. The dropper will automatically move across and dispense Benedict's reagent to the remaining tubes.

12. Click **Boil.** The entire tube rack will descend into the incubation unit and automatically boil the tube contents for a few moments.

13. Inspect the tubes for color change. A green-to-reddish color indicates that a reducing sugar is present; this is a positive sugar test. An orange-colored sample contains more sugar than a green sample. A reddish-brown color indicates even more sugar. A negative sugar test is indicated by no color change from the original bright blue. Click **Record Data** to display your results in the grid (and record your results in Chart 1).

After you complete the experiment, take the online **Post-lab Quiz** for Activity 1.

CHART 1	Salivary Amylase Digestion of Starch							
Tube No.	1	2	3	4	5	6	7	8
Additives	Amylase Starch pH 7.0 buffer	Amylase Starch pH 7.0 buffer	Amylase Starch pH 7.0 buffer	Amylase Deionized water pH 7.0 buffer	Deionized water Starch pH 7.0 buffer	Deionized water Maltose pH 7.0 buffer	Amylase Starch pH 2.0 buffer	Amylase Starch pH 9.0 buffer
Incubation condition	Boil first, then incubate at 37°C for 60 minutes	Freeze first, then incubate at 37°C for 60 minutes	37°C 60 minutes	37°C 60 minutes	37°C 60 minutes	37°C 60 minutes	37°C 60 minutes	37°C 60 minutes
IKI test								
Benedict's test								

Activity Questions

1. Describe the effect that boiling had on the activity of amylase. Why did boiling have this effect? How does the effect of freezing differ from the effect of boiling?

_____

_____

2. What is the purpose for including tube 3 and what can you conclude from the result?

_____

_____

3. Describe how you determined the optimal pH for amylase activity.

_____

_____

4. Judging from what you learned in this activity, suggest a reason why salivary amylase would be much less active in the stomach.

_____

_____

ACTIVITY 2

Exploring Amylase Substrate Specificity

OBJECTIVES

1. Explain how hydrolytic enzyme activity can be assessed with the IKI assay and the Benedict's assay.
2. Understand the specificity that enzymes have for their substrate.
3. Understand the difference between the substrates starch and cellulose.
4. Explain what would be the substrate specificity of peptidase.
5. Explain how bacteria might aid in digestion.

Introduction

In this activity you will investigate the specificity that enzymes have for their substrates. To do this you will hydrolyze starch to maltose and maltotriose using **salivary amylase,** the enzyme produced by the salivary glands and secreted into the mouth. To detect whether or not enzymatic action has occurred, you need to be able to identify the presence of the substrate and the product to determine to what extent hydrolysis has occurred. The **substrate** is the substance that the enzyme acts on. The enzyme has a pocket called the **active site,** which the substrate or substrates must fit into temporarily for catalysis to occur. The substrate is often held in the active site by non-covalent bonds (weak bonds), such as ionic bonds and hydrogen bonds.

With amylase activity, starch decreases and sugar increases as digestion proceeds according to the following equation.

$$\text{Starch} + \text{water} \xrightarrow{\text{amylase}} \text{maltose} + \text{maltotriose} + \text{starch}$$

Because the chemical changes that occur as starch is digested to maltose cannot be seen by the naked eye, you need to conduct an **enzyme assay,** the chemical method of detecting the presence of digested substances. You will perform two enzyme assays on each sample. The IKI assay detects the presence of starch or cellulose and the Benedict's assay tests for the presence of reducing sugars, such as glucose or maltose, which are the digestion products of starch. Normally a caramel-colored solution, IKI turns blue-black in the presence of starch or cellulose. Benedict's reagent is a bright blue solution that changes to green to orange to reddish brown with increasing amounts of maltose. It is important to understand that enzyme assays only indicate the presence or absence of substances. It is up to you to analyze the results of the experiments to decide whether enzymatic hydrolysis has occurred.

Starch is a polysaccharide found in plants, where it is used to store energy. Plants also have the polysaccharide **cellulose,** which provides rigidity to their cell walls. Both polysaccharides are polymers of glucose, but the glucose molecules are linked differently. You will be testing salivary amylase to determine whether it digests cellulose. Also, you will investigate to see whether a bacterial suspension can digest cellulose and whether **peptidase,** a pancreatic enzyme that digests peptides, can break down starch.

> **EQUIPMENT USED** The following equipment will be depicted on-screen: amylase—an enzyme that digests starch; starch—a polysaccharide; pH 7.0 buffer—a solution used to set the pH of the test tube solution; deionized water—used to adjust the test tube solution volume so it is the same for each reaction; glucose—a reducing sugar that is the monosaccharide subunit of both starch and cellulose; cellulose—a complex carbohydrate found in the cell wall of plants; peptidase—a pancreatic enzyme that breaks down peptides; bacteria—a suspension of live bacteria; test tubes—used as reaction vessels for the various tests; incubators—used for temperature treatments (37°C incubation); IKI—found in the assay cabinet; used to detect the presence of starch or cellulose; Benedict's reagent—found in the assay cabinet; used to detect the products of starch and cellulose digestion.

Experiment Instructions

Go to the home page in the PhysioEx software and click **Exercise 8: Chemical and Physical Processes of Digestion.** Click **Activity 2: Exploring Amylase Substrate Specificity,** and take the online **Pre-lab Quiz** for Activity 2.

After you take the online Pre-lab Quiz, click the **Experiment** tab and begin the experiment. The experiment instructions are reprinted here for your reference. The opening screen for the experiment is shown on the following page.

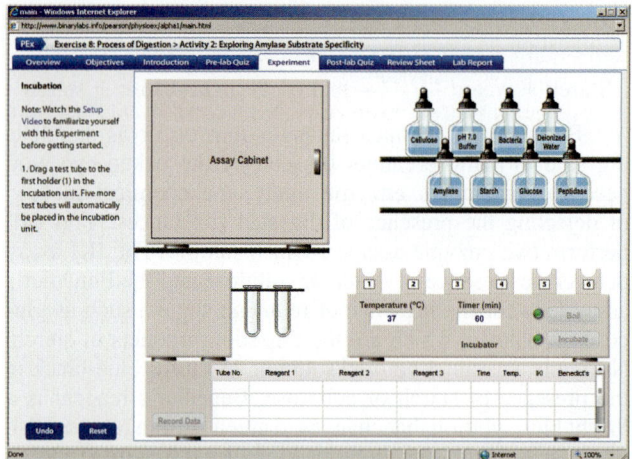

Incubation

1. Drag a test tube to the first holder (1) in the incubation unit. Five more test tubes will automatically be placed in the incubation unit.

2. Add the substances indicated below to tubes 1 through 6.

Tube 1: amylase, starch, pH 7.0 buffer

Tube 2: amylase, glucose, pH 7.0 buffer

Tube 3: amylase, cellulose, pH 7.0 buffer

Tube 4: cellulose, pH 7.0 buffer, deionized water

Tube 5: peptidase, starch, pH 7.0 buffer

Tube 6: bacteria, cellulose, pH 7.0 buffer

To add a substance to a test tube, drag the dropper cap of the bottle on the solutions shelf to the test tube.

3. Click **Incubate** to start the run. Note that the incubation temperature is set at 37°C and the timer is set at 60 min. The incubation unit will gently agitate the test tube rack, evenly mixing the contents of all test tubes throughout the incubation. The simulation compresses the 60-minute time period into 10 seconds of real time, so what would be a 60-minute incubation in real life will take only 10 seconds in the simulation. When the incubation time elapses, the test tube rack will automatically rise, and the doors to the assay cabinet will open.

? PREDICT Question 1
Do you think test tube 3 will show a positive Benedict's test?

_____

_____

Assays

After the assay cabinet doors open, notice the two reagents in the assay cabinet. IKI tests for the presence of starch and Benedict's reagent detects the presence of reducing sugars, such as glucose or maltose, which are the digestion products of starch. Below the reagents are seven small assay tubes into which you will dispense a portion of the incubated samples, plus a drop of IKI.

4. Drag the first tube in the incubation unit to the first small assay tube on the left side of the assay cabinet to decant approximately half of the contents in the test tube into the assay tube. The decanting step will automatically repeat for the remaining tubes in the incubation unit.

5. Drag the IKI dropper cap to the first assay tube to dispense a drop of IKI into the assay tube. The dropper will automatically dispense IKI into the remaining tubes.

6. Inspect the tubes for color change. A blue-black color indicates a positive starch test. If starch is not present, the mixture will look like diluted IKI, a negative starch test. Intermediate starch amounts result in a pale-gray color. Click **Record Data** to display your results in the grid (and record your results in Chart 2).

7. Drag the Benedict's reagent dropper cap to the test tube in the first holder (1) in the incubation unit to dispense five drops of Benedict's reagent into the tube. The dropper will automatically move across and dispense Benedict's reagent to the remaining tubes.

8. Click **Boil.** The entire tube rack will descend into the incubation unit and automatically boil the tube contents for a few moments.

9. Inspect the tubes for color change. A green-to-reddish color indicates that a reducing sugar is present; this is a positive sugar test. An orange-colored sample contains more sugar than a green sample. A reddish-brown color indicates even more sugar. A negative sugar test is indicated by no

CHART 2	Enzyme Digestion of Starch and Cellulose					
Tube No.	1	2	3	4	5	6
Additives	Amylase Starch pH 7.0 buffer	Amylase Glucose pH 7.0 buffer	Amylase Cellulose pH 7.0 buffer	Deionized water Cellulose pH 7.0 buffer	Peptidase Starch pH 7.0 buffer	Bacteria Cellulose pH 7.0 buffer
Incubation condition	37°C 60 minutes	37°C 60 minutes	37°C 60 minutes	37°C 60 minutes	37°C 60 minutes	37°C 60 minutes
IKI test						
Benedict's test						

color change from the original bright blue. Click **Record Data** to display your results in the grid (and record your results in Chart 2).

After you complete the Experiment, take the online **Post-lab Quiz** for Activity 2.

Activity Questions

1. Does amylase use cellulose as a substrate?

_____

_____

2. What effect did the addition of bacteria have on the digestion of cellulose?

_____

_____

3. What effect did the addition of peptidase to the starch have? Why?

_____

_____

4. What is the smallest subunit into which starch can be broken down?

_____

_____

ACTIVITY 3

Assessing Pepsin Digestion of Protein

OBJECTIVES

1. Explain how the enzyme activity of pepsin can be assessed with the BAPNA assay.
2. Identify the substrate specificity of pepsin.
3. Discuss the effects of temperature and pH on pepsin activity.
4. Understand the pH specificity of enzyme activity and how it relates to human physiology.

Introduction

In this activity, you will explore the digestion of protein (**peptides**). Peptides are two or more **amino acids** linked together by a peptide bond. A peptide chain containing 10 to 100 amino acids is typically called a **polypeptide.** Proteins can consist of a large peptide chain (more than 100 amino acids) or even multiple peptide chains.

During digestion, **chief cells** of the stomach glands secrete a protein-digesting enzyme called **pepsin.** Pepsin **hydrolyzes** peptide bonds. This activity breaks up ingested proteins and polypeptides into smaller peptide chains and free amino acids. In this activity, you will use **BAPNA** as a **substrate** to assess pepsin activity. BAPNA is a synthetic "peptide" that releases

a yellow dye **product** when hydrolyzed. BAPNA solutions turn yellow in the presence of an active peptidase, such as pepsin, but otherwise remain colorless.

To quantify the pepsin activity in each test solution, you will use a **spectrophotometer** to measure the amount of yellow dye produced. A spectrophotometer shines light through the sample and then measures how much light is absorbed. The fraction of light absorbed is expressed as the sample's **optical density.** Yellow solutions, where BAPNA has been hydrolyzed, will have optical densities greater than zero. The greater the optical density, the more hydrolysis has occurred. Colorless solutions, in contrast, do not absorb light and will have an optical density near zero.

Some negative controls are included in this activity. With negative controls, a negative result is expected. Negative results with negative controls validate the experiment. Negative controls are used to determine whether there are any contaminating substances in the reagents. So, when a positive result is produced but a negative result is expected, one or more contaminating substances are present to cause the change.

EQUIPMENT USED The following equipment will be depicted on-screen: pepsin—an enzyme that digests peptides; BAPNA—a synthetic "peptide"; pH buffers—solutions used to set the pH of the test tube solution; deionized water—used to adjust the test tube solution volume so it is the same for each reaction; test tubes—used as reaction vessels for the various tests; incubators—used for temperature treatments (boiling and 37°C incubation); spectrophotometer—found in the assay cabinet; used to measure the optical density of solutions.

Experiment Instructions

Go to the home page in the PhysioEx software and click **Exercise 8: Chemical and Physical Processes of Digestion.** Click **Activity 3: Assessing Pepsin Digestion of Protein,** and take the online **Pre-lab Quiz** for Activity 3.

After you take the online Pre-lab Quiz, click the **Experiment** tab and begin the experiment. The experiment instructions are reprinted here for your reference. The opening screen for the experiment is shown below.

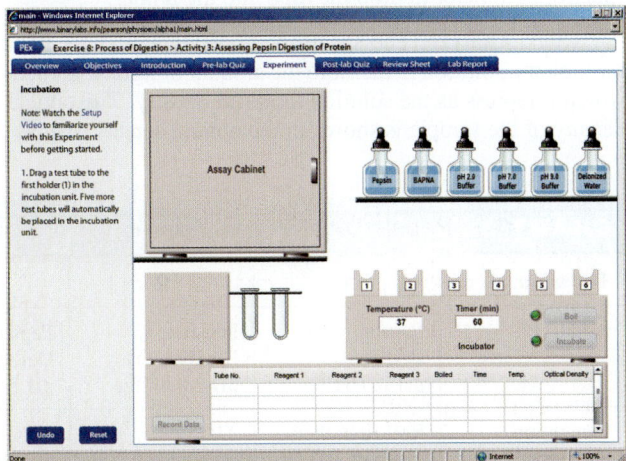

Incubation

1. Drag a test tube to the first holder (**1**) in the incubation unit. Five more test tubes will automatically be placed in the incubation unit.

2. Add the substances indicated below to tubes 1 through 6.

Tube 1: pepsin, BAPNA, pH 2.0 buffer

Tube 2: pepsin, BAPNA, pH 2.0 buffer

Tube 3: pepsin, deionized water, pH 2.0 buffer

Tube 4: deionized water, BAPNA, pH 2.0 buffer

Tube 5: pepsin, BAPNA, pH 7.0 buffer

Tube 6: pepsin, BAPNA, pH 9.0 buffer

To add a substance to a test tube, drag the dropper cap of the bottle on the solutions shelf to the test tube.

3. Click the number (1) under the first test tube. The tube will descend into the incubation unit. All other tubes should remain in the raised position.

4. Click **Boil** to boil tube 1. After boiling for a few moments, the tube will automatically rise.

5. Click **Incubate** to start the run. Note that the incubation temperature is set at 37°C and the timer is set at 60 min. The incubation unit will gently agitate the test tube rack, evenly mixing the contents of all test tubes throughout the incubation. The simulation compresses the 60-minute time period into 10 seconds of real time, so what would be a 60-minute incubation in real life will take only 10 seconds in the simulation. When the incubation time elapses, the test tube rack will automatically rise, and the doors to the assay cabinet will open. The spectrophotometer is in the assay cabinet.

> **?** **PREDICT Question 1**
> At which pH do you think pepsin will have the highest activity?
> _____
> _____

Assays

6. You will now use the spectrophotometer to measure how much yellow dye was liberated from BAPNA hydrolysis. Drag the first tube in the incubation unit to the holder in the spectrophotometer to drop the tube into the holder.

7. Click **Analyze.** The spectrophotometer will shine light through the solution to measure the amount of light absorbed, which it reports as the solution's optical density. The optical density of the sample is shown in the optical density display.

8. Click **Record Data** to display your results in the grid (and record your results in Chart 3).

9. Drag the tube to its original position in the incubation unit.

10. Analyze the remaining five tubes by repeating the following steps for each tube.

- Drag the tube to the holder in the spectrophotometer to drop the tube into the holder.

- Click **Analyze.**

- Drag the tube to its original position in the incubation unit.

After you have analyzed all five tubes, click **Record Data** to display your results in the grid (and record your results in Chart 3).

After you complete the experiment, take the online **Post-lab Quiz** for Activity 3.

Activity Questions

1. Describe the significance of the optimum pH for pepsin observed in the simulation and the secretion of pepsin by the chief cells of the gastric glands.

_____

_____

2. Would pepsin be active in the mouth? Explain your answer.

_____

_____

3. What are the subunit products of peptide digestion?

_____

_____

4 Describe the reason for including control tube 4.

_____

_____

CHART 3	Pepsin Digestion of Protein					
Tube No.	1	2	3	4	5	6
Additives	Pepsin BAPNA pH 2.0 buffer	Pepsin BAPNA pH 2.0 buffer	Pepsin Deionized water pH 2.0 buffer	Deionized water BAPNA pH 2.0 buffer	Pepsin BAPNA pH 7.0 buffer	Pepsin BAPNA pH 9.0 buffer
Incubation condition	Boil first, then incubate at 37°C for 60 minutes	37°C 60 minutes	37°C 60 minutes	37°C 60 minutes	37°C 60 minutes	37°C 60 minutes
Optical density						

Assessing Lipase Digestion of Fat

OBJECTIVES

1. Explain how the enzyme activity of pancreatic lipase can be assessed with a pH-based measurement.

2. Identify the hydrolysis products of fat digestion.

3. Understand the role that bile plays in fat digestion.

4. Understand the significance of pH specificity of lipase activity and how it relates to human physiology.

5. Discuss the difficulty of using pH to measure digestion when comparing the activity of lipase at various pHs.

Introduction

Fats and oils belong to a diverse class of molecules called lipids. **Triglycerides,** a type of lipid, make up both fats and oils. At room temperature, fats are solid and oils are liquid. Both are poorly soluble in water. This insolubility of triglycerides presents a challenge during digestion because they tend to clump together, leaving only the surface molecules exposed to **lipase** enzymes. To overcome this difficulty, **bile salts** are secreted into the small intestine during digestion to physically emulsify lipids. Bile salts act like a detergent, separating the lipid clumps and increasing the surface area accessible to lipase enzymes.

As a result, two reactions must occur. First,

$$\text{Triglyceride clumps} \xrightarrow[\text{(emulsification)}]{\text{bile}} \text{minute triglyceride droplets}$$

Then,

$$\text{Triglyceride} \xrightarrow{\text{lipase}} \text{monoglyceride} + \text{two fatty acids}$$

Lipase hydrolyzes each triglyceride to a monoglyceride and two fatty acids. In addition to the **pancreatic lipase** secreted into the small intestine, **lingual lipase** and **gastric lipase** are also secreted. Even though bile salts are not secreted in the mouth or the stomach, small amounts of lipids are digested by these other lipases.

Because some of the end products of fat digestion are acidic (that is, fatty acids), lipase activity can be easily measured by monitoring the solution's **pH.** A solution containing fatty acids liberated by lipase activity will have a lower pH than a solution without such fatty acid production. You will record pH in this activity with a **pH meter.**

> **EQUIPMENT USED** The following equipment will be depicted on-screen: lipase—an enzyme that digests triglycerides; vegetable oil—a mixture of triglycerides; bile salts—a solution that physically separates fats into smaller droplets; pH buffers—solutions used to set the pH of the test tube solution; deionized water—used to adjust the test tube solution volume so it is the same for each reaction; test tubes—used as reaction vessels for the various tests; incubators—used for temperature treatments (boiling and 37°C incubation); pH meter—found in the assay cabinet; used to measure pH.

Experiment Instructions

Go to the home page in the PhysioEx software and click **Exercise 8: Chemical and Physical Processes of Digestion.** Click **Activity 4: Assessing Lipase Digestion of Fat,** and take the online **Pre-lab Quiz** for Activity 4.

After you take the online Pre-lab Quiz, click the **Experiment** tab and begin the experiment. The experiment instructions are reprinted here for your reference. The opening screen for the experiment is shown below.

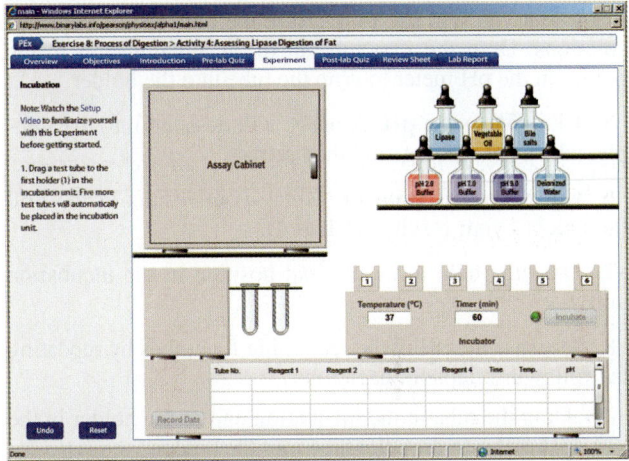

Incubation

1. Drag a test tube to the first holder **(1)** in the incubation unit. Five more test tubes will automatically be placed in the incubation unit.

2. Add the substances indicated below to tubes 1 through 6.

Tube 1: lipase, vegetable oil, bile salts, pH 7.0 buffer

Tube 2: lipase, vegetable oil, deionized water, pH 7.0 buffer

Tube 3: lipase, deionized water, bile salts, pH 9.0 buffer

Tube 4: deionized water, vegetable oil, bile salts, pH 7.0 buffer

Tube 5: lipase, vegetable oil, bile salts, pH 2.0 buffer

Tube 6: lipase, vegetable oil, bile salts, pH 9.0 buffer

To add a substance to a test tube, drag the dropper cap of the bottle on the solutions shelf to the test tube.

3. Click **Incubate** to start the run. Note that the incubation temperature is set at 37°C and the timer is set at 60 min. The incubation unit will gently agitate the test tube rack, evenly mixing the contents of all test tubes throughout the incubation. The simulation compresses the 60-minute time period into 10 seconds of real time, so what would be a 60-minute incubation in real life will take only 10 seconds in the simulation. When the incubation time elapses, the test tube rack will automatically rise, and the doors to the assay cabinet will open.

? PREDICT Question 1
Which tube do you think will have the highest lipase activity?

_____

_____

Assays

4. After the assay cabinet doors open, you will see a pH meter that you will use to measure the final pH of your test solutions. Drag the first tube in the incubation unit to the holder in the pH meter to drop the tube into the holder.

5. Click **Measure pH.** A probe will descend into the sample, take a pH reading, and then retract.

6. Click **Record Data** to display your results in the grid (and record your results in Chart 4).

7. Drag the tube to its original position in the incubation unit.

8. Measure the pH in the remaining five tubes by repeating the following steps for each tube.

- Drag the tube in the incubation unit to the holder in the pH meter to drop the tube into the holder.

- Click **Measure pH.**

- Drag the tube to its original position in the incubation unit.

After you have measured the pH in all five tubes, click **Record Data** to display your results in the grid (and record your results in Chart 4).

After you complete the experiment, take the online **Post-lab Quiz** for Activity 4.

Activity Questions

1. Describe how lipase activity is measured in the simulation.

_____

_____

2. Can you determine if fat hydrolysis occurred in tube 5? Why or why not?

_____

_____

3. Would pancreatic lipase be active in the mouth? Why or why not?

_____

_____

4. Describe the physical separation of fats by bile salts.

_____

_____

CHART 4	Pancreatic Lipase Digestion of Triglycerides and the Action of Bile					
Tube No.	1	2	3	4	5	6
Additives	Lipase Vegetable oil Bile salts pH 7.0 buffer	Lipase Vegetable oil Deionized water pH 7.0 buffer	Lipase Deionized water Bile salts pH 9.0 buffer	Deionized water Vegetable oil Bile salts pH 7.0 buffer	Lipase Vegetable oil Bile salts pH 2.0 buffer	Lipase Vegetable oil Bile salts pH 9.0 buffer
Incubation condition	37°C 60 minutes	37°C 60 minutes	37°C 60 minutes	37°C 60 minutes	37°C 60 minutes	37°C 60 minutes
pH						

NAME _____

LAB TIME/DATE _____

Chemical and Physical Processes of Digestion

ACTIVITY 1 **Assessing Starch Digestion by Salivary Amylase**

1. List the substrate and the subunit product of amylase. _____

_____

2. What effect did boiling and freezing have on enzyme activity? Why? How well did the results compare with your prediction?

_____

_____

3. At what pH was the amylase most active? Describe the significance of this result. _____

_____

4. Briefly describe the need for controls and give an example used in this activity. _____

_____

5. Describe the significance of using a 37°C incubation temperature to test salivary amylase activity. _____

_____

ACTIVITY 2 **Exploring Amylase Substrate Specificity**

1. Describe why the results in tube 1 and tube 2 are the same. _____

_____

2. Describe the result in tube 3. How well did the results compare with your prediction? _____

_____

3. Describe the usual substrate for peptidase. _____

_____

4. Explain how bacteria can aid in digestion. _____

_____

ACTIVITY 3 Assessing Pepsin Digestion of Protein

1. Describe the effect that boiling had on pepsin and how you could tell that it had that effect. _____

_____

2. Was your prediction correct about the optimal pH for pepsin activity? Discuss the physiological correlation behind your results.

_____

_____

3. What do you think would happen if you reduced the incubation time to 30 minutes for tube 5? _____

_____

ACTIVITY 4 Assessing Lipase Digestion of Fat

1. Explain why you can't fully test the lipase activity in tube 5. _____

_____

2. Which tube had the highest lipase activity? How well did the results compare with your prediction? Discuss possible reasons

why it may or may not have matched. _____

_____

_____

3. Explain why pancreatic lipase would be active in both the mouth and the intestine. _____

_____

_____

4. Describe the process of bile emulsification of lipids and how it improves lipase activity. _____

_____

_____

Renal System Physiology

Exercise Overview

The **kidney** is *both* an excretory and a regulatory organ. By filtering the water and solutes in the blood, the kidneys are able to *excrete* excess water, waste products, and even foreign materials from the body. However, the kidneys also *regulate* (1) plasma osmolarity (the concentration of a solution expressed as osmoles of solute per liter of solvent), (2) plasma volume, (3) the body's acid-base balance, and (4) the body's electrolyte balance. All these activities are extremely important for maintaining homeostasis in the body.

The paired kidneys are located between the posterior abdominal wall and the abdominal peritoneum. The right kidney is slightly lower than the left kidney. Each human kidney contains approximately one million **nephrons,** the functional units of the kidney.

Each nephron is composed of a **renal corpuscle** and a **renal tubule.** The renal corpuscle consists of a "ball" of capillaries, called the *glomerulus,* which is enclosed by a fluid-filled capsule, called *Bowman's capsule,* or the glomerular capsule. An **afferent arteriole** supplies blood to the glomerulus. As blood flows through the glomerular capillaries, protein-free plasma filters into the Bowman's capsule, a process called **glomerular filtration.** An **efferent arteriole** then drains the glomerulus of the remaining blood (view Figure 9.1).

The filtrate flows from Bowman's capsule into the start of the renal tubule, called the **proximal convoluted tubule,** then into the **loop of Henle,** a U-shaped

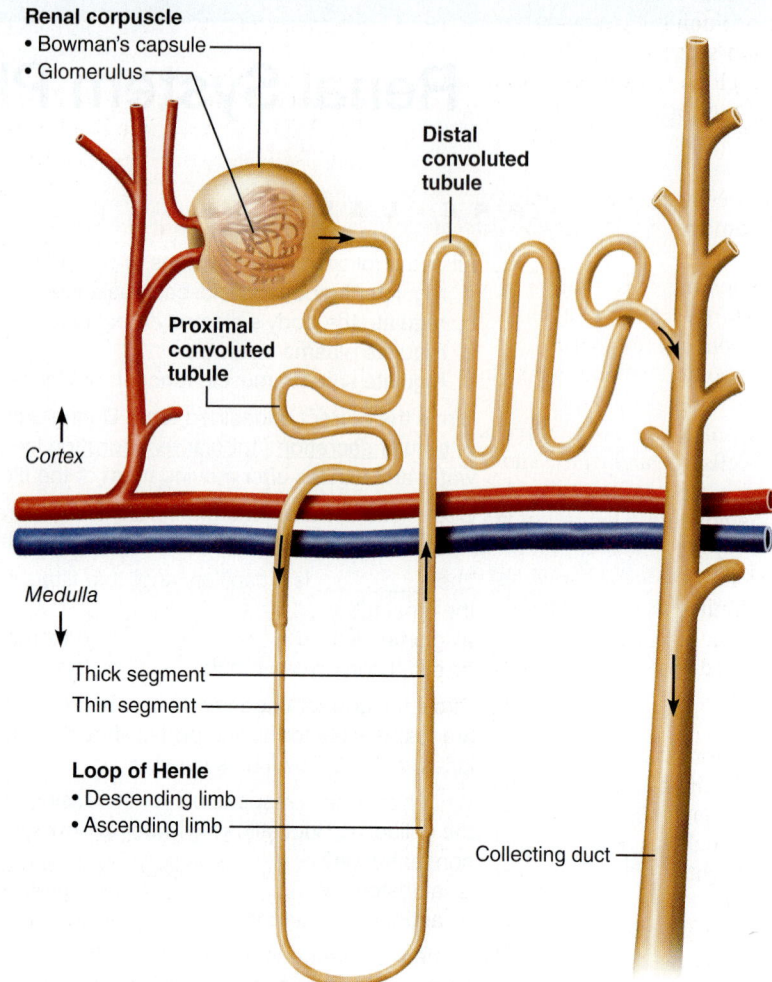

Renal corpuscle
• Bowman's capsule
• Glomerulus

Distal convoluted tubule

Proximal convoluted tubule

Cortex

Medulla

Thick segment
Thin segment

Loop of Henle
• Descending limb
• Ascending limb

Collecting duct

FIGURE 9.1 Location and structure of nephrons.

hairpin loop, and, finally, into the **distal convoluted tubule** before emptying into a **collecting duct.** From the collecting duct, the filtrate flows into, and collects in, the minor calyces.

The nephron performs three important functions that process blood into filtrate and urine: (1) glomerular filtration, (2) tubular reabsorption, and (3) tubular secretion. **Glomerular filtration** is a passive process in which fluid passes from the lumen of the glomerular capillary into the glomerular capsule of the renal tubule. **Tubular reabsorption** moves most of the filtrate back into the blood, leaving mainly salt water and the wastes in the lumen of the tubule. Some of the desirable, or needed, solutes are actively reabsorbed, and others move passively from the lumen of the tubule into the interstitial spaces. **Tubular secretion** is essentially the reverse of tubular reabsorption and is a process by which the kidneys can rid the blood of additional unwanted substances, such as creatinine and ammonia.

The reabsorbed solutes and water that move into the interstitial space between the nephrons need to be returned to the blood, or the kidneys will rapidly swell like balloons. The **peritubular capillaries** surrounding the renal tubule reclaim the reabsorbed substances and return them to general circulation. Peritubular capillaries arise from the efferent arteriole exiting the glomerulus and empty into the renal veins leaving the kidney.

The Effect of Arteriole Radius on Glomerular Filtration

OBJECTIVES

1. To understand the terms *nephron, glomerulus, glomerular capillaries, renal tubule, filtrate, Bowman's capsule, renal corpuscle, afferent arteriole, efferent arteriole, glomerular capillary pressure,* and *glomerular filtration rate.*

2. To understand how changes in afferent arteriole radius impact glomerular capillary pressure and filtration.

3. To understand how changes in efferent arteriole radius impact glomerular capillary pressure and filtration.

Introduction

Each of the million **nephrons** in each kidney contains two major parts: (1) a tubular component, the **renal tubule,** and (2) a vascular component, the **renal corpuscle** (view Figure 9.1). The **glomerulus** is a tangled capillary knot that filters fluid from the blood into the lumen of the renal tubule. The function of the renal tubule is to process the filtered fluid, also called the **filtrate.** The beginning of the renal tubule is an enlarged end called **Bowman's capsule** (or the glomerular capsule), which surrounds the glomerulus and serves to funnel the filtrate into the rest of the renal tubule. Collectively, the glomerulus and Bowman's capsule are called the renal corpuscle.

Two arterioles are associated with each glomerulus: an **afferent arteriole** feeds the **glomerular capillary** bed and an **efferent arteriole** drains it. These arterioles are responsible for blood flow through the glomerulus. The diameter of the efferent arteriole is smaller than the diameter of the afferent arteriole, restricting blood flow out of the glomerulus. Consequently, the pressure in the glomerular capillaries forces fluid through the endothelium of the capillaries into the lumen of the surrounding Bowman's capsule. In essence, everything in the blood except for the blood cells (red and white) and plasma proteins is filtered through the glomerular wall. From the Bowman's capsule, the filtrate moves into the rest of the renal tubule for processing. The job of the tubule is to reabsorb all the beneficial substances from its lumen and allow the wastes to travel down the tubule for elimination from the body.

During glomerular filtration, blood enters the glomerulus from the afferent arteriole and protein-free plasma flows from the blood across the walls of the glomerular capillaries and into the Bowman's capsule. The **glomerular filtration rate** is an index of kidney function. In humans, the filtration rate ranges from 80 to 140 ml/min, so that, in 24 hours, as much as 180 liters of filtrate is produced by the glomeruli. The filtrate formed is devoid of cellular debris, is essentially protein free, and contains a concentration of salts and organic molecules similar to that in blood.

The glomerular filtration rate can be altered by changing arteriole resistance or arteriole hydrostatic pressure. In this activity, you will explore the effect of arteriole radius on glomerular capillary pressure and filtration in a single nephron. You can apply the concepts you learn by studying a single nephron to understand the function of the kidney as a whole.

EQUIPMENT USED The following equipment will be depicted on-screen: source beaker for blood (first beaker on left side of screen)—simulates blood flow and pressure (mm Hg) from general circulation to the nephron; drain beaker for blood (second beaker on left side of screen)—simulates the renal vein; flow tube with adjustable radius—simulates the afferent arteriole and connects the blood supply to the glomerular capillaries; second flow tube with adjustable radius—simulates the efferent arteriole and drains the glomerular capillaries into the peritubular capillaries, which ultimately drain into the renal vein (drain beaker); simulated nephron (The filtrate forms in Bowman's capsule, flows through the renal tubule—the tubular components—and empties into a collecting duct, which in turn drains into the urinary bladder.); nephron tank; glomerulus—"ball" of capillaries that forms part of the filtration membrane; glomerular (Bowman's) capsule—forms part of the filtration membrane and a capsular space where the filtrate initially forms; proximal convoluted tubule; loop of Henle; distal convoluted tubule; collecting duct; drain beaker for filtrate (beaker on right side of screen)—simulates the urinary bladder.

Experiment Instructions

Go to the home page in the PhysioEx software and click **Exercise 9: Renal System Physiology.** Click **Activity 1: The Effect of Arteriole Radius on Glomerular Filtration,** and take the online **Pre-lab Quiz** for Activity 1.

After you take the online Pre-lab Quiz, click the **Experiment** tab and begin the experiment. The experiment instructions are reprinted here for your reference. The opening screen for the experiment is shown below.

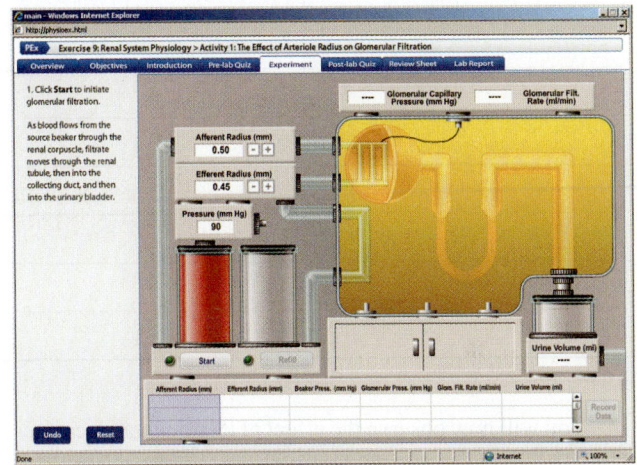

1. Click **Start** to initiate glomerular filtration. As blood flows from the source beaker through the renal corpuscle, filtrate moves through the renal tubule, then into the collecting duct, and then into the urinary bladder.

2. The glomerular capillary pressure display shows the hydrostatic blood pressure in the glomerular capillaries that promotes filtration, and the filtration rate display shows the flow rate of the fluid moving from the lumen of the glomerular capillaries into the lumen of Bowman's capsule. Click **Record Data** to display your results in the grid (and record your results in Chart 1).

3. Click **Refill** to replenish the source beaker and prepare the nephron for the next run.

? PREDICT Question 1
What will happen to the glomerular capillary pressure and filtration rate if you decrease the radius of the afferent arteriole?

---

4. Decrease the radius of the afferent arteriole to 0.45 mm by clicking the − button beside the afferent radius display. Click **Start** to initiate glomerular filtration.

5. Note the glomerular capillary pressure and glomerular filtration rate displays and click **Record Data** to display your results in the grid (and record your results in Chart 1).

6. Click **Refill** to replenish the source beaker and prepare the nephron for the next run.

CHART 1	Effect of Arteriole Radius on Glomerular Filtration		
Afferent arteriole radius (mm)	Efferent arteriole radius (mm)	Glomerular capillary pressure (mm Hg)	Glomerular filtration rate (ml/min)

7. You will now observe the effect of incremental decreases in the radius of the afferent arteriole.

- Decrease the radius of the afferent arteriole by 0.05 mm by clicking the − button beside the afferent radius display.
- Click **Start** to initiate glomerular filtration.
- Note the glomerular capillary pressure and glomerular filtration rate displays and click **Record Data** to display your results in the grid (and record your results in Chart 1).
- Click **Refill** to replenish the source beaker and prepare the nephron for the next run.

Repeat this step until you reach an afferent arteriole radius of 0.35 mm.

> **? PREDICT Question 2**
> What will happen to the glomerular capillary pressure and filtration rate if you increase the radius of the afferent arteriole?

8. Increase the radius of the afferent arteriole to 0.55 mm by clicking the + button beside the afferent radius display. Click **Start** to initiate glomerular filtration.

9. Note the glomerular capillary pressure and glomerular filtration rate displays and click **Record Data** to display your results in the grid (and record your results in Chart 1).

10. Click **Refill** to replenish the source beaker and prepare the nephron for the next run.

11. Increase the radius of the afferent arteriole to 0.60 mm. Click **Start** to initiate glomerular filtration.

12. Note the glomerular capillary pressure and glomerular filtration rate displays and click **Record Data** to display your results in the grid (and record your results in Chart 1).

13. Click **Refill** to replenish the source beaker and prepare the nephron for the next run.

> **? PREDICT Question 3**
> What will happen to the glomerular capillary pressure and filtration rate if you decrease the radius of the efferent arteriole?

14. Decrease the radius of the afferent arteriole to 0.50 mm by clicking the − button beside the afferent radius display. Click **Start** to initiate glomerular filtration.

15. Note the glomerular capillary pressure and glomerular filtration rate displays and click **Record Data** to display your results in the grid (and record your results in Chart 1).

16. Click **Refill** to replenish the source beaker and prepare the nephron for the next run.

17. You will now observe the effect of incremental decreases in the radius of the efferent arteriole.

- Decrease the radius of the efferent arteriole by 0.05 mm by clicking the − button beside the efferent radius display.
- Click **Start** to initiate glomerular filtration.

- Note the glomerular capillary pressure and glomerular filtration rate displays and click **Record Data** to display your results in the grid (and record your results in Chart 1).
- Click **Refill** to replenish the source beaker and prepare the nephron for the next run.

Repeat this step until you reach an efferent arteriole radius of 0.30 mm.

After you complete the experiment, take the online **Post-lab Quiz** for Activity 1.

Activity Questions

1. Activation of sympathetic nerves that innervate the kidney leads to a decreased urine production. Knowing that fact, what do you think the sympathetic nerves do to the afferent arteriole?

_____

_____

2. How is this effect of the sympathetic nervous system beneficial? Could this effect become harmful if it goes on too long?

_____

_____

ACTIVITY 2

The Effect of Pressure on Glomerular Filtration

OBJECTIVES

1. To understand the terms *glomerulus, glomerular capillaries, renal tubule, filtrate, Starling forces, Bowman's capsule, renal corpuscle, afferent arteriole, efferent arteriole, glomerular capillary pressure,* and *glomerular filtration rate.*
2. To understand how changes in glomerular capillary pressure affect glomerular filtration rate.
3. To understand how changes in renal tubule pressure affect glomerular filtration rate.

Introduction

Cellular metabolism produces a complex mixture of waste products that must be eliminated from the body. This excretory function is performed by a combination of organs, most importantly, the paired kidneys. Each kidney consists of approximately one million nephrons, which carry out three crucial processes: (1) glomerular filtration, (2) tubular reabsorption, and (3) tubular secretion.

Both the blood pressure in the **glomerular capillaries** and the **filtrate** pressure in the **renal tubule** can have a significant impact on the **glomerular filtration rate.** During glomerular filtration, blood enters the **glomerulus** from the

afferent arteriole. Starling forces (hydrostatic and osmotic pressure gradients) drive protein-free fluid between the blood in the glomerular capillaries and the filtrate in **Bowman's capsule.** The glomerular filtration rate is an index of kidney function. In humans, the filtration rate ranges from 80 to 140 ml/min, so that, in 24 hours, as much as 180 liters of filtrate is produced by the glomerular capillaries. The filtrate formed is devoid of blood cells, is essentially protein free, and contains a concentration of salts and organic molecules similar to that in blood.

Approximately 20% of the blood that enters the glomerular capillaries is normally filtered into Bowman's capsule, where it is then referred to as filtrate. The unusually high hydrostatic blood pressure in the glomerular capillaries promotes this filtration. Thus, the glomerular filtration rate can be altered by changing the afferent arteriole resistance (and, therefore, the hydrostatic pressure). In this activity you will explore the effect of blood pressure on the glomerular filtration rate in a single nephron. You can apply the concepts you learn by studying a single nephron to understand the function of the kidney as a whole.

EQUIPMENT USED The following equipment will be depicted on-screen: left source beaker (first beaker on left side of screen)—simulates blood flow and pressure (mm Hg) from general circulation to the nephron; drain beaker for blood (second beaker on left side of screen)—simulates the renal vein; flow tube with adjustable radius—simulates the afferent arteriole and connects the blood supply to the glomerular capillaries; second flow tube with adjustable radius—simulates the efferent arteriole and drains the glomerular capillaries into the peritubular capillaries, which ultimately drain into the renal vein (drain beaker); simulated nephron (The filtrate forms in Bowman's capsule, flows through the renal tubule—the tubular components—and empties into a collecting duct, which in turn drains into the urinary bladder.); nephron tank; glomerulus—"ball" of capillaries that forms part of the filtration membrane; glomerular (Bowman's) capsule—forms part of the filtration membrane and a capsular space where the filtrate initially forms; proximal convoluted tubule; loop of Henle; distal convoluted tubule; collecting duct; one-way valve between end of collecting tube (duct) and urinary bladder—used to restrict the flow of filtrate into the urinary bladder, increasing the volume and pressure in the renal tubule; drain beaker for filtrate (beaker on right side of screen)—simulates the urinary bladder.

Experiment Instructions

Go to the home page in the PhysioEx software and click **Exercise 9: Renal System Physiology.** Click **Activity 2: The Effect of Pressure on Glomerular Filtration,** and take the online **Pre-lab Quiz** for Activity 2.

After you take the online Pre-lab Quiz, click the **Experiment** tab and begin the experiment. The experiment instructions are reprinted here for your reference. The opening screen for the experiment is shown on the following page.

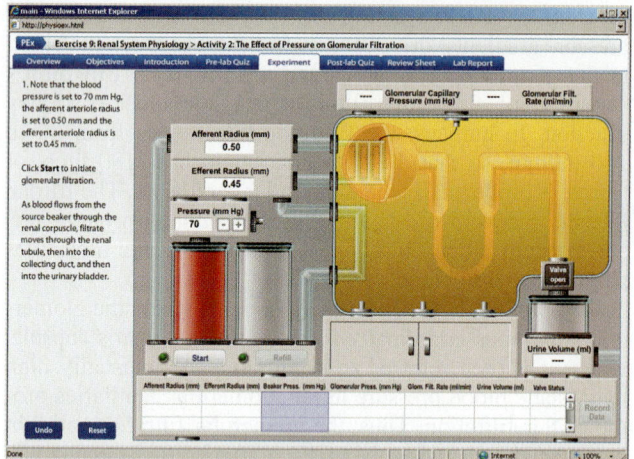

1. Note that the blood pressure is set to 70 mm Hg, the afferent arteriole radius is set to 0.50 mm, and the efferent arteriole radius is set to 0.45 mm. Click **Start** to initiate glomerular filtration. As blood flows from the source beaker through the renal corpuscle, filtrate moves through the renal tubule, then into the collecting duct, and then into the urinary bladder.

2. The glomerular capillary pressure display shows the hydrostatic blood pressure in the glomerular capillaries that promotes filtration, and the filtration rate display shows the flow rate of the fluid moving from the lumen of the glomerular capillaries into the lumen of Bowman's capsule. Click **Record Data** to display your results in the grid (and record your results in Chart 2.)

3. Click **Refill** to replenish the source beaker and prepare the nephron for the next run.

? PREDICT Question 1
What will happen to the glomerular capillary pressure and filtration rate if you increase the blood pressure in the left source beaker?

4. Increase the blood pressure to 80 mm Hg by clicking the + button beside the pressure display. Click **Start** to initiate glomerular filtration.

5. Note the glomerular capillary pressure and glomerular filtration rate displays and click **Record Data** to display your results in the grid (and record your results in Chart 2).

6. Click **Refill** to replenish the source beaker and prepare the nephron for the next run.

7. You will now observe the effect of further incremental increases in blood pressure.

- Increase the blood pressure by 10 mm Hg by clicking the + button beside the pressure display.

- Click **Start** to initiate glomerular filtration.

- Note the glomerular capillary pressure and glomerular filtration rate displays and click **Record Data** to display your results in the grid (and record your results in Chart 2).

- Click **Refill** to replenish the source beaker and prepare the nephron for the next run.

Repeat this step until you reach a blood pressure of 100 mm Hg.

? PREDICT Question 2
What will happen to the filtrate pressure in Bowman's capsule (not directly measured in this experiment) and the filtration rate if you close the one-way valve between the collecting duct and the urinary bladder?

8. Note that the valve between the collecting duct and the urinary bladder is open. Decrease the blood pressure to 70 mm Hg by clicking the button beside the pressure display. Click **Start** to initiate glomerular filtration.

CHART 2	Effect of Arteriole Radius on Glomerular Filtration			
Blood pressure (mm Hg)	Valve (open or closed)	Glomerular capillary pressure (mm Hg)	Glomerular filtration rate (ml/min)	Urine volume (ml)

9. Note the glomerular capillary pressure and glomerular filtration rate displays and click **Record Data** to display your results in the grid (and record your results in Chart 2).

10. Click **Refill** to replenish the source beaker and prepare the nephron for the next run.

11. Click the valve between the collecting duct and the urinary bladder to close it. Click **Start** to initiate glomerular filtration.

12. Note the glomerular capillary pressure and glomerular filtration rate displays and click **Record Data** to display your results in the grid (and record your results in Chart 2).

13. Click **Refill** to replenish the source beaker and prepare the nephron for the next run.

14. Increase the blood pressure to 100 mm Hg. Click **Start** to initiate glomerular filtration.

15. Note the glomerular capillary pressure and glomerular filtration rate displays and click **Record Data** to display your results in the grid (and record your results in Chart 2).

16. Click **Refill** to replenish the source beaker and prepare the nephron for the next run.

17. Click the valve between the collecting duct and the urinary bladder to open it. Click **Start** to initiate glomerular filtration.

18. Note the glomerular capillary pressure and glomerular filtration rate displays and click **Record Data** to display your results in the grid (and record your results in Chart 2).

After you complete the experiment, take the online **Post-lab Quiz** for Activity 2.

Activity Questions

1. Judging from the results in this laboratory activity, what *should be* the effect of blood pressure on glomerular filtration?

_____

_____

2. Persistent high blood pressure with inadequate glomerular filtration is now a frequent problem in Western cultures. Using the concepts in this activity, explain this health problem.

_____

_____ ▬

ACTIVITY 3

Renal Response to Altered Blood Pressure

OBJECTIVES

1. To understand the terms *nephron, renal tubule, filtrate, Bowman's capsule, blood pressure, afferent arteriole,* *efferent arteriole, glomerulus, glomerular filtration rate,* and *glomerular capillary pressure.*

2. To understand how blood pressure affects glomerular capillary pressure and glomerular filtration.

3. To observe which is more effective: changes in afferent or efferent arteriole radius when changes in blood pressure occur.

Introduction

In humans approximately 180 liters of filtrate flows into the **renal tubules** every day. As demonstrated in Activity 2, the **blood pressure** supplying the **nephron** can have a substantial impact on the **glomerular capillary pressure** and **glomerular filtration.** However, under most circumstances, glomerular capillary pressure and glomerular filtration remain relatively constant despite changes in blood pressure because the nephron has the capacity to alter its **afferent** and **efferent arteriole** radii.

During glomerular filtration, blood enters the **glomerulus** from the afferent arteriole. **Starling forces** (primarily hydrostatic pressure gradients) drive protein-free fluid out of the glomerular capillaries and into **Bowman's capsule.** Importantly for our body's homeostasis, a relatively constant glomerular filtration rate of 125 ml/min is maintained despite a wide range of blood pressures that occur throughout the day for an average human.

Activities 1 and 2 explored the independent effects of arteriole radii and blood pressure on glomerular capillary pressure and glomerular filtration. In the human body, these effects occur simultaneously. Therefore, in this activity, you will alter both variables to explore their combined effects on glomerular filtration and observe how changes in one variable can compensate for changes in the other to maintain an adequate glomerular filtration rate.

EQUIPMENT USED The following equipment will be depicted on-screen: left source beaker (first beaker on left side of screen)—simulates blood flow and pressure (mm Hg) from general circulation to the nephron; drain beaker for blood (second beaker on left side of screen)—simulates the renal vein; flow tube with adjustable radius—simulates the afferent arteriole and connects the blood supply to the glomerular capillaries; second flow tube with adjustable radius—simulates the efferent arteriole and drains the glomerular capillaries into the peritubular capillaries, which ultimately drain into the renal vein (drain beaker); simulated nephron (The filtrate forms in Bowman's capsule, flows through the renal tubule—the tubular components—and empties into a collecting duct, which in turn drains into the urinary bladder.); nephron tank; glomerulus—"ball" of capillaries that forms part of the filtration membrane; glomerular (Bowman's) capsule—forms part of the filtration membrane and a capsular space where the filtrate initially forms; proximal convoluted tubule; loop of Henle; distal convoluted tubule; collecting duct; one-way valve between end of collecting tube (duct) and urinary bladder—used to restrict the flow of filtrate into the urinary bladder, increasing the volume and pressure in the renal tubule; drain beaker for filtrate (beaker on right side of screen)—simulates the urinary bladder.

Experiment Instructions

Go to the home page in the PhysioEx software and click **Exercise 9: Renal System Physiology.** Click **Activity 3: Renal Response to Altered Blood Pressure,** and take the online **Pre-lab Quiz** for Activity 3.

After you take the online Pre-lab Quiz, click the **Experiment** tab and begin the experiment. The experiment instructions are reprinted here for your reference. The opening screen for the experiment is shown below.

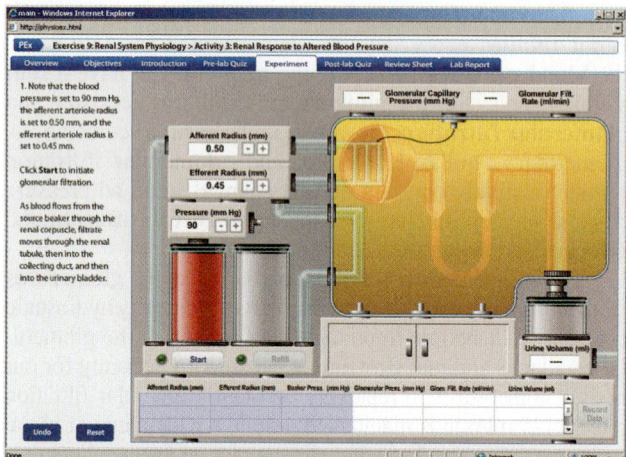

1. Note that the blood pressure is set to 90 mm Hg, the afferent arteriole radius is set to 0.50 mm, and the efferent arteriole radius is set to 0.45 mm. Click **Start** to initiate glomerular filtration. As blood flows from the source beaker through the renal corpuscle, filtrate moves through the renal tubule, then into the collecting duct, and then into the urinary bladder.

2. The glomerular capillary pressure display shows the hydrostatic blood pressure in the glomerular capillaries that promotes filtration, and the filtration rate display shows the flow rate of the fluid moving from the lumen of the glomerular capillaries into the lumen of Bowman's capsule. Click **Record Data** to display your results in the grid (and record your results in Chart 3).

3. Click **Refill** to replenish the source beaker and prepare the nephron for the next run.

4. You will now observe how the nephron might operate to keep the glomerular filtration rate relatively constant despite a large drop in blood pressure. Decrease the blood pressure to 70 mm Hg by clicking the − button beside the pressure display. Click **Start** to initiate glomerular filtration.

5. Note the glomerular capillary pressure and glomerular filtration rate displays and click **Record Data** to display your results in the grid (and record your results in Chart 3).

6. Click **Refill** to replenish the source beaker and prepare the nephron for the next run.

7. Increase the afferent arteriole radius to 0.60 mm by clicking the + button beside the afferent radius display. Click **Start** to initiate glomerular filtration.

8. Note the glomerular capillary pressure and glomerular filtration rate displays and click **Record Data** to display your results in the grid (and record your results in Chart 3).

9. Click **Refill** to replenish the source beaker and prepare the nephron for the next run.

10. Return the afferent arteriole radius to 0.50 mm by clicking the − button beside the afferent radius display and decrease the efferent radius to 0.35 mm by clicking the button beside the efferent radius display. Click **Start** to initiate glomerular filtration.

11. Note the glomerular capillary pressure and glomerular filtration rate displays and click **Record Data** to display your results in the grid (and record your results in Chart 3).

12. Click **Refill** to replenish the source beaker and prepare the nephron for the next run.

> **? PREDICT Question 1**
> What will happen to the glomerular capillary pressure and glomerular filtration rate if both of these arteriole radii changes are implemented simultaneously with the low blood pressure condition?

CHART 3	Renal Response to Altered Blood Pressure			
Afferent arteriole radius (mm)	**Efferent arteriole radius (mm)**	**Blood pressure (mm Hg)**	**Glomerular capillary pressure (mm Hg)**	**Glomerular filtration rate (ml/min)**

13. Set the afferent arteriole radius to 0.60 mm and keep the efferent arteriole radius at 0.35 mm. Click **Start** to initiate glomerular filtration.

14. Note the glomerular capillary pressure and glomerular filtration rate displays and click **Record Data** to display your results in the grid (and record your results in Chart 3).

After you complete the experiment, take the online **Post-lab Quiz** for Activity 3.

Activity Questions

1. How could an increased urine volume be viewed as beneficial to the body?

_____

_____

2. Diuretics are frequently given to people with persistent high blood pressure. Why?

_____

_____

Solute Gradients and Their Impact on Urine Concentration

OBJECTIVES

1. To understand the terms *antidiuretic hormone (ADH), reabsorption, loop of Henle, collecting duct, tubule lumen, interstitial space,* and *peritubular capillaries.*

2. To explain the process of water reabsorption in specific regions of the nephron.

3. To understand the role of ADH in water reabsorption by the nephron.

4. To describe how the kidneys can produce urine that is four times more concentrated than the blood.

Introduction

As filtrate moves through the tubules of a nephron, solutes and water move *from* the **tubule lumen** *into* the **interstitial spaces** of the nephron. This movement of solutes and water relies on the total solute concentration gradient in the interstitial spaces surrounding the tubule lumen. The interstitial fluid is comprised mostly of NaCl and urea. When the nephron is permeable to solutes or water, equilibrium will be reached between the interstitial fluid and the tubular fluid contents.

Antidiuretic hormone (ADH) increases the water permeability of the **collecting duct,** allowing water to flow to areas of higher solute concentration, from the tubule lumen into the surrounding interstitial spaces. **Reabsorption** describes this movement of filtered solutes and water from the lumen of the renal tubules back into the plasma. The reabsorbed solutes and water that move into the interstitial space need to be returned to the blood, or the kidneys will rapidly

swell like balloons. The **peritubular capillaries** surrounding the renal tubule reclaim the reabsorbed substances and return them to general circulation. Peritubular capillaries arise from the efferent arteriole exiting the glomerulus and empty into the renal veins leaving the kidney.

Without reabsorption, we would excrete the solutes and water that our bodies need to maintain homeostasis. In this activity you will examine the process of passive reabsorption that occurs while filtrate travels through a nephron and urine is formed. While completing the experiment, assume that when ADH is present, the conditions favor the formation of the most concentrated urine possible.

EQUIPMENT USED The following equipment will be depicted on-screen: simulated nephron surrounded by interstitial space between the nephron and peritubular capillaries (Reabsorbed solutes, such as glucose, will move from the lumen of the tubule into the interstitial space, and then into the peritubular capillaries that branch out from the efferent arteriole.); drain beaker for filtrate—simulates the urinary bladder; antidiuretic hormone (ADH).

Experiment Instructions

Go to the home page in the PhysioEx software and click **Exercise 9: Renal System Physiology.** Click **Activity 4: Solute Gradients and Their Impact on Urine Concentration,** and take the online **Pre-lab Quiz** for Activity 4.

After you take the online Pre-lab Quiz, click the **Experiment** tab and begin the experiment. The experiment instructions are reprinted here for your reference. The opening screen for the experiment is shown below.

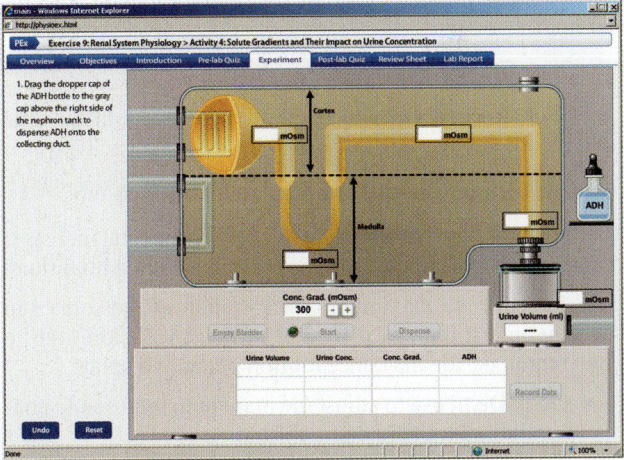

1. Drag the dropper cap of the ADH bottle to the gray cap above the right side of the nephron tank to dispense ADH onto the collecting duct.

2. Click **Dispense** beneath the concentration gradient display to adjust the maximum total solute concentration in the interstitial fluid to 300 mOsm. Because the blood solute concentration is also 300 mOsm, there is no osmotic difference between the lumen of the tubule and the surrounding interstitial fluid.

3. Click **Start** to initiate filtration. Filtrate will flow through the nephron, and solutes and water will move out of the tubules into the interstitial space. Fluid will also move

back into the peritubular capillaries, thus completing the process of reabsorption.

4. Click **Record Data** to display your results in the grid (and record your results in Chart 4).

CHART 4	Solute Gradients and Their Impact on Urine Concentration	
Urine volume (ml)	Urine concentration (mOsm)	Concentration gradient (mOsm)

5. Click **Empty Bladder** to prepare for the next run.

? PREDICT Question 1
What will happen to the urine volume and concentration as the solute gradient in the interstitial space is increased?

6. Increase the maximum concentration of the solutes in the interstitial space to 600 mOsm by clicking the + button beside the concentration gradient display. Click **Dispense** to adjust the maximum total solute concentration in the interstitial fluid.

7. Click **Start** to initiate filtration.

8. Click **Record Data** to display your results in the grid (and record your results in Chart 4).

9. Click **Empty Bladder** to prepare for the next run.

10. You will now observe the effect of incremental increases in maximum total solute concentration in the interstitial fluid.

- Increase the maximum concentration of the solutes in the interstitial space by 300 mOsm by clicking the + button beside the concentration gradient display.

- Click **Dispense** to adjust the maximum total solute concentration in the interstitial fluid.

- Click **Start** to initiate filtration.

- Click **Record Data** to display your results in the grid (and record your results in Chart 4).

- Click **Empty Bladder** to prepare for the next run.

Repeat this step until you reach the maximum total solute concentration in the interstitial fluid of 1200 mOsm.

After you complete the experiment, take the online **Post-lab Quiz** for Activity 4.

Activity Questions

1. From what you learned in this activity, speculate on ways that desert rats are able to concentrate their urine significantly more than humans.

_____

_____

2. Judging from this activity, what would be a reasonable mechanism for diuretics?

_____

_____

ACTIVITY 5

Reabsorption of Glucose via Carrier Proteins

OBJECTIVES

1. To understand the terms *reabsorption, carrier proteins, apical membrane, secondary active transport, facilitated diffusion,* and *basolateral membrane.*

2. To understand the role that glucose carrier proteins play in removing glucose from the filtrate.

3. To understand the concept of a glucose carrier transport maximum and why glucose is not normally present in the urine.

Introduction

Reabsorption is the movement of filtered solutes and water from the lumen of the renal tubules back into the plasma. Without reabsorption, we would excrete the solutes and water that our bodies require for homeostasis.

Glucose is not very large and is therefore easily filtered out of the plasma into Bowman's capsule as part of the filtrate. To ensure that glucose is reabsorbed into the body so that it can fuel cellular metabolism, glucose **carrier proteins** are present in the proximal tubule cells of the nephron. There are a finite number of these glucose carriers in each renal tubule cell. Therefore, if too much glucose is present in the filtrate, it will not all be reabsorbed and glucose will be inappropriately excreted into the urine.

Glucose is first absorbed by **secondary active transport** at the **apical membrane** of proximal tubule cells and then it leaves the tubule cell via **facilitated diffusion** along the **basolateral membrane.** Both types of carrier proteins that transport these molecules across the tubule membranes are transmembrane proteins. Because carrier proteins are needed to move glucose from the lumen of the nephron into the interstitial spaces, there is a limit to the amount of glucose that can be reabsorbed. When all glucose carriers are bound with the glucose they are transporting, excess glucose in the filtrate is eliminated in urine.

In this activity, you will examine the effect of varying the number of glucose transport proteins in the *proximal convoluted tubule.* It is important to note that, normally, the number of glucose carriers is constant in a human kidney and that it is the plasma glucose that varies during the day. Plasma glucose will be held constant in this activity, and the number of glucose carriers will be varied.

> **EQUIPMENT USED** The following equipment will be depicted on-screen: simulated nephron surrounded by interstitial space between the nephron and peritubular capillaries (Reabsorbed solutes, such as glucose, will move from the lumen of the tubule into the interstitial space, and then into the peritubular capillaries that branch out from the efferent arteriole.); drain beaker for filtrate—simulates the urinary bladder; glucose carrier protein control box—used to adjust the number of glucose carriers that will be inserted into the proximal tubule.

Experiment Instructions

Go to the home page in the PhysioEx software and click **Exercise 9: Renal System Physiology.** Click **Activity 5: Reabsorption of Glucose via Carrier Proteins,** and take the online **Pre-lab Quiz** for Activity 5.

After you take the online Pre-lab Quiz, click the **Experiment** tab and begin the experiment. The experiment instructions are reprinted here for your reference. The opening screen for the experiment is shown below.

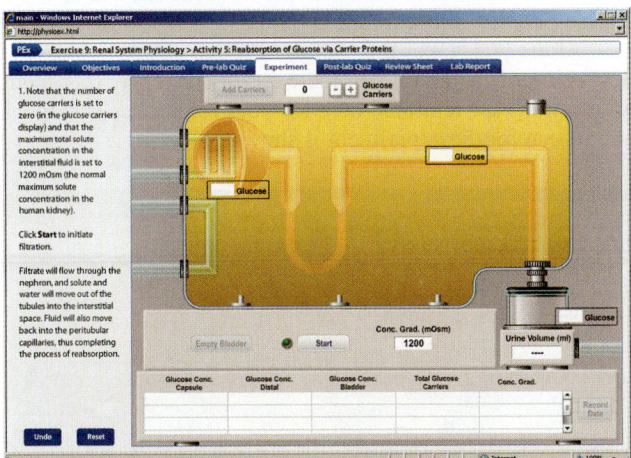

1. Note that the number of glucose carriers is set to zero (in the glucose carriers display) and that the maximum total solute concentration in the interstitial fluid is set to 1200 mOsm (the normal maximum solute concentration in the human kidney). Click **Start** to initiate filtration. Filtrate will flow through the nephron, and solute and water will move out of the tubules into the interstitial space. Fluid will also move back into the peritubular capillaries, thus completing the process of reabsorption.

2. Click **Record Data** to display your results in the grid (and record your results in Chart 5). The concentrations of glucose in Bowman's capsule, the distal convoluted tubule, and the urinary bladder will be displayed in the grid.

CHART 5	Reabsorption of Glucose via Carrier Proteins		
Glucose concentration (mM)			
Bowman's capsule	Distal convoluted tubule	Urinary bladder	Glucose carriers

3. Click **Empty Bladder** to prepare the nephron for the next run.

> **? PREDICT Question 1**
> What will happen to the glucose concentration in the urinary bladder as glucose carriers are added to the proximal tubule?

4. Increase the number of glucose carriers to 100 (an arbitrary number) by clicking the + button beside the glucose carriers display. Click **Add Carriers** to insert the specified number of glucose carrier proteins per unit area into the membrane of the proximal tubule.

5. Click **Start** to initiate filtration.

6. Click **Record Data** to display your results in the grid (and record your results in Chart 5).

7. Click **Empty Bladder** to prepare the nephron for the next run.

8. You will now observe the effect of incremental increases in the number of glucose carriers.

- Increase the number of glucose carriers by 100 by clicking the + button beside the glucose carriers display.

- Click **Add Carriers** to insert the specified number of glucose carrier proteins per unit area into the membrane of the proximal tubule.

- Click **Start** to initiate filtration.

- Click **Record Data** to display your results in the grid (and record your results in Chart 5).

- Click **Empty Bladder** to prepare the nephron for the next run.

Repeat this step until you have inserted 400 glucose carrier proteins per unit area into the membrane of the proximal tubule.

After you complete the experiment, take the online **Post-lab Quiz** for Activity 5.

Activity Questions

1. Why would your family physician at the turn of the twentieth century taste your urine?

_____

_____ ▬

The Effect of Hormones on Urine Formation

OBJECTIVES

1. To understand the terms *antidiuretic hormone (ADH), aldosterone, reabsorption, loop of Henle, distal convoluted tubule, collecting duct, tubule lumen,* and *interstitial space.*

2. To understand how the hormones aldosterone and ADH affect renal processes in a human kidney.

3. To understand the role of ADH in water reabsorption by the nephron.

4. To understand the role of aldosterone in solute reabsorption and secretion by the nephron.

Introduction

The concentration and volume of urine excreted by our kidneys will change depending on what our body needs for homeostasis. For example, if a person consumes a large quantity of water, the excess water will be eliminated as a large volume of dilute urine. On the other hand, when dehydration occurs, there is a clear benefit in being able to produce a small volume of concentrated urine to retain water. Activity 4 demonstrated how the total solute concentration gradient in the interstitial spaces surrounding the tubule lumen makes it possible to excrete concentrated urine.

Aldosterone is a hormone produced by the adrenal cortex under the control of the body's *renin-angiotensin system.* A decrease in blood pressure is detected by cells in the afferent arteriole, triggering the release of renin. Renin acts as a proteolytic enzyme, causing angiotensinogen to be converted into angiotensin I. Endothelial cells throughout the body possess a *converting enzyme* that converts angiotensin I into angiotensin II. Angiotensin II signals the adrenal cortex to secrete aldosterone. Aldosterone acts on the distal convoluted tubule cells in the nephron to promote the reabsorption of sodium from filtrate *into* the body and the secretion of potassium *from* the body. This electrolyte shift, coupled with the addition of **antidiuretic hormone (ADH),** also causes more water to be reabsorbed into the blood, resulting in increased blood pressure.

ADH is manufactured by the hypothalamus and stored in the posterior pituitary gland. ADH levels are influenced by the osmolality of body fluids and the volume and pressure of the cardiovascular system. A 1% change in body osmolality will cause this hormone to be secreted. The primary action of this hormone is to increase the permeability of the collecting duct to water so that more water is reabsorbed into the

body by inserting aquaporins, or water channels, in the apical membrane. Without this water reabsorption, the body would quickly dehydrate.

Thus, our kidneys tightly regulate the amount of water and solutes excreted to maintain water balance in the body. If water intake is down, or if there has been a fluid loss from the body, the kidneys work to conserve water by making the urine very hyperosmotic (having a relatively high solute concentration) to the blood. If there has been a large intake of fluid, the urine is more hypo-osmotic. In the normal individual, urine osmolarity varies from 50 to 1200 milliosmoles/kg of water.

> **EQUIPMENT USED** The following equipment will be depicted on-screen: simulated nephron surrounded by interstitial space between the nephron and peritubular capillaries (Reabsorbed solutes, such as glucose, will move from the lumen of the tubule into the interstitial space, and then into the peritubular capillaries that branch out from the efferent arteriole.); drain beaker for filtrate—simulates the urinary bladder; aldosterone; antidiuretic hormone (ADH).

Experiment Instructions

Go to the home page in the PhysioEx software and click **Exercise 9: Renal System Physiology.** Click **Activity 6: The Effect of Hormones on Urine Formation,** and take the online **Pre-lab Quiz** for Activity 6.

After you take the online Pre-lab Quiz, click the **Experiment** tab and begin the experiment. The experiment instructions are reprinted here for your reference. The opening screen for the experiment is shown below.

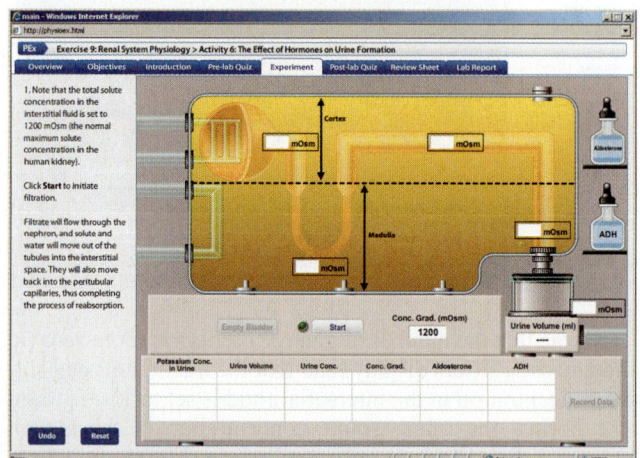

1. Note that the total solute concentration in the interstitial fluid is set to 1200 mOsm (the normal maximum solute concentration in the human kidney). Click **Start** to initiate filtration. Filtrate will flow through the nephron, and solute and water will move out of the tubules into the interstitial space. They will also move back into the peritubular capillaries, thus completing the process of reabsorption.

2. Click **Record Data** to display your results in the grid (and record your results in Chart 6). You will use this baseline data to compare the conditions of the filtrate and urine volume in the presence of the hormones aldosterone and ADH.

CHART 6	The Effect of Hormones on Urine Formation			
Potassium concentration (m*M*)	Urine volume (ml)	Urine concentration (mOsm)	Aldosterone	ADH

3. Click **Empty Bladder** to prepare the nephron for the next run.

PREDICT Question 1
What will happen to the urine volume (compared with baseline) when aldosterone is added to the distal tubule?

4. Drag the dropper cap of the aldosterone bottle to the gray cap above the right side of the nephron tank to dispense aldosterone into the tank surrounding the distal tubule and the collecting duct.

5. Click **Start** to initiate filtration.

6. Click **Record Data** to display your results in the grid (and record your results in Chart 6).

7. Click **Empty Bladder** to prepare the nephron for the next run.

8. Drag the dropper cap of the ADH bottle to the gray cap at the top right side of the nephron tank to dispense ADH into the tank surrounding the distal tubule and the collecting duct.

PREDICT Question 2
What will happen to the urine volume (compared with baseline) when ADH is added to the collecting duct?

9. Click **Start** to initiate filtration.

10. Click **Record Data** to display your results in the grid (and record your results in Chart 6).

11. Click **Empty Bladder** to prepare the nephron for the next run.

PREDICT Question 3
What will happen to the urine volume and the urine concentration (compared with baseline) in the presence of both aldosterone and ADH?

12. Drag the dropper cap of the aldosterone bottle and then the dropper cap of the ADH bottle to the gray cap above the right side of the nephron tank to dispense aldosterone and ADH into the tank surrounding the distal convoluted tubule and the collecting duct.

13. Click **Start** to initiate filtration.

14. Click **Record Data** to display your results in the grid (and record your results in Chart 6).

After you complete the experiment, take the online **Post-lab Quiz** for Activity 6.

Activity Questions

1. Why does ethanol consumption lead to a dramatic increase in urine production?

2. Why do angiotensin converting enzyme (ACE) inhibitors given to people with hypertension lead to increased urine production?

NAME _____

LAB TIME/DATE _____

Renal System Physiology

ACTIVITY 1   The Effect of Arteriole Radius on Glomerular Filtration

1. What are two primary functions of the kidney? _____

   _____

2. What are the components of the renal corpuscle? _____

   _____

3. Starting at the renal corpuscle, list the components of the renal tubule as they are encountered by filtrate. _____

   _____

   _____

4. Describe the effect of decreasing the afferent arteriole radius on glomerular capillary pressure and filtration rate. How well

   did the results compare with your prediction? _____

   _____

5. Describe the effect of increasing the afferent arteriole radius on glomerular capillary pressure and filtration rate. How well

   did the results compare with your prediction? _____

   _____

6. Describe the effect of decreasing the efferent arteriole radius on glomerular capillary pressure and filtration rate. How well

   did the results compare with your prediction? _____

   _____

7. Describe the effect of increasing the efferent radius on glomerular capillary pressure and filtration rate. _____

   _____

ACTIVITY 2   The Effect of Pressure on Glomerular Filtration

1. As blood pressure increased, what happened to the glomerular capillary pressure and the glomerular filtration rate? How

   well did the results compare with your prediction? _____

   _____

2. Compare the urine volume in your baseline data with the urine volume as you increased the blood pressure. How did the

   urine volume change? _____

   _____

3. How could the change in urine volume with the increase in blood pressure be viewed as being beneficial to the body?

   _____

   _____

4. When the one-way valve between the collecting duct and the urinary bladder was closed, what happened to the filtrate pressure in Bowman's capsule (this is not directly measured in this experiment) and the glomerular filtration rate? How well did

   the results compare with your prediction? _____

   _____

5. How did increasing the blood pressure alter the results when the valve was closed? _____

   _____

ACTIVITY 3 **Renal Response to Altered Blood Pressure**

1. List the several mechanisms you have explored that change the glomerular filtration rate. How does each mechanism specifi-

   cally alter the glomerular filtration rate? _____

   _____

2. Describe and explain what happened to the glomerular capillary pressure and glomerular filtration rate when *both* arteriole
   radii changes were implemented simultaneously with the low blood pressure condition. How well did the results compare

   with your prediction? _____

   _____

   _____

3. How could you adjust the afferent or efferent radius to compensate for the effect of reduced blood pressure on the glomerular

   filtration rate? _____

   _____

4. Which arteriole radius adjustment was more effective at compensating for the effect of low blood pressure on the glomerular

   filtration rate? Explain why you think this difference occurs. _____

   _____

5. In the body, how does a nephron maintain a near-constant glomerular filtration rate despite a constantly fluctuating blood

   pressure? _____

ACTIVITY 4 Solute Gradients and Their Impact on Urine Concentration

1. What happened to the urine concentration as the solute concentration in the interstitial space was increased? How well did the results compare to your prediction? _____

   _____

   _____

2. What happened to the volume of urine as the solute concentration in the interstitial space was increased? How well did the results compare to your prediction? _____

   _____

3. What do you think would happen to urine volume if you did not add ADH to the collecting duct? _____

   _____

   _____

4. Is most of the tubule filtrate reabsorbed into the body or excreted in urine? Explain. _____

   _____

   _____

5. Can the reabsorption of solutes influence water reabsorption from the tubule fluid? Explain. _____

   _____

ACTIVITY 5 Reabsorption of Glucose via Carrier Proteins

1. What happens to the concentration of glucose in the urinary bladder as the number of glucose carriers increases? _____

   _____

   _____

2. What types of transport are utilized during glucose reabsorption and where do they occur? _____

   _____

3. Why does the glucose concentration in the urinary bladder become zero in these experiments? _____

   _____

4. A person with type 1 diabetes cannot make insulin in the pancreas, and a person with untreated type 2 diabetes does not respond to the insulin that is made in the pancreas. In either case, why would you expect to find glucose in the person's urine?

   _____

   _____

ACTIVITY 6 The Effect of Hormones on Urine Formation

1. How did the addition of aldosterone affect urine volume (compared with baseline)? Can the reabsorption of solutes influence

   water reabsorption in the nephron? Explain. How well did the results compare with your prediction? _____

   _____

   _____

2. How did the addition of ADH affect urine volume (compared with baseline)? How well did the results compare with your
   prediction? Why did the addition of ADH also affect the concentration of potassium in the urine (compared with baseline)?

   _____

   _____

3. What is the principal determinant for the release of aldosterone from the adrenal cortex? _____

   _____

4. How did the addition of both aldosterone and ADH affect urine volume (compared with baseline)? How well did the results

   compare with your prediction? _____

   _____

   _____

5. What is the principal determinant for the release of ADH from the posterior pituitary gland? Does ADH favor the formation

   of dilute or concentrated urine? Explain why. _____

   _____

   _____

6. Which hormone (aldosterone or ADH) has the greater effect on urine volume? Why? _____

   _____

   _____

7. If ADH is not available, can the urine concentration still vary? Explain your answer. _____

   _____

   _____

8. Consider this situation: you want to reabsorb sodium ions but you do not want to increase the volume of the blood by
   reabsorbing large amounts of water from the filtrate. Assuming that aldosterone and ADH are both present, how would you

   adjust the hormones to accomplish the task? _____

   _____

Acid-Base Balance

Exercise Overview

pH denotes the hydrogen ion concentration, $[H^+]$, in a solution (such as body fluids). The reciprocal relationship between pH and $[H^+]$ is defined by the following equation.

$$pH = \log(1/[H^+])$$

Because the relationship is reciprocal, $[H^+]$ is higher at *lower* pH values (indicating higher acid levels) and lower at *higher* pH values (indicating lower acid levels).

The pH of a body's fluid is also referred to as its **acid-base balance.** An **acid** is a substance that releases H^+ in solution. A **base,** often a hydroxyl ion (OH^-) or bicarbonate ion (HCO_3^-), is a substance that binds, or buffers, the H^+. A **strong acid** completely dissociates in solution, releasing all of its hydrogen ions and, thus, lowering the solution's pH. A **weak acid** dissociates incompletely and does not release all of its hydrogen ions in solution, producing a lesser effect on the solution's pH. A **strong base** has a strong tendency to bind to H^+, raising the solution's pH. A **weak base** binds less of the H^+, producing a lesser effect on the solution's pH.

The pH of body fluids is very tightly regulated. Blood and tissue fluids normally have a pH between 7.35 and 7.45. Under pathological conditions, blood pH as low as 6.9 or as high as 7.8 has been recorded, but a higher or lower pH cannot sustain human life. The narrow range from 7.35 to 7.45 is remarkable when you consider the vast number of biochemical reactions that take place in the body. The human body normally produces a large amount of H^+ as the result of metabolic processes; ingested acids; and the products of fat, sugar, and amino acid metabolism. The regulation of a relatively constant internal pH is one of the major physiological functions of the body's organ systems.

To maintain pH homeostasis, the body utilizes both *chemical* and *physiological* buffering systems. Chemical buffers are composed of a mixture of weak acids and weak bases. They help regulate the body's pH levels by binding H^+ and removing it from solution as its concentration begins to rise or by releasing H^+ into solution as its concentration begins to fall. The body's three major chemical buffering systems are the *bicarbonate, phosphate,* and *protein buffer systems.* We will not focus on chemical buffering systems in this exercise, but keep in mind that chemical buffers are the fastest form of compensation and can return pH to normal within a fraction of a second.

The body's two major physiological buffering systems are the **renal system** and the **respiratory system.** The renal system is the slower of the two, taking hours to days to do its work. The respiratory system usually works within minutes, but cannot handle the amount of pH change that the renal system can. These physiological buffer systems help regulate body pH by controlling the output of acids, bases, or carbon dioxide (CO_2) from the body. For example, if there is too much acid in the body, the renal system may respond by excreting more H^+ from the body in urine. Similarly, if there is too much carbon dioxide in the blood, the respiratory system may respond by increasing ventilation to expel the excess carbon dioxide. Carbon dioxide levels have a direct effect on pH because the addition of carbon dioxide to the blood results in the generation of more H^+. The following equation shows what happens when carbon dioxide combines with water in the blood, producing carbonic acid.

$$H_2O + CO_2 \rightleftharpoons \underset{\substack{\text{carbonic}\\\text{acid}}}{H_2CO_3} \rightleftharpoons H^+ + \underset{\substack{\text{bicarbonate}\\\text{ion}}}{HCO_3^-}$$

ACTIVITY 1

Hyperventilation

OBJECTIVES

1. To introduce pH homeostasis in the body.
2. To understand the normal ranges for pH and P_{CO_2}.
3. To recognize respiratory alkalosis and its causes.
4. To interpret an oscilloscope tracing for hyperventilation and compare it with a tracing for normal breathing.

Introduction

Acid-base imbalances can have respiratory and metabolic causes. When diagnosing these disorders, two key signs are evaluated: the pH and the partial pressure of carbon dioxide in the blood (P_{CO_2}). The normal range for pH is between 7.35 and 7.45, and the normal range for P_{CO_2} is between 35 and 45 mm Hg. When the pH falls below 7.35, the body is said to be in a state of **acidosis.** When the pH rises above 7.45, the body is said to be in a state of **alkalosis.**

Respiratory alkalosis is the condition of too little carbon dioxide in the blood. Respiratory alkalosis commonly results from traveling to high altitude (where the air contains

less oxygen) or hyperventilation, which can be brought on by fever, panic attack, or anxiety. Hyperventilation, defined as an increase in the rate and depth of breathing, removes carbon dioxide from the blood faster than it is being produced by the cells of the body, reducing the amount of H^+ in the blood and, thus, increasing the blood's pH. The following equation shows the shift in the equilibrium that results in the increase in blood pH due to less carbon dioxide in the blood.

$$H_2O + CO_2 \leftarrow \underset{\substack{\text{carbonic}\\\text{acid}}}{H_2CO_3} \leftarrow H^+ + \underset{\substack{\text{bicarbonate}\\\text{ion}}}{HCO_3^-}$$

The renal system can compensate for alkalosis by retaining H^+ and excreting bicarbonate ions to lower the blood pH levels back to the normal range.

> **EQUIPMENT USED** The following equipment will be depicted on-screen: simulated lung chamber; pH meter; oscilloscope; two breathing patterns: normal and hyperventilation.

Experiment Instructions

Go to the home page in the PhysioEx software and click **Exercise 10: Acid-Base Balance.** Click **Activity 1: Hyperventilation,** and take the online **Pre-lab Quiz** for Activity 1.

After you take the online Pre-lab Quiz, click the **Experiment** tab and begin the experiment. The experiment instructions are reprinted here for your reference. The opening screen for the experiment is shown below.

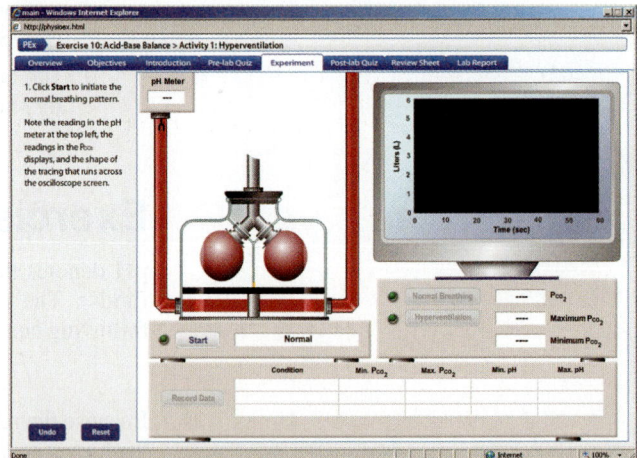

1. Click **Start** to initiate the normal breathing pattern. Note the reading in the pH meter at the top left, the readings in the P_{CO_2} displays, and the shape of the tracing that runs across the oscilloscope screen.

2. Click **Record Data** to display your results in the grid (and record your results in Chart 1).

> **? PREDICT Question 1**
> What do you think will happen to the pH and P_{CO_2} levels with hyperventilation?

CHART 1	Hyperventilation Breathing Patterns			
Condition	Minimum P_{CO_2}	Maximum P_{CO_2}	Minimum pH	Maximum pH

3. Click **Start** to initiate the normal breathing pattern. After the normal breathing tracing runs for 10 seconds, click **Hyperventilation** to initiate the hyperventilation breathing pattern. Note the reading in the pH meter at the top left, the readings in the P_{CO_2} displays, and the shape of the tracing that runs across the oscilloscope screen.

4. Click **Record Data** to display your results in the grid (and record your results in Chart 1).

5. Click **Start** to initiate the normal breathing pattern. After the normal breathing tracing runs for 10 seconds, click **Hyperventilation** to initiate the hyperventilation breathing pattern. After the hyperventilation tracing runs for 10 seconds, click **Normal Breathing** to return to the normal breathing pattern. Note the reading in the pH meter at the top left, the readings in the P_{CO_2} displays, and the shape of the tracing that runs across the oscilloscope screen.

6. Click **Record Data** to display your results in the grid (and record your results in Chart 1).

After you complete the experiment, take the online **Post-lab Quiz** for Activity 1.

Activity Questions

1. At what pH range is the body considered to be in a state of respiratory alkalosis?

_____

_____

2. How can the body compensate for respiratory alkalosis?

_____

_____

3. How did the tidal volume change with hyperventilation?

_____

_____

4. What might cause a person to hyperventilate?

_____

_____

ACTIVITY 2

Rebreathing

OBJECTIVES

1. To understand how rebreathing can simulate hypoventilation.
2. To observe the results of respiratory acidosis.
3. To describe the causes of respiratory acidosis.

Introduction

The body is said to be in a state of **acidosis** when the pH of the blood falls below 7.35 (although a pH of 7.35 is technically not acidic). Respiratory acidosis is the result of impaired respiration, or *hypoventilation*, which leads to the accumulation of too much carbon dioxide in the blood. The causes of impaired respiration include airway obstruction, depression of the respiratory center in the brain stem, lung disease (such as emphysema and chronic bronchitis), and drug overdose.

Recall that carbon dioxide contributes to the formation of carbonic acid when it combines with water through a reversible reaction catalyzed by carbonic anhydrase. The carbonic acid then dissociates into hydrogen ions and bicarbonate ions. Because hypoventilation results in elevated carbon dioxide levels in the blood, the equilibrium shifts, the H^+ levels increase, and the pH value of the blood decreases.

$$H_2O + CO_2 \rightarrow \underset{\substack{\text{carbonic}\\\text{acid}}}{H_2CO_3} \rightarrow H^+ + \underset{\substack{\text{bicarbonate}\\\text{ion}}}{HCO_3^-}$$

Rebreathing is the action of breathing in air that was just expelled from the lungs. Rebreathing results in the accumulation of carbon dioxide in the blood. Breathing into a paper bag is an example of rebreathing. (Note that breathing into a paper bag can deplete the body of oxygen and is therefore not the best therapy for hyperventilation because it can mask other life-threatening emergencies, such as a heart attack or asthma.) In this activity, you will observe what happens to pH and carbon dioxide levels in the blood during rebreathing. In the body, the kidneys regulate the acid-base balance by altering the amount of H^+ and HCO_3^- excreted in the urine.

> **EQUIPMENT USED** The following equipment will be depicted on-screen: simulated lung chamber; pH meter; oscilloscope; two breathing patterns: normal and rebreathing.

Experiment Instructions

Go to the home page in the PhysioEx software and click **Exercise 10: Acid-Base Balance.** Click **Activity 2: Rebreathing,** and take the online **Pre-lab Quiz** for Activity 2.

After you take the online Pre-lab Quiz, click the **Experiment** tab and begin the experiment. The experiment instructions are reprinted here for your reference. The opening screen for the experiment is shown below.

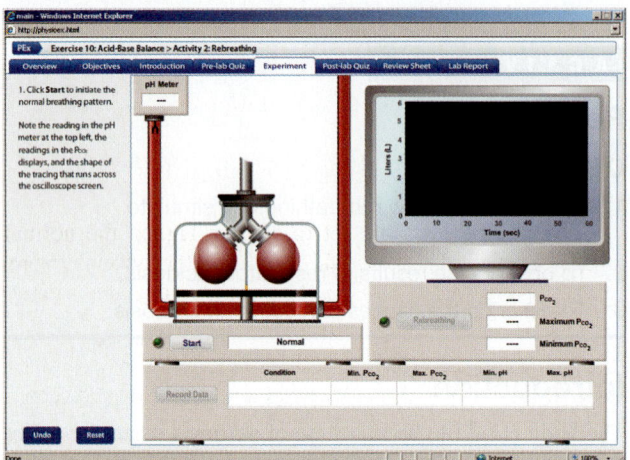

1. Click **Start** to initiate the normal breathing pattern. Note the reading in the pH meter at the top left, the readings in the P_{CO_2} displays, and the shape of the tracing that runs across the oscilloscope screen.

2. Click **Record Data** to display your results in the grid (and record your results in Chart 2).

> **? PREDICT Question 1**
> What do you think will happen to the pH and P_{CO_2} levels during rebreathing?

3. Click **Start** to initiate the normal breathing pattern. After the normal breathing tracing runs for 10 seconds, click **Rebreathing** to initiate the rebreathing pattern. Note the reading in the pH meter at the top left, the readings in the P_{CO_2} displays, and the shape of the tracing that runs across the oscilloscope screen.

4. Click **Record Data** to display your results in the grid (and record your results in Chart 2).

After you complete the experiment, take the online **Post-lab Quiz** for Activity 2.

Activity Questions

1. Did the pH level of the blood change at all with rebreathing? If so, how did it change?

_____

_____

2. What happens to the pH level of the blood when there is too much carbon dioxide remaining in the blood?

_____

_____

3. How did the tidal volumes change with rebreathing?

_____

_____

4. Describe two ways in which too much carbon dioxide might remain in the blood.

_____

_____

Renal Responses to Respiratory Acidosis and Respiratory Alkalosis

OBJECTIVES

1. To understand renal compensation mechanisms for respiratory acidosis and respiratory alkalosis.
2. To explore the functional unit of the kidneys that responds to acid-base balance.
3. To observe the changes in ion concentrations that occur with renal compensation.

Introduction

The kidneys play a major role in maintaining fluid, electrolyte, and acid-base balance in the body's internal environment. By regulating the amount of water lost in the urine, the kidneys defend the body against excessive hydration or dehydration. By regulating the acidity of urine and the rate of electrolyte excretion, the kidneys maintain plasma pH and electrolyte levels within normal limits.

CHART 2	Normal Breathing Patterns			
Condition	Minimum P_{CO_2}	Maximum P_{CO_2}	Minimum pH	Maximum pH

Renal compensation is the body's primary method of compensating for conditions of respiratory acidosis or respiratory alkalosis. The kidneys regulate the acid-base balance by altering the amount of H^+ and HCO_3^- excreted in the urine. If we revisit the equation for the dissociation of carbonic acid, a weak acid, we see that the conservation of bicarbonate ion (base) has the same net effect as the loss of acid, H^+.

$$H_2O + CO_2 \rightleftarrows \underset{\substack{\text{carbonic} \\ \text{acid}}}{H_2CO_3} \rightleftarrows H^+ + \underset{\substack{\text{bicarbonate} \\ \text{ion}}}{HCO_3^-}$$

In this activity you will examine how the renal system compensates for respiratory acidosis or respiratory alkalosis. Respiratory acidosis is generally caused by the accumulation of carbon dioxide in the blood from hypoventilation, but it can also be caused by rebreathing. Acidosis results in a lower-than-normal blood pH. Respiratory alkalosis is caused by a depletion of carbon dioxide, often caused by an episode of hyperventilation, and results in an elevated blood pH.

You will primarily be working with the variable P_{CO_2}. Recall that the normal range for pH is between 7.35 and 7.45 and the normal range for P_{CO_2} is between 35 and 45 mm Hg. You will observe how increases and decreases in P_{CO_2} affect the levels of H^+ and HCO_3^- that the kidneys excrete in urine. The functional unit for adjusting the plasma composition is the **nephron.** Remember that although the renal system can partially compensate for pH imbalances with a respiratory cause, the kidneys cannot fully compensate if respirations have not returned to normal because the carbon dioxide levels will still be abnormal.

EQUIPMENT USED The following equipment will be depicted on-screen: source beaker for blood (first beaker on left side of screen); drain beaker for blood (second beaker on left side of screen); simulated nephron (The filtrate forms in Bowman's capsule and flows through the renal tubule—the tubular components, and empties into a collecting duct which, in turn drains into the urinary bladder.); nephron tank; glomerulus—"ball" of capillaries that forms part of the filtration membrane; glomerular (Bowman's) capsule—forms part of the filtration membrane and a capsular space where the filtrate initially forms; proximal convoluted tubule; loop of Henle; distal convoluted tubule; collecting duct; drain beaker for filtrate (beaker on right side of screen)—simulates the urinary bladder.

Experiment Instructions

Go to the home page in the PhysioEx software and click **Exercise 10: Acid-Base Balance.** Click **Activity 3: Renal Responses to Respiratory Acidosis and Respiratory Alkalosis** and take the online **Pre-lab Quiz** for Activity 3.

After you take the online Pre-lab Quiz, click the **Experiment** tab and begin the experiment. The experiment instructions are reprinted here for your reference. The opening screen for the experiment is shown above.

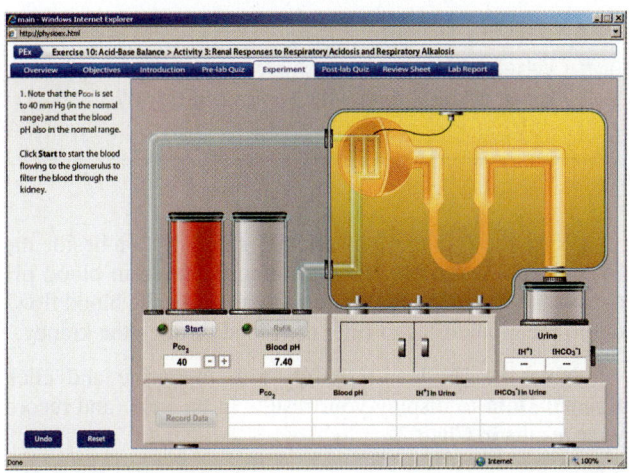

1. Note that the P_{CO_2} is set to 40 mm Hg (in the normal range) and that the blood pH is also in the normal range. Click **Start** to start the blood flowing to the glomerulus to filter the blood through the kidney.

2. Note the $[H^+]$ and $[HCO_3^-]$ in the urine and click **Record Data** to display your results in the grid (and record your results in Chart 3).

CHART 3	Renal Responses to Respiratory Acidosis and Respiratory Alkalosis		
P_{CO_2}	Blood pH	$[H^+]$ in urine	$[HCO_3^-]$ in urine

3. Click **Refill** to replenish the source beaker.

? PREDICT Question 1
What effect do you think lowering the P_{CO_2} will have on $[H^+]$ and $[HCO_3^-]$ in the urine?

---

4. Lower the P_{CO_2} to 30 by clicking the − button beside the P_{CO_2} display. Note the corresponding increase in blood pH (above the normal range). Click **Start** to start the blood flowing to the glomerulus to filter the blood through the kidney.

5. Note the $[H^+]$ and $[HCO_3^-]$ in the urine and click **Record Data** to display your results in the grid (and record your results in Chart 3).

6. Click **Refill** to replenish the source beaker.

? **PREDICT** Question 2
What effect do you think raising the P_{CO_2} will have on $[H^+]$ and $[HCO_3^-]$ in the urine?

_____

_____

7. Raise the P_{CO_2} to 60 by clicking the $+$ button beside the P_{CO_2} display. Note the corresponding decrease in blood pH (below the normal range). Click **Start** to start the blood flowing to the glomerulus to filter the blood through the kidney.

8. Note the $[H^+]$ and $[HCO_3^-]$ in the urine and click **Record Data** to display your results in the grid (and record your results in Chart 3).

After you complete the experiment, take the online **Post-lab Quiz** for Activity 3.

Activity Questions

1. Describe how the kidneys respond to respiratory acidosis.

_____

_____

2. What P_{CO_2} corresponded to respiratory acidosis?

_____

3. Describe how the kidneys respond to respiratory alkalosis.

_____

_____

4. What P_{CO_2} corresponded to respiratory alkalosis?

_____ ▬▬

Respiratory Responses to Metabolic Acidosis and Metabolic Alkalosis

OBJECTIVES

1. To understand the causes of metabolic acidosis and metabolic alkalosis.
2. To observe the physiological changes that occur with an increase and decrease in metabolic rate.
3. To explain how the respiratory system compensates for metabolic acidosis and alkalosis.

Introduction

Conditions of acidosis and alkalosis that do not have respiratory causes are termed *metabolic acidosis and metabolic alkalosis*. **Metabolic acidosis** is characterized by low plasma HCO_3^- and pH. The causes of metabolic acidosis include:

- **Ketoacidosis,** a buildup of keto acids that can result from diabetes mellitus
- **Salicylate poisoning,** a toxic condition resulting from ingestion of too much aspirin or oil of wintergreen (a substance often found in laboratories)
- The ingestion of too much alcohol, which metabolizes into acetic acid
- Diarrhea, which results in the loss of bicarbonate with the elimination of intestinal contents
- Strenuous exercise, which can cause a buildup of lactic acid from anaerobic muscle metabolism

Metabolic alkalosis is characterized by elevated plasma HCO_3^- and pH. The causes of metabolic alkalosis include:

- Ingestion of alkali, such as antacids or bicarbonate
- Vomiting, which can result in the loss of too much H^+
- Constipation, which may result in significant reabsorption of HCO_3^-

Increases or decreases in the body's normal metabolic rate can also result in metabolic acidosis or alkalosis. Recall that carbon dioxide—a waste product of metabolism—mixes with water in plasma to form carbonic acid, which in turn forms H^+.

$$H_2O + CO_2 \rightleftarrows \underset{\substack{\text{carbonic} \\ \text{acid}}}{H_2CO_3} \rightleftarrows H^+ + \underset{\substack{\text{bicarbonate} \\ \text{ion}}}{HCO_3^-}$$

An increase in the normal metabolic rate causes more carbon dioxide to form as a metabolic waste product, resulting in the formation of more H^+ and, therefore, lower plasma pH, potentially causing acidosis. Other acids that are also normal metabolic waste products (such as ketone bodies and phosphoric, uric, and lactic acids) would likewise accumulate with an increase in metabolic rate.

Conversely, a decrease in the normal metabolic rate causes less carbon dioxide to form as a metabolic waste product, resulting in the formation of less H^+ and, therefore, higher plasma pH, potentially causing alkalosis. Many factors can affect the rate of cell metabolism. For example, fever, stress, or the ingestion of food all cause the rate of cell metabolism to *increase*. Conversely, a fall in body temperature or a decrease in food intake causes the rate of cell metabolism to *decrease*.

The respiratory system compensates for metabolic acidosis or alkalosis by expelling or retaining carbon dioxide in the blood. During metabolic acidosis, respiration increases to expel carbon dioxide from the blood, thus decreasing $[H^+]$ and raising the pH. During metabolic alkalosis, respiration decreases to promote the accumulation of carbon dioxide in the blood, thus increasing $[H^+]$ and decreasing the pH.

The renal system also compensates for metabolic acidosis and alkalosis by conserving or excreting bicarbonate ions. Nevertheless, in this activity, you will focus on respiratory compensation of metabolic acidosis and alkalosis.

EQUIPMENT USED   The following equipment will be depicted on-screen: simulated heart pump; simulated lung chamber; oscilloscope.

Experiment Instructions

Go to the home page in the PhysioEx software and click **Exercise 10: Acid-Base Balance.** Click **Activity 4: Respiratory Responses to Metabolic Acidosis and Metabolic Alkalosis,** and take the online **Pre-lab Quiz** for Activity 4.

After you take the online Pre-lab Quiz, click the **Experiment** tab and begin the experiment. The experiment instructions are reprinted here for your reference. The opening screen for the experiment is shown below.

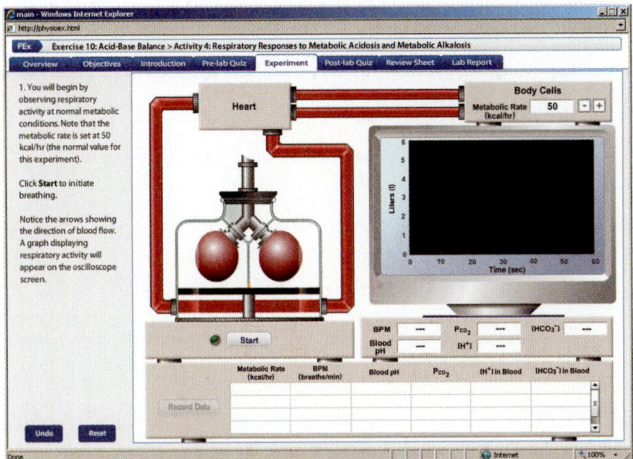

1. You will begin by observing respiratory activity at normal metabolic conditions. Note that the metabolic rate is set at 50 kcal/hr (the normal value for this experiment). Click **Start** to initiate breathing and blood flow. Notice the arrows showing the direction of blood flow. A graph displaying respiratory activity will appear on the oscilloscope screen.

2. Note the data in the displays below the oscilloscope screen and click **Record Data** to display your results in the grid (and record your results in Chart 4).

3. Increase the metabolic rate to 60 kcal/hr by clicking the + button beside the metabolic rate display. Click **Start** to initiate breathing and blood flow.

4. Note the data in the displays below the oscilloscope screen and click **Record Data** to display your results in the grid (and record your results in Chart 4).

5. Click **Clear Tracings** to clear the tracings on the oscilloscope.

? PREDICT Question 1
What do you think will happen when the metabolic rate is increased to 80 kcal/hr?

6. Increase the metabolic rate to 80 kcal/hr by clicking the + button beside the metabolic rate display. Click **Start** to initiate breathing and blood flow.

7. Note the data in the displays below the oscilloscope screen and click **Record Data** to display your results in the grid (and record your results in Chart 4).

8. Click **Clear Tracings** to clear the tracings on the oscilloscope.

9. Decrease the metabolic rate to 40 kcal/hr by clicking the − button beside the metabolic rate display. Click **Start** to initiate breathing and blood flow.

10. Note the data in the displays below the oscilloscope screen and click **Record Data** to display your results in the grid (and record your results in Chart 4).

11. Click **Clear Tracings** to clear the tracings on the oscilloscope.

? PREDICT Question 2
What do you think will happen when the metabolic rate is decreased to 20 kcal/hr?

CHART 4	**Respiratory Responses to Metabolic Acidosis and Metabolic Alkalosis**				
Metabolic rate	**BPM (breaths/min)**	**Blood pH**	**P_{CO_2}**	**[H$^+$] in blood**	**[HCO$_3^-$] in blood**

12. Decrease the metabolic rate to 20 kcal/hr by clicking the button beside the metabolic rate display. Click **Start** to initiate breathing and blood flow.

13. Note the data in the displays below the oscilloscope screen and click **Record Data** to display your results in the grid (and record your results in Chart 4).

After you complete the experiment, take the online **Post-lab Quiz** for Activity 4.

Activity Questions

1. Describe what happens to carbon dioxide and pH with increased metabolism.

_____

_____

2. Describe the respiratory response to metabolic acidosis.

_____

_____

3. When the respiratory system compensates for the metabolic acidosis, does the pH increase or decrease in value?

_____

_____

4. Describe the respiratory response to metabolic alkalosis.

_____

_____

NAME _____

LAB TIME/DATE _____

Acid-Base Balance

ACTIVITY 1 Hyperventilation

1. Describe the normal ranges for pH and carbon dioxide in the blood. _____

_____

_____

2. Describe what happened to the pH and the carbon dioxide levels with hyperventilation. How well did the results compare with your prediction? _____

_____

_____

3. Explain how returning to normal breathing after hyperventilation differed from hyperventilation without returning to normal breathing. _____

_____

_____

4. Describe some possible causes of respiratory alkalosis. _____

_____

_____

ACTIVITY 2 Rebreathing

1. Describe what happened to the pH and the carbon dioxide levels during rebreathing. How well did the results compare with your prediction? _____

_____

_____

2. Describe some possible causes of respiratory acidosis. _____

_____

_____

3. Explain how the renal system would compensate for respiratory acidosis. _____

_____

_____

ACTIVITY 3   Renal Responses to Respiratory Acidosis and Respiratory Alkalosis

1. Describe what happened to the concentration of ions in the urine when the P_{CO_2} was lowered. How well did the results

   compare with your prediction? _____

   _____

   _____

2. What condition was simulated when the P_{CO_2} was lowered? _____

   _____

   _____

3. Describe what happened to the concentration of ions in the urine when the P_{CO_2} was raised. How well did the results compare

   with your prediction? _____

   _____

   _____

4. What condition was simulated when the P_{CO_2} was raised? _____

   _____

   _____

ACTIVITY 4   Respiratory Responses to Metabolic Acidosis and Metabolic Alkalosis

1. Describe what happened to the blood pH when the metabolic rate was increased to 80 kcal/hr. What body system was

   compensating? How well did the results compare with your prediction? _____

   _____

   _____

2. List and describe some possible causes of metabolic acidosis. _____

   _____

   _____

3. Describe what happened to the blood pH when the metabolic rate was decreased to 20 kcal/hr. What body system was

    compensating? How well did the results compare with your prediction? _____

    _____

    _____

4. List and describe some possible causes of metabolic alkalosis. _____

    _____

    _____

Blood Analysis

P R E - L A B   Q U I Z

1. The percentage of erythrocytes in a sample of whole blood is measured by the:
   a. hemoglobin
   b. hematocrit
   c. ABO blood type
   d. erythropoietin

2. Circle the correct underlined term: The protein that transports oxygen from the lungs to the cells of the body is hemoglobin / hematocrit.

3. A lower-than-normal hematocrit is known as _____, in which insufficient oxygen is transported to the body cells.
   a. polycythemia
   b. hemophilia
   c. hemochromatosis
   d. anemia

4. The erythrocyte sedimentation rate (ESR) can be used to follow the progression of all of the following diseases or conditions *except:*
   a. anemia
   b. rheumatoid arthritis
   c. acute appendicitis (within 24 hours)
   d. myocardial infarction

5. Circle the correct underlined term: A rouleaux formation / hemoglobinometer will be used to compare a standard color value to an experimental sample to determine hemoglobin content.

6. Circle the correct underlined term: A person with type AB / O blood has two recessive alleles and has neither type A nor type B antigen.

7. Circle True or False: Blood transfusion reactions are of little consequence and occur when the recipient has antibodies that react with the antigens present on the transfused cells.

Exercise Overview

Blood transports soluble substances to and from all cells of the body. Laboratory analysis of our blood can reveal important information about how well this function is being achieved. The five activities in this exercise simulate common laboratory tests performed on blood: (1) *hematocrit* determination, (2) *erythrocyte sedimentation rate*, (3) *hemoglobin* determination, (4) *blood typing,* and (5) total *cholesterol* determination.

Hematocrit refers to the percentage of red blood cells (RBCs), or erythrocytes, in a sample of whole blood. A hematocrit of 48 means that 48% of the volume of blood consists of RBCs. RBCs transport oxygen to the cells of the body. Therefore, the higher the hematocrit, the more RBCs are present in the blood and the greater the oxygen-carrying potential of the blood. Males usually have higher hematocrit levels than females because males have higher levels of testosterone. In addition to promoting the male sex characteristics, testosterone is responsible for stimulating the release of erythropoietin from the kidneys. Erythropoietin (EPO) is a hormone that stimulates the synthesis of RBCs. Therefore, higher levels of testosterone lead to more EPO secretion and, thus, higher hematocrit levels.

The **erythrocyte sedimentation rate (ESR)** measures the settling of RBCs in a vertical, stationary tube of blood during one hour. In a healthy individual, RBCs do not settle very much in an hour. In some disease conditions, increased production of fibrinogen and immunoglobulins causes the RBCs to clump together, stack up, and form a column (called a *rouleaux formation*). RBCs in a rouleaux formation are heavier and settle faster (that is, they display an increase in the sedimentation rate.)

Hemoglobin (Hb), a protein found in RBCs, is necessary for the transport of oxygen from the lungs to the cells of the body. Four polypeptide chains of amino acids comprise the globin part of the molecule. Each polypeptide chain has a heme unit—a group of atoms that includes an atom of iron to which a molecule of oxygen binds. Each polypeptide chain, if it folds correctly, can bind a molecule of oxygen. Therefore, each hemoglobin molecule can carry four molecules of oxygen. Oxygen combined with hemoglobin forms oxyhemoglobin, which has a bright red color.

All of the cells in the human body, including RBCs, are surrounded by a plasma membrane that contains genetically determined glycoproteins, called antigens. On RBC membranes, there are certain antigens, called **agglutinogens,** that determine a person's blood type. Blood typing is used to identify the **ABO blood groups,** which are determined by the presence or absence of two antigens: **type A** and **type B.** Because these antigens are genetically determined, a person has two copies (alleles) of the gene for these antigens, one copy from each parent.

Cholesterol is a lipid substance that is essential for life—it is an important component of all cell membranes and is the base molecule of steroid hormones, vitamin D, and bile salts. Cholesterol is produced in the human liver and is present in some foods of animal origin, such as milk, meat, and eggs. Because cholesterol is a hydrophobic lipid, it needs to be wrapped in protein packages, called **lipoproteins,** to travel in the blood (which is mostly water) from the liver and digestive organs to the cells of the body.

Hematocrit Determination

OBJECTIVES

1. To understand the terms *hematocrit, red blood cells, hemoglobin, buffy coat, anemia,* and *polycythemia.*
2. To understand how the hematocrit (packed red blood cell volume) is determined.
3. To understand the implications of elevated or decreased hematocrit.
4. To understand the importance of proper disposal of laboratory material that comes in contact with blood.

Introduction

Hematocrit refers to the percentage of **red blood cells (RBCs),** or erythrocytes, in a sample of whole blood. A hematocrit of 48 means that 48% of the volume of blood consists of RBCs. RBCs transport oxygen to the cells of the body. Therefore, the higher the hematocrit, the more RBCs are present in the blood and the higher the oxygen-carrying potential of the blood. Hematocrit values are determined by spinning a microcapillary tube filled with a sample of whole blood in a special microhematocrit centrifuge. This procedure separates the blood cells from the blood plasma. A **buffy coat** layer

of white blood cells (WBCs) appears as a thin, white layer *between* the heavier RBC layer and the lighter, yellow plasma.

The hematocrit is determined after centrifuging by measuring the height of the RBC layer (in millimeters) and dividing that by the height of the total blood sample (in millimeters). This calculation gives the percentage of the total blood volume consisting of RBCs. The average hematocrit for males is 42–52%, and the average hematocrit for females is 37–47%. A lower-than-normal hematocrit indicates **anemia,** and a higher-than-normal hematocrit indicates **polycythemia.**

Anemia is a condition in which insufficient oxygen is transported to the body's cells. There are many possible causes for anemia, including inadequate numbers of RBCs, a decreased amount of the oxygen-carrying pigment **hemoglobin** in the RBCs, and abnormally shaped hemoglobin. The heme portion of a hemoglobin molecule contains an atom of iron to which a molecule of oxygen can bind. If adequate iron is not available, the body cannot manufacture hemoglobin, resulting in the condition *iron-deficiency anemia. Aplastic anemia* results from the failure of the bone marrow to produce adequate red blood cell numbers. *Sickle cell anemia* is an inherited condition in which the protein portion of hemoglobin molecules folds incorrectly when oxygen levels are low. As a result, oxygen molecules cannot bind to the misshapen hemoglobin, the RBCs develop a sickle shape, and anemia results. Regardless of the underlying cause, anemia causes a reduction in the blood's ability to transport oxygen to the cells of the body.

Polycythemia refers to an increase in RBCs, resulting in a higher-than-normal hematocrit. There are many possible causes of polycythemia, including living at high altitudes, strenuous athletic training, and tumors in the bone marrow. In this activity you will simulate the blood test used to determine hematocrit.

> **EQUIPMENT USED** The following equipment will be depicted on-screen: six heparinized capillary tubes (heparin keeps blood from clotting); blood samples from six individuals: sample 1: a healthy male living in Boston, sample 2: a healthy female living in Boston, sample 3: a healthy male living in Denver, sample 4: a healthy female living in Denver, sample 5: a male with aplastic anemia, sample 6: a female with iron-deficiency anemia; capillary tube sealer—a clay material (shown as an orange-yellow substance) used to seal the capillary tubes on one end so the blood sample can be centrifuged without having the blood spray out of the tube; microhematocrit centrifuge—used to centrifuge the samples (rotates at 14,500 revolutions per minute); metric ruler; biohazardous waste disposal—used to properly dispose of equipment that comes in contact with blood.

Experiment Instructions

Go to the home page in the PhysioEx software and click **Exercise 11: Blood Analysis.** Click **Activity 1: Hematocrit Determination,** and take the online **Pre-lab Quiz** for Activity 1.

After you take the online Pre-lab Quiz, click the **Experiment** tab and begin the experiment. The experiment

instructions are reprinted here for your reference. The opening screen for the experiment is shown below.

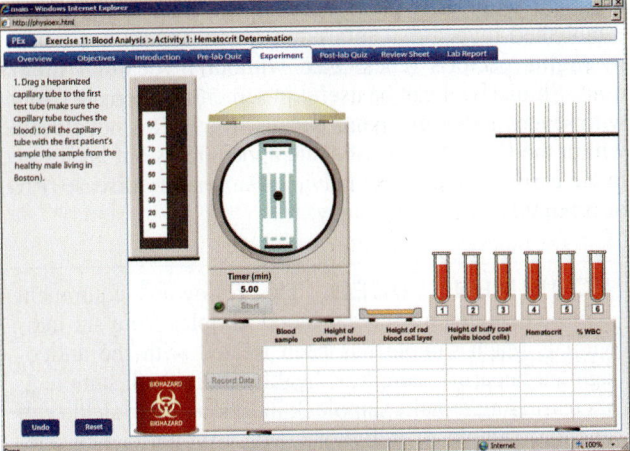

1. Drag a heparinized capillary tube to the first test tube (make sure the capillary tube touches the blood) to fill the capillary tube with the first patient's sample (the sample from the healthy male living in Boston).

2. Drag the capillary tube containing sample 1 to the container of capillary tube sealer to seal one end of the tube.

3. Drag the capillary tube to the microhematocrit centrifuge. The remaining samples will automatically be prepared for centrifugation.

4. Note that the timer is set to 5 minutes. Click **Start** to centrifuge the samples for 5 minutes at 14,500 revolutions per minute. The simulation compresses the 5-minute time period into 5 seconds of real time.

5. Drag capillary tube 1 from the centrifuge to the metric ruler to measure the height of the column of blood and the height of each layer.

6. Click **Record Data** to display your results in the grid (and record your results in Chart 1).

7. Drag capillary tube 1 to the biohazardous waste disposal.

> **? PREDICT Question 1**
> Predict how the hematocrits of the patients living in Denver, Colorado (approximately one mile above sea level), will compare with the hematocrit levels of the patients living in Boston, Massachusetts (at sea level).
>
> _____

8. You will now measure the column and layer heights of the remaining samples.

- Drag the next capillary tube from the centrifuge to the metric ruler.

- Click **Record data** to display your results in the grid (and record your results in Chart 1). The tube will automatically be placed in the biohazardous waste disposal.

Repeat this step for each of the remaining samples.

After you complete the experiment, take the online **Post-lab Quiz** for Activity 1.

CHART 1	Hematocrit Determination				
	Total height of column of blood (mm)	**Height of red blood cell layer (mm)**	**Height of buffy coat (mm)**	**Hematocrit**	**% WBC**
Sample 1 (healthy male living in Boston)					
Sample 2 (healthy female living in Boston)					
Sample 3 (healthy male living in Denver)					
Sample 4 (healthy female living in Denver)					
Sample 5 (male with aplastic anemia)					
Sample 6 (female with iron-deficiency anemia)					

Activity Questions

1. How do you calculate the hematocrit after you centrifuge the total blood sample? What does the result of this calculation indicate?

_____

_____

2. What is the significance of the "buffy coat" after you centrifuge the total blood sample?

_____

_____

3. As noted in the Exercise Overview, the average hematocrit for males is 42–52%, the average hematocrit for females is 37–47%, and erythropoietin is a hormone that is responsible for the synthesis of RBCs. Given this information, explain how a female could have a consistent hematocrit of 48, large, well-defined skeletal muscles, and an abnormally deep voice.

_____

_____

ACTIVITY 2

Erythrocyte Sedimentation Rate

OBJECTIVES

1. To understand *erythrocyte sedimentation rate (ESR)*, *red blood cells (RBCs)*, and *rouleaux formation*.

2. To learn how to perform an erythrocyte sedimentation rate blood test.

3. To understand the results (and their implications) from an erythrocyte sedimentation rate blood test.

4. To understand the importance of proper disposal of laboratory material that comes in contact with blood.

Introduction

The **erythrocyte sedimentation rate (ESR)** measures the settling of **red blood cells (RBCs)** in a vertical, stationary tube of whole blood during one hour. In a healthy individual, red blood cells do not settle very much in an hour. In some disease conditions, increased production of fibrinogen and immunoglobulins cause the RBCs to clump together, stack up, and form a dark red column (called a **rouleaux formation**). RBCs in a rouleaux formation are heavier and settle faster (that is, they exhibit an increase in the settling rate).

The ESR is neither very specific nor diagnostic, but it can be used to follow the progression of certain diseases, including sickle cell anemia, some cancers, and inflammatory diseases, such as rheumatoid arthritis. When the disease worsens, the ESR increases. When the disease improves, the ESR decreases.

The ESR can be elevated in iron-deficiency anemia, and menstruating females sometimes develop anemia and show an increase in ESR. The ESR can also be used to evaluate a patient with chest pains because the ESR is elevated in established myocardial infarction (heart attack) but normal in angina pectoris (chest pain without myocardial infarction). Similarly, it can be useful in screening a female patient with severe abdominal pains because the ESR is not elevated within the first 24 hours of acute appendicitis but is elevated in the early stage of acute pelvic inflammatory disease (PID) or ruptured ectopic pregnancy.

> **EQUIPMENT USED**   The following equipment will be depicted on-screen: blood samples from six individuals (each sample has been treated with the anticoagulant heparin): sample 1: healthy individual, sample 2: menstruating female, sample 3: individual with sickle cell anemia, sample 4: individual with iron-deficiency anemia, sample 5: individual suffering a myocardial infarction, sample 6: individual with angina pectoris; sodium citrate—used to bind with calcium and prevent the blood samples from clotting so they can be easily poured into the narrow sedimentation rate tubes; test tubes—used as reaction vessels for the tests; sedimentation tubes (contained in cabinet); magnifying chamber—used to help read the millimeter markings on the sedimentation tubes; biohazardous waste disposal—used to properly dispose of equipment that comes in contact with blood.

Experiment Instructions

Go to the home page in the PhysioEx software and click **Exercise 11: Blood Analysis.** Click **Activity 2: Erythrocyte Sedimentation Rate,** and take the online **Pre-lab Quiz** for Activity 2.

After you take the online Pre-lab Quiz, click the **Experiment** tab and begin the experiment. The experiment instructions are reprinted here for your reference. The opening screen for the experiment is shown below.

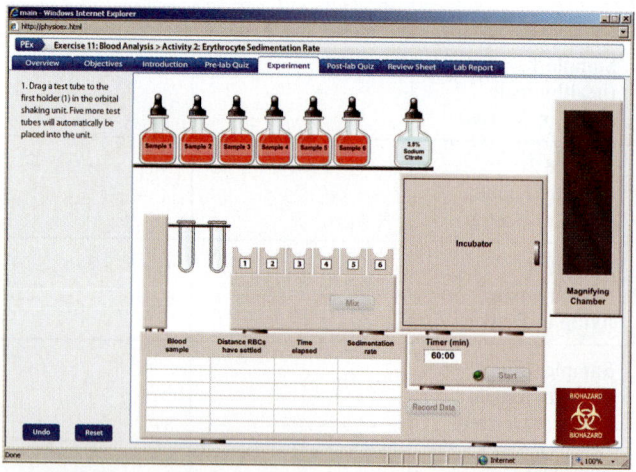

1. Drag a test tube to the first holder (1) in the orbital shaking unit. Five more test tubes will automatically be placed into the unit.

2. Drag the dropper cap of the sample 1 bottle (the sample from the healthy individual) to the first test tube (1) in the orbital shaking unit to dispense one milliliter of blood into the tube. The remaining five samples will be automatically dispensed.

3. Drag the dropper cap of the 3.8% sodium citrate bottle to the first test tube to dispense 0.5 milliliters of sodium citrate into each of the tubes.

4. Click **Mix** to mix the samples.

5. Drag the first test tube to the first sedimentation tube in the incubator to pour the contents of the test tube into the sedimentation tube.

6. Drag the now empty test tube to the biohazardous waste disposal. The contents of the remaining test tubes will automatically be poured into the sedimentation tubes, and the empty tubes will automatically be placed in the biohazardous waste disposal.

7. Note that the timer is set to 60 minutes. Click **Start** to incubate the sedimentation tubes for 60 minutes. The simulation compresses the 60-minute time period into 6 seconds of real time.

8. Drag the first sedimentation tube to the magnifying chamber to examine the tube. The tube is marked in millimeters (the distance between two marks is 5 mm).

9. Click **Record Data** to display your results in the grid (and record your results in Chart 2).

10. Drag the sedimentation tube to the biohazardous waste disposal.

> **? PREDICT Question 1**
> How will the sedimentation rate for sample 6 (unhealthy individual) compare with the sedimentation rate for sample 1 (healthy individual)?

11. You will now measure the sedimentation rate for the remaining samples.

- Drag the next sedimentation tube to the magnifying chamber to examine the tube.

- Click **Record Data** to display your results in the grid (and record your results in Chart 2). The tube will automatically be placed in the biohazardous waste disposal.

Repeat this step for each of the remaining samples.

After you complete the experiment, take the online **Post-lab Quiz** for Activity 2.

Activity Questions

1. Why is ESR useful, even though it is neither specific nor sensitive?

_____

_____

2. Describe the physical process underlying an accelerated erythrocyte sedimentation rate.

_____

_____

ACTIVITY 3

Hemoglobin Determination

OBJECTIVES

1. To understand the terms *hemoglobin (Hb), anemia, heme, oxyhemoglobin,* and *hemoglobinometer.*

2. To learn how to determine the amount of hemoglobin in a blood sample.

3. To understand the results and their implications when examining the amounts of hemoglobin present in a blood sample.

4. To understand the importance of proper disposal of laboratory material that comes in contact with blood.

CHART 2 Erythrocyte Sedimentation Rate			
Blood sample	**Distance RBCs have settled (mm)**	**Elapsed time**	**Sedimentation rate**
Sample 1 (healthy individual)			
Sample 2 (menstruating female)			
Sample 3 (individual with sickle cell anemia)			
Sample 4 (individual with iron-deficiency anemia)			
Sample 5 (individual suffering a myocardial infarction)			
Sample 6 (individual with angina pectoris)			

Introduction

Hemoglobin (Hb), a protein found in red blood cells, is necessary for the transport of oxygen from the lungs to the cells of the body. Four polypeptide chains of amino acids comprise the globin part of the molecule. Each polypeptide chain has a **heme** unit—a group of atoms that includes an atom of iron to which a molecule of oxygen binds. Each polypeptide chain, if it folds correctly, can bind a molecule of oxygen. Therefore, each hemoglobin molecule can carry four molecules of oxygen. Oxygen combined with hemoglobin forms **oxyhemoglobin,** which has a bright-red color. Anemia results when insufficient oxygen is carried in the blood.

A quantitative hemoglobin measurement is used to determine the classification and possible causes of anemia and also gives useful information on some other disease conditions. For example, a person can have anemia with a normal red blood cell count if there is inadequate hemoglobin in the red blood cells. Normal blood contains an average of 12–18 grams of hemoglobin per 100 milliliters of blood. A healthy male has 13.5–18 g/100 ml and a healthy female has 12–16 g/100 ml. Hemoglobin levels increase in patients with polycythemia, congestive heart failure, and chronic obstructive pulmonary disease (COPD). Hemoglobin levels also increase when dwelling at high altitudes. Hemoglobin levels decrease in patients with anemia, hyperthyroidism, cirrhosis of the liver, renal disease, systemic lupus erythematosus, and severe hemorrhage.

The hemoglobin level of a blood sample is determined by stirring the blood with a wooden stick to rupture, or lyse, the red blood cells. The color intensity of the hemolyzed blood reflects the amount of hemoglobin present. A **hemoglobinometer** transmits green light through the hemolyzed blood sample and then compares the amount of light that passes through the sample to standard color intensities to determine the hemoglobin content of the sample.

EQUIPMENT USED The following equipment will be depicted on-screen: blood samples from five individuals: sample 1: healthy male, sample 2: healthy female, sample 3: female with iron-deficiency anemia, sample 4: male with polycythemia, sample 5: female Olympic athlete; hemolysis sticks—used to stir the blood samples to lyse the red blood cells, thereby releasing their hemoglobin; blood chamber dispenser—used to dispense a blood chamber slide with a depression for the blood sample; hemoglobinometer—used to analyze the hemoglobin level in each sample; biohazardous waste disposal—used to properly dispose of equipment that comes in contact with blood.

Experiment Instructions

Go to the home page in the PhysioEx software and click **Exercise 11: Blood Analysis.** Click **Activity 3: Hemoglobin Determination,** and take the online **Pre-lab Quiz** for Activity 3.

After you take the online Pre-lab Quiz, click the **Experiment** tab and begin the experiment. The experiment instructions are reprinted here for your reference. The opening screen for the experiment is shown below.

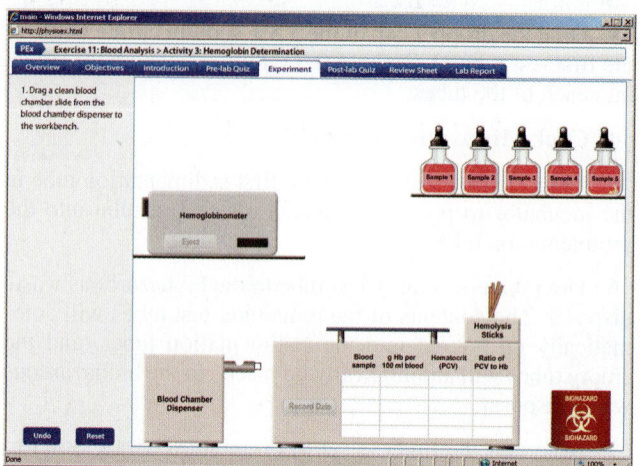

1. Drag a clean blood chamber slide from the blood chamber dispenser to the workbench.

2. Drag the bottle cap from the sample 1 bottle (the sample from the healthy male) to the depression in the blood chamber slide to dispense a drop of blood into the depression.

3. Drag a hemolysis stick to the drop of blood in the chamber to stir the blood sample for 45 seconds, lysing the red blood cells and releasing their hemoglobin.

4. Drag the hemolysis stick to the biohazardous waste disposal.

5. Drag the blood chamber slide to the dark rectangular slot on the hemoglobinometer to analyze the sample. After you insert the blood chamber slide into the hemoglobinometer, you will see a blowup of the inside of the hemoglobinometer.

6. The left half of the circular field shows the intensity of green light transmitted by blood sample 1. The right half of the circular field shows the intensity of green light for known levels of hemoglobin present in blood. Drag the lever on the right side of the hemoglobinometer down until the shade of green in the right half of the field matches the shade of green in the left half of the field and then click **Record Data** to display your results in the grid (and record your results in Chart 3).

7. Click **Eject** to remove the blood chamber slide from the hemoglobinometer.

8. Drag the blood chamber slide from the hemoglobinometer to the biohazardous waste disposal.

? PREDICT Question 1
How will the hemoglobin levels for the female Olympic athlete (sample 5) compare with the hemoglobin levels for the healthy female (sample 2)?

CHART 3	Hemoglobin Determination		
Blood sample	Hb in grams per 100 ml of blood	Hematocrit (PCV)	Ratio of PCV to Hb
Sample 1 (healthy male)		48	
Sample 2 (healthy female)		44	
Sample 3 (female with iron-deficiency anemia)		40	
Sample 4 (male with polycythemia)		60	
Sample 5 (female Olympic athlete)		60	

9. You will now measure the hemoglobin levels for each of the remaining samples.

- Drag a blood chamber slide to the workbench.

- Drag the bottle cap from the next sample bottle to the depression in the slide.

- Drag a hemolysis stick to the drop of blood in the chamber (after stirring the sample, the hemolysis stick will automatically be placed in the biohazardous waste disposal).

- Drag the blood chamber slide to the dark rectangular slot on the hemoglobinometer.

- Drag the lever on the right side of the hemoglobinometer down until the shade of green in the right half of the field matches the shade of green in the left half of the field and then click **Record Data** to display your results in the grid (and record your results in Chart 3).

- Click **Eject** to remove the blood chamber slide from the hemoglobinometer (the slide will automatically be placed in the biohazardous waste disposal).

Repeat this step until you analyze all five samples.

After you complete the experiment, take the online **Post-lab Quiz** for Activity 3.

Activity Questions

1. As mentioned in the introduction to this activity, hemoglobin levels increase for people living at high altitudes. Given that the atmospheric pressure of oxygen significantly declines as you ascend to higher elevations, why do you think hemoglobin levels would increase for those living at high altitudes?

2. Just by looking at the color of a freshly drawn blood sample, how could you distinguish between blood that is well oxygenated and blood that is poorly oxygenated?

ACTIVITY 4

Blood Typing

OBJECTIVES

1. To understand the terms *antigens, agglutinogens, ABO antigens, Rh antigens,* and *agglutinins.*

2. To learn how to perform a blood-typing assay.

3. To understand the results and their implications when examining agglutination reactions.

4. To understand the importance of proper disposal of laboratory material that comes in contact with blood.

Introduction

All of the cells in the human body, including red blood cells, are surrounded by a plasma membrane that contains genetically determined glycoproteins, called **antigens.** On red blood cell membranes, there are certain antigens, called **agglutinogens,** that determine a person's blood type. If a blood transfusion recipient has antibodies (called **agglutinins**) that react with the antigens present on the transfused cells, the red blood cells will become clumped together, or agglutinated, and then lysed, resulting in a potentially life-threatening blood transfusion reaction. It is therefore important to determine an individual's blood type before performing blood transfusions to avoid mixing incompatible blood. Although many different antigens are present on red blood cell membranes, the **ABO** and **Rh antigens** cause the most vigorous and potentially fatal transfusion reactions.

The ABO blood groups are determined by the presence or absence of two antigens: type A and type B. Because these antigens are genetically determined, a person has two copies (alleles) of the gene for these proteins, one copy from each parent. The presence of these antigens is due to a dominant allele, and their absence is due to a recessive allele.

- A person with type A blood can have two alleles for the type A antigen or one allele for the type A antigen and one allele for the absence of either the type A or type B antigen.

- A person with type B blood can have two alleles for the type B antigen or one allele for the type B antigen and one allele for the absence of either the type A or type B antigen.

- A person with type AB blood has one allele for the type A antigen and one allele for the type B antigen.

- A person with type O blood has two recessive alleles and has neither the type A nor type B antigen.

TABLE 11.1	ABO Blood Types	
Blood type	Antigens on RBCs	Antibodies present in plasma
A	A	anti-B
B	B	anti-A
AB	A and B	none
O	none	anti-A and anti-B

Antibodies against the A and B antigens are found preformed in the blood plasma. A person has antibodies only for the antigens not on his or her red blood cells, so a person with type A blood will have anti-B antibodies. View Table 11.1 for a summary of the antigens on red blood cells and the antibodies in the plasma for each blood type.

The Rh factor is another genetically determined protein that can be present on red blood cell membranes. Approximately 85% of the population is Rh positive (Rh^+), and their red blood cells have this protein on their surface. Antibodies against the Rh factor are not found preformed in the plasma. They are produced by an Rh negative (Rh^-) individual only after exposure to blood cells from someone who is Rh^+. Such exposure can occur during pregnancy when Rh^+ blood cells from the baby cross the placenta and expose the mother to the antigen.

To determine an individual's blood type, drops of an individual's blood sample are mixed separately with antiserum containing antibodies to either type A antigens, type B antigens, or Rh antigens. An agglutination reaction (showing clumping) indicates the presence of the agglutinogen.

EQUIPMENT USED The following equipment will be depicted on-screen: blood samples from six individuals with different blood types; anti-A serum (blue bottle), anti-B serum (yellow bottle), and anti-Rh serum (white bottle), containing antibodies to the A antigen, B antigen, and Rh antigen, respectively; blood-typing slide dispenser; color-coded stirring sticks—used to mix the blood sample and the serum (blue: used with anti-A serum, yellow: used with the anti-B serum, white: used with the anti-Rh serum); light box—used to view the blood type samples; biohazardous waste disposal—used to properly dispose of equipment that comes in contact with blood.

Experiment Instructions

Go to the home page in the PhysioEx software and click **Exercise 11, Blood Analysis.** Click **Activity 4, Blood Typing,** and take the online **Pre-lab Quiz** for Activity 4.

After you take the online Pre-lab Quiz, click the **Experiment** tab and begin the experiment. The experiment instructions are reprinted here for your reference. The opening screen for the experiment is shown above.

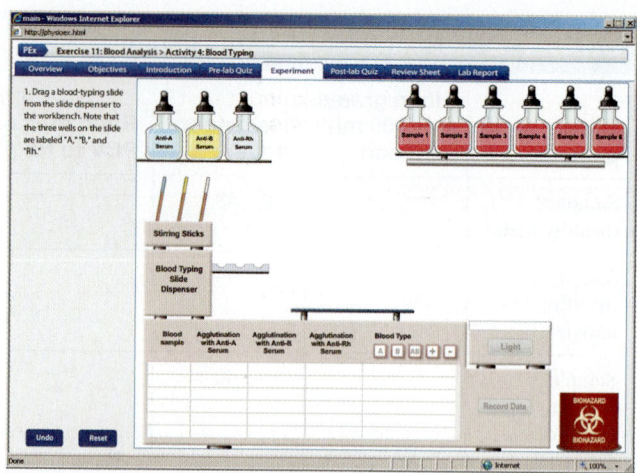

1. Drag a blood-typing slide from the slide dispenser to the workbench. Note that the three wells on the slide are labeled "A," "B," and "Rh."

2. Drag the dropper cap of the sample 1 bottle to well A on the blood-typing slide to dispense a drop of blood into each well.

3. Drag the dropper cap of the anti-A serum bottle to well A on the blood-typing slide to dispense a drop of anti-A serum into the well.

4. Drag the dropper cap of the anti-B serum bottle to well B on the blood-typing slide to dispense a drop of anti-B serum into the well.

5. Drag the dropper cap of the anti-Rh serum bottle to well Rh on the blood-typing slide to dispense a drop of anti-Rh serum into the well.

6. Drag a blue-tipped stirring stick to well A to mix the blood and anti-A serum.

7. Drag the stirring stick to the biohazardous waste disposal.

8. Drag a yellow-tipped stirring stick to well B to mix the blood and anti-B serum.

9. Drag the stirring stick to the biohazardous waste disposal.

10. Drag a white-tipped stirring stick to well Rh to mix the blood and anti-Rh serum.

11. Drag the stirring stick to the biohazardous waste disposal.

12. Drag the blood-typing slide to the light box and then click **Light** to analyze the slide.

13. Under each of the wells, click **Positive** if agglutination occurred (the sample shows clumping) or click **Negative** if agglutination did not occur (the sample looks smooth).

CHART 4	Blood Typing Results			
Blood sample	Agglutination with anti-A serum	Agglutination with anti-B serum	Agglutination with anti-Rh serum	Blood type
1				
2				
3				
4				
5				
6				

14. Click **Record Data** to display your results in the grid (and record your results in Chart 4).

15. Drag the blood-typing slide to the biohazardous waste disposal.

> **? PREDICT Question 1**
> If the patient's blood type is AB⁻, what would be the appearance of the A, B, and Rh samples?

16. You will now analyze the remaining samples.

- Drag a blood-typing slide from the slide dispenser to the workbench. The next sample will be added to each well on the slide, the appropriate antiserum will be added to each well, the sample and antisera will be mixed, and the slide will be placed in the light box.

- Under each of the wells, click **Positive** if agglutination occurred (the sample shows clumping) or click **Negative** if agglutination did not occur (the sample looks smooth).

- Click **Record Data** to display your results in the grid (and record your results in Chart 4).

  Repeat this step until you analyze all six samples.

17. You will now indicate the blood type for each sample and indicate whether the sample is Rh positive or Rh negative.

- Click the row for the sample in the grid (and record your results in Chart 4).

- Click A, B, AB, or O above the blood type column to indicate the blood type.

- Click the − button or the + button above the blood type column to indicate whether the sample is Rh negative or Rh positive.

  Repeat this step for all six samples. Record your results in Chart 4.

After you complete the experiment, take the online **Post-lab Quiz** for Activity 4.

Activity Questions

1. Antibodies against the A and B antigens are found in the plasma, and a person has antibodies only for the antigens that are not present on their red blood cells. Using this information, list the antigens found on red blood cells and the antibodies in the plasma for blood types 1) AB−, 2) O+, 3) B−, and 4) A+.

_____

_____

2. If an individual receives a bone marrow transplant from someone with a different ABO blood type, what happens to the recipient's ABO blood type?

_____

_____

ACTIVITY 5

Blood Cholesterol

OBJECTIVES

1. To understand the terms *cholesterol, lipoproteins, low-density lipoprotein (LDL), hypocholesterolemia, hypercholesterolemia,* and *atherosclerosis.*

2. To learn how to test for total blood cholesterol using a colorimetric assay.

3. To understand the results and their implications when examining total blood cholesterol.

4. To understand the importance of proper disposal of laboratory material that comes in contact with blood.

Introduction

Cholesterol is a lipid substance that is essential for life—it is an important component of all cell membranes and is the base molecule of steroid hormones, vitamin D, and bile salts. Cholesterol is produced in the human liver and is present in some foods of animal origin, such as milk, meat, and eggs. Because cholesterol is a water-insoluble lipid, it needs to be wrapped in protein packages, called **lipoproteins,** to travel in the blood (which is mostly water) from the liver and digestive organs to the cells of the body.

One type of lipoprotein package, called **low-density lipoprotein (LDL),** has been identified as a potential source of damage to the interior of arteries. LDLs can contribute to **atherosclerosis,** the buildup of plaque, in these blood vessels.

A total blood cholesterol determination does not measure the level of LDLs, but it does provide valuable information about the total amount of cholesterol in the blood.

Less than 200 milligrams of total cholesterol per deciliter of blood is considered desirable. Between 200 and 239 mg/dl is considered borderline high cholesterol. Over 240 mg/dl is considered high blood cholesterol (**hypercholesterolemia**) and is associated with an increased risk of cardiovascular disease. Abnormally low blood cholesterol levels (total cholesterol lower than 100 mg/dl) can also suggest a problem. Low levels may indicate hyperthyroidism (overactive thyroid gland), liver disease, inadequate absorption of nutrients from the intestine, or malnutrition. Other reports link **hypocholesterolemia** (low blood cholesterol) to depression, anxiety, and mood disturbances, which are thought to be controlled by the level of available serotonin, a neurotransmitter. There is evidence of a relationship between low levels of blood cholesterol and low levels of serotonin in the brain.

In this test for total blood cholesterol, a sample of blood is mixed with enzymes that produce a colored reaction with cholesterol. The intensity of the color indicates the amount of cholesterol present. The cholesterol tester compares the color of the sample to the colors of known levels of cholesterol (standard values).

EQUIPMENT USED The following equipment will be depicted on-screen: lancets—sharp, needlelike instruments used to prick the finger to obtain a drop of blood; four patients (represented by an extended finger); alcohol wipes—used to cleanse the patient's fingertip before it is punctured with the lancet; color wheel—divided into shades of green that correspond to total cholesterol levels; cholesterol strips—contain chemicals that convert, by a series of reactions, the cholesterol in the blood sample into a green-colored solution; biohazardous waste disposal—used to properly dispose of equipment that comes in contact with blood.

Experiment Instructions

Go to the home page in the PhysioEx software and click **Exercise 11: Blood Analysis.** Click **Activity 5: Blood Cholesterol,** and take the online **Pre-lab Quiz** for Activity 5.

After you take the online Pre-lab Quiz, click the **Experiment** tab and begin the experiment. The experiment instructions are reprinted here for your reference. The opening screen for the experiment is shown below.

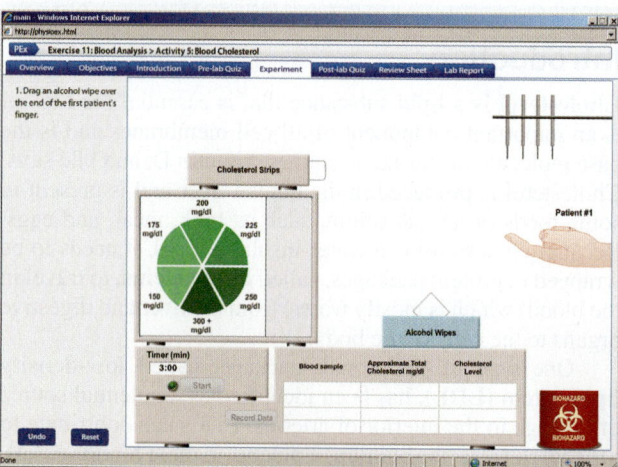

1. Drag an alcohol wipe over the end of the first patient's finger.

2. Drag the alcohol wipe to the biohazardous waste disposal.

3. Drag a lancet to the tip of the patient's finger to prick the finger and obtain a drop of blood.

4. Drag the lancet to the biohazardous waste disposal.

5. Drag a cholesterol strip to the finger to transfer a drop of blood from the patient's finger to the strip.

6. Drag the cholesterol strip to the rectangular box to the right of the color wheel.

7. Click **Start** to start the timer. It takes three minutes for the chemicals in the cholesterol strip to react with the blood. The simulation compresses the 3-minute time period into 3 seconds of real time.

8. Click the color on the color wheel that most closely matches the color on the cholesterol strip.

9. Click **Record Data** to display your results in the grid (and record your results in Chart 5).

CHART 5	Total Cholesterol Determination	
Blood sample	**Approximate total cholesterol (mg/dl)**	**Cholesterol level**
1		
2		
3		
4		

10. Drag the cholesterol test strip to the biohazardous waste disposal.

? PREDICT Question 1
Patient 4 prefers to cook all his meat in lard or bacon grease. Knowing this dietary preference, you anticipate his total cholesterol level to be:

_____

11. You will now test the total cholesterol levels for the remaining patients.

- Drag an alcohol wipe over the end of the patient's finger. The alcohol wipe will automatically be placed in the biohazardous waste disposal.

- Drag a lancet to the tip of the patient's finger to prick the finger and obtain a drop of blood. The lancet will automatically be placed in the biohazardous waste disposal.

- Drag a cholesterol strip to the finger to transfer a drop of blood from the patient's finger to the strip.

- Drag the cholesterol strip to the rectangular box to the right of the color wheel. The timer will automatically run for three minutes to allow the chemicals in the cholesterol strip to react with the blood.

- Click the color on the color wheel that most closely matches the color on the cholesterol strip.

- Click **Record Data** to display your results in the grid (and record your results in Chart 5). The cholesterol strip will automatically be placed in the biohazardous waste disposal.

Repeat this step until you determine the total cholesterol levels for all four patients.

After you complete the experiment, take the online **Post-lab Quiz** for Activity 5.

Activity Questions

1. Why do cholesterol plaques occur in arteries and not veins?

_____

_____

2. Phytosterols can alter absorption of certain molecules by the intestinal tract. Why would they be a beneficial dietary supplement for people with high LDL levels?

_____

_____

Blood Analysis

NAME _____

LAB TIME/DATE _____

ACTIVITY 1 **Hematocrit Determination**

1. List the hematocrits for the healthy male (sample 1) and female (sample 2) living in Boston (at sea level) and indicate whether they are normal or whether they indicate anemia or polycythemia.

   _____

   _____

2. Describe the difference between the hematocrits for the male and female living in Boston. Why does this difference between the sexes exist?

   _____

   _____

3. List the hematocrits for the healthy male and female living in Denver (approximately one mile above sea level) and indicate whether they are normal or whether they indicate anemia or polycythemia.

   _____

   _____

4. How did the hematocrit levels of the Denver residents differ from those of the Boston residents? Why? How well did the results compare with your prediction?

   _____

   _____

5. Describe how the kidneys respond to a chronic decrease in oxygen and what effect this has on hematocrit levels.

   _____

   _____

6. List the hematocrit for the male with aplastic anemia (sample 5) and indicate whether it is normal or abnormal. Explain your response.

   _____

   _____

7. List the hematocrit for the female with iron-deficiency anemia (sample 6) and indicate whether it is normal or abnormal. Explain your response.

_____

_____

ACTIVITY 2 **Erythrocyte Sedimentation Rate**

1. Describe the effect that sickle cell anemia has on the sedimentation rate (sample 3). Why do you think that it has this effect?

_____

_____

2. How did the sedimentation rate for the menstruating female (sample 2) compare with the sedimentation rate for the healthy individual (sample 1)? Why do you think this occurs?

_____

_____

3. How did the sedimentation rate for the individual with angina pectoris (sample 6) compare with the sedimentation rate for the healthy individual (sample 1)? Why? How well did the results compare with your prediction?

_____

_____

4. What effect does iron-deficiency anemia (sample 4) have on the sedimentation rate?

_____

_____

5. Compare the sedimentation rate for the individual suffering a myocardial infarction (sample 5) with the sedimentation rate for the individual with angina pectoris (sample 6). Explain how you might use this data to monitor heart conditions.

_____

_____

ACTIVITY 3 **Hemoglobin Determination**

1. Is the male with polycythemia (sample 4) deficient in hemoglobin? Why?

_____

_____

2. How did the hemoglobin levels for the female Olympic athlete (sample 5) compare with the hemoglobin levels for the healthy female (sample 2)? Is either person _deficient_ in hemoglobin? How well did the results compare with your prediction?

_____

_____

3. List conditions in which hemoglobin levels would be expected to decrease. Provide reasons for the change when possible.

_____

_____

4. List conditions in which hemoglobin levels would be expected to increase. Provide reasons for the change when possible.

_____

_____

5. Describe the ratio of hematocrit to hemoglobin for the healthy male (sample 1) and female (sample 2). (A normal ratio of hematocrit to grams of hemoglobin is approximately 3:1.) Discuss any differences between the two individuals.

_____

_____

6. Describe the ratio of hematocrit to hemoglobin for the female with iron-deficiency anemia (sample 3) and the female Olympic athlete (sample 5). (A normal ratio of hematocrit to grams of hemoglobin is approximately 3:1.) Discuss any differences between the two individuals.

_____

_____

ACTIVITY 4   Blood Typing

1. How did the appearance of the A, B, and Rh samples for the patient with AB⁻ blood type compare with your prediction?

_____

_____

2. Which blood sample contained the rarest blood type?

_____

3. Which blood sample contained the universal donor?

_____

4. Which blood sample contained the universal recipient?

_____

5. Which blood sample did not agglutinate with any of the antibodies tested? Why?

_____

_____

6. What antibodies would be found in the plasma of blood sample 1? _____

_____

7. When transfusing an individual with blood that is compatible but not the same type, it is important to separate packed cells from the plasma and administer only the packed cells. Why do you think this is done? (Hint: Think about what is *in plasma* versus what is *on RBCs*.)

_____

_____

8. List the blood samples in this activity that represent people who could donate blood to a person with type B$^+$ blood.

_____

_____

ACTIVITY 5 **Blood Cholesterol**

1. Which patient(s) had desirable cholesterol level(s)?

_____

2. Which patient(s) had elevated cholesterol level(s)?

_____

3. Describe the risks for the patient(s) you identified in question 2.

_____

_____

4. Was the cholesterol level for patient 4 low, desirable, or high? How well did the results compare with your prediction? What advice about diet and exercise would you give to this patient? Why?

_____

_____

5. Describe some reasons why a patient might have abnormally low blood cholesterol.

_____

_____

Serological Testing

Exercise Overview

Immunology, the study of the immune system, focuses on chemical interactions that are difficult to observe. A number of chemical techniques have been developed to visually represent antibodies and antigens in the **serum,** the fluid portion of the blood with the clotting factors removed. The study and use of these techniques is referred to as **serology.** These techniques are performed in vitro, outside of the body, and are primarily used as diagnostic tools to detect disease. Other applications include pregnancy testing and drug testing. These immunological techniques depend upon the principle that an antibody binds only to specific, corresponding antigens. The tests are relatively expensive to perform, so these activities will allow you to perform them without the sometimes cost-prohibitive supplies.

Antigens and Antibodies

The word **antigen** is derived from two words: *anti*body and *gen*erator. Antigens do not produce antibodies, but early scientists noted that when antigens were present, antibodies appeared. Plasma cells actually produce antibodies.

Antigens include proteins, polysaccharides, and various small molecules that stimulate antibody production. Antigens are often molecules that are described as **nonself,** or foreign to the body. There are also self-antigens that act as identifier tags, such as the proteins found on the surface of red blood cells. Most often,

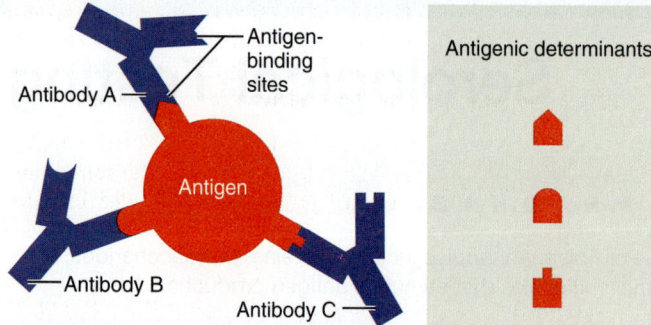

FIGURE 12.1 Antigen-antibody interaction with antigenic determinants.

antigens are a portion of an infectious agent, such as a bacterium or a virus, and the body produces antibodies in response to the presence of the infectious agent.

Antigens are often large and have multiple antigenic sites—locations that can bind to antibodies. We refer to these sites as **antigenic determinants,** or **epitopes.** The antibody has a corresponding antigen-binding site that has a "lock-and-key" recognition for the antigenic determinant on the antigen (view Figure 12.1). All of the simulated tests presented in this exercise take advantage of antigen-antibody specificity. These tests include direct fluorescent antibody technique, Ouchterlony technique, ELISA (enzyme-linked immunosorbent assay), and Western blotting technique.

Nonspecific Binding

The lock-and-key recognition that antigen and antibody have for each other is much like the specificity that an enzyme and its substrate have for one another. However, with antigen and antibody, **nonspecific binding** sometimes occurs. For this reason you will perform a number of washing steps in this exercise to remove any nonspecific binding.

Positive and Negative Controls

You will also use **positive** and **negative controls** to ensure that the test is working accurately. Positive controls include a substance that is known to react positively, thus giving you a standard against which to base your results. Negative controls include substances that should not react. A positive result with a negative control is a "false positive," which would invalidate all other results. Likewise, a negative result with a positive control is a "false negative," which would also invalidate your results.

ACTIVITY 1

Using Direct Fluorescent Antibody Technique to Test for Chlamydia

OBJECTIVES

1. To understand how fluorescent antibodies can be used diagnostically to detect the presence of a specific antigen.
2. To observe how to test for the sexually transmitted disease chlamydia.
3. To distinguish between antigens and antibodies.
4. To understand the terms *epitope* and *antigenic determinant*.
5. To observe nonspecific binding that can result between antigen and antibody.

Introduction

The direct fluorescent antibody technique uses antibodies to directly detect the presence of antigen. A fluorescent dye molecule attached to these antibodies acts as a visual signal for a positive result. This technique is typically used to test for antigens from infectious agents, such as bacteria or viruses. In this activity you will test for the presence of *Chlamydia trachomatis* (a bacterium that invades the cells of its host) using fluorescently labeled antibodies to detect the presence of the antigen and, therefore, the bacterium. *Chlamydia trachomatis* is an important infectious agent because it causes the sexually transmitted disease **chlamydia.** Left untreated, chlamydia can lead to sterility in men and women.

Chlamydia trachomatis is an obligate, intracellular bacterium, which means that it can only survive inside a host cell. The life cycle of the bacterium has two cellular types. The infectious cell type is the small, dense **elementary body,** which is capable of attaching to the host cell. The **reticulate body** is a larger, less-dense cell, which divides actively once inside the host cell. The reticulate body is also referred to as the vegetative form. The life cycle of *Chlamydia* begins when the elementary body enters the host cell and continues as the elementary body changes inside the host cell into a reticulate body. The reticulate body divides into more reticulate bodies and converts back to the elementary body form for release to infect other cells.

In this activity you will test three patient samples and two control samples for the *Chlamydia* infection. An epithelial scraping from the male urethra or from the cervix of the uterus is performed to collect squamous cells from the surface. The elementary bodies are measured by reacting antigen-specific antibodies to infected cells. The fluorescent dye attached to the antigen-specific antibodies makes the complex detectable. The sample is viewed with a fluorescent microscope. The presence of ten or more elementary bodies in a field of view with a diameter of 5 millimeters is considered a positive result. The elementary bodies will be stained green inside red host cells.

EQUIPMENT USED The following equipment will be depicted on-screen: five samples: patient A, patient B, patient C, a positive control, and a negative control; incubator; fluorescent microscope; 95% ethyl alcohol—used for fixing the sample to the microscope slide; chlamydia fluorescent antibody (Chlamydia FA)—antibodies specific for the *Chlamydia* antigen with a fluorescent dye attached; fluorescent antibody mounting media (FA mounting)—used to mount the prepared sample to the slide when ready for viewing under the microscope; phosphate buffered saline (PBS)—used to wash off excess antibodies and prevent nonspecific binding of the antigen and antibody; fluorescent antibody buffer (FA buffer)—used to remove excess ethyl alcohol; petri dishes—used for incubation of the slides to keep them moist; microscope slides—an incubation vessel where the antigen and antibody react; cotton-tipped applicators—used for application and mixing of the antibodies with the samples; filter paper—used to keep the samples moist in the petri dishes; biohazardous waste disposal.

Experiment Instructions

Go to the home page in the PhysioEx software and click **Exercise 12: Serological Testing.** Click **Activity 1: Using Direct Fluorescent Antibody Technique to Test for Chlamydia,** and take the online **Pre-lab Quiz** for Activity 1.

After you take the online Pre-lab Quiz, click the **Experiment** tab and begin the experiment. The experiment instructions are reprinted here for your reference. The opening screen for the experiment is shown below.

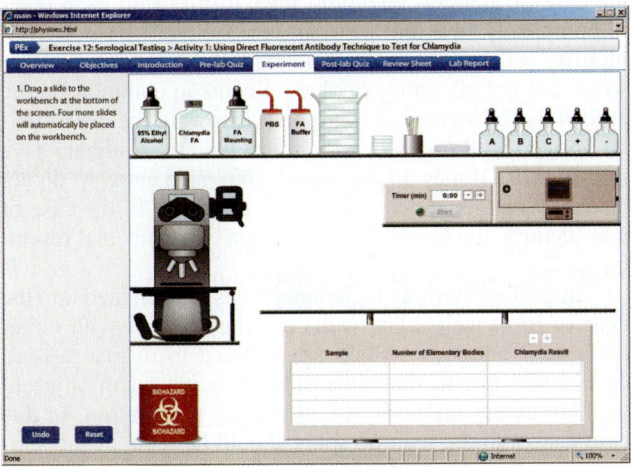

1. Drag a slide to the workbench at the bottom of the screen. Four more slides will automatically be placed on the workbench.

2. The patient samples have been suspended in a small amount of buffer and placed in dropper bottles for ease of dispensing. Drag the dropper cap of the patient A sample bottle to the first slide on the workbench to dispense a drop of the sample onto the slide. A drop from each sample will be placed on a separate slide.

3. Drag the dropper cap of the 95% ethyl alcohol bottle to the first slide on the workbench to dispense three drops of ethyl alcohol onto each slide.

4. Set the timer to 5 minutes by clicking the + button beside the timer display. Click **Start** to start the timer and allow the ethyl alcohol to fix the sample to the slide and prevent the sample from being washed off in the subsequent washing steps. The simulation compresses the 5-minute time period into 5 seconds of real time.

5. Drag the fluorescent antibody (FA) buffer squirt bottle to the first slide to rinse all five slides and remove excess ethyl alcohol.

6. Drag an applicator stick to the chlamydia fluorescent antibody (FA) bottle to soak its cotton tip with antibodies that are specific for *Chlamydia* and labeled with a fluorescent tag.

7. Drag the applicator stick to the first slide to apply the chlamydia fluorescent antibody. Separate applicator sticks will automatically be soaked in chlamydia fluorescent antibody and applied to each slide. Each applicator will automatically be placed in the biohazardous waste disposal.

8. Drag a petri dish to the workbench. A piece of filter paper will be placed into the petri dish. The filter paper has

been moistened with fluorescent antibody buffer to keep the samples from drying out during incubation. Four more petri dishes (and filter paper moistened with fluorescent antibody buffer) will automatically be placed on the workbench.

9. Drag the first slide into the first petri dish. The remaining four slides will automatically be placed into the remaining petri dishes and all five petri dishes will be loaded into the incubator.

10. Set the timer to 20 minutes by clicking the + button next to the timer display. Click **Start** to incubate the samples at 25°C. During incubation the antibodies will react with the corresponding antigens if they are present in the sample. The petri dishes will automatically be removed from the incubator when the time is complete. The simulation compresses the 20-minute incubation time period into 10 seconds of real time.

11. Drag the phosphate buffered saline (PBS) squirt bottle to the first petri dish to wash off excess antibodies and prevent nonspecific binding of the antigen and antibody. The timer will count down 10 minutes for a thorough washing.

12. Click the first petri dish to open the dish and remove the slide. The slides will automatically be removed from the remaining petri dishes.

13. Drag the first petri dish to the biohazardous waste disposal. The remaining petri dishes will automatically be placed in the biohazardous waste disposal.

14. Drag the dropper cap of the fluorescent antibody (FA) mounting media to the first slide to dispense a drop of mounting media onto each slide to mount the sample to the slide.

15. Drag the first slide (patient A) to the fluorescent microscope. Count the number of elementary bodies you see through the microscope (recall that elementary bodies stain green). Click **Submit** to display your results in the grid (and record your results in Chart 1). After you click **Submit,** the slide will automatically be placed in the biohazardous waste disposal.

CHART 1	Direct Fluorescent Antibody Technique Results	
Sample	Number of elementary bodies	Chlamydia result
Patient A		
Patient B		
Patient C		
Positive control		
Negative control		

16. Repeat step 15 for Patient B.

17. Repeat step 15 for Patient C.

18. Repeat step 15 for the Positive Control.

19. Repeat step 15 for the Negative Control.

20. You will now indicate whether each sample is negative or positive for *Chlamydia*. Click the row for the sample in the grid and then click the − button or the + button above the Chlamydia result column to indicate whether the sample is negative or positive for *Chlamydia*. Repeat this step for all five samples. Record your results in Chart 1.

After you complete the experiment, take the online **Post-lab Quiz** for Activity 1.

Activity Questions

1. With this technique, is the antigen or antibody found on the patient sample? Explain how you know this.

_____

_____

2. Explain the difference between an antigen and an epitope (antigenic determinant).

_____

_____

3. When a sample has a small number of elementary bodies but not enough to be a positive result, there appears to have been some nonspecific binding that was not removed by the washing steps. Which sample displayed this property?

_____

_____

Comparing Samples with Ouchterlony Double Diffusion

OBJECTIVES

1. To observe the precipitation reaction between antigen and antibody.

2. To distinguish between *epitope* and *antigen*.

3. To understand the specificity that antibodies have for their epitopes.

4. To observe how related proteins might share epitopes in common.

Introduction

The Ouchterlony technique is also known as double diffusion. In this technique antigen and antibody diffuse toward each other in a semisolid medium made up of clear, clarified agar. When the antigen and antibody are in optimal proportions, cross-linking of the antigen and antibody occurs, forming an insoluble precipitate, called a **precipitin line.** These lines can then be used to visually identify similarities between antigens. If optimum proportions have not been met—for example, if there is excess antigen or excess antibody—then no visible precipitate will form. This technique provides easily visible evidence of the binding between antigen and antibody, and sophisticated equipment is not needed to observe the antigen-antibody reaction.

The Ouchterlony technique is designed to determine whether antigens are identical, related, or unrelated. Antigens have **identity** if they are identical. Identical antigens have all their antigenic determinants, or epitopes, in common. In the case of identity, precipitin lines diffuse into each other to completely fuse and form an arc. Antigens have **partial identity** if they are similar or related. Related antigens have some, but not all, antigenic determinants in common. In the case of partial identity, a spur pointing toward the more similar antigen well forms in addition to the arc. Antigens have **non-identity** if they are unrelated. Unrelated antigens do not have any antigenic determinants in common. In the case of non-identity, the lines intersect to form two spurs that resemble an X.

In the Ouchterlony technique, holes are punched into the agar to form wells. The wells are then loaded with either antigen or antibody, which are allowed to diffuse toward each other. Often, the same antigen is placed in adjacent wells to assess the purity of an antigen preparation. In this case a smooth arc with no spurs should be seen, as the antigens are identical. Multiple antibodies can also be placed in a center well. The antibodies will diffuse out in all directions and react with the antigens that are placed in the surrounding wells.

In this activity you will use human and bovine (from cows) albumin as the antigens, and the antibodies will be made in goats against albumin from either humans or cows. The goals are to identify an unknown antigen and to observe the patterns produced by the various relationships: identity, partial identity, and non-identity.

> **EQUIPMENT USED** The following equipment will be depicted on-screen: goat anti–human albumin (Goat A-H)—an antiserum containing antibodies produced by goats against human albumin; goat anti–bovine albumin (Goat A-B)—an antiserum containing antibodies produced by goats against bovine (cow) albumin; bovine serum albumin (BSA); human serum albumin (HSA); unknown antigen; petri dishes filled with clear agar; well cutter.

Experiment Instructions

Go to the home page in the PhysioEx software and click **Exercise 12: Serological Testing.** Click **Activity 2: Comparing Samples with Ouchterlony Double Diffusion:** and take the online **Pre-lab Quiz** for Activity 2.

After you take the online Pre-lab Quiz, click the **Experiment** tab and begin the experiment. The experiment instructions are reprinted here for your reference. The opening screen for the experiment is shown on the following page.

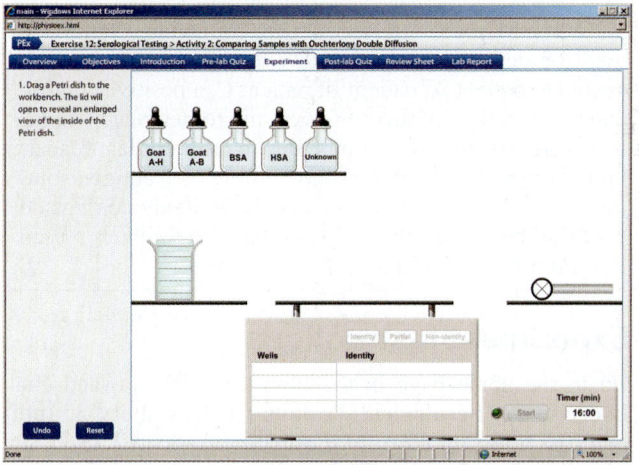

1. Drag a petri dish to the workbench. The lid will open to reveal an enlarged view of the inside of the petri dish.

2. Drag the well cutter to the middle of the enlarged view of the petri dish to punch a hole in the agar in the middle of the petri dish. Drag the well cutter to the upper left, upper right, lower left, and lower right of the petri dish to punch four more holes in the agar. After you punch all five wells into the agar, the wells will be labeled 1–5.

3. Drag the dropper cap of the goat anti–human albumin (Goat A-H) bottle to well 1 to fill it with a sample.

4. Drag the dropper cap of the goat anti–bovine albumin (Goat A-B) bottle to well 1 to fill it with a sample.

5. Drag the dropper cap of the bovine serum albumin (BSA) bottle to well 2 to fill it with a sample.

6. Drag the dropper cap of the bovine serum albumin (BSA) bottle to well 3 to fill it with a sample.

7. Drag the dropper cap of the human serum albumin (HSA) bottle to well 4 to fill it with a sample.

8. Drag the dropper cap of the unknown antigen bottle to well 5 to fill it with a sample.

> **? PREDICT Question 1**
> How do you think human serum albumin and bovine serum albumin will compare?

9. Note that the timer is set to 16 hours. Click **Start** to start the timer. The antigen and antibodies will diffuse toward each other and form a precipitate, detected as a precipitin line. The simulation compresses the 16-hour time period into 10 seconds of real time.

10. You will now examine the precipitin lines that formed and indicate the relationship between each pair of antigens. Click the row for the wells containing the antigens in the grid and then click **Identity, Partial,** or **Non-Identity** above the identity column to indicate whether the antigens have identity, partial identity, or non-identity. Repeat this step for the four pairs of antigens. Record your results in Chart 2.

CHART 2	Ouchterlony Double Diffusion Results
Wells	**Identity**
2 and 5	
2 and 3	
3 and 4	
4 and 5	

After you complete the Experiment, take the online **Post-lab Quiz** for Activity 2.

Activity Questions

1. Which type of identity was present between the samples in this activity? Describe this type of identity.

_____

_____

2. Describe the importance of what you place in the center well.

_____

_____

3. Why do you think it is important for the agar to be clear and clarified?

_____

_____

4. Describe the role that albumin plays in the blood.

_____

_____

ACTIVITY 3

Indirect Enzyme-Linked Immunosorbent Assay (ELISA)

OBJECTIVES

1. To understand how the enzyme-linked immunosorbent assay (ELISA) is used as a diagnostic test.
2. To distinguish between the direct and the indirect ELISA.
3. To describe the basic structure of antibodies.
4. To define *seroconversion*.
5. To understand how the indirect ELISA is used to detect antibodies against HIV.

Introduction

The **enzyme-linked immunosorbent assay (ELISA)** is used to test for the presence of an antigen or antibody. The assay is considered enzyme linked because an enzyme is chemically linked to an antibody in both the direct and indirect versions of the test. Immunosorbent refers to the fact that either antigens or antibodies are being adsorbed (stuck) to plastic. If the test is designed to detect an antigen or antigens, it is a **direct ELISA** because it is directly looking for the foreign substance. An **indirect ELISA** is designed to detect antibodies that the patient has made against the antigen. A positive result with the indirect ELISA requires **seroconversion.** Seroconversion occurs when a patient goes from testing negative for a specific antibody to testing positive for the same antibody.

In the direct ELISA, a 96-well microtiter plate is coated with homologous antibodies made against the antigen of interest. The number of wells makes it easy to test many samples at the same time. The patient serum sample is added to the plate to test for the presence of the antigen that binds to the antibody coating on the plate. ELISA takes advantage of the fact that protein sticks well to plastic. A secondary antibody is added to the plate after the patient serum sample is added. If the antigen is present, a "sandwich" of antibody, antigen, and secondary antibody will form. The secondary antibody is chemically linked to an enzyme. When the substrate is added, the enzyme converts the substrate from a colorless compound to a colored compound. The amount of color produced will be proportional to the amount of antigen binding to the antibodies and thus indicates whether the patient is positive for the antigen. If the antigen is not present, the secondary (enzyme-linked) antibodies will be rinsed away with the washing steps and the substrate will not be converted and will remain colorless. A common use of the direct ELISA is a home pregnancy test, which detects human chorionic gonadotropin (hCG), a hormone present in the urine of pregnant women.

In the indirect ELISA, a 96-well microtiter plate is coated with antigens. The patient serum sample is added to test for the presence of antibodies that bind to the antigens on the plate. The secondary antibody that is added has an enzyme linked to it that binds to the **constant region** of the primary antibody if it is present in the patient sample. The constant region of an antibody has the same sequence of amino acids within a class of antibodies (for example, all IgG antibodies have the same constant region). The **variable region** of an antibody provides the diversity of antibodies and is the site to which the antigen binds. The configuration that forms in the indirect ELISA is antigen, primary antibody, and secondary antibody. Just as in the direct ELISA, the addition of substrate is used to determine whether the sample is positive for the presence of antibody.

In this activity you will use the indirect ELISA to test for the presence of antibodies made against human immunodeficiency virus (HIV). You will use positive and negative controls to verify the results. You will note that an indeterminate result can be obtained if there is not enough color produced to warrant a positive result. The cause of an indeterminate result could be either nonspecific binding or that the individual has been recently infected and has not yet produced enough antibodies for a positive result. In either case, the individual would be retested.

EQUIPMENT USED The following equipment will be depicted on-screen: five samples in the samples cabinet: patient A, patient B, patient C, a positive control, and a negative control; 96-well microtiter plate; multichannel pipettor; 100-μl pipettor; microtiter plate reader; pipettor tip dispenser; washing buffer; HIV antigen solution; developing buffer—secondary antibody conjugated with an enzyme; substrate solution; paper towels—used for blotting; biohazardous waste disposal.

Experiment Instructions

Go to the home page in the PhysioEx software and click **Exercise 12: Serological Testing.** Click **Activity 3: Indirect Enzyme-Linked Immunosorbent Assay (ELISA),** and take the online **Pre-lab Quiz** for Activity 3.

After you take the online Pre-lab Quiz, click the **Experiment** tab and begin the experiment. The experiment instructions are reprinted here for your reference. The opening screen for the experiment is shown below.

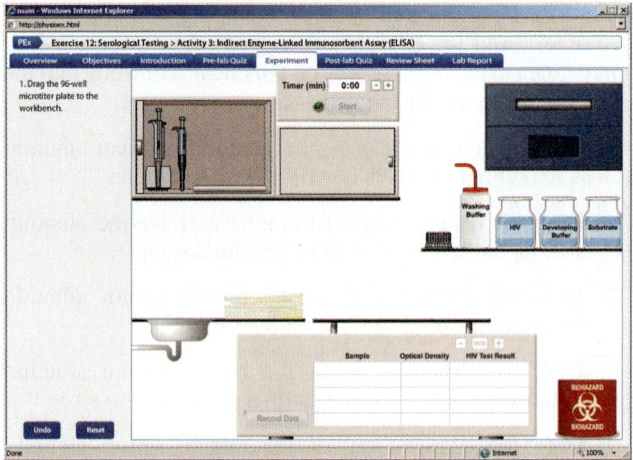

1. Drag the 96-well microtiter plate to the workbench.

2. Drag the multichannel pipettor to the pipette tip dispenser to insert the tips.

3. Drag the multichannel pipettor to the HIV antigens bottle to draw the antigen solution into the tips.

4. Drag the multichannel pipettor directly over the microtiter plate to dispense the liquid into the wells in one row of the plate.

5. Drag the multichannel pipettor to the biohazardous waste disposal for removal and disposal of the tips.

6. Set the timer to 14 hours by clicking the + button beside the timer display. This incubation time allows the antigens to stick to the plastic wells of the microtiter plate. Click **Start** to start the timer. The simulation compresses the 14-hour time period into 10 seconds of real time.

7. Drag the washing buffer squeeze bottle to the microtiter plate to remove excess antigens that are not adsorbed (stuck) to the plate.

8. Drag the microtiter plate to the sink to dump the contents of the tray into the sink to remove the washing buffer and excess antigens that are not stuck to the plastic.

9. Drag the microtiter plate to the paper towels. The plate will be pressed to the surface of the paper towels to remove the remaining liquid from the wells. In a typical ELISA, you would perform multiple washing steps to reduce any nonspecific binding. The number of washing steps in this simulation has been reduced for simplicity.

10. Drag the 100-µl pipettor to the tip dispenser to place a tip onto the pipettor.

11. Drag the 100-µl pipettor to the test tube containing the positive control sample (+) to draw the sample into the tip.

12. Drag the 100-µl pipettor to the microtiter plate to dispense the sample into the wells of the plate. The tip will automatically be removed and disposed of in the biohazardous waste disposal. Each of the remaining samples will automatically be dispensed into plate.

13. Set the timer to 1 hour by clicking the + button beside the timer display. This incubation time allows the antigens stuck to the plastic to bind to the antibodies present in the sample. Click **Start** to start the timer. The simulation compresses the 1-hour time period into 10 seconds of real time.

14. Drag the washing buffer squeeze bottle to the microtiter plate to wash off excess antibodies and prevent nonspecific binding of the antigen and antibody.

15. Drag the microtiter plate to the sink to dump washing buffer and unbound antibodies into the sink.

16. Drag the microtiter plate to the paper towels. The plate will be pressed to the surface of the paper towels to remove the remaining liquid from the wells.

17. Drag the multichannel pipettor to the pipette tip dispenser to insert the tips.

18. Drag the multichannel pipettor to the developing buffer bottle to draw the developing buffer into the tips. The developing buffer contains the conjugated secondary antibody.

19. Drag the multichannel pipettor to the microtiter plate to dispense the solution into the wells. The tips will automatically be removed and disposed of in the biohazardous waste disposal.

20. Set the timer to 1 hour and then click **Start** to start the timer and allow the conjugated secondary antibody to bind to the primary antibody if it is present in the sample.

21. Drag the washing buffer squeeze bottle to the microtiter plate to remove any nonspecific binding that occurred.

22. Drag the microtiter plate to the sink to dump the contents of the tray into the sink.

23. Drag the microtiter plate to the paper towels. The plate will be pressed to the surface of the paper towels to remove the remaining liquid from the wells.

24. Drag the multichannel pipettor to the pipette tip dispenser to insert the tips.

25. Drag the multichannel pipettor to the substrate bottle to draw the substrate into the tips.

26. Drag the multichannel pipettor to the microtiter plate to dispense the solution into the wells. The tips will automatically be removed and disposed of in the biohazardous waste disposal.

27. An enlargement of the wells will appear. The development will progress over time. To determine the optical density for each sample (the samples are in the first row, from top to bottom, of the microtiter plate):

- Click the well and the optical density will appear in the window of the microtiter plate reader.
- Click **Record Data** to display your results in the grid (and record your results in Chart 3).

CHART 3	Indirect ELISA Results	
Sample	**Optical density**	**HIV test result**
Patient A		
Patient B		
Patient C		
Positive control		
Negative control		

28. You will now indicate whether the result for each sample is negative, indeterminate, or positive for HIV.

- A result of <0.300 is read as negative for HIV-1.
- A result of 0.300–0.499 is read as indeterminate (need to retest).
- A result of >0.500 is read as positive for HIV-1.

Click the row for the sample in the grid and then click the − button, **IND,** or the + button above the HIV test result column to indicate whether the result for the sample is positive, indeterminate, or negative for HIV. Repeat this step for all five samples. Record your results in Chart 3.

After you complete the Experiment, take the online **Postlab Quiz** for Activity 3.

Activity Questions

1. Describe how you can tell that this test is the indirect ELISA rather than the direct ELISA.

_____

_____

2. Describe what the secondary antibody binds to in this activity and why.

_____

_____

3.   Define *seroconversion*. How can you tell that a sample has seroconverted?

_____

_____

Western Blotting Technique

OBJECTIVES

1. To compare the Western blotting technique to the ELISA.
2. To observe the use of the Western blotting technique to test for HIV.
3. To distinguish between antigens and antibodies.

Introduction

Southern blotting was developed by Ed Southern in 1975 to identify DNA. A variation of this technique, developed to identify RNA, was named Northern blotting, thus continuing the directional theme. Western blotting, another variation that identifies proteins, is named by the same convention.

Western blotting uses an electrical current to separate proteins on the basis of their size and charge. This technique uses **gel electrophoresis** to separate the proteins in a gel matrix. Because the resulting gel is fragile and would be difficult to use in further tests, the proteins are then transferred to a **nitrocellulose membrane.** The original Western blotting technique used blotting (diffusion) to transfer the proteins, but electricity is also used now for the transfer of the proteins to nitrocellulose strips. These strips are commercially available, eliminating the need for the electrophoresis and transfer equipment. In this activity you will begin the procedure after the HIV (human immunodeficiency virus) antigens have already been transferred to nitrocellulose and cut into strips.

Western blotting is also known as **immunoblotting** because the proteins that are transferred, or blotted, onto the membrane are later treated with antibodies—the same procedure used in the **indirect enzyme-linked immunosorbent assay (ELISA).** The ELISA is considered enzyme linked because an enzyme is chemically linked to an antibody in both the direct and indirect versions of the test. Immunosorbent refers to the fact that either antigens or antibodies are being adsorbed (stuck) to plastic. If the test is designed to detect an antigen or antigens, it is a **direct ELISA** because it is directly looking for the foreign substance. An **indirect ELISA** is designed to detect antibodies that the patient has made against the antigen.

Similar to the secondary antibodies used in the indirect ELISA technique, the secondary antibodies in the Western blot have an enzyme attached to them, allowing for the use of color to detect a particular protein. The secondary antibody binds to the constant region of the primary antibody found in the patient's sample. The main difference between these techniques is that the ELISA technique uses a well that corresponds to a mixture of antigens, and the Western blot has a discrete protein band that represents the specific antigen that the antibody is recognizing. Like HIV, Lyme disease can also be detected with the Western blot technique.

The initial test for HIV is the ELISA, which is less expensive and easier to perform than the Western blot. The Western blot is used as a confirmatory test after a positive ELISA because the ELISA is prone to false-positive results. The bands from a positive Western blot are from antibodies binding to specific proteins and glycoproteins from the human immunodeficiency virus. A positive result from the Western blot is determined by the presence of particular protein bands (view Table 12.1).

> **EQUIPMENT USED** The following equipment will be depicted on-screen: washing buffer; developing buffer—secondary antibody conjugated with an enzyme; substrate solution; five samples in the samples cabinet: patient A, patient B, patient C, positive control, and negative control; rocking apparatus; nitrocellulose strips; troughs; tray; biohazardous waste disposal.

Experiment Instructions

Go to the home page in the PhysioEx software and click **Exercise 12: Serological Testing.** Click **Activity 4: Western Blotting Technique,** and take the online **Pre-lab Quiz** for Activity 4.

After you take the online Pre-lab Quiz, click the **Experiment** tab and begin the experiment. The experiment instructions are reprinted here for your reference. The opening screen for the experiment is shown below.

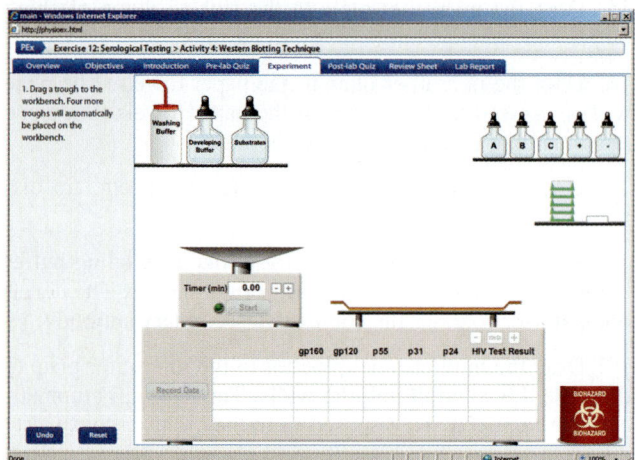

1. Drag a trough to the tray on the workbench. Four more troughs will automatically be placed on the tray.

2. Click the stack of nitrocellulose strips to place a nitrocellulose strip in each trough.

3. Drag the dropper cap of the patient A sample bottle to the first trough to dispense the antiserum from patient A to the nitrocellulose strip. A drop of antiserum for each patient will be dispensed into a separate trough.

4. Drag the tray holding the five troughs to the rocking apparatus.

5. Set the timer to 60 minutes by clicking the + button beside the timer display. Click **Start** to gently rock the samples and allow the antibodies to react with the antigens bound to the nitrocellulose. The tray will automatically be returned to the workbench and each trough will be drained into the biohazardous waste disposal when the time is complete. The simulation compresses the 60-minute time period into 10 seconds of real time.

6. Drag the washing buffer squirt bottle to the first trough to dispense washing buffer in each trough. Each trough will be automatically drained into the biohazardous waste container. The washing step removes any nonspecific binding of antibodies that occurred.

7. Drag the dropper cap of the developing buffer bottle to the first trough to dispense developing buffer to each trough.

8. Drag the tray holding the five troughs to the rocking apparatus.

9. Set the timer to 60 minutes by clicking the + button beside the timer display. Click **Start** to gently rock the samples and allow the antibodies to react with the antibodies bound to the nitrocellulose. The tray will automatically be returned to the workbench and each trough will be drained into the biohazardous waste disposal when the time is complete.

10. Drag the washing buffer squirt bottle to the first trough to add washing buffer to each trough. Each trough will be automatically drained into the biohazardous waste container. The washing step removes any nonspecific binding of secondary conjugated antibodies. Excess secondary conjugated antibodies could react erroneously with the substrate and give a false-positive result.

11. Drag the dropper cap of the substrates bottle to the first trough to dispense the substrates (tetramethyl benzidine and hydrogen peroxide) into each trough. The substrates are the chemicals that are being changed by the enzyme that is linked to the antibody.

12. Drag the tray holding the five troughs to the rocking apparatus.

13. Set the timer to 10 minutes by clicking the + button beside the timer display. Click **Start** to gently rock the samples and allow the enzyme to react with the substrates. The tray will automatically be returned to the workbench when the time is complete. The simulation compresses the 10-minute time period into 10 seconds of real time.

14. To determine the antigens present for each sample:

- Click the nitrocellulose strip inside the trough to visualize the results.
- Click **Record Data** to display your results in the grid (and record your results in Chart 4). The bands present

on the nitrocellulose strip represent the antibodies present in the sample that have reacted with the antigens (bands) on the strip (view Table 12.1).

Repeat this step for all five samples.

TABLE 12.1	**HIV Antigens**
Abbreviation	**Description**
gp160	Glycoprotein 160, a viral envelope precursor
gp120	Glycoprotein 120, a viral envelope protein that binds to CD4
p55	A precursor to the viral core protein p24
gp41	A final envelope glycoprotein
p31	Reverse transcriptase
p24	A viral core protein

15. You will now indicate whether the result for each sample is negative, indeterminate, or positive for HIV. The criteria for reporting a positive result varies slightly from agency to agency. The Centers for Disease Control and Prevention recommend the following criteria:

- If no bands are present, the result is negative.
- If bands are present but they do not match the criteria for a positive result, the result is indeterminate. Patients whose results are deemed indeterminate after multiple tests should be monitored and tested again at a later date.
- If either p31 or p24 is present *and* gp160 or gp120 is present, the result is positive. Click the row for the sample in the grid and then click the − button, **IND,** or the + button above the HIV test result column to indicate whether the result for the sample is positive, indeterminate, or negative for HIV.

Repeat this step for all five samples. Record your results in Chart 4.

After you complete the experiment, take the online **Post-lab Quiz** for Activity 4.

CHART 4	Western Blot Results					
Sample	gp160	gp120	p55	p31	p24	HIV test result
Patient A						
Patient B						
Patient C						
Positive control						
Negative control						

Activity Questions

1. Describe how gel electrophoresis is used to separate proteins.

_____

_____

_____

2. In a patient sample that is positive for HIV, would antibodies or antigens be present when using the Western blot technique? How do you know?

_____

_____

NAME _____

LAB TIME/DATE _____

Serological Testing

ACTIVITY 1 Using Direct Fluorescent Antibody Technique to Test for Chlamydia

1. Describe the importance of the washing steps in the direct antibody fluorescence test. _____
   _____
   _____

2. Explain where the epitope (antigenic determinant) is located. _____
   _____
   _____

3. Describe how a positive result is detected in this serological test. _____
   _____
   _____

4. How would the results be affected if a negative control gave a positive result? _____
   _____
   _____

ACTIVITY 2 Comparing Samples with Ouchterlony Double Diffusion

1. Describe how you were able to determine what antigen is in the unknown well. _____
   _____
   _____

2. Why does the precipitin line form? _____
   _____
   _____

3. Did you think human serum albumin and bovine serum albumin would have epitopes in common? How well did the results

compare with your prediction? _____

_____

_____

ACTIVITY 3 Indirect Enzyme-Linked Immunosorbent Assay (ELISA)

1. Describe how the direct and indirect ELISA are different. _____

_____

_____

2. Discuss why a patient might test indeterminate. _____

_____

_____

3. How would your results have been affected if your negative control had given an indeterminate result? _____

_____

_____

4. Briefly describe the basic structure of antibodies. _____

_____

_____

ACTIVITY 4 Western Blotting Technique

1. Describe why the HIV Western blot is a more specific test than the indirect ELISA for HIV. _____

_____

_____

2. Explain the procedure for a patient with an indeterminate HIV Western blot result. _____

_____

_____

3. Briefly describe how the nitrocellulose strips were prepared before the patient samples were added to them. _____

_____

_____

4. Describe the importance of the washing steps in the procedure. _____

_____

_____

The Metric System

Measurement	Unit and abbreviation	Metric equivalent	Metric to English conversion factor	English to metric conversion factor
Length	1 kilometer (km)	= 1000 (10^3) meters	1 km = 0.62 mile	1 mile = 1.61 km
	1 meter (m)	= 100 (10^2) centimeters	1 m = 1.09 yards	1 yard = 0.914 m
		= 1000 millimeters	1 m = 3.28 feet	1 foot = 0.305 m
			1 m = 39.37 inches	
	1 centimeter (cm)	= 0.01 (10^{-2}) meter	1 cm = 0.394 inch	1 foot = 30.5 cm
				1 inch = 2.54 cm
	1 millimeter (mm)	= 0.001 (10^{-3}) meter	1 mm = 0.039 inch	
	1 micrometer (μm) [formerly micron (μ)]	= 0.000001 (10^{-6}) meter		
	1 nanometer (nm) [formerly millimicron (mμ)]	= 0.000000001 (10^{-9}) meter		
	1 angstrom (Å)	= 0.0000000001 (10^{-10}) meter		
Area	1 square meter (m^2)	= 10,000 square centimeters	1 m^2 = 1.1960 square yards	1 square yard = 0.8361 m^2
			1 m^2 = 10.764 square feet	1 square foot = 0.0929 m^2
	1 square centimeter (cm^2)	= 100 square millimeters	1 cm^2 = 0.155 square inch	1 square inch = 6.4516 cm^2
Mass	1 metric ton (t)	= 1000 kilograms	1 t = 1.103 ton	1 ton = 0.907 t
	1 kilogram (kg)	= 1000 grams	1 kg = 2.205 pounds	1 pound = 0.4536 kg
	1 gram (g)	= 1000 milligrams	1 g = 0.0353 ounce	1 ounce = 28.35 g
			1 g = 15.432 grains	
	1 milligram (mg)	= 0.001 gram	1 mg = approx. 0.015 grain	
	1 microgram (μg)	= 0.000001 gram		
Volume (solids)	1 cubic meter (m^3)	= 1,000,000 cubic centimeters	1 m^3 = 1.3080 cubic yards	1 cubic yard = 0.7646 m^3
			1 m^3 = 35.315 cubic feet	1 cubic foot = 0.0283 m^3
	1 cubic centimeter (cm^3 or cc)	= 0.000001 cubic meter	1 cm^3 = 0.0610 cubic inch	1 cubic inch = 16.387 cm^3
		= 1 milliliter		
	1 cubic millimeter (mm^3)	= 0.000000001 cubic meter		
Volume (liquids and gases)	1 kiloliter (kl or kL)	= 1000 liters	1 kL = 264.17 gallons	1 gallon = 3.785 L
	1 liter (l or L)	= 1000 milliliters	1 L = 0.264 gallons	1 quart = 0.946 L
			1 L = 1.057 quarts	
	1 milliliter (ml or mL)	= 0.001 liter	1 ml = 0.034 fluid ounce	1 quart = 946 ml
		= 1 cubic centimeter	1 ml = approx. $\frac{1}{4}$ teaspoon	1 pint = 473 ml
			1 ml = approx. 15–16 drops (gtt.)	1 fluid ounce = 29.57 ml
				1 teaspoon = approx. 5 ml
	1 microliter (μl or μL)	= 0.000001 liter		
Time	1 second (s or sec)	= $\frac{1}{60}$ minute		
	1 millisecond (ms or msec)	= 0.001 second		
Temperature	Degrees Celsius (°C)		°F = $\frac{9}{5}$ (°C) + 32	°C = $\frac{5}{9}$ (°F − 32)

Credits

Illustrations

All illustrations are by Imagineering STA Media Services, except for Review Sheet art and as noted below.

Exercise 1
1.1: Imagineering STA Media Services/ Precision Graphics. 1.2, 1.4: Precision Graphics. 1.9: Adapted from Marieb and Mallatt, *Human Anatomy,* 3e, F1.10, © Benjamin Cummings, 2003.

Exercise 3
3.2–3.4, Activity 3: Precision Graphics.

Exercise 4
4.2: Imagineering STA Media Services/ Precision Graphics. 4.3, 4.RS1: Tomo Narashima.

Exercise 5
5.1: Precision Graphics.

Exercise 6
6.2: Precision Graphics.

Exercise 7
7.1, 7.2, 7.7: Electronic Publishing Services, Inc.

Exercise 9
9.1–9.3, 9.6, 9.8, 9.19: Nadine Sokol.

Exercise 12
12.4: Imagineering STA Media Services/ Precision Graphics. 12.5: Electronic Publishing Services, Inc.

Exercise 13
13.1: Imagineering STA Media Services/ Adapted from Martini, *Fundamentals of Anatomy & Physiology,* 4e, F11.1, Upper Saddle River, NJ: Prentice-Hall, © Frederic H. Martini, 1998.

Exercise 14
14.1–14.3: Precision Graphics. 14.7–14.9, 14.13–14.18, 14.U1: Biopac Systems.

Exercise 15
15.1: Imagineering STA Media Services/ Precision Graphics.

Exercise 16
16.2–16.4: Precision Graphics.

Exercise 17
17.1, 17.5, 17.7–17.10: Electronic Publishing Services, Inc. 17.11: Precision Graphics.

Exercise 18
18.4–18.7: Biopac Systems.

Exercise 19
19.2b: Electronic Publishing Services, Inc.

Exercise 20
20.7–20.13: Biopac Systems.

Exercise 21
21.1: Electronic Publishing Services, Inc. 21.8, 21.9: Biopac Systems.

Exercise 23
23.1, 23.3, 23.4, 23.6: Electronic Publishing Services, Inc.

Exercise 24
24.1: Shirley Bortoli. 24.4: Precision Graphics.

Exercise 25
25.1: Electronic Publishing Services, Inc./ Precision Graphics. 25.2, 25.3, 25.7, 25.9: Electronic Publishing Services, Inc.

Exercise 26
26.1, 26.2: Electronic Publishing Services, Inc.

Exercise 29
29.2, 29.5: Precision Graphics.

Exercise 30
30.1: Electronic Publishing Services, Inc./Precision Graphics. 30.2, 30.3, 30.6: Electronic Publishing Services, Inc.

Exercise 31
31.1: Electronic Publishing Services, Inc. 31.2, 31.4: Precsion Graphics. 31.7–31.12: Biopac Systems.

Exercise 32
32. 2–32.12, 32.14: Electronic Publishing Services, Inc.

Exercise 33
33.2, 33.9: Precision Graphics. 33.6–33.8: Biopac Systems.

Exercise 34
34.1, 34.2, 34.4: Precision Graphics. 34.5: Biopac Systems. 34.6: Electronic Publishing Services, Inc.

Exercise 35
35.8: Precision Graphics.

Exercise 36
36.1–36.5, 36.7: Electronic Publishing Services, Inc.

Exercise 37
37.5, 37.6: Precision Graphics. 37.11, 37.13, 37.14: Biopac Systems.

Exercise 38
38.1–38.5, 38.7, 38.8, 38.10, 38.15: Electronic Publishing Services, Inc. 38.16: Electronic Publishing Services, Inc./ Precision Graphics.

Exercise 39
39.A1–A3: Precision Graphics.

Exercise 40
40.1, 40.2: Electronic Publishing Services, Inc.

Exercise 41
41.1: Precision Graphics.

Exercise 42
42.1, 42.2, 42.7: Electronic Publishing Services, Inc.

Exercise 43
43.1: Precision Graphics. 43.2: Electronic Publishing Services, Inc.

Exercise 44
44.1: Electronic Publishing Services, Inc.

Exercise 46
46.17: Precision Graphics

Cat Dissection Exercises
1.1, 3.1, 3.3, 7.1: Precision Graphics. 2.1a, 2.2a, 3.2, 4.2, 4.4, 6.2, 7.2, 8.2, 9.1: Kristin Mount.

Fetal Pig Dissection Exercises
Kristin Mount.

PhysioEx Exercises
All illustrations by BinaryLabs, Inc. except as noted below.
2.1: Precision Graphics/Adapted from Stanfield, *Principles of Human Physiology,* 4e. F12.1, Upper Saddle River, NJ, © Pearson Science. 3.1, 12.1: Imagineering STA Media Services. 4.1, 8.1, 9.1: Electronic Publishing Services, Inc. 6.1: Precision Graphics/Adapted from Stanfield, *Principles of Human Physiology,* 4e. F13.13, Upper Saddle River, NJ, © Pearson Science. 7.1: Precision Graphics/Adapted from Stanfield, *Principles of Human Physiology,* 4e. F16.1, Upper Saddle River, NJ, © Pearson Science. 8.1: Electronic Publishing Services, Inc.

PHOTOGRAPHS

Exercise 1
1.3.1: Jenny Thomas, Pearson Education. 1.3a: Scott Camazine/Photo Researchers. 1.3b: James Cavallini/Photo Researchers. 1.3c: CNRI/Science Photo Library/Photo Researchers. 1.8a: John Wilson White, Pearson Education.

Exercise 2
2.1a–d, 2.2, 2.3a, 2.4a, 2.5b,c: Elena Dorfman, Pearson Education. 2.3b, 2.4b, 2.5a, 2.6a–c: From *A Stereoscopic Atlas of*

TRADEMARK ACKNOWLEDGMENTS

Index

Å. *See* Ångstrom
A bands, 186*f*, 187, 187*f*
Abdomen
  arteries of, 474–476, 475–476*f*
  muscles of, 198*f*, 206*t*, 207–208*f*, 685*f*, 686–687
    in breathing, **PEx-106, PEx-107**
    surface anatomy of, 685*f*, 686–687, 686*f*
  veins of, 477–478, 480*f*
Abdominal (term), **2**, 3*f*
Abdominal aorta, **472**, 474–476, 475*f*, 476*f*
Abdominal cavity, **6**, 7*f*, 18–20, 20*f*
Abdominal organs
  blood vessels of. *See* Abdomen, arteries of; Abdomen, veins of
  peritoneal attachments of, 580*f*
Abdominal reflex, 339
Abdominal wall muscles, 198*f*, 206*t*, 207–208*f*, 685*f*, 686–687
  in breathing, **PEx-106, PEx-107**
Abdominopelvic cavity, **6**, 7*f*, 18–21, 20*f*, 21*f*
  quadrants/regions of, **8**, 8*f*, 9*f*
Abducens nerve (cranial nerve VI), 286*t*, 288, 288*f*, 289*f*
Abduction (body movement), **173**, 175*f*
Abductor pollicis longus muscle, 214*t*, 215*f*
ABO antigens, 434*t*, 435*f*, PEx-162, **PEx-167**, PEx-168*t*. *See also* ABO blood group
ABO blood group, 434, 434*t*, 435*f*, PEx-162, **PEx-167**, PEx-168*t*
  inheritance of, 672, 672*t*, PEx-167
  typing for, 434–435, 434*t*, 435*f*
Absolute refractory period, 234, **267**, **PEx-45, PEx-94**
  cardiac, **PEx-94**
  muscle cell, **234**
  neuron, **267**, PEx-44 to **PEx-45**
Absorption, **574**
  in small intestine, 581–582, 581*f*
Accessory cutaneous organs, 94*f*, 97–100, 97*f*, 98*f*, 99*f*, 100*f*
Accessory digestive organs/ducts, **574**, 584–588, 587*f*. *See also specific organ*
Accessory ducts, male reproductive, 630–631*f*, 631–632
Accessory glands, male reproductive, 630–631*f*, 632
Accessory hemiazygos vein, **479**, 481*f*
Accessory nerve (cranial nerve XI), 287*t*, 288, 288*f*, 289*f*, 313*f*, **682**, 683*f*
Accommodation, **372**, 375
  near point of, **372**–373
Acetabular labrum, **176**, 176*f*
Acetabulum, **153**, 154*f*, 176
  in male versus female, 153, 155*t*
Acetylcholine, 324, **PEx-98**
  in muscle contraction, PEx-18
  in neuromuscular junction, 189, 189*f*
Achilles (calcaneal) tendon, 199*f*, 223*f*, 693*f*
Acid, **PEx-149**
  strong, **PEx-149**
  weak, 564, **PEx-149**
Acid ash foods, 622
Acid-base balance, 564, **PEx-149** to PEx-159
  renal system in, **PEx-150**, PEx-152 to PEx-154

respiratory system in, 564–565, 564*f*, **PEx-150**, PEx-150 to PEx-152, PEx-154 to PEx-156
Acid hydrolases, 43
Acidophil cells, **410**, 410*f*
Acidosis, **PEx-150, PEx-151**
  ketonuria and, 623
  metabolic, **PEx-154**
  respiratory, PEx-151, PEx-153
  respiratory and renal responses to, PEx-150, PEx-151, PEx-152 to PEx-154, PEx-154 to PEx-156
Acinar cells, pancreatic, 409*f*, **410**
Acne, **99**
Acoustic meatus
  external (external auditory canal/meatus), **120**, 121*f*, 122*f*, **384**, 384*f*, 680*f*, **681**
  internal, **120**, 123*f*
Acquired (learned) reflexes, 336
  reaction time of, 341–344, 343*f*
Acquired umbilical hernia, **686**
Acromegaly, **406**
Acromial (term), **2**, 3*f*
Acromial (lateral) end of clavicle, 148, 149*f*
Acromioclavicular joint, 148*f*, 152, 170*t*, **687**, 687*f*
Acromion, 148, 149*f*, 179*f*, **687**
Acrosome, **648**, 648*f*
ACTH (adrenocorticotropic hormone), **406**, 407*f*, **PEx-69** to PEx-70, PEx-69*t*
Actin, 43*t*, 44, **187**
Actin (thin) filaments, 43*t*, 44, 186*f*, **187**
  muscle contraction and, 234, 240
Action potential, 234, **267**, **PEx-35, PEx-41**
  cardiac, **PEx-94**, PEx-94*f*
  in muscle cells, **234**
  in neurons, 256, 266–**267**, 266*f*, **PEx-35, PEx-41**, PEx-50 to PEx-52. *See also* Nerve impulse
    chemical synaptic transmission/neurotransmitter release and, **PEx-49** to PEx-50
    compound, **267**, 270–271, 270*f*, 271*f*
    conduction velocity and, PEx-47 to PEx-49, **PEx-48**
    oscilloscope in study of, **270**–271, 270*f*, 271*f*
    refractory periods (absolute and relative) and, **267**, PEx-44 to PEx-45, **PEx-45**
    stimulus intensity coding and, PEx-46 to PEx-47
    threshold and, 266*f*, **267**, PEx-41 to **PEx-42**
    voltage-gated sodium channels and, **PEx-42** to PEx-44
  propagation of, 267
Active electrode, for EEG, 300
Active force, in muscle contraction, **PEx-26** to PEx-27
Active processes/transport, 54, 60–61, **PEx-3**, PEx-4, PEx-13
  active transport, 41, **60**, **PEx-4**, PEx-13 to PEx-14
    primary, 60
    secondary, 60, **PEx-140**
  vesicular transport, **60**–61, 61*f*, **PEx-4**
Active site, **PEx-123**
Acuity testing
  hearing, 387
  visual, 373–374
Adam's apple (laryngeal prominence), 538, 540*f*, **681**, 682*f*
Adaptation, **353**–354, PEx-46
  olfactory, 401–402, PEx-46
Adaptive immune system, **527**

Addison's disease, **97**, **PEx-69**, PEx-69*t*
Adduction (body movement), **173**, 175*f*
Adductor brevis muscle, 216*t*, 217*f*
Adductor longus muscle, 198*f*, 216*t*, 217*f*, 691*f*, 692*f*, **693**
Adductor magnus muscle, 199*f*, 216*t*, 217*f*, 219*f*
Adductor muscles of thigh, 216*t*, 217*f*, 692*f*, **693**
Adductor tubercle, **156**, 156*f*
Adenohypophysis (anterior pituitary gland), **406**, 407*f*, 410, 410*f*
  hormones of, 406–407, 407*f*, 418
Adequate stimulus, PEx-39, PEx-40
ADH. *See* Antidiuretic hormone
Adhesions, in joints, 180
Adipose tissue (fat), **75**, 77*f*, 82, 94*f*
Adjustment knobs (microscope), **28**, 29*f*
Adrenal cortex, 408, 410*f*, 411, PEx-69
Adrenal (suprarenal) glands, **20**, 21*f*, 408, 409*f*, PEx-60*f*
  microscopic anatomy of, 410*f*, 411
Adrenaline. *See* Epinephrine
Adrenal insufficiency, PEx-69, PEx-69*t*
Adrenal medulla, **408**, 410*f*, 411
Adrenergic fibers, 324
Adrenergic modifiers, **PEx-98**
Adrenocorticotropic hormone (ACTH), **406**, 407*f*, **PEx-69** to PEx-70, PEx-69*t*
Adventitia
  alimentary canal, **574**
  tracheal, 543*f*
Afferent arteriole, **613**, 614*f*, 615*f*, **PEx-131, PEx-133, PEx-135, PEx-137**
  radius of, glomerular filtration and, PEx-132 to PEx-135, **PEx-137** to PEx-139
Afferent (sensory) nerves, 258, **276**
Afferent (sensory) neurons, **258**, 259*f*, PEx-35, PEx-39, PEx-41, PEx-50 to PEx-52
  in reflex arc, 336, 336*f*
A fibers, conduction velocity of, PEx-48, PEx-48 to PEx-49
Afterimages, negative, temperature adaptation and, **354**
Afterload, **PEx-84, PEx-86**
Agar gel, diffusion through, 55–56, 55*f*
Agarose gel electrophoresis, in hemoglobin phenotyping, 672–673, 673*f*
Agglutinins, blood typing and, **434**, 434*t*, 435*f*, **PEx-167**
Agglutinogens, blood typing and, **434**, 434*t*, 435*f*, **PEx-162, PEx-167**
Agonists (chemical), **PEx-98**
Agonists (prime movers), **196**
Agranulocytes (agranular leukocytes), 426*t*, **428**
Air-conduction hearing, testing, 388, 388*f*
Ala(e)
  of nose, **681**, 681*f*
  sacral, **134**, 134*f*
Albumin, in urine (albuminuria), **622**–623, 624
Alcohol ingestion, metabolic acidosis and, PEx-154
Aldosterone, 408, **PEx-142**
  urine formation and, **PEx-142** to PEx-143
Alimentary canal (GI tract), **574**, PEx-119, PEx-120*f*. *See also specific organ*
  histology of, 574, 576*f*
Alkali ingestion, metabolic alkalosis and, PEx-154
Alkaline ash foods, 622
Alkalosis, **PEx-150**

metabolic, **PEx-154**
respiratory, **PEx-150**, PEx-152
respiratory and renal responses to, PEx-150, PEx-152 to PEx-154, PEx-154 to PEx-156
Allantois, **660**, 660*f*
Alleles, **668**
  codominant, 672
All-or-none phenomenon, action potential as, 267, PEx-46, PEx-48
Alpha block, **300**
Alpha (α) cells, pancreatic, 409*f*, **410**
Alpha waves/rhythm, **300**, 300*f*
Alveolar cells, 541*f*
Alveolar ducts, 540*f*, **541**, 544*f*
Alveolar pores, 541*f*, 544*f*
Alveolar process(es), 121*f*, **127**, 127*f*
Alveolar sacs, 70*f*, 540*f*, 541, 544*f*
Alveoli
  mammary gland, **637**, 637*f*
  respiratory, 70*f*, 540*f*, **541**, 541*f*, 544, 544*f*, **PEx-105**
Amacrine cells, retinal, 363*f*
Amino acids, **PEx-125**
Amnion, **660**, 660*f*, 662*f*
Amniotic sac, **661**
Amoeboid motion, by leukocytes, **428**
Amphiarthroses, **169**
Ampulla
  of ductus deferens, 630*f*, **631**, 631*f*
  hepatopancreatic, **581**, 587*f*
  of semicircular duct, **389**–390
  of uterine tube, 635*f*
Ampullary cupula, **390**
Amygdaloid body, 280
Amylase
  pancreatic, 597*f*
  salivary, 576, 585, **596**–598, 597*f*, **PEx-121, PEx-123**
Anabolism, 419, PEx-60
Anal canal, 575*f*, **583**, 583*f*
Anal sphincters
  external, 583, 583*f*
  internal, 583
Anaphase
  of meiosis, 646, 646*f*
  of mitosis, **45**, 47*f*
Anatomical crown of tooth, 585
Anatomical dead space, 541
Anatomical left/right, 8
Anatomical neck of humerus, 148, 150*f*
Anatomical position, **2**, 3*f*
Anatomical snuff box, **690**, 690*f*
Anatomical terminology, 1–14
Anconeus muscle, 212*f*, 212*t*, 215*f*
Androgens. *See also* Testosterone
  adrenal, 408
Anemia, 428, **429**, **PEx-162**
  blood viscosity and, PEx-78
  hemoglobin and, 428, PEx-166
Angina pectoris, referred pain and, 354
Angiotensin converting enzyme, PEx-142
Ångstrom (Å), 32*t*
Ankle, muscles acting on, 220*t*, 221*f*, 222–224*t*, 223*f*
Ankle-jerk (calcaneal tendon) reflex, **338**, 339*f*
Ankle joint, 171*t*
ANS. *See* Autonomic nervous system
Ansa cervicalis, 313*f*, 315*t*
Antacid ingestion, metabolic alkalosis and, PEx-154
Antagonists (chemical), **PEx-98**
Antagonists (muscle), **196**
Antebrachial (term), **2**, 3*f*
Antebrachial vein, median, **479**, 481*f*, 689*f*
Antebrachium. *See* Forearm
Antecubital (term), **2**, 3*f*

Anterior (term), **3**, 4*f*
Anterior axillary fold, 685*f*, **687**
Anterior border of tibia, **157**, 157*f*, 692*f*, **693**
Anterior cardiac vein, 444*f*, **448**, 448*f*
Anterior cerebellar lobe, 282, 282*f*
Anterior cerebral artery, **472**, 473*f*
Anterior chamber, **363**
Anterior circumflex humeral artery, **474**, 474*f*
Anterior commissure, 280*f*
Anterior communicating artery, **472**, 473*f*
Anterior cranial fossa, **120**, 123*f*
Anterior cruciate ligament, 177*f*
Anterior femoral cutaneous nerve, 316*f*, 317
Anterior (frontal) fontanelle, 137, 137*f*
Anterior (ventral) funiculus, **310**, 310*f*, 312*f*
Anterior gluteal line, 154*f*
Anterior (ventral) horns, 281*f*, 284*f*, **308**, 310*f*, 312*f*
Anterior inferior iliac spine, 153, 154*f*
Anterior intercostal artery, **474**, 474*f*
Anterior interventricular artery, 444*f*, **447**–448, 448*f*
Anterior muscle compartment
  forearm muscles and, 213*f*, 213–214*t*, 688
  lower limb muscles and
    foot and ankle, 220*t*, 221*f*, 692*f*
    thigh and leg, 217*f*, 218*t*, 692*f*
Anterior nasal spine, 127*f*
Anterior pituitary gland (adenohypophysis), **406**, 407*f*, 410, 410*f*
  hormones of, 406–407, 407*f*, 418
Anterior pole of eye, 362*f*
Anterior scalene muscle, 203*t*, 204*f*
Anterior segment, 362*f*, **363**
Anterior superior iliac spine, **153**, 154*f*, 157–158, 685*f*, **686**, 686*f*
Anterior tibial artery, **476**, 477*f*
Anterior tibial vein, **477**, 479*f*
Anterior triangle of neck, **682**, 683*f*
Anti-A antibodies, 434*t*, 435*f*, PEx-168, PEx-168*t*
Anti-B antibodies, 434*t*, 435*f*, PEx-168, PEx-168*t*
Antibodies, 529, **530**–532, 531*f*, PEx-177 to PEX-178, PEx-178*f*
  blood typing and, **434**, 434*t*, 435*f*, PEx-168, PEx-168*t*
  constant region of, 530, 531*f*, PEx-182
  fluorescent, in direct fluorescent antibody technique, PEx-178 to PEx-180
  variable region of, 530, 531*f*, PEx-182
Antibody monomers, **530**, 531*f*
Antidiuretic hormone (ADH), **407**, 407*f*, **PEx-139, PEx-142**
  urine concentration and, PEx-139 to PEx-140
  urine formation and, **PEx-139, PEx-142** to PEx-143
Antigen(s), 528, 529, 530, 531*f*, **PEx-177** to PEX-178, PEx-178*f*
  blood typing and, **434**, 434*t*, 435*f*, PEx-162, **PEx-167**, PEx-168*t*
  Ouchterlony technique in identification of, 531–532, 532*f*, 532*t*, PEx-180 to PEx-181
Antigen-antibody reaction, 531
Antigen-antibody specificity, PEx-178, PEx-178*f*
Antigen-binding site, **531**, 531*f*, PEx-178, PEx-178*f*
Antigenic determinants (epitopes), 531, 531*f*, **PEx-178**, PEx-178*f*
Antiporter, PEx-13
Antral (vesicular) follicle, 649*f*, **650**, 650*f*, **651**, 652*f*
Antrum, ovarian follicle, 650*f*, **651**
Anulospiral endings, 351*f*
Anulus fibrosus, 131
Anus, 20, 21*f*, 575*f*, **583**

Anvil (incus), 384*f*, **385**
Aorta, 445*f*, **446**, 448*f*, 450*f*, 451, **472**
  abdominal, **472**, 474–476, 475*f*
  ascending, 444*f*, **472**
  descending, **20**, 21*f*
  thoracic, 472–474, **472**, 474*f*
Aortic arch, 444*f*, 472–474, 482*f*
Aortic semilunar valve, 445*f*, **446**, 447*f*, 451
  auscultation of, 494, 494*f*
  stenosis of, PEx-86
Aperture selection dial (ophthalmoscope), **376**, 376*f*
Apex
  of heart, **444**, 444*f*, 445*f*, 450*f*, 452*f*
  of lung, 542*f*, **543**
  of nose, **681**, 681*f*
  of renal pyramid (papilla), 612, 612*f*
Apical foramen, **585**, 585*f*
Apical membrane, **PEx-140**
Apical pulse, 444, 444*f*, **497**
Apical-radial pulse, 497
Apical surface, of epithelium, 68, 68*f*
Aplastic anemia, PEx-162
Apocrine glands, **99**
Aponeuroses, **188**
Appendicitis, 583
Appendicular (term), **2**
Appendicular skeleton, 108*f*, **109**, **147**–165
  pectoral girdle/upper limb, 108*f*, **148**–152, 148*f*, 149*f*, 150*f*, 151*f*, 152*f*
  pelvic girdle/lower limb, 108*f*, **153**–158, 154*f*, 155*f*, 156*f*, 157*f*, 158*f*
Appendix, 527*f*, 575*f*, **583**, 583*f*
  appendectomy/appendicitis and, 686
Aqueous humor, 361, 362*f*, **363**
Arachnoid mater
  brain, **282**, 283*f*, 284*f*, 285
  spinal cord, 308, 309*f*, 310*f*
Arachnoid villi, **283**, 283*f*, 284*f*
Arbor vitae, 280*f*, **282**, 282*f*, 291*f*
Arches (fingerprint), 101, 101*f*
Arches of foot, 157, 158*f*
Arch of vertebra, **131**, 131*f*, 132*f*
Arcuate arteries
  of foot, **476**, 477*f*
  renal, 612*f*, **613**, 614*f*
Arcuate line, **153**, 154*f*
Arcuate popliteal ligament, 177*f*, **178**
Arcuate veins, 612*f*, **613**, 614*f*
Areola (breast), 637*f*
Areolar connective tissue, 74*f*, **75**, 76*f*, 82
Arm. *See also* Forearm; Hand; Upper limb
  bones of, 148–150, 150*f*
  deep artery of, **474**, 474*f*
  muscles of/muscles controlling, 198*f*, 199*f*, 211, 212*f*, 212*t*
  surface anatomy of, 152, 687–688, 687*f*, 688*f*
Arm (microscope), **28**, 29*f*
Arrector pili muscle, 94*f*, **98**, 98*f*
Arterial pressure, 497. *See also* Blood pressure
  mean (MAP), **499**
  measuring, 498–499, 498*f*
Arteries, 470, 470*f*, 471, 471*f*, 472–477. *See also specific structure supplied or specific named artery*
Arterioles, 519–520, 519*f*
  in nephron, **613**, 614*f*, 615*f*, PEx-131, PEx-133
  radius of, glomerular filtration and, PEx-132 to PEx-135, PEx-137 to PEx-139
  terminal, 519, 519*f*
Arteriosclerosis, **PEx-86**
Arthritis, 180
Articular capsule, 169, 169*f*, 171
  internal, **280**
Articular cartilage, **109**, 109*f*, **110**, 111*f*, 169*f*, 171
Articular discs, 171
Articular processes/facets, vertebral
  inferior, 132–133, **132**, 132*f*, 133*f*

superior, 131*f*, **132**, 132*f*, 133*f*, 134*f*, 134*t*
Articular tubercle, **179**, 180*f*
Articulations, 108, 167–184, 168*f*, **169**, 170–171*t*. *See also* Joint(s)
Arytenoid cartilages, **539**, 540*f*
Ascaris megalocephala, oogenesis in, 648–649
Ascending aorta, 444*f*, **472**
Ascending colon, 575*f*, **583**, 583*f*
Ascending lumbar vein, 479, 480*f*, 481*f*
Ascending (sensory) tracts, 310, 311*f*
Association areas of brain, 278*f*
Association neurons (interneurons), **258**, 259*f*, **PEx-35**, PEx-50
  in reflex arc, 336*f*, 337
Association tracts, 279
Asters, 46*f*
Asthma, **PEx-109**, PEx-111
  inhaler medication and, PEx-111
Astigmatism, 372, 374, 374*f*
Astrocytes, 254, 254*f*
Atelectasis, **PEx-112**
Atherosclerosis, 435–436, **PEx-86**, **PEx-169**
Atlantoaxial joint, 170*t*
Atlanto-occipital joint, 170*t*
Atlas (C$_1$), **132**, 132*f*
ATP
  for active processes, 54, 60, 61, PEx-3, PEx-4, PEx-13
  for muscle contraction, 234, 235
Atria, cardiac, 444*f*, 445*f*, **446**, 448*f*, 450*f*
  in frog, 513*f*
Atrioventricular (AV) bundle (bundle of His), **458**, 458*f*
Atrioventricular (AV) node, **458**, 458*f*
Atrioventricular (AV) valves, **446**, 447*f*
  auscultation of, 494, 494*f*
Atropine, heart rate and, 518, PEx-98 to PEx-99
Audiometry/audiometer, 388–389
Auditory association area, 278*f*
Auditory canal/meatus, external (external acoustic meatus), **120**, 121*f*, 122*f*, **384**, 384*f*, 680*f*, **681**
Auditory cortex, primary, 278*f*, 387
Auditory ossicles, 384*f*, **385**
Auditory (pharyngotympanic/eustachian) tube, **384**, **385**, 538, 539*f*
Auditory receptor (hair) cells, 386, 386*f*, 387
Auricle (pinna), 80*f*, **384**, 384*f*, **681**, 681*f*
Auricle(s), of heart, 445*f*, **450**, 450*f*
Auricular nerve, greater, 313*f*, 315*t*
Auricular surface of ilium, **153**, 154*f*
Auricular vein, posterior, 480*f*
Auscultation
  of heart sounds, 494, 494*f*
  of respiratory sounds, 551–552
  triangle of, 552, **683**, 684*f*
Autoimmunity, 528
Automaticity/autorhythmicity, cardiac muscle, **512**–513, **PEx-93**
Autonomic nervous system (ANS), **276**, **323**–333, 324*f*
  function of, 324–332
  in heart regulation, 458, **PEx-95**, PEx-98
  parasympathetic division of, **324**, 324*f*, **PEx-95**
  sympathetic division of, **324**, 324*f*, 325*f*, **PEx-95**
Autonomic (visceral) reflexes, **336**, 340–341
Autorhythmicity/automaticity, cardiac muscle, **512**–513, **PEx-93**
Autosomes, **669**
Avascularity, epithelial, 69
AV valves. *See* Atrioventricular (AV) valves
Axial (term), **2**
Axial skeleton, 108*f*, **109**, 119–145, **120**.
  See also specific region and specific bone
  skull, 108*f*, **120**–130, 121*f*, 122*f*, 123*f*, 124*f*, 125–126*f*, 127*f*, 128*f*, 129*f*

thoracic cage, 108*f*, **135**–136, 135*f*, 136*f*, 683, 685*f*
  vertebral column, 108*f*, **130**–135, 130*f*, 131*f*
Axilla, 685*f*, 687
Axillary (term), **2**, 3*f*
Axillary artery, 473*f*, **474**, 474*f*
Axillary (lateral) border of scapula, 148, 149*f*
Axillary folds, 685*f*, **687**
Axillary lymph nodes, 526*f*, 527, 687
Axillary nerve, **313**, 314*f*, 315*t*
Axillary vein, **479**, 481*f*
Axillary vessels, 687
Axis (C$_2$), **132**, 132*f*
Axon(s), 82*f*, 255*f*, **256**, 256*f*, 257*f*, 260*f*, **PEx-36**, PEx-36*f*, PEx-41, PEx-49
  of motor neurons, 241, 255*f*, 257*f*
Axon collaterals, **256**
Axon hillock, 255*f*, **256**, **PEx-41**
Axon terminals, 188, 189, 189*f*, 190*f*, **241**, 255*f*, **256**, PEx-36, PEx-36*f*, PEx-49
Azygos system, **479**
Azygos vein, **479**, 481*f*

Babinski's sign, 339
Back, surface anatomy (bones/muscles) of, 683, 684*f*
Balance tests, 391
Ball-and-socket joint, 172*f*, 173
BAPNA, 598–599, **PEx-125**, PEx-126
Barany test, 391–392
Basal lamina, 69
Basal layer
  of endometrium (stratum basalis), **636**
  of skin (stratum basale), 95*f*, **96**, 99*f*
Basal metabolic rate (BMR). *See also* Metabolic rate/metabolism
  determining, PEx-61 to PEx-62
  thyroid hormone affecting, 419, PEx-62
Basal nuclei (basal ganglia), 276*f*, **279**
Basal surface, of epithelium, 68, 68*f*
Base
  of axilla, 685*f*, **687**
  of heart, **444**, 450*f*
  of lung, 542*f*, **543**
  of metacarpal, 151
  of sacrum, 154*f*
Base (chemical), **PEx-149**
  strong, **PEx-149**
  weak, 564, **PEx-149**
Base (microscope), **28**, 29*f*
Basement membrane, **69**, 70*f*, 71*f*, 72*f*, 73*f*
Basilar artery, 473*f*, 473*f*
Basilar membrane, **386**, 386*f*, 387, 387*f*
Basilic vein, **479**, 481*f*, 688, 689*f*
Basolateral membrane, **PEx-140**
Basophil(s), 425*f*, 426*t*, **428**, 428*f*
Basophil cells, **410**, 410*f*
B cells (B lymphocytes), 428, 528–529, **528**
Beauchene skull, 120, 126*f*. *See also* Skull
Bedsores (decubitus ulcers), **96**, 96*f*
Benedict's assay, PEx-121, PEx-122, PEx-123, PEx-124
Beta (β) cells, pancreatic, 409*f*, **410**
Beta waves, **300**, 300*f*
B fibers, conduction velocity of, PEx-48, PEx-49
Biaxial joints/movement, 168*f*, 172*f*, 173
Bicarbonate (HCO$_3^-$)
  in acid-base balance, 564, 564*f*, PEx-149, PEx-150, PEx-151, PEx-153, PEx-154
  ingestion of, metabolic alkalosis and, PEx-154
Bicarbonate buffer system, PEx-150.
  See also Bicarbonate (HCO$_3^-$), in acid-base balance
Biceps (term), muscle name and, 197
Biceps brachii muscle, 196*f*, 198*f*, 205*f*, 211, 212*f*, 212*t*, 213*f*, **687**, 687*f*, 689*f*
Biceps femoris muscle, 199*f*, 218*t*, 219*f*, 222, **693**, 693*f*

Bicipital furrow, medial, **687**, 687*f*
Bicipital groove (intertubercular sulcus), **150**, 150*f*
Bicuspids (premolars), **584**, 584*f*
Bicuspid (mitral) valve, 445*f*, **446**, 447*f*
  auscultation of, 494, 494*f*
Bile, 581, 586, 587, 599–600, PEx-127
  emulsification action of, 586, 597*f*, 599–600
Bile canaliculi, **587**
Bile duct, **581**, **586**, 586*f*, 587, 587*f*, 588*f*
Bile pigments (bilirubin), in urine (bilirubinuria), **623**, 624–625
Bile salts, **PEx-127**. *See also* Bile
Bilirubin, in urine (bilirubinuria), **623**, 624–625
Binocular vision, 374–375, 374*f*
Biological action, hormone-receptor complex exerting, **PEx-59**
BIOPAC®
  for baseline heart activity, 514–515, 515–516, 515*f*, 516, 516*f*
  for ECG, 462–466, 462*f*, 463*f*, 464*f*, 465*f*
  for EEG, 301–304, 301*f*, 302*f*, 303*f*, 304*f*
  for EMG, **241**–248
  for galvanic skin response in polygraph, **326**–332, 327*f*, 328*f*, 329*f*, 330*f*
  for heart rate modifiers, 517–519
  for pulse, 495–497, 496*f*, 497*f*
  for reflex reaction time, 342–344, 343*f*
  for respiratory volumes, 559–562, 559*f*, 560*f*, 561*f*
Bipennate fascicles/muscles, 196*f*
Bipolar cells, 363*f*, **364**
Bipolar neurons, **257**, 258*f*
Birth, changes in fetal circulation at, 483*f*, 484
Bitter taste, 400
Blackheads, **99**
Bladder (urinary), **20**, 20*f*, 21*f*, 22*f*, 73*f*, 610*f*, **611**, 611*f*, **615**
  functional microscopic anatomy of, 615–616
  palpable, 687
Blastocyst, **659**, 659*f*, 660*f*
Blastomeres, **658**, 659*f*
Blastula, 48, **658**, 659
Blaze, genotype/phenotype and, **672**
Bleeding time, **433**
Blind spot (optic disc), **362**, 362*f*, 365*f*, 371–372, 372*f*
Blood, **75**, 81*f*, 423–441
  composition of, 424–429, 425*f*, 426*t*, 427*f*, 428*f*
  hematologic tests/analysis of, 429–436, 430*f*, 431*f*, 432*f*, 433*f*, 434*t*, 435*f*, PEx-161 to PEx-176. *See also specific test*
  pH of, 564, PEx-149. *See also* Acid-base balance
  precautions for handling, 424
  in urine, 623, 624
Blood air barrier (respiratory membrane), **541**, 541*f*
Blood cells (formed elements), 424–425, **424**, 425*f*, 426–429, 426*t*, 427*f*, 428*f*
  blood viscosity and, PEx-78
  microscopic appearance of, 426–429, 427*f*, 428*f*
  tissues forming (hematopoietic tissue), **75**
Blood (plasma) cholesterol, 435–436, **PEx-161**, **PEx-169** to PEx-171
Blood clotting, 433–434, 433*f*
Blood flow, **PEx-76**
  local, 519–520, 519*f*
  pressure and resistance and, 500, PEx-75 to PEx-76, PEx-76, PEx-76 to PEx-88, PEx-76 to PEx-88
    blood pressure and, 500, PEx-80 to PEx-82
    compensation/cardiovascular pathology and, PEx-86
    vessel length and, PEx-76, PEx-79 to PEx-80

vessel radius and, PEx-76, **PEx-76** to PEx-78
  viscosity and, PEx-76, PEx-78 to PEx-79
  skin color and, 97, 502–504
Blood pressure (BP), **497–501**, 498*f*
  arterial
    mean (MAP), **499**
    measuring, 498–499, 498*f*
  blood flow and, 500, PEx-80 to PEx-82
  cardiac cycle and, 493*f*
  factors affecting, 500–501
  glomerular filtration and, PEx-135 to PEx-137, **PEx-137** to PEx-139
  renal response to changes in, PEx-137 to PEx-139
  venous, 499
Blood smear, 427, 427*f*
Blood type, 434–435, 434*t*, 435*f*, PEx-161, PEx-162, PEx-167 to PEx-169, PEx-168*t*
  inheritance of, **672**, 672*t*, PEx-167
Blood vessel(s), 469–490, 470*f*. *See also* Circulation(s)
  arteries, 470, 470*f*, 471, 471*f*
  capillaries, 448*f*, 470, 470*f*, 472
  microscopic structure of, 470–472, 470*f*, 471*f*
  skin color affected by stimulation of, 504
  in special circulations, 479–484, 482*f*, 483*f*, 484*f*
  veins, 470, 470*f*, 472, 477–479, 478*f*
Blood vessel length, blood flow and, PEx-76, PEx-79
Blood vessel radius, **PEx-76**
  blood flow and, PEx-76, **PEx-76** to PEx-78
  glomerular filtration and, PEx-132 to PEx-135, PEx-137 to PEx-139
  pump activity and, PEx-82 to PEx-83
Blood vessel resistance, **PEx-75**, PEx-76
  blood flow and, PEx-75 to PEx-76, PEx-76, PEx-76 to PEx-88
  compensation/cardiovascular pathology and, PEx-86
  blood pressure and, 500, PEx-75 to PEx-76, PEx-76
Blood viscosity, PEx-76, PEx-78
  blood flow and, PEx-76, PEx-78 to PEx-79
BMR. *See* Basal metabolic rate
Body of bone
  ischium, 154*f*
  mandible, **127**, 127*f*
  sphenoid, 124*f*
  sternum, **135**, 135*f*, 683
  vertebra (centrum), **131**, 131*f*, 132*f*, 133*f*, 134*f*, 134*t*
Body cavities, 6–8, 7*f*, 8*f*, 9*f*
Body movements, 173–174, 173*f*, 174–175*f*
Body of nail, **97**, 97*f*
Body orientation/direction, 3–4, 4*f*
Body planes/sections, 4–6, 5*f*, 6*f*
Body of stomach, **577**
Body temperature regulation, PEx-97
  skin in, 96
Body tube (microscope), **28**
Body of uterus, **634**, 635*f*
Bolus, **585**
Bone(s)/bone (osseous) tissue, **75**, 81*f*, 108, 108*f*, 110–114
  chemical composition of, 111
  classification of, 110
  formation/growth of (ossification), **114**, 114*f*
  gross anatomy of, 110–111, 111*f*
  microscopic structure of, 113–114, 113*f*
Bone-conduction hearing, testing, 388, 388*f*
Bone density, loss of in osteoporosis, PEx-67
Bone markings, **110**, 112*f*
Bone marrow, **110**, 111*f*
Bone spurs, 180
Bone tissue. *See* Bone(s)

Bony labyrinth, **385**
Bony pelvis. *See* Pelvis
Bony thorax, 108*f*, 135–136, 135*f*, 136*f*, 683–684, 685*f*
Boutons, terminal (axon terminals), 188, 189, 189*f*, 190*f*, **241**, 255*f*, **256**, PEx-36, PEx-36*f*, PEx-49
Bowel sounds, 686–687
Bowman's (glomerular) capsule, **613**, 614*f*, 615*f*, PEx-131, PEx-132*f*, **PEx-133**, **PEx-135**, **PEx-137**
BP. *See* Blood pressure
Brachial (term), **2**, 3*f*
Brachial artery, 474, 474*f*, 687–688, **688**, 689*f*
  pulse at, 495, 495*f*, 688
Brachialis muscle, 198*f*, 212*f*, 212*t*, 689*f*
Brachial plexus, 313–316, 313*f*, 314*f*, 315*t*, **682**, 683*f*
Brachial vein, 479, 481*f*
Brachiocephalic artery/trunk, 444*f*, 450*f*, **451**, **472**, 473*f*, 474*f*
Brachiocephalic veins, 478, 480*f*, 481*f*
Brachioradialis muscle, 198*f*, 199*f*, 212*f*, 212*t*, 213*f*, 215*f*, **688**, 689*f*
Bradycardia, **460**
Brain, 276–285
  arterial supply of, 472–474, 473*f*
  cerebellum, 277*f*, **279**, 279*f*, 280*f*, 282, 282*f*, 288*f*
  cerebral hemispheres, **277–278**, 277–278*f*, 279–281, 280*f*, 281*f*, 288*f*
  cerebrospinal fluid and, 283–285, 284*f*
  development of, 276–277, 276*f*
  diencephalon, 276*f*, **278**, 279*f*, 281–282
  electrical activity of, recording (electroencephalography), 299–306, **300**, 300*f*, 301*f*, 302*f*, 303*f*, 304*f*
  meninges of, **282–283**, 283*f*
  sheep, dissection of, 285–292, 288*f*, 289–290*f*, 291*f*, 292*f*
  veins draining, 478–479, 480*f*
Brain stem, 276*f*, **278**, 279*f*, **282**, 282*f*, 288*f*
Brain vesicles, 276*f*
Brain waves, 299–306, 300*f*, 301*f*, 302*f*, 303*f*, 304*f*
Breast, 637, 637*f*. *See also under Mammary*
Breath holding, ECG during, 461–462
Breathing, 538, **550–572**, 551*f*, **PEx-105**, PEx-106*f*. *See also under Respiratory*
  acid-base balance and, 564–565, 564*f*, PEx-150, PEx-150 to PEx-152, PEx-154 to PEx-156
  factors affecting, 562–564
  frequency of, **PEx-106**
  mechanics of, 550–551, 551*f*, PEx-105 to PEx-117
  muscles of, 206*t*, 207*f*, 550, **PEx-106**, **PEx-107**
Breathing sounds, **551**
Brevis (term), muscle name and, 197
Bridge of nose, **681**, 681*f*
Broad ligament, 635*f*, **636**
Broca's area, **278**, 278*f*
Bronchi, **18**, **540**, 540*f*, 542*f*
Bronchial arteries, 474, **479**
Bronchial cartilages, 109
Bronchial sounds, **551**
Bronchial tree, **541**
Bronchioles, 540*f*, **541**, 544, 544*f*
Brush border, 71*f*, 581*f*. *See also* Microvilli
Brush border enzymes, **581**, 597*f*
BTPS factor, **555**
Buccal (term), **2**, 3*f*
Buccal phase, of deglutition (swallowing), 602, 602*f*
Buccinator muscle, 200*t*, 201*f*, 202*f*, 202*t*
Buffer(s)/buffer system(s), 564, **565**, PEx-150
  carbonic acid–bicarbonate (respiratory), 564, 564*f*, 565, **PEx-150**, PEx-150 to PEx-152. *See also* Respiratory buffering system
  chemical, PEx-150
  renal, **PEx-150**, PEx-153 to PEx-154

Buffy coat, PEx-162
Bulbar conjunctiva, 360, 360*f*
Bulbo-urethral glands, 21*f*, 630*f*, 631*f*, **632**
Bulbous corpuscle, 350*f*, **351**
Bundle branches, **458**, 458*f*
Bundle of His (atrioventricular/AV bundle), **458**, 458*f*
Bursa(e), 171
Buttocks, 690, 691*f*

Calcaneal (term), **2**, 3*f*
Calcaneal nerve, 318*t*
Calcaneal (Achilles) tendon, 199*f*, 223*f*, 693*f*
Calcaneal tendon (ankle-jerk) reflex, **338**, 339*f*
Calcaneus (heel bone), 108*f*, 157, 158, 158*f*
Calcitonin, **408**, PEx-67, PEx-68
Calcitriol, 408
Calcium
  automaticity/autorhythmicity and, 512, PEx-93
  in bone, 111
  calcitonin in regulation of, 408
  heart rate and, 518, PEx-99 to PEx-101, PEx-100*t*
  muscle contraction and, 234
  neurotransmitter release and, PEx-49
  parathyroid hormone in regulation of, 408
Calcium channel(s), fast, automaticity and, 512
Calcium channel blockers, PEx-99, PEx-100
Calvaria (cranial vault), **120**
cAMp, PEx-59 to PEx-60
Canaliculi
  bile, **587**
  in bone, 113*f*, **114**
  lacrimal, **360**, 360*f*
Cancellous (spongy) bone, **110**, 111*f*, 113*f*
Canines (eye teeth), **584**, 584*f*
Canthi, lateral and medial (lateral/medial commissures), **360**, 360*f*
Capillaries, 448*f*, 470, 470*f*, 472, 519–520, 519*f*
  glomerular, 613, 614*f*, 615*f*, **PEx-133**, **PEx-135**
  peritubular, **613**, 614*f*, 615*f*, **PEx-132**, **PEx-139**
  pulmonary, 448*f*, **479**
Capitate, 151, 152*f*
Capitulum, **150**, 150*f*
Capsule
  articular, 169, 169*f*, 171
  glomerular, **613**, 614*f*, 615*f*
  kidney, 610*f*, 611, **612**, 612*f*
  lymph node, 528*f*, **529**
  spleen, 529*f*
Carbohydrate (starch) digestion, 597–598, 597*f*, PEx-121 to PEx-123
  by salivary amylase, 576, 585, **596**–598, 597*f*, **PEx-121** to PEx-123
  substrate specificity and, PEx-123 to PEx-125
Carbon dioxide
  acid-base balance and, 564*f*, 565, PEx-150, PEx-151, PEx-153, PEx-154
  partial pressure of, PEx-106, PEx-150, PEx-153
  reaction of with water, 565
Carbonic acid, in acid-base balance, 564, 564*f*, PEx-150, PEx-151, PEx-153, PEx-154
Carbonic acid–bicarbonate (respiratory) buffer system, 564, 564*f*, 565, **PEx-150**, PEx-150 to PEx-152. *See also* Respiratory buffering system
Cardia (cardial part of stomach), **577**, 578*f*
Cardiac action potential, **PEx-94**, PEx-94*f*
  ion movement and, PEx-100*t*
Cardiac circulation, 446–449, 448*f*

Cardiac cycle, **492**, 493*f*, PEx-82
Cardiac muscle, **83**, 84*f*, 187, 449, 449*f*.
  See also Heart
  contraction of, PEx-84, PEx-93
  electrical properties of, 512–513
  microscopic anatomy of, 83, 84*f*, 449,
    449*f*
  refractory period of, 517, **PEx-94** to
    PEx-95
Cardiac muscle bundles, 449*f*, 451
Cardiac notch, 542*f*, **543**
Cardiac output, **PEx-76**, **PEx-82**, **PEx-84**
  blood pressure affected by, 500
Cardiac pacemaker, 458, 458*f*, PEx-93,
  PEx-96, PEx-98
  ectopic, **518**
  in frog, 513, 513*f*, 514
Cardiac skeleton, 446, 447*f*
Cardiac valves, 445*f*, 446, 447*f*, PEx-86
  murmurs and, 492
Cardiac veins, 444*f*, 445*f*, **448**, 448*f*
Cardiac ventricles, 444*f*, 445*f*,
    450–451, 450*f*
  in frog, 513*f*
  right versus left, 451–452, 452*f*
Cardial part of stomach (cardia), **577**,
    578*f*
Cardiovascular compensation
  pathological conditions and, PEx-86
    to PEx-88
  vessel radius affecting pump activity
    and, PEx-82 to PEx-83
Cardiovascular dynamics, PEx-75 to
    PEx-91
Cardiovascular physiology, PEx-93 to
    PEx-104
  frog, 511–524
  human, 491–510
Cardiovascular system, 16*t*, **444**. *See also*
    Blood vessel(s); Heart
Carotene, 97
Carotid artery, **682**, 683*f*
  common, **472**, 473*f*, 474*f*, 682
    pulse at, **495**, 495*f*
  external, **472**, 473*f*, **682**
    pulse at, 682
  internal, **472**, 473*f*
Carotid canal, **120**, 122*f*
Carotid pulse, 682
Carpal (term), **2**, 3*f*
Carpals, 108*f*, **151**, 152*f*
Carpometacarpal joint, 170*t*
Carpus, **151**, 152*f*
Carrier proteins, **PEx-140**
  active transport and, 60, PEx-4, PEx-13
  facilitated diffusion and, 55, PEx-3,
    PEx-6
  glucose transport/reabsorption and, 55,
    PEx-6 to PEx-8, **PEx-140** to PEx-142
Cartilages/cartilage tissue, **75**, 79*f*, 108,
  109–110, 109*f*. *See also* specific type
Cartilaginous joints, 168*f*, **169**, 170–171*t*
Caruncle, lacrimal, **360**, 360*f*
Casts, urinary, **623**, 626, 626*f*
Catabolism, 419, PEx-60
Catalysts, 596, PEx-120
Cataracts, 363
Catecholamine hormones, PEx-59
Cauda equina, **308**, 309*f*, 313*f*
Caudal (term), **3**, 4*f*
Caudate nucleus, 279–280, 281*f*, 292*f*
C (parafollicular) cells, **409**, 409*f*,
  PEx-67
Cecum, **18**, 20*f*, 575*f*, **583**, 583*f*
Celiac ganglia, 324
Celiac trunk, **474**, 475*f*, 476*f*
Cell(s), **15**, **40**
  anatomy of, 40–44, 40*f*, 41*f*, 42*f*, 43*t*
  cytoplasm/organelles of, **40**, 40*f*, **41–44**,
    41*f*, 42*f*, 43*t*
  differences/similarities in, 44–45
  division/life cycle of, **45–48**, 46–47*f*
  membrane of. *See* Plasma membrane
  microscopic examination of, 33–34,
    33*f*, 34*f*
  nucleus of, **40**, 40*f*, 42*f*
  permeability/transport mechanisms of,
    40–**41**, 53–65, **54**, PEx-3 to PEx-16

Cell body, neuron, 82, 82*f*, **254**, 255*f*,
    259*f*, **PEx-35**, PEx-36*f*
Cell division, **45–48**, 46–47*f*
Cell junctions, in epithelia, 69
Cellular immunity, **529**
Cellular respiration, 538
Cellulose, **PEx-123**
Cellulose digestion, PEx-123 to PEx-125
Cement (tooth covering), **585**, 585*f*
Centimeter, 32*t*
Central (Haversian) canal, 81*f*, 113*f*, **114**
Central canal of spinal cord, 276*f*, 277,
    308, 310*f*
Central nervous system (CNS), 254, **276**.
    See also Brain; Spinal cord
  supporting cells in, 254, 254*f*
Central retinal artery and vein, 362*f*
Central sulcus, **277**, 277*f*, 278*f*
Centrioles, 42*f*, 43*t*, **44**
  in sperm, 648
Centromere, 46*f*, 646
Centrosome, 42*f*, 44, 46*f*
Centrum (body), of vertebra, **131**, 131*f*,
    132*f*, 133*f*, 134*f*, 134*t*
Cephalad/cranial (term), **3**, 4*f*
Cephalic (term), **2**, 3*f*
Cephalic vein, **479**, 481*f*, 687*f*, **688**, 689*f*
Cerebellar cortex, 279, 282, 282*f*
Cerebellar hemispheres, 282
Cerebellar peduncles, 290
Cerebellum, 277*f*, **279**, 279*f*, 280*f*, 282,
    282*f*, 283*f*, 288*f*, 289*f*, 290, 290*f*,
    291, 291*f*
  development of, 276*f*, 277
Cerebral aqueduct, 276*f*, 280*f*, **282**, 284*f*,
    291
Cerebral arterial circle (circle of Willis),
    472, 473*f*, 474
Cerebral arteries, **472**, **473**, 473*f*
Cerebral cortex/cerebral gray matter,
    276*f*, 277*f*, **278**, 279, 281*f*
Cerebral fissures, **277**, 277*f*, 288*f*
Cerebral hemispheres (cerebrum),
    277–278, 277–278*f*, 279–281, 280*f*,
    281*f*, 288*f*, 289*f*, 290, 290*f*, 291*f*
  development of, 276*f*, 277
Cerebral peduncles, **278**, 279*f*, 287, 289*f*,
    291, 291*f*
Cerebral white matter, 276*f*, 277*f*, **278**,
    279, 281*f*
Cerebrospinal fluid, 283, 283–285, 284*f*, 308
Cerebrum. *See* Cerebral hemispheres
Ceruminous glands, **384**
Cervical (term), **2**, 3*f*
Cervical canal, 635*f*
Cervical curvature, 130*f*, 131
Cervical enlargement, 308, 309*f*, 313*f*
Cervical lymph nodes, 526*f*, 527, 682
Cervical plexus/spinal nerves, 309*f*, **313**,
    313*f*, 315*t*, **682**, 683*f*
Cervical vertebrae, 130*f*, 132, 132*f*, 133*f*,
    134*t*
Cervix, **634**, 635*f*
C fibers, conduction velocity of,
    PEx-48, PEx-49
Cheeks, **575**
Chemical buffers, PEx-150
Chemical neurotransmitters, **PEx-35**
Chemical stimulation, heart rate and,
    517–518, PEx-98 to PEx-99
Chemical synapses, 255*f*, **256**, **PEx-35**,
    PEx-49 to PEx-50, PEx-50
Chemoreceptors, 350, **398**
"Chenille stick" mitosis, 48
Chiasmata (crossovers), **646**, 646*f*
Chief cells, 578*f*, **579**
  gastric (zymogenic cells), **PEx-125**
Chlamydia/*Chlamydia trachomatis*,
  PEx-178
  direct fluorescent antibody technique in
    testing for, **PEx-178** to PEx-180
Chloride(s), in urine, 624
Chloride shift, 564, 564*f*
Cholesterol, 435–436, **PEx-161**,
  PEx-162, **PEx-169**
  blood/plasma concentration of, 435–
    436, **PEx-161**, **PEx-169** to PEx-171
  in plasma membrane, 40, 41*f*

Cholinergic fibers, 324
Cholinergic modifiers, **PEx-98**
Chondrocytes, 79*f*, 80*f*
Chordae tendinae, 445*f*, **446**, 447*f*, 452*f*
Chorion, **659**, 660*f*, 662*f*
Chorionic villi, **660**, 660*f*, 662*f*
Choroid, **361**, 362*f*
  pigmented coat of, **364**
Choroid plexus, 280*f*, **282**, **283**, 284*f*
Chromatids, 46*f*, 646
Chromatin, **40**, 40*f*, 42*f*, 46*f*
Chromatophilic substance (Nissl bodies),
    255*f*, **256**
Chromophobes, **410**, 410*f*
Chromosome(s), **40**, 46*f*
  homologous, **646**, 646*f*, 668
  sex, **669**–670, 669*f*
Chronic obstructive pulmonary disease,
    554. *See also* Obstructive lung
    disease
Chronotropic modifiers, **PEx-100**
Chyme, **577**
Cilia
  olfactory, **398**, 398*f*
  tracheal, 71*f*
Ciliary body, **361**, 362*f*, 363,
    364, 365*f*
Ciliary glands, **360**
Ciliary muscles, 361, 362*f*, 363
Ciliary processes, 361, 362*f*, **363**
Ciliary zonule, 362*f*, **363**
Ciliospinal reflex, **341**
Circle of Willis, **472**, 473*f*, 474
Circular fascicles/muscles, 196*f*
Circular folds, **581**–582, 581*f*
Circulation(s)
  coronary, 446–449, 448*f*
  fetal, 482–484, 483*f*
  hepatic portal, 484, 484*f*
  pulmonary, **446**, 448*f*
  systemic, **446**, 448*f*
    arterial, 471*f*, 472–477
    venous, 477–479, 478*f*
Circulatory dynamics, skin color
    indicating, 97, 502–504
Circumcision, 632
Circumduction (body movement), **173**,
    175*f*
Circumferential lamellae, 113*f*, 114
Circumflex artery, 444*f*, **448**, 448*f*
Circumflex femoral arteries, **476**, 477*f*
Circumflex humeral arteries, **474**, 474*f*
Cirrhosis, **587**
Cistern(s), Golgi apparatus, 43
Cisterna chyli, 526*f*, **527**
Clavicle (collarbone), 108*f*, **148**, 148*f*,
    149*f*, 152, **684**, 685*f*, **687**, 687*f*
Clavicular head, of sternocleidomastoid
    muscle, 682, 682*f*
Clavicular notch, 135*f*
Cleavage, 658–659, **658**
Cleavage furrow, 47*f*
Cleft palate, 538
Clinical crown of tooth, 585
Clitoris, **634**, 634*f*, 635*f*
Clonal selection, **528**–529
Clotting (blood), **433**–434, 433*f*
cm. *See* Centimeter
CNS. *See* Central nervous system
Coagulation, **433**–434, 433*f*
Coagulation time, 433–434
Coarse adjustment knob (microscope),
    28, 29*f*
Coccygeal nerve, 313*f*
Coccyx, 130, 130*f*, 134*f*, **135**, 683
  in male versus female, 155*t*
Cochlea, **385**, 385*f*, 386*f*
  microscopic anatomy of/hearing and,
    386–387, 386*f*, 387*f*
Cochlear duct, **385**, 385*f*, 386*f*, 387*f*
Cochlear nerve, 385*f*, 386*f*, 387, 387*f*
Codominant alleles, 672
Cold exposure, blood pressure/heart rate
    affected by, 501
Cold pressor test, **501**
Cold receptors, 352

Collagen fibers, 74*f*, **75**, 76*f*, 78*f*, 79*f*, 80*f*
  in skin, 96
Collarbone. *See* Clavicle
Collateral(s), axon, **256**
Collateral blood flow, skin color and,
    503–504
Collateral (prevertebral) ganglion, 324,
    325*f*
Collateral ligaments, of knee (fibular and
    tibial), 177*f*, **178**
Collateral vessels, **503**
Collecting ducts, **613**, 614*f*, **PEx-132**,
  PEx-132*f*, **PEx-139**
Collecting lymphatic vessels, 526*f*, **527**,
    527*f*
Colles' fracture, **688**
Colliculi, inferior and superior, **279**,
    280*f*, 291*f*
Collins spirometer, 552, 553*f*
Colloid, thyroid gland, 409, 409*f*, 410
Colon, 22*f*, 575*f*, **583**, 583*f*. *See also*
    Large intestine
Color blindness, **374**
  inheritance of, 670
Color vision, cones and, 364
Columnar epithelium, 68*f*, **69**
  pseudostratified, 71*f*
  simple, 71*f*
  stratified, 73*f*
Commissures, 279, 280*f*
  gray, **308**, 310*f*
  lateral and medial (lateral/medial
    canthi), **360**, 360*f*
Common carotid artery, **472**, 473*f*, 474*f*,
    682
  pulse at, **495**, 495*f*
Common fibular nerve, **317**–318, 317*f*,
    318*t*, 693
Common hepatic artery, **474**, 475*f*
Common hepatic duct, **586**, 587*f*
Common iliac arteries, 475*f*, **476**, 476*f*,
    477*f*
Common iliac veins, **477**, 479*f*, 480*f*
Common interosseous artery, 474*f*
Common tendinous ring, extrinsic eye
    muscles, 361*f*
Communicating arteries, **472**, 473*f*
Compact bone, **110**, 111*f*
  microscopic structure of, 113–114, 113*f*
Comparative spirometry, PEx-109 to
    PEx-112
Compensation, cardiovascular
  pathological conditions and, PEx-86
    to PEx-88
  vessel radius affecting pump activity
    and, PEx-82 to PEx-83
Compensatory pause, 516*f*, **517**, PEx-95
Complete heart block, **519**
Complete (fused) tetanus, 239*f*, **240**,
    PEx-24, **PEx-25**, PEx-94
Compound action potential, **267**,
    270–271, 270*f*, 271*f*
Compound microscope, **28**, 29*f*. *See also*
    Microscope(s)
Concentration gradient, 54–55, **PEx-4**
  in active transport, 60, PEx-4, PEx-13
  in diffusion, **54**–55, PEx-3, **PEx-4**,
    PEx-6
  in osmosis, 55, 59, 59*f*, PEx-8
  urine concentration and, PEx-139 to
    PEx-140
Concentric contraction, isotonic, **PEx-28**
Concentric lamellae, 114
Conchae, nasal (nasal turbinates)
  inferior, 125*f*, **128**, **538**, 539*f*
  middle, 124*f*, 125*f*, **127**, **538**, 539*f*
  superior, **127**, **538**, 539*f*
Condenser (microscope), **28**, 29*f*
Conductance, **PEx-36**
Conducting zone structures, **541**
Conduction, action potential, PEx-41,
  PEx-48
Conduction deafness, 388, 388*f*
Conduction system of heart, **458**, 458*f*
  disturbances of, 519
  in frog, 513–519, 514*f*, 515*f*, 516*f*
Conduction velocity, PEx-47 to
    PEx-49, **PEx-48**

Conductivity, neuron, 82, **266**, 266*f*
Condylar joint, 172*f*, 173
Condylar process, mandible, 121*f*, **127**, 127*f*, 681
Condyle (bone marking), 112*t*
Cones, 361–362, 363*f*, **364**
  color blindness and, 374
Congenital umbilical hernia, **686**
Conjunctiva, 360, 360*f*
Conjunctivitis, 360
Connective tissue, 74*f*, **75**–82, 76–81*f*, 83
  of hair, 98, 98*f*
  as nerve covering, 260, 260*f*
  of skeletal muscle, 188, 189*f*
Connective tissue fibers, 74*f*, **75**, 76*f*
Connective tissue proper, **75**, 76*f*, 77*f*, 78*f*, 79*f*
Conoid tubercle, **148**, 149*f*
Consensual reflex, **340**
Constant (C) region, immunoglobulin, 530, 531*f*, PEx-182
Constipation, **584**
  metabolic alkalosis and, PEx-154
Contractility (heart), **PEx-84**
  compensation and, PEx-86
Contraction, skeletal muscle, 234–248, PEx-17 to PEx-34. *See also* Muscle contraction
Contraction period/phase, of muscle twitch, **238**, 238*f*, PEx-18, PEx-20
Contralateral response, **340**
Controls, **PEx-121**, PEx-125, PEx-178
Conus medullaris, **308**, 309*f*
Convergence/convergence reflex, **375**
Convergent fascicles/muscles, 196*f*
Converting enzyme (angiotensin), PEx-142
COPD. *See* Chronic obstructive pulmonary disease
Coracoacromial ligament, 179*f*
Coracobrachialis muscle, 205*f*, 212*f*
Coracohumeral ligament, **178**, 179*f*
Coracoid process, **148**, 149*f*
Cord reflexes, superficial, **339**, 339*f*
Cords, brachial plexus, 313, 314*f*
Cornea, **361**, 362*f*, **364**, 365*f*
Corneal reflex, **340**
Corniculate cartilage, 540*f*
Cornua (horns)
  hyoid bone, **129**, 130*f*
  uterine, 21, 21*f*
Coronal (frontal) plane/section, **4**, 5*f*, 6*f*
Coronal suture, **120**, 121*f*
Corona radiata (cerebral projection fibers), **280**
Corona radiata (oocyte), 650*f*, **651**, 658
Coronary arteries, 444*f*, 445*f*, 447–448, 448*f*, **472**
Coronary circulation, 446–449, 448*f*
Coronary sinus, 445*f*, 446, **448**, 448*f*, 451, 452*f*
Coronary sulcus, 444*f*, 445*f*, **447**
Coronoid fossa, **150**, 150*f*
Coronoid process
  of mandible, 121*f*, **127**, 127*f*
  of ulna, **151**, 151*f*
Corpora cavernosa, 630*f*, 631*f*, **633**, 633*f*
Corpora quadrigemina, **279**, 280*f*, 290, 291*f*
Corpus albicans, **650**, 650*f*
Corpus callosum, **279**, 280*f*, 281*f*, 291*f*
Corpus luteum, 636, 649*f*, **650**, 650*f*, **651**, 652*f*
Corpus spongiosum, 630*f*, 631*f*, **633**, 633*f*
Corrugator supercilii muscle, 200*t*, 201*f*
Cortex
  adrenal, **408**
  cerebellar, 279, 282, 282*f*
  cerebral, 276*f*, 277*f*, **278**, 279, 281*f*
  of hair, 97, 98*f*
  lymph node, 528*f*, **529**
  renal, **612**, 612*f*
Cortical nephrons, **613**, 614*f*
Cortical radiate arteries, 612*f*, **613**, 614*f*, 615*f*
Cortical radiate veins, 612*f*, **613**, 614*f*
Corticosteroids, **408**, PEx-60
Corticosterone, 408

Corticotropin-releasing hormone (CRH), **PEx-69**
Cortisol (hydrocortisone), **408**, PEx-69 to PEx-70, PEx-69*t*
Cortisone, 408
Costal cartilages, 79*f*, **109**, 109*f*, 135, 135*f*, 136
Costal facets (demifacets), **132**, 132*f*, 133*f*, 136
Costal groove, 136*f*
Costal margin, 135*f*, **684**, 685*f*
Costal surface, **543**
Costocervical trunk, **472**, 473*f*, 474*f*
Costovertebral joint, 170*t*
Cough reflex, 538
Coupled transport, PEx-13
Cow eye, dissection of, 364, 365*f*
Coxal (term), 2, 3*f*
Coxal bones (ossa coxae/hip bones), **153**, 154*f*
Coxal (hip) joint, 171*t*, 176, 176*f*
Cranial/cephalad (term), **3**, 4*f*
Cranial base, **120**
Cranial cavity, 6, 7*f*
Cranial fossae, **120**, 123*f*
Cranial nerve reflex tests, 339–340
Cranial nerves, **285**, 286–287*t*, 288–290, 288*f*, 289–290*f*
Cranial sutures, **120**, 121*f*, 125*f*
Cranial vault (calvaria), **120**
Craniosacral (parasympathetic) division of autonomic nervous system, 324, 324*f*, **PEx-95**
  function of, 326
  in heart regulation, PEx-95, PEx-98
Cranium, 108*f*, 120–127, **120**, 121*f*, 122*f*, 123*f*, 124*f*, 125–126*f*, 127*f*, 128*f*. *See also specific bone*
  surface anatomy of, 680*f*, 681
C (constant) region, immunoglobulin, 530, 531*f*, PEx-182
Cremaster muscle, 339
Cremasteric reflex, 339
Crenation, **59**, 59*f*
Crest (bone marking), 112*t*
CRH. *See* Corticotropin-releasing hormone
Cribriform foramina, 123*f*
Cribriform plates, 123*f*, **126**–127
Cricoid cartilage, 109*f*, **538**, 539*f*, 540*f*, **682**, 682*f*
Cricothyroid ligament, 540*f*, **682**, 682*f*
Cricotracheal ligament, 540*f*
Crista(e), mitochondrial, 44
Crista ampullaris, 385*f*, **390**, 391*f*
Crista galli, 123*f*, 124*f*, **126**, 283*f*
Crossed-extensor reflex, **339**
Crossovers (chiasmata), 646, 646*f*
Cross section/transverse plane, 5*f*, **6**, 6*f*
Crown of tooth, **585**, 585*f*
Cruciate ligaments, 177*f*, **178**
Crural (term), **2**, 3*f*
Crypts
  intestinal, 581*f*, **582**, 582*f*
  of tonsils, **529**, 530*f*
CSF. *See* Cerebrospinal fluid
Cubital fossa, **688**
Cubital vein, median, 479*f*, 481*f*, **688**, 689*f*
Cuboid, 158*f*
Cuboidal epithelium, 68*f*, **69**
  simple, 70*f*
  stratified, 72*f*
Cuneiform cartilage, 540*f*
Cuneiforms, 158*f*
Cupula, ampullary, **390**
Curare, nerve function affected by, 269
Curvatures of spine, 130*f*, 131
Cushing's disease, **PEx-69**, PEx-69*t*
Cushing's syndrome, **PEx-69**, PEx-69*t*
Cutaneous femoral nerve, 316–317, 316*f*, 317*t*, 318*t*
Cutaneous glands, 98*f*, 99–100, 100*f*
Cutaneous membrane. *See* Skin
Cutaneous nerves of arm/forearm, medial, 314*f*
Cutaneous receptors/nerve endings, 94*f*, 95*f*, 96, 96–97, 350, 350*f*, 351, 352

Cutaneous vascular plexus, 94*f*
Cuticle
  hair, 97, 98*f*
  nail (eponychium), **97**, 97*f*
Cyanosis, 97
Cyclic adenosine monophosphate (cAMP), PEx-59 to PEx-60
Cystic duct, **586**, 587*f*
Cytokinesis, **45**, 47*f*
Cytoplasm, **40**, 40*f*, 41–44, 41*f*, 42*f*
  division of (cytokinesis), 45, 47*f*
Cytoplasmic inclusions, 44
Cytoskeletal elements/cytoskeleton, 41*f*, 42*f*, 43*t*, **44**
Cytosol, **41**, 42*f*
Cytotoxic T cells, 529

Daughter cells/chromosomes/nuclei
  in meiosis, 45, 646, 646*f*
  in mitosis, 45, 47*f*
DCT. *See* Distal convoluted tubule
Dead space, anatomical, 541
Deafness. *See also* Hearing
  testing for, 388, 388*f*
Death, absence of brain waves and, 300
Decidua basalis, **659**, 660*f*, 662*f*
Decidua capsularis, **659**, 660*f*, 662*f*
Deciduous (milk) teeth, **584**, 584*f*
Decubitus ulcers (bedsores), **96**, 96*f*
Decussation of pyramids, **278**, 279*f*
Deep/internal (term), **4**
Deep artery of arm, **474**, 474*f*
Deep artery of thigh (deep femoral artery), **476**, 477*f*
Deep breathing, ECG affected by, 463–464, 464*f*, 465–466
Deep fascia, **188**
Deep palmar arch, 474*f*
Deep palmar venous arch, 481*f*
Deep plantar arch, 479*f*
Deglutition (swallowing), **602**–603, 602*f*
Delta waves, **300**, 300*f*
Deltoid muscle, 196*f*, 197, 198*f*, 199*f*, 205*t*, 205*t*, 209*f*, 211, 212*f*, 684*f*, **687**, 687*f*
  for intramuscular injection, 687, 691*f*
Deltoid tuberosity, **150**, 150*f*
Demilunes, in salivary glands, 586, 586*f*
Dendrite(s), 82*f*, 255*f*, **256**, 259*f*, **PEx-35**, PEx-36*f*
Dendritic cells, epidermal (Langerhans' cells), 95*f*, **96**
Dens (odontoid process), **132**, 132*f*
Dense connective tissue, **75**, 78*f*, 79*f*
Dental formula, **584**
Denticulate ligaments, **308**, 309*f*
Dentin, **585**, 585*f*
Dentition (teeth), **584**–585, 584*f*, 585*f*
Depolarization, **234**, **266**, **PEx-39**
  cardiac, PEx-94
    ion movement and, PEx-100*t*
  muscle cell, 234, PEx-18
  neuron, 266–267, 266*f*, **PEx-39**, PEx-41
Depressor anguli oris muscle, 200*t*, 201*f*
Depressor labii inferioris muscle, 200*t*, 201*f*
Depth of field (microscope), **33**
Depth perception, 374–375
Dermal papillae, 94*f*, **96**
Dermatographism, **504**
Dermis, 94, 94*f*, 96–97, 96*f*, 99*f*
Dermography, 100–102, 101*f*. *See also* Fingerprints
Descending aorta, **20**, 21*f*
Descending colon, 575*f*, **583**, 583*f*
Descending (motor) tracts, 310, 311*f*
Desmosomes, in skin, 95*f*
Development, embryonic, 657–666, 659*f*, 660*f*
Diabetes insipidus, **407**
Diabetes mellitus, **408**, PEx-64 to PEx-67
  fasting plasma glucose levels and, PEx-65, PEx-66
  glycosuria in, **622**
  ketonuria in, 623
Dialysis, **PEx-3**, PEx-4 to PEx-6. *See also* Simple diffusion

Dialysis sacs, diffusion and osmosis through, 56–58
Diapedesis, **428**
Diaphragm, 7*f*, **18**, 19*f*, 206*t*, 207*t*, 542*f*, **543**, PEx-106, PEx-107
  during breathing, 550, 551*f*, PEx-107
Diaphysis, **110**, 111*f*
Diarrhea, **584**
  metabolic acidosis and, PEx-154
Diarthroses, 169. *See also* Synovial joints
Diastole, **446**, 492, 493*f*, **PEx-82**
Diastolic pressure, **497**
Dicrotic notch, 492, 493*f*
Diencephalon, 276*f*, **278**, 279*f*, 281–282
Differential (selective) permeability, 40–**41**, 53–65, **54**, PEx-3, PEx-4, PEx-8
Differential white blood cell count, **429**–430, 430*f*
Diffusion, 54, 54–**55**, 55–60, 55*f*, 59*f*, PEx-3, PEx-4
  facilitated, **55**, PEx-3, PEx-6 to PEx-8, PEx-140
  living membranes and, 58–60, 59*f*
  nonliving membranes and, 56–58
  simple, **55**, PEx-3, PEx-4 to PEx-6
Diffusion rates, molecular weight and, 55–56, 55*f*
Digastric muscle, 204*f*, 204*t*
Digestion, **574**, 597*f*, **PEx-119** to PEx-130
  chemical (enzymatic action), 596–599, 597*f*, PEx-120 to PEx-130, PEx-120*f*
  movements/sounds of, 602–603
  physical processes in (food propulsion/mixing), 602–603, 602*f*, 603*f*
Digestive system, 16*t*, 18, 20*f*, 573–594, **574**, 575*f*, **PEx-119**, PEx-120*f*. *See also specific organ*
  accessory organs of, **574**, 575*f*, 576*f*
  alimentary canal (GI tract), **574**, 575*f*
  chemical and physical processes of, 595–608, **PEx-119** to PEx-130. *See also* Digestion
  histological plan of, 574, 576*f*
Digital (term), **2**, 3*f*
Digital arteries, 474*f*
Digitalis, heart rate and, 518, PEx-98 to PEx-99
Digital veins
  in fingers, 481*f*
  in toes, 479*f*
Dilator pupillae, 361
Dimples, genotype/phenotype and, **671**
Diopter window (ophthalmoscope), **376**
Diploid chromosomal number (2*n*), 646, 668
Direct enzyme-linked immunosorbent assay (ELISA), **PEx-182**, PEx-184
Direct fluorescent antibody technique, PEx-178 to PEx-180
Dislocations, **180**
Distal (term), **4**, 4*f*
Distal convoluted tubule, **613**, 614*f*, **PEx-131**, PEx-132*f*
Distal phalanx
  finger, 152, 152*f*
  toe, 158*f*
Distal radioulnar joint, 151*f*, 170*t*
Divisions, brachial plexus, 313, 314*f*
Dominance, incomplete, **669**
Dominant arm, force measurement/fatigue and, **244**, 244*f*
Dominant gene, **668**
Dominant-recessive inheritance, 668
Dorsal (term), **3**–4, 4*f*
Dorsal body cavity, **6**, 7*f*
Dorsal (posterior) funiculus, **310**, 310*f*, 312*f*
Dorsal (posterior) horns, 284*f*, **308**, 310*f*, 312*f*
Dorsalis pedis artery, **476**, 477*f*, 692*f*, **694**
  pulse at, **495**, 495*f*, 692*f*, 694
Dorsalis pedis vein, **477**, 479*f*
Dorsal median sulcus, **308**, 309*f*, 310*f*, 312*f*
Dorsal metatarsal arteries, **476**, 477*f*
Dorsal metatarsal veins, 479*f*

Dorsal rami, **312,** 313*f*
Dorsal root, **308,** 309*f,* 310*f,* 313*f*
Dorsal root ganglion, 257–258, 258*f,* 259*f,* **308,** 310*f,* 313*f*
Dorsal scapular nerve, 314*f,* 315*t*
Dorsal venous arch, **477,** 479*f,* 692*f,* **694**
Dorsal venous network, **690,** 690*f*
Dorsiflexion (body movement), **174,** 175*f*
Dorsum (term), **2,** 3*f*
Dorsum nasi, **681,** 681*f*
Double-gel diffusion, Ouchterlony, 531–532, 532*f,* 532*t,* PEx-180 to PEx-181
Dry spirometers, 552, 552*f*
Dual X-ray absorptiometry (DXA), PEx-67
Ductus arteriosus, **451, 482,** 483*f*
Ductus (vas) deferens, **21,** 21*f,* **22,** 22*f,* 630*f,* **631,** 631*f*
Ductus venosus, **482,** 483*f,* 484
Duodenal glands, 581*f,* **582,** 582*f*
Duodenal papilla, major, **581,** 587*f*
Duodenum, 575*f,* 578*f,* **581, 582,** 582*f*
Dural sinuses, **479,** 480*f*
Dura mater
   brain, **282,** 283*f,* 284*f,* 285
   spinal cord, 308, 309*f,* 310*f*
Dwarfism, 406
DXA. *See* Dual X-ray absorptiometry
Dyads, 646, 648, 649
Dynamic equilibrium, **389**
Dynamometer, hand, **244,** 244*f,* 245, 245*f*
Dynamometry/dynagram, **244,** 246*f*

Ear, 383–396, 680*f,* 681
   cartilage in, 80*f,* 109*f*
   equilibrium and, 389–392, 390*f,* 391*f*
   gross anatomy of, 384–386, 384*f,* 385*f,* 386*f*
   hearing and, 386–389, 387*f,* 388*f*
   microscopic anatomy of
      of equilibrium apparatus, 389–391, 390*f,* 391*f*
      of spiral organ, 386–387, 386*f,* 387*f*
   otoscopic examination of, 385–386
Eardrum (tympanic membrane), **384,** 384*f*
Earlobe (lobule of ear), **384,** 384*f,* 680*f,* **681**
Early gastrula, 658
EC. *See* Expiratory capacity
Eccrine glands (merocrine sweat glands), 94*f,* **99,** 100, 100*f*
ECG. *See* Electrocardiography
Ectoderm, **658,** 659, 660*f*
Ectopic pacemakers, **518**
Ectopic pregnancy, 634
EDA. *See* Electrodermal activity
Edema, 527
EDV. *See* End diastolic volume
EEG. *See* Electroencephalography
Effector, in reflex arc, 336*f,* 337
Efferent arteriole, **613,** 614*f,* 615*f,* **PEx-131, PEx-133, PEx-137**
   radius of, glomerular filtration and, PEx-132 to PEx-135, **PEx-137** to PEx-139
Efferent (motor) nerves, **258, 276**
Efferent (motor) neurons, 241, 255*f,* **258,** 259*f,* **PEx-18, PEx-35**
   in reflex arc, 336*f,* 337
Eggs (ova), **650**
   development of (oogenesis), 648–651, **649,** 649*f*
   fertilized (zygote), **646,** 649, **658,** 659*f*
Einthoven's law, 460
Einthoven's triangle, 460, 460*f*
Ejaculation, **631**–632
Ejaculatory duct, 630*f,* **631,** 631*f*
EKG. *See* Electrocardiography
Elastic arteries, 471
Elastic cartilage, 80*f,* 109*f,* **110**
Elastic connective tissue, 78*f*
Elastic fibers, 74*f,* **75,** 76*f,* 78*f*
   in skin, 96
Elastic (titin) filaments, 186*f*
Elbow, 150*f,* 170*t,* 688, 688*f,* 689*f*
Electrical stimulation, muscle contraction and, 234, **PEx-17,** PEx-18

Electrocardiography/electrocardiogram (ECG), **458**–466, 459*f,* 459*t,* 460*f,* 462*f,* 463*f,* 464*f,* 465*f*
   with BIOPAC®, 462–466, 462*f,* 463*f,* 464*f,* 465*f*
   cardiac cycle and, 493*f*
   limb leads for, 460, 460*f,* 462, 462*f*
   with standard apparatus, 460–462
Electrochemical gradient, in active transport, 60
Electrodermal activity (EDA/galvanic skin response), **326**–332, 327*f,* 328*f,* 329*f,* 330*f*
Electrodes
   for ECG, 460, 460*f,* 462, 462*f*
   for EEG, 300, 300*f,* 301, 302, 302*f*
   for oscilloscope, 270, 271*f*
   for polygraph, 327, 327*f*
Electroencephalography/ electroencephalogram (EEG), 299–306, **300,** 300*f,* 301*f,* 302*f,* 303*f,* 304*f*
Electromyography/electromyogram (EMG), **241**–248
   force measurement/fatigue and, 244–248, 244*f,* 245*f,* 246*f,* 247*f*
   temporal/multiple motor unit summation and, 241–244, **242,** 242*f,* 243*f*
Electrophoresis, gel, **PEx-184**
   hemoglobin phenotyping using, 672–673, 673*f*
Elementary body (chlamydia), **PEx-178**
ELISA. *See* Enzyme-linked immunosorbent assay
Embryonic connective tissue (mesenchyme), 75, 76*f*
Embryonic development, 657–666, 659*f,* 660*f*
Embryonic disc, **659,** 660*f*
Embryonic membranes 660, 660*f*
EMG. *See* Electromyography
Emmetropic eye, **372,** 373*f*
Emphysema, 554, **PEx-109,** PEx-110
Emulsification, by bile, 586, 597*f,* 599–600
Enamel (tooth), **585,** 585*f*
Encapsulated nerve endings, **351**
Encephalitis, **283**
End diastolic volume (EDV), 493*f,* **PEx-82, PEx-84**
Endocardium, 445*f,* **446**
Endochondral ossification, **114**
Endocrine system/glands, 16*t,* **69,** 69*f,* 405–416, **406,** 407*f,* 409*f,* 409–410*f,* PEx-59, PEx-60*f. See also specific gland and hormone*
   epithelial cells forming, 69, 69*f*
   functional anatomy of, 405–416, 407*f,* 409*f,* 409–410*f*
   microscopic anatomy of, 409–411, 409–410*f*
   physiology of, 417–422, **PEx-59** to PEx-74
Endocytosis, **61,** 61*f*
Endoderm, **658,** 659, 660*f*
Endolymph, **385,** 386*f,* 389
Endometrium, 635*f,* **636,** 636*f*
   during menstrual cycle, 651, 651*f,* 652*f*
Endomysium, **188,** 189*f*
Endoneurium, **260,** 260*f*
Endoplasmic reticulum (ER), 40*f,* **42**–43, 43*t*
   rough, **42**–43, 42*f,* 43*t*
   smooth, **42,** 42*f,* 43, 43*t*
   muscle cell (sarcoplasmic reticulum), **187,** 187*f*
Endosteum, **110,** 111*f,* 113*f*
End-plate potential, **PEx-18**
End systolic volume (ESV), 493*f,* **PEx-82, PEx-84**
Energy
   for active processes, 54, 60, 61, PEx-3, PEx-4, PEx-13
   kinetic, **54,** 54*f,* **PEx-3,** PEx-4
Enteroendocrine cells, 581*f*
Envelope, nuclear, **40,** 40*f,* 42*f*

Enzyme(s), **596, PEx-120.** *See also specific type and* Enzyme substrates
   in digestion, 596–599, 597*f,* PEx-120 to PEx-130, PEx-120*f*
Enzyme assay, **PEx-121, PEx-123**
Enzyme-linked immunosorbent assay (ELISA), 531, **PEx-182**
   direct, **PEx-182, PEx-184**
   indirect, PEx-181 to PEx-184, **PEx-182, PEx-184**
Enzyme substrates, **596**
   specificity of, 596
   amylase, PEx-123 to PEx-125
Eosinophil(s), 425*f,* 426*t,* 427*f,* **428,** 428*f*
Ependymal cells, 254, 254*f*
Epicardium (visceral layer/pericardium), 7*f,* 445*f,* **446,** 449
Epicondyle (bone marking), 112*t. See also* Lateral epicondyle; Medial epicondyle
Epicranial aponeurosis, 200*f,* 201*f,* 681
Epicranius muscle, 197, 198*f,* 199*f,* 200*t,* 201*f,* 681, 681*f*
Epidermal dendritic cells (Langerhans' cells), 95*f,* **96**
Epidermal ridges (fingerprints), 96, 100–102, 101*f*
Epidermis, 94, 94–96, 94*f,* 95*f,* 99*f*
Epididymis, 630*f,* **631,** 631*f,* 633, 634*f*
Epidural space, 310*f*
Epigastric region, **8,** 9*f*
Epiglottis, **538,** 539*f,* 540*f,* 577*f*
Epimysium, **188,** 189*f*
Epinephrine, **408,** 520, **PEx-98**
   heart rate and, 518, **PEx-98** to PEx-99
Epineurium, **260,** 260*f*
Epiphyseal lines, **110,** 111*f*
Epiphyseal (growth) plate, **110,** 111, 114
Epiphysis, **110,** 111*f*
Epiploic appendages, 583*f,* 584
Epithalamus, 276*f,* 280*f,* **282**
Epithelial cells, cheek, 34, 34*f*
Epithelial tissues/epithelium, **68**–75, 68*f,* 69*f,* 70–73*f. See also specific type*
Epitopes (antigenic determinants), 531, 531*f,* **PEx-178,** PEx-178*f*
Eponychium, **97,** 97*f*
Equator (spindle), 47*f*
Equilibrium (balance), 389–392, 390*f,* 391*f. See also* Ear
Equilibrium (solution), **PEx-3**
ER. *See* Endoplasmic reticulum
Erection, 632
Erector spinae muscles, 210*f,* 210*t,* **683,** 684*f*
ERV. *See* Expiratory reserve volume
Erythrocyte(s) (red blood cells/RBCs), 81*f,* **424,** 425*f,* 426*t,* 427–428, 427*f*
   in hematocrit, PEx-161, PEx-162
   settling of (ESR), **PEx-161,** PEx-161 to PEx-162, **PEx-164** to PEx-165
   in urine (hematuria), 623
Erythrocyte (red blood cell) antigens, blood typing and, **434,** 434*t,* 435*f,* PEx-162, **PEx-167,** PEx-168*t*
Erythrocyte (red blood cell) count, total, **429**
Erythrocyte sedimentation rate (ESR), **PEx-161,** PEx-161 to PEx-162, **PEx-164** to PEx-165
Erythropoietin, PEx-161
Esophageal arteries, 474
Esophagus, **18,** 72*f,* 575*f,* **577,** 577*f*
ESR. *See* Erythrocyte sedimentation rate
Estrogen(s), **408,** 636, **PEx-67**
   adrenal, 408
   bone density and, PEx-67, PEx-68
   in menstrual cycle, 651, 652*f*
   in ovarian cycle, 650
Estrogen (hormone) replacement therapy, PEx-67 to PEx-69
ESV. *See* End systolic volume
Ether, nerve function affected by, 269
Ethmoidal air cells (sinuses), 124*f,* 129, 129*f,* 538
Ethmoid bone, 121*f,* 123*f,* 124*f,* 125*f,* **126**–127, 126*f,* 128*f*

Eustachian tube. *See* Pharyngotympanic (auditory/eustachian) tube
Eversion (body movement), **174,** 175*f*
Excitability, neuron, **266,** 266*f*
Excitation-contraction coupling, **PEx-18**
Excitatory postsynaptic potential, PEx-51
Excretion, by kidneys, PEx-131, PEx-135
Exercise
   blood pressure/heart rate affected by, 500
   breathing during, PEx-106, **PEx-109,** PEx-111
   ECG affected by, 461, 463–464, 464*f,* 465–466
   metabolic acidosis and, PEx-154
Exocrine glands, 61*f,* **69**
Exocytosis, 42*f,* **61,** 61*f*
Expiration, **550,** 551*f,* **PEx-106,** PEx-107
Expiratory capacity (EC), 559*f*
Expiratory muscles, 206*t,* 207*f*
Expiratory reserve volume (ERV), **552,** 552*f,* 554, 555–556, 556, 559*f,* **PEx-107**
Extension (body movement), **173,** 174*f*
Extensor carpi radialis brevis muscle, 214*t,* 215*f*
Extensor carpi radialis longus muscle, 199*f,* 213*f,* 214*t,* 215*f*
Extensor carpi ulnaris muscle, 199*f,* 214*t,* 215*f*
Extensor digiti minimi, 215*f*
Extensor digitorum brevis muscle, 221*f*
Extensor digitorum longus muscle, 196*f,* 198*f,* 220*t,* 221*f,* 692*f,* **694**
Extensor digitorum muscle, 199*f,* 211, 214*t,* 215*f,* **690,** 690*f*
Extensor hallucis brevis muscle, 221*f*
Extensor hallucis longus muscle, 220*t,* 221*f,* 692*f,* **694**
Extensor indicis muscle, 215*f*
Extensor pollicis longus and brevis muscles, 214*t,* 215*f,* **690,** 690*f*
Extensor retinacula, superior and inferior, 221*f*
External/superficial (term), **4**
External acoustic (auditory) meatus/canal, **120,** 121*f,* 122*f,* **384,** 384*f,* 680*f,* **681**
External anal sphincter, 583, 583*f*
External carotid artery, **472,** 473*f,* **682**
   pulse at, 682
External (outer) ear, **384,** 384*f*
   cartilage in, 80*f,* 109*f*
External genitalia, of female (vulva), **634,** 634*f*
External iliac artery, **476,** 477*f*
External iliac vein, **477,** 479*f,* 480*f*
External intercostal muscles, 206*t,* 207*f,* 550, **PEx-106, PEx-107**
External jugular vein, **479,** 480*f,* 481*f,* **682,** 683*f*
External oblique muscle, 198*f,* 206*t,* 207*f,* 208*f,* **686**
External occipital crest, 122*f,* **125,** 125*f*
External occipital protuberance, 122*f,* **125,** 125*f,* 130, 680*f,* **681**
External os, 635*f*
External respiration, **538,** PEx-106*f*
External urethral orifice, 21*f,* 611*f,* **612,** 630*f,* 631*f,* 634*f,* 635*f*
External urethral sphincter, 611*f,* **615**
Exteroceptors, **350**
Extracellular matrix, 74*f,* **75,** 79*f,* 80*f,* 82
Extrafusal fibers, 351*f*
Extrasystoles, 516*f,* **517, PEx-95**
Extrinsic eye muscles, **360,** 361*f,* 365*f*
   reflex activity of, 375
Eye, 359–363. *See also under* Visual and Vision
   cow/sheep, dissection of, 364, 365*f*
   emmetropic, **372,** 373*f*
   external anatomy/accessory structures of, 359–360, 360*f,* 361*f*
   extrinsic muscles of, **360,** 361*f,* 365*f*
      reflex activity of, 375
   hyperopic, **372,** 373*f*

internal anatomy of, 360–363, 362f, 363f
myopic, **372**, 373f
ophthalmoscopic examination of, 375–377, 376f, 377f
retinal anatomy and, 363f, 364
Eyelashes, **360**, 360f
Eyelids (palpebrae), **360**, 360f
Eye muscles
extrinsic, **360**, 361f, 365f
intrinsic, 361
reflex activity of, 375
Eyepiece/ocular (microscope), **28**, 29f
Eye reflexes, 375
Eye teeth (canines), **584**, 584f

Face. See also under Facial
bones of, 108f, **120**, 121f, 125–126f, **127**–128, 127f, 128f
muscles of, 198f, 200t, 201f, **681**
surface structures of, 681f, 681f
Facet (bone marking), 112t
Facial artery, **472**, 473f, 680f, **681**
pulse at, **495**, 495f, 680f
Facial bones, 108f, **120**, 121f, 125–126f, **127**–128, 127f, 128f.
See also specific bone
Facial expression, muscles controlling, 198f, 200t, 201f, **681**
Facial nerve (cranial nerve VII), 286t, 288, 288f, 289f, 399
Facial vein, **479**, 480f
Facilitated diffusion, 55, PEx-3, **PEx-6** to PEx-8, **PEx-140**
Falciform ligament, 20f, 580f, **586**, 586f
Fallopian (uterine) tubes, **22**, 22f, 635f, **636**, 637f
"False negative," PEx-178
False pelvis, **153**
"False positive," PEx-178
False ribs, 135f, 136
False vocal cords (vestibular folds), **539**, 539f, 540f
Falx cerebelli, **282**, 283f
Falx cerebri, **282**, 283f
Farsightedness (hyperopia), **372**, 373f
Fascia
deep, **188**
renal, 611
superficial (hypodermis), **94**, 94f
Fascicle(s)
muscle, **188**, 189f
muscle name and, 196f, 197
nerve, **260**, 260f
Fast calcium channels, automaticity and, 512
Fasting plasma glucose, PEx-65, PEx-66 to PEx-67
Fat (adipose tissue), **75**, 77f, 82, 94f
Fat cells, 74f, 77f, 82
Fat digestion, 597f, 599–601, PEx-127 to PEx-128
Fatigue, muscle, **240**, 241, **PEx-25** to PEx-26
inducing, 240
measurement of, 244–248, 244f, 245f, 246f, 247f
Fatty acids, PEx-127
Fauces
isthmus of, 539f
of uterine tube, 635f
Feedback mechanisms
negative, **PEx-60**
thyroid hormone secretion and, PEx-61
positive, PEx-60
Feet. See Foot
Female
pelvis in, male pelvis compared with, 153, 155t
reproductive system in, 16t, 21, 21f, 22, 22f, 634–637, 634f, 635f, 636f, 637f
urinary system in, 610f, 611f
X chromosome and, 669–670
Femoral (term), **2**, 3f
Femoral arteries, **476**, 477f, 693
circumflex, **476**, 477f
deep (deep artery of thigh), **476**, 477f
pulse at, **495**, 495f, 693

Femoral cutaneous nerve, 316–317, 316f, 317f, 317t, 318t
Femoral nerve, **316**–317, 316f, 317t, 318t
Femoral triangle, 691f, 692f, **693**
Femoral vein, **477**, 479f, 693
Femoropatellar joint, 171t, **176**
Femur, 108f, **156**, 692f, **693**. See also Thigh
ligament of head of (ligamentum teres), **176**, 176f
Fertilization, **658**, 659f. See also Zygote
Fertilization membrane, 648, 658
Fetal circulation, 482–484, 483f
Fetal skull, 136–137, 137f
Fetus, **660**, 661–662
circulation in, 482–484, 483f
FEV$_T$/FEV$_1$ (forced expiratory volume), **557**–559, 558f, **PEx-107**
Fibers (muscle). See Muscle fibers
Fibrillation, **460**
Fibrin/fibrin mesh, 433f, **434**
Fibrinogen, 433f, **434**
Fibroblast(s), 74f, 75, 76f, 78f, 79f
Fibrocartilage, 80f, 109f, **110**
Fibrous capsule of kidney, 610f, 611, **612**, 612f
Fibrous joints, 168f, **169**, 170–171t
Fibrous layer of eye, 360–361, 362f
Fibrous layer of synovial joint, 171
Fibrous pericardium, **446**
Fibrous sheath of hair, 98, 98f
Fibula, 108f, **157**, 157f, 692f, **693**
Fibular (term), **2**, 3f
Fibular artery, **476**, 477f
Fibular collateral ligament, 177f, **178**
Fibularis (peroneus) brevis muscle, 220t, 221f, 223f
Fibularis (peroneus) longus muscle, 198f, 199f, 220t, 221f, 223f
Fibularis (peroneus) tertius muscle, 220t, 221f
Fibular nerves, common, **317**–318, 317f, 318t, 693
Fibular retinaculum, 221f
Fibular vein, **477**, 479f
Field (microscope), **30**, 32–33, 32t
Field (visual), 374–375, 374f
Filter switch (ophthalmoscope), **376**, 376f
Filtrate, PEx-11, **PEx-133**, **PEx-135**
Filtrate pressure, PEx-135
Filtration, 54, 60, **613**, PEx-3, PEx-3 to PEx-4, **PEx-11** to PEx-13
glomerular, **613**, **PEx-131**, **PEx-132**
arteriole radius affecting, PEx-132 to PEx-135, PEx-137 to PEx-139
pressure affecting, PEx-135 to PEx-137, **PEx-137** to PEx-139
Filum terminale, **308**, 309f
Fimbriae, 635f, **636**
Fine adjustment knob (microscope), **28**, 29f
Finger(s)
bones of (phalanges), 108f, 151–152, 152f
joints of, 170t
muscles of/muscles acting on, 211, 213–215f, 213–214t, **688**–689, 689f
Finger hair, genotype/phenotype and, **671**, 672f
Fingerprints (epidermal ridges), 96, 100–102, 101f
First heart sound (S1), 492, 493f
First polar body, 648, 649f, **650**
Fissure(s) (bone marking), 112t
Fissure(s) (cerebral), **277**, 277f
Fixators (fixation muscles), **196**
Flare, mechanical stimulation of blood vessels causing, 504
Flat bones, **110**
Flexion (body movement), **173**, 174f
plantar, **174**, 175f
Flexion creases, **690**, 690f
Flexor carpi radialis muscle, 198f, 213f, 213t, **689**, 689f
Flexor carpi ulnaris muscle, 199f, 213f, 214t, 215f, **689**, 689f

Flexor digitorum longus muscle, 223f, 224f, 224t
Flexor digitorum profundus muscle, 213f, 214t
Flexor digitorum superficialis muscle, 213f, 214t, 689f
Flexor hallucis longus muscle, 221f, 223f, 224f, 224t
Flexor pollicis longus muscle, 213f, 214t
Flexor reflex, 336f, 337
Flexor retinaculum, 213f
Floating (vertebral) ribs, 135f, 136
Flocculonodular lobe, 282
Floor, of orbit, 128f
Flower spray endings, 351f
Fluid-phase endocytosis (pinocytosis), **61**, 61f
Fluorescent antibody technique, direct, PEx-178 to PEx-180
Flushing of skin, local metabolites and, 502–503
Foliate papillae, **399**, 399f, 400f
Follicle(s)
hair, 94f, **98**, 98f, 99f
lymphoid, 528f
ovarian, 636, 649, 649f, 650, 650f, 651, 652f
thyroid gland, **409**, 409f, 410
of tonsils, **529**, 530f
Follicle cells, ovarian, **649**, 649f
Follicle-stimulating hormone (FSH), **406**, 407f, **PEx-67**
oogenesis/ovarian/menstrual cycles affected by, 418, 650, 652f
in spermatogenesis, 647, 648
Fontanelles, **136**, 137f
Food, digestion of, 595–608, 597f, **PEx-119** to PEx-130
chemical (enzymatic action), 596–599, 597f, PEx-120 to PEx-130, PEx-120f
physical (food propulsion/mixing), 602–603, 602f, 603f
Foot. See also Lower limb
arches of, 157, 158f
bones of, 157, 158f
movements of, 174, 175f
muscles acting on, 220t, 221f, 222–224t, 223–224f
surface anatomy of, 3f, 158, 692f, 693–694
Footdrop, **318**
Foot processes, 613
Foramen (bone marking), 112t
Foramen lacerum, **120**, 122f, 123f
Foramen magnum, **120**, 122f, 123f
Foramen ovale (fetal heart), **451**, 482, 483f, 484
Foramen ovale (skull), 122f, 123f, 124f, **126**
Foramen rotundum, 123f, 124f, **126**
Foramen spinosum, 122f, 123f, 124f
Force
muscle, **PEx-17**, **PEx-20**
muscle contraction, **PEx-20**
Forced expiratory volume (FEV$_T$/FEV$_1$), **557**–559, 558f, **PEx-107**
Forced vital capacity (FVC), **557**–559, 558f, **PEx-107**
Force transducer, **PEx-17**
Forearm. See also Upper limb
bones of, 150–151, 151f
force measurement/fatigue and, 244–248, 244f, 245f, 246f, 247f
muscles of/muscles controlling, 198f, 199f, 211, 212f, 212t, 213–215f, 213–214t, **688**–689, 689f
surface anatomy of, 152, 688–690, 689f
Forebrain (prosencephalon), **276**, 276f
Foreskin (prepuce), 630f, 631f, **632**
Formed elements of blood, 424–425, **424**, 425f, 426–429, 426t, 427f, 428f. See also Blood cells
blood viscosity and, PEx-78
Fornix (cerebral), **279**, 280f, 290–291, 291f, 292f
Fornix (vaginal), 635f
Fossa (bone marking), 112t

Fossa ovalis, 445f, **451**, 452f, 483f, **484**
Fourth ventricle, 276f, 280f, 284f, 291f
Fovea capitis, **156**, 156f, **176**
Fovea centralis, **362**, 362f
FPG. See Fasting plasma glucose
Frank-Starling law of the heart, PEx-84
FRC. See Functional residual capacity
Freckle(s), 95
genotype/phenotype and, **671**, 672f
Free edge of nail, **97**, 97f
Free nerve endings, 94f, 96, 97, 350f, **351**
Frequency (stimulus), **PEx-22**, **PEx-24**, **PEx-25**
muscle contraction affected by, 239–240, 239f, PEx-22 to PEx-23
tetanus and, 239f, 240, **PEx-24**, **PEx-25**, PEx-94
Frequency of breathing, **PEx-106**
Frequency range of hearing, testing, 388
Friction ridges (fingerprints), 96, 100–102, 101f
Frog
cardiac anatomy in, 513, 513f
cardiovascular physiology in, 511–524, PEx-93 to PEx-104
gastrocnemius muscle in
contraction of, 236–241, 236f, 237f, 238f, 239f
dissection of, 237f, 238, 267–268, 267f
as poikilothermic animal, **PEx-97**
sciatic nerve in
dissection of, 267–268, 267f
inhibiting, 268–270
stimulating, 268, 270–271, 270f, 271f
Frontal (term), **2**, 3f
Frontal belly of epicranius muscle, 197, 198f, 200t, 201f, **681**, 681f
Frontal bone, **120**, 121f, 123f, 125f, 126f, 128f
in fetal skull, 137f
Frontal (anterior) fontanelle, 137, 137f
Frontal lobe, **277**, 277f, 279f
Frontal (coronal) plane/section, **4**, 5f, 6f
Frontal process, of maxilla, 127f, 128f
Frontal sinuses, 129, 129f, 538, 539f
Frontonasal suture, 125f
FSH. See Follicle-stimulating hormone
Functional layer (stratum functionalis) (endometrium), **636**
Functional residual capacity, 552f, 559f
Functional syncytium, myocardium as, **458**
Fundus
of eye, **362**
ophthalmoscopic examination of, 375–377, 376f, 377f
of stomach, **577**, 578f
of uterus, **634**, 635f
Fungiform papillae, **399**, 399f
Funiculi (dorsal/lateral/ventral), **310**, 310f, 312f
"Funny bone." See Medial epicondyle
Fused (complete) tetanus, 239f, 240, **PEx-24**, **PEx-25**, PEx-94
Fusiform fascicles/muscles, 196f
Fusion frequency (tetanus), **PEx-24** to PEx-25, PEx-25, PEx-94
FVC (forced vital capacity), **557**–559, 558f, **PEx-107**

Gag reflex, **340**
Gallbladder, 575f, **586**, 586f, 587f, 686
Galvanic skin potential (GSP), **326**
Galvanic skin resistance (GSR), **326**, 327f
Galvanic skin response, 326–332, 327f, 328f, 329f, 330f
Gametes, **630**, 646
meiosis in production of, **45**, **646**
Gametogenesis, **646**
oogenesis, 648–651, 649, 649f
spermatogenesis, **647**–648, 647f, 648f
Gamma globulins, 530
Ganglia (ganglion), **254**
basal (basal nuclei), 276f, **279**
dorsal root, 257–258, 258f, 259f
terminal (intramural), **324**
Ganglion cells, of retina, 363f, 364

Gas exchange, 541, 541f, PEx-105, PEx-106, PEx-106f
external respiration, **538**, PEx-106f
internal respiration, **538**, PEx-106f
Gastric arteries, **474**, 475f
Gastric glands, **577**, 579, 579f, PEx-120f, PEx-125
Gastric lipase, 597f, **PEx-127**
Gastric pits, 578f, 579, 579f
Gastric veins, **484**, 484f
Gastrocnemius muscle, 198f, 199f, 219f, 221f, 222, 222t, 223f, **693**, 693f
in frog
contraction of, 236–241, 236f, 237f, 238f, 239f
dissection of, 237f, 238
Gastroduodenal artery, **474**, 475f
Gastroepiploic arteries, **474**–475, 475f
Gastroepiploic vein, 484f
Gastroesophageal junction, **579**, 579f
Gastroesophageal sphincter, **577**
Gastrointestinal (GI) tract (alimentary canal), **574**, PEx-119, PEx-120f. *See also specific organ*
histology of, 574, 576f
Gastrula, **658**, 659
Gastrulation, **658**
Gated channels, PEx-36
voltage-gated potassium channels, refractory periods and, PEx-44
voltage-gated sodium channels, **PEx-42** to PEx-44
refractory periods and, PEx-44
Gel electrophoresis, **PEx-184**
hemoglobin phenotyping using, 672–673, 673f
Gender, sex chromosomes determining, 669–670
Gene(s), **668**. *See also* Inheritance
dominant, **668**
recessive, **668**
General sensation, 349–358, **350**. *See also under Sensory*
Genetics, 667–678. *See also* Inheritance
Geniculate body, lateral, **364**, 366f
Genitalia, external, of female (vulva), **634**, 634f
Genitofemoral nerve, 316f, 317t
Genotype, **668**
phenotype in determination of, 671–672, 672f, 672t
Germinal centers
lymph node, 528f, **529**
of tonsils, 529, 530f
Germinal epithelium, 650f, **651**
Germ layer, primary, **658**, 660f
GH. *See* Growth hormone
Gigantism, 406
Gingiva (gums), 576f, **585**, 585f
Gingival sulcus, 585, 585f
Glabella, **120**, 125f
Gland(s). *See also specific type*
epithelial cells forming, 69, 69f
Glans penis, 630f, 631f, **632**
Glassy membrane, 98, 98f
Glaucoma, **363**
Glenohumeral joint, 170t, **178**
Glenohumeral ligaments, **178**, 179f
Glenoid cavity, **148**, 149f, 179f
Glenoid labrum, **178**, 179f
Glial cells (neuroglia), **82**, 82f, **254**, 254f
myelination and, 254, 254f, 256–257, PEx-48
Globus pallidus, **280**, 281f
Glomerular capillaries, 613, 614f, 615f, **PEx-133, PEx-135**
Glomerular capillary pressure, PEx-133, PEx-135, **PEx-137**
glomerular filtration and, PEx-135 to PEx-137, PEx-137 to PEx-139
Glomerular (Bowman's) capsule, **613**, 614f, 615f, PEx-131, PEx-132f, **PEx-133, PEx-135, PEx-137**
Glomerular filtration, **613**, PEx-131, **PEx-132**
arteriole radius affecting, PEx-132 to PEx-135, PEx-137 to PEx-139

pressure affecting, PEx-135 to PEx-137, **PEx-137** to PEx-139
Glomerular filtration rate, **PEx-133, PEx-135**, PEx-137. *See also* Glomerular filtration
Glomerulus (glomeruli), **613**, 614f, 615f, PEx-131, PEx-132f, **PEx-132** to PEx-133, **PEx-135, PEx-137**
Glossopharyngeal nerve (cranial nerve IX), 287t, 288, 288f, 289f, 399
Glottis, **539**
Glucagon, **408, PEx-64**
Glucocorticoids, **408**
Glucose, **PEx-64** to PEx-67
in diabetes mellitus, 408
plasma, **PEx-64** to PEx-67
fasting levels of, PEx-65, PEx-66 to PEx-67
transport/reabsorption of, carrier proteins and, 55, PEx-6 to PEx-8, **PEx-140** to PEx-142
in urine (glycosuria), **622**, 624
Glucose carrier proteins, 55, PEx-6 to PEx-8, **PEx-140** to PEx-142. *See also* Carrier proteins
Glucose standard curve, **PEx-65**
developing, PEx-65 to PEx-66
Gluteal (term), **2**, 3f
Gluteal arteries, **476**, 477f
Gluteal (natal) cleft, **690**, 691f
Gluteal fold, **690**, 691f
Gluteal lines, anterior/posterior/inferior, 154f
Gluteal nerves, 317f, 318t
Gluteal prominences, 690
Gluteal region, 690–693, 691f
for intramuscular injection, 691–693, 691f
Gluteal tuberosity, **156**, 156f
Gluteus maximus muscle, 199f, 218t, 219f, 222, **690**, 691f
Gluteus medius muscle, 199f, 218t, 219f, **691**, 691f
for intramuscular injection, 691, 691f
Gluteus minimus muscle, 218t
Glycosuria, **622**, 624
Goblet cells, 71f
large intestine, 71f, 582f
small intestine, 581f
Goiter, **419, PEx-60**
Golgi apparatus, 42f, **43**, 43t
Gomphosis, **169**
Gonad(s), 408, **630**. *See also* Ovary; Testis
Gonadal arteries, **475**, 475f
Gonadal veins, **477**, 480f
Gonadocorticoids (sex hormones), **408**
Gonadotropins, **406**, 418
ovarian/menstrual cycles affected by, 418, 650, 652f
gp41, PEx-185t
gp120, PEx-185, PEx-185t
gp160, PEx-185, PEx-185t
Gracilis muscle, 198f, 216t, 217f, 219f
Graded depolarization, PEx-18
Graded muscle response (graded contractions), 238, **241, 244**
stimulus frequency increases and, 239–240, 239f
stimulus intensity increases and, 238–239
Graded potentials, PEx-39
Gradient, PEx-3
Granular (juxtaglomerular) cells, 613, 615f
Granulocytes, 426t, **428**
Granulosa cells, **649**, 650f, 658
Graves' disease, 419
Gray commissure, **308**, 310f
Gray matter
cerebellar, 279, 282, 282f
cerebral (cerebral cortex), 276f, 277f, **278**, 279, 281f
of spinal cord, **308**, 310f
Gray ramus communicans, **324**, 325f
Great cardiac vein, 444f, 445f, **448**, 448f
Greater auricular nerve, 313f, 315t
Greater curvature of stomach, **577**, 578f

Greater omentum, **18**, 20f, **577**, 580f
Greater sciatic notch, **153**, 154f
Greater trochanter, **156**, 156f, 158, **690**, 691f
Greater tubercle, **148**, 150f, **687**, 687f
Greater vestibular glands, **634**, 634f, 635f
Greater wings of sphenoid, 121f, 122f, 123f, 124f, 125f, **126**, 128f, 130
Great saphenous vein, **477**, 479f, 694
Groove (bone marking), 112t
Gross anatomy, **1**
Ground substance, 74f, **75**, 76f
Growth (ossification) centers, in fetal skull, **137**, 137f
Growth hormone (GH), **406**, 407f
Growth (epiphyseal) plate, **110**, 111, 114
GSP. *See* Galvanic skin potential
GSR. *See* Galvanic skin resistance
Gums (gingiva), 576f, **585**, 585f
Gustatory cortex, 278f, 399
Gustatory epithelial cells, **399**, 399f
Gustatory hairs, **399**, 399f
Gyri (gyrus), **277**, 277f

Hair(s), **97**–98, 98f
Hair bulb, **97**–98, 98f
Hair cells, 386, 386f
in equilibrium, **390**, 391f
in hearing, 386, 386f, 387
Hair follicle, 94f, **98**, 98f, 99f
Hair follicle receptor (root hair plexus), 94f, 97, 350f, **351**
Hair matrix, 98f
Hair papilla, **98**, 98f
Hair root, 94f, **97**, 98f
Hair shaft, 94f, **97**, 98f, 99f
Hallux (term), **2**, 3f
Hamate, **151**, 152f
Hammer (malleus), 384f, **385**
Hamstring muscles, 199f, 218–219t, **693**, 693f
Hand, **2**, 3f. *See also* Upper limb
bones of, 151–152, 152f
forearm muscles acting on, 213–215t, 213–214t, 688–689, 689f
surface anatomy of, 3f, 152, 689–690, 690f
Hand dynamometer, **244**, 244f, 245, 245f
Handheld dry spirometers, 552, 552f
Haploid (*n*) chromosomal number, **646**, 668
Hard palate, 122f, **538**, 539f, 575, 576f, 577f
Harvard step test, 500
Haustra, 583f, **584**
Haversian (central) canal, 81f, 113f, **114**
Haversian system (osteon), 113f, **114**
Hb. *See* Hemoglobin
HCO_3^-. *See* Bicarbonate
HDL. *See* High-density lipoprotein
Head
arteries of, 472, 473f
muscles of, 197, 198f, 200–202t, 201–202f
surface anatomy of, 680–682, 680f, 681f
veins of, 478–479, 480f
Head of bone, 112t
femur, 156f
ligament of, **176**, 176f
fibula, 157f, 692f, **693**
humerus, 148, 150f
metacarpals, 151
radius, 150, 150f, 151f, **688**, 688f
ulna, **151**, 151f, **688**, 688f, 689f
Head (microscope), **28**, 29f
Head of sperm, 648, 648f
Hearing, 386–389, 387f, 388f. *See also* Ear
cortical areas in, 278, 278f, 387
Hearing loss, testing for, 388, 388f
Heart, **18**, 19f, **444**, 684, PEx-82. *See also under Cardiac*
anatomy of, 443–456, 444f, 444–445f
in frog, 513, 513f
gross (human), 444–446, 444f, 444–445f

microscopic (cardiac muscle), **83**, 84f, 449, 449f
in sheep, 449–452, 450f, 452f
blood supply of, 446–449, 448f
chambers of, 446
conduction system of, **458**, 458f
disturbances of, 519
in frog, 513–519, 514f, 515f, 516f
contractility of, **PEx-84**
ECG in study of, **458**–466, 459f, 459t, 460f, 462f, 463f, 464f, 465f
epinephrine affecting, 518, PEx-98, **PEx-98** to PEx-99
fibrous skeleton of (cardiac skeleton), 446, 447f
Frank-Starling law of, PEx-84
in frog
anatomy of, 513, 513f
baseline activity of, 513–519, 514f, 515f, 516f
physiology of, 511–524
location of, 444, 444f, 684
nervous stimulation of, 458, PEx-95 to PEx-96
physiology of, PEx-93 to PEx-104
in frog, 511–524
in human, 491–510
pumping activity of, PEx-82
vessel radius affecting, PEx-82 to PEx-83
in sheep, dissection of, 449–452, 450f, 452f
valves of, 445f, 446, 447f, PEx-86
murmurs and, 492
Heart block, 519
Heartburn, referred pain and, 354
Heart murmurs, **492**
Heart rate, 460, 495. *See also* Pulse
in cardiac output, PEx-82, PEx-84
factors affecting, 500–501, 517–519
chemical modifiers, PEx-98 to PEx-99
epinephrine, **PEx-98**, PEx-98 to PEx-99
ions affecting, PEx-99 to PEx-101, PEx-100t
temperature affecting, PEx-96 to PEx-98
Heart sounds, 492–494, 493f, 494f
Heat, bone affected by, 111–112
Heat receptors, 352, PEx-40
Heavy chains, immunoglobulin, **530**, 531f
Heel bone (calcaneus), 108f, 157, 158, 158f
Helix of ear, 384f, 680f, **681**
Helper T cells, 529
Hematocrit, **430**–431, 431f, **PEx-76, PEx-161, PEx-162** to PEx-164
Hematologic tests, 429–436, 430f, 431f, 432f, 433f, 434t, 435f, PEx-161 to PEx-176. *See also specific test*
Hematopoietic tissue, **75**
Hematuria, **623**
Heme, **PEx-166**
Hemiazygos vein, **479**, 481f
Hemocytoblast, 428f
Hemoglobin, 431, **PEx-161, PEx-162**, PEx-165 to PEx-167, **PEx-166**
normal, electrophoresis in identification of, 672–673, 673f
sickle cell, 669
electrophoresis in identification of, 672–673, 673f
in urine (hemoglobinuria), **623**, 624
Hemoglobin concentration, 431–433, 432f
Hemoglobinometer, 432–433, 432f, **PEx-166**
Hemoglobin phenotype, 672–673
Hemoglobinuria, **623**, 624
Hemolysis, 59f, **60**
Hemophilia, inheritance of, 670
Hemostasis, 433–434, 433f
Henle, loop of (nephron loop), **613**, 614f, **PEx-131**, PEx-132f
Hepatic artery, 587
common, **474**, 475f

Hepatic artery proper, **474**, 475*f*
Hepatic duct, common, **586**, 587*f*
Hepatic (right colic) flexure, **583**, 583*f*
Hepatic (stellate) macrophages, **587**, 588*f*
Hepatic portal circulation, 484, 484*f*
Hepatic portal vein, 477–478, **484**, 484*f*, 587
Hepatic veins, **477–478**, 480*f*, 484*f*, 586*f*, 587
Hepatitis, **587**
Hepatocytes (liver cells), **587**, 588*f*
Hepatopancreatic ampulla, **581**, 587*f*
Hepatopancreatic sphincter, **581**, 587*f*
Heredity, 667–678. *See also* Inheritance
Hering-Breuer reflex, 562
Hernia
    inguinal, **686**, 686*f*
    umbilical, **686**
Herniated disc, 131
Hertz (Hz), **PEx-47**
Heterozygous individual, **668**
High-density lipoprotein (HDL), 436
High-power lens (microscope), **28**, 31
Hilum
    lung, **540**, 541, 542*f*
    lymph node, 528*f*, **529**
    renal, 610*f*, **611**
    spleen, 529*f*
Hindbrain (rhombencephalon), **276**, 276*f*, 282
Hinge joint, 172*f*, 173
Hip bones (coxal bones/ossa coxae), **153**, 154*f*
Hip (pelvic) girdle. *See also* Hip joint; Pelvis
    bones of, 108*f*, **153**, 154*f*, 155*t*
    surface anatomy of, 157–158
Hip joint, 171*t*, 176, 176*f*
Hip muscles, 199*f*, 216
Hirsutism, **408**
His (atrioventricular/AV) bundle, **458**, 458*f*
Histamine, microcirculation and, 520
Histology, **68**. *See also* Tissue(s)
HIV
    ELISA in testing for, 531, PEx-182 to PEx-184, **PEx-184**
    Western blotting in testing for, PEx-184 to PEx-186
HIV antigens, PEx-184, PEx-185*t*
Homeostasis, **PEx-59**
Homeothermic animals, humans as, PEx-60, **PEx-97**
Homologous chromosomes (homologues), **646**, 646*f*, 668
Homozygous individual, **668**
Horizontal cells, retinal, 363*f*
Horizontal plate
    of ethmoid bone, 127
    of palatine bone, 122*f*
Hormone(s), **406**, **PEx-59**. *See also specific hormone and* Endocrine system/glands
    functions of, 417–422
    metabolism and, **PEx-60** to PEx-64
    target organs of, **406**
    tropic, **406**, PEx-61, PEx-69
    urine formation and, PEx-142 to PEx-143
Hormone replacement therapy, PEx-67 to PEx-69
Horns (cornua)
    hyoid bone, **129**, 130*f*
    uterine, 21, 21*f*
Human(s)
    cardiovascular physiology in, 491–510
    embryonic developmental stages of, 658–661, 659*f*, 660*f*
    EMG in, 241–248. *See also* Electromyography
    muscle fatigue in, 241, **PEx-25** to PEx-26. *See also* Muscle fatigue
    oogenesis in, **649–651**, 649*f*
    reflex physiology in, 335–348
    reproductive systems in, 22, 22*f*.
        See also Reproductive system
        female, 22, 22*f*, 634–637, 634*f*, 635*f*, 636*f*, 637*f*

male, 22, 22*f*, 630–633, 630–631*f*, 632*f*, 633*f*, 634*f*
    torso of, 22, 23*f*
    urinary system in, 610*f*, 611–613, 611*f*, 612*f*
Human immunodeficiency virus (HIV)
    ELISA in testing for, 531, PEx-182 to PEx-184, **PEx-184**
    Western blotting in testing for, PEx-184 to PEx-186
Human immunodeficiency virus (HIV) antigens, PEx-184, PEx-185*t*
Humeral arteries, circumflex, **474**, 474*f*
Humerus, 108*f*, **148**–150, 150*f*, 152, 687, 687*f*, 688*f*
    muscles of acting on forearm, 212*f*, 212*t*
Humoral immunity, **529**. *See also* Antibodies
Hyaline cartilage, 79*f*, 109*f*, **110**, 169*f*, 171
Hydrocephalus, **285**
Hydrochloric acid, bone affected by, 111–112
Hydrocortisone (cortisol), 408, **PEx-69** to PEx-70, PEx-69*t*
Hydrogen ions, in acid-base balance, 564, 564*f*
Hydrolases, **596**, **PEx-120**
    acid, 43
Hydrolysis, **PEx-125**
    of fat, by lipase, 597*f*, 599–601, PEx-127 to PEx-128
    of proteins, by pepsin/trypsin, 597*f*, 598–599, **PEx-125** to PEx-126
    of starch, by salivary amylase, 596–598, 597*f*, PEx-121 to PEx-123
Hydrolysis test, positive/negative, **599**
Hydrostatic pressure, in filtration, 60
Hymen, **634**, 634*f*, 635*f*
Hyoid bone, 129, 130, 130*f*, **681**, 682*f*
Hypercholesterolemia, **PEx-170**
Hypercortisolism (Cushing's syndrome/ Cushing's disease), **PEx-69**, PEx-69*t*
Hyperextension (body movement), 173, 174*f*
Hyperinsulinism, 408, 418
Hyperkalemia, 518
Hyperopia (farsightedness), **372**, 373*f*
Hyperparathyroidism, 408
Hyperpolarization, **PEx-41**
Hyperreactors, in cold pressor test, **501**
Hyperthermia, **PEx-97**
Hyperthyroidism, 408, 419
Hypertonic solution, 58–59, 59*f*, **PEx-9**
Hyperventilation, acid-base balance and, 564, PEx-150 to PEx-151
Hypocholesterolemia, **PEx-170**
Hypochondriac regions, **8**, 9*f*
Hypocortisolism, PEx-69
Hypodermis (superficial fascia), **94**, 94*f*
Hypogastric ganglia, inferior, **324**
Hypogastric (pubic) region, **8**, 9*f*
Hypoglossal canal, 123*f*, **125**
Hypoglossal nerve (cranial nerve XII), 287*t*, 288, 288*f*, 289*f*, 313*f*
Hypoglycemia, **408**
Hyponychium, **97**, 97*f*
Hypoparathyroidism, 408
Hypophyseal fossa, 123*f*, 124*f*
Hypophyseal portal system, **407**, 407*f*
Hypophysectomy, PEx-61
Hypophysis (pituitary gland), **278**, 279*f*, 280*f*, 283*f*, **406**–407, 407*f*, 409*f*, PEx-60*f*
    hypothalamus relationship and, 406–407, 407*f*, PEx-61
    microscopic anatomy of, 410, 410*f*
    ovaries affected by hormones of, 418
    thyroxine production regulated by, 419, 419*f*, PEx-60
Hypopituitarism, PEx-69*t*
Hyporeactors, in cold pressor test, **501**
Hypotension, orthostatic, **326**

Hypothalamic-pituitary portal system, **PEx-61**
Hypothalamus, 276*f*, 280*f*, **282**, 291, 292*f*, 409*f*, **PEx-60**, PEx-60*f*
    pituitary gland relationship and, 406–407, 407*f*, PEx-61
    thyroid hormone/TSH production and, 419, 419*f*, **PEx-60**, PEx-61
Hypothenar eminence, **690**, 690*f*
Hypothermia, **PEx-97**
Hypothyroidism, 408, 419
Hypotonic solution, 59, 59*f*, **PEx-9**
Hypoventilation, acid-base balance and, 564, PEx-151
Hz. *See* Hertz
H zone, 186*f*, 187*f*

Iatrogenic, **PEx-69**
Iatrogenic Cushing's syndrome, **PEx-69**, PEx-69*t*
I bands, 186*f*, 187, 187*f*
IC. *See* Inspiratory capacity
ICM. *See* Inner cell mass
Identity (antigen), **PEx-180**
    partial, **PEx-180**
Igs. *See* Immunoglobulin(s)
IKI assay, PEx-121, PEx-122, PEx-123, PEx-124
Ileocecal valve, **581**, 583*f*
Ileocolic artery, 475, 476*f*
Ileum, 575*f*, **581**, 582, 582*f*, 583*f*
Iliac arteries
    common, 475*f*, **476**, 476*f*, 477*f*
    internal and external, **476**, 477*f*
Iliac crest, **153**, 154*f*, 157–158, **683**, 684*f*, 685*f*, 686, 691*f*
Iliac fossa, **153**, 154*f*
Iliac (inguinal) regions, **8**, 9*f*
Iliac spines
    anterior inferior, 153, 154*f*
    anterior superior, **153**, 154*f*, 157–158, 685*f*, **686**, 686*f*
    posterior inferior, 153, 154*f*
    posterior superior, **153**, 154*f*, 158, **691**, 691*f*
Iliacus muscle, 216*t*, 217*f*
Iliac veins
    common, **477**, 479*f*, 480*f*
    external and internal, **477**, 479*f*, 480*f*
Iliocostalis muscles, cervicis/lumborum/ thoracis, 210*f*, 210*t*
Iliofemoral ligament, **176**, 176*f*
Iliohypogastric nerve, 316*f*, 317*t*
Ilioinguinal nerve, 316*f*, 317*t*
Iliopsoas muscle, 198*f*, 216*t*, 217*f*
Iliotibial tract, 199*f*, 219*f*, 219*t*
Ilium, 108*f*, **153**, 154*f*
    in male versus female, 153
Image(s) (microscope), real and virtual, **29**, 30*f*
Image formation, 372–374, 373*f*, 374*f*
Immune response/immunity, 16*t*, **527**–529
Immune system, 15, 16*t*. *See also* Immune response; Lymphatic system
Immunoblotting (Western blotting), **PEx-184** to PEx-186
Immunocompetent cells, 528
Immunodeficiency, 529
Immunoglobulin(s), 529, **530**–532, 531*f*.
    See also Antibodies
Implantation, 659, 659*f*, 660*f*
Inborn (intrinsic) reflexes, 336
    reaction time of, 341–344, 343*f*
Incisive fossa, 122*f*, **127**
Incisors, **584**, 584*f*
Inclusions, cytoplasmic, **44**
Incomplete dominance, **669**
Incontinence, **616**
Incus (anvil), 384*f*, **385**
Index of physical fitness, 501
Indifferent electrode, for EEG, 300
Indirect enzyme-linked immunosorbent assay (ELISA), PEx-181 to PEx-184, **PEx-182**, **PEx-184**
Indirect pathway, 279
Inferior (term), **3**, 4*f*
Inferior angle of scapula, 148, 149*f*, **683**, 684*f*

Inferior articular process/facet, vertebral, 132–133, **132**, 132*f*, 133*f*
Inferior colliculi, **279**, 280*f*, 291*f*
Inferior gluteal artery, **476**
Inferior gluteal line, 154*f*
Inferior gluteal nerve, 317*f*, 318*t*
Inferior horns, 281*f*, 284*f*
Inferior hypogastric ganglia, **324**
Inferior mesenteric artery, **475**–476, 475*f*, 476*f*
Inferior mesenteric ganglia, **324**
Inferior mesenteric vein, **484**, 484*f*
Inferior nasal conchae/turbinates, 125*f*, **128**, **538**, 539*f*
Inferior nuchal line, 122*f*, 125*f*
Inferior oblique muscle of eye, 361*f*
Inferior orbital fissure, 125*f*, 128*f*
Inferior phrenic vein, 480*f*
Inferior rectus muscle of eye, 361*f*
Inferior vena cava, **20**, 21*f*, 444*f*, 445*f*, **446**, 448*f*, 450*f*, 480*f*, 481*f*
    veins draining into, 477–478, 479*f*, 480*f*, 481*f*
Inflammation, microcirculation and, 520
Infraglenoid tubercle, 149*f*
Infrahyoid muscles, **682**, 683*f*
Infraorbital foramen, 122*f*, 125*f*, **127**, 127*f*, 128*f*, 130
Infraorbital groove, 128*f*
Infraspinatus muscle, 178, 199*f*, 208*t*, 209*f*, 211, 212*f*
Infraspinous fossa, 148, 149*f*
Infrasternal angle, **684**, 685*f*
Infundibulum, **282**, 287, 288*f*, 289*f*, 291, **406**, 407*f*
    of uterine tube, 635*f*
Inguinal (term), **2**, 3*f*
Inguinal hernia, **686**, 686*f*
Inguinal ligament, 153, 207*f*, 208*f*, 685*f*, **686**, 686*f*, 691*f*, **693**
Inguinal lymph nodes, 526*f*, 527, 693
Inguinal (iliac) regions, **8**, 9*f*
Inguinal ring, superficial, **686**, 686*f*
Inhaler (asthma), PEx-111
Inheritance
    dominant-recessive, 668
    intermediate (incomplete dominance), **669**
    multiple-allele, **672**
    sex-linked, 669–670
Inhibition, reciprocal, 338
Initial segment, axon, 255*f*, 256, PEx-41
Inner cell mass (ICM), **659**, 659*f*
Inner ear. *See* Internal (inner) ear
Inotropic modifiers, **PEx-100**
Insertion (muscle), **173**, 173*f*, 188
    naming muscles and, 197
Inspiration, **550**, 551*f*, **PEx-106**, **PEx-106** to PEx-107
Inspiratory capacity (IC), 552*f*, 555, 556*f*, 559*f*
Inspiratory muscles, 206*t*, 207*f*, 550
Inspiratory reserve volume (IRV), **552**, 552*f*, 554, 555, 559*f*, **PEx-107**
Insula, **277**
Insulin, 408, 418, **PEx-64** to PEx-67
    hypersecretion of (hyperinsulinism), 408, 418
    hyposecretion of, in diabetes mellitus, 408
Integration center, in reflex arc, 336*f*, 337
Integumentary system, 16*t*, 93–107, **94**.
    See also Skin
Interatrial septum, **446**
Intercalated discs, 83, 84*f*, **449**, 449*f*
Intercarpal joints, 170*t*
Intercondylar eminence, **156**, 157*f*
Intercondylar fossa, **156**, 156*f*
Intercostal arteries, 474, 474*f*
Intercostal muscles, 198*f*, 206*t*, 207*f*
    external, 206*t*, 207*f*, 550, **PEx-106**, **PEx-107**
    internal, 206*t*, 207*f*, **PEx-106**, **PEx-107**
Intercostal nerves, **312**, 313*f*
Intercostal spaces, 135*f*
Intercostal veins, 481*f*
Interlobar arteries, 612*f*, **613**

Interlobar veins, 612f, **613**
Intermaxillary suture, 122f
Intermediate filaments, 42f, 43t, **44**
Intermediate inheritance (incomplete dominance), **669**
Intermediate mass (interthalamic adhesion), 280f, 291, 291f, 292f
Intermediate part of urethra, 611f, 612, 630f, 631f, **632**
Internal/deep (term), **4**
Internal acoustic meatus, **120**, 123f
Internal anal sphincter, 583
Internal capsule, **280**
Internal carotid artery, **472**, 473f
Internal (inner) ear, 384f, **385**
　equilibrium and, 389–392, 390f, 391f
　hearing and, 386–389, 387f, 388f
Internal iliac artery, **476**, 477f
Internal iliac vein, **477**, 479f, 480f
Internal intercostal muscles, 206t, 207f, **PEx-106, PEx-107**
Internal jugular vein, **478**–479, 480f, 481f, 682, 683f
Internal oblique muscle, 198f, 206t, 207f, 208f
Internal os, 635f
Internal pudendal artery, 476
Internal respiration, **538**, PEx-106f
Internal thoracic artery, 473f, **474**, 474f
Internal urethral sphincter, 611f, **615**
Interneurons (association neurons), **258**, 259f, **PEx-35**, PEx-50
　in reflex arc, 336f, 337, 337f
Interoceptors, **350**
Interossei muscles, 215f
Interosseous artery, common, 474f
Interosseous membrane, 151f, 157f
Interphalangeal joints
　of fingers, 170t
　of toes, 171t
Interphase, **45**, 46f
Interspike interval, PEx-47
Interstitial endocrine (Leydig) cells, **633, 648**
Interstitial lamellae, 113f, 114
Interstitial spaces, nephron, **PEx-139**
Intertarsal joint, 171t
Interthalamic adhesion (intermediate mass), 280f, 291, 291f, 292f
Intertrochanteric crest, **156**, 156f
Intertrochanteric line, **156**, 156f
Intertubercular sulcus (bicipital groove), **150**, 150f
Interventricular arteries, 444f, 445f, **447**–448
Interventricular foramen/foramina, 280f, **283**, 284f, 291
Interventricular septum, 445f, **446**, 447f
Intervertebral discs, 80f, **109**, 109f, **130**–131, 130f
　herniated, 131
Intervertebral foramina, 130f, **132**
Intervertebral joints, 170t
Intervillous spaces (lacunae), of placenta, 660f, 662, 662f
Intestinal arteries, 475, 476f
Intestinal crypts, 581f, **582**, 582f
Intrafusal fibers, 351f, **352**
Intramural (terminal) ganglion, **324**
Intramuscular injections, 687, 691, 691f
Intrapleural pressure, **PEx-112** to PEx-113
Intrapleural space, **PEx-112**
Intrapulmonary volume, 550
Intrinsic conduction system of heart, **458**, 458f
　disturbances of, 519
　in frog, 513–519, 514f, 515f, 516f
Intrinsic eye muscles, 361
　reflex activity of, 375
Intrinsic (inborn) reflexes, 336
　reaction time of, 341–344, 343f
Intrinsic nerve plexuses, in alimentary canal, 574, 576f
In utero development, 661–662
Inversion (body movement), **174**, 175f

Involuntary nervous system. See Autonomic nervous system
Ion(s), **PEx-36**
　heart rate and, 518, PEx-99 to PEx-101, PEx-100t
Ion channels, **PEx-36**
Ipsilateral response, **340**
Iris, **361**, 362f, **364**
Iris diaphragm lever (microscope), **28**, 29f
Iron deficiency anemia, PEx-162
Irregular bones, **110**
Irregular connective tissue, 79f, 82
IRV. See Inspiratory reserve volume
Ischial ramus, 154f
Ischial spine, **153**, 154f
　in male versus female, 153
Ischial tuberosity, **153**, 154f, **690**, 691f
Ischiofemoral ligament, **176**
Ischium, 108f, **153**, 154f
Ishihara's color plates, 374
ISI. See Interspike interval
Islets of Langerhans (pancreatic islets), 409f, **410**
Isometric contraction, PEx-20, **PEx-26** to PEx-28, **PEx-28**
Isometric length-tension relationship, **PEx-27**
Isotonic concentric contraction, **PEx-28**
Isotonic contraction, **PEx-28** to PEx-29
Isotonic solution, **59**, 59f, **PEx-9**
Isthmus
　of fauces, 539f
　of thyroid gland, 407, 682

Jaundice, **97**, 587
Jejunum, 575f, **581**
JGC. See Juxtaglomerular complex
JG cells. See Juxtaglomerular (granular) cells
Joint(s), 108, 167–184, 168f, **169**, 170–171t. See also specific joint
　articular cartilages in, 109f, **110**
　cartilaginous, 168f, **169**, 170–171t
　disorders of., 180
　fibrous, 168f, **169**, 170–171t
　synovial, 168f, **169**–180, 169f, 170–171t, 172f, 173f, 174–175f
Joint cavities. See Synovial cavities
Jugular foramen, **120**, 122f, 123f
Jugular notch, **135**, 135f, 148, **682**, 682f, 685f
Jugular veins, 478–479, 480f, 481f, **682**, 683f
　external, **479**, 480f, 481f, **682**, 683f
　internal, **478**–479, 480f, 481f, **682**, 683f
Juxtaglomerular (granular) cells, 613, 615f
Juxtaglomerular complex, **613**
Juxtamedullary nephrons, **613**, 614f

Karyotype, **669**, 669f
Keratin, **95**
Keratinocytes, **95**, 95f
Keratohyaline granules, 96
Ketoacidosis, **PEx-154**
Ketone bodies, in urine (ketonuria), 622, **623**, 624
Kidney(s), **20**, 21f, 610f, **611**, **PEx-131**. See also under Renal
　in acid-base balance, PEx-150, PEx-152 to PEx-154
　functional microscopic anatomy of, 613–615, 614f, 615f
　gross anatomy of, 611, 612–613, 612f
　physiology of, **PEx-131** to PEx-147
Kidney stones (renal calculi), **622**
Kinetic energy, **54**, 54f, **PEx-3**, PEx-4
Kinetochore(s), 46f
Kinetochore microtubules, 46f
Kinocilium, in equilibrium, 391f
Knee
　bones of, 156, 156f
　muscles of/muscles crossing, 216
　surface anatomy of, 692f, 693
Kneecap. See Patella

Knee-jerk (patellar) reflex, 336f, 337, 337f, **338**, 338f
Knee joint, 171t, 176–178, 177f
Knuckles (metacarpophalangeal joints), 151, 152, 170t
Korotkoff sounds, **498**, 498f
Kyphosis, 131, 131f

Labia (lips), **575**
Labial frenulum, 576f
Labia majora, **634**, 634f, 635f
Labia minora, **634**, 634f, 635f
Labyrinth, 384f, **385**
　bony, **385**
　membranous, **385**
Lacrimal apparatus, 359–360, 360f
Lacrimal bone, 121f, 125f, **128**, 128f
Lacrimal canaliculi, **360**, 360f
Lacrimal caruncle, **360**, 360f
Lacrimal fossa, 121f, **128**, **681**, 681f
Lacrimal glands, 359–360, 360f
Lacrimal puncta, **360**, 360f
Lacrimal sac, **360**, 360f
Lacteals, 581f
Lactiferous ducts, **637**, 637f
Lactiferous sinus, **637**, 637f
Lacunae
　in bone, 81f, 113f, **114**
　in connective tissue matrix, 75, 79f, 80f
Lacunae (intervillous spaces), of placenta, 660f, 662, 662f
Lambdoid suture, **120**, 121f, 125f
Lamella(e), 81f, 113f, **114**
　circumferential, 113f, **114**
　concentric, **114**
　interstitial, 113f
Lamellar (Pacinian) corpuscles (pressure receptors), 94f, 97, 350f, **351**, PEx-39, PEx-40
Lamellar granules, 96
Lamina(e), vertebral, 131f, 132f
Lamina propria, alimentary canal, 574, 576f
　gastric, 579f
　large intestine, 582f
　small intestine, 582f
Laminar flow, **PEx-76**
Langerhans, islets of (pancreatic islets), 409f, **410**
Langerhans' cells (dendritic cells), 95f, **96**
Large intestine, **18**, 20, 20f, 575f, 582f, **583–584**, 583f
Laryngeal cartilages, 109, 109f, 538, 539f, 540f
Laryngeal prominence (Adam's apple), 538, 540f, **681**, 682f
Laryngopharynx, 538, 539f, **577**, 577f
Larynx, 538, 539f, 540f
Latent period
　in isotonic concentric contraction, **PEx-28**
　of muscle twitch, 238, 238f, **PEx-18** to PEx-20
Lateral (term), **3**
Lateral angle of scapula, 148, 149f
Lateral apertures, 282, 284f
Lateral (axillary) border of scapula, 148, 149f
Lateral commissure (canthus), **360**, 360f
Lateral condyle
　of femur, **156**, 156f, 158, 692f, **693**
　of tibia, 156, 157f, 158, 692f, **693**
Lateral (acromial) end of clavicle, 148, 149f
Lateral epicondyle
　of femur, **156**, 156f
　of humerus, **150**, 150f, 152, **688**, 688f
Lateral femoral cutaneous nerve, 316f, 317f
Lateral funiculus, **310**, 310f, 312f
Lateral geniculate body, **364**, 366f
Lateral horn, **308**, 310f
Lateral ligament (temporomandibular joint), **179**, 180f
Lateral longitudinal arch of foot, 157, 158f

Lateral malleolus, **157**, 157f, 158, 692f, **693**
Lateral masses, 124f, **127**
Lateral muscle compartment, lower limb muscles and, 220t, 221f, 223f
Lateral patellar retinaculum, 178
Lateral plantar artery, **476**, 477f
Lateral plantar vein, **477**
Lateral pterygoid muscle, 202f, 202t
Lateral rectus muscle of eye, 361f
Lateral sacral crest, 134f
Lateral sulcus, **277**, 277f
Lateral supracondylar line, 156f
Lateral supracondylar ridge, 150f
Lateral thoracic artery, **474**, 474f
Lateral ventricle, 276f, 279, 280f, 281, 281f, 284f, 292f
Lateral wall of orbit, 128f
Latissimus dorsi muscle, 197, 199f, 208t, 209f, 212f, **683**, 684f
LDL. See Low-density lipoprotein
Leak channels, PEx-36
Learned (acquired) reflexes, 336
　reaction time of, 341–344, 343f
Left, anatomical, **8**
Left anterior descending (anterior interventricular) artery, 444f, **447**–448, 448f
Left ascending lumbar vein, **479**, 480f
Left atrium, 444f, 445f, 448f
　in frog, 513f
Left brachiocephalic vein, **478**
Left colic (splenic) flexure, **583**, 583f
Left common carotid artery, **472**
Left coronary artery, 444f, **447**, 448f, **472**
Left gastric artery, **474**, 475f
Left gastric vein, **484**
Left gastroepiploic artery, **474**–475, 475f
Left gonadal vein, **477**, 480f
Left pulmonary artery, **479**, 482f
Left subclavian artery, **472**, 474f
Left suprarenal vein, **477**, 480f
Left upper and lower abdominopelvic quadrants, 8, 8f
Left ventricle, 444f, 445f, 448f, 450, 450f, **452f**
　right ventricle compared with, 451–452, 452f
Leg. See also Lower limb
　bones of, 156–157, 157f
　muscles of/muscles acting on, 198f, 199f, 216, 216–218t, 217f, 218–219t, 219f
　surface anatomy of, 692f, 693–694, 693f
Length-tension relationship
　cardiac muscle, **PEx-84**
　skeletal muscle, PEx-26 to PEx-28
　　isometric, **PEx-27**
Lens of eye, 362f, **363**, 364, 365f
　accommodation and, 372, 373f
　astigmatism and, 372
　opacification of (cataracts), **363**
Lenses (microscope), **28**, 29f
Lens selection disc (ophthalmoscope), 376, 376f
Lesser curvature of stomach, **577**, 578f
Lesser occipital nerve, 313f, 315t
Lesser omentum, **577**, 580f
Lesser sciatic notch, **153**, 154f
Lesser trochanter, **156**, 156f
Lesser tubercle, 148, 150f
Lesser wings of sphenoid, 123f, 124f, **126**, 128f
Leukemia, **429**
Leukocyte(s) (white blood cells/WBCs), 81f, **424**–425, 425f, 426t, 427f, 428–429, 428f
　in urine (pyuria), 623, 624
Leukocyte count
　differential, **429**–430, 430f
　total, **429**
Leukocytosis, **429**
Leukopenia, **429**
Levator labii superioris muscle, 200t, 201f
Levator scapulae muscle, 209f, 209t
Leydig (interstitial endocrine) cells, **633, 648**

LH. *See* Luteinizing hormone
Lidocaine, PEx-42, PEx-43
Life cycle, cell, 45–48, 46–47*f*
Ligament of head of femur (ligamentum teres), **176**, 176*f*
Ligamentum arteriosum, 450*f*, **451**, 483*f*, **484**
Ligamentum nuchae, 210*f*
Ligamentum teres (ligament of head of femur), **176**, 176*f*
Ligamentum teres (round ligament), 483*f*, **484**, 586*f*
Ligamentum venosum, 483*f*, **484**
Light chains, immunoglobulin, **530**, 531*f*
Light receptors, PEx-40
Light reflex, pupillary, **340**
Light refraction in eye, 372–374, 373*f*, 374*f*
Limbic system, 280
Limb leads, for ECG, 460, 460*f*, 462, 462*f*
Line (bone marking), 112*t*
Linea alba, 207*f*, 685*f*, **686**
Linea aspera, **156**, 156*f*
Linea semilunaris, 685*f*, **686**
Lingual artery, 473*f*
Lingual frenulum, **575–576**, 576*f*
Lingual lipase, 597*f*, **PEx-127**
Lingual tonsil, 538, 539*f*, **576**, 577*f*
Lip(s) (labia), **575**, 576*f*
Lipase, **PEx-127**
 in fat digestion, 597*f*, 599–601, **PEx-127** to PEx-128
 gastric, 597*f*, **PEx-127**
 lingual, 597*f*, **PEx-127**
 pancreatic, 597*f*, **599–601**, **PEx-127**
Lipids, PEx-127
 digestion of, 597*f*, 599–601, PEx-127 to PEx-128
 in plasma membrane, 40, 41*f*
Lipoproteins, 436, **PEx-162**, **PEx-169**
Litmus blue, 599, 600
Liver, 19*f*, **20**, 20*f*, 575*f*, **586**, 586*f*, 587, 588*f*, PEx-120*f*. *See also under Hepatic*
 round ligament of (ligamentum teres), 483*f*, **484**, 586*f*
Liver cells (hepatocytes), **587**, 588*f*
Load, skeletal muscle affected by, 240–241, **PEx-28** to PEx-29
Load-velocity relationship, PEx-28 to PEx-29
Lobar arteries, of lung, 479, 482*f*
Lobe(s)
 of liver, 586, 586*f*
 of lung, 540*f*, 542*f*, **543**
 of mammary gland, **637**, 637*f*
 of pituitary gland, 406
 of thyroid gland, 682
Lobule(s)
 of ear (earlobe), 384, 384*f*, 680*f*, **681**
 of liver, **587**, 588*f*
 of mammary gland, **637**, 637*f*
Local potentials, PEx-39
Long bones, **110**
 gross anatomy of, 110–111, 111*f*
Longissimus muscles, capitis/cervicis/thoracis, 210*f*, 210*t*
Longitudinal arches of foot, 157, 158*f*
Longitudinal fissure, **277**, 277*f*
Long thoracic nerve, 314*f*, 315*t*
Longus (term), muscle name and, 197
Loop of Henle, **613**, 614*f*, **PEx-131**, PEx-132*f*
Loops (fingerprint), 101, 101*f*
Loose connective tissue, 76*f*, 77*f*
Lordosis, 131, 131*f*
Low-density lipoprotein (LDL), 436, **PEx-169** to PEx-170
Lower limb. *See also* Foot; Leg; Thigh
 blood vessels of
  arteries, 476, 477*f*
  veins, 477, 479*f*
 bones of, 108*f*, 156–157, 156*f*, 157*f*, 158*f*
 muscles of, 198*f*, 199*f*, 216–224, 216–218*t*, 217*f*, 218–219*t*, 219*f*, 220–224*t*, 221*f*, 223–224*f*

nerves of, **316**–318, 316*f*, 317*f*, 317*t*, 318*t*
 surface anatomy of, 3*f*, 157–158, 690–694, 691*f*, 692*f*, 693*f*
Lower respiratory system, 540–544, 540*f*, 541*f*, 542*f*, 543*f*, 544*f*
Low-power lens (microscope), **28**, 30
Lumbar (term), **2**, 3*f*
Lumbar arteries, 475*f*, **476**
Lumbar curvature, 130*f*, 131
Lumbar enlargement, 308, 309*f*, 313*f*
Lumbar plexus/spinal nerves, 309*f*, 313*f*, **316**–317, 316*f*, 317*t*
Lumbar puncture/tap, 133, 308, 683
Lumbar regions, **8**, 9*f*
Lumbar veins, **477**
 ascending, **479**, 480*f*, 481*f*
Lumbar vertebrae, 130*f*, 132–133, 133*f*, 134*f*
Lumbosacral plexus, **316**–318, 316*f*, 317*f*, 317*t*, 318*t*
Lumbosacral trunk, 316*f*, 317*f*
Lumbrical muscles, 213*f*
Lunate, 151, 152*f*
Lung(s), **18**, 19*f*, 540*f*, 541–543, 541*f*, 542*f*, 544, 544*f*, 684, 685*f*. *See also under Pulmonary and Respiratory*
 inflation of, 543
 mechanics of respiration and, 550–551, 551*f*, PEx-105 to PEx-117, PEx-106*f*
Lung capacity, total, 552*f*, 559*f*, **PEx-107**
Lung tissue, 544, 544*f*
Lunule, **97**, 97*f*
Luteinizing hormone (LH), **406**, 407*f*
 oogenesis/ovarian/menstrual cycles affected by, 418, 650, 652*f*
 in spermatogenesis, 648
Lymphatic capillaries, 526*f*, **527**
Lymphatic duct, 526*f*
 right, 526*f*, **527**
Lymphatic system, 16*t*, **526**–527, 526*f*, 527*f*
Lymphatic trunks, 526*f*, **527**
Lymphatic vessels (lymphatics), 526–527, 526*f*, 528*f*, 529
Lymph nodes, 526*f*, **527**, 528*f*, 529
 axillary, 526*f*, **527**, 687
 cervical, 526*f*, **527**, 682
 inguinal, 526*f*, **527**, 693
Lymphocyte(s), 74*f*, 77*f*, 81*f*, 425*f*, 426*t*, 427*f*, **428**–429, 428*f*. *See also* B cells; T cells
Lymphoid organs, 526, 527, 527*f*, **528**
Lymphoid tissues, **528**
 mucosa-associated, 576*f*
Lymph sinuses, 529
Lysosomes, 42*f*, **43**, 43*t*
Lysozyme, **360**

m. *See* Meter
mμ. *See* Nanometer
μm (μ). *See* Micrometer/micron
Macrophages, 74*f*, 75, **527**, 529
 in phagocytosis, 61
 stellate (hepatic), **587**, 588*f*
Macula(e), in vestibule, 385*f*, **390**, 391*f*
Macula densa, 613, 615*f*
Macula lutea, 362, 362*f*
Magnesium, muscle contraction and, 234, 235
Magnification, microscope, **29**, 30*f*
Main (primary) bronchi, **540**, 540*f*, 542*f*
Main pancreatic duct, 581, 587*f*
Major calyces, **612**, 612*f*
Major duodenal papilla, 581, 587*f*
Male
 pelvis in, female pelvis compared with, 153, 155*t*
 reproductive system in, 16*t*, 21, 21*f*, 22, 22*f*, 630–633, 630–631*f*, 632*f*, 633*f*, 634*f*
 urethra in, 22, 611*f*, 612, 630*f*, 631*f*, **632**, 633, 633*f*
 Y chromosome and, 669*f*, 670
Malleoli
 lateral, **157**, 157*f*, 158, 692*f*, **693**
 medial, **157**, 157*f*, 158, 692*f*, **693**

Malleus (hammer), 384*f*, **385**
Maltose/maltotriose, starch hydrolyzed to, 596–598, 597*f*, PEx-121 to PEx-123, PEx-123 to PEx-125
Mammary (term), **2**, 3*f*
Mammary glands, **637**, 637*f*
Mammillary bodies, 278, 279*f*, 280*f*, **282**, 287, 289*f*, 291
Mandible, 121*f*, 125*f*, 126*f*, **127**, 127*f*, 680*f*, **681**, 683*f*
 fetal, 137*f*
Mandibular angle, 121*f*, **127**, 127*f*, 130
Mandibular body, **127**, 127*f*
Mandibular foramen, **127**, 127*f*
Mandibular fossa, **120**, 122*f*, 127*f*, **179**, 180*f*
Mandibular notch, 121*f*, 127*f*
Mandibular ramus, 121*f*, **127**, 680*f*, **681**
Mandibular symphysis, 125*f*, **127**, 130
Manubrium, **135**, 135*f*, 683
Manus (term), **2**, 3*f*, 151. *See also* Hand
Marginal artery, right, 444*f*, **447**, 448*f*
Marrow (bone), **110**, 111*f*
Masseter muscle, 197, 198*f*, 201*f*, 202*f*, 202*t*, **681**
Mastication, 576
 muscles of, 202*f*, 202*t*
Mastoid fontanelle, 137, 137*f*
Mastoiditis, **120**
Mastoid process, **120**, 121*f*, 122*f*, 125*f*, 130, 680*f*, **681**
Matrix
 extracellular, 74*f*, **75**, 79*f*, 80*f*, 82
 hair, 98*f*
 nail, **97**, 97*f*
Maxilla(e), 121*f*, 122*f*, 125*f*, 126*f*, **127**–128, 127*f*, 128*f*
 fetal, 137*f*
Maxillary artery, **472**, 473*f*
Maxillary sinuses, 129, 129*f*, 538
Maximal shortening velocity, PEx-28
Maximal stimulus, muscle contraction and, **239**, **PEx-20**
Maximal tetanic tension, **PEx-24**, **PEx-25**
Maximal voltage, muscle contraction and, 239, **PEx-20**
Maximum force, 244, **245**
 tetanus and, 240, 244
Maximus (term), muscle name and, 197
McBurney's point, 685*f*, **686**
Mean arterial pressure (MAP), **499**
Meatus (bone marking), 112*t*
Mechanical stage (microscope), **28**, 29*f*
Mechanical stimulation, of blood vessels of skin, color change and, 504
Medial (term), **3**
Medial bicipital furrow, **687**, 687*f*
Medial (vertebral) border of scapula, 148, 149*f*, **683**, 684*f*
Medial commissure (canthus), **360**, 360*f*
Medial condyle
 of femur, **156**, 156*f*, 158, 692*f*, **693**
 of tibia, **156**, 157*f*, 158, 692*f*, **693**
Medial cutaneous nerves of arm/forearm, 314*f*
Medial (sternal) end of clavicle, 148, 149*f*
Medial epicondyle
 of femur, **156**, 156*f*
 of humerus, **150**, 150*f*, 152, 687*f*, **688**
Medial longitudinal arch of foot, 157, 158*f*
Medial malleolus, **157**, 157*f*, 158, 692*f*, **693**
Medial muscle compartment, lower limb muscles and, 216*t*, 217*f*
Medial patellar retinaculum, **178**
Medial plantar artery, 476, 477*f*
Medial plantar vein, **477**
Medial pterygoid muscle, 202*f*, 202*t*
Medial rectus muscle of eye, 361*f*
Medial supracondylar line, 156*f*
Medial supracondylar ridge, 150*f*
Medial surface of tibia, 692*f*, **693**
Medial umbilical ligaments, 483*f*, **484**

Medial wall of orbit, 128*f*
Median antebrachial vein, **479**, 481*f*, 689*f*
Median aperture, 282, 284*f*
Median cubital vein, **479**, 481*f*, **688**, 689*f*
Median femoral cutaneous nerve, 317
Median fissure, ventral, **308**, 310*f*, 312*f*
Median nerve, 314*f*, 315*t*, **316**, 689, 689*f*
Median palatine suture, 122*f*
Median (midsagittal) plane/section, **4**, 5*f*, 6*f*
Median sacral artery, 475*f*
Median sacral crest, **133**–134, 134*f*
Median sulcus, dorsal, **308**, 309*f*, 310*f*, 312*f*
Mediastinal arteries, 474
Mediastinum, 7*f*, 541, 542*f*
Medulla
 adrenal, **408**
 of hair, 97, 98*f*
 lymph node, 528*f*, **529**
 renal, **612**, 612*f*
Medulla oblongata, 276*f*, 277*f*, **278**, 279*f*, 280*f*, 282, 290*f*, 291, 291*f*
 respiratory centers in, 562
Medullary cavity, of bone, 110, 111*f*
Medullary (renal) pyramids, **612**, 612*f*
Megakaryocytes, **429**
Meiosis, **45**, 646, **646**–647, 646*f*
 in oogenesis, 648, 649*f*, 650
 in spermatogenesis, 647, 647*f*
Meiosis I, 646, 646*f*, 647, 647, 648, 649*f*, 650
Meiosis II, 646, 646*f*, 647, 647, 648, 649*f*, 650
Melanin, **95**, 95*f*, 97
Melanocytes, **95**, 95*f*, 98*f*
Melatonin, **407**
Membrane(s)
 cell. *See* Plasma membrane
 diffusion through, 54, 54–**55**, 55–60, 55*f*, 59*f*, **PEx-3**, **PEx-4**
  facilitated diffusion, **55**, PEx-3, **PEx-6** to PEx-8, **PEx-140**
  living membranes and, 58–60, 59*f*
  nonliving membranes and, 56–58
  simple diffusion, **55**, **PEx-3**, **PEx-4** to PEx-6
 filtration and, 60. *See also* Filtration
 synovial, 169*f*, 171
Membrane pores, in filtration, 60, PEx-11
Membrane potential, **PEx-36**
 resting, 41, **234**, **266**, **PEx-36**
  cardiac cell, PEx-99
   ion movement and, PEx-100*t*
  muscle cell, **234**
  neuron, **266**, 266*f*, **PEx-36** to PEx-39
Membrane proteins, 40, 41*f*
 neural, PEx-36, PEx-39
Membranous labyrinth, **385**
Membranous semicircular ducts, 385*f*, **389**
Membranous urethra (intermediate part of urethra), 611*f*, 612, 630*f*, 631*f*, **632**
Memory (cognitive), cortical areas in, 278*f*
Memory (immunological), **528**
Memory B cells, **529**
Mendel, Gregor, 668
Meningeal layer of dura mater, 282, 283*f*, 284*f*
Meninges
 brain, **282**–283, 283*f*
 spinal cord, 308, 309*f*, 310*f*
Meningitis, **283**
 mastoiditis and, **120**
Menisci of knee, 109*f*, **176**, 177*f*
Menses. *See* Menstruation
Menstrual (uterine) cycle, **651**, 651*f*, 652*f*
Menstruation (menses/menstrual phase), **636**, 651, 651*f*, 652*f*
Mental (term), **2**, 3*f*
Mental foramen, 121*f*, 125*f*, **127**, 127*f*
Mentalis muscle, 200*t*, 201*f*
Mercury, millimeters of, osmotic pressure measured in, PEx-9
Merkel (tactile) cells, 95*f*, **96**
Merkel (tactile) discs, 96, **351**
Merocrine sweat glands (eccrine glands), 94*f*, **99**, 100, 100*f*

Mesencephalon (midbrain), **276**, 276*f*, **278**, 279*f*, 280*f*, 290, 291, 291*f*
Mesenchyme, 75, 76*f*
Mesenteric artery
  inferior, **475–476**, 475*f*, 476*f*
  superior, **475**, 475*f*, 476*f*
Mesenteric ganglia, inferior and superior, **324**
Mesenteric vein
  inferior, **484**, 484*f*
  superior, **484**, 484*f*
Mesentery, **20**, 580*f*, **581**
Mesocolon, **577**, 580*f*, 583*f*
Mesoderm, **658**, 659–660, 660*f*
Mesometrium, 635*f*, **636**
Mesosalpinx, 635*f*, **636**
Mesovarium, 635*f*, **636**
Metabolic acidosis/metabolic alkalosis, **PEx-154**
  renal and respiratory response to, PEx-154 to PEx-156
Metabolic rate/metabolism, **419**, **PEx-60**
  acidosis/alkalosis and, PEx-154
  computing, PEx-61 to PEx-62
  thyroid hormone and, **419**, **PEx-60** to PEx-64
Metabolites, vasodilation/flushing and, 502–503
Metacarpals, 108*f*, **151**, 152*f*
Metacarpophalangeal joints (knuckles), 151, 152, 170*t*
Metaphase
  of meiosis, 646
  of mitosis, **45**, 47*f*
Metaphase plate, 47*f*, 646
Metarteriole–thoroughfare channels, 519, 519*f*
Metatarsal(s), 108*f*, **157**, 158*f*
Metatarsal arteries, dorsal, **476**, 477*f*
Metatarsal veins, dorsal, 479*f*
Metatarsophalangeal joints, 171*t*
Metencephalon, 276*f*
Meter, 32*t*
Metric system, 32*t*
Microcirculation, **519–520**, 519*f*
Microfilaments, 42*f*, 43*t*, **44**
Microglial cells, 254, 254*f*
Micrometer/micron, 32*t*
Microscope(s), 27–38
  care/structure of, 28, 29*f*
  depth of field and, **33**
  identifying parts of, 28, 29*f*
  magnification/resolution and, **29**, 30*f*
  viewing cells under, 33–34, 33*f*, 34*f*
  viewing objects through, 29–31, 31*f*
Microscope field, **30**, 32–33, 32*t*
Microscopic anatomy. *See* Histology
Microtubules, 42*f*, 43*t*, **44**
  kinetochore, 46*f*
  polar, 46*f*
Microvilli, **41**, 42*f*, 71*f*, **581**, 581*f*
Micturition (voiding), **615**
Micturition reflex, 615, 616
Midaxillary line, **684**, 685*f*
Midbrain (mesencephalon), **276**, 276*f*, **278**, 279*f*, 280*f*, 290, 291, 291*f*
Midclavicular line, **684**, 685*f*
Middle cardiac vein, 445*f*, **448**, 448*f*
Middle cerebellar peduncles, 290
Middle cerebral artery, **472**, 473*f*
Middle cranial fossa, **120**, 123*f*
Middle ear, **384–385**, 384*f*
Middle ear cavities, 8, 9*f*
Middle nasal conchae/turbinates, 124*f*, 125*f*, **127**, **538**, 539*f*
Middle phalanx
  finger, 152, 152*f*
  toe, 158*f*
Middle scalene muscle, 203*t*, 204*f*
Middle suprarenal arteries, **475**, 475*f*
Midpiece of sperm, 648, 648*f*
Midsagittal (median) plane/section, 4, 5*f*, 6*f*
Milk (deciduous) teeth, **584**, 584*f*
Millimeter, 32*t*
Millimicrometer/millimicron (nanometer), 32*t*
Millivolt(s), **PEx-36**

Mineralocorticoids, **408**
Minimus (term), muscle name and, 197
Minor calyces, **612–613**, 612*f*, PEx-132
Minute respiratory volume (MRV), **554**, **555**
Minute ventilation, **PEx-106**
Mirror (microscope), **28**
Mitochondria, 40*f*, 42*f*, 43*t*, **44**
  muscle fiber, 186*f*, 187*f*
Mitosis, **45**, 46–47*f*, 48, **646**
Mitotic spindle, 46*f*, 47*f*
Mitral (bicuspid) valve, 445*f*, **446**, 447*f*
  auscultation of, 494, 494*f*
Mixed nerves, **258**, **308**
M line, 186*f*, 187*f*
mm. *See* Millimeter
Moderator (septomarginal) band, **451**, 452*f*
Molars, **584**, 584*f*
Molecular weight, diffusion rates and, 55–56, 55*f*
Molecular weight cutoff (MWCO), PEx-4, PEx-11
Monocyte(s), 425*f*, 426*t*, 427*f*, 428*f*, **429**
Monomer(s), antibody, **530**, 531*f*
Monosynaptic reflex arc, 336*f*, 337
Mons pubis, **634**, 634*f*, 635*f*
Morula, **658**, 659*f*
Mother cell, 646
Motor areas of brain, 278, 278*f*
Motor cortex, primary, **278**, 278*f*
Motor end plate, **PEx-18**
Motor (efferent) nerves, **258**, **276**
Motor (efferent) neurons, 241, 255*f*, **258**, 259*f*, **PEx-18**, **PEx-35**
  in reflex arc, 336*f*, 337
Motor (descending) tracts, 310, 311*f*
Motor unit, **189**, 190*f*, **241**, **PEx-18**, **PEx-20**
  multiple, summation of (recruitment), 241–244, **242**, 242*f*, 243*f*, **PEx-20**
  temporal (wave) summation and, **239**–240, 239*f*, 241–244, **242**, 242*f*, 243*f*, **PEx-22**, **PEx-94**
Motor unit recruitment, **PEx-20**
Mouth (oral cavity), **8**, 9*f*, 17, **575–577**, 575*f*, 576*f*, 577*f*
Movement(s), at synovial joints (body movements), 173–174, 173*f*, 174–175*f*
MRV. *See* Minute respiratory volume
Mucin, in saliva, 585
Mucosa(e). *See* Mucous membrane(s)
Mucosa-associated lymphoid tissue, 576*f*, 577
Mucosal glands, gastric, 577
Mucous membrane(s)/mucosa(e)
  alimentary canal, **574**, 576*f*
    duodenal, **582**, 582*f*
    gastric, 579*f*
    tracheal, 543*f*
Multiaxial joints/movement, 168*f*, 172*f*, 173
Multifidus muscle, 210*f*
Multipennate fascicles/muscles, 196*f*
Multiple-allele inheritance, **672**
Multiple motor unit summation (recruitment), 241–244, **242**, 242*f*, 243*f*, **PEx-20**
Multipolar neurons, 257, 258*f*
Murmurs (heart), **492**
Muscle(s). *See* Muscle tissue; Muscular system
Muscle attachments, 173, 173*f*, 188
Muscle cells. *See* Muscle fibers
Muscle contraction, 234–248, **PEx-17** to PEx-34
  action potential and, **234**
  body movement and, 173, 173*f*
  electrical stimulation and, 234, **PEx-17**, PEx-18
  EMG in study of, **241**–248
    force measurement/fatigue and, 244–248, 244*f*, 245*f*, 246*f*, 247*f*
    temporal/multiple motor unit summation and, 241–244, **242**, 242*f*, 243*f*

fatigue and, **240**, 241, **PEx-25** to PEx-26. *See also* Muscle fatigue
  graded response and, 238, **241**, **244**
    stimulus frequency increases and, 239–240, 239*f*
    stimulus intensity increases and, 238–239
  isometric, PEx-20, **PEx-26** to PEx-28, PEx-28
  isotonic, **PEx-28** to PEx-29
  length-tension relationship in, PEx-26 to PEx-28
    isometric, **PEx-27**
  load affecting, 240–241, **PEx-28** to PEx-29
  muscle twitch and, **238**, 238*f*, 239, **PEx-18** to PEx-20, **PEx-20**
  recording activity and, 238, 238*f*
  stimulus frequency and, 239–240, 239*f*, **PEx-22** to PEx-23, PEx-24
  stimulus voltage and, 238–239, **PEx-20** to PEx-22
  tetanus and, 239*f*, 240, **PEx-24** to PEx-25, PEx-25, PEx-94
  threshold stimulus for, **238**, **PEx-20**
  treppe and, **PEx-22**
Muscle fatigue, **240**, 241, **PEx-25** to PEx-26
  inducing, 240
  measurement of, 244–248, 244*f*, 245*f*, 246*f*, 247*f*
Muscle fibers (muscle cells), 84*f*, 186*f*, **187–188**, 187*f*, 188*f*, 189*f*, **PEx-17**, **PEx-18**, PEx-18*f*
  action potential in, **234**
  contraction of, 234–248, PEx-17 to PEx-34. *See also* Muscle contraction
  load affecting, 240–241
  muscle name and, 196*f*, 197
  organization of into muscles, 188, 189*f*
Muscle spindles, 336*f*, 337, 337*f*, **351**, 351*f*
Muscle tension (muscle force), **PEx-17**, **PEx-20**
  muscle contraction and, **PEx-20**
Muscle tissue, **83–85**, 84–85*f*. *See also* Cardiac muscle; Skeletal muscle(s); Smooth (visceral) muscle
Muscle twitch, **238**, 238*f*, 239, **PEx-18** to PEx-20, **PEx-20**
Muscular arteries, 471
Muscularis externa, alimentary canal, **574**, 576*f*
  gastric, 577, 578*f*, 579, 579*f*
  small intestine, **582**, 582*f*
Muscularis mucosae, alimentary canal, 574, 576*f*
  gastric, 579*f*
  large intestine, 582*f*
  small intestine, 581*f*, 582*f*
Muscular system/muscles, 16*t*, **PEx-17**, PEx-18*f*
  gross anatomy of, 195–232, 196*f*, 198*f*, 199*f*. *See also* Skeletal muscle(s)
  microscopic anatomy/organization of, 185–193, 186*f*. *See also* Muscle fibers
  muscle fiber organization into, 188, 189*f*
  physiology of, 233–252, PEx-17 to PEx-34. *See also* Muscle contraction
Musculocutaneous nerve, 314*f*, 315*t*, **316**
mV. *See* Millivolt(s)
MWCO. *See* Molecular weight cutoff
Myelencephalon, 276*f*
Myelin, 256, 257*f*
Myelinated fibers, **256**, 256*f*, 257*f*, PEx-48
Myelination, 254, 254*f*, 256–257, 256*f*, **PEx-48**
Myelin sheath, 254*f*, **256**, 256*f*, 257*f*, 260*f*, PEx-48
Myelin sheath gaps (nodes of Ranvier), 255*f*, **256**, 257*f*, PEx-48
Myenteric plexus, 574, 576*f*
Mylohyoid muscle, 204*f*, 204*t*, 575
Myocardium, 445*f*, **446**

Myocytes, Purkinje, 458
Myofibrils, 186*f*, **187**, 187*f*
Myofilaments, **187**
  in muscle contraction, 234, 240
Myometrium, 635*f*, **636**, 636*f*
  oxytocin affecting, PEx-60
Myoneural (neuromuscular) junction, **188–190**, 189*f*, 190*f*, **PEx-18**
Myopia (nearsightedness), **372**, 373*f*
Myosin, **187**
Myosin (thick) filaments, 186*f*, **187**
  muscle contraction and, 234, 240
Myxedema, **408**, **419**

n (haploid chromosomal number), **646**, 668
2*n* (diploid chromosomal number), **646**, 668
Na⁺-K⁺ pump (sodium-potassium pump), PEx-13, PEx-36
Nail(s), 97, 97*f*
Nail bed, 97, 97*f*
Nail folds, **97**, 97*f*
Nail matrix, **97**, 97*f*
Nanometer/millimicrometer/millimicron, 32*t*
Nares (nostrils), **538**, 539*f*
Nasal (term), **2**, 3*f*
Nasal aperture, posterior, **538**, 539*f*
Nasal bones, 121*f*, 125*f*, 126*f*, **128**, 128*f*, 130
Nasal cartilages, **109**, 109*f*
Nasal cavity, **8**, 9*f*, **538**, 539*f*
Nasal conchae/turbinates
  inferior, 125*f*, **128**, **538**, 539*f*
  middle, 124*f*, 125*f*, **127**, **538**, 539*f*
  superior, **127**, **538**, 539*f*
Nasal septum, **538**
Nasal spine, anterior, 127*f*
Nasolacrimal duct, **360**, 360*f*
Nasopharynx, **538**, 539*f*, **577**
Natal (gluteal) cleft, **690**, 691*f*
Navel (umbilicus), 482, 483*f*, 686
Navicular, 158*f*
Near point of accommodation, **372**–373
Nearsightedness (myopia), **372**, 373*f*
Neck
  arteries of, 472, 473*f*
  bony landmarks of, 681–682, 682*f*
  muscles of, 197, 198*f*, 199*f*, 203–204*t*, 203–204*t*, 208–209*t*, 209–211*f*, 682, 682*f*, 683*f*
  nerves of, 309*f*, 313, 313*f*
  surface anatomy of, 681–682, 682*f*, 683*f*
  triangles of, 682, 683*f*
  veins of, 478–479, 480*f*
Neck of femur, 156*f*
Neck of fibula, 693
Neck of humerus, 148, 150*f*
Neck of radius, 150*f*, 151*f*
Neck of tooth, **585**, 585*f*
Negative afterimages, temperature adaptation and, **354**
Negative controls, PEx-121, PEx-125, **PEx-178**
Negative feedback mechanisms, **PEx-60**
  thyroid hormone secretion and, PEx-61
Negative hydrolysis test, **599**
Negative starch test, **598**, PEx-122, PEx-124
Negative sugar test, **598**, PEx-122, PEx-124
Nephron(s), **613–615**, 614*f*, 615*f*, **PEx-131**, **PEx-132**, **PEx-137**, **PEx-153**
  function of, 613, **PEx-131** to PEx-148
Nephron loop (loop of Henle), **613**, 614*f*, **PEx-131**, PEx-132*f*
Nerve(s), **256**, 260*f*, **PEx-41**
  cranial, **285**, 286–287*t*, 288–290, 288*f*, 289*f*
  mixed, **258**, **308**
  physiology of, 267–271, 267*f*, 270*f*, 271*f*, PEx-35 to PEx-55. *See also* Nerve impulse
  spinal, **308**, 309*f*
  structure of, 257*f*, 258–260, 260*f*

Nerve cell(s). *See* Neuron(s)
Nerve chamber, PEx-41
Nerve conduction velocity, PEx-47 to
  PEx-49, **PEx-48**
Nerve endings, in skin, 94f, 95f, 96,
  96–97, 350, 350f, 351, 352
Nerve fiber. *See* Axon(s)
Nerve impulse, 254, 255, 266f. *See also*
  Action potential
  neurophysiology of, 265–274,
    PEx-35 to PEx-55
Nerve plexuses, **312–313**. *See also*
  specific type
Nervous system, 16t, 276. *See also*
  specific division or structure and
  Nerve(s); Neurons(s)
  autonomic, 276, **323**–333, 324f, PEx-95
  brain, 276–285
  central, 254, **276**
  cranial nerves, 285, 286–287t,
    288–290, 288f, 289–290f
  electroencephalography and, 299–306,
    300, 300f, 301f, 302f, 303f, 304f
  nervous tissue histology and, 82–83,
    82f, 253–264, 254f, 255f
  peripheral, 254, **276**
  physiology/neurophysiology of,
    265–274, PEx-35 to PEx-55. *See also*
    Nerve impulse
  reflex physiology and, 335–348
  sensation and. *See also under* Sensory
    general sensation, 349–358, **350**
    special senses and, **350**
      hearing/equilibrium, 383–396
      olfaction/taste, 397–404
      vision
        visual system anatomy and,
          359–370
        visual tests/experiments and,
          371–381
  somatic, **276, 323**
  spinal cord, 276f, **308**–312, 309f, 310f,
    311f, 312f
  spinal nerves/nerve plexuses, **308**, 309f
  supporting cells and, 254, 254f
Nervous tissue, **82**–83, 82f, 253–264,
  254f, 255f. *See also* Neuroglia;
  Neuron(s)
Neural layer of retina, **361**–362
Neural tube, **276**
Neurofibril(s), **254**–256, 255f
Neuroglia (glial cells), **82**, 82f,
  254, 254f
  myelination and, 254, 254f, 256–257,
    PEx-48
Neurohormones, 407. *See also specific*
  hormone
Neurohypophysis (posterior pituitary
  gland), **406**, 407f, 410, 410f
  hormones of, 407, 407f
Neurolemmocytes (Schwann cells), 254,
  254f, 255f, **256**, 256f, 257f
  in myelination, 254, 254f, **256**, 256f,
    PEx-48
Neuromuscular (myoneural) junction,
  188–190, 189f, 190f, 256, **PEx-18**
Neuron(s), **82**–83, 82f, **254**–258, 255f,
  259f, **PEx-35** to PEx-36, PEx-36f.
  See also Motor (efferent) neurons;
  Sensory (afferent) neurons
  classification of, 257–258, 258f, 259f
  physiology/neurophysiology of,
    265–274, **PEx-35** to PEx-55. *See also*
    Nerve impulse
Neuron processes, **82**–83, 82f
  classification and, 257, 258f
Neurophysiology, 265–274, PEx-35 to
  PEx-55. *See also* Nerve impulse
Neurotransmitters, 324, **PEx-35**,
  PEx-49 to PEx-50
  neuron excitation/inhibition and, 256,
    267
Neutrophil(s), 74f, 81f, 426t, 427f, **428**, 428f
Newborn, circulation in, 483f, 484
Nipple, **637**, 637f, **684**, 685f
Nissl bodies (chromatophilic substance),
  255f, **256**
Nitrites, in urine, 623, 624

Nitrocellulose membrane, **PEx-184**
Nitrogen balance, negative, in diabetes,
  PEx-64
Nitrogenous wastes, in urine, 622
nm. *See* Nanometer
Nodal system. *See* Intrinsic conduction
  system of heart
Nodes of Ranvier (myelin sheath gaps),
  255f, **256**, 257f, PEx-48
Nonaxial joints, 172f, 173
Nondominant arm, force measurement/
  fatigue and, **244**, 246, 248
Nonencapsulated (free) nerve endings,
  94f, 96, 97, 350f, **351**
Non-identity (antigen), **PEx-180**
Nonself antigens, **PEx-177**
Nonspecific binding, **PEx-178**
Norepinephrine, 324, **408**, **PEx-98**
Nose, 681, 681f. *See also under* Nasal
Nosepiece (microscope), **28**, 29f
Nostrils (nares), **538**, 539f
Notch (bone marking), 112t
Noxious stimulus, blood pressure/heart
  rate affected by, 501
Nuchal line
  inferior, 122f, 125f
  superior, 122f, 125f, 681
Nuclear envelope, **40**, 40f, 42f
Nuclear pores, **40**, 42f
Nuclei (nucleus) (cell), **40**, 40f, 42f
Nuclei (nucleus) (neural), **254**, **279**
Nucleic acid digestion, 597f
Nucleoli (nucleolus), **40**, 40f, 42f
Nucleus pulposus, 131
Nutrient arteries, 111f
Nystagmus, **391**
  in equilibrium testing, 391–392

Objective lenses (microscope), **28**, 29f
Oblique (term), muscle name and, 197
Oblique muscles of abdomen, 198f, 206t,
  207f, 208f, **686**
Oblique muscles of eye, 361f
Oblique popliteal ligament, 177f, **178**
Obstructive lung disease, 554, 557,
  PEx-108
Obturator artery, **476**, 477f
Obturator foramen, **153**, 154f
Obturator nerve, 316f, 317t
Occipital (term), **2**, 3f
Occipital artery, **472**, 473f
Occipital belly of epicranius muscle,
  199f, 200t, 201f
Occipital bone, **120**–125, 121f, 122f,
  123f, 125f
  in fetal skull, 137f
Occipital condyles, 122f, **125**, 125f
Occipital crest, external, 122f, **125**, 125f
Occipital (posterior) fontanelle, 137, 137f
Occipital lobe, **277**, 277f, 283f
Occipital nerve, lesser, 313f, 315t
Occipital protuberance, external, 122f,
  125, 125f, 130, 680f, **681**
Occipital vein, 480f
Occipitomastoid suture, 121f, 125f
Ocular/eyepiece (microscope), **28**, 29f
Oculomotor nerve (cranial nerve III),
  286t, 287, 288f, 289f
Odontoblasts, **585**
Odontoid process (dens), **132**, 132f
Odor (olfactory) receptors. *See* Olfaction/
  olfactory receptors
Odor identification, taste and olfaction
  in, 401. *See also* Olfaction/olfactory
  receptors
Ohm(s), galvanic skin resistance recorded
  in, 326
Oil(s), PEx-127
Oil (sebaceous) glands, 94f, 98f, 99, 100f
Oil immersion lens (microscope), **28**, 31
Olecranal (term), **2**, 3f
Olecranon, 150f, **151**, 151f, 152, 687f, **688**
Olecranon fossa, **150**, 150f
Olfaction/olfactory receptors, 258f,
  398–399, 398f, **PEx-40**
  adaptation and, 401–402, PEx-46
  in odor identification, 401
  taste and, 400–401

Olfactory bulbs, **278**, 279f, 285, 288f,
  289f, 290f, 398, 398f
Olfactory cilia, **398**, 398f
Olfactory epithelium, **398**, 398f
Olfactory nerve (cranial nerve I), 286t,
  288f, 398, 398f
Olfactory sensory neurons, 258f, **398**, 398f
Olfactory tracts, **278**, 279f, 288f, 289f,
  398f
Oligodendrocytes, 254, 254f, **256**
  in myelination, 254, 254f, **256**–257,
    PEx-48
Omenta, 577
  greater, **18**, 20f, **577**, 580f
  lesser, **577**, 580f
Omohyoid muscle, 203f, 204f, 204t
Oocyte(s), 648, **649**f, 659f
  primary, 648, **649**–650, 649f
  secondary, 648, 649f, **650**, 650f
Oogenesis, **649**
  in *Ascaris megalocephala*, 648–649
  human, 649–651, 649f
Oogonia, 649, 649f
Ophthalmic artery, **472**, 473f
Ophthalmic vein, 480f
Ophthalmoscopic examination of eye,
  375–377, 376f, 377f
Optical density, **PEx-125**, PEx-126
  developing standard glucose curve and,
    PEx-65 to PEx-66
  measuring fasting plasma glucose and,
    PEx-66
Optic canals, 123f, 124f, 125f, **126**, 128f
Optic chiasma, **278**, 279f, 280f, 287, 288f,
  289f, 291, 291f, **364**, 366f
Optic disc (blind spot), 362, 362f, 365f,
  371–372, 372f
Optic nerve (cranial nerve II), **278**, 279f,
  286t, 287, 288f, 289f, 362f, **364**,
  365f, 366f
Optic radiation, **364**
Optic tracts, **278**, 279f, 287, 288f, 289f,
  364, 366f
Oral (term), **2**, 3f
Oral cavity (mouth), **8**, 9f, 17,
  575–577, 575f, 576f, 577f
Oral cavity proper, **576**
Oral vestibule, **576**, 576f
Ora serrata, 362f
Orbicularis oculi muscle, 197, 198f, 200t,
  201f
Orbicularis oris muscle, 196f, 197, 198f,
  200t, 201f, 202f
Orbital (term), **2**, 3f
Orbital cavities (orbits), **8**, 9f
  bones forming, 120, 126, 127, 128, 128f
Orbital fissures
  inferior, 125f, 128f
  superior, 124f, 125f, **126**, 128f
Orbital plates, 124f, 127, 128f
Organ(s), **15**, **68**
  target, **406**
Organelles, **41**–44, 42f, 43t
Organ systems, **15**–26, 16t. *See also*
  specific system
Origin (muscle), **173**, 173f, 188
  naming muscles and, 197
Oropharynx, **538**, 539f, 576f, **577**, 577f
Orthostatic hypotension, **326**
Oscilloscope, **270**
  action potential studied with,
    270–271, 270f, 271f
  brain waves studied with, 300–301
  membrane potential studied with,
    PEx-36
Osmometers, 58
Osmosis, **55**, **PEx-3**, **PEx-4**, **PEx-8** to
  PEx-10
  living membranes and, 58–60, 59f
  nonliving membranes and, 56–58
  tonicity and, 58–60, 59f, **PEx-9**
Osmotic pressure, PEx-8 to **PEx-10**
Ossa coxae (coxal bones/hip bones),
  153, 154f
Osseous (bone) tissue, **75**, 81f. *See also*
  Bone(s)
Ossicles, auditory, 384f, **385**
Ossification, **114**, 114f

Ossification (growth) centers, in fetal
  skull, **137**, 137f
Osteoblasts, 110
Osteoclasts, 110
Osteocytes, 113f, **114**
Osteogenic epiphyseal (growth) plate,
  110, 111, 114
Osteon (Haversian system), 113f, **114**
Osteoporosis, **PEx-67**
Otic (term), **2**, 3f
Otitis media, **385**, **538**
Otolith(s), **390**
Otolithic membrane, **390**, 391f
Otoscope, ear examination with, 385–386
Ouchterlony double-gel diffusion, 531–
  532, 532f, 532t, PEx-180 to PEx-181
Outer collar of perinuclear cytoplasm,
  256, 256f
Outer (external) ear, **384**, 384f
  cartilage in, 80f, 109f
Ova (eggs), **650**
  development of (oogenesis),
    648–651, **649**, 649f
  fertilized (zygote), **646**, 649, **658**, 659f
Oval window, **385**, 385f, 386, 387f
Ovarian (gonadal) artery, **475**, 475f
Ovarian cycle, 418, **649**–651, 649f, 652f
Ovarian follicles, 636, 649, 649f, 650,
  650f, 651, 652f
Ovarian (gonadal) vein, **477**, 480f
Ovarian ligament, 635f, **636**
Ovariectomy, PEx-67
Ovary/ovaries, **21**, 21f, **22**, 22f, 408, 409f,
  634, 635f, PEx-120f
  oogenesis in, 649–651, 649f, 650f
  pituitary hormones affecting, 418
Ovulation, 418, **636**, 649f, 652f, 659f
Ovum. *See* Ova
Oxidases, peroxisome, 43
Oxygen, partial pressure of, PEx-106
Oxygen consumption, thyroid hormone
  effect on metabolism and, PEx-61,
  PEx-61 to PEx-62
Oxyhemoglobin, **PEx-166**
Oxyphil cells, 409f, **410**
Oxytocin, **407**, 407f
  positive feedback regulating release of,
    PEx-60

p24, PEx-185, PEx-185t
p31, PEx-185, PEx-185t
p55, PEx-185t
Pacemaker, cardiac, 458, 458f, PEx-93,
  PEx-96, PEx-98
  ectopic, **518**
  in frog, 513, 513f, 514
Pacinian (lamellar) corpuscles (pressure
  receptors), 94f, 97, 350f, **351**,
  PEx-39, PEx-40
Packed cell volume (PCV/hematocrit),
  430–431, 431f, **PEx-76**, **PEx-161**,
  PEx-162 to PEx-164
Pain, referred, 354
Pain receptors, 96, 97, 350f, 351, 352
Palate, **575**, 576f, 577f
  cleft, **538**
  hard, 122f, **538**, 539f, **575**, 577f
  soft, **538**, 539f, **575**, 577f
Palatine bone, 122f, **128**, 128f
Palatine processes, 122f, **127**
Palatine raphe, 576f
Palatine suture, median, 122f
Palatine tonsils, 530f, 538, 539f, **576**,
  576f, 577f
Palatoglossal arch, **576**, 576f, 577f
Palatopharyngeal arch, **576**, 576f
Palm, superficial transverse ligament
  of, 213f
Palmar (term), **2**, 3f
Palmar aponeurosis, 213f
Palmar arches, 474f
  venous, 481f
Palmaris longus muscle, 198f, 213f, 213t,
  689, 689f
Palpation, **680**
Palpebrae (eyelids), **360**, 360f
Palpebral conjunctiva, 360, 360f

Pancreas, **20**, 408, 409f, 575f, **587**, 587f, PEx-60f, PEx-120f
 microscopic anatomy of, 409f, 410
Pancreatic amylase, 597f
Pancreatic duct, **581**, 587f
Pancreatic enzymes, 597f
Pancreatic islets (islets of Langerhans), 409f, **410**
Pancreatic lipase, 597f, **599**–601, **PEx-127**
Pancreatin, **599**
Panoramic vision, 374
Papilla(e)
 dermal, 94f, **96**
 hair, 98, 98f
 of renal pyramid, 612, 612f
 of tongue, 399, 399f, 400f
Papillary layer of dermis, 94f, **96**
Papillary muscles, 445f, **446**, 447f, 451, 452f
Parafollicular (C) cells, **409**, 409f, PEx-67
Parallel fascicles/muscles, 196f, 197
Paralysis, spinal cord injury and, 311
Paranasal sinuses, 129, 129f, **538**
Paraplegia, **311**
Parasympathetic (craniosacral) division of autonomic nervous system, **324**, 324f, **PEx-95**
 function of, 326
 in heart regulation, PEx-95, PEx-98
Parathyroid cells, 409f, **410**
Parathyroid glands, 408, 409f, PEx-60f
 microscopic anatomy of, 409f, 410
Parathyroid hormone (PTH), **408**
Paravertebral ganglia (sympathetic trunk ganglion), 313f, **324**, 325f
Parfocal microscope, **30**
Parietal bone, **120**, 121f, 122f, 123f, 125f, 126f
 in fetal skull, 137f
Parietal cells, 578f, **579**
Parietal layer/pericardium, 7f, **446**, 449
Parietal lobe, **277**, 277f
Parietal peritoneum, 7f, 580f
Parietal pleura, 7f, 542f, **543**
Parietal serosa, 6–7, 7f
Parieto-occipital sulcus, **277**, 277f
Parotid glands, 575f, **585**
Partial heart block, **519**
Partial identity (antigen), **PEx-180**
Partial pressure of gas, **PEx-106**
 carbon dioxide, PEx-106, PEx-150, PEx-153
 oxygen, PEx-106
Passive force, in muscle contraction, **PEx-26**
Passive processes/transport, 41, 54–60, **54**, **PEx-3**
 diffusion, 54, 54–**55**, 55–60, 55f, 59f, **PEx-3**, **PEx-4**
  facilitated diffusion, 55, **PEx-3**, **PEx-6** to PEx-8
  simple diffusion, 55, **PEx-3**, **PEx-4** to PEx-6
 filtration, 54, 60, **PEx-3**, **PEx-3** to PEx-4, **PEx-11** to PEx-13
Patella, 108f, **156**, 156f, 158, 177f, 692f, **693**
Patellar (term), 2, 3f
Patellar ligament, 177f, **178**, 217f, 692f, **693**
Patellar (knee-jerk) reflex, 336f, 337, 337f, **338**, 338f
Patellar retinaculi (medial and lateral), **178**
Patellar surface, **156**
Pattern area (fingerprint), 101
P$_{CO2}$ (partial pressure of carbon dioxide), PEx-106, PEx-150, PEx-153
PCT. See Proximal convoluted tubule
PCV. See Packed cell volume
Pectinate muscle, 445f, **451**
Pectineus muscle, 198f, 216t, 217f
Pectoral (shoulder) girdle. See also Shoulder joint
 bones of, 108f, **148**, 148f, 149f
 muscles of, 198f, 199f, 205f, 205–206t, 208–209t, 209–211f
 surface anatomy of, 152, 687f, 687f

Pectoralis major muscle, 196f, 197, 198f, 205f, 205t, 207f, 212f, **684**, 685f
Pectoralis minor muscle, 198f, 205f
Pectoral nerves, 314f, 315t
Pedal (term), **2**, 3f
Pedicle, vertebral, 131f, 132f
Pelvic (term), **2**, 3f
Pelvic articulations, 153
Pelvic cavity, **6**, 7f
Pelvic (hip) girdle. See also Hip joint; Pelvis
 bones of, 108f, **153**, 154f, 155t
 surface anatomy of, 157–158
Pelvic inflammatory disease (PID), **636**
Pelvic inlet (pelvic brim), **153**, 154f, 155t
Pelvic outlet, **153**, 155t
Pelvic splanchnic nerves, **324**
Pelvis (bony pelvis), **153**, 154f, 155t
 arteries of, 476, 477f
 false, **153**
 in male versus female, 153, 155t
 muscles originating on, 198f, 216t, 217f, 218t, 219f
 surface anatomy of, 157–158, 686, 687
 true, **153**
 veins of, 477, 479f
Pelvis (renal), **612**–613, 612f
Penis, **21**, 21f, **22**, 22f, **612**, 630f, 631f, **632**–633, 633, 633f
Pepsin, **PEx-125**
 protein digestion by, 597f, **PEx-125** to PEx-126
Peptidase, **PEx-123**
Peptide(s), PEx-59, **PEx-125**
 digestion of, **PEx-125** to PEx-126
Peptide hormones, PEx-59
Perception, 352
Perforating (Volkmann's) canals, 113f, **114**
Perforating (Sharpey's) fibers, **110**, 111f, 113f
Pericardial arteries, 474
Pericardial cavity, 7f
Pericarditis, **446**
Pericardium, **7**, 19f, 446
 fibrous, **446**
 parietal, 7f, **446**, 449
 serous, 446
 visceral (epicardium), 7f, 445f, **446**, 449
Perichondrium, 109
Perilymph, **385**, 386, 386f, 387f, 389
Perimetrium, 635f
Perimysium, **188**, 189f
Perineal (term), **2**, 3f
Perineum, **634**, 687
Perineurium, **260**, 260f
Perinuclear cytoplasm, outer collar of, **256**, 256f
Period (phase) of contraction, of muscle twitch, **238**, 238f, **PEx-18**, PEx-20
Period (phase) of relaxation, of muscle twitch, **238**, 238f, **PEx-18**, PEx-20
Periodontal ligament, **585**, 585f
Periosteal bud, 114
Periosteal layer of dura mater, 282, 283f, 284f
Periosteum, **110**, 111f, 113f
Peripheral nervous system (PNS), 254, **276**. See also Cranial nerves; Nerve(s); Spinal nerves
 supporting cells in, 254, 254f
Peripheral resistance, **PEx-75**, PEx-76
 blood flow and, PEx-75 to PEx-76, PEx-76, PEx-76 to PEx-88
 compensation/cardiovascular pathology and, PEx-86
 blood pressure and, 500, PEx-75 to PEx-76, PEx-76
Perirenal fat capsules, 610f, 611
Peristalsis (peristaltic movements), **603**, 603f
Peritoneum, **7**, 7f
 parietal, 7f, 580f
 visceral, 7f, 574, 580f
Peritubular capillary bed, **613**, 614f, 615f, **PEx-132, PEx-139**

Permanent teeth, **584**, 584f
Permeability (cell), 40–**41**, 53–65, **54**, PEx-3 to PEx-16
Permeability (ion channel), **PEx-36**
Peroneal (term), **2**, 3f
Peroneus (fibularis) brevis muscle, 220t, 221f, 223f
Peroneus (fibularis) longus muscle, 198f, 199f, 220t, 221f, 223f
Peroneus (fibularis) tertius muscle, 220t, 221f
Peroxisomes, 42f, **43**–44, 43t
Perpendicular plate of ethmoid bone, 124f, 125f, **127**
Petrous part, of temporal bone, 120, 122f, 123f
Peyer's patches, 527f, **582**, 582f
PF$_3$, **433**, 433f
pH, **PEx-127, PEx-149**. See also Acid-base balance
 blood, 564
  buffering systems in maintenance of, 564, 565, PEx-150
   carbonic acid–bicarbonate (respiratory), 564, 564f, 565, **PEx-150**, PEx-150 to PEx-152. See also Respiratory buffering system
   renal, **PEx-150**, PEx-153 to PEx-154
 lipase activity and, 599, 600, **PEx-127**
 normal, PEx-149, PEx-150, PEx-153
 plasma, 425
 urine, 622, 624
Phagocytosis, **61**, 61f
Phagosome, 61, 61f
Phalanges (phalanx)
 of fingers, 108f, **151**–152, 152f
 of toes, 108f, **157**, 158f
Pharyngeal-esophageal phase, of deglutition (swallowing), 602, 602f
Pharyngeal tonsil, 527f, 538, 539f
Pharyngotympanic (auditory/eustachian) tube, 384f, **385**, 538, 539f
Pharynx (throat), **538**, 539f, 575f, 577, 577f
Phenotype, **668**, 672f
 genotype determination and, 671–672, 672f, 672t
 hemoglobin, 672–673, 673f
Phipps and Bird wet spirometer, 552, 553f
pH meter, **PEx-127**
Phosphate(s), in urine, 624
Phosphate buffer system, PEx-150
Photopupillary reflex, 375. See also Pupillary reflexes
Photoreceptors (rods and cones), 361–362, 363f, **364**
Phrenic arteries, 474, 475f
Phrenic nerve, 313, 313f, 315t, **682**, 683f
Phrenic vein, inferior, 480f
Physical fitness index, 501
Physiograph
 baseline heart activity studied with, 514, 514f, 515, 516, 516f
 brain waves studied with, 300–301
 cardiac muscle refractory period studied with, 517
 heart rate modifiers studied with, 517–519
 muscle contraction studied with, 236–241, 236f
Physiological albuminuria, 623
Physiological buffering systems, PEx-150
Physiological fatigue, muscle, **240**, 241. See also Muscle fatigue
Pia mater
 brain, **282**, 283f, 285
 spinal cord, 308, 310f
PID. See Pelvic inflammatory disease
Pig kidney, anatomy of, 612–613, 612f
Pigmented choroid coat, 364
Pigmented layer of retina, 361, 363f, 364
Pilocarpine, heart rate and, 517–518, PEx-98 to PEx-99

Pineal gland, 280f, **282**, 290, 291, 291f, 407, 409f, PEx-60f
Pinna (auricle), 80f, **384**, 384f, **681**, 681f
Pinocytosis (fluid-phase endocytosis), **61**, 61f
Pisiform, 151, 152, 152f
Pituicytes, **410**, 410f
Pituitary gland (hypophysis), **278**, 279f, 280f, 283f, **406**–407, 407f, 409f, PEx-60f
 hypothalamus relationship and, 406–407, 407f, PEx-61
 microscopic anatomy of, 410, 410f
 ovaries affected by hormones of, 418
 thyroxine production regulated by, 419, 419f, PEx-60
Pivot joint, 172f, 173
Placenta, 482, 483f, **659**, **660**, **661**, 662, 662f
Plane(s) (body), **4**–6, 5f, 6f
Plane joint, 172f, 173
Plantar (term), **2**, 3f
Plantar arch
 arterial, 477f
 venous/deep, 479f
Plantar arteries, 476, 477f
Plantar flexion (body movement), **174**, 175f
Plantaris muscle, 223f
Plantar nerves, 317f, 318t
Plantar reflex, **339**, 339f
Plantar veins, **477**, 479f
Plasma, 81f, **424**, 425–426, 425f
 cholesterol concentration in, 435–436, **PEx-161, PEx-169** to PEx-171
Plasma cells, 428, **529**
Plasma glucose, **PEx-64** to PEx-67
 fasting, PEx-65, PEx-66 to PEx-67
Plasma membrane, **40**, **40**–41, 40f, 41f, 42f
 diffusion through, 55, 58–60, 59f, **PEx-3, PEx-4**
  facilitated diffusion, 55, **PEx-3**, **PEx-6** to PEx-8
  simple diffusion, 55, **PEx-3**, **PEx-4** to PEx-6
 filtration and, 60
 permeability of, 40–**41**, 53–65, **54**, PEx-3 to PEx-16
Plateau phase, in cardiac action potential, PEx-94, PEx-94f
 ion movement and, PEx-100t
Platelet(s), **425**, 425f, 426t, 427f, 429
 in blood clotting, 433, 433f
Platysma muscle, 198f, 201f, 203f, 203t
Pleura, **7**, 7f, 542f, **543**
Pleural cavity, 7f, 542f, **543**, 684
Pleural margin, 684
PMI. See Point of maximal intensity
Pneumothorax, 550, **PEx-112**
PNS. See Peripheral nervous system
Podocytes, **613**
Poikilothermic animal, frog as, **PEx-97**
Pointer (microscope), **28**
Point of maximal intensity (PMI), 444, 444f
Polar bodies, 648, 649f, **650**
Polarity, epithelial tissue, 68
Polarization
 cardiac, PEx-94, PEx-94f
 neuron, **PEx-36**
Polar microtubules, 46f
Pollex (thumb), **2**, 3f
 bones of, 151, 152f
Polycythemia, **429**, **PEx-162**
 blood viscosity and, PEx-78
Polygraph, **326**
 galvanic skin response in, **326**–332, 327f, 328f, 329f, 330f
Polypeptide, **PEx-125**
Polysynaptic reflex arc, 336f, 337
Pons, 276f, 277f, **278**, 279f, 280f, 282, 289f, 291, 291f
Popliteal (term), **2**, 3f
Popliteal artery, 476, 477f, 693
 pulse at, **495**, 495f, 693
Popliteal fossa, **693**, 693f

Popliteal ligaments, oblique and arcuate, 177f, **178**
Popliteal surface, **156**
Popliteal vein, **477**, 479f, 693
Popliteus muscle, 223f, 224f, 224t
Pore(s)
  membrane, in filtration, 60, PEx-11
  nuclear, **40**, 42f
  sweat (skin), 94f, 100f
Porta hepatis, 586f
Portal arteriole, 587, 588f
Portal triad, **587**, 588f
Portal vein (hepatic portal vein), 477–478, **484**, 484f, 587, **PEx-61**
Portal venule, 587, 588f
Positive controls, PEx-121, **PEx-178**
Positive feedback mechanisms, PEx-60
Positive hydrolysis test, **599**
Positive starch test, **598**, PEx-122, PEx-124
Positive sugar test, **598**, PEx-122, PEx-124
Postcentral gyrus, **277**, 277f
Posterior (term), **3**, 4f
Posterior auricular vein, 480f
Posterior axillary fold, 685f, **687**
Posterior cerebellar lobe, 282, 282f
Posterior cerebral artery, **473**, 473f
Posterior chamber, **363**
Posterior circumflex humeral artery, **474**, 474f
Posterior commissure, 280f
Posterior communicating artery, 472f, 473f
Posterior cranial fossa, **120**, 123f
Posterior cruciate ligament, 177f
Posterior femoral cutaneous nerve, 317f, 318t
Posterior (occipital) fontanelle, 137, 137f
Posterior (dorsal) funiculus, **310**, 310f, 312f
Posterior gluteal line, 154f
Posterior (dorsal) horns, 284f, **308**, 310f, 312f
Posterior inferior iliac spine, 153, 154f
Posterior intercostal artery, **474**
Posterior interventricular artery, 445f, **447**, 448f
Posterior median furrow, **683**, 684f
Posterior muscle compartment
  forearm muscles and, 214t, 215f, 688
  lower limb muscles and
    foot and ankle, 221f, 222–224t, 223–224f
    thigh and leg, 218–219t, 219f
Posterior nasal aperture, **538**, 539f
Posterior pituitary gland (neurohypophysis), **406**, 407f, 410, 410f
  hormones of, 407, 407f
Posterior pole of eye, 362f
Posterior scalene muscle, 203t, 204f
Posterior segment, 362f, **363**
Posterior superior iliac spine, **153**, 154f, 158, **691**, 691f
Posterior tibial artery, **476**, 477f, **693**
  pulse at, **495**, 495f, 693
Posterior tibial vein, **477**, 479f
Posterior triangle of neck, **682**, 683f
Postganglionic neuron, 323, 324
Postsynaptic neuron, 255f, 256
Postsynaptic potential, **PEx-49**, PEx-51
Posture, blood pressure/heart rate affected by, 500
Potassium
  heart rate and, 518, PEx-99 to PEx-101, PEx-100t
  muscle membrane potential/contraction and, 234, 235
  neuron membrane potential and, 266, 266f, 267
Potassium channels, PEx-36
  voltage-gated, refractory periods and, PEx-44
Precapillary sphincters, 519, 519f
Precentral gyrus, 277f, **278**
Precipitin line, **PEx-180**
  in Ouchterlony double-gel diffusion, 531, **PEx-180**
Predicted vital capacity, 556–557, **559**

Prefrontal cortex, **278**, 278f
Preganglionic neuron, 323, 324
Pregnancy, ectopic, **634**
Preload, **PEx-84**
Premolars (bicuspids), **584**, 584f
Premotor cortex, 278f
Prepuce
  of clitoris, **634**, 634f
  of penis (foreskin), 630f, 631f, **632**
Presbycusis, **387**
Presbyopia, **372**
Pressure. See also Blood pressure
  blood flow and, 500, PEx-80 to PEx-82
  glomerular filtration and, PEx-135 to PEx-137, **PEx-137** to PEx-139
  osmosis affecting, PEx-8
Pressure gradient
  blood flow and, **PEx-75**, PEx-81
  filtration and, 60, PEx-3 to PEx-4, PEx-11
Pressure receptors (lamellar/Pacinian) corpuscles, 94f, 97, 350f, **351**, PEx-39, PEx-40
Presynaptic neuron, 255f, 256
Prevertebral (collateral) ganglion, 324, 325f
Primary active transport, 60
Primary auditory cortex, 278f, 387
Primary (main) bronchi, **540**, 540f, 542f
Primary curvatures, 130f, 131
Primary follicles, 649f, **650**, 650f, **651**, 652f
Primary germ layer, 658, 660f
Primary lymphoid organs/tissues, 528
Primary motor cortex, **278**, 278f
Primary oocytes, 648, **649**–650, 649f
Primary ossification center, 114
Primary somatosensory cortex, **277**, 278f
Primary spermatocytes, **647**, 647f, 648
Primary visual cortex, 278f, **364**, 366f
Prime movers (agonists), **196**
Primordial follicles, **649**, 649f
P–R interval, 459, 459f, 459t
PRL. See Prolactin
Probability, heredity and, 670–671
Process (bone marking), 112t
Processes (neuron), 82–83, 82f
  classification of, 257, 258f
Progesterone, 408
  in menstrual cycle, 651, 652f
Projection/referred pain, **354**
Projection tracts, 279
Prolactin (PRL), **406**, 407f
Proliferative phase of menstrual cycle, 651, 651f, 652f
Prominences, gluteal, 690
Pronation (body movement), **173**, 175f
Pronator quadratus muscle, 213f, 214t
Pronator teres muscle, 198f, 213f, 213t, **688**, 689f
Pronuclei, 649
Propagation, action potential, 267, `PEx-41`, PEx-48
Prophase
  of meiosis, 646, 646f
  of mitosis, 45, 46f
Proprioceptors, 350, 351–352, 351f
Propylthiouracil, metabolic rate affected by, PEx-63
Prosencephalon (forebrain), **276**, 276f
Prostate gland, 21f, 22f, 611f, 630f, 631f, **632**
Prostatic urethra, 611f, 612, 630f, 631f, **632**
Protein(s), **PEx-125**
  membrane, 40, 41f
Protein buffer system, PEx-150
Protein digestion, 597f, 598–599, **PEx-125** to PEx-126
Prothrombin, 433f, **434**
Prothrombin activator, 433f, **434**
Proximal (term), **4**, 4f
Proximal convoluted tubule, **613**, 614f, **PEx-131**, PEx-132f
  glucose reabsorption in, PEx-140 to PEx-142
Proximal phalanx
  finger, 152, 152f
  toe, 158f

Proximal radioulnar joint, 151f, 170t
Pseudostratified epithelium, **69**, 71f
Pseudounipolar (unipolar) neurons, 257, 258f
Psoas major muscle, 216t, 217f
Psoas minor muscle, 217f
PTC taste, genotype/phenotype and, **671**
Pterygoid muscles, lateral and medial, 202f, 202t
Pterygoid processes, 124f, **126**
PTH. See Parathyroid hormone
Ptosis, **611**
Pubic (term), **2**, 3f
Pubic arch/angle, 153, 154f, 155t
Pubic bone (pubis), 22f, 108f, **153**, 154f, 686
Pubic crest, **153**, 154f, 685f, **686**
Pubic rami, **153**, 154f
Pubic (hypogastric) region, 8, 9f
Pubic symphysis, 109f, **153**, 154f, 171t, 685f, **686**
Pubic tubercle, 153, 154f, **686**, 686f
Pubis (pubic bone), 22f, 108f, **153**, 154f, 686
Pubofemoral ligament, 176, 176f
Pudendal artery, internal, 476
Pudendal nerve, 317f, 318t
Pulmonary arteries, 444f, 445f, 448f, 450f, **479**, 482f
Pulmonary capillaries, 448f, **479**
Pulmonary circulation, **446**, 448f, **479**–482, 482f
Pulmonary function tests, 557–559, 558f, **PEx-107**, PEx-108
  forced expiratory volume (FEV$_T$/FEV$_1$), **557**–559, 558f, **PEx-107**
  forced vital capacity (FVC), **557**–559, 558f, **PEx-107**
Pulmonary minute ventilation, **PEx-106**
Pulmonary semilunar valve, 445f, **446**, 447f, 451, 452f
  auscultation of, 494, 494f
Pulmonary trunk, 444f, 445f, **446**, 448f, 450f, 451, **479**, 482f
Pulmonary veins, 444f, 445f, **446**, 448f, 450f, **479**, 482f
Pulmonary ventilation (breathing), **538**, **550**–572, 551f, **PEx-105**, PEx-106, PEx-106f. See also under Respiratory
  acid-base balance and, 564–565, 564f, PEx-150, PEx-150 to PEx-152, PEx-154 to PEx-156
  factors affecting, 562–564
  frequency of, **PEx-106**
  mechanics of, 550–551, 551f, PEx-105 to PEx-117
  muscles of, 206t, 207f, 550, **PEx-106**, PEx-107
Pulp
  splenic, 529, 529f
  of tooth, 585
Pulp cavity, 585, 585f
Pulse, **494**–497, 495f, 496f, 497f
Pulse deficit, **497**
Pulse pressure, **495**, **499**
Pulse rate, 495. See also Heart rate; Pulse
Pump activity (heart), PEx-82
  stroke volume affecting, **PEx-84** to PEx-85
  vessel radius affecting, PEx-82 to PEx-83
Puncta, lacrimal, **360**, 360f
Punctate distribution, **352**
Punnett square, 668
Pupil, **361**, 362f
Pupillary light reflex, **340**
Pupillary reflexes, 340–341, 361, 375
Purkinje cells, 257–258, 258f, 259f
Purkinje fibers (subendocardial conducting network), **458**, 458f
Purkinje myocytes, 458
Putamen, **280**, 281f
P wave, 459f, 459t
Pyelonephritis, 622
Pyloric antrum, 577, 578f
Pyloric canal, 577, 578f

Pyloric part, of stomach, **577**
Pyloric sphincter/valve, **577**, 578f
Pylorus, 577, 578f
Pyramidal cells/neurons, 257–258, 258f, 259f, 279
Pyramids, decussation of, **278**, 279f
Pyuria (leukocytes in urine), **623**, 624

QRS complex, 459f, 459t, 460
Q–T interval, 459f, 459t, 460
Quadrants, abdominopelvic, **8**, 8f
Quadratus lumborum muscle, 210f, 211t, 217f
Quadriceps (term), muscle name and, 197
Quadriceps femoris muscles, 217f, 218t, 222, 692f, **693**
Quadriplegia, **311**
Quiescent period, **492**

Rack and pinion knob (microscope), 28, 29f
Radial-apical pulse, 497
Radial artery, **474**, 474f, **689**, 689f, 690
  pulse at, 152, **495**, 495f, 689, 689f, 690
Radial fossa, **150**, 150f
Radial groove, **150**, 150f
Radial nerve, **313**–316, 314f, 315t
Radial notch, 150f, **151**, 151f
Radial styloid process, 151f, 152, **688**, 689f
Radial tuberosity, **150**, 150f, 151f
Radial vein, **479**, 481f
Radiate arteries, cortical, 612f, **613**, 614f, 615f
Radiate veins, cortical, 612f, **613**, 614f
Radioulnar joint, 151f, 170t
Radius, 108f, **150**, 150f, 151f, 152f, 688, 688f, 689f
Rami
  communicantes (ramus communicans), **324**, 325f
  dorsal, **312**, 313f
  ventral, **312**, 313, 313f, 314f, 316, 316f, 317f
Ramus (bone marking), 112t
  mandibular, **127**, 127f, 680f, **681**
  of pubis, **153**, 154f
Ranvier, nodes of (myelin sheath gaps), 255f, **256**, 257f, PEx-48
Rat dissection, organ systems and, 16–22, 17f, 18f, 19f, 20f, 21f
RBCs. See Red blood cell(s)
Reabsorption
  tubular, **613**–614, **PEx-132**, **PEx-139**, PEx-140
    glucose carrier proteins and, **PEx-140** to PEx-142
    urine concentration and, 613–614, PEx-139 to PEx-140
Reaction time, reflex, 341–344, 343f
Real image, **372**
  in microscopy, 29, 30f
  refraction and, 372, 372f
Rebreathing, acid-base balance and, PEx-151 to PEx-152
Receptor(s). See also specific type
  hormone, **PEx-59**
  in reflex arc, 336, 336f
  sensory, 258, **350**. See also Sensory receptors
  special, 350. See also Special senses
Receptor-mediated endocytosis, **61**, 61f
Receptor potential, **PEx-35**, PEx-39 to PEx-41
Receptor proteins, PEx-39
Recessive gene, **668**
Reciprocal inhibition, 338
Recording electrodes, for oscilloscope, 270, 271f
Recruitment (multiple motor unit summation), 241–244, **242**, 242f, 243f, **PEx-20**
Rectal artery, superior, 476f
Rectouterine pouch, 635f
Rectum, **20**, 21f, 575f, **583**, 583f
Rectus (term), muscle name and, 197

Rectus abdominis muscle, 197, 198f, 206t, 207f, 208f, 685f, **686**
Rectus femoris muscle, 196f, 198f, 217f, 218t, 692f
Rectus muscles of eye, 361f
Red blood cell(s) (RBCs/erythrocytes), 81f, **424**, 425f, 426t, 427–428, 427f
 in hematocrit, PEx-161, PEx-162
 settling of (ESR), **PEx-161**, **PEx-161** to PEx-162, **PEx-164** to PEx-165
 in urine (hematuria), **623**
Red blood cell (erythrocyte) antigens, blood typing and, **434**, 434t, 435f, PEx-162, **PEx-167**, PEx-168t
Red blood cell (erythrocyte) count, total, **429**
Red blood cell (erythrocyte) sedimentation rate (ESR), **PEx-161**, **PEx-161** to PEx-162, **PEx-164** to PEx-165
Red marrow, **110**
Red pulp, **529**, 529f
Referred pain, **354**
Reflex arc, 336–337, 336f
Reflex physiology, 335–348
 autonomic reflexes and, **336**
 reaction time and, 341–344, 343f
 reflex arc and, 336–337, 336f
 somatic reflexes and, **336**, 337–340, 337f, 338f, 339f
Refraction, 372–374, 373f, 374f
Refractive index, 372
Refractory period
 cardiac muscle, 517, **PEx-94** to PEx-95
 muscle cell, 234
 neuron, 267, PEx-44 to PEx-45, **PEx-45**
Regeneration, epithelial, 69
Regions, abdominopelvic, 8, 9f
Regular connective tissue, 78f, 82
Regulatory T cells, 529
Relative refractory period, **234**, **267**, **PEx-94**
 cardiac, **PEx-94**
 muscle cell, **234**
 neuron, 267, PEx-44 to PEx-45
Relaxation phase, of muscle twitch, **238**, 238f, **PEx-18**, PEx-20
Renal arteries, **475**, 475f, 610f, **611**, 612f, **613**
Renal buffering system, **PEx-150**
 in respiratory acidosis/alkalosis, PEx-150, PEx-152 to PEx-154
Renal calculi (kidney stones), **622**
Renal calyces, **612–613**, 612f, PEx-132
Renal columns, **612**, 612f
Renal compensation, **PEx-153**
Renal corpuscle, **613**, 614f, **PEx-131**, **PEx-132**, PEx-132f, PEx-133
Renal cortex, **612**, 612f
Renal fascia, 611
Renal medulla, **612**, 612f
Renal pelvis, **612–613**, 612f
Renal physiology, PEx-131 to PEx-147
Renal (medullary) pyramids, **612**, 612f
Renal tubule, 70f, **613**, 614f, 615f, **PEx-131**, **PEx-132**, PEx-132f, **PEx-135**, **PEx-137**
 glucose reabsorption and, PEx-140 to PEx-142
 urine formation/concentration and, 613–614
Renal tubule lumen, **PEx-139**
Renal tubule pressure, PEx-135
 glomerular filtration and, PEx-135 to PEx-137, PEx-137 to PEx-139
Renal veins, **477**, 480f, 610f, **611**, 612f, **613**
Renin-angiotensin system, PEx-142
Repolarization, **234**, **267**
 cardiac, PEx-94, PEx-94f
 ion movement and, PEx-100t
 muscle cell, **234**
 neuron, 266f, **267**, PEx-44
Reproduction, 645–656

Reproductive system, 16t, 20–22, 21f, **630**. See also specific organ
 anatomy of, 629–644
  female
   gross anatomy, 16t, 21, 21f, 22, 22f, 634–636, 634f, 635f
   mammary glands, 637, 637f
   microscopic anatomy, 636, 636f, 637f
  male
   gross anatomy, 16t, 21, 21f, 22, 22f, 630–633, 630–631f
   microscopic anatomy, 632f, 633, 633f, 634f
 physiology of, 645–656
Reserve volume
 expiratory, **552**, 552f, 554, 555–556, 556, 559f, **PEx-107**
 inspiratory, **552**, 552f, 554, 555, 559f, **PEx-107**
Residual capacity, functional, 552f, 559f
Residual volume (RV), 552f, 557, **559**, 559f, **PEx-107**
Resistance, vascular, **PEx-75**, PEx-76
 blood flow and, PEx-75 to PEx-76, PEx-76, PEx-76 to PEx-88
 compensation/cardiovascular pathology and, PEx-86
 blood pressure and, 500, PEx-75 to PEx-76, PEx-76
Resolution/resolving power, microscope, **29**
Respiration, **538**, 549–572, **PEx-105**. See also Pulmonary ventilation
 cellular, 538
 external, **538**, PEx-106f
 factors affecting, 562–564
 internal, **538**, PEx-106f
 mechanics of, 550–551, 551f, PEx-105 to PEx-117, PEx-106f
 muscles of, 206t, 207f, 550, **PEx-106**, **PEx-107**
Respiratory acidosis, PEx-151, PEx-153
 renal response to, PEx-152 to PEx-154
 respiratory response to, PEx-151, PEx-152 to PEx-154
Respiratory alkalosis, **PEx-150**, PEx-153
 renal response to, PEx-150, PEx-152 to PEx-154
 respiratory response to, PEx-150
Respiratory bronchioles, 540f, 541, 544f. See also Bronchioles
Respiratory buffering system, 564–565, 564f, **PEx-150**
 in metabolic acidosis/metabolic alkalosis, PEx-154 to PEx-156
 in respiratory acidosis/respiratory alkalosis, PEx-150, PEx-151, PEx-152 to PEx-154
Respiratory capacities, 552–562, 552f, 559f, PEx-107
Respiratory gases transport of, **538**. See also Gas exchange
Respiratory membrane (blood air barrier), **541**, 541f
Respiratory rate/depth, factors affecting, 562–564
Respiratory sounds, 551–552
Respiratory system, 16t, **538**, **550**
 in acid-base balance, 564–565, 564f, **PEx-150**, PEx-150 to PEx-152, PEx-154 to PEx-156
 anatomy of, 537–548
  lower respiratory structures, 540–544, 540f, 541f, 542f, 543f, 544f
  upper respiratory structures, 538–539, 539f, 540f
 physiology/mechanics of, 549–572, PEx-105 to PEx-117
Respiratory volumes, 552–562, 552f, 559f, PEx-106 to PEx-109, PEx-107
 BIOPAC® in study of, 559–562, 559f, 560f, 561f

spirometry in study of, 552–557, 552f, 553f, 556f, PEx-109 to PEx-112
Respiratory zone structures, **541**
Resting membrane potential, 41, **234**, **266**, **PEx-36**
 cardiac cell, PEx-99
  ion movement and, PEx-100t
 muscle cell, **234**
 neuron, 266, 266f, **PEx-36** to PEx-39
Rest periods, muscle fatigue and, 241, **PEx-25**
Restrictive lung disease, 557, **PEx-108**
Rete testis, **633**
Reticular connective tissue, 77f
Reticular fibers, 74f, **75**, 77f
Reticular lamina, 69
Reticular layer of dermis, 94f, **96**
Reticulate body (chlamydia), **PEx-178**
Retina, 276f, **361–363**, 362f, 363f, **364**, 365f
 microscopic anatomy of, 363f, 364
Retinal cells, 258f
Rh antigens/blood group/factor, 434, **PEx-167**, PEx-168
 typing for, 434–435, **PEx-167** to **PEx-169**
Rheostat control (ophthalmoscope), 376, 376f
Rheostat lock button (ophthalmoscope), **376**, 376f
Rhombencephalon (hindbrain), **276**, 276f, 282
Rhomboid muscles, major and minor, 199f, 209f, 209t
Rhythmicity, cardiac muscle, **512**
Ribosomes, **42**, 42f, 43t
Ribs, 108f, 135f, **136**, 136f, 683
 during breathing, 550, 550f
Right, anatomical, **8**
Right atrium, 444f, 445f, 448f, 450f
 in frog, 513f
Right brachiocephalic vein, **478**
Right colic (hepatic) flexure, **583**, 583f
Right common carotid artery, **472**
Right coronary artery, 445f, **447**, 448f, **472**
Right gastric artery, **474**, 475f
Right gastroepiploic artery, 474–475, 475f
Right gonadal vein, **477**, 480f
Right lymphatic duct, 526f, **527**
Right marginal artery, 444f, **447**, 448f
Right pulmonary artery, 479f, 482f
Right subclavian artery, **472**, 474f
Right suprarenal vein, **477**, 480f
Right upper and lower abdominopelvic quadrants, 8, 8f
Right ventricle, 444f, 445f, 448f, 450–451, 450f, 452f
 left ventricle compared with, 451–452, 452f
Ringer's solution/irrigation, PEx-97
Rinne test, 388, 388f
Risorius muscle, 200t, 201f
Rods, 361–362, 363f, **364**
Romberg test, 392
Roof, of orbit, 128f
Root
 of hair, 94f, **97**, 98f
 of lung, **541**, 542f
 of nail, **97**, 97f
 of nose, **681**, 681f
 of tooth, **585**, 585f
Root canal, **585**, 585f
Root hair plexus (hair follicle receptor), 94f, 97, 350f, **351**
Rotation (body movement), **173**, 175f
Rotator cuff muscles, **178**, 179f
Rough endoplasmic reticulum, **42–43**, 42f, 43t
Rouleaux formation, PEx-162, **PEx-164**
Round ligament (ligamentum teres), 483f, **484**, 586f
Round ligament of uterus, 635f, **636**
Round window, **385**, 385f, 387f
R–R interval, 459t
Rugae
 of bladder, 611f
 gastric, 578f

Running in place, ECG during, 461, 463–464, 464f, 465–466
RV. See Residual volume

S₁ (first heart sound), 492, 493f
S₂ (second heart sound), 492, 493f
Saccule, 385f, **389**
Sacral (term), **2**, 3f
Sacral artery, median, 475f
Sacral canal, **134**, 134f
Sacral crests
 lateral, 134f
 median, **133–134**, 134f
Sacral curvature, 130f, 131
Sacral foramina, **134**, 134f
Sacral hiatus, **134**, 134f
Sacral plexus/spinal nerves, 309f, 313f, **317–318**, 317f, 318t
Sacral promontory, **134**, 134f, 154f
Sacroiliac joint, **153**, 154f, 171t, 691
Sacrum, 130, 130f, **133–134**, 134f, 683
 in male versus female, 153, 155t
Saddle joint, 172f, 173
Sagittal plane, **4**, 5f, 6f
Sagittal sinus, superior, **282**, 283f, 284f
Sagittal suture, **120**, 125f
Salicylate poisoning, metabolic acidosis and, **PEx-154**
Saliva, 576, 585
Salivary amylase, 576, 585, **596–598**, 597f, **PEx-121**, PEx-123
 substrate specificity of, PEx-123 to PEx-125
Salivary gland(s), 72f, 575f, 576, **585–586**, 586f, PEx-120f
Salivary reflex, **341**
Salty taste, 400
SA node. See Sinoatrial (SA) node
Saphenous nerve, 316f, 317
Saphenous veins, **477**, 479f, 694
Sarcolemma, 186f, 187, 187f, **PEx-18**
 action potential/muscle contraction and, 234
 in neuromuscular junction, 189, 189f
Sarcomeres, 186f, **187**
Sarcoplasmic reticulum, 187, 187f
Sartorius muscle, 196f, 198f, 216t, 217f, 219f, 691f, 692f, **693**
Satellite cells, 254, 254f, 259f
SCA. See Sickle cell anemia
Scala media, **385**, 386f
Scala tympani, **385**, 386, 386f, 387f
Scala vestibuli, **385**, 386f, 387f
Scalene muscles, 203t, 204f
Scalp, 283f, **681**
Scanning lens (microscope), **28**, 30
Scaphoid, 151, 152f
Scapula(e) (shoulder blades), 108f, **148**, 148f, 149f, 152, **683**, 684f
Scapular (term), **2**, 3f
Scapular nerve, dorsal, 314f, 315t
Schwann cells, 254, 254f, 255f, **256**, 256f, 257f
 in myelination, 254, 254f, **256**, 256f, PEx-48
Sciatica, **318**
Sciatic nerve, 267, **317–318**, 317f, 318t, 691, 691f
 in frog, 267–271
  dissection of, 267–268, 267f
  inhibiting, 268–270
  stimulating, 268, 270–271, 270f, 271f
Sciatic notches, greater and lesser, **153**, 154f
Sclera, **361**, 362f, 364, 365f
Scleral venous sinus, 362f, **363**
Scoliosis, 131, 131f
Scrotum, **21**, 21f, **22**, 408, 630f, **631**
SCT. See Sickle cell trait
Sea urchins, developmental stages of, 658
Sebaceous (oil) glands, 94f, 98f, 99, 100f
Sebum, 99
Secondary active transport, 60, **PEx-140**
Secondary curvatures, 130f, 131
Secondary follicles, 649f, 650f, **651**, 652f

Secondary lymphoid organs/tissues, 528
Secondary oocytes, 648, 649f, **650**, 650f
Secondary spermatocytes, 647f, 648
Second heart sound (S2), 492, 493f
Second polar body, 649, 649f, **650**
Secretory phase of menstrual cycle, 651, 651f, 652f
Secretory vesicle, 61
Sections (body), **4–6**, 5f, 6f
Sedimentation rate, **PEx-161**, **PEx-161** to PEx-162, **PEx-164** to PEx-165
Segmental arteries, 612f, **613**
Segmentation (segmental movements), **603**, 603f
Selective (differential) permeability, 40–**41**, 53–65, **54**, PEx-3, PEx-4, PEx-8
Self antigens, PEx-177
Self–nonself differentiation, **528**
Sella turcica (Turk's saddle), 123f, 124f, **126**
Semen, **632**
Semicircular canals, **385**, 385f, 389–390, 390f
Semicircular ducts, 385f, **389**
Semilunar cartilages (menisci) of knee, 109f, **176**, 177f
Semilunar valves, 445f, **446**, 447f, 451
auscultation of, 494, 494f
Semimembranosus muscle, 199f, 219f, 219t, **693**, 693f
Seminal fluid, **632**
Seminal glands (vesicles), 21f, 22f, 630f, 631f, **632**, 633, 633f
Seminiferous tubules, **633**
spermatogenesis in, 647, 647f, 648
Semipermeable membrane, PEx-4, PEx-8
Semispinalis muscle, capitis/cervicis/ thoracis, 209f, 210f
Semitendinosus muscle, 199f, 219f, 219t, **693**, 693f
Sensation, 352. *See also under Sensory*
general, 349–358, **350**
special, **350**
hearing/equilibrium, 383–396
olfaction/taste, 397–404
vision
visual system anatomy and, 359–370
visual tests/experiments and, 371–381
Sense organs, 350
Sensitivity, hormone receptor, **PEx-59**
Sensorineural deafness, 387, 388, 388f
Sensory areas of brain, 278, 278f
Sensory layer of eye, **361**–363, 362f, 363f. *See also* Retina
Sensory (afferent) nerve(s), **258**, **276**
Sensory nerve endings/fibers, in skin, 94f, 95f, 96, 96–97, 350, 350f, 351, 352
Sensory (afferent) neurons, **258**, 259f, PEx-35, PEx-39, PEx-41
olfactory, 258f, **398**, 398f
in reflex arc, 336, 336f
Sensory receptors, 258, **350**, **PEx-39**. *See also specific type*
general, 349–358, 350f, **351**
adaptation of, **353**–354, PEx-46
physiology of, 352–355
structure of, 350f, 351–352, 351f
special, 350. *See also* Special senses
Sensory (ascending) tracts, 310, 311f
Sensory transduction, **PEx-39**
Septomarginal (moderator) band, **451**, 452f
Septum pellucidum, **279**, 280f, 284f, 290
Seroconversion, **PEx-182**
Serological testing/serology, **PEx-177** to PEX-188. *See also specific test*
Serosa(e). *See* Serous membrane(s)
Serous membrane(s)/serosa(e), **6–7**, 7f
alimentary canal, **574**, 576f
duodenal, **582**
gastric, 578f
Serous pericardium, 446
Serratus anterior muscle, 198f, 205f, 205t, 207f, 208f, **684**, 685f
Sertoli cells (sustentocytes), 647f, **648**
Serum, **PEx-177**

Sesamoid bones, **110**, 152f
Sex (gender), chromosomes determining, 669–670
Sex chromosomes, **669**–670, 669f
Sex hormones (gonadocorticoids), **408**
Sex-linked inheritance, 669–670
Sexually transmitted infections/diseases, **636**
Shaft
of bone (diaphysis), **110**, 111f
of hair, 94f, **97**, 98f, 99f
Sharpey's (perforating) fibers, **110**, 111f, 113f
Sheep brain, dissection of, 285–292, 288f, 289–290f, 291f, 292f
Sheep eye, dissection of, 364, 365f
Sheep heart, dissection of, 449–452, 450f, 452f
Sheep kidney, anatomy of, 612–613, 612f
Sheep pluck, lung inflation in, 543
Shinbone. *See* Tibia
Short bones, **110**
Shortening velocity, **PEx-28**
Shoulder blades. *See* Scapula(e)
Shoulder (pectoral) girdle. *See also* Shoulder joint
bones of, 108f, **148**, 148f, 149f
muscles of, 198f, 199f, 205f, 205–206t, 208–209t, 209–211f
surface anatomy of, 152, 687, 687f
Shoulder joint, 170t, 178, 179f
Sickle cell anemia, 669, PEx-162
electrophoresis in identification of, 672–673, 673f
Sickle cell hemoglobin, 669
electrophoresis in identification of, 672–673, 673f
Sickle cell trait, 669
electrophoresis in identification of, 672–673, 673f
Sigmoidal arteries, 476f
Sigmoid colon, 575f, **583**, 583f
"Signet ring" cells, 82
Simple diffusion, **55**, **PEx-3**, **PEx-4** to PEx-6
Simple epithelium, 68f, **69**, 70–71f
columnar, 71f
cuboidal, 70f
squamous, 70f
Sinoatrial (SA) node, **458**, 458f, **PEx-95** to PEx-96, **PEx-98**
Sinus (bone marking), 112t
Sinuses (paranasal), 129, 129f, **538**
Sinusitis, 129
Sinusoids, liver, **587**, 588f
Sinus venosus (frog heart), **513**, 513f
Sister chromatids, 46f, 646
Skeletal muscle(s), **83**, 84f, **187**
action potential in, **234**
cells of, 84f, 186f, 187–188, 187f, 188f. *See also* Muscle fibers
classification/types/naming, 196–197, 196f
contraction of, PEx-17 to PEx-34. *See also* Muscle contraction
fiber organization into, 188, 189f
length-tension relationship in, PEx-26 to PEx-28
isometric, **PEx-27**
load affecting, 240–241, **PEx-28** to PEx-29
physiology of, 233–252, PEx-17 to PEx-34. *See also* Muscle contraction
structure of, 195–232, 196f, 198f, 199f, PEx-17, PEx-18f
microscopic anatomy/organization, 185–193, 186f. *See also* Muscle fibers
Skeletal muscle cells/fibers. *See* Muscle fibers
Skeletal muscle pump, venous return and, 472
Skeletal system/skeleton, 16t, 107–118, **108**, 108f
appendicular skeleton, 108f, **109**, 147–165
axial skeleton, 108f, **109**, 119–145, **120**
bones, 108, 108f, 110–114, 111f, 112t, 113f, 114f
cartilages, 108, 109–110, 109f

Skin (cutaneous membrane/integument), 16t, 79f, 93–107, **94**
accessory organs of, 94f, 97–100, 97f, 98f, 99f, 100f
color of, 97
circulatory dynamics and, 97, 502–504
structure of, 94–97, 94f, 95f
microscopic, 98, 99f
Skull, 108f, **120**–130, 121f, 122f, 123f, 124f, 125–126f, 127f, 128f, 129f, 170t. *See also* Cranium; Facial bones
fetal, 136–137, 137f
Sleep, brain waves during, 300, 300f
Sliding filaments, in muscle contraction, 235, 240
SL valves. *See* Semilunar valves
Small cardiac vein, 444f, **448**, 448f
Small intestine, **18**, 20, 20f, 71f, 575f, **581**–583, 581f, 582f, PEx-120f
Small saphenous vein, **477**, 479f, 694
Smell, sensation of, 398–399, 398f. *See also under Olfactory*
adaptation and, 401–402, PEx-46
cortical areas in, 278, 398
in odor identification, 401
taste and, 400–401
Smooth endoplasmic reticulum, **42**, 42f, 43, 43t
muscle cell (sarcoplasmic reticulum), **187**, 187f
Smooth (visceral) muscle, **85**, 85f, 187
Snellen eye chart, 373
Snuff box, anatomical, **690**, 690f
Sodium
autorhythmicity and, 512, PEx-93
heart rate and, 518, PEx-99 to PEx-101, PEx-100t
muscle cell membrane potential/ contraction and, 234
neuron membrane potential and, 266, 266f, 267, PEx-36
Sodium benzoate taste, genotype/ phenotype and, **671**
Sodium channels, PEx-36
voltage-gated, **PEx-42** to PEx-44
refractory periods and, PEx-44
Sodium-potassium pump, 234, 266, 266f, PEx-13, PEx-36
Soft palate, **538**, 539f, 575, 576f, 577f
Soleus muscle, 198f, 199f, 221f, 222t, 223f, 693f
Solute gradient, urine concentration and, PEx-139 to PEx-140
Solute particles, PEx-4
Solute pumps, **PEx-13**
Somatic nervous system, **276**, **323**. *See also* Brain; Spinal cord
Somatic reflexes, **336**, 337–340, 337f, 338f, 339f
Somatosensory association cortex, **277**–278, 278f
Somatosensory cortex, primary, **277**, 278f
Sound localization, 387–388
Sounds of Korotkoff, 498, **498**
Sound (traveling) waves, 384, 386, 387f
Sour taste, 400
Southern blotting, PEx-184
Special senses, **350**
cortical areas in, 278, 278f
hearing and equilibrium, 383–396
olfaction and taste, 397–404
vision
visual system anatomy and, 359–370
visual tests/experiments and, 371–381
Specific gravity, urine, 622, 624
Specificity, in immune response, **528**
Specificity (substrate), 596
amylase, PEx-123 to PEx-125
Spectrophotometer, **PEx-125**, PEx-126
developing standard glucose curve and, PEx-65 to PEx-66
measuring fasting plasma glucose and, PEx-66
in pepsin activity analysis, **PEx-125**, PEx-126
Sperm, **647**, 648, 648f

in epididymis, 631, 633, 634f
production of (spermatogenesis), **647**–648, 647f, 648f
Spermatic cord, **631**, 632
Spermatids, **647**, 647f, 648
Spermatocytes, **647**, 647f, 648
Spermatogenesis, **647**–648, 647f, 648f
Spermatogonia, **647**, 647f
Spermiogenesis, **647**, 647f
Sphenoidal fontanelle, 137, 137f
Sphenoidal sinuses, 129, 129f, 538, 539f
Sphenoid bone, 121f, 122f, 123f, 124f, **125**–126, 125f, 126f, 128f, 130
Sphincter pupillae, 361
Sphygmomanometer (blood pressure cuff), **498**–499, 498f
Spinal (vertebral) cavity, **6**, 7f
Spinal cord, 276f, **308**–312, 309f, 310f, 311f
dissection of, 311–312, 312f
Spinal curvatures, 130f, 131
Spinal (vertebral) foramen, **131**, 131f, 133f, 134t
Spinalis/spinalis thoracis muscles, 210f, 210t
Spinal nerves/nerve plexuses, **308**, 309f, 310f, **312**–318, 313f, 314f, 315t, 316f, 317f, 317t, 318t
Spinal reflexes, 337, 337–339, 337f, 338f, 339f
Spine (bone marking), 112t
of scapula, 149f, 152, **683**, 684f
Spine/spinal column. *See* Vertebral column
Spinous process, 130f, **131**, 131f, 132, 132f, 133f, 134t, 681, 683
Spiral organ, **385**, 385f, 386f
microscopic anatomy of/hearing and, 386–387, 386f, 387f
Spirometry/spirometer, **552**–557, 552f, 553f, 556f, PEx-109 to PEx-112
computer-generated spirogram and, 559, 559f
Splanchnic nerves, **324**, 325f
pelvic, **324**
Spleen, **20**, 20f, 77f, 527f, 529f, 575f
Splenic artery, **474**, 475f, 529f
Splenic cords, **529**, 529f
Splenic (left colic) flexure, **583**, 583f
Splenic pulp, **529**, 529f
Splenic vein, **484**, 484f, 529f
Splenius muscles
capitis, 201f, 211f, 211t
cervicis, 211f, 211t
Spongy (cancellous) bone, **110**, 111f, 113f
Spongy urethra, 611f, 612, 630f, 631f, **632**
Sprain, **180**
Spring clips (microscope), 28
Spurs (bone), 180
Squamous epithelium, 68f, **69**
simple, 70f
stratified, 72f
Squamous part
of frontal bone, 125f
of temporal bone, **120**
in fetal skull, 137f
Squamous suture, **120**, 121f
SR. *See* Sarcoplasmic reticulum
Stage (microscope), **28**, 29f
Standard limb leads, for ECG, 460, 460f, 462, 462f
Stapes (stirrup), 384f, **385**, 385f, 386, 387f
Starch, PEx-123
Starch digestion, 596–598, 597f, PEx-121 to PEx-123
by salivary amylase, 576, 585, **596**–598, 597f, **PEx-121** to PEx-123
substrate specificity and, PEx-123 to PEx-125
Starch test, positive/negative, **598**, PEx-122, PEx-124
Starling forces, **PEx-135**, **PEx-137**
Static equilibrium, **390**
STDs. *See* Sexually transmitted infections/diseases

Stellate (hepatic) macrophages, **587**, 588*f*
Stereocilia
  epididymal, 633, 634*f*
  equilibrium and, 391*f*
  in spiral organ, 386, 386*f*
Stereomicroscope, for study of muscle
  fiber contraction, 235
Sternal (term), **2**, 3*f*
Sternal angle, **135**–136, 135*f*, 540, **683**,
  685*f*
Sternal (medial) end of clavicle, 148, 149*f*
Sternal head, of sternocleidomastoid
  muscle, **682**, 682*f*
Sternoclavicular joint, 148, 170*t*
Sternocleidomastoid muscle, 197, 198*f*,
  199*f*, 201*f*, 203*f*, 203*t*, 204*f*, 205*f*,
  682, 682*f*, 683*f*
Sternocostal joints, 170*t*
Sternohyoid muscle, 198*f*, 203*f*, 204*f*,
  204*t*
Sternothyroid muscle, 203*f*, 204*f*, 204*t*
Sternum, 108*f*, **135**–136, 135*f*, 683
Steroid hormones, PEx-60
Stimulating electrodes, for oscilloscope,
  270, 271*f*
Stimuli, 350, 352
Stimulus artifact, 270, 270*f*
Stimulus frequency, **PEx-22, PEx-24,
  PEx-25**
  coding stimulus intensity as, PEx-46
    to PEx-47
  muscle contraction affected by,
    239–240, 239*f*, **PEx-22** to PEx-23
  tetanus and, 239*f*, 240, **PEx-24,
    PEx-25**, PEx-94
Stimulus intensity/stimulus voltage,
  PEx-20. *See also* Threshold voltage
  coding for, PEx-46 to PEx-47
  graded/local potentials and, PEx-39
  muscle contraction affected by,
    238–239, **PEx-20** to PEx-22
Stirrup (stapes), 384*f*, **385**, 385*f*, 386
STIs. *See* Sexually transmitted infections/
  diseases
Stomach, **18**, 20*f*, 575*f*, **577**, 578*f*, **579**,
  579*f*. *See also under* Gastric
Stomach (gastric) glands, PEx-120*f*,
  PEx-125
Straight sinus, 283*f*
Stratified epithelium, 68*f*, **69**, 72–73*f*
  columnar, 73*f*
  cuboidal, 72*f*
  squamous, 72*f*
Stratum basale (stratum germinativum)
  (skin), 95*f*, **96**, 99*f*
Stratum basalis (basal layer)
  (endometrium), **636**
Stratum corneum (skin), 95*f*, **96**, 99*f*
Stratum functionalis (functional layer)
  (endometrium), **636**
Stratum granulosum (skin), 95*f*, **96**, 99*f*
Stratum lucidum (skin), **96**, 99*f*
Stratum spinosum (skin), 95*f*, **96**, 99*f*
Stretch receptors, 350
Stretch reflexes, **337**–338, 337*f*, 338*f*
Striations, muscle tissue, 83, 83*f*. *See also*
  Skeletal muscle
Striatum, **281**, 281*f*
Stroke volume (SV), 460, 493*f*,
  PEx-82, PEx-84
  in cardiac output, PEx-82, PEx-84
  pump activity affected by, **PEx-84** to
    PEx-85
Strong acid, **PEx-149**
Strong base, **PEx-149**
S–T segment, 459*f*, 459*t*, 460
Sty, 360
Stylohyoid muscle, 204*f*, 204*t*
Styloid process
  of radius, 151*f*, 152, **688**, 689*f*
  of temporal bone, **120**, 121*f*, 122*f*
  of ulna, **151**, 151*f*, 152, **688**, 689*f*
Stylomastoid foramen, **120**, 122*f*
Subarachnoid space, **282**, 283*f*, 284*f*, 310*f*
Subcapsular sinus, lymph node, 528*f*
Subclavian arteries, **472**, 473*f*, 474*f*, **682**,
  682*f*, 683*f*
Subclavian vein, **479**, 480*f*, 481*f*

Subdural space, **282**, 283*f*, 310*f*
Subendocardial conducting network
  (Purkinje fibers), **458**, 458*f*
Sublingual glands, 575*f*, **585**
Submandibular glands, 575*f*, **585**, **682**,
  683*f*
Submucosa
  alimentary canal, **574**, 576*f*
    gastric, 579*f*
    large intestine, 582*f*
    small intestine, 581*f*, **582**
    tracheal, 543*f*
Submucosal plexus, 574, 576*f*
Subpapillary vascular plexus, 94*f*
Subscapular artery, **474**, 474*f*
Subscapular fossa, 148, 149*f*
Subscapularis muscle, 178, 205*f*, 211
  tendon of, 179*f*
Subscapular nerves, 314*f*, 315*t*
Substage light (microscope), **28**, 29*f*
Substrate(s), enzyme, 596, **PEx-120,
  PEx-123, PEx-125**
  specificity of, 596
    amylase, PEx-123 to PEx-125
Subthreshold stimuli, **238**
Sudoriferous (sweat) glands, 94*f*, 99–100
Sugar test, positive/negative, **598**,
  PEx-122, PEx-124
Sulci (sulcus), **277**, 277*f*
Sulfates, in urine, 624
Superciliary arches, 680*f*, **681**
Superficial/external (term), **4**
Superficial cord reflexes, **339**, 339*f*
Superficial fascia of skin (hypodermis),
  94, 94*f*
Superficial inguinal ring, **686**, 686*f*
Superficial palmar arch, 474*f*
Superficial palmar venous arch, 481*f*
Superficial temporal artery, **472**, 473*f*,
  680*f*, **681**
  pulse at, **495**, 495*f*, 680*f*
Superficial temporal vein, **478**–479, 480*f*
Superficial transverse ligament of palm,
  213*f*
Superior (term), **3**, 4*f*
Superior angle of scapula, 148, 149*f*
Superior articular process/facet, vertebral,
  131*f*, **132**, 132*f*, 133*f*, 134*f*, 134*t*
Superior border of scapula, 148, 149*f*
Superior cerebellar peduncles, 290
Superior colliculi, **279**, 280*f*, 291*f*
Superior gluteal artery, **476**, 477*f*
Superior gluteal nerve, 317*f*, 318*t*
Superior mesenteric artery, **475**, 475*f*,
  476*f*
Superior mesenteric ganglia, **324**
Superior mesenteric vein, **484**, 484*f*
Superior nasal conchae/turbinates, **127**,
  538, 539*f*
Superior nuchal line, 122*f*, 125*f*, 681
Superior oblique muscle of eye, 361*f*
Superior orbital fissure, 124*f*, 125*f*, **126**,
  128*f*
Superior rectal artery, 476*f*
Superior rectus muscle of eye, 361*f*
Superior sagittal sinus, **282**, 283*f*, 284*f*
Superior thyroid artery, 473*f*
Superior vena cava, 19*f*, 444*f*, 445*f*, **446**,
  448*f*, 450*f*, **477**, 480*f*, 481*f*
  veins draining into, 478–479, 480*f*, 481*f*
Supination (body movement),
  173–174, 175*f*
Supinator muscle, 213*f*, 214*t*, 215*f*
Supporting cells
  nervous tissue (neuroglia/glial cells),
    82, 82*f*, **254**, 254*f*
  olfactory epithelium, **398**, 398*f*
Supraclavicular nerves, 313*f*, 315*t*
Supracondylar lines, medial and lateral,
  156*f*
Supracondylar ridges, lateral and medial,
  150*f*
Supracristal line, **683**, 684*f*
Suprahyoid muscles, **682**, 683*f*
Supraorbital foramen (notch), **120**, 125*f*,
  130
Supraorbital margin, 125*f*
Suprarenal arteries, middle, **475**, 475*f*

Suprarenal glands. *See* Adrenal
  (suprarenal) glands
Suprarenal veins, **477**, 480*f*
Suprascapular artery, 474*f*
Suprascapular nerve, 314*f*, 315*t*
Suprascapular notch, **148**, 149*f*
Supraspinatus muscle, 178, 209*f*, 209*t*,
  211, 212*f*
Supraspinous fossa, 148, 149*f*
Sural (term), **2**, 3*f*
Sural nerve, 317*f*, 318*t*
Surface anatomy, 2, 3*f*, 679–695, **680**
  of abdomen, 685*f*, 686–687, 686*f*
  of head, 680–682, 680*f*, 681*f*
  of lower limb/pelvic girdle, 3*f*, 157–158
  of neck, 681–682, 682*f*, 683*f*
  terminology and, 2, 3*f*
  of trunk, 682–684, 684*f*, 685*f*
  of upper limb/pectoral girdle, 3*f*, 152
Surface tension, **PEx-112**
Surfactant, **PEx-112** to PEx-113
Surgical neck of humerus, 148, 150*f*
Suspensory ligament
  of breast, 637*f*
  of eye (ciliary zonule), 362*f*, **363**
  of ovary, 635*f*, **636**
Sustentocytes (Sertoli cells), 647*f*, **648**
Sutural bones, **110**, 125*f*
Sutures, 168*f*, **169**
  cranial, 120, 121*f*, 125*f*
SV. *See* Stroke volume
Swallowing (deglutition), **602**–603, 602*f*
Sweat (sudoriferous) glands, 94*f*, 99–100,
  100*f*
Sweat pore, 94*f*, 100*f*
Sweet taste, 400
Sympathetic (thoracolumbar) division of
  autonomic nervous system, 324, 324*f*,
  325*f*, **PEx-95**
  function of, 326
  in heart regulation, PEx-95, PEx-98
Sympathetic trunks/chains/sympathetic
  trunk ganglion, 313*f*, **324**, 325*f*
Symphyses, 168*f*, **169**
Symporter, PEx-13
Synapse(s), 255*f*, **256**, **PEx-35,
  PEx-49** to PEx-50, PEx-49 to PEx-50
Synapsis, in meiosis, **646**
Synaptic cleft/synaptic gap, **189**, 189*f*,
  255*f*, **256**, PEx-49
Synaptic potential, 256, **PEx-35** to
  PEx-36, PEx-51
Synaptic vesicles, 189*f*, 255*f*, 256,
  PEx-49
Synarthroses, **169**
Synchondroses, 168*f*, **169**
Syncytium, functional, myocardium as,
  458
Syndesmoses, 168*f*, **169**
Synergists, **196**
Synovial cavities, **8**, 9*f*, 169*f*
Synovial fluid, 169*f*, 171
Synovial joints, 168*f*, **169**–180, 169*f*,
  170–171*t*, 172*f*. *See also specific joint*
  movements at, 173–174, 173*f*,
    174–175*f*
  types of, 168*f*, 172–173, 172*f*
Synovial membranes, 169*f*, 171
Systemic circulation, **446**, 448*f*
  arterial, 471*f*, 472–477
  venous, 477–479, 478*f*
Systole, **446**, **492**, 493*f*, **PEx-82**
Systolic pressure, **497**

T₃ (triiodothyronine), **407**. *See also*
  Thyroid hormone
T₄ (thyroxine), **407**, PEx-60, **PEx-60**.
  See also Thyroid hormone
  metabolism/metabolic rate and,
    PEx-60 to PEx-64
Tachycardia, **460**
Tactile (Merkel) cells, 95*f*, **96**
Tactile corpuscles, 96, 350*f*, **351**
Tactile (Merkel) discs, 96, **351**
Tactile localization, **352**–353
Tail of sperm, 648, 648*f*
Tallquist method, 432
Talus, 108*f*, 157, 158*f*

Tapetum lucidum, **364**
Target cells, **PEx-59**
Target organs, **406**
Tarsal (term), **2**, 3*f*
Tarsal bones, 108*f*, **157**, 158*f*
Tarsal glands, **360**, 360*f*
Tarsometatarsal joint, 171*t*
Taste, 399–400, 399*f*
  cortical areas in, 278*f*, 399
  in odor identification, 401
  smell/texture/temperature affecting,
    400–401
Taste buds, **399**–400, 399*f*, 400*f*, 577
Taste pore, **399**, 399*f*
T cells (T lymphocytes), 429, 528–529,
  528
  thymus in development of, 408, 528
Tectorial membrane, **386**, 386*f*
Teeth, 584–585, 584*f*, 585*f*
Telencephalon, 276*f*
Telophase
  of meiosis, 646
  of mitosis, **45**, 47*f*
Temperature
  heart rate and, 517, PEx-96 to PEx-98
  microcirculation and, 520
  taste and, 401
Temperature receptors, 350*f*, **351**
  adaptation of, 353–354
Temporal artery, superficial, **472**, 473*f*,
  680*f*, **681**
  pulse at, **495**, 495*f*, 680*f*
Temporal bone, **120**, 121*f*, 122*f*, 123*f*,
  125*f*, 126*f*, 283*f*
  in fetal skull, 137*f*
Temporalis muscle, 197, 198*f*, 201*f*, 202*f*,
  202*t*
Temporal lobe, **277**, 277*f*, 279*f*
Temporal (wave) summation, **239**–240,
  239*f*, 241–244, **242**, 242*f*, 243*f*,
  PEx-22, PEx-94
Temporal vein, superficial, **478**–479, 480*f*
Temporomandibular joint (TMJ), 127*f*,
  130, 170*t*, **179**–180, 180*f*, 680*f*, **681**
Tendinous insertions (intersections/
  inscriptions), 207*f*, 685*f*, **686**
Tendon(s), **188**, 189*f*, **PEx-17, PEx-18*f***
Tendon organs, 351*f*, **352**
Teniae coli, 583*f*, **584**
Tension (muscle tension/force),
  PEx-17, PEx-20
  muscle contraction and, **PEx-20**
Tensor fasciae latae muscle, 198*f*, 217*f*,
  218*t*
Tentorium cerebelli, **282**, 283*f*
Teres major muscle, 199*f*, 208*t*, 209*f*, 212*f*
Teres minor muscle, 178, 208*t*, 209*f*, 212*f*
Terminal arterioles, 519, 519*f*
Terminal boutons (axon terminals), 188,
  189, 189*f*, 190*f*, **241**, 255*f*, **256**,
  PEx-36, PEx-36*f*, PEx-49
Terminal branches, 188, 190*f*, 255*f*
Terminal bronchioles, 540*f*
Terminal cisterns, **187**, 187*f*
Terminal (intramural) ganglion, **324**
Terminal web, 44
Terminology (anatomical), 1–14
Testicular (gonadal) artery, **475**, 475*f*
Testicular (gonadal) vein, **477**, 480*f*
Testis/testes, **21**, 21*f*, **22**, 22*f*, 408, 409*f*,
  630, 630*f*, 631, 631*f*, **633**, PEx-120*f*
Testosterone, **408**
Tetanus/tetany, 239*f*, 240, **408**, **PEx-24** to
  PEx-25, PEx-25, PEx-94
Tetrads, 646, **646**
Tetraiodothyronine. *See* Thyroxine
Tetrodotoxin, PEx-42, PEx-43
Texture, taste affected by, 400
TF. *See* Tissue factor
TH. *See* Thyroid hormone
Thalamus, 276*f*, 280*f*, 281*f*, 291
Theca folliculi, 650*f*, **651**
Thenar eminence/thenar muscles, 213*f*,
  690, 690*f*
Theta waves, **300**, 300*f*
Thick (myosin) filaments, 186*f*, **187**
  muscle contraction and, 234, 240
Thick skin, 97, 98, 99*f*

Thigh. *See also* Lower limb
bones of, 156, 156f
deep artery of, **476**, 477f
muscles of/muscles acting on, 198f, 199f, 216, 216–218t, 217f, 218–219t, 219f, 692f, 693, 693f
surface anatomy of, 692f, 693, 693f
Thin (actin) filaments, 43t, 44, 186f, **187**
muscle contraction and, 234, 240
Thin skin, 98, 99f
Third ventricle, 276f, 281, 281f, 282, 284f, 292f
Thoracic (term), **2**, 3f
Thoracic aorta, 472–474, **472**, 474f
Thoracic artery
internal, 473f, **474**, 474f
lateral, **474**, 474f
Thoracic cage, 108f, **135**–136, 135f, 136f, 683, 685f. *See also* Thorax
Thoracic cavity, **6**, 7f, 18, 19f
relationships of organs in, 542f
Thoracic curvature, 130f, 131
Thoracic duct, 526f, **527**
Thoracic nerves, 309f, 313f, 314f, 315t
Thoracic vertebrae, 130f, 132, 133f, 134t
Thoracoacromial artery, **474**, 474f
Thoracodorsal nerve, 314f
Thoracolumbar (sympathetic) division of autonomic nervous system, **324**, 324f, 325f, **PEx-95**
function of, 326
in heart regulation, PEx-95, PEx-98
Thorax, 683–684, 685f, **PEx-107**
arteries of, 474, 474f
bony, 108f, 135–136, 135f, 136f, 683–684, 685f
muscles of, 198f, 205f, 205–206t, 207f, 208–209t, 209–211f, 684, 685f
surface anatomy of, 683–684, 685f
veins of, 479, 481f
Thoroughfare channels, 519, 519f
Three-dimensional vision (depth perception), **374**–375
Threshold/threshold stimulus, **238**, **267**
muscle contraction and, **238**, PEx-20
nerve impulse/action potential and, 266f, **267**, PEx-41 to PEx-42
Threshold voltage, **PEx-42**
for muscle contraction, 238, **PEx-20**
for neuron action potential, 267, **PEx-42**
Throat (pharynx), **538**, 539f, 575f, 577
Thrombin, 433f, **434**
Thumb (pollex), **2**, 3f
bones of, 151, 152f
Thumb joint (carpometacarpal joint), 170t
Thymopoietins, **408**
Thymosins, **408**
Thymulin, **408**
Thymus, **18**, 19f, 408, 409f, 527f, PEx-60f
Thyrocervical trunk, **472**, 473f, 474f
Thyroglobulin, **409**
Thyrohyoid muscle, 204f, 204t
Thyroid artery, superior, 473f
Thyroid cartilage, 109f, **538**, 539f, 540f
Thyroidectomy, PEx-61
Thyroid gland, 407–408, 409f, 682, PEx-60, PEx-60f
isthmus of, 407, 682
metabolism and, 408, **PEx-60** to PEx-64
microscopic anatomy of, 409–410, 409f
Thyroid hormone (TH), 407–408, **419**, 419f, PEx-60, **PEx-60**
metabolism/metabolic rate and, **419**, **PEx-60** to PEx-64
Thyroid-stimulating hormone (TSH/thyrotropin), **406**, 407f, **PEx-60**
metabolic rate and, 419, PEx-62 to PEx-63
Thyroid veins, 480f
Thyrotropin (thyroid-stimulating hormone/TSH), **406**, 407f, **PEx-60**
metabolic rate and, 419, PEx-62 to PEx-63
Thyrotropin-releasing hormone (TRH), 419, **PEx-61**

Thyroxine (T$_4$), **407**, PEx-60, **PEx-60**. *See also* Thyroid hormone
metabolism/metabolic rate and, **PEx-60** to PEx-64
Tibia (shinbone), 108f, **156**, 157f, 177f, 692f, **693**
Tibial arteries
anterior, **476**, 477f
posterior, **476**, 477f, **693**
pulse at, **495**, 495f, 693
Tibial collateral ligament, 177f, **178**
Tibialis anterior muscle, 198f, 220t, 221f, 222, 692f
Tibialis posterior muscle, 223f, 224f, 224t
Tibial nerve, **317**–318, 317f, 318t
Tibial tuberosity, 156, 157f, 158, 692f, **693**
Tibial veins, anterior and posterior, **477**, 479f
Tibiofemoral joint, 171t, **176**
Tibiofibular joint, 157f, 171t
Tidal volume (TV), **552**, 552f, 554, 555, 556f, 559f, **562**, PEx-106, **PEx-107**
Tissue(s), **15**, **68**. *See also specific type*
classification of, 67–92
connective, 74f, **75**–82, 76–81f, 83
epithelial (epithelium), **68**–75, 68f, 69f, 70–73f
muscle, **83**–85, 84–85f
nervous, **82**–83, 82f
Tissue factor (TF), **433**, 433f
Titin (elastic) filaments, 186f
TLC. *See* Total lung capacity
TM. *See* Total magnification
TMJ. *See* Temporomandibular joint
Toe(s)
bones of (phalanges), 108f, **157**, 158f
joints of, 171t
Tongue, **575**–576, 575f, 576f, 577f
taste buds on, 399–400, 399f, 400f
Tonicity, 58–60, 59f, **PEx-9**
Tonsil(s), 527f, 529, 530f
lingual, 538, 539f, **576**, 577f
palatine, 530f, 538, 539f, **576**, 576f, 577f
pharyngeal, 527f, 538, 539f
tubal, 538, 539f
Tonsillar crypts, **529**, 530f
Tonsillitis, **576**
Tonus, 242
Total blood counts, **429**
Total force, muscle contraction, **PEx-27**. *See also* Force
Total heart block, **519**
Total lung capacity, 552f, 559f, **PEx-107**
Total magnification (TM), microscope, **29**
Total red blood cell count, **429**
Total white blood cell count, **429**
Touch receptors, 96, 350f, 351, 352. *See also under* Tactile
adaptation of, 353
T–P segment, 459t
Trabeculae
bone, 110, **113**
lymph node, 528f, **529**
spleen, 529f
Trabeculae carneae, 445f, **451**
Trachea, **18**, 19f, 71f, 539f, 540–541, **540**, 540f, 542f, 543, 543f
Tracheal cartilages, **109**, 109f, 540f, 541, 543, 543f
Trachealis muscle, 543f
Tracheal wall, 543, 543f
Tracts, **256**, 279
spinal cord, **310**, 311, 311f
Traits, alleles for, 668
Transducers
force, **PEx-17**
sensory receptors as, **352**
Transduction, sensory, **PEx-39**
Transfusion reactions, PEx-167
Transitional epithelium, **69**, 73f
Transmembrane proteins, in glucose reabsorption, PEx-140
Transport
cell, 53–65, **54**, PEx-3 to PEx-16
active processes in, 41, **54**, 60–61, **PEx-3**, PEx-4

passive processes in, 41, 54–60, **54**, **PEx-3**
of respiratory gases, **538**
Transverse (term), muscle name and, 197
Transverse arch of foot, 157, 158f
Transverse cerebral fissure, 277f, 288f
Transverse cervical nerve, 313f, 315t
Transverse colon, 575f, 583f
Transverse humeral ligament, 179f
Transverse mesocolon, 580f, 583f
Transverse plane/cross section, 5f, **6**, 6f
Transverse process, vertebral, 130f, **131**, 131f, 132, 132f, 133f, 134t
Transverse sinus, 283f
Transverse (T) tubules, **187**, 187f
Transversus abdominis muscle, 198f, 206t, 207f, 208f
Trapezium, 151, 152f, 201f
Trapezius muscles, 197, 198f, 199f, 208t, 209f, **683**, 683f, 684f
Trapezoid, 151, 152f
Traveling (sound) waves, 384, 386, 387f
Treppe, **PEx-22**
TRH. *See* Thyrotropin-releasing hormone
Triads, **187**, 187f
Triangle of auscultation, 552, **683**, 684f
Triceps (term), muscle name and, 197
Triceps brachii muscle, 198f, 199f, 211, 212f, 212t, 213f, **688**, 688f
Triceps surae muscle, 222t
Tricuspid valve, 445f, **446**, 447f, 451, 452f
auscultation of, 494, 494f
Trigeminal nerve (cranial nerve V), 286t, 288, 288f, 289f
Trigger zone, **PEx-41**
Triglycerides, **PEx-127**
Trigone, 611f
Triiodothyronine (T3), **407**. *See also* Thyroid hormone
Triquetrum, 151, 152f
Trochanter (bone marking), 112t
greater and lesser, **156**, 156f, 158, **690**
Trochlea
of humerus, **150**, 150f
of talus, 158f
Trochlear nerve (cranial nerve IV), 286t, 287, 288f, 289f
Trochlear notch, **151**, 151f
Trophoblast, **659**, 659f
Tropic hormones, **406**, PEx-61, PEx-69
Troponin, muscle contraction and, 234
True pelvis, **153**
True (vertebrosternal) ribs, 135f, 136
True vocal cords (vocal folds), 539, 539f, 540f
Trunk
muscles of, 197, 208–211t, 209–211f
surface anatomy of, 682–684, 684f, 685f
Trunks, brachial plexus, 313, 314f
Trypsin, **598**
protein digestion by, 597f, **598**–599
TSH. *See* Thyroid-stimulating hormone
T (transverse) tubules, **187**, 187f
TTX. *See* Tetrodotoxin
Tubal tonsils, 538, 539f
Tubercle (bone marking), 112t
Tuberosity (bone marking), 112t
Tubocurarine, nerve function affected by, 269
Tubular reabsorption, **613**–614, **PEx-132**, **PEx-139**, **PEx-140**
glucose carrier proteins and, **PEx-140** to PEx-142
urine concentration and, 613–614, PEx-139 to PEx-140
Tubular secretion, **614**, PEx-132
Tubulins, 44
Tunic(s)
alimentary canal, **574**
blood vessel, **470**–471, 470f
Tunica albuginea
ovarian, 650f
testicular, 631f, **633**, 633f
Tunica externa, 470f, **471**
Tunica intima, **470**, 470f

Tunica media, 470f, **471**
Tuning fork tests, of hearing, 388, 388f
Turbinates, nasal (nasal conchae)
inferior, 125f, **128**, **538**, 539f
middle, 124f, 125f, **127**, **538**, 539f
superior, **127**, **538**, 539f
Turk's saddle (sella turcica), 123f, 124f, **126**
TV. *See* Tidal volume
T wave, 459f, 459t
Twitch, muscle, **238**, 238f, 239, **PEx-18** to PEx-20, **PEx-20**
2n (diploid chromosomal number), **646**, 668
Two-point discrimination test, 352
Two-point threshold, **352**
Tympanic cavity, **384**–385
Tympanic membrane (eardrum), **384**, 384f
Tympanic part, of temporal bone, **120**
Type 1 diabetes mellitus, **PEx-64**
Type 2 diabetes mellitus, **PEx-64**
Type A blood/blood antigen, 434t, 435f, **PEx-162**, PEx-167, PEx-168t
antibodies and, 434t, 435f, PEx-168, PEx-168t
inheritance and, PEx-167
Type AB blood, 434t, 435f, PEx-167, PEx-168t
inheritance and, PEx-167
Type B blood/blood antigen, 434t, 435f, **PEx-162**, PEx-167, PEx-168t
antibodies and, 434t, 435f, PEx-168, PEx-168t
inheritance and, PEx-167
Type lines (fingerprint), 101
Type O blood, 434t, 435f, PEx-167, PEx-168t
inheritance and, PEx-167

Ulcers, decubitus (bedsores), **96**, 96f
Ulna, 108f, 150f, **151**, 151f, 152, 152f, **688**, 688f, 689f
Ulnar artery, **474**, 474f, **689**, 689f
pulse at, 689, 689f
Ulnar head, 151, 151f, **688**, 688f, 689f
Ulnar nerve, 314f, 315t, **316**
Ulnar notch, **150**, 151f
Ulnar styloid process, **151**, 151f, 152, **688**, 689f
Ulnar vein, **479**, 481f
Umami (taste), 400
Umbilical (term), **2**, 3f
Umbilical arteries, **482**, 483f, 662, 662f
Umbilical cord, 482, 483f, **660**, 660f, **661**, 662, 662f
Umbilical hernia, **686**
Umbilical ligaments, medial, 483f, **484**
Umbilical region, **8**, 9f
Umbilical vein, **482**, 483f, 662, 662f
Umbilicus (navel), 482, 483f, 686
Uncus, **278**, 280f
Unfused tetanus, **PEx-24**, PEx-25
Uniaxial joints/movement, 168f, 172f, 173
Unipennate fascicles/muscles, 196f
Unipolar neurons, **257**, 258f
Uniporter, PEx-13
Upper limb. *See also* Arm; Forearm; Hand
blood vessels of
arteries, 474, 474f
veins, 479, 481f
bones of, 108f, 148–152, 149f, 150f, 151f, 152f
muscles of, 198f, 199f, 211–215, 212f, 212t, 213–215f, 213–214f
nerves of, **313**–316, 313f, 314f, 315t
surface anatomy of, 3f, 152, 687–690, 687f, 688f, 689f, 690f
Upper respiratory system, 538–539, 539f, 540f
Urea, 622, 624
Ureter(s), **20**, 21f, 22f, 610f, **611**, 611f, 612f, 616f
Urethra, 22, 73f, 610f, **611**, 611f, 630f, 631f, **632**, 635f
Urethral orifice, external, 21f, 611f, **612**, 630f, 631f, 634f, 635f

Urethral sphincters, 611*f*, **615**
Uric acid, 622
Urinalysis, 621–628
Urinary bladder, **20**, 20*f*, 21*f*, 22*f*, 73*f*, 610*f*, **611**, 611*f*, **615**
  functional microscopic anatomy of, 615–616
  palpable, 687
Urinary casts, **623**, 626, 626*f*
Urinary system, 16*t*, **610**, 610*f*, 611*f*. *See also specific organ*
  anatomy of, 609–620
    functional microscopic anatomy of kidney/bladder, 613–616, 614*f*, 615*f*, 616*f*
    gross anatomy, 610*f*, 611–613, 611*f*, 612*f*
Urination (voiding/micturition), **615**
Urine
  abnormal constituents in, 622–623
  characteristics of, 622, 623–624
  formation of, 613–614
    hormones affecting, PEx-142 to PEx-143
  inorganic constituents in, 624
  organic constituents in, 624–625
  sample analysis and, 623–626
Urine concentration, 622
  solute gradient and, PEx-139 to PEx-140
Urine sediment, 626, 626*f*
Urobilinogen, 623, 624
Urochrome, 622
Uterine body, 635*f*, **636**
Uterine (menstrual) cycle, **651**, 651*f*, 652*f*
Uterine horns, 21, 21*f*
Uterine (Fallopian) tube(s), **22**, 22*f*, 635*f*, **636**, 637*f*
Uterosacral ligaments, 635*f*, **636**
Uterus, **21**, 21*f*, **22**, 22*f*, **634**, 635*f*, **636**, 636*f*
  oxytocin affecting, PEx-60
Utricle, 385*f*, **389**
Uvea (vascular layer of eye), **361**, 362*f*
Uvula, 539*f*, **575**, 576*f*, 577*f*

Vagal escape, **PEx-95**
Vagina, **21**, 21*f*, **22**, 22*f*, **634**, 635*f*
Vaginal orifice, **21**, 21*f*, **22**, 22*f*, 634*f*
Vagus nerve (cranial nerve X), 287*t*, 288, 288*f*, 289*f*, 399
  heart affected by stimulation of, 518–519, PEx-95 to PEx-96
Vallate papillae, **399**, 399*f*
Valves
  of collecting lymphatic vessels, 527, 527*f*
  of heart, 445*f*, **446**, 447*f*, PEx-86
    murmurs and, 482
  venous, 472
Variable (V) region, immunoglobulin, 530, 531*f*, PEx-182
Vasa recta, **613**, 614*f*
Vasa vasorum, **471**
Vascular layer of eye (uvea), **361**, 362*f*
Vas (ductus) deferens, **21**, 21*f*, **22**, 22*f*, 630*f*, **631**, 631*f*
Vasectomy, 632
Vasoconstriction, **PEx-76**
Vasodilation, **PEx-76**
  skin color change/local metabolites and, 502–503
Vastus intermedius muscle, 217*f*, 218*t*
Vastus lateralis muscle, 198*f*, 217*f*, 218*t*, 691*f*, 692*f*, **693**
  for intramuscular injections, 691*f*, 693
Vastus medialis muscle, 198*f*, 217*f*, 218*t*, 692*f*
VBD. *See* Vertebral bone density
VC. *See* Vital capacity
Veins, 470, 470*f*, 472, 477–479, 478*f*. *See also specific named vein*
Velocity, shortening, **PEx-28**

Venae cavae
  inferior, **20**, 21*f*, 444*f*, 445*f*, **446**, 448*f*, 450*f*, 480*f*, 481*f*
    veins draining into, 477–478, 479*f*, 480*f*, 481*f*
  superior, 19*f*, 444*f*, 445*f*, **446**, 448*f*, 450*f*, 480*f*, 481*f*
    veins draining into, 478–479, 480*f*, 481*f*
Venous arches
  deep palmar, 481*f*
  dorsal, **477**, 479*f*, 692*f*, **694**
  superficial palmar, 481*f*
Venous congestion, skin color and, 503
Venous pressure, 499
Venous valves, 472
Ventilation. *See* Breathing
Ventral (term), **3–4**, 4*f*
Ventral body cavity, **6–8**, 7*f*, 17–22
Ventral (anterior) funiculus, **310**, 310*f*, 312*f*
Ventral gluteal site, **691–693**, 691*f*
Ventral (anterior) horns, 281*f*, 284*f*, **308**, 310*f*, 312*f*
Ventral median fissure, **308**, 310*f*, 312*f*
Ventral rami, **312**, 313, 313*f*, 314*f*, 316, 316*f*, 317*f*
Ventral root, **308**, 310*f*, 313*f*
Ventricles
  brain, 276*f*, 277, 280*f*, 281*f*, 284*f*, 292*f*
    cerebrospinal fluid circulation and, 283–285, 284*f*
  cardiac, 444*f*, 445*f*, **446**, 448*f*, 450–451, 450*f*, 452*f*
    in frog, 513*f*
    right versus left, 451–452, 452*f*
Venules, 519–520, 519*f*
Vermis, **282**, 282*f*
Vertebra(e), 108*f*, **130**. *See also Vertebral column*
  cervical, 130*f*, 132, 132*f*, 133*f*, 134*t*
  lumbar, 130*f*, 132–133, 133*f*, 134*t*
  spinal cord relationship to, 310*f*
  structure of, 131–132, 131*f*, 134*t*
  thoracic, 130*f*, 132, 133*f*, 134*t*
Vertebral (term), **2**, 3*f*
Vertebral arch, **131**, 131*f*, 132*f*
Vertebral arteries, **472**, **473**, 473*f*, 474*f*
Vertebral bone density, PEx-67
Vertebral (medial) border of scapula, 148, 149*f*, **683**
Vertebral (spinal) cavity, **6**, 7*f*
Vertebral column, 108*f*, **130–135**, 130*f*, 131*f*. *See also Vertebra(e)*
  muscles associated with, 209–211*t*, 210–211*f*
Vertebral (spinal) foramen, **131**, 131*f*, 133*f*, 134*t*
Vertebral (floating) ribs, 135*f*, 136
Vertebral veins, **479**, 480*f*
Vertebra prominens, 130*f*, 132, 681
Vertebrochondral ribs, 136
Vertebrosternal (true) ribs, 135*f*, 136
Vertigo, **391**
  in equilibrium testing, 391–392
Vesicle(s), in transport, 60, 61, 61*f*
  secretory, 61
Vesicouterine pouch, 635*f*
Vesicular breathing sounds, **551**
Vesicular (antral) follicle, 649*f*, **650**, 650*f*, **651**, 652*f*
Vesicular transport, **60–61**, 61*f*, **PEx-4**
Vessel length, blood flow and, PEx-76, PEx-79
Vessel radius
  blood flow and, PEx-76, **PEx-76** to PEx-78
  glomerular filtration and, PEx-132 to PEx-135, PEx-137 to PEx-139
  pump activity and, PEx-82 to PEx-83
Vessel resistance, **PEx-75**, PEx-76
  blood flow and, PEx-75 to PEx-76, PEx-76, PEx-76 to PEx-88

compensation/cardiovascular pathology and, PEx-86
  blood pressure and, 500, PEx-75 to PEx-76, PEx-76
Vestibular apparatus, **389**
Vestibular folds (false vocal cords), **539**, 539*f*, 540*f*
Vestibular glands, greater, **634**, 634*f*, 635*f*
Vestibular membrane, **386**, 386*f*
Vestibule
  of ear, **385**, 385*f*
  oral, **576**, 576*f*
  of vagina, **634**, 634*f*
Vestibulocochlear nerve (cranial nerve VIII), 287*t*, 288, 288*f*, 289*f*
Viewing window (ophthalmoscope), **376**, 376*f*
Villi
  arachnoid, **283**, 283*f*, 284*f*
  chorionic, **660**, 660*f*, 662*f*
  intestinal, **581**, 581*f*, 582, 582*f*, 583
Virtual image, **29**, 30*f*
Visceral layer/pericardium (epicardium), 7*f*, 445*f*, **446**, 449
Visceral (smooth) muscle, **85**, 85*f*
Visceral peritoneum, 7*f*, **574**, 580*f*
  in duodenum, 582
Visceral pleura, 7*f*, 542*f*, **543**
Visceral (autonomic) reflexes, **336**, 340–341
Visceral serosa, **6–7**, 7*f*
Visceroceptors (interoceptors), **350**
Viscosity, PEx-78
  blood, PEx-76, PEx-78
    blood flow and, PEx-76, PEx-78 to PEx-79
Vision. *See also under Visual and Eye*
  binocular, **374–375**, 374*f*
  color, 364, 374
  cortical areas in, 278, 278*f*, 364, 366*f*
  equilibrium and, 392
  tests/experiments and, 371–381. *See also specific test*
  visual system anatomy and, 359–370. *See also Visual system anatomy*
Visual acuity, 373–374
Visual association area, 278*f*
Visual cortex, primary, 278*f*, **364**, 366*f*
Visual fields, **374–375**, 374*f*
Visual pathways, 364–366, 366*f*
Visual system anatomy, **359–370**
  external eye/accessory structures, 359–360, 360*f*, 361*f*
  internal eye, 360–363, 362*f*, 363*f*
  pathways to brain, 364–366, 366*f*
  retinal, 363*f*, 364
Visual tests/experiments, 371–381. *See also specific test*
Vital capacity (VC), **552**, 552*f*, 554, 556, 559, 559*f*, **PEx-107**
  forced, **557–559**, 558*f*, **PEx-107**
  predicted, 556–557, **559**
Vitreous humor/vitreous body, 362*f*, **363**
Vocal cords
  false (vestibular folds), **539**, 539*f*, 540*f*
  true (vocal folds), **539**, 539*f*, 540*f*
Vocal folds (true vocal cords), **539**, 539*f*, 540*f*
Vocal ligaments, 539
Voiding (micturition), **615**
Volkmann's (perforating) canals, 113*f*, **114**
Volt(s), galvanic skin potential measured in, 326
Voltage (stimulus). *See* Stimulus intensity/stimulus voltage; Threshold voltage
Voltage-gated potassium channels, refractory periods and, PEx-44
Voltage-gated sodium channels, **PEx-42** to PEx-44
  refractory periods and, PEx-44

Volume, osmosis affecting, PEx-8
Voluntary muscle, 187. *See also* Skeletal muscle(s)
Voluntary nervous system. *See* Somatic nervous system
Vomer, 122*f*, 125*f*, **128**
Vomiting, metabolic alkalosis and, PEx-154
V (variable) region, immunoglobulin, 530, 531*f*, PEx-182
Vulva (external genitalia), **634**, 634*f*

Water, diffusion through, 56
Wave (temporal) summation, **239–240**, 239*f*, 241–244, **242**, 242*f*, 243*f*, **PEx-22**, **PEx-94**
WBCs. *See* White blood cell(s)
Weak acid, 564, **PEx-149**
Weak base, 564, **PEx-149**
Weber test, 388, 388*f*
Wernicke's area, **278**, 278*f*
Western blotting, PEx-184 to PEx-186
Wet mount, **33–34**, 33*f*
Wet spirometers, 552, 553*f*
Wheal, 504
Wheel spirometer, 552, 552*f*
White blood cell(s) (leukocytes/WBCs), 81*f*, **424–425**, 425*f*, 426*t*, 427*f*, 428–429, 428*f*
  in urine (pyuria), **623**, 624
White blood cell count
  differential, **429–430**, 430*f*
  total, 429
White columns, **310**, 310*f*
White matter, 256
  cerebellar, 279, 282
  cerebral, 276*f*, 277*f*, **278**, 279, 281*f*
  of spinal cord, **308–311**
White pulp, **529**, 529*f*
White ramus communicans, **324**, 325*f*
Whorls (fingerprint), 101, 101*f*
Widow's peak, genotype/phenotype and, **671**, 672*f*
Willis, circle of, **472**, 473*f*, 474
Wisdom teeth, 584*f*, 585
Working distance (microscope), **30**, 31, 31*f*
Wright handheld dry spirometer, 552, 552*f*
Wrist
  bones of, 151, 152*f*
    fracture of, 688
  muscles acting on, 211, 213–215*f*, 213–214*t*, 689, 689*f*
Wrist joint, 170*t*

X chromosome, 669*f*, 670
Xiphisternal joint, 135*f*, **136**, 684, 685*f*
Xiphoid process, **135**, 135*f*, **683**, 685*f*
X-linked inheritance, 670

Y chromosome, 669*f*, 670
Yellow marrow, **110**, 111*f*
Yolk sac, **660**, 660*f*, 662*f*

Z discs, 186*f*, 187, 187*f*
Zona fasciculata, 410*f*, **411**
Zona glomerulosa, 410*f*, **411**
Zona pellucida, 650*f*, 658, 659*f*
Zona reticularis, 410*f*, **411**
Zygomatic arch, 120, 122*f*, 129, 680*f*, **681**
Zygomatic bone, 121*f*, 122*f*, 125*f*, 126*f*, **128**, 128*f*, 129
Zygomatic process, **120**, 121*f*, 122*f*, 127*f*, 128*f*
Zygomaticus muscles, 197, 198*f*, 200*t*, 201*f*
Zygote (fertilized egg), 646, 649, **658**, 659*f*
Zymogenic (chief) cells, **PEx-125**